Site Work & Landscape Costs with RSMeans data

Derrick Hale, PE, Senior Editor

2020

39th annual edition

Chief Data Officer
Noam Reininger

Vice President, Data
Tim Duggan

Principal Engineer
Bob Mewis (1, 4)

Contributing Editors
Brian Adams (21, 22)
Paul Cowan
Christopher Babbitt
Sam Babbitt
Stephen Bell
Michelle Curran
Antonio D'Aulerio (26, 27, 28, 48)

Matthew Doheny (8, 9, 10)
John Gomes, CCT (13, 41)
Derrick Hale, PE (2, 31, 32, 33, 34, 35, 44, 46)
Barry Hutchinson
Joseph Kelble (14, 23, 25)
Scott Keller (3, 5)
Charles Kibbee
Gerard Lafond, PE
Thomas Lane (6, 7)
Thomas Lyons
Jake MacDonald (11, 12)
John Melin, P.E.
Elisa Mello
Matthew Sorrentino

Kevin Souza
David Yazbek

Production Manager
Debbie Panarelli

Production
Jonathan Forgit
Sharon Larsen
Sheryl Rose
Janice Thalin

Data Quality Manager
Joseph Ingargiola

Innovation
Ray Diwakar
Kedar Gaikwad
Srini Narla

Cover Design
Blaire Collins

Data Analytics
David Byars
Ellen D'amico
Thomas Hauger
Cameron Jagoe
Matthew Kelliher-Gibson
Renee Rudicil

Numbers in italics are the divisional responsibilities for each editor. Please contact the designated editor directly with any questions.

RSMeans data from Gordian
Construction Publishers & Consultants
1099 Hingham Street, Suite 201
Rockland, MA 02370
United States of America
1.800.448.8182
RSMeans.com

Copyright 2019 by The Gordian Group Inc.
All rights reserved.
Cover photo © iStock.com/Gang Zhou

Printed in the United States of America
ISSN 1520-5320
ISBN 978-1-950656-17-2

0280

$319.99 per copy (in United States)
Price is subject to change without prior notice.

Related Data and Services

Our engineers recommend the following products and services to complement *Site Work & Landscape Costs with RSMeans data:*

Annual Cost Data Books
2020 Building Construction Costs with RSMeans data
2020 Heavy Construction Costs with RSMeans data

Reference Books
Landscape Estimating Methods
Estimating Building Costs
RSMeans Estimating Handbook
Green Building: Project Planning & Estimating
How to Estimate with Means Data and CostWorks
Plan Reading & Material Takeoff
Project Scheduling and Management for Construction

Seminars and In-House Training
Site Work & Heavy Construction Estimating
Training for our online estimating solution
Plan Reading & Material Takeoff
Scheduling with MSProject for Construction Professionals
Unit Price Estimating

RSMeans data
For access to the latest cost data, an intuitive search, and an easy-to-use estimate builder, take advantage of the time savings available from our online application.

To learn more visit: **RSMeans.com/online**

Enterprise Solutions
Building owners, facility managers, building product manufacturers, and attorneys across the public and private sectors engage with RSMeans data Enterprise to solve unique challenges where trusted construction cost data is critical.

To learn more visit: **RSMeans.com/Enterprise**

Custom Built Data Sets
Building and Space Models: Quickly plan construction costs across multiple locations based on geography, project size, building system component, product options, and other variables for precise budgeting and cost control.

Predictive Analytics: Accurately plan future builds with custom graphical interactive dashboards, negotiate future costs of tenant build-outs, and identify and compare national account pricing.

Consulting
Building Product Manufacturing Analytics: Validate your claims and assist with new product launches.

Third-Party Legal Resources: Used in cases of construction cost or estimate disputes, construction product failure vs. installation failure, eminent domain, class action construction product liability, and more.

API
For resellers or internal application integration, RSMeans data is offered via API. Deliver Unit, Assembly, and Square Foot Model data within your interface. To learn more about how you can provide your customers with the latest in localized construction cost data visit: **RSMeans.com/API**

Table of Contents

Foreword   iv

MasterFormat® Comparison Table   v

How the Cost Data Is Built: An Overview   vii

Estimating with RSMeans data: Unit Prices   ix

How to Use the Cost Data: The Details   xi

Unit Price Section   1

RSMeans data: Unit Prices—How They Work   4

Assemblies Section   569

RSMeans data: Assemblies—How They Work   572

Reference Section   727

     Construction Equipment Rental Costs   729

     Crew Listings   741

     Historical Cost Indexes   778

     City Cost Indexes   779

     Location Factors   822

     Reference Tables   828

     Change Orders   879

     Project Costs   883

     Abbreviations   888

Index   892

Other Data and Services   927

Labor Trade Rates including
Overhead & Profit   Inside Back Cover

Foreword

The Value of RSMeans data from Gordian

Since 1942, RSMeans data has been the industry-standard materials, labor, and equipment cost information database for contractors, facility owners and managers, architects, engineers, and anyone else that requires the latest localized construction cost information. More than 75 years later, the objective remains the same: to provide facility and construction professionals with the most current and comprehensive construction cost database possible.

With the constant influx of new construction methods and materials, in addition to ever-changing labor and material costs, last year's cost data is not reliable for today's designs, estimates, or budgets. Gordian's cost engineers apply real-world construction experience to identify and quantify new building products and methodologies, adjust productivity rates, and adjust costs to local market conditions across the nation. This adds up to more than 22,000 hours in cost research annually. This unparalleled construction cost expertise is why so many facility and construction professionals rely on RSMeans data year over year.

About Gordian

Gordian originated in the spirit of innovation and a strong commitment to helping clients reach and exceed their construction goals. In 1982, Gordian's chairman and founder, Harry H. Mellon, created Job Order Contracting while serving as chief engineer at the Supreme Headquarters Allied Powers Europe. Job Order Contracting is a unique indefinite delivery/indefinite quantity (IDIQ) process, which enables facility owners to complete a substantial number of repair, maintenance, and construction projects with a single, competitively awarded contract. Realizing facility and infrastructure owners across various industries could greatly benefit from the time and cost saving advantages of this innovative construction procurement solution, he established Gordian in 1990.

Continuing the commitment to providing the most relevant and accurate facility and construction data, software, and expertise in the industry, Gordian enhanced the fortitude of its data with the acquisition of RSMeans in 2014. And in an effort to expand its facility management capabilities, Gordian acquired Sightlines, the leading provider of facilities benchmarking data and analysis, in 2015.

Our Offerings

Gordian is the leader in facility and construction cost data, software, and expertise for all phases of the building life cycle. From planning to design, procurement, construction, and operations, Gordian's solutions help clients maximize efficiency, optimize cost savings, and increase building quality with its highly specialized data engineers, software, and unique proprietary data sets.

Our Commitment

At Gordian, we do more than talk about the quality of our data and the usefulness of its application. We stand behind all of our RSMeans data—from historical cost indexes to construction materials and techniques—to craft current costs and predict future trends. If you have any questions about our products or services, please call us toll-free at 800.448.8182 or visit our website at gordian.com.

MasterFormat® 2016/ MasterFormat® 2018 Comparison Table

This table compares the 2016 edition of the Construction Specifications Institute's MasterFormat® to the expanded 2018 edition. For your convenience, all revised 2016 numbers and titles are listed along with the corresponding 2018 numbers and titles.

CSI 2016 MF ID	CSI 2016 MF Description	CSI 2018 MF ID	CSI 2018 MF Description
015632	Temporary Security	015733	Temporary Security
019308	Facility Maintenance Equipment	019308	Facilities Maintenance, Equipment
024200	Removal and Salvage of Construction Materials	024200	Removal and Diversion of Construction Materials
040130	Unit Masonry Cleaning	04012052	Cleaning Masonry
068010	Composite Decking	067300	Composite Decking
072127	Reflective Insulation	072153	Reflective Insulation
072610	Above-Grade Vapor Retarders	072613	Above-Grade Vapor Retarders
074473	Metal Faced Panels	074433	Metal Faced Panels
075430	Ketone Ethylene Ester Roofing	075416	Ketone Ethylene Ester Roofing
077280	Vents	077280	Vent Options
087125	Weatherstripping	087125	Door Weatherstripping
087530	Weatherstripping	087530	Window Weatherstripping
096223	Bamboo Flooring	096436	Bamboo Flooring
099103	Paint Restoration	090190	Maintenance of Painting and Coating
102833	Laundry Accessories	102823	Laundry Accessories
117610	Operating Room Equipment	117610	Equipment for Operating Rooms
117710	Radiology Equipment	117710	Equipment for Radiology
122310	Wood Interior Shutters	122313	Wood Interior Shutters
123580	Commercial Kitchen Casework	123539	Commercial Kitchen Casework
124636	Desk Accessories	124113	Desk Accessories
141210	Dumbwaiters	141000	Dumbwaiters
211113	Facility Water Distribution Piping	211113	Facility Fire Suppression Piping
233715	Louvers	233715	Air Outlets and Inlets, HVAC Louvers
260580	Wiring Connections	260583	Wiring Connections
270110	Operation and Maintenance of Communications Systems	270110	Operation and Maintenance of Communication Systems
272123	Data Communications Switches and Hubs	272129	Data Communications Switches and Hubs
283149	Carbon-Monoxide Detection Sensors	284611.21	Carbon-Monoxide Detection Sensors
284621	Fire Alarm	284620	Fire Alarm
316233	Drilled Micropiles	316333	Drilled Micropiles
323420	Fabricated Pedestrian Bridges	323413	Fabricated Pedestrian Bridges
337543	Shunt Reactors	337253	Shunt Reactors
350100	Operation and Maint. of Waterway & Marine Construction	350100	Operation and Maintenance of Waterway and Marine Construction

How the Cost Data Is Built: An Overview

Unit Prices*

All cost data have been divided into 50 divisions according to the MasterFormat® system of classification and numbering.

Assemblies*

The cost data in this section have been organized in an "Assemblies" format. These assemblies are the functional elements of a building and are arranged according to the 7 elements of the UNIFORMAT II classification system. For a complete explanation of a typical "Assembly", see "RSMeans data: Assemblies—How They Work."

*Residential Models**

Model buildings for four classes of construction—economy, average, custom, and luxury—are developed and shown with complete costs per square foot.

*Commercial/Industrial/ Institutional Models**

This section contains complete costs for 77 typical model buildings expressed as costs per square foot.

*Green Commercial/Industrial/ Institutional Models**

This section contains complete costs for 25 green model buildings expressed as costs per square foot.

*References**

This section includes information on Equipment Rental Costs, Crew Listings, Historical Cost Indexes, City Cost Indexes, Location Factors, Reference Tables, and Change Orders, as well as a listing of abbreviations.

- **Equipment Rental Costs:** Included are the average costs to rent and operate hundreds of pieces of construction equipment.
- **Crew Listings:** This section lists all the crews referenced in the cost data. A crew is composed of more than one trade classification and/or the addition of power equipment to any trade classification. Power equipment is included in the cost of the crew. Costs are shown both with bare labor rates and with the installing contractor's overhead and profit added. For each, the total crew cost per eight-hour day and the composite cost per labor-hour are listed.

Unit Cost data

Assembly Cost data

Square Foot Models

- **Historical Cost Indexes:** These indexes provide you with data to adjust construction costs over time.
- **City Cost Indexes:** All costs in this data set are U.S. national averages. Costs vary by region. You can adjust for this by CSI Division to over 730 cities in 900+ 3-digit zip codes throughout the U.S. and Canada by using this data.
- **Location Factors:** You can adjust total project costs to over 730 cities in 900+ 3-digit zip codes throughout the U.S. and Canada by using the weighted number, which applies across all divisions.
- **Reference Tables:** At the beginning of selected major classifications in the Unit Prices are reference numbers indicators. These numbers refer you to related information in the Reference Section. In this section, you'll find reference tables, explanations, and estimating information that support how we develop the unit price data, technical data, and estimating procedures.
- **Change Orders:** This section includes information on the factors that influence the pricing of change orders.

- **Abbreviations:** A listing of abbreviations used throughout this information, along with the terms they represent, is included.

Index (printed versions only)

A comprehensive listing of all terms and subjects will help you quickly find what you need when you are not sure where it occurs in MasterFormat®.

Conclusion

This information is designed to be as comprehensive and easy to use as possible.

The Construction Specifications Institute (CSI) and Construction Specifications Canada (CSC) have produced the 2018 edition of MasterFormat®, a system of titles and numbers used extensively to organize construction information.

All unit prices in the RSMeans cost data are now arranged in the 50-division MasterFormat® 2018 system.

---

* Not all information is available in all data sets

*Note: The material prices in RSMeans cost data
are "contractor's prices." They are the prices that
contractors can expect to pay at the lumberyards,
suppliers'/distributors' warehouses, etc. Small
orders of specialty items would be higher than
the costs shown, while very large orders, such as
truckload lots, would be less. The variation would
depend on the size, timing, and negotiating power
of the contractor. The labor costs are primarily
for new construction or major renovation rather
than repairs or minor alterations. With reasonable
exercise of judgment, the figures can be used for
any building work.*

Estimating with RSMeans data: Unit Prices

Following these steps will allow you to complete an accurate estimate using RSMeans data: Unit Prices.

1. Scope Out the Project

- Think through the project and identify the CSI divisions needed in your estimate.
- Identify the individual work tasks that will need to be covered in your estimate.
- The Unit Price data have been divided into 50 divisions according to CSI MasterFormat® 2018.
- In printed versions, the Unit Price Section Table of Contents on page 1 may also be helpful when scoping out your project.
- Experienced estimators find it helpful to begin with Division 2 and continue through completion. Division 1 can be estimated after the full project scope is known.

2. Quantify

- Determine the number of units required for each work task that you identified.
- Experienced estimators include an allowance for waste in their quantities. (Waste is not included in our Unit Price line items unless otherwise stated.)

3. Price the Quantities

- Use the search tools available to locate individual Unit Price line items for your estimate.
- Reference Numbers indicated within a Unit Price section refer to additional information that you may find useful.
- The crew indicates who is performing the work for that task. Crew codes are expanded in the Crew Listings in the Reference Section to include all trades and equipment that comprise the crew.
- The Daily Output is the amount of work the crew is expected to complete in one day.
- The Labor-Hours value is the amount of time it will take for the crew to install one unit of work.
- The abbreviated Unit designation indicates the unit of measure upon which the crew, productivity, and prices are based.
- Bare Costs are shown for materials, labor, and equipment needed to complete the Unit Price line item. Bare costs do not include waste, project overhead, payroll insurance, payroll taxes, main office overhead, or profit.
- The Total Incl O&P cost is the billing rate or invoice amount of the installing contractor or subcontractor who performs the work for the Unit Price line item.

4. Multiply

- Multiply the total number of units needed for your project by the Total Incl O&P cost for each Unit Price line item.
- Be careful that your take off unit of measure matches the unit of measure in the Unit column.
- The price you calculate is an estimate for a completed item of work.
- Keep scoping individual tasks, determining the number of units required for those tasks, matching each task with individual Unit Price line items, and multiplying quantities by Total Incl O&P costs.
- An estimate completed in this manner is priced as if a subcontractor, or set of subcontractors, is performing the work. The estimate does not yet include Project Overhead or Estimate Summary components such as general contractor markups on subcontracted work, general contractor office overhead and profit, contingency, and location factors.

5. Project Overhead

- Include project overhead items from Division 1-General Requirements.
- These items are needed to make the job run. They are typically, but not always, provided by the general contractor. Items include, but are not limited to, field personnel, insurance, performance bond, permits, testing, temporary utilities, field office and storage facilities, temporary scaffolding and platforms, equipment mobilization and demobilization, temporary roads and sidewalks, winter protection, temporary barricades and fencing, temporary security, temporary signs, field engineering and layout, final cleaning, and commissioning.
- Each item should be quantified and matched to individual Unit Price line items in Division 1, then priced and added to your estimate.
- An alternate method of estimating project overhead costs is to apply a percentage of the total project cost—usually 5% to 15% with an average of 10% (see General Conditions).
- Include other project related expenses in your estimate such as:
  - Rented equipment not itemized in the Crew Listings
  - Rubbish handling throughout the project (see section 02 41 19.19)

6. Estimate Summary

- Include sales tax as required by laws of your state or county.
- Include the general contractor's markup on self-performed work, usually 5% to 15% with an average of 10%.
- Include the general contractor's markup on subcontracted work, usually 5% to 15% with an average of 10%.
- Include the general contractor's main office overhead and profit:
  - RSMeans data provides general guidelines on the general contractor's main office overhead (see section 01 31 13.60 and Reference Number R013113-50).
  - Markups will depend on the size of the general contractor's operations, projected annual revenue, the level of risk, and the level of competition in the local area and for this project in particular.
- Include a contingency, usually 3% to 5%, if appropriate.
- Adjust your estimate to the project's location by using the City Cost Indexes or the Location Factors in the Reference Section:
  - Look at the rules in "How to Use the City Cost Indexes" to see how to apply the Indexes for your location.
  - When the proper Index or Factor has been identified for the project's location, convert it to a multiplier by dividing it by 100, then multiply that multiplier by your estimated total cost. The original estimated total cost will now be adjusted up or down from the national average to a total that is appropriate for your location.

Editors' Note:
We urge you to spend time reading and understanding the supporting material. An accurate estimate requires experience, knowledge, and careful calculation. The more you know about how we at RSMeans developed the data, the more accurate your estimate will be. In addition, it is important to take into consideration the reference material such as Equipment Listings, Crew Listings, City Cost Indexes, Location Factors, and Reference Tables.

How to Use the Cost Data: The Details

What's Behind the Numbers? The Development of Cost Data

RSMeans data engineers continually monitor developments in the construction industry in order to ensure reliable, thorough, and up-to-date cost information. While overall construction costs may vary relative to general economic conditions, price fluctuations within the industry are dependent upon many factors. Individual price variations may, in fact, be opposite to overall economic trends. Therefore, costs are constantly tracked and complete updates are performed yearly. Also, new items are frequently added in response to changes in materials and methods.

Costs in U.S. Dollars

All costs represent U.S. national averages and are given in U.S. dollars. The City Cost Index (CCI) with RSMeans data can be used to adjust costs to a particular location. The CCI for Canada can be used to adjust U.S. national averages to local costs in Canadian dollars. No exchange rate conversion is necessary because it has already been factored in.

G The processes or products identified by the green symbol in our publications have been determined to be environmentally responsible and/or resource-efficient solely by RSMeans data engineering staff. The inclusion of the green symbol does not represent compliance with any specific industry association or standard.

Material Costs

RSMeans data engineers contact manufacturers, dealers, distributors, and contractors all across the U.S. and Canada to determine national average material costs. If you have access to current material costs for your specific location, you may wish to make adjustments to reflect differences from the national average. Included within material costs are fasteners for a normal installation. RSMeans data engineers use manufacturers' recommendations, written specifications, and/or standard construction practices for the sizing and spacing of fasteners. Adjustments to material costs may be required for your specific application or location. The manufacturer's warranty is assumed. Extended warranties are not included in the material costs. **Material costs do not include sales tax.**

Labor Costs

Labor costs are based upon a mathematical average of trade-specific wages in 30 major U.S. cities. The type of wage (union, open shop, or residential) is identified on the inside back cover of printed publications or selected by the estimator when using the electronic products. Markups for the wages can also be found on the inside back cover of printed publications and/or under the labor references found in the electronic products.

- If wage rates in your area vary from those used, or if rate increases are expected within a given year, labor costs should be adjusted accordingly.

Labor costs reflect productivity based on actual working conditions. In addition to actual installation, these figures include time spent during a normal weekday on tasks, such as material receiving and handling, mobilization at the site, site movement, breaks, and cleanup.

Productivity data is developed over an extended period so as not to be influenced by abnormal variations and reflects a typical average.

Equipment Costs

Equipment costs include not only rental but also operating costs for equipment under normal use. The operating costs include parts and labor for routine servicing, such as the repair and replacement of pumps, filters, and worn lines. Normal operating expendables, such as fuel, lubricants, tires, and electricity (where applicable), are also included. Extraordinary operating expendables with highly variable wear patterns, such as diamond bits and blades, are excluded. These costs are included under materials. Equipment rental rates are obtained from industry sources throughout North America—contractors, suppliers, dealers, manufacturers, and distributors.

Rental rates can also be treated as reimbursement costs for contractor-owned equipment. Owned equipment costs include depreciation, loan payments, interest, taxes, insurance, storage, and major repairs.

Equipment costs do not include operators' wages.

Equipment Cost/Day—The cost of equipment required for each crew is included in the Crew Listings in the Reference Section (small tools that are considered essential everyday tools are not listed out separately). The Crew Listings itemize specialized tools and heavy equipment along with labor trades. The daily cost of itemized equipment included in a crew is based on dividing the weekly bare rental rate by 5 (number of working days per week), then adding the hourly operating cost times 8 (the number of hours per day). This Equipment Cost/Day is shown in the last column of the Equipment Rental Costs in the Reference Section.

Mobilization, Demobilization—The cost to move construction equipment from an equipment yard or rental company to the job site and back again is not included in equipment costs. Mobilization (to the site) and demobilization (from the site) costs can be found in the Unit Price Section. If a piece of equipment is already at the job site, it is not appropriate to utilize mobilization or demobilization costs again in an estimate.

Overhead and Profit

Total Cost including O&P for the installing contractor is shown in the last column of the Unit Price and/or Assemblies. This figure is the sum of the bare material cost plus 10% for profit, the bare labor cost plus total overhead and profit, and the bare equipment cost plus 10% for profit. Details for the calculation of overhead and profit on labor are shown on the inside back cover of the printed product and in the Reference Section of the electronic product.

General Conditions

Cost data in this data set are presented in two ways: Bare Costs and Total Cost including O&P (Overhead and Profit). General Conditions, or General Requirements, of the contract should also be added to the Total Cost including O&P when applicable. Costs for General Conditions are listed in Division 1 of the Unit Price Section and in the Reference Section.

General Conditions for the installing contractor may range from 0% to 10% of the Total Cost including O&P. For the general or prime contractor, costs for General Conditions may range from 5% to 15% of the Total Cost including O&P, with a figure of 10% as the most typical allowance. If applicable, the Assemblies and Models sections use costs that include the installing contractor's overhead and profit (O&P).

Factors Affecting Costs

Costs can vary depending upon a number of variables. Here's a listing of some factors that affect costs and points to consider.

Quality—The prices for materials and the workmanship upon which productivity is based represent sound construction work. They are also in line with industry standard and manufacturer specifications and are frequently used by federal, state, and local governments.

Overtime—We have made no allowance for overtime. If you anticipate premium time or work beyond normal working hours, be sure to make an appropriate adjustment to your labor costs.

Productivity—The productivity, daily output, and labor-hour figures for each line item are based on an eight-hour work day in daylight hours in moderate temperatures and up to a 14' working height unless otherwise indicated. For work that extends beyond normal work hours or is performed under adverse conditions, productivity may decrease.

Size of Project—The size, scope of work, and type of construction project will have a significant impact on cost. Economies of scale can reduce costs for large projects. Unit costs can often run higher for small projects.

Location—Material prices are for metropolitan areas. However, in dense urban areas, traffic and site storage limitations may increase costs. Beyond a 20-mile radius of metropolitan areas, extra trucking or transportation charges may also increase the material costs slightly. On the other hand, lower wage rates may be in effect. Be sure to consider both of these factors when preparing an estimate, particularly if the job site is located in a central city or remote rural location. In addition, highly specialized subcontract items may require travel and per-diem expenses for mechanics.

Other Factors—

- season of year
- contractor management
- weather conditions
- local union restrictions
- building code requirements
- availability of:
  - adequate energy
  - skilled labor
  - building materials
- owner's special requirements/restrictions
- safety requirements
- environmental considerations
- access

Unpredictable Factors—General business conditions influence "in-place" costs of all items. Substitute materials and construction methods may have to be employed. These may affect the installed cost and/or life cycle costs. Such factors may be difficult to evaluate and cannot necessarily be predicted on the basis of the job's location in a particular section of the country. Thus, where these factors apply, you may find significant but unavoidable cost variations for which you will have to apply a measure of judgment to your estimate.

Rounding of Costs

In printed publications only, all unit prices in excess of $5.00 have been rounded to make them easier to use and still maintain adequate precision of the results.

How Subcontracted Items Affect Costs

A considerable portion of all large construction jobs is usually subcontracted. In fact, the percentage done by subcontractors is constantly increasing and may run over 90%. Since the workers employed by these companies do nothing else but install their particular products, they soon become experts in that line. As a result, installation by these firms is accomplished so efficiently that the total in-place cost, even with the general contractor's overhead and profit, is no more, and often less, than if the principal contractor had handled the installation. Companies that deal with construction specialties are anxious to have their products perform well and, consequently, the installation will be the best possible.

Contingencies

The allowance for contingencies generally provides for unforeseen construction difficulties. On alterations or repair jobs, 20% is not too much. If drawings are final and only field contingencies are being considered, 2% or 3% is probably sufficient and often nothing needs to be added. Contractually, changes in plans will be covered by extras. The contractor should consider inflationary price trends and possible material shortages during the course of the job. These escalation factors are dependent upon both economic conditions and the anticipated time between the estimate and actual construction. If drawings are not complete or approved, or a budget cost is wanted, it is wise to add 5% to 10%. Contingencies, then, are a matter of judgment.

Important Estimating Considerations

The productivity, or daily output, of each craftsman or crew assumes a well-managed job where tradesmen with the proper tools and equipment, along with the appropriate construction materials, are present. Included are daily set-up and cleanup time, break time, and plan layout time. Unless otherwise indicated, time for material movement on site (for items that can be transported by hand) of up to 200' into the building and to the first or second floor is also included. If material has to be transported by other means, over greater distances, or to higher floors, an additional allowance should be considered by the estimator.

While horizontal movement is typically a sole function of distances, vertical transport introduces other variables that can significantly impact productivity. In an occupied building, the use of elevators (assuming access, size, and required protective measures are acceptable) must be understood at the time of the estimate. For new construction, hoist wait and cycle times can easily be 15 minutes and may result in scheduled access extending beyond the normal work day. Finally, all vertical transport will impose strict weight limits likely to preclude the use of any motorized material handling.

The productivity, or daily output, also assumes installation that meets manufacturer/designer/standard specifications. A time allowance for quality control checks, minor adjustments, and any task required to ensure proper function or operation is also included. For items that require connections to services, time is included for positioning, leveling, securing the unit, and making all the necessary connections (and start up where applicable) to ensure a complete installation. Estimating of the services themselves (electrical, plumbing, water, steam, hydraulics, dust collection, etc.) is separate.

In some cases, the estimator must consider the use of a crane and an appropriate crew for the installation of large or heavy items. For those situations where a crane is not included in the assigned crew and as part of the line item cost,

then equipment rental costs, mobilization and demobilization costs, and operator and support personnel costs must be considered.

Labor-Hours

The labor-hours expressed in this publication are derived by dividing the total daily labor-hours for the crew by the daily output. Based on average installation time and the assumptions listed above, the labor-hours include: direct labor, indirect labor, and nonproductive time. A typical day for a craftsman might include but is not limited to:

- Direct Work
  - ☐ Measuring and layout
  - ☐ Preparing materials
  - ☐ Actual installation
  - ☐ Quality assurance/quality control
- Indirect Work
  - ☐ Reading plans or specifications
  - ☐ Preparing space
  - ☐ Receiving materials
  - ☐ Material movement
  - ☐ Giving or receiving instruction
  - ☐ Miscellaneous
- Non-Work
  - ☐ Chatting
  - ☐ Personal issues
  - ☐ Breaks
  - ☐ Interruptions (i.e., sickness, weather, material or equipment shortages, etc.)

If any of the items for a typical day do not apply to the particular work or project situation, the estimator should make any necessary adjustments.

Final Checklist

Estimating can be a straightforward process provided you remember the basics. Here's a checklist of some of the steps you should remember to complete before finalizing your estimate.

Did you remember to:

- factor in the City Cost Index for your locale?
- take into consideration which items have been marked up and by how much?
- mark up the entire estimate sufficiently for your purposes?
- read the background information on techniques and technical matters that could impact your project time span and cost?
- include all components of your project in the final estimate?
- double check your figures for accuracy?
- call RSMeans data engineers if you have any questions about your estimate or the data you've used? Remember, Gordian stands behind all of our products, including our extensive RSMeans data solutions. If you have any questions about your estimate, about the costs you've used from our data, or even about the technical aspects of the job that may affect your estimate, feel free to call the Gordian RSMeans editors at 1.800.448.8182.

Unit Price Section

Table of Contents

Sect. No.		Page
General Requirements		
01 11	Summary of Work	10
01 21	Allowances	10
01 31	Project Management and Coordination	12
01 32	Construction Progress Documentation	14
01 41	Regulatory Requirements	14
01 45	Quality Control	14
01 51	Temporary Utilities	16
01 52	Construction Facilities	18
01 54	Construction Aids	18
01 55	Vehicular Access and Parking	21
01 56	Temporary Barriers and Enclosures	21
01 57	Temporary Controls	23
01 58	Project Identification	23
01 66	Product Storage and Handling Requirements	24
01 71	Examination and Preparation	24
01 74	Cleaning and Waste Management	24
01 76	Protecting Installed Construction	24
01 91	Commissioning	24
Existing Conditions		
02 21	Surveys	26
02 32	Geotechnical Investigations	26
02 41	Demolition	27
02 42	Removal and Diversion of Construction Materials	44
02 43	Structure Moving	47
02 56	Site Containment	47
02 58	Snow Control	48
02 65	Underground Storage Tank Rmvl.	48
02 81	Transportation and Disposal of Hazardous Materials	49
02 91	Chemical Sampling, Testing and Analysis	49
Concrete		
03 01	Maintenance of Concrete	52
03 05	Common Work Results for Concrete	52
03 11	Concrete Forming	54
03 15	Concrete Accessories	63
03 21	Reinforcement Bars	71
03 22	Fabric and Grid Reinforcing	75
03 23	Stressed Tendon Reinforcing	75
03 24	Fibrous Reinforcing	76
03 30	Cast-In-Place Concrete	76
03 31	Structural Concrete	78
03 35	Concrete Finishing	81
03 37	Specialty Placed Concrete	83
03 39	Concrete Curing	84
03 41	Precast Structural Concrete	84
03 45	Precast Architectural Concrete	85
03 47	Site-Cast Concrete	86
03 48	Precast Concrete Specialties	86
03 53	Concrete Topping	86
03 54	Cast Underlayment	87
03 62	Non-Shrink Grouting	87
03 63	Epoxy Grouting	87
03 81	Concrete Cutting	88
03 82	Concrete Boring	88
Masonry		
04 01	Maintenance of Masonry	92
04 05	Common Work Results for Masonry	92
04 21	Clay Unit Masonry	97
04 22	Concrete Unit Masonry	99
04 27	Multiple-Wythe Unit Masonry	104
04 41	Dry-Placed Stone	104
04 43	Stone Masonry	104
04 54	Refractory Brick Masonry	108
04 57	Masonry Fireplaces	108
04 72	Cast Stone Masonry	109

Sect. No.		Page
Metals		
05 01	Maintenance of Metals	112
05 05	Common Work Results for Metals	112
05 12	Structural Steel Framing	116
05 15	Wire Rope Assemblies	120
05 21	Steel Joist Framing	123
05 31	Steel Decking	124
05 35	Raceway Decking Assemblies	125
05 51	Metal Stairs	125
05 52	Metal Railings	126
05 53	Metal Gratings	127
05 54	Metal Floor Plates	130
05 55	Metal Stair Treads and Nosings	131
05 56	Metal Castings	132
05 58	Formed Metal Fabrications	132
05 71	Decorative Metal Stairs	133
05 73	Decorative Metal Railings	134
Wood, Plastics & Composites		
06 05	Common Work Results for Wood, Plastics, and Composites	136
06 11	Wood Framing	136
06 13	Heavy Timber Construction	137
06 18	Glued-Laminated Construction	138
06 44	Ornamental Woodwork	139
Thermal & Moisture Protection		
07 01	Operation and Maint. of Thermal and Moisture Protection	142
07 11	Dampproofing	142
07 12	Built-up Bituminous Waterproofing	143
07 13	Sheet Waterproofing	143
07 16	Cementitious and Reactive Waterproofing	144
07 17	Bentonite Waterproofing	144
07 19	Water Repellents	144
07 21	Thermal Insulation	144
07 25	Weather Barriers	145
07 26	Vapor Retarders	146
07 31	Shingles and Shakes	146
07 32	Roof Tiles	147
07 41	Roof Panels	147
07 42	Wall Panels	148
07 46	Siding	149
07 51	Built-Up Bituminous Roofing	151
07 65	Flexible Flashing	152
07 71	Roof Specialties	152
07 72	Roof Accessories	155
07 76	Roof Pavers	155
07 81	Applied Fireproofing	155
07 91	Preformed Joint Seals	155
07 92	Joint Sealants	156
07 95	Expansion Control	157
Openings		
08 05	Common Work Results for Openings	160
08 12	Metal Frames	161
08 13	Metal Doors	161
08 14	Wood Doors	162
08 31	Access Doors and Panels	163
08 32	Sliding Glass Doors	163
08 36	Panel Doors	164
08 41	Entrances and Storefronts	164
08 42	Entrances	164
08 43	Storefronts	165
08 44	Curtain Wall and Glazed Assemblies	165
08 51	Metal Windows	165
08 52	Wood Windows	166
08 71	Door Hardware	166
08 75	Window Hardware	168
08 81	Glass Glazing	168
08 83	Mirrors	168
08 88	Special Function Glazing	168

Sect. No.		Page
Finishes		
09 05	Common Work Results for Finishes	170
09 21	Plaster and Gypsum Board Assemblies	171
09 23	Gypsum Plastering	171
09 24	Cement Plastering	172
09 26	Veneer Plastering	172
09 29	Gypsum Board	172
09 31	Thin-Set Tiling	173
09 32	Mortar-Bed Tiling	173
09 34	Waterproofing-Membrane Tiling	173
09 51	Acoustical Ceilings	174
09 63	Masonry Flooring	174
09 65	Resilient Flooring	175
09 66	Terrazzo Flooring	175
09 91	Painting	175
09 96	High-Performance Coatings	176
09 97	Special Coatings	177
Specialties		
10 05	Common Work Results for Specialties	180
10 13	Directories	180
10 14	Signage	180
10 22	Partitions	182
10 44	Fire Protection Specialties	182
10 57	Wardrobe and Closet Specialties	183
10 73	Protective Covers	183
10 74	Manufactured Exterior Specialties	184
10 75	Flagpoles	184
10 88	Scales	185
Equipment		
11 11	Vehicle Service Equipment	188
11 12	Parking Control Equipment	188
11 13	Loading Dock Equipment	189
11 14	Pedestrian Control Equipment	189
11 66	Athletic Equipment	190
11 68	Play Field Equipment and Structures	190
11 82	Facility Solid Waste Handling Equipment	192
Furnishings		
12 48	Rugs and Mats	194
12 92	Interior Planters and Artificial Plants	194
12 93	Interior Public Space Furnishings	197
Special Construction		
13 05	Common Work Results for Special Construction	200
13 11	Swimming Pools	201
13 12	Fountains	202
13 18	Ice Rinks	203
13 31	Fabric Structures	203
13 33	Geodesic Structures	205
13 34	Fabricated Engineered Structures	205
13 47	Facility Protection	209
13 53	Meteorological Instrumentation	210
Plumbing		
22 01	Operation and Maintenance of Plumbing	212
22 05	Common Work Results for Plumbing	213
22 11	Facility Water Distribution	217
22 12	Facility Potable-Water Storage Tanks	243
22 13	Facility Sanitary Sewerage	243
22 14	Facility Storm Drainage	253
22 41	Residential Plumbing Fixtures	255
22 47	Drinking Fountains and Water Coolers	255
22 51	Swimming Pool Plumbing Systems	256
22 52	Fountain Plumbing Systems	256

1

Table of Contents (cont.)

Sect. No.		Page
22 66	Chemical-Waste Systems for Lab. and Healthcare Facilities	256
Heating Ventilation Air Conditioning		
23 05	Common Work Results for HVAC	260
23 13	Facility Fuel-Storage Tanks	262
23 21	Hydronic Piping and Pumps	263
23 55	Fuel-Fired Heaters	264
Electrical		
26 05	Common Work Results for Electrical	266
26 12	Medium-Voltage Transformers	276
26 24	Switchboards and Panelboards	276
26 27	Low-Voltage Distribution Equipment	276
26 28	Low-Voltage Circuit Protective Devices	277
26 32	Packaged Generator Assemblies	278
26 36	Transfer Switches	279
26 51	Interior Lighting	279
26 56	Exterior Lighting	280
26 61	Lighting Systems and Accessories	285
Earthwork		
31 05	Common Work Results for Earthwork	288
31 06	Schedules for Earthwork	289
31 11	Clearing and Grubbing	289
31 13	Selective Tree and Shrub Removal and Trimming	290
31 14	Earth Stripping and Stockpiling	291
31 22	Grading	292
31 23	Excavation and Fill	292
31 25	Erosion and Sedimentation Controls	334
31 31	Soil Treatment	334
31 32	Soil Stabilization	335
31 36	Gabions	341
31 37	Riprap	341
31 41	Shoring	341
31 43	Concrete Raising	342
31 45	Vibroflotation and Densification	343
31 48	Underpinning	343
31 52	Cofferdams	343
31 56	Slurry Walls	344

Sect. No.		Page
31 62	Driven Piles	345
31 63	Bored Piles	355
Exterior Improvements		
32 01	Operation and Maintenance of Exterior Improvements	360
32 06	Schedules for Exterior Improvements	369
32 11	Base Courses	370
32 12	Flexible Paving	372
32 13	Rigid Paving	377
32 14	Unit Paving	378
32 16	Curbs, Gutters, Sidewalks, and Driveways	379
32 17	Paving Specialties	381
32 18	Athletic and Recreational Surfacing	384
32 31	Fences and Gates	385
32 32	Retaining Walls	395
32 33	Site Furnishings	397
32 34	Fabricated Bridges	398
32 35	Screening Devices	399
32 84	Planting Irrigation	399
32 91	Planting Preparation	402
32 92	Turf and Grasses	404
32 93	Plants	407
32 94	Planting Accessories	468
32 96	Transplanting	469
Utilities		
33 01	Operation and Maintenance of Utilities	472
33 05	Common Work Results for Utilities	477
33 11	Groundwater Sources	480
33 14	Water Utility Transmission and Distribution	482
33 16	Water Utility Storage Tanks	494
33 31	Sanitary Sewerage Piping	495
33 32	Sanitary Sewerage Equipment	499
33 34	Onsite Wastewater Disposal	499
33 41	Subdrainage	501
33 42	Stormwater Conveyance	503
33 46	Stormwater Management	510
33 52	Hydrocarbon Transmission and Distribution	510

Sect. No.		Page
33 59	Hydrocarbon Utility Metering	529
33 61	Hydronic Energy Distribution	530
33 63	Steam Energy Distribution	533
33 71	Electrical Utility Transmission and Distribution	535
33 81	Communications Structures	539
Transportation		
34 01	Operation and Maintenance of Transportation	542
34 11	Rail Tracks	543
34 41	Roadway Signaling and Control Equipment	544
34 71	Roadway Construction	544
34 72	Railway Construction	546
Waterway & Marine		
35 01	Operation and Maint. of Waterway & Marine Construction	550
35 22	Hydraulic Gates	553
35 24	Dredging	554
35 31	Shoreline Protection	555
35 49	Waterway Structures	555
35 51	Floating Construction	556
35 59	Marine Specialties	557
Material Processing & Handling Equipment		
41 21	Conveyors	560
41 22	Cranes and Hoists	560
Water and Wastewater Equipment		
46 07	Packaged Water and Wastewater Treatment Equipment	562
46 23	Grit Removal And Handling Equipment	562
46 25	Oil and Grease Separation and Removal Equipment	563
46 51	Air and Gas Diffusion Equipment	563
46 53	Biological Treatment Systems	567

RSMeans data: Unit Prices— How They Work

All RSMeans data: Unit Prices are organized in the same way.

03 30 Cast-In-Place Concrete

03 30 53 – Miscellaneous Cast-In-Place Concrete

03 30 53.40 Concrete In Place		Crew	Daily Output	Labor-Hours	Unit	Material	2020 Bare Costs Labor	Equipment	Total	Total Incl O&P
0010	**CONCRETE IN PLACE**	R033053-60								
0020	Including forms (4 uses), Grade 60 rebar, concrete (Portland cement									
0050	Type I), placement and finishing unless otherwise indicated	R033105-20								
0300	Beams (3500 psi), 5 kip/L.F., 10' span	R033105-70	C-14A	15.62	12.804	C.Y.	370	675	28	1,073	1,450
0350	25' span		"	18.55	10.782		390	570	23.50	983.50	1,300
0500	Chimney foundations (5000 psi), over 5 C.Y.		C-14C	32.22	3.476		178	175	.83	353.83	460
0510	(3500 psi), under 5 C.Y.		"	23.71	4.724		208	238	1.13	447.13	585
0700	Columns, square (4000 psi), 12" x 12", up to 1% reinforcing by area		C-14A	11.96	16.722		410	885	36.50	1,331.50	1,825
3540	Equipment pad (3000 psi), 3' x 3' x 6" thick		C-14H	45	1.067	Ea.	50.50	55	.60	106.10	139
3550	4' x 4' x 6" thick			30	1.600		78	82.50	.90	161.40	210
3560	5' x 5' x 8" thick			18	2.667		138	138	1.49	277.49	360
3570	6' x 6' x 8" thick			14	3.429		190	177	1.92	368.92	475
3580	8' x 8' x 10" thick			8	6		395	310	3.36	708.36	905
3590	10' x 10' x 12" thick			5	9.600		695	495	5.40	1,195.40	1,500

It is important to understand the structure of RSMeans data: Unit Prices so that you can find information easily and use it correctly.

1 Line Numbers

Line Numbers consist of 12 characters, which identify a unique location in the database for each task. The first 6 or 8 digits conform to the Construction Specifications Institute MasterFormat® 2018. The remainder of the digits are a further breakdown in order to arrange items in understandable groups of similar tasks. Line numbers are consistent across all of our publications, so a line number in any of our products will always refer to the same item of work.

2 Descriptions

Descriptions are shown in a hierarchical structure to make them readable. In order to read a complete description, read up through the indents to the top of the section. Include everything that is above and to the left that is not contradicted by information below. For instance, the complete description for line 03 30 53.40 3550 is "Concrete in place, including forms (4 uses), Grade 60 rebar, concrete (Portland cement Type 1), placement and finishing unless otherwise indicated; Equipment pad (3000 psi), 4' × 4' × 6" thick."

3 RSMeans data

When using **RSMeans data**, it is important to read through an entire section to ensure that you use the data that most closely matches your work. Note that sometimes there is additional information shown in the section that may improve your price. There are frequently lines that further describe, add to, or adjust data for specific situations.

4 Reference Information

Gordian's RSMeans engineers have created **reference** information to assist you in your estimate. **If** there is information that applies to a section, it will be indicated at the start of the section. The Reference Section is located in the back of the data set.

5 Crews

Crews include labor and/or equipment necessary to accomplish each task. In this case, Crew C-14H is used. Gordian's RSMeans staff selects a crew to represent the workers and equipment that are typically used for that task. In this case, Crew C-14H consists of one carpenter foreman (outside), two carpenters, one rodman, one laborer, one cement finisher, and one gas engine vibrator. Details of all crews can be found in the Reference Section.

Crews - Standard

Crew No.	Bare Costs		Incl. Subs O & P		Cost Per Labor-Hour	
Crew C-14H	Hr.	Daily	Hr.	Daily	Bare Costs	Incl. O&P
1 Carpenter Foreman (outside)	$55.15	$441.20	$82.85	$662.80	$51.65	$77.36
2 Carpenters	53.15	850.40	79.85	1277.60		
1 Rodman (reinf.)	56.40	451.20	84.90	679.20		
1 Laborer	42.10	336.80	63.25	506.00		
1 Cement Finisher	49.95	399.60	73.45	587.60		
1 Gas Engine Vibrator		26.85		29.54	.56	.62
48 L.H., Daily Totals		$2506.05		$3742.74	$52.21	$77.97

6 Daily Output
The **Daily Output** is the amount of work that the crew can do in a normal 8-hour workday, including mobilization, layout, movement of materials, and cleanup. In this case, crew C-14H can install thirty 4' × 4' × 6" thick concrete pads in a day. Daily output is variable and based on many factors, including the size of the job, location, and environmental conditions. RSMeans data represents work done in daylight (or adequate lighting) and temperate conditions.

7 Labor-Hours
The figure in the **Labor-Hours** column is the amount of labor required to perform one unit of work—in this case the amount of labor required to construct one 4' × 4' equipment pad. This figure is calculated by dividing the number of hours of labor in the crew by the daily output (48 labor-hours divided by 30 pads = 1.6 hours of labor per pad). Multiply 1.6 times 60 to see the value in minutes: 60 × 1.6 = 96

minutes. Note: the labor-hour figure is not dependent on the crew size. A change in crew size will result in a corresponding change in daily output, but the labor-hours per unit of work will not change.

8 Unit of Measure
All RSMeans data: Unit Prices include the typical **Unit of Measure** used for estimating that item. For concrete-in-place the typical unit is cubic yards (C.Y.) or each (Ea.). For installing broadloom carpet it is square yard and for gypsum board it is square foot. The estimator needs to take special care that the unit in the data matches the unit in the take-off. Unit conversions may be found in the Reference Section.

9 Bare Costs
Bare Costs are the costs of materials, labor, and equipment that the installing contractor pays. They represent the cost, in U.S. dollars, for one unit of work. They do not include any markups for profit or labor burden.

10 Bare Total
The **Total column** represents the total bare cost for the installing contractor in U.S. dollars. In this case, the sum of $78 for material + $82.50 for labor + $.90 for equipment is $161.40.

11 Total Incl O&P
The **Total Incl O&P column** is the total cost, including overhead and profit, that the installing contractor will charge the customer. This represents the cost of materials plus 10% profit, the cost of labor plus labor burden and 10% profit, and the cost of equipment plus 10% profit. It does not include the general contractor's overhead and profit. Note: See the inside back cover of the printed product or the Reference Section of the electronic product for details on how the labor burden is calculated.

National Average

The RSMeans data in our print publications represent a "national average" cost. This data should be modified to the project location using the **City Cost Indexes** *or* **Location Factors** *tables found in the Reference Section. Use the Location Factors to adjust estimate totals if the project covers multiple trades. Use the City Cost Indexes (CCI) for single trade*

projects or projects where a more detailed analysis is required. All figures in the two tables are derived from the same research. The last row of data in the CCI—the weighted average—is the same as the numbers reported for each location in the location factor table.

Project Name: Pre-Engineered Steel Building				Architect: As Shown				
Location:	**Anywhere, USA**						**01/01/20**	**STD**
Line Number	Description	Qty	Unit	Material	Labor	Equipment	SubContract	Estimate Total
03 30 53.40 3940	Strip footing, 12" x 24", reinforced	15	C.Y.	$2,565.00	$1,770.00	$8.40	$0.00	
03 30 53.40 3950	Strip footing, 12" x 36", reinforced	34	C.Y.	$5,610.00	$3,196.00	$15.30	$0.00	
03 11 13.65 3000	Concrete slab edge forms	500	L.F.	$165.00	$1,345.00	$0.00	$0.00	
03 22 11.10 0200	Welded wire fabric reinforcing	150	C.S.F.	$2,872.50	$4,350.00	$0.00	$0.00	
03 31 13.35 0300	Ready mix concrete, 4000 psi for slab on grade	278	C.Y.	$35,306.00	$0.00	$0.00	$0.00	
03 31 13.70 4300	Place, strike off & consolidate concrete slab	278	C.Y.	$0.00	$5,309.80	$136.22	$0.00	
03 35 13.30 0250	Machine float & trowel concrete slab	15,000	S.F.	$0.00	$9,900.00	$750.00	$0.00	
03 15 16.20 0140	Cut control joints in concrete slab	950	L.F.	$47.50	$418.00	$57.00	$0.00	
03 39 23.13 0300	Sprayed concrete curing membrane	150	C.S.F.	$1,867.50	$1,065.00	$0.00	$0.00	
Division 03	**Subtotal**			**$48,433.50**	**$27,353.80**	**$966.92**	**$0.00**	**$76,754.22**
08 36 13.10 2650	Manual 10' x 10' steel sectional overhead door	8	Ea.	$11,000.00	$3,760.00	$0.00	$0.00	
08 36 13.10 2860	Insulation and steel back panel for OH door	800	S.F.	$4,000.00	$0.00	$0.00	$0.00	
Division 08	**Subtotal**			**$15,000.00**	**$3,760.00**	**$0.00**	**$0.00**	**$18,760.00**
13 34 19.50 1100	Pre-Engineered Steel Building, 100' x 150' x 24'	15,000	SF Flr.	$0.00	$0.00	$0.00	$367,500.00	
13 34 19.50 6050	Framing for PESB door opening, 3' x 7'	4	Opng.	$0.00	$0.00	$0.00	$2,320.00	
13 34 19.50 6100	Framing for PESB door opening, 10' x 10'	8	Opng.	$0.00	$0.00	$0.00	$9,600.00	
13 34 19.50 6200	Framing for PESB window opening, 4' x 3'	6	Opng.	$0.00	$0.00	$0.00	$3,450.00	
13 34 19.50 5750	PESB door, 3' x 7', single leaf	4	Opng.	$2,920.00	$736.00	$0.00	$0.00	
13 34 19.50 7750	PESB sliding window, 4' x 3' with screen	6	Opng.	$2,940.00	$630.00	$67.80	$0.00	
13 34 19.50 6550	PESB gutter, eave type, 26 ga., painted	300	L.F.	$2,415.00	$864.00	$0.00	$0.00	
13 34 19.50 8650	PESB roof vent, 12" wide x 10' long	15	Ea.	$570.00	$3,465.00	$0.00	$0.00	
13 34 19.50 6900	PESB insulation, vinyl faced, 4" thick	27,400	S.F.	$11,782.00	$10,138.00	$0.00	$0.00	
Division 13	**Subtotal**			**$20,627.00**	**$15,833.00**	**$67.80**	**$382,870.00**	**$419,397.80**
			Subtotal	$84,060.50	$46,946.80	$1,034.72	$382,870.00	$514,912.02
Division 01	General Requirements @ 7%			5,884.24	3,286.28	72.43	26,800.90	
			Estimate Subtotal	$89,944.74	$50,233.08	$1,107.15	$409,670.90	$514,912.02
			Sales Tax @ 5%	4,497.24		55.36	10,241.77	
			Subtotal A	94,441.97	50,233.08	1,162.51	419,912.67	
			GC O & P	9,444.20	25,769.57	116.25	41,991.27	
			Subtotal B	103,886.17	76,002.64	1,278.76	461,903.94	$643,071.51
			Contingency @ 5%					32,153.58
			Subtotal C					$675,225.09
			Bond @ $12/1000 +10% O&P					8,912.97
			Subtotal D					$684,138.06
			Location Adjustment Factor		115.50			106,041.40
			Grand Total					**$790,179.46**

This estimate is based on an interactive spreadsheet. You are free to download it and adjust it to your methodology. A copy of this spreadsheet is available at **RSMeans.com/2020books**.

Sample Estimate

This sample demonstrates the elements of an estimate, including a tally of the RSMeans data lines and a summary of the markups on a contractor's work to arrive at a total cost to the owner. The Location Factor with RSMeans data is added at the bottom of the estimate to adjust the cost of the work to a specific location.

1 Work Performed

The body of the estimate shows the RSMeans data selected, including the line number, a brief description of each item, its take-off unit and quantity, and the bare costs of materials, labor, and equipment. This estimate also includes a column titled "SubContract." This data is taken from the column "Total Incl O&P" and represents the total that a subcontractor would charge a general contractor for the work, including the sub's markup for overhead and profit.

2 Division 1, General Requirements

This is the first division numerically but the last division estimated. Division 1 includes project-wide needs provided by the general contractor. These requirements vary by project but may include temporary facilities and utilities, security, testing, project cleanup, etc. For small projects a percentage can be used—typically between 5% and 15% of project cost. For large projects the costs may be itemized and priced individually.

3 Sales Tax

If the work is subject to state or local sales taxes, the amount must be added to the estimate. Sales tax may be added to material costs, equipment costs, and subcontracted work. In this case, sales tax was added in all three categories. It was assumed that approximately half the subcontracted work would be material cost, so the tax was applied to 50% of the subcontract total.

4 GC O&P

This entry represents the general contractor's markup on material, labor, equipment, and subcontractor costs. Our standard markup on materials, equipment, and subcontracted work is 10%. In this estimate, the markup on the labor performed by the GC's workers uses "Skilled Workers Average" shown in Column F on the table "Installing Contractor's Overhead & Profit," which can be found on the inside back cover of the printed product or in the Reference Section of the electronic product.

5 Contingency

A factor for contingency may be added to any estimate to represent the cost of unknowns that may occur between the time that the estimate is performed and the time the project is constructed. The amount of the allowance will depend on the stage of design at which the estimate is done and the contractor's assessment of the risk involved. Refer to section 01 21 16.50 for contingency allowances.

6 Bonds

Bond costs should be added to the estimate. The figures here represent a typical performance bond, ensuring the owner that if the general contractor does not complete the obligations in the construction contract the bonding company will pay the cost for completion of the work.

7 Location Adjustment

Published prices are based on national average costs. If necessary, adjust the total cost of the project using a location factor from the "Location Factor" table or the "City Cost Index" table. Use location factors if the work is general, covering multiple trades. If the work is by a single trade (e.g., masonry) use the more specific data found in the "City Cost Indexes."

Estimating Tips
01 20 00 Price and Payment Procedures

- Allowances that should be added to estimates to cover contingencies and job conditions that are not included in the national average material and labor costs are shown in Section 01 21.

- When estimating historic preservation projects (depending on the condition of the existing structure and the owner's requirements), a 15–20% contingency or allowance is recommended, regardless of the stage of the drawings.

01 30 00 Administrative Requirements

- Before determining a final cost estimate, it is good practice to review all the items listed in Subdivisions 01 31 and 01 32 to make final adjustments for items that may need customizing to specific job conditions.

- Requirements for initial and periodic submittals can represent a significant cost to the General Requirements of a job. Thoroughly check the submittal specifications when estimating a project to determine any costs that should be included.

01 40 00 Quality Requirements

- All projects will require some degree of quality control. This cost is not included in the unit cost of construction listed in each division. Depending upon the terms of the contract, the various costs of inspection and testing can be the responsibility of either the owner or the contractor. Be sure to include the required costs in your estimate.

01 50 00 Temporary Facilities and Controls

- Barricades, access roads, safety nets, scaffolding, security, and many more requirements for the execution of a safe project are elements of direct cost. These costs can easily be overlooked when preparing an estimate. When looking through the major classifications of this subdivision, determine which items apply to each division in your estimate.

- Construction equipment rental costs can be found in the Reference Section in Section 01 54 33. Operators' wages are not included in equipment rental costs.

- Equipment mobilization and demobilization costs are not included in equipment rental costs and must be considered separately.

- The cost of small tools provided by the installing contractor for his workers is covered in the "Overhead" column on the "Installing Contractor's Overhead and Profit" table that lists labor trades, base rates, and markups. Therefore, it is included in the "Total Incl. O&P" cost of any unit price line item.

01 70 00 Execution and Closeout Requirements

- When preparing an estimate, thoroughly read the specifications to determine the requirements for Contract Closeout. Final cleaning, record documentation, operation and maintenance data, warranties and bonds, and spare parts and maintenance materials can all be elements of cost for the completion of a contract. Do not overlook these in your estimate.

Reference Numbers

Reference numbers are shown at the beginning of some major classifications. These numbers refer to related items in the Reference Section. The reference information may be an estimating procedure, an alternate pricing method, or technical information.

Note: Not all subdivisions listed here necessarily appear. ■

Same Data. Simplified.

Enjoy the convenience and efficiency of accessing your costs anywhere:

- **Skip the multiplier** by setting your location
- **Quickly search,** edit, favorite and share costs
- **Stay on top of price changes** with automatic updates

Discover more at rsmeans.com/online

---

01 11 Summary of Work

01 11 31 – Professional Consultants

01 11 31.10 Architectural Fees		Crew	Daily Output	Labor-Hours	Unit	Material	2020 Bare Costs Labor	Equipment	Total	Total Incl O&P
0010	**ARCHITECTURAL FEES**	R011110-10								
0020	For new construction									
0060	Minimum				Project				4.90%	4.90%
0090	Maximum				"				16%	16%

01 11 31.20 Construction Management Fees										
0010	**CONSTRUCTION MANAGEMENT FEES**									
0020	$1,000,000 job, minimum				Project				4.50%	4.50%
0050	Maximum								7.50%	7.50%
0300	$50,000,000 job, minimum								2.50%	2.50%
0350	Maximum								4%	4%

01 11 31.30 Engineering Fees										
0010	**ENGINEERING FEES**	R011110-30								
0800	Landscaping & site development, minimum				Contrct				2.50%	2.50%
0900	Maximum				"				6%	6%

01 11 31.50 Models										
0010	**MODELS**									
0500	2 story building, scaled 100' x 200', simple materials and details				Ea.	4,750			4,750	5,225
0510	Elaborate materials and details				"	30,900			30,900	34,000

01 11 31.75 Renderings										
0010	**RENDERINGS** Color, matted, 20" x 30", eye level,									
0020	1 building, minimum				Ea.	2,250			2,250	2,475
0050	Average					3,200			3,200	3,525
0100	Maximum					5,150			5,150	5,675
1000	5 buildings, minimum					4,250			4,250	4,675
1100	Maximum					8,525			8,525	9,375
2000	Aerial perspective, color, 1 building, minimum					3,100			3,100	3,425
2100	Maximum					8,450			8,450	9,300
3000	5 buildings, minimum					6,050			6,050	6,650
3100	Maximum					12,100			12,100	13,400

01 21 Allowances

01 21 16 – Contingency Allowances

01 21 16.50 Contingencies

01 21 16.50 Contingencies		Crew	Daily Output	Labor-Hours	Unit	Material	Labor	Equipment	Total	Total Incl O&P
0010	**CONTINGENCIES**, Add to estimate									
0020	Conceptual stage				Project				20%	20%
0050	Schematic stage								15%	15%
0100	Preliminary working drawing stage (Design Dev.)								10%	10%
0150	Final working drawing stage								3%	3%

01 21 53 – Factors Allowance

01 21 53.50 Factors

01 21 53.50 Factors										
0010	**FACTORS** Cost adjustments									
0100	Add to construction costs for particular job requirements									
1100	Equipment usage curtailment, add, minimum				Costs	1%	1%			
1150	Maximum					3%	10%			
1400	Material handling & storage limitation, add, minimum					1%	1%			
1450	Maximum					6%	7%			
2300	Temporary shoring and bracing, add, minimum					2%	5%			
2350	Maximum					5%	12%			

01 21 53 – Factors Allowance

			Daily Output	Labor-Hours	Unit	Material	2020 Bare Costs Labor	Equipment	Total	Total Incl O&P
01 21 53.60 Security Factors		Crew								
0010	**SECURITY FACTORS**	R012153-60								
0100	Additional costs due to security requirements									
0110	Daily search of personnel, supplies, equipment and vehicles									
0120	Physical search, inventory and doc of assets, at entry					Costs	30%			
0130	At entry and exit						50%			
0140	Physical search, at entry						6.25%			
0150	At entry and exit						12.50%			
0160	Electronic scan search, at entry						2%			
0170	At entry and exit						4%			
0180	Visual inspection only, at entry						.25%			
0190	At entry and exit						.50%			
0200	ID card or display sticker only, at entry						.12%			
0210	At entry and exit						.25%			
0220	Day 1 as described below, then visual only for up to 5 day job duration									
0230	Physical search, inventory and doc of assets, at entry					Costs	5%			
0240	At entry and exit						10%			
0250	Physical search, at entry						1.25%			
0260	At entry and exit						2.50%			
0270	Electronic scan search, at entry						.42%			
0280	At entry and exit						.83%			
0290	Day 1 as described below, then visual only for 6-10 day job duration									
0300	Physical search, inventory and doc of assets, at entry					Costs	2.50%			
0310	At entry and exit						5%			
0320	Physical search, at entry						.63%			
0330	At entry and exit						1.25%			
0340	Electronic scan search, at entry						.21%			
0350	At entry and exit						.42%			
0360	Day 1 as described below, then visual only for 11-20 day job duration									
0370	Physical search, inventory and doc of assets, at entry					Costs	1.25%			
0380	At entry and exit						2.50%			
0390	Physical search, at entry						.31%			
0400	At entry and exit						.63%			
0410	Electronic scan search, at entry						.10%			
0420	At entry and exit						.21%			
0430	Beyond 20 days, costs are negligible									
0440	Escort required to be with tradesperson during work effort					Costs	6.25%			

01 21 57 – Overtime Allowance

			Daily Output	Labor-Hours	Unit	Material	Labor	Equipment	Total	Total Incl O&P
01 21 57.50 Overtime		Crew								
0010	**OVERTIME** for early completion of projects or where	R012909-90								
0020	labor shortages exist, add to usual labor, up to					Costs		100%			

01 21 63 – Taxes

			Daily Output	Labor-Hours	Unit	Material	Labor	Equipment	Total	Total Incl O&P
01 21 63.10 Taxes		Crew								
0010	**TAXES**	R012909-80								
0020	Sales tax, State, average					%	5.08%			
0050	Maximum	R012909-85					7.50%			
0200	Social Security, on first $118,500 of wages							7.65%		
0300	Unemployment, combined Federal and State, minimum	R012909-86						.60%		
0350	Average							9.60%		
0400	Maximum							12%		

For customer support on your Site Work & Landscape Costs with RSMeans data, call 800.448.8182.

11

01 31 Project Management and Coordination

01 31 13 – Project Coordination

01 31 13.20 Field Personnel

01 31 13.20 Field Personnel	Crew	Daily Output	Labor-Hours	Unit	Material	2020 Bare Costs Labor	2020 Bare Costs Equipment	Total	Total Incl O&P
0010 **FIELD PERSONNEL**									
0020 Clerk, average				Week		495		495	750
0100 Field engineer, junior engineer						1,241		1,241	1,877
0120 Engineer						1,825		1,825	2,775
0140 Senior engineer						2,400		2,400	3,625
0160 General purpose laborer, average	1 Clab	.20	40			1,675		1,675	2,525
0180 Project manager, minimum						2,175		2,175	3,300
0200 Average						2,500		2,500	3,800
0220 Maximum						2,850		2,850	4,325
0240 Superintendent, minimum						2,125		2,125	3,225
0260 Average						2,325		2,325	3,525
0280 Maximum						2,650		2,650	4,025
0290 Timekeeper, average						1,350		1,350	2,050

01 31 13.30 Insurance

01 31 13.30 Insurance				Unit	Material	Labor	Equipment	Total	Total Incl O&P
0010 **INSURANCE**	R013113-40								
0020 Builders risk, standard, minimum				Job				.24%	.24%
0050 Maximum	R013113-50							.64%	.64%
0200 All-risk type, minimum								.25%	.25%
0250 Maximum	R013113-60							.62%	.62%
0400 Contractor's equipment floater, minimum				Value				.50%	.50%
0450 Maximum				"				1.50%	1.50%
0600 Public liability, average				Job				2.02%	2.02%
0800 Workers' compensation & employer's liability, average									
0850 by trade, carpentry, general				Payroll		11.97%			
0900 Clerical						.38%			
0950 Concrete						10.84%			
1000 Electrical						4.91%			
1050 Excavation						7.81%			
1250 Masonry						13.40%			
1300 Painting & decorating						10.44%			
1350 Pile driving						12.06%			
1450 Plumbing						5.77%			
1600 Steel erection, structural						17.21%			
1650 Tile work, interior ceramic						8.10%			
1700 Waterproofing, brush or hand caulking						6.05%			
1800 Wrecking						15.48%			
2000 Range of 35 trades in 50 states, excl. wrecking & clerical, min.						1.37%			
2100 Average						10.60%			
2200 Maximum						120.29%			

01 31 13.40 Main Office Expense

01 31 13.40 Main Office Expense				Unit	Material	Labor	Equipment	Total	Total Incl O&P
0010 **MAIN OFFICE EXPENSE** Average for General Contractors	R013113-50								
0020 As a percentage of their annual volume									
0125 Annual volume under $1,000,000				% Vol.				17.50%	
0145 Up to $2,500,000								8%	
0150 Up to $4,000,000								6.80%	
0200 Up to $7,000,000								5.60%	
0250 Up to $10,000,000								5.10%	
0300 Over $10,000,000								3.90%	

01 31 13.50 General Contractor's Mark-Up

01 31 13.50 General Contractor's Mark-Up				Unit	Material	Labor	Equipment	Total	Total Incl O&P
0010 **GENERAL CONTRACTOR'S MARK-UP** on Change Orders									
0200 Extra work, by subcontractors, add				%				10%	10%
0250 By General Contractor, add								15%	15%

01 31 13 – Project Coordination

01 31 13.50 General Contractor's Mark-Up	Crew	Daily Output	Labor-Hours	Unit	Material	2020 Bare Costs Labor	Equipment	Total	Total Incl O&P
0400	Omitted work, by subcontractors, deduct all but				%				5%	5%
0450	By General Contractor, deduct all but								7.50%	7.50%
0600	Overtime work, by subcontractors, add								15%	15%
0650	By General Contractor, add								10%	10%

01 31 13.80 Overhead and Profit

		Crew	Daily Output	Labor-Hours	Unit	Material	Labor	Equipment	Total	Total Incl O&P
0010	**OVERHEAD & PROFIT** Allowance to add to items in this	R013113-50									
0020	book that do not include Subs O&P, average					%				25%	
0100	Allowance to add to items in this book that	R013113-55									
0110	do include Subs O&P, minimum					%				5%	5%
0150	Average									10%	10%
0200	Maximum									15%	15%
0300	Typical, by size of project, under $100,000									30%	
0350	$500,000 project									25%	
0400	$2,000,000 project									20%	
0450	Over $10,000,000 project									15%	

01 31 13.90 Performance Bond

		Crew	Daily Output	Labor-Hours	Unit	Material	Labor	Equipment	Total	Total Incl O&P
0010	**PERFORMANCE BOND**										
0020	For buildings, minimum				Job				.60%	.60%
0100	Maximum								2.50%	2.50%
0190	Highways & bridges, new construction, minimum								1%	1%
0200	Maximum								1.50%	1.50%
0300	Highways & bridges, resurfacing, minimum								.40%	.40%
0350	Maximum								.94%	.94%

01 31 14 – Facilities Services Coordination

01 31 14.20 Lock Out/Tag Out

		Crew	Daily Output	Labor-Hours	Unit	Material	Labor	Equipment	Total	Total Incl O&P
0010	**LOCK OUT / TAG OUT**									
0020	Miniature circuit breaker lock out device	1 Elec	220	.036	Ea.	24	2.23		26.23	30
0030	Miniature pin circuit breaker lock out device		220	.036		19.90	2.23		22.13	25.50
0040	Single circuit breaker lock out device		220	.036		21	2.23		23.23	26.50
0050	Multi-pole circuit breaker lock out device (15 to 225 Amp)		210	.038		19.90	2.34		22.24	25.50
0060	Large 3 pole circuit breaker lock out device (over 225 Amp)		210	.038		21	2.34		23.34	27
0080	Square D I-Line circuit breaker lock out device		210	.038		33	2.34		35.34	39.50
0090	Lock out disconnect switch, 30 to 100 Amp		330	.024		11	1.49		12.49	14.30
0100	100 to 400 Amp		330	.024		11	1.49		12.49	14.30
0110	Over 400 Amp		330	.024		11	1.49		12.49	14.30
0120	Lock out hasp for multiple lockout tags		200	.040		5.60	2.45		8.05	9.80
0130	Electrical cord plug lock out device		220	.036		10.50	2.23		12.73	14.85
0140	Electrical plug prong lock out device (3-wire grounding plug)		220	.036		6.95	2.23		9.18	10.95
0150	Wall switch lock out		200	.040		25	2.45		27.45	31.50
0160	Fire alarm pull station lock out	1 Stpi	200	.040		18.75	2.62		21.37	24.50
0170	Sprinkler valve tamper and flow switch lock out device	1 Skwk	220	.036		16.60	1.99		18.59	21.50
0180	Lock out sign		330	.024		18.90	1.33		20.23	23
0190	Lock out tag		440	.018		5	1		6	7

For customer support on your Site Work & Landscape Costs with RSMeans data, call 800.448.8182.

13

01 32 Construction Progress Documentation

01 32 13 – Scheduling of Work

01 32 13.50 Scheduling of Work	Crew	Daily Output	Labor-Hours	Unit	Material	2020 Bare Costs Labor	Equipment	Total	Total Incl O&P
0010 **SCHEDULING**									
0025 Scheduling, critical path, $50 million project, initial schedule				Ea.				25,750	28,325
0030 Monthly updates				"				3,348	3,682

01 32 33 – Photographic Documentation

01 32 33.50 Photographs

	Crew	Daily Output	Labor-Hours	Unit	Material	2020 Bare Costs Labor	Equipment	Total	Total Incl O&P
0010 **PHOTOGRAPHS**									
0020 8" x 10", 4 shots, 2 prints ea., std. mounting				Set	545			545	600
0100 Hinged linen mounts					550			550	605
0200 8" x 10", 4 shots, 2 prints each, in color					535			535	590
0300 For I.D. slugs, add to all above					5.10			5.10	5.60
0500 Aerial photos, initial fly-over, 5 shots, digital images					455			455	500
0550 10 shots, digital images, 1 print					515			515	570
0600 For each additional print from fly-over					252			252	278
0700 For full color prints, add					40%				
0750 Add for traffic control area				Ea.	355			355	390
0900 For over 30 miles from airport, add per				Mile	7.55			7.55	8.30
1500 Time lapse equipment, camera and projector, buy				Ea.	2,800			2,800	3,100
1550 Rent per month				"	1,325			1,325	1,450
1700 Cameraman and processing, black & white				Day	1,300			1,300	1,425
1720 Color				"	1,500			1,500	1,650

01 41 Regulatory Requirements

01 41 26 – Permit Requirements

01 41 26.50 Permits

	Crew	Daily Output	Labor-Hours	Unit	Material	2020 Bare Costs Labor	Equipment	Total	Total Incl O&P
0010 **PERMITS**									
0020 Rule of thumb, most cities, minimum				Job				.50%	.50%
0100 Maximum				"				2%	2%

01 45 Quality Control

01 45 23 – Testing and Inspecting Services

01 45 23.50 Testing

	Crew	Daily Output	Labor-Hours	Unit	Material	2020 Bare Costs Labor	Equipment	Total	Total Incl O&P
0010 **TESTING** and Inspecting Services									
0080 Supervision of Earthwork per day	1 Skwk	1	8	Day		440		440	665
0082 Quality Control of Earthwork per day	1 Clab	1	8	"		335		335	505
0200 Asphalt testing, compressive strength Marshall stability, set of 3				Ea.				545	600
0250 Extraction, individual tests on sample								236	260
0300 Penetration								91	100
0350 Mix design, 5 specimens								182	200
0360 Additional specimen								36	40
0400 Specific gravity								53	58
0420 Swell test								64	70
0450 Water effect and cohesion, set of 6								182	200
0470 Water effect and plastic flow								64	70
0600 Concrete testing, aggregates, abrasion, ASTM C 131								205	225
0650 Absorption, ASTM C 127								77	85
0800 Petrographic analysis, ASTM C 295								910	1,000
0900 Specific gravity, ASTM C 127								77	85
1000 Sieve analysis, washed, ASTM C 136								130	140
1050 Unwashed								130	140

01 45 23 – Testing and Inspecting Services

01 45 23.50 Testing	Crew	Daily Output	Labor-Hours	Unit	Material	2020 Bare Costs Labor	Equipment	Total	Total Incl O&P
1200	Sulfate soundness				Ea.				182	200
1300	Weight per cubic foot								80	88
1500	Cement, physical tests, ASTM C 150								320	350
1600	Chemical tests, ASTM C 150								245	270
1800	Compressive test, cylinder, delivered to lab, ASTM C 39								36	40
1900	Picked up by lab, 30 minute round trip, add to above	1 Skwk	16	.500			27.50		27.50	41.50
1950	1 hour round trip, add to above		8	1			55		55	83
2000	2 hour round trip, add to above		4	2			110		110	166
2200	Compressive strength, cores (not incl. drilling), ASTM C 42								95	105
2250	Core drilling, 4" diameter (plus technician), up to 6" thick	B-89A	14	1.143		.65	55.50	8.15	64.30	93
2260	Technician for core drilling				Hr.				50	55
2300	Patching core holes, to 12" diameter	1 Cefi	22	.364	Ea.	36.50	18.15		54.65	66.50
2400	Drying shrinkage at 28 days								236	260
2500	Flexural test beams, ASTM C 78								136	150
2600	Mix design, one batch mix								259	285
2650	Added trial batches								120	132
2800	Modulus of elasticity, ASTM C 469								195	215
2900	Tensile test, cylinders, ASTM C 496								52	58
3000	Water-Cement ratio curve, 3 batches								141	155
3100	4 batches								186	205
3300	Masonry testing, absorption, per 5 brick, ASTM C 67								72	80
3350	Chemical resistance, per 2 brick								50	55
3400	Compressive strength, per 5 brick, ASTM C 67								95	105
3420	Efflorescence, per 5 brick, ASTM C 67								101	112
3440	Imperviousness, per 5 brick								87	96
3470	Modulus of rupture, per 5 brick								86	95
3500	Moisture, block only								68	75
3550	Mortar, compressive strength, set of 3								32	35
4100	Reinforcing steel, bend test								68	75
4200	Tensile test, up to #8 bar								68	75
4220	#9 to #11 bar								114	125
4240	#14 bar and larger								182	200
4400	Soil testing, Atterberg limits, liquid and plastic limits								91	100
4510	Hydrometer analysis								155	170
4600	Sieve analysis, washed, ASTM D 422								136	150
4710	Consolidation test (ASTM D 2435), minimum								455	500
4750	Moisture content, ASTM D 2216								18	20
4780	Permeability test, double ring infiltrometer								500	550
4800	Permeability, var. or constant head, undist., ASTM D 2434								264	290
4850	Recompacted								250	275
4900	Proctor compaction, 4" standard mold, ASTM D 698								230	253
4950	6" modified mold								110	120
5100	Shear tests, triaxial, minimum								410	450
5150	Maximum								545	600
5300	Direct shear, minimum, ASTM D 3080								320	350
5350	Maximum								410	450
5550	Technician for inspection, per day, earthwork								556	612
5570	Concrete								556	612
5650	Bolting								556	612
5750	Roofing								556	612
5790	Welding								556	612
5820	Non-destructive metal testing, dye penetrant				Day				556	612
5840	Magnetic particle								556	612

01 45 Quality Control

01 45 23 – Testing and Inspecting Services

01 45 23.50 Testing

		Crew	Daily Output	Labor-Hours	Unit	Material	2020 Bare Costs Labor	Equipment	Total	Total Incl O&P
5860	Radiography				Day				556	612
5880	Ultrasonic				↓				556	612
6000	Welding certification, minimum				Ea.				91	100
6100	Maximum				"				364	400
7000	Underground storage tank									
7520	Volumetric tightness, <= 12,000 gal.				Ea.	470			470	520
7530	12,000 – 29,999 gal.					610			610	675
7540	>= 30,000 gal.				↓	665			665	730
7600	Vadose zone (soil gas) sampling, 10-40 samples, min.				Day				1,375	1,500
7610	Maximum				"				2,275	2,500
7700	Ground water monitoring incl. drilling 3 wells, min.				Total				4,550	5,000
7710	Maximum				"				6,375	7,000
8000	X-ray concrete slabs				Ea.				182	200

01 51 Temporary Utilities

01 51 13 – Temporary Electricity

01 51 13.50 Temporary Power Equip (Pro-Rated Per Job)

		Crew	Daily Output	Labor-Hours	Unit	Material	2020 Bare Costs Labor	Equipment	Total	Total Incl O&P
0010	**TEMPORARY POWER EQUIP (PRO-RATED PER JOB)**									
0020	Service, overhead feed, 3 use									
0030	100 Amp	1 Elec	1.25	6.400	Ea.	430	395		825	1,050
0040	200 Amp		1	8		470	490		960	1,250
0050	400 Amp		.75	10.667		830	655		1,485	1,900
0060	600 Amp	↓	.50	16	↓	1,200	980		2,180	2,775
0100	Underground feed, 3 use									
0110	100 Amp	1 Elec	2	4	Ea.	445	245		690	855
0120	200 Amp		1.15	6.957		520	425		945	1,200
0130	400 Amp		1	8		875	490		1,365	1,700
0140	600 Amp		.75	10.667		1,075	655		1,730	2,150
0150	800 Amp		.50	16		1,525	980		2,505	3,125
0160	1000 Amp		.35	22.857		1,700	1,400		3,100	3,950
0170	1200 Amp		.25	32		1,875	1,975		3,850	4,975
0180	2000 Amp	↓	.20	40	↓	2,250	2,450		4,700	6,125
0200	Transformers, 3 use									
0210	30 kVA	1 Elec	1	8	Ea.	1,925	490		2,415	2,850
0220	45 kVA		.75	10.667		2,450	655		3,105	3,675
0230	75 kVA		.50	16		3,725	980		4,705	5,550
0240	112.5 kVA	↓	.40	20	↓	4,100	1,225		5,325	6,325
0250	Feeder, PVC, CU wire in trench									
0260	60 Amp	1 Elec	96	.083	L.F.	2.73	5.10		7.83	10.60
0270	100 Amp		85	.094		4.96	5.75		10.71	14.05
0280	200 Amp		59	.136		12	8.30		20.30	25.50
0290	400 Amp	↓	42	.190	↓	32	11.70		43.70	52.50
0300	Feeder, PVC, aluminum wire in trench									
0310	60 Amp	1 Elec	96	.083	L.F.	2.67	5.10		7.77	10.55
0320	100 Amp		85	.094		3.19	5.75		8.94	12.10
0330	200 Amp		59	.136		8.65	8.30		16.95	22
0340	400 Amp	↓	42	.190	↓	15.75	11.70		27.45	34.50
0350	Feeder, EMT, CU wire									
0360	60 Amp	1 Elec	90	.089	L.F.	3.10	5.45		8.55	11.50
0370	100 Amp		80	.100		5.75	6.15		11.90	15.45
0380	200 Amp		60	.133		12.85	8.20		21.05	26.50

01 51 Temporary Utilities

01 51 13 – Temporary Electricity

01 51 13.50 Temporary Power Equip (Pro-Rated Per Job)	Crew	Daily Output	Labor-Hours	Unit	Material	2020 Bare Costs Labor	Equipment	Total	Total Incl O&P
0390 400 Amp	1 Elec	35	.229	L.F.	32.50	14		46.50	56.50
0400 Feeder, EMT, aluminum wire									
0410 60 Amp	1 Elec	90	.089	L.F.	4.59	5.45		10.04	13.15
0420 100 Amp		80	.100		5.05	6.15		11.20	14.70
0430 200 Amp		60	.133		9.25	8.20		17.45	22.50
0440 400 Amp		35	.229		17.30	14		31.30	40
0500 Equipment, 3 use									
0510 Spider box, 50 Amp	1 Elec	8	1	Ea.	785	61.50		846.50	955
0520 Lighting cord, 100'		8	1		126	61.50		187.50	231
0530 Light stanchion		8	1		72	61.50		133.50	171
0540 Temporary cords, 100', 3 use									
0550 Feeder cord, 50 Amp	1 Elec	16	.500	Ea.	237	30.50		267.50	305
0560 Feeder cord, 100 Amp		12	.667		1,200	41		1,241	1,375
0570 Tap cord, 50 Amp		12	.667		630	41		671	755
0580 Tap cord, 100 Amp		6	1.333		1,100	82		1,182	1,325
0590 Temporary cords, 50', 3 use									
0600 Feeder cord, 50 Amp	1 Elec	16	.500	Ea.	305	30.50		335.50	380
0610 Feeder cord, 100 Amp		12	.667		625	41		666	750
0620 Tap cord, 50 Amp		12	.667		305	41		346	395
0630 Tap cord, 100 Amp		6	1.333		615	82		697	800
0700 Connections									
0710 Compressor or pump									
0720 30 Amp	1 Elec	7	1.143	Ea.	17	70		87	123
0730 60 Amp		5.30	1.509		30	92.50		122.50	172
0740 100 Amp		4	2		80	123		203	271
0860 Office trailer									
0870 60 Amp	1 Elec	4.50	1.778	Ea.	55	109		164	223
0880 100 Amp		3	2.667		83.50	164		247.50	335
0890 200 Amp		2	4		292	245		537	685

01 51 13.80 Temporary Utilities

	Crew	Daily Output	Labor-Hours	Unit	Material	Labor	Equipment	Total	Total Incl O&P
0010 **TEMPORARY UTILITIES**									
0350 Lighting, lamps, wiring, outlets, 40,000 S.F. building, 8 strings	1 Elec	34	.235	CSF Flr	5.85	14.45		20.30	28
0360 16 strings	"	17	.471		11.75	29		40.75	56
0400 Power for temp lighting only, 6.6 KWH, per month								.92	1.01
0430 11.8 KWH, per month								1.65	1.82
0450 23.6 KWH, per month								3.30	3.63
0600 Power for job duration incl. elevator, etc., minimum								53	58
0650 Maximum								110	121
0675 Temporary cooling				Ea.	1,025			1,025	1,125
0700 Temporary construction water bill per month, average				Month	77			77	84.50

17

For customer support on your Site Work & Landscape Costs with RSMeans data, call 800.448.8182.

01 52 Construction Facilities

01 52 13 – Field Offices and Sheds

01 52 13.20 Office and Storage Space

		Crew	Daily Output	Labor-Hours	Unit	Material	2020 Bare Costs Labor	Equipment	Total	Total Incl O&P
0010	**OFFICE AND STORAGE SPACE**									
0020	Office trailer, furnished, no hookups, 20' x 8', buy	2 Skwk	1	16	Ea.	9,525	880		10,405	11,800
0250	Rent per month					195			195	214
0300	32' x 8', buy	2 Skwk	.70	22.857		15,300	1,250		16,550	18,700
0350	Rent per month					245			245	269
0400	50' x 10', buy	2 Skwk	.60	26.667		30,300	1,475		31,775	35,500
0450	Rent per month					355			355	390
0500	50' x 12', buy	2 Skwk	.50	32		25,800	1,750		27,550	31,100
0550	Rent per month					460			460	505
0700	For air conditioning, rent per month, add					51.50			51.50	56.50
0800	For delivery, add per mile				Mile	12.20			12.20	13.40
0900	Bunk house trailer, 8' x 40' duplex dorm with kitchen, no hookups, buy	2 Carp	1	16	Ea.	89,000	850		89,850	99,000
0910	9 man with kitchen and bath, no hookups, buy		1	16		91,000	850		91,850	101,500
0920	18 man sleeper with bath, no hookups, buy		1	16		98,000	850		98,850	109,500
1000	Portable buildings, prefab, on skids, economy, 8' x 8'		265	.060	S.F.	26.50	3.21		29.71	34
1100	Deluxe, 8' x 12'		150	.107	"	29.50	5.65		35.15	40.50
1200	Storage boxes, 20' x 8', buy	2 Skwk	1.80	8.889	Ea.	3,150	490		3,640	4,200
1250	Rent per month					86			86	94.50
1300	40' x 8', buy	2 Skwk	1.40	11.429		3,800	625		4,425	5,125
1350	Rent per month					126			126	139

01 52 13.40 Field Office Expense

		Crew	Daily Output	Labor-Hours	Unit	Material	2020 Bare Costs Labor	Equipment	Total	Total Incl O&P
0010	**FIELD OFFICE EXPENSE**									
0100	Office equipment rental average				Month	209			209	230
0120	Office supplies, average				"	87			87	96
0125	Office trailer rental, see Section 01 52 13.20									
0140	Telephone bill; avg. bill/month incl. long dist.				Month	87			87	96
0160	Lights & HVAC				"	163			163	179

01 54 Construction Aids

01 54 09 – Protection Equipment

01 54 09.60 Safety Nets

		Crew	Daily Output	Labor-Hours	Unit	Material	2020 Bare Costs Labor	Equipment	Total	Total Incl O&P
0010	**SAFETY NETS**									
0020	No supports, stock sizes, nylon, 3-1/2" mesh				S.F.	3.16			3.16	3.48
0100	Polypropylene, 6" mesh					1.64			1.64	1.80
0200	Small mesh debris nets, 1/4" mesh, stock sizes					.56			.56	.62
0220	Combined 3-1/2" mesh and 1/4" mesh, stock sizes					4.92			4.92	5.40
0300	Rental, 4" mesh, stock sizes, 3 months					.84			.84	.92
0320	6 month rental					1.03			1.03	1.13
0340	12 months					1.43			1.43	1.57
6220	Safety supplies and first aid kits				Month	28			28	30.50

01 54 16 – Temporary Hoists

01 54 16.50 Weekly Forklift Crew

		Crew	Daily Output	Labor-Hours	Unit	Material	2020 Bare Costs Labor	Equipment	Total	Total Incl O&P
0010	**WEEKLY FORKLIFT CREW**									
0100	All-terrain forklift, 45' lift, 35' reach, 9000 lb. capacity	A-3P	.20	40	Week		2,125	1,875	4,000	5,225

01 54 19 – Temporary Cranes

01 54 19.50 Daily Crane Crews

		Crew	Daily Output	Labor-Hours	Unit	Material	2020 Bare Costs Labor	Equipment	Total	Total Incl O&P
0010	**DAILY CRANE CREWS** for small jobs, portal to portal R015433-15									
0100	12-ton truck-mounted hydraulic crane	A-3H	1	8	Day		475	725	1,200	1,500
0200	25-ton	A-3I	1	8			475	800	1,275	1,600
0300	40-ton	A-3J	1	8			475	1,275	1,750	2,100

01 54 Construction Aids

01 54 19 – Temporary Cranes

01 54 19.50 Daily Crane Crews

		Crew	Daily Output	Labor-Hours	Unit	Material	2020 Bare Costs Labor	2020 Bare Costs Equipment	Total	Total Incl O&P
0400	55-ton	A-3K	1	16	Day		880	1,500	2,380	2,950
0500	80-ton	A-3L	1	16	↓		880	2,250	3,130	3,775
0900	If crane is needed on a Saturday, Sunday or Holiday									
0910	At time-and-a-half, add				Day		50%			
0920	At double time, add				"		100%			

01 54 19.60 Monthly Tower Crane Crew

		Crew	Daily Output	Labor-Hours	Unit	Material	Labor	Equipment	Total	Total Incl O&P
0010	**MONTHLY TOWER CRANE CREW**, excludes concrete footing									
0100	Static tower crane, 130' high, 106' jib, 6200 lb. capacity	A-3N	.05	176	Month		10,400	37,300	47,700	56,500

01 54 23 – Temporary Scaffolding and Platforms

01 54 23.70 Scaffolding

		Crew	Daily Output	Labor-Hours	Unit	Material	Labor	Equipment	Total	Total Incl O&P
0010	**SCAFFOLDING** R015423-10									
0015	Steel tube, regular, no plank, labor only to erect & dismantle									
0090	Building exterior, wall face, 1 to 5 stories, 6'-4" x 5' frames	3 Carp	8	3	C.S.F.		159		159	240
0460	Building interior, wall face area, up to 16' high	"	12	2			106		106	160
0906	Complete system for face of walls, no plank, material only rent/mo				↓	34			34	37.50
0908	Interior spaces, no plank, material only rent/mo				C.C.F.	3.88			3.88	4.27
0910	Steel tubular, heavy duty shoring, buy									
0920	Frames 5' high 2' wide				Ea.	89.50			89.50	98.50
0925	5' high 4' wide					106			106	116
0930	6' high 2' wide					107			107	118
0935	6' high 4' wide				↓	115			115	126
0940	Accessories									
0945	Cross braces				Ea.	17.05			17.05	18.80
0950	U-head, 8" x 8"					20.50			20.50	22.50
0955	J-head, 4" x 8"					14.90			14.90	16.40
0960	Base plate, 8" x 8"				↓	16.05			16.05	17.65
0965	Leveling jack				↓	35			35	38
1000	Steel tubular, regular, buy									
1100	Frames 3' high 5' wide				Ea.	92.50			92.50	102
1150	5' high 5' wide					108			108	119
1200	6'-4" high 5' wide					90.50			90.50	99.50
1350	7'-6" high 6' wide					151			151	166
1500	Accessories, cross braces					18.85			18.85	20.50
1550	Guardrail post					23.50			23.50	25.50
1600	Guardrail 7' section					8.20			8.20	9.05
1650	Screw jacks & plates					26			26	28.50
1700	Sidearm brackets					23.50			23.50	26
1750	8" casters					37.50			37.50	41
1800	Plank 2" x 10" x 16'-0"					64.50			64.50	71
1900	Stairway section					292			292	320
1910	Stairway starter bar					32.50			32.50	35.50
1920	Stairway inside handrail					54.50			54.50	60
1930	Stairway outside handrail					87.50			87.50	96
1940	Walk-thru frame guardrail				↓	42			42	46.50
2000	Steel tubular, regular, rent/mo.									
2100	Frames 3' high 5' wide				Ea.	4.50			4.50	4.95
2150	5' high 5' wide					4.50			4.50	4.95
2200	6'-4" high 5' wide					5.70			5.70	6.30
2250	7'-6" high 6' wide					9.90			9.90	10.90
2500	Accessories, cross braces					.90			.90	.99
2550	Guardrail post					.90			.90	.99
2600	Guardrail 7' section				↓	.90			.90	.99

01 54 23.70 Scaffolding

		Crew	Daily Output	Labor-Hours	Unit	Material	2020 Bare Costs Labor	Equipment	Total	Total Incl O&P
2650	Screw jacks & plates				Ea.	1.80			1.80	1.98
2700	Sidearm brackets					1.80			1.80	1.98
2750	8" casters					7.20			7.20	7.90
2800	Outrigger for rolling tower					2.70			2.70	2.97
2850	Plank 2" x 10" x 16'-0"					9.95			9.95	10.90
2900	Stairway section					31.50			31.50	35
2940	Walk-thru frame guardrail					2.25			2.25	2.48
3000	Steel tubular, heavy duty shoring, rent/mo.									
3250	5' high 2' & 4' wide				Ea.	8.45			8.45	9.30
3300	6' high 2' & 4' wide					8.45			8.45	9.30
3500	Accessories, cross braces					.90			.90	.99
3600	U-head, 8" x 8"					2.50			2.50	2.75
3650	J-head, 4" x 8"					2.50			2.50	2.75
3700	Base plate, 8" x 8"					.90			.90	.99
3750	Leveling jack					2.47			2.47	2.72
5700	Planks, 2" x 10" x 16'-0", labor only to erect & remove to 50' H	3 Carp	72	.333			17.70		17.70	26.50
5800	Over 50' high	4 Carp	80	.400			21.50		21.50	32
6000	Heavy duty shoring for elevated slab forms to 8'-2" high, floor area									
6100	Labor only to erect & dismantle	4 Carp	16	2	C.S.F.		106		106	160
6110	Materials only, rent/mo.				"	43.50			43.50	48
6500	To 14'-8" high									
6600	Labor only to erect & dismantle	4 Carp	10	3.200	C.S.F.		170		170	256
6610	Materials only, rent/mo				"	63.50			63.50	70

01 54 23.75 Scaffolding Specialties

		Crew	Daily Output	Labor-Hours	Unit	Material	2020 Bare Costs Labor	Equipment	Total	Total Incl O&P
0010	**SCAFFOLDING SPECIALTIES**									
1200	Sidewalk bridge, heavy duty steel posts & beams, including									
1210	parapet protection & waterproofing (material cost is rent/month)									
1220	8' to 10' wide, 2 posts	3 Carp	15	1.600	L.F.	46	85		131	179
1230	3 posts	"	10	2.400	"	71	128		199	270
1500	Sidewalk bridge using tubular steel scaffold frames including									
1510	planking (material cost is rent/month)	3 Carp	45	.533	L.F.	8.40	28.50		36.90	52
1600	For 2 uses per month, deduct from all above					50%				
1700	For 1 use every 2 months, add to all above					100%				
1900	Catwalks, 20" wide, no guardrails, 7' span, buy				Ea.	153			153	168
2000	10' span, buy					213			213	234
3800	Rolling ladders with handrails, 30" wide, buy, 2 step					320			320	355
4000	7 step					1,125			1,125	1,250
4050	10 step					1,625			1,625	1,775
4100	Rolling towers, buy, 5' wide, 7' long, 10' high					1,375			1,375	1,500
4200	For additional 5' high sections, to buy					255			255	280
4300	Complete incl. wheels, railings, outriggers,									
4350	21' high, to buy				Ea.	2,325			2,325	2,550
4400	Rent/month = 5% of purchase cost				"	116			116	128

01 54 36 – Equipment Mobilization

01 54 36.50 Mobilization

		Crew	Daily Output	Labor-Hours	Unit	Material	2020 Bare Costs Labor	Equipment	Total	Total Incl O&P
0010	**MOBILIZATION** (Use line item again for demobilization) R015436-50									
0015	Up to 25 mi. haul dist. (50 mi. RT for mob/demob crew)									
1200	Small equipment, placed in rear of, or towed by pickup truck	A-3A	4	2	Ea.		106	44	150	207
1300	Equipment hauled on 3-ton capacity towed trailer	A-3Q	2.67	3			159	92.50	251.50	340
1400	20-ton capacity	B-34U	2	8			410	221	631	855
1500	40-ton capacity	B-34N	2	8			425	340	765	1,000
1600	50-ton capacity	B-34V	1	24			1,300	985	2,285	3,000

For customer support on your Site Work & Landscape Costs with RSMeans data, call 800.448.8182.

01 54 Construction Aids

01 54 36 – Equipment Mobilization

	01 54 36.50 Mobilization	Crew	Daily Output	Labor-Hours	Unit	Material	2020 Bare Costs Labor	Equipment	Total	Total Incl O&P
1700	Crane, truck-mounted, up to 75 ton (driver only)	1 Eqhv	4	2	Ea.		118		118	177
1800	Over 75 ton (with chase vehicle)	A-3E	2.50	6.400			345	70.50	415.50	600
2400	Crane, large lattice boom, requiring assembly	B-34W	.50	144	↓		7,425	7,025	14,450	18,800
2500	For each additional 5 miles haul distance, add						10%	10%		
3000	For large pieces of equipment, allow for assembly/knockdown									
3001	For mob/demob of vibroflotation equip, see Section 31 45 13.10									
3100	For mob/demob of micro-tunneling equip, see Section 33 05 07.36									
3200	For mob/demob of pile driving equip, see Section 31 62 19.10									
3300	For mob/demob of caisson drilling equip, see Section 31 63 26.13									

01 55 Vehicular Access and Parking

01 55 23 – Temporary Roads

	01 55 23.50 Roads and Sidewalks	Crew	Daily Output	Labor-Hours	Unit	Material	2020 Bare Costs Labor	Equipment	Total	Total Incl O&P
0010	**ROADS AND SIDEWALKS** Temporary									
0050	Roads, gravel fill, no surfacing, 4" gravel depth	B-14	715	.067	S.Y.	3.38	2.97	.30	6.65	8.50
0100	8" gravel depth	"	615	.078	"	6.75	3.45	.35	10.55	13.05
1000	Ramp, 3/4" plywood on 2" x 6" joists, 16" OC	2 Carp	300	.053	S.F.	1.62	2.83		4.45	6.05
1100	On 2" x 10" joists, 16" OC	"	275	.058	"	2.25	3.09		5.34	7.10

01 56 Temporary Barriers and Enclosures

01 56 13 – Temporary Air Barriers

	01 56 13.60 Tarpaulins	Crew	Daily Output	Labor-Hours	Unit	Material	2020 Bare Costs Labor	Equipment	Total	Total Incl O&P
0010	**TARPAULINS**									
0020	Cotton duck, 10-13.13 oz./S.Y., 6' x 8'				S.F.	.75			.75	.83
0050	30' x 30'					.63			.63	.69
0100	Polyvinyl coated nylon, 14-18 oz., minimum					1.49			1.49	1.64
0150	Maximum					1.49			1.49	1.64
0200	Reinforced polyethylene 3 mils thick, white					.05			.05	.06
0300	4 mils thick, white, clear or black					.16			.16	.18
0400	5.5 mils thick, clear					.19			.19	.21
0500	White, fire retardant					.60			.60	.66
0600	12 mils, oil resistant, fire retardant					.50			.50	.55
0700	8.5 mils, black					.26			.26	.29
0710	Woven polyethylene, 6 mils thick					.08			.08	.09
0730	Polyester reinforced w/integral fastening system, 11 mils thick					.20			.20	.22
0740	Polyethylene, reflective, 23 mils thick					1.34			1.34	1.47

	01 56 13.90 Winter Protection	Crew	Daily Output	Labor-Hours	Unit	Material	2020 Bare Costs Labor	Equipment	Total	Total Incl O&P
0010	**WINTER PROTECTION**									
0100	Framing to close openings	2 Clab	500	.032	S.F.	.51	1.35		1.86	2.58
0200	Tarpaulins hung over scaffolding, 8 uses, not incl. scaffolding		1500	.011		.25	.45		.70	.95
0250	Tarpaulin polyester reinf. w/integral fastening system, 11 mils thick		1600	.010		.22	.42		.64	.87
0300	Prefab fiberglass panels, steel frame, 8 uses	↓	1200	.013	↓	2.78	.56		3.34	3.90

01 56 16 – Temporary Dust Barriers

	01 56 16.10 Dust Barriers, Temporary	Crew	Daily Output	Labor-Hours	Unit	Material	2020 Bare Costs Labor	Equipment	Total	Total Incl O&P
0010	**DUST BARRIERS, TEMPORARY**, erect and dismantle									
0020	Spring loaded telescoping pole & head, to 12', erect and dismantle	1 Clab	240	.033	Ea.		1.40		1.40	2.11
0025	Cost per day (based upon 250 days)				Day	.22			.22	.24
0030	To 21', erect and dismantle	1 Clab	240	.033	Ea.		1.40		1.40	2.11

01 56 Temporary Barriers and Enclosures

01 56 16 – Temporary Dust Barriers

01 56 16.10 Dust Barriers, Temporary

		Crew	Daily Output	Labor-Hours	Unit	Material	2020 Bare Costs Labor	Equipment	Total	Total Incl O&P
0035	Cost per day (based upon 250 days)				Day	.72			.72	.79
0040	Accessories, caution tape reel, erect and dismantle	1 Clab	480	.017	Ea.		.70		.70	1.05
0045	Cost per day (based upon 250 days)				Day	.33			.33	.37
0060	Foam rail and connector, erect and dismantle	1 Clab	240	.033	Ea.		1.40		1.40	2.11
0065	Cost per day (based upon 250 days)				Day	.10			.10	.11
0070	Caution tape	1 Clab	384	.021	C.L.F.	2.49	.88		3.37	4.06
0080	Zipper, standard duty		60	.133	Ea.	10.55	5.60		16.15	20
0090	Heavy duty		48	.167	"	10.65	7		17.65	22.50
0100	Polyethylene sheet, 4 mil		37	.216	Sq.	2.62	9.10		11.72	16.60
0110	6 mil		37	.216	"	4	9.10		13.10	18.10
1000	Dust partition, 6 mil polyethylene, 1" x 3" frame	2 Carp	2000	.008	S.F.	.33	.43		.76	1
1080	2" x 4" frame		2000	.008	"	.39	.43		.82	1.07
4000	Dust & infectious control partition, adj. to 10' high, obscured, 4' panel		90	.178	Ea.	580	9.45		589.45	655
4010	3' panel		90	.178		550	9.45		559.45	620
4020	2' panel		90	.178		430	9.45		439.45	490
4030	1' panel	1 Carp	90	.089		285	4.72		289.72	320
4040	6" panel	"	90	.089		265	4.72		269.72	299
4050	2' panel with HEPA filtered discharge port	2 Carp	90	.178		500	9.45		509.45	565
4060	3' panel with 32" door		90	.178		895	9.45		904.45	1,000
4070	4' panel with 36" door		90	.178		995	9.45		1,004.45	1,125
4080	4'-6" panel with 44" door		90	.178		1,200	9.45		1,209.45	1,350
4090	Hinged corner		80	.200		185	10.65		195.65	220
4100	Outside corner		80	.200		150	10.65		160.65	181
4110	T post		80	.200		150	10.65		160.65	181
4120	Accessories, ceiling grid clip		360	.044		7.45	2.36		9.81	11.75
4130	Panel locking clip		360	.044		5.15	2.36		7.51	9.20
4140	Panel joint closure strip		360	.044		8	2.36		10.36	12.35
4150	Screw jack		360	.044		6.65	2.36		9.01	10.85
4160	Digital pressure difference guage					275			275	305
4180	Combination lockset	1 Carp	13	.615		200	32.50		232.50	269
4185	Sealant tape, 2" wide	1 Clab	192	.042	C.L.F.	6.75	1.75		8.50	10.10
4190	System in place, including door and accessories									
4200	Based upon 25 uses	2 Carp	51	.314	L.F.	9.85	16.65		26.50	36
4210	Based upon 50 uses		51	.314		4.94	16.65		21.59	30.50
4230	Based upon 100 uses		51	.314		2.47	16.65		19.12	27.50

01 56 23 – Temporary Barricades

01 56 23.10 Barricades

		Crew	Daily Output	Labor-Hours	Unit	Material	Labor	Equipment	Total	Total Incl O&P
0010	**BARRICADES**									
0020	5' high, 3 rail @ 2" x 8", fixed	2 Carp	20	.800	L.F.	6.75	42.50		49.25	71.50
0150	Movable	"	30	.533	"	5.60	28.50		34.10	48.50
0300	Stock units, 58' high, 8' wide, reflective, buy				Ea.	211			211	232
0350	With reflective tape, buy				"	425			425	465
0400	Break-a-way 3" PVC pipe barricade									
0410	with 3 ea. 1' x 4' reflectorized panels, buy				Ea.	153			153	168
0500	Barricades, plastic, 8" x 24" wide, foldable					61			61	67
0800	Traffic cones, PVC, 18" high					10.95			10.95	12.05
0850	28" high					23.50			23.50	25.50
1000	Guardrail, wooden, 3' high, 1" x 6" on 2" x 4" posts	2 Carp	200	.080	L.F.	1.44	4.25		5.69	8
1100	2" x 6" on 4" x 4" posts	"	165	.097		2.80	5.15		7.95	10.85
1200	Portable metal with base pads, buy					14.45			14.45	15.90
1250	Typical installation, assume 10 reuses	2 Carp	600	.027		2.31	1.42		3.73	4.67
1300	Barricade tape, polyethylene, 7 mil, 3" wide x 300' long roll	1 Clab	128	.063	Ea.	7.50	2.63		10.13	12.20

22

01 56 Temporary Barriers and Enclosures

01 56 23 – Temporary Barricades

01 56 23.10 Barricades		Crew	Daily Output	Labor-Hours	Unit	Material	2020 Bare Costs Labor	Equipment	Total	Total Incl O&P
3000	Detour signs, set up and remove									
3010	Reflective aluminum, MUTCD, 24" x 24", post mounted	1 Clab	20	.400	Ea.	2.80	16.85		19.65	28.50

01 56 26 – Temporary Fencing

01 56 26.50 Temporary Fencing

		Crew	Daily Output	Labor-Hours	Unit	Material	Labor	Equipment	Total	Total Incl O&P
0010	TEMPORARY FENCING									
0020	Chain link, 11 ga., 4' high	2 Clab	400	.040	L.F.	2	1.68		3.68	4.73
0100	6' high		300	.053		4.96	2.25		7.21	8.80
0200	Rented chain link, 6' high, to 1000' (up to 12 mo.)		400	.040		3.37	1.68		5.05	6.25
0250	Over 1000' (up to 12 mo.)		300	.053		4.14	2.25		6.39	7.90
0350	Plywood, painted, 2" x 4" frame, 4' high	A-4	135	.178		6.90	8.95		15.85	21
0400	4" x 4" frame, 8' high	"	110	.218		13.45	10.95		24.40	31
0500	Wire mesh on 4" x 4" posts, 4' high	2 Carp	100	.160		11.15	8.50		19.65	25
0550	8' high	"	80	.200		16.60	10.65		27.25	34
0600	Plastic safety fence, light duty, 4' high, posts at 10'	B-1	500	.048		1.59	2.05		3.64	4.83
0610	Medium duty, 4' high, posts at 10'		500	.048		1.90	2.05		3.95	5.15
0620	Heavy duty, 4' high, posts at 10'		500	.048		2.51	2.05		4.56	5.85
0630	Reflective heavy duty, 4' high, posts at 10'		500	.048		4.96	2.05		7.01	8.55

01 56 29 – Temporary Protective Walkways

01 56 29.50 Protection

		Crew	Daily Output	Labor-Hours	Unit	Material	Labor	Equipment	Total	Total Incl O&P
0010	PROTECTION									
0020	Stair tread, 2" x 12" planks, 1 use	1 Carp	75	.107	Tread	5.15	5.65		10.80	14.20
0100	Exterior plywood, 1/2" thick, 1 use		65	.123		2.05	6.55		8.60	12.10
0200	3/4" thick, 1 use		60	.133		2.84	7.10		9.94	13.75
2200	Sidewalks, 2" x 12" planks, 2 uses		350	.023	S.F.	.86	1.22		2.08	2.78
2300	Exterior plywood, 2 uses, 1/2" thick		750	.011		.34	.57		.91	1.23
2400	5/8" thick		650	.012		.38	.65		1.03	1.40
2500	3/4" thick		600	.013		.47	.71		1.18	1.58

01 57 Temporary Controls

01 57 33 – Temporary Security

01 57 33.50 Watchman

		Crew	Daily Output	Labor-Hours	Unit	Material	Labor	Equipment	Total	Total Incl O&P
0010	WATCHMAN									
0020	Service, monthly basis, uniformed person, minimum				Hr.				27.65	30.40
0100	Maximum								56	61.62
0200	Person and command dog, minimum								31	34
0300	Maximum								60	65

01 58 Project Identification

01 58 13 – Temporary Project Signage

01 58 13.50 Signs

		Crew	Daily Output	Labor-Hours	Unit	Material	Labor	Equipment	Total	Total Incl O&P
0010	SIGNS									
0020	High intensity reflectorized, no posts, buy				Ea.	27			27	29.50

For customer support on your Site Work & Landscape Costs with RSMeans data, call 800.448.8182.

01 66 Product Storage and Handling Requirements

01 66 19 – Material Handling

01 66 19.10 Material Handling	Crew	Daily Output	Labor-Hours	Unit	Material	2020 Bare Costs Labor	2020 Bare Costs Equipment	Total	Total Incl O&P
0010 **MATERIAL HANDLING**									
0020 Above 2nd story, via stairs, per C.Y. of material per floor	2 Clab	145	.110	C.Y.		4.65		4.65	7
0030 Via elevator, per C.Y. of material		240	.067			2.81		2.81	4.22
0050 Distances greater than 200', per C.Y. of material per each addl 200'		300	.053			2.25		2.25	3.37

01 71 Examination and Preparation

01 71 23 – Field Engineering

01 71 23.13 Construction Layout

	Crew	Daily Output	Labor-Hours	Unit	Material	2020 Bare Costs Labor	2020 Bare Costs Equipment	Total	Total Incl O&P
0010 **CONSTRUCTION LAYOUT**									
1100 Crew for layout of building, trenching or pipe laying, 2 person crew	A-6	1	16	Day		845	31	876	1,300
1200 3 person crew	A-7	1	24			1,375	30.50	1,405.50	2,100
1400 Crew for roadway layout, 4 person crew	A-8	1	32			1,775	30.50	1,805.50	2,700

01 71 23.19 Surveyor Stakes

	Crew	Daily Output	Labor-Hours	Unit	Material	2020 Bare Costs Labor	2020 Bare Costs Equipment	Total	Total Incl O&P
0010 **SURVEYOR STAKES**									
0020 Hardwood, 1" x 1" x 48" long				C	71			71	78
0100 2" x 2" x 18" long					79			79	87
0150 2" x 2" x 24" long					155			155	170

01 74 Cleaning and Waste Management

01 74 13 – Progress Cleaning

01 74 13.20 Cleaning Up

	Crew	Daily Output	Labor-Hours	Unit	Material	2020 Bare Costs Labor	2020 Bare Costs Equipment	Total	Total Incl O&P
0010 **CLEANING UP**									
0020 After job completion, allow, minimum				Job				.30%	.30%
0040 Maximum				"				1%	1%
0200 Rubbish removal, see Section 02 41 19.19									

01 76 Protecting Installed Construction

01 76 13 – Temporary Protection of Installed Construction

01 76 13.20 Temporary Protection

	Crew	Daily Output	Labor-Hours	Unit	Material	2020 Bare Costs Labor	2020 Bare Costs Equipment	Total	Total Incl O&P
0010 **TEMPORARY PROTECTION**									
0020 Flooring, 1/8" tempered hardboard, taped seams	2 Carp	1500	.011	S.F.	.50	.57		1.07	1.40
0030 Peel away carpet protection	1 Clab	3200	.003	"	.13	.11		.24	.30

01 91 Commissioning

01 91 13 – General Commissioning Requirements

01 91 13.50 Building Commissioning

	Crew	Daily Output	Labor-Hours	Unit	Material	2020 Bare Costs Labor	2020 Bare Costs Equipment	Total	Total Incl O&P
0010 **BUILDING COMMISSIONING**									
0100 Systems operation and verification during turnover				%				.25%	.25%
0150 Including all systems subcontractors								.50%	.50%
0200 Systems design assistance, operation, verification and training								.50%	.50%
0250 Including all systems subcontractors								1%	1%

For customer support on your Site Work & Landscape Costs with RSMeans data, call 800.448.8182.

Estimating Tips

02 30 00 Subsurface Investigation

In preparing estimates on structures involving earthwork or foundations, all information concerning soil characteristics should be obtained. Look particularly for hazardous waste, evidence of prior dumping of debris, and previous stream beds.

02 40 00 Demolition and Structure Moving

The costs shown for selective demolition do not include rubbish handling or disposal. These items should be estimated separately using RSMeans data or other sources.

- Historic preservation often requires that the contractor remove materials from the existing structure, rehab them, and replace them. The estimator must be aware of any related measures and precautions that must be taken when doing selective demolition and cutting and patching. Requirements may include special handling and storage, as well as security.

- In addition to Subdivision 02 41 00, you can find selective demolition items in each division. Example: Roofing demolition is in Division 7.
- Absent of any other specific reference, an approximate demolish-in-place cost can be obtained by halving the new-install labor cost. To remove for reuse, allow the entire new-install labor figure.

02 40 00 Building Deconstruction

This section provides costs for the careful dismantling and recycling of most low-rise building materials.

02 50 00 Containment of Hazardous Waste

This section addresses on-site hazardous waste disposal costs.

02 80 00 Hazardous Material Disposal/Remediation

This subdivision includes information on hazardous waste handling, asbestos remediation, lead remediation, and mold remediation. See reference numbers

R028213-20 and R028319-60 for further guidance in using these unit price lines.

02 90 00 Monitoring Chemical Sampling, Testing Analysis

This section provides costs for on-site sampling and testing hazardous waste.

Reference Numbers

Reference numbers are shown at the beginning of some major classifications. These numbers refer to related items in the Reference Section. The reference information may be an estimating procedure, an alternate pricing method, or technical information.

Note: Not all subdivisions listed here necessarily appear. ■

Same Data. Simplified.

Enjoy the convenience and efficiency of accessing your costs anywhere:

- **Skip the multiplier** by setting your location
- **Quickly search,** edit, favorite and share costs
- **Stay on top of price changes** with automatic updates

Discover more at rsmeans.com/online

02 21 Surveys

02 21 13 – Site Surveys

02 21 13.09 Topographical Surveys	Crew	Daily Output	Labor-Hours	Unit	Material	2020 Bare Costs Labor	2020 Bare Costs Equipment	Total	Total Incl O&P
0010 **TOPOGRAPHICAL SURVEYS**									
0020 Topographical surveying, conventional, minimum	A-7	3.30	7.273	Acre	23.50	415	9.30	447.80	660
0100 Maximum	A-8	.60	53.333	"	63	2,975	51	3,089	4,600

02 21 13.13 Boundary and Survey Markers

	Crew	Daily Output	Labor-Hours	Unit	Material	Labor	Equipment	Total	Total Incl O&P
0010 **BOUNDARY AND SURVEY MARKERS**									
0300 Lot location and lines, large quantities, minimum	A-7	2	12	Acre	36	690	15.35	741.35	1,075
0320 Average	"	1.25	19.200		63.50	1,100	24.50	1,188	1,750
0400 Small quantities, maximum	A-8	1	32	↓	76.50	1,775	30.50	1,882	2,800
0600 Monuments, 3' long	A-7	10	2.400	Ea.	35.50	138	3.07	176.57	249
0800 Property lines, perimeter, cleared land	"	1000	.024	L.F.	.08	1.38	.03	1.49	2.19
0900 Wooded land	A-8	875	.037	"	.10	2.04	.04	2.18	3.21

02 21 13.16 Aerial Surveys

	Crew	Daily Output	Labor-Hours	Unit	Material	Labor	Equipment	Total	Total Incl O&P
0010 **AERIAL SURVEYS**									
1500 Aerial surveying, including ground control, minimum fee, 10 acres				Total				4,700	4,700
1510 100 acres								9,400	9,400
1550 From existing photography, deduct				↓				1,625	1,625
1600 2' contours, 10 acres				Acre				470	470
1850 100 acres								94	94
2000 1000 acres								90	90
2050 10,000 acres				↓				85	85

02 32 Geotechnical Investigations

02 32 13 – Subsurface Drilling and Sampling

02 32 13.10 Boring and Exploratory Drilling

	Crew	Daily Output	Labor-Hours	Unit	Material	Labor	Equipment	Total	Total Incl O&P
0010 **BORING AND EXPLORATORY DRILLING**									
0020 Borings, initial field stake out & determination of elevations	A-6	1	16	Day		845	31	876	1,300
0100 Drawings showing boring details				Total		335		335	425
0200 Report and recommendations from P.E.						775		775	970
0300 Mobilization and demobilization	B-55	4	6	↓		263	305	568	730
0350 For over 100 miles, per added mile		450	.053	Mile		2.34	2.69	5.03	6.45
0600 Auger holes in earth, no samples, 2-1/2" diameter		78.60	.305	L.F.		13.40	15.40	28.80	37
0650 4" diameter		67.50	.356			15.60	17.95	33.55	43.50
0800 Cased borings in earth, with samples, 2-1/2" diameter		55.50	.432		20.50	18.95	22	61.45	75
0850 4" diameter	↓	32.60	.736		29	32.50	37	98.50	122
1000 Drilling in rock, "BX" core, no sampling	B-56	34.90	.458			22	43.50	65.50	80.50
1050 With casing & sampling		31.70	.505		20.50	24	48	92.50	111
1200 "NX" core, no sampling		25.92	.617			29.50	58.50	88	109
1250 With casing and sampling	↓	25	.640	↓	21	30.50	60.50	112	136
1400 Borings, earth, drill rig and crew with truck mounted auger	B-55	1	24	Day		1,050	1,200	2,250	2,900
1450 Rock using crawler type drill	B-56	1	16	"		760	1,525	2,285	2,825
1500 For inner city borings, add, minimum								10%	10%
1510 Maximum								20%	20%

02 32 19 – Exploratory Excavations

02 32 19.10 Test Pits

	Crew	Daily Output	Labor-Hours	Unit	Material	Labor	Equipment	Total	Total Incl O&P
0010 **TEST PITS**									
0020 Hand digging, light soil	1 Clab	4.50	1.778	C.Y.		75		75	112
0100 Heavy soil	"	2.50	3.200			135		135	202
0120 Loader-backhoe, light soil	B-11M	28	.571			28.50	8.30	36.80	51.50
0130 Heavy soil	"	20	.800	↓		39.50	11.60	51.10	72
1000 Subsurface exploration, mobilization				Mile				6.75	8.40

02 32 Geotechnical Investigations

02 32 19 – Exploratory Excavations

02 32 19.10 Test Pits		Crew	Daily Output	Labor-Hours	Unit	Material	2020 Bare Costs Labor	Equipment	Total	Total Incl O&P
1010	Difficult access for rig, add				Hr.				260	320
1020	Auger borings, drill rig, incl. samples				L.F.				26.50	33
1030	Hand auger								31.50	40
1050	Drill and sample every 5', split spoon				↓				31.50	40
1060	Extra samples				Ea.				36	45.50

02 41 Demolition

02 41 13 – Selective Site Demolition

02 41 13.15 Hydrodemolition

		Crew	Daily Output	Labor-Hours	Unit	Material	Labor	Equipment	Total	Total Incl O&P
0010	**HYDRODEMOLITION**									
0015	Hydrodemolition, concrete pavement									
0120	Includes removal but exlcudes disposal of concrete									
0130	2" depth	B-5E	1000	.064	S.F.		3	2.75	5.75	7.55
0410	4" depth		800	.080			3.75	3.44	7.19	9.40
0420	6" depth		600	.107			5	4.59	9.59	12.55
0520	8" depth	↓	300	.213	↓		10	9.20	19.20	25

02 41 13.16 Hydroexcavation

		Crew	Daily Output	Labor-Hours	Unit	Material	Labor	Equipment	Total	Total Incl O&P
0010	**HYDROEXCAVATION**									
0015	Hydroexcavation									
0110	Mobilization or Demobilization	B-6D	8	2.500	Ea.		121	158	279	355
0130	Normal Conditions		48	.417	B.C.Y.		20	26.50	46.50	59
0140	Adverse Conditions		30	.667	"		32	42	74	95
0160	Minimum labor/equipment charge	↓	4	5	Ea.		242	315	557	705

02 41 13.17 Demolish, Remove Pavement and Curb

		Crew	Daily Output	Labor-Hours	Unit	Material	Labor	Equipment	Total	Total Incl O&P
0010	**DEMOLISH, REMOVE PAVEMENT AND CURB** R024119-10									
5010	Pavement removal, bituminous roads, up to 3" thick	B-38	690	.058	S.Y.		2.76	1.80	4.56	6.10
5050	4"-6" thick		420	.095			4.53	2.96	7.49	10.05
5100	Bituminous driveways		640	.063			2.98	1.94	4.92	6.60
5200	Concrete to 6" thick, hydraulic hammer, mesh reinforced		255	.157			7.45	4.87	12.32	16.55
5300	Rod reinforced		200	.200	↓		9.50	6.20	15.70	21
5400	Concrete, 7"-24" thick, plain		33	1.212	C.Y.		57.50	37.50	95	128
5500	Reinforced		24	1.667	"		79.50	51.50	131	176
5600	With hand held air equipment, bituminous, to 6" thick	B-39	1900	.025	S.F.		1.12	.19	1.31	1.89
5700	Concrete to 6" thick, no reinforcing		1600	.030			1.33	.22	1.55	2.24
5800	Mesh reinforced		1400	.034			1.52	.26	1.78	2.56
5900	Rod reinforced	↓	765	.063	↓		2.78	.47	3.25	4.68
6000	Curbs, concrete, plain	B-6	360	.067	L.F.		3.05	.59	3.64	5.20
6100	Reinforced		275	.087			3.99	.78	4.77	6.85
6200	Granite		360	.067			3.05	.59	3.64	5.20
6300	Bituminous		528	.045			2.08	.41	2.49	3.57
6400	Wood		570	.042			1.93	.38	2.31	3.30
6500	Site demo, berms under 4" in height, bituminous		528	.045			2.08	.41	2.49	3.57
6600	4" or over in height	↓	300	.080	↓		3.66	.71	4.37	6.30

02 41 13.20 Selective Demo, Highway Guard Rails & Barriers

		Crew	Daily Output	Labor-Hours	Unit	Material	Labor	Equipment	Total	Total Incl O&P
0010	**SELECTIVE DEMOLITION, HIGHWAY GUARD RAILS & BARRIERS**									
0100	Guard rail, corrugated steel	B-6	600	.040	L.F.		1.83	.36	2.19	3.13
0200	End sections		40	.600	Ea.		27.50	5.35	32.85	47
0300	Wrap around		40	.600	"		27.50	5.35	32.85	47
0400	Timber 4" x 8"		600	.040	L.F.		1.83	.36	2.19	3.13
0500	Three 3/4" cables	↓	600	.040	"		1.83	.36	2.19	3.13

27

02 41 13 – Selective Site Demolition

02 41 13.20 Selective Demo, Highway Guard Rails & Barriers	Crew	Daily Output	Labor-Hours	Unit	Material	2020 Bare Costs Labor	2020 Bare Costs Equipment	Total	Total Incl O&P
0600 Wood posts	B-6	240	.100	Ea.		4.57	.89	5.46	7.85
0700 Guide rail, 6" x 6" box beam	B-80B	120	.267	L.F.		11.95	1.97	13.92	20
0800 Median barrier, 6" x 8" box beam		240	.133			6	.98	6.98	10.05
0850 Precast concrete 3'-6" high x 2' wide		300	.107			4.78	.79	5.57	8
0900 Impact barrier, MUTCD, barrel type	B-16	60	.533	Ea.		23.50	9.55	33.05	46
1000 Resilient guide fence and light shield 6' high	"	120	.267	L.F.		11.80	4.77	16.57	23
1100 Concrete posts, 6'-5" triangular	B-6	200	.120	Ea.		5.50	1.07	6.57	9.45
1200 Speed bumps 10-1/2" x 2-1/4" x 48"		300	.080	L.F.		3.66	.71	4.37	6.30
1300 Pavement marking channelizing		200	.120	Ea.		5.50	1.07	6.57	9.45
1400 Barrier and curb delineators		300	.080			3.66	.71	4.37	6.30
1500 Rumble strips 24" x 3-1/2" x 1/2"		150	.160			7.30	1.43	8.73	12.50

02 41 13.23 Utility Line Removal

	Crew	Daily Output	Labor-Hours	Unit	Material	2020 Bare Costs Labor	2020 Bare Costs Equipment	Total	Total Incl O&P
0010 **UTILITY LINE REMOVAL**									
0015 No hauling, abandon catch basin or manhole	B-6	7	3.429	Ea.		157	30.50	187.50	269
0020 Remove existing catch basin or manhole, masonry		4	6			274	53.50	327.50	470
0030 Catch basin or manhole frames and covers, stored		13	1.846			84.50	16.45	100.95	145
0040 Remove and reset		7	3.429			157	30.50	187.50	269
0900 Hydrants, fire, remove only	B-21A	5	8			420	85.50	505.50	720
0950 Remove and reset	"	2	20			1,050	214	1,264	1,800
2900 Pipe removal, sewer/water, no excavation, 12" diameter	B-6	175	.137	L.F.		6.25	1.22	7.47	10.75
2930 15"-18" diameter	B-12Z	150	.160			7.65	9.75	17.40	22
2960 21"-24" diameter		120	.200			9.55	12.20	21.75	28
3000 27"-36" diameter		90	.267			12.75	16.30	29.05	37
3200 Steel, welded connections, 4" diameter	B-6	160	.150			6.85	1.34	8.19	11.75
3300 10" diameter	"	80	.300			13.70	2.67	16.37	23.50

02 41 13.30 Minor Site Demolition

	Crew	Daily Output	Labor-Hours	Unit	Material	2020 Bare Costs Labor	2020 Bare Costs Equipment	Total	Total Incl O&P
0010 **MINOR SITE DEMOLITION** R024119-10									
0100 Roadside delineators, remove only	B-80	175	.183	Ea.		8.50	5.85	14.35	19.20
0110 Remove and reset	"	100	.320	"		14.90	10.20	25.10	34
0800 Guiderail, corrugated steel, remove only	B-80A	100	.240	L.F.		10.10	8.20	18.30	24
0850 Remove and reset	"	40	.600	"		25.50	20.50	46	60.50
0860 Guide posts, remove only	B-80B	120	.267	Ea.		11.95	1.97	13.92	20
0870 Remove and reset	B-55	50	.480	"		21	24	45	58
1000 Masonry walls, block, solid	B-5	1800	.031	C.F.		1.45	.85	2.30	3.10
1200 Brick, solid		900	.062			2.90	1.70	4.60	6.20
1400 Stone, with mortar		900	.062			2.90	1.70	4.60	6.20
1500 Dry set		1500	.037			1.74	1.02	2.76	3.73
1600 Median barrier, precast concrete, remove and store	B-3	430	.112	L.F.		5.25	5.40	10.65	13.80
1610 Remove and reset	"	390	.123	"		5.80	5.95	11.75	15.25
4000 Sidewalk removal, bituminous, 2" thick	B-6	350	.069	S.Y.		3.14	.61	3.75	5.35
4010 2-1/2" thick		325	.074			3.38	.66	4.04	5.75
4050 Brick, set in mortar		185	.130			5.95	1.16	7.11	10.15
4060 Dry set		270	.089			4.06	.79	4.85	6.95
4100 Concrete, plain, 4"		160	.150			6.85	1.34	8.19	11.75
4110 Plain, 5"		140	.171			7.85	1.53	9.38	13.45
4120 Plain, 6"		120	.200			9.15	1.78	10.93	15.65
4200 Mesh reinforced, concrete, 4"		150	.160			7.30	1.43	8.73	12.50
4210 5" thick		131	.183			8.40	1.63	10.03	14.35
4220 6" thick		112	.214			9.80	1.91	11.71	16.80
4300 Slab on grade removal, plain	B-5	45	1.244	C.Y.		58	34	92	125
4310 Mesh reinforced		33	1.697			79	46	125	170
4320 Rod reinforced		25	2.240			104	61	165	223

02 41 Demolition

02 41 13 – Selective Site Demolition

		Daily Output	Labor-Hours	Unit	Material	2020 Bare Costs		Total	Total Incl O&P
	02 41 13.30 Minor Site Demolition	Crew				Labor	Equipment		
4400	For congested sites or small quantities, add up to			C.Y.				200%	200%
4450	For disposal on site, add	B-11A	232	.069			3.41	6.50	9.91	12.25
4500	To 5 miles, add	B-34D	76	.105			5.15	8.50	13.65	17.05

02 41 13.33 Railtrack Removal

		Crew	Daily Output	Labor-Hours	Unit	Material	Labor	Equipment	Total	Total Incl O&P
0010	**RAILTRACK REMOVAL**									
3500	Railroad track removal, ties and track	B-13	330	.170	L.F.		7.80	1.76	9.56	13.65
3600	Ballast	B-14	500	.096	C.Y.		4.25	.43	4.68	6.85
3700	Remove and re-install, ties & track using new bolts & spikes		50	.960	L.F.		42.50	4.27	46.77	68.50
3800	Turnouts using new bolts and spikes		1	48	Ea.		2,125	214	2,339	3,425

02 41 13.34 Selective Demolition, Utility Materials

		Crew	Daily Output	Labor-Hours	Unit	Material	Labor	Equipment	Total	Total Incl O&P
0010	**SELECTIVE DEMOLITION, UTILITY MATERIALS** R024119-10									
0015	Excludes excavation									
0020	See other utility items in Section 02 41 13.33									
0100	Fire hydrant extensions	B-20	14	1.714	Ea.		80.50		80.50	121
0200	Precast utility boxes up to 8' x 14' x 7'	B-13	2	28			1,300	290	1,590	2,250
0300	Handholes and meter pits	B-6	2	12			550	107	657	945
0400	Utility valves 4"-12"	B-20	4	6			282		282	425
0500	14"-24"	B-21	2	14			685	94.50	779.50	1,125

02 41 13.36 Selective Demolition, Utility Valves and Accessories

		Crew	Daily Output	Labor-Hours	Unit	Material	Labor	Equipment	Total	Total Incl O&P
0010	**SELECTIVE DEMOLITION, UTILITY VALVES & ACCESSORIES**									
0015	Excludes excavation									
0100	Utility valves 4"-12" diam.	B-20	4	6	Ea.		282		282	425
0200	14"-24" diam.	B-21	2	14			685	94.50	779.50	1,125
0300	Crosses 4"-12" diam.	B-20	8	3			141		141	212
0400	14"-24" diam.	B-21	4	7			340	47.50	387.50	565
0500	Utility cut-in valves 4"-12" diam.	B-20	20	1.200			56.50		56.50	85
0600	Curb boxes	"	20	1.200			56.50		56.50	85

02 41 13.38 Selective Demo., Water & Sewer Piping & Fittings

		Crew	Daily Output	Labor-Hours	Unit	Material	Labor	Equipment	Total	Total Incl O&P
0010	**SELECTIVE DEMOLITION, WATER & SEWER PIPING AND FITTINGS**									
0015	Excludes excavation									
0090	Concrete pipe 4"-10" diameter	B-6	250	.096	L.F.		4.39	.86	5.25	7.55
0100	42"-48" diameter	B-13B	96	.583			27	10.20	37.20	52
0200	60"-84" diameter	"	80	.700			32	12.25	44.25	62
0300	96" diameter	B-13C	80	.700			32	28.50	60.50	80
0400	108"-144" diameter	"	64	.875			40.50	35.50	76	100
0450	Concrete fittings 12" diameter	B-6	24	1	Ea.		45.50	8.90	54.40	78.50
0480	Concrete end pieces 12" diameter		200	.120	L.F.		5.50	1.07	6.57	9.45
0485	15" diameter		150	.160			7.30	1.43	8.73	12.50
0490	18" diameter		150	.160			7.30	1.43	8.73	12.50
0500	24"-36" diameter		100	.240			11	2.14	13.14	18.80
0600	Concrete fittings 24"-36" diameter		12	2	Ea.		91.50	17.80	109.30	157
0700	48"-84" diameter	B-13B	12	4.667			215	81.50	296.50	410
0800	96" diameter	"	8	7			320	123	443	620
0900	108"-144" diameter	B-13C	4	14			645	570	1,215	1,600
1000	Ductile iron pipe 4" diameter	B-21B	200	.200	L.F.		9.20	2.36	11.56	16.40
1100	6"-12" diameter		175	.229			10.50	2.69	13.19	18.70
1200	14"-24" diameter		120	.333			15.30	3.93	19.23	27.50
1300	Ductile iron fittings 4"-12" diameter		24	1.667	Ea.		76.50	19.65	96.15	137
1400	14"-16" diameter		18	2.222			102	26	128	182
1500	18"-24" diameter		12	3.333			153	39.50	192.50	273
1600	Plastic pipe 3/4"-4" diameter	B-6	700	.034	L.F.		1.57	.31	1.88	2.69
1700	6"-8" diameter		500	.048			2.20	.43	2.63	3.76

02 41 Demolition

02 41 13 – Selective Site Demolition

02 41 13.38 Selective Demo., Water & Sewer Piping & Fittings	Crew	Daily Output	Labor-Hours	Unit	Material	2020 Bare Costs Labor	Equipment	Total	Total Incl O&P
1800	10"-18" diameter	B-6	300	.080	L.F.		3.66	.71	4.37	6.30
1900	20"-36" diameter		200	.120			5.50	1.07	6.57	9.45
1910	42"-48" diameter		180	.133			6.10	1.19	7.29	10.45
1920	54"-60" diameter		160	.150			6.85	1.34	8.19	11.75
2000	Plastic fittings 4"-8" diameter		75	.320	Ea.		14.65	2.85	17.50	25
2100	10"-14" diameter		50	.480			22	4.28	26.28	37.50
2200	16"-24" diameter		20	1.200			55	10.70	65.70	94.50
2210	30"-36" diameter		15	1.600			73	14.25	87.25	126
2220	42"-48" diameter		12	2			91.50	17.80	109.30	157
2300	Copper pipe 3/4"-2" diameter	Q-1	500	.032	L.F.		1.86		1.86	2.78
2400	2-1/2"-3" diameter		300	.053			3.09		3.09	4.63
2500	4"-6" diameter		200	.080			4.64		4.64	6.95
2600	Copper fittings 3/4"-2" diameter		15	1.067	Ea.		62		62	92.50
2700	Cast iron pipe 4" diameter		200	.080	L.F.		4.64		4.64	6.95
2800	5"-6" diameter	Q-2	200	.120			7.20		7.20	10.80
2900	8"-12" diameter	Q-3	200	.160			9.80		9.80	14.70
3000	Cast iron fittings 4" diameter	Q-1	30	.533	Ea.		31		31	46.50
3100	5"-6" diameter	Q-2	30	.800			48		48	72
3200	8"-15" diameter	Q-3	30	1.067			65.50		65.50	98
3300	Vent cast iron pipe 4"-8" diameter	Q-1	200	.080	L.F.		4.64		4.64	6.95
3400	10"-15" diameter	Q-3	200	.160	"		9.80		9.80	14.70
3500	Vent cast iron fittings 4"-8" diameter	Q-2	30	.800	Ea.		48		48	72
3600	10"-15" diameter	"	20	1.200	"		72		72	108

02 41 13.40 Selective Demolition, Metal Drainage Piping

		Crew	Daily Output	Labor-Hours	Unit	Material	2020 Bare Costs Labor	Equipment	Total	Total Incl O&P
0010	**SELECTIVE DEMOLITION, METAL DRAINAGE PIPING**									
0015	Excludes excavation									
0100	CMP pipe, aluminum, 6"-10" diam.	B-11M	800	.020	L.F.		.99	.29	1.28	1.80
0110	12" diam.		600	.027			1.32	.39	1.71	2.40
0120	18" diam.		600	.027			1.32	.39	1.71	2.40
0140	Steel, 6"-10" diam.		800	.020			.99	.29	1.28	1.80
0150	12" diam.		600	.027			1.32	.39	1.71	2.40
0160	18" diam.		400	.040			1.98	.58	2.56	3.60
0170	24" diam.	B-13	300	.187			8.60	1.94	10.54	15.05
0180	30"-36" diam.		250	.224			10.30	2.32	12.62	18
0190	48"-60" diam.		200	.280			12.90	2.90	15.80	22.50
0200	72" diam.	B-13B	100	.560			26	9.80	35.80	49.50
0210	CMP end sections, steel, 10"-18" diam.	B-11M	40	.400	Ea.		19.75	5.80	25.55	36
0220	24"-36" diam.	B-13	30	1.867			86	19.35	105.35	151
0230	48" diam.		20	2.800			129	29	158	225
0240	60" diam.		10	5.600			258	58	316	450
0250	72" diam.	B-13B	10	5.600			258	98	356	495
0260	CMP fittings, 8"-12" diam.	B-11M	60	.267			13.20	3.87	17.07	24
0270	18" diam.	"	40	.400			19.75	5.80	25.55	36
0280	24"-48" diam.	B-13	30	1.867			86	19.35	105.35	151
0290	60" diam.	"	20	2.800			129	29	158	225
0300	72" diam.	B-13B	10	5.600			258	98	356	495
0310	Oval arch 17" x 13", 21" x 15", 15"-18" equivalent	B-11M	400	.040	L.F.		1.98	.58	2.56	3.60
0320	28" x 20", 24" equivalent	B-13	300	.187			8.60	1.94	10.54	15.05
0330	35" x 24", 42" x 29", 30"-36" equivalent		250	.224			10.30	2.32	12.62	18
0340	49" x 33", 57" x 38", 42"-48" equivalent		200	.280			12.90	2.90	15.80	22.50
0350	Oval arch 17" x 13" end piece, 15" equivalent	B-11M	40	.400	Ea.		19.75	5.80	25.55	36
0360	42" x 29" end piece, 36" equivalent	B-13	30	1.867	"		86	19.35	105.35	151

02 41 Demolition

02 41 13 – Selective Site Demolition

02 41 13.42 Selective Demolition, Manholes and Catch Basins	Crew	Daily Output	Labor-Hours	Unit	Material	2020 Bare Costs Labor	2020 Bare Costs Equipment	Total	Total Incl O&P
0010 **SELECTIVE DEMOLITION, MANHOLES & CATCH BASINS**									
0015 Excludes excavation									
0100 Manholes, precast or brick over 8' deep	B-6	8	3	V.L.F.		137	26.50	163.50	236
0200 Cast in place 4'-8' deep	B-9	127	.315	SF Face		13.40	2.81	16.21	23
0300 Over 8' deep	"	100	.400	"		17	3.57	20.57	29.50
0400 Top, precast, 8" thick, 4'-6' diam.	B-6	8	3	Ea.		137	26.50	163.50	236
0500 Steps	1 Clab	60	.133	"		5.60		5.60	8.45

02 41 13.43 Selective Demolition, Box Culvert

	Crew	Daily Output	Labor-Hours	Unit	Material	2020 Bare Costs Labor	2020 Bare Costs Equipment	Total	Total Incl O&P
0010 **SELECTIVE DEMOLITION, BOX CULVERT**									
0015 Excludes excavation									
0100 Box culvert 8' x 6' x 3' to 8' x 8' x 8'	B-69	300	.160	L.F.		7.45	4.95	12.40	16.65
0200 8' x 10' x 3' to 8' x 12' x 8'	"	200	.240	"		11.20	7.45	18.65	25

02 41 13.44 Selective Demolition, Septic Tanks and Related Components

	Crew	Daily Output	Labor-Hours	Unit	Material	2020 Bare Costs Labor	2020 Bare Costs Equipment	Total	Total Incl O&P
0010 **SELECTIVE DEMOLITION, SEPTIC TANKS & RELATED COMPONENTS**									
0020 Excludes excavation and disposal									
0100 Septic tanks, precast, 1000-1250 gal.	B-21	8	3.500	Ea.		171	23.50	194.50	283
0200 1500 gal.		7	4			195	27	222	325
0300 2000-2500 gal.		5	5.600			273	38	311	450
0400 4000 gal.		4	7			340	47.50	387.50	565
0500 Precast, 5000 gal., multiple sections	B-13	3	18.667			860	194	1,054	1,525
0600 15,000 gal.	B-13B	1.70	32.941			1,525	575	2,100	2,900
0800 40,000 gal.	"	.80	70			3,225	1,225	4,450	6,175
0900 Precast, 50,000 gal., 5 piece	B-13C	.60	93.333			4,300	3,800	8,100	10,700
1000 Cast-in-place, 75,000 gal.	B-13L	.50	32			1,900	4,150	6,050	7,400
1100 100,000 gal.	"	.30	53.333			3,150	6,925	10,075	12,400
1200 HDPE, 1000 gal.	B-21	9	3.111			152	21	173	251
1300 1500 gal.		8	3.500			171	23.50	194.50	283
1400 Galley, 4' x 4' x 4'		16	1.750			85.50	11.80	97.30	141
1500 Distribution boxes, concrete, 7 outlets	2 Clab	16	1			42		42	63.50
1600 9 outlets	"	8	2			84		84	127
1700 Leaching chambers 13' x 3'-7" x 1'-4", standard	B-13	16	3.500			161	36.50	197.50	282
1800 8' x 4' x 1'-6", heavy duty		14	4			184	41.50	225.50	320
1900 13' x 3'-9" x 1'-6"		12	4.667			215	48.50	263.50	375
2100 20' x 4' x 1'-6"		5	11.200			515	116	631	905
2200 Leaching pit 6'-6" x 6' deep	B-21	5	5.600			273	38	311	450
2300 6'-6" x 8' deep		4	7			340	47.50	387.50	565
2400 8' x 6' deep H20		4	7			340	47.50	387.50	565
2500 8' x 8' deep H20		3	9.333			455	63	518	755
2600 Velocity reducing pit, precast 6' x 3' deep		4.70	5.957			290	40	330	480

02 41 13.46 Selective Demolition, Steel Pipe With Insulation

	Crew	Daily Output	Labor-Hours	Unit	Material	2020 Bare Costs Labor	2020 Bare Costs Equipment	Total	Total Incl O&P
0010 **SELECTIVE DEMOLITION, STEEL PIPE WITH INSULATION**									
0020 Excludes excavation									
0100 Steel pipe, with insulation, 3/4"-4"	B-1A	400	.060	L.F.		2.57	.92	3.49	4.87
0200 5"-10"	B-1B	360	.089			4.17	2.33	6.50	8.80
0300 12"-16"		240	.133			6.25	3.50	9.75	13.20
0400 18"-24"		160	.200			9.40	5.25	14.65	19.80
0450 26"-36"		100	.320			15	8.40	23.40	32
0500 Steel gland seal, with insulation, 3/4"-4"	B-1A	100	.240	Ea.		10.25	3.69	13.94	19.45
0600 5"-10"	B-1B	75	.427			20	11.20	31.20	42.50
0700 12"-16"		60	.533			25	14	39	53
0800 18"-24"		50	.640			30	16.80	46.80	63.50

31

02 41 Demolition

02 41 13 – Selective Site Demolition

02 41 13.46 Selective Demolition, Steel Pipe With Insulation		Crew	Daily Output	Labor-Hours	Unit	Material	2020 Bare Costs		Total	Total Incl O&P
							Labor	Equipment		
0850	26"-36"	B-1B	40	.800	Ea.		37.50	21	58.50	79.50
0900	Demo steel fittings with insulation, 3/4"-4"	B-1A	60	.400			17.10	6.15	23.25	32.50
1000	5"-10"	B-1B	40	.800			37.50	21	58.50	79.50
1100	12"-16"		30	1.067			50	28	78	106
1200	18"-24"		20	1.600			75	42	117	159
1300	26"-36"		15	2.133			100	56	156	212
1400	Steel pipe anchors, 5"-10"		40	.800			37.50	21	58.50	79.50
1500	12"-16"		30	1.067			50	28	78	106
1600	18"-24"		20	1.600			75	42	117	159
1700	26"-36"	↓	15	2.133	↓		100	56	156	212

02 41 13.48 Selective Demolition, Gasoline Containment Piping

		Crew	Daily Output	Labor-Hours	Unit	Material	Labor	Equipment	Total	Total Incl O&P
0010	**SELECTIVE DEMOLITION, GASOLINE CONTAINMENT PIPING**									
0020	Excludes excavation									
0030	Excludes environmental site remediation									
0100	Gasoline plastic primary containment piping 2"-4"	Q-6	800	.030	L.F.		1.84		1.84	2.75
0200	Fittings 2"-4"		40	.600	Ea.		36.50		36.50	55
0300	Gasoline plastic secondary containment piping 3"-6"		800	.030	L.F.		1.84		1.84	2.75
0400	Fittings 3"-6"	↓	40	.600	Ea.		36.50		36.50	55

02 41 13.50 Selective Demolition, Natural Gas, PE Pipe

		Crew	Daily Output	Labor-Hours	Unit	Material	Labor	Equipment	Total	Total Incl O&P
0010	**SELECTIVE DEMOLITION, NATURAL GAS, PE PIPE**									
0020	Excludes excavation									
0100	Natural gas coils, PE, 1-1/4"-3"	Q-6	800	.030	L.F.		1.84		1.84	2.75
0200	Joints, 40', PE, 3"-4"		800	.030			1.84		1.84	2.75
0300	6"-8"	↓	600	.040	↓		2.45		2.45	3.66

02 41 13.51 Selective Demolition, Natural Gas, Steel Pipe

		Crew	Daily Output	Labor-Hours	Unit	Material	Labor	Equipment	Total	Total Incl O&P
0010	**SELECTIVE DEMOLITION, NATURAL GAS, STEEL PIPE**									
0020	Excludes excavation									
0100	Natural gas steel pipe 1"-4"	B-1A	800	.030	L.F.		1.28	.46	1.74	2.44
0200	5"-10"	B-1B	360	.089			4.17	2.33	6.50	8.80
0300	12"-16"		240	.133			6.25	3.50	9.75	13.20
0400	18"-24"	↓	160	.200	↓		9.40	5.25	14.65	19.80
0500	Natural gas steel fittings 1"-4"	B-1A	160	.150	Ea.		6.40	2.30	8.70	12.20
0600	5"-10"	B-1B	160	.200			9.40	5.25	14.65	19.80
0700	12"-16"		108	.296			13.90	7.75	21.65	29.50
0800	18"-24"	↓	70	.457			21.50	12	33.50	45

02 41 13.52 Selective Demo, Natural Gas, Valves, Fittings, Regulators

		Crew	Daily Output	Labor-Hours	Unit	Material	Labor	Equipment	Total	Total Incl O&P
0010	**SELECTIVE DEMO, NATURAL GAS, VALVES, FITTINGS, REGULATORS**									
0100	Gas stops 1-1/4"-2"	1 Plum	22	.364	Ea.		23.50		23.50	35
0200	Gas regulator 1-1/2"-2"	"	22	.364			23.50		23.50	35
0300	3"-4"	Q-1	22	.727			42		42	63
0400	Gas plug valve, 3/4"-2"	1 Plum	22	.364			23.50		23.50	35
0500	2-1/2"-3"	Q-1	10	1.600	↓		93		93	139

02 41 13.54 Selective Demolition, Electric Ducts and Fittings

		Crew	Daily Output	Labor-Hours	Unit	Material	Labor	Equipment	Total	Total Incl O&P
0010	**SELECTIVE DEMOLITION, ELECTRIC DUCTS & FITTINGS**									
0020	Excludes excavation									
0100	Plastic conduit, 1/2"-2"	1 Elec	600	.013	L.F.		.82		.82	1.22
0200	3"-6"	2 Elec	400	.040	"		2.45		2.45	3.65
0300	Fittings, 1/2"-2"	1 Elec	50	.160	Ea.		9.80		9.80	14.60
0400	3"-6"	"	40	.200	"		12.25		12.25	18.25

02 41 Demolition

02 41 13 - Selective Site Demolition

02 41 13.56 Selective Demolition, Electric Duct Banks	Crew	Daily Output	Labor-Hours	Unit	Material	2020 Bare Costs Labor	Equipment	Total	Total Incl O&P
0010 **SELECTIVE DEMOLITION, ELECTRIC DUCT BANKS**									
0020 Excludes excavation									
0100 Hand holes sized to 4' x 4' x 4'	R-3	7	2.857	Ea.		175	27	202	290
0200 Manholes sized to 6' x 10' x 7'	B-13	6	9.333	"		430	97	527	750
0300 Conduit 1 @ 2" diameter, EB plastic, no concrete	2 Elec	1000	.016	L.F.		.98		.98	1.46
0400 2 @ 2" diameter		500	.032			1.96		1.96	2.92
0500 4 @ 2" diameter		250	.064			3.93		3.93	5.85
0600 1 @ 3" diameter		800	.020			1.23		1.23	1.83
0700 2 @ 3" diameter		400	.040			2.45		2.45	3.65
0800 4 @ 3" diameter		200	.080			4.91		4.91	7.30
0900 1 @ 4" diameter		800	.020			1.23		1.23	1.83
1000 2 @ 4" diameter		400	.040			2.45		2.45	3.65
1100 4 @ 4" diameter		200	.080			4.91		4.91	7.30
1200 6 @ 4" diameter		100	.160			9.80		9.80	14.60
1300 1 @ 5" diameter		500	.032			1.96		1.96	2.92
1400 2 @ 5" diameter		250	.064			3.93		3.93	5.85
1500 4 @ 5" diameter		160	.100			6.15		6.15	9.15
1600 6 @ 5" diameter		100	.160			9.80		9.80	14.60
1700 1 @ 6" diameter		500	.032			1.96		1.96	2.92
1800 2 @ 6" diameter		250	.064			3.93		3.93	5.85
1900 4 @ 6" diameter		160	.100			6.15		6.15	9.15
2000 6 @ 6" diameter		100	.160			9.80		9.80	14.60
2100 Conduit 1 EB plastic, with concrete, 0.92 C.F./L.F.	B-9	150	.267			11.35	2.38	13.73	19.65
2200 2 EB plastic 1.52 C.F./L.F.		100	.400			17	3.57	20.57	29.50
2300 2 x 2 EB plastic 2.51 C.F./L.F.		80	.500			21.50	4.47	25.97	37
2400 2 x 3 EB plastic 3.56 C.F./L.F.		60	.667			28.50	5.95	34.45	49
2500 Conduit 2 @ 2" diameter, steel, no concrete	2 Elec	400	.040			2.45		2.45	3.65
2600 4 @ 2" diameter		200	.080			4.91		4.91	7.30
2700 2 @ 3" diameter		200	.080			4.91		4.91	7.30
2800 4 @ 3" diameter		100	.160			9.80		9.80	14.60
2900 2 @ 4" diameter		200	.080			4.91		4.91	7.30
3000 4 @ 4" diameter		100	.160			9.80		9.80	14.60
3100 6 @ 4" diameter		50	.320			19.65		19.65	29
3200 2 @ 5" diameter		120	.133			8.20		8.20	12.20
3300 4 @ 5" diameter		60	.267			16.35		16.35	24.50
3400 6 @ 5" diameter		40	.400			24.50		24.50	36.50
3500 2 @ 6" diameter		120	.133			8.20		8.20	12.20
3600 4 @ 6" diameter		60	.267			16.35		16.35	24.50
3700 6 @ 6" diameter		40	.400			24.50		24.50	36.50
3800 Conduit 2 steel, with concrete, 1.52 C.F./L.F.	B-9	80	.500			21.50	4.47	25.97	37
3900 2 x 2 EB steel 2.51 C.F./L.F.		60	.667			28.50	5.95	34.45	49
4000 2 x 3 steel 3.56 C.F./L.F.		50	.800			34	7.15	41.15	59
4100 Conduit fittings, PVC type EB, 2"-3"	1 Elec	30	.267	Ea.		16.35		16.35	24.50
4200 4"-6"	"	20	.400	"		24.50		24.50	36.50

02 41 13.60 Selective Demolition Fencing

	Crew	Daily Output	Labor-Hours	Unit	Material	Labor	Equipment	Total	Total Incl O&P
0010 **SELECTIVE DEMOLITION FENCING** R024119-10									
0700 Snow fence, 4' high	B-6	1000	.024	L.F.		1.10	.21	1.31	1.89
1600 Fencing, barbed wire, 3 strand	2 Clab	430	.037			1.57		1.57	2.35
1650 5 strand	"	280	.057			2.41		2.41	3.61
1700 Chain link, posts & fabric, 8'-10' high, remove only	B-6	445	.054			2.47	.48	2.95	4.23
1755 6' high, remove only	"	520	.046			2.11	.41	2.52	3.61
1760 Wood fence to 6', remove only, minimum	2 Clab	500	.032			1.35		1.35	2.02

For customer support on your Site Work & Landscape Costs with RSMeans data, call 800.448.8182.

33

02 41 13 – Selective Site Demolition

02 41 13.60 Selective Demolition Fencing	Crew	Daily Output	Labor-Hours	Unit	Material	2020 Bare Costs Labor	Equipment	Total	Total Incl O&P
1770	Maximum	2 Clab	250	.064	L.F.		2.69		2.69	4.05
1775	Fencing, wood, average all types, 4'-6' high		432	.037	↓		1.56		1.56	2.34
1780	Fence post, wood, 4" x 4", 6'-8' high		96	.167	Ea.		7		7	10.55
1785	Fence rail, 2" x 4", 8' long, remove only		7680	.002	L.F.		.09		.09	.13
1790	Remove and store	B-80	235	.136	"		6.35	4.34	10.69	14.30

02 41 13.62 Selective Demo., Chain Link Fences & Gates

		Crew	Daily Output	Labor-Hours	Unit	Material	Labor	Equipment	Total	Total Incl O&P
0010	**SELECTIVE DEMOLITION, CHAIN LINK FENCES & GATES**									
0100	Chain link, gates, 3'-4' width	B-6	30	.800	Ea.		36.50	7.15	43.65	63
0200	10'-12' width		16	1.500			68.50	13.35	81.85	118
0300	14' width		15	1.600			73	14.25	87.25	126
0400	20' width		10	2.400	↓		110	21.50	131.50	189
0500	18' width with overhead & cantilever		80	.300	L.F.		13.70	2.67	16.37	23.50
0510	Sliding		80	.300			13.70	2.67	16.37	23.50
0520	Cantilever to 40' wide		80	.300	↓		13.70	2.67	16.37	23.50
0530	Motor operators	2 Skwk	1	16	Ea.		880		880	1,325
0540	Transmitter systems	"	15	1.067	"		58.50		58.50	88.50
0600	Chain link, fence, 5' high	B-6	890	.027	L.F.		1.23	.24	1.47	2.11
0650	3'-4' high		1000	.024			1.10	.21	1.31	1.89
0675	12' high	↓	400	.060	↓		2.74	.53	3.27	4.70
0800	Chain link, fence, braces	B-1	2000	.012	Ea.		.51		.51	.77
0900	Privacy slats	"	2000	.012			.51		.51	.77
1000	Fence posts, steel, in concrete	B-6	80	.300	↓		13.70	2.67	16.37	23.50
1100	Fence fabric & accessories, fabric to 8' high		800	.030	L.F.		1.37	.27	1.64	2.35
1200	Barbed wire		5000	.005	"		.22	.04	.26	.38
1300	Extension arms & eye tops		300	.080	Ea.		3.66	.71	4.37	6.30
1400	Fence rails		2000	.012	L.F.		.55	.11	.66	.94
1500	Reinforcing wire	↓	5000	.005	"		.22	.04	.26	.38

02 41 13.64 Selective Demolition, Vinyl Fences and Gates

		Crew	Daily Output	Labor-Hours	Unit	Material	Labor	Equipment	Total	Total Incl O&P
0010	**SELECTIVE DEMOLITION, VINYL FENCES & GATES**									
0100	Vinyl fence up to 6' high	B-6	1000	.024	L.F.		1.10	.21	1.31	1.89
0200	Gates up to 6' high	"	40	.600	Ea.		27.50	5.35	32.85	47

02 41 13.66 Selective Demolition, Misc Metal Fences and Gates

		Crew	Daily Output	Labor-Hours	Unit	Material	Labor	Equipment	Total	Total Incl O&P
0010	**SELECTIVE DEMOLITION, MISC METAL FENCES & GATES**									
0100	Misc steel mesh fences, 4'-6' high	B-6	600	.040	L.F.		1.83	.36	2.19	3.13
0200	Kennels, 6'-12' long	2 Clab	8	2	Ea.		84		84	127
0300	Tops, 6'-12' long	"	30	.533	"		22.50		22.50	33.50
0400	Security fences, 12'-16' high	B-6	100	.240	L.F.		11	2.14	13.14	18.80
0500	Metal tubular picket fences, 4'-6' high		500	.048	"		2.20	.43	2.63	3.76
0600	Gates, 3'-4' wide	↓	20	1.200	Ea.		55	10.70	65.70	94.50

02 41 13.68 Selective Demolition, Wood Fences and Gates

		Crew	Daily Output	Labor-Hours	Unit	Material	Labor	Equipment	Total	Total Incl O&P
0010	**SELECTIVE DEMOLITION, WOOD FENCES & GATES**									
0100	Wood fence gates 3'-4' wide	2 Clab	20	.800	Ea.		33.50		33.50	50.50
0200	Wood fence, open rail, to 4' high		2560	.006	L.F.		.26		.26	.40
0300	to 8' high		368	.043	"		1.83		1.83	2.75
0400	Post, in concrete	↓	50	.320	Ea.		13.45		13.45	20

02 41 13.70 Selective Demolition, Rip-Rap and Rock Lining

		Crew	Daily Output	Labor-Hours	Unit	Material	Labor	Equipment	Total	Total Incl O&P
0010	**SELECTIVE DEMOLITION, RIP-RAP & ROCK LINING** R024119-10									
0100	Slope protection, broken stone	B-13	62	.903	C.Y.		41.50	9.35	50.85	73
0200	3/8 to 1/4 C.Y. pieces		60	.933	S.Y.		43	9.70	52.70	75
0300	18" depth		60	.933	"		43	9.70	52.70	75
0400	Dumped stone	↓	93	.602	Ton		27.50	6.25	33.75	48.50

02 41 Demolition

02 41 13 – Selective Site Demolition

02 41 13.70 Selective Demolition, Rip-Rap and Rock Lining		Crew	Daily Output	Labor-Hours	Unit	Material	2020 Bare Costs Labor	2020 Bare Costs Equipment	Total	Total Incl O&P
0500	Gabions, 6"-12" deep	B-13	60	.933	S.Y.		43	9.70	52.70	75
0600	18"-36" deep	↓	30	1.867	"		86	19.35	105.35	151

02 41 13.72 Selective Demo., Shore Protect/Mooring Struct.

0010	**SELECTIVE DEMOLITION, SHORE PROTECT/MOORING STRUCTURES**									
0100	Breakwaters, bulkheads, concrete, maximum [R024119-10]	B-9	12	3.333	L.F.		142	30	172	246
0200	Breakwaters, bulkheads, concrete, 12', minimum		10	4			170	35.50	205.50	295
0300	Maximum	↓	9	4.444			189	39.50	228.50	330
0400	Steel, from shore	B-40B	54	.889			41.50	37.50	79	104
0500	from barge	B-76A	30	2.133	↓		97	79	176	233
0600	Jetties, docks, floating	B-21B	600	.067	S.F.		3.06	.79	3.85	5.45
0700	Pier supported, 3"-4" decking		300	.133			6.10	1.57	7.67	10.95
0800	Floating, prefab, small boat, minimum		600	.067			3.06	.79	3.85	5.45
0900	Maximum		300	.133	↓		6.10	1.57	7.67	10.95
1000	Floating, prefab, per slip, minimum		3.20	12.500	Ea.		575	147	722	1,025
1010	Maximum	↓	2.80	14.286	"		655	168	823	1,175

02 41 13.74 Selective Demolition, Piles

0010	**SELECTIVE DEMOLITION, PILES**									
0100	Cast in place piles, corrugated, 8"-10"	B-19	600	.107	V.L.F.		5.90	3.35	9.25	12.70
0200	12"-14"		500	.128			7.05	4.02	11.07	15.25
0300	16"		400	.160			8.85	5	13.85	19.10
0400	fluted, 12"		600	.107			5.90	3.35	9.25	12.70
0500	14"-18"		500	.128			7.05	4.02	11.07	15.25
0600	end bearing, 12"		600	.107			5.90	3.35	9.25	12.70
0700	14"-18"		500	.128			7.05	4.02	11.07	15.25
0800	Precast prestressed piles, 12"-14" diam.		700	.091			5.05	2.87	7.92	10.90
0900	16"-24" diam.		600	.107			5.90	3.35	9.25	12.70
1000	36"-66" diam.		300	.213			11.80	6.70	18.50	25.50
1100	10"-14" thick	↓	600	.107			5.90	3.35	9.25	12.70
1200	16"-24" thick	B-38	500	.080			3.81	2.48	6.29	8.45
1300	Pressure grouted pile, 5"	B-19	150	.427			23.50	13.40	36.90	51
1400	Steel piles, 8"-12" tip		600	.107			5.90	3.35	9.25	12.70
1500	H sections HP8 to HP12		600	.107			5.90	3.35	9.25	12.70
1600	HP14	B-19A	600	.107			5.90	4.91	10.81	14.40
1700	Steel pipe piles, 8"-12"	B-19	600	.107			5.90	3.35	9.25	12.70
1800	14"-18" plain		500	.128			7.05	4.02	11.07	15.25
1900	concrete filled, 14"-18"		400	.160			8.85	5	13.85	19.10
2000	Timber piles to 14" diam.	↓	600	.107	↓		5.90	3.35	9.25	12.70

02 41 13.76 Selective Demolition, Water Wells

0010	**SELECTIVE DEMOLITION, WATER WELLS**									
0100	Well, 40' deep with casing & gravel pack, 24"-36" diam.	B-23	.25	160	Ea.		6,800	6,375	13,175	17,200
0200	Riser pipe, 1-1/4", for observation well	"	300	.133	L.F.		5.65	5.30	10.95	14.35
0300	Pump, 1/2 to 5 HP up to 100' depth	Q-1	3	5.333	Ea.		310		310	465
0400	Up to 150' well 25 HP pump	Q-22	1.50	10.667			620	315	935	1,275
0500	Up to 500' well 30 HP pump	"	1	16	↓		930	470	1,400	1,925
0600	Well screen 2"-8"	B-23	300	.133	V.L.F.		5.65	5.30	10.95	14.35
0700	10"-16"		200	.200			8.50	7.95	16.45	21.50
0800	18"-26"		150	.267			11.35	10.60	21.95	28.50
0900	Slotted PVC for wells 1-1/4"-8"		600	.067			2.83	2.65	5.48	7.20
1000	Well screen and casing 6"-16"		300	.133			5.65	5.30	10.95	14.35
1100	18"-26"		150	.267			11.35	10.60	21.95	28.50
1200	30"-36"	↓	100	.400	↓		17	15.90	32.90	43

For customer support on your Site Work & Landscape Costs with RSMeans data, call 800.448.8182.

35

02 41 Demolition

02 41 13 – Selective Site Demolition

02 41 13.78 Selective Demolition, Radio Towers	Crew	Daily Output	Labor-Hours	Unit	Material	2020 Bare Costs Labor	Equipment	Total	Total Incl O&P
0010 **SELECTIVE DEMOLITION, RADIO TOWERS**									
0100 Radio tower, guyed, 50'	2 Skwk	16	1	Ea.		55		55	83
0200 190', 40 lb. section	K-2	.70	34.286			1,875	1,175	3,050	4,200
0300 200', 70 lb. section		.70	34.286			1,875	1,175	3,050	4,200
0400 300', 70 lb. section		.40	60			3,300	2,050	5,350	7,350
0500 270', 90 lb. section		.40	60			3,300	2,050	5,350	7,350
0600 400'		.30	80			4,400	2,725	7,125	9,800
0700 Self supported, 60'		.90	26.667			1,475	910	2,385	3,275
0800 120'		.80	30			1,650	1,025	2,675	3,675
0900 190'		.40	60			3,300	2,050	5,350	7,350

02 41 13.80 Selective Demo., Utility Poles and Cross Arms

	Crew	Daily Output	Labor-Hours	Unit	Material	Labor	Equipment	Total	Total Incl O&P
0010 **SELECTIVE DEMOLITION, UTILITY POLES & CROSS ARMS**									
0100 Utility poles, wood, 20'-30' high	R-3	6	3.333	Ea.		204	31.50	235.50	340
0200 35'-45' high	"	5	4			244	38	282	405
0300 Cross arms, wood, 4'-6' long	1 Elec	5	1.600			98		98	146

02 41 13.82 Selective Removal, Pavement Lines and Markings

	Crew	Daily Output	Labor-Hours	Unit	Material	Labor	Equipment	Total	Total Incl O&P
0010 **SELECTIVE REMOVAL, PAVEMENT LINES & MARKINGS**									
0015 Does not include traffic control costs									
0020 See other items in Section 32 17 23.13									
0100 Remove permanent painted traffic lines and markings	B-78A	500	.016	C.L.F.		.85	1.83	2.68	3.29
0200 Temporary traffic line tape	2 Clab	1500	.011	L.F.		.45		.45	.67
0300 Thermoplastic traffic lines and markings	B-79A	500	.024	C.L.F.		1.27	2.78	4.05	4.96
0400 Painted pavement markings	B-78B	500	.036	S.F.		1.56	.81	2.37	3.23

02 41 13.84 Selective Demolition, Walks, Steps and Pavers

	Crew	Daily Output	Labor-Hours	Unit	Material	Labor	Equipment	Total	Total Incl O&P
0010 **SELECTIVE DEMOLITION, WALKS, STEPS AND PAVERS**									
0100 Splash blocks	1 Clab	300	.027	S.F.		1.12		1.12	1.69
0200 Tree grates	"	50	.160	Ea.		6.75		6.75	10.10
0300 Walks, limestone pavers	2 Clab	150	.107	S.F.		4.49		4.49	6.75
0400 Redwood sections		600	.027			1.12		1.12	1.69
0500 Redwood planks		480	.033			1.40		1.40	2.11
0600 Shale paver		300	.053			2.25		2.25	3.37
0700 Tile thinset paver		675	.024			1		1	1.50
0800 Wood round	B-1	350	.069	Ea.		2.93		2.93	4.41
0900 Asphalt block	2 Clab	450	.036	S.F.		1.50		1.50	2.25
1000 Bluestone		450	.036			1.50		1.50	2.25
1100 Slate, 1" or thinner		675	.024			1		1	1.50
1200 Granite blocks		300	.053			2.25		2.25	3.37
1300 Precast patio blocks		450	.036			1.50		1.50	2.25
1400 Planter blocks		600	.027			1.12		1.12	1.69
1500 Brick paving, dry set		300	.053			2.25		2.25	3.37
1600 Mortar set		180	.089			3.74		3.74	5.60
1700 Dry set on edge		240	.067			2.81		2.81	4.22
1800 Steps, brick		200	.080	L.F.		3.37		3.37	5.05
1900 Railroad tie		150	.107			4.49		4.49	6.75
2000 Bluestone		180	.089			3.74		3.74	5.60
2100 Wood/steel edging for steps		1000	.016			.67		.67	1.01
2200 Timber or railroad tie edging for steps		400	.040			1.68		1.68	2.53

02 41 13.86 Selective Demolition, Athletic Surfaces

	Crew	Daily Output	Labor-Hours	Unit	Material	Labor	Equipment	Total	Total Incl O&P
0010 **SELECTIVE DEMOLITION, ATHLETIC SURFACES**									
0100 Synthetic grass	2 Clab	2000	.008	S.F.		.34		.34	.51
0200 Surface coat latex rubber	"	2000	.008	"		.34		.34	.51

02 41 Demolition

02 41 13 – Selective Site Demolition

02 41 13.86 Selective Demolition, Athletic Surfaces	Crew	Daily Output	Labor-Hours	Unit	Material	2020 Bare Costs Labor	Equipment	Total	Total Incl O&P
0300	Tennis court posts	B-11C	16	1	Ea.		49.50	13.35	62.85	88.50

02 41 13.88 Selective Demolition, Lawn Sprinkler Systems

		Crew	Daily Output	Labor-Hours	Unit	Material	2020 Bare Costs Labor	Equipment	Total	Total Incl O&P
0010	**SELECTIVE DEMOLITION, LAWN SPRINKLER SYSTEMS**									
0100	Golf course sprinkler system, 9 hole	4 Clab	.10	320	Ea.		13,500		13,500	20,200
0200	Sprinkler system, 24' diam. @ 15' OC, per head	B-20	110	.218	Head		10.25		10.25	15.45
0300	60' diam. @ 24' OC, per head	"	52	.462	"		21.50		21.50	32.50
0400	Sprinkler heads, plastic	2 Clab	150	.107	Ea.		4.49		4.49	6.75
0500	Impact circle pattern, 28'-76' diam.		75	.213			9		9	13.50
0600	Pop-up, 42'-76' diam.		50	.320			13.45		13.45	20
0700	39'-99' diam.		50	.320			13.45		13.45	20
0800	Sprinkler valves		40	.400			16.85		16.85	25.50
0900	Valve boxes		40	.400			16.85		16.85	25.50
1000	Controls		2	8			335		335	505
1100	Backflow preventer		4	4			168		168	253
1200	Vacuum breaker		4	4			168		168	253

02 41 13.90 Selective Demolition, Retaining Walls

		Crew	Daily Output	Labor-Hours	Unit	Material	2020 Bare Costs Labor	Equipment	Total	Total Incl O&P
0010	**SELECTIVE DEMOLITION, RETAINING WALLS**									
0020	See other retaining wall items in Section 32 32									
0100	Concrete retaining wall, 6' high, no reinforcing	B-13K	200	.080	L.F.		4.74	10.15	14.89	18.25
0200	8' high		150	.107			6.30	13.55	19.85	24.50
0300	10' high		150	.107			6.30	13.55	19.85	24.50
0400	With reinforcing, 6' high		200	.080			4.74	10.15	14.89	18.25
0500	8' high		150	.107			6.30	13.55	19.85	24.50
0600	10' high		120	.133			7.90	16.90	24.80	30.50
0700	20' high		60	.267			15.80	34	49.80	60.50
0800	Concrete cribbing, 12' high, open/closed face		150	.107	S.F.		6.30	13.55	19.85	24.50
0900	Interlocking segmental retaining wall	B-62	800	.030			1.37	.22	1.59	2.30
1000	Wall caps	"	600	.040			1.83	.30	2.13	3.07
1100	Metal bin retaining wall, 10' wide, 4'-12' high	B-13	1200	.047			2.15	.48	2.63	3.75
1200	10' wide, 16'-28' high		1000	.056			2.58	.58	3.16	4.51
1300	Stone filled gabions, 6' x 3' x 1'		170	.329	Ea.		15.15	3.42	18.57	26.50
1400	6' x 3' x 1'-6"		75	.747			34.50	7.75	42.25	60
1500	6' x 3' x 3'		25	2.240			103	23	126	181
1600	9' x 3' x 1'		75	.747			34.50	7.75	42.25	60
1700	9' x 3' x 1'-6"		33	1.697			78	17.60	95.60	136
1800	9' x 3' x 3'		12	4.667			215	48.50	263.50	375
1900	12' x 3' x 1'		42	1.333			61.50	13.85	75.35	107
2000	12' x 3' x 1'-6"		20	2.800			129	29	158	225
2100	12' x 3' x 3'		6	9.333			430	97	527	750

02 41 13.92 Selective Demolition, Parking Appurtenances

		Crew	Daily Output	Labor-Hours	Unit	Material	2020 Bare Costs Labor	Equipment	Total	Total Incl O&P
0010	**SELECTIVE DEMOLITION, PARKING APPURTENANCES**									
0100	Bumper rails, garage, 6" wide	B-6	300	.080	L.F.		3.66	.71	4.37	6.30
0200	12" channel rail		300	.080			3.66	.71	4.37	6.30
0300	Parking bumper, timber		1000	.024			1.10	.21	1.31	1.89
0400	Folding, with locks	B-1	100	.240	Ea.		10.25		10.25	15.40
0500	Flexible fixed garage stanchion	B-6	150	.160			7.30	1.43	8.73	12.50
0600	Wheel stops, precast concrete		120	.200			9.15	1.78	10.93	15.65
0700	Thermoplastic		120	.200			9.15	1.78	10.93	15.65
0800	Pipe bollards, 6"-12" diam.		80	.300			13.70	2.67	16.37	23.50

02 41 Demolition

02 41 13 – Selective Site Demolition

02 41 13.93 Selective Demolition, Site Furnishings	Crew	Daily Output	Labor-Hours	Unit	Material	2020 Bare Costs Labor	Equipment	Total	Total Incl O&P
0010 **SELECTIVE DEMOLITION, SITE FURNISHINGS**									
0100 Benches, all types	2 Clab	10	1.600	Ea.		67.50		67.50	101
0200 Trash receptacles, all types		80	.200			8.40		8.40	12.65
0300 Trash enclosures, steel or wood	↓	10	1.600	↓		67.50		67.50	101

02 41 13.94 Selective Demolition, Athletic Screening

	Crew	Daily Output	Labor-Hours	Unit	Material	Labor	Equipment	Total	Total Incl O&P
0010 **SELECTIVE DEMOLITION, ATHLETIC SCREENING**									
0020 See other items in Sections 02 41 13.60 & 02 41 13.62									
0100 Baseball backstops	B-6	2	12	Ea.		550	107	657	945
0200 Basketball goal, single		6	4			183	35.50	218.50	315
0300 Double		4	6	↓		274	53.50	327.50	470
0400 Tennis wire mesh, pair ends		5	4.800	Set		220	43	263	375
0500 Enclosed court		3	8	Ea.		365	71.50	436.50	630
0600 Wood or masonry handball court	↓	1	24	"		1,100	214	1,314	1,875

02 41 13.95 Selective Demo., Athletic/Playground Equipment

	Crew	Daily Output	Labor-Hours	Unit	Material	Labor	Equipment	Total	Total Incl O&P
0010 **SELECTIVE DEMO., ATHLETIC/PLAYGROUND EQUIPMENT**									
0020 See other items in Section 02 41 13.96									
0100 Bike rack	B-6	32	.750	Ea.		34.50	6.70	41.20	59
0200 Climber arch		16	1.500			68.50	13.35	81.85	118
0300 Fitness trail, 9 to 10 stations wood or metal		.50	48			2,200	430	2,630	3,775
0400 16 to 20 stations wood or metal		.25	96			4,400	855	5,255	7,525
0500 Monkey bars 14' long		8	3			137	26.50	163.50	236
0600 Parallel bars 10' long		8	3			137	26.50	163.50	236
0700 Post tether ball		32	.750			34.50	6.70	41.20	59
0800 Poles 10'-6" long		32	.750			34.50	6.70	41.20	59
0900 Pole ground sockets		80	.300			13.70	2.67	16.37	23.50
1000 See-saw, 2 units		12	2			91.50	17.80	109.30	157
1100 4 units		10	2.400			110	21.50	131.50	189
1200 6 units		8	3			137	26.50	163.50	236
1300 Shelter, fiberglass, 3 person		10	2.400			110	21.50	131.50	189
1400 Slides, 12' long		8	3			137	26.50	163.50	236
1500 20' long		6	4			183	35.50	218.50	315
1600 Swings, 4 seats		6	4			183	35.50	218.50	315
1700 8 seats		4	6			274	53.50	327.50	470
1800 Whirlers, 8'-10' diameter		6	4	↓		183	35.50	218.50	315
1900 Football or football/soccer goal posts		3	8	Pair		365	71.50	436.50	630
2000 Soccer goal posts	↓	4	6	"		274	53.50	327.50	470
2100 Playground surfacing, 4" depth	B-62	2000	.012	S.F.		.55	.09	.64	.92
2200 2" topping	"	3500	.007	"		.31	.05	.36	.53
2300 Platform/paddle tennis court, alum. deck & frame	B-1	.20	120	Court		5,125		5,125	7,700
2400 Aluminum deck and wood frame		.25	96	"		4,100		4,100	6,175
2500 Heater		2	12	Ea.		515		515	770
2600 Wood deck & frame		.25	96	Court		4,100		4,100	6,175
2700 Wood deck & steel frame		.25	96			4,100		4,100	6,175
2800 Steel deck & wood frame	↓	.25	96	↓		4,100		4,100	6,175

02 41 13.96 Selective Demo., Modular Playground Equipment

	Crew	Daily Output	Labor-Hours	Unit	Material	Labor	Equipment	Total	Total Incl O&P
0010 **SELECTIVE DEMOLITION, MODULAR PLAYGROUND EQUIPMENT**									
0100 Modular playground, 48" deck, square or triangle shape	B-1	2	12	Ea.		515		515	770
0200 Various posts		18	1.333			57		57	85.50
0300 Roof 54" x 54"		18	1.333			57		57	85.50
0400 Wheel chair transfer module		6	4	↓		171		171	257
0500 Guardrail pipe	↓	90	.267	L.F.		11.40		11.40	17.15

02 41 Demolition

02 41 13 – Selective Site Demolition

02 41 13.96 Selective Demo., Modular Playground Equipment

		Crew	Daily Output	Labor-Hours	Unit	Material	2020 Bare Costs Labor	2020 Bare Costs Equipment	Total	Total Incl O&P
0600	Steps, 3 risers	B-1	16	1.500	Ea.		64		64	96.50
0700	Activity panel crawl through		8	3			128		128	193
0800	Alphabet/spelling panel		8	3			128		128	193
0900	Crawl tunnel, each 4' section		8	3			128		128	193
1000	Slide tunnel		16	1.500			64		64	96.50
1100	Spiral slide tunnel		12	2			85.50		85.50	129
1200	Ladder, horizontal or vertical		10	2.400			103		103	154
1300	Corkscrew climber		6	4			171		171	257
1400	Fire pole		12	2			85.50		85.50	129
1500	Bridge, ring chamber		8	3			128		128	193
1600	Bridge, suspension		8	3			128		128	193

02 41 13.97 Selective Demolition, Traffic Light Systems

		Crew	Daily Output	Labor-Hours	Unit	Material	2020 Bare Costs Labor	2020 Bare Costs Equipment	Total	Total Incl O&P
0010	**SELECTIVE DEMOLITION, TRAFFIC LIGHT SYSTEMS**									
0100	Traffic signal, pedestrian cross walk	R-2	.50	112	Total		6,450	755	7,205	10,500
0200	8 signal intersection, 2 each direction	"	.25	224			12,900	1,500	14,400	20,900
0300	Each traffic phase controller	L-9	3	12			575		575	870
0400	Each semi-actuated detector		1.50	24			1,150		1,150	1,750
0500	Each fully-actuated detector		1.50	24			1,150		1,150	1,750
0600	Pedestrian push button system		2	18			860		860	1,300
0650	Each optically programmed head		3	12			575		575	870
0700	School signal flashing system		1	36	Signal		1,725		1,725	2,625
0800	Complete traffic light intersection system, without lane control	R-2	.25	224	Ea.		12,900	1,500	14,400	20,900
0900	With lane control		.20	280			16,100	1,900	18,000	26,100
1000	Traffic light, each left turn protection system		.66	84.848			4,900	575	5,475	7,900

02 41 13.98 Select. Demo., Sod, Edging, Planters & Tree Guying

		Crew	Daily Output	Labor-Hours	Unit	Material	2020 Bare Costs Labor	2020 Bare Costs Equipment	Total	Total Incl O&P
0010	**SELECTIVE DEMOLITION, SOD, EDGING, PLANTERS & TREE GUYING**									
0100	Remove sod 1" deep, 3 M.S.F. or greater, level ground	B-63	40	1	M.S.F.		44.50	4.44	48.94	71.50
0200	Less than 3 M.S.F., level ground		27	1.481			65.50	6.60	72.10	106
0300	3 M.S.F. or greater, sloped ground		12	3.333			148	14.80	162.80	237
0400	Less than 3 M.S.F., sloped ground		8	5			221	22	243	355
0500	Edging, aluminum	B-1	800	.030	L.F.		1.28		1.28	1.93
0600	Brick horizontal		400	.060			2.57		2.57	3.86
0700	Brick vertical		200	.120			5.15		5.15	7.70
0800	Corrugated aluminum		1200	.020			.86		.86	1.29
0900	Granite 5" x 16"		600	.040			1.71		1.71	2.57
1000	Polyethylene grass barrier		1200	.020			.86		.86	1.29
1100	Precast 2" x 8" x 16"		800	.030			1.28		1.28	1.93
1200	Railroad ties		200	.120			5.15		5.15	7.70
1300	Wood		400	.060			2.57		2.57	3.86
1400	Steel strips including stakes		600	.040			1.71		1.71	2.57
1500	Planters, concrete, 48" diam.	2 Clab	30	.533	Ea.		22.50		22.50	33.50
1600	Concrete 7' diam.		20	.800			33.50		33.50	50.50
1700	Fiberglass 36" diam.		30	.533			22.50		22.50	33.50
1800	60" diam.		20	.800			33.50		33.50	50.50
1900	24" square		30	.533			22.50		22.50	33.50
2000	48" square		30	.533			22.50		22.50	33.50
2100	Wood 48" square		30	.533			22.50		22.50	33.50
2200	48" diam.		20	.800			33.50		33.50	50.50
2300	72" diam.		20	.800			33.50		33.50	50.50
2400	Planter/bench, fiberglass 72"-96" square		10	1.600			67.50		67.50	101
2500	Wood, 72" square		10	1.600			67.50		67.50	101
2600	Tree guying, stakes, 3"-4" caliper		40	.400			16.85		16.85	25.50

02 41 Demolition

02 41 13 – Selective Site Demolition

02 41 13.98 Select. Demo., Sod, Edging, Planters & Tree Guying

		Crew	Daily Output	Labor-Hours	Unit	Material	2020 Bare Costs Labor	Equipment	Total	Total Incl O&P
2700	anchors, 3" caliper	2 Clab	40	.400	Ea.		16.85		16.85	25.50
2800	3"-6" caliper		30	.533			22.50		22.50	33.50
2900	6" caliper		20	.800			33.50		33.50	50.50
3000	8" caliper	↓	15	1.067	↓		45		45	67.50

02 41 16 – Structure Demolition

02 41 16.13 Building Demolition

		Crew	Daily Output	Labor-Hours	Unit	Material	2020 Bare Costs Labor	Equipment	Total	Total Incl O&P
0010	**BUILDING DEMOLITION** Large urban projects, incl. 20 mi. haul R024119-10									
0011	No foundation or dump fees, C.F. is vol. of building standing									
0020	Steel	B-8	21500	.003	C.F.		.15	.13	.28	.37
0050	Concrete		15300	.004			.20	.19	.39	.52
0080	Masonry		20100	.003			.16	.14	.30	.39
0100	Mixture of types	↓	20100	.003			.16	.14	.30	.39
0500	Small bldgs, or single bldgs, no salvage included, steel	B-3	14800	.003			.15	.16	.31	.40
0600	Concrete		11300	.004			.20	.21	.41	.53
0650	Masonry		14800	.003			.15	.16	.31	.40
0700	Wood	↓	14800	.003			.15	.16	.31	.40
0750	For buildings with no interior walls, deduct				↓				30%	30%
1000	Demolition single family house, one story, wood 1600 S.F.	B-3	1	48	Ea.		2,275	2,325	4,600	5,950
1020	3200 S.F.		.50	96			4,525	4,625	9,150	11,900
1200	Demolition two family house, two story, wood 2400 S.F.		.67	71.964			3,400	3,475	6,875	8,925
1220	4200 S.F.		.38	128			6,025	6,175	12,200	15,900
1300	Demolition three family house, three story, wood 3200 S.F.		.50	96			4,525	4,625	9,150	11,900
1320	5400 S.F.	↓	.30	160			7,550	7,725	15,275	19,800
5000	For buildings with no interior walls, deduct				↓				30%	30%

02 41 16.15 Explosive/Implosive Demolition

		Crew	Daily Output	Labor-Hours	Unit	Material	2020 Bare Costs Labor	Equipment	Total	Total Incl O&P
0010	**EXPLOSIVE/IMPLOSIVE DEMOLITION** R024119-10									
0011	Large projects,									
0020	No disposal fee based on building volume, steel building	B-5B	16900	.003	C.F.		.15	.16	.31	.40
0100	Concrete building		16900	.003			.15	.16	.31	.40
0200	Masonry building	↓	16900	.003	↓		.15	.16	.31	.40
0400	Disposal of material, minimum	B-3	445	.108	C.Y.		5.10	5.20	10.30	13.35
0500	Maximum	"	365	.132	"		6.20	6.35	12.55	16.30

02 41 16.17 Building Demolition Footings and Foundations

		Crew	Daily Output	Labor-Hours	Unit	Material	2020 Bare Costs Labor	Equipment	Total	Total Incl O&P
0010	**BUILDING DEMOLITION FOOTINGS AND FOUNDATIONS** R024119-10									
0200	Floors, concrete slab on grade,									
0240	4" thick, plain concrete	B-13L	5000	.003	S.F.		.19	.42	.61	.74
0280	Reinforced, wire mesh		4000	.004			.24	.52	.76	.92
0300	Rods		4500	.004			.21	.46	.67	.83
0400	6" thick, plain concrete		4000	.004			.24	.52	.76	.92
0420	Reinforced, wire mesh		3200	.005			.30	.65	.95	1.15
0440	Rods		3600	.004	↓		.26	.58	.84	1.02
1000	Footings, concrete, 1' thick, 2' wide		300	.053	L.F.		3.16	6.95	10.11	12.30
1080	1'-6" thick, 2' wide		250	.064			3.79	8.30	12.09	14.80
1120	3' wide		200	.080			4.74	10.40	15.14	18.55
1140	2' thick, 3' wide	↓	175	.091			5.40	11.90	17.30	21
1200	Average reinforcing, add								10%	10%
1220	Heavy reinforcing, add				↓				20%	20%
2000	Walls, block, 4" thick	B-13L	8000	.002	S.F.		.12	.26	.38	.47
2040	6" thick		6000	.003			.16	.35	.51	.62
2080	8" thick		4000	.004			.24	.52	.76	.92
2100	12" thick	↓	3000	.005	↓		.32	.69	1.01	1.23
2200	For horizontal reinforcing, add								10%	10%

02 41 Demolition

02 41 16 – Structure Demolition

02 41 16.17 Building Demolition Footings and Foundations

		Crew	Daily Output	Labor-Hours	Unit	Material	2020 Bare Costs Labor	2020 Bare Costs Equipment	Total	Total Incl O&P
2220	For vertical reinforcing, add				S.F.				20%	20%
2400	Concrete, plain concrete, 6" thick	B-13L	4000	.004			.24	.52	.76	.92
2420	8" thick		3500	.005			.27	.59	.86	1.05
2440	10" thick		3000	.005			.32	.69	1.01	1.23
2500	12" thick	↓	2500	.006			.38	.83	1.21	1.48
2600	For average reinforcing, add								10%	10%
2620	For heavy reinforcing, add								20%	20%
4000	For congested sites or small quantities, add up to				↓				200%	200%
4200	Add for disposal, on site	B-11A	232	.069	C.Y.		3.41	6.50	9.91	12.25
4250	To five miles	B-30	220	.109	"		5.60	8.35	13.95	17.55

02 41 16.33 Bridge Demolition

		Crew	Daily Output	Labor-Hours	Unit	Material	2020 Bare Costs Labor	2020 Bare Costs Equipment	Total	Total Incl O&P
0010	**BRIDGE DEMOLITION**									
0100	Bridges, pedestrian, precast, 60'-150' long	B-21C	250	.224	S.F.		10.30	8.25	18.55	24.50
0200	Steel, 50'-160' long, 8'-10' wide	"	500	.112			5.15	4.13	9.28	12.30
0300	Laminated wood, 80'-130' long	C-12	300	.160	↓		8.40	1.57	9.97	14.40

02 41 19 – Selective Demolition

02 41 19.13 Selective Building Demolition

		Crew	Daily Output	Labor-Hours	Unit	Material	2020 Bare Costs Labor	2020 Bare Costs Equipment	Total	Total Incl O&P
0010	**SELECTIVE BUILDING DEMOLITION**									
0020	Costs related to selective demolition of specific building components									
0025	are included under Common Work Results (XX 05)									
0030	in the component's appropriate division.									

02 41 19.16 Selective Demolition, Cutout

		Crew	Daily Output	Labor-Hours	Unit	Material	2020 Bare Costs Labor	2020 Bare Costs Equipment	Total	Total Incl O&P
0010	**SELECTIVE DEMOLITION, CUTOUT** R024119-10									
0020	Concrete, elev. slab, light reinforcement, under 6 C.F.	B-9	65	.615	C.F.		26	5.50	31.50	45.50
0050	Light reinforcing, over 6 C.F.		75	.533	"		22.50	4.76	27.26	39.50
0200	Slab on grade to 6" thick, not reinforced, under 8 S.F.		85	.471	S.F.		20	4.20	24.20	34.50
0250	8-16 S.F.	↓	175	.229	"		9.70	2.04	11.74	16.85
0255	For over 16 S.F. see Line 02 41 16.17 0400									
0600	Walls, not reinforced, under 6 C.F.	B-9	60	.667	C.F.		28.50	5.95	34.45	49
0650	6-12 C.F.	"	80	.500	"		21.50	4.47	25.97	37
0655	For over 12 C.F. see Line 02 41 16.17 2500									
1000	Concrete, elevated slab, bar reinforced, under 6 C.F.	B-9	45	.889	C.F.		38	7.95	45.95	66
1050	Bar reinforced, over 6 C.F.		50	.800	"		34	7.15	41.15	59
1200	Slab on grade to 6" thick, bar reinforced, under 8 S.F.		75	.533	S.F.		22.50	4.76	27.26	39.50
1250	8-16 S.F.	↓	150	.267	"		11.35	2.38	13.73	19.65
1255	For over 16 S.F. see Line 02 41 16.17 0440									
1400	Walls, bar reinforced, under 6 C.F.	B-9	50	.800	C.F.		34	7.15	41.15	59
1450	6-12 C.F.	"	70	.571	"		24.50	5.10	29.60	42
1455	For over 12 C.F. see Lines 02 41 16.17 2500 and 2600									
2000	Brick, to 4 S.F. opening, not including toothing									
2040	4" thick	B-9	30	1.333	Ea.		56.50	11.90	68.40	98
2060	8" thick		18	2.222			94.50	19.85	114.35	164
2080	12" thick		10	4			170	35.50	205.50	295
2400	Concrete block, to 4 S.F. opening, 2" thick		35	1.143			48.50	10.20	58.70	84
2420	4" thick		30	1.333			56.50	11.90	68.40	98
2440	8" thick		27	1.481			63	13.25	76.25	109
2460	12" thick		24	1.667			71	14.90	85.90	122
2600	Gypsum block, to 4 S.F. opening, 2" thick		80	.500			21.50	4.47	25.97	37
2620	4" thick		70	.571			24.50	5.10	29.60	42
2640	8" thick		55	.727			31	6.50	37.50	53.50
2800	Terra cotta, to 4 S.F. opening, 4" thick		70	.571			24.50	5.10	29.60	42
2840	8" thick	↓	65	.615	↓		26	5.50	31.50	45.50

02 41 Demolition

02 41 19 – Selective Demolition

02 41 19.16 Selective Demolition, Cutout

		Crew	Daily Output	Labor-Hours	Unit	Material	2020 Bare Costs Labor	Equipment	Total	Total Incl O&P
2880	12" thick	B-9	50	.800	Ea.		34	7.15	41.15	59
3000	Toothing masonry cutouts, brick, soft old mortar	1 Brhe	40	.200	V.L.F.		8.35		8.35	12.65
3100	Hard mortar		30	.267			11.15		11.15	16.90
3200	Block, soft old mortar		70	.114			4.77		4.77	7.25
3400	Hard mortar	▼	50	.160	▼		6.70		6.70	10.15
6000	Walls, interior, not including re-framing,									
6010	openings to 5 S.F.									
6100	Drywall to 5/8" thick	1 Clab	24	.333	Ea.		14.05		14.05	21
6200	Paneling to 3/4" thick		20	.400			16.85		16.85	25.50
6300	Plaster, on gypsum lath		20	.400			16.85		16.85	25.50
6340	On wire lath	▼	14	.571	▼		24		24	36
7000	Wood frame, not including re-framing, openings to 5 S.F.									
7200	Floors, sheathing and flooring to 2" thick	1 Clab	5	1.600	Ea.		67.50		67.50	101
7310	Roofs, sheathing to 1" thick, not including roofing		6	1.333			56		56	84.50
7410	Walls, sheathing to 1" thick, not including siding	▼	7	1.143	▼		48		48	72.50

02 41 19.18 Selective Demolition, Disposal Only

		Crew	Daily Output	Labor-Hours	Unit	Material	2020 Bare Costs Labor	Equipment	Total	Total Incl O&P
0010	**SELECTIVE DEMOLITION, DISPOSAL ONLY** R024119-10									
0015	Urban bldg w/salvage value allowed									
0020	Including loading and 5 mile haul to dump									
0200	Steel frame	B-3	430	.112	C.Y.		5.25	5.40	10.65	13.80
0300	Concrete frame		365	.132			6.20	6.35	12.55	16.30
0400	Masonry construction		445	.108			5.10	5.20	10.30	13.35
0500	Wood frame	▼	247	.194	▼		9.15	9.35	18.50	24

02 41 19.19 Selective Demolition

		Crew	Daily Output	Labor-Hours	Unit	Material	2020 Bare Costs Labor	Equipment	Total	Total Incl O&P
0010	**SELECTIVE DEMOLITION**, Rubbish Handling R024119-10									
0020	The following are to be added to the demolition prices									
0050	The following are components for a complete chute system									
0100	Top chute circular steel, 4' long, 18" diameter R024119-30	B-1C	15	1.600	Ea.	285	68.50	20.50	374	440
0102	23" diameter		15	1.600		310	68.50	20.50	399	465
0104	27" diameter		15	1.600		330	68.50	20.50	419	490
0106	30" diameter		15	1.600		355	68.50	20.50	444	515
0108	33" diameter		15	1.600		380	68.50	20.50	469	545
0110	36" diameter		15	1.600		405	68.50	20.50	494	570
0112	Regular chute, 18" diameter		15	1.600		214	68.50	20.50	303	360
0114	23" diameter		15	1.600		238	68.50	20.50	327	390
0116	27" diameter		15	1.600		262	68.50	20.50	351	415
0118	30" diameter		15	1.600		268	68.50	20.50	357	420
0120	33" diameter		15	1.600		310	68.50	20.50	399	465
0122	36" diameter		15	1.600		330	68.50	20.50	419	490
0124	Control door chute, 18" diameter		15	1.600		405	68.50	20.50	494	570
0126	23" diameter		15	1.600		430	68.50	20.50	519	595
0128	27" diameter		15	1.600		450	68.50	20.50	539	620
0130	30" diameter		15	1.600		475	68.50	20.50	564	650
0132	33" diameter		15	1.600		500	68.50	20.50	589	675
0134	36" diameter		15	1.600		520	68.50	20.50	609	700
0136	Chute liners, 14 ga., 18"-30" diameter		15	1.600		228	68.50	20.50	317	375
0138	33"-36" diameter		15	1.600		284	68.50	20.50	373	440
0140	17% thinner chute, 30" diameter		15	1.600		233	68.50	20.50	322	385
0142	33% thinner chute, 30" diameter	▼	15	1.600		176	68.50	20.50	265	320
0144	Top chute cover	1 Clab	24	.333		155	14.05		169.05	192
0146	Door chute cover	"	24	.333		155	14.05		169.05	192
0148	Top chute trough	2 Clab	12	1.333		530	56		586	665

02 41 19.19 Selective Demolition	Crew	Daily Output	Labor-Hours	Unit	Material	2020 Bare Costs Labor	Equipment	Total	Total Incl O&P
0150 Bolt down frame & counter weights, 250 lb.	B-1	4	6	Ea.	4,325	257		4,582	5,125
0152 500 lb.		4	6		6,375	257		6,632	7,375
0154 750 lb.		4	6		11,100	257		11,357	12,600
0156 1000 lb.		2.67	8.989		12,000	385		12,385	13,800
0158 1500 lb.		2.67	8.989		14,000	385		14,385	16,000
0160 Chute warning light system, 5 stories	B-1C	4	6		8,525	257	77.50	8,859.50	9,850
0162 10 stories	"	2	12		14,800	515	155	15,470	17,200
0164 Dust control device for dumpsters	1 Clab	8	1		158	42		200	238
0166 Install or replace breakaway cord		8	1		28	42		70	94
0168 Install or replace warning sign		16	.500		11.10	21		32.10	43.50
0600 Dumpster, weekly rental, 1 dump/week, 6 C.Y. capacity (2 tons)				Week	415			415	455
0700 10 C.Y. capacity (3 tons)					480			480	530
0725 20 C.Y. capacity (5 tons) R024119-20					565			565	625
0800 30 C.Y. capacity (7 tons)					730			730	800
0840 40 C.Y. capacity (10 tons)					775			775	850
0900 Alternate pricing for dumpsters									
0910 Delivery, average for all sizes				Ea.	75			75	82.50
0920 Haul, average for all sizes					235			235	259
0930 Rent per day, average for all sizes					20			20	22
0940 Rent per month, average for all sizes					80			80	88
0950 Disposal fee per ton, average for all sizes				Ton	88			88	97
2000 Load, haul, dump and return, 0'-50' haul, hand carried	2 Clab	24	.667	C.Y.		28		28	42
2005 Wheeled		37	.432			18.20		18.20	27.50
2040 0'-100' haul, hand carried		16.50	.970			41		41	61.50
2045 Wheeled		25	.640			27		27	40.50
2050 Forklift	A-3R	25	.320			16.95	10.25	27.20	37
2080 Haul and return, add per each extra 100' haul, hand carried	2 Clab	35.50	.451			18.95		18.95	28.50
2085 Wheeled		54	.296			12.45		12.45	18.75
2120 For travel in elevators, up to 10 floors, add		140	.114			4.81		4.81	7.25
2130 0'-50' haul, incl. up to 5 riser stairs, hand carried		23	.696			29.50		29.50	44
2135 Wheeled		35	.457			19.25		19.25	29
2140 6-10 riser stairs, hand carried		22	.727			30.50		30.50	46
2145 Wheeled		34	.471			19.80		19.80	30
2150 11-20 riser stairs, hand carried		20	.800			33.50		33.50	50.50
2155 Wheeled		31	.516			21.50		21.50	32.50
2160 21-40 riser stairs, hand carried		16	1			42		42	63.50
2165 Wheeled		24	.667			28		28	42
2170 0'-100' haul, incl. 5 riser stairs, hand carried		15	1.067			45		45	67.50
2175 Wheeled		23	.696			29.50		29.50	44
2180 6-10 riser stairs, hand carried		14	1.143			48		48	72.50
2185 Wheeled		21	.762			32		32	48
2190 11-20 riser stairs, hand carried		12	1.333			56		56	84.50
2195 Wheeled		18	.889			37.50		37.50	56
2200 21-40 riser stairs, hand carried		8	2			84		84	127
2205 Wheeled		12	1.333			56		56	84.50
2210 Haul and return, add per each extra 100' haul, hand carried		35.50	.451			18.95		18.95	28.50
2215 Wheeled		54	.296			12.45		12.45	18.75
2220 For each additional flight of stairs, up to 5 risers, add		550	.029	Flight		1.22		1.22	1.84
2225 6-10 risers, add		275	.058			2.45		2.45	3.68
2230 11-20 risers, add		138	.116			4.88		4.88	7.35
2235 21-40 risers, add		69	.232			9.75		9.75	14.65
3000 Loading & trucking, including 2 mile haul, chute loaded	B-16	45	.711	C.Y.		31.50	12.75	44.25	61.50
3040 Hand loading truck, 50' haul	"	48	.667			29.50	11.95	41.45	57.50

02 41 Demolition

02 41 19 – Selective Demolition

02 41 19.19 Selective Demolition

02 41 19.19 Selective Demolition	Crew	Daily Output	Labor-Hours	Unit	Material	2020 Bare Costs Labor	2020 Bare Costs Equipment	Total	Total Incl O&P
3080	Machine loading truck	B-17	120	.267	C.Y.		12.40	5.35	17.75	24.50
5000	Haul, per mile, up to 8 C.Y. truck	B-34B	1165	.007			.34	.49	.83	1.04
5100	Over 8 C.Y. truck	"	1550	.005	↓		.25	.37	.62	.79

02 41 19.20 Selective Demolition, Dump Charges

		Crew	Daily Output	Labor-Hours	Unit	Material	Labor	Equipment	Total	Total Incl O&P
0010	**SELECTIVE DEMOLITION, DUMP CHARGES**	R024119-10								
0020	Dump charges, typical urban city, tipping fees only									
0100	Building construction materials				Ton	74			74	81
0200	Trees, brush, lumber					63			63	69.50
0300	Rubbish only					63			63	69.50
0500	Reclamation station, usual charge				↓	74			74	81

02 41 19.21 Selective Demolition, Gutting

		Crew	Daily Output	Labor-Hours	Unit	Material	Labor	Equipment	Total	Total Incl O&P
0010	**SELECTIVE DEMOLITION, GUTTING**	R024119-10								
0020	Building interior, including disposal, dumpster fees not included									
0500	Residential building									
0560	Minimum	B-16	400	.080	SF Flr.		3.54	1.43	4.97	6.90
0580	Maximum	"	360	.089	"		3.94	1.59	5.53	7.65
0900	Commercial building									
1000	Minimum	B-16	350	.091	SF Flr.		4.05	1.64	5.69	7.90
1020	Maximum	"	250	.128	"		5.65	2.29	7.94	11

02 41 19.25 Selective Demolition, Saw Cutting

		Crew	Daily Output	Labor-Hours	Unit	Material	Labor	Equipment	Total	Total Incl O&P
0010	**SELECTIVE DEMOLITION, SAW CUTTING**	R024119-10								
0015	Asphalt, up to 3" deep	B-89	1050	.015	L.F.	.20	.76	.98	1.94	2.44
0020	Each additional inch of depth	"	1800	.009		.07	.45	.57	1.09	1.37
1200	Masonry walls, hydraulic saw, brick, per inch of depth	B-89B	300	.053		.07	2.67	5	7.74	9.55
1220	Block walls, solid, per inch of depth	"	250	.064		.07	3.21	6	9.28	11.45
2000	Brick or masonry w/hand held saw, per inch of depth	A-1	125	.064		.05	2.69	.89	3.63	5.10
5000	Wood sheathing to 1" thick, on walls	1 Carp	200	.040			2.13		2.13	3.19
5020	On roof	"	250	.032	↓		1.70		1.70	2.56

02 41 19.27 Selective Demolition, Torch Cutting

		Crew	Daily Output	Labor-Hours	Unit	Material	Labor	Equipment	Total	Total Incl O&P
0010	**SELECTIVE DEMOLITION, TORCH CUTTING**	R024119-10								
0020	Steel, 1" thick plate	E-25	333	.024	L.F.	.90	1.43	.04	2.37	3.27
0040	1" diameter bar	"	600	.013	Ea.	.15	.79	.02	.96	1.43
1000	Oxygen lance cutting, reinforced concrete walls									
1040	12"-16" thick walls	1 Clab	10	.800	L.F.		33.50		33.50	50.50
1080	24" thick walls	"	6	1.333	"		56		56	84.50

02 42 Removal and Diversion of Construction Materials

02 42 10 – Building Deconstruction

02 42 10.10 Estimated Salvage Value or Savings

			Crew	Daily Output	Labor-Hours	Unit	Material	Labor	Equipment	Total	Total Incl O&P
0010	**ESTIMATED SALVAGE VALUE OR SAVINGS**										
0015	Excludes material handling, packaging, container costs and										
0020	transportation for salvage or disposal										
0050	All items in Section 02 42 10.10 are credit deducts and not costs										
0100	Copper wire salvage value	G				Lb.	1.70			1.70	1.87
0200	Brass salvage value	G					1.25			1.25	1.38
0250	Bronze						1.45			1.45	1.60
0300	Steel	G					.05			.05	.06
0304	304 Stainless Steel						.30			.30	.33
0316	316 Stainless Steel						.45			.45	.50
0400	Cast Iron	G				↓	.06			.06	.07

02 42 Removal and Diversion of Construction Materials

02 42 10 – Building Deconstruction

02 42 10.10 Estimated Salvage Value or Savings		Crew	Daily Output	Labor-Hours	Unit	Material	2020 Bare Costs Labor	Equipment	Total	Total Incl O&P
0600	Aluminum siding	G			Lb.	.33			.33	.36
0610	Aluminum wire				↓	.47			.47	.52

02 42 10.20 Deconstruction of Building Components		Crew	Daily Output	Labor-Hours	Unit	Material	2020 Bare Costs Labor	Equipment	Total	Total Incl O&P
0010	**DECONSTRUCTION OF BUILDING COMPONENTS**									
0012	Buildings one or two stories only									
0015	Excludes material handling, packaging, container costs and									
0020	transportation for salvage or disposal									
0050	Deconstruction of plumbing fixtures									
0100	Wall hung or countertop lavatory	G	2 Clab	16	1	Ea.		42		42	63.50
0110	Single or double compartment kitchen sink	G		14	1.143			48		48	72.50
0120	Wall hung urinal	G		14	1.143			48		48	72.50
0130	Floor mounted	G		8	2			84		84	127
0140	Floor mounted water closet	G		16	1			42		42	63.50
0150	Wall hung	G		14	1.143			48		48	72.50
0160	Water fountain, free standing	G		16	1			42		42	63.50
0170	Wall hung or deck mounted	G		12	1.333			56		56	84.50
0180	Bathtub, steel or fiberglass	G		10	1.600			67.50		67.50	101
0190	Cast iron	G		8	2			84		84	127
0200	Shower, single	G		6	2.667			112		112	169
0210	Group	G	↓	7	2.286	↓		96		96	145
0300	Deconstruction of electrical fixtures									
0310	Surface mounted incandescent fixtures	G	2 Clab	48	.333	Ea.		14.05		14.05	21
0320	Fluorescent, 2 lamp	G		32	.500			21		21	31.50
0330	4 lamp	G		24	.667			28		28	42
0340	Strip fluorescent, 1 lamp	G		40	.400			16.85		16.85	25.50
0350	2 lamp	G		32	.500			21		21	31.50
0400	Recessed drop-in fluorescent fixture, 2 lamp	G		27	.593			25		25	37.50
0410	4 lamp	G	↓	18	.889	↓		37.50		37.50	56
0500	Deconstruction of appliances									
0510	Cooking stoves	G	2 Clab	26	.615	Ea.		26		26	39
0520	Dishwashers	G	"	26	.615	"		26		26	39
0600	Deconstruction of millwork and trim									
0610	Cabinets, wood	G	2 Carp	40	.400	L.F.		21.50		21.50	32
0620	Countertops	G		100	.160	"		8.50		8.50	12.80
0630	Wall paneling, 1" thick	G		500	.032	S.F.		1.70		1.70	2.56
0640	Ceiling trim	G		500	.032	L.F.		1.70		1.70	2.56
0650	Wainscoting	G		500	.032	S.F.		1.70		1.70	2.56
0660	Base, 3/4"-1" thick	G	↓	600	.027	L.F.		1.42		1.42	2.13
0700	Deconstruction of doors and windows									
0710	Doors, wrap, interior, wood, single, no closers	G	2 Carp	21	.762	Ea.	4.75	40.50		45.25	66.50
0720	Double	G		13	1.231		9.50	65.50		75	109
0730	Solid core, single, exterior or interior	G		10	1.600		4.75	85		89.75	133
0740	Double	G	↓	8	2	↓	9.50	106		115.50	170
0810	Windows, wrap, wood, single									
0812	with no casement or cladding	G	2 Carp	21	.762	Ea.	4.75	40.50		45.25	66.50
0820	with casement and/or cladding	G	"	18	.889	"	4.75	47		51.75	76.50
0900	Deconstruction of interior finishes									
0910	Drywall for recycling	G	2 Clab	1775	.009	S.F.		.38		.38	.57
0920	Plaster wall, first floor	G		1775	.009			.38		.38	.57
0930	Second floor	G	↓	1330	.012	↓		.51		.51	.76
1000	Deconstruction of roofing and accessories									
1010	Built-up roofs	G	2 Clab	570	.028	S.F.		1.18		1.18	1.78

For customer support on your Site Work & Landscape Costs with RSMeans data, call 800.448.8182.

45

02 42 Removal and Diversion of Construction Materials

02 42 10 – Building Deconstruction

02 42 10.20 Deconstruction of Building Components

		Crew	Daily Output	Labor-Hours	Unit	Material	2020 Bare Costs Labor	2020 Bare Costs Equipment	Total	Total Incl O&P
1020	Gutters, fascia and rakes [G]	2 Clab	1140	.014	L.F.		.59		.59	.89
2000	Deconstruction of wood components									
2010	Roof sheeting [G]	2 Clab	570	.028	S.F.		1.18		1.18	1.78
2020	Main roof framing [G]		760	.021	L.F.		.89		.89	1.33
2030	Porch roof framing [G]		445	.036			1.51		1.51	2.27
2040	Beams 4" x 8" [G]	B-1	375	.064			2.74		2.74	4.11
2050	4" x 10" [G]		300	.080			3.42		3.42	5.15
2055	4" x 12" [G]		250	.096			4.11		4.11	6.15
2060	6" x 8" [G]		250	.096			4.11		4.11	6.15
2065	6" x 10" [G]		200	.120			5.15		5.15	7.70
2070	6" x 12" [G]		170	.141			6.05		6.05	9.05
2075	8" x 12" [G]		126	.190			8.15		8.15	12.25
2080	10" x 12" [G]		100	.240			10.25		10.25	15.40
2100	Ceiling joists [G]	2 Clab	800	.020			.84		.84	1.27
2150	Wall framing, interior [G]		1230	.013			.55		.55	.82
2160	Sub-floor [G]		2000	.008	S.F.		.34		.34	.51
2170	Floor joists [G]		2000	.008	L.F.		.34		.34	.51
2200	Wood siding (no lead or asbestos) [G]		1300	.012	S.F.		.52		.52	.78
2300	Wall framing, exterior [G]		1600	.010	L.F.		.42		.42	.63
2400	Stair risers [G]		53	.302	Ea.		12.70		12.70	19.10
2500	Posts [G]		800	.020	L.F.		.84		.84	1.27
3000	Deconstruction of exterior brick walls									
3010	Exterior brick walls, first floor [G]	2 Clab	200	.080	S.F.		3.37		3.37	5.05
3020	Second floor [G]		64	.250	"		10.55		10.55	15.80
3030	Brick chimney [G]		100	.160	C.F.		6.75		6.75	10.10
4000	Deconstruction of concrete									
4010	Slab on grade, 4" thick, plain concrete [G]	B-9	500	.080	S.F.		3.40	.71	4.11	5.90
4020	Wire mesh reinforced [G]		470	.085			3.62	.76	4.38	6.30
4030	Rod reinforced [G]		400	.100			4.25	.89	5.14	7.40
4110	Foundation wall, 6" thick, plain concrete [G]		160	.250			10.65	2.23	12.88	18.40
4120	8" thick [G]		140	.286			12.15	2.55	14.70	21
4130	10" thick [G]		120	.333			14.15	2.98	17.13	25
9000	Deconstruction process, support equipment as needed									
9010	Daily use, portal to portal, 12-ton truck-mounted hydraulic crane crew [G]	A-3H	1	8	Day		475	725	1,200	1,500
9020	Daily use, skid steer and operator [G]	A-3C	1	8			425	400	825	1,075
9030	Daily use, backhoe 48 HP, operator and labor [G]	"	1	8			425	400	825	1,075

02 42 10.30 Deconstruction Material Handling

		Crew	Daily Output	Labor-Hours	Unit	Material	2020 Bare Costs Labor	2020 Bare Costs Equipment	Total	Total Incl O&P
0010	**DECONSTRUCTION MATERIAL HANDLING**									
0012	Buildings one or two stories only									
0100	Clean and stack brick on pallet [G]	2 Clab	1200	.013	Ea.		.56		.56	.84
0200	Haul 50' and load rough lumber up to 2" x 8" size [G]		2000	.008	"		.34		.34	.51
0210	Lumber larger than 2" x 8" [G]		3200	.005	B.F.		.21		.21	.32
0300	Finish wood for recycling stack and wrap per pallet [G]		8	2	Ea.	38	84		122	169
0350	Light fixtures		6	2.667		68.50	112		180.50	244
0375	Windows		6	2.667		64.50	112		176.50	240
0400	Miscellaneous materials		8	2		19	84		103	148
1000	See Section 02 41 19.19 for bulk material handling									

02 43 Structure Moving

02 43 13 — Structure Relocation

02 43 13.13 Building Relocation

		Crew	Daily Output	Labor-Hours	Unit	Material	2020 Bare Costs Labor	Equipment	Total	Total Incl O&P
0010	**BUILDING RELOCATION**									
0011	One day move, up to 24' wide									
0020	Reset on existing foundation				Total				11,500	11,500
0040	Wood or steel frame bldg., based on ground floor area	G B-4	185	.259	S.F.		11.30	2.67	13.97	19.95
0060	Masonry bldg., based on ground floor area	G "	137	.350			15.25	3.60	18.85	27
0200	For 24'-42' wide, add						15%			15%

02 56 Site Containment

02 56 13 — Waste Containment

02 56 13.10 Containment of Hazardous Waste

		Crew	Daily Output	Labor-Hours	Unit	Material	2020 Bare Costs Labor	Equipment	Total	Total Incl O&P
0010	**CONTAINMENT OF HAZARDOUS WASTE**									
0020	OSHA Hazard level C									
0030	OSHA Hazard level D decrease labor and equipment, deduct						45%	45%		
0035	OSHA Hazard level B increase labor and equipment, add						22%	22%		
0040	OSHA Hazard level A increase labor and equipment, add						71%	71%		
0100	Excavation of contaminated soil & waste									
0105	Includes one respirator filter and two disposable suits per work day									
0110	3/4 C.Y. excavator to 10' deep	B-12F	51	.314	B.C.Y.	1.37	15.90	13.60	30.87	40.50
0120	Labor crew to 6' deep	B-2	19	2.105		9.20	89.50		98.70	144
0130	6'-12' deep	"	12	3.333		14.60	142		156.60	229
0200	Move contaminated soil/waste up to 150' on-site with 2.5 C.Y. loader	B-10T	300	.040	L.C.Y.	.23	2.07	2.34	4.64	5.95
0210	300'	"	186	.065	"	.38	3.35	3.78	7.51	9.55
0300	Secure burial cell construction									
0310	Various liner and cover materials									
0400	Very low density polyethylene (VLDPE)									
0410	50 mil top cover	B-47H	4000	.008	S.F.	.58	.44	.21	1.23	1.53
0420	80 mil liner	"	4000	.008	"	.85	.44	.21	1.50	1.84
0500	Chlorosulfonated polyethylene									
0510	36 mil hypalon top cover	B-47H	4000	.008	S.F.	2.41	.44	.21	3.06	3.55
0520	45 mil hypalon liner	"	4000	.008	"	2.89	.44	.21	3.54	4.08
0600	Polyvinyl chloride (PVC)									
0610	60 mil top cover	B-47H	4000	.008	S.F.	1	.44	.21	1.65	1.99
0620	80 mil liner	"	4000	.008	"	1.27	.44	.21	1.92	2.30
0700	Rough textured H.D. polyethylene (HDPE)									
0710	40 mil top cover	B-47H	4000	.008	S.F.	.53	.44	.21	1.18	1.48
0720	60 mil top cover		4000	.008		.66	.44	.21	1.31	1.62
0722	60 mil liner		4000	.008		.62	.44	.21	1.27	1.58
0730	80 mil liner		3800	.008		.77	.47	.22	1.46	1.79
1000	3/4" crushed stone, 6" deep ballast around liner	B-6	30	.800	L.C.Y.	27.50	36.50	7.15	71.15	93.50
1100	Hazardous waste, ballast cover with common borrow material	B-63	56	.714		12.60	31.50	3.17	47.27	65
1110	Mixture of common borrow & topsoil		56	.714		19.55	31.50	3.17	54.22	72.50
1120	Bank sand		56	.714		18.25	31.50	3.17	52.92	71
1130	Medium priced clay		44	.909		26	40.50	4.04	70.54	94
1140	Mixture of common borrow & medium priced clay		56	.714		19.40	31.50	3.17	54.07	72.50

02 58 Snow Control

02 58 13 – Snow Fencing

02 58 13.10 Snow Fencing System		Crew	Daily Output	Labor-Hours	Unit	Material	2020 Bare Costs Labor	Equipment	Total	Total Incl O&P
0010	**SNOW FENCING SYSTEM**									
7001	Snow fence on steel posts 10' OC, 4' high	B-1	500	.048	L.F.	.39	2.05		2.44	3.51

02 65 Underground Storage Tank Removal

02 65 10 – Underground Tank and Contaminated Soil Removal

02 65 10.30 Removal of Underground Storage Tanks

02 65 10.30 Removal of Underground Storage Tanks			Crew	Daily Output	Labor-Hours	Unit	Material	2020 Bare Costs Labor	Equipment	Total	Total Incl O&P
0010	**REMOVAL OF UNDERGROUND STORAGE TANKS**	R026510-20									
0011	Petroleum storage tanks, non-leaking										
0100	Excavate & load onto trailer										
0110	3,000 gal. to 5,000 gal. tank	G	B-14	4	12	Ea.		530	53.50	583.50	855
0120	6,000 gal. to 8,000 gal. tank	G	B-3A	3	13.333			600	229	829	1,150
0130	9,000 gal. to 12,000 gal. tank	G	"	2	20	↓		900	345	1,245	1,725
0190	Known leaking tank, add					%				100%	100%
0200	Remove sludge, water and remaining product from bottom										
0201	of tank with vacuum truck										
0300	3,000 gal. to 5,000 gal. tank	G	A-13	5	1.600	Ea.		85	148	233	289
0310	6,000 gal. to 8,000 gal. tank	G		4	2			106	184	290	360
0320	9,000 gal. to 12,000 gal. tank	G	↓	3	2.667	↓		141	246	387	480
0390	Dispose of sludge off-site, average					Gal.				6.25	6.80
0400	Insert inert solid CO_2 "dry ice" into tank										
0401	For cleaning/transporting tanks (1.5 lb./100 gal. cap)	G	1 Clab	500	.016	Lb.	1.22	.67		1.89	2.35
0503	Disconnect and remove piping	G	1 Plum	160	.050	L.F.		3.22		3.22	4.82
0603	Transfer liquids, 10% of volume	G	"	1600	.005	Gal.		.32		.32	.48
0703	Cut accessway into underground storage tank	G	1 Clab	5.33	1.501	Ea.		63		63	95
0813	Remove sludge, wash and wipe tank, 500 gal.	G	1 Plum	8	1			64.50		64.50	96.50
0823	3,000 gal.	G		6.67	1.199			77.50		77.50	116
0833	5,000 gal.	G		6.15	1.301			84		84	125
0843	8,000 gal.	G		5.33	1.501			96.50		96.50	145
0853	10,000 gal.	G		4.57	1.751			113		113	169
0863	12,000 gal.	G	↓	4.21	1.900	↓		122		122	183
1020	Haul tank to certified salvage dump, 100 miles round trip										
1023	3,000 gal. to 5,000 gal. tank					Ea.				760	830
1026	6,000 gal. to 8,000 gal. tank									880	960
1029	9,000 gal. to 12,000 gal. tank					↓				1,050	1,150
1100	Disposal of contaminated soil to landfill										
1110	Minimum					C.Y.				145	160
1111	Maximum					"				400	440
1120	Disposal of contaminated soil to										
1121	bituminous concrete batch plant										
1130	Minimum					C.Y.				80	88
1131	Maximum					"				115	125
1203	Excavate, pull, & load tank, backfill hole, 8,000 gal. +	G	B-12C	.50	32	Ea.		1,625	1,900	3,525	4,525
1213	Haul tank to certified dump, 100 miles rt, 8,000 gal. +	G	B-34K	1	8			390	855	1,245	1,525
1223	Excavate, pull, & load tank, backfill hole, 500 gal.	G	B-11C	1	16			790	214	1,004	1,400
1233	Excavate, pull, & load tank, backfill hole, 3,000-5,000 gal.	G	B-11M	.50	32			1,575	465	2,040	2,875
1243	Haul tank to certified dump, 100 miles rt, 500 gal.	G	B-34L	1	8			425	196	621	850
1253	Haul tank to certified dump, 100 miles rt, 3,000-5,000 gal.	G	B-34M	1	8	↓		425	820	1,245	1,525
2010	Decontamination of soil on site incl poly tarp on top/bottom										
2011	Soil containment berm and chemical treatment										
2020	Minimum	G	B-11C	100	.160	C.Y.	7.65	7.90	2.14	17.69	22.50
2021	Maximum	G	"	100	.160	↓	9.90	7.90	2.14	19.94	25

02 65 Underground Storage Tank Removal

02 65 10 – Underground Tank and Contaminated Soil Removal

02 65 10.30 Removal of Underground Storage Tanks	Crew	Daily Output	Labor-Hours	Unit	Material	2020 Bare Costs Labor	Equipment	Total	Total Incl O&P
2050	Disposal of decontaminated soil, minimum				C.Y.				135	150
2055	Maximum				↓				400	440

02 81 Transportation and Disposal of Hazardous Materials

02 81 20 – Hazardous Waste Handling

02 81 20.10 Hazardous Waste Cleanup/Pickup/Disposal

		Crew	Daily Output	Labor-Hours	Unit	Material	2020 Bare Costs Labor	Equipment	Total	Total Incl O&P
0010	**HAZARDOUS WASTE CLEANUP/PICKUP/DISPOSAL**									
0100	For contractor rental equipment, i.e., dozer,									
0110	Front end loader, dump truck, etc., see 01 54 33 Reference Section									
1000	Solid pickup									
1100	55 gal. drums				Ea.				240	265
1120	Bulk material, minimum				Ton				190	210
1130	Maximum				"				595	655
1200	Transportation to disposal site									
1220	Truckload = 80 drums or 25 C.Y. or 18 tons									
1260	Minimum				Mile				3.95	4.45
1270	Maximum				"				7.25	7.98
3000	Liquid pickup, vacuum truck, stainless steel tank									
3100	Minimum charge, 4 hours									
3110	1 compartment, 2200 gallon				Hr.				140	155
3120	2 compartment, 5000 gallon				"				200	225
3400	Transportation in 6900 gallon bulk truck				Mile				7.95	8.75
3410	In teflon lined truck				"				10.20	11.25
5000	Heavy sludge or dry vacuumable material				Hr.				140	155
6000	Dumpsite disposal charge, minimum				Ton				140	155
6020	Maximum				"				415	455

02 91 Chemical Sampling, Testing and Analysis

02 91 10 – Monitoring, Sampling, Testing and Analysis

02 91 10.10 Monitoring, Chem. Sampling, Testing & Analysis

		Crew	Daily Output	Labor-Hours	Unit	Material	2020 Bare Costs Labor	Equipment	Total	Total Incl O&P
0010	**MONITORING, CHEMICAL SAMPLING, TESTING AND ANALYSIS**									
0015	Field sampling of waste									
0100	Field samples, sample collection, sludge	1 Skwk	32	.250	Ea.		13.70		13.70	20.50
0110	Contaminated soils	"	32	.250	"		13.70		13.70	20.50
0200	Vials and bottles									
0210	32 oz. clear wide mouth jar (case of 12)				Ea.	41.50			41.50	46
0220	32 oz. Boston round bottle (case of 12)					37.50			37.50	41.50
0230	32 oz. HDPE bottle (case of 12)				↓	32.50			32.50	36
0300	Laboratory analytical services									
0310	Laboratory testing 13 metals				Ea.	224			224	247
0312	13 metals + mercury					217			217	239
0314	8 metals					167			167	183
0316	Mercury only					61.50			61.50	67.50
0318	Single metal (only Cs, Li, Sr, Ta)					47.50			47.50	52
0320	Single metal (excludes Hg, Cs, Li, Sr, Ta)					47.50			47.50	52.50
0400	Hydrocarbons standard					192			192	212
0410	Hydrocarbons fingerprint					243			243	267
0500	Radioactivity gross alpha					161			161	177
0510	Gross alpha & beta				↓	162			162	178

02 91 Chemical Sampling, Testing and Analysis

02 91 10 – Monitoring, Sampling, Testing and Analysis

02 91 10.10 Monitoring, Chem. Sampling, Testing & Analysis	Crew	Daily Output	Labor-Hours	Unit	Material	2020 Bare Costs Labor	Equipment	Total	Total Incl O&P
0520	Radium 226				Ea.	108			108	119
0530	Radium 228					128			128	141
0540	Radon					156			156	172
0550	Uranium					141			141	155
0600	Volatile organics without GC/MS					163			163	180
0610	Volatile organics including GC/MS					315			315	345
0630	Synthetic organic compounds					1,075			1,075	1,200
0640	Herbicides					221			221	243
0650	Pesticides					157			157	173
0660	PCB's					129			129	142

Estimating Tips

General

- Carefully check all the plans and specifications. Concrete often appears on drawings other than structural drawings, including mechanical and electrical drawings for equipment pads. The cost of cutting and patching is often difficult to estimate. See Subdivision 03 81 for Concrete Cutting, Subdivision 02 41 19.16 for Cutout Demolition, Subdivision 03 05 05.10 for Concrete Demolition, and Subdivision 02 41 19.19 for Rubbish Handling (handling, loading, and hauling of debris).

- Always obtain concrete prices from suppliers near the job site. A volume discount can often be negotiated, depending upon competition in the area. Remember to add for waste, particularly for slabs and footings on grade.

03 10 00 Concrete Forming and Accessories

- A primary cost for concrete construction is forming. Most jobs today are constructed with prefabricated forms. The selection of the forms best suited for the job and the total square feet of forms required for efficient concrete forming and placing are key elements in estimating concrete construction. Enough forms must be available for erection to make efficient use of the concrete placing equipment and crew.

- Concrete accessories for forming and placing depend upon the systems used. Study the plans and specifications to ensure that all special accessory requirements have been included in the cost estimate, such as anchor bolts, inserts, and hangers.

- Included within costs for forms-in-place are all necessary bracing and shoring.

03 20 00 Concrete Reinforcing

- Ascertain that the reinforcing steel supplier has included all accessories, cutting, bending, and an allowance for lapping, splicing, and waste. A good rule of thumb is 10% for lapping, splicing, and waste. Also, 10% waste should be allowed for welded wire fabric.

- The unit price items in the subdivisions for Reinforcing In Place, Glass Fiber Reinforcing, and Welded Wire Fabric include the labor to install accessories such as beam and slab bolsters, high chairs, and bar ties and tie wire. The material cost for these accessories is not included; they may be obtained from the Accessories Subdivisions.

03 30 00 Cast-In-Place Concrete

- When estimating structural concrete, pay particular attention to requirements for concrete additives, curing methods, and surface treatments. Special consideration for climate, hot or cold, must be included in your estimate. Be sure to include requirements for concrete placing equipment and concrete finishing.

- For accurate concrete estimating, the estimator must consider each of the following major components individually: forms, reinforcing steel, ready-mix concrete, placement of the concrete, and finishing of the top surface. For faster estimating, Subdivision 03 30 53.40 for Concrete-In-Place can be used; here, various items of concrete work are presented that include the costs of all five major components (unless specifically stated otherwise).

03 40 00 Precast Concrete
03 50 00 Cast Decks and Underlayment

- The cost of hauling precast concrete structural members is often an important factor. For this reason, it is important to get a quote from the nearest supplier. It may become economically feasible to set up precasting beds on the site if the hauling costs are prohibitive.

Reference Numbers

Reference numbers are shown at the beginning of some major classifications. These numbers refer to related items in the Reference Section. The reference information may be an estimating procedure, an alternate pricing method, or technical information.

Note: Not all subdivisions listed here necessarily appear. ■

Same Data. Simplified.

Enjoy the convenience and efficiency of accessing your costs anywhere:

- **Skip the multiplier** by setting your location
- **Quickly search**, edit, favorite and share costs
- **Stay on top of price changes** with automatic updates

Discover more at rsmeans.com/online

03 01 Maintenance of Concrete

03 01 30 – Maintenance of Cast-In-Place Concrete

03 01 30.62 Concrete Patching

03 01 30.62 Concrete Patching	Crew	Daily Output	Labor-Hours	Unit	Material	2020 Bare Costs Labor	2020 Bare Costs Equipment	Total	Total Incl O&P
0010 **CONCRETE PATCHING**									
0100 Floors, 1/4" thick, small areas, regular grout	1 Cefi	170	.047	S.F.	1.61	2.35		3.96	5.25
0150 Epoxy grout	"	100	.080	"	9.90	4		13.90	16.80
2000 Walls, including chipping, cleaning and epoxy grout									
2100 1/4" deep	1 Cefi	65	.123	S.F.	7.80	6.15		13.95	17.65
2150 1/2" deep		50	.160		15.60	8		23.60	29
2200 3/4" deep	↓	40	.200	↓	23.50	10		33.50	40
2400 Walls, including chipping or sand blasting,									
2410 Priming, and two part polymer mix, 1/4" deep	1 Cefi	80	.100	S.F.	2.41	5		7.41	10
2420 1/2" deep		60	.133		4.82	6.65		11.47	15.10
2430 3/4" deep	↓	40	.200	↓	7.20	10		17.20	22.50

03 05 Common Work Results for Concrete

03 05 05 – Selective Demolition for Concrete

03 05 05.10 Selective Demolition, Concrete

03 05 05.10 Selective Demolition, Concrete	Crew	Daily Output	Labor-Hours	Unit	Material	2020 Bare Costs Labor	2020 Bare Costs Equipment	Total	Total Incl O&P
0010 **SELECTIVE DEMOLITION, CONCRETE** R024119-10									
0012 Excludes saw cutting, torch cutting, loading or hauling									
0050 Break into small pieces, reinf. less than 1% of cross-sectional area	B-9	24	1.667	C.Y.		71	14.90	85.90	122
0060 Reinforcing 1% to 2% of cross-sectional area		16	2.500			106	22.50	128.50	185
0070 Reinforcing more than 2% of cross-sectional area	↓	8	5	↓		213	44.50	257.50	370
0150 Remove whole pieces, up to 2 tons per piece	E-18	36	1.111	Ea.		64.50	41.50	106	146
0160 2-5 tons per piece		30	1.333			77	49.50	126.50	175
0170 5-10 tons per piece		24	1.667			96.50	62	158.50	218
0180 10-15 tons per piece		18	2.222			129	82.50	211.50	291
0250 Precast unit embedded in masonry, up to 1 C.F.	D-1	16	1			47		47	71
0260 1-2 C.F.		12	1.333			62.50		62.50	95
0270 2-5 C.F.		10	1.600			75		75	114
0280 5-10 C.F.	↓	8	2	↓		94		94	142
0990 For hydrodemolition see Section 02 41 13.15									

03 05 13 – Basic Concrete Materials

03 05 13.20 Concrete Admixtures and Surface Treatments

03 05 13.20 Concrete Admixtures and Surface Treatments	Crew	Daily Output	Labor-Hours	Unit	Material	2020 Bare Costs Labor	2020 Bare Costs Equipment	Total	Total Incl O&P
0010 **CONCRETE ADMIXTURES AND SURFACE TREATMENTS**									
0040 Abrasives, aluminum oxide, over 20 tons				Lb.	2.51			2.51	2.76
0050 1 to 20 tons					2.61			2.61	2.87
0070 Under 1 ton					2.71			2.71	2.98
0100 Silicon carbide, black, over 20 tons					4.36			4.36	4.80
0110 1 to 20 tons					4.53			4.53	4.99
0120 Under 1 ton				↓	4.71			4.71	5.20
0200 Air entraining agent, .7 to 1.5 oz. per bag, 55 gallon drum				Gal.	24.50			24.50	27
0220 5 gallon pail					36.50			36.50	40
0300 Bonding agent, acrylic latex, 250 S.F./gallon, 5 gallon pail					28			28	31
0320 Epoxy resin, 80 S.F./gallon, 4 gallon case				↓	69			69	76
0400 Calcium chloride, 50 lb. bags, T.L. lots				Ton	1,525			1,525	1,675
0420 Less than truckload lots				Bag	24.50			24.50	27
0500 Carbon black, liquid, 2 to 8 lb. per bag of cement				Lb.	17			17	18.70
0600 Concrete admixture, integral colors, dry pigment, 5 lb. bag				Ea.	31			31	34.50
0610 10 lb. bag					38.50			38.50	42
0620 25 lb. bag				↓	78.50			78.50	86
0920 Dustproofing compound, 250 S.F./gal., 5 gallon pail				Gal.	7.70			7.70	8.50
1010 Epoxy based, 125 S.F./gal., 5 gallon pail				"	63.50			63.50	69.50

03 05 Common Work Results for Concrete

03 05 13 – Basic Concrete Materials

03 05 13.20 Concrete Admixtures and Surface Treatments

		Crew	Daily Output	Labor-Hours	Unit	Material	2020 Bare Costs Labor	Equipment	Total	Total Incl O&P
1100	Hardeners, metallic, 55 lb. bags, natural (grey)				Lb.	.92			.92	1.01
1200	Colors					1.43			1.43	1.57
1300	Non-metallic, 55 lb. bags, natural grey					.74			.74	.82
1320	Colors				▼	.91			.91	1
1550	Release agent, for tilt slabs, 5 gallon pail				Gal.	18.05			18.05	19.90
1570	For forms, 5 gallon pail					13.35			13.35	14.65
1590	Concrete release agent for forms, 100% biodegradable, zero VOC, 5 gal. pail [G]					23.50			23.50	26
1595	55 gallon drum [G]					19.30			19.30	21
1600	Sealer, hardener and dustproofer, epoxy-based, 125 S.F./gal., 5 gallon unit					63.50			63.50	69.50
1620	3 gallon unit					69.50			69.50	76.50
1630	Sealer, solvent-based, 250 S.F./gal., 55 gallon drum					23			23	25.50
1640	5 gallon pail					33			33	36
1650	Sealer, water based, 350 S.F., 55 gallon drum					23.50			23.50	26
1660	5 gallon pail					26			26	29
1900	Set retarder, 100 S.F./gal., 1 gallon pail				▼	6.95			6.95	7.65
2000	Waterproofing, integral 1 lb. per bag of cement				Lb.	4.02			4.02	4.42
2100	Powdered metallic, 40 lbs. per 100 S.F., standard colors					4.77			4.77	5.25
2120	Premium colors				▼	6.70			6.70	7.35
3000	For colored ready-mix concrete, add to prices in section 03 31 13.35									
3100	Subtle shades, 5 lb. dry pigment per C.Y., add				C.Y.	31			31	34.50
3400	Medium shades, 10 lb. dry pigment per C.Y., add					38.50			38.50	42
3700	Deep shades, 25 lb. dry pigment per C.Y., add				▼	78.50			78.50	86
6000	Concrete ready mix additives, recycled coal fly ash, mixed at plant [G]				Ton	64			64	70
6010	Recycled blast furnace slag, mixed at plant [G]				"	90.50			90.50	99.50

03 05 13.25 Aggregate

		Crew	Daily Output	Labor-Hours	Unit	Material	2020 Bare Costs Labor	Equipment	Total	Total Incl O&P
0010	**AGGREGATE** R033105-20									
0100	Lightweight vermiculite or perlite, 4 C.F. bag, C.L. lots [G]				Bag	22			22	24.50
0150	L.C.L. lots [G]				"	24.50			24.50	27
0250	Sand & stone, loaded at pit, crushed bank gravel				Ton	19.85			19.85	22
0350	Sand, washed, for concrete					21.50			21.50	24
0400	For plaster or brick					21.50			21.50	24
0450	Stone, 3/4" to 1-1/2"					20.50			20.50	22.50
0470	Round, river stone					41			41	45
0500	3/8" roofing stone & 1/2" pea stone					31.50			31.50	34.50
0550	For trucking 10-mile round trip, add to the above	B-34B	117	.068			3.35	4.90	8.25	10.40
0600	For trucking 30-mile round trip, add to the above	"	72	.111	▼		5.45	7.95	13.40	16.90
0850	Sand & stone, loaded at pit, crushed bank gravel				C.Y.	28			28	30.50
0950	Sand, washed, for concrete					30			30	33
1000	For plaster or brick					30			30	33
1050	Stone, 3/4" to 1-1/2"					39			39	43
1055	Round, river stone					44			44	48.50
1100	3/8" roofing stone & 1/2" pea stone					30			30	33
1150	For trucking 10-mile round trip, add to the above	B-34B	78	.103			5	7.35	12.35	15.60
1200	For trucking 30-mile round trip, add to the above	"	48	.167	▼		8.15	11.95	20.10	25.50
1310	Onyx chips, 50 lb. bags				Cwt.	34			34	37
1330	Quartz chips, 50 lb. bags					34.50			34.50	37.50
1410	White marble, 3/8" to 1/2", 50 lb. bags				▼	9.55			9.55	10.50
1430	3/4", bulk				Ton	189			189	207

03 05 13.30 Cement

		Crew	Daily Output	Labor-Hours	Unit	Material	2020 Bare Costs Labor	Equipment	Total	Total Incl O&P
0010	**CEMENT** R033105-20									
0240	Portland, Type I/II, T.L. lots, 94 lb. bags				Bag	14.45			14.45	15.85
0250	L.T.L./L.C.L. lots				"	15.95			15.95	17.55

For customer support on your Site Work & Landscape Costs with RSMeans data, call 800.448.8182.

53

03 05 Common Work Results for Concrete

03 05 13 – Basic Concrete Materials

03 05 13.30 Cement

		Crew	Daily Output	Labor-Hours	Unit	Material	2020 Bare Costs Labor	2020 Bare Costs Equipment	Total	Total Incl O&P
0300	Trucked in bulk, per cwt.				Cwt.	8.70			8.70	9.55
0400	Type III, high early strength, T.L. lots, 94 lb. bags				Bag	13.85			13.85	15.25
0420	L.T.L. or L.C.L. lots					19.30			19.30	21.50
0500	White, type III, high early strength, T.L. or C.L. lots, bags					27.50			27.50	30.50
0520	L.T.L. or L.C.L. lots					38.50			38.50	42.50
0600	White, type I, T.L. or C.L. lots, bags					30			30	33.50
0620	L.T.L. or L.C.L. lots					32			32	35

03 05 13.80 Waterproofing and Dampproofing

		Crew	Daily Output	Labor-Hours	Unit	Material	2020 Bare Costs Labor	2020 Bare Costs Equipment	Total	Total Incl O&P
0010	**WATERPROOFING AND DAMPPROOFING**									
0050	Integral waterproofing, add to cost of regular concrete				C.Y.	24			24	26.50

03 05 13.85 Winter Protection

		Crew	Daily Output	Labor-Hours	Unit	Material	2020 Bare Costs Labor	2020 Bare Costs Equipment	Total	Total Incl O&P
0010	**WINTER PROTECTION**									
0012	For heated ready mix, add				C.Y.	5.35			5.35	5.90
0100	Temporary heat to protect concrete, 24 hours	2 Clab	50	.320	M.S.F.	203	13.45		216.45	243
0200	Temporary shelter for slab on grade, wood frame/polyethylene sheeting									
0201	Build or remove, light framing for short spans	2 Carp	10	1.600	M.S.F.	345	85		430	510
0210	Large framing for long spans	"	3	5.333	"	460	283		743	935
0500	Electrically heated pads, 110 volts, 15 watts/S.F., buy				S.F.	11.60			11.60	12.80
0600	20 watts/S.F., buy					15.45			15.45	17
0710	Electrically heated pads, 15 watts/S.F., 20 uses					.58			.58	.64

03 11 Concrete Forming

03 11 13 – Structural Cast-In-Place Concrete Forming

03 11 13.20 Forms In Place, Beams and Girders

		Crew	Daily Output	Labor-Hours	Unit	Material	2020 Bare Costs Labor	2020 Bare Costs Equipment	Total	Total Incl O&P
0010	**FORMS IN PLACE, BEAMS AND GIRDERS** R031113-40 R031113-60									
3500	Bottoms only, to 30" wide, job-built plywood, 1 use	C-2	230	.209	SFCA	4.49	10.80		15.29	21
3550	2 use		265	.181		2.51	9.35		11.86	16.80
3600	3 use		280	.171		1.79	8.85		10.64	15.25
3650	4 use		290	.166		1.46	8.55		10.01	14.45
4000	Sides only, vertical, 36" high, job-built plywood, 1 use		335	.143		5.95	7.40		13.35	17.65
4050	2 use		405	.119		3.28	6.10		9.38	12.80
4100	3 use		430	.112		2.39	5.75		8.14	11.30
4150	4 use		445	.108		1.94	5.55		7.49	10.50
4500	Sloped sides, 36" high, 1 use		305	.157		5.70	8.15		13.85	18.45
4550	2 use		370	.130		3.17	6.70		9.87	13.55
4600	3 use		405	.119		2.28	6.10		8.38	11.70
4650	4 use		425	.113		1.85	5.85		7.70	10.80
5000	Upstanding beams, 36" high, 1 use		225	.213		7.45	11		18.45	25
5050	2 use		255	.188		4.13	9.70		13.83	19.15
5100	3 use		275	.175		3	9		12	16.85
5150	4 use		280	.171		2.43	8.85		11.28	16

03 11 13.25 Forms In Place, Columns

		Crew	Daily Output	Labor-Hours	Unit	Material	2020 Bare Costs Labor	2020 Bare Costs Equipment	Total	Total Incl O&P
0010	**FORMS IN PLACE, COLUMNS** R031113-40									
0500	Round fiberglass, 4 use per mo., rent, 12" diameter	C-1	160	.200	L.F.	9.95	10.10		20.05	26
0550	16" diameter R031113-60		150	.213		11.85	10.75		22.60	29
0600	18" diameter		140	.229		13.25	11.50		24.75	32
0650	24" diameter		135	.237		16.50	11.95		28.45	36
0700	28" diameter		130	.246		18.45	12.40		30.85	39
0800	30" diameter		125	.256		19.25	12.90		32.15	40.50
0850	36" diameter		120	.267		25.50	13.45		38.95	48

03 11 Concrete Forming

03 11 13 – Structural Cast-In-Place Concrete Forming

03 11 13.25 Forms In Place, Columns

		Crew	Daily Output	Labor-Hours	Unit	Material	2020 Bare Costs Labor	Equipment	Total	Total Incl O&P
1500	Round fiber tube, recycled paper, 1 use, 8" diameter	**G** C-1	155	.206	L.F.	2.96	10.40		13.36	18.90
1550	10" diameter	**G**	155	.206		3.91	10.40		14.31	19.95
1600	12" diameter	**G**	150	.213		4.54	10.75		15.29	21
1650	14" diameter	**G**	145	.221		4.54	11.10		15.64	21.50
1700	16" diameter	**G**	140	.229		5.65	11.50		17.15	23.50
1720	18" diameter	**G**	140	.229		6.60	11.50		18.10	24.50
1750	20" diameter	**G**	135	.237		7.90	11.95		19.85	26.50
1800	24" diameter	**G**	130	.246		11.40	12.40		23.80	31
1850	30" diameter	**G**	125	.256		14.80	12.90		27.70	35.50
1900	36" diameter	**G**	115	.278		20.50	14		34.50	43.50
1950	42" diameter	**G**	100	.320		44.50	16.10		60.60	73
2000	48" diameter	**G**	85	.376		55.50	18.95		74.45	89.50
2200	For seamless type, add					15%				
5000	Job-built plywood, 8" x 8" columns, 1 use	C-1	165	.194	SFCA	2.88	9.75		12.63	17.85
5050	2 use		195	.164		1.65	8.25		9.90	14.20
5100	3 use		210	.152		1.15	7.70		8.85	12.80
5150	4 use		215	.149		.95	7.50		8.45	12.30
5500	12" x 12" columns, 1 use		180	.178		2.80	8.95		11.75	16.55
5550	2 use		210	.152		1.54	7.70		9.24	13.25
5600	3 use		220	.145		1.12	7.35		8.47	12.25
5650	4 use		225	.142		.91	7.15		8.06	11.75
6000	16" x 16" columns, 1 use		185	.173		2.81	8.70		11.51	16.20
6050	2 use		215	.149		1.50	7.50		9	12.90
6100	3 use		230	.139		1.13	7		8.13	11.80
6150	4 use		235	.136		.92	6.85		7.77	11.30
6500	24" x 24" columns, 1 use		190	.168		3.19	8.50		11.69	16.25
6550	2 use		216	.148		1.75	7.45		9.20	13.15
6600	3 use		230	.139		1.27	7		8.27	11.95
6650	4 use		238	.134		1.04	6.75		7.79	11.35
7000	36" x 36" columns, 1 use		200	.160		2.33	8.05		10.38	14.65
7050	2 use		230	.139		1.32	7		8.32	12
7100	3 use		245	.131		.93	6.60		7.53	10.90
7150	4 use		250	.128		.76	6.45		7.21	10.55
7400	Steel framed plywood, based on 50 uses of purchased									
7420	forms, and 4 uses of bracing lumber									
7500	8" x 8" column	C-1	340	.094	SFCA	2.22	4.74		6.96	9.55
7550	10" x 10"		350	.091		1.98	4.61		6.59	9.05
7600	12" x 12"		370	.086		1.68	4.36		6.04	8.40
7650	16" x 16"		400	.080		1.30	4.03		5.33	7.50
7700	20" x 20"		420	.076		1.16	3.84		5	7
7750	24" x 24"		440	.073		.81	3.66		4.47	6.40
7755	30" x 30"		440	.073		1.04	3.66		4.70	6.65
7760	36" x 36"		460	.070		.92	3.51		4.43	6.25

03 11 13.30 Forms In Place, Culvert

			Crew	Daily Output	Labor-Hours	Unit	Material	2020 Bare Costs Labor	Equipment	Total	Total Incl O&P
0010	**FORMS IN PLACE, CULVERT**	R031113-40									
0015	5' to 8' square or rectangular, 1 use		C-1	170	.188	SFCA	4.25	9.50		13.75	18.95
0050	2 use	R031113-60		180	.178		2.57	8.95		11.52	16.25
0100	3 use			190	.168		2.01	8.50		10.51	14.95
0150	4 use			200	.160		1.73	8.05		9.78	14

For customer support on your Site Work & Landscape Costs with RSMeans data, call 800.448.8182.

55

03 11 Concrete Forming

03 11 13 – Structural Cast-In-Place Concrete Forming

03 11 13.35 Forms In Place, Elevated Slabs

		Crew	Daily Output	Labor-Hours	Unit	Material	2020 Bare Costs Labor	2020 Bare Costs Equipment	Total	Total Incl O&P
0010	**FORMS IN PLACE, ELEVATED SLABS** R031113-40									
1000	Flat plate, job-built plywood, to 15' high, 1 use R031113-60	C-2	470	.102	S.F.	4.09	5.25		9.34	12.40
1050	2 use		520	.092		2.25	4.77		7.02	9.65
1100	3 use		545	.088		1.64	4.55		6.19	8.65
1150	4 use		560	.086		1.33	4.43		5.76	8.10
1500	15' to 20' high ceilings, 4 use		495	.097		1.32	5		6.32	8.95
1600	21' to 35' high ceilings, 4 use		450	.107		1.65	5.50		7.15	10.10
2000	Flat slab, drop panels, job-built plywood, to 15' high, 1 use		449	.107		4.85	5.50		10.35	13.65
2050	2 use		509	.094		2.67	4.87		7.54	10.25
2100	3 use		532	.090		1.94	4.66		6.60	9.15
2150	4 use		544	.088		1.58	4.56		6.14	8.60
2250	15' to 20' high ceilings, 4 use		480	.100		2.70	5.15		7.85	10.70
2350	20' to 35' high ceilings, 4 use		435	.110		3.03	5.70		8.73	11.90
3000	Floor slab hung from steel beams, 1 use		485	.099		2.66	5.10		7.76	10.65
3050	2 use		535	.090		2.08	4.63		6.71	9.25
3100	3 use		550	.087		1.89	4.51		6.40	8.80
3150	4 use		565	.085		1.79	4.39		6.18	8.55
5000	Box out for slab openings, over 16" deep, 1 use		190	.253	SFCA	3.75	13.05		16.80	23.50
5050	2 use		240	.200	"	2.06	10.35		12.41	17.75
5500	Shallow slab box outs, to 10 S.F.		42	1.143	Ea.	12.60	59		71.60	102
5550	Over 10 S.F. (use perimeter)		600	.080	L.F.	1.69	4.13		5.82	8.05
6000	Bulkhead forms for slab, with keyway, 1 use, 2 piece		500	.096		1.80	4.96		6.76	9.45
6100	3 piece (see also edge forms)		460	.104		2.06	5.40		7.46	10.35
6200	Slab bulkhead form, 4-1/2" high, exp metal, w/keyway & stakes G	C-1	1200	.027		.94	1.34		2.28	3.05
6210	5-1/2" high G		1100	.029		1.19	1.47		2.66	3.51
6215	7-1/2" high G		960	.033		1.40	1.68		3.08	4.06
6220	9-1/2" high G		840	.038		1.56	1.92		3.48	4.60
6500	Curb forms, wood, 6" to 12" high, on elevated slabs, 1 use		180	.178	SFCA	1.80	8.95		10.75	15.45
6550	2 use		205	.156		.99	7.85		8.84	12.90
6600	3 use		220	.145		.72	7.35		8.07	11.80
6650	4 use		225	.142		.58	7.15		7.73	11.40
7000	Edge forms to 6" high, on elevated slab, 4 use		500	.064	L.F.	.22	3.23		3.45	5.10
7070	7" to 12" high, 1 use		162	.198	SFCA	1.33	9.95		11.28	16.40
7080	2 use		198	.162		.73	8.15		8.88	13.05
7090	3 use		222	.144		.53	7.25		7.78	11.50
7101	4 use		350	.091		.22	4.61		4.83	7.15
7500	Depressed area forms to 12" high, 4 use		300	.107	L.F.	.83	5.40		6.23	8.95
7550	12" to 24" high, 4 use		175	.183		1.13	9.20		10.33	15.10
8000	Perimeter deck and rail for elevated slabs, straight		90	.356		12.70	17.90		30.60	41
8050	Curved		65	.492		17.45	25		42.45	56.50
8500	Void forms, round plastic, 8" high x 3" diameter G		450	.071	Ea.	2.05	3.58		5.63	7.65
8550	4" diameter G		425	.075		2.32	3.79		6.11	8.25
8600	6" diameter G		400	.080		4.05	4.03		8.08	10.50
8650	8" diameter G		375	.085		7	4.30		11.30	14.15

03 11 13.40 Forms In Place, Equipment Foundations

		Crew	Daily Output	Labor-Hours	Unit	Material	2020 Bare Costs Labor	2020 Bare Costs Equipment	Total	Total Incl O&P
0010	**FORMS IN PLACE, EQUIPMENT FOUNDATIONS** R031113-40									
0020	1 use	C-2	160	.300	SFCA	2.95	15.50		18.45	27
0050	2 use R031113-60		190	.253		1.62	13.05		14.67	21.50
0100	3 use		200	.240		1.18	12.40		13.58	19.90
0150	4 use		205	.234		.96	12.10		13.06	19.20

03 11 Concrete Forming

03 11 13 – Structural Cast-In-Place Concrete Forming

03 11 13.45 Forms In Place, Footings

		Crew	Daily Output	Labor-Hours	Unit	Material	2020 Bare Costs Labor	Equipment	Total	Total Incl O&P
0010	**FORMS IN PLACE, FOOTINGS** R031113-40	C-1	375	.085	SFCA	6.85	4.30		11.15	14
0020	Continuous wall, plywood, 1 use		375	.085	SFCA	6.85	4.30		11.15	14
0050	2 use R031113-60		440	.073		3.77	3.66		7.43	9.65
0100	3 use		470	.068		2.74	3.43		6.17	8.15
0150	4 use		485	.066		2.23	3.32		5.55	7.45
0500	Dowel supports for footings or beams, 1 use		500	.064	L.F.	1.05	3.23		4.28	6
1000	Integral starter wall, to 4" high, 1 use		400	.080		1.01	4.03		5.04	7.15
1500	Keyway, 4 use, tapered wood, 2" x 4"	1 Carp	530	.015		.25	.80		1.05	1.48
1550	2" x 6"		500	.016		.34	.85		1.19	1.65
2000	Tapered plastic		530	.015		1.53	.80		2.33	2.88
2250	For keyway hung from supports, add		150	.053		1.05	2.83		3.88	5.40
3000	Pile cap, square or rectangular, job-built plywood, 1 use	C-1	290	.110	SFCA	3.16	5.55		8.71	11.85
3050	2 use		346	.092		1.74	4.66		6.40	8.90
3100	3 use		371	.086		1.26	4.35		5.61	7.95
3150	4 use		383	.084		1.03	4.21		5.24	7.45
4000	Triangular or hexagonal, 1 use		225	.142		3.70	7.15		10.85	14.80
4050	2 use		280	.114		2.03	5.75		7.78	10.90
4100	3 use		305	.105		1.48	5.30		6.78	9.60
4150	4 use		315	.102		1.20	5.10		6.30	9
5000	Spread footings, job-built lumber, 1 use		305	.105		2.28	5.30		7.58	10.45
5050	2 use		371	.086		1.27	4.35		5.62	7.95
5100	3 use		401	.080		.91	4.02		4.93	7.05
5150	4 use		414	.077		.74	3.89		4.63	6.65
6000	Supports for dowels, plinths or templates, 2' x 2' footing		25	1.280	Ea.	6.50	64.50		71	104
6050	4' x 4' footing		22	1.455		12.95	73.50		86.45	124
6100	8' x 8' footing		20	1.600		26	80.50		106.50	150
6150	12' x 12' footing		17	1.882		31	95		126	177
7000	Plinths, job-built plywood, 1 use		250	.128	SFCA	3.25	6.45		9.70	13.25
7100	4 use		270	.119	"	1.06	5.95		7.01	10.10

03 11 13.47 Forms In Place, Gas Station Forms

		Crew	Daily Output	Labor-Hours	Unit	Material	2020 Bare Costs Labor	Equipment	Total	Total Incl O&P
0010	**FORMS IN PLACE, GAS STATION FORMS**									
0050	Curb fascia, with template, 12 ga. steel, left in place, 9" high [G]	1 Carp	50	.160	L.F.	15.05	8.50		23.55	29.50
1000	Sign or light bases, 18" diameter, 9" high [G]		9	.889	Ea.	95	47		142	175
1050	30" diameter, 13" high [G]		8	1		151	53		204	246
2000	Island forms, 10' long, 9" high, 3'-6" wide [G]	C-1	10	3.200		420	161		581	705
2050	4' wide [G]		9	3.556		435	179		614	750
2500	20' long, 9" high, 4' wide [G]		6	5.333		700	269		969	1,175
2550	5' wide [G]		5	6.400		725	325		1,050	1,275

03 11 13.50 Forms In Place, Grade Beam

		Crew	Daily Output	Labor-Hours	Unit	Material	2020 Bare Costs Labor	Equipment	Total	Total Incl O&P
0010	**FORMS IN PLACE, GRADE BEAM** R031113-40									
0020	Job-built plywood, 1 use	C-2	530	.091	SFCA	3.15	4.68		7.83	10.50
0050	2 use R031113-60		580	.083		1.73	4.27		6	8.30
0100	3 use		600	.080		1.26	4.13		5.39	7.60
0150	4 use		605	.079		1.02	4.10		5.12	7.30

03 11 13.55 Forms In Place, Mat Foundation

		Crew	Daily Output	Labor-Hours	Unit	Material	2020 Bare Costs Labor	Equipment	Total	Total Incl O&P
0010	**FORMS IN PLACE, MAT FOUNDATION** R031113-40									
0020	Job-built plywood, 1 use	C-2	290	.166	SFCA	3.16	8.55		11.71	16.35
0050	2 use R031113-60		310	.155		1.26	8		9.26	13.40
0100	3 use		330	.145		.81	7.50		8.31	12.20
0120	4 use		350	.137		.74	7.10		7.84	11.45

For customer support on your Site Work & Landscape Costs with RSMeans data, call 800.448.8182.

57

03 11 Concrete Forming

03 11 13 – Structural Cast-In-Place Concrete Forming

03 11 13.65 Forms In Place, Slab On Grade

		Crew	Daily Output	Labor-Hours	Unit	Material	2020 Bare Costs Labor	2020 Bare Costs Equipment	Total	Total Incl O&P
0010	**FORMS IN PLACE, SLAB ON GRADE** R031113-40									
1000	Bulkhead forms w/keyway, wood, 6" high, 1 use	C-1	510	.063	L.F.	1.10	3.16		4.26	5.95
1050	2 uses R031113-60		400	.080		.60	4.03		4.63	6.70
1100	4 uses		350	.091		.36	4.61		4.97	7.30
1400	Bulkhead form for slab, 4-1/2" high, exp metal, incl. keyway & stakes G		1200	.027		.94	1.34		2.28	3.05
1410	5-1/2" high G		1100	.029		1.19	1.47		2.66	3.51
1420	7-1/2" high G		960	.033		1.40	1.68		3.08	4.06
1430	9-1/2" high G		840	.038		1.56	1.92		3.48	4.60
2000	Curb forms, wood, 6" to 12" high, on grade, 1 use		215	.149	SFCA	2.19	7.50		9.69	13.65
2050	2 use		250	.128		1.21	6.45		7.66	11.05
2100	3 use		265	.121		.88	6.10		6.98	10.10
2150	4 use		275	.116		.71	5.85		6.56	9.60
3000	Edge forms, wood, 4 use, on grade, to 6" high		600	.053	L.F.	.33	2.69		3.02	4.40
3050	7" to 12" high		435	.074	SFCA	.75	3.71		4.46	6.40
3060	Over 12"		350	.091	"	.97	4.61		5.58	7.95
3500	For depressed slabs, 4 use, to 12" high		300	.107	L.F.	.81	5.40		6.21	8.95
3550	To 24" high		175	.183		1.04	9.20		10.24	15
4000	For slab blockouts, to 12" high, 1 use		200	.160		.82	8.05		8.87	13
4050	To 24" high, 1 use		120	.267		1.04	13.45		14.49	21
4100	Plastic (extruded), to 6" high, multiple use, on grade		800	.040		6.50	2.02		8.52	10.20
5020	Wood, incl. wood stakes, 1" x 3"		900	.036		.87	1.79		2.66	3.65
5050	2" x 4"		900	.036		.90	1.79		2.69	3.68
6000	Trench forms in floor, wood, 1 use		160	.200	SFCA	2.05	10.10		12.15	17.40
6050	2 use		175	.183		1.13	9.20		10.33	15.10
6100	3 use		180	.178		.82	8.95		9.77	14.35
6150	4 use		185	.173		.67	8.70		9.37	13.85
8760	Void form, corrugated fiberboard, 4" x 12", 4' long G		3000	.011	S.F.	3.64	.54		4.18	4.81
8770	6" x 12", 4' long		3000	.011		4.47	.54		5.01	5.75
8780	1/4" thick hardboard protective cover for void form	2 Carp	1500	.011		.72	.57		1.29	1.64

03 11 13.85 Forms In Place, Walls

		Crew	Daily Output	Labor-Hours	Unit	Material	2020 Bare Costs Labor	2020 Bare Costs Equipment	Total	Total Incl O&P
0010	**FORMS IN PLACE, WALLS** R031113-10									
0100	Box out for wall openings, to 16" thick, to 10 S.F.	C-2	24	2	Ea.	28	103		131	186
0150	Over 10 S.F. (use perimeter) R031113-40	"	280	.171	L.F.	2.45	8.85		11.30	16
0250	Brick shelf, 4" w, add to wall forms, use wall area above shelf									
0260	1 use R031113-60	C-2	240	.200	SFCA	2.70	10.35		13.05	18.45
0300	2 use		275	.175		1.48	9		10.48	15.20
0350	4 use		300	.160		1.08	8.25		9.33	13.60
0500	Bulkhead, wood with keyway, 1 use, 2 piece		265	.181	L.F.	2.19	9.35		11.54	16.45
0600	Bulkhead forms with keyway, 1 piece expanded metal, 8" wall G	C-1	1000	.032		1.40	1.61		3.01	3.96
0610	10" wall G		800	.040		1.56	2.02		3.58	4.75
0620	12" wall G		525	.061		1.87	3.07		4.94	6.65
0700	Buttress, to 8' high, 1 use	C-2	350	.137	SFCA	4.85	7.10		11.95	16
0750	2 use		430	.112		2.67	5.75		8.42	11.60
0800	3 use		460	.104		1.95	5.40		7.35	10.25
0850	4 use		480	.100		1.60	5.15		6.75	9.50
1000	Corbel or haunch, to 12" wide, add to wall forms, 1 use		150	.320	L.F.	2.59	16.50		19.09	28
1050	2 use		170	.282		1.42	14.60		16.02	23.50
1100	3 use		175	.274		1.03	14.15		15.18	22.50
1150	4 use		180	.267		.84	13.75		14.59	21.50
2000	Wall, job-built plywood, to 8' high, 1 use		370	.130	SFCA	2.99	6.70		9.69	13.35
2050	2 use		435	.110		1.91	5.70		7.61	10.65
2100	3 use		495	.097		1.39	5		6.39	9.05

58
For customer support on your Site Work & Landscape Costs with RSMeans data, call 800.448.8182.

03 11 Concrete Forming

03 11 13 – Structural Cast-In-Place Concrete Forming

03 11 13.85 Forms In Place, Walls		Crew	Daily Output	Labor-Hours	Unit	Material	2020 Bare Costs Labor	Equipment	Total	Total Incl O&P
2150	4 use	C-2	505	.095	SFCA	1.13	4.91		6.04	8.60
2400	Over 8' to 16' high, 1 use		280	.171		3.32	8.85		12.17	16.95
2450	2 use		345	.139		1.44	7.20		8.64	12.40
2500	3 use		375	.128		1.03	6.60		7.63	11.10
2550	4 use		395	.122		.84	6.30		7.14	10.40
2700	Over 16' high, 1 use		235	.204		2.94	10.55		13.49	19.10
2750	2 use		290	.166		1.62	8.55		10.17	14.65
2800	3 use		315	.152		1.18	7.85		9.03	13.10
2850	4 use		330	.145		.96	7.50		8.46	12.35
4000	Radial, smooth curved, job-built plywood, 1 use		245	.196		2.65	10.10		12.75	18.10
4050	2 use		300	.160		1.46	8.25		9.71	14
4100	3 use		325	.148		1.06	7.65		8.71	12.60
4150	4 use		335	.143		.86	7.40		8.26	12.05
4200	Below grade, job-built plywood, 1 use		225	.213		2.86	11		13.86	19.70
4210	2 use		225	.213		1.58	11		12.58	18.30
4220	3 use		225	.213		1.31	11		12.31	18
4230	4 use		225	.213		.93	11		11.93	17.60
4300	Curved, 2' chords, job-built plywood, to 8' high, 1 use		290	.166		2.32	8.55		10.87	15.40
4350	2 use		355	.135		1.27	7		8.27	11.90
4400	3 use		385	.125		.93	6.45		7.38	10.65
4450	4 use		400	.120		.75	6.20		6.95	10.15
4500	Over 8' to 16' high, 1 use		290	.166		.95	8.55		9.50	13.90
4525	2 use		355	.135		.52	7		7.52	11.05
4550	3 use		385	.125		.38	6.45		6.83	10.05
4575	4 use		400	.120		.31	6.20		6.51	9.65
4600	Retaining wall, battered, job-built plyw'd, to 8' high, 1 use		300	.160		2.18	8.25		10.43	14.80
4650	2 use		355	.135		1.20	7		8.20	11.80
4700	3 use		375	.128		.87	6.60		7.47	10.90
4750	4 use		390	.123		.71	6.35		7.06	10.35
4900	Over 8' to 16' high, 1 use		240	.200		2.39	10.35		12.74	18.10
4950	2 use		295	.163		1.31	8.40		9.71	14.05
5000	3 use		305	.157		.95	8.15		9.10	13.25
5050	4 use		320	.150		.78	7.75		8.53	12.50
5100	Retaining wall form, plywood, smooth curve, 1 use		200	.240		3.39	12.40		15.79	22.50
5120	2 use		235	.204		1.86	10.55		12.41	17.90
5130	3 use		250	.192		1.35	9.90		11.25	16.40
5140	4 use		260	.185		1.11	9.55		10.66	15.50
5500	For gang wall forming, 192 S.F. sections, deduct					10%	10%			
5550	384 S.F. sections, deduct					20%	20%			
7500	Lintel or sill forms, 1 use	1 Carp	30	.267		3.51	14.15		17.66	25.50
7520	2 use		34	.235		1.93	12.50		14.43	21
7540	3 use		36	.222		1.40	11.80		13.20	19.30
7560	4 use		37	.216		1.14	11.50		12.64	18.50
7800	Modular prefabricated plywood, based on 20 uses of purchased									
7820	forms, and 4 uses of bracing lumber									
7860	To 8' high	C-2	800	.060	SFCA	1.15	3.10		4.25	5.90
8060	Over 8' to 16' high		600	.080		1.20	4.13		5.33	7.50
8600	Pilasters, 1 use		270	.178		3.49	9.20		12.69	17.65
8620	2 use		330	.145		1.92	7.50		9.42	13.40
8640	3 use		370	.130		1.40	6.70		8.10	11.60
8660	4 use		385	.125		1.13	6.45		7.58	10.90
9010	Steel framed plywood, based on 50 uses of purchased									
9020	forms, and 4 uses of bracing lumber									

03 11 Concrete Forming

03 11 13 – Structural Cast-In-Place Concrete Forming

03 11 13.85 Forms In Place, Walls

	Crew	Daily Output	Labor-Hours	Unit	Material	2020 Bare Costs Labor	2020 Bare Costs Equipment	Total	Total Incl O&P
9060	To 8' high	C-2	600	.080	SFCA	.79	4.13		4.92	7.05
9260	Over 8' to 16' high		450	.107		.79	5.50		6.29	9.15
9460	Over 16' to 20' high	↓	400	.120	↓	.79	6.20		6.99	10.15
9480	For battered walls, 1 side battered, add					10%	10%			
9485	For battered walls, 2 sides battered, add					15%	15%			

03 11 16 – Architectural Cast-in-Place Concrete Forming

03 11 16.13 Concrete Form Liners

		Crew	Daily Output	Labor-Hours	Unit	Material	2020 Bare Costs Labor	2020 Bare Costs Equipment	Total	Total Incl O&P
0010	**CONCRETE FORM LINERS**									
5750	Liners for forms (add to wall forms), ABS plastic									
5800	Aged wood, 4" wide, 1 use	1 Carp	256	.031	SFCA	3.34	1.66		5	6.15
5820	2 use		256	.031		1.84	1.66		3.50	4.52
5830	3 use		256	.031		1.34	1.66		3	3.97
5840	4 use		256	.031		1.09	1.66		2.75	3.69
5900	Fractured rope rib, 1 use		192	.042		4.91	2.21		7.12	8.75
5925	2 use		192	.042		2.70	2.21		4.91	6.30
5950	3 use		192	.042		1.96	2.21		4.17	5.50
6000	4 use		192	.042		1.60	2.21		3.81	5.10
6100	Ribbed, 3/4" deep x 1-1/2" OC, 1 use		224	.036		5.05	1.90		6.95	8.45
6125	2 use		224	.036		2.79	1.90		4.69	5.90
6150	3 use		224	.036		2.03	1.90		3.93	5.10
6200	4 use		224	.036		1.65	1.90		3.55	4.66
6300	Rustic brick pattern, 1 use		224	.036		3.47	1.90		5.37	6.65
6325	2 use		224	.036		1.91	1.90		3.81	4.95
6350	3 use		224	.036		1.39	1.90		3.29	4.38
6400	4 use		224	.036		1.13	1.90		3.03	4.09
6500	3/8" striated, random, 1 use		224	.036		3.54	1.90		5.44	6.75
6525	2 use		224	.036		1.95	1.90		3.85	4.99
6550	3 use		224	.036		1.42	1.90		3.32	4.41
6600	4 use		224	.036		1.15	1.90		3.05	4.12
6850	Random vertical rustication, 1 use		384	.021		6.60	1.11		7.71	8.90
6900	2 use		384	.021		3.62	1.11		4.73	5.65
6925	3 use		384	.021		2.63	1.11		3.74	4.56
6950	4 use		384	.021		2.14	1.11		3.25	4.01
7050	Wood, beveled edge, 3/4" deep, 1 use		384	.021	L.F.	.18	1.11		1.29	1.86
7100	1" deep, 1 use		384	.021	"	.29	1.11		1.40	1.98
7200	4" wide aged cedar, 1 use		256	.031	SFCA	3.47	1.66		5.13	6.30
7300	4" variable depth rough cedar	↓	224	.036	"	4.85	1.90		6.75	8.20

03 11 19 – Insulating Concrete Forming

03 11 19.10 Insulating Forms, Left In Place

		Crew	Daily Output	Labor-Hours	Unit	Material	2020 Bare Costs Labor	2020 Bare Costs Equipment	Total	Total Incl O&P
0010	**INSULATING FORMS, LEFT IN PLACE**									
0020	Forms include layout, exclude rebar, embedments, bucks for openings,									
0030	scaffolding, wall bracing, concrete, and concrete placing.									
0040	S.F. is for exterior face but includes forms for both faces of wall									
0100	Straight blocks or panels, molded, walls up to 4' high									
0110	4" core wall	4 Carp	1984	.016	S.F.	3.62	.86		4.48	5.25
0120	6" core wall		1808	.018		3.64	.94		4.58	5.40
0130	8" core wall		1536	.021		3.77	1.11		4.88	5.80
0140	10" core wall		1152	.028		4.30	1.48		5.78	6.95
0150	12" core wall		992	.032	↓	4.89	1.71		6.60	8
0200	90 degree corner blocks or panels, molded, walls up to 4' high									
0210	4" core wall	4 Carp	1880	.017	S.F.	3.73	.90		4.63	5.45
0220	6" core wall	↓	1708	.019	↓	3.75	1		4.75	5.65

60

03 11 19.10 Insulating Forms, Left In Place	Crew	Daily Output	Labor-Hours	Unit	Material	2020 Bare Costs Labor	Equipment	Total	Total Incl O&P
0230 8" core wall	4 Carp	1324	.024	S.F.	3.88	1.28		5.16	6.20
0240 10" core wall		987	.032		4.49	1.72		6.21	7.55
0250 12" core wall	↓	884	.036	↓	4.81	1.92		6.73	8.20
0300 45 degree corner blocks or panels, molded, walls up to 4' high									
0310 4" core wall	4 Carp	1880	.017	S.F.	3.93	.90		4.83	5.70
0320 6" core wall		1712	.019		4.06	.99		5.05	5.95
0330 8" core wall	↓	1324	.024	↓	4.17	1.28		5.45	6.50
0400 T blocks or panels, molded, walls up to 4' high									
0420 6" core wall	4 Carp	1540	.021	S.F.	4.79	1.10		5.89	6.90
0430 8" core wall		1325	.024		4.84	1.28		6.12	7.25
0440 Non-standard corners or Ts requiring trimming & strapping	↓	192	.167	↓	4.89	8.85		13.74	18.70
0500 Radius blocks or panels, molded, walls up to 4' high, 6" core wall									
0520 5' to 10' diameter, molded blocks or panels	4 Carp	2400	.013	S.F.	5.90	.71		6.61	7.55
0530 10' to 15' diameter, requiring trimming and strapping, add		500	.064		4.44	3.40		7.84	10
0540 15'-1" to 30' diameter, requiring trimming and strapping, add		1200	.027		4.44	1.42		5.86	7
0550 30'-1" to 60' diameter, requiring trimming and strapping, add	↓	1600	.020		4.44	1.06		5.50	6.50
0560 60'-1" to 100' diameter, requiring trimming and strapping, add	↓	2800	.011	↓	4.44	.61		5.05	5.80
0600 Additional labor for blocks/panels in higher walls (excludes scaffolding)									
0610 4'-1" to 9'-4" high, add						10%			
0620 9'-5" to 12'-0" high, add						20%			
0630 12'-1" to 20'-0" high, add						35%			
0640 Over 20'-0" high, add						55%			
0700 Taper block or panels, molded, single course									
0720 6" core wall	4 Carp	1600	.020	S.F.	3.91	1.06		4.97	5.90
0730 8" core wall	"	1392	.023	"	3.96	1.22		5.18	6.20
0800 ICF brick ledge (corbel) block or panels, molded, single course									
0820 6" core wall	4 Carp	1200	.027	S.F.	4.31	1.42		5.73	6.85
0830 8" core wall	"	1152	.028	"	4.41	1.48		5.89	7.05
0900 ICF curb (shelf) block or panels, molded, single course									
0930 8" core wall	4 Carp	688	.047	S.F.	3.72	2.47		6.19	7.80
0940 10" core wall	"	544	.059	"	4.25	3.13		7.38	9.40
0950 Wood form to hold back concrete to form shelf, 8" high	2 Carp	400	.040	L.F.	.87	2.13		3	4.14
1000 ICF half height block or panels, molded, single course									
1010 4" core wall	4 Carp	1248	.026	S.F.	4.72	1.36		6.08	7.25
1020 6" core wall		1152	.028		4.74	1.48		6.22	7.40
1030 8" core wall		942	.034		4.79	1.81		6.60	7.95
1040 10" core wall		752	.043		5.30	2.26		7.56	9.20
1050 12" core wall	↓	648	.049	↓	5.20	2.62		7.82	9.70
1100 ICF half height block/panels, made by field sawing full height block/panels									
1110 4" core wall	4 Carp	800	.040	S.F.	1.81	2.13		3.94	5.20
1120 6" core wall		752	.043		1.82	2.26		4.08	5.40
1130 8" core wall		600	.053		1.89	2.83		4.72	6.35
1140 10" core wall		496	.065		2.15	3.43		5.58	7.50
1150 12" core wall	↓	400	.080		2.45	4.25		6.70	9.10
1200 Additional insulation inserted into forms between ties									
1210 1 layer (2" thick)	4 Carp	14000	.002	S.F.	1.03	.12		1.15	1.31
1220 2 layers (4" thick)		7000	.005		2.06	.24		2.30	2.63
1230 3 layers (6" thick)	↓	4622	.007	↓	3.09	.37		3.46	3.95
1300 EPS window/door bucks, molded, permanent									
1310 4" core wall (9" wide)	2 Carp	200	.080	L.F.	3.47	4.25		7.72	10.20
1320 6" core wall (11" wide)		200	.080		3.56	4.25		7.81	10.30
1330 8" core wall (13" wide)		176	.091		3.73	4.83		8.56	11.35
1340 10" core wall (15" wide)		152	.105		4.86	5.60		10.46	13.75

For customer support on your Site Work & Landscape Costs with RSMeans data, call 800.448.8182.

61

03 11 Concrete Forming

03 11 19 – Insulating Concrete Forming

03 11 19.10 Insulating Forms, Left In Place

		Crew	Daily Output	Labor-Hours	Unit	Material	2020 Bare Costs Labor	Equipment	Total	Total Incl O&P
1350	12" core wall (17" wide)	2 Carp	152	.105	L.F.	5.45	5.60		11.05	14.35
1360	2" x 6" temporary buck bracing (includes installing and removing)	↓	400	.040	↓	.68	2.13		2.81	3.93
1400	Wood window/door bucks (instead of EPS bucks), permanent									
1410	4" core wall (9" wide)	2 Carp	400	.040	L.F.	1.31	2.13		3.44	4.63
1420	6" core wall (11" wide)		400	.040		1.72	2.13		3.85	5.10
1430	8" core wall (13" wide)		350	.046		7.35	2.43		9.78	11.70
1440	10" core wall (15" wide)		300	.053		10.15	2.83		12.98	15.45
1450	12" core wall (17" wide)		300	.053		10.15	2.83		12.98	15.45
1460	2" x 6" temporary buck bracing (includes installing and removing)	↓	800	.020	↓	.68	1.06		1.74	2.34
1500	ICF alignment brace (incl. stiff-back, diagonal kick-back, work platform									
1501	bracket & guard rail post), fastened to one face of wall forms @ 6' O.C.									
1510	1st tier up to 10' tall									
1520	Rental of ICF alignment brace set, per set				Week	10.05			10.05	11.05
1530	Labor (includes installing & removing)	2 Carp	30	.533	Ea.		28.50		28.50	42.50
1560	2nd tier from 10' to 20' tall (excludes mason's scaffolding up to 10' high)									
1570	Rental of ICF alignment brace set, per set				Week	10.05			10.05	11.05
1580	Labor (includes installing & removing)	4 Carp	30	1.067	Ea.		56.50		56.50	85
1600	2" x 10" wood plank for work platform, 16' long									
1610	Plank material cost pro-rated over 20 uses				Ea.	1.05			1.05	1.15
1620	Labor (includes installing & removing)	2 Carp	48	.333	"		17.70		17.70	26.50
1700	2" x 4" lumber for top & middle rails for work platform									
1710	Railing material cost pro-rated over 20 uses				Ea.	.02			.02	.03
1720	Labor (includes installing & removing)	2 Carp	2400	.007	L.F.		.35		.35	.53
1800	ICF accessories									
1810	Wire clip to secure forms in place	2 Carp	2100	.008	Ea.	.39	.41		.80	1.04
1820	Masonry anchor embedment (excludes ties by mason)		1600	.010		4.24	.53		4.77	5.45
1830	Ledger anchor embedment (excludes timber hanger & screws)	↓	128	.125	↓	7.60	6.65		14.25	18.35
1900	See section 01 54 23.70 for mason's scaffolding components									
1910	See section 03 15 19.05 for anchor bolt sleeves									
1920	See section 03 15 19.10 for anchor bolts									
1930	See section 03 15 19.20 for dovetail anchor components									
1940	See section 03 15 19.30 for embedded inserts									
1950	See section 03 21 05.10 for rebar accessories									
1960	See section 03 21 11.60 for reinforcing bars in place									
1970	See section 03 31 13.35 for ready-mix concrete material									
1980	See section 03 31 13.70 for placement and consolidation of concrete									
1990	See section 06 05 23.60 for timber connectors									

03 11 23 – Permanent Stair Forming

03 11 23.75 Forms In Place, Stairs

		Crew	Daily Output	Labor-Hours	Unit	Material	2020 Bare Costs Labor	Equipment	Total	Total Incl O&P
0010	**FORMS IN PLACE, STAIRS** R031113-40									
0015	(Slant length x width), 1 use	C-2	165	.291	S.F.	6.15	15		21.15	29.50
0050	2 use R031113-60		170	.282		3.50	14.60		18.10	26
0100	3 use		180	.267		2.62	13.75		16.37	23.50
0150	4 use		190	.253	↓	2.17	13.05		15.22	22
1000	Alternate pricing method (1.0 L.F./S.F.), 1 use		100	.480	LF Rsr	6.15	25		31.15	44
1050	2 use		105	.457		3.50	23.50		27	39.50
1100	3 use		110	.436		2.62	22.50		25.12	37
1150	4 use		115	.417	↓	2.17	21.50		23.67	35
2000	Stairs, cast on sloping ground (length x width), 1 use		220	.218	S.F.	2.57	11.25		13.82	19.80
2025	2 use		232	.207		1.42	10.70		12.12	17.60
2050	3 use		244	.197		1.03	10.15		11.18	16.40
2100	4 use	↓	256	.188	↓	.84	9.70		10.54	15.45

62

For customer support on your Site Work & Landscape Costs with RSMeans data, call 800.448.8182.

03 15 Concrete Accessories

03 15 05 – Concrete Forming Accessories

03 15 05.12 Chamfer Strips

		Crew	Daily Output	Labor-Hours	Unit	Material	2020 Bare Costs Labor	Equipment	Total	Total Incl O&P
0010	**CHAMFER STRIPS**									
2000	Polyvinyl chloride, 1/2" wide with leg	1 Carp	535	.015	L.F.	.75	.79		1.54	2.02
2200	3/4" wide with leg		525	.015		.77	.81		1.58	2.07
2400	1" radius with leg		515	.016		.81	.83		1.64	2.13
2800	2" radius with leg		500	.016		1.79	.85		2.64	3.25
5000	Wood, 1/2" wide		535	.015		.14	.79		.93	1.34
5200	3/4" wide		525	.015		.18	.81		.99	1.42
5400	1" wide		515	.016		.29	.83		1.12	1.56

03 15 05.30 Hangers

					Unit	Material	Labor	Equipment	Total	Total Incl O&P
0010	**HANGERS**	G								
8500	Wire, black annealed, 15 gauge				Cwt.	187			187	206
8600	16 gauge				"	197			197	217

03 15 05.70 Shores

		Crew	Daily Output	Labor-Hours	Unit	Material	Labor	Equipment	Total	Total Incl O&P
0010	**SHORES**									
0020	Erect and strip, by hand, horizontal members									
0500	Aluminum joists and stringers	G	2 Carp	60	.267	Ea.		14.15		14.15	21.50
0600	Steel, adjustable beams	G		45	.356			18.90		18.90	28.50
0700	Wood joists		50	.320			17		17	25.50
0800	Wood stringers		30	.533			28.50		28.50	42.50
1000	Vertical members to 10' high	G	55	.291			15.45		15.45	23
1050	To 13' high	G	50	.320			17		17	25.50
1100	To 16' high	G	45	.356			18.90		18.90	28.50
1500	Reshoring	G	1400	.011	S.F.	.63	.61		1.24	1.60
1600	Flying truss system	G	C-17D	9600	.009	SFCA		.49	.08	.57	.82
1760	Horizontal, aluminum joists, 6-1/4" high x 5' to 21' span, buy	G			L.F.	16.40			16.40	18.05
1770	Beams, 7-1/4" high x 4' to 30' span	G			"	19.70			19.70	21.50
1810	Horizontal, steel beam, W8x10, 7' span, buy	G			Ea.	64.50			64.50	71
1830	10' span	G				75			75	82.50
1920	15' span	G				129			129	142
1940	20' span	G				181			181	199
1970	Steel stringer, W8x10, 4' to 16' span, buy	G			L.F.	7.45			7.45	8.20
3000	Rent for job duration, aluminum joist @ 2' OC, per mo.	G			SF Flr.	.41			.41	.45
3050	Steel W8x10	G				.19			.19	.21
3060	Steel adjustable	G				.19			.19	.21
3500	#1 post shore, steel, 5'-7" to 9'-6" high, 10,000# cap., buy	G			Ea.	161			161	178
3550	#2 post shore, 7'-3" to 12'-10" high, 7800# capacity	G				186			186	205
3600	#3 post shore, 8'-10" to 16'-1" high, 3800# capacity	G				204			204	224
5010	Frame shoring systems, steel, 12,000#/leg, buy									
5040	Frame, 2' wide x 6' high	G			Ea.	107			107	118
5250	X-brace	G				17.05			17.05	18.80
5550	Base plate	G				16.05			16.05	17.65
5600	Screw jack	G				35			35	38
5650	U-head, 8" x 8"	G				20.50			20.50	22.50

03 15 05.75 Sleeves and Chases

		Crew	Daily Output	Labor-Hours	Unit	Material	Labor	Equipment	Total	Total Incl O&P
0010	**SLEEVES AND CHASES**									
0100	Plastic, 1 use, 12" long, 2" diameter	1 Carp	100	.080	Ea.	2.32	4.25		6.57	8.95
0150	4" diameter		90	.089		4.97	4.72		9.69	12.55
0200	6" diameter		75	.107		10.85	5.65		16.50	20.50
0250	12" diameter		60	.133		28.50	7.10		35.60	41.50
5000	Sheet metal, 2" diameter	G	100	.080		1.54	4.25		5.79	8.10
5100	4" diameter	G	90	.089		1.93	4.72		6.65	9.20
5150	6" diameter	G	75	.107		2.03	5.65		7.68	10.75

03 15 Concrete Accessories

03 15 05 – Concrete Forming Accessories

03 15 05.75 Sleeves and Chases

		Crew	Daily Output	Labor-Hours	Unit	Material	2020 Bare Costs Labor	Equipment	Total	Total Incl O&P
5200	12" diameter	G	1 Carp	60	.133	Ea.	4.24	7.10		11.34	15.30
6000	Steel pipe, 2" diameter	G		100	.080		3.26	4.25		7.51	10
6100	4" diameter	G		90	.089		17.20	4.72		21.92	26
6150	6" diameter	G		75	.107		36.50	5.65		42.15	48.50
6200	12" diameter	G		60	.133		85	7.10		92.10	104

03 15 05.80 Snap Ties

			Crew	Daily Output	Labor-Hours	Unit	Material	Labor	Equipment	Total	Total Incl O&P
0010	**SNAP TIES**, 8-1/4" L&W (Lumber and wedge)										
0100	2250 lb., w/flat washer, 8" wall	G				C	97.50			97.50	107
0150	10" wall	G					163			163	179
0200	12" wall	G					166			166	183
0250	16" wall	G					178			178	196
0300	18" wall	G					185			185	203
0500	With plastic cone, 8" wall	G					86			86	94.50
0550	10" wall	G					88			88	97
0600	12" wall	G					96.50			96.50	106
0650	16" wall	G					106			106	116
0700	18" wall	G					109			109	120
1000	3350 lb., w/flat washer, 8" wall	G					204			204	225
1100	10" wall	G					224			224	246
1150	12" wall	G					204			204	224
1200	16" wall	G					262			262	289
1250	18" wall	G					274			274	300
1500	With plastic cone, 8" wall	G					164			164	180
1550	10" wall	G					180			180	197
1600	12" wall	G					163			163	179
1650	16" wall	G					211			211	232
1700	18" wall	G					213			213	234

03 15 05.85 Stair Tread Inserts

		Crew	Daily Output	Labor-Hours	Unit	Material	Labor	Equipment	Total	Total Incl O&P
0010	**STAIR TREAD INSERTS**									
0105	Cast nosing insert, abrasive surface, pre-drilled, includes screws									
0110	Aluminum, 3" wide x 3' long	1 Cefi	32	.250	Ea.	58	12.50		70.50	82
0120	4' long		31	.258		75	12.90		87.90	101
0130	5' long		30	.267		90	13.30		103.30	119
0135	Extruded nosing insert, black abrasive strips, continuous anchor									
0140	Aluminum, 3" wide x 3' long	1 Cefi	64	.125	Ea.	35.50	6.25		41.75	48
0150	4' long		60	.133		48	6.65		54.65	63
0160	5' long		56	.143		62.50	7.15		69.65	79.50
0165	Extruded nosing insert, black abrasive strips, pre-drilled, incl. screws									
0170	Aluminum, 3" wide x 3' long	1 Cefi	32	.250	Ea.	45	12.50		57.50	68
0180	4' long		31	.258		59.50	12.90		72.40	84.50
0190	5' long		30	.267		83.50	13.30		96.80	112

03 15 05.95 Wall and Foundation Form Accessories

			Crew	Daily Output	Labor-Hours	Unit	Material	Labor	Equipment	Total	Total Incl O&P
0010	**WALL AND FOUNDATION FORM ACCESSORIES**										
0020	Coil tie system										
0050	Coil ties 1/2", 6000 lb., to 8"	G				C	259			259	285
0150	24"	G					435			435	480
0300	3/4", 12,000 lb., to 8"	G					500			500	550
0400	24"	G					740			740	815
0900	Coil bolts, 1/2" diameter x 3" long	G					153			153	168
0940	12" long	G					440			440	480
1000	3/4" diameter x 3" long	G					335			335	365
1040	12" long	G					920			920	1,000

03 15 Concrete Accessories

03 15 05 – Concrete Forming Accessories

03 15 05.95 Wall and Foundation Form Accessories		Crew	Daily Output	Labor-Hours	Unit	Material	2020 Bare Costs Labor	Equipment	Total	Total Incl O&P
1300	Adjustable coil tie, 3/4" diameter, 20" long	G			C	1,275			1,275	1,400
1350	3/4" diameter, 36" long	G				1,550			1,550	1,725
1400	Tie cones, plastic, 1" setback length, 1/2" bolt diameter					89			89	98
1420	3/4" bolt diameter					182			182	200
1500	Welding coil tie, 1/2" diameter x 4" long	G				239			239	263
1550	3/4" diameter x 6" long	G				325			325	355
1700	Waler holders, 1/2" diameter	G				230			230	253
1750	3/4" diameter	G				257			257	282
1900	Flat washers, 4" x 5" x 1/4" for 3/4" diameter	G				1,125			1,125	1,250
2000	Footings, turnbuckle form aligner	G			Ea.	17.40			17.40	19.10
2050	Spreaders for footer, adjustable	G			"	25.50			25.50	28
2100	Lagstud, threaded, 1/2" diameter	G			C.L.F.	136			136	150
2200	1" diameter	G			"	495			495	545
2300	Lagnuts, 1/2" diameter	G			C	35.50			35.50	39
2400	1" diameter	G				248			248	273
2600	Lagnuts with handle, 1/2" diameter	G				325			325	360
2700	1" diameter	G				910			910	1,000
2750	Plastic set back plugs for 3/4" diameter					100			100	110
2800	Rock anchors, 1/2" diameter	G				1,225			1,225	1,350
2900	1" diameter	G				1,750			1,750	1,925
2950	Batter washer, 1/2" diameter	G				705			705	780
3000	Form oil, up to 1200 S.F./gallon coverage				Gal.	15.95			15.95	17.55
3050	Up to 800 S.F./gallon				"	23.50			23.50	25.50
3500	Form patches, 1-3/4" diameter				C	27			27	29.50
3550	2-3/4" diameter				"	45.50			45.50	50.50
4000	Nail stakes, 3/4" diameter, 18" long	G			Ea.	2.70			2.70	2.97
4050	24" long	G				2.68			2.68	2.95
4200	30" long	G				3.38			3.38	3.72
4250	36" long	G				4.25			4.25	4.68
5000	Pencil rods, 1/4" diameter	G			C.L.F.	56.50			56.50	62
5200	Clamps for 1/4" pencil rods	G			Ea.	5.65			5.65	6.20
5300	Clamping jacks for 1/4" pencil rod clamp	G			"	101			101	111
6000	She-bolts, 7/8" x 20"	G			C	5,350			5,350	5,875
6150	1-1/4" x 20"	G				6,250			6,250	6,900
6300	Inside rods, threaded, 1/2" diameter x 6"	G				182			182	201
6350	1/2" diameter x 12"	G				285			285	315
6500	3/4" diameter x 6"	G				420			420	465
6550	3/4" diameter x 12"	G				630			630	695
6700	For wing nuts, see taper ties									
7000	Taper tie system									
7100	Taper ties, 3/4" to 1/2" diameter x 30"	G			Ea.	61			61	67
7200	1-1/4" to 1" diameter x 30"	G			"	80.50			80.50	88.50
7300	Wing nuts, 1/2" diameter	G			C	740			740	815
7550	1-1/4" diameter	G				1,125			1,125	1,225
7700	Flat washers, 1/4" x 3" x 4", for 1/2" diam. bolt	G				268			268	295
7800	7/16" x 5" x 5", for 1" diam. bolt	G				1,475			1,475	1,625
9000	Wood form accessories									
9100	Tie plate	G			C	630			630	690
9200	Corner washer	G				1,250			1,250	1,375
9300	Panel bolt	G				970			970	1,075
9400	Panel wedge	G				172			172	190
9500	Stud clamp	G				700			700	770

For customer support on your Site Work & Landscape Costs with RSMeans data, call 800.448.8182.

65

03 15 Concrete Accessories

03 15 13 – Waterstops

03 15 13.50 Waterstops	Crew	Daily Output	Labor-Hours	Unit	Material	2020 Bare Costs Labor	2020 Bare Costs Equipment	Total	Total Incl O&P
0010 **WATERSTOPS**, PVC and Rubber									
0020 PVC, ribbed 3/16" thick, 4" wide	1 Carp	155	.052	L.F.	1.55	2.74		4.29	5.85
0050 6" wide		145	.055		2.78	2.93		5.71	7.45
0500 With center bulb, 6" wide, 3/16" thick		135	.059		2.68	3.15		5.83	7.70
0550 3/8" thick		130	.062		4.85	3.27		8.12	10.25
0600 9" wide x 3/8" thick		125	.064		8.15	3.40		11.55	14.05
0800 Dumbbell type, 6" wide, 3/16" thick		150	.053		4.31	2.83		7.14	9
0850 3/8" thick		145	.055		4.13	2.93		7.06	8.95
1000 9" wide, 3/8" thick, plain		130	.062		6.60	3.27		9.87	12.15
1050 Center bulb		130	.062		9.15	3.27		12.42	14.95
1250 Ribbed type, split, 3/16" thick, 6" wide		145	.055		2.21	2.93		5.14	6.85
1300 3/8" thick		130	.062		5.10	3.27		8.37	10.50
2000 Rubber, flat dumbbell, 3/8" thick, 6" wide		145	.055		5.65	2.93		8.58	10.60
2050 9" wide		135	.059		10.40	3.15		13.55	16.15
2500 Flat dumbbell split, 3/8" thick, 6" wide		145	.055		2.21	2.93		5.14	6.85
2550 9" wide		135	.059		5.10	3.15		8.25	10.35
3000 Center bulb, 1/4" thick, 6" wide		145	.055		6.95	2.93		9.88	12.05
3050 9" wide		135	.059		14.85	3.15		18	21
3500 Center bulb split, 3/8" thick, 6" wide		145	.055		6.10	2.93		9.03	11.10
3550 9" wide		135	.059		10.25	3.15		13.40	16
5000 Waterstop fittings, rubber, flat									
5010 Dumbbell or center bulb, 3/8" thick,									
5200 Field union, 6" wide	1 Carp	50	.160	Ea.	31.50	8.50		40	47.50
5250 9" wide		50	.160		33.50	8.50		42	50
5500 Flat cross, 6" wide		30	.267		52.50	14.15		66.65	79.50
5550 9" wide		30	.267		78	14.15		92.15	108
6000 Flat tee, 6" wide		30	.267		50.50	14.15		64.65	77
6050 9" wide		30	.267		66	14.15		80.15	94.50
6500 Flat ell, 6" wide		40	.200		49.50	10.65		60.15	70.50
6550 9" wide		40	.200		61.50	10.65		72.15	83.50
7000 Vertical tee, 6" wide		25	.320		21	17		38	48.50
7050 9" wide		25	.320		28.50	17		45.50	57
7500 Vertical ell, 6" wide		35	.229		20.50	12.15		32.65	41
7550 9" wide		35	.229		39	12.15		51.15	61.50

03 15 16 – Concrete Construction Joints

03 15 16.20 Control Joints, Saw Cut

	Crew	Daily Output	Labor-Hours	Unit	Material	2020 Bare Costs Labor	2020 Bare Costs Equipment	Total	Total Incl O&P
0010 **CONTROL JOINTS, SAW CUT**									
0100 Sawcut control joints in green concrete									
0120 1" depth	C-27	2000	.008	L.F.	.03	.40	.06	.49	.69
0140 1-1/2" depth		1800	.009		.05	.44	.06	.55	.78
0160 2" depth		1600	.010		.07	.50	.07	.64	.88
0180 Sawcut joint reservoir in cured concrete									
0182 3/8" wide x 3/4" deep, with single saw blade	C-27	1000	.016	L.F.	.05	.80	.11	.96	1.36
0184 1/2" wide x 1" deep, with double saw blades		900	.018		.10	.89	.12	1.11	1.56
0186 3/4" wide x 1-1/2" deep, with double saw blades		800	.020		.20	1	.14	1.34	1.84
0190 Water blast joint to wash away laitance, 2 passes	C-29	2500	.003			.13	.04	.17	.24
0200 Air blast joint to blow out debris and air dry, 2 passes	C-28	2000	.004			.20	.02	.22	.31
0300 For backer rod, see Section 07 91 23.10									
0340 For joint sealant, see Section 03 15 16.30 or 07 92 13.20									
0900 For replacement of joint sealant, see Section 07 01 90.81									

03 15 16 – Concrete Construction Joints

03 15 16.30 Expansion Joints		Crew	Daily Output	Labor-Hours	Unit	Material	2020 Bare Costs Labor	Equipment	Total	Total Incl O&P
0010	**EXPANSION JOINTS**									
0020	Keyed, cold, 24 ga., incl. stakes, 3-1/2" high	G	1 Carp	200	.040	L.F.	.89	2.13		3.02	4.17
0050	4-1/2" high	G		200	.040		.94	2.13		3.07	4.22
0100	5-1/2" high	G		195	.041		1.19	2.18		3.37	4.59
0150	7-1/2" high	G		190	.042		1.40	2.24		3.64	4.90
0160	9-1/2" high	G		185	.043		1.56	2.30		3.86	5.15
0300	Poured asphalt, plain, 1/2" x 1"		1 Clab	450	.018		.24	.75		.99	1.38
0350	1" x 2"			400	.020		.97	.84		1.81	2.34
0500	Neoprene, liquid, cold applied, 1/2" x 1"			450	.018		2.53	.75		3.28	3.90
0550	1" x 2"			400	.020		9.95	.84		10.79	12.20
0700	Polyurethane, poured, 2 part, 1/2" x 1"			400	.020		1.40	.84		2.24	2.81
0750	1" x 2"			350	.023		5.60	.96		6.56	7.65
0900	Rubberized asphalt, hot or cold applied, 1/2" x 1"			450	.018		.32	.75		1.07	1.47
0950	1" x 2"			400	.020		1.27	.84		2.11	2.67
1100	Hot applied, fuel resistant, 1/2" x 1"			450	.018		.48	.75		1.23	1.65
1150	1" x 2"			400	.020		1.91	.84		2.75	3.37
2000	Premolded, bituminous fiber, 1/2" x 6"		1 Carp	375	.021		.44	1.13		1.57	2.18
2050	1" x 12"			300	.027		2.03	1.42		3.45	4.36
2140	Concrete expansion joint, recycled paper and fiber, 1/2" x 6"	G		390	.021		.43	1.09		1.52	2.11
2150	1/2" x 12"	G		360	.022		.86	1.18		2.04	2.72
2250	Cork with resin binder, 1/2" x 6"			375	.021		1.36	1.13		2.49	3.20
2300	1" x 12"			300	.027		3.50	1.42		4.92	6
2500	Neoprene sponge, closed cell, 1/2" x 6"			375	.021		2.43	1.13		3.56	4.37
2550	1" x 12"			300	.027		8.90	1.42		10.32	11.90
2750	Polyethylene foam, 1/2" x 6"			375	.021		.59	1.13		1.72	2.35
2800	1" x 12"			300	.027		3.11	1.42		4.53	5.55
3000	Polyethylene backer rod, 3/8" diameter			460	.017		.03	.92		.95	1.43
3050	3/4" diameter			460	.017		.07	.92		.99	1.46
3100	1" diameter			460	.017		.15	.92		1.07	1.55
3500	Polyurethane foam, with polybutylene, 1/2" x 1/2"			475	.017		1.25	.90		2.15	2.72
3550	1" x 1"			450	.018		3.19	.95		4.14	4.93
3750	Polyurethane foam, regular, closed cell, 1/2" x 6"			375	.021		.93	1.13		2.06	2.72
3800	1" x 12"			300	.027		3.32	1.42		4.74	5.80
4000	Polyvinyl chloride foam, closed cell, 1/2" x 6"			375	.021		2.53	1.13		3.66	4.48
4050	1" x 12"			300	.027		8.70	1.42		10.12	11.70
4100	Rod, 3/8" polyethylene/rubberized asphalt sealer, 1/4" x 3/4"		2 Clab	600	.027		.19	1.12		1.31	1.90
4250	Rubber, gray sponge, 1/2" x 6"		1 Carp	375	.021		2.12	1.13		3.25	4.03
4300	1" x 12"			300	.027		7.65	1.42		9.07	10.55
4400	Redwood heartwood, 1" x 4"			400	.020		1.18	1.06		2.24	2.89
4450	1" x 6"			375	.021		1.79	1.13		2.92	3.67
5000	For installation in walls, add							75%			
5250	For installation in boxouts, add							25%			

03 15 19 – Cast-In Concrete Anchors

03 15 19.05 Anchor Bolt Accessories

			Crew	Daily Output	Labor-Hours	Unit	Material	Labor	Equipment	Total	Total Incl O&P
0010	**ANCHOR BOLT ACCESSORIES**										
0015	For anchor bolts set in fresh concrete, see Section 03 15 19.10										
8150	Anchor bolt sleeve, plastic, 1" diameter bolts		1 Carp	60	.133	Ea.	11.60	7.10		18.70	23.50
8500	1-1/2" diameter			28	.286		21.50	15.20		36.70	46.50
8600	2" diameter			24	.333		19.10	17.70		36.80	47.50
8650	3" diameter			20	.400		35	21.50		56.50	70.50
8800	Templates, steel, 8" bolt spacing	G	2 Carp	16	1		11.70	53		64.70	93
8850	12" bolt spacing	G		15	1.067		12.20	56.50		68.70	98.50

03 15 19 – Cast-In Concrete Anchors

03 15 19.05 Anchor Bolt Accessories

		Crew	Daily Output	Labor-Hours	Unit	Material	2020 Bare Costs Labor	Equipment	Total	Total Incl O&P
8900	16" bolt spacing	G 2 Carp	14	1.143	Ea.	14.65	60.50		75.15	108
8950	24" bolt spacing	G	12	1.333		19.50	71		90.50	128
9100	Wood, 8" bolt spacing		16	1		.90	53		53.90	81
9150	12" bolt spacing		15	1.067		1.19	56.50		57.69	86.50
9200	16" bolt spacing		14	1.143		1.49	60.50		61.99	93
9250	24" bolt spacing		16	1		2.08	53		55.08	82.50

03 15 19.10 Anchor Bolts

		Crew	Daily Output	Labor-Hours	Unit	Material	2020 Bare Costs Labor	Equipment	Total	Total Incl O&P
0010	**ANCHOR BOLTS**									
0015	Made from recycled materials									
0025	Single bolts installed in fresh concrete, no templates									
0030	Hooked w/nut and washer, 1/2" diameter, 8" long	G 1 Carp	132	.061	Ea.	1.62	3.22		4.84	6.60
0040	12" long	G	131	.061		1.80	3.25		5.05	6.85
0050	5/8" diameter, 8" long	G	129	.062		3.39	3.30		6.69	8.70
0060	12" long	G	127	.063		4.18	3.35		7.53	9.65
0070	3/4" diameter, 8" long	G	127	.063		4.18	3.35		7.53	9.65
0080	12" long	G	125	.064		5.20	3.40		8.60	10.85
0090	2-bolt pattern, including job-built 2-hole template, per set									
0100	J-type, incl. hex nut & washer, 1/2" diameter x 6" long	G 1 Carp	21	.381	Set	5.60	20.50		26.10	36.50
0110	12" long	G	21	.381		6.35	20.50		26.85	37.50
0120	18" long	G	21	.381		7.40	20.50		27.90	38.50
0130	3/4" diameter x 8" long	G	20	.400		11.10	21.50		32.60	44
0140	12" long	G	20	.400		13.15	21.50		34.65	46.50
0150	18" long	G	20	.400		16.30	21.50		37.80	50
0160	1" diameter x 12" long	G	19	.421		26.50	22.50		49	62.50
0170	18" long	G	19	.421		31	22.50		53.50	68
0180	24" long	G	19	.421		37.50	22.50		60	74.50
0190	36" long	G	18	.444		50.50	23.50		74	91
0200	1-1/2" diameter x 18" long	G	17	.471		47.50	25		72.50	89.50
0210	24" long	G	16	.500		56	26.50		82.50	102
0300	L-type, incl. hex nut & washer, 3/4" diameter x 12" long	G	20	.400		18.20	21.50		39.70	52
0310	18" long	G	20	.400		22.50	21.50		44	57
0320	24" long	G	20	.400		27	21.50		48.50	61.50
0330	30" long	G	20	.400		33.50	21.50		55	69
0340	36" long	G	20	.400		38	21.50		59.50	73.50
0350	1" diameter x 12" long	G	19	.421		24.50	22.50		47	60.50
0360	18" long	G	19	.421		29.50	22.50		52	66
0370	24" long	G	19	.421		36	22.50		58.50	73
0380	30" long	G	19	.421		42	22.50		64.50	80
0390	36" long	G	18	.444		47.50	23.50		71	88
0400	42" long	G	18	.444		57.50	23.50		81	98.50
0410	48" long	G	18	.444		64	23.50		87.50	106
0420	1-1/4" diameter x 18" long	G	18	.444		43.50	23.50		67	83.50
0430	24" long	G	18	.444		51.50	23.50		75	92
0440	30" long	G	17	.471		59	25		84	103
0450	36" long	G	17	.471		66.50	25		91.50	111
0460	42" long	G 2 Carp	32	.500		75	26.50		101.50	123
0470	48" long	G	32	.500		85	26.50		111.50	134
0480	54" long	G	31	.516		100	27.50		127.50	151
0490	60" long	G	31	.516		109	27.50		136.50	161
0500	1-1/2" diameter x 18" long	G	33	.485		47	26		73	90
0510	24" long	G	32	.500		54	26.50		80.50	99.50
0520	30" long	G	31	.516		61	27.50		88.50	108

For customer support on your Site Work & Landscape Costs with RSMeans data, call 800.448.8182.

03 15 19 – Cast-In Concrete Anchors

03 15 19.10 Anchor Bolts		Crew	Daily Output	Labor-Hours	Unit	Material	2020 Bare Costs Labor	Equipment	Total	Total Incl O&P
0530	36" long	G 2 Carp	30	.533	Set	69.50	28.50		98	119
0540	42" long	G	30	.533		79	28.50		107.50	129
0550	48" long	G	29	.552		88	29.50		117.50	141
0560	54" long	G	28	.571		107	30.50		137.50	163
0570	60" long	G	28	.571		117	30.50		147.50	174
0580	1-3/4" diameter x 18" long	G	31	.516		71.50	27.50		99	120
0590	24" long	G	30	.533		83.50	28.50		112	135
0600	30" long	G	29	.552		96.50	29.50		126	150
0610	36" long	G	28	.571		110	30.50		140.50	167
0620	42" long	G	27	.593		123	31.50		154.50	184
0630	48" long	G	26	.615		135	32.50		167.50	198
0640	54" long	G	26	.615		167	32.50		199.50	233
0650	60" long	G	25	.640		180	34		214	249
0660	2" diameter x 24" long	G	27	.593		135	31.50		166.50	197
0670	30" long	G	27	.593		152	31.50		183.50	215
0680	36" long	G	26	.615		167	32.50		199.50	232
0690	42" long	G	25	.640		185	34		219	255
0700	48" long	G	24	.667		213	35.50		248.50	287
0710	54" long	G	23	.696		253	37		290	335
0720	60" long	G	23	.696		272	37		309	355
0730	66" long	G	22	.727		291	38.50		329.50	380
0740	72" long	G	21	.762		320	40.50		360.50	410
1000	4-bolt pattern, including job-built 4-hole template, per set									
1100	J-type, incl. hex nut & washer, 1/2" diameter x 6" long	G 1 Carp	19	.421	Set	8.50	22.50		31	43
1110	12" long	G	19	.421		9.95	22.50		32.45	44.50
1120	18" long	G	18	.444		12.10	23.50		35.60	49
1130	3/4" diameter x 8" long	G	17	.471		19.45	25		44.45	59
1140	12" long	G	17	.471		23.50	25		48.50	63.50
1150	18" long	G	17	.471		30	25		55	70.50
1160	1" diameter x 12" long	G	16	.500		50	26.50		76.50	95
1170	18" long	G	15	.533		59.50	28.50		88	108
1180	24" long	G	15	.533		72	28.50		100.50	122
1190	36" long	G	15	.533		98	28.50		126.50	151
1200	1-1/2" diameter x 18" long	G	13	.615		92	32.50		124.50	150
1210	24" long	G	12	.667		109	35.50		144.50	173
1300	L-type, incl. hex nut & washer, 3/4" diameter x 12" long	G	17	.471		33.50	25		58.50	74.50
1310	18" long	G	17	.471		42.50	25		67.50	84
1320	24" long	G	17	.471		51	25		76	93.50
1330	30" long	G	16	.500		64	26.50		90.50	111
1340	36" long	G	16	.500		73	26.50		99.50	120
1350	1" diameter x 12" long	G	16	.500		46	26.50		72.50	90.50
1360	18" long	G	15	.533		56.50	28.50		85	105
1370	24" long	G	15	.533		69.50	28.50		98	119
1380	30" long	G	15	.533		81.50	28.50		110	132
1390	36" long	G	15	.533		92.50	28.50		121	145
1400	42" long	G	14	.571		112	30.50		142.50	169
1410	48" long	G	14	.571		125	30.50		155.50	184
1420	1-1/4" diameter x 18" long	G	14	.571		84.50	30.50		115	139
1430	24" long	G	14	.571		100	30.50		130.50	156
1440	30" long	G	13	.615		115	32.50		147.50	176
1450	36" long	G	13	.615		131	32.50		163.50	193
1460	42" long	G 2 Carp	25	.640		147	34		181	213
1470	48" long	G	24	.667		168	35.50		203.50	237

03 15 19 – Cast-In Concrete Anchors

03 15 19.10 Anchor Bolts

			Crew	Daily Output	Labor-Hours	Unit	Material	2020 Bare Costs Labor	Equipment	Total	Total Incl O&P
1480	54" long	G	2 Carp	23	.696	Set	197	37		234	273
1490	60" long	G		23	.696		216	37		253	294
1500	1-1/2" diameter x 18" long	G		25	.640		91	34		125	151
1510	24" long	G		24	.667		105	35.50		140.50	169
1520	30" long	G		23	.696		119	37		156	187
1530	36" long	G		22	.727		136	38.50		174.50	208
1540	42" long	G		22	.727		155	38.50		193.50	228
1550	48" long	G		21	.762		173	40.50		213.50	252
1560	54" long	G		20	.800		211	42.50		253.50	296
1570	60" long	G		20	.800		231	42.50		273.50	320
1580	1-3/4" diameter x 18" long	G		22	.727		140	38.50		178.50	212
1590	24" long	G		21	.762		164	40.50		204.50	242
1600	30" long	G		21	.762		191	40.50		231.50	271
1610	36" long	G		20	.800		217	42.50		259.50	305
1620	42" long	G		19	.842		244	45		289	335
1630	48" long	G		18	.889		267	47		314	365
1640	54" long	G		18	.889		330	47		377	435
1650	60" long	G		17	.941		360	50		410	470
1660	2" diameter x 24" long	G		19	.842		267	45		312	360
1670	30" long	G		18	.889		300	47		347	400
1680	36" long	G		18	.889		330	47		377	435
1690	42" long	G		17	.941		370	50		420	480
1700	48" long	G		16	1		425	53		478	545
1710	54" long	G		15	1.067		505	56.50		561.50	640
1720	60" long	G		15	1.067		540	56.50		596.50	680
1730	66" long	G		14	1.143		580	60.50		640.50	725
1740	72" long	G		14	1.143		635	60.50		695.50	785
1990	For galvanized, add					Ea.	75%				

03 15 19.20 Dovetail Anchor System

			Crew	Daily Output	Labor-Hours	Unit	Material	2020 Bare Costs Labor	Equipment	Total	Total Incl O&P
0010	**DOVETAIL ANCHOR SYSTEM**										
0500	Dovetail anchor slot, galvanized, foam-filled, 26 ga.	G	1 Carp	425	.019	L.F.	1.38	1		2.38	3.02
0600	24 ga.	G		400	.020		1.67	1.06		2.73	3.44
0625	22 ga.	G		400	.020		2.10	1.06		3.16	3.91
0900	Stainless steel, foam-filled, 26 ga.	G		375	.021		2.21	1.13		3.34	4.13
1200	Dovetail brick anchor, corrugated, galvanized, 3-1/2" long, 16 ga.	G	1 Bric	10.50	.762	C	44.50	39.50		84	109
1300	12 ga.	G		10.50	.762		67	39.50		106.50	134
1500	Seismic, galvanized, 3-1/2" long, 16 ga.	G		10.50	.762		75.50	39.50		115	143
1600	12 ga.	G		10.50	.762		86	39.50		125.50	155
6000	Dovetail stone panel anchors, galvanized, 1/8" x 1" wide, 3-1/2" long	G		10.50	.762		103	39.50		142.50	173
6100	1/4" x 1" wide	G		10.50	.762		183	39.50		222.50	262

03 15 19.30 Inserts

			Crew	Daily Output	Labor-Hours	Unit	Material	2020 Bare Costs Labor	Equipment	Total	Total Incl O&P
0010	**INSERTS**										
6000	Thin slab, ferrule type										
6100	1/2" diameter bolt	G	1 Carp	60	.133	Ea.	4.05	7.10		11.15	15.10
6150	5/8" diameter bolt	G		60	.133		5	7.10		12.10	16.15
6200	3/4" diameter bolt	G		60	.133		5.80	7.10		12.90	17
6250	1" diameter bolt	G		60	.133		9.35	7.10		16.45	21
9950	For galvanized inserts, add						30%				

03 15 Concrete Accessories

03 15 19 – Cast-In Concrete Anchors

03 15 19.45 Machinery Anchors

		Crew	Daily Output	Labor-Hours	Unit	Material	2020 Bare Costs Labor	Equipment	Total	Total Incl O&P
0010	**MACHINERY ANCHORS**, heavy duty, incl. sleeve, floating base nut,									
0020	lower stud & coupling nut, fiber plug, connecting stud, washer & nut.									
0030	For flush mounted embedment in poured concrete heavy equip. pads.									
0200	Stud & bolt, 1/2" diameter G	E-16	40	.400	Ea.	61.50	23.50	3.68	88.68	109
0300	5/8" diameter G		35	.457		67	26.50	4.20	97.70	120
0500	3/4" diameter G		30	.533		79.50	31	4.90	115.40	142
0600	7/8" diameter G		25	.640		93.50	37.50	5.90	136.90	168
0800	1" diameter G		20	.800		102	47	7.35	156.35	194
0900	1-1/4" diameter G	↓	15	1.067	↓	133	62.50	9.80	205.30	256

03 21 Reinforcement Bars

03 21 05 – Reinforcing Steel Accessories

03 21 05.10 Rebar Accessories

		Crew	Daily Output	Labor-Hours	Unit	Material	2020 Bare Costs Labor	Equipment	Total	Total Incl O&P
0010	**REBAR ACCESSORIES** R032110-70									
0030	Steel & plastic made from recycled materials									
0700	Bag ties, 16 ga., plain, 4" long G				C	4.30			4.30	4.73
0710	5" long G					5.55			5.55	6.10
0720	6" long G					3.93			3.93	4.32
0730	7" long G					5.85			5.85	6.45
1200	High chairs, individual (HC), 3" high, plain steel G					66.50			66.50	73.50
1202	Galvanized G					95.50			95.50	105
1204	Stainless tipped legs G					530			530	585
1206	Plastic tipped legs G					70.50			70.50	77.50
1210	5" high, plain G					97			97	107
1212	Galvanized G					124			124	136
1214	Stainless tipped legs G					570			570	625
1216	Plastic tipped legs G					105			105	116
1220	8" high, plain G					141			141	155
1222	Galvanized G					194			194	214
1224	Stainless tipped legs G					645			645	705
1226	Plastic tipped legs G					152			152	168
1230	12" high, plain G					340			340	375
1232	Galvanized G					385			385	425
1234	Stainless tipped legs G					700			700	770
1236	Plastic tipped legs G					360			360	400
1400	Individual high chairs, with plate (HCP), 5" high G					192			192	211
1410	8" high G					265			265	292
1500	Bar chair (BC), 1-1/2" high, plain steel G					51			51	56
1520	Galvanized G					45.50			45.50	50
1530	Stainless tipped legs G					480			480	525
1540	Plastic tipped legs G				↓	44			44	48.50
1700	Continuous high chairs (CHC), legs 8" OC, 4" high, plain steel G				C.L.F.	57			57	63
1705	Galvanized G					67.50			67.50	74.50
1710	Stainless tipped legs G					540			540	590
1715	Plastic tipped legs G					78.50			78.50	86
1718	Epoxy dipped G					95			95	104
1720	6" high, plain G					75			75	82.50
1725	Galvanized G					109			109	119
1730	Stainless tipped legs G					560			560	615
1735	Plastic tipped legs G				↓	101			101	111

71

03 21 05 – Reinforcing Steel Accessories

03 21 05.10 Rebar Accessories

			Crew	Daily Output	Labor-Hours	Unit	Material	2020 Bare Costs Labor	Equipment	Total	Total Incl O&P
1738	Epoxy dipped	G				C.L.F.	129			129	142
1740	8" high, plain	G					132			132	145
1745	Galvanized	G					134			134	148
1750	Stainless tipped legs	G					630			630	695
1755	Plastic tipped legs	G					132			132	145
1758	Epoxy dipped	G					168			168	185
1900	For continuous bottom wire runners, add	G					32			32	35.50
1940	For continuous bottom plate, add	G				↓	191			191	210
2200	Screed chair base, 1/2" coil thread diam., 2-1/2" high, plain steel	G				C	345			345	380
2210	Galvanized	G					385			385	420
2220	5-1/2" high, plain	G					420			420	460
2250	Galvanized	G					460			460	510
2300	3/4" coil thread diam., 2-1/2" high, plain steel	G					420			420	460
2310	Galvanized	G					485			485	530
2320	5-1/2" high, plain steel	G					535			535	585
2350	Galvanized	G					665			665	730
2400	Screed holder, 1/2" coil thread diam. for pipe screed, plain steel, 6" long	G					445			445	490
2420	12" long	G					635			635	700
2500	3/4" coil thread diam. for pipe screed, plain steel, 6" long	G					590			590	645
2520	12" long	G					940			940	1,025
2700	Screw anchor for bolts, plain steel, 3/4" diameter x 4" long	G					535			535	585
2720	1" diameter x 6" long	G					1,000			1,000	1,100
2740	1-1/2" diameter x 8" long	G					1,225			1,225	1,350
2800	Screw anchor eye bolts, 3/4" x 3" long	G					3,125			3,125	3,450
2820	1" x 3-1/2" long	G					3,700			3,700	4,075
2840	1-1/2" x 6" long	G					12,400			12,400	13,600
2900	Screw anchor bolts, 3/4" x 9" long	G					1,775			1,775	1,950
2920	1" x 12" long	G					3,425			3,425	3,750
3800	Subgrade chairs, #4 bar head, 3-1/2" high	G					44			44	48
3850	12" high	G					53			53	58
3900	#6 bar head, 3-1/2" high	G					44			44	48
3950	12" high	G					53			53	58
4200	Subgrade stakes, no nail holes, 3/4" diameter, 12" long	G					143			143	157
4250	24" long	G					274			274	300
4300	7/8" diameter, 12" long	G					420			420	465
4350	24" long	G				↓	680			680	745
4500	Tie wire, 16 ga. annealed steel	G				Cwt.	197			197	217

03 21 11 – Plain Steel Reinforcement Bars

03 21 11.50 Reinforcing Steel, Mill Base Plus Extras

			Crew	Daily Output	Labor-Hours	Unit	Material	2020 Bare Costs Labor	Equipment	Total	Total Incl O&P
0010	**REINFORCING STEEL, MILL BASE PLUS EXTRAS**	R032110-10									
0150	Reinforcing, A615 grade 40, mill base	G				Ton	790			790	870
0200	Detailed, cut, bent, and delivered	G					1,100			1,100	1,200
0650	Reinforcing steel, A615 grade 60, mill base	G					795			795	875
0700	Detailed, cut, bent, and delivered	G				↓	1,100			1,100	1,200
1000	Reinforcing steel, extras, included in delivered price										
1005	Mill extra, added for delivery to shop					Ton	40.50			40.50	44.50
1010	Shop extra, added for handling & storage						49.50			49.50	54.50
1020	Shop extra, added for bending, limited percent of bars						38			38	41.50
1030	Average percent of bars						75.50			75.50	83
1050	Large percent of bars	R032110-70					151			151	166
1200	Shop extra, added for detailing, under 50 tons						57			57	62.50
1250	50 to 150 tons	R032110-80				↓	42.50			42.50	47

For customer support on your Site Work & Landscape Costs with RSMeans data, call 800.448.8182.

03 21 Reinforcement Bars

03 21 11 – Plain Steel Reinforcement Bars

03 21 11.50 Reinforcing Steel, Mill Base Plus Extras

		Crew	Daily Output	Labor-Hours	Unit	Material	2020 Bare Costs Labor	Equipment	Total	Total Incl O&P
1300	150 to 500 tons				Ton	40.50			40.50	44.50
1350	Over 500 tons					38.50			38.50	42.50
1700	Shop extra, added for listing					6.10			6.10	6.70
2000	Mill extra, added for quantity, under 20 tons					24			24	26.50
2100	Shop extra, added for quantity, under 20 tons					37.50			37.50	41.50
2200	20 to 49 tons					28			28	31
2250	50 to 99 tons					18.75			18.75	20.50
2300	100 to 300 tons					11.25			11.25	12.40
2500	Shop extra, added for size, #3					168			168	185
2550	#4					84			84	92.50
2600	#5					42			42	46
2650	#6					38			38	41.50
2700	#7 to #11					50.50			50.50	55.50
2750	#14					63			63	69.50
2800	#18					71.50			71.50	78.50
2900	Shop extra, added for delivery to job				▼	19.10			19.10	21

03 21 11.60 Reinforcing In Place

			Crew	Daily Output	Labor-Hours	Unit	Material	2020 Bare Costs Labor	Equipment	Total	Total Incl O&P
0010	**REINFORCING IN PLACE**, 50-60 ton lots, A615 Grade 60	R032110-10									
0020	Includes labor, but not material cost, to install accessories										
0030	Made from recycled materials										
0100	Beams & Girders, #3 to #7	G	4 Rodm	1.60	20	Ton	1,125	1,125		2,250	2,950
0150	#8 to #18	G		2.70	11.852		1,125	670		1,795	2,250
0200	Columns, #3 to #7	G		1.50	21.333		1,125	1,200		2,325	3,050
0250	#8 to #18	G		2.30	13.913		1,125	785		1,910	2,425
0300	Spirals, hot rolled, 8" to 15" diameter	G		2.20	14.545		1,675	820		2,495	3,050
0320	15" to 24" diameter	G		2.20	14.545		1,600	820		2,420	2,975
0330	24" to 36" diameter	G		2.30	13.913		1,525	785		2,310	2,850
0340	36" to 48" diameter	G		2.40	13.333		1,450	750		2,200	2,700
0360	48" to 64" diameter	G		2.50	12.800		1,600	720		2,320	2,825
0380	64" to 84" diameter	R032110-70	G		2.60	12.308		1,675	695		2,370	2,875
0390	84" to 96" diameter		G		2.70	11.852		1,750	670		2,420	2,925
0400	Elevated slabs, #4 to #7	R032110-80	G		2.90	11.034		1,125	620		1,745	2,175
0500	Footings, #4 to #7		G		2.10	15.238		1,125	860		1,985	2,550
0550	#8 to #18		G		3.60	8.889		1,125	500		1,625	2,000
0600	Slab on grade, #3 to #7		G		2.30	13.913		1,125	785		1,910	2,425
0700	Walls, #3 to #7		G		3	10.667		1,125	600		1,725	2,150
0750	#8 to #18		G	▼	4	8	▼	1,125	450		1,575	1,925
0900	For other than 50-60 ton lots										
1000	Under 10 ton job, #3 to #7, add							25%	10%		
1010	#8 to #18, add							20%	10%		
1050	10-50 ton job, #3 to #7, add							10%			
1060	#8 to #18, add							5%			
1100	60-100 ton job, #3 to #7, deduct							5%			
1110	#8 to #18, deduct							10%			
1150	Over 100 ton job, #3 to #7, deduct							10%			
1160	#8 to #18, deduct							15%			
1200	Reinforcing in place, A615 Grade 75, add		G				Ton	102			102	112
1220	Grade 90, add							140			140	154
2000	Unloading & sorting, add to above			C-5	100	.560			31.50	5.80	37.30	54
2200	Crane cost for handling, 90 picks/day, up to 1.5 tons/bundle, add to above				135	.415			23.50	4.30	27.80	39.50
2210	1.0 ton/bundle				92	.609			34	6.30	40.30	58.50
2220	0.5 ton/bundle				35	1.600			90	16.60	106.60	153

03 21 Reinforcement Bars

03 21 11 – Plain Steel Reinforcement Bars

03 21 11.60 Reinforcing In Place

		Crew	Daily Output	Labor-Hours	Unit	Material	2020 Bare Costs Labor	Equipment	Total	Total Incl O&P
2400	Dowels, 2 feet long, deformed, #3 [G]	2 Rodm	520	.031	Ea.	.47	1.74		2.21	3.12
2410	#4 [G]		480	.033		.83	1.88		2.71	3.74
2420	#5 [G]		435	.037		1.30	2.07		3.37	4.55
2430	#6 [G]		360	.044		1.87	2.51		4.38	5.80
2450	Longer and heavier dowels, add [G]		725	.022	Lb.	.62	1.24		1.86	2.55
2500	Smooth dowels, 12" long, 1/4" or 3/8" diameter [G]		140	.114	Ea.	.87	6.45		7.32	10.65
2520	5/8" diameter [G]		125	.128		1.53	7.20		8.73	12.55
2530	3/4" diameter [G]		110	.145		1.89	8.20		10.09	14.45
2600	Dowel sleeves for CIP concrete, 2-part system									
2610	Sleeve base, plastic, for 5/8" smooth dowel sleeve, fasten to edge form	1 Rodm	200	.040	Ea.	.54	2.26		2.80	3.99
2615	Sleeve, plastic, 12" long, for 5/8" smooth dowel, snap onto base		400	.020		1.54	1.13		2.67	3.39
2620	Sleeve base, for 3/4" smooth dowel sleeve		175	.046		.54	2.58		3.12	4.47
2625	Sleeve, 12" long, for 3/4" smooth dowel		350	.023		1.22	1.29		2.51	3.28
2630	Sleeve base, for 1" smooth dowel sleeve		150	.053		.71	3.01		3.72	5.30
2635	Sleeve, 12" long, for 1" smooth dowel		300	.027		1.67	1.50		3.17	4.10
2700	Dowel caps, visual warning only, plastic, #3 to #8	2 Rodm	800	.020		.37	1.13		1.50	2.11
2720	#8 to #18		750	.021		.89	1.20		2.09	2.79
2750	Impalement protective, plastic, #4 to #9		800	.020		1.18	1.13		2.31	3

03 21 13 – Galvanized Reinforcement Steel Bars

03 21 13.10 Galvanized Reinforcing

		Crew	Daily Output	Labor-Hours	Unit	Material	2020 Bare Costs Labor	Equipment	Total	Total Incl O&P
0010	**GALVANIZED REINFORCING**									
0150	Add to plain steel rebar pricing for galvanized rebar				Ton	510			510	560

03 21 16 – Epoxy-Coated Reinforcement Steel Bars

03 21 16.10 Epoxy-Coated Reinforcing

		Crew	Daily Output	Labor-Hours	Unit	Material	2020 Bare Costs Labor	Equipment	Total	Total Incl O&P
0010	**EPOXY-COATED REINFORCING**									
0100	Add to plain steel rebar pricing for epoxy-coated rebar				Ton	1,075			1,075	1,200

03 21 21 – Composite Reinforcement Bars

03 21 21.11 Glass Fiber-Reinforced Polymer Reinf. Bars

		Crew	Daily Output	Labor-Hours	Unit	Material	2020 Bare Costs Labor	Equipment	Total	Total Incl O&P
0010	**GLASS FIBER-REINFORCED POLYMER REINFORCEMENT BARS**									
0020	Includes labor, but not material cost, to install accessories									
0050	#2 bar, .043 lb./L.F.	4 Rodm	9500	.003	L.F.	.41	.19		.60	.74
0100	#3 bar, .092 lb./L.F.		9300	.003		.65	.19		.84	1.01
0150	#4 bar, .160 lb./L.F.		9100	.004		.94	.20		1.14	1.33
0200	#5 bar, .258 lb./L.F.		8700	.004		1.14	.21		1.35	1.56
0250	#6 bar, .372 lb./L.F.		8300	.004		1.92	.22		2.14	2.44
0300	#7 bar, .497 lb./L.F.		7900	.004		2.59	.23		2.82	3.19
0350	#8 bar, .620 lb./L.F.		7400	.004		2.45	.24		2.69	3.07
0400	#9 bar, .800 lb./L.F.		6800	.005		4.22	.27		4.49	5.05
0450	#10 bar, 1.08 lb./L.F.		5800	.006		3.83	.31		4.14	4.68
0500	For bends, add per bend				Ea.	1.64			1.64	1.80

03 22 Fabric and Grid Reinforcing

03 22 11 – Plain Welded Wire Fabric Reinforcing

03 22 11.10 Plain Welded Wire Fabric

		Crew	Daily Output	Labor-Hours	Unit	Material	2020 Bare Costs Labor	2020 Bare Costs Equipment	Total	Total Incl O&P
0010	**PLAIN WELDED WIRE FABRIC** ASTM A185 R032205-30									
0020	Includes labor, but not material cost, to install accessories									
0030	Made from recycled materials									
0050	Sheets									
0100	6 x 6 - W1.4 x W1.4 (10 x 10) 21 lb./C.S.F.	[G] 2 Rodm	35	.457	C.S.F.	16.25	26		42.25	57
0200	6 x 6 - W2.1 x W2.1 (8 x 8) 30 lb./C.S.F.	[G]	31	.516		19.15	29		48.15	65
0300	6 x 6 - W2.9 x W2.9 (6 x 6) 42 lb./C.S.F.	[G]	29	.552		27.50	31		58.50	77.50
0400	6 x 6 - W4 x W4 (4 x 4) 58 lb./C.S.F.	[G]	27	.593		38.50	33.50		72	93
0500	4 x 4 - W1.4 x W1.4 (10 x 10) 31 lb./C.S.F.	[G]	31	.516		24.50	29		53.50	70.50
0600	4 x 4 - W2.1 x W2.1 (8 x 8) 44 lb./C.S.F.	[G]	29	.552		30.50	31		61.50	80.50
0650	4 x 4 - W2.9 x W2.9 (6 x 6) 61 lb./C.S.F.	[G]	27	.593		48.50	33.50		82	104
0700	4 x 4 - W4 x W4 (4 x 4) 85 lb./C.S.F.	[G]	25	.640		60.50	36		96.50	122
0750	Rolls									
0800	2 x 2 - #14 galv., 21 lb./C.S.F., beam & column wrap	[G] 2 Rodm	6.50	2.462	C.S.F.	48.50	139		187.50	263
0900	2 x 2 - #12 galv. for gunite reinforcing	[G] "	6.50	2.462	"	73	139		212	290

03 22 13 – Galvanized Welded Wire Fabric Reinforcing

03 22 13.10 Galvanized Welded Wire Fabric

		Crew	Daily Output	Labor-Hours	Unit	Material	Labor	Equipment	Total	Total Incl O&P
0010	**GALVANIZED WELDED WIRE FABRIC**									
0100	Add to plain welded wire pricing for galvanized welded wire				Lb.	.26			.26	.28

03 22 16 – Epoxy-Coated Welded Wire Fabric Reinforcing

03 22 16.10 Epoxy-Coated Welded Wire Fabric

		Crew	Daily Output	Labor-Hours	Unit	Material	Labor	Equipment	Total	Total Incl O&P
0010	**EPOXY-COATED WELDED WIRE FABRIC**									
0100	Add to plain welded wire pricing for epoxy-coated welded wire				Lb.	.54			.54	.60

03 23 Stressed Tendon Reinforcing

03 23 05 – Prestressing Tendons

03 23 05.50 Prestressing Steel

		Crew	Daily Output	Labor-Hours	Unit	Material	Labor	Equipment	Total	Total Incl O&P
0010	**PRESTRESSING STEEL**									
3000	Slabs on grade, 0.5-inch diam. non-bonded strands, HDPE sheathed,									
3050	attached dead-end anchors, loose stressing-end anchors									
3100	25' x 30' slab, strands @ 36" OC, placing	2 Rodm	2940	.005	S.F.	.68	.31		.99	1.21
3105	Stressing	C-4A	3750	.004			.24	.02	.26	.38
3110	42" OC, placing	2 Rodm	3200	.005		.60	.28		.88	1.08
3115	Stressing	C-4A	4040	.004			.22	.02	.24	.36
3120	48" OC, placing	2 Rodm	3510	.005		.53	.26		.79	.97
3125	Stressing	C-4A	4390	.004			.21	.02	.23	.33
3150	25' x 40' slab, strands @ 36" OC, placing	2 Rodm	3370	.005		.66	.27		.93	1.12
3155	Stressing	C-4A	4360	.004			.21	.02	.23	.33
3160	42" OC, placing	2 Rodm	3760	.004		.57	.24		.81	.99
3165	Stressing	C-4A	4820	.003			.19	.02	.21	.30
3170	48" OC, placing	2 Rodm	4090	.004		.51	.22		.73	.89
3175	Stressing	C-4A	5190	.003			.17	.01	.18	.28
3200	30' x 30' slab, strands @ 36" OC, placing	2 Rodm	3260	.005		.66	.28		.94	1.14
3205	Stressing	C-4A	4190	.004			.22	.02	.24	.34
3210	42" OC, placing	2 Rodm	3530	.005		.59	.26		.85	1.03
3215	Stressing	C-4A	4500	.004			.20	.02	.22	.32
3220	48" OC, placing	2 Rodm	3840	.004		.52	.24		.76	.93
3225	Stressing	C-4A	4850	.003			.19	.02	.21	.30
3230	30' x 40' slab, strands @ 36" OC, placing	2 Rodm	3780	.004		.63	.24		.87	1.05
3235	Stressing	C-4A	4920	.003			.18	.02	.20	.30

75

03 23 Stressed Tendon Reinforcing

03 23 05 - Prestressing Tendons

03 23 05.50 Prestressing Steel

		Crew	Daily Output	Labor-Hours	Unit	Material	2020 Bare Costs Labor	Equipment	Total	Total Incl O&P
3240	42" OC, placing	2 Rodm	4190	.004	S.F.	.55	.22		.77	.93
3245	Stressing	C-4A	5410	.003			.17	.01	.18	.27
3250	48" OC, placing	2 Rodm	4520	.004		.50	.20		.70	.85
3255	Stressing	C-4A	5790	.003			.16	.01	.17	.24
3260	30' x 50' slab, strands @ 36" OC, placing	2 Rodm	4300	.004		.60	.21		.81	.97
3265	Stressing	C-4A	5650	.003			.16	.01	.17	.25
3270	42" OC, placing	2 Rodm	4720	.003		.53	.19		.72	.87
3275	Stressing	C-4A	6150	.003			.15	.01	.16	.23
3280	48" OC, placing	2 Rodm	5240	.003		.46	.17		.63	.77
3285	Stressing	C-4A	6760	.002	▼		.13	.01	.14	.21

03 24 Fibrous Reinforcing

03 24 05 - Reinforcing Fibers

03 24 05.30 Synthetic Fibers

					Unit	Material	Labor	Equipment	Total	Total Incl O&P
0010	**SYNTHETIC FIBERS**									
0100	Synthetic fibers, add to concrete				Lb.	5.05			5.05	5.55
0110	1-1/2 lb./C.Y.				C.Y.	7.80			7.80	8.55

03 24 05.70 Steel Fibers

						Unit	Material	Labor	Equipment	Total	Total Incl O&P
0010	**STEEL FIBERS**										
0140	ASTM A850, Type V, continuously deformed, 1-1/2" long x 0.045" diam.										
0150	Add to price of ready mix concrete	G				Lb.	1.23			1.23	1.35
0205	Alternate pricing, dosing at 5 lb./C.Y., add to price of RMC	G				C.Y.	6.15			6.15	6.75
0210	10 lb./C.Y.	G					12.30			12.30	13.55
0215	15 lb./C.Y.	G					18.45			18.45	20.50
0220	20 lb./C.Y.	G					24.50			24.50	27
0225	25 lb./C.Y.	G					31			31	34
0230	30 lb./C.Y.	G					37			37	40.50
0235	35 lb./C.Y.	G					43			43	47.50
0240	40 lb./C.Y.	G					49			49	54
0250	50 lb./C.Y.	G					61.50			61.50	67.50
0275	75 lb./C.Y.	G					92.50			92.50	101
0300	100 lb./C.Y.	G			▼		123			123	135

03 30 Cast-In-Place Concrete

03 30 53 - Miscellaneous Cast-In-Place Concrete

03 30 53.40 Concrete In Place

			Crew	Daily Output	Labor-Hours	Unit	Material	Labor	Equipment	Total	Total Incl O&P
0010	**CONCRETE IN PLACE**	R033053-60									
0020	Including forms (4 uses), Grade 60 rebar, concrete (Portland cement										
0050	Type I), placement and finishing unless otherwise indicated	R033105-20									
0300	Beams (3500 psi), 5 kip/L.F., 10' span	R033105-70	C-14A	15.62	12.804	C.Y.	370	675	28	1,073	1,450
0350	25' span		"	18.55	10.782		390	570	23.50	983.50	1,300
0500	Chimney foundations (5000 psi), over 5 C.Y.		C-14C	32.22	3.476		178	175	.83	353.83	460
0510	(3500 psi), under 5 C.Y.		"	23.71	4.724		208	238	1.13	447.13	585
0700	Columns, square (4000 psi), 12" x 12", up to 1% reinforcing by area		C-14A	11.96	16.722		410	885	36.50	1,331.50	1,825
0800	16" x 16", up to 1% reinforcing by area			16.22	12.330		325	650	27	1,002	1,375
0900	24" x 24", up to 1% reinforcing by area			23.66	8.453		278	445	18.50	741.50	995
1000	36" x 36", up to 1% reinforcing by area			33.69	5.936		247	315	13	575	755
1100	Columns, round (4000 psi), tied, 12" diameter, up to 1% reinforcing by area			20.97	9.537		415	505	21	941	1,225
1200	16" diameter, up to 1% reinforcing by area		▼	31.49	6.351	▼	340	335	13.90	688.90	890

03 30 53 – Miscellaneous Cast-In-Place Concrete

03 30 53.40 Concrete In Place	Crew	Daily Output	Labor-Hours	Unit	Material	2020 Bare Costs Labor	2020 Bare Costs Equipment	Total	Total Incl O&P
1300 20" diameter, up to 1% reinforcing by area	C-14A	41.04	4.873	C.Y.	325	258	10.65	593.65	755
1400 24" diameter, up to 1% reinforcing by area		51.85	3.857		315	204	8.45	527.45	665
1500 36" diameter, up to 1% reinforcing by area		75.04	2.665		315	141	5.85	461.85	565
3100 Elevated slabs, flat plate, including finish, not									
3110 including forms or reinforcing									
3150 Regular concrete (4000 psi), 4" slab	C-8	2613	.021	S.F.	1.69	1	.16	2.85	3.52
3200 6" slab		2585	.022		2.48	1.01	.16	3.65	4.40
3250 2-1/2" thick floor fill		2685	.021		1.10	.97	.15	2.22	2.84
3300 Lightweight, 110 #/C.F., 2-1/2" thick floor fill		2585	.022		1.31	1.01	.16	2.48	3.13
3400 Cellular concrete, 1-5/8" fill, under 5000 S.F.		2000	.028		.90	1.31	.21	2.42	3.17
3450 Over 10,000 S.F.		2200	.025		.86	1.19	.19	2.24	2.92
3540 Equipment pad (3000 psi), 3' x 3' x 6" thick	C-14H	45	1.067	Ea.	50.50	55	.60	106.10	139
3550 4' x 4' x 6" thick		30	1.600		78	82.50	.90	161.40	210
3560 5' x 5' x 8" thick		18	2.667		138	138	1.49	277.49	360
3570 6' x 6' x 8" thick		14	3.429		190	177	1.92	368.92	475
3580 8' x 8' x 10" thick		8	6		395	310	3.36	708.36	905
3590 10' x 10' x 12" thick		5	9.600		695	495	5.40	1,195.40	1,500
3600 Flexural concrete on grade, direct chute, 500 psi, no forms, reinf, finish	C-8A	150	.320	C.Y.	129	14.40		143.40	164
3610 650 psi		150	.320		135	14.40		149.40	171
3620 750 psi		150	.320		203	14.40		217.40	245
3650 Pumped, 500 psi	C-8	70	.800		129	37.50	5.85	172.35	204
3660 650 psi		70	.800		135	37.50	5.85	178.35	211
3670 750 psi		70	.800		203	37.50	5.85	246.35	285
3800 Footings (3000 psi), spread under 1 C.Y.	C-14C	28	4		203	201	.96	404.96	525
3813 Install new concrete (3000 psi) light pole base, 24" diam. x 8'	C-1	2.66	12.030		325	605		930	1,275
3825 1 C.Y. to 5 C.Y.	C-14C	43	2.605		240	131	.63	371.63	465
3850 Over 5 C.Y.	"	75	1.493		220	75	.36	295.36	355
3900 Footings, strip (3000 psi), 18" x 9", unreinforced	C-14L	40	2.400		154	118	.67	272.67	350
3920 18" x 9", reinforced	C-14C	35	3.200		181	161	.77	342.77	440
3925 20" x 10", unreinforced	C-14L	45	2.133		150	105	.60	255.60	325
3930 20" x 10", reinforced	C-14C	40	2.800		172	141	.67	313.67	400
3935 24" x 12", unreinforced	C-14L	55	1.745		148	86	.49	234.49	293
3940 24" x 12", reinforced	C-14C	48	2.333		171	118	.56	289.56	365
3945 36" x 12", unreinforced	C-14L	70	1.371		144	67.50	.38	211.88	261
3950 36" x 12", reinforced	C-14C	60	1.867		165	94	.45	259.45	320
4000 Foundation mat (3000 psi), under 10 C.Y.		38.67	2.896		246	146	.70	392.70	490
4050 Over 20 C.Y.		56.40	1.986		217	100	.48	317.48	390
4200 Wall, free-standing (3000 psi), 8" thick, 8' high	C-14D	45.83	4.364		197	230	9.55	436.55	570
4250 14' high		27.26	7.337		232	385	16.05	633.05	855
4260 12" thick, 8' high		64.32	3.109		179	164	6.80	349.80	450
4270 14' high		40.01	4.999		190	263	10.95	463.95	615
4300 15" thick, 8' high		80.02	2.499		173	132	5.45	310.45	395
4350 12' high		51.26	3.902		173	205	8.55	386.55	510
4500 18' high		48.85	4.094		194	215	8.95	417.95	550
4520 Handicap access ramp (4000 psi), railing both sides, 3' wide	C-14H	14.58	3.292	L.F.	390	170	1.84	561.84	685
4525 5' wide		12.22	3.928		400	203	2.20	605.20	745
4530 With 6" curb and rails both sides, 3' wide		8.55	5.614		400	290	3.14	693.14	880
4535 5' wide		7.31	6.566		405	340	3.68	748.68	965
4650 Slab on grade (3500 psi), not including finish, 4" thick	C-14E	60.75	1.449	C.Y.	147	74	.45	221.45	273
4700 6" thick	"	92	.957	"	141	49	.30	190.30	230
4701 Thickened slab edge (3500 psi), for slab on grade poured									
4702 monolithically with slab; depth is in addition to slab thickness;									
4703 formed vertical outside edge, earthen bottom and inside slope									

For customer support on your Site Work & Landscape Costs with RSMeans data, call 800.448.8182.

03 30 Cast-In-Place Concrete

03 30 53 – Miscellaneous Cast-In-Place Concrete

03 30 53.40 Concrete In Place

		Crew	Daily Output	Labor-Hours	Unit	Material	2020 Bare Costs Labor	Equipment	Total	Total Incl O&P
4705	8" deep x 8" wide bottom, unreinforced	C-14L	2190	.044	L.F.	4.07	2.16	.01	6.24	7.75
4710	8" x 8", reinforced	C-14C	1670	.067		6.70	3.38	.02	10.10	12.45
4715	12" deep x 12" wide bottom, unreinforced	C-14L	1800	.053		8.30	2.63	.01	10.94	13.10
4720	12" x 12", reinforced	C-14C	1310	.086		13.15	4.31	.02	17.48	21
4725	16" deep x 16" wide bottom, unreinforced	C-14L	1440	.067		14	3.29	.02	17.31	20.50
4730	16" x 16", reinforced	C-14C	1120	.100		19.75	5.05	.02	24.82	29.50
4735	20" deep x 20" wide bottom, unreinforced	C-14L	1150	.083		21	4.12	.02	25.14	29.50
4740	20" x 20", reinforced	C-14C	920	.122		28.50	6.15	.03	34.68	40.50
4745	24" deep x 24" wide bottom, unreinforced	C-14L	930	.103		30	5.10	.03	35.13	40.50
4750	24" x 24", reinforced	C-14C	740	.151	↓	39.50	7.60	.04	47.14	55
4751	Slab on grade (3500 psi), incl. troweled finish, not incl. forms									
4760	or reinforcing, over 10,000 S.F., 4" thick	C-14F	3425	.021	S.F.	1.62	1	.01	2.63	3.27
4820	6" thick		3350	.021		2.36	1.02	.01	3.39	4.12
4840	8" thick		3184	.023		3.24	1.08	.01	4.33	5.15
4900	12" thick		2734	.026		4.85	1.25	.01	6.11	7.20
4950	15" thick	↓	2505	.029	↓	6.10	1.37	.01	7.48	8.75
5000	Slab on grade (3000 psi), incl. broom finish, not incl. forms									
5001	or reinforcing, 4" thick	C-14G	2873	.019	S.F.	1.67	.91	.01	2.59	3.20
5010	6" thick		2590	.022		2.62	1.01	.01	3.64	4.39
5020	8" thick	↓	2320	.024	↓	3.41	1.13	.01	4.55	5.45
5200	Lift slab in place above the foundation, incl. forms, reinforcing,									
5210	concrete (4000 psi) and columns, over 20,000 S.F./floor	C-14B	2113	.098	S.F.	7.75	5.20	.21	13.16	16.55
5900	Pile caps (3000 psi), incl. forms and reinf., sq. or rect., under 10 C.Y.	C-14C	54.14	2.069	C.Y.	200	104	.50	304.50	375
5950	Over 10 C.Y.		75	1.493		189	75	.36	264.36	320
6000	Triangular or hexagonal, under 10 C.Y.		53	2.113		153	106	.51	259.51	330
6050	Over 10 C.Y.		85	1.318		171	66.50	.32	237.82	288
6200	Retaining walls (3000 psi), gravity, 4' high see Section 32 32	C-14D	66.20	3.021		173	159	6.60	338.60	435
6250	10' high		125	1.600		166	84	3.50	253.50	315
6300	Cantilever, level backfill loading, 8' high		70	2.857		184	150	6.25	340.25	435
6350	16' high	↓	91	2.198	↓	179	116	4.81	299.81	375
6800	Stairs (3500 psi), not including safety treads, free standing, 3'-6" wide	C-14H	83	.578	LF Nose	6.40	30	.32	36.72	52
6850	Cast on ground		125	.384	"	5.40	19.85	.22	25.47	35.50
7000	Stair landings, free standing		200	.240	S.F.	5.10	12.40	.13	17.63	24.50
7050	Cast on ground	↓	475	.101	"	4.15	5.20	.06	9.41	12.45

03 31 Structural Concrete

03 31 13 – Heavyweight Structural Concrete

03 31 13.25 Concrete, Hand Mix

		Crew	Daily Output	Labor-Hours	Unit	Material	2020 Bare Costs Labor	Equipment	Total	Total Incl O&P
0010	**CONCRETE, HAND MIX** for small quantities or remote areas									
0050	Includes bulk local aggregate, bulk sand, bagged Portland									
0060	cement (Type I) and water, using gas powered cement mixer									
0125	2500 psi	C-30	135	.059	C.F.	4.33	2.49	1.04	7.86	9.65
0130	3000 psi		135	.059		4.68	2.49	1.04	8.21	10.05
0135	3500 psi		135	.059		4.88	2.49	1.04	8.41	10.25
0140	4000 psi		135	.059		5.10	2.49	1.04	8.63	10.55
0145	4500 psi		135	.059		5.40	2.49	1.04	8.93	10.85
0150	5000 psi	↓	135	.059	↓	5.80	2.49	1.04	9.33	11.25
0300	Using pre-bagged dry mix and wheelbarrow (80-lb. bag = 0.6 C.F.)									
0340	4000 psi	1 Clab	48	.167	C.F.	8.50	7		15.50	19.90

03 31 Structural Concrete

03 31 13 – Heavyweight Structural Concrete

03 31 13.30 Concrete, Volumetric Site-Mixed

		Crew	Daily Output	Labor-Hours	Unit	Material	2020 Bare Costs Labor	Equipment	Total	Total Incl O&P
0010	**CONCRETE, VOLUMETRIC SITE-MIXED**									
0015	Mixed on-site in volumetric truck									
0020	Includes local aggregate, sand, Portland cement (Type I) and water									
0025	Excludes all additives and treatments									
0100	3000 psi, 1 C.Y. mixed and discharged				C.Y.	213			213	234
0110	2 C.Y.					164			164	180
0120	3 C.Y.					143			143	157
0130	4 C.Y.					125			125	138
0140	5 C.Y.					113			113	125
0200	For truck holding/waiting time past first 2 on-site hours, add				Hr.	91.50			91.50	101
0210	For trip charge beyond first 20 miles, each way, add				Mile	3.65			3.65	4.02
0220	For each additional increase of 500 psi, add				Ea.	4.64			4.64	5.10

03 31 13.35 Heavyweight Concrete, Ready Mix

		Crew	Daily Output	Labor-Hours	Unit	Material	2020 Bare Costs Labor	Equipment	Total	Total Incl O&P
0010	**HEAVYWEIGHT CONCRETE, READY MIX**, delivered									
0012	Includes local aggregate, sand, Portland cement (Type I) and water									
0015	Excludes all additives and treatments R033105-20									
0020	2000 psi				C.Y.	107			107	117
0100	2500 psi					110			110	121
0150	3000 psi					129			129	142
0200	3500 psi					124			124	137
0300	4000 psi					127			127	140
0350	4500 psi					131			131	144
0400	5000 psi					135			135	149
0411	6000 psi					139			139	153
0412	8000 psi					146			146	161
0413	10,000 psi					154			154	169
0414	12,000 psi					161			161	177
1000	For high early strength (Portland cement Type III), add					10%				
1010	For structural lightweight with regular sand, add					25%				
1300	For winter concrete (hot water), add					5.35			5.35	5.90
1410	For mid-range water reducer, add					4.18			4.18	4.60
1420	For high-range water reducer/superplasticizer, add					6.40			6.40	7
1430	For retarder, add					3.45			3.45	3.80
1440	For non-Chloride accelerator, add					6.80			6.80	7.45
1450	For Chloride accelerator, per 1%, add					4.10			4.10	4.51
1460	For fiber reinforcing, synthetic (1 lb./C.Y.), add					8.10			8.10	8.90
1500	For Saturday delivery, add					9.20			9.20	10.10
1510	For truck holding/waiting time past 1st hour per load, add				Hr.	109			109	120
1520	For short load (less than 4 C.Y.), add per load				Ea.	85			85	93.50
2000	For all lightweight aggregate, add				C.Y.	45%				
4000	Flowable fill: ash, cement, aggregate, water									
4100	40-80 psi				C.Y.	78.50			78.50	86.50
4150	Structural: ash, cement, aggregate, water & sand									
4200	50 psi				C.Y.	78.50			78.50	86.50
4250	140 psi					79.50			79.50	87
4300	500 psi					81.50			81.50	90
4350	1000 psi					85			85	93.50

03 31 13.70 Placing Concrete

		Crew	Daily Output	Labor-Hours	Unit	Material	2020 Bare Costs Labor	Equipment	Total	Total Incl O&P
0010	**PLACING CONCRETE** R033105-70									
0020	Includes labor and equipment to place, level (strike off) and consolidate									
0050	Beams, elevated, small beams, pumped	C-20	60	1.067	C.Y.		48	7.75	55.75	80.50
0100	With crane and bucket	C-7	45	1.600			73	24	97	137

03 31 Structural Concrete

03 31 13 – Heavyweight Structural Concrete

03 31 13.70 Placing Concrete	Crew	Daily Output	Labor-Hours	Unit	Material	2020 Bare Costs Labor	2020 Bare Costs Equipment	Total	Total Incl O&P
0200	Large beams, pumped	C-20	90	.711	C.Y.		32	5.15	37.15	53.50
0250	With crane and bucket	C-7	65	1.108			50.50	16.60	67.10	94.50
0400	Columns, square or round, 12" thick, pumped	C-20	60	1.067			48	7.75	55.75	80.50
0450	With crane and bucket	C-7	40	1.800			82.50	27	109.50	153
0600	18" thick, pumped	C-20	90	.711			32	5.15	37.15	53.50
0650	With crane and bucket	C-7	55	1.309			60	19.65	79.65	111
0800	24" thick, pumped	C-20	92	.696			31.50	5.05	36.55	52.50
0850	With crane and bucket	C-7	70	1.029			47	15.45	62.45	87.50
1000	36" thick, pumped	C-20	140	.457			20.50	3.31	23.81	34.50
1050	With crane and bucket	C-7	100	.720			33	10.80	43.80	61.50
1200	Duct bank, direct chute	C-6	155	.310			13.55	.35	13.90	21
1400	Elevated slabs, less than 6" thick, pumped	C-20	140	.457			20.50	3.31	23.81	34.50
1450	With crane and bucket	C-7	95	.758			34.50	11.35	45.85	64.50
1500	6" to 10" thick, pumped	C-20	160	.400			18.05	2.90	20.95	30
1550	With crane and bucket	C-7	110	.655			30	9.80	39.80	56
1600	Slabs over 10" thick, pumped	C-20	180	.356			16.05	2.58	18.63	27
1650	With crane and bucket	C-7	130	.554			25.50	8.30	33.80	47
1900	Footings, continuous, shallow, direct chute	C-6	120	.400			17.50	.45	17.95	26.50
1950	Pumped	C-20	150	.427			19.25	3.09	22.34	32.50
2000	With crane and bucket	C-7	90	.800			36.50	12	48.50	68
2100	Footings, continuous, deep, direct chute	C-6	140	.343			15	.38	15.38	23
2150	Pumped	C-20	160	.400			18.05	2.90	20.95	30
2200	With crane and bucket	C-7	110	.655			30	9.80	39.80	56
2400	Footings, spread, under 1 C.Y., direct chute	C-6	55	.873			38	.98	38.98	58
2450	Pumped	C-20	65	.985			44.50	7.15	51.65	74.50
2500	With crane and bucket	C-7	45	1.600			73	24	97	137
2600	Over 5 C.Y., direct chute	C-6	120	.400			17.50	.45	17.95	26.50
2650	Pumped	C-20	150	.427			19.25	3.09	22.34	32.50
2700	With crane and bucket	C-7	100	.720			33	10.80	43.80	61.50
2900	Foundation mats, over 20 C.Y., direct chute	C-6	350	.137			6	.15	6.15	9.15
2950	Pumped	C-20	400	.160			7.25	1.16	8.41	12.10
3000	With crane and bucket	C-7	300	.240			11	3.60	14.60	20.50
3200	Grade beams, direct chute	C-6	150	.320			14	.36	14.36	21.50
3250	Pumped	C-20	180	.356			16.05	2.58	18.63	27
3300	With crane and bucket	C-7	120	.600			27.50	9	36.50	51
3700	Pile caps, under 5 C.Y., direct chute	C-6	90	.533			23.50	.60	24.10	35.50
3750	Pumped	C-20	110	.582			26.50	4.22	30.72	44
3800	With crane and bucket	C-7	80	.900			41	13.50	54.50	76.50
3850	Pile cap, 5 C.Y. to 10 C.Y., direct chute	C-6	175	.274			12	.31	12.31	18.30
3900	Pumped	C-20	200	.320			14.45	2.32	16.77	24
3950	With crane and bucket	C-7	150	.480			22	7.20	29.20	41
4000	Over 10 C.Y., direct chute	C-6	215	.223			9.75	.25	10	14.85
4050	Pumped	C-20	240	.267			12.05	1.93	13.98	20
4100	With crane and bucket	C-7	185	.389			17.80	5.85	23.65	33
4300	Slab on grade, up to 6" thick, direct chute R033105-70	C-6	110	.436			19.10	.49	19.59	29
4350	Pumped	C-20	130	.492			22	3.57	25.57	37.50
4400	With crane and bucket	C-7	110	.655			30	9.80	39.80	56
4600	Over 6" thick, direct chute	C-6	165	.291			12.70	.33	13.03	19.40
4650	Pumped	C-20	185	.346			15.60	2.51	18.11	26.50
4700	With crane and bucket	C-7	145	.497			22.50	7.45	29.95	42
4900	Walls, 8" thick, direct chute	C-6	90	.533			23.50	.60	24.10	35.50
4950	Pumped	C-20	100	.640			29	4.64	33.64	48.50
5000	With crane and bucket	C-7	80	.900			41	13.50	54.50	76.50

03 31 Structural Concrete

03 31 13 – Heavyweight Structural Concrete

03 31 13.70 Placing Concrete

		Crew	Daily Output	Labor-Hours	Unit	Material	2020 Bare Costs Labor	2020 Bare Costs Equipment	Total	Total Incl O&P
5050	12" thick, direct chute	C-6	100	.480	C.Y.		21	.54	21.54	32
5100	Pumped	C-20	110	.582			26.50	4.22	30.72	44
5200	With crane and bucket	C-7	90	.800			36.50	12	48.50	68
5300	15" thick, direct chute	C-6	105	.457			20	.51	20.51	30.50
5350	Pumped	C-20	120	.533			24	3.87	27.87	40.50
5400	With crane and bucket	C-7	95	.758	▼		34.50	11.35	45.85	64.50
5600	Wheeled concrete dumping, add to placing costs above									
5610	Walking cart, 50' haul, add	C-18	32	.281	C.Y.		11.90	4	15.90	22.50
5620	150' haul, add		24	.375			15.85	5.35	21.20	30
5700	250' haul, add	▼	18	.500			21	7.10	28.10	40
5800	Riding cart, 50' haul, add	C-19	80	.113			4.76	1.92	6.68	9.25
5810	150' haul, add		60	.150			6.35	2.56	8.91	12.35
5900	250' haul, add	▼	45	.200	▼		8.45	3.41	11.86	16.45
6000	Concrete in-fill for pan-type metal stairs and landings. Manual placement									
6010	includes up to 50' horizontal haul from point of concrete discharge.									
6100	Stair pan treads, 2" deep									
6110	Flights in 1st floor level up/down from discharge point	C-8A	3200	.015	S.F.		.68		.68	1.01
6120	2nd floor level		2500	.019			.87		.87	1.29
6130	3rd floor level		2000	.024			1.08		1.08	1.61
6140	4th floor level	▼	1800	.027	▼		1.20		1.20	1.79
6200	Intermediate stair landings, pan-type 4" deep									
6210	Flights in 1st floor level up/down from discharge point	C-8A	2000	.024	S.F.		1.08		1.08	1.61
6220	2nd floor level		1500	.032			1.44		1.44	2.15
6230	3rd floor level		1200	.040			1.80		1.80	2.69
6240	4th floor level	▼	1000	.048	▼		2.16		2.16	3.22

03 35 Concrete Finishing

03 35 13 – High-Tolerance Concrete Floor Finishing

03 35 13.30 Finishing Floors, High Tolerance

		Crew	Daily Output	Labor-Hours	Unit	Material	2020 Bare Costs Labor	2020 Bare Costs Equipment	Total	Total Incl O&P
0010	**FINISHING FLOORS, HIGH TOLERANCE**									
0012	Finishing of fresh concrete flatwork requires that concrete									
0013	first be placed, struck off & consolidated									
0015	Basic finishing for various unspecified flatwork									
0100	Bull float only	C-10	4000	.006	S.F.		.28		.28	.42
0125	Bull float & manual float		2000	.012			.57		.57	.84
0150	Bull float, manual float & broom finish, w/edging & joints		1850	.013			.61		.61	.91
0200	Bull float, manual float & manual steel trowel	▼	1265	.019	▼		.90		.90	1.33
0210	For specified Random Access Floors in ACI Classes 1, 2, 3 and 4 to achieve									
0215	Composite Overall Floor Flatness and Levelness values up to FF35/FL25									
0250	Bull float, machine float & machine trowel (walk-behind)	C-10C	1715	.014	S.F.		.66	.05	.71	1.04
0300	Power screed, bull float, machine float & trowel (walk-behind)	C-10D	2400	.010			.47	.07	.54	.77
0350	Power screed, bull float, machine float & trowel (ride-on)	C-10E	4000	.006	▼		.28	.06	.34	.49
0352	For specified Random Access Floors in ACI Classes 5, 6, 7 and 8 to achieve									
0354	Composite Overall Floor Flatness and Levelness values up to FF50/FL50									
0356	Add for two-dimensional restraightening after power float	C-10	6000	.004	S.F.		.19		.19	.28
0358	For specified Random or Defined Access Floors in ACI Class 9 to achieve									
0360	Composite Overall Floor Flatness and Levelness values up to FF100/FL100									
0362	Add for two-dimensional restraightening after bull float & power float	C-10	3000	.008	S.F.		.38		.38	.56
0364	For specified Superflat Defined Access Floors in ACI Class 9 to achieve									
0366	Minimum Floor Flatness and Levelness values of FF100/FL100									
0368	Add for 2-dim'l restraightening after bull float, power float, power trowel	C-10	2000	.012	S.F.		.57		.57	.84

03 35 Concrete Finishing

03 35 16 – Heavy-Duty Concrete Floor Finishing

03 35 16.30 Finishing Floors, Heavy-Duty

		Crew	Daily Output	Labor-Hours	Unit	Material	2020 Bare Costs Labor	2020 Bare Costs Equipment	Total	Total Incl O&P
0010	**FINISHING FLOORS, HEAVY-DUTY**									
1800	Floor abrasives, dry shake on fresh concrete, .25 psf, aluminum oxide	1 Cefi	850	.009	S.F.	.60	.47		1.07	1.34
1850	Silicon carbide		850	.009		.84	.47		1.31	1.62
2000	Floor hardeners, dry shake, metallic, light service, .50 psf		850	.009		.50	.47		.97	1.24
2050	Medium service, .75 psf		750	.011		.75	.53		1.28	1.61
2100	Heavy service, 1.0 psf		650	.012		1	.61		1.61	2
2150	Extra heavy, 1.5 psf		575	.014		1.50	.69		2.19	2.67
2300	Non-metallic, light service, .50 psf		850	.009		.18	.47		.65	.88
2350	Medium service, .75 psf		750	.011		.26	.53		.79	1.07
2400	Heavy service, 1.0 psf		650	.012		.35	.61		.96	1.29
2450	Extra heavy, 1.5 psf		575	.014		.53	.69		1.22	1.60
2800	Trap rock wearing surface, dry shake, for monolithic floors									
2810	2.0 psf	C-10B	1250	.032	S.F.	.02	1.45	.25	1.72	2.46
3800	Dustproofing, liquid, for cured concrete, solvent-based, 1 coat	1 Cefi	1900	.004		.18	.21		.39	.51
3850	2 coats		1300	.006		.64	.31		.95	1.16
4000	Epoxy-based, 1 coat		1500	.005		.15	.27		.42	.56
4050	2 coats		1500	.005		.31	.27		.58	.73

03 35 19 – Colored Concrete Finishing

03 35 19.30 Finishing Floors, Colored

		Crew	Daily Output	Labor-Hours	Unit	Material	2020 Bare Costs Labor	2020 Bare Costs Equipment	Total	Total Incl O&P
0010	**FINISHING FLOORS, COLORED**									
3000	Floor coloring, dry shake on fresh concrete (0.6 psf)	1 Cefi	1300	.006	S.F.	.45	.31		.76	.94
3050	(1.0 psf)	"	625	.013	"	.74	.64		1.38	1.76
3100	Colored dry shake powder only				Lb.	.74			.74	.82
3600	1/2" topping using 0.6 psf dry shake powdered color	C-10B	590	.068	S.F.	4.62	3.07	.53	8.22	10.25
3650	1.0 psf dry shake powdered color	"	590	.068	"	4.92	3.07	.53	8.52	10.55

03 35 23 – Exposed Aggregate Concrete Finishing

03 35 23.30 Finishing Floors, Exposed Aggregate

		Crew	Daily Output	Labor-Hours	Unit	Material	2020 Bare Costs Labor	2020 Bare Costs Equipment	Total	Total Incl O&P
0010	**FINISHING FLOORS, EXPOSED AGGREGATE**									
1600	Exposed local aggregate finish, seeded on fresh concrete, 3 lb./S.F.	1 Cefi	625	.013	S.F.	.21	.64		.85	1.18
1650	4 lb./S.F.	"	465	.017	"	.36	.86		1.22	1.66

03 35 29 – Tooled Concrete Finishing

03 35 29.30 Finishing Floors, Tooled

		Crew	Daily Output	Labor-Hours	Unit	Material	2020 Bare Costs Labor	2020 Bare Costs Equipment	Total	Total Incl O&P
0010	**FINISHING FLOORS, TOOLED**									
4400	Stair finish, fresh concrete, float finish	1 Cefi	275	.029	S.F.		1.45		1.45	2.14
4500	Steel trowel finish		200	.040			2		2	2.94
4600	Silicon carbide finish, dry shake on fresh concrete, .25 psf		150	.053		.60	2.66		3.26	4.57

03 35 29.60 Finishing Walls

		Crew	Daily Output	Labor-Hours	Unit	Material	2020 Bare Costs Labor	2020 Bare Costs Equipment	Total	Total Incl O&P
0010	**FINISHING WALLS**									
0020	Break ties and patch voids	1 Cefi	540	.015	S.F.	.04	.74		.78	1.14
0050	Burlap rub with grout		450	.018		.04	.89		.93	1.36
0100	Carborundum rub, dry		270	.030			1.48		1.48	2.18
0150	Wet rub		175	.046			2.28		2.28	3.36
0300	Bush hammer, green concrete	B-39	1000	.048			2.12	.36	2.48	3.58
0350	Cured concrete	"	650	.074			3.27	.55	3.82	5.50
0500	Acid etch	1 Cefi	575	.014		.14	.69		.83	1.17
0600	Float finish, 1/16" thick	"	300	.027		.40	1.33		1.73	2.40
0700	Sandblast, light penetration	E-11	1100	.029		.51	1.36	.27	2.14	2.99
0750	Heavy penetration	"	375	.085		1.03	3.98	.80	5.81	8.20
0850	Grind form fins flush	1 Clab	700	.011	L.F.		.48		.48	.72

03 35 Concrete Finishing

03 35 33 – Stamped Concrete Finishing

03 35 33.50 Slab Texture Stamping	Crew	Daily Output	Labor-Hours	Unit	Material	2020 Bare Costs Labor	2020 Bare Costs Equipment	Total	Total Incl O&P
0010 **SLAB TEXTURE STAMPING**									
0050 Stamping requires that concrete first be placed, struck off, consolidated,									
0060 bull floated and free of bleed water. Decorative stamping tasks include:									
0100 Step 1 - first application of dry shake colored hardener	1 Cefi	6400	.001	S.F.	.43	.06		.49	.56
0110 Step 2 - bull float		6400	.001			.06		.06	.09
0130 Step 3 - second application of dry shake colored hardener		6400	.001		.21	.06		.27	.32
0140 Step 4 - bull float, manual float & steel trowel	3 Cefi	1280	.019			.94		.94	1.38
0150 Step 5 - application of dry shake colored release agent	1 Cefi	6400	.001		.10	.06		.16	.20
0160 Step 6 - place, tamp & remove mats	3 Cefi	2400	.010		.87	.50		1.37	1.68
0170 Step 7 - touch up edges, mat joints & simulated grout lines	1 Cefi	1280	.006			.31		.31	.46
0300 Alternate stamping estimating method includes all tasks above	4 Cefi	800	.040		1.60	2		3.60	4.70
0400 Step 8 - pressure wash @ 3000 psi after 24 hours	1 Cefi	1600	.005			.25		.25	.37
0500 Step 9 - roll 2 coats cure/seal compound when dry	"	800	.010		.69	.50		1.19	1.49

03 37 Specialty Placed Concrete

03 37 13 – Shotcrete

03 37 13.30 Gunite (Dry-Mix)

	Crew	Daily Output	Labor-Hours	Unit	Material	2020 Bare Costs Labor	2020 Bare Costs Equipment	Total	Total Incl O&P
0010 **GUNITE (DRY-MIX)**									
0020 Typical in place, 1" layers, no mesh included	C-16	2000	.028	S.F.	.42	1.31	.20	1.93	2.63
0100 Mesh for gunite 2 x 2, #12	2 Rodm	800	.020		.73	1.13		1.86	2.50
0150 #4 reinforcing bars @ 6" each way	"	500	.032		1.88	1.80		3.68	4.79
0300 Typical in place, including mesh, 2" thick, flat surfaces	C-16	1000	.056		1.57	2.62	.39	4.58	6.05
0350 Curved surfaces		500	.112		1.57	5.25	.78	7.60	10.40
0500 4" thick, flat surfaces		750	.075		2.41	3.49	.52	6.42	8.40
0550 Curved surfaces		350	.160		2.41	7.50	1.12	11.03	15.05
0900 Prepare old walls, no scaffolding, good condition	C-10	1000	.024			1.14		1.14	1.68
0950 Poor condition	"	275	.087			4.13		4.13	6.10
1100 For high finish requirement or close tolerance, add						50%			
1150 Very high						110%			

03 37 13.60 Shotcrete (Wet-Mix)

	Crew	Daily Output	Labor-Hours	Unit	Material	2020 Bare Costs Labor	2020 Bare Costs Equipment	Total	Total Incl O&P
0010 **SHOTCRETE (WET-MIX)**									
0020 Wet mix, placed @ up to 12 C.Y./hour, 3000 psi	C-8C	80	.600	C.Y.	136	27.50	7.45	170.95	199
0100 Up to 35 C.Y./hour	C-8E	240	.200	"	122	9.10	2.70	133.80	151
1010 Fiber reinforced, 1" thick	C-8C	1740	.028	S.F.	1	1.27	.34	2.61	3.38
1020 2" thick		900	.053		1.99	2.46	.66	5.11	6.60
1030 3" thick		825	.058		2.99	2.69	.72	6.40	8.10
1040 4" thick		750	.064		3.98	2.96	.79	7.73	9.65

03 37 23 – Roller-Compacted Concrete

03 37 23.50 Concrete, Roller-Compacted

	Crew	Daily Output	Labor-Hours	Unit	Material	2020 Bare Costs Labor	2020 Bare Costs Equipment	Total	Total Incl O&P
0010 **CONCRETE, ROLLER-COMPACTED**									
0100 Mass placement, 1' lift, 1' layer	B-10C	1280	.009	C.Y.		.49	1.58	2.07	2.47
0200 2' lift, 6" layer	"	1600	.008			.39	1.26	1.65	1.97
0210 Vertical face, formed, 1' lift	B-11V	400	.060			2.53	.42	2.95	4.26
0220 6" lift	"	200	.120			5.05	.83	5.88	8.50
0300 Sloped face, nonformed, 1' lift	B-11L	384	.042			2.06	2.77	4.83	6.15
0360 6" lift	"	192	.083			4.12	5.55	9.67	12.25
0400 Surface preparation, vacuum truck	B-6A	3280	.006	S.Y.		.30	.11	.41	.56
0450 Water clean	B-9A	3000	.008			.36	.16	.52	.70
0460 Water blast	B-9B	800	.030			1.33	.71	2.04	2.78
0500 Joint bedding placement, 1" thick	B-11C	975	.016			.81	.22	1.03	1.46

03 37 Specialty Placed Concrete

03 37 23 – Roller-Compacted Concrete

03 37 23.50 Concrete, Roller-Compacted

		Crew	Daily Output	Labor-Hours	Unit	Material	2020 Bare Costs Labor	Equipment	Total	Total Incl O&P
0510	Conveyance of materials, 18 C.Y. truck, 5 min. cycle	B-34F	2048	.004	C.Y.		.19	.46	.65	.79
0520	10 min. cycle		1024	.008			.38	.91	1.29	1.57
0540	15 min. cycle	↓	680	.012			.58	1.37	1.95	2.37
0550	With crane and bucket	C-23A	1600	.025			1.19	1.63	2.82	3.57
0560	With 4 C.Y. loader, 4 min. cycle	B-10U	480	.025			1.30	1.99	3.29	4.13
0570	8 min. cycle	↓	240	.050			2.59	3.99	6.58	8.25
0580	12 min. cycle		160	.075			3.89	6	9.89	12.40
0590	With belt conveyor	C-7D	600	.093			4.15	.34	4.49	6.60
0600	With 17 C.Y. scraper, 5 min. cycle	B-33J	1440	.006			.32	1.66	1.98	2.30
0610	10 min. cycle		720	.011			.63	3.33	3.96	4.60
0620	15 min. cycle		480	.017			.95	4.99	5.94	6.90
0630	20 min. cycle	↓	360	.022	↓		1.26	6.65	7.91	9.20
0640	Water cure, small job, < 500 C.Y.	B-94C	8	1	Hr.		42	11.55	53.55	76
0650	Large job, over 500 C.Y.	B-59A	8	3	"		133	57.50	190.50	263
0660	RCC paving, with asphalt paver including material	B-25C	1000	.048	C.Y.	73	2.27	2.35	77.62	86.50
0670	8" thick layers		4200	.011	S.Y.	19.05	.54	.56	20.15	22.50
0680	12" thick layers	↓	2800	.017	"	28.50	.81	.84	30.15	33

03 39 Concrete Curing

03 39 13 – Water Concrete Curing

03 39 13.50 Water Curing

		Crew	Daily Output	Labor-Hours	Unit	Material	2020 Bare Costs Labor	Equipment	Total	Total Incl O&P
0010	**WATER CURING**									
0015	With burlap, 4 uses assumed, 7.5 oz.	2 Clab	55	.291	C.S.F.	15.65	12.25		27.90	35.50
0100	10 oz.	"	55	.291	"	28	12.25		40.25	49.50
0400	Curing blankets, 1" to 2" thick, buy				S.F.	.26			.26	.29

03 39 23 – Membrane Concrete Curing

03 39 23.13 Chemical Compound Membrane Concrete Curing

		Crew	Daily Output	Labor-Hours	Unit	Material	2020 Bare Costs Labor	Equipment	Total	Total Incl O&P
0010	**CHEMICAL COMPOUND MEMBRANE CONCRETE CURING**									
0300	Sprayed membrane curing compound	2 Clab	95	.168	C.S.F.	12.45	7.10		19.55	24.50
0700	Curing compound, solvent based, 400 S.F./gal., 55 gallon lots				Gal.	23			23	25.50
0720	5 gallon lots					33			33	36
0800	Curing compound, water based, 250 S.F./gal., 55 gallon lots					23.50			23.50	26
0820	5 gallon lots					26			26	29

03 39 23.23 Sheet Membrane Concrete Curing

		Crew	Daily Output	Labor-Hours	Unit	Material	2020 Bare Costs Labor	Equipment	Total	Total Incl O&P
0010	**SHEET MEMBRANE CONCRETE CURING**									
0200	Curing blanket, burlap/poly, 2-ply	2 Clab	70	.229	C.S.F.	20.50	9.60		30.10	37

03 41 Precast Structural Concrete

03 41 13 – Precast Concrete Hollow Core Planks

03 41 13.50 Precast Slab Planks

		Crew	Daily Output	Labor-Hours	Unit	Material	2020 Bare Costs Labor	Equipment	Total	Total Incl O&P
0010	**PRECAST SLAB PLANKS**									
0020	Prestressed roof/floor members, grouted, solid, 4" thick	C-11	2400	.030	S.F.	8.55	1.72	.95	11.22	13.15
0050	6" thick		2800	.026		8.55	1.47	.81	10.83	12.60
0100	Hollow, 8" thick		3200	.023		11.90	1.29	.71	13.90	15.85
0150	10" thick		3600	.020		9.95	1.15	.63	11.73	13.35
0200	12" thick	↓	4000	.018	↓	11.60	1.03	.57	13.20	15

03 41 Precast Structural Concrete

03 41 23 – Precast Concrete Stairs

03 41 23.50 Precast Stairs

		Crew	Daily Output	Labor-Hours	Unit	Material	2020 Bare Costs Labor	Equipment	Total	Total Incl O&P
0010	**PRECAST STAIRS**									
0020	Precast concrete treads on steel stringers, 3' wide	C-12	75	.640	Riser	149	33.50	6.30	188.80	221
0200	Front entrance 4' wide with 48" platform, 2 risers		17	2.824	Flight	605	149	27.50	781.50	920
0250	5 risers		13	3.692		960	194	36	1,190	1,375
0300	Front entrance, 5' wide with 48" platform, 2 risers		16	3		675	158	29.50	862.50	1,025
1200	Basement entrance stairwell, 6 steps, incl. steel bulkhead door	B-51	22	2.182		1,875	94.50	8.90	1,978.40	2,200
1250	14 steps	"	11	4.364		3,350	189	17.85	3,556.85	3,975

03 41 33 – Precast Structural Pretensioned Concrete

03 41 33.10 Precast Beams

		Crew	Daily Output	Labor-Hours	Unit	Material	2020 Bare Costs Labor	Equipment	Total	Total Incl O&P
0010	**PRECAST BEAMS**									
0011	L-shaped, 20' span, 12" x 20"	C-11	32	2.250	Ea.	4,500	129	71	4,700	5,225
1200	Rectangular, 20' span, 12" x 20"		32	2.250		3,925	129	71	4,125	4,575
1250	18" x 36"		24	3		5,800	172	94.50	6,066.50	6,750
1300	24" x 44"		22	3.273		6,850	187	103	7,140	7,950
1400	30' span, 12" x 36"		24	3		6,475	172	94.50	6,741.50	7,500
1450	18" x 44"		20	3.600		9,000	206	113	9,319	10,300
1500	24" x 52"		16	4.500		11,200	258	142	11,600	12,900
1600	40' span, 12" x 52"		20	3.600		10,800	206	113	11,119	12,200
1650	18" x 52"		16	4.500		13,000	258	142	13,400	14,900
1700	24" x 52"		12	6		14,900	345	189	15,434	17,100
2000	"T" shaped, 20' span, 12" x 20"		32	2.250		5,575	129	71	5,775	6,400
2050	18" x 36"		24	3		7,175	172	94.50	7,441.50	8,250
2100	24" x 44"		22	3.273		8,075	187	103	8,365	9,275
2200	30' span, 12" x 36"		24	3		9,950	172	94.50	10,216.50	11,300
2250	18" x 44"		20	3.600		12,500	206	113	12,819	14,200
2300	24" x 52"		16	4.500		13,300	258	142	13,700	15,300
2500	40' span, 12" x 52"		20	3.600		16,200	206	113	16,519	18,200
2550	18" x 52"		16	4.500		16,700	258	142	17,100	18,900
2600	24" x 52"		12	6		17,800	345	189	18,334	20,300

03 45 Precast Architectural Concrete

03 45 13 – Faced Architectural Precast Concrete

03 45 13.50 Precast Wall Panels

		Crew	Daily Output	Labor-Hours	Unit	Material	2020 Bare Costs Labor	Equipment	Total	Total Incl O&P
0010	**PRECAST WALL PANELS**									
0050	Uninsulated, smooth gray									
0150	Low rise, 4' x 8' x 4" thick	C-11	320	.225	S.F.	31.50	12.90	7.10	51.50	62.50
0210	8' x 8', 4" thick		576	.125		31	7.15	3.94	42.09	50
0250	8' x 16' 4" thick		1024	.070		31	4.03	2.22	37.25	42.50
0600	High rise, 4' x 8' x 4" thick		288	.250		31.50	14.30	7.90	53.70	65
0650	8' x 8' x 4" thick		512	.141		31	8.05	4.43	43.48	52
0700	8' x 16' x 4" thick		768	.094		31	5.35	2.95	39.30	45.50
0750	10' x 20', 6" thick		1400	.051		52.50	2.94	1.62	57.06	64.50
0800	Insulated panel, 2" polystyrene, add					1.22			1.22	1.34
0850	2" urethane, add					.74			.74	.81
1200	Finishes, white, add					3.54			3.54	3.89
1250	Exposed aggregate, add					.77			.77	.84
1300	Granite faced, domestic, add					30.50			30.50	33.50
1350	Brick faced, modular, red, add					4.83			4.83	5.30
2200	Fiberglass reinforced cement with urethane core									
2210	R20, 8' x 8', 5" plain finish	E-2	750	.075	S.F.	29.50	4.27	2.26	36.03	41.50

85

03 45 Precast Architectural Concrete

03 45 13 – Faced Architectural Precast Concrete

03 45 13.50 Precast Wall Panels	Crew	Daily Output	Labor-Hours	Unit	Material	2020 Bare Costs Labor	Equipment	Total	Total Incl O&P
2220 Exposed aggregate or brick finish	E-2	600	.093	S.F.	48	5.35	2.83	56.18	64

03 47 Site-Cast Concrete

03 47 13 – Tilt-Up Concrete

03 47 13.50 Tilt-Up Wall Panels

	Crew	Daily Output	Labor-Hours	Unit	Material	Labor	Equipment	Total	Total Incl O&P
0010 **TILT-UP WALL PANELS**									
0015 Wall panel construction, walls only, 5-1/2" thick	C-14	1600	.090	S.F.	6.40	4.62	.93	11.95	15
0100 7-1/2" thick	"	1550	.093	"	7.95	4.77	.96	13.68	16.95

03 48 Precast Concrete Specialties

03 48 43 – Precast Concrete Trim

03 48 43.40 Precast Lintels

	Crew	Daily Output	Labor-Hours	Unit	Material	Labor	Equipment	Total	Total Incl O&P
0010 **PRECAST LINTELS**, smooth gray, prestressed, stock units only									
0800 4" wide x 8" high x 4' long	D-10	28	1.143	Ea.	35.50	59	15.30	109.80	145
0850 8' long		24	1.333		81	69	17.85	167.85	213
1000 6" wide x 8" high x 4' long		26	1.231		58.50	63.50	16.45	138.45	179
1050 10' long		22	1.455		149	75.50	19.45	243.95	300
1200 8" wide x 8" high x 4' long		24	1.333		61	69	17.85	147.85	191
1250 12' long		20	1.600		203	83	21.50	307.50	370
1275 For custom sizes, types, colors, or finishes of precast lintels, add					150%				

03 53 Concrete Topping

03 53 16 – Iron-Aggregate Concrete Topping

03 53 16.50 Floor Topping

	Crew	Daily Output	Labor-Hours	Unit	Material	Labor	Equipment	Total	Total Incl O&P
0010 **FLOOR TOPPING**									
0400 Integral topping/finish, on fresh concrete, using 1:1:2 mix, 3/16" thick	C-10B	1000	.040	S.F.	.14	1.81	.31	2.26	3.19
0450 1/2" thick		950	.042		.37	1.91	.33	2.61	3.61
0500 3/4" thick		850	.047		.56	2.13	.37	3.06	4.19
0600 1" thick		750	.053		.74	2.41	.42	3.57	4.87
0800 Granolithic topping, on fresh or cured concrete, 1:1:1-1/2 mix, 1/2" thick		590	.068		.41	3.07	.53	4.01	5.60
0820 3/4" thick		580	.069		.62	3.12	.54	4.28	5.90
0850 1" thick		575	.070		.82	3.15	.55	4.52	6.20
0950 2" thick		500	.080		1.65	3.62	.63	5.90	7.90
1200 Heavy duty, 1:1:2, 3/4" thick, preshrunk, gray, 20 M.S.F.		320	.125		.95	5.65	.98	7.58	10.55
1300 100 M.S.F.		380	.105		.56	4.76	.83	6.15	8.60

03 54 Cast Underlayment

03 54 16 – Hydraulic Cement Underlayment

03 54 16.50 Cement Underlayment		Crew	Daily Output	Labor-Hours	Unit	Material	2020 Bare Costs Labor	Equipment	Total	Total Incl O&P
0010	**CEMENT UNDERLAYMENT**									
2510	Underlayment, P.C. based, self-leveling, 4100 psi, pumped, 1/4" thick	C-8	20000	.003	S.F.	1.75	.13	.02	1.90	2.15
2520	1/2" thick		19000	.003		3.50	.14	.02	3.66	4.09
2530	3/4" thick		18000	.003		5.25	.15	.02	5.42	6.05
2540	1" thick		17000	.003		7	.15	.02	7.17	7.95
2550	1-1/2" thick	↓	15000	.004		10.50	.17	.03	10.70	11.85
2560	Hand placed, 1/2" thick	C-18	450	.020		3.50	.85	.28	4.63	5.45
2610	Topping, P.C. based, self-leveling, 6100 psi, pumped, 1/4" thick	C-8	20000	.003		2.09	.13	.02	2.24	2.52
2620	1/2" thick		19000	.003		4.18	.14	.02	4.34	4.83
2630	3/4" thick		18000	.003		6.25	.15	.02	6.42	7.15
2660	1" thick		17000	.003		8.35	.15	.02	8.52	9.45
2670	1-1/2" thick	↓	15000	.004		12.55	.17	.03	12.75	14.10
2680	Hand placed, 1/2" thick	C-18	450	.020	↓	4.18	.85	.28	5.31	6.20

03 62 Non-Shrink Grouting

03 62 13 – Non-Metallic Non-Shrink Grouting

03 62 13.50 Grout, Non-Metallic Non-Shrink		Crew	Daily Output	Labor-Hours	Unit	Material	2020 Bare Costs Labor	Equipment	Total	Total Incl O&P
0010	**GROUT, NON-METALLIC NON-SHRINK**									
0300	Non-shrink, non-metallic, 1" deep	1 Cefi	35	.229	S.F.	7.60	11.40		19	25
0350	2" deep	"	25	.320	"	15.25	16		31.25	40.50

03 62 16 – Metallic Non-Shrink Grouting

03 62 16.50 Grout, Metallic Non-Shrink		Crew	Daily Output	Labor-Hours	Unit	Material	2020 Bare Costs Labor	Equipment	Total	Total Incl O&P
0010	**GROUT, METALLIC NON-SHRINK**									
0020	Column & machine bases, non-shrink, metallic, 1" deep	1 Cefi	35	.229	S.F.	12.05	11.40		23.45	30
0050	2" deep	"	25	.320	"	24	16		40	50

03 63 Epoxy Grouting

03 63 05 – Grouting of Dowels and Fasteners

03 63 05.10 Epoxy Only		Crew	Daily Output	Labor-Hours	Unit	Material	2020 Bare Costs Labor	Equipment	Total	Total Incl O&P
0010	**EPOXY ONLY**									
1500	Chemical anchoring, epoxy cartridge, excludes layout, drilling, fastener									
1530	For fastener 3/4" diam. x 6" embedment	2 Skwk	72	.222	Ea.	4.78	12.20		16.98	23.50
1535	1" diam. x 8" embedment		66	.242		7.15	13.30		20.45	28
1540	1-1/4" diam. x 10" embedment		60	.267		14.35	14.65		29	38
1545	1-3/4" diam. x 12" embedment		54	.296		24	16.25		40.25	51
1550	14" embedment		48	.333		28.50	18.30		46.80	59
1555	2" diam. x 12" embedment		42	.381		38	21		59	73.50
1560	18" embedment	↓	32	.500	↓	48	27.50		75.50	94

03 81 Concrete Cutting

03 81 13 – Flat Concrete Sawing

03 81 13.50 Concrete Floor/Slab Cutting	Crew	Daily Output	Labor-Hours	Unit	Material	2020 Bare Costs Labor	Equipment	Total	Total Incl O&P
0010 **CONCRETE FLOOR/SLAB CUTTING**									
0050 Includes blade cost, layout and set-up time									
0300 Saw cut concrete slabs, plain, up to 3" deep	B-89	1060	.015	L.F.	.12	.76	.98	1.86	2.33
0320 Each additional inch of depth		3180	.005		.04	.25	.33	.62	.78
0400 Mesh reinforced, up to 3" deep		980	.016		.13	.82	1.06	2.01	2.53
0420 Each additional inch of depth		2940	.005		.04	.27	.35	.66	.85
0500 Rod reinforced, up to 3" deep		800	.020		.17	1	1.29	2.46	3.10
0520 Each additional inch of depth		2400	.007		.06	.33	.43	.82	1.03

03 81 13.75 Concrete Saw Blades

	Crew	Daily Output	Labor-Hours	Unit	Material	2020 Bare Costs Labor	Equipment	Total	Total Incl O&P
0010 **CONCRETE SAW BLADES**									
3000 Blades for saw cutting, included in cutting line items									
3020 Diamond, 12" diameter				Ea.	249			249	274
3040 18" diameter					405			405	445
3080 24" diameter					585			585	645
3120 30" diameter					865			865	950
3160 36" diameter					1,150			1,150	1,250
3200 42" diameter					2,450			2,450	2,700

03 81 16 – Track Mounted Concrete Wall Sawing

03 81 16.50 Concrete Wall Cutting

	Crew	Daily Output	Labor-Hours	Unit	Material	2020 Bare Costs Labor	Equipment	Total	Total Incl O&P
0010 **CONCRETE WALL CUTTING**									
0750 Includes blade cost, layout and set-up time									
0800 Concrete walls, hydraulic saw, plain, per inch of depth	B-89B	250	.064	L.F.	.04	3.21	6	9.25	11.45
0820 Rod reinforcing, per inch of depth	"	150	.107	"	.06	5.35	10.05	15.46	19.10

03 82 Concrete Boring

03 82 13 – Concrete Core Drilling

03 82 13.10 Core Drilling

	Crew	Daily Output	Labor-Hours	Unit	Material	2020 Bare Costs Labor	Equipment	Total	Total Incl O&P
0010 **CORE DRILLING**									
0015 Includes bit cost, layout and set-up time									
0020 Reinforced concrete slab, up to 6" thick									
0100 1" diameter core	B-89A	17	.941	Ea.	.20	45.50	6.70	52.40	76.50
0150 For each additional inch of slab thickness in same hole, add		1440	.011		.03	.54	.08	.65	.94
0200 2" diameter core		16.50	.970		.26	47	6.90	54.16	79
0250 For each additional inch of slab thickness in same hole, add		1080	.015		.04	.72	.11	.87	1.25
0300 3" diameter core		16	1		.40	48.50	7.15	56.05	81.50
0350 For each additional inch of slab thickness in same hole, add		720	.022		.07	1.08	.16	1.31	1.86
0500 4" diameter core		15	1.067		.45	51.50	7.60	59.55	87
0550 For each additional inch of slab thickness in same hole, add		480	.033		.07	1.62	.24	1.93	2.78
0700 6" diameter core		14	1.143		.65	55.50	8.15	64.30	93
0750 For each additional inch of slab thickness in same hole, add		360	.044		.11	2.15	.32	2.58	3.72
0900 8" diameter core		13	1.231		1.07	59.50	8.80	69.37	101
0950 For each additional inch of slab thickness in same hole, add		288	.056		.18	2.69	.40	3.27	4.70
1100 10" diameter core		12	1.333		1.20	64.50	9.50	75.20	109
1150 For each additional inch of slab thickness in same hole, add		240	.067		.20	3.23	.48	3.91	5.60
1300 12" diameter core		11	1.455		1.84	70.50	10.40	82.74	119
1350 For each additional inch of slab thickness in same hole, add		206	.078		.31	3.77	.55	4.63	6.65
1500 14" diameter core		10	1.600		1.84	77.50	11.40	90.74	132
1550 For each additional inch of slab thickness in same hole, add		180	.089		.31	4.31	.63	5.25	7.55
1700 18" diameter core		9	1.778		3.06	86	12.70	101.76	147
1750 For each additional inch of slab thickness in same hole, add		144	.111		.51	5.40	.79	6.70	9.55

For customer support on your Site Work & Landscape Costs with RSMeans data, call 800.448.8182.

03 82 Concrete Boring

03 82 13 – Concrete Core Drilling

	03 82 13.10 Core Drilling	Crew	Daily Output	Labor-Hours	Unit	Material	2020 Bare Costs Labor	Equipment	Total	Total Incl O&P
1754	24" diameter core	B-89A	8	2	Ea.	4.15	97	14.30	115.45	166
1756	For each additional inch of slab thickness in same hole, add		120	.133		.69	6.45	.95	8.09	11.55
1760	For horizontal holes, add to above						20%	20%		
1770	Prestressed hollow core plank, 8" thick									
1780	1" diameter core	B-89A	17.50	.914	Ea.	.27	44.50	6.55	51.32	74.50
1790	For each additional inch of plank thickness in same hole, add		3840	.004		.03	.20	.03	.26	.37
1794	2" diameter core		17.25	.928		.35	45	6.60	51.95	75.50
1796	For each additional inch of plank thickness in same hole, add		2880	.006		.04	.27	.04	.35	.50
1800	3" diameter core		17	.941		.53	45.50	6.70	52.73	77
1810	For each additional inch of plank thickness in same hole, add		1920	.008		.07	.40	.06	.53	.75
1820	4" diameter core		16.50	.970		.60	47	6.90	54.50	79.50
1830	For each additional inch of plank thickness in same hole, add		1280	.013		.07	.61	.09	.77	1.09
1840	6" diameter core		15.50	1.032		.87	50	7.35	58.22	84.50
1850	For each additional inch of plank thickness in same hole, add		960	.017		.11	.81	.12	1.04	1.47
1860	8" diameter core		15	1.067		1.43	51.50	7.60	60.53	88
1870	For each additional inch of plank thickness in same hole, add		768	.021		.18	1.01	.15	1.34	1.88
1880	10" diameter core		14	1.143		1.60	55.50	8.15	65.25	94.50
1890	For each additional inch of plank thickness in same hole, add		640	.025		.20	1.21	.18	1.59	2.25
1900	12" diameter core		13.50	1.185		2.45	57.50	8.45	68.40	98.50
1910	For each additional inch of plank thickness in same hole, add		548	.029		.31	1.42	.21	1.94	2.70
3000	Bits for core drilling, included in drilling line items									
3010	Diamond, premium, 1" diameter				Ea.	80.50			80.50	89
3020	2" diameter					104			104	114
3030	3" diameter					159			159	175
3040	4" diameter					180			180	198
3060	6" diameter					262			262	288
3080	8" diameter					430			430	470
3110	10" diameter					480			480	530
3120	12" diameter					735			735	810
3140	14" diameter					735			735	810
3180	18" diameter					1,225			1,225	1,350
3240	24" diameter					1,650			1,650	1,825

03 82 16 – Concrete Drilling

03 82 16.10 Concrete Impact Drilling

	03 82 16.10 Concrete Impact Drilling	Crew	Daily Output	Labor-Hours	Unit	Material	2020 Bare Costs Labor	Equipment	Total	Total Incl O&P
0010	**CONCRETE IMPACT DRILLING**									
0020	Includes bit cost, layout and set-up time, no anchors									
0050	Up to 4" deep in concrete/brick floors/walls									
0100	Holes, 1/4" diameter	1 Carp	75	.107	Ea.	.07	5.65		5.72	8.60
0150	For each additional inch of depth in same hole, add		430	.019		.02	.99		1.01	1.51
0200	3/8" diameter		63	.127		.05	6.75		6.80	10.20
0250	For each additional inch of depth in same hole, add		340	.024		.01	1.25		1.26	1.89
0300	1/2" diameter		50	.160		.05	8.50		8.55	12.85
0350	For each additional inch of depth in same hole, add		250	.032		.01	1.70		1.71	2.57
0400	5/8" diameter		48	.167		.10	8.85		8.95	13.40
0450	For each additional inch of depth in same hole, add		240	.033		.02	1.77		1.79	2.69
0500	3/4" diameter		45	.178		.11	9.45		9.56	14.30
0550	For each additional inch of depth in same hole, add		220	.036		.03	1.93		1.96	2.93
0600	7/8" diameter		43	.186		.17	9.90		10.07	15.05
0650	For each additional inch of depth in same hole, add		210	.038		.04	2.03		2.07	3.09
0700	1" diameter		40	.200		.18	10.65		10.83	16.15
0750	For each additional inch of depth in same hole, add		190	.042		.04	2.24		2.28	3.41
0800	1-1/4" diameter		38	.211		.33	11.20		11.53	17.15

03 82 Concrete Boring

03 82 16 – Concrete Drilling

03 82 16.10 Concrete Impact Drilling	Crew	Daily Output	Labor-Hours	Unit	Material	2020 Bare Costs Labor	Equipment	Total	Total Incl O&P
0850	For each additional inch of depth in same hole, add	1 Carp	180	.044	Ea.	.08	2.36		2.44	3.64
0900	1-1/2" diameter		35	.229		.43	12.15		12.58	18.70
0950	For each additional inch of depth in same hole, add	↓	165	.048	↓	.11	2.58		2.69	3.99
1000	For ceiling installations, add						40%			

Estimating Tips
04 05 00 Common Work Results for Masonry

- The terms mortar and grout are often used interchangeably—and incorrectly. Mortar is used to bed masonry units, seal the entry of air and moisture, provide architectural appearance, and allow for size variations in the units. Grout is used primarily in reinforced masonry construction and to bond the masonry to the reinforcing steel. Common mortar types are M (2500 psi), S (1800 psi), N (750 psi), and O (350 psi), and they conform to ASTM C270. Grout is either fine or coarse and conforms to ASTM C476, and in-place strengths generally exceed 2500 psi. Mortar and grout are different components of masonry construction and are placed by entirely different methods. An estimator should be aware of their unique uses and costs.

- Mortar is included in all assembled masonry line items. The mortar cost, part of the assembled masonry material cost, includes all ingredients, all labor, and all equipment required. Please see reference number R040513-10.

- Waste, specifically the loss/droppings of mortar and the breakage of brick and block, is included in all unit cost lines that include mortar and masonry units in this division. A factor of 25% is added for mortar and 3% for brick and concrete masonry units.

- Scaffolding or staging is not included in any of the Division 4 costs. Refer to Subdivision 01 54 23 for scaffolding and staging costs.

04 20 00 Unit Masonry

- The most common types of unit masonry are brick and concrete masonry. The major classifications of brick are building brick (ASTM C62), facing brick (ASTM C216), glazed brick, fire brick, and pavers. Many varieties of texture and appearance can exist within these classifications, and the estimator would be wise to check local custom and availability within the project area. For repair and remodeling jobs, matching the existing brick may be the most important criteria.

- Brick and concrete block are priced by the piece and then converted into a price per square foot of wall. Openings less than two square feet are generally ignored by the estimator because any savings in units used are offset by the cutting and trimming required.

- It is often difficult and expensive to find and purchase small lots of historic brick. Costs can vary widely. Many design issues affect costs, selection of mortar mix, and repairs or replacement of masonry materials. Cleaning techniques must be reflected in the estimate.

- All masonry walls, whether interior or exterior, require bracing. The cost of bracing walls during construction should be included by the estimator, and this bracing must remain in place until permanent bracing is complete. Permanent bracing of masonry walls is accomplished by masonry itself, in the form of pilasters or abutting wall corners, or by anchoring the walls to the structural frame. Accessories in the form of anchors, anchor slots, and ties are used, but their supply and installation can be by different trades. For instance, anchor slots on spandrel beams and columns are supplied and welded in place by the steel fabricator, but the ties from the slots into the masonry are installed by the bricklayer. Regardless of the installation method, the estimator must be certain that these accessories are accounted for in pricing.

Reference Numbers
Reference numbers are shown at the beginning of some major classifications. These numbers refer to related items in the Reference Section. The reference information may be an estimating procedure, an alternate pricing method, or technical information.

Note: Not all subdivisions listed here necessarily appear. ∎

Same Data. Simplified.

Enjoy the convenience and efficiency of accessing your costs anywhere:

- **Skip the multiplier** by setting your location
- **Quickly search,** edit, favorite and share costs
- **Stay on top of price changes** with automatic updates

Discover more at rsmeans.com/online

04 01 20.10 Patching Masonry

		Crew	Daily Output	Labor-Hours	Unit	Material	2020 Bare Costs Labor	Equipment	Total	Total Incl O&P
0010	**PATCHING MASONRY**									
0500	CMU patching, includes chipping, cleaning and epoxy									
0520	1/4" deep	1 Cefi	65	.123	S.F.	8.75	6.15		14.90	18.65
0540	3/8" deep		50	.160		13.10	8		21.10	26
0580	1/2" deep		40	.200		17.50	10		27.50	34

04 01 20.20 Pointing Masonry

		Crew	Daily Output	Labor-Hours	Unit	Material	2020 Bare Costs Labor	Equipment	Total	Total Incl O&P
0010	**POINTING MASONRY**									
0300	Cut and repoint brick, hard mortar, running bond	1 Bric	80	.100	S.F.	.57	5.20		5.77	8.55
0320	Common bond		77	.104		.57	5.40		5.97	8.85
0360	Flemish bond		70	.114		.60	5.95		6.55	9.65
0400	English bond		65	.123		.60	6.40		7	10.35
0600	Soft old mortar, running bond		100	.080		.57	4.16		4.73	6.95
0620	Common bond		96	.083		.57	4.34		4.91	7.20
0640	Flemish bond		90	.089		.60	4.63		5.23	7.65
0680	English bond		82	.098		.60	5.10		5.70	8.35
0700	Stonework, hard mortar		140	.057	L.F.	.76	2.97		3.73	5.35
0720	Soft old mortar		160	.050	"	.76	2.60		3.36	4.78
1000	Repoint, mask and grout method, running bond		95	.084	S.F.	.76	4.38		5.14	7.50
1020	Common bond		90	.089		.76	4.63		5.39	7.85
1040	Flemish bond		86	.093		.80	4.84		5.64	8.25
1060	English bond		77	.104		.80	5.40		6.20	9.10
2000	Scrub coat, sand grout on walls, thin mix, brushed		120	.067		3.59	3.47		7.06	9.20
2020	Troweled		98	.082		4.99	4.25		9.24	11.95

04 05 05.10 Selective Demolition

		Crew	Daily Output	Labor-Hours	Unit	Material	2020 Bare Costs Labor	Equipment	Total	Total Incl O&P
0010	**SELECTIVE DEMOLITION** R024119-10									
1000	Chimney, 16" x 16", soft old mortar	1 Clab	55	.145	C.F.		6.10		6.10	9.20
1020	Hard mortar		40	.200			8.40		8.40	12.65
1030	16" x 20", soft old mortar		55	.145			6.10		6.10	9.20
1040	Hard mortar		40	.200			8.40		8.40	12.65
1050	16" x 24", soft old mortar		55	.145			6.10		6.10	9.20
1060	Hard mortar		40	.200			8.40		8.40	12.65
1080	20" x 20", soft old mortar		55	.145			6.10		6.10	9.20
1100	Hard mortar		40	.200			8.40		8.40	12.65
1110	20" x 24", soft old mortar		55	.145			6.10		6.10	9.20
1120	Hard mortar		40	.200			8.40		8.40	12.65
1140	20" x 32", soft old mortar		55	.145			6.10		6.10	9.20
1160	Hard mortar		40	.200			8.40		8.40	12.65
1200	48" x 48", soft old mortar		55	.145			6.10		6.10	9.20
1220	Hard mortar		40	.200			8.40		8.40	12.65
1250	Metal, high temp steel jacket, 24" diameter	E-2	130	.431	V.L.F.		24.50	13.05	37.55	52.50
1260	60" diameter	"	60	.933			53.50	28.50	82	114
1280	Flue lining, up to 12" x 12"	1 Clab	200	.040			1.68		1.68	2.53
1282	Up to 24" x 24"	"	150	.053			2.25		2.25	3.37
4000	Fireplace, brick, 30" x 24" opening									
4020	Soft old mortar	1 Clab	2	4	Ea.		168		168	253
4040	Hard mortar		1.25	6.400			269		269	405
4100	Stone, soft old mortar		1.50	5.333			225		225	335
4120	Hard mortar		1	8			335		335	505

04 05 Common Work Results for Masonry

04 05 13 – Masonry Mortaring

04 05 13.10 Cement

		Crew	Daily Output	Labor-Hours	Unit	Material	2020 Bare Costs Labor	Equipment	Total	Total Incl O&P
0010	**CEMENT**									
0100	Masonry, 70 lb. bag, T.L. lots				Bag	11.40			11.40	12.50
0150	L.T.L. lots					12.05			12.05	13.25
0200	White, 70 lb. bag, T.L. lots					18.65			18.65	20.50
0250	L.T.L. lots				↓	22.50			22.50	24.50

04 05 13.20 Lime

		Crew	Daily Output	Labor-Hours	Unit	Material	2020 Bare Costs Labor	Equipment	Total	Total Incl O&P
0010	**LIME**									
0020	Masons, hydrated, 50 lb. bag, T.L. lots				Bag	11.25			11.25	12.40
0050	L.T.L. lots					12.40			12.40	13.65
0200	Finish, double hydrated, 50 lb. bag, T.L. lots					11.90			11.90	13.10
0250	L.T.L. lots				↓	13.10			13.10	14.40

04 05 13.23 Surface Bonding Masonry Mortaring

		Crew	Daily Output	Labor-Hours	Unit	Material	2020 Bare Costs Labor	Equipment	Total	Total Incl O&P
0010	**SURFACE BONDING MASONRY MORTARING**									
0020	Gray or white colors, not incl. block work	1 Bric	540	.015	S.F.	.20	.77		.97	1.39

04 05 13.30 Mortar

		Crew	Daily Output	Labor-Hours	Unit	Material	2020 Bare Costs Labor	Equipment	Total	Total Incl O&P
0010	**MORTAR** R040513-10									
0020	With masonry cement									
0100	Type M, 1:1:6 mix	1 Brhe	143	.056	C.F.	6.15	2.34		8.49	10.30
0200	Type N, 1:3 mix		143	.056		5.35	2.34		7.69	9.45
0300	Type O, 1:3 mix		143	.056		5.20	2.34		7.54	9.25
0400	Type PM, 1:1:6 mix, 2500 psi		143	.056		6.20	2.34		8.54	10.35
0500	Type S, 1/2:1:4 mix	↓	143	.056	↓	6.55	2.34		8.89	10.75
2000	With Portland cement and lime									
2100	Type M, 1:1/4:3 mix	1 Brhe	143	.056	C.F.	9.10	2.34		11.44	13.55
2200	Type N, 1:1:6 mix, 750 psi		143	.056		7.30	2.34		9.64	11.60
2300	Type O, 1:2:9 mix (Pointing Mortar)		143	.056		8.75	2.34		11.09	13.20
2400	Type PL, 1:1/2:4 mix, 2500 psi		143	.056		6.35	2.34		8.69	10.55
2600	Type S, 1:1/2:4 mix, 1800 psi	↓	143	.056		8.55	2.34		10.89	13
2650	Pre-mixed, type S or N					6.25			6.25	6.90
2700	Mortar for glass block	1 Brhe	143	.056	↓	13.25	2.34		15.59	18.10
2900	Mortar for fire brick, dry mix, 10 lb. pail				Ea.	25			25	28

04 05 13.91 Masonry Restoration Mortaring

		Crew	Daily Output	Labor-Hours	Unit	Material	2020 Bare Costs Labor	Equipment	Total	Total Incl O&P
0010	**MASONRY RESTORATION MORTARING**									
0020	Masonry restoration mix				Lb.	.85			.85	.94
0050	White				"	1.40			1.40	1.54

04 05 13.93 Mortar Pigments

		Crew	Daily Output	Labor-Hours	Unit	Material	2020 Bare Costs Labor	Equipment	Total	Total Incl O&P
0010	**MORTAR PIGMENTS**, 50 lb. bags (2 bags per M bricks) R040513-10									
0020	Color admixture, range 2 to 10 lb. per bag of cement, light colors				Lb.	6.30			6.30	6.95
0050	Medium colors					7.05			7.05	7.75
0100	Dark colors				↓	15.95			15.95	17.55

04 05 13.95 Sand

		Crew	Daily Output	Labor-Hours	Unit	Material	2020 Bare Costs Labor	Equipment	Total	Total Incl O&P
0010	**SAND**, screened and washed at pit									
0020	For mortar, per ton				Ton	21.50			21.50	24
0050	With 10 mile haul					42			42	46
0100	With 30 mile haul				↓	67			67	74
0200	Screened and washed, at the pit				C.Y.	30			30	33
0250	With 10 mile haul					58.50			58.50	64
0300	With 30 mile haul				↓	93.50			93.50	103

For customer support on your Site Work & Landscape Costs with RSMeans data, call 800.448.8182.

93

04 05 13 – Masonry Mortaring

04 05 13.98 Mortar Admixtures	Crew	Daily Output	Labor-Hours	Unit	Material	2020 Bare Costs Labor	Equipment	Total	Total Incl O&P
0010 **MORTAR ADMIXTURES**									
0020 Waterproofing admixture, per quart (1 qt. to 2 bags of masonry cement)				Qt.	4.10			4.10	4.51

04 05 16 – Masonry Grouting

04 05 16.30 Grouting

	Crew	Daily Output	Labor-Hours	Unit	Material	2020 Bare Costs Labor	Equipment	Total	Total Incl O&P
0010 **GROUTING**									
0011 Bond beams & lintels, 8" deep, 6" thick, 0.15 C.F./L.F.	D-4	1480	.022	L.F.	.80	1.02	.13	1.95	2.56
0020 8" thick, 0.2 C.F./L.F.		1400	.023		1.29	1.08	.13	2.50	3.20
0050 10" thick, 0.25 C.F./L.F.		1200	.027		1.34	1.26	.16	2.76	3.54
0060 12" thick, 0.3 C.F./L.F.		1040	.031		1.61	1.45	.18	3.24	4.16
0200 Concrete block cores, solid, 4" thk., by hand, 0.067 C.F./S.F. of wall	D-8	1100	.036	S.F.	.36	1.74		2.10	3.03
0210 6" thick, pumped, 0.175 C.F./S.F.	D-4	720	.044		.94	2.09	.26	3.29	4.48
0250 8" thick, pumped, 0.258 C.F./S.F.		680	.047		1.38	2.22	.28	3.88	5.15
0300 10" thick, pumped, 0.340 C.F./S.F.		660	.048		1.82	2.29	.29	4.40	5.75
0350 12" thick, pumped, 0.422 C.F./S.F.		640	.050		2.26	2.36	.29	4.91	6.35
0500 Cavity walls, 2" space, pumped, 0.167 C.F./S.F. of wall		1700	.019		.89	.89	.11	1.89	2.44
0550 3" space, 0.250 C.F./S.F.		1200	.027		1.34	1.26	.16	2.76	3.54
0600 4" space, 0.333 C.F./S.F.		1150	.028		1.78	1.31	.16	3.25	4.12
0700 6" space, 0.500 C.F./S.F.		800	.040		2.68	1.89	.24	4.81	6.05
0800 Door frames, 3' x 7' opening, 2.5 C.F. per opening		60	.533	Opng.	13.40	25	3.14	41.54	56
0850 6' x 7' opening, 3.5 C.F. per opening		45	.711	"	18.75	33.50	4.19	56.44	75.50
2000 Grout, C476, for bond beams, lintels and CMU cores		350	.091	C.F.	5.35	4.31	.54	10.20	13

04 05 19 – Masonry Anchorage and Reinforcing

04 05 19.05 Anchor Bolts

	Crew	Daily Output	Labor-Hours	Unit	Material	2020 Bare Costs Labor	Equipment	Total	Total Incl O&P
0010 **ANCHOR BOLTS**									
0015 Installed in fresh grout in CMU bond beams or filled cores, no templates									
0020 Hooked, with nut and washer, 1/2" diameter, 8" long	1 Bric	132	.061	Ea.	1.62	3.15		4.77	6.55
0030 12" long		131	.061		1.80	3.18		4.98	6.80
0040 5/8" diameter, 8" long		129	.062		3.39	3.23		6.62	8.60
0050 12" long		127	.063		4.18	3.28		7.46	9.55
0060 3/4" diameter, 8" long		127	.063		4.18	3.28		7.46	9.55
0070 12" long		125	.064		5.20	3.33		8.53	10.80

04 05 19.16 Masonry Anchors

	Crew	Daily Output	Labor-Hours	Unit	Material	2020 Bare Costs Labor	Equipment	Total	Total Incl O&P
0010 **MASONRY ANCHORS**									
0020 For brick veneer, galv., corrugated, 7/8" x 7", 22 ga.	1 Bric	10.50	.762	C	16.70	39.50		56.20	78.50
0100 24 ga.		10.50	.762		10.35	39.50		49.85	71.50
0150 16 ga.		10.50	.762		31.50	39.50		71	95
0200 Buck anchors, galv., corrugated, 16 ga., 2" bend, 8" x 2"		10.50	.762		65	39.50		104.50	132
0250 8" x 3"		10.50	.762		69.50	39.50		109	137
0300 Adjustable, rectangular, 4-1/8" wide									
0350 Anchor and tie, 3/16" wire, mill galv.									
0400 2-3/4" eye, 3-1/4" tie	1 Bric	1.05	7.619	M	475	395		870	1,125
0500 4-3/4" tie		1.05	7.619		515	395		910	1,175
0520 5-1/2" tie		1.05	7.619		555	395		950	1,225
0550 4-3/4" eye, 3-1/4" tie		1.05	7.619		520	395		915	1,175
0570 4-3/4" tie		1.05	7.619		550	395		945	1,200
0580 5-1/2" tie		1.05	7.619		610	395		1,005	1,275
0660 Cavity wall, Z-type, galvanized, 6" long, 1/8" diam.		10.50	.762	C	25	39.50		64.50	87.50
0670 3/16" diameter		10.50	.762		31.50	39.50		71	94.50
0680 1/4" diameter		10.50	.762		40.50	39.50		80	105
0850 8" long, 3/16" diameter		10.50	.762		27	39.50		66.50	90
0855 1/4" diameter		10.50	.762		49	39.50		88.50	114

04 05 19 – Masonry Anchorage and Reinforcing

04 05 19.16 Masonry Anchors	Crew	Daily Output	Labor-Hours	Unit	Material	2020 Bare Costs Labor	Equipment	Total	Total Incl O&P
1000	Rectangular type, galvanized, 1/4" diameter, 2" x 6"	1 Bric	10.50	.762	C	80	39.50		119.50	148
1050	4" x 6"		10.50	.762		91.50	39.50		131	160
1100	3/16" diameter, 2" x 6"		10.50	.762		54.50	39.50		94	120
1150	4" x 6"		10.50	.762		55	39.50		94.50	121
1200	Mesh wall tie, 1/2" mesh, hot dip galvanized									
1400	16 ga., 12" long, 3" wide	1 Bric	9	.889	C	92.50	46.50		139	172
1420	6" wide		9	.889		135	46.50		181.50	219
1440	12" wide		8.50	.941		214	49		263	310
1500	Rigid partition anchors, plain, 8" long, 1" x 1/8"		10.50	.762		241	39.50		280.50	325
1550	1" x 1/4"		10.50	.762		325	39.50		364.50	415
1580	1-1/2" x 1/8"		10.50	.762		261	39.50		300.50	345
1600	1-1/2" x 1/4"		10.50	.762		340	39.50		379.50	435
1650	2" x 1/8"		10.50	.762		330	39.50		369.50	425
1700	2" x 1/4"		10.50	.762		445	39.50		484.50	550
2000	Column flange ties, wire, galvanized									
2300	3/16" diameter, up to 3" wide	1 Bric	10.50	.762	C	88	39.50		127.50	157
2350	To 5" wide		10.50	.762		96	39.50		135.50	166
2400	To 7" wide		10.50	.762		104	39.50		143.50	174
2600	To 9" wide		10.50	.762		110	39.50		149.50	181
2650	1/4" diameter, up to 3" wide		10.50	.762		112	39.50		151.50	183
2700	To 5" wide		10.50	.762		133	39.50		172.50	207
2800	To 7" wide		10.50	.762		150	39.50		189.50	225
2850	To 9" wide		10.50	.762		162	39.50		201.50	239
2900	For hot dip galvanized, add					35%				
4000	Channel slots, 1-3/8" x 1/2" x 8"									
4100	12 ga., plain	1 Bric	10.50	.762	C	244	39.50		283.50	330
4150	16 ga., galvanized	"	10.50	.762	"	155	39.50		194.50	230
4200	Channel slot anchors									
4300	16 ga., galvanized, 1-1/4" x 3-1/2"				C	61.50			61.50	67.50
4350	1-1/4" x 5-1/2"					70.50			70.50	77.50
4400	1-1/4" x 7-1/2"					80			80	88
4500	1/8" plain, 1-1/4" x 3-1/2"					160			160	176
4550	1-1/4" x 5-1/2"					171			171	188
4600	1-1/4" x 7-1/2"					186			186	205
4700	For corrugation, add					86			86	94.50
4750	For hot dip galvanized, add					35%				
5000	Dowels									
5100	Plain, 1/4" diameter, 3" long				C	54.50			54.50	60
5150	4" long					60.50			60.50	66.50
5200	6" long					74			74	81
5300	3/8" diameter, 3" long					72			72	79
5350	4" long					89.50			89.50	98.50
5400	6" long					103			103	113
5500	1/2" diameter, 3" long					105			105	115
5550	4" long					124			124	137
5600	6" long					161			161	177
5700	5/8" diameter, 3" long					144			144	158
5750	4" long					177			177	195
5800	6" long					243			243	267
6000	3/4" diameter, 3" long					179			179	197
6100	4" long					227			227	249
6150	6" long					325			325	355
6300	For hot dip galvanized, add					35%				

04 05 Common Work Results for Masonry

04 05 19 – Masonry Anchorage and Reinforcing

04 05 19.26 Masonry Reinforcing Bars

04 05 19.26 Masonry Reinforcing Bars	Crew	Daily Output	Labor-Hours	Unit	Material	2020 Bare Costs Labor	Equipment	Total	Total Incl O&P
0010 **MASONRY REINFORCING BARS** R040519-50									
0015 Steel bars A615, placed horiz., #3 & #4 bars	1 Bric	450	.018	Lb.	.56	.93		1.49	2.02
0020 #5 & #6 bars		800	.010		.56	.52		1.08	1.41
0050 Placed vertical, #3 & #4 bars		350	.023		.56	1.19		1.75	2.42
0060 #5 & #6 bars		650	.012		.56	.64		1.20	1.59
0200 Joint reinforcing, regular truss, to 6" wide, mill std galvanized		30	.267	C.L.F.	23	13.90		36.90	46.50
0250 12" wide		20	.400		29.50	21		50.50	64
0400 Cavity truss with drip section, to 6" wide		30	.267		23	13.90		36.90	46.50
0450 12" wide		20	.400		26.50	21		47.50	60.50
0500 Joint reinforcing, ladder type, mill std galvanized									
0600 9 ga. sides, 9 ga. ties, 4" wall	1 Bric	30	.267	C.L.F.	24	13.90		37.90	47
0650 6" wall		30	.267		22	13.90		35.90	45
0700 8" wall		25	.320		26.50	16.65		43.15	54.50
0750 10" wall		20	.400		24.50	21		45.50	58.50
0800 12" wall		20	.400		26	21		47	60
1000 Truss type									
1100 9 ga. sides, 9 ga. ties, 4" wall	1 Bric	30	.267	C.L.F.	24.50	13.90		38.40	48
1150 6" wall		30	.267		26	13.90		39.90	49.50
1200 8" wall		25	.320		28	16.65		44.65	56.50
1250 10" wall		20	.400		24.50	21		45.50	58
1300 12" wall		20	.400		25	21		46	59
1500 3/16" sides, 9 ga. ties, 4" wall		30	.267		28	13.90		41.90	52
1550 6" wall		30	.267		36	13.90		49.90	60.50
1600 8" wall		25	.320		32.50	16.65		49.15	61
1650 10" wall		20	.400		39	21		60	74.50
1700 12" wall		20	.400		38.50	21		59.50	73.50
2000 3/16" sides, 3/16" ties, 4" wall		30	.267		40	13.90		53.90	65
2050 6" wall		30	.267		41.50	13.90		55.40	66.50
2100 8" wall		25	.320		40	16.65		56.65	69.50
2150 10" wall		20	.400		45	21		66	81
2200 12" wall		20	.400		46.50	21		67.50	83
2500 Cavity truss type, galvanized									
2600 9 ga. sides, 9 ga. ties, 4" wall	1 Bric	25	.320	C.L.F.	37	16.65		53.65	66.50
2650 6" wall		25	.320		36	16.65		52.65	65
2700 8" wall		20	.400		51	21		72	87.50
2750 10" wall		15	.533		43.50	28		71.50	90
2800 12" wall		15	.533		49	28		77	96
3000 3/16" sides, 9 ga. ties, 4" wall		25	.320		36.50	16.65		53.15	65.50
3050 6" wall		25	.320		36.50	16.65		53.15	66
3100 8" wall		20	.400		56	21		77	93
3150 10" wall		15	.533		40	28		68	86
3200 12" wall		15	.533		41.50	28		69.50	87.50
3500 For hot dip galvanizing, add				Ton	510			510	560

04 05 23 – Masonry Accessories

04 05 23.13 Masonry Control and Expansion Joints

04 05 23.13 Masonry Control and Expansion Joints	Crew	Daily Output	Labor-Hours	Unit	Material	Labor	Equipment	Total	Total Incl O&P
0010 **MASONRY CONTROL AND EXPANSION JOINTS**									
0020 Rubber, for double wythe 8" minimum wall (Brick/CMU)	1 Bric	400	.020	L.F.	2.46	1.04		3.50	4.29
0025 "T" shaped		320	.025		1.22	1.30		2.52	3.31
0030 Cross-shaped for CMU units		280	.029		1.66	1.49		3.15	4.08
0050 PVC, for double wythe 8" minimum wall (Brick/CMU)		400	.020		1.70	1.04		2.74	3.45
0120 "T" shaped		320	.025		.92	1.30		2.22	2.98
0160 Cross-shaped for CMU units		280	.029		1.14	1.49		2.63	3.50

96

04 05 Common Work Results for Masonry

04 05 23 – Masonry Accessories

	04 05 23.95 Wall Plugs	Crew	Daily Output	Labor-Hours	Unit	Material	2020 Bare Costs Labor	Equipment	Total	Total Incl O&P
0010	**WALL PLUGS** (for nailing to brickwork)									
0020	25 ga., galvanized, plain	1 Bric	10.50	.762	C	26.50	39.50		66	89
0050	Wood filled	"	10.50	.762	"	73	39.50		112.50	141

04 21 Clay Unit Masonry

04 21 13 – Brick Masonry

04 21 13.13 Brick Veneer Masonry

		Crew	Daily Output	Labor-Hours	Unit	Material	2020 Bare Costs Labor	Equipment	Total	Total Incl O&P
0010	**BRICK VENEER MASONRY**, T.L. lots, excl. scaff., grout & reinforcing R042110-20									
0015	Material costs incl. 3% brick and 25% mortar waste									
0020	Standard, select common, 4" x 2-2/3" x 8" (6.75/S.F.)	D-8	1.50	26.667	M	670	1,275		1,945	2,675
0050	Red, 4" x 2-2/3" x 8", running bond		1.50	26.667		720	1,275		1,995	2,750
0100	Full header every 6th course (7.88/S.F.) R042110-50		1.45	27.586		720	1,325		2,045	2,800
0150	English, full header every 2nd course (10.13/S.F.)		1.40	28.571		715	1,375		2,090	2,875
0200	Flemish, alternate header every course (9.00/S.F.)		1.40	28.571		715	1,375		2,090	2,875
0250	Flemish, alt. header every 6th course (7.13/S.F.)		1.45	27.586		720	1,325		2,045	2,800
0300	Full headers throughout (13.50/S.F.)		1.40	28.571		715	1,375		2,090	2,850
0350	Rowlock course (13.50/S.F.)		1.35	29.630		715	1,425		2,140	2,925
0400	Rowlock stretcher (4.50/S.F.)		1.40	28.571		725	1,375		2,100	2,875
0450	Soldier course (6.75/S.F.)		1.40	28.571		720	1,375		2,095	2,875
0500	Sailor course (4.50/S.F.)		1.30	30.769		725	1,475		2,200	3,025
0601	Buff or gray face, running bond (6.75/S.F.)		1.50	26.667		720	1,275		1,995	2,750
0700	Glazed face, 4" x 2-2/3" x 8", running bond		1.40	28.571		2,175	1,375		3,550	4,475
0750	Full header every 6th course (7.88/S.F.)		1.35	29.630		2,075	1,425		3,500	4,425
1000	Jumbo, 6" x 4" x 12" (3.00/S.F.)		1.30	30.769		1,950	1,475		3,425	4,375
1051	Norman, 4" x 2-2/3" x 12" (4.50/S.F.)		1.45	27.586		1,225	1,325		2,550	3,350
1100	Norwegian, 4" x 3-1/5" x 12" (3.75/S.F.)		1.40	28.571		1,650	1,375		3,025	3,900
1150	Economy, 4" x 4" x 8" (4.50/S.F.)		1.40	28.571		1,050	1,375		2,425	3,225
1201	Engineer, 4" x 3-1/5" x 8" (5.63/S.F.)		1.45	27.586		695	1,325		2,020	2,775
1251	Roman, 4" x 2" x 12" (6.00/S.F.)		1.50	26.667		1,300	1,275		2,575	3,375
1300	S.C.R., 6" x 2-2/3" x 12" (4.50/S.F.)		1.40	28.571		1,425	1,375		2,800	3,625
1350	Utility, 4" x 4" x 12" (3.00/S.F.)		1.08	37.037		1,850	1,775		3,625	4,725
1360	For less than truck load lots, add					15%				
1400	For battered walls, add						30%			
1450	For corbels, add						75%			
1500	For curved walls, add						30%			
1550	For pits and trenches, deduct						20%			
1999	Alternate method of figuring by square foot									
2000	Standard, sel. common, 4" x 2-2/3" x 8" (6.75/S.F.)	D-8	230	.174	S.F.	4.52	8.35		12.87	17.65
2020	Red, 4" x 2-2/3" x 8", running bond		220	.182		4.86	8.70		13.56	18.55
2050	Full header every 6th course (7.88/S.F.)		185	.216		5.65	10.35		16	22
2100	English, full header every 2nd course (10.13/S.F.)		140	.286		7.25	13.70		20.95	29
2150	Flemish, alternate header every course (9.00/S.F.)		150	.267		6.45	12.80		19.25	26.50
2200	Flemish, alt. header every 6th course (7.13/S.F.)		205	.195		5.15	9.35		14.50	19.85
2250	Full headers throughout (13.50/S.F.)		105	.381		9.65	18.25		27.90	38
2300	Rowlock course (13.50/S.F.)		100	.400		9.65	19.15		28.80	39.50
2350	Rowlock stretcher (4.50/S.F.)		310	.129		3.27	6.20		9.47	13
2400	Soldier course (6.75/S.F.)		200	.200		4.86	9.60		14.46	19.90
2450	Sailor course (4.50/S.F.)		290	.138		3.27	6.60		9.87	13.60
2600	Buff or gray face, running bond (6.75/S.F.)		220	.182		5.15	8.70		13.85	18.85
2700	Glazed face brick, running bond		210	.190		14	9.15		23.15	29.50

04 21 13.13 Brick Veneer Masonry	Crew	Daily Output	Labor-Hours	Unit	Material	2020 Bare Costs Labor	Equipment	Total	Total Incl O&P
2750	Full header every 6th course (7.88/S.F.)	D-8	170	.235	S.F.	16.30	11.30		27.60	35
3000	Jumbo, 6" x 4" x 12" running bond (3.00/S.F.)		435	.092		5.40	4.41		9.81	12.60
3050	Norman, 4" x 2-2/3" x 12" running bond (4.5/S.F.)		320	.125		6.65	6		12.65	16.45
3100	Norwegian, 4" x 3-1/5" x 12" (3.75/S.F.)		375	.107		6.05	5.10		11.15	14.40
3150	Economy, 4" x 4" x 8" (4.50/S.F.)		310	.129		4.63	6.20		10.83	14.50
3200	Engineer, 4" x 3-1/5" x 8" (5.63/S.F.)		260	.154		3.89	7.35		11.24	15.50
3250	Roman, 4" x 2" x 12" (6.00/S.F.)		250	.160		7.65	7.65		15.30	20
3300	S.C.R., 6" x 2-2/3" x 12" (4.50/S.F.)		310	.129		6.30	6.20		12.50	16.35
3350	Utility, 4" x 4" x 12" (3.00/S.F.)		360	.111		5.40	5.35		10.75	14
3360	For less than truck load lots, add					.10%				
3400	For cavity wall construction, add						15%			
3450	For stacked bond, add						10%			
3500	For interior veneer construction, add						15%			
3550	For curved walls, add						30%			

04 21 13.14 Thin Brick Veneer

		Crew	Daily Output	Labor-Hours	Unit	Material	Labor	Equipment	Total	Total Incl O&P
0010	**THIN BRICK VENEER**									
0015	Material costs incl. 3% brick and 25% mortar waste									
0020	On & incl. metal panel support sys, modular, 2-2/3" x 5/8" x 8", red	D-7	92	.174	S.F.	9.70	7.70		17.40	22
0100	Closure, 4" x 5/8" x 8"		110	.145		9.40	6.40		15.80	19.75
0110	Norman, 2-2/3" x 5/8" x 12"		110	.145		9.65	6.40		16.05	20
0120	Utility, 4" x 5/8" x 12"		125	.128		9.10	5.65		14.75	18.30
0130	Emperor, 4" x 3/4" x 16"		175	.091		10.30	4.04		14.34	17.20
0140	Super emperor, 8" x 3/4" x 16"		195	.082		11.15	3.62		14.77	17.55
0150	For L shaped corners with 4" return, add				L.F.	9.25			9.25	10.20
0200	On masonry/plaster back-up, modular, 2-2/3" x 5/8" x 8", red	D-7	137	.117	S.F.	4.66	5.15		9.81	12.70
0210	Closure, 4" x 5/8" x 8"		165	.097		4.37	4.28		8.65	11.10
0220	Norman, 2-2/3" x 5/8" x 12"		165	.097		4.61	4.28		8.89	11.35
0230	Utility, 4" x 5/8" x 12"		185	.086		4.09	3.82		7.91	10.10
0240	Emperor, 4" x 3/4" x 16"		260	.062		5.25	2.72		7.97	9.80
0250	Super emperor, 8" x 3/4" x 16"		285	.056		6.15	2.48		8.63	10.40
0260	For L shaped corners with 4" return, add				L.F.	10			10	11
0270	For embedment into pre-cast concrete panels, add				S.F.	14.40			14.40	15.85

04 21 13.18 Columns

		Crew	Daily Output	Labor-Hours	Unit	Material	Labor	Equipment	Total	Total Incl O&P
0010	**COLUMNS**, solid, excludes scaffolding, grout and reinforcing									
0050	Brick, 8" x 8", 9 brick/V.L.F.	D-1	56	.286	V.L.F.	6.30	13.40		19.70	27.50
0100	12" x 8", 13.5 brick/V.L.F.		37	.432		9.45	20.50		29.95	41.50
0200	12" x 12", 20 brick/V.L.F.		25	.640		14	30		44	61
0300	16" x 12", 27 brick/V.L.F.		19	.842		18.90	39.50		58.40	81
0400	16" x 16", 36 brick/V.L.F.		14	1.143		25	53.50		78.50	109
0500	20" x 16", 45 brick/V.L.F.		11	1.455		31.50	68		99.50	138
0600	20" x 20", 56 brick/V.L.F.		9	1.778		39	83.50		122.50	169
0700	24" x 20", 68 brick/V.L.F.		7	2.286		47.50	107		154.50	216
0800	24" x 24", 81 brick/V.L.F.		6	2.667		56.50	125		181.50	253
1000	36" x 36", 182 brick/V.L.F.		3	5.333		127	250		377	520

04 21 13.35 Common Building Brick

		Crew	Daily Output	Labor-Hours	Unit	Material	Labor	Equipment	Total	Total Incl O&P
0010	**COMMON BUILDING BRICK**, C62, T.L. lots, material only									
0020	Standard				M	590			590	650
0050	Select				"	555			555	615

04 21 13.45 Face Brick

		Crew	Daily Output	Labor-Hours	Unit	Material	Labor	Equipment	Total	Total Incl O&P
0010	**FACE BRICK** Material Only, C216, T.L. lots									
0300	Standard modular, 4" x 2-2/3" x 8"				M	605			605	665
0450	Economy, 4" x 4" x 8"					900			900	990

04 21 Clay Unit Masonry

04 21 13 – Brick Masonry

04 21 13.45 Face Brick	Crew	Daily Output	Labor-Hours	Unit	Material	2020 Bare Costs Labor	Equipment	Total	Total Incl O&P
0510	Economy, 4" x 4" x 12"				M	1,425			1,425	1,550
0550	Jumbo, 6" x 4" x 12"					1,625			1,625	1,800
0610	Jumbo, 8" x 4" x 12"					1,625			1,625	1,800
0650	Norwegian, 4" x 3-1/5" x 12"					1,475			1,475	1,600
0710	Norwegian, 6" x 3-1/5" x 12"					1,725			1,725	1,900
0850	Standard glazed, plain colors, 4" x 2-2/3" x 8"					1,925			1,925	2,100
1000	Deep trim shades, 4" x 2-2/3" x 8"					2,350			2,350	2,600
1080	Jumbo utility, 4" x 4" x 12"					1,650			1,650	1,800
1120	4" x 8" x 8"					1,975			1,975	2,175
1140	4" x 8" x 16"					6,250			6,250	6,875
1260	Engineer, 4" x 3-1/5" x 8"					575			575	635
1350	King, 4" x 2-3/4" x 10"					600			600	660
1770	Standard modular, double glazed, 4" x 2-2/3" x 8"					2,950			2,950	3,250
1850	Jumbo, colored glazed ceramic, 6" x 4" x 12"					3,050			3,050	3,350
2050	Jumbo utility, glazed, 4" x 4" x 12"					6,125			6,125	6,725
2100	4" x 8" x 8"					7,200			7,200	7,925
2150	4" x 16" x 8"					7,875			7,875	8,675
3050	Used brick					480			480	530
3150	Add for brick to match existing work, minimum					5%				
3200	Maximum					50%				

04 22 Concrete Unit Masonry

04 22 10 – Concrete Masonry Units

04 22 10.10 Concrete Block

		Crew	Daily Output	Labor-Hours	Unit	Material	2020 Bare Costs Labor	Equipment	Total	Total Incl O&P
0010	**CONCRETE BLOCK** Material Only R042210-20									
0020	2" x 8" x 16" solid, normal-weight, 2,000 psi				Ea.	1.19			1.19	1.31
0050	3,500 psi					1.31			1.31	1.44
0100	5,000 psi					1.40			1.40	1.54
0150	Lightweight, std.					1.43			1.43	1.57
0300	3" x 8" x 16" solid, normal-weight, 2,000 psi					1.16			1.16	1.28
0350	3,500 psi					1.45			1.45	1.60
0400	5,000 psi					1.68			1.68	1.85
0450	Lightweight, std.					1.48			1.48	1.63
0600	4" x 8" x 16" hollow, normal-weight, 2,000 psi					1.50			1.50	1.65
0650	3,500 psi					1.38			1.38	1.52
0700	5,000 psi					1.73			1.73	1.90
0750	Lightweight, std.					1.44			1.44	1.58
1300	Solid, normal-weight, 2,000 psi					1.64			1.64	1.80
1350	3,500 psi					1.77			1.77	1.95
1400	5,000 psi					1.57			1.57	1.73
1450	Lightweight, std.					1.44			1.44	1.58
1600	6" x 8" x 16" hollow, normal-weight, 2,000 psi					1.87			1.87	2.06
1650	3,500 psi					1.77			1.77	1.95
1700	5,000 psi					2.08			2.08	2.29
1750	Lightweight, std.					2			2	2.20
2300	Solid, normal-weight, 2,000 psi					2.25			2.25	2.48
2350	3,500 psi					1.67			1.67	1.84
2400	5,000 psi					2.22			2.22	2.44
2450	Lightweight, std.					2.61			2.61	2.87
2600	8" x 8" x 16" hollow, normal-weight, 2,000 psi					1.75			1.75	1.93
2650	3,500 psi					1.98			1.98	2.18

04 22 10.10 Concrete Block

	Crew	Daily Output	Labor-Hours	Unit	Material	2020 Bare Costs Labor	2020 Bare Costs Equipment	Total	Total Incl O&P
2700	5,000 psi				Ea.	2.54			2.54	2.79
2750	Lightweight, std.					2.74			2.74	3.01
3200	Solid, normal-weight, 2,000 psi					2.77			2.77	3.05
3250	3,500 psi					2.93			2.93	3.22
3300	5,000 psi					3.21			3.21	3.53
3350	Lightweight, std.					2.80			2.80	3.08
3400	10" x 8" x 16" hollow, normal-weight, 2,000 psi					1.75			1.75	1.93
3410	3,500 psi					1.98			1.98	2.18
3420	5,000 psi					2.54			2.54	2.79
3430	Lightweight, std.					2.74			2.74	3.01
3480	Solid, normal-weight, 2,000 psi					2.77			2.77	3.05
3490	3,500 psi					2.93			2.93	3.22
3500	5,000 psi					3.21			3.21	3.53
3510	Lightweight, std.					2.80			2.80	3.08
3600	12" x 8" x 16" hollow, normal-weight, 2,000 psi					3.27			3.27	3.60
3650	3,500 psi					3.13			3.13	3.44
3700	5,000 psi					3.40			3.40	3.74
3750	Lightweight, std.					2.97			2.97	3.27
4300	Solid, normal-weight, 2,000 psi					4.37			4.37	4.81
4350	3,500 psi					3.72			3.72	4.09
4400	5,000 psi					3.81			3.81	4.19
4500	Lightweight, std.					3.18			3.18	3.50

04 22 10.14 Concrete Block, Back-Up

	Crew	Daily Output	Labor-Hours	Unit	Material	2020 Bare Costs Labor	2020 Bare Costs Equipment	Total	Total Incl O&P
0010	**CONCRETE BLOCK, BACK-UP**, C90, 2000 psi									
0020	Normal weight, 8" x 16" units, tooled joint 1 side									
0050	Not-reinforced, 2000 psi, 2" thick	D-8	475	.084	S.F.	1.68	4.04		5.72	7.95
0200	4" thick		460	.087		2.16	4.17		6.33	8.70
0300	6" thick		440	.091		2.72	4.36		7.08	9.60
0350	8" thick		400	.100		2.73	4.79		7.52	10.25
0400	10" thick		330	.121		3.25	5.80		9.05	12.35
0450	12" thick	D-9	310	.155		4.72	7.25		11.97	16.20

04 22 10.16 Concrete Block, Bond Beam

	Crew	Daily Output	Labor-Hours	Unit	Material	2020 Bare Costs Labor	2020 Bare Costs Equipment	Total	Total Incl O&P
0010	**CONCRETE BLOCK, BOND BEAM**, C90, 2000 psi									
0020	Not including grout or reinforcing									
0125	Regular block, 6" thick	D-8	584	.068	L.F.	2.77	3.28		6.05	8
0130	8" high, 8" thick	"	565	.071		2.90	3.39		6.29	8.35
0150	12" thick	D-9	510	.094		4.09	4.41		8.50	11.20
0525	Lightweight, 6" thick	D-8	592	.068		2.90	3.24		6.14	8.10
0530	8" high, 8" thick	"	575	.070		3.41	3.33		6.74	8.80
0550	12" thick	D-9	520	.092		4.63	4.33		8.96	11.65
2000	Including grout and 2 #5 bars									
2100	Regular block, 8" high, 8" thick	D-8	300	.133	L.F.	5.35	6.40		11.75	15.60
2150	12" thick	D-9	250	.192		7.25	9		16.25	21.50
2500	Lightweight, 8" high, 8" thick	D-8	305	.131		5.90	6.30		12.20	16
2550	12" thick	D-9	255	.188		7.80	8.85		16.65	22

04 22 10.19 Concrete Block, Insulation Inserts

	Crew	Daily Output	Labor-Hours	Unit	Material	2020 Bare Costs Labor	2020 Bare Costs Equipment	Total	Total Incl O&P
0010	**CONCRETE BLOCK, INSULATION INSERTS**									
0100	Styrofoam, plant installed, add to block prices									
0200	8" x 16" units, 6" thick				S.F.	1.25			1.25	1.38
0250	8" thick					1.57			1.57	1.73
0300	10" thick					1.47			1.47	1.62
0350	12" thick					1.85			1.85	2.04

04 22 10.19 Concrete Block, Insulation Inserts	Crew	Daily Output	Labor-Hours	Unit	Material	2020 Bare Costs Labor	Equipment	Total	Total Incl O&P
0500	8" x 8" units, 8" thick				S.F.	1.40			1.40	1.54
0550	12" thick				↓	1.47			1.47	1.62

04 22 10.23 Concrete Block, Decorative		Crew	Daily Output	Labor-Hours	Unit	Material	2020 Bare Costs Labor	Equipment	Total	Total Incl O&P
0010	**CONCRETE BLOCK, DECORATIVE**, C90, 2000 psi									
0020	Embossed, simulated brick face									
0100	8" x 16" units, 4" thick	D-8	400	.100	S.F.	3.23	4.79		8.02	10.80
0200	8" thick		340	.118		3.25	5.65		8.90	12.15
0250	12" thick	↓	300	.133	↓	5.60	6.40		12	15.85
0400	Embossed both sides									
0500	8" thick	D-8	300	.133	S.F.	4.88	6.40		11.28	15.05
0550	12" thick	"	275	.145	"	6.10	6.95		13.05	17.25
1000	Fluted high strength									
1100	8" x 16" x 4" thick, flutes 1 side,	D-8	345	.116	S.F.	4.11	5.55		9.66	12.90
1150	Flutes 2 sides		335	.119		5.40	5.70		11.10	14.65
1200	8" thick	↓	300	.133		5.85	6.40		12.25	16.15
1250	For special colors, add				↓	.67			.67	.73
1400	Deep grooved, smooth face									
1450	8" x 16" x 4" thick	D-8	345	.116	S.F.	2.97	5.55		8.52	11.65
1500	8" thick	"	300	.133	"	4.48	6.40		10.88	14.65
2000	Formblock, incl. inserts & reinforcing									
2100	8" x 16" x 8" thick	D-8	345	.116	S.F.	4.22	5.55		9.77	13.05
2150	12" thick	"	310	.129	"	5.50	6.20		11.70	15.45
2500	Ground face									
2600	8" x 16" x 4" thick	D-8	345	.116	S.F.	4.86	5.55		10.41	13.75
2650	6" thick		325	.123		4.90	5.90		10.80	14.35
2700	8" thick	↓	300	.133		5.25	6.40		11.65	15.45
2750	12" thick	D-9	265	.181	↓	6.15	8.50		14.65	19.65
2900	For special colors, add, minimum					15%				
2950	For special colors, add, maximum					45%				
4000	Slump block									
4100	4" face height x 16" x 4" thick	D-1	165	.097	S.F.	4.76	4.55		9.31	12.15
4150	6" thick		160	.100		6.25	4.69		10.94	14
4200	8" thick		155	.103		8	4.84		12.84	16.15
4250	10" thick		140	.114		14.10	5.35		19.45	23.50
4300	12" thick		130	.123		12.55	5.75		18.30	22.50
4400	6" face height x 16" x 6" thick		155	.103		5.80	4.84		10.64	13.75
4450	8" thick		150	.107		8.75	5		13.75	17.25
4500	10" thick		130	.123		14.90	5.75		20.65	25
4550	12" thick	↓	120	.133	↓	14.50	6.25		20.75	25.50
5000	Split rib profile units, 1" deep ribs, 8 ribs									
5100	8" x 16" x 4" thick	D-8	345	.116	S.F.	4.61	5.55		10.16	13.50
5150	6" thick		325	.123		5.25	5.90		11.15	14.75
5200	8" thick	↓	300	.133		6.55	6.40		12.95	16.95
5250	12" thick	D-9	275	.175		7.50	8.20		15.70	20.50
5400	For special deeper colors, 4" thick, add					1.32			1.32	1.45
5450	12" thick, add					1.41			1.41	1.55
5600	For white, 4" thick, add					1.32			1.32	1.45
5650	6" thick, add					1.39			1.39	1.53
5700	8" thick, add					1.48			1.48	1.63
5750	12" thick, add				↓	1.47			1.47	1.62
6000	Split face									
6100	8" x 16" x 4" thick	D-8	350	.114	S.F.	3.90	5.50		9.40	12.60

04 22 Concrete Unit Masonry

04 22 10 – Concrete Masonry Units

04 22 10.23 Concrete Block, Decorative

		Crew	Daily Output	Labor-Hours	Unit	Material	2020 Bare Costs Labor	2020 Bare Costs Equipment	Total	Total Incl O&P
6150	6" thick	D-8	325	.123	S.F.	4.40	5.90		10.30	13.80
6200	8" thick	↓	300	.133		5.60	6.40		12	15.90
6250	12" thick	D-9	270	.178		6.25	8.35		14.60	19.55
6300	For scored, add					.39			.39	.43
6400	For special deeper colors, 4" thick, add					.63			.63	.70
6450	6" thick, add					.77			.77	.85
6500	8" thick, add					.81			.81	.89
6550	12" thick, add					.86			.86	.94
6650	For white, 4" thick, add					1.32			1.32	1.45
6700	6" thick, add					1.29			1.29	1.42
6750	8" thick, add					1.41			1.41	1.55
6800	12" thick, add				↓	1.41			1.41	1.55
7000	Scored ground face, 2 to 5 scores									
7100	8" x 16" x 4" thick	D-8	340	.118	S.F.	9	5.65		14.65	18.50
7150	6" thick	↓	310	.129		10.10	6.20		16.30	20.50
7200	8" thick	↓	290	.138		11.45	6.60		18.05	22.50
7250	12" thick	D-9	265	.181	↓	15.65	8.50		24.15	30
8000	Hexagonal face profile units, 8" x 16" units									
8100	4" thick, hollow	D-8	340	.118	S.F.	4.06	5.65		9.71	13
8200	Solid		340	.118		5.15	5.65		10.80	14.25
8300	6" thick, hollow		310	.129		4.37	6.20		10.57	14.20
8350	8" thick, hollow	↓	290	.138		4.74	6.60		11.34	15.20
8500	For stacked bond, add				↓		26%			
8550	For high rise construction, add per story	D-8	67.80	.590	M.S.F.		28.50		28.50	43
8600	For scored block, add					10%				
8650	For honed or ground face, per face, add				Ea.	1.23			1.23	1.35
8700	For honed or ground end, per end, add				"	1.36			1.36	1.49
8750	For bullnose block, add					10%				
8800	For special color, add					13%				

04 22 10.24 Concrete Block, Exterior

		Crew	Daily Output	Labor-Hours	Unit	Material	2020 Bare Costs Labor	2020 Bare Costs Equipment	Total	Total Incl O&P
0010	**CONCRETE BLOCK, EXTERIOR**, C90, 2000 psi R042210-20									
0020	Reinforced alt courses, tooled joints 2 sides									
0100	Normal weight, 8" x 16" x 6" thick	D-8	395	.101	S.F.	2.56	4.85		7.41	10.15
0200	8" thick		360	.111		4.16	5.35		9.51	12.65
0250	10" thick	↓	290	.138		4.50	6.60		11.10	14.95
0300	12" thick	D-9	250	.192		5.30	9		14.30	19.50
0500	Lightweight, 8" x 16" x 6" thick	D-8	450	.089		3.62	4.26		7.88	10.45
0600	8" thick		430	.093		4.01	4.46		8.47	11.15
0650	10" thick	↓	395	.101		4.50	4.85		9.35	12.30
0700	12" thick	D-9	350	.137	↓	4.71	6.45		11.16	14.95

04 22 10.26 Concrete Block Foundation Wall

		Crew	Daily Output	Labor-Hours	Unit	Material	2020 Bare Costs Labor	2020 Bare Costs Equipment	Total	Total Incl O&P
0010	**CONCRETE BLOCK FOUNDATION WALL**, C90/C145									
0050	Normal-weight, cut joints, horiz joint reinf, no vert reinf.									
0200	Hollow, 8" x 16" x 6" thick	D-8	455	.088	S.F.	3.20	4.21		7.41	9.90
0250	8" thick		425	.094		3.23	4.51		7.74	10.40
0300	10" thick	↓	350	.114		3.73	5.50		9.23	12.40
0350	12" thick	D-9	300	.160		5.20	7.50		12.70	17.15
0500	Solid, 8" x 16" block, 6" thick	D-8	440	.091		3.63	4.36		7.99	10.60
0550	8" thick	"	415	.096		4.38	4.62		9	11.80
0600	12" thick	D-9	350	.137	↓	6.45	6.45		12.90	16.85
1000	Reinforced, #4 vert @ 48"									
1100	Hollow, 8" x 16" block, 4" thick	D-8	455	.088	S.F.	3.29	4.21		7.50	10

102

04 22 10.26 Concrete Block Foundation Wall

		Crew	Daily Output	Labor-Hours	Unit	Material	2020 Bare Costs Labor	Equipment	Total	Total Incl O&P
1125	6" thick	D-8	445	.090	S.F.	4.25	4.31		8.56	11.25
1150	8" thick	↓	415	.096		4.72	4.62		9.34	12.20
1200	10" thick	↓	340	.118		5.65	5.65		11.30	14.80
1250	12" thick	D-9	290	.166		7.60	7.75		15.35	20
1500	Solid, 8" x 16" block, 6" thick	D-8	430	.093		3.64	4.46		8.10	10.75
1600	8" thick	"	405	.099		4.38	4.73		9.11	12
1650	12" thick	D-9	340	.141	↓	6.45	6.60		13.05	17.15

04 22 10.32 Concrete Block, Lintels

		Crew	Daily Output	Labor-Hours	Unit	Material	2020 Bare Costs Labor	Equipment	Total	Total Incl O&P
0010	**CONCRETE BLOCK, LINTELS**, C90, normal weight									
0100	Including grout and horizontal reinforcing									
0200	8" x 8" x 8", 1 #4 bar	D-4	300	.107	L.F.	4.54	5.05	.63	10.22	13.30
0250	2 #4 bars		295	.108		4.79	5.10	.64	10.53	13.65
0400	8" x 16" x 8", 1 #4 bar		275	.116		4.24	5.50	.69	10.43	13.70
0450	2 #4 bars		270	.119		4.49	5.60	.70	10.79	14.15
1000	12" x 8" x 8", 1 #4 bar		275	.116		6.05	5.50	.69	12.24	15.70
1100	2 #4 bars		270	.119		6.30	5.60	.70	12.60	16.10
1150	2 #5 bars		270	.119		6.55	5.60	.70	12.85	16.45
1200	2 #6 bars		265	.121		6.90	5.70	.71	13.31	17
1500	12" x 16" x 8", 1 #4 bar		250	.128		7.05	6.05	.75	13.85	17.70
1600	2 #3 bars		245	.131		7.10	6.15	.77	14.02	17.95
1650	2 #4 bars		245	.131		7.30	6.15	.77	14.22	18.20
1700	2 #5 bars	↓	240	.133	↓	7.60	6.30	.79	14.69	18.70

04 22 10.34 Concrete Block, Partitions

		Crew	Daily Output	Labor-Hours	Unit	Material	2020 Bare Costs Labor	Equipment	Total	Total Incl O&P
0010	**CONCRETE BLOCK, PARTITIONS**, excludes scaffolding R042210-20									
1000	Lightweight block, tooled joints, 2 sides, hollow									
1100	Not reinforced, 8" x 16" x 4" thick	D-8	440	.091	S.F.	2	4.36		6.36	8.80
1150	6" thick		410	.098		2.77	4.68		7.45	10.15
1200	8" thick		385	.104		3.75	4.98		8.73	11.65
1250	10" thick	↓	370	.108		4.18	5.20		9.38	12.45
1300	12" thick	D-9	350	.137		4.28	6.45		10.73	14.45
1500	Reinforced alternate courses, 4" thick	D-8	435	.092		2.18	4.41		6.59	9.10
1600	6" thick		405	.099		2.93	4.73		7.66	10.40
1650	8" thick		380	.105		3.95	5.05		9	12
1700	10" thick	↓	365	.110		4.36	5.25		9.61	12.75
1750	12" thick	D-9	345	.139	↓	4.48	6.55		11.03	14.80
4000	Regular block, tooled joints, 2 sides, hollow									
4100	Not reinforced, 8" x 16" x 4" thick	D-8	430	.093	S.F.	2.06	4.46		6.52	9
4150	6" thick		400	.100		2.62	4.79		7.41	10.15
4200	8" thick		375	.107		2.63	5.10		7.73	10.65
4250	10" thick	↓	360	.111		3.15	5.35		8.50	11.50
4300	12" thick	D-9	340	.141		4.62	6.60		11.22	15.15
4500	Reinforced alternate courses, 8" x 16" x 4" thick	D-8	425	.094		2.25	4.51		6.76	9.30
4550	6" thick		395	.101		2.81	4.85		7.66	10.45
4600	8" thick		370	.108		2.84	5.20		8.04	10.95
4650	10" thick	↓	355	.113		4.67	5.40		10.07	13.35
4700	12" thick	D-9	335	.143	↓	4.81	6.70		11.51	15.50

For customer support on your Site Work & Landscape Costs with RSMeans data, call 800.448.8182.

103

04 27 Multiple-Wythe Unit Masonry

04 27 10 – Multiple-Wythe Masonry

04 27 10.30 Brick Walls

		Crew	Daily Output	Labor-Hours	Unit	Material	2020 Bare Costs Labor	Equipment	Total	Total Incl O&P
0010	**BRICK WALLS**, including mortar, excludes scaffolding R042110-20									
0020	Estimating by number of brick									
0140	Face brick, 4" thick wall, 6.75 brick/S.F.	D-8	1.45	27.586	M	710	1,325		2,035	2,775
0150	Common brick, 4" thick wall, 6.75 brick/S.F. R042110-50		1.60	25		690	1,200		1,890	2,575
0204	8" thick, 13.50 brick/S.F.		1.80	22.222		710	1,075		1,785	2,400
0250	12" thick, 20.25 brick/S.F.		1.90	21.053		715	1,000		1,715	2,300
0304	16" thick, 27.00 brick/S.F.		2	20		720	960		1,680	2,250
0500	Reinforced, face brick, 4" thick wall, 6.75 brick/S.F.		1.40	28.571		735	1,375		2,110	2,875
0520	Common brick, 4" thick wall, 6.75 brick/S.F.		1.55	25.806		720	1,225		1,945	2,675
0550	8" thick, 13.50 brick/S.F.		1.75	22.857		740	1,100		1,840	2,450
0600	12" thick, 20.25 brick/S.F.		1.85	21.622		740	1,025		1,765	2,400
0650	16" thick, 27.00 brick/S.F.		1.95	20.513		750	985		1,735	2,325
0790	Alternate method of figuring by square foot									
0800	Face brick, 4" thick wall, 6.75 brick/S.F.	D-8	215	.186	S.F.	4.78	8.90		13.68	18.75
0850	Common brick, 4" thick wall, 6.75 brick/S.F.		240	.167		4.67	8		12.67	17.25
0900	8" thick, 13.50 brick/S.F.		135	.296		9.60	14.20		23.80	32
1000	12" thick, 20.25 brick/S.F.		95	.421		14.45	20		34.45	46.50
1050	16" thick, 27.00 brick/S.F.		75	.533		19.50	25.50		45	60.50
1200	Reinforced, face brick, 4" thick wall, 6.75 brick/S.F.		210	.190		4.96	9.15		14.11	19.30
1220	Common brick, 4" thick wall, 6.75 brick/S.F.		235	.170		4.86	8.15		13.01	17.70
1250	8" thick, 13.50 brick/S.F.		130	.308		9.95	14.75		24.70	33.50
1260	8" thick, 2.25 brick/S.F.		130	.308		1.79	14.75		16.54	24.50
1300	12" thick, 20.25 brick/S.F.		90	.444		15	21.50		36.50	49
1350	16" thick, 27.00 brick/S.F.		70	.571		20	27.50		47.50	63.50

04 27 10.40 Steps

		Crew	Daily Output	Labor-Hours	Unit	Material	2020 Bare Costs Labor	Equipment	Total	Total Incl O&P
0010	**STEPS**									
0012	Entry steps, select common brick	D-1	.30	53.333	M	555	2,500		3,055	4,425

04 41 Dry-Placed Stone

04 41 10 – Dry Placed Stone

04 41 10.10 Rough Stone Wall

			Crew	Daily Output	Labor-Hours	Unit	Material	2020 Bare Costs Labor	Equipment	Total	Total Incl O&P
0011	**ROUGH STONE WALL**, Dry										
0012	Dry laid (no mortar), under 18" thick	G	D-1	60	.267	C.F.	14.15	12.50		26.65	34.50
0100	Random fieldstone, under 18" thick	G	D-12	60	.533		14.15	25.50		39.65	54
0150	Over 18" thick	G	"	63	.508		17	24		41	55
0500	Field stone veneer	G	D-8	120	.333	S.F.	12.95	16		28.95	38.50

04 43 Stone Masonry

04 43 10 – Masonry with Natural and Processed Stone

04 43 10.05 Ashlar Veneer

		Crew	Daily Output	Labor-Hours	Unit	Material	2020 Bare Costs Labor	Equipment	Total	Total Incl O&P
0011	**ASHLAR VENEER** +/- 4" thk, random or random rectangular									
0150	Sawn face, split joints, low priced stone	D-8	140	.286	S.F.	12.15	13.70		25.85	34.50
0200	Medium priced stone		130	.308		15.40	14.75		30.15	39.50
0300	High priced stone		120	.333		18.45	16		34.45	44.50
0600	Seam face, split joints, medium price stone		125	.320		18.20	15.35		33.55	43.50
0700	High price stone		120	.333		18.10	16		34.10	44
1000	Split or rock face, split joints, medium price stone		125	.320		10.95	15.35		26.30	35.50
1100	High price stone		120	.333		17.25	16		33.25	43

04 43 10 – Masonry with Natural and Processed Stone

04 43 10.10 Bluestone	Crew	Daily Output	Labor-Hours	Unit	Material	2020 Bare Costs Labor	Equipment	Total	Total Incl O&P
0010 **BLUESTONE**, cut to size									
0100 Paving, natural cleft, to 4', 1" thick	D-8	150	.267	S.F.	7.95	12.80		20.75	28
0150 1-1/2" thick		145	.276		7.90	13.20		21.10	28.50
0200 Smooth finish, 1" thick		150	.267		7.60	12.80		20.40	28
0250 1-1/2" thick		145	.276		8.90	13.20		22.10	30
0300 Thermal finish, 1" thick		150	.267		7.55	12.80		20.35	27.50
0350 1-1/2" thick		145	.276		8.50	13.20		21.70	29.50
0500 Sills, natural cleft, 10" wide to 6' long, 1-1/2" thick	D-11	70	.343	L.F.	13.50	16.90		30.40	40.50
0550 2" thick		63	.381		14.80	18.75		33.55	45
0600 Smooth finish, 1-1/2" thick		70	.343		14.45	16.90		31.35	41.50
0650 2" thick		63	.381		18.10	18.75		36.85	48.50
0800 Thermal finish, 1-1/2" thick		70	.343		15.35	16.90		32.25	42.50
0850 2" thick		63	.381		12.20	18.75		30.95	42
1000 Stair treads, natural cleft, 12" wide, 6' long, 1-1/2" thick	D-10	115	.278		14.05	14.40	3.72	32.17	41.50
1050 2" thick		105	.305		15.30	15.75	4.08	35.13	45.50
1100 Smooth finish, 1-1/2" thick		115	.278		13.45	14.40	3.72	31.57	41
1150 2" thick		105	.305		14.55	15.75	4.08	34.38	44.50
1300 Thermal finish, 1-1/2" thick		115	.278		13.80	14.40	3.72	31.92	41.50
1350 2" thick		105	.305		13.50	15.75	4.08	33.33	43.50
2000 Coping, finished top & 2 sides, 12" to 6'									
2100 Natural cleft, 1-1/2" thick	D-10	115	.278	L.F.	15.05	14.40	3.72	33.17	42.50
2150 2" thick		105	.305		17.50	15.75	4.08	37.33	47.50
2200 Smooth finish, 1-1/2" thick		115	.278		15.05	14.40	3.72	33.17	42.50
2250 2" thick		105	.305		17.50	15.75	4.08	37.33	47.50
2300 Thermal finish, 1-1/2" thick		115	.278		15.20	14.40	3.72	33.32	43
2350 2" thick		105	.305		17.70	15.75	4.08	37.53	48

04 43 10.45 Granite	Crew	Daily Output	Labor-Hours	Unit	Material	2020 Bare Costs Labor	Equipment	Total	Total Incl O&P
0010 **GRANITE**, cut to size									
0050 Veneer, polished face, 3/4" to 1-1/2" thick									
0150 Low price, gray, light gray, etc.	D-10	130	.246	S.F.	27	12.75	3.29	43.04	52.50
0180 Medium price, pink, brown, etc.		130	.246		30.50	12.75	3.29	46.54	56.50
0220 High price, red, black, etc.		130	.246		42.50	12.75	3.29	58.54	70
0300 1-1/2" to 2-1/2" thick, veneer									
0350 Low price, gray, light gray, etc.	D-10	130	.246	S.F.	29	12.75	3.29	45.04	55
0500 Medium price, pink, brown, etc.		130	.246		34	12.75	3.29	50.04	60
0550 High price, red, black, etc.		130	.246		52.50	12.75	3.29	68.54	81
0700 2-1/2" to 4" thick, veneer									
0750 Low price, gray, light gray, etc.	D-10	110	.291	S.F.	39	15.05	3.89	57.94	70
0850 Medium price, pink, brown, etc.		110	.291		44.50	15.05	3.89	63.44	76
0950 High price, red, black, etc.		110	.291		64	15.05	3.89	82.94	97.50
1000 For bush hammered finish, deduct					5%				
1050 Coarse rubbed finish, deduct					10%				
1100 Honed finish, deduct					5%				
1150 Thermal finish, deduct					18%				
1800 Carving or bas-relief, from templates or plaster molds									
1850 Low price, gray, light gray, etc.	D-10	80	.400	C.F.	181	20.50	5.35	206.85	236
1875 Medium price, pink, brown, etc.		80	.400		355	20.50	5.35	380.85	425
1900 High price, red, black, etc.		80	.400		525	20.50	5.35	550.85	615
2000 Intricate or hand finished pieces									
2010 Mouldings, radius cuts, bullnose edges, etc.									
2050 Add for low price gray, light gray, etc.					30%				
2075 Add for medium price, pink, brown, etc.					165%				

For customer support on your Site Work & Landscape Costs with RSMeans data, call 800.448.8182.

105

04 43 10 – Masonry with Natural and Processed Stone

04 43 10.45 Granite		Crew	Daily Output	Labor-Hours	Unit	Material	2020 Bare Costs Labor	2020 Bare Costs Equipment	Total	Total Incl O&P
2100	Add for high price red, black, etc.					300%				
2450	For radius under 5', add				L.F.	100%				
2500	Steps, copings, etc., finished on more than one surface									
2550	Low price, gray, light gray, etc.	D-10	50	.640	C.F.	92	33	8.55	133.55	160
2575	Medium price, pink, brown, etc.		50	.640		120	33	8.55	161.55	191
2600	High price, red, black, etc.		50	.640		147	33	8.55	188.55	221
2700	Pavers, Belgian block, 8"-13" long, 4"-6" wide, 4"-6" deep	D-11	120	.200	S.F.	27	9.85		36.85	45
2800	Pavers, 4" x 4" x 4" blocks, split face and joints									
2850	Low price, gray, light gray, etc.	D-11	80	.300	S.F.	13.30	14.80		28.10	37
2875	Medium price, pink, brown, etc.		80	.300		21.50	14.80		36.30	46
2900	High price, red, black, etc.		80	.300		29.50	14.80		44.30	55
3000	Pavers, 4" x 4" x 4", thermal face, sawn joints									
3050	Low price, gray, light gray, etc.	D-11	65	.369	S.F.	24.50	18.20		42.70	54.50
3075	Medium price, pink, brown, etc.		65	.369		28.50	18.20		46.70	59
3100	High price, red, black, etc.		65	.369		32.50	18.20		50.70	63.50
4000	Soffits, 2" thick, low price, gray, light gray	D-13	35	1.371		38	69	12.25	119.25	159
4050	Medium price, pink, brown, etc.		35	1.371		65.50	69	12.25	146.75	189
4100	High price, red, black, etc.		35	1.371		92.50	69	12.25	173.75	219
4200	Low price, gray, light gray, etc.		35	1.371		63.50	69	12.25	144.75	187
4250	Medium price, pink, brown, etc.		35	1.371		92	69	12.25	173.25	218
4300	High price, red, black, etc.		35	1.371		120	69	12.25	201.25	249
5000	Reclaimed or antique									
5010	Treads, up to 12" wide	D-10	100	.320	L.F.	42	16.55	4.28	62.83	75.50
5020	Up to 18" wide		100	.320		45.50	16.55	4.28	66.33	79.50
5030	Capstone, size varies		50	.640		26.50	33	8.55	68.05	88.50
5040	Posts		30	1.067	V.L.F.	31.50	55	14.25	100.75	134

04 43 10.50 Lightweight Natural Stone

04 43 10.50 Lightweight Natural Stone		Crew	Daily Output	Labor-Hours	Unit	Material	2020 Bare Costs Labor	2020 Bare Costs Equipment	Total	Total Incl O&P
0011	**LIGHTWEIGHT NATURAL STONE** Lava type									
0100	Veneer, rubble face, sawed back, irregular shapes G	D-10	130	.246	S.F.	9.10	12.75	3.29	25.14	33
0200	Sawed face and back, irregular shapes G		130	.246	"	9.10	12.75	3.29	25.14	33
1000	Reclaimed or antique, barn or foundation stone		1	32	Ton	385	1,650	430	2,465	3,400

04 43 10.55 Limestone

04 43 10.55 Limestone		Crew	Daily Output	Labor-Hours	Unit	Material	2020 Bare Costs Labor	2020 Bare Costs Equipment	Total	Total Incl O&P
0010	**LIMESTONE**, cut to size									
0020	Veneer facing panels									
0500	Texture finish, light stick, 4-1/2" thick, 5' x 12'	D-4	300	.107	S.F.	22.50	5.05	.63	28.18	33.50
1400	Sugarcube, textured finish, 4-1/2" thick, 5' x 12'	D-10	275	.116		35.50	6	1.56	43.06	50
1450	5" thick, 5' x 14' panels		275	.116		35.50	6	1.56	43.06	50
2000	Coping, sugarcube finish, top & 2 sides		30	1.067	C.F.	66.50	55	14.25	135.75	173
2100	Sills, lintels, jambs, trim, stops, sugarcube finish, simple		20	1.600		66.50	83	21.50	171	222
2150	Detailed		20	1.600		66.50	83	21.50	171	222
2300	Steps, extra hard, 14" wide, 6" rise		50	.640	L.F.	29	33	8.55	70.55	91.50
3000	Quoins, plain finish, 6" x 12" x 12"	D-12	25	1.280	Ea.	44.50	60.50		105	141
3050	6" x 16" x 24"	"	25	1.280	"	59.50	60.50		120	158

04 43 10.60 Marble

04 43 10.60 Marble		Crew	Daily Output	Labor-Hours	Unit	Material	2020 Bare Costs Labor	2020 Bare Costs Equipment	Total	Total Incl O&P
0011	**MARBLE**, ashlar, split face, +/- 4" thick, random									
0040	Lengths 1' to 4' & heights 2" to 7-1/2", average	D-8	175	.229	S.F.	17.60	10.95		28.55	36
1000	Facing, polished finish, cut to size, 3/4" to 7/8" thick									
1050	Carrara or equal	D-10	130	.246	S.F.	20.50	12.75	3.29	36.54	46
1100	Arabescato or equal		130	.246		38	12.75	3.29	54.04	64.50
1300	1-1/4" thick, Botticino Classico or equal		125	.256		24.50	13.25	3.43	41.18	51
1350	Statuarietto or equal		125	.256		39	13.25	3.43	55.68	67
1500	2" thick, Crema Marfil or equal		120	.267		48.50	13.80	3.57	65.87	78.50

04 43 Stone Masonry

04 43 10 – Masonry with Natural and Processed Stone

04 43 10.60 Marble

		Crew	Daily Output	Labor-Hours	Unit	Material	2020 Bare Costs Labor	2020 Bare Costs Equipment	Total	Total Incl O&P
1550	Cafe Pinta or equal	D-10	120	.267	S.F.	65	13.80	3.57	82.37	96.50
1700	Rubbed finish, cut to size, 4" thick									
1740	Average	D-10	100	.320	S.F.	42.50	16.55	4.28	63.33	76
1780	Maximum	"	100	.320	"	71	16.55	4.28	91.83	108
2500	Flooring, polished tiles, 12" x 12" x 3/8" thick									
2510	Thin set, Giallo Solare or equal	D-11	90	.267	S.F.	15.65	13.15		28.80	37
2600	Sky Blue or equal		90	.267		15.05	13.15		28.20	36.50
2700	Mortar bed, Giallo Solare or equal		65	.369		15.75	18.20		33.95	45
2740	Sky Blue or equal		65	.369		15.05	18.20		33.25	44
3210	Stairs, risers, 7/8" thick x 6" high	D-10	115	.278	L.F.	15.15	14.40	3.72	33.27	43
3360	Treads, 12" wide x 1-1/4" thick	"	115	.278	"	50	14.40	3.72	68.12	81
3500	Thresholds, 3' long, 7/8" thick, 4" to 5" wide, plain	D-12	24	1.333	Ea.	36	63		99	136
3550	Beveled	"	24	1.333	"	76	63		139	180

04 43 10.75 Sandstone or Brownstone

		Crew	Daily Output	Labor-Hours	Unit	Material	2020 Bare Costs Labor	2020 Bare Costs Equipment	Total	Total Incl O&P
0011	**SANDSTONE OR BROWNSTONE**									
0100	Sawed face veneer, 2-1/2" thick, to 2' x 4' panels	D-10	130	.246	S.F.	22	12.75	3.29	38.04	47
0150	4" thick, to 3'-6" x 8' panels		100	.320		22	16.55	4.28	42.83	53.50
0300	Split face, random sizes		100	.320		14.50	16.55	4.28	35.33	45.50

04 43 10.80 Slate

		Crew	Daily Output	Labor-Hours	Unit	Material	2020 Bare Costs Labor	2020 Bare Costs Equipment	Total	Total Incl O&P
0010	**SLATE**									
0040	Pennsylvania - blue gray to black									
0050	Vermont - unfading green, mottled green & purple, gray & purple									
0100	Virginia - blue black									
0200	Exterior paving, natural cleft, 1" thick									
0250	6" x 6" Pennsylvania	D-12	100	.320	S.F.	7.05	15.15		22.20	31
0300	Vermont		100	.320		11.20	15.15		26.35	35.50
0350	Virginia		100	.320		14.70	15.15		29.85	39
0500	24" x 24", Pennsylvania		120	.267		13.55	12.65		26.20	34
0550	Vermont		120	.267		28	12.65		40.65	49.50
0600	Virginia		120	.267		21.50	12.65		34.15	42.50
0700	18" x 30" Pennsylvania		120	.267		15.35	12.65		28	36
0750	Vermont		120	.267		28	12.65		40.65	49.50
0800	Virginia		120	.267		19	12.65		31.65	40
1000	Interior flooring, natural cleft, 1/2" thick									
1100	6" x 6" Pennsylvania	D-12	100	.320	S.F.	4.16	15.15		19.31	27.50
1150	Vermont		100	.320		9.85	15.15		25	34
1200	Virginia		100	.320		11.60	15.15		26.75	36
1300	24" x 24" Pennsylvania		120	.267		8.10	12.65		20.75	28
1350	Vermont		120	.267		22.50	12.65		35.15	43.50
1400	Virginia		120	.267		15.35	12.65		28	36
1500	18" x 24" Pennsylvania		120	.267		8.10	12.65		20.75	28
1550	Vermont		120	.267		16.95	12.65		29.60	38
1600	Virginia		120	.267		15.55	12.65		28.20	36.50
2000	Facing panels, 1-1/4" thick, to 4' x 4' panels									
2100	Natural cleft finish, Pennsylvania	D-10	180	.178	S.F.	35	9.20	2.38	46.58	55
2110	Vermont		180	.178		28.50	9.20	2.38	40.08	47.50
2120	Virginia		180	.178		34.50	9.20	2.38	46.08	54.50
2150	Sand rubbed finish, surface, add					10.60			10.60	11.65
2200	Honed finish, add					7.65			7.65	8.45
2500	Ribbon, natural cleft finish, 1" thick, to 9 S.F.	D-10	80	.400		13.65	20.50	5.35	39.50	52.50
2550	Sand rubbed finish		80	.400		18.50	20.50	5.35	44.35	58
2600	Honed finish		80	.400		17.20	20.50	5.35	43.05	56.50

107

04 43 Stone Masonry

04 43 10 – Masonry with Natural and Processed Stone

04 43 10.80 Slate	Crew	Daily Output	Labor-Hours	Unit	Material	2020 Bare Costs Labor	Equipment	Total	Total Incl O&P
2700	1-1/2" thick	D-10	78	.410	S.F.	17.75	21	5.50	44.25	57.50
2750	Sand rubbed finish		78	.410		23.50	21	5.50	50	64
2800	Honed finish		78	.410		22	21	5.50	48.50	62.50
2850	2" thick		76	.421		21.50	22	5.65	49.15	62.50
2900	Sand rubbed finish		76	.421		29.50	22	5.65	57.15	71.50
2950	Honed finish	↓	76	.421		27	22	5.65	54.65	69
3100	Stair landings, 1" thick, black, clear	D-1	65	.246		21	11.55		32.55	40.50
3200	Ribbon	"	65	.246	↓	23	11.55		34.55	42.50
3500	Stair treads, sand finish, 1" thick x 12" wide									
3550	Under 3 L.F.	D-10	85	.376	L.F.	23	19.50	5.05	47.55	60
3600	3 L.F. to 6 L.F.	"	120	.267	"	25	13.80	3.57	42.37	52
3700	Ribbon, sand finish, 1" thick x 12" wide									
3750	To 6 L.F.	D-10	120	.267	L.F.	21	13.80	3.57	38.37	48
4000	Stools or sills, sand finish, 1" thick, 6" wide	D-12	160	.200		12	9.50		21.50	27.50
4100	Honed finish		160	.200		11.45	9.50		20.95	27
4200	10" wide		90	.356		18.50	16.85		35.35	46
4250	Honed finish		90	.356		17.20	16.85		34.05	44.50
4400	2" thick, 6" wide		140	.229		19.30	10.85		30.15	37.50
4450	Honed finish		140	.229		18.20	10.85		29.05	36.50
4600	10" wide		90	.356		30	16.85		46.85	58.50
4650	Honed finish	↓	90	.356		28.50	16.85		45.35	57
4800	For lengths over 3', add				↓	25%				

04 54 Refractory Brick Masonry

04 54 10 – Refractory Brick Work

04 54 10.10 Fire Brick

		Crew	Daily Output	Labor-Hours	Unit	Material	2020 Bare Costs Labor	Equipment	Total	Total Incl O&P
0010	**FIRE BRICK**									
0012	Low duty, 2000°F, 9" x 2-1/2" x 4-1/2"	D-1	.60	26.667	M	1,675	1,250		2,925	3,750
0050	High duty, 3000°F	"	.60	26.667	"	2,800	1,250		4,050	5,000

04 54 10.20 Fire Clay

		Crew	Daily Output	Labor-Hours	Unit	Material	2020 Bare Costs Labor	Equipment	Total	Total Incl O&P
0010	**FIRE CLAY**									
0020	Gray, high duty, 100 lb. bag				Bag	37			37	40.50
0050	100 lb. drum, premixed (400 brick per drum)				Drum	48.50			48.50	53.50

04 57 Masonry Fireplaces

04 57 10 – Brick or Stone Fireplaces

04 57 10.10 Fireplace

		Crew	Daily Output	Labor-Hours	Unit	Material	2020 Bare Costs Labor	Equipment	Total	Total Incl O&P
0010	**FIREPLACE**									
0100	Brick fireplace, not incl. foundations or chimneys									
0110	30" x 29" opening, incl. chamber, plain brickwork	D-1	.40	40	Ea.	625	1,875		2,500	3,525
0200	Fireplace box only (110 brick)	"	2	8	"	172	375		547	760
0300	For elaborate brickwork and details, add					35%	35%			
0400	For hearth, brick & stone, add	D-1	2	8	Ea.	215	375		590	805

04 72 Cast Stone Masonry

04 72 10 – Cast Stone Masonry Features

04 72 10.10 Coping

		Crew	Daily Output	Labor- Hours	Unit	Material	2020 Bare Costs Labor	Equipment	Total	Total Incl O&P
0010	**COPING**, stock units									
0050	Precast concrete, 10" wide, 4" tapers to 3-1/2", 8" wall	D-1	75	.213	L.F.	25.50	10		35.50	43
0100	12" wide, 3-1/2" tapers to 3", 10" wall		70	.229		27.50	10.70		38.20	46.50
0110	14" wide, 4" tapers to 3-1/2", 12" wall		65	.246		31.50	11.55		43.05	52
0150	16" wide, 4" tapers to 3-1/2", 14" wall		60	.267		34	12.50		46.50	56
0250	Precast concrete corners		40	.400	Ea.	44	18.75		62.75	76.50
0300	Limestone for 12" wall, 4" thick		90	.178	L.F.	16.80	8.35		25.15	31
0350	6" thick		80	.200		24	9.40		33.40	40
0500	Marble, to 4" thick, no wash, 9" wide		90	.178		12.30	8.35		20.65	26
0550	12" wide		80	.200		19.90	9.40		29.30	36
0700	Terra cotta, 9" wide		90	.178		8.15	8.35		16.50	21.50
0750	12" wide		80	.200		8.60	9.40		18	23.50
0800	Aluminum, for 12" wall		80	.200		9.45	9.40		18.85	24.50

04 72 20 – Cultured Stone Veneer

04 72 20.10 Cultured Stone Veneer Components

		Crew	Daily Output	Labor- Hours	Unit	Material	2020 Bare Costs Labor	Equipment	Total	Total Incl O&P
0010	**CULTURED STONE VENEER COMPONENTS**									
0110	On wood frame and sheathing substrate, random sized cobbles, corner stones	D-8	70	.571	V.L.F.	10.55	27.50		38.05	53
0120	Field stones		140	.286	S.F.	7.45	13.70		21.15	29
0130	Random sized flats, corner stones		70	.571	V.L.F.	10.55	27.50		38.05	53
0140	Field stones		140	.286	S.F.	8.85	13.70		22.55	30.50
0150	Horizontal lined ledgestones, corner stones		75	.533	V.L.F.	10.35	25.50		35.85	50.50
0160	Field stones		150	.267	S.F.	7.80	12.80		20.60	28
0170	Random shaped flats, corner stones		65	.615	V.L.F.	10.35	29.50		39.85	56
0180	Field stones		150	.267	S.F.	7.75	12.80		20.55	28
0190	Random shaped/textured face, corner stones		65	.615	V.L.F.	10.55	29.50		40.05	56
0200	Field stones		130	.308	S.F.	7.70	14.75		22.45	31
0210	Random shaped river rock, corner stones		65	.615	V.L.F.	10.55	29.50		40.05	56
0220	Field stones		130	.308	S.F.	7.70	14.75		22.45	31
0240	On concrete or CMU substrate, random sized cobbles, corner stones		70	.571	V.L.F.	9.70	27.50		37.20	52
0250	Field stones		140	.286	S.F.	7.05	13.70		20.75	29
0260	Random sized flats, corner stones		70	.571	V.L.F.	9.70	27.50		37.20	52
0270	Field stones		140	.286	S.F.	8.40	13.70		22.10	30.50
0280	Horizontal lined ledgestones, corner stones		75	.533	V.L.F.	9.50	25.50		35	49.50
0290	Field stones		150	.267	S.F.	7.35	12.80		20.15	27.50
0300	Random shaped flats, corner stones		70	.571	V.L.F.	9.50	27.50		37	52
0310	Field stones		140	.286	S.F.	7.35	13.70		21.05	29
0320	Random shaped/textured face, corner stones		65	.615	V.L.F.	9.70	29.50		39.20	55
0330	Field stones		130	.308	S.F.	7.30	14.75		22.05	30.50
0340	Random shaped river rock, corner stones		65	.615	V.L.F.	9.70	29.50		39.20	55
0350	Field stones		130	.308	S.F.	7.30	14.75		22.05	30.50
0360	Cultured stone veneer, #15 felt weather resistant barrier	1 Clab	3700	.002	Sq.	5.25	.09		5.34	5.95
0370	Expanded metal lath, diamond, 2.5 lb./S.Y., galvanized	1 Lath	85	.094	S.Y.	3.80	4.90		8.70	11.40
0390	Water table or window sill, 18" long	1 Bric	80	.100	Ea.	9.95	5.20		15.15	18.85

For customer support on your Site Work & Landscape Costs with RSMeans data, call 800.448.8182.

109

Division Notes

	CREW	DAILY OUTPUT	LABOR-HOURS	UNIT	BARE COSTS				TOTAL INCL O&P
					MAT.	LABOR	EQUIP.	TOTAL	

Estimating Tips

05 05 00 Common Work Results for Metals

- Nuts, bolts, washers, connection angles, and plates can add a significant amount to both the tonnage of a structural steel job and the estimated cost. As a rule of thumb, add 10% to the total weight to account for these accessories.

- Type 2 steel construction, commonly referred to as "simple construction," consists generally of field-bolted connections with lateral bracing supplied by other elements of the building, such as masonry walls or x-bracing. The estimator should be aware, however, that shop connections may be accomplished by welding or bolting. The method may be particular to the fabrication shop and may have an impact on the estimated cost.

05 10 00 Structural Steel

- Steel items can be obtained from two sources: a fabrication shop or a metals service center. Fabrication shops can fabricate items under more controlled conditions than crews in the field can. They are also more efficient and can produce items more economically. Metal service centers serve as a source of long mill shapes to both fabrication shops and contractors.

- Most line items in this structural steel subdivision, and most items in 05 50 00 Metal Fabrications, are indicated as being shop fabricated. The bare material cost for these shop fabricated items is the "Invoice Cost" from the shop and includes the mill base price of steel plus mill extras, transportation to the shop, shop drawings and detailing where warranted, shop fabrication and handling, sandblasting and a shop coat of primer paint, all necessary structural bolts, and delivery to the job site. The bare labor cost and bare equipment cost for these shop fabricated items are for field installation or erection.

- Line items in Subdivision 05 12 23.40 Lightweight Framing, and other items scattered in Division 5, are indicated as being field fabricated. The bare material cost for these field fabricated items is the "Invoice Cost" from the metals service center and includes the mill base price of steel plus mill extras, transportation to the metals service center, material handling, and delivery of long lengths of mill shapes to the job site. Material costs for structural bolts and welding rods should be added to the estimate. The bare labor cost and bare equipment cost for these items are for both field fabrication and field installation or erection, and include time for cutting, welding, and drilling in the fabricated metal items. Drilling into concrete and fasteners to fasten field fabricated items to other work is not included and should be added to the estimate.

05 20 00 Steel Joist Framing

- In any given project the total weight of open web steel joists is determined by the loads to be supported and the design. However, economies can be realized in minimizing the amount of labor used to place the joists. This is done by maximizing the joist spacing and therefore minimizing the number of joists required to be installed on the job. Certain spacings and locations may be required by the design, but in other cases maximizing the spacing and keeping it as uniform as possible will keep the costs down.

05 30 00 Steel Decking

- The takeoff and estimating of a metal deck involve more than the area of the floor or roof and the type of deck specified or shown on the drawings. Many different sizes and types of openings may exist. Small openings for individual pipes or conduits may be drilled after the floor/roof is installed, but larger openings may require special deck lengths as well as reinforcing or structural support. The estimator should determine who will be supplying this reinforcing. Additionally, some deck terminations are part of the deck package, such as screed angles and pour stops, and others will be part of the steel contract, such as angles attached to structural members and cast-in-place angles and plates. The estimator must ensure that all pieces are accounted for in the complete estimate.

05 50 00 Metal Fabrications

- The most economical steel stairs are those that use common materials, standard details, and most importantly, a uniform and relatively simple method of field assembly. Commonly available A36/A992 channels and plates are very good choices for the main stringers of the stairs, as are angles and tees for the carrier members. Risers and treads are usually made by specialty shops, and it is most economical to use a typical detail in as many places as possible. The stairs should be pre-assembled and shipped directly to the site. The field connections should be simple and straightforward enough to be accomplished efficiently, and with minimum equipment and labor.

Reference Numbers

Reference numbers are shown at the beginning of some major classifications. These numbers refer to related items in the Reference Section. The reference information may be an estimating procedure, an alternate pricing method, or technical information.

Note: Not all subdivisions listed here necessarily appear. ∎

Same Data. Simplified.

Enjoy the convenience and efficiency of accessing your costs anywhere:

- **Skip the multiplier** by setting your location
- **Quickly search**, edit, favorite and share costs
- **Stay on top of price changes** with automatic updates

Discover more at rsmeans.com/online

---

05 01 Maintenance of Metals

05 01 10 – Maintenance of Structural Metal Framing

05 01 10.51 Cleaning of Structural Metal Framing	Crew	Daily Output	Labor-Hours	Unit	Material	2020 Bare Costs Labor	Equipment	Total	Total Incl O&P
0010 **CLEANING OF STRUCTURAL METAL FRAMING**									
6125 Steel surface treatments, PDCA guidelines									
6170 Wire brush, hand (SSPC-SP2)	1 Psst	400	.020	S.F.	.02	.92		.94	1.51
6180 Power tool (SSPC-SP3)	"	700	.011		.07	.52		.59	.93
6215 Pressure washing, up to 5,000 psi, 5,000-15,000 S.F./day	1 Pord	10000	.001			.04		.04	.05
6220 Steam cleaning, 600 psi @ 300 degree F, 1,250-2,500 S.F./day		2000	.004			.18		.18	.26
6225 Water blasting, up to 25,000 psi, 1,750-3,500 S.F./day		2500	.003			.14		.14	.21
6230 Brush-off blast (SSPC-SP7)	E-11	1750	.018		.17	.85	.17	1.19	1.71
6235 Com'l blast (SSPC-SP6), loose scale, fine pwder rust, 2.0#/S.F. sand		1200	.027		.34	1.25	.25	1.84	2.60
6240 Tight mill scale, little/no rust, 3.0#/S.F. sand		1000	.032		.51	1.49	.30	2.30	3.23
6245 Exist coat blistered/pitted, 4.0#/S.F. sand		875	.037		.69	1.71	.34	2.74	3.79
6250 Exist coat badly pitted/nodules, 6.7#/S.F. sand		825	.039		1.15	1.81	.36	3.32	4.49
6255 Near white blast (SSPC-SP10), loose scale, fine rust, 5.6#/S.F. sand		450	.071		.96	3.32	.67	4.95	7
6260 Tight mill scale, little/no rust, 6.9#/S.F. sand		325	.098		1.18	4.60	.93	6.71	9.45
6265 Exist coat blistered/pitted, 9.0#/S.F. sand		225	.142		1.54	6.65	1.34	9.53	13.50
6270 Exist coat badly pitted/nodules, 11.3#/S.F. sand		150	.213		1.94	9.95	2.01	13.90	19.90

05 05 Common Work Results for Metals

05 05 05 – Selective Demolition for Metals

05 05 05.10 Selective Demolition, Metals

	Crew	Daily Output	Labor-Hours	Unit	Material	2020 Bare Costs Labor	Equipment	Total	Total Incl O&P
0010 **SELECTIVE DEMOLITION, METALS** R024119-10									
0015 Excludes shores, bracing, cutting, loading, hauling, dumping									
0020 Remove nuts only up to 3/4" diameter	1 Sswk	480	.017	Ea.		.96		.96	1.51
0030 7/8" to 1-1/4" diameter		240	.033			1.92		1.92	3.01
0040 1-3/8" to 2" diameter		160	.050			2.88		2.88	4.52
0060 Unbolt and remove structural bolts up to 3/4" diameter		240	.033			1.92		1.92	3.01
0070 7/8" to 2" diameter		160	.050			2.88		2.88	4.52
0140 Light weight framing members, remove whole or cut up, up to 20 lb.		240	.033			1.92		1.92	3.01
0150 21-40 lb.	2 Sswk	210	.076			4.39		4.39	6.90
0160 41-80 lb.	3 Sswk	180	.133			7.70		7.70	12.05
0170 81-120 lb.	4 Sswk	150	.213			12.30		12.30	19.30
0230 Structural members, remove whole or cut up, up to 500 lb.	E-19	48	.500			28.50	31	59.50	78
0240 1/4-2 tons	E-18	36	1.111			64.50	41.50	106	146
0250 2-5 tons	E-24	30	1.067			61.50	19.35	80.85	117
0260 5-10 tons	E-20	24	2.667			153	84.50	237.50	330
0270 10-15 tons	E-2	18	3.111			178	94	272	380
0340 Fabricated item, remove whole or cut up, up to 20 lb.	1 Sswk	96	.083			4.80		4.80	7.55
0350 21-40 lb.	2 Sswk	84	.190			11		11	17.25
0360 41-80 lb.	3 Sswk	72	.333			19.20		19.20	30
0370 81-120 lb.	4 Sswk	60	.533			31		31	48
0380 121-500 lb.	E-19	48	.500			28.50	31	59.50	78
0390 501-1000 lb.	"	36	.667			38	41.50	79.50	104
0500 Steel roof decking, uncovered, bare	B-2	5000	.008	S.F.		.34		.34	.51

05 05 13 – Shop-Applied Coatings for Metal

05 05 13.50 Paints and Protective Coatings

	Crew	Daily Output	Labor-Hours	Unit	Material	2020 Bare Costs Labor	Equipment	Total	Total Incl O&P
0010 **PAINTS AND PROTECTIVE COATINGS**									
5900 Galvanizing structural steel in shop, under 1 ton				Ton	580			580	640
5950 1 ton to 20 tons					570			570	625
6000 Over 20 tons					535			535	590

05 05 19 – Post-Installed Concrete Anchors

		Crew	Daily Output	Labor-Hours	Unit	Material	2020 Bare Costs Labor	Equipment	Total	Total Incl O&P
05 05 19.10	**Chemical Anchors**									
0010	**CHEMICAL ANCHORS**									
0020	Includes layout & drilling									
1430	Chemical anchor, w/rod & epoxy cartridge, 3/4" diameter x 9-1/2" long	B-89A	27	.593	Ea.	10.15	28.50	4.23	42.88	59.50
1435	1" diameter x 11-3/4" long		24	.667		20.50	32.50	4.76	57.76	76.50
1440	1-1/4" diameter x 14" long		21	.762		39	37	5.45	81.45	105
1445	1-3/4" diameter x 15" long		20	.800		68.50	39	5.70	113.20	140
1450	18" long		17	.941		82.50	45.50	6.70	134.70	167
1455	2" diameter x 18" long		16	1		113	48.50	7.15	168.65	206
1460	24" long		15	1.067		148	51.50	7.60	207.10	249
05 05 19.20	**Expansion Anchors**									
0010	**EXPANSION ANCHORS**									
0100	Anchors for concrete, brick or stone, no layout and drilling									
0200	Expansion shields, zinc, 1/4" diameter, 1-5/16" long, single	G 1 Carp	90	.089	Ea.	.46	4.72		5.18	7.60
0300	1-3/8" long, double	G	85	.094		.66	5		5.66	8.25
0600	1/2" diameter, 2-1/16" long, single	G	80	.100		1.28	5.30		6.58	9.40
1000	3/4" diameter, 2-3/4" long, single	G	70	.114		3.66	6.05		9.71	13.20
1100	3-15/16" long, double	G	65	.123		5.65	6.55		12.20	16.10
3000	Toggle bolts, bright steel, 1/8" diameter, 2" long	G	85	.094		.21	5		5.21	7.75
3400	1/4" diameter, 3" long	G	75	.107		.44	5.65		6.09	9
3600	3/8" diameter, 3" long	G	70	.114		.88	6.05		6.93	10.10
5000	Screw anchors for concrete, masonry,									
5100	stone & tile, no layout or drilling included									
6600	Lead, #6 & #8, 3/4" long	G 1 Carp	260	.031	Ea.	.18	1.64		1.82	2.66
6700	#10 - #14, 1-1/2" long	G	200	.040		.48	2.13		2.61	3.72
6800	#16 & #18, 1-1/2" long	G	160	.050		.42	2.66		3.08	4.45
6900	Plastic, #6 & #8, 3/4" long	G	260	.031		.05	1.64		1.69	2.52
8000	Wedge anchors, not including layout or drilling									
8050	Carbon steel, 1/4" diameter, 1-3/4" long	G 1 Carp	150	.053	Ea.	.62	2.83		3.45	4.94
8100	3-1/4" long	G	140	.057		.82	3.04		3.86	5.45
8150	3/8" diameter, 2-1/4" long	G	145	.055		.44	2.93		3.37	4.90
8200	5" long	G	140	.057		.78	3.04		3.82	5.40
8250	1/2" diameter, 2-3/4" long	G	140	.057		.99	3.04		4.03	5.65
8300	7" long	G	125	.064		1.70	3.40		5.10	6.95
8350	5/8" diameter, 3-1/2" long	G	130	.062		2.05	3.27		5.32	7.15
8400	8-1/2" long	G	115	.070		4.36	3.70		8.06	10.35
8450	3/4" diameter, 4-1/4" long	G	115	.070		2.97	3.70		6.67	8.80
8500	10" long	G	95	.084		6.75	4.48		11.23	14.15
8550	1" diameter, 6" long	G	100	.080		8.80	4.25		13.05	16.05
8575	9" long	G	85	.094		11.40	5		16.40	20
8600	12" long	G	75	.107		12.30	5.65		17.95	22
8650	1-1/4" diameter, 9" long	G	70	.114		23.50	6.05		29.55	34.50
8700	12" long	G	60	.133		30	7.10		37.10	43.50
8750	For type 303 stainless steel, add					350%				
8800	For type 316 stainless steel, add					450%				
8950	Self-drilling concrete screw, hex washer head, 3/16" diam. x 1-3/4" long	G 1 Carp	300	.027	Ea.	.18	1.42		1.60	2.33
8960	2-1/4" long	G	250	.032		.24	1.70		1.94	2.82
8970	Phillips flat head, 3/16" diam. x 1-3/4" long	G	300	.027		.19	1.42		1.61	2.34
8980	2-1/4" long	G	250	.032		.23	1.70		1.93	2.81

113

05 05 21 – Fastening Methods for Metal

05 05 21.10 Cutting Steel

		Crew	Daily Output	Labor-Hours	Unit	Material	2020 Bare Costs Labor	2020 Bare Costs Equipment	Total	Total Incl O&P
0010	**CUTTING STEEL**									
0020	Hand burning, incl. preparation, torch cutting & grinding, no staging									
0050	Steel to 1/4" thick	E-25	400	.020	L.F.	.21	1.19	.03	1.43	2.14
0100	1/2" thick		320	.025		.39	1.49	.04	1.92	2.80
0150	3/4" thick		260	.031		.63	1.83	.05	2.51	3.61
0200	1" thick		200	.040		.90	2.38	.06	3.34	4.79

05 05 21.15 Drilling Steel

		Crew	Daily Output	Labor-Hours	Unit	Material	2020 Bare Costs Labor	2020 Bare Costs Equipment	Total	Total Incl O&P
0010	**DRILLING STEEL**									
1910	Drilling & layout for steel, up to 1/4" deep, no anchor									
1920	Holes, 1/4" diameter	1 Sswk	112	.071	Ea.	.10	4.12		4.22	6.55
1925	For each additional 1/4" depth, add		336	.024		.10	1.37		1.47	2.26
1930	3/8" diameter		104	.077		.08	4.43		4.51	7.05
1935	For each additional 1/4" depth, add		312	.026		.08	1.48		1.56	2.41
1940	1/2" diameter		96	.083		.08	4.80		4.88	7.65
1945	For each additional 1/4" depth, add		288	.028		.08	1.60		1.68	2.60
1950	5/8" diameter		88	.091		.16	5.25		5.41	8.35
1955	For each additional 1/4" depth, add		264	.030		.16	1.75		1.91	2.91
1960	3/4" diameter		80	.100		.17	5.75		5.92	9.25
1965	For each additional 1/4" depth, add		240	.033		.17	1.92		2.09	3.19
1970	7/8" diameter		72	.111		.24	6.40		6.64	10.30
1975	For each additional 1/4" depth, add		216	.037		.24	2.14		2.38	3.62
1980	1" diameter		64	.125		.26	7.20		7.46	11.60
1985	For each additional 1/4" depth, add		192	.042		.26	2.40		2.66	4.05
1990	For drilling up, add						40%			

05 05 21.90 Welding Steel

		Crew	Daily Output	Labor-Hours	Unit	Material	2020 Bare Costs Labor	2020 Bare Costs Equipment	Total	Total Incl O&P
0010	**WELDING STEEL**, Structural R050521-20									
0020	Field welding, 1/8" E6011, cost per welder, no operating engineer	E-14	8	1	Hr.	5.55	59.50	18.40	83.45	120
0200	With 1/2 operating engineer	E-13	8	1.500		5.55	86	18.40	109.95	159
0300	With 1 operating engineer	E-12	8	2		5.55	112	18.40	135.95	198
0500	With no operating engineer, 2# weld rod per ton	E-14	8	1	Ton	5.55	59.50	18.40	83.45	120
0600	8# E6011 per ton	"	2	4		22	238	73.50	333.50	480
0800	With one operating engineer per welder, 2# E6011 per ton	E-12	8	2		5.55	112	18.40	135.95	198
0900	8# E6011 per ton	"	2	8		22	450	73.50	545.50	795
1200	Continuous fillet, down welding									
1300	Single pass, 1/8" thick, 0.1#/L.F.	E-14	150	.053	L.F.	.28	3.17	.98	4.43	6.35
1400	3/16" thick, 0.2#/L.F.		75	.107		.55	6.35	1.96	8.86	12.70
1500	1/4" thick, 0.3#/L.F.		50	.160		.83	9.50	2.94	13.27	19.05
1610	5/16" thick, 0.4#/L.F.		38	.211		1.11	12.50	3.87	17.48	25
1800	3 passes, 3/8" thick, 0.5#/L.F.		30	.267		1.38	15.85	4.90	22.13	32
2010	4 passes, 1/2" thick, 0.7#/L.F.		22	.364		1.93	21.50	6.70	30.13	43.50
2600	For vertical joint welding, add						20%			
2700	Overhead joint welding, add						300%			
2900	For semi-automatic welding, obstructed joints, deduct						5%			
3000	Exposed joints, deduct						15%			
4000	Cleaning and welding plates, bars, or rods									
4010	to existing beams, columns, or trusses	E-14	12	.667	L.F.	1.38	39.50	12.25	53.13	77

05 05 23 – Metal Fastenings

05 05 23.10 Bolts and Hex Nuts

			Crew	Daily Output	Labor-Hours	Unit	Material	2020 Bare Costs Labor	2020 Bare Costs Equipment	Total	Total Incl O&P
0010	**BOLTS & HEX NUTS**, Steel, A307										
0100	1/4" diameter, 1/2" long	G	1 Sswk	140	.057	Ea.	.06	3.29		3.35	5.20
0200	1" long	G		140	.057		.07	3.29		3.36	5.25

05 05 Common Work Results for Metals

05 05 23 – Metal Fastenings

		Crew	Daily Output	Labor-Hours	Unit	Material	2020 Bare Costs Labor	Equipment	Total	Total Incl O&P
05 05 23.10 Bolts and Hex Nuts										
0400	3" long	G 1 Sswk	130	.062	Ea.	.15	3.55		3.70	5.70
0600	3/8" diameter, 1" long	G	130	.062		.14	3.55		3.69	5.70
0800	3" long	G	120	.067		.23	3.84		4.07	6.30
1100	1/2" diameter, 1-1/2" long	G	120	.067		.41	3.84		4.25	6.50
1400	6" long	G	110	.073		1.08	4.19		5.27	7.80
1600	5/8" diameter, 1-1/2" long	G	120	.067		.88	3.84		4.72	7
2000	8" long	G	105	.076		2.56	4.39		6.95	9.70
2200	3/4" diameter, 2" long	G	120	.067		1.22	3.84		5.06	7.40
2600	10" long	G	85	.094		4.62	5.45		10.07	13.60
2800	1" diameter, 3" long	G	105	.076		3.02	4.39		7.41	10.20
3000	12" long	G	75	.107		8.20	6.15		14.35	18.70
3100	For galvanized, add					75%				
3200	For stainless, add					350%				
05 05 23.30 Lag Screws										
0010	**LAG SCREWS**									
0020	Steel, 1/4" diameter, 2" long	G 1 Carp	200	.040	Ea.	.10	2.13		2.23	3.30
0100	3/8" diameter, 3" long	G	150	.053		.31	2.83		3.14	4.60
0200	1/2" diameter, 3" long	G	130	.062		.72	3.27		3.99	5.70
0300	5/8" diameter, 3" long	G	120	.067		1.26	3.54		4.80	6.70
05 05 23.35 Machine Screws										
0010	**MACHINE SCREWS**									
0020	Steel, round head, #8 x 1" long	G 1 Carp	4.80	1.667	C	3.98	88.50		92.48	137
0110	#8 x 2" long	G	2.40	3.333		8.80	177		185.80	276
0200	#10 x 1" long	G	4	2		4.89	106		110.89	165
0300	#10 x 2" long	G	2	4		9.80	213		222.80	330
05 05 23.50 Powder Actuated Tools and Fasteners										
0010	**POWDER ACTUATED TOOLS & FASTENERS**									
0020	Stud driver, .22 caliber, single shot				Ea.	159			159	175
0100	.27 caliber, semi automatic, strip				"	460			460	505
0300	Powder load, single shot, .22 cal, power level 2, brown				C	6.25			6.25	6.85
0400	Strip, .27 cal, power level 4, red					10.30			10.30	11.35
0600	Drive pin, .300 x 3/4" long	G 1 Carp	4.80	1.667		4.48	88.50		92.98	138
0700	.300 x 3" long with washer	G "	4	2		11.95	106		117.95	173
05 05 23.80 Vibration and Bearing Pads										
0010	**VIBRATION & BEARING PADS**									
0300	Laminated synthetic rubber impregnated cotton duck, 1/2" thick	2 Sswk	24	.667	S.F.	74.50	38.50		113	143
0400	1" thick		20	.800		156	46		202	245
0600	Neoprene bearing pads, 1/2" thick		24	.667		30	38.50		68.50	93.50
0700	1" thick		20	.800		60.50	46		106.50	139
0900	Fabric reinforced neoprene, 5000 psi, 1/2" thick		24	.667		12.30	38.50		50.80	74
1000	1" thick		20	.800		24.50	46		70.50	99.50
1200	Felt surfaced vinyl pads, cork and sisal, 5/8" thick		24	.667		6.35	38.50		44.85	67.50
1300	1" thick		20	.800		11.40	46		57.40	85
1500	Teflon bonded to 10 ga. carbon steel, 1/32" layer		24	.667		59	38.50		97.50	126
1600	3/32" layer		24	.667		88.50	38.50		127	158
1800	Bonded to 10 ga. stainless steel, 1/32" layer		24	.667		105	38.50		143.50	176
1900	3/32" layer		24	.667		124	38.50		162.50	198
2100	Circular machine leveling pad & stud				Kip	7.55			7.55	8.30

For customer support on your Site Work & Landscape Costs with RSMeans data, call 800.448.8182.

115

05 05 Common Work Results for Metals

05 05 23 – Metal Fastenings

05 05 23.90 Welding Rod		Crew	Daily Output	Labor-Hours	Unit	Material	2020 Bare Costs Labor	2020 Bare Costs Equipment	Total	Total Incl O&P
0010	**WELDING ROD**									
0020	Steel, type 6011, 1/8" diam., less than 500#				Lb.	2.76			2.76	3.04
0100	500# to 2,000#					2.49			2.49	2.74
0200	2,000# to 5,000#					2.34			2.34	2.57
0300	5/32" diam., less than 500#					2.46			2.46	2.71
0310	500# to 2,000#					2.22			2.22	2.44
0320	2,000# to 5,000#					2.09			2.09	2.30
0400	3/16" diam., less than 500#					2.75			2.75	3.03
0500	500# to 2,000#					2.48			2.48	2.73
0600	2,000# to 5,000#					2.33			2.33	2.56
0620	Steel, type 6010, 1/8" diam., less than 500#					2.49			2.49	2.74
0630	500# to 2,000#					2.24			2.24	2.46
0640	2,000# to 5,000#					2.11			2.11	2.32
0650	Steel, type 7018 Low Hydrogen, 1/8" diam., less than 500#					2.94			2.94	3.24
0660	500# to 2,000#					2.65			2.65	2.92
0670	2,000# to 5,000#					2.49			2.49	2.74
0700	Steel, type 7024 Jet Weld, 1/8" diam., less than 500#					2.55			2.55	2.81
0710	500# to 2,000#					2.30			2.30	2.53
0720	2,000# to 5,000#					2.16			2.16	2.38
1550	Aluminum, type 4043 TIG, 1/8" diam., less than 10#					5.20			5.20	5.75
1560	10# to 60#					4.70			4.70	5.15
1570	Over 60#					4.42			4.42	4.86
1600	Aluminum, type 5356 TIG, 1/8" diam., less than 10#					5.55			5.55	6.10
1610	10# to 60#					4.99			4.99	5.50
1620	Over 60#					4.69			4.69	5.15
1900	Cast iron, type 8 Nickel, 1/8" diam., less than 500#					22.50			22.50	24.50
1910	500# to 1,000#					20			20	22
1920	Over 1,000#					18.90			18.90	21
2000	Stainless steel, type 316/316L, 1/8" diam., less than 500#					7.15			7.15	7.85
2100	500# to 1,000#					6.45			6.45	7.10
2220	Over 1,000#					6.05			6.05	6.65

05 12 Structural Steel Framing

05 12 23 – Structural Steel for Buildings

05 12 23.15 Columns, Lightweight

05 12 23.15 Columns, Lightweight		Crew	Daily Output	Labor-Hours	Unit	Material	2020 Bare Costs Labor	2020 Bare Costs Equipment	Total	Total Incl O&P
0010	**COLUMNS, LIGHTWEIGHT**									
1000	Lightweight units (lally), 3-1/2" diameter	E-2	780	.072	L.F.	6.30	4.10	2.17	12.57	15.70
1050	4" diameter	"	900	.062	"	9.40	3.56	1.88	14.84	17.85
5800	Adjustable jack post, 8' maximum height, 2-3/4" diameter [G]				Ea.	52			52	57
5850	4" diameter [G]				"	83			83	91.50

05 12 23.17 Columns, Structural

05 12 23.17 Columns, Structural		Crew	Daily Output	Labor-Hours	Unit	Material	2020 Bare Costs Labor	2020 Bare Costs Equipment	Total	Total Incl O&P
0010	**COLUMNS, STRUCTURAL** R051223-10									
0015	Made from recycled materials									
0020	Shop fab'd for 100-ton, 1-2 story project, bolted connections									
0800	Steel, concrete filled, extra strong pipe, 3-1/2" diameter	E-2	660	.085	L.F.	47.50	4.85	2.57	54.92	62.50
0830	4" diameter		780	.072		53	4.10	2.17	59.27	66.50
0890	5" diameter		1020	.055		63	3.14	1.66	67.80	75.50
0930	6" diameter		1200	.047		83.50	2.67	1.41	87.58	97.50
0940	8" diameter		1100	.051		83.50	2.91	1.54	87.95	98
1100	For galvanizing, add				Lb.	.29			.29	.31

05 12 23 – Structural Steel for Buildings

05 12 23.17 Columns, Structural

		Crew	Daily Output	Labor-Hours	Unit	Material	2020 Bare Costs Labor	Equipment	Total	Total Incl O&P
1500	Steel pipe, extra strong, no concrete, 3" to 5" diameter	G E-2	16000	.004	Lb.	1.44	.20	.11	1.75	2.01
1600	6" to 12" diameter	G	14000	.004		1.44	.23	.12	1.79	2.06
3300	Structural tubing, square, A500GrB, 4" to 6" square, light section	G	11270	.005		1.44	.28	.15	1.87	2.19
3600	Heavy section	G	32000	.002		1.44	.10	.05	1.59	1.79
5100	Structural tubing, rect., 5" to 6" wide, light section	G	8000	.007		1.44	.40	.21	2.05	2.43
5200	Heavy section	G	12000	.005		1.44	.27	.14	1.85	2.15
5300	7" to 10" wide, light section	G	15000	.004		1.44	.21	.11	1.76	2.03
5400	Heavy section	G	18000	.003		1.44	.18	.09	1.71	1.96
6800	W Shape, A992 steel, 2 tier, W8 x 24	G	1080	.052	L.F.	38	2.96	1.57	42.53	48
6900	W8 x 48	G	1032	.054		76	3.10	1.64	80.74	90
7050	W10 x 68	G	984	.057		107	3.25	1.72	111.97	125
7100	W10 x 112	G	960	.058		177	3.33	1.77	182.10	202
7200	W12 x 87	G	984	.057		138	3.25	1.72	142.97	158
7250	W12 x 120	G	960	.058		190	3.33	1.77	195.10	216
7350	W14 x 74	G	984	.057		117	3.25	1.72	121.97	136
7450	W14 x 176	G	912	.061		278	3.51	1.86	283.37	315
8090	For projects 75 to 99 tons, add				%	10%				
8092	50 to 74 tons, add					20%				
8094	25 to 49 tons, add					30%	10%			
8096	10 to 24 tons, add					50%	25%			
8098	2 to 9 tons, add					75%	50%			
8099	Less than 2 tons, add					100%	100%			

05 12 23.18 Corner Guards

		Crew	Daily Output	Labor-Hours	Unit	Material	2020 Bare Costs Labor	Equipment	Total	Total Incl O&P
0010	**CORNER GUARDS**									
0020	Steel angle w/anchors, 1" x 1" x 1/4", 1.5#/L.F.	2 Carp	160	.100	L.F.	7.95	5.30		13.25	16.70
0100	2" x 2" x 1/4" angles, 3.2#/L.F.		150	.107		10.60	5.65		16.25	20
0200	3" x 3" x 5/16" angles, 6.1#/L.F.		140	.114		17.95	6.05		24	29
0300	4" x 4" x 5/16" angles, 8.2#/L.F.		120	.133		18.10	7.10		25.20	30.50
0350	For angles drilled and anchored to masonry, add					15%	120%			
0370	Drilled and anchored to concrete, add					20%	170%			
0400	For galvanized angles, add					35%				
0450	For stainless steel angles, add					100%				

05 12 23.20 Curb Edging

		Crew	Daily Output	Labor-Hours	Unit	Material	2020 Bare Costs Labor	Equipment	Total	Total Incl O&P
0010	**CURB EDGING**									
0020	Steel angle w/anchors, shop fabricated, on forms, 1" x 1", 0.8#/L.F.	G E-4	350	.091	L.F.	1.74	5.30	.42	7.46	10.70
0100	2" x 2" angles, 3.92#/L.F.	G	330	.097		7.10	5.65	.45	13.20	17.20
0200	3" x 3" angles, 6.1#/L.F.	G	300	.107		11.70	6.20	.49	18.39	23
0300	4" x 4" angles, 8.2#/L.F.	G	275	.116		15.30	6.75	.53	22.58	28
1000	6" x 4" angles, 12.3#/L.F.	G	250	.128		22.50	7.45	.59	30.54	37
1050	Steel channels with anchors, on forms, 3" channel, 5#/L.F.	G	290	.110		9	6.40	.51	15.91	20.50
1100	4" channel, 5.4#/L.F.	G	270	.119		9.65	6.90	.54	17.09	22
1200	6" channel, 8.2#/L.F.	G	255	.125		15.30	7.30	.58	23.18	29
1300	8" channel, 11.5#/L.F.	G	225	.142		21	8.25	.65	29.90	36.50
1400	10" channel, 15.3#/L.F.	G	180	.178		27.50	10.35	.82	38.67	47.50
1500	12" channel, 20.7#/L.F.	G	140	.229		37	13.30	1.05	51.35	62.50
2000	For curved edging, add					35%	10%			

05 12 23.40 Lightweight Framing

		Crew	Daily Output	Labor-Hours	Unit	Material	2020 Bare Costs Labor	Equipment	Total	Total Incl O&P
0010	**LIGHTWEIGHT FRAMING** R051223-35									
0015	Made from recycled materials									
0400	Angle framing, field fabricated, 4" and larger	G E-3	440	.055	Lb.	.83	3.18	.33	4.34	6.25
0450	Less than 4" angles R051223-45	G	265	.091		.86	5.30	.56	6.72	9.85
0600	Channel framing, field fabricated, 8" and larger	G	500	.048		.86	2.80	.29	3.95	5.65

117

For customer support on your Site Work & Landscape Costs with RSMeans data, call 800.448.8182.

05 12 23.40 Lightweight Framing

			Crew	Daily Output	Labor-Hours	Unit	Material	2020 Bare Costs Labor	Equipment	Total	Total Incl O&P
0650	Less than 8" channels	G	E-3	335	.072	Lb.	.86	4.17	.44	5.47	8
1000	Continuous slotted channel framing system, shop fab, simple framing	G	2 Sswk	2400	.007		4.45	.38		4.83	5.50
1200	Complex framing	G	"	1600	.010		5.05	.58		5.63	6.45
1300	Cross bracing, rods, shop fabricated, 3/4" diameter	G	E-3	700	.034		1.72	2	.21	3.93	5.25
1310	7/8" diameter	G		850	.028		1.72	1.65	.17	3.54	4.67
1320	1" diameter	G		1000	.024		1.72	1.40	.15	3.27	4.25
1330	Angle, 5" x 5" x 3/8"	G		2800	.009		1.72	.50	.05	2.27	2.74
1350	Hanging lintels, shop fabricated	G	↓	850	.028		1.72	1.65	.17	3.54	4.67
1380	Roof frames, shop fabricated, 3'-0" square, 5' span	G	E-2	4200	.013		1.72	.76	.40	2.88	3.52
1400	Tie rod, not upset, 1-1/2" to 4" diameter, with turnbuckle	G	2 Sswk	800	.020		1.87	1.15		3.02	3.86
1420	No turnbuckle	G		700	.023		1.80	1.32		3.12	4.05
1500	Upset, 1-3/4" to 4" diameter, with turnbuckle	G		800	.020		1.87	1.15		3.02	3.86
1520	No turnbuckle	G	↓	700	.023	↓	1.80	1.32		3.12	4.05

05 12 23.45 Lintels

			Crew	Daily Output	Labor-Hours	Unit	Material	2020 Bare Costs Labor	Equipment	Total	Total Incl O&P
0010	**LINTELS**										
0015	Made from recycled materials										
0020	Plain steel angles, shop fabricated, under 500 lb.	G	1 Bric	550	.015	Lb.	1.11	.76		1.87	2.37
0100	500 to 1,000 lb.	G		640	.013		1.08	.65		1.73	2.18
0200	1,000 to 2,000 lb.	G		640	.013		1.05	.65		1.70	2.14
0300	2,000 to 4,000 lb.	G	↓	640	.013		1.02	.65		1.67	2.11
0500	For built-up angles and plates, add to above	G					1.44			1.44	1.58
0700	For engineering, add to above						.14			.14	.16
0900	For galvanizing, add to above, under 500 lb.						.32			.32	.35
0950	500 to 2,000 lb.						.29			.29	.32
1000	Over 2,000 lb.					↓	.29			.29	.31
2000	Steel angles, 3-1/2" x 3", 1/4" thick, 2'-6" long	G	1 Bric	47	.170	Ea.	15.50	8.85		24.35	30.50
2100	4'-6" long	G		26	.308		28	16		44	55
2600	4" x 3-1/2", 1/4" thick, 5'-0" long	G		21	.381		35.50	19.85		55.35	69
2700	9'-0" long	G	↓	12	.667	↓	64	34.50		98.50	123

05 12 23.60 Pipe Support Framing

			Crew	Daily Output	Labor-Hours	Unit	Material	2020 Bare Costs Labor	Equipment	Total	Total Incl O&P
0010	**PIPE SUPPORT FRAMING**										
0020	Under 10#/L.F., shop fabricated	G	E-4	3900	.008	Lb.	1.93	.48	.04	2.45	2.91
0200	10.1 to 15#/L.F.	G		4300	.007		1.90	.43	.03	2.36	2.81
0400	15.1 to 20#/L.F.	G		4800	.007		1.87	.39	.03	2.29	2.69
0600	Over 20#/L.F.	G	↓	5400	.006	↓	1.84	.34	.03	2.21	2.59

05 12 23.65 Plates

			Crew	Daily Output	Labor-Hours	Unit	Material	2020 Bare Costs Labor	Equipment	Total	Total Incl O&P
0010	**PLATES**	R051223-80									
0015	Made from recycled materials										
0020	For connections & stiffener plates, shop fabricated										
0050	1/8" thick (5.1 lb./S.F.)	G				S.F.	7.35			7.35	8.05
0100	1/4" thick (10.2 lb./S.F.)	G					14.65			14.65	16.10
0300	3/8" thick (15.3 lb./S.F.)	G					22			22	24
0400	1/2" thick (20.4 lb./S.F.)	G					29.50			29.50	32
0450	3/4" thick (30.6 lb./S.F.)	G					44			44	48.50
0500	1" thick (40.8 lb./S.F.)	G				↓	58.50			58.50	64.50
2000	Steel plate, warehouse prices, no shop fabrication										
2100	1/4" thick (10.2 lb./S.F.)	G				S.F.	7.20			7.20	7.90

05 12 Structural Steel Framing

05 12 23 – Structural Steel for Buildings

05 12 23.75 Structural Steel Members

		Crew	Daily Output	Labor-Hours	Unit	Material	2020 Bare Costs Labor	2020 Bare Costs Equipment	Total	Total Incl O&P
0010	**STRUCTURAL STEEL MEMBERS** R051223-10									
0015	Made from recycled materials									
0020	Shop fab'd for 100-ton, 1-2 story project, bolted connections									
0300	W 8 x 10 [G]	E-2	600	.093	L.F.	15.80	5.35	2.83	23.98	29
0500	x 31 [G]		550	.102		49	5.80	3.08	57.88	66.50
0600	W 10 x 12 [G]		600	.093		18.95	5.35	2.83	27.13	32.50
0700	x 22 [G]		600	.093		35	5.35	2.83	43.18	50
0900	x 49 [G]		550	.102		77.50	5.80	3.08	86.38	97.50
1100	W 12 x 16 [G]		880	.064		25.50	3.64	1.93	31.07	36
1700	x 72 [G]		640	.088		114	5	2.65	121.65	136
1900	W 14 x 26 [G]		990	.057		41	3.23	1.71	45.94	52
2500	x 120 [G]		720	.078		190	4.44	2.36	196.80	218
3300	W 18 x 35 [G]	E-5	960	.083		55.50	4.79	1.92	62.21	70.50
4100	W 21 x 44 [G]		1064	.075		69.50	4.32	1.73	75.55	85
4900	W 24 x 55 [G]		1110	.072		87	4.14	1.66	92.80	104
5900	x 94 [G]		1190	.067		149	3.86	1.55	154.41	171
6100	W 30 x 99 [G]		1200	.067		156	3.83	1.54	161.37	180
6700	W 33 x 118 [G]		1176	.068		187	3.91	1.57	192.48	213
7300	W 36 x 135 [G]		1170	.068		213	3.93	1.58	218.51	243
8490	For projects 75 to 99 tons, add					10%				
8492	50 to 74 tons, add					20%				
8494	25 to 49 tons, add					30%	10%			
8496	10 to 24 tons, add					50%	25%			
8498	2 to 9 tons, add					75%	50%			
8499	Less than 2 tons, add					100%	100%			

05 12 23.77 Structural Steel Projects

		Crew	Daily Output	Labor-Hours	Unit	Material	2020 Bare Costs Labor	2020 Bare Costs Equipment	Total	Total Incl O&P
0010	**STRUCTURAL STEEL PROJECTS** R050516-30									
0015	Made from recycled materials									
0020	Shop fab'd for 100-ton, 1-2 story project, bolted connections									
0200	Apartments, nursing homes, etc., 1 to 2 stories R050523-10 [G]	E-5	10.30	7.767	Ton	2,875	445	179	3,499	4,050
1300	Industrial bldgs., 1 story, beams & girders, steel bearing [G]		12.90	6.202		2,875	355	143	3,373	3,850
1400	Masonry bearing R051223-10 [G]		10	8		2,875	460	184	3,519	4,075
2800	Power stations, fossil fuels, simple connections [G]	E-6	11	11.636		2,875	670	195	3,740	4,425
2900	Moment/composite connections R051223-20 [G]	"	5.70	22.456		4,300	1,300	375	5,975	7,175
5390	For projects 75 to 99 tons, add					10%				
5392	50 to 74 tons, add R051223-25					20%				
5394	25 to 49 tons, add					30%	10%			
5396	10 to 24 tons, add R051223-30					50%	25%			
5398	2 to 9 tons, add					75%	50%			
5399	Less than 2 tons, add					100%	100%			

05 12 23.78 Structural Steel Secondary Members

		Crew	Daily Output	Labor-Hours	Unit	Material	2020 Bare Costs Labor	2020 Bare Costs Equipment	Total	Total Incl O&P
0010	**STRUCTURAL STEEL SECONDARY MEMBERS**									
0015	Made from recycled materials									
0020	Shop fabricated for 20-ton girt/purlin framing package, materials only									
0100	Girts/purlins, C/Z-shapes, includes clips and bolts									
0110	6" x 2-1/2" x 2-1/2", 16 ga., 3.0 lb./L.F.				L.F.	3.88			3.88	4.27
0115	14 ga., 3.5 lb./L.F.					4.53			4.53	4.98
0120	8" x 2-3/4" x 2-3/4", 16 ga., 3.4 lb./L.F.					4.40			4.40	4.84
0125	14 ga., 4.1 lb./L.F.					5.30			5.30	5.85
0130	12 ga., 5.6 lb./L.F.					7.25			7.25	7.95
0135	10" x 3-1/2" x 3-1/2", 14 ga., 4.7 lb./L.F.					6.10			6.10	6.70
0140	12 ga., 6.7 lb./L.F.					8.65			8.65	9.55

119

05 12 Structural Steel Framing

05 12 23 – Structural Steel for Buildings

05 12 23.78 Structural Steel Secondary Members	Crew	Daily Output	Labor-Hours	Unit	Material	2020 Bare Costs Labor	Equipment	Total	Total Incl O&P
0145	12" x 3-1/2" x 3-1/2", 14 ga., 5.3 lb./L.F.				L.F.	6.85			6.85	7.55
0150	12 ga., 7.4 lb./L.F.				↓	9.55			9.55	10.55
0200	Eave struts, C-shape, includes clips and bolts									
0210	6" x 4" x 3", 16 ga., 3.1 lb./L.F.				L.F.	4.01			4.01	4.41
0215	14 ga., 3.9 lb./L.F.					5.05			5.05	5.55
0220	8" x 4" x 3", 16 ga., 3.5 lb./L.F.					4.53			4.53	4.98
0225	14 ga., 4.4 lb./L.F.					5.70			5.70	6.25
0230	12 ga., 6.2 lb./L.F.					8			8	8.80
0235	10" x 5" x 3", 14 ga., 5.2 lb./L.F.					6.70			6.70	7.40
0240	12 ga., 7.3 lb./L.F.					9.45			9.45	10.40
0245	12" x 5" x 4", 14 ga., 6.0 lb./L.F.					7.75			7.75	8.55
0250	12 ga., 8.4 lb./L.F.				↓	10.85			10.85	11.95
0300	Rake/base angle, excludes concrete drilling and expansion anchors									
0310	2" x 2", 14 ga., 1.0 lb./L.F.	2 Sswk	640	.025	L.F.	1.29	1.44		2.73	3.68
0315	3" x 2", 14 ga., 1.3 lb./L.F.		535	.030		1.68	1.72		3.40	4.56
0320	3" x 3", 14 ga., 1.6 lb./L.F.		500	.032		2.07	1.84		3.91	5.15
0325	4" x 3", 14 ga., 1.8 lb./L.F.	↓	480	.033	↓	2.33	1.92		4.25	5.55
0600	Installation of secondary members, erection only									
0610	Girts, purlins, eave struts, 16 ga., 6" deep	E-18	100	.400	Ea.		23	14.90	37.90	52.50
0615	8" deep		80	.500			29	18.60	47.60	65.50
0620	14 ga., 6" deep		80	.500			29	18.60	47.60	65.50
0625	8" deep		65	.615			35.50	23	58.50	80.50
0630	10" deep		55	.727			42	27	69	95.50
0635	12" deep		50	.800			46.50	30	76.50	105
0640	12 ga., 8" deep		50	.800			46.50	30	76.50	105
0645	10" deep		45	.889			51.50	33	84.50	117
0650	12" deep	↓	40	1	↓		58	37	95	131
0900	For less than 20-ton job lots									
0905	For 15 to 19 tons, add				%	10%				
0910	For 10 to 14 tons, add					25%				
0915	For 5 to 9 tons, add					50%	50%	50%		
0920	For 1 to 4 tons, add					75%	75%	75%		
0925	For less than 1 ton, add				↓	100%	100%	100%		

05 15 Wire Rope Assemblies

05 15 16 – Steel Wire Rope Assemblies

05 15 16.05 Accessories for Steel Wire Rope

			Crew	Daily Output	Labor-Hours	Unit	Material	2020 Bare Costs Labor	Equipment	Total	Total Incl O&P
0010	**ACCESSORIES FOR STEEL WIRE ROPE**										
0015	Made from recycled materials										
1500	Thimbles, heavy duty, 1/4"	G	E-17	160	.100	Ea.	.51	5.85		6.36	9.75
1510	1/2"	G		160	.100		2.24	5.85		8.09	11.65
1520	3/4"	G		105	.152		5.10	8.95		14.05	19.60
1530	1"	G		52	.308		10.15	18.05		28.20	39.50
1540	1-1/4"	G		38	.421		15.65	24.50		40.15	56
1550	1-1/2"	G		13	1.231		44	72		116	162
1560	1-3/4"	G		8	2		91	117		208	284
1570	2"	G		6	2.667		132	156		288	390
1580	2-1/4"	G		4	4		179	235		414	565
1600	Clips, 1/4" diameter	G		160	.100		1.81	5.85		7.66	11.20
1610	3/8" diameter	G		160	.100		1.98	5.85		7.83	11.40
1620	1/2" diameter	G	↓	160	.100	↓	3.19	5.85		9.04	12.70

05 15 Wire Rope Assemblies

05 15 16 – Steel Wire Rope Assemblies

05 15 16.05 Accessories for Steel Wire Rope		Crew	Daily Output	Labor-Hours	Unit	Material	2020 Bare Costs Labor	Equipment	Total	Total Incl O&P
1630	3/4" diameter	G	E-17	102	.157	Ea.	5.20	9.20		14.40	20
1640	1" diameter	G		64	.250		8.65	14.65		23.30	32.50
1650	1-1/4" diameter	G		35	.457		14.15	27		41.15	57.50
1670	1-1/2" diameter	G		26	.615		19.05	36		55.05	77.50
1680	1-3/4" diameter	G		16	1		44.50	58.50		103	141
1690	2" diameter	G		12	1.333		49.50	78		127.50	178
1700	2-1/4" diameter	G		10	1.600		73	94		167	227
1800	Sockets, open swage, 1/4" diameter	G		160	.100		31.50	5.85		37.35	44
1810	1/2" diameter	G		77	.208		45.50	12.20		57.70	69
1820	3/4" diameter	G		19	.842		71	49.50		120.50	156
1830	1" diameter	G		9	1.778		127	104		231	305
1840	1-1/4" diameter	G		5	3.200		176	188		364	485
1850	1-1/2" diameter	G		3	5.333		385	315		700	915
1860	1-3/4" diameter	G		3	5.333		685	315		1,000	1,250
1870	2" diameter	G		1.50	10.667		1,025	625		1,650	2,125
1900	Closed swage, 1/4" diameter	G		160	.100		19	5.85		24.85	30
1910	1/2" diameter	G		104	.154		33	9		42	50
1920	3/4" diameter	G		32	.500		49	29.50		78.50	100
1930	1" diameter	G		15	1.067		86.50	62.50		149	193
1940	1-1/4" diameter	G		7	2.286		129	134		263	350
1950	1-1/2" diameter	G		4	4		235	235		470	630
1960	1-3/4" diameter	G		3	5.333		345	315		660	870
1970	2" diameter	G		2	8		675	470		1,145	1,475
2000	Open spelter, galv., 1/4" diameter	G		160	.100		42	5.85		47.85	55
2010	1/2" diameter	G		70	.229		43.50	13.40		56.90	69
2020	3/4" diameter	G		26	.615		65.50	36		101.50	129
2030	1" diameter	G		10	1.600		182	94		276	345
2040	1-1/4" diameter	G		5	3.200		260	188		448	580
2050	1-1/2" diameter	G		4	4		550	235		785	975
2060	1-3/4" diameter	G		2	8		960	470		1,430	1,775
2070	2" diameter	G		1.20	13.333		1,100	780		1,880	2,450
2080	2-1/2" diameter	G		1	16		2,025	940		2,965	3,725
2100	Closed spelter, galv., 1/4" diameter	G		160	.100		33	5.85		38.85	45
2110	1/2" diameter	G		88	.182		35	10.65		45.65	55.50
2120	3/4" diameter	G		30	.533		53	31.50		84.50	108
2130	1" diameter	G		13	1.231		113	72		185	238
2140	1-1/4" diameter	G		7	2.286		181	134		315	410
2150	1-1/2" diameter	G		6	2.667		390	156		546	675
2160	1-3/4" diameter	G		2.80	5.714		520	335		855	1,100
2170	2" diameter	G		2	8		640	470		1,110	1,450
2200	Jaw & jaw turnbuckles, 1/4" x 4"	G		160	.100		9.65	5.85		15.50	19.80
2250	1/2" x 6"	G		96	.167		12.20	9.80		22	29
2260	1/2" x 9"	G		77	.208		16.25	12.20		28.45	37
2270	1/2" x 12"	G		66	.242		18.30	14.20		32.50	42.50
2300	3/4" x 6"	G		38	.421		24	24.50		48.50	65
2310	3/4" x 9"	G		30	.533		26.50	31.50		58	78
2320	3/4" x 12"	G		28	.571		34	33.50		67.50	90
2330	3/4" x 18"	G		23	.696		40.50	41		81.50	109
2350	1" x 6"	G		17	.941		46	55		101	138
2360	1" x 12"	G		13	1.231		51	72		123	169
2370	1" x 18"	G		10	1.600		76	94		170	231
2380	1" x 24"	G		9	1.778		84	104		188	256
2400	1-1/4" x 12"	G		7	2.286		85.50	134		219.50	305

121

For customer support on your Site Work & Landscape Costs with RSMeans data, call 800.448.8182.

05 15 Wire Rope Assemblies

05 15 16 – Steel Wire Rope Assemblies

05 15 16.05 Accessories for Steel Wire Rope

			Crew	Daily Output	Labor-Hours	Unit	Material	2020 Bare Costs Labor	Equipment	Total	Total Incl O&P
2410	1-1/4" x 18"	G	E-17	6.50	2.462	Ea.	106	144		250	340
2420	1-1/4" x 24"	G		5.60	2.857		142	168		310	420
2450	1-1/2" x 12"	G		5.20	3.077		238	180		418	545
2460	1-1/2" x 18"	G		4	4		254	235		489	650
2470	1-1/2" x 24"	G		3.20	5		340	293		633	835
2500	1-3/4" x 18"	G		3.20	5		515	293		808	1,025
2510	1-3/4" x 24"	G		2.80	5.714		585	335		920	1,175
2550	2" x 24"	G		1.60	10		795	585		1,380	1,800

05 15 16.50 Steel Wire Rope

			Crew	Daily Output	Labor-Hours	Unit	Material	2020 Bare Costs Labor	Equipment	Total	Total Incl O&P
0010	**STEEL WIRE ROPE**										
0015	Made from recycled materials										
0020	6 x 19, bright, fiber core, 5000' rolls, 1/2" diameter	G				L.F.	.70			.70	.77
0050	Steel core	G					.92			.92	1.01
0100	Fiber core, 1" diameter	G					2.36			2.36	2.59
0150	Steel core	G					2.69			2.69	2.96
0300	6 x 19, galvanized, fiber core, 1/2" diameter	G					1.03			1.03	1.13
0350	Steel core	G					1.18			1.18	1.30
0400	Fiber core, 1" diameter	G					3.02			3.02	3.32
0450	Steel core	G					3.16			3.16	3.48
0500	6 x 7, bright, IPS, fiber core, <500 L.F. w/acc., 1/4" diameter	G	E-17	6400	.003		1.20	.15		1.35	1.55
0510	1/2" diameter	G		2100	.008		2.93	.45		3.38	3.92
0520	3/4" diameter	G		960	.017		5.30	.98		6.28	7.40
0550	6 x 19, bright, IPS, IWRC, <500 L.F. w/acc., 1/4" diameter	G		5760	.003		.77	.16		.93	1.11
0560	1/2" diameter	G		1730	.009		1.25	.54		1.79	2.23
0570	3/4" diameter	G		770	.021		2.17	1.22		3.39	4.30
0580	1" diameter	G		420	.038		3.68	2.23		5.91	7.55
0590	1-1/4" diameter	G		290	.055		6.10	3.24		9.34	11.80
0600	1-1/2" diameter	G		192	.083		7.50	4.89		12.39	15.90
0610	1-3/4" diameter	G	E-18	240	.167		11.95	9.65	6.20	27.80	35
0620	2" diameter	G		160	.250		15.35	14.45	9.30	39.10	49.50
0630	2-1/4" diameter	G		160	.250		20.50	14.45	9.30	44.25	55.50
0650	6 x 37, bright, IPS, IWRC, <500 L.F. w/acc., 1/4" diameter	G	E-17	6400	.003		1.12	.15		1.27	1.46
0660	1/2" diameter	G		1730	.009		1.90	.54		2.44	2.94
0670	3/4" diameter	G		770	.021		3.07	1.22		4.29	5.30
0680	1" diameter	G		430	.037		4.88	2.18		7.06	8.75
0690	1-1/4" diameter	G		290	.055		7.35	3.24		10.59	13.20
0700	1-1/2" diameter	G		190	.084		10.55	4.94		15.49	19.35
0710	1-3/4" diameter	G	E-18	260	.154		16.75	8.90	5.70	31.35	38.50
0720	2" diameter	G		200	.200		21.50	11.55	7.45	40.50	50
0730	2-1/4" diameter	G		160	.250		28.50	14.45	9.30	52.25	64.50
0800	6 x 19 & 6 x 37, swaged, 1/2" diameter	G	E-17	1220	.013		3.91	.77		4.68	5.50
0810	9/16" diameter	G		1120	.014		4.55	.84		5.39	6.30
0820	5/8" diameter	G		930	.017		5.40	1.01		6.41	7.55
0830	3/4" diameter	G		640	.025		6.85	1.47		8.32	9.85
0840	7/8" diameter	G		480	.033		8.65	1.95		10.60	12.60
0850	1" diameter	G		350	.046		10.55	2.68		13.23	15.85
0860	1-1/8" diameter	G		288	.056		13	3.26		16.26	19.40
0870	1-1/4" diameter	G		230	.070		15.75	4.08		19.83	23.50
0880	1-3/8" diameter	G		192	.083		18.20	4.89		23.09	27.50
0890	1-1/2" diameter	G	E-18	300	.133		22	7.70	4.96	34.66	42

122

05 15 Wire Rope Assemblies

05 15 16 – Steel Wire Rope Assemblies

05 15 16.60 Galvanized Steel Wire Rope and Accessories		Crew	Daily Output	Labor-Hours	Unit	Material	2020 Bare Costs Labor	2020 Bare Costs Equipment	Total	Total Incl O&P
0010	**GALVANIZED STEEL WIRE ROPE & ACCESSORIES**									
0015	Made from recycled materials									
3000	Aircraft cable, galvanized, 7 x 7 x 1/8"	G	E-17	5000	.003	L.F.	.18	.19		.37	.49
3100	Clamps, 1/8"	G	"	125	.128	Ea.	1.29	7.50		8.79	13.20

05 15 16.70 Temporary Cable Safety Railing

		Crew	Daily Output	Labor-Hours	Unit	Material	2020 Bare Costs Labor	2020 Bare Costs Equipment	Total	Total Incl O&P
0010	**TEMPORARY CABLE SAFETY RAILING**, Each 100' strand incl.									
0020	2 eyebolts, 1 turnbuckle, 100' cable, 2 thimbles, 6 clips									
0025	Made from recycled materials									
0100	One strand using 1/4" cable & accessories	G	2 Sswk	4	4	C.L.F.	158	231		389	535
0200	1/2" cable & accessories	G	"	2	8	"	345	460		805	1,100

05 21 Steel Joist Framing

05 21 13 – Deep Longspan Steel Joist Framing

05 21 13.50 Deep Longspan Joists

		Crew	Daily Output	Labor-Hours	Unit	Material	2020 Bare Costs Labor	2020 Bare Costs Equipment	Total	Total Incl O&P
0010	**DEEP LONGSPAN JOISTS**									
3015	Made from recycled materials									
3500	For less than 40-ton job lots									
3502	For 30 to 39 tons, add					%	10%			
3504	20 to 29 tons, add						20%			
3506	10 to 19 tons, add						30%			
3507	5 to 9 tons, add						50%	25%		
3508	1 to 4 tons, add						75%	50%		
3509	Less than 1 ton, add						100%	100%		
4010	SLH series, 40-ton job lots, bolted cross bridging, shop primer									
4040	Spans to 200' (shipped in 3 pieces)	G	E-7	13	6.154	Ton	2,350	355	153	2,858	3,325
6100	For less than 40-ton job lots									
6102	For 30 to 39 tons, add					%	10%			
6104	20 to 29 tons, add						20%			
6106	10 to 19 tons, add						30%			
6107	5 to 9 tons, add						50%	25%		
6108	1 to 4 tons, add						75%	50%		
6109	Less than 1 ton, add						100%	100%		

05 21 16 – Longspan Steel Joist Framing

05 21 16.50 Longspan Joists

		Crew	Daily Output	Labor-Hours	Unit	Material	2020 Bare Costs Labor	2020 Bare Costs Equipment	Total	Total Incl O&P
0010	**LONGSPAN JOISTS**									
2000	LH series, 40-ton job lots, bolted cross bridging, shop primer									
2015	Made from recycled materials									
2600	For less than 40-ton job lots									
2602	For 30 to 39 tons, add					%	10%			
2604	20 to 29 tons, add						20%			
2606	10 to 19 tons, add						30%			
2607	5 to 9 tons, add						50%	25%		
2608	1 to 4 tons, add						75%	50%		
2609	Less than 1 ton, add						100%	100%		

05 21 Steel Joist Framing

05 21 19 – Open Web Steel Joist Framing

05 21 19.10 Open Web Joists

05 21 19.10 Open Web Joists		Crew	Daily Output	Labor-Hours	Unit	Material	2020 Bare Costs Labor	Equipment	Total	Total Incl O&P
0010	**OPEN WEB JOISTS**									
0015	Made from recycled materials									
0440	K series, 30' to 50' spans	E-7	17	4.706	Ton	1,950	270	117	2,337	2,700
0800	For less than 40-ton job lots									
0802	For 30 to 39 tons, add				%	10%				
0804	20 to 29 tons, add					20%				
0806	10 to 19 tons, add					30%				
0807	5 to 9 tons, add					50%	25%			
0808	1 to 4 tons, add					75%	50%			
0809	Less than 1 ton, add					100%	100%			
1010	CS series, 40-ton job lots, horizontal bridging, shop primer									
1040	Spans to 30'	E-7	12	6.667	Ton	2,000	385	166	2,551	2,975
1500	For less than 40-ton job lots									
1502	For 30 to 39 tons, add				%	10%				
1504	20 to 29 tons, add					20%				
1506	10 to 19 tons, add					30%				
1507	5 to 9 tons, add					50%	25%			
1508	1 to 4 tons, add					75%	50%			
1509	Less than 1 ton, add					100%	100%			

05 31 Steel Decking

05 31 13 – Steel Floor Decking

05 31 13.50 Floor Decking

05 31 13.50 Floor Decking		Crew	Daily Output	Labor-Hours	Unit	Material	2020 Bare Costs Labor	Equipment	Total	Total Incl O&P
0010	**FLOOR DECKING** R053100-10									
0015	Made from recycled materials									
5100	Non-cellular composite decking, galvanized, 1-1/2" deep, 16 ga.	E-4	3500	.009	S.F.	4.82	.53	.04	5.39	6.20
5120	18 ga.		3650	.009		3	.51	.04	3.55	4.13
5140	20 ga.		3800	.008		3.52	.49	.04	4.05	4.68
5200	2" deep, 22 ga.		3860	.008		2.38	.48	.04	2.90	3.42
5300	20 ga.		3600	.009		3.38	.52	.04	3.94	4.57
5400	18 ga.		3380	.009		3.12	.55	.04	3.71	4.34
5500	16 ga.		3200	.010		4.77	.58	.05	5.40	6.20
5700	3" deep, 22 ga.		3200	.010		2.59	.58	.05	3.22	3.81
5800	20 ga.		3000	.011		3.72	.62	.05	4.39	5.10
5900	18 ga.		2850	.011		3.12	.65	.05	3.82	4.51
6000	16 ga.		2700	.012		5.40	.69	.05	6.14	7.10

05 31 23 – Steel Roof Decking

05 31 23.50 Roof Decking

05 31 23.50 Roof Decking		Crew	Daily Output	Labor-Hours	Unit	Material	2020 Bare Costs Labor	Equipment	Total	Total Incl O&P
0010	**ROOF DECKING**									
0015	Made from recycled materials									
2100	Open type, 1-1/2" deep, Type B, wide rib, galv., 22 ga., under 50 sq.	E-4	4500	.007	S.F.	2.68	.41	.03	3.12	3.64
2400	Over 500 squares		5100	.006		1.93	.36	.03	2.32	2.72
2600	20 ga., under 50 squares		3865	.008		2.99	.48	.04	3.51	4.09
2700	Over 500 squares		4300	.007		2.15	.43	.03	2.61	3.09
2900	18 ga., under 50 squares		3800	.008		3.84	.49	.04	4.37	5.05
3000	Over 500 squares		4300	.007		2.76	.43	.03	3.22	3.76
3050	16 ga., under 50 squares		3700	.009		5.20	.50	.04	5.74	6.55
3100	Over 500 squares		4200	.008		3.74	.44	.04	4.22	4.85
3200	3" deep, Type N, 22 ga., under 50 squares		3600	.009		3.72	.52	.04	4.28	4.94
3260	over 500 squares		4000	.008		2.68	.47	.04	3.19	3.72

05 31 Steel Decking

05 31 23 – Steel Roof Decking

05 31 23.50 Roof Decking

		Crew	Daily Output	Labor-Hours	Unit	Material	2020 Bare Costs Labor	Equipment	Total	Total Incl O&P
3300	20 ga., under 50 squares	G	E-4	3400	.009	S.F.	4.04	.55	.04	4.63	5.35
3360	over 500 squares	G		3800	.008		2.91	.49	.04	3.44	4.01
3400	18 ga., under 50 squares	G		3200	.010		5.25	.58	.05	5.88	6.70
3460	over 500 squares	G		3600	.009		3.77	.52	.04	4.33	5
3500	16 ga., under 50 squares	G		3000	.011		6.95	.62	.05	7.62	8.60
3560	over 500 squares	G		3400	.009		4.99	.55	.04	5.58	6.40
3700	4-1/2" deep, Type J, 20 ga., over 50 squares	G		2700	.012		4.47	.69	.05	5.21	6.05
4100	6" deep, Type H, 18 ga., over 50 squares	G		2000	.016		6.55	.93	.07	7.55	8.75
4500	7-1/2" deep, Type H, 18 ga., over 50 squares	G		1690	.019		7.75	1.10	.09	8.94	10.40

05 31 33 – Steel Form Decking

05 31 33.50 Form Decking

		Crew	Daily Output	Labor-Hours	Unit	Material	2020 Bare Costs Labor	Equipment	Total	Total Incl O&P
0010	**FORM DECKING**										
0015	Made from recycled materials										
6100	Slab form, steel, 28 ga., 9/16" deep, Type UFS, uncoated	G	E-4	4000	.008	S.F.	1.90	.47	.04	2.41	2.86
6200	Galvanized	G	"	4000	.008	"	1.68	.47	.04	2.19	2.62

05 35 Raceway Decking Assemblies

05 35 13 – Steel Cellular Decking

05 35 13.50 Cellular Decking

		Crew	Daily Output	Labor-Hours	Unit	Material	2020 Bare Costs Labor	Equipment	Total	Total Incl O&P
0010	**CELLULAR DECKING**										
0015	Made from recycled materials										
0200	Cellular units, galv, 1-1/2" deep, Type BC, 20-20 ga., over 15 squares	G	E-4	1460	.022	S.F.	11.55	1.27	.10	12.92	14.80
0250	18-20 ga.	G		1420	.023		10.25	1.31	.10	11.66	13.45
0300	18-18 ga.	G		1390	.023		10.55	1.34	.11	12	13.80
0320	16-18 ga.	G		1360	.024		16.05	1.37	.11	17.53	19.90
0340	16-16 ga.	G		1330	.024		17.85	1.40	.11	19.36	22
0400	3" deep, Type NC, galvanized, 20-20 ga.	G		1375	.023		12.70	1.35	.11	14.16	16.25

05 51 Metal Stairs

05 51 13 – Metal Pan Stairs

05 51 13.50 Pan Stairs

		Crew	Daily Output	Labor-Hours	Unit	Material	2020 Bare Costs Labor	Equipment	Total	Total Incl O&P
0010	**PAN STAIRS**, shop fabricated, steel stringers										
0015	Made from recycled materials										
0200	Metal pan tread for concrete in-fill, picket rail, 3'-6" wide	G	E-4	35	.914	Riser	600	53	4.20	657.20	750
0300	4'-0" wide	G		30	1.067		590	62	4.90	656.90	755
0350	Wall rail, both sides, 3'-6" wide	G		53	.604		460	35	2.77	497.77	565
1500	Landing, steel pan, conventional	G		160	.200	S.F.	79.50	11.65	.92	92.07	107
1600	Pre-erected	G		255	.125	"	138	7.30	.58	145.88	164
1700	Pre-erected, steel pan tread, 3'-6" wide, 2 line pipe rail	G	E-2	87	.644	Riser	585	37	19.50	641.50	725

05 51 16 – Metal Floor Plate Stairs

05 51 16.50 Floor Plate Stairs

		Crew	Daily Output	Labor-Hours	Unit	Material	2020 Bare Costs Labor	Equipment	Total	Total Incl O&P
0010	**FLOOR PLATE STAIRS**, shop fabricated, steel stringers										
0015	Made from recycled materials										
0400	Cast iron tread and pipe rail, 3'-6" wide	G	E-4	35	.914	Riser	570	53	4.20	627.20	715
0500	Checkered plate tread, industrial, 3'-6" wide	G		28	1.143		385	66.50	5.25	456.75	535
0550	Circular, for tanks, 3'-0" wide	G		33	.970		435	56.50	4.45	495.95	575
0600	For isolated stairs, add							100%			
0800	Custom steel stairs, 3'-6" wide, economy	G	E-4	35	.914		585	53	4.20	642.20	735

125

05 51 Metal Stairs

05 51 16 – Metal Floor Plate Stairs

05 51 16.50 Floor Plate Stairs

		Crew	Daily Output	Labor-Hours	Unit	Material	2020 Bare Costs Labor	Equipment	Total	Total Incl O&P
0810	Medium priced	G E-4	30	1.067	Riser	775	62	4.90	841.90	960
0900	Deluxe	G	20	1.600		970	93	7.35	1,070.35	1,225
1100	For 4' wide stairs, add					5%	5%			
1300	For 5' wide stairs, add					10%	10%			

05 51 19 – Metal Grating Stairs

05 51 19.50 Grating Stairs

		Crew	Daily Output	Labor-Hours	Unit	Material	Labor	Equipment	Total	Total Incl O&P
0010	**GRATING STAIRS**, shop fabricated, steel stringers, safety nosing on treads									
0015	Made from recycled materials									
0020	Grating tread and pipe railing, 3'-6" wide	G E-4	35	.914	Riser	390	53	4.20	447.20	520
0100	4'-0" wide	G	30	1.067	"	455	62	4.90	521.90	605

05 51 33 – Metal Ladders

05 51 33.13 Vertical Metal Ladders

		Crew	Daily Output	Labor-Hours	Unit	Material	Labor	Equipment	Total	Total Incl O&P
0010	**VERTICAL METAL LADDERS**, shop fabricated									
0015	Made from recycled materials									
0020	Steel, 20" wide, bolted to concrete, with cage	G E-4	50	.640	V.L.F.	92	37	2.94	131.94	163
0100	Without cage	G	85	.376		42	22	1.73	65.73	83
0300	Aluminum, bolted to concrete, with cage	G	50	.640		130	37	2.94	169.94	205
0400	Without cage	G	85	.376		51	22	1.73	74.73	92.50

05 51 33.16 Inclined Metal Ladders

		Crew	Daily Output	Labor-Hours	Unit	Material	Labor	Equipment	Total	Total Incl O&P
0010	**INCLINED METAL LADDERS**, shop fabricated									
0015	Made from recycled materials									
3900	Industrial ships ladder, steel, 24" W, grating treads, 2 line pipe rail	G E-4	30	1.067	Riser	205	62	4.90	271.90	330
4000	Aluminum	G "	30	1.067	"	300	62	4.90	366.90	435

05 52 Metal Railings

05 52 13 – Pipe and Tube Railings

05 52 13.50 Railings, Pipe

		Crew	Daily Output	Labor-Hours	Unit	Material	Labor	Equipment	Total	Total Incl O&P
0010	**RAILINGS, PIPE**, shop fab'd, 3'-6" high, posts @ 5' OC									
0015	Made from recycled materials									
0020	Aluminum, 2 rail, satin finish, 1-1/4" diameter	G E-4	160	.200	L.F.	51.50	11.65	.92	64.07	76
0030	Clear anodized	G	160	.200		61.50	11.65	.92	74.07	87
0040	Dark anodized	G	160	.200		68	11.65	.92	80.57	94.50
0080	1-1/2" diameter, satin finish	G	160	.200		58.50	11.65	.92	71.07	84
0090	Clear anodized	G	160	.200		66	11.65	.92	78.57	92
0100	Dark anodized	G	160	.200		72.50	11.65	.92	85.07	99.50
0140	Aluminum, 3 rail, 1-1/4" diam., satin finish	G	137	.234		67	13.60	1.07	81.67	96.50
0150	Clear anodized	G	137	.234		83.50	13.60	1.07	98.17	115
0160	Dark anodized	G	137	.234		92	13.60	1.07	106.67	124
0200	1-1/2" diameter, satin finish	G	137	.234		79.50	13.60	1.07	94.17	110
0210	Clear anodized	G	137	.234		90.50	13.60	1.07	105.17	122
0220	Dark anodized	G	137	.234		99.50	13.60	1.07	114.17	132
0500	Steel, 2 rail, on stairs, primed, 1-1/4" diameter	G	160	.200		32	11.65	.92	44.57	55
0520	1-1/2" diameter	G	160	.200		31	11.65	.92	43.57	53.50
0540	Galvanized, 1-1/4" diameter	G	160	.200		41.50	11.65	.92	54.07	65.50
0560	1-1/2" diameter	G	160	.200		49	11.65	.92	61.57	73.50
0580	Steel, 3 rail, primed, 1-1/4" diameter	G	137	.234		46	13.60	1.07	60.67	73
0600	1-1/2" diameter	G	137	.234		50.50	13.60	1.07	65.17	78
0620	Galvanized, 1-1/4" diameter	G	137	.234		64.50	13.60	1.07	79.17	93.50
0640	1-1/2" diameter	G	137	.234		77.50	13.60	1.07	92.17	108
0700	Stainless steel, 2 rail, 1-1/4" diam., #4 finish	G	137	.234		141	13.60	1.07	155.67	178

05 52 Metal Railings

05 52 13 – Pipe and Tube Railings

05 52 13.50 Railings, Pipe

		Crew	Daily Output	Labor-Hours	Unit	Material	2020 Bare Costs Labor	Equipment	Total	Total Incl O&P
0720	High polish	G E-4	137	.234	L.F.	228	13.60	1.07	242.67	274
0740	Mirror polish	G	137	.234		285	13.60	1.07	299.67	340
0760	Stainless steel, 3 rail, 1-1/2" diam., #4 finish	G	120	.267		212	15.50	1.22	228.72	259
0770	High polish	G	120	.267		350	15.50	1.22	366.72	410
0780	Mirror finish	G	120	.267		430	15.50	1.22	446.72	495
0900	Wall rail, alum. pipe, 1-1/4" diam., satin finish	G	213	.150		25	8.75	.69	34.44	42
0905	Clear anodized	G	213	.150		31	8.75	.69	40.44	48.50
0910	Dark anodized	G	213	.150		36.50	8.75	.69	45.94	55
0915	1-1/2" diameter, satin finish	G	213	.150		27.50	8.75	.69	36.94	45
0920	Clear anodized	G	213	.150		34.50	8.75	.69	43.94	52.50
0925	Dark anodized	G	213	.150		43.50	8.75	.69	52.94	62.50
0930	Steel pipe, 1-1/4" diameter, primed	G	213	.150		18.50	8.75	.69	27.94	35
0935	Galvanized	G	213	.150		27	8.75	.69	36.44	44
0940	1-1/2" diameter	G	176	.182		17.75	10.55	.83	29.13	37
0945	Galvanized	G	213	.150		27	8.75	.69	36.44	44
0955	Stainless steel pipe, 1-1/2" diam., #4 finish	G	107	.299		113	17.40	1.37	131.77	153
0960	High polish	G	107	.299		229	17.40	1.37	247.77	281
0965	Mirror polish	G	107	.299		271	17.40	1.37	289.77	325
2000	2-line pipe rail (1-1/2" T&B) with 1/2" pickets @ 4-1/2" OC,									
2005	attached handrail on brackets									
2010	42" high aluminum, satin finish, straight & level	G E-4	120	.267	L.F.	355	15.50	1.22	371.72	415
2050	42" high steel, primed, straight & level	G "	120	.267		149	15.50	1.22	165.72	190
4000	For curved and level rails, add						10%	10%		
4100	For sloped rails for stairs, add						30%	30%		

05 52 16 – Industrial Railings

05 52 16.50 Railings, Industrial

		Crew	Daily Output	Labor-Hours	Unit	Material	2020 Bare Costs Labor	Equipment	Total	Total Incl O&P
0010	**RAILINGS, INDUSTRIAL**, shop fab'd, 3'-6" high, posts @ 5' OC									
0020	2 rail, 3'-6" high, 1-1/2" pipe	G E-4	255	.125	L.F.	39	7.30	.58	46.88	55
0100	2" angle rail	G "	255	.125		35	7.30	.58	42.88	50.50
0200	For 4" high kick plate, 10 ga., add	G				6.20			6.20	6.80
0300	1/4" thick, add	G				8.15			8.15	8.95
0500	For curved level rails, add						10%	10%		
0550	For sloped rails for stairs, add						30%	30%		

05 53 Metal Gratings

05 53 13 – Bar Gratings

05 53 13.10 Floor Grating, Aluminum

		Crew	Daily Output	Labor-Hours	Unit	Material	2020 Bare Costs Labor	Equipment	Total	Total Incl O&P
0010	**FLOOR GRATING, ALUMINUM**, field fabricated from panels									
0015	Made from recycled materials									
0110	Bearing bars @ 1-3/16" OC, cross bars @ 4" OC,									
0111	Up to 300 S.F., 1" x 1/8" bar	G E-4	900	.036	S.F.	22	2.07	.16	24.23	28
0112	Over 300 S.F.	G	850	.038		20	2.19	.17	22.36	25.50
0113	1-1/4" x 1/8" bar, up to 300 S.F.	G	800	.040		19.10	2.33	.18	21.61	25
0114	Over 300 S.F.	G	1000	.032		17.35	1.86	.15	19.36	22
0122	1-1/4" x 3/16" bar, up to 300 S.F.	G	750	.043		28	2.48	.20	30.68	35
0124	Over 300 S.F.	G	1000	.032		25.50	1.86	.15	27.51	31
0132	1-1/2" x 3/16" bar, up to 300 S.F.	G	700	.046		41	2.66	.21	43.87	49.50
0134	Over 300 S.F.	G	1000	.032		37.50	1.86	.15	39.51	44
0136	1-3/4" x 3/16" bar, up to 300 S.F.	G	500	.064		43.50	3.72	.29	47.51	53.50
0138	Over 300 S.F.	G	1000	.032		39.50	1.86	.15	41.51	46.50

05 53 Metal Gratings

05 53 13 – Bar Gratings

05 53 13.10 Floor Grating, Aluminum

		Crew	Daily Output	Labor-Hours	Unit	Material	2020 Bare Costs Labor	Equipment	Total	Total Incl O&P
0146	2-1/4" x 3/16" bar, up to 300 S.F. [G]	E-4	600	.053	S.F.	59	3.10	.24	62.34	69.50
0148	Over 300 S.F. [G]		1000	.032		53.50	1.86	.15	55.51	62
0162	Cross bars @ 2" OC, 1" x 1/8", up to 300 S.F. [G]		600	.053		31.50	3.10	.24	34.84	39.50
0164	Over 300 S.F. [G]		1000	.032		28.50	1.86	.15	30.51	34.50
0172	1-1/4" x 3/16" bar, up to 300 S.F. [G]		600	.053		50.50	3.10	.24	53.84	60.50
0174	Over 300 S.F. [G]		1000	.032		46	1.86	.15	48.01	53.50
0182	1-1/2" x 3/16" bar, up to 300 S.F. [G]		600	.053		61.50	3.10	.24	64.84	73
0184	Over 300 S.F. [G]		1000	.032		56	1.86	.15	58.01	64.50
0186	1-3/4" x 3/16" bar, up to 300 S.F. [G]		600	.053		63	3.10	.24	66.34	74.50
0188	Over 300 S.F. [G]		1000	.032		57.50	1.86	.15	59.51	66
0200	For straight cuts, add				L.F.	4.53			4.53	4.98
0212	Bearing bars @ 15/16" OC, 1" x 1/8", up to 300 S.F. [G]	E-4	520	.062	S.F.	32	3.58	.28	35.86	41.50
0214	Over 300 S.F. [G]		920	.035		29	2.02	.16	31.18	35.50
0222	1-1/4" x 3/16", up to 300 S.F. [G]		520	.062		51	3.58	.28	54.86	62
0224	Over 300 S.F. [G]		920	.035		46.50	2.02	.16	48.68	54.50
0232	1-1/2" x 3/16", up to 300 S.F. [G]		520	.062		57	3.58	.28	60.86	68.50
0234	Over 300 S.F. [G]		920	.035		51.50	2.02	.16	53.68	60.50
0300	For curved cuts, add				L.F.	5.55			5.55	6.15
0400	For straight banding, add [G]					5.80			5.80	6.35
0500	For curved banding, add [G]					7.10			7.10	7.85
0600	For aluminum checkered plate nosings, add [G]					7.50			7.50	8.25
0700	For straight toe plate, add [G]					11.35			11.35	12.50
0800	For curved toe plate, add [G]					13.25			13.25	14.55
1000	For cast aluminum abrasive nosings, add [G]					11.05			11.05	12.15
1400	Extruded I bars are 10% less than 3/16" bars									
1600	Heavy duty, all extruded plank, 3/4" deep, 1.8 #/S.F. [G]	E-4	1100	.029	S.F.	27.50	1.69	.13	29.32	33
1700	1-1/4" deep, 2.9 #/S.F. [G]		1000	.032		34.50	1.86	.15	36.51	41
1800	1-3/4" deep, 4.2 #/S.F. [G]		925	.035		43.50	2.01	.16	45.67	51.50
1900	2-1/4" deep, 5.0 #/S.F. [G]		875	.037		67.50	2.13	.17	69.80	78
2100	For safety serrated surface, add					15%				

05 53 13.70 Floor Grating, Steel

		Crew	Daily Output	Labor-Hours	Unit	Material	2020 Bare Costs Labor	Equipment	Total	Total Incl O&P
0010	**FLOOR GRATING, STEEL**, field fabricated from panels									
0015	Made from recycled materials									
0300	Platforms, to 12' high, rectangular [G]	E-4	3150	.010	Lb.	3.34	.59	.05	3.98	4.65
0400	Circular [G]	"	2300	.014	"	4.17	.81	.06	5.04	5.95
0410	Painted bearing bars @ 1-3/16"									
0412	Cross bars @ 4" OC, 3/4" x 1/8" bar, up to 300 S.F. [G]	E-2	500	.112	S.F.	8.35	6.40	3.39	18.14	23
0414	Over 300 S.F. [G]		750	.075		7.60	4.27	2.26	14.13	17.45
0422	1-1/4" x 3/16", up to 300 S.F. [G]		400	.140		13.30	8	4.24	25.54	31.50
0424	Over 300 S.F. [G]		600	.093		12.10	5.35	2.83	20.28	24.50
0432	1-1/2" x 3/16", up to 300 S.F. [G]		400	.140		15.45	8	4.24	27.69	34
0434	Over 300 S.F. [G]		600	.093		14.05	5.35	2.83	22.23	27
0436	1-3/4" x 3/16", up to 300 S.F. [G]		400	.140		20.50	8	4.24	32.74	39.50
0438	Over 300 S.F. [G]		600	.093		18.50	5.35	2.83	26.68	32
0452	2-1/4" x 3/16", up to 300 S.F. [G]		300	.187		24	10.65	5.65	40.30	48.50
0454	Over 300 S.F. [G]		450	.124		21.50	7.10	3.77	32.37	39
0462	Cross bars @ 2" OC, 3/4" x 1/8", up to 300 S.F. [G]		500	.112		15.20	6.40	3.39	24.99	30.50
0464	Over 300 S.F. [G]		750	.075		12.65	4.27	2.26	19.18	23
0472	1-1/4" x 3/16", up to 300 S.F. [G]		400	.140		19.90	8	4.24	32.14	39
0474	Over 300 S.F. [G]		600	.093		16.55	5.35	2.83	24.73	29.50
0482	1-1/2" x 3/16", up to 300 S.F. [G]		400	.140		22.50	8	4.24	34.74	41.50
0484	Over 300 S.F. [G]		600	.093		18.65	5.35	2.83	26.83	32

05 53 Metal Gratings

05 53 13 – Bar Gratings

05 53 13.70 Floor Grating, Steel

		Crew	Daily Output	Labor-Hours	Unit	Material	2020 Bare Costs Labor	Equipment	Total	Total Incl O&P
0486	1-3/4" x 3/16", up to 300 S.F.	G E-2	400	.140	S.F.	33.50	8	4.24	45.74	54
0488	Over 300 S.F.	G	600	.093		28	5.35	2.83	36.18	42.50
0502	2-1/4" x 3/16", up to 300 S.F.	G	300	.187		31	10.65	5.65	47.30	56.50
0504	Over 300 S.F.	G	450	.124		26	7.10	3.77	36.87	43.50
0601	Painted bearing bars @ 15/16" OC, cross bars @ 4" OC,									
0612	Up to 300 S.F., 3/4" x 3/16" bars	G E-4	850	.038	S.F.	12.45	2.19	.17	14.81	17.30
0622	1-1/4" x 3/16" bars	G	600	.053		16.90	3.10	.24	20.24	23.50
0632	1-1/2" x 3/16" bars	G	550	.058		19.25	3.38	.27	22.90	26.50
0636	1-3/4" x 3/16" bars	G	450	.071		24.50	4.14	.33	28.97	34
0652	2-1/4" x 3/16" bars	G E-2	300	.187		26	10.65	5.65	42.30	51
0662	Cross bars @ 2" OC, up to 300 S.F., 3/4" x 3/16"	G	500	.112		15.80	6.40	3.39	25.59	31
0672	1-1/4" x 3/16" bars	G	400	.140		20.50	8	4.24	32.74	39.50
0682	1-1/2" x 3/16" bars	G	400	.140		23	8	4.24	35.24	42
0686	1-3/4" x 3/16" bars	G	300	.187		29	10.65	5.65	45.30	54.50
0690	For galvanized grating, add					25%				
0800	For straight cuts, add				L.F.	6.45			6.45	7.10
0900	For curved cuts, add					8.20			8.20	9.05
1000	For straight banding, add	G				7.10			7.10	7.85
1100	For curved banding, add	G				9.15			9.15	10.05
1200	For checkered plate nosings, add	G				8.05			8.05	8.85
1300	For straight toe or kick plate, add	G				14.40			14.40	15.85
1400	For curved toe or kick plate, add	G				16.30			16.30	17.95
1500	For abrasive nosings, add	G				10.65			10.65	11.75
1600	For safety serrated surface, bearing bars @ 1-3/16" OC, add					15%				
1700	Bearing bars @ 15/16" OC, add					25%				
2000	Stainless steel gratings, close spaced, 1" x 1/8" bars, up to 300 S.F.	G E-4	450	.071	S.F.	76.50	4.14	.33	80.97	91
2100	Standard spacing, 3/4" x 1/8" bars	G	500	.064		96.50	3.72	.29	100.51	112
2200	1-1/4" x 3/16" bars	G	400	.080		118	4.65	.37	123.02	137

05 53 16 – Plank Gratings

05 53 16.50 Grating Planks

		Crew	Daily Output	Labor-Hours	Unit	Material	2020 Bare Costs Labor	Equipment	Total	Total Incl O&P
0010	**GRATING PLANKS**, field fabricated from planks									
0020	Aluminum, 9-1/2" wide, 14 ga., 2" rib	G E-4	950	.034	L.F.	36	1.96	.15	38.11	42.50
0200	Galvanized steel, 9-1/2" wide, 14 ga., 2-1/2" rib	G	950	.034		16.95	1.96	.15	19.06	22
0300	4" rib	G	950	.034		18.20	1.96	.15	20.31	23
0500	12 ga., 2-1/2" rib	G	950	.034		19.80	1.96	.15	21.91	25
0600	3" rib	G	950	.034		21.50	1.96	.15	23.61	27
0800	Stainless steel, type 304, 16 ga., 2" rib	G	950	.034		42	1.96	.15	44.11	49
0900	Type 316	G	950	.034		56.50	1.96	.15	58.61	65

05 53 19 – Expanded Metal Gratings

05 53 19.10 Expanded Grating, Aluminum

		Crew	Daily Output	Labor-Hours	Unit	Material	2020 Bare Costs Labor	Equipment	Total	Total Incl O&P
0010	**EXPANDED GRATING, ALUMINUM**									
1200	Expanded aluminum, .65 #/S.F.	G E-4	1050	.030	S.F.	22	1.77	.14	23.91	27

05 53 19.20 Expanded Grating, Steel

		Crew	Daily Output	Labor-Hours	Unit	Material	2020 Bare Costs Labor	Equipment	Total	Total Incl O&P
0010	**EXPANDED GRATING, STEEL**									
2400	Expanded steel grating, at ground, 3.0 #/S.F.	G E-4	900	.036	S.F.	8.10	2.07	.16	10.33	12.30
2500	3.14 #/S.F.	G	900	.036		7.35	2.07	.16	9.58	11.50
2600	4.0 #/S.F.	G	850	.038		8.85	2.19	.17	11.21	13.35
2650	4.27 #/S.F.	G	850	.038		9.70	2.19	.17	12.06	14.30
2700	5.0 #/S.F.	G	800	.040		13.80	2.33	.18	16.31	19.05
2800	6.25 #/S.F.	G	750	.043		17.85	2.48	.20	20.53	24
2900	7.0 #/S.F.	G	700	.046		19.70	2.66	.21	22.57	26

129

For customer support on your Site Work & Landscape Costs with RSMeans data, call 800.448.8182.

05 53 Metal Gratings

05 53 19 – Expanded Metal Gratings

05 53 19.20 Expanded Grating, Steel		Crew	Daily Output	Labor-Hours	Unit	Material	2020 Bare Costs Labor	Equipment	Total	Total Incl O&P
3100	For flattened expanded steel grating, add				S.F.	8%				
3300	For elevated installation above 15', add				↓		15%			

05 53 19.30 Grating Frame

0010	GRATING FRAME, field fabricated										
0020	Aluminum, for gratings 1" to 1-1/2" deep	G	1 Sswk	70	.114	L.F.	3.82	6.60		10.42	14.55
0100	For each corner, add	G				Ea.	5.70			5.70	6.25

05 54 Metal Floor Plates

05 54 13 – Floor Plates

05 54 13.20 Checkered Plates

			Crew	Daily Output	Labor-Hours	Unit	Material	Labor	Equipment	Total	Total Incl O&P
0010	CHECKERED PLATES, steel, field fabricated										
0015	Made from recycled materials										
0020	1/4" & 3/8", 2000 to 5000 S.F., bolted	G	E-4	2900	.011	Lb.	.80	.64	.05	1.49	1.95
0100	Welded	G		4400	.007	"	.91	.42	.03	1.36	1.70
0300	Pit or trench cover and frame, 1/4" plate, 2' to 3' wide	G		100	.320	S.F.	10.95	18.60	1.47	31.02	42.50
0400	For galvanizing, add	G				Lb.	.32			.32	.35
0500	Platforms, 1/4" plate, no handrails included, rectangular	G	E-4	4200	.008		3.55	.44	.04	4.03	4.65
0600	Circular	G	"	2500	.013		4.44	.74	.06	5.24	6.10

05 54 13.70 Trench Covers

			Crew	Daily Output	Labor-Hours	Unit	Material	Labor	Equipment	Total	Total Incl O&P
0010	TRENCH COVERS, field fabricated										
0020	Cast iron grating with bar stops and angle frame, to 18" wide	G	1 Sswk	20	.400	L.F.	212	23		235	269
0100	Frame only (both sides of trench), 1" grating	G		45	.178		1.41	10.25		11.66	17.65
0150	2" grating	G		35	.229		3.23	13.20		16.43	24
0200	Aluminum, stock units, including frames and										
0210	3/8" plain cover plate, 4" opening	G	E-4	205	.156	L.F.	15.85	9.10	.72	25.67	32.50
0300	6" opening	G		185	.173		19.95	10.05	.79	30.79	38.50
0400	10" opening	G		170	.188		30	10.95	.86	41.81	51
0500	16" opening	G		155	.206		43.50	12	.95	56.45	67.50
0700	Add per inch for additional widths to 24"	G					1.87			1.87	2.05
0900	For custom fabrication, add						50%				
1100	For 1/4" plain cover plate, deduct						12%				
1500	For cover recessed for tile, 1/4" thick, deduct						12%				
1600	3/8" thick, add						5%				
1800	For checkered plate cover, 1/4" thick, deduct						12%				
1900	3/8" thick, add						2%				
2100	For slotted or round holes in cover, 1/4" thick, add						3%				
2200	3/8" thick, add						4%				
2300	For abrasive cover, add					↓	12%				

05 55 Metal Stair Treads and Nosings

05 55 13 – Metal Stair Treads

05 55 13.50 Stair Treads		Crew	Daily Output	Labor-Hours	Unit	Material	2020 Bare Costs Labor	Equipment	Total	Total Incl O&P
0010	**STAIR TREADS**, stringers and bolts not included									
3000	Diamond plate treads, steel, 1/8" thick									
3005	Open riser, black enamel									
3010	9" deep x 36" long	G	2 Sswk	48	.333	Ea.	112	19.20		131.20	153
3020	42" long	G		48	.333		118	19.20		137.20	160
3030	48" long	G		48	.333		123	19.20		142.20	165
3040	11" deep x 36" long	G		44	.364		118	21		139	163
3050	42" long	G		44	.364		125	21		146	170
3060	48" long	G		44	.364		133	21		154	179
3110	Galvanized, 9" deep x 36" long	G		48	.333		177	19.20		196.20	224
3120	42" long	G		48	.333		188	19.20		207.20	237
3130	48" long	G		48	.333		199	19.20		218.20	249
3140	11" deep x 36" long	G		44	.364		183	21		204	234
3150	42" long	G		44	.364		199	21		220	252
3160	48" long	G	↓	44	.364	↓	207	21		228	261
3200	Closed riser, black enamel									
3210	12" deep x 36" long	G	2 Sswk	40	.400	Ea.	135	23		158	185
3220	42" long	G		40	.400		147	23		170	197
3230	48" long	G		40	.400		155	23		178	206
3240	Galvanized, 12" deep x 36" long	G		40	.400		213	23		236	271
3250	42" long	G		40	.400		233	23		256	292
3260	48" long	G	↓	40	.400	↓	244	23		267	305
4000	Bar grating treads									
4005	Steel, 1-1/4" x 3/16" bars, anti-skid nosing, black enamel									
4010	8-5/8" deep x 30" long	G	2 Sswk	48	.333	Ea.	56	19.20		75.20	91.50
4020	36" long	G		48	.333		65.50	19.20		84.70	102
4030	48" long	G		48	.333		101	19.20		120.20	141
4040	10-15/16" deep x 36" long	G		44	.364		72.50	21		93.50	113
4050	48" long	G		44	.364		104	21		125	147
4060	Galvanized, 8-5/8" deep x 30" long	G		48	.333		64	19.20		83.20	101
4070	36" long	G		48	.333		76	19.20		95.20	114
4080	48" long	G		48	.333		112	19.20		131.20	153
4090	10-15/16" deep x 36" long	G		44	.364		89	21		110	131
4100	48" long	G	↓	44	.364	↓	115	21		136	160
4200	Aluminum, 1-1/4" x 3/16" bars, serrated, with nosing									
4210	7-5/8" deep x 18" long	G	2 Sswk	52	.308	Ea.	53	17.75		70.75	86.50
4220	24" long	G		52	.308		63	17.75		80.75	97
4230	30" long	G		52	.308		73	17.75		90.75	108
4240	36" long	G		52	.308		188	17.75		205.75	235
4250	8-13/16" deep x 18" long	G		48	.333		71.50	19.20		90.70	109
4260	24" long	G		48	.333		106	19.20		125.20	146
4270	30" long	G		48	.333		122	19.20		141.20	164
4280	36" long	G		48	.333		207	19.20		226.20	258
4290	10" deep x 18" long	G		44	.364		139	21		160	185
4300	30" long	G		44	.364		184	21		205	236
4310	36" long	G	↓	44	.364	↓	224	21		245	279
5000	Channel grating treads									
5005	Steel, 14 ga., 2-1/2" thick, galvanized									
5010	9" deep x 36" long	G	2 Sswk	48	.333	Ea.	106	19.20		125.20	147
5020	48" long	G	"	48	.333	"	143	19.20		162.20	188

05 55 Metal Stair Treads and Nosings

05 55 19 – Metal Stair Tread Covers

05 55 19.50 Stair Tread Covers for Renovation

	05 55 19.50 Stair Tread Covers for Renovation	Crew	Daily Output	Labor-Hours	Unit	Material	2020 Bare Costs Labor	Equipment	Total	Total Incl O&P
0010	**STAIR TREAD COVERS FOR RENOVATION**									
0205	Extruded tread cover with nosing, pre-drilled, includes screws									
0210	Aluminum with black abrasive strips, 9" wide x 3' long	1 Carp	24	.333	Ea.	104	17.70		121.70	142
0220	4' long		22	.364		136	19.35		155.35	179
0230	5' long		20	.400		186	21.50		207.50	236
0240	11" wide x 3' long		24	.333		147	17.70		164.70	188
0250	4' long		22	.364		199	19.35		218.35	247
0260	5' long		20	.400		238	21.50		259.50	293
0305	Black abrasive strips with yellow front strips									
0310	Aluminum, 9" wide x 3' long	1 Carp	24	.333	Ea.	124	17.70		141.70	164
0320	4' long		22	.364		166	19.35		185.35	211
0330	5' long		20	.400		213	21.50		234.50	266
0340	11" wide x 3' long		24	.333		161	17.70		178.70	204
0350	4' long		22	.364		204	19.35		223.35	254
0360	5' long		20	.400		267	21.50		288.50	325
0405	Black abrasive strips with photoluminescent front strips									
0410	Aluminum, 9" wide x 3' long	1 Carp	24	.333	Ea.	157	17.70		174.70	200
0420	4' long		22	.364		178	19.35		197.35	225
0430	5' long		20	.400		223	21.50		244.50	277
0440	11" wide x 3' long		24	.333		162	17.70		179.70	205
0450	4' long		22	.364		216	19.35		235.35	266
0460	5' long		20	.400		270	21.50		291.50	330

05 56 Metal Castings

05 56 13 – Metal Construction Castings

05 56 13.50 Construction Castings

	05 56 13.50 Construction Castings		Crew	Daily Output	Labor-Hours	Unit	Material	2020 Bare Costs Labor	Equipment	Total	Total Incl O&P
0010	**CONSTRUCTION CASTINGS**										
0020	Manhole covers and frames, see Section 33 44 13.13										
0100	Column bases, cast iron, 16" x 16", approx. 65 lb.	G	E-4	46	.696	Ea.	140	40.50	3.19	183.69	221
0200	32" x 32", approx. 256 lb.	G		23	1.391	"	525	81	6.40	612.40	710
0600	Miscellaneous C.I. castings, light sections, less than 150 lb.	G		3200	.010	Lb.	9.10	.58	.05	9.73	11
1100	Heavy sections, more than 150 lb.	G		4200	.008		5.30	.44	.04	5.78	6.60
1300	Special low volume items	G		3200	.010		11.05	.58	.05	11.68	13.10
1500	For ductile iron, add						100%				

05 58 Formed Metal Fabrications

05 58 21 – Formed Chain

05 58 21.05 Alloy Steel Chain

	05 58 21.05 Alloy Steel Chain		Crew	Daily Output	Labor-Hours	Unit	Material	2020 Bare Costs Labor	Equipment	Total	Total Incl O&P
0010	**ALLOY STEEL CHAIN**, Grade 80, for lifting										
0015	Self-colored, cut lengths, 1/4"	G	E-17	4	4	C.L.F.	920	235		1,155	1,400
0020	3/8"	G		2	8		1,400	470		1,870	2,275
0030	1/2"	G		1.20	13.333		2,300	780		3,080	3,750
0040	5/8"	G		.72	22.222		3,800	1,300		5,100	6,225
0050	3/4"	G	E-18	.48	83.333		3,750	4,825	3,100	11,675	15,000
0060	7/8"	G		.40	100		7,425	5,775	3,725	16,925	21,300
0070	1"	G		.35	114		9,200	6,625	4,250	20,075	25,100
0080	1-1/4"	G		.24	167		13,600	9,650	6,200	29,450	36,700
0110	Hook, Grade 80, Clevis slip, 1/4"	G				Ea.	32.50			32.50	35.50
0120	3/8"	G					51.50			51.50	56.50

05 58 Formed Metal Fabrications

05 58 21 – Formed Chain

05 58 21.05 Alloy Steel Chain

		Crew	Daily Output	Labor-Hours	Unit	Material	2020 Bare Costs Labor	Equipment	Total	Total Incl O&P
0130	1/2"	G			Ea.	75			75	82.50
0140	5/8"	G				107			107	117
0150	3/4"	G				123			123	135
0160	Hook, Grade 80, eye/sling w/hammerlock coupling, 15 ton	G				425			425	470
0170	22 ton	G				945			945	1,050
0180	37 ton	G				3,050			3,050	3,350

05 58 25 – Formed Lamp Posts

05 58 25.40 Lamp Posts

		Crew	Daily Output	Labor-Hours	Unit	Material	2020 Bare Costs Labor	Equipment	Total	Total Incl O&P
0010	**LAMP POSTS**									
0020	Aluminum, 7' high, stock units, post only	G	1 Carp	16	.500	Ea.	84	26.50	110.50	132
0100	Mild steel, plain	G	"	16	.500	"	86	26.50	112.50	135

05 71 Decorative Metal Stairs

05 71 13 – Fabricated Metal Spiral Stairs

05 71 13.50 Spiral Stairs

		Crew	Daily Output	Labor-Hours	Unit	Material	2020 Bare Costs Labor	Equipment	Total	Total Incl O&P
0010	**SPIRAL STAIRS**									
1805	Shop fabricated, custom ordered									
1810	Aluminum, 5'-0" diameter, plain units	G	E-4	45	.711	Riser	720	41.50	3.26	764.76	860
1900	Cast iron, 4'-0" diameter, plain units	G		45	.711		715	41.50	3.26	759.76	855
1920	Fancy units	G		25	1.280		1,400	74.50	5.90	1,480.40	1,675
2000	Steel, industrial checkered plate, 4' diameter	G		45	.711		700	41.50	3.26	744.76	835
2200	6' diameter	G		40	.800		715	46.50	3.67	765.17	860
3100	Spiral stair kits, 12 stacking risers to fit exact floor height									
3110	Steel, flat metal treads, primed, 3'-6" diameter	G	2 Carp	1.60	10	Flight	1,300	530		1,830	2,225
3120	4'-0" diameter	G		1.45	11.034		1,500	585		2,085	2,525
3130	4'-6" diameter	G		1.35	11.852		1,650	630		2,280	2,775
3140	5'-0" diameter	G		1.25	12.800		1,850	680		2,530	3,050
3210	Galvanized, 3'-6" diameter	G		1.60	10		1,875	530		2,405	2,850
3220	4'-0" diameter	G		1.45	11.034		2,150	585		2,735	3,225
3230	4'-6" diameter	G		1.35	11.852		2,325	630		2,955	3,500
3240	5'-0" diameter	G		1.25	12.800		2,575	680		3,255	3,850
3310	Checkered plate tread, primed, 3'-6" diameter	G		1.45	11.034		1,575	585		2,160	2,625
3320	4'-0" diameter	G		1.35	11.852		1,725	630		2,355	2,850
3330	4'-6" diameter	G		1.25	12.800		1,875	680		2,555	3,075
3340	5'-0" diameter	G		1.15	13.913		1,950	740		2,690	3,250
3410	Galvanized, 3'-6" diameter	G		1.45	11.034		2,150	585		2,735	3,250
3420	4'-0" diameter	G		1.35	11.852		2,425	630		3,055	3,600
3430	4'-6" diameter	G		1.25	12.800		2,625	680		3,305	3,900
3440	5'-0" diameter	G		1.15	13.913		2,925	740		3,665	4,300
3510	Red oak covers on flat metal treads, 3'-6" diameter			1.35	11.852		2,375	630		3,005	3,575
3520	4'-0" diameter			1.25	12.800		2,850	680		3,530	4,175
3530	4'-6" diameter			1.15	13.913		3,075	740		3,815	4,500
3540	5'-0" diameter			1.05	15.238		3,300	810		4,110	4,850

For customer support on your Site Work & Landscape Costs with RSMeans data, call 800.448.8182.

133

05 73 Decorative Metal Railings

05 73 16 – Wire Rope Decorative Metal Railings

05 73 16.10 Cable Railings		Crew	Daily Output	Labor-Hours	Unit	Material	2020 Bare Costs Labor	2020 Bare Costs Equipment	Total	Total Incl O&P
0010	**CABLE RAILINGS**, with 316 stainless steel 1 x 19 cable, 3/16" diameter									
0015	Made from recycled materials									
0100	1-3/4" diameter stainless steel posts x 42" high, cables 4" OC	G	2 Sswk	25	.640	L.F.	43.50	37		80.50	106

05 73 23 – Ornamental Railings

05 73 23.50 Railings, Ornamental		Crew	Daily Output	Labor-Hours	Unit	Material	2020 Bare Costs Labor	2020 Bare Costs Equipment	Total	Total Incl O&P
0010	**RAILINGS, ORNAMENTAL**, 3'-6" high, posts @ 6' OC									
0020	Bronze or stainless, hand forged, plain	G	2 Sswk	24	.667	L.F.	265	38.50		303.50	355
0100	Fancy	G		18	.889		525	51		576	660
0200	Aluminum, panelized, plain	G		24	.667		13.35	38.50		51.85	75
0300	Fancy	G		18	.889		25.50	51		76.50	109
0400	Wrought iron, hand forged, plain	G		24	.667		96.50	38.50		135	167
0500	Fancy	G		18	.889		231	51		282	335
0550	Steel, panelized, plain	G		24	.667		21	38.50		59.50	84
0560	Fancy	G		18	.889		33	51		84	117
0600	Composite metal/wood/glass, plain			18	.889		145	51		196	240
0700	Fancy			12	1.333		290	77		367	440

Estimating Tips
06 05 00 Common Work Results for Wood, Plastics, and Composites

- Common to any wood-framed structure are the accessory connector items such as screws, nails, adhesives, hangers, connector plates, straps, angles, and hold-downs. For typical wood-framed buildings, such as residential projects, the aggregate total for these items can be significant, especially in areas where seismic loading is a concern. For floor and wall framing, the material cost is based on 10 to 25 lbs. of accessory connectors per MBF. Hold-downs, hangers, and other connectors should be taken off by the piece.

  Included with material costs are fasteners for a normal installation. Gordian's RSMeans engineers use manufacturers' recommendations, written specifications, and/or standard construction practice for the sizing and spacing of fasteners. Prices for various fasteners are shown for informational purposes only. Adjustments should be made if unusual fastening conditions exist.

06 10 00 Carpentry

- Lumber is a traded commodity and therefore sensitive to supply and demand in the marketplace. Even with "budgetary" estimating of wood-framed projects, it is advisable to call local suppliers for the latest market pricing.

- The common quantity unit for wood-framed projects is "thousand board feet" (MBF). A board foot is a volume of wood—1" x 1' x 1' or 144 cubic inches. Board-foot quantities are generally calculated using nominal material dimensions—dressed sizes are ignored. Board foot per lineal foot of any stick of lumber can be calculated by dividing the nominal cross-sectional area by 12. As an example, 2,000 lineal feet of 2 x 12 equates to 4 MBF by dividing the nominal area, 2 x 12, by 12, which equals 2, and multiplying that by 2,000 to give 4,000 board feet. This simple rule applies to all nominal dimensioned lumber.

- Waste is an issue of concern at the quantity takeoff for any area of construction. Framing lumber is sold in even foot lengths, i.e., 8', 10', 12', 14', 16', and depending on spans, wall heights, and the grade of lumber, waste is inevitable. A rule of thumb for lumber waste is 5–10% depending on material quality and the complexity of the framing.

- Wood in various forms and shapes is used in many projects, even where the main structural framing is steel, concrete, or masonry. Plywood as a back-up partition material and 2x boards used as blocking and cant strips around roof edges are two common examples. The estimator should ensure that the costs of all wood materials are included in the final estimate.

06 20 00 Finish Carpentry

- It is necessary to consider the grade of workmanship when estimating labor costs for erecting millwork and an interior finish. In practice, there are three grades: premium, custom, and economy. The RSMeans daily output for base and case moldings is in the range of 200 to 250 L.F. per carpenter per day. This is appropriate for most average custom-grade projects. For premium projects, an adjustment to productivity of 25–50% should be made, depending on the complexity of the job.

Reference Numbers

Reference numbers are shown at the beginning of some major classifications. These numbers refer to related items in the Reference Section. The reference information may be an estimating procedure, an alternate pricing method, or technical information.

Note: Not all subdivisions listed here necessarily appear. ■

Same Data. Simplified.

Enjoy the convenience and efficiency of accessing your costs anywhere:

- **Skip the multiplier** by setting your location
- **Quickly search,** edit, favorite and share costs
- **Stay on top of price changes** with automatic updates

Discover more at rsmeans.com/online

06 05 Common Work Results for Wood, Plastics, and Composites

06 05 05 – Selective Demolition for Wood, Plastics, and Composites

06 05 05.10 Selective Demolition Wood Framing

		Crew	Daily Output	Labor-Hours	Unit	Material	2020 Bare Costs Labor	Equipment	Total	Total Incl O&P
0010	**SELECTIVE DEMOLITION WOOD FRAMING** R024119-10									
0100	Timber connector, nailed, small	1 Clab	96	.083	Ea.		3.51		3.51	5.25
0110	Medium		60	.133			5.60		5.60	8.45
0120	Large		48	.167			7		7	10.55
0130	Bolted, small		48	.167			7		7	10.55
0140	Medium		32	.250			10.55		10.55	15.80
0150	Large		24	.333			14.05		14.05	21
3162	Alternate pricing method	B-1	1.10	21.818	M.B.F.		935		935	1,400

06 05 23 – Wood, Plastic, and Composite Fastenings

06 05 23.60 Timber Connectors

		Crew	Daily Output	Labor-Hours	Unit	Material	2020 Bare Costs Labor	Equipment	Total	Total Incl O&P
0010	**TIMBER CONNECTORS**									
0020	Add up cost of each part for total cost of connection									
0100	Connector plates, steel, with bolts, straight	2 Carp	75	.213	Ea.	27.50	11.35		38.85	47.50
0110	Tee, 7 ga.		50	.320		41.50	17		58.50	71
0120	T- Strap, 14 ga., 12" x 8" x 2"		50	.320		41.50	17		58.50	71
0150	Anchor plates, 7 ga., 9" x 7"		75	.213		27.50	11.35		38.85	47.50
0200	Bolts, machine, sq. hd. with nut & washer, 1/2" diameter, 4" long	1 Carp	140	.057		.77	3.04		3.81	5.40
0300	7-1/2" long		130	.062		1.41	3.27		4.68	6.45
0500	3/4" diameter, 7-1/2" long		130	.062		3.51	3.27		6.78	8.75
0610	Machine bolts, w/nut, washer, 3/4" diameter, 15" L, HD's & beam hangers		95	.084		6.60	4.48		11.08	13.95
0800	Drilling bolt holes in timber, 1/2" diameter		450	.018	Inch		.95		.95	1.42
0900	1" diameter		350	.023	"		1.22		1.22	1.83

06 11 Wood Framing

06 11 10 – Framing with Dimensional, Engineered or Composite Lumber

06 11 10.14 Posts and Columns

		Crew	Daily Output	Labor-Hours	Unit	Material	2020 Bare Costs Labor	Equipment	Total	Total Incl O&P
0010	**POSTS AND COLUMNS**									
0100	4" x 4"	2 Carp	390	.041	L.F.	2.04	2.18		4.22	5.50
0150	4" x 6"		275	.058		3.06	3.09		6.15	8
0200	4" x 8"		220	.073		4.33	3.87		8.20	10.55
0250	6" x 6"		215	.074		5.35	3.96		9.31	11.85
0300	6" x 8"		175	.091		8.35	4.86		13.21	16.45
0350	6" x 10"		150	.107		7.10	5.65		12.75	16.30

06 11 10.20 Framing, Light

		Crew	Daily Output	Labor-Hours	Unit	Material	2020 Bare Costs Labor	Equipment	Total	Total Incl O&P
0010	**FRAMING, LIGHT**									
0020	Average cost for all light framing	2 Carp	1.05	15.238	M.B.F.	760	810		1,570	2,075
2870	Redwood joists, 2" x 6"		1.50	10.667		3,175	565		3,740	4,325
2880	2" x 8"		1.75	9.143		3,825	485		4,310	4,925
2890	2" x 10"		1.75	9.143		4,150	485		4,635	5,300
4000	Platform framing, 2" x 4"		.70	22.857		710	1,225		1,935	2,600
4100	2" x 6"		.76	21.053		685	1,125		1,810	2,425
4200	Pressure treated 6" x 6"		.80	20		1,325	1,075		2,400	3,050
4220	8" x 8"		.60	26.667		1,875	1,425		3,300	4,175
4500	Cedar post, columns & girts, 4" x 4"		.52	30.769		3,075	1,625		4,700	5,850
4510	4" x 6"		.55	29.091		3,300	1,550		4,850	5,950
4520	6" x 6"		.65	24.615		4,725	1,300		6,025	7,175
4530	8" x 8"		.70	22.857		4,675	1,225		5,900	6,950
4540	Redwood post, columns & girts, 4" x 4"		.52	30.769		4,950	1,625		6,575	7,900
4550	4" x 6"		.55	29.091		6,400	1,550		7,950	9,375
4560	6" x 6"		.65	24.615		8,400	1,300		9,700	11,200

06 11 Wood Framing

06 11 10 − Framing with Dimensional, Engineered or Composite Lumber

06 11 10.20 Framing, Light	Crew	Daily Output	Labor-Hours	Unit	Material	2020 Bare Costs Labor	Equipment	Total	Total Incl O&P
4570	8" x 8"	2 Carp	.70	22.857	M.B.F.	9,000	1,225		10,225	11,700

06 11 10.34 Sleepers

		Crew	Daily Output	Labor-Hours	Unit	Material	Labor	Equipment	Total	Total Incl O&P
0010	**SLEEPERS**									
0100	On concrete, treated, 1" x 2"	2 Carp	2350	.007	L.F.	.31	.36		.67	.88
0150	1" x 3"		2000	.008		.51	.43		.94	1.20
0200	2" x 4"		1500	.011		.61	.57		1.18	1.52
0250	2" x 6"	↓	1300	.012	↓	.84	.65		1.49	1.91

06 11 10.38 Treated Lumber Framing Material

		Crew	Daily Output	Labor-Hours	Unit	Material	Labor	Equipment	Total	Total Incl O&P
0010	**TREATED LUMBER FRAMING MATERIAL**									
0100	2" x 4"				M.B.F.	850			850	935
0110	2" x 6"					755			755	830
0120	2" x 8"					915			915	1,000
0130	2" x 10"					845			845	930
0140	2" x 12"					1,150			1,150	1,275
0200	4" x 4"					1,025			1,025	1,125
0210	4" x 6"					1,075			1,075	1,200
0220	4" x 8"					1,575			1,575	1,725

06 11 10.42 Furring

		Crew	Daily Output	Labor-Hours	Unit	Material	Labor	Equipment	Total	Total Incl O&P
0010	**FURRING**									
0012	Wood strips, 1" x 2", on walls, on wood	1 Carp	550	.015	L.F.	.29	.77		1.06	1.48
0015	On wood, pneumatic nailed		710	.011		.29	.60		.89	1.22
0300	On masonry		495	.016		.31	.86		1.17	1.64
0400	On concrete		260	.031		.31	1.64		1.95	2.81
0600	1" x 3", on walls, on wood		550	.015		.44	.77		1.21	1.64
0605	On wood, pneumatic nailed		710	.011		.44	.60		1.04	1.38
0700	On masonry		495	.016		.47	.86		1.33	1.81
0800	On concrete	↓	260	.031	↓	.47	1.64		2.11	2.98

06 13 Heavy Timber Construction

06 13 33 − Heavy Timber Pier Construction

06 13 33.50 Jetties, Docks, Fixed

		Crew	Daily Output	Labor-Hours	Unit	Material	Labor	Equipment	Total	Total Incl O&P
0010	**JETTIES, DOCKS, FIXED**									
0020	Pile supported, treated wood,									
0030	5' x 20' platform, freshwater	F-3	115	.348	S.F.	19.10	18.90	4.10	42.10	54
0060	Saltwater		115	.348		41	18.90	4.10	64	78.50
0100	6' x 20' platform, freshwater		110	.364		16.70	19.75	4.28	40.73	52.50
0160	Saltwater		110	.364		36.50	19.75	4.28	60.53	74
0200	8' x 20' platform, freshwater		105	.381		13.90	20.50	4.49	38.89	51
0260	Saltwater		105	.381		31.50	20.50	4.49	56.49	70.50
0420	5' x 30' platform, freshwater		100	.400		18.35	21.50	4.71	44.56	57.50
0460	Saltwater	↓	100	.400	↓	55	21.50	4.71	81.21	98.50
0500	For Greenhart lumber, add					40%				
0550	Diagonal planking, add								25%	25%

06 13 33.52 Jetties, Piers

		Crew	Daily Output	Labor-Hours	Unit	Material	Labor	Equipment	Total	Total Incl O&P
0010	**JETTIES, PIERS**, Municipal with 3" x 12" framing and 3" decking,									
0020	wood piles and cross bracing, alternate bents battered	B-76	60	1.200	S.F.	79	66	61.50	206.50	255
0200	Treated piles, not including mobilization									
0210	50' long, CCA treated, shore driven	B-19	540	.119	V.L.F.	20.50	6.55	3.72	30.77	36.50
0220	Barge driven	B-76	320	.225		20.50	12.40	11.55	44.45	54
0230	ACZA treated, shore driven	B-19	540	.119	↓	16.90	6.55	3.72	27.17	32.50

06 13 Heavy Timber Construction

06 13 33 – Heavy Timber Pier Construction

06 13 33.52 Jetties, Piers

		Crew	Daily Output	Labor-Hours	Unit	Material	2020 Bare Costs Labor	Equipment	Total	Total Incl O&P
0240	Barge driven	B-76	320	.225	V.L.F.	16.90	12.40	11.55	40.85	50.50
0250	30' long, CCA treated, shore driven	B-19	540	.119		18.90	6.55	3.72	29.17	35
0260	Barge driven	B-76	320	.225		18.90	12.40	11.55	42.85	52.50
0270	ACZA treated, shore driven	B-19	540	.119		7.35	6.55	3.72	17.62	22
0280	Barge driven	B-76	320	.225	↓	7.35	12.40	11.55	31.30	40
0300	Mobilization, barge, by tug boat	B-83	25	.640	Mile		31.50	28.50	60	79
0350	Standby time for shore pile driving crew	B-19	8	8	Hr.		440	251	691	950
0360	Standby time for barge driving rig	B-76	8	9	"		495	460	955	1,275

06 18 Glued-Laminated Construction

06 18 13 – Glued-Laminated Beams

06 18 13.20 Laminated Framing

		Crew	Daily Output	Labor-Hours	Unit	Material	2020 Bare Costs Labor	Equipment	Total	Total Incl O&P
0010	**LAMINATED FRAMING**									
0200	Straight roof beams, 20' clear span, beams 8' OC	F-3	2560	.016	SF Flr.	2.27	.85	.18	3.30	3.98
0500	40' clear span, beams 8' OC	"	3200	.013		4.31	.68	.15	5.14	5.90
0800	60' clear span, beams 8' OC	F-4	2880	.017		7.40	.90	.34	8.64	9.80
1700	Radial arches, 60' clear span, frames 8' OC		1920	.025		9.70	1.34	.51	11.55	13.20
2000	100' clear span, frames 8' OC		1600	.030		10	1.61	.61	12.22	14.10
2300	120' clear span, frames 8' OC	↓	1440	.033		13.25	1.79	.68	15.72	18.05
2600	Bowstring trusses, 20' OC, 40' clear span	F-3	2400	.017		6	.91	.20	7.11	8.20
2700	60' clear span	F-4	3600	.013		5.40	.72	.27	6.39	7.30
2800	100' clear span	"	4000	.012	↓	7.65	.64	.25	8.54	9.65
4300	Alternate pricing method: (use nominal footage of									
4310	components). Straight beams, camber less than 6"	F-3	3.50	11.429	M.B.F.	3,075	620	135	3,830	4,450
4600	Curved members, radius over 32'		2.50	16		3,375	870	188	4,433	5,200
4700	Radius 10' to 32'	↓	3	13.333	↓	3,350	725	157	4,232	4,950
6000	Laminated veneer members, southern pine or western species									
6050	1-3/4" wide x 5-1/2" deep	2 Carp	480	.033	L.F.	3.36	1.77		5.13	6.35
6100	9-1/2" deep		480	.033		4.56	1.77		6.33	7.65
6150	14" deep		450	.036		7.35	1.89		9.24	10.90
6200	18" deep	↓	450	.036	↓	10.15	1.89		12.04	14.05
6300	Parallel strand members, southern pine or western species									
6350	1-3/4" wide x 9-1/4" deep	2 Carp	480	.033	L.F.	5.05	1.77		6.82	8.20
6400	11-1/4" deep		450	.036		5.65	1.89		7.54	9.10
6450	14" deep		400	.040		8.35	2.13		10.48	12.40
6500	3-1/2" wide x 9-1/4" deep		480	.033		18.10	1.77		19.87	22.50
6550	11-1/4" deep		450	.036		19.70	1.89		21.59	24.50
6600	14" deep		400	.040		22.50	2.13		24.63	28
6650	7" wide x 9-1/4" deep		450	.036		33	1.89		34.89	39.50
6700	11-1/4" deep		420	.038		42	2.03		44.03	49
6750	14" deep	↓	400	.040	↓	50	2.13		52.13	58

06 44 Ornamental Woodwork

06 44 39 – Wood Posts and Columns

06 44 39.20 Columns	Crew	Daily Output	Labor-Hours	Unit	Material	2020 Bare Costs Labor	Equipment	Total	Total Incl O&P
0010 **COLUMNS**									
4000 Rough sawn cedar posts, 4" x 4"	2 Carp	250	.064	V.L.F.	4.61	3.40		8.01	10.15
4100 4" x 6"		235	.068		8.55	3.62		12.17	14.85
4200 6" x 6"		220	.073		13.10	3.87		16.97	20
4300 8" x 8"		200	.080		21	4.25		25.25	29.50

For customer support on your Site Work & Landscape Costs with RSMeans data, call 800.448.8182.

139

Division Notes

	CREW	DAILY OUTPUT	LABOR-HOURS	UNIT	BARE COSTS				TOTAL INCL O&P
					MAT.	LABOR	EQUIP.	TOTAL	

Estimating Tips
07 10 00 Dampproofing and Waterproofing
- Be sure of the job specifications before pricing this subdivision. The difference in cost between waterproofing and dampproofing can be great. Waterproofing will hold back standing water. Dampproofing prevents the transmission of water vapor. Also included in this section are vapor retarding membranes.

07 20 00 Thermal Protection
- Insulation and fireproofing products are measured by area, thickness, volume, or R-value. Specifications may give only what the specific R-value should be in a certain situation. The estimator may need to choose the type of insulation to meet that R-value.

07 30 00 Steep Slope Roofing
07 40 00 Roofing and Siding Panels
- Many roofing and siding products are bought and sold by the square. One square is equal to an area that measures 100 square feet.

  This simple change in unit of measure could create a large error if the estimator is not observant. Accessories necessary for a complete installation must be figured into any calculations for both material and labor.

07 50 00 Membrane Roofing
07 60 00 Flashing and Sheet Metal
07 70 00 Roofing and Wall Specialties and Accessories
- The items in these subdivisions compose a roofing system. No one component completes the installation, and all must be estimated. Built-up or single-ply membrane roofing systems are made up of many products and installation trades. Wood blocking at roof perimeters or penetrations, parapet coverings, reglets, roof drains, gutters, downspouts, sheet metal flashing, skylights, smoke vents, and roof hatches all need to be considered along with the roofing material. Several different installation trades will need to work together on the roofing system. Inherent difficulties in the scheduling and coordination of various trades must be accounted for when estimating labor costs.

07 90 00 Joint Protection
- To complete the weather-tight shell, the sealants and caulkings must be estimated. Where different materials meet—at expansion joints, at flashing penetrations, and at hundreds of other locations throughout a construction project—caulking and sealants provide another line of defense against water penetration. Often, an entire system is based on the proper location and placement of caulking or sealants. The detailed drawings that are included as part of a set of architectural plans show typical locations for these materials. When caulking or sealants are shown at typical locations, this means the estimator must include them for all the locations where this detail is applicable. Be careful to keep different types of sealants separate, and remember to consider backer rods and primers if necessary.

Reference Numbers
Reference numbers are shown at the beginning of some major classifications. These numbers refer to related items in the Reference Section. The reference information may be an estimating procedure, an alternate pricing method, or technical information.

Note: Not all subdivisions listed here necessarily appear. ■

Same Data. Simplified.

Enjoy the convenience and efficiency of accessing your costs anywhere:
- **Skip the multiplier** by setting your location
- **Quickly search,** edit, favorite and share costs
- **Stay on top of price changes** with automatic updates

Discover more at rsmeans.com/online

07 01 Operation and Maint. of Thermal and Moisture Protection

07 01 90 – Maintenance of Joint Protection

07 01 90.81 Joint Sealant Replacement

07 01 90.81 Joint Sealant Replacement	Crew	Daily Output	Labor-Hours	Unit	Material	2020 Bare Costs Labor	Equipment	Total	Total Incl O&P
0010 **JOINT SEALANT REPLACEMENT**									
0050 Control joints in concrete floors/slabs									
0100 Option 1 for joints with hard dry sealant									
0110 Step 1: Sawcut to remove 95% of old sealant									
0112 1/4" wide x 1/2" deep, with single saw blade	C-27	4800	.003	L.F.	.02	.17	.02	.21	.29
0114 3/8" wide x 3/4" deep, with single saw blade		4000	.004		.03	.20	.03	.26	.35
0116 1/2" wide x 1" deep, with double saw blades		3600	.004		.06	.22	.03	.31	.42
0118 3/4" wide x 1-1/2" deep, with double saw blades		3200	.005		.11	.25	.03	.39	.53
0120 Step 2: Water blast joint faces and edges	C-29	2500	.003			.13	.04	.17	.24
0130 Step 3: Air blast joint faces and edges	C-28	2000	.004			.20	.02	.22	.31
0140 Step 4: Sand blast joint faces and edges	E-11	2000	.016			.75	.15	.90	1.34
0150 Step 5: Air blast joint faces and edges	C-28	2000	.004			.20	.02	.22	.31
0200 Option 2 for joints with soft pliable sealant									
0210 Step 1: Plow joint with rectangular blade	B-62	2600	.009	L.F.		.42	.07	.49	.71
0220 Step 2: Sawcut to re-face joint faces									
0222 1/4" wide x 1/2" deep, with single saw blade	C-27	2400	.007	L.F.	.02	.33	.05	.40	.56
0224 3/8" wide x 3/4" deep, with single saw blade		2000	.008		.04	.40	.06	.50	.69
0226 1/2" wide x 1" deep, with double saw blades		1800	.009		.07	.44	.06	.57	.80
0228 3/4" wide x 1-1/2" deep, with double saw blades		1600	.010		.15	.50	.07	.72	.98
0230 Step 3: Water blast joint faces and edges	C-29	2500	.003			.13	.04	.17	.24
0240 Step 4: Air blast joint faces and edges	C-28	2000	.004			.20	.02	.22	.31
0250 Step 5: Sand blast joint faces and edges	E-11	2000	.016			.75	.15	.90	1.34
0260 Step 6: Air blast joint faces and edges	C-28	2000	.004			.20	.02	.22	.31
0290 For saw cutting new control joints, see Section 03 15 16.20									
8910 For backer rod, see Section 07 91 23.10									
8920 For joint sealant, see Section 03 15 16.30 or 07 92 13.20									

07 11 Dampproofing

07 11 13 – Bituminous Dampproofing

07 11 13.10 Bituminous Asphalt Coating

07 11 13.10 Bituminous Asphalt Coating	Crew	Daily Output	Labor-Hours	Unit	Material	Labor	Equipment	Total	Total Incl O&P
0010 **BITUMINOUS ASPHALT COATING**									
0030 Brushed on, below grade, 1 coat	1 Rofc	665	.012	S.F.	.24	.56		.80	1.18
0100 2 coat		500	.016		.48	.74		1.22	1.74
0300 Sprayed on, below grade, 1 coat		830	.010		.24	.45		.69	1
0400 2 coat		500	.016		.46	.74		1.20	1.73
0500 Asphalt coating, with fibers				Gal.	9.40			9.40	10.35
0600 Troweled on, asphalt with fibers, 1/16" thick	1 Rofc	500	.016	S.F.	.41	.74		1.15	1.67
0700 1/8" thick		400	.020		.72	.92		1.64	2.32
1000 1/2" thick		350	.023		2.35	1.06		3.41	4.32

07 11 16 – Cementitious Dampproofing

07 11 16.20 Cementitious Parging

07 11 16.20 Cementitious Parging	Crew	Daily Output	Labor-Hours	Unit	Material	Labor	Equipment	Total	Total Incl O&P
0010 **CEMENTITIOUS PARGING**									
0020 Portland cement, 2 coats, 1/2" thick	D-1	250	.064	S.F.	.42	3		3.42	5
0100 Waterproofed Portland cement, 1/2" thick, 2 coats	"	250	.064	"	7.25	3		10.25	12.55

07 12 Built-up Bituminous Waterproofing

07 12 13 – Built-Up Asphalt Waterproofing

07 12 13.20 Membrane Waterproofing	Crew	Daily Output	Labor-Hours	Unit	Material	2020 Bare Costs Labor	Equipment	Total	Total Incl O&P
0010 **MEMBRANE WATERPROOFING**									
0012 On slabs, 1 ply, felt, mopped	G-1	3000	.019	S.F.	.40	.81	.19	1.40	1.98
0015 On walls, 1 ply, felt, mopped		3000	.019		.40	.81	.19	1.40	1.98
0100 On slabs, 1 ply, glass fiber fabric, mopped		2100	.027		.45	1.15	.27	1.87	2.69
0105 On walls, 1 ply, glass fiber fabric, mopped		2100	.027		.45	1.15	.27	1.87	2.69
0300 On slabs, 2 ply, felt, mopped		2500	.022		.80	.97	.23	2	2.73
0305 On walls, 2 ply, felt, mopped		2500	.022		.80	.97	.23	2	2.73
0400 On slabs, 2 ply, glass fiber fabric, mopped		1650	.034		.99	1.47	.34	2.80	3.89
0405 On walls, 2 ply, glass fiber fabric, mopped		1650	.034		.99	1.47	.34	2.80	3.89
0600 On slabs, 3 ply, felt, mopped		2100	.027		1.20	1.15	.27	2.62	3.52
0605 On walls, 3 ply, felt, mopped		2100	.027		1.20	1.15	.27	2.62	3.52
0700 On slabs, 3 ply, glass fiber fabric, mopped		1550	.036		1.35	1.56	.37	3.28	4.45
0705 On walls, 3 ply, glass fiber fabric, mopped	▼	1550	.036		1.35	1.56	.37	3.28	4.45
0710 Asphaltic hardboard protection board, 1/8" thick	2 Rofc	500	.032		.76	1.48		2.24	3.28
0715 1/4" thick		450	.036		1.51	1.64		3.15	4.37
1000 EPS membrane protection board, 1/4" thick		3500	.005		.33	.21		.54	.71
1050 3/8" thick		3500	.005		.36	.21		.57	.75
1060 1/2" thick		3500	.005	▼	.40	.21		.61	.79
1070 Fiberglass fabric, black, 20/10 mesh	▼	116	.138	Sq.	15.10	6.35		21.45	27

07 13 Sheet Waterproofing

07 13 53 – Elastomeric Sheet Waterproofing

07 13 53.10 Elastomeric Sheet Waterproofing and Access.	Crew	Daily Output	Labor-Hours	Unit	Material	2020 Bare Costs Labor	Equipment	Total	Total Incl O&P
0010 **ELASTOMERIC SHEET WATERPROOFING AND ACCESS.**									
0090 EPDM, plain, 45 mils thick	2 Rofc	580	.028	S.F.	1.50	1.27		2.77	3.75
0100 60 mils thick		570	.028		1.72	1.30		3.02	4.03
0300 Nylon reinforced sheets, 45 mils thick		580	.028		1.62	1.27		2.89	3.88
0400 60 mils thick	▼	570	.028	▼	1.69	1.30		2.99	4
0600 Vulcanizing splicing tape for above, 2" wide				C.L.F.	61.50			61.50	67.50
0700 4" wide				"	124			124	136
0900 Adhesive, bonding, 60 S.F./gal.				Gal.	28			28	31
1000 Splicing, 75 S.F./gal.				"	37.50			37.50	41.50
1200 Neoprene sheets, plain, 45 mils thick	2 Rofc	580	.028	S.F.	2.17	1.27		3.44	4.49
1300 60 mils thick		570	.028		2.26	1.30		3.56	4.63
1500 Nylon reinforced, 45 mils thick		580	.028		2.40	1.27		3.67	4.74
1600 60 mils thick		570	.028		2.56	1.30		3.86	4.96
1800 120 mils thick	▼	500	.032	▼	5.25	1.48		6.73	8.25
1900 Adhesive, splicing, 150 S.F./gal. per coat				Gal.	37.50			37.50	41.50
2100 Fiberglass reinforced, fluid applied, 1/8" thick	2 Rofc	500	.032	S.F.	1.72	1.48		3.20	4.33
2200 Polyethylene and rubberized asphalt sheets, 60 mils thick		550	.029		.96	1.34		2.30	3.28
2400 Polyvinyl chloride sheets, plain, 10 mils thick		580	.028		.15	1.27		1.42	2.27
2500 20 mils thick		570	.028		.22	1.30		1.52	2.38
2700 30 mils thick	▼	560	.029	▼	.24	1.32		1.56	2.44
3000 Adhesives, trowel grade, 40-100 S.F./gal.				Gal.	23.50			23.50	26
3100 Brush grade, 100-250 S.F./gal.				"	21			21	23.50
3300 Bitumen modified polyurethane, fluid applied, 55 mils thick	2 Rofc	665	.024	S.F.	1.12	1.11		2.23	3.06

07 16 Cementitious and Reactive Waterproofing

07 16 16 – Crystalline Waterproofing

07 16 16.20 Cementitious Waterproofing	Crew	Daily Output	Labor-Hours	Unit	Material	2020 Bare Costs Labor	Equipment	Total	Total Incl O&P
0010 **CEMENTITIOUS WATERPROOFING**									
0020 1/8" application, sprayed on	G-2A	1000	.024	S.F.	.71	.98	.63	2.32	3.05
0050 4 coat cementitious metallic slurry	1 Cefi	1.20	6.667	C.S.F.	75	335		410	575

07 17 Bentonite Waterproofing

07 17 13 – Bentonite Panel Waterproofing

07 17 13.10 Bentonite

07 17 13.10 Bentonite	Crew	Daily Output	Labor-Hours	Unit	Material	2020 Bare Costs Labor	Equipment	Total	Total Incl O&P
0010 **BENTONITE**									
0020 Panels, 4' x 4', 3/16" thick	1 Rofc	625	.013	S.F.	1.83	.59		2.42	2.99
0100 Rolls, 3/8" thick, with geotextile fabric both sides	"	550	.015	"	1.78	.67		2.45	3.07
0300 Granular bentonite, 50 lb. bags (.625 C.F.)				Bag	17.65			17.65	19.40
0400 3/8" thick, troweled on	1 Rofc	475	.017	S.F.	.88	.78		1.66	2.25
0500 Drain board, expanded polystyrene, 1-1/2" thick	1 Rohe	1600	.005		.44	.17		.61	.77
0510 2" thick		1600	.005		.58	.17		.75	.93
0520 3" thick		1600	.005		.87	.17		1.04	1.25
0530 4" thick		1600	.005		1.16	.17		1.33	1.57
0600 With filter fabric, 1-1/2" thick		1600	.005		.50	.17		.67	.84
0625 2" thick		1600	.005		.65	.17		.82	1
0650 3" thick		1600	.005		.94	.17		1.11	1.32
0675 4" thick		1600	.005		1.23	.17		1.40	1.64

07 19 Water Repellents

07 19 19 – Silicone Water Repellents

07 19 19.10 Silicone Based Water Repellents

07 19 19.10 Silicone Based Water Repellents	Crew	Daily Output	Labor-Hours	Unit	Material	2020 Bare Costs Labor	Equipment	Total	Total Incl O&P
0010 **SILICONE BASED WATER REPELLENTS**									
0020 Water base liquid, roller applied	2 Rofc	7000	.002	S.F.	.43	.11		.54	.64
0200 Silicone or stearate, sprayed on CMU, 1 coat	1 Rofc	4000	.002		.40	.09		.49	.59
0300 2 coats	"	3000	.003		.79	.12		.91	1.07

07 21 Thermal Insulation

07 21 13 – Board Insulation

07 21 13.10 Rigid Insulation

07 21 13.10 Rigid Insulation		Crew	Daily Output	Labor-Hours	Unit	Material	2020 Bare Costs Labor	Equipment	Total	Total Incl O&P
0010 **RIGID INSULATION**, for walls										
0040 Fiberglass, 1.5#/C.F., unfaced, 1" thick, R4.1	G	1 Carp	1000	.008	S.F.	.33	.43		.76	1
0060 1-1/2" thick, R6.2	G		1000	.008		.43	.43		.86	1.11
0080 2" thick, R8.3	G		1000	.008		.49	.43		.92	1.18
0120 3" thick, R12.4	G		800	.010		.56	.53		1.09	1.42
0370 3#/C.F., unfaced, 1" thick, R4.3	G		1000	.008		.57	.43		1	1.27
0390 1-1/2" thick, R6.5	G		1000	.008		.78	.43		1.21	1.50
0400 2" thick, R8.7	G		890	.009		1.09	.48		1.57	1.92
0420 2-1/2" thick, R10.9	G		800	.010		1.12	.53		1.65	2.03
0440 3" thick, R13	G		800	.010		1.65	.53		2.18	2.62
0520 Foil faced, 1" thick, R4.3	G		1000	.008		.85	.43		1.28	1.58
0540 1-1/2" thick, R6.5	G		1000	.008		1.28	.43		1.71	2.05
0560 2" thick, R8.7	G		890	.009		1.62	.48		2.10	2.50
0580 2-1/2" thick, R10.9	G		800	.010		1.91	.53		2.44	2.90
0600 3" thick, R13	G		800	.010		2.01	.53		2.54	3.01

07 21 Thermal Insulation

07 21 13 – Board Insulation

07 21 13.10 Rigid Insulation		Crew	Daily Output	Labor-Hours	Unit	Material	2020 Bare Costs Labor	Equipment	Total	Total Incl O&P
1600	Isocyanurate, 4' x 8' sheet, foil faced, both sides									
1610	1/2" thick	G	1 Carp	800	.010	S.F.	.31	.53		.84	1.14
1620	5/8" thick	G		800	.010		.62	.53		1.15	1.48
1630	3/4" thick	G		800	.010		.43	.53		.96	1.27
1640	1" thick	G		800	.010		.62	.53		1.15	1.48
1650	1-1/2" thick	G		730	.011		.76	.58		1.34	1.72
1660	2" thick	G		730	.011		.74	.58		1.32	1.69
1670	3" thick	G		730	.011		3.11	.58		3.69	4.30
1680	4" thick	G		730	.011		2.58	.58		3.16	3.72
1700	Perlite, 1" thick, R2.77	G		800	.010		.47	.53		1	1.32
1750	2" thick, R5.55	G		730	.011		.86	.58		1.44	1.83
1900	Extruded polystyrene, 25 psi compressive strength, 1" thick, R5	G		800	.010		.62	.53		1.15	1.48
1940	2" thick, R10	G		730	.011		1.22	.58		1.80	2.22
1960	3" thick, R15	G		730	.011		1.67	.58		2.25	2.72
2100	Expanded polystyrene, 1" thick, R3.85	G		800	.010		.29	.53		.82	1.12
2120	2" thick, R7.69	G		730	.011		.58	.58		1.16	1.52
2140	3" thick, R11.49	G		730	.011		.87	.58		1.45	1.84

07 21 13.13 Foam Board Insulation

		Crew	Daily Output	Labor-Hours	Unit	Material	2020 Bare Costs Labor	Equipment	Total	Total Incl O&P
0010	**FOAM BOARD INSULATION**										
0600	Polystyrene, expanded, 1" thick, R4	G	1 Carp	680	.012	S.F.	.29	.63		.92	1.26
0700	2" thick, R8	G	"	675	.012	"	.58	.63		1.21	1.59

07 21 23 – Loose-Fill Insulation

07 21 23.20 Masonry Loose-Fill Insulation

		Crew	Daily Output	Labor-Hours	Unit	Material	2020 Bare Costs Labor	Equipment	Total	Total Incl O&P
0010	**MASONRY LOOSE-FILL INSULATION**, vermiculite or perlite										
0100	In cores of concrete block, 4" thick wall, .115 C.F./S.F.	G	D-1	4800	.003	S.F.	.61	.16		.77	.91
0200	6" thick wall, .175 C.F./S.F.	G		3000	.005		.93	.25		1.18	1.40
0300	8" thick wall, .258 C.F./S.F.	G		2400	.007		1.36	.31		1.67	1.97
0400	10" thick wall, .340 C.F./S.F.	G		1850	.009		1.80	.41		2.21	2.60
0500	12" thick wall, .422 C.F./S.F.	G		1200	.013		2.23	.63		2.86	3.41
0600	Poured cavity wall, vermiculite or perlite, water repellent	G		250	.064	C.F.	5.30	3		8.30	10.35
0700	Foamed in place, urethane in 2-5/8" cavity	G	G-2A	1035	.023	S.F.	1.39	.95	.61	2.95	3.72
0800	For each 1" added thickness, add	G	"	2372	.010	"	.53	.41	.27	1.21	1.53

07 25 Weather Barriers

07 25 10 – Weather Barriers or Wraps

07 25 10.10 Weather Barriers

		Crew	Daily Output	Labor-Hours	Unit	Material	2020 Bare Costs Labor	Equipment	Total	Total Incl O&P
0010	**WEATHER BARRIERS**										
0400	Asphalt felt paper, #15		1 Carp	37	.216	Sq.	5.25	11.50		16.75	23
0401	Per square foot		"	3700	.002	S.F.	.05	.11		.16	.23
0450	Housewrap, exterior, spun bonded polypropylene										
0470	Small roll		1 Carp	3800	.002	S.F.	.15	.11		.26	.34
0480	Large roll		"	4000	.002	"	.16	.11		.27	.34
2100	Asphalt felt roof deck vapor barrier, class 1 metal decks		1 Rofc	37	.216	Sq.	21	10		31	40
2200	For all other decks		"	37	.216		15.90	10		25.90	34
2800	Asphalt felt, 50% recycled content, 15 lb., 4 sq./roll		1 Carp	36	.222		6.10	11.80		17.90	24.50
2810	30 lb., 2 sq./roll		"	36	.222		10.45	11.80		22.25	29.50
3000	Building wrap, spun bonded polyethylene		2 Carp	8000	.002	S.F.	.14	.11		.25	.31

07 26 Vapor Retarders

07 26 13 – Above-Grade Vapor Retarders

07 26 13.10 Vapor Retarders

07 26 13.10 Vapor Retarders		Crew	Daily Output	Labor-Hours	Unit	Material	2020 Bare Costs Labor	Equipment	Total	Total Incl O&P
0010	**VAPOR RETARDERS**									
0020	Aluminum and kraft laminated, foil 1 side G	1 Carp	37	.216	Sq.	16.05	11.50		27.55	35
0100	Foil 2 sides G		37	.216		15.50	11.50		27	34.50
0600	Polyethylene vapor barrier, standard, 2 mil G		37	.216		1.63	11.50		13.13	19.05
0700	4 mil G		37	.216		2.62	11.50		14.12	20
0900	6 mil G		37	.216		4	11.50		15.50	21.50
1200	10 mil G		37	.216		9.70	11.50		21.20	28
1300	Clear reinforced, fire retardant, 8 mil G		37	.216		28	11.50		39.50	48.50
1350	Cross laminated type, 3 mil G		37	.216		12.90	11.50		24.40	31.50
1400	4 mil G		37	.216		14.15	11.50		25.65	33
1800	Reinf. waterproof, 2 mil polyethylene backing, 1 side		37	.216		11.35	11.50		22.85	30
1900	2 sides		37	.216		14.35	11.50		25.85	33
2400	Waterproofed kraft with sisal or fiberglass fibers		37	.216		21	11.50		32.50	41

07 31 Shingles and Shakes

07 31 13 – Asphalt Shingles

07 31 13.10 Asphalt Roof Shingles

07 31 13.10 Asphalt Roof Shingles		Crew	Daily Output	Labor-Hours	Unit	Material	2020 Bare Costs Labor	Equipment	Total	Total Incl O&P
0010	**ASPHALT ROOF SHINGLES**									
0100	Standard strip shingles									
0150	Inorganic, class A, 25 year	1 Rofc	5.50	1.455	Sq.	75	67		142	194
0155	Pneumatic nailed		7	1.143		75	53		128	170
0200	30 year		5	1.600		111	74		185	244
0205	Pneumatic nailed		6.25	1.280		111	59		170	220
0250	Standard laminated multi-layered shingles									
0300	Class A, 240-260 lb./square	1 Rofc	4.50	1.778	Sq.	103	82		185	249
0305	Pneumatic nailed		5.63	1.422		103	65.50		168.50	221
0350	Class A, 250-270 lb./square		4	2		103	92.50		195.50	266
0355	Pneumatic nailed		5	1.600		103	74		177	235
0400	Premium, laminated multi-layered shingles									
0450	Class A, 260-300 lb./square	1 Rofc	3.50	2.286	Sq.	148	106		254	335
0455	Pneumatic nailed		4.37	1.831		148	84.50		232.50	305
0500	Class A, 300-385 lb./square		3	2.667		310	123		433	545
0505	Pneumatic nailed		3.75	2.133		310	98.50		408.50	505
0800	#15 felt underlayment		64	.125		5.25	5.80		11.05	15.35
0825	#30 felt underlayment		58	.138		10.20	6.35		16.55	21.50
0850	Self adhering polyethylene and rubberized asphalt underlayment		22	.364		76	16.80		92.80	111
0900	Ridge shingles		330	.024	L.F.	2.31	1.12		3.43	4.39
0905	Pneumatic nailed		412.50	.019	"	2.31	.90		3.21	4.02
1000	For steep roofs (7 to 12 pitch or greater), add						50%			

07 31 26 – Slate Shingles

07 31 26.10 Slate Roof Shingles

07 31 26.10 Slate Roof Shingles		Crew	Daily Output	Labor-Hours	Unit	Material	2020 Bare Costs Labor	Equipment	Total	Total Incl O&P
0010	**SLATE ROOF SHINGLES**									
0100	Buckingham Virginia black, 3/16" - 1/4" thick G	1 Rots	1.75	4.571	Sq.	560	211		771	965

146

For customer support on your Site Work & Landscape Costs with RSMeans data, call 800.448.8182.

07 32 Roof Tiles

07 32 13 – Clay Roof Tiles

07 32 13.10 Clay Tiles	Crew	Daily Output	Labor-Hours	Unit	Material	2020 Bare Costs Labor	Equipment	Total	Total Incl O&P
0010 **CLAY TILES**, including accessories									
0300 Flat shingle, interlocking, 15", 166 pcs./sq., fireflashed blend	3 Rots	6	4	Sq.	535	185		720	895
0500 Terra cotta red		6	4		545	185		730	905
0600 Roman pan and top, 18", 102 pcs./sq., fireflashed blend	↓	5.50	4.364		495	202		697	880
0640 Terra cotta red	1 Rots	2.40	3.333		585	154		739	900
1100 Barrel mission tile, 18", 166 pcs./sq., fireflashed blend	3 Rots	5.50	4.364		460	202		662	840
1140 Terra cotta red		5.50	4.364		465	202		667	845
1700 Scalloped edge flat shingle, 14", 145 pcs./sq., fireflashed blend		6	4		1,125	185		1,310	1,525
1800 Terra cotta red	↓	6	4		1,050	185		1,235	1,450
3010 #15 felt underlayment	1 Rofc	64	.125		5.25	5.80		11.05	15.35
3020 #30 felt underlayment		58	.138		10.20	6.35		16.55	21.50
3040 Polyethylene and rubberized asph. underlayment	↓	22	.364	↓	76	16.80		92.80	111

07 32 16 – Concrete Roof Tiles

07 32 16.10 Concrete Tiles	Crew	Daily Output	Labor-Hours	Unit	Material	2020 Bare Costs Labor	Equipment	Total	Total Incl O&P
0010 **CONCRETE TILES**									
0020 Corrugated, 13" x 16-1/2", 90 per sq., 950 lb./sq.									
0050 Earthtone colors, nailed to wood deck	1 Rots	1.35	5.926	Sq.	107	274		381	565
0150 Blues		1.35	5.926		106	274		380	565
0200 Greens		1.35	5.926		117	274		391	580
0250 Premium colors	↓	1.35	5.926	↓	117	274		391	580
0500 Shakes, 13" x 16-1/2", 90 per sq., 950 lb./sq.									
0600 All colors, nailed to wood deck	1 Rots	1.50	5.333	Sq.	153	247		400	575
1500 Accessory pieces, ridge & hip, 10" x 16-1/2", 8 lb. each	"	120	.067	Ea.	3.72	3.08		6.80	9.20
1700 Rake, 6-1/2" x 16-3/4", 9 lb. each					3.72			3.72	4.09
1800 Mansard hip, 10" x 16-1/2", 9.2 lb. each					3.72			3.72	4.09
1900 Hip starter, 10" x 16-1/2", 10.5 lb. each					10.65			10.65	11.75
2000 3 or 4 way apex, 10" each side, 11.5 lb. each				↓	12.20			12.20	13.40

07 32 19 – Metal Roof Tiles

07 32 19.10 Metal Roof Tiles	Crew	Daily Output	Labor-Hours	Unit	Material	2020 Bare Costs Labor	Equipment	Total	Total Incl O&P
0010 **METAL ROOF TILES**									
0020 Accessories included, .032" thick aluminum, mission tile	1 Carp	2.50	3.200	Sq.	830	170		1,000	1,175
0200 Spanish tiles	"	3	2.667	"	550	142		692	820

07 41 Roof Panels

07 41 13 – Metal Roof Panels

07 41 13.10 Aluminum Roof Panels	Crew	Daily Output	Labor-Hours	Unit	Material	2020 Bare Costs Labor	Equipment	Total	Total Incl O&P
0010 **ALUMINUM ROOF PANELS**									
0020 Corrugated or ribbed, .0155" thick, natural	G-3	1200	.027	S.F.	1.02	1.39		2.41	3.22
0300 Painted		1200	.027		1.50	1.39		2.89	3.75
0400 Corrugated, .018" thick, on steel frame, natural finish		1200	.027		1.25	1.39		2.64	3.48
0600 Painted		1200	.027		1.57	1.39		2.96	3.83
0700 Corrugated, on steel frame, natural, .024" thick		1200	.027		1.82	1.39		3.21	4.10
0800 Painted		1200	.027		2.20	1.39		3.59	4.52
0900 .032" thick, natural		1200	.027		3.15	1.39		4.54	5.55
1200 Painted		1200	.027		3.41	1.39		4.80	5.85
1300 V-Beam, on steel frame construction, .032" thick, natural		1200	.027		2.73	1.39		4.12	5.10
1500 Painted		1200	.027		4.03	1.39		5.42	6.55
1600 .040" thick, natural		1200	.027		3.98	1.39		5.37	6.50
1800 Painted		1200	.027		4.79	1.39		6.18	7.35
1900 .050" thick, natural	↓	1200	.027	↓	4.11	1.39		5.50	6.60

07 41 Roof Panels

07 41 13 - Metal Roof Panels

07 41 13.10 Aluminum Roof Panels

07 41 13.10 Aluminum Roof Panels		Crew	Daily Output	Labor-Hours	Unit	Material	2020 Bare Costs Labor	Equipment	Total	Total Incl O&P
2100	Painted	G-3	1200	.027	S.F.	4.94	1.39		6.33	7.55
2200	For roofing on wood frame, deduct		4600	.007		.08	.36		.44	.64
2400	Ridge cap, .032" thick, natural		800	.040	L.F.	3.36	2.09		5.45	6.85

07 41 13.20 Steel Roofing Panels

	07 41 13.20 Steel Roofing Panels	Crew	Daily Output	Labor-Hours	Unit	Material	2020 Bare Costs Labor	Equipment	Total	Total Incl O&P
0010	**STEEL ROOFING PANELS**									
0012	Corrugated or ribbed, on steel framing, 30 ga. galv	G-3	1100	.029	S.F.	1.80	1.52		3.32	4.27
0100	28 ga.		1050	.030		1.63	1.59		3.22	4.19
0300	26 ga.		1000	.032		2	1.67		3.67	4.72
0400	24 ga.		950	.034		3.06	1.76		4.82	6.05
0600	Colored, 28 ga.		1050	.030		1.90	1.59		3.49	4.49
0700	26 ga.		1000	.032		2.15	1.67		3.82	4.89

07 41 33 - Plastic Roof Panels

07 41 33.10 Fiberglass Panels

	07 41 33.10 Fiberglass Panels	Crew	Daily Output	Labor-Hours	Unit	Material	2020 Bare Costs Labor	Equipment	Total	Total Incl O&P
0010	**FIBERGLASS PANELS**									
0012	Corrugated panels, roofing, 8 oz./S.F.	G-3	1000	.032	S.F.	2.55	1.67		4.22	5.35
0100	12 oz./S.F.		1000	.032		4.68	1.67		6.35	7.65
0300	Corrugated siding, 6 oz./S.F.		880	.036		2.02	1.90		3.92	5.10
0400	8 oz./S.F.		880	.036		2.55	1.90		4.45	5.70
0500	Fire retardant		880	.036		4.02	1.90		5.92	7.30
0600	12 oz. siding, textured		880	.036		3.78	1.90		5.68	7.05
0700	Fire retardant		880	.036		4.74	1.90		6.64	8.05
0900	Flat panels, 6 oz./S.F., clear or colors		880	.036		2.78	1.90		4.68	5.95
1100	Fire retardant, class A		880	.036		3.64	1.90		5.54	6.85
1300	8 oz./S.F., clear or colors		880	.036		2.41	1.90		4.31	5.50
1700	Sandwich panels, fiberglass, 1-9/16" thick, panels to 20 S.F.		180	.178		36.50	9.30		45.80	54
1900	As above, but 2-3/4" thick, panels to 100 S.F.		265	.121		25.50	6.30		31.80	37.50

07 42 Wall Panels

07 42 13 - Metal Wall Panels

07 42 13.20 Aluminum Siding Panels

	07 42 13.20 Aluminum Siding Panels	Crew	Daily Output	Labor-Hours	Unit	Material	2020 Bare Costs Labor	Equipment	Total	Total Incl O&P
0010	**ALUMINUM SIDING PANELS**									
0012	Corrugated, on steel framing, .019" thick, natural finish	G-3	775	.041	S.F.	1.72	2.16		3.88	5.15
0100	Painted		775	.041		1.86	2.16		4.02	5.30
0400	Farm type, .021" thick on steel frame, natural		775	.041		1.76	2.16		3.92	5.20
0600	Painted		775	.041		1.86	2.16		4.02	5.30
0700	Industrial type, corrugated, on steel, .024" thick, mill		775	.041		2.42	2.16		4.58	5.90
0900	Painted		775	.041		2.57	2.16		4.73	6.10
1000	.032" thick, mill		775	.041		2.65	2.16		4.81	6.20
1200	Painted		775	.041		3.28	2.16		5.44	6.85
2750	.050" thick, natural		775	.041		3.52	2.16		5.68	7.15
2760	Painted		775	.041		4.06	2.16		6.22	7.75
3300	For siding on wood frame, deduct from above		2800	.011		.09	.60		.69	1
3400	Screw fasteners, aluminum, self tapping, neoprene washer, 1"				M	219			219	241
3600	Stitch screws, self tapping, with neoprene washer, 5/8"				"	165			165	181
3630	Flashing, sidewall, .032" thick	G-3	800	.040	L.F.	3.11	2.09		5.20	6.60
3650	End wall, .040" thick		800	.040		3.64	2.09		5.73	7.15
3670	Closure strips, corrugated, .032" thick		800	.040		.94	2.09		3.03	4.19
3680	Ribbed, 4" or 8", .032" thick		800	.040		.99	2.09		3.08	4.25
3690	V-beam, .040" thick		800	.040		1.45	2.09		3.54	4.76
3800	Horizontal, colored clapboard, 8" wide, plain	2 Carp	515	.031	S.F.	2.72	1.65		4.37	5.45

07 42 Wall Panels

07 42 13 – Metal Wall Panels

07 42 13.20 Aluminum Siding Panels	Crew	Daily Output	Labor-Hours	Unit	Material	2020 Bare Costs Labor	Equipment	Total	Total Incl O&P
3810	Insulated	2 Carp	515	.031	S.F.	3.17	1.65		4.82	5.95

07 42 13.30 Steel Siding

		Crew	Daily Output	Labor-Hours	Unit	Material	Labor	Equipment	Total	Total Incl O&P
0010	**STEEL SIDING**									
0020	Beveled, vinyl coated, 8" wide	1 Carp	265	.030	S.F.	1.87	1.60		3.47	4.47
0050	10" wide	"	275	.029		1.95	1.55		3.50	4.47
0080	Galv, corrugated or ribbed, on steel frame, 30 ga.	G-3	800	.040		1.20	2.09		3.29	4.48
0100	28 ga.		795	.040		1.30	2.10		3.40	4.60
0300	26 ga.		790	.041		1.69	2.11		3.80	5.05
0400	24 ga.		785	.041		2.22	2.13		4.35	5.65
0600	22 ga.		770	.042		2.31	2.17		4.48	5.80
0700	Colored, corrugated/ribbed, on steel frame, 10 yr. finish, 28 ga.		800	.040		2.33	2.09		4.42	5.70
0900	26 ga.		795	.040		2.01	2.10		4.11	5.40
1000	24 ga.		790	.041		2.24	2.11		4.35	5.65
1020	20 ga.		785	.041		2.85	2.13		4.98	6.35

07 46 Siding

07 46 23 – Wood Siding

07 46 23.10 Wood Board Siding

		Crew	Daily Output	Labor-Hours	Unit	Material	Labor	Equipment	Total	Total Incl O&P
0010	**WOOD BOARD SIDING**									
3200	Wood, cedar bevel, A grade, 1/2" x 6"	1 Carp	295	.027	S.F.	4.46	1.44		5.90	7.10
3300	1/2" x 8"		330	.024		8.50	1.29		9.79	11.30
3500	3/4" x 10", clear grade		375	.021		8.45	1.13		9.58	11
3600	"B" grade		375	.021		4.16	1.13		5.29	6.30
3800	Cedar, rough sawn, 1" x 4", A grade, natural		220	.036		7.90	1.93		9.83	11.60
3900	Stained		220	.036		8	1.93		9.93	11.70
4100	1" x 12", board & batten, #3 & Btr., natural		420	.019		4.77	1.01		5.78	6.75
4200	Stained		420	.019		5.20	1.01		6.21	7.20
4400	1" x 8" channel siding, #3 & Btr., natural		330	.024		5.15	1.29		6.44	7.60
4500	Stained		330	.024		5.05	1.29		6.34	7.50
4700	Redwood, clear, beveled, vertical grain, 1/2" x 4"		220	.036		5	1.93		6.93	8.40
4750	1/2" x 6"		295	.027		5.35	1.44		6.79	8
4800	1/2" x 8"		330	.024		5.45	1.29		6.74	7.90
5000	3/4" x 10"		375	.021		5.35	1.13		6.48	7.55
5400	White pine, rough sawn, 1" x 8", natural		330	.024		2.55	1.29		3.84	4.75

07 46 29 – Plywood Siding

07 46 29.10 Plywood Siding Options

		Crew	Daily Output	Labor-Hours	Unit	Material	Labor	Equipment	Total	Total Incl O&P
0010	**PLYWOOD SIDING OPTIONS**									
0900	Plywood, medium density overlaid, 3/8" thick	2 Carp	750	.021	S.F.	1.35	1.13		2.48	3.19
1000	1/2" thick		700	.023		1.58	1.22		2.80	3.57
1100	3/4" thick		650	.025		2.29	1.31		3.60	4.49
1600	Texture 1-11, cedar, 5/8" thick, natural		675	.024		2.59	1.26		3.85	4.74
1900	Texture 1-11, fir, 5/8" thick, natural		675	.024		1.21	1.26		2.47	3.22
2050	Texture 1-11, S.Y.P., 5/8" thick, natural		675	.024		1.44	1.26		2.70	3.47
2200	Rough sawn cedar, 3/8" thick, natural		675	.024		1.28	1.26		2.54	3.30
2500	Rough sawn fir, 3/8" thick, natural		675	.024		.97	1.26		2.23	2.96

07 46 33 – Plastic Siding

07 46 33.10 Vinyl Siding

		Crew	Daily Output	Labor-Hours	Unit	Material	Labor	Equipment	Total	Total Incl O&P
0010	**VINYL SIDING**									
3995	Clapboard profile, woodgrain texture, .048 thick, double 4	2 Carp	495	.032	S.F.	1.09	1.72		2.81	3.78
4000	Double 5		550	.029		1.09	1.55		2.64	3.52

07 46 Siding

07 46 33 – Plastic Siding

07 46 33.10 Vinyl Siding	Crew	Daily Output	Labor-Hours	Unit	Material	2020 Bare Costs Labor	Equipment	Total	Total Incl O&P
4005	Single 8	2 Carp	495	.032	S.F.	1.65	1.72		3.37	4.40
4010	Single 10		550	.029		1.98	1.55		3.53	4.50
4015	.044 thick, double 4		495	.032		1.08	1.72		2.80	3.77
4020	Double 5		550	.029		1.10	1.55		2.65	3.54
4025	.042 thick, double 4		495	.032		1.08	1.72		2.80	3.77
4030	Double 5		550	.029		1.08	1.55		2.63	3.51
4035	Cross sawn texture, .040 thick, double 4		495	.032		.72	1.72		2.44	3.38
4040	Double 5		550	.029		.65	1.55		2.20	3.04
4045	Smooth texture, .042 thick, double 4		495	.032		.91	1.72		2.63	3.59
4050	Double 5		550	.029		.80	1.55		2.35	3.21
4055	Single 8		495	.032		.80	1.72		2.52	3.47
4060	Cedar texture, .044 thick, double 4		495	.032		1.18	1.72		2.90	3.88
4065	Double 6		600	.027		1.31	1.42		2.73	3.58
4070	Dutch lap profile, woodgrain texture, .048 thick, double 5		550	.029		1.15	1.55		2.70	3.59
4075	.044 thick, double 4.5		525	.030		1.09	1.62		2.71	3.63
4080	.042 thick, double 4.5		525	.030		.92	1.62		2.54	3.45
4085	.040 thick, double 4.5		525	.030		.72	1.62		2.34	3.23
4100	Shake profile, 10" wide		400	.040		4.04	2.13		6.17	7.65
4105	Vertical pattern, .046 thick, double 5		550	.029		1.77	1.55		3.32	4.27
4110	.044 thick, triple 3		550	.029		1.77	1.55		3.32	4.27
4115	.040 thick, triple 4		550	.029		1.72	1.55		3.27	4.22
4120	.040 thick, triple 2.66		550	.029		1.88	1.55		3.43	4.39
4125	Insulation, fan folded extruded polystyrene, 1/4"		2000	.008		.29	.43		.72	.96
4130	3/8"		2000	.008		.32	.43		.75	1
4135	Accessories, J channel, 5/8" pocket		700	.023	L.F.	.57	1.22		1.79	2.45
4140	3/4" pocket		695	.023		.63	1.22		1.85	2.53
4145	1-1/4" pocket		680	.024		1.15	1.25		2.40	3.14
4150	Flexible, 3/4" pocket		600	.027		2.85	1.42		4.27	5.25
4155	Under sill finish trim		500	.032		.59	1.70		2.29	3.21
4160	Vinyl starter strip		700	.023		.73	1.22		1.95	2.63
4165	Aluminum starter strip		700	.023		.31	1.22		1.53	2.17
4170	Window casing, 2-1/2" wide, 3/4" pocket		510	.031		1.78	1.67		3.45	4.46
4175	Outside corner, woodgrain finish, 4" face, 3/4" pocket		700	.023		2.35	1.22		3.57	4.42
4180	5/8" pocket		700	.023		2.29	1.22		3.51	4.35
4185	Smooth finish, 4" face, 3/4" pocket		700	.023		2.36	1.22		3.58	4.43
4190	7/8" pocket		690	.023		1.91	1.23		3.14	3.96
4195	1-1/4" pocket		700	.023		1.35	1.22		2.57	3.32
4200	Soffit and fascia, 1' overhang, solid		120	.133		5.15	7.10		12.25	16.30
4205	Vented		120	.133		5.15	7.10		12.25	16.30
4207	18" overhang, solid		110	.145		6.05	7.75		13.80	18.25
4208	Vented		110	.145		6.05	7.75		13.80	18.25
4210	2' overhang, solid		100	.160		6.90	8.50		15.40	20.50
4215	Vented		100	.160		6.90	8.50		15.40	20.50
4217	3' overhang, solid		100	.160		8.70	8.50		17.20	22.50
4218	Vented		100	.160		8.70	8.50		17.20	22.50
4220	Colors for siding and soffits, add				S.F.	.15			.15	.17
4225	Colors for accessories and trim, add				L.F.	.31			.31	.34

07 46 33.20 Polypropylene Siding

		Crew	Daily Output	Labor-Hours	Unit	Material	Labor	Equipment	Total	Total Incl O&P
0010	**POLYPROPYLENE SIDING**									
4090	Shingle profile, random grooves, double 7	2 Carp	400	.040	S.F.	3.51	2.13		5.64	7.05
4092	Cornerpost for above	1 Carp	365	.022	L.F.	12.85	1.17		14.02	15.90
4095	Triple 5	2 Carp	400	.040	S.F.	3.72	2.13		5.85	7.30

07 46 Siding

07 46 33 – Plastic Siding

07 46 33.20 Polypropylene Siding	Crew	Daily Output	Labor-Hours	Unit	Material	2020 Bare Costs Labor	Equipment	Total	Total Incl O&P
4097	Cornerpost for above	1 Carp	365	.022	L.F.	12.40	1.17		13.57	15.40
5000	Staggered butt, double 7"	2 Carp	400	.040	S.F.	3.64	2.13		5.77	7.20
5002	Cornerpost for above	1 Carp	365	.022	L.F.	13.35	1.17		14.52	16.45
5010	Half round, double 6-1/4"	2 Carp	360	.044	S.F.	4.27	2.36		6.63	8.25
5020	Shake profile, staggered butt, double 9"	"	510	.031	"	3.80	1.67		5.47	6.70
5022	Cornerpost for above	1 Carp	365	.022	L.F.	10	1.17		11.17	12.75
5030	Straight butt, double 7"	2 Carp	400	.040	S.F.	4.27	2.13		6.40	7.90
5032	Cornerpost for above	1 Carp	365	.022	L.F.	14.55	1.17		15.72	17.75
6000	Accessories, J channel, 5/8" pocket	2 Carp	700	.023		.57	1.22		1.79	2.45
6010	3/4" pocket		695	.023		.63	1.22		1.85	2.53
6020	1-1/4" pocket		680	.024		1.15	1.25		2.40	3.14
6030	Aluminum starter strip		700	.023		.31	1.22		1.53	2.17

07 51 Built-Up Bituminous Roofing

07 51 13 – Built-Up Asphalt Roofing

07 51 13.10 Built-Up Roofing Components

		Crew	Daily Output	Labor-Hours	Unit	Material	Labor	Equipment	Total	Total Incl O&P
0010	**BUILT-UP ROOFING COMPONENTS**									
0012	Asphalt saturated felt, #30, 2 sq./roll	1 Rofc	58	.138	Sq.	10.20	6.35		16.55	21.50
0200	#15, 4 sq./roll, plain or perforated, not mopped		58	.138		5.25	6.35		11.60	16.30
0300	Roll roofing, smooth, #65		15	.533		10.30	24.50		34.80	52
0500	#90		12	.667		37.50	31		68.50	92
0520	Mineralized		12	.667		36	31		67	90.50
0540	D.C. (double coverage), 19" selvage edge		10	.800		48	37		85	114
0580	Adhesive (lap cement)				Gal.	8.50			8.50	9.35
0800	Steep, flat or dead level asphalt, 10 ton lots, packaged				Ton	885			885	975

07 51 13.20 Built-Up Roofing Systems

		Crew	Daily Output	Labor-Hours	Unit	Material	Labor	Equipment	Total	Total Incl O&P
0010	**BUILT-UP ROOFING SYSTEMS**									
0120	Asphalt flood coat with gravel/slag surfacing, not including									
0140	Insulation, flashing or wood nailers									
0200	Asphalt base sheet, 3 plies #15 asphalt felt, mopped	G-1	22	2.545	Sq.	108	110	26	244	330
0350	On nailable decks		21	2.667		100	115	27	242	330
0500	4 plies #15 asphalt felt, mopped		20	2.800		133	121	28.50	282.50	375
0550	On nailable decks		19	2.947		118	127	30	275	375
0700	Coated glass base sheet, 2 plies glass (type IV), mopped		22	2.545		108	110	26	244	330
0850	3 plies glass, mopped		20	2.800		130	121	28.50	279.50	375
0950	On nailable decks		19	2.947		123	127	30	280	380

07 51 13.30 Cants

		Crew	Daily Output	Labor-Hours	Unit	Material	Labor	Equipment	Total	Total Incl O&P
0010	**CANTS**									
0012	Lumber, treated, 4" x 4" cut diagonally	1 Rofc	325	.025	L.F.	1.89	1.14		3.03	3.96
0300	Mineral or fiber, trapezoidal, 1" x 4" x 48"		325	.025		.31	1.14		1.45	2.22
0400	1-1/2" x 5-5/8" x 48"		325	.025		.48	1.14		1.62	2.41

07 65 Flexible Flashing

07 65 10 – Sheet Metal Flashing

07 65 10.10 Sheet Metal Flashing and Counter Flashing	Crew	Daily Output	Labor-Hours	Unit	Material	2020 Bare Costs Labor	2020 Bare Costs Equipment	Total	Total Incl O&P
0010 **SHEET METAL FLASHING AND COUNTER FLASHING**									
0011 Including up to 4 bends									
0020 Aluminum, mill finish, .013" thick	1 Rofc	145	.055	S.F.	1.01	2.55		3.56	5.30
0030 .016" thick		145	.055		1.15	2.55		3.70	5.50
0060 .019" thick		145	.055		1.50	2.55		4.05	5.85
0100 .032" thick		145	.055		1.38	2.55		3.93	5.75
0200 .040" thick		145	.055		2.33	2.55		4.88	6.75
0300 .050" thick		145	.055		3.22	2.55		5.77	7.75
0400 Painted finish, add					.34			.34	.37

07 65 12 – Fabric and Mastic Flashings

07 65 12.10 Fabric and Mastic Flashing and Counter Flashing

	Crew	Daily Output	Labor-Hours	Unit	Material	Labor	Equipment	Total	Total Incl O&P
0010 **FABRIC AND MASTIC FLASHING AND COUNTER FLASHING**									
1300 Asphalt flashing cement, 5 gallon				Gal.	11.35			11.35	12.50
4900 Fabric, asphalt-saturated cotton, specification grade	1 Rofc	35	.229	S.Y.	3.26	10.55		13.81	21
5000 Utility grade		35	.229		1.55	10.55		12.10	19.15
5300 Close-mesh fabric, saturated, 17 oz./S.Y.		35	.229		2.21	10.55		12.76	19.90
5500 Fiberglass, resin-coated		35	.229		1.24	10.55		11.79	18.80

07 71 Roof Specialties

07 71 19 – Manufactured Gravel Stops and Fasciae

07 71 19.10 Gravel Stop

	Crew	Daily Output	Labor-Hours	Unit	Material	Labor	Equipment	Total	Total Incl O&P
0010 **GRAVEL STOP**									
0020 Aluminum, .050" thick, 4" face height, mill finish	1 Shee	145	.055	L.F.	6.95	3.44		10.39	12.85
0080 Duranodic finish	"	145	.055	"	8.25	3.44		11.69	14.25

07 71 23 – Manufactured Gutters and Downspouts

07 71 23.10 Downspouts

	Crew	Daily Output	Labor-Hours	Unit	Material	Labor	Equipment	Total	Total Incl O&P
0010 **DOWNSPOUTS**									
0020 Aluminum, embossed, .020" thick, 2" x 3"	1 Shee	190	.042	L.F.	.95	2.62		3.57	5.05
0100 Enameled		190	.042		1.37	2.62		3.99	5.50
0300 .024" thick, 2" x 3"		180	.044		2.17	2.77		4.94	6.60
0400 3" x 4"		140	.057		2.29	3.56		5.85	7.90
0600 Round, corrugated aluminum, 3" diameter, .020" thick		190	.042		2.13	2.62		4.75	6.30
0700 4" diameter, .025" thick		140	.057		3.21	3.56		6.77	8.95
4800 Steel, galvanized, round, corrugated, 2" or 3" diameter, 28 ga.		190	.042		2.30	2.62		4.92	6.50
4900 4" diameter, 28 ga.		145	.055		2.19	3.44		5.63	7.60
5100 5" diameter, 26 ga.		130	.062		3.86	3.83		7.69	10.05
5400 6" diameter, 28 ga.		105	.076		3.72	4.75		8.47	11.30
5700 Rectangular, corrugated, 28 ga., 2" x 3"		190	.042		2.38	2.62		5	6.60
5800 3" x 4"		145	.055		1.96	3.44		5.40	7.35
6000 Rectangular, plain, 28 ga., galvanized, 2" x 3"		190	.042		3.98	2.62		6.60	8.35
6100 3" x 4"		145	.055		4.83	3.44		8.27	10.50

07 71 23.30 Gutters

	Crew	Daily Output	Labor-Hours	Unit	Material	Labor	Equipment	Total	Total Incl O&P
0010 **GUTTERS**									
0012 Aluminum, stock units, 5" K type, .027" thick, plain	1 Shee	125	.064	L.F.	2.85	3.99		6.84	9.20
0100 Enameled		125	.064		2.94	3.99		6.93	9.30
0300 5" K type, .032" thick, plain		125	.064		3.79	3.99		7.78	10.20
0400 Enameled		125	.064		3.58	3.99		7.57	10
0700 Copper, half round, 16 oz., stock units, 4" wide		125	.064		8.60	3.99		12.59	15.55
0900 5" wide		125	.064		7.65	3.99		11.64	14.50
1000 6" wide		118	.068		10.90	4.22		15.12	18.35

07 71 Roof Specialties

07 71 23 – Manufactured Gutters and Downspouts

07 71 23.30 Gutters		Crew	Daily Output	Labor-Hours	Unit	Material	2020 Bare Costs Labor	Equipment	Total	Total Incl O&P
1200	K type, 16 oz., stock, 5" wide	1 Shee	125	.064	L.F.	8.75	3.99		12.74	15.65
1300	6" wide		125	.064		8.45	3.99		12.44	15.35
1500	Lead coated copper, 16 oz., half round, stock, 4" wide		125	.064		17.55	3.99		21.54	25.50
1600	6" wide		118	.068		20	4.22		24.22	28.50
1800	K type, stock, 5" wide		125	.064		18.55	3.99		22.54	26.50
1900	6" wide		125	.064		21	3.99		24.99	29
2100	Copper clad stainless steel, K type, 5" wide		125	.064		7.60	3.99		11.59	14.40
2200	6" wide		125	.064		10.30	3.99		14.29	17.40
2400	Steel, galv, half round or box, 28 ga., 5" wide, plain		125	.064		2.40	3.99		6.39	8.70
2500	Enameled		125	.064		2.24	3.99		6.23	8.50
2700	26 ga., stock, 5" wide		125	.064		2.66	3.99		6.65	9
2800	6" wide		125	.064		2.64	3.99		6.63	8.95
3000	Vinyl, O.G., 4" wide	1 Carp	115	.070		1.39	3.70		5.09	7.10
3100	5" wide		115	.070		1.51	3.70		5.21	7.20
3200	4" half round, stock units		115	.070		1.42	3.70		5.12	7.10
3250	Joint connectors				Ea.	3.10			3.10	3.41
3300	Wood, clear treated cedar, fir or hemlock, 3" x 4"	1 Carp	100	.080	L.F.	12.70	4.25		16.95	20.50
3400	4" x 5"	"	100	.080	"	24.50	4.25		28.75	33.50
5000	Accessories, end cap, K type, aluminum 5"	1 Shee	625	.013	Ea.	.74	.80		1.54	2.02
5010	6"		625	.013		1.49	.80		2.29	2.85
5020	Copper, 5"		625	.013		3.41	.80		4.21	4.96
5030	6"		625	.013		3.59	.80		4.39	5.15
5040	Lead coated copper, 5"		625	.013		13.20	.80		14	15.70
5050	6"		625	.013		14.10	.80		14.90	16.70
5060	Copper clad stainless steel, 5"		625	.013		3.58	.80		4.38	5.15
5070	6"		625	.013		3.58	.80		4.38	5.15
5080	Galvanized steel, 5"		625	.013		1.57	.80		2.37	2.94
5090	6"		625	.013		2.72	.80		3.52	4.20
5100	Vinyl, 4"	1 Carp	625	.013		6.50	.68		7.18	8.15
5110	5"	"	625	.013		6.85	.68		7.53	8.55
5120	Half round, copper, 4"	1 Shee	625	.013		4.65	.80		5.45	6.30
5130	5"		625	.013		5.25	.80		6.05	6.95
5140	6"		625	.013		8.50	.80		9.30	10.55
5150	Lead coated copper, 5"		625	.013		14.85	.80		15.65	17.55
5160	6"		625	.013		22	.80		22.80	25.50
5170	Copper clad stainless steel, 5"		625	.013		5	.80		5.80	6.70
5180	6"		625	.013		4.62	.80		5.42	6.30
5190	Galvanized steel, 5"		625	.013		2.80	.80		3.60	4.29
5200	6"		625	.013		3.47	.80		4.27	5.05
5210	Outlet, aluminum, 2" x 3"		420	.019		.72	1.19		1.91	2.59
5220	3" x 4"		420	.019		1.18	1.19		2.37	3.10
5230	2-3/8" round		420	.019		.66	1.19		1.85	2.53
5240	Copper, 2" x 3"		420	.019		7.80	1.19		8.99	10.40
5250	3" x 4"		420	.019		8.80	1.19		9.99	11.50
5260	2-3/8" round		420	.019		4.87	1.19		6.06	7.15
5270	Lead coated copper, 2" x 3"		420	.019		26	1.19		27.19	31
5280	3" x 4"		420	.019		30	1.19		31.19	35
5290	2-3/8" round		420	.019		26.50	1.19		27.69	31
5300	Copper clad stainless steel, 2" x 3"		420	.019		7.30	1.19		8.49	9.85
5310	3" x 4"		420	.019		8.35	1.19		9.54	10.95
5320	2-3/8" round		420	.019		4.87	1.19		6.06	7.15
5330	Galvanized steel, 2" x 3"		420	.019		3.81	1.19		5	6
5340	3" x 4"		420	.019		6.10	1.19		7.29	8.50

For customer support on your Site Work & Landscape Costs with RSMeans data, call 800.448.8182.

153

07 71 Roof Specialties

07 71 23 – Manufactured Gutters and Downspouts

07 71 23.30 Gutters

		Crew	Daily Output	Labor-Hours	Unit	Material	2020 Bare Costs Labor	Equipment	Total	Total Incl O&P
5350	2-3/8" round	1 Shee	420	.019	Ea.	4.95	1.19		6.14	7.25
5360	K type mitres, aluminum		65	.123		4.94	7.65		12.59	17.10
5370	Copper		65	.123		17.45	7.65		25.10	31
5380	Lead coated copper		65	.123		56	7.65		63.65	73
5390	Copper clad stainless steel		65	.123		28	7.65		35.65	42
5400	Galvanized steel		65	.123		28	7.65		35.65	42.50
5420	Half round mitres, copper		65	.123		70.50	7.65		78.15	89
5430	Lead coated copper		65	.123		91.50	7.65		99.15	113
5440	Copper clad stainless steel		65	.123		57.50	7.65		65.15	74.50
5450	Galvanized steel		65	.123		33.50	7.65		41.15	48.50
5460	Vinyl mitres and outlets		65	.123		11	7.65		18.65	24
5470	Sealant		940	.009	L.F.	.01	.53		.54	.81
5480	Soldering		96	.083	"	.29	5.20		5.49	8.15

07 71 29 – Manufactured Roof Expansion Joints

07 71 29.10 Expansion Joints

		Crew	Daily Output	Labor-Hours	Unit	Material	2020 Bare Costs Labor	Equipment	Total	Total Incl O&P
0010	**EXPANSION JOINTS**									
0300	Butyl or neoprene center with foam insulation, metal flanges									
0400	Aluminum, .032" thick for openings to 2-1/2"	1 Rofc	165	.048	L.F.	12.85	2.24		15.09	17.85
0600	For joint openings to 3-1/2"		165	.048		12.40	2.24		14.64	17.35
0610	For joint openings to 5"		165	.048		14.75	2.24		16.99	19.95
0620	For joint openings to 8"		165	.048		16.70	2.24		18.94	22
0700	Copper, 16 oz. for openings to 2-1/2"		165	.048		21.50	2.24		23.74	27.50
0900	For joint openings to 3-1/2"		165	.048		19.70	2.24		21.94	25
0910	For joint openings to 5"		165	.048		22.50	2.24		24.74	28.50
0920	For joint openings to 8"		165	.048		25	2.24		27.24	30.50
1000	Galvanized steel, 26 ga. for openings to 2-1/2"		165	.048		10.80	2.24		13.04	15.60
1200	For joint openings to 3-1/2"		165	.048		10.55	2.24		12.79	15.30
1210	For joint openings to 5"		165	.048		14.10	2.24		16.34	19.25
1220	For joint openings to 8"		165	.048		15.20	2.24		17.44	20.50
1810	For joint openings to 5"		165	.048		15.25	2.24		17.49	20.50
1820	For joint openings to 8"		165	.048		23.50	2.24		25.74	29
1900	Neoprene, double-seal type with thick center, 4-1/2" wide		125	.064		14.95	2.96		17.91	21.50
1950	Polyethylene bellows, with galv steel flat flanges		100	.080		7.20	3.70		10.90	14
1960	With galvanized angle flanges		100	.080		7.75	3.70		11.45	14.65
2000	Roof joint with extruded aluminum cover, 2"	1 Shee	115	.070		31.50	4.33		35.83	41
2700	Wall joint, closed cell foam on PVC cover, 9" wide	1 Rofc	125	.064		6.10	2.96		9.06	11.65
2800	12" wide	"	115	.070		7	3.21		10.21	13

07 71 43 – Drip Edge

07 71 43.10 Drip Edge, Rake Edge, Ice Belts

		Crew	Daily Output	Labor-Hours	Unit	Material	2020 Bare Costs Labor	Equipment	Total	Total Incl O&P
0010	**DRIP EDGE, RAKE EDGE, ICE BELTS**									
0020	Aluminum, .016" thick, 5" wide, mill finish	1 Carp	400	.020	L.F.	.59	1.06		1.65	2.25
0100	White finish		400	.020		.65	1.06		1.71	2.32
0200	8" wide, mill finish		400	.020		1.48	1.06		2.54	3.23
0300	Ice belt, 28" wide, mill finish		100	.080		8	4.25		12.25	15.20
0400	Galvanized, 5" wide		400	.020		.62	1.06		1.68	2.28
0500	8" wide, mill finish		400	.020		.84	1.06		1.90	2.52

07 72 Roof Accessories

07 72 53 – Snow Guards

07 72 53.10 Snow Guard Options

		Crew	Daily Output	Labor-Hours	Unit	Material	2020 Bare Costs Labor	Equipment	Total	Total Incl O&P
0010	**SNOW GUARD OPTIONS**									
0100	Slate & asphalt shingle roofs, fastened with nails	1 Rofc	160	.050	Ea.	12.45	2.31		14.76	17.50
0200	Standing seam metal roofs, fastened with set screws		48	.167		17.75	7.70		25.45	32.50
0300	Surface mount for metal roofs, fastened with solder		48	.167		7.75	7.70		15.45	21.50
0400	Double rail pipe type, including pipe		130	.062	L.F.	35	2.84		37.84	43

07 76 Roof Pavers

07 76 16 – Roof Decking Pavers

07 76 16.10 Roof Pavers and Supports

		Crew	Daily Output	Labor-Hours	Unit	Material	2020 Bare Costs Labor	Equipment	Total	Total Incl O&P
0010	**ROOF PAVERS AND SUPPORTS**									
1000	Roof decking pavers, concrete blocks, 2" thick, natural	1 Clab	115	.070	S.F.	3.58	2.93		6.51	8.35
1100	Colors		115	.070	"	3.71	2.93		6.64	8.50
1200	Support pedestal, bottom cap		960	.008	Ea.	2.66	.35		3.01	3.46
1300	Top cap		960	.008		4.88	.35		5.23	5.90
1400	Leveling shims, 1/16"		1920	.004		1.22	.18		1.40	1.60
1500	1/8"		1920	.004		1.22	.18		1.40	1.60
1600	Buffer pad		960	.008		2.54	.35		2.89	3.32
1700	PVC legs (4" SDR 35)		2880	.003	Inch	.14	.12		.26	.33
2000	Alternate pricing method, system in place		101	.079	S.F.	7.20	3.33		10.53	12.90

07 81 Applied Fireproofing

07 81 16 – Cementitious Fireproofing

07 81 16.10 Sprayed Cementitious Fireproofing

		Crew	Daily Output	Labor-Hours	Unit	Material	2020 Bare Costs Labor	Equipment	Total	Total Incl O&P
0010	**SPRAYED CEMENTITIOUS FIREPROOFING**									
0050	Not including canvas protection, normal density									
0100	Per 1" thick, on flat plate steel	G-2	3000	.008	S.F.	.59	.36	.06	1.01	1.25
0200	Flat decking		2400	.010		.59	.44	.08	1.11	1.40
0400	Beams		1500	.016		.59	.71	.13	1.43	1.85
0500	Corrugated or fluted decks		1250	.019		.88	.85	.15	1.88	2.41
0700	Columns, 1-1/8" thick		1100	.022		.66	.97	.17	1.80	2.37

07 91 Preformed Joint Seals

07 91 23 – Backer Rods

07 91 23.10 Backer Rods

		Crew	Daily Output	Labor-Hours	Unit	Material	2020 Bare Costs Labor	Equipment	Total	Total Incl O&P
0010	**BACKER RODS**									
0030	Backer rod, polyethylene, 1/4" diameter	1 Bric	4.60	1.739	C.L.F.	2.78	90.50		93.28	140
0050	1/2" diameter		4.60	1.739		4.10	90.50		94.60	142
0070	3/4" diameter		4.60	1.739		6.70	90.50		97.20	144
0090	1" diameter		4.60	1.739		14.80	90.50		105.30	153

07 91 26 – Joint Fillers

07 91 26.10 Joint Fillers

		Crew	Daily Output	Labor-Hours	Unit	Material	2020 Bare Costs Labor	Equipment	Total	Total Incl O&P
0010	**JOINT FILLERS**									
4360	Butyl rubber filler, 1/4" x 1/4"	1 Bric	290	.028	L.F.	.23	1.44		1.67	2.43
4365	1/2" x 1/2"		250	.032		.92	1.67		2.59	3.53
4370	1/2" x 3/4"		210	.038		1.38	1.98		3.36	4.52
4375	3/4" x 3/4"		230	.035		2.06	1.81		3.87	5
4380	1" x 1"		180	.044		2.75	2.31		5.06	6.55

155

07 91 Preformed Joint Seals

07 91 26 – Joint Fillers

07 91 26.10 Joint Fillers	Crew	Daily Output	Labor-Hours	Unit	Material	2020 Bare Costs Labor	Equipment	Total	Total Incl O&P
4390 For coloring, add					12%				
4980 Polyethylene joint backing, 1/4" x 2"	1 Bric	2.08	3.846	C.L.F.	14.25	200		214.25	320
4990 1/4" x 6"		1.28	6.250	"	32.50	325		357.50	530
5600 Silicone, room temp vulcanizing foam seal, 1/4" x 1/2"		1312	.006	L.F.	.46	.32		.78	.98
5610 1/2" x 1/2"		656	.012		.92	.64		1.56	1.97
5620 1/2" x 3/4"		442	.018		1.37	.94		2.31	2.94
5630 3/4" x 3/4"		328	.024		2.06	1.27		3.33	4.19
5640 1/8" x 1"		1312	.006		.46	.32		.78	.98
5650 1/8" x 3"		442	.018		1.37	.94		2.31	2.94
5670 1/4" x 3"		295	.027		2.75	1.41		4.16	5.15
5680 1/4" x 6"		148	.054		5.50	2.81		8.31	10.30
5690 1/2" x 6"		82	.098		11	5.10		16.10	19.80
5700 1/2" x 9"		52.50	.152		16.50	7.95		24.45	30
5710 1/2" x 12"		33	.242		22	12.60		34.60	43

07 92 Joint Sealants

07 92 13 – Elastomeric Joint Sealants

07 92 13.20 Caulking and Sealant Options

	Crew	Daily Output	Labor-Hours	Unit	Material	2020 Bare Costs Labor	Equipment	Total	Total Incl O&P
0010 **CAULKING AND SEALANT OPTIONS**									
0050 Latex acrylic based, bulk				Gal.	31			31	34
0055 Bulk in place 1/4" x 1/4" bead	1 Bric	300	.027	L.F.	.10	1.39		1.49	2.21
0060 1/4" x 3/8"		294	.027		.16	1.42		1.58	2.33
0065 1/4" x 1/2"		288	.028		.22	1.45		1.67	2.43
0075 3/8" x 3/8"		284	.028		.24	1.47		1.71	2.49
0080 3/8" x 1/2"		280	.029		.32	1.49		1.81	2.61
0085 3/8" x 5/8"		276	.029		.40	1.51		1.91	2.74
0095 3/8" x 3/4"		272	.029		.48	1.53		2.01	2.85
0100 1/2" x 1/2"		275	.029		.43	1.51		1.94	2.78
0105 1/2" x 5/8"		269	.030		.54	1.55		2.09	2.94
0110 1/2" x 3/4"		263	.030		.65	1.58		2.23	3.11
0115 1/2" x 7/8"		256	.031		.76	1.63		2.39	3.30
0120 1/2" x 1"		250	.032		.87	1.67		2.54	3.47
0125 3/4" x 3/4"		244	.033		.97	1.71		2.68	3.66
0130 3/4" x 1"		225	.036		1.30	1.85		3.15	4.24
0135 1" x 1"		200	.040		1.73	2.08		3.81	5.05
0190 Cartridges				Gal.	33.50			33.50	37
0200 11 fl. oz. cartridge				Ea.	2.89			2.89	3.18
0400 1/4" x 1/4"	1 Bric	300	.027	L.F.	.12	1.39		1.51	2.23
0410 1/4" x 3/8"		294	.027		.18	1.42		1.60	2.34
0500 1/4" x 1/2"		288	.028		.24	1.45		1.69	2.45
0510 3/8" x 3/8"		284	.028		.26	1.47		1.73	2.51
0520 3/8" x 1/2"		280	.029		.35	1.49		1.84	2.64
0530 3/8" x 5/8"		276	.029		.44	1.51		1.95	2.78
0540 3/8" x 3/4"		272	.029		.53	1.53		2.06	2.90
0600 1/2" x 1/2"		275	.029		.47	1.51		1.98	2.82
0610 1/2" x 5/8"		269	.030		.59	1.55		2.14	3
0620 1/2" x 3/4"		263	.030		.71	1.58		2.29	3.18
0630 1/2" x 7/8"		256	.031		.83	1.63		2.46	3.38
0640 1/2" x 1"		250	.032		.94	1.67		2.61	3.56
0800 3/4" x 3/4"		244	.033		1.06	1.71		2.77	3.76
0900 3/4" x 1"		225	.036		1.42	1.85		3.27	4.37

07 92 Joint Sealants

07 92 13 – Elastomeric Joint Sealants

07 92 13.20 Caulking and Sealant Options		Crew	Daily Output	Labor-Hours	Unit	Material	2020 Bare Costs Labor	Equipment	Total	Total Incl O&P
1000	1" x 1"	1 Bric	200	.040	L.F.	1.77	2.08		3.85	5.10
3200	Polyurethane, 1 or 2 component				Gal.	54			54	59.50
3300	Cartridges				"	67.50			67.50	74.50
3500	Bulk, in place, 1/4" x 1/4"	1 Bric	300	.027	L.F.	.18	1.39		1.57	2.29
3510	1/4" x 3/8"		294	.027		.26	1.42		1.68	2.44
3520	3/8" x 3/8"		284	.028		.39	1.47		1.86	2.65
3530	3/8" x 1/2"		280	.029		.52	1.49		2.01	2.83
3540	3/8" x 5/8"		276	.029		.66	1.51		2.17	3.02
3550	3/8" x 3/4"		272	.029		.80	1.53		2.33	3.19
3560	1/2" x 1/2"		275	.029		.70	1.51		2.21	3.07
3570	1/2" x 5/8"		269	.030		.87	1.55		2.42	3.31
3580	1/2" x 3/4"		263	.030		1.06	1.58		2.64	3.57
3590	1/2" x 7/8"		256	.031		1.23	1.63		2.86	3.82
3655	1/2" x 1/4"		288	.028		.35	1.45		1.80	2.58
3800	3/4" x 3/8"		272	.029		.80	1.53		2.33	3.19
3850	3/4" x 3/4"		244	.033		1.59	1.71		3.30	4.34
3875	3/4" x 1"		225	.036		2.08	1.85		3.93	5.10
3900	1" x 1/2"		250	.032		1.40	1.67		3.07	4.06
4000	For coloring, add					15%				

07 95 Expansion Control

07 95 13 – Expansion Joint Cover Assemblies

07 95 13.50 Expansion Joint Assemblies

		Crew	Daily Output	Labor-Hours	Unit	Material	2020 Bare Costs Labor	Equipment	Total	Total Incl O&P
0010	**EXPANSION JOINT ASSEMBLIES**									
0200	Floor cover assemblies, 1" space, aluminum	1 Sswk	38	.211	L.F.	26	12.15		38.15	47.50
0300	Bronze		38	.211		52.50	12.15		64.65	77
2000	Roof closures, aluminum, flat roof, low profile, 1" space		57	.140		24	8.10		32.10	39
2300	Roof to wall, low profile, 1" space		57	.140		23.50	8.10		31.60	38.50

Division Notes

	CREW	DAILY OUTPUT	LABOR-HOURS	UNIT	BARE COSTS				TOTAL INCL O&P
					MAT.	LABOR	EQUIP.	TOTAL	

Estimating Tips
08 10 00 Doors and Frames

All exterior doors should be addressed for their energy conservation (insulation and seals).

- Most metal doors and frames look alike, but there may be significant differences among them. When estimating these items, be sure to choose the line item that most closely compares to the specification or door schedule requirements regarding:
  - □ type of metal
  - □ metal gauge
  - □ door core material
  - □ fire rating
  - □ finish
- Wood and plastic doors vary considerably in price. The primary determinant is the veneer material. Lauan, birch, and oak are the most common veneers. Other variables include the following:
  - □ hollow or solid core
  - □ fire rating
  - □ flush or raised panel
  - □ finish
- Door pricing includes bore for cylindrical locksets and mortise for hinges.

08 30 00 Specialty Doors and Frames

- There are many varieties of special doors, and they are usually priced per each. Add frames, hardware, or operators required for a complete installation.

08 40 00 Entrances, Storefronts, and Curtain Walls

- Glazed curtain walls consist of the metal tube framing and the glazing material. The cost data in this subdivision is presented for the metal tube framing alone or the composite wall. If your estimate requires a detailed takeoff of the framing, be sure to add the glazing cost and any tints.

08 50 00 Windows

- Steel windows are unglazed and aluminum can be glazed or unglazed. Some metal windows are priced without glass. Refer to 08 80 00 Glazing for glass pricing. The grade C indicates commercial grade windows, usually ASTM C-35.
- All wood windows and vinyl are priced preglazed. The glazing is insulating glass. Add the cost of screens and grills if required and not already included.

08 70 00 Hardware

- Hardware costs add considerably to the cost of a door. The most efficient method to determine the hardware requirements for a project is to review the door and hardware schedule together. One type of door may have different hardware, depending on the door usage.
- Door hinges are priced by the pair, with most doors requiring 1-1/2 pairs per door. The hinge prices do not include installation labor because it is included in door installation.

Hinges are classified according to the frequency of use, base material, and finish.

08 80 00 Glazing

- Different openings require different types of glass. The most common types are:
  - □ float
  - □ tempered
  - □ insulating
  - □ impact-resistant
  - □ ballistic-resistant
- Most exterior windows are glazed with insulating glass. Entrance doors and window walls, where the glass is less than 18" from the floor, are generally glazed with tempered glass. Interior windows and some residential windows are glazed with float glass.
- Coastal communities require the use of impact-resistant glass, dependent on wind speed.
- The insulation or 'u' value is a strong consideration, along with solar heat gain, to determine total energy efficiency.

Reference Numbers

Reference numbers are shown at the beginning of some major classifications. These numbers refer to related items in the Reference Section. The reference information may be an estimating procedure, an alternate pricing method, or technical information.

Note: Not all subdivisions listed here necessarily appear. ■

08 05 05.10 Selective Demolition Doors

		Crew	Daily Output	Labor-Hours	Unit	Material	2020 Bare Costs Labor	2020 Bare Costs Equipment	Total	Total Incl O&P
0010	**SELECTIVE DEMOLITION DOORS** R024119-10									
0200	Doors, exterior, 1-3/4" thick, single, 3' x 7' high	1 Clab	16	.500	Ea.		21		21	31.50
0220	Double, 6' x 7' high		12	.667			28		28	42
0500	Interior, 1-3/8" thick, single, 3' x 7' high		20	.400			16.85		16.85	25.50
0520	Double, 6' x 7' high		16	.500			21		21	31.50
0700	Bi-folding, 3' x 6'-8" high		20	.400			16.85		16.85	25.50
0720	6' x 6'-8" high		18	.444			18.70		18.70	28
0900	Bi-passing, 3' x 6'-8" high		16	.500			21		21	31.50
0940	6' x 6'-8" high		14	.571			24		24	36
1100	Door demo, floor door	2 Sswk	5	3.200			184		184	289
1500	Remove and reset, hollow core	1 Carp	8	1			53		53	80
1520	Solid		6	1.333			71		71	106
2000	Frames, including trim, metal		8	1			53		53	80
2200	Wood	2 Carp	32	.500			26.50		26.50	40
3000	Special doors, counter doors		6	2.667			142		142	213
3100	Double acting		10	1.600			85		85	128
3200	Floor door (trap type), or access type		8	2			106		106	160
3300	Glass, sliding, including frames		12	1.333			71		71	106
3400	Overhead, commercial, 12' x 12' high		4	4			213		213	320
3440	up to 20' x 16' high		3	5.333			283		283	425
3445	up to 35' x 30' high		1	16			850		850	1,275
3500	Residential, 9' x 7' high		8	2			106		106	160
3540	16' x 7' high		7	2.286			121		121	183
3620	Large		2.50	6.400			340		340	510
3700	Roll-up grille		5	3.200			170		170	256
3800	Revolving door, no wire or connections		2	8			425		425	640
3900	Storefront swing door		3	5.333			283		283	425
7550	Hangar door demo	2 Sswk	330	.048	S.F.		2.79		2.79	4.39

08 05 05.20 Selective Demolition of Windows

		Crew	Daily Output	Labor-Hours	Unit	Material	2020 Bare Costs Labor	2020 Bare Costs Equipment	Total	Total Incl O&P
0010	**SELECTIVE DEMOLITION OF WINDOWS** R024119-10									
0200	Aluminum, including trim, to 12 S.F.	1 Clab	16	.500	Ea.		21		21	31.50
0240	To 25 S.F.		11	.727			30.50		30.50	46
0280	To 50 S.F.		5	1.600			67.50		67.50	101
0600	Glass, up to 10 S.F./window		200	.040	S.F.		1.68		1.68	2.53
0620	Over 10 S.F./window		150	.053	"		2.25		2.25	3.37
1000	Steel, including trim, to 12 S.F.		13	.615	Ea.		26		26	39
1020	To 25 S.F.		9	.889			37.50		37.50	56
1040	To 50 S.F.		4	2			84		84	127
2000	Wood, including trim, to 12 S.F.		22	.364			15.30		15.30	23
2020	To 25 S.F.		18	.444			18.70		18.70	28
2060	To 50 S.F.		13	.615			26		26	39
2065	To 180 S.F.		8	1			42		42	63.50
5083	Remove unit skylight	1 Carp	1.54	5.202			276		276	415

08 12 Metal Frames

08 12 13 – Hollow Metal Frames

08 12 13.13 Standard Hollow Metal Frames

		Crew	Daily Output	Labor-Hours	Unit	Material	2020 Bare Costs Labor	Equipment	Total	Total Incl O&P
0010	**STANDARD HOLLOW METAL FRAMES**									
0020	16 ga., up to 5-3/4" jamb depth									
0025	3'-0" x 6'-8" single	G 2 Carp	16	1	Ea.	217	53		270	320
0028	3'-6" wide, single	G	16	1		270	53		323	375
0030	4'-0" wide, single	G	16	1		305	53		358	415
0040	6'-0" wide, double	G	14	1.143		256	60.50		316.50	375
0045	8'-0" wide, double	G	14	1.143		248	60.50		308.50	365
3600	up to 5-3/4" jamb depth, 4'-0" x 7'-0" single		15	1.067		256	56.50		312.50	365
3620	6'-0" wide, double		12	1.333		201	71		272	325
3640	8'-0" wide, double		12	1.333		300	71		371	435
4400	8-3/4" deep, 4'-0" x 7'-0", single		15	1.067		430	56.50		486.50	560
4440	8'-0" wide, double		12	1.333		455	71		526	605
4900	For welded frames, add					59.50			59.50	65.50
6200	8-3/4" deep, 4'-0" x 7'-0" single	2 Carp	15	1.067		269	56.50		325.50	380
6300	For "A" label use same price as "B" label									

08 13 Metal Doors

08 13 13 – Hollow Metal Doors

08 13 13.13 Standard Hollow Metal Doors

		Crew	Daily Output	Labor-Hours	Unit	Material	2020 Bare Costs Labor	Equipment	Total	Total Incl O&P
0010	**STANDARD HOLLOW METAL DOORS** R081313-20									
0015	Flush, full panel, hollow core									
0017	When noted doors are prepared but do not include glass or louvers									
0020	1-3/8" thick, 20 ga., 2'-0" x 6'-8"	G 2 Carp	20	.800	Ea.	345	42.50		387.50	445
0060	3'-0" x 6'-8"	G	17	.941		360	50		410	470
0100	3'-0" x 7'-0"	G	17	.941		375	50		425	485
0120	For vision lite, add					111			111	122
0140	For narrow lite, add					115			115	126
0320	Half glass, 20 ga., 2'-0" x 6'-8"	G 2 Carp	20	.800		695	42.50		737.50	830
0360	3'-0" x 6'-8"	G	17	.941		605	50		655	740
0400	3'-0" x 7'-0"	G	17	.941		475	50		525	600
0410	1-3/8" thick, 18 ga., 2'-0" x 6'-8"	G	20	.800		465	42.50		507.50	580
0420	3'-0" x 6'-8"	G	17	.941		470	50		520	590
0425	3'-0" x 7'-0"	G	17	.941		460	50		510	580
0450	For vision lite, add					111			111	122
0452	For narrow lite, add					115			115	126
0460	Half glass, 18 ga., 2'-0" x 6'-8"	G 2 Carp	20	.800		635	42.50		677.50	765
0470	3'-0" x 6'-8"	G	17	.941		645	50		695	785
0475	3'-0" x 7'-0"	G	17	.941		605	50		655	740
0500	Hollow core, 1-3/4" thick, full panel, 20 ga., 2'-8" x 6'-8"	G	18	.889		495	47		542	615
0520	3'-0" x 6'-8"	G	17	.941		495	50		545	620
0640	3'-0" x 7'-0"	G	17	.941		515	50		565	640
1000	18 ga., 2'-8" x 6'-8"	G	17	.941		525	50		575	650
1020	3'-0" x 6'-8"	G	16	1		510	53		563	640
1120	3'-0" x 7'-0"	G	17	.941		505	50		555	630
1212	For vision lite, add					111			111	122
1214	For narrow lite, add					115			115	126
1230	Half glass, 20 ga., 2'-8" x 6'-8"	G 2 Carp	20	.800		575	42.50		617.50	695
1240	3'-0" x 6'-8"	G	18	.889		570	47		617	695
1280	Embossed panel, 1-3/4" thick, poly core, 20 ga., 3'-0" x 7'-0"	G	18	.889		425	47		472	540
1290	Half glass, 1-3/4" thick, poly core, 20 ga., 3'-0" x 7'-0"	G	18	.889		625	47		672	755
1320	18 ga., 2'-8" x 6'-8"	G	18	.889		645	47		692	780

08 13 Metal Doors

08 13 13 – Hollow Metal Doors

08 13 13.13 Standard Hollow Metal Doors

		Crew	Daily Output	Labor-Hours	Unit	Material	2020 Bare Costs Labor	Equipment	Total	Total Incl O&P
1340	3'-0" x 6'-8" [G]	2 Carp	17	.941	Ea.	640	50		690	780
1720	Insulated, 1-3/4" thick, full panel, 18 ga., 3'-0" x 6'-8" [G]		15	1.067		500	56.50		556.50	640
1760	3'-0" x 7'-0" [G]		15	1.067		520	56.50		576.50	660
1820	Half glass, 18 ga., 3'-0" x 6'-8" [G]		16	1		630	53		683	775
1860	3'-0" x 7'-0" [G]	↓	16	1		690	53		743	840
2000	For vision lite, add					111			111	122
2010	For narrow lite, add					115			115	126
8100	For bottom louver, add				↓	243			243	267

08 13 13.15 Metal Fire Doors

		Crew	Daily Output	Labor-Hours	Unit	Material	2020 Bare Costs Labor	Equipment	Total	Total Incl O&P
0010	**METAL FIRE DOORS**									
0015	Steel, flush, "B" label, 90 minutes									
0020	Full panel, 20 ga., 2'-0" x 6'-8"	2 Carp	20	.800	Ea.	420	42.50		462.50	525
0060	3'-0" x 6'-8"		17	.941		435	50		485	555
0080	3'-0" x 7'-0"		17	.941		450	50		500	570
0140	18 ga., 3'-0" x 6'-8"	↓	16	1	↓	530	53		583	665
0220	For "A" label, 3 hour, 18 ga., use same price as "B" label									
0240	For vision lite, add				Ea.	183			183	201
0520	Flush, "B" label, 90 minutes, egress core, 20 ga., 2'-0" x 6'-8"	2 Carp	18	.889		680	47		727	820
0580	3'-0" x 7'-0"		16	1		715	53		768	865
0640	Flush, "A" label, 3 hour, egress core, 18 ga., 3'-0" x 6'-8"		15	1.067		740	56.50		796.50	900
0680	3'-0" x 7'-0"	↓	15	1.067	↓	410	56.50		466.50	540

08 14 Wood Doors

08 14 16 – Flush Wood Doors

08 14 16.09 Smooth Wood Doors

		Crew	Daily Output	Labor-Hours	Unit	Material	2020 Bare Costs Labor	Equipment	Total	Total Incl O&P
0010	**SMOOTH WOOD DOORS**									
0015	Flush, interior, hollow core									
0180	3'-0" x 6'-8"	2 Carp	17	.941	Ea.	101	50		151	186
0202	1-3/4", 2'-0" x 6'-8"	"	17	.941		69	50		119	151
0430	For 7'-0" high, add					31.50			31.50	35
0480	For prefinishing, clear, add					57			57	63
1720	H.P. plastic laminate, 1-3/8", 2'-0" x 6'-8"	2 Carp	16	1		286	53		339	395
1780	3'-0" x 6'-8"		15	1.067		305	56.50		361.50	420
2020	Particle core, lauan face, 1-3/8", 2'-6" x 6'-8"		15	1.067		105	56.50		161.50	201
2080	3'-0" x 7'-0"		13	1.231		128	65.50		193.50	240
2110	1-3/4", 3'-0" x 7'-0"	↓	13	1.231	↓	226	65.50		291.50	350

08 14 16.20 Wood Fire Doors

		Crew	Daily Output	Labor-Hours	Unit	Material	2020 Bare Costs Labor	Equipment	Total	Total Incl O&P
0010	**WOOD FIRE DOORS**									
0020	Particle core, 7 face plys, "B" label,									
0040	1 hour, birch face, 1-3/4" x 2'-6" x 6'-8"	2 Carp	14	1.143	Ea.	465	60.50		525.50	605
0080	3'-0" x 6'-8"		13	1.231		540	65.50		605.50	695
0090	3'-0" x 7'-0"		12	1.333		500	71		571	655
0440	M.D. overlay on hardboard, 2'-6" x 6'-8"		15	1.067		370	56.50		426.50	495
0480	3'-0" x 6'-8"		14	1.143		460	60.50		520.50	595
0740	90 minutes, birch face, 1-3/4" x 2'-6" x 6'-8"		14	1.143		345	60.50		405.50	465
0780	3'-0" x 6'-8"		13	1.231		410	65.50		475.50	550
0790	3'-0" x 7'-0"	↓	12	1.333	↓	460	71		531	610
2200	Custom architectural "B" label, flush, 1-3/4" thick, birch,									
2260	3'-0" x 7'-0"	2 Carp	14	1.143	Ea.	370	60.50		430.50	495

08 14 Wood Doors

08 14 33 – Stile and Rail Wood Doors

08 14 33.20 Wood Doors Residential		Crew	Daily Output	Labor-Hours	Unit	Material	2020 Bare Costs Labor	Equipment	Total	Total Incl O&P
0010	**WOOD DOORS RESIDENTIAL**									
1000	Entrance door, colonial, 1-3/4" x 6'-8" x 2'-8" wide	2 Carp	16	1	Ea.	625	53		678	770
1300	Flush, birch, solid core, 1-3/4" x 6'-8" x 2'-8" wide	"	16	1	"	169	53		222	266
7310	Passage doors, flush, no frame included									
7700	Birch, hollow core, 1-3/8" x 6'-8" x 1'-6" wide	2 Carp	18	.889	Ea.	78	47		125	157
7740	2'-6" wide	"	18	.889	"	87.50	47		134.50	168

08 31 Access Doors and Panels

08 31 13 – Access Doors and Frames

08 31 13.20 Bulkhead/Cellar Doors

		Crew	Daily Output	Labor-Hours	Unit	Material	2020 Bare Costs Labor	Equipment	Total	Total Incl O&P
0010	**BULKHEAD/CELLAR DOORS**									
0020	Steel, not incl. sides, 44" x 62"	1 Carp	5.50	1.455	Ea.	705	77.50		782.50	890
0100	52" x 73"		5.10	1.569		865	83.50		948.50	1,075
0500	With sides and foundation plates, 57" x 45" x 24"		4.70	1.702		895	90.50		985.50	1,125
0600	42" x 49" x 51"	↓	4.30	1.860	↓	600	99		699	810

08 31 13.30 Commercial Floor Doors

		Crew	Daily Output	Labor-Hours	Unit	Material	2020 Bare Costs Labor	Equipment	Total	Total Incl O&P
0010	**COMMERCIAL FLOOR DOORS**									
0020	Aluminum tile, steel frame, one leaf, 2' x 2' opng.	2 Sswk	3.50	4.571	Opng.	725	264		989	1,200
0050	3'-6" x 3'-6" opening	"	3.50	4.571	"	1,325	264		1,589	1,900

08 32 Sliding Glass Doors

08 32 19 – Sliding Wood-Framed Glass Doors

08 32 19.10 Sliding Wood Doors

		Crew	Daily Output	Labor-Hours	Unit	Material	2020 Bare Costs Labor	Equipment	Total	Total Incl O&P
0010	**SLIDING WOOD DOORS**									
0020	Wood, tempered insul. glass, 6' wide, premium	2 Carp	4	4	Ea.	1,525	213		1,738	2,000
0100	Economy		4	4		1,250	213		1,463	1,700
0150	8' wide, wood, premium		3	5.333		1,950	283		2,233	2,575
0200	Economy		3	5.333		1,600	283		1,883	2,175
0235	10' wide, wood, premium		2.50	6.400		2,875	340		3,215	3,650
0240	Economy		2.50	6.400		2,450	340		2,790	3,175
0250	12' wide, wood, premium		2.50	6.400		3,275	340		3,615	4,100
0300	Economy	↓	2.50	6.400	↓	2,675	340		3,015	3,425

08 32 19.15 Sliding Glass Vinyl-Clad Wood Doors

			Crew	Daily Output	Labor-Hours	Unit	Material	2020 Bare Costs Labor	Equipment	Total	Total Incl O&P
0010	**SLIDING GLASS VINYL-CLAD WOOD DOORS**										
0020	Glass, sliding vinyl-clad, insul. glass, 6'-0" x 6'-8"	G	2 Carp	4	4	Opng.	1,575	213		1,788	2,075
0025	6'-0" x 6'-10" high	G		4	4		1,775	213		1,988	2,275
0030	6'-0" x 8'-0" high	G		4	4		2,175	213		2,388	2,725
0100	8'-0" x 6'-10" high	G		4	4		2,125	213		2,338	2,650
0500	4 leaf, 9'-0" x 6'-10" high	G		3	5.333		3,675	283		3,958	4,475
0600	12'-0" x 6'-10" high	G	↓	3	5.333	↓	4,350	283		4,633	5,200

For customer support on your Site Work & Landscape Costs with RSMeans data, call 800.448.8182.

163

08 36 Panel Doors

08 36 13 – Sectional Doors

08 36 13.10 Overhead Commercial Doors

	08 36 13.10 Overhead Commercial Doors	Crew	Daily Output	Labor-Hours	Unit	Material	2020 Bare Costs Labor	Equipment	Total	Total Incl O&P
0010	**OVERHEAD COMMERCIAL DOORS**									
1000	Stock, sectional, heavy duty, wood, 1-3/4" thick, 8' x 8' high	2 Carp	2	8	Ea.	1,250	425		1,675	2,025
1200	12' x 12' high		1.50	10.667		2,775	565		3,340	3,900
1300	Chain hoist, 14' x 14' high		1.30	12.308		3,825	655		4,480	5,175
1500	20' x 8' high		1.30	12.270		2,950	650		3,600	4,225
2300	Fiberglass and aluminum, heavy duty, sectional, 12' x 12' high		1.50	10.667		3,425	565		3,990	4,600
2600	Steel, 24 ga. sectional, manual, 8' x 8' high		2	8		1,125	425		1,550	1,900
2800	Chain hoist, 20' x 14' high		.70	22.857		3,725	1,225		4,950	5,925
2850	For 1-1/4" rigid insulation and 26 ga. galv.									
2860	back panel, add				S.F.	5			5	5.50
2900	For electric trolley operator, 1/3 HP, to 12' x 12', add	1 Carp	2	4	Ea.	1,125	213		1,338	1,550
2950	Over 12' x 12', 1/2 HP, add	"	1	8	"	1,250	425		1,675	2,025

08 41 Entrances and Storefronts

08 41 13 – Aluminum-Framed Entrances and Storefronts

08 41 13.20 Tube Framing

	08 41 13.20 Tube Framing	Crew	Daily Output	Labor-Hours	Unit	Material	2020 Bare Costs Labor	Equipment	Total	Total Incl O&P
0010	**TUBE FRAMING**, For window walls and storefronts, aluminum stock									
1000	Flush tube frame, mill finish, 1/4" glass, 1-3/4" x 4", open header	2 Glaz	80	.200	L.F.	14.90	10.20		25.10	31.50
1050	Open sill		82	.195		12	9.95		21.95	28
1100	Closed back header		83	.193		21	9.85		30.85	38.50
1200	Vertical mullion, one piece		75	.213		21.50	10.90		32.40	40
1300	90° or 180° vertical corner post		75	.213		34.50	10.90		45.40	54.50
5000	Flush tube frame, mill fin., thermal brk., 2-1/4" x 4-1/2", open header		74	.216		16.50	11.05		27.55	34.50
5050	Open sill		75	.213		14.25	10.90		25.15	32
5100	Vertical mullion, one piece		69	.232		19.35	11.85		31.20	39
5200	90° or 180° vertical corner post		69	.232		18.80	11.85		30.65	38

08 41 26 – All-Glass Entrances and Storefronts

08 41 26.10 Window Walls Aluminum, Stock

	08 41 26.10 Window Walls Aluminum, Stock	Crew	Daily Output	Labor-Hours	Unit	Material	2020 Bare Costs Labor	Equipment	Total	Total Incl O&P
0010	**WINDOW WALLS ALUMINUM, STOCK**, including glazing									
0020	Minimum	H-2	160	.150	S.F.	51.50	7.20		58.70	67.50
0050	Average		140	.171		71.50	8.25		79.75	91
0100	Maximum		110	.218		188	10.50		198.50	223

08 42 Entrances

08 42 26 – All-Glass Entrances

08 42 26.10 Swinging Glass Doors

	08 42 26.10 Swinging Glass Doors	Crew	Daily Output	Labor-Hours	Unit	Material	2020 Bare Costs Labor	Equipment	Total	Total Incl O&P
0010	**SWINGING GLASS DOORS**									
0020	Including hardware, 1/2" thick, tempered, 3' x 7' opening	2 Glaz	2	8	Opng.	2,400	410		2,810	3,225
0100	6' x 7' opening	"	1.40	11.429	"	4,650	585		5,235	6,000

08 43 Storefronts

08 43 13 – Aluminum-Framed Storefronts

08 43 13.20 Storefront Systems

		Crew	Daily Output	Labor-Hours	Unit	Material	2020 Bare Costs Labor	Equipment	Total	Total Incl O&P
0010	**STOREFRONT SYSTEMS**, aluminum frame clear 3/8" plate glass									
0020	incl. 3' x 7' door with hardware (400 sq. ft. max. wall)									
0500	Wall height to 12' high, commercial grade	2 Glaz	150	.107	S.F.	27.50	5.45		32.95	38
0700	Monumental grade		115	.139		84.50	7.10		91.60	104
1000	6' x 7' door with hardware, commercial grade		135	.119		76	6.05		82.05	92.50
1200	Monumental grade		100	.160		117	8.15		125.15	140

08 44 Curtain Wall and Glazed Assemblies

08 44 13 – Glazed Aluminum Curtain Walls

08 44 13.10 Glazed Curtain Walls

		Crew	Daily Output	Labor-Hours	Unit	Material	2020 Bare Costs Labor	Equipment	Total	Total Incl O&P
0010	**GLAZED CURTAIN WALLS**, aluminum, stock, including glazing									
0020	Minimum	H-1	205	.156	S.F.	43.50	8.50		52	61
0150	Average, double glazed	"	180	.178	"	81	9.65		90.65	104

08 51 Metal Windows

08 51 13 – Aluminum Windows

08 51 13.10 Aluminum Sash

		Crew	Daily Output	Labor-Hours	Unit	Material	2020 Bare Costs Labor	Equipment	Total	Total Incl O&P
0010	**ALUMINUM SASH**									
0020	Stock, grade C, glaze & trim not incl., casement	2 Sswk	200	.080	S.F.	43	4.61		47.61	54.50
0100	Fixed casement	"	200	.080	"	18.45	4.61		23.06	28

08 51 13.20 Aluminum Windows

		Crew	Daily Output	Labor-Hours	Unit	Material	2020 Bare Costs Labor	Equipment	Total	Total Incl O&P
0010	**ALUMINUM WINDOWS**, incl. frame and glazing, commercial grade									
1000	Stock units, casement, 3'-1" x 3'-2" opening	2 Sswk	10	1.600	Ea.	395	92		487	575
3000	Single-hung, 2' x 3' opening, enameled, standard glazed		10	1.600		219	92		311	385
3100	Insulating glass		10	1.600		265	92		357	435
3890	Awning type, 3' x 3' opening, standard glass		14	1.143		435	66		501	585
4000	Sliding aluminum, 3' x 2' opening, standard glazed		10	1.600		218	92		310	385
5500	Sliding, with thermal barrier and screen, 6' x 4', 2 track		8	2		725	115		840	980
5700	4 track		8	2		915	115		1,030	1,175

08 51 23 – Steel Windows

08 51 23.10 Steel Sash

		Crew	Daily Output	Labor-Hours	Unit	Material	2020 Bare Costs Labor	Equipment	Total	Total Incl O&P
0010	**STEEL SASH** Custom units, glazing and trim not included									
0100	Casement, 100% vented	2 Sswk	200	.080	S.F.	69.50	4.61		74.11	83.50
0200	50% vented		200	.080		56	4.61		60.61	69
0300	Fixed		200	.080		29	4.61		33.61	39
2000	Industrial security sash, 50% vented		200	.080		61.50	4.61		66.11	75
2100	Fixed		200	.080		49	4.61		53.61	61.50

08 51 66 – Metal Window Screens

08 51 66.10 Screens

		Crew	Daily Output	Labor-Hours	Unit	Material	2020 Bare Costs Labor	Equipment	Total	Total Incl O&P
0010	**SCREENS**									
0020	For metal sash, aluminum or bronze mesh, flat screen	2 Sswk	1200	.013	S.F.	4.63	.77		5.40	6.30
0800	Security screen, aluminum frame with stainless steel cloth	"	1200	.013	"	24.50	.77		25.27	28

08 52 Wood Windows

08 52 10 – Plain Wood Windows

08 52 10.40 Casement Window

		Crew	Daily Output	Labor-Hours	Unit	Material	2020 Bare Costs Labor	Equipment	Total	Total Incl O&P
0010	**CASEMENT WINDOW**, including frame, screen and grilles									
0100	2'-0" x 3'-0" H, double insulated glass [G]	1 Carp	10	.800	Ea.	275	42.50		317.50	365
0200	2'-0" x 4'-6" high, double insulated glass [G]		9	.889		390	47		437	500
0522	Vinyl-clad, premium, double insulated glass, 2'-0" x 3'-0" [G]		10	.800		281	42.50		323.50	375
0525	2'-0" x 5'-0" [G]	↓	8	1	↓	365	53		418	480

08 52 10.50 Double-Hung

		Crew	Daily Output	Labor-Hours	Unit	Material	2020 Bare Costs Labor	Equipment	Total	Total Incl O&P
0010	**DOUBLE-HUNG**, Including frame, screens and grilles									
0100	2'-0" x 3'-0" high, low E insul. glass [G]	1 Carp	10	.800	Ea.	193	42.50		235.50	276
0300	4'-0" x 4'-6" high, low E insulated glass [G]	"	8	1	"	325	53		378	440

08 52 10.70 Sliding Windows

		Crew	Daily Output	Labor-Hours	Unit	Material	2020 Bare Costs Labor	Equipment	Total	Total Incl O&P
0010	**SLIDING WINDOWS**									
0100	3'-0" x 3'-0" high, double insulated [G]	1 Carp	10	.800	Ea.	286	42.50		328.50	380
0200	4'-0" x 3'-6" high, double insulated [G]	"	9	.889	"	425	47		472	535

08 52 13 – Metal-Clad Wood Windows

08 52 13.10 Awning Windows, Metal-Clad

		Crew	Daily Output	Labor-Hours	Unit	Material	2020 Bare Costs Labor	Equipment	Total	Total Incl O&P
0010	**AWNING WINDOWS, METAL-CLAD**									
2000	Metal-clad, awning deluxe, double insulated glass, 34" x 22"	1 Carp	9	.889	Ea.	251	47		298	350
2100	40" x 22"	"	9	.889	"	297	47		344	395

08 52 13.20 Casement Windows, Metal-Clad

		Crew	Daily Output	Labor-Hours	Unit	Material	2020 Bare Costs Labor	Equipment	Total	Total Incl O&P
0010	**CASEMENT WINDOWS, METAL-CLAD**									
0100	Metal-clad, deluxe, dbl. insul. glass, 2'-0" x 3'-0" high [G]	1 Carp	10	.800	Ea.	300	42.50		342.50	395
0130	2'-0" x 5'-0" high [G]	"	8	1	"	360	53		413	475

08 52 13.30 Double-Hung Windows, Metal-Clad

		Crew	Daily Output	Labor-Hours	Unit	Material	2020 Bare Costs Labor	Equipment	Total	Total Incl O&P
0010	**DOUBLE-HUNG WINDOWS, METAL-CLAD**									
0100	Metal-clad, deluxe, dbl. insul. glass, 2'-6" x 3'-0" high [G]	1 Carp	10	.800	Ea.	294	42.50		336.50	390
0160	3'-0" x 4'-6" high [G]	"	9	.889	"	365	47		412	470

08 52 13.35 Picture and Sliding Windows Metal-Clad

		Crew	Daily Output	Labor-Hours	Unit	Material	2020 Bare Costs Labor	Equipment	Total	Total Incl O&P
0010	**PICTURE AND SLIDING WINDOWS METAL-CLAD**									
2400	Metal-clad, dlx sliding, dbl. insul. glass, 3'-0" x 3'-0" high [G]	1 Carp	10	.800	Ea.	330	42.50		372.50	425
2420	4'-0" x 3'-6" high [G]	"	9	.889	"	400	47		447	510

08 52 16 – Plastic-Clad Wood Windows

08 52 16.70 Vinyl-Clad, Premium, DBL. Insul. Glass

		Crew	Daily Output	Labor-Hours	Unit	Material	2020 Bare Costs Labor	Equipment	Total	Total Incl O&P
0010	**VINYL-CLAD, PREMIUM, DBL. INSUL. GLASS**									
1000	Sliding, 3'-0" x 3'-0" [G]	1 Carp	10	.800	Ea.	645	42.50		687.50	775
1100	5'-0" x 4'-0" [G]	"	9	.889	"	940	47		987	1,100

08 71 Door Hardware

08 71 20.40 Lockset

		Crew	Daily Output	Labor-Hours	Unit	Material	2020 Bare Costs Labor	Equipment	Total	Total Incl O&P
0010	**LOCKSET**, Standard duty									
0020	Non-keyed, passage, w/sect. trim	1 Carp	12	.667	Ea.	81	35.50		116.50	142
0100	Privacy		12	.667		77.50	35.50		113	139
0400	Keyed, single cylinder function		10	.800		163	42.50		205.50	243
0500	Lever handled, keyed, single cylinder function	↓	10	.800	↓	159	42.50		201.50	239

08 71 20.41 Dead Locks

		Crew	Daily Output	Labor-Hours	Unit	Material	2020 Bare Costs Labor	Equipment	Total	Total Incl O&P
0010	**DEAD LOCKS**									
0011	Mortise heavy duty outside key (security item)	1 Carp	9	.889	Ea.	193	47		240	284
0020	Double cylinder		9	.889		193	47		240	283
1000	Tubular, standard duty, outside key	↓	10	.800		54	42.50		96.50	124

08 71 Door Hardware

08 52 16 – Plastic-Clad Wood Windows

08 71 20.41 Dead Locks		Crew	Daily Output	Labor-Hours	Unit	Material	2020 Bare Costs Labor	Equipment	Total	Total Incl O&P
1010	Double cylinder	1 Carp	10	.800	Ea.	71.50	42.50		114	143
08 71 20.50 Door Stops										
0010	**DOOR STOPS**									
0020	Holder & bumper, floor or wall	1 Carp	32	.250	Ea.	36.50	13.30		49.80	60
1300	Wall bumper, 4" diameter, with rubber pad, aluminum		32	.250		13.55	13.30		26.85	35
1600	Door bumper, floor type, aluminum		32	.250		3.94	13.30		17.24	24.50
1900	Plunger type, door mounted		32	.250		32	13.30		45.30	55.50
08 71 20.65 Thresholds										
0010	**THRESHOLDS**									
0011	Threshold 3' long saddles aluminum	1 Carp	48	.167	L.F.	11.10	8.85		19.95	25.50
0100	Aluminum, 8" wide, 1/2" thick		12	.667	Ea.	56.50	35.50		92	115
0700	Rubber, 1/2" thick, 5-1/2" wide		20	.400		43	21.50		64.50	79.50
0800	2-3/4" wide		20	.400		54.50	21.50		76	91.50
08 71 20.90 Hinges										
0010	**HINGES**									
0012	Full mortise, avg. freq., steel base, USP, 4-1/2" x 4-1/2"				Pr.	47.50			47.50	52.50
0080	US10A					60			60	66
0400	Brass base, 4-1/2" x 4-1/2", US10					73.50			73.50	81
0480	US10B					73			73	80.50
0950	Full mortise, high frequency, steel base, 3-1/2" x 3-1/2", US26D					31.50			31.50	34.50
1000	4-1/2" x 4-1/2", USP					67.50			67.50	74
1400	Brass base, 3-1/2" x 3-1/2", US4					53.50			53.50	58.50
1430	4-1/2" x 4-1/2", US10					74.50			74.50	82
3000	Half surface, half mortise, or full surface, average frequency									
3010	Steel base, 4-1/2" x 4-1/2", USP				Pr.	54			54	59.50
3400	Brass base, 4-1/2" x 4-1/2", US10				"	185			185	203
08 71 20.91 Special Hinges										
0010	**SPECIAL HINGES**									
1500	Non template, full mortise, average frequency									
1510	Steel base, 4" x 4", USP				Pr.	32			32	35
4600	Spring hinge, single acting, 6" flange, steel				Ea.	49.50			49.50	54.50
4700	Brass				"	96			96	106

08 71 21 – Astragals

08 71 21.10 Exterior Mouldings, Astragals

		Crew	Daily Output	Labor-Hours	Unit	Material	Labor	Equipment	Total	Total Incl O&P
0010	**EXTERIOR MOULDINGS, ASTRAGALS**									
2000	"L" extrusion, neoprene bulbs	1 Carp	75	.107	L.F.	4.51	5.65		10.16	13.45
2400	Spring hinged security seal, with cam	"	75	.107	"	6.75	5.65		12.40	15.90
3000	One piece stile protection									
3020	Neoprene fabric loop, nail on aluminum strips	1 Carp	60	.133	L.F.	1.25	7.10		8.35	12.05
3600	Spring bronze strip, nail on type		105	.076		2.16	4.05		6.21	8.50
4000	Extruded aluminum retainer, flush mount, pile insert		105	.076		2.49	4.05		6.54	8.85
5000	Two piece overlapping astragal, extruded aluminum retainer									
5010	Pile insert	1 Carp	60	.133	L.F.	3.26	7.10		10.36	14.25
5500	Interlocking aluminum, 5/8" x 1" neoprene bulb insert	"	45	.178	"	5.30	9.45		14.75	20

08 71 25 – Door Weatherstripping

08 71 25.10 Mechanical Seals, Weatherstripping

		Crew	Daily Output	Labor-Hours	Unit	Material	Labor	Equipment	Total	Total Incl O&P
0010	**MECHANICAL SEALS, WEATHERSTRIPPING**									
1000	Doors, wood frame, interlocking, for 3' x 7' door, zinc	1 Carp	3	2.667	Opng.	58.50	142		200.50	278
1300	6' x 7' opening, zinc	"	2	4	"	60.50	213		273.50	385
1700	Wood frame, spring type, bronze									

08 71 Door Hardware

08 71 25 – Door Weatherstripping

	08 71 25.10 Mechanical Seals, Weatherstripping	Crew	Daily Output	Labor-Hours	Unit	Material	2020 Bare Costs Labor	2020 Bare Costs Equipment	Total	Total Incl O&P
1800	3' x 7' door	1 Carp	7.60	1.053	Opng.	24.50	56		80.50	111

08 75 Window Hardware

08 75 30 – Window Weatherstripping

08 75 30.10 Mechanical Weather Seals

		Crew	Daily Output	Labor-Hours	Unit	Material	2020 Bare Costs Labor	2020 Bare Costs Equipment	Total	Total Incl O&P
0010	**MECHANICAL WEATHER SEALS**, Window, double-hung, 3' x 5'									
0020	Zinc	1 Carp	7.20	1.111	Opng.	22.50	59		81.50	114
0500	As above but heavy duty, zinc	"	4.60	1.739	"	21	92.50		113.50	162

08 81 Glass Glazing

08 81 10 – Float Glass

08 81 10.10 Various Types and Thickness of Float Glass

		Crew	Daily Output	Labor-Hours	Unit	Material	2020 Bare Costs Labor	2020 Bare Costs Equipment	Total	Total Incl O&P
0010	**VARIOUS TYPES AND THICKNESS OF FLOAT GLASS**									
0020	3/16" plain	2 Glaz	130	.123	S.F.	7.10	6.30		13.40	17.25
0200	Tempered, clear	"	130	.123	"	7.50	6.30		13.80	17.65

08 81 30 – Insulating Glass

08 81 30.10 Reduce Heat Transfer Glass

			Crew	Daily Output	Labor-Hours	Unit	Material	2020 Bare Costs Labor	2020 Bare Costs Equipment	Total	Total Incl O&P
0010	**REDUCE HEAT TRANSFER GLASS**										
0015	2 lites 1/8" float, 1/2" thk under 15 S.F.										
0020	Clear	G	2 Glaz	95	.168	S.F.	10.60	8.60		19.20	24.50
0200	2 lites 3/16" float, for 5/8" thk unit, 15 to 30 S.F., clear	G		90	.178		14.25	9.05		23.30	29.50
0400	1" thk, dbl. glazed, 1/4" float, 30 to 70 S.F., clear	G		75	.213		17.10	10.90		28	35

08 83 Mirrors

08 83 13 – Mirrored Glass Glazing

08 83 13.10 Mirrors

		Crew	Daily Output	Labor-Hours	Unit	Material	2020 Bare Costs Labor	2020 Bare Costs Equipment	Total	Total Incl O&P
0010	**MIRRORS**, No frames, wall type, 1/4" plate glass, polished edge									
0100	Up to 5 S.F.	2 Glaz	125	.128	S.F.	9.85	6.55		16.40	20.50
1000	Float glass, up to 10 S.F., 1/8" thick		160	.100		6.75	5.10		11.85	15.05
1100	3/16" thick		150	.107		7.85	5.45		13.30	16.80

08 88 Special Function Glazing

08 88 56 – Ballistics-Resistant Glazing

08 88 56.10 Laminated Glass

		Crew	Daily Output	Labor-Hours	Unit	Material	2020 Bare Costs Labor	2020 Bare Costs Equipment	Total	Total Incl O&P
0010	**LAMINATED GLASS**									
0020	Clear float .03" vinyl 1/4" thick	2 Glaz	90	.178	S.F.	13.30	9.05		22.35	28.50
0200	.06" vinyl, 1/2" thick	"	65	.246	"	28	12.55		40.55	50

Estimating Tips

General

- Room Finish Schedule: A complete set of plans should contain a room finish schedule. If one is not available, it would be well worth the time and effort to obtain one.

09 20 00 Plaster and Gypsum Board

- Lath is estimated by the square yard plus a 5% allowance for waste. Furring, channels, and accessories are measured by the linear foot. An extra foot should be allowed for each accessory miter or stop.

- Plaster is also estimated by the square yard. Deductions for openings vary by preference, from zero deduction to 50% of all openings over 2 feet in width. The estimator should allow one extra square foot for each linear foot of horizontal interior or exterior angle located below the ceiling level. Also, double the areas of small radius work.

- Drywall accessories, studs, track, and acoustical caulking are all measured by the linear foot. Drywall taping is figured by the square foot. Gypsum wallboard is estimated by the square foot. No material deductions should be made for door or window openings under 32 S.F.

09 60 00 Flooring

- Tile and terrazzo areas are taken off on a square foot basis. Trim and base materials are measured by the linear foot. Accent tiles are listed per each. Two basic methods of installation are used. Mud set is approximately 30% more expensive than thin set.

The cost of grout is included with tile unit price lines unless otherwise noted. In terrazzo work, be sure to include the linear footage of embedded decorative strips, grounds, machine rubbing, and power cleanup.

- Wood flooring is available in strip, parquet, or block configuration. The latter two types are set in adhesives with quantities estimated by the square foot. The laying pattern will influence labor costs and material waste. In addition to the material and labor for laying wood floors, the estimator must make allowances for sanding and finishing these areas, unless the flooring is prefinished.

- Sheet flooring is measured by the square yard. Roll widths vary, so consideration should be given to use the most economical width, as waste must be figured into the total quantity. Consider also the installation methods available—direct glue down or stretched. Direct glue-down installation is assumed with sheet carpet unit price lines unless otherwise noted.

09 70 00 Wall Finishes

- Wall coverings are estimated by the square foot. The area to be covered is measured—length by height of the wall above the baseboards—to calculate the square footage of each wall. This figure is divided by the number of square feet in the single roll which is being used. Deduct, in full, the areas of openings such as doors and windows. Where a pattern match is required allow 25–30% waste.

09 80 00 Acoustic Treatment

- Acoustical systems fall into several categories. The takeoff of these materials should be by the square foot of area with a 5% allowance for waste. Do not forget about scaffolding, if applicable, when estimating these systems.

09 90 00 Painting and Coating

- New line items created for cut-ins with reference diagram.

- A major portion of the work in painting involves surface preparation. Be sure to include cleaning, sanding, filling, and masking costs in the estimate.

- Protection of adjacent surfaces is not included in painting costs. When considering the method of paint application, an important factor is the amount of protection and masking required. These must be estimated separately and may be the determining factor in choosing the method of application.

Reference Numbers

Reference numbers are shown at the beginning of some major classifications. These numbers refer to related items in the Reference Section. The reference information may be an estimating procedure, an alternate pricing method, or technical information.

Note: Not all subdivisions listed here necessarily appear. ∎

Same Data. Simplified.

Enjoy the convenience and efficiency of accessing your costs anywhere:

- **Skip the multiplier** by setting your location
- **Quickly search**, edit, favorite and share costs
- **Stay on top of price changes** with automatic updates

Discover more at rsmeans.com/online

09 05 Common Work Results for Finishes

09 05 05 – Selective Demolition for Finishes

09 05 05.10 Selective Demolition, Ceilings

		Crew	Daily Output	Labor-Hours	Unit	Material	2020 Bare Costs Labor	Equipment	Total	Total Incl O&P
0010	**SELECTIVE DEMOLITION, CEILINGS** R024119-10									
0200	Ceiling, gypsum wall board, furred and nailed or screwed	2 Clab	800	.020	S.F.		.84		.84	1.27
0220	On metal frame		760	.021			.89		.89	1.33
0240	On suspension system, including system		720	.022			.94		.94	1.41
1000	Plaster, lime and horse hair, on wood lath, incl. lath		700	.023			.96		.96	1.45
1020	On metal lath		570	.028			1.18		1.18	1.78
1100	Gypsum, on gypsum lath		720	.022			.94		.94	1.41
1120	On metal lath		500	.032			1.35		1.35	2.02
1200	Suspended ceiling, mineral fiber, 2' x 2' or 2' x 4'		1500	.011			.45		.45	.67
1250	On suspension system, incl. system		1200	.013			.56		.56	.84
1500	Tile, wood fiber, 12" x 12", glued		900	.018			.75		.75	1.12
1540	Stapled		1500	.011			.45		.45	.67
1580	On suspension system, incl. system		760	.021			.89		.89	1.33
2000	Wood, tongue and groove, 1" x 4"		1000	.016			.67		.67	1.01
2040	1" x 8"		1100	.015			.61		.61	.92
2400	Plywood or wood fiberboard, 4' x 8' sheets		1200	.013			.56		.56	.84
2500	Remove & refinish textured ceiling	1 Plas	222	.036		.04	1.75		1.79	2.65

09 05 05.20 Selective Demolition, Flooring

		Crew	Daily Output	Labor-Hours	Unit	Material	2020 Bare Costs Labor	Equipment	Total	Total Incl O&P
0010	**SELECTIVE DEMOLITION, FLOORING** R024119-10									
0200	Brick with mortar	2 Clab	475	.034	S.F.		1.42		1.42	2.13
0400	Carpet, bonded, including surface scraping		2000	.008			.34		.34	.51
0440	Scrim applied		8000	.002			.08		.08	.13
0480	Tackless		9000	.002			.07		.07	.11
0500	Carpet pad		3500	.005	S.Y.		.19		.19	.29
0550	Carpet tile, releasable adhesive		5000	.003	S.F.		.13		.13	.20
0560	Permanent adhesive		1850	.009			.36		.36	.55
0600	Composition, acrylic or epoxy		400	.040			1.68		1.68	2.53
0700	Concrete, scarify skin	A-1A	225	.036			1.95	.92	2.87	3.96
0800	Resilient, sheet goods	2 Clab	1400	.011			.48		.48	.72
0820	For gym floors	"	900	.018			.75		.75	1.12
0850	Vinyl or rubber cove base	1 Clab	1000	.008	L.F.		.34		.34	.51
0860	Vinyl or rubber cove base, molded corner	"	1000	.008	Ea.		.34		.34	.51
0870	For glued and caulked installation, add to labor						50%			
0900	Vinyl composition tile, 12" x 12"	2 Clab	1000	.016	S.F.		.67		.67	1.01
2000	Tile, ceramic, thin set		675	.024			1		1	1.50
2020	Mud set		625	.026			1.08		1.08	1.62
2200	Marble, slate, thin set		675	.024			1		1	1.50
2220	Mud set		625	.026			1.08		1.08	1.62
2600	Terrazzo, thin set		450	.036			1.50		1.50	2.25
2620	Mud set		425	.038			1.59		1.59	2.38
2640	Terrazzo, cast in place		300	.053			2.25		2.25	3.37
8000	Remove flooring, bead blast, simple floor plan	A-1A	1000	.008		.05	.44	.21	.70	.95
8100	Complex floor plan		400	.020		.05	1.10	.52	1.67	2.29
8150	Mastic only		1500	.005		.05	.29	.14	.48	.65

09 05 05.30 Selective Demolition, Walls and Partitions

		Crew	Daily Output	Labor-Hours	Unit	Material	2020 Bare Costs Labor	Equipment	Total	Total Incl O&P
0010	**SELECTIVE DEMOLITION, WALLS AND PARTITIONS** R024119-10									
0100	Brick, 4" to 12" thick	B-9	220	.182	C.F.		7.75	1.62	9.37	13.40
0200	Concrete block, 4" thick		1150	.035	S.F.		1.48	.31	1.79	2.56
0280	8" thick		1050	.038			1.62	.34	1.96	2.80
1000	Gypsum wallboard, nailed or screwed	1 Clab	1000	.008			.34		.34	.51
1010	2 layers		400	.020			.84		.84	1.27
1020	Glued and nailed		900	.009			.37		.37	.56

170

170

For customer support on your **Site Work & Landscape Costs** with RSMeans data, call 800.448.8182.

09 05 Common Work Results for Finishes

09 05 05 – Selective Demolition for Finishes

09 05 05.30 Selective Demolition, Walls and Partitions

		Crew	Daily Output	Labor-Hours	Unit	Material	2020 Bare Costs Labor	Equipment	Total	Total Incl O&P
2200	Metal or wood studs, finish 2 sides, fiberboard	B-1	520	.046	S.F.		1.97		1.97	2.97
2250	Lath and plaster		260	.092			3.95		3.95	5.95
2300	Gypsum wallboard		520	.046			1.97		1.97	2.97
2350	Plywood		450	.053			2.28		2.28	3.43
3760	Tile, ceramic, on walls, thin set	1 Clab	300	.027			1.12		1.12	1.69
3800	Toilet partitions, slate or marble		5	1.600	Ea.		67.50		67.50	101
3820	Metal or plastic		8	1	"		42		42	63.50
5000	Wallcovering, vinyl	1 Pape	700	.011	S.F.		.51		.51	.76
5010	With release agent		1500	.005			.24		.24	.35
5025	Wallpaper, 2 layers or less, by hand		250	.032			1.43		1.43	2.12
5035	3 layers or more		165	.048			2.16		2.16	3.21

09 21 Plaster and Gypsum Board Assemblies

09 21 13 – Plaster Assemblies

09 21 13.10 Plaster Partition Wall

		Crew	Daily Output	Labor-Hours	Unit	Material	2020 Bare Costs Labor	Equipment	Total	Total Incl O&P
0010	**PLASTER PARTITION WALL**									
0400	Stud walls, 3.4 lb. metal lath, 3 coat gypsum plaster, 2 sides									
0600	2" x 4" wood studs, 16" OC	J-2	315	.152	S.F.	3.01	7.20	.37	10.58	14.40
0700	2-1/2" metal studs, 25 ga., 12" OC		325	.148		2.71	6.95	.36	10.02	13.75
0800	3-5/8" metal studs, 25 ga., 16" OC		320	.150		2.68	7.05	.37	10.10	13.85
0900	Gypsum lath, 2 coat vermiculite plaster, 2 sides									
1000	2" x 4" wood studs, 16" OC	J-2	355	.135	S.F.	2.63	6.35	.33	9.31	12.75
1200	2-1/2" metal studs, 25 ga., 12" OC		365	.132		2.54	6.20	.32	9.06	12.35
1300	3-5/8" metal studs, 25 ga., 16" OC		360	.133		2.51	6.30	.33	9.14	12.45

09 21 16 – Gypsum Board Assemblies

09 21 16.33 Partition Wall

		Crew	Daily Output	Labor-Hours	Unit	Material	2020 Bare Costs Labor	Equipment	Total	Total Incl O&P
0010	**PARTITION WALL** Stud wall, 8' to 12' high									
3200	5/8", interior, gypsum board, standard, tape & finish 2 sides									
3400	Installed on and including 2" x 4" wood studs, 16" OC	2 Carp	300	.053	S.F.	1.29	2.83		4.12	5.70
3800	Metal studs, NLB, 25 ga., 16" OC, 3-5/8" wide		340	.047		1.17	2.50		3.67	5.05
4800	Water resistant, on 2" x 4" wood studs, 16" OC		300	.053		1.51	2.83		4.34	5.90
5200	Metal studs, NLB, 25 ga. 16" OC, 3-5/8" wide		340	.047		1.39	2.50		3.89	5.30
6000	Fire resistant, 2 layers, 2 hr., on 2" x 4" wood studs, 16" OC		205	.078		2.15	4.15		6.30	8.60
6400	Metal studs, NLB, 25 ga., 16" OC, 3-5/8" wide		245	.065		2.06	3.47		5.53	7.45
9600	Partitions, for work over 8' high, add		1530	.010			.56		.56	.84

09 23 Gypsum Plastering

09 23 20 – Gypsum Plaster

09 23 20.10 Gypsum Plaster On Walls and Ceilings

		Crew	Daily Output	Labor-Hours	Unit	Material	2020 Bare Costs Labor	Equipment	Total	Total Incl O&P
0010	**GYPSUM PLASTER ON WALLS AND CEILINGS**									
0020	80# bag, less than 1 ton				Bag	15.90			15.90	17.50
0100	Over 1 ton				"	13.70			13.70	15.05
0300	2 coats, no lath included, on walls	J-1	105	.381	S.Y.	3.90	17.55	1.12	22.57	31.50
0600	On and incl. 3/8" gypsum lath on steel, on walls	J-2	97	.495		7.40	23.50	1.21	32.11	44
0900	3 coats, no lath included, on walls	J-1	87	.460		5.55	21	1.35	27.90	39
1200	On and including painted metal lath, on wood studs	J-2	86	.558		10.70	26.50	1.36	38.56	52.50

For customer support on your Site Work & Landscape Costs with RSMeans data, call 800.448.8182.

09 24 Cement Plastering

09 24 23 – Cement Stucco

09 24 23.40 Stucco		Crew	Daily Output	Labor-Hours	Unit	Material	2020 Bare Costs Labor	Equipment	Total	Total Incl O&P
0010	**STUCCO**									
0015	3 coats 7/8" thick, float finish, with mesh, on wood frame	J-2	63	.762	S.Y.	7.65	36	1.86	45.51	64
0100	On masonry construction, no mesh incl.	J-1	67	.597		3.02	27.50	1.75	32.27	46.50
0300	For trowel finish, add	1 Plas	170	.047			2.29		2.29	3.41
1000	Stucco, with bonding agent, 3 coats, on walls, no mesh incl.	J-1	200	.200	↓	4.23	9.20	.59	14.02	19.05

09 26 Veneer Plastering

09 26 13 – Gypsum Veneer Plastering

09 26 13.80 Thin Coat Plaster

		Crew	Daily Output	Labor-Hours	Unit	Material	2020 Bare Costs Labor	Equipment	Total	Total Incl O&P
0010	**THIN COAT PLASTER**									
0012	1 coat veneer, not incl. lath	J-1	3600	.011	S.F.	.10	.51	.03	.64	.91
1000	In 50 lb. bags				Bag	14			14	15.40

09 29 Gypsum Board

09 29 10 – Gypsum Board Panels

09 29 10.30 Gypsum Board

		Crew	Daily Output	Labor-Hours	Unit	Material	2020 Bare Costs Labor	Equipment	Total	Total Incl O&P
0010	**GYPSUM BOARD** on walls & ceilings R092910-10									
0150	3/8" thick, on walls, standard, no finish included	2 Carp	2000	.008	S.F.	.35	.43		.78	1.03
0300	1/2" thick, on walls, standard, no finish included		2000	.008		.33	.43		.76	1
0350	Taped and finished (level 4 finish)		965	.017		.38	.88		1.26	1.74
0390	With compound skim coat (level 5 finish)		775	.021		.44	1.10		1.54	2.13
2000	5/8" thick, on walls, standard, no finish included		2000	.008		.33	.43		.76	1
2090	With compound skim coat (level 5 finish)		775	.021		.44	1.10		1.54	2.13
2200	Water resistant, no finish included		2000	.008		.44	.43		.87	1.12
2290	With compound skim coat (level 5 finish)		775	.021		.55	1.10		1.65	2.25
2510	Mold resistant, no finish included		2000	.008		.54	.43		.97	1.23
2530	With compound skim coat (level 5 finish)		775	.021		.65	1.10		1.75	2.36
4000	Fireproofing, beams or columns, 2 layers, 1/2" thick, incl finish		330	.048		.86	2.58		3.44	4.82
4010	Mold resistant		330	.048		.98	2.58		3.56	4.95
4050	5/8" thick		300	.053		.86	2.83		3.69	5.20
4060	Mold resistant		300	.053		1.18	2.83		4.01	5.55
6010	1/2" thick on walls, multi-layer, lightweight, no finish included		1500	.011		2.25	.57		2.82	3.33
6015	Taped and finished (level 4 finish)		725	.022		2.30	1.17		3.47	4.29
6025	5/8" thick on walls, for wood studs, no finish included		1500	.011		2.46	.57		3.03	3.56
6035	With compound skim coat (level 5 finish)		580	.028		2.57	1.47		4.04	5.05
6040	For metal stud, no finish included		1500	.011		2.22	.57		2.79	3.29
6050	With compound skim coat (level 5 finish)		580	.028		2.33	1.47		3.80	4.76
6055	Abuse resist, no finish included		1500	.011		4.23	.57		4.80	5.50
6065	With compound skim coat (level 5 finish)		580	.028		4.34	1.47		5.81	6.95
6070	Shear rated, no finish included		1500	.011		5.60	.57		6.17	7.05
6080	With compound skim coat (level 5 finish)		580	.028		5.75	1.47		7.22	8.50
6085	For SCIF applications, no finish included		1500	.011		5.30	.57		5.87	6.70
6095	With compound skim coat (level 5 finish)		580	.028		5.40	1.47		6.87	8.15
6100	1-3/8" thick on walls, THX certified, no finish included		1500	.011		9.35	.57		9.92	11.15
6110	With compound skim coat (level 5 finish)		580	.028		9.50	1.47		10.97	12.65
6115	5/8" thick on walls, score & snap installation, no finish included		2000	.008		1.99	.43		2.42	2.83
6125	With compound skim coat (level 5 finish)		775	.021		2.10	1.10		3.20	3.96
7020	5/8" thick on ceilings, for wood joists, no finish included		1200	.013		2.46	.71		3.17	3.77
7030	With compound skim coat (level 5 finish)		410	.039		2.57	2.07		4.64	5.95

09 29 Gypsum Board

09 29 10 – Gypsum Board Panels

09 29 10.30 Gypsum Board		Crew	Daily Output	Labor-Hours	Unit	Material	2020 Bare Costs Labor	Equipment	Total	Total Incl O&P
7035	For metal joists, no finish included	2 Carp	1200	.013	S.F.	2.22	.71		2.93	3.50
7045	With compound skim coat (level 5 finish)		410	.039		2.33	2.07		4.40	5.70
7050	Abuse resist, no finish included		1200	.013		4.23	.71		4.94	5.70
7060	With compound skim coat (level 5 finish)		410	.039		4.34	2.07		6.41	7.90
7065	Shear rated, no finish included		1200	.013		5.60	.71		6.31	7.25
7075	With compound skim coat (level 5 finish)		410	.039		5.75	2.07		7.82	9.40
7080	For SCIF applications, no finish included		1200	.013		5.30	.71		6.01	6.90
7090	With compound skim coat (level 5 finish)		410	.039		5.40	2.07		7.47	9.05
8010	5/8" thick on ceilings, score & snap installation, no finish included		1600	.010		1.99	.53		2.52	2.99
8020	With compound skim coat (level 5 finish)	↓	545	.029	↓	2.10	1.56		3.66	4.65

09 31 Thin-Set Tiling

09 31 13 – Thin-Set Ceramic Tiling

09 31 13.10 Thin-Set Ceramic Tile		Crew	Daily Output	Labor-Hours	Unit	Material	2020 Bare Costs Labor	Equipment	Total	Total Incl O&P
0010	THIN-SET CERAMIC TILE									
3300	Ceramic tile, porcelain type, 1 color, color group 2, 1" x 1"	D-7	183	.087	S.F.	6.25	3.86		10.11	12.50
3310	2" x 2" or 2" x 1"		190	.084		6.35	3.72		10.07	12.45
5400	Walls, interior, 4-1/4" x 4-1/4" tile		190	.084		3.10	3.72		6.82	8.85
5700	8-1/2" x 4-1/4" tile	↓	190	.084	↓	5.95	3.72		9.67	12

09 32 Mortar-Bed Tiling

09 32 13 – Mortar-Bed Ceramic Tiling

09 32 13.10 Ceramic Tile		Crew	Daily Output	Labor-Hours	Unit	Material	2020 Bare Costs Labor	Equipment	Total	Total Incl O&P
0010	CERAMIC TILE									
6300	Exterior walls, frostproof, 4-1/4" x 4-1/4"	D-7	102	.157	S.F.	7.25	6.95		14.20	18.15
6400	1-3/8" x 1-3/8"	"	93	.172	"	6.60	7.60		14.20	18.40

09 32 16 – Mortar-Bed Quarry Tiling

09 32 16.10 Quarry Tile		Crew	Daily Output	Labor-Hours	Unit	Material	2020 Bare Costs Labor	Equipment	Total	Total Incl O&P
0010	QUARRY TILE									
0100	Base, cove or sanitary, to 5" high, 1/2" thick	D-7	110	.145	L.F.	6.35	6.40		12.75	16.35
0700	Floors, 1,000 S.F. lots, red, 4" x 4" x 1/2" thick		120	.133	S.F.	8.55	5.90		14.45	18.05
0900	6" x 6" x 1/2" thick	↓	140	.114	"	8.70	5.05		13.75	17

09 34 Waterproofing-Membrane Tiling

09 34 13 – Waterproofing-Membrane Ceramic Tiling

09 34 13.10 Ceramic Tile Waterproofing Membrane		Crew	Daily Output	Labor-Hours	Unit	Material	2020 Bare Costs Labor	Equipment	Total	Total Incl O&P
0010	CERAMIC TILE WATERPROOFING MEMBRANE									
0020	On floors, including thinset									
0030	Fleece laminated polyethylene grid, 1/8" thick	D-7	250	.064	S.F.	2.13	2.83		4.96	6.50
0040	5/16" thick	"	250	.064	"	2.61	2.83		5.44	7
0050	On walls, including thinset									
0060	Fleece laminated polyethylene sheet, 8 mil thick	D-7	480	.033	S.F.	2.30	1.47		3.77	4.69
0070	Accessories, including thinset									
0080	Joint and corner sheet, 4 mils thick, 5" wide	1 Tilf	240	.033	L.F.	1.35	1.65		3	3.90
0090	7-1/4" wide		180	.044		1.72	2.20		3.92	5.10
0100	10" wide		120	.067	↓	2.09	3.30		5.39	7.15
0110	Pre-formed corners, inside		32	.250	Ea.	7.85	12.40		20.25	27

173

For customer support on your Site Work & Landscape Costs with RSMeans data, call 800.448.8182.

09 34 Waterproofing-Membrane Tiling

09 34 13 – Waterproofing-Membrane Ceramic Tiling

09 34 13.10 Ceramic Tile Waterproofing Membrane

	09 34 13.10 Ceramic Tile Waterproofing Membrane	Crew	Daily Output	Labor-Hours	Unit	Material	2020 Bare Costs Labor	Equipment	Total	Total Incl O&P
0120	Outside	1 Tilf	32	.250	Ea.	7.70	12.40		20.10	26.50
0130	2" flanged floor drain with 6" stainless steel grate		16	.500	↓	390	25		415	465
0140	EPS, sloped shower floor		480	.017	S.F.	5.55	.83		6.38	7.30
0150	Curb	↓	32	.250	L.F.	14.10	12.40		26.50	33.50

09 51 Acoustical Ceilings

09 51 23 – Acoustical Tile Ceilings

09 51 23.30 Suspended Ceilings, Complete

		Crew	Daily Output	Labor-Hours	Unit	Material	2020 Bare Costs Labor	Equipment	Total	Total Incl O&P
0010	**SUSPENDED CEILINGS, COMPLETE**, incl. standard									
0100	suspension system but not incl. 1-1/2" carrier channels									
0600	Fiberglass ceiling board, 2' x 4' x 5/8", plain faced	1 Carp	500	.016	S.F.	2.27	.85		3.12	3.78
1800	Tile, Z bar suspension, 5/8" mineral fiber tile	"	150	.053	"	2.15	2.83		4.98	6.65

09 63 Masonry Flooring

09 63 13 – Brick Flooring

09 63 13.10 Miscellaneous Brick Flooring

		Crew	Daily Output	Labor-Hours	Unit	Material	2020 Bare Costs Labor	Equipment	Total	Total Incl O&P
0010	**MISCELLANEOUS BRICK FLOORING**									
0020	Acid-proof shales, red, 8" x 3-3/4" x 1-1/4" thick	D-7	.43	37.209	M	725	1,650		2,375	3,200
0050	2-1/4" thick	D-1	.40	40		1,075	1,875		2,950	4,050
0200	Acid-proof clay brick, 8" x 3-3/4" x 2-1/4" thick G	"	.40	40	↓	1,025	1,875		2,900	3,975
0260	Cast ceramic, pressed, 4" x 8" x 1/2", unglazed	D-7	100	.160	S.F.	7.55	7.05		14.60	18.70
0270	Glazed		100	.160		10.10	7.05		17.15	21.50
0280	Hand molded flooring, 4" x 8" x 3/4", unglazed		95	.168		10	7.45		17.45	22
0290	Glazed		95	.168		11.40	7.45		18.85	23.50
0300	8" hexagonal, 3/4" thick, unglazed		85	.188		9.95	8.30		18.25	23
0310	Glazed	↓	85	.188		18	8.30		26.30	32
0400	Heavy duty industrial, cement mortar bed, 2" thick, not incl. brick	D-1	80	.200		1.18	9.40		10.58	15.50
0450	Acid-proof joints, 1/4" wide	"	65	.246		1.74	11.55		13.29	19.40
0500	Pavers, 8" x 4", 1" to 1-1/4" thick, red	D-7	95	.168		4.39	7.45		11.84	15.75
0510	Ironspot	"	95	.168		5.65	7.45		13.10	17.10
0540	1-3/8" to 1-3/4" thick, red	D-1	95	.168		4.24	7.90		12.14	16.60
0560	Ironspot		95	.168		5.60	7.90		13.50	18.10
0580	2-1/4" thick, red		90	.178		4.32	8.35		12.67	17.40
0590	Ironspot	↓	90	.178		6.70	8.35		15.05	20
0700	Paver, adobe brick, 6" x 12", 1/2" joint G	↓	42	.381		1.47	17.85		19.32	28.50
0710	Mexican red, 12" x 12" G	1 Tilf	48	.167		2.08	8.25		10.33	14.40
0720	Saltillo, 12" x 12" G	"	48	.167	↓	1.66	8.25		9.91	13.95
0800	For sidewalks and patios with pavers, see Section 32 14 16.10									
0870	For epoxy joints, add	D-1	600	.027	S.F.	3.15	1.25		4.40	5.35
0880	For Furan underlayment, add	"	600	.027		2.60	1.25		3.85	4.76
0890	For waxed surface, steam cleaned, add	A-1H	1000	.008	↓	.21	.34	.08	.63	.82

09 65 Resilient Flooring

09 65 13 – Resilient Base and Accessories

09 65 13.13 Resilient Base

		Crew	Daily Output	Labor-Hours	Unit	Material	2020 Bare Costs Labor	Equipment	Total	Total Incl O&P
0010	**RESILIENT BASE**									
0800	1/8" rubber base, 2-1/2" H, straight or cove, standard colors	1 Tilf	315	.025	L.F.	1.20	1.26		2.46	3.16
1100	4" high	"	315	.025	"	1.43	1.26		2.69	3.41

09 65 13.23 Resilient Stair Treads and Risers

		Crew	Daily Output	Labor-Hours	Unit	Material	2020 Bare Costs Labor	Equipment	Total	Total Incl O&P
0010	**RESILIENT STAIR TREADS AND RISERS**									
0300	Rubber, molded tread, 12" wide, 5/16" thick, black	1 Tilf	115	.070	L.F.	14.35	3.44		17.79	21
1800	Risers, 7" high, 1/8" thick, flat		250	.032		8.55	1.58		10.13	11.70
2100	Vinyl, molded tread, 12" wide, colors, 1/8" thick		115	.070		6.05	3.44		9.49	11.70
2400	Riser, 7" high, 1/8" thick, coved	↓	175	.046	↓	3.39	2.26		5.65	7.05

09 65 19 – Resilient Tile Flooring

09 65 19.19 Vinyl Composition Tile Flooring

		Crew	Daily Output	Labor-Hours	Unit	Material	2020 Bare Costs Labor	Equipment	Total	Total Incl O&P
0010	**VINYL COMPOSITION TILE FLOORING**									
7000	Vinyl composition tile, 12" x 12", 1/16" thick	1 Tilf	500	.016	S.F.	1.23	.79		2.02	2.51
7350	1/8" thick, marbleized	"	500	.016	"	2.47	.79		3.26	3.88

09 65 19.23 Vinyl Tile Flooring

		Crew	Daily Output	Labor-Hours	Unit	Material	2020 Bare Costs Labor	Equipment	Total	Total Incl O&P
0010	**VINYL TILE FLOORING**									
7500	Vinyl tile, 12" x 12", 3/32" thick, standard colors/patterns	1 Tilf	500	.016	S.F.	3.58	.79		4.37	5.10
7550	1/8" thick, standard colors/patterns	"	500	.016	"	5.25	.79		6.04	6.90

09 66 Terrazzo Flooring

09 66 16 – Terrazzo Floor Tile

09 66 16.10 Tile or Terrazzo Base

		Crew	Daily Output	Labor-Hours	Unit	Material	2020 Bare Costs Labor	Equipment	Total	Total Incl O&P
0010	**TILE OR TERRAZZO BASE**									
0020	Scratch coat only	1 Mstz	150	.053	S.F.	.58	2.63		3.21	4.49
0500	Scratch and brown coat only	"	75	.107	"	1.15	5.25		6.40	9

09 91 Painting

09 91 13 – Exterior Painting

09 91 13.60 Siding Exterior

		Crew	Daily Output	Labor-Hours	Unit	Material	2020 Bare Costs Labor	Equipment	Total	Total Incl O&P
0010	**SIDING EXTERIOR**, Alkyd (oil base)									
0450	Steel siding, oil base, paint 1 coat, brushwork	2 Pord	2015	.008	S.F.	.11	.35		.46	.64
0800	Paint 2 coats, brushwork		1300	.012		.22	.55		.77	1.05
1200	Stucco, rough, oil base, paint 2 coats, brushwork		1300	.012		.22	.55		.77	1.05
1400	Roller		1625	.010		.23	.44		.67	.90
1800	Texture 1-11 or clapboard, oil base, primer coat, brushwork		1300	.012		.14	.55		.69	.96
2400	Paint 2 coats, brushwork	↓	810	.020	↓	.32	.88		1.20	1.65
8000	For latex paint, deduct					10%				

09 91 23 – Interior Painting

09 91 23.52 Miscellaneous, Interior

		Crew	Daily Output	Labor-Hours	Unit	Material	2020 Bare Costs Labor	Equipment	Total	Total Incl O&P
0010	**MISCELLANEOUS, INTERIOR**									
2400	Floors, conc./wood, oil base, primer/sealer coat, brushwork	2 Pord	1950	.008	S.F.	.09	.36		.45	.64
2450	Roller	"	5200	.003		.09	.14		.23	.30
3800	Grilles, per side, oil base, primer coat, brushwork	1 Pord	520	.015		.14	.68		.82	1.17
3900	Spray	"	1140	.007	↓	.28	.31		.59	.77
5000	Pipe, 1"-4" diameter, primer or sealer coat, oil base, brushwork	2 Pord	1250	.013	L.F.	.09	.57		.66	.95
5350	Paint 2 coats, brushwork		775	.021		.23	.92		1.15	1.61
6300	13"-16" diameter, primer or sealer coat, brushwork		310	.052		.37	2.29		2.66	3.81
6500	Paint 2 coats, brushwork	↓	195	.082	↓	.91	3.64		4.55	6.40

175

09 91 Painting

09 91 23 – Interior Painting

09 91 23.52 Miscellaneous, Interior

	Crew	Daily Output	Labor-Hours	Unit	Material	2020 Bare Costs Labor	Equipment	Total	Total Incl O&P
8900 Trusses and wood frames, primer coat, oil base, brushwork	1 Pord	800	.010	S.F.	.07	.44		.51	.74
9220 Paint 2 coats, brushwork		500	.016		.25	.71		.96	1.33
9260 Stain, brushwork, wipe off		600	.013		.10	.59		.69	.99
9280 Varnish, 3 coats, brushwork		275	.029		.29	1.29		1.58	2.24

09 91 23.72 Walls and Ceilings, Interior

	Crew	Daily Output	Labor-Hours	Unit	Material	2020 Bare Costs Labor	Equipment	Total	Total Incl O&P
0010 **WALLS AND CEILINGS, INTERIOR**									
0100 Concrete, drywall or plaster, latex, primer or sealer coat									
0150 Smooth finish, cut-in by brush	1 Pord	1150	.007	L.F.	.02	.31		.33	.48
0200 Brushwork		1150	.007	S.F.	.06	.31		.37	.53
0590 Paint 2 coats, smooth finish, cut-in by brush		680	.012	L.F.	.04	.52		.56	.82
0800 Brushwork		680	.012	S.F.	.15	.52		.67	.95
0880 Spray		1625	.005	"	.14	.22		.36	.49

09 91 23.74 Walls and Ceilings, Interior, Zero VOC Latex

	Crew	Daily Output	Labor-Hours	Unit	Material	2020 Bare Costs Labor	Equipment	Total	Total Incl O&P
0010 **WALLS AND CEILINGS, INTERIOR, ZERO VOC LATEX**									
0100 Concrete, dry wall or plaster, latex, primer or sealer coat									
0190 Smooth finish, cut-in by brush	1 Pord	1150	.007	L.F.	.02	.31		.33	.48
0200 Brushwork G		1150	.007	S.F.	.07	.31		.38	.54
0790 Paint 2 coats, smooth finish, cut-in by brush		680	.012	L.F.	.04	.52		.56	.82
0800 Brushwork G		680	.012	S.F.	.04	.52		.56	.82
0880 Spray G		1625	.005	"	.14	.22		.36	.49

09 91 23.75 Dry Fall Painting

	Crew	Daily Output	Labor-Hours	Unit	Material	2020 Bare Costs Labor	Equipment	Total	Total Incl O&P
0010 **DRY FALL PAINTING**									
0100 Sprayed on walls, gypsum board or plaster									
0220 One coat	1 Pord	2600	.003	S.F.	.08	.14		.22	.29
0250 Two coats		1560	.005		.16	.23		.39	.52
0280 Concrete or textured plaster, one coat		1560	.005		.08	.23		.31	.43
0310 Two coats		1300	.006		.16	.27		.43	.59
0340 Concrete block, one coat		1560	.005		.08	.23		.31	.43
0370 Two coats		1300	.006		.16	.27		.43	.59
0400 Wood, one coat		877	.009		.08	.40		.48	.69
0430 Two coats		650	.012		.16	.55		.71	.99
0440 On ceilings, gypsum board or plaster									
0470 One coat	1 Pord	1560	.005	S.F.	.08	.23		.31	.43
0500 Two coats		1300	.006		.16	.27		.43	.59
0530 Concrete or textured plaster, one coat		1560	.005		.08	.23		.31	.43
0560 Two coats		1300	.006		.16	.27		.43	.59
0570 Structural steel, bar joists or metal deck, one coat		1560	.005		.08	.23		.31	.43
0580 Two coats		1040	.008		.16	.34		.50	.69

09 96 High-Performance Coatings

09 96 23 – Graffiti-Resistant Coatings

09 96 23.10 Graffiti-Resistant Treatments

	Crew	Daily Output	Labor-Hours	Unit	Material	2020 Bare Costs Labor	Equipment	Total	Total Incl O&P
0010 **GRAFFITI-RESISTANT TREATMENTS**, sprayed on walls									
0100 Non-sacrificial, permanent non-stick coating, clear, on metals	1 Pord	2000	.004	S.F.	2.05	.18		2.23	2.52
0200 Concrete		2000	.004		2.33	.18		2.51	2.82
0300 Concrete block		2000	.004		3.02	.18		3.20	3.58
0400 Brick		2000	.004		3.42	.18		3.60	4.02
0500 Stone		2000	.004		3.42	.18		3.60	4.02
0600 Unpainted wood		2000	.004		3.94	.18		4.12	4.60
2000 Semi-permanent cross linking polymer primer, on metals		2000	.004		.62	.18		.80	.94

09 96 High-Performance Coatings

09 96 23 – Graffiti-Resistant Coatings

09 96 23.10 Graffiti-Resistant Treatments		Crew	Daily Output	Labor-Hours	Unit	Material	2020 Bare Costs Labor	Equipment	Total	Total Incl O&P
2100	Concrete	1 Pord	2000	.004	S.F.	.74	.18		.92	1.07
2200	Concrete block		2000	.004		.92	.18		1.10	1.28
2300	Brick		2000	.004		.74	.18		.92	1.07
2400	Stone		2000	.004		.74	.18		.92	1.07
2500	Unpainted wood		2000	.004		1.03	.18		1.21	1.39
3000	Top coat, on metals		2000	.004		.56	.18		.74	.87
3100	Concrete		2000	.004		.64	.18		.82	.96
3200	Concrete block		2000	.004		.89	.18		1.07	1.24
3300	Brick		2000	.004		.74	.18		.92	1.08
3400	Stone		2000	.004		.74	.18		.92	1.08
3500	Unpainted wood		2000	.004		.89	.18		1.07	1.24
5000	Sacrificial, water based, on metal		2000	.004		.34	.18		.52	.63
5100	Concrete		2000	.004		.34	.18		.52	.63
5200	Concrete block		2000	.004		.34	.18		.52	.63
5300	Brick		2000	.004		.34	.18		.52	.63
5400	Stone		2000	.004		.34	.18		.52	.63
5500	Unpainted wood		2000	.004		.34	.18		.52	.63
8000	Cleaner for use after treatment									
8100	Towels or wipes, per package of 30				Ea.	.72			.72	.79
8200	Aerosol spray, 24 oz. can				"	19			19	21

09 96 46 – Intumescent Painting

09 96 46.10 Coatings, Intumescent

		Crew	Daily Output	Labor-Hours	Unit	Material	Labor	Equipment	Total	Total Incl O&P
0010	**COATINGS, INTUMESCENT**, spray applied									
0100	On exterior structural steel, 0.25" d.f.t.	1 Pord	475	.017	S.F.	.50	.75		1.25	1.66
0150	0.51" d.f.t.		350	.023		.50	1.02		1.52	2.06
0200	0.98" d.f.t.		280	.029		.50	1.27		1.77	2.44
0300	On interior structural steel, 0.108" d.f.t.		300	.027		.49	1.18		1.67	2.30
0350	0.310" d.f.t.		150	.053		.49	2.37		2.86	4.06
0400	0.670" d.f.t.		100	.080		.49	3.55		4.04	5.85

09 97 Special Coatings

09 97 10 – Coatings and Paints

09 97 10.10 Coatings and Paints

		Crew	Daily Output	Labor-Hours	Unit	Material	Labor	Equipment	Total	Total Incl O&P
0010	**COATINGS & PAINTS** in 5 gallon lots									
0050	For 100 gallons or more, deduct					10%				
0100	Paint, exterior alkyd (oil base)									
0200	Flat				Gal.	48			48	53
0300	Gloss					47.50			47.50	52
0400	Primer					35			35	38.50
0500	Latex (water base)									
0600	Acrylic stain				Gal.	41			41	45
0700	Gloss enamel					34.50			34.50	38
0800	Flat					28.50			28.50	31.50
0900	Primer					26			26	28.50
1000	Semi-gloss					35			35	38.50
2400	Masonry, exterior									
2500	Alkali-resistant primer				Gal.	24.50			24.50	27
2600	Block filler, epoxy					30.50			30.50	33.50
2700	Latex					18.25			18.25	20
2800	Latex, flat					25.50			25.50	28.50

For customer support on your Site Work & Landscape Costs with RSMeans data, call 800.448.8182.

09 97 10.10 Coatings and Paints

		Crew	Daily Output	Labor-Hours	Unit	Material	2020 Bare Costs Labor	Equipment	Total	Total Incl O&P
2900	Semi-gloss				Gal.	33			33	36.50
4000	Metal									
4100	Galvanizing paint				Gal.	92			92	101
4200	High heat					67			67	73.50
4300	Heat-resistant					40.50			40.50	44.50
4400	Machinery enamel, alkyd					54			54	59
4500	Metal pretreatment (polyvinyl butyral)					117			117	128
4600	Rust inhibitor, ferrous metal					40.50			40.50	44.50
4700	Zinc chromate					150			150	165
4800	Zinc rich primer					170			170	187
5500	Coatings									
5600	Heavy duty									
5700	Acrylic urethane				Gal.	59.50			59.50	65.50
5800	Chlorinated rubber					70.50			70.50	77.50
5900	Coal tar epoxy					75			75	82.50
6000	Polyamide epoxy, finish					72.50			72.50	80
6100	Primer					52			52	57.50
6200	Silicone alkyd					62.50			62.50	68.50
6300	2 component solvent based acrylic epoxy					113			113	124
6400	Polyester epoxy					100			100	110
6500	Vinyl					37			37	40.50
6600	Special/Miscellaneous									
6700	Aluminum				Gal.	47.50			47.50	52
6900	Dry fall out, flat					24.50			24.50	27
7000	Fire retardant, intumescent					50			50	55
7100	Linseed oil					26			26	28.50
7200	Shellac					52.50			52.50	58
7300	Swimming pool, epoxy or urethane base					54			54	59
7400	Rubber base					74			74	81.50
7500	Texture paint					27.50			27.50	30.50
7600	Turpentine					37.50			37.50	41.50
7700	Water repellent, 5% silicone					36			36	39.50

09 97 13 – Steel Coatings

09 97 13.23 Exterior Steel Coatings

		Crew	Daily Output	Labor-Hours	Unit	Material	2020 Bare Costs Labor	Equipment	Total	Total Incl O&P
0010	**EXTERIOR STEEL COATINGS**									
6100	Cold galvanizing, brush in field	1 Psst	1100	.007	S.F.	.23	.33		.56	.79
6510	Paints & protective coatings, sprayed in field									
6520	Alkyds, primer	2 Psst	3600	.004	S.F.	.09	.20		.29	.43
6540	Gloss topcoats		3200	.005		.09	.23		.32	.47
6560	Silicone alkyd		3200	.005		.18	.23		.41	.57
6610	Epoxy, primer		3000	.005		.26	.24		.50	.69
6630	Intermediate or topcoat		2800	.006		.29	.26		.55	.75
6650	Enamel coat		2800	.006		.36	.26		.62	.83
6810	Latex primer		3600	.004		.07	.20		.27	.40
6830	Topcoats		3200	.005		.08	.23		.31	.46
6910	Universal primers, one part, phenolic, modified alkyd		2000	.008		.39	.37		.76	1.03
6940	Two part, epoxy spray		2000	.008		.45	.37		.82	1.10
7000	Zinc rich primers, self cure, spray, inorganic		1800	.009		.85	.41		1.26	1.60
7010	Epoxy, spray, organic		1800	.009		.29	.41		.70	.98
7020	Above one story, spray painting simple structures, add						25%			
7030	Intricate structures, add						50%			

178

Estimating Tips
General

- The items in this division are usually priced per square foot or each.
- Many items in Division 10 require some type of support system or special anchors that are not usually furnished with the item. The required anchors must be added to the estimate in the appropriate division.
- Some items in Division 10, such as lockers, may require assembly before installation. Verify the amount of assembly required. Assembly can often exceed installation time.

10 20 00 Interior Specialties

- Support angles and blocking are not included in the installation of toilet compartments, shower/dressing compartments, or cubicles. Appropriate line items from Division 5 or 6 may need to be added to support the installations.
- Toilet partitions are priced by the stall. A stall consists of a side wall, pilaster, and door with hardware. Toilet tissue holders and grab bars are extra.
- The required acoustical rating of a folding partition can have a significant impact on costs. Verify the sound transmission coefficient rating of the panel priced against the specification requirements.

- Grab bar installation does not include supplemental blocking or backing to support the required load. When grab bars are installed at an existing facility, provisions must be made to attach the grab bars to a solid structure.

Reference Numbers

Reference numbers are shown at the beginning of some major classifications. These numbers refer to related items in the Reference Section. The reference information may be an estimating procedure, an alternate pricing method, or technical information.

Note: Not all subdivisions listed here necessarily appear. ■

Same Data. Simplified.

Enjoy the convenience and efficiency of accessing your costs anywhere:

- **Skip the multiplier** by setting your location
- **Quickly search,** edit, favorite and share costs
- **Stay on top of price changes** with automatic updates

Discover more at rsmeans.com/online

---

10 05 Common Work Results for Specialties

10 05 05 – Selective Demolition for Specialties

10 05 05.10 Selective Demolition, Specialties	Crew	Daily Output	Labor-Hours	Unit	Material	2020 Bare Costs Labor	Equipment	Total	Total Incl O&P
0010 **SELECTIVE DEMOLITION, SPECIALTIES**									
3500 Flagpole, groundset, to 70' high, excluding base/foundation	K-1	1	16	Ea.		805	820	1,625	2,100
3555 To 30' high	"	2.50	6.400	"		320	330	650	840
4000 Removal of traffic signs, including supports									
4020 To 10 S.F.	B-80B	16	2	Ea.		89.50	14.75	104.25	150
4030 11 S.F. to 20 S.F.	"	5	6.400			287	47	334	480
4040 21 S.F. to 40 S.F.	B-14	1.80	26.667			1,175	119	1,294	1,900
4050 41 S.F. to 100 S.F.	B-13	1.30	43.077			1,975	445	2,420	3,475
4070 Remove traffic posts to 12'-0" high	B-6	100	.240			11	2.14	13.14	18.80
4310 Signs, street, reflective aluminum, including post and bracket	1 Clab	60	.133	▼		5.60		5.60	8.45

10 13 Directories

10 13 10 – Building Directories

10 13 10.10 Directory Boards

	Crew	Daily Output	Labor-Hours	Unit	Material	2020 Bare Costs Labor	Equipment	Total	Total Incl O&P
0010 **DIRECTORY BOARDS**									
0900 Outdoor, weatherproof, black plastic, 36" x 24"	2 Carp	2	8	Ea.	700	425		1,125	1,400
1000 36" x 36"	"	1.50	10.667	"	815	565		1,380	1,750

10 14 Signage

10 14 19 – Dimensional Letter Signage

10 14 19.10 Exterior Signs

	Crew	Daily Output	Labor-Hours	Unit	Material	2020 Bare Costs Labor	Equipment	Total	Total Incl O&P
0010 **EXTERIOR SIGNS**									
0020 Letters, 2" high, 3/8" deep, cast bronze	1 Carp	24	.333	Ea.	33	17.70		50.70	63
0140 1/2" deep, cast aluminum		18	.444		33	23.50		56.50	72
0160 Cast bronze		32	.250		34.50	13.30		47.80	58
0300 6" high, 5/8" deep, cast aluminum		24	.333		30	17.70		47.70	59.50
0400 Cast bronze		24	.333		60.50	17.70		78.20	93
0600 8" high, 3/4" deep, cast aluminum		14	.571		33	30.50		63.50	82
0700 Cast bronze		20	.400		100	21.50		121.50	142
0900 10" high, 1" deep, cast aluminum		18	.444		64.50	23.50		88	107
1000 Cast bronze		18	.444		102	23.50		125.50	148
1200 12" high, 1-1/4" deep, cast aluminum		12	.667		59.50	35.50		95	119
1500 Cast bronze		18	.444		163	23.50		186.50	215
1600 14" high, 2-5/16" deep, cast aluminum		12	.667		110	35.50		145.50	174
1800 Fabricated stainless steel, 6" high, 2" deep		20	.400		45.50	21.50		67	82
1900 12" high, 3" deep		18	.444		85	23.50		108.50	129
2100 18" high, 3" deep		12	.667		101	35.50		136.50	164
2200 24" high, 4" deep		10	.800		198	42.50		240.50	282
2700 Acrylic, on high density foam, 12" high, 2" deep		20	.400		18.65	21.50		40.15	52.50
2800 18" high, 2" deep		18	.444		38	23.50		61.50	77.50
3900 Plaques, custom, 20" x 30", for up to 450 letters, cast aluminum	2 Carp	4	4		1,875	213		2,088	2,375
4000 Cast bronze		4	4		1,950	213		2,163	2,475
4200 30" x 36", up to 900 letters, cast aluminum		3	5.333		2,925	283		3,208	3,650
4300 Cast bronze		3	5.333		4,200	283		4,483	5,050
4500 36" x 48", for up to 1300 letters, cast bronze		2	8		4,500	425		4,925	5,600
4800 Signs, reflective alum. directional signs, dbl. face, 2-way, w/bracket		30	.533		107	28.50		135.50	161
4900 4-way		30	.533		197	28.50		225.50	260
5100 Exit signs, 24 ga. alum., 14" x 12" surface mounted	1 Carp	30	.267		52	14.15		66.15	78.50
5200 10" x 7"		20	.400	▼	32.50	21.50		54	68

For customer support on your Site Work & Landscape Costs with RSMeans data, call 800.448.8182.

10 14 Signage

10 14 19 – Dimensional Letter Signage

10 14 19.10 Exterior Signs

		Crew	Daily Output	Labor-Hours	Unit	Material	2020 Bare Costs Labor	2020 Bare Costs Equipment	Total	Total Incl O&P
5400	Bracket mounted, double face, 12" x 10"	1 Carp	30	.267	Ea.	59.50	14.15		73.65	87
5500	Sticky back, stock decals, 14" x 10"	1 Clab	50	.160		24.50	6.75		31.25	36.50
6400	Replacement sign faces, 6" or 8"	"	50	.160	↓	66.50	6.75		73.25	83.50
8000	Internally illuminated, custom									
8100	On pedestal, 84" x 30" x 12"	L-7	.50	56	Ea.	11,500	2,875		14,375	17,000

10 14 26 – Post and Panel/Pylon Signage

10 14 26.10 Post and Panel Signage

		Crew	Daily Output	Labor-Hours	Unit	Material	2020 Bare Costs Labor	2020 Bare Costs Equipment	Total	Total Incl O&P
0010	POST AND PANEL SIGNAGE, Alum., incl. 2 posts,									
0011	2-sided panel (blank), concrete footings per post									
0025	24" W x 18" H	B-1	25	.960	Ea.	905	41		946	1,050
0050	36" W x 24" H		25	.960		950	41		991	1,100
0100	57" W x 36" H		25	.960		2,150	41		2,191	2,400
0150	81" W x 46" H	↓	25	.960	↓	2,225	41		2,266	2,500

10 14 43 – Photoluminescent Signage

10 14 43.10 Photoluminescent Signage

		Crew	Daily Output	Labor-Hours	Unit	Material	2020 Bare Costs Labor	2020 Bare Costs Equipment	Total	Total Incl O&P
0010	PHOTOLUMINESCENT SIGNAGE									
0011	Photoluminescent exit sign, plastic, 7"x10"	1 Carp	32	.250	Ea.	24	13.30		37.30	46.50
0022	Polyester		32	.250		15.20	13.30		28.50	36.50
0033	Aluminum		32	.250		53.50	13.30		66.80	78.50
0044	Plastic, 10"x14"		32	.250		59.50	13.30		72.80	85.50
0055	Polyester		32	.250		35.50	13.30		48.80	59
0066	Aluminum		32	.250		70.50	13.30		83.80	97.50
0111	Directional exit sign, plastic, 10"x14"		32	.250		58.50	13.30		71.80	84
0122	Polyester		32	.250		46	13.30		59.30	70.50
0133	Aluminum		32	.250		68.50	13.30		81.80	95
0153	Kick plate exit sign, 10"x34"		28	.286	↓	278	15.20		293.20	330
0173	Photoluminescent tape, 1"x60'		300	.027	L.F.	.50	1.42		1.92	2.68
0183	2"x60'		290	.028	"	1.01	1.47		2.48	3.31
0211	Photoluminescent directional arrows	↓	42	.190	Ea.	1.30	10.10		11.40	16.65

10 14 53 – Traffic Signage

10 14 53.20 Traffic Signs

		Crew	Daily Output	Labor-Hours	Unit	Material	2020 Bare Costs Labor	2020 Bare Costs Equipment	Total	Total Incl O&P
0010	TRAFFIC SIGNS									
0012	Stock, 24" x 24", no posts, .080" alum. reflectorized	B-80	70	.457	Ea.	93.50	21.50	14.60	129.60	151
0100	High intensity		70	.457		93.50	21.50	14.60	129.60	151
0300	30" x 30", reflectorized		70	.457		65	21.50	14.60	101.10	120
0400	High intensity		70	.457		172	21.50	14.60	208.10	237
0600	Guide and directional signs, 12" x 18", reflectorized		70	.457		34	21.50	14.60	70.10	85
0700	High intensity		70	.457		65.50	21.50	14.60	101.60	120
0900	18" x 24", stock signs, reflectorized		70	.457		48.50	21.50	14.60	84.60	101
1000	High intensity		70	.457		53.50	21.50	14.60	89.60	107
1200	24" x 24", stock signs, reflectorized		70	.457		59	21.50	14.60	95.10	113
1300	High intensity		70	.457		63.50	21.50	14.60	99.60	118
1500	Add to above for steel posts, galvanized, 10'-0" upright, bolted		200	.160		44	7.45	5.10	56.55	65.50
1600	12'-0" upright, bolted		140	.229	↓	38.50	10.65	7.30	56.45	66.50
1800	Highway road signs, aluminum, over 20 S.F., reflectorized		350	.091	S.F.	16.45	4.26	2.92	23.63	27.50
2000	High intensity		350	.091		16.45	4.26	2.92	23.63	27.50
2200	Highway, suspended over road, 80 S.F. min., reflectorized		165	.194		19.35	9.05	6.20	34.60	42
2300	High intensity		165	.194	↓	19.20	9.05	6.20	34.45	41.50
2350	Roadway delineators and reference markers		500	.064	Ea.	15.80	2.98	2.04	20.82	24
2360	Delineator post only, 6'		500	.064		18.90	2.98	2.04	23.92	27.50
2400	Highway sign bridge structure, 45' to 80'	↓			↓				33,900	33,900

For customer support on your Site Work & Landscape Costs with RSMeans data, call 800.448.8182.

181

10 14 Signage

10 14 53 – Traffic Signage

10 14 53.20 Traffic Signs		Crew	Daily Output	Labor-Hours	Unit	Material	2020 Bare Costs Labor	Equipment	Total	Total Incl O&P
2410	Cantilever structure, add				Ea.				7,000	7,000
5200	Remove and relocate signs, including supports									
5210	To 10 S.F.	B-80B	5	6.400	Ea.	395	287	47	729	915
5220	11 S.F. to 20 S.F.	"	1.70	18.824		885	845	139	1,869	2,400
5230	21 S.F. to 40 S.F.	B-14	.56	85.714		935	3,800	380	5,115	7,150
5240	41 S.F. to 100 S.F.	B-13	.32	175		1,550	8,050	1,825	11,425	15,800
8000	For temporary barricades and lights, see Section 01 56 23.10									
9000	Replace directional sign	B-80	6	5.333	Ea.	172	249	170	591	750

10 22 Partitions

10 22 16 – Folding Gates

10 22 16.10 Security Gates

0010	SECURITY GATES									
0300	Scissors type folding gate, ptd. steel, single, 6-1/2' high, 5-1/2' wide	2 Sswk	4	4	Opng.	235	231		466	620
0400	7-1/2' wide		4	4		223	231		454	605
0600	Double gate, 8' high, 8' wide		2.50	6.400		385	370		755	1,000
0750	14' wide		2	8		625	460		1,085	1,400

10 44 Fire Protection Specialties

10 44 16 – Fire Extinguishers

10 44 16.13 Portable Fire Extinguishers

0010	PORTABLE FIRE EXTINGUISHERS									
0140	CO_2, with hose and "H" horn, 10 lb.				Ea.	300			300	330
0160	15 lb.				"	385			385	425
1000	Dry chemical, pressurized									
1040	Standard type, portable, painted, 2-1/2 lb.				Ea.	44			44	48.50
1060	5 lb.					62			62	68
1080	10 lb.					96			96	106
1100	20 lb.					138			138	152
1120	30 lb.					430			430	470
1300	Standard type, wheeled, 150 lb.					1,500			1,500	1,650
2000	ABC all purpose type, portable, 2-1/2 lb.					23.50			23.50	26
2060	5 lb.					28.50			28.50	31.50
2080	9-1/2 lb.					52			52	57
2100	20 lb.					91			91	100
5000	Pressurized water, 2-1/2 gallon, stainless steel					105			105	115
5060	With anti-freeze					118			118	130
9400	Installation of extinguishers, 12 or more, on nailable surface	1 Carp	30	.267			14.15		14.15	21.50
9420	On masonry or concrete	"	15	.533			28.50		28.50	42.50

10 44 16.16 Wheeled Fire Extinguisher Units

0010	WHEELED FIRE EXTINGUISHER UNITS									
0350	CO_2, portable, with swivel horn									
0360	Wheeled type, cart mounted, 50 lb.				Ea.	1,375			1,375	1,525
0400	100 lb.				"	4,250			4,250	4,675
2200	ABC all purpose type									
2300	Wheeled, 45 lb.				Ea.	775			775	850
2360	150 lb.				"	1,975			1,975	2,150

10 57 Wardrobe and Closet Specialties

10 57 23 – Closet and Utility Shelving

10 57 23.19 Wood Closet and Utility Shelving	Crew	Daily Output	Labor-Hours	Unit	Material	2020 Bare Costs Labor	2020 Bare Costs Equipment	Total	Total Incl O&P
0010 **WOOD CLOSET AND UTILITY SHELVING**									
0020 Pine, clear grade, no edge band, 1" x 8"	1 Carp	115	.070	L.F.	3.35	3.70		7.05	9.25
0100 1" x 10"		110	.073		4.17	3.87		8.04	10.40
0200 1" x 12"		105	.076		5.05	4.05		9.10	11.65

10 73 Protective Covers

10 73 13 – Awnings

10 73 13.10 Awnings, Fabric

	Crew	Daily Output	Labor-Hours	Unit	Material	2020 Bare Costs Labor	2020 Bare Costs Equipment	Total	Total Incl O&P
0010 **AWNINGS, FABRIC**									
0020 Including acrylic canvas and frame, standard design									
0100 Door and window, slope, 3' high, 4' wide	1 Carp	4.50	1.778	Ea.	755	94.50		849.50	970
0110 6' wide		3.50	2.286		970	121		1,091	1,250
0120 8' wide		3	2.667		1,175	142		1,317	1,525
0200 Quarter round convex, 4' wide		3	2.667		1,175	142		1,317	1,525
0210 6' wide		2.25	3.556		1,525	189		1,714	1,950
0220 8' wide		1.80	4.444		1,875	236		2,111	2,400
0300 Dome, 4' wide		7.50	1.067		455	56.50		511.50	585
0310 6' wide		3.50	2.286		1,025	121		1,146	1,300
0320 8' wide		2	4		1,800	213		2,013	2,325
0350 Elongated dome, 4' wide		1.33	6.015		1,700	320		2,020	2,350
0360 6' wide		1.11	7.207		2,025	385		2,410	2,800
0370 8' wide		1	8		2,375	425		2,800	3,250
1000 Entry or walkway, peak, 12' long, 4' wide	2 Carp	.90	17.778		5,375	945		6,320	7,350
1010 6' wide		.60	26.667		8,300	1,425		9,725	11,300
1020 8' wide		.40	40		11,500	2,125		13,625	15,800
1100 Radius with dome end, 4' wide		1.10	14.545		4,075	775		4,850	5,650
1110 6' wide		.70	22.857		6,575	1,225		7,800	9,050
1120 8' wide		.50	32		9,325	1,700		11,025	12,900
2000 Retractable lateral arm awning, manual									
2010 To 12' wide, 8'-6" projection	2 Carp	1.70	9.412	Ea.	1,225	500		1,725	2,100
2020 To 14' wide, 8'-6" projection		1.10	14.545		1,425	775		2,200	2,725
2030 To 19' wide, 8'-6" projection		.85	18.824		1,925	1,000		2,925	3,625
2040 To 24' wide, 8'-6" projection		.67	23.881		2,425	1,275		3,700	4,575
2050 Motor for above, add	1 Carp	2.67	3		1,050	159		1,209	1,400
3000 Patio/deck canopy with frame									
3010 12' wide, 12' projection	2 Carp	2	8	Ea.	1,725	425		2,150	2,550
3020 16' wide, 14' projection	"	1.20	13.333		2,675	710		3,385	4,025
9000 For fire retardant canvas, add					7%				
9010 For lettering or graphics, add					35%				
9020 For painted or coated acrylic canvas, deduct					8%				
9030 For translucent or opaque vinyl canvas, add					10%				
9040 For 6 or more units, deduct					20%	15%			

10 73 16 – Canopies

10 73 16.20 Metal Canopies

	Crew	Daily Output	Labor-Hours	Unit	Material	2020 Bare Costs Labor	2020 Bare Costs Equipment	Total	Total Incl O&P
0010 **METAL CANOPIES**									
0020 Wall hung, .032", aluminum, prefinished, 8' x 10'	K-2	1.30	18.462	Ea.	2,500	1,025	630	4,155	5,025
0300 8' x 20'		1.10	21.818		4,200	1,200	745	6,145	7,275
0500 10' x 10'		1.30	18.462		3,075	1,025	630	4,730	5,650
0700 10' x 20'		1.10	21.818		4,725	1,200	745	6,670	7,875
1000 12' x 20'		1	24		5,825	1,325	820	7,970	9,350

For customer support on your Site Work & Landscape Costs with RSMeans data, call 800.448.8182.

183

10 73 Protective Covers

10 73 16 – Canopies

10 73 16.20 Metal Canopies	Crew	Daily Output	Labor-Hours	Unit	Material	2020 Bare Costs Labor	Equipment	Total	Total Incl O&P
1360	12' x 30'	K-2	.80	30	Ea.	8,775	1,650	1,025	11,450	13,400
1700	12' x 40'	↓	.60	40		11,700	2,200	1,375	15,275	17,800
1900	For free standing units, add				↓	20%	10%			
7000	Carport, baked vinyl finish, .032", 20' x 10', no foundations, flat panel	K-2	4	6	Car	4,125	330	205	4,660	5,275
7250	Insulated flat panel		2	12	"	4,950	660	410	6,020	6,925
7500	Walkway cover, to 12' wide, stl., vinyl finish, .032", no fndtns., flat		250	.096	S.F.	22	5.25	3.28	30.53	36
7750	Arched	↓	200	.120	"	53.50	6.60	4.10	64.20	73.50

10 74 Manufactured Exterior Specialties

10 74 46 – Window Wells

10 74 46.10 Area Window Wells

		Crew	Daily Output	Labor-Hours	Unit	Material	2020 Bare Costs Labor	Equipment	Total	Total Incl O&P
0010	**AREA WINDOW WELLS**, Galvanized steel									
0020	20 ga., 3'-2" wide, 1' deep	1 Sswk	29	.276	Ea.	18.70	15.90		34.60	45.50
0100	2' deep		23	.348		30	20		50	64.50
0300	16 ga., 3'-2" wide, 1' deep		29	.276		25.50	15.90		41.40	53
0400	3' deep		23	.348		48.50	20		68.50	85
0600	Welded grating for above, 15 lb., painted		45	.178		103	10.25		113.25	129
0700	Galvanized	↓	45	.178	↓	126	10.25		136.25	154

10 75 Flagpoles

10 75 16 – Ground-Set Flagpoles

10 75 16.10 Flagpoles

		Crew	Daily Output	Labor-Hours	Unit	Material	2020 Bare Costs Labor	Equipment	Total	Total Incl O&P
0010	**FLAGPOLES**, ground set									
0050	Not including base or foundation									
0100	Aluminum, tapered, ground set 20' high	K-1	2	8	Ea.	1,175	400	410	1,985	2,325
0200	25' high		1.70	9.412		1,275	470	485	2,230	2,650
0300	30' high		1.50	10.667		1,250	535	545	2,330	2,775
0400	35' high		1.40	11.429		1,800	575	585	2,960	3,475
0500	40' high		1.20	13.333		3,100	670	685	4,455	5,175
0600	50' high		1	16		3,950	805	820	5,575	6,425
0700	60' high		.90	17.778		4,900	890	910	6,700	7,750
0800	70' high		.80	20		7,725	1,000	1,025	9,750	11,100
1100	Counterbalanced, internal halyard, 20' high		1.80	8.889		3,125	445	455	4,025	4,600
1200	30' high		1.50	10.667		2,900	535	545	3,980	4,575
1300	40' high		1.30	12.308		7,350	620	630	8,600	9,725
1400	50' high		1	16		10,200	805	820	11,825	13,400
2820	Aluminum, electronically operated, 30' high		1.40	11.429		4,525	575	585	5,685	6,475
2840	35' high		1.30	12.308		5,450	620	630	6,700	7,600
2860	39' high		1.10	14.545		6,775	730	745	8,250	9,375
2880	45' high		1	16		6,250	805	820	7,875	8,975
2900	50' high		.90	17.778		9,100	890	910	10,900	12,400
3000	Fiberglass, tapered, ground set, 23' high		2	8		785	400	410	1,595	1,925
3100	29'-7" high		1.50	10.667		1,300	535	545	2,380	2,825
3200	36'-1" high		1.40	11.429		1,825	575	585	2,985	3,500
3300	39'-5" high		1.20	13.333		2,050	670	685	3,405	4,000
3400	49'-2" high		1	16		4,825	805	820	6,450	7,400
3500	59' high	↓	.90	17.778	↓	5,075	890	910	6,875	7,950
4300	Steel, direct imbedded installation									
4400	Internal halyard, 20' high	K-1	2.50	6.400	Ea.	1,575	320	330	2,225	2,575

10 75 Flagpoles

10 75 16 – Ground-Set Flagpoles

10 75 16.10 Flagpoles

		Crew	Daily Output	Labor-Hours	Unit	Material	2020 Bare Costs Labor	Equipment	Total	Total Incl O&P
4500	25' high	K-1	2.50	6.400	Ea.	2,150	320	330	2,800	3,200
4600	30' high		2.30	6.957		2,475	350	355	3,180	3,650
4700	40' high		2.10	7.619		3,525	380	390	4,295	4,875
4800	50' high		1.90	8.421		4,500	425	430	5,355	6,050
5000	60' high		1.80	8.889		7,150	445	455	8,050	9,025
5100	70' high		1.60	10		8,500	500	515	9,515	10,700
5200	80' high		1.40	11.429		11,100	575	585	12,260	13,700
5300	90' high		1.20	13.333		17,100	670	685	18,455	20,600
5500	100' high		1	16		18,600	805	820	20,225	22,600
6400	Wood poles, tapered, clear vertical grain fir with tilting									
6410	base, not incl. foundation, 4" butt, 25' high	K-1	1.90	8.421	Ea.	1,875	425	430	2,730	3,150
6800	6" butt, 30' high	"	1.30	12.308	"	3,275	620	630	4,525	5,225
7300	Foundations for flagpoles, including									
7400	excavation and concrete, to 35' high poles	C-1	10	3.200	Ea.	730	161		891	1,050
7600	40' to 50' high		3.50	9.143		1,350	460		1,810	2,175
7700	Over 60' high		2	16		1,675	805		2,480	3,050

10 75 23 – Wall-Mounted Flagpoles

10 75 23.10 Flagpoles

		Crew	Daily Output	Labor-Hours	Unit	Material	2020 Bare Costs Labor	Equipment	Total	Total Incl O&P
0010	**FLAGPOLES**, structure mounted									
0100	Fiberglass, vertical wall set, 19'-8" long	K-1	1.50	10.667	Ea.	1,125	535	545	2,205	2,625
0200	23' long		1.40	11.429		1,750	575	585	2,910	3,425
0300	26'-3" long		1.30	12.308		1,625	620	630	2,875	3,400
0800	19'-8" long outrigger		1.30	12.308		1,500	620	630	2,750	3,275
1300	Aluminum, vertical wall set, tapered, with base, 20' high		1.20	13.333		1,300	670	685	2,655	3,175
1400	29'-6" high		1	16		3,350	805	820	4,975	5,775
2400	Outrigger poles with base, 12' long		1.30	12.308		1,375	620	630	2,625	3,150
2500	14' long		1	16		1,575	805	820	3,200	3,825

10 88 Scales

10 88 05 – Commercial Scales

10 88 05.10 Scales

		Crew	Daily Output	Labor-Hours	Unit	Material	2020 Bare Costs Labor	Equipment	Total	Total Incl O&P
0010	**SCALES**									
0700	Truck scales, incl. steel weigh bridge,									
0800	not including foundation, pits									
1550	Digital, electronic, 100 ton capacity, steel deck 12' x 10' platform	3 Carp	.20	120	Ea.	14,900	6,375		21,275	26,000
1600	40' x 10' platform		.14	171		29,700	9,100		38,800	46,400
1640	60' x 10' platform		.13	185		43,700	9,800		53,500	63,000
1680	70' x 10' platform		.12	200		41,800	10,600		52,400	62,000
2000	For standard automatic printing device, add					1,550			1,550	1,700
2100	For remote reading electronic system, add					2,925			2,925	3,225
2300	Concrete foundation pits for above, 8' x 6', 5 C.Y. required	C-1	.50	64		1,125	3,225		4,350	6,100
2400	14' x 6' platform, 10 C.Y. required		.35	91.429		1,675	4,600		6,275	8,750
2600	50' x 10' platform, 30 C.Y. required		.25	128		2,250	6,450		8,700	12,200
2700	70' x 10' platform, 40 C.Y. required		.15	213		4,900	10,700		15,600	21,500
2750	Crane scales, dial, 1 ton capacity					1,175			1,175	1,275
2780	5 ton capacity					1,625			1,625	1,775
2800	Digital, 1 ton capacity					1,925			1,925	2,100
2850	10 ton capacity					4,775			4,775	5,250
3800	Portable, beam type, capacity 1000 lb., platform 18" x 24"					875			875	965
3900	Dial type, capacity 2000 lb., platform 24" x 24"					1,625			1,625	1,800

10 88 Scales

10 88 05 – Commercial Scales

10 88 05.10 Scales		Crew	Daily Output	Labor-Hours	Unit	Material	2020 Bare Costs Labor	Equipment	Total	Total Incl O&P
4000	Digital type, capacity 1000 lb., platform 24" x 30"				Ea.	2,600			2,600	2,875
4100	Portable contractor truck scales, 50 ton cap., 40' x 10' platform					29,700			29,700	32,700
4200	60' x 10' platform					36,300			36,300	39,900
4400	Heavy-Duty Steel Deck Truck Scales 20' x 10'	3 Carp	.20	120		18,200	6,375		24,575	29,700
4500	Heavy-Duty Steel Deck Truck Scales 80' x 10'		.10	240		48,100	12,800		60,900	72,000
4600	Heavy-Duty Steel Deck Truck Scales 160' x 10'		.04	600		105,500	31,900		137,400	164,000
4700	Heavy-Duty Steel Deck Truck Scales 20' x 12'		.18	137		21,300	7,300		28,600	34,500
4800	Heavy-Duty Steel Deck Truck Scales 80' x 12'		.09	267		61,000	14,200		75,200	88,500
4900	Heavy-Duty Steel Deck Truck Scales 160' x 12'		.04	600		123,000	31,900		154,900	183,500
5000	Heavy-Duty Steel Deck Truck Scales 20' x 14'		.16	150		28,300	7,975		36,275	43,200
5100	Heavy-Duty Steel Deck Truck Scales 80' x 14'		.06	400		67,000	21,300		88,300	105,500
5200	Heavy-Duty Steel Deck Truck Scales 160' x 14'		.03	800		131,000	42,500		173,500	208,000

For customer support on your Site Work & Landscape Costs with RSMeans data, call 800.448.8182.

Estimating Tips
General

- The items in this division are usually priced per square foot or each. Many of these items are purchased by the owner for installation by the contractor. Check the specifications for responsibilities and include time for receiving, storage, installation, and mechanical and electrical hookups in the appropriate divisions.

- Many items in Division 11 require some type of support system that is not usually furnished with the item. Examples of these systems include blocking for the attachment of casework and support angles for ceiling-hung projection screens. The required blocking or supports must be added to the estimate in the appropriate division.

- Some items in Division 11 may require assembly or electrical hookups. Verify the amount of assembly required or the need for a hard electrical connection and add the appropriate costs.

Reference Numbers

Reference numbers are shown at the beginning of some major classifications. These numbers refer to related items in the Reference Section. The reference information may be an estimating procedure, an alternate pricing method, or technical information.

Same Data. Simplified.

Enjoy the convenience and efficiency of accessing your costs anywhere:

- **Skip the multiplier** by setting your location
- **Quickly search,** edit, favorite and share costs
- **Stay on top of price changes** with automatic updates

Discover more at rsmeans.com/online

11 11 Vehicle Service Equipment

11 11 13 – Compressed-Air Vehicle Service Equipment

11 11 13.10 Compressed Air Equipment

		Crew	Daily Output	Labor-Hours	Unit	Material	2020 Bare Costs Labor	Equipment	Total	Total Incl O&P
0010	**COMPRESSED AIR EQUIPMENT**									
0030	Compressors, electric, 1-1/2 HP, standard controls	L-4	1.50	16	Ea.	505	800		1,305	1,750
0550	Dual controls		1.50	16		1,050	800		1,850	2,350
0600	5 HP, 115/230 volt, standard controls		1	24		1,725	1,200		2,925	3,725
0650	Dual controls		1	24		3,225	1,200		4,425	5,375

11 11 19 – Vehicle Lubrication Equipment

11 11 19.10 Lubrication Equipment

		Crew	Daily Output	Labor-Hours	Unit	Material	2020 Bare Costs Labor	Equipment	Total	Total Incl O&P
0010	**LUBRICATION EQUIPMENT**									
3000	Lube equipment, 3 reel type, with pumps, not including piping	L-4	.50	48	Set	10,400	2,400		12,800	15,100
3100	Hose reel, including hose, oil/lube, 1000 psi	2 Sswk	2	8	Ea.	480	460		940	1,250
3200	Grease, 5000 psi		2	8		585	460		1,045	1,375
3300	Air, 50', 160 psi		2	8		605	460		1,065	1,400
3350	25', 160 psi		2	8		365	460		825	1,125

11 11 33 – Vehicle Spray Painting Equipment

11 11 33.10 Spray Painting Equipment

		Crew	Daily Output	Labor-Hours	Unit	Material	2020 Bare Costs Labor	Equipment	Total	Total Incl O&P
0010	**SPRAY PAINTING EQUIPMENT**									
4000	Spray painting booth, 26' long, complete	L-4	.40	60	Ea.	11,600	3,000		14,600	17,200

11 12 Parking Control Equipment

11 12 13 – Parking Key and Card Control Units

11 12 13.10 Parking Control Units

		Crew	Daily Output	Labor-Hours	Unit	Material	2020 Bare Costs Labor	Equipment	Total	Total Incl O&P
0010	**PARKING CONTROL UNITS**									
5100	Card reader	1 Elec	2	4	Ea.	1,475	245		1,720	2,000
5120	Proximity with customer display	2 Elec	1	16		6,025	980		7,005	8,075
6000	Parking control software, basic functionality	1 Elec	.50	16		24,000	980		24,980	27,900
6020	Multi-function	"	.20	40		96,000	2,450		98,450	109,000

11 12 16 – Parking Ticket Dispensers

11 12 16.10 Ticket Dispensers

		Crew	Daily Output	Labor-Hours	Unit	Material	2020 Bare Costs Labor	Equipment	Total	Total Incl O&P
0010	**TICKET DISPENSERS**									
5900	Ticket spitter with time/date stamp, standard	2 Elec	2	8	Ea.	6,625	490		7,115	8,000
5920	Mag stripe encoding	"	2	8	"	18,900	490		19,390	21,500

11 12 26 – Parking Fee Collection Equipment

11 12 26.13 Parking Fee Coin Collection Equipment

		Crew	Daily Output	Labor-Hours	Unit	Material	2020 Bare Costs Labor	Equipment	Total	Total Incl O&P
0010	**PARKING FEE COIN COLLECTION EQUIPMENT**									
5200	Cashier booth, average	B-22	1	30	Ea.	10,600	1,475	284	12,359	14,200
5300	Collector station, pay on foot	2 Elec	.20	80		111,500	4,900		116,400	130,500
5320	Credit card only	"	.50	32		20,600	1,975		22,575	25,600

11 12 26.23 Fee Equipment

		Crew	Daily Output	Labor-Hours	Unit	Material	2020 Bare Costs Labor	Equipment	Total	Total Incl O&P
0010	**FEE EQUIPMENT**									
5600	Fee computer	1 Elec	1.50	5.333	Ea.	13,000	325		13,325	14,800

11 12 33 – Parking Gates

11 12 33.13 Lift Arm Parking Gates

		Crew	Daily Output	Labor-Hours	Unit	Material	2020 Bare Costs Labor	Equipment	Total	Total Incl O&P
0010	**LIFT ARM PARKING GATES**									
5000	Barrier gate with programmable controller	2 Elec	3	5.333	Ea.	3,300	325		3,625	4,100
5020	Industrial		3	5.333		5,225	325		5,550	6,225
5050	Non-programmable, with reader and 12' arm		3	5.333		1,700	325		2,025	2,350
5500	Exit verifier		1	16		19,600	980		20,580	23,100
5700	Full sign, 4" letters	1 Elec	2	4		1,250	245		1,495	1,725

11 12 Parking Control Equipment

11 12 33 – Parking Gates

11 12 33.13 Lift Arm Parking Gates	Crew	Daily Output	Labor-Hours	Unit	Material	2020 Bare Costs Labor	Equipment	Total	Total Incl O&P
5800	Inductive loop	2 Elec	4	4	Ea.	209	245		454	595
5950	Vehicle detector, microprocessor based	1 Elec	3	2.667	↓	470	164		634	760

11 13 Loading Dock Equipment

11 13 13 – Loading Dock Bumpers

11 13 13.10 Dock Bumpers

		Crew	Daily Output	Labor-Hours	Unit	Material	Labor	Equipment	Total	Total Incl O&P
0010	**DOCK BUMPERS** Bolts not included									
0020	2" x 6" to 4" x 8", average	1 Carp	.30	26.667	M.B.F.	1,425	1,425		2,850	3,700
0050	Bumpers, lam. rubber blocks 4-1/2" thick, 10" high, 14" long		26	.308	Ea.	60.50	16.35		76.85	91
0300	36" long		17	.471		340	25		365	415
0500	12" high, 14" long		25	.320		81	17		98	115
0600	36" long		15	.533		127	28.50		155.50	183
0800	Laminated rubber blocks 6" thick, 10" high, 14" long		22	.364		72.50	19.35		91.85	109
0900	36" long		13	.615		212	32.50		244.50	282
0920	Extruded rubber bumpers, T section, 22" x 22" x 3" thick		41	.195		73.50	10.35		83.85	96.50
0940	Molded rubber bumpers, 24" x 12" x 3" thick		20	.400	↓	66.50	21.50		88	105

11 14 Pedestrian Control Equipment

11 14 13 – Pedestrian Gates

11 14 13.19 Turnstiles

		Crew	Daily Output	Labor-Hours	Unit	Material	Labor	Equipment	Total	Total Incl O&P
0010	**TURNSTILES**									
0020	One way, 4 arm, 46" diameter, economy, manual	2 Carp	5	3.200	Ea.	2,250	170		2,420	2,725
0100	Electric		1.20	13.333		2,375	710		3,085	3,700
0420	Three arm, 24" opening, light duty, manual		2	8		3,100	425		3,525	4,050
0450	Heavy duty		1.50	10.667		4,925	565		5,490	6,250
0460	Manual, with registering & controls, light duty		2	8		4,600	425		5,025	5,700
0470	Heavy duty	↓	1.50	10.667	↓	4,900	565		5,465	6,250

11 14 19 – Portable Posts and Railings

11 14 19.13 Portable Posts and Railings

		Crew	Daily Output	Labor-Hours	Unit	Material	Labor	Equipment	Total	Total Incl O&P
0010	**PORTABLE POSTS AND RAILINGS**									
0020	Portable for pedestrian traffic control, standard				Ea.	166			166	182
0300	Deluxe posts				"	239			239	263
0600	Ropes for above posts, plastic covered, 1-1/2" diameter				L.F.	20			20	22
0700	Chain core				"	14.65			14.65	16.10
1500	Portable security or safety barrier, black with 7' yellow strap				Ea.	206			206	227
1510	12' yellow strap				↓	251			251	276
1550	Sign holder, standard design					86			86	95

For customer support on your Site Work & Landscape Costs with RSMeans data, call 800.448.8182.

189

11 66 Athletic Equipment

11 66 43 – Interior Scoreboards

11 66 43.10 Scoreboards		Crew	Daily Output	Labor-Hours	Unit	Material	2020 Bare Costs Labor	Equipment	Total	Total Incl O&P
0010	**SCOREBOARDS**									
7000	Baseball, minimum	R-3	1.30	15.385	Ea.	4,125	940	145	5,210	6,100
7200	Maximum		.05	400		19,800	24,400	3,775	47,975	62,500
7300	Football, minimum		.86	23.256		5,225	1,425	220	6,870	8,125
7400	Maximum		.20	100		20,200	6,100	945	27,245	32,400

11 68 Play Field Equipment and Structures

11 68 13 – Playground Equipment

11 68 13.10 Free-Standing Playground Equipment

			Crew	Daily Output	Labor-Hours	Unit	Material	2020 Bare Costs Labor	Equipment	Total	Total Incl O&P
0010	**FREE-STANDING PLAYGROUND EQUIPMENT** See also individual items										
0200	Bike rack, 10' long, permanent	G	B-1	12	2	Ea.	540	85.50		625.50	725
0240	Climber, arch, 6' high, 12' long, 5' wide			4	6		810	257		1,067	1,275
0260	Fitness trail, with signs, 9 to 10 stations, treated pine, minimum			.25	96		7,725	4,100		11,825	14,700
0270	Maximum			.17	141		23,100	6,050		29,150	34,500
0280	Metal, minimum			.25	96		11,700	4,100		15,800	19,100
0285	Maximum			.17	141		24,600	6,050		30,650	36,100
0300	Redwood, minimum			.25	96		11,800	4,100		15,900	19,200
0310	Maximum			.17	141		12,600	6,050		18,650	22,900
0320	16 to 20 stations, treated pine, minimum			.17	141		15,300	6,050		21,350	26,000
0330	Maximum			.13	185		16,300	7,900		24,200	29,900
0340	Metal, minimum			.17	141		13,200	6,050		19,250	23,700
0350	Maximum			.13	185		20,300	7,900		28,200	34,200
0360	Redwood, minimum			.17	141		17,800	6,050		23,850	28,700
0370	Maximum			.13	185		32,300	7,900		40,200	47,400
0392	Upper body warm-up station			2.60	9.231		3,475	395		3,870	4,400
0394	Bench stepper station			2.60	9.231		1,750	395		2,145	2,525
0396	Standing push up station			2.60	9.231		1,200	395		1,595	1,925
0398	Upper body stretch station			2.60	9.231		2,450	395		2,845	3,300
0400	Horizontal monkey ladder, 14' long, 6' high			4	6		1,200	257		1,457	1,700
0590	Parallel bars, 10' long			4	6		635	257		892	1,075
0600	Posts, tether ball set, 2-3/8" OD			12	2		430	85.50		515.50	605
0800	Poles, multiple purpose, 10'-6" long			12	2	Pr.	259	85.50		344.50	415
1000	Ground socket for movable posts, 2-3/8" post			10	2.400		114	103		217	280
1100	3-1/2" post			10	2.400		225	103		328	400
1300	See-saw, spring, steel, 2 units			6	4	Ea.	730	171		901	1,050
1400	4 units			4	6		1,275	257		1,532	1,775
1500	6 units			3	8		2,000	340		2,340	2,725
1700	Shelter, fiberglass golf tee, 3 person			4.60	5.217		4,375	223		4,598	5,150
1900	Slides, stainless steel bed, 12' long, 6' high			3	8		4,400	340		4,740	5,375
2000	20' long, 10' high			2	12		8,250	515		8,765	9,850
2200	Swings, plain seats, 8' high, 4 seats			2	12		1,675	515		2,190	2,625
2300	8 seats			1.30	18.462		2,925	790		3,715	4,375
2500	12' high, 4 seats			2	12		2,150	515		2,665	3,125
2600	8 seats			1.30	18.462		4,825	790		5,615	6,500
2800	Whirlers, 8' diameter			3	8		3,175	340		3,515	4,000
2900	10' diameter			3	8		6,625	340		6,965	7,825

11 68 13.20 Modular Playground

			Crew	Daily Output	Labor-Hours	Unit	Material	2020 Bare Costs Labor	Equipment	Total	Total Incl O&P
0010	**MODULAR PLAYGROUND** Basic components										
0100	Deck, square, steel, 48" x 48"		B-1	1	24	Ea.	695	1,025		1,720	2,325
0110	Recycled polyurethane			1	24		690	1,025		1,715	2,300
0120	Triangular, steel, 48" side			1	24		815	1,025		1,840	2,450

11 68 Play Field Equipment and Structures

11 68 13 – Playground Equipment

11 68 13.20 Modular Playground

		Crew	Daily Output	Labor-Hours	Unit	Material	2020 Bare Costs Labor	2020 Bare Costs Equipment	Total	Total Incl O&P
0130	Post, steel, 5" square	B-1	18	1.333	L.F.	51.50	57		108.50	142
0140	Aluminum, 2-3/8" square		20	1.200		59.50	51.50		111	142
0150	5" square		18	1.333	↓	50.50	57		107.50	141
0160	Roof, square poly, 54" side		18	1.333	Ea.	1,975	57		2,032	2,250
0170	Wheelchair transfer module, for 3' high deck		3	8	"	3,425	340		3,765	4,300
0180	Guardrail, pipe, 36" high		60	.400	L.F.	294	17.10		311.10	350
0190	Steps, deck-to-deck, three 8" steps		8	3	Ea.	1,375	128		1,503	1,700
0200	Activity panel, crawl through panel		2	12		580	515		1,095	1,400
0210	Alphabet/spelling panel		2	12		660	515		1,175	1,500
0360	With guardrails		3	8		2,000	340		2,340	2,725
0370	Crawl tunnel, straight, 56" long		4	6		1,375	257		1,632	1,900
0380	90°, 4' long		4	6		1,575	257		1,832	2,125
1200	Slide, tunnel, for 56" high deck		8	3		2,300	128		2,428	2,750
1210	Straight, poly		8	3		520	128		648	765
1220	Stainless steel, 54" high deck		6	4		995	171		1,166	1,350
1230	Curved, poly, 40" high deck		6	4		940	171		1,111	1,275
1240	Spiral slide, 56"-72" high		5	4.800		4,900	205		5,105	5,700
1300	Ladder, vertical, for 24"-72" high deck		5	4.800		545	205		750	910
1310	Horizontal, 8' long		5	4.800		1,050	205		1,255	1,475
1320	Corkscrew climber, 6' high		3	8		1,200	340		1,540	1,850
1330	Fire pole for 72" high deck		6	4		360	171		531	650
1340	Bridge, ring climber, 8' long		4	6	↓	2,775	257		3,032	3,425
1350	Suspension	↓	4	6	L.F.	375	257		632	795

11 68 16 – Play Structures

11 68 16.10 Handball/Squash Court

		Crew	Daily Output	Labor-Hours	Unit	Material	2020 Bare Costs Labor	2020 Bare Costs Equipment	Total	Total Incl O&P
0010	**HANDBALL/SQUASH COURT**, outdoor									
0900	Handball or squash court, outdoor, wood	2 Carp	.50	32	Ea.	5,075	1,700		6,775	8,150
1000	Masonry handball/squash court	D-1	.30	53.333	"	26,500	2,500		29,000	33,000

11 68 16.30 Platform/Paddle Tennis Court

		Crew	Daily Output	Labor-Hours	Unit	Material	2020 Bare Costs Labor	2020 Bare Costs Equipment	Total	Total Incl O&P
0010	**PLATFORM/PADDLE TENNIS COURT** Complete with lighting, etc.									
0100	Aluminum slat deck with aluminum frame	B-1	.08	300	Court	68,500	12,800		81,300	95,000
0500	Aluminum slat deck with wood frame	C-1	.12	267		80,500	13,400		93,900	108,500
0800	Aluminum deck heater, add	B-1	1.18	20.339		2,925	870		3,795	4,525
0900	Douglas fir planking with wood frame 2" x 6" x 30'	C-1	.12	267		75,500	13,400		88,900	103,000
1000	Plywood deck with steel frame		.12	267		75,500	13,400		88,900	103,000
1100	Steel slat deck with wood frame	↓	.12	267	↓	45,600	13,400		59,000	70,000

11 68 33 – Athletic Field Equipment

11 68 33.13 Football Field Equipment

		Crew	Daily Output	Labor-Hours	Unit	Material	2020 Bare Costs Labor	2020 Bare Costs Equipment	Total	Total Incl O&P
0010	**FOOTBALL FIELD EQUIPMENT**									
0020	Goal posts, steel, football, double post	B-1	1.50	16	Pr.	4,600	685		5,285	6,100
0100	Deluxe, single post		1.50	16		3,525	685		4,210	4,925
0300	Football, convertible to soccer		1.50	16		3,150	685		3,835	4,500
0500	Soccer, regulation	↓	2	12	↓	2,075	515		2,590	3,050

11 82 Facility Solid Waste Handling Equipment

11 82 19 – Packaged Incinerators

11 82 19.10 Packaged Gas Fired Incinerators	Crew	Daily Output	Labor-Hours	Unit	Material	2020 Bare Costs Labor	Equipment	Total	Total Incl O&P
0010 **PACKAGED GAS FIRED INCINERATORS**									
4750 Large municipal incinerators, incl. stack, minimum	Q-3	.25	128	Ton/day	21,700	7,850		29,550	35,700
4850 Maximum	"	.10	320	"	58,000	19,600		77,600	93,000

11 82 26 – Facility Waste Compactors

11 82 26.10 Compactors

	Crew	Daily Output	Labor-Hours	Unit	Material	2020 Bare Costs Labor	Equipment	Total	Total Incl O&P
0010 **COMPACTORS**									
0020 Compactors, 115 volt, 250 lbs./hr., chute fed	L-4	1	24	Ea.	12,600	1,200		13,800	15,600
1400 For handling hazardous waste materials, 55 gallon drum packer, std.					21,700			21,700	23,900
1410 55 gallon drum packer w/HEPA filter					26,400			26,400	29,100
1420 55 gallon drum packer w/charcoal & HEPA filter					35,200			35,200	38,800
1430 All of the above made explosion proof, add					1,725			1,725	1,900
6000 Transfer station compactor, with power unit									
6050 and pedestal, not including pit, 50 tons/hour				Ea.	239,000			239,000	263,000

Estimating Tips
General

- The items in this division are usually priced per square foot or each. Most of these items are purchased by the owner and installed by the contractor. Do not assume the items in Division 12 will be purchased and installed by the contractor. Check the specifications for responsibilities and include receiving, storage, installation, and mechanical and electrical hookups in the appropriate divisions.

- Some items in this division require some type of support system that is not usually furnished with the item. Examples of these systems include blocking for the attachment of casework and heavy drapery rods. The required blocking must be added to the estimate in the appropriate division.

Reference Numbers

Reference numbers are shown at the beginning of some major classifications. These numbers refer to related items in the Reference Section. The reference information may be an estimating procedure, an alternate pricing method, or technical information.

Same Data. Simplified.

Enjoy the convenience and efficiency of accessing your costs anywhere:

- **Skip the multiplier** by setting your location
- **Quickly search,** edit, favorite and share costs
- **Stay on top of price changes** with automatic updates

Discover more at rsmeans.com/online

---

12 48 Rugs and Mats

12 48 13 – Entrance Floor Mats and Frames

12 48 13.13 Entrance Floor Mats		Crew	Daily Output	Labor-Hours	Unit	Material	2020 Bare Costs Labor	2020 Bare Costs Equipment	Total	Total Incl O&P
0010	**ENTRANCE FLOOR MATS**									
0020	Recessed, black rubber, 3/8" thick, solid	1 Clab	155	.052	S.F.	27.50	2.17		29.67	34

12 92 Interior Planters and Artificial Plants

12 92 13 – Interior Artificial Plants

12 92 13.10 Plants

		Crew	Daily Output	Labor-Hours	Unit	Material	2020 Bare Costs Labor	2020 Bare Costs Equipment	Total	Total Incl O&P
0010	**PLANTS** Permanent only, weighted									
0020	Preserved or polyester leaf, natural wood									
0030	Or molded trunk. For pots see Section 12 92 33.10									
0100	Plants, acuba, 5' high				Ea.	360			360	395
0200	Apidistra, 4' high					315			315	345
0300	Beech, variegated, 5' high					310			310	340
0400	Birds nest fern, 6' high					550			550	610
0500	Croton, 4' high					135			135	148
0600	Diffenbachia, 3' high					124			124	136
0700	Ficus Benjamina, 3' high					147			147	162
0720	6' high					370			370	410
0760	Nitida, 3' high					330			330	365
0780	6' high					580			580	635
0800	Helicona					310			310	340
1200	Palm, green date fan, 5' high					320			320	350
1220	7' high					405			405	445
1280	Green chamdora date, 5' high					290			290	320
1300	7' high					440			440	485
1400	Green giant, 7' high					685			685	755
1600	Rubber plant, 5' high					233			233	257
1800	Schefflera, 3' high					160			160	176
1820	4' high					192			192	211
1840	5' high					305			305	335
1860	6' high					415			415	460
1880	7' high					625			625	685
1900	Spathiphyllum, 3' high					222			222	244
4000	Trees, polyester or preserved, with									
4020	Natural trunks									
4100	Acuba, 10' high				Ea.	1,750			1,750	1,925
4120	12' high					2,450			2,450	2,675
4140	14' high					3,400			3,400	3,750
4400	Bamboo, 10' high					760			760	835
4420	12' high					880			880	965
4800	Beech, 10' high					2,325			2,325	2,575
4820	12' high					2,675			2,675	2,950
4840	14' high					3,625			3,625	3,975
5000	Birch, 10' high					2,325			2,325	2,575
5020	12' high					2,675			2,675	2,950
5040	14' high					2,925			2,925	3,200
5060	16' high					3,625			3,625	3,975
5080	18' high					4,075			4,075	4,500
5500	Ficus Benjamina, 10' high					1,725			1,725	1,900
5520	12' high					2,200			2,200	2,425
5540	14' high					2,450			2,450	2,700
5560	16' high					3,425			3,425	3,775

12 92 Interior Planters and Artificial Plants

12 92 13 – Interior Artificial Plants

12 92 13.10 Plants		Crew	Daily Output	Labor-Hours	Unit	Material	2020 Bare Costs Labor	Equipment	Total	Total Incl O&P
5580	18' high				Ea.	5,150			5,150	5,675
6000	Magnolia, 10' high					2,525			2,525	2,775
6020	12' high					3,575			3,575	3,925
6040	14' high					4,875			4,875	5,375
6500	Maple, 10' high					270			270	297
6520	12' high					405			405	445
6540	14' high					500			500	550
7000	Palms, chamadora, 10' high					490			490	540
7020	12' high					625			625	690
7040	Date, 10' high					675			675	745
7060	12' high					775			775	850

12 92 33 – Interior Planters

12 92 33.10 Planters		Crew	Daily Output	Labor-Hours	Unit	Material	2020 Bare Costs Labor	Equipment	Total	Total Incl O&P
0010	**PLANTERS**									
1000	Fiberglass, hanging, 12" diameter, 7" high				Ea.	109			109	120
1100	15" diameter, 7" high					158			158	174
1200	36" diameter, 8" high					218			218	240
1500	Rectangular, 48" long, 16" high, 15" wide					615			615	675
1550	16" high, 24" wide					670			670	735
1600	24" high, 24" wide					980			980	1,075
1650	60" long, 30" high, 28" wide					1,025			1,025	1,150
1700	72" long, 16" high, 15" wide					845			845	930
1750	21" high, 24" wide					1,075			1,075	1,200
1800	30" high, 24" wide					1,200			1,200	1,300
2000	Round, 12" diameter, 13" high					203			203	223
2050	25" high					241			241	265
2150	14" diameter, 15" high					188			188	207
2200	16" diameter, 16" high					198			198	217
2250	18" diameter, 19" high					239			239	262
2300	23" high					370			370	410
2350	20" diameter, 16" high					223			223	245
2400	18" high					245			245	270
2450	21" high					315			315	350
2500	22" diameter, 10" high					240			240	264
2550	24" diameter, 16" high					288			288	315
2600	19" high					345			345	380
2650	25" high					395			395	435
2700	36" high					620			620	680
2750	48" high					865			865	950
2800	30" diameter, 16" high					340			340	375
2850	18" high					345			345	375
2900	21" high					385			385	420
3000	24" high					395			395	435
3350	27" high					435			435	480
3400	36" diameter, 16" high					425			425	470
3450	18" high					445			445	485
3500	21" high					475			475	525
3550	24" high					510			510	560
3600	27" high					555			555	610
3650	30" high					600			600	660
3700	48" diameter, 16" high					745			745	820
3750	21" high					805			805	885

195

12 92 33.10 Planters		Crew	Daily Output	Labor-Hours	Unit	Material	2020 Bare Costs Labor	Equipment	Total	Total Incl O&P
3800	24" high				Ea.	900			900	990
3850	27" high					945			945	1,050
3900	30" high					1,025			1,025	1,125
3950	36" high					1,075			1,075	1,175
4000	60" diameter, 16" high					985			985	1,075
4100	21" high					1,125			1,125	1,225
4150	27" high					1,250			1,250	1,375
4200	30" high					1,375			1,375	1,525
4250	33" high					1,575			1,575	1,725
4300	36" high					1,925			1,925	2,100
4400	39" high					2,075			2,075	2,275
5000	Square, 10" side, 20" high					206			206	227
5100	14" side, 15" high					238			238	262
5200	18" side, 19" high					250			250	274
5300	20" side, 16" high					305			305	335
5320	18" high					475			475	525
5340	21" high					395			395	435
5400	24" side, 16" high					385			385	425
5420	21" high					580			580	640
5440	25" high					540			540	595
5460	30" side, 16" high					515			515	565
5480	24" high					815			815	895
5490	27" high					600			600	660
5500	Round, 36" diameter, 16" high					580			580	640
5510	18" high					805			805	890
5520	21" high					890			890	975
5530	24" high					985			985	1,075
5540	27" high					1,125			1,125	1,225
5550	30" high					1,550			1,550	1,700
5800	48" diameter, 16" high					820			820	900
5820	21" high					915			915	1,000
5840	24" high					1,125			1,125	1,250
5860	27" high					1,225			1,225	1,350
5880	30" high					325			325	360
5900	60" diameter, 16" high					985			985	1,075
5920	21" high					1,150			1,150	1,250
5940	27" high					1,250			1,250	1,375
5960	30" high					1,400			1,400	1,550
5980	36" high					1,525			1,525	1,675
6000	Metal bowl, 32" diameter, 8" high, minimum					580			580	640
6050	Maximum					860			860	945
6100	Rectangle, 30" long x 12" wide, 6" high, minimum					435			435	475
6200	Maximum					610			610	670
6300	36" long x 12" wide, 6" high, minimum					855			855	940
6400	Maximum					560			560	620
6500	Square, 15" side, minimum					635			635	700
6600	Maximum					940			940	1,025
6700	20" side, minimum					1,350			1,350	1,475
6800	Maximum					650			650	715
6900	Round, 6" diameter x 6" high, minimum					62.50			62.50	68.50
7000	Maximum					74.50			74.50	81.50
7100	8" diameter x 8" high, minimum					83			83	91.50
7200	Maximum					89.50			89.50	98

12 92 Interior Planters and Artificial Plants

12 92 33 – Interior Planters

12 92 33.10 Planters		Crew	Daily Output	Labor-Hours	Unit	Material	2020 Bare Costs Labor	Equipment	Total	Total Incl O&P
7300	10" diameter x 11" high, minimum				Ea.	107			107	117
7400	Maximum					157			157	173
7420	12" diameter x 13" high, minimum					98.50			98.50	109
7440	Maximum					227			227	250
7500	14" diameter x 15" high, minimum					133			133	146
7550	Maximum					233			233	257
7580	16" diameter x 17" high, minimum					144			144	158
7600	Maximum					255			255	281
7620	18" diameter x 19" high, minimum					174			174	191
7640	Maximum					310			310	345
7680	22" diameter x 20" high, minimum					219			219	241
7700	Maximum					400			400	440
7750	24" diameter x 21" high, minimum					285			285	315
7800	Maximum					535			535	585
7850	31" diameter x 18" high, minimum					620			620	680
7900	Maximum					1,575			1,575	1,750
7950	38" diameter x 24" high, minimum					1,050			1,050	1,150
8000	Maximum					2,650			2,650	2,900
8050	48" diameter x 24" high, minimum					1,375			1,375	1,525
8150	Maximum					2,725			2,725	3,000
8750	Wood, fiberglass liner, square									
8780	14" square, 15" high, minimum				Ea.	430			430	470
8800	Maximum					525			525	580
8820	24" square, 16" high, minimum					525			525	580
8840	Maximum					695			695	765
8860	36" square, 21" high, minimum					675			675	745
8880	Maximum					1,025			1,025	1,125
9000	Rectangle, 36" long x 12" wide, 10" high, minimum					480			480	525
9050	Maximum					620			620	685
9100	48" long x 12" wide, 10" high, minimum					515			515	570
9120	Maximum					650			650	715
9200	48" long x 12" wide, 24" high, minimum					620			620	685
9300	Maximum					915			915	1,000
9400	Plastic cylinder, molded, 10" diameter, 10" high					47.50			47.50	52.50
9500	11" diameter, 11" high					90.50			90.50	99.50
9600	13" diameter, 12" high					148			148	162
9700	16" diameter, 14" high					172			172	189

12 93 Interior Public Space Furnishings

12 93 13 – Bicycle Racks

12 93 13.10 Bicycle Racks		Crew	Daily Output	Labor-Hours	Unit	Material	2020 Bare Costs Labor	Equipment	Total	Total Incl O&P
0010	**BICYCLE RACKS**									
0020	Single side, grid, 1-5/8" OD stl. pipe, w/1/2" bars, galv, 5 bike cap	2 Clab	10	1.600	Ea.	282	67.50		349.50	410
0025	Powder coat finish		10	1.600		288	67.50		355.50	415
0030	Single side, grid, 1-5/8" OD stl. pipe, w/1/2" bars, galv, 9 bike cap		8	2		400	84		484	565
0035	Powder coat finish		8	2		430	84		514	595
0040	Single side, grid, 1-5/8" OD stl. pipe, w/1/2" bars, galv, 18 bike cap		4	4		520	168		688	830
0045	Powder coat finish		4	4		535	168		703	845
0050	S curve, 1-7/8" OD stl. pipe, 11 ga., galv, 5 bike cap		10	1.600		160	67.50		227.50	277
0055	Powder coat finish		10	1.600		160	67.50		227.50	277
0060	S curve, 1-7/8" OD stl. pipe, 11 ga., galv, 7 bike cap		9	1.778		203	75		278	335

12 93 Interior Public Space Furnishings

12 93 13 – Bicycle Racks

12 93 13.10 Bicycle Racks	Crew	Daily Output	Labor-Hours	Unit	Material	2020 Bare Costs Labor	Equipment	Total	Total Incl O&P
0065 Powder coat finish	2 Clab	9	1.778	Ea.	219	75		294	355
0070 S curve, 1-7/8" OD stl. pipe, 11 ga., galv, 9 bike cap		8	2		320	84		404	475
0075 Powder coat finish		8	2		310	84		394	465
0080 S curve, 1-7/8" OD stl. pipe, 11 ga., galv, 11 bike cap		6	2.667		425	112		537	635
0085 Powder coat finish		6	2.667		395	112		507	605

12 93 23 – Trash and Litter Receptacles

12 93 23.10 Trash Receptacles

	Crew	Daily Output	Labor-Hours	Unit	Material	2020 Bare Costs Labor	Equipment	Total	Total Incl O&P
0010 **TRASH RECEPTACLES**									
0020 Fiberglass, 2' square, 18" high	2 Clab	30	.533	Ea.	610	22.50		632.50	710
0100 2' square, 2'-6" high		30	.533		975	22.50		997.50	1,100
0300 Circular, 2' diameter, 18" high		30	.533		575	22.50		597.50	670
0400 2' diameter, 2'-6" high		30	.533		615	22.50		637.50	710
1000 Alum. frame, hardboard panels, steel drum base,									
1020 30 gal. capacity, silk screen on plastic finish	2 Clab	25	.640	Ea.	515	27		542	605
1040 Aggregate finish		25	.640		655	27		682	760
1100 50 gal. capacity, silk screen on plastic finish		20	.800		770	33.50		803.50	900
1140 Aggregate finish		20	.800		825	33.50		858.50	960
1200 Formed plastic liner, 14 gal., silk screen on plastic finish		40	.400		430	16.85		446.85	500
1240 Aggregate finish		40	.400		470	16.85		486.85	540
1300 30 gal. capacity, silk screen on plastic finish		35	.457		555	19.25		574.25	640
1340 Aggregate finish		35	.457		605	19.25		624.25	695
1400 Redwood slats, plastic liner, leg base, 14 gal. capacity,									
1420 Varnish w/routed message	2 Clab	40	.400	Ea.	450	16.85		466.85	520
2000 Concrete, precast, 2' to 2-1/2' wide, 3' high, sandblasted	"	15	1.067	"	590	45		635	720
3000 Galv. steel frame and panels, leg base, poly bag retainer,									
3020 40 gal. capacity, silk screen on enamel finish	2 Clab	25	.640	Ea.	630	27		657	735
3040 Aggregate finish		25	.640		725	27		752	835
3200 Formed plastic liner, 50 gal., silk screen on enamel finish		20	.800		1,050	33.50		1,083.50	1,200
3240 Aggregate finish		20	.800		1,125	33.50		1,158.50	1,275
4000 Perforated steel, pole mounted, 12" diam., 10 gal., painted		25	.640		166	27		193	224
4040 Redwood slats		25	.640		330	27		357	405
4100 22 gal. capacity, painted		25	.640		217	27		244	280
4140 Redwood slats		25	.640		620	27		647	720
4500 Galvanized steel street basket, 52 gal. capacity, unpainted		40	.400		390	16.85		406.85	450

12 93 23.20 Trash Closure

	Crew	Daily Output	Labor-Hours	Unit	Material	2020 Bare Costs Labor	Equipment	Total	Total Incl O&P
0010 **TRASH CLOSURE**									
0020 Steel with pullover cover, 2'-3" wide, 4'-7" high, 6'-2" long	2 Clab	5	3.200	Ea.	2,050	135		2,185	2,450
0100 10'-1" long		4	4		2,600	168		2,768	3,100
0300 Wood, 10' wide, 6' high, 10' long		1.20	13.333		2,325	560		2,885	3,400

Estimating Tips
General

- The items and systems in this division are usually estimated, purchased, supplied, and installed as a unit by one or more subcontractors. The estimator must ensure that all parties are operating from the same set of specifications and assumptions, and that all necessary items are estimated and will be provided. Many times the complex items and systems are covered, but the more common ones, such as excavation or a crane, are overlooked for the very reason that everyone assumes nobody could miss them. The estimator should be the central focus and be able to ensure that all systems are complete.

- It is important to consider factors such as site conditions, weather, shape and size of building, as well as labor availability as they may impact the overall cost of erecting special structures and systems included in this division.

- Another area where problems can develop in this division is at the interface between systems.

The estimator must ensure, for instance, that anchor bolts, nuts, and washers are estimated and included for the air-supported structures and pre-engineered buildings to be bolted to their foundations. Utility supply is a common area where essential items or pieces of equipment can be missed or overlooked because each subcontractor may feel it is another's responsibility. The estimator should also be aware of certain items which may be supplied as part of a package but installed by others, and ensure that the installing contractor's estimate includes the cost of installation. Conversely, the estimator must also ensure that items are not costed by two different subcontractors, resulting in an inflated overall estimate.

13 30 00 Special Structures

- The foundations and floor slab, as well as rough mechanical and electrical, should be estimated, as this work is required for the assembly and erection of the structure. Generally, as noted in the data set, the pre-engineered building comes

as a shell. Pricing is based on the size and structural design parameters stated in the reference section. Additional features, such as windows and doors with their related structural framing, must also be included by the estimator. Here again, the estimator must have a clear understanding of the scope of each portion of the work and all the necessary interfaces.

Reference Numbers

Reference numbers are shown at the beginning of some major classifications. These numbers refer to related items in the Reference Section. The reference information may be an estimating procedure, an alternate pricing method, or technical information.

Note: Not all subdivisions listed here necessarily appear. ■

Same Data. Simplified.

Enjoy the convenience and efficiency of accessing your costs anywhere:

- **Skip the multiplier** by setting your location
- **Quickly search,** edit, favorite and share costs
- **Stay on top of price changes** with automatic updates

Discover more at rsmeans.com/online

13 05 05 – Selective Demolition for Special Construction

13 05 05.25 Selective Demolition, Geodesic Domes

		Crew	Daily Output	Labor-Hours	Unit	Material	2020 Bare Costs Labor	Equipment	Total	Total Incl O&P
0010	**SELECTIVE DEMOLITION, GEODESIC DOMES**									
0050	Shell only, interlocking plywood panels, 30' diameter	F-5	3.20	10	Ea.		535		535	805
0060	34' diameter		2.30	13.913			745		745	1,125
0070	39' diameter		2	16			860		860	1,300
0080	45' diameter	F-3	2.20	18.182			990	214	1,204	1,700
0090	55' diameter		2	20			1,075	236	1,311	1,875
0100	60' diameter		2	20			1,075	236	1,311	1,875
0110	65' diameter		1.60	25			1,350	295	1,645	2,375

13 05 05.45 Selective Demolition, Lightning Protection

		Crew	Daily Output	Labor-Hours	Unit	Material	2020 Bare Costs Labor	Equipment	Total	Total Incl O&P
0010	**SELECTIVE DEMOLITION, LIGHTNING PROTECTION**									
0020	Air terminal & base, copper, 3/8" diam. x 10", to 75' H	1 Clab	16	.500	Ea.		21		21	31.50
0030	1/2" diam. x 12", over 75' H		16	.500			21		21	31.50
0050	Aluminum, 1/2" diam. x 12", to 75' H		16	.500			21		21	31.50
0060	5/8" diam. x 12", over 75' H		16	.500			21		21	31.50
0070	Cable, copper, 220 lb. per thousand feet, to 75' H		640	.013	L.F.		.53		.53	.79
0080	375 lb. per thousand feet, over 75' H		460	.017			.73		.73	1.10
0090	Aluminum, 101 lb. per thousand feet, to 75' H		560	.014			.60		.60	.90
0100	199 lb. per thousand feet, over 75' H		480	.017			.70		.70	1.05
0110	Arrester, 175 V AC, to ground		16	.500	Ea.		21		21	31.50
0120	650 V AC, to ground		13	.615	"		26		26	39

13 05 05.50 Selective Demolition, Pre-Engineered Steel Buildings

		Crew	Daily Output	Labor-Hours	Unit	Material	2020 Bare Costs Labor	Equipment	Total	Total Incl O&P
0010	**SELECTIVE DEMOLITION, PRE-ENGINEERED STEEL BUILDINGS**									
0500	Pre-engd. steel bldgs., rigid frame, clear span & multi post, excl. salvage									
0550	3,500 to 7,500 S.F.	L-10	1000	.024	SF Flr.		1.41	.47	1.88	2.70
0600	7,501 to 12,500 S.F.		1500	.016			.94	.31	1.25	1.80
0650	12,501 S.F. or greater		1650	.015			.86	.29	1.15	1.63
0700	Pre-engd. steel building components									
0710	Entrance canopy, including frame 4' x 4'	E-24	8	4	Ea.		230	72.50	302.50	435
0720	4' x 8'	"	7	4.571			263	83	346	495
0730	HM doors, self framing, single leaf	2 Skwk	8	2			110		110	166
0740	Double leaf		5	3.200			176		176	265
0760	Gutter, eave type		600	.027	L.F.		1.46		1.46	2.21
0770	Sash, single slide, double slide or fixed		24	.667	Ea.		36.50		36.50	55.50
0780	Skylight, fiberglass, to 30 S.F.		16	1			55		55	83
0785	Roof vents, circular, 12" to 24" diameter		12	1.333			73		73	111
0790	Continuous, 10' long		8	2			110		110	166
0900	Shelters, aluminum frame									
0910	Acrylic glazing, 3' x 9' x 8' high	2 Skwk	2	8	Ea.		440		440	665
0920	9' x 12' x 8' high	"	1.50	10.667	"		585		585	885

13 05 05.60 Selective Demolition, Silos

		Crew	Daily Output	Labor-Hours	Unit	Material	2020 Bare Costs Labor	Equipment	Total	Total Incl O&P
0010	**SELECTIVE DEMOLITION, SILOS**									
0020	Conc stave, indstrl, conical/sloping bott, excl fndtn, 12' diam., 35' H	E-24	.18	178	Ea.		10,200	3,225	13,425	19,400
0030	16' diam., 45' H		.12	267			15,300	4,850	20,150	29,000
0040	25' diam., 75' H		.08	400			23,000	7,250	30,250	43,600
0050	Steel, factory fabricated, 30,000 gal. cap, painted or epoxy lined	L-5	2	28			1,625	290	1,915	2,850

13 05 05.75 Selective Demolition, Storage Tanks

		Crew	Daily Output	Labor-Hours	Unit	Material	2020 Bare Costs Labor	Equipment	Total	Total Incl O&P
0010	**SELECTIVE DEMOLITION, STORAGE TANKS**									
0500	Steel tank, single wall, above ground, not incl. fdn., pumps or piping									
0510	Single wall, 275 gallon R024119-10	Q-1	3	5.333	Ea.		310		310	465
0520	550 thru 2,000 gallon	B-34P	2	12			680	515	1,195	1,600
0530	5,000 thru 10,000 gallon	B-34Q	2	12			690	570	1,260	1,650
0540	15,000 thru 30,000 gallon	B-34S	2	16			955	1,550	2,505	3,150

200

For customer support on your Site Work & Landscape Costs with RSMeans data, call 800.448.8182.

13 05 Common Work Results for Special Construction

13 05 05 – Selective Demolition for Special Construction

13 05 05.75 Selective Demolition, Storage Tanks

		Crew	Daily Output	Labor-Hours	Unit	Material	2020 Bare Costs Labor	2020 Bare Costs Equipment	Total	Total Incl O&P
0600	Steel tank, double wall, above ground not incl. fdn., pumps & piping									
0620	500 thru 2,000 gallon	B-34P	2	12	Ea.		680	515	1,195	1,600

13 05 05.90 Selective Demolition, Tension Structures

		Crew	Daily Output	Labor-Hours	Unit	Material	2020 Bare Costs Labor	2020 Bare Costs Equipment	Total	Total Incl O&P
0010	**SELECTIVE DEMOLITION, TENSION STRUCTURES**									
0020	Steel/alum. frame, fabric shell, 60' clear span, 6,000 S.F.	B-41	2000	.022	SF Flr.		.96	.15	1.11	1.61
0030	12,000 S.F.		2200	.020			.87	.14	1.01	1.46
0040	80' clear span, 20,800 S.F.	↓	2440	.018			.79	.12	.91	1.32
0050	100' clear span, 10,000 S.F.	L-5	4350	.013			.75	.13	.88	1.32
0060	26,000 S.F.		4600	.012			.71	.13	.84	1.24
0070	36,000 S.F.		5000	.011			.65	.12	.77	1.14
0080	120' clear span, 24,000 S.F.		6000	.009			.54	.10	.64	.96
0090	150' clear span, 30,000 S.F.	↓	7000	.008			.47	.08	.55	.81
0100	200' clear span, 40,000 S.F.	E-6	16000	.008	↓		.46	.13	.59	.86
0110	For roll-up door, 12' x 14'	L-2	2	8	Ea.		370		370	565

13 11 Swimming Pools

13 11 13 – Below-Grade Swimming Pools

13 11 13.50 Swimming Pools

		Crew	Daily Output	Labor-Hours	Unit	Material	2020 Bare Costs Labor	2020 Bare Costs Equipment	Total	Total Incl O&P
0010	**SWIMMING POOLS** Residential in-ground, vinyl lined									
0020	Concrete sides, w/equip, sand bottom	B-52	300	.187	SF Surf	28.50	9.10	1.95	39.55	47
0100	Metal or polystyrene sides	B-14	410	.117		24	5.20	.52	29.72	35
0200	Add for vermiculite bottom				↓	1.82			1.82	2
0500	Gunite bottom and sides, white plaster finish									
0600	12' x 30' pool	B-52	145	.386	SF Surf	53	18.80	4.03	75.83	91
0720	16' x 32' pool		155	.361		48	17.60	3.77	69.37	83
0750	20' x 40' pool	↓	250	.224	↓	42.50	10.90	2.34	55.74	66
0810	Concrete bottom and sides, tile finish									
0820	12' x 30' pool	B-52	80	.700	SF Surf	53.50	34	7.30	94.80	118
0830	16' x 32' pool		95	.589		44.50	28.50	6.15	79.15	99
0840	20' x 40' pool	↓	130	.431	↓	35.50	21	4.50	61	75.50
1100	Motel, gunite with plaster finish, incl. medium									
1150	capacity filtration & chlorination	B-52	115	.487	SF Surf	65.50	23.50	5.10	94.10	113
1200	Municipal, gunite with plaster finish, incl. high									
1250	capacity filtration & chlorination	B-52	100	.560	SF Surf	84.50	27.50	5.85	117.85	140
1350	Add for formed gutters				L.F.	124			124	137
1360	Add for stainless steel gutters				"	370			370	405
1700	Filtration and deck equipment only, as % of total				Total				20%	20%
1800	Automatic vacuum, hand tools, etc., 20' x 40' pool				SF Pool				.56	.62
1900	5,000 S.F. pool				"				.11	.12
3000	Painting pools, preparation + 3 coats, 20' x 40' pool, epoxy	2 Pord	.33	48.485	Total	1,450	2,150		3,600	4,800
3100	Rubber base paint, 18 gallons	"	.33	48.485		1,250	2,150		3,400	4,575
3500	42' x 82' pool, 75 gallons, epoxy paint	3 Pord	.14	171		6,150	7,600		13,750	18,100
3600	Rubber base paint	"	.14	171	↓	5,150	7,600		12,750	17,000

13 11 23 – On-Grade Swimming Pools

13 11 23.50 Swimming Pools

		Crew	Daily Output	Labor-Hours	Unit	Material	2020 Bare Costs Labor	2020 Bare Costs Equipment	Total	Total Incl O&P
0010	**SWIMMING POOLS** Residential above ground, steel construction									
0100	Round, 15' diam.	B-80A	3	8	Ea.	895	335	273	1,503	1,800
0120	18' diam.		2.50	9.600		1,025	405	330	1,760	2,100
0140	21' diam.		2	12		1,175	505	410	2,090	2,500
0160	24' diam.	↓	1.80	13.333	↓	1,350	560	455	2,365	2,850

13 11 Swimming Pools

13 11 23 – On-Grade Swimming Pools

13 11 23.50 Swimming Pools		Crew	Daily Output	Labor-Hours	Unit	Material	2020 Bare Costs Labor	Equipment	Total	Total Incl O&P
0180	27' diam.	B-80A	1.50	16	Ea.	1,550	675	545	2,770	3,325
0200	30' diam.		1	24		1,575	1,000	820	3,395	4,175
0220	Oval, 12' x 24'		2.30	10.435		1,700	440	355	2,495	2,900
0240	15' x 30'		1.80	13.333		1,925	560	455	2,940	3,475
0260	18' x 33'		1	24		2,150	1,000	820	3,970	4,775

13 11 46 – Swimming Pool Accessories

13 11 46.50 Swimming Pool Equipment

13 11 46.50 Swimming Pool Equipment		Crew	Daily Output	Labor-Hours	Unit	Material	2020 Bare Costs Labor	Equipment	Total	Total Incl O&P
0010	**SWIMMING POOL EQUIPMENT**									
0020	Diving stand, stainless steel, 3 meter	2 Carp	.40	40	Ea.	18,000	2,125		20,125	23,000
0300	1 meter		2.70	5.926		10,900	315		11,215	12,500
0600	Diving boards, 16' long, aluminum		2.70	5.926		4,275	315		4,590	5,175
0700	Fiberglass		2.70	5.926		3,475	315		3,790	4,275
0800	14' long, aluminum		2.70	5.926		3,925	315		4,240	4,775
0850	Fiberglass		2.70	5.926		3,675	315		3,990	4,525
1100	Bulkhead, movable, PVC, 8'-2" wide	2 Clab	8	2		3,075	84		3,159	3,525
1120	7'-9" wide		8	2		2,975	84		3,059	3,400
1140	7'-3" wide		8	2		2,300	84		2,384	2,650
1160	6'-9" wide		8	2		2,300	84		2,384	2,650
1200	Ladders, heavy duty, stainless steel, 2 tread	2 Carp	7	2.286		960	121		1,081	1,225
1500	4 tread		6	2.667		910	142		1,052	1,225
1800	Lifeguard chair, stainless steel, fixed		2.70	5.926		3,150	315		3,465	3,950
1900	Portable					3,050			3,050	3,350
2100	Lights, underwater, 12 volt, with transformer, 300 watt	1 Elec	1	8		365	490		855	1,125
2200	110 volt, 500 watt, standard		1	8		291	490		781	1,050
2400	Low water cutoff type		1	8		310	490		800	1,075
3000	Pool covers, reinforced vinyl	3 Clab	1800	.013	S.F.	1.28	.56		1.84	2.25
3050	Automatic, electric								9.15	10.10
3100	Vinyl, for winter, 400 S.F. max pool surface	3 Clab	3200	.008		.23	.32		.55	.72
3200	With water tubes, 400 S.F. max pool surface	"	3000	.008		.34	.34		.68	.88
3250	Sealed air bubble polyethylene solar blanket, 16 mils					.35			.35	.39
3300	Slides, tubular, fiberglass, aluminum handrails & ladder, 5'-0", straight	2 Carp	1.60	10	Ea.	3,850	530		4,380	5,025
3320	8'-0", curved		3	5.333		8,000	283		8,283	9,225
3400	10'-0", curved		1	16		23,800	850		24,650	27,500
3420	12'-0", straight with platform		1.20	13.333		16,600	710		17,310	19,300

13 12 Fountains

13 12 13 – Exterior Fountains

13 12 13.10 Outdoor Fountains

13 12 13.10 Outdoor Fountains		Crew	Daily Output	Labor-Hours	Unit	Material	2020 Bare Costs Labor	Equipment	Total	Total Incl O&P
0010	**OUTDOOR FOUNTAINS**									
0100	Outdoor fountain, 48" high with bowl and figures	2 Clab	2	8	Ea.	435	335		770	985
0200	Commercial, concrete or cast stone, 40-60" H, simple		2	8		955	335		1,290	1,550
0220	Average		2	8		1,850	335		2,185	2,525
0240	Ornate		2	8		4,200	335		4,535	5,125
0260	Metal, 72" high		2	8		1,500	335		1,835	2,150
0280	90" high		2	8		2,250	335		2,585	2,975
0300	120" high		2	8		5,275	335		5,610	6,300
0320	Resin or fiberglass, 40-60" H, wall type		2	8		675	335		1,010	1,250
0340	Waterfall type		2	8		1,150	335		1,485	1,750

13 18 Ice Rinks

13 18 13 – Ice Rink Floor Systems

13 18 13.50 Ice Skating

		Crew	Daily Output	Labor-Hours	Unit	Material	2020 Bare Costs Labor	Equipment	Total	Total Incl O&P
0010	**ICE SKATING** Equipment incl. refrigeration, plumbing & cooling									
0020	coils & concrete slab, 85' x 200' rink									
0300	55° system, 5 mos., 100 ton				Total	591,000			591,000	650,000
0700	90° system, 12 mos., 135 ton				"	683,000			683,000	751,500
1200	Subsoil heating system (recycled from compressor), 85' x 200'	Q-7	.27	119	Ea.	41,400	7,400		48,800	56,500
1300	Subsoil insulation, 2 lb. polystyrene with vapor barrier, 85' x 200'	2 Carp	.14	114	"	30,900	6,075		36,975	43,100

13 18 16 – Ice Rink Dasher Boards

13 18 16.50 Ice Rink Dasher Boards

		Crew	Daily Output	Labor-Hours	Unit	Material	2020 Bare Costs Labor	Equipment	Total	Total Incl O&P
0010	**ICE RINK DASHER BOARDS**									
1000	Dasher boards, 1/2" H.D. polyethylene faced steel frame, 3' acrylic									
1020	screen at sides, 5' acrylic ends, 85' x 200'	F-5	.06	533	Ea.	144,500	28,600		173,100	202,000
1100	Fiberglass & aluminum construction, same sides and ends	"	.06	533	"	162,000	28,600		190,600	221,000

13 31 Fabric Structures

13 31 13 – Air-Supported Fabric Structures

13 31 13.09 Air Supported Tank Covers

		Crew	Daily Output	Labor-Hours	Unit	Material	2020 Bare Costs Labor	Equipment	Total	Total Incl O&P
0010	**AIR SUPPORTED TANK COVERS**, vinyl polyester									
0100	Scrim, double layer, with hardware, blower, standby & controls									
0200	Round, 75' diameter	B-2	4500	.009	S.F.	12.70	.38		13.08	14.55
0300	100' diameter		5000	.008		11.55	.34		11.89	13.20
0400	150' diameter		5000	.008		9.15	.34		9.49	10.55
0500	Rectangular, 20' x 20'		4500	.009		24.50	.38		24.88	27.50
0600	30' x 40'		4500	.009		24.50	.38		24.88	27.50
0700	50' x 60'		4500	.009		24.50	.38		24.88	27.50
0800	For single wall construction, deduct, minimum					.86			.86	.95
0900	Maximum					2.48			2.48	2.73
1000	For maximum resistance to atmosphere or cold, add					1.24			1.24	1.36
1100	For average shipping charges, add				Total	2,125			2,125	2,350

13 31 13.13 Single-Walled Air-Supported Structures

		Crew	Daily Output	Labor-Hours	Unit	Material	2020 Bare Costs Labor	Equipment	Total	Total Incl O&P
0010	**SINGLE-WALLED AIR-SUPPORTED STRUCTURES**									
0020	Site preparation, incl. anchor placement and utilities	B-11B	1000	.016	SF Flr.	1.25	.76	.40	2.41	2.96
0030	For concrete, see Section 03 30 53.40									
0050	Warehouse, polyester/vinyl fabric, 28 oz., over 10 yr. life, welded									
0060	Seams, tension cables, primary & auxiliary inflation system,									
0070	airlock, personnel doors and liner									
0100	5,000 S.F.	4 Clab	5000	.006	SF Flr.	29.50	.27		29.77	33
0250	12,000 S.F.	"	6000	.005		20	.22		20.22	22.50
0400	24,000 S.F.	8 Clab	12000	.005		15	.22		15.22	16.85
0500	50,000 S.F.	"	12500	.005		13.20	.22		13.42	14.80
0700	12 oz. reinforced vinyl fabric, 5 yr. life, sewn seams,									
0710	accordion door, including liner									
0750	3,000 S.F.	4 Clab	3000	.011	SF Flr.	13.75	.45		14.20	15.80
0800	12,000 S.F.	"	6000	.005		11.70	.22		11.92	13.25
0850	24,000 S.F.	8 Clab	12000	.005		9.90	.22		10.12	11.25
0950	Deduct for single layer					1.12			1.12	1.23
1000	Add for welded seams					1.56			1.56	1.72
1050	Add for double layer, welded seams included					3.13			3.13	3.44
1250	Tedlar/vinyl fabric, 28 oz., with liner, over 10 yr. life,									
1260	incl. overhead and personnel doors									
1300	3,000 S.F.	4 Clab	3000	.011	SF Flr.	26	.45		26.45	29

13 31 Fabric Structures

13 31 13 – Air-Supported Fabric Structures

13 31 13.13 Single-Walled Air-Supported Structures

		Crew	Daily Output	Labor-Hours	Unit	Material	2020 Bare Costs Labor	Equipment	Total	Total Incl O&P
1450	12,000 S.F.	4 Clab	6000	.005	SF Flr.	17.50	.22		17.72	19.60
1550	24,000 S.F.	8 Clab	12000	.005		14.05	.22		14.27	15.80
1700	Deduct for single layer				↓	2.08			2.08	2.29
2250	Greenhouse/shelter, woven polyethylene with liner, 2 yr. life,									
2260	sewn seams, including doors									
2300	3,000 S.F.	4 Clab	3000	.011	SF Flr.	16.80	.45		17.25	19.15
2350	12,000 S.F.	"	6000	.005		14.70	.22		14.92	16.50
2450	24,000 S.F.	8 Clab	12000	.005		12.60	.22		12.82	14.20
2550	Deduct for single layer				↓	1.03			1.03	1.13
2600	Tennis/gymnasium, polyester/vinyl fabric, 28 oz., over 10 yr. life,									
2610	including thermal liner, heat and lights									
2650	7,200 S.F.	4 Clab	6000	.005	SF Flr.	24	.22		24.22	26.50
2750	13,000 S.F.	"	6500	.005		19	.21		19.21	21.50
2850	Over 24,000 S.F.	8 Clab	12000	.005		17.35	.22		17.57	19.45
2860	For low temperature conditions, add				↓	1.23			1.23	1.35
2870	For average shipping charges, add				Total	5,825			5,825	6,425
2900	Thermal liner, translucent reinforced vinyl				SF Flr.	1.23			1.23	1.35
2950	Metalized mylar fabric and mesh, double liner				"	2.47			2.47	2.72
3050	Stadium/convention center, teflon coated fiberglass, heavy weight,									
3060	over 20 yr. life, incl. thermal liner and heating system									
3100	Minimum	9 Clab	26000	.003	SF Flr.	60.50	.12		60.62	66.50
3110	Maximum	"	19000	.004	"	74.50	.16		74.66	82
3400	Doors, air lock, 15' long, 10' x 10'	2 Carp	.80	20	Ea.	21,600	1,075		22,675	25,400
3600	15' x 15'	"	.50	32		33,200	1,700		34,900	39,100
3700	For each added 5' length, add					5,925			5,925	6,525
3900	Revolving personnel door, 6' diameter, 6'-6" high	2 Carp	.80	20	↓	16,600	1,075		17,675	19,800

13 31 23 – Tensioned Fabric Structures

13 31 23.50 Tension Structures

		Crew	Daily Output	Labor-Hours	Unit	Material	2020 Bare Costs Labor	Equipment	Total	Total Incl O&P
0010	**TENSION STRUCTURES** Rigid steel/alum. frame, vinyl coated poly									
0100	Fabric shell, 60' clear span, not incl. foundations or floors									
0200	6,000 S.F.	B-41	1000	.044	SF Flr.	17.10	1.92	.30	19.32	22
0300	12,000 S.F.		1100	.040		19.35	1.75	.27	21.37	24.50
0400	80' to 99' clear span, 20,800 S.F.	↓	1220	.036		18.60	1.57	.25	20.42	23
0410	100' to 119' clear span, 10,000 S.F.	L-5	2175	.026		16.80	1.50	.27	18.57	21
0430	26,000 S.F.		2300	.024		17.20	1.42	.25	18.87	21.50
0450	36,000 S.F.		2500	.022		17.25	1.30	.23	18.78	21
0460	120' to 149' clear span, 24,000 S.F.		3000	.019		17.05	1.09	.19	18.33	20.50
0470	150' to 199' clear span, 30,000 S.F.	↓	6000	.009		19.75	.54	.10	20.39	22.50
0480	200' clear span, 40,000 S.F.	E-6	8000	.016	↓	25	.92	.27	26.19	29
0500	For roll-up door, 12' x 14', add	L-2	1	16	Ea.	6,925	745		7,670	8,750

13 33 Geodesic Structures

13 33 13 – Geodesic Domes

13 33 13.35 Geodesic Domes

		Crew	Daily Output	Labor-Hours	Unit	Material	2020 Bare Costs Labor	Equipment	Total	Total Incl O&P
0010	**GEODESIC DOMES** Shell only, interlocking plywood panels R133423-30									
0400	30' diameter	F-5	1.60	20	Ea.	27,000	1,075		28,075	31,300
0500	33' diameter		1.14	28.070		26,700	1,500		28,200	31,700
0600	40' diameter	↓	1	32		30,500	1,725		32,225	36,100
0700	45' diameter	F-3	1.13	35.556		32,100	1,925	420	34,445	38,700
0750	56' diameter		1	40		58,500	2,175	470	61,145	68,000
0800	60' diameter		1	40		61,000	2,175	470	63,645	71,500
0850	67' diameter	↓	.80	50	↓	94,000	2,725	590	97,315	108,000
1100	Aluminum panel, with 6" insulation									
1200	100' diameter				SF Flr.	25			25	27.50
1300	500' diameter				"	24.50			24.50	27
1600	Aluminum framed, plexiglass closure panels									
1700	40' diameter				SF Flr.	61.50			61.50	67.50
1800	200' diameter				"	58			58	63.50
2100	Aluminum framed, aluminum closure panels									
2200	40' diameter				SF Flr.	20			20	22
2300	100' diameter					24			24	26.50
2400	200' diameter				↓	20.50			20.50	22.50
2700	Aluminum framed, fiberglass sandwich panel closure									
2800	6' diameter	2 Carp	150	.107	SF Flr.	27	5.65		32.65	38.50
2900	28' diameter	"	350	.046	"	25	2.43		27.43	30.50

13 34 Fabricated Engineered Structures

13 34 13 – Glazed Structures

13 34 13.13 Greenhouses

		Crew	Daily Output	Labor-Hours	Unit	Material	2020 Bare Costs Labor	Equipment	Total	Total Incl O&P
0010	**GREENHOUSES**, Shell only, stock units, not incl. 2' stub walls,									
0020	foundation, floors, heat or compartments									
0300	Residential type, free standing, 8'-6" long x 7'-6" wide	2 Carp	59	.271	SF Flr.	24	14.40		38.40	47.50
0400	10'-6" wide		85	.188		42.50	10		52.50	62
0600	13'-6" wide		108	.148		45	7.85		52.85	61.50
0700	17'-0" wide		160	.100		43.50	5.30		48.80	55.50
0900	Lean-to type, 3'-10" wide		34	.471		43.50	25		68.50	85
1000	6'-10" wide		58	.276		28	14.65		42.65	53
1050	8'-0" wide	↓	60	.267	↓	73	14.15		87.15	102
1100	Wall mounted to existing window, 3' x 3'	1 Carp	4	2	Ea.	2,225	106		2,331	2,600
1120	4' x 5'	"	3	2.667	"	2,600	142		2,742	3,100
1200	Deluxe quality, free standing, 7'-6" wide	2 Carp	55	.291	SF Flr.	87.50	15.45		102.95	120
1220	10'-6" wide		81	.198		65	10.50		75.50	87.50
1240	13'-6" wide		104	.154		54	8.20		62.20	72
1260	17'-0" wide		150	.107		51.50	5.65		57.15	65
1400	Lean-to type, 3'-10" wide		31	.516		107	27.50		134.50	158
1420	6'-10" wide		55	.291		77	15.45		92.45	108
1440	8'-0" wide	↓	97	.165	↓	64.50	8.75		73.25	84
1500	Commercial, custom, truss frame, incl. equip., plumbing, elec.,									
1550	benches and controls, under 2,000 S.F.				SF Flr.	13.50			13.50	14.85
1700	Over 5,000 S.F.				"	11.95			11.95	13.15
2000	Institutional, custom, rigid frame, including compartments and									
2050	multi-controls, under 500 S.F.				SF Flr.	29.50			29.50	32.50
2150	Over 2,000 S.F.				"	10.50			10.50	11.55
3700	For 1/4" tempered glass, add				SF Surf	1.75			1.75	1.93
3900	Cooling, 1,200 CFM exhaust fan, add				Ea.	325			325	355

205

13 34 Fabricated Engineered Structures

13 34 13 – Glazed Structures

13 34 13.13 Greenhouses

13 34 13.13 Greenhouses		Crew	Daily Output	Labor-Hours	Unit	Material	2020 Bare Costs Labor	Equipment	Total	Total Incl O&P
4000	7,850 CFM				Ea.	1,025			1,025	1,150
4200	For heaters, 10 MBH, add					246			246	270
4300	60 MBH, add					845			845	925
4500	For benches, 2' x 8', add					188			188	207
4600	4' x 10', add					227			227	249
4800	For ventilation & humidity control w/4 integrated outlets, add				Total	273			273	300
4900	For environmental controls and automation, 8 outputs, 9 stages, add				"	1,200			1,200	1,325
5100	For humidification equipment, add				Ea.	325			325	355
5200	For vinyl shading, add				S.F.	.44			.44	.48
6000	Geodesic hemisphere, 1/8" plexiglass glazing									
6050	8' diameter	2 Carp	2	8	Ea.	5,950	425		6,375	7,200
6150	24' diameter		.35	45.714		16,000	2,425		18,425	21,300
6250	48' diameter		.20	80		34,700	4,250		38,950	44,600

13 34 13.19 Swimming Pool Enclosures

		Crew	Daily Output	Labor-Hours	Unit	Material	Labor	Equipment	Total	Total Incl O&P
0010	**SWIMMING POOL ENCLOSURES** Translucent, free standing R131113-20									
0020	not including foundations, heat or light									
0200	Economy	2 Carp	200	.080	SF Hor.	46.50	4.25		50.75	57.50
0600	Deluxe	"	70	.229		103	12.15		115.15	131
0700	For motorized roof, 40% opening, solid roof, add					23			23	25.50
0800	Skylight type roof, add					12.95			12.95	14.25

13 34 16 – Grandstands and Bleachers

13 34 16.13 Grandstands

		Crew	Daily Output	Labor-Hours	Unit	Material	Labor	Equipment	Total	Total Incl O&P
0010	**GRANDSTANDS** Permanent, municipal, including foundation									
0300	Steel, economy				Seat	27			27	30
0400	Steel, deluxe					30.50			30.50	34
0900	Composite, steel, wood and plastic, stock design, economy					38			38	42
1000	Deluxe					91.50			91.50	101

13 34 16.53 Bleachers

		Crew	Daily Output	Labor-Hours	Unit	Material	Labor	Equipment	Total	Total Incl O&P
0010	**BLEACHERS**									
0020	Bleachers, outdoor, portable, 5 tiers, 42 seats	2 Sswk	120	.133	Seat	71.50	7.70		79.20	90.50
0100	5 tiers, 54 seats		80	.200		70	11.55		81.55	95
0200	10 tiers, 104 seats		120	.133		87.50	7.70		95.20	108
0300	10 tiers, 144 seats		80	.200		64.50	11.55		76.05	89
0500	Permanent bleachers, aluminum seat, steel frame, 24" row									
0600	8 tiers, 80 seats	2 Sswk	60	.267	Seat	74	15.35		89.35	105
0700	8 tiers, 160 seats		48	.333		66.50	19.20		85.70	104
0925	15 tiers, 154 to 165 seats		60	.267		74	15.35		89.35	105
0975	15 tiers, 214 to 225 seats		60	.267		66	15.35		81.35	96.50
1050	15 tiers, 274 to 285 seats		60	.267		68.50	15.35		83.85	99
1200	Seat backs only, 30" row, fiberglass		160	.100		25	5.75		30.75	36.50
1300	Steel and wood		160	.100		29	5.75		34.75	41
1400	NOTE: average seating is 1.5' in width									

13 34 19 – Metal Building Systems

13 34 19.50 Pre-Engineered Steel Buildings

		Crew	Daily Output	Labor-Hours	Unit	Material	Labor	Equipment	Total	Total Incl O&P
0010	**PRE-ENGINEERED STEEL BUILDINGS** R133419-10									
0100	Clear span rigid frame, 26 ga. colored roofing and siding									
0150	20' to 29' wide, 10' eave height	E-2	425	.132	SF Flr.	9.50	7.55	3.99	21.04	26.50
0160	14' eave height		350	.160		10.25	9.15	4.85	24.25	31
0170	16' eave height		320	.175		10.85	10	5.30	26.15	33.50
0180	20' eave height		275	.204		11.95	11.65	6.15	29.75	38
0190	24' eave height		240	.233		13.30	13.35	7.05	33.70	43

13 34 19 – Metal Building Systems

13 34 19.50 Pre-Engineered Steel Buildings	Crew	Daily Output	Labor-Hours	Unit	Material	2020 Bare Costs Labor	Equipment	Total	Total Incl O&P
0200	30' to 49' wide, 10' eave height	E-2	535	.105	SF Flr.	7	6	3.17	16.17	20.50
0300	14' eave height		450	.124		7.55	7.10	3.77	18.42	23.50
0400	16' eave height		415	.135		8.05	7.70	4.09	19.84	25.50
0500	20' eave height		360	.156		8.70	8.90	4.71	22.31	28.50
0600	24' eave height		320	.175		9.60	10	5.30	24.90	32
0700	50' to 100' wide, 10' eave height		770	.073		5.95	4.16	2.20	12.31	15.35
0900	16' eave height		600	.093		6.85	5.35	2.83	15.03	18.85
1000	20' eave height		490	.114		7.40	6.55	3.46	17.41	22
1100	24' eave height	▼	435	.129	▼	8.15	7.35	3.90	19.40	24.50
1200	Clear span tapered beam frame, 26 ga. colored roofing/siding									
1300	30' to 39' wide, 10' eave height	E-2	535	.105	SF Flr.	7.90	6	3.17	17.07	21.50
1400	14' eave height		450	.124		8.75	7.10	3.77	19.62	25
1500	16' eave height		415	.135		9.15	7.70	4.09	20.94	26.50
1600	20' eave height		360	.156		10.15	8.90	4.71	23.76	30
1700	40' wide, 10' eave height		600	.093		7.05	5.35	2.83	15.23	19.10
1800	14' eave height		510	.110		7.80	6.25	3.33	17.38	22
1900	16' eave height		475	.118		8.15	6.75	3.57	18.47	23.50
2000	20' eave height		415	.135		8.95	7.70	4.09	20.74	26.50
2100	50' to 79' wide, 10' eave height		770	.073		6.55	4.16	2.20	12.91	16.10
2200	14' eave height		675	.083		7.10	4.74	2.51	14.35	17.95
2300	16' eave height		635	.088		7.40	5.05	2.67	15.12	18.90
2400	20' eave height		490	.114		8.70	6.55	3.46	18.71	23.50
2410	80' to 100' wide, 10' eave height		935	.060		5.85	3.42	1.81	11.08	13.70
2420	14' eave height		750	.075		6.40	4.27	2.26	12.93	16.15
2430	16' eave height		685	.082		6.70	4.67	2.48	13.85	17.35
2440	20' eave height		560	.100		7.15	5.70	3.03	15.88	20
2460	101' to 120' wide, 10' eave height		950	.059		5.35	3.37	1.79	10.51	13.05
2470	14' eave height		770	.073		5.95	4.16	2.20	12.31	15.40
2480	16' eave height		675	.083		6.30	4.74	2.51	13.55	17.05
2490	20' eave height	▼	560	.100	▼	6.75	5.70	3.03	15.48	19.65
2500	Single post 2-span frame, 26 ga. colored roofing and siding									
2600	80' wide, 14' eave height	E-2	740	.076	SF Flr.	5.90	4.32	2.29	12.51	15.70
2700	16' eave height		695	.081		6.30	4.60	2.44	13.34	16.80
2800	20' eave height		625	.090		6.85	5.10	2.71	14.66	18.45
2900	24' eave height		570	.098		7.45	5.60	2.98	16.03	20
3000	100' wide, 14' eave height		835	.067		5.75	3.83	2.03	11.61	14.55
3100	16' eave height		795	.070		5.35	4.02	2.13	11.50	14.50
3200	20' eave height		730	.077		6.50	4.38	2.32	13.20	16.50
3300	24' eave height		670	.084		7.20	4.78	2.53	14.51	18.10
3400	120' wide, 14' eave height		870	.064		7.60	3.68	1.95	13.23	16.20
3500	16' eave height		830	.067		5.95	3.86	2.04	11.85	14.75
3600	20' eave height		765	.073		6.90	4.18	2.22	13.30	16.55
3700	24' eave height	▼	705	.079	▼	7.55	4.54	2.41	14.50	18
3800	Double post 3-span frame, 26 ga. colored roofing and siding									
3900	150' wide, 14' eave height	E-2	925	.061	SF Flr.	4.71	3.46	1.83	10	12.55
4000	16' eave height		890	.063		4.90	3.60	1.91	10.41	13.05
4100	20' eave height		820	.068		5.35	3.90	2.07	11.32	14.15
4200	24' eave height	▼	765	.073	▼	5.90	4.18	2.22	12.30	15.45
4300	Triple post 4-span frame, 26 ga. colored roofing and siding									
4400	160' wide, 14' eave height	E-2	970	.058	SF Flr.	4.58	3.30	1.75	9.63	12.05
4500	16' eave height		930	.060		4.81	3.44	1.82	10.07	12.65
4600	20' eave height		870	.064		4.73	3.68	1.95	10.36	13.05
4700	24' eave height	▼	815	.069		5.35	3.93	2.08	11.36	14.30

13 34 19 – Metal Building Systems

13 34 19.50 Pre-Engineered Steel Buildings

		Crew	Daily Output	Labor-Hours	Unit	Material	2020 Bare Costs Labor	Equipment	Total	Total Incl O&P
4800	200' wide, 14' eave height	E-2	1030	.054	SF Flr.	4.23	3.11	1.65	8.99	11.25
4900	16' eave height	↓	995	.056		4.38	3.22	1.70	9.30	11.65
5000	20' eave height		935	.060		4.82	3.42	1.81	10.05	12.60
5100	24' eave height	↓	885	.063	↓	5.40	3.62	1.92	10.94	13.65
7635	Insulation installation, over the purlin, second layer, up to 4" thick, add						90%			
7640	Insulation installation, between the purlins, up to 4" thick, add						100%			

13 34 23 – Fabricated Structures

13 34 23.15 Domes

		Crew	Daily Output	Labor-Hours	Unit	Material	2020 Bare Costs Labor	Equipment	Total	Total Incl O&P
0010	**DOMES**									
1500	Domes, bulk storage, shell only, dual radius hemisphere, arch, steel									
1600	framing, corrugated steel covering, 150' diameter	E-2	550	.102	SF Flr.	34	5.80	3.08	42.88	49.50
1700	400' diameter	"	720	.078		27.50	4.44	2.36	34.30	40
1800	Wood framing, wood decking, to 400' diameter	F-4	400	.120	↓	35	6.45	2.45	43.90	51
1900	Radial framed wood (2" x 6"), 1/2" thick									
2000	plywood, asphalt shingles, 50' diameter	F-3	2000	.020	SF Flr.	69.50	1.09	.24	70.83	78.50
2100	60' diameter		1900	.021		59.50	1.14	.25	60.89	67.50
2200	72' diameter		1800	.022		49.50	1.21	.26	50.97	56.50
2300	116' diameter		1730	.023		33.50	1.26	.27	35.03	39
2400	150' diameter	↓	1500	.027	↓	36	1.45	.31	37.76	42.50

13 34 23.16 Fabricated Control Booths

		Crew	Daily Output	Labor-Hours	Unit	Material	2020 Bare Costs Labor	Equipment	Total	Total Incl O&P
0010	**FABRICATED CONTROL BOOTHS**									
0100	Guard House, prefab conc. w/bullet resistant doors & windows, roof & wiring									
0110	8' x 8', Level III	L-10	1	24	Ea.	54,500	1,400	470	56,370	62,500
0120	8' x 8', Level IV	"	1	24	"	72,000	1,400	470	73,870	81,500

13 34 23.25 Garage Costs

		Crew	Daily Output	Labor-Hours	Unit	Material	2020 Bare Costs Labor	Equipment	Total	Total Incl O&P
0010	**GARAGE COSTS**									
0300	Residential, wood, 12' x 20', one car prefab shell, stock, economy	2 Carp	1	16	Total	6,350	850		7,200	8,275
0350	Custom		.67	23.881		7,125	1,275		8,400	9,725
0400	Two car, 24' x 20', economy		.67	23.881		12,900	1,275		14,175	16,100
0450	Custom	↓	.50	32	↓	14,100	1,700		15,800	18,100

13 34 23.30 Garden House

		Crew	Daily Output	Labor-Hours	Unit	Material	2020 Bare Costs Labor	Equipment	Total	Total Incl O&P
0010	**GARDEN HOUSE** Prefab wood, no floors or foundations									
0100	6' x 6'	2 Carp	200	.080	SF Flr.	52.50	4.25		56.75	64.50
0300	8' x 12'	"	48	.333	"	30	17.70		47.70	59.50

13 34 23.45 Kiosks

		Crew	Daily Output	Labor-Hours	Unit	Material	2020 Bare Costs Labor	Equipment	Total	Total Incl O&P
0010	**KIOSKS**									
0020	Round, advertising type, 5' diameter, 7' high, aluminum wall, illuminated				Ea.	23,500			23,500	25,900
0100	Aluminum wall, non-illuminated					22,500			22,500	24,800
0500	Rectangular, 5' x 9', 7'-6" high, aluminum wall, illuminated					25,500			25,500	28,100
0600	Aluminum wall, non-illuminated				↓	23,800			23,800	26,200

13 34 23.60 Portable Booths

		Crew	Daily Output	Labor-Hours	Unit	Material	2020 Bare Costs Labor	Equipment	Total	Total Incl O&P
0010	**PORTABLE BOOTHS** Prefab. aluminum with doors, windows, ext. roof									
0100	lights wiring & insulation, 15 S.F. building, OD, painted				S.F.	282			282	310
0300	30 S.F. building					259			259	285
0400	50 S.F. building					189			189	208
0600	80 S.F. building					161			161	177
0700	100 S.F. building				↓	147			147	162
0900	Acoustical booth, 27 Db @ 1,000 Hz, 15 S.F. floor				Ea.	4,250			4,250	4,700
1000	7' x 7'-6", including light & ventilation					8,775			8,775	9,650
1200	Ticket booth, galv. steel, not incl. foundations., 4' x 4'					5,275			5,275	5,800
1300	4' x 6'				↓	8,200			8,200	9,025

13 34 Fabricated Engineered Structures

13 34 23 – Fabricated Structures

13 34 23.70 Shelters	Crew	Daily Output	Labor-Hours	Unit	Material	2020 Bare Costs Labor	Equipment	Total	Total Incl O&P
0010 **SHELTERS**									
0020　Aluminum frame, acrylic glazing, 3' x 9' x 8' high	2 Sswk	1.14	14.035	Ea.	3,450	810		4,260	5,075
0100　　9' x 12' x 8' high	"	.73	21.918	"	7,575	1,275		8,850	10,300

13 34 43 – Aircraft Hangars

13 34 43.50 Hangars	Crew	Daily Output	Labor-Hours	Unit	Material	2020 Bare Costs Labor	Equipment	Total	Total Incl O&P
0010 **HANGARS** Prefabricated steel T hangars, galv. steel roof &									
0100　walls, incl. electric bi-folding doors									
0110　　not including floors or foundations, 4 unit	E-2	1275	.044	SF Flr.	14.85	2.51	1.33	18.69	21.50
0130　　　8 unit	"	1063	.053	"	13.05	3.01	1.60	17.66	21
1200　Alternate pricing method:									
1300　　Galv. roof and walls, electric bi-folding doors, 4 plane	E-2	1.06	52.830	Plane	19,600	3,025	1,600	24,225	27,900
1500　　　8 plane		.91	61.538		13,900	3,525	1,875	19,300	22,800
1600　　With bottom rolling doors, 4 plane		1.25	44.800		19,200	2,550	1,350	23,100	26,600
1800　　　8 plane		.97	57.732		14,300	3,300	1,750	19,350	22,800

13 34 53 – Agricultural Structures

13 34 53.50 Silos	Crew	Daily Output	Labor-Hours	Unit	Material	2020 Bare Costs Labor	Equipment	Total	Total Incl O&P
0010 **SILOS**									
0500　Steel, factory fab., 30,000 gallon cap., painted, economy	L-5	1	56	Ea.	26,300	3,250	580	30,130	34,600
0700　　Deluxe	"	.50	112	"	41,800	6,525	1,150	49,475	57,500

13 47 Facility Protection

13 47 13 – Cathodic Protection

13 47 13.16 Cathodic Prot. for Underground Storage Tanks	Crew	Daily Output	Labor-Hours	Unit	Material	2020 Bare Costs Labor	Equipment	Total	Total Incl O&P
0010 **CATHODIC PROTECTION FOR UNDERGROUND STORAGE TANKS**									
1000　Anodes, magnesium type, 9 #	R-15	18.50	2.595	Ea.	38.50	156	15.05	209.55	291
1010　　17 #		13	3.692		81.50	222	21.50	325	445
1020　　32 #		10	4.800		123	288	28	439	595
1030　　48 #		7.20	6.667		160	400	38.50	598.50	815
1100　Graphite type w/epoxy cap, 3" x 60" (32 #)	R-22	8.40	4.438		148	249		397	535
1110　　4" x 80" (68 #)		6	6.213		246	350		596	790
1120　　6" x 72" (80 #)		5.20	7.169		1,525	405		1,930	2,275
1130　　6" x 36" (45 #)		9.60	3.883		770	218		988	1,175
2000　Rectifiers, silicon type, air cooled, 28 V/10 A	R-19	3.50	5.714		2,325	350		2,675	3,075
2010　　20 V/20 A		3.50	5.714		2,325	350		2,675	3,100
2100　　Oil immersed, 28 V/10 A		3	6.667		2,875	410		3,285	3,775
2110　　20 V/20 A		3	6.667		3,250	410		3,660	4,175
3000　Anode backfill, coke breeze	R-22	3850	.010	Lb.	.30	.54		.84	1.14
4000　Cable, HMWPE, No. 8		2.40	15.533	M.L.F.	605	870		1,475	1,975
4010　　No. 6		2.40	15.533		820	870		1,690	2,200
4020　　No. 4		2.40	15.533		1,225	870		2,095	2,650
4030　　No. 2		2.40	15.533		1,900	870		2,770	3,400
4040　　No. 1		2.20	16.945		2,575	950		3,525	4,275
4050　　No. 1/0		2.20	16.945		3,350	950		4,300	5,100
4060　　No. 2/0		2.20	16.945		5,250	950		6,200	7,200
4070　　No. 4/0		2	18.640		7,225	1,050		8,275	9,500
5000　Test station, 7 terminal box, flush curb type w/lockable cover	R-19	12	1.667	Ea.	79.50	102		181.50	241
5010　　Reference cell, 2" diam. PVC conduit, cplg., plug, set flush	"	4.80	4.167	"	158	256		414	555

13 53 Meteorological Instrumentation

13 53 09 – Weather Instrumentation

13 53 09.50 Weather Station	Crew	Daily Output	Labor-Hours	Unit	Material	2020 Bare Costs Labor	Equipment	Total	Total Incl O&P
0010 **WEATHER STATION**									
0020 Remote recording, solar powered, with rain gauge & display, 400' range				Ea.	850			850	935
0100 1 mile range				"	1,950			1,950	2,150

Estimating Tips
22 10 00 Plumbing Piping and Pumps

This subdivision is primarily basic pipe and related materials. The pipe may be used by any of the mechanical disciplines, i.e., plumbing, fire protection, heating, and air conditioning.

Note: CPVC plastic piping approved for fire protection is located in 21 11 13.

- The labor adjustment factors listed in Subdivision 22 01 02.20 apply throughout Divisions 21, 22, and 23. CAUTION: the correct percentage may vary for the same items. For example, the percentage add for the basic pipe installation should be based on the maximum height that the installer must install for that particular section. If the pipe is to be located 14' above the floor but it is suspended on threaded rod from beams, the bottom flange of which is 18' high (4' rods), then the height is actually 18' and the add is 20%. The pipe cover, however, does not have to go above the 14' and so the add should be 10%.

- Most pipe is priced first as straight pipe with a joint (coupling, weld, etc.) every 10' and a hanger usually every 10'. There are exceptions with hanger spacing such as for cast iron pipe (5')

and plastic pipe (3 per 10'). Following each type of pipe there are several lines listing sizes and the amount to be subtracted to delete couplings and hangers. This is for pipe that is to be buried or supported together on trapeze hangers. The reason that the couplings are deleted is that these runs are usually long, and frequently longer lengths of pipe are used. By deleting the couplings, the estimator is expected to look up and add back the correct reduced number of couplings.

- When preparing an estimate, it may be necessary to approximate the fittings. Fittings usually run between 25% and 50% of the cost of the pipe. The lower percentage is for simpler runs, and the higher number is for complex areas, such as mechanical rooms.

- For historic restoration projects, the systems must be as invisible as possible, and pathways must be sought for pipes, conduit, and ductwork. While installations in accessible spaces (such as basements and attics) are relatively straightforward to estimate, labor costs may be more difficult to determine when delivery systems must be concealed.

22 40 00 Plumbing Fixtures

- Plumbing fixture costs usually require two lines: the fixture itself and its "rough-in, supply, and waste."

- In the Assemblies Section (Plumbing D2010) for the desired fixture, the System Components Group at the center of the page shows the fixture on the first line. The rest of the list (fittings, pipe, tubing, etc.) will total up to what we refer to in the Unit Price section as "Rough-in, supply, waste, and vent." Note that for most fixtures we allow a nominal 5' of tubing to reach from the fixture to a main or riser.

- Remember that gas- and oil-fired units need venting.

Reference Numbers

Reference numbers are shown at the beginning of some major classifications. These numbers refer to related items in the Reference Section. The reference information may be an estimating procedure, an alternate pricing method, or technical information.

Note: Not all subdivisions listed here necessarily appear. ∎

Same Data. Simplified.

Enjoy the convenience and efficiency of accessing your costs anywhere:

- **Skip the multiplier** by setting your location
- **Quickly search,** edit, favorite and share costs
- **Stay on top of price changes** with automatic updates

Discover more at rsmeans.com/online

---

22 01 02.20 Labor Adjustment Factors	Crew	Daily Output	Labor-Hours	Unit	Material	2020 Bare Costs Labor	Equipment	Total	Total Incl O&P
0010 LABOR ADJUSTMENT FACTORS (For Div. 21, 22 and 23) R220102-20									
0100 Labor factors: The below are reasonable suggestions, but									
0110 each project must be evaluated for its own peculiarities, and									
0120 the adjustments be increased or decreased depending on the									
0130 severity of the special conditions.									
1000 Add to labor for elevated installation (Above floor level)									
1080 10' to 14.5' high						10%			
1100 15' to 19.5' high						20%			
1120 20' to 24.5' high						25%			
1140 25' to 29.5' high						35%			
1160 30' to 34.5' high						40%			
1180 35' to 39.5' high						50%			
1200 40' and higher						55%			
2000 Add to labor for crawl space									
2100 3' high						40%			
2140 4' high						30%			
3000 Add to labor for multi-story building									
3010 For new construction (No elevator available)									
3100 Add for floors 3 thru 10						5%			
3110 Add for floors 11 thru 15						10%			
3120 Add for floors 16 thru 20						15%			
3130 Add for floors 21 thru 30						20%			
3140 Add for floors 31 and up						30%			
3170 For existing structure (Elevator available)									
3180 Add for work on floor 3 and above						2%			
4000 Add to labor for working in existing occupied buildings									
4100 Hospital						35%			
4140 Office building						25%			
4180 School						20%			
4220 Factory or warehouse						15%			
4260 Multi dwelling						15%			
5000 Add to labor, miscellaneous									
5100 Cramped shaft						35%			
5140 Congested area						15%			
5180 Excessive heat or cold						30%			
9000 Labor factors: The above are reasonable suggestions, but									
9010 each project should be evaluated for its own peculiarities.									
9100 Other factors to be considered are:									
9140 Movement of material and equipment through finished areas									
9180 Equipment room									
9220 Attic space									
9260 No service road									
9300 Poor unloading/storage area									
9340 Congested site area/heavy traffic									

22 05 Common Work Results for Plumbing

22 05 05 – Selective Demolition for Plumbing

22 05 05.10 Plumbing Demolition

		Crew	Daily Output	Labor-Hours	Unit	Material	2020 Bare Costs Labor	Equipment	Total	Total Incl O&P
0010	**PLUMBING DEMOLITION**									
1020	Fixtures, including 10' piping									
1100	Bathtubs, cast iron	1 Plum	4	2	Ea.		129		129	193
1120	Fiberglass		6	1.333			86		86	129
1140	Steel		5	1.600			103		103	154
1200	Lavatory, wall hung		10	.800			51.50		51.50	77
1220	Counter top		8	1			64.50		64.50	96.50
1300	Sink, single compartment		8	1			64.50		64.50	96.50
1320	Double compartment		7	1.143			73.50		73.50	110
1400	Water closet, floor mounted		8	1			64.50		64.50	96.50
1420	Wall mounted		7	1.143			73.50		73.50	110
1500	Urinal, floor mounted		4	2			129		129	193
1520	Wall mounted		7	1.143			73.50		73.50	110
1600	Water fountains, free standing		8	1			64.50		64.50	96.50
1620	Wall or deck mounted		6	1.333			86		86	129
2000	Piping, metal, up thru 1-1/2" diameter		200	.040	L.F.		2.58		2.58	3.86
2050	2" thru 3-1/2" diameter		150	.053			3.44		3.44	5.15
2100	4" thru 6" diameter	2 Plum	100	.160			10.30		10.30	15.45
2150	8" thru 14" diameter	"	60	.267			17.20		17.20	25.50
2153	16" thru 20" diameter	Q-18	70	.343			21	1.52	22.52	33
2155	24" thru 26" diameter		55	.436			26.50	1.93	28.43	42
2156	30" thru 36" diameter		40	.600			36.50	2.66	39.16	58
2160	Plastic pipe with fittings, up thru 1-1/2" diameter	1 Plum	250	.032			2.06		2.06	3.09
2162	2" thru 3" diameter	"	200	.040			2.58		2.58	3.86
2164	4" thru 6" diameter	Q-1	200	.080			4.64		4.64	6.95
2166	8" thru 14" diameter		150	.107			6.20		6.20	9.25
2168	16" diameter		100	.160			9.30		9.30	13.90
2212	Deduct for salvage, aluminum scrap				Ton				635	695
2214	Brass scrap								2,700	2,975
2216	Copper scrap								4,475	4,925
2218	Lead scrap								850	935
2220	Steel scrap								204	224
2250	Water heater, 40 gal.	1 Plum	6	1.333	Ea.		86		86	129

22 05 23 – General-Duty Valves for Plumbing Piping

22 05 23.20 Valves, Bronze

		Crew	Daily Output	Labor-Hours	Unit	Material	2020 Bare Costs Labor	Equipment	Total	Total Incl O&P
0010	**VALVES, BRONZE** R220523-80									
1020	Angle, 150 lb., rising stem, threaded									
1030	1/8"	1 Plum	24	.333	Ea.	142	21.50		163.50	188
1040	1/4"		24	.333		162	21.50		183.50	210
1050	3/8"		24	.333		165	21.50		186.50	213
1060	1/2"		22	.364		169	23.50		192.50	221
1070	3/4"		20	.400		230	26		256	291
1080	1"		19	.421		315	27		342	385
1100	1-1/2"		13	.615		525	39.50		564.50	635
1102	Soldered same price as threaded									
1110	2"	1 Plum	11	.727	Ea.	845	47		892	1,000
1750	Check, swing, class 150, regrinding disc, threaded									
1800	1/8"	1 Plum	24	.333	Ea.	82.50	21.50		104	123
1830	1/4"		24	.333		76.50	21.50		98	117
1840	3/8"		24	.333		71	21.50		92.50	111
1850	1/2"		24	.333		77	21.50		98.50	117
1860	3/4"		20	.400		94	26		120	142

213

22 05 Common Work Results for Plumbing

22 05 23 – General-Duty Valves for Plumbing Piping

22 05 23.20 Valves, Bronze		Crew	Daily Output	Labor-Hours	Unit	Material	2020 Bare Costs Labor	Equipment	Total	Total Incl O&P
1870	1"	1 Plum	19	.421	Ea.	178	27		205	237
1880	1-1/4"		15	.533		210	34.50		244.50	283
1890	1-1/2"		13	.615		300	39.50		339.50	390
1900	2"		11	.727		380	47		427	490
1910	2-1/2"	Q-1	15	1.067		925	62		987	1,125
2000	For 200 lb., add					5%	10%			
2040	For 300 lb., add					15%	15%			
2850	Gate, N.R.S., soldered, 125 psi									
2900	3/8"	1 Plum	24	.333	Ea.	75.50	21.50		97	115
2920	1/2"		24	.333		60.50	21.50		82	98.50
2940	3/4"		20	.400		74	26		100	120
2950	1"		19	.421		78.50	27		105.50	127
2960	1-1/4"		15	.533		152	34.50		186.50	219
2970	1-1/2"		13	.615		184	39.50		223.50	263
2980	2"		11	.727		213	47		260	305
2990	2-1/2"	Q-1	15	1.067		515	62		577	660
3000	3"	"	13	1.231		570	71.50		641.50	735
3850	Rising stem, soldered, 300 psi									
3950	1"	1 Plum	19	.421	Ea.	229	27		256	293
3980	2"	"	11	.727		615	47		662	750
4000	3"	Q-1	13	1.231		2,025	71.50		2,096.50	2,350
4250	Threaded, class 150									
4310	1/4"	1 Plum	24	.333	Ea.	90	21.50		111.50	131
4320	3/8"		24	.333		90	21.50		111.50	131
4330	1/2"		24	.333		69.50	21.50		91	109
4340	3/4"		20	.400		81.50	26		107.50	128
4350	1"		19	.421		125	27		152	179
4360	1-1/4"		15	.533		132	34.50		166.50	197
4370	1-1/2"		13	.615		167	39.50		206.50	243
4380	2"		11	.727		224	47		271	315
4390	2-1/2"	Q-1	15	1.067		670	62		732	835
4400	3"	"	13	1.231		935	71.50		1,006.50	1,125
4500	For 300 psi, threaded, add					100%	15%			
4540	For chain operated type, add					15%				
4850	Globe, class 150, rising stem, threaded									
4920	1/4"	1 Plum	24	.333	Ea.	123	21.50		144.50	168
4940	3/8"		24	.333		121	21.50		142.50	166
4950	1/2"		24	.333		106	21.50		127.50	149
4960	3/4"		20	.400		172	26		198	228
4970	1"		19	.421		239	27		266	305
4980	1-1/4"		15	.533		292	34.50		326.50	370
4990	1-1/2"		13	.615		460	39.50		499.50	565
5000	2"		11	.727		690	47		737	830
5010	2-1/2"	Q-1	15	1.067		1,475	62		1,537	1,725
5020	3"	"	13	1.231		2,125	71.50		2,196.50	2,425
5120	For 300 lb. threaded, add					50%	15%			
5600	Relief, pressure & temperature, self-closing, ASME, threaded									
5640	3/4"	1 Plum	28	.286	Ea.	257	18.40		275.40	310
5650	1"		24	.333		425	21.50		446.50	495
5660	1-1/4"		20	.400		805	26		831	925
5670	1-1/2"		18	.444		1,550	28.50		1,578.50	1,750
5680	2"		16	.500		1,675	32		1,707	1,900
5950	Pressure, poppet type, threaded									

22 05 23 – General-Duty Valves for Plumbing Piping

22 05 23.20 Valves, Bronze		Crew	Daily Output	Labor-Hours	Unit	Material	2020 Bare Costs Labor	Equipment	Total	Total Incl O&P
6000	1/2"	1 Plum	30	.267	Ea.	90	17.20		107.20	125
6040	3/4"	"	28	.286	"	104	18.40		122.40	143
6400	Pressure, water, ASME, threaded									
6440	3/4"	1 Plum	28	.286	Ea.	119	18.40		137.40	159
6450	1"		24	.333		340	21.50		361.50	405
6460	1-1/4"		20	.400		485	26		511	575
6470	1-1/2"		18	.444		750	28.50		778.50	865
6480	2"		16	.500		1,075	32		1,107	1,250
6490	2-1/2"		15	.533		4,400	34.50		4,434.50	4,900
6900	Reducing, water pressure									
6920	300 psi to 25-75 psi, threaded or sweat									
6940	1/2"	1 Plum	24	.333	Ea.	470	21.50		491.50	550
6950	3/4"		20	.400		550	26		576	645
6960	1"		19	.421		850	27		877	975
6970	1-1/4"		15	.533		1,475	34.50		1,509.50	1,675
6980	1-1/2"		13	.615		2,225	39.50		2,264.50	2,500
8350	Tempering, water, sweat connections									
8400	1/2"	1 Plum	24	.333	Ea.	118	21.50		139.50	162
8440	3/4"	"	20	.400	"	167	26		193	222
8650	Threaded connections									
8700	1/2"	1 Plum	24	.333	Ea.	162	21.50		183.50	210
8740	3/4"		20	.400		1,050	26		1,076	1,225
8750	1"		19	.421		1,200	27		1,227	1,350
8760	1-1/4"		15	.533		1,975	34.50		2,009.50	2,225
8770	1-1/2"		13	.615		2,025	39.50		2,064.50	2,275
8780	2"		11	.727		3,000	47		3,047	3,375

22 05 23.60 Valves, Plastic		Crew	Daily Output	Labor-Hours	Unit	Material	2020 Bare Costs Labor	Equipment	Total	Total Incl O&P
0010	**VALVES, PLASTIC**									
1150	Ball, PVC, socket or threaded, true union									
1230	1/2"	1 Plum	26	.308	Ea.	39	19.85		58.85	72.50
1240	3/4"		25	.320		50	20.50		70.50	86
1250	1"		23	.348		55.50	22.50		78	94.50
1260	1-1/4"		21	.381		94	24.50		118.50	140
1270	1-1/2"		20	.400		86	26		112	134
1280	2"		17	.471		136	30.50		166.50	195
1290	2-1/2"	Q-1	26	.615		236	35.50		271.50	315
1300	3"		24	.667		272	38.50		310.50	355
1310	4"		20	.800		455	46.50		501.50	575
1360	For PVC, flanged, add					100%	15%			
3150	Ball check, PVC, socket or threaded									
3200	1/4"	1 Plum	26	.308	Ea.	48	19.85		67.85	82.50
3220	3/8"		26	.308		48	19.85		67.85	82.50
3240	1/2"		26	.308		46.50	19.85		66.35	80.50
3250	3/4"		25	.320		52	20.50		72.50	88
3260	1"		23	.348		65	22.50		87.50	105
3270	1-1/4"		21	.381		124	24.50		148.50	173
3280	1-1/2"		20	.400		109	26		135	159
3290	2"		17	.471		149	30.50		179.50	209
3310	3"	Q-1	24	.667		415	38.50		453.50	515
3320	4"	"	20	.800		585	46.50		631.50	710
3360	For PVC, flanged, add					50%	15%			

22 05 76.10 Cleanouts		Crew	Daily Output	Labor-Hours	Unit	Material	2020 Bare Costs Labor	Equipment	Total	Total Incl O&P
0010	**CLEANOUTS**									
0060	Floor type									
0080	Round or square, scoriated nickel bronze top									
0100	2" pipe size	1 Plum	10	.800	Ea.	183	51.50		234.50	279
0120	3" pipe size		8	1		264	64.50		328.50	385
0140	4" pipe size		6	1.333		284	86		370	440

22 05 76.20 Cleanout Tees

22 05 76.20 Cleanout Tees		Crew	Daily Output	Labor-Hours	Unit	Material	2020 Bare Costs Labor	Equipment	Total	Total Incl O&P
0010	**CLEANOUT TEES**									
0100	Cast iron, B&S, with countersunk plug									
0200	2" pipe size	1 Plum	4	2	Ea.	143	129		272	350
0220	3" pipe size		3.60	2.222		181	143		324	415
0240	4" pipe size		3.30	2.424		275	156		431	535
0260	5" pipe size	Q-1	5.50	2.909		545	169		714	855
0280	6" pipe size	"	5	3.200		790	186		976	1,150
0300	8" pipe size	Q-3	5	6.400		1,100	395		1,495	1,825
0500	For round smooth access cover, same price									
0600	For round scoriated access cover, same price									
0700	For square smooth access cover, add				Ea.	60%				
4000	Plastic, tees and adapters. Add plugs									
4010	ABS, DWV									
4020	Cleanout tee, 1-1/2" pipe size	1 Plum	15	.533	Ea.	14.15	34.50		48.65	67
4030	2" pipe size	Q-1	27	.593		12.65	34.50		47.15	65.50
4040	3" pipe size		21	.762		25.50	44		69.50	94
4050	4" pipe size		16	1		64.50	58		122.50	158
4100	Cleanout plug, 1-1/2" pipe size	1 Plum	32	.250		2.72	16.10		18.82	27
4110	2" pipe size	Q-1	56	.286		3.28	16.55		19.83	28.50
4120	3" pipe size		36	.444		5.35	26		31.35	44.50
4130	4" pipe size		30	.533		9.15	31		40.15	56.50
4180	Cleanout adapter fitting, 1-1/2" pipe size	1 Plum	32	.250		4.18	16.10		20.28	28.50
4190	2" pipe size	Q-1	56	.286		6.10	16.55		22.65	31.50
4200	3" pipe size		36	.444		15.60	26		41.60	55.50
4210	4" pipe size		30	.533		29	31		60	78.50
5000	PVC, DWV									
5010	Cleanout tee, 1-1/2" pipe size	1 Plum	15	.533	Ea.	11.90	34.50		46.40	64.50
5020	2" pipe size	Q-1	27	.593		9.15	34.50		43.65	61.50
5030	3" pipe size		21	.762		27	44		71	96
5040	4" pipe size		16	1		36.50	58		94.50	128
5090	Cleanout plug, 1-1/2" pipe size	1 Plum	32	.250		1.87	16.10		17.97	26
5100	2" pipe size	Q-1	56	.286		2.09	16.55		18.64	27.50
5110	3" pipe size		36	.444		3.73	26		29.73	42.50
5120	4" pipe size		30	.533		5.50	31		36.50	52.50
5130	6" pipe size		24	.667		17.75	38.50		56.25	77.50
5170	Cleanout adapter fitting, 1-1/2" pipe size	1 Plum	32	.250		2.51	16.10		18.61	27
5180	2" pipe size	Q-1	56	.286		3.08	16.55		19.63	28.50
5190	3" pipe size		36	.444		8.45	26		34.45	48
5200	4" pipe size		30	.533		13.90	31		44.90	62
5210	6" pipe size		24	.667		40	38.50		78.50	102

22 11 Facility Water Distribution

22 11 13 – Facility Water Distribution Piping

22 11 13.23 Pipe/Tube, Copper

		Crew	Daily Output	Labor-Hours	Unit	Material	2020 Bare Costs Labor	Equipment	Total	Total Incl O&P
0010	**PIPE/TUBE, COPPER**, Solder joints									
1000	Type K tubing, couplings & clevis hanger assemblies 10' OC									
1100	1/4" diameter	1 Plum	84	.095	L.F.	4.25	6.15		10.40	13.85
1120	3/8" diameter		82	.098		4.76	6.30		11.06	14.65
1140	1/2" diameter		78	.103		5.30	6.60		11.90	15.70
1160	5/8" diameter		77	.104		6.75	6.70		13.45	17.40
1180	3/4" diameter		74	.108		8.65	6.95		15.60	19.95
1200	1" diameter		66	.121		12.80	7.80		20.60	26
1220	1-1/4" diameter		56	.143		15.30	9.20		24.50	30.50
1240	1-1/2" diameter		50	.160		18.60	10.30		28.90	36
1260	2" diameter		40	.200		27	12.90		39.90	49.50
1280	2-1/2" diameter	Q-1	60	.267		44	15.45		59.45	71.50
1300	3" diameter	"	54	.296		58	17.20		75.20	89
1950	To delete cplgs. & hngrs., 1/4"-1" pipe, subtract					27%	60%			
1960	1-1/4"-3" pipe, subtract					14%	52%			

22 11 13.25 Pipe/Tube Fittings, Copper

		Crew	Daily Output	Labor-Hours	Unit	Material	2020 Bare Costs Labor	Equipment	Total	Total Incl O&P
0010	**PIPE/TUBE FITTINGS, COPPER**, Wrought unless otherwise noted									
0040	Solder joints, copper x copper									
0070	90° elbow, 1/4"	1 Plum	22	.364	Ea.	3.88	23.50		27.38	39.50
0090	3/8"		22	.364		4.15	23.50		27.65	39.50
0100	1/2"		20	.400		1.26	26		27.26	40
0110	5/8"		19	.421		2.77	27		29.77	43.50
0120	3/4"		19	.421		2.68	27		29.68	43.50
0130	1"		16	.500		7.05	32		39.05	56
0140	1-1/4"		15	.533		11.65	34.50		46.15	64.50
0150	1-1/2"		13	.615		16.10	39.50		55.60	77
0160	2"		11	.727		29.50	47		76.50	103
0180	3"	Q-1	11	1.455		86.50	84.50		171	221
0200	4"	"	9	1.778		236	103		339	415
0220	6"	Q-2	9	2.667		1,025	160		1,185	1,375
0250	45° elbow, 1/4"	1 Plum	22	.364		7.75	23.50		31.25	43.50
0270	3/8"		22	.364		6.80	23.50		30.30	42.50
0280	1/2"		20	.400		2.75	26		28.75	41.50
0290	5/8"		19	.421		11.45	27		38.45	53
0300	3/4"		19	.421		4.46	27		31.46	45.50
0310	1"		16	.500		11.20	32		43.20	60.50
0320	1-1/4"		15	.533		16	34.50		50.50	69
0340	2"		11	.727		30.50	47		77.50	104
0450	Tee, 1/4"		14	.571		8.45	37		45.45	64.50
0470	3/8"		14	.571		6.85	37		43.85	62.50
0480	1/2"		13	.615		2.44	39.50		41.94	62
0490	5/8"		12	.667		16.25	43		59.25	82.50
0500	3/4"		12	.667		6.05	43		49.05	71
0510	1"		10	.800		17.05	51.50		68.55	96
0520	1-1/4"		9	.889		24.50	57.50		82	113
0530	1-1/2"		8	1		36.50	64.50		101	137
0540	2"		7	1.143		58	73.50		131.50	174
0560	3"	Q-1	7	2.286		158	133		291	370
0580	4"	"	5	3.200		365	186		551	680
0600	6"	Q-2	6	4		1,475	241		1,716	1,975
0612	Tee, reducing on the outlet, 1/4"	1 Plum	15	.533		17.05	34.50		51.55	70.50
0613	3/8"		15	.533		15.55	34.50		50.05	68.50

22 11 13 – Facility Water Distribution Piping

22 11 13.25 Pipe/Tube Fittings, Copper

		Crew	Daily Output	Labor-Hours	Unit	Material	2020 Bare Costs Labor	Equipment	Total	Total Incl O&P
0614	1/2"	1 Plum	14	.571	Ea.	15.60	37		52.60	72
0615	5/8"		13	.615		30.50	39.50		70	93
0616	3/4"		12	.667		8	43		51	73.50
0617	1"		11	.727		29	47		76	102
0618	1-1/4"		10	.800		32.50	51.50		84	113
0619	1-1/2"		9	.889		32	57.50		89.50	121
0620	2"		8	1		64.50	64.50		129	168
0621	2-1/2"	Q-1	9	1.778		137	103		240	305
0622	3"		8	2		163	116		279	355
0623	4"		6	2.667		299	155		454	560
0624	5"		5	3.200		1,725	186		1,911	2,175
0625	6"	Q-2	7	3.429		2,150	206		2,356	2,675
0626	8"	"	6	4		10,800	241		11,041	12,300
0630	Tee, reducing on the run, 1/4"	1 Plum	15	.533		23.50	34.50		58	77.50
0631	3/8"		15	.533		31	34.50		65.50	85.50
0632	1/2"		14	.571		20.50	37		57.50	77.50
0633	5/8"		13	.615		29	39.50		68.50	91.50
0634	3/4"		12	.667		21	43		64	87.50
0635	1"		11	.727		26	47		73	98.50
0636	1-1/4"		10	.800		41	51.50		92.50	122
0637	1-1/2"		9	.889		72	57.50		129.50	165
0638	2"		8	1		82.50	64.50		147	188
0639	2-1/2"	Q-1	9	1.778		177	103		280	350
0640	3"		8	2		320	116		436	530
0641	4"		6	2.667		585	155		740	875
0642	5"		5	3.200		1,950	186		2,136	2,425
0643	6"	Q-2	7	3.429		2,950	206		3,156	3,550
0644	8"	"	6	4		10,200	241		10,441	11,600
0650	Coupling, 1/4"	1 Plum	24	.333		1.07	21.50		22.57	33
0670	3/8"		24	.333		1.53	21.50		23.03	33.50
0680	1/2"		22	.364		.96	23.50		24.46	36
0690	5/8"		21	.381		4.23	24.50		28.73	41
0700	3/4"		21	.381		2.71	24.50		27.21	39.50
0710	1"		18	.444		5.30	28.50		33.80	49
0715	1-1/4"		17	.471		7.80	30.50		38.30	54
0716	1-1/2"		15	.533		10.80	34.50		45.30	63.50
0718	2"		13	.615		15.30	39.50		54.80	76.50

22 11 13.44 Pipe, Steel

		Crew	Daily Output	Labor-Hours	Unit	Material	2020 Bare Costs Labor	Equipment	Total	Total Incl O&P
0010	**PIPE, STEEL**		R221113-50							
0020	All pipe sizes are to Spec. A-53 unless noted otherwise									
0050	Schedule 40, threaded, with couplings, and clevis hanger									
0060	assemblies sized for covering, 10' OC									
0540	Black, 1/4" diameter	1 Plum	66	.121	L.F.	6.15	7.80		13.95	18.50
0550	3/8" diameter		65	.123		6.85	7.95		14.80	19.40
0560	1/2" diameter		63	.127		4.03	8.20		12.23	16.70
0570	3/4" diameter		61	.131		4.40	8.45		12.85	17.50
0580	1" diameter		53	.151		7.80	9.75		17.55	23
0590	1-1/4" diameter	Q-1	89	.180		8.20	10.45		18.65	24.50
0600	1-1/2" diameter		80	.200		8.75	11.60		20.35	27
0610	2" diameter		64	.250		6.90	14.50		21.40	29
0620	2-1/2" diameter		50	.320		11.40	18.55		29.95	40.50
0630	3" diameter		43	.372		14.35	21.50		35.85	48.50

22 11 13 – Facility Water Distribution Piping

22 11 13.44 Pipe, Steel

		Crew	Daily Output	Labor-Hours	Unit	Material	2020 Bare Costs Labor	Equipment	Total	Total Incl O&P
0640	3-1/2" diameter	Q-1	40	.400	L.F.	21	23		44	58
0650	4" diameter	↓	36	.444	↓	23	26		49	64
1220	To delete coupling & hanger, subtract									
1230	1/4" diam. to 3/4" diam.					31%	56%			
1240	1" diam. to 1-1/2" diam.					23%	51%			
1250	2" diam. to 4" diam.					23%	41%			
1280	All pipe sizes are to Spec. A-53 unless noted otherwise									
1281	Schedule 40, threaded, with couplings and clevis hanger									
1282	assemblies sized for covering, 10' OC									
1290	Galvanized, 1/4" diameter	1 Plum	66	.121	L.F.	8.10	7.80		15.90	20.50
1300	3/8" diameter		65	.123		8.35	7.95		16.30	21
1310	1/2" diameter		63	.127		4.10	8.20		12.30	16.75
1320	3/4" diameter		61	.131		5	8.45		13.45	18.15
1330	1" diameter	↓	53	.151		7.90	9.75		17.65	23.50
1340	1-1/4" diameter	Q-1	89	.180		8.85	10.45		19.30	25.50
1350	1-1/2" diameter		80	.200		9.45	11.60		21.05	28
1360	2" diameter		64	.250		7.90	14.50		22.40	30
1370	2-1/2" diameter		50	.320		13.15	18.55		31.70	42.50
1380	3" diameter		43	.372		16	21.50		37.50	50
1390	3-1/2" diameter		40	.400		23	23		46	59.50
1400	4" diameter	↓	36	.444	↓	25.50	26		51.50	66.50
1750	To delete coupling & hanger, subtract									
1760	1/4" diam. to 3/4" diam.					31%	56%			
1770	1" diam. to 1-1/2" diam.					23%	51%			
1780	2" diam. to 4" diam.					23%	41%			

22 11 13.45 Pipe Fittings, Steel, Threaded

		Crew	Daily Output	Labor-Hours	Unit	Material	2020 Bare Costs Labor	Equipment	Total	Total Incl O&P
0010	**PIPE FITTINGS, STEEL, THREADED**									
0020	Cast iron									
1300	Extra heavy weight, black									
1310	Couplings, steel straight									
1320	1/4"	1 Plum	19	.421	Ea.	5.55	27		32.55	46.50
1330	3/8"		19	.421		6.05	27		33.05	47
1340	1/2"		19	.421		8.20	27		35.20	49.50
1350	3/4"		18	.444		8.75	28.50		37.25	52.50
1360	1"	↓	15	.533		11.15	34.50		45.65	64
1370	1-1/4"	Q-1	26	.615		17.85	35.50		53.35	73
1380	1-1/2"		24	.667		17.85	38.50		56.35	77.50
1390	2"		21	.762		27	44		71	96
1400	2-1/2"		18	.889		40.50	51.50		92	122
1410	3"		14	1.143		48	66.50		114.50	152
1420	3-1/2"		12	1.333		64.50	77.50		142	187
1430	4"	↓	10	1.600	↓	76	93		169	223
1510	90° elbow, straight									
1520	1/2"	1 Plum	15	.533	Ea.	39	34.50		73.50	94.50
1530	3/4"		14	.571		39.50	37		76.50	98.50
1540	1"	↓	13	.615		48	39.50		87.50	113
1550	1-1/4"	Q-1	22	.727		71.50	42		113.50	142
1560	1-1/2"		20	.800		88.50	46.50		135	167
1580	2"		18	.889		109	51.50		160.50	197
1590	2-1/2"		14	1.143		266	66.50		332.50	390
1600	3"		10	1.600		360	93		453	535
1610	4"	↓	6	2.667		785	155		940	1,100

22 11 13.45 Pipe Fittings, Steel, Threaded		Crew	Daily Output	Labor-Hours	Unit	Material	2020 Bare Costs Labor	Equipment	Total	Total Incl O&P
1650	45° elbow, straight									
1660	1/2"	1 Plum	15	.533	Ea.	55.50	34.50		90	113
1670	3/4"		14	.571		53.50	37		90.50	114
1680	1"		13	.615		64.50	39.50		104	131
1690	1-1/4"	Q-1	22	.727		106	42		148	180
1700	1-1/2"		20	.800		117	46.50		163.50	199
1710	2"		18	.889		166	51.50		217.50	260
1720	2-1/2"		14	1.143		287	66.50		353.50	415
1800	Tee, straight									
1810	1/2"	1 Plum	9	.889	Ea.	61	57.50		118.50	153
1820	3/4"		9	.889		61	57.50		118.50	153
1830	1"		8	1		73.50	64.50		138	178
1840	1-1/4"	Q-1	14	1.143		110	66.50		176.50	220
1850	1-1/2"		13	1.231		141	71.50		212.50	262
1860	2"		11	1.455		175	84.50		259.50	320
1870	2-1/2"		9	1.778		370	103		473	560
1880	3"		6	2.667		525	155		680	805
1890	4"		4	4		1,025	232		1,257	1,475
4000	Standard weight, black									
4010	Couplings, steel straight, merchants									
4030	1/4"	1 Plum	19	.421	Ea.	1.48	27		28.48	42
4040	3/8"		19	.421		1.79	27		28.79	42.50
4050	1/2"		19	.421		1.91	27		28.91	42.50
4060	3/4"		18	.444		2.42	28.50		30.92	45.50
4070	1"		15	.533		3.39	34.50		37.89	55
4080	1-1/4"	Q-1	26	.615		4.32	35.50		39.82	58.50
4090	1-1/2"		24	.667		5.50	38.50		44	64
4100	2"		21	.762		7.85	44		51.85	74.50
4110	2-1/2"		18	.889		25.50	51.50		77	105
4120	3"		14	1.143		36	66.50		102.50	139
4130	3-1/2"		12	1.333		64	77.50		141.50	186
4140	4"		10	1.600		64	93		157	209
4200	Standard weight, galvanized									
4210	Couplings, steel straight, merchants									
4230	1/4"	1 Plum	19	.421	Ea.	1.72	27		28.72	42.50
4240	3/8"		19	.421		2.20	27		29.20	43
4250	1/2"		19	.421		2.33	27		29.33	43
4260	3/4"		18	.444		2.91	28.50		31.41	46
4270	1"		15	.533		4.07	34.50		38.57	56
4280	1-1/4"	Q-1	26	.615		5.20	35.50		40.70	59.50
4290	1-1/2"		24	.667		6.45	38.50		44.95	65
4300	2"		21	.762		9.65	44		53.65	76.50
4310	2-1/2"		18	.889		27	51.50		78.50	107
4320	3"		14	1.143		36	66.50		102.50	139
4330	3-1/2"		12	1.333		73.50	77.50		151	197
4340	4"		10	1.600		73.50	93		166.50	220
5000	Malleable iron, 150 lb.									
5020	Black									
5040	90° elbow, straight									
5060	1/4"	1 Plum	16	.500	Ea.	5.90	32		37.90	54.50
5070	3/8"		16	.500		5.90	32		37.90	54.50
5080	1/2"		15	.533		4	34.50		38.50	56
5090	3/4"		14	.571		4.12	37		41.12	59.50

220

For customer support on your Site Work & Landscape Costs with RSMeans data, call 800.448.8182.

22 11 13.45 Pipe Fittings, Steel, Threaded

		Crew	Daily Output	Labor-Hours	Unit	Material	2020 Bare Costs Labor	Equipment	Total	Total Incl O&P
5100	1"	1 Plum	13	.615	Ea.	8.35	39.50		47.85	68.50
5110	1-1/4"	Q-1	22	.727		13.85	42		55.85	78.50
5120	1-1/2"		20	.800		14.60	46.50		61.10	85.50
5130	2"		18	.889		25.50	51.50		77	105
5140	2-1/2"		14	1.143		71	66.50		137.50	178
5150	3"		10	1.600		104	93		197	254
5160	3-1/2"		8	2		287	116		403	490
5170	4"		6	2.667		196	155		351	445
5250	45° elbow, straight									
5270	1/4"	1 Plum	16	.500	Ea.	8.85	32		40.85	58
5280	3/8"		16	.500		8.85	32		40.85	58
5290	1/2"		15	.533		6.75	34.50		41.25	59
5300	3/4"		14	.571		8.35	37		45.35	64
5310	1"		13	.615		10.50	39.50		50	71
5320	1-1/4"	Q-1	22	.727		18.55	42		60.55	83.50
5330	1-1/2"		20	.800		23	46.50		69.50	95
5340	2"		18	.889		34.50	51.50		86	115
5350	2-1/2"		14	1.143		100	66.50		166.50	209
5360	3"		10	1.600		131	93		224	283
5370	3-1/2"		8	2		261	116		377	460
5380	4"		6	2.667		256	155		411	510
5450	Tee, straight									
5470	1/4"	1 Plum	10	.800	Ea.	8.55	51.50		60.05	86.50
5480	3/8"		10	.800		8.55	51.50		60.05	86.50
5490	1/2"		9	.889		5.45	57.50		62.95	91.50
5500	3/4"		9	.889		7.85	57.50		65.35	94
5510	1"		8	1		13.40	64.50		77.90	111
5520	1-1/4"	Q-1	14	1.143		21.50	66.50		88	123
5530	1-1/2"		13	1.231		27	71.50		98.50	137
5540	2"		11	1.455		46	84.50		130.50	177
5550	2-1/2"		9	1.778		99	103		202	263
5560	3"		6	2.667		146	155		301	390
5570	3-1/2"		5	3.200		340	186		526	655
5580	4"		4	4		355	232		587	735
5650	Coupling									
5670	1/4"	1 Plum	19	.421	Ea.	7.30	27		34.30	48.50
5680	3/8"		19	.421		7.30	27		34.30	48.50
5690	1/2"		19	.421		5.65	27		32.65	46.50
5700	3/4"		18	.444		6.60	28.50		35.10	50.50
5710	1"		15	.533		9.90	34.50		44.40	62.50
5720	1-1/4"	Q-1	26	.615		11.95	35.50		47.45	66.50
5730	1-1/2"		24	.667		15.80	38.50		54.30	75.50
5740	2"		21	.762		23.50	44		67.50	91.50
5750	2-1/2"		18	.889		64.50	51.50		116	148
5760	3"		14	1.143		87.50	66.50		154	195
5770	3-1/2"		12	1.333		161	77.50		238.50	293
5780	4"		10	1.600		176	93		269	330
6000	For galvanized elbows, tees, and couplings, add					20%				

22 11 13.47 Pipe Fittings, Steel

		Crew	Daily Output	Labor-Hours	Unit	Material	2020 Bare Costs Labor	Equipment	Total	Total Incl O&P
0010	**PIPE FITTINGS, STEEL**, flanged, welded & special									
0620	Gasket and bolt set, 150 lb., 1/2" pipe size	1 Plum	20	.400	Ea.	3.83	26		29.83	42.50
0622	3/4" pipe size		19	.421		4.13	27		31.13	45

221

22 11 13.47 Pipe Fittings, Steel

		Crew	Daily Output	Labor-Hours	Unit	Material	2020 Bare Costs Labor	Equipment	Total	Total Incl O&P
0624	1" pipe size	1 Plum	18	.444	Ea.	4.62	28.50		33.12	48
0626	1-1/4" pipe size		17	.471		4.36	30.50		34.86	50.50
0628	1-1/2" pipe size		15	.533		4.38	34.50		38.88	56.50
0630	2" pipe size		13	.615		8.90	39.50		48.40	69.50
0640	2-1/2" pipe size		12	.667		8.45	43		51.45	74
0650	3" pipe size		11	.727		9.10	47		56.10	80
0660	3-1/2" pipe size		9	.889		20	57.50		77.50	108
0670	4" pipe size		8	1		17.90	64.50		82.40	116
0680	5" pipe size		7	1.143		27	73.50		100.50	140
0690	6" pipe size		6	1.333		28.50	86		114.50	160
0700	8" pipe size		5	1.600		31	103		134	188
0710	10" pipe size		4.50	1.778		57	115		172	234
0720	12" pipe size		4.20	1.905		59	123		182	249
0730	14" pipe size		4	2		54.50	129		183.50	253
0740	16" pipe size		3	2.667		65.50	172		237.50	330
0750	18" pipe size		2.70	2.963		127	191		318	425
0760	20" pipe size		2.30	3.478		206	224		430	560
0780	24" pipe size	▼	1.90	4.211	▼	257	271		528	690

22 11 13.48 Pipe, Fittings and Valves, Steel, Grooved-Joint

		Crew	Daily Output	Labor-Hours	Unit	Material	2020 Bare Costs Labor	Equipment	Total	Total Incl O&P
0010	**PIPE, FITTINGS AND VALVES, STEEL, GROOVED-JOINT**									
0012	Fittings are ductile iron. Steel fittings noted.									
0020	Pipe includes coupling & clevis type hanger assemblies, 10' OC									
1000	Schedule 40, black									
1040	3/4" diameter	1 Plum	71	.113	L.F.	6.55	7.25		13.80	18.05
1050	1" diameter		63	.127		6.45	8.20		14.65	19.35
1060	1-1/4" diameter		58	.138		7.70	8.90		16.60	22
1070	1-1/2" diameter		51	.157		8.40	10.10		18.50	24.50
1080	2" diameter	▼	40	.200		9.70	12.90		22.60	30
1090	2-1/2" diameter	Q-1	57	.281		16.50	16.30		32.80	42.50
1100	3" diameter		50	.320		22	18.55		40.55	52
1110	4" diameter		45	.356		26	20.50		46.50	59.50
1120	5" diameter	▼	37	.432		41	25		66	82.50
1130	6" diameter	Q-2	42	.571		52.50	34.50		87	109
1140	8" diameter		37	.649		84.50	39		123.50	152
1150	10" diameter		31	.774		114	46.50		160.50	195
1160	12" diameter	▼	27	.889	▼	128	53.50		181.50	220
1740	To delete coupling & hanger, subtract									
1750	3/4" diam. to 2" diam.					65%	27%			
1760	2-1/2" diam. to 5" diam.					41%	18%			
1770	6" diam. to 12" diam.					31%	13%			
1780	14" diam. to 24" diam.					35%	10%			
1800	Galvanized									
1840	3/4" diameter	1 Plum	71	.113	L.F.	7.15	7.25		14.40	18.70
1850	1" diameter		63	.127		7.10	8.20		15.30	20
1860	1-1/4" diameter		58	.138		8.50	8.90		17.40	22.50
1870	1-1/2" diameter		51	.157		9.25	10.10		19.35	25.50
1880	2" diameter	▼	40	.200		10.80	12.90		23.70	31
1890	2-1/2" diameter	Q-1	57	.281		16.70	16.30		33	43
1900	3" diameter		50	.320		19.65	18.55		38.20	49.50
1910	4" diameter		45	.356		23	20.50		43.50	56.50
1920	5" diameter	▼	37	.432		30.50	25		55.50	71
1930	6" diameter	Q-2	42	.571		39.50	34.50		74	95

22 11 Facility Water Distribution

22 11 13 – Facility Water Distribution Piping

22 11 13.48 Pipe, Fittings and Valves, Steel, Grooved-Joint

		Crew	Daily Output	Labor-Hours	Unit	Material	2020 Bare Costs Labor	Equipment	Total	Total Incl O&P
1940	8" diameter	Q-2	37	.649	L.F.	58.50	39		97.50	123
1950	10" diameter		31	.774		114	46.50		160.50	195
1960	12" diameter		27	.889		137	53.50		190.50	231
2540	To delete coupling & hanger, subtract									
2550	3/4" diam. to 2" diam.					36%	27%			
2560	2-1/2" diam. to 5" diam.					19%	18%			
2570	6" diam. to 12" diam.					14%	13%			
4690	Tee, painted									
4700	3/4" diameter	1 Plum	38	.211	Ea.	88.50	13.55		102.05	118
4740	1" diameter		33	.242		68.50	15.60		84.10	98.50
4750	1-1/4" diameter		27	.296		68.50	19.10		87.60	104
4760	1-1/2" diameter		22	.364		68.50	23.50		92	110
4770	2" diameter		17	.471		68.50	30.50		99	121
4780	2-1/2" diameter	Q-1	27	.593		68.50	34.50		103	127
4790	3" diameter		22	.727		93.50	42		135.50	166
4800	4" diameter		17	.941		142	54.50		196.50	238
4810	5" diameter		13	1.231		330	71.50		401.50	470
4820	6" diameter	Q-2	17	1.412		380	85		465	545
4830	8" diameter		14	1.714		835	103		938	1,075
4840	10" diameter		12	2		1,050	120		1,170	1,350
4850	12" diameter		10	2.400		1,375	144		1,519	1,750
4900	For galvanized tees, add					24%				
4939	Couplings									
4940	Flexible, standard, painted									
4950	3/4" diameter	1 Plum	100	.080	Ea.	25	5.15		30.15	35
4960	1" diameter		100	.080		25	5.15		30.15	35
4970	1-1/4" diameter		80	.100		32.50	6.45		38.95	45
4980	1-1/2" diameter		67	.119		35	7.70		42.70	50
4990	2" diameter		50	.160		38	10.30		48.30	57
5000	2-1/2" diameter	Q-1	80	.200		44	11.60		55.60	66
5010	3" diameter		67	.239		48.50	13.85		62.35	74
5020	3-1/2" diameter		57	.281		69.50	16.30		85.80	101
5030	4" diameter		50	.320		70	18.55		88.55	105
5040	5" diameter		40	.400		105	23		128	151
5050	6" diameter	Q-2	50	.480		125	29		154	180
5070	8" diameter		42	.571		202	34.50		236.50	274
5090	10" diameter		35	.686		330	41.50		371.50	420
5110	12" diameter		32	.750		375	45		420	480
5200	For galvanized couplings, add					33%				
5750	Flange, w/groove gasket, black steel									
5754	See Line 22 11 13.47 0620 for gasket & bolt set									
5760	ANSI class 125 and 150, painted									
5780	2" pipe size	1 Plum	23	.348	Ea.	145	22.50		167.50	194
5790	2-1/2" pipe size	Q-1	37	.432		181	25		206	237
5800	3" pipe size		31	.516		195	30		225	259
5820	4" pipe size		23	.696		260	40.50		300.50	345
5830	5" pipe size		19	.842		300	49		349	405
5840	6" pipe size	Q-2	23	1.043		330	63		393	455
5850	8" pipe size		17	1.412		370	85		455	535
5860	10" pipe size		14	1.714		585	103		688	800
5870	12" pipe size		12	2		765	120		885	1,025
8000	Butterfly valve, 2 position handle, with standard trim									
8010	1-1/2" pipe size	1 Plum	30	.267	Ea.	350	17.20		367.20	410

For customer support on your Site Work & Landscape Costs with RSMeans data, call 800.448.8182.

223

22 11 13 – Facility Water Distribution Piping

22 11 13.48 Pipe, Fittings and Valves, Steel, Grooved-Joint

		Crew	Daily Output	Labor-Hours	Unit	Material	2020 Bare Costs Labor	2020 Bare Costs Equipment	Total	Total Incl O&P
8020	2" pipe size	1 Plum	23	.348	Ea.	350	22.50		372.50	420
8030	3" pipe size	Q-1	30	.533		505	31		536	600
8050	4" pipe size	"	22	.727		555	42		597	675
8070	6" pipe size	Q-2	23	1.043		1,125	63		1,188	1,325
8080	8" pipe size		18	1.333		1,450	80		1,530	1,725
8090	10" pipe size		15	1.600		2,000	96		2,096	2,350
8200	With stainless steel trim									
8240	1-1/2" pipe size	1 Plum	30	.267	Ea.	445	17.20		462.20	515
8250	2" pipe size	"	23	.348		445	22.50		467.50	525
8270	3" pipe size	Q-1	30	.533		600	31		631	705
8280	4" pipe size	"	22	.727		650	42		692	780
8300	6" pipe size	Q-2	23	1.043		1,200	63		1,263	1,425
8310	8" pipe size		18	1.333		3,000	80		3,080	3,425
8320	10" pipe size		15	1.600		4,800	96		4,896	5,425
9000	Cut one groove, labor									
9010	3/4" pipe size	Q-1	152	.105	Ea.		6.10		6.10	9.15
9020	1" pipe size		140	.114			6.65		6.65	9.90
9030	1-1/4" pipe size		124	.129			7.50		7.50	11.20
9040	1-1/2" pipe size		114	.140			8.15		8.15	12.20
9050	2" pipe size		104	.154			8.90		8.90	13.35
9060	2-1/2" pipe size		96	.167			9.65		9.65	14.45
9070	3" pipe size		88	.182			10.55		10.55	15.80
9080	3-1/2" pipe size		83	.193			11.20		11.20	16.75
9090	4" pipe size		78	.205			11.90		11.90	17.80
9100	5" pipe size		72	.222			12.90		12.90	19.30
9110	6" pipe size		70	.229			13.25		13.25	19.85
9120	8" pipe size		54	.296			17.20		17.20	25.50
9130	10" pipe size		38	.421			24.50		24.50	36.50
9140	12" pipe size		30	.533			31		31	46.50
9210	Roll one groove									
9220	3/4" pipe size	Q-1	266	.060	Ea.		3.49		3.49	5.20
9230	1" pipe size		228	.070			4.07		4.07	6.10
9240	1-1/4" pipe size		200	.080			4.64		4.64	6.95
9250	1-1/2" pipe size		178	.090			5.20		5.20	7.80
9260	2" pipe size		116	.138			8		8	11.95
9270	2-1/2" pipe size		110	.145			8.45		8.45	12.65
9280	3" pipe size		100	.160			9.30		9.30	13.90
9290	3-1/2" pipe size		94	.170			9.85		9.85	14.75
9300	4" pipe size		86	.186			10.80		10.80	16.15
9310	5" pipe size		84	.190			11.05		11.05	16.55
9320	6" pipe size		80	.200			11.60		11.60	17.35
9330	8" pipe size		66	.242			14.05		14.05	21
9340	10" pipe size		58	.276			16		16	24
9350	12" pipe size		46	.348			20		20	30

22 11 13.74 Pipe, Plastic

		Crew	Daily Output	Labor-Hours	Unit	Material	2020 Bare Costs Labor	2020 Bare Costs Equipment	Total	Total Incl O&P
0010	**PIPE, PLASTIC**									
0020	Fiberglass reinforced, couplings 10' OC, clevis hanger assy's, 3 per 10'									
0080	General service									
0120	2" diameter	Q-1	59	.271	L.F.	17.50	15.75		33.25	43
0140	3" diameter		52	.308		27.50	17.85		45.35	57
0150	4" diameter		48	.333		23	19.35		42.35	54.50
0160	6" diameter		39	.410		43	24		67	83

224

For customer support on your Site Work & Landscape Costs with RSMeans data, call 800.448.8182.

22 11 13 – Facility Water Distribution Piping

22 11 13.74 Pipe, Plastic		Crew	Daily Output	Labor-Hours	Unit	Material	2020 Bare Costs Labor	Equipment	Total	Total Incl O&P
0170	8" diameter	Q-2	49	.490	L.F.	67.50	29.50		97	118
0180	10" diameter		41	.585		95.50	35		130.50	158
0190	12" diameter		36	.667		125	40		165	198
1800	PVC, couplings 10' OC, clevis hanger assemblies, 3 per 10'									
1820	Schedule 40									
1860	1/2" diameter	1 Plum	54	.148	L.F.	4.90	9.55		14.45	19.70
1870	3/4" diameter		51	.157		5.40	10.10		15.50	21
1880	1" diameter		46	.174		9.45	11.20		20.65	27
1890	1-1/4" diameter		42	.190		9.95	12.30		22.25	29.50
1900	1-1/2" diameter		36	.222		10.25	14.30		24.55	33
1910	2" diameter	Q-1	59	.271		12.15	15.75		27.90	37
1920	2-1/2" diameter		56	.286		20	16.55		36.55	47.50
1930	3" diameter		53	.302		22	17.50		39.50	50.50
1940	4" diameter		48	.333		14.80	19.35		34.15	45.50
1950	5" diameter		43	.372		23.50	21.50		45	58.50
1960	6" diameter		39	.410		30.50	24		54.50	69
2340	To delete coupling & hangers, subtract									
2360	1/2" diam. to 1-1/4" diam.					65%	74%			
2370	1-1/2" diam. to 6" diam.					44%	57%			
4100	DWV type, schedule 40, couplings 10' OC, clevis hanger assy's, 3 per 10'									
4210	ABS, schedule 40, foam core type									
4212	Plain end black									
4214	1-1/2" diameter	1 Plum	39	.205	L.F.	8.60	13.20		21.80	29.50
4216	2" diameter	Q-1	62	.258		9.25	14.95		24.20	32.50
4218	3" diameter		56	.286		17.60	16.55		34.15	44.50
4220	4" diameter		51	.314		9.80	18.20		28	38
4222	6" diameter		42	.381		21.50	22		43.50	56.50
4240	To delete coupling & hangers, subtract									
4244	1-1/2" diam. to 6" diam.					43%	48%			
4400	PVC									
4410	1-1/4" diameter	1 Plum	42	.190	L.F.	9.10	12.30		21.40	28.50
4420	1-1/2" diameter	"	36	.222		8.25	14.30		22.55	30.50
4460	2" diameter	Q-1	59	.271		9.15	15.75		24.90	33.50
4470	3" diameter		53	.302		17.45	17.50		34.95	45
4480	4" diameter		48	.333		19.20	19.35		38.55	50
4490	6" diameter		39	.410		18.60	24		42.60	56
4510	To delete coupling & hangers, subtract									
4520	1-1/4" diam. to 1-1/2" diam.					48%	60%			
4530	2" diam. to 8" diam.					42%	54%			
4532	to delete hangers, 2" diam. to 8" diam.	Q-1	50	.320	L.F.	5.50	18.55		24.05	34
5300	CPVC, socket joint, couplings 10' OC, clevis hanger assemblies, 3 per 10'									
5302	Schedule 40									
5304	1/2" diameter	1 Plum	54	.148	L.F.	5.80	9.55		15.35	20.50
5305	3/4" diameter		51	.157		6.95	10.10		17.05	23
5306	1" diameter		46	.174		11.35	11.20		22.55	29.50
5307	1-1/4" diameter		42	.190		12.85	12.30		25.15	32.50
5308	1-1/2" diameter		36	.222		12.45	14.30		26.75	35
5309	2" diameter	Q-1	59	.271		15.70	15.75		31.45	41
5310	2-1/2" diameter		56	.286		28	16.55		44.55	56
5311	3" diameter		53	.302		32	17.50		49.50	61
5318	To delete coupling & hangers, subtract									
5319	1/2" diam. to 3/4" diam.					37%	77%			
5320	1" diam. to 1-1/4" diam.					27%	70%			

225

22 11 13.74 Pipe, Plastic

		Crew	Daily Output	Labor-Hours	Unit	Material	2020 Bare Costs Labor	Equipment	Total	Total Incl O&P
5321	1-1/2" diam. to 3" diam.					21%	57%			
5360	CPVC, threaded, couplings 10' OC, clevis hanger assemblies, 3 per 10'									
5380	Schedule 40									
5460	1/2" diameter	1 Plum	54	.148	L.F.	6.65	9.55		16.20	21.50
5470	3/4" diameter		51	.157		8.45	10.10		18.55	24.50
5480	1" diameter		46	.174		12.90	11.20		24.10	31
5490	1-1/4" diameter		42	.190		14.05	12.30		26.35	34
5500	1-1/2" diameter		36	.222		13.45	14.30		27.75	36.50
5510	2" diameter	Q-1	59	.271		16.90	15.75		32.65	42
5520	2-1/2" diameter		56	.286		29.50	16.55		46.05	57.50
5530	3" diameter		53	.302		33.50	17.50		51	63
5730	To delete coupling & hangers, subtract									
5740	1/2" diam. to 3/4" diam.					37%	77%			
5750	1" diam. to 1-1/4" diam.					27%	70%			
5760	1-1/2" diam. to 3" diam.					21%	57%			
7280	PEX, flexible, no couplings or hangers									
7282	Note: For labor costs add 25% to the couplings and fittings labor total.									
7300	Non-barrier type, hot/cold tubing rolls									
7310	1/4" diameter x 100'				L.F.	.56			.56	.62
7350	3/8" diameter x 100'					.55			.55	.61
7360	1/2" diameter x 100'					.76			.76	.84
7370	1/2" diameter x 500'					.74			.74	.81
7380	1/2" diameter x 1000'					.72			.72	.79
7400	3/4" diameter x 100'					1.04			1.04	1.14
7410	3/4" diameter x 500'					1.17			1.17	1.29
7420	3/4" diameter x 1000'					1.17			1.17	1.29
7460	1" diameter x 100'					2.02			2.02	2.22
7470	1" diameter x 300'					2.02			2.02	2.22
7480	1" diameter x 500'					2.03			2.03	2.23
7500	1-1/4" diameter x 100'					3.44			3.44	3.78
7510	1-1/4" diameter x 300'					3.44			3.44	3.78
7540	1-1/2" diameter x 100'					4.71			4.71	5.20
7550	1-1/2" diameter x 300'					4.72			4.72	5.20
7596	Most sizes available in red or blue									

22 11 13.76 Pipe Fittings, Plastic

		Crew	Daily Output	Labor-Hours	Unit	Material	2020 Bare Costs Labor	Equipment	Total	Total Incl O&P
0010	**PIPE FITTINGS, PLASTIC**									
0030	Epoxy resin, fiberglass reinforced, general service									
0100	3"	Q-1	20.80	.769	Ea.	118	44.50		162.50	196
0110	4"		16.50	.970		121	56		177	217
0120	6"		10.10	1.584		234	92		326	395
0130	8"	Q-2	9.30	2.581		430	155		585	705
0140	10"		8.50	2.824		540	170		710	850
0150	12"		7.60	3.158		775	190		965	1,150
0380	Couplings									
0410	2"	Q-1	33.10	.483	Ea.	27	28		55	71.50
0420	3"		20.80	.769		31.50	44.50		76	102
0430	4"		16.50	.970		43.50	56		99.50	132
0440	6"		10.10	1.584		101	92		193	249
0450	8"	Q-2	9.30	2.581		171	155		326	420
0460	10"		8.50	2.824		233	170		403	510
0470	12"		7.60	3.158		335	190		525	655
0473	High corrosion resistant couplings, add					30%				

22 11 13.76 Pipe Fittings, Plastic	Crew	Daily Output	Labor-Hours	Unit	Material	2020 Bare Costs Labor	Equipment	Total	Total Incl O&P
2100	PVC schedule 80, socket joint									
2110	90° elbow, 1/2"	1 Plum	30.30	.264	Ea.	2.67	17		19.67	28.50
2130	3/4"		26	.308		3.42	19.85		23.27	33.50
2140	1"		22.70	.352		5.55	22.50		28.05	40
2150	1-1/4"		20.20	.396		7.35	25.50		32.85	46
2160	1-1/2"		18.20	.440		7.90	28.50		36.40	51
2170	2"	Q-1	33.10	.483		9.55	28		37.55	52.50
2180	3"		20.80	.769		25	44.50		69.50	94.50
2190	4"		16.50	.970		38	56		94	126
2200	6"		10.10	1.584		109	92		201	257
2210	8"	Q-2	9.30	2.581		299	155		454	560
2250	45° elbow, 1/2"	1 Plum	30.30	.264		5.05	17		22.05	31
2270	3/4"		26	.308		7.70	19.85		27.55	38
2280	1"		22.70	.352		11.55	22.50		34.05	46.50
2290	1-1/4"		20.20	.396		14.60	25.50		40.10	54
2300	1-1/2"		18.20	.440		17.35	28.50		45.85	61.50
2310	2"	Q-1	33.10	.483		22.50	28		50.50	67
2320	3"		20.80	.769		57	44.50		101.50	130
2330	4"		16.50	.970		104	56		160	198
2340	6"		10.10	1.584		130	92		222	282
2350	8"	Q-2	9.30	2.581		283	155		438	540
2400	Tee, 1/2"	1 Plum	20.20	.396		7.55	25.50		33.05	46.50
2420	3/4"		17.30	.462		7.90	30		37.90	53
2430	1"		15.20	.526		9.90	34		43.90	62
2440	1-1/4"		13.50	.593		27	38		65	86.50
2450	1-1/2"		12.10	.661		27	42.50		69.50	94
2460	2"	Q-1	20	.800		34	46.50		80.50	107
2470	3"		13.90	1.151		46	67		113	151
2480	4"		11	1.455		53.50	84.50		138	185
2490	6"		6.70	2.388		183	139		322	410
2500	8"	Q-2	6.20	3.871		420	233		653	810
2510	Flange, socket, 150 lb., 1/2"	1 Plum	55.60	.144		14.55	9.25		23.80	30
2514	3/4"		47.60	.168		15.55	10.85		26.40	33.50
2518	1"		41.70	.192		17.35	12.35		29.70	37.50
2522	1-1/2"		33.30	.240		18.30	15.50		33.80	43
2526	2"	Q-1	60.60	.264		24.50	15.30		39.80	49.50
2530	4"		30.30	.528		52.50	30.50		83	104
2534	6"		18.50	.865		82.50	50		132.50	166
2538	8"	Q-2	17.10	1.404		147	84.50		231.50	288
2550	Coupling, 1/2"	1 Plum	30.30	.264		5.10	17		22.10	31
2570	3/4"		26	.308		7.55	19.85		27.40	38
2580	1"		22.70	.352		7.45	22.50		29.95	42
2590	1-1/4"		20.20	.396		10.15	25.50		35.65	49
2600	1-1/2"		18.20	.440		11.40	28.50		39.90	55
2610	2"	Q-1	33.10	.483		12.25	28		40.25	55.50
2620	3"		20.80	.769		33.50	44.50		78	104
2630	4"		16.50	.970		43.50	56		99.50	132
2640	6"		10.10	1.584		93	92		185	241
2650	8"	Q-2	9.30	2.581		122	155		277	365
2660	10"		8.50	2.824		420	170		590	715
2670	12"		7.60	3.158		485	190		675	815
2700	PVC (white), schedule 40, socket joints									
2810	2"	Q-1	36.40	.440	Ea.	3.11	25.50		28.61	41.50

For customer support on your Site Work & Landscape Costs with RSMeans data, call 800.448.8182.

227

22 11 13.76 Pipe Fittings, Plastic	Crew	Daily Output	Labor-Hours	Unit	Material	2020 Bare Costs Labor	Equipment	Total	Total Incl O&P
2840	4"	Q-1	18.20	.879	Ea.	20.50	51		71.50	99
4500	DWV, ABS, non pressure, socket joints									
4540	1/4 bend, 1-1/4"	1 Plum	20.20	.396	Ea.	4.48	25.50		29.98	43
4560	1-1/2"	"	18.20	.440		3.44	28.50		31.94	46.50
4570	2"	Q-1	33.10	.483		5.30	28		33.30	48
4580	3"	"	20.80	.769		13.45	44.50		57.95	82
4650	1/8 bend, same as 1/4 bend									
4800	Tee, sanitary									
4820	1-1/4"	1 Plum	13.50	.593	Ea.	5.95	38		43.95	63.50
4830	1-1/2"	"	12.10	.661		5.15	42.50		47.65	69.50
4840	2"	Q-1	20	.800		7.90	46.50		54.40	78
4850	3"		13.90	1.151		21.50	67		88.50	124
4860	4"		11	1.455		38.50	84.50		123	169
4862	Tee, sanitary, reducing, 2" x 1-1/2"		22	.727		7	42		49	70.50
4864	3" x 2"		15.30	1.046		13.40	60.50		73.90	106
4868	4" x 3"		12.10	1.322		37.50	76.50		114	157
4870	Combination Y and 1/8 bend									
4872	1-1/2"	1 Plum	12.10	.661	Ea.	12.25	42.50		54.75	77.50
4874	2"	Q-1	20	.800		13.70	46.50		60.20	84.50
4876	3"		13.90	1.151		32	67		99	136
4878	4"		11	1.455		62	84.50		146.50	194
4880	3" x 1-1/2"		15.50	1.032		32.50	60		92.50	126
4882	4" x 3"		12.10	1.322		49.50	76.50		126	170
4900	Wye, 1-1/4"	1 Plum	13.50	.593		6.85	38		44.85	64.50
4902	1-1/2"	"	12.10	.661		7.85	42.50		50.35	72.50
4904	2"	Q-1	20	.800		10.20	46.50		56.70	80.50
4906	3"		13.90	1.151		24	67		91	127
4908	4"		11	1.455		49	84.50		133.50	180
4910	6"		6.70	2.388		146	139		285	370
4918	3" x 1-1/2"		15.50	1.032		19.40	60		79.40	111
4920	4" x 3"		12.10	1.322		38.50	76.50		115	158
4922	6" x 4"		6.90	2.319		116	134		250	330
4930	Double wye, 1-1/2"	1 Plum	9.10	.879		25	56.50		81.50	113
4932	2"	Q-1	16.60	.964		30	56		86	117
4934	3"		10.40	1.538		70.50	89		159.50	212
4936	4"		8.25	1.939		138	112		250	320
4940	2" x 1-1/2"		16.80	.952		27.50	55		82.50	113
4942	3" x 2"		10.60	1.509		49	87.50		136.50	185
4944	4" x 3"		8.45	1.893		109	110		219	284
4946	6" x 4"		7.25	2.207		156	128		284	365
4950	Reducer bushing, 2" x 1-1/2"		36.40	.440		2.72	25.50		28.22	41
4952	3" x 1-1/2"		27.30	.586		11.95	34		45.95	64
4954	4" x 2"		18.20	.879		22.50	51		73.50	101
4956	6" x 4"		11.10	1.441		62.50	83.50		146	194
4960	Couplings, 1-1/2"	1 Plum	18.20	.440		1.66	28.50		30.16	44.50
4962	2"	Q-1	33.10	.483		2.22	28		30.22	44.50
4963	3"		20.80	.769		6.30	44.50		50.80	74
4964	4"		16.50	.970		11.35	56		67.35	96.50
4966	6"		10.10	1.584		47.50	92		139.50	190
4970	2" x 1-1/2"		33.30	.480		4.80	28		32.80	47
4972	3" x 1-1/2"		21	.762		13.85	44		57.85	81.50
4974	4" x 3"		16.70	.958		21	55.50		76.50	107
4978	Closet flange, 4"	1 Plum	32	.250		11.60	16.10		27.70	37

22 11 Facility Water Distribution

22 11 13 – Facility Water Distribution Piping

22 11 13.76 Pipe Fittings, Plastic		Crew	Daily Output	Labor-Hours	Unit	Material	2020 Bare Costs Labor	Equipment	Total	Total Incl O&P
4980	4" x 3"	1 Plum	34	.235	Ea.	14.10	15.15		29.25	38
5000	DWV, PVC, schedule 40, socket joints									
5040	1/4 bend, 1-1/4"	1 Plum	20.20	.396	Ea.	8.80	25.50		34.30	47.50
5060	1-1/2"	"	18.20	.440		2.48	28.50		30.98	45
5070	2"	Q-1	33.10	.483		3.91	28		31.91	46.50
5080	3"		20.80	.769		11.65	44.50		56.15	80
5090	4"		16.50	.970		23	56		79	109
5100	6"		10.10	1.584		79.50	92		171.50	226
5105	8"	Q-2	9.30	2.581		122	155		277	365
5110	1/4 bend, long sweep, 1-1/2"	1 Plum	18.20	.440		5.85	28.50		34.35	49
5112	2"	Q-1	33.10	.483		6.50	28		34.50	49
5114	3"		20.80	.769		15.05	44.50		59.55	83.50
5116	4"		16.50	.970		28.50	56		84.50	116
5150	1/8 bend, 1-1/4"	1 Plum	20.20	.396		6.05	25.50		31.55	44.50
5215	8"	Q-2	9.30	2.581		117	155		272	360
5250	Tee, sanitary 1-1/4"	1 Plum	13.50	.593		9.45	38		47.45	67.50
5254	1-1/2"	"	12.10	.661		4.41	42.50		46.91	69
5255	2"	Q-1	20	.800		6.50	46.50		53	76.50
5256	3"	"	13.90	1.151		17.05	67		84.05	119
5374	Coupling, 1-1/4"	1 Plum	20.20	.396		5.70	25.50		31.20	44.50
5376	1-1/2"	"	18.20	.440		1.19	28.50		29.69	44
5378	2"	Q-1	33.10	.483		1.62	28		29.62	44
5380	3"		20.80	.769		5.65	44.50		50.15	73
5390	4"		16.50	.970		9.65	56		65.65	94.50
5450	Solvent cement for PVC, industrial grade, per quart				Qt.	31			31	34
5500	CPVC, Schedule 80, threaded joints									
5540	90° elbow, 1/4"	1 Plum	32	.250	Ea.	13	16.10		29.10	38.50
5560	1/2"		30.30	.264		7.55	17		24.55	34
5570	3/4"		26	.308		11.30	19.85		31.15	42
5580	1"		22.70	.352		15.85	22.50		38.35	51.50
5590	1-1/4"		20.20	.396		30.50	25.50		56	71.50
5600	1-1/2"		18.20	.440		33	28.50		61.50	78.50
5610	2"	Q-1	33.10	.483		44	28		72	90.50
5620	2-1/2"		24.20	.661		137	38.50		175.50	208
5630	3"		20.80	.769		147	44.50		191.50	228
5660	45° elbow same as 90° elbow									
5700	Tee, 1/4"	1 Plum	22	.364	Ea.	25.50	23.50		49	63
5704	3/4"		17.30	.462		36.50	30		66.50	84.50
5706	1"		15.20	.526		39.50	34		73.50	94
5708	1-1/4"		13.50	.593		39.50	38		77.50	101
5710	1-1/2"		12.10	.661		41.50	42.50		84	110
5712	2"	Q-1	20	.800		46	46.50		92.50	120
5714	2-1/2"		16.20	.988		225	57.50		282.50	335
5716	3"		13.90	1.151		263	67		330	390
5730	Coupling, 1/4"	1 Plum	32	.250		16.65	16.10		32.75	42.50
5732	1/2"		30.30	.264		13.75	17		30.75	40.50
5734	3/4"		26	.308		22	19.85		41.85	54
5736	1"		22.70	.352		25	22.50		47.50	61.50
5738	1-1/4"		20.20	.396		26.50	25.50		52	67
5740	1-1/2"		18.20	.440		28.50	28.50		57	74
5742	2"	Q-1	33.10	.483		33.50	28		61.50	79
5744	2-1/2"		24.20	.661		60.50	38.50		99	124
5746	3"		20.80	.769		70	44.50		114.50	145

22 11 13.76 Pipe Fittings, Plastic	Crew	Daily Output	Labor-Hours	Unit	Material	2020 Bare Costs Labor	Equipment	Total	Total Incl O&P
5900	CPVC, Schedule 80, socket joints									
5904	90° elbow, 1/4"	1 Plum	32	.250	Ea.	12.45	16.10		28.55	37.50
5906	1/2"		30.30	.264		4.88	17		21.88	31
5908	3/4"		26	.308		6.25	19.85		26.10	36.50
5910	1"		22.70	.352		9.90	22.50		32.40	45
5912	1-1/4"		20.20	.396		21.50	25.50		47	61.50
5914	1-1/2"	↓	18.20	.440		24	28.50		52.50	68.50
5916	2"	Q-1	33.10	.483		29	28		57	73.50
5918	2-1/2"		24.20	.661		66.50	38.50		105	131
5920	3"	↓	20.80	.769		75	44.50		119.50	150
5930	45° elbow, 1/4"	1 Plum	32	.250		18.50	16.10		34.60	44.50
5932	1/2"		30.30	.264		5.95	17		22.95	32
5934	3/4"		26	.308		7.80	19.85		27.65	38
5936	1"		22.70	.352		13.70	22.50		36.20	49
5938	1-1/4"		20.20	.396		27	25.50		52.50	67.50
5940	1-1/2"	↓	18.20	.440		27.50	28.50		56	73
5942	2"	Q-1	33.10	.483		27.50	28		55.50	72
5944	2-1/2"		24.20	.661		63.50	38.50		102	127
5946	3"	↓	20.80	.769		81.50	44.50		126	157
5960	Tee, 1/4"	1 Plum	22	.364		11.45	23.50		34.95	47.50
5964	3/4"		17.30	.462		11.65	30		41.65	57.50
5966	1"		15.20	.526		14.25	34		48.25	66.50
5968	1-1/4"		13.50	.593		30	38		68	90
5970	1-1/2"	↓	12.10	.661		34.50	42.50		77	102
5972	2"	Q-1	20	.800		34	46.50		80.50	107
5974	2-1/2"		16.20	.988		97.50	57.50		155	193
5976	3"	↓	13.90	1.151		97.50	67		164.50	207
5990	Coupling, 1/4"	1 Plum	32	.250		13.25	16.10		29.35	38.50
5992	1/2"		30.30	.264		5.15	17		22.15	31
5994	3/4"		26	.308		7.20	19.85		27.05	37.50
5996	1"		22.70	.352		9.70	22.50		32.20	44.50
5998	1-1/4"		20.20	.396		14.55	25.50		40.05	54
6000	1-1/2"	↓	18.20	.440		18.25	28.50		46.75	62.50
6002	2"	Q-1	33.10	.483		21.50	28		49.50	65.50
6004	2-1/2"		24.20	.661		47.50	38.50		86	110
6006	3"	↓	20.80	.769		51.50	44.50		96	124
6360	Solvent cement for CPVC, commercial grade, per quart				Qt.	51.50			51.50	56.50
7340	PVC flange, slip-on, Sch 80 std., 1/2"	1 Plum	22	.364	Ea.	14.65	23.50		38.15	51
7350	3/4"		21	.381		15.65	24.50		40.15	54
7360	1"		18	.444		17.45	28.50		45.95	62
7370	1-1/4"		17	.471		17.95	30.50		48.45	65.50
7380	1-1/2"	↓	16	.500		18.35	32		50.35	68
7390	2"	Q-1	26	.615		24.50	35.50		60	80.50
7400	2-1/2"		24	.667		37.50	38.50		76	99.50
7410	3"		18	.889		41.50	51.50		93	123
7420	4"		15	1.067		52.50	62		114.50	151
7430	6"	↓	10	1.600		83	93		176	230
7440	8"	Q-2	11	2.182		148	131		279	360
7550	Union, schedule 40, socket joints, 1/2"	1 Plum	19	.421		5.20	27		32.20	46.50
7560	3/4"		18	.444		5.80	28.50		34.30	49.50
7570	1"		15	.533		6	34.50		40.50	58
7580	1-1/4"		14	.571		17.85	37		54.85	74.50
7590	1-1/2"	↓	13	.615		20	39.50		59.50	81.50

22 11 Facility Water Distribution

22 11 13 – Facility Water Distribution Piping

22 11 13.76 Pipe Fittings, Plastic	Crew	Daily Output	Labor-Hours	Unit	Material	2020 Bare Costs Labor	Equipment	Total	Total Incl O&P
7600	2"	Q-1	20	.800	Ea.	27	46.50		73.50	99.50

22 11 13.78 Pipe, High Density Polyethylene Plastic (HDPE)

		Crew	Daily Output	Labor-Hours	Unit	Material	Labor	Equipment	Total	Total Incl O&P
0010	**PIPE, HIGH DENSITY POLYETHYLENE PLASTIC (HDPE)**									
0020	Not incl. hangers, trenching, backfill, hoisting or digging equipment.									
0030	Standard length is 40', add a weld for each joint									
0035	For HDPE weld machine see 015433401685 in equipment rental									
0040	Single wall									
0050	Straight									
0054	1" diameter DR 11				L.F.	.89			.89	.98
0058	1-1/2" diameter DR 11					1.14			1.14	1.25
0062	2" diameter DR 11					1.89			1.89	2.08
0066	3" diameter DR 11					2.29			2.29	2.52
0070	3" diameter DR 17					1.83			1.83	2.01
0074	4" diameter DR 11					3.83			3.83	4.21
0078	4" diameter DR 17					3.83			3.83	4.21
0082	6" diameter DR 11					9.50			9.50	10.45
0086	6" diameter DR 17					6.30			6.30	6.90
0090	8" diameter DR 11					15.80			15.80	17.40
0094	8" diameter DR 26					7.40			7.40	8.15
0098	10" diameter DR 11					25			25	27
0102	10" diameter DR 26					11.45			11.45	12.60
0106	12" diameter DR 11					36			36	40
0110	12" diameter DR 26					17.15			17.15	18.85
0114	16" diameter DR 11					55			55	61
0118	16" diameter DR 26					25			25	27
0122	18" diameter DR 11					70.50			70.50	77.50
0126	18" diameter DR 26					32.50			32.50	35.50
0130	20" diameter DR 11					85.50			85.50	94.50
0134	20" diameter DR 26					38			38	42
0138	22" diameter DR 11					105			105	115
0142	22" diameter DR 26					47.50			47.50	52.50
0146	24" diameter DR 11					124			124	136
0150	24" diameter DR 26					55			55	61
0154	28" diameter DR 17					114			114	126
0158	28" diameter DR 26					76			76	84
0162	30" diameter DR 21					107			107	117
0166	30" diameter DR 26					87.50			87.50	96.50
0170	36" diameter DR 26					87.50			87.50	96.50
0174	42" diameter DR 26					170			170	186
0178	48" diameter DR 26					223			223	245
0182	54" diameter DR 26				↓	280			280	310
0300	90° elbow									
0304	1" diameter DR 11				Ea.	5.50			5.50	6.05
0308	1-1/2" diameter DR 11					7.30			7.30	8.05
0312	2" diameter DR 11					7.30			7.30	8.05
0316	3" diameter DR 11					14.60			14.60	16.05
0320	3" diameter DR 17					14.60			14.60	16.05
0324	4" diameter DR 11					20.50			20.50	22.50
0328	4" diameter DR 17					20.50			20.50	22.50
0332	6" diameter DR 11					46.50			46.50	51.50
0336	6" diameter DR 17					46.50			46.50	51.50
0340	8" diameter DR 11				↓	115			115	127

22 11 13 – Facility Water Distribution Piping

22 11 13.78 Pipe, High Density Polyethylene Plastic (HDPE)	Crew	Daily Output	Labor-Hours	Unit	Material	2020 Bare Costs Labor	Equipment	Total	Total Incl O&P
0344	8" diameter DR 26				Ea.	102			102	112
0348	10" diameter DR 11					430			430	475
0352	10" diameter DR 26					400			400	440
0356	12" diameter DR 11					455			455	500
0360	12" diameter DR 26					410			410	450
0364	16" diameter DR 11					540			540	590
0368	16" diameter DR 26					530			530	585
0372	18" diameter DR 11					675			675	745
0376	18" diameter DR 26					635			635	700
0380	20" diameter DR 11					795			795	875
0384	20" diameter DR 26					770			770	845
0388	22" diameter DR 11					820			820	905
0392	22" diameter DR 26					795			795	875
0396	24" diameter DR 11					930			930	1,025
0400	24" diameter DR 26					900			900	990
0404	28" diameter DR 17					1,125			1,125	1,250
0408	28" diameter DR 26					1,050			1,050	1,175
0412	30" diameter DR 17					1,600			1,600	1,750
0416	30" diameter DR 26					1,450			1,450	1,600
0420	36" diameter DR 26					1,850			1,850	2,050
0424	42" diameter DR 26					2,400			2,400	2,625
0428	48" diameter DR 26					2,775			2,775	3,075
0432	54" diameter DR 26					6,625			6,625	7,300
0500	45° elbow									
0512	2" diameter DR 11				Ea.	5.85			5.85	6.45
0516	3" diameter DR 11					14.60			14.60	16.05
0520	3" diameter DR 17					14.60			14.60	16.05
0524	4" diameter DR 11					20.50			20.50	22.50
0528	4" diameter DR 17					20.50			20.50	22.50
0532	6" diameter DR 11					46.50			46.50	51.50
0536	6" diameter DR 17					46.50			46.50	51.50
0540	8" diameter DR 11					115			115	127
0544	8" diameter DR 26					67			67	74
0548	10" diameter DR 11					430			430	475
0552	10" diameter DR 26					400			400	440
0556	12" diameter DR 11					455			455	500
0560	12" diameter DR 26					410			410	450
0564	16" diameter DR 11					239			239	263
0568	16" diameter DR 26					226			226	248
0572	18" diameter DR 11					253			253	279
0576	18" diameter DR 26					232			232	255
0580	20" diameter DR 11					385			385	425
0584	20" diameter DR 26					365			365	400
0588	22" diameter DR 11					530			530	585
0592	22" diameter DR 26					505			505	555
0596	24" diameter DR 11					650			650	715
0600	24" diameter DR 26					625			625	685
0604	28" diameter DR 17					740			740	810
0608	28" diameter DR 26					715			715	790
0612	30" diameter DR 17					915			915	1,000
0616	30" diameter DR 26					890			890	980
0620	36" diameter DR 26					1,125			1,125	1,250
0624	42" diameter DR 26					1,450			1,450	1,600

22 11 13.78 Pipe, High Density Polyethylene Plastic (HDPE)	Crew	Daily Output	Labor-Hours	Unit	Material	2020 Bare Costs Labor	Equipment	Total	Total Incl O&P
0628	48" diameter DR 26				Ea.	1,575			1,575	1,725
0632	54" diameter DR 26				▼	2,100			2,100	2,325
0700	Tee									
0704	1" diameter DR 11				Ea.	7.60			7.60	8.40
0708	1-1/2" diameter DR 11					10.70			10.70	11.75
0712	2" diameter DR 11					9.15			9.15	10.10
0716	3" diameter DR 11					16.75			16.75	18.45
0720	3" diameter DR 17					16.75			16.75	18.45
0724	4" diameter DR 11					24.50			24.50	27
0728	4" diameter DR 17					24.50			24.50	27
0732	6" diameter DR 11					61			61	67
0736	6" diameter DR 17					61			61	67
0740	8" diameter DR 11					151			151	166
0744	8" diameter DR 17					151			151	166
0748	10" diameter DR 11					450			450	495
0752	10" diameter DR 17					450			450	495
0756	12" diameter DR 11					600			600	660
0760	12" diameter DR 17					600			600	660
0764	16" diameter DR 11					315			315	350
0768	16" diameter DR 17					261			261	287
0772	18" diameter DR 11					445			445	490
0776	18" diameter DR 17					365			365	405
0780	20" diameter DR 11					545			545	600
0784	20" diameter DR 17					445			445	490
0788	22" diameter DR 11					700			700	770
0792	22" diameter DR 17					555			555	610
0796	24" diameter DR 11					870			870	955
0800	24" diameter DR 17					730			730	805
0804	28" diameter DR 17					1,425			1,425	1,550
0812	30" diameter DR 17					1,625			1,625	1,775
0820	36" diameter DR 17					2,675			2,675	2,950
0824	42" diameter DR 26					2,975			2,975	3,275
0828	48" diameter DR 26				▼	3,200			3,200	3,525
1000	Flange adptr, w/back-up ring and 1/2 cost of plated bolt set									
1004	1" diameter DR 11				Ea.	29			29	32
1008	1-1/2" diameter DR 11					29			29	32
1012	2" diameter DR 11					18.30			18.30	20
1016	3" diameter DR 11					21.50			21.50	23.50
1020	3" diameter DR 17					21.50			21.50	23.50
1024	4" diameter DR 11					29			29	32
1028	4" diameter DR 17					29			29	32
1032	6" diameter DR 11					41			41	45.50
1036	6" diameter DR 17					41			41	45.50
1040	8" diameter DR 11					59.50			59.50	65.50
1044	8" diameter DR 26					59.50			59.50	65.50
1048	10" diameter DR 11					94.50			94.50	104
1052	10" diameter DR 26					94.50			94.50	104
1056	12" diameter DR 11					139			139	153
1060	12" diameter DR 26					139			139	153
1064	16" diameter DR 11					297			297	325
1068	16" diameter DR 26					297			297	325
1072	18" diameter DR 11					385			385	425
1076	18" diameter DR 26					385			385	425

22 11 13.78 Pipe, High Density Polyethylene Plastic (HDPE)	Crew	Daily Output	Labor-Hours	Unit	Material	2020 Bare Costs Labor	Equipment	Total	Total Incl O&P
1080	20" diameter DR 11				Ea.	535			535	590
1084	20" diameter DR 26					535			535	590
1088	22" diameter DR 11					580			580	640
1092	22" diameter DR 26					580			580	640
1096	24" diameter DR 17					630			630	695
1100	24" diameter DR 32.5					630			630	695
1104	28" diameter DR 15.5					855			855	940
1108	28" diameter DR 32.5					855			855	940
1112	30" diameter DR 11					990			990	1,100
1116	30" diameter DR 21					990			990	1,100
1120	36" diameter DR 26					1,075			1,075	1,200
1124	42" diameter DR 26					1,225			1,225	1,350
1128	48" diameter DR 26					1,500			1,500	1,650
1132	54" diameter DR 26				↓	1,825			1,825	2,025
1200	Reducer									
1208	2" x 1-1/2" diameter DR 11				Ea.	8.75			8.75	9.65
1212	3" x 2" diameter DR 11					8.75			8.75	9.65
1216	4" x 2" diameter DR 11					10.20			10.20	11.25
1220	4" x 3" diameter DR 11					13.15			13.15	14.45
1224	6" x 4" diameter DR 11					30.50			30.50	33.50
1228	8" x 6" diameter DR 11					46.50			46.50	51.50
1232	10" x 8" diameter DR 11					80.50			80.50	88.50
1236	12" x 8" diameter DR 11					131			131	144
1240	12" x 10" diameter DR 11					105			105	116
1244	14" x 12" diameter DR 11					117			117	128
1248	16" x 14" diameter DR 11					149			149	164
1252	18" x 16" diameter DR 11					181			181	199
1256	20" x 18" diameter DR 11					360			360	395
1260	22" x 20" diameter DR 11					440			440	485
1264	24" x 22" diameter DR 11					495			495	545
1268	26" x 24" diameter DR 11					585			585	640
1272	28" x 24" diameter DR 11					745			745	820
1276	32" x 28" diameter DR 17					965			965	1,050
1280	36" x 32" diameter DR 17				↓	1,325			1,325	1,450
4000	Welding labor per joint, not including welding machine									
4010	Pipe joint size (cost based on thickest wall for each diam.)									
4030	1" pipe size	4 Skwk	273	.117	Ea.		6.45		6.45	9.70
4040	1-1/2" pipe size		175	.183			10.05		10.05	15.15
4050	2" pipe size		128	.250			13.70		13.70	20.50
4060	3" pipe size		100	.320			17.55		17.55	26.50
4070	4" pipe size	↓	77	.416			23		23	34.50
4080	6" pipe size	5 Skwk	63	.635			35		35	52.50
4090	8" pipe size		48	.833			45.50		45.50	69
4100	10" pipe size	↓	40	1			55		55	83
4110	12" pipe size	6 Skwk	41	1.171			64		64	97
4120	16" pipe size		34	1.412			77.50		77.50	117
4130	18" pipe size	↓	32	1.500			82.50		82.50	124
4140	20" pipe size	8 Skwk	37	1.730			95		95	143
4150	22" pipe size		35	1.829			100		100	152
4160	24" pipe size		34	1.882			103		103	156
4170	28" pipe size		33	1.939			106		106	161
4180	30" pipe size		32	2			110		110	166
4190	36" pipe size		31	2.065	↓		113		113	171

22 11 Facility Water Distribution

22 11 13 – Facility Water Distribution Piping

22 11 13.78 Pipe, High Density Polyethylene Plastic (HDPE)	Crew	Daily Output	Labor-Hours	Unit	Material	2020 Bare Costs Labor	Equipment	Total	Total Incl O&P
4200	42" pipe size	8 Skwk	30	2.133	Ea.		117		117	177
4210	48" pipe size	9 Skwk	33	2.182			120		120	181
4220	54" pipe size	"	31	2.323			127		127	193
5000	Dual wall contained pipe									
5040	Straight									
5054	1" DR 11 x 3" DR 11				L.F.	5.35			5.35	5.90
5058	1" DR 11 x 4" DR 11					5.65			5.65	6.25
5062	1-1/2" DR 11 x 4" DR 17					6.25			6.25	6.90
5066	2" DR 11 x 4" DR 17					6.60			6.60	7.25
5070	2" DR 11 x 6" DR 17					10			10	11
5074	3" DR 11 x 6" DR 17					11.20			11.20	12.30
5078	3" DR 11 x 6" DR 26					9.55			9.55	10.50
5086	3" DR 17 x 8" DR 17					12.15			12.15	13.40
5090	4" DR 11 x 8" DR 17					16.85			16.85	18.55
5094	4" DR 17 x 8" DR 26					13.20			13.20	14.50
5098	6" DR 11 x 10" DR 17					26			26	29
5102	6" DR 17 x 10" DR 26					17.80			17.80	19.60
5106	6" DR 26 x 10" DR 26					16.70			16.70	18.35
5110	8" DR 17 x 12" DR 26					25.50			25.50	28.50
5114	8" DR 26 x 12" DR 32.5					23.50			23.50	26
5118	10" DR 17 x 14" DR 26					38.50			38.50	42
5122	10" DR 17 x 16" DR 26					27.50			27.50	30.50
5126	10" DR 26 x 16" DR 26					36.50			36.50	40
5130	12" DR 26 x 16" DR 26					38.50			38.50	42.50
5134	12" DR 17 x 18" DR 26					53.50			53.50	58.50
5138	12" DR 26 x 18" DR 26					50.50			50.50	56
5142	14" DR 26 x 20" DR 32.5					51.50			51.50	57
5146	16" DR 26 x 22" DR 32.5					57			57	63
5150	18" DR 26 x 24" DR 32.5					71			71	78.50
5154	20" DR 32.5 x 28" DR 32.5					80.50			80.50	88.50
5158	22" DR 32.5 x 30" DR 32.5					80.50			80.50	88.50
5162	24" DR 32.5 x 32" DR 32.5					83.50			83.50	92
5166	36" DR 32.5 x 42" DR 32.5					97.50			97.50	107
5300	Force transfer coupling									
5354	1" DR 11 x 3" DR 11				Ea.	188			188	207
5358	1" DR 11 x 4" DR 17					221			221	243
5362	1-1/2" DR 11 x 4" DR 17					212			212	233
5366	2" DR 11 x 4" DR 17					212			212	233
5370	2" DR 11 x 6" DR 17					365			365	400
5374	3" DR 11 x 6" DR 17					390			390	425
5378	3" DR 11 x 6" DR 26					390			390	425
5382	3" DR 11 x 8" DR 11					390			390	430
5386	3" DR 11 x 8" DR 17					425			425	465
5390	4" DR 11 x 8" DR 17					435			435	475
5394	4" DR 17 x 8" DR 26					440			440	480
5398	6" DR 11 x 10" DR 17					520			520	570
5402	6" DR 17 x 10" DR 26					495			495	545
5406	6" DR 26 x 10" DR 26					675			675	745
5410	8" DR 17 x 12" DR 26					720			720	790
5414	8" DR 26 x 12" DR 32.5					830			830	910
5418	10" DR 17 x 14" DR 26					980			980	1,075
5422	10" DR 17 x 16" DR 26					990			990	1,100
5426	10" DR 26 x 16" DR 26					1,100			1,100	1,200

22 11 13.78 Pipe, High Density Polyethylene Plastic (HDPE)	Crew	Daily Output	Labor-Hours	Unit	Material	2020 Bare Costs Labor	Equipment	Total	Total Incl O&P
5430	12" DR 26 x 16" DR 26				Ea.	1,350			1,350	1,475
5434	12" DR 17 x 18" DR 26					1,350			1,350	1,475
5438	12" DR 26 x 18" DR 26					1,475			1,475	1,625
5442	14" DR 26 x 20" DR 32.5					1,650			1,650	1,800
5446	16" DR 26 x 22" DR 32.5					1,825			1,825	2,000
5450	18" DR 26 x 24" DR 32.5					2,000			2,000	2,200
5454	20" DR 32.5 x 28" DR 32.5					2,200			2,200	2,425
5458	22" DR 32.5 x 30" DR 32.5					2,425			2,425	2,675
5462	24" DR 32.5 x 32" DR 32.5					2,675			2,675	2,950
5466	36" DR 32.5 x 42" DR 32.5					2,875			2,875	3,175
5600	90° elbow									
5654	1" DR 11 x 3" DR 11				Ea.	206			206	226
5658	1" DR 11 x 4" DR 17					243			243	267
5662	1-1/2" DR 11 x 4" DR 17					250			250	275
5666	2" DR 11 x 4" DR 17					244			244	269
5670	2" DR 11 x 6" DR 17					355			355	390
5674	3" DR 11 x 6" DR 17					380			380	420
5678	3" DR 17 x 6" DR 26					390			390	425
5682	3" DR 17 x 8" DR 11					520			520	575
5686	3" DR 17 x 8" DR 17					555			555	610
5690	4" DR 11 x 8" DR 17					470			470	520
5694	4" DR 17 x 8" DR 26					450			450	495
5698	6" DR 11 x 10" DR 17					820			820	905
5702	6" DR 17 x 10" DR 26					835			835	920
5706	6" DR 26 x 10" DR 26					815			815	900
5710	8" DR 17 x 12" DR 26					1,100			1,100	1,200
5714	8" DR 26 x 12" DR 32.5					905			905	995
5718	10" DR 17 x 14" DR 26					1,075			1,075	1,175
5722	10" DR 17 x 16" DR 26					1,050			1,050	1,150
5726	10" DR 26 x 16" DR 26					1,025			1,025	1,150
5730	12" DR 26 x 16" DR 26					1,050			1,050	1,175
5734	12" DR 17 x 18" DR 26					1,150			1,150	1,275
5738	12" DR 26 x 18" DR 26					1,175			1,175	1,300
5742	14" DR 26 x 20" DR 32.5					1,275			1,275	1,400
5746	16" DR 26 x 22" DR 32.5					1,325			1,325	1,450
5750	18" DR 26 x 24" DR 32.5					1,375			1,375	1,500
5754	20" DR 32.5 x 28" DR 32.5					1,425			1,425	1,550
5758	22" DR 32.5 x 30" DR 32.5					1,475			1,475	1,600
5762	24" DR 32.5 x 32" DR 32.5					1,525			1,525	1,675
5766	36" DR 32.5 x 42" DR 32.5					1,575			1,575	1,725
5800	45° elbow									
5804	1" DR 11 x 3" DR 11				Ea.	222			222	244
5808	1" DR 11 x 4" DR 17					195			195	215
5812	1-1/2" DR 11 x 4" DR 17					199			199	219
5816	2" DR 11 x 4" DR 17					190			190	209
5820	2" DR 11 x 6" DR 17					272			272	299
5824	3" DR 11 x 6" DR 17					292			292	320
5828	3" DR 17 x 6" DR 26					295			295	325
5832	3" DR 17 x 8" DR 11					375			375	410
5836	3" DR 17 x 8" DR 17					420			420	465
5840	4" DR 11 x 8" DR 17					345			345	380
5844	4" DR 17 x 8" DR 26					350			350	385
5848	6" DR 11 x 10" DR 17					545			545	600

22 11 13.78 Pipe, High Density Polyethylene Plastic (HDPE)	Crew	Daily Output	Labor-Hours	Unit	Material	2020 Bare Costs Labor	Equipment	Total	Total Incl O&P
5852	6" DR 17 x 10" DR 26				Ea.	555			555	610
5856	6" DR 26 x 10" DR 26					625			625	685
5860	8" DR 17 x 12" DR 26					740			740	815
5864	8" DR 26 x 12" DR 32.5					695			695	765
5868	10" DR 17 x 14" DR 26					785			785	860
5872	10" DR 17 x 16" DR 26					755			755	830
5876	10" DR 26 x 16" DR 26					800			800	880
5880	12" DR 26 x 16" DR 26					720			720	795
5884	12" DR 17 x 18" DR 26					1,025			1,025	1,125
5888	12" DR 26 x 18" DR 26					910			910	1,000
5892	14" DR 26 x 20" DR 32.5					990			990	1,100
5896	16" DR 26 x 22" DR 32.5					1,025			1,025	1,125
5900	18" DR 26 x 24" DR 32.5					1,100			1,100	1,200
5904	20" DR 32.5 x 28" DR 32.5					1,075			1,075	1,200
5908	22" DR 32.5 x 30" DR 32.5					1,125			1,125	1,225
5912	24" DR 32.5 x 32" DR 32.5					1,150			1,150	1,275
5916	36" DR 32.5 x 42" DR 32.5					1,200			1,200	1,325
6000	Access port with 4" riser									
6050	1" DR 11 x 4" DR 17				Ea.	284			284	310
6054	1-1/2" DR 11 x 4" DR 17					286			286	315
6058	2" DR 11 x 6" DR 17					350			350	385
6062	3" DR 11 x 6" DR 17					355			355	390
6066	3" DR 17 x 6" DR 26					345			345	380
6070	3" DR 17 x 8" DR 11					355			355	390
6074	3" DR 17 x 8" DR 17					370			370	405
6078	4" DR 11 x 8" DR 17					380			380	415
6082	4" DR 17 x 8" DR 26					360			360	400
6086	6" DR 11 x 10" DR 17					440			440	485
6090	6" DR 17 x 10" DR 26					405			405	445
6094	6" DR 26 x 10" DR 26					425			425	465
6098	8" DR 17 x 12" DR 26					455			455	500
6102	8" DR 26 x 12" DR 32.5					445			445	490
6200	End termination with vent plug									
6204	1" DR 11 x 3" DR 11				Ea.	116			116	128
6208	1" DR 11 x 4" DR 17					141			141	155
6212	1-1/2" DR 11 x 4" DR 17					141			141	155
6216	2" DR 11 x 4" DR 17					141			141	155
6220	2" DR 11 x 6" DR 17					286			286	315
6224	3" DR 11 x 6" DR 17					286			286	315
6228	3" DR 17 x 6" DR 26					286			286	315
6232	3" DR 17 x 8" DR 11					276			276	305
6236	3" DR 17 x 8" DR 17					310			310	340
6240	4" DR 11 x 8" DR 17					370			370	405
6244	4" DR 17 x 8" DR 26					345			345	380
6248	6" DR 11 x 10" DR 17					400			400	440
6252	6" DR 17 x 10" DR 26					370			370	410
6256	6" DR 26 x 10" DR 26					540			540	595
6260	8" DR 17 x 12" DR 26					565			565	625
6264	8" DR 26 x 12" DR 32.5					680			680	745
6268	10" DR 17 x 14" DR 26					900			900	990
6272	10" DR 17 x 16" DR 26					850			850	935
6276	10" DR 26 x 16" DR 26					955			955	1,050
6280	12" DR 26 x 16" DR 26					1,050			1,050	1,175

For customer support on your Site Work & Landscape Costs with RSMeans data, call 800.448.8182.

237

22 11 13.78 Pipe, High Density Polyethylene Plastic (HDPE)	Crew	Daily Output	Labor-Hours	Unit	Material	2020 Bare Costs Labor	Equipment	Total	Total Incl O&P
6284	12" DR 17 x 18" DR 26				Ea.	1,225			1,225	1,350
6288	12" DR 26 x 18" DR 26					1,350			1,350	1,475
6292	14" DR 26 x 20" DR 32.5					1,525			1,525	1,675
6296	16" DR 26 x 22" DR 32.5					1,700			1,700	1,850
6300	18" DR 26 x 24" DR 32.5					1,900			1,900	2,100
6304	20" DR 32.5 x 28" DR 32.5					2,125			2,125	2,325
6308	22" DR 32.5 x 30" DR 32.5					2,375			2,375	2,600
6312	24" DR 32.5 x 32" DR 32.5					2,650			2,650	2,925
6316	36" DR 32.5 x 42" DR 32.5					2,925			2,925	3,225
6600	Tee									
6604	1" DR 11 x 3" DR 11				Ea.	239			239	263
6608	1" DR 11 x 4" DR 17					260			260	286
6612	1-1/2" DR 11 x 4" DR 17					269			269	296
6616	2" DR 11 x 4" DR 17					244			244	268
6620	2" DR 11 x 6" DR 17					390			390	430
6624	3" DR 11 x 6" DR 17					460			460	505
6628	3" DR 17 x 6" DR 26					435			435	480
6632	3" DR 17 x 8" DR 11					535			535	590
6636	3" DR 17 x 8" DR 17					580			580	640
6640	4" DR 11 x 8" DR 17					700			700	770
6644	4" DR 17 x 8" DR 26					700			700	770
6648	6" DR 11 x 10" DR 17					910			910	1,000
6652	6" DR 17 x 10" DR 26					915			915	1,000
6656	6" DR 26 x 10" DR 26					1,125			1,125	1,225
6660	8" DR 17 x 12" DR 26					1,125			1,125	1,225
6664	8" DR 26 x 12" DR 32.5					1,450			1,450	1,600
6668	10" DR 17 x 14" DR 26					1,675			1,675	1,850
6672	10" DR 17 x 16" DR 26					1,875			1,875	2,050
6676	10" DR 26 x 16" DR 26					1,875			1,875	2,050
6680	12" DR 26 x 16" DR 26					2,125			2,125	2,350
6684	12" DR 17 x 18" DR 26					2,125			2,125	2,350
6688	12" DR 26 x 18" DR 26					2,425			2,425	2,650
6692	14" DR 26 x 20" DR 32.5					2,725			2,725	3,000
6696	16" DR 26 x 22" DR 32.5					3,075			3,075	3,400
6700	18" DR 26 x 24" DR 32.5					3,500			3,500	3,850
6704	20" DR 32.5 x 28" DR 32.5					3,950			3,950	4,350
6708	22" DR 32.5 x 30" DR 32.5					4,475			4,475	4,925
6712	24" DR 32.5 x 32" DR 32.5					5,075			5,075	5,575
6716	36" DR 32.5 x 42" DR 32.5					5,825			5,825	6,400
6800	Wye									
6816	2" DR 11 x 4" DR 17				Ea.	595			595	655
6820	2" DR 11 x 6" DR 17					735			735	810
6824	3" DR 11 x 6" DR 17					780			780	855
6828	3" DR 17 x 6" DR 26					780			780	855
6832	3" DR 17 x 8" DR 11					830			830	910
6836	3" DR 17 x 8" DR 17					900			900	990
6840	4" DR 11 x 8" DR 17					990			990	1,100
6844	4" DR 17 x 8" DR 26					625			625	685
6848	6" DR 11 x 10" DR 17					1,525			1,525	1,675
6852	6" DR 17 x 10" DR 26					950			950	1,050
6856	6" DR 26 x 10" DR 26					1,325			1,325	1,450
6860	8" DR 17 x 12" DR 26					1,850			1,850	2,025
6864	8" DR 26 x 12" DR 32.5					1,525			1,525	1,675

22 11 13 – Facility Water Distribution Piping

22 11 13.78 Pipe, High Density Polyethylene Plastic (HDPE)	Crew	Daily Output	Labor-Hours	Unit	Material	2020 Bare Costs Labor	2020 Bare Costs Equipment	Total	Total Incl O&P
6868	10" DR 17 x 14" DR 26				Ea.	1,625			1,625	1,800
6872	10" DR 17 x 16" DR 26					1,750			1,750	1,925
6876	10" DR 26 x 16" DR 26					1,900			1,900	2,100
6880	12" DR 26 x 16" DR 26					2,050			2,050	2,250
6884	12" DR 17 x 18" DR 26					2,200			2,200	2,425
6888	12" DR 26 x 18" DR 26					2,400			2,400	2,625
6892	14" DR 26 x 20" DR 32.5					2,550			2,550	2,825
6896	16" DR 26 x 22" DR 32.5					2,700			2,700	2,975
6900	18" DR 26 x 24" DR 32.5					3,100			3,100	3,400
6904	20" DR 32.5 x 28" DR 32.5					3,225			3,225	3,550
6908	22" DR 32.5 x 30" DR 32.5					3,500			3,500	3,850
6912	24" DR 32.5 x 32" DR 32.5					3,775			3,775	4,150
9000	Welding labor per joint, not including welding machine									
9010	Pipe joint size, outer pipe (cost based on the thickest walls)									
9020	Straight pipe									
9050	3" pipe size	4 Skwk	96	.333	Ea.		18.30		18.30	27.50
9060	4" pipe size	"	77	.416			23		23	34.50
9070	6" pipe size	5 Skwk	60	.667			36.50		36.50	55.50
9080	8" pipe size	"	40	1			55		55	83
9090	10" pipe size	6 Skwk	41	1.171			64		64	97
9100	12" pipe size		39	1.231			67.50		67.50	102
9110	14" pipe size		38	1.263			69.50		69.50	105
9120	16" pipe size		35	1.371			75		75	114
9130	18" pipe size	8 Skwk	45	1.422			78		78	118
9140	20" pipe size		42	1.524			83.50		83.50	126
9150	22" pipe size		40	1.600			88		88	133
9160	24" pipe size		38	1.684			92.50		92.50	140
9170	28" pipe size		37	1.730			95		95	143
9180	30" pipe size		36	1.778			97.50		97.50	147
9190	32" pipe size		35	1.829			100		100	152
9200	42" pipe size		32	2			110		110	166

22 11 19 – Domestic Water Piping Specialties

22 11 19.38 Water Supply Meters

		Crew	Daily Output	Labor-Hours	Unit	Material	2020 Bare Costs Labor	2020 Bare Costs Equipment	Total	Total Incl O&P
0010	**WATER SUPPLY METERS**									
1000	Detector, serves dual systems such as fire and domestic or									
1020	process water, wide range cap., UL and FM approved									
1100	3" mainline x 2" by-pass, 400 GPM	Q-1	3.60	4.444	Ea.	8,025	258		8,283	9,200
1140	4" mainline x 2" by-pass, 700 GPM	"	2.50	6.400		8,025	370		8,395	9,375
1180	6" mainline x 3" by-pass, 1,600 GPM	Q-2	2.60	9.231		12,300	555		12,855	14,300
1220	8" mainline x 4" by-pass, 2,800 GPM		2.10	11.429		18,200	685		18,885	21,000
1260	10" mainline x 6" by-pass, 4,400 GPM		2	12		26,000	720		26,720	29,700
1300	10" x 12" mainlines x 6" by-pass, 5,400 GPM		1.70	14.118		35,300	850		36,150	40,100
2000	Domestic/commercial, bronze									
2020	Threaded									
2060	5/8" diameter, to 20 GPM	1 Plum	16	.500	Ea.	54	32		86	108
2080	3/4" diameter, to 30 GPM		14	.571		98.50	37		135.50	163
2100	1" diameter, to 50 GPM		12	.667		149	43		192	229
2300	Threaded/flanged									
2340	1-1/2" diameter, to 100 GPM	1 Plum	8	1	Ea.	365	64.50		429.50	495
2360	2" diameter, to 160 GPM	"	6	1.333	"	495	86		581	675
2600	Flanged, compound									
2640	3" diameter, 320 GPM	Q-1	3	5.333	Ea.	2,650	310		2,960	3,400

239

For customer support on your Site Work & Landscape Costs with RSMeans data, call 800.448.8182.

22 11 Facility Water Distribution

22 11 19 – Domestic Water Piping Specialties

22 11 19.38 Water Supply Meters

		Crew	Daily Output	Labor-Hours	Unit	Material	2020 Bare Costs Labor	Equipment	Total	Total Incl O&P
2660	4" diameter, to 500 GPM	Q-1	1.50	10.667	Ea.	4,250	620		4,870	5,600
2680	6" diameter, to 1,000 GPM		1	16		6,850	930		7,780	8,950
2700	8" diameter, to 1,800 GPM	↓	.80	20	↓	10,700	1,150		11,850	13,500
7000	Turbine									
7260	Flanged									
7300	2" diameter, to 160 GPM	1 Plum	7	1.143	Ea.	615	73.50		688.50	785
7320	3" diameter, to 450 GPM	Q-1	3.60	4.444		1,100	258		1,358	1,575
7340	4" diameter, to 650 GPM	"	2.50	6.400		1,825	370		2,195	2,550
7360	6" diameter, to 1,800 GPM	Q-2	2.60	9.231		3,000	555		3,555	4,125
7380	8" diameter, to 2,500 GPM		2.10	11.429		5,150	685		5,835	6,700
7400	10" diameter, to 5,500 GPM	↓	1.70	14.118	↓	6,925	850		7,775	8,900

22 11 19.42 Backflow Preventers

		Crew	Daily Output	Labor-Hours	Unit	Material	2020 Bare Costs Labor	Equipment	Total	Total Incl O&P
0010	**BACKFLOW PREVENTERS**, Includes valves									
0020	and four test cocks, corrosion resistant, automatic operation									
1000	Double check principle									
1010	Threaded, with ball valves									
1020	3/4" pipe size	1 Plum	16	.500	Ea.	241	32		273	315
1030	1" pipe size		14	.571		253	37		290	335
1040	1-1/2" pipe size		10	.800		620	51.50		671.50	760
1050	2" pipe size	↓	7	1.143	↓	665	73.50		738.50	845
1080	Threaded, with gate valves									
1100	3/4" pipe size	1 Plum	16	.500	Ea.	1,150	32		1,182	1,300
1120	1" pipe size		14	.571		1,150	37		1,187	1,325
1140	1-1/2" pipe size		10	.800		1,225	51.50		1,276.50	1,425
1160	2" pipe size	↓	7	1.143	↓	1,825	73.50		1,898.50	2,100
1200	Flanged, valves are gate									
1210	3" pipe size	Q-1	4.50	3.556	Ea.	2,975	206		3,181	3,575
1220	4" pipe size	"	3	5.333		3,000	310		3,310	3,775
1230	6" pipe size	Q-2	3	8		4,675	480		5,155	5,850
1240	8" pipe size		2	12		7,975	720		8,695	9,850
1250	10" pipe size	↓	1	24		12,000	1,450		13,450	15,400
1300	Flanged, valves are OS&Y									
1380	3" pipe size	Q-1	4.50	3.556	Ea.	2,850	206		3,056	3,450
1400	4" pipe size	"	3	5.333		4,675	310		4,985	5,600
1420	6" pipe size	Q-2	3	8	↓	6,700	480		7,180	8,100
4000	Reduced pressure principle									
4100	Threaded, bronze, valves are ball									
4120	3/4" pipe size	1 Plum	16	.500	Ea.	530	32		562	635
4140	1" pipe size		14	.571		560	37		597	670
4150	1-1/4" pipe size		12	.667		1,125	43		1,168	1,300
4160	1-1/2" pipe size		10	.800		1,150	51.50		1,201.50	1,350
4180	2" pipe size	↓	7	1.143	↓	1,325	73.50		1,398.50	1,575
5000	Flanged, bronze, valves are OS&Y									
5060	2-1/2" pipe size	Q-1	5	3.200	Ea.	5,475	186		5,661	6,300
5080	3" pipe size		4.50	3.556		6,600	206		6,806	7,550
5100	4" pipe size	↓	3	5.333		7,725	310		8,035	8,975
5120	6" pipe size	Q-2	3	8	↓	12,000	480		12,480	13,900
5200	Flanged, iron, valves are gate									
5210	2-1/2" pipe size	Q-1	5	3.200	Ea.	3,200	186		3,386	3,775
5220	3" pipe size		4.50	3.556		3,300	206		3,506	3,950
5230	4" pipe size	↓	3	5.333		3,775	310		4,085	4,650
5240	6" pipe size	Q-2	3	8		5,675	480		6,155	6,975

22 11 19 – Domestic Water Piping Specialties

22 11 19.42 Backflow Preventers		Crew	Daily Output	Labor-Hours	Unit	Material	2020 Bare Costs Labor	Equipment	Total	Total Incl O&P
5250	8" pipe size	Q-2	2	12	Ea.	9,575	720		10,295	11,600
5260	10" pipe size	↓	1	24	↓	16,200	1,450		17,650	20,100
5600	Flanged, iron, valves are OS&Y									
5660	2-1/2" pipe size	Q-1	5	3.200	Ea.	3,300	186		3,486	3,900
5680	3" pipe size		4.50	3.556		3,275	206		3,481	3,900
5700	4" pipe size	↓	3	5.333		4,725	310		5,035	5,675
5720	6" pipe size	Q-2	3	8		5,825	480		6,305	7,150
5740	8" pipe size		2	12		10,300	720		11,020	12,400
5760	10" pipe size	↓	1	24	↓	19,800	1,450		21,250	24,000

22 11 19.54 Water Hammer Arresters/Shock Absorbers		Crew	Daily Output	Labor-Hours	Unit	Material	2020 Bare Costs Labor	Equipment	Total	Total Incl O&P
0010	**WATER HAMMER ARRESTERS/SHOCK ABSORBERS**									
0490	Copper									
0500	3/4" male IPS for 1 to 11 fixtures	1 Plum	12	.667	Ea.	31	43		74	98.50
0600	1" male IPS for 12 to 32 fixtures		8	1		51.50	64.50		116	153
0700	1-1/4" male IPS for 33 to 60 fixtures		8	1		50.50	64.50		115	153
0800	1-1/2" male IPS for 61 to 113 fixtures		8	1		73	64.50		137.50	177
0900	2" male IPS for 114 to 154 fixtures		8	1		108	64.50		172.50	215
1000	2-1/2" male IPS for 155 to 330 fixtures	↓	4	2	↓	330	129		459	560

22 11 19.64 Hydrants		Crew	Daily Output	Labor-Hours	Unit	Material	2020 Bare Costs Labor	Equipment	Total	Total Incl O&P
0010	**HYDRANTS**									
0050	Wall type, moderate climate, bronze, encased									
0200	3/4" IPS connection	1 Plum	16	.500	Ea.	1,025	32		1,057	1,175
0300	1" IPS connection		14	.571		1,125	37		1,162	1,300
0500	Anti-siphon type, 3/4" connection	↓	16	.500	↓	815	32		847	945
1000	Non-freeze, bronze, exposed									
1100	3/4" IPS connection, 4" to 9" thick wall	1 Plum	14	.571	Ea.	470	37		507	570
1120	10" to 14" thick wall		12	.667		445	43		488	555
1140	15" to 19" thick wall		12	.667		550	43		593	670
1160	20" to 24" thick wall	↓	10	.800	↓	830	51.50		881.50	985
1200	For 1" IPS connection, add					15%	10%			
1240	For 3/4" adapter type vacuum breaker, add				Ea.	83			83	91.50
1280	For anti-siphon type, add				"	176			176	194
2000	Non-freeze bronze, encased, anti-siphon type									
2100	3/4" IPS connection, 5" to 9" thick wall	1 Plum	14	.571	Ea.	1,275	37		1,312	1,450
2120	10" to 14" thick wall		12	.667		1,625	43		1,668	1,875
2140	15" to 19" thick wall		12	.667		1,725	43		1,768	1,975
2160	20" to 24" thick wall	↓	10	.800	↓	1,800	51.50		1,851.50	2,050
2200	For 1" IPS connection, add					10%	10%			
3000	Ground box type, bronze frame, 3/4" IPS connection									
3080	Non-freeze, all bronze, polished face, set flush									
3100	2' depth of bury	1 Plum	8	1	Ea.	1,325	64.50		1,389.50	1,550
3120	3' depth of bury		8	1		1,425	64.50		1,489.50	1,675
3140	4' depth of bury		8	1		1,525	64.50		1,589.50	1,775
3160	5' depth of bury		7	1.143		1,625	73.50		1,698.50	1,875
3180	6' depth of bury		7	1.143		1,700	73.50		1,773.50	1,950
3200	7' depth of bury		6	1.333		1,675	86		1,761	1,975
3220	8' depth of bury		5	1.600		1,775	103		1,878	2,100
3240	9' depth of bury		4	2		1,875	129		2,004	2,275
3260	10' depth of bury	↓	4	2		1,975	129		2,104	2,350
3400	For 1" IPS connection, add					15%	10%			
3450	For 1-1/4" IPS connection, add					325%	14%			
3500	For 1-1/2" IPS connection, add					370%	18%			

For customer support on your Site Work & Landscape Costs with RSMeans data, call 800.448.8182.

241

22 11 Facility Water Distribution

22 11 19 – Domestic Water Piping Specialties

22 11 19.64 Hydrants	Crew	Daily Output	Labor-Hours	Unit	Material	2020 Bare Costs Labor	Equipment	Total	Total Incl O&P
3550	For 2" IPS connection, add				Ea.	445%	24%			
3600	For tapped drain port in box, add				↓	107			107	118
4000	Non-freeze, CI body, bronze frame & scoriated cover									
4010	with hose storage									
4100	2' depth of bury	1 Plum	7	1.143	Ea.	2,200	73.50		2,273.50	2,525
4120	3' depth of bury		7	1.143		2,300	73.50		2,373.50	2,625
4140	4' depth of bury		7	1.143		2,375	73.50		2,448.50	2,700
4160	5' depth of bury		6.50	1.231		2,425	79.50		2,504.50	2,775
4180	6' depth of bury		6	1.333		2,450	86		2,536	2,825
4200	7' depth of bury		5.50	1.455		2,550	94		2,644	2,950
4220	8' depth of bury		5	1.600		2,625	103		2,728	3,025
4240	9' depth of bury		4.50	1.778		2,825	115		2,940	3,300
4260	10' depth of bury	↓	4	2		2,825	129		2,954	3,300
4280	For 1" IPS connection, add					505			505	555
4300	For tapped drain port in box, add				↓	113			113	124
5000	Moderate climate, all bronze, polished face									
5020	and scoriated cover, set flush									
5100	3/4" IPS connection	1 Plum	16	.500	Ea.	830	32		862	965
5120	1" IPS connection	"	14	.571		2,000	37		2,037	2,275
5200	For tapped drain port in box, add				↓	113			113	124
6000	Ground post type, all non-freeze, all bronze, aluminum casing									
6010	guard, exposed head, 3/4" IPS connection									
6100	2' depth of bury	1 Plum	8	1	Ea.	1,275	64.50		1,339.50	1,500
6120	3' depth of bury		8	1		1,300	64.50		1,364.50	1,550
6140	4' depth of bury		8	1		1,475	64.50		1,539.50	1,725
6160	5' depth of bury		7	1.143		1,450	73.50		1,523.50	1,700
6180	6' depth of bury		7	1.143		1,775	73.50		1,848.50	2,050
6200	7' depth of bury		6	1.333		1,675	86		1,761	1,950
6220	8' depth of bury		5	1.600		1,775	103		1,878	2,100
6240	9' depth of bury		4	2		1,875	129		2,004	2,275
6260	10' depth of bury	↓	4	2		2,000	129		2,129	2,375
6300	For 1" IPS connection, add					40%	10%			
6350	For 1-1/4" IPS connection, add					140%	14%			
6400	For 1-1/2" IPS connection, add					225%	18%			
6450	For 2" IPS connection, add					315%	24%			

22 11 23 – Domestic Water Pumps

22 11 23.10 General Utility Pumps

		Crew	Daily Output	Labor-Hours	Unit	Material	2020 Bare Costs Labor	Equipment	Total	Total Incl O&P
0010	**GENERAL UTILITY PUMPS**									
2000	Single stage									
3000	Double suction,									
3190	75 HP, to 2,500 GPM	Q-3	.28	114	Ea.	16,500	7,000		23,500	28,600
3220	100 HP, to 3,000 GPM		.26	123		20,300	7,550		27,850	33,600
3240	150 HP, to 4,000 GPM	↓	.24	133	↓	23,600	8,175		31,775	38,200

242

For customer support on your Site Work & Landscape Costs with RSMeans data, call 800.448.8182.

22 12 Facility Potable-Water Storage Tanks

22 12 21 – Facility Underground Potable-Water Storage Tanks

22 12 21.13 Fiberglass, Undrgrnd Pot.-Water Storage Tanks	Crew	Daily Output	Labor-Hours	Unit	Material	2020 Bare Costs Labor	Equipment	Total	Total Incl O&P
0010 **FIBERGLASS, UNDERGROUND POTABLE-WATER STORAGE TANKS**									
0020 Excludes excavation, backfill & piping									
0030 Single wall									
2000 600 gallon capacity	B-21B	3.75	10.667	Ea.	4,125	490	126	4,741	5,425
2010 1,000 gallon capacity		3.50	11.429		5,275	525	135	5,935	6,725
2020 2,000 gallon capacity		3.25	12.308		7,500	565	145	8,210	9,250
2030 4,000 gallon capacity		3	13.333		10,100	610	157	10,867	12,200
2040 6,000 gallon capacity		2.65	15.094		11,400	695	178	12,273	13,800
2050 8,000 gallon capacity		2.30	17.391		13,900	800	205	14,905	16,700
2060 10,000 gallon capacity		2	20		15,400	920	236	16,556	18,600
2070 12,000 gallon capacity		1.50	26.667		21,600	1,225	315	23,140	25,900
2080 15,000 gallon capacity		1	40		24,500	1,825	470	26,795	30,300
2090 20,000 gallon capacity		.75	53.333		31,800	2,450	630	34,880	39,400
2100 25,000 gallon capacity		.50	80		47,400	3,675	940	52,015	58,500
2110 30,000 gallon capacity		.35	114		57,000	5,250	1,350	63,600	72,000
2120 40,000 gallon capacity		.30	133		80,000	6,125	1,575	87,700	99,000

22 13 Facility Sanitary Sewerage

22 13 16 – Sanitary Waste and Vent Piping

22 13 16.20 Pipe, Cast Iron

	Crew	Daily Output	Labor-Hours	Unit	Material	2020 Bare Costs Labor	Equipment	Total	Total Incl O&P
0010 **PIPE, CAST IRON**, Soil, on clevis hanger assemblies, 5' OC									
0020 Single hub, service wt., lead & oakum joints 10' OC									
2120 2" diameter	Q-1	63	.254	L.F.	17.10	14.75		31.85	41
2140 3" diameter		60	.267		26.50	15.45		41.95	52
2160 4" diameter		55	.291		24.50	16.85		41.35	52.50
2180 5" diameter	Q-2	76	.316		31	19		50	62.50
2200 6" diameter	"	73	.329		42.50	19.80		62.30	76
2220 8" diameter	Q-3	59	.542		64.50	33.50		98	121
2240 10" diameter		54	.593		102	36.50		138.50	168
2260 12" diameter		48	.667		145	41		186	221
2320 For service weight, double hub, add					10%				
2340 For extra heavy, single hub, add					48%	4%			
2360 For extra heavy, double hub, add					71%	4%			
2400 Lead for caulking (1#/diam. in.)	Q-1	160	.100	Lb.	1.01	5.80		6.81	9.80
2420 Oakum for caulking (1/8#/diam. in.)	"	40	.400	"	4.37	23		27.37	39.50
2960 To delete hangers, subtract									
2970 2" diam. to 4" diam.					16%	19%			
2980 5" diam. to 8" diam.					14%	14%			
2990 10" diam. to 15" diam.					13%	19%			
3000 Single hub, service wt., push-on gasket joints 10' OC									
3010 2" diameter	Q-1	66	.242	L.F.	18.30	14.05		32.35	41
3020 3" diameter		63	.254		28	14.75		42.75	53
3030 4" diameter		57	.281		26.50	16.30		42.80	53.50
3040 5" diameter	Q-2	79	.304		34	18.25		52.25	65
3050 6" diameter	"	75	.320		45.50	19.25		64.75	79
3060 8" diameter	Q-3	62	.516		71	31.50		102.50	126
3070 10" diameter		56	.571		114	35		149	178
3080 12" diameter		49	.653		160	40		200	236
3082 15" diameter		40	.800		191	49		240	284
3100 For service weight, double hub, add					65%				
3110 For extra heavy, single hub, add					48%	4%			

22 13 16.20 Pipe, Cast Iron

		Crew	Daily Output	Labor-Hours	Unit	Material	2020 Bare Costs Labor	Equipment	Total	Total Incl O&P
3120	For extra heavy, double hub, add					29%	4%			
3130	To delete hangers, subtract									
3140	2" diam. to 4" diam.					12%	21%			
3150	5" diam. to 8" diam.					10%	16%			
3160	10" diam. to 15" diam.					9%	21%			
4000	No hub, couplings 10' OC									
4100	1-1/2" diameter	Q-1	71	.225	L.F.	16.85	13.05		29.90	38
4120	2" diameter		67	.239		20	13.85		33.85	42.50
4140	3" diameter		64	.250		27	14.50		41.50	51
4160	4" diameter		58	.276		25	16		41	51.50
4180	5" diameter	Q-2	83	.289		36.50	17.40		53.90	66.50
4200	6" diameter	"	79	.304		43.50	18.25		61.75	75.50
4220	8" diameter	Q-3	69	.464		71	28.50		99.50	121
4240	10" diameter		61	.525		117	32		149	177
4244	12" diameter		58	.552		135	34		169	200
4248	15" diameter		52	.615		207	38		245	285
4280	To delete hangers, subtract									
4290	1-1/2" diam. to 6" diam.					22%	47%			
4300	8" diam. to 10" diam.					21%	44%			
4310	12" diam. to 15" diam.					19%	40%			

22 13 16.30 Pipe Fittings, Cast Iron

		Crew	Daily Output	Labor-Hours	Unit	Material	2020 Bare Costs Labor	Equipment	Total	Total Incl O&P
0010	**PIPE FITTINGS, CAST IRON**, Soil									
0040	Hub and spigot, service weight, lead & oakum joints									
0080	1/4 bend, 2"	Q-1	16	1	Ea.	23.50	58		81.50	113
0120	3"		14	1.143		31	66.50		97.50	133
0140	4"		13	1.231		49	71.50		120.50	161
0160	5"	Q-2	18	1.333		68	80		148	195
0180	6"	"	17	1.412		85	85		170	221
0200	8"	Q-3	11	2.909		256	178		434	550
0220	10"		10	3.200		370	196		566	700
0224	12"		9	3.556		510	218		728	885
0266	Closet bend, 3" diameter with flange 10" x 16"	Q-1	14	1.143		128	66.50		194.50	240
0268	16" x 16"		12	1.333		139	77.50		216.50	269
0270	Closet bend, 4" diameter, 2-1/2" x 4" ring, 6" x 16"		13	1.231		110	71.50		181.50	228
0280	8" x 16"		13	1.231		97.50	71.50		169	214
0290	10" x 12"		12	1.333		129	77.50		206.50	258
0300	10" x 18"		11	1.455		130	84.50		214.50	269
0310	12" x 16"		11	1.455		110	84.50		194.50	247
0330	16" x 16"		10	1.600		143	93		236	297
0340	1/8 bend, 2"		16	1		16.65	58		74.65	105
0350	3"		14	1.143		26	66.50		92.50	128
0360	4"		13	1.231		38	71.50		109.50	149
0380	5"	Q-2	18	1.333		54	80		134	180
0400	6"	"	17	1.412		64.50	85		149.50	198
0420	8"	Q-3	11	2.909		193	178		371	480
0440	10"		10	3.200		277	196		473	600
0460	12"		9	3.556		525	218		743	905
0500	Sanitary tee, 2"	Q-1	10	1.600		32.50	93		125.50	175
0540	3"		9	1.778		53	103		156	212
0620	4"		8	2		65.50	116		181.50	246
0700	5"	Q-2	12	2		129	120		249	320
0800	6"	"	11	2.182		146	131		277	355

22 13 16.30 Pipe Fittings, Cast Iron		Crew	Daily Output	Labor-Hours	Unit	Material	2020 Bare Costs Labor	Equipment	Total	Total Incl O&P
0880	8"	Q-3	7	4.571	Ea.	385	280		665	845
1000	Tee, 2"	Q-1	10	1.600		47	93		140	191
1060	3"		9	1.778		70.50	103		173.50	232
1120	4"		8	2		88.50	116		204.50	271
1200	5"	Q-2	12	2		187	120		307	385
1300	6"	"	11	2.182		186	131		317	400
1380	8"	Q-3	7	4.571		370	280		650	825
1400	Combination Y and 1/8 bend									
1420	2"	Q-1	10	1.600	Ea.	41	93		134	184
1460	3"		9	1.778		62	103		165	222
1520	4"		8	2		85.50	116		201.50	268
1540	5"	Q-2	12	2		162	120		282	360
1560	6"		11	2.182		205	131		336	420
1580	8"		7	3.429		505	206		711	870
1582	12"	Q-3	6	5.333		1,025	325		1,350	1,625
1600	Double Y, 2"	Q-1	8	2		72	116		188	253
1610	3"		7	2.286		90	133		223	297
1620	4"		6.50	2.462		118	143		261	345
1630	5"	Q-2	9	2.667		210	160		370	470
1640	6"	"	8	3		310	180		490	610
1650	8"	Q-3	5.50	5.818		745	355		1,100	1,350
1660	10"		5	6.400		1,675	395		2,070	2,450
1670	12"		4.50	7.111		1,925	435		2,360	2,775
1740	Reducer, 3" x 2"	Q-1	15	1.067		23	62		85	118
1750	4" x 2"		14.50	1.103		26	64		90	125
1760	4" x 3"		14	1.143		29.50	66.50		96	132
1770	5" x 2"		14	1.143		61.50	66.50		128	167
1780	5" x 3"		13.50	1.185		66	68.50		134.50	176
1790	5" x 4"		13	1.231		38.50	71.50		110	149
1800	6" x 2"		13.50	1.185		59	68.50		127.50	168
1810	6" x 3"		13	1.231		61	71.50		132.50	174
1830	6" x 4"		12.50	1.280		60	74		134	177
1840	6" x 5"		11	1.455		64.50	84.50		149	197
1880	8" x 3"	Q-2	13.50	1.778		116	107		223	288
1900	8" x 4"		13	1.846		100	111		211	276
1920	8" x 5"		12	2		106	120		226	297
1940	8" x 6"		12	2		100	120		220	290
1960	Increaser, 2" x 3"	Q-1	15	1.067		55	62		117	153
1980	2" x 4"		14	1.143		55	66.50		121.50	159
2000	2" x 5"		13	1.231		62.50	71.50		134	176
2020	3" x 4"		13	1.231		60	71.50		131.50	173
2040	3" x 5"		13	1.231		62.50	71.50		134	176
2060	3" x 6"		12	1.333		80.50	77.50		158	205
2070	4" x 5"		13	1.231		71	71.50		142.50	185
2080	4" x 6"		12	1.333		81.50	77.50		159	206
2090	4" x 8"	Q-2	13	1.846		170	111		281	355
2100	5" x 6"	Q-1	11	1.455		118	84.50		202.50	255
2110	5" x 8"	Q-2	12	2		183	120		303	380
2120	6" x 8"		12	2		195	120		315	395
2130	6" x 10"		8	3		350	180		530	655
2140	8" x 10"		6.50	3.692		360	222		582	730
2150	10" x 12"		5.50	4.364		640	262		902	1,100
2500	Y, 2"	Q-1	10	1.600		29.50	93		122.50	172

22 13 16.30 Pipe Fittings, Cast Iron		Crew	Daily Output	Labor-Hours	Unit	Material	2020 Bare Costs Labor	Equipment	Total	Total Incl O&P
2510	3"	Q-1	9	1.778	Ea.	55	103		158	215
2520	4"		8	2		72	116		188	253
2530	5"	Q-2	12	2		128	120		248	320
2540	6"	"	11	2.182		170	131		301	385
2550	8"	Q-3	7	4.571		415	280		695	875
2560	10"		6	5.333		670	325		995	1,225
2570	12"		5	6.400		1,550	395		1,945	2,325
2580	15"		4	8		3,350	490		3,840	4,425
3000	For extra heavy, add					44%	4%			
3600	Hub and spigot, service weight gasket joint									
3605	Note: gaskets and joint labor have									
3606	been included with all listed fittings.									
3610	1/4 bend, 2"	Q-1	20	.800	Ea.	35	46.50		81.50	108
3620	3"		17	.941		46.50	54.50		101	133
3630	4"		15	1.067		68	62		130	168
3640	5"	Q-2	21	1.143		97.50	68.50		166	210
3650	6"	"	19	1.263		116	76		192	241
3660	8"	Q-3	12	2.667		325	164		489	600
3670	10"		11	2.909		485	178		663	795
3680	12"		10	3.200		655	196		851	1,025
3700	Closet bend, 3" diameter with ring 10" x 16"	Q-1	17	.941		143	54.50		197.50	239
3710	16" x 16"		15	1.067		155	62		217	263
3730	Closet bend, 4" diameter, 1" x 4" ring, 6" x 16"		15	1.067		129	62		191	235
3740	8" x 16"		15	1.067		117	62		179	221
3750	10" x 12"		14	1.143		148	66.50		214.50	262
3760	10" x 18"		13	1.231		149	71.50		220.50	271
3770	12" x 16"		13	1.231		129	71.50		200.50	249
3780	16" x 16"		12	1.333		162	77.50		239.50	295
3800	1/8 bend, 2"		20	.800		28.50	46.50		75	101
3810	3"		17	.941		41	54.50		95.50	127
3820	4"		15	1.067		57	62		119	156
3830	5"	Q-2	21	1.143		84	68.50		152.50	196
3840	6"	"	19	1.263		95.50	76		171.50	219
3850	8"	Q-3	12	2.667		261	164		425	530
3860	10"		11	2.909		390	178		568	695
3870	12"		10	3.200		670	196		866	1,025
3900	Sanitary tee, 2"	Q-1	12	1.333		56	77.50		133.50	178
3910	3"		10	1.600		83	93		176	231
3920	4"		9	1.778		103	103		206	268
3930	5"	Q-2	13	1.846		188	111		299	375
3940	6"	"	11	2.182		208	131		339	425
3950	8"	Q-3	8.50	3.765		520	231		751	920
3980	Tee, 2"	Q-1	12	1.333		70.50	77.50		148	194
3990	3"		10	1.600		101	93		194	250
4000	4"		9	1.778		126	103		229	293
4010	5"	Q-2	13	1.846		246	111		357	435
4020	6"	"	11	2.182		248	131		379	470
4030	8"	Q-3	8	4		505	245		750	920
4060	Combination Y and 1/8 bend									
4070	2"	Q-1	12	1.333	Ea.	64.50	77.50		142	187
4080	3"		10	1.600		92.50	93		185.50	241
4090	4"		9	1.778		124	103		227	290
4100	5"	Q-2	13	1.846		222	111		333	410

246

22 13 Facility Sanitary Sewerage

22 13 16 – Sanitary Waste and Vent Piping

22 13 16.30 Pipe Fittings, Cast Iron		Crew	Daily Output	Labor-Hours	Unit	Material	2020 Bare Costs Labor	Equipment	Total	Total Incl O&P
4110	6"	Q-2	11	2.182	Ea.	267	131		398	490
4120	8"	Q-3	8	4		640	245		885	1,075
4121	12"	"	7	4.571		1,325	280		1,605	1,875
4160	Double Y, 2"	Q-1	10	1.600		107	93		200	257
4170	3"		8	2		135	116		251	325
4180	4"		7	2.286		175	133		308	390
4190	5"	Q-2	10	2.400		300	144		444	545
4200	6"	"	9	2.667		405	160		565	685
4210	8"	Q-3	6	5.333		950	325		1,275	1,550
4220	10"		5	6.400		2,025	395		2,420	2,825
4230	12"		4.50	7.111		2,375	435		2,810	3,250
4260	Reducer, 3" x 2"	Q-1	17	.941		50	54.50		104.50	137
4270	4" x 2"		16.50	.970		56.50	56		112.50	147
4280	4" x 3"		16	1		64	58		122	158
4290	5" x 2"		16	1		103	58		161	201
4300	5" x 3"		15.50	1.032		111	60		171	212
4310	5" x 4"		15	1.067		87	62		149	189
4320	6" x 2"		15.50	1.032		102	60		162	202
4330	6" x 3"		15	1.067		107	62		169	211
4336	6" x 4"		14	1.143		110	66.50		176.50	220
4340	6" x 5"		13	1.231		125	71.50		196.50	245
4360	8" x 3"	Q-2	15	1.600		199	96		295	365
4370	8" x 4"		15	1.600		187	96		283	350
4380	8" x 5"		14	1.714		204	103		307	380
4390	8" x 6"		14	1.714		199	103		302	375
4430	Increaser, 2" x 3"	Q-1	17	.941		70	54.50		124.50	159
4440	2" x 4"		16	1		74	58		132	168
4450	2" x 5"		15	1.067		92.50	62		154.50	195
4460	3" x 4"		15	1.067		79	62		141	180
4470	3" x 5"		15	1.067		92.50	62		154.50	195
4480	3" x 6"		14	1.143		111	66.50		177.50	222
4490	4" x 5"		15	1.067		101	62		163	204
4500	4" x 6"		14	1.143		112	66.50		178.50	223
4510	4" x 8"	Q-2	15	1.600		237	96		333	405
4520	5" x 6"	Q-1	13	1.231		149	71.50		220.50	270
4530	5" x 8"	Q-2	14	1.714		250	103		353	430
4540	6" x 8"		14	1.714		263	103		366	445
4550	6" x 10"		10	2.400		465	144		609	725
4560	8" x 10"		8.50	2.824		475	170		645	780
4570	10" x 12"		7.50	3.200		785	192		977	1,150
4600	Y, 2"	Q-1	12	1.333		53.50	77.50		131	175
4610	3"		10	1.600		85.50	93		178.50	233
4620	4"		9	1.778		110	103		213	275
4630	5"	Q-2	13	1.846		188	111		299	375
4640	6"	"	11	2.182		232	131		363	450
4650	8"	Q-3	8	4		550	245		795	970
4660	10"		7	4.571		900	280		1,180	1,400
4670	12"		6	5.333		1,850	325		2,175	2,525
4672	15"		5	6.400		3,700	395		4,095	4,675
4900	For extra heavy, add					44%	4%			
4940	Gasket and making push-on joint									
4950	2"	Q-1	40	.400	Ea.	11.75	23		34.75	47.50
4960	3"		35	.457		15.15	26.50		41.65	56

247

For customer support on your Site Work & Landscape Costs with RSMeans data, call 800.448.8182.

22 13 16.30 Pipe Fittings, Cast Iron		Crew	Daily Output	Labor-Hours	Unit	Material	2020 Bare Costs Labor	2020 Bare Costs Equipment	Total	Total Incl O&P
4970	4"	Q-1	32	.500	Ea.	19.05	29		48.05	64.50
4980	5"	Q-2	43	.558		30	33.50		63.50	83
4990	6"	"	40	.600		31	36		67	88
5000	8"	Q-3	32	1		67.50	61.50		129	167
5010	10"		29	1.103		114	67.50		181.50	227
5020	12"		25	1.280		146	78.50		224.50	279
5022	15"	↓	21	1.524	↓	175	93.50		268.50	330
5030	Note: gaskets and joint labor have									
5040	been included with all listed fittings.									
5990	No hub									
6000	Cplg. & labor required at joints not incl. in fitting									
6010	price. Add 1 coupling per joint for installed price									
6020	1/4 bend, 1-1/2"				Ea.	11.65			11.65	12.80
6060	2"					12.60			12.60	13.85
6080	3"					17.65			17.65	19.40
6120	4"					26			26	28.50
6140	5"					65			65	71.50
6160	6"					63			63	69.50
6180	8"					178			178	195
6184	1/4 bend, long sweep, 1-1/2"					29.50			29.50	32.50
6186	2"					28			28	30.50
6188	3"					33.50			33.50	37
6189	4"					53			53	58.50
6190	5"					103			103	114
6191	6"					118			118	129
6192	8"					310			310	345
6193	10"					655			655	725
6200	1/8 bend, 1-1/2"					9.70			9.70	10.70
6210	2"					10.90			10.90	12
6212	3"					14.55			14.55	16
6214	4"					19.15			19.15	21
6216	5"					40.50			40.50	44.50
6218	6"					42.50			42.50	46.50
6220	8"					122			122	134
6222	10"					233			233	256
6380	Sanitary tee, tapped, 1-1/2"					23.50			23.50	26
6382	2" x 1-1/2"					21			21	23
6384	2"					22			22	24
6386	3" x 2"					33.50			33.50	36.50
6388	3"					56			56	62
6390	4" x 1-1/2"					30			30	32.50
6392	4" x 2"					34			34	37
6393	4"					34			34	37
6394	6" x 1-1/2"					76.50			76.50	84
6396	6" x 2"					77.50			77.50	85
6459	Sanitary tee, 1-1/2"					16.25			16.25	17.85
6460	2"					17.40			17.40	19.15
6470	3"					21.50			21.50	23.50
6472	4"					41			41	45
6474	5"					97			97	107
6476	6"					97			97	106
6478	8"					395			395	435
6730	Y, 1-1/2"				↓	16.45			16.45	18.10

For customer support on your Site Work & Landscape Costs with RSMeans data, call 800.448.8182.

22 13 16.30 Pipe Fittings, Cast Iron		Crew	Daily Output	Labor-Hours	Unit	Material	2020 Bare Costs Labor	Equipment	Total	Total Incl O&P
6740	2"				Ea.	16.10			16.10	17.70
6750	3"					23.50			23.50	26
6760	4"					37.50			37.50	41.50
6762	5"					90.50			90.50	99.50
6764	6"					100			100	110
6768	8"					239			239	263
6769	10"					540			540	595
6770	12"					1,025			1,025	1,150
6771	15"					2,325			2,325	2,550
6791	Y, reducing, 3" x 2"					17.40			17.40	19.15
6792	4" x 2"					25			25	27.50
6793	5" x 2"					56			56	61.50
6794	6" x 2"					62.50			62.50	68.50
6795	6" x 4"					79.50			79.50	87.50
6796	8" x 4"					139			139	153
6797	8" x 6"					168			168	185
6798	10" x 6"					385			385	425
6799	10" x 8"					455			455	500
6800	Double Y, 2"					26			26	28.50
6920	3"					48			48	52.50
7000	4"					96			96	106
7100	6"					173			173	190
7120	8"					490			490	540
7200	Combination Y and 1/8 bend									
7220	1-1/2"				Ea.	17.80			17.80	19.60
7260	2"					18.40			18.40	20
7320	3"					29.50			29.50	32.50
7400	4"					56.50			56.50	62.50
7480	5"					116			116	127
7500	6"					153			153	169
7520	8"					360			360	395
7800	Reducer, 3" x 2"					9.05			9.05	9.95
7820	4" x 2"					13.70			13.70	15.05
7840	4" x 3"					13.70			13.70	15.05
7842	6" x 3"					37			37	41
7844	6" x 4"					37.50			37.50	41
7846	6" x 5"					38			38	41.50
7848	8" x 2"					58.50			58.50	64.50
7850	8" x 3"					54.50			54.50	60
7852	8" x 4"					57			57	63
7854	8" x 5"					64.50			64.50	71
7856	8" x 6"					63.50			63.50	70
7858	10" x 4"					112			112	123
7860	10" x 6"					119			119	131
7862	10" x 8"					139			139	153
7864	12" x 4"					235			235	259
7866	12" x 6"					250			250	275
7868	12" x 8"					257			257	283
7870	12" x 10"					262			262	289
7872	15" x 4"					490			490	540
7874	15" x 6"					465			465	510
7876	15" x 8"					550			550	605
7878	15" x 10"					550			550	605

22 13 16.30 Pipe Fittings, Cast Iron	Crew	Daily Output	Labor-Hours	Unit	Material	2020 Bare Costs Labor	Equipment	Total	Total Incl O&P
7880	15" x 12"				Ea.	555			555	610
8000	Coupling, standard (by CISPI Mfrs.)									
8020	1-1/2"	Q-1	48	.333	Ea.	16.70	19.35		36.05	47.50
8040	2"		44	.364		17.80	21		38.80	51
8080	3"		38	.421		20	24.50		44.50	58.50
8120	4"		33	.485		23	28		51	67.50
8160	5"	Q-2	44	.545		58	33		91	113
8180	6"	"	40	.600		60	36		96	120
8200	8"	Q-3	33	.970		111	59.50		170.50	211
8220	10"	"	26	1.231		149	75.50		224.50	277
8300	Coupling, cast iron clamp & neoprene gasket (by MG)									
8310	1-1/2"	Q-1	48	.333	Ea.	9.30	19.35		28.65	39.50
8320	2"		44	.364		12.90	21		33.90	45.50
8330	3"		38	.421		11.55	24.50		36.05	49
8340	4"		33	.485		18.85	28		46.85	62.50
8350	5"	Q-2	44	.545		33	33		66	85
8360	6"	"	40	.600		36	36		72	93.50
8380	8"	Q-3	33	.970		110	59.50		169.50	210
8400	10"	"	26	1.231		181	75.50		256.50	315
8600	Coupling, stainless steel, heavy duty									
8620	1-1/2"	Q-1	48	.333	Ea.	5.90	19.35		25.25	35.50
8630	2"		44	.364		6.40	21		27.40	38.50
8640	2" x 1-1/2"		44	.364		11.45	21		32.45	44
8650	3"		38	.421		6.70	24.50		31.20	44
8660	4"		33	.485		7.55	28		35.55	50.50
8670	4" x 3"		33	.485		18.85	28		46.85	62.50
8680	5"	Q-2	44	.545		16.20	33		49.20	67
8690	6"	"	40	.600		18.15	36		54.15	74
8700	8"	Q-3	33	.970		30.50	59.50		90	123
8710	10"		26	1.231		39	75.50		114.50	156
8712	12"		22	1.455		69.50	89		158.50	211
8715	15"		18	1.778		111	109		220	285

22 13 19.13 Sanitary Drains

		Crew	Daily Output	Labor-Hours	Unit	Material	Labor	Equipment	Total	Total Incl O&P
0010	**SANITARY DRAINS**									
0400	Deck, auto park, CI, 13" top									
0440	3", 4", 5", and 6" pipe size	Q-1	8	2	Ea.	1,875	116		1,991	2,225
0480	For galvanized body, add				"	1,125			1,125	1,250
0800	Promenade, heelproof grate, CI, 14" top									
0840	2", 3", and 4" pipe size	Q-1	10	1.600	Ea.	750	93		843	965
0860	5" and 6" pipe size		9	1.778		935	103		1,038	1,175
0880	8" pipe size		8	2		1,050	116		1,166	1,325
0940	For galvanized body, add					590			590	645
0960	With polished bronze top, 2"-3"-4" diam.					1,325			1,325	1,450
1200	Promenade, heelproof grate, CI, lateral, 14" top									
1240	2", 3" and 4" pipe size	Q-1	10	1.600	Ea.	985	93		1,078	1,225
1260	5" and 6" pipe size		9	1.778		1,100	103		1,203	1,375
1280	8" pipe size		8	2		1,275	116		1,391	1,575
1340	For galvanized body, add					645			645	710
1360	For polished bronze top, add					1,000			1,000	1,100
1500	Promenade, slotted grate, CI, 11" top									
1540	2", 3", 4", 5", and 6" pipe size	Q-1	12	1.333	Ea.	555	77.50		632.50	725

250

For customer support on your Site Work & Landscape Costs with RSMeans data, call 800.448.8182.

22 13 Facility Sanitary Sewerage

22 13 19 – Sanitary Waste Piping Specialties

22 13 19.13 Sanitary Drains

		Crew	Daily Output	Labor-Hours	Unit	Material	2020 Bare Costs Labor	Equipment	Total	Total Incl O&P
1600	For galvanized body, add				Ea.	335			335	365
1640	With polished bronze top				↓	920			920	1,000
2000	Floor, medium duty, CI, deep flange, 7" diam. top	Q-1	12	1.333	Ea.	227	77.50		304.50	365
2040	2" and 3" pipe size									
2080	For galvanized body, add					140			140	154
2120	With polished bronze top				↓	375			375	410
2400	Heavy duty, with sediment bucket, CI, 12" diam. loose grate	Q-1	9	1.778	Ea.	935	103		1,038	1,175
2420	2", 3", 4", 5", and 6" pipe size				"					
2460	With polished bronze top					1,150			1,150	1,250
2500	Heavy duty, cleanout & trap w/bucket, CI, 15" top									
2540	2", 3", and 4" pipe size	Q-1	6	2.667	Ea.	8,125	155		8,280	9,175
2560	For galvanized body, add					2,425			2,425	2,675
2580	With polished bronze top				↓	9,700			9,700	10,700

22 13 23 – Sanitary Waste Interceptors

22 13 23.10 Interceptors

		Crew	Daily Output	Labor-Hours	Unit	Material	2020 Bare Costs Labor	Equipment	Total	Total Incl O&P
0010	**INTERCEPTORS**									
0150	Grease, fabricated steel, 4 GPM, 8 lb. fat capacity	1 Plum	4	2	Ea.	1,575	129		1,704	1,925
0200	7 GPM, 14 lb. fat capacity		4	2		2,050	129		2,179	2,475
1000	10 GPM, 20 lb. fat capacity		4	2		2,575	129		2,704	3,025
1040	15 GPM, 30 lb. fat capacity		4	2		3,575	129		3,704	4,150
1060	20 GPM, 40 lb. fat capacity	↓	3	2.667		4,650	172		4,822	5,375
1120	50 GPM, 100 lb. fat capacity	Q-1	2	8		8,075	465		8,540	9,575
1160	100 GPM, 200 lb. fat capacity	"	2	8		20,100	465		20,565	22,900
1580	For seepage pan, add					7%				
3000	Hair, cast iron, 1-1/4" and 1-1/2" pipe connection	1 Plum	8	1	Ea.	560	64.50		624.50	710
3100	For chrome-plated cast iron, add					375			375	410
4000	Oil, fabricated steel, 10 GPM, 2" pipe size	1 Plum	4	2		3,300	129		3,429	3,850
4100	15 GPM, 2" or 3" pipe size		4	2		4,525	129		4,654	5,200
4120	20 GPM, 2" or 3" pipe size	↓	3	2.667		6,150	172		6,322	7,000
4220	100 GPM, 3" pipe size	Q-1	2	8		18,300	465		18,765	20,800
6000	Solids, precious metals recovery, CI, 1-1/4" to 2" pipe	1 Plum	4	2		755	129		884	1,025
6100	Dental lab., large, CI, 1-1/2" to 2" pipe	"	3	2.667	↓	2,650	172		2,822	3,150

22 13 29 – Sanitary Sewerage Pumps

22 13 29.13 Wet-Pit-Mounted, Vertical Sewerage Pumps

		Crew	Daily Output	Labor-Hours	Unit	Material	2020 Bare Costs Labor	Equipment	Total	Total Incl O&P
0010	**WET-PIT-MOUNTED, VERTICAL SEWERAGE PUMPS**									
0020	Controls incl. alarm/disconnect panel w/wire. Excavation not included									
0260	Simplex, 9 GPM at 60 PSIG, 91 gal. tank				Ea.	3,525			3,525	3,900
0300	Unit with manway, 26" ID, 18" high					4,150			4,150	4,575
0340	26" ID, 36" high					4,075			4,075	4,500
0380	43" ID, 4' high					4,250			4,250	4,675
0600	Simplex, 9 GPM at 60 PSIG, 150 gal. tank, indoor					4,275			4,275	4,700
0700	Unit with manway, 26" ID, 36" high					4,975			4,975	5,475
0740	26" ID, 4' high					5,025			5,025	5,525
2000	Duplex, 18 GPM at 60 PSIG, 150 gal. tank, indoor					8,175			8,175	9,000
2060	Unit with manway, 43" ID, 4' high					8,975			8,975	9,875
2400	For core only				↓	2,050			2,050	2,250
3000	Indoor residential type installation									
3020	Simplex, 9 GPM at 60 PSIG, 91 gal. HDPE tank				Ea.	3,625			3,625	3,975

For customer support on your Site Work & Landscape Costs with RSMeans data, call 800.448.8182.

22 13 29.14 Sewage Ejector Pumps	Crew	Daily Output	Labor-Hours	Unit	Material	2020 Bare Costs Labor	Equipment	Total	Total Incl O&P
0010 **SEWAGE EJECTOR PUMPS**, With operating and level controls									
0100 Simplex system incl. tank, cover, pump 15' head									
0500 37 gal. PE tank, 12 GPM, 1/2 HP, 2" discharge	Q-1	3.20	5	Ea.	505	290		795	990
0510 3" discharge		3.10	5.161		550	299		849	1,050
0530 87 GPM, .7 HP, 2" discharge		3.20	5		770	290		1,060	1,275
0540 3" discharge		3.10	5.161		835	299		1,134	1,375
0600 45 gal. coated stl. tank, 12 GPM, 1/2 HP, 2" discharge		3	5.333		900	310		1,210	1,450
0610 3" discharge		2.90	5.517		935	320		1,255	1,500
0630 87 GPM, .7 HP, 2" discharge		3	5.333		1,150	310		1,460	1,750
0640 3" discharge		2.90	5.517		1,225	320		1,545	1,825
0660 134 GPM, 1 HP, 2" discharge		2.80	5.714		1,250	330		1,580	1,875
0680 3" discharge		2.70	5.926		1,325	345		1,670	1,975
0700 70 gal. PE tank, 12 GPM, 1/2 HP, 2" discharge		2.60	6.154		970	355		1,325	1,600
0710 3" discharge		2.40	6.667		1,050	385		1,435	1,725
0730 87 GPM, .7 HP, 2" discharge		2.50	6.400		1,275	370		1,645	1,975
0740 3" discharge		2.30	6.957		1,325	405		1,730	2,075
0760 134 GPM, 1 HP, 2" discharge		2.20	7.273		1,375	420		1,795	2,125
0770 3" discharge		2	8		1,450	465		1,915	2,300
0800 75 gal. coated stl. tank, 12 GPM, 1/2 HP, 2" discharge		2.40	6.667		1,075	385		1,460	1,750
0810 3" discharge		2.20	7.273		1,125	420		1,545	1,850
0830 87 GPM, .7 HP, 2" discharge		2.30	6.957		1,350	405		1,755	2,100
0840 3" discharge		2.10	7.619		1,425	440		1,865	2,225
0860 134 GPM, 1 HP, 2" discharge		2	8		1,450	465		1,915	2,300
0880 3" discharge		1.80	8.889		1,525	515		2,040	2,450
1040 Duplex system incl. tank, covers, pumps									
1060 110 gal. fiberglass tank, 24 GPM, 1/2 HP, 2" discharge	Q-1	1.60	10	Ea.	1,950	580		2,530	3,025
1080 3" discharge		1.40	11.429		2,050	665		2,715	3,250
1100 174 GPM, .7 HP, 2" discharge		1.50	10.667		2,525	620		3,145	3,700
1120 3" discharge		1.30	12.308		2,625	715		3,340	3,950
1140 268 GPM, 1 HP, 2" discharge		1.20	13.333		2,725	775		3,500	4,150
1160 3" discharge		1	16		2,825	930		3,755	4,500
1260 135 gal. coated stl. tank, 24 GPM, 1/2 HP, 2" discharge	Q-2	1.70	14.118		2,000	850		2,850	3,475
2000 3" discharge		1.60	15		2,125	900		3,025	3,700
2640 174 GPM, .7 HP, 2" discharge		1.60	15		2,625	900		3,525	4,225
2660 3" discharge		1.50	16		2,775	960		3,735	4,500
2700 268 GPM, 1 HP, 2" discharge		1.30	18.462		2,850	1,100		3,950	4,775
3040 3" discharge		1.10	21.818		3,025	1,300		4,325	5,300
3060 275 gal. coated stl. tank, 24 GPM, 1/2 HP, 2" discharge		1.50	16		2,500	960		3,460	4,200
3080 3" discharge		1.40	17.143		2,550	1,025		3,575	4,350
3100 174 GPM, .7 HP, 2" discharge		1.40	17.143		3,225	1,025		4,250	5,100
3120 3" discharge		1.30	18.462		3,475	1,100		4,575	5,475
3140 268 GPM, 1 HP, 2" discharge		1.10	21.818		3,550	1,300		4,850	5,875
3160 3" discharge		.90	26.667		3,750	1,600		5,350	6,500
3260 Pump system accessories, add									
3300 Alarm horn and lights, 115 V mercury switch	Q-1	8	2	Ea.	102	116		218	286
3340 Switch, mag. contactor, alarm bell, light, 3 level control		5	3.200		495	186		681	825
3380 Alternator, mercury switch activated		4	4		895	232		1,127	1,325

22 13 63.10 Graywater Recovery Systems

	Crew	Daily Output	Labor-Hours	Unit	Material	Labor	Equipment	Total	Total Incl O&P
0010 **GRAYWATER RECOVERY SYSTEMS**, for ext. irrigation									
2010 Residential system, sink to toilet	Q-1	2	8	Ea.	310	465		775	1,025
2020 Toilet top sink	1 Plum	8	1		165	64.50		229.50	279
3010 Prepackaged residential system, 6 people	Q-2	5	4.800		2,225	289		2,514	2,875

22 13 Facility Sanitary Sewerage

22 13 29 – Sanitary Sewerage Pumps

22 13 63.10 Graywater Recovery Systems		Crew	Daily Output	Labor-Hours	Unit	Material	2020 Bare Costs Labor	Equipment	Total	Total Incl O&P
3020	9 people	Q-2	4	6	Ea.	2,475	360		2,835	3,275
3030	12 people	↓	3.50	6.857		2,700	410		3,110	3,575
3040	for irrigation supply	1 Plum	3	2.667		730	172		902	1,050
4010	Prepackaged commercial system, 1,530 gal.	Q-3	4.25	7.529		30,400	460		30,860	34,200
4020	3,060 gal.	↓	3.50	9.143		51,500	560		52,060	58,000
4030	4,590 gal.		2.50	12.800		64,500	785		65,285	72,000
5010	Lift station 75 GPM	Q-1	3	5.333	↓	1,500	310		1,810	2,125
6010	System components									

22 14 Facility Storm Drainage

22 14 23 – Storm Drainage Piping Specialties

22 14 23.33 Backwater Valves

		Crew	Daily Output	Labor-Hours	Unit	Material	2020 Bare Costs Labor	Equipment	Total	Total Incl O&P
0010	**BACKWATER VALVES**, CI Body									
6980	Bronze gate and automatic flapper valves									
7000	3" and 4" pipe size	Q-1	13	1.231	Ea.	2,525	71.50		2,596.50	2,875
7100	5" and 6" pipe size	"	13	1.231	"	3,850	71.50		3,921.50	4,350
7240	Bronze flapper valve, bolted cover									
7260	2" pipe size	Q-1	16	1	Ea.	735	58		793	895
7280	3" pipe size		14.50	1.103		1,150	64		1,214	1,375
7300	4" pipe size	↓	13	1.231		1,300	71.50		1,371.50	1,525
7320	5" pipe size	Q-2	18	1.333		1,600	80		1,680	1,875
7340	6" pipe size	"	17	1.412		2,275	85		2,360	2,625
7360	8" pipe size	Q-3	10	3.200		3,225	196		3,421	3,825
7380	10" pipe size	"	9	3.556	↓	5,250	218		5,468	6,100
7500	For threaded cover, same cost									
7540	Revolving disk type, same cost as flapper type									

22 14 26 – Facility Storm Drains

22 14 26.19 Facility Trench Drains

		Crew	Daily Output	Labor-Hours	Unit	Material	2020 Bare Costs Labor	Equipment	Total	Total Incl O&P
0010	**FACILITY TRENCH DRAINS**									
5980	Trench, floor, heavy duty, modular, CI, 12" x 12" top									
6000	2", 3", 4", 5" & 6" pipe size	Q-1	8	2	Ea.	1,100	116		1,216	1,400
6100	For unit with polished bronze top		8	2		1,625	116		1,741	1,975
6200	For 12" extension section, CI top		8	2		1,100	116		1,216	1,400
6240	For 12" extension section, polished bronze top	↓	8	2	↓	1,825	116		1,941	2,175
6600	Trench, floor, for cement concrete encasement									
6610	Not including trenching or concrete									
6640	Polyester polymer concrete									
6650	4" internal width, with grate									
6660	Light duty steel grate	Q-1	120	.133	L.F.	68.50	7.75		76.25	86.50
6670	Medium duty steel grate		115	.139		129	8.05		137.05	154
6680	Heavy duty iron grate	↓	110	.145	↓	122	8.45		130.45	147
6700	12" internal width, with grate									
6770	Heavy duty galvanized grate	Q-1	80	.200	L.F.	171	11.60		182.60	205
6800	Fiberglass									
6810	8" internal width, with grate									
6820	Medium duty galvanized grate	Q-1	115	.139	L.F.	122	8.05		130.05	146
6830	Heavy duty iron grate	"	110	.145	"	201	8.45		209.45	234

For customer support on your Site Work & Landscape Costs with RSMeans data, call 800.448.8182.

253

22 14 Facility Storm Drainage

22 14 29 – Sump Pumps

22 14 29.13 Wet-Pit-Mounted, Vertical Sump Pumps	Crew	Daily Output	Labor-Hours	Unit	Material	2020 Bare Costs Labor	Equipment	Total	Total Incl O&P
0010 **WET-PIT-MOUNTED, VERTICAL SUMP PUMPS**									
0400 Molded PVC base, 21 GPM at 15' head, 1/3 HP	1 Plum	5	1.600	Ea.	141	103		244	310
0800 Iron base, 21 GPM at 15' head, 1/3 HP		5	1.600		155	103		258	325
1200 Solid brass, 21 GPM at 15' head, 1/3 HP		5	1.600		243	103		346	420
2000 Sump pump, single stage									
2010 25 GPM, 1 HP, 1-1/2" discharge	Q-1	1.80	8.889	Ea.	3,825	515		4,340	5,000
2020 75 GPM, 1-1/2 HP, 2" discharge		1.50	10.667		4,050	620		4,670	5,375
2030 100 GPM, 2 HP, 2-1/2" discharge		1.30	12.308		4,125	715		4,840	5,625
2040 150 GPM, 3 HP, 3" discharge		1.10	14.545		4,125	845		4,970	5,825
2050 200 GPM, 3 HP, 3" discharge		1	16		4,375	930		5,305	6,225
2060 300 GPM, 10 HP, 4" discharge	Q-2	1.20	20		4,725	1,200		5,925	7,000
2070 500 GPM, 15 HP, 5" discharge		1.10	21.818		5,375	1,300		6,675	7,875
2080 800 GPM, 20 HP, 6" discharge		1	24		6,350	1,450		7,800	9,125
2090 1,000 GPM, 30 HP, 6" discharge		.85	28.235		6,975	1,700		8,675	10,200
2100 1,600 GPM, 50 HP, 8" discharge		.72	33.333		10,900	2,000		12,900	15,000
2110 2,000 GPM, 60 HP, 8" discharge	Q-3	.85	37.647		11,100	2,300		13,400	15,700
2202 For general purpose float switch, copper coated float, add	Q-1	5	3.200		107	186		293	395

22 14 29.16 Submersible Sump Pumps

	Crew	Daily Output	Labor-Hours	Unit	Material	2020 Bare Costs Labor	Equipment	Total	Total Incl O&P
0010 **SUBMERSIBLE SUMP PUMPS**									
7000 Sump pump, automatic									
7100 Plastic, 1-1/4" discharge, 1/4 HP	1 Plum	6.40	1.250	Ea.	168	80.50		248.50	305
7140 1/3 HP		6	1.333		224	86		310	375
7160 1/2 HP		5.40	1.481		245	95.50		340.50	415
7180 1-1/2" discharge, 1/2 HP		5.20	1.538		325	99		424	505
7500 Cast iron, 1-1/4" discharge, 1/4 HP		6	1.333		225	86		311	375
7540 1/3 HP		6	1.333		264	86		350	420
7560 1/2 HP		5	1.600		320	103		423	505

22 14 53 – Rainwater Storage Tanks

22 14 53.13 Fiberglass, Rainwater Storage Tank

	Crew	Daily Output	Labor-Hours	Unit	Material	2020 Bare Costs Labor	Equipment	Total	Total Incl O&P
0010 **FIBERGLASS, RAINWATER STORAGE TANK**									
2000 600 gallon	B-21B	3.75	10.667	Ea.	4,125	490	126	4,741	5,425
2010 1,000 gallon		3.50	11.429		5,275	525	135	5,935	6,725
2020 2,000 gallon		3.25	12.308		7,500	565	145	8,210	9,250
2030 4,000 gallon		3	13.333		10,100	610	157	10,867	12,200
2040 6,000 gallon		2.65	15.094		11,400	695	178	12,273	13,800
2050 8,000 gallon		2.30	17.391		13,900	800	205	14,905	16,700
2060 10,000 gallon		2	20		15,400	920	236	16,556	18,600
2070 12,000 gallon		1.50	26.667		21,600	1,225	315	23,140	25,900
2080 15,000 gallon		1	40		24,500	1,825	470	26,795	30,300
2090 20,000 gallon		.75	53.333		31,800	2,450	630	34,880	39,400
2100 25,000 gallon		.50	80		47,400	3,675	940	52,015	58,500
2110 30,000 gallon		.35	114		57,000	5,250	1,350	63,600	72,000
2120 40,000 gallon		.30	133		80,000	6,125	1,575	87,700	99,000

22 41 Residential Plumbing Fixtures

22 41 39 – Residential Faucets, Supplies and Trim

22 41 39.10 Faucets and Fittings	Crew	Daily Output	Labor-Hours	Unit	Material	2020 Bare Costs Labor	Equipment	Total	Total Incl O&P
0010 **FAUCETS AND FITTINGS**									
5000 Sillcock, compact, brass, IPS or copper to hose	1 Plum	24	.333	Ea.	12.05	21.50		33.55	45.50

22 47 Drinking Fountains and Water Coolers

22 47 13 – Drinking Fountains

22 47 13.10 Drinking Water Fountains

	22 47 13.10 Drinking Water Fountains	Crew	Daily Output	Labor-Hours	Unit	Material	2020 Bare Costs Labor	Equipment	Total	Total Incl O&P
0010	**DRINKING WATER FOUNTAINS**, For connection to cold water supply									
1000	Wall mounted, non-recessed									
1400	Bronze, with no back	1 Plum	4	2	Ea.	1,125	129		1,254	1,425
1800	Cast aluminum, enameled, for correctional institutions		4	2		1,400	129		1,529	1,750
2000	Fiberglass, 12" back, single bubbler unit		4	2		2,000	129		2,129	2,425
2040	Dual bubbler		3.20	2.500		2,575	161		2,736	3,075
2400	Precast stone, no back		4	2		1,050	129		1,179	1,350
2700	Stainless steel, single bubbler, no back		4	2		915	129		1,044	1,200
2740	With back		4	2		1,100	129		1,229	1,400
2780	Dual handle, ADA compliant		4	2		760	129		889	1,025
2820	Dual level, ADA compliant		3.20	2.500		1,600	161		1,761	2,000
3300	Vitreous china									
3340	7" back	1 Plum	4	2	Ea.	575	129		704	830
3940	For vandal-resistant bottom plate, add					75			75	82.50
3960	For freeze-proof valve system, add	1 Plum	2	4		770	258		1,028	1,225
3980	For rough-in, supply and waste, add	"	2.21	3.620		229	233		462	600
4000	Wall mounted, semi-recessed									
4200	Poly-marble, single bubbler	1 Plum	4	2	Ea.	960	129		1,089	1,250
4600	Stainless steel, satin finish, single bubbler		4	2		1,400	129		1,529	1,750
4900	Vitreous china, single bubbler		4	2		935	129		1,064	1,225
5980	For rough-in, supply and waste, add		1.83	4.372		229	282		511	670
6000	Wall mounted, fully recessed									
6400	Poly-marble, single bubbler	1 Plum	4	2	Ea.	1,775	129		1,904	2,150
6800	Stainless steel, single bubbler		4	2		1,750	129		1,879	2,125
7560	For freeze-proof valve system, add		2	4		985	258		1,243	1,450
7580	For rough-in, supply and waste, add		1.83	4.372		229	282		511	670
7600	Floor mounted, pedestal type									
7780	ADA compliant unit	1 Plum	2	4	Ea.	1,675	258		1,933	2,225
8400	Stainless steel, architectural style		2	4		2,150	258		2,408	2,750
8600	Enameled iron, heavy duty service, 2 bubblers		2	4		3,675	258		3,933	4,425
8660	4 bubblers		2	4		4,650	258		4,908	5,475
8880	For freeze-proof valve system, add		2	4		700	258		958	1,150
8900	For rough-in, supply and waste, add		1.83	4.372		229	282		511	670
9100	Deck mounted									
9500	Stainless steel, circular receptor	1 Plum	4	2	Ea.	455	129		584	700
9760	White enameled steel, 14" x 9" receptor		4	2		400	129		529	635
9860	White enameled cast iron, 24" x 16" receptor		3	2.667		525	172		697	835
9980	For rough-in, supply and waste, add		1.83	4.372		229	282		511	670

For customer support on your Site Work & Landscape Costs with RSMeans data, call 800.448.8182.

22 51 Swimming Pool Plumbing Systems

22 51 19 – Swimming Pool Water Treatment Equipment

22 51 19.50 Swimming Pool Filtration Equipment	Crew	Daily Output	Labor-Hours	Unit	Material	2020 Bare Costs Labor	Equipment	Total	Total Incl O&P
0010 **SWIMMING POOL FILTRATION EQUIPMENT**									
0900 Filter system, sand or diatomite type, incl. pump, 6,000 gal./hr.	2 Plum	1.80	8.889	Total	2,375	575		2,950	3,475
1020 Add for chlorination system, 800 S.F. pool		3	5.333	Ea.	235	345		580	775
1040 5,000 S.F. pool		3	5.333	"	1,950	345		2,295	2,650

22 52 Fountain Plumbing Systems

22 52 16 – Fountain Pumps

22 52 16.10 Fountain Water Pumps

22 52 16.10 Fountain Water Pumps	Crew	Daily Output	Labor-Hours	Unit	Material	2020 Bare Costs Labor	Equipment	Total	Total Incl O&P
0010 **FOUNTAIN WATER PUMPS**									
0100 Pump w/controls									
0200 Single phase, 100' cord, 1/2 HP pump	2 Skwk	4.40	3.636	Ea.	1,325	199		1,524	1,750
0300 3/4 HP pump		4.30	3.721		1,375	204		1,579	1,825
0400 1 HP pump		4.20	3.810		1,925	209		2,134	2,450
0500 1-1/2 HP pump		4.10	3.902		2,575	214		2,789	3,175
0600 2 HP pump		4	4		4,350	219		4,569	5,100
0700 Three phase, 200' cord, 5 HP pump		3.90	4.103		6,250	225		6,475	7,225
0800 7-1/2 HP pump		3.80	4.211		12,600	231		12,831	14,200
0900 10 HP pump		3.70	4.324		15,200	237		15,437	17,200
1000 15 HP pump		3.60	4.444		21,500	244		21,744	24,000
2000 DESIGN NOTE: Use two horsepower per surface acre.									

22 52 33 – Fountain Ancillary

22 52 33.10 Fountain Miscellaneous

22 52 33.10 Fountain Miscellaneous	Crew	Daily Output	Labor-Hours	Unit	Material	2020 Bare Costs Labor	Equipment	Total	Total Incl O&P
0010 **FOUNTAIN MISCELLANEOUS**									
1300 Lights w/mounting kits, 200 watt	2 Skwk	18	.889	Ea.	1,200	49		1,249	1,400
1400 300 watt		18	.889		1,350	49		1,399	1,575
1500 500 watt		18	.889		1,525	49		1,574	1,750
1600 Color blender		12	1.333		600	73		673	770

22 66 Chemical-Waste Systems for Lab. and Healthcare Facilities

22 66 53 – Laboratory Chemical-Waste and Vent Piping

22 66 53.30 Glass Pipe

22 66 53.30 Glass Pipe	Crew	Daily Output	Labor-Hours	Unit	Material	2020 Bare Costs Labor	Equipment	Total	Total Incl O&P
0010 **GLASS PIPE**, Borosilicate, couplings & clevis hanger assemblies, 10' OC									
0020 Drainage									
1100 1-1/2" diameter	Q-1	52	.308	L.F.	15.30	17.85		33.15	43.50
1120 2" diameter		44	.364		19.50	21		40.50	53
1140 3" diameter		39	.410		26	24		50	64
1160 4" diameter		30	.533		42.50	31		73.50	93.50
1180 6" diameter		26	.615		75	35.50		110.50	136

22 66 53.40 Pipe Fittings, Glass

22 66 53.40 Pipe Fittings, Glass	Crew	Daily Output	Labor-Hours	Unit	Material	2020 Bare Costs Labor	Equipment	Total	Total Incl O&P
0010 **PIPE FITTINGS, GLASS**									
0020 Drainage, beaded ends									
0040 Coupling & labor required at joints not incl. in fitting									
0050 price. Add 1 per joint for installed price									
0070 90° bend or sweep, 1-1/2"				Ea.	36.50			36.50	40
0110 4"				"	121			121	134

22 66 53.60 Corrosion Resistant Pipe

		Crew	Daily Output	Labor-Hours	Unit	Material	2020 Bare Costs Labor	2020 Bare Costs Equipment	Total	Total Incl O&P
0010	**CORROSION RESISTANT PIPE**, No couplings or hangers									
0020	Iron alloy, drain, mechanical joint									
1000	1-1/2" diameter	Q-1	70	.229	L.F.	84.50	13.25		97.75	113
1100	2" diameter		66	.242		87	14.05		101.05	117
1120	3" diameter		60	.267		95	15.45		110.45	128
1140	4" diameter		52	.308		117	17.85		134.85	156
1980	Iron alloy, drain, B&S joint									
2000	2" diameter	Q-1	54	.296	L.F.	85.50	17.20		102.70	120
2100	3" diameter		52	.308		82.50	17.85		100.35	117
2120	4" diameter		48	.333		101	19.35		120.35	140
2140	6" diameter	Q-2	59	.407		152	24.50		176.50	204
2160	8" diameter	"	54	.444		315	26.50		341.50	385
2980	Plastic, epoxy, fiberglass filament wound, B&S joint									
3000	2" diameter	Q-1	62	.258	L.F.	12.35	14.95		27.30	36
3100	3" diameter		51	.314		14.40	18.20		32.60	43
3120	4" diameter		45	.356		20.50	20.50		41	53.50
3140	6" diameter		32	.500		29	29		58	75
3160	8" diameter	Q-2	38	.632		45	38		83	107
3980	Polyester, fiberglass filament wound, B&S joint									
4000	2" diameter	Q-1	62	.258	L.F.	13.40	14.95		28.35	37.50
4100	3" diameter		51	.314		17.55	18.20		35.75	46.50
4120	4" diameter		45	.356		25.50	20.50		46	59.50
4140	6" diameter		32	.500		37	29		66	84.50
4160	8" diameter	Q-2	38	.632		88	38		126	154
4980	Polypropylene, acid resistant, fire retardant, Schedule 40									
5000	1-1/2" diameter	Q-1	68	.235	L.F.	9.05	13.65		22.70	30.50
5100	2" diameter		62	.258		14.55	14.95		29.50	38.50
5120	3" diameter		51	.314		25	18.20		43.20	54.50
5140	4" diameter		45	.356		32	20.50		52.50	66
5160	6" diameter		32	.500		63.50	29		92.50	114
5980	Proxylene, fire retardant, Schedule 40									
6000	1-1/2" diameter	Q-1	68	.235	L.F.	14	13.65		27.65	36
6100	2" diameter		62	.258		19.20	14.95		34.15	43.50
6120	3" diameter		51	.314		34.50	18.20		52.70	65
6140	4" diameter		45	.356		49	20.50		69.50	85
6820	For Schedule 80, add					35%	2%			

22 66 53.70 Pipe Fittings, Corrosion Resistant

		Crew	Daily Output	Labor-Hours	Unit	Material	2020 Bare Costs Labor	2020 Bare Costs Equipment	Total	Total Incl O&P
0010	**PIPE FITTINGS, CORROSION RESISTANT**									
0030	Iron alloy									
0050	Mechanical joint									
0060	1/4 bend, 1-1/2"	Q-1	12	1.333	Ea.	125	77.50		202.50	253
0080	2"		10	1.600		204	93		297	365
0090	3"		9	1.778		245	103		348	425
0100	4"		8	2		282	116		398	485
0160	Tee and Y, sanitary, straight									
0170	1-1/2"	Q-1	8	2	Ea.	136	116		252	325
0180	2"		7	2.286		182	133		315	400
0190	3"		6	2.667		282	155		437	540
0200	4"		5	3.200		520	186		706	850
3000	Epoxy, filament wound									
3030	Quick-lock joint									
3040	90° elbow, 2"	Q-1	28	.571	Ea.	100	33		133	161

For customer support on your Site Work & Landscape Costs with RSMeans data, call 800.448.8182.

22 66 53.70 Pipe Fittings, Corrosion Resistant		Crew	Daily Output	Labor-Hours	Unit	Material	2020 Bare Costs Labor	2020 Bare Costs Equipment	Total	Total Incl O&P
3060	3"	Q-1	16	1	Ea.	116	58		174	214
3070	4"	▼	13	1.231	▼	158	71.50		229.50	280
4000	Polypropylene, acid resistant									
4020	Non-pressure, electrofusion joints									
4060	2"	Q-1	28	.571	Ea.	33	33		66	85.50
4090	4"	"	14	1.143	"	58	66.50		124.50	163

Estimating Tips

The labor adjustment factors listed in Subdivision 22 01 02.20 also apply to Division 23.

23 10 00 Facility Fuel Systems

- The prices in this subdivision for above- and below-ground storage tanks do not include foundations or hold-down slabs, unless noted. The estimator should refer to Divisions 3 and 31 for foundation system pricing. In addition to the foundations, required tank accessories, such as tank gauges, leak detection devices, and additional manholes and piping, must be added to the tank prices.

23 50 00 Central Heating Equipment

- When estimating the cost of an HVAC system, check to see who is responsible for providing and installing the temperature control system. It is possible to overlook controls, assuming that they would be included in the electrical estimate.
- When looking up a boiler, be careful on specified capacity. Some manufacturers rate their products on output while others use input.
- Include HVAC insulation for pipe, boiler, and duct (wrap and liner).
- Be careful when looking up mechanical items to get the correct pressure rating and connection type (thread, weld, flange).

23 70 00 Central HVAC Equipment

- Combination heating and cooling units are sized by the air conditioning requirements. (See Reference No. R236000-20 for the preliminary sizing guide.)
- A ton of air conditioning is nominally 400 CFM.
- Rectangular duct is taken off by the linear foot for each size, but its cost is usually estimated by the pound. Remember that SMACNA standards now base duct on internal pressure.
- Prefabricated duct is estimated and purchased like pipe: straight sections and fittings.
- Note that cranes or other lifting equipment are not included on any lines in Division 23. For example, if a crane is required to lift a heavy piece of pipe into place high above a gym floor, or to put a rooftop unit on the roof of a four-story building, etc., it must be added. Due to the potential for extreme variation—from nothing additional required to a major crane or helicopter—we feel that including a nominal amount for "lifting contingency" would be useless and detract from the accuracy of the estimate. When using equipment rental cost data from RSMeans, do not forget to include the cost of the operator(s).

Reference Numbers

Reference numbers are shown at the beginning of some major classifications. These numbers refer to related items in the Reference Section. The reference information may be an estimating procedure, an alternate pricing method, or technical information.

Note: Not all subdivisions listed here necessarily appear. ■

---

23 05 05 – Selective Demolition for HVAC

23 05 05.10 HVAC Demolition

		Crew	Daily Output	Labor-Hours	Unit	Material	2020 Bare Costs Labor	2020 Bare Costs Equipment	Total	Total Incl O&P
0010	**HVAC DEMOLITION**									
0100	Air conditioner, split unit, 3 ton	Q-5	2	8	Ea.		470		470	705
0150	Package unit, 3 ton	Q-6	3	8	"		490		490	735
0298	Boilers									
0300	Electric, up thru 148 kW	Q-19	2	12	Ea.		715		715	1,075
0310	150 thru 518 kW	"	1	24			1,425		1,425	2,150
0320	550 thru 2,000 kW	Q-21	.40	80			4,900		4,900	7,325
0330	2,070 kW and up	"	.30	107			6,525		6,525	9,750
0340	Gas and/or oil, up thru 150 MBH	Q-7	2.20	14.545			910		910	1,350
0350	160 thru 2,000 MBH		.80	40			2,500		2,500	3,725
0360	2,100 thru 4,500 MBH		.50	64			4,000		4,000	5,975
0370	4,600 thru 7,000 MBH		.30	107			6,650		6,650	9,950
0380	7,100 thru 12,000 MBH		.16	200			12,500		12,500	18,700
0390	12,200 thru 25,000 MBH		.12	267			16,600		16,600	24,900
1000	Ductwork, 4" high, 8" wide	1 Clab	200	.040	L.F.		1.68		1.68	2.53
1100	6" high, 8" wide		165	.048			2.04		2.04	3.07
1200	10" high, 12" wide		125	.064			2.69		2.69	4.05
1300	12"-14" high, 16"-18" wide		85	.094			3.96		3.96	5.95
1400	18" high, 24" wide		67	.119			5.05		5.05	7.55
1500	30" high, 36" wide		56	.143			6		6	9.05
1540	72" wide		50	.160			6.75		6.75	10.10
3000	Mechanical equipment, light items. Unit is weight, not cooling.	Q-5	.90	17.778	Ton		1,050		1,050	1,575
3600	Heavy items	"	1.10	14.545	"		860		860	1,275
5090	Remove refrigerant from system	1 Stpi	40	.200	Lb.		13.10		13.10	19.60

23 05 23 – General-Duty Valves for HVAC Piping

23 05 23.30 Valves, Iron Body

		Crew	Daily Output	Labor-Hours	Unit	Material	2020 Bare Costs Labor	2020 Bare Costs Equipment	Total	Total Incl O&P
0010	**VALVES, IRON BODY** R220523-80									
1020	Butterfly, wafer type, gear actuator, 200 lb.									
1030	2"	1 Plum	14	.571	Ea.	118	37		155	185
1040	2-1/2"	Q-1	9	1.778		125	103		228	291
1050	3"		8	2		124	116		240	310
1060	4"		5	3.200		139	186		325	430
1070	5"	Q-2	5	4.800		157	289		446	600
1080	6"	"	5	4.800		177	289		466	625
1650	Gate, 125 lb., N.R.S.									
2150	Flanged									
2200	2"	1 Plum	5	1.600	Ea.	610	103		713	830
2240	2-1/2"	Q-1	5	3.200		630	186		816	970
2260	3"		4.50	3.556		710	206		916	1,100
2280	4"		3	5.333		1,000	310		1,310	1,575
2300	6"	Q-2	3	8		1,725	480		2,205	2,625
3550	OS&Y, 125 lb., flanged									
3600	2"	1 Plum	5	1.600	Ea.	400	103		503	595
3660	3"	Q-1	4.50	3.556		460	206		666	820
3680	4"	"	3	5.333		685	310		995	1,225
3700	6"	Q-2	3	8		1,075	480		1,555	1,925
3900	For 175 lb., flanged, add					200%	10%			
5450	Swing check, 125 lb., threaded									
5500	2"	1 Plum	11	.727	Ea.	575	47		622	700
5540	2-1/2"	Q-1	15	1.067		710	62		772	875
5550	3"		13	1.231		770	71.50		841.50	955
5560	4"		10	1.600		1,225	93		1,318	1,500

23 05 Common Work Results for HVAC

23 05 23 – General-Duty Valves for HVAC Piping

23 05 23.30 Valves, Iron Body		Crew	Daily Output	Labor-Hours	Unit	Material	2020 Bare Costs Labor	Equipment	Total	Total Incl O&P
5950	Flanged									
6000	2"	1 Plum	5	1.600	Ea.	510	103		613	720
6040	2-1/2"	Q-1	5	3.200		365	186		551	680
6050	3"		4.50	3.556		490	206		696	850
6060	4"	↓	3	5.333		785	310		1,095	1,325
6070	6"	Q-2	3	8	↓	1,325	480		1,805	2,175

23 05 23.80 Valves, Steel

	23 05 23.80 Valves, Steel	Crew	Daily Output	Labor-Hours	Unit	Material	2020 Bare Costs Labor	Equipment	Total	Total Incl O&P
0010	**VALVES, STEEL** R220523-80									
0800	Cast									
1350	Check valve, swing type, 150 lb., flanged									
1370	1"	1 Plum	10	.800	Ea.	360	51.50		411.50	475
1400	2"	"	8	1		765	64.50		829.50	935
1440	2-1/2"	Q-1	5	3.200		885	186		1,071	1,250
1450	3"		4.50	3.556		900	206		1,106	1,300
1460	4"	↓	3	5.333		1,100	310		1,410	1,700
1540	For 300 lb., flanged, add					50%	15%			
1548	For 600 lb., flanged, add				↓	110%	20%			
1950	Gate valve, 150 lb., flanged									
2000	2"	1 Plum	8	1	Ea.	645	64.50		709.50	800
2040	2-1/2"	Q-1	5	3.200		1,075	186		1,261	1,450
2050	3"		4.50	3.556		910	206		1,116	1,300
2060	4"	↓	3	5.333		1,150	310		1,460	1,725
2070	6"	Q-2	3	8	↓	1,825	480		2,305	2,725
3650	Globe valve, 150 lb., flanged									
3700	2"	1 Plum	8	1	Ea.	950	64.50		1,014.50	1,150
3740	2-1/2"	Q-1	5	3.200		1,200	186		1,386	1,600
3750	3"		4.50	3.556		1,200	206		1,406	1,625
3760	4"	↓	3	5.333		1,750	310		2,060	2,400
3770	6"	Q-2	3	8	↓	2,750	480		3,230	3,750
5150	Forged									
5650	Check valve, class 800, horizontal									
5698	Threaded									
5700	1/4"	1 Plum	24	.333	Ea.	86	21.50		107.50	127
5720	3/8"		24	.333		86	21.50		107.50	127
5730	1/2"		24	.333		86	21.50		107.50	127
5740	3/4"		20	.400		92	26		118	140
5750	1"		19	.421		108	27		135	160
5760	1-1/4"	↓	15	.533	↓	212	34.50		246.50	285

23 05 93 – Testing, Adjusting, and Balancing for HVAC

23 05 93.50 Piping, Testing

	23 05 93.50 Piping, Testing	Crew	Daily Output	Labor-Hours	Unit	Material	2020 Bare Costs Labor	Equipment	Total	Total Incl O&P
0010	**PIPING, TESTING**									
0100	Nondestructive testing									
0120	1" - 4" pipe									
0140	0-250 L.F.	1 Stpi	1.33	6.015	Ea.		395		395	590
0160	250-500 L.F.	"	.80	10			655		655	980
0180	500-1000 L.F.	Q-5	1.14	14.035			830		830	1,250
0200	1000-2000 L.F.		.80	20			1,175		1,175	1,775
0320	0-250 L.F.		1	16			945		945	1,425
0340	250-500 L.F.		.73	21.918			1,300		1,300	1,925
0360	500-1000 L.F.		.53	30.189			1,775		1,775	2,675
0380	1000-2000 L.F.	↓	.38	42.105	↓		2,475		2,475	3,725
1000	Pneumatic pressure test, includes soaping joints									

For customer support on your Site Work & Landscape Costs with RSMeans data, call 800.448.8182.

23 05 Common Work Results for HVAC

23 05 93 – Testing, Adjusting, and Balancing for HVAC

23 05 93.50 Piping, Testing		Crew	Daily Output	Labor-Hours	Unit	Material	2020 Bare Costs Labor	Equipment	Total	Total Incl O&P
1120	1" - 4" pipe									
1140	0-250 L.F.	Q-5	2.67	5.993	Ea.	12.30	355		367.30	545
1160	250-500 L.F.		1.33	12.030		24.50	710		734.50	1,075
1180	500-1000 L.F.		.80	20		37	1,175		1,212	1,825
1200	1000-2000 L.F.		.50	32		49	1,900		1,949	2,875
1300	6" - 10" pipe									
1320	0-250 L.F.	Q-5	1.33	12.030	Ea.	12.30	710		722.30	1,075
1340	250-500 L.F.		.67	23.881		24.50	1,400		1,424.50	2,125
1360	500-1000 L.F.		.40	40		49	2,350		2,399	3,575
1380	1000-2000 L.F.		.25	64		61.50	3,775		3,836.50	5,725
2000	X-Ray of welds									
2110	2" diam.	1 Stpi	8	1	Ea.	17.25	65.50		82.75	117
2120	3" diam.		8	1		17.25	65.50		82.75	117
2130	4" diam.		8	1		26	65.50		91.50	127
2140	6" diam.		8	1		26	65.50		91.50	127
2150	8" diam.		6.60	1.212		26	79.50		105.50	148
2160	10" diam.		6	1.333		34.50	87.50		122	169
3000	Liquid penetration of welds									
3110	2" diam.	1 Stpi	14	.571	Ea.	13.05	37.50		50.55	70.50
3120	3" diam.		13.60	.588		13.05	38.50		51.55	72
3130	4" diam.		13.40	.597		13.05	39		52.05	73
3140	6" diam.		13.20	.606		13.05	39.50		52.55	74
3150	8" diam.		13	.615		19.60	40.50		60.10	82
3160	10" diam.		12.80	.625		19.60	41		60.60	83

23 13 Facility Fuel-Storage Tanks

23 13 13 – Facility Underground Fuel-Oil, Storage Tanks

23 13 13.09 Single-Wall Steel Fuel-Oil Tanks

		Crew	Daily Output	Labor-Hours	Unit	Material	Labor	Equipment	Total	Total Incl O&P
0010	**SINGLE-WALL STEEL FUEL-OIL TANKS**									
5000	Tanks, steel ugnd., sti-p3, not incl. hold-down bars									
5500	Excavation, pad, pumps and piping not included									
5510	Single wall, 500 gallon capacity, 7 ga. shell	Q-5	2.70	5.926	Ea.	1,875	350		2,225	2,600
5520	1,000 gallon capacity, 7 ga. shell	"	2.50	6.400		2,925	380		3,305	3,800
5530	2,000 gallon capacity, 1/4" thick shell	Q-7	4.60	6.957		3,725	435		4,160	4,750
5535	2,500 gallon capacity, 7 ga. shell	Q-5	3	5.333		5,675	315		5,990	6,700
5540	5,000 gallon capacity, 1/4" thick shell	Q-7	3.20	10		13,400	625		14,025	15,600
5560	10,000 gallon capacity, 1/4" thick shell		2	16		10,800	1,000		11,800	13,400
5580	15,000 gallon capacity, 5/16" thick shell		1.70	18.824		11,200	1,175		12,375	14,100
5600	20,000 gallon capacity, 5/16" thick shell		1.50	21.333		22,600	1,325		23,925	26,900
5610	25,000 gallon capacity, 3/8" thick shell		1.30	24.615		29,600	1,525		31,125	34,900
5620	30,000 gallon capacity, 3/8" thick shell		1.10	29.091		35,000	1,825		36,825	41,200
5630	40,000 gallon capacity, 3/8" thick shell		.90	35.556		44,500	2,225		46,725	52,500
5640	50,000 gallon capacity, 3/8" thick shell		.80	40		49,800	2,500		52,300	58,000

23 13 13.23 Glass-Fiber-Reinfcd-Plastic, Fuel-Oil, Storage

		Crew	Daily Output	Labor-Hours	Unit	Material	Labor	Equipment	Total	Total Incl O&P
0010	**GLASS-FIBER-REINFCD-PLASTIC, UNDERGRND. FUEL-OIL, STORAGE**									
0210	Fiberglass, underground, single wall, UL listed, not including									
0220	manway or hold-down strap									
0225	550 gallon capacity	Q-5	2.67	5.993	Ea.	4,075	355		4,430	5,025
0230	1,000 gallon capacity	"	2.46	6.504		5,275	385		5,660	6,375
0240	2,000 gallon capacity	Q-7	4.57	7.002		8,275	435		8,710	9,750
0245	3,000 gallon capacity		3.90	8.205		8,400	510		8,910	10,000

262

For customer support on your Site Work & Landscape Costs with RSMeans data, call 800.448.8182.

23 13 Facility Fuel-Storage Tanks

23 13 13 – Facility Underground Fuel-Oil, Storage Tanks

23 13 13.23 Glass-Fiber-Reinfcd-Plastic, Fuel-Oil, Storage		Crew	Daily Output	Labor-Hours	Unit	Material	2020 Bare Costs Labor	Equipment	Total	Total Incl O&P
0250	4,000 gallon capacity	Q-7	3.55	9.014	Ea.	11,100	560		11,660	13,000
0255	5,000 gallon capacity		3.20	10		10,400	625		11,025	12,300
0260	6,000 gallon capacity		2.67	11.985		11,800	750		12,550	14,100
0270	8,000 gallon capacity		2.29	13.974		14,900	870		15,770	17,700
0280	10,000 gallon capacity		2	16		15,600	1,000		16,600	18,700
0282	12,000 gallon capacity		1.88	17.021		20,900	1,050		21,950	24,600
0284	15,000 gallon capacity		1.68	19.048		27,900	1,200		29,100	32,500
0290	20,000 gallon capacity		1.45	22.069		33,200	1,375		34,575	38,600
0300	25,000 gallon capacity		1.28	25		58,500	1,550		60,050	67,000
0320	30,000 gallon capacity		1.14	28.070		104,500	1,750		106,250	117,500
0340	40,000 gallon capacity		.89	35.955		114,000	2,250		116,250	129,000
0360	48,000 gallon capacity	▼	.81	39.506		177,000	2,475		179,475	198,000
0500	For manway, fittings and hold-downs, add					20%	15%			
1000	For helical heating coil, add	Q-5	2.50	6.400	▼	5,450	380		5,830	6,575
2210	Fiberglass, underground, single wall, UL listed, including									
2220	hold-down straps, no manways									
2225	550 gallon capacity	Q-5	2	8	Ea.	4,600	470		5,070	5,750
2230	1,000 gallon capacity	"	1.88	8.511		5,775	500		6,275	7,100
2240	2,000 gallon capacity	Q-7	3.55	9.014		8,775	560		9,335	10,500
2250	4,000 gallon capacity		2.90	11.034		11,600	690		12,290	13,800
2260	6,000 gallon capacity		2	16		12,800	1,000		13,800	15,600
2270	8,000 gallon capacity		1.78	17.978		15,900	1,125		17,025	19,200
2280	10,000 gallon capacity		1.60	20		16,600	1,250		17,850	20,200
2282	12,000 gallon capacity		1.52	21.053		21,900	1,325		23,225	26,100
2284	15,000 gallon capacity		1.39	23.022		28,900	1,425		30,325	34,000
2290	20,000 gallon capacity		1.14	28.070		34,700	1,750		36,450	40,800
2300	25,000 gallon capacity		.96	33.333		60,500	2,075		62,575	69,500
2320	30,000 gallon capacity	▼	.80	40	▼	107,500	2,500		110,000	121,500

23 21 Hydronic Piping and Pumps

23 21 23 – Hydronic Pumps

23 21 23.13 In-Line Centrifugal Hydronic Pumps

		Crew	Daily Output	Labor-Hours	Unit	Material	2020 Bare Costs Labor	Equipment	Total	Total Incl O&P
0010	**IN-LINE CENTRIFUGAL HYDRONIC PUMPS**									
0600	Bronze, sweat connections, 1/40 HP, in line									
0640	3/4" size	Q-1	16	1	Ea.	275	58		333	385
1000	Flange connection, 3/4" to 1-1/2" size									
1040	1/12 HP	Q-1	6	2.667	Ea.	710	155		865	1,025
1060	1/8 HP		6	2.667		1,250	155		1,405	1,600
1100	1/3 HP		6	2.667		1,400	155		1,555	1,750
1140	2" size, 1/6 HP		5	3.200		1,750	186		1,936	2,200
1180	2-1/2" size, 1/4 HP	▼	5	3.200	▼	2,200	186		2,386	2,700
2000	Cast iron, flange connection									
2040	3/4" to 1-1/2" size, in line, 1/12 HP	Q-1	6	2.667	Ea.	470	155		625	750
2100	1/3 HP		6	2.667		870	155		1,025	1,175
2140	2" size, 1/6 HP		5	3.200		955	186		1,141	1,325
2180	2-1/2" size, 1/4 HP		5	3.200		1,150	186		1,336	1,525
2220	3" size, 1/4 HP	▼	4	4	▼	1,150	232		1,382	1,625
2600	For nonferrous impeller, add					3%				

23 55 Fuel-Fired Heaters

23 55 13 – Fuel-Fired Duct Heaters

23 55 13.16 Gas-Fired Duct Heaters	Crew	Daily Output	Labor-Hours	Unit	Material	2020 Bare Costs Labor	Equipment	Total	Total Incl O&P
0010 **GAS-FIRED DUCT HEATERS**, Includes burner, controls, stainless steel									
0020 heat exchanger. Gas fired, electric ignition									
1000 Outdoor installation, with power venter									
1020 75 MBH output	Q-5	4	4	Ea.	3,550	236		3,786	4,250
1060 120 MBH output		4	4		4,075	236		4,311	4,850
1100 187 MBH output		3	5.333		5,500	315		5,815	6,525
1140 300 MBH output		1.80	8.889		6,375	525		6,900	7,775
1180 450 MBH output		1.40	11.429		8,300	675		8,975	10,100

Estimating Tips
26 05 00 Common Work Results for Electrical

- Conduit should be taken off in three main categories—power distribution, branch power, and branch lighting—so the estimator can concentrate on systems and components, therefore making it easier to ensure all items have been accounted for.

- For cost modifications for elevated conduit installation, add the percentages to labor according to the height of installation and only to the quantities exceeding the different height levels, not to the total conduit quantities. Refer to subdivision 26 01 02.20 for labor adjustment factors.

- Remember that aluminum wiring of equal ampacity is larger in diameter than copper and may require larger conduit.

- If more than three wires at a time are being pulled, deduct percentages from the labor hours of that grouping of wires.

- When taking off grounding systems, identify separately the type and size of wire, and list each unique type of ground connection.

- The estimator should take the weights of materials into consideration when completing a takeoff. Topics to consider include: How will the materials be supported? What methods of support are available? How high will the support structure have to reach? Will the final support structure be able to withstand the total burden? Is the support material included or separate from the fixture, equipment, and material specified?

- Do not overlook the costs for equipment used in the installation. If scaffolding or highlifts are available in the field, contractors may use them in lieu of the proposed ladders and rolling staging.

26 20 00 Low-Voltage Electrical Transmission

- Supports and concrete pads may be shown on drawings for the larger equipment, or the support system may be only a piece of plywood for the back of a panelboard. In either case, they must be included in the costs.

26 40 00 Electrical and Cathodic Protection

- When taking off cathodic protection systems, identify the type and size of cable, and list each unique type of anode connection.

26 50 00 Lighting

- Fixtures should be taken off room by room using the fixture schedule, specifications, and the ceiling plan. For large concentrations of lighting fixtures in the same area, deduct the percentages from labor hours.

Reference Numbers

Reference numbers are shown at the beginning of some major classifications. These numbers refer to related items in the Reference Section. The reference information may be an estimating procedure, an alternate pricing method, or technical information.

Note: Not all subdivisions listed here necessarily appear. ■

Same Data. Simplified.

Enjoy the convenience and efficiency of accessing your costs anywhere:

- **Skip the multiplier** by setting your location
- **Quickly search,** edit, favorite and share costs
- **Stay on top of price changes** with automatic updates

Discover more at rsmeans.com/online

Note: "Powered in part by CINX™, based on liscensed proprietary information of Harrison Publishing House, Inc."

Note: The following companies, in part, have been used as a reference source for some of the material prices used in Division 26:

Electriflex

Trade Service

26 05 05.10 Electrical Demolition

		Crew	Daily Output	Labor-Hours	Unit	Material	2020 Bare Costs Labor	Equipment	Total	Total Incl O&P
0010	**ELECTRICAL DEMOLITION**									
0020	Electrical demolition, conduit to 10' high, incl. fittings & hangers									
0100	Rigid galvanized steel, 1/2" to 1" diameter	1 Elec	242	.033	L.F.		2.03		2.03	3.02
0120	1-1/4" to 2"	"	200	.040			2.45		2.45	3.65
0140	2-1/2" to 3-1/2"	2 Elec	302	.053			3.25		3.25	4.84
0160	4" to 6"	"	160	.100			6.15		6.15	9.15
0200	Electric metallic tubing (EMT), 1/2" to 1"	1 Elec	394	.020			1.25		1.25	1.85
0220	1-1/4" to 1-1/2"		326	.025			1.51		1.51	2.24
0240	2" to 3"		236	.034			2.08		2.08	3.10
0260	3-1/2" to 4"	2 Elec	310	.052			3.17		3.17	4.71
1300	Transformer, dry type, 1 phase, incl. removal of									
1320	supports, wire & conduit terminations									
1340	1 kVA	1 Elec	7.70	1.039	Ea.		63.50		63.50	95
1360	5 kVA		4.70	1.702			104		104	155
1380	10 kVA		3.60	2.222			136		136	203
1400	37.5 kVA	2 Elec	3	5.333			325		325	485
1420	75 kVA	"	2.50	6.400			395		395	585
1440	3 phase to 600 V, primary									
1460	3 kVA	1 Elec	3.87	2.067	Ea.		127		127	189
1520	75 kVA	2 Elec	2.69	5.948			365		365	545
1550	300 kVA	R-3	1.80	11.111			680	105	785	1,125
1570	750 kVA	"	1.10	18.182			1,100	172	1,272	1,850

26 05 13 – Medium-Voltage Cables

26 05 13.16 Medium-Voltage, Single Cable

		Crew	Daily Output	Labor-Hours	Unit	Material	2020 Bare Costs Labor	Equipment	Total	Total Incl O&P
0010	**MEDIUM-VOLTAGE, SINGLE CABLE** Splicing & terminations not included									
0040	Copper, XLP shielding, 5 kV, #6 R260533-22	2 Elec	4.40	3.636	C.L.F.	171	223		394	520
0050	#4		4.40	3.636		222	223		445	575
0100	#2		4	4		254	245		499	645
0200	#1		4	4		340	245		585	740
0400	1/0		3.80	4.211		335	258		593	750
0600	2/0		3.60	4.444		410	273		683	855
0800	4/0		3.20	5		545	305		850	1,050
1000	250 kcmil	3 Elec	4.50	5.333		655	325		980	1,200
1200	350 kcmil		3.90	6.154		850	380		1,230	1,500
1400	500 kcmil		3.60	6.667		1,100	410		1,510	1,800
1600	15 kV, ungrounded neutral, #1	2 Elec	4	4		365	245		610	770
1800	1/0		3.80	4.211		440	258		698	865
2000	2/0		3.60	4.444		500	273		773	955
2200	4/0		3.20	5		660	305		965	1,175
2400	250 kcmil	3 Elec	4.50	5.333		735	325		1,060	1,300
2600	350 kcmil		3.90	6.154		925	380		1,305	1,575
2800	500 kcmil		3.60	6.667		1,050	410		1,460	1,750
3000	25 kV, grounded neutral, 1/0	2 Elec	3.60	4.444		605	273		878	1,075
3200	2/0		3.40	4.706		670	289		959	1,175
3400	4/0		3	5.333		840	325		1,165	1,400
3600	250 kcmil	3 Elec	4.20	5.714		1,050	350		1,400	1,675
3800	350 kcmil		3.60	6.667		1,225	410		1,635	1,950
3900	500 kcmil		3.30	7.273		1,425	445		1,870	2,250
4000	35 kV, grounded neutral, 1/0	2 Elec	3.40	4.706		595	289		884	1,075
4200	2/0		3.20	5		755	305		1,060	1,275
4400	4/0		2.80	5.714		950	350		1,300	1,575
4600	250 kcmil	3 Elec	3.90	6.154		1,125	380		1,505	1,775

For customer support on your Site Work & Landscape Costs with RSMeans data, call 800.448.8182.

26 05 13 – Medium-Voltage Cables

26 05 13.16 Medium-Voltage, Single Cable		Crew	Daily Output	Labor-Hours	Unit	Material	2020 Bare Costs		Total	Total Incl O&P
							Labor	Equipment		
4800	350 kcmil	3 Elec	3.30	7.273	C.L.F.	1,350	445		1,795	2,150
5000	500 kcmil	↓	3	8		1,575	490		2,065	2,450
5050	Aluminum, XLP shielding, 5 kV, #2	2 Elec	5	3.200		242	196		438	560
5070	#1		4.40	3.636		251	223		474	605
5090	1/0		4	4		290	245		535	685
5100	2/0		3.80	4.211		325	258		583	745
5150	4/0	↓	3.60	4.444		385	273		658	830
5200	250 kcmil	3 Elec	4.80	5		465	305		770	970
5220	350 kcmil		4.50	5.333		550	325		875	1,100
5240	500 kcmil		3.90	6.154		705	380		1,085	1,325
5260	750 kcmil	↓	3.60	6.667		930	410		1,340	1,625
5300	15 kV aluminum, XLP, #1	2 Elec	4.40	3.636		310	223		533	670
5320	1/0		4	4		320	245		565	720
5340	2/0		3.80	4.211		345	258		603	765
5360	4/0	↓	3.60	4.444		425	273		698	870
5380	250 kcmil	3 Elec	4.80	5		510	305		815	1,025
5400	350 kcmil		4.50	5.333		570	325		895	1,100
5420	500 kcmil		3.90	6.154		785	380		1,165	1,425
5440	750 kcmil	↓	3.60	6.667	↓	1,050	410		1,460	1,750
6010	Copper, XLP shielding, 5 kV, buried in trench, #6	R-15	2.90	16.552	M.L.F.	1,700	995	96	2,791	3,450
6020	#4		2.90	16.552		2,225	995	96	3,316	4,025
6030	#2		2.90	16.552		2,550	995	96	3,641	4,375
6040	1/0		2.60	18.462		3,325	1,100	107	4,532	5,450
6050	2/0		2.60	18.462		4,100	1,100	107	5,307	6,275
6060	3/0		2.40	20		4,725	1,200	116	6,041	7,125
6070	4/0		2.40	20		5,475	1,200	116	6,791	7,950
6080	250 kcmil		2.20	21.818		6,550	1,300	127	7,977	9,300
6090	350 kcmil		2.15	22.326		8,475	1,350	130	9,955	11,500
6100	500 kcmil	↓	2.05	23.415		10,900	1,400	136	12,436	14,300
6200	Installed on poles, #6	R-16	5.50	11.636		1,700	700	50.50	2,450.50	2,975
6210	#4		5.10	12.549		2,225	760	54.50	3,039.50	3,625
6220	#2		4.05	15.802		2,550	955	69	3,574	4,300
6230	1/0		3.35	19.104		3,325	1,150	83.50	4,558.50	5,500
6240	2/0		2.85	22.456		4,100	1,350	98	5,548	6,625
6250	3/0		2.50	25.600		4,725	1,550	112	6,387	7,625
6260	4/0		2.15	29.767		5,475	1,800	130	7,405	8,850
6270	250 kcmil		1.70	37.647		6,550	2,275	164	8,989	10,800
6280	350 kcmil		1.35	47.407		8,475	2,850	207	11,532	13,800
6290	500 kcmil	↓	1.05	60.952		10,900	3,675	266	14,841	17,800
6300	15 kV, pulled in duct, #1	R-15	1.35	35.556		3,675	2,125	207	6,007	7,425
6310	1/0		1.35	35.556		4,375	2,125	207	6,707	8,225
6320	2/0		1.35	35.556		4,975	2,125	207	7,307	8,875
6330	3/0		1.30	36.923		5,400	2,225	215	7,840	9,450
6340	4/0		1.30	36.923		6,625	2,225	215	9,065	10,800
6350	250 kcmil		1.25	38.400		7,325	2,300	223	9,848	11,700
6360	350 kcmil		1.20	40		9,250	2,400	232	11,882	14,000
6370	500 kcmil		1.20	40		10,400	2,400	232	13,032	15,200
6400	Direct burial, #1		2.40	20		3,675	1,200	116	4,991	5,950
6410	1/0		2.40	20		4,375	1,200	116	5,691	6,750
6420	2/0		2.40	20		4,975	1,200	116	6,291	7,400
6430	3/0		2.20	21.818		5,400	1,300	127	6,827	8,025
6440	4/0		2.20	21.818		6,625	1,300	127	8,052	9,375
6450	250 kcmil		2	24		7,325	1,450	139	8,914	10,400

267

26 05 13 – Medium-Voltage Cables

26 05 13.16 Medium-Voltage, Single Cable

		Crew	Daily Output	Labor-Hours	Unit	Material	2020 Bare Costs Labor	Equipment	Total	Total Incl O&P
6460	350 kcmil	R-15	1.90	25.263	M.L.F.	9,250	1,525	147	10,922	12,600
6470	500 kcmil	↓	1.85	25.946		10,400	1,550	151	12,101	13,900
6500	Installed on poles, #1	R-16	3.85	16.623		3,675	1,000	72.50	4,747.50	5,600
6510	1/0		3.35	19.104		4,375	1,150	83.50	5,608.50	6,650
6520	2/0		2.85	22.456		4,975	1,350	98	6,423	7,600
6530	3/0		2.50	25.600		5,400	1,550	112	7,062	8,350
6540	4/0		2.15	29.767		6,625	1,800	130	8,555	10,100
6550	250 kcmil		1.85	34.595		7,325	2,100	151	9,576	11,300
6560	350 kcmil		1.60	40		9,250	2,425	174	11,849	14,000
6570	500 kcmil	↓	1.35	47.407	↓	10,400	2,850	207	13,457	15,900
6600	URD, 15 kV, 3/C, direct burial, #2	R-22	1500	.025	L.F.	12.35	1.40		13.75	15.70
7000	Copper, shielded, 35 kV, 1/0	R-16	1.10	58.182	M.L.F.	5,950	3,500	254	9,704	12,100
7020	3/0		1.05	60.952		7,100	3,675	266	11,041	13,600
7100	69 kV, 500 kcmil	↓	.90	71.111	↓	26,700	4,300	310	31,310	36,100

26 05 19 – Low-Voltage Electrical Power Conductors and Cables

26 05 19.55 Non-Metallic Sheathed Cable

		Crew	Daily Output	Labor-Hours	Unit	Material	2020 Bare Costs Labor	Equipment	Total	Total Incl O&P
0010	**NON-METALLIC SHEATHED CABLE** 600 volt R260533-22									
0100	Copper with ground wire (Romex)									
0150	#14, 2 conductor	1 Elec	2.70	2.963	C.L.F.	18.40	182		200.40	291
0200	3 conductor		2.40	3.333		26.50	205		231.50	335
0220	4 conductor		2.20	3.636		42	223		265	375
0250	#12, 2 conductor		2.50	3.200		26.50	196		222.50	320
0300	3 conductor		2.20	3.636		41	223		264	375
0320	4 conductor		2	4		65	245		310	435
0350	#10, 2 conductor		2.20	3.636		45	223		268	380
0400	3 conductor		1.80	4.444		66.50	273		339.50	480
0420	4 conductor		1.60	5		105	305		410	570
0430	#8, 2 conductor		1.60	5		87.50	305		392.50	550
0450	3 conductor		1.50	5.333		117	325		442	615
0500	#6, 3 conductor	↓	1.40	5.714		179	350		529	715
0520	#4, 3 conductor	2 Elec	2.40	6.667		400	410		810	1,050
0540	#2, 3 conductor	"	2.20	7.273	↓	910	445		1,355	1,675
0550	SE type SER aluminum cable, 3 RHW and									
0600	1 bare neutral, 3 #8 & 1 #8	1 Elec	1.60	5	C.L.F.	63	305		368	525
0650	3 #6 & 1 #6	"	1.40	5.714		70.50	350		420.50	600
0700	3 #4 & 1 #6	2 Elec	2.40	6.667		78.50	410		488.50	695
0750	3 #2 & 1 #4		2.20	7.273		143	445		588	825
0800	3 #1/0 & 1 #2		2	8		167	490		657	915
0850	3 #2/0 & 1 #1		1.80	8.889		199	545		744	1,025
0900	3 #4/0 & 1 #2/0		1.60	10		292	615		907	1,225
1000	URD - triplex underground distribution cable, alum. 2 #4 + #4 neutral		2.80	5.714		84	350		434	615
1010	2 #2 + #4 neutral		2.65	6.038		106	370		476	665
1020	2 #2 + #2 neutral		2.55	6.275		105	385		490	690
1030	2 1/0 + #2 neutral		2.40	6.667		130	410		540	755
1040	2 1/0 + 1/0 neutral		2.30	6.957		147	425		572	795
1050	2 2/0 + #1 neutral		2.20	7.273		146	445		591	825
1060	2 2/0 + 2/0 neutral		2.10	7.619		154	465		619	865
1070	2 3/0 + 1/0 neutral		1.95	8.205		194	505		699	965
1080	2 3/0 + 3/0 neutral		1.95	8.205		198	505		703	970
1090	2 4/0 + 2/0 neutral		1.85	8.649		188	530		718	995
1100	2 4/0 + 4/0 neutral	↓	1.85	8.649		226	530		756	1,050
1450	UF underground feeder cable, copper with ground, #14, 2 conductor	1 Elec	4	2		36	123		159	223

26 05 19 – Low-Voltage Electrical Power Conductors and Cables

26 05 19.55 Non-Metallic Sheathed Cable

		Crew	Daily Output	Labor-Hours	Unit	Material	2020 Bare Costs Labor	Equipment	Total	Total Incl O&P
1500	#12, 2 conductor	1 Elec	3.50	2.286	C.L.F.	33.50	140		173.50	246
1550	#10, 2 conductor		3	2.667		53.50	164		217.50	305
1600	#14, 3 conductor		3.50	2.286		36.50	140		176.50	250
1650	#12, 3 conductor		3	2.667		49.50	164		213.50	299
1700	#10, 3 conductor		2.50	3.200		80	196		276	380
1710	#8, 3 conductor		2	4		139	245		384	520
1720	#6, 3 conductor		1.80	4.444		213	273		486	640
2400	SEU service entrance cable, copper 2 conductors, #8 + #8 neutral		1.50	5.333		98.50	325		423.50	595
2600	#6 + #8 neutral		1.30	6.154		139	380		519	710
2800	#6 + #6 neutral		1.30	6.154		156	380		536	730
3000	#4 + #6 neutral	2 Elec	2.20	7.273		222	445		667	910
3200	#4 + #4 neutral		2.20	7.273		234	445		679	920
3400	#3 + #5 neutral		2.10	7.619		320	465		785	1,050
3600	#3 + #3 neutral		2.10	7.619		305	465		770	1,025
3800	#2 + #4 neutral		2	8		360	490		850	1,125
4000	#1 + #1 neutral		1.90	8.421		375	515		890	1,175
4200	1/0 + 1/0 neutral		1.80	8.889		850	545		1,395	1,750
4400	2/0 + 2/0 neutral		1.70	9.412		810	575		1,385	1,750
4600	3/0 + 3/0 neutral		1.60	10		905	615		1,520	1,900
4620	4/0 + 4/0 neutral		1.45	11.034		1,775	675		2,450	2,950
4800	Aluminum 2 conductors, #8 + #8 neutral	1 Elec	1.60	5		56	305		361	515
5000	#6 + #6 neutral	"	1.40	5.714		57	350		407	585
5100	#4 + #6 neutral	2 Elec	2.50	6.400		70	395		465	660
5200	#4 + #4 neutral		2.40	6.667		81	410		491	700
5300	#2 + #4 neutral		2.30	6.957		105	425		530	750
5400	#2 + #2 neutral		2.20	7.273		108	445		553	785
5450	1/0 + #2 neutral		2.10	7.619		160	465		625	870
5500	1/0 + 1/0 neutral		2	8		170	490		660	915
5550	2/0 + #1 neutral		1.90	8.421		180	515		695	965
5600	2/0 + 2/0 neutral		1.80	8.889		180	545		725	1,000
5800	3/0 + 1/0 neutral		1.70	9.412		242	575		817	1,125
6000	3/0 + 3/0 neutral		1.70	9.412		198	575		773	1,075
6200	4/0 + 2/0 neutral		1.60	10		199	615		814	1,125
6400	4/0 + 4/0 neutral		1.60	10		220	615		835	1,150
6500	Service entrance cap for copper SEU									
6600	100 amp	1 Elec	12	.667	Ea.	7.85	41		48.85	69.50
6700	150 amp	"	10	.800	"	11.75	49		60.75	86

26 05 26 – Grounding and Bonding for Electrical Systems

26 05 26.80 Grounding

		Crew	Daily Output	Labor-Hours	Unit	Material	2020 Bare Costs Labor	Equipment	Total	Total Incl O&P
0010	**GROUNDING**									
0030	Rod, copper clad, 8' long, 1/2" diameter	1 Elec	5.50	1.455	Ea.	22.50	89		111.50	158
0040	5/8" diameter		5.50	1.455		22.50	89		111.50	158
0050	3/4" diameter		5.30	1.509		36	92.50		128.50	178
0080	10' long, 1/2" diameter		4.80	1.667		24.50	102		126.50	179
0090	5/8" diameter		4.60	1.739		22.50	107		129.50	184
0100	3/4" diameter		4.40	1.818		53	112		165	225
0130	15' long, 3/4" diameter		4	2		55	123		178	244
0150	Coupling, bronze, 1/2" diameter					5.05			5.05	5.55
0160	5/8" diameter					5.45			5.45	6
0170	3/4" diameter					14.75			14.75	16.20
0190	Drive studs, 1/2" diameter					12.75			12.75	14.05
0210	5/8" diameter					20.50			20.50	22.50

26 05 26.80 Grounding		Crew	Daily Output	Labor-Hours	Unit	Material	2020 Bare Costs Labor	Equipment	Total	Total Incl O&P
0220	3/4" diameter				Ea.	23			23	25
0230	Clamp, bronze, 1/2" diameter	1 Elec	32	.250		5.25	15.35		20.60	29
0240	5/8" diameter		32	.250		4.64	15.35		19.99	28
0250	3/4" diameter		32	.250		5.65	15.35		21	29.50
0260	Wire ground bare armored, #8-1 conductor		2	4	C.L.F.	74.50	245		319.50	445
0270	#6-1 conductor		1.80	4.444		84	273		357	500
0280	#4-1 conductor		1.60	5		146	305		451	615
0320	Bare copper wire, #14 solid		14	.571		7.10	35		42.10	60
0330	#12		13	.615		12.85	38		50.85	70
0340	#10		12	.667		17.70	41		58.70	80.50
0350	#8		11	.727		25	44.50		69.50	94
0360	#6		10	.800		39.50	49		88.50	117
0370	#4		8	1		98	61.50		159.50	200
0380	#2		5	1.600		127	98		225	285
0390	Bare copper wire, stranded, #8		11	.727		40	44.50		84.50	111
0400	#6		10	.800		42	49		91	120
0450	#4	2 Elec	16	1		85.50	61.50		147	186
0600	#2		10	1.600		89	98		187	244
0650	#1		9	1.778		130	109		239	305
0700	1/0		8	2		156	123		279	355
0750	2/0		7.20	2.222		253	136		389	480
0800	3/0		6.60	2.424		305	149		454	555
1000	4/0		5.70	2.807		420	172		592	720
1200	250 kcmil	3 Elec	7.20	3.333		440	205		645	785
1210	300 kcmil		6.60	3.636		560	223		783	950
1220	350 kcmil		6	4		550	245		795	970
1230	400 kcmil		5.70	4.211		605	258		863	1,050
1240	500 kcmil		5.10	4.706		785	289		1,074	1,300
1260	750 kcmil		3.60	6.667		1,300	410		1,710	2,025
1270	1,000 kcmil		3	8		1,575	490		2,065	2,450
1360	Bare aluminum, stranded, #6	1 Elec	9	.889		14.55	54.50		69.05	97
1370	#4	2 Elec	16	1		31.50	61.50		93	126
1380	#2		13	1.231		45.50	75.50		121	162
1390	#1		10.60	1.509		40	92.50		132.50	182
1400	1/0		9	1.778		50.50	109		159.50	218
1410	2/0		8	2		77.50	123		200.50	269
1420	3/0		7.20	2.222		86	136		222	298
1430	4/0		6.60	2.424		109	149		258	340
1440	250 kcmil	3 Elec	9.30	2.581		115	158		273	360
1450	300 kcmil		8.70	2.759		121	169		290	385
1460	400 kcmil		7.50	3.200		159	196		355	465
1470	500 kcmil		6.90	3.478		193	213		406	530
1480	600 kcmil		6	4		230	245		475	620
1490	700 kcmil		5.70	4.211		247	258		505	655
1500	750 kcmil		5.10	4.706		273	289		562	730
1510	1,000 kcmil		4.80	5		350	305		655	840
1800	Water pipe ground clamps, heavy duty									
2000	Bronze, 1/2" to 1" diameter	1 Elec	8	1	Ea.	29	61.50		90.50	123
2100	1-1/4" to 2" diameter		8	1		31	61.50		92.50	126
2200	2-1/2" to 3" diameter		6	1.333		60.50	82		142.50	189
2740	4/0 wire to building steel		7	1.143		9.45	70		79.45	114
2750	4/0 wire to motor frame		7	1.143		9.45	70		79.45	114
2760	4/0 wire to 4/0 wire		7	1.143		9.45	70		79.45	114

26 05 26 – Grounding and Bonding for Electrical Systems

26 05 26.80 Grounding		Crew	Daily Output	Labor-Hours	Unit	Material	2020 Bare Costs Labor	2020 Bare Costs Equipment	Total	Total Incl O&P
2770	4/0 wire to #4 wire	1 Elec	7	1.143	Ea.	9.45	70		79.45	114
2780	4/0 wire to #8 wire		7	1.143		9.45	70		79.45	114
2790	Mold, reusable, for above					141			141	155
2800	Brazed connections, #6 wire	1 Elec	12	.667		16.15	41		57.15	79
3000	#2 wire		10	.800		21.50	49		70.50	97
3100	3/0 wire		8	1		32	61.50		93.50	127
3200	4/0 wire		7	1.143		36.50	70		106.50	145
3400	250 kcmil wire		5	1.600		43	98		141	193
3600	500 kcmil wire		4	2		52.50	123		175.50	241
3700	Insulated ground wire, copper #14		13	.615	C.L.F.	7.70	38		45.70	64.50
3710	#12		11	.727		10.60	44.50		55.10	78
3720	#10		10	.800		20.50	49		69.50	95.50
3730	#8		8	1		27	61.50		88.50	122
3740	#6		6.50	1.231		44	75.50		119.50	161
3750	#4	2 Elec	10.60	1.509		70	92.50		162.50	215
3770	#2		9	1.778		109	109		218	281
3780	#1		8	2		138	123		261	335
3790	1/0		6.60	2.424		200	149		349	440
3800	2/0		5.80	2.759		208	169		377	480
3810	3/0		5	3.200		259	196		455	575
3820	4/0		4.40	3.636		330	223		553	690
3830	250 kcmil	3 Elec	6	4		380	245		625	785
3840	300 kcmil		5.70	4.211		515	258		773	950
3850	350 kcmil		5.40	4.444		535	273		808	995
3860	400 kcmil		5.10	4.706		605	289		894	1,100
3870	500 kcmil		4.80	5		775	305		1,080	1,300
3880	600 kcmil		3.90	6.154		1,275	380		1,655	1,950
3890	750 kcmil		3.30	7.273		1,150	445		1,595	1,950
3900	1,000 kcmil		2.70	8.889		1,875	545		2,420	2,850
3960	Insulated ground wire, aluminum, #6	1 Elec	8	1		20.50	61.50		82	114
3970	#4	2 Elec	13	1.231		25	75.50		100.50	140
3980	#2		10.60	1.509		48	92.50		140.50	191
3990	#1		9	1.778		57	109		166	225
4000	1/0		8	2		80	123		203	272
4010	2/0		7.20	2.222		82	136		218	293
4020	3/0		6.60	2.424		103	149		252	335
4030	4/0		6.20	2.581		131	158		289	380
4040	250 kcmil	3 Elec	8.70	2.759		137	169		306	405
4050	300 kcmil		8.10	2.963		190	182		372	480
4060	350 kcmil		7.50	3.200		192	196		388	505
4070	400 kcmil		6.90	3.478		219	213		432	560
4080	500 kcmil		6	4		246	245		491	635
4090	600 kcmil		5.70	4.211		315	258		573	730
4100	700 kcmil		5.10	4.706		385	289		674	855
4110	750 kcmil		4.80	5		375	305		680	870
5000	Copper electrolytic ground rod system									
5010	Includes augering hole, mixing bentonite clay,									
5020	Installing rod, and terminating ground wire									
5100	Straight vertical type, 2" diam.									
5120	8.5' long, clamp connection	1 Elec	2.67	2.996	Ea.	720	184		904	1,075
5130	With exothermic weld connection		1.95	4.103		730	252		982	1,175
5140	10' long		2.35	3.404		710	209		919	1,100
5150	With exothermic weld connection		1.78	4.494		890	276		1,166	1,400

For customer support on your Site Work & Landscape Costs with RSMeans data, call 800.448.8182.

271

26 05 26 – Grounding and Bonding for Electrical Systems

26 05 26.80 Grounding		Crew	Daily Output	Labor-Hours	Unit	Material	2020 Bare Costs Labor	Equipment	Total	Total Incl O&P
5160	12' long	1 Elec	2.16	3.704	Ea.	1,025	227		1,252	1,475
5170	With exothermic weld connection		1.67	4.790		1,125	294		1,419	1,675
5180	20' long		1.74	4.598		1,400	282		1,682	1,950
5190	With exothermic weld connection		1.40	5.714		1,375	350		1,725	2,050
5195	40' long with exothermic weld connection	2 Elec	2	8		3,050	490		3,540	4,100
5200	L-shaped, 2" diam.									
5220	4' vert. x 10' horiz., clamp connection	1 Elec	5.33	1.501	Ea.	1,325	92		1,417	1,600
5230	With exothermic weld connection	"	3.08	2.597	"	1,325	159		1,484	1,675
5300	Protective box at grade level, with breather slots									
5320	Round 12" long, fiberlyte	1 Elec	32	.250	Ea.	70	15.35		85.35	100
5330	Concrete	"	16	.500		112	30.50		142.50	169
5400	Bentonite clay, 50# bag, 1 per 10' of rod					56			56	61.50
5500	Equipotential earthing bar	1 Elec	2	4		138	245		383	515

26 05 33 – Raceway and Boxes for Electrical Systems

26 05 33.13 Conduit

26 05 33.13 Conduit		Crew	Daily Output	Labor-Hours	Unit	Material	2020 Bare Costs Labor	Equipment	Total	Total Incl O&P
0010	**CONDUIT** To 10' high, includes 2 terminations, 2 elbows,									
0020	11 beam clamps, and 11 couplings per 100 L.F.									
9100	PVC, schedule 40, 1/2" diameter	1 Elec	190	.042	L.F.	.87	2.58		3.45	4.80
9110	3/4" diameter		145	.055		1.03	3.38		4.41	6.20
9120	1" diameter		125	.064		1.54	3.93		5.47	7.55
9130	1-1/4" diameter		110	.073		1.99	4.46		6.45	8.85
9140	1-1/2" diameter		100	.080		2.39	4.91		7.30	9.95
9150	2" diameter		90	.089		3.36	5.45		8.81	11.80
9160	2-1/2" diameter		65	.123		4.60	7.55		12.15	16.30
9170	3" diameter	2 Elec	110	.145		4.84	8.90		13.74	18.65
9180	3-1/2" diameter		100	.160		6.40	9.80		16.20	21.50
9190	4" diameter		90	.178		5.75	10.90		16.65	22.50
9200	5" diameter		70	.229		8.35	14		22.35	30
9210	6" diameter		60	.267		12.50	16.35		28.85	38.50
9220	Elbows, 1/2" diameter	1 Elec	50	.160	Ea.	.58	9.80		10.38	15.25
9225	3/4" diameter		42	.190		.60	11.70		12.30	18.05
9230	1" diameter		35	.229		1.21	14		15.21	22.50
9235	1-1/4" diameter		28	.286		2.72	17.55		20.27	29
9240	1-1/2" diameter		20	.400		2.71	24.50		27.21	39.50
9245	2" diameter	R-1A	36.40	.440		3.54	24.50		28.04	40
9250	2-1/2" diameter		26.70	.599		7.75	33		40.75	58
9255	3" diameter		22.90	.699		12.25	38.50		50.75	71
9260	3-1/2" diameter		22.20	.721		18.65	40		58.65	80
9265	4" diameter		18.20	.879		23	48.50		71.50	97.50
9270	5" diameter		12.10	1.322		34.50	73		107.50	147
9275	6" diameter		11.10	1.441		35	79.50		114.50	158
9312	Couplings, 1/2" diameter	1 Elec	50	.160		.11	9.80		9.91	14.70
9314	3/4" diameter		42	.190		.31	11.70		12.01	17.75
9316	1" diameter		35	.229		.25	14		14.25	21.50
9318	1-1/4" diameter		28	.286		.35	17.55		17.90	26.50
9320	1-1/2" diameter		20	.400		.54	24.50		25.04	37
9322	2" diameter	R-1A	36.40	.440		1.26	24.50		25.76	37.50
9324	2-1/2" diameter		26.70	.599		1.80	33		34.80	51.50
9326	3" diameter		22.90	.699		1.48	38.50		39.98	59
9328	3-1/2" diameter		22.20	.721		2.05	40		42.05	62
9330	4" diameter		18.20	.879		2.65	48.50		51.15	75.50
9332	5" diameter		12.10	1.322		7	73		80	117

26 05 33.13 Conduit		Crew	Daily Output	Labor-Hours	Unit	Material	2020 Bare Costs Labor	Equipment	Total	Total Incl O&P
9334	6" diameter	R-1A	11.10	1.441	Ea.	10.20	79.50		89.70	130
9335	See note on line 26 05 33.13 9995									
9340	Field bends, 45° & 90°, 1/2" diameter	1 Elec	45	.178	Ea.		10.90		10.90	16.25
9350	3/4" diameter		40	.200			12.25		12.25	18.25
9360	1" diameter		35	.229			14		14	21
9370	1-1/4" diameter		32	.250			15.35		15.35	23
9380	1-1/2" diameter		27	.296			18.20		18.20	27
9390	2" diameter		20	.400			24.50		24.50	36.50
9400	2-1/2" diameter		16	.500			30.50		30.50	45.50
9410	3" diameter		13	.615			38		38	56
9420	3-1/2" diameter		12	.667			41		41	61
9430	4" diameter		10	.800			49		49	73
9440	5" diameter		9	.889			54.50		54.50	81
9450	6" diameter		8	1			61.50		61.50	91.50
9460	PVC adapters, 1/2" diameter		50	.160		.26	9.80		10.06	14.90
9470	3/4" diameter		42	.190		.37	11.70		12.07	17.80
9480	1" diameter		38	.211		.39	12.90		13.29	19.70
9490	1-1/4" diameter		35	.229		.76	14		14.76	22
9500	1-1/2" diameter		32	.250		.71	15.35		16.06	24
9510	2" diameter		27	.296		1.39	18.20		19.59	28.50
9520	2-1/2" diameter		23	.348		2.01	21.50		23.51	34
9530	3" diameter		18	.444		2.85	27.50		30.35	43.50
9540	3-1/2" diameter		13	.615		5.20	38		43.20	61.50
9550	4" diameter		11	.727		4.80	44.50		49.30	72
9560	5" diameter		8	1		9.70	61.50		71.20	102
9570	6" diameter		6	1.333		18.90	82		100.90	143
9580	PVC-LB, LR or LL fittings & covers									
9590	1/2" diameter	1 Elec	20	.400	Ea.	2.35	24.50		26.85	39
9600	3/4" diameter		16	.500		4.78	30.50		35.28	51
9610	1" diameter		12	.667		4.63	41		45.63	66
9620	1-1/4" diameter		9	.889		8.65	54.50		63.15	90.50
9630	1-1/2" diameter		7	1.143		10.25	70		80.25	115
9640	2" diameter		6	1.333		14.85	82		96.85	138
9650	2-1/2" diameter		6	1.333		80.50	82		162.50	211
9660	3" diameter		5	1.600		42.50	98		140.50	193
9670	3-1/2" diameter		4	2		56.50	123		179.50	246
9680	4" diameter		3	2.667		62	164		226	310
9690	PVC-tee fitting & cover									
9700	1/2"	1 Elec	14	.571	Ea.	5.30	35		40.30	58
9710	3/4"		13	.615		5.70	38		43.70	62.50
9720	1"		10	.800		5.35	49		54.35	79
9730	1-1/4"		9	.889		10.95	54.50		65.45	93
9740	1-1/2"		8	1		13.40	61.50		74.90	106
9750	2"		7	1.143		18.10	70		88.10	124
9760	PVC-reducers, 3/4" x 1/2" diameter					1.01			1.01	1.11
9770	1" x 1/2" diameter					3.19			3.19	3.51
9780	1" x 3/4" diameter					2.80			2.80	3.08
9790	1-1/4" x 3/4" diameter					2.40			2.40	2.64
9800	1-1/4" x 1" diameter					4.32			4.32	4.75
9810	1-1/2" x 1-1/4" diameter					3.44			3.44	3.78
9820	2" x 1-1/4" diameter					4.51			4.51	4.96
9830	2-1/2" x 2" diameter					18.30			18.30	20
9840	3" x 2" diameter					21			21	23

273

26 05 Common Work Results for Electrical

26 05 33 – Raceway and Boxes for Electrical Systems

26 05 33.13 Conduit

		Crew	Daily Output	Labor-Hours	Unit	Material	2020 Bare Costs Labor	Equipment	Total	Total Incl O&P
9850	4" x 3" diameter				Ea.	23			23	25.50
9860	Cement, quart					15.55			15.55	17.10
9870	Gallon					71			71	78
9880	Heat bender, to 6" diameter					1,500			1,500	1,650
9900	Add to labor for higher elevated installation									
9905	10' to 14.5' high, add						10%			
9910	15' to 20' high, add						20%			
9920	20' to 25' high, add						25%			
9930	25' to 30' high, add						35%			
9940	30' to 35' high, add						40%			
9950	35' to 40' high, add						50%			
9960	Over 40' high, add						55%			
9995	Do not include labor when adding couplings to a fitting installation									

26 05 33.18 Pull Boxes

		Crew	Daily Output	Labor-Hours	Unit	Material	2020 Bare Costs Labor	Equipment	Total	Total Incl O&P
0010	**PULL BOXES**									
2100	Pull box, NEMA 3R, type SC, raintight & weatherproof									
2150	6" L x 6" W x 6" D	1 Elec	10	.800	Ea.	17.25	49		66.25	92
2200	8" L x 6" W x 6" D		8	1		25	61.50		86.50	119
2250	10" L x 6" W x 6" D		7	1.143		32	70		102	139
2300	12" L x 12" W x 6" D		5	1.600		67.50	98		165.50	220
2350	16" L x 16" W x 6" D		4.50	1.778		92	109		201	263
2400	20" L x 20" W x 6" D		4	2		86	123		209	278
2450	24" L x 18" W x 8" D		3	2.667		153	164		317	415
2500	24" L x 24" W x 10" D		2.50	3.200		288	196		484	605
2550	30" L x 24" W x 12" D		2	4		400	245		645	800
2600	36" L x 36" W x 12" D		1.50	5.333		695	325		1,020	1,250
2800	Cast iron, pull boxes for surface mounting									
3000	NEMA 4, watertight & dust tight									
3050	6" L x 6" W x 6" D	1 Elec	4	2	Ea.	276	123		399	490
3100	8" L x 6" W x 6" D		3.20	2.500		420	153		573	690
3150	10" L x 6" W x 6" D		2.50	3.200		440	196		636	775
3200	12" L x 12" W x 6" D		2.30	3.478		745	213		958	1,150
3250	16" L x 16" W x 6" D		1.30	6.154		840	380		1,220	1,475
3300	20" L x 20" W x 6" D		.80	10		965	615		1,580	1,975
3350	24" L x 18" W x 8" D		.70	11.429		2,600	700		3,300	3,900
3400	24" L x 24" W x 10" D		.50	16		4,725	980		5,705	6,650
3450	30" L x 24" W x 12" D		.40	20		6,725	1,225		7,950	9,225
3500	36" L x 36" W x 12" D		.20	40		8,275	2,450		10,725	12,800

26 05 39 – Underfloor Raceways for Electrical Systems

26 05 39.30 Conduit In Concrete Slab

		Crew	Daily Output	Labor-Hours	Unit	Material	2020 Bare Costs Labor	Equipment	Total	Total Incl O&P
0010	**CONDUIT IN CONCRETE SLAB** Including terminations,	R337119-30								
0020	fittings and supports									
3230	PVC, schedule 40, 1/2" diameter	1 Elec	270	.030	L.F.	.55	1.82		2.37	3.31
3250	3/4" diameter		230	.035		.59	2.13		2.72	3.83
3270	1" diameter		200	.040		.78	2.45		3.23	4.50
3300	1-1/4" diameter		170	.047		1.05	2.89		3.94	5.45
3330	1-1/2" diameter		140	.057		1.27	3.51		4.78	6.60
3350	2" diameter		120	.067		1.59	4.09		5.68	7.85
3370	2-1/2" diameter		90	.089		2.62	5.45		8.07	11
3400	3" diameter	2 Elec	160	.100		3.47	6.15		9.62	12.95
3430	3-1/2" diameter		120	.133		4.50	8.20		12.70	17.15
3440	4" diameter		100	.160		4.76	9.80		14.56	19.85

26 05 39 – Underfloor Raceways for Electrical Systems

26 05 39.30 Conduit In Concrete Slab

		Crew	Daily Output	Labor-Hours	Unit	Material	2020 Bare Costs Labor	Equipment	Total	Total Incl O&P
3450	5" diameter	2 Elec	80	.200	L.F.	7.45	12.25		19.70	26.50
3460	6" diameter	↓	60	.267	↓	9.65	16.35		26	35
3530	Sweeps, 1" diameter, 30" radius	1 Elec	32	.250	Ea.	9.55	15.35		24.90	33.50
3550	1-1/4" diameter		24	.333		8.55	20.50		29.05	40
3570	1-1/2" diameter		21	.381		9.35	23.50		32.85	45.50
3600	2" diameter		18	.444		11.25	27.50		38.75	53
3630	2-1/2" diameter		14	.571		80	35		115	140
3650	3" diameter		10	.800		68	49		117	148
3670	3-1/2" diameter		8	1		52	61.50		113.50	149
3700	4" diameter		7	1.143		25	70		95	132
3710	5" diameter	↓	6	1.333		55	82		137	183
3730	Couplings, 1/2" diameter					.12			.12	.13
3750	3/4" diameter					.15			.15	.17
3770	1" diameter					.23			.23	.25
3800	1-1/4" diameter					.38			.38	.42
3830	1-1/2" diameter					.43			.43	.47
3850	2" diameter					.58			.58	.64
3870	2-1/2" diameter					1.11			1.11	1.22
3900	3" diameter					1.76			1.76	1.94
3930	3-1/2" diameter					2.45			2.45	2.70
3950	4" diameter					2.74			2.74	3.01
3960	5" diameter					5.65			5.65	6.20
3970	6" diameter					8.85			8.85	9.75
4030	End bells, 1" diameter, PVC	1 Elec	60	.133		2.68	8.20		10.88	15.15
4050	1-1/4" diameter		53	.151		4.54	9.25		13.79	18.80
4100	1-1/2" diameter		48	.167		3.10	10.25		13.35	18.65
4150	2" diameter		34	.235		4.36	14.45		18.81	26.50
4170	2-1/2" diameter		27	.296		4.05	18.20		22.25	31.50
4200	3" diameter		20	.400		6.30	24.50		30.80	43.50
4250	3-1/2" diameter		16	.500		6.90	30.50		37.40	53
4300	4" diameter		14	.571		7.75	35		42.75	60.50
4310	5" diameter		12	.667		12.05	41		53.05	74.50
4320	6" diameter		9	.889	↓	12.35	54.50		66.85	94.50
4350	Rigid galvanized steel, 1/2" diameter		200	.040	L.F.	2.34	2.45		4.79	6.20
4400	3/4" diameter		170	.047		4.45	2.89		7.34	9.20
4450	1" diameter		130	.062		6.90	3.78		10.68	13.20
4500	1-1/4" diameter		110	.073		4.53	4.46		8.99	11.65
4600	1-1/2" diameter		100	.080		8.40	4.91		13.31	16.55
4800	2" diameter	↓	90	.089	↓	8.20	5.45		13.65	17.15

26 05 39.40 Conduit In Trench

		Crew	Daily Output	Labor-Hours	Unit	Material	2020 Bare Costs Labor	Equipment	Total	Total Incl O&P
0010	**CONDUIT IN TRENCH** Includes terminations and fittings	R337119-30								
0020	Does not include excavation or backfill, see Section 31 23 16									
0200	Rigid galvanized steel, 2" diameter	1 Elec	150	.053	L.F.	7.80	3.27		11.07	13.45
0400	2-1/2" diameter	"	100	.080		13	4.91		17.91	21.50
0600	3" diameter	2 Elec	160	.100		14.55	6.15		20.70	25
0800	3-1/2" diameter		140	.114		19.20	7		26.20	31.50
1000	4" diameter		100	.160		23.50	9.80		33.30	40
1200	5" diameter		80	.200		37.50	12.25		49.75	59.50
1400	6" diameter	↓	60	.267	↓	57.50	16.35		73.85	88

26 12 Medium-Voltage Transformers

26 12 19 – Pad-Mounted, Liquid-Filled, Medium-Voltage Transformers

26 12 19.10 Transformer, Oil-Filled	Crew	Daily Output	Labor-Hours	Unit	Material	2020 Bare Costs Labor	Equipment	Total	Total Incl O&P
0010 **TRANSFORMER, OIL-FILLED** primary delta or Y,									
0050 Pad mounted 5 kV or 15 kV, with taps, 277/480 V secondary, 3 phase									
0100 150 kVA	R-3	.65	30.769	Ea.	9,875	1,875	291	12,041	14,000
0200 300 kVA	"	.45	44.444	"	14,800	2,725	420	17,945	20,700

26 24 Switchboards and Panelboards

26 24 16 – Panelboards

26 24 16.30 Panelboards Commercial Applications

	Crew	Daily Output	Labor-Hours	Unit	Material	2020 Bare Costs Labor	Equipment	Total	Total Incl O&P
0010 **PANELBOARDS COMMERCIAL APPLICATIONS**									
0050 NQOD, w/20 amp 1 pole bolt-on circuit breakers									
0100 3 wire, 120/240 volts, 100 amp main lugs									
0150 10 circuits	1 Elec	1	8	Ea.	935	490		1,425	1,750
0200 14 circuits		.88	9.091		1,050	560		1,610	1,975
0250 18 circuits		.75	10.667		1,150	655		1,805	2,225
0300 20 circuits		.65	12.308		1,275	755		2,030	2,550
0350 225 amp main lugs, 24 circuits	2 Elec	1.20	13.333		1,450	820		2,270	2,825
0400 30 circuits		.90	17.778		1,675	1,100		2,775	3,475
0450 36 circuits		.80	20		1,925	1,225		3,150	3,925
0500 38 circuits		.72	22.222		2,075	1,375		3,450	4,325
0550 42 circuits		.66	24.242		2,150	1,475		3,625	4,575
0600 4 wire, 120/208 volts, 100 amp main lugs, 12 circuits	1 Elec	1	8		1,025	490		1,515	1,850
0650 16 circuits		.75	10.667		1,175	655		1,830	2,250
0700 20 circuits		.65	12.308		1,350	755		2,105	2,600
0750 24 circuits		.60	13.333		1,275	820		2,095	2,625
0800 30 circuits		.53	15.094		1,675	925		2,600	3,225
0850 225 amp main lugs, 32 circuits	2 Elec	.90	17.778		1,875	1,100		2,975	3,700
0900 34 circuits		.84	19.048		1,925	1,175		3,100	3,850
0950 36 circuits		.80	20		1,975	1,225		3,200	3,975
1000 42 circuits		.68	23.529		2,200	1,450		3,650	4,575
1010 400 amp main lugs, 42 circs		.68	23.529		2,200	1,450		3,650	4,575

26 27 Low-Voltage Distribution Equipment

26 27 13 – Electricity Metering

26 27 13.10 Meter Centers and Sockets

	Crew	Daily Output	Labor-Hours	Unit	Material	2020 Bare Costs Labor	Equipment	Total	Total Incl O&P
0010 **METER CENTERS AND SOCKETS**									
0100 Sockets, single position, 4 terminal, 100 amp	1 Elec	3.20	2.500	Ea.	52.50	153		205.50	286
0200 150 amp		2.30	3.478		62.50	213		275.50	390
0300 200 amp		1.90	4.211		110	258		368	505
0500 Double position, 4 terminal, 100 amp		2.80	2.857		240	175		415	525
0600 150 amp		2.10	3.810		290	234		524	670
0700 200 amp		1.70	4.706		550	289		839	1,025
1100 Meter centers and sockets, three phase, single pos, 7 terminal, 100 amp		2.80	2.857		136	175		311	410
1200 200 amp		2.10	3.810		239	234		473	615
1400 400 amp		1.70	4.706		825	289		1,114	1,350
2000 Meter center, main fusible switch, 1P 3W 120/240 V									
2030 400 amp	2 Elec	1.60	10	Ea.	855	615		1,470	1,850
2040 600 amp		1.10	14.545		1,300	890		2,190	2,775
2050 800 amp		.90	17.778		5,000	1,100		6,100	7,125
2060 Rainproof 1P 3W 120/240 V, 400 A		1.60	10		1,950	615		2,565	3,075

26 27 Low-Voltage Distribution Equipment

26 27 13 – Electricity Metering

26 27 13.10 Meter Centers and Sockets

		Crew	Daily Output	Labor-Hours	Unit	Material	2020 Bare Costs Labor	2020 Bare Costs Equipment	Total	Total Incl O&P
2070	600 amp	2 Elec	1.10	14.545	Ea.	3,400	890		4,290	5,050
2080	800 amp		.90	17.778		5,325	1,100		6,425	7,475
2100	3P 4W 120/208 V, 400 amp		1.60	10		865	615		1,480	1,875
2110	600 amp		1.10	14.545		1,475	890		2,365	2,925
2120	800 amp		.90	17.778		2,300	1,100		3,400	4,150
2130	Rainproof 3P 4W 120/208 V, 400 amp		1.60	10		2,225	615		2,840	3,375
2140	600 amp		1.10	14.545		3,575	890		4,465	5,275
2150	800 amp		.90	17.778		7,750	1,100		8,850	10,200
2170	Main circuit breaker, 1P 3W 120/240 V									
2180	400 amp	2 Elec	1.60	10	Ea.	1,475	615		2,090	2,550
2190	600 amp		1.10	14.545		1,800	890		2,690	3,300
2200	800 amp		.90	17.778		2,775	1,100		3,875	4,700
2210	1,000 amp		.80	20		3,200	1,225		4,425	5,350
2220	1,200 amp		.76	21.053		4,300	1,300		5,600	6,650
2230	1,600 amp		.68	23.529		18,800	1,450		20,250	22,900
2240	Rainproof 1P 3W 120/240 V, 400 amp		1.60	10		3,000	615		3,615	4,225
2250	600 amp		1.10	14.545		4,700	890		5,590	6,500
2260	800 amp		.90	17.778		5,475	1,100		6,575	7,650
2270	1,000 amp		.80	20		7,550	1,225		8,775	10,100
2280	1,200 amp		.76	21.053		10,200	1,300		11,500	13,100
2300	3P 4W 120/208 V, 400 amp		1.60	10		3,350	615		3,965	4,625
2310	600 amp		1.10	14.545		5,600	890		6,490	7,475
2320	800 amp		.90	17.778		6,650	1,100		7,750	8,950
2330	1,000 amp		.80	20		8,725	1,225		9,950	11,400
2340	1,200 amp		.76	21.053		11,200	1,300		12,500	14,200
2350	1,600 amp		.68	23.529		22,900	1,450		24,350	27,400
2360	Rainproof 3P 4W 120/208 V, 400 amp		1.60	10		3,975	615		4,590	5,300
2370	600 amp		1.10	14.545		5,600	890		6,490	7,475
2380	800 amp		.90	17.778		6,650	1,100		7,750	8,950
2390	1,000 amp		.76	21.053		8,725	1,300		10,025	11,500

26 28 Low-Voltage Circuit Protective Devices

26 28 16 – Enclosed Switches and Circuit Breakers

26 28 16.10 Circuit Breakers

		Crew	Daily Output	Labor-Hours	Unit	Material	2020 Bare Costs Labor	2020 Bare Costs Equipment	Total	Total Incl O&P
0010	**CIRCUIT BREAKERS** (in enclosure)									
2300	Enclosed (NEMA 7), explosion proof, 600 volt 3 pole, 50 amp	1 Elec	2.30	3.478	Ea.	1,925	213		2,138	2,450
2350	100 amp		1.50	5.333		2,000	325		2,325	2,675
2400	150 amp		1	8		4,825	490		5,315	6,050
2450	250 amp	2 Elec	1.60	10		6,050	615		6,665	7,575
2500	400 amp	"	1.20	13.333		6,700	820		7,520	8,600

26 28 16.20 Safety Switches

		Crew	Daily Output	Labor-Hours	Unit	Material	2020 Bare Costs Labor	2020 Bare Costs Equipment	Total	Total Incl O&P
0010	**SAFETY SWITCHES**									
0100	General duty 240 volt, 3 pole NEMA 1, fusible, 30 amp	1 Elec	3.20	2.500	Ea.	59.50	153		212.50	294
0200	60 amp		2.30	3.478		102	213		315	430
0300	100 amp		1.90	4.211		175	258		433	575
0400	200 amp		1.30	6.154		375	380		755	970
0500	400 amp	2 Elec	1.80	8.889		985	545		1,530	1,875
0600	600 amp	"	1.20	13.333		1,925	820		2,745	3,325
2900	Heavy duty, 240 volt, 3 pole NEMA 1 fusible									
2910	30 amp	1 Elec	3.20	2.500	Ea.	102	153		255	340
3000	60 amp		2.30	3.478		170	213		383	510

277

26 28 Low-Voltage Circuit Protective Devices

26 28 16 – Enclosed Switches and Circuit Breakers

26 28 16.20 Safety Switches		Crew	Daily Output	Labor-Hours	Unit	Material	2020 Bare Costs Labor	Equipment	Total	Total Incl O&P
3300	100 amp	1 Elec	1.90	4.211	Ea.	263	258		521	675
3500	200 amp	↓	1.30	6.154		455	380		835	1,050
3700	400 amp	2 Elec	1.80	8.889		1,225	545		1,770	2,150
3900	600 amp	"	1.20	13.333	↓	1,900	820		2,720	3,300

26 32 Packaged Generator Assemblies

26 32 13 – Engine Generators

26 32 13.13 Diesel-Engine-Driven Generator Sets

		Crew	Daily Output	Labor-Hours	Unit	Material	2020 Bare Costs Labor	Equipment	Total	Total Incl O&P
0010	**DIESEL-ENGINE-DRIVEN GENERATOR SETS**									
2000	Diesel engine, including battery, charger,									
2010	muffler, & day tank, 30 kW	R-3	.55	36.364	Ea.	11,200	2,225	345	13,770	16,000
2100	50 kW		.42	47.619		21,800	2,900	450	25,150	28,800
2200	75 kW		.35	57.143		25,300	3,500	540	29,340	33,600
2300	100 kW		.31	64.516		27,600	3,950	610	32,160	36,900
2400	125 kW		.29	68.966		29,000	4,225	650	33,875	38,900
2500	150 kW		.26	76.923		45,800	4,700	725	51,225	58,500
2501	Generator set, dsl eng in alum encl, incl btry, chgr, muf & day tank, 150 kW		.26	76.923		41,500	4,700	725	46,925	53,500
2600	175 kW		.25	80		48,900	4,900	755	54,555	61,500
2700	200 kW		.24	83.333		49,900	5,100	790	55,790	63,500
2800	250 kW		.23	86.957		52,500	5,325	820	58,645	67,000
2900	300 kW		.22	90.909		58,000	5,550	860	64,410	72,500
3000	350 kW		.20	100		67,000	6,100	945	74,045	84,000
3100	400 kW		.19	105		78,000	6,425	995	85,420	96,500
3200	500 kW		.18	111		106,500	6,800	1,050	114,350	128,500
3220	600 kW	↓	.17	118		128,500	7,200	1,100	136,800	153,500
3240	750 kW	R-13	.38	111	↓	160,000	6,550	620	167,170	186,500

26 32 13.16 Gas-Engine-Driven Generator Sets

		Crew	Daily Output	Labor-Hours	Unit	Material	2020 Bare Costs Labor	Equipment	Total	Total Incl O&P
0010	**GAS-ENGINE-DRIVEN GENERATOR SETS**									
0020	Gas or gasoline operated, includes battery,									
0050	charger, & muffler									
0200	7.5 kW	R-3	.83	24.096	Ea.	8,750	1,475	228	10,453	12,100
0300	11.5 kW		.71	28.169		12,400	1,725	266	14,391	16,500
0400	20 kW		.63	31.746		14,600	1,950	300	16,850	19,300
0500	35 kW		.55	36.364		17,400	2,225	345	19,970	22,800
0520	60 kW	↓	.50	40		22,900	2,450	380	25,730	29,300
0600	80 kW	R-13	.40	105		28,500	6,225	590	35,315	41,300
0700	100 kW		.33	127		31,200	7,550	715	39,465	46,400
0800	125 kW		.28	150		64,000	8,900	845	73,745	84,500
0900	185 kW	↓	.25	168	↓	84,500	9,950	945	95,395	109,000

26 36 Transfer Switches

26 36 13 – Manual Transfer Switches

26 36 13.10 Non-Automatic Transfer Switches

		Crew	Daily Output	Labor-Hours	Unit	Material	2020 Bare Costs Labor	2020 Bare Costs Equipment	Total	Total Incl O&P
0010	**NON-AUTOMATIC TRANSFER SWITCHES** enclosed									
0100	Manual operated, 480 volt 3 pole, 30 amp	1 Elec	2.30	3.478	Ea.	1,100	213		1,313	1,550
0200	100 amp	"	1.30	6.154		3,525	380		3,905	4,450
0250	200 amp	2 Elec	2	8		4,450	490		4,940	5,625
0300	400 amp	"	1.60	10		8,350	615		8,965	10,100
1000	250 volt 3 pole, 30 amp	1 Elec	2.30	3.478		1,050	213		1,263	1,475
1150	100 amp	"	1.30	6.154		2,875	380		3,255	3,725
1200	200 amp	2 Elec	2	8		4,150	490		4,640	5,275
1500	Electrically operated, 480 volt 3 pole, 60 amp	1 Elec	1.90	4.211		1,650	258		1,908	2,200
1600	100 amp	"	1.30	6.154		1,650	380		2,030	2,375
1650	200 amp	2 Elec	2	8		2,725	490		3,215	3,700
1700	400 amp	"	1.60	10		3,800	615		4,415	5,100
2000	250 volt 3 pole, 30 amp	1 Elec	2.30	3.478		1,800	213		2,013	2,300
2050	60 amp	"	1.90	4.211		2,150	258		2,408	2,725
2150	200 amp	2 Elec	2	8		3,525	490		4,015	4,600
2200	400 amp	"	1.60	10		4,925	615		5,540	6,325
2500	NEMA 3R, 480 volt 3 pole, 60 amp	1 Elec	1.80	4.444		2,550	273		2,823	3,200
2550	100 amp	"	1.20	6.667		3,875	410		4,285	4,850
2650	400 amp	2 Elec	1.40	11.429		4,150	700		4,850	5,625

26 51 Interior Lighting

26 51 13 – Interior Lighting Fixtures, Lamps, and Ballasts

26 51 13.70 Residential Fixtures

		Crew	Daily Output	Labor-Hours	Unit	Material	2020 Bare Costs Labor	2020 Bare Costs Equipment	Total	Total Incl O&P
0010	**RESIDENTIAL FIXTURES**									
2000	Incandescent, exterior lantern, wall mounted, 60 watt	1 Elec	16	.500	Ea.	61.50	30.50		92	113
2100	Post light, 150 W, with 7' post		4	2		291	123		414	505
2500	Lamp holder, weatherproof with 150 W PAR		16	.500		35.50	30.50		66	84.50
2550	With reflector and guard		12	.667		66	41		107	134

26 51 13.90 Ballast, Replacement HID

		Crew	Daily Output	Labor-Hours	Unit	Material	2020 Bare Costs Labor	2020 Bare Costs Equipment	Total	Total Incl O&P
0010	**BALLAST, REPLACEMENT HID**									
7510	Multi-tap 120/208/240/277 V									
7550	High pressure sodium, 70 watt	1 Elec	10	.800	Ea.	183	49		232	274
7560	100 watt		9.40	.851		130	52		182	221
7570	150 watt		9	.889		205	54.50		259.50	305
7580	250 watt		8.50	.941		305	57.50		362.50	420
7590	400 watt		7	1.143		269	70		339	400
7600	1,000 watt		6	1.333		287	82		369	435
7610	Metal halide, 175 watt		8	1		97	61.50		158.50	199
7620	250 watt		8	1		128	61.50		189.50	233
7630	400 watt		7	1.143		158	70		228	278
7640	1,000 watt		6	1.333		258	82		340	405
7650	1,500 watt		5	1.600		245	98		343	415

26 56 Exterior Lighting

26 56 13 – Lighting Poles and Standards

26 56 13.10 Lighting Poles	Crew	Daily Output	Labor-Hours	Unit	Material	2020 Bare Costs Labor	2020 Bare Costs Equipment	Total	Total Incl O&P
0010 **LIGHTING POLES**									
2800 Light poles, anchor base									
2820 not including concrete bases									
2840 Aluminum pole, 8' high	1 Elec	4	2	Ea.	810	123		933	1,075
2850 10' high		4	2		855	123		978	1,125
2860 12' high		3.80	2.105		890	129		1,019	1,175
2870 14' high		3.40	2.353		920	144		1,064	1,225
2880 16' high		3	2.667		1,025	164		1,189	1,375
3000 20' high	R-3	2.90	6.897		1,100	420	65	1,585	1,900
3200 30' high		2.60	7.692		2,075	470	72.50	2,617.50	3,050
3400 35' high		2.30	8.696		2,275	530	82	2,887	3,375
3600 40' high		2	10		2,750	610	94.50	3,454.50	4,050
3800 Bracket arms, 1 arm	1 Elec	8	1		141	61.50		202.50	247
4000 2 arms		8	1		280	61.50		341.50	400
4200 3 arms		5.30	1.509		425	92.50		517.50	605
4400 4 arms		4.80	1.667		565	102		667	775
4500 Steel pole, galvanized, 8' high		3.80	2.105		675	129		804	930
4510 10' high		3.70	2.162		700	133		833	975
4520 12' high		3.40	2.353		755	144		899	1,050
4530 14' high		3.10	2.581		805	158		963	1,125
4540 16' high		2.90	2.759		855	169		1,024	1,200
4550 18' high		2.70	2.963		900	182		1,082	1,250
4600 20' high	R-3	2.60	7.692		1,275	470	72.50	1,817.50	2,175
4800 30' high		2.30	8.696		1,275	530	82	1,887	2,275
5000 35' high		2.20	9.091		1,550	555	86	2,191	2,650
5200 40' high		1.70	11.765		1,725	720	111	2,556	3,075
5400 Bracket arms, 1 arm	1 Elec	8	1		214	61.50		275.50	325
5600 2 arms		8	1		310	61.50		371.50	430
5800 3 arms		5.30	1.509		215	92.50		307.50	375
6000 4 arms		5.30	1.509		350	92.50		442.50	525
6100 Fiberglass pole, 1 or 2 fixtures, 20' high	R-3	4	5		860	305	47.50	1,212.50	1,450
6200 30' high		3.60	5.556		980	340	52.50	1,372.50	1,650
6300 35' high		3.20	6.250		1,675	380	59	2,114	2,475
6400 40' high		2.80	7.143		1,850	435	67.50	2,352.50	2,750
6420 Wood pole, 4-1/2" x 5-1/8", 8' high	1 Elec	6	1.333		385	82		467	545
6430 10' high		6	1.333		450	82		532	620
6440 12' high		5.70	1.404		570	86		656	755
6450 15' high		5	1.600		665	98		763	875
6460 20' high		4	2		810	123		933	1,075
6461 Light poles, anchor base, w/o conc base, pwdr ct stl, 16' H	2 Elec	3.10	5.161		855	315		1,170	1,400
6462 20' high	R-3	2.90	6.897		1,275	420	65	1,760	2,100
6463 30' high		2.30	8.696		1,275	530	82	1,887	2,275
6464 35' high		2.40	8.333		1,550	510	79	2,139	2,575
6465 25' high		2.70	7.407		1,275	455	70	1,800	2,150
6470 Light pole conc base, max 6' buried, 2' exposed, 18" diam., average cost	C-6	6	8		190	350	8.95	548.95	745
7300 Transformer bases, not including concrete bases									
7320 Maximum pole size, steel, 40' high	1 Elec	2	4	Ea.	1,600	245		1,845	2,150
7340 Cast aluminum, 30' high		3	2.667		850	164		1,014	1,175
7350 40' high		2.50	3.200		1,300	196		1,496	1,725
8000 Line cover protective devices									
8010 line cover									
8015 Refer to 26 01 02.20 for labor adjustment factors as they apply									
8100 MVLC 1" W x 1-1/2" H, 14/5'	1 Elec	60	.133	Ea.	4.70	8.20		12.90	17.35

280

26 56 Exterior Lighting

26 56 13 – Lighting Poles and Standards

26 56 13.10 Lighting Poles		Crew	Daily Output	Labor-Hours	Unit	Material	2020 Bare Costs Labor	2020 Bare Costs Equipment	Total	Total Incl O&P
8110	8'	1 Elec	58	.138	Ea.	5.25	8.45		13.70	18.35
8120	MVLC 1" W x 1-1/2" H, 18/5'		60	.133		5.95	8.20		14.15	18.75
8130	8'		58	.138		6.65	8.45		15.10	19.90
8140	MVLC 1" W x 1-1/2" H, 38/5'		60	.133		10.60	8.20		18.80	24
8150	8'		58	.138		12.90	8.45		21.35	27

26 56 19 – LED Exterior Lighting

26 56 19.55 Roadway LED Luminaire

			Crew	Daily Output	Labor-Hours	Unit	Material	Labor	Equipment	Total	Total Incl O&P
0010	**ROADWAY LED LUMINAIRE**										
0100	LED fixture, 72 LEDs, 120 V AC or 12 V DC, equal to 60 watt	G	1 Elec	2.70	2.963	Ea.	585	182		767	910
0110	108 LEDs, 120 V AC or 12 V DC, equal to 90 watt	G		2.70	2.963		750	182		932	1,100
0120	144 LEDs, 120 V AC or 12 V DC, equal to 120 watt	G		2.70	2.963		920	182		1,102	1,300
0130	252 LEDs, 120 V AC or 12 V DC, equal to 210 watt	G	2 Elec	4.40	3.636		1,275	223		1,498	1,725
0140	Replaces high pressure sodium fixture, 75 watt	G	1 Elec	2.70	2.963		405	182		587	715
0150	125 watt	G		2.70	2.963		460	182		642	780
0160	150 watt	G		2.70	2.963		540	182		722	865
0170	175 watt	G		2.70	2.963		795	182		977	1,150
0180	200 watt	G		2.70	2.963		760	182		942	1,100
0190	250 watt	G	2 Elec	4.40	3.636		870	223		1,093	1,275
0200	320 watt	G	"	4.40	3.636		950	223		1,173	1,375

26 56 19.60 Parking LED Lighting

			Crew	Daily Output	Labor-Hours	Unit	Material	Labor	Equipment	Total	Total Incl O&P
0010	**PARKING LED LIGHTING**										
0100	Round pole mounting, 88 lamp watts	G	1 Elec	2	4	Ea.	1,200	245		1,445	1,700

26 56 21 – HID Exterior Lighting

26 56 21.20 Roadway Luminaire

		Crew	Daily Output	Labor-Hours	Unit	Material	Labor	Equipment	Total	Total Incl O&P
0010	**ROADWAY LUMINAIRE**									
2650	Roadway area luminaire, low pressure sodium, 135 watt	1 Elec	2	4	Ea.	835	245		1,080	1,275
2700	180 watt	"	2	4		995	245		1,240	1,475
2750	Metal halide, 400 watt	2 Elec	4.40	3.636		690	223		913	1,100
2760	1,000 watt		4	4		775	245		1,020	1,225
2780	High pressure sodium, 400 watt		4.40	3.636		825	223		1,048	1,250
2790	1,000 watt		4	4		940	245		1,185	1,400

26 56 23 – Area Lighting

26 56 23.10 Exterior Fixtures

		Crew	Daily Output	Labor-Hours	Unit	Material	Labor	Equipment	Total	Total Incl O&P
0010	**EXTERIOR FIXTURES** With lamps									
0200	Wall mounted, incandescent, 100 watt	1 Elec	8	1	Ea.	51.50	61.50		113	148
0400	Quartz, 500 watt		5.30	1.509		62	92.50		154.50	207
0420	1,500 watt		4.20	1.905		108	117		225	292
1100	Wall pack, low pressure sodium, 35 watt		4	2		205	123		328	410
1150	55 watt		4	2		241	123		364	450
1160	High pressure sodium, 70 watt		4	2		197	123		320	400
1170	150 watt		4	2		218	123		341	420
1175	High pressure sodium, 250 watt		4	2		218	123		341	420
1180	Metal halide, 175 watt		4	2		221	123		344	425
1190	250 watt		4	2		251	123		374	460
1195	400 watt		4	2		410	123		533	640
1250	Induction lamp, 40 watt		4	2		495	123		618	730
1260	80 watt		4	2		590	123		713	830
1278	LED, poly lens, 26 watt		4	2		310	123		433	525
1280	110 watt		4	2		830	123		953	1,100
1500	LED, glass lens, 13 watt		4	2		310	123		433	525

26 56 23 – Area Lighting

26 56 23.55 Exterior LED Fixtures

		Crew	Daily Output	Labor-Hours	Unit	Material	2020 Bare Costs Labor	2020 Bare Costs Equipment	Total	Total Incl O&P
0010	**EXTERIOR LED FIXTURES**									
0100	Wall mounted, indoor/outdoor, 12 watt G	1 Elec	10	.800	Ea.	199	49		248	292
0110	32 watt G		10	.800		455	49		504	575
0120	66 watt G		10	.800		480	49		529	600
0200	outdoor, 110 watt G		10	.800		1,150	49		1,199	1,350
0210	220 watt G		10	.800		1,775	49		1,824	2,025
0300	modular, type IV, 120 V, 50 lamp watts G		9	.889		1,325	54.50		1,379.50	1,550
0310	101 lamp watts G		9	.889		1,325	54.50		1,379.50	1,525
0320	126 lamp watts G		9	.889		1,650	54.50		1,704.50	1,875
0330	202 lamp watts G		9	.889		1,900	54.50		1,954.50	2,175
0340	240 V, 50 lamp watts G		8	1		1,225	61.50		1,286.50	1,450
0350	101 lamp watts G		8	1		1,375	61.50		1,436.50	1,600
0360	126 lamp watts G		8	1		1,275	61.50		1,336.50	1,500
0370	202 lamp watts G		8	1		1,925	61.50		1,986.50	2,200
0400	wall pack, glass, 13 lamp watts G		4	2		450	123		573	680
0410	poly w/photocell, 26 lamp watts G		4	2		330	123		453	550
0420	50 lamp watts G		4	2		705	123		828	960
0430	replacement, 40 watts G		4	2		385	123		508	610
0440	60 watts G		4	2		435	123		558	665

26 56 26 – Landscape Lighting

26 56 26.20 Landscape Fixtures

		Crew	Daily Output	Labor-Hours	Unit	Material	2020 Bare Costs Labor	2020 Bare Costs Equipment	Total	Total Incl O&P
0010	**LANDSCAPE FIXTURES**									
0012	Incl. conduit, wire, trench									
0030	Bollards									
0040	Incandescent, 24"	1 Elec	2.50	3.200	Ea.	370	196		566	695
0050	36"		2	4		485	245		730	900
0060	42"		2	4		515	245		760	935
0070	H.I.D., 24"		2.50	3.200		530	196		726	870
0080	36"		2	4		585	245		830	1,000
0090	42"		2	4		850	245		1,095	1,300
0100	Concrete, 18" diam.		1.20	6.667		1,675	410		2,085	2,425
0110	24" diam.		.75	10.667		2,100	655		2,755	3,275
0120	Dry niche									
0130	300 W 120 Volt	1 Elec	4	2	Ea.	1,100	123		1,223	1,375
0140	1000 W 120 Volt		2	4		1,450	245		1,695	1,975
0150	300 W 12 Volt		4	2		1,100	123		1,223	1,375
0160	Low voltage									
0170	Recessed uplight	1 Elec	2.20	3.636	Ea.	435	223		658	810
0180	Walkway		4	2		400	123		523	625
0190	Malibu - 5 light set		3	2.667		289	164		453	560
0200	Mushroom 24" pier		4	2		296	123		419	510
0210	Recessed, adjustable									
0220	Incandescent, 150 W	1 Elec	2.50	3.200	Ea.	680	196		876	1,025
0230	300 W	"	2.50	3.200	"	865	196		1,061	1,250
0250	Recessed uplight									
0260	Incandescent, 50 W	1 Elec	2.50	3.200	Ea.	550	196		746	895
0270	150 W		2.50	3.200		575	196		771	920
0280	300 W		2.50	3.200		640	196		836	995
0310	Quartz 500 W		2.50	3.200		575	196		771	925
0400	Recessed wall light									
0410	Incandescent 100 W	1 Elec	4	2	Ea.	213	123		336	420
0420	Fluorescent		4	2		194	123		317	395

For customer support on your Site Work & Landscape Costs with RSMeans data, call 800.448.8182.

26 56 Exterior Lighting

26 56 26 – Landscape Lighting

26 56 26.20 Landscape Fixtures

		Crew	Daily Output	Labor-Hours	Unit	Material	2020 Bare Costs Labor	Equipment	Total	Total Incl O&P
0430	H.I.D. 100 W	1 Elec	3	2.667	Ea.	400	164		564	685
0500	Step lights									
0510	Incandescent	1 Elec	5	1.600	Ea.	146	98		244	305
0520	Fluorescent	"	5	1.600	"	152	98		250	315
0600	Tree lights, surface adjustable									
0610	Incandescent 50 W	1 Elec	3	2.667	Ea.	297	164		461	570
0620	Incandescent 100 W		3	2.667		181	164		345	445
0630	Incandescent 150 W	↓	2	4	↓	415	245		660	825
0700	Underwater lights									
0710	150 W 120 Volt	1 Elec	6	1.333	Ea.	1,025	82		1,107	1,250
0720	300 W 120 Volt		6	1.333		1,325	82		1,407	1,575
0730	1000 W 120 Volt		4	2		1,675	123		1,798	2,000
0740	50 W 12 Volt		6	1.333		1,250	82		1,332	1,500
0750	300 W 12 Volt	↓	6	1.333		1,125	82		1,207	1,375
0800	Walkway, adjustable									
0810	Fluorescent, 2'	1 Elec	3	2.667	Ea.	455	164		619	745
0820	Fluorescent, 4'		3	2.667		485	164		649	775
0830	Fluorescent, 8'		2	4		945	245		1,190	1,425
0840	Incandescent, 50 W		4	2		430	123		553	660
0850	150 W	↓	4	2	↓	460	123		583	690
0900	Wet niche									
0910	300 W 120 Volt	1 Elec	2.50	3.200	Ea.	940	196		1,136	1,325
0920	1000 W 120 Volt		1.50	5.333		1,125	325		1,450	1,725
0930	300 W 12 Volt	↓	2.50	3.200	↓	1,075	196		1,271	1,475
0960	Landscape lights, solar powered, 6" diam. x 12" H, one piece w/stake [G]	1 Clab	80	.100		30.50	4.21		34.71	40
7380	Landscape recessed uplight, incl. housing, ballast, transformer									
7390	& reflector, not incl. conduit, wire, trench									
7420	Incandescent, 250 watt	1 Elec	5	1.600	Ea.	665	98		763	875
7440	Quartz, 250 watt		5	1.600		630	98		728	840
7460	500 watt	↓	4	2		650	123		773	900

26 56 26.50 Landscape LED Fixtures

		Crew	Daily Output	Labor-Hours	Unit	Material	2020 Bare Costs Labor	Equipment	Total	Total Incl O&P
0010	**LANDSCAPE LED FIXTURES**									
0100	12 volt alum bullet hooded-BLK	1 Elec	5	1.600	Ea.	97.50	98		195.50	253
0200	12 volt alum bullet hooded-BRZ		5	1.600		97.50	98		195.50	253
0300	12 volt alum bullet hooded-GRN		5	1.600		97.50	98		195.50	253
1000	12 volt alum large bullet hooded-BLK		5	1.600		72	98		170	225
1100	12 volt alum large bullet hooded-BRZ		5	1.600		72	98		170	225
1200	12 volt alum large bullet hooded-GRN		5	1.600		72	98		170	225
2000	12 volt large bullet landscape light fixture		5	1.600		72	98		170	225
2100	12 volt alum light large bullet		5	1.600		72	98		170	225
2200	12 volt alum bullet light		5	1.600	↓	72	98		170	225

26 56 33 – Walkway Lighting

26 56 33.10 Walkway Luminaire

		Crew	Daily Output	Labor-Hours	Unit	Material	2020 Bare Costs Labor	Equipment	Total	Total Incl O&P
0010	**WALKWAY LUMINAIRE**									
6500	Bollard light, lamp & ballast, 42" high with polycarbonate lens									
6800	Metal halide, 175 watt	1 Elec	3	2.667	Ea.	985	164		1,149	1,325
6900	High pressure sodium, 70 watt		3	2.667		950	164		1,114	1,300
7000	100 watt		3	2.667		950	164		1,114	1,300
7100	150 watt		3	2.667		925	164		1,089	1,275
7200	Incandescent, 150 watt		3	2.667		675	164		839	985
7810	Walkway luminaire, square 16", metal halide 250 watt		2.70	2.963		770	182		952	1,125
7820	High pressure sodium, 70 watt	↓	3	2.667	↓	880	164		1,044	1,225

283

26 56 33.10 Walkway Luminaire

		Crew	Daily Output	Labor-Hours	Unit	Material	2020 Bare Costs Labor	Equipment	Total	Total Incl O&P
7830	100 watt	1 Elec	3	2.667	Ea.	895	164		1,059	1,225
7840	150 watt		3	2.667		895	164		1,059	1,225
7850	200 watt		3	2.667		900	164		1,064	1,225
7910	Round 19", metal halide, 250 watt		2.70	2.963		1,250	182		1,432	1,650
7920	High pressure sodium, 70 watt		3	2.667		1,375	164		1,539	1,750
7930	100 watt		3	2.667		1,375	164		1,539	1,750
7940	150 watt		3	2.667		1,375	164		1,539	1,750
7950	250 watt		2.70	2.963		1,300	182		1,482	1,700
8000	Sphere 14" opal, incandescent, 200 watt		4	2		355	123		478	575
8020	Sphere 18" opal, incandescent, 300 watt		3.50	2.286		430	140		570	680
8040	Sphere 16" clear, high pressure sodium, 70 watt		3	2.667		740	164		904	1,050
8050	100 watt		3	2.667		790	164		954	1,125
8100	Cube 16" opal, incandescent, 300 watt		3.50	2.286		470	140		610	725
8120	High pressure sodium, 70 watt		3	2.667		685	164		849	1,000
8130	100 watt		3	2.667		705	164		869	1,025
8230	Lantern, high pressure sodium, 70 watt		3	2.667		615	164		779	920
8240	100 watt		3	2.667		660	164		824	975
8250	150 watt		3	2.667		620	164		784	925
8260	250 watt		2.70	2.963		865	182		1,047	1,225
8270	Incandescent, 300 watt		3.50	2.286		455	140		595	715
8330	Reflector 22" w/globe, high pressure sodium, 70 watt		3	2.667		555	164		719	855
8340	100 watt		3	2.667		565	164		729	865
8350	150 watt		3	2.667		570	164		734	875
8360	250 watt		2.70	2.963		730	182		912	1,075

26 56 33.55 Walkway LED Luminaire

			Crew	Daily Output	Labor-Hours	Unit	Material	2020 Bare Costs Labor	Equipment	Total	Total Incl O&P
0010	**WALKWAY LED LUMINAIRE**										
0100	Pole mounted, 86 watts, 4,350 lumens	G	1 Elec	3	2.667	Ea.	1,350	164		1,514	1,725
0110	4,630 lumens	G		3	2.667		2,200	164		2,364	2,650
0120	80 watts, 4,000 lumens	G		3	2.667		2,075	164		2,239	2,525

26 56 36 – Flood Lighting

26 56 36.20 Floodlights

			Crew	Daily Output	Labor-Hours	Unit	Material	2020 Bare Costs Labor	Equipment	Total	Total Incl O&P
0010	**FLOODLIGHTS** with ballast and lamp,										
1290	floor mtd, mount with swivel bracket										
1300	Induction lamp, 40 watt		1 Elec	3	2.667	Ea.	565	164		729	865
1310	80 watt			3	2.667		750	164		914	1,075
1320	150 watt			3	2.667		1,350	164		1,514	1,750
1400	Pole mounted, pole not included										
1950	Metal halide, 175 watt		1 Elec	2.70	2.963	Ea.	213	182		395	505
2000	400 watt		2 Elec	4.40	3.636		219	223		442	570
2200	1,000 watt			4	4		900	245		1,145	1,350
2210	1,500 watt			3.70	4.324		425	265		690	860
2250	Low pressure sodium, 55 watt		1 Elec	2.70	2.963		460	182		642	775
2270	90 watt			2	4		620	245		865	1,050
2290	180 watt			2	4		680	245		925	1,100
2340	High pressure sodium, 70 watt			2.70	2.963		254	182		436	550
2360	100 watt			2.70	2.963		261	182		443	560
2380	150 watt			2.70	2.963		293	182		475	595
2400	400 watt		2 Elec	4.40	3.636		320	223		543	685
2600	1,000 watt		"	4	4		635	245		880	1,050
9005	Solar powered floodlight, w/motion det, incl. batt pack for cloudy days	G	1 Elec	8	1		71	61.50		132.50	170
9020	Replacement battery pack	G	"	8	1		20	61.50		81.50	114

26 56 Exterior Lighting

26 56 36 – Flood Lighting

26 56 36.55 LED Floodlights		Crew	Daily Output	Labor-Hours	Unit	Material	2020 Bare Costs Labor	Equipment	Total	Total Incl O&P
0010	**LED FLOODLIGHTS** with ballast and lamp,									
0020	Pole mounted, pole not included									
0100	11 watt	G	1 Elec	4	2	Ea.	440	123		563	670
0110	46 watt	G		4	2		1,500	123		1,623	1,825
0120	90 watt	G		4	2		2,175	123		2,298	2,575
0130	288 watt	G		4	2		2,000	123		2,123	2,375

26 61 Lighting Systems and Accessories

26 61 23 – Lamps Applications

26 61 23.10 Lamps

26 61 23.10 Lamps		Crew	Daily Output	Labor-Hours	Unit	Material	2020 Bare Costs Labor	Equipment	Total	Total Incl O&P
0010	**LAMPS**									
0520	Very high output, 4' long, 110 watt		1 Elec	.90	8.889	C	1,600	545		2,145	2,575
0525	8' long, 195 watt energy saver	G		.70	11.429		1,375	700		2,075	2,575
0550	8' long, 215 watt			.70	11.429		1,325	700		2,025	2,500
0600	Mercury vapor, mogul base, deluxe white, 100 watt			.30	26.667		5,150	1,625		6,775	8,100
0650	175 watt			.30	26.667		2,725	1,625		4,350	5,425
0700	250 watt			.30	26.667		5,000	1,625		6,625	7,925
0800	400 watt			.30	26.667		4,575	1,625		6,200	7,475
0900	1,000 watt			.20	40		12,700	2,450		15,150	17,700
1000	Metal halide, mogul base, 175 watt			.30	26.667		1,050	1,625		2,675	3,575
1100	250 watt			.30	26.667		1,650	1,625		3,275	4,225
1200	400 watt			.30	26.667		2,000	1,625		3,625	4,625
1300	1,000 watt			.20	40		3,925	2,450		6,375	7,950
1320	1,000 watt, 125,000 initial lumens			.20	40		16,100	2,450		18,550	21,400
1330	1,500 watt			.20	40		3,575	2,450		6,025	7,575
1350	High pressure sodium, 70 watt			.30	26.667		1,700	1,625		3,325	4,300
1360	100 watt			.30	26.667		1,750	1,625		3,375	4,350
1370	150 watt			.30	26.667		1,625	1,625		3,250	4,225
1380	250 watt			.30	26.667		1,950	1,625		3,575	4,575
1400	400 watt			.30	26.667		1,650	1,625		3,275	4,225
1450	1,000 watt			.20	40		4,500	2,450		6,950	8,600
1500	Low pressure sodium, 35 watt			.30	26.667		15,900	1,625		17,525	19,800
1550	55 watt			.30	26.667		18,100	1,625		19,725	22,300
1600	90 watt			.30	26.667		6,375	1,625		8,000	9,425
1650	135 watt			.20	40		26,300	2,450		28,750	32,600
1700	180 watt			.20	40		39,400	2,450		41,850	47,000
1750	Quartz line, clear, 500 watt			1.10	7.273		745	445		1,190	1,475
1760	1,500 watt			.20	40		2,125	2,450		4,575	5,975
1762	Spot, MR 16, 50 watt			1.30	6.154		1,050	380		1,430	1,700
1800	Incandescent, interior, A21, 100 watt			1.60	5		2,700	305		3,005	3,425
2500	Exterior, PAR 38, 75 watt			1.30	6.154		1,825	380		2,205	2,550
2600	PAR 38, 150 watt			1.30	6.154		2,050	380		2,430	2,800
2700	PAR 46, 200 watt			1.10	7.273		3,725	445		4,170	4,775
2800	PAR 56, 300 watt			1.10	7.273		2,550	445		2,995	3,475

285

For customer support on your Site Work & Landscape Costs with RSMeans data, call 800.448.8182.

Division Notes

	CREW	DAILY OUTPUT	LABOR-HOURS	UNIT	BARE COSTS				TOTAL INCL O&P
					MAT.	LABOR	EQUIP.	TOTAL	

Estimating Tips
31 05 00 Common Work Results for Earthwork

- Estimating the actual cost of performing earthwork requires careful consideration of the variables involved. This includes items such as type of soil, whether water will be encountered, dewatering, whether banks need bracing, disposal of excavated earth, and length of haul to fill or spoil sites, etc. If the project has large quantities of cut or fill, consider raising or lowering the site to reduce costs, while paying close attention to the effect on site drainage and utilities.

- If the project has large quantities of fill, creating a borrow pit on the site can significantly lower the costs.

- It is very important to consider what time of year the project is scheduled for completion. Bad weather can create large cost overruns from dewatering, site repair, and lost productivity from cold weather.

Reference Numbers

Reference numbers are shown at the beginning of some major classifications. These numbers refer to related items in the Reference Section. The reference information may be an estimating procedure, an alternate pricing method, or technical information.

Note: Not all subdivisions listed here necessarily appear. ■

Same Data. Simplified.

Enjoy the convenience and efficiency of accessing your costs anywhere:

- **Skip the multiplier** by setting your location
- **Quickly search,** edit, favorite and share costs
- **Stay on top of price changes** with automatic updates

Discover more at rsmeans.com/online

31 05 Common Work Results for Earthwork

31 05 13 – Soils for Earthwork

31 05 13.10 Borrow

		Crew	Daily Output	Labor-Hours	Unit	Material	2020 Bare Costs Labor	2020 Bare Costs Equipment	Total	Total Incl O&P
0010	**BORROW**									
0020	Spread, 200 HP dozer, no compaction, 2 mile RT haul									
0200	Common borrow	B-15	600	.047	C.Y.	12.60	2.34	4.42	19.36	22
0700	Screened loam		600	.047		29	2.34	4.42	35.76	40.50
0800	Topsoil, weed free		600	.047		26.50	2.34	4.42	33.26	37.50
0900	For 5 mile haul, add	B-34B	200	.040			1.96	2.86	4.82	6.10

31 05 16 – Aggregates for Earthwork

31 05 16.10 Borrow

		Crew	Daily Output	Labor-Hours	Unit	Material	2020 Bare Costs Labor	2020 Bare Costs Equipment	Total	Total Incl O&P
0010	**BORROW**									
0020	Spread, with 200 HP dozer, no compaction, 2 mile RT haul									
0100	Bank run gravel	B-15	600	.047	L.C.Y.	18.60	2.34	4.42	25.36	29
0300	Crushed stone (1.40 tons per C.Y.), 1-1/2"		600	.047		27.50	2.34	4.42	34.26	39
0320	3/4"		600	.047		27.50	2.34	4.42	34.26	39
0340	1/2"		600	.047		27	2.34	4.42	33.76	38
0360	3/8"		600	.047		32	2.34	4.42	38.76	44
0400	Sand, washed, concrete		600	.047		35.50	2.34	4.42	42.26	47.50
0500	Dead or bank sand		600	.047		18.25	2.34	4.42	25.01	28.50
0600	Select structural fill		600	.047		24.50	2.34	4.42	31.26	35.50
0610	Import 10 mile RT select structural fill		408	.069		24.50	3.45	6.50	34.45	39.50
0900	For 5 mile haul, add	B-34B	200	.040			1.96	2.86	4.82	6.10
1000	For flowable fill, see Section 03 31 13.35									

31 05 19 – Geosynthetics for Earthwork

31 05 19.53 Reservoir Liners

		Crew	Daily Output	Labor-Hours	Unit	Material	2020 Bare Costs Labor	2020 Bare Costs Equipment	Total	Total Incl O&P
0010	**RESERVOIR LINERS**									
0011	Membrane lining									
1100	30 mil, LLDPE	3 Skwk	1850	.013	S.F.	.38	.71		1.09	1.50
1200	60 mil, HDPE		1600	.015		.83	.82		1.65	2.15
1300	120 mil thick		1440	.017		1.41	.91		2.32	2.93

31 05 23 – Cement and Concrete for Earthwork

31 05 23.30 Plant Mixed Bituminous Concrete

		Crew	Daily Output	Labor-Hours	Unit	Material	2020 Bare Costs Labor	2020 Bare Costs Equipment	Total	Total Incl O&P
0010	**PLANT MIXED BITUMINOUS CONCRETE**									
0020	Asphaltic concrete plant mix (145 lb./C.F.)				Ton	68			68	74.50
0040	Asphaltic concrete less than 300 tons add trucking costs									
0200	All weather patching mix, hot				Ton	66			66	73
0250	Cold patch					69.50			69.50	76.50
0300	Berm mix					68			68	75
0400	Base mix					68			68	74.50
0500	Binder mix					68			68	74.50
0600	Sand or sheet mix					65			65	71.50

31 05 23.40 Recycled Plant Mixed Bituminous Concrete

			Crew	Daily Output	Labor-Hours	Unit	Material	2020 Bare Costs Labor	2020 Bare Costs Equipment	Total	Total Incl O&P
0010	**RECYCLED PLANT MIXED BITUMINOUS CONCRETE**										
0200	Reclaimed pavement in stockpile	G				Ton	26			26	29
0400	Recycled pavement, at plant, ratio old:new, 70:30	G					38			38	42
0600	Ratio old:new, 30:70	G					54			54	59.50

31 06 Schedules for Earthwork

31 06 60 – Schedules for Special Foundations and Load Bearing Elements

31 06 60.14 Piling Special Costs

		Crew	Daily Output	Labor-Hours	Unit	Material	2020 Bare Costs Labor	Equipment	Total	Total Incl O&P
0010	**PILING SPECIAL COSTS**									
0011	Piling special costs, pile caps, see Section 03 30 53.40									
0500	Cutoffs, concrete piles, plain	1 Pile	5.50	1.455	Ea.		79		79	122
0600	With steel thin shell, add		38	.211			11.40		11.40	17.75
0700	Steel pile or "H" piles		19	.421			23		23	35.50
0800	Wood piles		38	.211			11.40		11.40	17.75
0900	Pre-augering up to 30' deep, average soil, 24" diameter	B-43	180	.267	L.F.		12.45	4.28	16.73	23.50
0920	36" diameter		115	.417			19.50	6.70	26.20	36.50
0960	48" diameter		70	.686			32	11	43	60
0980	60" diameter		50	.960			45	15.40	60.40	84
1000	Testing, any type piles, test load is twice the design load									
1050	50 ton design load, 100 ton test				Ea.				14,000	15,500
1100	100 ton design load, 200 ton test								20,000	22,000
1150	150 ton design load, 300 ton test								26,000	28,500
1200	200 ton design load, 400 ton test								28,000	31,000
1250	400 ton design load, 800 ton test								32,000	35,000
1500	Wet conditions, soft damp ground									
1600	Requiring mats for crane, add								40%	40%
1700	Barge mounted driving rig, add								30%	30%

31 06 60.15 Mobilization

		Crew	Daily Output	Labor-Hours	Unit	Material	2020 Bare Costs Labor	Equipment	Total	Total Incl O&P
0010	**MOBILIZATION**									
0020	Set up & remove, air compressor, 600 CFM	A-5	3.30	5.455	Ea.		233	14.85	247.85	365
0100	1,200 CFM	"	2.20	8.182			350	22.50	372.50	550
0200	Crane, with pile leads and pile hammer, 75 ton	B-19	.60	107			5,900	3,350	9,250	12,700
0300	150 ton	"	.36	178			9,825	5,575	15,400	21,200
0500	Drill rig, for caissons, to 36", minimum	B-43	2	24			1,125	385	1,510	2,100
0520	Maximum		.50	96			4,475	1,550	6,025	8,425
0600	Up to 84"		1	48			2,250	770	3,020	4,200
0800	Auxiliary boiler, for steam small	A-5	1.66	10.843			465	29.50	494.50	730
0900	Large	"	.83	21.687			925	59	984	1,475
1100	Rule of thumb: complete pile driving set up, small	B-19	.45	142			7,850	4,475	12,325	16,900
1200	Large	"	.27	237			13,100	7,450	20,550	28,200
1300	Mobilization by water for barge driving rig									
1310	Minimum				Ea.				7,000	7,700
1320	Maximum				"				45,000	49,500
1500	Mobilization, barge, by tug boat	B-83	25	.640	Mile		31.50	28.50	60	79
1600	Standby time for shore pile driving crew	B-19	8	8	Hr.		440	251	691	950
1700	Standby time for barge driving rig	B-76	8	9	"		495	460	955	1,275

31 11 Clearing and Grubbing

31 11 10 – Clearing and Grubbing Land

31 11 10.10 Clear and Grub Site

		Crew	Daily Output	Labor-Hours	Unit	Material	2020 Bare Costs Labor	Equipment	Total	Total Incl O&P
0010	**CLEAR AND GRUB SITE**									
0020	Cut & chip light trees to 6" diam.	B-7	1	48	Acre		2,150	1,650	3,800	5,025
0150	Grub stumps and remove	B-30	2	12			620	915	1,535	1,925
0160	Clear & grub brush including stumps	"	.58	41.379			2,125	3,150	5,275	6,675
0200	Cut & chip medium trees to 12" diam.	B-7	.70	68.571			3,075	2,350	5,425	7,200
0250	Grub stumps and remove	B-30	1	24			1,225	1,825	3,050	3,875
0260	Clear & grub dense brush including stumps	"	.47	51.064			2,625	3,900	6,525	8,250
0300	Cut & chip heavy trees to 24" diam.	B-7	.30	160			7,175	5,475	12,650	16,800
0350	Grub stumps and remove	B-30	.50	48			2,475	3,675	6,150	7,725

289

31 11 Clearing and Grubbing

31 11 10 – Clearing and Grubbing Land

31 11 10.10 Clear and Grub Site

		Crew	Daily Output	Labor-Hours	Unit	Material	2020 Bare Costs Labor	2020 Bare Costs Equipment	Total	Total Incl O&P
0400	If burning is allowed, deduct cut & chip				Acre				40%	40%
3000	Chipping stumps, to 18" deep, 12" diam.	B-86	20	.400	Ea.		22.50	9.70	32.20	44.50
3040	18" diameter		16	.500			28.50	12.15	40.65	56
3080	24" diameter		14	.571			32.50	13.90	46.40	64
3100	30" diameter		12	.667			38	16.20	54.20	74.50
3120	36" diameter		10	.800			45.50	19.45	64.95	89.50
3160	48" diameter		8	1			57	24.50	81.50	112
5000	Tree thinning, feller buncher, conifer									
5080	Up to 8" diameter	B-93	240	.033	Ea.		1.89	2.45	4.34	5.55
5120	12" diameter		160	.050			2.84	3.68	6.52	8.30
5240	Hardwood, up to 4" diameter		240	.033			1.89	2.45	4.34	5.55
5280	8" diameter		180	.044			2.52	3.27	5.79	7.35
5320	12" diameter		120	.067			3.78	4.91	8.69	11.05
7000	Tree removal, congested area, aerial lift truck									
7040	8" diameter	B-85	7	5.714	Ea.		265	115	380	525
7080	12" diameter		6	6.667			310	134	444	610
7120	18" diameter		5	8			370	161	531	730
7160	24" diameter		4	10			465	201	666	915
7240	36" diameter		3	13.333			620	268	888	1,225
7280	48" diameter		2	20			930	400	1,330	1,850

31 13 Selective Tree and Shrub Removal and Trimming

31 13 13 – Selective Tree and Shrub Removal

31 13 13.10 Selective Clearing

		Crew	Daily Output	Labor-Hours	Unit	Material	2020 Bare Costs Labor	2020 Bare Costs Equipment	Total	Total Incl O&P
0010	**SELECTIVE CLEARING**									
0020	Clearing brush with brush saw	A-1C	.25	32	Acre		1,350	208	1,558	2,250
0100	By hand	1 Clab	.12	66.667			2,800		2,800	4,225
0300	With dozer, ball and chain, light clearing	B-11A	2	8			395	750	1,145	1,425
0400	Medium clearing		1.50	10.667			525	1,000	1,525	1,900
0500	With dozer and brush rake, light		10	1.600			79	150	229	283
0550	Medium brush to 4" diameter		8	2			99	188	287	355
0600	Heavy brush to 4" diameter		6.40	2.500			124	235	359	445
1000	Brush mowing, tractor w/rotary mower, no removal									
1020	Light density	B-84	2	4	Acre		227	183	410	540
1040	Medium density		1.50	5.333			305	244	549	725
1080	Heavy density		1	8			455	365	820	1,075

31 13 13.20 Selective Tree Removal

		Crew	Daily Output	Labor-Hours	Unit	Material	2020 Bare Costs Labor	2020 Bare Costs Equipment	Total	Total Incl O&P
0010	**SELECTIVE TREE REMOVAL**									
0011	With tractor, large tract, firm									
0020	level terrain, no boulders, less than 12" diam. trees									
0300	300 HP dozer, up to 400 trees/acre, up to 25% hardwoods	B-10M	.75	16	Acre		830	2,350	3,180	3,825
0340	25% to 50% hardwoods		.60	20			1,025	2,950	3,975	4,775
0370	75% to 100% hardwoods		.45	26.667			1,375	3,925	5,300	6,375
0400	500 trees/acre, up to 25% hardwoods		.60	20			1,025	2,950	3,975	4,775
0440	25% to 50% hardwoods		.48	25			1,300	3,675	4,975	6,000
0470	75% to 100% hardwoods		.36	33.333			1,725	4,900	6,625	8,000
0500	More than 600 trees/acre, up to 25% hardwoods		.52	23.077			1,200	3,400	4,600	5,525
0540	25% to 50% hardwoods		.42	28.571			1,475	4,200	5,675	6,850
0570	75% to 100% hardwoods		.31	38.710			2,000	5,700	7,700	9,250
0900	Large tract clearing per tree									
1500	300 HP dozer, to 12" diameter, softwood	B-10M	320	.038	Ea.		1.95	5.50	7.45	8.95

290

31 13 Selective Tree and Shrub Removal and Trimming

31 13 13 – Selective Tree and Shrub Removal

31 13 13.20 Selective Tree Removal		Crew	Daily Output	Labor-Hours	Unit	Material	2020 Bare Costs		Total	Total Incl O&P
							Labor	Equipment		
1550	Hardwood	B-10M	100	.120	Ea.		6.20	17.65	23.85	28.50
1600	12" to 24" diameter, softwood		200	.060			3.11	8.80	11.91	14.35
1650	Hardwood		80	.150			7.80	22	29.80	36
1700	24" to 36" diameter, softwood		100	.120			6.20	17.65	23.85	28.50
1750	Hardwood		50	.240			12.45	35.50	47.95	57.50
1800	36" to 48" diameter, softwood		70	.171			8.90	25	33.90	41
1850	Hardwood	↓	35	.343	↓		17.80	50.50	68.30	82
2000	Stump removal on site by hydraulic backhoe, 1-1/2 C.Y.									
2040	4" to 6" diameter	B-17	60	.533	Ea.		25	10.70	35.70	49
2050	8" to 12" diameter	B-30	33	.727			37.50	55.50	93	117
2100	14" to 24" diameter		25	.960			49.50	73.50	123	155
2150	26" to 36" diameter	↓	16	1.500	↓		77.50	115	192.50	242
3000	Remove selective trees, on site using chain saws and chipper,									
3050	not incl. stumps, up to 6" diameter	B-7	18	2.667	Ea.		120	91.50	211.50	280
3100	8" to 12" diameter		12	4			180	137	317	420
3150	14" to 24" diameter		10	4.800			215	164	379	505
3200	26" to 36" diameter	↓	8	6			269	206	475	630
3300	Machine load, 2 mile haul to dump, 12" diam. tree	A-3B	8	2	↓		106	159	265	335

31 14 Earth Stripping and Stockpiling

31 14 13 – Soil Stripping and Stockpiling

31 14 13.23 Topsoil Stripping and Stockpiling

		Crew	Daily Output	Labor-Hours	Unit	Material	Labor	Equipment	Total	Total Incl O&P
0010	**TOPSOIL STRIPPING AND STOCKPILING**									
0020	200 HP dozer, ideal conditions	B-10B	2300	.005	C.Y.		.27	.65	.92	1.13
0100	Adverse conditions	"	1150	.010			.54	1.31	1.85	2.25
0200	300 HP dozer, ideal conditions	B-10M	3000	.004			.21	.59	.80	.96
0300	Adverse conditions	"	1650	.007			.38	1.07	1.45	1.74
0400	400 HP dozer, ideal conditions	B-10X	3900	.003			.16	.71	.87	1.02
0500	Adverse conditions	"	2000	.006			.31	1.39	1.70	2
0600	Clay, dry and soft, 200 HP dozer, ideal conditions	B-10B	1600	.008			.39	.94	1.33	1.61
0700	Adverse conditions	"	800	.015			.78	1.88	2.66	3.23
1000	Medium hard, 300 HP dozer, ideal conditions	B-10M	2000	.006			.31	.88	1.19	1.44
1100	Adverse conditions	"	1100	.011			.57	1.60	2.17	2.61
1200	Very hard, 400 HP dozer, ideal conditions	B-10X	2600	.005			.24	1.07	1.31	1.54
1300	Adverse conditions	"	1340	.009	↓		.46	2.07	2.53	2.98
1400	Loam or topsoil, remove and stockpile on site									
1420	6" deep, 200' haul	B-10B	865	.014	C.Y.		.72	1.74	2.46	2.99
1430	300' haul		520	.023			1.20	2.89	4.09	4.97
1440	500' haul		225	.053	↓		2.77	6.70	9.47	11.50
1450	Alternate method: 6" deep, 200' haul		5090	.002	S.Y.		.12	.30	.42	.51
1460	500' haul	↓	1325	.009	"		.47	1.14	1.61	1.95
1500	Loam or topsoil, remove/stockpile on site									
1510	By hand, 6" deep, 50' haul, less than 100 S.Y.	B-1	100	.240	S.Y.		10.25		10.25	15.40
1520	By skid steer, 6" deep, 100' haul, 101-500 S.Y.	B-62	500	.048			2.20	.36	2.56	3.68
1530	100' haul, 501-900 S.Y.	"	900	.027			1.22	.20	1.42	2.05
1540	200' haul, 901-1,100 S.Y.	B-63	1000	.040			1.77	.18	1.95	2.86
1550	By dozer, 200' haul, 1,101-4,000 S.Y.	B-10B	4000	.003	↓		.16	.38	.54	.64

For customer support on your Site Work & Landscape Costs with RSMeans data, call 800.448.8182.

291

31 22 Grading

31 22 13 – Rough Grading

31 22 13.20 Rough Grading Sites

		Crew	Daily Output	Labor-Hours	Unit	Material	2020 Bare Costs Labor	2020 Bare Costs Equipment	Total	Total Incl O&P
0010	**ROUGH GRADING SITES**									
0100	Rough grade sites 400 S.F. or less, hand labor	B-1	2	12	Ea.		515		515	770
0120	410-1,000 S.F.	"	1	24			1,025		1,025	1,550
0130	1,100-3,000 S.F., skid steer & labor	B-62	1.50	16			730	118	848	1,225
0140	3,100-5,000 S.F.	"	1	24			1,100	178	1,278	1,850
0150	5,100-8,000 S.F.	B-63	1	40			1,775	178	1,953	2,850
0160	8,100-10,000 S.F.	"	.75	53.333			2,350	237	2,587	3,800
0170	8,100-10,000 S.F., dozer	B-10L	1	12			620	400	1,020	1,375
0200	Rough grade open sites 10,000-20,000 S.F., grader	B-11L	1.80	8.889			440	590	1,030	1,300
0210	20,100-25,000 S.F.		1.40	11.429			565	760	1,325	1,675
0220	25,100-30,000 S.F.		1.20	13.333			660	885	1,545	1,950
0230	30,100-35,000 S.F.		1	16			790	1,050	1,840	2,350
0240	35,100-40,000 S.F.		.90	17.778			880	1,175	2,055	2,625
0250	40,100-45,000 S.F.		.80	20			990	1,325	2,315	2,925
0260	45,100-50,000 S.F.		.72	22.222			1,100	1,475	2,575	3,275
0270	50,100-75,000 S.F.		.50	32			1,575	2,125	3,700	4,700
0280	75,100-100,000 S.F.	▼	.36	44.444	▼		2,200	2,950	5,150	6,550

31 22 16 – Fine Grading

31 22 16.10 Finish Grading

		Crew	Daily Output	Labor-Hours	Unit	Material	2020 Bare Costs Labor	2020 Bare Costs Equipment	Total	Total Incl O&P
0010	**FINISH GRADING**									
0011	Finish grading granular subbase for highway paving	B-32C	8000	.006	S.Y.		.30	.35	.65	.84
0012	Finish grading area to be paved with grader, small area	B-11L	400	.040			1.98	2.66	4.64	5.90
0100	Large area		2000	.008			.40	.53	.93	1.17
0200	Grade subgrade for base course, roadways	▼	3500	.005			.23	.30	.53	.67
1020	For large parking lots	B-32C	5000	.010			.48	.56	1.04	1.34
1050	For small irregular areas	"	2000	.024			1.19	1.40	2.59	3.33
1100	Fine grade for slab on grade, machine	B-11L	1040	.015			.76	1.02	1.78	2.26
1150	Hand grading	B-18	700	.034			1.47	.24	1.71	2.46
1200	Fine grade granular base for sidewalks and bikeways	B-62	1200	.020	▼		.91	.15	1.06	1.53
2550	Hand grade select gravel	2 Clab	60	.267	C.S.F.		11.25		11.25	16.85
3000	Hand grade select gravel, including compaction, 4" deep	B-18	555	.043	S.Y.		1.85	.30	2.15	3.11
3100	6" deep		400	.060			2.57	.41	2.98	4.31
3120	8" deep	▼	300	.080			3.42	.55	3.97	5.75
3300	Finishing grading slopes, gentle	B-11L	8900	.002			.09	.12	.21	.26
3310	Steep slopes		7100	.002	▼		.11	.15	.26	.33
3312	Steep slopes, large quantities		64	.250	M.S.F.		12.35	16.60	28.95	37
3500	Finish grading lagoon bottoms		4	4			198	266	464	590
3600	Fine grade, top of lagoon banks for compaction	▼	30	.533	▼		26.50	35.50	62	78.50

31 23 Excavation and Fill

31 23 16 – Excavation

31 23 16.13 Excavating, Trench

			Crew	Daily Output	Labor-Hours	Unit	Material	2020 Bare Costs Labor	2020 Bare Costs Equipment	Total	Total Incl O&P
0010	**EXCAVATING, TRENCH**	R312316-40									
0011	Or continuous footing										
0020	Common earth with no sheeting or dewatering included										
0050	1' to 4' deep, 3/8 C.Y. excavator		B-11C	150	.107	B.C.Y.		5.25	1.43	6.68	9.45
0060	1/2 C.Y. excavator		B-11M	200	.080			3.95	1.16	5.11	7.20
0062	3/4 C.Y. excavator		B-12F	270	.059			3	2.57	5.57	7.35
0090	4' to 6' deep, 1/2 C.Y. excavator		B-11M	200	.080			3.95	1.16	5.11	7.20
0100	5/8 C.Y. excavator		B-12Q	250	.064	▼		3.24	2.39	5.63	7.50

31 23 16 – Excavation

31 23 16.13 Excavating, Trench

		Crew	Daily Output	Labor-Hours	Unit	Material	2020 Bare Costs Labor	Equipment	Total	Total Incl O&P
0110	3/4 C.Y. excavator	B-12F	300	.053	B.C.Y.		2.70	2.31	5.01	6.60
0120	1 C.Y. hydraulic excavator	B-12A	400	.040			2.03	1.99	4.02	5.25
0130	1-1/2 C.Y. excavator	B-12B	540	.030			1.50	1.27	2.77	3.65
0300	1/2 C.Y. excavator, truck mounted	B-12J	200	.080			4.05	4.23	8.28	10.70
0500	6' to 10' deep, 3/4 C.Y. excavator	B-12F	225	.071			3.60	3.09	6.69	8.80
0510	1 C.Y. excavator	B-12A	400	.040			2.03	1.99	4.02	5.25
0600	1 C.Y. excavator, truck mounted	B-12K	400	.040			2.03	2.43	4.46	5.70
0610	1-1/2 C.Y. excavator	B-12B	600	.027			1.35	1.15	2.50	3.28
0620	2-1/2 C.Y. excavator	B-12S	1000	.016			.81	1.47	2.28	2.82
0900	10' to 14' deep, 3/4 C.Y. excavator	B-12F	200	.080			4.05	3.47	7.52	9.85
0910	1 C.Y. excavator	B-12A	360	.044			2.25	2.21	4.46	5.80
1000	1-1/2 C.Y. excavator	B-12B	540	.030			1.50	1.27	2.77	3.65
1020	2-1/2 C.Y. excavator	B-12S	900	.018			.90	1.63	2.53	3.14
1030	3 C.Y. excavator	B-12D	1400	.011			.58	1.54	2.12	2.57
1040	3-1/2 C.Y. excavator	"	1800	.009			.45	1.20	1.65	1.99
1300	14' to 20' deep, 1 C.Y. excavator	B-12A	320	.050			2.53	2.49	5.02	6.50
1310	1-1/2 C.Y. excavator	B-12B	480	.033			1.69	1.43	3.12	4.11
1320	2-1/2 C.Y. excavator	B-12S	765	.021			1.06	1.92	2.98	3.70
1330	3 C.Y. excavator	B-12D	1000	.016			.81	2.16	2.97	3.58
1335	3-1/2 C.Y. excavator	"	1150	.014			.70	1.88	2.58	3.12
1340	20' to 24' deep, 1 C.Y. excavator	B-12A	288	.056			2.81	2.76	5.57	7.25
1342	1-1/2 C.Y. excavator	B-12B	432	.037			1.88	1.59	3.47	4.56
1344	2-1/2 C.Y. excavator	B-12S	685	.023			1.18	2.14	3.32	4.12
1346	3 C.Y. excavator	B-12D	900	.018			.90	2.40	3.30	3.99
1348	3-1/2 C.Y. excavator	"	1050	.015			.77	2.06	2.83	3.42
1352	4' to 6' deep, 1/2 C.Y. excavator w/trench box	B-13H	188	.085			4.31	5.15	9.46	12.10
1354	5/8 C.Y. excavator	"	235	.068			3.45	4.10	7.55	9.65
1356	3/4 C.Y. excavator	B-13G	282	.057			2.87	2.88	5.75	7.50
1358	1 C.Y. excavator	B-13D	376	.043			2.16	2.43	4.59	5.90
1360	1-1/2 C.Y. excavator	B-13E	508	.032			1.60	1.59	3.19	4.13
1362	6' to 10' deep, 3/4 C.Y. excavator w/trench box	B-13G	212	.075			3.82	3.83	7.65	9.95
1370	1 C.Y. excavator	B-13D	376	.043			2.16	2.43	4.59	5.90
1371	1-1/2 C.Y. excavator	B-13E	564	.028			1.44	1.43	2.87	3.72
1372	2-1/2 C.Y. excavator	B-13J	940	.017			.86	1.68	2.54	3.14
1374	10' to 14' deep, 3/4 C.Y. excavator w/trench box	B-13G	188	.085			4.31	4.32	8.63	11.20
1375	1 C.Y. excavator	B-13D	338	.047			2.40	2.70	5.10	6.55
1376	1-1/2 C.Y. excavator	B-13E	508	.032			1.60	1.59	3.19	4.13
1377	2-1/2 C.Y. excavator	B-13J	845	.019			.96	1.87	2.83	3.50
1378	3 C.Y. excavator	B-13F	1316	.012			.62	1.73	2.35	2.82
1380	3-1/2 C.Y. excavator	"	1692	.009			.48	1.35	1.83	2.20
1381	14' to 20' deep, 1 C.Y. excavator w/trench box	B-13D	301	.053			2.69	3.03	5.72	7.35
1382	1-1/2 C.Y. excavator	B-13E	451	.035			1.80	1.79	3.59	4.65
1383	2-1/2 C.Y. excavator	B-13J	720	.022			1.13	2.20	3.33	4.11
1384	3 C.Y. excavator	B-13F	940	.017			.86	2.42	3.28	3.95
1385	3-1/2 C.Y. excavator	"	1081	.015			.75	2.11	2.86	3.44
1386	20' to 24' deep, 1 C.Y. excavator w/trench box	B-13D	271	.059			2.99	3.37	6.36	8.20
1387	1-1/2 C.Y. excavator	B-13E	406	.039			2	1.98	3.98	5.15
1388	2-1/2 C.Y. excavator	B-13J	645	.025			1.26	2.45	3.71	4.58
1389	3 C.Y. excavator	B-13F	846	.019			.96	2.69	3.65	4.39
1390	3-1/2 C.Y. excavator	"	987	.016			.82	2.31	3.13	3.77
1391	Shoring by S.F./day trench wall protected loose mat., 4' W	B-6	3200	.008	SF Wall	.51	.34	.07	.92	1.14
1392	Rent shoring per week per S.F. wall protected, loose mat., 4' W					1.40			1.40	1.54
1395	Hydraulic shoring, S.F. trench wall protected stable mat., 4' W	2 Clab	2700	.006		.20	.25		.45	.60

293

31 23 16.13 Excavating, Trench	Crew	Daily Output	Labor-Hours	Unit	Material	2020 Bare Costs Labor	Equipment	Total	Total Incl O&P
1397 semi-stable material, 4' W	2 Clab	2400	.007	SF Wall	.28	.28		.56	.73
1398 Rent hydraulic shoring per day/S.F. wall, stable mat., 4' W					.32			.32	.36
1399 semi-stable material				↓	.41			.41	.45
1400 By hand with pick and shovel 2' to 6' deep, light soil	1 Clab	8	1	B.C.Y.		42		42	63.50
1500 Heavy soil	"	4	2	"		84		84	127
1700 For tamping backfilled trenches, air tamp, add	A-1G	100	.080	E.C.Y.		3.37	.54	3.91	5.65
1900 Vibrating plate, add	B-18	180	.133	"		5.70	.92	6.62	9.55
2100 Trim sides and bottom for concrete pours, common earth		1500	.016	S.F.		.68	.11	.79	1.15
2300 Hardpan	↓	600	.040	"		1.71	.28	1.99	2.87
2400 Pier and spread footing excavation, add to above				B.C.Y.				30%	30%
3000 Backfill trench, F.E. loader, wheel mtd., 1 C.Y. bucket									
3020 Minimal haul	B-10R	400	.030	L.C.Y.		1.56	.76	2.32	3.16
3040 100' haul		200	.060			3.11	1.51	4.62	6.30
3060 200' haul	↓	100	.120			6.20	3.02	9.22	12.60
3080 2-1/4 C.Y. bucket, minimum haul	B-10T	600	.020			1.04	1.17	2.21	2.84
3090 100' haul		300	.040			2.07	2.34	4.41	5.70
3100 200' haul	↓	150	.080	↓		4.15	4.68	8.83	11.35
4000 For backfill with dozer, see Section 31 23 23.14									
4010 For compaction of backfill, see Section 31 23 23.23									
5020 Loam & sandy clay with no sheeting or dewatering included									
5050 1' to 4' deep, 3/8 C.Y. tractor loader/backhoe	B-11C	162	.099	B.C.Y.		4.88	1.32	6.20	8.75
5060 1/2 C.Y. excavator	B-11M	216	.074			3.66	1.08	4.74	6.70
5070 3/4 C.Y. excavator	B-12F	292	.055			2.78	2.38	5.16	6.80
5080 4' to 6' deep, 1/2 C.Y. excavator	B-11M	216	.074			3.66	1.08	4.74	6.70
5090 5/8 C.Y. excavator	B-12Q	276	.058			2.94	2.17	5.11	6.80
5100 3/4 C.Y. excavator	B-12F	324	.049			2.50	2.14	4.64	6.10
5110 1 C.Y. excavator	B-12A	432	.037			1.88	1.84	3.72	4.84
5120 1-1/2 C.Y. excavator	B-12B	583	.027			1.39	1.18	2.57	3.38
5130 1/2 C.Y. excavator, truck mounted	B-12J	216	.074			3.75	3.92	7.67	9.90
5140 6' to 10' deep, 3/4 C.Y. excavator	B-12F	243	.066			3.33	2.86	6.19	8.15
5150 1 C.Y. excavator	B-12A	432	.037			1.88	1.84	3.72	4.84
5160 1 C.Y. excavator, truck mounted	B-12K	432	.037			1.88	2.25	4.13	5.30
5170 1-1/2 C.Y. excavator	B-12B	648	.025			1.25	1.06	2.31	3.04
5180 2-1/2 C.Y. excavator	B-12S	1080	.015			.75	1.36	2.11	2.61
5190 10' to 14' deep, 3/4 C.Y. excavator	B-12F	216	.074			3.75	3.21	6.96	9.15
5200 1 C.Y. excavator	B-12A	389	.041			2.08	2.04	4.12	5.35
5210 1-1/2 C.Y. excavator	B-12B	583	.027			1.39	1.18	2.57	3.38
5220 2-1/2 C.Y. hydraulic backhoe	B-12S	970	.016			.84	1.51	2.35	2.91
5230 3 C.Y. excavator	B-12D	1512	.011			.54	1.43	1.97	2.37
5240 3-1/2 C.Y. excavator	"	1944	.008			.42	1.11	1.53	1.84
5250 14' to 20' deep, 1 C.Y. excavator	B-12A	346	.046			2.34	2.30	4.64	6.05
5260 1-1/2 C.Y. excavator	B-12B	518	.031			1.56	1.33	2.89	3.80
5270 2-1/2 C.Y. excavator	B-12S	826	.019			.98	1.77	2.75	3.42
5280 3 C.Y. excavator	B-12D	1080	.015			.75	2	2.75	3.32
5290 3-1/2 C.Y. excavator	"	1242	.013			.65	1.74	2.39	2.89
5300 20' to 24' deep, 1 C.Y. excavator	B-12A	311	.051			2.61	2.56	5.17	6.70
5310 1-1/2 C.Y. excavator	B-12B	467	.034			1.74	1.47	3.21	4.22
5320 2-1/2 C.Y. excavator	B-12S	740	.022			1.10	1.98	3.08	3.82
5330 3 C.Y. excavator	B-12D	972	.016			.83	2.22	3.05	3.69
5340 3-1/2 C.Y. excavator	"	1134	.014			.71	1.90	2.61	3.16
5352 4' to 6' deep, 1/2 C.Y. excavator w/trench box	B-13H	205	.078			3.95	4.70	8.65	11.05
5354 5/8 C.Y. excavator	"	257	.062			3.15	3.75	6.90	8.85
5356 3/4 C.Y. excavator	B-13G	308	.052	↓		2.63	2.64	5.27	6.85

31 23 16.13 Excavating, Trench

		Crew	Daily Output	Labor-Hours	Unit	Material	2020 Bare Costs Labor	2020 Bare Costs Equipment	Total	Total Incl O&P
5358	1 C.Y. excavator	B-13D	410	.039	B.C.Y.		1.98	2.23	4.21	5.40
5360	1-1/2 C.Y. excavator	B-13E	554	.029			1.46	1.45	2.91	3.79
5362	6' to 10' deep, 3/4 C.Y. excavator w/trench box	B-13G	231	.069			3.51	3.51	7.02	9.10
5364	1 C.Y. excavator	B-13D	410	.039			1.98	2.23	4.21	5.40
5366	1-1/2 C.Y. excavator	B-13E	616	.026			1.32	1.31	2.63	3.41
5368	2-1/2 C.Y. excavator	B-13J	1015	.016			.80	1.56	2.36	2.91
5370	10' to 14' deep, 3/4 C.Y. excavator w/trench box	B-13G	205	.078			3.95	3.96	7.91	10.25
5372	1 C.Y. excavator	B-13D	370	.043			2.19	2.47	4.66	6
5374	1-1/2 C.Y. excavator	B-13E	554	.029			1.46	1.45	2.91	3.79
5376	2-1/2 C.Y. excavator	B-13J	914	.018			.89	1.73	2.62	3.24
5378	3 C.Y. excavator	B-13F	1436	.011			.56	1.58	2.14	2.59
5380	3-1/2 C.Y. excavator	"	1847	.009			.44	1.23	1.67	2.01
5382	14' to 20' deep, 1 C.Y. excavator w/trench box	B-13D	329	.049			2.46	2.78	5.24	6.75
5384	1-1/2 C.Y. excavator	B-13E	492	.033			1.65	1.64	3.29	4.27
5386	2-1/2 C.Y. excavator	B-13J	780	.021			1.04	2.03	3.07	3.79
5388	3 C.Y. excavator	B-13F	1026	.016			.79	2.22	3.01	3.62
5390	3-1/2 C.Y. excavator	"	1180	.014			.69	1.93	2.62	3.15
5392	20' to 24' deep, 1 C.Y. excavator w/trench box	B-13D	295	.054			2.75	3.10	5.85	7.50
5394	1-1/2 C.Y. excavator	B-13E	444	.036			1.83	1.81	3.64	4.73
5396	2-1/2 C.Y. excavator	B-13J	695	.023			1.17	2.28	3.45	4.25
5398	3 C.Y. excavator	B-13F	923	.017			.88	2.46	3.34	4.03
5400	3-1/2 C.Y. excavator	"	1077	.015	↓		.75	2.11	2.86	3.46
6020	Sand & gravel with no sheeting or dewatering included									
6050	1' to 4' deep, 3/8 C.Y. excavator	B-11C	165	.097	B.C.Y.		4.79	1.30	6.09	8.65
6060	1/2 C.Y. excavator	B-11M	220	.073			3.60	1.06	4.66	6.55
6070	3/4 C.Y. excavator	B-12F	297	.054			2.73	2.34	5.07	6.65
6080	4' to 6' deep, 1/2 C.Y. excavator	B-11M	220	.073			3.60	1.06	4.66	6.55
6090	5/8 C.Y. excavator	B-12Q	275	.058			2.95	2.18	5.13	6.80
6100	3/4 C.Y. excavator	B-12F	330	.048			2.46	2.10	4.56	6
6110	1 C.Y. excavator	B-12A	440	.036			1.84	1.81	3.65	4.75
6120	1-1/2 C.Y. excavator	B-12B	594	.027			1.36	1.16	2.52	3.31
6130	1/2 C.Y. excavator, truck mounted	B-12J	220	.073			3.68	3.85	7.53	9.75
6140	6' to 10' deep, 3/4 C.Y. excavator	B-12F	248	.065			3.27	2.80	6.07	8
6150	1 C.Y. excavator	B-12A	440	.036			1.84	1.81	3.65	4.75
6160	1 C.Y. excavator, truck mounted	B-12K	440	.036			1.84	2.21	4.05	5.20
6170	1-1/2 C.Y. excavator	B-12B	660	.024			1.23	1.04	2.27	2.99
6180	2-1/2 C.Y. excavator	B-12S	1100	.015			.74	1.33	2.07	2.57
6190	10' to 14' deep, 3/4 C.Y. excavator	B-12F	220	.073			3.68	3.16	6.84	8.95
6200	1 C.Y. excavator	B-12A	396	.040			2.05	2.01	4.06	5.30
6210	1-1/2 C.Y. excavator	B-12B	594	.027			1.36	1.16	2.52	3.31
6220	2-1/2 C.Y. excavator	B-12S	990	.016			.82	1.48	2.30	2.86
6230	3 C.Y. excavator	B-12D	1540	.010			.53	1.40	1.93	2.33
6240	3-1/2 C.Y. excavator	"	1980	.008			.41	1.09	1.50	1.81
6250	14' to 20' deep, 1 C.Y. excavator	B-12A	352	.045			2.30	2.26	4.56	5.95
6260	1-1/2 C.Y. excavator	B-12B	528	.030			1.53	1.30	2.83	3.73
6270	2-1/2 C.Y. excavator	B-12S	840	.019			.96	1.74	2.70	3.37
6280	3 C.Y. excavator	B-12D	1100	.015			.74	1.96	2.70	3.26
6290	3-1/2 C.Y. excavator	"	1265	.013			.64	1.71	2.35	2.84
6300	20' to 24' deep, 1 C.Y. excavator	B-12A	317	.050			2.56	2.51	5.07	6.60
6310	1-1/2 C.Y. excavator	B-12B	475	.034			1.71	1.45	3.16	4.15
6320	2-1/2 C.Y. excavator	B-12S	755	.021			1.07	1.94	3.01	3.74
6330	3 C.Y. excavator	B-12D	990	.016			.82	2.18	3	3.63
6340	3-1/2 C.Y. excavator	"	1155	.014	↓		.70	1.87	2.57	3.10

31 23 16 – Excavation

31 23 16.13 Excavating, Trench	Crew	Daily Output	Labor-Hours	Unit	Material	2020 Bare Costs Labor	Equipment	Total	Total Incl O&P
6352	4' to 6' deep, 1/2 C.Y. excavator w/trench box	B-13H	209	.077	B.C.Y.		3.88	4.61	8.49	10.90
6354	5/8 C.Y. excavator	"	261	.061			3.10	3.69	6.79	8.70
6356	3/4 C.Y. excavator	B-13G	314	.051			2.58	2.59	5.17	6.70
6358	1 C.Y. excavator	B-13D	418	.038			1.94	2.18	4.12	5.30
6360	1-1/2 C.Y. excavator	B-13E	564	.028			1.44	1.43	2.87	3.72
6362	6' to 10' deep, 3/4 C.Y. excavator w/trench box	B-13G	236	.068			3.43	3.44	6.87	8.95
6364	1 C.Y. excavator	B-13D	418	.038			1.94	2.18	4.12	5.30
6366	1-1/2 C.Y. excavator	B-13E	627	.026			1.29	1.28	2.57	3.35
6368	2-1/2 C.Y. excavator	B-13J	1035	.015			.78	1.53	2.31	2.85
6370	10' to 14' deep, 3/4 C.Y. excavator w/trench box	B-13G	209	.077			3.88	3.89	7.77	10.05
6372	1 C.Y. excavator	B-13D	376	.043			2.16	2.43	4.59	5.90
6374	1-1/2 C.Y. excavator	B-13E	564	.028			1.44	1.43	2.87	3.72
6376	2-1/2 C.Y. excavator	B-13J	930	.017			.87	1.70	2.57	3.18
6378	3 C.Y. excavator	B-13F	1463	.011			.55	1.56	2.11	2.54
6380	3-1/2 C.Y. excavator	"	1881	.009			.43	1.21	1.64	1.98
6382	14' to 20' deep, 1 C.Y. excavator w/trench box	B-13D	334	.048			2.43	2.73	5.16	6.65
6384	1-1/2 C.Y. excavator	B-13E	502	.032			1.61	1.60	3.21	4.18
6386	2-1/2 C.Y. excavator	B-13J	790	.020			1.03	2	3.03	3.74
6388	3 C.Y. excavator	B-13F	1045	.015			.78	2.18	2.96	3.56
6390	3-1/2 C.Y. excavator	"	1202	.013			.67	1.89	2.56	3.09
6392	20' to 24' deep, 1 C.Y. excavator w/trench box	B-13D	301	.053			2.69	3.03	5.72	7.35
6394	1-1/2 C.Y. excavator	B-13E	452	.035			1.79	1.78	3.57	4.65
6396	2-1/2 C.Y. excavator	B-13J	710	.023			1.14	2.23	3.37	4.16
6398	3 C.Y. excavator	B-13F	941	.017			.86	2.42	3.28	3.95
6400	3-1/2 C.Y. excavator	"	1097	.015			.74	2.08	2.82	3.39
7020	Dense hard clay with no sheeting or dewatering included									
7050	1' to 4' deep, 3/8 C.Y. excavator	B-11C	132	.121	B.C.Y.		6	1.62	7.62	10.80
7060	1/2 C.Y. excavator	B-11M	176	.091			4.49	1.32	5.81	8.20
7070	3/4 C.Y. excavator	B-12F	238	.067			3.41	2.92	6.33	8.30
7080	4' to 6' deep, 1/2 C.Y. excavator	B-11M	176	.091			4.49	1.32	5.81	8.20
7090	5/8 C.Y. excavator	B-12Q	220	.073			3.68	2.72	6.40	8.50
7100	3/4 C.Y. excavator	B-12F	264	.061			3.07	2.63	5.70	7.50
7110	1 C.Y. excavator	B-12A	352	.045			2.30	2.26	4.56	5.95
7120	1-1/2 C.Y. excavator	B-12B	475	.034			1.71	1.45	3.16	4.15
7130	1/2 C.Y. excavator, truck mounted	B-12J	176	.091			4.60	4.81	9.41	12.20
7140	6' to 10' deep, 3/4 C.Y. excavator	B-12F	198	.081			4.09	3.51	7.60	10
7150	1 C.Y. excavator	B-12A	352	.045			2.30	2.26	4.56	5.95
7160	1 C.Y. excavator, truck mounted	B-12K	352	.045			2.30	2.77	5.07	6.50
7170	1-1/2 C.Y. excavator	B-12B	528	.030			1.53	1.30	2.83	3.73
7180	2-1/2 C.Y. excavator	B-12S	880	.018			.92	1.66	2.58	3.21
7190	10' to 14' deep, 3/4 C.Y. excavator	B-12F	176	.091			4.60	3.94	8.54	11.25
7200	1 C.Y. excavator	B-12A	317	.050			2.56	2.51	5.07	6.60
7210	1-1/2 C.Y. excavator	B-12B	475	.034			1.71	1.45	3.16	4.15
7220	2-1/2 C.Y. excavator	B-12S	790	.020			1.03	1.85	2.88	3.58
7230	3 C.Y. excavator	B-12D	1232	.013			.66	1.75	2.41	2.92
7240	3-1/2 C.Y. excavator	"	1584	.010			.51	1.36	1.87	2.27
7250	14' to 20' deep, 1 C.Y. excavator	B-12A	282	.057			2.87	2.82	5.69	7.40
7260	1-1/2 C.Y. excavator	B-12B	422	.038			1.92	1.63	3.55	4.67
7270	2-1/2 C.Y. excavator	B-12S	675	.024			1.20	2.17	3.37	4.19
7280	3 C.Y. excavator	B-12D	880	.018			.92	2.45	3.37	4.08
7290	3-1/2 C.Y. excavator	"	1012	.016			.80	2.13	2.93	3.55
7300	20' to 24' deep, 1 C.Y. excavator	B-12A	254	.063			3.19	3.13	6.32	8.20
7310	1-1/2 C.Y. excavator	B-12B	380	.042			2.13	1.81	3.94	5.20

31 23 Excavation and Fill

31 23 16 – Excavation

31 23 16.13 Excavating, Trench	Crew	Daily Output	Labor-Hours	Unit	Material	2020 Bare Costs Labor	2020 Bare Costs Equipment	Total	Total Incl O&P
7320	2-1/2 C.Y. excavator	B-12S	605	.026	B.C.Y.		1.34	2.42	3.76	4.67
7330	3 C.Y. excavator	B-12D	792	.020			1.02	2.72	3.74	4.53
7340	3-1/2 C.Y. excavator	"	924	.017	↓		.88	2.34	3.22	3.88

31 23 16.14 Excavating, Utility Trench

		Crew	Daily Output	Labor-Hours	Unit	Material	Labor	Equipment	Total	Total Incl O&P
0010	**EXCAVATING, UTILITY TRENCH**									
0011	Common earth									
0050	Trenching with chain trencher, 12 HP, operator walking									
0100	4" wide trench, 12" deep	B-53	800	.010	L.F.		.53	.20	.73	1.01
0150	18" deep		750	.011			.57	.21	.78	1.08
0200	24" deep		700	.011			.61	.22	.83	1.16
0300	6" wide trench, 12" deep		650	.012			.65	.24	.89	1.25
0350	18" deep		600	.013			.71	.26	.97	1.35
0400	24" deep		550	.015			.77	.29	1.06	1.46
0450	36" deep		450	.018			.94	.35	1.29	1.79
0600	8" wide trench, 12" deep		475	.017			.89	.33	1.22	1.69
0650	18" deep		400	.020			1.06	.39	1.45	2.01
0700	24" deep		350	.023			1.21	.45	1.66	2.30
0750	36" deep	↓	300	.027	↓		1.41	.52	1.93	2.69
0830	Fly wheel trencher, 18" wide trench, 6' deep, light soil	B-54A	1992	.005	B.C.Y.		.26	.57	.83	1.01
0840	Medium soil		1594	.006			.32	.71	1.03	1.26
0850	Heavy soil	↓	1295	.007			.40	.87	1.27	1.55
0860	24" wide trench, 9' deep, light soil	B-54B	4981	.002			.11	.27	.38	.45
0870	Medium soil		4000	.003			.14	.33	.47	.56
0880	Heavy soil	↓	3237	.003	↓		.17	.41	.58	.70
1000	Backfill by hand including compaction, add									
1050	4" wide trench, 12" deep	A-1G	800	.010	L.F.		.42	.07	.49	.70
1100	18" deep		530	.015			.64	.10	.74	1.06
1150	24" deep		400	.020			.84	.14	.98	1.42
1300	6" wide trench, 12" deep		540	.015			.62	.10	.72	1.05
1350	18" deep		405	.020			.83	.13	.96	1.40
1400	24" deep		270	.030			1.25	.20	1.45	2.09
1450	36" deep		180	.044			1.87	.30	2.17	3.14
1600	8" wide trench, 12" deep		400	.020			.84	.14	.98	1.42
1650	18" deep		265	.030			1.27	.20	1.47	2.13
1700	24" deep		200	.040			1.68	.27	1.95	2.83
1750	36" deep	↓	135	.059	↓		2.49	.40	2.89	4.19
2000	Chain trencher, 40 HP operator riding									
2050	6" wide trench and backfill, 12" deep	B-54	1200	.007	L.F.		.35	.34	.69	.90
2100	18" deep		1000	.008			.42	.40	.82	1.07
2150	24" deep		975	.008			.44	.41	.85	1.10
2200	36" deep		900	.009			.47	.45	.92	1.19
2250	48" deep		750	.011			.57	.54	1.11	1.44
2300	60" deep		650	.012			.65	.62	1.27	1.66
2400	8" wide trench and backfill, 12" deep		1000	.008			.42	.40	.82	1.07
2450	18" deep		950	.008			.45	.42	.87	1.14
2500	24" deep		900	.009			.47	.45	.92	1.19
2550	36" deep		800	.010			.53	.50	1.03	1.34
2600	48" deep		650	.012			.65	.62	1.27	1.66
2700	12" wide trench and backfill, 12" deep		975	.008			.44	.41	.85	1.10
2750	18" deep		860	.009			.49	.47	.96	1.25
2800	24" deep		800	.010			.53	.50	1.03	1.34
2850	36" deep	↓	725	.011	↓		.58	.55	1.13	1.48

31 23 Excavation and Fill

31 23 16 – Excavation

31 23 16.14 Excavating, Utility Trench

		Crew	Daily Output	Labor-Hours	Unit	Material	2020 Bare Costs Labor	Equipment	Total	Total Incl O&P
3000	16" wide trench and backfill, 12" deep	B-54	835	.010	L.F.		.51	.48	.99	1.29
3050	18" deep		750	.011			.57	.54	1.11	1.44
3100	24" deep	↓	700	.011	↓		.61	.57	1.18	1.54
3200	Compaction with vibratory plate, add								35%	35%
5100	Hand excavate and trim for pipe bells after trench excavation									
5200	8" pipe	1 Clab	155	.052	L.F.		2.17		2.17	3.26
5300	18" pipe	"	130	.062	"		2.59		2.59	3.89
5400	For clay or till, add up to								150%	150%
6000	Excavation utility trench, rock material									
6050	Note: assumption teeth change every 100' of trench									
6100	Chain trencher, 6" wide, 30" depth, rock saw	B-54D	400	.040	L.F.	20.50	1.98	1.07	23.55	26.50
6200	Chain trencher, 18" wide, 8' depth, rock trencher	B-54E	280	.057	"	32.50	2.82	3.59	38.91	43.50

31 23 16.15 Excavating, Utility Trench, Plow

		Crew	Daily Output	Labor-Hours	Unit	Material	2020 Bare Costs Labor	Equipment	Total	Total Incl O&P
0010	**EXCAVATING, UTILITY TRENCH, PLOW**									
0100	Single cable, plowed into fine material	B-54C	3800	.004	L.F.		.21	.30	.51	.64
0200	Two cable		3200	.005			.25	.35	.60	.76
0300	Single cable, plowed into coarse material	↓	2000	.008	↓		.40	.56	.96	1.21

31 23 16.16 Structural Excavation for Minor Structures

		Crew	Daily Output	Labor-Hours	Unit	Material	2020 Bare Costs Labor	Equipment	Total	Total Incl O&P
0010	**STRUCTURAL EXCAVATION FOR MINOR STRUCTURES** R312316-40									
0015	Hand, pits to 6' deep, sandy soil	1 Clab	8	1	B.C.Y.		42		42	63.50
0020	Normal soil	B-2	16	2.500			106		106	160
0030	Medium clay		12	3.333			142		142	213
0040	Heavy clay		8	5			213		213	320
0050	Loose rock	↓	6	6.667			283		283	425
0100	Heavy soil or clay	1 Clab	4	2			84		84	127
0200	Pits to 2' deep, normal soil	B-2	24	1.667			71		71	106
0210	Sand and gravel		24	1.667			71		71	106
0220	Medium clay		18	2.222			94.50		94.50	142
0230	Heavy clay		12	3.333			142		142	213
0300	Pits 6' to 12' deep, sandy soil	1 Clab	5	1.600			67.50		67.50	101
0500	Heavy soil or clay		3	2.667			112		112	169
0700	Pits 12' to 18' deep, sandy soil		4	2			84		84	127
0900	Heavy soil or clay	↓	2	4	↓		168		168	253
1000	Hand trimming, bottom of excavation	B-2	2400	.017	S.F.		.71		.71	1.06
1010	Slopes and sides		2400	.017	"		.71		.71	1.06
1030	Around obstructions	↓	8	5	B.C.Y.		213		213	320
1100	Hand loading trucks from stock pile, sandy soil	1 Clab	12	.667			28		28	42
1300	Heavy soil or clay	"	8	1			42		42	63.50
1500	For wet or muck hand excavation, add to above								50%	50%
1550	Excavation rock by hand/air tool	B-9	3.40	11.765	B.C.Y.		500	105	605	865
6000	Machine excavation, for spread and mat footings, elevator pits,									
6001	and small building foundations									
6030	Common earth, hydraulic backhoe, 1/2 C.Y. bucket	B-12E	55	.291	B.C.Y.		14.75	8.20	22.95	31
6035	3/4 C.Y. bucket	B-12F	90	.178			9	7.70	16.70	22
6040	1 C.Y. bucket	B-12A	108	.148			7.50	7.35	14.85	19.35
6050	1-1/2 C.Y. bucket	B-12B	144	.111			5.65	4.77	10.42	13.70
6060	2 C.Y. bucket	B-12C	200	.080			4.05	4.76	8.81	11.30
6070	Sand and gravel, 3/4 C.Y. bucket	B-12F	100	.160			8.10	6.95	15.05	19.80
6080	1 C.Y. bucket	B-12A	120	.133			6.75	6.65	13.40	17.40
6090	1-1/2 C.Y. bucket	B-12B	160	.100			5.05	4.30	9.35	12.35
6100	2 C.Y. bucket	B-12C	220	.073			3.68	4.32	8	10.25
6110	Clay, till, or blasted rock, 3/4 C.Y. bucket	B-12F	80	.200			10.15	8.70	18.85	25

298

31 23 Excavation and Fill

31 23 16 – Excavation

31 23 16.16 Structural Excavation for Minor Structures	Crew	Daily Output	Labor-Hours	Unit	Material	2020 Bare Costs Labor	2020 Bare Costs Equipment	Total	Total Incl O&P
6120 ~ 1 C.Y. bucket	B-12A	95	.168	B.C.Y.		8.55	8.35	16.90	22
6130 1-1/2 C.Y. bucket	B-12B	130	.123			6.25	5.30	11.55	15.15
6140 2 C.Y. bucket	B-12C	175	.091			4.63	5.45	10.08	12.95
6230 Sandy clay & loam, hydraulic backhoe, 1/2 C.Y. bucket	B-12E	60	.267			13.50	7.55	21.05	28.50
6235 3/4 C.Y. bucket	B-12F	98	.163			8.25	7.10	15.35	20
6240 1 C.Y. bucket	B-12A	116	.138			7	6.85	13.85	18
6250 1-1/2 C.Y. bucket	B-12B	156	.103			5.20	4.41	9.61	12.65
6260 2 C.Y. bucket	B-12C	216	.074			3.75	4.40	8.15	10.45
6280 3-1/2 C.Y. bucket	B-12D	300	.053	↓		2.70	7.20	9.90	11.95
9010 For mobilization or demobilization, see Section 01 54 36.50									
9020 For dewatering, see Section 31 23 19.20									
9022 For larger structures, see Bulk Excavation, Section 31 23 16.42									
9024 For loading onto trucks, add								15%	15%
9026 For hauling, see Section 31 23 23.20									
9030 For sheeting or soldier bms/lagging, see Section 31 52 16.10									
9040 For trench excavation of strip ftgs, see Section 31 23 16.13									

31 23 16.26 Rock Removal

	Crew	Daily Output	Labor-Hours	Unit	Material	Labor	Equipment	Total	Total Incl O&P
0010 **ROCK REMOVAL**									
0015 Drilling only rock, 2" hole for rock bolts	B-47	316	.076	L.F.		3.52	4.92	8.44	10.70
0800 2-1/2" hole for pre-splitting		250	.096			4.45	6.20	10.65	13.55
4600 Quarry operations, 2-1/2" to 3-1/2" diameter	↓	240	.100	↓		4.64	6.50	11.14	14.10

31 23 16.30 Drilling and Blasting Rock

	Crew	Daily Output	Labor-Hours	Unit	Material	Labor	Equipment	Total	Total Incl O&P
0010 **DRILLING AND BLASTING ROCK**									
0020 Rock, open face, under 1,500 C.Y.	B-47	225	.107	B.C.Y.	4.54	4.95	6.90	16.39	20
0100 Over 1,500 C.Y.		300	.080		4.54	3.71	5.20	13.45	16.25
0200 Areas where blasting mats are required, under 1,500 C.Y.		175	.137		4.54	6.35	8.90	19.79	24.50
0250 Over 1,500 C.Y.	↓	250	.096		4.54	4.45	6.20	15.19	18.55
0300 Bulk drilling and blasting, can vary greatly, average								9.65	12.20
0500 Pits, average								25.50	31.50
1300 Deep hole method, up to 1,500 C.Y.	B-47	50	.480		4.54	22.50	31	58.04	72.50
1400 Over 1,500 C.Y.		66	.364		4.54	16.85	23.50	44.89	56.50
1900 Restricted areas, up to 1,500 C.Y.		13	1.846		4.54	85.50	120	210.04	265
2000 Over 1,500 C.Y.		20	1.200		4.54	55.50	78	138.04	174
2200 Trenches, up to 1,500 C.Y.		22	1.091		13.15	50.50	70.50	134.15	169
2300 Over 1,500 C.Y.		26	.923		13.15	43	60	116.15	145
2500 Pier holes, up to 1,500 C.Y.		22	1.091		4.54	50.50	70.50	125.54	159
2600 Over 1,500 C.Y.	↓	31	.774		4.54	36	50	90.54	114
2800 Boulders under 1/2 C.Y., loaded on truck, no hauling	B-100	80	.150			7.80	12	19.80	25
2900 Boulders, drilled, blasted	B-47	100	.240	↓	4.54	11.15	15.55	31.24	39
3100 Jackhammer operators with foreman compressor, air tools	B-9	1	40	Day		1,700	355	2,055	2,950
3300 Track drill, compressor, operator and foreman	B-47	1	24	"		1,125	1,550	2,675	3,375
3500 Blasting caps				Ea.	6.65			6.65	7.30
3700 Explosives					.54			.54	.59
3800 Blasting mats, for purchase, no mobilization, 10' x 15' x 12"					1,275			1,275	1,400
3900 Blasting mats, rent, for first day					211			211	232
4000 Per added day					65.50			65.50	72
4200 Preblast survey for 6 room house, individual lot, minimum	A-6	2.40	6.667			355	12.85	367.85	545
4300 Maximum	"	1.35	11.852	↓		625	23	648	970
4500 City block within zone of influence, minimum	A-8	25200	.001	S.F.		.07		.07	.11
4600 Maximum	"	15100	.002	"		.12		.12	.18
5000 Excavate and load boulders, less than 0.5 C.Y.	B-10T	80	.150	B.C.Y.		7.80	8.80	16.60	21.50
5020 0.5 C.Y. to 1 C.Y.	B-10U	100	.120	↓		6.20	9.55	15.75	19.85

For customer support on your Site Work & Landscape Costs with RSMeans data, call 800.448.8182.

299

31 23 Excavation and Fill

31 23 16 – Excavation

31 23 16.30 Drilling and Blasting Rock

		Crew	Daily Output	Labor-Hours	Unit	Material	2020 Bare Costs Labor	Equipment	Total	Total Incl O&P
5200	Excavate and load blasted rock, 3 C.Y. power shovel	B-12T	1530	.010	B.C.Y.		.53	1.38	1.91	2.31
5400	Haul boulders, 25 ton off-highway dump, 1 mile round trip	B-34E	330	.024			1.19	4.18	5.37	6.40
5420	2 mile round trip		275	.029			1.42	5	6.42	7.65
5440	3 mile round trip		225	.036			1.74	6.15	7.89	9.35
5460	4 mile round trip		200	.040			1.96	6.90	8.86	10.55
5600	Bury boulders on site, less than 0.5 C.Y., 300 HP dozer									
5620	150' haul	B-10M	310	.039	B.C.Y.		2.01	5.70	7.71	9.25
5640	300' haul		210	.057			2.96	8.40	11.36	13.70
5800	0.5 to 1 C.Y., 300 HP dozer, 150' haul		300	.040			2.07	5.90	7.97	9.55
5820	300' haul		200	.060			3.11	8.80	11.91	14.35

31 23 16.32 Ripping

		Crew	Daily Output	Labor-Hours	Unit	Material	2020 Bare Costs Labor	Equipment	Total	Total Incl O&P
0010	**RIPPING** R312316-40									
0020	Ripping, trap rock, soft, 300 HP dozer, ideal conditions	B-11S	700	.017	B.C.Y.		.89	2.65	3.54	4.24
1500	Adverse conditions		660	.018			.94	2.81	3.75	4.50
1600	Medium hard, 300 HP dozer, ideal conditions		600	.020			1.04	3.09	4.13	4.95
1700	Adverse conditions		540	.022			1.15	3.43	4.58	5.50
2000	Very hard, 410 HP dozer, ideal conditions	B-11T	350	.034			1.78	8.35	10.13	11.80
2100	Adverse conditions	"	310	.039			2.01	9.40	11.41	13.35
2200	Shale, soft, 300 HP dozer, ideal conditions	B-11S	1500	.008			.42	1.24	1.66	1.98
2300	Adverse conditions	"	1350	.009			.46	1.37	1.83	2.20
2310	Grader rear ripper, 180 HP ideal conditions	B-11J	740	.022			1.07	1.56	2.63	3.31
2320	Adverse conditions	"	630	.025			1.26	1.83	3.09	3.89
2400	Medium hard, 300 HP dozer, ideal conditions	B-11S	1200	.010			.52	1.55	2.07	2.48
2500	Adverse conditions	"	1080	.011			.58	1.72	2.30	2.75
2510	Grader rear ripper, 180 HP ideal conditions	B-11J	625	.026			1.27	1.84	3.11	3.93
2520	Adverse conditions	"	530	.030			1.49	2.17	3.66	4.63
2600	Very hard, 410 HP dozer, ideal conditions	B-11T	800	.015			.78	3.64	4.42	5.15
2700	Adverse conditions	"	720	.017			.86	4.05	4.91	5.75
2800	Till, boulder clay/hardpan, soft, 300 HP dozer, ideal conditions	B-11S	7000	.002			.09	.26	.35	.42
2810	Adverse conditions	"	6300	.002			.10	.29	.39	.47
2815	Grader rear ripper, 180 HP ideal conditions	B-11J	1500	.011			.53	.77	1.30	1.64
2816	Adverse conditions	"	1275	.013			.62	.90	1.52	1.92
2820	Medium hard, 300 HP dozer, ideal conditions	B-11S	6000	.002			.10	.31	.41	.50
2830	Adverse conditions	"	5400	.002			.12	.34	.46	.55
2835	Grader rear ripper, 180 HP ideal conditions	B-11J	740	.022			1.07	1.56	2.63	3.31
2836	Adverse conditions	"	630	.025			1.26	1.83	3.09	3.89
2840	Very hard, 410 HP dozer, ideal conditions	B-11T	5000	.002			.12	.58	.70	.83
2850	Adverse conditions	"	4500	.003			.14	.65	.79	.92
3000	Dozing ripped material, 200 HP, 100' haul	B-10B	700	.017			.89	2.15	3.04	3.69
3050	300' haul	"	250	.048			2.49	6	8.49	10.35
3200	300 HP, 100' haul	B-10M	1150	.010			.54	1.53	2.07	2.50
3250	300' haul	"	400	.030			1.56	4.41	5.97	7.20
3400	410 HP, 100' haul	B-10X	1680	.007			.37	1.65	2.02	2.37
3450	300' haul	"	600	.020			1.04	4.63	5.67	6.65

31 23 16.35 Hydraulic Rock Breaking and Loading

		Crew	Daily Output	Labor-Hours	Unit	Material	2020 Bare Costs Labor	Equipment	Total	Total Incl O&P
0010	**HYDRAULIC ROCK BREAKING AND LOADING**									
0020	Solid rock mass excavation									
0022	Excavator/breaker and excavator/bucket (into trucks)									
0024	Double costs for trenching									
0026	Mobilization costs to be repeated for demobilization									
0100	110 HP excavator w/4,000 ft lb. breaker									
0120	Mobilization 2/20 ton									

300

31 23 16 – Excavation

31 23 16.35 Hydraulic Rock Breaking and Loading

		Crew	Daily Output	Labor-Hours	Unit	Material	2020 Bare Costs Labor	2020 Bare Costs Equipment	Total	Total Incl O&P
0130	6,000 psi rock	B-13K	640	.025	B.C.Y.		1.48	3.17	4.65	5.70
0140	12,000 psi rock		240	.067			3.95	8.45	12.40	15.20
0150	18,000 psi rock		80	.200			11.85	25.50	37.35	45.50
0200	170 HP excavator w/5,000 ft lb. breaker									
0220	Mobilization 30 ton & 20 ton									
0240	12,000 psi rock	B-13L	640	.025	B.C.Y.		1.48	3.25	4.73	5.80
0250	18,000 psi rock		240	.067			3.95	8.65	12.60	15.45
0260	24,000 psi rock		80	.200			11.85	26	37.85	46
0300	270 HP excavator w/8,000 ft lb. breaker									
0320	Mobilization 40 ton & 30 ton									
0350	18,000 psi rock	B-13M	480	.033	B.C.Y.		1.97	6.40	8.37	9.95
0360	24,000 psi rock		240	.067			3.95	12.75	16.70	19.95
0370	30,000 psi rock		80	.200			11.85	38.50	50.35	59.50
0400	350 HP excavator w/12,000 ft lb. breaker									
0420	Mobilization 50 ton & 30 ton (may require disassembly & third hauler)									
0450	18,000 psi rock	B-13N	640	.025	B.C.Y.		1.48	5.80	7.28	8.60
0460	24,000 psi rock		400	.040			2.37	9.30	11.67	13.75
0470	30,000 psi rock		240	.067			3.95	15.50	19.45	23

31 23 16.42 Excavating, Bulk Bank Measure

		Crew	Daily Output	Labor-Hours	Unit	Material	2020 Bare Costs Labor	2020 Bare Costs Equipment	Total	Total Incl O&P
0010	**EXCAVATING, BULK BANK MEASURE** R312316-40									
0011	Common earth piled									
0020	For loading onto trucks, add								15%	15%
0050	For mobilization and demobilization, see Section 01 54 36.50 R312316-45									
0100	For hauling, see Section 31 23 23.20									
0200	Excavator, hydraulic, crawler mtd., 1 C.Y. cap. = 100 C.Y./hr.	B-12A	800	.020	B.C.Y.		1.01	.99	2	2.61
0250	1-1/2 C.Y. cap. = 125 C.Y./hr.	B-12B	1000	.016			.81	.69	1.50	1.97
0260	2 C.Y. cap. = 165 C.Y./hr.	B-12C	1320	.012			.61	.72	1.33	1.71
0300	3 C.Y. cap. = 260 C.Y./hr.	B-12D	2080	.008			.39	1.04	1.43	1.72
0305	3-1/2 C.Y. cap. = 300 C.Y./hr.	"	2400	.007			.34	.90	1.24	1.50
0310	Wheel mounted, 1/2 C.Y. cap. = 40 C.Y./hr.	B-12E	320	.050			2.53	1.41	3.94	5.35
0360	3/4 C.Y. cap. = 60 C.Y./hr.	B-12F	480	.033			1.69	1.45	3.14	4.12
0500	Clamshell, 1/2 C.Y. cap. = 20 C.Y./hr.	B-12G	160	.100			5.05	5.35	10.40	13.50
0550	1 C.Y. cap. = 35 C.Y./hr.	B-12H	280	.057			2.89	4.28	7.17	9.05
0950	Dragline, 1/2 C.Y. cap. = 30 C.Y./hr.	B-12I	240	.067			3.38	4.26	7.64	9.75
1000	3/4 C.Y. cap. = 35 C.Y./hr.	"	280	.057			2.89	3.65	6.54	8.35
1050	1-1/2 C.Y. cap. = 65 C.Y./hr.	B-12P	520	.031			1.56	2.44	4	5
1100	3 C.Y. cap. = 112 C.Y./hr.	B-12V	900	.018			.90	2.30	3.20	3.88
1200	Front end loader, track mtd., 1-1/2 C.Y. cap. = 70 C.Y./hr.	B-10N	560	.021			1.11	1.01	2.12	2.77
1250	2-1/2 C.Y. cap. = 95 C.Y./hr.	B-10O	760	.016			.82	1.26	2.08	2.62
1300	3 C.Y. cap. = 130 C.Y./hr.	B-10P	1040	.012			.60	1.12	1.72	2.14
1350	5 C.Y. cap. = 160 C.Y./hr.	B-10Q	1280	.009			.49	1.16	1.65	2.01
1500	Wheel mounted, 3/4 C.Y. cap. = 45 C.Y./hr.	B-10R	360	.033			1.73	.84	2.57	3.51
1550	1-1/2 C.Y. cap. = 80 C.Y./hr.	B-10S	640	.019			.97	.66	1.63	2.19
1600	2-1/4 C.Y. cap. = 100 C.Y./hr.	B-10T	800	.015			.78	.88	1.66	2.13
1601	3 C.Y. cap. = 140 C.Y./hr.	"	1120	.011			.56	.63	1.19	1.52
1650	5 C.Y. cap. = 185 C.Y./hr.	B-10U	1480	.008			.42	.65	1.07	1.34
1800	Hydraulic excavator, truck mtd. 1/2 C.Y. = 30 C.Y./hr.	B-12J	240	.067			3.38	3.53	6.91	8.95
1850	48" bucket, 1 C.Y. = 45 C.Y./hr.	B-12K	360	.044			2.25	2.70	4.95	6.35
3700	Shovel, 1/2 C.Y. cap. = 55 C.Y./hr.	B-12L	440	.036			1.84	1.95	3.79	4.90
3750	3/4 C.Y. cap. = 85 C.Y./hr.	B-12M	680	.024			1.19	1.52	2.71	3.46
3800	1 C.Y. cap. = 120 C.Y./hr.	B-12N	960	.017			.84	1.26	2.10	2.64
3850	1-1/2 C.Y. cap. = 160 C.Y./hr.	B-12O	1280	.013			.63	1.01	1.64	2.06

For customer support on your Site Work & Landscape Costs with RSMeans data, call 800.448.8182.

301

31 23 16.42 Excavating, Bulk Bank Measure

		Crew	Daily Output	Labor-Hours	Unit	Material	2020 Bare Costs Labor	2020 Bare Costs Equipment	Total	Total Incl O&P
3900	3 C.Y. cap. = 250 C.Y./hr.	B-12T	2000	.008	B.C.Y.		.41	1.06	1.47	1.77
4000	For soft soil or sand, deduct								15%	15%
4100	For heavy soil or stiff clay, add								60%	60%
4200	For wet excavation with clamshell or dragline, add								100%	100%
4250	All other equipment, add								50%	50%
4400	Clamshell in sheeting or cofferdam, minimum	B-12H	160	.100			5.05	7.50	12.55	15.85
4450	Maximum	"	60	.267	▼		13.50	19.95	33.45	42
5000	Excavating, bulk bank measure, sandy clay & loam piled									
5020	For loading onto trucks, add								15%	15%
5100	Excavator, hydraulic, crawler mtd., 1 C.Y. cap. = 120 C.Y./hr.	B-12A	960	.017	B.C.Y.		.84	.83	1.67	2.17
5150	1-1/2 C.Y. cap. = 150 C.Y./hr.	B-12B	1200	.013			.68	.57	1.25	1.64
5300	2 C.Y. cap. = 195 C.Y./hr.	B-12C	1560	.010			.52	.61	1.13	1.45
5400	3 C.Y. cap. = 300 C.Y./hr.	B-12D	2400	.007			.34	.90	1.24	1.50
5500	3-1/2 C.Y. cap. = 350 C.Y./hr.	"	2800	.006			.29	.77	1.06	1.28
5610	Wheel mounted, 1/2 C.Y. cap. = 44 C.Y./hr.	B-12E	352	.045			2.30	1.28	3.58	4.86
5660	3/4 C.Y. cap. = 66 C.Y./hr.	B-12F	528	.030	▼		1.53	1.31	2.84	3.75
8000	For hauling excavated material, see Section 31 23 23.20									

31 23 16.43 Excavating, Large Volume Projects

		Crew	Daily Output	Labor-Hours	Unit	Material	2020 Bare Costs Labor	2020 Bare Costs Equipment	Total	Total Incl O&P
0010	**EXCAVATING, LARGE VOLUME PROJECTS**, various materials									
0050	Minimum project size 200,000 B.C.Y.									
0080	For hauling, see Section 31 23 23.20									
0100	Unrestricted means continuous loading into hopper or railroad cars									
0200	Loader, 8 C.Y. bucket, 110% fill factor, unrestricted operation	B-14J	5280	.002	L.C.Y.		.12	.43	.55	.65
0250	10 C.Y. bucket	B-14K	7040	.002			.09	.38	.47	.55
0300	8 C.Y. bucket, 105% fill factor	B-14J	5040	.002			.12	.45	.57	.67
0350	10 C.Y. bucket	B-14K	6720	.002			.09	.40	.49	.58
0400	8 C.Y. bucket, 100% fill factor	B-14J	4800	.003			.13	.47	.60	.71
0450	10 C.Y. bucket	B-14K	6400	.002			.10	.42	.52	.61
0500	8 C.Y. bucket, 95% fill factor	B-14J	4560	.003			.14	.50	.64	.75
0550	10 C.Y. bucket	B-14K	6080	.002			.10	.44	.54	.63
0600	8 C.Y. bucket, 90% fill factor	B-14J	4320	.003			.14	.52	.66	.80
0650	10 C.Y. bucket	B-14K	5760	.002			.11	.46	.57	.67
0700	8 C.Y. bucket, 85% fill factor	B-14J	4080	.003			.15	.55	.70	.84
0750	10 C.Y. bucket	B-14K	5440	.002			.11	.49	.60	.71
0800	8 C.Y. bucket, 80% fill factor	B-14J	3840	.003			.16	.59	.75	.89
0850	10 C.Y. bucket	B-14K	5120	.002			.12	.52	.64	.75
0900	8 C.Y. bucket, 75% fill factor	B-14J	3600	.003			.17	.63	.80	.95
0950	10 C.Y. bucket	B-14K	4800	.003			.13	.56	.69	.80
1000	8 C.Y. bucket, 70% fill factor	B-14J	3360	.004			.19	.67	.86	1.02
1050	10 C.Y. bucket	B-14K	4480	.003			.14	.60	.74	.87
1100	8 C.Y. bucket, 65% fill factor	B-14J	3120	.004			.20	.73	.93	1.10
1150	10 C.Y. bucket	B-14K	4160	.003			.15	.64	.79	.93
1200	8 C.Y. bucket, 60% fill factor	B-14J	2880	.004			.22	.79	1.01	1.18
1250	10 C.Y. bucket	B-14K	3840	.003	▼		.16	.70	.86	1.01
2100	Restricted loading trucks									
2200	Loader, 8 C.Y. bucket, 110% fill factor, loading trucks	B-14J	4400	.003	L.C.Y.		.14	.51	.65	.78
2250	10 C.Y. bucket	B-14K	5940	.002			.10	.45	.55	.66
2300	8 C.Y. bucket, 105% fill factor	B-14J	4200	.003			.15	.54	.69	.81
2350	10 C.Y. bucket	B-14K	5670	.002			.11	.47	.58	.68
2400	8 C.Y. bucket, 100% fill factor	B-14J	4000	.003			.16	.57	.73	.85
2450	10 C.Y. bucket	B-14K	5400	.002			.12	.49	.61	.71
2500	8 C.Y. bucket, 95% fill factor	B-14J	3800	.003			.16	.60	.76	.90

31 23 16 – Excavation

31 23 16.43 Excavating, Large Volume Projects

		Crew	Daily Output	Labor-Hours	Unit	Material	2020 Bare Costs Labor	2020 Bare Costs Equipment	Total	Total Incl O&P
2550	10 C.Y. bucket	B-14K	5130	.002	L.C.Y.		.12	.52	.64	.75
2600	8 C.Y. bucket, 90% fill factor	B-14J	3600	.003			.17	.63	.80	.95
2650	10 C.Y. bucket	B-14K	4860	.002			.13	.55	.68	.80
2700	8 C.Y. bucket, 85% fill factor	B-14J	3400	.004			.18	.67	.85	1
2750	10 C.Y. bucket	B-14K	4590	.003			.14	.58	.72	.84
2800	8 C.Y. bucket, 80% fill factor	B-14J	3200	.004			.19	.71	.90	1.07
2850	10 C.Y. bucket	B-14K	4320	.003			.14	.62	.76	.90
2900	8 C.Y. bucket, 75% fill factor	B-14J	3000	.004			.21	.75	.96	1.14
2950	10 C.Y. bucket	B-14K	4050	.003			.15	.66	.81	.96
4000	8 C.Y. bucket, 70% fill factor	B-14J	2800	.004			.22	.81	1.03	1.22
4050	10 C.Y. bucket	B-14K	3780	.003			.16	.71	.87	1.03
4100	8 C.Y. bucket, 65% fill factor	B-14J	2600	.005			.24	.87	1.11	1.32
4150	10 C.Y. bucket	B-14K	3510	.003			.18	.76	.94	1.11
4200	8 C.Y. bucket, 60% fill factor	B-14J	2400	.005			.26	.94	1.20	1.43
4250	10 C.Y. bucket	B-14K	3240	.004			.19	.82	1.01	1.20
4290	Continuous excavation with no truck loading									
4300	Excavator, 4.5 C.Y. bucket, 100% fill factor, no truck loading	B-14A	4000	.003	B.C.Y.		.16	.85	1.01	1.18
4320	6 C.Y. bucket	B-14B	5900	.002			.11	.59	.70	.80
4330	7 C.Y. bucket	B-14C	6900	.002			.09	.50	.59	.69
4340	Shovel, 7 C.Y. bucket	B-14F	8900	.001			.07	.46	.53	.62
4360	12 C.Y. bucket	B-14G	14800	.001			.04	.40	.44	.50
4400	Excavator, 4.5 C.Y. bucket, 95% fill factor	B-14A	3800	.003			.17	.90	1.07	1.24
4420	6 C.Y. bucket	B-14B	5605	.002			.11	.62	.73	.85
4430	7 C.Y. bucket	B-14C	6555	.002			.10	.52	.62	.73
4440	Shovel, 7 C.Y. bucket	B-14F	8455	.001			.08	.49	.57	.64
4460	12 C.Y. bucket	B-14G	14060	.001			.05	.42	.47	.53
4500	Excavator, 4.5 C.Y. bucket, 90% fill factor	B-14A	3600	.003			.18	.95	1.13	1.31
4520	6 C.Y. bucket	B-14B	5310	.002			.12	.65	.77	.90
4530	7 C.Y. bucket	B-14C	6210	.002			.10	.55	.65	.76
4540	Shovel, 7 C.Y. bucket	B-14F	8010	.002			.08	.51	.59	.68
4560	12 C.Y. bucket	B-14G	13320	.001			.05	.45	.50	.56
4600	Excavator, 4.5 C.Y. bucket, 85% fill factor	B-14A	3400	.004			.19	1	1.19	1.38
4620	6 C.Y. bucket	B-14B	5015	.002			.13	.69	.82	.95
4630	7 C.Y. bucket	B-14C	5865	.002			.11	.59	.70	.81
4640	Shovel, 7 C.Y. bucket	B-14F	7565	.002			.09	.54	.63	.73
4660	12 C.Y. bucket	B-14G	12580	.001			.05	.47	.52	.60
4700	Excavator, 4.5 C.Y. bucket, 80% fill factor	B-14A	3200	.004			.20	1.07	1.27	1.47
4720	6 C.Y. bucket	B-14B	4720	.003			.14	.73	.87	1.01
4730	7 C.Y. bucket	B-14C	5520	.002			.12	.62	.74	.85
4740	Shovel, 7 C.Y. bucket	B-14F	7120	.002			.09	.58	.67	.78
4760	12 C.Y. bucket	B-14G	11840	.001			.05	.50	.55	.63
5290	Excavation with truck loading									
5300	Excavator, 4.5 C.Y. bucket, 100% fill factor, with truck loading	B-14A	3400	.004	B.C.Y.		.19	1	1.19	1.38
5320	6 C.Y. bucket	B-14B	5000	.002			.13	.69	.82	.95
5330	7 C.Y. bucket	B-14C	5800	.002			.11	.59	.70	.82
5340	Shovel, 7 C.Y. bucket	B-14F	7500	.002			.09	.55	.64	.73
5360	12 C.Y. bucket	B-14G	12500	.001			.05	.48	.53	.60
5400	Excavator, 4.5 C.Y. bucket, 95% fill factor	B-14A	3230	.004			.20	1.06	1.26	1.46
5420	6 C.Y. bucket	B-14B	4750	.003			.14	.73	.87	1
5430	7 C.Y. bucket	B-14C	5510	.002			.12	.62	.74	.86
5440	Shovel, 7 C.Y. bucket	B-14F	7125	.002			.09	.57	.66	.76
5460	12 C.Y. bucket	B-14G	11875	.001			.05	.50	.55	.63
5500	Excavator, 4.5 C.Y. bucket, 90% fill factor	B-14A	3060	.004			.21	1.11	1.32	1.54

303

31 23 16 – Excavation

31 23 16.43 Excavating, Large Volume Projects		Crew	Daily Output	Labor-Hours	Unit	Material	2020 Bare Costs Labor	Equipment	Total	Total Incl O&P
5520	6 C.Y. bucket	B-14B	4500	.003	B.C.Y.		.14	.77	.91	1.06
5530	7 C.Y. bucket	B-14C	5220	.002			.12	.66	.78	.90
5540	Shovel, 7 C.Y. bucket	B-14F	6750	.002			.10	.61	.71	.81
5560	12 C.Y. bucket	B-14G	11250	.001			.06	.53	.59	.67
5600	Excavator, 4.5 C.Y. bucket, 85% fill factor	B-14A	2890	.004			.22	1.18	1.40	1.63
5620	6 C.Y. bucket	B-14B	4250	.003			.15	.81	.96	1.13
5630	7 C.Y. bucket	B-14C	4930	.002			.13	.70	.83	.96
5640	Shovel, 7 C.Y. bucket	B-14F	6375	.002			.10	.64	.74	.86
5660	12 C.Y. bucket	B-14G	10625	.001			.06	.56	.62	.71
5700	Excavator, 4.5 C.Y. bucket, 80% fill factor	B-14A	2720	.004			.24	1.25	1.49	1.73
5720	6 C.Y. bucket	B-14B	4000	.003			.16	.87	1.03	1.19
5740	Shovel, 7 C.Y. bucket	B-14F	6000	.002			.11	.68	.79	.91
5760	12 C.Y. bucket	B-14G	10000	.001			.06	.60	.66	.75
6000	Self propelled scraper, 1/4 push dozer, average productivity									
6100	31 C.Y., 1,500' haul	B-33K	2046	.007	B.C.Y.		.36	2.13	2.49	2.88
6200	44 C.Y.	B-33H	2640	.005	"		.28	2.02	2.30	2.64

31 23 16.46 Excavating, Bulk, Dozer

		Crew	Daily Output	Labor-Hours	Unit	Material	2020 Bare Costs Labor	Equipment	Total	Total Incl O&P
0010	**EXCAVATING, BULK, DOZER**									
0011	Open site									
2000	80 HP, 50' haul, sand & gravel	B-10L	460	.026	B.C.Y.		1.35	.87	2.22	2.99
2010	Sandy clay & loam		440	.027			1.41	.91	2.32	3.12
2020	Common earth		400	.030			1.56	1	2.56	3.43
2040	Clay		250	.048			2.49	1.61	4.10	5.50
2200	150' haul, sand & gravel		230	.052			2.71	1.75	4.46	5.95
2210	Sandy clay & loam		220	.055			2.83	1.82	4.65	6.25
2220	Common earth		200	.060			3.11	2.01	5.12	6.85
2240	Clay		125	.096			4.98	3.21	8.19	11
2400	300' haul, sand & gravel		120	.100			5.20	3.35	8.55	11.45
2410	Sandy clay & loam		115	.104			5.40	3.49	8.89	11.95
2420	Common earth		100	.120			6.20	4.01	10.21	13.70
2440	Clay		65	.185			9.60	6.20	15.80	21
3000	105 HP, 50' haul, sand & gravel	B-10W	700	.017			.89	.91	1.80	2.33
3010	Sandy clay & loam		680	.018			.92	.93	1.85	2.40
3020	Common earth		610	.020			1.02	1.04	2.06	2.67
3040	Clay		385	.031			1.62	1.65	3.27	4.23
3200	150' haul, sand & gravel		310	.039			2.01	2.04	4.05	5.25
3210	Sandy clay & loam		300	.040			2.07	2.11	4.18	5.45
3220	Common earth		270	.044			2.31	2.35	4.66	6.05
3240	Clay		170	.071			3.66	3.73	7.39	9.60
3300	300' haul, sand & gravel		140	.086			4.45	4.53	8.98	11.65
3310	Sandy clay & loam		135	.089			4.61	4.70	9.31	12.05
3320	Common earth		120	.100			5.20	5.30	10.50	13.55
3340	Clay		100	.120			6.20	6.35	12.55	16.25
4000	200 HP, 50' haul, sand & gravel	B-10B	1400	.009			.44	1.07	1.51	1.85
4010	Sandy clay & loam		1360	.009			.46	1.11	1.57	1.90
4020	Common earth		1230	.010			.51	1.22	1.73	2.11
4040	Clay		770	.016			.81	1.95	2.76	3.36
4200	150' haul, sand & gravel		595	.020			1.05	2.53	3.58	4.35
4210	Sandy clay & loam		580	.021			1.07	2.59	3.66	4.46
4220	Common earth		516	.023			1.21	2.92	4.13	5
4240	Clay		325	.037			1.92	4.63	6.55	7.95
4400	300' haul, sand & gravel		310	.039			2.01	4.85	6.86	8.35

31 23 16.46 Excavating, Bulk, Dozer

		Crew	Daily Output	Labor-Hours	Unit	Material	2020 Bare Costs Labor	2020 Bare Costs Equipment	Total	Total Incl O&P
4410	Sandy clay & loam	B-10B	300	.040	B.C.Y.		2.07	5	7.07	8.60
4420	Common earth		270	.044			2.31	5.55	7.86	9.60
4440	Clay		170	.071			3.66	8.85	12.51	15.25
5000	300 HP, 50' haul, sand & gravel	B-10M	1900	.006			.33	.93	1.26	1.51
5010	Sandy clay & loam		1850	.006			.34	.95	1.29	1.55
5020	Common earth		1650	.007			.38	1.07	1.45	1.74
5040	Clay		1025	.012			.61	1.72	2.33	2.80
5200	150' haul, sand & gravel		920	.013			.68	1.92	2.60	3.12
5210	Sandy clay & loam		895	.013			.70	1.97	2.67	3.21
5220	Common earth		800	.015			.78	2.21	2.99	3.59
5240	Clay		500	.024			1.24	3.53	4.77	5.75
5400	300' haul, sand & gravel		470	.026			1.32	3.75	5.07	6.10
5410	Sandy clay & loam		455	.026			1.37	3.88	5.25	6.30
5420	Common earth		410	.029			1.52	4.30	5.82	7
5440	Clay		250	.048			2.49	7.05	9.54	11.50
5500	460 HP, 50' haul, sand & gravel	B-10X	1930	.006			.32	1.44	1.76	2.06
5506	Sandy clay & loam		1880	.006			.33	1.48	1.81	2.12
5510	Common earth		1680	.007			.37	1.65	2.02	2.37
5520	Clay		1050	.011			.59	2.65	3.24	3.80
5530	150' haul, sand & gravel		1290	.009			.48	2.15	2.63	3.09
5535	Sandy clay & loam		1250	.010			.50	2.22	2.72	3.19
5540	Common earth		1120	.011			.56	2.48	3.04	3.56
5550	Clay		700	.017			.89	3.97	4.86	5.70
5560	300' haul, sand & gravel		660	.018			.94	4.21	5.15	6.05
5565	Sandy clay & loam		640	.019			.97	4.34	5.31	6.25
5570	Common earth		575	.021			1.08	4.83	5.91	6.90
5580	Clay		350	.034			1.78	7.95	9.73	11.40
6000	700 HP, 50' haul, sand & gravel	B-10V	3500	.003			.18	1.46	1.64	1.88
6006	Sandy clay & loam		3400	.004			.18	1.51	1.69	1.93
6010	Common earth		3035	.004			.20	1.69	1.89	2.16
6020	Clay		1925	.006			.32	2.66	2.98	3.40
6030	150' haul, sand & gravel		2025	.006			.31	2.53	2.84	3.24
6035	Sandy clay & loam		1960	.006			.32	2.61	2.93	3.35
6040	Common earth		1750	.007			.36	2.93	3.29	3.75
6050	Clay		1100	.011			.57	4.66	5.23	5.95
6060	300' haul, sand & gravel		1030	.012			.60	4.97	5.57	6.35
6065	Sandy clay & loam		1005	.012			.62	5.10	5.72	6.55
6070	Common earth		900	.013			.69	5.70	6.39	7.30
6080	Clay		550	.022			1.13	9.30	10.43	11.95
6090	For dozer with ripper, see Section 31 23 16.32									

31 23 16.48 Excavation, Bulk, Dragline

		Crew	Daily Output	Labor-Hours	Unit	Material	2020 Bare Costs Labor	2020 Bare Costs Equipment	Total	Total Incl O&P
0010	**EXCAVATION, BULK, DRAGLINE**									
0011	Excavate and load on truck, bank measure									
0012	Bucket drag line, 3/4 C.Y., sand/gravel	B-12I	440	.036	B.C.Y.		1.84	2.32	4.16	5.30
0100	Light clay		310	.052			2.61	3.30	5.91	7.55
0110	Heavy clay		280	.057			2.89	3.65	6.54	8.35
0120	Unclassified soil		250	.064			3.24	4.09	7.33	9.35
0200	1-1/2 C.Y. bucket, sand/gravel	B-12P	575	.028			1.41	2.21	3.62	4.54
0210	Light clay		440	.036			1.84	2.89	4.73	5.95
0220	Heavy clay		352	.045			2.30	3.61	5.91	7.40
0230	Unclassified soil		300	.053			2.70	4.24	6.94	8.70
0300	3 C.Y., sand/gravel	B-12V	720	.022			1.13	2.88	4.01	4.86

For customer support on your Site Work & Landscape Costs with RSMeans data, call 800.448.8182.

305

31 23 Excavation and Fill

31 23 16 – Excavation

31 23 16.48 Excavation, Bulk, Dragline		Crew	Daily Output	Labor-Hours	Unit	Material	2020 Bare Costs Labor	Equipment	Total	Total Incl O&P
0310	Light clay	B-12V	700	.023	B.C.Y.		1.16	2.96	4.12	4.99
0320	Heavy clay		600	.027			1.35	3.46	4.81	5.80
0330	Unclassified soil		550	.029			1.47	3.77	5.24	6.35

31 23 16.50 Excavation, Bulk, Scrapers

31 23 16.50		Crew	Daily Output	Labor-Hours	Unit	Material	Labor	Equipment	Total	Total Incl O&P
0010	**EXCAVATION, BULK, SCRAPERS** R312316-40									
0100	Elev. scraper 11 C.Y., sand & gravel 1,500' haul, 1/4 dozer	B-33F	690	.020	B.C.Y.		1.07	2.28	3.35	4.11
0150	3,000' haul R312316-45		610	.023			1.21	2.58	3.79	4.65
0200	5,000' haul		505	.028			1.46	3.12	4.58	5.60
0300	Common earth, 1,500' haul		600	.023			1.23	2.62	3.85	4.73
0350	3,000' haul		530	.026			1.39	2.97	4.36	5.35
0400	5,000' haul		440	.032			1.67	3.58	5.25	6.45
0410	Sandy clay & loam, 1,500' haul		648	.022			1.14	2.43	3.57	4.37
0420	3,000' haul		572	.024			1.29	2.75	4.04	4.96
0430	5,000' haul		475	.029			1.55	3.31	4.86	5.95
0500	Clay, 1,500' haul		375	.037			1.96	4.20	6.16	7.55
0550	3,000' haul		330	.042			2.23	4.77	7	8.60
0600	5,000' haul		275	.051			2.68	5.70	8.38	10.30
1000	Self propelled scraper, 14 C.Y., 1/4 push dozer									
1050	Sand and gravel, 1,500' haul	B-33D	920	.015	B.C.Y.		.80	3.08	3.88	4.59
1100	3,000' haul		805	.017			.91	3.52	4.43	5.25
1200	5,000' haul		645	.022			1.14	4.40	5.54	6.55
1300	Common earth, 1,500' haul		800	.018			.92	3.55	4.47	5.30
1350	3,000' haul		700	.020			1.05	4.05	5.10	6.05
1400	5,000' haul		560	.025			1.31	5.05	6.36	7.50
1420	Sandy clay & loam, 1,500' haul		864	.016			.85	3.28	4.13	4.88
1430	3,000' haul		786	.018			.94	3.61	4.55	5.35
1440	5,000' haul		605	.023			1.22	4.69	5.91	6.95
1500	Clay, 1,500' haul		500	.028			1.47	5.65	7.12	8.45
1550	3,000' haul		440	.032			1.67	6.45	8.12	9.60
1600	5,000' haul		350	.040			2.10	8.10	10.20	12.05
2000	21 C.Y., 1/4 push dozer, sand & gravel, 1,500' haul	B-33E	1180	.012			.62	2.60	3.22	3.79
2100	3,000' haul		910	.015			.81	3.37	4.18	4.92
2200	5,000' haul		750	.019			.98	4.09	5.07	5.95
2300	Common earth, 1,500' haul		1030	.014			.71	2.98	3.69	4.35
2350	3,000' haul		790	.018			.93	3.88	4.81	5.65
2400	5,000' haul		650	.022			1.13	4.72	5.85	6.90
2420	Sandy clay & loam, 1,500' haul		1112	.013			.66	2.76	3.42	4.03
2430	3,000' haul		854	.016			.86	3.59	4.45	5.25
2440	5,000' haul		702	.020			1.05	4.37	5.42	6.40
2500	Clay, 1,500' haul		645	.022			1.14	4.76	5.90	6.95
2550	3,000' haul		495	.028			1.49	6.20	7.69	9.05
2600	5,000' haul		405	.035			1.82	7.60	9.42	11.05
2700	Towed, 10 C.Y., 1/4 push dozer, sand & gravel, 1,500' haul	B-33B	560	.025			1.31	4.22	5.53	6.60
2720	3,000' haul		450	.031			1.64	5.25	6.89	8.25
2730	5,000' haul		365	.038			2.02	6.50	8.52	10.15
2750	Common earth, 1,500' haul		420	.033			1.75	5.65	7.40	8.80
2770	3,000' haul		400	.035			1.84	5.90	7.74	9.25
2780	5,000' haul		310	.045			2.37	7.65	10.02	11.95
2785	Sandy clay & loam, 1,500' haul		454	.031			1.62	5.20	6.82	8.20
2790	3,000' haul		432	.032			1.70	5.45	7.15	8.55
2795	5,000' haul		340	.041			2.16	6.95	9.11	10.90
2800	Clay, 1,500' haul		315	.044			2.34	7.50	9.84	11.75

For customer support on your Site Work & Landscape Costs with RSMeans data, call 800.448.8182.

31 23 Excavation and Fill

31 23 16 – Excavation

31 23 16.50 Excavation, Bulk, Scrapers		Crew	Daily Output	Labor-Hours	Unit	Material	2020 Bare Costs Labor	2020 Bare Costs Equipment	Total	Total Incl O&P
2820	3,000' haul	B-33B	300	.047	B.C.Y.		2.45	7.90	10.35	12.30
2840	5,000' haul	↓	225	.062			3.27	10.50	13.77	16.45
2900	15 C.Y., 1/4 push dozer, sand & gravel, 1,500' haul	B-33C	800	.018			.92	2.98	3.90	4.66
2920	3,000' haul		640	.022			1.15	3.72	4.87	5.80
2940	5,000' haul		520	.027			1.41	4.58	5.99	7.15
2960	Common earth, 1,500' haul		600	.023			1.23	3.97	5.20	6.20
2980	3,000' haul		560	.025			1.31	4.25	5.56	6.65
3000	5,000' haul		440	.032			1.67	5.40	7.07	8.45
3005	Sandy clay & loam, 1,500' haul		648	.022			1.14	3.67	4.81	5.75
3010	3,000' haul		605	.023			1.22	3.94	5.16	6.15
3015	5,000' haul		475	.029			1.55	5	6.55	7.80
3020	Clay, 1,500' haul		450	.031			1.64	5.30	6.94	8.25
3040	3,000' haul		420	.033			1.75	5.65	7.40	8.85
3060	5,000' haul	↓	320	.044	↓		2.30	7.45	9.75	11.65

31 23 19 – Dewatering

31 23 19.10 Cut Drainage Ditch

		Crew	Daily Output	Labor-Hours	Unit	Material	2020 Bare Costs Labor	2020 Bare Costs Equipment	Total	Total Incl O&P
0010	**CUT DRAINAGE DITCH**									
0020	Cut drainage ditch common earth, 30' wide x 1' deep	B-11L	6000	.003	L.F.		.13	.18	.31	.39
0200	Clay and till		4200	.004			.19	.25	.44	.56
0250	Clean wet drainage ditch, 30' wide	↓	10000	.002	↓		.08	.11	.19	.24

31 23 19.20 Dewatering Systems

		Crew	Daily Output	Labor-Hours	Unit	Material	2020 Bare Costs Labor	2020 Bare Costs Equipment	Total	Total Incl O&P
0010	**DEWATERING SYSTEMS**									
0020	Excavate drainage trench, 2' wide, 2' deep	B-11C	90	.178	C.Y.		8.80	2.38	11.18	15.75
0100	2' wide, 3' deep, with backhoe loader	"	135	.119			5.85	1.58	7.43	10.55
0200	Excavate sump pits by hand, light soil	1 Clab	7.10	1.127			47.50		47.50	71.50
0300	Heavy soil	"	3.50	2.286	↓		96		96	145
0500	Pumping 8 hrs., attended 2 hrs./day, incl. 20 L.F.									
0550	of suction hose & 100 L.F. discharge hose									
0600	2" diaphragm pump used for 8 hrs.	B-10H	4	3	Day		156	24.50	180.50	260
0620	Add per additional pump							72	72	79
0650	4" diaphragm pump used for 8 hrs.	B-10I	4	3			156	37	193	274
0670	Add per additional pump							116	116	127
0800	8 hrs. attended, 2" diaphragm pump	B-10H	1	12			620	98.50	718.50	1,050
0820	Add per additional pump							72	72	79
0900	3" centrifugal pump	B-10J	1	12			620	92	712	1,025
0920	Add per additional pump							79	79	86.50
1000	4" diaphragm pump	B-10I	1	12			620	148	768	1,100
1020	Add per additional pump							116	116	127
1100	6" centrifugal pump	B-10K	1	12			620	294	914	1,250
1120	Add per additional pump				↓			340	340	375
1300	CMP, incl. excavation 3' deep, 12" diameter	B-6	115	.209	L.F.	12.40	9.55	1.86	23.81	30
1400	18" diameter		100	.240	"	18.80	11	2.14	31.94	39.50
1600	Sump hole construction, incl. excavation and gravel, pit		1250	.019	C.F.	1.10	.88	.17	2.15	2.72
1700	With 12" gravel collar, 12" pipe, corrugated, 16 ga.		70	.343	L.F.	23	15.70	3.05	41.75	52.50
1800	15" pipe, corrugated, 16 ga.		55	.436		30	19.95	3.89	53.84	67.50
1900	18" pipe, corrugated, 16 ga.		50	.480		35	22	4.28	61.28	76
2000	24" pipe, corrugated, 14 ga.		40	.600	↓	42	27.50	5.35	74.85	93
2200	Wood lining, up to 4' x 4', add	↓	300	.080	SFCA	16.40	3.66	.71	20.77	24.50
9950	See Section 31 23 19.40 for wellpoints									
9960	See Section 31 23 19.30 for deep well systems									
9970	See Section 22 11 23 for pumps									

31 23 19 – Dewatering

31 23 19.30 Wells

		Crew	Daily Output	Labor-Hours	Unit	Material	2020 Bare Costs Labor	2020 Bare Costs Equipment	Total	Total Incl O&P
0010	**WELLS**									
0011	For dewatering 10' to 20' deep, 2' diameter									
0020	with steel casing, minimum	B-6	165	.145	V.L.F.	37.50	6.65	1.30	45.45	53
0050	Average		98	.245		44.50	11.20	2.18	57.88	68
0100	Maximum	↓	49	.490	↓	48	22.50	4.36	74.86	91
0300	For dewatering pumps see 01 54 33 in Reference Section									
0500	For domestic water wells, see Section 33 21 13.10									

31 23 19.40 Wellpoints

		Crew	Daily Output	Labor-Hours	Unit	Material	2020 Bare Costs Labor	2020 Bare Costs Equipment	Total	Total Incl O&P
0010	**WELLPOINTS** R312319-90									
0011	For equipment rental, see 01 54 33 in Reference Section									
0100	Installation and removal of single stage system									
0110	Labor only, 0.75 labor-hours per L.F.	1 Clab	10.70	.748	LF Hdr		31.50		31.50	47.50
0200	2.0 labor-hours per L.F.	"	4	2	"		84		84	127
0400	Pump operation, 4 @ 6 hr. shifts									
0410	Per 24 hr. day	4 Eqlt	1.27	25.197	Day		1,325		1,325	2,000
0500	Per 168 hr. week, 160 hr. straight, 8 hr. double time		.18	178	Week		9,425		9,425	14,100
0550	Per 4.3 week month	↓	.04	800	Month		42,400		42,400	63,500
0600	Complete installation, operation, equipment rental, fuel &									
0610	removal of system with 2" wellpoints 5' OC									
0700	100' long header, 6" diameter, first month	4 Eqlt	3.23	9.907	LF Hdr	160	525		685	960
0800	Thereafter, per month		4.13	7.748		128	410		538	755
1000	200' long header, 8" diameter, first month		6	5.333		154	283		437	590
1100	Thereafter, per month		8.39	3.814		72	202		274	380
1300	500' long header, 8" diameter, first month		10.63	3.010		56	160		216	300
1400	Thereafter, per month		20.91	1.530		40	81		121	165
1600	1,000' long header, 10" diameter, first month		11.62	2.754		48	146		194	271
1700	Thereafter, per month	↓	41.81	.765	↓	24	40.50		64.50	87
1900	Note: above figures include pumping 168 hrs. per week,									
1910	the pump operator, and one stand-by pump.									

31 23 23 – Fill

31 23 23.13 Backfill

		Crew	Daily Output	Labor-Hours	Unit	Material	2020 Bare Costs Labor	2020 Bare Costs Equipment	Total	Total Incl O&P
0010	**BACKFILL** R312323-30									
0015	By hand, no compaction, light soil	1 Clab	14	.571	L.C.Y.		24		24	36
0100	Heavy soil		11	.727	"		30.50		30.50	46
0300	Compaction in 6" layers, hand tamp, add to above	↓	20.60	.388	E.C.Y.		16.35		16.35	24.50
0400	Roller compaction operator walking, add	B-10A	100	.120			6.20	1.66	7.86	11.15
0500	Air tamp, add	B-9D	190	.211			8.95	1.67	10.62	15.30
0600	Vibrating plate, add	A-1D	60	.133			5.60	.53	6.13	9.05
0800	Compaction in 12" layers, hand tamp, add to above	1 Clab	34	.235			9.90		9.90	14.90
0900	Roller compaction operator walking, add	B-10A	150	.080			4.15	1.11	5.26	7.40
1000	Air tamp, add	B-9	285	.140			5.95	1.25	7.20	10.35
1100	Vibrating plate, add	A-1E	90	.089	↓		3.74	1.84	5.58	7.60

31 23 23.14 Backfill, Structural

		Crew	Daily Output	Labor-Hours	Unit	Material	2020 Bare Costs Labor	2020 Bare Costs Equipment	Total	Total Incl O&P
0010	**BACKFILL, STRUCTURAL**									
0011	Dozer or F.E. loader									
0020	From existing stockpile, no compaction									
1000	55 HP wheeled loader, 50' haul, common earth	B-11C	200	.080	L.C.Y.		3.95	1.07	5.02	7.10
2000	80 HP, 50' haul, sand & gravel	B-10L	1100	.011			.57	.36	.93	1.25
2010	Sandy clay & loam		1070	.011			.58	.38	.96	1.28
2020	Common earth		975	.012			.64	.41	1.05	1.41
2040	Clay	↓	850	.014			.73	.47	1.20	1.62

31 23 23 – Fill

31 23 23.14 Backfill, Structural	Crew	Daily Output	Labor-Hours	Unit	Material	2020 Bare Costs		Total	Total Incl O&P
						Labor	Equipment		
2200 150' haul, sand & gravel	B-10L	550	.022	L.C.Y.		1.13	.73	1.86	2.49
2210 Sandy clay & loam		535	.022			1.16	.75	1.91	2.57
2220 Common earth		490	.024			1.27	.82	2.09	2.80
2240 Clay		425	.028			1.46	.94	2.40	3.23
2400 300' haul, sand & gravel		370	.032			1.68	1.08	2.76	3.71
2410 Sandy clay & loam		360	.033			1.73	1.11	2.84	3.82
2420 Common earth		330	.036			1.89	1.22	3.11	4.16
2440 Clay		290	.041			2.15	1.38	3.53	4.73
3000 105 HP, 50' haul, sand & gravel	B-10W	1350	.009			.46	.47	.93	1.21
3010 Sandy clay & loam		1325	.009			.47	.48	.95	1.23
3020 Common earth		1225	.010			.51	.52	1.03	1.33
3040 Clay		1100	.011			.57	.58	1.15	1.48
3200 150' haul, sand & gravel		670	.018			.93	.95	1.88	2.43
3210 Sandy clay & loam		655	.018			.95	.97	1.92	2.48
3220 Common earth		610	.020			1.02	1.04	2.06	2.67
3240 Clay		550	.022			1.13	1.15	2.28	2.96
3300 300' haul, sand & gravel		465	.026			1.34	1.36	2.70	3.50
3310 Sandy clay & loam		455	.026			1.37	1.39	2.76	3.58
3320 Common earth		415	.029			1.50	1.53	3.03	3.93
3340 Clay		370	.032			1.68	1.71	3.39	4.40
4000 200 HP, 50' haul, sand & gravel	B-10B	2500	.005			.25	.60	.85	1.03
4010 Sandy clay & loam		2435	.005			.26	.62	.88	1.06
4020 Common earth		2200	.005			.28	.68	.96	1.17
4040 Clay		1950	.006			.32	.77	1.09	1.33
4200 150' haul, sand & gravel		1225	.010			.51	1.23	1.74	2.11
4210 Sandy clay & loam		1200	.010			.52	1.25	1.77	2.16
4220 Common earth		1100	.011			.57	1.37	1.94	2.35
4240 Clay		975	.012			.64	1.54	2.18	2.66
4400 300' haul, sand & gravel		805	.015			.77	1.87	2.64	3.22
4410 Sandy clay & loam		790	.015			.79	1.90	2.69	3.27
4420 Common earth		735	.016			.85	2.05	2.90	3.52
4440 Clay		660	.018			.94	2.28	3.22	3.92
5000 300 HP, 50' haul, sand & gravel	B-10M	3170	.004			.20	.56	.76	.90
5010 Sandy clay & loam		3110	.004			.20	.57	.77	.92
5020 Common earth		2900	.004			.21	.61	.82	.99
5040 Clay		2700	.004			.23	.65	.88	1.06
5200 150' haul, sand & gravel		2200	.005			.28	.80	1.08	1.30
5210 Sandy clay & loam		2150	.006			.29	.82	1.11	1.33
5220 Common earth		1950	.006			.32	.90	1.22	1.47
5240 Clay		1700	.007			.37	1.04	1.41	1.69
5400 300' haul, sand & gravel		1500	.008			.42	1.18	1.60	1.91
5410 Sandy clay & loam		1470	.008			.42	1.20	1.62	1.95
5420 Common earth		1350	.009			.46	1.31	1.77	2.13
5440 Clay		1225	.010			.51	1.44	1.95	2.34
6000 For compaction, see Section 31 23 23.23									
6010 For trench backfill, see Sections 31 23 16.13 and 31 23 16.14									

31 23 23.15 Borrow, Loading and/or Spreading

		Crew	Daily Output	Labor-Hours	Unit	Material	2020 Bare Costs		Total	Total Incl O&P
							Labor	Equipment		
0010	**BORROW, LOADING AND/OR SPREADING**									
4000	Common earth, shovel, 1 C.Y. bucket	B-12N	840	.019	B.C.Y.	17	.96	1.44	19.40	21.50
4010	1-1/2 C.Y. bucket	B-12O	1135	.014		17	.71	1.14	18.85	21
4020	3 C.Y. bucket	B-12T	1800	.009		17	.45	1.18	18.63	20.50
4030	Front end loader, wheel mounted									

31 23 23 – Fill

31 23 23.15 Borrow, Loading and/or Spreading

	Crew	Daily Output	Labor-Hours	Unit	Material	2020 Bare Costs Labor	Equipment	Total	Total Incl O&P
4050	3/4 C.Y. bucket	B-10R	550	.022	B.C.Y.	17	1.13	.55	18.68	21
4060	1-1/2 C.Y. bucket	B-10S	970	.012		17	.64	.44	18.08	20
4070	3 C.Y. bucket	B-10T	1575	.008		17	.40	.45	17.85	19.80
4080	5 C.Y. bucket	B-10U	2600	.005		17	.24	.37	17.61	19.45
5000	Select granular fill, shovel, 1 C.Y. bucket	B-12N	925	.017		24.50	.88	1.30	26.68	29.50
5010	1-1/2 C.Y. bucket	B-12O	1250	.013		24.50	.65	1.03	26.18	29
5020	3 C.Y. bucket	B-12T	1980	.008		24.50	.41	1.07	25.98	29
5030	Front end loader, wheel mounted									
5050	3/4 C.Y. bucket	B-10R	800	.015	B.C.Y.	24.50	.78	.38	25.66	28.50
5060	1-1/2 C.Y. bucket	B-10S	1065	.011		24.50	.58	.40	25.48	28.50
5070	3 C.Y. bucket	B-10T	1735	.007		24.50	.36	.40	25.26	28
5080	5 C.Y. bucket	B-10U	2850	.004		24.50	.22	.34	25.06	27.50
6000	Clay, till, or blasted rock, shovel, 1 C.Y. bucket	B-12N	715	.022		12.60	1.13	1.69	15.42	17.40
6010	1-1/2 C.Y. bucket	B-12O	965	.017		12.60	.84	1.34	14.78	16.60
6020	3 C.Y. bucket	B-12T	1530	.010		12.60	.53	1.38	14.51	16.15
6030	Front end loader, wheel mounted									
6035	3/4 C.Y. bucket	B-10R	465	.026	B.C.Y.	12.60	1.34	.65	14.59	16.55
6040	1-1/2 C.Y. bucket	B-10S	825	.015		12.60	.75	.51	13.86	15.55
6045	3 C.Y. bucket	B-10T	1340	.009		12.60	.46	.52	13.58	15.15
6050	5 C.Y. bucket	B-10U	2200	.005		12.60	.28	.43	13.31	14.75
6060	Front end loader, track mounted									
6065	1-1/2 C.Y. bucket	B-10N	715	.017	B.C.Y.	12.60	.87	.79	14.26	16
6070	3 C.Y. bucket	B-10P	1190	.010		12.60	.52	.98	14.10	15.70
6075	5 C.Y. bucket	B-10Q	1835	.007		12.60	.34	.81	13.75	15.25
7000	Topsoil or loam from stockpile, shovel, 1 C.Y. bucket	B-12N	840	.019		26.50	.96	1.44	28.90	32
7010	1-1/2 C.Y. bucket	B-12O	1135	.014		26.50	.71	1.14	28.35	31.50
7020	3 C.Y. bucket	B-12T	1800	.009		26.50	.45	1.18	28.13	31
7030	Front end loader, wheel mounted									
7050	3/4 C.Y. bucket	B-10R	550	.022	B.C.Y.	26.50	1.13	.55	28.18	31.50
7060	1-1/2 C.Y. bucket	B-10S	970	.012		26.50	.64	.44	27.58	30.50
7070	3 C.Y. bucket	B-10T	1575	.008		26.50	.40	.45	27.35	30
7080	5 C.Y. bucket	B-10U	2600	.005		26.50	.24	.37	27.11	30
9000	Hauling only, excavated or borrow material, see Section 31 23 23.20									

31 23 23.16 Fill By Borrow and Utility Bedding

		Crew	Daily Output	Labor-Hours	Unit	Material	2020 Bare Costs Labor	Equipment	Total	Total Incl O&P
0010	**FILL BY BORROW AND UTILITY BEDDING**									
0049	Utility bedding, for pipe & conduit, not incl. compaction									
0050	Crushed or screened bank run gravel	B-6	150	.160	L.C.Y.	21	7.30	1.43	29.73	35.50
0100	Crushed stone 3/4" to 1/2"		150	.160		27.50	7.30	1.43	36.23	43
0200	Sand, dead or bank		150	.160		18.25	7.30	1.43	26.98	32.50
0500	Compacting bedding in trench	A-1D	90	.089	E.C.Y.		3.74	.35	4.09	6
0600	If material source exceeds 2 miles, add for extra mileage.									
0610	See Section 31 23 23.20 for hauling mileage add.									

31 23 23.17 General Fill

		Crew	Daily Output	Labor-Hours	Unit	Material	2020 Bare Costs Labor	Equipment	Total	Total Incl O&P
0010	**GENERAL FILL**									
0011	Spread dumped material, no compaction									
0020	By dozer	B-10B	1000	.012	L.C.Y.		.62	1.50	2.12	2.58
0100	By hand	1 Clab	12	.667	"		28		28	42
0150	Spread fill, from stockpile with 2-1/2 C.Y. F.E. loader									
0170	130 HP, 300' haul	B-10P	600	.020	L.C.Y.		1.04	1.95	2.99	3.69
0190	With dozer 300 HP, 300' haul	B-10M	600	.020	"		1.04	2.94	3.98	4.78
0400	For compaction of embankment, see Section 31 23 23.23									
0500	Gravel fill, compacted, under floor slabs, 4" deep	B-37	10000	.005	S.F.	.49	.21	.03	.73	.89

31 23 23 – Fill

		Crew	Daily Output	Labor-Hours	Unit	Material	2020 Bare Costs Labor	Equipment	Total	Total Incl O&P
31 23 23.17 General Fill										
0600	6" deep	B-37	8600	.006	S.F.	.74	.25	.03	1.02	1.21
0700	9" deep		7200	.007		1.23	.30	.04	1.57	1.83
0800	12" deep		6000	.008	↓	1.72	.35	.04	2.11	2.47
1000	Alternate pricing method, 4" deep		120	.400	E.C.Y.	37	17.70	2.13	56.83	69.50
1100	6" deep		160	.300		37	13.30	1.60	51.90	62
1200	9" deep		200	.240		37	10.60	1.28	48.88	58
1300	12" deep	↓	220	.218	↓	37	9.65	1.16	47.81	56.50
1500	For fill under exterior paving, see Section 32 11 23.23									
1600	For flowable fill, see Section 03 31 13.35									
31 23 23.20 Hauling										
0010	**HAULING**									
0011	Excavated or borrow, loose cubic yards									
0012	no loading equipment, including hauling, waiting, loading/dumping									
0013	time per cycle (wait, load, travel, unload or dump & return)									
0014	8 C.Y. truck, 15 MPH avg., cycle 0.5 miles, 10 min. wait/ld./uld.	B-34A	320	.025	L.C.Y.		1.22	1.34	2.56	3.30
0016	cycle 1 mile		272	.029			1.44	1.57	3.01	3.89
0018	cycle 2 miles		208	.038			1.88	2.06	3.94	5.10
0020	cycle 4 miles		144	.056			2.72	2.97	5.69	7.35
0022	cycle 6 miles		112	.071			3.50	3.82	7.32	9.45
0024	cycle 8 miles		88	.091			4.45	4.86	9.31	12
0026	20 MPH avg., cycle 0.5 mile		336	.024			1.17	1.27	2.44	3.15
0028	cycle 1 mile		296	.027			1.32	1.45	2.77	3.57
0030	cycle 2 miles		240	.033			1.63	1.78	3.41	4.40
0032	cycle 4 miles		176	.045			2.22	2.43	4.65	6
0034	cycle 6 miles		136	.059			2.88	3.15	6.03	7.75
0036	cycle 8 miles		112	.071			3.50	3.82	7.32	9.45
0044	25 MPH avg., cycle 4 miles		192	.042			2.04	2.23	4.27	5.50
0046	cycle 6 miles		160	.050			2.45	2.68	5.13	6.60
0048	cycle 8 miles		128	.063			3.06	3.34	6.40	8.25
0050	30 MPH avg., cycle 4 miles		216	.037			1.81	1.98	3.79	4.90
0052	cycle 6 miles		176	.045			2.22	2.43	4.65	6
0054	cycle 8 miles		144	.056			2.72	2.97	5.69	7.35
0114	15 MPH avg., cycle 0.5 mile, 15 min. wait/ld./uld.		224	.036			1.75	1.91	3.66	4.72
0116	cycle 1 mile		200	.040			1.96	2.14	4.10	5.30
0118	cycle 2 miles		168	.048			2.33	2.55	4.88	6.30
0120	cycle 4 miles		120	.067			3.26	3.57	6.83	8.80
0122	cycle 6 miles		96	.083			4.08	4.46	8.54	11
0124	cycle 8 miles		80	.100			4.90	5.35	10.25	13.25
0126	20 MPH avg., cycle 0.5 mile		232	.034			1.69	1.85	3.54	4.56
0128	cycle 1 mile		208	.038			1.88	2.06	3.94	5.10
0130	cycle 2 miles		184	.043			2.13	2.33	4.46	5.75
0132	cycle 4 miles		144	.056			2.72	2.97	5.69	7.35
0134	cycle 6 miles		112	.071			3.50	3.82	7.32	9.45
0136	cycle 8 miles		96	.083			4.08	4.46	8.54	11
0144	25 MPH avg., cycle 4 miles		152	.053			2.58	2.82	5.40	6.95
0146	cycle 6 miles		128	.063			3.06	3.34	6.40	8.25
0148	cycle 8 miles		112	.071			3.50	3.82	7.32	9.45
0150	30 MPH avg., cycle 4 miles		168	.048			2.33	2.55	4.88	6.30
0152	cycle 6 miles		144	.056			2.72	2.97	5.69	7.35
0154	cycle 8 miles		120	.067			3.26	3.57	6.83	8.80
0214	15 MPH avg., cycle 0.5 mile, 20 min. wait/ld./uld.		176	.045			2.22	2.43	4.65	6
0216	cycle 1 mile		160	.050			2.45	2.68	5.13	6.60

311

31 23 23.20 Hauling		Crew	Daily Output	Labor-Hours	Unit	Material	2020 Bare Costs		Total	Total Incl O&P
							Labor	Equipment		
0218	cycle 2 miles	B-34A	136	.059	L.C.Y.		2.88	3.15	6.03	7.75
0220	cycle 4 miles		104	.077			3.77	4.12	7.89	10.20
0222	cycle 6 miles		88	.091			4.45	4.86	9.31	12
0224	cycle 8 miles		72	.111			5.45	5.95	11.40	14.70
0226	20 MPH avg., cycle 0.5 mile		176	.045			2.22	2.43	4.65	6
0228	cycle 1 mile		168	.048			2.33	2.55	4.88	6.30
0230	cycle 2 miles		144	.056			2.72	2.97	5.69	7.35
0232	cycle 4 miles		120	.067			3.26	3.57	6.83	8.80
0234	cycle 6 miles		96	.083			4.08	4.46	8.54	11
0236	cycle 8 miles		88	.091			4.45	4.86	9.31	12
0244	25 MPH avg., cycle 4 miles		128	.063			3.06	3.34	6.40	8.25
0246	cycle 6 miles		112	.071			3.50	3.82	7.32	9.45
0248	cycle 8 miles		96	.083			4.08	4.46	8.54	11
0250	30 MPH avg., cycle 4 miles		136	.059			2.88	3.15	6.03	7.75
0252	cycle 6 miles		120	.067			3.26	3.57	6.83	8.80
0254	cycle 8 miles		104	.077			3.77	4.12	7.89	10.20
0314	15 MPH avg., cycle 0.5 mile, 25 min. wait/ld./uld.		144	.056			2.72	2.97	5.69	7.35
0316	cycle 1 mile		128	.063			3.06	3.34	6.40	8.25
0318	cycle 2 miles		112	.071			3.50	3.82	7.32	9.45
0320	cycle 4 miles		96	.083			4.08	4.46	8.54	11
0322	cycle 6 miles		80	.100			4.90	5.35	10.25	13.25
0324	cycle 8 miles		64	.125			6.10	6.70	12.80	16.50
0326	20 MPH avg., cycle 0.5 mile		144	.056			2.72	2.97	5.69	7.35
0328	cycle 1 mile		136	.059			2.88	3.15	6.03	7.75
0330	cycle 2 miles		120	.067			3.26	3.57	6.83	8.80
0332	cycle 4 miles		104	.077			3.77	4.12	7.89	10.20
0334	cycle 6 miles		88	.091			4.45	4.86	9.31	12
0336	cycle 8 miles		80	.100			4.90	5.35	10.25	13.25
0344	25 MPH avg., cycle 4 miles		112	.071			3.50	3.82	7.32	9.45
0346	cycle 6 miles		96	.083			4.08	4.46	8.54	11
0348	cycle 8 miles		88	.091			4.45	4.86	9.31	12
0350	30 MPH avg., cycle 4 miles		112	.071			3.50	3.82	7.32	9.45
0352	cycle 6 miles		104	.077			3.77	4.12	7.89	10.20
0354	cycle 8 miles		96	.083			4.08	4.46	8.54	11
0414	15 MPH avg., cycle 0.5 mile, 30 min. wait/ld./uld.		120	.067			3.26	3.57	6.83	8.80
0416	cycle 1 mile		112	.071			3.50	3.82	7.32	9.45
0418	cycle 2 miles		96	.083			4.08	4.46	8.54	11
0420	cycle 4 miles		80	.100			4.90	5.35	10.25	13.25
0422	cycle 6 miles		72	.111			5.45	5.95	11.40	14.70
0424	cycle 8 miles		64	.125			6.10	6.70	12.80	16.50
0426	20 MPH avg., cycle 0.5 mile		120	.067			3.26	3.57	6.83	8.80
0428	cycle 1 mile		112	.071			3.50	3.82	7.32	9.45
0430	cycle 2 miles		104	.077			3.77	4.12	7.89	10.20
0432	cycle 4 miles		88	.091			4.45	4.86	9.31	12
0434	cycle 6 miles		80	.100			4.90	5.35	10.25	13.25
0436	cycle 8 miles		72	.111			5.45	5.95	11.40	14.70
0444	25 MPH avg., cycle 4 miles		96	.083			4.08	4.46	8.54	11
0446	cycle 6 miles		88	.091			4.45	4.86	9.31	12
0448	cycle 8 miles		80	.100			4.90	5.35	10.25	13.25
0450	30 MPH avg., cycle 4 miles		96	.083			4.08	4.46	8.54	11
0452	cycle 6 miles		88	.091			4.45	4.86	9.31	12
0454	cycle 8 miles		80	.100			4.90	5.35	10.25	13.25
0514	15 MPH avg., cycle 0.5 mile, 35 min. wait/ld./uld.		104	.077			3.77	4.12	7.89	10.20

31 23 23 – Fill

31 23 23.20 Hauling		Crew	Daily Output	Labor-Hours	Unit	Material	2020 Bare Costs Labor	2020 Bare Costs Equipment	Total	Total Incl O&P
0516	cycle 1 mile	B-34A	96	.083	L.C.Y.		4.08	4.46	8.54	11
0518	cycle 2 miles		88	.091			4.45	4.86	9.31	12
0520	cycle 4 miles		72	.111			5.45	5.95	11.40	14.70
0522	cycle 6 miles		64	.125			6.10	6.70	12.80	16.50
0524	cycle 8 miles		56	.143			7	7.65	14.65	18.90
0526	20 MPH avg., cycle 0.5 mile		104	.077			3.77	4.12	7.89	10.20
0528	cycle 1 mile		96	.083			4.08	4.46	8.54	11
0530	cycle 2 miles		96	.083			4.08	4.46	8.54	11
0532	cycle 4 miles		80	.100			4.90	5.35	10.25	13.25
0534	cycle 6 miles		72	.111			5.45	5.95	11.40	14.70
0536	cycle 8 miles		64	.125			6.10	6.70	12.80	16.50
0544	25 MPH avg., cycle 4 miles		88	.091			4.45	4.86	9.31	12
0546	cycle 6 miles		80	.100			4.90	5.35	10.25	13.25
0548	cycle 8 miles		72	.111			5.45	5.95	11.40	14.70
0550	30 MPH avg., cycle 4 miles		88	.091			4.45	4.86	9.31	12
0552	cycle 6 miles		80	.100			4.90	5.35	10.25	13.25
0554	cycle 8 miles		72	.111			5.45	5.95	11.40	14.70
1014	12 C.Y. truck, cycle 0.5 mile, 15 MPH avg., 15 min. wait/ld./uld.	B-34B	336	.024			1.17	1.71	2.88	3.63
1016	cycle 1 mile		300	.027			1.31	1.91	3.22	4.06
1018	cycle 2 miles		252	.032			1.55	2.27	3.82	4.83
1020	cycle 4 miles		180	.044			2.18	3.18	5.36	6.75
1022	cycle 6 miles		144	.056			2.72	3.98	6.70	8.45
1024	cycle 8 miles		120	.067			3.26	4.77	8.03	10.15
1025	cycle 10 miles		96	.083			4.08	5.95	10.03	12.65
1026	20 MPH avg., cycle 0.5 mile		348	.023			1.13	1.65	2.78	3.50
1028	cycle 1 mile		312	.026			1.26	1.84	3.10	3.90
1030	cycle 2 miles		276	.029			1.42	2.08	3.50	4.41
1032	cycle 4 miles		216	.037			1.81	2.65	4.46	5.65
1034	cycle 6 miles		168	.048			2.33	3.41	5.74	7.25
1036	cycle 8 miles		144	.056			2.72	3.98	6.70	8.45
1038	cycle 10 miles		120	.067			3.26	4.77	8.03	10.15
1040	25 MPH avg., cycle 4 miles		228	.035			1.72	2.51	4.23	5.35
1042	cycle 6 miles		192	.042			2.04	2.98	5.02	6.35
1044	cycle 8 miles		168	.048			2.33	3.41	5.74	7.25
1046	cycle 10 miles		144	.056			2.72	3.98	6.70	8.45
1050	30 MPH avg., cycle 4 miles		252	.032			1.55	2.27	3.82	4.83
1052	cycle 6 miles		216	.037			1.81	2.65	4.46	5.65
1054	cycle 8 miles		180	.044			2.18	3.18	5.36	6.75
1056	cycle 10 miles		156	.051			2.51	3.67	6.18	7.80
1060	35 MPH avg., cycle 4 miles		264	.030			1.48	2.17	3.65	4.61
1062	cycle 6 miles		228	.035			1.72	2.51	4.23	5.35
1064	cycle 8 miles		204	.039			1.92	2.81	4.73	5.95
1066	cycle 10 miles		180	.044			2.18	3.18	5.36	6.75
1068	cycle 20 miles		120	.067			3.26	4.77	8.03	10.15
1069	cycle 30 miles		84	.095			4.66	6.80	11.46	14.50
1070	cycle 40 miles		72	.111			5.45	7.95	13.40	16.90
1072	40 MPH avg., cycle 6 miles		240	.033			1.63	2.39	4.02	5.05
1074	cycle 8 miles		216	.037			1.81	2.65	4.46	5.65
1076	cycle 10 miles		192	.042			2.04	2.98	5.02	6.35
1078	cycle 20 miles		120	.067			3.26	4.77	8.03	10.15
1080	cycle 30 miles		96	.083			4.08	5.95	10.03	12.65
1082	cycle 40 miles		72	.111			5.45	7.95	13.40	16.90
1084	cycle 50 miles		60	.133			6.55	9.55	16.10	20.50

For customer support on your Site Work & Landscape Costs with RSMeans data, call 800.448.8182.

313

31 23 Excavation and Fill

31 23 23 – Fill

31 23 23.20 Hauling		Crew	Daily Output	Labor-Hours	Unit	Material	2020 Bare Costs Labor	Equipment	Total	Total Incl O&P
1094	45 MPH avg., cycle 8 miles	B-34B	216	.037	L.C.Y.		1.81	2.65	4.46	5.65
1096	cycle 10 miles		204	.039			1.92	2.81	4.73	5.95
1098	cycle 20 miles		132	.061			2.97	4.34	7.31	9.20
1100	cycle 30 miles		108	.074			3.63	5.30	8.93	11.30
1102	cycle 40 miles		84	.095			4.66	6.80	11.46	14.50
1104	cycle 50 miles		72	.111			5.45	7.95	13.40	16.90
1106	50 MPH avg., cycle 10 miles		216	.037			1.81	2.65	4.46	5.65
1108	cycle 20 miles		144	.056			2.72	3.98	6.70	8.45
1110	cycle 30 miles		108	.074			3.63	5.30	8.93	11.30
1112	cycle 40 miles		84	.095			4.66	6.80	11.46	14.50
1114	cycle 50 miles		72	.111			5.45	7.95	13.40	16.90
1214	15 MPH avg., cycle 0.5 mile, 20 min. wait/ld./uld.		264	.030			1.48	2.17	3.65	4.61
1216	cycle 1 mile		240	.033			1.63	2.39	4.02	5.05
1218	cycle 2 miles		204	.039			1.92	2.81	4.73	5.95
1220	cycle 4 miles		156	.051			2.51	3.67	6.18	7.80
1222	cycle 6 miles		132	.061			2.97	4.34	7.31	9.20
1224	cycle 8 miles		108	.074			3.63	5.30	8.93	11.30
1225	cycle 10 miles		96	.083			4.08	5.95	10.03	12.65
1226	20 MPH avg., cycle 0.5 mile		264	.030			1.48	2.17	3.65	4.61
1228	cycle 1 mile		252	.032			1.55	2.27	3.82	4.83
1230	cycle 2 miles		216	.037			1.81	2.65	4.46	5.65
1232	cycle 4 miles		180	.044			2.18	3.18	5.36	6.75
1234	cycle 6 miles		144	.056			2.72	3.98	6.70	8.45
1236	cycle 8 miles		132	.061			2.97	4.34	7.31	9.20
1238	cycle 10 miles		108	.074			3.63	5.30	8.93	11.30
1240	25 MPH avg., cycle 4 miles		192	.042			2.04	2.98	5.02	6.35
1242	cycle 6 miles		168	.048			2.33	3.41	5.74	7.25
1244	cycle 8 miles		144	.056			2.72	3.98	6.70	8.45
1246	cycle 10 miles		132	.061			2.97	4.34	7.31	9.20
1250	30 MPH avg., cycle 4 miles		204	.039			1.92	2.81	4.73	5.95
1252	cycle 6 miles		180	.044			2.18	3.18	5.36	6.75
1254	cycle 8 miles		156	.051			2.51	3.67	6.18	7.80
1256	cycle 10 miles		144	.056			2.72	3.98	6.70	8.45
1260	35 MPH avg., cycle 4 miles		216	.037			1.81	2.65	4.46	5.65
1262	cycle 6 miles		192	.042			2.04	2.98	5.02	6.35
1264	cycle 8 miles		168	.048			2.33	3.41	5.74	7.25
1266	cycle 10 miles		156	.051			2.51	3.67	6.18	7.80
1268	cycle 20 miles		108	.074			3.63	5.30	8.93	11.30
1269	cycle 30 miles		72	.111			5.45	7.95	13.40	16.90
1270	cycle 40 miles		60	.133			6.55	9.55	16.10	20.50
1272	40 MPH avg., cycle 6 miles		192	.042			2.04	2.98	5.02	6.35
1274	cycle 8 miles		180	.044			2.18	3.18	5.36	6.75
1276	cycle 10 miles		156	.051			2.51	3.67	6.18	7.80
1278	cycle 20 miles		108	.074			3.63	5.30	8.93	11.30
1280	cycle 30 miles		84	.095			4.66	6.80	11.46	14.50
1282	cycle 40 miles		72	.111			5.45	7.95	13.40	16.90
1284	cycle 50 miles		60	.133			6.55	9.55	16.10	20.50
1294	45 MPH avg., cycle 8 miles		180	.044			2.18	3.18	5.36	6.75
1296	cycle 10 miles		168	.048			2.33	3.41	5.74	7.25
1298	cycle 20 miles		120	.067			3.26	4.77	8.03	10.15
1300	cycle 30 miles		96	.083			4.08	5.95	10.03	12.65
1302	cycle 40 miles		72	.111			5.45	7.95	13.40	16.90
1304	cycle 50 miles		60	.133			6.55	9.55	16.10	20.50

31 23 Excavation and Fill

31 23 23 – Fill

31 23 23.20 Hauling

		Crew	Daily Output	Labor-Hours	Unit	Material	2020 Bare Costs Labor	2020 Bare Costs Equipment	Total	Total Incl O&P
1306	50 MPH avg., cycle 10 miles	B-34B	180	.044	L.C.Y.		2.18	3.18	5.36	6.75
1308	cycle 20 miles		132	.061			2.97	4.34	7.31	9.20
1310	cycle 30 miles		96	.083			4.08	5.95	10.03	12.65
1312	cycle 40 miles		84	.095			4.66	6.80	11.46	14.50
1314	cycle 50 miles		72	.111			5.45	7.95	13.40	16.90
1414	15 MPH avg., cycle 0.5 mile, 25 min. wait/ld./uld.		204	.039			1.92	2.81	4.73	5.95
1416	cycle 1 mile		192	.042			2.04	2.98	5.02	6.35
1418	cycle 2 miles		168	.048			2.33	3.41	5.74	7.25
1420	cycle 4 miles		132	.061			2.97	4.34	7.31	9.20
1422	cycle 6 miles		120	.067			3.26	4.77	8.03	10.15
1424	cycle 8 miles		96	.083			4.08	5.95	10.03	12.65
1425	cycle 10 miles		84	.095			4.66	6.80	11.46	14.50
1426	20 MPH avg., cycle 0.5 mile		216	.037			1.81	2.65	4.46	5.65
1428	cycle 1 mile		204	.039			1.92	2.81	4.73	5.95
1430	cycle 2 miles		180	.044			2.18	3.18	5.36	6.75
1432	cycle 4 miles		156	.051			2.51	3.67	6.18	7.80
1434	cycle 6 miles		132	.061			2.97	4.34	7.31	9.20
1436	cycle 8 miles		120	.067			3.26	4.77	8.03	10.15
1438	cycle 10 miles		96	.083			4.08	5.95	10.03	12.65
1440	25 MPH avg., cycle 4 miles		168	.048			2.33	3.41	5.74	7.25
1442	cycle 6 miles		144	.056			2.72	3.98	6.70	8.45
1444	cycle 8 miles		132	.061			2.97	4.34	7.31	9.20
1446	cycle 10 miles		108	.074			3.63	5.30	8.93	11.30
1450	30 MPH avg., cycle 4 miles		168	.048			2.33	3.41	5.74	7.25
1452	cycle 6 miles		156	.051			2.51	3.67	6.18	7.80
1454	cycle 8 miles		132	.061			2.97	4.34	7.31	9.20
1456	cycle 10 miles		120	.067			3.26	4.77	8.03	10.15
1460	35 MPH avg., cycle 4 miles		180	.044			2.18	3.18	5.36	6.75
1462	cycle 6 miles		156	.051			2.51	3.67	6.18	7.80
1464	cycle 8 miles		144	.056			2.72	3.98	6.70	8.45
1466	cycle 10 miles		132	.061			2.97	4.34	7.31	9.20
1468	cycle 20 miles		96	.083			4.08	5.95	10.03	12.65
1469	cycle 30 miles		72	.111			5.45	7.95	13.40	16.90
1470	cycle 40 miles		60	.133			6.55	9.55	16.10	20.50
1472	40 MPH avg., cycle 6 miles		168	.048			2.33	3.41	5.74	7.25
1474	cycle 8 miles		156	.051			2.51	3.67	6.18	7.80
1476	cycle 10 miles		144	.056			2.72	3.98	6.70	8.45
1478	cycle 20 miles		96	.083			4.08	5.95	10.03	12.65
1480	cycle 30 miles		84	.095			4.66	6.80	11.46	14.50
1482	cycle 40 miles		60	.133			6.55	9.55	16.10	20.50
1484	cycle 50 miles		60	.133			6.55	9.55	16.10	20.50
1494	45 MPH avg., cycle 8 miles		156	.051			2.51	3.67	6.18	7.80
1496	cycle 10 miles		144	.056			2.72	3.98	6.70	8.45
1498	cycle 20 miles		108	.074			3.63	5.30	8.93	11.30
1500	cycle 30 miles		84	.095			4.66	6.80	11.46	14.50
1502	cycle 40 miles		72	.111			5.45	7.95	13.40	16.90
1504	cycle 50 miles		60	.133			6.55	9.55	16.10	20.50
1506	50 MPH avg., cycle 10 miles		156	.051			2.51	3.67	6.18	7.80
1508	cycle 20 miles		120	.067			3.26	4.77	8.03	10.15
1510	cycle 30 miles		96	.083			4.08	5.95	10.03	12.65
1512	cycle 40 miles		72	.111			5.45	7.95	13.40	16.90
1514	cycle 50 miles		60	.133			6.55	9.55	16.10	20.50
1614	15 MPH avg., cycle 0.5 mile, 30 min. wait/ld./uld.		180	.044			2.18	3.18	5.36	6.75

31 23 Excavation and Fill

31 23 23 – Fill

	31 23 23.20 Hauling	Crew	Daily Output	Labor-Hours	Unit	Material	2020 Bare Costs Labor	Equipment	Total	Total Incl O&P
1616	cycle 1 mile	B-34B	168	.048	L.C.Y.		2.33	3.41	5.74	7.25
1618	cycle 2 miles		144	.056			2.72	3.98	6.70	8.45
1620	cycle 4 miles		120	.067			3.26	4.77	8.03	10.15
1622	cycle 6 miles		108	.074			3.63	5.30	8.93	11.30
1624	cycle 8 miles		84	.095			4.66	6.80	11.46	14.50
1625	cycle 10 miles		84	.095			4.66	6.80	11.46	14.50
1626	20 MPH avg., cycle 0.5 mile		180	.044			2.18	3.18	5.36	6.75
1628	cycle 1 mile		168	.048			2.33	3.41	5.74	7.25
1630	cycle 2 miles		156	.051			2.51	3.67	6.18	7.80
1632	cycle 4 miles		132	.061			2.97	4.34	7.31	9.20
1634	cycle 6 miles		120	.067			3.26	4.77	8.03	10.15
1636	cycle 8 miles		108	.074			3.63	5.30	8.93	11.30
1638	cycle 10 miles		96	.083			4.08	5.95	10.03	12.65
1640	25 MPH avg., cycle 4 miles		144	.056			2.72	3.98	6.70	8.45
1642	cycle 6 miles		132	.061			2.97	4.34	7.31	9.20
1644	cycle 8 miles		108	.074			3.63	5.30	8.93	11.30
1646	cycle 10 miles		108	.074			3.63	5.30	8.93	11.30
1650	30 MPH avg., cycle 4 miles		144	.056			2.72	3.98	6.70	8.45
1652	cycle 6 miles		132	.061			2.97	4.34	7.31	9.20
1654	cycle 8 miles		120	.067			3.26	4.77	8.03	10.15
1656	cycle 10 miles		108	.074			3.63	5.30	8.93	11.30
1660	35 MPH avg., cycle 4 miles		156	.051			2.51	3.67	6.18	7.80
1662	cycle 6 miles		144	.056			2.72	3.98	6.70	8.45
1664	cycle 8 miles		132	.061			2.97	4.34	7.31	9.20
1666	cycle 10 miles		120	.067			3.26	4.77	8.03	10.15
1668	cycle 20 miles		84	.095			4.66	6.80	11.46	14.50
1669	cycle 30 miles		72	.111			5.45	7.95	13.40	16.90
1670	cycle 40 miles		60	.133			6.55	9.55	16.10	20.50
1672	40 MPH avg., cycle 6 miles		144	.056			2.72	3.98	6.70	8.45
1674	cycle 8 miles		132	.061			2.97	4.34	7.31	9.20
1676	cycle 10 miles		120	.067			3.26	4.77	8.03	10.15
1678	cycle 20 miles		96	.083			4.08	5.95	10.03	12.65
1680	cycle 30 miles		72	.111			5.45	7.95	13.40	16.90
1682	cycle 40 miles		60	.133			6.55	9.55	16.10	20.50
1684	cycle 50 miles		48	.167			8.15	11.95	20.10	25.50
1694	45 MPH avg., cycle 8 miles		144	.056			2.72	3.98	6.70	8.45
1696	cycle 10 miles		132	.061			2.97	4.34	7.31	9.20
1698	cycle 20 miles		96	.083			4.08	5.95	10.03	12.65
1700	cycle 30 miles		84	.095			4.66	6.80	11.46	14.50
1702	cycle 40 miles		60	.133			6.55	9.55	16.10	20.50
1704	cycle 50 miles		60	.133			6.55	9.55	16.10	20.50
1706	50 MPH avg., cycle 10 miles		132	.061			2.97	4.34	7.31	9.20
1708	cycle 20 miles		108	.074			3.63	5.30	8.93	11.30
1710	cycle 30 miles		84	.095			4.66	6.80	11.46	14.50
1712	cycle 40 miles		72	.111			5.45	7.95	13.40	16.90
1714	cycle 50 miles		60	.133			6.55	9.55	16.10	20.50
2000	Hauling, 8 C.Y. truck, small project cost per hour	B-34A	8	1	Hr.		49	53.50	102.50	133
2100	12 C.Y. truck	B-34B	8	1			49	71.50	120.50	153
2150	16.5 C.Y. truck	B-34C	8	1			49	78.50	127.50	160
2175	18 C.Y. 8 wheel truck	B-34I	8	1			49	89	138	172
2200	20 C.Y. truck	B-34D	8	1			49	80.50	129.50	162
2300	Grading at dump, or embankment if required, by dozer	B-10B	1000	.012	L.C.Y.		.62	1.50	2.12	2.58
2310	Spotter at fill or cut, if required	1 Clab	8	1	Hr.		42		42	63.50

316

31 23 Excavation and Fill

31 23 23 – Fill

31 23 23.20 Hauling		Crew	Daily Output	Labor-Hours	Unit	Material	2020 Bare Costs Labor	Equipment	Total	Total Incl O&P
2500	Dust control, light	B-59	1	8	Day		390	460	850	1,100
2510	Heavy	"	.50	16			785	920	1,705	2,175
2600	Haul road maintenance	B-86A	1	8	↓		455	1,050	1,505	1,850
3014	16.5 C.Y. truck, 15 min. wait/ld./uld., 15 MPH, cycle 0.5 mile	B-34C	462	.017	L.C.Y.		.85	1.36	2.21	2.77
3016	cycle 1 mile		413	.019			.95	1.53	2.48	3.10
3018	cycle 2 miles		347	.023			1.13	1.82	2.95	3.69
3020	cycle 4 miles		248	.032			1.58	2.54	4.12	5.15
3022	cycle 6 miles		198	.040			1.98	3.18	5.16	6.45
3024	cycle 8 miles		165	.048			2.37	3.82	6.19	7.75
3025	cycle 10 miles		132	.061			2.97	4.77	7.74	9.70
3026	20 MPH avg., cycle 0.5 mile		479	.017			.82	1.32	2.14	2.67
3028	cycle 1 mile		429	.019			.91	1.47	2.38	2.99
3030	cycle 2 miles		380	.021			1.03	1.66	2.69	3.36
3032	cycle 4 miles		281	.028			1.39	2.24	3.63	4.56
3034	cycle 6 miles		231	.035			1.70	2.73	4.43	5.55
3036	cycle 8 miles		198	.040			1.98	3.18	5.16	6.45
3038	cycle 10 miles		165	.048			2.37	3.82	6.19	7.75
3040	25 MPH avg., cycle 4 miles		314	.025			1.25	2.01	3.26	4.08
3042	cycle 6 miles		264	.030			1.48	2.39	3.87	4.84
3044	cycle 8 miles		231	.035			1.70	2.73	4.43	5.55
3046	cycle 10 miles		198	.040			1.98	3.18	5.16	6.45
3050	30 MPH avg., cycle 4 miles		347	.023			1.13	1.82	2.95	3.69
3052	cycle 6 miles		281	.028			1.39	2.24	3.63	4.56
3054	cycle 8 miles		248	.032			1.58	2.54	4.12	5.15
3056	cycle 10 miles		215	.037			1.82	2.93	4.75	5.95
3060	35 MPH avg., cycle 4 miles		363	.022			1.08	1.74	2.82	3.53
3062	cycle 6 miles		314	.025			1.25	2.01	3.26	4.08
3064	cycle 8 miles		264	.030			1.48	2.39	3.87	4.84
3066	cycle 10 miles		248	.032			1.58	2.54	4.12	5.15
3068	cycle 20 miles		149	.054			2.63	4.23	6.86	8.60
3070	cycle 30 miles		116	.069			3.38	5.45	8.83	11
3072	cycle 40 miles		83	.096			4.72	7.60	12.32	15.40
3074	40 MPH avg., cycle 6 miles		330	.024			1.19	1.91	3.10	3.88
3076	cycle 8 miles		281	.028			1.39	2.24	3.63	4.56
3078	cycle 10 miles		264	.030			1.48	2.39	3.87	4.84
3080	cycle 20 miles		165	.048			2.37	3.82	6.19	7.75
3082	cycle 30 miles		132	.061			2.97	4.77	7.74	9.70
3084	cycle 40 miles		99	.081			3.96	6.35	10.31	12.95
3086	cycle 50 miles		83	.096			4.72	7.60	12.32	15.40
3094	45 MPH avg., cycle 8 miles		297	.027			1.32	2.12	3.44	4.31
3096	cycle 10 miles		281	.028			1.39	2.24	3.63	4.56
3098	cycle 20 miles		182	.044			2.15	3.46	5.61	7.05
3100	cycle 30 miles		132	.061			2.97	4.77	7.74	9.70
3102	cycle 40 miles		116	.069			3.38	5.45	8.83	11
3104	cycle 50 miles		99	.081			3.96	6.35	10.31	12.95
3106	50 MPH avg., cycle 10 miles		281	.028			1.39	2.24	3.63	4.56
3108	cycle 20 miles		198	.040			1.98	3.18	5.16	6.45
3110	cycle 30 miles		149	.054			2.63	4.23	6.86	8.60
3112	cycle 40 miles		116	.069			3.38	5.45	8.83	11
3114	cycle 50 miles		99	.081			3.96	6.35	10.31	12.95
3214	20 min. wait/ld./uld., 15 MPH, cycle 0.5 mile		363	.022			1.08	1.74	2.82	3.53
3216	cycle 1 mile		330	.024			1.19	1.91	3.10	3.88
3218	cycle 2 miles		281	.028	↓		1.39	2.24	3.63	4.56

317

31 23 23.20 Hauling		Crew	Daily Output	Labor-Hours	Unit	Material	2020 Bare Costs Labor	2020 Bare Costs Equipment	Total	Total Incl O&P
3220	cycle 4 miles	B-34C	215	.037	L.C.Y.		1.82	2.93	4.75	5.95
3222	cycle 6 miles		182	.044			2.15	3.46	5.61	7.05
3224	cycle 8 miles		149	.054			2.63	4.23	6.86	8.60
3225	cycle 10 miles		132	.061			2.97	4.77	7.74	9.70
3226	20 MPH avg., cycle 0.5 mile		363	.022			1.08	1.74	2.82	3.53
3228	cycle 1 mile		347	.023			1.13	1.82	2.95	3.69
3230	cycle 2 miles		297	.027			1.32	2.12	3.44	4.31
3232	cycle 4 miles		248	.032			1.58	2.54	4.12	5.15
3234	cycle 6 miles		198	.040			1.98	3.18	5.16	6.45
3236	cycle 8 miles		182	.044			2.15	3.46	5.61	7.05
3238	cycle 10 miles		149	.054			2.63	4.23	6.86	8.60
3240	25 MPH avg., cycle 4 miles		264	.030			1.48	2.39	3.87	4.84
3242	cycle 6 miles		231	.035			1.70	2.73	4.43	5.55
3244	cycle 8 miles		198	.040			1.98	3.18	5.16	6.45
3246	cycle 10 miles		182	.044			2.15	3.46	5.61	7.05
3250	30 MPH avg., cycle 4 miles		281	.028			1.39	2.24	3.63	4.56
3252	cycle 6 miles		248	.032			1.58	2.54	4.12	5.15
3254	cycle 8 miles		215	.037			1.82	2.93	4.75	5.95
3256	cycle 10 miles		198	.040			1.98	3.18	5.16	6.45
3260	35 MPH avg., cycle 4 miles		297	.027			1.32	2.12	3.44	4.31
3262	cycle 6 miles		264	.030			1.48	2.39	3.87	4.84
3264	cycle 8 miles		231	.035			1.70	2.73	4.43	5.55
3266	cycle 10 miles		215	.037			1.82	2.93	4.75	5.95
3268	cycle 20 miles		149	.054			2.63	4.23	6.86	8.60
3270	cycle 30 miles		99	.081			3.96	6.35	10.31	12.95
3272	cycle 40 miles		83	.096			4.72	7.60	12.32	15.40
3274	40 MPH avg., cycle 6 miles		264	.030			1.48	2.39	3.87	4.84
3276	cycle 8 miles		248	.032			1.58	2.54	4.12	5.15
3278	cycle 10 miles		215	.037			1.82	2.93	4.75	5.95
3280	cycle 20 miles		149	.054			2.63	4.23	6.86	8.60
3282	cycle 30 miles		116	.069			3.38	5.45	8.83	11
3284	cycle 40 miles		99	.081			3.96	6.35	10.31	12.95
3286	cycle 50 miles		83	.096			4.72	7.60	12.32	15.40
3294	45 MPH avg., cycle 8 miles		248	.032			1.58	2.54	4.12	5.15
3296	cycle 10 miles		231	.035			1.70	2.73	4.43	5.55
3298	cycle 20 miles		165	.048			2.37	3.82	6.19	7.75
3300	cycle 30 miles		132	.061			2.97	4.77	7.74	9.70
3302	cycle 40 miles		99	.081			3.96	6.35	10.31	12.95
3304	cycle 50 miles		83	.096			4.72	7.60	12.32	15.40
3306	50 MPH avg., cycle 10 miles		248	.032			1.58	2.54	4.12	5.15
3308	cycle 20 miles		182	.044			2.15	3.46	5.61	7.05
3310	cycle 30 miles		132	.061			2.97	4.77	7.74	9.70
3312	cycle 40 miles		116	.069			3.38	5.45	8.83	11
3314	cycle 50 miles		99	.081			3.96	6.35	10.31	12.95
3414	25 min. wait/ld./uld., 15 MPH, cycle 0.5 mile		281	.028			1.39	2.24	3.63	4.56
3416	cycle 1 mile		264	.030			1.48	2.39	3.87	4.84
3418	cycle 2 miles		231	.035			1.70	2.73	4.43	5.55
3420	cycle 4 miles		182	.044			2.15	3.46	5.61	7.05
3422	cycle 6 miles		165	.048			2.37	3.82	6.19	7.75
3424	cycle 8 miles		132	.061			2.97	4.77	7.74	9.70
3425	cycle 10 miles		116	.069			3.38	5.45	8.83	11
3426	20 MPH avg., cycle 0.5 mile		297	.027			1.32	2.12	3.44	4.31
3428	cycle 1 mile		281	.028			1.39	2.24	3.63	4.56

31 23 23 – Fill

31 23 23.20 Hauling		Crew	Daily Output	Labor-Hours	Unit	Material	2020 Bare Costs Labor	Equipment	Total	Total Incl O&P
3430	cycle 2 miles	B-34C	248	.032	L.C.Y.		1.58	2.54	4.12	5.15
3432	cycle 4 miles		215	.037			1.82	2.93	4.75	5.95
3434	cycle 6 miles		182	.044			2.15	3.46	5.61	7.05
3436	cycle 8 miles		165	.048			2.37	3.82	6.19	7.75
3438	cycle 10 miles		132	.061			2.97	4.77	7.74	9.70
3440	25 MPH avg., cycle 4 miles		231	.035			1.70	2.73	4.43	5.55
3442	cycle 6 miles		198	.040			1.98	3.18	5.16	6.45
3444	cycle 8 miles		182	.044			2.15	3.46	5.61	7.05
3446	cycle 10 miles		165	.048			2.37	3.82	6.19	7.75
3450	30 MPH avg., cycle 4 miles		231	.035			1.70	2.73	4.43	5.55
3452	cycle 6 miles		215	.037			1.82	2.93	4.75	5.95
3454	cycle 8 miles		182	.044			2.15	3.46	5.61	7.05
3456	cycle 10 miles		165	.048			2.37	3.82	6.19	7.75
3460	35 MPH avg., cycle 4 miles		248	.032			1.58	2.54	4.12	5.15
3462	cycle 6 miles		215	.037			1.82	2.93	4.75	5.95
3464	cycle 8 miles		198	.040			1.98	3.18	5.16	6.45
3466	cycle 10 miles		182	.044			2.15	3.46	5.61	7.05
3468	cycle 20 miles		132	.061			2.97	4.77	7.74	9.70
3470	cycle 30 miles		99	.081			3.96	6.35	10.31	12.95
3472	cycle 40 miles		83	.096			4.72	7.60	12.32	15.40
3474	40 MPH, cycle 6 miles		231	.035			1.70	2.73	4.43	5.55
3476	cycle 8 miles		215	.037			1.82	2.93	4.75	5.95
3478	cycle 10 miles		198	.040			1.98	3.18	5.16	6.45
3480	cycle 20 miles		132	.061			2.97	4.77	7.74	9.70
3482	cycle 30 miles		116	.069			3.38	5.45	8.83	11
3484	cycle 40 miles		83	.096			4.72	7.60	12.32	15.40
3486	cycle 50 miles		83	.096			4.72	7.60	12.32	15.40
3494	45 MPH avg., cycle 8 miles		215	.037			1.82	2.93	4.75	5.95
3496	cycle 10 miles		198	.040			1.98	3.18	5.16	6.45
3498	cycle 20 miles		149	.054			2.63	4.23	6.86	8.60
3500	cycle 30 miles		116	.069			3.38	5.45	8.83	11
3502	cycle 40 miles		99	.081			3.96	6.35	10.31	12.95
3504	cycle 50 miles		83	.096			4.72	7.60	12.32	15.40
3506	50 MPH avg., cycle 10 miles		215	.037			1.82	2.93	4.75	5.95
3508	cycle 20 miles		165	.048			2.37	3.82	6.19	7.75
3510	cycle 30 miles		132	.061			2.97	4.77	7.74	9.70
3512	cycle 40 miles		99	.081			3.96	6.35	10.31	12.95
3514	cycle 50 miles		83	.096			4.72	7.60	12.32	15.40
3614	30 min. wait/ld./uld., 15 MPH, cycle 0.5 mile		248	.032			1.58	2.54	4.12	5.15
3616	cycle 1 mile		231	.035			1.70	2.73	4.43	5.55
3618	cycle 2 miles		198	.040			1.98	3.18	5.16	6.45
3620	cycle 4 miles		165	.048			2.37	3.82	6.19	7.75
3622	cycle 6 miles		149	.054			2.63	4.23	6.86	8.60
3624	cycle 8 miles		116	.069			3.38	5.45	8.83	11
3625	cycle 10 miles		116	.069			3.38	5.45	8.83	11
3626	20 MPH avg., cycle 0.5 mile		248	.032			1.58	2.54	4.12	5.15
3628	cycle 1 mile		231	.035			1.70	2.73	4.43	5.55
3630	cycle 2 miles		215	.037			1.82	2.93	4.75	5.95
3632	cycle 4 miles		182	.044			2.15	3.46	5.61	7.05
3634	cycle 6 miles		165	.048			2.37	3.82	6.19	7.75
3636	cycle 8 miles		149	.054			2.63	4.23	6.86	8.60
3638	cycle 10 miles		132	.061			2.97	4.77	7.74	9.70
3640	25 MPH avg., cycle 4 miles		198	.040			1.98	3.18	5.16	6.45

319

31 23 23.20 Hauling		Crew	Daily Output	Labor-Hours	Unit	Material	2020 Bare Costs Labor	2020 Bare Costs Equipment	Total	Total Incl O&P
3642	cycle 6 miles	B-34C	182	.044	L.C.Y.		2.15	3.46	5.61	7.05
3644	cycle 8 miles		149	.054			2.63	4.23	6.86	8.60
3646	cycle 10 miles		149	.054			2.63	4.23	6.86	8.60
3650	30 MPH avg., cycle 4 miles		198	.040			1.98	3.18	5.16	6.45
3652	cycle 6 miles		182	.044			2.15	3.46	5.61	7.05
3654	cycle 8 miles		165	.048			2.37	3.82	6.19	7.75
3656	cycle 10 miles		149	.054			2.63	4.23	6.86	8.60
3660	35 MPH avg., cycle 4 miles		215	.037			1.82	2.93	4.75	5.95
3662	cycle 6 miles		198	.040			1.98	3.18	5.16	6.45
3664	cycle 8 miles		182	.044			2.15	3.46	5.61	7.05
3666	cycle 10 miles		165	.048			2.37	3.82	6.19	7.75
3668	cycle 20 miles		116	.069			3.38	5.45	8.83	11
3670	cycle 30 miles		99	.081			3.96	6.35	10.31	12.95
3672	cycle 40 miles		83	.096			4.72	7.60	12.32	15.40
3674	40 MPH, cycle 6 miles		198	.040			1.98	3.18	5.16	6.45
3676	cycle 8 miles		182	.044			2.15	3.46	5.61	7.05
3678	cycle 10 miles		165	.048			2.37	3.82	6.19	7.75
3680	cycle 20 miles		132	.061			2.97	4.77	7.74	9.70
3682	cycle 30 miles		99	.081			3.96	6.35	10.31	12.95
3684	cycle 40 miles		83	.096			4.72	7.60	12.32	15.40
3686	cycle 50 miles		66	.121			5.95	9.55	15.50	19.40
3694	45 MPH avg., cycle 8 miles		198	.040			1.98	3.18	5.16	6.45
3696	cycle 10 miles		182	.044			2.15	3.46	5.61	7.05
3698	cycle 20 miles		132	.061			2.97	4.77	7.74	9.70
3700	cycle 30 miles		116	.069			3.38	5.45	8.83	11
3702	cycle 40 miles		83	.096			4.72	7.60	12.32	15.40
3704	cycle 50 miles		83	.096			4.72	7.60	12.32	15.40
3706	50 MPH avg., cycle 10 miles		182	.044			2.15	3.46	5.61	7.05
3708	cycle 20 miles		149	.054			2.63	4.23	6.86	8.60
3710	cycle 30 miles		116	.069			3.38	5.45	8.83	11
3712	cycle 40 miles		99	.081			3.96	6.35	10.31	12.95
3714	cycle 50 miles		83	.096			4.72	7.60	12.32	15.40
4014	20 C.Y. truck, 15 min. wait/ld./uld., 15 MPH, cycle 0.5 mile	B-34D	560	.014			.70	1.15	1.85	2.32
4016	cycle 1 mile		500	.016			.78	1.29	2.07	2.59
4018	cycle 2 miles		420	.019			.93	1.54	2.47	3.09
4020	cycle 4 miles		300	.027			1.31	2.15	3.46	4.33
4022	cycle 6 miles		240	.033			1.63	2.69	4.32	5.40
4024	cycle 8 miles		200	.040			1.96	3.22	5.18	6.50
4025	cycle 10 miles		160	.050			2.45	4.03	6.48	8.10
4026	20 MPH avg., cycle 0.5 mile		580	.014			.68	1.11	1.79	2.23
4028	cycle 1 mile		520	.015			.75	1.24	1.99	2.49
4030	cycle 2 miles		460	.017			.85	1.40	2.25	2.82
4032	cycle 4 miles		340	.024			1.15	1.90	3.05	3.82
4034	cycle 6 miles		280	.029			1.40	2.30	3.70	4.63
4036	cycle 8 miles		240	.033			1.63	2.69	4.32	5.40
4038	cycle 10 miles		200	.040			1.96	3.22	5.18	6.50
4040	25 MPH avg., cycle 4 miles		380	.021			1.03	1.70	2.73	3.41
4042	cycle 6 miles		320	.025			1.22	2.02	3.24	4.05
4044	cycle 8 miles		280	.029			1.40	2.30	3.70	4.63
4046	cycle 10 miles		240	.033			1.63	2.69	4.32	5.40
4050	30 MPH avg., cycle 4 miles		420	.019			.93	1.54	2.47	3.09
4052	cycle 6 miles		340	.024			1.15	1.90	3.05	3.82
4054	cycle 8 miles		300	.027			1.31	2.15	3.46	4.33

31 23 23 – Fill

31 23 23.20 Hauling		Crew	Daily Output	Labor-Hours	Unit	Material	2020 Bare Costs Labor	Equipment	Total	Total Incl O&P
4056	cycle 10 miles	B-34D	260	.031	L.C.Y.		1.51	2.48	3.99	4.99
4060	35 MPH avg., cycle 4 miles		440	.018			.89	1.47	2.36	2.94
4062	cycle 6 miles		380	.021			1.03	1.70	2.73	3.41
4064	cycle 8 miles		320	.025			1.22	2.02	3.24	4.05
4066	cycle 10 miles		300	.027			1.31	2.15	3.46	4.33
4068	cycle 20 miles		180	.044			2.18	3.58	5.76	7.20
4070	cycle 30 miles		140	.057			2.80	4.61	7.41	9.25
4072	cycle 40 miles		100	.080			3.92	6.45	10.37	12.95
4074	40 MPH avg., cycle 6 miles		400	.020			.98	1.61	2.59	3.24
4076	cycle 8 miles		340	.024			1.15	1.90	3.05	3.82
4078	cycle 10 miles		320	.025			1.22	2.02	3.24	4.05
4080	cycle 20 miles		200	.040			1.96	3.22	5.18	6.50
4082	cycle 30 miles		160	.050			2.45	4.03	6.48	8.10
4084	cycle 40 miles		120	.067			3.26	5.35	8.61	10.80
4086	cycle 50 miles		100	.080			3.92	6.45	10.37	12.95
4094	45 MPH avg., cycle 8 miles		360	.022			1.09	1.79	2.88	3.60
4096	cycle 10 miles		340	.024			1.15	1.90	3.05	3.82
4098	cycle 20 miles		220	.036			1.78	2.93	4.71	5.90
4100	cycle 30 miles		160	.050			2.45	4.03	6.48	8.10
4102	cycle 40 miles		140	.057			2.80	4.61	7.41	9.25
4104	cycle 50 miles		120	.067			3.26	5.35	8.61	10.80
4106	50 MPH avg., cycle 10 miles		340	.024			1.15	1.90	3.05	3.82
4108	cycle 20 miles		240	.033			1.63	2.69	4.32	5.40
4110	cycle 30 miles		180	.044			2.18	3.58	5.76	7.20
4112	cycle 40 miles		140	.057			2.80	4.61	7.41	9.25
4114	cycle 50 miles		120	.067			3.26	5.35	8.61	10.80
4214	20 min. wait/ld./uld., 15 MPH, cycle 0.5 mile		440	.018			.89	1.47	2.36	2.94
4216	cycle 1 mile		400	.020			.98	1.61	2.59	3.24
4218	cycle 2 miles		340	.024			1.15	1.90	3.05	3.82
4220	cycle 4 miles		260	.031			1.51	2.48	3.99	4.99
4222	cycle 6 miles		220	.036			1.78	2.93	4.71	5.90
4224	cycle 8 miles		180	.044			2.18	3.58	5.76	7.20
4225	cycle 10 miles		160	.050			2.45	4.03	6.48	8.10
4226	20 MPH avg., cycle 0.5 mile		440	.018			.89	1.47	2.36	2.94
4228	cycle 1 mile		420	.019			.93	1.54	2.47	3.09
4230	cycle 2 miles		360	.022			1.09	1.79	2.88	3.60
4232	cycle 4 miles		300	.027			1.31	2.15	3.46	4.33
4234	cycle 6 miles		240	.033			1.63	2.69	4.32	5.40
4236	cycle 8 miles		220	.036			1.78	2.93	4.71	5.90
4238	cycle 10 miles		180	.044			2.18	3.58	5.76	7.20
4240	25 MPH avg., cycle 4 miles		320	.025			1.22	2.02	3.24	4.05
4242	cycle 6 miles		280	.029			1.40	2.30	3.70	4.63
4244	cycle 8 miles		240	.033			1.63	2.69	4.32	5.40
4246	cycle 10 miles		220	.036			1.78	2.93	4.71	5.90
4250	30 MPH avg., cycle 4 miles		340	.024			1.15	1.90	3.05	3.82
4252	cycle 6 miles		300	.027			1.31	2.15	3.46	4.33
4254	cycle 8 miles		260	.031			1.51	2.48	3.99	4.99
4256	cycle 10 miles		240	.033			1.63	2.69	4.32	5.40
4260	35 MPH avg., cycle 4 miles		360	.022			1.09	1.79	2.88	3.60
4262	cycle 6 miles		320	.025			1.22	2.02	3.24	4.05
4264	cycle 8 miles		280	.029			1.40	2.30	3.70	4.63
4266	cycle 10 miles		260	.031			1.51	2.48	3.99	4.99
4268	cycle 20 miles		180	.044			2.18	3.58	5.76	7.20

31 23 23.20 Hauling		Crew	Daily Output	Labor-Hours	Unit	Material	2020 Bare Costs Labor	2020 Bare Costs Equipment	Total	Total Incl O&P
4270	cycle 30 miles	B-34D	120	.067	L.C.Y.		3.26	5.35	8.61	10.80
4272	cycle 40 miles		100	.080			3.92	6.45	10.37	12.95
4274	40 MPH avg., cycle 6 miles		320	.025			1.22	2.02	3.24	4.05
4276	cycle 8 miles		300	.027			1.31	2.15	3.46	4.33
4278	cycle 10 miles		260	.031			1.51	2.48	3.99	4.99
4280	cycle 20 miles		180	.044			2.18	3.58	5.76	7.20
4282	cycle 30 miles		140	.057			2.80	4.61	7.41	9.25
4284	cycle 40 miles		120	.067			3.26	5.35	8.61	10.80
4286	cycle 50 miles		100	.080			3.92	6.45	10.37	12.95
4294	45 MPH avg., cycle 8 miles		300	.027			1.31	2.15	3.46	4.33
4296	cycle 10 miles		280	.029			1.40	2.30	3.70	4.63
4298	cycle 20 miles		200	.040			1.96	3.22	5.18	6.50
4300	cycle 30 miles		160	.050			2.45	4.03	6.48	8.10
4302	cycle 40 miles		120	.067			3.26	5.35	8.61	10.80
4304	cycle 50 miles		100	.080			3.92	6.45	10.37	12.95
4306	50 MPH avg., cycle 10 miles		300	.027			1.31	2.15	3.46	4.33
4308	cycle 20 miles		220	.036			1.78	2.93	4.71	5.90
4310	cycle 30 miles		180	.044			2.18	3.58	5.76	7.20
4312	cycle 40 miles		140	.057			2.80	4.61	7.41	9.25
4314	cycle 50 miles		120	.067			3.26	5.35	8.61	10.80
4414	25 min. wait/ld./uld., 15 MPH, cycle 0.5 mile		340	.024			1.15	1.90	3.05	3.82
4416	cycle 1 mile		320	.025			1.22	2.02	3.24	4.05
4418	cycle 2 miles		280	.029			1.40	2.30	3.70	4.63
4420	cycle 4 miles		220	.036			1.78	2.93	4.71	5.90
4422	cycle 6 miles		200	.040			1.96	3.22	5.18	6.50
4424	cycle 8 miles		160	.050			2.45	4.03	6.48	8.10
4425	cycle 10 miles		140	.057			2.80	4.61	7.41	9.25
4426	20 MPH avg., cycle 0.5 mile		360	.022			1.09	1.79	2.88	3.60
4428	cycle 1 mile		340	.024			1.15	1.90	3.05	3.82
4430	cycle 2 miles		300	.027			1.31	2.15	3.46	4.33
4432	cycle 4 miles		260	.031			1.51	2.48	3.99	4.99
4434	cycle 6 miles		220	.036			1.78	2.93	4.71	5.90
4436	cycle 8 miles		200	.040			1.96	3.22	5.18	6.50
4438	cycle 10 miles		160	.050			2.45	4.03	6.48	8.10
4440	25 MPH avg., cycle 4 miles		280	.029			1.40	2.30	3.70	4.63
4442	cycle 6 miles		240	.033			1.63	2.69	4.32	5.40
4444	cycle 8 miles		220	.036			1.78	2.93	4.71	5.90
4446	cycle 10 miles		200	.040			1.96	3.22	5.18	6.50
4450	30 MPH avg., cycle 4 miles		280	.029			1.40	2.30	3.70	4.63
4452	cycle 6 miles		260	.031			1.51	2.48	3.99	4.99
4454	cycle 8 miles		220	.036			1.78	2.93	4.71	5.90
4456	cycle 10 miles		200	.040			1.96	3.22	5.18	6.50
4460	35 MPH avg., cycle 4 miles		300	.027			1.31	2.15	3.46	4.33
4462	cycle 6 miles		260	.031			1.51	2.48	3.99	4.99
4464	cycle 8 miles		240	.033			1.63	2.69	4.32	5.40
4466	cycle 10 miles		220	.036			1.78	2.93	4.71	5.90
4468	cycle 20 miles		160	.050			2.45	4.03	6.48	8.10
4470	cycle 30 miles		120	.067			3.26	5.35	8.61	10.80
4472	cycle 40 miles		100	.080			3.92	6.45	10.37	12.95
4474	40 MPH, cycle 6 miles		280	.029			1.40	2.30	3.70	4.63
4476	cycle 8 miles		260	.031			1.51	2.48	3.99	4.99
4478	cycle 10 miles		240	.033			1.63	2.69	4.32	5.40
4480	cycle 20 miles		160	.050			2.45	4.03	6.48	8.10

31 23 23.20 Hauling

		Crew	Daily Output	Labor-Hours	Unit	Material	2020 Bare Costs Labor	2020 Bare Costs Equipment	Total	Total Incl O&P
4482	cycle 30 miles	B-34D	140	.057	L.C.Y.		2.80	4.61	7.41	9.25
4484	cycle 40 miles		100	.080			3.92	6.45	10.37	12.95
4486	cycle 50 miles		100	.080			3.92	6.45	10.37	12.95
4494	45 MPH avg., cycle 8 miles		260	.031			1.51	2.48	3.99	4.99
4496	cycle 10 miles		240	.033			1.63	2.69	4.32	5.40
4498	cycle 20 miles		180	.044			2.18	3.58	5.76	7.20
4500	cycle 30 miles		140	.057			2.80	4.61	7.41	9.25
4502	cycle 40 miles		120	.067			3.26	5.35	8.61	10.80
4504	cycle 50 miles		100	.080			3.92	6.45	10.37	12.95
4506	50 MPH avg., cycle 10 miles		260	.031			1.51	2.48	3.99	4.99
4508	cycle 20 miles		200	.040			1.96	3.22	5.18	6.50
4510	cycle 30 miles		160	.050			2.45	4.03	6.48	8.10
4512	cycle 40 miles		120	.067			3.26	5.35	8.61	10.80
4514	cycle 50 miles		100	.080			3.92	6.45	10.37	12.95
4614	30 min. wait/ld./uld., 15 MPH, cycle 0.5 mile		300	.027			1.31	2.15	3.46	4.33
4616	cycle 1 mile		280	.029			1.40	2.30	3.70	4.63
4618	cycle 2 miles		240	.033			1.63	2.69	4.32	5.40
4620	cycle 4 miles		200	.040			1.96	3.22	5.18	6.50
4622	cycle 6 miles		180	.044			2.18	3.58	5.76	7.20
4624	cycle 8 miles		140	.057			2.80	4.61	7.41	9.25
4625	cycle 10 miles		140	.057			2.80	4.61	7.41	9.25
4626	20 MPH avg., cycle 0.5 mile		300	.027			1.31	2.15	3.46	4.33
4628	cycle 1 mile		280	.029			1.40	2.30	3.70	4.63
4630	cycle 2 miles		260	.031			1.51	2.48	3.99	4.99
4632	cycle 4 miles		220	.036			1.78	2.93	4.71	5.90
4634	cycle 6 miles		200	.040			1.96	3.22	5.18	6.50
4636	cycle 8 miles		180	.044			2.18	3.58	5.76	7.20
4638	cycle 10 miles		160	.050			2.45	4.03	6.48	8.10
4640	25 MPH avg., cycle 4 miles		240	.033			1.63	2.69	4.32	5.40
4642	cycle 6 miles		220	.036			1.78	2.93	4.71	5.90
4644	cycle 8 miles		180	.044			2.18	3.58	5.76	7.20
4646	cycle 10 miles		180	.044			2.18	3.58	5.76	7.20
4650	30 MPH avg., cycle 4 miles		240	.033			1.63	2.69	4.32	5.40
4652	cycle 6 miles		220	.036			1.78	2.93	4.71	5.90
4654	cycle 8 miles		200	.040			1.96	3.22	5.18	6.50
4656	cycle 10 miles		180	.044			2.18	3.58	5.76	7.20
4660	35 MPH avg., cycle 4 miles		260	.031			1.51	2.48	3.99	4.99
4662	cycle 6 miles		240	.033			1.63	2.69	4.32	5.40
4664	cycle 8 miles		220	.036			1.78	2.93	4.71	5.90
4666	cycle 10 miles		200	.040			1.96	3.22	5.18	6.50
4668	cycle 20 miles		140	.057			2.80	4.61	7.41	9.25
4670	cycle 30 miles		120	.067			3.26	5.35	8.61	10.80
4672	cycle 40 miles		100	.080			3.92	6.45	10.37	12.95
4674	40 MPH, cycle 6 miles		240	.033			1.63	2.69	4.32	5.40
4676	cycle 8 miles		220	.036			1.78	2.93	4.71	5.90
4678	cycle 10 miles		200	.040			1.96	3.22	5.18	6.50
4680	cycle 20 miles		160	.050			2.45	4.03	6.48	8.10
4682	cycle 30 miles		120	.067			3.26	5.35	8.61	10.80
4684	cycle 40 miles		100	.080			3.92	6.45	10.37	12.95
4686	cycle 50 miles		80	.100			4.90	8.05	12.95	16.20
4694	45 MPH avg., cycle 8 miles		220	.036			1.78	2.93	4.71	5.90
4696	cycle 10 miles		220	.036			1.78	2.93	4.71	5.90
4698	cycle 20 miles		160	.050			2.45	4.03	6.48	8.10

323

31 23 23.20 Hauling		Crew	Daily Output	Labor-Hours	Unit	Material	2020 Bare Costs Labor	Equipment	Total	Total Incl O&P
4700	cycle 30 miles	B-34D	140	.057	L.C.Y.		2.80	4.61	7.41	9.25
4702	cycle 40 miles		100	.080			3.92	6.45	10.37	12.95
4704	cycle 50 miles		100	.080			3.92	6.45	10.37	12.95
4706	50 MPH avg., cycle 10 miles		220	.036			1.78	2.93	4.71	5.90
4708	cycle 20 miles		180	.044			2.18	3.58	5.76	7.20
4710	cycle 30 miles		140	.057			2.80	4.61	7.41	9.25
4712	cycle 40 miles		120	.067			3.26	5.35	8.61	10.80
4714	cycle 50 miles		100	.080			3.92	6.45	10.37	12.95
5000	22 C.Y. off-road, 15 min. wait/ld./uld., 5 MPH, cycle 2,000'	B-34F	528	.015			.74	1.77	2.51	3.06
5010	cycle 3,000'		484	.017			.81	1.93	2.74	3.34
5020	cycle 4,000'		440	.018			.89	2.13	3.02	3.67
5030	cycle 0.5 mile		506	.016			.77	1.85	2.62	3.19
5040	cycle 1 mile		374	.021			1.05	2.50	3.55	4.32
5050	cycle 2 miles		264	.030			1.48	3.54	5.02	6.10
5060	10 MPH, cycle 2,000'		594	.013			.66	1.57	2.23	2.72
5070	cycle 3,000'		572	.014			.68	1.64	2.32	2.83
5080	cycle 4,000'		528	.015			.74	1.77	2.51	3.06
5090	cycle 0.5 mile		572	.014			.68	1.64	2.32	2.83
5100	cycle 1 mile		506	.016			.77	1.85	2.62	3.19
5110	cycle 2 miles		374	.021			1.05	2.50	3.55	4.32
5120	cycle 4 miles		264	.030			1.48	3.54	5.02	6.10
5130	15 MPH, cycle 2,000'		638	.013			.61	1.47	2.08	2.53
5140	cycle 3,000'		594	.013			.66	1.57	2.23	2.72
5150	cycle 4,000'		572	.014			.68	1.64	2.32	2.83
5160	cycle 0.5 mile		616	.013			.64	1.52	2.16	2.62
5170	cycle 1 mile		550	.015			.71	1.70	2.41	2.94
5180	cycle 2 miles		462	.017			.85	2.02	2.87	3.50
5190	cycle 4 miles		330	.024			1.19	2.83	4.02	4.90
5200	20 MPH, cycle 2 miles		506	.016			.77	1.85	2.62	3.19
5210	cycle 4 miles		374	.021			1.05	2.50	3.55	4.32
5220	25 MPH, cycle 2 miles		528	.015			.74	1.77	2.51	3.06
5230	cycle 4 miles		418	.019			.94	2.24	3.18	3.86
5300	20 min. wait/ld./uld., 5 MPH, cycle 2,000'		418	.019			.94	2.24	3.18	3.86
5310	cycle 3,000'		396	.020			.99	2.36	3.35	4.08
5320	cycle 4,000'		352	.023			1.11	2.66	3.77	4.59
5330	cycle 0.5 mile		396	.020			.99	2.36	3.35	4.08
5340	cycle 1 mile		330	.024			1.19	2.83	4.02	4.90
5350	cycle 2 miles		242	.033			1.62	3.86	5.48	6.70
5360	10 MPH, cycle 2,000'		462	.017			.85	2.02	2.87	3.50
5370	cycle 3,000'		440	.018			.89	2.13	3.02	3.67
5380	cycle 4,000'		418	.019			.94	2.24	3.18	3.86
5390	cycle 0.5 mile		462	.017			.85	2.02	2.87	3.50
5400	cycle 1 mile		396	.020			.99	2.36	3.35	4.08
5410	cycle 2 miles		330	.024			1.19	2.83	4.02	4.90
5420	cycle 4 miles		242	.033			1.62	3.86	5.48	6.70
5430	15 MPH, cycle 2,000'		484	.017			.81	1.93	2.74	3.34
5440	cycle 3,000'		462	.017			.85	2.02	2.87	3.50
5450	cycle 4,000'		462	.017			.85	2.02	2.87	3.50
5460	cycle 0.5 mile		484	.017			.81	1.93	2.74	3.34
5470	cycle 1 mile		440	.018			.89	2.13	3.02	3.67
5480	cycle 2 miles		374	.021			1.05	2.50	3.55	4.32
5490	cycle 4 miles		286	.028			1.37	3.27	4.64	5.65
5500	20 MPH, cycle 2 miles		396	.020			.99	2.36	3.35	4.08

31 23 23.20 Hauling	Crew	Daily Output	Labor-Hours	Unit	Material	2020 Bare Costs Labor	Equipment	Total	Total Incl O&P
5510	cycle 4 miles	B-34F	330	.024	L.C.Y.		1.19	2.83	4.02	4.90
5520	25 MPH, cycle 2 miles		418	.019			.94	2.24	3.18	3.86
5530	cycle 4 miles		352	.023			1.11	2.66	3.77	4.59
5600	25 min. wait/ld./uld., 5 MPH, cycle 2,000'		352	.023			1.11	2.66	3.77	4.59
5610	cycle 3,000'		330	.024			1.19	2.83	4.02	4.90
5620	cycle 4,000'		308	.026			1.27	3.04	4.31	5.25
5630	cycle 0.5 mile		330	.024			1.19	2.83	4.02	4.90
5640	cycle 1 mile		286	.028			1.37	3.27	4.64	5.65
5650	cycle 2 miles		220	.036			1.78	4.25	6.03	7.35
5660	10 MPH, cycle 2,000'		374	.021			1.05	2.50	3.55	4.32
5670	cycle 3,000'		374	.021			1.05	2.50	3.55	4.32
5680	cycle 4,000'		352	.023			1.11	2.66	3.77	4.59
5690	cycle 0.5 mile		374	.021			1.05	2.50	3.55	4.32
5700	cycle 1 mile		330	.024			1.19	2.83	4.02	4.90
5710	cycle 2 miles		286	.028			1.37	3.27	4.64	5.65
5720	cycle 4 miles		220	.036			1.78	4.25	6.03	7.35
5730	15 MPH, cycle 2,000'		396	.020			.99	2.36	3.35	4.08
5740	cycle 3,000'		374	.021			1.05	2.50	3.55	4.32
5750	cycle 4,000'		374	.021			1.05	2.50	3.55	4.32
5760	cycle 0.5 mile		374	.021			1.05	2.50	3.55	4.32
5770	cycle 1 mile		352	.023			1.11	2.66	3.77	4.59
5780	cycle 2 miles		308	.026			1.27	3.04	4.31	5.25
5790	cycle 4 miles		242	.033			1.62	3.86	5.48	6.70
5800	20 MPH, cycle 2 miles		330	.024			1.19	2.83	4.02	4.90
5810	cycle 4 miles		286	.028			1.37	3.27	4.64	5.65
5820	25 MPH, cycle 2 miles		352	.023			1.11	2.66	3.77	4.59
5830	cycle 4 miles		308	.026			1.27	3.04	4.31	5.25
6000	34 C.Y. off-road, 15 min. wait/ld./uld., 5 MPH, cycle 2,000'	B-34G	816	.010			.48	2.17	2.65	3.11
6010	cycle 3,000'		748	.011			.52	2.37	2.89	3.39
6020	cycle 4,000'		680	.012			.58	2.60	3.18	3.72
6030	cycle 0.5 mile		782	.010			.50	2.26	2.76	3.24
6040	cycle 1 mile		578	.014			.68	3.06	3.74	4.39
6050	cycle 2 miles		408	.020			.96	4.34	5.30	6.20
6060	10 MPH, cycle 2,000'		918	.009			.43	1.93	2.36	2.76
6070	cycle 3,000'		884	.009			.44	2	2.44	2.86
6080	cycle 4,000'		816	.010			.48	2.17	2.65	3.11
6090	cycle 0.5 mile		884	.009			.44	2	2.44	2.86
6100	cycle 1 mile		782	.010			.50	2.26	2.76	3.24
6110	cycle 2 miles		578	.014			.68	3.06	3.74	4.39
6120	cycle 4 miles		408	.020			.96	4.34	5.30	6.20
6130	15 MPH, cycle 2,000'		986	.008			.40	1.80	2.20	2.56
6140	cycle 3,000'		918	.009			.43	1.93	2.36	2.76
6150	cycle 4,000'		884	.009			.44	2	2.44	2.86
6160	cycle 0.5 mile		952	.008			.41	1.86	2.27	2.67
6170	cycle 1 mile		850	.009			.46	2.08	2.54	2.98
6180	cycle 2 miles		714	.011			.55	2.48	3.03	3.55
6190	cycle 4 miles		510	.016			.77	3.47	4.24	4.97
6200	20 MPH, cycle 2 miles		782	.010			.50	2.26	2.76	3.24
6210	cycle 4 miles		578	.014			.68	3.06	3.74	4.39
6220	25 MPH, cycle 2 miles		816	.010			.48	2.17	2.65	3.11
6230	cycle 4 miles		646	.012			.61	2.74	3.35	3.92
6300	20 min. wait/ld./uld., 5 MPH, cycle 2,000'		646	.012			.61	2.74	3.35	3.92
6310	cycle 3,000'		612	.013			.64	2.89	3.53	4.14

31 23 23.20 Hauling		Crew	Daily Output	Labor-Hours	Unit	Material	2020 Bare Costs Labor	Equipment	Total	Total Incl O&P
6320	cycle 4,000'	B-34G	544	.015	L.C.Y.		.72	3.26	3.98	4.66
6330	cycle 0.5 mile		612	.013			.64	2.89	3.53	4.14
6340	cycle 1 mile		510	.016			.77	3.47	4.24	4.97
6350	cycle 2 miles		374	.021			1.05	4.74	5.79	6.75
6360	10 MPH, cycle 2,000'		714	.011			.55	2.48	3.03	3.55
6370	cycle 3,000'		680	.012			.58	2.60	3.18	3.72
6380	cycle 4,000'		646	.012			.61	2.74	3.35	3.92
6390	cycle 0.5 mile		714	.011			.55	2.48	3.03	3.55
6400	cycle 1 mile		612	.013			.64	2.89	3.53	4.14
6410	cycle 2 miles		510	.016			.77	3.47	4.24	4.97
6420	cycle 4 miles		374	.021			1.05	4.74	5.79	6.75
6430	15 MPH, cycle 2,000'		748	.011			.52	2.37	2.89	3.39
6440	cycle 3,000'		714	.011			.55	2.48	3.03	3.55
6450	cycle 4,000'		714	.011			.55	2.48	3.03	3.55
6460	cycle 0.5 mile		748	.011			.52	2.37	2.89	3.39
6470	cycle 1 mile		680	.012			.58	2.60	3.18	3.72
6480	cycle 2 miles		578	.014			.68	3.06	3.74	4.39
6490	cycle 4 miles		442	.018			.89	4.01	4.90	5.75
6500	20 MPH, cycle 2 miles		612	.013			.64	2.89	3.53	4.14
6510	cycle 4 miles		510	.016			.77	3.47	4.24	4.97
6520	25 MPH, cycle 2 miles		646	.012			.61	2.74	3.35	3.92
6530	cycle 4 miles		544	.015			.72	3.26	3.98	4.66
6600	25 min. wait/ld./uld., 5 MPH, cycle 2,000'		544	.015			.72	3.26	3.98	4.66
6610	cycle 3,000'		510	.016			.77	3.47	4.24	4.97
6620	cycle 4,000'		476	.017			.82	3.72	4.54	5.30
6630	cycle 0.5 mile		510	.016			.77	3.47	4.24	4.97
6640	cycle 1 mile		442	.018			.89	4.01	4.90	5.75
6650	cycle 2 miles		340	.024			1.15	5.20	6.35	7.50
6660	10 MPH, cycle 2,000'		578	.014			.68	3.06	3.74	4.39
6670	cycle 3,000'		578	.014			.68	3.06	3.74	4.39
6680	cycle 4,000'		544	.015			.72	3.26	3.98	4.66
6690	cycle 0.5 mile		578	.014			.68	3.06	3.74	4.39
6700	cycle 1 mile		510	.016			.77	3.47	4.24	4.97
6710	cycle 2 miles		442	.018			.89	4.01	4.90	5.75
6720	cycle 4 miles		340	.024			1.15	5.20	6.35	7.50
6730	15 MPH, cycle 2,000'		612	.013			.64	2.89	3.53	4.14
6740	cycle 3,000'		578	.014			.68	3.06	3.74	4.39
6750	cycle 4,000'		578	.014			.68	3.06	3.74	4.39
6760	cycle 0.5 mile		612	.013			.64	2.89	3.53	4.14
6770	cycle 1 mile		544	.015			.72	3.26	3.98	4.66
6780	cycle 2 miles		476	.017			.82	3.72	4.54	5.30
6790	cycle 4 miles		374	.021			1.05	4.74	5.79	6.75
6800	20 MPH, cycle 2 miles		510	.016			.77	3.47	4.24	4.97
6810	cycle 4 miles		442	.018			.89	4.01	4.90	5.75
6820	25 MPH, cycle 2 miles		544	.015			.72	3.26	3.98	4.66
6830	cycle 4 miles		476	.017			.82	3.72	4.54	5.30
7000	42 C.Y. off-road, 20 min. wait/ld./uld., 5 MPH, cycle 2,000'	B-34H	798	.010			.49	2.40	2.89	3.38
7010	cycle 3,000'		756	.011			.52	2.53	3.05	3.57
7020	cycle 4,000'		672	.012			.58	2.85	3.43	4
7030	cycle 0.5 mile		756	.011			.52	2.53	3.05	3.57
7040	cycle 1 mile		630	.013			.62	3.04	3.66	4.27
7050	cycle 2 miles		462	.017			.85	4.15	5	5.85
7060	10 MPH, cycle 2,000'		882	.009			.44	2.17	2.61	3.06

31 23 Excavation and Fill

31 23 23 – Fill

31 23 23.20 Hauling		Crew	Daily Output	Labor-Hours	Unit	Material	2020 Bare Costs Labor	Equipment	Total	Total Incl O&P
7070	cycle 3,000'	B-34H	840	.010	L.C.Y.		.47	2.28	2.75	3.21
7080	cycle 4,000'		798	.010			.49	2.40	2.89	3.38
7090	cycle 0.5 mile		882	.009			.44	2.17	2.61	3.06
7100	cycle 1 mile		798	.010			.49	2.40	2.89	3.38
7110	cycle 2 miles		630	.013			.62	3.04	3.66	4.27
7120	cycle 4 miles		462	.017			.85	4.15	5	5.85
7130	15 MPH, cycle 2,000'		924	.009			.42	2.07	2.49	2.92
7140	cycle 3,000'		882	.009			.44	2.17	2.61	3.06
7150	cycle 4,000'		882	.009			.44	2.17	2.61	3.06
7160	cycle 0.5 mile		882	.009			.44	2.17	2.61	3.06
7170	cycle 1 mile		840	.010			.47	2.28	2.75	3.21
7180	cycle 2 miles		714	.011			.55	2.68	3.23	3.77
7190	cycle 4 miles		546	.015			.72	3.51	4.23	4.93
7200	20 MPH, cycle 2 miles		756	.011			.52	2.53	3.05	3.57
7210	cycle 4 miles		630	.013			.62	3.04	3.66	4.27
7220	25 MPH, cycle 2 miles		798	.010			.49	2.40	2.89	3.38
7230	cycle 4 miles		672	.012			.58	2.85	3.43	4
7300	25 min. wait/ld./uld., 5 MPH, cycle 2,000'		672	.012			.58	2.85	3.43	4
7310	cycle 3,000'		630	.013			.62	3.04	3.66	4.27
7320	cycle 4,000'		588	.014			.67	3.26	3.93	4.58
7330	cycle 0.5 mile		630	.013			.62	3.04	3.66	4.27
7340	cycle 1 mile		546	.015			.72	3.51	4.23	4.93
7350	cycle 2 miles		378	.021			1.04	5.05	6.09	7.10
7360	10 MPH, cycle 2,000'		714	.011			.55	2.68	3.23	3.77
7370	cycle 3,000'		714	.011			.55	2.68	3.23	3.77
7380	cycle 4,000'		672	.012			.58	2.85	3.43	4
7390	cycle 0.5 mile		714	.011			.55	2.68	3.23	3.77
7400	cycle 1 mile		630	.013			.62	3.04	3.66	4.27
7410	cycle 2 miles		546	.015			.72	3.51	4.23	4.93
7420	cycle 4 miles		378	.021			1.04	5.05	6.09	7.10
7430	15 MPH, cycle 2,000'		756	.011			.52	2.53	3.05	3.57
7440	cycle 3,000'		714	.011			.55	2.68	3.23	3.77
7450	cycle 4,000'		714	.011			.55	2.68	3.23	3.77
7460	cycle 0.5 mile		714	.011			.55	2.68	3.23	3.77
7470	cycle 1 mile		672	.012			.58	2.85	3.43	4
7480	cycle 2 miles		588	.014			.67	3.26	3.93	4.58
7490	cycle 4 miles		462	.017			.85	4.15	5	5.85
7500	20 MPH, cycle 2 miles		630	.013			.62	3.04	3.66	4.27
7510	cycle 4 miles		546	.015			.72	3.51	4.23	4.93
7520	25 MPH, cycle 2 miles		672	.012			.58	2.85	3.43	4
7530	cycle 4 miles		588	.014			.67	3.26	3.93	4.58
8000	60 C.Y. off-road, 20 min. wait/ld./uld., 5 MPH, cycle 2,000'	B-34J	1140	.007			.34	2.40	2.74	3.15
8010	cycle 3,000'		1080	.007			.36	2.53	2.89	3.33
8020	cycle 4,000'		960	.008			.41	2.85	3.26	3.74
8030	cycle 0.5 mile		1080	.007			.36	2.53	2.89	3.33
8040	cycle 1 mile		900	.009			.44	3.04	3.48	3.99
8050	cycle 2 miles		660	.012			.59	4.15	4.74	5.45
8060	10 MPH, cycle 2,000'		1260	.006			.31	2.17	2.48	2.86
8070	cycle 3,000'		1200	.007			.33	2.28	2.61	3
8080	cycle 4,000'		1140	.007			.34	2.40	2.74	3.15
8090	cycle 0.5 mile		1260	.006			.31	2.17	2.48	2.86
8100	cycle 1 mile		1080	.007			.36	2.53	2.89	3.33
8110	cycle 2 miles		900	.009			.44	3.04	3.48	3.99

For customer support on your Site Work & Landscape Costs with RSMeans data, call 800.448.8182.

327

31 23 23 – Fill

31 23 23.20 Hauling		Crew	Daily Output	Labor-Hours	Unit	Material	2020 Bare Costs Labor	2020 Bare Costs Equipment	Total	Total Incl O&P
8120	cycle 4 miles	B-34J	660	.012	L.C.Y.		.59	4.15	4.74	5.45
8130	15 MPH, cycle 2,000'		1320	.006			.30	2.07	2.37	2.72
8140	cycle 3,000'		1260	.006			.31	2.17	2.48	2.86
8150	cycle 4,000'		1260	.006			.31	2.17	2.48	2.86
8160	cycle 0.5 mile		1320	.006			.30	2.07	2.37	2.72
8170	cycle 1 mile		1200	.007			.33	2.28	2.61	3
8180	cycle 2 miles		1020	.008			.38	2.68	3.06	3.53
8190	cycle 4 miles		780	.010			.50	3.51	4.01	4.61
8200	20 MPH, cycle 2 miles		1080	.007			.36	2.53	2.89	3.33
8210	cycle 4 miles		900	.009			.44	3.04	3.48	3.99
8220	25 MPH, cycle 2 miles		1140	.007			.34	2.40	2.74	3.15
8230	cycle 4 miles		960	.008			.41	2.85	3.26	3.74
8300	25 min. wait/ld./uld., 5 MPH, cycle 2,000'		960	.008			.41	2.85	3.26	3.74
8310	cycle 3,000'		900	.009			.44	3.04	3.48	3.99
8320	cycle 4,000'		840	.010			.47	3.26	3.73	4.28
8330	cycle 0.5 mile		900	.009			.44	3.04	3.48	3.99
8340	cycle 1 mile		780	.010			.50	3.51	4.01	4.61
8350	cycle 2 miles		600	.013			.65	4.56	5.21	6
8360	10 MPH, cycle 2,000'		1020	.008			.38	2.68	3.06	3.53
8370	cycle 3,000'		1020	.008			.38	2.68	3.06	3.53
8380	cycle 4,000'		960	.008			.41	2.85	3.26	3.74
8390	cycle 0.5 mile		1020	.008			.38	2.68	3.06	3.53
8400	cycle 1 mile		900	.009			.44	3.04	3.48	3.99
8410	cycle 2 miles		780	.010			.50	3.51	4.01	4.61
8420	cycle 4 miles		600	.013			.65	4.56	5.21	6
8430	15 MPH, cycle 2,000'		1080	.007			.36	2.53	2.89	3.33
8440	cycle 3,000'		1020	.008			.38	2.68	3.06	3.53
8450	cycle 4,000'		1020	.008			.38	2.68	3.06	3.53
8460	cycle 0.5 mile		1080	.007			.36	2.53	2.89	3.33
8470	cycle 1 mile		960	.008			.41	2.85	3.26	3.74
8480	cycle 2 miles		840	.010			.47	3.26	3.73	4.28
8490	cycle 4 miles		660	.012			.59	4.15	4.74	5.45
8500	20 MPH, cycle 2 miles		900	.009			.44	3.04	3.48	3.99
8510	cycle 4 miles		780	.010			.50	3.51	4.01	4.61
8520	25 MPH, cycle 2 miles		960	.008			.41	2.85	3.26	3.74
8530	cycle 4 miles		840	.010			.47	3.26	3.73	4.28
9014	18 C.Y. truck, 8 wheels,15 min. wait/ld./uld.,15 MPH, cycle 0.5 mi.	B-34I	504	.016			.78	1.42	2.20	2.72
9016	cycle 1 mile		450	.018			.87	1.59	2.46	3.05
9018	cycle 2 miles		378	.021			1.04	1.89	2.93	3.63
9020	cycle 4 miles		270	.030			1.45	2.64	4.09	5.10
9022	cycle 6 miles		216	.037			1.81	3.31	5.12	6.35
9024	cycle 8 miles		180	.044			2.18	3.97	6.15	7.60
9025	cycle 10 miles		144	.056			2.72	4.96	7.68	9.55
9026	20 MPH avg., cycle 0.5 mile		522	.015			.75	1.37	2.12	2.62
9028	cycle 1 mile		468	.017			.84	1.52	2.36	2.93
9030	cycle 2 miles		414	.019			.95	1.72	2.67	3.32
9032	cycle 4 miles		324	.025			1.21	2.20	3.41	4.23
9034	cycle 6 miles		252	.032			1.55	2.83	4.38	5.45
9036	cycle 8 miles		216	.037			1.81	3.31	5.12	6.35
9038	cycle 10 miles		180	.044			2.18	3.97	6.15	7.60
9040	25 MPH avg., cycle 4 miles		342	.023			1.14	2.09	3.23	4.02
9042	cycle 6 miles		288	.028			1.36	2.48	3.84	4.77
9044	cycle 8 miles		252	.032			1.55	2.83	4.38	5.45

31 23 23 – Fill

31 23 23.20 Hauling		Crew	Daily Output	Labor-Hours	Unit	Material	2020 Bare Costs Labor	2020 Bare Costs Equipment	Total	Total Incl O&P
9046	cycle 10 miles	B-34I	216	.037	L.C.Y.		1.81	3.31	5.12	6.35
9050	30 MPH avg., cycle 4 miles		378	.021			1.04	1.89	2.93	3.63
9052	cycle 6 miles		324	.025			1.21	2.20	3.41	4.23
9054	cycle 8 miles		270	.030			1.45	2.64	4.09	5.10
9056	cycle 10 miles		234	.034			1.67	3.05	4.72	5.85
9060	35 MPH avg., cycle 4 miles		396	.020			.99	1.80	2.79	3.46
9062	cycle 6 miles		342	.023			1.14	2.09	3.23	4.02
9064	cycle 8 miles		288	.028			1.36	2.48	3.84	4.77
9066	cycle 10 miles		270	.030			1.45	2.64	4.09	5.10
9068	cycle 20 miles		162	.049			2.42	4.41	6.83	8.45
9070	cycle 30 miles		126	.063			3.11	5.65	8.76	10.90
9072	cycle 40 miles		90	.089			4.35	7.95	12.30	15.20
9074	40 MPH avg., cycle 6 miles		360	.022			1.09	1.98	3.07	3.81
9076	cycle 8 miles		324	.025			1.21	2.20	3.41	4.23
9078	cycle 10 miles		288	.028			1.36	2.48	3.84	4.77
9080	cycle 20 miles		180	.044			2.18	3.97	6.15	7.60
9082	cycle 30 miles		144	.056			2.72	4.96	7.68	9.55
9084	cycle 40 miles		108	.074			3.63	6.60	10.23	12.70
9086	cycle 50 miles		90	.089			4.35	7.95	12.30	15.20
9094	45 MPH avg., cycle 8 miles		324	.025			1.21	2.20	3.41	4.23
9096	cycle 10 miles		306	.026			1.28	2.33	3.61	4.49
9098	cycle 20 miles		198	.040			1.98	3.60	5.58	6.95
9100	cycle 30 miles		144	.056			2.72	4.96	7.68	9.55
9102	cycle 40 miles		126	.063			3.11	5.65	8.76	10.90
9104	cycle 50 miles		108	.074			3.63	6.60	10.23	12.70
9106	50 MPH avg., cycle 10 miles		324	.025			1.21	2.20	3.41	4.23
9108	cycle 20 miles		216	.037			1.81	3.31	5.12	6.35
9110	cycle 30 miles		162	.049			2.42	4.41	6.83	8.45
9112	cycle 40 miles		126	.063			3.11	5.65	8.76	10.90
9114	cycle 50 miles		108	.074			3.63	6.60	10.23	12.70
9214	20 min. wait/ld./uld.,15 MPH, cycle 0.5 mi.		396	.020			.99	1.80	2.79	3.46
9216	cycle 1 mile		360	.022			1.09	1.98	3.07	3.81
9218	cycle 2 miles		306	.026			1.28	2.33	3.61	4.49
9220	cycle 4 miles		234	.034			1.67	3.05	4.72	5.85
9222	cycle 6 miles		198	.040			1.98	3.60	5.58	6.95
9224	cycle 8 miles		162	.049			2.42	4.41	6.83	8.45
9225	cycle 10 miles		144	.056			2.72	4.96	7.68	9.55
9226	20 MPH avg., cycle 0.5 mile		396	.020			.99	1.80	2.79	3.46
9228	cycle 1 mile		378	.021			1.04	1.89	2.93	3.63
9230	cycle 2 miles		324	.025			1.21	2.20	3.41	4.23
9232	cycle 4 miles		270	.030			1.45	2.64	4.09	5.10
9234	cycle 6 miles		216	.037			1.81	3.31	5.12	6.35
9236	cycle 8 miles		198	.040			1.98	3.60	5.58	6.95
9238	cycle 10 miles		162	.049			2.42	4.41	6.83	8.45
9240	25 MPH avg., cycle 4 miles		288	.028			1.36	2.48	3.84	4.77
9242	cycle 6 miles		252	.032			1.55	2.83	4.38	5.45
9244	cycle 8 miles		216	.037			1.81	3.31	5.12	6.35
9246	cycle 10 miles		198	.040			1.98	3.60	5.58	6.95
9250	30 MPH avg., cycle 4 miles		306	.026			1.28	2.33	3.61	4.49
9252	cycle 6 miles		270	.030			1.45	2.64	4.09	5.10
9254	cycle 8 miles		234	.034			1.67	3.05	4.72	5.85
9256	cycle 10 miles		216	.037			1.81	3.31	5.12	6.35
9260	35 MPH avg., cycle 4 miles		324	.025			1.21	2.20	3.41	4.23

31 23 23.20 Hauling		Crew	Daily Output	Labor-Hours	Unit	Material	2020 Bare Costs		Total	Total Incl O&P
							Labor	Equipment		
9262	cycle 6 miles	B-34I	288	.028	L.C.Y.		1.36	2.48	3.84	4.77
9264	cycle 8 miles		252	.032			1.55	2.83	4.38	5.45
9266	cycle 10 miles		234	.034			1.67	3.05	4.72	5.85
9268	cycle 20 miles		162	.049			2.42	4.41	6.83	8.45
9270	cycle 30 miles		108	.074			3.63	6.60	10.23	12.70
9272	cycle 40 miles		90	.089			4.35	7.95	12.30	15.20
9274	40 MPH avg., cycle 6 miles		288	.028			1.36	2.48	3.84	4.77
9276	cycle 8 miles		270	.030			1.45	2.64	4.09	5.10
9278	cycle 10 miles		234	.034			1.67	3.05	4.72	5.85
9280	cycle 20 miles		162	.049			2.42	4.41	6.83	8.45
9282	cycle 30 miles		126	.063			3.11	5.65	8.76	10.90
9284	cycle 40 miles		108	.074			3.63	6.60	10.23	12.70
9286	cycle 50 miles		90	.089			4.35	7.95	12.30	15.20
9294	45 MPH avg., cycle 8 miles		270	.030			1.45	2.64	4.09	5.10
9296	cycle 10 miles		252	.032			1.55	2.83	4.38	5.45
9298	cycle 20 miles		180	.044			2.18	3.97	6.15	7.60
9300	cycle 30 miles		144	.056			2.72	4.96	7.68	9.55
9302	cycle 40 miles		108	.074			3.63	6.60	10.23	12.70
9304	cycle 50 miles		90	.089			4.35	7.95	12.30	15.20
9306	50 MPH avg., cycle 10 miles		270	.030			1.45	2.64	4.09	5.10
9308	cycle 20 miles		198	.040			1.98	3.60	5.58	6.95
9310	cycle 30 miles		144	.056			2.72	4.96	7.68	9.55
9312	cycle 40 miles		126	.063			3.11	5.65	8.76	10.90
9314	cycle 50 miles		108	.074			3.63	6.60	10.23	12.70
9414	25 min. wait/ld./uld.,15 MPH, cycle 0.5 mi.		306	.026			1.28	2.33	3.61	4.49
9416	cycle 1 mile		288	.028			1.36	2.48	3.84	4.77
9418	cycle 2 miles		252	.032			1.55	2.83	4.38	5.45
9420	cycle 4 miles		198	.040			1.98	3.60	5.58	6.95
9422	cycle 6 miles		180	.044			2.18	3.97	6.15	7.60
9424	cycle 8 miles		144	.056			2.72	4.96	7.68	9.55
9425	cycle 10 miles		126	.063			3.11	5.65	8.76	10.90
9426	20 MPH avg., cycle 0.5 mile		324	.025			1.21	2.20	3.41	4.23
9428	cycle 1 mile		306	.026			1.28	2.33	3.61	4.49
9430	cycle 2 miles		270	.030			1.45	2.64	4.09	5.10
9432	cycle 4 miles		234	.034			1.67	3.05	4.72	5.85
9434	cycle 6 miles		198	.040			1.98	3.60	5.58	6.95
9436	cycle 8 miles		180	.044			2.18	3.97	6.15	7.60
9438	cycle 10 miles		144	.056			2.72	4.96	7.68	9.55
9440	25 MPH avg., cycle 4 miles		252	.032			1.55	2.83	4.38	5.45
9442	cycle 6 miles		216	.037			1.81	3.31	5.12	6.35
9444	cycle 8 miles		198	.040			1.98	3.60	5.58	6.95
9446	cycle 10 miles		180	.044			2.18	3.97	6.15	7.60
9450	30 MPH avg., cycle 4 miles		252	.032			1.55	2.83	4.38	5.45
9452	cycle 6 miles		234	.034			1.67	3.05	4.72	5.85
9454	cycle 8 miles		198	.040			1.98	3.60	5.58	6.95
9456	cycle 10 miles		180	.044			2.18	3.97	6.15	7.60
9460	35 MPH avg., cycle 4 miles		270	.030			1.45	2.64	4.09	5.10
9462	cycle 6 miles		234	.034			1.67	3.05	4.72	5.85
9464	cycle 8 miles		216	.037			1.81	3.31	5.12	6.35
9466	cycle 10 miles		198	.040			1.98	3.60	5.58	6.95
9468	cycle 20 miles		144	.056			2.72	4.96	7.68	9.55
9470	cycle 30 miles		108	.074			3.63	6.60	10.23	12.70
9472	cycle 40 miles		90	.089			4.35	7.95	12.30	15.20

31 23 Excavation and Fill

31 23 23 – Fill

31 23 23.20 Hauling		Crew	Daily Output	Labor-Hours	Unit	Material	2020 Bare Costs Labor	Equipment	Total	Total Incl O&P
9474	40 MPH avg., cycle 6 miles	B-34I	252	.032	L.C.Y.		1.55	2.83	4.38	5.45
9476	cycle 8 miles		234	.034			1.67	3.05	4.72	5.85
9478	cycle 10 miles		216	.037			1.81	3.31	5.12	6.35
9480	cycle 20 miles		144	.056			2.72	4.96	7.68	9.55
9482	cycle 30 miles		126	.063			3.11	5.65	8.76	10.90
9484	cycle 40 miles		90	.089			4.35	7.95	12.30	15.20
9486	cycle 50 miles		90	.089			4.35	7.95	12.30	15.20
9494	45 MPH avg., cycle 8 miles		234	.034			1.67	3.05	4.72	5.85
9496	cycle 10 miles		216	.037			1.81	3.31	5.12	6.35
9498	cycle 20 miles		162	.049			2.42	4.41	6.83	8.45
9500	cycle 30 miles		126	.063			3.11	5.65	8.76	10.90
9502	cycle 40 miles		108	.074			3.63	6.60	10.23	12.70
9504	cycle 50 miles		90	.089			4.35	7.95	12.30	15.20
9506	50 MPH avg., cycle 10 miles		234	.034			1.67	3.05	4.72	5.85
9508	cycle 20 miles		180	.044			2.18	3.97	6.15	7.60
9510	cycle 30 miles		144	.056			2.72	4.96	7.68	9.55
9512	cycle 40 miles		108	.074			3.63	6.60	10.23	12.70
9514	cycle 50 miles		90	.089			4.35	7.95	12.30	15.20
9614	30 min. wait/ld./uld.,15 MPH, cycle 0.5 mi.		270	.030			1.45	2.64	4.09	5.10
9616	cycle 1 mile		252	.032			1.55	2.83	4.38	5.45
9618	cycle 2 miles		216	.037			1.81	3.31	5.12	6.35
9620	cycle 4 miles		180	.044			2.18	3.97	6.15	7.60
9622	cycle 6 miles		162	.049			2.42	4.41	6.83	8.45
9624	cycle 8 miles		126	.063			3.11	5.65	8.76	10.90
9625	cycle 10 miles		126	.063			3.11	5.65	8.76	10.90
9626	20 MPH avg., cycle 0.5 mile		270	.030			1.45	2.64	4.09	5.10
9628	cycle 1 mile		252	.032			1.55	2.83	4.38	5.45
9630	cycle 2 miles		234	.034			1.67	3.05	4.72	5.85
9632	cycle 4 miles		198	.040			1.98	3.60	5.58	6.95
9634	cycle 6 miles		180	.044			2.18	3.97	6.15	7.60
9636	cycle 8 miles		162	.049			2.42	4.41	6.83	8.45
9638	cycle 10 miles		144	.056			2.72	4.96	7.68	9.55
9640	25 MPH avg., cycle 4 miles		216	.037			1.81	3.31	5.12	6.35
9642	cycle 6 miles		198	.040			1.98	3.60	5.58	6.95
9644	cycle 8 miles		180	.044			2.18	3.97	6.15	7.60
9646	cycle 10 miles		162	.049			2.42	4.41	6.83	8.45
9650	30 MPH avg., cycle 4 miles		216	.037			1.81	3.31	5.12	6.35
9652	cycle 6 miles		198	.040			1.98	3.60	5.58	6.95
9654	cycle 8 miles		180	.044			2.18	3.97	6.15	7.60
9656	cycle 10 miles		162	.049			2.42	4.41	6.83	8.45
9660	35 MPH avg., cycle 4 miles		234	.034			1.67	3.05	4.72	5.85
9662	cycle 6 miles		216	.037			1.81	3.31	5.12	6.35
9664	cycle 8 miles		198	.040			1.98	3.60	5.58	6.95
9666	cycle 10 miles		180	.044			2.18	3.97	6.15	7.60
9668	cycle 20 miles		126	.063			3.11	5.65	8.76	10.90
9670	cycle 30 miles		108	.074			3.63	6.60	10.23	12.70
9672	cycle 40 miles		90	.089			4.35	7.95	12.30	15.20
9674	40 MPH avg., cycle 6 miles		216	.037			1.81	3.31	5.12	6.35
9676	cycle 8 miles		198	.040			1.98	3.60	5.58	6.95
9678	cycle 10 miles		180	.044			2.18	3.97	6.15	7.60
9680	cycle 20 miles		144	.056			2.72	4.96	7.68	9.55
9682	cycle 30 miles		108	.074			3.63	6.60	10.23	12.70
9684	cycle 40 miles		90	.089			4.35	7.95	12.30	15.20

331

31 23 23.20 Hauling		Crew	Daily Output	Labor-Hours	Unit	Material	2020 Bare Costs Labor	Equipment	Total	Total Incl O&P
9686	cycle 50 miles	B-34I	72	.111	L.C.Y.		5.45	9.90	15.35	19.05
9694	45 MPH avg., cycle 8 miles		216	.037			1.81	3.31	5.12	6.35
9696	cycle 10 miles		198	.040			1.98	3.60	5.58	6.95
9698	cycle 20 miles		144	.056			2.72	4.96	7.68	9.55
9700	cycle 30 miles		126	.063			3.11	5.65	8.76	10.90
9702	cycle 40 miles		108	.074			3.63	6.60	10.23	12.70
9704	cycle 50 miles		90	.089			4.35	7.95	12.30	15.20
9706	50 MPH avg., cycle 10 miles		198	.040			1.98	3.60	5.58	6.95
9708	cycle 20 miles		162	.049			2.42	4.41	6.83	8.45
9710	cycle 30 miles		126	.063			3.11	5.65	8.76	10.90
9712	cycle 40 miles		108	.074			3.63	6.60	10.23	12.70
9714	cycle 50 miles		90	.089			4.35	7.95	12.30	15.20

31 23 23.23 Compaction

		Crew	Daily Output	Labor-Hours	Unit	Material	2020 Bare Costs Labor	Equipment	Total	Total Incl O&P
0010	**COMPACTION** R312323-30									
5000	Riding, vibrating roller, 6" lifts, 2 passes	B-10Y	3000	.004	E.C.Y.		.21	.19	.40	.52
5020	3 passes		2300	.005			.27	.25	.52	.69
5040	4 passes		1900	.006			.33	.30	.63	.83
5050	8" lifts, 2 passes		4100	.003			.15	.14	.29	.39
5060	12" lifts, 2 passes		5200	.002			.12	.11	.23	.30
5080	3 passes		3500	.003			.18	.17	.35	.45
5100	4 passes		2600	.005			.24	.22	.46	.61
5600	Sheepsfoot or wobbly wheel roller, 6" lifts, 2 passes	B-10G	2400	.005			.26	.56	.82	1.01
5620	3 passes		1735	.007			.36	.78	1.14	1.40
5640	4 passes		1300	.009			.48	1.04	1.52	1.86
5680	12" lifts, 2 passes		5200	.002			.12	.26	.38	.47
5700	3 passes		3500	.003			.18	.39	.57	.69
5720	4 passes		2600	.005			.24	.52	.76	.93
6000	Towed sheepsfoot or wobbly wheel roller, 6" lifts, 2 passes	B-10D	10000	.001			.06	.19	.25	.30
6020	3 passes		2000	.006			.31	.97	1.28	1.54
6030	4 passes		1500	.008			.42	1.29	1.71	2.04
6050	12" lifts, 2 passes		6000	.002			.10	.32	.42	.52
6060	3 passes		4000	.003			.16	.49	.65	.76
6070	4 passes		3000	.004			.21	.65	.86	1.02
6200	Vibrating roller, 6" lifts, 2 passes	B-10C	2600	.005			.24	.78	1.02	1.21
6210	3 passes		1735	.007			.36	1.16	1.52	1.82
6220	4 passes		1300	.009			.48	1.55	2.03	2.43
6250	12" lifts, 2 passes		5200	.002			.12	.39	.51	.61
6260	3 passes		3465	.003			.18	.58	.76	.91
6270	4 passes		2600	.005			.24	.78	1.02	1.21
7000	Walk behind, vibrating plate 18" wide, 6" lifts, 2 passes	A-1D	200	.040			1.68	.16	1.84	2.70
7020	3 passes		185	.043			1.82	.17	1.99	2.92
7040	4 passes		140	.057			2.41	.23	2.64	3.86
7200	12" lifts, 2 passes, 21" wide	A-1E	560	.014			.60	.30	.90	1.23
7220	3 passes		375	.021			.90	.44	1.34	1.84
7240	4 passes		280	.029			1.20	.59	1.79	2.46
7500	Vibrating roller 24" wide, 6" lifts, 2 passes	B-10A	420	.029			1.48	.40	1.88	2.66
7520	3 passes		280	.043			2.22	.59	2.81	3.98
7540	4 passes		210	.057			2.96	.79	3.75	5.30
7600	12" lifts, 2 passes		840	.014			.74	.20	.94	1.33
7620	3 passes		560	.021			1.11	.30	1.41	1.99
7640	4 passes		420	.029			1.48	.40	1.88	2.66
8000	Rammer tamper, 6" to 11", 4" lifts, 2 passes	A-1F	130	.062			2.59	.36	2.95	4.28

31 23 Excavation and Fill

31 23 23 – Fill

31 23 23.23 Compaction

		Crew	Daily Output	Labor-Hours	Unit	Material	2020 Bare Costs Labor	2020 Bare Costs Equipment	Total	Total Incl O&P
8050	3 passes	A-1F	97	.082	E.C.Y.		3.47	.48	3.95	5.75
8100	4 passes		65	.123			5.20	.72	5.92	8.60
8200	8" lifts, 2 passes		260	.031			1.30	.18	1.48	2.15
8250	3 passes		195	.041			1.73	.24	1.97	2.86
8300	4 passes		130	.062			2.59	.36	2.95	4.28
8400	13" to 18", 4" lifts, 2 passes	A-1G	390	.021			.86	.14	1	1.45
8450	3 passes		290	.028			1.16	.19	1.35	1.96
8500	4 passes		195	.041			1.73	.28	2.01	2.90
8600	8" lifts, 2 passes		780	.010			.43	.07	.50	.73
8650	3 passes		585	.014			.58	.09	.67	.97
8700	4 passes		390	.021			.86	.14	1	1.45
9000	Water, 3,000 gal. truck, 3 mile haul	B-45	1888	.008		1.22	.45	.44	2.11	2.49
9010	6 mile haul		1444	.011		1.22	.59	.57	2.38	2.85
9020	12 mile haul		1000	.016		1.22	.85	.82	2.89	3.52
9030	6,000 gal. wagon, 3 mile haul	B-59	2000	.004		1.22	.20	.23	1.65	1.88
9040	6 mile haul	"	1600	.005		1.22	.24	.29	1.75	2.03

31 23 23.24 Compaction, Structural

		Crew	Daily Output	Labor-Hours	Unit	Material	2020 Bare Costs Labor	2020 Bare Costs Equipment	Total	Total Incl O&P
0010	**COMPACTION, STRUCTURAL**									
0020	Steel wheel tandem roller, 5 tons	B-10E	8	1.500	Hr.		78	32	110	151
0050	Air tamp, 6" to 8" lifts, common fill	B-9	250	.160	E.C.Y.		6.80	1.43	8.23	11.75
0060	Select fill	"	300	.133	"		5.65	1.19	6.84	9.80
0100	10 tons	B-10F	8	1.500	Hr.		78	29.50	107.50	149
0300	Sheepsfoot or wobbly wheel roller, 8" lifts, common fill	B-10G	1300	.009	E.C.Y.		.48	1.04	1.52	1.86
0400	Select fill	"	1500	.008	"		.42	.90	1.32	1.61
0420	Static roller, riding 33" diam.	B-10A	2800	.004	S.Y.		.22	.06	.28	.40
0600	Vibratory plate, 8" lifts, common fill	A-1D	200	.040	E.C.Y.		1.68	.16	1.84	2.70
0700	Select fill	"	216	.037	"		1.56	.15	1.71	2.50

31 23 23.25 Compaction, Airports

		Crew	Daily Output	Labor-Hours	Unit	Material	2020 Bare Costs Labor	2020 Bare Costs Equipment	Total	Total Incl O&P
0010	**COMPACTION, AIRPORTS**									
0100	Airport subgrade, compaction									
0110	non cohesive soils, 85% proctor, 12" depth	B-10G	15600	.001	S.Y.		.04	.09	.13	.16
0200	24" depth		7800	.002			.08	.17	.25	.31
0300	36" depth		5200	.002			.12	.26	.38	.47
0400	48" depth		3900	.003			.16	.35	.51	.62
0500	60" depth		3120	.004			.20	.43	.63	.78
0600	90% proctor, 12" depth		7200	.002			.09	.19	.28	.34
0700	24" depth		3600	.003			.17	.37	.54	.67
0800	36" depth		2400	.005			.26	.56	.82	1.01
0900	48" depth		1800	.007			.35	.75	1.10	1.34
1000	60" depth		1440	.008			.43	.94	1.37	1.68
1100	95% proctor, 12" depth	B-10D	6000	.002			.10	.32	.42	.52
1200	24" depth		3000	.004			.21	.65	.86	1.02
1300	36" depth		2000	.006			.31	.97	1.28	1.54
1400	42" depth		1715	.007			.36	1.13	1.49	1.79
1500	100% proctor, 12" depth		4500	.003			.14	.43	.57	.69
1550	18" depth		3000	.004			.21	.65	.86	1.02
1600	24" depth		2250	.005			.28	.86	1.14	1.36
1700	cohesive soils, 80% proctor, 12" depth	B-10G	7200	.002			.09	.19	.28	.34
1800	24" depth	B-10D	15000	.001			.04	.13	.17	.20
1900	36" depth		10000	.001			.06	.19	.25	.30
2000	85% proctor, 12" depth		6000	.002			.10	.32	.42	.52
2100	18" depth		4000	.003			.16	.49	.65	.76

For customer support on your Site Work & Landscape Costs with RSMeans data, call 800.448.8182.

333

31 23 Excavation and Fill

31 23 23 – Fill

31 23 23.25 Compaction, Airports		Crew	Daily Output	Labor-Hours	Unit	Material	2020 Bare Costs Labor	Equipment	Total	Total Incl O&P
2200	24" depth	B-10D	3000	.004	S.Y.		.21	.65	.86	1.02
2300	27" depth		2670	.004			.23	.73	.96	1.15
2400	90% proctor, 6" depth		10400	.001			.06	.19	.25	.29
2500	9" depth		6935	.002			.09	.28	.37	.44
2600	12" depth		5200	.002			.12	.37	.49	.59
2700	15" depth		4160	.003			.15	.47	.62	.73
2800	18" depth		3470	.003			.18	.56	.74	.89
2900	95% proctor, 6" depth		1500	.008			.42	1.29	1.71	2.04
3000	7" depth		1285	.009			.48	1.51	1.99	2.39
3100	8" depth		1125	.011			.55	1.73	2.28	2.73
3200	9" depth		1000	.012			.62	1.94	2.56	3.07

31 25 Erosion and Sedimentation Controls

31 25 14 – Stabilization Measures for Erosion and Sedimentation Control

31 25 14.16 Rolled Erosion Control Mats and Blankets

			Crew	Daily Output	Labor-Hours	Unit	Material	2020 Bare Costs Labor	Equipment	Total	Total Incl O&P
0010	**ROLLED EROSION CONTROL MATS AND BLANKETS**										
0020	Jute mesh, 100 S.Y. per roll, 4' wide, stapled	G	B-80A	2400	.010	S.Y.	1.03	.42	.34	1.79	2.14
0060	Polyethylene 3 dimensional geomatrix, 50 mil thick	G		700	.034		3.28	1.44	1.17	5.89	7.05
0062	120 mil thick	G		515	.047		6.20	1.96	1.59	9.75	11.55
0068	Slope stakes (placed @ 3' - 5' intervals)	G				Ea.	.15			.15	.17
0070	Paper biodegradable mesh	G	B-1	2500	.010	S.Y.	.18	.41		.59	.82
0080	Paper mulch	G	B-64	20000	.001		.16	.04	.02	.22	.25
0100	Plastic netting, stapled, 2" x 1" mesh, 20 mil	G	B-1	2500	.010		.27	.41		.68	.92
0120	Revegetation mat, webbed	G	2 Clab	1000	.016		2.96	.67		3.63	4.27
0160	Underdrain fabric, 18" x 100' roll	G	"	32	.500	Roll	28.50	21		49.50	63
0200	Polypropylene mesh, stapled, 6.5 oz./S.Y.	G	B-1	2500	.010	S.Y.	1.70	.41		2.11	2.49
0300	Tobacco netting, or jute mesh #2, stapled	G	"	2500	.010		.31	.41		.72	.96
0400	Soil sealant, liquid sprayed from truck	G	B-81	5000	.005		.38	.24	.12	.74	.90
0600	Straw in polymeric netting, biodegradable log		A-2	1000	.024	L.F.	6.45	1.05	.20	7.70	8.90
0705	Sediment Log, Filter Sock, 9"			1000	.024		3.50	1.05	.20	4.75	5.65
0710	Sediment Log, Filter Sock, 12"			1000	.024		4.95	1.05	.20	6.20	7.25
1000	Silt fence, install and maintain, remove	G	B-62	1300	.018		.45	.84	.14	1.43	1.92
1305	For less than 3 to 1 slope, add							15%			
1310	For greater than 3 to 1 slope, add							25%			
1400	Barriers, w/degradable component, 3' H, incl. wd stakes	G	2 Clab	1600	.010	L.F.	.79	.42		1.21	1.50

31 31 Soil Treatment

31 31 16 – Termite Control

31 31 16.13 Chemical Termite Control

			Crew	Daily Output	Labor-Hours	Unit	Material	2020 Bare Costs Labor	Equipment	Total	Total Incl O&P
0010	**CHEMICAL TERMITE CONTROL**										
0020	Slab and walls, residential		1 Skwk	1200	.007	SF Flr.	.33	.37		.70	.91
0030	SS mesh, no chemicals, avg 1,400 S.F. home, min	G		1000	.008		.33	.44		.77	1.02
0040	Max	G		500	.016		.33	.88		1.21	1.69
0100	Commercial, minimum			2496	.003		.35	.18		.53	.66
0200	Maximum			1645	.005		.53	.27		.80	.98
0400	Insecticides for termite control, minimum			14.20	.563	Gal.	70.50	31		101.50	124
0500	Maximum			11	.727	"	120	40		160	193
3000	Soil poisoning (sterilization)		1 Clab	4496	.002	S.F.	.36	.07		.43	.51
3100	Herbicide application from truck		B-59	19000	.001	S.Y.		.02	.02	.04	.06

31 32 13 – Soil Mixing Stabilization

31 32 13.13 Asphalt Soil Stabilization

	Crew	Daily Output	Labor-Hours	Unit	Material	2020 Bare Costs Labor	Equipment	Total	Total Incl O&P
0010 **ASPHALT SOIL STABILIZATION**									
0011 Including scarifying and compaction									
0020 Asphalt, 1-1/2" deep, 1/2 gal./S.Y.	B-75	4000	.014	S.Y.	1.11	.72	1.35	3.18	3.79
0040 3/4 gal./S.Y.		4000	.014		1.67	.72	1.35	3.74	4.40
0100 3" deep, 1 gal./S.Y.		3500	.016		2.22	.83	1.55	4.60	5.40
0140 1-1/2 gal./S.Y.		3500	.016		3.33	.83	1.55	5.71	6.60
0200 6" deep, 2 gal./S.Y.		3000	.019		4.44	.97	1.81	7.22	8.30
0240 3 gal./S.Y.		3000	.019		6.65	.97	1.81	9.43	10.80
0300 8" deep, 2-2/3 gal./S.Y.		2800	.020		5.95	1.03	1.94	8.92	10.20
0340 4 gal./S.Y.		2800	.020		8.90	1.03	1.94	11.87	13.45
0500 12" deep, 4 gal./S.Y.		5000	.011		8.90	.58	1.08	10.56	11.80
0540 6 gal./S.Y.		2600	.022		13.30	1.11	2.08	16.49	18.60

31 32 13.16 Cement Soil Stabilization

	Crew	Daily Output	Labor-Hours	Unit	Material	2020 Bare Costs Labor	Equipment	Total	Total Incl O&P
0010 **CEMENT SOIL STABILIZATION**									
0011 Including scarifying and compaction									
1020 Cement, 4% mix, by volume, 6" deep	B-74	1100	.058	S.Y.	1.88	2.99	5.40	10.27	12.50
1030 8" deep		1050	.061		2.45	3.13	5.65	11.23	13.65
1060 12" deep		960	.067		3.68	3.43	6.20	13.31	16.05
1100 6% mix, 6" deep		1100	.058		2.70	2.99	5.40	11.09	13.40
1120 8" deep		1050	.061		3.51	3.13	5.65	12.29	14.80
1160 12" deep		960	.067		5.30	3.43	6.20	14.93	17.85
1200 9% mix, 6" deep		1100	.058		4.09	2.99	5.40	12.48	14.90
1220 8" deep		1050	.061		5.30	3.13	5.65	14.08	16.80
1260 12" deep		960	.067		8	3.43	6.20	17.63	21
1300 12% mix, 6" deep		1100	.058		5.30	2.99	5.40	13.69	16.30
1320 8" deep		1050	.061		7.10	3.13	5.65	15.88	18.75
1360 12" deep		960	.067		10.60	3.43	6.20	20.23	23.50

31 32 13.19 Lime Soil Stabilization

	Crew	Daily Output	Labor-Hours	Unit	Material	2020 Bare Costs Labor	Equipment	Total	Total Incl O&P
0010 **LIME SOIL STABILIZATION**									
0011 Including scarifying and compaction									
2020 Hydrated lime, for base, 2% mix by weight, 6" deep	B-74	1800	.036	S.Y.	.76	1.83	3.31	5.90	7.20
2030 8" deep		1700	.038		1.01	1.93	3.50	6.44	7.85
2060 12" deep		1550	.041		1.51	2.12	3.84	7.47	9.05
2100 4% mix, 6" deep		1800	.036		1.51	1.83	3.31	6.65	8.05
2120 8" deep		1700	.038		2.03	1.93	3.50	7.46	9
2160 12" deep		1550	.041		3.02	2.12	3.84	8.98	10.75
2200 6% mix, 6" deep		1800	.036		2.27	1.83	3.31	7.41	8.85
2220 8" deep		1700	.038		3.04	1.93	3.50	8.47	10.10
2260 12" deep		1550	.041		4.54	2.12	3.84	10.50	12.40

31 32 13.30 Calcium Chloride

	Crew	Daily Output	Labor-Hours	Unit	Material	2020 Bare Costs Labor	Equipment	Total	Total Incl O&P
0010 **CALCIUM CHLORIDE**									
0020 Calcium chloride, delivered, 100 lb. bags, truckload lots				Ton	635			635	700
0030 Solution, 4 lb. flake per gallon, tank truck delivery				Gal.	1.67			1.67	1.84

31 32 19 – Geosynthetic Soil Stabilization and Layer Separation

31 32 19.16 Geotextile Soil Stabilization

	Crew	Daily Output	Labor-Hours	Unit	Material	2020 Bare Costs Labor	Equipment	Total	Total Incl O&P
0010 **GEOTEXTILE SOIL STABILIZATION**									
1500 Geotextile fabric, woven, 200 lb. tensile strength	2 Clab	2500	.006	S.Y.	.96	.27		1.23	1.46
1510 Heavy duty, 600 lb. tensile strength		2400	.007		1.80	.28		2.08	2.40
1550 Non-woven, 120 lb. tensile strength		2500	.006		.86	.27		1.13	1.35

31 32 36.16 Grouted Soil Nailing	Crew	Daily Output	Labor-Hours	Unit	Material	2020 Bare Costs Labor	Equipment	Total	Total Incl O&P
0010 **GROUTED SOIL NAILING**									
0020 Soil nailing does not include guniting of surfaces									
0030 See Section 03 37 13.30 for guniting of surfaces									
0035 Layout and vertical and horizontal control per day	A-6	1	16	Day		845	31	876	1,300
0038 Layout and vertical and horizontal control average holes per day	"	60	.267	Ea.		14.10	.51	14.61	22
0050 For grade 150 soil nail add $1.37/L.F. to base material cost									
0060 Material delivery add $1.83 to $2.00 per truck mile for shipping									
0090 Average soil nailing, grade 75, 15 min. setup per hole & 80'/hr. drilling									
0100 Soil nailing, drill hole, install #8 nail, grout 20' depth average	B-47G	16	2	Ea.	165	90.50	117	372.50	445
0110 25' depth average		14.20	2.254		206	102	132	440	525
0120 30' depth average		12.80	2.500		375	113	146	634	740
0130 35' depth average		11.60	2.759		435	125	162	722	845
0140 40' depth average		10.70	2.991		500	136	175	811	945
0150 45' depth average		9.90	3.232		560	147	189	896	1,050
0160 50' depth average		9.10	3.516		645	159	206	1,010	1,175
0170 55' depth average		8.50	3.765		705	171	221	1,097	1,275
0180 60' depth average		8	4		765	181	234	1,180	1,375
0190 65' depth average		7.50	4.267		830	193	250	1,273	1,475
0200 70' depth average		7.10	4.507		890	204	264	1,358	1,575
0210 75' depth average	▼	6.70	4.776	▼	950	217	280	1,447	1,675
0290 Average soil nailing, grade 75, 15 min. setup per hole & 90'/hr. drilling									
0300 Soil nailing, drill hole, install #8 nail, grout 20' depth average	B-47G	16.60	1.928	Ea.	165	87.50	113	365.50	435
0310 25' depth average		15	2.133		206	96.50	125	427.50	510
0320 30' depth average		13.70	2.336		375	106	137	618	720
0330 35' depth average		12.30	2.602		435	118	152	705	825
0340 40' depth average		11.40	2.807		500	127	164	791	920
0350 45' depth average		10.70	2.991		560	136	175	871	1,000
0360 50' depth average		9.80	3.265		645	148	191	984	1,125
0370 55' depth average		9.20	3.478		705	158	204	1,067	1,225
0380 60' depth average		8.70	3.678		765	167	215	1,147	1,325
0390 65' depth average		8.10	3.951		830	179	231	1,240	1,425
0400 70' depth average		7.70	4.156		890	188	243	1,321	1,525
0410 75' depth average	▼	7.40	4.324	▼	950	196	253	1,399	1,625
0490 Average soil nailing, grade 75, 15 min. setup per hole & 100'/hr. drilling									
0500 Soil nailing, drill hole, install #8 nail, grout 20' depth average	B-47G	17.80	1.798	Ea.	165	81.50	105	351.50	420
0510 25' depth average		16	2		206	90.50	117	413.50	490
0520 30' depth average		14.60	2.192		375	99.50	128	602.50	700
0530 35' depth average		13.30	2.406		435	109	141	685	800
0540 40' depth average		12.30	2.602		500	118	152	770	895
0550 45' depth average		11.40	2.807		560	127	164	851	985
0560 50' depth average		10.70	2.991		645	136	175	956	1,100
0570 55' depth average		10	3.200		705	145	187	1,037	1,200
0580 60' depth average		9.40	3.404		765	154	199	1,118	1,300
0590 65' depth average		8.90	3.596		830	163	211	1,204	1,375
0600 70' depth average		8.40	3.810		890	173	223	1,286	1,475
0610 75' depth average	▼	8	4	▼	950	181	234	1,365	1,575
0690 Average soil nailing, grade 75, 15 min. setup per hole & 110'/hr. drilling									
0700 Soil nailing, drill hole, install #8 nail, grout 20' depth average	B-47G	18.50	1.730	Ea.	165	78.50	101	344.50	410
0710 25' depth average		16.60	1.928		206	87.50	113	406.50	480
0720 30' depth average		15.50	2.065		375	93.50	121	589.50	685
0730 35' depth average		14.10	2.270		435	103	133	671	780
0740 40' depth average	▼	13	2.462	▼	500	112	144	756	875

31 32 36 – Soil Nailing

31 32 36.16 Grouted Soil Nailing		Crew	Daily Output	Labor-Hours	Unit	Material	2020 Bare Costs		Total	Total Incl O&P
							Labor	Equipment		
0750	45' depth average	B-47G	12	2.667	Ea.	560	121	156	837	970
0760	50' depth average		11.40	2.807		645	127	164	936	1,075
0770	55' depth average		10.70	2.991		705	136	175	1,016	1,175
0780	60' depth average		10	3.200		765	145	187	1,097	1,275
0790	65' depth average		9.40	3.404		830	154	199	1,183	1,350
0800	70' depth average		9.10	3.516		890	159	206	1,255	1,450
0810	75' depth average		8.60	3.721		950	169	218	1,337	1,550
0890	Average soil nailing, grade 75, 15 min. setup per hole & 120'/hr. drilling									
0900	Soil nailing, drill hole, install #8 nail, grout 20' depth average	B-47G	19.20	1.667	Ea.	165	75.50	97.50	338	400
0910	25' depth average		17.10	1.871		206	85	110	401	475
0920	30' depth average		16	2		375	90.50	117	582.50	675
0930	35' depth average		14.60	2.192		435	99.50	128	662.50	770
0940	40' depth average		13.70	2.336		500	106	137	743	860
0950	45' depth average		12.60	2.540		560	115	149	824	950
0960	50' depth average		12	2.667		645	121	156	922	1,050
0970	55' depth average		11.20	2.857		705	130	167	1,002	1,150
0980	60' depth average		10.70	2.991		765	136	175	1,076	1,250
0990	65' depth average		10	3.200		830	145	187	1,162	1,325
1000	70' depth average		9.60	3.333		890	151	195	1,236	1,425
1010	75' depth average		9.10	3.516		950	159	206	1,315	1,525
1190	Difficult soil nailing, grade 75, 20 min. setup per hole & 80'/hr. drilling									
1200	Soil nailing, drill hole, install #8 nail, grout 20' depth difficult	B-47G	13.70	2.336	Ea.	165	106	137	408	490
1210	25' depth difficult		12.30	2.602		206	118	152	476	570
1220	30' depth difficult		11.20	2.857		375	130	167	672	790
1230	35' depth difficult		10.40	3.077		435	139	180	754	885
1240	40' depth difficult		9.60	3.333		500	151	195	846	990
1250	45' depth difficult		8.90	3.596		560	163	211	934	1,100
1260	50' depth difficult		8.30	3.855		645	175	226	1,046	1,225
1270	55' depth difficult		7.90	4.051		705	184	237	1,126	1,300
1280	60' depth difficult		7.40	4.324		765	196	253	1,214	1,425
1290	65' depth difficult		7	4.571		830	207	268	1,305	1,525
1300	70' depth difficult		6.60	4.848		890	220	284	1,394	1,625
1310	75' depth difficult		6.30	5.079		950	230	298	1,478	1,725
1390	Difficult soil nailing, grade 75, 20 min. setup per hole & 90'/hr. drilling									
1400	Soil nailing, drill hole, install #8 nail, grout 20' depth difficult	B-47G	14.10	2.270	Ea.	165	103	133	401	480
1410	25' depth difficult		13	2.462		206	112	144	462	550
1420	30' depth difficult		12	2.667		375	121	156	652	765
1430	35' depth difficult		10.90	2.936		435	133	172	740	870
1440	40' depth difficult		10.20	3.137		500	142	184	826	965
1450	45' depth difficult		9.60	3.333		560	151	195	906	1,050
1460	50' depth difficult		8.90	3.596		645	163	211	1,019	1,175
1470	55' depth difficult		8.40	3.810		705	173	223	1,101	1,275
1480	60' depth difficult		8	4		765	181	234	1,180	1,375
1490	65' depth difficult		7.50	4.267		830	193	250	1,273	1,475
1500	70' depth difficult		7.20	4.444		890	201	260	1,351	1,575
1510	75' depth difficult		6.90	4.638		950	210	272	1,432	1,675
1590	Difficult soil nailing, grade 75, 20 min. setup per hole & 100'/hr drilling									
1600	Soil nailing, drill hole, install #8 nail, grout 20' depth difficult	B-47G	15	2.133	Ea.	165	96.50	125	386.50	465
1610	25' depth difficult		13.70	2.336		206	106	137	449	535
1620	30' depth difficult		12.60	2.540		375	115	149	639	745
1630	35' depth difficult		11.70	2.735		435	124	160	719	840
1640	40' depth difficult		10.90	2.936		500	133	172	805	940
1650	45' depth difficult		10.20	3.137		560	142	184	886	1,025

337

For customer support on your Site Work & Landscape Costs with RSMeans data, call 800.448.8182.

31 32 36.16 Grouted Soil Nailing		Crew	Daily Output	Labor-Hours	Unit	Material	2020 Bare Costs			Total	Total Incl O&P
							Labor	Equipment		Total	
1660	50' depth difficult	B-47G	9.60	3.333	Ea.	645	151	195		991	1,150
1670	55' depth difficult		9.10	3.516		705	159	206		1,070	1,250
1680	60' depth difficult		8.60	3.721		765	169	218		1,152	1,350
1690	65' depth difficult		8.10	3.951		830	179	231		1,240	1,425
1700	70' depth difficult		7.70	4.156		890	188	243		1,321	1,525
1710	75' depth difficult		7.40	4.324		950	196	253		1,399	1,625
1790	Difficult soil nailing, grade 75, 20 min. setup per hole & 110'/hr drilling										
1800	Soil nailing, drill hole, install #8 nail, grout 20' depth difficult	B-47G	15.50	2.065	Ea.	165	93.50	121		379.50	455
1810	25' depth difficult		14.10	2.270		206	103	133		442	525
1820	30' depth difficult		13.30	2.406		375	109	141		625	730
1830	35' depth difficult		12.30	2.602		435	118	152		705	825
1840	40' depth difficult		11.40	2.807		500	127	164		791	920
1850	45' depth difficult		10.70	2.991		560	136	175		871	1,000
1860	50' depth difficult		10.20	3.137		645	142	184		971	1,125
1870	55' depth difficult		9.60	3.333		705	151	195		1,051	1,225
1880	60' depth difficult		9.10	3.516		765	159	206		1,130	1,300
1890	65' depth difficult		8.60	3.721		830	169	218		1,217	1,400
1900	70' depth difficult		8.30	3.855		890	175	226		1,291	1,500
1910	75' depth difficult		7.90	4.051		950	184	237		1,371	1,575
1990	Difficult soil nailing, grade 75, 20 min. setup per hole & 120'/hr drilling										
2000	Soil nailing, drill hole, install #8 nail, grout 20' depth difficult	B-47G	16	2	Ea.	165	90.50	117		372.50	445
2010	25' depth difficult		14.60	2.192		206	99.50	128		433.50	515
2020	30' depth difficult		13.70	2.336		375	106	137		618	720
2030	35' depth difficult		12.60	2.540		435	115	149		699	815
2040	40' depth difficult		12	2.667		500	121	156		777	905
2050	45' depth difficult		11.20	2.857		560	130	167		857	995
2060	50' depth difficult		10.70	2.991		645	136	175		956	1,100
2070	55' depth difficult		10	3.200		705	145	187		1,037	1,200
2080	60' depth difficult		9.60	3.333		765	151	195		1,111	1,275
2090	65' depth difficult		9.10	3.516		830	159	206		1,195	1,375
2100	70' depth difficult		8.70	3.678		890	167	215		1,272	1,475
2110	75' depth difficult		8.30	3.855		950	175	226		1,351	1,550
2990	Very difficult soil nailing, grade 75, 25 min. setup & 80'/hr. drilling										
3000	Soil nailing, drill, install #8 nail, grout 20' depth very difficult	B-47G	12	2.667	Ea.	165	121	156		442	535
3010	25' depth very difficult		10.90	2.936		206	133	172		511	615
3020	30' depth very difficult		10	3.200		375	145	187		707	835
3030	35' depth very difficult		9.40	3.404		435	154	199		788	930
3040	40' depth very difficult		8.70	3.678		500	167	215		882	1,025
3050	45' depth very difficult		8.10	3.951		560	179	231		970	1,150
3060	50' depth very difficult		7.60	4.211		645	191	247		1,083	1,250
3070	55' depth very difficult		7.30	4.384		705	199	257		1,161	1,350
3080	60' depth very difficult		6.90	4.638		765	210	272		1,247	1,450
3090	65' depth very difficult		6.50	4.923		830	223	288		1,341	1,550
3100	70' depth very difficult		6.20	5.161		890	234	300		1,424	1,650
3110	75' depth very difficult		5.90	5.424		950	246	320		1,516	1,775
3190	Very difficult soil nailing, grade 75, 25 min. setup & 90'/hr. drilling										
3200	Soil nailing, drill, install #8 nail, grout 20' depth very difficult	B-47G	12.30	2.602	Ea.	165	118	152		435	525
3210	25' depth very difficult		11.40	2.807		206	127	164		497	600
3220	30' depth very difficult		10.70	2.991		375	136	175		686	805
3230	35' depth very difficult		9.80	3.265		435	148	191		774	910
3240	40' depth very difficult		9.20	3.478		500	158	204		862	1,000
3250	45' depth very difficult		8.70	3.678		560	167	215		942	1,100
3260	50' depth very difficult		8.10	3.951		645	179	231		1,055	1,225

31 32 36.16 Grouted Soil Nailing	Crew	Daily Output	Labor-Hours	Unit	Material	2020 Bare Costs Labor	Equipment	Total	Total Incl O&P
3270	55' depth very difficult	B-47G	7.70	4.156	Ea.	705	188	243	1,136	1,325
3280	60' depth very difficult		7.40	4.324		765	196	253	1,214	1,425
3290	65' depth very difficult		7	4.571		830	207	268	1,305	1,525
3300	70' depth very difficult		6.70	4.776		890	217	280	1,387	1,625
3310	75' depth very difficult		6.40	5		950	227	293	1,470	1,700
3390	Very difficult soil nailing, grade 75, 25 min. setup & 100'/hr. drilling									
3400	Soil nailing, drill, install #8 nail, grout 20' depth very difficult	B-47G	13	2.462	Ea.	165	112	144	421	505
3410	25' depth very difficult		12	2.667		206	121	156	483	580
3420	30' depth very difficult		11.20	2.857		375	130	167	672	790
3430	35' depth very difficult		10.40	3.077		435	139	180	754	885
3440	40' depth very difficult		9.80	3.265		500	148	191	839	980
3450	45' depth very difficult		9.20	3.478		560	158	204	922	1,075
3460	50' depth very difficult		8.70	3.678		645	167	215	1,027	1,200
3470	55' depth very difficult		8.30	3.855		705	175	226	1,106	1,275
3480	60' depth very difficult		7.90	4.051		765	184	237	1,186	1,375
3490	65' depth very difficult		7.50	4.267		830	193	250	1,273	1,475
3500	70' depth very difficult		7.20	4.444		890	201	260	1,351	1,575
3510	75' depth very difficult		6.90	4.638		950	210	272	1,432	1,675
3590	Very difficult soil nailing, grade 75, 25 min. setup & 110'/hr. drilling									
3600	Soil nailing, drill, install #8 nail, grout 20' depth very difficult	B-47G	13.30	2.406	Ea.	165	109	141	415	500
3610	25' depth very difficult		13.70	2.336		206	106	137	449	535
3620	30' depth very difficult		11.70	2.735		375	124	160	659	770
3630	35' depth very difficult		10.90	2.936		435	133	172	740	870
3640	40' depth very difficult		10.20	3.137		500	142	184	826	965
3650	45' depth very difficult		9.60	3.333		560	151	195	906	1,050
3660	50' depth very difficult		9.20	3.478		645	158	204	1,007	1,175
3670	55' depth very difficult		8.70	3.678		705	167	215	1,087	1,250
3680	60' depth very difficult		8.30	3.855		765	175	226	1,166	1,350
3690	65' depth very difficult		7.90	4.051		830	184	237	1,251	1,450
3700	70' depth very difficult		7.60	4.211		890	191	247	1,328	1,525
3710	75' depth very difficult		7.30	4.384		950	199	257	1,406	1,625
3790	Very difficult soil nailing, grade 75, 25 min. setup & 120'/hr. drilling									
3800	Soil nailing, drill, install #8 nail, grout 20' depth very difficult	B-47G	13.70	2.336	Ea.	165	106	137	408	490
3810	25' depth very difficult		12.60	2.540		206	115	149	470	565
3820	30' depth very difficult		12	2.667		375	121	156	652	765
3830	35' depth very difficult		11.20	2.857		435	130	167	732	860
3840	40' depth very difficult		10.70	2.991		500	136	175	811	945
3850	45' depth very difficult		10	3.200		560	145	187	892	1,050
3860	50' depth very difficult		9.60	3.333		645	151	195	991	1,150
3870	55' depth very difficult		9.10	3.516		705	159	206	1,070	1,250
3880	60' depth very difficult		8.70	3.678		765	167	215	1,147	1,325
3890	65' depth very difficult		8.30	3.855		830	175	226	1,231	1,425
3900	70' depth very difficult		8	4		890	181	234	1,305	1,500
3910	75' depth very difficult		7.60	4.211		950	191	247	1,388	1,600
3990	Severe soil nailing, grade 75, 30 min. setup per hole & 80'/hr. drilling									
4000	Soil nailing, drill hole, install #8 nail, grout 20' depth severe	B-47G	10.70	2.991	Ea.	165	136	175	476	575
4010	25' depth severe		9.80	3.265		206	148	191	545	660
4020	30' depth severe		9.10	3.516		375	159	206	740	875
4030	35' depth severe		8.60	3.721		435	169	218	822	975
4040	40' depth severe		8	4		500	181	234	915	1,075
4050	45' depth severe		7.50	4.267		560	193	250	1,003	1,175
4060	50' depth severe		7.10	4.507		645	204	264	1,113	1,300
4070	55' depth severe		6.80	4.706		705	213	276	1,194	1,400

31 32 36.16 Grouted Soil Nailing		Crew	Daily Output	Labor-Hours	Unit	Material	2020 Bare Costs Labor	Equipment	Total	Total Incl O&P
4080	60' depth severe	B-47G	6.40	5	Ea.	765	227	293	1,285	1,500
4090	65' depth severe		6.10	5.246		830	238	305	1,373	1,600
4100	70' depth severe		5.80	5.517		890	250	325	1,465	1,700
4110	75' depth severe		5.60	5.714		950	259	335	1,544	1,800
4190	Severe soil nailing, grade 75, 30 min. setup per hole & 90'/hr. drilling									
4200	Soil nailing, drill hole, install #8 nail, grout 20' depth severe	B-47G	10.90	2.936	Ea.	165	133	172	470	570
4210	25' depth severe		10.20	3.137		206	142	184	532	640
4220	30' depth severe		9.60	3.333		375	151	195	721	850
4230	35' depth severe		8.90	3.596		435	163	211	809	955
4240	40' depth severe		8.40	3.810		500	173	223	896	1,050
4250	45' depth severe		8	4		560	181	234	975	1,150
4260	50' depth severe		7.50	4.267		645	193	250	1,088	1,275
4270	55' depth severe		7.20	4.444		705	201	260	1,166	1,350
4280	60' depth severe		6.90	4.638		765	210	272	1,247	1,450
4290	65' depth severe		6.50	4.923		830	223	288	1,341	1,550
4300	70' depth severe		6.20	5.161		890	234	300	1,424	1,650
4310	75' depth severe		6	5.333		950	242	310	1,502	1,750
4390	Severe soil nailing, grade 75, 30 min. setup per hole & 100'/hr. drilling									
4400	Soil nailing, drill hole, install #8 nail, grout 20' depth severe	B-47G	11.40	2.807	Ea.	165	127	164	456	555
4410	25' depth severe		10.70	2.991		206	136	175	517	620
4420	30' depth severe		10	3.200		375	145	187	707	835
4430	35' depth severe		9.40	3.404		435	154	199	788	930
4440	40' depth severe		8.90	3.596		500	163	211	874	1,025
4450	45' depth severe		8.40	3.810		560	173	223	956	1,125
4460	50' depth severe		8	4		645	181	234	1,060	1,225
4470	55' depth severe		7.60	4.211		705	191	247	1,143	1,325
4480	60' depth severe		7.30	4.384		765	199	257	1,221	1,425
4490	65' depth severe		7	4.571		830	207	268	1,305	1,525
4500	70' depth severe		6.70	4.776		890	217	280	1,387	1,625
4510	75' depth severe		6.40	5		950	227	293	1,470	1,700
4590	Severe soil nailing, grade 75, 30 min. setup per hole & 110'/hr. drilling									
4600	Soil nailing, drill hole, install #8 nail, grout 20' depth severe	B-47G	11.70	2.735	Ea.	165	124	160	449	545
4610	25' depth severe		10.90	2.936		206	133	172	511	615
4620	30' depth severe		10.40	3.077		375	139	180	694	815
4630	35' depth severe		9.80	3.265		435	148	191	774	910
4640	40' depth severe		9.20	3.478		500	158	204	862	1,000
4650	45' depth severe		8.70	3.678		560	167	215	942	1,100
4660	50' depth severe		8.40	3.810		645	173	223	1,041	1,200
4670	55' depth severe		8	4		705	181	234	1,120	1,300
4680	60' depth severe		7.60	4.211		765	191	247	1,203	1,400
4690	65' depth severe		7.30	4.384		830	199	257	1,286	1,500
4700	70' depth severe		7.10	4.507		890	204	264	1,358	1,575
4710	75' depth severe		6.80	4.706		950	213	276	1,439	1,675
4790	Severe soil nailing, grade 75, 30 min. setup per hole & 120'/hr. drilling									
4800	Soil nailing, drill hole, install #8 nail, grout 20' depth severe	B-47G	12	2.667	Ea.	165	121	156	442	535
4810	25' depth severe		11.20	2.857		206	130	167	503	605
4820	30' depth severe		10.70	2.991		375	136	175	686	805
4830	35' depth severe		10	3.200		435	145	187	767	905
4840	40' depth severe		9.60	3.333		500	151	195	846	990
4850	45' depth severe		9.10	3.516		560	159	206	925	1,075
4860	50' depth severe		8.70	3.678		645	167	215	1,027	1,200
4870	55' depth severe		8.30	3.855		705	175	226	1,106	1,275
4880	60' depth severe		8	4		765	181	234	1,180	1,375

For customer support on your Site Work & Landscape Costs with RSMeans data, call 800.448.8182.

31 32 Soil Stabilization

31 32 36 – Soil Nailing

31 32 36.16 Grouted Soil Nailing	Crew	Daily Output	Labor-Hours	Unit	Material	2020 Bare Costs Labor	Equipment	Total	Total Incl O&P
4890	65' depth severe	B-47G	7.60	4.211	Ea.	830	191	247	1,268	1,475
4900	70' depth severe		7.40	4.324		890	196	253	1,339	1,550
4910	75' depth severe	↓	7.10	4.507	↓	950	204	264	1,418	1,650

31 36 Gabions

31 36 13 – Gabion Boxes

31 36 13.10 Gabion Box Systems

		Crew	Daily Output	Labor-Hours	Unit	Material	Labor	Equipment	Total	Total Incl O&P
0010	**GABION BOX SYSTEMS**									
0400	Gabions, galvanized steel mesh mats or boxes, stone filled, 6" deep	B-13	200	.280	S.Y.	18.30	12.90	2.90	34.10	42.50
0500	9" deep		163	.344		25.50	15.80	3.56	44.86	56
0600	12" deep		153	.366		37	16.85	3.80	57.65	70.50
0700	18" deep	↓	102	.549		47.50	25.50	5.70	78.70	96.50
0800	36" deep	↓	60	.933	↓	80	43	9.70	132.70	163

31 37 Riprap

31 37 13 – Machined Riprap

31 37 13.10 Riprap and Rock Lining

		Crew	Daily Output	Labor-Hours	Unit	Material	Labor	Equipment	Total	Total Incl O&P
0010	**RIPRAP AND ROCK LINING**									
0011	Random, broken stone									
0100	Machine placed for slope protection	B-12G	62	.258	L.C.Y.	32	13.05	13.85	58.90	70
0110	3/8 to 1/4 C.Y. pieces, grouted	B-13	80	.700	S.Y.	64.50	32	7.25	103.75	128
0200	18" minimum thickness, not grouted	"	53	1.057	"	19.85	48.50	10.95	79.30	107
0300	Dumped, 50 lb. average	B-11A	800	.020	Ton	28.50	.99	1.88	31.37	35
0350	100 lb. average		700	.023		28.50	1.13	2.15	31.78	35.50
0370	300 lb. average	↓	600	.027	↓	28.50	1.32	2.51	32.33	36

31 41 Shoring

31 41 13 – Timber Shoring

31 41 13.10 Building Shoring

		Crew	Daily Output	Labor-Hours	Unit	Material	Labor	Equipment	Total	Total Incl O&P
0010	**BUILDING SHORING**									
0020	Shoring, existing building, with timber, no salvage allowance	B-51	2.20	21.818	M.B.F.	860	945	89	1,894	2,475
1000	On cribbing with 35 ton screw jacks, per box and jack	"	3.60	13.333	Jack	65	575	54.50	694.50	995

31 41 16 – Sheet Piling

31 41 16.10 Sheet Piling Systems

			Crew	Daily Output	Labor-Hours	Unit	Material	Labor	Equipment	Total	Total Incl O&P
0010	**SHEET PILING SYSTEMS**										
0020	Sheet piling, 50,000 psi steel, not incl. wales, 22 psf, left in place		B-40	10.81	5.920	Ton	1,875	325	320	2,520	2,900
0100	Drive, extract & salvage	R314116-40		6	10.667		525	590	580	1,695	2,100
0300	20' deep excavation, 27 psf, left in place			12.95	4.942		1,875	273	269	2,417	2,775
0400	Drive, extract & salvage	R314116-45		6.55	9.771		525	540	530	1,595	1,975
0600	25' deep excavation, 38 psf, left in place			19	3.368		1,875	186	183	2,244	2,525
0700	Drive, extract & salvage			10.50	6.095		525	335	330	1,190	1,450
0900	40' deep excavation, 38 psf, left in place			21.20	3.019		1,875	167	164	2,206	2,475
1000	Drive, extract & salvage			12.25	5.224	↓	525	289	284	1,098	1,325
1200	15' deep excavation, 22 psf, left in place			983	.065	S.F.	21.50	3.60	3.54	28.64	33.50
1300	Drive, extract & salvage			545	.117		5.90	6.50	6.40	18.80	23.50
1500	20' deep excavation, 27 psf, left in place			960	.067		27.50	3.68	3.62	34.80	39.50
1600	Drive, extract & salvage		↓	485	.132	↓	7.65	7.30	7.15	22.10	27.50

341

31 41 Shoring

31 41 16 – Sheet Piling

31 41 16.10 Sheet Piling Systems	Crew	Daily Output	Labor-Hours	Unit	Material	2020 Bare Costs Labor	Equipment	Total	Total Incl O&P
1800	25' deep excavation, 38 psf, left in place	B-40	1000	.064	S.F.	40	3.54	3.48	47.02	53
1900	Drive, extract & salvage	↓	553	.116	↓	10.50	6.40	6.30	23.20	28.50
2100	Rent steel sheet piling and wales, first month				Ton	335			335	370
2200	Per added month					33.50			33.50	37
2300	Rental piling left in place, add to rental					1,200			1,200	1,325
2500	Wales, connections & struts, 2/3 salvage					535			535	585
2700	High strength piling, 60,000 psi, add					187			187	206
2800	65,000 psi, add					280			280	310
3000	Tie rod, not upset, 1-1/2" to 4" diameter with turnbuckle					2,275			2,275	2,500
3100	No turnbuckle					1,650			1,650	1,800
3300	Upset, 1-3/4" to 4" diameter with turnbuckle					2,400			2,400	2,650
3400	No turnbuckle				▼	2,125			2,125	2,325
3600	Lightweight, 18" to 28" wide, 7 ga., 9.22 psf, and									
3610	9 ga., 8.6 psf, minimum				Lb.	.73			.73	.80
3700	Average					.92			.92	1.01
3750	Maximum				▼	1.02			1.02	1.12
3900	Wood, solid sheeting, incl. wales, braces and spacers,									
3910	drive, extract & salvage, 8' deep excavation	B-31	330	.121	S.F.	2.06	5.40	.78	8.24	11.25
4000	10' deep, 50 S.F./hr. in & 150 S.F./hr. out		300	.133		2.12	5.95	.86	8.93	12.20
4100	12' deep, 45 S.F./hr. in & 135 S.F./hr. out		270	.148		2.18	6.60	.95	9.73	13.40
4200	14' deep, 42 S.F./hr. in & 126 S.F./hr. out		250	.160		2.24	7.15	1.03	10.42	14.35
4300	16' deep, 40 S.F./hr. in & 120 S.F./hr. out		240	.167		2.31	7.45	1.07	10.83	14.90
4400	18' deep, 38 S.F./hr. in & 114 S.F./hr. out		230	.174		2.39	7.80	1.12	11.31	15.55
4500	20' deep, 35 S.F./hr. in & 105 S.F./hr. out		210	.190		2.47	8.50	1.22	12.19	16.85
4520	Left in place, 8' deep, 55 S.F./hr.		440	.091		3.70	4.06	.58	8.34	10.80
4540	10' deep, 50 S.F./hr.		400	.100		3.90	4.47	.64	9.01	11.70
4560	12' deep, 45 S.F./hr.		360	.111		4.11	4.97	.71	9.79	12.75
4565	14' deep, 42 S.F./hr.		335	.119		4.35	5.35	.77	10.47	13.65
4570	16' deep, 40 S.F./hr.		320	.125		4.63	5.60	.80	11.03	14.40
4580	18' deep, 38 S.F./hr.		305	.131		4.94	5.85	.84	11.63	15.20
4590	20' deep, 35 S.F./hr.	▼	280	.143	▼	5.30	6.40	.92	12.62	16.40
4700	Alternate pricing, left in place, 8' deep		1.76	22.727	M.B.F.	830	1,025	146	2,001	2,600
4800	Drive, extract and salvage, 8' deep	↓	1.32	30.303	"	740	1,350	195	2,285	3,050
5000	For treated lumber add cost of treatment to lumber									

31 43 Concrete Raising

31 43 13 – Pressure Grouting

31 43 13.13 Concrete Pressure Grouting

		Crew	Daily Output	Labor-Hours	Unit	Material	2020 Bare Costs Labor	Equipment	Total	Total Incl O&P
0010	**CONCRETE PRESSURE GROUTING**									
0020	Grouting, pressure, cement & sand, 1:1 mix, minimum	B-61	124	.323	Bag	19.30	14.40	2.49	36.19	45.50
0100	Maximum		51	.784	"	19.30	35	6.05	60.35	80.50
0200	Cement and sand, 1:1 mix, minimum		250	.160	C.F.	38.50	7.15	1.23	46.88	54.50
0300	Maximum		100	.400		58	17.85	3.08	78.93	94.50
0310	Grouting, cement and sand, 1:2 mix, geothermal wells		250	.160		15.45	7.15	1.23	23.83	29
0320	Bentonite mix 1:4		250	.160		9.50	7.15	1.23	17.88	22.50
0400	Epoxy cement grout, minimum		137	.292		785	13.05	2.25	800.30	885
0500	Maximum	▼	57	.702	▼	785	31.50	5.40	821.90	920
0700	Alternate pricing method: (Add for materials)									
0710	5 person crew and equipment	B-61	1	40	Day		1,775	310	2,085	3,025

31 43 Concrete Raising

31 43 19 – Mechanical Jacking

31 43 19.10 Slabjacking	Crew	Daily Output	Labor-Hours	Unit	Material	2020 Bare Costs Labor	Equipment	Total	Total Incl O&P
0010 **SLABJACKING**									
0100 4" thick slab	D-4	1500	.021	S.F.	.38	1.01	.13	1.52	2.08
0150 6" thick slab		1200	.027		.53	1.26	.16	1.95	2.65
0200 8" thick slab		1000	.032		.63	1.51	.19	2.33	3.19
0250 10" thick slab		900	.036		.67	1.68	.21	2.56	3.50
0300 12" thick slab	↓	850	.038	↓	.74	1.77	.22	2.73	3.73

31 45 Vibroflotation and Densification

31 45 13 – Vibroflotation

31 45 13.10 Vibroflotation Densification	Crew	Daily Output	Labor-Hours	Unit	Material	2020 Bare Costs Labor	Equipment	Total	Total Incl O&P
0010 **VIBROFLOTATION DENSIFICATION**									
0900 Vibroflotation compacted sand cylinder, minimum	B-60	750	.075	V.L.F.		3.67	3	6.67	8.80
0950 Maximum		325	.172			8.45	6.90	15.35	20.50
1100 Vibro replacement compacted stone cylinder, minimum		500	.112			5.50	4.49	9.99	13.20
1150 Maximum		250	.224	↓		11	9	20	26.50
1300 Mobilization and demobilization, minimum		.47	119	Total		5,850	4,775	10,625	14,000
1400 Maximum	↓	.14	400	"		19,700	16,000	35,700	47,000

31 48 Underpinning

31 48 13 – Underpinning Piers

31 48 13.10 Underpinning Foundations	Crew	Daily Output	Labor-Hours	Unit	Material	2020 Bare Costs Labor	Equipment	Total	Total Incl O&P
0010 **UNDERPINNING FOUNDATIONS**									
0011 Including excavation,									
0020 forming, reinforcing, concrete and equipment									
0100 5' to 16' below grade, 100 to 500 C.Y.	B-52	2.30	24.348	C.Y.	340	1,175	254	1,769	2,425
0200 Over 500 C.Y.		2.50	22.400		305	1,100	234	1,639	2,225
0400 16' to 25' below grade, 100 to 500 C.Y.		2	28		370	1,375	292	2,037	2,775
0500 Over 500 C.Y.		2.10	26.667		350	1,300	278	1,928	2,650
0700 26' to 40' below grade, 100 to 500 C.Y.		1.60	35		405	1,700	365	2,470	3,400
0800 Over 500 C.Y.	↓	1.80	31.111		370	1,525	325	2,220	3,050
0900 For under 50 C.Y., add					10%	40%			
1000 For 50 C.Y. to 100 C.Y., add				↓	5%	20%			

31 52 Cofferdams

31 52 16 – Timber Cofferdams

31 52 16.10 Cofferdams	Crew	Daily Output	Labor-Hours	Unit	Material	2020 Bare Costs Labor	Equipment	Total	Total Incl O&P
0010 **COFFERDAMS**									
0011 Incl. mobilization and temporary sheeting									
0020 Shore driven	B-40	960	.067	S.F.	26.50	3.68	3.62	33.80	39
0060 Barge driven	"	550	.116	"	26.50	6.45	6.30	39.25	46.50
0080 Soldier beams & lagging H-piles with 3" wood sheeting									
0090 horizontal between piles, including removal of wales & braces									
0100 No hydrostatic head, 15' deep, 1 line of braces, minimum	B-50	545	.206	S.F.	8.60	10.75	4.72	24.07	31
0200 Maximum		495	.226		9.55	11.85	5.20	26.60	34.50
0400 15' to 22' deep with 2 lines of braces, 10" H, minimum		360	.311		10.15	16.30	7.15	33.60	44
0500 Maximum		330	.339		11.50	17.75	7.80	37.05	48
0700 23' to 35' deep with 3 lines of braces, 12" H, minimum		325	.345		13.25	18.05	7.90	39.20	51

31 52 Cofferdams

31 52 16 – Timber Cofferdams

31 52 16.10 Cofferdams		Crew	Daily Output	Labor-Hours	Unit	Material	2020 Bare Costs Labor	Equipment	Total	Total Incl O&P
0800	Maximum	B-50	295	.380	S.F.	14.35	19.90	8.70	42.95	56
1000	36' to 45' deep with 4 lines of braces, 14" H, minimum		290	.386		14.85	20	8.85	43.70	57
1100	Maximum		265	.423		15.65	22	9.70	47.35	62
1300	No hydrostatic head, left in place, 15' deep, 1 line of braces, min.		635	.176		11.50	9.25	4.05	24.80	31
1400	Maximum		575	.195		12.30	10.20	4.47	26.97	34
1600	15' to 22' deep with 2 lines of braces, minimum		455	.246		17.20	12.90	5.65	35.75	45
1700	Maximum		415	.270		19.15	14.15	6.20	39.50	49.50
1900	23' to 35' deep with 3 lines of braces, minimum		420	.267		20.50	13.95	6.10	40.55	51
2000	Maximum		380	.295		22.50	15.45	6.75	44.70	56
2200	36' to 45' deep with 4 lines of braces, minimum		385	.291		24.50	15.25	6.70	46.45	58
2300	Maximum	↓	350	.320		28.50	16.75	7.35	52.60	65
2350	Lagging only, 3" thick wood between piles 8' OC, minimum	B-46	400	.120		1.91	5.80	.10	7.81	11.10
2370	Maximum		250	.192		2.87	9.30	.17	12.34	17.60
2400	Open sheeting no bracing, for trenches to 10' deep, min.		1736	.028		.86	1.34	.02	2.22	3.03
2450	Maximum	↓	1510	.032		.96	1.54	.03	2.53	3.44
2500	Tie-back method, add to open sheeting, add, minimum								20%	20%
2550	Maximum				↓				60%	60%
2700	Tie-backs only, based on tie-backs total length, minimum	B-46	86.80	.553	L.F.	17.25	27	.48	44.73	60.50
2750	Maximum		38.50	1.247	"	30.50	60.50	1.07	92.07	127
3500	Tie-backs only, typical average, 25' long		2	24	Ea.	760	1,175	20.50	1,955.50	2,625
3600	35' long	↓	1.58	30.380	"	1,025	1,475	26	2,526	3,400
4500	Trench box, 7' deep, 16' x 8', see 01 54 33 in Reference Section				Day				191	210
4600	20' x 10', see 01 54 33 in Reference Section				"				240	265
5200	Wood sheeting, in trench, jacks at 4' OC, 8' deep	B-1	800	.030	S.F.	.66	1.28		1.94	2.66
5250	12' deep		700	.034		.78	1.47		2.25	3.06
5300	15' deep	↓	600	.040	↓	1.08	1.71		2.79	3.75
6000	See also Section 31 41 16.10									

31 56 Slurry Walls

31 56 23 – Lean Concrete Slurry Walls

31 56 23.20 Slurry Trench

		Crew	Daily Output	Labor-Hours	Unit	Material	2020 Bare Costs Labor	Equipment	Total	Total Incl O&P
0010	**SLURRY TRENCH**									
0011	Excavated slurry trench in wet soils									
0020	backfilled with 3,000 psi concrete, no reinforcing steel									
0050	Minimum	C-7	333	.216	C.F.	9.55	9.90	3.24	22.69	29
0100	Maximum		200	.360	"	16	16.45	5.40	37.85	48
0200	Alternate pricing method, minimum		150	.480	S.F.	19.10	22	7.20	48.30	62
0300	Maximum	↓	120	.600		28.50	27.50	9	65	82.50
0500	Reinforced slurry trench, minimum	B-48	177	.316		14.30	15.05	5.90	35.25	45
0600	Maximum	"	69	.812	↓	47.50	38.50	15.20	101.20	127
0800	Haul for disposal, 2 mile haul, excavated material, add	B-34B	99	.081	C.Y.		3.96	5.80	9.76	12.30
0900	Haul bentonite castings for disposal, add	"	40	.200	"		9.80	14.30	24.10	30.50

31 62 Driven Piles

31 62 13 – Concrete Piles

31 62 13.23 Prestressed Concrete Piles	Crew	Daily Output	Labor-Hours	Unit	Material	2020 Bare Costs Labor	Equipment	Total	Total Incl O&P
0010 **PRESTRESSED CONCRETE PILES**, 200 piles									
0020 Unless specified otherwise, not incl. pile caps or mobilization									
2200 Precast, prestressed, 50' long, cylinder, 12" diam., 2-3/8" wall	B-19	720	.089	V.L.F.	29.50	4.91	2.79	37.20	42.50
2300 14" diameter, 2-1/2" wall		680	.094		38	5.20	2.96	46.16	53
2500 16" diameter, 3" wall	↓	640	.100		50.50	5.50	3.14	59.14	67.50
2600 18" diameter, 3-1/2" wall	B-19A	600	.107		63.50	5.90	4.91	74.31	84.50
2800 20" diameter, 4" wall		560	.114		55.50	6.30	5.25	67.05	76.50
2900 24" diameter, 5" wall	↓	520	.123		75.50	6.80	5.65	87.95	99.50
2920 36" diameter, 5-1/2" wall	B-19	400	.160		99.50	8.85	5	113.35	128
2940 54" diameter, 6" wall		340	.188		236	10.40	5.90	252.30	282
2950 60" diameter, 6" wall		280	.229		251	12.65	7.20	270.85	305
2960 66" diameter, 6-1/2" wall		220	.291		340	16.05	9.15	365.20	405
3100 Precast, prestressed, 40' long, 10" thick, square		700	.091		14.10	5.05	2.87	22.02	26.50
3200 12" thick, square	↓	680	.094		26	5.20	2.96	34.16	39.50
3275 Shipping for 60' long concrete piles, add to material cost					2.06			2.06	2.27
3400 14" thick, square	B-19	600	.107		29	5.90	3.35	38.25	44.50
3500 Octagonal		640	.100		28	5.50	3.14	36.64	43
3700 16" thick, square		560	.114		30	6.30	3.59	39.89	46.50
3800 Octagonal	↓	600	.107		33.50	5.90	3.35	42.75	49.50
4000 18" thick, square	B-19A	520	.123		52	6.80	5.65	64.45	73.50
4100 Octagonal	B-19	560	.114		45	6.30	3.59	54.89	63
4300 20" thick, square	B-19A	480	.133		47.50	7.35	6.15	61	70.50
4400 Octagonal	B-19	520	.123		50.50	6.80	3.86	61.16	70
4600 24" thick, square	B-19A	440	.145		57.50	8.05	6.70	72.25	82.50
4700 Octagonal	B-19	480	.133		59.50	7.35	4.19	71.04	81.50
4730 Precast, prestressed, 60' long, 10" thick, square		700	.091		14.85	5.05	2.87	22.77	27
4740 12" thick, square (60' long)		680	.094		27	5.20	2.96	35.16	40.50
4750 Mobilization for 10,000 L.F. pile job, add		3300	.019			1.07	.61	1.68	2.31
4800 25,000 L.F. pile job, add	↓	8500	.008	↓		.42	.24	.66	.90
4850 Mobilization by water for barge driving rig, add								100%	100%

31 62 16 – Steel Piles

31 62 16.13 Steel Piles

	Crew	Daily Output	Labor-Hours	Unit	Material	2020 Bare Costs Labor	Equipment	Total	Total Incl O&P
0010 **STEEL PILES**									
0100 Step tapered, round, concrete filled									
0110 8" tip, 12" butt, 60 ton capacity, 30' depth	B-19	760	.084	V.L.F.	18.85	4.65	2.64	26.14	30.50
0120 60' depth with extension		740	.086		36.50	4.78	2.72	44	51
0130 80' depth with extensions		700	.091		57	5.05	2.87	64.92	73.50
0250 "H" Sections, 50' long, HP8 x 36		640	.100		18.35	5.50	3.14	26.99	32
0400 HP10 x 42		610	.105		20.50	5.80	3.29	29.59	35
0500 HP10 x 57		610	.105		28	5.80	3.29	37.09	43.50
0700 HP12 x 53	↓	590	.108		27.50	6	3.41	36.91	43
0800 HP12 x 74	B-19A	590	.108		35	6	4.99	45.99	53
1000 HP14 x 73		540	.119		37	6.55	5.45	49	56.50
1100 HP14 x 89		540	.119		47.50	6.55	5.45	59.50	68
1300 HP14 x 102		510	.125		54	6.95	5.80	66.75	76.50
1400 HP14 x 117	↓	510	.125	↓	62.50	6.95	5.80	75.25	85.50
1600 Splice on standard points, not in leads, 8" or 10"	1 Sswl	5	1.600	Ea.	111	92		203	266
1700 12" or 14"		4	2		155	115		270	350
1900 Heavy duty points, not in leads, 10" wide		4	2		186	115		301	385
2100 14" wide	↓	3.50	2.286		228	131		359	455

31 62 Driven Piles

31 62 19 – Timber Piles

31 62 19.10 Wood Piles

		Crew	Daily Output	Labor-Hours	Unit	Material	2020 Bare Costs Labor	Equipment	Total	Total Incl O&P
0010	**WOOD PILES**									
0011	Friction or end bearing, not including									
0050	mobilization or demobilization									
0100	ACZA Treated piles, 1.0 lb/C.F., up to 30' long, 12" butts, 8" points	B-19	625	.102	V.L.F.	13.70	5.65	3.22	22.57	27
0200	30' to 39' long, 12" butts, 8" points		700	.091		14.45	5.05	2.87	22.37	27
0300	40' to 49' long, 12" butts, 7" points		720	.089		15.20	4.91	2.79	22.90	27.50
0400	50' to 59' long, 13" butts, 7" points		800	.080		21.50	4.42	2.51	28.43	33
0500	60' to 69' long, 13" butts, 7" points		840	.076		24.50	4.21	2.39	31.10	36
0600	70' to 80' long, 13" butts, 6" points		840	.076		25	4.21	2.39	31.60	36
0800	ACZA Treated piles, 1.5 lb./C.F.									
0810	friction or end bearing, ASTM class B									
1000	Up to 30' long, 12" butts, 8" points	B-19	625	.102	V.L.F.	18.20	5.65	3.22	27.07	32
1100	30' to 39' long, 12" butts, 8" points		700	.091		18.90	5.05	2.87	26.82	32
1200	40' to 49' long, 12" butts, 7" points		720	.089		20.50	4.91	2.79	28.20	33
1300	50' to 59' long, 13" butts, 7" points		800	.080		23.50	4.42	2.51	30.43	35.50
1400	60' to 69' long, 13" butts, 6" points	B-19A	840	.076		25.50	4.21	3.51	33.22	38.50
1500	70' to 80' long, 13" butts, 6" points	"	840	.076		28	4.21	3.51	35.72	41
1600	ACZA Treated piles, 2.5 lb./C.F.									
1610	8" butts, 10' long	B-19	400	.160	V.L.F.	7.35	8.85	5	21.20	27
1620	11' to 16' long		500	.128		7.35	7.05	4.02	18.42	23.50
1630	17' to 20' long		575	.111		7.35	6.15	3.49	16.99	21.50
1640	10" butts, 10' to 16' long		500	.128		22	7.05	4.02	33.07	39.50
1650	17' to 20' long		575	.111		22	6.15	3.49	31.64	37.50
1660	21' to 40' long		700	.091		22	5.05	2.87	29.92	35
1670	12" butts, 10' to 20' long		575	.111		23	6.15	3.49	32.64	39
1680	21' to 35' long		650	.098		23	5.45	3.09	31.54	37.50
1690	36' to 40' long		700	.091		23.50	5.05	2.87	31.42	37
1695	14" butts, to 40' long		700	.091		16.90	5.05	2.87	24.82	29.50
1700	Boot for pile tip, minimum	1 Pile	27	.296	Ea.	45.50	16.05		61.55	75
1800	Maximum		21	.381		136	20.50		156.50	181
2000	Point for pile tip, minimum		20	.400		45.50	21.50		67	83.50
2100	Maximum		15	.533		163	29		192	224
2300	Splice for piles over 50' long, minimum	B-46	35	1.371		59	66.50	1.18	126.68	168
2400	Maximum		20	2.400		75.50	116	2.06	193.56	263
2600	Concrete encasement with wire mesh and tube		331	.145	V.L.F.	75.50	7.05	.12	82.67	94
2700	Mobilization for 10,000 L.F. pile job, add	B-19	3300	.019			1.07	.61	1.68	2.31
2800	25,000 L.F. pile job, add	"	8500	.008			.42	.24	.66	.90
2900	Mobilization by water for barge driving rig, add								100%	100%

31 62 19.30 Marine Wood Piles

		Crew	Daily Output	Labor-Hours	Unit	Material	2020 Bare Costs Labor	Equipment	Total	Total Incl O&P
0010	**MARINE WOOD PILES**									
0040	Friction or end bearing, not including									
0050	mobilization or demobilization									
0055	Marine piles include production adjusted									
0060	due to thickness of silt layers									
0090	Shore driven piles with silt layer									
1000	ACZA treated 1.0lbs/C.F., 50', 13" butts, 7" points, 10' silt	B-19	952	.067	V.L.F.	17.85	3.71	2.11	23.67	27.50
1050	55', 7" points		936	.068		20.50	3.78	2.15	26.43	30.50
1075	60', 7" points		969	.066		21.50	3.65	2.07	27.22	31.50
1100	65', 7" points		958	.067		24	3.69	2.10	29.79	34.50
1125	70', 6" points		948	.068		24.50	3.73	2.12	30.35	35
1150	75', 6" points		940	.068		25	3.76	2.14	30.90	35
1175	80', 6" points		933	.069		27	3.79	2.15	32.94	37.50

346

31 62 19.30 Marine Wood Piles		Crew	Daily Output	Labor-Hours	Unit	Material	2020 Bare Costs Labor	Equipment	Total	Total Incl O&P
1200	ACZA treated 1.5lbs/C.F., 50', 7" points,10' silt	B-19	952	.067	V.L.F.	21	3.71	2.11	26.82	31
1225	55', 7" points		936	.068		23.50	3.78	2.15	29.43	34
1250	60', 7" points		969	.066		24	3.65	2.07	29.72	34.50
1275	65', 7" points		958	.067		27.50	3.69	2.10	33.29	38
1300	70', 6" points		948	.068		27.50	3.73	2.12	33.35	38
1325	75', 6" points		940	.068		27.50	3.76	2.14	33.40	38.50
1350	80', 6" points		933	.069		30	3.79	2.15	35.94	41
1400	Shore driven, ACZA treated 1.0lbs/C.F., 50', 13" butts, 7" points, 15' silt		1053	.061		18.45	3.36	1.91	23.72	28
1450	55', 7" points		1023	.063		20.50	3.46	1.96	25.92	30
1475	60', 7" points		1050	.061		21.50	3.37	1.91	26.78	31
1500	65', 7" points		1030	.062		24	3.43	1.95	29.38	34
1525	70', 6" points		1013	.063		24.50	3.49	1.98	29.97	34.50
1550	75', 6" points		1000	.064		25	3.54	2.01	30.55	34.50
1575	80', 6" points		988	.065		27	3.58	2.03	32.61	37
1600	ACZA treated 1.5lbs/C.F., 50', 7" points, 15' silt		1053	.061		21	3.36	1.91	26.27	30.50
1625	55', 7" points		1023	.063		23.50	3.46	1.96	28.92	33.50
1650	60', 7" points		1050	.061		24	3.37	1.91	29.28	34
1675	65', 7" points		1030	.062		27.50	3.43	1.95	32.88	37.50
1700	70', 6" points		1014	.063		27.50	3.49	1.98	32.97	37.50
1725	75', 6" points		1000	.064		27.50	3.54	2.01	33.05	38
1750	80', 6" points		988	.065		30	3.58	2.03	35.61	40.50
1800	Shore driven, ACZA treated 1.0lbs/C.F., 50', 13" butts, 7" points, 20' silt		1177	.054		18.45	3	1.71	23.16	27
1850	55', 7" points		1128	.057		20.50	3.13	1.78	25.41	29.50
1875	60', 7" points		1145	.056		21.50	3.09	1.76	26.35	30
1900	65', 7" points		1114	.057		24	3.17	1.80	28.97	33.50
1925	70', 6" points		1089	.059		24.50	3.25	1.85	29.60	34
1950	75', 6" points		1068	.060		25	3.31	1.88	30.19	34
1975	80', 6" points		1050	.061		27	3.37	1.91	32.28	37
2000	ACZA treated 1.5lbs/C.F., 50', 7" points		1177	.054		21	3	1.71	25.71	29.50
2025	55', 7" points		1128	.057		23.50	3.13	1.78	28.41	33
2050	60', 7" points		1145	.056		24	3.09	1.76	28.85	33
2075	65', 7" points		1114	.057		27.50	3.17	1.80	32.47	37
2100	70', 6" points		1089	.059		27.50	3.25	1.85	32.60	37
2125	75', 6" points		1068	.060		27.50	3.31	1.88	32.69	37.50
2150	80', 6" points		1050	.061		30	3.37	1.91	35.28	40.50
2200	Shore driven, ACZA treated 1.0bs/C.F., 60', 13" butts, 7" points, 25' silt		1260	.051		21.50	2.81	1.59	25.90	29.50
2225	65', 7" points		1213	.053		24	2.91	1.66	28.57	33
2250	70', 6" points		1176	.054		24.50	3.01	1.71	29.22	33.50
2275	75', 6" points		1146	.056		25	3.09	1.75	29.84	33.50
2300	80', 6" points		1120	.057		27	3.16	1.79	31.95	36.50
2350	ACZA treated 1.5lbs/C.F., 60', 6" points		1260	.051		24	2.81	1.59	28.40	32.50
2375	65', 6" points		1213	.053		27.50	2.91	1.66	32.07	36.50
2400	70', 6" points		1176	.054		27.50	3.01	1.71	32.22	36.50
2425	75', 6" points		1146	.056		27.50	3.09	1.75	32.34	37
2450	80', 6" points		1120	.057		30	3.16	1.79	34.95	40
2500	Shore driven, ACZA treated 1.0lbs/C.F., 70', 13" butts, 6" points, 30' sil		1278	.050		24.50	2.77	1.57	28.84	33
2525	75', 6" points		1235	.052		25	2.86	1.63	29.49	33
2550	80', 6" points		1200	.053		27	2.95	1.67	31.62	36
2600	ACZA treated 1.5lbs/C.F., 70', 6" points		1278	.050		27.50	2.77	1.57	31.84	36
2625	75', 6" points		1235	.052		27.50	2.86	1.63	31.99	36.50
2650	80', 6" points		1200	.053		30	2.95	1.67	34.62	39.50
2700	Shore driven, ACZA treated 1.0lbs/C.F., 70', 13" butts, 6" points, 35' silt		1400	.046		24.50	2.53	1.44	28.47	32.50
2725	75', 6" points		1340	.048		25	2.64	1.50	29.14	32.50

For customer support on your Site Work & Landscape Costs with RSMeans data, call 800.448.8182.

347

31 62 Driven Piles

31 62 19 – Timber Piles

31 62 19.30 Marine Wood Piles

		Crew	Daily Output	Labor-Hours	Unit	Material	2020 Bare Costs Labor	2020 Bare Costs Equipment	Total	Total Incl O&P
2750	80', 6" points	B-19	1292	.050	V.L.F.	27	2.74	1.56	31.30	35.50
2800	ACZA treated 1.5lbs/C.F., 70', 6" points		1400	.046		27.50	2.53	1.44	31.47	35.50
2825	75', 6" points		1340	.048		27.50	2.64	1.50	31.64	36
2850	80', 6" points		1292	.050		30	2.74	1.56	34.30	39
2875	Shore driven, ACZA treated 1.0lbs/C.F., 75', 13" butts, 6" points, 40' silt		1465	.044		25	2.41	1.37	28.78	32
2900	80', 6" points		1400	.046		27	2.53	1.44	30.97	35
2925	ACZA treated 1.5lbs/C.F., 75', 6" points		1465	.044		27.50	2.41	1.37	31.28	35.50
2950	80', 6" points	▼	1400	.046	▼	30	2.53	1.44	33.97	38.50
2990	Barge driven piles with silt layer									
3000	ACZA treated 1.5lbs/C.F. 50', 13" butts, 7" points, 10' silt	B-19B	762	.084	V.L.F.	21	4.64	3.76	29.40	34
3050	55', 7" points		749	.085		23.50	4.72	3.83	32.05	37.50
3075	60', 7" points		775	.083		24	4.56	3.70	32.26	37.50
3100	65', 7" points		766	.084		27.50	4.62	3.74	35.86	41
3125	70', 6" points		759	.084		27.50	4.66	3.78	35.94	41.50
3150	75', 6" points		752	.085		27.50	4.70	3.81	36.01	42
3175	80', 6" points		747	.086		30	4.73	3.84	38.57	44.50
3200	ACZA treated 2.5lbs/C.F., 50', 7" points		762	.084		24.50	4.64	3.76	32.90	38
3225	55', 7" points		749	.085		25	4.72	3.83	33.55	39
3250	60', 7" points		775	.083		25.50	4.56	3.70	33.76	39
3275	65', 7" points		766	.084		31	4.62	3.74	39.36	45
3300	70', 6" points		759	.084		31	4.66	3.78	39.44	46
3325	75', 6" points		752	.085		31.50	4.70	3.81	40.01	46
3350	80', 6" points		747	.086		36.50	4.73	3.84	45.07	51.50
3400	Barge driven, ACZA treated 1.5lbs/C.F. 50', 13" butts, 7" points, 15' silt		842	.076		21	4.20	3.41	28.61	33
3450	55', 7" points		819	.078		23.50	4.32	3.50	31.32	36.50
3475	60', 7" points		840	.076		24	4.21	3.41	31.62	36.50
3500	65', 7" points		824	.078		27.50	4.29	3.48	35.27	40.50
3525	70', 6" points		811	.079		27.50	4.36	3.54	35.40	40.50
3550	75', 6" points		800	.080		27.50	4.42	3.59	35.51	41
3575	80', 6" points		791	.081		30	4.47	3.63	38.10	44
3600	ACZA treated 2.5lbs/C.F., 50', 7" points		842	.076		24.50	4.20	3.41	32.11	37
3625	55', 7" points		819	.078		25	4.32	3.50	32.82	38
3650	60', 7" points		840	.076		25.50	4.21	3.41	33.12	38
3675	65', 7" points		824	.078		31	4.29	3.48	38.77	44.50
3700	70', 6" points		811	.079		31	4.36	3.54	38.90	45
3725	75', 6" points		800	.080		31.50	4.42	3.59	39.51	45
3750	80', 6" points		791	.081		36.50	4.47	3.63	44.60	51
3800	Barge driven, ACZA treated 1.5lbs/C.F. 50', 13" butts, 7" points, 20' silt		941	.068		21	3.76	3.05	27.81	32
3850	55', 7" points		903	.071		23.50	3.91	3.18	30.59	35.50
3875	60', 7" points		916	.070		24	3.86	3.13	30.99	36
3900	65', 7" points		891	.072		27.50	3.97	3.22	34.69	39.50
3925	70', 6" points		871	.073		27.50	4.06	3.29	34.85	40
3950	75', 6" points		854	.075		27.50	4.14	3.36	35	40.50
3975	80', 6" points		840	.076		30	4.21	3.41	37.62	43
4000	ACZA treated 2.5lbs/C.F., 50', 7" points		941	.068		24.50	3.76	3.05	31.31	36
4025	55', 7" points		903	.071		25	3.91	3.18	32.09	37
4050	60', 7" points		916	.070		25.50	3.86	3.13	32.49	37.50
4075	65', 7" points		891	.072		31	3.97	3.22	38.19	43.50
4100	70', 6" points		871	.073		31	4.06	3.29	38.35	44.50
4125	75', 6" points		854	.075		31.50	4.14	3.36	39	44.50
4150	80', 6" points		840	.076		36.50	4.21	3.41	44.12	50
4200	Barge driven, ACZA treated 1.5lbs/C.F. 60', 3" butts, 7" points, 25' silt		1008	.063		24	3.51	2.85	30.36	35
4225	65', 7" point	▼	971	.066	▼	27.50	3.64	2.95	34.09	39

348

31 62 Driven Piles

31 62 19 – Timber Piles

31 62 19.30 Marine Wood Piles		Crew	Daily Output	Labor-Hours	Unit	Material	2020 Bare Costs Labor	Equipment	Total	Total Incl O&P
4250	70', 6" points	B-19B	941	.068	V.L.F.	27.50	3.76	3.05	34.31	39
4275	75', 6" points		916	.070		27.50	3.86	3.13	34.49	40
4300	80', 6" points		896	.071		30	3.95	3.20	37.15	42.50
4350	ACZA treated 2.5lbs/C.F., 60', 7" points		1008	.063		25.50	3.51	2.85	31.86	36.50
4375	65', 7" points		971	.066		31	3.64	2.95	37.59	43
4400	70', 6" points		941	.068		31	3.76	3.05	37.81	43.50
4425	75', 6" points		916	.070		31.50	3.86	3.13	38.49	44
4450	80', 6" points		896	.071		36.50	3.95	3.20	43.65	49.50
4500	Barge driven, ACZA treated 1.5lbs/C.F. 70', 13" butts, 6" points, 30' silt		1023	.063		27.50	3.46	2.80	33.76	38.50
4525	75', 6" points		988	.065		27.50	3.58	2.90	33.98	39
4550	80', 6" points		960	.067		30	3.68	2.99	36.67	42
4600	ACZA treated 2.5lbs/C.F., 70', 6" points		1023	.063		31	3.46	2.80	37.26	43
4625	75', 6" points		988	.065		31.50	3.58	2.90	37.98	43
4650	80', 6" points		960	.067		36.50	3.68	2.99	43.17	49
4700	Barge driven, ACZA treated 1.5lbs/C.F. 70', 13" butts, 6" points, 35' silt		1120	.057		27.50	3.16	2.56	33.22	37.50
4725	75', 6" points		1072	.060		27.50	3.30	2.68	33.48	38.50
4750	80', 6" points		1034	.062		30	3.42	2.77	36.19	41.50
4800	ACZA treated 2.5lbs/C.F., 70', 6" points		1120	.057		31	3.16	2.56	36.72	42
4825	75', 6" points		1072	.060		31.50	3.30	2.68	37.48	42.50
4850	80', 6" points		1034	.062		36.50	3.42	2.77	42.69	48.50
4875	Barge driven, ACZA treated 1.5lbs/C.F. 75', 13" butts, 6" points, 40' silt		1172	.055		27.50	3.02	2.45	32.97	38
4900	80' untreated pile, 6" points		1120	.057		30	3.16	2.56	35.72	40.50
4925	ACZA treated 2.5lbs/C.F., 75', 6" points		1172	.055		31.50	3.02	2.45	36.97	42
4950	80' treated pile, 6" points		1120	.057		36.50	3.16	2.56	42.22	47.50
5000	Piles, wood, treated, added undriven length, 13" butt					31.50			31.50	34.50
5020	Untreated pile, 13" butts					26			26	28.50

31 62 23 – Composite Piles

31 62 23.10 Recycled Plastic and Fiberglass Piles

			Crew	Daily Output	Labor-Hours	Unit	Material	Labor	Equipment	Total	Total Incl O&P
0010	RECYCLED PLASTIC AND FIBERGLASS PILES, 200 PILES										
5000	Marine pilings, recycled plastic w/fiberglass reinf, up to 90' L, 8" diam.	G	B-19	500	.128	V.L.F.	32.50	7.05	4.02	43.57	51
5010	10" diam.	G		500	.128		36.50	7.05	4.02	47.57	56
5020	13" diam.	G		400	.160		55.50	8.85	5	69.35	80
5030	16" diam.	G		400	.160		87.50	8.85	5	101.35	115
5040	20" diam.	G		350	.183		88.50	10.10	5.75	104.35	119
5050	24" diam.	G		350	.183		108	10.10	5.75	123.85	141

31 62 23.12 Marine Recycled Plastic and Fiberglass Piles

			Crew	Daily Output	Labor-Hours	Unit	Material	Labor	Equipment	Total	Total Incl O&P
0010	MARINE RECYCLED PLASTIC AND FIBERGLASS PILES, 200 PILES										
0040	Friction or end bearing, not including										
0050	mobilization or demobilization										
0055	Marine piles include production adjusted										
0060	due to thickness of silt layers										
0090	Shore driven piles with silt layer										
0100	Marine pilings, recycled plastic w/fiberglass, 50' long, 8" diam.,10' silt	G	B-19	595	.108	V.L.F.	32.50	5.95	3.38	41.83	48.50
0110	50' long, 10" diam.	G		595	.108		36.50	5.95	3.38	45.83	53.50
0120	50' long, 13" diam.	G		476	.134		55.50	7.45	4.22	67.17	77
0130	50' long, 16" diam.	G		476	.134		87.50	7.45	4.22	99.17	112
0140	50' long, 20" diam.	G		417	.153		88.50	8.50	4.82	101.82	115
0150	50' long, 24" diam.	G		417	.153		108	8.50	4.82	121.32	137
0210	55' long, 8" diam.	G		585	.109		32.50	6.05	3.44	41.99	48.50
0220	55' long, 10" diam.	G		585	.109		36.50	6.05	3.44	45.99	53.50
0230	55' long, 13" diam.	G		468	.137		55.50	7.55	4.29	67.34	77.50
0240	55' long, 16" diam.	G		468	.137		87.50	7.55	4.29	99.34	112

For customer support on your Site Work & Landscape Costs with RSMeans data, call 800.448.8182.

31 62 23.12 Marine Recycled Plastic and Fiberglass Piles		Crew	Daily Output	Labor-Hours	Unit	Material	2020 Bare Costs Labor	Equipment	Total	Total Incl O&P
0250	55' long, 20" diam. G	B-19	410	.156	V.L.F.	88.50	8.60	4.90	102	116
0260	55' long, 24" diam. G		410	.156		108	8.60	4.90	121.50	138
0310	60' long, 8" diam. G		576	.111		32.50	6.15	3.49	42.14	48.50
0320	60' long, 10" diam. G		577	.111		36.50	6.15	3.48	46.13	53.50
0330	60' long, 13" diam. G		462	.139		55.50	7.65	4.35	67.50	77.50
0340	60' long, 16" diam. G		462	.139		87.50	7.65	4.35	99.50	112
0350	60' long, 20" diam. G		404	.158		88.50	8.75	4.97	102.22	116
0360	60' long, 24" diam. G		404	.158		108	8.75	4.97	121.72	138
0410	65' long, 8" diam. G		570	.112		32.50	6.20	3.53	42.23	49
0420	65' long, 10" diam. G		570	.112		36.50	6.20	3.53	46.23	54
0430	65' long, 13" diam. G		456	.140		55.50	7.75	4.41	67.66	77.50
0440	65' long, 16" diam. G		456	.140		87.50	7.75	4.41	99.66	113
0450	65' long, 20" diam. G		399	.160		88.50	8.85	5.05	102.40	116
0460	65' long, 24" diam. G		399	.160		108	8.85	5.05	121.90	138
0510	70' long, 8" diam. G		565	.113		32.50	6.25	3.56	42.31	49
0520	70' long, 10" diam. G		565	.113		36.50	6.25	3.56	46.31	54
0530	70' long, 13" diam. G		452	.142		55.50	7.80	4.45	67.75	78
0540	70' long, 16" diam. G		452	.142		87.50	7.80	4.45	99.75	113
0550	70' long, 20" diam. G		395	.162		88.50	8.95	5.10	102.55	116
0560	70' long, 24" diam. G		399	.160		108	8.85	5.05	121.90	138
0610	75' long, 8" diam. G		560	.114		32.50	6.30	3.59	42.39	49
0620	75' long, 10" diam. G		560	.114		36.50	6.30	3.59	46.39	54
0630	75' long, 13" diam. G		448	.143		55.50	7.90	4.49	67.89	78
0640	75' long, 16" diam. G		448	.143		87.50	7.90	4.49	99.89	113
0650	75' long, 20" diam. G		392	.163		88.50	9	5.15	102.65	116
0660	75' long, 24" diam. G		392	.163		108	9	5.15	122.15	138
0710	80' long, 8" diam. G		556	.115		32.50	6.35	3.61	42.46	49
0720	80' long, 10" diam. G		556	.115		36.50	6.35	3.61	46.46	54
0730	80' long, 13" diam. G		444	.144		55.50	7.95	4.53	67.98	78
0740	80' long, 16" diam. G		444	.144		87.50	7.95	4.53	99.98	113
0750	80' long, 20" diam. G		389	.165		88.50	9.10	5.15	102.75	117
0760	80' long, 24" diam. G		389	.165		108	9.10	5.15	122.25	139
1000	Marine pilings, recycled plastic w/fiberglass, 50' long, 8" diam.,15' silt G		658	.097		32.50	5.35	3.05	40.90	47
1010	50' long, 10" diam. G		658	.097		36.50	5.35	3.05	44.90	52
1020	50' long, 13" diam. G		526	.122		55.50	6.70	3.82	66.02	75.50
1030	50' long, 16" diam. G		526	.122		87.50	6.70	3.82	98.02	111
1040	50' long, 20" diam. G		461	.139		88.50	7.65	4.36	100.51	114
1050	50' long, 24" diam. G		461	.139		108	7.65	4.36	120.01	136
1210	55' long, 8" diam. G		640	.100		32.50	5.50	3.14	41.14	47.50
1220	55' long, 10" diam. G		640	.100		36.50	5.50	3.14	45.14	52.50
1230	55' long, 13" diam. G		512	.125		55.50	6.90	3.93	66.33	76
1240	55' long, 16" diam. G		512	.125		87.50	6.90	3.93	98.33	111
1250	55' long, 20" diam. G		448	.143		88.50	7.90	4.49	100.89	114
1260	55' long, 24" diam. G		448	.143		108	7.90	4.49	120.39	136
1310	60' long, 8" diam. G		625	.102		32.50	5.65	3.22	41.37	47.50
1320	60' long, 10" diam. G		625	.102		36.50	5.65	3.22	45.37	52.50
1330	60' long, 13" diam. G		500	.128		55.50	7.05	4.02	66.57	76.50
1340	60' long, 16" diam. G		500	.128		87.50	7.05	4.02	98.57	111
1350	60' long, 20" diam. G		438	.146		88.50	8.05	4.59	101.14	114
1360	60' long, 24" diam. G		438	.146		108	8.05	4.59	120.64	136
1410	65' long, 8" diam. G		613	.104		32.50	5.75	3.28	41.53	48
1420	65' long, 10" diam. G		613	.104		36.50	5.75	3.28	45.53	53
1430	65' long, 13" diam. G		491	.130		55.50	7.20	4.09	66.79	76.50

350

31 62 Driven Piles

31 62 23 – Composite Piles

31 62 23.12 Marine Recycled Plastic and Fiberglass Piles		Crew	Daily Output	Labor-Hours	Unit	Material	2020 Bare Costs Labor	Equipment	Total	Total Incl O&P
1440	65' long, 16" diam.	G	B-19	491	.130	V.L.F.	87.50	7.20	4.09	98.79	112
1450	65' long, 20" diam.	G		429	.149		88.50	8.25	4.68	101.43	115
1460	65' long, 24" diam.	G		429	.149		108	8.25	4.68	120.93	137
1510	70' long, 8" diam.	G		603	.106		32.50	5.85	3.33	41.68	48
1520	70' long, 10" diam.	G		603	.106		36.50	5.85	3.33	45.68	53
1530	70' long, 13" diam.	G		483	.133		55.50	7.30	4.16	66.96	77
1540	70' long, 16" diam.	G		483	.133		87.50	7.30	4.16	98.96	112
1550	70' long, 20" diam.	G		422	.152		88.50	8.40	4.76	101.66	115
1560	70' long, 24" diam.	G		422	.152		108	8.40	4.76	121.16	137
1610	75' long, 8" diam.	G		595	.108		32.50	5.95	3.38	41.83	48.50
1620	75' long, 10" diam.	G		595	.108		36.50	5.95	3.38	45.83	53.50
1630	75' long, 13" diam.	G		476	.134		55.50	7.45	4.22	67.17	77
1640	75' long, 16" diam.	G		476	.134		87.50	7.45	4.22	99.17	112
1650	75' long, 20" diam.	G		417	.153		88.50	8.50	4.82	101.82	115
1660	75' long, 24" diam.	G		417	.153		108	8.50	4.82	121.32	137
1710	80' long, 8" diam.	G		588	.109		32.50	6	3.42	41.92	48.50
1720	80' long, 10" diam.	G		588	.109		36.50	6	3.42	45.92	53.50
1730	80' long, 13" diam.	G		471	.136		55.50	7.50	4.27	67.27	77
1740	80' long, 16" diam.	G		471	.136		87.50	7.50	4.27	99.27	112
1750	80' long, 20" diam.	G		412	.155		88.50	8.60	4.88	101.98	116
1760	80' long, 24" diam.	G		412	.155		108	8.60	4.88	121.48	138
2000	Marine pilings, recycled plastic w/fiberglass, 50' long, 8" diam., 20' silt	G		735	.087		32.50	4.81	2.73	40.04	46
2010	50' long, 10" diam.	G		735	.087		36.50	4.81	2.73	44.04	51
2020	50' long, 13" diam.	G		588	.109		55.50	6	3.42	64.92	74
2030	50' long, 16" diam.	G		588	.109		87.50	6	3.42	96.92	109
2040	50' long, 20" diam.	G		515	.124		88.50	6.85	3.90	99.25	112
2050	50' long, 24" diam.	G		515	.124		108	6.85	3.90	118.75	134
2210	55' long, 8" diam.	G		705	.091		32.50	5	2.85	40.35	46.50
2220	55' long, 10" diam.	G		705	.091		36.50	5	2.85	44.35	51.50
2230	55' long, 13" diam.	G		564	.113		55.50	6.25	3.56	65.31	74.50
2240	55' long, 16" diam.	G		564	.113		87.50	6.25	3.56	97.31	110
2250	55' long, 20" diam.	G		494	.130		88.50	7.15	4.07	99.72	112
2260	55' long, 24" diam.	G		494	.130		108	7.15	4.07	119.22	134
2310	60' long, 8" diam.	G		682	.094		32.50	5.20	2.95	40.65	46.50
2320	60' long, 10" diam.	G		682	.094		36.50	5.20	2.95	44.65	51.50
2330	60' long, 13" diam.	G		546	.117		55.50	6.50	3.68	65.68	75
2340	60' long, 16" diam.	G		546	.117		87.50	6.50	3.68	97.68	110
2350	60' long, 20" diam.	G		477	.134		88.50	7.40	4.21	100.11	113
2360	60' long, 24" diam.	G		477	.134		108	7.40	4.21	119.61	135
2410	65' long, 8" diam.	G		663	.097		32.50	5.35	3.03	40.88	47
2420	65' long, 10" diam.	G		663	.097		36.50	5.35	3.03	44.88	52
2430	65' long, 13" diam.	G		531	.121		55.50	6.65	3.78	65.93	75.50
2440	65' long, 16" diam.	G		531	.121		87.50	6.65	3.78	97.93	110
2450	65' long, 20" diam.	G		464	.138		88.50	7.60	4.33	100.43	113
2460	65' long, 24" diam.	G		464	.138		108	7.60	4.33	119.93	135
2510	70' long, 8" diam.	G		648	.099		32.50	5.45	3.10	41.05	47.50
2520	70' long, 10" diam.	G		648	.099		36.50	5.45	3.10	45.05	52.50
2530	70' long, 13" diam.	G		518	.124		55.50	6.80	3.88	66.18	75.50
2540	70' long, 16" diam.	G		518	.124		87.50	6.80	3.88	98.18	111
2550	70' long, 20" diam.	G		454	.141		88.50	7.80	4.43	100.73	114
2560	70' long, 24" diam.	G		454	.141		108	7.80	4.43	120.23	136
2610	75' long, 8" diam.	G		636	.101		32.50	5.55	3.16	41.21	47.50
2620	75' long, 10" diam.	G		636	.101		36.50	5.55	3.16	45.21	52.50

31 62 23.12 Marine Recycled Plastic and Fiberglass Piles		Crew	Daily Output	Labor-Hours	Unit	Material	2020 Bare Costs Labor	2020 Bare Costs Equipment	Total	Total Incl O&P
2630	75' long, 13" diam. G	B-19	508	.126	V.L.F.	55.50	6.95	3.96	66.41	76
2640	75' long, 16" diam. G		508	.126		87.50	6.95	3.96	98.41	111
2650	75' long, 20" diam. G		445	.144		88.50	7.95	4.52	100.97	114
2660	75' long, 24" diam. G		445	.144		108	7.95	4.52	120.47	136
2710	80' long, 8" diam. G		625	.102		32.50	5.65	3.22	41.37	47.50
2720	80' long, 10" diam. G		625	.102		36.50	5.65	3.22	45.37	52.50
2730	80' long, 13" diam. G		500	.128		55.50	7.05	4.02	66.57	76.50
2740	80' long, 16" diam. G		500	.128		87.50	7.05	4.02	98.57	111
2750	80' long, 20" diam. G		437	.146		88.50	8.10	4.60	101.20	114
2760	80' long, 24" diam. G		437	.146		108	8.10	4.60	120.70	136
4990	Barge driven piles with silt layer									
5100	Marine pilings, recycled plastic w/fiberglass, 50' long, 8" diam., 10' silt G	B-19B	476	.134	V.L.F.	32.50	7.45	6.05	46	53.50
5110	50' long, 10" diam. G		476	.134		36.50	7.45	6.05	50	58.50
5120	50' long, 13" diam. G		381	.168		55.50	9.30	7.55	72.35	83.50
5130	50' long, 16" diam. G		381	.168		87.50	9.30	7.55	104.35	119
5140	50' long, 20" diam. G		333	.192		88.50	10.60	8.60	107.70	123
5150	50' long, 24" diam. G		333	.192		108	10.60	8.60	127.20	145
5210	55' long, 8" diam. G		468	.137		32.50	7.55	6.15	46.20	54
5220	55' long, 10" diam. G		468	.137		36.50	7.55	6.15	50.20	59
5230	55' long, 13" diam. G		374	.171		55.50	9.45	7.65	72.60	84
5240	55' long, 16" diam. G		374	.171		87.50	9.45	7.65	104.60	119
5250	55' long, 20" diam. G		328	.195		88.50	10.80	8.75	108.05	123
5260	55' long, 24" diam. G		328	.195		108	10.80	8.75	127.55	145
5310	60' long, 8" diam. G		461	.139		32.50	7.65	6.20	46.35	54
5320	60' long, 10" diam. G		461	.139		36.50	7.65	6.20	50.35	59
5330	60' long, 13" diam. G		369	.173		55.50	9.60	7.75	72.85	84
5340	60' long, 16" diam. G		369	.173		87.50	9.60	7.75	104.85	119
5350	60' long, 20" diam. G		323	.198		88.50	10.95	8.90	108.35	124
5360	60' long, 24" diam. G		323	.198		108	10.95	8.90	127.85	146
5410	65' long, 8" diam. G		456	.140		32.50	7.75	6.30	46.55	54.50
5420	65' long, 10" diam. G		456	.140		36.50	7.75	6.30	50.55	59.50
5430	65' long, 13" diam. G		365	.175		55.50	9.70	7.85	73.05	84.50
5440	65' long, 16" diam. G		365	.175		87.50	9.70	7.85	105.05	120
5450	65' long, 20" diam. G		319	.201		88.50	11.10	9	108.60	124
5460	65' long, 24" diam. G		319	.201		108	11.10	9	128.10	146
5510	70' long, 8" diam. G		452	.142		32.50	7.80	6.35	46.65	54.50
5520	70' long, 10" diam. G		452	.142		36.50	7.80	6.35	50.65	59.50
5530	70' long, 13" diam. G		361	.177		55.50	9.80	7.95	73.25	85
5540	70' long, 16" diam. G		361	.177		87.50	9.80	7.95	105.25	120
5550	70' long, 20" diam. G		316	.203		88.50	11.20	9.10	108.80	124
5560	70' long, 24" diam. G		316	.203		108	11.20	9.10	128.30	146
5610	75' long, 8" diam. G		448	.143		32.50	7.90	6.40	46.80	54.50
5620	75' long, 10" diam. G		448	.143		36.50	7.90	6.40	50.80	59.50
5630	75' long, 13" diam. G		358	.179		55.50	9.90	8	73.40	85
5640	75' long, 16" diam. G		358	.179		87.50	9.90	8	105.40	120
5650	75' long, 20" diam. G		313	.204		88.50	11.30	9.15	108.95	124
5660	75' long, 24" diam. G		313	.204		108	11.30	9.15	128.45	146
5710	80' long, 8" diam. G		444	.144		32.50	7.95	6.45	46.90	55
5720	80' long, 10" diam. G		444	.144		36.50	7.95	6.45	50.90	60
5730	80' long, 13" diam. G		356	.180		55.50	9.95	8.05	73.50	85
5740	80' long, 16" diam. G		356	.180		87.50	9.95	8.05	105.50	120
5750	80' long, 20" diam. G		311	.206		88.50	11.35	9.20	109.05	125
5760	80' long, 24" diam. G		311	.206		108	11.35	9.20	128.55	147

352

For customer support on your Site Work & Landscape Costs with RSMeans data, call 800.448.8182.

31 62 23.12 Marine Recycled Plastic and Fiberglass Piles		Crew	Daily Output	Labor-Hours	Unit	Material	2020 Bare Costs Labor	Equipment	Total	Total Incl O&P
6100	Marine pilings, recycled plastic w/fiberglass, 50' long, 8" diam.,15' silt	G	B-19B	526	.122	V.L.F.	32.50	6.70	5.45	44.65	52
6110	50' long, 10" diam.	G		526	.122		36.50	6.70	5.45	48.65	57
6120	50' long, 13" diam.	G		421	.152		55.50	8.40	6.80	70.70	81.50
6130	50' long, 16" diam.	G		421	.152		87.50	8.40	6.80	102.70	116
6140	50' long, 20" diam.	G		368	.174		88.50	9.60	7.80	105.90	120
6150	50' long, 24" diam.	G		368	.174		108	9.60	7.80	125.40	142
6210	55' long, 8" diam.	G		512	.125		32.50	6.90	5.60	45	52
6220	55' long, 10" diam.	G		512	.125		36.50	6.90	5.60	49	57
6230	55' long, 13" diam.	G		409	.156		55.50	8.65	7	71.15	82
6240	55' long, 16" diam.	G		409	.156		87.50	8.65	7	103.15	117
6250	55' long, 20" diam.	G		358	.179		88.50	9.90	8	106.40	121
6260	55' long, 24" diam.	G		358	.179		108	9.90	8	125.90	143
6310	60' long, 8" diam.	G		500	.128		32.50	7.05	5.75	45.30	52.50
6320	60' long, 10" diam.	G		500	.128		36.50	7.05	5.75	49.30	57.50
6330	60' long, 13" diam.	G		400	.160		55.50	8.85	7.15	71.50	82.50
6340	60' long, 16" diam.	G		400	.160		87.50	8.85	7.15	103.50	117
6350	60' long, 20" diam.	G		350	.183		88.50	10.10	8.20	106.80	121
6360	60' long, 24" diam.	G		350	.183		108	10.10	8.20	126.30	143
6410	65' long, 8" diam.	G		491	.130		32.50	7.20	5.85	45.55	53
6420	65' long, 10" diam.	G		491	.130		36.50	7.20	5.85	49.55	58
6430	65' long, 13" diam.	G		392	.163		55.50	9	7.30	71.80	83
6440	65' long, 16" diam.	G		392	.163		87.50	9	7.30	103.80	118
6450	65' long, 20" diam.	G		343	.187		88.50	10.30	8.35	107.15	122
6460	65' long, 24" diam.	G		343	.187		108	10.30	8.35	126.65	144
6510	70' long, 8" diam.	G		483	.133		32.50	7.30	5.95	45.75	53.50
6520	70' long, 10" diam.	G		483	.133		36.50	7.30	5.95	49.75	58.50
6530	70' long, 13" diam.	G		386	.166		55.50	9.15	7.45	72.10	83
6540	70' long, 16" diam.	G		386	.166		87.50	9.15	7.45	104.10	118
6550	70' long, 20" diam.	G		338	.189		88.50	10.45	8.50	107.45	122
6560	70' long, 24" diam.	G		338	.189		108	10.45	8.50	126.95	144
6610	75' long, 8" diam.	G		476	.134		32.50	7.45	6.05	46	53.50
6620	75' long, 10" diam.	G		476	.134		36.50	7.45	6.05	50	58.50
6630	75' long, 13" diam.	G		381	.168		55.50	9.30	7.55	72.35	83.50
6640	75' long, 16" diam.	G		381	.168		87.50	9.30	7.55	104.35	119
6650	75' long, 20" diam.	G		333	.192		88.50	10.60	8.60	107.70	123
6660	75' long, 24" diam.	G		333	.192		108	10.60	8.60	127.20	145
6710	80' long, 8" diam.	G		471	.136		32.50	7.50	6.10	46.10	53.50
6720	80' long, 10" diam.	G		471	.136		36.50	7.50	6.10	50.10	58.50
6730	80' long, 13" diam.	G		376	.170		55.50	9.40	7.65	72.55	84
6740	80' long, 16" diam.	G		376	.170		87.50	9.40	7.65	104.55	119
6750	80' long, 20" diam.	G		329	.195		88.50	10.75	8.70	107.95	123
6760	80' long, 24" diam.	G		329	.195		108	10.75	8.70	127.45	145
7100	Marine pilings, recycled plastic w/fiberglass, 50' long, 8" diam., 20' silt	G		588	.109		32.50	6	4.88	43.38	50
7110	50' long, 10" diam.	G		588	.109		36.50	6	4.88	47.38	55
7120	50' long, 13" diam.	G		471	.136		55.50	7.50	6.10	69.10	79
7130	50' long, 16" diam.	G		471	.136		87.50	7.50	6.10	101.10	114
7140	50' long, 20" diam.	G		412	.155		88.50	8.60	6.95	104.05	118
7150	50' long, 24" diam.	G		412	.155		108	8.60	6.95	123.55	140
7210	55' long, 8" diam.	G		564	.113		32.50	6.25	5.10	43.85	50.50
7220	55' long, 10" diam.	G		564	.113		36.50	6.25	5.10	47.85	55.50
7230	55' long, 13" diam.	G		451	.142		55.50	7.85	6.35	69.70	80
7240	55' long, 16" diam.	G		451	.142		87.50	7.85	6.35	101.70	115
7250	55' long, 20" diam.	G		395	.162		88.50	8.95	7.25	104.70	119

For customer support on your Site Work & Landscape Costs with RSMeans data, call 800.448.8182.

353

31 62 Driven Piles

31 62 23 – Composite Piles

31 62 23.12 Marine Recycled Plastic and Fiberglass Piles		Crew	Daily Output	Labor-Hours	Unit	Material	2020 Bare Costs Labor	2020 Bare Costs Equipment	Total	Total Incl O&P
7260	55' long, 24" diam.	G	B-19B	395	.162	V.L.F.	108	8.95	7.25	124.20	141
7310	60' long, 8" diam.	G		545	.117		32.50	6.50	5.25	44.25	51.50
7320	60' long, 10" diam.	G		545	.117		36.50	6.50	5.25	48.25	56.50
7330	60' long, 13" diam.	G		436	.147		55.50	8.10	6.60	70.20	80.50
7340	60' long, 16" diam.	G		436	.147		87.50	8.10	6.60	102.20	116
7350	60' long, 20" diam.	G		382	.168		88.50	9.25	7.50	105.25	119
7360	60' long, 24" diam.	G		382	.168		108	9.25	7.50	124.75	141
7410	65' long, 8" diam.	G		531	.121		32.50	6.65	5.40	44.55	51.50
7420	65' long, 10" diam.	G		531	.121		36.50	6.65	5.40	48.55	56.50
7430	65' long, 13" diam.	G		424	.151		55.50	8.35	6.75	70.60	81
7440	65' long, 16" diam.	G		424	.151		87.50	8.35	6.75	102.60	116
7450	65' long, 20" diam.	G		371	.173		88.50	9.55	7.75	105.80	120
7460	65' long, 24" diam.	G		371	.173		108	9.55	7.75	125.30	142
7510	70' long, 8" diam.	G		518	.124		32.50	6.80	5.55	44.85	52
7520	70' long, 10" diam.	G		518	.124		36.50	6.80	5.55	48.85	57
7530	70' long, 13" diam.	G		415	.154		55.50	8.50	6.90	70.90	81.50
7540	70' long, 16" diam.	G		415	.154		87.50	8.50	6.90	102.90	117
7550	70' long, 20" diam.	G		363	.176		88.50	9.75	7.90	106.15	121
7560	70' long, 24" diam.	G		363	.176		108	9.75	7.90	125.65	143
7610	75' long, 8" diam.	G		508	.126		32.50	6.95	5.65	45.10	52.50
7620	75' long, 10" diam.	G		508	.126		36.50	6.95	5.65	49.10	57.50
7630	75' long, 13" diam.	G		407	.157		55.50	8.70	7.05	71.25	82
7640	75' long, 16" diam.	G		407	.157		87.50	8.70	7.05	103.25	117
7650	75' long, 20" diam.	G		356	.180		88.50	9.95	8.05	106.50	121
7660	75' long, 24" diam.	G		356	.180		108	9.95	8.05	126	143
7710	80' long, 8" diam.	G		500	.128		32.50	7.05	5.75	45.30	52.50
7720	80' long, 10" diam.	G		500	.128		36.50	7.05	5.75	49.30	57.50
7730	80' long, 13" diam.	G		400	.160		55.50	8.85	7.15	71.50	82.50
7740	80' long, 16" diam.	G		400	.160		87.50	8.85	7.15	103.50	117
7750	80' long, 20" diam.	G		350	.183		88.50	10.10	8.20	106.80	121
7760	80' long, 24" diam.	G		350	.183		108	10.10	8.20	126.30	143

31 62 23.13 Concrete-Filled Steel Piles

0010	CONCRETE-FILLED STEEL PILES no mobilization or demobilization		Crew	Daily Output	Labor-Hours	Unit	Material	Labor	Equipment	Total	Total Incl O&P
2600	Pipe piles, 50' L, 8" diam., 29 lb./L.F., no concrete		B-19	500	.128	V.L.F.	21.50	7.05	4.02	32.57	39
2700	Concrete filled			460	.139		25.50	7.70	4.37	37.57	44.50
2900	10" diameter, 34 lb./L.F., no concrete			500	.128		24	7.05	4.02	35.07	42
3000	Concrete filled			450	.142		30	7.85	4.47	42.32	50
3200	12" diameter, 44 lb./L.F., no concrete			475	.135		31.50	7.45	4.23	43.18	50.50
3300	Concrete filled			415	.154		35.50	8.50	4.84	48.84	57.50
3500	14" diameter, 46 lb./L.F., no concrete			430	.149		33.50	8.20	4.67	46.37	55
3600	Concrete filled			355	.180		41.50	9.95	5.65	57.10	67
3800	16" diameter, 52 lb./L.F., no concrete			385	.166		38	9.20	5.20	52.40	61.50
3900	Concrete filled			335	.191		52	10.55	6	68.55	80.50
4100	18" diameter, 59 lb./L.F., no concrete			355	.180		47	9.95	5.65	62.60	73
4200	Concrete filled			310	.206		61.50	11.40	6.50	79.40	92.50
4400	Splices for pipe piles, stl., not in leads, 8" diameter		1 Sswl	5	1.600	Ea.	81	92		173	233
4410	10" diameter			4.75	1.684		88.50	97		185.50	249
4430	12" diameter			4.50	1.778		106	102		208	276
4500	14" diameter			4.25	1.882		146	108		254	330
4600	16" diameter			4	2		165	115		280	360
4650	18" diameter			3.75	2.133		274	123		397	490
4710	Steel pipe pile backing rings, w/spacer, 8" diameter			12	.667		9.10	38.50		47.60	70

31 62 Driven Piles

31 62 23 – Composite Piles

31 62 23.13 Concrete-Filled Steel Piles

		Crew	Daily Output	Labor-Hours	Unit	Material	2020 Bare Costs Labor	Equipment	Total	Total Incl O&P
4720	10" diameter	1 Sswl	12	.667	Ea.	11.90	38.50		50.40	73
4730	12" diameter		10	.800		14.35	46		60.35	88
4740	14" diameter		9	.889		18.35	51		69.35	100
4750	16" diameter		8	1		21.50	57.50		79	114
4760	18" diameter		6	1.333		24	76.50		100.50	146
4800	Points, standard, 8" diameter		4.61	1.735		93	99.50		192.50	258
4840	10" diameter		4.45	1.798		123	103		226	297
4880	12" diameter		4.25	1.882		164	108		272	350
4900	14" diameter		4.05	1.975		209	113		322	410
5000	16" diameter		3.37	2.374		296	136		432	540
5050	18" diameter		3.50	2.286		385	131		516	625
5200	Points, heavy duty, 10" diameter		2.90	2.759		242	158		400	515
5240	12" diameter		2.95	2.712		305	156		461	580
5260	14" diameter		2.95	2.712		305	156		461	580
5280	16" diameter		2.95	2.712		350	156		506	630
5290	18" diameter		2.80	2.857		425	164		589	725
5500	For reinforcing steel, add		1150	.007	Lb.	.79	.40		1.19	1.50
5700	For thick wall sections, add				"	.68			.68	.75
6020	Steel pipe pile end plates, 8" diameter	1 Sswl	14	.571	Ea.	31	33		64	85.50
6050	10" diameter		14	.571		40.50	33		73.50	96
6100	12" diameter		12	.667		52	38.50		90.50	117
6150	14" diameter		10	.800		60.50	46		106.50	139
6200	16" diameter		9	.889		76	51		127	164
6250	18" diameter		8	1		100	57.50		157.50	200
6300	Steel pipe pile shoes, 8" diameter		12	.667		65.50	38.50		104	132
6350	10" diameter		12	.667		82	38.50		120.50	150
6400	12" diameter		10	.800		97	46		143	179
6450	14" diameter		9	.889		143	51		194	237
6500	16" diameter		8	1		149	57.50		206.50	254
6550	18" diameter		6	1.333		217	76.50		293.50	360

31 63 Bored Piles

31 63 26 – Drilled Caissons

31 63 26.13 Fixed End Caisson Piles

		Crew	Daily Output	Labor-Hours	Unit	Material	2020 Bare Costs Labor	Equipment	Total	Total Incl O&P
0010	**FIXED END CAISSON PILES** R316326-60									
0015	Including excavation, concrete, 50 lb. reinforcing									
0020	per C.Y., not incl. mobilization, boulder removal, disposal									
0100	Open style, machine drilled, to 50' deep, in stable ground, no									
0110	casings or ground water, 18" diam., 0.065 C.Y./L.F.	B-43	200	.240	V.L.F.	10.20	11.20	3.85	25.25	32
0200	24" diameter, 0.116 C.Y./L.F.		190	.253		18.20	11.80	4.06	34.06	42
0300	30" diameter, 0.182 C.Y./L.F.		150	.320		28.50	14.95	5.15	48.60	59.50
0400	36" diameter, 0.262 C.Y./L.F.		125	.384		41	17.95	6.15	65.10	79
0500	48" diameter, 0.465 C.Y./L.F.		100	.480		73	22.50	7.70	103.20	123
0600	60" diameter, 0.727 C.Y./L.F.		90	.533		114	25	8.55	147.55	173
0700	72" diameter, 1.05 C.Y./L.F.		80	.600		165	28	9.65	202.65	234
0800	84" diameter, 1.43 C.Y./L.F.		75	.640		225	30	10.30	265.30	305
1000	For bell excavation and concrete, add									
1020	4' bell diameter, 24" shaft, 0.444 C.Y.	B-43	20	2.400	Ea.	57	112	38.50	207.50	274
1040	6' bell diameter, 30" shaft, 1.57 C.Y.		5.70	8.421		202	395	135	732	960
1060	8' bell diameter, 36" shaft, 3.72 C.Y.		2.40	20		480	935	320	1,735	2,275
1080	9' bell diameter, 48" shaft, 4.48 C.Y.		2	24		575	1,125	385	2,085	2,725

31 63 Bored Piles

31 63 26 – Drilled Caissons

31 63 26.13 Fixed End Caisson Piles	Crew	Daily Output	Labor-Hours	Unit	Material	2020 Bare Costs Labor	Equipment	Total	Total Incl O&P
1100	10' bell diameter, 60" shaft, 5.24 C.Y.	B-43	1.70	28.235	Ea.	675	1,325	455	2,455	3,225
1120	12' bell diameter, 72" shaft, 8.74 C.Y.		1	48		1,125	2,250	770	4,145	5,450
1140	14' bell diameter, 84" shaft, 13.6 C.Y.	↓	.70	68.571	↓	1,750	3,200	1,100	6,050	7,925
1200	Open style, machine drilled, to 50' deep, in wet ground, pulled									
1300	casing and pumping, 18" diameter, 0.065 C.Y./L.F.	B-48	160	.350	V.L.F.	10.20	16.65	6.55	33.40	43.50
1400	24" diameter, 0.116 C.Y./L.F.		125	.448		18.20	21.50	8.40	48.10	61
1500	30" diameter, 0.182 C.Y./L.F.		85	.659		28.50	31.50	12.30	72.30	92
1600	36" diameter, 0.262 C.Y./L.F.	↓	60	.933		41	44.50	17.45	102.95	131
1700	48" diameter, 0.465 C.Y./L.F.	B-49	55	1.600		73	80	29.50	182.50	234
1800	60" diameter, 0.727 C.Y./L.F.		35	2.514		114	126	46.50	286.50	365
1900	72" diameter, 1.05 C.Y./L.F.		30	2.933		165	147	54.50	366.50	465
2000	84" diameter, 1.43 C.Y./L.F.	↓	25	3.520	↓	225	176	65	466	585
2100	For bell excavation and concrete, add									
2120	4' bell diameter, 24" shaft, 0.444 C.Y.	B-48	19.80	2.828	Ea.	57	135	53	245	325
2140	6' bell diameter, 30" shaft, 1.57 C.Y.		5.70	9.825		202	470	184	856	1,125
2160	8' bell diameter, 36" shaft, 3.72 C.Y.	↓	2.40	23.333		480	1,100	435	2,015	2,675
2180	9' bell diameter, 48" shaft, 4.48 C.Y.	B-49	3.30	26.667		575	1,325	495	2,395	3,200
2200	10' bell diameter, 60" shaft, 5.24 C.Y.		2.80	31.429		675	1,575	580	2,830	3,750
2220	12' bell diameter, 72" shaft, 8.74 C.Y.		1.60	55		1,125	2,750	1,025	4,900	6,525
2240	14' bell diameter, 84" shaft, 13.6 C.Y.	↓	1	88	↓	1,750	4,400	1,625	7,775	10,400
2300	Open style, machine drilled, to 50' deep, in soft rocks and									
2400	medium hard shales, 18" diameter, 0.065 C.Y./L.F.	B-49	50	1.760	V.L.F.	10.20	88	32.50	130.70	180
2500	24" diameter, 0.116 C.Y./L.F.		30	2.933		18.20	147	54.50	219.70	300
2600	30" diameter, 0.182 C.Y./L.F.		20	4.400		28.50	221	81.50	331	455
2700	36" diameter, 0.262 C.Y./L.F.		15	5.867		41	294	109	444	610
2800	48" diameter, 0.465 C.Y./L.F.		10	8.800		73	440	163	676	925
2900	60" diameter, 0.727 C.Y./L.F.		7	12.571		114	630	233	977	1,325
3000	72" diameter, 1.05 C.Y./L.F.		6	14.667		165	735	271	1,171	1,575
3100	84" diameter, 1.43 C.Y./L.F.	↓	5	17.600	↓	225	880	325	1,430	1,925
3200	For bell excavation and concrete, add									
3220	4' bell diameter, 24" shaft, 0.444 C.Y.	B-49	10.90	8.073	Ea.	57	405	149	611	835
3240	6' bell diameter, 30" shaft, 1.57 C.Y.		3.10	28.387		202	1,425	525	2,152	2,950
3260	8' bell diameter, 36" shaft, 3.72 C.Y.		1.30	67.692		480	3,400	1,250	5,130	7,025
3280	9' bell diameter, 48" shaft, 4.48 C.Y.		1.10	80		575	4,000	1,475	6,050	8,300
3300	10' bell diameter, 60" shaft, 5.24 C.Y.		.90	97.778		675	4,900	1,800	7,375	10,100
3320	12' bell diameter, 72" shaft, 8.74 C.Y.		.60	147		1,125	7,350	2,725	11,200	15,300
3340	14' bell diameter, 84" shaft, 13.6 C.Y.		.40	220	↓	1,750	11,000	4,075	16,825	23,000
3600	For rock excavation, sockets, add, minimum		120	.733	C.F.		37	13.55	50.55	70.50
3650	Average		95	.926			46.50	17.15	63.65	89
3700	Maximum	↓	48	1.833	↓		92	34	126	177
3900	For 50' to 100' deep, add				V.L.F.				7%	7%
4000	For 100' to 150' deep, add								25%	25%
4100	For 150' to 200' deep, add				↓				30%	30%
4200	For casings left in place, add				Lb.	1.38			1.38	1.52
4300	For other than 50 lb. reinf. per C.Y., add or deduct				"	1.29			1.29	1.42
4400	For steel I-beam cores, add	B-49	8.30	10.602	Ton	2,300	530	196	3,026	3,550
4500	Load and haul excess excavation, 2 miles	B-34B	178	.045	L.C.Y.		2.20	3.22	5.42	6.85
4600	For mobilization, 50 mile radius, rig to 36"	B-43	2	24	Ea.		1,125	385	1,510	2,100
4650	Rig to 84"	B-48	1.75	32			1,525	600	2,125	2,925
4700	For low headroom, add								50%	50%
4750	For difficult access, add								25%	25%
5000	Bottom inspection	1 Skwk	1.20	6.667	↓		365		365	555

31 63 Bored Piles

31 63 26 – Drilled Caissons

31 63 26.16 Concrete Caissons for Marine Construction

31 63 26.16 Concrete Caissons for Marine Construction	Crew	Daily Output	Labor-Hours	Unit	Material	2020 Bare Costs Labor	Equipment	Total	Total Incl O&P
0010 **CONCRETE CAISSONS FOR MARINE CONSTRUCTION**									
0100 Caissons, incl. mobilization and demobilization, up to 50 miles									
0200 Uncased shafts, 30 to 80 tons cap., 17" diam., 10' depth	B-44	88	.727	V.L.F.	26	39.50	23	88.50	115
0300 25' depth		165	.388		18.40	21	12.30	51.70	65.50
0400 80 to 150 ton capacity, 22" diameter, 10' depth		80	.800		32	43.50	25.50	101	130
0500 20' depth		130	.492		26	26.50	15.65	68.15	86.50
0700 Cased shafts, 10 to 30 ton capacity, 10-5/8" diam., 20' depth		175	.366		18.40	19.80	11.60	49.80	63.50
0800 30' depth		240	.267		17.15	14.45	8.45	40.05	50
0850 30 to 60 ton capacity, 12" diameter, 20' depth		160	.400		26	21.50	12.70	60.20	75.50
0900 40' depth		230	.278		19.80	15.10	8.85	43.75	54.50
1000 80 to 100 ton capacity, 16" diameter, 20' depth		160	.400		37	21.50	12.70	71.20	87.50
1100 40' depth		230	.278		34.50	15.10	8.85	58.45	70.50
1200 110 to 140 ton capacity, 17-5/8" diameter, 20' depth		160	.400		39.50	21.50	12.70	73.70	90.50
1300 40' depth		230	.278		37	15.10	8.85	60.95	73
1400 140 to 175 ton capacity, 19" diameter, 20' depth		130	.492		43	26.50	15.65	85.15	105
1500 40' depth	↓	210	.305	↓	39.50	16.50	9.70	65.70	79.50
1700 Over 30' long, L.F. cost tends to be lower									
1900 Maximum depth is about 90'									

31 63 29 – Drilled Concrete Piers and Shafts

31 63 29.13 Uncased Drilled Concrete Piers

31 63 29.13 Uncased Drilled Concrete Piers	Crew	Daily Output	Labor-Hours	Unit	Material	2020 Bare Costs Labor	Equipment	Total	Total Incl O&P
0010 **UNCASED DRILLED CONCRETE PIERS**									
0020 Unless specified otherwise, not incl. pile caps or mobilization									
0050 Cast in place augered piles, no casing or reinforcing									
0060 8" diameter	B-43	540	.089	V.L.F.	4.23	4.15	1.43	9.81	12.40
0065 10" diameter		480	.100		6.75	4.67	1.61	13.03	16.15
0070 12" diameter		420	.114		9.50	5.35	1.84	16.69	20.50
0075 14" diameter		360	.133		12.75	6.25	2.14	21.14	26
0080 16" diameter		300	.160		17.20	7.45	2.57	27.22	33
0085 18" diameter	↓	240	.200	↓	21.50	9.35	3.21	34.06	41
0100 Cast in place, thin wall shell pile, straight sided,									
0110 not incl. reinforcing, 8" diam., 16 ga., 5.8 lb./L.F.	B-19	700	.091	V.L.F.	9.50	5.05	2.87	17.42	21.50
0200 10" diameter, 16 ga. corrugated, 7.3 lb./L.F.		650	.098		12.45	5.45	3.09	20.99	25.50
0300 12" diameter, 16 ga. corrugated, 8.7 lb./L.F.		600	.107		16.20	5.90	3.35	25.45	30.50
0400 14" diameter, 16 ga. corrugated, 10.0 lb./L.F.		550	.116		19.05	6.45	3.65	29.15	35
0500 16" diameter, 16 ga. corrugated, 11.6 lb./L.F.	↓	500	.128	↓	23.50	7.05	4.02	34.57	41
0800 Cast in place friction pile, 50' long, fluted,									
0810 tapered steel, 4,000 psi concrete, no reinforcing									
0900 12" diameter, 7 ga.	B-19	600	.107	V.L.F.	29.50	5.90	3.35	38.75	45
1000 14" diameter, 7 ga.		560	.114		32	6.30	3.59	41.89	49
1100 16" diameter, 7 ga.		520	.123		38	6.80	3.86	48.66	56
1200 18" diameter, 7 ga.	↓	480	.133	↓	44.50	7.35	4.19	56.04	64.50
1300 End bearing, fluted, constant diameter,									
1320 4,000 psi concrete, no reinforcing									
1340 12" diameter, 7 ga.	B-19	600	.107	V.L.F.	31	5.90	3.35	40.25	46.50
1360 14" diameter, 7 ga.		560	.114		39	6.30	3.59	48.89	56
1380 16" diameter, 7 ga.		520	.123		45	6.80	3.86	55.66	64
1400 18" diameter, 7 ga.	↓	480	.133	↓	49.50	7.35	4.19	61.04	70.50

31 63 29.20 Cast In Place Piles, Adds

31 63 29.20 Cast In Place Piles, Adds	Crew	Daily Output	Labor-Hours	Unit	Material	2020 Bare Costs Labor	Equipment	Total	Total Incl O&P
0010 **CAST IN PLACE PILES, ADDS**									
1500 For reinforcing steel, add				Lb.	1.13			1.13	1.24
1700 For ball or pedestal end, add	B-19	11	5.818	C.Y.	150	320	183	653	855

357

31 63 Bored Piles

31 63 29 – Drilled Concrete Piers and Shafts

31 63 29.20 Cast In Place Piles, Adds	Crew	Daily Output	Labor-Hours	Unit	Material	2020 Bare Costs Labor	Equipment	Total	Total Incl O&P
1900	For lengths above 60', concrete, add	B-19	11	5.818	C.Y.	157	320	183	660	865
2000	For steel thin shell, pipe only				Lb.	1.67			1.67	1.84

31 63 33 – Drilled Micropiles

31 63 33.10 Drilled Micropiles Metal Pipe

		Crew	Daily Output	Labor-Hours	Unit	Material	2020 Bare Costs Labor	Equipment	Total	Total Incl O&P
0010	**DRILLED MICROPILES METAL PIPE**									
0011	No mobilization or demobilization									
5000	Pressure grouted pin pile, 5" diam., cased, up to 50 ton,									
5040	End bearing, less than 20'	B-48	160	.350	V.L.F.	41	16.65	6.55	64.20	77.50
5080	More than 40'		240	.233		39	11.10	4.36	54.46	64.50
5120	Friction, loose sand and gravel		240	.233		41	11.10	4.36	56.46	67
5160	Dense sand and gravel		240	.233		41	11.10	4.36	56.46	67
5200	Uncased, up to 10 ton capacity, 20'		200	.280		22	13.35	5.25	40.60	50

Estimating Tips

32 01 00 Operations and Maintenance of Exterior Improvements

- Recycling of asphalt pavement is becoming very popular and is an alternative to removal and replacement. It can be a good value engineering proposal if removed pavement can be recycled, either at the project site or at another site that is reasonably close to the project site. Sections on repair of flexible and rigid pavement are included.

32 10 00 Bases, Ballasts, and Paving

- When estimating paving, keep in mind the project schedule. Also note that prices for asphalt and concrete are generally higher in the cold seasons. Lines for pavement markings, including tactile warning systems and fence lines, are included.

32 90 00 Planting

- The timing of planting and guarantee specifications often dictate the costs for establishing tree and shrub growth and a stand of grass or ground cover. Establish the work performance schedule to coincide with the local planting season. Maintenance and growth guarantees can add 20–100% to the total landscaping cost and can be contractually cumbersome. The cost to replace trees and shrubs can be as high as 5% of the total cost, depending on the planting zone, soil conditions, and time of year.

Reference Numbers

Reference numbers are shown at the beginning of some major classifications. These numbers refer to related items in the Reference Section. The reference information may be an estimating procedure, an alternate pricing method, or technical information.

Note: Not all subdivisions listed here necessarily appear. ■

Same Data. Simplified.

Enjoy the convenience and efficiency of accessing your costs anywhere:

- **Skip the multiplier** by setting your location
- **Quickly search,** edit, favorite and share costs
- **Stay on top of price changes** with automatic updates

Discover more at rsmeans.com/online

---

32 01 11 – Paving Cleaning

32 01 11.51 Rubber and Paint Removal From Paving

		Crew	Daily Output	Labor-Hours	Unit	Material	2020 Bare Costs Labor	Equipment	Total	Total Incl O&P
0010	**RUBBER AND PAINT REMOVAL FROM PAVING**									
0015	Does not include traffic control costs									
0020	See other items in Section 32 17 23.13									
0100	Remove permanent painted traffic lines and markings	B-78A	500	.016	C.L.F.		.85	1.83	2.68	3.29
0200	Temporary traffic line tape	2 Clab	1500	.011	L.F.		.45		.45	.67
0300	Thermoplastic traffic lines and markings	B-79A	500	.024	C.L.F.		1.27	2.78	4.05	4.96
0400	Painted pavement markings	B-78B	500	.036	S.F.		1.56	.81	2.37	3.23
1000	Prepare and clean, 25,000 S.Y.	B-91A	7.36	2.174	Mile		103	112	215	278
2000	Minimum labor/equipment charge	1 Clab	2	4	Job		168		168	253

32 01 13 – Flexible Paving Surface Treatment

32 01 13.61 Slurry Seal (Latex Modified)

		Crew	Daily Output	Labor-Hours	Unit	Material	Labor	Equipment	Total	Total Incl O&P
0010	**SLURRY SEAL (LATEX MODIFIED)**									
0011	Chip seal, slurry seal, and microsurfacing, see section 32 12 36									

32 01 13.62 Asphalt Surface Treatment

		Crew	Daily Output	Labor-Hours	Unit	Material	Labor	Equipment	Total	Total Incl O&P
0010	**ASPHALT SURFACE TREATMENT**									
3000	Pavement overlay, polypropylene									
3040	6 oz./S.Y., ideal conditions	B-63	10000	.004	S.Y.	1.09	.18	.02	1.29	1.49
3190	Prime coat, bituminous, 0.28 gallon/S.Y.	B-45	2400	.007	C.S.F.	8.20	.35	.34	8.89	9.95
3200	Tack coat, emulsion, 0.05 gal./S.Y., 1,000 S.Y.		2500	.006	S.Y.	.35	.34	.33	1.02	1.26
3240	10,000 S.Y.		10000	.002		.28	.08	.08	.44	.53
3275	10,000 S.Y.		10000	.002		.53	.08	.08	.69	.80
3280	0.15 gal./S.Y., 1,000 S.Y.		2500	.006		.96	.34	.33	1.63	1.93
3320	10,000 S.Y.		10000	.002		.77	.08	.08	.93	1.07
3780	Rubberized asphalt (latex) seal		5000	.003		1.39	.17	.16	1.72	1.96
5400	Thermoplastic coal-tar, Type I, small or irregular area	B-90	2400	.027		4.32	1.25	.92	6.49	7.65
5450	Roadway or large area		8000	.008		4.32	.37	.27	4.96	5.60
5500	Type II, small or irregular area		2400	.027		5.50	1.25	.92	7.67	8.95
6000	Gravel surfacing on asphalt, screened and rolled	B-11L	160	.100	C.Y.	23	4.94	6.65	34.59	40
7000	For subbase treatment, see Section 31 32 13									

32 01 13.64 Sand Seal

		Crew	Daily Output	Labor-Hours	Unit	Material	Labor	Equipment	Total	Total Incl O&P
0010	**SAND SEAL**									
2080	Sand sealing, sharp sand, asphalt emulsion, small area	B-91	10000	.006	S.Y.	1.48	.32	.23	2.03	2.36
2120	Roadway or large area	"	18000	.004	"	1.27	.18	.13	1.58	1.81
3000	Sealing random cracks, min. 1/2" wide, to 1-1/2", 1,000 L.F.	B-77	2800	.014	L.F.	1.65	.62	.37	2.64	3.15
3040	10,000 L.F.		4000	.010	"	1.14	.44	.26	1.84	2.19
3080	Alternate method, 1,000 L.F.		200	.200	Gal.	45	8.70	5.15	58.85	68
3120	10,000 L.F.		325	.123	"	36	5.35	3.16	44.51	51.50
3200	Multi-cracks (flooding), 1 coat, small area	B-92	460	.070	S.Y.	2.55	2.96	2.57	8.08	10.10
3320	Large area		1425	.022	"	14.35	.96	.83	16.14	18.15
3360	Alternate method, small area		115	.278	Gal.	15.55	11.85	10.30	37.70	46
3400	Large area		715	.045	"	14.60	1.91	1.65	18.16	20.50
3500	Sealing, roads, resealing joints in concrete	B-77	525	.076	L.F.	.39	3.32	1.96	5.67	7.55

32 01 13.66 Fog Seal

		Crew	Daily Output	Labor-Hours	Unit	Material	Labor	Equipment	Total	Total Incl O&P
0010	**FOG SEAL**									
0012	Sealcoating, 2 coat coal tar pitch emulsion over 10,000 S.Y.	B-45	5000	.003	S.Y.	.84	.17	.16	1.17	1.35
0030	1,000 to 10,000 S.Y.	"	3000	.005		.84	.28	.27	1.39	1.64
0100	Under 1,000 S.Y.	B-1	1050	.023		.84	.98		1.82	2.39
0300	Petroleum resistant, over 10,000 S.Y.	B-45	5000	.003		1.40	.17	.16	1.73	1.97
0320	1,000 to 10,000 S.Y.	"	3000	.005		1.40	.28	.27	1.95	2.26
0400	Under 1,000 S.Y.	B-1	1050	.023		1.40	.98		2.38	3.01
0600	Non-skid pavement renewal, over 10,000 S.Y.	B-45	5000	.003		1.37	.17	.16	1.70	1.94

360

For customer support on your Site Work & Landscape Costs with RSMeans data, call 800.448.8182.

32 01 13 – Flexible Paving Surface Treatment

32 01 13.66 Fog Seal

		Crew	Daily Output	Labor-Hours	Unit	Material	2020 Bare Costs Labor	2020 Bare Costs Equipment	Total	Total Incl O&P
0620	1,000 to 10,000 S.Y.	B-45	3000	.005	S.Y.	1.37	.28	.27	1.92	2.23
0700	Under 1,000 S.Y.	B-1	1050	.023		1.37	.98		2.35	2.98
0800	Prepare and clean surface for above	A-2	8545	.003	▼		.12	.02	.14	.21
1000	Hand seal asphalt curbing	B-1	4420	.005	L.F.	.59	.23		.82	1
1900	Asphalt surface treatment, single course, small area									
1901	0.30 gal./S.Y. asphalt material, 20 lb./S.Y. aggregate	B-91	5000	.013	S.Y.	1.41	.65	.45	2.51	3.02
1910	Roadway or large area		10000	.006	"	1.30	.32	.23	1.85	2.16
1920	Bituminous surface treatment, 12' wide roadway, single course		7	9.143	Mile	7,525	460	320	8,305	9,325
1950	Asphalt surface treatment, dbl. course for small area		3000	.021	S.Y.	2.97	1.08	.75	4.80	5.70
1960	Roadway or large area		6000	.011		2.67	.54	.38	3.59	4.16
1980	Asphalt surface treatment, single course, for shoulders	▼	7500	.009	▼	1.46	.43	.30	2.19	2.59

32 01 16 – Flexible Paving Rehabilitation

32 01 16.71 Cold Milling Asphalt Paving

		Crew	Daily Output	Labor-Hours	Unit	Material	2020 Bare Costs Labor	2020 Bare Costs Equipment	Total	Total Incl O&P
0010	**COLD MILLING ASPHALT PAVING**									
5200	Cold planing & cleaning, 1" to 3" asphalt pavmt., over 25,000 S.Y.	B-71	6000	.009	S.Y.		.45	.76	1.21	1.52
5280	5,000 S.Y. to 10,000 S.Y.		4000	.014	"		.68	1.15	1.83	2.28
5285	5,000 S.Y. to 10,000 S.Y.	▼	36000	.002	S.F.		.08	.13	.21	.25
5300	Asphalt pavement removal from conc. base, no haul									
5320	Rip, load & sweep 1" to 3"	B-70	8000	.007	S.Y.		.34	.29	.63	.83
5330	3" to 6" deep	"	5000	.011			.55	.46	1.01	1.32
5340	Profile grooving, asphalt pavement load & sweep, 1" deep	B-71	12500	.004			.22	.37	.59	.73
5350	3" deep		9000	.006			.30	.51	.81	1.01
5360	6" deep	▼	5000	.011			.55	.92	1.47	1.83
5400	Mixing material in windrow, 180 HP grader	B-11L	9400	.002	C.Y.		.08	.11	.19	.25
5450	For cold laid asphalt pavement, see Section 32 12 16.19									
7400	(Over 3") 5,000-25,000 S.F.	B-71	7500	.007	S.F.		.36	.61	.97	1.21
7410	2,5001-100,000 S.F.		27000	.002			.10	.17	.27	.34
7420	More than 100,000 S.F.	▼	50000	.001	▼		.05	.09	.14	.18

32 01 16.73 In Place Cold Reused Asphalt Paving

		Crew	Daily Output	Labor-Hours	Unit	Material	2020 Bare Costs Labor	2020 Bare Costs Equipment	Total	Total Incl O&P
0010	**IN PLACE COLD REUSED ASPHALT PAVING**									
5000	Reclamation, pulverizing and blending with existing base									
5040	Aggregate base, 4" thick pavement, over 15,000 S.Y.	B-73	2400	.027	S.Y.		1.37	1.87	3.24	4.12
5080	5,000 S.Y. to 15,000 S.Y.		2200	.029			1.50	2.04	3.54	4.48
5120	8" thick pavement, over 15,000 S.Y.		2200	.029			1.50	2.04	3.54	4.48
5160	5,000 S.Y. to 15,000 S.Y.	▼	2000	.032	▼		1.65	2.24	3.89	4.94

32 01 16.74 In Place Hot Reused Asphalt Paving

			Crew	Daily Output	Labor-Hours	Unit	Material	2020 Bare Costs Labor	2020 Bare Costs Equipment	Total	Total Incl O&P
0010	**IN PLACE HOT REUSED ASPHALT PAVING**										
5500	Recycle asphalt pavement at site										
5520	Remove, rejuvenate and spread 4" deep	G	B-72	2500	.026	S.Y.	4.58	1.27	3.29	9.14	10.60
5521	6" deep	G	"	2000	.032	"	6.70	1.59	4.11	12.40	14.30

32 01 17 – Flexible Paving Repair

32 01 17.10 Repair of Asphalt Pavement Holes

		Crew	Daily Output	Labor-Hours	Unit	Material	2020 Bare Costs Labor	2020 Bare Costs Equipment	Total	Total Incl O&P
0010	**REPAIR OF ASPHALT PAVEMENT HOLES** (cold patch)									
0100	Flexible pavement repair holes, roadway, light traffic, 1 C.F. size each	B-37A	24	1	Ea.	9.50	44	14.70	68.20	92.50
0150	Group of two, 1 C.F. size each		16	1.500	Set	19	65.50	22	106.50	144
0200	Group of three, 1 C.F. size each		12	2	"	28.50	87.50	29.50	145.50	196
0300	Medium traffic, 1 C.F. size each	B-37B	24	1.333	Ea.	9.50	58	14.70	82.20	114
0350	Group of two, 1 C.F. size each		16	2	Set	19	87	22	128	175
0400	Group of three, 1 C.F. size each	▼	12	2.667	"	28.50	116	29.50	174	238
0500	Highway/heavy traffic, 1 C.F. size each	B-37C	24	1.333	Ea.	9.50	59.50	23	92	125
0550	Group of two, 1 C.F. size each	▼	16	2	Set	19	89.50	34.50	143	193

32 01 17.10 Repair of Asphalt Pavement Holes

		Crew	Daily Output	Labor-Hours	Unit	Material	2020 Bare Costs Labor	Equipment	Total	Total Incl O&P
0600	Group of three, 1 C.F. size each	B-37C	12	2.667	Set	28.50	119	45.50	193	261
0700	Add police officer and car for traffic control				Hr.	54.50			54.50	60
1000	Flexible pavement repair holes, parking lot, bag material, 1 C.F. size each	B-37D	18	.889	Ea.	59	39.50	6.15	104.65	131
1010	Economy bag material, 1 C.F. size each		18	.889	"	37	39.50	6.15	82.65	107
1100	Group of two, 1 C.F. size each		12	1.333	Set	118	59.50	9.25	186.75	229
1110	Group of two, economy bag, 1 C.F. size each		12	1.333		73.50	59.50	9.25	142.25	181
1200	Group of three, 1 C.F. size each		10	1.600		176	71.50	11.10	258.60	315
1210	Economy bag, group of three, 1 C.F. size each		10	1.600		110	71.50	11.10	192.60	240
1300	Flexible pavement repair holes, parking lot, 1 C.F. single hole	A-3A	4	2	Ea.	59	106	44	209	271
1310	Economy material, 1 C.F. single hole		4	2	"	37	106	44	187	247
1400	Flexible pavement repair holes, parking lot, 1 C.F. four holes		2	4	Set	235	212	88	535	670
1410	Economy material, 1 C.F. four holes		2	4	"	147	212	88	447	575
1500	Flexible pavement repair holes, large parking lot, bulk matl, 1 C.F. size	B-37A	32	.750	Ea.	9.50	33	11	53.50	72

32 01 17.20 Repair of Asphalt Pavement Patches

		Crew	Daily Output	Labor-Hours	Unit	Material	2020 Bare Costs Labor	Equipment	Total	Total Incl O&P
0010	**REPAIR OF ASPHALT PAVEMENT PATCHES**									
0100	Flexible pavement patches, roadway, light traffic, sawcut 10-25 S.F.	B-89	12	1.333	Ea.		67	86	153	195
0150	Sawcut 26-60 S.F.		10	1.600			80	103	183	234
0200	Sawcut 61-100 S.F.		8	2			100	129	229	292
0210	Sawcut groups of small size patches		24	.667			33.50	43	76.50	97.50
0220	Large size patches		16	1			50	64.50	114.50	146
0300	Flexible pavement patches, roadway, light traffic, digout 10-25 S.F.	B-6	16	1.500			68.50	13.35	81.85	118
0350	Digout 26-60 S.F.		12	2			91.50	17.80	109.30	157
0400	Digout 61-100 S.F.		8	3			137	26.50	163.50	236
0450	Add 8 C.Y. truck, small project debris haulaway	B-34A	8	1	Hr.		49	53.50	102.50	133
0460	Add 12 C.Y. truck, small project debris haulaway	B-34B	8	1			49	71.50	120.50	153
0480	Add flagger for non-intersection medium traffic	1 Clab	8	1			42		42	63.50
0490	Add flasher truck for intersection medium traffic or heavy traffic	A-2B	8	1			47.50	24.50	72	98
0500	Flexible pavement patches, roadway, repave, cold, 15 S.F., 4" D	B-37	8	6	Ea.	45	266	32	343	485
0510	6" depth		8	6		68	266	32	366	510
0520	Repave, cold, 20 S.F., 4" depth		8	6		60	266	32	358	500
0530	6" depth		8	6		90.50	266	32	388.50	535
0540	Repave, cold, 25 S.F., 4" depth		8	6		74.50	266	32	372.50	515
0550	6" depth		8	6		113	266	32	411	560
0600	Repave, cold, 30 S.F., 4" depth		8	6		89.50	266	32	387.50	535
0610	6" depth		8	6		136	266	32	434	585
0640	Repave, cold, 40 S.F., 4" depth		8	6		119	266	32	417	565
0650	6" depth		8	6		181	266	32	479	635
0680	Repave, cold, 50 S.F., 4" depth		8	6		149	266	32	447	600
0690	6" depth		8	6		226	266	32	524	685
0720	Repave, cold, 60 S.F., 4" depth		8	6		179	266	32	477	630
0730	6" depth		8	6		271	266	32	569	735
0800	Repave, cold, 70 S.F., 4" depth		7	6.857		209	305	36.50	550.50	725
0810	6" depth		7	6.857		315	305	36.50	656.50	845
0820	Repave, cold, 75 S.F., 4" depth		7	6.857		222	305	36.50	563.50	740
0830	6" depth		7	6.857		340	305	36.50	681.50	870
0900	Repave, cold, 80 S.F., 4" depth		6	8		239	355	42.50	636.50	840
0910	6" depth		6	8		360	355	42.50	757.50	975
0940	Repave, cold, 90 S.F., 4" depth		6	8		268	355	42.50	665.50	870
0950	6" depth		6	8		405	355	42.50	802.50	1,025
0980	Repave, cold, 100 S.F., 4" depth		6	8		298	355	42.50	695.50	905
0990	6" depth		6	8		450	355	42.50	847.50	1,075
1000	Add flasher truck for paving operations in medium or heavy traffic	A-2B	8	1	Hr.		47.50	24.50	72	98

362

32 01 17 – Flexible Paving Repair

32 01 17.20 Repair of Asphalt Pavement Patches

		Crew	Daily Output	Labor-Hours	Unit	Material	2020 Bare Costs Labor	2020 Bare Costs Equipment	Total	Total Incl O&P
1100	Prime coat for repair 15-40 S.F., 25% overspray	B-37A	48	.500	Ea.	8.75	22	7.35	38.10	51
1150	41-60 S.F.		40	.600		17.50	26.50	8.80	52.80	68.50
1175	61-80 S.F.		32	.750		23.50	33	11	67.50	87.50
1200	81-100 S.F.		32	.750		29	33	11	73	93.50
1210	Groups of patches w/25% overspray		3600	.007	S.F.	.29	.29	.10	.68	.87
1300	Flexible pavement repair patches, street repave, hot, 60 S.F., 4" D	B-37	8	6	Ea.	97	266	32	395	540
1310	6" depth		8	6		147	266	32	445	595
1320	Repave, hot, 70 S.F., 4" depth		7	6.857		113	305	36.50	454.50	620
1330	6" depth		7	6.857		172	305	36.50	513.50	685
1340	Repave, hot, 75 S.F., 4" depth		7	6.857		121	305	36.50	462.50	630
1350	6" depth		7	6.857		184	305	36.50	525.50	695
1360	Repave, hot, 80 S.F., 4" depth		6	8		130	355	42.50	527.50	720
1370	6" depth		6	8		196	355	42.50	593.50	795
1380	Repave, hot, 90 S.F., 4" depth		6	8		146	355	42.50	543.50	735
1390	6" depth		6	8		221	355	42.50	618.50	820
1400	Repave, hot, 100 S.F., 4" depth		6	8		162	355	42.50	559.50	755
1410	6" depth		6	8		245	355	42.50	642.50	845
1420	Pave hot groups of patches, 4" depth		900	.053	S.F.	1.62	2.36	.28	4.26	5.65
1430	6" depth		900	.053	"	2.46	2.36	.28	5.10	6.55
1500	Add 8 C.Y. truck for hot asphalt paving operations	B-34A	1	8	Day		390	430	820	1,050
1550	Add flasher truck for hot paving operations in med/heavy traffic	A-2B	8	1	Hr.		47.50	24.50	72	98
2000	Add police officer and car for traffic control				"	54.50			54.50	60

32 01 17.61 Sealing Cracks In Asphalt Paving

		Crew	Daily Output	Labor-Hours	Unit	Material	2020 Bare Costs Labor	2020 Bare Costs Equipment	Total	Total Incl O&P
0010	**SEALING CRACKS IN ASPHALT PAVING**									
0100	Sealing cracks in asphalt paving, 1/8" wide x 1/2" depth, slow set	B-37A	2000	.012	L.F.	.17	.53	.18	.88	1.17
0110	1/4" wide x 1/2" depth		2000	.012		.20	.53	.18	.91	1.20
0130	3/8" wide x 1/2" depth		2000	.012		.22	.53	.18	.93	1.23
0140	1/2" wide x 1/2" depth		1800	.013		.25	.58	.20	1.03	1.38
0150	3/4" wide x 1/2" depth		1600	.015		.31	.66	.22	1.19	1.57
0160	1" wide x 1/2" depth		1600	.015		.37	.66	.22	1.25	1.63
0165	1/8" wide x 1" depth, rapid set	B-37F	2000	.016		.18	.69	.26	1.13	1.53
0170	1/4" wide x 1" depth		2000	.016		.36	.69	.26	1.31	1.72
0175	3/8" wide x 1" depth		1800	.018		.53	.77	.29	1.59	2.07
0180	1/2" wide x 1" depth		1800	.018		.71	.77	.29	1.77	2.26
0185	5/8" wide x 1" depth		1800	.018		.89	.77	.29	1.95	2.46
0190	3/4" wide x 1" depth		1600	.020		1.07	.87	.33	2.27	2.84
0195	1" wide x 1" depth		1600	.020		1.42	.87	.33	2.62	3.23
0200	1/8" wide x 2" depth, rapid set		2000	.016		.36	.69	.26	1.31	1.72
0210	1/4" wide x 2" depth		2000	.016		.71	.69	.26	1.66	2.11
0220	3/8" wide x 2" depth		1800	.018		1.07	.77	.29	2.13	2.66
0230	1/2" wide x 2" depth		1800	.018		1.42	.77	.29	2.48	3.05
0240	5/8" wide x 2" depth		1800	.018		1.78	.77	.29	2.84	3.44
0250	3/4" wide x 2" depth		1600	.020		2.14	.87	.33	3.34	4.01
0260	1" wide x 2" depth		1600	.020		2.85	.87	.33	4.05	4.80
0300	Add flagger for non-intersection medium traffic	1 Clab	8	1	Hr.		42		42	63.50
0400	Add flasher truck for intersection medium traffic or heavy traffic	A-2B	8	1	"		47.50	24.50	72	98

32 01 19 – Rigid Paving Surface Treatment

32 01 19.61 Sealing of Joints In Rigid Paving

		Crew	Daily Output	Labor-Hours	Unit	Material	2020 Bare Costs Labor	2020 Bare Costs Equipment	Total	Total Incl O&P
0010	**SEALING OF JOINTS IN RIGID PAVING**									
1000	Sealing, concrete pavement, preformed elastomeric	B-77	500	.080	L.F.	3.70	3.48	2.06	9.24	11.60
2000	Waterproofing, membrane, tar and fabric, small area	B-63	233	.172	S.Y.	15.40	7.60	.76	23.76	29
2500	Large area		1435	.028		13.60	1.23	.12	14.95	16.95

363

32 01 19 – Rigid Paving Surface Treatment

32 01 19.61 Sealing of Joints In Rigid Paving	Crew	Daily Output	Labor-Hours	Unit	Material	2020 Bare Costs Labor	Equipment	Total	Total Incl O&P
3000	Preformed rubberized asphalt, small area	B-63	100	.400	S.Y.	19.90	17.70	1.78	39.38	50.50
3500	Large area	↓	367	.109	↓	18.10	4.83	.48	23.41	27.50
9000	Minimum labor/equipment charge	1 Clab	2	4	Job		168		168	253

32 01 29 – Rigid Paving Repair

32 01 29.61 Partial Depth Patching of Rigid Pavement

		Crew	Daily Output	Labor-Hours	Unit	Material	Labor	Equipment	Total	Total Incl O&P
0010	**PARTIAL DEPTH PATCHING OF RIGID PAVEMENT**									
0100	Rigid pavement repair, roadway, light traffic, 25% pitting 15 S.F.	B-37F	16	2	Ea.	24	87	32.50	143.50	192
0110	Pitting 20 S.F.		16	2		31.50	87	32.50	151	201
0120	Pitting 25 S.F.		16	2		39.50	87	32.50	159	210
0130	Pitting 30 S.F.		16	2		47.50	87	32.50	167	219
0140	Pitting 40 S.F.		16	2		63.50	87	32.50	183	236
0150	Pitting 50 S.F.		12	2.667		79	116	43.50	238.50	310
0160	Pitting 60 S.F.		12	2.667		95	116	43.50	254.50	325
0170	Pitting 70 S.F.		12	2.667		111	116	43.50	270.50	345
0180	Pitting 75 S.F.		12	2.667		119	116	43.50	278.50	355
0190	Pitting 80 S.F.		8	4		127	174	65	366	470
0200	Pitting 90 S.F.		8	4		143	174	65	382	490
0210	Pitting 100 S.F.		8	4		158	174	65	397	505
0300	Roadway, light traffic, 50% pitting 15 S.F.		12	2.667		47.50	116	43.50	207	275
0310	Pitting 20 S.F.		12	2.667		63.50	116	43.50	223	292
0320	Pitting 25 S.F.		12	2.667		79	116	43.50	238.50	310
0330	Pitting 30 S.F.		12	2.667		95	116	43.50	254.50	325
0340	Pitting 40 S.F.		12	2.667		127	116	43.50	286.50	360
0350	Pitting 50 S.F.		8	4		158	174	65	397	505
0360	Pitting 60 S.F.		8	4		190	174	65	429	540
0370	Pitting 70 S.F.		8	4		222	174	65	461	575
0380	Pitting 75 S.F.		8	4		238	174	65	477	595
0390	Pitting 80 S.F.		6	5.333		253	231	87	571	720
0400	Pitting 90 S.F.		6	5.333		285	231	87	603	755
0410	Pitting 100 S.F.		6	5.333		315	231	87	633	790
1000	Rigid pavement repair, light traffic, surface patch, 2" deep, 15 S.F.		8	4		95	174	65	334	440
1010	Surface patch, 2" deep, 20 S.F.		8	4		127	174	65	366	470
1020	Surface patch, 2" deep, 25 S.F.		8	4		158	174	65	397	505
1030	Surface patch, 2" deep, 30 S.F.		8	4		190	174	65	429	540
1040	Surface patch, 2" deep, 40 S.F.		8	4		253	174	65	492	610
1050	Surface patch, 2" deep, 50 S.F.		6	5.333		315	231	87	633	790
1060	Surface patch, 2" deep, 60 S.F.		6	5.333		380	231	87	698	860
1070	Surface patch, 2" deep, 70 S.F.		6	5.333		445	231	87	763	930
1080	Surface patch, 2" deep, 75 S.F.		6	5.333		475	231	87	793	965
1090	Surface patch, 2" deep, 80 S.F.		5	6.400		505	278	104	887	1,100
1100	Surface patch, 2" deep, 90 S.F.		5	6.400		570	278	104	952	1,150
1200	Surface patch, 2" deep, 100 S.F.	↓	5	6.400	↓	635	278	104	1,017	1,225
2000	Add flagger for non-intersection medium traffic	1 Clab	8	1	Hr.		42		42	63.50
2100	Add flasher truck for intersection medium traffic or heavy traffic	A-2B	8	1			47.50	24.50	72	98
2200	Add police officer and car for traffic control				↓	54.50			54.50	60

32 01 29.70 Full Depth Patching of Rigid Pavement

		Crew	Daily Output	Labor-Hours	Unit	Material	Labor	Equipment	Total	Total Incl O&P
0010	**FULL DEPTH PATCHING OF RIGID PAVEMENT**									
0015	Pavement preparation includes sawcut, remove pavement and replace									
0020	6" of subbase and one layer of reinforcement									
0030	Trucking and haulaway of debris excluded									
0100	Rigid pavement replace, light traffic, prep, 15 S.F., 6" depth	B-37E	16	3.500	Ea.	23	165	58	246	335
0110	Replacement preparation 20 S.F., 6" depth	↓	16	3.500	↓	30.50	165	58	253.50	345

364

For customer support on your Site Work & Landscape Costs with RSMeans data, call 800.448.8182.

32 01 29 – Rigid Paving Repair

32 01 29.70 Full Depth Patching of Rigid Pavement		Crew	Daily Output	Labor-Hours	Unit	Material	2020 Bare Costs Labor	2020 Bare Costs Equipment	Total	Total Incl O&P
0115	25 S.F., 6" depth	B-37E	16	3.500	Ea.	38	165	58	261	355
0120	30 S.F., 6" depth		14	4		46	189	66	301	405
0125	35 S.F., 6" depth		14	4		53.50	189	66	308.50	415
0130	40 S.F., 6" depth		14	4		61	189	66	316	425
0135	45 S.F., 6" depth		14	4		69	189	66	324	430
0140	50 S.F., 6" depth		12	4.667		76.50	220	77	373.50	500
0145	55 S.F., 6" depth		12	4.667		84	220	77	381	510
0150	60 S.F., 6" depth		10	5.600		92	264	92.50	448.50	600
0155	65 S.F., 6" depth		10	5.600		99.50	264	92.50	456	605
0160	70 S.F., 6" depth		10	5.600		107	264	92.50	463.50	615
0165	75 S.F., 6" depth		10	5.600		115	264	92.50	471.50	625
0170	80 S.F., 6" depth		8	7		122	330	116	568	755
0175	85 S.F., 6" depth		8	7		130	330	116	576	765
0180	90 S.F., 6" depth		8	7		138	330	116	584	775
0185	95 S.F., 6" depth		8	7		145	330	116	591	780
0190	100 S.F., 6" depth		8	7		153	330	116	599	790
0200	Pavement preparation for 8" rigid paving same as 6"									
0290	Pavement preparation for 9" rigid paving same as 10"									
0300	Rigid pavement replace, light traffic, prep, 15 S.F., 10" depth	B-37E	16	3.500	Ea.	38	165	58	261	355
0302	Pavement preparation 10" includes sawcut, remove pavement and									
0304	replace 6" of subbase and two layers of reinforcement									
0306	Trucking and haulaway of debris excluded									
0310	Replacement preparation 20 S.F., 10" depth	B-37E	16	3.500	Ea.	51	165	58	274	370
0315	25 S.F., 10" depth		16	3.500		63.50	165	58	286.50	380
0320	30 S.F., 10" depth		14	4		76.50	189	66	331.50	440
0325	35 S.F., 10" depth		14	4		58.50	189	66	313.50	420
0330	40 S.F., 10" depth		14	4		102	189	66	357	470
0335	45 S.F., 10" depth		12	4.667		115	220	77	412	540
0340	50 S.F., 10" depth		12	4.667		127	220	77	424	555
0345	55 S.F., 10" depth		12	4.667		140	220	77	437	570
0350	60 S.F., 10" depth		10	5.600		153	264	92.50	509.50	665
0355	65 S.F., 10" depth		10	5.600		166	264	92.50	522.50	680
0360	70 S.F., 10" depth		10	5.600		178	264	92.50	534.50	695
0365	75 S.F., 10" depth		8	7		191	330	116	637	830
0370	80 S.F., 10" depth		8	7		204	330	116	650	845
0375	85 S.F., 10" depth		8	7		216	330	116	662	860
0380	90 S.F., 10" depth		8	7		229	330	116	675	875
0385	95 S.F., 10" depth		8	7		242	330	116	688	890
0390	100 S.F., 10" depth		8	7		255	330	116	701	900
0500	Add 8 C.Y. truck, small project debris haulaway	B-34A	8	1	Hr.		49	53.50	102.50	133
0550	Add 12 C.Y. truck, small project debris haulaway	B-34B	8	1			49	71.50	120.50	153
0600	Add flagger for non-intersection medium traffic	1 Clab	8	1			42		42	63.50
0650	Add flasher truck for intersection medium traffic or heavy traffic	A-2B	8	1			47.50	24.50	72	98
0700	Add police officer and car for traffic control					54.50			54.50	60
0990	Concrete will be replaced using quick set mix with aggregate									
1000	Rigid pavement replace, light traffic, repour, 15 S.F., 6" depth	B-37F	18	1.778	Ea.	174	77	29	280	340
1010	20 S.F., 6" depth		18	1.778		232	77	29	338	405
1015	25 S.F., 6" depth		18	1.778		290	77	29	396	470
1020	30 S.F., 6" depth		16	2		350	87	32.50	469.50	550
1025	35 S.F., 6" depth		16	2		405	87	32.50	524.50	610
1030	40 S.F., 6" depth		16	2		465	87	32.50	584.50	675
1035	45 S.F., 6" depth		16	2		525	87	32.50	644.50	740
1040	50 S.F., 6" depth		16	2		580	87	32.50	699.50	805

For customer support on your Site Work & Landscape Costs with RSMeans data, call 800.448.8182.

365

32 01 29 – Rigid Paving Repair

32 01 29.70 Full Depth Patching of Rigid Pavement

		Crew	Daily Output	Labor-Hours	Unit	Material	2020 Bare Costs Labor	2020 Bare Costs Equipment	Total	Total Incl O&P
1045	55 S.F., 6" depth	B-37F	12	2.667	Ea.	640	116	43.50	799.50	925
1050	60 S.F., 6" depth		12	2.667		695	116	43.50	854.50	985
1055	65 S.F., 6" depth		12	2.667		755	116	43.50	914.50	1,050
1060	70 S.F., 6" depth		12	2.667		815	116	43.50	974.50	1,125
1065	75 S.F., 6" depth		10	3.200		870	139	52	1,061	1,225
1070	80 S.F., 6" depth		10	3.200		930	139	52	1,121	1,300
1075	85 S.F., 6" depth		8	4		990	174	65	1,229	1,400
1080	90 S.F., 6" depth		8	4		1,050	174	65	1,289	1,475
1085	95 S.F., 6" depth		8	4		1,100	174	65	1,339	1,550
1090	100 S.F., 6" depth	↓	8	4	↓	1,150	174	65	1,389	1,600
1099	Concrete will be replaced using 4,500 psi concrete ready mix									
1100	Rigid pavement replace, light traffic, repour, 15 S.F., 6" depth	A-2	12	2	Ea.	345	87.50	16.35	448.85	530
1110	20-50 S.F., 6" depth		12	2		345	87.50	16.35	448.85	530
1120	55-65 S.F., 6" depth		10	2.400		380	105	19.60	504.60	600
1130	70-80 S.F., 6" depth		10	2.400		415	105	19.60	539.60	635
1140	85-100 S.F., 6" depth		8	3		485	131	24.50	640.50	760
1200	Repour 15-40 S.F., 8" depth		12	2		345	87.50	16.35	448.85	530
1210	45-50 S.F., 8" depth		12	2		380	87.50	16.35	483.85	570
1220	55-60 S.F., 8" depth		10	2.400		415	105	19.60	539.60	635
1230	65-80 S.F., 8" depth		10	2.400		485	105	19.60	609.60	715
1240	85-90 S.F., 8" depth		8	3		525	131	24.50	680.50	800
1250	95-100 S.F., 8" depth		8	3		555	131	24.50	710.50	835
1300	Repour 15-30 S.F., 10" depth		12	2		345	87.50	16.35	448.85	530
1310	35-40 S.F., 10" depth		12	2		380	87.50	16.35	483.85	570
1320	45 S.F., 10" depth		12	2		415	87.50	16.35	518.85	605
1330	50-65 S.F., 10" depth		10	2.400		485	105	19.60	609.60	715
1340	70 S.F., 10" depth		10	2.400		525	105	19.60	649.60	755
1350	75-80 S.F., 10" depth		10	2.400		555	105	19.60	679.60	790
1360	85-95 S.F., 10" depth		8	3		625	131	24.50	780.50	915
1370	100 S.F., 10" depth	↓	8	3	↓	665	131	24.50	820.50	955

32 01 30 – Operation and Maintenance of Site Improvements

32 01 30.10 Site Maintenance

		Crew	Daily Output	Labor-Hours	Unit	Material	2020 Bare Costs Labor	2020 Bare Costs Equipment	Total	Total Incl O&P
0010	**SITE MAINTENANCE**									
0800	Flower bed maintenance									
0810	Cultivate bed, no mulch	1 Clab	14	.571	M.S.F.		24		24	36
0830	Fall clean-up of flower bed, including pick-up mulch for re-use		1	8			335		335	505
0840	Fertilize flower bed, dry granular 3 lb./M.S.F.		85	.094		1.23	3.96		5.19	7.30
1130	Police, hand pickup		30	.267			11.25		11.25	16.85
1140	Vacuum (outside)		48	.167			7		7	10.55
1200	Spring prepare		2	4			168		168	253
1300	Weed mulched bed		20	.400			16.85		16.85	25.50
1310	Unmulched bed	↓	8	1	↓		42		42	63.50
3000	Lawn maintenance									
3010	Aerate lawn, 18" cultivating width, walk behind	A-1K	95	.084	M.S.F.		3.55	1.07	4.62	6.50
3040	48" cultivating width	B-66	750	.011			.57	.32	.89	1.20
3060	72" cultivating width	"	1100	.007			.39	.22	.61	.82
3100	Edge lawn, by hand at walks	1 Clab	16	.500	C.L.F.		21		21	31.50
3150	At planting beds		7	1.143			48		48	72.50
3200	Using gas powered edger at walks		88	.091			3.83		3.83	5.75
3250	At planting beds		24	.333	↓		14.05		14.05	21
3260	Vacuum, 30" gas, outdoors with hose		96	.083	M.L.F.		3.51		3.51	5.25
3400	Weed lawn, by hand	↓	3	2.667	M.S.F.		112		112	169

32 01 30.10 Site Maintenance

		Crew	Daily Output	Labor-Hours	Unit	Material	2020 Bare Costs Labor	2020 Bare Costs Equipment	Total	Total Incl O&P
3800	Lawn bed preparation see Section 32 91 13.23									
4510	Power rake	1 Clab	45	.178	M.S.F.		7.50		7.50	11.25
4700	Seeding lawn, see Section 32 92 19.14									
4750	Sodding, see Section 32 92 23.10									
5900	Road & walk maintenance									
5910	Asphaltic concrete paving, cold patch, 2" thick	B-37	350	.137	S.Y.	13.60	6.05	.73	20.38	25
5913	3" thick	"	260	.185	"	23.50	8.15	.98	32.63	39.50
5915	De-icing roads and walks									
5920	Calcium chloride in truckload lots see Section 31 32 13.30									
6000	Ice melting comp., 90% calc. chlor., effec. to -30°F									
6010	50-80 lb. poly bags, med. applic. 19 lb./M.S.F., by hand	1 Clab	60	.133	M.S.F.	197	5.60		202.60	224
6050	With hand operated rotary spreader		110	.073		197	3.06		200.06	221
6100	Rock salt, med. applic. on road & walkway, by hand		60	.133		108	5.60		113.60	127
6110	With hand operated rotary spreader		110	.073		108	3.06		111.06	124
6130	Hosing, sidewalks & other paved areas		30	.267			11.25		11.25	16.85
6260	Sidewalk, brick pavers, steam cleaning	A-1H	950	.008	S.F.	.11	.35	.08	.54	.74
6400	Sweep walk by hand	1 Clab	15	.533	M.S.F.		22.50		22.50	33.50
6410	Power vacuum		100	.080			3.37		3.37	5.05
6420	Drives & parking areas with power vacuum		120	.067			2.81		2.81	4.22
6600	Shrub maintenance									
6640	Shrub bed fertilize dry granular 3 lb./M.S.F.	1 Clab	85	.094	M.S.F.	1.23	3.96		5.19	7.30
6800	Weed, by handhoe		8	1			42		42	63.50
6810	Spray out		32	.250			10.55		10.55	15.80
6820	Spray after mulch		48	.167			7		7	10.55
7100	Tree maintenance									
7200	Fertilize, tablets, slow release, 30 gram/tree	1 Clab	100	.080	Ea.	2.30	3.37		5.67	7.60
7420	Pest control, spray		24	.333		28.50	14.05		42.55	52.50
7430	Systemic		48	.167		29	7		36	42

32 01 30.20 Snow Removal

		Crew	Daily Output	Labor-Hours	Unit	Material	2020 Bare Costs Labor	2020 Bare Costs Equipment	Total	Total Incl O&P
0010	**SNOW REMOVAL**									
0020	Plowing, 12 ton truck, 2"-4" deep	B-34A	250	.032	M.S.F.		1.57	1.71	3.28	4.23
0040	4"-10" deep		200	.040			1.96	2.14	4.10	5.30
0060	10"-15" deep		150	.053			2.61	2.85	5.46	7.05
0080	Pickup truck, 2"-4" deep	A-3A	175	.046			2.42	1	3.42	4.73
0100	4"-10" deep		130	.062			3.26	1.35	4.61	6.35
0120	10"-15" deep		75	.107			5.65	2.34	7.99	11.05
0140	Load and haul snow, 1 mile round trip	A-3B	230	.070	C.Y.		3.68	5.55	9.23	11.60
0160	2 mile round trip		175	.091			4.83	7.30	12.13	15.25
0180	3 mile round trip		150	.107			5.65	8.50	14.15	17.80
0200	4 mile round trip		130	.123			6.50	9.80	16.30	20.50
0220	5 mile round trip		120	.133			7.05	10.65	17.70	22.50
0240	Clearing with wheeled skid steer loader, 1 C.Y.	A-3C	240	.033			1.77	1.67	3.44	4.47
0260	Spread sand and salt mix	B-34A	375	.021	M.S.F.	7.70	1.04	1.14	9.88	11.25
0280	Sidewalks and drives, by hand	1 Clab	1200	.007	C.F.		.28		.28	.42
0300	Power, 24" blower	A-1M	7000	.001	"		.05	.01	.06	.08
0320	2"-4" deep, single driveway (10' x 50')		16	.500	Ea.		21	4.22	25.22	36
0340	Double driveway (20' x 50')		16	.500			21	4.22	25.22	36
0360	4"-10" deep, single driveway		16	.500			21	4.22	25.22	36
0380	Double driveway		16	.500			21	4.22	25.22	36
0400	10"-15" deep, single driveway		12	.667			28	5.65	33.65	48
0420	Double driveway		12	.667			28	5.65	33.65	48
0440	For heavy wet snow, add						20%			20%

For customer support on your Site Work & Landscape Costs with RSMeans data, call 800.448.8182.

32 01 90.13 Fertilizing

		Crew	Daily Output	Labor-Hours	Unit	Material	2020 Bare Costs Labor	2020 Bare Costs Equipment	Total	Total Incl O&P
0010	**FERTILIZING**									
0100	Dry granular, 4 lb./M.S.F., hand spread	1 Clab	24	.333	M.S.F.	2.32	14.05		16.37	23.50
0110	Push rotary	"	140	.057		2.32	2.41		4.73	6.15
0120	Tractor towed spreader, 8'	B-66	500	.016		2.32	.85	.48	3.65	4.35
0130	12' spread		800	.010		2.32	.53	.30	3.15	3.67
0140	Truck whirlwind spreader		1200	.007		2.32	.35	.20	2.87	3.30
0180	Water soluble, hydro spread, 1.5 lb./M.S.F.	B-64	600	.027		2.44	1.19	.70	4.33	5.25
0190	Add for weed control					.30			.30	.33

32 01 90.19 Mowing

		Crew	Daily Output	Labor-Hours	Unit	Material	2020 Bare Costs Labor	2020 Bare Costs Equipment	Total	Total Incl O&P
0010	**MOWING**									
1650	Mowing brush, tractor with rotary mower									
1660	Light density	B-84	22	.364	M.S.F.		20.50	16.65	37.15	49.50
1670	Medium density		13	.615			35	28	63	83
1680	Heavy density		9	.889			50.50	41	91.50	121
2000	Mowing, brush/grass, tractor, rotary mower, highway/airport median		13	.615			35	28	63	83
2010	Traffic safety flashing truck for highway/airport median mowing	A-2B	1	8	Day		380	196	576	780
4050	Lawn mowing, power mower, 18" - 22"	1 Clab	65	.123	M.S.F.		5.20		5.20	7.80
4100	22" - 30"		110	.073			3.06		3.06	4.60
4150	30" - 32"		140	.057			2.41		2.41	3.61
4160	Riding mower, 36" - 44"	B-66	300	.027			1.41	.80	2.21	3
4170	48" - 58"	"	480	.017			.88	.50	1.38	1.87
4175	Mowing with tractor & attachments									
4180	3 gang reel, 7'	B-66	930	.009	M.S.F.		.46	.26	.72	.97
4190	5 gang reel, 12'		1200	.007			.35	.20	.55	.75
4200	Cutter or sickle-bar, 5', rough terrain		210	.038			2.02	1.15	3.17	4.28
4210	Cutter or sickle-bar, 5', smooth terrain		340	.024			1.25	.71	1.96	2.64
4220	Drainage channel, 5' sickle bar		5	1.600	Mile		85	48.50	133.50	180
4250	Lawn mower, rotary type, sharpen (all sizes)	1 Clab	10	.800	Ea.		33.50		33.50	50.50
4260	Repair or replace part		7	1.143	"		48		48	72.50
5000	Edge trimming with weed whacker		5760	.001	L.F.		.06		.06	.09

32 01 90.23 Pruning

		Crew	Daily Output	Labor-Hours	Unit	Material	2020 Bare Costs Labor	2020 Bare Costs Equipment	Total	Total Incl O&P
0010	**PRUNING**									
0020	1-1/2" caliper	1 Clab	84	.095	Ea.		4.01		4.01	6
0030	2" caliper		70	.114			4.81		4.81	7.25
0040	2-1/2" caliper		50	.160			6.75		6.75	10.10
0050	3" caliper		30	.267			11.25		11.25	16.85
0060	4" caliper, by hand	2 Clab	21	.762			32		32	48
0070	Aerial lift equipment	B-85	38	1.053			49	21	70	96.50
0100	6" caliper, by hand	2 Clab	12	1.333			56		56	84.50
0110	Aerial lift equipment	B-85	20	2			93	40	133	183
0200	9" caliper, by hand	2 Clab	7.50	2.133			90		90	135
0210	Aerial lift equipment	B-85	12.50	3.200			148	64	212	294
0300	12" caliper, by hand	2 Clab	6.50	2.462			104		104	156
0310	Aerial lift equipment	B-85	10.80	3.704			172	74.50	246.50	340
0400	18" caliper by hand	2 Clab	5.60	2.857			120		120	181
0410	Aerial lift equipment	B-85	9.30	4.301			200	86.50	286.50	395
0500	24" caliper, by hand	2 Clab	4.60	3.478			146		146	220
0510	Aerial lift equipment	B-85	7.70	5.195			241	104	345	475
0600	30" caliper, by hand	2 Clab	3.70	4.324			182		182	274
0610	Aerial lift equipment	B-85	6.20	6.452			299	129	428	590
0700	36" caliper, by hand	2 Clab	2.70	5.926			249		249	375
0710	Aerial lift equipment	B-85	4.50	8.889			410	178	588	815

32 01 Operation and Maintenance of Exterior Improvements

32 01 90 – Operation and Maintenance of Planting

32 01 90.23 Pruning

		Crew	Daily Output	Labor-Hours	Unit	Material	2020 Bare Costs Labor	Equipment	Total	Total Incl O&P
0800	48" caliper, by hand	2 Clab	1.70	9.412	Ea.		395		395	595
0810	Aerial lift equipment	B-85	2.80	14.286	↓		665	287	952	1,300

32 01 90.24 Shrub Pruning

		Crew	Daily Output	Labor-Hours	Unit	Material	2020 Bare Costs Labor	Equipment	Total	Total Incl O&P
0010	**SHRUB PRUNING**									
6700	Prune, shrub bed	1 Clab	7	1.143	M.S.F.		48		48	72.50
6710	Shrub under 3' height		190	.042	Ea.		1.77		1.77	2.66
6720	4' height		90	.089			3.74		3.74	5.60
6730	Over 6'		50	.160			6.75		6.75	10.10
7350	Prune trees from ground		20	.400			16.85		16.85	25.50
7360	High work	▼	8	1	↓		42		42	63.50

32 01 90.26 Watering

		Crew	Daily Output	Labor-Hours	Unit	Material	2020 Bare Costs Labor	Equipment	Total	Total Incl O&P
0010	**WATERING**									
4900	Water lawn or planting bed with hose, 1" of water	1 Clab	16	.500	M.S.F.		21		21	31.50
4910	50' soaker hoses, in place		82	.098			4.11		4.11	6.15
4920	60' soaker hoses, in place		89	.090	↓		3.78		3.78	5.70
7500	Water trees or shrubs, under 1" caliper		32	.250	Ea.		10.55		10.55	15.80
7550	1" - 3" caliper		17	.471			19.80		19.80	30
7600	3" - 4" caliper		12	.667			28		28	42
7650	Over 4" caliper	▼	10	.800	↓		33.50		33.50	50.50
9000	For sprinkler irrigation systems, see Section 32 84 23.10									

32 01 90.29 Topsoil Preservation

		Crew	Daily Output	Labor-Hours	Unit	Material	2020 Bare Costs Labor	Equipment	Total	Total Incl O&P
0010	**TOPSOIL PRESERVATION**									
0100	Weed planting bed	1 Clab	800	.010	S.Y.		.42		.42	.63

32 06 Schedules for Exterior Improvements

32 06 10 – Schedules for Bases, Ballasts, and Paving

32 06 10.10 Sidewalks, Driveways and Patios

		Crew	Daily Output	Labor-Hours	Unit	Material	2020 Bare Costs Labor	Equipment	Total	Total Incl O&P
0010	**SIDEWALKS, DRIVEWAYS AND PATIOS** No base									
0020	Asphaltic concrete, 2" thick	B-37	720	.067	S.Y.	7.20	2.95	.36	10.51	12.70
0100	2-1/2" thick	"	660	.073	"	9.10	3.22	.39	12.71	15.30
0110	Bedding for brick or stone, mortar, 1" thick	D-1	300	.053	S.F.	.82	2.50		3.32	4.69
0120	2" thick	"	200	.080		2.05	3.75		5.80	7.95
0130	Sand, 2" thick	B-18	8000	.003		.34	.13	.02	.49	.59
0140	4" thick	"	4000	.006	↓	.68	.26	.04	.98	1.19
0300	Concrete, 3,000 psi, CIP, 6 x 6 - W1.4 x W1.4 mesh,									
0310	broomed finish, no base, 4" thick	B-24	600	.040	S.F.	2.31	1.94		4.25	5.45
0350	5" thick		545	.044		2.86	2.13		4.99	6.35
0400	6" thick	▼	510	.047	↓	3.34	2.28		5.62	7.05
0440	For other finishes, see Section 03 35 29.30									
0450	For bank run gravel base, 4" thick, add	B-18	2500	.010	S.F.	.43	.41	.07	.91	1.17
0520	8" thick, add	"	1600	.015		.87	.64	.10	1.61	2.03
0550	Exposed aggregate finish, add to above, minimum	B-24	1875	.013		.13	.62		.75	1.06
0600	Maximum		455	.053		.43	2.55		2.98	4.28
0700	Patterned surface, add to above, minimum		1200	.020			.97		.97	1.44
0710	Maximum	▼	500	.048	↓		2.32		2.32	3.46
0850	Splash block, precast concrete	1 Clab	150	.053	Ea.	11.95	2.25		14.20	16.45
0950	Concrete tree grate, 5' square	B-6	25	.960		440	44	8.55	492.55	560
0955	Tree well & cover, concrete, 3' square		25	.960		245	44	8.55	297.55	345
0960	Cast iron tree grate with frame, 2 piece, round, 5' diameter		25	.960		1,075	44	8.55	1,127.55	1,250
0980	Square, 5' side	▼	25	.960		1,075	44	8.55	1,127.55	1,250

For customer support on your Site Work & Landscape Costs with RSMeans data, call 800.448.8182.

369

32 06 10.10 Sidewalks, Driveways and Patios	Crew	Daily Output	Labor-Hours	Unit	Material	2020 Bare Costs Labor	Equipment	Total	Total Incl O&P
1000	Crushed stone, 1" thick, white marble	2 Clab	1700	.009	S.F.	.49	.40		.89	1.13
1050	Bluestone		1700	.009		.18	.40		.58	.80
1070	Granite chips		1700	.009		.24	.40		.64	.86
1200	For 2" asphaltic conc. base and tack coat, add to above	B-37	7200	.007		.98	.30	.04	1.32	1.56
1660	Limestone pavers, 3" thick	D-1	72	.222		10.75	10.40		21.15	27.50
1670	4" thick		70	.229		14.25	10.70		24.95	32
1680	5" thick		68	.235		17.85	11.05		28.90	36.50
1700	Redwood, prefabricated, 4' x 4' sections	2 Carp	316	.051		4.89	2.69		7.58	9.40
1750	Redwood planks, 1" thick, on sleepers	"	240	.067		4.89	3.54		8.43	10.65
1830	1-1/2" thick	B-28	167	.144		6.05	7.10		13.15	17.35
1840	2" thick		167	.144		7.75	7.10		14.85	19.25
1850	3" thick		150	.160		11.45	7.90		19.35	24.50
1860	4" thick		150	.160		15.05	7.90		22.95	28.50
1870	5" thick		150	.160		19.10	7.90		27	33
2100	River or beach stone, stock	B-1	18	1.333	Ton	33	57		90	122
2150	Quarried	"	18	1.333	"	58	57		115	150
2160	Load, dump, and spread stone with skid steer, 100' haul	B-62	24	1	C.Y.		45.50	7.40	52.90	76.50
2165	200' haul		18	1.333			61	9.85	70.85	102
2168	300' haul		12	2			91.50	14.80	106.30	153
2170	Shale paver, 2-1/4" thick	D-1	200	.080	S.F.	3.78	3.75		7.53	9.85
2200	Coarse washed sand bed, 1" thick	B-62	1350	.018	S.Y.	1.03	.81	.13	1.97	2.50
2250	Stone dust, 4" thick	"	900	.027	"	5.25	1.22	.20	6.67	7.85
2300	Tile thinset pavers, 3/8" thick	D-1	300	.053	S.F.	3.74	2.50		6.24	7.90
2350	3/4" thick	"	280	.057	"	5.95	2.68		8.63	10.60
2400	Wood rounds, cypress	B-1	175	.137	Ea.	10.20	5.85		16.05	20

32 06 10.20 Steps

		Crew	Daily Output	Labor-Hours	Unit	Material	2020 Bare Costs Labor	Equipment	Total	Total Incl O&P
0010	**STEPS**									
0011	Incl. excav., borrow & concrete base as required									
0100	Brick steps	B-24	35	.686	LF Riser	17.95	33		50.95	69.50
0200	Railroad ties	2 Clab	25	.640		3.75	27		30.75	44.50
0300	Bluestone treads, 12" x 2" or 12" x 1-1/2"	B-24	30	.800		44.50	38.50		83	107
0500	Concrete, cast in place, see Section 03 30 53.40									
0600	Precast concrete, see Section 03 41 23.50									
4000	Edging, redwood, 2" x 4"	2 Carp	330	.048	L.F.	3.06	2.58		5.64	7.25
4025	Steel edge strips, incl. stakes, 1/4" x 5"	B-1	390	.062		4.58	2.63		7.21	9
4050	Edging, landscape timber or railroad ties, 6" x 8"	2 Carp	170	.094		2.48	5		7.48	10.25

32 11 Base Courses
32 11 23 – Aggregate Base Courses
32 11 23.23 Base Course Drainage Layers

		Crew	Daily Output	Labor-Hours	Unit	Material	2020 Bare Costs Labor	Equipment	Total	Total Incl O&P
0010	**BASE COURSE DRAINAGE LAYERS**									
0011	For Soil Stabilization, see Section 31 32									
0012	For roadways and large areas									
0050	Crushed 3/4" stone base, compacted, 3" deep	B-36C	5200	.008	S.Y.	2.70	.41	.82	3.93	4.47
0100	6" deep		5000	.008		5.40	.42	.85	6.67	7.50
0200	9" deep		4600	.009		8.10	.46	.92	9.48	10.60
0300	12" deep		4200	.010		10.80	.50	1.01	12.31	13.75
0301	Crushed 1-1/2" stone base, compacted to 4" deep	B-36B	6000	.011		4.87	.54	.82	6.23	7.05
0302	6" deep		5400	.012		7.30	.60	.91	8.81	9.95
0303	8" deep		4500	.014		9.75	.72	1.09	11.56	13
0304	12" deep		3800	.017		14.60	.85	1.29	16.74	18.80

For customer support on your Site Work & Landscape Costs with RSMeans data, call 800.448.8182.

32 11 Base Courses

32 11 23 - Aggregate Base Courses

32 11 23.23 Base Course Drainage Layers

		Crew	Daily Output	Labor-Hours	Unit	Material	2020 Bare Costs Labor	Equipment	Total	Total Incl O&P
0350	Bank run gravel, spread and compacted									
0370	6" deep	B-32	6000	.005	S.Y.	3.62	.28	.47	4.37	4.91
0390	9" deep		4900	.007		5.45	.35	.57	6.37	7.10
0400	12" deep	↓	4200	.008	↓	7.25	.40	.67	8.32	9.30
0600	Cold laid asphalt pavement, see Section 32 12 16.19									
1500	Alternate method to figure base course									
1510	Crushed stone, 3/4", compacted, 3" deep	B-36C	435	.092	E.C.Y.	27.50	4.84	9.75	42.09	48.50
1511	6" deep	B-36B	835	.077		27.50	3.87	5.85	37.22	43
1512	9" deep		1150	.056		27.50	2.81	4.26	34.57	39.50
1513	12" deep		1400	.046		27.50	2.31	3.50	33.31	38
1520	Crushed stone, 1-1/2", compacted, 4" deep		665	.096		27.50	4.86	7.35	39.71	46
1521	6" deep		900	.071		27.50	3.59	5.45	36.54	42
1522	8" deep		1000	.064		27.50	3.23	4.90	35.63	40.50
1523	12" deep	↓	1265	.051		27.50	2.56	3.88	33.94	38.50
1530	Gravel, bank run, compacted, 6" deep	B-36C	835	.048		18.60	2.52	5.10	26.22	30
1531	9" deep		1150	.035		18.60	1.83	3.69	24.12	27.50
1532	12" deep	↓	1400	.029	↓	18.60	1.50	3.03	23.13	26
2010	Crushed stone, 3/4" maximum size, 3" deep	B-36	540	.074	Ton	18	3.58	3.35	24.93	29
2011	6" deep		1625	.025		18	1.19	1.11	20.30	23
2012	9" deep		1785	.022		18	1.08	1.01	20.09	22.50
2013	12" deep		1950	.021		18	.99	.93	19.92	22.50
2020	Crushed stone, 1-1/2" maximum size, 4" deep		720	.056		18	2.69	2.51	23.20	26.50
2021	6" deep		815	.049		18	2.37	2.22	22.59	26
2022	8" deep		835	.048		18	2.32	2.16	22.48	25.50
2023	12" deep	↓	975	.041		18	1.98	1.85	21.83	25
2030	Bank run gravel, 6" deep	B-32A	875	.027		12.50	1.42	1.97	15.89	18.05
2031	9" deep		970	.025		12.50	1.28	1.78	15.56	17.65
2032	12" deep	↓	1060	.023	↓	12.50	1.17	1.63	15.30	17.30
6000	Stabilization fabric, polypropylene, 6 oz./S.Y.	B-6	10000	.002	S.Y.	.91	.11	.02	1.04	1.18
6900	For small and irregular areas, add						50%	50%		
7000	Prepare and roll sub-base, small areas to 2,500 S.Y.	B-32A	1500	.016	S.Y.		.83	1.15	1.98	2.51
8000	Large areas over 2,500 S.Y.	"	3500	.007			.36	.49	.85	1.07
8050	For roadways	B-32	4000	.008	↓		.42	.70	1.12	1.41

32 11 26 - Asphaltic Base Courses

32 11 26.13 Plant Mix Asphaltic Base Courses

		Crew	Daily Output	Labor-Hours	Unit	Material	2020 Bare Costs Labor	Equipment	Total	Total Incl O&P
0010	**PLANT MIX ASPHALTIC BASE COURSES**									
0011	For roadways and large paved areas									
0500	Bituminous concrete, 4" thick	B-25	4545	.019	S.Y.	15.30	.90	.59	16.79	18.80
0550	6" thick		3700	.024		22.50	1.10	.73	24.33	27
0560	8" thick		3000	.029		30	1.36	.90	32.26	36
0570	10" thick	↓	2545	.035	↓	37	1.60	1.06	39.66	44
1600	Macadam base, crushed stone or slag, dry-bound	B-36D	1400	.023	E.C.Y.	49	1.23	2.49	52.72	58
1610	Water-bound	B-36C	1400	.029	"	48.50	1.50	3.03	53.03	59
2000	Alternate method to figure base course									
2005	Bituminous concrete, 4" thick	B-25	1000	.088	Ton	68	4.07	2.70	74.77	83.50
2006	6" thick		1220	.072		68	3.34	2.21	73.55	82
2007	8" thick		1320	.067		68	3.09	2.05	73.14	81.50
2008	10" thick	↓	1400	.063	↓	68	2.91	1.93	72.84	81
8900	For small and irregular areas, add						50%	50%		

32 11 Base Courses

32 11 26 – Asphaltic Base Courses

32 11 26.19 Bituminous-Stabilized Base Courses

		Crew	Daily Output	Labor-Hours	Unit	Material	2020 Bare Costs Labor	2020 Bare Costs Equipment	Total	Total Incl O&P
0010	**BITUMINOUS-STABILIZED BASE COURSES**									
0020	For roadways and large paved areas									
0700	Liquid application to gravel base, asphalt emulsion	B-45	6000	.003	Gal.	4.84	.14	.14	5.12	5.65
0800	Prime and seal, cut back asphalt		6000	.003	"	5.70	.14	.14	5.98	6.65
1000	Macadam penetration crushed stone, 2 gal./S.Y., 4" thick		6000	.003	S.Y.	9.70	.14	.14	9.98	11
1100	6" thick, 3 gal./S.Y.		4000	.004		14.50	.21	.21	14.92	16.50
1200	8" thick, 4 gal./S.Y.		3000	.005		19.35	.28	.27	19.90	22
8900	For small and irregular areas, add						50%	50%		

32 12 Flexible Paving

32 12 16 – Asphalt Paving

32 12 16.13 Plant-Mix Asphalt Paving

		Crew	Daily Output	Labor-Hours	Unit	Material	2020 Bare Costs Labor	2020 Bare Costs Equipment	Total	Total Incl O&P
0010	**PLANT-MIX ASPHALT PAVING**									
0020	For highways and large paved areas, exludes hauling									
0025	See Section 31 23 23.20 for hauling costs									
0080	Binder course, 1-1/2" thick	B-25	7725	.011	S.Y.	5.55	.53	.35	6.43	7.25
0120	2" thick		6345	.014		7.40	.64	.43	8.47	9.55
0130	2-1/2" thick		5620	.016		9.25	.72	.48	10.45	11.75
0160	3" thick		4905	.018		11.05	.83	.55	12.43	14.05
0170	3-1/2" thick		4520	.019		12.90	.90	.60	14.40	16.20
0200	4" thick		4140	.021		14.75	.98	.65	16.38	18.45
0300	Wearing course, 1" thick	B-25B	10575	.009		3.66	.43	.28	4.37	4.98
0340	1-1/2" thick		7725	.012		6.15	.59	.38	7.12	8.05
0380	2" thick		6345	.015		8.25	.71	.46	9.42	10.70
0420	2-1/2" thick		5480	.018		10.20	.83	.54	11.57	13.05
0460	3" thick		4900	.020		12.15	.92	.60	13.67	15.40
0470	3-1/2" thick		4520	.021		14.25	1	.65	15.90	17.90
0480	4" thick		4140	.023		16.30	1.09	.71	18.10	20.50
0500	Open graded friction course	B-25C	5000	.010		2.55	.45	.47	3.47	4
0800	Alternate method of figuring paving costs									
0810	Binder course, 1-1/2" thick	B-25	630	.140	Ton	68	6.45	4.29	78.74	89
0811	2" thick		690	.128		68	5.90	3.91	77.81	87.50
0812	3" thick		800	.110		68	5.10	3.37	76.47	86
0813	4" thick		900	.098		68	4.53	3	75.53	84.50
0850	Wearing course, 1" thick	B-25B	575	.167		73.50	7.85	5.10	86.45	98.50
0851	1-1/2" thick		630	.152		73.50	7.20	4.66	85.36	97
0852	2" thick		690	.139		73.50	6.55	4.26	84.31	95.50
0853	2-1/2" thick		765	.125		73.50	5.90	3.84	83.24	94
0854	3" thick		800	.120		73.50	5.65	3.67	82.82	93.50
1000	Pavement replacement over trench, 2" thick	B-37	90	.533	S.Y.	7.60	23.50	2.84	33.94	47
1050	4" thick		70	.686		15.10	30.50	3.65	49.25	66
1080	6" thick		55	.873		24	38.50	4.65	67.15	89.50
1100	Turnouts and driveway entrances to highways									
1110	Binder course, 1-1/2" thick	B-25	315	.279	Ton	68	12.95	8.55	89.50	103
1120	2" thick		345	.255		68	11.80	7.85	87.65	101
1130	3" thick		400	.220		68	10.20	6.75	84.95	97
1140	4" thick		450	.196		68	9.05	6	83.05	94.50
1150	Binder course, 1-1/2" thick		3860	.023	S.Y.	5.55	1.06	.70	7.31	8.45
1160	2" thick		3175	.028		7.40	1.28	.85	9.53	10.95
1170	2-1/2" thick		2810	.031		9.25	1.45	.96	11.66	13.40

32 12 Flexible Paving

32 12 16 – Asphalt Paving

32 12 16.13 Plant-Mix Asphalt Paving

		Crew	Daily Output	Labor-Hours	Unit	Material	2020 Bare Costs Labor	2020 Bare Costs Equipment	Total	Total Incl O&P
1180	3" thick	B-25	2455	.036	S.Y.	11.05	1.66	1.10	13.81	15.90
1190	3-1/2" thick		2260	.039		12.90	1.80	1.19	15.89	18.20
1195	4" thick	↓	2070	.043	↓	14.75	1.97	1.30	18.02	20.50
1200	Wearing course, 1" thick	B-25B	290	.331	Ton	73.50	15.60	10.15	99.25	116
1210	1-1/2" thick		315	.305		73.50	14.35	9.30	97.15	113
1220	2" thick		345	.278		73.50	13.10	8.50	95.10	110
1230	2-1/2" thick		385	.249		73.50	11.75	7.65	92.90	107
1240	3" thick		400	.240	↓	73.50	11.30	7.35	92.15	106
1250	Wearing course, 1" thick		5290	.018	S.Y.	3.66	.86	.56	5.08	5.90
1260	1-1/2" thick		3860	.025		6.15	1.17	.76	8.08	9.35
1270	2" thick		3175	.030		8.25	1.43	.93	10.61	12.25
1280	2-1/2" thick		2740	.035		10.20	1.65	1.07	12.92	14.85
1285	3" thick		2450	.039		12.15	1.85	1.20	15.20	17.45
1290	3-1/2" thick		2260	.042		14.25	2	1.30	17.55	20
1295	4" thick	↓	2070	.046	↓	16.30	2.19	1.42	19.91	23
1400	Bridge approach less than 300 tons									
1410	Binder course, 1-1/2" thick	B-25	330	.267	Ton	68	12.35	8.20	88.55	102
1420	2" thick		360	.244		68	11.30	7.50	86.80	99.50
1430	3" thick		420	.210		68	9.70	6.45	84.15	96
1440	4" thick	↓	470	.187	↓	68	8.65	5.75	82.40	94
1450	Binder course, 1-1/2" thick		4055	.022	S.Y.	5.55	1	.67	7.22	8.35
1460	2" thick		3330	.026		7.40	1.22	.81	9.43	10.80
1470	2-1/2" thick		2950	.030		9.25	1.38	.92	11.55	13.25
1480	3" thick		2575	.034		11.05	1.58	1.05	13.68	15.70
1490	3-1/2" thick		2370	.037		12.90	1.72	1.14	15.76	18.05
1495	4" thick	↓	2175	.040	↓	14.75	1.87	1.24	17.86	20.50
1500	Wearing course, 1" thick	B-25B	300	.320	Ton	73.50	15.10	9.80	98.40	114
1510	1-1/2" thick		330	.291		73.50	13.70	8.90	96.10	111
1520	2" thick		360	.267		73.50	12.55	8.15	94.20	109
1530	2-1/2" thick		400	.240		73.50	11.30	7.35	92.15	106
1540	3" thick		420	.229	↓	73.50	10.80	7	91.30	105
1550	Wearing course, 1" thick		5555	.017	S.Y.	3.66	.81	.53	5	5.85
1560	1-1/2" thick		4055	.024		6.15	1.12	.72	7.99	9.20
1570	2" thick		3330	.029		8.25	1.36	.88	10.49	12.10
1580	2-1/2" thick		2875	.033		10.20	1.57	1.02	12.79	14.70
1585	3" thick		2570	.037		12.15	1.76	1.14	15.05	17.25
1590	3-1/2" thick		2375	.040		14.25	1.91	1.24	17.40	19.90
1595	4" thick	↓	2175	.044	↓	16.30	2.08	1.35	19.73	22.50
1600	Bridge approach 300-800 tons									
1610	Binder course, 1-1/2" thick	B-25	390	.226	Ton	68	10.45	6.90	85.35	98
1620	2" thick		430	.205		68	9.45	6.30	83.75	95.50
1630	3" thick		500	.176		68	8.15	5.40	81.55	92.50
1640	4" thick	↓	560	.157	↓	68	7.25	4.82	80.07	90.50
1650	Binder course, 1-1/2" thick		4815	.018	S.Y.	5.55	.85	.56	6.96	8
1660	2" thick		3955	.022		7.40	1.03	.68	9.11	10.40
1670	2-1/2" thick		3500	.025		9.25	1.16	.77	11.18	12.75
1680	3" thick		3060	.029		11.05	1.33	.88	13.26	15.15
1690	3-1/2" thick		2820	.031		12.90	1.44	.96	15.30	17.40
1695	4" thick	↓	2580	.034	↓	14.75	1.58	1.05	17.38	19.75
1700	Wearing course, 1" thick	B-25B	360	.267	Ton	73.50	12.55	8.15	94.20	109
1710	1-1/2" thick		390	.246		73.50	11.60	7.55	92.65	107
1720	2" thick		430	.223		73.50	10.55	6.85	90.90	104
1730	2-1/2" thick	↓	475	.202	↓	73.50	9.55	6.20	89.25	102

For customer support on your Site Work & Landscape Costs with RSMeans data, call 800.448.8182.

373

32 12 16.13 Plant-Mix Asphalt Paving		Crew	Daily Output	Labor-Hours	Unit	Material	2020 Bare Costs Labor	Equipment	Total	Total Incl O&P
1740	3" thick	B-25B	500	.192	Ton	73.50	9.05	5.85	88.40	101
1750	Wearing course, 1" thick		6590	.015	S.Y.	3.66	.69	.45	4.80	5.55
1760	1-1/2" thick		4815	.020		6.15	.94	.61	7.70	8.85
1770	2" thick		3955	.024		8.25	1.14	.74	10.13	11.65
1780	2-1/2" thick		3415	.028		10.20	1.33	.86	12.39	14.15
1785	3" thick		3055	.031		12.15	1.48	.96	14.59	16.65
1790	3-1/2" thick		2820	.034		14.25	1.61	1.04	16.90	19.25
1795	4" thick		2580	.037		16.30	1.75	1.14	19.19	22
1800	Bridge approach over 800 tons									
1810	Binder course, 1-1/2" thick	B-25	525	.168	Ton	68	7.75	5.15	80.90	92
1820	2" thick		575	.153		68	7.10	4.70	79.80	90.50
1830	3" thick		670	.131		68	6.10	4.03	78.13	88
1840	4" thick		750	.117		68	5.45	3.60	77.05	86.50
1850	Binder course, 1-1/2" thick		6450	.014	S.Y.	5.55	.63	.42	6.60	7.50
1860	2" thick		5300	.017		7.40	.77	.51	8.68	9.80
1870	2-1/2" thick		4690	.019		9.25	.87	.58	10.70	12.10
1880	3" thick		4095	.021		11.05	.99	.66	12.70	14.40
1890	3-1/2" thick		3775	.023		12.90	1.08	.72	14.70	16.60
1895	4" thick		3455	.025		14.75	1.18	.78	16.71	18.90
1900	Wearing course, 1" thick	B-25B	480	.200	Ton	73.50	9.45	6.10	89.05	102
1910	1-1/2" thick		525	.183		73.50	8.60	5.60	87.70	100
1920	2" thick		575	.167		73.50	7.85	5.10	86.45	98.50
1930	2-1/2" thick		640	.150		73.50	7.05	4.59	85.14	96.50
1940	3" thick		670	.143		73.50	6.75	4.38	84.63	96
1950	Wearing course, 1" thick		8830	.011	S.Y.	3.66	.51	.33	4.50	5.15
1960	1-1/2" thick		6450	.015		6.15	.70	.46	7.31	8.30
1970	2" thick		5300	.018		8.25	.85	.55	9.65	11
1980	2-1/2" thick		4575	.021		10.20	.99	.64	11.83	13.40
1985	3" thick		4090	.023		12.15	1.11	.72	13.98	15.80
1990	3-1/2" thick		3775	.025		14.25	1.20	.78	16.23	18.35
1995	4" thick		3455	.028		16.30	1.31	.85	18.46	21
2000	Intersections									
2010	Binder course, 1-1/2" thick	B-25	440	.200	Ton	68	9.25	6.15	83.40	95
2020	2" thick		485	.181		68	8.40	5.55	81.95	93
2030	3" thick		560	.157		68	7.25	4.82	80.07	90.50
2040	4" thick		630	.140		68	6.45	4.29	78.74	89
2050	Binder course, 1-1/2" thick		5410	.016	S.Y.	5.55	.75	.50	6.80	7.80
2060	2" thick		4440	.020		7.40	.92	.61	8.93	10.15
2070	2-1/2" thick		3935	.022		9.25	1.03	.69	10.97	12.45
2080	3" thick		3435	.026		11.05	1.19	.79	13.03	14.85
2090	3-1/2" thick		3165	.028		12.90	1.29	.85	15.04	17.05
2095	4" thick		2900	.030		14.75	1.40	.93	17.08	19.40
2100	Wearing course, 1" thick	B-25B	400	.240	Ton	73.50	11.30	7.35	92.15	106
2110	1-1/2" thick		440	.218		73.50	10.30	6.65	90.45	104
2120	2" thick		485	.198		73.50	9.35	6.05	88.90	102
2130	2-1/2" thick		535	.179		73.50	8.45	5.50	87.45	100
2140	3" thick		560	.171		73.50	8.10	5.25	86.85	99
2150	Wearing course, 1" thick		7400	.013	S.Y.	3.66	.61	.40	4.67	5.40
2160	1-1/2" thick		5410	.018		6.15	.84	.54	7.53	8.60
2170	2" thick		4440	.022		8.25	1.02	.66	9.93	11.35
2180	2-1/2" thick		3835	.025		10.20	1.18	.77	12.15	13.80
2185	3" thick		3430	.028		12.15	1.32	.86	14.33	16.25
2190	3-1/2" thick		3165	.030		14.25	1.43	.93	16.61	18.85

32 12 Flexible Paving

32 12 16 – Asphalt Paving

32 12 16.13 Plant-Mix Asphalt Paving

		Crew	Daily Output	Labor-Hours	Unit	Material	2020 Bare Costs Labor	Equipment	Total	Total Incl O&P
2195	4" thick	B-25B	2900	.033	S.Y.	16.30	1.56	1.01	18.87	21.50
3000	Prime coat, emulsion, 0.30 gal./S.Y., 1000 S.Y.	B-45	2500	.006		2.10	.34	.33	2.77	3.18
3100	Tack coat, emulsion, 0.10 gal./S.Y., 1000 S.Y.	"	2500	.006	↓	.70	.34	.33	1.37	1.64

32 12 16.14 Asphaltic Concrete Paving

		Crew	Daily Output	Labor-Hours	Unit	Material	2020 Bare Costs Labor	Equipment	Total	Total Incl O&P
0011	**ASPHALTIC CONCRETE PAVING**, parking lots & driveways									
0015	No asphalt hauling included									
0018	Use 6.05 C.Y. per inch per M.S.F. for hauling									
0020	6" stone base, 2" binder course, 1" topping	B-25C	9000	.005	S.F.	1.92	.25	.26	2.43	2.78
0025	2" binder course, 2" topping		9000	.005		2.36	.25	.26	2.87	3.27
0030	3" binder course, 2" topping		9000	.005		2.78	.25	.26	3.29	3.73
0035	4" binder course, 2" topping		9000	.005		3.19	.25	.26	3.70	4.17
0040	1-1/2" binder course, 1" topping		9000	.005		1.72	.25	.26	2.23	2.56
0042	3" binder course, 1" topping		9000	.005		2.33	.25	.26	2.84	3.24
0045	3" binder course, 3" topping		9000	.005		3.23	.25	.26	3.74	4.22
0050	4" binder course, 3" topping		9000	.005		3.63	.25	.26	4.14	4.67
0055	4" binder course, 4" topping		9000	.005		4.08	.25	.26	4.59	5.15
0300	Binder course, 1-1/2" thick		35000	.001		.62	.06	.07	.75	.85
0400	2" thick		25000	.002		.80	.09	.09	.98	1.12
0500	3" thick		15000	.003		1.23	.15	.16	1.54	1.76
0600	4" thick		10800	.004		1.62	.21	.22	2.05	2.34
0800	Sand finish course, 3/4" thick		41000	.001		.31	.06	.06	.43	.48
0900	1" thick	↓	34000	.001		.38	.07	.07	.52	.60
1000	Fill pot holes, hot mix, 2" thick	B-16	4200	.008		.82	.34	.14	1.30	1.56
1100	4" thick		3500	.009		1.20	.41	.16	1.77	2.11
1120	6" thick	↓	3100	.010		1.61	.46	.18	2.25	2.66
1140	Cold patch, 2" thick	B-51	3000	.016		.87	.69	.07	1.63	2.07
1160	4" thick		2700	.018		1.66	.77	.07	2.50	3.07
1180	6" thick	↓	1900	.025	↓	2.58	1.09	.10	3.77	4.59
2000	From 60% recycled content, base course 3" thick **G**	B-25	800	.110	Ton	44	5.10	3.37	52.47	59.50
3000	Prime coat, emulsion, 0.30 gal./S.Y., 1000 S.Y.	B-45	2500	.006	S.Y.	2.10	.34	.33	2.77	3.18
3100	Tack coat, emulsion, 0.10 gal./S.Y., 1000 S.Y.	"	2500	.006	"	.70	.34	.33	1.37	1.64

32 12 16.19 Cold-Mix Asphalt Paving

		Crew	Daily Output	Labor-Hours	Unit	Material	2020 Bare Costs Labor	Equipment	Total	Total Incl O&P
0010	**COLD-MIX ASPHALT PAVING** 0.5 gal. asphalt/S.Y. per in depth									
0020	Well graded granular aggregate									
0100	Blade mixed in windrows, spread & compacted 4" course	B-90A	1600	.035	S.Y.	22	1.78	1.69	25.47	29
0200	Traveling plant mixed in windrows, compacted 4" course	B-90B	3000	.016		22	.80	.75	23.55	26.50
0300	Rotary plant mixed in place, compacted 4" course	"	3500	.014		22	.68	.64	23.32	26
0400	Central stationary plant, mixed, compacted 4" course	B-36	7200	.006	↓	44.50	.27	.25	45.02	49.50

32 12 36 – Seal Coats

32 12 36.13 Chip Seal

		Crew	Daily Output	Labor-Hours	Unit	Material	2020 Bare Costs Labor	Equipment	Total	Total Incl O&P
0010	**CHIP SEAL**									
0011	Excludes crack repair and flush coat									
1000	Fine - PMCRS-2h (20lbs/sy, 1/4" (No.10), .30gal/sy app. rate)									
1010	Small, irregular areas	B-91	5000	.013	S.Y.	1.10	.65	.45	2.20	2.68
1020	Parking Lot	B-91D	15000	.007		1.10	.34	.22	1.66	1.97
1030	Roadway	"	30000	.003		1.10	.17	.11	1.38	1.58
1090	For Each .5% Latex Additive, Add				↓	.23			.23	.25
1100	Medium Fine - PMCRS-2h (25lbs/sy, 5/16" (No.8), .35gal/sy app. rate)									
1110	Small, irregular areas	B-91	4000	.016	S.Y.	1.29	.81	.56	2.66	3.24
1120	Parking Lot	B-91D	12000	.009		1.29	.42	.28	1.99	2.36
1130	Roadway	"	24000	.004		1.29	.21	.14	1.64	1.88
1190	For Each .5% Latex Additive, Add					.28			.28	.31

For customer support on your Site Work & Landscape Costs with RSMeans data, call 800.448.8182.

375

32 12 Flexible Paving

32 12 36 – Seal Coats

32 12 36.13 Chip Seal

		Crew	Daily Output	Labor-Hours	Unit	Material	2020 Bare Costs Labor	Equipment	Total	Total Incl O&P
1200	Medium - PMCRS-2h (30lbs/sy, 3/8" (No.6), .40gal/sy app. rate)									
1210	Small, irregular areas	B-91	3330	.019	S.Y.	1.41	.97	.68	3.06	3.74
1220	Parking Lot	B-91D	10000	.010		1.41	.51	.34	2.26	2.68
1230	Roadway	"	20000	.005		1.41	.25	.17	1.83	2.11
1290	For Each .5% Latex Additive, Add					.34			.34	.37
1300	Course - PMCRS-2h (30lbs/sy, 1/2" (No.4), .40gal/sy app. rate)									
1310	Small, irregular areas	B-91	2500	.026	S.Y.	1.39	1.29	.90	3.58	4.45
1320	Parking Lot	B-91D	7500	.014		1.39	.68	.45	2.52	3.03
1330	Roadway	"	15000	.007		1.39	.34	.22	1.95	2.28
1390	For Each .5% Latex Additive, Add					.34			.34	.37
1400	Double - PMCRS-2h (Course Base w/ Fine Top)									
1410	Small, irregular areas	B-91	2000	.032	S.Y.	2.36	1.62	1.13	5.11	6.25
1420	Parking Lot	B-91D	6000	.017		2.36	.85	.56	3.77	4.49
1430	Roadway	"	12000	.009		2.36	.42	.28	3.06	3.55
1490	For Each .5% Latex Additive, Add					.56			.56	.62

32 12 36.14 Flush Coat

		Crew	Daily Output	Labor-Hours	Unit	Material	2020 Bare Costs Labor	Equipment	Total	Total Incl O&P
0010	**FLUSH COAT**									
0011	Fog Seal w/ Sand Cover.18 gal/sy, 6lbs/sy									
1010	Small, irregular areas	B-91	2000	.032	S.Y.	.52	1.62	1.13	3.27	4.24
1020	Parking lot		6000	.011		.52	.54	.38	1.44	1.80
1030	Roadway		12000	.005		.52	.27	.19	.98	1.19

32 12 36.33 Slurry Seal

		Crew	Daily Output	Labor-Hours	Unit	Material	2020 Bare Costs Labor	Equipment	Total	Total Incl O&P
0010	**SLURRY SEAL**									
0011	Includes sweeping and cleaning of area									
1000	Type I-PMCQS-1h-EAS (12lbs/sy, 1/8", 20% asphalt emulsion)									
1010	Small, irregular areas	B-90	8000	.008	S.Y.	.98	.37	.27	1.62	1.93
1020	Parking Lot		25000	.003		.98	.12	.09	1.19	1.35
1030	Roadway		50000	.001		.98	.06	.04	1.08	1.21
1090	For Each .5% Latex Additive, Add					.15			.15	.16
1100	Type II-PMCQS-1h-EAS (15lbs/sy, 1/4", 18% asphalt emulsion)									
1110	Small, irregular areas	B-90	6000	.011	S.Y.	1.10	.50	.37	1.97	2.36
1120	Parking lot		20000	.003		1.10	.15	.11	1.36	1.55
1130	Roadway		40000	.002		1.10	.07	.05	1.22	1.38
1200	Type III-PMCQS-1h-EAS (25lbs/sy, 3/8", 15% asphalt emulsion)									
1210	Small, irregular areas	B-90	4000	.016	S.Y.	1.51	.75	.55	2.81	3.38
1220	Parking lot		12000	.005		1.51	.25	.18	1.94	2.23
1230	Roadway		24000	.003		1.51	.12	.09	1.72	1.95
1290	For Each .5% Latex Additive, Add					.29			.29	.32

32 12 36.36 Microsurfacing

		Crew	Daily Output	Labor-Hours	Unit	Material	2020 Bare Costs Labor	Equipment	Total	Total Incl O&P
0010	**MICROSURFACING**									
1100	Type II-MSE (20lbs/sy, 1/4", 18% microsurfacing emulsion)									
1110	Small, irregular areas	B-90	5000	.013	S.Y.	1.65	.60	.44	2.69	3.19
1120	Parking lot		15000	.004		1.65	.20	.15	2	2.27
1130	Roadway		30000	.002		1.65	.10	.07	1.82	2.04
1200	Type IIIa-MSE (32lbs/sy, 3/8", 15% microsurfacing emulsion)									
1210	Small, irregular areas	B-90	3000	.021	S.Y.	2.29	1	.73	4.02	4.83
1220	Parking lot		9000	.007		2.29	.33	.24	2.86	3.29
1230	Roadway		18000	.004		2.29	.17	.12	2.58	2.90

32 13 13 – Concrete Paving

32 13 13.25 Concrete Pavement, Highways	Crew	Daily Output	Labor-Hours	Unit	Material	2020 Bare Costs Labor	Equipment	Total	Total Incl O&P
0010 **CONCRETE PAVEMENT, HIGHWAYS**									
0015 Including joints, finishing and curing									
0020 Fixed form, 12' pass, unreinforced, 6" thick	B-26	3000	.029	S.Y.	26	1.38	1.18	28.56	32
0030 7" thick		2850	.031		31.50	1.45	1.24	34.19	38
0100 8" thick		2750	.032		35.50	1.50	1.29	38.29	42.50
0110 8" thick, small area		1375	.064		35.50	3.01	2.57	41.08	46.50
0200 9" thick		2500	.035		40	1.65	1.41	43.06	48.50
0300 10" thick		2100	.042		44	1.97	1.68	47.65	53.50
0310 10" thick, small area		1050	.084		44	3.94	3.37	51.31	58
0400 12" thick		1800	.049		50.50	2.30	1.96	54.76	61.50
0410 Conc. pavement, w/jt., fnsh.& curing, fix form, 24' pass, unreinforced, 6"T		6000	.015		24.50	.69	.59	25.78	28.50
0420 7" thick		5700	.015		29.50	.73	.62	30.85	34.50
0430 8" thick		5500	.016		33.50	.75	.64	34.89	39
0440 9" thick		5000	.018		38.50	.83	.71	40.04	44
0450 10" thick		4200	.021		42.50	.98	.84	44.32	49
0460 12" thick		3600	.024		49	1.15	.98	51.13	56.50
0470 15" thick		3000	.029		64	1.38	1.18	66.56	73.50
0500 Fixed form 12' pass, 15" thick		1500	.059		64.50	2.76	2.36	69.62	77.50
0510 For small irregular areas, add				%	10%	100%	100%		
0520 Welded wire fabric, sheets for rigid paving 2.33 lb./S.Y.	2 Rodm	389	.041	S.Y.	1.46	2.32		3.78	5.10
0530 Reinforcing steel for rigid paving 12 lb./S.Y.		666	.024		7.15	1.35		8.50	9.90
0540 Reinforcing steel for rigid paving 18 lb./S.Y.		444	.036		10.70	2.03		12.73	14.80
0610 For under 10' pass, add				%	10%	100%	100%		
0620 Slip form, 12' pass, unreinforced, 6" thick	B-26A	5600	.016	S.Y.	25	.74	.65	26.39	29.50
0622 7" thick		5600	.016		30.50	.74	.65	31.89	35.50
0624 8" thick		5300	.017		34.50	.78	.69	35.97	39.50
0626 9" thick		4820	.018		39	.86	.76	40.62	45
0628 10" thick		4050	.022		43	1.02	.90	44.92	50
0630 12" thick		3470	.025		49.50	1.19	1.05	51.74	57.50
0632 15" thick		2890	.030		62.50	1.43	1.26	65.19	72.50
0640 Slip form, 24' pass, unreinforced, 6" thick		11200	.008		24.50	.37	.33	25.20	28
0642 7" thick		11200	.008		29	.37	.33	29.70	33
0644 8" thick		10600	.008		33	.39	.34	33.73	37
0646 9" thick		9640	.009		38	.43	.38	38.81	42.50
0648 10" thick		8100	.011		41.50	.51	.45	42.46	47.50
0650 12" thick		6940	.013		48.50	.60	.53	49.63	54.50
0652 15" thick		5780	.015		60.50	.71	.63	61.84	69
0700 Finishing, broom finish small areas	2 Cefi	120	.133			6.65		6.65	9.80
0730 Transverse expansion joints, incl. premolded bit. jt. filler	C-1	150	.213	L.F.	2.06	10.75		12.81	18.40
0740 Transverse construction joint using bulkhead	"	73	.438	"	2.99	22		24.99	36.50
0750 Longitudinal joint tie bars, grouted	B-23	70	.571	Ea.	4.64	24.50	22.50	51.64	66.50
1000 Curing, with sprayed membrane by hand	2 Clab	1500	.011	S.Y.	1.12	.45		1.57	1.90

377

For customer support on your **Site Work & Landscape Costs** with RSMeans data, call 800.448.8182.

32 14 Unit Paving

32 14 13 – Precast Concrete Unit Paving

32 14 13.13 Interlocking Precast Concrete Unit Paving

		Crew	Daily Output	Labor-Hours	Unit	Material	2020 Bare Costs Labor	2020 Bare Costs Equipment	Total	Total Incl O&P
0010	**INTERLOCKING PRECAST CONCRETE UNIT PAVING**									
0020	"V" blocks for retaining soil	D-1	205	.078	S.F.	11.15	3.66		14.81	17.85

32 14 13.16 Precast Concrete Unit Paving Slabs

		Crew	Daily Output	Labor-Hours	Unit	Material	2020 Bare Costs Labor	2020 Bare Costs Equipment	Total	Total Incl O&P
0010	**PRECAST CONCRETE UNIT PAVING SLABS**									
0710	Precast concrete patio blocks, 2-3/8" thick, colors, 8" x 16"	D-1	265	.060	S.F.	10.65	2.83		13.48	16.05
0715	12" x 12"		300	.053		13.85	2.50		16.35	19.05
0720	16" x 16"		335	.048		15.25	2.24		17.49	20
0730	24" x 24"		510	.031		19	1.47		20.47	23
0740	Green, 8" x 16"		265	.060		13.75	2.83		16.58	19.45
0750	Exposed local aggregate, natural	2 Bric	250	.064		9.20	3.33		12.53	15.15
0800	Colors		250	.064		9.40	3.33		12.73	15.40
0850	Exposed granite or limestone aggregate		250	.064		9.10	3.33		12.43	15.05
0900	Exposed white tumblestone aggregate		250	.064		10.65	3.33		13.98	16.80

32 14 13.18 Precast Concrete Plantable Pavers

		Crew	Daily Output	Labor-Hours	Unit	Material	2020 Bare Costs Labor	2020 Bare Costs Equipment	Total	Total Incl O&P
0010	**PRECAST CONCRETE PLANTABLE PAVERS** (50% grass)									
0015	Subgrade preparation and grass planting not included									
0100	Precast concrete plantable pavers with topsoil, 24" x 16"	B-63	800	.050	S.F.	4.52	2.21	.22	6.95	8.55
0200	Less than 600 S.F. or irregular area	"	500	.080	"	4.52	3.54	.36	8.42	10.65
0300	3/4" crushed stone base for plantable pavers, 6" depth	B-62	1000	.024	S.Y.	4.20	1.10	.18	5.48	6.45
0400	8" depth		900	.027		5.60	1.22	.20	7.02	8.20
0500	10" depth		800	.030		7	1.37	.22	8.59	10
0600	12" depth		700	.034		8.40	1.57	.25	10.22	11.90
0700	Hydro seeding plantable pavers	B-81A	20	.800	M.S.F.	12.10	35.50	47	94.60	118
0800	Apply fertilizer and seed to plantable pavers	1 Clab	8	1	"	46	42		88	115

32 14 16 – Brick Unit Paving

32 14 16.10 Brick Paving

		Crew	Daily Output	Labor-Hours	Unit	Material	2020 Bare Costs Labor	2020 Bare Costs Equipment	Total	Total Incl O&P
0010	**BRICK PAVING**									
0012	4" x 8" x 1-1/2", without joints (4.5 bricks/S.F.)	D-1	110	.145	S.F.	2.66	6.80		9.46	13.25
0100	Grouted, 3/8" joint (3.9 bricks/S.F.)		90	.178		2.17	8.35		10.52	15.05
0200	4" x 8" x 2-1/4", without joints (4.5 bricks/S.F.)		110	.145		2.51	6.80		9.31	13.10
0300	Grouted, 3/8" joint (3.9 bricks/S.F.)		90	.178		2.17	8.35		10.52	15.05
0400	4" x 8" x 2-1/4", dry, on edge (8/S.F.)		140	.114		4.46	5.35		9.81	13.05
0450	Grouted, 3/8" joint (6.5/S.F.)		85	.188		3.62	8.85		12.47	17.40
0455	Pervious brick paving, 4" x 8" x 3-1/4", without joints (4.5 bricks/S.F.)		110	.145		3.70	6.80		10.50	14.40
0500	Bedding, asphalt, 3/4" thick	B-25	5130	.017		.68	.79	.53	2	2.52
0540	Course washed sand bed, 1" thick	B-18	5000	.005		.38	.21	.03	.62	.77
0580	Mortar, 1" thick	D-1	300	.053		.68	2.50		3.18	4.54
0620	2" thick		200	.080		1.37	3.75		5.12	7.20
2000	Brick pavers, laid on edge, 7.2/S.F.		70	.229		4.61	10.70		15.31	21.50
2500	For 4" thick concrete bed and joints, add		595	.027		1.31	1.26		2.57	3.36
2800	For steam cleaning, add	A-1H	950	.008		.12	.35	.08	.55	.75

32 14 23 – Asphalt Unit Paving

32 14 23.10 Asphalt Blocks

		Crew	Daily Output	Labor-Hours	Unit	Material	2020 Bare Costs Labor	2020 Bare Costs Equipment	Total	Total Incl O&P
0010	**ASPHALT BLOCKS**									
0020	Rectangular, 6" x 12" x 1-1/4", w/bed & neopr. adhesive	D-1	135	.119	S.F.	10.20	5.55		15.75	19.65
0100	3" thick		130	.123		14.30	5.75		20.05	24.50
0300	Hexagonal tile, 8" wide, 1-1/4" thick		135	.119		10.20	5.55		15.75	19.65
0400	2" thick		130	.123		14.30	5.75		20.05	24.50
0500	Square, 8" x 8", 1-1/4" thick		135	.119		10.20	5.55		15.75	19.65
0600	2" thick		130	.123		14.30	5.75		20.05	24.50
0900	For exposed aggregate (ground finish), add					.58			.58	.64

32 14 Unit Paving

32 14 23 – Asphalt Unit Paving

32 14 23.10 Asphalt Blocks	Crew	Daily Output	Labor-Hours	Unit	Material	2020 Bare Costs Labor	Equipment	Total	Total Incl O&P
0910	For colors, add				S.F.	.58			.58	.64

32 14 40 – Stone Paving

32 14 40.10 Stone Pavers

	32 14 40.10 Stone Pavers	Crew	Daily Output	Labor-Hours	Unit	Material	2020 Bare Costs Labor	Equipment	Total	Total Incl O&P
0010	**STONE PAVERS**									
1100	Flagging, bluestone, irregular, 1" thick,	D-1	81	.198	S.F.	10.35	9.25		19.60	25.50
1110	1-1/2" thick		90	.178		12.20	8.35		20.55	26
1120	Pavers, 1/2" thick		110	.145		17.25	6.80		24.05	29.50
1130	3/4" thick		95	.168		22	7.90		29.90	36
1140	1" thick		81	.198		23.50	9.25		32.75	40
1150	Snapped random rectangular, 1" thick		92	.174		15.70	8.15		23.85	29.50
1200	1-1/2" thick		85	.188		18.85	8.85		27.70	34
1250	2" thick		83	.193		22	9.05		31.05	37.50
1300	Slate, natural cleft, irregular, 3/4" thick		92	.174		9.75	8.15		17.90	23
1310	1" thick		85	.188		11.35	8.85		20.20	26
1350	Random rectangular, gauged, 1/2" thick		105	.152		21	7.15		28.15	34
1400	Random rectangular, butt joint, gauged, 1/4" thick	▼	150	.107		22.50	5		27.50	32.50
1450	For sand rubbed finish, add				▼	9.65			9.65	10.65
1500	For interior setting, add					25%			25%	25%
1550	Granite blocks, 3-1/2" x 3-1/2" x 3-1/2"	D-1	92	.174	S.F.	21.50	8.15		29.65	36
1560	4" x 4" x 4"		95	.168		22.50	7.90		30.40	37
1600	4" to 12" long, 3" to 5" wide, 3" to 5" thick		98	.163		17.95	7.65		25.60	31.50
1650	6" to 15" long, 3" to 6" wide, 3" to 5" thick	▼	105	.152		9.60	7.15		16.75	21.50
2000	Granite paver, sawn with thermal finish, white, 4" x 4" x 2"	D-2	80	.550		23.50	26.50		50	66.50
2005	4" x 4" x 3"		80	.550		36	26.50		62.50	80
2010	4" x 4" x 4"		80	.550		37.50	26.50		64	82
2015	4" x 8" x 2"		170	.259		27	12.55		39.55	48.50
2020	4" x 8" x 3"		170	.259		34	12.55		46.55	56.50
2025	4" x 8" x 4"		170	.259		37.50	12.55		50.05	60.50
2030	8" x 8" x 2"		300	.147		27	7.10		34.10	41
2035	8" x 8" x 3"		300	.147		36	7.10		43.10	51
2040	8" x 8" x 4"		300	.147		33.50	7.10		40.60	48
2045	12" x 12" x 2"		725	.061		24	2.94		26.94	30.50
2050	12" x 12" x 3"		725	.061		29	2.94		31.94	36.50
2055	12" x 12" x 4"	▼	725	.061	▼	34.50	2.94		37.44	42.50

32 16 Curbs, Gutters, Sidewalks, and Driveways

32 16 13 – Curbs and Gutters

32 16 13.13 Cast-in-Place Concrete Curbs and Gutters

	32 16 13.13 Cast-in-Place Concrete Curbs and Gutters	Crew	Daily Output	Labor-Hours	Unit	Material	2020 Bare Costs Labor	Equipment	Total	Total Incl O&P
0010	**CAST-IN-PLACE CONCRETE CURBS AND GUTTERS**									
0290	Forms only, no concrete									
0300	Concrete, wood forms, 6" x 18", straight	C-2	500	.096	L.F.	3.16	4.96		8.12	10.90
0400	6" x 18", radius	"	200	.240	"	3.30	12.40		15.70	22
0402	Forms and concrete complete									
0404	Concrete, wood forms, 6" x 18", straight & concrete	C-2A	500	.096	L.F.	6.75	4.91		11.66	14.80
0406	6" x 18", radius		200	.240		6.90	12.25		19.15	26
0410	Steel forms, 6" x 18", straight		700	.069		7.10	3.50		10.60	13.05
0411	6" x 18", radius	▼	400	.120		7.55	6.15		13.70	17.50
0415	Machine formed, 6" x 18", straight	B-69A	2000	.024		5.15	1.11	.61	6.87	8
0416	6" x 18", radius	"	900	.053	▼	5.20	2.46	1.35	9.01	10.90
0421	Curb and gutter, straight									

32 16 13.13 Cast-in-Place Concrete Curbs and Gutters

	32 16 13.13 Cast-in-Place Concrete Curbs and Gutters	Crew	Daily Output	Labor-Hours	Unit	Material	2020 Bare Costs Labor	Equipment	Total	Total Incl O&P
0422	with 6" high curb and 6" thick gutter, wood forms									
0430	24" wide, 0.055 C.Y./L.F.	C-2A	375	.128	L.F.	17.35	6.55		23.90	29
0435	30" wide, 0.066 C.Y./L.F.		340	.141		19.20	7.20		26.40	32
0440	Steel forms, 24" wide, straight		700	.069		9.75	3.50		13.25	15.95
0441	Radius		500	.096		9.25	4.91		14.16	17.50
0442	30" wide, straight		700	.069		11.40	3.50		14.90	17.80
0443	Radius		500	.096		10.40	4.91		15.31	18.80
0445	Machine formed, 24" wide, straight	B-69A	2000	.024		7.20	1.11	.61	8.92	10.30
0446	Radius		900	.053		7.20	2.46	1.35	11.01	13.10
0447	30" wide, straight		2000	.024		8.35	1.11	.61	10.07	11.55
0448	Radius		900	.053		8.35	2.46	1.35	12.16	14.35
0451	Median mall, machine formed, 2' x 9" high, straight		2200	.022		7.20	1.01	.55	8.76	10.05
0452	Radius	B-69B	900	.053		7.20	2.46	.88	10.54	12.60
0453	4' x 9" high, straight		2000	.024		14.40	1.11	.40	15.91	17.95
0454	Radius		800	.060		14.40	2.77	.99	18.16	21

32 16 13.23 Precast Concrete Curbs and Gutters

		Crew	Daily Output	Labor-Hours	Unit	Material	Labor	Equipment	Total	Total Incl O&P
0010	**PRECAST CONCRETE CURBS AND GUTTERS**									
0550	Precast, 6" x 18", straight	B-29	700	.080	L.F.	9.15	3.68	1.21	14.04	16.95
0600	6" x 18", radius	"	325	.172	"	12.20	7.95	2.61	22.76	28

32 16 13.33 Asphalt Curbs

		Crew	Daily Output	Labor-Hours	Unit	Material	Labor	Equipment	Total	Total Incl O&P
0010	**ASPHALT CURBS**									
0012	Curbs, asphaltic, machine formed, 8" wide, 6" high, 40 L.F./ton	B-27	1000	.032	L.F.	1.64	1.36	.27	3.27	4.15
0100	8" wide, 8" high, 30 L.F./ton		900	.036		2.19	1.51	.30	4	5
0150	Asphaltic berm, 12" W, 3" to 6" H, 35 L.F./ton, before pavement		700	.046		.04	1.95	.39	2.38	3.41
0200	12" W, 1-1/2" to 4" H, 60 L.F./ton, laid with pavement	B-2	1050	.038		.02	1.62		1.64	2.46

32 16 13.43 Stone Curbs

		Crew	Daily Output	Labor-Hours	Unit	Material	Labor	Equipment	Total	Total Incl O&P
0010	**STONE CURBS**									
1000	Granite, split face, straight, 5" x 16"	D-13	275	.175	L.F.	16.50	8.80	1.56	26.86	33
1100	6" x 18"	"	250	.192		21.50	9.65	1.71	32.86	40.50
1300	Radius curbing, 6" x 18", over 10' radius	B-29	260	.215		26.50	9.90	3.26	39.66	47.50
1400	Corners, 2' radius	"	80	.700	Ea.	89	32	10.60	131.60	158
1600	Edging, 4-1/2" x 12", straight	D-13	300	.160	L.F.	8.25	8.05	1.43	17.73	23
1800	Curb inlets (guttermouth) straight	B-29	41	1.366	Ea.	198	63	20.50	281.50	335
2000	Indian granite (Belgian block)									
2100	Jumbo, 10-1/2" x 7-1/2" x 4", grey	D-1	150	.107	L.F.	9.65	5		14.65	18.20
2150	Pink		150	.107		9.55	5		14.55	18.10
2200	Regular, 9" x 4-1/2" x 4-1/2", grey		160	.100		4.64	4.69		9.33	12.20
2250	Pink		160	.100		6.75	4.69		11.44	14.55
2300	Cubes, 4" x 4" x 4", grey		175	.091		3.34	4.29		7.63	10.15
2350	Pink		175	.091		3.99	4.29		8.28	10.90
2400	6" x 6" x 6", pink		155	.103		12.60	4.84		17.44	21.50
2500	Alternate pricing method for Indian granite									
2550	Jumbo, 10-1/2" x 7-1/2" x 4" (30 lb.), grey				Ton	550			550	605
2600	Pink					555			555	610
2650	Regular, 9" x 4-1/2" x 4-1/2" (20 lb.), grey					325			325	360
2700	Pink					475			475	520
2750	Cubes, 4" x 4" x 4" (5 lb.), grey					405			405	445
2800	Pink					515			515	565
2850	6" x 6" x 6" (25 lb.), pink					490			490	540
2900	For pallets, add					22			22	24

32 17 13 – Parking Bumpers

32 17 13.13 Metal Parking Bumpers

		Crew	Daily Output	Labor-Hours	Unit	Material	2020 Bare Costs Labor	Equipment	Total	Total Incl O&P
0010	**METAL PARKING BUMPERS**									
0015	Bumper rails for garages, 12 ga. rail, 6" wide, with steel									
0020	posts 12'-6" OC, minimum	E-4	190	.168	L.F.	22	9.80	.77	32.57	40
0030	Average		165	.194		27.50	11.30	.89	39.69	48.50
0100	Maximum		140	.229		33	13.30	1.05	47.35	58
0300	12" channel rail, minimum		160	.200		27.50	11.65	.92	40.07	49.50
0400	Maximum		120	.267		41	15.50	1.22	57.72	71.50
1300	Pipe bollards, conc. filled/paint, 8' L x 4' D hole, 6" diam.	B-6	20	1.200	Ea.	635	55	10.70	700.70	795
1400	8" diam.		15	1.600		740	73	14.25	827.25	940
1500	12" diam.		12	2		1,025	91.50	17.80	1,134.30	1,275
2030	Folding with individual padlocks	B-2	50	.800		205	34		239	276

32 17 13.16 Plastic Parking Bumpers

		Crew	Daily Output	Labor-Hours	Unit	Material	Labor	Equipment	Total	Total Incl O&P
0010	**PLASTIC PARKING BUMPERS**									
1200	Thermoplastic, 6" x 10" x 6'-0"	B-2	120	.333	Ea.	57	14.15		71.15	84

32 17 13.19 Precast Concrete Parking Bumpers

		Crew	Daily Output	Labor-Hours	Unit	Material	Labor	Equipment	Total	Total Incl O&P
0010	**PRECAST CONCRETE PARKING BUMPERS**									
1000	Wheel stops, precast concrete incl. dowels, 6" x 10" x 6'-0"	B-2	120	.333	Ea.	63.50	14.15		77.65	91
1100	8" x 13" x 6'-0"	"	120	.333	"	74	14.15		88.15	103

32 17 13.26 Wood Parking Bumpers

		Crew	Daily Output	Labor-Hours	Unit	Material	Labor	Equipment	Total	Total Incl O&P
0010	**WOOD PARKING BUMPERS**									
0020	Parking barriers, timber w/saddles, treated type									
0100	4" x 4" for cars	B-2	520	.077	L.F.	3.16	3.27		6.43	8.40
0200	6" x 6" for trucks		520	.077	"	6.55	3.27		9.82	12.10
0600	Flexible fixed stanchion, 2' high, 3" diameter		100	.400	Ea.	44.50	17		61.50	74.50

32 17 23 – Pavement Markings

32 17 23.13 Painted Pavement Markings

		Crew	Daily Output	Labor-Hours	Unit	Material	Labor	Equipment	Total	Total Incl O&P
0010	**PAINTED PAVEMENT MARKINGS**									
0020	Acrylic waterborne, white or yellow, 4" wide, less than 3,000 L.F.	B-78	20000	.002	L.F.	.11	.10	.05	.26	.34
0030	3,000-16,000 L.F.		20000	.002		.13	.10	.05	.28	.36
0040	over 16,000 L.F.		20000	.002		.11	.10	.05	.26	.34
0200	6" wide, less than 3,000 L.F.		11000	.004		.16	.19	.10	.45	.57
0220	3,000-16,000 L.F.		11000	.004		.19	.19	.10	.48	.60
0230	over 16,000 L.F.		11000	.004		.16	.19	.10	.45	.57
0500	8" wide, less than 3,000 L.F.		10000	.005		.21	.21	.11	.53	.67
0520	3,000-16,000 L.F.		10000	.005		.25	.21	.11	.57	.71
0530	over 16,000 L.F.		10000	.005		.21	.21	.11	.53	.67
0600	12" wide, less than 3,000 L.F.		4000	.012		.32	.52	.26	1.10	1.42
0604	3,000-16,000 L.F.		4000	.012		.38	.52	.26	1.16	1.49
0608	over 16,000 L.F.		4000	.012		.32	.52	.26	1.10	1.42
0620	Arrows or gore lines		2300	.021	S.F.	.22	.90	.46	1.58	2.11
0640	Temporary paint, white or yellow, less than 3,000 L.F.		15000	.003	L.F.	.12	.14	.07	.33	.42
0642	3,000-16,000 L.F.		15000	.003		.09	.14	.07	.30	.39
0644	Over 16,000 L.F.		15000	.003		.09	.14	.07	.30	.39
0660	Removal	1 Clab	300	.027			1.12		1.12	1.69
0680	Temporary tape	2 Clab	1500	.011		.53	.45		.98	1.25
0710	Thermoplastic, white or yellow, 4" wide, less than 6,000 L.F.	B-79	15000	.003		.42	.12	.12	.66	.76
0715	6,000 L.F. or more		15000	.003		.32	.12	.12	.56	.65
0730	6" wide, less than 6,000 L.F.		14000	.003		.63	.12	.12	.87	1.02
0735	6,000 L.F. or more		14000	.003		.48	.12	.12	.72	.86
0740	8" wide, less than 6,000 L.F.		12000	.003		.84	.15	.14	1.13	1.30
0745	6,000 L.F. or more		12000	.003		.64	.15	.14	.93	1.08

For customer support on your Site Work & Landscape Costs with RSMeans data, call 800.448.8182.

381

32 17 23 – Pavement Markings

32 17 23.13 Painted Pavement Markings

		Crew	Daily Output	Labor-Hours	Unit	Material	2020 Bare Costs Labor	Equipment	Total	Total Incl O&P
0750	12" wide, less than 6,000 L.F.	B-79	6000	.007	L.F.	1.24	.29	.29	1.82	2.12
0755	6,000 L.F. or more		6000	.007	↓	.94	.29	.29	1.52	1.80
0760	Arrows		660	.061	S.F.	.69	2.64	2.63	5.96	7.60
0770	Gore lines		2500	.016		.69	.70	.69	2.08	2.57
0780	Letters	↓	660	.061	↓	.70	2.64	2.63	5.97	7.60
0782	Thermoplastic material, small users				Ton	2,050			2,050	2,250
0784	Glass beads, highway use, add				Lb.	.68			.68	.75
0786	Thermoplastic material, highway departments				Ton	1,750			1,750	1,925
1000	Airport painted markings									
1050	Traffic safety flashing truck for airport painting	A-2B	1	8	Day		380	196	576	780
1100	Painting, white or yellow, taxiway markings	B-78	4000	.012	S.F.	.38	.52	.26	1.16	1.49
1110	with 12 lb. beads per 100 S.F.		4000	.012		.46	.52	.26	1.24	1.58
1200	Runway markings		3500	.014		.38	.59	.30	1.27	1.64
1210	with 12 lb. beads per 100 S.F.		3500	.014		.46	.59	.30	1.35	1.73
1300	Pavement location or direction signs		2500	.019		.38	.83	.42	1.63	2.14
1310	with 12 lb. beads per 100 S.F.		2500	.019	↓	.46	.83	.42	1.71	2.23
1350	Mobilization airport pavement painting		4	12	Ea.		520	265	785	1,075
1400	Paint markings or pavement signs removal daytime	B-78B	400	.045	S.F.		1.95	1.01	2.96	4.04
1500	Removal nighttime		335	.054	"		2.33	1.20	3.53	4.81
1600	Mobilization pavement paint removal	↓	4	4.500	Ea.		195	101	296	405

32 17 23.14 Pavement Parking Markings

		Crew	Daily Output	Labor-Hours	Unit	Material	2020 Bare Costs Labor	Equipment	Total	Total Incl O&P
0010	**PAVEMENT PARKING MARKINGS**									
0790	Layout of pavement marking	A-2	25000	.001	L.F.		.04	.01	.05	.07
0800	Lines on pvmt., parking stall, paint, white, 4" wide	B-78B	400	.045	Stall	5.05	1.95	1.01	8.01	9.60
0825	Parking stall, small quantities	2 Pord	80	.200		10.10	8.90		19	24.50
0830	Lines on pvmt., parking stall, thermoplastic, white, 4" wide	B-79	300	.133	↓	20	5.80	5.80	31.60	37
1000	Street letters and numbers	B-78B	1600	.011	S.F.	.82	.49	.25	1.56	1.91
1100	Pavement marking letter, 6"	2 Pord	400	.040	Ea.	12.05	1.78		13.83	15.90
1110	12" letter		272	.059		15.15	2.61		17.76	20.50
1120	24" letter		160	.100		35.50	4.44		39.94	46
1130	36" letter		84	.190		43	8.45		51.45	60
1140	42" letter		84	.190		52	8.45		60.45	70
1150	72" letter		40	.400		46	17.75		63.75	77.50
1200	Handicap symbol	↓	40	.400	↓	34	17.75		51.75	63.50
1210	Handicap parking sign 12" x 18" and post	A-2	12	2	↓	135	87.50	16.35	238.85	299
1300	Pavement marking, thermoplastic tape including layout, 4" width	B-79B	320	.025	L.F.	2.51	1.05	.54	4.10	4.93
1310	12" width		192	.042	"	9.15	1.75	.89	11.79	13.65
1320	Letters including layout, 4"		240	.033	Ea.	16.30	1.40	.71	18.41	21
1330	6"		160	.050		18.35	2.11	1.07	21.53	24.50
1340	12"		120	.067		21	2.81	1.43	25.24	29
1350	48"		64	.125		79	5.25	2.68	86.93	98
1360	96"		32	.250		100	10.55	5.35	115.90	132
1380	4 letter words, 8' tall	↓	8	1	↓	370	42	21.50	433.50	490
1385	Sample words: BUMP, FIRE, LANE									

32 17 23.33 Plastic Pavement Markings

		Crew	Daily Output	Labor-Hours	Unit	Material	2020 Bare Costs Labor	Equipment	Total	Total Incl O&P
0010	**PLASTIC PAVEMENT MARKINGS**									
0020	Thermoplastic markings with glass beads									
0100	Thermoplastic striping, 6 lb. beads per 100 S.F., white or yellow, 4" wide	B-79	15000	.003	L.F.	.43	.12	.12	.67	.78
0110	8 lb. beads per 100 S.F.		15000	.003		.44	.12	.12	.68	.78
0120	10 lb. beads per 100 S.F.		15000	.003		.44	.12	.12	.68	.79
0130	12 lb. beads per 100 S.F.		15000	.003		.45	.12	.12	.69	.79
0200	6 lb. beads per 100 S.F., white or yellow, 6" wide		14000	.003		.65	.12	.12	.89	1.05

382

For customer support on your Site Work & Landscape Costs with RSMeans data, call 800.448.8182.

32 17 Paving Specialties

32 17 23 – Pavement Markings

32 17 23.33 Plastic Pavement Markings

		Crew	Daily Output	Labor-Hours	Unit	Material	2020 Bare Costs Labor	Equipment	Total	Total Incl O&P
0210	8 lb. beads per 100 S.F.	B-79	14000	.003	L.F.	.66	.12	.12	.90	1.05
0220	10 lb. beads per 100 S.F.		14000	.003		.66	.12	.12	.90	1.06
0230	12 lb. beads per 100 S.F.		14000	.003		.67	.12	.12	.91	1.07
0300	6 lb. beads per 100 S.F., white or yellow, 8" wide		12000	.003		.87	.15	.14	1.16	1.33
0310	8 lb. beads per 100 S.F.		12000	.003		.88	.15	.14	1.17	1.34
0320	10 lb. beads per 100 S.F.		12000	.003		.89	.15	.14	1.18	1.35
0330	12 lb. beads per 100 S.F.		12000	.003		.89	.15	.14	1.18	1.36
0400	6 lb. beads per 100 S.F., white or yellow, 12" wide		6000	.007		1.30	.29	.29	1.88	2.19
0410	8 lb. beads per 100 S.F.		6000	.007		1.31	.29	.29	1.89	2.21
0420	10 lb. beads per 100 S.F.		6000	.007		1.33	.29	.29	1.91	2.22
0430	12 lb. beads per 100 S.F.		6000	.007		1.34	.29	.29	1.92	2.24
0500	Gore lines with 6 lb. beads per 100 S.F.		2500	.016	S.F.	.73	.70	.69	2.12	2.61
0510	8 lb. beads per 100 S.F.		2500	.016		.74	.70	.69	2.13	2.63
0520	10 lb. beads per 100 S.F.		2500	.016		.76	.70	.69	2.15	2.64
0530	12 lb. beads per 100 S.F.		2500	.016		.77	.70	.69	2.16	2.66
0600	Arrows with 6 lb. beads per 100 S.F.		660	.061		.73	2.64	2.63	6	7.65
0610	8 lb. beads per 100 S.F.		660	.061		.74	2.64	2.63	6.01	7.65
0620	10 lb. beads per 100 S.F.		660	.061		.76	2.64	2.63	6.03	7.70
0630	12 lb. beads per 100 S.F.		660	.061		.77	2.64	2.63	6.04	7.70
0700	Letters with 6 lb. beads per 100 S.F.		660	.061		.74	2.64	2.63	6.01	7.65
0710	8 lb. beads per 100 S.F.		660	.061		.75	2.64	2.63	6.02	7.70
0720	10 lb. beads per 100 S.F.		660	.061		.77	2.64	2.63	6.04	7.70
0730	12 lb. beads per 100 S.F.		660	.061		.78	2.64	2.63	6.05	7.70
1000	Airport thermoplastic markings									
1050	Traffic safety flashing truck for airport markings	A-2B	1	8	Day		380	196	576	780
1100	Thermoplastic taxiway markings with 24 lb. beads per 100 S.F.	B-79	2500	.016	S.F.	.81	.70	.69	2.20	2.70
1200	Runway markings		2100	.019		.81	.83	.83	2.47	3.05
1250	Location or direction signs		1500	.027		.81	1.16	1.16	3.13	3.90
1350	Mobilization airport pavement marking		4	10	Ea.		435	435	870	1,125
1400	Thermostatic markings or pavement signs removal daytime	B-78B	400	.045	S.F.		1.95	1.01	2.96	4.04
1500	Removal nighttime		335	.054	"		2.33	1.20	3.53	4.81
1600	Mobilization thermostatic marking removal		4	4.500	Ea.		195	101	296	405
1700	Thermoplastic road marking material, small users				Ton	2,050			2,050	2,250
1710	Highway departments					1,750			1,750	1,925
1720	Thermoplastic airport marking material, small users					1,950			1,950	2,150
1730	Large users					1,700			1,700	1,875
1800	Glass beads material for highway use, type II				Lb.	.68			.68	.75
1900	Glass beads material for airport use, type III				"	.68			.68	.75

32 17 26 – Tactile Warning Surfacing

32 17 26.10 Tactile Warning Surfacing

		Crew	Daily Output	Labor-Hours	Unit	Material	2020 Bare Costs Labor	Equipment	Total	Total Incl O&P
0010	**TACTILE WARNING SURFACING**									
0100	Tactile warning tiles S.F.	2 Clab	400	.040	S.F.	18.35	1.68		20.03	22.50

For customer support on your Site Work & Landscape Costs with RSMeans data, call 800.448.8182.

383

32 18 Athletic and Recreational Surfacing

32 18 13 – Synthetic Grass Surfacing

32 18 13.10 Artificial Grass Surfacing

		Crew	Daily Output	Labor-Hours	Unit	Material	2020 Bare Costs Labor	Equipment	Total	Total Incl O&P
0010	**ARTIFICIAL GRASS SURFACING**									
0015	Not including asphalt base or drainage,									
0020	but including cushion pad, over 50,000 S.F.									
0200	1/2" pile and 5/16" cushion pad, standard	C-17	3200	.025	S.F.	6.50	1.38		7.88	9.25
0300	Deluxe		2560	.031		6.30	1.73		8.03	9.55
0500	1/2" pile and 5/8" cushion pad, standard		2844	.028		6.20	1.55		7.75	9.15
0600	Deluxe		2327	.034		6.30	1.90		8.20	9.75
0700	Acrylic emulsion texture coat, sand filled	B-1	5200	.005	S.Y.	3.19	.20		3.39	3.81
0710	Rubber filled	"	5200	.005	"	3.29	.20		3.49	3.92
0720	Acrylic emulsion color coat									
0730	Brown, red, or tan	B-1	6000	.004	S.Y.	3.19	.17		3.36	3.77
0740	Green		6000	.004		3.19	.17		3.36	3.77
0750	Blue		6000	.004		3.08	.17		3.25	3.65
0760	Rubber acrylic cushion base coat		2600	.009		4.45	.39		4.84	5.50
0800	For asphaltic concrete base, 2-1/2" thick,									
0900	with 6" crushed stone sub-base, add	B-25	12000	.007	S.F.	1.73	.34	.22	2.29	2.66

32 18 16 – Synthetic Resilient Surfacing

32 18 16.13 Playground Protective Surfacing

		Crew	Daily Output	Labor-Hours	Unit	Material	2020 Bare Costs Labor	Equipment	Total	Total Incl O&P
0010	**PLAYGROUND PROTECTIVE SURFACING**									
0100	Resilient rubber surface, 4" thick, black	2 Skwk	300	.053	S.F.	14.95	2.93		17.88	21
0150	2" thick topping, colors	"	2800	.006		7.55	.31		7.86	8.75
0200	Wood chip mulch, 6" deep	1 Clab	300	.027		.91	1.12		2.03	2.69

32 18 16.16 Epoxy Acrylic Coatings

		Crew	Daily Output	Labor-Hours	Unit	Material	2020 Bare Costs Labor	Equipment	Total	Total Incl O&P
0010	**EPOXY ACRYLIC COATINGS**									
0100	Surface Prep, Sand blasting to SSPC-SP6, 2.0#/S.F. sand	E-11A	1200	.027	S.F.	.34	1.25	.51	2.10	2.88
0150	Layout and line striping by others									
0200	Painting epoxy, Foundation/Prime coat, large area (100 S.F. per Gal)	E-11B	2400	.010	S.F.	.37	.45	.17	.99	1.29
0210	Epoxy intermediate coat		2800	.009		.37	.38	.14	.89	1.17
0220	Epoxy top coat		2800	.009		.37	.38	.14	.89	1.17
0300	Painting epoxy, Foundation/Prime coat, small areas (100 S.F. per Gal)		1400	.017		.37	.76	.28	1.41	1.92
0310	Epoxy intermediate coat		1500	.016		.37	.71	.27	1.35	1.82
0320	Epoxy top coat		1500	.016		.37	.71	.27	1.35	1.82
0330	Adhesive Promoter		1400	.017		.27	.76	.28	1.31	1.82
0340	Sealer Concentrate		2400	.010		.15	.45	.17	.77	1.05
0400	Install debris tarp for construction work under 600 S.F.	2 Clab	2100	.008		.76	.32		1.08	1.32
0410	600 S.F. or more	"	2400	.007		.61	.28		.89	1.09

32 18 23 – Athletic Surfacing

32 18 23.33 Running Track Surfacing

		Crew	Daily Output	Labor-Hours	Unit	Material	2020 Bare Costs Labor	Equipment	Total	Total Incl O&P
0010	**RUNNING TRACK SURFACING**									
0020	Running track, asphalt concrete pavement, 2-1/2"	B-37	300	.160	S.Y.	15.95	7.10	.85	23.90	29
0102	Surface, latex rubber system, 1/2" thick, black	B-20	115	.209		53.50	9.80		63.30	73.50
0152	Colors		115	.209		65.50	9.80		75.30	87
0302	Urethane rubber system, 1/2" thick, black		110	.218		41	10.25		51.25	61
0402	Color coating		110	.218		51	10.25		61.25	71.50

32 18 23.53 Tennis Court Surfacing

		Crew	Daily Output	Labor-Hours	Unit	Material	2020 Bare Costs Labor	Equipment	Total	Total Incl O&P
0010	**TENNIS COURT SURFACING**									
0020	Tennis court, asphalt, incl. base, 2-1/2" thick, one court	B-37	450	.107	S.Y.	33.50	4.72	.57	38.79	44
0200	Two courts		675	.071		19.35	3.15	.38	22.88	26.50
0300	Clay courts		360	.133		46	5.90	.71	52.61	60
0400	Pulverized natural greenstone with 4" base, fast dry		250	.192		39	8.50	1.02	48.52	56.50
0800	Rubber-acrylic base resilient pavement		600	.080		67	3.54	.43	70.97	80

32 18 Athletic and Recreational Surfacing

32 18 23 – Athletic Surfacing

32 18 23.53 Tennis Court Surfacing

		Crew	Daily Output	Labor-Hours	Unit	Material	2020 Bare Costs Labor	Equipment	Total	Total Incl O&P
1000	Colored sealer, acrylic emulsion, 3 coats	2 Clab	800	.020	S.Y.	7.05	.84		7.89	9
1100	3 coats, 2 colors	"	900	.018		9.80	.75		10.55	11.90
1200	For preparing old courts, add	1 Clab	825	.010			.41		.41	.61
1400	Posts for nets, 3-1/2" diameter with eye bolts	B-1	3.40	7.059	Pr.	282	300		582	765
1500	With pulley & reel		3.40	7.059	"	900	300		1,200	1,450
1700	Net, 42' long, nylon thread with binder		50	.480	Ea.	275	20.50		295.50	335
1800	All metal		6.50	3.692	"	560	158		718	850
2000	Paint markings on asphalt, 2 coats	1 Pord	1.78	4.494	Court	252	200		452	575
2200	Complete court with fence, etc., asphaltic conc., minimum	B-37	.20	240		34,500	10,600	1,275	46,375	55,000
2300	Maximum		.16	300		82,500	13,300	1,600	97,400	112,000
2800	Clay courts, minimum		.20	240		38,000	10,600	1,275	49,875	59,000
2900	Maximum		.16	300		69,500	13,300	1,600	84,400	98,000

32 31 Fences and Gates

32 31 11 – Gate Operators

32 31 11.10 Gate Operators

		Crew	Daily Output	Labor-Hours	Unit	Material	2020 Bare Costs Labor	Equipment	Total	Total Incl O&P
0010	**GATE OPERATORS**									
7810	Motor operators for gates (no elec wiring), 3' wide swing	2 Skwk	.50	32	Ea.	1,150	1,750		2,900	3,925
7815	Up to 20' wide swing		.50	32		1,450	1,750		3,200	4,250
7820	Up to 45' sliding		.50	32		3,300	1,750		5,050	6,300
7825	Overhead gate, 6' to 18' wide, sliding/cantilever		45	.356	L.F.	340	19.50		359.50	405
7830	Gate operators, digital receiver		7	2.286	Ea.	87	125		212	286
7835	Two button transmitter		24	.667		28	36.50		64.50	86.50
7840	3 button station		14	1.143		43	62.50		105.50	143
7845	Master slave system		4	4		189	219		408	540

32 31 13 – Chain Link Fences and Gates

32 31 13.10 Chain Link Gates and Fences

		Crew	Daily Output	Labor-Hours	Unit	Material	2020 Bare Costs Labor	Equipment	Total	Total Incl O&P
0010	**CHAIN LINK GATES AND FENCES**									
4750	Gate, transom for 10' fence, galv. steel, single, 3' x 7'	B-80A	52	.462	Ea.	400	19.45	15.80	435.25	485
4752	4' x 7'		10	2.400		430	101	82	613	710
4754	3' x 10'		8	3		365	126	103	594	705
4756	4' x 10'		10	2.400		390	101	82	573	670
4758	Double transom, 10' x 7'	B-80B	10	3.200		690	143	23.50	856.50	1,000
4760	12' x 7'		6	5.333		730	239	39.50	1,008.50	1,200
4762	14' x 7'		5	6.400		795	287	47	1,129	1,350
4764	10' x 10'		4	8		990	360	59	1,409	1,700
4766	12' x 10'		7	4.571		1,100	205	33.50	1,338.50	1,550
4768	14' x 10'		7	4.571		1,175	205	33.50	1,413.50	1,625
4780	Vinyl-clad, single transom, 3' x 7'		10	3.200		525	143	23.50	691.50	820
4782	4' x 7'		10	3.200		570	143	23.50	736.50	870
4784	3' x 10'		8	4		575	179	29.50	783.50	935
4786	4' x 10'		8	4		650	179	29.50	858.50	1,025
4788	Double transom, 10' x 7'		10	3.200		895	143	23.50	1,061.50	1,225
4790	12' x 7'		6	5.333		920	239	39.50	1,198.50	1,425
4792	14' x 7'		5	6.400		1,025	287	47	1,359	1,600
4794	10' x 10'		4	8		995	360	59	1,414	1,700
4798	12' x 12'		7	4.571		1,425	205	33.50	1,663.50	1,925
4799	14' x 14'		7	4.571		1,900	205	33.50	2,138.50	2,425

32 31 Fences and Gates

32 31 13 – Chain Link Fences and Gates

32 31 13.20 Fence, Chain Link Industrial

		Crew	Daily Output	Labor-Hours	Unit	Material	2020 Bare Costs Labor	Equipment	Total	Total Incl O&P
0010	**FENCE, CHAIN LINK INDUSTRIAL**									
0011	Schedule 40, including concrete									
0020	3 strands barb wire, 2" post @ 10' OC, set in concrete, 6' H									
0200	9 ga. wire, galv. steel, in concrete	B-80C	240	.100	L.F.	19.95	4.38	1.04	25.37	30
0248	Fence, add for vinyl coated fabric				S.F.	.71			.71	.78
0300	Aluminized steel	B-80C	240	.100	L.F.	22.50	4.38	1.04	27.92	33
0301	Fence, wrought iron		240	.100		43	4.38	1.04	48.42	55
0303	Fence, commercial 4' high		240	.100		32.50	4.38	1.04	37.92	43.50
0304	Fence, commercial 6' high		240	.100		58	4.38	1.04	63.42	72
0500	6 ga. wire, galv. steel		240	.100		23.50	4.38	1.04	28.92	34
0600	Aluminized steel		240	.100		31	4.38	1.04	36.42	42
0800	6 ga. wire, 6' high but omit barbed wire, galv. steel		250	.096		21.50	4.21	1	26.71	31
0900	Aluminized steel, in concrete		250	.096		28.50	4.21	1	33.71	38.50
0920	8' H, 6 ga. wire, 2-1/2" line post, galv. steel, in concrete		180	.133		33.50	5.85	1.39	40.74	47.50
0940	Aluminized steel, in concrete		180	.133	↓	41	5.85	1.39	48.24	55.50
1400	Gate for 6' high fence, 1-5/8" frame, 3' wide, galv. steel		10	2.400	Ea.	209	105	25	339	415
1500	Aluminized steel, in concrete	↓	10	2.400	"	218	105	25	348	425
2000	5'-0" high fence, 9 ga., no barbed wire, 2" line post, in concrete									
2010	10' OC, 1-5/8" top rail, in concrete									
2100	Galvanized steel, in concrete	B-80C	300	.080	L.F.	22.50	3.51	.83	26.84	31
2200	Aluminized steel, in concrete		300	.080	"	21	3.51	.83	25.34	29
2400	Gate, 4' wide, 5' high, 2" frame, galv. steel, in concrete		10	2.400	Ea.	229	105	25	359	440
2500	Aluminized steel, in concrete		10	2.400	"	191	105	25	321	395
3100	Overhead slide gate, chain link, 6' high, to 18' wide, in concrete	↓	38	.632	L.F.	95	27.50	6.60	129.10	154
3105	8' high, in concrete	B-80	30	1.067		96.50	49.50	34	180	218
3108	10' high, in concrete		24	1.333		191	62	42.50	295.50	350
3110	Cantilever type, in concrete		48	.667		168	31	21.50	220.50	255
3120	8' high, in concrete		24	1.333		176	62	42.50	280.50	335
3130	10' high, in concrete	↓	18	1.778	↓	218	83	56.50	357.50	425
5000	Double swing gates, incl. posts & hardware, in concrete									
5010	5' high, 12' opening, in concrete	B-80C	3.40	7.059	Opng.	535	310	73.50	918.50	1,125
5020	20' opening, in concrete		2.80	8.571		595	375	89.50	1,059.50	1,325
5060	6' high, 12' opening, in concrete		3.20	7.500		465	330	78	873	1,100
5070	20' opening, in concrete	↓	2.60	9.231		710	405	96	1,211	1,500
5080	8' high, 12' opening, in concrete	B-80	2.13	15.002		545	700	480	1,725	2,175
5090	20' opening, in concrete		1.45	22.069		845	1,025	705	2,575	3,250
5100	10' high, 12' opening, in concrete		1.31	24.427		830	1,150	780	2,760	3,475
5110	20' opening, in concrete		1.03	31.068		880	1,450	990	3,320	4,250
5120	12' high, 12' opening, in concrete		1.05	30.476		1,300	1,425	970	3,695	4,625
5130	20' opening, in concrete	↓	.85	37.647	↓	1,325	1,750	1,200	4,275	5,400
5190	For aluminized steel, add					20%				
7055	Braces, galv. steel	B-80A	960	.025	L.F.	2.72	1.05	.85	4.62	5.50
7056	Aluminized steel	"	960	.025	"	3.26	1.05	.85	5.16	6.10
7071	Privacy slats, vertical, vinyl	1 Clab	500	.016	S.F.	1.76	.67		2.43	2.95
7072	Redwood		450	.018		1.51	.75		2.26	2.78
7073	Diagonal, aluminum	↓	300	.027	↓	4.86	1.12		5.98	7.05

32 31 13.25 Fence, Chain Link Residential

		Crew	Daily Output	Labor-Hours	Unit	Material	2020 Bare Costs Labor	Equipment	Total	Total Incl O&P
0010	**FENCE, CHAIN LINK RESIDENTIAL**									
0011	Schedule 20, 11 ga. wire, 1-5/8" post									
0020	10' OC, 1-3/8" top rail, 2" corner post, galv. stl. 3' high	B-80C	500	.048	L.F.	4.95	2.10	.50	7.55	9.15
0050	4' high		400	.060		5.60	2.63	.63	8.86	10.80
0100	6' high	↓	200	.120	↓	7.25	5.25	1.25	13.75	17.30

32 31 Fences and Gates

32 31 13 – Chain Link Fences and Gates

32 31 13.25 Fence, Chain Link Residential

		Crew	Daily Output	Labor-Hours	Unit	Material	2020 Bare Costs Labor	Equipment	Total	Total Incl O&P
0150	Add for gate 3' wide, 1-3/8" frame, 3' high	B-80C	12	2	Ea.	84.50	87.50	21	193	248
0170	4' high		10	2.400		91.50	105	25	221.50	287
0190	6' high		10	2.400		104	105	25	234	300
0200	Add for gate 4' wide, 1-3/8" frame, 3' high		9	2.667		96	117	28	241	310
0220	4' high		9	2.667		103	117	28	248	320
0240	6' high		8	3		118	131	31.50	280.50	360
0350	Aluminized steel, 11 ga. wire, 3' high		500	.048	L.F.	8.60	2.10	.50	11.20	13.20
0380	4' high		400	.060		7.85	2.63	.63	11.11	13.30
0400	6' high		200	.120		11.15	5.25	1.25	17.65	21.50
0450	Add for gate 3' wide, 1-3/8" frame, 3' high		12	2	Ea.	104	87.50	21	212.50	269
0470	4' high		10	2.400		97	105	25	227	293
0490	6' high		10	2.400		137	105	25	267	335
0500	Add for gate 4' wide, 1-3/8" frame, 3' high		10	2.400		108	105	25	238	305
0520	4' high		9	2.667		119	117	28	264	335
0540	6' high		8	3		133	131	31.50	295.50	380
0620	Vinyl covered, 9 ga. wire, 3' high		500	.048	L.F.	7.70	2.10	.50	10.30	12.15
0640	4' high		400	.060		8.20	2.63	.63	11.46	13.65
0660	6' high		200	.120		9.35	5.25	1.25	15.85	19.60
0720	Add for gate 3' wide, 1-3/8" frame, 3' high		12	2	Ea.	88	87.50	21	196.50	252
0740	4' high		10	2.400		95.50	105	25	225.50	291
0760	6' high		10	2.400		132	105	25	262	330
0780	Add for gate 4' wide, 1-3/8" frame, 3' high		10	2.400		109	105	25	239	305
0800	4' high		9	2.667		100	117	28	245	315
0820	6' high		8	3		138	131	31.50	300.50	385
7076	Fence, for small jobs 100 L.F. fence or less w/or w/o gate, add				L.F.	20%				

32 31 13.26 Tennis Court Fences and Gates

		Crew	Daily Output	Labor-Hours	Unit	Material	2020 Bare Costs Labor	Equipment	Total	Total Incl O&P
0010	**TENNIS COURT FENCES AND GATES**									
0860	Tennis courts, 11 ga. wire, 2-1/2" post set									
0870	in concrete, 10' OC, 1-5/8" top rail									
0900	10' high	B-80	190	.168	L.F.	22	7.85	5.35	35.20	41.50
0920	12' high		170	.188	"	23	8.75	6	37.75	45
1000	Add for gate 4' wide, 1-5/8" frame 7' high		10	3.200	Ea.	231	149	102	482	590
1040	Aluminized steel, 11 ga. wire 10' high		190	.168	L.F.	20.50	7.85	5.35	33.70	40
1100	12' high		170	.188	"	25.50	8.75	6	40.25	48
1140	Add for gate 4' wide, 1-5/8" frame, 7' high		10	3.200	Ea.	238	149	102	489	595
1250	Vinyl covered, 9 ga. wire, 10' high		190	.168	L.F.	21.50	7.85	5.35	34.70	41
1300	12' high		170	.188	"	25	8.75	6	39.75	47.50
1310	Fence, CL, tennis court, transom gate, single, galv., 4' x 7'	B-80A	8.72	2.752	Ea.	350	116	94	560	665
1400	Add for gate 4' wide, 1-5/8" frame, 7' high	B-80	10	3.200	"	315	149	102	566	680

32 31 13.30 Fence, Chain Link, Gates and Posts

		Crew	Daily Output	Labor-Hours	Unit	Material	2020 Bare Costs Labor	Equipment	Total	Total Incl O&P
0010	**FENCE, CHAIN LINK, GATES & POSTS**									
0011	(1/3 post length in ground)									
0013	For Concrete, See Section 03 31									
6580	Line posts, galvanized, 2-1/2" OD, set in conc., 4'	B-80	80	.400	Ea.	49.50	18.65	12.75	80.90	96.50
6585	5'		76	.421		55	19.65	13.45	88.10	105
6590	6'		74	.432		62.50	20	13.80	96.30	114
6595	7'		72	.444		68.50	20.50	14.20	103.20	122
6600	8'		69	.464		80.50	21.50	14.80	116.80	137
6635	Vinyl coated, 2-1/2" OD, set in conc., 4'		79	.405		58.50	18.90	12.90	90.30	107
6640	5'		77	.416		62.50	19.35	13.25	95.10	113
6645	6'		74	.432		66	20	13.80	99.80	118
6650	7'		72	.444		78	20.50	14.20	112.70	133

32 31 Fences and Gates

32 31 13 – Chain Link Fences and Gates

32 31 13.30 Fence, Chain Link, Gates and Posts

		Crew	Daily Output	Labor-Hours	Unit	Material	2020 Bare Costs Labor	Equipment	Total	Total Incl O&P
6655	8'	B-80	69	.464	Ea.	99.50	21.50	14.80	135.80	159
6660	End gate post, steel, 3" OD, set in conc., 4'		68	.471		55	22	15	92	110
6665	5'		65	.492		54.50	23	15.70	93.20	112
6670	6'		63	.508		56.50	23.50	16.20	96.20	115
6675	7'		61	.525		64.50	24.50	16.75	105.75	126
6680	8'		59	.542		77.50	25.50	17.30	120.30	142
6685	Vinyl, 4'		68	.471		35.50	22	15	72.50	88.50
6690	5'		65	.492		46	23	15.70	84.70	102
6695	6'		63	.508		54	23.50	16.20	93.70	113
6700	7'		61	.525		55.50	24.50	16.75	96.75	116
6705	8'		59	.542		65.50	25.50	17.30	108.30	129
6710	Corner post, galv. steel, 4" OD, set in conc., 4'		65	.492		70.50	23	15.70	109.20	129
6715	6'		63	.508		109	23.50	16.20	148.70	173
6720	7'		61	.525		131	24.50	16.75	172.25	200
6725	8'		65	.492		202	23	15.70	240.70	274
6730	Vinyl, 5'		65	.492		50	23	15.70	88.70	107
6735	6'		63	.508		91	23.50	16.20	130.70	153
6740	7'		61	.525		73.50	24.50	16.75	114.75	136
6745	8'		59	.542		96.50	25.50	17.30	139.30	163
7031	For corner, end & pull post bracing, add					20%	15%			
7770	Gates, sliding w/overhead support, 4' high	B-80B	35	.914	L.F.	142	41	6.75	189.75	226
7775	5' high		32	1		160	45	7.40	212.40	251
7780	6' high		28	1.143		150	51	8.45	209.45	251
7785	7' high		25	1.280		185	57.50	9.45	251.95	300
7790	8' high		23	1.391		193	62.50	10.25	265.75	320
7795	Cantilever, manual, exp. roller (pr), 40' wide x 8' high	B-22	1	30	Ea.	5,950	1,475	284	7,709	9,075
7800	30' wide x 8' high		1	30		4,575	1,475	284	6,334	7,550
7805	24' wide x 8' high		1	30		3,075	1,475	284	4,834	5,900
7900	Auger fence post hole, 3' deep, medium soil, by hand	1 Clab	30	.267			11.25		11.25	16.85
7925	By machine	B-80	175	.183			8.50	5.85	14.35	19.20
7950	Rock, with jackhammer	B-9	32	1.250			53	11.15	64.15	92.50
7975	With rock drill	B-47C	65	.246			11.70	25.50	37.20	46

32 31 13.33 Chain Link Backstops

		Crew	Daily Output	Labor-Hours	Unit	Material	2020 Bare Costs Labor	Equipment	Total	Total Incl O&P
0010	**CHAIN LINK BACKSTOPS**									
0015	Backstops, baseball, prefabricated, 30' wide, 12' high & 1 overhang	B-1	1	24	Ea.	2,775	1,025		3,800	4,625
0100	40' wide, 12' high & 2 overhangs	"	.75	32		7,075	1,375		8,450	9,825
0300	Basketball, steel, single goal	B-13	3.04	18.421		1,775	850	191	2,816	3,425
0400	Double goal	"	1.92	29.167		2,550	1,350	300	4,200	5,150
0600	Tennis, wire mesh with pair of ends	B-1	2.48	9.677	Set	2,775	415		3,190	3,675
0700	Enclosed court	"	1.30	18.462	Ea.	9,200	790		9,990	11,300

32 31 13.40 Fence, Fabric and Accessories

		Crew	Daily Output	Labor-Hours	Unit	Material	2020 Bare Costs Labor	Equipment	Total	Total Incl O&P
0010	**FENCE, FABRIC & ACCESSORIES**									
1000	Fabric, 9 ga., galv., 1.2 oz. coat, 2" chain link, 4'	B-80A	304	.079	L.F.	3.56	3.32	2.70	9.58	11.90
1150	5'		285	.084		4.30	3.55	2.88	10.73	13.25
1200	6'		266	.090		9.20	3.80	3.08	16.08	19.25
1250	7'		247	.097		10.05	4.09	3.32	17.46	21
1300	8'		228	.105		12.80	4.43	3.60	20.83	24.50
1400	9 ga., fused, 4'		304	.079		4.25	3.32	2.70	10.27	12.65
1450	5'		285	.084		4.81	3.55	2.88	11.24	13.80
1500	6'		266	.090		4.80	3.80	3.08	11.68	14.40
1550	7'		247	.097		6.05	4.09	3.32	13.46	16.50
1600	8'		228	.105		11	4.43	3.60	19.03	22.50

388

32 31 Fences and Gates

32 31 13 – Chain Link Fences and Gates

	32 31 13.40 Fence, Fabric and Accessories	Crew	Daily Output	Labor-Hours	Unit	Material	2020 Bare Costs Labor	Equipment	Total	Total Incl O&P
1650	Barbed wire, galv., cost per strand	B-80A	2280	.011	L.F.	.15	.44	.36	.95	1.24
1700	Vinyl coated		2280	.011	↓	.17	.44	.36	.97	1.26
1750	Extension arms, 3 strands		143	.168	Ea.	4.18	7.05	5.75	16.98	21.50
1800	6 strands, 2-3/8"		119	.202		11.15	8.50	6.90	26.55	32.50
1850	Eye tops, 2-3/8"		143	.168	↓	1.73	7.05	5.75	14.53	18.80
1900	Top rail, incl. tie wires, 1-5/8", galv.		912	.026	L.F.	4.86	1.11	.90	6.87	8
1950	Vinyl coated		912	.026		6.35	1.11	.90	8.36	9.60
2100	Rail, middle/bottom, w/tie wire, 1-5/8", galv.		912	.026		5.15	1.11	.90	7.16	8.35
2150	Vinyl coated		912	.026		5.45	1.11	.90	7.46	8.65
2200	Reinforcing wire, coiled spring, 7 ga. galv.		2279	.011		.11	.44	.36	.91	1.19
2250	9 ga., vinyl coated		2282	.011	↓	.58	.44	.36	1.38	1.71
2300	Steel T-post, galvanized with clips, 5', common earth, flat		200	.120	Ea.	10.05	5.05	4.10	19.20	23
2310	Clay		176	.136		10.05	5.75	4.66	20.46	25
2320	Soil & rock		144	.167		10.05	7	5.70	22.75	28
2330	5-1/2', common earth, flat		200	.120		10.70	5.05	4.10	19.85	24
2340	Clay		176	.136		10.70	5.75	4.66	21.11	25.50
2350	Soil & rock		144	.167		10.70	7	5.70	23.40	28.50
2360	6', common earth, flat		200	.120		10.75	5.05	4.10	19.90	24
2370	Clay		176	.136		10.75	5.75	4.66	21.16	25.50
2375	Soil & rock		144	.167		10.75	7	5.70	23.45	28.50
2385	6-1/2', common earth, flat		200	.120		12.65	5.05	4.10	21.80	26
2390	Clay		176	.136		12.65	5.75	4.66	23.06	27.50
2395	Soil & rock		144	.167		12.65	7	5.70	25.35	31
2400	7', common earth, flat		184	.130		12.40	5.50	4.46	22.36	27
2410	Clay		160	.150		12.40	6.30	5.15	23.85	29
2420	Soil & rock		128	.188		12.40	7.90	6.40	26.70	32.50
2430	8', common earth, flat		184	.130		14.05	5.50	4.46	24.01	28.50
2440	Clay		160	.150		14.05	6.30	5.15	25.50	30.50
2450	Soil & rock		128	.188		14.05	7.90	6.40	28.35	34.50
2460	9', common earth, flat		184	.130		17.10	5.50	4.46	27.06	32
2470	Clay		160	.150		17.10	6.30	5.15	28.55	34
2480	Soil & rock		128	.188		17.10	7.90	6.40	31.40	38
2485	10', common earth, flat		168	.143		17.30	6	4.88	28.18	33.50
2490	Clay		144	.167		17.30	7	5.70	30	36
2495	Soil & rock		112	.214		17.30	9	7.30	33.60	40.50
2500	11', common earth, flat		168	.143		20.50	6	4.88	31.38	37
2510	Clay		144	.167		20.50	7	5.70	33.20	39.50
2530	Soil & rock		112	.214		20.50	9	7.30	36.80	44
2540	12', common earth, flat		168	.143		21	6	4.88	31.88	38
2550	Clay		144	.167		21	7	5.70	33.70	40.50
2560	Soil & rock		112	.214		21	9	7.30	37.30	45
2600	Steel T-post, galvanized with clips, 5', common earth, hills		180	.133		10.05	5.60	4.56	20.21	24.50
2610	Clay		160	.150		10.05	6.30	5.15	21.50	26
2620	Soil & rock		130	.185		10.05	7.75	6.30	24.10	29.50
2630	5-1/2', common earth, hills		180	.133		10.70	5.60	4.56	20.86	25.50
2640	Clay		160	.150		10.70	6.30	5.15	22.15	27
2650	Soil & rock		130	.185		10.70	7.75	6.30	24.75	30.50
2660	6', common earth, hills		180	.133		10.75	5.60	4.56	20.91	25.50
2670	Clay		160	.150		10.75	6.30	5.15	22.20	27
2675	Soil & rock		130	.185		10.75	7.75	6.30	24.80	30.50
2685	6-1/2', common earth, hills		180	.133		12.65	5.60	4.56	22.81	27.50
2690	Clay		160	.150		12.65	6.30	5.15	24.10	29
2695	Soil & rock	↓	130	.185	↓	12.65	7.75	6.30	26.70	32.50

32 31 Fences and Gates

32 31 13 – Chain Link Fences and Gates

32 31 13.40 Fence, Fabric and Accessories

		Crew	Daily Output	Labor-Hours	Unit	Material	2020 Bare Costs Labor	Equipment	Total	Total Incl O&P
2700	7', common earth, hills	B-80A	166	.145	Ea.	12.40	6.10	4.94	23.44	28.50
2710	Clay		144	.167		12.40	7	5.70	25.10	30.50
2720	Soil & rock		116	.207		12.40	8.70	7.05	28.15	34.50
2730	8', common earth, hills		166	.145		14.05	6.10	4.94	25.09	30
2740	Clay		144	.167		14.05	7	5.70	26.75	32.50
2750	Soil & rock		116	.207		14.05	8.70	7.05	29.80	36.50
2760	9', common earth, hills		166	.145		17.10	6.10	4.94	28.14	33.50
2770	Clay		144	.167		17.10	7	5.70	29.80	35.50
2780	Soil & rock		116	.207		17.10	8.70	7.05	32.85	40
2785	10', common earth, hills		152	.158		17.30	6.65	5.40	29.35	35
2790	Clay		130	.185		17.30	7.75	6.30	31.35	37.50
2795	Soil & rock		101	.238		17.30	10	8.10	35.40	43
2800	11', common earth, hills		152	.158		20.50	6.65	5.40	32.55	38.50
2810	Clay		130	.185		20.50	7.75	6.30	34.55	41
2830	Soil & rock		101	.238		20.50	10	8.10	38.60	46.50
2840	12', common earth, hills		152	.158		21	6.65	5.40	33.05	39.50
2850	Clay		130	.185		21	7.75	6.30	35.05	42
2860	Soil & rock		101	.238		21	10	8.10	39.10	47.50

32 31 13.53 High-Security Chain Link Fences, Gates and Sys.

		Crew	Daily Output	Labor-Hours	Unit	Material	2020 Bare Costs Labor	Equipment	Total	Total Incl O&P
0010	**HIGH-SECURITY CHAIN LINK FENCES, GATES AND SYSTEMS**									
0100	Fence, chain link, security, 7' H, standard FE-7, incl. excavation & posts	B-80C	480	.050	L.F.	46.50	2.19	.52	49.21	55
0200	Fence, barbed wire, security, 7' H, with 3 wire barbed wire arm	"	400	.060	"	8.20	2.63	.63	11.46	13.65
0300	Complete systems, including material and installation									
0310	Taunt wire fence detection system				M.L.F.				25,100	27,600
0410	Microwave fence detection system								41,300	45,400
0510	Passive magnetic fence detection system								19,500	21,400
0610	Infrared fence detection system								12,900	14,400
0710	Strain relief fence detection system								25,100	27,600
0810	Electro-shock fence detection system								35,900	39,500
0910	Photo-electric fence detection system								16,300	18,000

32 31 13.64 Chain Link Terminal Post

		Crew	Daily Output	Labor-Hours	Unit	Material	2020 Bare Costs Labor	Equipment	Total	Total Incl O&P
0010	**CHAIN LINK TERMINAL POST**									
0110	16 ga., steel, 2-1/2" x 6' x 0.065 wall, incl. post cap, excavation	B-80C	80	.300	Ea.	13.45	13.15	3.13	29.73	38
0120	2-1/2" x 7'-6" x 0.065 wall		80	.300		17.70	13.15	3.13	33.98	42.50
0130	2-1/2" x 8'-6" x 0.095 wall		80	.300		26	13.15	3.13	42.28	51.50
0210	16 ga., steel, 2-1/2" x 6' x 0.065 wall, incl. floor flange	B-80A	80	.300		32.50	12.65	10.25	55.40	66.50
0220	2-1/2" x 8' x 0.065 wall		80	.300		37	12.65	10.25	59.90	71
0230	4" x 10' x 0.160 wall		80	.300		118	12.65	10.25	140.90	160
0240	4" x 12' x 0.160 wall		80	.300		130	12.65	10.25	152.90	173
0310	16 ga., steel, 4" x 11' x 0.226 wall, incl. post cap, excavation	B-80C	80	.300		192	13.15	3.13	208.28	234
0320	4" x 13'-6" x 0.226 wall		80	.300		223	13.15	3.13	239.28	269
0330	4" x 21' x 0.226 wall		80	.300		305	13.15	3.13	321.28	360

32 31 13.65 Chain Link Line Post

		Crew	Daily Output	Labor-Hours	Unit	Material	2020 Bare Costs Labor	Equipment	Total	Total Incl O&P
0010	**CHAIN LINK LINE POST**									
0110	16 ga., steel, 1-5/8" x 6' x 0.065 wall, incl. post cap and excavation	B-80C	80	.300	Ea.	9.50	13.15	3.13	25.78	33.50
0120	1-5/8" x 7'-6" x 0.065 wall		80	.300		12.40	13.15	3.13	28.68	37
0130	2" x 8'-6" x 0.095 wall		80	.300		19.75	13.15	3.13	36.03	44.50
0210	16 ga., steel, 2" x 6' x 0.065 wall, incl. post top and floor flange	B-80A	80	.300		28.50	12.65	10.25	51.40	62
0220	2" x 8' x 0.065 wall		80	.300		32	12.65	10.25	54.90	65.50
0230	3" x 10' x 0.160 wall		80	.300		82.50	12.65	10.25	105.40	121
0240	3" x 12' x 0.160 wall		80	.300		94	12.65	10.25	116.90	133
0410	16 ga., steel, 3" x 11' x 0.203 wall, incl. post cap and excavation	B-80C	80	.300		122	13.15	3.13	138.28	158

32 31 Fences and Gates

32 31 13 – Chain Link Fences and Gates

		Crew	Daily Output	Labor-Hours	Unit	Material	2020 Bare Costs Labor	Equipment	Total	Total Incl O&P
32 31 13.65 Chain Link Line Post										
0420	3" x 13'-6" x 0.203 wall	B-80C	80	.300	Ea.	143	13.15	3.13	159.28	180
0430	3" x 21' x 0.203 wall	↓	80	.300	↓	186	13.15	3.13	202.28	227
32 31 13.66 Chain Link Top Rail										
0010	**CHAIN LINK TOP RAIL**									
0110	Fence, rail tubing, 16 ga., 1-3/8" x 21' x 0.065 wall incl. hardware	B-80A	266	.090	L.F.	1.18	3.80	3.08	8.06	10.40
0120	1-5/8" x 21' x 0.065 wall, swedge		266	.090		1.49	3.80	3.08	8.37	10.75
0130	1-5/8" x 21' x 0.111 wall		266	.090		2.25	3.80	3.08	9.13	11.55
0140	2" x 18' x 0.145 wall	↓	266	.090	↓	4.88	3.80	3.08	11.76	14.45
32 31 13.68 Chain Link Fabric										
0010	**CHAIN LINK FABRIC**									
0110	Fence, fabric, steel, galv., 11-1/2 ga., 2-1/4" mesh, 4' H, incl. hardware	B-80A	266	.090	L.F.	2.60	3.80	3.08	9.48	11.95
0120	5' H		266	.090		2.95	3.80	3.08	9.83	12.35
0130	6' H		266	.090		3.24	3.80	3.08	10.12	12.65
0210	Fence, fabric, steel, galv., 11 ga., 2" mesh, 6' H, incl. hardware		266	.090		3.95	3.80	3.08	10.83	13.45
0220	8' H		266	.090		5.20	3.80	3.08	12.08	14.80
0230	10' H		266	.090		9.75	3.80	3.08	16.63	19.85
0240	12' H		266	.090		11.55	3.80	3.08	18.43	22
0310	Fence, fabric, steel, galv., 9 ga., 2" mesh, 8' H, incl. hardware		266	.090		5.25	3.80	3.08	12.13	14.85
0320	10' H		266	.090		6.10	3.80	3.08	12.98	15.80
0330	12' H		266	.090		7.15	3.80	3.08	14.03	16.95
0340	14' H		266	.090		7.25	3.80	3.08	14.13	17.05
0410	Fence, fabric, steel, galv., 9 ga., 1" mesh, 8' H, incl. hardware		266	.090		11.40	3.80	3.08	18.28	21.50
0420	10' H		266	.090		18.35	3.80	3.08	25.23	29
0430	12' H		266	.090		22	3.80	3.08	28.88	33
0440	14' H	↓	266	.090	↓	25.50	3.80	3.08	32.38	37
32 31 13.70 Chain Link Barbed Wire										
0010	**CHAIN LINK BARBED WIRE**									
0110	Fence, barbed wire, 12-1/2 ga., 1320' roll incl. barb arm and brace band	B-80A	266	.090	L.F.	.63	3.80	3.08	7.51	9.80
32 31 13.80 Residential Chain Link Gate										
0010	**RESIDENTIAL CHAIN LINK GATE**									
0110	Residential 4' gate, single incl. hardware and concrete	B-80C	10	2.400	Ea.	128	105	25	258	325
0120	5'		10	2.400		138	105	25	268	340
0130	6'		10	2.400		149	105	25	279	350
0510	Residential 4' gate, double incl. hardware and concrete		10	2.400		228	105	25	358	435
0520	5'		10	2.400		244	105	25	374	455
0530	6'	↓	10	2.400	↓	284	105	25	414	495
32 31 13.82 Internal Chain Link Gate										
0010	**INTERNAL CHAIN LINK GATE**									
0110	Internal 6' gate, single incl. post flange, hardware and concrete	B-80C	10	2.400	Ea.	279	105	25	409	490
0120	8'		10	2.400		320	105	25	450	535
0130	10'		10	2.400		435	105	25	565	665
0510	Internal 6' gate, double incl. post flange, hardware and concrete		10	2.400		510	105	25	640	745
0520	8'		10	2.400		585	105	25	715	830
0530	10'	↓	10	2.400	↓	740	105	25	870	1,000
32 31 13.84 Industrial Chain Link Gate										
0010	**INDUSTRIAL CHAIN LINK GATE**									
0110	Industrial 8' gate, single incl. hardware and concrete	B-80C	10	2.400	Ea.	475	105	25	605	710
0120	10'		10	2.400		540	105	25	670	780
0510	Industrial 8' gate, double incl. hardware and concrete		10	2.400		740	105	25	870	1,000
0520	10'	↓	10	2.400	↓	845	105	25	975	1,125

32 31 Fences and Gates

32 31 13 – Chain Link Fences and Gates

32 31 13.88 Chain Link Transom

		Crew	Daily Output	Labor-Hours	Unit	Material	2020 Bare Costs Labor	2020 Bare Costs Equipment	Total	Total Incl O&P
0010	**CHAIN LINK TRANSOM**									
0110	Add for, single transom, 3' wide, incl. components & hardware	B-80C	10	2.400	Ea.	118	105	25	248	315
0120	Add for, double transom, 6' wide, incl. components & hardware	"	10	2.400	"	127	105	25	257	325

32 31 19 – Decorative Metal Fences and Gates

32 31 19.10 Decorative Fence

		Crew	Daily Output	Labor-Hours	Unit	Material	2020 Bare Costs Labor	2020 Bare Costs Equipment	Total	Total Incl O&P
0010	**DECORATIVE FENCE**									
5300	Tubular picket, steel, 6' sections, 1-9/16" posts, 4' high	B-80C	300	.080	L.F.	37	3.51	.83	41.34	47
5400	2" posts, 5' high		240	.100		44	4.38	1.04	49.42	56
5600	2" posts, 6' high		200	.120		53	5.25	1.25	59.50	68
5700	Staggered picket, 1-9/16" posts, 4' high		300	.080		39.50	3.51	.83	43.84	49.50
5800	2" posts, 5' high		240	.100		45	4.38	1.04	50.42	57.50
5900	2" posts, 6' high	↓	200	.120	↓	53	5.25	1.25	59.50	68
6200	Gates, 4' high, 3' wide	B-1	10	2.400	Ea.	335	103		438	525
6300	5' high, 3' wide		10	2.400		245	103		348	425
6400	6' high, 3' wide		10	2.400		288	103		391	470
6500	4' wide	↓	10	2.400	↓	345	103		448	535

32 31 23 – Plastic Fences and Gates

32 31 23.10 Fence, Vinyl

		Crew	Daily Output	Labor-Hours	Unit	Material	2020 Bare Costs Labor	2020 Bare Costs Equipment	Total	Total Incl O&P
0010	**FENCE, VINYL**									
0011	White, steel reinforced, stainless steel fasteners									
0020	Picket, 4" x 4" posts @ 6'-0 OC, 3' high	B-1	140	.171	L.F.	24.50	7.35		31.85	38
0030	4' high		130	.185		26.50	7.90		34.40	41.50
0040	5' high		120	.200		34	8.55		42.55	50.50
0100	Board (semi-privacy), 5" x 5" posts @ 7'-6" OC, 5' high		130	.185		27.50	7.90		35.40	42
0120	6' high		125	.192		30	8.20		38.20	45.50
0200	Basket weave, 5" x 5" posts @ 7'-6" OC, 5' high		160	.150		25.50	6.40		31.90	38
0220	6' high		150	.160		29	6.85		35.85	42.50
0300	Privacy, 5" x 5" posts @ 7'-6" OC, 5' high		130	.185		26.50	7.90		34.40	41
0320	6' high		150	.160	↓	30	6.85		36.85	43.50
0350	Gate, 5' high		9	2.667	Ea.	330	114		444	530
0360	6' high		9	2.667		375	114		489	580
0400	For posts set in concrete, add		25	.960	↓	9.30	41		50.30	71.50
0500	Post and rail fence, 2 rail		150	.160	L.F.	6.25	6.85		13.10	17.20
0510	3 rail		150	.160		8.10	6.85		14.95	19.20
0515	4 rail	↓	150	.160	↓	11.05	6.85		17.90	22.50

32 31 23.20 Fence, Recycled Plastic

			Crew	Daily Output	Labor-Hours	Unit	Material	2020 Bare Costs Labor	2020 Bare Costs Equipment	Total	Total Incl O&P
0010	**FENCE, RECYCLED PLASTIC**										
9015	Fence rail, made from recycled plastic, various colors, 2 rail	G	B-1	150	.160	L.F.	9.45	6.85		16.30	20.50
9018	3 rail	G		150	.160		12.15	6.85		19	23.50
9020	4 rail	G		150	.160	↓	15.85	6.85		22.70	27.50
9030	Fence pole, made from recycled plastic, various colors, 7'	G	↓	96	.250	Ea.	50.50	10.70		61.20	71.50
9040	Stockade fence, made from recycled plastic, various colors, 4' high	G	B-80C	160	.150	L.F.	35.50	6.55	1.56	43.61	50.50
9050	6' high	G		160	.150	"	36.50	6.55	1.56	44.61	52
9060	6' pole	G		96	.250	Ea.	39	10.95	2.61	52.56	62.50
9070	9' pole	G		96	.250	"	49	10.95	2.61	62.56	73
9080	Picket fence, made from recycled plastic, various colors, 3' high	G		160	.150	L.F.	24	6.55	1.56	32.11	38
9090	4' high	G		160	.150	"	35	6.55	1.56	43.11	50
9100	3' high gate	G		8	3	Ea.	26.50	131	31.50	189	261
9110	4' high gate	G		8	3		34.50	131	31.50	197	269
9120	5' high pole	G		96	.250		36.50	10.95	2.61	50.06	60
9130	6' high pole	G	↓	96	.250		42	10.95	2.61	55.56	66

32 31 Fences and Gates

32 31 23 – Plastic Fences and Gates

32 31 23.20 Fence, Recycled Plastic

		Crew	Daily Output	Labor-Hours	Unit	Material	2020 Bare Costs Labor	2020 Bare Costs Equipment	Total	Total Incl O&P
9140	Pole cap only [G]				Ea.	6.05			6.05	6.65
9150	Keeper pins only [G]					.43			.43	.47

32 31 26 – Wire Fences and Gates

32 31 26.10 Fences, Misc. Metal

		Crew	Daily Output	Labor-Hours	Unit	Material	Labor	Equipment	Total	Total Incl O&P
0010	**FENCES, MISC. METAL**									
0012	Chicken wire, posts @ 4', 1" mesh, 4' high	B-80C	410	.059	L.F.	3.53	2.57	.61	6.71	8.40
0100	2" mesh, 6' high		350	.069		4.01	3	.71	7.72	9.70
0200	Galv. steel, 12 ga., 2" x 4" mesh, posts 5' OC, 3' high		300	.080		2.80	3.51	.83	7.14	9.25
0300	5' high		300	.080		3.37	3.51	.83	7.71	9.90
0400	14 ga., 1" x 2" mesh, 3' high		300	.080		3.48	3.51	.83	7.82	10
0500	5' high		300	.080		4.50	3.51	.83	8.84	11.10
1000	Kennel fencing, 1-1/2" mesh, 6' long, 3'-6" wide, 6'-2" high	2 Clab	4	4	Ea.	495	168		663	800
1050	12' long		4	4		700	168		868	1,025
1200	Top covers, 1-1/2" mesh, 6' long		15	1.067		141	45		186	223
1250	12' long		12	1.333		210	56		266	315
4492	Security fence, prison grade, barbed wire, set in concrete, 10' high	B-80	22	1.455	L.F.	64	68	46.50	178.50	224
4494	Security fence, prison grade, razor wire, set in concrete, 10' high		18	1.778		65.50	83	56.50	205	259
4500	Security fence, prison grade, set in concrete, 12' high		25	1.280		60	59.50	41	160.50	201
4600	16' high		20	1.600		79.50	74.50	51	205	256
4990	Security fence, prison grade, set in concrete, 10' high		25	1.280		60	59.50	41	160.50	201

32 31 26.20 Wire Fencing, General

		Crew	Daily Output	Labor-Hours	Unit	Material	Labor	Equipment	Total	Total Incl O&P
0010	**WIRE FENCING, GENERAL**									
0015	Barbed wire, galvanized, domestic steel, hi-tensile 15-1/2 ga.				M.L.F.	149			149	164
0020	Standard, 12-3/4 ga.					164			164	180
0210	Barbless wire, 2-strand galvanized, 12-1/2 ga.					164			164	180
0500	Helical razor ribbon, stainless steel, 18" diam. x 18" spacing				C.L.F.	195			195	215
0600	Hardware cloth galv., 1/4" mesh, 23 ga., 2' wide				C.S.F.	43.50			43.50	48
0700	3' wide					27.50			27.50	30.50
0900	1/2" mesh, 19 ga., 2' wide					36.50			36.50	40
1000	4' wide					50			50	55
1200	Chain link fabric, steel, 2" mesh, 6 ga., galvanized					61.50			61.50	67.50
1300	9 ga., galvanized					60			60	66
1350	Vinyl coated					32			32	35
1360	Aluminized					228			228	251
1400	2-1/4" mesh, 11-1/2 ga., galvanized					48			48	53
1600	1-3/4" mesh (tennis courts), 11-1/2 ga. (core), vinyl coated					56			56	61.50
1700	9 ga., galvanized					91			91	100
2100	Welded wire fabric, galvanized, 1" x 2", 14 ga.	2 Carp	1600	.010	S.F.	.67	.53		1.20	1.54
2200	2" x 4", 12-1/2 ga.				C.S.F.	39			39	43

32 31 29 – Wood Fences and Gates

32 31 29.10 Fence, Wood

		Crew	Daily Output	Labor-Hours	Unit	Material	Labor	Equipment	Total	Total Incl O&P
0010	**FENCE, WOOD**									
0011	Basket weave, 3/8" x 4" boards, 2" x 4"									
0020	stringers on spreaders, 4" x 4" posts									
0050	No. 1 cedar, 6' high	B-80C	160	.150	L.F.	26	6.55	1.56	34.11	40
0070	Treated pine, 6' high	"	150	.160	"	38	7	1.67	46.67	54
0200	Board fence, 1" x 4" boards, 2" x 4" rails, 4" x 4" post									
0220	Preservative treated, 2 rail, 3' high	B-80C	145	.166	L.F.	11.80	7.25	1.72	20.77	26
0240	4' high		135	.178		13.45	7.80	1.85	23.10	28.50
0260	3 rail, 5' high		130	.185		12.70	8.10	1.92	22.72	28.50
0300	6' high		125	.192		16.30	8.40	2	26.70	33

32 31 29 – Wood Fences and Gates

32 31 29.10 Fence, Wood

32 31 29.10 Fence, Wood	Crew	Daily Output	Labor-Hours	Unit	Material	2020 Bare Costs Labor	Equipment	Total	Total Incl O&P
0320	No. 2 grade western cedar, 2 rail, 3' high	B-80C	145	.166	L.F.	12.75	7.25	1.72	21.72	27
0340	4' high		135	.178		12.40	7.80	1.85	22.05	27.50
0360	3 rail, 5' high		130	.185		14.65	8.10	1.92	24.67	30.50
0400	6' high		125	.192		15.50	8.40	2	25.90	32
0420	No. 1 grade cedar, 2 rail, 3' high		145	.166		13.90	7.25	1.72	22.87	28
0440	4' high		135	.178		15.20	7.80	1.85	24.85	30.50
0460	3 rail, 5' high		130	.185		18.25	8.10	1.92	28.27	34.50
0500	6' high		125	.192		21.50	8.40	2	31.90	39
0540	Shadow box, 1" x 6" board, 2" x 4" rail, 4" x 4" post									
0560	Pine, pressure treated, 3 rail, 6' high	B-80C	150	.160	L.F.	28	7	1.67	36.67	43.50
0600	Gate, 3'-6" wide		8	3	Ea.	136	131	31.50	298.50	380
0620	No. 1 cedar, 3 rail, 4' high		130	.185	L.F.	19.15	8.10	1.92	29.17	35.50
0640	6' high		125	.192		26.50	8.40	2	36.90	44
0860	Open rail fence, split rails, 2 rail, 3' high, no. 1 cedar		160	.150		9.25	6.55	1.56	17.36	21.50
0870	No. 2 cedar		160	.150		8.65	6.55	1.56	16.76	21
0880	3 rail, 4' high, no. 1 cedar		150	.160		13.30	7	1.67	21.97	27
0890	No. 2 cedar		150	.160		9	7	1.67	17.67	22.50
0920	Rustic rails, 2 rail, 3' high, no. 1 cedar		160	.150		13.85	6.55	1.56	21.96	27
0930	No. 2 cedar		160	.150		12.70	6.55	1.56	20.81	25.50
0940	3 rail, 4' high		150	.160		13.85	7	1.67	22.52	27.50
0950	No. 2 cedar		150	.160		8	7	1.67	16.67	21
0960	Picket fence, gothic, pressure treated pine									
1000	2 rail, 3' high	B-80C	140	.171	L.F.	8.90	7.50	1.79	18.19	23
1020	3 rail, 4' high		130	.185	"	10.95	8.10	1.92	20.97	26.50
1040	Gate, 3'-6" wide		9	2.667	Ea.	77.50	117	28	222.50	291
1060	No. 2 cedar, 2 rail, 3' high		140	.171	L.F.	10.30	7.50	1.79	19.59	24.50
1100	3 rail, 4' high		130	.185	"	10.40	8.10	1.92	20.42	25.50
1120	Gate, 3'-6" wide		9	2.667	Ea.	82.50	117	28	227.50	296
1140	No. 1 cedar, 2 rail, 3' high		140	.171	L.F.	14.45	7.50	1.79	23.74	29
1160	3 rail, 4' high		130	.185	"	20.50	8.10	1.92	30.52	37
1170	Gate, 3'-6" wide		9	2.667	Ea.	272	117	28	417	505
1200	Rustic picket, molded pine, 2 rail, 3' high		140	.171	L.F.	9.65	7.50	1.79	18.94	24
1220	No. 1 cedar, 2 rail, 3' high		140	.171		11.60	7.50	1.79	20.89	26
1240	Stockade fence, no. 1 cedar, 3-1/4" rails, 6' high		160	.150		14.05	6.55	1.56	22.16	27
1260	8' high		155	.155		20.50	6.80	1.61	28.91	34.50
1270	Gate, 3'-6" wide		9	2.667	Ea.	277	117	28	422	510
1300	No. 2 cedar, treated wood rails, 6' high		160	.150	L.F.	14.75	6.55	1.56	22.86	28
1320	Gate, 3'-6" wide		8	3	Ea.	95.50	131	31.50	258	335
1360	Treated pine, treated rails, 6' high		160	.150	L.F.	14.85	6.55	1.56	22.96	28
1400	8' high		150	.160	"	20.50	7	1.67	29.17	35
1420	Gate, 3'-6" wide		9	2.667	Ea.	103	117	28	248	320

32 31 29.20 Fence, Wood Rail

32 31 29.20 Fence, Wood Rail	Crew	Daily Output	Labor-Hours	Unit	Material	2020 Bare Costs Labor	Equipment	Total	Total Incl O&P
0010	**FENCE, WOOD RAIL**									
0012	Picket, No. 2 cedar, Gothic, 2 rail, 3' high	B-1	160	.150	L.F.	8.60	6.40		15	19.10
0050	Gate, 3'-6" wide	B-80C	9	2.667	Ea.	75	117	28	220	288
0400	3 rail, 4' high		150	.160	L.F.	9.65	7	1.67	18.32	23
0500	Gate, 3'-6" wide		9	2.667	Ea.	90	117	28	235	305
5000	Fence rail, redwood, 2" x 4", merch. grade, 8'	B-1	2400	.010	L.F.	2.76	.43		3.19	3.68
5050	Select grade, 8'		2400	.010	"	6.10	.43		6.53	7.40
6000	Fence post, select redwood, earth packed & treated, 4" x 4" x 6'		96	.250	Ea.	14.70	10.70		25.40	32.50
6010	4" x 4" x 8'		96	.250		22	10.70		32.70	40
6020	Set in concrete, 4" x 4" x 6'		50	.480		23	20.50		43.50	56.50

32 31 Fences and Gates

32 31 29 – Wood Fences and Gates

32 31 29.20 Fence, Wood Rail

	Crew	Daily Output	Labor-Hours	Unit	Material	2020 Bare Costs Labor	Equipment	Total	Total Incl O&P
6030	4" x 4" x 8'	B-1	50	.480	Ea.	24.50	20.50		45	58
6040	Wood post, 4' high, set in concrete, incl. concrete		50	.480		15.50	20.50		36	48
6050	Earth packed		96	.250		17.40	10.70		28.10	35
6060	6' high, set in concrete, incl. concrete		50	.480		20	20.50		40.50	53
6070	Earth packed		96	.250		13.20	10.70		23.90	30.50

32 32 Retaining Walls

32 32 13 – Cast-in-Place Concrete Retaining Walls

32 32 13.10 Retaining Walls, Cast Concrete

		Crew	Daily Output	Labor-Hours	Unit	Material	2020 Bare Costs Labor	Equipment	Total	Total Incl O&P
0010	**RETAINING WALLS, CAST CONCRETE**									
1800	Concrete gravity wall with vertical face including excavation & backfill									
1850	No reinforcing									
1900	6' high, level embankment	C-17C	36	2.306	L.F.	82.50	128	15.45	225.95	300
2000	33° slope embankment		32	2.594		109	144	17.40	270.40	355
2200	8' high, no surcharge		27	3.074		117	170	20.50	307.50	410
2300	33° slope embankment		24	3.458		142	192	23	357	470
2500	10' high, level embankment		19	4.368		167	242	29.50	438.50	580
2600	33° slope embankment		18	4.611		232	255	31	518	675
2800	Reinforced concrete cantilever, incl. excavation, backfill & reinf.									
2900	6' high, 33° slope embankment	C-17C	35	2.371	L.F.	86	131	15.90	232.90	310
3000	8' high, 33° slope embankment		29	2.862		99	159	19.20	277.20	370
3100	10' high, 33° slope embankment		20	4.150		129	230	28	387	525
3200	20' high, 500 lb./L.F. surcharge		7.50	11.067		385	615	74.50	1,074.50	1,425
3500	Concrete cribbing, incl. excavation and backfill									
3700	12' high, open face	B-13	210	.267	S.F.	43	12.30	2.77	58.07	68.50
3900	Closed face	"	210	.267	"	40.50	12.30	2.77	55.57	66
4100	Concrete filled slurry trench, see Section 31 56 23.20									

32 32 23 – Segmental Retaining Walls

32 32 23.13 Segmental Conc. Unit Masonry Retaining Walls

		Crew	Daily Output	Labor-Hours	Unit	Material	2020 Bare Costs Labor	Equipment	Total	Total Incl O&P
0010	**SEGMENTAL CONC. UNIT MASONRY RETAINING WALLS**									
7100	Segmental retaining wall system, incl. pins and void fill									
7120	base and backfill not included									
7140	Large unit, 8" high x 18" wide x 20" deep, 3 plane split	B-62	300	.080	S.F.	14.70	3.66	.59	18.95	22.50
7150	Straight split		300	.080		14.60	3.66	.59	18.85	22.50
7160	Medium, lt. wt., 8" high x 18" wide x 12" deep, 3 plane split		400	.060		7.50	2.74	.44	10.68	12.85
7170	Straight split		400	.060		10.60	2.74	.44	13.78	16.30
7180	Small unit, 4" x 18" x 10" deep, 3 plane split		400	.060		17.25	2.74	.44	20.43	23.50
7190	Straight split		400	.060		12.45	2.74	.44	15.63	18.30
7200	Cap unit, 3 plane split		300	.080		14.70	3.66	.59	18.95	22.50
7210	Cap unit, straight split		300	.080		14.70	3.66	.59	18.95	22.50
7250	Geo-grid soil reinforcement 4' x 50'	2 Clab	22500	.001		.80	.03		.83	.92
7255	Geo-grid soil reinforcement 6' x 150'	"	22500	.001		.62	.03		.65	.72

32 32 26 – Metal Crib Retaining Walls

32 32 26.10 Metal Bin Retaining Walls

		Crew	Daily Output	Labor-Hours	Unit	Material	2020 Bare Costs Labor	Equipment	Total	Total Incl O&P
0010	**METAL BIN RETAINING WALLS**									
0011	Aluminized steel bin, excavation									
0020	and backfill not included, 10' wide									
0100	4' high, 5.5' deep	B-13	650	.086	S.F.	29.50	3.97	.89	34.36	39
0200	8' high, 5.5' deep		615	.091		33.50	4.19	.94	38.63	44.50
0300	10' high, 7.7' deep		580	.097		37.50	4.45	1	42.95	49

395

32 32 26 – Metal Crib Retaining Walls

32 32 26.10 Metal Bin Retaining Walls

		Crew	Daily Output	Labor-Hours	Unit	Material	2020 Bare Costs Labor	Equipment	Total	Total Incl O&P
0400	12' high, 7.7' deep	B-13	530	.106	S.F.	40.50	4.86	1.10	46.46	53
0500	16' high, 7.7' deep		515	.109		42.50	5	1.13	48.63	55
0600	16' high, 9.9' deep		500	.112		45.50	5.15	1.16	51.81	59
0700	20' high, 9.9' deep		470	.119		51	5.50	1.24	57.74	65.50
0800	20' high, 12.1' deep		460	.122		45	5.60	1.26	51.86	59.50
0900	24' high, 12.1' deep		455	.123		47.50	5.65	1.28	54.43	62.50
1000	24' high, 14.3' deep		450	.124		57.50	5.75	1.29	64.54	73.50
1100	28' high, 14.3' deep	▼	440	.127	▼	60	5.85	1.32	67.17	76.50
1300	For plain galvanized bin type walls, deduct					10%				

32 32 29 – Timber Retaining Walls

32 32 29.10 Landscape Timber Retaining Walls

		Crew	Daily Output	Labor-Hours	Unit	Material	2020 Bare Costs Labor	Equipment	Total	Total Incl O&P
0010	**LANDSCAPE TIMBER RETAINING WALLS**									
0100	Treated timbers, 6" x 6"	1 Clab	265	.030	L.F.	2.76	1.27		4.03	4.95
0110	6" x 8'	"	200	.040	"	6.65	1.68		8.33	9.85
0120	Drilling holes in timbers for fastening, 1/2"	1 Carp	450	.018	Inch		.95		.95	1.42
0130	5/8"	"	450	.018	"		.95		.95	1.42
0140	Reinforcing rods for fastening, 1/2"	1 Clab	312	.026	L.F.	.41	1.08		1.49	2.08
0150	5/8"	"	312	.026	"	.65	1.08		1.73	2.33
0160	Reinforcing fabric	2 Clab	2500	.006	S.Y.	2.22	.27		2.49	2.84
0170	Gravel backfill		28	.571	C.Y.	16.90	24		40.90	54.50
0180	Perforated pipe, 4" diameter with silt sock	▼	1200	.013	L.F.	1.05	.56		1.61	2
0190	Galvanized 60d common nails	1 Clab	625	.013	Ea.	.21	.54		.75	1.04
0200	20d common nails	"	3800	.002	"	.05	.09		.14	.18

32 32 36 – Gabion Retaining Walls

32 32 36.10 Stone Gabion Retaining Walls

		Crew	Daily Output	Labor-Hours	Unit	Material	2020 Bare Costs Labor	Equipment	Total	Total Incl O&P
0010	**STONE GABION RETAINING WALLS**									
4300	Stone filled gabions, not incl. excavation,									
4310	Stone, delivered, 3' wide									
4340	Galvanized, 6' long, 1' high	B-13	113	.496	Ea.	84	23	5.15	112.15	132
4400	1'-6" high		50	1.120		105	51.50	11.60	168.10	205
4490	3'-0" high		13	4.308		173	198	44.50	415.50	535
4590	9' long, 1' high		50	1.120		195	51.50	11.60	258.10	305
4650	1'-6" high		22	2.545		146	117	26.50	289.50	365
4690	3'-0" high		6	9.333		370	430	97	897	1,150
4890	12' long, 1' high		28	2		165	92	20.50	277.50	340
4950	1'-6" high		13	4.308		195	198	44.50	437.50	560
4990	3'-0" high		3	18.667		435	860	194	1,489	2,000
5200	PVC coated, 6' long, 1' high		113	.496		95	23	5.15	123.15	145
5250	1'-6" high		50	1.120		116	51.50	11.60	179.10	217
5300	3' high		13	4.308		173	198	44.50	415.50	535
5500	9' long, 1' high		50	1.120		124	51.50	11.60	187.10	227
5550	1'-6" high		22	2.545		153	117	26.50	296.50	375
5600	3' high		6	9.333		273	430	97	800	1,050
5800	12' long, 1' high		28	2		194	92	20.50	306.50	375
5850	1'-6" high		13	4.308		232	198	44.50	474.50	600
5900	3' high		3	18.667	▼	330	860	194	1,384	1,875
6000	Galvanized, 6' long, 1' high		75	.747	C.Y.	77.50	34.50	7.75	119.75	146
6010	1'-6" high		50	1.120		105	51.50	11.60	168.10	205
6020	3'-0" high		25	2.240		86.50	103	23	212.50	276
6030	9' long, 1' high		50	1.120		195	51.50	11.60	258.10	305
6040	1'-6" high		33.30	1.682		97.50	77.50	17.45	192.45	242
6050	3'-0" high	▼	16.70	3.353	▼	123	154	35	312	405

396

For customer support on your Site Work & Landscape Costs with RSMeans data, call 800.448.8182.

32 32 Retaining Walls

32 32 36 – Gabion Retaining Walls

32 32 36.10 Stone Gabion Retaining Walls		Crew	Daily Output	Labor-Hours	Unit	Material	2020 Bare Costs Labor	Equipment	Total	Total Incl O&P
6060	12' long, 1' high	B-13	37.50	1.493	C.Y.	123	69	15.50	207.50	256
6070	1'-6" high		25	2.240		97.50	103	23	223.50	288
6080	3'-0" high		12.50	4.480		109	206	46.50	361.50	480
6100	PVC coated, 6' long, 1' high		75	.747		143	34.50	7.75	185.25	217
6110	1'-6" high		50	1.120		116	51.50	11.60	179.10	217
6120	3' high		25	2.240		86.50	103	23	212.50	276
6130	9' long, 1' high		50	1.120		124	51.50	11.60	187.10	227
6140	1'-6" high		33.30	1.682		102	77.50	17.45	196.95	247
6150	3' high		16.67	3.359		91	155	35	281	370
6160	12' long, 1' high		37.50	1.493		145	69	15.50	229.50	280
6170	1'-6" high		25	2.240		104	103	23	230	295
6180	3' high		12.50	4.480		83	206	46.50	335.50	450

32 32 53 – Stone Retaining Walls

32 32 53.10 Retaining Walls, Stone		Crew	Daily Output	Labor-Hours	Unit	Material	2020 Bare Costs Labor	Equipment	Total	Total Incl O&P
0010	**RETAINING WALLS, STONE**									
0015	Including excavation, concrete footing and									
0020	stone 3' below grade. Price is exposed face area.									
0200	Decorative random stone, to 6' high, 1'-6" thick, dry set	D-1	35	.457	S.F.	77	21.50		98.50	118
0300	Mortar set		40	.400		78.50	18.75		97.25	115
0500	Cut stone, to 6' high, 1'-6" thick, dry set		35	.457		80	21.50		101.50	121
0600	Mortar set		40	.400		80.50	18.75		99.25	117
0800	Random stone, 6' to 10' high, 2' thick, dry set		45	.356		80	16.70		96.70	114
0900	Mortar set		50	.320		83	15		98	115
1100	Cut stone, 6' to 10' high, 2' thick, dry set		45	.356		80.50	16.70		97.20	114
1200	Mortar set		50	.320		83.50	15		98.50	115
5100	Setting stone, dry		100	.160	C.F.		7.50		7.50	11.40
5600	With mortar		120	.133	"		6.25		6.25	9.50

32 33 Site Furnishings

32 33 33 – Site Manufactured Planters

32 33 33.10 Planters		Crew	Daily Output	Labor-Hours	Unit	Material	2020 Bare Costs Labor	Equipment	Total	Total Incl O&P
0010	**PLANTERS**									
0012	Concrete, sandblasted, precast, 48" diameter, 24" high	2 Clab	15	1.067	Ea.	640	45		685	775
0100	Fluted, precast, 7' diameter, 36" high		10	1.600		1,625	67.50		1,692.50	1,875
0300	Fiberglass, circular, 36" diameter, 24" high		15	1.067		745	45		790	890
0400	60" diameter, 24" high		10	1.600		1,175	67.50		1,242.50	1,375
0600	Square, 24" side, 36" high		15	1.067		680	45		725	815
0700	48" side, 36" high		15	1.067		1,125	45		1,170	1,300
0900	Planter/bench, 72" square, 36" high		5	3.200		2,950	135		3,085	3,450
1000	96" square, 27" high		5	3.200		3,525	135		3,660	4,075
1200	Wood, square, 48" side, 24" high		15	1.067		1,625	45		1,670	1,850
1300	Circular, 48" diameter, 30" high		10	1.600		1,125	67.50		1,192.50	1,325
1500	72" diameter, 30" high		10	1.600		2,000	67.50		2,067.50	2,300
1600	Planter/bench, 72"		5	3.200		3,850	135		3,985	4,425

32 33 43 – Site Seating and Tables

32 33 43.13 Site Seating		Crew	Daily Output	Labor-Hours	Unit	Material	2020 Bare Costs Labor	Equipment	Total	Total Incl O&P
0010	**SITE SEATING**									
0012	Seating, benches, park, precast conc., w/backs, wood rails, 4' long	2 Clab	5	3.200	Ea.	610	135		745	870
0100	8' long		4	4		1,125	168		1,293	1,500
0300	Fiberglass, without back, one piece, 4' long		10	1.600		1,725	67.50		1,792.50	2,000

32 33 Site Furnishings

32 33 43 – Site Seating and Tables

32 33 43.13 Site Seating		Crew	Daily Output	Labor-Hours	Unit	Material	2020 Bare Costs Labor	Equipment	Total	Total Incl O&P
0400	8' long	2 Clab	7	2.286	Ea.	2,300	96		2,396	2,675
0500	Steel barstock pedestals w/backs, 2" x 3" wood rails, 4' long		10	1.600		1,375	67.50		1,442.50	1,625
0510	8' long		7	2.286		1,725	96		1,821	2,050
0520	3" x 8" wood plank, 4' long		10	1.600		1,525	67.50		1,592.50	1,775
0530	8' long		7	2.286		1,550	96		1,646	1,875
0540	Backless, 4" x 4" wood plank, 4' square		10	1.600		1,275	67.50		1,342.50	1,500
0550	8' long		7	2.286		1,175	96		1,271	1,450
0600	Aluminum pedestals, with backs, aluminum slats, 8' long		8	2		480	84		564	655
0610	15' long		5	3.200		1,025	135		1,160	1,325
0620	Portable, aluminum slats, 8' long		8	2		470	84		554	645
0630	15' long		5	3.200		505	135		640	760
0800	Cast iron pedestals, back & arms, wood slats, 4' long		8	2		470	84		554	645
0820	8' long		5	3.200		1,050	135		1,185	1,350
0840	Backless, wood slats, 4' long		8	2		715	84		799	910
0860	8' long		5	3.200		1,200	135		1,335	1,525
1700	Steel frame, fir seat, 10' long		10	1.600		410	67.50		477.50	550

32 34 Fabricated Bridges

32 34 10 – Fabricated Highway Bridges

32 34 10.10 Bridges, Highway		Crew	Daily Output	Labor-Hours	Unit	Material	2020 Bare Costs Labor	Equipment	Total	Total Incl O&P
0010	**BRIDGES, HIGHWAY**									
0020	Structural steel, rolled beams	E-5	8.50	9.412	Ton	2,875	540	217	3,632	4,225
0500	Built up, plate girders	E-6	10.50	12.190	"	3,400	700	205	4,305	5,075
1000	Concrete in place, no reinforcing, abutment footings	C-17B	30	2.733	C.Y.	273	151	13.10	437.10	545
1100	Walls, stems and wing walls		20	4.100		370	227	19.70	616.70	770
1200	Decks		17	4.824		555	267	23	845	1,050
1300	Sidewalks and parapets		11	7.455		970	415	36	1,421	1,750
2000	Reinforcing, in place	4 Rodm	3	10.667	Ton	2,275	600		2,875	3,400
2050	Galvanized coated		3	10.667		3,000	600		3,600	4,200
2100	Epoxy coated		3	10.667		2,850	600		3,450	4,025
3000	Expansion dams, steel, double upset 4" x 8" angles welded to									
3010	double 8" x 8" angles, 1-3/4" compression seal	C-22	30	1.400	L.F.	620	79.50	2.42	701.92	805
3040	Double 8" x 8" angles only, 1-3/4" compression seal		35	1.200		465	68	2.08	535.08	615
3050	Galvanized		35	1.200		575	68	2.08	645.08	735
3060	Double 8" x 6" angles only, 1-3/4" compression seal		35	1.200		355	68	2.08	425.08	495
3100	Double 10" channels, 1-3/4" compression seal		35	1.200		325	68	2.08	395.08	465
3420	For 3" compression seal, add					62.50			62.50	69
3440	For double slotted extrusions with seal strip, add					310			310	345
3490	For galvanizing, add				Lb.	.45			.45	.50
4000	Approach railings, steel, galv. pipe, 2 line	C-22	140	.300	L.F.	238	17	.52	255.52	288
4200	Bridge railings, steel, galv. pipe, 3 line w/screen		85	.494		320	28	.85	348.85	395
4220	4 line w/screen		75	.560		375	32	.97	407.97	460
4300	Aluminum, pipe, 3 line w/screen		95	.442		245	25	.76	270.76	310
8000	For structural excavation, see Section 31 23 16.16									
8010	For dewatering, see Section 31 23 19.20									

32 34 Fabricated Bridges

32 34 13 – Fabricated Pedestrian Bridges

32 34 13.10 Bridges, Pedestrian

32 34 13.10 Bridges, Pedestrian	Crew	Daily Output	Labor-Hours	Unit	Material	2020 Bare Costs Labor	2020 Bare Costs Equipment	Total	Total Incl O&P
0010 **BRIDGES, PEDESTRIAN**									
0011　Spans over streams, roadways, etc.									
0020　　including erection, not including foundations									
0050　Precast concrete, complete in place, 8' wide, 60' span	E-2	215	.260	S.F.	161	14.90	7.90	183.80	209
0100　　100' span		185	.303		177	17.30	9.15	203.45	231
0150　　120' span		160	.350		192	20	10.60	222.60	254
0200　　150' span		145	.386		200	22	11.70	233.70	267
0300　Steel, trussed or arch spans, compl. in place, 8' wide, 40' span		320	.175		131	10	5.30	146.30	165
0400　　50' span		395	.142		118	8.10	4.29	130.39	146
0500　　60' span		465	.120		118	6.90	3.65	128.55	144
0600　　80' span		570	.098		140	5.60	2.98	148.58	166
0700　　100' span		465	.120		197	6.90	3.65	207.55	232
0800　　120' span		365	.153		249	8.75	4.65	262.40	293
0900　　150' span		310	.181		265	10.30	5.45	280.75	315
1000　　160' span		255	.220		265	12.55	6.65	284.20	320
1100　　10' wide, 80' span		640	.088		131	5	2.65	138.65	155
1200　　120' span		415	.135		170	7.70	4.09	181.79	203
1300　　150' span		445	.126		191	7.20	3.81	202.01	225
1400　　200' span	▼	205	.273		204	15.60	8.25	227.85	257
1600　Wood, laminated type, complete in place, 80' span	C-12	203	.236		98	12.45	2.32	112.77	129
1700　　130' span	"	153	.314	▼	102	16.50	3.08	121.58	140

32 35 Screening Devices

32 35 16 – Sound Barriers

32 35 16.10 Traffic Barriers, Highway Sound Barriers

32 35 16.10 Traffic Barriers, Highway Sound Barriers	Crew	Daily Output	Labor-Hours	Unit	Material	2020 Bare Costs Labor	2020 Bare Costs Equipment	Total	Total Incl O&P
0010 **TRAFFIC BARRIERS, HIGHWAY SOUND BARRIERS**									
0020　Highway sound barriers, not including footing									
0100　Precast concrete, concrete columns @ 30' OC, 8" T, 8' H	C-12	400	.120	L.F.	175	6.30	1.18	182.48	204
0110　　12' H		265	.181		263	9.55	1.78	274.33	305
0120　　16' H		200	.240		350	12.65	2.35	365	405
0130　　20' H	▼	160	.300		440	15.80	2.94	458.74	505
0400　Lt. wt. composite panel, cementitious face, st. posts @ 12' OC, 8' H	B-80B	190	.168		179	7.55	1.24	187.79	210
0410　　12' H		125	.256		269	11.50	1.89	282.39	315
0420　　16' H		95	.337		360	15.10	2.49	377.59	420
0430　　20' H	▼	75	.427	▼	450	19.15	3.15	472.30	525

32 84 Planting Irrigation

32 84 13 – Drip Irrigation

32 84 13.10 Subsurface Drip Irrigation

32 84 13.10 Subsurface Drip Irrigation		Crew	Daily Output	Labor-Hours	Unit	Material	2020 Bare Costs Labor	2020 Bare Costs Equipment	Total	Total Incl O&P
0010 **SUBSURFACE DRIP IRRIGATION**										
0011　Looped grid, pressure compensating										
0100　Preinserted PE emitter, line, hand bury, irregular area, small	G	3 Skwk	1200	.020	L.F.	.26	1.10		1.36	1.95
0150　　Medium	G		1800	.013		.26	.73		.99	1.40
0200　　Large	G		2520	.010		.26	.52		.78	1.08
0250　　Rectangular area, small	G		2040	.012		.26	.65		.91	1.27
0300　　Medium	G		2640	.009		.26	.50		.76	1.04
0350　　Large	G		3600	.007		.26	.37		.63	.84
0400　　Install in trench, irregular area, small	G		4050	.006		.26	.33		.59	.78
0450　　Medium	G		7488	.003		.26	.18		.44	.56

32 84 13 – Drip Irrigation

32 84 13.10 Subsurface Drip Irrigation	Crew	Daily Output	Labor-Hours	Unit	Material	2020 Bare Costs Labor	2020 Bare Costs Equipment	Total	Total Incl O&P
0500 Large ⒢	3 Skwk	16560	.001	L.F.	.26	.08		.34	.41
0550 Rectangular area, small ⒢		8100	.003		.26	.16		.42	.54
0600 Medium		21960	.001		.26	.06		.32	.38
0650 Large ⒢		33264	.001		.26	.04		.30	.35
0700 Trenching and backfill ⒢	B-53	500	.016			.85	.31	1.16	1.62
0800 Vinyl tubing, 1/4", material only ⒢					.08			.08	.09
0850 Supply tubing, 1/2", material only, 100' coil ⒢					.13			.13	.14
0900 500' coil ⒢					.12			.12	.13
0950 Compression fittings ⒢	1 Skwk	90	.089	Ea.	1.71	4.88		6.59	9.25
1000 Barbed fittings, 1/4" ⒢		360	.022		.12	1.22		1.34	1.97
1100 Flush risers ⒢		60	.133		3.76	7.30		11.06	15.20
1150 Flush ends, figure eight ⒢		180	.044		.55	2.44		2.99	4.30
1300 Auto flush, spring loaded ⒢		90	.089		2.30	4.88		7.18	9.90
1350 Volumetric ⒢		90	.089		8	4.88		12.88	16.15
1400 Air relief valve, inline with compensation tee, 1/2" ⒢		45	.178		18.70	9.75		28.45	35.50
1450 1" ⒢		30	.267		17.45	14.65		32.10	41
1500 Round box for flush ends, 6" ⒢		30	.267		8.10	14.65		22.75	31
1600 Screen filter, 3/4" screen ⒢		12	.667		10.20	36.50		46.70	66.50
1650 1" disk ⒢		8	1		28	55		83	114
1700 1-1/2" disk ⒢		4	2		105	110		215	281
1750 2" disk ⒢		3	2.667		147	146		293	385
1800 Typical installation 18" OC, small, minimum				S.F.				1.53	2.35
1850 Maximum								1.78	2.75
1900 Large, minimum								1.48	2.28
2000 Maximum								1.71	2.64
2100 For non-pressure compensating systems, deduct								10%	10%
2150 For supply piping, see Section 33 11 13.45									
2680 Install PVC or hose bib to PE drip adapter fittings	1 Skwk	40	.200	Ea.	1.70	10.95		12.65	18.45
2690 Install three outlet manual manifold		4	2		58	110		168	230
2700 Install four zone control module		4	2		51.50	110		161.50	223
2710 Install six zone control module		4	2		100	110		210	276

32 84 23 – Underground Sprinklers

32 84 23.10 Sprinkler Irrigation System

	Crew	Daily Output	Labor-Hours	Unit	Material	2020 Bare Costs Labor	2020 Bare Costs Equipment	Total	Total Incl O&P
0010 **SPRINKLER IRRIGATION SYSTEM**									
0011 For lawns									
0800 Residential system, custom, 1" supply	B-20	2000	.012	S.F.	.27	.56		.83	1.15
0900 1-1/2" supply	"	1800	.013	"	.42	.63		1.05	1.40
1020 Pop up spray head w/risers, hi-pop, full circle pattern, 4"	2 Skwk	76	.211	Ea.	4.85	11.55		16.40	23
1030 1/2 circle pattern, 4"		76	.211		7	11.55		18.55	25
1040 6", full circle pattern		76	.211		12.40	11.55		23.95	31
1050 1/2 circle pattern, 6"		76	.211		12.95	11.55		24.50	31.50
1060 12", full circle pattern		76	.211		15.75	11.55		27.30	35
1070 1/2 circle pattern, 12"		76	.211		15.95	11.55		27.50	35
1080 Pop up bubbler head w/risers, hi-pop bubbler head, 4"		76	.211		5.60	11.55		17.15	23.50
1110 Impact full/part circle sprinklers, 28'-54' @ 25-60 psi		37	.432		21.50	23.50		45	59.50
1120 Spaced 37'-49' @ 25-50 psi		37	.432		22	23.50		45.50	60
1130 Spaced 43'-61' @ 30-60 psi		37	.432		67.50	23.50		91	110
1140 Spaced 54'-78' @ 40-80 psi		37	.432		124	23.50		147.50	172
1145 Impact rotor pop-up full/part commercial circle sprinklers									
1150 Spaced 42'-65' @ 35-80 psi	2 Skwk	25	.640	Ea.	17.70	35		52.70	72.50
1160 Spaced 48'-76' @ 45-85 psi	"	25	.640	"	19.45	35		54.45	74.50
1165 Impact rotor pop-up part. circle comm., 53'-75', 55-100 psi, w/accessories									

32 84 23.10 Sprinkler Irrigation System	Crew	Daily Output	Labor-Hours	Unit	Material	2020 Bare Costs Labor	Equipment	Total	Total Incl O&P
1180 Sprinkler, premium, pop-up rotator, 50'-100'	2 Skwk	25	.640	Ea.	96.50	35		131.50	159
1250 Plastic case, 2 nozzle, metal cover		25	.640		105	35		140	169
1260 Rubber cover		25	.640		104	35		139	167
1270 Iron case, 2 nozzle, metal cover		22	.727		149	40		189	225
1280 Rubber cover		22	.727		148	40		188	224
1282 Impact rotor pop-up full circle commercial, 39'-99', 30-100 psi									
1284 Plastic case, metal cover	2 Skwk	25	.640	Ea.	74	35		109	135
1286 Rubber cover		25	.640		108	35		143	172
1288 Iron case, metal cover		22	.727		143	40		183	218
1290 Rubber cover		22	.727		168	40		208	246
1292 Plastic case, 2 nozzle, metal cover		22	.727		107	40		147	178
1294 Rubber cover		22	.727		109	40		149	181
1296 Iron case, 2 nozzle, metal cover		20	.800		142	44		186	224
1298 Rubber cover		20	.800		148	44		192	230
1305 Electric remote control valve, plastic, 3/4"		18	.889		20.50	49		69.50	96
1310 1"		18	.889		25	49		74	101
1320 1-1/2"		18	.889		99	49		148	183
1330 2"		18	.889		126	49		175	212
1335 Quick coupling valves, brass, locking cover									
1340 Inlet coupling valve, 3/4"	2 Skwk	18.75	.853	Ea.	24.50	47		71.50	98
1350 1"		18.75	.853		33.50	47		80.50	108
1360 Controller valve boxes, 6" round boxes		18.75	.853		7.55	47		54.55	79.50
1370 10" round boxes		14.25	1.123		17.95	61.50		79.45	113
1380 12" square box		9.75	1.641		19.10	90		109.10	157
1388 Electromech. control, 14 day 3-60 min., auto start to 23/day									
1390 4 station	2 Skwk	1.04	15.385	Ea.	82	845		927	1,375
1400 7 station		.64	25		156	1,375		1,531	2,250
1410 12 station		.40	40		184	2,200		2,384	3,525
1420 Dual programs, 18 station		.24	66.667		217	3,650		3,867	5,775
1430 23 station		.16	100		267	5,475		5,742	8,600
1435 Backflow preventer, bronze, 0-175 psi, w/valves, test cocks									
1440 3/4"	2 Skwk	6	2.667	Ea.	93.50	146		239.50	325
1450 1"		6	2.667		106	146		252	340
1460 1-1/2"		6	2.667		269	146		415	515
1470 2"		6	2.667		395	146		541	655
1475 Pressure vacuum breaker, brass, 15-150 psi									
1480 3/4"	2 Skwk	6	2.667	Ea.	24	146		170	248
1490 1"		6	2.667		69	146		215	297
1500 1-1/2"		6	2.667		85	146		231	315
1510 2"		6	2.667		157	146		303	395

For customer support on your **Site Work & Landscape Costs** with RSMeans data, call 800.448.8182.

401

32 91 13.16 Mulching

	Crew	Daily Output	Labor-Hours	Unit	Material	2020 Bare Costs Labor	2020 Bare Costs Equipment	Total	Total Incl O&P
0010 MULCHING									
0100 Aged barks, 3" deep, hand spread	1 Clab	100	.080	S.Y.	4.12	3.37		7.49	9.60
0150 Skid steer loader	B-63	13.50	2.963	M.S.F.	460	131	13.15	604.15	715
0160 Skid steer loader	"	1500	.027	S.Y.	4.12	1.18	.12	5.42	6.45
0200 Hay, 1" deep, hand spread	1 Clab	475	.017	"	.51	.71		1.22	1.63
0250 Power mulcher, small	B-64	180	.089	M.S.F.	56.50	3.97	2.32	62.79	71
0350 Large	B-65	530	.030	"	56.50	1.35	.99	58.84	65.50
0370 Fiber mulch recycled newsprint hand spread	1 Clab	500	.016	S.Y.	.24	.67		.91	1.27
0380 Power mulcher small	B-64	200	.080	M.S.F.	26.50	3.57	2.09	32.16	37
0390 Power mulcher large	B-65	600	.027	"	26.50	1.19	.87	28.56	32.50
0400 Humus peat, 1" deep, hand spread	1 Clab	700	.011	S.Y.	3.27	.48		3.75	4.32
0450 Push spreader	"	2500	.003	"	3.27	.13		3.40	3.80
0550 Tractor spreader	B-66	700	.011	M.S.F.	365	.61	.35	365.96	400
0600 Oat straw, 1" deep, hand spread	1 Clab	475	.017	S.Y.	.58	.71		1.29	1.71
0650 Power mulcher, small	B-64	180	.089	M.S.F.	64.50	3.97	2.32	70.79	79.50
0700 Large	B-65	530	.030	"	64.50	1.35	.99	66.84	74
0750 Add for asphaltic emulsion	B-45	1770	.009	Gal.	5.90	.48	.47	6.85	7.75
0760 Pine straw, 1" deep, hand spread	1 Clab	1950	.004	S.F.	.13	.17		.30	.40
0800 Peat moss, 1" deep, hand spread		900	.009	S.Y.	5.15	.37		5.52	6.20
0850 Push spreader		2500	.003	"	5.15	.13		5.28	5.85
0950 Tractor spreader	B-66	700	.011	M.S.F.	570	.61	.35	570.96	625
1000 Polyethylene film, 6 mil	2 Clab	2000	.008	S.Y.	.51	.34		.85	1.07
1010 4 mil		2300	.007		.45	.29		.74	.94
1020 1-1/2 mil		2500	.006		.26	.27		.53	.69
1050 Filter fabric weed barrier		2000	.008		.86	.34		1.20	1.46
1100 Redwood nuggets, 3" deep, hand spread	1 Clab	150	.053		3.33	2.25		5.58	7.05
1150 Skid steer loader	B-63	13.50	2.963	M.S.F.	370	131	13.15	514.15	615
1200 Stone mulch, hand spread, ceramic chips, economy	1 Clab	125	.064	S.Y.	7.25	2.69		9.94	12.05
1250 Deluxe	"	95	.084	"	11.40	3.55		14.95	17.90
1300 Granite chips	B-1	10	2.400	C.Y.	80	103		183	242
1400 Marble chips		10	2.400		245	103		348	425
1600 Pea gravel		28	.857		115	36.50		151.50	181
1700 Quartz		10	2.400		197	103		300	370
1760 Landscape mulch, 100% recycld tires, var colors, hand spread, 3" deep		10	2.400		305	103		408	490
1800 Tar paper, 15 lb. felt	1 Clab	800	.010	S.Y.	.47	.42		.89	1.15
1900 Wood chips, 2" deep, hand spread	"	220	.036	"	1.68	1.53		3.21	4.15
1950 Skid steer loader	B-63	20.30	1.970	M.S.F.	187	87.50	8.75	283.25	345

32 91 13.23 Structural Soil Mixing

	Crew	Daily Output	Labor-Hours	Unit	Material	2020 Bare Costs Labor	2020 Bare Costs Equipment	Total	Total Incl O&P
0010 STRUCTURAL SOIL MIXING									
0100 Rake topsoil, site material, harley rock rake, ideal	B-6	33	.727	M.S.F.		33.50	6.50	40	57
0200 Adverse	"	7	3.429			157	30.50	187.50	269
0250 By hand (raking)	1 Clab	10	.800			33.50		33.50	50.50
0300 Screened loam, york rake and finish, ideal	B-62	24	1			45.50	7.40	52.90	76.50
0400 Adverse	"	20	1.200			55	8.90	63.90	92.50
0450 By hand (raking)	1 Clab	15	.533			22.50		22.50	33.50
1000 Remove topsoil & stock pile on site, 75 HP dozer, 6" deep, 50' haul	B-10L	30	.400			21	13.40	34.40	45.50
1050 300' haul		6.10	1.967			102	66	168	226
1100 12" deep, 50' haul		15.50	.774			40	26	66	88.50
1150 300' haul		3.10	3.871			201	129	330	440
1200 200 HP dozer, 6" deep, 50' haul	B-10B	125	.096			4.98	12.05	17.03	20.50
1250 300' haul		30.70	.391			20.50	49	69.50	84.50
1300 12" deep, 50' haul		62	.194			10.05	24.50	34.55	41.50

32 91 Planting Preparation

32 91 13 – Soil Preparation

32 91 13.23 Structural Soil Mixing		Crew	Daily Output	Labor-Hours	Unit	Material	2020 Bare Costs Labor	Equipment	Total	Total Incl O&P
1350	300' haul	B-10B	15.40	.779	M.S.F.		40.50	97.50	138	168
1400	Alternate method, 75 HP dozer, 50' haul	B-10L	860	.014	C.Y.		.72	.47	1.19	1.59
1450	300' haul	"	114	.105			5.45	3.52	8.97	12
1500	200 HP dozer, 50' haul	B-10B	2660	.005			.23	.57	.80	.97
1600	300' haul	"	570	.021			1.09	2.64	3.73	4.53
1800	Rolling topsoil, hand push roller	1 Clab	3200	.003	S.F.		.11		.11	.16
1850	Tractor drawn roller	B-66	10666	.001	"		.04	.02	.06	.08
1900	Remove rocks & debris from grade, by hand	B-62	80	.300	M.S.F.		13.70	2.22	15.92	23
1920	With rock picker	B-10S	140	.086			4.45	3.02	7.47	10
2000	Root raking and loading, residential, no boulders	B-6	53.30	.450			20.50	4.01	24.51	35.50
2100	With boulders		32	.750			34.50	6.70	41.20	59
2200	Municipal, no boulders		200	.120			5.50	1.07	6.57	9.45
2300	With boulders		120	.200			9.15	1.78	10.93	15.65
2400	Large commercial, no boulders	B-10B	400	.030			1.56	3.76	5.32	6.45
2500	With boulders	"	240	.050			2.59	6.25	8.84	10.80
2600	Rough grade & scarify subsoil to receive topsoil, common earth									
2610	200 HP dozer with scarifier	B-11A	80	.200	M.S.F.		9.90	18.80	28.70	35.50
2620	180 HP grader with scarifier	B-11L	110	.145			7.20	9.65	16.85	21.50
2700	Clay and till, 200 HP dozer with scarifier	B-11A	50	.320			15.80	30	45.80	56.50
2710	180 HP grader with scarifier	B-11L	40	.400			19.75	26.50	46.25	58.50
3000	Scarify subsoil, residential, skid steer loader w/scarifiers, 50 HP	B-66	32	.250			13.25	7.55	20.80	28
3050	Municipal, skid steer loader w/scarifiers, 50 HP	"	120	.067			3.53	2.01	5.54	7.50
3100	Large commercial, 75 HP, dozer w/scarifier	B-10L	240	.050			2.59	1.67	4.26	5.70
3200	Grader with scarifier, 135 HP	B-11L	280	.057			2.82	3.79	6.61	8.40
3500	Screen topsoil from stockpile, vibrating screen, wet material (organic)	B-10P	200	.060	C.Y.		3.11	5.85	8.96	11.10
3550	Dry material	"	300	.040			2.07	3.90	5.97	7.40
3600	Mixing with conditioners, manure and peat	B-10R	550	.022			1.13	.55	1.68	2.29
3650	Mobilization add for 2 days or less operation	B-34K	3	2.667	Job		131	286	417	510
3800	Spread conditioned topsoil, 6" deep, by hand	B-1	360	.067	S.Y.	5.90	2.85		8.75	10.80
3850	300 HP dozer	B-10M	27	.444	M.S.F.	635	23	65.50	723.50	805
3900	4" deep, by hand	B-1	470	.051	S.Y.	5.30	2.18		7.48	9.15
3920	300 HP dozer	B-10M	34	.353	M.S.F.	475	18.30	52	545.30	610
3940	180 HP grader	B-11L	37	.432	"	475	21.50	28.50	525	590
4000	Spread soil conditioners, alum. sulfate, 1#/S.Y., hand push spreader	1 Clab	17500	.001	S.Y.	23	.02		23.02	25.50
4050	Tractor spreader	B-66	700	.011	M.S.F.	2,575	.61	.35	2,575.96	2,825
4100	Fertilizer, 0.2#/S.Y., push spreader	1 Clab	17500	.001	S.Y.	.08	.02		.10	.12
4150	Tractor spreader	B-66	700	.011	M.S.F.	9.10	.61	.35	10.06	11.30
4200	Ground limestone, 1#/S.Y., push spreader	1 Clab	17500	.001	S.Y.	.14	.02		.16	.18
4250	Tractor spreader	B-66	700	.011	M.S.F.	15.55	.61	.35	16.51	18.40
4400	Manure, 18#/S.Y., push spreader	1 Clab	2500	.003	S.Y.	8.10	.13		8.23	9.10
4450	Tractor spreader	B-66	280	.029	M.S.F.	900	1.51	.86	902.37	995
4500	Perlite, 1" deep, push spreader	1 Clab	17500	.001	S.Y.	11.15	.02		11.17	12.30
4550	Tractor spreader	B-66	700	.011	M.S.F.	1,225	.61	.35	1,225.96	1,350
4600	Vermiculite, push spreader	1 Clab	17500	.001	S.Y.	8.45	.02		8.47	9.35
4650	Tractor spreader	B-66	700	.011	M.S.F.	940	.61	.35	940.96	1,025
5000	Spread topsoil, skid steer loader and hand dress	B-62	270	.089	C.Y.	26.50	4.06	.66	31.22	36
5100	Articulated loader and hand dress	B-100	320	.038		26.50	1.95	3	31.45	35
5200	Articulated loader and 75 HP dozer	B-10M	500	.024		26.50	1.24	3.53	31.27	34.50
5300	Road grader and hand dress	B-11L	1000	.016		26.50	.79	1.06	28.35	31.50
6000	Tilling topsoil, 20 HP tractor, disk harrow, 2" deep	B-66	450	.018	M.S.F.		.94	.54	1.48	2
6050	4" deep		360	.022			1.18	.67	1.85	2.50
6100	6" deep		270	.030			1.57	.89	2.46	3.33
6150	26" rototiller, 2" deep	A-1J	1250	.006	S.Y.		.27	.04	.31	.45

403

32 91 Planting Preparation

32 91 13 – Soil Preparation

32 91 13.23 Structural Soil Mixing

		Crew	Daily Output	Labor-Hours	Unit	Material	2020 Bare Costs Labor	Equipment	Total	Total Incl O&P
6200	4" deep	A-1J	1000	.008	S.Y.		.34	.05	.39	.57
6250	6" deep	↓	750	.011	↓		.45	.07	.52	.75

32 91 13.26 Planting Beds

		Crew	Daily Output	Labor-Hours	Unit	Material	2020 Bare Costs Labor	Equipment	Total	Total Incl O&P
0010	**PLANTING BEDS**									
0100	Backfill planting pit, by hand, on site topsoil	2 Clab	18	.889	C.Y.		37.50		37.50	56
0200	Prepared planting mix, by hand	"	24	.667			28		28	42
0300	Skid steer loader, on site topsoil	B-62	340	.071			3.23	.52	3.75	5.40
0400	Prepared planting mix	"	410	.059			2.68	.43	3.11	4.49
1000	Excavate planting pit, by hand, sandy soil	2 Clab	16	1			42		42	63.50
1100	Heavy soil or clay	"	8	2			84		84	127
1200	1/2 C.Y. backhoe, sandy soil	B-11C	150	.107			5.25	1.43	6.68	9.45
1300	Heavy soil or clay	"	115	.139			6.90	1.86	8.76	12.35
2000	Mix planting soil, incl. loam, manure, peat, by hand	2 Clab	60	.267		46.50	11.25		57.75	68
2100	Skid steer loader	B-62	150	.160	↓	46.50	7.30	1.18	54.98	63.50
3000	Pile sod, skid steer loader	"	2800	.009	S.Y.		.39	.06	.45	.66
3100	By hand	2 Clab	400	.040			1.68		1.68	2.53
4000	Remove sod, F.E. loader	B-10S	2000	.006			.31	.21	.52	.70
4100	Sod cutter	B-12K	3200	.005			.25	.30	.55	.71
4200	By hand	2 Clab	240	.067	↓		2.81		2.81	4.22
6000	For planting bed edging, see Section 32 94 13.20									

32 91 19 – Landscape Grading

32 91 19.13 Topsoil Placement and Grading

		Crew	Daily Output	Labor-Hours	Unit	Material	2020 Bare Costs Labor	Equipment	Total	Total Incl O&P
0010	**TOPSOIL PLACEMENT AND GRADING**									
0300	Fine grade, base course for paving, see Section 32 11 23.23									
0400	Spread from pile to rough finish grade, F.E. loader, 1-1/2 C.Y.	B-10S	200	.060	C.Y.		3.11	2.12	5.23	7
0500	Up to 200' radius, by hand	1 Clab	14	.571			24		24	36
0600	Top dress by hand, 1 C.Y. for 600 S.F.	"	11.50	.696	↓	29	29.50		58.50	76
0700	Furnish and place, truck dumped, screened, 4" deep	B-10S	1300	.009	S.Y.	3.65	.48	.33	4.46	5.10
0800	6" deep	"	820	.015	"	4.68	.76	.52	5.96	6.85
0900	Fine grading and seeding, incl. lime, fertilizer & seed,									
1000	With equipment	B-14	1000	.048	S.Y.	.49	2.12	.21	2.82	3.97

32 92 Turf and Grasses

32 92 19 – Seeding

32 92 19.13 Mechanical Seeding

		Crew	Daily Output	Labor-Hours	Unit	Material	2020 Bare Costs Labor	Equipment	Total	Total Incl O&P
0010	**MECHANICAL SEEDING**									
0020	Mechanical seeding, 215 lb./acre	B-66	1.50	5.333	Acre	595	283	161	1,039	1,250
0100	44 lb./M.S.Y.	"	2500	.003	S.Y.	.23	.17	.10	.50	.61
0101	44 lb./M.S.Y.	1 Clab	13950	.001	S.F.	.03	.02		.05	.07
0300	Fine grading and seeding incl. lime, fertilizer & seed,									
0310	with equipment	B-14	1000	.048	S.Y.	.47	2.12	.21	2.80	3.95
0400	Fertilizer hand push spreader, 35 lb./M.S.F.	1 Clab	200	.040	M.S.F.	10.70	1.68		12.38	14.30
0600	Limestone hand push spreader, 50 lb./M.S.F.		180	.044		5.60	1.87		7.47	8.95
0800	Grass seed hand push spreader, 4.5 lb./M.S.F.	↓	180	.044	↓	25.50	1.87		27.37	31
1000	Hydro or air seeding for large areas, incl. seed and fertilizer	B-81	8900	.003	S.Y.	.64	.13	.07	.84	.97
1100	With wood fiber mulch added	"	8900	.003	"	2.07	.13	.07	2.27	2.55
1300	Seed only, over 100 lb., field seed, minimum				Lb.	2.08			2.08	2.29
1400	Maximum					1.97			1.97	2.17
1500	Lawn seed, minimum					1.42			1.42	1.56
1600	Maximum				↓	2.70			2.70	2.97

32 92 Turf and Grasses

32 92 19 – Seeding

32 92 19.13 Mechanical Seeding

		Crew	Daily Output	Labor-Hours	Unit	Material	2020 Bare Costs Labor	2020 Bare Costs Equipment	Total	Total Incl O&P
1800	Aerial operations, seeding only, field seed	B-58	50	.480	Acre	675	22	46	743	830
1900	Lawn seed		50	.480		460	22	46	528	595
2100	Seed and liquid fertilizer, field seed		50	.480		760	22	46	828	920
2200	Lawn seed	↓	50	.480	↓	545	22	46	613	685

32 92 19.14 Seeding, Athletic Fields

		Crew	Daily Output	Labor-Hours	Unit	Material	2020 Bare Costs Labor	2020 Bare Costs Equipment	Total	Total Incl O&P
0010	**SEEDING, ATHLETIC FIELDS** R329219-50									
0020	Seeding, athletic fields, athletic field mix, 8#/M.S.F. push spreader	1 Clab	8	1	M.S.F.	11	42		53	75.50
0100	Tractor spreader	B-66	52	.154		11	8.15	4.64	23.79	29.50
0200	Hydro or air seeding, with mulch and fertilizer	B-81	80	.300		12.10	14.80	7.30	34.20	43.50
0400	Birdsfoot trefoil, 0.45#/M.S.F., push spreader	1 Clab	8	1		4.84	42		46.84	69
0500	Tractor spreader	B-66	52	.154		4.84	8.15	4.64	17.63	22.50
0600	Hydro or air seeding, with mulch and fertilizer	B-81	80	.300		9.30	14.80	7.30	31.40	40.50
0800	Bluegrass, 4#/M.S.F., common, push spreader	1 Clab	8	1		15.40	42		57.40	80.50
0900	Tractor spreader	B-66	52	.154		15.40	8.15	4.64	28.19	34.50
1000	Hydro or air seeding, with mulch and fertilizer	B-81	80	.300		25.50	14.80	7.30	47.60	58
1100	Baron, push spreader	1 Clab	8	1		15.50	42		57.50	80.50
1200	Tractor spreader	B-66	52	.154		15.50	8.15	4.64	28.29	34.50
1300	Hydro or air seeding, with mulch and fertilizer	B-81	80	.300		21.50	14.80	7.30	43.60	53.50
1500	Clover, 0.67#/M.S.F., white, push spreader	1 Clab	8	1		3.14	42		45.14	67
1600	Tractor spreader	B-66	52	.154		3.14	8.15	4.64	15.93	21
1700	Hydro or air seeding, with mulch and fertilizer	B-81	80	.300		17.25	14.80	7.30	39.35	49
1800	Ladino, push spreader	1 Clab	8	1		3.62	42		45.62	67.50
1900	Tractor spreader	B-66	52	.154		3.62	8.15	4.64	16.41	21.50
2000	Hydro or air seeding, with mulch and fertilizer	B-81	80	.300		15.95	14.80	7.30	38.05	47.50
2200	Fescue 5.5#/M.S.F., tall, push spreader	1 Clab	8	1		9.75	42		51.75	74
2300	Tractor spreader	B-66	52	.154		9.75	8.15	4.64	22.54	28
2400	Hydro or air seeding, with mulch and fertilizer	B-81	80	.300		32	14.80	7.30	54.10	65.50
2500	Chewing, push spreader	1 Clab	8	1		8.95	42		50.95	73.50
2600	Tractor spreader	B-66	52	.154		8.95	8.15	4.64	21.74	27
2700	Hydro or air seeding, with mulch and fertilizer	B-81	80	.300		29.50	14.80	7.30	51.60	62.50
2800	Creeping, push spreader	1 Clab	8	1		7.60	42		49.60	72
2810	Tractor spreader	B-66	26	.308		7.60	16.30	9.30	33.20	43
2820	Hydro or air seeding, with mulch and fertilizer	B-81	80	.300		25	14.80	7.30	47.10	57.50
2900	Crown vetch, 4#/M.S.F., push spreader	1 Clab	8	1		128	42		170	205
3000	Tractor spreader	B-66	52	.154		128	8.15	4.64	140.79	158
3100	Hydro or air seeding, with mulch and fertilizer	B-81	80	.300		176	14.80	7.30	198.10	224
3300	Rye, 10#/M.S.F., annual, push spreader	1 Clab	8	1		17.25	42		59.25	82.50
3400	Tractor spreader	B-66	52	.154		17.25	8.15	4.64	30.04	36.50
3500	Hydro or air seeding, with mulch and fertilizer	B-81	80	.300		38	14.80	7.30	60.10	72
3600	Fine textured, push spreader	1 Clab	8	1		16.30	42		58.30	81.50
3700	Tractor spreader	B-66	52	.154		16.30	8.15	4.64	29.09	35.50
3800	Hydro or air seeding, with mulch and fertilizer	B-81	80	.300		36	14.80	7.30	58.10	69.50
4000	Shade mix, 6#/M.S.F., push spreader	1 Clab	8	1		10.65	42		52.65	75
4100	Tractor spreader	B-66	52	.154		10.65	8.15	4.64	23.44	29
4200	Hydro or air seeding, with mulch and fertilizer	B-81	80	.300		23.50	14.80	7.30	45.60	56
4400	Slope mix, 6#/M.S.F., push spreader	1 Clab	8	1		11.95	42		53.95	76.50
4500	Tractor spreader	B-66	52	.154		11.95	8.15	4.64	24.74	30.50
4600	Hydro or air seeding, with mulch and fertilizer	B-81	80	.300		30	14.80	7.30	52.10	63
4800	Turf mix, 4#/M.S.F., push spreader	1 Clab	8	1		13.55	42		55.55	78.50
4900	Tractor spreader	B-66	52	.154		13.55	8.15	4.64	26.34	32.50
5000	Hydro or air seeding, with mulch and fertilizer	B-81	80	.300		34	14.80	7.30	56.10	67.50
5200	Utility mix, 7#/M.S.F., push spreader	1 Clab	8	1	↓	12.45	42		54.45	77

32 92 Turf and Grasses

32 92 19 – Seeding

32 92 19.14 Seeding, Athletic Fields

		Crew	Daily Output	Labor-Hours	Unit	Material	2020 Bare Costs Labor	Equipment	Total	Total Incl O&P
5300	Tractor spreader	B-66	52	.154	M.S.F.	12.45	8.15	4.64	25.24	31
5400	Hydro or air seeding, with mulch and fertilizer	B-81	80	.300		46.50	14.80	7.30	68.60	81.50
5600	Wildflower, 0.10#/M.S.F., push spreader	1 Clab	8	1		1.80	42		43.80	65.50
5700	Tractor spreader	B-66	52	.154		1.80	8.15	4.64	14.59	19.30
5800	Hydro or air seeding, with mulch and fertilizer	B-81	80	.300		9.90	14.80	7.30	32	41
7000	Apply fertilizer, 800 lb./acre	B-66	4	2	Ton	925	106	60.50	1,091.50	1,250
7025	Fertilizer, mechanical spread	1 Clab	1.75	4.571	Acre	5.60	192		197.60	295
7050	Apply limestone, 800 lb./acre	B-66	4.25	1.882	Ton	203	100	57	360	435
7060	Limestone, mechanical spread	1 Clab	1.74	4.598	Acre	89.50	194		283.50	390
7100	Apply mulch, see Section 32 91 13.16									

32 92 23 – Sodding

32 92 23.10 Sodding Systems

		Crew	Daily Output	Labor-Hours	Unit	Material	2020 Bare Costs Labor	Equipment	Total	Total Incl O&P
0010	**SODDING SYSTEMS**									
0020	Sodding, 1" deep, bluegrass sod, on level ground, over 8 M.S.F.	B-63	22	1.818	M.S.F.	223	80.50	8.05	311.55	375
0200	4 M.S.F.		17	2.353		286	104	10.45	400.45	485
0300	1,000 S.F.		13.50	2.963		335	131	13.15	479.15	580
0500	Sloped ground, over 8 M.S.F.		6	6.667		223	295	29.50	547.50	725
0600	4 M.S.F.		5	8		286	355	35.50	676.50	885
0700	1,000 S.F.		4	10		335	445	44.50	824.50	1,075
1000	Bent grass sod, on level ground, over 6 M.S.F.		20	2		305	88.50	8.90	402.40	480
1100	3 M.S.F.		18	2.222		320	98.50	9.85	428.35	515
1200	Sodding 1,000 S.F. or less		14	2.857		345	127	12.70	484.70	585
1500	Sloped ground, over 6 M.S.F.		15	2.667		305	118	11.85	434.85	525
1600	3 M.S.F.		13.50	2.963		320	131	13.15	464.15	565
1700	1,000 S.F.		12	3.333		345	148	14.80	507.80	615

32 92 26 – Sprigging

32 92 26.13 Stolonizing

		Crew	Daily Output	Labor-Hours	Unit	Material	2020 Bare Costs Labor	Equipment	Total	Total Incl O&P
0010	**STOLONIZING**									
0100	6" OC, by hand	1 Clab	4	2	M.S.F.	56.50	84		140.50	189
0110	Walk behind sprig planter	"	80	.100		56.50	4.21		60.71	68.50
0120	Towed sprig planter	B-66	350	.023		56.50	1.21	.69	58.40	64.50
0130	9" OC, by hand	1 Clab	5.20	1.538		26.50	65		91.50	127
0140	Walk behind sprig planter	"	92	.087		26.50	3.66		30.16	34.50
0150	Towed sprig planter	B-66	420	.019		26.50	1.01	.57	28.08	31
0160	12" OC, by hand	1 Clab	6	1.333		13.55	56		69.55	99.50
0170	Walk behind sprig planter	"	110	.073		13.55	3.06		16.61	19.50
0180	Towed sprig planter	B-66	500	.016		13.55	.85	.48	14.88	16.70
0200	Broadcast, by hand, 2 Bu. per M.S.F.	1 Clab	15	.533		6	22.50		28.50	40
0210	4 Bu. per M.S.F.		10	.800		10.75	33.50		44.25	62.50
0220	6 Bu. per M.S.F.		6.50	1.231		15.40	52		67.40	95
0300	Hydro planter, 6 Bu. per M.S.F.	B-64	100	.160		15.40	7.15	4.17	26.72	32
0320	Manure spreader planting 6 Bu. per M.S.F.	B-66	200	.040		15.40	2.12	1.21	18.73	21.50

32 93 13.30 Ground Covers	Crew	Daily Output	Labor-Hours	Unit	Material	2020 Bare Costs Labor	Equipment	Total	Total Incl O&P
0010 **GROUND COVERS**									
0011 Vines and climbing plants									
1000 Achillea tomentosa (Woolly Yarrow), Z4, cont									
1010 1 qt.				Ea.	4.48			4.48	4.93
1020 2 gal.				"	7.80			7.80	8.55
1100 Aegopodium podagaria (Snow-on-the-Mountain Goutweed), Z3, cont									
1110 1 qt.				Ea.	9.35			9.35	10.30
1120 1 gal.					11.70			11.70	12.90
1130 Flat/24 plants					44			44	48
1200 Ajuga reptans (Carpet Bugle), Z4, cont									
1210 4" pot				Ea.	4.81			4.81	5.30
1220 1 qt.					7.65			7.65	8.45
1230 1 gal.					14.10			14.10	15.50
1240 Flat/24 plants					58			58	63.50
1300 Akebia quinata (Five-Leaf Akebia), Z4, cont									
1310 1 gal.				Ea.	14.55			14.55	16
1320 5 gal.				"	52			52	57
1400 Alchemilla vulgaris (Common Lady's Mantle), Z3, cont									
1410 1 qt.				Ea.	8.10			8.10	8.95
1420 1 gal.				"	33			33	36.50
1500 Ampelopsis brevipedunculata elegans (Ampelopsis), Z4, cont									
1510 1 gal.				Ea.	19.30			19.30	21.50
1520 3 gal.				"	35.50			35.50	39.50
1600 Arctostaphylos uva-ursi (Bearberry), Z2, cont									
1610 3" pot				Ea.	11.50			11.50	12.65
1620 1 gal.					16.40			16.40	18.05
1630 5 gal.					24.50			24.50	27
1700 Aristolochia durior (Dutchman's Pipe), Z5, cont									
1710 2 gal.				Ea.	30.50			30.50	33.50
1720 5 gal.				"	31			31	34
1800 Armeria maritima (Thrift, Sea Pink), Z6, cont									
1810 1 qt.				Ea.	13			13	14.30
1820 1 gal.				"	14.15			14.15	15.60
1900 Artemisia schmitdiana (Silver Mound/Satiny Wormwood), Z3, cont									
1910 1 qt.				Ea.	4.18			4.18	4.60
1920 1 gal.				"	11.20			11.20	12.30
2000 Artemisia stellerana (Dusty Miller), Z2, cont									
2010 1 qt.				Ea.	9			9	9.90
2020 1 gal.				"	33			33	36
2100 Asarum europaeum (European Wild Ginger), Z4, cont									
2110 4" pot				Ea.	11.65			11.65	12.85
2120 1 gal.				"	27.50			27.50	30.50
2200 Baccharis pilularis (Dwarf Coyote Bush), Z7, cont									
2210 1 qt.				Ea.	6.50			6.50	7.15
2220 1 gal.				"	8.05			8.05	8.85
2300 Bergenia cordifolia (Heartleaf Bergenia), Z3, cont									
2310 1 qt.				Ea.	17.75			17.75	19.55
2320 1 gal.				"	27.50			27.50	30.50
2400 Bougainvillea glabra, Z10, cont									
2410 1 gal.				Ea.	18.70			18.70	20.50
2420 5 gal.				"	26			26	28.50
2500 Calluna vulgaris cultivars (Scotch Heather), Z4, cont									
2510 3" pot				Ea.	8.30			8.30	9.10

407

32 93 13 – Ground Covers

32 93 13.30 Ground Covers	Crew	Daily Output	Labor-Hours	Unit	Material	2020 Bare Costs Labor	Equipment	Total	Total Incl O&P
2520 1 qt.				Ea.	16.35			16.35	18
2530 1 gal.					33.50			33.50	37
2540 2 gal.				↓	63			63	69
2600 Campsis radicans (Trumpet Creeper), Z5, cont									
2610 1 gal.				Ea.	16.90			16.90	18.60
2620 2 gal.					21			21	23
2630 5 gal.				↓	31.50			31.50	34.50
2700 Celastrus scandens (American Bittersweet), Z2, cont									
2710 1 gal.				Ea.	14.10			14.10	15.50
2720 2 gal.				"	25			25	27.50
2800 Cerastium tomentosum (Snow-in-Summer), Z2, cont									
2810 1 qt.				Ea.	21.50			21.50	23.50
2820 1 gal.				"	43.50			43.50	48
2900 Ceratostigma (Dwarf Plumbago), Z6, cont									
2910 1 qt.				Ea.	15.20			15.20	16.75
2920 1 gal.				"	25			25	27.50
3000 Cissus antartica (Kangaroo Vine), Z8, cont									
3010 1 gal.				Ea.	6.10			6.10	6.70
3020 5 gal.				"	22			22	24.50
3100 Cistus crispus (Wrinkleleaf Rock Rose), Z7, cont									
3110 1 gal.				Ea.	8.85			8.85	9.70
3120 5 gal.				"	16.35			16.35	18
3200 Clematis (Hybrids & Clones), Z6, cont									
3210 1 gal.				Ea.	13.75			13.75	15.15
3220 2 gal.					20.50			20.50	22.50
3230 5 gal.				↓	31.50			31.50	34.50
3300 Cocculus laurifolius (Laurel-leaf Snailseed), Z8, cont									
3310 1 gal.				Ea.	4.88			4.88	5.35
3320 5 gal.				"	14.25			14.25	15.70
3400 Conuallaria majalis (Lily-of-the-Valley), Z3, cont									
3410 1 qt.				Ea.	9.85			9.85	10.80
3420 1 gal.				"	12.70			12.70	14
3500 Coprosma prostrata (Prostrate Coprosma), Z7, cont									
3510 1 gal.				Ea.	5.15			5.15	5.65
3520 5 gal.				"	15.15			15.15	16.65
3600 Coronilla varia (Crown Vetch), Z3, cont									
3610 2-1/4" pot				Ea.	5.60			5.60	6.15
3620 Flat/50 plants				"	85			85	93.50
3700 Cotoneaster dammeri (Bearberry Cotoneaster), Z5, cont									
3710 1 gal.				Ea.	5.50			5.50	6.05
3720 2 gal.					15.45			15.45	17
3730 5 gal.				↓	14.50			14.50	15.95
3800 Cytisus decumbens (Prostrate Broom), Z5, cont									
3810 1 gal.				Ea.	21.50			21.50	23.50
3820 2 gal.				"	15.55			15.55	17.10
3900 Dianthus plumarius (Grass Pinks), Z4, cont									
3910 1 qt.				Ea.	11.10			11.10	12.20
3920 1 gal.				"	40			40	44
4000 Dicentra formosa (Western Bleeding Heart), Z3, cont									
4010 2 qt.				Ea.	12.55			12.55	13.85
4020 1 gal.				"	14.90			14.90	16.35
4100 Distictis buccinatoria (Blood Red Trumpet Vine), Z9, cont									
4110 1 gal.				Ea.	11.70			11.70	12.85

32 93 13.30 Ground Covers	Crew	Daily Output	Labor-Hours	Unit	Material	2020 Bare Costs Labor	Equipment	Total	Total Incl O&P
4120	5 gal.				Ea.	29.50			29.50	32.50
4200	Epimedium grandiflorum (Barrenwort, Bishopshat), Z3, cont									
4210	1 qt.				Ea.	7.50			7.50	8.25
4220	1 gal.				"	13.40			13.40	14.75
4300	Euonymus fortunei colorata (Purple-Leaf Wintercreeper), Z5, cont									
4310	1 qt.				Ea.	3.11			3.11	3.42
4320	1 gal.					14.85			14.85	16.35
4330	Flat/24 plants				↓	37.50			37.50	41
4400	Euonymus fortunei kewensis (Euonymus), Z5, cont									
4410	1 gal.				Ea.	10.70			10.70	11.75
4420	2 gal.				"	22.50			22.50	24.50
4500	Fragaria vesca (Alpine Strawberry), Z6, cont									
4510	4" pot				Ea.	8.70			8.70	9.55
4520	1 qt.					12.80			12.80	14.05
4530	1 gal.				↓	16.90			16.90	18.60
4600	Galium odoratum (Sweet Woodruff), Z4, cont									
4610	4" pot				Ea.	10.85			10.85	11.95
4620	1 gal.				"	30.50			30.50	33.50
4700	Gardenia jasminoides "Radicans" (Creeping Gardenia), Z8, cont									
4710	1 gal.				Ea.	10.40			10.40	11.45
4720	5 gal.				"	23			23	25.50
4800	Gelsemium sempervirens (Carolina Jessamine), Z7, cont									
4810	1 gal.				Ea.	10.70			10.70	11.80
4820	5 gal.				"	23			23	25
4900	Genista pilosa (Silky-leaf Woadwaxen), Z5, cont									
4910	1 gal.				Ea.	13.15			13.15	14.45
4920	2 gal.				"	27.50			27.50	30.50
5000	Hardenbergia violacea (Happy Wanderer), Z9, cont									
5010	1 gal.				Ea.	10.75			10.75	11.85
5020	5 gal.				"	24.50			24.50	27
5100	Hedera helix varieties (English Ivy), Z5, cont									
5110	3" pot				Ea.	5.15			5.15	5.65
5120	5 gal.					35			35	38.50
5130	Flat/50 plants				↓	54.50			54.50	59.50
5200	Helianthemum nummularium (Sun-Rose), Z5, cont									
5210	1 qt.				Ea.	8.25			8.25	9.05
5220	1 gal.				"	10.30			10.30	11.35
5300	Hemerocallis hybrids (Assorted Daylillies), Z4, cont									
5310	2 qt.				Ea.	10.85			10.85	11.90
5320	1 gal.					16.05			16.05	17.70
5330	5 gal.				↓	23			23	25.50
5400	Hosta assorted (Plantain Lily), Z4, cont									
5410	2 qt.				Ea.	20.50			20.50	22.50
5420	1 gal.					33			33	36.50
5430	2 gal.				↓	47			47	51.50
5500	Hydrangea petiolaris (Climbing Hydrangea), Z4, cont									
5510	1 gal.				Ea.	29			29	32
5520	2 gal.					60.50			60.50	66.50
5530	5 gal.				↓	73			73	80
5600	Hypericum calycinum (St. John's Wort), Z6, cont									
5610	1 qt.				Ea.	10.85			10.85	11.95
5620	1 gal.					48.50			48.50	53.50
5630	5 gal.				↓	59			59	65

409

32 93 13.30 Ground Covers	Crew	Daily Output	Labor-Hours	Unit	Material	2020 Bare Costs Labor	Equipment	Total	Total Incl O&P
5700	Iberis sempervirens (Candytuft), Z4, cont								
5710	1 qt.				Ea.	8.10			8.10	8.95
5720	1 gal.				"	10.95			10.95	12.05
5800	Jasminum nudiflorum (Winter Jasmine), Z5, cont								
5810	1 gal.				Ea.	11.55			11.55	12.70
5820	2 gal.					28			28	31
5830	5 gal.					37.50			37.50	41.50
5900	Juniperus horizontalis varieties (Prostrate Juniper), Z2, cont								
5910	1 gal.				Ea.	24.50			24.50	27
5920	2 gal.					35			35	38.50
5930	5 gal.					69			69	76
6000	Lamium maculatum (Dead Nettle), Z6, cont								
6010	1 qt.				Ea.	7.45			7.45	8.15
6020	1 gal.					10.70			10.70	11.75
6030	Flat/24 plants					71.50			71.50	78.50
6100	Lantana camara (Common Lantana), Z8, cont								
6110	3" pot				Ea.	12.35			12.35	13.55
6120	1 gal.				"	21.50			21.50	23.50
6200	Lantana montevidensis (Trailing Lantana), Z10, cont								
6210	1 gal.				Ea.	4.75			4.75	5.25
6220	5 gal.				"	17.35			17.35	19.10
6300	Lavandula angustifolia (English Lavender), Z5, cont								
6310	4" pot				Ea.	4.83			4.83	5.30
6320	1 qt.					3.59			3.59	3.95
6330	1 gal.					17.50			17.50	19.25
6400	Liriope spicata (Creeping Lilyturf), Z6, cont								
6410	2 qt.				Ea.	9.40			9.40	10.35
6420	1 gal.					10.55			10.55	11.60
6430	5 gal.					20.50			20.50	22.50
6500	Lonicera heckrotti (Gold Flame Honeysuckle), Z5, cont								
6510	1 gal.				Ea.	26			26	29
6520	2 gal.				"	48.50			48.50	53.50
6600	Lonicera japonica "Halliana" (Halls Honeysuckle), Z4, cont								
6610	3" pot				Ea.	14.30			14.30	15.75
6620	1 gal.					38			38	42
6630	2 gal.					58			58	63.50
6700	Lysimachia nummularia aurea (Creeping Jenny), Z3, cont								
6710	2 qt.				Ea.	9.75			9.75	10.75
6720	1 gal.				"	10.45			10.45	11.50
6800	Macfadyena unguis-cali (Cat Claw Vine), Z8, cont								
6810	1 gal.				Ea.	8.10			8.10	8.90
6820	5 gal.				"	21			21	23.50
6900	Myoporum parvifolium (Prostrate Myoporum), Z9, cont								
6910	1 gal.				Ea.	2.56			2.56	2.82
6920	5 gal.				"	7.55			7.55	8.30
7000	Nepeta (Persian Catmint), Z5, cont								
7010	1 qt.				Ea.	4.61			4.61	5.05
7020	1 gal.				"	32			32	35
7100	Ophiupogon japonicus (Mondo-Grass), Z7, cont								
7110	4" pot				Ea.	5.25			5.25	5.80
7120	1 gal.					10.60			10.60	11.65
7130	5 gal.					15.90			15.90	17.50
7200	Pachysandra terminalis (Japanese Spurge), Z5, cont								

32 93 Plants

32 93 13 - Ground Covers

32 93 13.30 Ground Covers	Crew	Daily Output	Labor-Hours	Unit	Material	2020 Bare Costs Labor	Equipment	Total	Total Incl O&P
7210	2-1/2" pot				Ea.	9.45			9.45	10.40
7220	1 gal.					31.50			31.50	34.50
7230	Tray/50 plants					56.50			56.50	62
7240	Tray/100 plants				↓	111			111	122
7300	Parthenocissus quinquefolia (Virginia Creeper), Z3, cont									
7310	3" pot				Ea.	6.90			6.90	7.60
7320	1 gal.				"	7.05			7.05	7.75
7400	Parthenocissus tricuspidata veitchi (Boston Ivy), Z4, cont									
7410	3" pot				Ea.	10.90			10.90	12
7420	1 gal.				↓	34			34	37
7430	5 gal.				↓	55.50			55.50	61
7500	Phlox subulata (Moss Phlox, Creeping Phlox), Z4, cont									
7510	1 qt.				Ea.	18.70			18.70	20.50
7520	1 gal.				"	29.50			29.50	32.50
7600	Polygonum auberti (Silver Fleece Vine), Z4, cont									
7610	1 qt.				Ea.	16.50			16.50	18.15
7620	1 gal.				"	27.50			27.50	30
7700	Polygonum cuspidatum compactum (Fleece Flower), Z4, cont									
7710	1 qt.				Ea.	20			20	22
7720	1 gal.				"	29			29	32
7800	Potentilla verna (Spring Cinquefoil), Z6, cont									
7810	1 qt.				Ea.	5.50			5.50	6.05
7820	1 gal.				"	8.45			8.45	9.25
7900	Rosmarinus officinalis "Prostratus" (Creeping Rosemary), Z8, cont									
7910	3" pot				Ea.	11.25			11.25	12.35
7920	1 gal.				"	38			38	42
8000	Sagina subulata (Irish Moss, Scotch Moss), Z6, cont									
8010	3" pot				Ea.	8.85			8.85	9.70
8020	4" pot					10.10			10.10	11.10
8030	1 qt.				↓	23.50			23.50	25.50
8100	Sedum (Stonecrop), Z4, cont									
8110	1 qt.				Ea.	7.90			7.90	8.70
8120	1 gal.				"	17.80			17.80	19.55
8200	Solanum jasminoides (Jasmine Nightshade), Z9, cont									
8210	1 gal.				Ea.	18.30			18.30	20
8220	5 gal.				"	108			108	118
8300	Stachys byzantina (Lamb's ears), Z4, cont									
8310	1 qt.				Ea.	22.50			22.50	25
8320	1 gal.				"	28			28	31
8400	Thymus pseudolanuginosus (Wooly Thyme), Z5, cont									
8410	3" pot				Ea.	10.60			10.60	11.65
8420	1 qt.					15.10			15.10	16.60
8430	1 gal.				↓	18			18	19.80
8500	Vaccinium crassifolium procumbent (Creeping Blueberry), Z7, cont									
8510	1 gal.				Ea.	20			20	22
8520	3 gal.				"	52.50			52.50	57.50
8600	Verbena peruviana "St. Paul", Z8, cont									
8610	2 qt.				Ea.	9.70			9.70	10.65
8620	1 gal.				"	15.30			15.30	16.85
8700	Veronica repens (Creeping Speedwell), Z5, cont									
8710	1 qt.				Ea.	18.50			18.50	20.50
8720	1 gal.				"	24.50			24.50	27
8800	Vinca major variegata (Variegated Greater Periwinkle), Z7, cont									

32 93 13.30 Ground Covers

		Crew	Daily Output	Labor-Hours	Unit	Material	2020 Bare Costs Labor	Equipment	Total	Total Incl O&P
8810	2-1/4" pot				Ea.	7.35			7.35	8.10
8820	4" pot					13.70			13.70	15.05
8830	Bare Root				↓	1.66			1.66	1.83
8900	Vinca minor (Periwinkle), Z4, cont									
8910	2-1/4" pot				Ea.	7.20			7.20	7.90
8920	4" pot					10.25			10.25	11.30
8930	Bare Root				↓	1.85			1.85	2.04
9000	Waldsteinia sibirica (Barren Strawberry), Z4, cont									
9010	1 qt.				Ea.	12.20			12.20	13.45
9020	1 gal.				"	14.85			14.85	16.30
9100	Wisteria sinensis (Chinese Wisteria), Z5, cont									
9110	1 gal.				Ea.	36.50			36.50	40
9120	3 gal.				"	55.50			55.50	61

32 93 13.40 Ornamental Grasses

		Crew	Daily Output	Labor-Hours	Unit	Material	2020 Bare Costs Labor	Equipment	Total	Total Incl O&P
0010	**ORNAMENTAL GRASSES**									
2000	Acorus gramineus (Japanese Sweet Flag), Z7, cont									
2010	1 gal.				Ea.	7.95			7.95	8.75
2100	Arundo donax (Giant Reed), Z7, cont									
2110	1 qt.				Ea.	8.90			8.90	9.75
2120	1 gal.				"	44			44	48.50
2200	Calamagrostis acutiflora stricta (Feather Reed Grass), Z5, cont									
2210	1 qt.				Ea.	8.50			8.50	9.35
2220	1 gal.					19.50			19.50	21.50
2230	2 gal.				↓	36			36	39.50
2300	Carex elata "Bowles Golden" (Variegated Sedge), Z5, cont									
2310	1 qt.				Ea.	5.05			5.05	5.60
2320	1 gal.				"	16.40			16.40	18.05
2400	Carex morrowii "aureo variegata" (Japanese Sedge), Z5, cont									
2410	1 qt.				Ea.	6.60			6.60	7.25
2420	1 gal.					15.50			15.50	17.05
2430	2 gal.				↓	23.50			23.50	26
2500	Cortaderia selloana (Pampas Grass), Z8, cont									
2510	1 gal.				Ea.	9.85			9.85	10.80
2520	2 gal.					18.25			18.25	20
2530	3 gal.					33			33	36.50
2540	5 gal.				↓	60			60	66
2600	Deschampsia caespitosa (Tufted Hair Grass), Z4, cont									
2610	1 qt.				Ea.	7.10			7.10	7.80
2620	1 gal.				"	9			9	9.90
2700	Elymus glaucus (Blue Wild Rye), Z4, cont									
2710	1 qt.				Ea.	7.95			7.95	8.70
2720	1 gal.				"	11.95			11.95	13.15
2800	Erianthus ravennae (Ravenna Grass), Z6, cont									
2810	1 gal.				Ea.	24			24	26.50
2820	2 gal.					30.50			30.50	33.50
2830	7 gal.				↓	48			48	53
2900	Festuca cineria "Sea Urchin" (Sea Urchin Blue Fescue), Z6, cont									
2910	1 qt.				Ea.	7.15			7.15	7.85
2920	1 gal.					11			11	12.10
2930	2 gal.				↓	12.85			12.85	14.15
3000	Glyceria maxima variegata (Manna Grass), Z5, cont									
3010	1 qt.				Ea.	15.75			15.75	17.35

32 93 Plants

32 93 13 – Ground Covers

32 93 13.40 Ornamental Grasses		Crew	Daily Output	Labor-Hours	Unit	Material	2020 Bare Costs Labor	Equipment	Total	Total Incl O&P
3020	1 gal.				Ea.	24			24	26
3100	Hakonechloa macra aureola (Variegated Hakonechloa), Z7, cont									
3110	1 qt.				Ea.	15.55			15.55	17.10
3120	1 gal.				"	23			23	25
3200	Helictotrichon sempervirens (Blue Oat Grass), Z4, cont									
3210	1 qt.				Ea.	6.75			6.75	7.45
3220	1 gal.					7.20			7.20	7.90
3230	2 gal.				↓	18.40			18.40	20.50
3300	Imperata cylindrica rubra (Japanese Blood Grass), Z7, cont									
3310	1 qt.				Ea.	8.25			8.25	9.05
3320	1 gal.					10.25			10.25	11.25
3330	2 gal.				↓	9.85			9.85	10.80
3400	Juncus effusus spiralis (Corkscrew Rush), Z4, cont									
3410	1 qt.				Ea.	7.15			7.15	7.90
3420	2 gal.				"	13.50			13.50	14.85
3500	Koeleria glauca (Blue Hair Grass), Z6, cont									
3510	1 qt.				Ea.	7.60			7.60	8.35
3520	1 gal.					8.50			8.50	9.35
3530	3 gal.				↓	23			23	25.50
3600	Miscanthus sinensis gracillimus (Maiden Grass), Z6, cont									
3610	1 gal.				Ea.	9.30			9.30	10.25
3620	2 gal.					14.05			14.05	15.45
3630	5 gal.				↓	24			24	26.50
3700	Miscanthus sinensis zebrinus (Zebra Grass), Z6, cont									
3710	1 gal.				Ea.	9.10			9.10	10
3720	2 gal.					12.95			12.95	14.25
3730	5 gal.				↓	21.50			21.50	24
3800	Molinia caerulea arundinacea (Tall Purple Moor Grass), Z5, cont									
3810	1 gal.				Ea.	22.50			22.50	25
3820	3 gal.				"	32.50			32.50	35.50
3900	Molinia caerulea variegata (Variegated Moor Grass), Z5, cont									
3910	1 qt.				Ea.	12.35			12.35	13.55
3920	1 gal.				"	22.50			22.50	25
4000	Pennisetum alopecuroides (Fountain Grass), Z6, cont									
4010	1 qt.				Ea.	8.45			8.45	9.30
4020	1 gal.					23.50			23.50	25.50
4030	2 gal.				↓	31			31	34
4100	Phalaris arundinacea "Picta" (Ribbon Grass), Z4, cont									
4110	1 qt.				Ea.	7.30			7.30	8
4120	1 gal.					11.05			11.05	12.20
4130	2 gal.				↓	13.10			13.10	14.45
4200	Sesleria autumnalis (Autumn Moor Grass), Z5, cont									
4210	1 qt.				Ea.	8.70			8.70	9.55
4220	1 gal.				"	14.20			14.20	15.65
4300	Spartina pectinata "Aureomarginata" (Cord Grass), Z5, cont									
4310	1 gal.				Ea.	13.90			13.90	15.30
4320	2 gal.				"	22			22	24.50
4400	Stipa gigantea (Giant Feather Grass), Z7, cont									
4410	1 gal.				Ea.	10.40			10.40	11.45
4420	2 gal.				"	13.95			13.95	15.30

32 93 23.10 Perennials		Crew	Daily Output	Labor-Hours	Unit	Material	2020 Bare Costs Labor	Equipment	Total	Total Incl O&P
0010	**PERENNIALS**									
0100	Achillea filipendulina (Fernleaf Yarrow), Z2, cont									
0110	1 gal.				Ea.	12.95			12.95	14.25
0120	2 gal.					14.40			14.40	15.85
0130	3 gal.				↓	18.80			18.80	20.50
0200	Agapanthus africanus (Lily-of-the-Nile), Z9, cont									
0210	1 gal.				Ea.	11.80			11.80	13
0220	5 gal.				"	21			21	23.50
0300	Amsonia ciliata (Blue Star), Z7, cont									
0310	1 qt.				Ea.	25.50			25.50	28
0320	1 gal.				"	34.50			34.50	37.50
0400	Amsonia tabernaemontana (Willow Amsonia), Z3, cont									
0410	1 qt.				Ea.	18.30			18.30	20
0420	1 gal.				"	21			21	23
0500	Anchusa azurea (Italian Bugloss), Z3, cont									
0510	4" pot				Ea.	15.80			15.80	17.35
0520	1 qt.					23.50			23.50	26
0530	1 gal.				↓	34.50			34.50	38
0600	Anemone x hybrida (Japanese Anemone), Z5, cont									
0610	4" pot				Ea.	10.10			10.10	11.10
0620	1 qt.					12.20			12.20	13.40
0630	2 gal.				↓	18.55			18.55	20.50
0700	Anthemis tinctoria (Golden Marguerite), Z3, cont									
0710	4" pot				Ea.	5.70			5.70	6.30
0720	1 qt.					9.65			9.65	10.60
0730	1 gal.				↓	16.25			16.25	17.90
0800	Aonitum napellus (Aconite Monkshood), Z2, cont									
0810	4" pot				Ea.	6.85			6.85	7.55
0820	1 qt.					11.45			11.45	12.60
0830	1 gal.				↓	19.75			19.75	21.50
0900	Aquilegia species and hybrids (Columbine), Z2, cont									
0910	2" pot				Ea.	4.79			4.79	5.25
0920	1 qt.					10.10			10.10	11.10
0930	1 gal.				↓	11.45			11.45	12.60
1000	Aruncus dioicus (Goatsbeard), Z4, cont									
1010	1 qt.				Ea.	13.25			13.25	14.60
1020	1 gal.					16.50			16.50	18.15
1030	2 gal.				↓	17.80			17.80	19.60
1100	Asclepias tuberosa (Butterfly Flower), Z4, cont									
1110	1 gal.				Ea.	19.95			19.95	22
1120	2 gal.				"	28			28	30.50
1200	Aster species and hybrids (Aster, Michaelemas Daisy), Z4, cont									
1210	1 gal.				Ea.	18.35			18.35	20
1220	2 gal.					25.50			25.50	28.50
1230	3 gal.				↓	33.50			33.50	36.50
1300	Astilbe species and hybrids (False Spirea), Z4, cont									
1310	4" pot				Ea.	6.50			6.50	7.15
1320	1 gal.					11.60			11.60	12.75
1330	2 gal.				↓	12.35			12.35	13.55
1400	Aubrieta deltoidea (Purple Rock Cress), Z4, cont									
1410	4" pot				Ea.	7.50			7.50	8.25
1420	1 qt.				↓	8.90			8.90	9.80

For customer support on your Site Work & Landscape Costs with RSMeans data, call 800.448.8182.

32 93 23 – Plants and Bulbs

32 93 23.10 Perennials	Crew	Daily Output	Labor-Hours	Unit	Material	2020 Bare Costs Labor	Equipment	Total	Total Incl O&P
1430	1 gal.				Ea.	19.95			19.95	22
1500	Aurinia saxatilis (Basket of Gold), Z3, cont									
1510	4" pot				Ea.	10.90			10.90	12
1520	1 qt.					22.50			22.50	25
1530	1 gal.				↓	28			28	31
1600	Baptisia australis (False Indigo), Z3, cont									
1610	1 qt.				Ea.	8.30			8.30	9.15
1620	1 gal.				"	11.35			11.35	12.45
1700	Begonia grandis (Evans Begonia), Z6, cont									
1710	1 qt.				Ea.	14.35			14.35	15.75
1720	1 gal.				"	28			28	31
1800	Belamcanda chinensis (Blackberry Lily), Z5, cont									
1810	1 qt.				Ea.	12			12	13.20
1820	1 gal.				"	26			26	28.50
1900	Boltonia asteroides "Snowbank" (White Boltonia), Z3, cont									
1910	1 qt.				Ea.	14.45			14.45	15.85
1920	1 gal.				"	28			28	30.50
2000	Brunnera macrophylla (Siberian Bugloss), Z3, cont									
2010	1 qt.				Ea.	8.60			8.60	9.45
2020	1 gal.				"	10.85			10.85	11.95
2100	Caltha palustris (Marsh Marigold), Z3, cont									
2110	1 qt.				Ea.	14.05			14.05	15.50
2120	1 gal.				"	26.50			26.50	29
2200	Campanula carpatica "Blue Chips" (Bellflower), Z3, cont									
2210	1 gal.				Ea.	11.10			11.10	12.20
2220	2 gal.				"	15.10			15.10	16.60
2300	Ceratostigma plumbaginoides (Blue Plumbago), Z6, cont									
2310	1 qt.				Ea.	17.40			17.40	19.15
2320	1 gal.				"	31			31	34
2400	Chrysanthemum hybrids (Hardy Chrysanthemum), Z5, cont									
2410	1 qt.				Ea.	11.90			11.90	13.10
2420	6" to 8" pot					18.70			18.70	20.50
2430	1 gal.				↓	20			20	22
2500	Chrysogonum virginianum (Golden Star), Z6, cont									
2510	1 gal.				Ea.	19.10			19.10	21
2520	2 gal.				"	25.50			25.50	28
2600	Cimicifuga racemosa (Bugbane), Z5, cont									
2610	4" pot				Ea.	10.45			10.45	11.50
2620	1 qt.					16.70			16.70	18.35
2630	1 gal.				↓	29			29	31.50
2700	Coreopsis verticillata (Threadleaf Coreopsis), Z5, cont									
2710	2" pot				Ea.	10.60			10.60	11.65
2720	1 gal.				"	14.75			14.75	16.25
2800	Delphinium elatum and hybrids (Delphinium), Z2, cont									
2810	1 qt.				Ea.	8.95			8.95	9.85
2820	1 gal.					33.50			33.50	36.50
2830	2 gal.				↓	12.90			12.90	14.20
2900	Dianthus species and hybrids (Cottage Pink/Scotch Pink), Z3, cont									
2910	1 gal.				Ea.	11.80			11.80	12.95
2920	2 gal.				"	14.45			14.45	15.90
3000	Dicentra spectabilis (Bleeding Heart), Z3, cont									
3010	2 qt.				Ea.	16.90			16.90	18.55
3020	1 gal.				↓	17.15			17.15	18.85

415

32 93 23.10 Perennials		Crew	Daily Output	Labor-Hours	Unit	Material	2020 Bare Costs Labor	Equipment	Total	Total Incl O&P
3030	2 gal.				Ea.	24			24	26.50
3100	Dictamnus (Gas Plant), Z3, cont									
3110	1 gal.				Ea.	34			34	37
3120	2 gal.					42.50			42.50	47
3130	3 gal.				↓	55			55	60.50
3200	Digitalis purpurea (Foxglove), Z4, cont									
3210	1 gal.				Ea.	20.50			20.50	22.50
3220	2 gal.					64			64	70.50
3230	3 gal.				↓	72.50			72.50	80
3300	Doronicum cordatun (Leopard's Bane), Z4, cont									
3310	2 qt.				Ea.	9.80			9.80	10.80
3320	1 gal.				"	16.65			16.65	18.35
3400	Echinacea purpurea (Purple Coneflower), Z3, cont									
3410	2 qt.				Ea.	11.15			11.15	12.30
3420	1 gal.					18.85			18.85	20.50
3430	2 gal.				↓	20			20	22
3500	Echinops exaltatus (Globe Thistle), Z3, cont									
3510	1 qt.				Ea.	9.55			9.55	10.50
3520	1 gal.				"	18.35			18.35	20
3600	Erigeron speciosus (Oregon Fleabane), Z3, cont									
3610	1 qt.				Ea.	8.90			8.90	9.80
3620	1 gal.				"	16.80			16.80	18.50
3700	Euphorbia epithymoides (Cushion Spurge), Z4, cont									
3710	1 qt.				Ea.	10			10	11
3720	1 gal.				"	12.10			12.10	13.30
3800	Filipendula hexapetala (Meadowsweet), Z5, cont									
3810	1 qt.				Ea.	9.85			9.85	10.85
3820	1 gal.				"	10.70			10.70	11.75
3900	Gaillardia aristata (Blanket Flower), Z2, cont									
3910	1 qt.				Ea.	10.40			10.40	11.45
3920	1 gal.					9.80			9.80	10.75
3930	2 gal.				↓	16.05			16.05	17.70
4000	Geranium sanguineum (Bloodred Geranium), Z3, cont									
4010	2" pot				Ea.	5.15			5.15	5.70
4020	1 gal.				"	11.60			11.60	12.75
4100	Geum x borisii (Boris Avens), Z3, cont									
4110	1 qt.				Ea.	8.25			8.25	9.10
4120	1 gal.				"	11.25			11.25	12.35
4200	Gypsophila paniculata (Baby's Breath), Z3, cont									
4210	2 qt.				Ea.	9.90			9.90	10.90
4220	1 gal.					12.75			12.75	14.05
4230	3 gal.				↓	43.50			43.50	47.50
4300	Helenium autumnale (Sneezeweed), Z3, cont									
4310	2 qt.				Ea.	10.50			10.50	11.55
4320	1 gal.					13.05			13.05	14.35
4330	2 gal.				↓	17.30			17.30	19.05
4400	Helianthus salicifolius (Willow Leaf Sunflower), Z5, cont									
4410	1 qt.				Ea.	7.85			7.85	8.60
4420	1 gal.					16.95			16.95	18.60
4430	2 gal.				↓	22.50			22.50	24.50
4500	Heliopsis helianthoides (False Sunflower), Z3, cont									
4510	2 qt.				Ea.	10.20			10.20	11.20
4520	1 gal.				↓	10.95			10.95	12.05

32 93 23.10 Perennials		Crew	Daily Output	Labor-Hours	Unit	Material	2020 Bare Costs Labor	Equipment	Total	Total Incl O&P
4530	2 gal.				Ea.	17.30			17.30	19.05
4600	Helleborus niger (Christmas Rose), Z3, cont									
4610	2 qt.				Ea.	14.75			14.75	16.20
4620	1 gal.				"	23.50			23.50	26
4700	Hemerocallis hybrids (Diploid) (Day Lilly), Z3, cont									
4710	1 qt.				Ea.	8.70			8.70	9.60
4720	1 gal.					11.20			11.20	12.35
4730	3 gal.				↓	19.20			19.20	21
4800	Hemerocallis hybrids (Tetraploid), Z3, cont									
4810	1 qt.				Ea.	8.15			8.15	9
4820	1 gal.					15.30			15.30	16.85
4830	3 gal.				↓	22.50			22.50	24.50
4900	Hibiscus moscheutos (Rose Mallow), Z5, cont									
4910	1 qt.				Ea.	8.25			8.25	9.05
4920	2 gal.					42			42	46
4930	3 gal.				↓	57.50			57.50	63
5000	Hosta species and hybrids (Plantain Lily), Z3, cont									
5010	1 gal.				Ea.	12.15			12.15	13.35
5020	2 gal.					12.95			12.95	14.25
5030	3 gal.				↓	22			22	24
5100	Iris hybrids (Pacific Coast Iris), Z6, cont									
5110	1 gal.				Ea.	12			12	13.20
5120	3 gal.				"	17			17	18.70
5200	Iris sibiricu and hybrids (Siberian Iris), Z3, cont									
5210	1 gal.				Ea.	10.60			10.60	11.65
5220	4 gal.					20			20	22
5230	3" pot				↓	6			6	6.60
5300	Kniphofia uvaria and hybrids (Torch Lily), Z7, cont									
5310	1 qt.				Ea.	8.95			8.95	9.85
5320	1 gal.				"	10.60			10.60	11.65
5400	Liatris spicata (Blazing Star, Gayfeather), Z3, cont									
5410	1 gal.				Ea.	10.65			10.65	11.75
5420	2 gal.				"	14.85			14.85	16.35
5500	Ligularia dentata (Golden Groundsel), Z5, cont									
5510	2 qt.				Ea.	10.15			10.15	11.15
5520	2 gal.				"	13.90			13.90	15.30
5600	Linum perenne (Perennial Flax), Z4, cont									
5610	1 qt.				Ea.	9.35			9.35	10.30
5620	1 gal.				"	11.85			11.85	13.05
5700	Lobelia cardinalis (Cardinal Flower), Z3, cont									
5710	1 qt.				Ea.	9.25			9.25	10.15
5720	1 gal.				"	10.90			10.90	12
5800	Lupinus "Rusell Hybrids" (Russell Lupines), Z3, cont									
5810	1 gal.				Ea.	11.85			11.85	13.05
5820	2 gal.					14.15			14.15	15.60
5830	3 gal.				↓	17.70			17.70	19.50
5900	Lychnis chalcedonica (Maltese Cross), Z3, cont									
5910	1 qt.				Ea.	8.15			8.15	8.95
5920	1 gal.				"	11.75			11.75	12.90
6000	Lysimachia clethroides (Gooseneck Loosestrife), Z3, cont									
6010	1 qt.				Ea.	8.25			8.25	9.10
6020	1 gal.				"	10.75			10.75	11.85
6100	Lythrum salicaria (Purple Loosestrife), Z3, cont									

417

32 93 23.10 Perennials		Crew	Daily Output	Labor-Hours	Unit	Material	2020 Bare Costs Labor	Equipment	Total	Total Incl O&P
6110	1 qt.				Ea.	8.25			8.25	9.05
6120	1 gal.				"	12.40			12.40	13.65
6200	Macleaya cordata (Plume Poppy), Z3, cont									
6210	1 qt.				Ea.	8.25			8.25	9.05
6220	1 gal.				"	12.55			12.55	13.80
6300	Mertensia virginica (Virginia Bluebells), Z3, cont									
6310	1 qt.				Ea.	7.30			7.30	8
6320	1 gal.				"	14.45			14.45	15.85
6400	Paeonia hybrids (Herbaceous Peony), Z5, cont									
6410	1 gal.				Ea.	14.85			14.85	16.35
6420	2 gal.					17.55			17.55	19.30
6430	3 gal.				↓	109			109	120
6500	Papaver orientalis (Oriental Poppy), Z3, cont									
6510	1 qt.				Ea.	8.70			8.70	9.55
6520	1 gal.				"	10.80			10.80	11.90
6600	Penstemon azureus (Azure Penstemon), Z8, cont									
6610	2" pot				Ea.	5.45			5.45	6
6620	1 gal.				"	10.65			10.65	11.70
6700	Phlox paniculata (Garden Phlox), Z4, cont									
6710	1 qt.				Ea.	8.90			8.90	9.80
6720	1 gal.				"	18.45			18.45	20.50
6800	Physostegia virginiana (False Dragonhead), Z3, cont									
6810	4" pot				Ea.	5.35			5.35	5.90
6820	1 gal.				"	10.75			10.75	11.80
6900	Platycodon grandiflorus (Balloon Flower), Z3, cont									
6910	2" pot				Ea.	4.69			4.69	5.15
6920	1 gal.				"	12.40			12.40	13.65
7000	Pulmonaria saccharata (Bethlehem Sage), Z4, cont									
7010	4" pot				Ea.	5.60			5.60	6.20
7020	1 qt.					9.50			9.50	10.45
7030	1 gal.				↓	14.20			14.20	15.60
7100	Rudbeckia fulgida "Goldstrum" (Black-eyed Susan), Z4, cont									
7110	1 qt.				Ea.	9.15			9.15	10.05
7120	1 gal.					11.70			11.70	12.85
7130	2 gal.				↓	14.45			14.45	15.90
7200	Salvia x superba (Perennial Salvia), Z5, cont									
7210	4" pot				Ea.	5.75			5.75	6.30
7220	1 qt.					10.10			10.10	11.15
7230	1 gal.				↓	11.40			11.40	12.55
7300	Scabiosa caucasica (Pincushion Flower), Z3, cont									
7310	1 qt.				Ea.	9.30			9.30	10.25
7320	1 gal.				"	11.20			11.20	12.30
7400	Sedum spectabile (Showy Sedum), Z3, cont									
7410	4" pot				Ea.	6.35			6.35	7
7420	2" pot					6.55			6.55	7.20
7430	1 gal.				↓	10.95			10.95	12
7500	Solidago hybrids (Goldenrod), Z3, cont									
7510	1 qt.				Ea.	8.75			8.75	9.60
7520	1 gal.				"	10.60			10.60	11.70
7700	Trollius europaeus (Common Globeflower), Z4, cont									
7710	1 qt.				Ea.	9.05			9.05	9.95
7720	1 gal.				"	15.20			15.20	16.70
7800	Veronica hybrids (Speedwell), Z4, cont									

32 93 Plants

32 93 23 – Plants and Bulbs

32 93 23.10 Perennials		Crew	Daily Output	Labor-Hours	Unit	Material	2020 Bare Costs Labor	2020 Bare Costs Equipment	Total	Total Incl O&P
7810	1 gal.				Ea.	11.95			11.95	13.15
7820	2 gal.				"	19.60			19.60	21.50

32 93 33 – Shrubs

32 93 33.40 Shrubs

		Crew	Daily Output	Labor-Hours	Unit	Material	Labor	Equipment	Total	Total Incl O&P
0010	**SHRUBS**, Temperate Z2 to 6									
1000	Abelia grandiflora (Abelia), Z6, cont									
1002	15" to 18"				Ea.	33.50			33.50	37
1003	18" to 21"				"	56.50			56.50	62
1050	Acanthopanax sieboldianus (Five-Leaved Azalea), Z5, cont/B&B									
1051	15" to 18"				Ea.	35.50			35.50	39.50
1052	3' to 4'					43			43	47.50
1053	4' to 5'					41			41	45
1100	Aronia arbutifolia brilliantissima (Brilliant Chokeberry), Z5, B&B									
1101	2' to 3'				Ea.	33			33	36
1102	3' to 4'					38.50			38.50	42
1103	4' to 5'					52			52	57
1104	5' to 6'					61.50			61.50	67.50
1150	Aronia melancarpa "Elata" (Elata Black Chokeberry), Z3, B&B									
1151	2' to 3'				Ea.	16.75			16.75	18.45
1152	3' to 4'				"	19.05			19.05	21
1200	Azalea arborescens (Sweet Azalea), Z5, cont									
1201	15" to 18"				Ea.	17.55			17.55	19.30
1202	18" to 24"				"	39			39	43
1250	Azalea Evergreen Hybrids (Azalea Evergreen), Z6, cont/B&B									
1251	1 gal.				Ea.	7.20			7.20	7.90
1252	2 gal.					10.85			10.85	11.90
1253	3 gal.					16.70			16.70	18.40
1254	5 gal.					21			21	23
1255	15" to 18"					16			16	17.60
1256	18" to 24"					22			22	24.50
1257	2' to 2-1/2'					42			42	46.50
1258	2-1/2' to 3'					53			53	58.50
1300	Azalea exbury, Ilam Hybrids (Azalea exbury), Z5, cont/B&B									
1301	1 gal.				Ea.	7.95			7.95	8.70
1302	2 gal.					11.50			11.50	12.65
1303	15" to 18"					18.40			18.40	20
1304	18" to 24"					21.50			21.50	23.50
1305	2' to 2-1/2'					50			50	55
1306	2-1/2' to 3'					45			45	49.50
1307	4' to 5'					77			77	84.50
1350	Azalea gandavensis (Ghent Azalea), Z4, B&B									
1351	15" to 18"				Ea.	33			33	36.50
1352	18" to 24"					36			36	39.50
1353	2' to 3'					41.50			41.50	45.50
1354	3' to 4'					50.50			50.50	55.50
1355	4' to 5'					68.50			68.50	75
1356	5' to 6'					86.50			86.50	95.50
1400	Azalea kaempferi (Torch Azalea), Z6, cont/B&B									
1401	12" to 15"				Ea.	10.35			10.35	11.40
1402	15" to 18"					15.40			15.40	16.90
1403	18" to 24"					23			23	25.50
1404	2' to 3'					39			39	43

32 93 33.40 Shrubs	Crew	Daily Output	Labor-Hours	Unit	Material	2020 Bare Costs Labor	Equipment	Total	Total Incl O&P
1450 Berberis julianae "Nana" (Dwarf Wintergreen Barberry), Z6, B&B									
1451 18" to 24"				Ea.	22.50			22.50	25
1452 2' to 2-1/2'					25			25	27.50
1453 2-1/2' to 3'					29.50			29.50	32.50
1500 Berberis mentorensis (Mentor Barberry), Z6, cont/B&B									
1501 1 gal.				Ea.	9.05			9.05	10
1502 5 gal.					18.55			18.55	20.50
1503 2' to 3'					23			23	25.50
1504 3' to 4'					29			29	32
1550 Berberis thumbergii (Crimson Pygmy Barberry), Z5, cont									
1551 10" to 12"				Ea.	9.05			9.05	9.95
1552 12" to 15"					14.90			14.90	16.40
1553 15" to 18"					21.50			21.50	23.50
1600 Berberis thunbergi (Japanese Green Barberry), Z5, cont/B&B									
1601 9" to 12"				Ea.	4.28			4.28	4.71
1602 12" to 15"					14.15			14.15	15.60
1603 15" to 18"					14			14	15.40
1604 18" to 24"					26			26	28.50
1605 2' to 3'					36			36	39.50
1650 Berberis verruculosa (Barberry), Z5, cont/B&B									
1651 15" to 18"				Ea.	22			22	24
1652 18" to 24"				"	30			30	33
1700 Buddleia davidii (Orange-Eye Butterfly Bush), Z5, cont									
1701 9" to 12"				Ea.	12.05			12.05	13.30
1702 15" to 18"				"	16.50			16.50	18.15
1750 Buxus microphylla japonica (Japanese Boxwood), Z5, cont/B&B									
1751 1 gal.				Ea.	5.35			5.35	5.90
1752 8" to 10"					10.25			10.25	11.30
1753 12" to 15"					13.80			13.80	15.15
1754 15" to 18"					41			41	45
1755 18" to 24"					48			48	53
1756 24" to 30"					92			92	101
1757 30" to 36"					86			86	94.50
1800 Buxus sempervirens (Common Boxwood), Z5, cont/B&B									
1801 10" to 12"				Ea.	9.70			9.70	10.65
1802 15" to 18"					22			22	24
1803 18" to 24"					44.50			44.50	49
1804 24" to 30"					58			58	64
1850 Callicarpa japonica (Japanese Beautyberry), Z6, B&B									
1851 2' to 3'				Ea.	20.50			20.50	22.50
1852 3' to 4'					27.50			27.50	30
1853 4' to 5'					33.50			33.50	36.50
1900 Caragana arborescens (Siberian Peashrub), Z2, B&B									
1901 18" to 24"				Ea.	14.15			14.15	15.55
1902 2' to 3'					28			28	31
1903 5' to 6'					61.50			61.50	67.50
1904 6' to 8'					82			82	90
1950 Caragana arborescens (Walkers Weeping Peashrub), Z3, B&B									
1951 1-1/2" to 1-3/4" Cal.				Ea.	136			136	150
1952 1-3/4" to 2" Cal.				"	173			173	190
2000 Caryopteris x clandonensis (Bluebeard), Z5, cont									
2001 12" to 15"				Ea.	12.80			12.80	14.05
2002 1 gal.					21.50			21.50	24

For customer support on your Site Work & Landscape Costs with RSMeans data, call 800.448.8182.

32 93 33 – Shrubs

32 93 33.40 Shrubs	Crew	Daily Output	Labor-Hours	Unit	Material	2020 Bare Costs Labor	Equipment	Total	Total Incl O&P
2003	18" to 24"				Ea.	32			32	35
2050	Chaenomeles speciosa (Flowering Quince), Z5, cont/B&B									
2051	10" to 12"				Ea.	13.60			13.60	14.95
2052	2' to 3'					18.60			18.60	20.50
2053	3' to 4'					23			23	25.50
2100	Clethra alnifolia rosea (Pink Summersweet), Z4, B&B									
2101	2' to 3'				Ea.	24.50			24.50	27
2102	3' to 4'					32.50			32.50	35.50
2103	4' to 5'					35.50			35.50	39.50
2150	Cornus alba siberica (Siberian Dogwood), Z3, B&B									
2151	3' to 4'				Ea.	26.50			26.50	29.50
2152	4' to 5'					35.50			35.50	39
2153	5' to 6'					59			59	64.50
2154	6' to 7'					56			56	61.50
2200	Cornus alternifolia (Alternate Leaved Dogwood), Z3, B&B									
2201	4' to 5'				Ea.	74			74	81.50
2202	5' to 6'					91.50			91.50	101
2203	6' to 8'					95.50			95.50	105
2250	Cornus mas (Cornelian Cherry), Z5, B&B									
2251	2' to 3'				Ea.	73.50			73.50	80.50
2252	3' to 4'					91			91	100
2253	4' to 5'					94			94	103
2254	5' to 6'					111			111	123
2300	Cornus racemosa (Gray Dogwood), Z4, B&B									
2301	18" to 24"				Ea.	26.50			26.50	29
2302	2' to 3'					36			36	40
2303	3' to 4'					46.50			46.50	51.50
2350	Cornus stolonifera flaviramea (Goldentwig Dogwood), Z3, cont/B&B									
2351	2' to 3'				Ea.	29			29	32
2352	3' to 4'					42			42	46.50
2353	4' to 5'					46.50			46.50	51
2354	5' to 6'					56			56	61.50
2355	6' to 7'					67			67	73.50
2400	Corylus americana (Filbert, American Hazelnut), Z4, B&B									
2401	2-1/2' to 3'				Ea.	17.25			17.25	18.95
2402	3' to 4'					22.50			22.50	24.50
2403	4' to 5'					58.50			58.50	64.50
2450	Cotinus coggygria (Smoke Tree), Z5, B&B									
2451	5 gal.				Ea.	49			49	54
2452	6' to 8'					65			65	71.50
2453	8' to 10'					105			105	115
2500	Cotoneaster acutifolius (Peking Cotoneaster), Z4, B&B									
2501	2' to 3'				Ea.	13.10			13.10	14.45
2502	3' to 4'					20.50			20.50	22.50
2503	4' to 5'					26			26	29
2550	Cotoneaster adpressa (Creeping Cotoneaster), Z5, cont									
2551	15" to 18" spread				Ea.	14.20			14.20	15.65
2600	Cotoneaster adpressa praecox (Early Cotoneaster), Z5, cont									
2601	12" to 15"				Ea.	20.50			20.50	22.50
2602	15" to 18"					34.50			34.50	38
2603	18" to 24"					24.50			24.50	27
2650	Cotoneaster divaricatus (Spreading Cotoneaster), Z5, cont/B&B									
2651	1 gal.				Ea.	23.50			23.50	26

32 93 33.40 Shrubs		Crew	Daily Output	Labor-Hours	Unit	Material	2020 Bare Costs Labor	Equipment	Total	Total Incl O&P
2652	2 gal.				Ea.	26.50			26.50	29
2653	2' to 3'					24			24	26.50
2654	3' to 4'					29.50			29.50	32.50
2655	4' to 5'				↓	35.50			35.50	39
2700	Cotoneaster horizontalis (Rock Cotoneaster), Z5, cont									
2701	12" to 15"				Ea.	14.30			14.30	15.70
2702	15" to 18"					33.50			33.50	37
2703	18" to 21"				↓	40.50			40.50	44.50
2750	Cytisus praecox (Warminster Broom), Z6, cont									
2751	1 gal.				Ea.	5.50			5.50	6.05
2752	2 gal.					12			12	13.20
2753	5 gal.				↓	22			22	24
2800	Daphne burkwoodi somerset (Burkwood Daphne), Z6, cont									
2801	15" to 18"				Ea.	25			25	27.50
2802	18" to 24"				"	40			40	44
2850	Deutzia gracilis (Slender Deutzia), Z5, cont									
2851	1 gal.				Ea.	11.25			11.25	12.40
2852	2 gal.				"	15.35			15.35	16.90
2900	Diervilla sessilifolia (Southern Bush Honeysuckle), Z5, B&B									
2901	18" to 24"				Ea.	29			29	32
2902	2' to 3'					27.50			27.50	30.50
2903	3' to 4'				↓	40			40	44
2950	Elaeagnus angustifolia (Russian Olive), Z3, B&B									
2951	5 gal.				Ea.	29			29	32
3000	Elaeagnus umbellata (Autumn Olive), Z3, B&B									
3001	3' to 4'				Ea.	44			44	48.50
3002	4' to 5'					101			101	111
3003	5' to 6'				↓	170			170	187
3050	Enkianthus campanulatus (Bellflowertree), Z5, cont/B&B									
3051	2' to 3'				Ea.	46.50			46.50	51
3052	3' to 4'					54.50			54.50	60
3053	4' to 5'					61.50			61.50	68
3054	5' to 6'				↓	74			74	81.50
3100	Euonymus alatus (Winged Burning Bush), Z4, B&B									
3101	18" to 24"				Ea.	22.50			22.50	25
3102	2' to 2-1/2'					26.50			26.50	29
3103	2-1/2' to 3'					44			44	48
3104	3' to 3-1/2'					50.50			50.50	55.50
3105	3-1/2' to 4'					64			64	70
3106	4' to 5'					68			68	75
3107	5' to 6'					82			82	90.50
3108	6' to 8'				↓	85			85	93.50
3150	Euonymus alatus compacta (Dwarf Winged Burning Bush), Z4, B&B									
3151	15" to 18"				Ea.	29			29	32
3152	18" to 24"					32			32	35.50
3153	2' to 2-1/2'					37.50			37.50	41.50
3154	2-1/2' to 3'					48.50			48.50	53
3155	3' to 3-1/2'					46			46	51
3156	3-1/2' to 4'					54.50			54.50	60
3157	4' to 5'					69.50			69.50	76.50
3158	5' to 6'				↓	73.50			73.50	80.50
3200	Euonymus europaeus aldenhamensis (Spindletree), Z4, B&B									
3201	3' to 4'				Ea.	18.35			18.35	20

32 93 33.40 Shrubs		Crew	Daily Output	Labor-Hours	Unit	Material	2020 Bare Costs Labor	Equipment	Total	Total Incl O&P
3202	4' to 5'				Ea.	23.50			23.50	26
3203	5' to 6'					35.50			35.50	39
3204	6' to 8'				↓	58.50			58.50	64.50
3250	Euonymus fortunei (Wintercreeper), Z5, cont/B&B									
3251	1 gal.				Ea.	11.75			11.75	12.90
3252	2 gal.					16.10			16.10	17.70
3253	5 gal.					28			28	30.50
3254	24" to 30"				↓	26			26	28.50
3300	Euonymus fortunei "Emerald 'n Gold" (Wintercreeper), Z5, cont									
3301	10" to 12"				Ea.	4.89			4.89	5.40
3302	12" to 15"					10.70			10.70	11.80
3303	15" to 18"					24			24	26.50
3304	5 gal.				↓	22			22	24
3350	Euonymus fortunei emerald gaiety (Gaiety Wintercreeper), Z5, cont									
3351	10" to 12"				Ea.	7.50			7.50	8.25
3352	12" to 15"					14.65			14.65	16.10
3353	15" to 18"					17.15			17.15	18.90
3354	18" to 21"				↓	29.50			29.50	32.50
3400	Forsythia intermedia (Border Goldenbells), Z5, cont/B&B									
3401	5 gal.				Ea.	18.10			18.10	19.90
3402	2' to 3'					14.70			14.70	16.20
3403	3' to 4'					17.95			17.95	19.75
3404	4' to 5'					38.50			38.50	42.50
3405	5' to 6'					58			58	63.50
3406	6' to 8'				↓	55			55	60.50
3450	Forsythia ovata robusta (Korean Forsythia), Z5, B&B									
3451	2' to 3'				Ea.	23.50			23.50	25.50
3452	3' to 4'					23			23	25.50
3453	4' to 5'				↓	31			31	34
3500	Hamamelis vernalis (Vernal Witch-Hazel), Z4, B&B									
3501	2' to 3'				Ea.	42.50			42.50	47
3502	3' to 4'				↓	53			53	58
3503	4' to 5'				↓	71			71	78
3550	Hibiscus syriacus (Rose of Sharon), Z6, cont/B&B									
3551	18" to 24"				Ea.	23.50			23.50	26
3552	2' to 3'					24			24	26.50
3553	3' to 4'					25.50			25.50	28
3554	4' to 5'					49			49	54
3555	5' to 6'				↓	56.50			56.50	62
3600	Hydrangea arborescens (Smooth Hydrangea), Z4, cont									
3601	15" to 18"				Ea.	28			28	30.50
3602	18" to 24"				"	21			21	23
3650	Hydrangea macrophylla nikko blue (Blue Hydrangea), Z6, cont									
3651	18" to 24"				Ea.	23.50			23.50	26
3652	24" to 30"				"	59.50			59.50	65.50
3700	Hydrangea paniculata grandiflora (Peegee Hydrangea), Z5, B&B									
3701	2' to 3'				Ea.	30.50			30.50	33.50
3702	3' to 4'				"	51.50			51.50	57
3750	Hydrangea quercifolia (Oakleaf Hydrangea), Z6, cont									
3751	18" to 24"				Ea.	31			31	34
3800	Hypericum hidcote (St. John's Wort), Z6, cont									
3801	12" to 15"				Ea.	29			29	32
3850	Ilex cornuta "Carissa" (Carissa Holly), Z7, cont									

For customer support on your Site Work & Landscape Costs with RSMeans data, call 800.448.8182.

423

32 93 Plants

32 93 33 – Shrubs

32 93 33.40 Shrubs		Crew	Daily Output	Labor-Hours	Unit	Material	2020 Bare Costs Labor	Equipment	Total	Total Incl O&P
3851	1 gal.				Ea.	24.50			24.50	27
3852	2 gal.					28.50			28.50	31.50
3853	5 gal.					51			51	56
3900	Ilex cornuta rotunda (Dwarf Chinese Holly), Z6, cont									
3901	1 gal.				Ea.	23			23	25
3902	2 gal.					35.50			35.50	39
3903	5 gal.					54			54	59.50
3950	Ilex crenata "Helleri" (Hellers Japanese Holly), Z6, cont									
3951	24" to 30"				Ea.	23.50			23.50	26
3952	30" to 36"					40.50			40.50	45
3953	3' to 4'					49			49	54
4000	Ilex crenata "Hetzi" (Hetzi Japanese Holly), Z6, cont/B&B									
4001	2 gal.				Ea.	21			21	23
4002	3 gal.					30			30	33
4003	2' to 3'					76.50			76.50	84.50
4004	3' to 4'					103			103	113
4050	Ilex glabra compacta (Compact Inkberry), Z4, B&B									
4051	15" to 18"				Ea.	23.50			23.50	26
4052	18" to 24"					32			32	35
4053	2' to 2-1/2'					39			39	43
4100	Ilex meserve blue holly hybrids (Blue Holly), Z5, cont/B&B									
4101	15" to 18"				Ea.	21			21	23.50
4102	18" to 24"					22			22	24
4103	24" to 30"					44.50			44.50	49
4104	2-1/2' to 3'					54.50			54.50	60
4105	3-1/2' to 4'					56.50			56.50	62
4106	4' to 5'					86.50			86.50	95.50
4107	5' to 6'					131			131	144
4150	Ilex opaca (American Holly), Z6, B&B									
4151	2' to 3'				Ea.	39			39	42.50
4152	3' to 4'					64			64	70.50
4153	4' to 5'					88			88	97
4154	5' to 6'					161			161	177
4155	6' to 7'					218			218	240
4156	7' to 8'					294			294	325
4157	8' to 10'					410			410	450
4158	10' to 12'					760			760	835
4159	12' to 14'					860			860	950
4200	Ilex verticillata female (Winterberry), Z4, B&B									
4201	2' to 3'				Ea.	29.50			29.50	32.50
4202	3' to 4'				"	41			41	45
4250	Ilex verticillata male (Winterberry), Z4, B&B									
4251	2' to 3'				Ea.	33.50			33.50	37
4252	3' to 4'					41			41	45
4253	4' to 5'					58.50			58.50	64.50
4300	Ilex x attenuata "Fosteri" (Foster Holly), Z6, B&B									
4301	4' to 5'				Ea.	164			164	181
4302	5' to 6'					221			221	243
4303	6' to 7'					296			296	325
4304	7' to 8'					545			545	600
4350	Itea virginica (Virginia Sweetspire), Z6, B&B									
4351	21" to 30"				Ea.	29.50			29.50	32.50
4400	Kalmia latifolia (Mountain Laurel), Z5, cont/B&B									

For customer support on your Site Work & Landscape Costs with RSMeans data, call 800.448.8182.

32 93 33.40 Shrubs		Crew	Daily Output	Labor-Hours	Unit	Material	2020 Bare Costs Labor	Equipment	Total	Total Incl O&P
4401	15" to 18"				Ea.	45.50			45.50	50
4402	18" to 24"					64			64	70.50
4403	2' to 3'					79.50			79.50	87.50
4404	4' to 5'					131			131	144
4405	5' to 6'					181			181	199
4406	6' to 8'				↓	233			233	257
4450	Kerria japonica (Kerria), Z6, cont									
4451	12" to 15"				Ea.	17.60			17.60	19.35
4452	5 gal.				"	28.50			28.50	31.50
4500	Kolkwitzia amabilis (Beautybush), Z5, cont/B&B									
4501	2 gal.				Ea.	18.65			18.65	20.50
4502	15" to 18"					23.50			23.50	25.50
4503	2' to 3'					30			30	33
4504	3' to 4'					39.50			39.50	43.50
4505	4' to 5'					45			45	49.50
4506	5' to 6'					75.50			75.50	83
4507	6' to 8'				↓	84.50			84.50	93
4550	Leucothoe axillaris (Coast Leucothoe), Z6, cont									
4551	15" to 18"				Ea.	16.50			16.50	18.15
4552	18" to 24"					22.50			22.50	25
4553	3 gal.				↓	41			41	45
4600	Leucothoe fontanesiana (Drooping Leucothoe), Z6, cont/B&B									
4601	2 gal.				Ea.	29.50			29.50	32.50
4602	12" to 15"				"	25			25	27.50
4650	Ligustrum obtusifolium regelianum (Regal Privet), Z4, cont/B&B									
4651	18" to 24"				Ea.	26			26	28.50
4652	2' to 3'					16.10			16.10	17.70
4653	Bare root, 18" to 24"					8.80			8.80	9.70
4654	2' to 2-1/2'				↓	10.30			10.30	11.35
4700	Ligustrum amurense (Amur Privet), Z4, Bare Root									
4701	2' to 3'				Ea.	15.45			15.45	17
4702	3' to 4'				"	28.50			28.50	31.50
4750	Lindera benzoin (Spicebush), Z5, cont/B&B									
4751	18" to 24"				Ea.	21			21	23
4752	24" to 30"				"	38.50			38.50	42.50
4800	Lonicera fragrantissima (Winter Honeysuckle), Z6, cont/B&B									
4801	2' to 3'				Ea.	30.50			30.50	33.50
4802	3' to 4'				"	39.50			39.50	43.50
4850	Lonicera xylosteum (Emerald Mound Honeysuckle), Z4, cont									
4851	15" to 18"				Ea.	22.50			22.50	25
4852	3 gal.					28.50			28.50	31.50
4853	5 gal.				↓	41.50			41.50	45.50
4900	Mahonia aquifolium (Oregon Grape Holly), Z6, cont/B&B									
4901	1 gal.				Ea.	5.70			5.70	6.25
4902	2 gal.					18.10			18.10	19.90
4903	5 gal.				↓	17.25			17.25	19
4950	Myrica pensylvanica (Northern Bayberry), Z2, B&B									
4951	15" to 18"				Ea.	20			20	22
4952	2' to 2-1/2'					28.50			28.50	31.50
4953	2-1/2' to 3'				↓	48			48	52.50
5000	Osmanthus heterophyllus "Gulftide" (Gulftide Sweet Holly), Z6, B&B									
5001	18" to 24"				Ea.	21.50			21.50	23.50
5002	2' to 2-1/2'				↓	44			44	48

32 93 33.40 Shrubs		Crew	Daily Output	Labor-Hours	Unit	Material	2020 Bare Costs Labor	Equipment	Total	Total Incl O&P
5003	2-1/2' to 3'				Ea.	48			48	52.50
5004	3' to 3-1/2'				↓	55.50			55.50	61
5050	Paxistima canbyi (Canby Paxistima), Z5, B&B									
5051	6" to 9"				Ea.	8.80			8.80	9.65
5052	9" to 12"					11.70			11.70	12.90
5053	12" to 15"				↓	14.85			14.85	16.35
5100	Philadelphus coronarius (Sweet Mockorange), Z5, cont/B&B									
5101	2' to 3'				Ea.	27.50			27.50	30.50
5102	3' to 4'					54			54	59
5103	4' to 5'					77			77	85
5104	5' to 6'				↓	106			106	117
5150	Philadelphus virginalis (Hybrid Mockorange), Z6, cont/B&B									
5151	18" to 24"				Ea.	19.45			19.45	21.50
5152	2' to 3'					26.50			26.50	29.50
5153	3' to 4'					48.50			48.50	53.50
5154	4' to 5'				↓	78			78	85.50
5200	Physocarpus opulifolius luteus (Golden Ninebark), Z3, cont/B&B									
5201	5 gal.				Ea.	22.50			22.50	24.50
5202	3' to 4'					19.45			19.45	21.50
5203	4' to 5'				↓	26			26	28.50
5250	Pieris floribunda (Mountain Andromeda), Z5, cont									
5251	12" to 15" spread				Ea.	18.50			18.50	20.50
5252	15" to 18" spread				"	31			31	34.50
5300	Pieris japonica (Japanese Andromeda), Z6, cont/B&B									
5301	9" to 12"				Ea.	14.40			14.40	15.85
5302	12" to 15"					33.50			33.50	37
5303	18" to 24"					56			56	61.50
5304	24" to 30"					56.50			56.50	62
5305	2-1/2' to 3'				↓	80			80	88
5350	Potentilla fruticosa (Shrubby Cinquefoil), Z2, cont/B&B									
5351	1 gal.				Ea.	12.60			12.60	13.85
5352	10" to 12"					13.35			13.35	14.70
5353	12" to 15"					5.75			5.75	6.35
5354	15" to 18"					15.35			15.35	16.90
5355	18" to 24"					16.10			16.10	17.75
5356	2' to 3'				↓	47			47	51.50
5400	Prunus cistena (Purple-Leaf Sand Cherry), Z3, cont/B&B									
5401	5 gal.				Ea.	13.95			13.95	15.35
5402	18" to 24"					16.95			16.95	18.65
5403	2' to 3'					20.50			20.50	22.50
5404	3' to 4'					26.50			26.50	29
5405	4' to 5'					33			33	36
5406	5' to 6'				↓	46.50			46.50	51
5450	Prunus glandulosa albo plena (Flowering Almond), Z6, cont/B&B									
5451	15" to 18"				Ea.	31			31	34
5452	18" to 24"					47			47	52
5453	2' to 3'					58			58	64
5454	3' to 4'				↓	89.50			89.50	98.50
5500	Prunus glandulosa sinensis (Pink Flowering Almond), Z6, cont/B&B									
5501	5 gal.				Ea.	25			25	27.50
5502	18" to 24"					32.50			32.50	36
5503	2' to 3'				↓	25.50			25.50	28
5550	Prunus laurocerasus schiphaewsis (Cherry Laurel), Z5, B&B									

For customer support on your Site Work & Landscape Costs with RSMeans data, call 800.448.8182.

32 93 33.40 Shrubs		Crew	Daily Output	Labor-Hours	Unit	Material	2020 Bare Costs Labor	Equipment	Total	Total Incl O&P
5551	5 gal.				Ea.	19.60			19.60	21.50
5552	18" to 24"					43.50			43.50	47.50
5553	24" to 30"				↓	53.50			53.50	59
5600	Prunus maritima (Beach Plum), Z4, cont/B&B									
5601	1 gal.				Ea.	24			24	26.50
5602	18" to 24"					54			54	59
5603	2' to 3'				↓	73.50			73.50	81
5650	Pyracantha coccinea (LaLande Firethorn), Z6, cont/B&B									
5651	1 gal.				Ea.	44			44	48.50
5652	5 gal.					81.50			81.50	90
5653	5 gal. Espalier					146			146	161
5654	10 gal. Espalier				↓	213			213	234
5700	Rhamus frangula columnaris (Tallhedge), Z2, B&B									
5701	2' to 3'				Ea.	24			24	26.50
5702	3' to 4'					42.50			42.50	47
5703	4' to 5'					34.50			34.50	37.50
5704	5' to 6'				↓	45			45	49.50
5750	Rhododendron carolinianum (Carolina Rhododendron), Z5, cont/B&B									
5751	15" to 18"				Ea.	22.50			22.50	24.50
5752	18" to 24"					38.50			38.50	42
5753	24" to 30"				↓	52			52	57.50
5800	Rhododendron catawbiense hybrids (Rhododendron), Z5, cont/B&B									
5801	1 gal.				Ea.	10.75			10.75	11.85
5802	2 gal.					22			22	24.50
5803	5 gal.					35			35	38.50
5804	12" to 15"					37			37	40.50
5805	15" to 18"					38			38	42
5806	18" to 24"					47.50			47.50	52
5807	24" to 30"					63.50			63.50	70
5808	30" to 36"					66			66	73
5809	36" to 42"				↓	86.50			86.50	95.50
5850	Rhododendron PJM (PJM Rhododendron), Z4, cont/B&B									
5851	15" to 18"				Ea.	28.50			28.50	31.50
5852	18" to 24"					36.50			36.50	40
5853	2' to 2-1/2'					50			50	55
5854	2-1/2' to 3'					68			68	75
5855	3' to 3-1/2'				↓	84			84	92.50
5900	Rhus aromatica "Gro-Low" (Fragrant Sumac), Z3, cont									
5901	15" to 18"				Ea.	19.70			19.70	21.50
5902	3 gal.					17.55			17.55	19.30
5903	5 gal.					25.50			25.50	28
5950	Rhus glabra cismontana (Dwarf Smooth Sumac), Z2, cont									
5951	1 gal.				Ea.	15.55			15.55	17.10
5952	5 gal.				"	37			37	41
6000	Rhus typhina lacianata (Cut-Leaf Staghorn Sumac), Z4, cont/B&B									
6001	5 gal.				Ea.	26.50			26.50	29.50
6002	7 gal.					32			32	35.50
6003	6' to 7'				↓	59.50			59.50	65.50
6050	Ribes alpinum (Alpine Currant), Z3, cont/B&B									
6051	18" to 24"				Ea.	13.75			13.75	15.15
6052	2' to 3'				"	21			21	23
6100	Rosa hugonis (Father Hugo Rose), Z6, cont									
6101	1 gal.				Ea.	10.65			10.65	11.70

32 93 33.40 Shrubs	Crew	Daily Output	Labor-Hours	Unit	Material	2020 Bare Costs Labor	Equipment	Total	Total Incl O&P
6102 2 gal.				Ea.	15.40			15.40	16.95
6150 Rosa meidiland hybrids, Z6, cont									
6151 9" to 12"				Ea.	20			20	22
6152 12" to 15"				"	24			24	26.50
6200 Rosa multiflora (Japanese Rose), Z6, cont									
6201 1 gal.				Ea.	7.15			7.15	7.90
6202 2 gal.				"	12.80			12.80	14.05
6250 Rosa rugosa and hybrids, Z4, cont									
6251 18" to 24"				Ea.	24.50			24.50	27
6252 2' to 3'				"	21.50			21.50	23.50
6300 Rosa: hybrid teas, grandifloras and climbers, Zna, cont									
6301 2 gal. Non Pat.				Ea.	19.30			19.30	21
6302 2 gal. Patented				"	16.05			16.05	17.70
6350 Salix caprea (French Pussy Willow), Z5, B&B									
6351 4' to 5'				Ea.	36.50			36.50	40
6352 5' to 6'				"	39.50			39.50	43.50
6400 Salix discolor (Pussy Willow), Z2, cont									
6401 2' to 3'				Ea.	26.50			26.50	29
6450 Sambucus canadensis (American Elder), Z4, B&B									
6451 2' to 3'				Ea.	26			26	29
6500 Skimmia japonica (Japanese Skimmia), Z6, cont									
6501 1 gal.				Ea.	9.60			9.60	10.55
6502 2 gal.					17.40			17.40	19.15
6503 5 gal.				↓	25			25	27.50
6550 Spiraea arguta compacta (Garland Spirea), Z5, B&B									
6551 3' to 4'				Ea.	29			29	32
6552 4' to 5'				"	48.50			48.50	53.50
6600 Spirea bumalda (Bumalda Spirea), Z5, cont/B&B									
6601 1 gal.				Ea.	9			9	9.90
6602 5 gal.					29.50			29.50	32.50
6603 15" to 18"					17.25			17.25	19
6604 18" to 24"				↓	18.30			18.30	20
6650 Spirea japonica (Japanese Spirea), Z5, cont									
6651 15" to 18"				Ea.	16.75			16.75	18.40
6652 18" to 24"					34			34	37.50
6653 3 gal.					17.20			17.20	18.90
6654 5 gal.				↓	23.50			23.50	25.50
6700 Spirea nipponica snowmound (Snowmound Spirea), Z5, cont/B&B									
6701 15" to 18"				Ea.	15.35			15.35	16.85
6702 18" to 24"					25			25	27.50
6703 2' to 3'					38.50			38.50	42.50
6704 3' to 4'					55.50			55.50	61
6705 4' to 5'				↓	64.50			64.50	71
6750 Spirea vanhouttei (Van Houtte Spirea), Z4, cont/B&B									
6751 15" to 18"				Ea.	18.90			18.90	21
6752 2' to 3'					33.50			33.50	37
6753 3' to 4'					43.50			43.50	47.50
6754 4' to 5'					39.50			39.50	43
6755 5' to 6'				↓	56.50			56.50	62.50
6800 Symphoricarpos albus (Common Snowberry), Z4, cont/B&B									
6801 18" to 24"				Ea.	12.45			12.45	13.70
6802 2' to 3'					16.15			16.15	17.75
6803 3' to 4'				↓	18.45			18.45	20.50

32 93 33 – Shrubs

32 93 33.40 Shrubs	Crew	Daily Output	Labor-Hours	Unit	Material	2020 Bare Costs Labor	Equipment	Total	Total Incl O&P
6850 Syringa chinensis (Roven Lilac), Z6, B&B									
6851 3' to 4'				Ea.	31			31	34
6852 4' to 5'					45.50			45.50	50
6853 5' to 6'				↓	76.50			76.50	84
6900 Syringa hugo koster (Hugo Koster Lilac), Z4, B&B									
6901 2' to 3'				Ea.	29.50			29.50	32.50
6902 3' to 4'					35.50			35.50	39.50
6903 4' to 5'					48			48	52.50
6904 5' to 6'				↓	55			55	60.50
6950 Syringa meyeri (Meyer Lilac), Z6, cont/B&B									
6951 12" to 15"				Ea.	17.75			17.75	19.55
6952 15" to 18"					47			47	52
6953 18" to 24"					63.50			63.50	69.50
6954 2' to 3'					69.50			69.50	76.50
6955 3' to 4'				↓	109			109	120
7000 Syringa persica (Persian Lilac), Z5, cont									
7001 5 gal.				Ea.	55.50			55.50	61
7050 Syringa prestonae hybrids (Preston Lilac), Z3, B&B									
7051 2' to 3'				Ea.	38			38	41.50
7052 3' to 4'					78			78	85.50
7053 4' to 5'					128			128	141
7054 5' to 6'					191			191	210
7055 6' to 7'				↓	173			173	191
7100 Syringa vulgaris and Mixed Hybrids (Common Lilac), Z4, cont/B&B									
7101 5 gal.				Ea.	30			30	33
7102 3' to 4'					38.50			38.50	42.50
7103 4' to 5'					55.50			55.50	61
7104 5' to 6'					96.50			96.50	106
7105 6' to 8'				↓	103			103	114
7200 Vaccinium corymbosum (Highbush Blueberry), Z4, cont/B&B									
7201 15" to 18"				Ea.	22.50			22.50	25
7202 18" to 24"					31			31	34
7203 2' to 3'					32.50			32.50	35.50
7204 3' to 4'				↓	39.50			39.50	43.50
7250 Viburnum bodnantense (Bodnant Viburnum), Z5, cont/B&B									
7251 18" to 24"				Ea.	12.75			12.75	14
7252 3' to 4'					18.10			18.10	19.90
7253 4' to 5'					50			50	55
7254 5' to 6'				↓	67.50			67.50	74
7300 Viburnum carlcephalum (Fragrant Snowball), Z6, cont/B&B									
7301 2' to 3'				Ea.	28			28	31
7302 3' to 4'					21			21	23
7303 4' to 5'				↓	49.50			49.50	54
7350 Viburnum carlesi (Korean Spice Viburnum), Z4, B&B									
7351 18" to 24"				Ea.	25			25	27.50
7352 2' to 2-1/2'					28.50			28.50	31.50
7353 2-1/2' to 3'					33			33	36.50
7354 3' to 4'				↓	34			34	37
7400 Viburnum dilatatum (Linden Viburnum), Z6, B&B									
7401 3' to 4'				Ea.	31.50			31.50	34.50
7402 4' to 5'					35.50			35.50	39
7403 5' to 6'				↓	49.50			49.50	54
7450 Viburnum juddii (Judd Viburnum), Z5, B&B									

For customer support on your Site Work & Landscape Costs with RSMeans data, call 800.448.8182.

429

32 93 33.40 Shrubs		Crew	Daily Output	Labor-Hours	Unit	Material	2020 Bare Costs Labor	Equipment	Total	Total Incl O&P
7451	18" to 24"				Ea.	18.55			18.55	20.50
7452	2' to 2-1/2'					28			28	31
7453	2-1/2' to 3'					38.50			38.50	42
7454	3' to 4'				↓	45.50			45.50	50
7500	Viburnum opulus sterile (Common Snowball), Z4, cont/B&B									
7501	2 gal.				Ea.	18.50			18.50	20.50
7502	5 gal.					22.50			22.50	24.50
7503	2' to 3'					34			34	37.50
7504	3' to 4'				↓	42.50			42.50	47
7550	Viburnum plicatum (Japanese Snowball), Z6, cont/B&B									
7551	18" to 21"				Ea.	27.50			27.50	30.50
7552	5' to 6'				"	64.50			64.50	71
7600	Viburnum plicatum var. tomentosum (Viburnum), Z5, cont/B&B									
7601	18" to 21"				Ea.	21.50			21.50	23.50
7602	2' to 3'					29			29	31.50
7603	3' to 4'					45			45	49.50
7604	4' to 5'				↓	46			46	50.50
7650	Viburnum prunifolium (Blackhaw), Z4, B&B									
7651	3' to 4'				Ea.	28			28	30.50
7652	4' to 5'					72			72	79
7653	5' to 6'					77.50			77.50	85.50
7654	6' to 7'				↓	153			153	168
7700	Viburnum rhytidophyllum (Leatherleaf Viburnum), Z6, B&B									
7701	2' to 3'				Ea.	28.50			28.50	31.50
7702	3' to 4'					29			29	32
7703	4' to 5'				↓	37.50			37.50	41.50
7750	Viburnum sieboldi (Siebold Viburnum), Z5, B&B									
7751	3' to 4'				Ea.	95			95	105
7752	4' to 5'					62.50			62.50	68.50
7753	5' to 6'					97.50			97.50	107
7754	6' to 7'				↓	87.50			87.50	96
7800	Viburnum trilobum (American Cranberry Bush), Z2, cont/B&B									
7801	2' to 3'				Ea.	22.50			22.50	25
7802	3' to 4'					27.50			27.50	30.50
7803	5' to 6'					32.50			32.50	35.50
7804	6' to 8'					45.50			45.50	50.50
7805	8' to 10'				↓	61			61	67
7850	Weigela florida and Hybrids (Old Fashioned Weigela), Z6, cont/B&B									
7851	2' to 3'				Ea.	18.10			18.10	19.90
7852	3' to 4'					27			27	29.50
7853	4' to 5'					39.50			39.50	43.50
7854	5' to 6'				↓	49			49	54
7900	Xanthorhiza simplicissima (Yellowroot), Z4, cont									
7901	4" pot				Ea.	7.15			7.15	7.90
7902	1 gal.				"	15.65			15.65	17.20
7950	Yucca filamentosa (Adams Needle), Z4, cont									
7951	9" to 12"				Ea.	11			11	12.10
7952	15" to 18"					23.50			23.50	26
7953	18" to 24"				↓	24			24	26.50

32 93 33.50 Shrubs	Crew	Daily Output	Labor- Hours	Unit	Material	2020 Bare Costs Labor	Equipment	Total	Total Incl O&P
0010 **SHRUBS**, Warm temperate									
0011 & subtropical, Z7 to 10									
1000 Aspidistra elatior (Cast Iron Plant), Z7, cont									
1010 1 gal.				Ea.	7.60			7.60	8.35
1020 2 gal.				"	19.65			19.65	21.50
1100 Aucuba japonica "Variegata" (Gold Dust Plant), Z7, cont									
1110 1 gal.				Ea.	9.55			9.55	10.55
1120 3 gal.					17.30			17.30	19
1130 5 gal.				↓	29.50			29.50	32
1200 Azara microphylla (Boxleaf Azara), Z8, cont									
1210 5 gal.				Ea.	30			30	33
1220 15 gal.				"	68			68	74.50
1300 Boronia megastigma (Sweet Boronia), Z9, cont									
1310 1 gal.				Ea.	8.60			8.60	9.50
1320 5 gal.				"	23.50			23.50	26
1400 Brunfelsia pauciflora calycina (Brazil Raisin-tree), Z10, cont									
1410 1 gal.				Ea.	8.20			8.20	9
1420 5 gal.				"	25			25	27.50
1500 Calliandra haematocephala (Pink Powerpuff), Z10, cont									
1510 1 gal.				Ea.	7.50			7.50	8.25
1520 5 gal.					16.25			16.25	17.85
1530 15 gal.				↓	64			64	70.50
1600 Callistemon citrinus (Lemon Bottlebrush), Z9, cont									
1610 1 gal.				Ea.	7.40			7.40	8.10
1620 5 gal.					24.50			24.50	27
1630 10 gal.					51.50			51.50	56.50
1640 20 gal.				↓	79.50			79.50	87
1700 Camellia japonica (Japanese Camellia), Z7, cont									
1710 1 gal.				Ea.	9.10			9.10	10
1720 3 gal.					12.80			12.80	14.10
1730 5 gal.				↓	33			33	36
1800 Camellia sasanqua (Sasanqua Camellia), Z7, cont									
1810 1 gal.				Ea.	11.25			11.25	12.35
1820 3 gal.					12.25			12.25	13.45
1830 5 gal.				↓	30.50			30.50	33.50
1900 Carissa grandiflora (Natal Plum), Z9, cont									
1910 1 gal.				Ea.	5.55			5.55	6.15
1920 5 gal.					11.95			11.95	13.15
1930 15 gal.				↓	59.50			59.50	65
2000 Carpenteria californica (Evergreen Mockorange), Z8, cont									
2010 1 gal.				Ea.	9.50			9.50	10.45
2020 5 gal.				"	30.50			30.50	33.50
2100 Cassia artemisioides (Feathery Cassia), Z9, cont									
2110 1 gal.				Ea.	6.20			6.20	6.80
2120 5 gal.					18.05			18.05	19.90
2130 15 gal.				↓	66			66	73
2200 Ceanothus concha (Concha Wild Lilac), Z8, cont									
2210 1 gal.				Ea.	5.65			5.65	6.20
2220 5 gal.				"	19.25			19.25	21
2300 Choisa ternata (Mexican Orange), Z7, cont									
2310 1 gal.				Ea.	11.45			11.45	12.55
2320 5 gal.				"	25			25	27.50

32 93 33.50 Shrubs	Crew	Daily Output	Labor-Hours	Unit	Material	2020 Bare Costs Labor	Equipment	Total	Total Incl O&P
2400 Cistus Hybridus (White Rockrose), Z7, cont									
2410 1 gal.				Ea.	6.05			6.05	6.65
2420 5 gal.				"	13.45			13.45	14.80
2500 Coprosma repens (Mirror Plant), Z9, cont									
2510 1 gal.				Ea.	5.55			5.55	6.10
2520 5 gal.				"	16.90			16.90	18.60
2600 Correa pulchella "Carminebells" (Australian Fuchsia), Z9, cont									
2610 1 gal.				Ea.	7.10			7.10	7.80
2620 5 gal.				"	17.20			17.20	18.90
2700 Cyrtomium falcatum (Japanese Holly Fern), Z10, cont									
2710 1 gal.				Ea.	10.05			10.05	11.05
2720 5 gal.				"	21			21	23
2800 Daphne odora "Marginata" (Variegated Winter Daphne), Z7, cont									
2810 2 gal.				Ea.	33.50			33.50	37
2820 5 gal.					69.50			69.50	76.50
2830 15 gal.				↓	99.50			99.50	109
2900 Dizygotheca elegantissima (False Aralia), Z10, cont									
2910 1 gal.				Ea.	6.30			6.30	6.95
2920 5 gal.					15.90			15.90	17.50
2930 15 gal.				↓	62.50			62.50	69
3000 Dodonaea viscosa (Akeake), Z10, cont									
3010 1 gal.				Ea.	5.40			5.40	5.95
3020 5 gal.					11.70			11.70	12.90
3030 15 gal.				↓	52			52	57
3100 Eleagnus ebbingei "Gilt Edge" (Variegated Silverberry), Z7, cont									
3110 1 gal.				Ea.	4.66			4.66	5.15
3120 5 gal.					13.40			13.40	14.75
3130 15 gal.				↓	49			49	54
3200 Eleagnus pungens (Silverberry), Z7, cont									
3210 1 gal.				Ea.	4.69			4.69	5.15
3220 3 gal.					9.65			9.65	10.60
3230 5 gal.				↓	13.35			13.35	14.70
3300 Escallonia rubra (Red Escallonia), Z8, cont									
3310 1 gal.				Ea.	5.30			5.30	5.85
3320 5 gal.					13.85			13.85	15.25
3330 15 gal.				↓	44.50			44.50	49
3400 Euonymus japonica "Aureo-Variegata" (Gold Euonymus), Z7, cont									
3410 1 gal.				Ea.	5.90			5.90	6.45
3420 2 gal.					19.10			19.10	21
3430 5 gal.				↓	31			31	34
3500 Fatsia japonica (Japanese Fatsia), Z8, cont									
3510 1 gal.				Ea.	4.98			4.98	5.50
3520 3 gal.					13.75			13.75	15.10
3530 5 gal.					13.55			13.55	14.90
3540 15 gal.				↓	54.50			54.50	60
3600 Fremontodendron californicum (Pacific "Fremontia"), Z9, cont									
3610 2 gal.				Ea.	22			22	24
3620 5 gal.					42.50			42.50	47
3630 15 gal.				↓	127			127	140
3700 Gardenia jasminoides (Cape Jasmine), Z8, cont									
3710 1 gal.				Ea.	8.40			8.40	9.25
3720 3 gal.					16.55			16.55	18.20
3730 5 gal.				↓	21			21	23.50

432

32 93 33.50 Shrubs	Crew	Daily Output	Labor-Hours	Unit	Material	2020 Bare Costs Labor	Equipment	Total	Total Incl O&P
3800 Garrya elliptica (Silk Tassel), Z8, cont									
3810 1 gal.				Ea.	10.35			10.35	11.35
3820 5 gal.				"	29.50			29.50	32.50
3900 Grewia caffra (Lavender Starflower), Z9, cont									
3910 5 gal.				Ea.	14.60			14.60	16.05
3920 15 gal.				"	50			50	55.50
4000 Hebe buxifolia (Boxleaf Hebe), Z7, cont									
4010 1 gal.				Ea.	4.87			4.87	5.35
4020 5 gal.				"	15.40			15.40	16.95
4100 Heteromeles arbutifolia (Toyon), Z9, cont									
4110 1 gal.				Ea.	11.35			11.35	12.50
4120 5 gal.					29			29	32
4130 15 gal.				↓	100			100	110
4200 Hibiscus rosa-sinensis (Chinese Hibiscus), Z9, cont									
4210 1 gal.				Ea.	6.75			6.75	7.45
4220 5 gal.					20.50			20.50	22.50
4230 15 gal.				↓	49			49	54
4300 Ilex vomitoria (Yaupon Holly), Z7, cont									
4310 1 gal.				Ea.	10.15			10.15	11.20
4320 3 gal.					36.50			36.50	40.50
4330 10 gal.				↓	95.50			95.50	105
4400 Illicium parvifolium (Florida Anise Tree), Z8, cont									
4410 1 gal.				Ea.	10.10			10.10	11.10
4420 5 gal.				"	25			25	27.50
4500 Lagerstroemia indica (Crapemyrtle), Z7, cont									
4510 1 gal.				Ea.	6			6	6.60
4520 5 gal.					19.35			19.35	21.50
4530 3' to 4'					70			70	77
4540 4' to 5'					78.50			78.50	86
4550 5' to 6'				↓	93			93	102
4600 Ligustrum japonicum texanum (Wax Leaf Privet), Z7, cont									
4610 1 gal.				Ea.	7.45			7.45	8.20
4620 2 gal.					15.90			15.90	17.45
4630 5 gal.				↓	22			22	24
4700 Ligustrum lucidum (Glossy Privet), Z7, cont									
4710 1 gal.				Ea.	7.10			7.10	7.80
4720 5 gal.					16			16	17.60
4730 15 gal.				↓	80			80	88
4800 Ligustrum sinensis "Variegatum" (Variegated Privet), Z7, cont									
4810 1 gal.				Ea.	7.40			7.40	8.15
4820 2 gal.					15.65			15.65	17.25
4830 3 gal.				↓	20			20	22
4900 Mahonia lomariifolia (Burmese Grape), Z8, cont									
4910 1 gal.				Ea.	7.25			7.25	8
4920 5 gal.					20			20	22
4930 15 gal.				↓	58.50			58.50	64.50
5000 Myoporum laetum (Coast Sandalwood), Z9, cont									
5010 1 gal.				Ea.	6.95			6.95	7.65
5020 5 gal.					21.50			21.50	23.50
5030 15 gal.				↓	62.50			62.50	68.50
5100 Myrica californica (Pacific Wax Myrtle), Z7, cont									
5110 1 gal.				Ea.	11.30			11.30	12.40
5120 5 gal.				"	31.50			31.50	34.50

433

32 93 33.50 Shrubs	Crew	Daily Output	Labor-Hours	Unit	Material	2020 Bare Costs Labor	Equipment	Total	Total Incl O&P
5200 Myrtus communis "Compacta" (Dwarf Roman Myrtle), Z8, cont									
5210 1 gal.				Ea.	10.50			10.50	11.55
5220 5 gal.				"	36			36	40
5300 Nandina domestica (Heavenly Bamboo), Z7, cont									
5310 1 gal.				Ea.	4.45			4.45	4.90
5320 2 gal.					9.20			9.20	10.10
5330 5 gal.				▼	14.45			14.45	15.90
5400 Nerium oleander (Oleander), Z8, cont									
5410 1 gal.				Ea.	5.15			5.15	5.70
5420 3 gal.					12.30			12.30	13.55
5430 5 gal.				▼	19.35			19.35	21.50
5500 Osmanthus fragrans (Fragrant Tea Olive), Z8, cont									
5510 1 gal.				Ea.	4.49			4.49	4.94
5520 3 gal.				"	13.65			13.65	15.05
5600 Philodendron selloum (Selloum), Z10, cont									
5610 1 gal.				Ea.	5.60			5.60	6.15
5620 5 gal.					17.20			17.20	18.90
5630 15 gal.				▼	53			53	58
5700 Photinia fraseri (Christmas Berry), Z7, cont									
5710 1 gal.				Ea.	10.45			10.45	11.50
5720 2 gal.					16.95			16.95	18.65
5730 5 gal.					35.50			35.50	39
5740 15 gal.				▼	94			94	104
5800 Phyllostachys aurea (Golden Bamboo), Z7, cont									
5810 1 gal.				Ea.	7.10			7.10	7.80
5820 5 gal.					27			27	29.50
5830 15 gal.				▼	62			62	68
5900 Pittosporum tobira (Japanese Pittosporum), Z8, cont									
5910 1 gal.				Ea.	4.95			4.95	5.45
5920 3 gal.					8.70			8.70	9.55
5930 5 gal.				▼	13.80			13.80	15.20
6000 Plumbago auriculata (Cape Plumbago), Z9, cont									
6010 1 gal.				Ea.	4.93			4.93	5.40
6020 5 gal.					13.05			13.05	14.35
6030 15 gal.				▼	54.50			54.50	60
6100 Prunus carolina (Carolina Cherry Laurel), Z7, cont									
6110 1 gal.				Ea.	6.20			6.20	6.80
6120 3 gal.					14.25			14.25	15.70
6130 15 gal.				▼	63			63	69.50
6200 Punica granatum "nana" (Dwarf Pomegranate), Z7, cont									
6210 1 gal.				Ea.	5.85			5.85	6.40
6220 5 gal.					23			23	25.50
6230 7 gal.				▼	37			37	41
6300 Pyracantha koidzumii "Santa Cruz" (Pyracantha), Z8, cont									
6310 1 gal.				Ea.	5.70			5.70	6.30
6320 5 gal.					13.90			13.90	15.30
6330 5 gal. Espalier				▼	18.10			18.10	19.90
6400 Raphiolepis indica (India Hawthorne), Z8, cont									
6410 1 gal.				Ea.	15			15	16.50
6420 2 gal.					46.50			46.50	51
6430 5 gal.				▼	55			55	60.50
6500 Rhamnus alaternus (Italian Buckthorn), Z7, cont									
6510 1 gal.				Ea.	6.30			6.30	6.95

32 93 33 – Shrubs

32 93 33.50 Shrubs	Crew	Daily Output	Labor-Hours	Unit	Material	2020 Bare Costs Labor	Equipment	Total	Total Incl O&P
6520	5 gal.				Ea.	18.65			18.65	20.50
6530	15 gal.				▼	62			62	68
6600	Sarcococca hookerana humilis (Himalayan Sarcococca), Z8, cont									
6610	1 gal.				Ea.	6.80			6.80	7.50
6620	2 gal.				"	13.15			13.15	14.50
6700	Sarcococca ruscifolia (Fragrant Sarcococca), Z7, cont									
6710	1 gal.				Ea.	7.90			7.90	8.70
6720	5 gal.				"	14.35			14.35	15.80
6800	Solanum rantonnetii (Paraguay Nightshade), Z9, cont									
6810	1 gal.				Ea.	6.85			6.85	7.55
6820	5 gal.				"	19.20			19.20	21
6900	Ternstroemia gymnanthera (Japanese Cleyera), Z7, cont									
6910	1 gal.				Ea.	8.05			8.05	8.90
6920	3 gal.					25.50			25.50	28.50
6930	5 gal.				▼	32.50			32.50	35.50
7000	Viburnum davidii (David Viburnum), Z8, cont									
7010	1 gal.				Ea.	7.15			7.15	7.90
7020	3 gal.				"	13.75			13.75	15.15
7100	Viburnum japonicum (Japanese Viburnum), Z8, cont									
7110	1 gal.				Ea.	8.35			8.35	9.15
7120	3 gal.				"	14.05			14.05	15.45
7200	Viburnum tinus (Spring Bouquet Laurustinis), Z7, cont									
7210	1 gal.				Ea.	5.05			5.05	5.55
7220	5 gal.				"	13.60			13.60	14.95
7300	Vitex agnus castus (Chastetree), Z7, cont									
7310	5 gal.				Ea.	14.10			14.10	15.50
7400	Xylosma congestum (Shiny Xylosma), Z8, cont									
7410	1 gal.				Ea.	5.05			5.05	5.55
7420	5 gal.				↓	17.15			17.15	18.90
7430	15 gal.				▼	51.50			51.50	57
7500	Yucca aloifolia (Spanish Bayonet), Z7, cont									
7510	1 gal.				Ea.	8			8	8.80
7520	3 gal.				"	16.50			16.50	18.15

32 93 33.60 Cacti

		Crew	Daily Output	Labor-Hours	Unit	Material	2020 Bare Costs Labor	Equipment	Total	Total Incl O&P
0010	**CACTI**									
1000	Column cactus, Cereus species, Z8, cont									
1020	5 gal.				Ea.	31.50			31.50	34.50
1030	regular size, 15 gal.					78			78	86
1032	large size, 15 gal.				▼	132			132	145
1100	Golden barrel cactus, Echinocactus grusonii, Z8, cont									
1130	14"				Ea.	68			68	75
1200	Orchid cactus, Epiphyllum varieties, Z8, cont									
1210	1 gal.				Ea.	10.40			10.40	11.45
1212	2 gal.				"	21			21	23
1300	Cluster cactus, Mammillaria species, Z8, cont									
1310	1 gal.				Ea.	8.80			8.80	9.65
1312	2 gal.					18.30			18.30	20
1320	5 gal.				▼	25.50			25.50	28.50
1400	Burbank spineless Prickly Pear, Opuntia fiscus-indica, Z8, cont									
1420	5 gal.				Ea.	20			20	22
1430	15 gal.				"	51.50			51.50	56.50
1500	Bunny ears, Opuntia microdasys, Z8, cont									

For customer support on your Site Work & Landscape Costs with RSMeans data, call 800.448.8182.

435

32 93 Plants

32 93 33 – Shrubs

32 93 33.60 Cacti		Crew	Daily Output	Labor-Hours	Unit	Material	2020 Bare Costs Labor	Equipment	Total	Total Incl O&P
1510	1 gal.				Ea.	4.69			4.69	5.15
1520	5 gal.				"	18.75			18.75	20.50
1600	Prickly Pear, Opuntia species, Z8, cont									
1610	1 gal.				Ea.	6.55			6.55	7.20
1620	5 gal.					20			20	22
1630	15 gal.				↓	52			52	57.50
1700	Giant Saguaro Cactus, Carnegia gigantea, Z8, cont									
1713	3 gal.				Ea.	43			43	47.50
1720	5 gal.				"	70.50			70.50	78

32 93 43 – Trees

32 93 43.10 Planting

		Crew	Daily Output	Labor-Hours	Unit	Material	2020 Bare Costs Labor	Equipment	Total	Total Incl O&P
0010	**PLANTING**									
0011	Trees, shrubs and ground cover									
0100	Light soil									
0110	Bare root seedlings, 3" to 5" height	1 Clab	960	.008	Ea.		.35		.35	.53
0120	6" to 10"		520	.015			.65		.65	.97
0130	11" to 16"		370	.022			.91		.91	1.37
0140	17" to 24"		210	.038			1.60		1.60	2.41
0200	Potted, 2-1/4" diameter		840	.010			.40		.40	.60
0210	3" diameter		700	.011			.48		.48	.72
0220	4" diameter		620	.013			.54		.54	.82
0300	Container, 1 gallon	2 Clab	84	.190			8		8	12.05
0310	2 gallon		52	.308			12.95		12.95	19.45
0320	3 gallon		40	.400			16.85		16.85	25.50
0330	5 gallon		29	.552			23		23	35
0400	Bagged and burlapped, 12" diameter ball, by hand	↓	19	.842			35.50		35.50	53.50
0410	Backhoe/loader, 48 HP	B-6	40	.600			27.50	5.35	32.85	47
0415	15" diameter, by hand	2 Clab	16	1			42		42	63.50
0416	Backhoe/loader, 48 HP	B-6	30	.800			36.50	7.15	43.65	63
0420	18" diameter by hand	2 Clab	12	1.333			56		56	84.50
0430	Backhoe/loader, 48 HP	B-6	27	.889			40.50	7.90	48.40	69.50
0440	24" diameter by hand	2 Clab	9	1.778			75		75	112
0450	Backhoe/loader, 48 HP	B-6	21	1.143			52.50	10.20	62.70	89.50
0470	36" diameter, backhoe/loader, 48 HP	"	17	1.412	↓		64.50	12.60	77.10	111
0550	Medium soil									
0560	Bare root seedlings, 3" to 5"	1 Clab	672	.012	Ea.		.50		.50	.75
0561	6" to 10"		364	.022			.93		.93	1.39
0562	11" to 16"		260	.031			1.30		1.30	1.95
0563	17" to 24"		145	.055			2.32		2.32	3.49
0570	Potted, 2-1/4" diameter		590	.014			.57		.57	.86
0572	3" diameter		490	.016			.69		.69	1.03
0574	4" diameter		435	.018			.77		.77	1.16
0590	Container, 1 gallon	2 Clab	59	.271			11.40		11.40	17.15
0592	2 gallon		36	.444			18.70		18.70	28
0594	3 gallon		28	.571			24		24	36
0595	5 gallon		20	.800			33.50		33.50	50.50
0600	Bagged and burlapped, 12" diameter ball, by hand	↓	13	1.231			52		52	78
0605	Backhoe/loader, 48 HP	B-6	28	.857			39	7.65	46.65	67.50
0607	15" diameter, by hand	2 Clab	11.20	1.429			60		60	90.50
0608	Backhoe/loader, 48 HP	B-6	21	1.143			52.50	10.20	62.70	89.50
0610	18" diameter, by hand	2 Clab	8.50	1.882			79.50		79.50	119
0615	Backhoe/loader, 48 HP	B-6	19	1.263	↓		58	11.25	69.25	99

436

32 93 Plants

32 93 43 – Trees

32 93 43.10 Planting		Crew	Daily Output	Labor-Hours	Unit	Material	2020 Bare Costs Labor	Equipment	Total	Total Incl O&P
0620	24" diameter, by hand	2 Clab	6.30	2.540	Ea.		107		107	161
0625	Backhoe/loader, 48 HP	B-6	14.70	1.633			74.50	14.55	89.05	128
0630	36" diameter, backhoe/loader, 48 HP	"	12	2	↓		91.50	17.80	109.30	157
0700	Heavy or stoney soil									
0710	Bare root seedlings, 3" to 5"	1 Clab	470	.017	Ea.		.72		.72	1.08
0711	6" to 10"		255	.031			1.32		1.32	1.98
0712	11" to 16"		182	.044			1.85		1.85	2.78
0713	17" to 24"		101	.079			3.33		3.33	5
0720	Potted, 2-1/4" diameter		360	.022			.94		.94	1.41
0722	3" diameter		343	.023			.98		.98	1.48
0724	4" diameter	↓	305	.026			1.10		1.10	1.66
0730	Container, 1 gallon	2 Clab	41.30	.387			16.30		16.30	24.50
0732	2 gallon		25.20	.635			26.50		26.50	40
0734	3 gallon		19.60	.816			34.50		34.50	51.50
0735	5 gallon		14	1.143			48		48	72.50
0750	Bagged and burlapped, 12" diameter ball, by hand	↓	9.10	1.758			74		74	111
0751	Backhoe/loader	B-6	19.60	1.224			56	10.90	66.90	96
0752	15" diameter, by hand	2 Clab	7.80	2.051			86.50		86.50	130
0753	Backhoe/loader, 48 HP	B-6	14.70	1.633			74.50	14.55	89.05	128
0754	18" diameter, by hand	2 Clab	5.60	2.857			120		120	181
0755	Backhoe/loader, 48 HP	B-6	13.30	1.805			82.50	16.10	98.60	142
0756	24" diameter, by hand	2 Clab	4.40	3.636			153		153	230
0757	Backhoe/loader, 48 HP	B-6	10.30	2.330			107	21	128	183
0758	36" diameter backhoe/loader, 48 HP	"	8.40	2.857			131	25.50	156.50	224
2000	Stake out tree and shrub locations	2 Clab	220	.073	↓		3.06		3.06	4.60

32 93 43.30 Trees, Deciduous

		Crew	Daily Output	Labor-Hours	Unit	Material	Labor	Equipment	Total	Total Incl O&P
0010	**TREES, DECIDUOUS**									
0011	Z2 to 6									
0100	Acer campestre (Hedge Maple), Z4, B&B									
0110	4' to 5'				Ea.	57			57	62.50
0120	5' to 6'					85			85	93.50
0130	1-1/2" to 2" Cal.					155			155	171
0140	2" to 2-1/2" Cal.					272			272	299
0150	2-1/2" to 3" Cal.				↓	320			320	350
0200	Acer ginnala (Amur Maple), Z2, cont/BB									
0210	8' to 10'				Ea.	146			146	161
0220	10' to 12'					243			243	267
0230	12' to 14'				↓	225			225	247
0300	Acer griseum (Paperbark Maple), Z5, B&B									
0310	1-1/2" to 1-3/4" Cal.				Ea.	260			260	286
0320	1-3/4" to 2" Cal.				"	276			276	305
0400	Acer palmatum (Japanese Maple), Z6, cont/BB									
0410	2-1/2' to 3'				Ea.	83.50			83.50	92
0420	4' to 5'					111			111	122
0430	5' to 6'					146			146	161
0440	6' to 7'					153			153	169
0450	7' to 8'					182			182	200
0460	8' to 10'					255			255	280
0470	10' to 12'				↓	300			300	330
0500	Acer palmatum atropurpureum (Bloodgood Japan Maple), Z5, B&B									
0510	3' to 3-1/2'				Ea.	86			86	94.50
0520	3-1/2' to 4'					86.50			86.50	95

For customer support on your Site Work & Landscape Costs with RSMeans data, call 800.448.8182.

437

32 93 43.30 Trees, Deciduous	Crew	Daily Output	Labor-Hours	Unit	Material	2020 Bare Costs Labor	Equipment	Total	Total Incl O&P
0530 4' to 4-1/2'				Ea.	122			122	134
0540 4-1/2' to 5'					154			154	170
0550 5' to 6'				↓	143			143	157
0600 Acer platanoides (Norway Maple), Z4, B&B									
0610 8' to 10'				Ea.	200			200	220
0620 1-1/2" to 2" Cal.					128			128	141
0630 2" to 2-1/2" Cal.					140			140	154
0640 2-1/2" to 3" Cal.					186			186	204
0650 3" to 3-1/2" Cal.					211			211	232
0660 Bare root, 8' to 10'					315			315	345
0670 10' to 12'					335			335	365
0680 12' to 14'				↓	340			340	375
0700 Acer platanoides columnare (Column maple), Z4, B&B									
0710 2" to 2-1/2" Cal.				Ea.	186			186	204
0720 2-1/2" to 3" Cal.					193			193	212
0730 3" to 3-1/2" Cal.					246			246	270
0740 3-1/2" to 4" Cal.					272			272	299
0750 4" to 4-1/2" Cal.					298			298	330
0760 4-1/2" to 5" Cal.					400			400	440
0770 5" to 5-1/2" Cal.				↓	485			485	530
0800 Acer rubrum (Red Maple), Z4, B&B									
0810 1-1/2" to 2" Cal.				Ea.	127			127	140
0820 2" to 2-1/2" Cal.					191			191	210
0830 2-1/2" to 3" Cal.					230			230	253
0840 Bare Root, 8' to 10'					219			219	241
0850 10' to 12'				↓	315			315	345
0900 Acer saccharum (Sugar Maple), Z3, B&B									
0910 1-1/2" to 2" Cal.				Ea.	158			158	173
0920 2" to 2-1/2" Cal.					209			209	230
0930 2-1/2" to 3" Cal.					213			213	234
0940 3" to 3-1/2" Cal.					225			225	248
0950 3-1/2" to 4" Cal.					253			253	278
0960 Bare Root, 8' to 10'					146			146	160
0970 10' to 12'					236			236	260
0980 12' to 14'				↓	320			320	355
1000 Aesculus carnea biroti (Ruby Horsechestnut), Z3, B&B									
1010 2' to 3'				Ea.	82			82	90
1020 3' to 4'					114			114	126
1030 4' to 5'					165			165	181
1040 5' to 6'				↓	162			162	179
1100 Amelanchier canadensis (shadblow), Z4, B&B									
1110 3' to 4'				Ea.	50			50	55
1120 4' to 5'					92.50			92.50	102
1130 5' to 6'					118			118	129
1140 6' to 8'					123			123	135
1150 8' to 10'				↓	170			170	187
1200 Betula maximowicziana (monarch birch), Z5, B&B									
1210 6' to 8'				Ea.	40			40	44
1220 1-1/2" to 2" Cal.					175			175	193
1230 2" to 2-1/2" Cal.				↓	215			215	237
1300 Betula nigra (river birch), Z5, B&B									
1310 1-1/2" to 2" Cal.				Ea.	83			83	91.50
1320 2" to 2-1/2" Cal.				↓	118			118	130

438

32 93 43.30 Trees, Deciduous	Crew	Daily Output	Labor-Hours	Unit	Material	2020 Bare Costs Labor	Equipment	Total	Total Incl O&P
1330	2-1/2" to 3" Cal.				Ea.	181			181	199
1340	3" to 3-1/2" Cal.				↓	298			298	325
1400	Betula papyrifera (canoe or paper birch), Z2, B&B									
1410	4' to 5'				Ea.	78			78	86
1420	1-1/2" to 2" Cal.					109			109	120
1430	2" to 2-1/2" Cal.					125			125	137
1440	2-1/2" to 3" Cal.					148			148	163
1450	3" to 3-1/2" Cal.					155			155	170
1460	3-1/2" to 4" Cal.					195			195	214
1470	4" to 4-1/2" Cal.					213			213	234
1480	4-1/2" to 5" Cal.				↓	262			262	289
1500	Castanea mollissima (Chinese Chestnut), Z5, B&B									
1510	2' to 3'				Ea.	18.70			18.70	20.50
1520	6' to 8'					31			31	34.50
1530	8' to 10'					117			117	128
1540	10' to 12'				↓	144			144	159
1600	Catalpa speciosa (northern catalpa), Z5, B&B									
1610	6' to 8'				Ea.	81			81	89
1620	2" Cal.				"	246			246	270
1700	Cercidiphyllum japonica (katsura tree), Z5, B&B									
1710	6' to 8'				Ea.	100			100	110
1720	1-1/2" to 2" Cal.					113			113	125
1730	2" to 2-1/2" Cal.					103			103	114
1740	2-1/2" to 3" Cal.					134			134	147
1750	3-1/2" to 4" Cal.					151			151	166
1760	4" to 4-1/2" Cal.				↓	152			152	168
1800	Cercis canadensis (eastern redbud), Z5, B&B									
1810	5 gal.				Ea.	71.50			71.50	78.50
1820	7 gal.					85			85	93.50
1830	4' to 5'					71			71	78
1840	5' to 6'					65			65	71.50
1850	6' to 8'				↓	84.50			84.50	93
1900	Chionanthus virginicus (fringetree), Z4, B&B									
1910	5 gal.				Ea.	39.50			39.50	43.50
1920	5' to 6'					85.50			85.50	94
1930	6' to 8'				↓	113			113	124
2000	Cladrastis lutea (yellowwood), Z4, B&B									
2010	1-1/2" to 2" Cal.				Ea.	128			128	141
2020	2" to 2-1/2" Cal.					147			147	162
2030	2-1/2" to 3" Cal.					199			199	219
2040	3" to 3-1/2" Cal.					268			268	295
2050	3-1/2" to 4" Cal.				↓	365			365	400
2100	Cornus florida (white flowering dogwood), Z5, B&B									
2110	4' to 5'				Ea.	47.50			47.50	52
2120	5' to 6'					78.50			78.50	86.50
2130	6' to 7'				↓	91			91	100
2200	Cornus florida rubra (pink flowering dogwood), Z6, B&B									
2210	4' to 5'				Ea.	60.50			60.50	66.50
2220	5' to 6'					108			108	119
2230	6' to 7'					145			145	159
2240	7' to 8'				↓	158			158	174
2300	Cornus kousa (Kousa Dogwood), Z5, B&B									
2310	3' to 4'				Ea.	55.50			55.50	61

32 93 43.30 Trees, Deciduous	Crew	Daily Output	Labor-Hours	Unit	Material	2020 Bare Costs Labor	Equipment	Total	Total Incl O&P
2320 4' to 5'				Ea.	55.50			55.50	61
2330 5' to 6'					85.50			85.50	94
2340 6' to 8'					180			180	198
2350 8' to 10'				↓	227			227	250
2400 Crataegus crus-galli (Cockspur Hawthorn), Z5, B&B									
2410 55 gal.				Ea.	142			142	156
2420 5' to 6'				"	101			101	111
2500 Crataegus oxyacantha superba (Crimson Hawthorn), Z5, B&B									
2510 5' to 6'				Ea.	78.50			78.50	86
2520 6' to 8'					85			85	93.50
2530 1-1/2" to 2" Cal.					93.50			93.50	103
2540 2" to 2-1/2" Cal.					82.50			82.50	91
2550 2-1/2" to 3" Cal.					99.50			99.50	110
2560 3" to 3-1/2" Cal.					239			239	263
2570 3-1/2" to 4" Cal.					257			257	283
2580 4" to 5" Cal.				↓	260			260	286
2600 Crataegus phaenopyrum (Washington Hawthorn), Z5, B&B									
2610 5' to 6'				Ea.	162			162	178
2620 6' to 7'					208			208	229
2630 1-3/4" to 2" Cal.					130			130	142
2640 2" to 2-1/2" Cal.					156			156	171
2650 2-1/2" to 3" Cal.					223			223	245
2660 3" to 3-1/2" Cal.				↓	296			296	325
2700 Cydonia oblonga orange (Orange Quince), Z5, B&B									
2710 5' to 6'				Ea.	59.50			59.50	65.50
2720 6' to 8'				"	91			91	100
2800 Diospyros virginiana (Persimmon), Z4, B&B									
2810 3' to 4'				Ea.	25.50			25.50	28
2820 4' to 5'					26			26	28.50
2830 1-1/2" to 2" Cal.					139			139	153
2840 2" to 2-1/2" Cal.					285			285	315
2850 2-1/2" to 3" Cal.				↓	350			350	380
2900 Eleagnus angustifolia (Russian Olive, tree form), Z3, B&B									
2910 1-1/2" to 2" Cal.				Ea.	188			188	206
2920 2" to 2-1/2" Cal.					420			420	460
2930 2-1/2" to 3" Cal.				↓	620			620	680
3000 Eucommia ulmoides (Hardy Rubber Tree), Z4, B&B									
3010 6' to 8'				Ea.	111			111	122
3020 8' to 10'					128			128	141
3030 1-1/2" to 2" Cal.					125			125	137
3040 2" to 2-1/2" Cal.					144			144	158
3050 2-1/2" to 3" Cal.				↓	183			183	202
3100 Fagus grandiflora (American Beech), Z3, B&B									
3110 1-1/2" to 1-3/4" Cal.				Ea.	206			206	227
3120 1-3/4" to 2" Cal.					256			256	281
3130 2" to 2-1/2" Cal.				↓	350			350	385
3200 Fagus sylvatica (European Beech), Z5, B&B									
3210 6' to 8'				Ea.	213			213	234
3220 1-1/2" to 2" Cal.					370			370	410
3230 2" to 2-1/2" Cal.				↓	251			251	277
3300 Fagus sylvatica pendula (Weeping European Beech), Z4, B&B									
3310 1-3/4" to 2" cal.				Ea.	99			99	109
3320 2" to 2-1/2" Cal.				↓	118			118	129

32 93 43.30 Trees, Deciduous	Crew	Daily Output	Labor-Hours	Unit	Material	2020 Bare Costs Labor	Equipment	Total	Total Incl O&P
3330	2-1/2" to 3" Cal.				Ea.	155			155	170
3340	3" to 3-1/2" Cal.				↓	385			385	425
3400	Franklinia alatamaha (Franklin Tree), Z5, B&B									
3410	3 gal.				Ea.	30.50			30.50	33.50
3420	7 gal.					75			75	83
3430	15 gal.					147			147	161
3440	5' to 6'					224			224	246
3450	6' to 8'				↓	158			158	174
3500	Fraxinus americana (Autumn Purple Ash), Z3, B&B									
3510	5 gal.				Ea.	46.50			46.50	51.50
3520	1-1/2" to 2" Cal.					160			160	176
3530	2" to 2-1/2" Cal.				↓	199			199	219
3600	Fraxinus pensylvanica (Seedless Green Ash), Z3, B&B									
3610	5 gal.				Ea.	33			33	36.50
3615	20 gal.					103			103	113
3620	55 gal.					136			136	149
3625	6' to 8'					78			78	86
3630	1-1/2" to 2" Cal.					130			130	143
3635	2" to 2-1/2" Cal.					165			165	181
3640	2-1/2" to 3" Cal.					182			182	200
3645	3" to 3-1/2" Cal.					212			212	233
3650	3-1/2" to 4" Cal.					276			276	305
3655	4" to 4-1/2" Cal.					292			292	320
3660	4-1/2" to 5" Cal.					455			455	500
3665	Bare root, 8' to 10'					88.50			88.50	97
3670	10' to 12'					130			130	143
3675	12' to 14'				↓	154			154	169
3700	Ginkgo biloba (Maidenhair Tree), Z5, B&B									
3710	5 gal.				Ea.	66.50			66.50	73.50
3720	1-1/2" to 2" Cal.					178			178	196
3730	2" to 2-1/2" Cal.					276			276	305
3740	2-1/2" to 3" Cal.					273			273	300
3750	3" to 3-1/2" Cal.				↓	365			365	405
3800	Gleditsia triacanthos inermis (Thornless Honeylocust), Z5, B&B									
3810	20 gal.				Ea.	64			64	70.50
3815	55 gal.					95			95	104
3820	8' to 10'					137			137	150
3825	1-1/2" to 2" Cal.					186			186	205
3830	2" to 2-1/2" Cal.					240			240	264
3835	2-1/2" to 3" Cal.					271			271	298
3840	3" to 3-1/2" Cal.					310			310	340
3845	3-1/2" to 4" Cal.					345			345	380
3850	4" to 4-1/2" Cal.					435			435	480
3855	4-1/2" to 5" Cal.					440			440	480
3860	Bare root, 10' to 12'					248			248	273
3865	12' to 14'					270			270	297
3870	14' to 16'				↓	345			345	380
3900	Gymnocladus dioicus (Kentucky Coffee Tree), Z5, B&B									
3910	1-3/4" Cal.				Ea.	151			151	166
3920	2" Cal.					235			235	258
3930	2-1/2" Cal.					284			284	315
4000	Halesia carolina (Silverbell), Z6, B&B									
4010	4' to 5'				Ea.	75.50			75.50	83

32 93 43.30 Trees, Deciduous	Crew	Daily Output	Labor-Hours	Unit	Material	2020 Bare Costs Labor	Equipment	Total	Total Incl O&P
4020	5' to 6'				Ea.	82			82	90.50
4030	6' to 8'					96			96	106
4040	8' to 10'					109			109	120
4050	10' to 12'				↓	160			160	176
4100	Hippophae rhamnoides (Common Sea Buckthorn), Z3, B&B									
4110	3 gal.				Ea.	18.30			18.30	20
4120	5' to 6'					96.50			96.50	106
4130	6' to 8'				↓	105			105	116
4200	Juglans cinera (Butternut), Z4, B&B									
4210	4' to 5'				Ea.	29.50			29.50	32.50
4220	5' to 6'					43.50			43.50	48
4230	6' to 8'				↓	72			72	79.50
4300	Juglans nigra (Black Walnut), Z5, B&B									
4310	4' to 5'				Ea.	25			25	28
4320	5' to 6'					57			57	62.50
4330	1-1/2" to 2" Cal.					153			153	168
4340	2" to 2-1/2" Cal.					182			182	200
4350	2-1/2" to 3" Cal.				↓	229			229	252
4400	Koelreuteria paniculata (Goldenrain Tree), Z6, B&B									
4410	5 gal.				Ea.	21			21	23
4420	15 gal.					55.50			55.50	61
4430	20 gal.					184			184	203
4440	55 gal.					220			220	242
4450	1-1/2" to 2" Cal.					275			275	305
4460	2" to 2-1/2" Cal.					300			300	330
4470	2-1/2" to 3" Cal.				↓	450			450	495
4500	Laburnum x watereri (Goldenchain tree), Z5, B&B									
4510	6' to 8'				Ea.	65.50			65.50	72.50
4520	8' to 9'					56.50			56.50	62.50
4530	1-3/4" to 2" Cal.					123			123	135
4540	2" to 2-1/2" Cal.					126			126	139
4550	2-1/2" to 3" Cal.				↓	187			187	205
4600	Larix decidua pendula (Weeping European Larch), Z2, B&B									
4610	3' to 4'				Ea.	95			95	105
4620	4' to 5'					93			93	102
4630	5' to 6'					152			152	168
4640	6' to 7'				↓	263			263	289
4700	Larix kaempferi (Japanese Larch), Z4, B&B									
4710	5' to 6'				Ea.	161			161	177
4720	6' to 8'					162			162	178
4730	8' to 10'				↓	184			184	202
4800	Liquidambar styraciflua (Sweetgum), Z6, B&B									
4810	6' to 8'				Ea.	82.50			82.50	90.50
4820	8' to 10'					82			82	90
4830	1-1/2" to 2" Cal.					118			118	130
4840	2" to 2-1/2" Cal.					132			132	145
4850	2-1/2" to 3" Cal.				↓	153			153	168
4900	Liriodendron tulipifera (Tuliptree), Z5, B&B									
4910	7' to 8'				Ea.	56			56	61.50
4920	1-1/2" to 2" Cal.					162			162	179
4930	2" to 2-1/2" Cal.					184			184	202
4940	2-1/2" to 3" Cal.				↓	249			249	274
5000	Maclura pomifera (Osage Orange), Z5, B&B									

32 93 43.30 Trees, Deciduous	Crew	Daily Output	Labor-Hours	Unit	Material	2020 Bare Costs Labor	Equipment	Total	Total Incl O&P
5010	3' to 4'				Ea.	33.50			33.50	37
5020	4' to 5'					59			59	64.50
5030	6' to 8'				▼	75.50			75.50	83
5100	Magnolia loebneri merrill (Merrill Magnolia), Z5, B&B									
5110	4' to 5'				Ea.	76			76	83.50
5120	5' to 6'					115			115	126
5130	6' to 7'					144			144	159
5140	7' to 8'				▼	180			180	198
5200	Magnolia soulangena (Saucer Magnolia), Z6, B&B									
5210	3' to 4'				Ea.	60.50			60.50	66.50
5220	4' to 5'					73			73	80
5230	5' to 6'					93.50			93.50	103
5240	6' to 7'				▼	183			183	202
5300	Magnolia stellata (Star Magnolia), Z6, B&B									
5310	2' to 3'				Ea.	33			33	36
5320	3' to 4'					42			42	46
5330	4' to 5'					80.50			80.50	88.50
5340	5' to 6'				▼	110			110	121
5400	Magnolia virginiana (Sweetbay Magnolia), Z5, B&B									
5410	2' to 3'				Ea.	41			41	45.50
5420	3' to 4'					44.50			44.50	49
5430	4' to 5'					82			82	90
5440	5' to 6'				▼	114			114	126
5500	Magnolia x galaxy (Galaxy Magnolia), Z6, B&B									
5510	5 gal.				Ea.	76			76	84
5520	7 gal.					99			99	109
5530	2-1/2" to 3" Cal.				▼	178			178	196
5600	Malus (Crabapple), Z5, B&B									
5610	5 gal.				Ea.	18.70			18.70	20.50
5615	15 gal.					36			36	39.50
5620	20 gal.					54.50			54.50	60
5625	55 gal.					111			111	122
5630	6' to 8'					86			86	94.50
5635	1-1/2" to 2" Cal.					131			131	144
5640	2" to 2-1/2" Cal.					162			162	178
5645	2-1/2" to 3" Cal.					206			206	227
5650	3" to 3-1/2" Cal.					233			233	256
5655	3-1/2" to 4" Cal.					265			265	292
5660	4" to 5" Cal.					375			375	415
5665	Bare root, 5' to 6'				▼	66			66	72.50
5700	Metasequoia glyptostroboides (Dawn Redwood), Z5, B&B									
5710	5' to 6'				Ea.	139			139	153
5720	6' to 8'					213			213	234
5730	8' to 10'				▼	276			276	305
5800	Morus alba (White Mulberry), Z4, B&B									
5810	5 gal.				Ea.	25			25	27.50
5900	Morus alba tatarica (Russian Mulberry), Z4, B&B									
5910	6' to 8'				Ea.	69.50			69.50	76.50
5920	10' to 12'					159			159	175
5930	2-1/2" to 3" Cal.				▼	355			355	390
6000	Nyssa sylvatica (Tupelo), Z4, B&B									
6010	5' to 6'				Ea.	81.50			81.50	89.50
6020	1-1/2" to 2" Cal.				▼	187			187	206

For customer support on your Site Work & Landscape Costs with RSMeans data, call 800.448.8182.

443

32 93 43.30 Trees, Deciduous	Crew	Daily Output	Labor-Hours	Unit	Material	2020 Bare Costs Labor	Equipment	Total	Total Incl O&P
6030	2" to 2-1/2" Cal.				Ea.	241			241	265
6040	2-1/2" to 3" Cal.				▼	370			370	405
6100	Ostrya virginiana (Hop Hornbeam), Z5, B&B									
6110	2" to 2-1/2" Cal.				Ea.	191			191	210
6120	2-1/2" to 3" Cal.				"	240			240	265
6200	Oxydendron arboreum (Sourwood), Z4, B&B									
6210	4' to 5'				Ea.	103			103	114
6220	5' to 6'					93.50			93.50	103
6230	6' to 7'					158			158	174
6240	7' to 8'				▼	166			166	183
6300	Parrotia persica (Persian Parrotia), Z6, B&B									
6310	5 gal.				Ea.	32.50			32.50	36
6320	3' to 4'				"	97.50			97.50	107
6400	Phellodendron amurense (Amur Corktree), Z3, B&B									
6410	6' to 8'				Ea.	82.50			82.50	90.50
6420	8' to 10'					126			126	138
6430	2" to 2-1/2" Cal.					178			178	196
6440	2-1/2" to 3" Cal.					222			222	244
6450	3" to 3-1/2" Cal.				▼	275			275	305
6500	Platanus acerifolia (London Plane Tree), Z5, B&B									
6510	15 gal.				Ea.	60			60	66
6520	20 gal.					76.50			76.50	84.50
6530	8' to 10'					94.50			94.50	104
6540	1-1/2" to 2" Cal.					148			148	163
6550	2" to 2-1/2" Cal.					184			184	202
6560	2-1/2" to 3" Cal.					223			223	245
6570	3" to 3-1/2" Cal.				▼	251			251	276
6600	Platanus occidentalis (Buttonwood, Sycamore), Z4, B&B									
6610	30" box				Ea.	288			288	315
6620	36" box					365			365	405
6630	42" box					845			845	930
6640	1-1/2" to 2" Cal.					159			159	175
6650	2" to 2-1/2" Cal.					188			188	207
6660	2-1/2" to 3" Cal.					219			219	241
6670	3" to 3-1/2" Cal.				▼	251			251	276
6700	Populus alba pyramidalis (Bolleana Poplar), Z3, cont/BB									
6710	5 gal.				Ea.	35			35	38.50
6720	15 gal.					54.50			54.50	60
6730	20 gal.				▼	68			68	75
6800	Populus canadensis (Hybrid Black Poplar), Z5, B&B									
6810	6' to 8'				Ea.	162			162	178
6820	1-1/2" to 2" Cal.					225			225	247
6830	2" to 2-1/2" Cal.				▼	315			315	345
6900	Populus nigra italica (Lombardy Poplar), Z4, cont									
6910	5 gal.				Ea.	22.50			22.50	24.50
6920	20 gal.					73.50			73.50	80.50
6930	6' to 8' potted				▼	88.50			88.50	97.50
7000	Prunus cerasifera pissardi (Flowering Plum), Z5, cont/BB									
7010	5 gal.				Ea.	34.50			34.50	37.50
7020	15 gal.					117			117	129
7030	20 gal.					176			176	194
7040	5' to 6'					92			92	101
7050	6' to 8'				▼	156			156	172

For customer support on your Site Work & Landscape Costs with RSMeans data, call 800.448.8182.

32 93 Plants

32 93 43 – Trees

32 93 43.30 Trees, Deciduous	Crew	Daily Output	Labor-Hours	Unit	Material	2020 Bare Costs Labor	Equipment	Total	Total Incl O&P
7060 1-1/2" to 2" Cal.				Ea.	224			224	247
7070 2" to 2-1/2" Cal.				↓	345			345	380
7100 Prunus sargenti columnaris (Columnar Cherry), Z4, cont/BB									
7110 6' to 8'				Ea.	87			87	95.50
7120 1-1/2" to 2" Cal.					214			214	235
7130 2" to 2-1/2" Cal.					247			247	272
7140 2-1/2" to 3" Cal.				↓	485			485	535
7200 Prunus serrulata kwanzan (Flowering Cherry), Z6, cont/BB									
7210 20 gal.				Ea.	163			163	179
7220 1-1/2" to 2" Cal.					152			152	167
7230 2" to 2-1/2" Cal.					214			214	235
7240 3" to 3-1/2" Cal.					275			275	300
7250 3-1/2" to 4" Cal.					430			430	470
7260 4" to 5" Cal.				↓	400			400	440
7300 Prunus subhirtella pendula (Weeping Cherry), Z5, B&B									
7310 1-1/4" to 1-1/2" Cal.				Ea.	85.50			85.50	94
7320 1-1/2" to 1-3/4" Cal.					139			139	153
7330 1-3/4" to 2" Cal.					144			144	158
7340 2" to 2-1/2" Cal.					174			174	192
7350 2-1/2" to 3" Cal.				↓	207			207	228
7400 Prunus yedoensis (Yoshino Cherry), Z5, B&B									
7410 5 gal.				Ea.	149			149	164
7420 1-1/2" to 2" Cal.					253			253	278
7430 2" to 2-1/2" Cal.					210			210	231
7440 2-1/2" to 3" Cal.				↓	330			330	365
7500 Pyrus calleryana aristocrat (Aristocrat Flwrng Pear), Z5, cont/BB									
7510 15 gal.				Ea.	40			40	44
7520 20 gal.					69			69	76
7530 6' to 8'					63			63	69.50
7540 1-1/2" to 2" Cal.					129			129	142
7550 2" to 2-1/2" Cal.					160			160	176
7560 2-1/2" to 3" Cal.					251			251	277
7570 3" to 3-1/2" Cal.					226			226	249
7580 3-1/2" to 4" Cal.				↓	258			258	284
7600 Pyrus calleryana bradford (Bradford Flowering Pear), Z5, cont/BB									
7610 5 gal.				Ea.	21			21	23
7620 8' to 9'					78.50			78.50	86.50
7630 1-1/2" to 2" Cal.					143			143	157
7640 2" to 2-1/2" Cal.					218			218	240
7650 2-1/2" to 3" Cal.					310			310	345
7660 3" to 3-1/2" Cal.					455			455	500
7670 3-1/2" to 4" Cal.				↓	675			675	745
7700 Quercus acutissima (Sawtooth Oak), Z5, B&B									
7710 6' to 8'				Ea.	60			60	66
7720 8' to 10'					73			73	80.50
7730 10' to 12'					104			104	114
7740 1-1/2" to 2" Cal.					88			88	96.50
7750 2" to 2-1/2" Cal.				↓	103			103	114
7800 Quercus coccinea (Scarlet Oak), Z4, B&B									
7810 1-1/2" to 2" Cal.				Ea.	188			188	206
7820 2" to 2-1/2" Cal.				"	210			210	231
7900 Quercus palustris (Pin Oak), Z5, B&B									
7910 5 gal.				Ea.	39			39	43

445

32 93 43 – Trees

32 93 43.30 Trees, Deciduous	Crew	Daily Output	Labor-Hours	Unit	Material	2020 Bare Costs Labor	Equipment	Total	Total Incl O&P
7920 1-1/2" to 2" Cal.				Ea.	130			130	143
7930 2" to 2-1/2" Cal.					179			179	197
7940 2-1/2" to 3" Cal.					231			231	254
7950 3" to 3-1/2" Cal.					276			276	305
7960 3-1/2" to 4" Cal.					310			310	340
7970 4" to 4-1/2" Cal.					445			445	490
7980 4-1/2" to 5" Cal.					355			355	390
7985 6" to 7" Cal.					750			750	825
7990 8" to 9" Cal.				▼	895			895	985
8000 Quercus prinus (Chesnut Oak), Z4, B&B									
8010 6' to 8'				Ea.	77			77	84.50
8020 8' to 10'				"	193			193	212
8100 Quercus robur fastigiata (Columnar English Oak), Z4, B&B									
8110 6' to 8'				Ea.	70.50			70.50	78
8120 8' to 10'					106			106	117
8130 1-1/2" to 2" Cal.					150			150	165
8140 2" to 2-1/2" Cal.					232			232	255
8150 2-1/2" to 3" Cal.				▼	268			268	294
8200 Quercus rubra (Red Oak), Z4, cont/BB									
8210 5 gal.				Ea.	12			12	13.20
8220 1-1/2" to 2" Cal.					112			112	124
8230 2" to 2-1/2" Cal.					156			156	171
8240 2-1/2" to 3" Cal.				▼	215			215	237
8300 Robinia pseudoacacia (Black Locust), Z5, cont									
8310 24" box				Ea.	209			209	229
8320 30" box					272			272	299
8330 36" box				▼	460			460	505
8400 Salix babylonica (Weeping Willow), Z5, cont/BB									
8410 24" box				Ea.	133			133	147
8420 36" box					425			425	465
8430 1-1/2" to 2" Cal.					111			111	123
8440 2" to 2-1/2" Cal.					185			185	204
8450 2-1/2" to 3" Cal.				▼	200			200	220
8500 Sophora japonica "Regent" (Regent Scholartree), Z4, B&B									
8510 1-3/4" to 2" Cal.				Ea.	320			320	355
8520 2" to 2-1/2" Cal.					395			395	435
8530 2-1/2" to 3" Cal.				▼	450			450	495
8600 Sorbus alnifolia (Korean Mountain Ash), Z5, B&B									
8610 8' to 10'				Ea.	103			103	113
8620 1-1/2" to 2" Cal.					133			133	147
8630 2" to 2-1/2" Cal.					165			165	182
8640 2-1/2" to 3" Cal.				▼	185			185	204
8700 Sorbus aucuparia (European Mountain Ash), Z2, cont/BB									
8710 15 gal.				Ea.	58.50			58.50	64.50
8715 20 gal.					179			179	197
8720 1-1/2" to 2" Cal.					152			152	167
8725 2" to 2-1/2" Cal.					276			276	305
8730 2-1/2" to 3" Cal.					370			370	405
8735 3" to 3-1/2" Cal.					490			490	540
8740 3-1/2" to 4" Cal.					540			540	590
8745 4" to 5" Cal.					705			705	775
8750 5" to 6" Cal.					885			885	970
8755 Bare root, 8' to 10'				▼	91			91	100

32 93 43.30 Trees, Deciduous	Crew	Daily Output	Labor-Hours	Unit	Material	2020 Bare Costs Labor	Equipment	Total	Total Incl O&P
8760	10' to 12'				Ea.	121			121	133
8800	Stewartia pseudocamellia (Japanese Stewartia), Z5, B&B									
8810	2' to 3'				Ea.	94			94	103
8820	3' to 4'					133			133	146
8830	6' to 8'					510			510	560
8900	Styrax japonica (Japanese Snowbell), Z6, B&B									
8910	4' to 5'				Ea.	141			141	155
8920	5' to 6'					221			221	244
8930	6' to 8'					283			283	310
9000	Syringa japonica (Japanese Tree, Lilac), Z4, B&B									
9010	5' to 6'				Ea.	90.50			90.50	99.50
9020	6' to 8'					105			105	115
9030	8' to 10'					256			256	282
9040	10' to 12'					405			405	450
9100	Taxodium distichum (Bald Cypress), Z4, B&B									
9110	5' to 6'				Ea.	100			100	110
9120	6' to 7'					134			134	148
9130	7' to 8'					163			163	179
9140	8' to 10'					229			229	252
9200	Tilia americana "Redmond" (Redmond Linden), Z3, B&B									
9210	8' to 10'				Ea.	170			170	187
9220	10' to 12'					206			206	227
9230	1-1/2" to 2" Cal.					192			192	211
9240	2" to 2-1/2" Cal.					204			204	225
9250	2-1/2" to 3" Cal.					222			222	245
9300	Tilia cordata greenspire (Littleleaf Linden), Z4, B&B									
9310	1-1/2" to 2" Cal.				Ea.	200			200	220
9315	2" to 2-1/2" Cal.					238			238	262
9320	2-1/2" to 3" Cal.					305			305	335
9325	3" to 3-1/2" Cal.					340			340	370
9330	3-1/2" to 4" Cal.					385			385	420
9335	4" to 4-1/2" Cal.					430			430	470
9340	4-1/2" to 5" Cal.					525			525	580
9345	5" to 5-1/2" Cal.					505			505	555
9350	Bare root, 10' to 12'					213			213	235
9355	12' to 14'					247			247	272
9400	Tilia tomentosa (Silver Linden), Z4, B&B									
9410	6' to 8'				Ea.	143			143	157
9420	8' to 10'					166			166	182
9430	10' to 12'					200			200	220
9440	2" to 2-1/2" Cal.					230			230	253
9450	2-1/2" to 3" Cal.					288			288	315
9500	Ulmus americana (American Elm), Z3, B&B									
9510	6' to 8'				Ea.	72.50			72.50	80
9600	Zeklova serrata (Japanese Keaki Tree), Z4, B&B									
9610	1-1/2" to 2" Cal.				Ea.	167			167	184
9620	2" to 2-1/2" Cal.					179			179	197
9630	2-1/2" to 3" Cal.					200			200	220

32 93 43.40 Trees, Conifers	Crew	Daily Output	Labor-Hours	Unit	Material	2020 Bare Costs Labor	Equipment	Total	Total Incl O&P
0010 TREES, CONIFERS									
0011 Z2 to 7									
1000 Abies balsamea nana (Dwarf Balsam Fir), Z3, cont									
1001 1 gal.				Ea.	18.15			18.15	19.95
1002 15" to 18"					41			41	45
1003 18" to 24"				↓	82			82	90.50
1050 Abies concolor (White Fir), Z4, B&B									
1051 2' to 3'				Ea.	47.50			47.50	52.50
1052 3' to 4'					57			57	62.50
1053 4' to 5'					118			118	129
1054 5' to 6'					150			150	165
1055 6' to 8'				↓	199			199	219
1100 Abies concolor violacea, Z4, cont/BB									
1101 5' to 6'				Ea.	185			185	203
1102 6' to 7'				"	232			232	255
1150 Abies fraseri (Fraser Balsam Fir), Z5, B&B									
1151 5' to 6'				Ea.	156			156	171
1152 6' to 7'				"	211			211	232
1200 Abies iasiocarpa arizonica (Cork Fir), Z5, cont/BB									
1201 2' to 3'				Ea.	51			51	56
1202 3' to 4'					73			73	80.50
1203 4' to 5'					128			128	141
1204 5' to 6'					148			148	163
1205 6' to 7'					206			206	227
1206 8' to 10'					237			237	261
1207 10' to 12'				↓	395			395	435
1250 Abies iasiocarpa compacta (Dwarf Alpine Fir), Z5, cont/BB									
1251 18" to 24"				Ea.	125			125	138
1252 2' to 3'					117			117	129
1253 3' to 4'					144			144	159
1254 4' to 5'					208			208	229
1255 5' to 6'				↓	251			251	277
1300 Abies koreana (Korean Fir), Z5, B&B									
1301 10' to 12'				Ea.	455			455	500
1302 12' to 14'				"	555			555	610
1350 Abies pinsapo glauca (Blue Spanish Fir), Z6, cont									
1351 2' to 3' spread				Ea.	93			93	102
1352 3' to 4'					144			144	158
1353 4' to 5'				↓	175			175	193
1400 Abies procera (nobilis) glauca (Noble Fir), Z5, B&B									
1401 15" to 18" spread				Ea.	48.50			48.50	53.50
1402 18" to 24" spread					71.50			71.50	79
1403 2' to 3'					89			89	98
1404 3' to 4'				↓	141			141	155
1450 Abies veitchi (Veitch Fir), Z3, cont/BB									
1451 4' to 5'				Ea.	130			130	142
1452 5' to 6'				"	157			157	172
1500 Cedrus atlantica glauca (Blue Atlas Cedar), Z6, cont/BB									
1501 2' to 3'				Ea.	102			102	112
1502 3' to 4'					143			143	157
1503 4' to 5'					204			204	224
1504 5' to 6'				↓	249			249	273

For customer support on your Site Work & Landscape Costs with RSMeans data, call 800.448.8182.

32 93 43 – Trees

32 93 43.40 Trees, Conifers	Crew	Daily Output	Labor-Hours	Unit	Material	2020 Bare Costs Labor	Equipment	Total	Total Incl O&P
1505 7' to 8'				Ea.	325			325	355
1550 Cedrus deodara (Indian Cedar), Z7, cont/BB									
1551 5 gal.				Ea.	25.50			25.50	28
1552 6' to 8'					293			293	325
1553 8' to 10'					305			305	340
1554 10' to 12'					350			350	380
1555 12' to 14'					640			640	700
1556 14' to 16'					615			615	675
1600 Cedrus deodara kashmir (Hardy Deodar Cedar), Z6, cont/BB									
1601 3' to 4'				Ea.	174			174	191
1602 4' to 5'					243			243	267
1603 5' to 6'					288			288	315
1650 Cedrus deodara shalimar (Hardy Deodar Cedar), Z6, cont/BB									
1651 2' to 3'				Ea.	136			136	150
1652 3' to 4'				"	154			154	169
1700 Cedrus libani sargenti (Weeping Cedar of Lebanon), Z6, cont									
1701 2' to 3'				Ea.	117			117	128
1702 3' to 4'				"	174			174	191
1750 Chamaecyparis lawsoniana (Dwarf Lawson's Cypress), Z6, cont									
1751 18" to 24"				Ea.	93			93	103
1752 2' to 2-1/2'				"	123			123	135
1800 Chamaecyparis nootkatensis glauca (Alaska Cedar), Z6, cont/BB									
1801 2' to 3'				Ea.	109			109	120
1802 3' to 4'					153			153	168
1803 4' to 5'					203			203	223
1804 5' to 6'					254			254	280
1850 Chamaecyparis nootkatensis lutea (Gold Cedar), Z6, cont/BB									
1851 3' to 4'				Ea.	125			125	137
1852 4' to 5'					151			151	166
1853 5' to 6'					163			163	180
1900 Chamaecyparis nootkatensis pendula (Alaska Cedar), Z6, cont/BB									
1901 2' to 3'				Ea.	86			86	94.50
1902 4' to 5'					122			122	134
1903 5' to 6'					148			148	163
1904 6' to 7'					186			186	205
1950 Chamaecyparis obtusa coralliformis (Torulosa), Z5, cont/BB									
1951 1 gal.				Ea.	5.70			5.70	6.25
1952 2 gal.					12.55			12.55	13.80
1953 5 gal.					16.65			16.65	18.30
1954 7 gal.					28			28	30.50
1955 4' to 5'					66.50			66.50	73.50
1956 7' to 8'					131			131	144
2000 Chamaecyparis obtusa crippsi (Gold Hinoki Cypress), Z6, cont/BB									
2001 1 gal.				Ea.	23.50			23.50	26
2002 5 gal.					42.50			42.50	47
2003 2' to 3'					65			65	71.50
2004 3' to 4'					73.50			73.50	80.50
2005 4' to 5'					84			84	92.50
2050 Chamaecyparis obtusa filiciodes (Fernspray Cypress), Z5, cont/BB									
2051 5 gal.				Ea.	27			27	30
2052 2' to 3'					98			98	108
2053 3' to 4'					128			128	141
2100 Chamaecyparis obtusa gracilis (Hinoki Cypress), Z5, cont/BB									

449

32 93 43.40 Trees, Conifers		Crew	Daily Output	Labor-Hours	Unit	Material	2020 Bare Costs Labor	Equipment	Total	Total Incl O&P
2101	2' to 3'				Ea.	76.50			76.50	84
2102	3' to 4'					132			132	145
2103	10' to 12'				▼	300			300	330
2150	Chamaecyparis obtusa gracilis compacta (Cypress), Z5, cont/BB									
2151	15" to 18"				Ea.	46.50			46.50	51
2152	18" to 24"					65.50			65.50	72
2153	2' to 3'					102			102	113
2154	3' to 4'				▼	175			175	193
2200	Chamaecyparis obtusa gracilis nana (Hinoki Cypress), Z5, cont/BB									
2201	10" to 12"				Ea.	23.50			23.50	26
2202	12" to 15"					32			32	35.50
2203	15" to 18"					72.50			72.50	79.50
2204	18" to 24"				▼	80.50			80.50	88.50
2250	Chamaecyparis obtusa lycopodioides (Cypress), Z5, cont/BB									
2251	18" to 24"				Ea.	30			30	33
2252	2' to 3'					79.50			79.50	87
2253	3' to 4'					126			126	139
2254	4' to 5'					171			171	188
2255	5' to 6'					187			187	205
2256	6' to 7'					183			183	201
2257	7' to 8'				▼	219			219	241
2300	Chamaecyparis obtusa magnifica (Hinoki Cypress), Z5, cont/BB									
2301	2' to 3'				Ea.	69			69	76
2302	3' to 4'					67			67	74
2303	4' to 5'					166			166	183
2304	5' to 6'					194			194	213
2305	6' to 7'				▼	217			217	239
2350	Chamaecyparis pisifera (Sawara Cypress), Z5, cont/BB									
2351	4' to 5'				Ea.	49			49	54
2352	5' to 6'					65.50			65.50	72
2353	6' to 7'					135			135	149
2354	7' to 8'				▼	143			143	158
2400	Chamaecyparis pisifera aurea (Sawara Cypress), Z4, cont/BB									
2401	4' to 5'				Ea.	66			66	72.50
2402	5' to 6'					77			77	84.50
2403	6' to 7'					151			151	166
2404	7' to 8'				▼	178			178	196
2450	Chamaecyparis pisifera boulevard (Moss Cypress), Z5, cont/BB									
2451	18" to 24"				Ea.	23			23	25
2452	2' to 3'					67			67	74
2453	3' to 4'					90			90	99
2454	4' to 5'					109			109	120
2455	5' to 6'					161			161	177
2456	6' to 7'				▼	202			202	223
2500	Chamaecyparis pisifera compacta (Sawara Cypress), Z4, cont									
2501	15" to 18" spread				Ea.	30.50			30.50	34
2502	18" to 24" spread					46.50			46.50	51
2503	2' to 2-1/2' spread					67			67	74
2504	2-1/2' to 3' spread				▼	116			116	127
2550	Chamaecyparis pisifera filifera (Thread Cypress), Z4, cont/BB									
2551	18" to 24"				Ea.	37			37	40.50
2552	2' to 3'					76.50			76.50	84.50
2553	3' to 4'				▼	118			118	130

32 93 43.40 Trees, Conifers	Crew	Daily Output	Labor-Hours	Unit	Material	2020 Bare Costs Labor	Equipment	Total	Total Incl O&P
2554 4' to 5'				Ea.	143			143	157
2555 5' to 6'					158			158	174
2556 6' to 7'				↓	240			240	264
2600 Chamaecyparis pisifera filifera aurea (Thread Cypress), Z4, cont									
2601 8" to 10"				Ea.	18.70			18.70	20.50
2602 10" to 12"					25.50			25.50	28
2603 15" to 18"				↓	29.50			29.50	32.50
2650 Chamaecyparis pisifera filifera nana (Thread Cypress), Z5, cont									
2651 15" to 18"				Ea.	40.50			40.50	44.50
2652 18" to 24"					56			56	62
2653 2' to 2-1/2'					81			81	89
2654 2-1/2' to 3'					92.50			92.50	102
2655 3' to 4'					129			129	141
2671 Gold spangle, 12" to 15"					34.50			34.50	38
2672 18" to 24"					50			50	55
2673 Bare root, 2' to 3'				↓	67.50			67.50	74.50
2700 Chamaecyparis pisifera plumosa aurea (Plumed Cypress), Z6, B&B									
2701 18" to 24"				Ea.	61.50			61.50	67.50
2702 3' to 4'					91.50			91.50	101
2703 4' to 5'				↓	90			90	99
2750 Chamaecyparis pisifera plumosa compress., Z6, cont/BB									
2751 12" to 15"				Ea.	28			28	30.50
2752 15" to 18"					36.50			36.50	40
2753 18" to 24"					48.50			48.50	53.50
2754 2' to 2-1/2'					62			62	68.50
2755 2-1/2' to 3'				↓	71.50			71.50	78.50
2800 Chamaecyparis pisifera squarrosa (Cypress), Z5, cont/BB									
2801 1 gal.				Ea.	8.85			8.85	9.70
2802 2 gal.					17.75			17.75	19.55
2803 4' to 5'					23			23	25
2804 5' to 6'				↓	75.50			75.50	83
2850 Chamaecyparis pisifera squarrosa pygmaea (Cypress), Z6, cont									
2851 12" to 15"				Ea.	28			28	31
2852 15" to 18"					36.50			36.50	40
2853 18" to 24"				↓	46.50			46.50	51
2900 Chamaecyparis thyoides andelyensis (White Cedar), Z5, cont/BB									
2901 2' to 3'				Ea.	10.05			10.05	11.05
2902 5 gal.					77.50			77.50	85.50
2903 4' to 5'					116			116	128
2904 5' to 6'					157			157	173
2905 6' to 7'				↓	169			169	186
2950 Cryptomeria japonica cristata (Japanese Cedar), Z6, B&B									
2951 5' to 6'				Ea.	115			115	126
2952 6' to 8'				"	180			180	198
3000 Cryptomeria japonica lobbi (Lobb Cryptomeria), Z6, B&B									
3001 6' to 8'				Ea.	213			213	235
3002 8' to 10'					265			265	291
3003 10' to 12'				↓	320			320	350
3050 Cupressocyparis leylandi blue (Leylandi Cypress), Z6, B&B									
3051 5' to 6'				Ea.	171			171	188
3052 6' to 7'				"	176			176	194
3100 Cupressocyparis leylandii (Leylandi Cypress), Z6, cont/BB									
3101 1 gal.				Ea.	9.35			9.35	10.30

For customer support on your Site Work & Landscape Costs with RSMeans data, call 800.448.8182.

451

32 93 43.40 Trees, Conifers		Crew	Daily Output	Labor-Hours	Unit	Material	2020 Bare Costs Labor	Equipment	Total	Total Incl O&P
3102	5 gal.				Ea.	25.50			25.50	28.50
3103	4' to 5'					108			108	118
3104	5' to 6'					171			171	188
3105	6' to 7'					169			169	186
3106	Box, 24"				↓	17.55			17.55	19.30
3150	Cupressus arizonica (Arizona Cypress), Z6, cont									
3151	5 gal.				Ea.	44			44	48.50
3152	7 gal.				"	63.50			63.50	69.50
3200	Cupressus macrocarpa (Monterey Cypress), Z8, cont									
3201	1 gal.				Ea.	7			7	7.70
3202	5 gal.				↓	21			21	23
3203	15 gal.					68.50			68.50	75.50
3250	Cupressus sempervirens (Italian Cypress), Z7, cont									
3251	5 gal.				Ea.	29.50			29.50	32.50
3252	15 gal.				"	66.50			66.50	73
3300	Juniperus chinensis (Green Clmnr Chinese Juniper), Z4, cont/BB									
3301	24" to 30"				Ea.	40.50			40.50	45
3302	30" to 36"					45			45	50
3303	3' to 4'					72.50			72.50	79.50
3304	4' to 5'					82			82	90
3305	5' to 6'				↓	99			99	109
3350	Juniperus chinensis "Armstrongii" (Armstrong Juniper), Z4, cont									
3351	1 gal.				Ea.	5.95			5.95	6.50
3352	2 gal.					19.05			19.05	21
3353	5 gal.				↓	27.50			27.50	30
3400	Juniperus chinensis "Blue Point" (Blue Point Juniper), Z4, cont/BB									
3401	1 gal.				Ea.	9.35			9.35	10.30
3402	5 gal.					21.50			21.50	24
3403	4' to 5'					59.50			59.50	65.50
3404	5' to 6'					119			119	130
3405	6' to 8'				↓	135			135	149
3450	Juniperus chinensis "Keteleeri" (Keteleer Juniper), Z5, cont/BB									
3451	3-1/2' to 4'				Ea.	62			62	68.50
3452	4' to 4-1/2'				"	147			147	162
3500	Juniperus chinensis "Sargenti Glauca" (Sargent Juniper), Z4, cont									
3501	12" to 15"				Ea.	18.65			18.65	20.50
3502	15" to 18"					19.80			19.80	22
3503	2' to 2-1/2'					48			48	52.50
3504	2-1/2' to 3'				↓	48			48	53
3550	Juniperus chinensis "Sea Green" (Sea Green Juniper), Z4, cont									
3555	15" to 18"				Ea.	19.10			19.10	21
3556	18" to 24"					29			29	32
3557	2' to 2-1/2'					46			46	51
3558	2-1/2' to 3'				↓	47			47	51.50
3600	Juniperus chinensis blaauwi (Blaauwi Juniper), Z4, B&B									
3601	18" to 24"				Ea.	35.50			35.50	39
3602	2' to 3'					45			45	49
3603	3' to 4'					63			63	69.50
3604	4' to 5'					125			125	137
3605	5' to 6'				↓	176			176	193
3650	Juniperus chinensis columnaris (Chinese Juniper), Z4, B&B									
3651	4' to 5'				Ea.	131			131	144
3652	5' to 6'				↓	147			147	161

32 93 43.40 Trees, Conifers	Crew	Daily Output	Labor-Hours	Unit	Material	2020 Bare Costs Labor	Equipment	Total	Total Incl O&P
3653 6' to 7'				Ea.	169			169	185
3700 Juniperus chinensis densaerecta (Spartan Juniper), Z5, cont									
3701 1 gal.				Ea.	5.60			5.60	6.20
3702 5 gal.					25			25	27.50
3703 15 gal.					51.50			51.50	56.50
3704 3' to 4'					56.50			56.50	62
3705 4' to 5'					126			126	138
3706 5' to 6'					177			177	195
3707 6' to 8'				↓	168			168	185
3750 Juniperus chinensis Hetzii (Hetz Juniper), Z4, cont/BB									
3751 2-1/2' to 3'				Ea.	49.50			49.50	54.50
3752 3' to 3-1/2'					64			64	70.50
3753 2' to 2-1/2'					69			69	76
3754 4' to 4-1/2'					116			116	127
3755 4-1/2' to 5'				↓	140			140	155
3800 Juniperus chinensis iowa (Chinese Juniper), Z4, cont/BB									
3801 24" to 30"				Ea.	51			51	56
3802 3' to 4'					68.50			68.50	75
3803 4' to 5'					129			129	142
3804 5' to 6'				↓	151			151	166
3850 Juniperus chinensis mountbatten (Chinese Juniper), Z4, cont/BB									
3851 2' to 3'				Ea.	50.50			50.50	55.50
3852 3' to 4'					60.50			60.50	66.50
3853 4' to 5'					114			114	125
3854 5' to 6'					149			149	164
3855 6' to 7'					178			178	195
3856 7' to 8'					191			191	210
3857 8' to 10'				↓	234			234	258
3900 Juniperus chinensis obelisk (Chinese Juniper), Z5, cont/BB									
3901 2' to 3'				Ea.	55.50			55.50	61.50
3902 3' to 4'					68.50			68.50	75
3903 4' to 5'					116			116	128
3904 5' to 6'					170			170	187
3905 6' to 7'				↓	162			162	178
3950 Juniperus chinensis pfitzeriana (Pfitzer Juniper), Z4, cont/BB									
3951 1 gal.				Ea.	6.45			6.45	7.10
3952 2 gal.					17.55			17.55	19.30
3953 5 gal.					28			28	31
3954 9" to 12"					12.10			12.10	13.30
3955 15" to 18"					19			19	21
3956 18" to 24"					25			25	27
3957 2' to 2-1/2'					36			36	39.50
3958 2-1/2' to 3'					56			56	61.50
3959 3' to 3-1/2'				↓	48.50			48.50	53
3970 Juniperus chinensis pfitzeriana aurea (Pfitzer Juniper), Z4, cont									
3971 1 gal.				Ea.	9.30			9.30	10.20
3972 2 gal.					23.50			23.50	26
3973 5 gal.				↓	28			28	30.50
4000 Juniperus chinensis pfitzeriana compact (Pfitzer Juniper), Z4, cont									
4001 5 gal.				Ea.	28			28	31
4002 15" to 18"					22.50			22.50	25
4003 18" to 24"					26			26	28.50
4004 2' to 2-1/2'				↓	38.50			38.50	42.50

32 93 43.40 Trees, Conifers	Crew	Daily Output	Labor-Hours	Unit	Material	2020 Bare Costs Labor	Equipment	Total	Total Incl O&P
4005 2-1/2' to 3'				Ea.	42			42	46
4050 Juniperus chinensis robusta green (Green Juniper), Z5, cont/BB									
4051 15" to 18"				Ea.	21.50			21.50	23.50
4052 2' to 3'					39			39	43
4053 3' to 4'					56			56	62
4054 4' to 5'					153			153	168
4055 5' to 6'					166			166	183
4056 6' to 7'					165			165	182
4057 7' to 8'					197			197	217
4058 8' to 10'					187			187	206
4100 Juniperus chinensis torulosa (Hollywood Juniper), Z6, cont/BB									
4101 1 gal.				Ea.	6.20			6.20	6.80
4102 5 gal.					28.50			28.50	31.50
4103 15 gal.					61			61	67
4104 4' to 5'					103			103	113
4105 5' to 6'					176			176	194
4150 Juniperus horizontalis "Bar Harbor" (Bar Harbor Juniper), Z4, cont									
4151 9" to 12"				Ea.	17.10			17.10	18.80
4152 15" to 18"					25.50			25.50	28
4153 18" to 21"					30.50			30.50	33.50
4154 5 gal.					27.50			27.50	30
4200 Juniperus horizontalis "Wiltoni" (Blue Rug Juniper), Z3, cont									
4201 12" to 15"				Ea.	19.10			19.10	21
4202 15" to 18"					25			25	27.50
4203 18" to 24"					24.50			24.50	26.50
4204 24" to 30"					35			35	39
4250 Juniperus horizontalis Plumosa Compacta (Juniper), Z3, cont									
4251 18" to 24"				Ea.	24.50			24.50	27
4252 24" to 30"					33			33	36
4253 30" to 36"					49			49	53.50
4254 3' to 3-1/2'					88			88	96.50
4300 Juniperus procumbens "Nana" (Procumbens Juniper), Z4, cont									
4301 10" to 12"				Ea.	14.35			14.35	15.75
4302 12" to 15"					22.50			22.50	25
4303 15" to 18"					28.50			28.50	31.50
4304 18" to 24"					26.50			26.50	29.50
4350 Juniperus sabina "Broadmoor" (Broadmoor Juniper), Z3, cont									
4351 10" to 12"				Ea.	18.05			18.05	19.85
4352 12" to 15"					22.50			22.50	24.50
4353 15" to 18"					31.50			31.50	34.50
4354 18" to 24"					35.50			35.50	39
4400 Juniperus sabina Tamariscifolia New Blue (Tam Juniper), Z4, cont									
4401 9" to 12"				Ea.	8.50			8.50	9.35
4402 15" to 18"					17.35			17.35	19.10
4403 18" to 21"					18.90			18.90	21
4450 Juniperus scopulorum "Gray Gleam" (Gray Gleam Juniper), Z3, cont									
4451 3-1/2' to 4'				Ea.	78			78	85.50
4452 4' to 4-1/2'				"	103			103	114
4500 Juniperus scopulorum blue heaven (Blue Heaven Juniper), Z4, B&B									
4501 4' to 5'				Ea.	114			114	126
4502 5' to 6'					146			146	161
4503 6' to 7'					171			171	188
4504 7' to 8'					211			211	232

32 93 43.40 Trees, Conifers	Crew	Daily Output	Labor-Hours	Unit	2020 Bare Costs Material	Labor	Equipment	Total	Total Incl O&P
4550	Juniperus scopulorum cologreen (Cologreen Juniper), Z5, cont									
4551	5 gal.				Ea.	43.50			43.50	47.50
4600	Juniperus scopulorum hillburn's silver globe (Juniper), Z5, BB									
4601	3' to 4'				Ea.	73.50			73.50	81
4602	4' to 5'				"	97.50			97.50	107
4650	Juniperus scopulorum pathfinder (Pathfinder Juniper), Z5, B&B									
4651	4' to 5'				Ea.	113			113	125
4652	5' to 6'					128			128	141
4653	7' to 8'				↓	171			171	188
4700	Juniperus scopulorum (Rocky Mountain Juniper), Z5, cont/BB									
4701	20 gal.				Ea.	125			125	138
4702	3' to 4'					39.50			39.50	43.50
4703	4' to 5'					85			85	93.50
4704	5' to 6'				↓	115			115	126
4750	Juniperus scopulorum welchii (Welchi Juniper), Z3, cont									
4751	5 gal.				Ea.	78.50			78.50	86.50
4752	15 gal.				"	104			104	114
4800	Juniperus scopulorum wichita blue (Wichita Blue Juniper), Z5, cont									
4801	3' to 4'				Ea.	84			84	92.50
4802	4' to 5'				"	132			132	145
4850	Juniperus squamata "Parsonii" (Parson's Juniper), Z4, cont									
4851	1 gal.				Ea.	8.25			8.25	9.05
4852	2 gal.					15.55			15.55	17.10
4853	3 gal.					19.40			19.40	21.50
4854	5 gal.				↓	24.50			24.50	27
4900	Juniperus squamata Blue Star (Blue Juniper), Z4, cont/BB									
4901	12" to 15"				Ea.	21			21	23.50
4902	15" to 18"				"	25.50			25.50	28
4950	Juniperus squamata meyeri (Meyer's Juniper), Z5, cont/BB									
4951	3 gal.				Ea.	23			23	25.50
4952	5 gal.					34			34	37.50
4953	2' to 3'				↓	44			44	48.50
5000	Juniperus virginiana (Eastern Red Cedar), Z2, B&B									
5001	2' to 3'				Ea.	43			43	47.50
5002	3' to 4'					46.50			46.50	51
5003	4' to 5'					90			90	99
5004	5' to 6'				↓	130			130	143
5050	Juniperus virginiana burkii (Burki Juniper), Z2, cont									
5051	4' to 4-1/2'				Ea.	113			113	125
5052	4-1/2' to 5'				"	122			122	134
5100	Juniperus virginiana canaerti (Canaert Red Cedar), Z4, B&B									
5101	2' to 3'				Ea.	45			45	49.50
5102	3' to 4'					53.50			53.50	59
5103	4' to 5'					67			67	73.50
5104	5' to 6'				↓	122			122	134
5150	Juniperus virginiana glauca (Silver Cedar), Z4, cont/BB									
5151	2' to 3'				Ea.	41.50			41.50	45.50
5152	3' to 4'					42.50			42.50	46.50
5153	4' to 5'				↓	79			79	87
5200	Juniperus virginiana globusa (Globe Red Cedar), Z4, cont/BB									
5201	2' to 3'				Ea.	42.50			42.50	46.50
5250	Juniperus virginiana gray owl (Gray Owl Juniper), Z4, cont									
5251	18" to 24"				Ea.	26			26	29

32 93 43.40 Trees, Conifers	Crew	Daily Output	Labor-Hours	Unit	Material	2020 Bare Costs Labor	Equipment	Total	Total Incl O&P
5252	2' to 2-1/2'				Ea.	53			53	58.50
5300	Juniperus virginiana skyrocket (Skyrocket Juniper), Z5, cont/BB									
5301	5 gal.				Ea.	78.50			78.50	86.50
5302	15 gal.					88			88	96.50
5303	2' to 3'					78.50			78.50	86.50
5304	3' to 4'					86			86	94.50
5305	4' to 5'					96.50			96.50	106
5306	5' to 6'					113			113	125
5307	6' to 8'					157			157	173
5350	Picea abies (Norway Spruce), Z2, cont/BB									
5351	2' to 3'				Ea.	39			39	42.50
5352	3' to 4'					74			74	81.50
5353	4' to 5'					100			100	110
5354	5' to 6'					158			158	174
5355	6' to 7'					174			174	191
5356	7' to 8'					196			196	216
5400	Picea abies clanbrasiliana (Barry Spruce), Z4, cont/BB									
5401	2' to 3'				Ea.	40.50			40.50	44.50
5402	3' to 4'				"	65			65	71.50
5450	Picea abies gregoryana (Gregory's Dwarf Spruce), Z4, cont/BB									
5451	18" to 24"				Ea.	90.50			90.50	99.50
5452	2' to 2-1/2'				"	128			128	141
5500	Picea abies nidiformis (Nest Spruce), Z4, cont/BB									
5501	12" to 15"				Ea.	21.50			21.50	23.50
5502	15" to 18"					31			31	34
5503	18" to 24"					39.50			39.50	43.50
5504	2' to 3'					49.50			49.50	54.50
5505	3' to 4'					72			72	79
5506	4' to 5'					118			118	130
5550	Picea abies pendula (Weeping Norway Spruce), Z4, cont/BB									
5551	2' to 3'				Ea.	56			56	61.50
5552	3' to 4'					83.50			83.50	92
5553	4' to 5'					121			121	133
5554	6' to 7'					204			204	225
5555	7' to 8'					247			247	272
5600	Picea abies remonti (Remont Spruce), Z4, cont/BB									
5601	2' to 3'				Ea.	77.50			77.50	85.50
5650	Picea glauca (White or Canadian Spruce), Z3, cont/BB									
5651	3' to 4'				Ea.	70			70	77
5652	4' to 5'					130			130	143
5653	5' to 6'					158			158	174
5654	6' to 7'					193			193	212
5655	7' to 8'					206			206	227
5700	Picea glauca conica (Dwarf Alberta Spruce), Z3, cont/BB									
5701	9" to 12"				Ea.	19.50			19.50	21.50
5702	15" to 18"					38.50			38.50	42.50
5703	18" to 24"					57			57	63
5704	2' to 2-1/2'					77.50			77.50	85
5705	2-1/2' to 3'					99			99	109
5750	Picea omorika (Serbian Spruce), Z4, cont/BB									
5751	2' to 3'				Ea.	53.50			53.50	59
5752	3' to 4'					99.50			99.50	109
5753	4' to 5'					150			150	165

32 93 43 – Trees

32 93 43.40 Trees, Conifers		Crew	Daily Output	Labor-Hours	Unit	Material	2020 Bare Costs Labor	Equipment	Total	Total Incl O&P
5754	5' to 6'				Ea.	184			184	202
5755	6' to 7'				↓	231			231	255
5800	Picea omorika nana (Dwarf Serbian Spruce), Z4, cont									
5801	9" to 12"				Ea.	24.50			24.50	26.50
5802	12" to 15"					41.50			41.50	45.50
5803	18" to 24"				↓	48.50			48.50	53.50
5850	Picea omorika pendula (Weeping Serbian Spruce), Z4, cont/BB									
5851	3' to 4'				Ea.	98.50			98.50	108
5900	Picea pungens (Colorado Spruce), Z3, cont/BB									
5901	18" to 24"				Ea.	49.50			49.50	54.50
5902	2' to 3'					63.50			63.50	69.50
5903	3' to 4'					90			90	99
5904	4' to 5'					133			133	146
5905	5' to 6'					180			180	197
5906	6' to 7'					188			188	206
5907	7' to 8'				↓	204			204	225
5950	Picea pungens glauca (Blue Colorado Spruce), Z3, cont/BB									
5951	18" to 24"				Ea.	44.50			44.50	49
5952	2' to 3'					72			72	79
5953	3' to 4'					91.50			91.50	101
5954	4' to 5'					131			131	145
5955	5' to 6'					174			174	192
5956	6' to 7'				↓	249			249	274
6000	Picea pungens hoopsi (Hoops Blue Spruce), Z3, cont/BB									
6001	2' to 3'				Ea.	81.50			81.50	90
6002	3' to 4'					75			75	82.50
6003	4' to 5'					153			153	168
6004	5' to 6'					155			155	170
6005	14' to 16'					248			248	273
6006	16' to 18'					410			410	450
6007	18' to 20'				↓	735			735	810
6050	Picea pungens hunnewelliana (Dwarf Blue Spruce), Z4, cont/BB									
6051	12" to 15"				Ea.	33			33	36.50
6052	15" to 18"					52.50			52.50	58
6053	18" to 24"					60			60	66.50
6054	2' to 2-1/2'				↓	107			107	118
6100	Picea pungens kosteriana (Koster's Blue Spruce), Z3, cont/BB									
6101	2' to 3'				Ea.	73			73	80
6102	3' to 4'					78			78	86
6103	4' to 5'					165			165	182
6104	5' to 6'					134			134	148
6105	6' to 7'					244			244	268
6106	7' to 8'				↓	279			279	305
6150	Picea pungens montgomery (Dwarf Blue Spruce), Z4, cont/BB									
6151	12" to 15"				Ea.	38			38	42
6152	15" to 18"					56			56	62
6153	18" to 24"					74.50			74.50	82
6154	4' to 5'					231			231	255
6155	5' to 6'				↓	625			625	685
6200	Pinus albicaulis (White Bark Pine), Z4, cont/BB									
6201	2' to 3'				Ea.	44.50			44.50	49
6202	3' to 4'					81			81	89
6203	5' to 6'				↓	180			180	198

32 93 43.40 Trees, Conifers	Crew	Daily Output	Labor-Hours	Unit	Material	2020 Bare Costs Labor	Equipment	Total	Total Incl O&P
6204	6' to 8'				Ea.	271			271	298
6205	8' to 10'					460			460	505
6206	10' to 12'					510			510	560
6207	12' to 14'					500			500	550
6208	14' to 16'				↓	625			625	690
6250	Pinus aristata (Bristlecone Pine), Z5, cont/BB									
6251	15" to 18"				Ea.	45.50			45.50	50
6252	18" to 24"					64			64	70.50
6253	2' to 3'					65.50			65.50	72
6254	3' to 4'					181			181	199
6255	4' to 5'				↓	350			350	385
6300	Pinus banksiana (Jack Pine), Z3, cont/BB									
6301	5' to 6'				Ea.	42			42	46
6302	6' to 8'					52.50			52.50	57.50
6303	8' to 10'				↓	63			63	69
6350	Pinus bungeana (Lacebark Pine), Z5, cont/BB									
6351	2' to 3'				Ea.	43.50			43.50	48
6352	3' to 4'					76.50			76.50	84
6353	4' to 5'					188			188	207
6354	5' to 6'					186			186	205
6355	6' to 7'					290			290	320
6356	7' to 8'				↓	365			365	400
6400	Pinus canariensis (Canary Island Pine), Z8, cont									
6401	1 gal.				Ea.	6.85			6.85	7.55
6402	5 gal.					14.80			14.80	16.30
6403	15 gal.				↓	67			67	74
6450	Pinus cembra (Swiss Stone Pine), Z3, cont/BB									
6451	1 gal.				Ea.	24.50			24.50	26.50
6452	2 gal.					39			39	43
6453	18" to 24"					79			79	86.50
6454	2' to 3'				↓	159			159	175
6500	Pinus densiflora umbraculifera (Tanyosho Pine), Z5, cont/BB									
6501	18" to 24"				Ea.	53.50			53.50	58.50
6502	2' to 3'					204			204	225
6503	3' to 4'					267			267	293
6504	4' to 5'					305			305	335
6505	5' to 6'					300			300	330
6506	7' to 8'				↓	420			420	460
6550	Pinus flexilis (Limber Pine), Z4, B&B									
6551	5 gal.				Ea.	70.50			70.50	77.50
6552	4' to 5'					156			156	172
6553	5' to 6'				↓	196			196	216
6600	Pinus koraiensis (Korean Pine), Z4, cont/BB									
6601	3' to 4'				Ea.	250			250	275
6602	4' to 5'					320			320	350
6603	5' to 6'				↓	405			405	450
6650	Pinus lambertiana (Sugar Pine), Z6, cont/BB									
6651	3' to 4'				Ea.	54			54	59.50
6652	4' to 5'					61			61	67
6653	7' to 8'					146			146	160
6654	8' to 10'				↓	226			226	249
6700	Pinus monticola (Western White Pine), Z6, B&B									
6701	5' to 6'				Ea.	40			40	44

32 93 43.40 Trees, Conifers	Crew	Daily Output	Labor-Hours	Unit	Material	2020 Bare Costs Labor	Equipment	Total	Total Incl O&P
6702	6' to 7'				Ea.	50			50	55
6703	7' to 8'					60			60	66
6704	8' to 10'				↓	100			100	110
6750	Pinus mugo var. mugo (Mugho Pine), Z3, cont/BB									
6751	12" to 15"				Ea.	22.50			22.50	24.50
6752	15" to 18"					26			26	28.50
6753	18" to 24"					27			27	30
6754	2' to 2-1/2'					30.50			30.50	33.50
6755	2-1/2' to 3'					41			41	45.50
6756	3' to 3-1/2' spread					52			52	57
6757	3-1/2' to 4'					63			63	69
6758	1 gal.					20			20	22
6759	2 gal.				↓	29.50			29.50	32.50
6800	Pinus nigra (Austrian Pine), Z5, cont/BB									
6801	5 gal.				Ea.	25			25	27.50
6802	3' to 4'					57			57	63
6803	4' to 5'					73			73	80
6804	5' to 6'					163			163	179
6805	6' to 7'					207			207	228
6806	7' to 8'					220			220	242
6807	8' to 9'					233			233	257
6808	9' to 10'					340			340	370
6809	10' to 12'				↓	470			470	515
6850	Pinus parviflora (Japanese White Pine), Z5, cont									
6851	12' to 14'				Ea.	585			585	645
6900	Pinus parviflora glauca (Silver Japanese White Pine), Z5, cont/BB									
6901	5 gal.				Ea.	67			67	74
6902	3' to 4'					85			85	93.50
6903	4' to 5'					236			236	259
6904	5' to 6'					325			325	355
6905	6' to 7'				↓	345			345	380
6950	Pinus ponderosa (Western Yellow Pine), Z6, B&B									
6951	16' to 20'				Ea.	1,025			1,025	1,125
7000	Pinus radiata (Monterey Pine), Z8, cont									
7001	1 gal.				Ea.	9.85			9.85	10.80
7002	5 gal.					25			25	27.50
7003	15 gal.					98			98	108
7004	24" box				↓	310			310	340
7050	Pinus resinosa (Red Pine), Z3, cont/BB									
7051	2' to 3'				Ea.	42			42	46.50
7052	3' to 4'					61			61	67
7053	4' to 5'					79.50			79.50	87.50
7054	5' to 6'					151			151	166
7055	6' to 8'					175			175	193
7056	8' to 10'					246			246	271
7057	10' to 12'					380			380	415
7058	12' to 14'				↓	415			415	455
7100	Pinus rigida (Pitch Pine), Z5, cont/BB									
7101	2' to 3'				Ea.	42			42	46.50
7150	Pinus strobus (White Pine), Z3, cont/BB									
7151	2' to 3'				Ea.	22			22	24
7152	3' to 4'					52			52	57
7153	4' to 5'				↓	75.50			75.50	83

For customer support on your Site Work & Landscape Costs with RSMeans data, call 800.448.8182.

459

32 93 43.40 Trees, Conifers		Crew	Daily Output	Labor-Hours	Unit	Material	2020 Bare Costs Labor	Equipment	Total	Total Incl O&P
7154	5' to 6'				Ea.	125			125	138
7155	6' to 8'					140			140	154
7156	8' to 10'					253			253	278
7157	10' to 12'					405			405	445
7158	12' to 14'					495			495	545
7200	Pinus strobus fastigiata (Upright White Pine), Z3, B&B									
7201	5' to 6'				Ea.	141			141	155
7250	Pinus strobus nana (Dwarf White Pine), Z3, cont/BB									
7251	5 gal.				Ea.	57.50			57.50	63.50
7252	15" to 18" spread					60.50			60.50	67
7253	18" to 24" spread					72.50			72.50	79.50
7254	2' to 2-1/2' spread					76.50			76.50	84
7255	2-1/2' to 3' spread					77			77	84.50
7256	3' to 4' spread					121			121	133
7257	4' to 5' spread					325			325	355
7258	5' to 6' spread					420			420	465
7300	Pinus strobus pendula (Weeping White Pine), Z3, cont/BB									
7301	3' to 4'				Ea.	74.50			74.50	82
7302	4' to 5'					88			88	97
7303	6' to 8'					207			207	228
7304	8' to 10'					320			320	355
7305	10' to 12'					435			435	480
7350	Pinus sylvestris (Scotch Pine), Z3, cont/BB									
7351	Seedlings				Ea.	6.70			6.70	7.40
7352	3' to 4'					69			69	76
7353	4' to 5'					89			89	98
7354	5' to 6'					140			140	154
7355	6' to 7'					179			179	197
7356	7' to 8'					188			188	207
7400	Pinus sylvestris beauvronensis (Mini Scotch Pine), Z4, cont/BB									
7401	12" to 15"				Ea.	53.50			53.50	59
7402	15" to 18"					64.50			64.50	70.50
7403	2' to 2-1/2'					119			119	130
7450	Pinus sylvestris fastigiata (Pyramidal Scotch Pine), Z3, cont/BB									
7451	5 gal.				Ea.	66.50			66.50	73
7452	7 gal.				"	103			103	114
7500	Pinus thumbergi (Japanese Black Pine), Z5, cont/BB									
7501	1 gal.				Ea.	7.45			7.45	8.15
7502	5 gal.					22.50			22.50	25
7503	7 gal.					44			44	48.50
7504	2' to 3'					51			51	56
7505	3' to 4'					84			84	92.50
7506	4' to 5'					101			101	111
7507	5' to 6'					174			174	191
7508	6' to 7'					233			233	256
7509	7' to 8'					204			204	224
7550	Podocarpus macrophyllus "Maki" (Podocarpus), Z7, cont/BB									
7551	1 gal.				Ea.	8.75			8.75	9.60
7552	5 gal.					27			27	29.50
7553	7 gal.					48.50			48.50	53.50
7554	15 gal.					66.50			66.50	73.50
7555	4' to 5'					72.50			72.50	80
7600	Pseudotsuga menziesii (Douglas Fir), Z5, cont/BB									

For customer support on your Site Work & Landscape Costs with RSMeans data, call 800.448.8182.

32 93 43.40 Trees, Conifers	Crew	Daily Output	Labor-Hours	Unit	2020 Bare Costs Material	2020 Bare Costs Labor	2020 Bare Costs Equipment	Total	Total Incl O&P
7601	5 gal.				Ea.	31.50			31.50	34.50
7602	3' to 4'					58.50			58.50	64
7603	4' to 5'					86			86	94.50
7604	5' to 6'					129			129	142
7605	6' to 7'					180			180	198
7650	Sciadopitys verticillata (Umbrella Pine), Z6, B&B									
7651	2 gal.				Ea.	47.50			47.50	52
7652	2' to 2-1/2'					74			74	81
7653	6' to 7'					294			294	325
7654	7' to 8'					405			405	450
7700	Sequoia sempervirens (Coast Redwood), Z7, cont									
7701	5 gal.				Ea.	24			24	26.50
7702	15 gal.					57.50			57.50	63.50
7703	24" box					168			168	184
7750	Taxus baccata adpressa fowle (Midget Boxleaf Yew), Z5, cont/BB									
7751	15" to 18"				Ea.	26			26	28.50
7752	18" to 24"					27.50			27.50	30
7753	2' to 2-1/2'					44			44	48.50
7754	2-1/2' to 3'					64			64	70.50
7755	3' to 4'					71.50			71.50	78.50
7756	4' to 5'					148			148	162
7800	Taxus baccata repandens (Spreading English Yew), Z5, B&B									
7801	12" to 15"				Ea.	24.50			24.50	27
7802	15" to 18"					46.50			46.50	51.50
7803	2' to 2-1/2'					85			85	93.50
7804	2-1/2' to 3'					106			106	117
7805	3' to 3-1/2'					136			136	149
7850	Taxus cuspidata (Spreading Japanese Yew), Z4, cont/BB									
7851	15" to 18"				Ea.	47.50			47.50	52
7852	18" to 24"					50			50	55
7853	2' to 2-1/2'					84			84	92.50
7854	2-1/2' to 3'					103			103	113
7855	3' to 3-1/2'					138			138	152
7900	Taxus cuspidata capitata (Upright Japanese Yew), Z4, cont/BB									
7901	15" to 18"				Ea.	44.50			44.50	49
7902	18" to 24"					53.50			53.50	59
7903	2' to 2-1/2'					79			79	87
7904	2-1/2' to 3'					95			95	105
7905	3' to 4'					124			124	136
7906	4' to 5'					188			188	207
7907	5' to 6'					370			370	410
7908	6' to 7'					385			385	420
7909	7' to 8'					440			440	485
7950	Taxus cuspidata densiformis (Dense Spreading Yew), Z5, cont/BB									
7951	15" to 18"				Ea.	48.50			48.50	53.50
7952	18" to 24"					55.50			55.50	61
7953	2' to 2-1/2'					69.50			69.50	76.50
7954	2-1/2' to 3'					79			79	87
7955	3' to 3-1/2'					90.50			90.50	100
8000	Taxus cuspidata hicksi (Hick's Yew), Z4, cont/BB									
8001	15" to 18"				Ea.	42			42	46
8002	18" to 24"					72.50			72.50	79.50
8003	2' to 2-1/2'					88.50			88.50	97.50

32 93 43.40 Trees, Conifers	Crew	Daily Output	Labor-Hours	Unit	Material	2020 Bare Costs Labor	Equipment	Total	Total Incl O&P
8004 2-1/2' to 3'				Ea.	97.50			97.50	107
8050 Taxus cuspidata nana (Dwarf Yew), Z4, cont/BB									
8051 15" to 18"				Ea.	36.50			36.50	40
8052 18" to 24"					72.50			72.50	79.50
8053 3' to 4'					98			98	108
8054 4' to 5'					116			116	128
8055 5' to 6'					138			138	152
8100 Thuja occidentalis "Smaragd" (Emerald Arborvitae), Z3, cont/BB									
8101 3' to 3-1/2'				Ea.	53.50			53.50	59
8102 3-1/2' to 4'					52			52	57
8103 4' to 4-1/2'					77.50			77.50	85.50
8104 4-1/2' to 5'					78			78	85.50
8105 5' to 6'					97.50			97.50	107
8150 Thuja occidentalis "Woodwardii" (Globe Arborvitae), Z2, B&B									
8151 15" to 18"				Ea.	23			23	25.50
8152 18" to 24"					28.50			28.50	31
8153 2' to 2-1/2'					33.50			33.50	37
8154 2-1/2' to 3'					38.50			38.50	42.50
8200 Thuja occidentalis douglasi pyramidalis (Arborvitae), Z4, cont/BB									
8201 2' to 3'				Ea.	50.50			50.50	55.50
8202 3' to 4'					62.50			62.50	69
8203 4' to 5'					90.50			90.50	99.50
8204 5' to 6'					92			92	101
8205 6' to 7'					148			148	163
8206 7' to 8'					152			152	168
8250 Thuja occidentalis nigra (Dark American Arborvitae), Z4, cont/BB									
8251 2' to 3'				Ea.	30			30	33
8252 3' to 4'					42			42	46.50
8253 4' to 5'					57.50			57.50	63.50
8254 5' to 6'					96			96	106
8255 6' to 8'					139			139	153
8256 8' to 10'					183			183	201
8300 Thuja occidentalis techney (Mission Arborvitae), Z4, cont/BB									
8301 2' to 3'				Ea.	40.50			40.50	44.50
8302 3' to 4'					49.50			49.50	54
8303 4' to 5'					72.50			72.50	79.50
8304 5' to 6'					104			104	115
8305 6' to 7'					143			143	157
8306 7' to 8'					185			185	203
8350 Thuja orientalis "aurea nana"' (Gold Arborvitae), Z4, cont									
8351 1 gal.				Ea.	8.80			8.80	9.70
8352 5 gal.				"	23			23	25
8400 Thuja orientalis "blue cone" (Blue Cone Arborvitae), Z6, cont									
8401 1 gal.				Ea.	7.45			7.45	8.20
8402 5 gal.				"	23			23	25
8450 Tsuga canadensis (Canadian Hemlock), Z4, cont/BB									
8451 2' to 2-1/2'				Ea.	61.50			61.50	67.50
8452 2-1/2' to 3'					65.50			65.50	72
8453 3' to 3-1/2'					86.50			86.50	95.50
8454 3-1/2' to 4'					94			94	103
8455 4' to 5'					132			132	145
8456 5' to 5-1/2'					163			163	179
8457 5-1/2' to 6'					173			173	191

32 93 43.40 Trees, Conifers	Crew	Daily Output	Labor-Hours	Unit	Material	2020 Bare Costs Labor	Equipment	Total	Total Incl O&P
8458	6' to 7'				Ea.	194			194	214
8459	7' to 8'				▼	240			240	264
8500	Tsuga canadensis sargentii (Sargents Weeping Hemlock), Z8, cont									
8501	1 gal.				Ea.	9.50			9.50	10.45
8502	3 gal.					23.50			23.50	26
8503	30" to 36"					97.50			97.50	107
8504	3' to 4'					116			116	127
8505	4' to 5'					157			157	172
8506	5' to 6'				▼	163			163	179
8550	Tsuga caroliniana (Caroline Hemlock), Z5, B&B									
8551	2-1/2' to 3'				Ea.	90			90	99
8552	3' to 3-1/2'					118			118	130
8553	3-1/2' to 4'					120			120	132
8554	4' to 5'				▼	167			167	184
8600	Tsuga diversifolia (Northern Japanese Hemlock), Z6, cont/BB									
8601	3' to 4'				Ea.	133			133	147
8602	4' to 5'				"	150			150	165

32 93 43.50 Trees and Palms

32 93 43.50 Trees and Palms	Crew	Daily Output	Labor-Hours	Unit	Material	2020 Bare Costs Labor	Equipment	Total	Total Incl O&P
0010	**TREES & PALMS**									
0011	Warm temperate and subtropical									
0100	Acacia baileyana (Bailey Acacia), Z9, cont									
0110	1 gal.				Ea.	7.40			7.40	8.15
0120	5 gal.					27.50			27.50	30
0130	15 gal.				▼	59.50			59.50	65.50
0200	Acacia latifolia (Broadleaf Acacia (multistem)), Z10, cont									
0210	1 gal.				Ea.	8.60			8.60	9.45
0220	5 gal.					26			26	28.50
0230	15 gal.				▼	66			66	72.50
0300	Acacia melanoxylon (Blackwood Acacia), Z9, cont									
0310	1 gal.				Ea.	8.60			8.60	9.45
0320	5 gal.					27			27	29.50
0330	15 gal.				▼	68			68	75
0400	Agonis flexuosa (Willow Myrtle), Z9, cont									
0410	5 gal.				Ea.	26			26	29
0420	15 gal.				"	63.50			63.50	69.50
0500	Albizzia julibrissin (Silk Tree), Z7, cont									
0510	5 gal.				Ea.	23.50			23.50	25.50
0520	15 gal.					48			48	52.50
0530	20 gal.					87.50			87.50	96
0540	24" box				▼	171			171	188
0600	Arbutus menziesii (Pacific Madrone), Z7, cont									
0610	1 gal.				Ea.	9.40			9.40	10.35
0620	5 gal.					27.50			27.50	30.50
0630	15 gal.				▼	56.50			56.50	62
0700	Arbutus unedo (Strawberry Tree), Z7, cont									
0710	1 gal.				Ea.	7.45			7.45	8.20
0720	5 gal.					30.50			30.50	33.50
0730	15 gal.				▼	54.50			54.50	60
0800	Archontophoenix cunninghamiana (King Palm), Z10, cont									
0810	5 gal.				Ea.	33.50			33.50	37
0820	15 gal.					130			130	143
0830	24" box				▼	305			305	335

32 93 43.50 Trees and Palms	Crew	Daily Output	Labor-Hours	Unit	Material	2020 Bare Costs Labor	Equipment	Total	Total Incl O&P
0900	Arecastrum romanzoffianum (Queen Palm), Z10, cont								
0910	5 gal.				Ea.	19.25			19.25	21
0920	15 gal.					57			57	62.50
0930	24" box				↓	229			229	252
1000	Bauhinia variegata (Orchid Tree), Z9, cont								
1010	1 gal.				Ea.	6.75			6.75	7.40
1020	5 gal.					21			21	23.50
1030	24" box				↓	181			181	199
1100	Brachychiton populneus (Kurrajong), Z9, cont								
1110	5 gal.				Ea.	24			24	26
1120	15 gal.					65.50			65.50	72
1130	24" box				↓	188			188	207
1200	Callistemon citroides (Lemon Bottlebrush (Tree Form)), Z9, cont								
1210	15 gal.				Ea.	54.50			54.50	60
1220	24" box					176			176	194
1230	36" box				↓	595			595	655
1300	Casuarina stricta (Mountain She-Oak), Z9, cont								
1310	5 gal.				Ea.	23			23	25
1320	15 gal.				"	69			69	76
1400	Ceratoma siliqua (Carob), Z10, cont								
1410	5 gal.				Ea.	19.05			19.05	21
1420	15 gal.					63			63	69.50
1430	24" box				↓	210			210	231
1500	Chamaerops humilis (Mediterranean Fan Palm), Z9, cont								
1510	5 gal.				Ea.	21			21	23
1520	15 gal.					83			83	91.50
1530	24" box				↓	231			231	254
1600	Chrysobalanus icaco (Coco-plum), Z10, cont								
1610	1 gal.				Ea.	3.84			3.84	4.22
1620	3 gal.					4.83			4.83	5.30
1630	10 gal.					22.50			22.50	24.50
1640	20 gal.				↓	59			59	64.50
1700	Cinnamonum camphora (Camphor Tree), Z9, cont								
1710	5 gal.				Ea.	27.50			27.50	30.50
1720	15 gal.					60			60	66
1730	24" box				↓	221			221	243
1800	Cornus nattali (California Laurel), Z7, B&B								
1810	5' to 6'				Ea.	29			29	31.50
1820	6' to 8'					78.50			78.50	86.50
1830	8' to 10'				↓	136			136	149
2000	Eriobotrya japonica (Loquat), Z8, cont								
2010	3 gal.				Ea.	26			26	28.50
2020	10 gal.					38.50			38.50	42.50
2030	15 gal.					83.50			83.50	92
2040	15 gal. Espalier				↓	138			138	152
2100	Erythrina caffra (Kaffirboom Coral Tree), Z10, cont								
2110	5 gal.				Ea.	17.75			17.75	19.50
2120	15 gal.					56			56	62
2130	24" box				↓	185			185	203
2200	Eucalyptus camaldulensis (Red Gum), Z9, cont								
2210	5 gal.				Ea.	22			22	24.50
2220	15 gal.					48.50			48.50	53.50
2230	24" box				↓	163			163	180

32 93 Plants

32 93 43 – Trees

32 93 43.50 Trees and Palms	Crew	Daily Output	Labor-Hours	Unit	Material	2020 Bare Costs Labor	Equipment	Total	Total Incl O&P
2300 Eucalyptus ficifolia (Flaming Gum), Z10, cont									
2310 5 gal.				Ea.	19.05			19.05	21
2320 15 gal.					46			46	50.50
2330 24" box				↓	154			154	169
2400 Feijoa sellowiana (Guava), Z8, cont									
2410 1 gal.				Ea.	4.74			4.74	5.20
2420 5 gal.					13.50			13.50	14.85
2430 15 gal.				↓	50			50	54.50
2500 Ficus benjamina (Benjamin Fig), Z10, cont									
2510 1 gal.				Ea.	6.70			6.70	7.40
2520 5 gal.					21.50			21.50	23.50
2530 24" box				↓	212			212	233
2600 Ficus retusa nitida (India Laurel Fig), Z10, cont									
2610 5 gal.				Ea.	15.90			15.90	17.50
2620 15 gal.					53.50			53.50	59
2630 24" box				↓	199			199	219
2700 Fraxinus uhdei (Shamel Ash), Z9, cont									
2710 5 gal.				Ea.	17.20			17.20	18.90
2720 15 gal.					60			60	66
2730 24" box					185			185	203
2740 36" box				↓	545			545	600
2800 Fraxinus velutina (Velvet Ash), Z7, cont									
2810 5 gal.				Ea.	18.10			18.10	19.90
2820 15 gal.					58			58	63.50
2830 24" box				↓	169			169	186
2900 Gordonia lasianthus (Loblolly Bay Gordonia), Z8, cont									
2910 5 gal.				Ea.	34			34	37
2920 7 gal.				"	59			59	65
3000 Grevillea robusta (Silk Oak), Z10, cont									
3010 5 gal.				Ea.	18.70			18.70	20.50
3020 15 gal.				"	51			51	56
3100 Hakea suaveolens (Sea Urchin Tree), Z10, cont									
3110 1 gal.				Ea.	5.05			5.05	5.55
3120 5 gal.					14.60			14.60	16.05
3130 15 gal.				↓	40			40	44
3200 Harpephyllum caffrum (Kaffir Plum), Z10, cont									
3210 5 gal.				Ea.	19.25			19.25	21
3220 15 gal.					51			51	56
3230 24" box				↓	198			198	218
3300 Hymenosporum flavum (Sweetshade), Z10, cont									
3310 5 gal.				Ea.	19.40			19.40	21.50
3320 15 gal.					50			50	55
3330 24" box				↓	147			147	162
3400 Jacaranda mimosifolia (Jacaranda), Z10, cont									
3410 5 gal.				Ea.	17.55			17.55	19.30
3420 15 gal.					61.50			61.50	68
3430 24" box				↓	206			206	227
3500 Jatropha curcas (Barbados Nut), Z10, cont/BB									
3510 3 gal.				Ea.	7.65			7.65	8.40
3520 10 gal.					35.50			35.50	39
3530 6' to 7'				↓	50			50	55
3600 Laurus nobilis (Sweet Bay), Z8, cont									
3610 1 gal.				Ea.	5.95			5.95	6.55

For customer support on your Site Work & Landscape Costs with RSMeans data, call 800.448.8182.

465

32 93 43.50 Trees and Palms	Crew	Daily Output	Labor-Hours	Unit	Material	2020 Bare Costs Labor	Equipment	Total	Total Incl O&P
3620	5 gal.				Ea.	15.90			15.90	17.50
3630	15 gal.					56			56	61.50
3640	24" box				↓	205			205	225
3700	Leptospermum laevigatum (Australian Tea Tree), Z9, cont									
3710	1 gal.				Ea.	5.55			5.55	6.10
3720	5 gal.					24.50			24.50	27
3730	15 gal.				↓	61.50			61.50	68
3800	Leptospermum scoparium Ruby Glow (Tea Tree), Z9, cont									
3810	1 gal.				Ea.	5.70			5.70	6.25
3820	5 gal.				"	14.90			14.90	16.40
3900	Magnolia grandiflora (Southern Magnolia), Z7, B&B									
3910	4' to 5'				Ea.	36.50			36.50	40
3920	5' to 6'					55			55	60.50
3930	6' to 7'					89.50			89.50	98.50
3940	7' to 8'				↓	135			135	148
4000	Melaleuca linariifolia (Flaxleaf Paperbark), Z9, cont									
4010	5 gal.				Ea.	19.85			19.85	22
4020	15 gal.					60.50			60.50	66.50
4030	24" box				↓	209			209	230
4100	Maytenus boaria (Mayten Tree), Z9, cont									
4110	5 gal.				Ea.	28.50			28.50	31.50
4120	15 gal.				"	82.50			82.50	91
4200	Metrosideros excelsus (New Zealand Christmas Tree), Z10, cont									
4210	5 gal.				Ea.	22			22	24
4220	15 gal.					47.50			47.50	52
4230	24" box				↓	188			188	207
4300	Olea europea (Olive), Z9, cont									
4310	1 gal.				Ea.	10.85			10.85	11.95
4320	5 gal.					25.50			25.50	28
4330	15 gal.				↓	63			63	69.50
4400	Parkinsonia aculeata (Jerusalem Thorn), Z9, cont									
4410	5 gal.				Ea.	21.50			21.50	23.50
4420	15 gal.					62			62	68
4430	24" box				↓	170			170	187
4500	Pistachia chinensis (Chinese Pistache), Z9, cont									
4510	5 gal.				Ea.	23.50			23.50	25.50
4520	15 gal.					58.50			58.50	64.50
4530	24" box				↓	149			149	164
4600	Pyrus Kawakamii (Evergreen Pear), Z8, cont									
4610	5 gal.				Ea.	30			30	33
4620	15 gal.					64.50			64.50	71
4630	15 gal. Espalier					68			68	74.50
4640	24" box				↓	179			179	196
4700	Quercus agrifolia (Coast Live Oak), Z9, cont									
4710	5 gal.				Ea.	18.35			18.35	20
4720	15 gal.					57.50			57.50	63.50
4730	20 gal.				↓	202			202	222
4800	Sapium sebiferum (Chinese Tallow Tree), Z8, cont									
4810	5 gal.				Ea.	20.50			20.50	22.50
4820	15 gal.					51.50			51.50	57
4830	24" box				↓	163			163	179
4900	Schinus molle (Pepper Tree), Z9, cont									
4910	5 gal.				Ea.	25.50			25.50	28

32 93 Plants

32 93 43 – Trees

32 93 43.50 Trees and Palms	Crew	Daily Output	Labor-Hours	Unit	Material	2020 Bare Costs Labor	Equipment	Total	Total Incl O&P
4920	15 gal.				Ea.	64.50			64.50	71
4930	24" box				▼	205			205	225
5000	Strelitzia nicolai (Giant Bird of Paradise), Z10, cont									
5010	5 gal.				Ea.	15.75			15.75	17.35
5020	15 gal.					69			69	76
5030	24" box				▼	234			234	257
5100	Syzygium paniculatum (Brush Cherry), Z9, cont									
5110	1 gal.				Ea.	6.75			6.75	7.40
5120	5 gal.					14.50			14.50	15.95
5130	15 gal.				▼	53			53	58.50
5200	Tabebuia chrysotricha (Golden Trumpet Tree), Z10, cont									
5210	5 gal.				Ea.	18.70			18.70	20.50
5220	15 gal.					54.50			54.50	59.50
5230	24" box				▼	200			200	221
5300	Thevetia peruviana (Yellow Oleander), Z10, cont									
5310	1 gal.				Ea.	9.20			9.20	10.10
5320	5 gal.					15.75			15.75	17.35
5330	15 gal.				▼	58			58	64
5400	Tipuana tipi (Common Tiputree), Z10, cont									
5410	1 gal.				Ea.	6.20			6.20	6.80
5420	5 gal.					18.95			18.95	21
5430	15 gal.					51.50			51.50	56.50
5440	24" box				▼	149			149	164
5500	Tristania conferta (Brisbane Box), Z10, cont									
5510	5 gal.				Ea.	18.95			18.95	21
5520	15 gal.				"	51.50			51.50	56.50
5600	Umbellularia californica (California Laurel), Z7, cont									
5610	1 gal.				Ea.	7.75			7.75	8.50
5620	5 gal.					19.90			19.90	22
5630	15 gal.				▼	61			61	67

32 93 43.60 Trees, Fruits and Nuts

		Crew	Daily Output	Labor-Hours	Unit	Material	Labor	Equipment	Total	Total Incl O&P
0010	**TREES, FRUITS AND NUTS**									
1000	Apple, different varieties, Z4, cont									
1010	5 gal.				Ea.	29			29	32
1020	Standard, 2-3 yr.					35			35	38.50
1030	Semidwarf, 2-3 yr.				▼	31.50			31.50	34.50
1500	Apricot, Z5, cont/BB									
1510	5 gal.				Ea.	23			23	25.50
1520	Standard, 2-3 yr.					31			31	34.50
1530	Semidwarf, 2-3 yr.				▼	35.50			35.50	39.50
2000	Carya ovata, Z4, bare root									
2010	Bare root, 2 yr.				Ea.	112			112	123
2500	Castanea mollissima (Chinese Chestnut), Z4, bare root									
2510	Bare root, 2 yr.				Ea.	58			58	63.50
3000	Cherry, Z4, B&B									
3010	Standard, 2-3 yr.				Ea.	32			32	35.50
3020	Semidwarf, 2-3 yr.				"	34			34	37.50
3500	Corylus americana (Filbert, American Hazelnut), Z5, B&B									
3510	2' to 3'				Ea.	13.40			13.40	14.75
3520	3' to 4'				"	22			22	24.50
4500	Grape different varieties, Z5, potted									
4510	Potted				Ea.	14.60			14.60	16.10

For customer support on your Site Work & Landscape Costs with RSMeans data, call 800.448.8182.

467

32 93 Plants

32 93 43 – Trees

32 93 43.60 Trees, Fruits and Nuts	Crew	Daily Output	Labor-Hours	Unit	Material	2020 Bare Costs Labor	Equipment	Total	Total Incl O&P
5000 Juglans cinera (Butternut), Z4, bare root									
5010 Bare root, 2 yr.				Ea.	37			37	40.50
5500 Juglans nigra (Black Walnut), Z4, bare root									
5510 Bare root, 2'				Ea.	31.50			31.50	34.50
6000 Peach, Z6, cont/BB									
6010 5 gal.				Ea.	25.50			25.50	28
6020 Standard, 2-3 yr.					28			28	30.50
6030 Semidwarf, 2-3 yr.				↓	35.50			35.50	39
6500 Pear, Z5, B&B									
6510 Standard, 2-3 yr.				Ea.	20			20	22
6520 Semidwarf, 2-3 yr.				"	28.50			28.50	31.50
7000 Plum, Z5, cont or B&B									
7010 5 gal.				Ea.	25.50			25.50	28
7020 Standard, 2-3 yr.					25.50			25.50	28
7030 Semidwarf, 2-3 yr.				↓	31			31	34
7500 Raspberry (Everbearing), Z4									
7510 package of 5				Ea.	23			23	25.50
8000 Strawberry (Hybrid), Z5									
8010 25 per box				Ea.	13.10			13.10	14.45
8020 50 per box				"	30			30	33

32 94 Planting Accessories

32 94 13 – Landscape Edging

32 94 13.20 Edging

		Crew	Daily Output	Labor-Hours	Unit	Material	2020 Bare Costs Labor	Equipment	Total	Total Incl O&P
0010	**EDGING**									
0050	Aluminum alloy, including stakes, 1/8" x 4", mill finish	B-1	390	.062	L.F.	2.15	2.63		4.78	6.30
0051	Black paint		390	.062		2.49	2.63		5.12	6.70
0052	Black anodized		390	.062		2.88	2.63		5.51	7.10
0060	3/16" x 4", mill finish		380	.063		3.10	2.70		5.80	7.45
0061	Black paint		380	.063		3.57	2.70		6.27	8
0062	Black anodized		380	.063		4.17	2.70		6.87	8.65
0070	1/8" x 5-1/2" mill finish		370	.065		3.12	2.77		5.89	7.60
0071	Black paint		370	.065		3.70	2.77		6.47	8.25
0072	Black anodized		370	.065		4.24	2.77		7.01	8.85
0080	3/16" x 5-1/2" mill finish		360	.067		4.17	2.85		7.02	8.85
0081	Black paint		360	.067		4.71	2.85		7.56	9.50
0082	Black anodized	↓	360	.067		5.45	2.85		8.30	10.30
0100	Brick, set horizontally, 1-1/2 bricks per L.F.	D-1	370	.043		1.70	2.03		3.73	4.94
0150	Set vertically, 3 bricks per L.F.	"	135	.119		3.80	5.55		9.35	12.65
0200	Corrugated aluminum, roll, 4" wide	1 Carp	650	.012		2.25	.65		2.90	3.46
0250	6" wide	"	550	.015	↓	2.81	.77		3.58	4.25
0300	Concrete, cast in place, see Section 03 30 53.40									
0350	Granite, 5" x 16", straight	B-29	300	.187	L.F.	16.50	8.60	2.82	27.92	34
0400	Polyethylene grass barrier, 5" x 1/8"	D-1	400	.040		1.47	1.88		3.35	4.46
0410	5" x 1/4"		400	.040		1.69	1.88		3.57	4.70
0420	5" x 5/32"		400	.040		1.96	1.88		3.84	4.99
0430	6" x 3/32"		400	.040		1.57	1.88		3.45	4.57
0500	Precast scallops, green, 2" x 8" x 16"		400	.040		2.18	1.88		4.06	5.25
0550	2" x 8" x 16" other than green		400	.040		1.74	1.88		3.62	4.76
0600	Railroad ties, 6" x 8"	2 Carp	170	.094		2.48	5		7.48	10.25
0650	7" x 9"	↓	136	.118	↓	2.75	6.25		9	12.45

32 94 Planting Accessories

32 94 13 – Landscape Edging

32 94 13.20 Edging

		Crew	Daily Output	Labor-Hours	Unit	Material	2020 Bare Costs Labor	Equipment	Total	Total Incl O&P
0750	Redwood 2" x 4"	2 Carp	330	.048	L.F.	2.31	2.58		4.89	6.40
0780	Landscape timbers, 100% recycled plastic, var colors, 4" x 4" x 8' G		250	.064		6	3.40		9.40	11.70
0790	6" x 6" x 8' G		250	.064		14.20	3.40		17.60	20.50
0800	Steel edge strips, incl. stakes, 1/4" x 5"	B-1	390	.062		4.58	2.63		7.21	9
0850	3/16" x 4"	"	390	.062		3.62	2.63		6.25	7.95
0900	Hardwood, pressure treated, 4" x 6"	2 Carp	250	.064		2.37	3.40		5.77	7.70
0940	6" x 6"		200	.080		3.39	4.25		7.64	10.15
0980	6" x 8"		170	.094		3.86	5		8.86	11.75
1000	Pine, pressure treated, 1" x 4"		500	.032		.68	1.70		2.38	3.31
1040	2" x 4"		330	.048		1.14	2.58		3.72	5.15
1080	4" x 6"		250	.064		3.81	3.40		7.21	9.30
1100	6" x 6"		200	.080		5.70	4.25		9.95	12.70
1140	6" x 8"		170	.094		7.50	5		12.50	15.75
1200	Edging, lawn, made from recycled tires, black, 1" x 6"	B-1	390	.062		1.24	2.63		3.87	5.30

32 94 50 – Tree Guying

32 94 50.10 Tree Guying Systems

		Crew	Daily Output	Labor-Hours	Unit	Material	2020 Bare Costs Labor	Equipment	Total	Total Incl O&P
0010	**TREE GUYING SYSTEMS**									
0015	Tree guying including stakes, guy wire and wrap									
0100	Less than 3" caliper, 2 stakes	2 Clab	35	.457	Ea.	14.15	19.25		33.40	44.50
0200	3" to 4" caliper, 3 stakes	"	21	.762		20	32		52	70
0300	6" to 8" caliper, 3 stakes	B-1	8	3		40.50	128		168.50	238
1000	Including arrowhead anchor, cable, turnbuckles and wrap									
1100	Less than 3" caliper, 3" anchors	2 Clab	20	.800	Ea.	23.50	33.50		57	76
1200	3" to 6" caliper, 4" anchors		15	1.067		34	45		79	105
1300	6" caliper, 6" anchors		12	1.333		27.50	56		83.50	115
1400	8" caliper, 8" anchors		9	1.778		115	75		190	238
2000	Tree guard, preformed plastic, 36" high		168	.095		2.63	4.01		6.64	8.90
2010	Snow fence		140	.114		31.50	4.81		36.31	42.50

32 96 Transplanting

32 96 23 – Plant and Bulb Transplanting

32 96 23.23 Planting

		Crew	Daily Output	Labor-Hours	Unit	Material	2020 Bare Costs Labor	Equipment	Total	Total Incl O&P
0010	**PLANTING**									
0012	Moving shrubs on site, 12" ball	B-62	28	.857	Ea.		39	6.35	45.35	66
0100	24" ball	"	22	1.091	"		50	8.05	58.05	84

32 96 23.43 Moving Trees

		Crew	Daily Output	Labor-Hours	Unit	Material	2020 Bare Costs Labor	Equipment	Total	Total Incl O&P
0010	**MOVING TREES**, On site									
0300	Moving trees on site, 36" ball	B-6	3.75	6.400	Ea.		293	57	350	505
0400	60" ball	"	1	24	"		1,100	214	1,314	1,875

32 96 43 – Tree Transplanting

32 96 43.20 Tree Removal

		Crew	Daily Output	Labor-Hours	Unit	Material	2020 Bare Costs Labor	Equipment	Total	Total Incl O&P
0010	**TREE REMOVAL**									
0100	Dig & lace, shrubs, broadleaf evergreen, 18"-24" high	B-1	55	.436	Ea.		18.65		18.65	28
0200	2'-3'	"	35	.686			29.50		29.50	44
0300	3'-4'	B-6	30	.800			36.50	7.15	43.65	63
0400	4'-5'	"	20	1.200			55	10.70	65.70	94.50
1000	Deciduous, 12"-15"	B-1	110	.218			9.35		9.35	14
1100	18"-24"		65	.369			15.80		15.80	23.50
1200	2'-3'		55	.436			18.65		18.65	28
1300	3'-4'	B-6	50	.480			22	4.28	26.28	37.50

For customer support on your Site Work & Landscape Costs with RSMeans data, call 800.448.8182.

469

32 96 43.20 Tree Removal		Crew	Daily Output	Labor-Hours	Unit	Material	2020 Bare Costs Labor	Equipment	Total	Total Incl O&P
2000	Evergreen, 18"-24"	B-1	55	.436	Ea.		18.65		18.65	28
2100	2'-0" to 2'-6"		50	.480			20.50		20.50	31
2200	2'-6" to 3'-0"		35	.686			29.50		29.50	44
2300	3'-0" to 3'-6"		20	1.200			51.50		51.50	77
3000	Trees, deciduous, small, 2'-3'		55	.436			18.65		18.65	28
3100	3'-4'	B-6	50	.480			22	4.28	26.28	37.50
3200	4'-5'		35	.686			31.50	6.10	37.60	53.50
3300	5'-6'		30	.800			36.50	7.15	43.65	63
4000	Shade, 5'-6'		50	.480			22	4.28	26.28	37.50
4100	6'-8'		35	.686			31.50	6.10	37.60	53.50
4200	8'-10'		25	.960			44	8.55	52.55	75.50
4300	2" caliper		12	2			91.50	17.80	109.30	157
5000	Evergreen, 4'-5'		35	.686			31.50	6.10	37.60	53.50
5100	5'-6'		25	.960			44	8.55	52.55	75.50
5200	6'-7'		19	1.263			58	11.25	69.25	99
5300	7'-8'		15	1.600			73	14.25	87.25	126
5400	8'-10'		11	2.182			100	19.45	119.45	172

Estimating Tips
33 10 00 Water Utilities
33 30 00 Sanitary Sewerage Utilities
33 40 00 Storm Drainage Utilities

- Never assume that the water, sewer, and drainage lines will go in at the early stages of the project. Consider the site access needs before dividing the site in half with open trenches, loose pipe, and machinery obstructions. Always inspect the site to establish that the site drawings are complete. Check off all existing utilities on your drawings as you locate them. Be especially careful with underground utilities because appurtenances are sometimes buried during regrading or repaving operations. If you find any discrepancies, mark up the site plan for further research. Differing site conditions can be very costly if discovered later in the project.

- See also Section 33 01 00 for restoration of pipe where removal/replacement may be undesirable. Use of new types of piping materials can reduce the overall project cost. Owners/design engineers should consider the installing contractor as a valuable source of current information on utility products and local conditions that could lead to significant cost savings.

Reference Numbers
Reference numbers are shown at the beginning of some major classifications. These numbers refer to related items in the Reference Section. The reference information may be an estimating procedure, an alternate pricing method, or technical information.

Note: Not all subdivisions listed here necessarily appear. ■

Same Data. Simplified.

Enjoy the convenience and efficiency of accessing your costs anywhere:

- **Skip the multiplier** by setting your location
- **Quickly search,** edit, favorite and share costs
- **Stay on top of price changes** with automatic updates

Discover more at rsmeans.com/online

---

33 01 10.10 Corrosion Resistance

		Crew	Daily Output	Labor-Hours	Unit	Material	2020 Bare Costs Labor	Equipment	Total	Total Incl O&P
0010	**CORROSION RESISTANCE**									
0012	Wrap & coat, add to pipe, 4" diameter				L.F.	2.38			2.38	2.62
0020	5" diameter					3.04			3.04	3.34
0040	6" diameter					3.67			3.67	4.04
0060	8" diameter					5			5	5.50
0080	10" diameter					6.30			6.30	6.95
0100	12" diameter					7.30			7.30	8
0120	14" diameter					8.35			8.35	9.20
0140	16" diameter					10.55			10.55	11.60
0160	18" diameter					11			11	12.10
0180	20" diameter					12.25			12.25	13.50
0200	24" diameter					14.75			14.75	16.20
0220	Small diameter pipe, 1" diameter, add					1.01			1.01	1.11
0240	2" diameter					1.28			1.28	1.41
0260	2-1/2" diameter					1.61			1.61	1.77
0280	3" diameter					1.93			1.93	2.12
0300	Fittings, field covered, add				S.F.	9.40			9.40	10.35
0500	Coating, bituminous, per diameter inch, 1 coat, add				L.F.	.21			.21	.23
0540	3 coat					.57			.57	.63
0560	Coal tar epoxy, per diameter inch, 1 coat, add					.21			.21	.23
0600	3 coat					.66			.66	.73
1000	Polyethylene H.D. extruded, 0.025" thk., 1/2" diameter, add					.10			.10	.11
1020	3/4" diameter					.15			.15	.17
1040	1" diameter					.19			.19	.21
1060	1-1/4" diameter					.25			.25	.28
1080	1-1/2" diameter					.28			.28	.31
1100	0.030" thk., 2" diameter					.39			.39	.43
1120	2-1/2" diameter					.50			.50	.55
1140	0.035" thk., 3" diameter					.56			.56	.62
1160	3-1/2" diameter					.65			.65	.72
1180	4" diameter					.81			.81	.89
1200	0.040" thk., 5" diameter					.93			.93	1.02
1220	6" diameter					1.14			1.14	1.25
1240	8" diameter					1.46			1.46	1.61
1260	10" diameter					1.79			1.79	1.97
1280	12" diameter					2.15			2.15	2.37
1300	0.060" thk., 14" diameter					2.76			2.76	3.04
1320	16" diameter					3.16			3.16	3.48
1340	18" diameter					3.30			3.30	3.63
1360	20" diameter					3.67			3.67	4.04
1380	Fittings, field wrapped, add				S.F.	4.30			4.30	4.73

33 01 10.20 Pipe Repair

		Crew	Daily Output	Labor-Hours	Unit	Material	2020 Bare Costs Labor	Equipment	Total	Total Incl O&P
0010	**PIPE REPAIR**									
0020	Not including excavation or backfill									
0100	Clamp, stainless steel, lightweight, for steel pipe									
0110	3" long, 1/2" diameter pipe	1 Plum	34	.235	Ea.	11.15	15.15		26.30	35
0120	3/4" diameter pipe		32	.250		14.10	16.10		30.20	39.50
0130	1" diameter pipe		30	.267		12.50	17.20		29.70	39.50
0140	1-1/4" diameter pipe		28	.286		13.15	18.40		31.55	42
0150	1-1/2" diameter pipe		26	.308		16.85	19.85		36.70	48
0160	2" diameter pipe		24	.333		18.15	21.50		39.65	52
0170	2-1/2" diameter pipe		23	.348		17.40	22.50		39.90	52.50

472

For customer support on your Site Work & Landscape Costs with RSMeans data, call 800.448.8182.

33 01 10 – Operation and Maintenance of Water Utilities

33 01 10.20 Pipe Repair	Crew	Daily Output	Labor-Hours	Unit	Material	2020 Bare Costs Labor	Equipment	Total	Total Incl O&P
0180	3" diameter pipe	1 Plum	22	.364	Ea.	24.50	23.50		48	62
0190	3-1/2" diameter pipe	↓	21	.381		25	24.50		49.50	64
0200	4" diameter pipe	B-20	44	.545		23	25.50		48.50	63.50
0210	5" diameter pipe		42	.571		26.50	27		53.50	69.50
0220	6" diameter pipe		38	.632		29.50	29.50		59	77
0230	8" diameter pipe		30	.800		35.50	37.50		73	96
0240	10" diameter pipe		28	.857		120	40.50		160.50	193
0250	12" diameter pipe		24	1		131	47		178	216
0260	14" diameter pipe		22	1.091		166	51.50		217.50	261
0270	16" diameter pipe		20	1.200		179	56.50		235.50	282
0280	18" diameter pipe		18	1.333		206	62.50		268.50	320
0290	20" diameter pipe		16	1.500		234	70.50		304.50	365
0300	24" diameter pipe	↓	14	1.714		238	80.50		318.50	385
0360	For 6" long, add					100%	40%			
0370	For 9" long, add					200%	100%			
0380	For 12" long, add					300%	150%			
0390	For 18" long, add			↓		500%	200%			
0400	Pipe freezing for live repairs of systems 3/8" to 6"									
0410	Note: Pipe freezing can also be used to install a valve into a live system									
0420	Pipe freezing each side 3/8"	2 Skwk	8	2	Ea.	550	110		660	770
0425	Pipe freezing each side 3/8", second location same kit		8	2		32	110		142	202
0430	Pipe freezing each side 3/4"		8	2		515	110		625	735
0435	Pipe freezing each side 3/4", second location same kit		8	2		32	110		142	202
0440	Pipe freezing each side 1-1/2"		6	2.667		515	146		661	790
0445	Pipe freezing each side 1-1/2", second location same kit		6	2.667		32	146		178	257
0450	Pipe freezing each side 2"		6	2.667		945	146		1,091	1,275
0455	Pipe freezing each side 2", second location same kit		6	2.667		32	146		178	257
0460	Pipe freezing each side 2-1/2" to 3"		6	2.667		955	146		1,101	1,275
0465	Pipe freezing each side 2-1/2" to 3", second location same kit		6	2.667		64	146		210	292
0470	Pipe freezing each side 4"		4	4		1,575	219		1,794	2,050
0475	Pipe freezing each side 4", second location same kit		4	4		73	219		292	410
0480	Pipe freezing each side 5" to 6"		4	4		4,400	219		4,619	5,175
0485	Pipe freezing each side 5" to 6", second location same kit	↓	4	4		218	219		437	570
0490	Pipe freezing extra 20 lb. CO_2 cylinders (3/8" to 2" - 1 ea, 3" - 2 ea)					231			231	254
0500	Pipe freezing extra 50 lb. CO_2 cylinders (4" - 2 ea, 5"-6" - 6 ea)				↓	530			530	585
1000	Clamp, stainless steel, with threaded service tap									
1040	Full seal for iron, steel, PVC pipe									
1100	6" long, 2" diameter pipe	1 Plum	17	.471	Ea.	115	30.50		145.50	173
1110	2-1/2" diameter pipe		16	.500		114	32		146	174
1120	3" diameter pipe		15.60	.513		112	33		145	173
1130	3-1/2" diameter pipe	↓	15	.533		135	34.50		169.50	200
1140	4" diameter pipe	B-20	32	.750		125	35.50		160.50	190
1150	6" diameter pipe		28	.857		164	40.50		204.50	241
1160	8" diameter pipe		21	1.143		176	53.50		229.50	275
1170	10" diameter pipe		20	1.200		232	56.50		288.50	340
1180	12" diameter pipe	↓	17	1.412		266	66.50		332.50	390
1200	8" long, 2" diameter pipe	1 Plum	11.72	.683		117	44		161	195
1210	2-1/2" diameter pipe		11	.727		127	47		174	210
1220	3" diameter pipe		10.75	.744		125	48		173	210
1230	3-1/2" diameter pipe	↓	10.34	.774		132	50		182	220
1240	4" diameter pipe	B-20	22	1.091		158	51.50		209.50	252
1250	6" diameter pipe		19.31	1.243		162	58.50		220.50	267
1260	8" diameter pipe	↓	14.48	1.657	↓	211	78		289	350

33 01 10.20 Pipe Repair	Crew	Daily Output	Labor-Hours	Unit	Material	Labor	2020 Bare Costs Equipment	Total	Total Incl O&P
1270	10" diameter pipe	B-20	13.80	1.739	Ea.	243	82		325	390
1280	12" diameter pipe	↓	11.72	2.048		305	96.50		401.50	485
1300	12" long, 2" diameter pipe	1 Plum	9.44	.847		248	54.50		302.50	355
1310	2-1/2" diameter pipe		8.89	.900		204	58		262	310
1320	3" diameter pipe		8.67	.923		191	59.50		250.50	299
1330	3-1/2" diameter pipe	↓	8.33	.960		228	62		290	345
1340	4" diameter pipe	B-20	17.78	1.350		237	63.50		300.50	355
1350	6" diameter pipe		15.56	1.542		243	72.50		315.50	375
1360	8" diameter pipe		11.67	2.057		288	96.50		384.50	460
1370	10" diameter pipe		11.11	2.160		470	102		572	675
1380	12" diameter pipe	↓	9.44	2.542		475	120		595	705
1400	20" long, 2" diameter pipe	1 Plum	8.10	.988		231	63.50		294.50	350
1410	2-1/2" diameter pipe		7.62	1.050		272	67.50		339.50	400
1420	3" diameter pipe		7.43	1.077		345	69.50		414.50	485
1430	3-1/2" diameter pipe	↓	7.14	1.120		310	72		382	450
1440	4" diameter pipe	B-20	15.24	1.575		340	74		414	485
1450	6" diameter pipe		13.33	1.800		405	84.50		489.50	575
1460	8" diameter pipe		10	2.400		495	113		608	715
1470	10" diameter pipe		9.52	2.521		555	119		674	795
1480	12" diameter pipe	↓	8.10	2.963	↓	660	139		799	935
1600	Clamp, stainless steel, single section									
1640	Full seal for iron, steel, PVC pipe									
1700	6" long, 2" diameter pipe	1 Plum	17	.471	Ea.	91	30.50		121.50	146
1710	2-1/2" diameter pipe		16	.500		94	32		126	151
1720	3" diameter pipe		15.60	.513		86.50	33		119.50	145
1730	3-1/2" diameter pipe	↓	15	.533		113	34.50		147.50	176
1740	4" diameter pipe	B-20	32	.750		123	35.50		158.50	188
1750	6" diameter pipe		27	.889		152	42		194	231
1760	8" diameter pipe		21	1.143		161	53.50		214.50	258
1770	10" diameter pipe		20	1.200		227	56.50		283.50	335
1780	12" diameter pipe	↓	17	1.412		218	66.50		284.50	340
1800	8" long, 2" diameter pipe	1 Plum	11.72	.683		126	44		170	205
1805	2-1/2" diameter pipe		11.03	.725		129	46.50		175.50	212
1810	3" diameter pipe		10.76	.743		136	48		184	222
1815	3-1/2" diameter pipe	↓	10.34	.774		124	50		174	212
1820	4" diameter pipe	B-20	22.07	1.087		147	51		198	239
1825	6" diameter pipe		19.31	1.243		187	58.50		245.50	294
1830	8" diameter pipe		14.48	1.657		223	78		301	365
1835	10" diameter pipe		13.79	1.740		300	82		382	460
1840	12" diameter pipe	↓	11.72	2.048		335	96.50		431.50	515
1850	12" long, 2" diameter pipe	1 Plum	9.44	.847		190	54.50		244.50	290
1855	2-1/2" diameter pipe		8.89	.900		196	58		254	305
1860	3" diameter pipe		8.67	.923		215	59.50		274.50	325
1865	3-1/2" diameter pipe	↓	8.33	.960		197	62		259	310
1870	4" diameter pipe	B-20	17.78	1.350		224	63.50		287.50	345
1875	6" diameter pipe		15.56	1.542		305	72.50		377.50	445
1880	8" diameter pipe		11.67	2.057		330	96.50		426.50	510
1885	10" diameter pipe		11.11	2.160		450	102		552	645
1890	12" diameter pipe	↓	9.44	2.542		535	120		655	770
1900	20" long, 2" diameter pipe	1 Plum	8.10	.988		197	63.50		260.50	315
1905	2-1/2" diameter pipe		7.62	1.050		218	67.50		285.50	340
1910	3" diameter pipe		7.43	1.077		250	69.50		319.50	380
1915	3-1/2" diameter pipe	↓	7.14	1.120	↓	320	72		392	465

33 01 10 – Operation and Maintenance of Water Utilities

33 01 10.20 Pipe Repair

33 01 10.20 Pipe Repair		Crew	Daily Output	Labor-Hours	Unit	Material	2020 Bare Costs Labor	2020 Bare Costs Equipment	Total	Total Incl O&P
1920	4" diameter pipe	B-20	15.24	1.575	Ea.	425	74		499	575
1925	6" diameter pipe		13.33	1.800		465	84.50		549.50	640
1930	8" diameter pipe		10	2.400		555	113		668	780
1935	10" diameter pipe		9.52	2.521		700	119		819	950
1940	12" diameter pipe		8.10	2.963		830	139		969	1,125
2000	Clamp, stainless steel, two section									
2040	Full seal, for iron, steel, PVC pipe									
2100	8" long, 4" diameter pipe	B-20	24	1	Ea.	230	47		277	325
2110	6" diameter pipe		20	1.200		264	56.50		320.50	375
2120	8" diameter pipe		13	1.846		305	87		392	470
2130	10" diameter pipe		12	2		293	94		387	460
2140	12" diameter pipe		10	2.400		395	113		508	605
2200	10" long, 4" diameter pipe		16	1.500		300	70.50		370.50	435
2210	6" diameter pipe		13	1.846		340	87		427	505
2220	8" diameter pipe		9	2.667		380	125		505	605
2230	10" diameter pipe		8	3		490	141		631	750
2240	12" diameter pipe		7	3.429		565	161		726	870
2242	Clamp, stainless steel, three section									
2244	Full seal, for iron, steel, PVC pipe									
2250	10" long, 14" diameter pipe	B-20	6.40	3.750	Ea.	655	176		831	985
2260	16" diameter pipe		6	4		695	188		883	1,050
2270	18" diameter pipe		5	4.800		795	226		1,021	1,225
2280	20" diameter pipe		4.60	5.217		905	245		1,150	1,375
2290	24" diameter pipe		4	6		1,250	282		1,532	1,800
2320	For 12" long, add to 10"					15%	25%			
2330	For 20" long, add to 10"					70%	55%			
8000	For internal cleaning and inspection, see Section 33 01 30.11									
8100	For pipe testing, see Section 23 05 93.50									

33 01 30 – Operation and Maintenance of Sewer Utilities

33 01 30.11 Television Inspection of Sewers

		Crew	Daily Output	Labor-Hours	Unit	Material	2020 Bare Costs Labor	2020 Bare Costs Equipment	Total	Total Incl O&P
0010	**TELEVISION INSPECTION OF SEWERS**									
0100	Pipe internal cleaning & inspection, cleaning, pressure pipe systems									
0120	Pig method, lengths 1000' to 10,000'									
0140	4" diameter thru 24" diameter, minimum				L.F.				3.60	4.14
0160	Maximum				"				18	21
6000	Sewage/sanitary systems									
6100	Power rodder with header & cutters									
6110	Mobilization charge, minimum				Total				695	800
6120	Mobilization charge, maximum				"				9,125	10,600
6140	Cleaning 4"-12" diameter				L.F.				6	6.60
6190	14"-24" diameter								8	8.80
6240	30" diameter								9.60	10.50
6250	36" diameter								12	13.20
6260	48" diameter								16	17.60
6270	60" diameter								24	26.50
6280	72" diameter								48	53
9000	Inspection, television camera with video									
9060	up to 500 linear feet				Total				715	820

For customer support on your Site Work & Landscape Costs with RSMeans data, call 800.448.8182.

475

33 01 30.23 Pipe Bursting

		Crew	Daily Output	Labor-Hours	Unit	Material	2020 Bare Costs Labor	2020 Bare Costs Equipment	Total	Total Incl O&P
0010	**PIPE BURSTING**									
0011	300' runs, replace with HDPE pipe									
0020	Not including excavation, backfill, shoring, or dewatering									
0100	6" to 15" diameter, minimum				L.F.				200	220
0200	Maximum								550	605
0300	18" to 36" diameter, minimum								650	715
0400	Maximum								1,050	1,150
0500	Mobilize and demobilize, minimum				Job				3,000	3,300
0600	Maximum				"				32,100	35,400

33 01 30.72 Cured-In-Place Pipe Lining

		Crew	Daily Output	Labor-Hours	Unit	Material	2020 Bare Costs Labor	2020 Bare Costs Equipment	Total	Total Incl O&P
0010	**CURED-IN-PLACE PIPE LINING**									
0011	Not incl. bypass or cleaning									
0020	Less than 10,000 L.F., urban, 6" to 10"	C-17E	130	.615	L.F.	9.40	34	.28	43.68	62
0050	10" to 12"		125	.640		11.60	35.50	.29	47.39	66.50
0070	12" to 16"		115	.696		11.85	38.50	.31	50.66	71.50
0100	16" to 20"		95	.842		13.95	46.50	.38	60.83	86
0200	24" to 36"		90	.889		15.30	49	.40	64.70	91.50
0300	48" to 72"		80	1		24	55.50	.45	79.95	110
0500	Rural, 6" to 10"		180	.444		9.40	24.50	.20	34.10	47.50
0550	10" to 12"		175	.457		11.60	25.50	.21	37.31	51
0570	12" to 16"		160	.500		12.15	27.50	.23	39.88	55.50
0600	16" to 20"		135	.593		14.95	32.50	.27	47.72	66
0700	24" to 36"		125	.640		15.35	35.50	.29	51.14	70.50
0800	48" to 72"		100	.800		24	44	.36	68.36	94
1000	Greater than 10,000 L.F., urban, 6" to 10"		160	.500		9.40	27.50	.23	37.13	52.50
1050	10" to 12"		155	.516		11.55	28.50	.23	40.28	56
1070	12" to 16"		140	.571		11.85	31.50	.26	43.61	61.50
1100	16" to 20"		120	.667		14.95	37	.30	52.25	72
1200	24" to 36"		115	.696		15.35	38.50	.31	54.16	75
1300	48" to 72"		95	.842		24	46.50	.38	70.88	97.50
1500	Rural, 6" to 10"		215	.372		9.40	20.50	.17	30.07	41.50
1550	10" to 12"		210	.381		11.60	21	.17	32.77	45
1570	12" to 16"		185	.432		11.85	24	.19	36.04	49.50
1600	16" to 20"		150	.533		14.95	29.50	.24	44.69	61
1700	24" to 36"		140	.571		15.35	31.50	.26	47.11	65
1800	48" to 72"		120	.667		24.50	37	.30	61.80	83
2000	Cured in place pipe, non-pressure, flexible felt resin									
2100	6" diameter				L.F.				25.50	28
2200	8" diameter								26.50	29
2201	8" diameter									
2300	10" diameter								29	32
2400	12" diameter								34.50	38
2500	15" diameter								59	65
2600	18" diameter								82	90
2700	21" diameter								105	115
2800	24" diameter								177	195
2900	30" diameter								195	215
3000	36" diameter								205	225
3100	48" diameter								218	240

33 01 Operation and Maintenance of Utilities

33 01 30 – Operation and Maintenance of Sewer Utilities

33 01 30.74 Sliplining, Excludes Cleaning

		Crew	Daily Output	Labor-Hours	Unit	Material	2020 Bare Costs Labor	2020 Bare Costs Equipment	Total	Total Incl O&P
0010	**SLIPLINING, excludes cleaning** and video inspection									
0020	Pipe relined with one pipe size smaller than original (4" for 6")									
0100	6" diameter, original size	B-6B	600	.080	L.F.	5.60	3.42	1.67	10.69	13.20
0150	8" diameter, original size		600	.080		9.60	3.42	1.67	14.69	17.55
0200	10" diameter, original size		600	.080		11.25	3.42	1.67	16.34	19.40
0250	12" diameter, original size		400	.120		16	5.15	2.50	23.65	28
0300	14" diameter, original size	↓	400	.120		15.60	5.15	2.50	23.25	27.50
0350	16" diameter, original size	B-6C	300	.160		18.30	6.85	7.75	32.90	39
0400	18" diameter, original size	"	300	.160	↓	28	6.85	7.75	42.60	50
1000	Pipe HDPE lining, make service line taps	B-6	4	6	Ea.	87	274	53.50	414.50	565

33 05 Common Work Results for Utilities

33 05 07 – Trenchless Installation of Utility Piping

33 05 07.13 Utility Directional Drilling

		Crew	Daily Output	Labor-Hours	Unit	Material	2020 Bare Costs Labor	2020 Bare Costs Equipment	Total	Total Incl O&P
0010	**UTILITY DIRECTIONAL DRILLING**									
0011	Excluding access and splice pits (if required) and conduit (required)									
0012	Drilled hole diameters shown should be 50% larger than conduit									
0013	Assume access to H2O & removal of spoil/drilling mud as non-hazard material									
0014	Actual production rates adj. to account for risk factors for each soil type									
0100	Sand, silt, clay, common earth									
0110	Mobilization or demobilization	B-82A	1	32	Ea.		1,525	2,475	4,000	5,000
0120	6" diameter		180	.178	L.F.		8.45	13.80	22.25	28
0130	12" diameter		70	.457			21.50	35.50	57	71.50
0140	18" diameter		50	.640			30.50	49.50	80	100
0150	24" diameter	↓	35	.914	↓		43.50	71	114.50	143
0200	Hard clay, cobble, random boulders									
0210	Mobilization or demobilization	B-82B	1	32	Ea.		1,525	2,600	4,125	5,125
0220	6" diameter		125	.256	L.F.		12.15	20.50	32.65	41.50
0230	12" diameter		50	.640			30.50	52	82.50	103
0240	18" diameter		30	1.067			50.50	86.50	137	171
0250	24" diameter		25	1.280			61	104	165	205
0260	30" diameter		20	1.600			76	129	205	256
0270	36" diameter	↓	15	2.133	↓		101	173	274	340
0300	Hard rock (solid bed, 24,000+ psi)									
0310	Mobilization or demobilization	B-82C	1	32	Ea.		1,525	2,775	4,300	5,325
0320	6" diameter		30	1.067	L.F.		50.50	92	142.50	177
0330	12" diameter		12	2.667			127	230	357	445
0340	18" diameter		8	4			190	345	535	665
0344	20" diameter		7.50	4.267			203	370	573	710
0350	24" diameter		7	4.571			217	395	612	760
0360	30" diameter		6	5.333			254	460	714	885
0370	36" diameter	↓	4.50	7.111			340	615	955	1,175

33 05 07.23 Utility Boring and Jacking

		Crew	Daily Output	Labor-Hours	Unit	Material	2020 Bare Costs Labor	2020 Bare Costs Equipment	Total	Total Incl O&P
0010	**UTILITY BORING AND JACKING**									
0011	Casing only, 100' minimum,									
0020	not incl. jacking pits or dewatering									
0100	Roadwork, 1/2" thick wall, 24" diameter casing	B-42	20	3.200	L.F.	122	152	52	326	420
0200	36" diameter		16	4		215	190	65	470	595
0300	48" diameter		15	4.267		300	203	69.50	572.50	710
0500	Railroad work, 24" diameter	↓	15	4.267	↓	122	203	69.50	394.50	515

33 05 07 – Trenchless Installation of Utility Piping

33 05 07.23 Utility Boring and Jacking

		Crew	Daily Output	Labor-Hours	Unit	Material	2020 Bare Costs Labor	Equipment	Total	Total Incl O&P
0600	36" diameter	B-42	14	4.571	L.F.	215	217	74.50	506.50	650
0700	48" diameter	↓	12	5.333		300	253	87	640	805
0900	For ledge, add								20%	20%
1000	Small diameter boring, 3", sandy soil	B-82	900	.018		22	.85	.20	23.05	26
1040	Rocky soil	"	500	.032	↓	22	1.52	.36	23.88	27
1100	Prepare jacking pits, incl. mobilization & demobilization, minimum				Ea.				3,225	3,700
1101	Maximum				"				22,000	25,500

33 05 07.36 Microtunneling

		Crew	Daily Output	Labor-Hours	Unit	Material	2020 Bare Costs Labor	Equipment	Total	Total Incl O&P
0010	**MICROTUNNELING**									
0011	Not including excavation, backfill, shoring,									
0020	or dewatering, average 50'/day, slurry method									
0100	24" to 48" outside diameter, minimum				L.F.				965	965
0110	Adverse conditions, add				"				500	500
1000	Rent microtunneling machine, average monthly lease				Month				97,500	107,000
1010	Operating technician				Day				630	705
1100	Mobilization and demobilization, minimum				Job				41,200	45,900
1110	Maximum				"				445,500	490,500

33 05 61 – Concrete Manholes

33 05 61.10 Storm Drainage Manholes, Frames and Covers

		Crew	Daily Output	Labor-Hours	Unit	Material	2020 Bare Costs Labor	Equipment	Total	Total Incl O&P
0010	**STORM DRAINAGE MANHOLES, FRAMES & COVERS**									
0020	Excludes footing, excavation, backfill (See line items for frame & cover)									
0050	Brick, 4' inside diameter, 4' deep	D-1	1	16	Ea.	615	750		1,365	1,825
0100	6' deep		.70	22.857		870	1,075		1,945	2,575
0150	8' deep		.50	32	↓	1,125	1,500		2,625	3,500
0200	For depths over 8', add		4	4	V.L.F.	95.50	188		283.50	390
0400	Concrete blocks (radial), 4' ID, 4' deep		1.50	10.667	Ea.	430	500		930	1,225
0500	6' deep		1	16		565	750		1,315	1,775
0600	8' deep		.70	22.857	↓	705	1,075		1,780	2,400
0700	For depths over 8', add		5.50	2.909	V.L.F.	71.50	136		207.50	286
0800	Concrete, cast in place, 4' x 4', 8" thick, 4' deep	C-14H	2	24	Ea.	580	1,250	13.45	1,843.45	2,500
0900	6' deep		1.50	32		830	1,650	17.90	2,497.90	3,400
1000	8' deep		1	48	↓	1,200	2,475	27	3,702	5,075
1100	For depths over 8', add	↓	8	6	V.L.F.	133	310	3.36	446.36	615
1110	Precast, 4' ID, 4' deep	B-22	4.10	7.317	Ea.	880	360	69	1,309	1,600
1120	6' deep		3	10		1,025	495	94.50	1,614.50	1,975
1130	8' deep		2	15	↓	1,200	740	142	2,082	2,600
1140	For depths over 8', add	↓	16	1.875	V.L.F.	134	92.50	17.70	244.20	305
1150	5' ID, 4' deep	B-6	3	8	Ea.	1,725	365	71.50	2,161.50	2,525
1160	6' deep		2	12		2,075	550	107	2,732	3,225
1170	8' deep		1.50	16	↓	2,625	730	143	3,498	4,125
1180	For depths over 8', add		12	2	V.L.F.	325	91.50	17.80	434.30	515
1190	6' ID, 4' deep		2	12	Ea.	2,600	550	107	3,257	3,800
1200	6' deep		1.50	16		2,975	730	143	3,848	4,525
1210	8' deep		1	24	↓	3,675	1,100	214	4,989	5,925
1220	For depths over 8', add		8	3	V.L.F.	410	137	26.50	573.50	685
1250	Slab tops, precast, 8" thick									
1300	4' diameter manhole	B-6	8	3	Ea.	282	137	26.50	445.50	545
1400	5' diameter manhole		7.50	3.200		440	146	28.50	614.50	735
1500	6' diameter manhole	↓	7	3.429		705	157	30.50	892.50	1,050
3800	Steps, heavyweight cast iron, 7" x 9"	1 Bric	40	.200		18.05	10.40		28.45	35.50
3900	8" x 9"		40	.200		21.50	10.40		31.90	40
3928	12" x 10-1/2"	↓	40	.200	↓	29.50	10.40		39.90	48.50

33 05 Common Work Results for Utilities

33 05 61 – Concrete Manholes

33 05 61.10 Storm Drainage Manholes, Frames and Covers

	Crew	Daily Output	Labor-Hours	Unit	Material	2020 Bare Costs Labor	Equipment	Total	Total Incl O&P
4000	Standard sizes, galvanized steel	1 Bric	40	.200	Ea.	26.50	10.40		36.90	45
4100	Aluminum		40	.200		30	10.40		40.40	49
4150	Polyethylene		40	.200		32.50	10.40		42.90	52
4210	Rubber boot 6" diam. or smaller	1 Clab	32	.250		104	10.55		114.55	130
4215	8" diam.		24	.333		118	14.05		132.05	151
4220	10" diam.		19	.421		135	17.75		152.75	176
4225	12" diam.		16	.500		172	21		193	221
4230	16" diam.		15	.533		226	22.50		248.50	282
4235	18" diam.		15	.533		262	22.50		284.50	320
4240	24" diam.		14	.571		286	24		310	350
4245	30" diam.		12	.667		375	28		403	455

33 05 63 – Concrete Vaults and Chambers

33 05 63.13 Precast Concrete Utility Structures

	Crew	Daily Output	Labor-Hours	Unit	Material	2020 Bare Costs Labor	Equipment	Total	Total Incl O&P
0010	**PRECAST CONCRETE UTILITY STRUCTURES**, 6" thick									
0040	4' x 6' x 6' high, ID	B-13	2	28	Ea.	1,500	1,300	290	3,090	3,900
0050	5' x 10' x 6' high, ID		2	28		1,850	1,300	290	3,440	4,300
0100	6' x 10' x 6' high, ID		2	28		1,925	1,300	290	3,515	4,375
0150	5' x 12' x 6' high, ID		2	28		2,050	1,300	290	3,640	4,500
0200	6' x 12' x 6' high, ID		1.80	31.111		2,275	1,425	325	4,025	5,025
0250	6' x 13' x 6' high, ID		1.50	37.333		3,000	1,725	385	5,110	6,300
0300	8' x 14' x 7' high, ID		1	56		3,250	2,575	580	6,405	8,100
0350	Hand hole, precast concrete, 1-1/2" thick									
0400	1'-0" x 2'-0" x 1'-9", ID, light duty	B-1	4	6	Ea.	540	257		797	980
0450	4'-6" x 3'-2" x 2'-0", OD, heavy duty	B-6	3	8		1,625	365	71.50	2,061.50	2,400
0460	Meter pit, 4' x 4', 4' deep		2	12		1,600	550	107	2,257	2,700
0470	6' deep		1.60	15		2,275	685	134	3,094	3,675
0480	8' deep		1.40	17.143		3,000	785	153	3,938	4,650
0490	10' deep		1.20	20		3,800	915	178	4,893	5,775
0500	15' deep		1	24		5,575	1,100	214	6,889	8,000
0510	6' x 6', 4' deep		1.40	17.143		2,725	785	153	3,663	4,350
0520	6' deep		1.20	20		4,100	915	178	5,193	6,075
0530	8' deep		1	24		5,450	1,100	214	6,764	7,875
0540	10' deep		.80	30		6,825	1,375	267	8,467	9,850
0550	15' deep		.60	40		10,400	1,825	355	12,580	14,500

33 05 97 – Identification and Signage for Utilities

33 05 97.05 Utility Connection

	Crew	Daily Output	Labor-Hours	Unit	Material	2020 Bare Costs Labor	Equipment	Total	Total Incl O&P
0010	**UTILITY CONNECTION**									
0020	Water, sanitary, stormwater, gas, single connection	B-14	1	48	Ea.	3,050	2,125	214	5,389	6,775
0030	Telecommunication	"	3	16	"	400	710	71	1,181	1,600

33 05 97.10 Utility Accessories

	Crew	Daily Output	Labor-Hours	Unit	Material	2020 Bare Costs Labor	Equipment	Total	Total Incl O&P
0010	**UTILITY ACCESSORIES**									
0400	Underground tape, detectable, reinforced, alum. foil core, 2"	1 Clab	150	.053	C.L.F.	9	2.25		11.25	13.25
0500	6"	"	140	.057	"	38	2.41		40.41	45

33 05 97.15 Pipe Casing

	Crew	Daily Output	Labor-Hours	Unit	Material	2020 Bare Costs Labor	Equipment	Total	Total Incl O&P
0010	**PIPE CASING**									
1000	Spacers									
1010	Poly, std config, std runners, for hot/cold carrier pipe, 8" diameter band	2 Clab	60	.267	Ea.	28.50	11.25		39.75	48
1020	12" diameter band		60	.267		39	11.25		50.25	60
1030	16" diameter band		60	.267		68	11.25		79.25	92
1040	20" diameter band		40	.400		92	16.85		108.85	127
1050	24" diameter band		40	.400		116	16.85		132.85	154

33 05 Common Work Results for Utilities

33 05 97 – Identification and Signage for Utilities

33 05 97.15 Pipe Casing

		Crew	Daily Output	Labor-Hours	Unit	Material	2020 Bare Costs Labor	Equipment	Total	Total Incl O&P
1060	30" diameter band	2 Clab	30	.533	Ea.	140	22.50		162.50	188
1070	36" diameter band		30	.533		165	22.50		187.50	215
1110	Carbon steel, 14 ga., poly coat, std config, 4 runners, 8" diameter band		60	.267		77.50	11.25		88.75	102
1120	12" diameter band		60	.267		87	11.25		98.25	113
1130	6 runners, 16" diameter band		60	.267		103	11.25		114.25	131
1140	20" diameter band		40	.400		141	16.85		157.85	181
1150	24" diameter band		40	.400		187	16.85		203.85	232
1160	30" diameter band		30	.533		265	22.50		287.50	325
1170	36" diameter band		30	.533		370	22.50		392.50	440
1210	Stainless steel, 14 ga., grade 304, std config, 4 runners, 8" diameter band		60	.267		97.50	11.25		108.75	124
1220	12" diameter band		60	.267		115	11.25		126.25	143
1230	6 runners, 16" diameter band		60	.267		139	11.25		150.25	169
1240	20" diameter band		40	.400		188	16.85		204.85	232
1250	24" diameter band		40	.400		247	16.85		263.85	298
1260	30" diameter band		30	.533		305	22.50		327.50	375
1270	36" diameter band	↓	30	.533	↓	465	22.50		487.50	545
2000	End seals									
2010	Pull-on, synthetic rubber, stainless steel bands, 8" carrier pipe	2 Clab	60	.267	Ea.	76	11.25		87.25	100
2020	12" carrier pipe		60	.267		104	11.25		115.25	132
2030	16" carrier pipe		60	.267		144	11.25		155.25	176
2040	20" carrier pipe		40	.400		194	16.85		210.85	239
2050	24" carrier pipe		40	.400		237	16.85		253.85	287
2060	30" carrier pipe		30	.533		320	22.50		342.50	385
2070	36" carrier pipe	↓	30	.533	↓	360	22.50		382.50	430

33 11 Groundwater Sources

33 11 13 – Potable Water Supply Wells

33 11 13.10 Wells and Accessories

		Crew	Daily Output	Labor-Hours	Unit	Material	2020 Bare Costs Labor	Equipment	Total	Total Incl O&P
0010	**WELLS & ACCESSORIES**									
0011	Domestic									
0100	Drilled, 4" to 6" diameter	B-23	120	.333	L.F.		14.15	13.25	27.40	36
0200	8" diameter	"	95.20	.420	"		17.85	16.70	34.55	45.50
0400	Gravel pack well, 40' deep, incl. gravel & casing, complete									
0500	24" diameter casing x 18" diameter screen	B-23	.13	308	Total	38,500	13,100	12,200	63,800	75,500
0600	36" diameter casing x 18" diameter screen	↓	.12	333	"	40,800	14,200	13,300	68,300	81,000
0800	Observation wells, 1-1/4" riser pipe		163	.245	V.L.F.	19.90	10.45	9.75	40.10	48.50
0900	For flush Buffalo roadway box, add	1 Skwk	16.60	.482	Ea.	53	26.50		79.50	98
1200	Test well, 2-1/2" diameter, up to 50' deep (15 to 50 GPM)	B-23	1.51	26.490	"	830	1,125	1,050	3,005	3,775
1300	Over 50' deep, add	"	121.80	.328	L.F.	22	13.95	13.05	49	60
1400	Remove & reset pump, minimum	B-21	4	7	Ea.		340	47.50	387.50	565
1420	Maximum	"	2	14	"		685	94.50	779.50	1,125
1500	Pumps, installed in wells to 100' deep, 4" submersible									
1510	1/2 HP	Q-1	3.22	4.969	Ea.	660	288		948	1,150
1520	3/4 HP		2.66	6.015		795	350		1,145	1,400
1600	1 HP		2.29	6.987		975	405		1,380	1,675
1700	1-1/2 HP	Q-22	1.60	10		2,150	580	294	3,024	3,575
1800	2 HP		1.33	12.030		1,800	700	355	2,855	3,450
1900	3 HP		1.14	14.035		2,425	815	415	3,655	4,350
2000	5 HP		1.14	14.035		2,850	815	415	4,080	4,800
2050	Remove and install motor only, 4 HP		1.14	14.035		1,450	815	415	2,680	3,275
3000	Pump, 6" submersible, 25' to 150' deep, 25 HP, 103 to 400 GPM	↓	.89	17.978	↓	9,325	1,050	530	10,905	12,400

480

33 11 13 – Potable Water Supply Wells

33 11 13.10 Wells and Accessories		Crew	Daily Output	Labor-Hours	Unit	Material	2020 Bare Costs Labor	Equipment	Total	Total Incl O&P
3100	25' to 500' deep, 30 HP, 104 to 400 GPM	Q-22	.73	21.918	Ea.	11,000	1,275	645	12,920	14,700
5000	Wells to 180' deep, 4" submersible, 1 HP	B-21	1.10	25.455		2,425	1,250	172	3,847	4,725
5500	2 HP		1.10	25.455		3,875	1,250	172	5,297	6,325
6000	3 HP		1	28		4,500	1,375	189	6,064	7,200
7000	5 HP		.90	31.111		7,100	1,525	210	8,835	10,300
8000	Steel well casing	B-23A	3020	.008	Lb.	1.29	.38	.29	1.96	2.31
8110	Well screen assembly, stainless steel, 2" diameter		273	.088	L.F.	81.50	4.19	3.23	88.92	100
8120	3" diameter		253	.095		143	4.52	3.49	151.01	168
8130	4" diameter		200	.120		178	5.70	4.41	188.11	209
8140	5" diameter		168	.143		184	6.80	5.25	196.05	218
8150	6" diameter		126	.190		204	9.10	7	220.10	246
8160	8" diameter		98.50	.244		274	11.60	8.95	294.55	325
8170	10" diameter		73	.329		340	15.65	12.10	367.75	410
8180	12" diameter		62.50	.384		395	18.30	14.10	427.40	480
8190	14" diameter		54.30	.442		445	21	16.25	482.25	540
8200	16" diameter		48.30	.497		490	23.50	18.25	531.75	595
8210	18" diameter		39.20	.612		620	29	22.50	671.50	755
8220	20" diameter		31.20	.769		700	36.50	28.50	765	855
8230	24" diameter		23.80	1.008		855	48	37	940	1,050
8240	26" diameter		21	1.143		960	54.50	42	1,056.50	1,175
8244	Well casing or drop pipe, PVC, 1/2" diameter		550	.044		1.31	2.08	1.60	4.99	6.30
8245	3/4" diameter		550	.044		1.32	2.08	1.60	5	6.35
8246	1" diameter		550	.044		1.36	2.08	1.60	5.04	6.40
8247	1-1/4" diameter		520	.046		1.71	2.20	1.70	5.61	7.05
8248	1-1/2" diameter		490	.049		1.81	2.33	1.80	5.94	7.45
8249	1-3/4" diameter		380	.063		1.85	3.01	2.32	7.18	9.10
8250	2" diameter		280	.086		2.12	4.08	3.15	9.35	11.90
8252	3" diameter		260	.092		4.16	4.40	3.39	11.95	14.90
8254	4" diameter		205	.117		4.46	5.60	4.30	14.36	18
8255	5" diameter		170	.141		4.48	6.75	5.20	16.43	20.50
8256	6" diameter		130	.185		8.60	8.80	6.80	24.20	30
8258	8" diameter		100	.240		13	11.45	8.80	33.25	41
8260	10" diameter		73	.329		19.90	15.65	12.10	47.65	59
8262	12" diameter		62	.387		24	18.45	14.25	56.70	69
8300	Slotted PVC, 1-1/4" diameter		521	.046		2.89	2.20	1.69	6.78	8.35
8310	1-1/2" diameter		488	.049		3.27	2.34	1.81	7.42	9.10
8320	2" diameter		273	.088		4.09	4.19	3.23	11.51	14.35
8330	3" diameter		253	.095		6.75	4.52	3.49	14.76	18.05
8340	4" diameter		200	.120		8.35	5.70	4.41	18.46	22.50
8350	5" diameter		168	.143		15.05	6.80	5.25	27.10	32.50
8360	6" diameter		126	.190		16.30	9.10	7	32.40	39
8370	8" diameter		98.50	.244		24	11.60	8.95	44.55	53.50
8400	Artificial gravel pack, 2" screen, 6" casing	B-23B	174	.138		4.68	6.55	6.40	17.63	22
8405	8" casing		111	.216		6.30	10.30	10.05	26.65	33.50
8410	10" casing		74.50	.322		8.80	15.35	14.95	39.10	49
8415	12" casing		60	.400		14.05	19.05	18.60	51.70	64.50
8420	14" casing		50.20	.478		14.65	23	22	59.65	74.50
8425	16" casing		40.70	.590		18.55	28	27.50	74.05	92.50
8430	18" casing		36	.667		22.50	32	31	85.50	106
8435	20" casing		29.50	.814		26	39	38	103	128
8440	24" casing		25.70	.934		30	44.50	43.50	118	147
8445	26" casing		24.60	.976		33	46.50	45.50	125	156
8450	30" casing		20	1.200		38	57	55.50	150.50	189

33 11 Groundwater Sources

33 11 13 – Potable Water Supply Wells

33 11 13.10 Wells and Accessories

	33 11 13.10 Wells and Accessories	Crew	Daily Output	Labor-Hours	Unit	Material	2020 Bare Costs Labor	2020 Bare Costs Equipment	Total	Total Incl O&P
8455	36" casing	B-23B	16.40	1.463	L.F.	40.50	69.50	68	178	225
8500	Develop well		8	3	Hr.	300	143	139	582	695
8550	Pump test well	↓	8	3		96.50	143	139	378.50	475
8560	Standby well	B-23A	8	3		105	143	110	358	450
8570	Standby, drill rig		8	3			143	110	253	335
8580	Surface seal well, concrete filled	↓	1	24	Ea.	1,025	1,150	880	3,055	3,850
8590	Well test pump, install & remove	B-23	1	40			1,700	1,600	3,300	4,300
8600	Well sterilization, chlorine	2 Clab	1	16		141	675		816	1,150
8610	Well water pressure switch	1 Clab	12	.667		98	28		126	150
8630	Well water pressure switch with manual reset	"	12	.667	↓	130	28		158	185
9950	See Section 31 23 19.40 for wellpoints									
9960	See Section 31 23 19.30 for drainage wells									

33 11 13.20 Water Supply Wells, Pumps

	33 11 13.20 Water Supply Wells, Pumps	Crew	Daily Output	Labor-Hours	Unit	Material	2020 Bare Costs Labor	2020 Bare Costs Equipment	Total	Total Incl O&P
0010	**WATER SUPPLY WELLS, PUMPS**									
0011	With pressure control									
1000	Deep well, jet, 42 gal. galvanized tank									
1040	3/4 HP	1 Plum	.80	10	Ea.	1,125	645		1,770	2,225
3000	Shallow well, jet, 30 gal. galvanized tank									
3040	1/2 HP	1 Plum	2	4	Ea.	915	258		1,173	1,375

33 14 Water Utility Transmission and Distribution

33 14 13 – Public Water Utility Distribution Piping

33 14 13.10 Water Supply, Concrete Pipe

	33 14 13.10 Water Supply, Concrete Pipe	Crew	Daily Output	Labor-Hours	Unit	Material	2020 Bare Costs Labor	2020 Bare Costs Equipment	Total	Total Incl O&P
0010	**WATER SUPPLY, CONCRETE PIPE**									
0020	Not including excavation or backfill									
3000	Prestressed Concrete Pipe (PCCP), 150 psi, 12" diameter	B-13	192	.292	L.F.	83	13.45	3.02	99.47	115
3010	24" diameter	"	128	.438		83	20	4.54	107.54	126
3040	36" diameter	B-13B	96	.583		129	27	10.20	166.20	194
3050	48" diameter		64	.875		184	40.50	15.30	239.80	280
3070	72" diameter		60	.933		360	43	16.35	419.35	475
3080	84" diameter	↓	40	1.400		465	64.50	24.50	554	635
3090	96" diameter	B-13C	40	1.400		680	64.50	57	801.50	910
3100	108" diameter		32	1.750		965	80.50	71.50	1,117	1,275
3102	120" diameter		16	3.500		1,800	161	143	2,104	2,400
3104	144" diameter	↓	16	3.500	↓	1,700	161	143	2,004	2,275
3110	Prestressed Concrete Pipe (PCCP), 150 psi, elbow, 90°, 12" diameter	B-13	24	2.333	Ea.	955	107	24	1,086	1,250
3140	24" diameter	"	6	9.333		1,950	430	97	2,477	2,875
3150	36" diameter	B-13B	4	14		3,650	645	245	4,540	5,225
3160	48" diameter		3	18.667		7,200	860	325	8,385	9,575
3180	72" diameter		1.60	35		20,500	1,600	615	22,715	25,700
3190	84" diameter		1.30	43.077		31,700	1,975	755	34,430	38,700
3200	96" diameter	↓	1	56		38,600	2,575	980	42,155	47,500
3210	108" diameter	B-13C	.66	84.848		50,000	3,900	3,475	57,375	64,500
3220	120" diameter		.40	140		56,000	6,450	5,725	68,175	77,500
3225	144" diameter	↓	.30	184		79,500	8,475	7,525	95,500	108,500
3230	Prestressed Concrete Pipe (PCCP), 150 psi, elbow, 45°, 12" diameter	B-13	24	2.333		595	107	24	726	845
3250	24" diameter	"	6	9.333		1,250	430	97	1,777	2,125
3260	36" diameter	B-13B	4	14		2,175	645	245	3,065	3,625
3270	48" diameter		3	18.667		4,150	860	325	5,335	6,225
3290	72" diameter		1.60	35		12,700	1,600	615	14,915	17,100
3300	84" diameter	↓	1.30	42.945	↓	19,500	1,975	750	22,225	25,200

33 14 Water Utility Transmission and Distribution

33 14 13 – Public Water Utility Distribution Piping

33 14 13.10 Water Supply, Concrete Pipe

		Crew	Daily Output	Labor-Hours	Unit	Material	2020 Bare Costs Labor	2020 Bare Costs Equipment	Total	Total Incl O&P
3310	96" diameter	B-13B	1	56	Ea.	24,100	2,575	980	27,655	31,500
3320	108" diameter	B-13C	.66	84.337		30,400	3,875	3,450	37,725	43,000
3330	120" diameter		.40	140		34,500	6,450	5,725	46,675	54,000
3340	144" diameter		.30	184		49,200	8,475	7,525	65,200	75,000

33 14 13.15 Water Supply, Ductile Iron Pipe

		Crew	Daily Output	Labor-Hours	Unit	Material	2020 Bare Costs Labor	2020 Bare Costs Equipment	Total	Total Incl O&P
0010	**WATER SUPPLY, DUCTILE IRON PIPE**									
0011	Cement lined									
0020	Not including excavation or backfill									
2000	Pipe, class 50 water piping, 18' lengths									
2020	Mechanical joint, 4" diameter	B-21A	200	.200	L.F.	42.50	10.45	2.14	55.09	64.50
2040	6" diameter		160	.250		51	13.05	2.68	66.73	78.50
2060	8" diameter		133.33	.300		53.50	15.70	3.21	72.41	86
2080	10" diameter		114.29	.350		71	18.30	3.75	93.05	110
2100	12" diameter		105.26	.380		90.50	19.85	4.07	114.42	134
2120	14" diameter		100	.400		105	21	4.28	130.28	152
2140	16" diameter		72.73	.550		112	29	5.90	146.90	173
2160	18" diameter		68.97	.580		129	30.50	6.20	165.70	194
2170	20" diameter		57.14	.700		132	36.50	7.50	176	208
2180	24" diameter		47.06	.850		168	44.50	9.10	221.60	262
2260	30" diameter		66.67	.600		130	31.50	6.45	167.95	197
2270	36" diameter		58.82	.680		153	35.50	7.30	195.80	231
2300	Mechanical restrained joint, no fittings, 4" diameter		173.91	.230		37.50	12	2.46	51.96	62
3000	Push-on joint, 4" diameter		400	.100		23	5.25	1.07	29.32	34.50
3020	6" diameter		333.33	.120		23.50	6.25	1.29	31.04	37
3040	8" diameter		200	.200		32.50	10.45	2.14	45.09	54
3060	10" diameter		181.82	.220		50	11.50	2.36	63.86	75
3080	12" diameter		160	.250		53	13.05	2.68	68.73	80.50
3100	14" diameter		133.33	.300		53	15.70	3.21	71.91	85
3120	16" diameter		114.29	.350		55	18.30	3.75	77.05	92
3140	18" diameter		100	.400		61	21	4.28	86.28	103
3160	20" diameter		88.89	.450		63.50	23.50	4.82	91.82	110
3180	24" diameter		76.92	.520		86.50	27	5.55	119.05	142
6170	Cap, 4" diameter	B-20A	32	1	Ea.	84	50.50		134.50	169
6180	6" diameter		25.60	1.250		131	63		194	239
6190	8" diameter		21.33	1.500		203	76		279	335
6198	10" diameter		16.84	1.900		370	96		466	550
6200	12" diameter		16.84	1.900		370	96		466	550
6205	15" diameter		11	2.909		805	147		952	1,100
6206	16" diameter		11	2.909		805	147		952	1,100
6210	18" diameter		11	2.909		975	147		1,122	1,300
6220	24" diameter	B-21A	9.41	4.251		1,925	222	45.50	2,192.50	2,500
6230	30" diameter		8	5		3,700	261	53.50	4,014.50	4,525
6240	36" diameter		6.90	5.797		5,000	305	62	5,367	6,025
8000	Piping, fittings, mechanical joint, AWWA C110									
8006	90° bend, 4" diameter	B-20A	16	2	Ea.	176	101		277	345
8020	6" diameter		12.80	2.500		256	126		382	470
8040	8" diameter		10.67	2.999		500	152		652	775
8060	10" diameter	B-21A	11.43	3.500		730	183	37.50	950.50	1,125
8080	12" diameter		10.53	3.799		960	199	40.50	1,199.50	1,400
8100	14" diameter		10	4		1,375	209	43	1,627	1,875
8120	16" diameter		7.27	5.502		1,750	288	59	2,097	2,425
8140	18" diameter		6.90	5.797		2,400	305	62	2,767	3,175

For customer support on your Site Work & Landscape Costs with RSMeans data, call 800.448.8182.

483

33 14 13 – Public Water Utility Distribution Piping

33 14 13.15 Water Supply, Ductile Iron Pipe

		Crew	Daily Output	Labor-Hours	Unit	Material	2020 Bare Costs Labor	Equipment	Total	Total Incl O&P
8160	20" diameter	B-21A	5.71	7.005	Ea.	2,875	365	75	3,315	3,775
8180	24" diameter	↓	4.70	8.511		4,975	445	91	5,511	6,250
8200	Wye or tee, 4" diameter	B-20A	10.67	2.999		390	152		542	650
8220	6" diameter		8.53	3.751		585	190		775	925
8240	8" diameter	↓	7.11	4.501		890	228		1,118	1,325
8260	10" diameter	B-21A	7.62	5.249		1,175	274	56	1,505	1,775
8280	12" diameter		7.02	5.698		1,975	298	61	2,334	2,675
8300	14" diameter		6.67	5.997		2,425	315	64	2,804	3,225
8320	16" diameter		4.85	8.247		2,750	430	88.50	3,268.50	3,775
8340	18" diameter		4.60	8.696		4,500	455	93	5,048	5,725
8360	20" diameter		3.81	10.499		6,525	550	112	7,187	8,125
8380	24" diameter	↓	3.14	12.739		8,800	665	136	9,601	10,800
8398	45° bend, 4" diameter	B-20A	16	2		206	101		307	380
8400	6" diameter		12.80	2.500		295	126		421	515
8405	8" diameter	↓	10.67	2.999		430	152		582	700
8410	12" diameter	B-21A	10.53	3.799		875	199	40.50	1,114.50	1,300
8420	16" diameter		7.27	5.502		1,725	288	59	2,072	2,400
8430	20" diameter		5.71	7.005		2,450	365	75	2,890	3,325
8440	24" diameter	↓	4.70	8.511		3,450	445	91	3,986	4,575
8450	Decreaser, 6" x 4" diameter	B-20A	14.22	2.250		262	114		376	460
8460	8" x 6" diameter	"	11.64	2.749		350	139		489	595
8470	10" x 6" diameter	B-21A	13.33	3.001		425	157	32	614	740
8480	12" x 6" diameter		12.70	3.150		620	165	33.50	818.50	965
8490	16" x 6" diameter		10	4		1,075	209	43	1,327	1,525
8500	20" x 6" diameter	↓	8.42	4.751	↓	1,800	248	51	2,099	2,425
8552	For water utility valves see Section 33 14 19									
8700	Joint restraint, ductile iron mechanical joints									
8710	4" diameter	B-20A	32	1	Ea.	32.50	50.50		83	112
8720	6" diameter		25.60	1.250		41	63		104	140
8730	8" diameter		21.33	1.500		59	76		135	179
8740	10" diameter		18.28	1.751		102	88.50		190.50	246
8750	12" diameter		16.84	1.900		117	96		213	273
8760	14" diameter		16	2		177	101		278	345
8770	16" diameter		11.64	2.749		190	139		329	415
8780	18" diameter		11.03	2.901		267	147		414	515
8785	20" diameter		9.14	3.501		330	177		507	630
8790	24" diameter	↓	7.53	4.250		450	215		665	815
9600	Steel sleeve with tap, 4" diameter	B-20	3	8		505	375		880	1,125
9620	6" diameter		2	12		555	565		1,120	1,475
9630	8" diameter	↓	2	12	↓	720	565		1,285	1,650

33 14 13.20 Water Supply, Polyethylene Pipe, C901

		Crew	Daily Output	Labor-Hours	Unit	Material	2020 Bare Costs Labor	Equipment	Total	Total Incl O&P
0010	**WATER SUPPLY, POLYETHYLENE PIPE, C901**									
0020	Not including excavation or backfill									
1000	Piping, 160 psi, 3/4" diameter	Q-1A	525	.019	L.F.	.40	1.24		1.64	2.29
1120	1" diameter		485	.021		.58	1.34		1.92	2.64
1140	1-1/2" diameter		450	.022		.96	1.44		2.40	3.22
1160	2" diameter	↓	365	.027	↓	1.44	1.78		3.22	4.24
2000	Fittings, insert type, nylon, 160 & 250 psi, cold water									
2220	Clamp ring, stainless steel, 3/4" diameter	Q-1A	345	.029	Ea.	1.23	1.88		3.11	4.16
2240	1" diameter		321	.031		1.44	2.02		3.46	4.60
2260	1-1/2" diameter		285	.035		1.91	2.28		4.19	5.50
2280	2" diameter	↓	255	.039		2.63	2.54		5.17	6.70

33 14 13.20 Water Supply, Polyethylene Pipe, C901

		Crew	Daily Output	Labor-Hours	Unit	Material	2020 Bare Costs Labor	2020 Bare Costs Equipment	Total	Total Incl O&P
2300	Coupling, 3/4" diameter	Q-1A	66	.152	Ea.	1	9.85		10.85	15.80
2320	1" diameter		57	.175		1.41	11.40		12.81	18.60
2340	1-1/2" diameter		51	.196		3.18	12.70		15.88	22.50
2360	2" diameter		48	.208		4.38	13.50		17.88	25
2400	Elbow, 90°, 3/4" diameter		66	.152		1.76	9.85		11.61	16.65
2420	1" diameter		57	.175		2.09	11.40		13.49	19.35
2440	1-1/2" diameter		51	.196		4.41	12.70		17.11	24
2460	2" diameter		48	.208		9.55	13.50		23.05	30.50
2500	Tee, 3/4" diameter		42	.238		2.03	15.45		17.48	25
2520	1" diameter		39	.256		3.19	16.65		19.84	28.50
2540	1-1/2" diameter		33	.303		6.65	19.65		26.30	37
2560	2" diameter		30	.333		11.35	21.50		32.85	45

33 14 13.25 Water Supply, Polyvinyl Chloride Pipe

		Crew	Daily Output	Labor-Hours	Unit	Material	2020 Bare Costs Labor	2020 Bare Costs Equipment	Total	Total Incl O&P
0010	**WATER SUPPLY, POLYVINYL CHLORIDE PIPE**									
0020	Not including excavation or backfill, unless specified									
2100	PVC pipe, Class 150, 1-1/2" diameter	Q-1A	750	.013	L.F.	.61	.86		1.47	1.96
2120	2" diameter		686	.015		.88	.95		1.83	2.39
2140	2-1/2" diameter		500	.020		1.18	1.30		2.48	3.24
2160	3" diameter	B-20	430	.056		1.59	2.62		4.21	5.70
2180	4" diameter		375	.064		3.03	3.01		6.04	7.85
2200	6" diameter		316	.076		5	3.57		8.57	10.90
2210	8" diameter		260	.092		8.60	4.34		12.94	16
3010	AWWA C905, PR 100, DR 25									
3030	14" diameter	B-20A	213	.150	L.F.	14.20	7.60		21.80	27
3040	16" diameter		200	.160		17.30	8.10		25.40	31
3050	18" diameter		160	.200		27.50	10.10		37.60	45.50
3060	20" diameter		133	.241		30.50	12.15		42.65	52
3070	24" diameter		107	.299		45	15.10		60.10	72
3080	30" diameter		80	.400		74.50	20		94.50	113
3090	36" diameter		80	.400		115	20		135	158
3100	42" diameter		60	.533		153	27		180	209
3200	48" diameter		60	.533		200	27		227	261
3960	Pressure pipe, class 200, ASTM 2241, SDR 21, 3/4" diameter	Q-1A	1000	.010		.35	.65		1	1.36
3980	1" diameter		900	.011		.35	.72		1.07	1.47
4000	1-1/2" diameter		750	.013		1.12	.86		1.98	2.52
4010	2" diameter		686	.015		1.24	.95		2.19	2.78
4020	2-1/2" diameter		500	.020		2.32	1.30		3.62	4.49
4030	3" diameter	B-20A	430	.074		2.40	3.76		6.16	8.30
4040	4" diameter		375	.085		3.48	4.31		7.79	10.30
4050	6" diameter		316	.101		7.25	5.10		12.35	15.60
4060	8" diameter		260	.123		8.80	6.20		15	19
4090	Including trenching to 3' deep, 3/4" diameter	Q-1C	300	.080		.35	4.61	6.25	11.21	14.15
4100	1" diameter		280	.086		.35	4.94	6.70	11.99	15.15
4110	1-1/2" diameter		260	.092		1.12	5.30	7.20	13.62	17.10
4120	2" diameter		220	.109		1.24	6.30	8.50	16.04	20
4130	2-1/2" diameter		200	.120		2.32	6.90	9.35	18.57	23
4140	3" diameter		175	.137		2.40	7.90	10.70	21	26
4150	4" diameter		150	.160		3.48	9.20	12.50	25.18	31.50
4160	6" diameter		125	.192		7.25	11.05	15	33.30	41
4165	Fittings									
4170	Elbow, 90°, 3/4"	Q-1A	114	.088	Ea.	1.22	5.70		6.92	9.85
4180	1"		100	.100		3.58	6.50		10.08	13.65

For customer support on your Site Work & Landscape Costs with RSMeans data, call 800.448.8182.

485

33 14 Water Utility Transmission and Distribution

33 14 13 – Public Water Utility Distribution Piping

33 14 13.25 Water Supply, Polyvinyl Chloride Pipe	Crew	Daily Output	Labor-Hours	Unit	Material	2020 Bare Costs Labor	Equipment	Total	Total Incl O&P
4190	1-1/2"	Q-1A	80	.125	Ea.	15.40	8.10		23.50	29
4200	2"	▼	72	.139		19.25	9		28.25	34.50
4210	3"	B-20A	46	.696		25.50	35		60.50	80.50
4220	4"		36	.889		40.50	45		85.50	112
4230	6"		24	1.333		71	67.50		138.50	179
4240	8"	▼	14	2.286		153	116		269	340
4250	Elbow, 45°, 3/4"	Q-1A	114	.088		2.49	5.70		8.19	11.25
4260	1"		100	.100		5.70	6.50		12.20	16
4270	1-1/2"		80	.125		15.15	8.10		23.25	29
4280	2"		72	.139		17.25	9		26.25	32.50
4290	2-1/2"	▼	54	.185		19.55	12		31.55	39.50
4300	3"	B-20A	46	.696		22	35		57	77
4310	4"		36	.889		36	45		81	107
4320	6"		24	1.333		76	67.50		143.50	185
4330	8"	▼	14	2.286		161	116		277	350
4340	Tee, 3/4"	Q-1A	76	.132		3.08	8.55		11.63	16.15
4350	1"		66	.152		11.90	9.85		21.75	28
4360	1-1/2"		54	.185		25	12		37	45.50
4370	2"		48	.208		24	13.50		37.50	46.50
4380	2-1/2"	▼	36	.278		29	18		47	59
4390	3"	B-20A	30	1.067		34	54		88	118
4400	4"		24	1.333		47	67.50		114.50	153
4410	6"		14.80	2.162		109	109		218	284
4420	8"	▼	9	3.556		250	180		430	545
4430	Coupling, 3/4"	Q-1A	114	.088		1.79	5.70		7.49	10.45
4440	1"		100	.100		7.05	6.50		13.55	17.45
4450	1-1/2"		80	.125		9.80	8.10		17.90	23
4460	2"		72	.139		11.05	9		20.05	25.50
4470	2-1/2"	▼	54	.185		11.40	12		23.40	30.50
4480	3"	B-20A	46	.696		17.40	35		52.40	71.50
4490	4"		36	.889		29.50	45		74.50	100
4500	6"		24	1.333		44.50	67.50		112	150
4510	8"		14	2.286	▼	94	116		210	276
4520	Pressure pipe Class 150, SDR 18, AWWA C900, 4" diameter		380	.084	L.F.	3.03	4.26		7.29	9.75
4530	6" diameter		316	.101		5	5.10		10.10	13.15
4540	8" diameter		264	.121		8.60	6.15		14.75	18.65
4550	10" diameter		220	.145		12.65	7.35		20	25
4560	12" diameter	▼	186	.172	▼	17.75	8.70		26.45	32.50
8000	Fittings with rubber gasket									
8003	Class 150, DR 18									
8006	90° bend , 4" diameter	B-20	100	.240	Ea.	42.50	11.30		53.80	63.50
8020	6" diameter		90	.267		74	12.55		86.55	100
8040	8" diameter		80	.300		139	14.10		153.10	175
8060	10" diameter		50	.480		365	22.50		387.50	435
8080	12" diameter		30	.800		445	37.50		482.50	545
8100	Tee, 4" diameter		90	.267		63	12.55		75.55	88
8120	6" diameter		80	.300		139	14.10		153.10	175
8140	8" diameter		70	.343		199	16.10		215.10	243
8160	10" diameter		40	.600		660	28		688	770
8180	12" diameter		20	1.200		905	56.50		961.50	1,075
8200	45° bend, 4" diameter		100	.240		41.50	11.30		52.80	62.50
8220	6" diameter		90	.267		77	12.55		89.55	103
8240	8" diameter	▼	50	.480		137	22.50		159.50	184

33 14 13.25 Water Supply, Polyvinyl Chloride Pipe		Crew	Daily Output	Labor-Hours	Unit	Material	2020 Bare Costs Labor	Equipment	Total	Total Incl O&P
8260	10" diameter	B-20	50	.480	Ea.	305	22.50		327.50	370
8280	12" diameter		30	.800		400	37.50		437.50	495
8300	Reducing tee 6" x 4"		100	.240		126	11.30		137.30	156
8320	8" x 6"		90	.267		202	12.55		214.55	241
8330	10" x 6"		90	.267		340	12.55		352.55	395
8340	10" x 8"		90	.267		365	12.55		377.55	420
8350	12" x 6"		90	.267		420	12.55		432.55	480
8360	12" x 8"		90	.267		440	12.55		452.55	505
8400	Tapped service tee (threaded type) 6" x 6" x 3/4"		100	.240		96	11.30		107.30	123
8430	6" x 6" x 1"		90	.267		96	12.55		108.55	125
8440	6" x 6" x 1-1/2"		90	.267		90.50	12.55		103.05	118
8450	6" x 6" x 2"		90	.267		96	12.55		108.55	125
8460	8" x 8" x 3/4"		90	.267		126	12.55		138.55	158
8470	8" x 8" x 1"		90	.267		126	12.55		138.55	158
8480	8" x 8" x 1-1/2"		90	.267		126	12.55		138.55	158
8490	8" x 8" x 2"		90	.267		126	12.55		138.55	158
8500	Repair coupling 4"		100	.240		23.50	11.30		34.80	43
8520	6" diameter		90	.267		36.50	12.55		49.05	59.50
8540	8" diameter		50	.480		90	22.50		112.50	133
8560	10" diameter		50	.480		175	22.50		197.50	227
8580	12" diameter		50	.480		265	22.50		287.50	325
8600	Plug end 4"		100	.240		21	11.30		32.30	40
8620	6" diameter		90	.267		43	12.55		55.55	66
8640	8" diameter		50	.480		63.50	22.50		86	104
8660	10" diameter		50	.480		97.50	22.50		120	141
8680	12" diameter	↓	50	.480	↓	120	22.50		142.50	166
8700	PVC pipe, joint restraint									
8710	4" diameter	B-20A	32	1	Ea.	46.50	50.50		97	128
8720	6" diameter		25.60	1.250		58	63		121	159
8730	8" diameter		21.33	1.500		84.50	76		160.50	207
8740	10" diameter		18.28	1.751		139	88.50		227.50	286
8750	12" diameter		16.84	1.900		146	96		242	305
8760	14" diameter		16	2		201	101		302	375
8770	16" diameter		11.64	2.749		270	139		409	505
8780	18" diameter		11.03	2.901		335	147		482	585
8785	20" diameter		9.14	3.501		445	177		622	755
8790	24" diameter	↓	7.53	4.250	↓	515	215		730	885

33 14 13.35 Water Supply, HDPE

		Crew	Daily Output	Labor-Hours	Unit	Material	2020 Bare Costs Labor	Equipment	Total	Total Incl O&P
0010	**WATER SUPPLY, HDPE**									
0011	Butt fusion joints, SDR 21 40' lengths not including excavation or backfill									
0100	4" diameter	B-22A	400	.100	L.F.	2.59	4.85	2.01	9.45	12.35
0200	6" diameter		380	.105		5.60	5.10	2.12	12.82	16.20
0300	8" diameter		320	.125		9.60	6.05	2.52	18.17	22.50
0400	10" diameter		300	.133		11.25	6.45	2.68	20.38	25
0500	12" diameter	↓	260	.154		16	7.45	3.10	26.55	32
0600	14" diameter	B-22B	220	.182		15.60	8.80	6.60	31	37.50
0700	16" diameter		180	.222		18.30	10.75	8.10	37.15	45
0800	18" diameter		140	.286		28	13.85	10.40	52.25	63.50
0850	20" diameter		130	.308		34	14.90	11.20	60.10	72.50
0900	24" diameter	↓	100	.400	↓	55.50	19.40	14.55	89.45	106
1000	Fittings									
1100	Elbows, 90 degrees									

For customer support on your Site Work & Landscape Costs with RSMeans data, call 800.448.8182.

487

33 14 13 – Public Water Utility Distribution Piping

33 14 13.35 Water Supply, HDPE		Crew	Daily Output	Labor-Hours	Unit	Material	2020 Bare Costs Labor	Equipment	Total	Total Incl O&P
1200	4" diameter	B-22A	32	1.250	Ea.	17.25	60.50	25	102.75	138
1300	6" diameter		28	1.429		47.50	69	28.50	145	188
1400	8" diameter		24	1.667		113	81	33.50	227.50	282
1500	10" diameter		18	2.222		254	108	44.50	406.50	490
1600	12" diameter		12	3.333		297	162	67	526	640
1700	14" diameter	B-22B	9	4.444		630	215	162	1,007	1,200
1800	16" diameter		6	6.667		935	325	242	1,502	1,775
1900	18" diameter		4	10		1,050	485	365	1,900	2,275
2000	24" diameter		3	13.333		1,825	645	485	2,955	3,500
2100	Tees									
2200	4" diameter	B-22A	30	1.333	Ea.	22.50	64.50	27	114	152
2300	6" diameter		26	1.538		74	74.50	31	179.50	227
2400	8" diameter		22	1.818		130	88	36.50	254.50	315
2500	10" diameter		15	2.667		172	129	53.50	354.50	440
2600	12" diameter		10	4		400	194	80.50	674.50	820
2700	14" diameter	B-22B	8	5		475	242	182	899	1,075
2800	16" diameter		6	6.667		555	325	242	1,122	1,350
2900	18" diameter		4	10		670	485	365	1,520	1,875
3000	24" diameter		2	20		895	970	725	2,590	3,225
4100	Caps									
4110	4" diameter	B-22A	34	1.176	Ea.	15.40	57	23.50	95.90	128
4120	6" diameter		30	1.333		33.50	64.50	27	125	164
4130	8" diameter		26	1.538		57	74.50	31	162.50	209
4150	10" diameter		20	2		139	97	40	276	345
4160	12" diameter		14	2.857		168	138	57.50	363.50	455

33 14 13.40 Water Supply, Black Steel Pipe

33 14 13.40 Water Supply, Black Steel Pipe		Crew	Daily Output	Labor-Hours	Unit	Material	2020 Bare Costs Labor	Equipment	Total	Total Incl O&P
0010	**WATER SUPPLY, BLACK STEEL PIPE**									
0011	Not including excavation or backfill									
1000	Pipe, black steel, plain end, welded, 1/4" wall thk, 8" diam.	B-35A	208	.269	L.F.	30.50	13.75	10.35	54.60	65.50
1010	10" diameter		204	.275		39	14	10.55	63.55	75
1020	12" diameter		195	.287		45.50	14.65	11	71.15	84
1030	18" diameter		175	.320		72	16.35	12.30	100.65	117
1040	5/16" wall thickness, 12" diameter		195	.287		65.50	14.65	11	91.15	106
1050	18" diameter		175	.320		102	16.35	12.30	130.65	151
1060	36" diameter		28.96	1.934		149	98.50	74	321.50	395
1070	3/8" wall thickness, 18" diameter		43.20	1.296		120	66	50	236	286
1080	24" diameter		36	1.556		141	79.50	59.50	280	340
1090	30" diameter		30.40	1.842		161	94	70.50	325.50	395
1100	1/2" wall thickness, 36" diameter		26.08	2.147		305	110	82.50	497.50	595
1110	48" diameter		21.68	2.583		540	132	99	771	900
1135	7/16" wall thickness, 48" diameter		20.80	2.692		365	137	103	605	720
1140	5/8" wall thickness, 48" diameter		21.68	2.583		485	132	99	716	835

33 14 13.45 Water Supply, Copper Pipe

33 14 13.45 Water Supply, Copper Pipe		Crew	Daily Output	Labor-Hours	Unit	Material	2020 Bare Costs Labor	Equipment	Total	Total Incl O&P
0010	**WATER SUPPLY, COPPER PIPE**									
0020	Not including excavation or backfill									
2000	Tubing, type K, 20' joints, 3/4" diameter	Q-1	400	.040	L.F.	7.05	2.32		9.37	11.20
2200	1" diameter		320	.050		9.70	2.90		12.60	15
3000	1-1/2" diameter		265	.060		14.95	3.50		18.45	21.50
3020	2" diameter		230	.070		23	4.04		27.04	31.50
3040	2-1/2" diameter		146	.110		35	6.35		41.35	48
3060	3" diameter		134	.119		48.50	6.95		55.45	63.50
4012	4" diameter		95	.168		80	9.75		89.75	103

33 14 Water Utility Transmission and Distribution

33 14 13 – Public Water Utility Distribution Piping

33 14 13.45 Water Supply, Copper Pipe

		Crew	Daily Output	Labor-Hours	Unit	Material	2020 Bare Costs Labor	Equipment	Total	Total Incl O&P
4014	5" diameter	Q-1	80	.200	L.F.	115	11.60		126.60	144
4016	6" diameter	Q-2	80	.300		134	18.05		152.05	175
4018	8" diameter	"	80	.300	↓	325	18.05		343.05	380
5000	Tubing, type L									
5108	2" diameter	Q-1	230	.070	L.F.	14.70	4.04		18.74	22
6010	3" diameter		134	.119		36.50	6.95		43.45	51
6012	4" diameter	↓	95	.168		48	9.75		57.75	67.50
6016	6" diameter	Q-2	80	.300	↓	112	18.05		130.05	150
7165	Fittings, brass, corporation stops, no lead, 3/4" diameter	1 Plum	19	.421	Ea.	76.50	27		103.50	125
7166	1" diameter		16	.500		98	32		130	156
7167	1-1/2" diameter		13	.615		208	39.50		247.50	289
7168	2" diameter		11	.727		330	47		377	430
7170	Curb stops, no lead, 3/4" diameter		19	.421		96	27		123	146
7171	1" diameter		16	.500		142	32		174	204
7172	1-1/2" diameter		13	.615		269	39.50		308.50	355
7173	2" diameter		11	.727		325	47		372	425
7180	Curb box, cast iron, 1/2" to 1" curb stops		12	.667		62	43		105	133
7200	1-1/4" to 2" curb stops	↓	8	1		116	64.50		180.50	225
7220	Saddles, 3/4" & 1" diameter, add					76.50			76.50	84
7240	1-1/2" to 2" diameter, add				↓	92			92	101
7250	For copper fittings, see Section 22 11 13.25									

33 14 13.90 Water Supply, Thrust Blocks

		Crew	Daily Output	Labor-Hours	Unit	Material	2020 Bare Costs Labor	Equipment	Total	Total Incl O&P
0010	**WATER SUPPLY, THRUST BLOCKS**									
0015	Piping, not including excavation or backfill									
0110	Thrust block for 90 degree elbow, 4" diameter	C-30	41	.195	Ea.	21.50	8.20	3.42	33.12	40
0115	6" diameter		23	.348		37.50	14.65	6.10	58.25	69.50
0120	8" diameter		14	.571		58	24	10	92	111
0125	10" diameter		9	.889		83.50	37.50	15.60	136.60	165
0130	12" diameter		7	1.143		114	48	20	182	220
0135	14" diameter		5	1.600		154	67.50	28	249.50	300
0140	16" diameter		4	2		194	84	35	313	380
0145	18" diameter		3	2.667		240	112	47	399	485
0150	20" diameter		2.50	3.200		290	135	56	481	585
0155	24" diameter		2	4		415	168	70	653	790
0210	Thrust block for tee or deadend, 4" diameter		65	.123		14.75	5.20	2.16	22.11	26.50
0215	6" diameter		35	.229		26.50	9.60	4.01	40.11	48
0220	8" diameter		21	.381		41	16.05	6.70	63.75	77
0225	10" diameter		14	.571		60	24	10	94	113
0230	12" diameter		10	.800		82	33.50	14.05	129.55	156
0235	14" diameter		7	1.143		110	48	20	178	216
0240	16" diameter		5	1.600		139	67.50	28	234.50	285
0245	18" diameter		4	2		172	84	35	291	355
0250	20" diameter		3.50	2.286		208	96	40	344	420
0255	24" diameter	↓	2.50	3.200	↓	295	135	56	486	590

33 14 17 – Site Water Utility Service Laterals

33 14 17.15 Tapping, Crosses and Sleeves

		Crew	Daily Output	Labor-Hours	Unit	Material	2020 Bare Costs Labor	Equipment	Total	Total Incl O&P
0010	**TAPPING, CROSSES AND SLEEVES**									
4000	Drill and tap pressurized main (labor only)									
4100	6" main, 1" to 2" service	Q-1	3	5.333	Ea.		310		310	465
4150	8" main, 1" to 2" service	"	2.75	5.818	"		335		335	505
4500	Tap and insert gate valve									
4600	8" main, 4" branch	B-21	3.20	8.750	Ea.		425	59	484	705

For customer support on your Site Work & Landscape Costs with RSMeans data, call 800.448.8182.

489

33 14 17.15 Tapping, Crosses and Sleeves	Crew	Daily Output	Labor-Hours	Unit	Material	2020 Bare Costs Labor	2020 Bare Costs Equipment	Total	Total Incl O&P
4650	6" branch	B-21	2.70	10.370	Ea.		505	70	575	835
4700	10" main, 4" branch		2.70	10.370			505	70	575	835
4750	6" branch		2.35	11.915			580	80.50	660.50	965
4800	12" main, 6" branch		2.35	11.915			580	80.50	660.50	965
4850	8" branch		2.35	11.915			580	80.50	660.50	965
7000	Tapping crosses, sleeves, valves; with rubber gaskets									
7020	Crosses, 4" x 4"	B-21	37	.757	Ea.	1,450	37	5.10	1,492.10	1,625
7030	6" x 4"		25	1.120		1,650	54.50	7.55	1,712.05	1,925
7040	6" x 6"		25	1.120		1,725	54.50	7.55	1,787.05	2,000
7060	8" x 6"		21	1.333		2,050	65	9	2,124	2,350
7080	8" x 8"		21	1.333		2,175	65	9	2,249	2,500
7100	10" x 6"		21	1.333		4,150	65	9	4,224	4,675
7120	10" x 10"		21	1.333		4,475	65	9	4,549	5,025
7140	12" x 6"		18	1.556		4,125	76	10.50	4,211.50	4,675
7160	12" x 12"		18	1.556		5,300	76	10.50	5,386.50	5,950
7180	14" x 6"		16	1.750		10,200	85.50	11.80	10,297.30	11,300
7200	14" x 14"		16	1.750		10,800	85.50	11.80	10,897.30	12,000
7220	16" x 6"		14	2		10,900	97.50	13.50	11,011	12,200
7240	16" x 10"		14	2		11,100	97.50	13.50	11,211	12,400
7260	16" x 16"		14	2		11,700	97.50	13.50	11,811	13,000
7280	18" x 6"		10	2.800		16,400	137	18.90	16,555.90	18,200
7300	18" x 12"		10	2.800		16,500	137	18.90	16,655.90	18,400
7320	18" x 18"		10	2.800		17,100	137	18.90	17,255.90	19,000
7340	20" x 6"		8	3.500		13,200	171	23.50	13,394.50	14,900
7360	20" x 12"		8	3.500		14,200	171	23.50	14,394.50	15,900
7380	20" x 20"		8	3.500		21,000	171	23.50	21,194.50	23,400
7400	24" x 6"		6	4.667		17,100	228	31.50	17,359.50	19,200
7420	24" x 12"		6	4.667		17,400	228	31.50	17,659.50	19,600
7440	24" x 18"		6	4.667		27,100	228	31.50	27,359.50	30,200
7460	24" x 24"		6	4.667		27,300	228	31.50	27,559.50	30,500
7600	Cut-in sleeves with rubber gaskets, 4"		18	1.556		560	76	10.50	646.50	740
7620	6"		12	2.333		705	114	15.75	834.75	965
7640	8"		10	2.800		895	137	18.90	1,050.90	1,200
7660	10"		10	2.800		1,250	137	18.90	1,405.90	1,600
7680	12"		9	3.111		1,800	152	21	1,973	2,225
7800	Cut-in valves with rubber gaskets, 4"		18	1.556		560	76	10.50	646.50	740
7820	6"		12	2.333		715	114	15.75	844.75	975
7840	8"		10	2.800		1,000	137	18.90	1,155.90	1,325
7860	10"		10	2.800		1,775	137	18.90	1,930.90	2,175
7880	12"		9	3.111		2,350	152	21	2,523	2,825
7900	Tapping valve 4", MJ, ductile iron		18	1.556		455	76	10.50	541.50	625
7920	6", MJ, ductile iron		12	2.333		620	114	15.75	749.75	870
7940	Tapping valve 8", MJ, ductile iron		10	2.800		840	137	18.90	995.90	1,150
7960	Tapping valve 10", MJ, ductile iron		10	2.800		1,175	137	18.90	1,330.90	1,525
7980	Tapping valve 12", MJ, ductile iron		8	3.500		1,875	171	23.50	2,069.50	2,325
8000	Sleeves with rubber gaskets, 4" x 4"		37	.757		1,050	37	5.10	1,092.10	1,200
8010	6" x 4"		25	1.120		985	54.50	7.55	1,047.05	1,175
8020	6" x 6"		25	1.120		1,300	54.50	7.55	1,362.05	1,525
8030	8" x 4"		21	1.333		1,025	65	9	1,099	1,225
8040	8" x 6"		21	1.333		910	65	9	984	1,100
8060	8" x 8"		21	1.333		1,350	65	9	1,424	1,575
8070	10" x 4"		21	1.333		830	65	9	904	1,025
8080	10" x 6"		21	1.333		1,025	65	9	1,099	1,225

490

33 14 Water Utility Transmission and Distribution

33 14 17 – Site Water Utility Service Laterals

33 14 17.15 Tapping, Crosses and Sleeves

		Crew	Daily Output	Labor-Hours	Unit	Material	2020 Bare Costs Labor	Equipment	Total	Total Incl O&P
8090	10" x 8"	B-21	21	1.333	Ea.	1,525	65	9	1,599	1,800
8100	10" x 10"		21	1.333		2,225	65	9	2,299	2,550
8110	12" x 4"		18	1.556		1,375	76	10.50	1,461.50	1,625
8120	12" x 6"		18	1.556		1,425	76	10.50	1,511.50	1,675
8130	12" x 8"		18	1.556		1,500	76	10.50	1,586.50	1,775
8135	12" x 10"		18	1.556		2,600	76	10.50	2,686.50	2,975
8140	12" x 12"		18	1.556		2,725	76	10.50	2,811.50	3,125
8160	14" x 6"		16	1.750		1,600	85.50	11.80	1,697.30	1,900
8180	14" x 14"		16	1.750		2,575	85.50	11.80	2,672.30	3,000
8200	16" x 6"		14	2		1,875	97.50	13.50	1,986	2,225
8220	16" x 10"		14	2		3,750	97.50	13.50	3,861	4,275
8240	16" x 16"		14	2		9,700	97.50	13.50	9,811	10,900
8260	18" x 6"		10	2.800		1,150	137	18.90	1,305.90	1,500
8280	18" x 12"		10	2.800		2,925	137	18.90	3,080.90	3,425
8300	18" x 18"		10	2.800		8,650	137	18.90	8,805.90	9,750
8320	20" x 6"		8	3.500		2,425	171	23.50	2,619.50	2,950
8340	20" x 12"		8	3.500		4,750	171	23.50	4,944.50	5,500
8360	20" x 20"		8	3.500		10,600	171	23.50	10,794.50	11,900
8380	24" x 6"		6	4.667		2,325	228	31.50	2,584.50	2,925
8400	24" x 12"		6	4.667		6,350	228	31.50	6,609.50	7,350
8420	24" x 18"		6	4.667		11,300	228	31.50	11,559.50	12,800
8440	24" x 24"	↓	6	4.667		13,300	228	31.50	13,559.50	15,000
8800	Hydrant valve box, 6' long	B-20	20	1.200		230	56.50		286.50	340
8820	8' long		18	1.333		305	62.50		367.50	430
8830	Valve box w/lid 4' deep		14	1.714		96.50	80.50		177	227
8840	Valve box and large base w/lid	↓	14	1.714	↓	325	80.50		405.50	480

33 14 19 – Valves and Hydrants for Water Utility Service

33 14 19.10 Valves

		Crew	Daily Output	Labor-Hours	Unit	Material	2020 Bare Costs Labor	Equipment	Total	Total Incl O&P
0010	**VALVES**, water distribution									
0011	See Sections 22 05 23.20 and 22 05 23.60									
3000	Butterfly valves with boxes, cast iron, mechanical joint									
3100	4" diameter	B-6	6	4	Ea.	550	183	35.50	768.50	925
3140	6" diameter		6	4		725	183	35.50	943.50	1,100
3180	8" diameter		6	4		910	183	35.50	1,128.50	1,325
3300	10" diameter		6	4		1,475	183	35.50	1,693.50	1,925
3340	12" diameter		6	4		1,650	183	35.50	1,868.50	2,150
3400	14" diameter		4	6		2,825	274	53.50	3,152.50	3,575
3440	16" diameter		4	6		4,150	274	53.50	4,477.50	5,050
3460	18" diameter		4	6		4,950	274	53.50	5,277.50	5,925
3480	20" diameter		4	6		6,525	274	53.50	6,852.50	7,650
3500	24" diameter		4	6		10,800	274	53.50	11,127.50	12,400
3510	30" diameter		4	6		11,600	274	53.50	11,927.50	13,300
3520	36" diameter		4	6		14,400	274	53.50	14,727.50	16,300
3530	42" diameter		4	6		18,900	274	53.50	19,227.50	21,300
3540	48" diameter	↓	4	6	↓	24,300	274	53.50	24,627.50	27,200
3600	With lever operator									
3610	4" diameter	B-6	6	4	Ea.	455	183	35.50	673.50	815
3614	6" diameter		6	4		625	183	35.50	843.50	1,000
3616	8" diameter		6	4		815	183	35.50	1,033.50	1,200
3618	10" diameter		6	4		1,375	183	35.50	1,593.50	1,825
3620	12" diameter		6	4		1,550	183	35.50	1,768.50	2,050
3622	14" diameter		4	6		2,500	274	53.50	2,827.50	3,225

491

33 14 19 – Valves and Hydrants for Water Utility Service

33 14 19.10 Valves		Crew	Daily Output	Labor-Hours	Unit	Material	2020 Bare Costs Labor	Equipment	Total	Total Incl O&P
3624	16" diameter	B-6	4	6	Ea.	3,825	274	53.50	4,152.50	4,675
3626	18" diameter		4	6		4,625	274	53.50	4,952.50	5,575
3628	20" diameter		4	6		6,200	274	53.50	6,527.50	7,300
3630	24" diameter		4	6		10,500	274	53.50	10,827.50	12,100
3700	Check valves, flanged									
3710	4" diameter	B-6	6	4	Ea.	785	183	35.50	1,003.50	1,175
3712	5" diameter		6	4		1,000	183	35.50	1,218.50	1,425
3714	6" diameter		6	4		1,325	183	35.50	1,543.50	1,775
3716	8" diameter		6	4		2,525	183	35.50	2,743.50	3,100
3718	10" diameter		6	4		4,275	183	35.50	4,493.50	5,025
3720	12" diameter		6	4		7,000	183	35.50	7,218.50	8,025
3726	18" diameter		4	6		25,900	274	53.50	26,227.50	29,000
3730	24" diameter		4	6		66,000	274	53.50	66,327.50	73,000
3800	Gate valves, C.I., 125 psi, mechanical joint, w/boxes									
3810	4" diameter	B-6	6	4	Ea.	780	183	35.50	998.50	1,175
3814	6" diameter		6	4		1,175	183	35.50	1,393.50	1,625
3816	8" diameter		6	4		2,025	183	35.50	2,243.50	2,550
3818	10" diameter		6	4		3,625	183	35.50	3,843.50	4,300
3820	12" diameter		6	4		4,900	183	35.50	5,118.50	5,700
3822	14" diameter		4	6		10,700	274	53.50	11,027.50	12,300
3824	16" diameter		4	6		15,100	274	53.50	15,427.50	17,100
3826	18" diameter		4	6		24,500	274	53.50	24,827.50	27,500
3828	20" diameter		4	6		25,600	274	53.50	25,927.50	28,700
3830	24" diameter		4	6		38,300	274	53.50	38,627.50	42,700
3831	30" diameter		4	6		59,000	274	53.50	59,327.50	65,500
3832	36" diameter		4	6		86,000	274	53.50	86,327.50	95,000
3880	Sleeve, for tapping mains, 8" x 4", add					1,025			1,025	1,125
3884	10" x 6", add					1,025			1,025	1,125
3888	12" x 6", add					1,425			1,425	1,550
3892	12" x 8", add					1,500			1,500	1,650

33 14 19.20 Valves

		Crew	Daily Output	Labor-Hours	Unit	Material	2020 Bare Costs Labor	Equipment	Total	Total Incl O&P
0010	**VALVES**									
0011	Special trim or use									
0150	Altitude valve, single acting, modulating, 2-1/2" diameter	B-6	6	4	Ea.	3,500	183	35.50	3,718.50	4,175
0155	3" diameter		6	4		3,575	183	35.50	3,793.50	4,250
0160	4" diameter		6	4		4,200	183	35.50	4,418.50	4,950
0165	6" diameter		6	4		5,150	183	35.50	5,368.50	5,975
0170	8" diameter		6	4		7,525	183	35.50	7,743.50	8,600
0175	10" diameter		6	4		8,225	183	35.50	8,443.50	9,375
0180	12" diameter		6	4		12,600	183	35.50	12,818.50	14,100
0250	Altitude valve, single acting, non-modulating, 2-1/2" diameter		6	4		3,525	183	35.50	3,743.50	4,200
0255	3" diameter		6	4		3,625	183	35.50	3,843.50	4,325
0260	4" diameter		6	4		4,325	183	35.50	4,543.50	5,075
0265	6" diameter		6	4		5,400	183	35.50	5,618.50	6,275
0270	8" diameter		6	4		7,550	183	35.50	7,768.50	8,650
0275	10" diameter		6	4		8,350	183	35.50	8,568.50	9,525
0280	12" diameter		6	4		12,900	183	35.50	13,118.50	14,500
0350	Altitude valve, double acting, non-modulating, 2-1/2" diameter		6	4		4,275	183	35.50	4,493.50	5,025
0355	3" diameter		6	4		4,450	183	35.50	4,668.50	5,225
0360	4" diameter		6	4		5,250	183	35.50	5,468.50	6,075
0365	6" diameter		6	4		5,900	183	35.50	6,118.50	6,825
0370	8" diameter		6	4		7,500	183	35.50	7,718.50	8,575

33 14 19 – Valves and Hydrants for Water Utility Service

33 14 19.20 Valves

		Crew	Daily Output	Labor-Hours	Unit	Material	2020 Bare Costs Labor	Equipment	Total	Total Incl O&P
0375	10" diameter	B-6	6	4	Ea.	10,200	183	35.50	10,418.50	11,500
0380	12" diameter	↓	6	4		15,500	183	35.50	15,718.50	17,300
1000	Air release valve for water, 1/2" inlet	1 Plum	16	.500		83.50	32		115.50	140
1005	3/4" inlet		16	.500		83.50	32		115.50	140
1010	1" inlet		14	.571		83.50	37		120.50	147
1020	2" inlet		9	.889		240	57.50		297.50	350
1100	Air release & vacuum valve for water, 1/2" inlet		16	.500		345	32		377	430
1105	3/4" inlet		16	.500		365	32		397	450
1110	1" inlet	↓	14	.571		485	37		522	585
1120	2" inlet		9	.889		820	57.50		877.50	990
1130	3" inlet	Q-1	8	2		2,650	116		2,766	3,100
1140	4" inlet		5	3.200		2,500	186		2,686	3,025
1150	6" inlet		5	3.200		4,625	186		4,811	5,350
1160	8" inlet		5	3.200		11,400	186		11,586	12,900
1170	10" inlet	↓	5	3.200		12,800	186		12,986	14,400
9000	Valves, gate valve, N.R.S. PIV with post, 4" diameter	B-6	6	4		2,000	183	35.50	2,218.50	2,525
9020	6" diameter		6	4		2,400	183	35.50	2,618.50	2,975
9040	8" diameter		6	4		3,250	183	35.50	3,468.50	3,875
9060	10" diameter		6	4		4,850	183	35.50	5,068.50	5,650
9080	12" diameter		6	4		6,125	183	35.50	6,343.50	7,050
9100	14" diameter		6	4		11,700	183	35.50	11,918.50	13,200
9120	OS&Y, 4" diameter		6	4		920	183	35.50	1,138.50	1,325
9140	6" diameter		6	4		1,275	183	35.50	1,493.50	1,725
9160	8" diameter		6	4		1,650	183	35.50	1,868.50	2,150
9180	10" diameter		6	4		2,575	183	35.50	2,793.50	3,175
9200	12" diameter		6	4		4,025	183	35.50	4,243.50	4,750
9220	14" diameter		4	6		6,150	274	53.50	6,477.50	7,250
9400	Check valves, rubber disc, 2-1/2" diameter		6	4		475	183	35.50	693.50	840
9420	3" diameter		6	4		595	183	35.50	813.50	970
9440	4" diameter		6	4		785	183	35.50	1,003.50	1,175
9480	6" diameter		6	4		1,325	183	35.50	1,543.50	1,775
9500	8" diameter		6	4		2,525	183	35.50	2,743.50	3,100
9520	10" diameter		6	4		4,275	183	35.50	4,493.50	5,025
9540	12" diameter		6	4		7,000	183	35.50	7,218.50	8,025
9542	14" diameter		4	6		6,750	274	53.50	7,077.50	7,900
9700	Detector check valves, reducing, 4" diameter		6	4		1,125	183	35.50	1,343.50	1,575
9720	6" diameter		6	4		1,925	183	35.50	2,143.50	2,450
9740	8" diameter		6	4		2,875	183	35.50	3,093.50	3,500
9760	10" diameter		6	4		5,025	183	35.50	5,243.50	5,850
9800	Galvanized, 4" diameter		6	4		2,025	183	35.50	2,243.50	2,550
9820	6" diameter		6	4		2,700	183	35.50	2,918.50	3,275
9840	8" diameter		6	4		4,025	183	35.50	4,243.50	4,750
9860	10" diameter	↓	6	4	↓	5,425	183	35.50	5,643.50	6,275

33 14 19.30 Fire Hydrants

		Crew	Daily Output	Labor-Hours	Unit	Material	2020 Bare Costs Labor	Equipment	Total	Total Incl O&P
0010	**FIRE HYDRANTS**									
0020	Mechanical joints unless otherwise noted									
1000	Fire hydrants, two way; excavation and backfill not incl.									
1100	4-1/2" valve size, depth 2'-0"	B-21	10	2.800	Ea.	2,900	137	18.90	3,055.90	3,400
1120	2'-6"		10	2.800		2,925	137	18.90	3,080.90	3,450
1140	3'-0"		10	2.800		3,000	137	18.90	3,155.90	3,525
1160	3'-6"		9	3.111		2,850	152	21	3,023	3,400
1170	4'-0"	↓	9	3.111	↓	2,925	152	21	3,098	3,450

For customer support on your Site Work & Landscape Costs with RSMeans data, call 800.448.8182.

493

33 14 Water Utility Transmission and Distribution

33 14 19 – Valves and Hydrants for Water Utility Service

33 14 19.30 Fire Hydrants		Crew	Daily Output	Labor-Hours	Unit	Material	2020 Bare Costs Labor	Equipment	Total	Total Incl O&P
1200	4'-6"	B-21	9	3.111	Ea.	3,000	152	21	3,173	3,550
1220	5'-0"		8	3.500		3,350	171	23.50	3,544.50	3,975
1240	5'-6"		8	3.500		3,425	171	23.50	3,619.50	4,025
1260	6'-0"		7	4		3,400	195	27	3,622	4,075
1280	6'-6"		7	4		3,525	195	27	3,747	4,225
1300	7'-0"		6	4.667		3,475	228	31.50	3,734.50	4,200
1340	8'-0"		6	4.667		2,775	228	31.50	3,034.50	3,425
1420	10'-0"		5	5.600		4,000	273	38	4,311	4,875
2000	5-1/4" valve size, depth 2'-0"		10	2.800		2,375	137	18.90	2,530.90	2,850
2080	4'-0"		9	3.111		2,700	152	21	2,873	3,225
2160	6'-0"		7	4		2,900	195	27	3,122	3,500
2240	8'-0"		6	4.667		3,325	228	31.50	3,584.50	4,050
2320	10'-0"	↓	5	5.600		3,700	273	38	4,011	4,525
2350	For threeway valves, add					7%				
2400	Lower barrel extensions with stems, 1'-0"	B-20	14	1.714		360	80.50		440.50	515
2440	2'-0"		13	1.846		440	87		527	615
2480	3'-0"		12	2		915	94		1,009	1,150
2520	4'-0"	↓	10	2.400		835	113		948	1,075
3200	Yard hydrant, flush, non freeze, 4' depth of bury, 3/4" conn.	2 Plum	4	4		140	258		398	540
4000	Above ground, 1" connection	1 Plum	8	1	↓	151	64.50		215.50	263
5000	Indicator post									
5020	Adjustable, valve size 4" to 14", 4' bury	B-21	10	2.800	Ea.	1,475	137	18.90	1,630.90	1,850
5060	8' bury		7	4		1,325	195	27	1,547	1,775
5080	10' bury		6	4.667		1,775	228	31.50	2,034.50	2,325
5100	12' bury		5	5.600		1,900	273	38	2,211	2,550
5120	14' bury		4	7		1,825	340	47.50	2,212.50	2,575
5500	Non-adjustable, valve size 4" to 14", 3' bury		10	2.800		925	137	18.90	1,080.90	1,250
5520	3'-6" bury		10	2.800		925	137	18.90	1,080.90	1,250
5540	4' bury	↓	9	3.111	↓	925	152	21	1,098	1,275

33 16 Water Utility Storage Tanks

33 16 11 – Elevated Composite Water Storage Tanks

33 16 11.50 Elevated Water Storage Tanks

		Crew	Daily Output	Labor-Hours	Unit	Material	Labor	Equipment	Total	Total Incl O&P
0010	**ELEVATED WATER STORAGE TANKS**									
0011	Not incl. pipe, pumps or foundation									
3000	Elevated water tanks, 100' to bottom capacity line, incl. painting									
3010	50,000 gallons				Ea.				185,000	204,000
3300	100,000 gallons								280,000	307,500
3400	250,000 gallons								751,500	826,500
3600	500,000 gallons								1,336,000	1,470,000
3700	750,000 gallons								1,622,000	1,783,500
3900	1,000,000 gallons				↓				2,322,000	2,556,000

33 16 23 – Ground-Level Steel Water Storage Tanks

33 16 23.13 Steel Water Storage Tanks

		Crew	Daily Output	Labor-Hours	Unit	Material	Labor	Equipment	Total	Total Incl O&P
0010	**STEEL WATER STORAGE TANKS**									
0910	Steel, ground level, ht./diam. less than 1, not incl. fdn., 100,000 gallons				Ea.				202,000	244,500
1000	250,000 gallons								295,500	324,000
1200	500,000 gallons								417,000	458,500
1250	750,000 gallons								538,000	591,500
1300	1,000,000 gallons				↓				558,000	725,500

33 16 Water Utility Storage Tanks

33 16 23 – Ground-Level Steel Water Storage Tanks

33 16 23.13 Steel Water Storage Tanks

		Crew	Daily Output	Labor-Hours	Unit	Material	2020 Bare Costs Labor	Equipment	Total	Total Incl O&P
1500	2,000,000 gallons				Ea.				1,043,000	1,148,000
1600	4,000,000 gallons								2,121,000	2,333,000
1800	6,000,000 gallons								3,095,000	3,405,000
1850	8,000,000 gallons								4,068,000	4,475,000
1910	10,000,000 gallons								5,050,000	5,554,500
2100	Steel standpipes, ht./diam. more than 1,100' to overflow, no fdn.									
2200	500,000 gallons				Ea.				546,500	600,500
2400	750,000 gallons								722,500	794,500
2500	1,000,000 gallons								1,060,500	1,167,000
2700	1,500,000 gallons								1,749,000	1,923,000
2800	2,000,000 gallons								2,327,000	2,559,000

33 16 36 – Ground-Level Reinforced Concrete Water Storage Tanks

33 16 36.16 Prestressed Conc. Water Storage Tanks

		Crew	Daily Output	Labor-Hours	Unit	Material	2020 Bare Costs Labor	Equipment	Total	Total Incl O&P
0010	**PRESTRESSED CONC. WATER STORAGE TANKS**									
0020	Not including fdn., pipe or pumps, 250,000 gallons				Ea.				299,000	329,500
0100	500,000 gallons								487,000	536,000
0300	1,000,000 gallons								707,000	807,500
0400	2,000,000 gallons								1,072,000	1,179,000
0600	4,000,000 gallons								1,706,000	1,877,000
0700	6,000,000 gallons								2,266,000	2,493,000
0750	8,000,000 gallons								2,924,000	3,216,000
0800	10,000,000 gallons								3,533,000	3,886,000

33 16 56 – Ground-Level Plastic Water Storage Cisterns

33 16 56.23 Plastic-Coated Fabric Pillow Water Tanks

		Crew	Daily Output	Labor-Hours	Unit	Material	2020 Bare Costs Labor	Equipment	Total	Total Incl O&P
0010	**PLASTIC-COATED FABRIC PILLOW WATER TANKS**									
7000	Water tanks, vinyl coated fabric pillow tanks, freestanding, 5,000 gallons	4 Clab	4	8	Ea.	3,800	335		4,135	4,675
7100	Supporting embankment not included, 25,000 gallons	6 Clab	2	24		13,500	1,000		14,500	16,300
7200	50,000 gallons	8 Clab	1.50	42.667		20,500	1,800		22,300	25,300
7300	100,000 gallons	9 Clab	.90	80		37,700	3,375		41,075	46,500
7400	150,000 gallons		.50	144		44,400	6,050		50,450	58,000
7500	200,000 gallons		.40	180		84,500	7,575		92,075	104,500
7600	250,000 gallons		.30	240		89,000	10,100		99,100	113,000

33 31 Sanitary Sewerage Piping

33 31 11 – Public Sanitary Sewerage Gravity Piping

33 31 11.10 Sewage Collection, Valves

		Crew	Daily Output	Labor-Hours	Unit	Material	2020 Bare Costs Labor	Equipment	Total	Total Incl O&P
0010	**SEWAGE COLLECTION, VALVES**									
1300	PVC swing check backwater valve									
1310	3" size	2 Skwk	15	1.067	Ea.	59.50	58.50		118	154
1315	4" size		15	1.067		72.50	58.50		131	169
1320	6" size		13	1.231		293	67.50		360.50	420
2000	Sewer relief valve									
2010	Plug style popper for outside cleanout									
2025	2" size	2 Skwk	56	.286	Ea.	3.01	15.65		18.66	27
2030	3" size		36	.444		3.97	24.50		28.47	41.50
2040	4" size		30	.533		4.80	29.50		34.30	49.50
3000	Air release valves for sewer pipes									
3020	2" size inlet connection	1 Plum	9	.889	Ea.	485	57.50		542.50	615
3030	3" size inlet connection	Q-1	8	2		675	116		791	915
3040	4" size inlet connection	"	5	3.200		860	186		1,046	1,225

495

33 31 11 – Public Sanitary Sewerage Gravity Piping

33 31 11.10 Sewage Collection, Valves	Crew	Daily Output	Labor-Hours	Unit	Material	2020 Bare Costs Labor	2020 Bare Costs Equipment	Total	Total Incl O&P
3100	Air release valves for sewer pipes with backwash valve and fittings									
3120	2" size inlet connection	1 Plum	9	.889	Ea.	800	57.50		857.50	965
3130	3" size inlet connection	Q-1	8	2		1,050	116		1,166	1,325
3140	4" size inlet connection	"	5	3.200	↓	1,575	186		1,761	2,000
3200	Air release & vacuum valves for sewer pipes									
3220	2" size inlet connection	1 Plum	9	.889	Ea.	1,400	57.50		1,457.50	1,625
3230	3" size inlet connection	Q-1	8	2	"	1,300	116		1,416	1,600

33 31 11.13 Sewage Collection, Vent Cast Iron Pipe

		Crew	Daily Output	Labor-Hours	Unit	Material	2020 Bare Costs Labor	2020 Bare Costs Equipment	Total	Total Incl O&P
0010	**SEWAGE COLLECTION, VENT CAST IRON PIPE**									
0020	Not including excavation or backfill									
2022	Sewage vent cast iron, B&S, 4" diameter	Q-1	66	.242	L.F.	23	14.05		37.05	46
2024	5" diameter	Q-2	88	.273		30	16.40		46.40	57.50
2026	6" diameter	"	84	.286		39	17.20		56.20	68
2028	8" diameter	Q-3	70	.457		62.50	28		90.50	111
2030	10" diameter		66	.485		104	30		134	159
2032	12" diameter		57	.561		149	34.50		183.50	216
2034	15" diameter	↓	49	.653	↓	170	40		210	247
8001	Fittings, bends and elbows									
8110	4" diameter	Q-1	13	1.231	Ea.	68	71.50		139.50	182
8112	5" diameter	Q-2	18	1.333		97.50	80		177.50	227
8114	6" diameter	"	17	1.412		116	85		201	254
8116	8" diameter	Q-3	11	2.909		325	178		503	620
8118	10" diameter		10	3.200		485	196		681	825
8120	12" diameter		9	3.556		655	218		873	1,050
8122	15" diameter	↓	7	4.571	↓	1,925	280		2,205	2,550
8500	Wyes and tees									
8510	4" diameter	Q-1	8	2	Ea.	110	116		226	295
8512	5" diameter	Q-2	12	2		188	120		308	385
8514	6" diameter	"	11	2.182		232	131		363	450
8516	8" diameter	Q-3	7	4.571		550	280		830	1,025
8518	10" diameter		6	5.333		900	325		1,225	1,475
8520	12" diameter		4	8		1,850	490		2,340	2,750
8522	15" diameter	↓	3	10.667	↓	3,700	655		4,355	5,050

33 31 11.15 Sewage Collection, Concrete Pipe

0010	**SEWAGE COLLECTION, CONCRETE PIPE**									
0020	See Section 33 42 11.60 for sewage/drainage collection, concrete pipe									

33 31 11.20 Sewage Collection, Plastic Pipe

		Crew	Daily Output	Labor-Hours	Unit	Material	2020 Bare Costs Labor	2020 Bare Costs Equipment	Total	Total Incl O&P
0010	**SEWAGE COLLECTION, PLASTIC PIPE**									
0020	Not including excavation & backfill									
3000	Piping, HDPE Corrugated Type S with watertight gaskets, 4" diameter	B-20	425	.056	L.F.	1.08	2.66		3.74	5.20
3020	6" diameter		400	.060		3.13	2.82		5.95	7.70
3040	8" diameter		380	.063		4.12	2.97		7.09	9
3060	10" diameter		370	.065		5.50	3.05		8.55	10.65
3080	12" diameter		340	.071		7.90	3.32		11.22	13.70
3100	15" diameter	↓	300	.080		11.65	3.76		15.41	18.50
3120	18" diameter	B-21	275	.102		14.35	4.96	.69	20	24
3130	21" diameter		250	.112		23	5.45	.76	29.21	34
3140	24" diameter		250	.112		23	5.45	.76	29.21	34
3160	30" diameter		200	.140		29	6.85	.95	36.80	43.50
3180	36" diameter		180	.156		42	7.60	1.05	50.65	59
3200	42" diameter		175	.160		49.50	7.80	1.08	58.38	67
3220	48" diameter	↓	170	.165	↓	69.50	8.05	1.11	78.66	90

33 31 11 – Public Sanitary Sewerage Gravity Piping

33 31 11.20 Sewage Collection, Plastic Pipe

		Crew	Daily Output	Labor-Hours	Unit	Material	2020 Bare Costs Labor	Equipment	Total	Total Incl O&P
3240	54" diameter	B-21	160	.175	L.F.	90.50	8.55	1.18	100.23	114
3260	60" diameter		150	.187		167	9.10	1.26	177.36	198
3300	Watertight elbows 12" diameter	B-20	11	2.182	Ea.	82.50	103		185.50	246
3320	15" diameter	"	9	2.667		125	125		250	325
3340	18" diameter	B-21	9	3.111		206	152	21	379	475
3350	21" diameter		9	3.111		297	152	21	470	575
3360	24" diameter		9	3.111		435	152	21	608	725
3380	30" diameter		8	3.500		690	171	23.50	884.50	1,050
3400	36" diameter		8	3.500		890	171	23.50	1,084.50	1,250
3420	42" diameter		6	4.667		1,125	228	31.50	1,384.50	1,600
3440	48" diameter		6	4.667		2,000	228	31.50	2,259.50	2,575
3460	Watertight tee 12" diameter	B-20	7	3.429		109	161		270	365
3480	15" diameter	"	6	4		296	188		484	610
3500	18" diameter	B-21	6	4.667		430	228	31.50	689.50	850
3520	24" diameter		5	5.600		795	273	38	1,106	1,325
3540	30" diameter		5	5.600		1,350	273	38	1,661	1,950
3560	36" diameter		4	7		1,300	340	47.50	1,687.50	2,025
3580	42" diameter		4	7		1,500	340	47.50	1,887.50	2,225
3600	48" diameter		4	7		1,600	340	47.50	1,987.50	2,325

33 31 11.25 Sewage Collection, Polyvinyl Chloride Pipe

		Crew	Daily Output	Labor-Hours	Unit	Material	2020 Bare Costs Labor	Equipment	Total	Total Incl O&P
0010	**SEWAGE COLLECTION, POLYVINYL CHLORIDE PIPE**									
0020	Not including excavation or backfill									
2000	20' lengths, SDR 35, B&S, 4" diameter	B-20	375	.064	L.F.	1.67	3.01		4.68	6.35
2040	6" diameter		350	.069		4.05	3.22		7.27	9.30
2080	13' lengths, SDR 35, B&S, 8" diameter		335	.072		6.70	3.37		10.07	12.40
2120	10" diameter	B-21	330	.085		11.30	4.14	.57	16.01	19.30
2160	12" diameter		320	.088		15.90	4.27	.59	20.76	24.50
2170	14" diameter		280	.100		19.45	4.88	.68	25.01	29.50
2200	15" diameter		240	.117		22	5.70	.79	28.49	33.50
2250	16" diameter		220	.127		29	6.20	.86	36.06	42
2300	18" diameter		200	.140		34.50	6.85	.95	42.30	49.50
2350	20" diameter		195	.144		40.50	7	.97	48.47	56
2400	21" diameter		190	.147		48.50	7.20	.99	56.69	65.50
2500	24" diameter		180	.156		63	7.60	1.05	71.65	82
2550	27" diameter		170	.165		80	8.05	1.11	89.16	101
2600	30" diameter		150	.187		104	9.10	1.26	114.36	129
2650	36" diameter		130	.215		151	10.50	1.45	162.95	183
3040	Fittings, bends or elbows, 4" diameter	2 Skwk	24	.667	Ea.	11.30	36.50		47.80	68
3080	6" diameter		24	.667		36.50	36.50		73	95.50
3120	Tees, 4" diameter		16	1		18.85	55		73.85	104
3160	6" diameter		16	1		59.50	55		114.50	149
3200	Wyes, 4" diameter		16	1		19.05	55		74.05	104
3240	6" diameter		16	1		52	55		107	140
3661	Cap, 4"		48	.333		6.85	18.30		25.15	35
3670	6"		48	.333		23	18.30		41.30	52.50
3671	8"		40	.400		26	22		48	61.50
3672	10"		24	.667		84	36.50		120.50	148
3673	12"		20	.800		117	44		161	196
3674	15"		18	.889		190	49		239	283
3675	18"		18	.889		279	49		328	380
3676	21"		18	.889		605	49		654	740
3677	24"		18	.889		770	49		819	920

For customer support on your Site Work & Landscape Costs with RSMeans data, call 800.448.8182.

497

33 31 11 – Public Sanitary Sewerage Gravity Piping

33 31 11.25 Sewage Collection, Polyvinyl Chloride Pipe	Crew	Daily Output	Labor-Hours	Unit	Material	2020 Bare Costs Labor	2020 Bare Costs Equipment	Total	Total Incl O&P
3745	Elbow, 90 degree, 8" diameter	2 Skwk	20	.800	Ea.	48	44		92	120
3750	10" diameter		12	1.333		174	73		247	300
3755	12" diameter		10	1.600		213	88		301	365
3760	15" diameter		9	1.778		525	97.50		622.50	725
3765	18" diameter		9	1.778		890	97.50		987.50	1,125
3770	21" diameter		9	1.778		1,525	97.50		1,622.50	1,825
3775	24" diameter		9	1.778		2,075	97.50		2,172.50	2,425
3795	Elbow, 45 degree, 8" diameter		20	.800		37.50	44		81.50	108
3805	12" diameter		10	1.600		186	88		274	340
3810	15" diameter		9	1.778		390	97.50		487.50	575
3815	18" diameter		9	1.778		650	97.50		747.50	860
3820	21" diameter		9	1.778		1,025	97.50		1,122.50	1,275
3825	24" diameter		9	1.778		1,400	97.50		1,497.50	1,675
3875	Elbow, 22-1/2 degree, 8" diameter		20	.800		53.50	44		97.50	125
3880	10" diameter		12	1.333		116	73		189	238
3885	12" diameter		12	1.333		150	73		223	276
3890	15" diameter		9	1.778		455	97.50		552.50	645
3895	18" diameter		9	1.778		710	97.50		807.50	925
3900	21" diameter		9	1.778		870	97.50		967.50	1,100
3905	24" diameter		9	1.778		1,450	97.50		1,547.50	1,750
3910	Tee, 8" diameter		13	1.231		92	67.50		159.50	203
3915	10" diameter		8	2		282	110		392	475
3920	12" diameter		7	2.286		340	125		465	560
3925	15" diameter		6	2.667		615	146		761	895
3930	18" diameter		6	2.667		1,125	146		1,271	1,450
3935	21" diameter		6	2.667		2,350	146		2,496	2,800
3940	24" diameter		6	2.667		4,075	146		4,221	4,725
4000	Piping, DWV PVC, no exc./bkfill., 10' L, Sch 40, 4" diameter	B-20	375	.064	L.F.	3.83	3.01		6.84	8.75
4010	6" diameter		350	.069		8.25	3.22		11.47	13.90
4020	8" diameter		335	.072		13.40	3.37		16.77	19.80
4030	Fittings, 1/4 bend DWV PVC, 4" diameter		19	1.263	Ea.	23	59.50		82.50	115
4040	6" diameter		12	2		122	94		216	276
4050	8" diameter		11	2.182		122	103		225	289
4060	1/8 bend DWV PVC, 4" diameter		19	1.263		18.95	59.50		78.45	111
4070	6" diameter		12	2		71.50	94		165.50	221
4080	8" diameter		11	2.182		117	103		220	283
4090	Tee DWV PVC, 4" diameter		12	2		31	94		125	177
4100	6" diameter		10	2.400		126	113		239	310
4110	8" diameter		9	2.667		293	125		418	515

33 31 11.30 Piping, Drainage and Sewage, Vitrified Clay C700

		Crew	Daily Output	Labor-Hours	Unit	Material	2020 Bare Costs Labor	2020 Bare Costs Equipment	Total	Total Incl O&P
0010	**PIPING, DRAINAGE & SEWAGE, VITRIFIED CLAY C700**									
0020	Not including excavation or backfill,									
0025	Piping									
4030	Extra strength, compression joints, C425									
5000	4" diameter x 4' long	B-20	265	.091	L.F.	5.35	4.26		9.61	12.25
5020	6" diameter x 5' long	"	200	.120		6.70	5.65		12.35	15.85
5040	8" diameter x 5' long	B-21	200	.140		9.05	6.85	.95	16.85	21.50
5060	10" diameter x 5' long		190	.147		15	7.20	.99	23.19	28.50
5080	12" diameter x 6' long		150	.187		15.95	9.10	1.26	26.31	32.50
5100	15" diameter x 7' long		110	.255		30	12.40	1.72	44.12	53.50
5120	18" diameter x 7' long		88	.318		42.50	15.50	2.15	60.15	73
5140	24" diameter x 7' long		45	.622		79.50	30.50	4.20	114.20	138

33 31 Sanitary Sewerage Piping

33 31 11 – Public Sanitary Sewerage Gravity Piping

33 31 11.30 Piping, Drainage and Sewage, Vitrified Clay C700	Crew	Daily Output	Labor-Hours	Unit	Material	2020 Bare Costs Labor	Equipment	Total	Total Incl O&P
5160	30" diameter x 7' long	B-22	31	.968	L.F.	96.50	48	9.15	153.65	188
5180	36" diameter x 7' long	"	20	1.500	↓	127	74	14.20	215.20	268

33 32 Sanitary Sewerage Equipment

33 32 13 – Packaged Wastewater Pumping Stations

33 32 13.13 Packaged Sewage Lift Stations

		Crew	Daily Output	Labor-Hours	Unit	Material	2020 Bare Costs Labor	Equipment	Total	Total Incl O&P
0010	**PACKAGED SEWAGE LIFT STATIONS**									
2500	Sewage lift station, 200,000 GPD	E-8	.20	520	Ea.	238,000	29,700	11,400	279,100	320,500
2510	500,000 GPD		.15	684		271,500	39,100	15,000	325,600	375,500
2520	800,000 GPD	↓	.13	813	↓	327,500	46,400	17,800	391,700	451,500

33 34 Onsite Wastewater Disposal

33 34 13 – Septic Tanks

33 34 13.13 Concrete Septic Tanks

		Crew	Daily Output	Labor-Hours	Unit	Material	2020 Bare Costs Labor	Equipment	Total	Total Incl O&P
0010	**CONCRETE SEPTIC TANKS**									
0011	Not including excavation or piping									
0015	Septic tanks, precast, 1,000 gallon	B-21	8	3.500	Ea.	1,050	171	23.50	1,244.50	1,450
0020	1,250 gallon		8	3.500		1,425	171	23.50	1,619.50	1,850
0060	1,500 gallon		7	4		1,625	195	27	1,847	2,125
0100	2,000 gallon		5	5.600		2,500	273	38	2,811	3,200
0140	2,500 gallon		5	5.600		2,375	273	38	2,686	3,050
0180	4,000 gallon	↓	4	7		5,850	340	47.50	6,237.50	7,025
0200	5,000 gallon	B-13	3.50	16		6,325	735	166	7,226	8,225
0220	5,000 gallon, 4 piece	"	3	18.667		9,025	860	194	10,079	11,400
0300	15,000 gallon, 4 piece	B-13B	1.70	32.941		23,700	1,525	575	25,800	29,000
0400	25,000 gallon, 4 piece		1.10	50.909		47,900	2,350	890	51,140	57,000
0500	40,000 gallon, 4 piece	↓	.80	70		53,500	3,225	1,225	57,950	64,500
0520	50,000 gallon, 5 piece	B-13C	.60	93.333		61,500	4,300	3,800	69,600	78,000
0640	75,000 gallon, cast in place	C-14C	.25	448		74,500	22,600	108	97,208	116,000
0660	100,000 gallon	"	.15	747		92,500	37,600	179	130,279	158,000
0800	Septic tanks, precast, 2 compartment, 1,000 gallon	B-21	8	3.500		695	171	23.50	889.50	1,050
0805	1,250 gallon		8	3.500		1,050	171	23.50	1,244.50	1,425
0810	1,500 gallon		7	4		1,550	195	27	1,772	2,050
0815	2,000 gallon		5	5.600		2,525	273	38	2,836	3,225
0820	2,500 gallon	↓	5	5.600	↓	3,225	273	38	3,536	4,000
0890	Septic tanks, butyl rope sealant				L.F.	.42			.42	.46
0900	Concrete riser 24" x 12" with standard lid	1 Clab	6	1.333	Ea.	87.50	56		143.50	181
0905	24" x 12" with heavy duty lid		6	1.333		111	56		167	207
0910	24" x 8" with standard lid		6	1.333		82	56		138	175
0915	24" x 8" with heavy duty lid		6	1.333		106	56		162	201
0917	24" x 12" extension		12	.667		59	28		87	107
0920	24" x 8" extension		12	.667		53.50	28		81.50	101
0950	HDPE riser 20" x 12" with standard lid		8	1		135	42		177	212
0951	HDPE 20" x 12" with heavy duty lid		8	1		146	42		188	225
0955	HDPE 24" x 12" with standard lid		8	1		159	42		201	238
0956	HDPE 24" x 12" with heavy duty lid		8	1		164	42		206	245
0960	HDPE 20" x 6" extension		48	.167		29.50	7		36.50	43
0962	HDPE 20" x 12" extension		48	.167		51	7		58	66.50
0965	HDPE 24" x 6" extension	↓	48	.167	↓	37.50	7		44.50	52

33 34 Onsite Wastewater Disposal

33 34 13 – Septic Tanks

33 34 13.13 Concrete Septic Tanks

	33 34 13.13 Concrete Septic Tanks	Crew	Daily Output	Labor-Hours	Unit	Material	2020 Bare Costs Labor	Equipment	Total	Total Incl O&P
0967	HDPE 24" x 12" extension	1 Clab	48	.167	Ea.	60	7		67	76.50
1150	Leaching field chambers, 13' x 3'-7" x 1'-4", standard	B-13	16	3.500		500	161	36.50	697.50	830
1200	Heavy duty, 8' x 4' x 1'-6"		14	4		310	184	41.50	535.50	660
1300	13' x 3'-9" x 1'-6"		12	4.667		1,350	215	48.50	1,613.50	1,850
1350	20' x 4' x 1'-6"		5	11.200		1,225	515	116	1,856	2,250
1400	Leaching pit, precast concrete, 3' diameter, 3' deep	B-21	8	3.500		735	171	23.50	929.50	1,100
1500	6' diameter, 3' section		4.70	5.957		940	290	40	1,270	1,500
1600	Leaching pit, 6'-6" diameter, 6' deep		5	5.600		1,075	273	38	1,386	1,625
1620	8' deep		4	7		1,175	340	47.50	1,562.50	1,875
1700	8' diameter, H-20 load, 6' deep		4	7		1,475	340	47.50	1,862.50	2,175
1720	8' deep		3	9.333		2,700	455	63	3,218	3,725
2000	Velocity reducing pit, precast conc., 6' diameter, 3' deep		4.70	5.957		1,825	290	40	2,155	2,500

33 34 13.33 Polyethylene Septic Tanks

	33 34 13.33 Polyethylene Septic Tanks	Crew	Daily Output	Labor-Hours	Unit	Material	2020 Bare Costs Labor	Equipment	Total	Total Incl O&P
0010	**POLYETHYLENE SEPTIC TANKS**									
0015	High density polyethylene, 1,000 gallon	B-21	8	3.500	Ea.	1,475	171	23.50	1,669.50	1,900
0020	1,250 gallon		8	3.500		1,325	171	23.50	1,519.50	1,725
0025	1,500 gallon		7	4		1,375	195	27	1,597	1,850
1015	Septic tanks, HDPE, 2 compartment, 1,000 gallon		8	3.500		1,475	171	23.50	1,669.50	1,900
1020	1,500 gallon		7	4		1,650	195	27	1,872	2,125

33 34 16 – Septic Tank Effluent Filters

33 34 16.13 Septic Tank Gravity Effluent Filters

	33 34 16.13 Septic Tank Gravity Effluent Filters	Crew	Daily Output	Labor-Hours	Unit	Material	2020 Bare Costs Labor	Equipment	Total	Total Incl O&P
0010	**SEPTIC TANK GRAVITY EFFLUENT FILTERS**									
3000	Effluent filter, 4" diameter	1 Skwk	8	1	Ea.	44	55		99	131
3020	6" diameter		7	1.143		42.50	62.50		105	142
3030	8" diameter		7	1.143		252	62.50		314.50	370
3040	8" diameter, very fine		7	1.143		500	62.50		562.50	645
3050	10" diameter, very fine		6	1.333		237	73		310	370
3060	10" diameter		6	1.333		276	73		349	415
3080	12" diameter		6	1.333		655	73		728	830
3090	15" diameter		5	1.600		1,075	88		1,163	1,325

33 34 51 – Drainage Field Systems

33 34 51.10 Drainage Field Excavation and Fill

	33 34 51.10 Drainage Field Excavation and Fill	Crew	Daily Output	Labor-Hours	Unit	Material	2020 Bare Costs Labor	Equipment	Total	Total Incl O&P
0010	**DRAINAGE FIELD EXCAVATION AND FILL**									
2200	Septic tank & drainage field excavation with 3/4 C.Y. backhoe	B-12F	145	.110	C.Y.		5.60	4.79	10.39	13.60
2400	4' trench for disposal field, 3/4 C.Y. backhoe	"	335	.048	L.F.		2.42	2.07	4.49	5.90
2600	Gravel fill, run of bank	B-6	150	.160	C.Y.	16.90	7.30	1.43	25.63	31
2800	Crushed stone, 3/4"	"	150	.160	"	42	7.30	1.43	50.73	58.50

33 34 51.13 Utility Septic Tank Tile Drainage Field

	33 34 51.13 Utility Septic Tank Tile Drainage Field	Crew	Daily Output	Labor-Hours	Unit	Material	2020 Bare Costs Labor	Equipment	Total	Total Incl O&P
0010	**UTILITY SEPTIC TANK TILE DRAINAGE FIELD**									
0015	Distribution box, concrete, 5 outlets	2 Clab	20	.800	Ea.	93.50	33.50		127	154
0020	7 outlets		16	1		103	42		145	178
0025	9 outlets		8	2		545	84		629	725
0115	Distribution boxes, HDPE, 5 outlets		20	.800		76.50	33.50		110	135
0117	6 outlets		15	1.067		81	45		126	157
0118	7 outlets		15	1.067		75	45		120	151
0120	8 outlets		10	1.600		79	67.50		146.50	188
0240	Distribution boxes, outlet flow leveler	1 Clab	50	.160		2.53	6.75		9.28	12.90
0260	Bull run valve 4" diam.	2 Skwk	16	1		66	55		121	156
0261	Bull run valve key					15.60			15.60	17.15
0300	Precast concrete, galley, 4' x 4' x 4'	B-21	16	1.750		246	85.50	11.80	343.30	410
0350	HDPE infiltration chamber 12" H x 15" W	2 Clab	300	.053	L.F.	7	2.25		9.25	11.05

33 34 Onsite Wastewater Disposal

33 34 51 – Drainage Field Systems

33 34 51.13 Utility Septic Tank Tile Drainage Field

		Crew	Daily Output	Labor-Hours	Unit	Material	2020 Bare Costs Labor	Equipment	Total	Total Incl O&P
0351	12" H x 15" W end cap	1 Clab	32	.250	Ea.	18.90	10.55		29.45	37
0355	chamber 12" H x 22" W	2 Clab	300	.053	L.F.	6.65	2.25		8.90	10.70
0356	12" H x 22" W end cap	1 Clab	32	.250	Ea.	16.60	10.55		27.15	34
0360	chamber 13" H x 34" W	2 Clab	300	.053	L.F.	14.30	2.25		16.55	19.10
0361	13" H x 34" W end cap	1 Clab	32	.250	Ea.	50.50	10.55		61.05	72
0365	chamber 16" H x 34" W	2 Clab	300	.053	L.F.	19.25	2.25		21.50	24.50
0366	16" H x 34" W end cap	1 Clab	32	.250	Ea.	16.55	10.55		27.10	34
0370	chamber 8" H x 16" W	2 Clab	300	.053	L.F.	9.90	2.25		12.15	14.25
0371	8" H x 16" W end cap	1 Clab	32	.250	Ea.	11.10	10.55		21.65	28

33 41 Subdrainage

33 41 16 – Subdrainage Piping

33 41 16.10 Piping, Subdrainage, Vitrified Clay

		Crew	Daily Output	Labor-Hours	Unit	Material	2020 Bare Costs Labor	Equipment	Total	Total Incl O&P
0010	**PIPING, SUBDRAINAGE, VITRIFIED CLAY**									
0020	Not including excavation and backfill									
1000	Foundation drain, 6" diameter	B-14	315	.152	L.F.	6.25	6.75	.68	13.68	17.75
1010	8" diameter		290	.166		9.30	7.30	.74	17.34	22
1020	12" diameter		275	.175		20	7.70	.78	28.48	34.50
3000	Perforated, 5' lengths, C700, 4" diameter		400	.120		4.70	5.30	.53	10.53	13.70
3020	6" diameter		315	.152		6.65	6.75	.68	14.08	18.20
3040	8" diameter		290	.166		7.25	7.30	.74	15.29	19.80
3060	12" diameter		275	.175		20.50	7.70	.78	28.98	35

33 41 16.25 Piping, Subdrainage, Corrugated Metal

		Crew	Daily Output	Labor-Hours	Unit	Material	2020 Bare Costs Labor	Equipment	Total	Total Incl O&P
0010	**PIPING, SUBDRAINAGE, CORRUGATED METAL**									
0021	Not including excavation and backfill									
2010	Aluminum, perforated									
2020	6" diameter, 18 ga.	B-20	380	.063	L.F.	6.40	2.97		9.37	11.50
2200	8" diameter, 16 ga.	"	370	.065		8.80	3.05		11.85	14.30
2220	10" diameter, 16 ga.	B-21	360	.078		11	3.79	.53	15.32	18.40
2240	12" diameter, 16 ga.		285	.098		12.30	4.79	.66	17.75	21.50
2260	18" diameter, 16 ga.		205	.137		18.50	6.65	.92	26.07	31.50
3000	Uncoated galvanized, perforated									
3020	6" diameter, 18 ga.	B-20	380	.063	L.F.	6.65	2.97		9.62	11.75
3200	8" diameter, 16 ga.	"	370	.065		8.10	3.05		11.15	13.50
3220	10" diameter, 16 ga.	B-21	360	.078		8.60	3.79	.53	12.92	15.75
3240	12" diameter, 16 ga.		285	.098		9.55	4.79	.66	15	18.50
3260	18" diameter, 16 ga.		205	.137		14.65	6.65	.92	22.22	27
4000	Steel, perforated, asphalt coated									
4020	6" diameter, 18 ga.	B-20	380	.063	L.F.	6.35	2.97		9.32	11.45
4030	8" diameter, 18 ga.	"	370	.065		9	3.05		12.05	14.50
4040	10" diameter, 16 ga.	B-21	360	.078		10.05	3.79	.53	14.37	17.35
4050	12" diameter, 16 ga.		285	.098		11.30	4.79	.66	16.75	20.50
4060	18" diameter, 16 ga.		205	.137		17.80	6.65	.92	25.37	30.50

33 41 16.30 Piping, Subdrainage, Plastic

		Crew	Daily Output	Labor-Hours	Unit	Material	2020 Bare Costs Labor	Equipment	Total	Total Incl O&P
0010	**PIPING, SUBDRAINAGE, PLASTIC**									
0020	Not including excavation and backfill									
2100	Perforated PVC, 4" diameter	B-14	314	.153	L.F.	1.67	6.75	.68	9.10	12.75
2110	6" diameter		300	.160		4.05	7.10	.71	11.86	15.90
2120	8" diameter		290	.166		6.10	7.30	.74	14.14	18.50
2130	10" diameter		280	.171		10.45	7.60	.76	18.81	23.50

For customer support on your Site Work & Landscape Costs with RSMeans data, call 800.448.8182.

501

33 41 Subdrainage

33 41 16 – Subdrainage Piping

33 41 16.30 Piping, Subdrainage, Plastic		Crew	Daily Output	Labor-Hours	Unit	Material	2020 Bare Costs Labor	Equipment	Total	Total Incl O&P
2140	12" diameter	B-14	270	.178	L.F.	14.75	7.85	.79	23.39	29

33 41 16.35 Piping, Subdrain., Corr. Plas. Tubing, Perf. or Plain

		Crew	Daily Output	Labor-Hours	Unit	Material	2020 Bare Costs Labor	Equipment	Total	Total Incl O&P
0010	**PIPING, SUBDRAINAGE, CORR. PLASTIC TUBING, PERF. OR PLAIN**									
0020	In rolls, not including excavation and backfill									
0030	3" diameter	2 Clab	1200	.013	L.F.	.62	.56		1.18	1.52
0040	4" diameter		1200	.013		.75	.56		1.31	1.67
0041	With silt sock		1200	.013		1.05	.56		1.61	2
0050	5" diameter		900	.018		1.29	.75		2.04	2.54
0060	6" diameter		900	.018		2.09	.75		2.84	3.42
0080	8" diameter		700	.023		4.11	.96		5.07	5.95
0200	Fittings									
0230	Elbows, 3" diameter	1 Clab	32	.250	Ea.	5.20	10.55		15.75	21.50
0240	4" diameter		32	.250		5.25	10.55		15.80	21.50
0250	5" diameter		32	.250		6.20	10.55		16.75	22.50
0260	6" diameter		32	.250		9	10.55		19.55	25.50
0280	8" diameter		32	.250		30	10.55		40.55	49.50
0330	Tees, 3" diameter		27	.296		4.01	12.45		16.46	23
0340	4" diameter		27	.296		5.40	12.45		17.85	24.50
0350	5" diameter		27	.296		7.40	12.45		19.85	27
0360	6" diameter		27	.296		11.80	12.45		24.25	31.50
0370	6" x 6" x 4"		27	.296		52.50	12.45		64.95	76.50
0380	8" diameter		27	.296		35.50	12.45		47.95	58
0390	8" x 8" x 6"		27	.296		80	12.45		92.45	107
0430	End cap, 3" diameter		32	.250		1.98	10.55		12.53	18
0440	4" diameter		32	.250		2.37	10.55		12.92	18.40
0460	6" diameter		32	.250		4.95	10.55		15.50	21.50
0480	8" diameter		32	.250		8.30	10.55		18.85	25
0530	Coupler, 3" diameter		32	.250		2.49	10.55		13.04	18.55
0540	4" diameter		32	.250		2.04	10.55		12.59	18.05
0550	5" diameter		32	.250		3.63	10.55		14.18	19.80
0560	6" diameter		32	.250		3.78	10.55		14.33	19.95
0580	8" diameter		32	.250		6.60	10.55		17.15	23
0590	Heavy duty highway type, add					10%				
0660	Reducer, 6" to 4"	1 Clab	32	.250		5.80	10.55		16.35	22
0680	8" to 6"		32	.250		13.85	10.55		24.40	31
0730	"Y" fitting, 3" diameter		27	.296		4.93	12.45		17.38	24
0740	4" diameter		27	.296		8.70	12.45		21.15	28.50
0750	5" diameter		27	.296		9.45	12.45		21.90	29
0760	6" diameter		27	.296		13.20	12.45		25.65	33.50
0780	8" diameter		27	.296		35.50	12.45		47.95	58.50
0860	Silt sock only for above tubing, 6" diameter				L.F.	3.85			3.85	4.24
0880	8" diameter				"	6.35			6.35	7

33 41 23 – Drainage Layers

33 41 23.19 Geosynthetic Drainage Layers

		Crew	Daily Output	Labor-Hours	Unit	Material	2020 Bare Costs Labor	Equipment	Total	Total Incl O&P
0010	**GEOSYNTHETIC DRAINAGE LAYERS**									
0100	Fabric, laid in trench, polypropylene, ideal conditions	2 Clab	2400	.007	S.Y.	1.57	.28		1.85	2.15
0110	Adverse conditions		1600	.010	"	1.57	.42		1.99	2.36
0170	Fabric ply bonded to 3 dimen. nylon mat, 0.4" thick, ideal conditions		2000	.008	S.F.	.27	.34		.61	.80
0180	Adverse conditions		1200	.013	"	.33	.56		.89	1.21
0185	Soil drainage mat on vertical wall, 0.44" thick		265	.060	S.Y.	1.80	2.54		4.34	5.80
0188	0.25" thick		300	.053	"	.96	2.25		3.21	4.43
0190	0.8" thick, ideal conditions		2400	.007	S.F.	.28	.28		.56	.73

502

33 41 Subdrainage

33 41 23 – Drainage Layers

33 41 23.19 Geosynthetic Drainage Layers	Crew	Daily Output	Labor-Hours	Unit	Material	2020 Bare Costs Labor	Equipment	Total	Total Incl O&P
0200	Adverse conditions	2 Clab	1600	.010	S.F.	.48	.42		.90	1.16
0300	Drainage material, 3/4" gravel fill in trench	B-6	260	.092	C.Y.	20.50	4.22	.82	25.54	30
0400	Pea stone	"	260	.092	"	20.50	4.22	.82	25.54	30

33 42 Stormwater Conveyance

33 42 11 – Stormwater Gravity Piping

33 42 11.40 Piping, Storm Drainage, Corrugated Metal

		Crew	Daily Output	Labor-Hours	Unit	Material	2020 Bare Costs Labor	Equipment	Total	Total Incl O&P
0010	**PIPING, STORM DRAINAGE, CORRUGATED METAL**									
0020	Not including excavation or backfill									
2000	Corrugated metal pipe, galvanized									
2020	Bituminous coated with paved invert, 20' lengths									
2040	8" diameter, 16 ga.	B-14	330	.145	L.F.	8.35	6.45	.65	15.45	19.50
2060	10" diameter, 16 ga.		260	.185		8.65	8.15	.82	17.62	22.50
2080	12" diameter, 16 ga.		210	.229		12.40	10.10	1.02	23.52	30
2100	15" diameter, 16 ga.		200	.240		14.60	10.60	1.07	26.27	33
2120	18" diameter, 16 ga.		190	.253		18.80	11.20	1.12	31.12	38.50
2140	24" diameter, 14 ga.		160	.300		24.50	13.30	1.34	39.14	48.50
2160	30" diameter, 14 ga.	B-13	120	.467		29	21.50	4.84	55.34	69.50
2180	36" diameter, 12 ga.		120	.467		35	21.50	4.84	61.34	76
2200	48" diameter, 12 ga.		100	.560		46.50	26	5.80	78.30	96
2220	60" diameter, 10 ga.	B-13B	75	.747		80	34.50	13.05	127.55	153
2240	72" diameter, 8 ga.	"	45	1.244		84.50	57.50	22	164	203
2250	End sections, 8" diameter, 16 ga.	B-14	20	2.400	Ea.	42	106	10.70	158.70	217
2255	10" diameter, 16 ga.		20	2.400		52	106	10.70	168.70	228
2260	12" diameter, 16 ga.		18	2.667		107	118	11.85	236.85	305
2265	15" diameter, 16 ga.		18	2.667		206	118	11.85	335.85	415
2270	18" diameter, 16 ga.		16	3		240	133	13.35	386.35	480
2275	24" diameter, 16 ga.	B-13	16	3.500		315	161	36.50	512.50	630
2280	30" diameter, 16 ga.		14	4		520	184	41.50	745.50	890
2285	36" diameter, 14 ga.		14	4		665	184	41.50	890.50	1,050
2290	48" diameter, 14 ga.		10	5.600		1,850	258	58	2,166	2,500
2292	60" diameter, 14 ga.		6	9.333		1,875	430	97	2,402	2,825
2294	72" diameter, 14 ga.	B-13B	5	11.200		2,925	515	196	3,636	4,225
2300	Bends or elbows, 8" diameter	B-14	28	1.714		110	76	7.65	193.65	243
2320	10" diameter		25	1.920		137	85	8.55	230.55	287
2340	12" diameter, 16 ga.		23	2.087		162	92.50	9.30	263.80	325
2342	18" diameter, 16 ga.		20	2.400		231	106	10.70	347.70	425
2344	24" diameter, 14 ga.		16	3		360	133	13.35	506.35	610
2346	30" diameter, 14 ga.		15	3.200		430	142	14.25	586.25	705
2348	36" diameter, 14 ga.	B-13	15	3.733		590	172	38.50	800.50	950
2350	48" diameter, 12 ga.	"	12	4.667		760	215	48.50	1,023.50	1,200
2352	60" diameter, 10 ga.	B-13B	10	5.600		1,025	258	98	1,381	1,625
2354	72" diameter, 10 ga.	"	6	9.333		1,275	430	163	1,868	2,225
2360	Wyes or tees, 8" diameter	B-14	25	1.920		156	85	8.55	249.55	310
2380	10" diameter		21	2.286		194	101	10.15	305.15	375
2400	12" diameter, 16 ga.		19	2.526		222	112	11.25	345.25	425
2410	18" diameter, 16 ga.		16	3		305	133	13.35	451.35	555
2412	24" diameter, 14 ga.		16	3		525	133	13.35	671.35	790
2414	30" diameter, 14 ga.	B-13	12	4.667		665	215	48.50	928.50	1,100
2416	36" diameter, 14 ga.		11	5.091		815	234	53	1,102	1,300
2418	48" diameter, 12 ga.		10	5.600		1,100	258	58	1,416	1,650

33 42 11.40 Piping, Storm Drainage, Corrugated Metal	Crew	Daily Output	Labor-Hours	Unit	Material	2020 Bare Costs Labor	Equipment	Total	Total Incl O&P
2420	60" diameter, 10 ga.	B-13B	8	7	Ea.	1,600	320	123	2,043	2,400
2422	72" diameter, 10 ga.	"	5	11.200	↓	1,925	515	196	2,636	3,125
2500	Galvanized, uncoated, 20' lengths									
2520	8" diameter, 16 ga.	B-14	355	.135	L.F.	7.85	6	.60	14.45	18.25
2540	10" diameter, 16 ga.		280	.171		8.40	7.60	.76	16.76	21.50
2560	12" diameter, 16 ga.		220	.218		9.75	9.65	.97	20.37	26.50
2580	15" diameter, 16 ga.		220	.218		11.05	9.65	.97	21.67	28
2600	18" diameter, 16 ga.		205	.234		13.50	10.35	1.04	24.89	31.50
2610	21" diameter, 16 ga.		205	.234		19.50	10.35	1.04	30.89	38
2620	24" diameter, 14 ga.		175	.274		23	12.15	1.22	36.37	45
2630	27" diameter, 14 ga.	▼	170	.282		28.50	12.50	1.26	42.26	51
2640	30" diameter, 14 ga.	B-13	130	.431		32	19.85	4.47	56.32	70
2660	36" diameter, 12 ga.		130	.431		39	19.85	4.47	63.32	77
2680	48" diameter, 12 ga.	▼	110	.509		54	23.50	5.30	82.80	100
2690	60" diameter, 10 ga.	B-13B	78	.718		84	33	12.55	129.55	155
2695	72" diameter, 10 ga.	"	60	.933	↓	109	43	16.35	168.35	202
2711	Bends or elbows, 12" diameter, 16 ga.	B-14	30	1.600	Ea.	130	71	7.10	208.10	257
2712	15" diameter, 16 ga.		25.04	1.917		167	85	8.55	260.55	320
2714	18" diameter, 16 ga.		20	2.400		187	106	10.70	303.70	375
2716	24" diameter, 14 ga.		16	3		281	133	13.35	427.35	525
2718	30" diameter, 14 ga.	▼	15	3.200		350	142	14.25	506.25	615
2720	36" diameter, 14 ga.	B-13	15	3.733		485	172	38.50	695.50	835
2722	48" diameter, 12 ga.		12	4.667		655	215	48.50	918.50	1,100
2724	60" diameter, 10 ga.		10	5.600		1,025	258	58	1,341	1,575
2726	72" diameter, 10 ga.	▼	6	9.333		1,325	430	97	1,852	2,200
2728	Wyes or tees, 12" diameter, 16 ga.	B-14	22.48	2.135		168	94.50	9.50	272	335
2730	18" diameter, 16 ga.		15	3.200		261	142	14.25	417.25	515
2732	24" diameter, 14 ga.		15	3.200		435	142	14.25	591.25	710
2734	30" diameter, 14 ga.		14	3.429		565	152	15.25	732.25	865
2736	36" diameter, 14 ga.	B-13	14	4		715	184	41.50	940.50	1,100
2738	48" diameter, 12 ga.		12	4.667		1,050	215	48.50	1,313.50	1,525
2740	60" diameter, 10 ga.		10	5.600		1,475	258	58	1,791	2,075
2742	72" diameter, 10 ga.	▼	6	9.333		1,800	430	97	2,327	2,725
2780	End sections, 8" diameter	B-14	35	1.371		78.50	60.50	6.10	145.10	184
2785	10" diameter		35	1.371		87	60.50	6.10	153.60	193
2790	12" diameter		35	1.371		100	60.50	6.10	166.60	208
2800	18" diameter	▼	30	1.600		117	71	7.10	195.10	243
2810	24" diameter	B-13	25	2.240		197	103	23	323	395
2820	30" diameter		25	2.240		350	103	23	476	565
2825	36" diameter		20	2.800		470	129	29	628	740
2830	48" diameter	▼	10	5.600		940	258	58	1,256	1,475
2835	60" diameter	B-13B	5	11.200		1,725	515	196	2,436	2,900
2840	72" diameter	"	4	14		2,175	645	245	3,065	3,625
2850	Couplings, 12" diameter					16.50			16.50	18.15
2855	18" diameter					21.50			21.50	24
2860	24" diameter					28			28	30.50
2865	30" diameter					29			29	32
2870	36" diameter					38.50			38.50	42
2875	48" diameter					56.50			56.50	62
2880	60" diameter					62.50			62.50	69
2885	72" diameter					71			71	78

33 42 Stormwater Conveyance

33 42 11 – Stormwater Gravity Piping

33 42 11.50 Piping, Drainage & Sewage, Corrug. HDPE Type S	Crew	Daily Output	Labor-Hours	Unit	Material	2020 Bare Costs Labor	Equipment	Total	Total Incl O&P
0010 **PIPING, DRAINAGE & SEWAGE, CORRUGATED HDPE TYPE S**									
0020 Not including excavation & backfill, bell & spigot									
1000 With gaskets, 4" diameter	B-20	425	.056	L.F.	.94	2.66		3.60	5.05
1010 6" diameter		400	.060		2.72	2.82		5.54	7.25
1020 8" diameter		380	.063		3.58	2.97		6.55	8.40
1030 10" diameter		370	.065		4.77	3.05		7.82	9.85
1040 12" diameter		340	.071		6.85	3.32		10.17	12.55
1050 15" diameter		300	.080		10.15	3.76		13.91	16.80
1060 18" diameter	B-21	275	.102		12.45	4.96	.69	18.10	22
1070 24" diameter		250	.112		19.90	5.45	.76	26.11	31
1080 30" diameter		200	.140		25.50	6.85	.95	33.30	39.50
1090 36" diameter		180	.156		36.50	7.60	1.05	45.15	53
1100 42" diameter		175	.160		43	7.80	1.08	51.88	60
1110 48" diameter		170	.165		60.50	8.05	1.11	69.66	80
1120 54" diameter		160	.175		78.50	8.55	1.18	88.23	101
1130 60" diameter		150	.187		145	9.10	1.26	155.36	174
1135 Add 15% to material pipe cost for water tight connection bell & spigot									
1140 HDPE type S, elbows 12" diameter	B-20	11	2.182	Ea.	72	103		175	234
1150 15" diameter	"	9	2.667		109	125		234	310
1160 18" diameter	B-21	9	3.111		179	152	21	352	450
1170 24" diameter		9	3.111		375	152	21	548	665
1180 30" diameter		8	3.500		600	171	23.50	794.50	945
1190 36" diameter		8	3.500		775	171	23.50	969.50	1,125
1200 42" diameter		6	4.667		975	228	31.50	1,234.50	1,450
1220 48" diameter		6	4.667		1,750	228	31.50	2,009.50	2,300
1240 HDPE type S, Tee 12" diameter	B-20	7	3.429		95	161		256	350
1260 15" diameter	"	6	4		257	188		445	565
1280 18" diameter	B-21	6	4.667		375	228	31.50	634.50	790
1300 24" diameter		5	5.600		690	273	38	1,001	1,200
1320 30" diameter		5	5.600		1,175	273	38	1,486	1,750
1340 36" diameter		4	7		1,150	340	47.50	1,537.50	1,825
1360 42" diameter		4	7		1,300	340	47.50	1,687.50	2,000
1380 48" diameter		4	7		1,375	340	47.50	1,762.50	2,100
1400 Add to basic installation cost for each split coupling joint									
1402 HDPE type S, split coupling, 12" diameter	B-20	17	1.412	Ea.	10.65	66.50		77.15	112
1420 15" diameter		15	1.600		19.90	75		94.90	135
1440 18" diameter		13	1.846		27.50	87		114.50	162
1460 24" diameter		12	2		40.50	94		134.50	187
1480 30" diameter		10	2.400		87.50	113		200.50	267
1500 36" diameter		9	2.667		141	125		266	345
1520 42" diameter		8	3		168	141		309	395
1540 48" diameter		8	3		198	141		339	430

33 42 11.60 Sewage/Drainage Collection, Concrete Pipe

	Crew	Daily Output	Labor-Hours	Unit	Material	2020 Bare Costs Labor	Equipment	Total	Total Incl O&P
0010 **SEWAGE/DRAINAGE COLLECTION, CONCRETE PIPE**									
0020 Not including excavation or backfill									
0050 Box culvert, cast in place, 6' x 6'	C-15	16	4.500	L.F.	240	222		462	595
0060 8' x 8'		14	5.143		355	254		609	770
0070 12' x 12'		10	7.200		700	355		1,055	1,300
0100 Box culvert, precast, base price, 8' long, 6' x 3'	B-69	140	.343		265	16	10.60	291.60	330
0150 6' x 7'		125	.384		320	17.95	11.90	349.85	390
0200 8' x 3'		113	.425		350	19.85	13.15	383	430
0250 8' x 8'		100	.480		415	22.50	14.85	452.35	505

33 42 Stormwater Conveyance

33 42 11 – Stormwater Gravity Piping

33 42 11.60 Sewage/Drainage Collection, Concrete Pipe	Crew	Daily Output	Labor-Hours	Unit	Material	2020 Bare Costs Labor	Equipment	Total	Total Incl O&P
0300	10' x 3'	B-69	110	.436	L.F.	405	20.50	13.50	439	490
0350	10' x 8'		80	.600		605	28	18.60	651.60	730
0400	12' x 3'		100	.480		825	22.50	14.85	862.35	960
0450	12' x 8'	↓	67	.716	↓	940	33.50	22	995.50	1,100
0500	Set up charge at plant, add to base price				Job	6,225			6,225	6,825
0510	Inserts and keyway, add				Ea.	640			640	705
0520	Sloped or skewed end, add				"	895			895	985
1000	Non-reinforced pipe, extra strength, B&S or T&G joints									
1010	6" diameter	B-14	265.04	.181	L.F.	7.80	8	.81	16.61	21.50
1020	8" diameter		224	.214		8.55	9.50	.95	19	24.50
1030	10" diameter		216	.222		9.50	9.85	.99	20.34	26.50
1040	12" diameter		200	.240		10.05	10.60	1.07	21.72	28
1050	15" diameter		180	.267		14.40	11.80	1.19	27.39	35
1060	18" diameter		144	.333		18.05	14.75	1.48	34.28	43.50
1070	21" diameter		112	.429		18.95	18.95	1.91	39.81	51.50
1080	24" diameter	↓	100	.480	↓	22	21	2.14	45.14	58.50
1560	Reinforced culvert, class 2, no gaskets									
1590	27" diameter	B-21	88	.318	L.F.	48	15.50	2.15	65.65	79
1592	30" diameter	B-13	80	.700		49.50	32	7.25	88.75	111
1594	36" diameter	"	72	.778	↓	79.50	36	8.05	123.55	150
2000	Reinforced culvert, class 3, no gaskets									
2010	12" diameter	B-14	150	.320	L.F.	16.15	14.15	1.42	31.72	41
2020	15" diameter		150	.320		22	14.15	1.42	37.57	47
2030	18" diameter		132	.364		24.50	16.10	1.62	42.22	53
2035	21" diameter		120	.400		28.50	17.70	1.78	47.98	60
2040	24" diameter		100	.480		28.50	21	2.14	51.64	66
2045	27" diameter	B-13	92	.609		47.50	28	6.30	81.80	101
2050	30" diameter		88	.636		52	29.50	6.60	88.10	108
2060	36" diameter	↓	72	.778		83	36	8.05	127.05	153
2070	42" diameter	B-13B	68	.824		99.50	38	14.40	151.90	182
2080	48" diameter		64	.875		100	40.50	15.30	155.80	187
2085	54" diameter		56	1		147	46	17.50	210.50	249
2090	60" diameter		48	1.167		181	53.50	20.50	255	300
2100	72" diameter		40	1.400		273	64.50	24.50	362	425
2120	84" diameter		32	1.750		405	80.50	30.50	516	600
2140	96" diameter	↓	24	2.333		485	107	41	633	740
2200	With gaskets, class 3, 12" diameter	B-21	168	.167		17.80	8.15	1.13	27.08	33
2220	15" diameter		160	.175		24	8.55	1.18	33.73	40.50
2230	18" diameter		152	.184		27	9	1.24	37.24	45
2235	21" diameter		152	.184		31.50	9	1.24	41.74	49.50
2240	24" diameter	↓	136	.206		35.50	10.05	1.39	46.94	55.50
2260	30" diameter	B-13	88	.636		60	29.50	6.60	96.10	117
2270	36" diameter	"	72	.778		92.50	36	8.05	136.55	164
2290	48" diameter	B-13B	64	.875		114	40.50	15.30	169.80	202
2310	72" diameter	"	40	1.400	↓	294	64.50	24.50	383	450
2330	Flared ends, 12" diameter	B-21	31	.903	Ea.	305	44	6.10	355.10	410
2340	15" diameter		25	1.120		350	54.50	7.55	412.05	475
2400	18" diameter		20	1.400		400	68.50	9.45	477.95	555
2420	24" diameter	↓	14	2		455	97.50	13.50	566	665
2440	36" diameter	B-13	10	5.600	↓	990	258	58	1,306	1,550
2500	Class 4									
2510	12" diameter	B-21	168	.167	L.F.	18.80	8.15	1.13	28.08	34
2512	15" diameter	↓	160	.175	↓	21	8.55	1.18	30.73	37

33 42 11 – Stormwater Gravity Piping

33 42 11.60 Sewage/Drainage Collection, Concrete Pipe		Crew	Daily Output	Labor-Hours	Unit	Material	2020 Bare Costs Labor	Equipment	Total	Total Incl O&P
2514	18" diameter	B-21	152	.184	L.F.	21.50	9	1.24	31.74	38.50
2516	21" diameter		144	.194		25.50	9.50	1.31	36.31	43.50
2518	24" diameter		136	.206		30	10.05	1.39	41.44	49.50
2520	27" diameter		120	.233		32	11.40	1.58	44.98	54
2522	30" diameter	B-13	88	.636		40	29.50	6.60	76.10	95.50
2524	36" diameter		72	.778		57.50	36	8.05	101.55	125
2528	42" diameter		68	.824		70	38	8.55	116.55	143
2600	Class 5									
2610	12" diameter	B-21	168	.167	L.F.	17.65	8.15	1.13	26.93	33
2612	15" diameter		160	.175		22.50	8.55	1.18	32.23	39
2614	18" diameter		152	.184		26.50	9	1.24	36.74	44.50
2616	21" diameter		144	.194		35.50	9.50	1.31	46.31	55
2618	24" diameter		136	.206		32	10.05	1.39	43.44	52
2620	27" diameter		120	.233		54	11.40	1.58	66.98	78.50
2622	30" diameter	B-13	88	.636		52.50	29.50	6.60	88.60	109
2624	36" diameter	"	72	.778		90	36	8.05	134.05	161
2800	Add for rubber joints 12" to 36" diameter					12%				
3080	Radius pipe, add to pipe prices, 12" to 60" diameter					50%				
3090	Over 60" diameter, add					20%				
3500	Reinforced elliptical, 8' lengths, C507 class 3									
3520	14" x 23" inside, round equivalent 18" diameter	B-21	82	.341	L.F.	40.50	16.65	2.30	59.45	72
3530	24" x 38" inside, round equivalent 30" diameter	B-13	58	.966		69.50	44.50	10	124	154
3540	29" x 45" inside, round equivalent 36" diameter		52	1.077		106	49.50	11.15	166.65	204
3550	38" x 60" inside, round equivalent 48" diameter		38	1.474		171	68	15.30	254.30	305
3560	48" x 76" inside, round equivalent 60" diameter		26	2.154		215	99	22.50	336.50	410
3570	58" x 91" inside, round equivalent 72" diameter		22	2.545		330	117	26.50	473.50	565
3780	Concrete slotted pipe, class 4 mortar joint									
3800	12" diameter	B-21	168	.167	L.F.	33	8.15	1.13	42.28	49.50
3840	18" diameter	"	152	.184	"	37	9	1.24	47.24	55.50
3900	Concrete slotted pipe, Class 4 O-ring joint									
3940	12" diameter	B-21	168	.167	L.F.	30.50	8.15	1.13	39.78	47
3960	18" diameter	"	152	.184	"	32	9	1.24	42.24	50
6200	Gasket, conc. pipe joint, 12"				Ea.	4.21			4.21	4.63
6220	24"					6.90			6.90	7.60
6240	36"					9.80			9.80	10.75
6260	48"					13.50			13.50	14.85
6270	60"					16.35			16.35	17.95
6280	72"					20.50			20.50	22.50

33 42 13 – Stormwater Culverts

33 42 13.13 Public Pipe Culverts

		Crew	Daily Output	Labor-Hours	Unit	Material	2020 Bare Costs Labor	Equipment	Total	Total Incl O&P
0010	**PUBLIC PIPE CULVERTS**									
0020	Headwall, concrete									
0520	Precast, 12" diameter pipe	B-6	12	2	Ea.	1,900	91.50	17.80	2,009.30	2,250
0530	18" diameter pipe		10	2.400		1,675	110	21.50	1,806.50	2,050
0540	24" diameter pipe		10	2.400		2,500	110	21.50	2,631.50	2,950
0550	30" diameter pipe		8	3		2,700	137	26.50	2,863.50	3,200
0560	36" diameter pipe		8	3		2,725	137	26.50	2,888.50	3,225
0570	48" diameter pipe	B-69	7	6.857		7,400	320	212	7,932	8,875
0580	60" diameter pipe	"	6	8		11,600	375	248	12,223	13,600

33 42 13 – Stormwater Culverts

33 42 13.15 Oval Arch Culverts

33 42 13.15 Oval Arch Culverts	Crew	Daily Output	Labor-Hours	Unit	Material	2020 Bare Costs Labor	Equipment	Total	Total Incl O&P
0010 **OVAL ARCH CULVERTS**									
3000 Corrugated galvanized or aluminum, coated & paved									
3020 17" x 13", 16 ga., 15" equivalent	B-14	200	.240	L.F.	14.80	10.60	1.07	26.47	33.50
3040 21" x 15", 16 ga., 18" equivalent		150	.320		19.25	14.15	1.42	34.82	44
3060 28" x 20", 14 ga., 24" equivalent		125	.384		23	17	1.71	41.71	52.50
3080 35" x 24", 14 ga., 30" equivalent		100	.480		30.50	21	2.14	53.64	68
3100 42" x 29", 12 ga., 36" equivalent	B-13	100	.560		36	26	5.80	67.80	84.50
3120 49" x 33", 12 ga., 42" equivalent		90	.622		43.50	28.50	6.45	78.45	98
3140 57" x 38", 12 ga., 48" equivalent		75	.747		56	34.50	7.75	98.25	122
3160 Steel, plain oval arch culverts, plain									
3180 17" x 13", 16 ga., 15" equivalent	B-14	225	.213	L.F.	13.25	9.45	.95	23.65	30
3200 21" x 15", 16 ga., 18" equivalent		175	.274		17.25	12.15	1.22	30.62	38.50
3220 28" x 20", 14 ga., 24" equivalent		150	.320		27	14.15	1.42	42.57	52.50
3240 35" x 24", 14 ga., 30" equivalent	B-13	108	.519		32	24	5.40	61.40	77
3260 42" x 29", 12 ga., 36" equivalent		108	.519		40	24	5.40	69.40	86
3280 49" x 33", 12 ga., 42" equivalent		92	.609		59.50	28	6.30	93.80	114
3300 57" x 38", 12 ga., 48" equivalent		75	.747		69	34.50	7.75	111.25	136
3320 End sections, 17" x 13"		22	2.545	Ea.	139	117	26.50	282.50	360
3340 42" x 29"		17	3.294	"	425	152	34	611	735
3360 Multi-plate arch, steel	B-20	1690	.014	Lb.	1.39	.67		2.06	2.54
8000 Bends or elbows, 6" diameter		30	.800	Ea.	71.50	37.50		109	135
8020 8" diameter		28	.857		92.50	40.50		133	163
8040 10" diameter		25	.960		110	45		155	189
8060 12" diameter	B-21	20	1.400		144	68.50	9.45	221.95	271
8100 Wyes or tees, 6" diameter	B-20	27	.889		115	42		157	189
8120 8" diameter		25	.960		145	45		190	227
8140 10" diameter		21	1.143		171	53.50		224.50	269
8160 12" diameter	B-21	18	1.556		184	76	10.50	270.50	330

33 42 33 – Stormwater Curbside Drains and Inlets

33 42 33.13 Catch Basins

33 42 33.13 Catch Basins	Crew	Daily Output	Labor-Hours	Unit	Material	2020 Bare Costs Labor	Equipment	Total	Total Incl O&P
0010 **CATCH BASINS**									
0011 Not including footing & excavation									
1510 Grates only, for pipe bells, plastic, 4" diameter pipe	1 Clab	50	.160	Ea.	3.84	6.75		10.59	14.30
1514 6" diameter pipe		50	.160		11.30	6.75		18.05	22.50
1516 8" diameter pipe		50	.160		15.05	6.75		21.80	26.50
1520 For beehive type grate, add					60%				
1550 Gray iron, 4" diam. pipe, light duty	1 Clab	50	.160		26	6.75		32.75	39
1551 Heavy duty		50	.160		44	6.75		50.75	58.50
1552 6" diameter pipe, light duty		50	.160		31	6.75		37.75	44
1553 Heavy duty		50	.160		23.50	6.75		30.25	35.50
1554 8" diameter pipe, light duty		50	.160		52.50	6.75		59.25	67.50
1555 Heavy duty		50	.160		43.50	6.75		50.25	58
1556 10" diameter pipe, light duty		40	.200		64	8.40		72.40	83
1557 Heavy duty		40	.200		56	8.40		64.40	74
1558 12" diameter pipe, light duty		40	.200		104	8.40		112.40	127
1559 Heavy duty		40	.200		91	8.40		99.40	113
1560 15" diameter pipe, light duty		32	.250		126	10.55		136.55	154
1561 Heavy duty		32	.250		128	10.55		138.55	156
1562 18" diameter pipe, light duty		32	.250		149	10.55		159.55	180
1563 Heavy duty		32	.250		147	10.55		157.55	178
1568 For beehive type grate, add					60%				
1570 Covers only for pipe bells, gray iron, med. duty									

508

33 42 Stormwater Conveyance

33 42 33 – Stormwater Curbside Drains and Inlets

33 42 33.13 Catch Basins

		Crew	Daily Output	Labor-Hours	Unit	Material	2020 Bare Costs Labor	Equipment	Total	Total Incl O&P
1571	4" diameter pipe	1 Clab	50	.160	Ea.	74	6.75		80.75	91
1572	6" diameter pipe		50	.160		93	6.75		99.75	112
1573	8" diameter pipe		50	.160		88	6.75		94.75	107
1574	10" diameter pipe		40	.200		222	8.40		230.40	257
1575	12" diameter pipe		40	.200		101	8.40		109.40	125
1576	15" diameter pipe		32	.250		115	10.55		125.55	142
1577	18" diameter pipe		32	.250		182	10.55		192.55	217
1578	24" diameter pipe	2 Clab	24	.667		305	28		333	380
1579	36" diameter pipe	"	24	.667		340	28		368	415
1580	Curb inlet frame, grate, and curb box									
1582	Large 24" x 36" heavy duty	B-24	2	12	Ea.	800	580		1,380	1,750
1590	Small 10" x 21" medium duty	"	2	12		510	580		1,090	1,425
1600	Frames & grates, C.I., 24" square, 500 lb.	B-6	7.80	3.077		375	141	27.50	543.50	650
1700	26" D shape, 600 lb.		7	3.429		720	157	30.50	907.50	1,050
1800	Light traffic, 18" diameter, 100 lb.		10	2.400		215	110	21.50	346.50	425
1900	24" diameter, 300 lb.		8.70	2.759		199	126	24.50	349.50	435
2000	36" diameter, 900 lb.		5.80	4.138		765	189	37	991	1,175
2100	Heavy traffic, 24" diameter, 400 lb.		7.80	3.077		295	141	27.50	463.50	565
2200	36" diameter, 1,150 lb.		3	8		935	365	71.50	1,371.50	1,650
2300	Mass. State standard, 26" diameter, 475 lb.		7	3.429		750	157	30.50	937.50	1,100
2400	30" diameter, 620 lb.		7	3.429		340	157	30.50	527.50	645
2500	Watertight, 24" diameter, 350 lb.		7.80	3.077		415	141	27.50	583.50	695
2600	26" diameter, 500 lb.		7	3.429		450	157	30.50	637.50	765
2700	32" diameter, 575 lb.		6	4		960	183	35.50	1,178.50	1,375
2800	3 piece cover & frame, 10" deep,									
2900	1,200 lb., for heavy equipment	B-6	3	8	Ea.	1,200	365	71.50	1,636.50	1,950
3000	Raised for paving 1-1/4" to 2" high									
3100	4 piece expansion ring									
3200	20" to 26" diameter	1 Clab	3	2.667	Ea.	213	112		325	405
3300	30" to 36" diameter	"	3	2.667	"	295	112		407	495
3320	Frames and covers, existing, raised for paving, 2", including									
3340	row of brick, concrete collar, up to 12" wide frame	B-6	18	1.333	Ea.	52	61	11.90	124.90	162
3360	20" to 26" wide frame		11	2.182		78.50	100	19.45	197.95	258
3380	30" to 36" wide frame		9	2.667		97	122	24	243	315
3400	Inverts, single channel brick	D-1	3	5.333		112	250		362	505
3500	Concrete		5	3.200		124	150		274	365
3600	Triple channel, brick		2	8		182	375		557	770
3700	Concrete		3	5.333		156	250		406	550

33 42 33.15 Remove and Replace Catch Basin Cover

		Crew	Daily Output	Labor-Hours	Unit	Material	2020 Bare Costs Labor	Equipment	Total	Total Incl O&P
0010	**REMOVE AND REPLACE CATCH BASIN COVER**									
0023	Remove catch basin cover	1 Clab	80	.100	Ea.		4.21		4.21	6.35
0033	Replace catch basin cover	"	80	.100	"		4.21		4.21	6.35

33 42 33.50 Stormwater Management

		Crew	Daily Output	Labor-Hours	Unit	Material	2020 Bare Costs Labor	Equipment	Total	Total Incl O&P
0010	**STORMWATER MANAGEMENT**									
0030	Add per S.F. of impervious surface	B-37	6000	.008	S.F.	2.63	.35	.04	3.02	3.48

For customer support on your Site Work & Landscape Costs with RSMeans data, call 800.448.8182.

509

33 46 Stormwater Management

33 46 23 – Modular Buried Stormwater Storage Units

33 46 23.10 HDPE Storm Water Infiltration Chamber		Crew	Daily Output	Labor-Hours	Unit	Material	2020 Bare Costs Labor	Equipment	Total	Total Incl O&P
0010	**HDPE STORM WATER INFILTRATION CHAMBER**									
0020	Not including excavation or backfill									
0100	11" H x 16" W	2 Clab	300	.053	L.F.	6.65	2.25		8.90	10.70
0110	12" H x 22" W		300	.053		9.05	2.25		11.30	13.30
0120	12" H x 34" W		300	.053		10.70	2.25		12.95	15.15
0130	16" H x 34" W		300	.053		23.50	2.25		25.75	29.50
0140	30" H x 52" W	↓	270	.059	↓	36.50	2.49		38.99	44.50

33 52 Hydrocarbon Transmission and Distribution

33 52 13 – Liquid Hydrocarbon Piping

33 52 13.16 Gasoline Piping

		Crew	Daily Output	Labor-Hours	Unit	Material	2020 Bare Costs Labor	Equipment	Total	Total Incl O&P
0010	**GASOLINE PIPING**									
0020	Primary containment pipe, fiberglass-reinforced									
0030	Plastic pipe 15' & 30' lengths									
0040	2" diameter	Q-6	425	.056	L.F.	7	3.45		10.45	12.85
0050	3" diameter		400	.060		10.65	3.67		14.32	17.20
0060	4" diameter	↓	375	.064	↓	13.85	3.92		17.77	21
0100	Fittings									
0110	Elbows, 90° & 45°, bell ends, 2"	Q-6	24	1	Ea.	40.50	61		101.50	137
0120	3" diameter		22	1.091		56	66.50		122.50	162
0130	4" diameter		20	1.200		70	73.50		143.50	188
0200	Tees, bell ends, 2"		21	1.143		57	70		127	168
0210	3" diameter		18	1.333		68	81.50		149.50	197
0220	4" diameter		15	1.600		81.50	98		179.50	237
0230	Flanges bell ends, 2"		24	1		31.50	61		92.50	126
0240	3" diameter		22	1.091		38.50	66.50		105	142
0250	4" diameter		20	1.200		43.50	73.50		117	158
0260	Sleeve couplings, 2"		21	1.143		12.05	70		82.05	118
0270	3" diameter		18	1.333		18.05	81.50		99.55	142
0280	4" diameter		15	1.600		22.50	98		120.50	172
0290	Threaded adapters, 2"		21	1.143		19.25	70		89.25	126
0300	3" diameter		18	1.333		32	81.50		113.50	158
0310	4" diameter		15	1.600		35.50	98		133.50	186
0320	Reducers, 2"		27	.889		27	54.50		81.50	111
0330	3" diameter		22	1.091		27	66.50		93.50	130
0340	4" diameter	↓	20	1.200	↓	36	73.50		109.50	150
1010	Gas station product line for secondary containment (double wall)									
1100	Fiberglass reinforced plastic pipe 25' lengths									
1120	Pipe, plain end, 3" diameter	Q-6	375	.064	L.F.	29.50	3.92		33.42	38.50
1130	4" diameter		350	.069		33.50	4.20		37.70	43.50
1140	5" diameter		325	.074		35.50	4.52		40.02	46.50
1150	6" diameter	↓	300	.080	↓	39.50	4.89		44.39	51
1200	Fittings									
1230	Elbows, 90° & 45°, 3" diameter	Q-6	18	1.333	Ea.	131	81.50		212.50	266
1240	4" diameter		16	1.500		163	92		255	315
1250	5" diameter		14	1.714		170	105		275	345
1260	6" diameter		12	2		211	122		333	415
1270	Tees, 3" diameter		15	1.600		157	98		255	320
1280	4" diameter		12	2		184	122		306	385
1290	5" diameter		9	2.667		299	163		462	575
1300	6" diameter	↓	6	4		340	245		585	740

510

33 52 Hydrocarbon Transmission and Distribution

33 52 13 – Liquid Hydrocarbon Piping

33 52 13.16 Gasoline Piping

		Crew	Daily Output	Labor-Hours	Unit	Material	2020 Bare Costs Labor	Equipment	Total	Total Incl O&P
1310	Couplings, 3" diameter	Q-6	18	1.333	Ea.	54.50	81.50		136	182
1320	4" diameter		16	1.500		117	92		209	266
1330	5" diameter		14	1.714		218	105		323	395
1340	6" diameter		12	2		320	122		442	535
1350	Cross-over nipples, 3" diameter		18	1.333		10.30	81.50		91.80	133
1360	4" diameter		16	1.500		12.55	92		104.55	151
1370	5" diameter		14	1.714		16.30	105		121.30	175
1380	6" diameter		12	2		17.55	122		139.55	202
1400	Telescoping, reducers, concentric 4" x 3"		18	1.333		44.50	81.50		126	171
1410	5" x 4"		17	1.412		94.50	86.50		181	233
1420	6" x 5"	▼	16	1.500	▼	237	92		329	400

33 52 16 – Gas Hydrocarbon Piping

33 52 16.13 Steel Natural Gas Piping

		Crew	Daily Output	Labor-Hours	Unit	Material	2020 Bare Costs Labor	Equipment	Total	Total Incl O&P
0010	**STEEL NATURAL GAS PIPING**									
0020	Not including excavation or backfill, tar coated and wrapped									
4000	Pipe schedule 40, plain end									
4040	1" diameter	Q-4	300	.107	L.F.	5.10	6.55	.36	12.01	15.80
4080	2" diameter		280	.114		8	7	.38	15.38	19.65
4120	3" diameter	▼	260	.123		13.20	7.55	.41	21.16	26.50
4160	4" diameter	B-35	255	.188		17.15	9.90	3.14	30.19	37
4200	5" diameter		220	.218		25	11.45	3.64	40.09	48.50
4240	6" diameter		180	.267		30.50	14	4.45	48.95	59.50
4280	8" diameter		140	.343		48.50	18	5.70	72.20	86.50
4320	10" diameter		100	.480		114	25	8	147	173
4360	12" diameter		80	.600		127	31.50	10	168.50	199
4400	14" diameter		75	.640		136	33.50	10.70	180.20	211
4440	16" diameter		70	.686		148	36	11.45	195.45	230
4480	18" diameter		65	.738		191	39	12.30	242.30	282
4520	20" diameter		60	.800		297	42	13.35	352.35	405
4560	24" diameter	▼	50	.960	▼	340	50.50	16	406.50	470
6000	Schedule 80, plain end									
6002	4" diameter	B-35	144	.333	L.F.	40	17.50	5.55	63.05	76.50
6004	5" diameter		140	.343		52.50	18	5.70	76.20	91
6006	6" diameter		126	.381		85.50	20	6.35	111.85	131
6008	8" diameter		108	.444		114	23.50	7.40	144.90	168
6010	10" diameter		90	.533		171	28	8.90	207.90	240
6012	12" diameter	▼	72	.667	▼	228	35	11.10	274.10	315
8008	Elbow, weld joint, standard weight									
8020	4" diameter	Q-16	6.80	3.529	Ea.	113	212	15.65	340.65	460
8022	5" diameter		5.40	4.444		223	267	19.70	509.70	670
8024	6" diameter		4.50	5.333		251	320	23.50	594.50	785
8026	8" diameter		3.40	7.059		475	425	31.50	931.50	1,200
8028	10" diameter		2.70	8.889		705	535	39.50	1,279.50	1,625
8030	12" diameter		2.30	10.435		1,000	630	46	1,676	2,100
8032	14" diameter		1.80	13.333		1,525	800	59	2,384	2,950
8034	16" diameter		1.50	16		2,100	960	71	3,131	3,850
8036	18" diameter		1.40	17.143		2,625	1,025	76	3,726	4,525
8038	20" diameter		1.20	20		3,625	1,200	88.50	4,913.50	5,875
8040	24" diameter	▼	1.02	23.529	▼	5,450	1,425	104	6,979	8,250
8100	Extra heavy									
8102	4" diameter	Q-16	5.30	4.528	Ea.	226	272	20	518	680
8104	5" diameter	▼	4.20	5.714	▼	445	345	25.50	815.50	1,025

For customer support on your Site Work & Landscape Costs with RSMeans data, call 800.448.8182.

511

33 52 16 – Gas Hydrocarbon Piping

33 52 16.13 Steel Natural Gas Piping	Crew	Daily Output	Labor-Hours	Unit	Material	2020 Bare Costs Labor	Equipment	Total	Total Incl O&P
8106	6" diameter	Q-16	3.50	6.857	Ea.	505	410	30.50	945.50	1,200
8108	8" diameter		2.60	9.231		710	555	41	1,306	1,650
8110	10" diameter		2.10	11.429		940	685	50.50	1,675.50	2,100
8112	12" diameter		1.80	13.333		1,325	800	59	2,184	2,750
8114	14" diameter		1.40	17.143		2,050	1,025	76	3,151	3,875
8116	16" diameter		1.20	20		2,800	1,200	88.50	4,088.50	5,000
8118	18" diameter		1.10	21.818		3,500	1,300	96.50	4,896.50	5,925
8120	20" diameter		.94	25.532		4,825	1,525	113	6,463	7,750
8122	24" diameter		.80	30		7,275	1,800	133	9,208	10,800
8200	Malleable, standard weight									
8202	4" diameter	B-20	12	2	Ea.	745	94		839	960
8204	5" diameter		9	2.667		860	125		985	1,125
8206	6" diameter		8	3		1,075	141		1,216	1,400
8208	8" diameter		6	4		2,025	188		2,213	2,525
8210	10" diameter		5	4.800		1,950	226		2,176	2,500
8212	12" diameter		4	6		2,175	282		2,457	2,800
8300	Extra heavy									
8302	4" diameter	B-20	12	2	Ea.	1,500	94		1,594	1,800
8304	5" diameter	B-21	9	3.111		1,725	152	21	1,898	2,125
8306	6" diameter		8	3.500		2,050	171	23.50	2,244.50	2,525
8308	8" diameter		6	4.667		2,550	228	31.50	2,809.50	3,175
8310	10" diameter		5	5.600		3,100	273	38	3,411	3,850
8312	12" diameter		4	7		3,450	340	47.50	3,837.50	4,375
8500	Tee weld, standard weight									
8510	4" diameter	Q-16	4.50	5.333	Ea.	186	320	23.50	529.50	710
8512	5" diameter		3.60	6.667		305	400	29.50	734.50	975
8514	6" diameter		3	8		320	480	35.50	835.50	1,100
8516	8" diameter		2.30	10.435		555	630	46	1,231	1,600
8518	10" diameter		1.80	13.333		1,100	800	59	1,959	2,475
8520	12" diameter		1.50	16		1,525	960	71	2,556	3,200
8522	14" diameter		1.20	20		2,675	1,200	88.50	3,963.50	4,850
8524	16" diameter		1	24		3,000	1,450	106	4,556	5,575
8526	18" diameter		.90	26.667		4,750	1,600	118	6,468	7,750
8528	20" diameter		.80	30		7,500	1,800	133	9,433	11,100
8530	24" diameter		.70	34.286		9,675	2,050	152	11,877	13,800
8810	Malleable, standard weight									
8812	4" diameter	B-20	8	3	Ea.	1,275	141		1,416	1,600
8814	5" diameter	B-21	6	4.667		1,725	228	31.50	1,984.50	2,275
8816	6" diameter		5.30	5.283		1,500	258	35.50	1,793.50	2,075
8818	8" diameter		4	7		2,175	340	47.50	2,562.50	2,975
8820	10" diameter		3.30	8.485		3,425	415	57.50	3,897.50	4,450
8822	12" diameter		2.70	10.370		4,000	505	70	4,575	5,225
8900	Extra heavy									
8902	4" diameter	B-20	8	3	Ea.	1,900	141		2,041	2,300
8904	5" diameter	B-21	6	4.667		2,600	228	31.50	2,859.50	3,250
8906	6" diameter		5.30	5.283		2,275	258	35.50	2,568.50	2,925
8908	8" diameter		4	7		2,725	340	47.50	3,112.50	3,575
8910	10" diameter		3.30	8.485		3,425	415	57.50	3,897.50	4,450
8912	12" diameter		2.70	10.370		4,000	505	70	4,575	5,225

33 52 16.20 Piping, Gas Service and Distribution, P.E.	Crew	Daily Output	Labor-Hours	Unit	Material	2020 Bare Costs Labor	Equipment	Total	Total Incl O&P
0010 **PIPING, GAS SERVICE AND DISTRIBUTION, POLYETHYLENE**									
0020 Not including excavation or backfill									
1000 60 psi coils, compression coupling @ 100', 1/2" diameter, SDR 11	B-20A	608	.053	L.F.	.49	2.66		3.15	4.53
1010 1" diameter, SDR 11		544	.059		1.14	2.97		4.11	5.70
1040 1-1/4" diameter, SDR 11		544	.059		1.64	2.97		4.61	6.25
1100 2" diameter, SDR 11		488	.066		2.48	3.31		5.79	7.70
1160 3" diameter, SDR 11		408	.078		6.20	3.96		10.16	12.75
1500 60 psi 40' joints with coupling, 3" diameter, SDR 11	B-21A	408	.098		7.35	5.15	1.05	13.55	16.95
1540 4" diameter, SDR 11		352	.114		13.35	5.95	1.22	20.52	25
1600 6" diameter, SDR 11		328	.122		35.50	6.40	1.31	43.21	50
1640 8" diameter, SDR 11		272	.147		54	7.70	1.58	63.28	72.50

33 52 16.23 Medium Density Polyethylene Piping

	Crew	Daily Output	Labor-Hours	Unit	Material	2020 Bare Costs Labor	Equipment	Total	Total Incl O&P
2010 **MEDIUM DENSITY POLYETHYLENE PIPING**									
2020 ASTM D2513, not including excavation or backfill									
2200 Butt fused pipe									
2205 80psi coils, butt fusion joint @ 100', 1/2" CTS diameter, SDR 7	B-22C	2050	.008	L.F.	.18	.38	.08	.64	.85
2210 1" CTS diameter, SDR 11.5		1900	.008		.25	.41	.08	.74	.99
2215 80psi coils, butt fusion joint @ 100', IPS 1/2" diameter, SDR 9.3		1950	.008		.26	.40	.08	.74	.98
2220 3/4" diameter, SDR 11		1950	.008		.33	.40	.08	.81	1.05
2225 1" diameter, SDR 11		1850	.009		.80	.42	.08	1.30	1.60
2230 1-1/4" diameter, SDR 11		1850	.009		.80	.42	.08	1.30	1.60
2235 1-1/2" diameter, SDR 11		1750	.009		.98	.44	.09	1.51	1.85
2240 2" diameter, SDR 11		1750	.009		1.33	.44	.09	1.86	2.23
2245 80psi 40' lengths, butt fusion joint, IPS 3" diameter, SDR 11.5	B-22A	660	.061		2.55	2.94	1.22	6.71	8.55
2250 4" diameter, SDR 11.5	"	500	.080		4.02	3.88	1.61	9.51	12.05
2400 Socket fused pipe									
2405 80psi coils, socket fusion coupling @ 100', 1/2" CTS diameter, SDR 7	B-20	1050	.023	L.F.	.20	1.07		1.27	1.84
2410 1" CTS diameter, SDR 11.5		1000	.024		.28	1.13		1.41	2.01
2415 80psi coils, socket fusion coupling @ 100', IPS 1/2" diameter, SDR 9.3		1000	.024		.28	1.13		1.41	2.01
2420 3/4" diameter, SDR 11		1000	.024		.35	1.13		1.48	2.09
2425 1" diameter, SDR 11		950	.025		.82	1.19		2.01	2.70
2430 1-1/4" diameter, SDR 11		950	.025		.82	1.19		2.01	2.70
2435 1-1/2" diameter, SDR 11		900	.027		1.01	1.25		2.26	3
2440 2" diameter, SDR 11		850	.028		1.36	1.33		2.69	3.49
2445 80psi 40' lengths, socket fusion coupling, IPS 3" diameter, SDR 11.5	B-21A	340	.118		3.03	6.15	1.26	10.44	13.90
2450 4" diameter, SDR 11.5	"	260	.154		4.91	8.05	1.65	14.61	19.25
2600 Compression coupled pipe									
2605 80psi coils, compression coupling @ 100', 1/2" CTS diameter, SDR 7	B-20	2250	.011	L.F.	.31	.50		.81	1.10
2610 1" CTS diameter, SDR 11.5		2175	.011		.50	.52		1.02	1.33
2615 80psi coils, compression coupling @ 100', IPS 1/2" diameter, SDR 9.3		2175	.011		.50	.52		1.02	1.33
2620 3/4" diameter, SDR 11		2100	.011		.57	.54		1.11	1.43
2625 1" diameter, SDR 11		2025	.012		1.13	.56		1.69	2.08
3000 Fittings, butt fusion									
3010 SDR 11, IPS unless noted CTS									
3100 Caps, 3/4" diameter	B-22C	28.50	.561	Ea.	3.36	27	5.45	35.81	50.50
3105 1" diameter		27	.593		3.06	28.50	5.80	37.36	53
3110 1-1/4" diameter		27	.593		4.71	28.50	5.80	39.01	55
3115 1-1/2" diameter		25.50	.627		3.67	30.50	6.10	40.27	57
3120 2" diameter		24	.667		6.05	32.50	6.50	45.05	62.50
3125 3" diameter		24	.667		11.50	32.50	6.50	50.50	68.50
3130 4" diameter		18	.889		24	43	8.65	75.65	101
3200 Reducers, 1" x 3/4" diameters		13.50	1.185		6.85	57.50	11.55	75.90	107

513

For customer support on your Site Work & Landscape Costs with RSMeans data, call 800.448.8182.

33 52 16.23 Medium Density Polyethylene Piping	Crew	Daily Output	Labor-Hours	Unit	Material	2020 Bare Costs Labor	Equipment	Total	Total Incl O&P
3205	1-1/4" x 1" diameters	B-22C	13.50	1.185	Ea.	7	57.50	11.55	76.05	107
3210	1-1/2" x 3/4" diameters		12.75	1.255		7.15	61	12.25	80.40	113
3215	1-1/2" x 1" diameters		12.75	1.255		7.15	61	12.25	80.40	113
3220	1-1/2" x 1-1/4" diameters		12.75	1.255		7.15	61	12.25	80.40	113
3225	2" x 1" diameters		12	1.333		7.70	64.50	13	85.20	120
3230	2" x 1-1/4" diameters		12	1.333		10.85	64.50	13	88.35	124
3235	2" x 1-1/2" diameters		12	1.333		11.95	64.50	13	89.45	125
3240	3" x 2" diameters		12	1.333		15.10	64.50	13	92.60	128
3245	4" x 2" diameters		9	1.778		22.50	86	17.35	125.85	174
3250	4" x 3" diameters		9	1.778		32	86	17.35	135.35	184
3300	Elbows, 90°, 3/4" diameter		14.25	1.123		4.92	54.50	10.95	70.37	99.50
3302	1" diameter		13.50	1.185		5.10	57.50	11.55	74.15	105
3304	1-1/4" diameter		13.50	1.185		9.75	57.50	11.55	78.80	110
3306	1-1/2" diameter		12.75	1.255		8.60	61	12.25	81.85	114
3308	2" diameter		12	1.333		11.40	64.50	13	88.90	124
3310	3" diameter		12	1.333		25.50	64.50	13	103	140
3312	4" diameter		9	1.778		34.50	86	17.35	137.85	187
3350	45°, 3" diameter		12	1.333		25.50	64.50	13	103	140
3352	4" diameter		9	1.778		34.50	86	17.35	137.85	187
3400	Tees, 3/4" diameter		9.50	1.684		16.95	81.50	16.40	114.85	160
3405	1" diameter		9	1.778		5.10	86	17.35	108.45	155
3410	1-1/4" diameter		9	1.778		10.25	86	17.35	113.60	160
3415	1-1/2" diameter		8.50	1.882		8.70	91.50	18.35	118.55	168
3420	2" diameter		8	2		12.70	97	19.50	129.20	182
3425	3" diameter		8	2		25	97	19.50	141.50	195
3430	4" diameter		6	2.667		34.50	129	26	189.50	261
3500	Tapping tees, high volume, butt fusion outlets									
3505	1-1/2" punch, 2" x 2" outlet	B-22C	11	1.455	Ea.	92.50	70.50	14.20	177.20	224
3510	1-7/8" punch, 3" x 2" outlet		11	1.455		93	70.50	14.20	177.70	224
3515	4" x 2" outlet		9	1.778		94.50	86	17.35	197.85	253
3550	For protective sleeves, add					2.92			2.92	3.21
3600	Service saddles, saddle contour x outlet, butt fusion outlets									
3602	1-1/4" x 3/4" outlet	B-22C	17	.941	Ea.	6.65	45.50	9.20	61.35	86.50
3604	1-1/4" x 1" outlet		17	.941		6.65	45.50	9.20	61.35	86.50
3606	1-1/4" x 1-1/4" outlet		16	1		6.65	48.50	9.75	64.90	91
3608	1-1/2" x 3/4" outlet		16	1		6.65	48.50	9.75	64.90	91
3610	1-1/2" x 1-1/4" outlet		15	1.067		6.65	51.50	10.40	68.55	97
3612	2" x 3/4" outlet		12	1.333		6.65	64.50	13	84.15	119
3614	2" x 1" outlet		12	1.333		6.65	64.50	13	84.15	119
3616	2" x 1-1/4" outlet		11	1.455		6.65	70.50	14.20	91.35	129
3618	3" x 3/4" outlet		12	1.333		6.65	64.50	13	84.15	119
3620	3" x 1" outlet		12	1.333		6.65	64.50	13	84.15	119
3622	3" x 1-1/4" outlet		11	1.455		6.65	70.50	14.20	91.35	129
3624	4" x 3/4" outlet		10	1.600		6.65	77.50	15.60	99.75	141
3626	4" x 1" outlet		10	1.600		6.65	77.50	15.60	99.75	141
3628	4" x 1-1/4" outlet		9	1.778		6.65	86	17.35	110	156
3685	For protective sleeves, 3/4" diameter outlets, add					.86			.86	.95
3690	1" diameter outlets, add					1.45			1.45	1.60
3695	1-1/4" diameter outlets, add					2.49			2.49	2.74
3700	Branch saddles, contour x outlet, butt outlets, round base									
3702	2" x 2" outlet	B-22C	11	1.455	Ea.	17.30	70.50	14.20	102	141
3704	3" x 2" outlet		11	1.455		18.80	70.50	14.20	103.50	142
3706	3" x 3" outlet		11	1.455		27	70.50	14.20	111.70	151

For customer support on your Site Work & Landscape Costs with RSMeans data, call 800.448.8182.

33 52 Hydrocarbon Transmission and Distribution

33 52 16 - Gas Hydrocarbon Piping

33 52 16.23 **Medium Density Polyethylene Piping**	Crew	Daily Output	Labor-Hours	Unit	Material	2020 Bare Costs Labor	Equipment	Total	Total Incl O&P
3708	4" x 2" outlet	B-22C	9	1.778	Ea.	21.50	86	17.35	124.85	173
3710	4" x 3" outlet		9	1.778		27	86	17.35	130.35	179
3712	4" x 4" outlet		9	1.778		36	86	17.35	139.35	189
3750	Rectangular base, 2" x 2" outlet		11	1.455		16.90	70.50	14.20	101.60	140
3752	3" x 2" outlet		11	1.455		18.35	70.50	14.20	103.05	142
3754	3" x 3" outlet		11	1.455		26.50	70.50	14.20	111.20	151
3756	4" x 2" outlet		9	1.778		21	86	17.35	124.35	172
3758	4" x 3" outlet		9	1.778		25.50	86	17.35	128.85	177
3760	4" x 4" outlet		9	1.778		31	86	17.35	134.35	184
4000	Fittings, socket fusion									
4010	SDR 11, IPS unless noted CTS									
4100	Caps, 1/2" CTS diameter	B-20	30	.800	Ea.	2.57	37.50		40.07	59.50
4105	1/2" diameter		28.50	.842		2.43	39.50		41.93	62
4110	3/4" diameter		28.50	.842		3.05	39.50		42.55	63
4115	1" CTS diameter		28	.857		6.90	40.50		47.40	68
4120	1" diameter		27	.889		3.65	42		45.65	67
4125	1-1/4" diameter		27	.889		4.02	42		46.02	67.50
4130	1-1/2" diameter		25.50	.941		5.05	44.50		49.55	72
4135	2" diameter		24	1		4.84	47		51.84	76.50
4140	3" diameter		24	1		19.30	47		66.30	92
4145	4" diameter		18	1.333		33	62.50		95.50	131
4200	Reducers, 1/2" x 1/2" CTS diameters		14.25	1.684		6.40	79		85.40	126
4202	3/4" x 1/2" CTS diameters		14.25	1.684		5.70	79		84.70	125
4204	3/4" x 1/2" diameters		14.25	1.684		6.30	79		85.30	126
4206	1" CTS x 1/2" diameters		14	1.714		5.80	80.50		86.30	127
4208	1" CTS x 3/4" diameters		13.50	1.778		5.55	83.50		89.05	132
4210	1" x 1/2" CTS diameters		13.50	1.778		7.45	83.50		90.95	134
4212	1" x 1/2" diameters		13.50	1.778		6.15	83.50		89.65	133
4214	1" x 3/4" diameters		13.50	1.778		9	83.50		92.50	136
4216	1" x 1" CTS diameters		13.50	1.778		7.85	83.50		91.35	135
4218	1-1/4" x 1/2" CTS diameters		13.50	1.778		16.35	83.50		99.85	144
4220	1-1/4" x 1/2" diameters		13.50	1.778		8.65	83.50		92.15	136
4222	1-1/4" x 3/4" diameters		13.50	1.778		12.80	83.50		96.30	140
4224	1-1/4" x 1" CTS diameters		13.50	1.778		9.20	83.50		92.70	136
4226	1-1/4" x 1" diameters		13.50	1.778		10.90	83.50		94.40	138
4228	1-1/2" x 3/4" diameters		12.75	1.882		9.50	88.50		98	143
4230	1-1/2" x 1" diameters		12.75	1.882		8.25	88.50		96.75	142
4232	1-1/2" x 1-1/4" diameters		12.75	1.882		9.35	88.50		97.85	143
4234	2" x 3/4" diameters		12	2		10.30	94		104.30	153
4236	2" x 1" CTS diameters		12	2		9.75	94		103.75	153
4238	2" x 1" diameters		12	2		11.10	94		105.10	154
4240	2" x 1-1/4" diameters		12	2		10.10	94		104.10	153
4242	2" x 1-1/2" diameters		12	2		9.80	94		103.80	153
4244	3" x 2" diameters		12	2		11.75	94		105.75	155
4246	4" x 2" diameters		9	2.667		26.50	125		151.50	218
4248	4" x 3" diameters		9	2.667		30	125		155	222
4300	Couplings, 1/2" CTS diameter		15	1.600		1.86	75		76.86	115
4305	1/2" diameter		14.25	1.684		1.79	79		80.79	121
4310	3/4" diameter		14.25	1.684		2.25	79		81.25	121
4315	1" CTS diameter		14	1.714		3.08	80.50		83.58	124
4320	1" diameter		13.50	1.778		2.28	83.50		85.78	129
4325	1-1/4" diameter		13.50	1.778		2.34	83.50		85.84	129
4330	1-1/2" diameter		12.75	1.882		2.65	88.50		91.15	136

For customer support on your Site Work & Landscape Costs with RSMeans data, call 800.448.8182.

515

33 52 16.23 Medium Density Polyethylene Piping	Crew	Daily Output	Labor-Hours	Unit	Material	2020 Bare Costs Labor	Equipment	Total	Total Incl O&P
4335 2" diameter	B-20	12	2	Ea.	2.86	94		96.86	145
4340 3" diameter		12	2		19.25	94		113.25	163
4345 4" diameter		9	2.667		35.50	125		160.50	228
4400 Elbows, 90°, 1/2" CTS diameter		15	1.600		3.41	75		78.41	117
4405 1/2" diameter		14.25	1.684		3.42	79		82.42	123
4410 3/4" diameter		14.25	1.684		4.07	79		83.07	123
4415 1" CTS diameter		14	1.714		3.37	80.50		83.87	125
4420 1" diameter		13.50	1.778		3.57	83.50		87.07	130
4425 1-1/4" diameter		13.50	1.778		4.04	83.50		87.54	130
4430 1-1/2" diameter		12.75	1.882		6.80	88.50		95.30	141
4435 2" diameter		12	2		7.40	94		101.40	150
4440 3" diameter		12	2		25	94		119	170
4445 4" diameter		9	2.667		71.50	125		196.50	268
4460 45°, 1" diameter		13.50	1.778		7	83.50		90.50	134
4465 2" diameter		12	2		9.45	94		103.45	152
4500 Tees, 1/2" CTS diameter		10	2.400		3.26	113		116.26	174
4505 1/2" diameter		9.50	2.526		3.07	119		122.07	182
4510 3/4" diameter		9.50	2.526		3.89	119		122.89	183
4515 1" CTS diameter		9.33	2.571		3.81	121		124.81	186
4520 1" diameter		9	2.667		4.55	125		129.55	194
4525 1-1/4" diameter		9	2.667		5.20	125		130.20	195
4530 1-1/2" diameter		8.50	2.824		8.95	133		141.95	210
4535 2" diameter		8	3		10.45	141		151.45	223
4540 3" diameter		8	3		38	141		179	254
4545 4" diameter		6	4		74.50	188		262.50	365
4600 Tapping tees, type II, 3/4" punch, socket fusion outlets									
4601 Saddle contour x outlet diameter, IPS unless noted CTS									
4602 1-1/4" x 1/2" CTS outlet	B-20	17	1.412	Ea.	14.95	66.50		81.45	116
4604 1-1/4" x 1/2" outlet		17	1.412		14.95	66.50		81.45	116
4606 1-1/4" x 3/4" outlet		17	1.412		14.95	66.50		81.45	116
4608 1-1/4" x 1" CTS outlet		17	1.412		15.55	66.50		82.05	117
4610 1-1/4" x 1" outlet		17	1.412		15.55	66.50		82.05	117
4612 1-1/4" x 1-1/4" outlet		16	1.500		22	70.50		92.50	130
4614 1-1/2" x 1/2" CTS outlet		16	1.500		11.95	70.50		82.45	119
4616 1-1/2" x 1/2" outlet		16	1.500		11.95	70.50		82.45	119
4618 1-1/2" x 3/4" outlet		16	1.500		11.95	70.50		82.45	119
4620 1-1/2" x 1" CTS outlet		16	1.500		13.05	70.50		83.55	120
4622 1-1/2" x 1" outlet		16	1.500		13.05	70.50		83.55	120
4624 1-1/2" x 1-1/4" outlet		15	1.600		22	75		97	137
4626 2" x 1/2" CTS outlet		12	2		14	94		108	157
4628 2" x 1/2" outlet		12	2		14	94		108	157
4630 2" x 3/4" outlet		12	2		14	94		108	157
4632 2" x 1" CTS outlet		12	2		14.60	94		108.60	158
4634 2" x 1" outlet		12	2		14.60	94		108.60	158
4636 2" x 1-1/4" outlet		11	2.182		22.50	103		125.50	180
4638 3" x 1/2" CTS outlet		12	2		14.50	94		108.50	158
4640 3" x 1/2" outlet		12	2		14.50	94		108.50	158
4642 3" x 3/4" outlet		12	2		14.50	94		108.50	158
4644 3" x 1" CTS outlet		12	2		15.55	94		109.55	159
4646 3" x 1" outlet		12	2		15.10	94		109.10	159
4648 3" x 1-1/4" outlet		11	2.182		23.50	103		126.50	181
4650 4" x 1/2" CTS outlet		10	2.400		12.05	113		125.05	183
4652 4" x 1/2" outlet		10	2.400		12.05	113		125.05	183

For customer support on your Site Work & Landscape Costs with RSMeans data, call 800.448.8182.

33 52 16 – Gas Hydrocarbon Piping

33 52 16.23 Medium Density Polyethylene Piping	Crew	Daily Output	Labor-Hours	Unit	Material	2020 Bare Costs Labor	Equipment	Total	Total Incl O&P
4654	4" x 3/4" outlet	B-20	10	2.400	Ea.	12.05	113		125.05	183
4656	4" x 1" CTS outlet		10	2.400		13.15	113		126.15	184
4658	4" x 1" outlet		10	2.400		17.50	113		130.50	189
4660	4" x 1-1/4" outlet		9	2.667		25.50	125		150.50	218
4685	For protective sleeves, 1/2" CTS to 3/4" diameter outlets, add					.86			.86	.95
4690	1" CTS & IPS diameter outlets, add					1.45			1.45	1.60
4695	1-1/4" diameter outlets, add					2.49			2.49	2.74
4700	Tapping tees, high volume, socket fusion outlets									
4705	1-1/2" punch, 2" x 1-1/4" outlet	B-20	11	2.182	Ea.	111	103		214	277
4710	1-7/8" punch, 3" x 1-1/4" outlet		11	2.182		111	103		214	277
4715	4" x 1-1/4" outlet		9	2.667		111	125		236	310
4750	For protective sleeves, add					2.92			2.92	3.21
4800	Service saddles, saddle contour x outlet, socket fusion outlets									
4801	IPS unless noted CTS									
4802	1-1/4" x 1/2" CTS outlet	B-20	17	1.412	Ea.	3.73	66.50		70.23	104
4804	1-1/4" x 1/2" outlet		17	1.412		3.73	66.50		70.23	104
4806	1-1/4" x 3/4" outlet		17	1.412		3.73	66.50		70.23	104
4808	1-1/4" x 1" CTS outlet		17	1.412		6.15	66.50		72.65	107
4810	1-1/4" x 1" outlet		17	1.412		6.15	66.50		72.65	107
4812	1-1/4" x 1-1/4" outlet		16	1.500		9.80	70.50		80.30	117
4814	1-1/2" x 1/2" CTS outlet		16	1.500		3.70	70.50		74.20	110
4816	1-1/2" x 1/2" outlet		16	1.500		3.70	70.50		74.20	110
4818	1-1/2" x 3/4" outlet		16	1.500		3.70	70.50		74.20	110
4820	1-1/2" x 1-1/4" outlet		15	1.600		9.80	75		84.80	124
4822	2" x 1/2" CTS outlet		12	2		3.73	94		97.73	146
4824	2" x 1/2" outlet		12	2		3.73	94		97.73	146
4826	2" x 3/4" outlet		12	2		3.73	94		97.73	146
4828	2" x 1" CTS outlet		12	2		6.15	94		100.15	149
4830	2" x 1" outlet		12	2		6.15	94		100.15	149
4832	2" x 1-1/4" outlet		11	2.182		9.80	103		112.80	166
4834	3" x 1/2" CTS outlet		12	2		3.70	94		97.70	146
4836	3" x 1/2" outlet		12	2		3.70	94		97.70	146
4838	3" x 3/4" outlet		12	2		3.73	94		97.73	146
4840	3" x 1" CTS outlet		12	2		6.10	94		100.10	149
4842	3" x 1" outlet		12	2		6.15	94		100.15	149
4844	3" x 1-1/4" outlet		11	2.182		9.90	103		112.90	166
4846	4" x 1/2" CTS outlet		10	2.400		3.70	113		116.70	174
4848	4" x 1/2" outlet		10	2.400		3.70	113		116.70	174
4850	4" x 3/4" outlet		10	2.400		3.73	113		116.73	174
4852	4" x 1" CTS outlet		10	2.400		6.10	113		119.10	177
4854	4" x 1" outlet		10	2.400		6.15	113		119.15	177
4856	4" x 1-1/4" outlet		9	2.667		9.90	125		134.90	200
4885	For protective sleeves, 1/2" CTS to 3/4" diameter outlets, add					.86			.86	.95
4890	1" CTS & IPS diameter outlets, add					1.45			1.45	1.60
4895	1-1/4" diameter outlets, add					2.49			2.49	2.74
4900	Spigot fittings, tees, SDR 7, 1/2" CTS diameter	B-20	10	2.400		25.50	113		138.50	198
4901	SDR 9.3, 1/2" diameter		9.50	2.526		11.30	119		130.30	191
4902	SDR 10, 1-1/4" diameter		9	2.667		11.30	125		136.30	201
4903	SDR 11, 3/4" diameter		9.50	2.526		23.50	119		142.50	205
4904	2" diameter		8	3		21.50	141		162.50	236
4906	SDR 11.5, 1" CTS diameter		9.33	2.571		60	121		181	248
4907	3" diameter		8	3		61	141		202	279
4908	4" diameter		6	4		157	188		345	455

33 52 Hydrocarbon Transmission and Distribution

33 52 16 – Gas Hydrocarbon Piping

33 52 16.23 Medium Density Polyethylene Piping	Crew	Daily Output	Labor-Hours	Unit	Material	2020 Bare Costs Labor	Equipment	Total	Total Incl O&P
4921	90° elbows, SDR 7, 1/2" CTS diameter	B-20	15	1.600	Ea.	13.45	75		88.45	128
4922	SDR 9.3, 1/2" diameter		14.25	1.684		5.80	79		84.80	125
4923	SDR 10, 1-1/4" diameter		13.50	1.778		5.90	83.50		89.40	133
4924	SDR 11, 3/4" diameter		14.25	1.684		6.60	79		85.60	126
4925	2" diameter		12	2		25	94		119	170
4927	SDR 11.5, 1" CTS diameter		14	1.714		33	80.50		113.50	158
4928	3" diameter		12	2		56.50	94		150.50	204
4929	4" diameter		9	2.667		121	125		246	320
4935	Caps, SDR 7, 1/2" CTS diameter		30	.800		4.46	37.50		41.96	61.50
4936	SDR 9.3, 1/2" diameter		28.50	.842		4.46	39.50		43.96	64.50
4937	SDR 10, 1-1/4" diameter		27	.889		4.65	42		46.65	68
4938	SDR 11, 3/4" diameter		28.50	.842		4.77	39.50		44.27	65
4939	1" diameter		27	.889		5.30	42		47.30	69
4940	2" diameter		24	1		5.70	47		52.70	77.50
4942	SDR 11.5, 1" CTS diameter		28	.857		18.70	40.50		59.20	81
4943	3" diameter		24	1		30.50	47		77.50	105
4944	4" diameter		18	1.333		29	62.50		91.50	127
4946	SDR 13.5, 4" diameter		18	1.333		73.50	62.50		136	175
4950	Reducers, SDR 10 x SDR 11, 1-1/4" x 3/4" diameters		13.50	1.778		15.25	83.50		98.75	143
4951	1-1/4" x 1" diameters		13.50	1.778		15.25	83.50		98.75	143
4952	SDR 10 x SDR 11.5, 1-1/4" x 1" CTS diameters		13.50	1.778		15.25	83.50		98.75	143
4953	SDR 11 x SDR 7, 3/4" x 1/2" CTS diameters		14.25	1.684		6.80	79		85.80	127
4954	SDR 11 x SDR 9.3, 3/4" x 1/2" diameters		14.25	1.684		6.80	79		85.80	127
4955	SDR 11 x SDR 10, 2" x 1-1/4" diameters		12	2		15.80	94		109.80	159
4956	SDR 11 x SDR 11, 1" x 3/4" diameters		13.50	1.778		7.85	83.50		91.35	135
4957	2" x 3/4" diameters		12	2		15.60	94		109.60	159
4958	2" x 1" diameters		12	2		15.70	94		109.70	159
4959	SDR 11 x SDR 11.5, 1" x 1" CTS diameters		13.50	1.778		8	83.50		91.50	135
4960	2" x 1" CTS diameters		12	2		15.60	94		109.60	159
4962	SDR 11.5 x SDR 11, 3" x 2" diameters		12	2		21	94		115	165
4963	4" x 2" diameters		9	2.667		32	125		157	224
4964	SDR 11.5 x SDR 11.5, 4" x 3" diameters		9	2.667		34.50	125		159.50	227
6100	Fittings, compression									
6101	MDPE gas pipe, ASTM D2513/ASTM F1924-98									
6102	Caps, SDR 7, 1/2" CTS diameter	B-20	60	.400	Ea.	12.85	18.80		31.65	42.50
6104	SDR 9.3, 1/2" IPS diameter		58	.414		23	19.45		42.45	55
6106	SDR 10, 1/2" CTS diameter		60	.400		11.35	18.80		30.15	41
6108	1-1/4" IPS diameter		54	.444		74.50	21		95.50	114
6110	SDR 11, 3/4" IPS diameter		58	.414		20	19.45		39.45	51.50
6112	1" IPS diameter		54	.444		32.50	21		53.50	67
6114	1-1/4" IPS diameter		54	.444		78	21		99	118
6116	1-1/2" IPS diameter		52	.462		98.50	21.50		120	141
6118	2" IPS diameter		50	.480		86	22.50		108.50	129
6120	SDR 11.5, 1" CTS diameter		56	.429		21	20		41	53.50
6122	SDR 12.5, 1" CTS diameter		56	.429		24	20		44	57
6202	Reducers, SDR 7 x SDR 10, 1/2" CTS x 1/2" CTS diameters		30	.800		17.35	37.50		54.85	75.50
6204	SDR 9.3 x SDR 7, 1/2" IPS x 1/2" CTS diameters		29	.828		51.50	39		90.50	115
6206	SDR 11 x SDR 7, 3/4" IPS x 1/2" CTS diameters		29	.828		51.50	39		90.50	115
6208	1" IPS x 1/2" CTS diameters		27	.889		60	42		102	129
6210	SDR 11 x SDR 9.3, 3/4" IPS x 1/2" IPS diameters		29	.828		50	39		89	114
6212	1" IPS x 1/2" IPS diameters		27	.889		64.50	42		106.50	134
6214	SDR 11 x SDR 10, 2" IPS x 1-1/4" IPS diameters		25	.960		95.50	45		140.50	173
6216	SDR 11 x SDR 11, 1" IPS x 3/4" IPS diameters		27	.889		60	42		102	129

518

33 52 Hydrocarbon Transmission and Distribution

33 52 16 – Gas Hydrocarbon Piping

33 52 16.23 Medium Density Polyethylene Piping	Crew	Daily Output	Labor-Hours	Unit	Material	2020 Bare Costs Labor	Equipment	Total	Total Incl O&P
6218	1-1/4" IPS x 1" IPS diameters	B-20	27	.889	Ea.	58.50	42		100.50	128
6220	2" IPS x 1-1/4" IPS diameters		25	.960		95.50	45		140.50	173
6222	SDR 11 x SDR 11.5, 1" IPS x 1" CTS diameters		27	.889		46	42		88	114
6224	SDR 11 x SDR 12.5, 1" IPS x 1" CTS diameters		27	.889		56	42		98	125
6226	SDR 11.5 x SDR 7, 1" CTS x 1/2" CTS diameters		28	.857		34	40.50		74.50	98
6228	SDR 11.5 x SDR 9.3, 1" CTS x 1/2" IPS diameters		28	.857		39.50	40.50		80	104
6230	SDR 11.5 x SDR 11, 1" CTS x 3/4" IPS diameters		28	.857		55	40.50		95.50	121
6232	SDR 12.5 x SDR 7, 1" CTS x 1/2" CTS diameters		28	.857		36	40.50		76.50	100
6234	SDR 12.5 x SDR 9.3, 1" CTS x 1/2" IPS diameters		28	.857		40.50	40.50		81	105
6236	SDR 12.5 x SDR 11, 1" CTS x 3/4" IPS diameters		28	.857		34.50	40.50		75	98.50
6302	Couplings, SDR 7, 1/2" CTS diameter		30	.800		13.10	37.50		50.60	71
6304	SDR 9.3, 1/2" IPS diameter		29	.828		23.50	39		62.50	84.50
6306	SDR 10, 1/2" CTS diameter		30	.800		12.40	37.50		49.90	70
6308	SDR 11, 3/4" IPS diameter		29	.828		24	39		63	84.50
6310	1" IPS diameter		27	.889		32.50	42		74.50	99
6312	SDR 11.5, 1" CTS diameter		28	.857		25	40.50		65.50	88.50
6314	SDR 12.5, 1" CTS diameter		28	.857		28	40.50		68.50	91
6402	Repair couplings, SDR 10, 1-1/4" IPS diameter		27	.889		71.50	42		113.50	142
6404	SDR 11, 1-1/4" IPS diameter		27	.889		76	42		118	147
6406	1-1/2" IPS diameter		26	.923		81.50	43.50		125	155
6408	2" IPS diameter		25	.960		95	45		140	172
6502	Elbows, 90°, SDR 7, 1/2" CTS diameter		30	.800		22.50	37.50		60	81.50
6504	SDR 9.3, 1/2" IPS diameter		29	.828		41.50	39		80.50	104
6506	SDR 10, 1-1/4" IPS diameter		27	.889		102	42		144	175
6508	SDR 11, 3/4" IPS diameter		29	.828		37.50	39		76.50	99.50
6510	1" IPS diameter		27	.889		35	42		77	102
6512	1-1/4" IPS diameter		27	.889		101	42		143	174
6514	1-1/2" IPS diameter		26	.923		151	43.50		194.50	232
6516	2" IPS diameter		25	.960		124	45		169	205
6518	SDR 11.5, 1" CTS diameter		28	.857		37	40.50		77.50	101
6520	SDR 12.5, 1" CTS diameter		28	.857		53.50	40.50		94	120
6602	Tees, SDR 7, 1/2" CTS diameter		20	1.200		42	56.50		98.50	131
6604	SDR 9.3, 1/2" IPS diameter		19.33	1.241		46	58.50		104.50	139
6606	SDR 10, 1-1/4" IPS diameter		18	1.333		127	62.50		189.50	235
6608	SDR 11, 3/4" IPS diameter		19.33	1.241		58.50	58.50		117	153
6610	1" IPS diameter		18	1.333		64.50	62.50		127	165
6612	1-1/4" IPS diameter		18	1.333		107	62.50		169.50	213
6614	1-1/2" IPS diameter		17.33	1.385		224	65		289	345
6616	2" IPS diameter		16.67	1.440		143	67.50		210.50	260
6618	SDR 11.5, 1" CTS diameter		18.67	1.286		59	60.50		119.50	156
6620	SDR 12.5, 1" CTS diameter		18.67	1.286		59	60.50		119.50	156
8100	Fittings, accessories									
8105	SDR 11, IPS unless noted CTS									
8110	Protective sleeves, for high volume tapping tees, butt fusion outlets	1 Skwk	16	.500	Ea.	5.45	27.50		32.95	47.50
8120	For tapping tees, socket outlets, CTS, 1/2" diameter		18	.444		1.42	24.50		25.92	38.50
8122	1" diameter		18	.444		2.51	24.50		27.01	40
8124	IPS, 1/2" diameter		18	.444		1.96	24.50		26.46	39
8126	3/4" diameter		18	.444		2.51	24.50		27.01	40
8128	1" diameter		18	.444		3.42	24.50		27.92	41
8130	1-1/4" diameter		18	.444		4.31	24.50		28.81	41.50
8205	Tapping tee test caps, type I, yellow PE cap, "aldyl style"		45	.178		44	9.75		53.75	63.50
8210	Type II, yellow polyethylene cap		45	.178		44	9.75		53.75	63.50
8215	High volume, yellow polyethylene cap		45	.178		56.50	9.75		66.25	77.50

For customer support on your Site Work & Landscape Costs with RSMeans data, call 800.448.8182.

33 52 16.23 Medium Density Polyethylene Piping

		Crew	Daily Output	Labor-Hours	Unit	Material	2020 Bare Costs Labor	Equipment	Total	Total Incl O&P
8250	Quick connector, female x female inlets, 1/4" N.P.T.	1 Skwk	50	.160	Ea.	18.45	8.80		27.25	34
8255	Test hose, 24" length, 3/8" ID, male outlets, 1/4" N.P.T.		50	.160		25	8.80		33.80	41
8260	Quick connector and test hose assembly		50	.160		43.50	8.80		52.30	61.50
8305	Purge point caps, butt fusion, SDR 10, 1-1/4" diameter	B-22C	27	.593		32	28.50	5.80	66.30	85
8310	SDR 11, 1-1/4" diameter		27	.593		32	28.50	5.80	66.30	85
8315	2" diameter		24	.667		32.50	32.50	6.50	71.50	91
8320	3" diameter		24	.667		46	32.50	6.50	85	106
8325	4" diameter		18	.889		45	43	8.65	96.65	124
8340	Socket fusion, SDR 11, 1-1/4" diameter		27	.593		8.20	28.50	5.80	42.50	59
8345	2" diameter		24	.667		9.85	32.50	6.50	48.85	66.50
8350	3" diameter		24	.667		60.50	32.50	6.50	99.50	122
8355	4" diameter		18	.889		44.50	43	8.65	96.15	124
8360	Purge test quick connector, female x female inlets, 1/4" N.P.T.	1 Skwk	50	.160		18.45	8.80		27.25	34
8365	Purge test hose, 24" length, 3/8" ID, male outlets, 1/4" N.P.T.		50	.160		25	8.80		33.80	41
8370	Purge test quick connector and test hose assembly		50	.160		43.50	8.80		52.30	61.50
8405	Transition fittings, MDPE x zinc plated steel, SDR 7, 1/2" CTS x 1/2" MPT	B-22C	30	.533		25	26	5.20	56.20	72
8410	SDR 9.3, 1/2" IPS x 3/4" MPT		28.50	.561		39.50	27	5.45	71.95	90.50
8415	SDR 10, 1-1/4" IPS x 1-1/4" MPT		27	.593		37.50	28.50	5.80	71.80	91
8420	SDR 11, 3/4" IPS x 3/4" MPT		28.50	.561		20	27	5.45	52.45	69
8425	1" IPS x 1" MPT		27	.593		24.50	28.50	5.80	58.80	77
8430	1-1/4" IPS x 1-1/4" MPT		17	.941		46	45.50	9.20	100.70	130
8435	1-1/2" IPS x 1-1/2" MPT		25.50	.627		47	30.50	6.10	83.60	104
8440	2" IPS x 2" MPT		24	.667		53.50	32.50	6.50	92.50	115
8445	SDR 11.5, 1" CTS x 1" MPT		28	.571		33	27.50	5.55	66.05	84.50

33 52 16.26 High Density Polyethylene Piping

		Crew	Daily Output	Labor-Hours	Unit	Material	2020 Bare Costs Labor	Equipment	Total	Total Incl O&P
2010	**HIGH DENSITY POLYETHYLENE PIPING**									
2020	ASTM D2513, not including excavation or backfill									
2200	Butt fused pipe									
2205	125psi coils, butt fusion joint @ 100', 1/2" CTS diameter, SDR 7	B-22C	2050	.008	L.F.	.13	.38	.08	.59	.79
2210	160psi coils, butt fusion joint @ 100', IPS 1/2" diameter, SDR 9		1950	.008		.18	.40	.08	.66	.89
2215	3/4" diameter, SDR 11		1950	.008		.26	.40	.08	.74	.98
2220	1" diameter, SDR 11		1850	.009		.42	.42	.08	.92	1.18
2225	1-1/4" diameter, SDR 11		1850	.009		.62	.42	.08	1.12	1.40
2230	1-1/2" diameter, SDR 11		1750	.009		1.17	.44	.09	1.70	2.06
2235	2" diameter, SDR 11		1750	.009		.94	.44	.09	1.47	1.80
2240	3" diameter, SDR 11		1650	.010		1.55	.47	.09	2.11	2.52
2245	160psi 40' lengths, butt fusion joint, IPS 3" diameter, SDR 11	B-22A	660	.061		1.69	2.94	1.22	5.85	7.60
2250	4" diameter, SDR 11		500	.080		2.78	3.88	1.61	8.27	10.70
2255	6" diameter, SDR 11		500	.080		6.05	3.88	1.61	11.54	14.25
2260	8" diameter, SDR 11		420	.095		10.20	4.62	1.92	16.74	20.50
2265	10" diameter, SDR 11		340	.118		24	5.70	2.37	32.07	37
2270	12" diameter, SDR 11		300	.133		33	6.45	2.68	42.13	49
2400	Socket fused pipe									
2405	125psi coils, socket fusion coupling @ 100', 1/2" CTS diameter, SDR 7	B-20	1050	.023	L.F.	.15	1.07		1.22	1.79
2410	160psi coils, socket fusion coupling @ 100', IPS 1/2" diameter, SDR 9		1000	.024		.20	1.13		1.33	1.92
2415	3/4" diameter, SDR 11		1000	.024		.28	1.13		1.41	2
2420	1" diameter, SDR 11		950	.025		.44	1.19		1.63	2.27
2425	1-1/4" diameter, SDR 11		950	.025		.64	1.19		1.83	2.50
2430	1-1/2" diameter, SDR 11		900	.027		1.19	1.25		2.44	3.20
2435	2" diameter, SDR 11		850	.028		.97	1.33		2.30	3.06
2440	3" diameter, SDR 11		850	.028		1.66	1.33		2.99	3.83
2445	160psi 40' lengths, socket fusion coupling, IPS 3" diameter, SDR 11	B-21A	340	.118		1.98	6.15	1.26	9.39	12.75

33 52 16 – Gas Hydrocarbon Piping

33 52 16.26 High Density Polyethylene Piping	Crew	Daily Output	Labor-Hours	Unit	Material	2020 Bare Costs Labor	Equipment	Total	Total Incl O&P
2450	4" diameter, SDR 11	B-21A	260	.154	L.F.	3.34	8.05	1.65	13.04	17.55
2600	Compression coupled pipe									
2610	160psi coils, compression coupling @ 100', IPS 1/2" diameter, SDR 9	B-20	1000	.024	L.F.	.42	1.13		1.55	2.16
2615	3/4" diameter, SDR 11		1000	.024		.50	1.13		1.63	2.25
2620	1" diameter, SDR 11		950	.025		.75	1.19		1.94	2.61
2625	1-1/4" diameter, SDR 11		950	.025		1.38	1.19		2.57	3.31
2630	1-1/2" diameter, SDR 11		900	.027		1.99	1.25		3.24	4.07
2635	2" diameter, SDR 11		850	.028		1.89	1.33		3.22	4.08
3000	Fittings, butt fusion									
3010	SDR 11, IPS unless noted CTS									
3105	Caps, 1/2" diameter	B-22C	28.50	.561	Ea.	6.60	27	5.45	39.05	54.50
3110	3/4" diameter		28.50	.561		3.53	27	5.45	35.98	51
3115	1" diameter		27	.593		3.62	28.50	5.80	37.92	54
3120	1-1/4" diameter		27	.593		4.79	28.50	5.80	39.09	55
3125	1-1/2" diameter		25.50	.627		4.42	30.50	6.10	41.02	57.50
3130	2" diameter		24	.667		4.73	32.50	6.50	43.73	61
3135	3" diameter		24	.667		7.15	32.50	6.50	46.15	63.50
3140	4" diameter		18	.889		15.05	43	8.65	66.70	91
3145	6" diameter	B-22A	18	2.222		38	108	44.50	190.50	253
3150	8" diameter		15	2.667		62.50	129	53.50	245	320
3155	10" diameter		12	3.333		340	162	67	569	690
3160	12" diameter		10.50	3.810		365	185	76.50	626.50	760
3205	Reducers, 1/2" x 1/2" CTS diameters	B-22C	14.25	1.123		12.25	54.50	10.95	77.70	108
3210	3/4" x 1/2" CTS diameters		14.25	1.123		8.80	54.50	10.95	74.25	104
3215	1" x 1/2" CTS diameters		13.50	1.185		13.65	57.50	11.55	82.70	114
3220	1" x 1/2" diameters		13.50	1.185		12.25	57.50	11.55	81.30	113
3225	1" x 3/4" diameters		13.50	1.185		10.65	57.50	11.55	79.70	111
3230	1-1/4" x 1" diameters		13.50	1.185		10.80	57.50	11.55	79.85	111
3232	1-1/2" x 3/4" diameters		12.75	1.255		7.80	61	12.25	81.05	114
3233	1-1/2" x 1" diameters		12.75	1.255		7.80	61	12.25	81.05	114
3234	1-1/2" x 1-1/4" diameters		12.75	1.255		7.80	61	12.25	81.05	114
3235	2" x 1" diameters		12	1.333		8.60	64.50	13	86.10	121
3240	2" x 1-1/4" diameters		12	1.333		8.20	64.50	13	85.70	121
3245	2" x 1-1/2" diameters		12	1.333		9.05	64.50	13	86.55	122
3250	3" x 2" diameters		12	1.333		10.40	64.50	13	87.90	123
3255	4" x 2" diameters		9	1.778		14.40	86	17.35	117.75	165
3260	4" x 3" diameters		9	1.778		15.40	86	17.35	118.75	166
3262	6" x 3" diameters	B-22A	9	4.444		35.50	215	89.50	340	465
3265	6" x 4" diameters		9	4.444		48.50	215	89.50	353	475
3270	8" x 6" diameters		7.50	5.333		81	259	107	447	595
3275	10" x 8" diameters		6	6.667		174	325	134	633	825
3280	12" x 8" diameters		5.25	7.619		257	370	153	780	1,000
3285	12" x 10" diameters		5.25	7.619		120	370	153	643	855
3305	Elbows, 90°, 3/4" diameter	B-22C	14.25	1.123		5.85	54.50	10.95	71.30	101
3310	1" diameter		13.50	1.185		5.95	57.50	11.55	75	106
3315	1-1/4" diameter		13.50	1.185		6.60	57.50	11.55	75.65	106
3320	1-1/2" diameter		12.75	1.255		7.90	61	12.25	81.15	114
3325	2" diameter		12	1.333		8.25	64.50	13	85.75	121
3330	3" diameter		12	1.333		16.25	64.50	13	93.75	130
3335	4" diameter		9	1.778		20.50	86	17.35	123.85	172
3340	6" diameter	B-22A	9	4.444		55	215	89.50	359.50	485
3345	8" diameter		7.50	5.333		164	259	107	530	690
3350	10" diameter		6	6.667		410	325	134	869	1,075

For customer support on your Site Work & Landscape Costs with RSMeans data, call 800.448.8182.

521

33 52 16.26 High Density Polyethylene Piping	Crew	Daily Output	Labor-Hours	Unit	Material	2020 Bare Costs Labor	Equipment	Total	Total Incl O&P
3355	12" diameter	B-22A	5.25	7.619	Ea.	380	370	153	903	1,150
3380	45°, 3/4" diameter	B-22C	14.25	1.123		6.70	54.50	10.95	72.15	101
3385	1" diameter		13.50	1.185		6.70	57.50	11.55	75.75	107
3390	1-1/4" diameter		13.50	1.185		6.90	57.50	11.55	75.95	107
3395	1-1/2" diameter		12.75	1.255		9	61	12.25	82.25	115
3400	2" diameter		12	1.333		9.95	64.50	13	87.45	123
3405	3" diameter		12	1.333		16.45	64.50	13	93.95	130
3410	4" diameter		9	1.778		20.50	86	17.35	123.85	172
3415	6" diameter	B-22A	9	4.444		55	215	89.50	359.50	485
3420	8" diameter		7.50	5.333		166	259	107	532	690
3425	10" diameter		6	6.667		390	325	134	849	1,075
3430	12" diameter		5.25	7.619		760	370	153	1,283	1,550
3505	Tees, 1/2" CTS diameter	B-22C	10	1.600		8.05	77.50	15.60	101.15	143
3510	1/2" diameter		9.50	1.684		8.90	81.50	16.40	106.80	151
3515	3/4" diameter		9.50	1.684		6.70	81.50	16.40	104.60	148
3520	1" diameter		9	1.778		7.25	86	17.35	110.60	157
3525	1-1/4" diameter		9	1.778		7.35	86	17.35	110.70	157
3530	1-1/2" diameter		8.50	1.882		12.65	91.50	18.35	122.50	172
3535	2" diameter		8	2		10	97	19.50	126.50	179
3540	3" diameter		8	2		17.85	97	19.50	134.35	187
3545	4" diameter		6	2.667		27	129	26	182	253
3550	6" diameter	B-22A	6	6.667		66	325	134	525	705
3555	8" diameter		5	8		264	390	161	815	1,050
3560	10" diameter		4	10		505	485	201	1,191	1,500
3565	12" diameter		3.50	11.429		670	555	230	1,455	1,825
3600	Tapping tees, type II, 3/4" punch, butt fusion outlets									
3601	Saddle contour x outlet diameter, IPS unless noted CTS									
3602	1-1/4" x 1/2" CTS outlet	B-22C	17	.941	Ea.	11.20	45.50	9.20	65.90	91.50
3604	1-1/4" x 1/2" outlet		17	.941		11.20	45.50	9.20	65.90	91.50
3606	1-1/4" x 3/4" outlet		17	.941		11.20	45.50	9.20	65.90	91.50
3608	1-1/4" x 1" outlet		17	.941		13.05	45.50	9.20	67.75	93.50
3610	1-1/4" x 1-1/4" outlet		16	1		23.50	48.50	9.75	81.75	109
3612	1-1/2" x 1/2" CTS outlet		16	1		11.20	48.50	9.75	69.45	96
3614	1-1/2" x 1/2" outlet		16	1		11.20	48.50	9.75	69.45	96
3616	1-1/2" x 3/4" outlet		16	1		11.20	48.50	9.75	69.45	96
3618	1-1/2" x 1" outlet		16	1		13.05	48.50	9.75	71.30	98
3620	1-1/2" x 1-1/4" outlet		15	1.067		23.50	51.50	10.40	85.40	115
3622	2" x 1/2" CTS outlet		12	1.333		11.20	64.50	13	88.70	124
3624	2" x 1/2" outlet		12	1.333		11.20	64.50	13	88.70	124
3626	2" x 3/4" outlet		12	1.333		11.20	64.50	13	88.70	124
3628	2" x 1" outlet		12	1.333		13.05	64.50	13	90.55	126
3630	2" x 1-1/4" outlet		11	1.455		23.50	70.50	14.20	108.20	147
3632	3" x 1/2" CTS outlet		12	1.333		11.20	64.50	13	88.70	124
3634	3" x 1/2" outlet		12	1.333		11.20	64.50	13	88.70	124
3636	3" x 3/4" outlet		12	1.333		11.20	64.50	13	88.70	124
3638	3" x 1" outlet		12	1.333		13.15	64.50	13	90.65	126
3640	3" x 1-1/4" outlet		11	1.455		23.50	70.50	14.20	108.20	148
3642	4" x 1/2" CTS outlet		10	1.600		11.20	77.50	15.60	104.30	146
3644	4" x 1/2" outlet		10	1.600		11.20	77.50	15.60	104.30	146
3646	4" x 3/4" outlet		10	1.600		11.20	77.50	15.60	104.30	146
3648	4" x 1" outlet		10	1.600		13.05	77.50	15.60	106.15	149
3650	4" x 1-1/4" outlet		9	1.778		23.50	86	17.35	126.85	175
3652	6" x 1/2" CTS outlet	B-22A	10	4		11.20	194	80.50	285.70	390

33 52 16.26 **High Density Polyethylene Piping**		Crew	Daily Output	Labor-Hours	Unit	Material	2020 Bare Costs Labor	Equipment	Total	Total Incl O&P
3654	6" x 1/2" outlet	B-22A	10	4	Ea.	11.20	194	80.50	285.70	390
3656	6" x 3/4" outlet		10	4		11.20	194	80.50	285.70	390
3658	6" x 1" outlet		10	4		13.05	194	80.50	287.55	395
3660	6" x 1-1/4" outlet		9	4.444		23.50	215	89.50	328	450
3662	8" x 1/2" CTS outlet		8	5		11.20	242	101	354.20	490
3664	8" x 1/2" outlet		8	5		11.20	242	101	354.20	490
3666	8" x 3/4" outlet		8	5		11.20	242	101	354.20	490
3668	8" x 1" outlet		8	5		13.05	242	101	356.05	490
3670	8" x 1-1/4" outlet		7	5.714		23.50	277	115	415.50	570
3675	For protective sleeves, 1/2" to 3/4" diameter outlets, add					.86			.86	.95
3680	1" diameter outlets, add					1.45			1.45	1.60
3685	1-1/4" diameter outlets, add					2.49			2.49	2.74
3700	Tapping tees, high volume, butt fusion outlets									
3705	1-1/2" punch, 2" x 2" outlet	B-22C	11	1.455	Ea.	110	70.50	14.20	194.70	243
3710	1-7/8" punch, 3" x 2" outlet		11	1.455		110	70.50	14.20	194.70	243
3715	4" x 2" outlet		9	1.778		110	86	17.35	213.35	270
3720	6" x 2" outlet	B-22A	9	4.444		110	215	89.50	414.50	545
3725	8" x 2" outlet		7	5.714		110	277	115	502	665
3730	10" x 2" outlet		6	6.667		110	325	134	569	755
3735	12" x 2" outlet		5	8		110	390	161	661	885
3750	For protective sleeves, add					2.92			2.92	3.21
3800	Service saddles, saddle contour x outlet, butt fusion outlets									
3801	IPS unless noted CTS									
3802	1-1/4" x 3/4" outlet	B-22C	17	.941	Ea.	6.65	45.50	9.20	61.35	86.50
3804	1-1/4" x 1" outlet		16	1		6.65	48.50	9.75	64.90	91
3806	1-1/4" x 1-1/4" outlet		16	1		6.95	48.50	9.75	65.20	91.50
3808	1-1/2" x 3/4" outlet		16	1		6.65	48.50	9.75	64.90	91
3810	1-1/2" x 1-1/4" outlet		15	1.067		6.95	51.50	10.40	68.85	97
3812	2" x 3/4" outlet		12	1.333		6.65	64.50	13	84.15	119
3814	2" x 1" outlet		11	1.455		6.65	70.50	14.20	91.35	129
3816	2" x 1-1/4" outlet		11	1.455		6.95	70.50	14.20	91.65	129
3818	3" x 3/4" outlet		12	1.333		6.65	64.50	13	84.15	119
3820	3" x 1" outlet		11	1.455		6.65	70.50	14.20	91.35	129
3822	3" x 1-1/4" outlet		11	1.455		6.95	70.50	14.20	91.65	129
3824	4" x 3/4" outlet		10	1.600		6.65	77.50	15.60	99.75	141
3826	4" x 1" outlet		9	1.778		6.65	86	17.35	110	156
3828	4" x 1-1/4" outlet		9	1.778		6.95	86	17.35	110.30	157
3830	6" x 3/4" outlet	B-22A	10	4		6.65	194	80.50	281.15	385
3832	6" x 1" outlet		9	4.444		6.65	215	89.50	311.15	430
3834	6" x 1-1/4" outlet		9	4.444		6.95	215	89.50	311.45	430
3836	8" x 3/4" outlet		8	5		6.65	242	101	349.65	485
3838	8" x 1" outlet		7	5.714		6.65	277	115	398.65	550
3840	8" x 1-1/4" outlet		7	5.714		6.95	277	115	398.95	550
3880	For protective sleeves, 3/4" diameter outlets, add					.86			.86	.95
3885	1" diameter outlets, add					1.45			1.45	1.60
3890	1-1/4" diameter outlets, add					2.49			2.49	2.74
3905	Branch saddles, contour x outlet, 2" x 2" outlet	B-22C	11	1.455		18.75	70.50	14.20	103.45	142
3910	3" x 2" outlet		11	1.455		18.75	70.50	14.20	103.45	142
3915	3" x 3" outlet		11	1.455		25	70.50	14.20	109.70	149
3920	4" x 2" outlet		9	1.778		18.75	86	17.35	122.10	170
3925	4" x 3" outlet		9	1.778		25	86	17.35	128.35	177
3930	4" x 4" outlet		9	1.778		34.50	86	17.35	137.85	187
3935	6" x 2" outlet	B-22A	9	4.444		18.75	215	89.50	323.25	445

33 52 16.26 High Density Polyethylene Piping	Crew	Daily Output	Labor-Hours	Unit	Material	2020 Bare Costs Labor	Equipment	Total	Total Incl O&P
3940	6" x 3" outlet	B-22A	9	4.444	Ea.	25	215	89.50	329.50	450
3945	6" x 4" outlet		9	4.444		34.50	215	89.50	339	460
3950	6" x 6" outlet		9	4.444		82	215	89.50	386.50	515
3955	8" x 2" outlet		7	5.714		18.75	277	115	410.75	565
3960	8" x 3" outlet		7	5.714		25	277	115	417	570
3965	8" x 4" outlet		7	5.714		34.50	277	115	426.50	580
3970	8" x 6" outlet		7	5.714		82	277	115	474	635
3980	Ball valves, full port, 3/4" diameter	B-22C	14.25	1.123		59	54.50	10.95	124.45	159
3982	1" diameter		13.50	1.185		59	57.50	11.55	128.05	164
3984	1-1/4" diameter		13.50	1.185		59.50	57.50	11.55	128.55	165
3986	1-1/2" diameter		12.75	1.255		86.50	61	12.25	159.75	200
3988	2" diameter		12	1.333		135	64.50	13	212.50	261
3990	3" diameter		12	1.333		320	64.50	13	397.50	460
3992	4" diameter		9	1.778		420	86	17.35	523.35	610
3994	6" diameter	B-22A	9	4.444		1,125	215	89.50	1,429.50	1,650
3996	8" diameter	"	7.50	5.333		1,800	259	107	2,166	2,500
4000	Fittings, socket fusion									
4010	SDR 11, IPS unless noted CTS									
4105	Caps, 1/2" CTS diameter	B-20	30	.800	Ea.	2.39	37.50		39.89	59
4110	1/2" diameter		28.50	.842		2.44	39.50		41.94	62
4115	3/4" diameter		28.50	.842		2.60	39.50		42.10	62.50
4120	1" diameter		27	.889		3.38	42		45.38	66.50
4125	1-1/4" diameter		27	.889		3.53	42		45.53	67
4130	1-1/2" diameter		25.50	.941		3.91	44.50		48.41	71
4135	2" diameter		24	1		4.01	47		51.01	75.50
4140	3" diameter		24	1		24.50	47		71.50	98
4145	4" diameter		18	1.333		41	62.50		103.50	140
4205	Reducers, 1/2" x 1/2" CTS diameters		14.25	1.684		6.05	79		85.05	126
4210	3/4" x 1/2" CTS diameters		14.25	1.684		6.15	79		85.15	126
4215	3/4" x 1/2" diameters		14.25	1.684		6.15	79		85.15	126
4220	1" x 1/2" CTS diameters		13.50	1.778		6	83.50		89.50	133
4225	1" x 1/2" diameters		13.50	1.778		6.35	83.50		89.85	133
4230	1" x 3/4" diameters		13.50	1.778		3.99	83.50		87.49	130
4235	1-1/4" x 1/2" CTS diameters		13.50	1.778		6	83.50		89.50	133
4240	1-1/4" x 3/4" diameters		13.50	1.778		6.60	83.50		90.10	133
4245	1-1/4" x 1" diameters		13.50	1.778		5.80	83.50		89.30	132
4250	1-1/2" x 3/4" diameters		12.75	1.882		8.30	88.50		96.80	142
4252	1-1/2" x 1" diameters		12.75	1.882		8.25	88.50		96.75	142
4255	1-1/2" x 1-1/4" diameters		12.75	1.882		8.45	88.50		96.95	142
4260	2" x 3/4" diameters		12	2		8.85	94		102.85	152
4265	2" x 1" diameters		12	2		9	94		103	152
4270	2" x 1-1/4" diameters		12	2		9.10	94		103.10	152
4275	3" x 2" diameters		12	2		12.15	94		106.15	155
4280	4" x 2" diameters		9	2.667		28	125		153	220
4285	4" x 3" diameters		9	2.667		26	125		151	218
4305	Couplings, 1/2" CTS diameter		15	1.600		2.02	75		77.02	115
4310	1/2" diameter		14.25	1.684		2.02	79		81.02	121
4315	3/4" diameter		14.25	1.684		1.54	79		80.54	121
4320	1" diameter		13.50	1.778		1.54	83.50		85.04	128
4325	1-1/4" diameter		13.50	1.778		2.22	83.50		85.72	128
4330	1-1/2" diameter		12.75	1.882		2.49	88.50		90.99	136
4335	2" diameter		12	2		2.59	94		96.59	145
4340	3" diameter		12	2		11.45	94		105.45	155

For customer support on your Site Work & Landscape Costs with RSMeans data, call 800.448.8182.

33 52 16 – Gas Hydrocarbon Piping

33 52 16.26 High Density Polyethylene Piping	Crew	Daily Output	Labor-Hours	Unit	Material	2020 Bare Costs Labor	2020 Bare Costs Equipment	Total	Total Incl O&P
4345	4" diameter	B-20	9	2.667	Ea.	22.50	125		147.50	214
4405	Elbows, 90°, 1/2" CTS diameter		15	1.600		3.46	75		78.46	117
4410	1/2" diameter		14.25	1.684		3.48	79		82.48	123
4415	3/4" diameter		14.25	1.684		2.73	79		81.73	122
4420	1" diameter		13.50	1.778		3.36	83.50		86.86	130
4425	1-1/4" diameter		13.50	1.778		4.12	83.50		87.62	131
4430	1-1/2" diameter		12.75	1.882		6.85	88.50		95.35	141
4435	2" diameter		12	2		7	94		101	150
4440	3" diameter		12	2		25	94		119	170
4445	4" diameter		9	2.667		102	125		227	300
4450	45°, 3/4" diameter		14.25	1.684		7.05	79		86.05	127
4455	1" diameter		13.50	1.778		7.85	83.50		91.35	135
4460	1-1/4" diameter		13.50	1.778		6.60	83.50		90.10	133
4465	1-1/2" diameter		12.75	1.882		10.35	88.50		98.85	144
4470	2" diameter		12	2		7.70	94		101.70	150
4475	3" diameter		12	2		42.50	94		136.50	189
4480	4" diameter		9	2.667		55	125		180	250
4505	Tees, 1/2" CTS diameter		10	2.400		3.14	113		116.14	173
4510	1/2" diameter		9.50	2.526		2.95	119		121.95	182
4515	3/4" diameter		9.50	2.526		2.94	119		121.94	182
4520	1" diameter		9	2.667		3.62	125		128.62	193
4525	1-1/4" diameter		9	2.667		5.05	125		130.05	195
4530	1-1/2" diameter		8.50	2.824		8.45	133		141.45	209
4535	2" diameter		8	3		9.20	141		150.20	222
4540	3" diameter		8	3		37.50	141		178.50	253
4545	4" diameter		6	4		72.50	188		260.50	365
4600	Tapping tees, type II, 3/4" punch, socket fusion outlets									
4601	Saddle countour x outlet diameter, IPS unless noted CTS									
4602	1-1/4" x 1/2" CTS outlet	B-20	17	1.412	Ea.	11.25	66.50		77.75	112
4604	1-1/4" x 1/2" outlet		17	1.412		11.25	66.50		77.75	112
4606	1-1/4" x 3/4" outlet		17	1.412		11.25	66.50		77.75	112
4608	1-1/4" x 1" outlet		17	1.412		13.15	66.50		79.65	114
4610	1-1/4" x 1-1/4" outlet		16	1.500		23.50	70.50		94	132
4612	1-1/2" x 1/2" CTS outlet		16	1.500		11.25	70.50		81.75	118
4614	1-1/2" x 1/2" outlet		16	1.500		11.25	70.50		81.75	118
4616	1-1/2" x 3/4" outlet		16	1.500		11.25	70.50		81.75	118
4618	1-1/2" x 1" outlet		16	1.500		13.15	70.50		83.65	120
4620	1-1/2" x 1-1/4" outlet		15	1.600		23.50	75		98.50	139
4622	2" x 1/2" CTS outlet		12	2		11.25	94		105.25	154
4624	2" x 1/2" outlet		12	2		11.25	94		105.25	154
4626	2" x 3/4" outlet		12	2		11.25	94		105.25	154
4628	2" x 1" outlet		12	2		13.15	94		107.15	156
4630	2" x 1-1/4" outlet		11	2.182		23.50	103		126.50	181
4632	3" x 1/2" CTS outlet		12	2		11.25	94		105.25	154
4634	3" x 1/2" outlet		12	2		11.25	94		105.25	154
4636	3" x 3/4" outlet		12	2		11.25	94		105.25	154
4638	3" x 1" outlet		12	2		13.15	94		107.15	156
4640	3" x 1-1/4" outlet		11	2.182		23.50	103		126.50	181
4642	4" x 1/2" CTS outlet		10	2.400		11.25	113		124.25	182
4644	4" x 1/2" outlet		10	2.400		11.25	113		124.25	182
4646	4" x 3/4" outlet		10	2.400		11.25	113		124.25	182
4648	4" x 1" outlet		10	2.400		13.15	113		126.15	184
4650	4" x 1-1/4" outlet		9	2.667		23.50	125		148.50	215

33 52 16.26 High Density Polyethylene Piping	Crew	Daily Output	Labor-Hours	Unit	Material	2020 Bare Costs Labor	Equipment	Total	Total Incl O&P
4652	6" x 1/2" CTS outlet	B-21A	10	4	Ea.	11.25	209	43	263.25	375
4654	6" x 1/2" outlet	B-20	10	2.400		11.25	113		124.25	182
4656	6" x 3/4" outlet	B-21A	10	4		11.25	209	43	263.25	375
4658	6" x 1" outlet		10	4		13.15	209	43	265.15	375
4660	6" x 1-1/4" outlet		9	4.444		23.50	232	47.50	303	430
4662	8" x 1/2" CTS outlet		8	5		11.25	261	53.50	325.75	460
4664	8" x 1/2" outlet		8	5		11.25	261	53.50	325.75	460
4666	8" x 3/4" outlet		8	5		11.25	261	53.50	325.75	460
4668	8" x 1" outlet		8	5		13.15	261	53.50	327.65	465
4670	8" x 1-1/4" outlet		7	5.714		23.50	299	61	383.50	545
4675	For protective sleeves, 1/2" to 3/4" diameter outlets, add					.86			.86	.95
4680	1" diameter outlets, add					1.45			1.45	1.60
4685	1-1/4" diameter outlets, add					2.49			2.49	2.74
4700	Tapping tees, high volume, socket fusion outlets									
4705	1-1/2" punch, 2" x 1-1/4" outlet	B-20	11	2.182	Ea.	111	103		214	277
4710	1-7/8" punch, 3" x 1-1/4" outlet		11	2.182		111	103		214	277
4715	4" x 1-1/4" outlet		9	2.667		111	125		236	310
4720	6" x 1-1/4" outlet	B-21A	9	4.444		111	232	47.50	390.50	525
4725	8" x 1-1/4" outlet		7	5.714		111	299	61	471	640
4730	10" x 1-1/4" outlet		6	6.667		111	350	71.50	532.50	720
4735	12" x 1-1/4" outlet		5	8		111	420	85.50	616.50	840
4750	For protective sleeves, add					2.92			2.92	3.21
4800	Service saddles, saddle contour x outlet, socket fusion outlets									
4801	IPS unless noted CTS									
4802	1-1/4" x 1/2" CTS outlet	B-20	17	1.412	Ea.	5.95	66.50		72.45	107
4804	1-1/4" x 1/2" outlet		17	1.412		5.95	66.50		72.45	107
4806	1-1/4" x 3/4" outlet		17	1.412		5.95	66.50		72.45	107
4808	1-1/4" x 1" outlet		16	1.500		7	70.50		77.50	114
4810	1-1/4" x 1-1/4" outlet		16	1.500		9.90	70.50		80.40	117
4812	1-1/2" x 1/2" CTS outlet		16	1.500		5.95	70.50		76.45	113
4814	1-1/2" x 1/2" outlet		16	1.500		5.95	70.50		76.45	113
4816	1-1/2" x 3/4" outlet		16	1.500		5.95	70.50		76.45	113
4818	1-1/2" x 1-1/4" outlet		15	1.600		9.90	75		84.90	124
4820	2" x 1/2" CTS outlet		12	2		5.95	94		99.95	149
4822	2" x 1/2" outlet		12	2		5.95	94		99.95	149
4824	2" x 3/4" outlet		12	2		5.95	94		99.95	149
4826	2" x 1" outlet		11	2.182		7	103		110	163
4828	2" x 1-1/4" outlet		11	2.182		9.90	103		112.90	166
4830	3" x 1/2" CTS outlet		12	2		5.95	94		99.95	149
4832	3" x 1/2" outlet		12	2		5.95	94		99.95	149
4834	3" x 3/4" outlet		12	2		5.95	94		99.95	149
4836	3" x 1" outlet		11	2.182		7	103		110	163
4838	3" x 1-1/4" outlet		11	2.182		9.90	103		112.90	166
4840	4" x 1/2" CTS outlet		10	2.400		5.95	113		118.95	177
4842	4" x 1/2" outlet		10	2.400		5.95	113		118.95	177
4844	4" x 3/4" outlet		10	2.400		5.95	113		118.95	177
4846	4" x 1" outlet		9	2.667		7	125		132	197
4848	4" x 1-1/4" outlet		9	2.667		9.90	125		134.90	200
4850	6" x 1/2" CTS outlet	B-21A	10	4		5.95	209	43	257.95	370
4852	6" x 1/2" outlet		10	4		5.95	209	43	257.95	370
4854	6" x 3/4" outlet		10	4		5.95	209	43	257.95	370
4856	6" x 1" outlet		9	4.444		7	232	47.50	286.50	410
4858	6" x 1-1/4" outlet		9	4.444		9.90	232	47.50	289.40	415

33 52 16.26 High Density Polyethylene Piping	Crew	Daily Output	Labor-Hours	Unit	Material	2020 Bare Costs Labor	2020 Bare Costs Equipment	Total	Total Incl O&P
4860	8" x 1/2" CTS outlet	B-21A	8	5	Ea.	5.95	261	53.50	320.45	455
4862	8" x 1/2" outlet		8	5		5.95	261	53.50	320.45	455
4864	8" x 3/4" outlet		8	5		5.95	261	53.50	320.45	455
4866	8" x 1" outlet		7	5.714		7	299	61	367	525
4868	8" x 1-1/4" outlet		7	5.714		9.90	299	61	369.90	530
4880	For protective sleeves, 1/2" to 3/4" diameter outlets, add					.86			.86	.95
4885	1" diameter outlets, add					1.45			1.45	1.60
4890	1-1/4" diameter outlets, add					2.49			2.49	2.74
4900	Reducer tees, 1-1/4" x 3/4" x 3/4" diameters	B-20	9	2.667		7.35	125		132.35	197
4902	1-1/4" x 3/4" x 1" diameters		9	2.667		9.55	125		134.55	200
4904	1-1/4" x 3/4" x 1-1/4" diameters		9	2.667		9.55	125		134.55	200
4906	1-1/4" x 1" x 3/4" diameters		9	2.667		9.55	125		134.55	200
4908	1-1/4" x 1" x 1" diameters		9	2.667		7.60	125		132.60	197
4910	1-1/4" x 1" x 1-1/4" diameters		9	2.667		9.55	125		134.55	200
4912	1-1/4" x 1-1/4" x 3/4" diameters		9	2.667		7.20	125		132.20	197
4914	1-1/4" x 1-1/4" x 1" diameters		9	2.667		7.30	125		132.30	197
4916	1-1/2" x 3/4" x 3/4" diameters		8.50	2.824		12.15	133		145.15	213
4918	1-1/2" x 3/4" x 1" diameters		8.50	2.824		12.15	133		145.15	213
4920	1-1/2" x 3/4" x 1-1/4" diameters		8.50	2.824		12.15	133		145.15	213
4922	1-1/2" x 3/4" x 1-1/2" diameters		8.50	2.824		12.15	133		145.15	213
4924	1-1/2" x 1" x 3/4" diameters		8.50	2.824		12.15	133		145.15	213
4926	1-1/2" x 1" x 1" diameters		8.50	2.824		12.15	133		145.15	213
4928	1-1/2" x 1" x 1-1/4" diameters		8.50	2.824		12.15	133		145.15	213
4930	1-1/2" x 1" x 1-1/2" diameters		8.50	2.824		12.15	133		145.15	213
4932	1-1/2" x 1-1/4" x 3/4" diameters		8.50	2.824		10.15	133		143.15	211
4934	1-1/2" x 1-1/4" x 1" diameters		8.50	2.824		10.15	133		143.15	211
4936	1-1/2" x 1-1/4" x 1-1/4" diameters		8.50	2.824		10.15	133		143.15	211
4938	1-1/2" x 1-1/4" x 1-1/2" diameters		8.50	2.824		12.15	133		145.15	213
4940	1-1/2" x 1-1/2" x 3/4" diameters		8.50	2.824		9.75	133		142.75	211
4942	1-1/2" x 1-1/2" x 1" diameters		8.50	2.824		9.75	133		142.75	211
4944	1-1/2" x 1-1/2" x 1-1/4" diameters		8.50	2.824		9.75	133		142.75	211
4946	2" x 1-1/4" x 3/4" diameters		8	3		11	141		152	224
4948	2" x 1-1/4" x 1" diameters		8	3		11	141		152	224
4950	2" x 1-1/4" x 1-1/4" diameters		8	3		11.45	141		152.45	225
4952	2" x 1-1/2" x 3/4" diameters		8	3		11	141		152	224
4954	2" x 1-1/2" x 1" diameters		8	3		11	141		152	224
4956	2" x 1-1/2" x 1-1/4" diameters		8	3		11	141		152	224
4958	2" x 2" x 3/4" diameters		8	3		10.95	141		151.95	224
4960	2" x 2" x 1" diameters		8	3		11.10	141		152.10	224
4962	2" x 2" x 1-1/4" diameters		8	3		11.45	141		152.45	225
8100	Fittings, accessories									
8105	SDR 11, IPS unless noted CTS									
8110	Flange adapters, 2" x 6" long	1 Skwk	32	.250	Ea.	14.55	13.70		28.25	36.50
8115	3" x 6" long		32	.250		17.65	13.70		31.35	40
8120	4" x 6" long		24	.333		22	18.30		40.30	51.50
8125	6" x 8" long	2 Skwk	24	.667		33.50	36.50		70	92
8130	8" x 9" long	"	20	.800		48	44		92	120
8135	Backup flanges, 2" diameter	1 Skwk	32	.250		19.55	13.70		33.25	42
8140	3" diameter		32	.250		26.50	13.70		40.20	49.50
8145	4" diameter		24	.333		51.50	18.30		69.80	84.50
8150	6" diameter	2 Skwk	24	.667		53	36.50		89.50	114
8155	8" diameter	"	20	.800		91	44		135	167
8200	Tapping tees, test caps									

527

For customer support on your Site Work & Landscape Costs with RSMeans data, call 800.448.8182.

33 52 16.26 High Density Polyethylene Piping	Crew	Daily Output	Labor-Hours	Unit	Material	2020 Bare Costs Labor	Equipment	Total	Total Incl O&P
8205	Type II, yellow polyethylene cap	1 Skwk	45	.178	Ea.	44	9.75		53.75	63.50
8210	High volume, black polyethylene cap		45	.178		56.50	9.75		66.25	77.50
8215	Quick connector, female x female inlets, 1/4" N.P.T.		50	.160		18.45	8.80		27.25	34
8220	Test hose, 24" length, 3/8" ID, male outlets, 1/4" N.P.T.		50	.160		25	8.80		33.80	41
8225	Quick connector and test hose assembly		50	.160		43.50	8.80		52.30	61.50
8300	Threaded transition fittings									
8302	HDPE x MPT zinc plated steel, SDR 7, 1/2" CTS x 1/2" MPT	B-22C	30	.533	Ea.	31.50	26	5.20	62.70	79
8304	SDR 9.3, 1/2" IPS x 3/4" MPT		28.50	.561		41.50	27	5.45	73.95	93
8306	SDR 11, 3/4" IPS x 3/4" MPT		28.50	.561		20.50	27	5.45	52.95	69.50
8308	1" IPS x 1" MPT		27	.593		26.50	28.50	5.80	60.80	79
8310	1-1/4" IPS x 1-1/4" MPT		27	.593		42.50	28.50	5.80	76.80	97
8312	1-1/2" IPS x 1-1/2" MPT		25.50	.627		49.50	30.50	6.10	86.10	107
8314	2" IPS x 2" MPT		24	.667		56.50	32.50	6.50	95.50	118
8322	HDPE x MPT 316 stainless steel, SDR 11, 3/4" IPS x 3/4" MPT		28.50	.561		27	27	5.45	59.45	77
8324	1" IPS x 1" MPT		27	.593		28	28.50	5.80	62.30	81
8326	1-1/4" IPS x 1-1/4" MPT		27	.593		34.50	28.50	5.80	68.80	88
8328	2" IPS x 2" MPT		24	.667		40.50	32.50	6.50	79.50	100
8330	3" IPS x 3" MPT		24	.667		82.50	32.50	6.50	121.50	147
8332	4" IPS x 4" MPT		18	.889		113	43	8.65	164.65	200
8334	6" IPS x 6" MPT	B-22A	18	2.222		208	108	44.50	360.50	440
8342	HDPE x FPT 316 stainless steel, SDR 11, 3/4" IPS x 3/4" FPT	B-22C	28.50	.561		35.50	27	5.45	67.95	86
8344	1" IPS x 1" FPT		27	.593		47.50	28.50	5.80	81.80	102
8346	1-1/4" IPS x 1-1/4" FPT		27	.593		76.50	28.50	5.80	110.80	134
8348	1-1/2" IPS x 1-1/2" FPT		25.50	.627		72.50	30.50	6.10	109.10	132
8350	2" IPS x 2" FPT		24	.667		96.50	32.50	6.50	135.50	162
8352	3" IPS x 3" FPT		24	.667		187	32.50	6.50	226	262
8354	4" IPS x 4" FPT		18	.889		269	43	8.65	320.65	370
8362	HDPE x MPT epoxy carbon steel, SDR 11, 3/4" IPS x 3/4" MPT		28.50	.561		20.50	27	5.45	52.95	69.50
8364	1" IPS x 1" MPT		27	.593		22.50	28.50	5.80	56.80	75
8366	1-1/4" IPS x 1-1/4" MPT		27	.593		24	28.50	5.80	58.30	76.50
8368	1-1/2" IPS x 1-1/2" MPT		25.50	.627		26	30.50	6.10	62.60	82
8370	2" IPS x 2" MPT		24	.667		28.50	32.50	6.50	67.50	86.50
8372	3" IPS x 3" MPT		24	.667		43.50	32.50	6.50	82.50	104
8374	4" IPS x 4" MPT		18	.889		59	43	8.65	110.65	140
8382	HDPE x FPT epoxy carbon steel, SDR 11, 3/4" IPS x 3/4" FPT		28.50	.561		25	27	5.45	57.45	74.50
8384	1" IPS x 1" FPT		27	.593		27	28.50	5.80	61.30	80
8386	1-1/4" IPS x 1-1/4" FPT		27	.593		58.50	28.50	5.80	92.80	114
8388	1-1/2" IPS x 1-1/2" FPT		25.50	.627		58.50	30.50	6.10	95.10	117
8390	2" IPS x 2" FPT		24	.667		63	32.50	6.50	102	125
8392	3" IPS x 3" FPT		24	.667		107	32.50	6.50	146	174
8394	4" IPS x 4" FPT		18	.889		134	43	8.65	185.65	222
8402	HDPE x MPT poly coated carbon steel, SDR 11, 1" IPS x 1" MPT		27	.593		22	28.50	5.80	56.30	74
8404	1-1/4" IPS x 1-1/4" MPT		27	.593		28.50	28.50	5.80	62.80	81.50
8406	1-1/2" IPS x 1-1/2" MPT		25.50	.627		34	30.50	6.10	70.60	90.50
8408	2" IPS x 2" MPT		24	.667		38.50	32.50	6.50	77.50	98
8410	3" IPS x 3" MPT		24	.667		75	32.50	6.50	114	138
8412	4" IPS x 4" MPT		18	.889		116	43	8.65	167.65	202
8414	6" IPS x 6" MPT	B-22A	18	2.222		270	108	44.50	422.50	510
8602	Socket fused HDPE x MPT brass, SDR 11, 3/4" IPS x 3/4" MPT	B-20	28.50	.842		12	39.50		51.50	72.50
8604	1" IPS x 3/4" MPT		27	.889		14.50	42		56.50	79
8606	1" IPS x 1" MPT		27	.889		14	42		56	78.50
8608	1-1/4" IPS x 3/4" MPT		27	.889		15	42		57	79.50
8610	1-1/4" IPS x 1" MPT		27	.889		15	42		57	79.50

For customer support on your Site Work & Landscape Costs with RSMeans data, call 800.448.8182.

33 52 Hydrocarbon Transmission and Distribution

33 52 16 – Gas Hydrocarbon Piping

33 52 16.26 High Density Polyethylene Piping	Crew	Daily Output	Labor-Hours	Unit	Material	2020 Bare Costs Labor	Equipment	Total	Total Incl O&P
8612	1-1/4" IPS x 1-1/4" MPT	B-20	27	.889	Ea.	21.50	42		63.50	86.50
8614	1-1/2" IPS x 1-1/2" MPT		25.50	.941		29.50	44.50		74	99
8616	2" IPS x 2" MPT		24	1		30	47		77	104
8618	Socket fused HDPE x FPT brass, SDR 11, 3/4" IPS x 3/4" FPT		28.50	.842		12	39.50		51.50	72.50
8620	1" IPS x 1/2" FPT		27	.889		12	42		54	76
8622	1" IPS x 3/4" FPT		27	.889		14.50	42		56.50	79
8624	1" IPS x 1" FPT		27	.889		14	42		56	78.50
8626	1-1/4" IPS x 3/4" FPT		27	.889		15	42		57	79.50
8628	1-1/4" IPS x 1" FPT		27	.889		14.50	42		56.50	79
8630	1-1/4" IPS x 1-1/4" FPT		27	.889		21.50	42		63.50	86.50
8632	1-1/2" IPS x 1-1/2" FPT		25.50	.941		30	44.50		74.50	99.50
8634	2" IPS x 1-1/2" FPT		24	1		30.50	47		77.50	105
8636	2" IPS x 2" FPT		24	1		32	47		79	106

33 59 Hydrocarbon Utility Metering

33 59 33 – Natural-Gas Metering

33 59 33.10 Piping, Valves and Meters, Gas Distribution	Crew	Daily Output	Labor-Hours	Unit	Material	2020 Bare Costs Labor	Equipment	Total	Total Incl O&P
0010	PIPING, VALVES & METERS, GAS DISTRIBUTION									
0020	Not including excavation or backfill									
0100	Gas stops, with or without checks									
0140	1-1/4" size	1 Plum	12	.667	Ea.	70.50	43		113.50	142
0180	1-1/2" size		10	.800		91.50	51.50		143	178
0200	2" size		8	1		143	64.50		207.50	255
0600	Pressure regulator valves, iron and bronze									
0640	1-1/2" diameter	1 Plum	13	.615	Ea.	174	39.50		213.50	252
0680	2" diameter	"	11	.727		281	47		328	380
0700	3" diameter	Q-1	13	1.231		650	71.50		721.50	820
0740	4" diameter	"	8	2		2,200	116		2,316	2,600
1000	Meter risers, ASTM D2513, 125 psi									
1002	MDPE x MPT epoxy coated carbon steel, fixed, anodeless									
1005	SDR 7, 1/2" CTS x 1/2" MPT, 22" x 24"	B-22C	30	.533	Ea.	69	26	5.20	100.20	121
1010	1/2" CTS x 1/2" MPT, 30" x 15"		30	.533		66.50	26	5.20	97.70	118
1015	1/2" CTS x 3/4" MPT, 22" x 24"		30	.533		36	26	5.20	67.20	84
1020	1/2" CTS x 3/4" MPT, 30" x 17"		30	.533		39	26	5.20	70.20	87.50
1025	1/2" CTS x 3/4" MPT, 30" x 28"		30	.533		25	26	5.20	56.20	72
1030	SDR 9.3, 1/2" IPS x 3/4" MPT, 30" x 17"		28.50	.561		41	27	5.45	73.45	92
1035	1/2" IPS x 3/4" MPT, 30" x 28"		28.50	.561		78.50	27	5.45	110.95	134
1040	SDR 10, 1-1/4" IPS x 1-1/4" MPT, 30" x 22"		27	.593		109	28.50	5.80	143.30	170
1045	SDR 11, 3/4" IPS x 3/4" MPT, 30" x 16"		28.50	.561		56.50	27	5.45	88.95	109
1050	3/4" IPS x 3/4" MPT, 36" x 23"		28.50	.561		61	27	5.45	93.45	115
1055	1" IPS x 1" MPT, 30" x 16"		27	.593		64.50	28.50	5.80	98.80	121
1060	1-1/4" IPS x 1-1/4" MPT, 30" x 22"		27	.593		111	28.50	5.80	145.30	172
1065	1-1/2" IPS x 1-1/2" MPT, 36" x 26"		25.50	.627		155	30.50	6.10	191.60	224
1070	2" IPS x 2" MPT, 36" x 36"		24	.667		202	32.50	6.50	241	278
1075	SDR 11.5, 1" CTS x 3/4" MPT, 30" x 16"		28	.571		60	27.50	5.55	93.05	114
1100	Compression joint x MPT stainless steel, HDPE or MDPE, flexible									
1105	SDR 7, 1/2" CTS x 1/2" MPT, 60" long	B-20	60	.400	Ea.	42	18.80		60.80	74.50
1110	1/2" CTS x 3/4" MPT, 60" long		60	.400		39.50	18.80		58.30	72
1115	SDR 11, 3/4" IPS x 3/4" MPT, 60" long		58	.414		63.50	19.45		82.95	99
1120	1" IPS x 1" MPT, 60" long		54	.444		139	21		160	185
1125	SDR 11.5, 1" CTS x 1" MPT, 60" long		56	.429		81.50	20		101.50	120

33 59 Hydrocarbon Utility Metering

33 59 33 – Natural-Gas Metering

33 59 33.10 Piping, Valves and Meters, Gas Distribution	Crew	Daily Output	Labor-Hours	Unit	Material	2020 Bare Costs Labor	Equipment	Total	Total Incl O&P
2000 Lubricated semi-steel plug valve									
2040 3/4" diameter	1 Plum	16	.500	Ea.	125	32		157	186
2080 1" diameter		14	.571		135	37		172	203
2100 1-1/4" diameter		12	.667		159	43		202	240
2140 1-1/2" diameter		11	.727		154	47		201	240
2180 2" diameter	↓	8	1		208	64.50		272.50	325
2300 2-1/2" diameter	Q-1	5	3.200		299	186		485	610
2340 3" diameter	"	4.50	3.556	↓	425	206		631	780

33 61 Hydronic Energy Distribution

33 61 13 – Underground Hydronic Energy Distribution

33 61 13.10 Chilled/HVAC Hot Water Distribution

		Crew	Daily Output	Labor-Hours	Unit	Material	2020 Bare Costs Labor	Equipment	Total	Total Incl O&P
0010	**CHILLED/HVAC HOT WATER DISTRIBUTION**									
1005	Pipe, black steel w/2" polyurethane insul., 20' lengths									
1010	Align & tackweld on sleepers (NIC), 1-1/4" diameter	B-35	864	.056	L.F.	34	2.92	.93	37.85	43
1020	1-1/2"		824	.058		36.50	3.06	.97	40.53	45.50
1030	2"		680	.071		46	3.71	1.18	50.89	58
1040	2-1/2"		560	.086		51	4.50	1.43	56.93	64.50
1050	3"		528	.091		48.50	4.78	1.52	54.80	62.50
1060	4"		384	.125		53.50	6.55	2.09	62.14	71
1070	5"		360	.133		63.50	7	2.22	72.72	83
1080	6"		296	.162		86	8.50	2.70	97.20	110
1090	8"		264	.182		114	9.55	3.03	126.58	143
1100	12"		216	.222		177	11.70	3.71	192.41	217
1110	On trench bottom, 18" diameter	↓	176.13	.273		285	14.30	4.55	303.85	340
1120	24"	B-35A	145	.386		405	19.70	14.80	439.50	490
1130	30"		121	.463		655	23.50	17.75	696.25	775
1140	36"	↓	100	.560	↓	905	28.50	21.50	955	1,050
1150	Elbows, on sleepers, 1-1/2" diameter	Q-17	21	.762	Ea.	970	45	5.05	1,020.05	1,150
1160	3"		9.36	1.709		1,250	101	11.35	1,362.35	1,550
1170	4"		8	2		1,500	118	13.30	1,631.30	1,850
1180	6"		6	2.667		1,850	157	17.75	2,024.75	2,275
1190	8"		4.64	3.448		2,425	203	23	2,651	3,000
1200	Tees, 1-1/2" diameter		17	.941		1,775	55.50	6.25	1,836.75	2,050
1210	3"		8.50	1.882		2,225	111	12.50	2,348.50	2,625
1220	4"	↓	6	2.667		2,425	157	17.75	2,599.75	2,925
1230	6"	Q-17A	6.72	3.571		3,000	211	86	3,297	3,700
1240	8"	"	6.40	3.750		3,800	222	90	4,112	4,625
1250	Reducer, 3" diameter	Q-17	16	1		525	59	6.65	590.65	675
1260	4"		12	1.333		620	78.50	8.85	707.35	815
1270	6"		12	1.333		900	78.50	8.85	987.35	1,125
1280	8"	↓	10	1.600		1,175	94.50	10.65	1,280.15	1,450
1290	Anchor, 4" diameter	Q-17A	12	2		1,350	118	48	1,516	1,725
1300	6"		10.50	2.286		1,650	135	55	1,840	2,075
1310	8"	↓	10	2.400		1,975	142	57.50	2,174.50	2,450
1320	Cap, 1-1/2" diameter	Q-17	42	.381		167	22.50	2.53	192.03	219
1330	3"		14.64	1.093		221	64.50	7.25	292.75	350
1340	4"		16	1		254	59	6.65	319.65	375
1350	6"		16	1		440	59	6.65	505.65	580
1360	8"		13.50	1.185		560	70	7.90	637.90	730
1365	12"	↓	11	1.455	↓	845	86	9.65	940.65	1,075

33 61 13 – Underground Hydronic Energy Distribution

33 61 13.10 Chilled/HVAC Hot Water Distribution

		Crew	Daily Output	Labor-Hours	Unit	Material	2020 Bare Costs Labor	Equipment	Total	Total Incl O&P
1370	Elbow, fittings on trench bottom, 12" diameter	Q-17A	12	2	Ea.	3,600	118	48	3,766	4,175
1380	18"		8	3		6,975	177	72	7,224	8,025
1390	24"		6	4		9,850	236	96	10,182	11,300
1400	30"		5.36	4.478		14,300	264	108	14,672	16,200
1410	36"		4	6		22,400	355	144	22,899	25,400
1420	Tee, 12" diameter		6.72	3.571		4,600	211	86	4,897	5,450
1430	18"		6	4		8,375	236	96	8,707	9,650
1440	24"		4.64	5.172		13,300	305	124	13,729	15,200
1450	30"		4.16	5.769		21,500	340	139	21,979	24,400
1460	36"		3.36	7.143		36,700	420	172	37,292	41,200
1470	Reducer, 12" diameter		10.64	2.256		1,475	133	54.50	1,662.50	1,875
1480	18"		9.68	2.479		2,525	146	59.50	2,730.50	3,050
1490	24"		8	3		3,950	177	72	4,199	4,700
1500	30"		7.04	3.409		6,500	201	82	6,783	7,550
1510	36"		5.36	4.478		10,800	264	108	11,172	12,400
1520	Anchor, 12" diameter		11	2.182		1,925	129	52.50	2,106.50	2,350
1530	18"		6.72	3.571		1,925	211	86	2,222	2,525
1540	24"		6.32	3.797		2,725	224	91.50	3,040.50	3,425
1550	30"		4.64	5.172		3,825	305	124	4,254	4,825
1560	36"		4	6		5,425	355	144	5,924	6,675
1565	Weld in place and install shrink collar									
1570	On sleepers, 1-1/2" diameter	Q-17A	18.50	1.297	Ea.	31.50	76.50	31	139	185
1580	3"		6.72	3.571		54	211	86	351	470
1590	4"		5.36	4.478		61	264	108	433	580
1600	6"		4	6		63.50	355	144	562.50	760
1610	8"		3.36	7.143		98	420	172	690	925
1620	12"		2.64	9.091		127	535	219	881	1,175
1630	On trench bottom, 18" diameter		2	12		191	710	289	1,190	1,575
1640	24"		1.36	17.647		254	1,050	425	1,729	2,300
1650	30"		1.04	23.077		320	1,375	555	2,250	3,000
1660	36"		1	24		380	1,425	575	2,380	3,175

33 61 13.20 Pipe Conduit, Prefabricated/Preinsulated

		Crew	Daily Output	Labor-Hours	Unit	Material	2020 Bare Costs Labor	Equipment	Total	Total Incl O&P
0010	**PIPE CONDUIT, PREFABRICATED/PREINSULATED**									
0020	Does not include trenching, fittings or crane.									
0300	For cathodic protection, add 12 to 14%									
0310	of total built-up price (casing plus service pipe)									
0580	Polyurethane insulated system, 250°F max. temp.									
0620	Black steel service pipe, standard wt., 1/2" insulation									
0660	3/4" diam. pipe size	Q-17	54	.296	L.F.	88.50	17.50	1.97	107.97	126
0670	1" diam. pipe size		50	.320		97.50	18.90	2.13	118.53	138
0680	1-1/4" diam. pipe size		47	.340		109	20	2.26	131.26	152
0690	1-1/2" diam. pipe size		45	.356		118	21	2.36	141.36	164
0700	2" diam. pipe size		42	.381		122	22.50	2.53	147.03	171
0710	2-1/2" diam. pipe size		34	.471		124	28	3.13	155.13	182
0720	3" diam. pipe size		28	.571		144	33.50	3.80	181.30	213
0730	4" diam. pipe size		22	.727		182	43	4.84	229.84	269
0740	5" diam. pipe size		18	.889		232	52.50	5.90	290.40	340
0750	6" diam. pipe size	Q-18	23	1.043		269	64	4.62	337.62	395
0760	8" diam. pipe size		19	1.263		395	77.50	5.60	478.10	555
0770	10" diam. pipe size		16	1.500		500	92	6.65	598.65	695
0780	12" diam. pipe size		13	1.846		625	113	8.20	746.20	865
0790	14" diam. pipe size		11	2.182		695	133	9.65	837.65	975

33 61 13.20 Pipe Conduit, Prefabricated/Preinsulated	Crew	Daily Output	Labor-Hours	Unit	Material	2020 Bare Costs Labor	Equipment	Total	Total Incl O&P
0800	16" diam. pipe size	Q-18	10	2.400	L.F.	800	147	10.65	957.65	1,100
0810	18" diam. pipe size		8	3		920	184	13.30	1,117.30	1,300
0820	20" diam. pipe size		7	3.429		1,025	210	15.20	1,250.20	1,450
0830	24" diam. pipe size		6	4		1,250	245	17.70	1,512.70	1,750
0900	For 1" thick insulation, add					10%				
0940	For 1-1/2" thick insulation, add					13%				
0980	For 2" thick insulation, add					20%				
1500	Gland seal for system, 3/4" diam. pipe size	Q-17	32	.500	Ea.	1,025	29.50	3.33	1,057.83	1,175
1510	1" diam. pipe size		32	.500		1,025	29.50	3.33	1,057.83	1,175
1540	1-1/4" diam. pipe size		30	.533		1,100	31.50	3.55	1,135.05	1,250
1550	1-1/2" diam. pipe size		30	.533		1,100	31.50	3.55	1,135.05	1,250
1560	2" diam. pipe size		28	.571		1,300	33.50	3.80	1,337.30	1,475
1570	2-1/2" diam. pipe size		26	.615		1,400	36.50	4.09	1,440.59	1,600
1580	3" diam. pipe size		24	.667		1,500	39.50	4.43	1,543.93	1,725
1590	4" diam. pipe size		22	.727		1,775	43	4.84	1,822.84	2,025
1600	5" diam. pipe size		19	.842		2,200	49.50	5.60	2,255.10	2,475
1610	6" diam. pipe size	Q-18	26	.923		2,325	56.50	4.09	2,385.59	2,650
1620	8" diam. pipe size		25	.960		2,700	58.50	4.25	2,762.75	3,075
1630	10" diam. pipe size		23	1.043		3,175	64	4.62	3,243.62	3,600
1640	12" diam. pipe size		21	1.143		3,500	70	5.05	3,575.05	3,950
1650	14" diam. pipe size		19	1.263		3,925	77.50	5.60	4,008.10	4,450
1660	16" diam. pipe size		18	1.333		4,650	81.50	5.90	4,737.40	5,225
1670	18" diam. pipe size		16	1.500		4,925	92	6.65	5,023.65	5,575
1680	20" diam. pipe size		14	1.714		5,600	105	7.60	5,712.60	6,325
1690	24" diam. pipe size		12	2		6,200	122	8.85	6,330.85	7,025
2000	Elbow, 45° for system									
2020	3/4" diam. pipe size	Q-17	14	1.143	Ea.	675	67.50	7.60	750.10	850
2040	1" diam. pipe size		13	1.231		690	72.50	8.20	770.70	880
2050	1-1/4" diam. pipe size		11	1.455		785	86	9.65	880.65	1,000
2060	1-1/2" diam. pipe size		9	1.778		790	105	11.80	906.80	1,050
2070	2" diam. pipe size		6	2.667		845	157	17.75	1,019.75	1,175
2080	2-1/2" diam. pipe size		4	4		930	236	26.50	1,192.50	1,400
2090	3" diam. pipe size		3.50	4.571		1,075	270	30.50	1,375.50	1,625
2100	4" diam. pipe size		3	5.333		1,250	315	35.50	1,600.50	1,875
2110	5" diam. pipe size		2.80	5.714		1,600	335	38	1,973	2,325
2120	6" diam. pipe size	Q-18	4	6		1,825	365	26.50	2,216.50	2,575
2130	8" diam. pipe size		3	8		2,650	490	35.50	3,175.50	3,675
2140	10" diam. pipe size		2.40	10		3,250	610	44.50	3,904.50	4,550
2150	12" diam. pipe size		2	12		4,275	735	53	5,063	5,850
2160	14" diam. pipe size		1.80	13.333		5,325	815	59	6,199	7,150
2170	16" diam. pipe size		1.60	15		6,325	920	66.50	7,311.50	8,400
2180	18" diam. pipe size		1.30	18.462		7,925	1,125	82	9,132	10,500
2190	20" diam. pipe size		1	24		10,000	1,475	106	11,581	13,300
2200	24" diam. pipe size		.70	34.286		12,600	2,100	152	14,852	17,100
2260	For elbow, 90°, add					25%				
2300	For tee, straight, add					85%	30%			
2340	For tee, reducing, add					170%	30%			
2380	For weldolet, straight, add					50%				

33 63 13.10 Calcium Silicate Insulated System	Crew	Daily Output	Labor-Hours	Unit	Material	2020 Bare Costs Labor	Equipment	Total	Total Incl O&P
0010 **CALCIUM SILICATE INSULATED SYSTEM**									
0011 High temp. (1200 degrees F)									
2840 Steel casing with protective exterior coating									
2850 6-5/8" diameter	Q-18	52	.462	L.F.	130	28	2.04	160.04	188
2860 8-5/8" diameter		50	.480		141	29.50	2.13	172.63	201
2870 10-3/4" diameter		47	.511		165	31	2.26	198.26	231
2880 12-3/4" diameter		44	.545		180	33.50	2.42	215.92	251
2890 14" diameter		41	.585		203	36	2.59	241.59	280
2900 16" diameter		39	.615		218	37.50	2.73	258.23	300
2910 18" diameter		36	.667		240	41	2.95	283.95	330
2920 20" diameter		34	.706		271	43	3.13	317.13	365
2930 22" diameter		32	.750		375	46	3.32	424.32	480
2940 24" diameter		29	.828		430	50.50	3.67	484.17	550
2950 26" diameter		26	.923		490	56.50	4.09	550.59	630
2960 28" diameter		23	1.043		605	64	4.62	673.62	765
2970 30" diameter		21	1.143		650	70	5.05	725.05	825
2980 32" diameter		19	1.263		730	77.50	5.60	813.10	920
2990 34" diameter		18	1.333		735	81.50	5.90	822.40	940
3000 36" diameter		16	1.500		790	92	6.65	888.65	1,025
3040 For multi-pipe casings, add					10%				
3060 For oversize casings, add					2%				
3400 Steel casing gland seal, single pipe									
3420 6-5/8" diameter	Q-18	25	.960	Ea.	1,975	58.50	4.25	2,037.75	2,275
3440 8-5/8" diameter		23	1.043		2,325	64	4.62	2,393.62	2,650
3450 10-3/4" diameter		21	1.143		2,575	70	5.05	2,650.05	2,950
3460 12-3/4" diameter		19	1.263		3,050	77.50	5.60	3,133.10	3,475
3470 14" diameter		17	1.412		3,375	86.50	6.25	3,467.75	3,850
3480 16" diameter		16	1.500		3,950	92	6.65	4,048.65	4,500
3490 18" diameter		15	1.600		4,400	98	7.10	4,505.10	4,975
3500 20" diameter		13	1.846		4,850	113	8.20	4,971.20	5,525
3510 22" diameter		12	2		5,450	122	8.85	5,580.85	6,200
3520 24" diameter		11	2.182		6,125	133	9.65	6,267.65	6,950
3530 26" diameter		10	2.400		7,000	147	10.65	7,157.65	7,925
3540 28" diameter		9.50	2.526		7,900	155	11.20	8,066.20	8,950
3550 30" diameter		9	2.667		8,050	163	11.80	8,224.80	9,100
3560 32" diameter		8.50	2.824		9,050	173	12.50	9,235.50	10,200
3570 34" diameter		8	3		9,850	184	13.30	10,047.30	11,100
3580 36" diameter		7	3.429		10,500	210	15.20	10,725.20	11,800
3620 For multi-pipe casings, add					5%				
4000 Steel casing anchors, single pipe									
4020 6-5/8" diameter	Q-18	8	3	Ea.	1,775	184	13.30	1,972.30	2,250
4040 8-5/8" diameter		7.50	3.200		1,850	196	14.20	2,060.20	2,325
4050 10-3/4" diameter		7	3.429		2,400	210	15.20	2,625.20	2,975
4060 12-3/4" diameter		6.50	3.692		2,575	226	16.35	2,817.35	3,200
4070 14" diameter		6	4		3,050	245	17.70	3,312.70	3,725
4080 16" diameter		5.50	4.364		3,525	267	19.35	3,811.35	4,300
4090 18" diameter		5	4.800		3,975	294	21.50	4,290.50	4,850
4100 20" diameter		4.50	5.333		4,400	325	23.50	4,748.50	5,350
4110 22" diameter		4	6		4,900	365	26.50	5,291.50	5,950
4120 24" diameter		3.50	6.857		5,325	420	30.50	5,775.50	6,525
4130 26" diameter		3	8		6,025	490	35.50	6,550.50	7,400
4140 28" diameter		2.50	9.600		6,625	585	42.50	7,252.50	8,225
4150 30" diameter		2	12		7,050	735	53	7,838	8,900

33 63 13 – Underground Steam and Condensate Distribution Piping

33 63 13.10 Calcium Silicate Insulated System	Crew	Daily Output	Labor-Hours	Unit	Material	2020 Bare Costs Labor	Equipment	Total	Total Incl O&P
4160	32" diameter	Q-18	1.50	16	Ea.	8,400	980	71	9,451	10,800
4170	34" diameter		1	24		9,450	1,475	106	11,031	12,700
4180	36" diameter		1	24		10,300	1,475	106	11,881	13,600
4220	For multi-pipe, add					5%	20%			
4800	Steel casing elbow									
4820	6-5/8" diameter	Q-18	15	1.600	Ea.	2,525	98	7.10	2,630.10	2,925
4830	8-5/8" diameter		15	1.600		2,700	98	7.10	2,805.10	3,100
4850	10-3/4" diameter		14	1.714		3,250	105	7.60	3,362.60	3,750
4860	12-3/4" diameter		13	1.846		3,850	113	8.20	3,971.20	4,400
4870	14" diameter		12	2		4,125	122	8.85	4,255.85	4,750
4880	16" diameter		11	2.182		4,500	133	9.65	4,642.65	5,150
4890	18" diameter		10	2.400		5,150	147	10.65	5,307.65	5,875
4900	20" diameter		9	2.667		5,500	163	11.80	5,674.80	6,300
4910	22" diameter		8	3		5,900	184	13.30	6,097.30	6,800
4920	24" diameter		7	3.429		6,500	210	15.20	6,725.20	7,475
4930	26" diameter		6	4		7,125	245	17.70	7,387.70	8,225
4940	28" diameter		5	4.800		7,700	294	21.50	8,015.50	8,925
4950	30" diameter		4	6		7,725	365	26.50	8,116.50	9,075
4960	32" diameter		3	8		8,775	490	35.50	9,300.50	10,400
4970	34" diameter		2	12		9,650	735	53	10,438	11,800
4980	36" diameter		2	12		10,300	735	53	11,088	12,500
5500	Black steel service pipe, std. wt., 1" thick insulation									
5510	3/4" diameter pipe size	Q-17	54	.296	L.F.	79.50	17.50	1.97	98.97	115
5540	1" diameter pipe size		50	.320		81.50	18.90	2.13	102.53	121
5550	1-1/4" diameter pipe size		47	.340		104	20	2.26	126.26	146
5560	1-1/2" diameter pipe size		45	.356		109	21	2.36	132.36	153
5570	2" diameter pipe size		42	.381		118	22.50	2.53	143.03	166
5580	2-1/2" diameter pipe size		34	.471		124	28	3.13	155.13	181
5590	3" diameter pipe size		28	.571		108	33.50	3.80	145.30	174
5600	4" diameter pipe size		22	.727		121	43	4.84	168.84	202
5610	5" diameter pipe size		18	.889		136	52.50	5.90	194.40	235
5620	6" diameter pipe size	Q-18	23	1.043		147	64	4.62	215.62	262
6000	Black steel service pipe, std. wt., 1-1/2" thick insul.									
6010	3/4" diameter pipe size	Q-17	54	.296	L.F.	94	17.50	1.97	113.47	131
6040	1" diameter pipe size		50	.320		99	18.90	2.13	120.03	140
6050	1-1/4" diameter pipe size		47	.340		107	20	2.26	129.26	149
6060	1-1/2" diameter pipe size		45	.356		106	21	2.36	129.36	150
6070	2" diameter pipe size		42	.381		113	22.50	2.53	138.03	161
6080	2-1/2" diameter pipe size		34	.471		104	28	3.13	135.13	159
6090	3" diameter pipe size		28	.571		115	33.50	3.80	152.30	182
6100	4" diameter pipe size		22	.727		132	43	4.84	179.84	214
6110	5" diameter pipe size		18	.889		157	52.50	5.90	215.40	258
6120	6" diameter pipe size	Q-18	23	1.043		164	64	4.62	232.62	281
6130	8" diameter pipe size		19	1.263		208	77.50	5.60	291.10	350
6140	10" diameter pipe size		16	1.500		278	92	6.65	376.65	450
6150	12" diameter pipe size		13	1.846		370	113	8.20	491.20	585
6190	For 2" thick insulation, add					15%				
6220	For 2-1/2" thick insulation, add					25%				
6260	For 3" thick insulation, add					30%				
6800	Black steel service pipe, ex. hvy. wt., 1" thick insul.									
6820	3/4" diameter pipe size	Q-17	50	.320	L.F.	95	18.90	2.13	116.03	136
6840	1" diameter pipe size		47	.340		101	20	2.26	123.26	143
6850	1-1/4" diameter pipe size		44	.364		112	21.50	2.42	135.92	158

33 63 Steam Energy Distribution

33 63 13 – Underground Steam and Condensate Distribution Piping

33 63 13.10 Calcium Silicate Insulated System	Crew	Daily Output	Labor-Hours	Unit	Material	2020 Bare Costs Labor	Equipment	Total	Total Incl O&P
6860	1-1/2" diameter pipe size	Q-17	42	.381	L.F.	114	22.50	2.53	139.03	161
6870	2" diameter pipe size		40	.400		125	23.50	2.66	151.16	175
6880	2-1/2" diameter pipe size		31	.516		102	30.50	3.43	135.93	162
6890	3" diameter pipe size		27	.593		118	35	3.94	156.94	187
6900	4" diameter pipe size		21	.762		129	45	5.05	179.05	215
6910	5" diameter pipe size		17	.941		156	55.50	6.25	217.75	262
6920	6" diameter pipe size	Q-18	22	1.091		166	66.50	4.83	237.33	288
7400	Black steel service pipe, ex. hvy. wt., 1-1/2" thick insul.									
7420	3/4" diameter pipe size	Q-17	50	.320	L.F.	70	18.90	2.13	91.03	108
7440	1" diameter pipe size		47	.340		88.50	20	2.26	110.76	129
7450	1-1/4" diameter pipe size		44	.364		97.50	21.50	2.42	121.42	142
7460	1-1/2" diameter pipe size		42	.381		96	22.50	2.53	121.03	142
7470	2" diameter pipe size		40	.400		93	23.50	2.66	119.16	140
7480	2-1/2" diameter pipe size		31	.516		95.50	30.50	3.43	129.43	154
7490	3" diameter pipe size		27	.593		105	35	3.94	143.94	173
7500	4" diameter pipe size		21	.762		134	45	5.05	184.05	220
7510	5" diameter pipe size		17	.941		185	55.50	6.25	246.75	293
7520	6" diameter pipe size	Q-18	22	1.091		197	66.50	4.83	268.33	320
7530	8" diameter pipe size		18	1.333		270	81.50	5.90	357.40	425
7540	10" diameter pipe size		15	1.600		300	98	7.10	405.10	490
7550	12" diameter pipe size		13	1.846		360	113	8.20	481.20	580
7590	For 2" thick insulation, add					13%				
7640	For 2-1/2" thick insulation, add					18%				
7680	For 3" thick insulation, add					24%				

33 71 Electrical Utility Transmission and Distribution

33 71 16 – Electrical Utility Poles

33 71 16.20 Line Poles and Fixtures

		Crew	Daily Output	Labor-Hours	Unit	Material	2020 Bare Costs Labor	Equipment	Total	Total Incl O&P
0010	**LINE POLES & FIXTURES**									
0200	Wood poles, material handling and spotting	R-7	6.49	7.396	Ea.		320	25.50	345.50	520
0220	Erect wood poles & backfill holes, in earth	R-5	6.77	12.999	"	1,850	695	175	2,720	3,275
0300	Crossarms for wood pole structure									
0310	Material handling and spotting	R-7	14.55	3.299	Ea.		144	11.40	155.40	231
0320	Install crossarms	R-5	11	8	"	605	430	108	1,143	1,425

33 71 16.23 Steel Electrical Utility Poles

		Crew	Daily Output	Labor-Hours	Unit	Material	2020 Bare Costs Labor	Equipment	Total	Total Incl O&P
0010	**STEEL ELECTRICAL UTILITY POLES**									
6000	Digging holes in earth, average	R-5	25.14	3.500	Ea.		188	47	235	335
6010	In rock, average	"	4.51	19.512	"		1,050	262	1,312	1,875
6020	Formed plate pole structure									
6030	Material handling and spotting	R-7	2.40	20	Ea.		870	69	939	1,400
6040	Erect steel plate pole	R-5	1.95	45.128		11,400	2,425	605	14,430	16,900
6050	Guys, anchors and hardware for pole, in earth		7.04	12.500		695	670	168	1,533	1,950
6060	In rock		17.96	4.900		830	263	66	1,159	1,375
6070	Foundations for line poles									
6080	Excavation, in earth	R-5	135.38	.650	C.Y.		35	8.75	43.75	61.50
6090	In rock		20	4.400			236	59	295	420
6110	Concrete foundations		11	8		164	430	108	702	945

33 71 16 – Electrical Utility Poles

33 71 16.33 Wood Electrical Utility Poles	Crew	Daily Output	Labor-Hours	Unit	Material	2020 Bare Costs Labor	Equipment	Total	Total Incl O&P
0010 **WOOD ELECTRICAL UTILITY POLES**									
0011 Excludes excavation, backfill and cast-in-place concrete									
6200 Wood, class 3 Douglas Fir, penta-treated, 20'	R-3	3.10	6.452	Ea.	256	395	61	712	940
6600 30'		2.60	7.692		320	470	72.50	862.50	1,125
7000 40'		2.30	8.696		630	530	82	1,242	1,575
7200 45'		1.70	11.765		900	720	111	1,731	2,175
7400 Cross arms with hardware & insulators									
7600 4' long	1 Elec	2.50	3.200	Ea.	156	196		352	465
7800 5' long		2.40	3.333		168	205		373	490
8000 6' long		2.20	3.636		190	223		413	540
9000 Disposal of pole & hardware surplus material	R-7	20.87	2.300	Mile		100	7.95	107.95	161
9100 Disposal of crossarms & hardware surplus material	"	40	1.200	"		52.50	4.15	56.65	84

33 71 19 – Electrical Underground Ducts and Manholes

33 71 19.15 Underground Ducts and Manholes

	Crew	Daily Output	Labor-Hours	Unit	Material	2020 Bare Costs Labor	Equipment	Total	Total Incl O&P
0010 **UNDERGROUND DUCTS AND MANHOLES**									
0011 Not incl. excavation, backfill, or concrete in slab and duct bank									
1000 Direct burial									
1010 PVC, schedule 40, w/coupling, 1/2" diameter	1 Elec	340	.024	L.F.	.34	1.44		1.78	2.52
1020 3/4" diameter		290	.028		.41	1.69		2.10	2.97
1030 1" diameter		260	.031		.67	1.89		2.56	3.55
1040 1-1/2" diameter		210	.038		.99	2.34		3.33	4.57
1050 2" diameter		180	.044		1.29	2.73		4.02	5.50
1060 3" diameter	2 Elec	240	.067		2.38	4.09		6.47	8.70
1070 4" diameter		160	.100		3.16	6.15		9.31	12.65
1080 5" diameter		120	.133		4.69	8.20		12.89	17.35
1090 6" diameter		90	.178		6.20	10.90		17.10	23
1110 Elbows, 1/2" diameter	1 Elec	48	.167	Ea.	.58	10.25		10.83	15.90
1120 3/4" diameter		38	.211		.60	12.90		13.50	19.90
1130 1" diameter		32	.250		1.21	15.35		16.56	24.50
1140 1-1/2" diameter		21	.381		2.71	23.50		26.21	38
1150 2" diameter		16	.500		3.54	30.50		34.04	49.50
1160 3" diameter		12	.667		12.25	41		53.25	74.50
1170 4" diameter		9	.889		23	54.50		77.50	106
1180 5" diameter		8	1		34.50	61.50		96	130
1190 6" diameter		5	1.600		35	98		133	185
1210 Adapters, 1/2" diameter		52	.154		.26	9.45		9.71	14.35
1220 3/4" diameter		43	.186		.37	11.40		11.77	17.40
1230 1" diameter		39	.205		.39	12.60		12.99	19.20
1240 1-1/2" diameter		35	.229		.71	14		14.71	22
1250 2" diameter		26	.308		1.39	18.90		20.29	29.50
1260 3" diameter		20	.400		2.85	24.50		27.35	39.50
1270 4" diameter		14	.571		4.80	35		39.80	57.50
1280 5" diameter		12	.667		9.70	41		50.70	71.50
1290 6" diameter		9	.889		18.90	54.50		73.40	102
1340 Bell end & cap, 1-1/2" diameter		35	.229		4.69	14		18.69	26
1350 Bell end & plug, 2" diameter		26	.308		5.90	18.90		24.80	34.50
1360 3" diameter		20	.400		8.25	24.50		32.75	45.50
1370 4" diameter		14	.571		9.90	35		44.90	63
1380 5" diameter		12	.667		15.95	41		56.95	78.50
1390 6" diameter		9	.889		17.05	54.50		71.55	100
1450 Base spacer, 2" diameter		56	.143		1.37	8.75		10.12	14.55
1460 3" diameter		46	.174		1.78	10.65		12.43	17.85

33 71 19.15 Underground Ducts and Manholes	Crew	Daily Output	Labor-Hours	Unit	Material	2020 Bare Costs Labor	Equipment	Total	Total Incl O&P
1470	4" diameter	1 Elec	41	.195	Ea.	1.78	11.95		13.73	19.75
1480	5" diameter		37	.216		1.99	13.25		15.24	22
1490	6" diameter		34	.235		2.48	14.45		16.93	24
1550	Intermediate spacer, 2" diameter		60	.133		1.25	8.20		9.45	13.60
1560	3" diameter		46	.174		1.76	10.65		12.41	17.85
1570	4" diameter		41	.195		1.59	11.95		13.54	19.55
1580	5" diameter		37	.216		1.94	13.25		15.19	22
1590	6" diameter		34	.235	↓	2.58	14.45		17.03	24.50
4010	PVC, schedule 80, w/coupling, 1/2" diameter		215	.037	L.F.	1.03	2.28		3.31	4.53
4020	3/4" diameter		180	.044		1.41	2.73		4.14	5.60
4030	1" diameter		145	.055		1.91	3.38		5.29	7.15
4040	1-1/2" diameter		120	.067		3.15	4.09		7.24	9.55
4050	2" diameter	↓	100	.080		4.28	4.91		9.19	12
4060	3" diameter	2 Elec	130	.123		8.95	7.55		16.50	21
4070	4" diameter		90	.178		12.85	10.90		23.75	30.50
4080	5" diameter		70	.229		18.05	14		32.05	41
4090	6" diameter	↓	50	.320	↓	25.50	19.65		45.15	57
4110	Elbows, 1/2" diameter	1 Elec	29	.276	Ea.	2.47	16.90		19.37	27.50
4120	3/4" diameter		23	.348		5.70	21.50		27.20	38.50
4130	1" diameter		20	.400		6.80	24.50		31.30	44
4140	1-1/2" diameter		16	.500		11.40	30.50		41.90	58
4150	2" diameter		12	.667		15.25	41		56.25	78
4160	3" diameter		9	.889		38	54.50		92.50	123
4170	4" diameter		7	1.143		67	70		137	178
4180	5" diameter		6	1.333		168	82		250	305
4190	6" diameter		4	2		72.50	123		195.50	263
4210	Adapter, 1/2" diameter		39	.205		.26	12.60		12.86	19.05
4220	3/4" diameter		33	.242		.37	14.85		15.22	22.50
4230	1" diameter		29	.276		.39	16.90		17.29	25.50
4240	1-1/2" diameter		26	.308		.71	18.90		19.61	29
4250	2" diameter		23	.348		1.39	21.50		22.89	33.50
4260	3" diameter		18	.444		2.85	27.50		30.35	43.50
4270	4" diameter		13	.615		4.80	38		42.80	61.50
4280	5" diameter		11	.727		9.70	44.50		54.20	77
4290	6" diameter		8	1		18.90	61.50		80.40	113
4310	Bell end & cap, 1-1/2" diameter		26	.308		4.69	18.90		23.59	33
4320	Bell end & plug, 2" diameter		23	.348		5.90	21.50		27.40	38.50
4330	3" diameter		18	.444		8.25	27.50		35.75	49.50
4340	4" diameter		13	.615		9.90	38		47.90	67
4350	5" diameter		11	.727		15.95	44.50		60.45	84
4360	6" diameter		8	1		17.05	61.50		78.55	110
4370	Base spacer, 2" diameter		42	.190		1.37	11.70		13.07	18.90
4380	3" diameter		33	.242		1.78	14.85		16.63	24
4390	4" diameter		29	.276		1.78	16.90		18.68	27
4400	5" diameter		26	.308		1.99	18.90		20.89	30
4410	6" diameter		25	.320		2.48	19.65		22.13	31.50
4420	Intermediate spacer, 2" diameter		45	.178		1.25	10.90		12.15	17.65
4430	3" diameter		34	.235		1.76	14.45		16.21	23.50
4440	4" diameter		31	.258		1.59	15.85		17.44	25.50
4450	5" diameter		28	.286		1.94	17.55		19.49	28
4460	6" diameter	↓	25	.320	↓	2.58	19.65		22.23	32

For customer support on your Site Work & Landscape Costs with RSMeans data, call 800.448.8182.

537

33 71 19.17 Electric and Telephone Underground	Crew	Daily Output	Labor-Hours	Unit	Material	2020 Bare Costs Labor	2020 Bare Costs Equipment	Total	Total Incl O&P
0010 **ELECTRIC AND TELEPHONE UNDERGROUND**									
0011 Not including excavation									
0200 backfill and cast in place concrete									
0250 For bedding, see Section 31 23 23.16									
0400 Hand holes, precast concrete, with concrete cover									
0600 2' x 2' x 3' deep	R-3	2.40	8.333	Ea.	545	510	79	1,134	1,450
0800 3' x 3' x 3' deep		1.90	10.526		705	645	99.50	1,449.50	1,850
1000 4' x 4' x 4' deep		1.40	14.286		1,600	875	135	2,610	3,200
1200 Manholes, precast with iron racks & pulling irons, C.I. frame									
1400 and cover, 4' x 6' x 7' deep	B-13	2	28	Ea.	3,000	1,300	290	4,590	5,550
1600 6' x 8' x 7' deep		1.90	29.474		3,375	1,350	305	5,030	6,050
1800 6' x 10' x 7' deep		1.80	31.111		3,775	1,425	325	5,525	6,650
4200 Underground duct, banks ready for concrete fill, min. of 7.5"									
4400 between conduits, center to center									
4580 PVC, type EB, 1 @ 2" diameter	2 Elec	480	.033	L.F.	1.12	2.04		3.16	4.27
4600 2 @ 2" diameter		240	.067		2.23	4.09		6.32	8.55
4800 4 @ 2" diameter		120	.133		4.47	8.20		12.67	17.10
4900 1 @ 3" diameter		400	.040		1.62	2.45		4.07	5.45
5000 2 @ 3" diameter		200	.080		3.24	4.91		8.15	10.85
5200 4 @ 3" diameter		100	.160		6.45	9.80		16.25	21.50
5300 1 @ 4" diameter		320	.050		1.86	3.07		4.93	6.60
5400 2 @ 4" diameter		160	.100		3.72	6.15		9.87	13.25
5600 4 @ 4" diameter		80	.200		7.45	12.25		19.70	26.50
5800 6 @ 4" diameter		54	.296		11.15	18.20		29.35	39.50
5810 1 @ 5" diameter		260	.062		2.09	3.78		5.87	7.90
5820 2 @ 5" diameter		130	.123		4.18	7.55		11.73	15.85
5840 4 @ 5" diameter		70	.229		8.35	14		22.35	30
5860 6 @ 5" diameter		50	.320		12.55	19.65		32.20	43
5870 1 @ 6" diameter		200	.080		3.59	4.91		8.50	11.25
5880 2 @ 6" diameter		100	.160		7.20	9.80		17	22.50
5900 4 @ 6" diameter		50	.320		14.35	19.65		34	45
5920 6 @ 6" diameter		30	.533		21.50	32.50		54	72
6200 Rigid galvanized steel, 2 @ 2" diameter		180	.089		15.75	5.45		21.20	25.50
6400 4 @ 2" diameter		90	.178		31.50	10.90		42.40	51
6800 2 @ 3" diameter		100	.160		29	9.80		38.80	46
7000 4 @ 3" diameter		50	.320		57.50	19.65		77.15	92.50
7200 2 @ 4" diameter		70	.229		45.50	14		59.50	71
7400 4 @ 4" diameter		34	.471		91.50	29		120.50	143
7600 6 @ 4" diameter		22	.727		137	44.50		181.50	218
7620 2 @ 5" diameter		60	.267		72	16.35		88.35	104
7640 4 @ 5" diameter		30	.533		144	32.50		176.50	208
7660 6 @ 5" diameter		18	.889		216	54.50		270.50	320
7680 2 @ 6" diameter		40	.400		110	24.50		134.50	158
7700 4 @ 6" diameter		20	.800		219	49		268	315
7720 6 @ 6" diameter		14	1.143		330	70		400	465
7800 For cast-in-place concrete, add									
7801 For cable, see Section 26 05 19.55									
7810 Under 1 C.Y.	C-6	16	3	C.Y.	206	131	3.36	340.36	425
7820 1 C.Y. to 5 C.Y.		19.20	2.500		187	109	2.80	298.80	370
7830 Over 5 C.Y.		24	2		155	87.50	2.24	244.74	305
7850 For reinforcing rods, add									
7860 #4 to #7	2 Rodm	1.10	14.545	Ton	1,125	820		1,945	2,475

33 71 Electrical Utility Transmission and Distribution

33 71 19 – Electrical Underground Ducts and Manholes

33 71 19.17 Electric and Telephone Underground	Crew	Daily Output	Labor-Hours	Unit	Material	2020 Bare Costs Labor	Equipment	Total	Total Incl O&P
7870	#8 to #14	2 Rodm	1.50	10.667	Ton	1,125	600		1,725	2,150
8000	Fittings, PVC type EB, elbow, 2" diameter	1 Elec	16	.500	Ea.	15.85	30.50		46.35	63
8200	3" diameter		14	.571		18.35	35		53.35	72
8400	4" diameter		12	.667		40	41		81	105
8420	5" diameter		10	.800		123	49		172	208
8440	6" diameter		9	.889		191	54.50		245.50	291
8500	Coupling, 2" diameter					.85			.85	.94
8600	3" diameter					4.14			4.14	4.55
8700	4" diameter					4.48			4.48	4.93
8720	5" diameter					8.55			8.55	9.40
8740	6" diameter					24			24	26
8800	Adapter, 2" diameter	1 Elec	26	.308		1.01	18.90		19.91	29
9000	3" diameter		20	.400		3.07	24.50		27.57	40
9200	4" diameter		16	.500		3.36	30.50		33.86	49
9220	5" diameter		13	.615		8.55	38		46.55	65.50
9240	6" diameter		10	.800		12.50	49		61.50	87
9400	End bell, 2" diameter		16	.500		2.30	30.50		32.80	48
9600	3" diameter		14	.571		4.08	35		39.08	56.50
9800	4" diameter		12	.667		4.63	41		45.63	66
9810	5" diameter		10	.800		8.10	49		57.10	82
9820	6" diameter		8	1		10.15	61.50		71.65	103
9830	5° angle coupling, 2" diameter		26	.308		7.80	18.90		26.70	36.50
9840	3" diameter		20	.400		26	24.50		50.50	65
9850	4" diameter		16	.500		19.15	30.50		49.65	66.50
9860	5" diameter		13	.615		16.70	38		54.70	74.50
9870	6" diameter		10	.800		26	49		75	102
9880	Expansion joint, 2" diameter		16	.500		33.50	30.50		64	82.50
9890	3" diameter		18	.444		70	27.50		97.50	118
9900	4" diameter		12	.667		102	41		143	174
9910	5" diameter		10	.800		203	49		252	296
9920	6" diameter		8	1		176	61.50		237.50	286
9930	Heat bender, 2" diameter					490			490	540
9940	6" diameter					1,500			1,500	1,650
9950	Cement, quart					15.55			15.55	17.10

33 81 Communications Structures

33 81 13 – Communications Transmission Towers

33 81 13.10 Radio Towers

0010	**RADIO TOWERS**									
0020	Guyed, 50' H, 40 lb. section, 70 MPH basic wind speed	2 Sswk	1	16	Ea.	2,875	920		3,795	4,625
0100	Wind load 90 MPH basic wind speed	"	1	16		3,325	920		4,245	5,100
0300	190' high, 40 lb. section, wind load 70 MPH basic wind speed	K-2	.33	72.727		9,575	4,000	2,475	16,050	19,400
0400	200' high, 70 lb. section, wind load 90 MPH basic wind speed		.33	72.727		15,100	4,000	2,475	21,575	25,500
0600	300' high, 70 lb. section, wind load 70 MPH basic wind speed		.20	120		24,900	6,575	4,100	35,575	42,100
0700	270' high, 90 lb. section, wind load 90 MPH basic wind speed		.20	120		28,100	6,575	4,100	38,775	45,600
0800	400' high, 100 lb. section, wind load 70 MPH basic wind speed		.14	171		37,600	9,400	5,850	52,850	62,500
0900	Self-supporting, 60' high, wind load 70 MPH basic wind speed		.80	30		4,750	1,650	1,025	7,425	8,900
0910	60' high, wind load 90 MPH basic wind speed		.45	53.333		5,575	2,925	1,825	10,325	12,700
1000	120' high, wind load 70 MPH basic wind speed		.40	60		10,100	3,300	2,050	15,450	18,500
1200	190' high, wind load 90 MPH basic wind speed		.20	120		29,000	6,575	4,100	39,675	46,500
2000	For states west of Rocky Mountains, add for shipping					10%				

For customer support on your Site Work & Landscape Costs with RSMeans data, call 800.448.8182.

539

Division Notes

	CREW	DAILY OUTPUT	LABOR-HOURS	UNIT	BARE COSTS				TOTAL INCL O&P
					MAT.	LABOR	EQUIP.	TOTAL	

Estimating Tips
34 11 00 Rail Tracks
This subdivision includes items that may involve either repair of existing or construction of new railroad tracks. Additional preparation work, such as the roadbed earthwork, would be found in Division 31. Additional new construction siding and turnouts are found in Subdivision 34 72. Maintenance of railroads is found under 34 01 23 Operation and Maintenance of Railways.

34 40 00 Traffic Signals
This subdivision includes traffic signal systems. Other traffic control devices such as traffic signs are found in Subdivision 10 14 53 Traffic Signage.

34 70 00 Vehicle Barriers
This subdivision includes security vehicle barriers, guide and guard rails, crash barriers, and delineators. The actual maintenance and construction of concrete and asphalt pavement are found in Division 32.

Reference Numbers
Reference numbers are shown at the beginning of some major classifications. These numbers refer to related items in the Reference Section. The reference information may be an estimating procedure, an alternate pricing method, or technical information.

Note: Not all subdivisions listed here necessarily appear. ∎

Same Data. Simplified.

Enjoy the convenience and efficiency of accessing your costs anywhere:

- **Skip the multiplier** by setting your location
- **Quickly search,** edit, favorite and share costs
- **Stay on top of price changes** with automatic updates

Discover more at rsmeans.com/online

34 01 Operation and Maintenance of Transportation

34 01 13 – Operation and Maintenance of Roadways

34 01 13.10 Maintenance Grading of Roadways		Crew	Daily Output	Labor-Hours	Unit	Material	2020 Bare Costs		Total	Total Incl O&P
							Labor	Equipment		
0010	**MAINTENANCE GRADING OF ROADWAYS**									
0100	Maintenance grading mob/demob of equipment per hour	B-11L	8	2	Hr.		99	133	232	294
0200	Maintenance grading of ditches one ditch slope only 1.0 MPH average		5	3.200	Mile		158	212	370	470
0210	1.5 MPH		7.50	2.133			105	142	247	315
0220	2.0 MPH		10	1.600			79	106	185	235
0230	2.5 MPH		12.50	1.280			63.50	85	148.50	189
0240	3.0 MPH		15	1.067			52.50	71	123.50	157
0300	Maintenance grading of roadway, 3 passes, 3.0 MPH average		5	3.200			158	212	370	470
0310	4 passes		3.80	4.211			208	280	488	615
0320	3 passes, 3.5 MPH average		5.80	2.759			136	183	319	405
0330	4 passes		4.40	3.636			180	241	421	535
0340	3 passes, 4.0 MPH average		6.70	2.388			118	159	277	350
0350	4 passes		5	3.200			158	212	370	470
0360	3 passes, 4.5 MPH average		7.50	2.133			105	142	247	315
0370	4 passes		5.60	2.857			141	190	331	420
0380	3 passes, 5.0 MPH average		8.30	1.928			95.50	128	223.50	284
0390	4 passes		6.30	2.540			126	169	295	375
0400	3 passes, 5.5 MPH average		9.20	1.739			86	115	201	256
0410	4 passes		6.90	2.319			115	154	269	340
0420	3 passes, 6.0 MPH average		10	1.600			79	106	185	235
0430	4 passes		7.50	2.133			105	142	247	315
0440	3 passes, 6.5 MPH average		10.80	1.481			73	98.50	171.50	218
0450	4 passes		8.10	1.975			97.50	131	228.50	290
0460	3 passes, 7.0 MPH average		11.70	1.368			67.50	91	158.50	201
0470	4 passes		8.80	1.818			90	121	211	268
0480	3 passes, 7.5 MPH average		12.50	1.280			63.50	85	148.50	189
0490	4 passes		9.40	1.702			84	113	197	250
0500	3 passes, 8.0 MPH average		13.30	1.203			59.50	80	139.50	177
0510	4 passes		10	1.600			79	106	185	235
0520	3 passes, 8.5 MPH average		14.20	1.127			55.50	75	130.50	166
0530	4 passes		10.60	1.509			74.50	100	174.50	222
0540	3 passes, 9.0 MPH average		15	1.067			52.50	71	123.50	157
0550	4 passes		11.30	1.416			70	94	164	208
0560	3 passes, 9.5 MPH average		15.80	1.013			50	67	117	149
0570	4 passes		11.90	1.345			66.50	89.50	156	198
0600	Maintenance grading of roadway, 8' wide pass, 3.0 MPH average		15	1.067			52.50	71	123.50	157
0610	3.5 MPH average		17.50	.914			45	60.50	105.50	135
0620	4.0 MPH average		20	.800			39.50	53	92.50	118
0630	4.5 MPH average		22.50	.711			35	47	82	105
0640	5.0 MPH average		25	.640			31.50	42.50	74	94
0650	5.5 MPH average		27.50	.582			29	38.50	67.50	85.50
0660	6.0 MPH average		30	.533			26.50	35.50	62	78.50
0670	6.5 MPH average		32.50	.492			24.50	32.50	57	72.50
0680	7.0 MPH average		35	.457			22.50	30.50	53	67.50
0690	7.5 MPH average		37.50	.427			21	28.50	49.50	62.50
0700	8.0 MPH average		40	.400			19.75	26.50	46.25	58.50
0710	8.5 MPH average		42.50	.376			18.60	25	43.60	55.50
0720	9.0 MPH average		45	.356			17.60	23.50	41.10	52.50
0730	9.5 MPH average		47.50	.337			16.65	22.50	39.15	49.50

34 01 Operation and Maintenance of Transportation

34 01 23 – Operation and Maintenance of Railways

34 01 23.51 Maintenance of Railroads	Crew	Daily Output	Labor-Hours	Unit	Material	2020 Bare Costs Labor	2020 Bare Costs Equipment	Total	Total Incl O&P
0010 **MAINTENANCE OF RAILROADS**									
0400 Resurface and realign existing track	B-14	200	.240	L.F.		10.60	1.07	11.67	17.15
0600 For crushed stone ballast, add	"	500	.096	"	14.65	4.25	.43	19.33	23

34 11 Rail Tracks

34 11 13 – Track Rails

34 11 13.23 Heavy Rail Track

	Crew	Daily Output	Labor-Hours	Unit	Material	2020 Bare Costs Labor	2020 Bare Costs Equipment	Total	Total Incl O&P
0010 **HEAVY RAIL TRACK** R347216-10									
1000 Rail, 100 lb. prime grade				L.F.	33			33	36
1500 Relay rail				"	16.40			16.40	18.05

34 11 33 – Track Cross Ties

34 11 33.13 Concrete Track Cross Ties

	Crew	Daily Output	Labor-Hours	Unit	Material	2020 Bare Costs Labor	2020 Bare Costs Equipment	Total	Total Incl O&P
0010 **CONCRETE TRACK CROSS TIES**									
1400 Ties, concrete, 8'-6" long, 30" OC	B-14	80	.600	Ea.	153	26.50	2.67	182.17	211

34 11 33.16 Timber Track Cross Ties

	Crew	Daily Output	Labor-Hours	Unit	Material	2020 Bare Costs Labor	2020 Bare Costs Equipment	Total	Total Incl O&P
0010 **TIMBER TRACK CROSS TIES**									
1600 Wood, pressure treated, 6" x 8" x 8'-6", C.L. lots	B-14	90	.533	Ea.	56	23.50	2.37	81.87	99.50
1700 L.C.L. lots		90	.533		59	23.50	2.37	84.87	103
1900 Heavy duty, 7" x 9" x 8'-6", C.L. lots		70	.686		61.50	30.50	3.05	95.05	117
2000 L.C.L. lots	↓	70	.686	↓	61.50	30.50	3.05	95.05	117

34 11 33.17 Timber Switch Ties

	Crew	Daily Output	Labor-Hours	Unit	Material	2020 Bare Costs Labor	2020 Bare Costs Equipment	Total	Total Incl O&P
0010 **TIMBER SWITCH TIES**									
1200 Switch timber, for a #8 switch, pressure treated	B-14	3.70	12.973	M.B.F.	3,375	575	57.50	4,007.50	4,625
1300 Complete set of timbers, 3.7 MBF for #8 switch	"	1	48	Total	13,000	2,125	214	15,339	17,700

34 11 93 – Track Appurtenances and Accessories

34 11 93.50 Track Accessories

	Crew	Daily Output	Labor-Hours	Unit	Material	2020 Bare Costs Labor	2020 Bare Costs Equipment	Total	Total Incl O&P
0010 **TRACK ACCESSORIES**									
0020 Car bumpers, test	B-14	2	24	Ea.	3,650	1,050	107	4,807	5,750
0100 Heavy duty		2	24		6,925	1,050	107	8,082	9,325
0200 Derails hand throw (sliding)		10	4.800		1,525	212	21.50	1,758.50	2,025
0300 Hand throw with standard timbers, open stand & target		8	6	↓	1,650	266	26.50	1,942.50	2,225
2400 Wheel stops, fixed		18	2.667	Pr.	920	118	11.85	1,049.85	1,225
2450 Hinged	↓	14	3.429	"	1,250	152	15.25	1,417.25	1,625

34 11 93.60 Track Material

	Crew	Daily Output	Labor-Hours	Unit	Material	2020 Bare Costs Labor	2020 Bare Costs Equipment	Total	Total Incl O&P
0010 **TRACK MATERIAL**									
0020 Track bolts				Ea.	3.64			3.64	4
0100 Joint bars				Pr.	90			90	98.50
0200 Spikes				Ea.	2.37			2.37	2.61
0300 Tie plates				"	15			15	16.50

34 41 Roadway Signaling and Control Equipment

34 41 13 – Traffic Signals

34 41 13.10 Traffic Signals Systems

		Crew	Daily Output	Labor-Hours	Unit	Material	2020 Bare Costs Labor	Equipment	Total	Total Incl O&P
0010	**TRAFFIC SIGNALS SYSTEMS**									
0020	Component costs									
0600	Crew employs crane/directional driller as required									
1000	Vertical mast with foundation									
1010	Mast sized for single arm to 40'; no lighting or power function	R-11	.50	112	Signal	10,700	6,525	1,275	18,500	23,000
1100	Horizontal arm									
1110	Per linear foot of arm	R-11	50	1.120	Signal	230	65.50	12.75	308.25	365
1200	Traffic signal									
1210	Includes signal, bracket, sensor, and wiring	R-11	2.50	22.400	Signal	1,150	1,300	255	2,705	3,500
1300	Pedestrian signals and callers									
1310	Includes four signals with brackets and two call buttons	R-11	2.50	22.400	Signal	3,775	1,300	255	5,330	6,375
1400	Controller, design, and underground conduit									
1410	Includes miscellaneous signage and adjacent surface work	R-11	.25	224	Signal	25,700	13,100	2,550	41,350	50,500

34 71 Roadway Construction

34 71 13 – Vehicle Barriers

34 71 13.17 Security Vehicle Barriers

		Crew	Daily Output	Labor-Hours	Unit	Material	2020 Bare Costs Labor	Equipment	Total	Total Incl O&P
0010	**SECURITY VEHICLE BARRIERS**									
0020	Security planters excludes filling material									
0100	Concrete security planter, exposed aggregate 36" diam. x 30" high	B-11M	8	2	Ea.	555	99	29	683	790
0200	48" diam. x 36" high		8	2		790	99	29	918	1,050
0300	53" diam. x 18" high		8	2		720	99	29	848	970
0400	72" diam. x 18" high with seats		8	2		1,975	99	29	2,103	2,350
0450	84" diam. x 36" high		8	2		2,075	99	29	2,203	2,475
0500	36" x 36" x 24" high square		8	2		460	99	29	588	685
0600	36" x 36" x 30" high square		8	2		580	99	29	708	815
0700	48" L x 24" W x 30" H rectangle		8	2		485	99	29	613	710
0800	72" L x 24" W x 30" H rectangle		8	2		580	99	29	708	815
0900	96" L x 24" W x 30" H rectangle		8	2		745	99	29	873	1,000
0950	Decorative geometric concrete barrier, 96" L x 24" W x 36" H		8	2		820	99	29	948	1,075
1000	Concrete security planter, filling with washed sand or gravel <1 C.Y.		8	2		34	99	29	162	217
1050	2 C.Y. or less		6	2.667		67.50	132	39	238.50	315
1200	Jersey concrete barrier, 10' L x 2' by 0.5' W x 30" H	B-21B	16	2.500		395	115	29.50	539.50	640
1300	10 or more same site		24	1.667		395	76.50	19.65	491.15	570
1400	10' L x 2' by 0.5' W x 32" H		16	2.500		820	115	29.50	964.50	1,100
1500	10 or more same site		24	1.667		820	76.50	19.65	916.15	1,025
1600	20' L x 2' by 0.5' W x 30" H		12	3.333		790	153	39.50	982.50	1,150
1700	10 or more same site		18	2.222		790	102	26	918	1,050
1800	20' L x 2' by 0.5' W x 32" H		12	3.333		695	153	39.50	887.50	1,050
1900	10 or more same site		18	2.222		695	102	26	823	945
2000	GFRC decorative security barrier per 10' section including concrete		4	10		3,625	460	118	4,203	4,825
2100	Per 12' section including concrete		4	10		4,350	460	118	4,928	5,625
2210	GFRC decorative security barrier will stop 30 MPH, 4,000 lb. vehicle									
2300	High security barrier base prep per 12' section on bare ground	B-11C	4	4	Ea.	25	198	53.50	276.50	385
2310	GFRC barrier base prep does not include haulaway of excavated matl.									
2400	GFRC decorative high security barrier per 12' section w/concrete	B-21B	3	13.333	Ea.	5,850	610	157	6,617	7,525
2410	GFRC decorative high security barrier will stop 50 MPH, 15,000 lb. vehicle									
2500	GFRC decorative impaler security barrier per 10' section w/prep	B-6	4	6	Ea.	2,275	274	53.50	2,602.50	2,975
2600	Per 12' section w/prep	"	4	6	"	2,750	274	53.50	3,077.50	3,500
2610	Impaler barrier should stop 50 MPH, 15,000 lb. vehicle w/some penetr.									
2700	Pipe bollards, steel, concrete filled/painted, 8' L x 4' D hole, 8" diam.	B-6	10	2.400	Ea.	865	110	21.50	996.50	1,150

544

34 71 Roadway Construction

34 71 13 – Vehicle Barriers

34 71 13.17 Security Vehicle Barriers

		Crew	Daily Output	Labor-Hours	Unit	Material	2020 Bare Costs Labor	Equipment	Total	Total Incl O&P
2710	Schedule 80 concrete bollards will stop 4,000 lb. vehicle @ 30 MPH									
2800	GFRC decorative jersey barrier cover per 10' section excludes soil	B-6	8	3	Ea.	1,000	137	26.50	1,163.50	1,325
2900	Per 12' section excludes soil		8	3		1,225	137	26.50	1,388.50	1,550
3000	GFRC decorative 8" diameter bollard cover		12	2		390	91.50	17.80	499.30	585
3100	GFRC decorative barrier face 10' section excludes earth backing		10	2.400		980	110	21.50	1,111.50	1,275
3200	Drop arm crash barrier, 15,000 lb. vehicle @ 50 MPH									
3205	Includes all material, labor for complete installation									
3210	12' width				Ea.				64,500	71,000
3310	24' width								86,000	95,000
3410	Wedge crash barrier, 10' width								98,000	108,000
3510	12.5' width								108,000	119,000
3520	15' width								118,500	130,000
3610	Sliding crash barrier, 12' width								178,500	196,000
3710	Sliding roller crash barrier, 20' width								198,000	217,500
3810	Sliding cantilever crash barrier, 20' width								198,000	217,500
3890	Note: Raised bollard crash barriers should be used w/tire shredders									
3910	Raised bollard crash barrier, 10' width				Ea.				38,800	42,900
4010	12' width								44,900	49,000
4110	Raised bollard crash barrier, 10' width, solar powered								46,900	52,000
4210	12' width								53,000	58,000
4310	In ground tire shredder, 16' width								43,900	47,900

34 71 13.26 Vehicle Guide Rails

		Crew	Daily Output	Labor-Hours	Unit	Material	2020 Bare Costs Labor	Equipment	Total	Total Incl O&P
0010	**VEHICLE GUIDE RAILS**									
0012	Corrugated stl., galv. stl. posts, 6'-3" OC	B-80	850	.038	L.F.	27.50	1.75	1.20	30.45	34
0100	Double face		570	.056	"	37	2.62	1.79	41.41	47
0200	End sections, galvanized, flared		50	.640	Ea.	99	30	20.50	149.50	176
0300	Wrap around end		50	.640		141	30	20.50	191.50	223
0350	Anchorage units		15	2.133		1,225	99.50	68	1,392.50	1,575
0365	End section, flared	B-80A	78	.308		46	12.95	10.50	69.45	82
0370	End section, wrap-around, corrugated steel	"	78	.308		58.50	12.95	10.50	81.95	95.50
0400	Timber guide rail, 4" x 8" with 6" x 8" wood posts, treated	B-80	960	.033	L.F.	13.45	1.55	1.06	16.06	18.25
0600	Cable guide rail, 3 at 3/4" cables, steel posts, single face		900	.036		11.75	1.66	1.13	14.54	16.70
0650	Double face		635	.050		24	2.35	1.61	27.96	31.50
0700	Wood posts		950	.034		13.30	1.57	1.07	15.94	18.15
0750	Double face		650	.049		23.50	2.29	1.57	27.36	31
0760	Breakaway wood posts		195	.164	Ea.	360	7.65	5.25	372.90	410
0800	Anchorage units, breakaway		15	2.133	"	1,500	99.50	68	1,667.50	1,875
0900	Guide rail, steel box beam, 6" x 6"		120	.267	L.F.	37.50	12.45	8.50	58.45	69
0950	End assembly	B-80A	48	.500	Ea.	1,500	21	17.10	1,538.10	1,725
1100	Median barrier, steel box beam, 6" x 8"	B-80	215	.149	L.F.	52	6.95	4.75	63.70	72.50
1120	Shop curved	B-80A	92	.261	"	67	11	8.90	86.90	100
1140	End assembly		48	.500	Ea.	2,525	21	17.10	2,563.10	2,825
1150	Corrugated beam		400	.060	L.F.	49	2.53	2.05	53.58	60
1400	Resilient guide fence and light shield, 6' high	B-2	130	.308	"	22	13.10		35.10	44
1500	Concrete posts, individual, 6'-5", triangular	B-80	110	.291	Ea.	71.50	13.55	9.30	94.35	110
1550	Square		110	.291		77	13.55	9.30	99.85	116
1600	Wood guide posts		150	.213		46	9.95	6.80	62.75	73
2000	Median, precast concrete, 3'-6" high, 2' wide, single face	B-29	380	.147	L.F.	60.50	6.80	2.23	69.53	79
2200	Double face	"	340	.165		69	7.60	2.49	79.09	90
2300	Cast in place, steel forms	C-2	170	.282		52.50	14.60		67.10	80
2320	Slipformed	C-7	352	.205		43.50	9.35	3.07	55.92	65.50
2400	Speed bumps, thermoplastic, 10-1/2" x 2-1/4" x 48" long	B-2	120	.333	Ea.	112	14.15		126.15	145

For customer support on your Site Work & Landscape Costs with RSMeans data, call 800.448.8182.

545

34 71 Roadway Construction

34 71 13 – Vehicle Barriers

34 71 13.26 Vehicle Guide Rails

		Crew	Daily Output	Labor-Hours	Unit	Material	2020 Bare Costs Labor	Equipment	Total	Total Incl O&P
3000	Energy absorbing terminal, 10 bay, 3' wide	B-80B	.10	320	Ea.	40,300	14,300	2,350	56,950	68,500
3010	7 bay, 2'-6" wide		.20	160		32,100	7,175	1,175	40,450	47,400
3020	5 bay, 2' wide	↓	.20	160		24,600	7,175	1,175	32,950	39,100
3030	Impact barrier, MUTCD, barrel type	B-16	30	1.067		405	47.50	19.10	471.60	535
3100	Wide hazard protection, foam sandwich, 7 bay, 7'-6" wide	B-80B	.14	229		33,400	10,200	1,675	45,275	54,000
3110	5' wide		.15	213		26,200	9,575	1,575	37,350	44,800
3120	3' wide	↓	.18	178	↓	25,000	7,975	1,300	34,275	41,000
5000	For bridge railing, see Section 32 34 10.10									

34 71 19 – Vehicle Delineators

34 71 19.13 Fixed Vehicle Delineators

		Crew	Daily Output	Labor-Hours	Unit	Material	2020 Bare Costs Labor	Equipment	Total	Total Incl O&P
0010	**FIXED VEHICLE DELINEATORS**									
0020	Crash barriers									
0100	Traffic channelizing pavement markers, layout only	A-7	2000	.012	Ea.		.69	.02	.71	1.05
0110	13" x 7-1/2" x 2-1/2" high, non-plowable, install	2 Clab	96	.167		29.50	7		36.50	43
0200	8" x 8" x 3-1/4" high, non-plowable, install		96	.167		29	7		36	42.50
0230	4" x 4" x 3/4" high, non-plowable, install	↓	120	.133		2.68	5.60		8.28	11.40
0240	9-1/4" x 5-7/8" x 1/4" high, plowable, concrete pavmt.	A-2A	70	.343		4.78	15	4.40	24.18	32.50
0250	9-1/4" x 5-7/8" x 1/4" high, plowable, asphalt pavmt.	"	120	.200		3.93	8.75	2.56	15.24	20.50
0300	Barrier and curb delineators, reflectorized, 4" x 6"	2 Clab	150	.107		3.66	4.49		8.15	10.80
0310	3" x 5"	"	150	.107	↓	5	4.49		9.49	12.25
0500	Rumble strip, polycarbonate									
0510	24" x 3-1/2" x 1/2" high	2 Clab	50	.320	Ea.	10.25	13.45		23.70	31.50

34 72 Railway Construction

34 72 16 – Railway Siding

34 72 16.50 Railroad Sidings

		Crew	Daily Output	Labor-Hours	Unit	Material	2020 Bare Costs Labor	Equipment	Total	Total Incl O&P
0010	**RAILROAD SIDINGS** R347216-10									
0800	Siding, yard spur, level grade									
0808	Wood ties and ballast, 80 lb. new rail	B-14	57	.842	L.F.	116	37.50	3.75	157.25	188
0809	80 lb. relay rail		57	.842		85.50	37.50	3.75	126.75	154
0812	90 lb. new rail		57	.842		124	37.50	3.75	165.25	197
0813	90 lb. relay rail		57	.842		86.50	37.50	3.75	127.75	155
0820	100 lb. new rail		57	.842		132	37.50	3.75	173.25	205
0822	100 lb. relay rail		57	.842		90.50	37.50	3.75	131.75	160
0830	110 lb. new rail		57	.842		140	37.50	3.75	181.25	214
0832	110 lb. relay rail	↓	57	.842	↓	95.50	37.50	3.75	136.75	165
1002	Steel ties in concrete, incl. fasteners & plates									
1003	80 lb. new rail	B-14	22	2.182	L.F.	181	96.50	9.70	287.20	355
1005	80 lb. relay rail		22	2.182		153	96.50	9.70	259.20	325
1012	90 lb. new rail		22	2.182		187	96.50	9.70	293.20	360
1015	90 lb. relay rail		22	2.182		156	96.50	9.70	262.20	330
1020	100 lb. new rail		22	2.182		193	96.50	9.70	299.20	370
1025	100 lb. relay rail		22	2.182		159	96.50	9.70	265.20	330
1030	110 lb. new rail		22	2.182		202	96.50	9.70	308.20	380
1035	110 lb. relay rail	↓	22	2.182	↓	164	96.50	9.70	270.20	335

34 72 Railway Construction

34 72 16 – Railway Siding

34 72 16.60 Railroad Turnouts	Crew	Daily Output	Labor-Hours	Unit	Material	2020 Bare Costs Labor	Equipment	Total	Total Incl O&P
0010 **RAILROAD TURNOUTS**									
2200 Turnout, #8 complete, w/rails, plates, bars, frog, switch point,									
2250 timbers, and ballast to 6" below bottom of ties									
2280 90 lb. rails	B-13	.25	224	Ea.	40,900	10,300	2,325	53,525	63,000
2290 90 lb. relay rails		.25	224		27,500	10,300	2,325	40,125	48,400
2300 100 lb. rails		.25	224		45,600	10,300	2,325	58,225	68,000
2310 100 lb. relay rails		.25	224		30,700	10,300	2,325	43,325	52,000
2320 110 lb. rails		.25	224		50,500	10,300	2,325	63,125	73,500
2330 110 lb. relay rails		.25	224		33,800	10,300	2,325	46,425	55,500
2340 115 lb. rails		.25	224		55,000	10,300	2,325	67,625	78,500
2350 115 lb. relay rails		.25	224		35,800	10,300	2,325	48,425	57,500
2360 132 lb. rails		.25	224		63,000	10,300	2,325	75,625	87,000
2370 132 lb. relay rails		.25	224		41,700	10,300	2,325	54,325	64,000

Division Notes

		CREW	DAILY OUTPUT	LABOR-HOURS	UNIT	BARE COSTS				TOTAL INCL O&P
						MAT.	LABOR	EQUIP.	TOTAL	

Estimating Tips

35 01 50 Operation and Maintenance of Marine Construction

Includes unit price lines for pile cleaning and pile wrapping for protection.

35 20 16 Hydraulic Gates

This subdivision includes various types of gates that are commonly used in waterway and canal construction. Various earthwork and structural support items are found in Division 31, and concrete work is found in Division 3.

35 20 23 Dredging

This subdivision includes barge and shore dredging systems for rivers, canals, and channels.

35 31 00 Shoreline Protection

This subdivision includes breakwaters, bulkheads, and revetments for ocean and river inlets. Additional earthwork may be required from Division 31, as well as concrete work from Division 3.

35 41 00 Levees

Contains information on levee construction, including the estimated cost of clay cone material.

35 49 00 Waterway Structures

This subdivision includes breakwaters and bulkheads for canals.

35 51 00 Floating Construction

This section includes floating piers, docks, and dock accessories. Fixed Pier Timber Construction is found in Section 06 13 33. Driven piles are found in Division 31, as are sheet piling, cofferdams, and riprap.

Reference Numbers

Reference numbers are shown at the beginning of some major classifications. These numbers refer to related items in the Reference Section. The reference information may be an estimating procedure, an alternate pricing method, or technical information.

Note: Not all subdivisions listed here necessarily appear. ■

Same Data. Simplified.

Enjoy the convenience and efficiency of accessing your costs anywhere:

- **Skip the multiplier** by setting your location
- **Quickly search,** edit, favorite and share costs
- **Stay on top of price changes** with automatic updates

Discover more at rsmeans.com/online

---

35 01 50.10 Cleaning of Marine Pier Piles		Crew	Daily Output	Labor-Hours	Unit	Material	2020 Bare Costs Labor	Equipment	Total	Total Incl O&P
0010	**CLEANING OF MARINE PIER PILES**									
0015	Exposed piles pressure washing from boat									
0100	Clean wood piles from boat, 8" diam., 5' length	G	B-1D	17	.941	Ea.		39.50	12.35	51.85	73
0110	6-10' long	G		9	1.778			75	23.50	98.50	138
0120	11-15' long	G		6.50	2.462			104	32.50	136.50	192
0130	10" diam., 5' length	G		13.75	1.164			49	15.30	64.30	90.50
0140	6-10' long	G		7.25	2.207			93	29	122	172
0150	11-15' long	G		5	3.200			135	42	177	248
0160	12" diam., 5' length	G		11.50	1.391			58.50	18.25	76.75	108
0170	6-10' long	G		6	2.667			112	35	147	208
0180	11-15' long	G		4.25	3.765			158	49.50	207.50	293
0190	13" diam., 5' length	G		10.50	1.524			64	20	84	119
0200	6-10' long	G		5.50	2.909			122	38	160	226
0210	11-15' long	G		4	4			168	52.50	220.50	310
0220	14" diam., 5' length	G		9.75	1.641			69	21.50	90.50	128
0230	6-10' long	G		5.25	3.048			128	40	168	237
0240	11-15' long	G		3.75	4.267			180	56	236	330
0300	Clean concrete piles from boat, 12" diam., 5' length	G		14	1.143			48	15	63	89
0310	6-10' long	G		7.25	2.207			93	29	122	172
0320	11-15' long	G		5	3.200			135	42	177	248
0330	14" diam., 5' length	G		12	1.333			56	17.50	73.50	104
0340	6-10' long	G		6.25	2.560			108	33.50	141.50	199
0350	11-15' long	G		4.50	3.556			150	46.50	196.50	277
0360	16" diam., 5' length	G		10.50	1.524			64	20	84	119
0370	6-10' long	G		5.50	2.909			122	38	160	226
0380	11-15' long	G		4	4			168	52.50	220.50	310
0390	18" diam., 5' length	G		9.25	1.730			73	22.50	95.50	134
0400	6-10' long	G		5	3.200			135	42	177	248
0410	11-15' long	G		3.50	4.571			192	60	252	355
0420	20" diam., 5' length	G		8.50	1.882			79.50	24.50	104	146
0430	6-10' long	G		4.50	3.556			150	46.50	196.50	277
0440	11-15' long	G		3	5.333			225	70	295	410
0450	24" diam., 5' length	G		7	2.286			96	30	126	178
0460	6-10' long	G		3.75	4.267			180	56	236	330
0470	11-15' long	G		2.50	6.400			269	84	353	500
0480	36" diam., 5' length	G		4.75	3.368			142	44	186	262
0490	6-10' long	G		2.50	6.400			269	84	353	500
0500	11-15' long	G		1.75	9.143			385	120	505	710
0510	54" diam., 5' length	G		3	5.333			225	70	295	410
0520	6-10' long	G		1.75	9.143			385	120	505	710
0530	11-15' long	G		1.25	12.800			540	168	708	995
0540	66" diam., 5' length	G		2.50	6.400			269	84	353	500
0550	6-10' long	G		1.50	10.667			450	140	590	830
0560	11-15' long	G		1	16			675	210	885	1,225
0600	10" square pile, 5' length	G		13.25	1.208			51	15.85	66.85	94
0610	6-10' long	G		7	2.286			96	30	126	178
0620	11-15' long	G		4.75	3.368			142	44	186	262
0630	12" square pile, 5' length	G		11	1.455			61	19.10	80.10	113
0640	6-10' long	G		5.75	2.783			117	36.50	153.50	216
0650	11-15' long	G		4	4			168	52.50	220.50	310
0660	14" square pile, 5' length	G		9.50	1.684			71	22	93	132
0670	6-10' long	G		5	3.200			135	42	177	248
0680	11-15' long	G		3.50	4.571			192	60	252	355

35 01 50.10 Cleaning of Marine Pier Piles		Crew	Daily Output	Labor-Hours	Unit	Material	2020 Bare Costs Labor	Equipment	Total	Total Incl O&P
0690	16" square pile, 5' length	G	B-1D	8.25	1.939	Ea.		81.50	25.50	107	151
0700	6-10' long	G		4.25	3.765			158	49.50	207.50	293
0710	11-15' long	G		3	5.333			225	70	295	410
0720	18" square pile, 5' length	G		7.25	2.207			93	29	122	172
0730	6-10' long	G		3.75	4.267			180	56	236	330
0740	11-15' long	G		2.75	5.818			245	76.50	321.50	455
0750	20" square pile, 5' length	G		6.50	2.462			104	32.50	136.50	192
0760	6-10' long	G		3.50	4.571			192	60	252	355
0770	11-15' long	G		2.50	6.400			269	84	353	500
0780	24" square pile, 5' length	G		5.50	2.909			122	38	160	226
0790	6-10' long	G		3	5.333			225	70	295	410
0800	11-15' long	G		2	8			335	105	440	620
0810	14" diameter octagonal piles, 5' length	G		12.50	1.280			54	16.80	70.80	99.50
0820	6-10' long	G		6.50	2.462			104	32.50	136.50	192
0830	11-15' long	G		4.50	3.556			150	46.50	196.50	277
0840	16" diameter octagonal piles, 5' length	G		11	1.455			61	19.10	80.10	113
0850	6-10' long	G		5.75	2.783			117	36.50	153.50	216
0860	11-15' long	G		4	4			168	52.50	220.50	310
0870	18" diameter octagonal piles, 5' length	G		9.75	1.641			69	21.50	90.50	128
0880	6-10' long	G		5	3.200			135	42	177	248
0890	11-15' long	G		3.50	4.571			192	60	252	355
0900	20" diameter octagonal piles, 5' length	G		8.75	1.829			77	24	101	143
0910	6-10' long	G		4.50	3.556			150	46.50	196.50	277
0920	11-15' long	G		3.25	4.923			207	64.50	271.50	380
0930	24" diameter octagonal piles, 5' length	G		7.25	2.207			93	29	122	172
0940	6-10' long	G		3.75	4.267			180	56	236	330
0950	11-15' long	G		2.75	5.818			245	76.50	321.50	455
1000	Clean steel piles from boat, 8" diam., 5' length	G		14.50	1.103			46.50	14.50	61	86
1010	6-10' long	G		7.75	2.065			87	27	114	161
1020	11-15' long	G		5.50	2.909			122	38	160	226
1030	10" diam., 5' length	G		11.50	1.391			58.50	18.25	76.75	108
1040	6-10' long	G		6	2.667			112	35	147	208
1050	11-15' long	G		4.50	3.556			150	46.50	196.50	277
1060	12" diam., 5' length	G		9.50	1.684			71	22	93	132
1070	6-10' long	G		5	3.200			135	42	177	248
1080	11-15' long	G		3.75	4.267			180	56	236	330
1100	HP 8 x 36, 5' length	G		7.50	2.133			90	28	118	166
1110	6-10' long	G		4	4			168	52.50	220.50	310
1120	11-15' long	G		3	5.333			225	70	295	410
1130	HP 10 x 42, 5' length	G		6.25	2.560			108	33.50	141.50	199
1140	6-10' long	G		3.25	4.923			207	64.50	271.50	380
1150	11-15' long	G		2.30	6.957			293	91.50	384.50	540
1160	HP 10 x 57, 5' length	G		6	2.667			112	35	147	208
1170	6-10' long	G		3.25	4.923			207	64.50	271.50	380
1180	11-15' long	G		2.30	6.957			293	91.50	384.50	540
1200	HP 12 x 53, 5' length	G		5	3.200			135	42	177	248
1210	6-10' long	G		2.75	5.818			245	76.50	321.50	455
1220	11-15' long	G		2	8			335	105	440	620
1230	HP 12 x 74, 5' length	G		5	3.200			135	42	177	248
1240	6-10' long	G		2.75	5.818			245	76.50	321.50	455
1250	11-15' long	G		1.90	8.421			355	111	466	655
1300	HP 14 x 73, 5' length	G		4.25	3.765			158	49.50	207.50	293
1310	6-10' long	G		2.33	6.867			289	90	379	535

For customer support on your Site Work & Landscape Costs with RSMeans data, call 800.448.8182.

551

35 01 Operation and Maint. of Waterway & Marine Construction

35 01 50 – Operation and Maintenance of Marine Construction

35 01 50.10 Cleaning of Marine Pier Piles

			Crew	Daily Output	Labor-Hours	Unit	Material	2020 Bare Costs Labor	Equipment	Total	Total Incl O&P
1320	11-15' long	G	B-1D	1.60	10	Ea.		420	131	551	780
1330	HP 14 x 89, 5' length	G		4.25	3.765			158	49.50	207.50	293
1340	6-10' long	G		2.33	6.867			289	90	379	535
1350	11-15' long	G		1.60	10			420	131	551	780
1360	HP 14 x 102, 5' length	G		4.25	3.765			158	49.50	207.50	293
1370	6-10' long	G		2.33	6.867			289	90	379	535
1380	11-15' long	G		1.60	10			420	131	551	780
1385	HP 14 x 117, 5' length	G		4.25	3.765			158	49.50	207.50	293
1390	6-10' long	G		2.33	6.867			289	90	379	535
1395	11-15' long	G		1.60	10			420	131	551	780

35 01 50.20 Protective Wrapping of Marine Pier Piles

			Crew	Daily Output	Labor-Hours	Unit	Material	2020 Bare Costs Labor	Equipment	Total	Total Incl O&P
0010	**PROTECTIVE WRAPPING OF MARINE PIER PILES**										
0015	Exposed piles wrapped using boat										
0020	Note: piles must be cleaned before wrapping										
0100	Wrap protective material on wood piles with nails 8" diameter	G	B-1G	162	.099	V.L.F.	18.55	4.16	.74	23.45	27.50
0110	10" diameter	G		130	.123		23.50	5.20	.92	29.62	34.50
0120	12" diameter	G		108	.148		28	6.25	1.11	35.36	41
0130	13" diameter	G		99	.162		30	6.80	1.21	38.01	44.50
0140	14" diameter	G		93	.172		32.50	7.25	1.29	41.04	48.50
0150	Wrap protective material on wood piles, straps, 8" diameter	G		180	.089		28	3.74	.66	32.40	37.50
0160	10" diameter	G		144	.111		35	4.68	.83	40.51	46.50
0170	12" diameter	G		120	.133		42	5.60	1	48.60	56
0180	13" diameter	G		110	.145		45.50	6.10	1.09	52.69	60.50
0190	14" diameter	G		103	.155		49	6.55	1.16	56.71	65
0200	Wrap protective material on concrete piles, straps, 12" diameter	G		120	.133		42	5.60	1	48.60	56
0210	14" diameter	G		103	.155		49	6.55	1.16	56.71	65
0220	16" diameter	G		90	.178		56	7.50	1.33	64.83	74
0230	18" diameter	G		80	.200		63	8.40	1.49	72.89	84
0240	20" diameter	G		72	.222		70	9.35	1.66	81.01	93
0250	24" diameter	G		60	.267		84	11.25	1.99	97.24	112
0260	36" diameter	G		40	.400		126	16.85	2.99	145.84	168
0270	54" diameter	G		27	.593		189	25	4.43	218.43	250
0280	66" diameter	G		22	.727		231	30.50	5.45	266.95	305
0300	Wrap protective material on concrete piles, straps, 10" square	G		113	.142		44.50	5.95	1.06	51.51	59
0310	12" square	G		94	.170		53.50	7.15	1.27	61.92	71
0320	14" square	G		81	.198		62.50	8.30	1.48	72.28	82.50
0330	16" square	G		71	.225		71.50	9.50	1.68	82.68	94.50
0340	18" square	G		63	.254		80.50	10.70	1.90	93.10	107
0350	20" square	G		63	.254		89	10.70	1.90	101.60	116
0360	24" square	G		47	.340		107	14.35	2.54	123.89	142
0400	Wrap protective material, concrete piles, straps, 14" octagon	G		107	.150		47	6.30	1.12	54.42	62.50
0410	16" octagon	G		94	.170		54	7.15	1.27	62.42	71.50
0420	18" octagon	G		83	.193		60.50	8.10	1.44	70.04	80.50
0430	20" octagon	G		75	.213		67	9	1.59	77.59	89.50
0440	24" octagon	G		62	.258		80.50	10.85	1.93	93.28	107
0500	Wrap protective material, steel piles, straps, HP 8 x 36	G		90	.178		71	7.50	1.33	79.83	90.50
0510	HP 10 x 42	G		72	.222		87.50	9.35	1.66	98.51	112
0520	HP 10 x 57	G		72	.222		89	9.35	1.66	100.01	114
0530	HP 12 x 53	G		60	.267		105	11.25	1.99	118.24	135
0540	HP 12 x 74	G		60	.267		107	11.25	1.99	120.24	137
0550	HP 14 x 73	G		52	.308		125	12.95	2.30	140.25	159
0560	HP 14 x 89	G		52	.308		126	12.95	2.30	141.25	161

For customer support on your Site Work & Landscape Costs with RSMeans data, call 800.448.8182.

35 01 Operation and Maint. of Waterway & Marine Construction

35 01 50 – Operation and Maintenance of Marine Construction

35 01 50.20 Protective Wrapping of Marine Pier Piles		Crew	Daily Output	Labor-Hours	Unit	Material	2020 Bare Costs Labor	Equipment	Total	Total Incl O&P
0570	HP 14 x 102	B-1G [G]	52	.308	V.L.F.	126	12.95	2.30	141.25	161
0580	HP 14 x 117	[G]	52	.308		128	12.95	2.30	143.25	163

35 22 Hydraulic Gates

35 22 26 – Sluice Gates

35 22 26.16 Hydraulic Sluice Gates

		Crew	Daily Output	Labor-Hours	Unit	Material	2020 Bare Costs Labor	Equipment	Total	Total Incl O&P
0010	**HYDRAULIC SLUICE GATES**									
0100	Heavy duty, self contained w/crank oper. gate, 18" x 18"	L-5A	1.70	18.824	Ea.	7,625	1,100	675	9,400	10,800
0110	24" x 24"		1.20	26.667		11,200	1,550	955	13,705	15,800
0120	30" x 30"		1	32		12,400	1,875	1,150	15,425	17,800
0130	36" x 36"		.90	35.556		15,300	2,075	1,275	18,650	21,400
0140	42" x 42"		.80	40		17,500	2,350	1,425	21,275	24,400
0150	48" x 48"		.50	64		18,100	3,750	2,300	24,150	28,200
0160	54" x 54"		.40	80		29,500	4,675	2,850	37,025	42,800
0170	60" x 60"		.30	107		26,900	6,250	3,825	36,975	43,500
0180	66" x 66"		.30	107		31,600	6,250	3,825	41,675	48,700
0190	72" x 72"		.20	160		35,100	9,375	5,725	50,200	59,500
0200	78" x 78"		.20	160		47,600	9,375	5,725	62,700	73,500
0210	84" x 84"		.10	320		48,900	18,700	11,400	79,000	95,000
0220	90" x 90"	E-20	.30	213		58,000	12,200	6,750	76,950	90,500
0230	96" x 96"		.30	213		62,000	12,200	6,750	80,950	94,500
0240	108" x 108"		.20	320		70,500	18,300	10,100	98,900	117,000
0250	120" x 120"		.10	640		83,000	36,600	20,300	139,900	171,000
0260	132" x 132"		.10	640		97,500	36,600	20,300	154,400	187,000

35 22 63 – Through-Levee Access Gates

35 22 63.16 Canal Gates

		Crew	Daily Output	Labor-Hours	Unit	Material	2020 Bare Costs Labor	Equipment	Total	Total Incl O&P
0010	**CANAL GATES**									
0011	Cast iron body, fabricated frame									
0100	12" diameter	L-5A	4.60	6.957	Ea.	1,200	405	249	1,854	2,225
0110	18" diameter		4	8		1,425	470	286	2,181	2,625
0120	24" diameter		3.50	9.143		2,075	535	325	2,935	3,475
0130	30" diameter		2.80	11.429		2,700	670	410	3,780	4,425
0140	36" diameter		2.30	13.913		5,075	815	495	6,385	7,375
0150	42" diameter		1.70	18.824		9,375	1,100	675	11,150	12,700
0160	48" diameter		1.20	26.667		10,500	1,550	955	13,005	15,100
0170	54" diameter		.90	35.556		12,000	2,075	1,275	15,350	17,900
0180	60" diameter		.50	64		12,600	3,750	2,300	18,650	22,200
0190	66" diameter		.50	64		17,100	3,750	2,300	23,150	27,100
0200	72" diameter		.40	80		21,300	4,675	2,850	28,825	33,800

35 22 66 – Flap Gates, Hydraulic

35 22 66.16 Flap Gates

		Crew	Daily Output	Labor-Hours	Unit	Material	2020 Bare Costs Labor	Equipment	Total	Total Incl O&P
0010	**FLAP GATES**									
0100	Aluminum, 18" diameter	L-5A	5	6.400	Ea.	2,400	375	229	3,004	3,475
0110	24" diameter		4	8		2,800	470	286	3,556	4,125
0120	30" diameter		3.50	9.143		3,900	535	325	4,760	5,475
0130	36" diameter		2.80	11.429		4,775	670	410	5,855	6,725
0140	42" diameter		2.30	13.913		4,625	815	495	5,935	6,900
0150	48" diameter		1.70	18.824		5,425	1,100	675	7,200	8,425
0160	54" diameter		1.20	26.667		6,000	1,550	955	8,505	10,100
0170	60" diameter		.80	40		6,425	2,350	1,425	10,200	12,300

For customer support on your Site Work & Landscape Costs with RSMeans data, call 800.448.8182.

553

35 22 Hydraulic Gates

35 22 66 – Flap Gates, Hydraulic

35 22 66.16 Flap Gates		Crew	Daily Output	Labor-Hours	Unit	Material	2020 Bare Costs Labor	2020 Bare Costs Equipment	Total	Total Incl O&P
0180	66" diameter	L-5A	.50	64	Ea.	6,875	3,750	2,300	12,925	15,900
0190	72" diameter		.40	80		8,000	4,675	2,850	15,525	19,200

35 22 69 – Knife Gates, Hydraulic

35 22 69.16 Knife Gates

		Crew	Daily Output	Labor-Hours	Unit	Material	Labor	Equipment	Total	Total Incl O&P
0010	**KNIFE GATES**									
0100	Incl. handwheel operator for hub, 6" diameter	Q-23	7.70	3.117	Ea.	1,050	195	193	1,438	1,650
0110	8" diameter		7.20	3.333		1,575	209	207	1,991	2,250
0120	10" diameter		4.80	5		2,400	315	310	3,025	3,425
0130	12" diameter		3.60	6.667		3,725	415	415	4,555	5,175
0140	14" diameter		3.40	7.059		4,425	440	440	5,305	6,025
0150	16" diameter		3.20	7.500		5,450	470	465	6,385	7,175
0160	18" diameter		3	8		10,300	500	495	11,295	12,600
0170	20" diameter		2.70	8.889		9,625	555	550	10,730	12,000
0180	24" diameter		2.40	10		15,000	625	620	16,245	18,100
0190	30" diameter		1.80	13.333		29,900	835	825	31,560	35,000
0200	36" diameter		1.20	20		45,100	1,250	1,250	47,600	53,000

35 22 73 – Slide Gates, Hydraulic

35 22 73.16 Slide Gates

		Crew	Daily Output	Labor-Hours	Unit	Material	Labor	Equipment	Total	Total Incl O&P
0010	**SLIDE GATES**									
0100	Steel, self contained incl. anchor bolts and grout, 12" x 12"	L-5A	4.60	6.957	Ea.	4,700	405	249	5,354	6,075
0110	18" x 18"		4	8		5,250	470	286	6,006	6,825
0120	24" x 24"		3.50	9.143		5,575	535	325	6,435	7,325
0130	30" x 30"		2.80	11.429		6,550	670	410	7,630	8,700
0140	36" x 36"		2.30	13.913		7,000	815	495	8,310	9,500
0150	42" x 42"		1.70	18.824		7,350	1,100	675	9,125	10,500
0160	48" x 48"		1.20	26.667		7,650	1,550	955	10,155	11,900
0170	54" x 54"		.90	35.556		7,700	2,075	1,275	11,050	13,100
0180	60" x 60"		.55	58.182		9,025	3,400	2,075	14,500	17,500
0190	72" x 72"		.36	88.889		11,200	5,200	3,175	19,575	23,900

35 24 Dredging

35 24 13 – Suction Dredging

35 24 13.13 Cutter Suction Dredging

		Crew	Daily Output	Labor-Hours	Unit	Material	Labor	Equipment	Total	Total Incl O&P
0010	**CUTTER SUCTION DREDGING**									
0015	Add, Marine Equipment Rental, See Section 01 54 33.80									
1000	Hydraulic method, pumped 1,000' to shore dump, minimum	B-57	460	.104	B.C.Y.		5.05	3.94	8.99	11.95
1100	Maximum		310	.155			7.50	5.85	13.35	17.70
1400	Into scows dumped 20 miles, minimum		425	.113			5.50	4.27	9.77	12.90
1500	Maximum		243	.198			9.60	7.45	17.05	22.50
1600	For inland rivers and canals in South, deduct								30%	30%

35 24 23 – Clamshell Dredging

35 24 23.13 Mechanical Dredging

		Crew	Daily Output	Labor-Hours	Unit	Material	Labor	Equipment	Total	Total Incl O&P
0010	**MECHANICAL DREDGING**									
0015	Add, Marine Equipment Rental, See Section 01 54 33.80									
0020	Dredging mobilization and demobilization, add to below, minimum	B-8	.53	121	Total		5,900	5,475	11,375	14,800
0100	Maximum	"	.10	640	"		31,200	29,000	60,200	78,500
0300	Barge mounted clamshell excavation into scows									
0310	Dumped 20 miles at sea, minimum	B-57	310	.155	B.C.Y.		7.50	5.85	13.35	17.70
0400	Maximum	"	213	.225	"		10.95	8.50	19.45	26

35 24 Dredging

35 24 23 – Clamshell Dredging

35 24 23.13 Mechanical Dredging

		Crew	Daily Output	Labor-Hours	Unit	Material	2020 Bare Costs Labor	2020 Bare Costs Equipment	Total	Total Incl O&P
0500	Barge mounted dragline or clamshell, hopper dumped,									
0510	pumped 1,000' to shore dump, minimum	B-57	340	.141	B.C.Y.		6.85	5.35	12.20	16.10
0525	All pumping uses 2,000 gallons of water per cubic yard									
0600	Maximum	B-57	243	.198	B.C.Y.		9.60	7.45	17.05	22.50

35 31 Shoreline Protection

35 31 16 – Seawalls

35 31 16.13 Concrete Seawalls

		Crew	Daily Output	Labor-Hours	Unit	Material	2020 Bare Costs Labor	2020 Bare Costs Equipment	Total	Total Incl O&P
0010	**CONCRETE SEAWALLS**									
0011	Reinforced concrete									
0015	include footing and tie-backs									
0020	Up to 6' high, minimum	C-17C	28	2.964	L.F.	64.50	164	19.90	248.40	340
0060	Maximum		24.25	3.423		103	190	23	316	425
0100	12' high, minimum		20	4.150		167	230	28	425	565
0160	Maximum		18.50	4.486		193	249	30	472	620
0180	Precast bulkhead, complete, including									
0190	vertical and battered piles, face panels, and cap									
0195	Using 16' vertical piles				L.F.				445	515
0196	Using 20' vertical piles				"				475	550

35 31 16.19 Steel Sheet Piling Seawalls

		Crew	Daily Output	Labor-Hours	Unit	Material	2020 Bare Costs Labor	2020 Bare Costs Equipment	Total	Total Incl O&P
0010	**STEEL SHEET PILING SEAWALLS**									
0200	Steel sheeting, with 4' x 4' x 8" concrete deadmen, @ 10' OC									
0210	12' high, shore driven	B-40	27	2.370	L.F.	125	131	129	385	480
0260	Barge driven	B-76	15	4.800	"	187	265	246	698	880
6000	Crushed stone placed behind bulkhead by clam bucket	B-12H	120	.133	L.C.Y.	22	6.75	10	38.75	45

35 31 19 – Revetments

35 31 19.18 Revetments, Concrete

		Crew	Daily Output	Labor-Hours	Unit	Material	2020 Bare Costs Labor	2020 Bare Costs Equipment	Total	Total Incl O&P
0010	**REVETMENTS, CONCRETE**									
0100	Concrete revetment matt 8' x 20' x 4-1/2" excluding site preparation									
0110	Includes all labor, material and equip. for complete installation	B-57	5	9.600	Ea.	2,650	465	365	3,480	4,025

35 49 Waterway Structures

35 49 13 – Floodwalls

35 49 13.30 Breakwaters, Bulkheads, Residential Canal

		Crew	Daily Output	Labor-Hours	Unit	Material	2020 Bare Costs Labor	2020 Bare Costs Equipment	Total	Total Incl O&P
0010	**BREAKWATERS, BULKHEADS, RESIDENTIAL CANAL**									
0020	Aluminum panel sheeting, incl. concrete cap and anchor									
0030	Coarse compact sand, 4'-0" high, 2'-0" embedment	B-40	200	.320	L.F.	65	17.70	17.40	100.10	118
0040	3'-6" embedment		140	.457		76.50	25.50	25	127	151
0060	6'-0" embedment		90	.711		97.50	39.50	38.50	175.50	210
0120	6'-0" high, 2'-6" embedment		170	.376		87.50	21	20.50	129	151
0140	4'-0" embedment		125	.512		103	28.50	28	159.50	188
0160	5'-6" embedment		95	.674		136	37	36.50	209.50	248
0220	8'-0" high, 3'-6" embedment		140	.457		118	25.50	25	168.50	196
0240	5'-0" embedment		100	.640		118	35.50	35	188.50	223
0420	Medium compact sand, 3'-0" high, 2'-0" embedment		235	.272		154	15.05	14.80	183.85	208
0440	4'-0" embedment		150	.427		189	23.50	23	235.50	270
0460	5'-6" embedment		115	.557		229	30.50	30	289.50	335
0520	5'-0" high, 3'-6" embedment		165	.388		202	21.50	21	244.50	278

35 49 Waterway Structures

35 49 13 – Floodwalls

35 49 13.30 Breakwaters, Bulkheads, Residential Canal	Crew	Daily Output	Labor-Hours	Unit	Material	2020 Bare Costs Labor	Equipment	Total	Total Incl O&P
0540	5'-0" embedment	B-40	120	.533	L.F.	233	29.50	29	291.50	335
0560	6'-6" embedment		105	.610		305	33.50	33	371.50	425
0620	7'-0" high, 4'-6" embedment		135	.474		289	26	26	341	390
0640	6'-0" embedment		110	.582		320	32	31.50	383.50	435
0720	Loose silty sand, 3'-0" high, 3'-0" embedment		205	.312		152	17.25	16.95	186.20	212
0740	4'-6" embedment		155	.413		183	23	22.50	228.50	262
0760	6'-0" embedment		125	.512		214	28.50	28	270.50	310
0820	4'-6" high, 4'-6" embedment		155	.413		196	23	22.50	241.50	275
0840	6'-0" embedment		125	.512		227	28.50	28	283.50	325
0860	7'-0" embedment		115	.557		274	30.50	30	334.50	380
0920	6'-0" high, 5'-6" embedment		130	.492		296	27	27	350	395
0940	7'-0" embedment		115	.557		325	30.50	30	385.50	440

35 51 Floating Construction

35 51 13 – Floating Piers

35 51 13.23 Floating Wood Piers

		Crew	Daily Output	Labor-Hours	Unit	Material	2020 Bare Costs Labor	Equipment	Total	Total Incl O&P
0010	**FLOATING WOOD PIERS**									
0020	Polyethylene encased polystyrene, no pilings included	F-3	330	.121	S.F.	30.50	6.60	1.43	38.53	45
0030	See Section 06 13 33.50 or Section 06 13 33.52 for fixed docks									
0200	Pile supported, shore constructed, minimum	F-3	130	.308	S.F.	29.50	16.75	3.62	49.87	61.50
0250	Maximum		120	.333		32.50	18.10	3.93	54.53	67
0400	Floating, small boat, prefab, no shore facilities, minimum		250	.160		27.50	8.70	1.88	38.08	45.50
0500	Maximum		150	.267		59.50	14.50	3.14	77.14	91
0700	Per slip, minimum (180 S.F. each)		1.59	25.157	Ea.	5,850	1,375	296	7,521	8,800
0800	Maximum		1.40	28.571	"	9,100	1,550	335	10,985	12,700

35 51 13.24 Jetties, Docks

		Crew	Daily Output	Labor-Hours	Unit	Material	2020 Bare Costs Labor	Equipment	Total	Total Incl O&P
0010	**JETTIES, DOCKS**									
0011	Floating including anchors									
0030	See Section 06 13 33.50 or Section 06 13 33.52 for fixed docks									
1000	Polystyrene flotation, minimum	F-3	200	.200	S.F.	29.50	10.85	2.36	42.71	51.50
1040	Maximum		135	.296	"	38.50	16.10	3.49	58.09	70.50
1100	Alternate method of figuring, minimum		1.13	35.398	Slip	5,950	1,925	415	8,290	9,875
1140	Maximum		.70	57.143	"	8,525	3,100	675	12,300	14,800
1200	Galv. steel frame and wood deck, 3' wide, minimum		320	.125	S.F.	18.25	6.80	1.47	26.52	32
1240	Maximum		200	.200		29.50	10.85	2.36	42.71	51.50
1300	4' wide, minimum		320	.125		21	6.80	1.47	29.27	35.50
1340	Maximum		200	.200		34	10.85	2.36	47.21	56.50
1500	8' wide, minimum		250	.160		22	8.70	1.88	32.58	39.50
1540	Maximum		160	.250		33.50	13.60	2.95	50.05	60.50
1700	Treated wood frames and deck, 3' wide, minimum		250	.160		24	8.70	1.88	34.58	41.50
1740	Maximum		125	.320		83	17.40	3.77	104.17	121
2000	Polyethylene drums, treated wood frame and deck									
2100	6' wide, minimum	F-3	250	.160	S.F.	30.50	8.70	1.88	41.08	48.50
2140	Maximum		125	.320		40.50	17.40	3.77	61.67	74.50
2200	8' wide, minimum		233	.172		30.50	9.35	2.02	41.87	50
2240	Maximum		120	.333		42	18.10	3.93	64.03	78
2300	10' wide, minimum		200	.200		34	10.85	2.36	47.21	56.50
2340	Maximum		110	.364		47.50	19.75	4.28	71.53	86
2400	Concrete pontoons, treated wood frame and deck									
2500	Breakwater, concrete pontoon, 50' L x 8' W x 6.5' D	F-4	4	12	Ea.	51,000	645	245	51,890	57,000
2600	Inland breakwater, concrete pontoon, 50' L x 8' W x 4' D		4	12	"	47,200	645	245	48,090	53,000

556

35 51 Floating Construction

35 51 13 – Floating Piers

35 51 13.24 Jetties, Docks	Crew	Daily Output	Labor-Hours	Unit	Material	2020 Bare Costs Labor	Equipment	Total	Total Incl O&P
2700	Docks, concrete pontoon w/treated wood deck	F-4	2400	.020	S.F.	105	1.07	.41	106.48	118
2800	Docks, concrete pontoons docks 400 S.F. minimum order									

35 51 13.28 Jetties, Floating Dock Accessories

		Crew	Daily Output	Labor-Hours	Unit	Material	Labor	Equipment	Total	Total Incl O&P
0010	**JETTIES, FLOATING DOCK ACCESSORIES**									
0200	Dock connectors, stressed cables with rubber spacers									
0220	25" long, 3' wide dock	1 Clab	2	4	Joint	244	168		412	520
0240	5' wide dock		2	4		284	168		452	565
0400	38" long, 4' wide dock		1.75	4.571		267	192		459	585
0440	6' wide dock		1.45	5.517		305	232		537	685
1000	Gangway, aluminum, one end rolling, no hand rails									
1020	3' wide, minimum	1 Clab	67	.119	L.F.	226	5.05		231.05	256
1040	Maximum		32	.250		259	10.55		269.55	300
1100	4' wide, minimum		40	.200		239	8.40		247.40	276
1140	Maximum		24	.333		275	14.05		289.05	325
1180	For handrails, add					75.50			75.50	83.50
2000	Pile guides, beads on stainless cable	1 Clab	4	2	Ea.	87	84		171	223
2020	Rod type, 8" diameter piles, minimum		4	2		72	84		156	206
2040	Maximum		2	4		105	168		273	370
2100	10" to 14" diameter piles, minimum		3.20	2.500		82.50	105		187.50	249
2140	Maximum		1.75	4.571		144	192		336	450
2200	Roller type, 4 rollers, minimum		4	2		190	84		274	335
2240	Maximum		1.75	4.571		271	192		463	585

35 59 Marine Specialties

35 59 33 – Marine Bollards and Cleats

35 59 33.50 Jetties, Dock Accessories

		Crew	Daily Output	Labor-Hours	Unit	Material	Labor	Equipment	Total	Total Incl O&P
0010	**JETTIES, DOCK ACCESSORIES**									
0100	Cleats, aluminum, "S" type, 12" long	1 Clab	8	1	Ea.	27	42		69	93.50
0140	10" long		6.70	1.194		66	50.50		116.50	148
0180	15" long		6	1.333		87	56		143	180
0400	Dock wheel for corners and piles, vinyl, 12" diameter		4	2		89.50	84		173.50	226
1000	Electrical receptacle with circuit breaker,									
1020	Pile mounted, double 30 amp, 125 volt	1 Elec	2	4	Unit	865	245		1,110	1,325
1060	Double 50 amp, 125/240 volt		1.60	5		1,175	305		1,480	1,725
1120	Free standing, add		4	2		210	123		333	415
1140	Double free standing, add		2.70	2.963		216	182		398	510
1160	Light, 2 louvered, with photo electric switch, add		8	1		205	61.50		266.50	315
1180	Telephone jack on stanchion		8	1		207	61.50		268.50	320
1300	Fender, vinyl, 4" high	1 Clab	160	.050	L.F.	14.65	2.11		16.76	19.30
1380	Corner piece		80	.100	Ea.	14.05	4.21		18.26	22
1400	Hose holder, cast aluminum		16	.500	"	48.50	21		69.50	84.50
1500	Ladder, aluminum, heavy duty									
1520	Crown top, 5 to 7 step, minimum	1 Clab	5.30	1.509	Ea.	195	63.50		258.50	310
1560	Maximum		2	4		280	168		448	565
1580	Bracket for portable clamp mounting		8	1		9.45	42		51.45	74
1800	Line holder, treated wood, small		16	.500		13.45	21		34.45	46.50
1840	Large		13.30	.602		23	25.50		48.50	63
2000	Mooring whip, fiberglass bolted to dock,									
2020	1,200 lb. boat	1 Clab	8.80	.909	Pr.	390	38.50		428.50	490
2040	10,000 lb. boat		6.70	1.194		790	50.50		840.50	945
2080	60,000 lb. boat		4	2		925	84		1,009	1,150

For customer support on your Site Work & Landscape Costs with RSMeans data, call 800.448.8182.

557

35 59 Marine Specialties

35 59 33 – Marine Bollards and Cleats

35 59 33.50 Jetties, Dock Accessories	Crew	Daily Output	Labor-Hours	Unit	Material	2020 Bare Costs Labor	Equipment	Total	Total Incl O&P
2400	Shock absorbing tubing, vertical bumpers									
2420	3" diameter, vinyl, white	1 Clab	80	.100	L.F.	5.95	4.21		10.16	12.90
2440	Polybutyl, clear		80	.100	"	6.60	4.21		10.81	13.60
2480	Mounts, polybutyl		20	.400	Ea.	6.95	16.85		23.80	33
2490	Deluxe		20	.400	"	15.55	16.85		32.40	42.50

Estimating Tips

Products such as conveyors, material handling cranes and hoists, and other items specified in this division require trained installers. The general contractor may not have any choice as to who will perform the installation or when it will be performed. Long lead times are often required for these products, making early decisions in purchasing and scheduling necessary. The installation of this type of equipment may require the embedment of mounting hardware during the construction of floors, structural walls, or interior walls/partitions. Electrical connections will require coordination with the electrical contractor.

Reference Numbers

Reference numbers are shown at the beginning of some major classifications. These numbers refer to related items in the Reference Section. The reference information may be an estimating procedure, an alternate pricing method, or technical information.

Same Data. Simplified.

Enjoy the convenience and efficiency of accessing your costs anywhere:

- **Skip the multiplier** by setting your location
- **Quickly search,** edit, favorite and share costs
- **Stay on top of price changes** with automatic updates

Discover more at rsmeans.com/online

41 21 Conveyors

41 21 23 – Piece Material Conveyors

41 21 23.16 Container Piece Material Conveyors	Crew	Daily Output	Labor-Hours	Unit	Material	2020 Bare Costs Labor	Equipment	Total	Total Incl O&P
0010 **CONTAINER PIECE MATERIAL CONVEYORS**									
0020 Gravity fed, 2" rollers, 3" OC									
0050 10' sections with 2 supports, 600 lb. capacity, 18" wide				Ea.	530			530	580
0350 Horizontal belt, center drive and takeup, 60 fpm									
0400 16" belt, 26.5' length	2 Mill	.50	32	Ea.	3,675	1,800		5,475	6,625
0600 Inclined belt, 10' rise with horizontal loader and									
0620 End idler assembly, 27.5' length, 18" belt	2 Mill	.30	53.333	Ea.	7,925	2,975		10,900	13,000

41 22 Cranes and Hoists

41 22 13 – Cranes

41 22 13.10 Crane Rail

	Crew	Daily Output	Labor-Hours	Unit	Material	Labor	Equipment	Total	Total Incl O&P
0010 **CRANE RAIL**									
0020 Box beam bridge, no equipment included	E-4	3400	.009	Lb.	1.44	.55	.04	2.03	2.49
0200 Running track only, 104 lb. per yard	"	5600	.006	"	.72	.33	.03	1.08	1.34

41 22 23 – Hoists

41 22 23.10 Material Handling

	Crew	Daily Output	Labor-Hours	Unit	Material	Labor	Equipment	Total	Total Incl O&P
0010 **MATERIAL HANDLING**, cranes, hoists and lifts									
2100 Hoists, electric overhead, chain, hook hung, 15' lift, 1 ton cap.				Ea.	2,825			2,825	3,100
2200 3 ton capacity					3,800			3,800	4,200
2500 5 ton capacity					8,350			8,350	9,200

Estimating Tips

This division contains information about water and wastewater equipment and systems, which was formerly located in Division 44. The main areas of focus are total wastewater treatment plants and components of wastewater treatment plants. Also included in this section are oil/water separators for wastewater treatment.

Reference Numbers

Reference numbers are shown at the beginning of some major classifications. These numbers refer to related items in the Reference Section. The reference information may be an estimating procedure, an alternate pricing method, or technical information.

Same Data. Simplified.

Enjoy the convenience and efficiency of accessing your costs anywhere:

- **Skip the multiplier** by setting your location
- **Quickly search,** edit, favorite and share costs
- **Stay on top of price changes** with automatic updates

Discover more at rsmeans.com/online

46 07 Packaged Water and Wastewater Treatment Equipment

46 07 53 – Packaged Wastewater Treatment Equipment

46 07 53.10 Biological Pkg. Wastewater Treatment Plants	Crew	Daily Output	Labor-Hours	Unit	Material	2020 Bare Costs Labor	Equipment	Total	Total Incl O&P
0010 **BIOLOGICAL PACKAGED WASTEWATER TREATMENT PLANTS**									
0011 Not including fencing or external piping									
0020 Steel packaged, blown air aeration plants									
0100 1,000 GPD				Gal.				55	60.50
0200 5,000 GPD								22	24
0300 15,000 GPD								22	24
0400 30,000 GPD								15.40	16.95
0500 50,000 GPD								11	12.10
0600 100,000 GPD								9.90	10.90
0700 200,000 GPD								8.80	9.70
0800 500,000 GPD								7.70	8.45
1000 Concrete, extended aeration, primary and secondary treatment									
1010 10,000 GPD				Gal.				22	24
1100 30,000 GPD								15.40	16.95
1200 50,000 GPD								11	12.10
1400 100,000 GPD								9.90	10.90
1500 500,000 GPD								7.70	8.45
1700 Municipal wastewater treatment facility									
1720 1.0 MGD				Gal.				11	12.10
1740 1.5 MGD								10.60	11.65
1760 2.0 MGD								10	11
1780 3.0 MGD								7.80	8.60
1800 5.0 MGD								5.80	6.70
2000 Holding tank system, not incl. excavation or backfill									
2010 Recirculating chemical water closet	2 Plum	4	4	Ea.	550	258		808	990
2100 For voltage converter, add	"	16	1		320	64.50		384.50	445
2200 For high level alarm, add	1 Plum	7.80	1.026		123	66		189	235

46 07 53.20 Wastewater Treatment System

	Crew	Daily Output	Labor-Hours	Unit	Material	Labor	Equipment	Total	Total Incl O&P
0010 **WASTEWATER TREATMENT SYSTEM**									
0020 Fiberglass, 1,000 gallon	B-21	1.29	21.705	Ea.	4,450	1,050	147	5,647	6,650
0100 1,500 gallon	"	1.03	27.184	"	9,700	1,325	184	11,209	12,900

46 23 Grit Removal And Handling Equipment

46 23 23 – Vortex Grit Removal Equipment

46 23 23.10 Rainwater Filters

	Crew	Daily Output	Labor-Hours	Unit	Material	Labor	Equipment	Total	Total Incl O&P
0010 **RAINWATER FILTERS**									
0100 42 gal./min	B-21	3.50	8	Ea.	40,600	390	54	41,044	45,300
0200 65 gal./min		3.50	8		40,700	390	54	41,144	45,300
0300 208 gal./min		3.50	8		40,700	390	54	41,144	45,400

46 25 Oil and Grease Separation and Removal Equipment

46 25 13 – Coalescing Oil-Water Separators

46 25 13.20 Oil/Water Separators

46 25 13.20 Oil/Water Separators	Crew	Daily Output	Labor-Hours	Unit	Material	2020 Bare Costs Labor	Equipment	Total	Total Incl O&P
0010 **OIL/WATER SEPARATORS**									
0020 Underground, tank only									
0030 Excludes excavation, backfill & piping									
0100 200 GPM	B-21	3.50	8	Ea.	40,400	390	54	40,844	45,000
0110 400 GPM		3.25	8.615		85,500	420	58	85,978	94,500
0120 600 GPM		2.75	10.182		90,000	495	68.50	90,563.50	100,500
0130 800 GPM		2.50	11.200		108,000	545	75.50	108,620.50	119,500
0140 1,000 GPM		2	14		122,000	685	94.50	122,779.50	135,500
0150 1,200 GPM		1.50	18.667		134,000	910	126	135,036	148,500
0160 1,500 GPM		1	28		154,000	1,375	189	155,564	171,500

46 25 16 – API Oil-Water Separators

46 25 16.10 Oil/Water Separators

46 25 16.10 Oil/Water Separators	Crew	Daily Output	Labor-Hours	Unit	Material	2020 Bare Costs Labor	Equipment	Total	Total Incl O&P
0010 **OIL/WATER SEPARATORS**									
0020 Complete system, not including excavation and backfill									
0100 Treated capacity 0.2 C.F./second	B-22	1	30	Ea.	9,225	1,475	284	10,984	12,600
0200 0.5 C.F./second	B-13	.75	74.667		11,800	3,450	775	16,025	19,000
0300 1.0 C.F./second		.60	93.333		17,000	4,300	970	22,270	26,200
0400 2.4 to 3.0 C.F./second		.30	187		24,800	8,600	1,925	35,325	42,300
0500 11 C.F./second		.17	329		29,100	15,200	3,425	47,725	58,500
0600 22 C.F./second		.10	560		48,500	25,800	5,800	80,100	98,500

46 51 Air and Gas Diffusion Equipment

46 51 13 – Floating Mechanical Aerators

46 51 13.10 Aeration Equipment

46 51 13.10 Aeration Equipment	Crew	Daily Output	Labor-Hours	Unit	Material	2020 Bare Costs Labor	Equipment	Total	Total Incl O&P
0010 **AERATION EQUIPMENT**									
0020 Aeration equipment includes floats and excludes power supply and anchorage									
4320 Surface aerator, 50 lb. oxygen/hr, 30 HP, 900 RPM, inc floats, no anchoring	B-21B	1	40	Ea.	33,200	1,825	470	35,495	39,900
4322 Anchoring aerators per cell, 6 anchors per cell	B-6	.50	48		3,125	2,200	430	5,755	7,200
4324 Anchoring aerators per cell, 8 anchors per cell	"	.33	72.727		4,275	3,325	650	8,250	10,400

46 51 20 – Air and Gas Handling Equipment

46 51 20.10 Blowers and System Components

46 51 20.10 Blowers and System Components	Crew	Daily Output	Labor-Hours	Unit	Material	2020 Bare Costs Labor	Equipment	Total	Total Incl O&P
0010 **BLOWERS AND SYSTEM COMPONENTS**									
0020 Rotary lobe blowers									
0030 Medium pressure (7 to 14 PSIG)									
0120 38 CFM, 2.1 BHP	Q-2	2.50	9.600	Ea.	1,350	575		1,925	2,350
0130 125 CFM, 5.5 BHP		2.40	10		1,625	600		2,225	2,675
0140 245 CFM, 10.2 BHP		2	12		1,925	720		2,645	3,200
0150 363 CFM, 14.5 BHP		1.80	13.333		2,900	800		3,700	4,375
0160 622 CFM, 24.5 BHP		1.40	17.143		4,800	1,025		5,825	6,825
0170 1,125 CFM, 42.8 BHP		1.10	21.818		7,525	1,300		8,825	10,300
0180 1,224 CFM, 47.4 BHP		1	24		8,675	1,450		10,125	11,700
0400 Prepackaged medium pressure (7 to 14 PSIG) blower									
0405 incl. 3 ph. motor, filter, silencer, valves, check valve, press. gage									
0420 38 CFM, 2.1 BHP	Q-2	2.50	9.600	Ea.	3,050	575		3,625	4,225
0430 125 CFM, 5.5 BHP		2.40	10		3,850	600		4,450	5,150
0440 245 CFM, 10.2 BHP		2	12		5,000	720		5,720	6,575
0450 363 CFM, 14.5 BHP		1.80	13.333		6,975	800		7,775	8,850
0460 521 CFM, 20 BHP		1.50	16		7,625	960		8,585	9,825
1010 Filters and silencers									
1100 Silencer with paper filter									

For customer support on your Site Work & Landscape Costs with RSMeans data, call 800.448.8182.

563

46 51 20 – Air and Gas Handling Equipment

46 51 20.10 Blowers and System Components		Crew	Daily Output	Labor-Hours	Unit	Material	2020 Bare Costs Labor	Equipment	Total	Total Incl O&P
1105	1" connection	1 Plum	14	.571	Ea.	35.50	37		72.50	94
1110	1.5" connection		11	.727		39	47		86	113
1115	2" connection		9	.889		142	57.50		199.50	242
1120	2.5" connection		8	1		143	64.50		207.50	254
1125	3" connection	Q-1	8	2		147	116		263	335
1130	4" connection	"	5	3.200		300	186		486	610
1135	5" connection	Q-2	5	4.800		340	289		629	805
1140	6" connection	"	5	4.800		370	289		659	835
1300	Silencer with polyester filter									
1305	1" connection	1 Plum	14	.571	Ea.	41	37		78	100
1310	1.5" connection		11	.727		42.50	47		89.50	117
1315	2" connection		9	.889		159	57.50		216.50	261
1320	2.5" connection		8	1		148	64.50		212.50	260
1325	3" connection	Q-1	8	2		164	116		280	355
1330	4" connection	"	5	3.200		310	186		496	625
1335	5" connection	Q-2	5	4.800		300	289		589	760
1340	6" connection	"	5	4.800		405	289		694	875
1500	Chamber silencers									
1505	1" connection	1 Plum	14	.571	Ea.	81.50	37		118.50	145
1510	1.5" connection		11	.727		102	47		149	182
1515	2" connection		9	.889		127	57.50		184.50	225
1520	2.5" connection		8	1		204	64.50		268.50	320
1525	3" connection	Q-1	8	2		289	116		405	495
1530	4" connection	"	5	3.200		375	186		561	690
1700	Blower couplings									
1701	Blower flexible coupling									
1710	1.5" connection	Q-1	71	.225	Ea.	30	13.05		43.05	52.50
1715	2" connection		67	.239		33	13.85		46.85	56.50
1720	2.5" connection		65	.246		38	14.30		52.30	63
1725	3" connection		64	.250		51	14.50		65.50	77.50
1730	4" connection		58	.276		61	16		77	91
1735	5" connection	Q-2	83	.289		66	17.40		83.40	99
1740	6" connection		79	.304		91	18.25		109.25	128
1745	8" connection		69	.348		150	21		171	197

46 51 20.20 Aeration System Air Process Piping

		Crew	Daily Output	Labor-Hours	Unit	Material	2020 Bare Costs Labor	Equipment	Total	Total Incl O&P
0010	**AERATION SYSTEM AIR PROCESS PIPING**									
0100	Blower pressure relief valves - adjustable									
0105	1" diameter	1 Plum	14	.571	Ea.	115	37		152	181
0110	2" diameter		9	.889		125	57.50		182.50	223
0115	2.5" diameter		8	1		174	64.50		238.50	289
0120	3" diameter	Q-1	8	2		180	116		296	370
0125	4" diameter	"	5	3.200		180	186		366	475
0200	Blower pressure relief valves - weight loaded									
0205	1" diameter	1 Plum	14	.571	Ea.	178	37		215	251
0210	2" diameter	"	9	.889		210	57.50		267.50	315
0220	3" diameter	Q-1	8	2		315	116		431	520
0225	4" diameter	"	5	3.200		395	186		581	715
0300	Pressure relief valves - preset high flow									
0310	2" diameter	1 Plum	9	.889	Ea.	315	57.50		372.50	435
0320	3" diameter	Q-1	8	2	"	685	116		801	930
1000	Check valves, wafer style									
1110	2" diameter	1 Plum	9	.889	Ea.	133	57.50		190.50	232

46 51 20 - Air and Gas Handling Equipment

46 51 20.20 Aeration System Air Process Piping		Crew	Daily Output	Labor-Hours	Unit	Material	2020 Bare Costs Labor	2020 Bare Costs Equipment	Total	Total Incl O&P
1120	3" diameter	Q-1	8	2	Ea.	168	116		284	360
1125	4" diameter	"	5	3.200		220	186		406	520
1130	5" diameter	Q-2	6	4		298	241		539	685
1135	6" diameter		5	4.800		370	289		659	835
1140	8" diameter		4.50	5.333		525	320		845	1,050
1200	Check valves, flanged steel									
1210	2" diameter	1 Plum	8	1	Ea.	297	64.50		361.50	420
1220	3" diameter	Q-1	4.50	3.556		335	206		541	680
1225	4" diameter	"	3	5.333		440	310		750	945
1230	5" diameter	Q-2	3	8		585	480		1,065	1,375
1235	6" diameter		3	8		735	480		1,215	1,525
1240	8" diameter		2.50	9.600		1,075	575		1,650	2,075
2100	Butterfly valves, lever operated - wafer style									
2110	2" diameter	1 Plum	14	.571	Ea.	54.50	37		91.50	115
2120	3" diameter	Q-1	8	2		59	116		175	239
2125	4" diameter	"	5	3.200		83.50	186		269.50	370
2130	5" diameter	Q-2	6	4		108	241		349	480
2135	6" diameter		5	4.800		126	289		415	570
2140	8" diameter		4.50	5.333		186	320		506	685
2145	10" diameter		4	6		255	360		615	820
2200	Butterfly valves, gear operated - wafer style									
2210	2" diameter	1 Plum	14	.571	Ea.	99	37		136	164
2220	3" diameter	Q-1	8	2		115	116		231	300
2225	4" diameter	"	5	3.200		129	186		315	420
2230	5" diameter	Q-2	6	4		159	241		400	535
2235	6" diameter		5	4.800		159	289		448	605
2240	8" diameter		4.50	5.333		243	320		563	745
2245	10" diameter		4	6		295	360		655	865
2250	12" diameter		3	8		590	480		1,070	1,375

46 51 20.30 Aeration System Blower Control Panels

		Crew	Daily Output	Labor-Hours	Unit	Material	2020 Bare Costs Labor	2020 Bare Costs Equipment	Total	Total Incl O&P
0010	**AERATION SYSTEM BLOWER CONTROL PANELS**									
0020	Single phase simplex									
0030	7 to 10 overload amp range	1 Elec	2.70	2.963	Ea.	780	182		962	1,125
0040	9 to 13 overload amp range		2.70	2.963		845	182		1,027	1,200
0050	12 to 18 overload amp range		2.70	2.963		810	182		992	1,150
0060	16 to 24 overload amp range		2.70	2.963		840	182		1,022	1,200
0070	23 to 32 overload amp range		2.70	2.963		945	182		1,127	1,325
0080	30 to 40 overload amp range		2.70	2.963		1,100	182		1,282	1,500
0090	Single phase duplex									
0100	7 to 10 overload amp range	1 Elec	2	4	Ea.	1,350	245		1,595	1,850
0110	9 to 13 overload amp range		2	4		1,275	245		1,520	1,775
0120	12 to 18 overload amp range		2	4		1,275	245		1,520	1,775
0130	16 to 24 overload amp range		2	4		1,375	245		1,620	1,900
0140	23 to 32 overload amp range		2	4		1,450	245		1,695	1,975
0150	30 to 40 overload amp range		2	4		1,300	245		1,545	1,800
0160	Three phase simplex									
0170	1.6 to 2.5 overload amp range	1 Elec	2.50	3.200	Ea.	975	196		1,171	1,375
0180	2.5 to 4 overload amp range		2.50	3.200		1,050	196		1,246	1,475
0190	4 to 6.3 overload amp range		2.50	3.200		1,050	196		1,246	1,475
0200	6 to 10 overload amp range		2.50	3.200		1,050	196		1,246	1,475
0210	9 to 14 overload amp range		2.50	3.200		1,150	196		1,346	1,575
0220	13 to 18 overload amp range		2.50	3.200		1,100	196		1,296	1,500

46 51 Air and Gas Diffusion Equipment

46 51 20 – Air and Gas Handling Equipment

46 51 20.30 Aeration System Blower Control Panels

		Crew	Daily Output	Labor-Hours	Unit	Material	2020 Bare Costs Labor	Equipment	Total	Total Incl O&P
0230	17 to 23 overload amp range	1 Elec	2.50	3.200	Ea.	1,025	196		1,221	1,425
0240	20 to 25 overload amp range		2.50	3.200		1,025	196		1,221	1,425
0250	23 to 32 overload amp range		2.50	3.200		1,175	196		1,371	1,575
0260	37 to 50 overload amp range		2.50	3.200		1,425	196		1,621	1,875
0270	Three phase duplex									
0280	1.6 to 2.5 overload amp range	1 Elec	1.90	4.211	Ea.	1,325	258		1,583	1,825
0290	2.5 to 4 overload amp range		1.90	4.211		1,325	258		1,583	1,825
0300	4 to 6.3 overload amp range		1.90	4.211		1,325	258		1,583	1,825
0310	6 to 10 overload amp range		1.90	4.211		1,325	258		1,583	1,825
0320	9 to 14 overload amp range		1.90	4.211		1,400	258		1,658	1,900
0330	13 to 18 overload amp range		1.90	4.211		1,425	258		1,683	1,950
0340	17 to 23 overload amp range		1.90	4.211		1,450	258		1,708	1,975
0350	20 to 25 overload amp range		1.90	4.211		1,450	258		1,708	1,975
0360	23 to 32 overload amp range		1.90	4.211		1,525	258		1,783	2,050
0370	37 to 50 overload amp range		1.90	4.211		2,250	258		2,508	2,850

46 51 36 – Ceramic Disc Fine Bubble Diffusers

46 51 36.10 Ceramic Disc Air Diffuser Systems

		Crew	Daily Output	Labor-Hours	Unit	Material	2020 Bare Costs Labor	Equipment	Total	Total Incl O&P
0010	**CERAMIC DISC AIR DIFFUSER SYSTEMS**									
0020	Price for air diffuser pipe system by cell size excluding concrete work									
0030	Price for air diffuser pipe system by cell size excluding air supply									
0040	Depth of 12' is the waste depth, not the cell dimensions									
0100	Ceramic disc air diffuser system, cell size 20' x 20' x 12'	2 Plum	.38	42.667	Ea.	10,800	2,750		13,550	15,900
0120	20' x 30' x 12'		.26	61.326		15,900	3,950		19,850	23,400
0140	20' x 40' x 12'		.20	80		21,100	5,150		26,250	30,900
0160	20' x 50' x 12'		.16	98.644		26,300	6,350		32,650	38,400
0180	20' x 60' x 12'		.14	117		31,400	7,550		38,950	45,900
0200	20' x 70' x 12'		.12	136		36,600	8,775		45,375	53,500
0220	20' x 80' x 12'		.10	155		41,800	9,975		51,775	61,000
0240	20' x 90' x 12'		.09	173		46,900	11,200		58,100	68,000
0260	20' x 100' x 12'		.08	192		52,000	12,400		64,400	76,000
0280	20' x 120' x 12'		.07	229		62,500	14,800		77,300	90,500
0300	20' x 140' x 12'		.06	267		73,000	17,200		90,200	105,500
0320	20' x 160' x 12'		.05	304		83,000	19,600		102,600	121,000
0340	20' x 180' x 12'		.05	341		93,500	22,000		115,500	136,000
0360	20' x 200' x 12'		.04	378		104,000	24,400		128,400	150,500
0380	20' x 250' x 12'		.03	472		129,500	30,400		159,900	188,000
0400	20' x 300' x 12'		.03	565		155,500	36,400		191,900	225,500
0420	20' x 350' x 12'		.02	658		181,500	42,400		223,900	263,000
0440	20' x 400' x 12'		.02	751		207,000	48,400		255,400	300,500
0460	20' x 450' x 12'		.02	847		233,000	54,500		287,500	337,500
0480	20' x 500' x 12'		.02	941		259,000	60,500		319,500	375,500

46 53 Biological Treatment Systems

46 53 17 – Activated Sludge Treatment

46 53 17.10 Activated Sludge Treatment Cells	Crew	Daily Output	Labor-Hours	Unit	Material	2020 Bare Costs Labor	Equipment	Total	Total Incl O&P
0010 **ACTIVATED SLUDGE TREATMENT CELLS**									
0015 Price for cell construction excluding aerator piping & drain									
0020 Cell construction by dimensions									
0100 Treatment cell, end or single, 20' x 20' x 14' high (2' freeboard)	C-14D	.46	434	Ea.	21,200	22,900	950	45,050	58,500
0120 20' x 30' x 14' high		.35	568		27,600	29,900	1,250	58,750	76,500
0140 20' x 40' x 14' high		.29	702		33,900	36,900	1,525	72,325	94,500
0160 20' x 50' x 14' high		.24	835		40,200	44,000	1,825	86,025	112,500
0180 20' x 60' x 14' high		.21	969		46,600	51,000	2,125	99,725	130,000
0200 20' x 70' x 14' high		.18	1103		53,000	58,000	2,425	113,425	147,500
0220 20' x 80' x 14' high		.16	1237		59,500	65,000	2,700	127,200	165,500
0240 20' x 90' x 14' high		.15	1371		65,500	72,000	3,000	140,500	184,000
0260 20' x 100' x 14' high		.13	1504		72,000	79,000	3,300	154,300	201,500
0280 20' x 120' x 14' high		.11	1771		84,500	93,000	3,875	181,375	237,500
0300 20' x 140' x 14' high		.10	2039		97,500	107,500	4,475	209,475	273,000
0320 20' x 160' x 14' high		.09	2307		110,000	121,500	5,050	236,550	308,500
0340 20' x 180' x 14' high		.08	2574		122,500	135,500	5,625	263,625	344,500
0360 20' x 200' x 14' high		.07	2841		135,500	149,500	6,225	291,225	380,500
0380 20' x 250' x 14' high		.06	3509		167,000	184,500	7,675	359,175	469,500
0400 20' x 300' x 14' high		.05	4175		199,000	219,500	9,150	427,650	559,000
0420 20' x 350' x 14' high		.04	4843		230,500	255,000	10,600	496,100	647,500
0440 20' x 400' x 14' high		.04	5510		262,500	290,000	12,100	564,600	736,500
0460 20' x 450' x 14' high		.03	6192		294,000	326,000	13,600	633,600	827,500
0480 20' x 500' x 14' high		.03	6849		326,000	360,500	15,000	701,500	916,000
0500 Treatment cell, end or single, 20' x 20' x 12' high (2' freeboard)		.48	414		20,100	21,800	905	42,805	56,000
0520 20' x 30' x 12' high		.37	543		26,100	28,600	1,200	55,900	73,000
0540 20' x 40' x 12' high		.30	672		32,200	35,300	1,475	68,975	90,000
0560 20' x 50' x 12' high		.25	800		38,200	42,100	1,750	82,050	107,000
0580 20' x 60' x 12' high		.22	929		44,300	48,900	2,025	95,225	124,500
0600 20' x 70' x 12' high		.19	1058		50,500	55,500	2,325	108,325	141,500
0620 20' x 80' x 12' high		.17	1186		56,500	62,500	2,600	121,600	158,500
0640 20' x 90' x 12' high		.15	1315		62,500	69,000	2,875	134,375	175,500
0660 20' x 100' x 12' high		.14	1444		68,500	76,000	3,150	147,650	193,000
0680 20' x 120' x 12' high		.12	1701		80,500	89,500	3,725	173,725	227,000
0700 20' x 140' x 12' high		.10	1959		93,000	103,000	4,300	200,300	261,000
0720 20' x 160' x 12' high		.09	2215		105,000	116,500	4,850	226,350	296,000
0740 20' x 180' x 12' high		.08	2472		117,000	130,000	5,425	252,425	330,000
0760 20' x 200' x 12' high		.07	2732		129,000	144,000	5,975	278,975	364,500
0780 20' x 250' x 12' high		.06	3373		159,500	177,500	7,375	344,375	450,000
0800 20' x 300' x 12' high		.05	4016		189,500	211,500	8,800	409,800	535,000
0820 20' x 350' x 12' high		.04	4662		220,000	245,500	10,200	475,700	621,500
0840 20' x 400' x 12' high		.04	5305		250,500	279,000	11,600	541,100	707,000
0860 20' x 450' x 12' high		.03	5952		280,500	313,000	13,000	606,500	793,000
0880 20' x 500' x 12' high		.03	6601		311,000	347,500	14,500	673,000	879,500
1100 Treatment cell, connecting, 20' x 20' x 14' high (2' freeboard)		.57	351		16,900	18,500	770	36,170	47,100
1120 20' x 30' x 14' high		.45	443		21,200	23,300	970	45,470	59,500
1140 20' x 40' x 14' high		.37	535		25,400	28,100	1,175	54,675	71,500
1160 20' x 50' x 14' high		.32	627		29,700	33,000	1,375	64,075	83,500
1180 20' x 60' x 14' high		.28	719		33,800	37,800	1,575	73,175	96,000
1200 20' x 70' x 14' high		.25	811		38,000	42,700	1,775	82,475	108,000
1220 20' x 80' x 14' high		.22	903		42,300	47,500	1,975	91,775	120,000
1240 20' x 90' x 14' high		.20	995		46,500	52,500	2,175	101,175	132,000
1260 20' x 100' x 14' high		.18	1086		50,500	57,000	2,375	109,875	144,500
1280 20' x 120' x 14' high		.16	1271		59,000	67,000	2,775	128,775	168,500

46 53 17.10 Activated Sludge Treatment Cells	Crew	Daily Output	Labor-Hours	Unit	Material	2020 Bare Costs Labor	Equipment	Total	Total Incl O&P
1300	20' x 140' x 14' high	C-14D	.14	1455	Ea.	67,500	76,500	3,175	147,175	193,000
1320	20' x 160' x 14' high		.12	1638		76,000	86,000	3,575	165,575	217,000
1340	20' x 180' x 14' high		.11	1823		84,500	96,000	4,000	184,500	241,500
1360	20' x 200' x 14' high		.10	2006		93,000	105,500	4,400	202,900	265,500
1380	20' x 250' x 14' high		.08	2466		114,000	130,000	5,400	249,400	326,500
1400	20' x 300' x 14' high		.07	2924		135,000	154,000	6,400	295,400	386,500
1420	20' x 350' x 14' high		.06	3384		156,000	178,000	7,400	341,400	447,500
1440	20' x 400' x 14' high		.05	3846		177,500	202,500	8,425	388,425	508,000
1460	20' x 450' x 14' high		.05	4301		198,500	226,500	9,425	434,425	568,000
1480	20' x 500' x 14' high		.04	4762		219,500	250,500	10,400	480,400	629,000
1500	Treatment cell, connecting, 20' x 20' x 12' high (2' freeboard)		.60	336		16,100	17,700	735	34,535	45,000
1520	20' x 30' x 12' high		.47	425		20,200	22,400	930	43,530	57,000
1540	20' x 40' x 12' high		.39	515		24,200	27,100	1,125	52,425	68,500
1560	20' x 50' x 12' high		.33	604		28,300	31,800	1,325	61,425	80,500
1580	20' x 60' x 12' high		.29	694		32,400	36,500	1,525	70,425	92,500
1600	20' x 70' x 12' high		.26	783		36,500	41,200	1,725	79,425	104,000
1620	20' x 80' x 12' high		.23	873		40,500	45,900	1,900	88,300	115,500
1640	20' x 90' x 12' high		.21	962		44,600	50,500	2,100	97,200	127,500
1660	20' x 100' x 12' high		.19	1052		48,700	55,500	2,300	106,500	139,000
1680	20' x 120' x 12' high		.16	1230		57,000	64,500	2,700	124,200	162,500
1700	20' x 140' x 12' high		.14	1409		65,000	74,000	3,075	142,075	186,500
1720	20' x 160' x 12' high		.13	1589		73,000	83,500	3,475	159,975	210,000
1740	20' x 180' x 12' high		.11	1767		81,500	93,000	3,875	178,375	233,500
1760	20' x 200' x 12' high		.10	1946		89,500	102,500	4,250	196,250	256,500
1780	20' x 250' x 12' high		.08	2392		110,000	126,000	5,250	241,250	316,000
1800	20' x 300' x 12' high		.07	2841		130,000	149,500	6,225	285,725	374,500
1820	20' x 350' x 12' high		.06	3289		150,500	173,000	7,200	330,700	433,500
1840	20' x 400' x 12' high		.05	3738		171,000	196,500	8,175	375,675	492,500
1860	20' x 450' x 12' high		.05	4184		191,500	220,000	9,175	420,675	551,000
1880	20' x 500' x 12' high		.04	5571		211,500	293,000	12,200	516,700	686,500
1996	Price for aerator curb per cell excluding aerator piping & drain									
2000	Treatment cell, 20' x 20' aerator curb	F-7	1.88	17.067	Ea.	665	815		1,480	1,950
2010	20' x 30'		1.25	25.600		995	1,225		2,220	2,925
2020	20' x 40'		.94	34.133		1,325	1,625		2,950	3,900
2030	20' x 50'		.75	42.667		1,650	2,025		3,675	4,875
2040	20' x 60'		.63	51.200		2,000	2,450		4,450	5,850
2050	20' x 70'		.54	59.735		2,325	2,850		5,175	6,825
2060	20' x 80'		.47	68.259		2,650	3,250		5,900	7,800
2070	20' x 90'		.42	76.794		2,975	3,650		6,625	8,775
2080	20' x 100'		.38	85.333		3,325	4,075		7,400	9,750
2090	20' x 120'		.31	102		3,975	4,875		8,850	11,700
2100	20' x 140'		.27	119		4,650	5,700		10,350	13,700
2110	20' x 160'		.23	137		5,300	6,500		11,800	15,600
2120	20' x 180'		.21	154		5,975	7,325		13,300	17,600
2130	20' x 200'		.19	171		6,625	8,125		14,750	19,500
2160	20' x 250'		.15	213		8,275	10,200		18,475	24,400
2190	20' x 300'		.13	256		9,950	12,200		22,150	29,200
2220	20' x 350'		.11	299		11,600	14,200		25,800	34,200
2250	20' x 400'		.09	341		13,300	16,200		29,500	39,000
2280	20' x 450'		.08	384		14,900	18,300		33,200	43,900
2310	20' x 500'		.08	427		16,600	20,300		36,900	48,700

Assemblies Section

Table of Contents

Table No.   **Page**

A SUBSTRUCTURE

A1010   Standard Foundations
A1010 105 Wall Foundations 578
A1010 110 Strip Footings 580
A1010 210 Spread Footings 581
A1010 250 Pile Caps 582
A1010 310 Foundation Underdrain 584
A1010 320 Foundation Dampproofing 585

A1020   Special Foundations
A1020 110 C.I.P. Concrete Piles 586
A1020 120 Precast Concrete Piles 588
A1020 130 Steel Pipe Piles 590
A1020 140 Steel H Piles 592
A1020 150 Step-Tapered Steel Piles 594
A1020 160 Treated Wood Piles 595
A1020 210 Grade Beams 597
A1020 310 Caissons 598
A1020 710 Pressure Injected Footings 599

A1030   Slab on Grade
A1030 120 Plain & Reinforced 600

A2010   Basement Excavation
A2010 110 Building Excavation & Backfill 601

A2020   Basement Walls
A2020 110 Walls, Cast in Place 603
A2020 150 Wood Wall Foundations 605

D SERVICES

D3050   Terminal & Package Units
D3050 248 Geothermal Heat Pump System 609

G BUILDING SITEWORK

G1010   Site Clearing
G1010 120 Clear & Grub Site 612
G1010 122 Strip Topsoil 613

G1020   Site Demolitions and Relocations
G1020 205 Remove Underground Water Pipe Including Earthwork 614
G1020 206 Remove Underground Sewer Pipe Including Earthwork 617
G1020 210 Demolition Bituminous Sidewalk 619
G1020 212 Demolition Concrete Sidewalk 619

G1030   Site Earthwork
G1030 105 Cut & Fill Gravel 620
G1030 110 Excavate and Haul Sand & Gravel 622
G1030 115 Cut & Fill Common Earth 624
G1030 120 Excavate and Haul Common Earth 626
G1030 125 Cut & Fill Clay 628
G1030 130 Excavate and Haul Clay 630
G1030 150 Load & Haul Rock 632

Table No.   **Page**

G1030 160 Excavate and Haul Sandy Clay/Loam 633
G1030 205 Gravel Backfill 634
G1030 210 Common Earth Backfill 636
G1030 215 Clay Backfill 638
G1030 220 Sandy Clay/Loam Backfill 640
G1030 805 Trenching Common Earth 642
G1030 806 Trenching Loam & Sandy Clay 644
G1030 807 Trenching Sand & Gravel 646
G1030 815 Pipe Bedding 648

G2010   Roadways
G2010 230 Bituminous Roadways Gravel Base 650
G2010 232 Bituminous Roadways Crushed Stone 653

G2020   Parking Lots
G2020 210 Parking Lots Gravel Base 656
G2020 212 Parking Lots Crushed Stone 657
G2020 214 Parking Lots With Handicap & Lighting 658

G2030   Pedestrian Paving
G2030 110 Bituminous Sidewalks 662
G2030 120 Concrete Sidewalks 663
G2030 150 Brick & Tile Plazas 664
G2030 310 Stairs 666

G2040   Site Development
G2040 210 Concrete Retaining Walls 668
G2040 220 Masonry Retaining Walls 669
G2040 230 Wood Post Retaining Walls 670
G2040 240 Post & Board Retaining Walls 671
G2040 250 Wood Tie Retaining Walls 672
G2040 260 Stone Retaining Walls 673
G2040 270 Gabion Retaining Walls 675
G2040 910 Wood Decks 677
G2040 920 Swimming Pools 678
G2040 990 Site Development Components for Buildings 679

G2050   Landscaping
G2050 410 Lawns & Ground Cover 681
G2050 710 Site Irrigation 683
G2050 720 Site Irrigation 686
G2050 910 Tree Pits 687

G3010   Water Supply
G3010 121 Water Service, Lead Free 689
G3010 122 Waterline (Common Earth Excavation) 691
G3010 124 Waterline (Loam & Sandy Clay Excavation) 694
G3010 410 Fire Hydrants 697

G3020   Sanitary Sewer
G3020 112 Sewerline (Common Earth Excavation) 699
G3020 114 Sewerline (loam & sandy clay excavation) 701
G3020 302 Septic Systems 703
G3020 730 Secondary Sewage Lagoon 705
G3020 740 Blower System for Wastewater Aeration 707

Table of Contents

Table No.	Page
G3030 Storm Sewer	
G3030 210 Manholes & Catch Basins	708
G3030 310 Headwalls	710
G3030 610 Stormwater Management (costs per CF of stormwater)	712
G3060 Fuel Distribution	
G3060 112 Gasline (Common Earth Excavation)	713
G4010 Electrical Distribution	
G4010 312 Underground Power Feed	716
G4010 320 Underground Electrical Conduit	717

Table No.	Page
G4020 Site Lighting	
G4020 110 Site Lighting	720
G4020 210 Light Pole (Installed)	722
G9020 Site Repair & Maintenance	
G9020 100 Clean and Wrap Marine Piles	723

RSMeans data: Assemblies—How They Work

Assemblies estimating provides a fast and reasonably accurate way to develop construction costs. An assembly is the grouping of individual work items—with appropriate quantities—to provide a cost for a major construction component in a convenient unit of measure.

An assemblies estimate is often used during early stages of design development to compare the cost impact of various design alternatives on total building cost.

Assemblies estimates are also used as an efficient tool to verify construction estimates.

Assemblies estimates do not require a completed design or detailed drawings. Instead, they are based on the general size of the structure and other known parameters of the project. The degree of accuracy of an assemblies estimate is generally within +/- 15%.

Most assemblies consist of three major elements: a graphic, the system components, and the cost data itself. The **Graphic** is a visual representation showing the typical appearance of the assembly

1 Unique 12-character Identifier

Our assemblies are identified by a **unique 12-character identifier**. The assemblies are numbered using UNIFORMAT II, ASTM Standard E1557. The first 5 characters represent this system to Level 3. The last 7 characters represent further breakdown in order to arrange items in understandable groups of similar tasks. Line numbers are consistent across all of our publications, so a line number in any assemblies data set will always refer to the same item.

2 Narrative Descriptions

Our assemblies descriptions appear in two formats: narrative and table. **Narrative descriptions** are shown in a hierarchical structure to make them readable. In order to read a complete description, read up through the indents to the top of the section. Include everything that is above and to the left that is not contradicted by information below.

Narrative Format

G20 Site Improvements

G2040 Site Development

There are four basic types of Concrete Retaining Wall Systems: reinforced concrete with level backfill; reinforced concrete with sloped backfill or surcharge; unreinforced with level backfill; and unreinforced with sloped backfill or surcharge. System elements include: all necessary forms (4 uses); 3,000 p.s.i. concrete with an 8" chute; all necessary reinforcing steel; and underdrain. Exposed concrete is patched and rubbed.

The Expanded System Listing shows walls that range in thickness from 10" to 18" for reinforced concrete walls with level backfill and 12" to 24" for reinforced walls with sloped backfill. Walls range from a height of 4' to 20'. Unreinforced level and sloped backfill walls range from a height of 3' to 10'.

System Components

SYSTEM G2040 210 1000
CONC. RETAIN. WALL REINFORCED, LEVEL BACKFILL, 4' HIGH

System Components	QUANTITY	UNIT	COST PER L.F.		
			MAT.	INST.	TOTAL
Forms in place, cont. wall footing & keyway, 4 uses	2.000	S.F.	4.92	9.98	14.90
Forms in place, retaining wall forms, battered to 8' high, 4 uses	8.000	SFCA	6.24	76.40	82.64
Reinforcing in place, walls, #3 to #7	.004	Ton	5	3.62	8.62
Concrete ready mix, regular weight, 3000 psi	.204	C.Y.	28.97		28.97
Placing concrete and vibrating footing con., shallow direct chute	.074	C.Y.		1.96	1.96
Placing concrete and vibrating walls, 8" thick, direct chute	.130	C.Y.		4.64	4.64
Pipe bedding, crushed or screened bank run gravel	1.000	L.F.	2.53	1.37	3.90
Pipe, subdrainage, corrugated plastic, 4" diameter	1.000	L.F.	.83	.84	1.67
Finish walls and break ties, patch walls	4.000	S.F.	.20	4.36	4.56
TOTAL			48.69	103.17	151.86

G2040 210	Concrete Retaining Walls	COST PER L.F.		
		MAT.	INST.	TOTAL
1000	Conc. retain. wall, reinforced, level backfill, 4' high x 2'-2" base,10" th	48.50	103	151.50
1200	6' high x 3'-3" base, 10" thick	70	149	219
1400	8' high x 4'-3" base, 10" thick	91	196	287
1600	10' high x 5'-4" base, 13" thick	118	287	405
2200	16' high x 8'-6" base, 16" thick	224	465	689
2600	20' high x 10'-5" base, 18" thick	330	600	930
3000	Sloped backfill, 4' high x 3'-2" base, 12" thick	59.50	107	166.50
3200	6' high x 4'-6" base, 12" thick	84	153	237
3400	8' high x 5'-11" base, 12" thick	112	202	314
3600	10' high x 7'-5" base, 16" thick	159	300	459
3800	12' high x 8'-10" base, 18" thick	208	365	573
4200	16' high x 11'-10" base, 21" thick	350	515	865
4600	20' high x 15'-0" base, 24" thick	545	690	1,235
5000	Unreinforced, level backfill, 3'-0" high x 1'-6" base	26	69.50	95.50
5200	4'-0" high x 2'-0" base	40	93	133
5400	6'-0" high x 3'-0" base	75.50	143	218.50
5600	8'-0" high x 4'-0" base	121	191	312
5800	10'-0" high x 5'-0" base	180	296	476
7000	Sloped backfill, 3'-0" high x 2'-0" base	31	72	103
7200	4'-0" high x 3'-0" base	52.50	96.50	149
7400	6'-0" high x 5'-0" base	110	153	263
7600	8'-0" high x 7'-0" base	188	207	395
7800	10'-0" high x 9'-0" base	290	320	610

For supplemental customizable square foot estimating forms, visit: **RSMeans.com/2020books** in question. It is frequently accompanied by additional explanatory technical information describing the class of items. The **System Components** is a listing of the individual tasks that make up the assembly, including the quantity and unit of measure for each item, along with the cost of material and installation. The **Assemblies**

data below lists prices for other similar systems with dimensional and/or size variations.

All of our assemblies costs represent the cost for the installing contractor. An allowance for profit has been added to all material, labor, and equipment rental costs. A markup for labor burdens, including workers' compensation, fixed

overhead, and business overhead, is included with installation costs.

The information in RSMeans cost data represents a "national average" cost. This data should be modified to the project location using the **City Cost Indexes** or **Location Factors** tables found in the Reference Section.

Table Format

A10 Foundations

A1010 Standard Foundations

The Foundation Bearing Wall System includes: forms up to 6' high (four uses); 3,000 p.s.i. concrete placed and vibrated; and form removal with breaking form ties and patching walls. The wall systems list walls from 6" to 16" thick and are designed with minimum reinforcement.

Excavation and backfill are not included.

Please see the reference section for further design and cost information.

System Components	QUANTITY	UNIT	MAT.	INST.	TOTAL
SYSTEM A1010 105 1500					
FOUNDATION WALL, CAST IN PLACE, DIRECT CHUTE, 4' HIGH, 6" THICK					
Formwork	8.000	SFCA	6.96	49.60	56.56
Reinforcing	3.300	Lb.	2.06	1.49	3.55
Unloading & sorting reinforcing	3.300	Lb.		.09	.09
Concrete, 3,000 psi	.074	C.Y.	10.51		10.51
Place concrete, direct chute	.074	C.Y.		2.64	2.64
Finish walls, break ties and patch voids, one side	4.000	S.F.	.20	4.36	4.56
TOTAL			19.73	58.18	77.91

COST PER L.F. MAT., INST., TOTAL

A1010 105		Wall Foundations						
	WALL HEIGHT (FT.)	PLACING METHOD	CONCRETE (C.Y. per L.F.)	REINFORCING (LBS. per L.F.)	WALL THICKNESS (IN.)	MAT.	INST.	TOTAL
1500	4'	direct chute	.074	3.3	6	19.75	58	77.75
1520			.099	4.8	8	24	59.50	83.50
1540			.123	6.0	10	28.50	60.50	89
1560			.148	7.2	12	32.50	62	94.50
1580			.173	8.1	14	37	63	100
1600			.197	9.44	16	41	64.50	105.50
1700	4'	pumped	.074	3.3	6	19.75	59.50	79.25
1720			.099	4.8	8	24	61	85
1740			.123	6.0	10	28.50	62	90.50
1760			.148	7.2	12	32.50	63.50	96
1780			.173	8.1	14	37	65	102
1800			.197	9.44	16	41	66.50	107.50

COST PER L.F. (column group: MAT., INST., TOTAL)

③ Unit of Measure

All RSMeans data: Assemblies include a typical **Unit of Measure** used for estimating that item. For instance, for continuous footings or foundation walls the unit is linear feet (L.F.). For spread footings the unit is each (Ea.). The estimator needs to take special care that the unit in the data matches the unit in the takeoff. Abbreviations and unit conversions can be found in the Reference Section.

④ System Components

System components are listed separately to detail what is included in the development of the total system price.

⑤ Table Descriptions

Table descriptions work similar to Narrative Descriptions, except that if there is a blank in the column at a particular line number, read up to the description above in the same column.

RSMeans data: Assemblies— How They Work (Continued)

Sample Estimate

This sample demonstrates the elements of an estimate, including a tally of the RSMeans data lines. Published assemblies costs include all markups for labor burden and profit for the installing contractor. This estimate adds a summary of the markups applied by a general contractor on the installing contractor's work. These figures represent the total cost to the owner. The location factor with RSMeans data is applied at the bottom of the estimate to adjust the cost of the work to a specific location.

Project Name:	Interior Fit-out, ABC Office				
Location:	**Anywhere, USA**		**Date: 1/1/2020**		**STD**
Assembly Number	**Description**	**Qty.**	**Unit**		**Subtotal**
① C1010 124 1200	Wood partition, 2 x 4 @ 16" OC w/5/8" FR gypsum board	560	S.F.		$2,996.00
C1020 114 1800	Metal door & frame, flush hollow core, 3'-0" x 7'-0"	2	Ea.		$2,620.00
C3010 230 0080	Painting, brushwork, primer & 2 coats	1,120	S.F.		$1,467.20
C3020 410 0140	Carpet, tufted, nylon, roll goods, 12' wide, 26 oz	240	S.F.		$928.80
C3030 210 6000	Acoustic ceilings, 24" x 48" tile, tee grid suspension	200	S.F.		$1,326.00
D5020 125 0560	Receptacles incl plate, box, conduit, wire, 20 A duplex	8	Ea.		$2,428.00
D5020 125 0720	Light switch incl plate, box, conduit, wire, 20 A single pole	2	Ea.		$591.00
D5020 210 0560	Fluorescent fixtures, recess mounted, 20 per 1000 SF	200	S.F.		$2,320.00
	Assembly Subtotal				**$14,677.00**
	Sales Tax @ ②		5 %		$ 366.93
	General Requirements @ ③		7 %		$ 1,027.39
	Subtotal A				**$16,071.32**
	GC Overhead @ ④		5 %		$ 803.57
	Subtotal B				**$16,874.88**
	GC Profit @ ⑤		5 %		$ 843.74
	Subtotal C				**$17,718.62**
	Adjusted by Location Factor ⑥	115.5			$ 20,465.01
	Architects Fee @ ⑦		8 %		$ 1,637.20
	Contingency @ ⑧		15 %		$ 3,069.75
	Project Total Cost				**$ 25,171.96**

This estimate is based on an interactive spreadsheet. You are free to download it and adjust it to your methodology.
A copy of this spreadsheet is available at **RSMeans.com/2020books.**

① Work Performed

The body of the estimate shows the RSMeans data selected, including line numbers, a brief description of each item, its takeoff quantity and unit, and the total installed cost, including the installing contractor's overhead and profit.

② Sales Tax

If the work is subject to state or local sales taxes, the amount must be added to the estimate. In a conceptual estimate it can be assumed that one half of the total represents material costs. Therefore, apply the sales tax rate to 50% of the assembly subtotal.

③ General Requirements

This item covers project-wide needs provided by the general contractor. These items vary by project but may include temporary facilities and utilities, security, testing, project cleanup, etc. In assemblies estimates a percentage is used—typically between 5% and 15% of project cost.

④ General Contractor Overhead

This entry represents the general contractor's markup on all work to cover project administration costs.

⑤ General Contractor Profit

This entry represents the GC's profit on all work performed. The value included here can vary widely by project and is influenced by the GC's perception of the project's financial risk and market conditions.

⑥ Location Factor

RSMeans published data are based on national average costs. If necessary, adjust the total cost of the project using a location factor from the "Location Factor" table or the "City Cost Indexes" table found in the Reference Section. Use location factors if the work is general, covering the work of multiple trades. If the work is by a single trade (e.g., masonry) use the more specific data found in the City Cost Indexes.

To adjust costs by location factors, multiply the base cost by the factor and divide by 100.

⑦ Architect's Fee

If appropriate, add the design cost to the project estimate. These fees vary based on project complexity and size. Typical design and engineering fees can be found in the Reference Section.

⑧ Contingency

A factor for contingency may be added to any estimate to represent the cost of unknowns that may occur between the time that the estimate is performed and the time the project is constructed. The amount of the allowance will depend on the stage of design at which the estimate is done, as well as the contractor's assessment of the risk involved.

Same Data. Simplified.

Enjoy the convenience and efficiency of accessing your costs anywhere:

- **Skip the multiplier** by setting your location
- **Quickly search,** edit, favorite and share costs
- **Stay on top of price changes** with automatic updates

Discover more at rsmeans.com/online

A1010   Standard Foundations

The Foundation Bearing Wall System includes: forms up to 6' high (four uses); 3,000 p.s.i. concrete placed and vibrated; and form removal with breaking form ties and patching walls. The wall systems list walls from 6" to 16" thick and are designed with minimum reinforcement.

Excavation and backfill are not included.

Please see the reference section for further design and cost information.

System Components	QUANTITY	UNIT	COST PER L.F.		
			MAT.	INST.	TOTAL
SYSTEM A1010 105 1500					
FOUNDATION WALL, CAST IN PLACE, DIRECT CHUTE, 4' HIGH, 6" THICK					
Formwork	8.000	SFCA	6.96	49.60	56.56
Reinforcing	3.300	Lb.	2.06	1.49	3.55
Unloading & sorting reinforcing	3.300	Lb.		.09	.09
Concrete, 3,000 psi	.074	C.Y.	10.51		10.51
Place concrete, direct chute	.074	C.Y.		2.64	2.64
Finish walls, break ties and patch voids, one side	4.000	S.F.	.20	4.36	4.56
TOTAL			19.73	58.18	77.91

A1010 105			Wall Foundations					
	WALL HEIGHT (FT.)	PLACING METHOD	CONCRETE (C.Y. per L.F.)	REINFORCING (LBS. per L.F.)	WALL THICKNESS (IN.)	COST PER L.F.		
						MAT.	INST.	TOTAL
1500	4'	direct chute	.074	3.3	6	19.75	58	77.75
1520			.099	4.8	8	24	59.50	83.50
1540			.123	6.0	10	28.50	60.50	89
1560			.148	7.2	12	32.50	62	94.50
1580			.173	8.1	14	37	63	100
1600			.197	9.44	16	41	64.50	105.50
1700	4'	pumped	.074	3.3	6	19.75	59.50	79.25
1720			.099	4.8	8	24	61	85
1740			.123	6.0	10	28.50	62	90.50
1760			.148	7.2	12	32.50	63.50	96
1780			.173	8.1	14	37	65	102
1800			.197	9.44	16	41	66.50	107.50

A10 Foundations

A1010 Standard Foundations

A1010 105		Wall Foundations						
	WALL HEIGHT (FT.)	PLACING METHOD	CONCRETE (C.Y. per L.F.)	REINFORCING (LBS. per L.F.)	WALL THICKNESS (IN.)	COST PER L.F.		
						MAT.	INST.	TOTAL
3000	6'	direct chute	.111	4.95	6	29.50	87	116.50
3020			.149	7.20	8	36.50	89.50	126
3040			.184	9.00	10	42.50	91	133.50
3060			.222	10.8	12	49	93	142
3080			.260	12.15	14	55.50	94.50	150
3100			.300	14.39	16	62.50	97	159.50
3200	6'	pumped	.111	4.95	6	29.50	88.50	118
3220			.149	7.20	8	36.50	92	128.50
3240			.184	9.00	10	42.50	93.50	136
3260			.222	10.8	12	49	96	145
3280			.260	12.15	14	55.50	97	152.50
3300			.300	14.39	16	62.50	100	162.50

For customer support on your Site Work & Landscape Costs with RSMeans data, call 800.448.8182.

A1010 Standard Foundations

The Strip Footing System includes: excavation; hand trim; all forms needed for footing placement; forms for 2″ x 6″ keyway (four uses); dowels; and 3,000 p.s.i. concrete.

The footing size required varies for different soils. Soil bearing capacities are listed for 3 KSF and 6 KSF. Depths of the system range from 8″ and deeper. Widths range from 16″ and wider. Smaller strip footings may not require reinforcement.

Please see the reference section for further design and cost information.

System Components	QUANTITY	UNIT	COST PER L.F.		
			MAT.	INST.	TOTAL
SYSTEM A1010 110 2500					
STRIP FOOTING, LOAD 5.1 KLF, SOIL CAP. 3 KSF, 24″ WIDE X 12″ DEEP, REINF.					
Trench excavation	.148	C.Y.		1.40	1.40
Hand trim	2.000	S.F.		2.30	2.30
Compacted backfill	.074	C.Y.		.33	.33
Formwork, 4 uses	2.000	S.F.	4.92	9.98	14.90
Keyway form, 4 uses	1.000	L.F.	.37	1.28	1.65
Reinforcing, fy = 60000 psi	3.000	Lb.	1.95	1.95	3.90
Dowels	2.000	Ea.	1.82	5.66	7.48
Concrete, f'c = 3000 psi	.074	C.Y.	10.51		10.51
Place concrete, direct chute	.074	C.Y.		1.96	1.96
Screed finish	2.000	S.F.		.84	.84
TOTAL			19.57	25.70	45.27

A1010 110	Strip Footings	COST PER L.F.		
		MAT.	INST.	TOTAL
2100	Strip footing, load 2.6 KLF, soil capacity 3 KSF, 16″ wide x 8″ deep, plain	8.35	12.05	20.40
2300	Load 3.9 KLF, soil capacity 3 KSF, 24″ wide x 8″ deep, plain	10.75	13.60	24.35
2500	Load 5.1 KLF, soil capacity 3 KSF, 24″ wide x 12″ deep, reinf.	19.55	25.50	45.05
2700	Load 11.1 KLF, soil capacity 6 KSF, 24″ wide x 12″ deep, reinf.	19.55	25.50	45.05
2900	Load 6.8 KLF, soil capacity 3 KSF, 32″ wide x 12″ deep, reinf.	24	28.50	52.50
3100	Load 14.8 KLF, soil capacity 6 KSF, 32″ wide x 12″ deep, reinf.	24	28	52
3300	Load 9.3 KLF, soil capacity 3 KSF, 40″ wide x 12″ deep, reinf.	28	30.50	58.50
3500	Load 18.4 KLF, soil capacity 6 KSF, 40″ wide x 12″ deep, reinf.	28	31	59
3700	Load 10.1 KLF, soil capacity 3 KSF, 48″ wide x 12″ deep, reinf.	31.50	32.50	64
3900	Load 22.1 KLF, soil capacity 6 KSF, 48″ wide x 12″ deep, reinf.	33.50	34.50	68
4100	Load 11.8 KLF, soil capacity 3 KSF, 56″ wide x 12″ deep, reinf.	36.50	36	72.50
4300	Load 25.8 KLF, soil capacity 6 KSF, 56″ wide x 12″ deep, reinf.	39.50	39	78.50
4500	Load 10 KLF, soil capacity 3 KSF, 48″ wide x 16″ deep, reinf.	40	38.50	78.50
4700	Load 22 KLF, soil capacity 6 KSF, 48″ wide, 16″ deep, reinf.	41	39	80
4900	Load 11.6 KLF, soil capacity 3 KSF, 56″ wide x 16″ deep, reinf.	45.50	54.50	100
5100	Load 25.6 KLF, soil capacity 6 KSF, 56″ wide x 16″ deep, reinf.	47.50	56.50	104
5300	Load 13.3 KLF, soil capacity 3 KSF, 64″ wide x 16″ deep, reinf.	52	45	97
5500	Load 29.3 KLF, soil capacity 6 KSF, 64″ wide x 16″ deep, reinf.	55	48.50	103.50
5700	Load 15 KLF, soil capacity 3 KSF, 72″ wide x 20″ deep, reinf.	69.50	54.50	124
5900	Load 33 KLF, soil capacity 6 KSF, 72″ wide x 20″ deep, reinf.	73	58	131
6100	Load 18.3 KLF, soil capacity 3 KSF, 88″ wide x 24″ deep, reinf.	98	69	167
6300	Load 40.3 KLF, soil capacity 6 KSF, 88″ wide x 24″ deep, reinf.	106	76	182
6500	Load 20 KLF, soil capacity 3 KSF, 96″ wide x 24″ deep, reinf.	107	73.50	180.50
6700	Load 44 KLF, soil capacity 6 KSF, 96″ wide x 24″ deep, reinf.	112	79	191

A1010 Standard Foundations

The Spread Footing System includes: excavation; backfill; forms (four uses); all reinforcement; 3,000 p.s.i. concrete (chute placed); and float finish.

Footing systems are priced per individual unit. The Expanded System Listing at the bottom shows various footing sizes. It is assumed that excavation is done by a truck mounted hydraulic excavator with an operator and oiler.

Backfill is with a dozer, and compaction by air tamp. The excavation and backfill equipment is assumed to operate at 30 C.Y. per hour.

Please see the reference section for further design and cost information.

System Components	QUANTITY	UNIT	COST EACH MAT.	COST EACH INST.	COST EACH TOTAL
SYSTEM A1010 210 2200					
SPREAD FOOTINGS, 3'-0" SQ. X 12" DEEP, 3 KSF SOIL CAP., 25K LOAD					
Excavation, Gradall, 30" bucket, 30 CY/Hr	.590	C.Y.		5.27	5.27
Trim sides & bottom	9.000	S.F.		10.35	10.35
Dozer backfill & roller compaction	.260	C.Y.		1.17	1.17
Forms in place, spread footing, 4 uses	12.000	S.F.	9.84	70.20	80.04
Reinforcing in place footings, #4 to #7	.006	Ton	7.50	7.80	15.30
Supports for dowels in footings, walls, and beams	6.000	L.F.	6.96	29.04	36
Concrete ready mix, regular weight, 3000 psi	.330	C.Y.	46.86		46.86
Place and vibrate concrete for spread footings over 5 CY, direct chute	.330	C.Y.		8.74	8.74
Finishing floor, bull float only	9.000	S.F.		3.78	3.78
TOTAL			71.16	136.35	207.51

A1010 210	Spread Footings	COST EACH MAT.	COST EACH INST.	COST EACH TOTAL
2200	Spread footings, 3'-0 square x 12" deep, 3 KSF soil cap., 25K load	71	136	207
2210	4'-0" sq. x 12" deep, 6 KSF soil cap., 75K load	124	201	325
2400	4'-6" sq. x 12" deep, 3 KSF soil capacity, 50K load	153	234	387
2600	15" deep, 6 KSF soil capacity, 100K load	188	276	464
2800	5'-0" sq. x 12" deep, 3 KSF soil capacity, 62K load	187	287	474
3000	16" deep, 6 KSF soil capacity, 125K load	272	345	617
3200	5'-6" sq. x 13" deep, 3 KSF soil capacity, 75K load	244	330	574
3400	18" deep, 6 KSF soil capacity, 150K load	325	415	740
3600	6'-0" sq. x 14" deep, 3 KSF soil capacity, 100K load	310	395	705
3800	20" deep, 6 KSF soil capacity, 200K load	425	515	940
4000	7'-6" sq. x 18" deep, 3 KSF soil capacity, 150K load	595	660	1,255
4200	25" deep, 6 KSF soil capacity, 300K load	810	845	1,655
4400	8'-6" sq. x 20" deep, 3 KSF soil capacity, 200K load	845	870	1,715
4600	27" deep, 6 KSF soil capacity, 400K load	1,125	1,100	2,225
4800	9'-6" sq. x 24" deep, 3 KSF soil capacity, 250K load	1,225	1,200	2,425
5000	30" deep, 6 KSF soil capacity, 500K load	1,550	1,425	2,975
5200	10'-6" sq. x 26" deep, 3 KSF soil capacity, 300K load	1,550	1,400	2,950
5400	33" deep, 6 KSF soil capacity, 600K load	2,100	1,825	3,925
5600	12'-0" sq. x 28" deep, 3 KSF soil capacity, 400K load	2,525	2,100	4,625
5800	38" deep, 6 KSF soil capacity, 800K load	3,025	2,475	5,500
6200	13'-6" sq. x 30" deep, 3 KSF soil capacity, 500K load	3,125	2,500	5,625
6400	42" deep, 6 KSF soil capacity, 1000K load	4,225	3,300	7,525

A10 Foundations

A1010 Standard Foundations

These pile cap systems include excavation with a truck mounted hydraulic excavator, hand trimming, compacted backfill, forms for concrete, templates for dowels or anchor bolts, reinforcing steel and concrete placed and floated.

Pile embedment is assumed as 6″. Design is consistent with the Concrete Reinforcing Steel Institute Handbook f'c = 3000 psi, fy = 60,000.

Please see the reference section for further design and cost information.

System Components	QUANTITY	UNIT	COST EACH		
			MAT.	INST.	TOTAL
SYSTEM A1010 250 5100					
CAP FOR 2 PILES, 6′-6″X3′-6″X20″, 15 TON PILE, 8″ MIN. COL., 45K COL. LOAD					
Excavation, bulk, hyd excavator, truck mtd. 30″ bucket 1/2 CY	2.890	C.Y.		25.80	25.80
Trim sides and bottom of trench, regular soil	23.000	S.F.		26.45	26.45
Dozer backfill & roller compaction	1.500	C.Y.		6.76	6.76
Forms in place pile cap, square or rectangular, 4 uses	33.000	SFCA	37.29	207.90	245.19
Templates for dowels or anchor bolts	8.000	Ea.	9.28	38.72	48
Reinforcing in place footings, #8 to #14	.025	Ton	31.25	18.88	50.13
Concrete ready mix, regular weight, 3000 psi	1.400	C.Y.	198.80		198.80
Place and vibrate concrete for pile caps, under 5 CY, direct chute	1.400	C.Y.		49.92	49.92
Float finish	23.000	S.F.		9.66	9.66
TOTAL			276.62	384.09	660.71

A1010 250			Pile Caps					
	NO. PILES	SIZE FT-IN X FT-IN X IN	PILE CAPACITY (TON)	COLUMN SIZE (IN)	COLUMN LOAD (K)	COST EACH		
						MAT.	INST.	TOTAL
5100	2	6-6x3-6x20	15	8	45	277	385	662
5150		26	40	8	155	340	470	810
5200		34	80	11	314	460	600	1,060
5250		37	120	14	473	495	645	1,140
5300	3	5-6x5-1x23	15	8	75	330	440	770
5350		28	40	10	232	370	500	870
5400		32	80	14	471	425	565	990
5450		38	120	17	709	490	645	1,135
5500	4	5-6x5-6x18	15	10	103	375	440	815
5550		30	40	11	308	545	635	1,180
5600		36	80	16	626	640	745	1,385
5650		38	120	19	945	680	775	1,455
5700	6	8-6x5-6x18	15	12	156	625	635	1,260
5750		37	40	14	458	1,025	945	1,970
5800		40	80	19	936	1,150	1,050	2,200
5850		45	120	24	1413	1,300	1,175	2,475
5900	8	8-6x7-9x19	15	12	205	945	860	1,805
5950		36	40	16	610	1,375	1,150	2,525
6000		44	80	22	1243	1,700	1,375	3,075
6050		47	120	27	1881	1,825	1,450	3,275

A1010 Standard Foundations

A1010 250	Pile Caps

	NO. PILES	SIZE FT-IN X FT-IN X IN	PILE CAPACITY (TON)	COLUMN SIZE (IN)	COLUMN LOAD (K)	COST EACH		
						MAT.	INST.	TOTAL
6100	10	11-6x7-9x21	15	14	250	1,325	1,050	2,375
6150		39	40	17	756	2,050	1,525	3,575
6200		47	80	25	1547	2,475	1,800	4,275
6250		49	120	31	2345	2,625	1,875	4,500
6300	12	11-6x8-6x22	15	15	316	1,700	1,250	2,950
6350		49	40	19	900	2,700	1,925	4,625
6400		52	80	27	1856	2,975	2,100	5,075
6450		55	120	34	2812	3,225	2,225	5,450
6500	14	11-6x10-9x24	15	16	345	2,200	1,550	3,750
6550		41	40	21	1056	2,900	1,950	4,850
6600		55	80	29	2155	3,800	2,475	6,275
6700	16	11-6x11-6x26	15	18	400	2,650	1,800	4,450
6750		48	40	22	1200	3,575	2,325	5,900
6800		60	80	31	2460	4,450	2,825	7,275
6900	18	13-0x11-6x28	15	20	450	3,025	1,975	5,000
6950		49	40	23	1349	4,150	2,600	6,750
7000		56	80	33	2776	4,900	3,050	7,950
7100	20	14-6x11-6x30	15	20	510	3,700	2,375	6,075
7150		52	40	24	1491	5,000	3,050	8,050

A1010   Standard Foundations

General: Footing drains can be placed either inside or outside of foundation walls depending upon the source of water to be intercepted. If the source of subsurface water is principally from grade or a subsurface stream above the bottom of the footing, outside drains should be used. For high water tables, use inside drains or both inside and outside.

The effectiveness of underdrains depends on good waterproofing. This must be carefully installed and protected during construction.

Costs below include the labor and materials for the pipe and 6″ of crushed stone around pipe. Excavation and backfill are not included.

System Components	QUANTITY	UNIT	COST PER L.F.		
			MAT.	INST.	TOTAL
SYSTEM A1010 310 1000					
FOUNDATION UNDERDRAIN, OUTSIDE ONLY, PVC, 4″ DIAM.					
PVC pipe 4″ diam. S.D.R. 35	1.000	L.F.	1.84	4.53	6.37
Crushed stone 3/4″ to 1/2″	.070	C.Y.	2.14	.88	3.02
TOTAL			3.98	5.41	9.39

A1010 310	Foundation Underdrain	COST PER L.F.		
		MAT.	INST.	TOTAL
1000	Foundation underdrain, outside only, PVC, 4″ diameter	3.98	5.40	9.38
1100	6″ diameter	7.20	6	13.20
1400	Perforated HDPE, 6″ diameter	5.05	2.25	7.30
1450	8″ diameter	7.90	2.82	10.72
1500	12″ diameter	12.45	7	19.45
1600	Corrugated metal, 16 ga. asphalt coated, 6″ diameter	9.80	5.60	15.40
1650	8″ diameter	13.05	5.95	19
1700	10″ diameter	16.05	7.90	23.95
3000	Outside and inside, PVC, 4″ diameter	7.95	10.80	18.75
3100	6″ diameter	14.40	12	26.40
3400	Perforated HDPE, 6″ diameter	10.10	4.49	14.59
3450	8″ diameter	15.75	5.65	21.40
3500	12″ diameter	25	14	39
3600	Corrugated metal, 16 ga., asphalt coated, 6″ diameter	19.60	11.20	30.80
3650	8″ diameter	26	11.95	37.95
3700	10″ diameter	32	15.80	47.80

A1010 Standard Foundations

General: Apply foundation wall dampproofing over clean concrete giving particular attention to the joint between the wall and the footing. Use care in backfilling to prevent damage to the dampproofing.

Costs for four types of dampproofing are listed below.

System Components			COST PER L.F.		
	QUANTITY	UNIT	MAT.	INST.	TOTAL
SYSTEM A1010 320 1000					
FOUNDATION DAMPPROOFING, BITUMINOUS, 1 COAT, 4' HIGH					
Bituminous asphalt dampproofing brushed on below grade, 1 coat	4.000	S.F.	1.04	3.68	4.72
Labor for protection of dampproofing during backfilling	4.000	S.F.		1.52	1.52
TOTAL			1.04	5.20	6.24

A1010 320	Foundation Dampproofing	COST PER L.F.		
		MAT.	INST.	TOTAL
1000	Foundation dampproofing, bituminous, 1 coat, 4' high	1.04	5.20	6.24
1400	8' high	2.08	10.40	12.48
1800	12' high	3.12	16.10	19.22
2000	2 coats, 4' high	2.08	6.40	8.48
2400	8' high	4.16	12.80	16.96
2800	12' high	6.25	19.70	25.95
3000	Asphalt with fibers, 1/16" thick, 4' high	1.80	6.40	8.20
3400	8' high	3.60	12.80	16.40
3800	12' high	5.40	19.70	25.10
4000	1/8" thick, 4' high	3.16	7.65	10.81
4400	8' high	6.30	15.25	21.55
4800	12' high	9.50	23.50	33
5000	Asphalt coated board and mastic, 1/4" thick, 4' high	5.35	6.95	12.30
5400	8' high	10.70	13.90	24.60
5800	12' high	16.10	21.50	37.60
6000	1/2" thick, 4' high	8.05	9.70	17.75
6400	8' high	16.10	19.35	35.45
6800	12' high	24	29.50	53.50
7000	Cementitious coating, on walls, 1/8" thick coating, 4' high	3.12	9.10	12.22
7400	8' high	6.25	18.15	24.40
7800	12' high	9.35	27.50	36.85
8000	Cementitious/metallic slurry, 4 coat, 1/2"thick, 2' high	1.65	9.80	11.45
8400	4' high	3.30	19.60	22.90
8800	6' high	4.95	29.50	34.45

The Cast-in-Place Concrete Pile System includes: a defined number of 4,000 p.s.i. concrete piles with thin-wall, straight-sided, steel shells that have a standard steel plate driving point. An allowance for cutoffs is included.

The Expanded System Listing shows costs per cluster of piles. Clusters range from one pile to twenty piles. Loads vary from 50 Kips to 1,600 Kips. Both end-bearing and friction-type piles are shown.

Please see the reference section for cost of mobilization of the pile driving equipment and other design and cost information.

System Components			COST EACH		
	QUANTITY	UNIT	MAT.	INST.	TOTAL
SYSTEM A1020 110 2220					
CIP SHELL CONCRETE PILE, 25′ LONG, 50K LOAD, END BEARING, 1 PILE					
7 Ga. shell, 12″ diam.	27.000	V.L.F.	918	342.63	1,260.63
Steel pipe pile standard point, 12″ or 14″ diameter pile	1.000	Ea.	230	178	408
Pile cutoff, conc. pile with thin steel shell	1.000	Ea.		17.75	17.75
TOTAL			1,148	538.38	1,686.38

A1020 110	C.I.P. Concrete Piles	COST EACH		
		MAT.	INST.	TOTAL
2220	CIP shell concrete pile, 25′ long, 50K load, end bearing, 1 pile	1,150	540	1,690
2240	100K load, end bearing, 2 pile cluster	2,075	900	2,975
2260	200K load, end bearing, 4 pile cluster	4,125	1,800	5,925
2280	400K load, end bearing, 7 pile cluster	7,225	3,150	10,375
2300	10 pile cluster	10,300	4,500	14,800
2320	800K load, end bearing, 13 pile cluster	22,000	7,400	29,400
2340	17 pile cluster	28,800	9,675	38,475
2360	1200K load, end bearing, 14 pile cluster	23,700	7,975	31,675
2380	19 pile cluster	32,200	10,800	43,000
2400	1600K load, end bearing, 19 pile cluster	32,200	10,800	43,000
2420	50′ long, 50K load, end bearing, 1 pile	1,925	780	2,705
2440	Friction type, 2 pile cluster	3,675	1,575	5,250
2460	3 pile cluster	5,525	2,325	7,850
2480	100K load, end bearing, 2 pile cluster	3,825	1,575	5,400
2500	Friction type, 4 pile cluster	7,350	3,100	10,450
2520	6 pile cluster	11,000	4,675	15,675
2540	200K load, end bearing, 4 pile cluster	7,675	3,100	10,775
2560	Friction type, 8 pile cluster	14,700	6,250	20,950
2580	10 pile cluster	18,400	7,800	26,200
2600	400K load, end bearing, 7 pile cluster	13,400	5,450	18,850
2620	Friction type, 16 pile cluster	29,400	12,500	41,900
2640	19 pile cluster	34,900	14,800	49,700
2660	800K load, end bearing, 14 pile cluster	43,600	13,700	57,300
2680	20 pile cluster	62,000	19,700	81,700
2700	1200K load, end bearing, 15 pile cluster	46,700	14,800	61,500
2720	1600K load, end bearing, 20 pile cluster	62,000	19,700	81,700
3740	75′ long, 50K load, end bearing, 1 pile	2,950	1,275	4,225
3760	Friction type, 2 pile cluster	5,675	2,550	8,225

A10 Foundations

A1020 Special Foundations

A1020 110	C.I.P. Concrete Piles	COST EACH		
		MAT.	INST.	TOTAL
3780	3 pile cluster	8,525	3,850	12,375
3800	100K load, end bearing, 2 pile cluster	5,925	2,550	8,475
3820	Friction type, 3 pile cluster	8,525	3,850	12,375
3840	5 pile cluster	14,200	6,400	20,600
3860	200K load, end bearing, 4 pile cluster	11,800	5,125	16,925
3880	6 pile cluster	17,800	7,675	25,475
3900	Friction type, 6 pile cluster	17,100	7,675	24,775
3910	7 pile cluster	19,900	8,975	28,875
3920	400K load, end bearing, 7 pile cluster	20,700	8,975	29,675
3930	11 pile cluster	32,600	14,100	46,700
3940	Friction type, 12 pile cluster	34,100	15,400	49,500
3950	14 pile cluster	39,800	17,900	57,700
3960	800K load, end bearing, 15 pile cluster	70,500	23,700	94,200
3970	20 pile cluster	94,000	31,500	125,500
3980	1200K load, end bearing, 17 pile cluster	80,000	26,800	106,800

A10 Foundations

A1020   Special Foundations

The Precast Concrete Pile System includes: pre-stressed concrete piles; standard steel driving point; and an allowance for cutoffs.

The Expanded System Listing shows costs per cluster of piles. Clusters range from one pile to twenty piles. Loads vary from 50 Kips to 1,600 Kips. Both end-bearing and friction type piles are listed.

Please see the reference section for cost of mobilization of the pile driving equipment and other design and cost information.

System Components	QUANTITY	UNIT	COST EACH		
			MAT.	INST.	TOTAL
SYSTEM A1020 120 2220					
PRECAST CONCRETE PILE, 50′ LONG, 50K LOAD, END BEARING, 1 PILE					
Precast, prestressed conc. piles, 10″ square, no mobil.	53.000	V.L.F.	824.15	578.23	1,402.38
Steel pipe pile standard point, 8″ to 10″ diameter	1.000	Ea.	51	78	129
Piling special costs cutoffs concrete piles plain	1.000	Ea.		122	122
TOTAL			875.15	778.23	1,653.38

A1020 120	Precast Concrete Piles	COST EACH		
		MAT.	INST.	TOTAL
2220	Precast conc pile, 50′ long, 50K load, end bearing, 1 pile	875	775	1,650
2240	Friction type, 2 pile cluster	3,250	1,625	4,875
2260	4 pile cluster	6,500	3,225	9,725
2280	100K load, end bearing, 2 pile cluster	1,750	1,550	3,300
2300	Friction type, 2 pile cluster	3,250	1,625	4,875
2320	4 pile cluster	6,500	3,225	9,725
2340	7 pile cluster	11,400	5,625	17,025
2360	200K load, end bearing, 3 pile cluster	2,625	2,325	4,950
2380	4 pile cluster	3,500	3,125	6,625
2400	Friction type, 8 pile cluster	13,000	6,425	19,425
2420	9 pile cluster	14,600	7,250	21,850
2440	14 pile cluster	22,800	11,300	34,100
2460	400K load, end bearing, 6 pile cluster	5,250	4,675	9,925
2480	8 pile cluster	7,000	6,225	13,225
2500	Friction type, 14 pile cluster	22,800	11,300	34,100
2520	16 pile cluster	26,000	12,900	38,900
2540	18 pile cluster	29,300	14,500	43,800
2560	800K load, end bearing, 12 pile cluster	11,400	10,100	21,500
2580	16 pile cluster	14,000	12,500	26,500
2600	1200K load, end bearing, 19 pile cluster	53,000	18,100	71,100
2620	20 pile cluster	55,500	19,000	74,500
2640	1600K load, end bearing, 19 pile cluster	53,000	18,100	71,100
4660	100′ long, 50K load, end bearing, 1 pile	1,675	1,350	3,025
4680	Friction type, 1 pile	3,100	1,400	4,500

A1020 Special Foundations

A1020 120	Precast Concrete Piles	COST EACH		
		MAT.	INST.	TOTAL
4700	2 pile cluster	6,225	2,775	9,000
4720	100K load, end bearing, 2 pile cluster	3,375	2,700	6,075
4740	Friction type, 2 pile cluster	6,225	2,775	9,000
4760	3 pile cluster	9,325	4,150	13,475
4780	4 pile cluster	12,400	5,550	17,950
4800	200K load, end bearing, 3 pile cluster	5,050	4,050	9,100
4820	4 pile cluster	6,725	5,375	12,100
4840	Friction type, 3 pile cluster	9,325	4,150	13,475
4860	5 pile cluster	15,500	6,925	22,425
4880	400K load, end bearing, 6 pile cluster	10,100	8,075	18,175
4900	8 pile cluster	13,500	10,800	24,300
4910	Friction type, 8 pile cluster	24,900	11,100	36,000
4920	10 pile cluster	31,100	13,900	45,000
4930	800K load, end bearing, 13 pile cluster	21,900	17,500	39,400
4940	16 pile cluster	26,900	21,500	48,400
4950	1200K load, end bearing, 19 pile cluster	102,000	31,500	133,500
4960	20 pile cluster	107,000	33,100	140,100
4970	1600K load, end bearing, 19 pile cluster	102,000	31,500	133,500

589

A10 Foundations

A1020 Special Foundations

The Steel Pipe Pile System includes: steel pipe sections filled with 4,000 p.s.i. concrete; a standard steel driving point; splices when required and an allowance for cutoffs.

The Expanded System Listing shows costs per cluster of piles. Clusters range from one pile to twenty piles. Loads vary from 50 Kips to 1,600 Kips. Both end-bearing and friction-type piles are shown.

Please see the reference section for cost of mobilization of the pile driving equipment and other design and cost information.

System Components	QUANTITY	UNIT	COST EACH		
			MAT.	INST.	TOTAL
SYSTEM A1020 130 2220					
CONC. FILL STEEL PIPE PILE, 50′ LONG, 50K LOAD, END BEARING, 1 PILE					
Piles, steel, pipe, conc. filled, 12″ diameter	53.000	V.L.F.	2,067	975.20	3,042.20
Steel pipe pile, standard point, for 12″ or 14″ diameter pipe	1.000	Ea.	230	178	408
Pile cut off, concrete pile, thin steel shell	1.000	Ea.		17.75	17.75
TOTAL			2,297	1,170.95	3,467.95

A1020 130	Steel Pipe Piles	COST EACH		
		MAT.	INST.	TOTAL
2220	Conc. fill steel pipe pile, 50′ long, 50K load, end bearing, 1 pile	2,300	1,175	3,475
2240	Friction type, 2 pile cluster	4,600	2,350	6,950
2250	100K load, end bearing, 2 pile cluster	4,600	2,350	6,950
2260	3 pile cluster	6,900	3,500	10,400
2300	Friction type, 4 pile cluster	9,200	4,675	13,875
2320	5 pile cluster	11,500	5,850	17,350
2340	10 pile cluster	23,000	11,700	34,700
2360	200K load, end bearing, 3 pile cluster	6,900	3,500	10,400
2380	4 pile cluster	9,200	4,675	13,875
2400	Friction type, 4 pile cluster	9,200	4,675	13,875
2420	8 pile cluster	18,400	9,375	27,775
2440	9 pile cluster	20,700	10,500	31,200
2460	400K load, end bearing, 6 pile cluster	13,800	7,025	20,825
2480	7 pile cluster	16,100	8,175	24,275
2500	Friction type, 9 pile cluster	20,700	10,500	31,200
2520	16 pile cluster	36,800	18,700	55,500
2540	19 pile cluster	43,600	22,300	65,900
2560	800K load, end bearing, 11 pile cluster	25,300	12,900	38,200
2580	14 pile cluster	32,200	16,400	48,600
2600	15 pile cluster	34,500	17,600	52,100
2620	Friction type, 17 pile cluster	39,000	19,900	58,900
2640	1200K load, end bearing, 16 pile cluster	36,800	18,700	55,500
2660	20 pile cluster	45,900	23,400	69,300
2680	1600K load, end bearing, 17 pile cluster	39,000	19,900	58,900
3700	100′ long, 50K load, end bearing, 1 pile	4,475	2,275	6,750
3720	Friction type, 1 pile	4,475	2,275	6,750

A1020 Special Foundations

A1020 130	Steel Pipe Piles	COST EACH		
		MAT.	INST.	TOTAL
3740	2 pile cluster	8,975	4,600	13,575
3760	100K load, end bearing, 2 pile cluster	8,975	4,600	13,575
3780	Friction type, 2 pile cluster	8,975	4,600	13,575
3800	3 pile cluster	13,500	6,875	20,375
3820	200K load, end bearing, 3 pile cluster	13,500	6,875	20,375
3840	4 pile cluster	17,900	9,200	27,100
3860	Friction type, 3 pile cluster	13,500	6,875	20,375
3880	4 pile cluster	17,900	9,200	27,100
3900	400K load, end bearing, 6 pile cluster	26,900	13,800	40,700
3910	7 pile cluster	31,400	16,100	47,500
3920	Friction type, 5 pile cluster	22,400	11,500	33,900
3930	8 pile cluster	35,900	18,400	54,300
3940	800K load, end bearing, 11 pile cluster	49,300	25,300	74,600
3950	14 pile cluster	63,000	32,200	95,200
3960	15 pile cluster	67,500	34,400	101,900
3970	1200K load, end bearing, 16 pile cluster	72,000	36,800	108,800
3980	20 pile cluster	89,500	45,900	135,400
3990	1600K load, end bearing, 17 pile cluster	76,500	39,100	115,600

For customer support on your Site Work & Landscape Costs with RSMeans data, call 800.448.8182.

A10 Foundations

A1020 Special Foundations

H

A Steel "H" Pile System includes: steel H sections; heavy duty driving point; splices where applicable and allowance for cutoffs.

The Expanded System Listing shows costs per cluster of piles. Clusters range from one pile to eighteen piles. Loads vary from 50 Kips to 2,000 Kips. All loads for Steel H Pile systems are given in terms of end bearing capacity.

Steel sections range from 10" x 10" to 14" x 14" in the Expanded System Listing. The 14" x 14" steel section is used for all H piles used in applications requiring a working load over 800 Kips.

Please see the reference section for cost of mobilization of the pile driving equipment and other design and cost information.

System Components	QUANTITY	UNIT	COST EACH		
			MAT.	INST.	TOTAL
SYSTEM A1020 140 2220					
STEEL H PILES, 50' LONG, 100K LOAD, END BEARING, 1 PILE					
Steel H piles 10" x 10", 42 #/L.F.	53.000	V.L.F.	1,192.50	661.44	1,853.94
Heavy duty point, 10"	1.000	Ea.	205	180	385
Pile cut off, steel pipe or H piles	1.000	Ea.		35.50	35.50
TOTAL			1,397.50	876.94	2,274.44

A1020 140	Steel H Piles	COST EACH		
		MAT.	INST.	TOTAL
2220	Steel H piles, 50' long, 100K load, end bearing, 1 pile	1,400	875	2,275
2260	2 pile cluster	2,800	1,750	4,550
2280	200K load, end bearing, 2 pile cluster	2,800	1,750	4,550
2300	3 pile cluster	4,200	2,625	6,825
2320	400K load, end bearing, 3 pile cluster	4,200	2,625	6,825
2340	4 pile cluster	5,600	3,525	9,125
2360	6 pile cluster	8,375	5,250	13,625
2380	800K load, end bearing, 5 pile cluster	7,000	4,375	11,375
2400	7 pile cluster	9,775	6,150	15,925
2420	12 pile cluster	16,800	10,500	27,300
2440	1200K load, end bearing, 8 pile cluster	11,200	7,025	18,225
2460	11 pile cluster	15,400	9,650	25,050
2480	17 pile cluster	23,800	14,900	38,700
2500	1600K load, end bearing, 10 pile cluster	18,200	9,175	27,375
2520	14 pile cluster	25,400	12,900	38,300
2540	2000K load, end bearing, 12 pile cluster	21,800	11,000	32,800
2560	18 pile cluster	32,700	16,500	49,200
3580	100' long, 50K load, end bearing, 1 pile	4,675	2,100	6,775
3600	100K load, end bearing, 1 pile	4,675	2,100	6,775
3620	2 pile cluster	9,350	4,200	13,550
3640	200K load, end bearing, 2 pile cluster	9,350	4,200	13,550
3660	3 pile cluster	14,000	6,325	20,325
3680	400K load, end bearing, 3 pile cluster	14,000	6,325	20,325
3700	4 pile cluster	18,700	8,400	27,100
3720	6 pile cluster	28,000	12,600	40,600
3740	800K load, end bearing, 5 pile cluster	23,400	10,500	33,900

A1020 Special Foundations

A1020 140	Steel H Piles	COST EACH		
		MAT.	INST.	TOTAL
3760	7 pile cluster	32,700	14,700	47,400
3780	12 pile cluster	56,000	25,300	81,300
3800	1200K load, end bearing, 8 pile cluster	37,400	16,900	54,300
3820	11 pile cluster	51,500	23,100	74,600
3840	17 pile cluster	79,500	35,700	115,200
3860	1600K load, end bearing, 10 pile cluster	46,700	21,000	67,700
3880	14 pile cluster	65,500	29,400	94,900
3900	2000K load, end bearing, 12 pile cluster	56,000	25,300	81,300
3920	18 pile cluster	84,000	37,800	121,800

A10 Foundations

A1020 Special Foundations

The Step Tapered Steel Pile System includes: step tapered piles filled with 4,000 p.s.i. concrete. The cost for splices and pile cutoffs is included.

The Expanded System Listing shows costs per cluster of piles. Clusters range from one pile to twenty-four piles. Both end bearing piles and friction piles are listed. Loads vary from 50 Kips to 1,600 Kips.

Please see the reference section for cost of mobilization of the pile driving equipment and other design and cost information.

System Components	QUANTITY	UNIT	COST EACH		
			MAT.	INST.	TOTAL
SYSTEM A1020 150 1000					
STEEL PILE, STEP TAPERED, 50' LONG, 50K LOAD, END BEARING, 1 PILE					
Steel shell step tapered conc. filled piles, 8" tip 60 ton capacity to 60'	53.000	V.L.F.	2,146.50	545.37	2,691.87
Pile cutoff, steel pipe or H piles	1.000	Ea.		35.50	35.50
TOTAL			2,146.50	580.87	2,727.37

A1020 150	Step-Tapered Steel Piles	COST EACH		
		MAT.	INST.	TOTAL
1000	Steel pile, step tapered, 50' long, 50K load, end bearing, 1 pile	2,150	580	2,730
1200	Friction type, 3 pile cluster	6,450	1,750	8,200
1400	100K load, end bearing, 2 pile cluster	4,300	1,150	5,450
1600	Friction type, 4 pile cluster	8,575	2,325	10,900
1800	200K load, end bearing, 4 pile cluster	8,575	2,325	10,900
2000	Friction type, 6 pile cluster	12,900	3,475	16,375
2200	400K load, end bearing, 7 pile cluster	15,000	4,050	19,050
2400	Friction type, 10 pile cluster	21,500	5,800	27,300
2600	800K load, end bearing, 14 pile cluster	30,100	8,150	38,250
2800	Friction type, 18 pile cluster	38,600	10,500	49,100
3000	1200K load, end bearing, 16 pile cluster	18,200	9,950	28,150
3200	Friction type, 21 pile cluster	23,900	13,000	36,900
3400	1600K load, end bearing, 18 pile cluster	20,500	11,200	31,700
3600	Friction type, 24 pile cluster	27,300	14,900	42,200
5000	100' long, 50K load, end bearing, 1 pile	6,450	1,150	7,600
5200	Friction type, 2 pile cluster	12,900	2,325	15,225
5400	100K load, end bearing, 2 pile cluster	12,900	2,325	15,225
5600	Friction type, 3 pile cluster	19,300	3,475	22,775
5800	200K load, end bearing, 4 pile cluster	25,800	4,625	30,425
6000	Friction type, 5 pile cluster	32,200	5,800	38,000
6200	400K load, end bearing, 7 pile cluster	45,100	8,100	53,200
6400	Friction type, 8 pile cluster	51,500	9,275	60,775
6600	800K load, end bearing, 15 pile cluster	96,500	17,400	113,900
6800	Friction type, 16 pile cluster	103,000	18,500	121,500
7000	1200K load, end bearing, 17 pile cluster	61,000	30,300	91,300
7200	Friction type, 19 pile cluster	46,000	23,000	69,000
7400	1600K load, end bearing, 20 pile cluster	48,400	24,200	72,600
7600	Friction type, 22 pile cluster	53,500	26,600	80,100

594

The Treated Wood Pile System includes: creosoted wood piles; a standard steel driving point; and an allowance for cutoffs.

The Expanded System Listing shows costs per cluster of piles. Clusters range from two piles to twenty piles. Loads vary from 50 Kips to 400 Kips. Both end-bearing and friction type piles are listed.

Please see the reference section for cost of mobilization of the pile driving equipment and other design and cost information.

System Components	QUANTITY	UNIT	COST EACH		
			MAT.	INST.	TOTAL
SYSTEM A1020 160 2220					
WOOD PILES, 25' LONG, 50K LOAD, END BEARING, 3 PILE CLUSTER					
Wood piles, treated, 12" butt, 8" tip, up to 30' long	81.000	V.L.F.	1,620	987.39	2,607.39
Point for driving wood piles	3.000	Ea.	150	100.50	250.50
Pile cutoff, wood piles	3.000	Ea.		53.25	53.25
TOTAL			1,770	1,141.14	2,911.14

A1020 160	Treated Wood Piles	COST EACH		
		MAT.	INST.	TOTAL
2220	Wood piles, 25' long, 50K load, end bearing, 3 pile cluster	1,775	1,150	2,925
2240	Friction type, 3 pile cluster	1,625	1,050	2,675
2260	5 pile cluster	2,700	1,725	4,425
2280	100K load, end bearing, 4 pile cluster	2,350	1,525	3,875
2300	5 pile cluster	2,950	1,900	4,850
2320	6 pile cluster	3,550	2,275	5,825
2340	Friction type, 5 pile cluster	2,700	1,725	4,425
2360	6 pile cluster	3,250	2,075	5,325
2380	10 pile cluster	5,400	3,475	8,875
2400	200K load, end bearing, 8 pile cluster	4,725	3,050	7,775
2420	10 pile cluster	5,900	3,800	9,700
2440	12 pile cluster	7,075	4,575	11,650
2460	Friction type, 10 pile cluster	5,400	3,475	8,875
2480	400K load, end bearing, 16 pile cluster	9,450	6,075	15,525
2500	20 pile cluster	11,800	7,600	19,400
4520	50' long, 50K load, end bearing, 3 pile cluster	4,275	1,675	5,950
4540	4 pile cluster	5,700	2,250	7,950
4560	Friction type, 2 pile cluster	2,700	1,025	3,725
4580	3 pile cluster	4,050	1,525	5,575
4600	100K load, end bearing, 5 pile cluster	7,125	2,825	9,950
4620	8 pile cluster	11,400	4,500	15,900
4640	Friction type, 3 pile cluster	4,050	1,525	5,575
4660	5 pile cluster	6,750	2,575	9,325
4680	200K load, end bearing, 9 pile cluster	12,800	5,075	17,875
4700	10 pile cluster	14,300	5,625	19,925
4720	15 pile cluster	21,400	8,425	29,825
4740	Friction type, 5 pile cluster	6,750	2,575	9,325
4760	6 pile cluster	8,100	3,075	11,175

A10 Foundations

A1020 Special Foundations

A1020 160	Treated Wood Piles	COST EACH		
		MAT.	INST.	TOTAL
4780	10 pile cluster	13,500	5,125	18,625
4800	400K load, end bearing, 18 pile cluster	25,700	10,100	35,800
4820	20 pile cluster	28,500	11,300	39,800
4840	Friction type, 9 pile cluster	12,200	4,625	16,825
4860	10 pile cluster	13,500	5,125	18,625
9000	Add for boot for driving tip, each pile	50	25	75

A1020 Special Foundations

The Grade Beam System includes: excavation with a truck mounted backhoe; hand trim; backfill; forms (four uses); reinforcing steel; and 3,000 p.s.i. concrete placed from chute.

Superimposed loads vary in the listing from 1 Kip per linear foot (KLF) and above. In the Expanded System Listing, the span of the beams varies from 15' to 40'. Depth varies from 28" to 52". Width varies from 12" and wider.

Please see the reference section for further design and cost information.

System Components	QUANTITY	UNIT	COST PER L.F.		
			MAT.	INST.	TOTAL
SYSTEM A1020 210 2220					
GRADE BEAM, 15' SPAN, 28" DEEP, 12" WIDE, 8 KLF LOAD					
Excavation, trench, hydraulic backhoe, 3/8 CY bucket	.260	C.Y.		2.46	2.46
Trim sides and bottom of trench, regular soil	2.000	S.F.		2.30	2.30
Backfill, by hand, compaction in 6" layers, using vibrating plate	.170	C.Y.		1.54	1.54
Forms in place, grade beam, 4 uses	4.700	SFCA	5.31	28.91	34.22
Reinforcing in place, beams & girders, #8 to #14	.019	Ton	23.75	19	42.75
Concrete ready mix, regular weight, 3000 psi	.090	C.Y.	12.78		12.78
Place and vibrate conc. for grade beam, direct chute	.090	C.Y.		1.93	1.93
TOTAL			41.84	56.14	97.98

A1020 210	Grade Beams	COST PER L.F.		
		MAT.	INST.	TOTAL
2220	Grade beam, 15' span, 28" deep, 12" wide, 8 KLF load	42	56.50	98.50
2240	14" wide, 12 KLF load	43.50	57	100.50
2260	40" deep, 12" wide, 16 KLF load	43.50	70.50	114
2280	20 KLF load	49.50	75.50	125
2300	52" deep, 12" wide, 30 KLF load	57.50	94.50	152
2320	40 KLF load	70	104	174
3360	20' span, 28" deep, 12" wide, 2 KLF load	27	44.50	71.50
3380	16" wide, 4 KLF load	35.50	49	84.50
3400	40" deep, 12" wide, 8 KLF load	43.50	70.50	114
3440	14" wide, 16 KLF load	66	87	153
3460	52" deep, 12" wide, 20 KLF load	71.50	105	176.50
3480	14" wide, 30 KLF load	98	124	222
4540	30' span, 28" deep, 12" wide, 1 KLF load	28	45.50	73.50
4560	14" wide, 2 KLF load	44	61.50	105.50
4580	40" deep, 12" wide, 4 KLF load	53	74	127
4600	18" wide, 8 KLF load	74.50	89.50	164
4620	52" deep, 14" wide, 12 KLF load	94.50	121	215.50
4640	20" wide, 16 KLF load	116	132	248
4660	24" wide, 20 KLF load	143	150	293
5720	40' span, 40" deep, 12" wide, 1 KLF load	38.50	66.50	105
5740	2 KLF load	47	73.50	120.50
5760	52" deep, 12" wide, 4 KLF load	70	104	174
5780	20" wide, 8 KLF load	112	129	241
5800	28" wide, 12 KLF load	154	155	309

597

A10 Foundations

A1020 Special Foundations

Caisson Systems are listed for three applications: stable ground, wet ground and soft rock. Concrete used is 3,000 p.s.i. placed from chute. Included are a bell at the bottom of the caisson shaft (if applicable) along with required excavation and disposal of excess excavated material up to two miles from job site.

The Expanded System lists cost per caisson. End-bearing loads vary from 200 Kips to 3,200 Kips. The dimensions of the caissons range from 2' x 50' to 7' x 200'.

Please see the reference section for further design and cost information.

System Components	QUANTITY	UNIT	COST EACH		
			MAT.	INST.	TOTAL
SYSTEM A1020 310 2200					
CAISSON, STABLE GROUND, 3000 PSI CONC., 10 KSF BRNG, 200K LOAD, 2'X50'					
Caissons, drilled, to 50', 24" shaft diameter, .116 C.Y./L.F.	50.000	V.L.F.	1,000	1,108	2,108
Reinforcing in place, columns, #3 to #7	.060	Ton	75	108	183
4' bell diameter, 24" shaft, 0.444 C.Y.	1.000	Ea.	63	210.50	273.50
Load & haul excess excavation, 2 miles	6.240	C.Y.		42.68	42.68
TOTAL			1,138	1,469.18	2,607.18

A1020 310	Caissons	COST EACH		
		MAT.	INST.	TOTAL
2200	Caisson, stable ground, 3000 PSI conc, 10 KSF brng, 200K load, 2'-0"x50'-0	1,150	1,475	2,625
2400	400K load, 2'-6"x50'-0"	1,875	2,350	4,225
2600	800K load, 3'-0"x100'-0"	5,175	5,550	10,725
2800	1200K load, 4'-0"x100'-0"	8,950	7,025	15,975
3000	1600K load, 5'-0"x150'-0"	20,200	11,100	31,300
3200	2400K load, 6'-0"x150'-0"	29,300	14,500	43,800
3400	3200K load, 7'-0"x200'-0"	52,500	21,300	73,800
5000	Wet ground, 3000 PSI conc., 10 KSF brng, 200K load, 2'-0"x50'-0"	990	2,175	3,165
5200	400K load, 2'-6"x50'-0"	1,650	3,700	5,350
5400	800K load, 3'-0"x100'-0"	4,475	9,850	14,325
5600	1200K load, 4'-0"x100'-0"	7,750	16,400	24,150
5800	1600K load, 5'-0"x150'-0"	17,300	35,300	52,600
6000	2400K load, 6'-0"x150'-0"	25,200	43,600	68,800
6200	3200K load, 7'-0"x200'-0"	45,300	70,000	115,300
7800	Soft rock, 3000 PSI conc., 10 KSF brng, 200K load, 2'-0"x50'-0"	990	18,400	19,390
8000	400K load, 2'-6"x50'-0"	1,650	21,000	22,650
8200	800K load, 3'-0"x100'-0"	4,475	55,000	59,475
8400	1200K load, 4'-0"x100'-0"	7,750	80,000	87,750
8600	1600K load, 5'-0"x150'-0"	17,300	164,500	181,800
8800	2400K load, 6'-0"x150'-0"	25,200	194,500	219,700
9000	3200K load, 7'-0"x200'-0"	45,300	311,500	356,800

A10 Foundations

A1020 Special Foundations

Pressure Injected Piles are usually uncased up to 25' and cased over 25' depending on soil conditions.

These costs include excavation and hauling of excess materials; steel casing over 25'; reinforcement; 3,000 p.s.i. concrete; plus mobilization and demobilization of equipment for a distance of up to fifty miles to and from the job site.

The Expanded System lists cost per cluster of piles. Clusters range from one pile to eight piles. End-bearing loads range from 50 Kips to 1,600 Kips.

Please see the reference section for further design and cost information.

System Components	QUANTITY	UNIT	COST EACH		
			MAT.	INST.	TOTAL
SYSTEM A1020 710 4200					
PRESSURE INJECTED FOOTING, END BEARING, 50' LONG, 50K LOAD, 1 PILE					
Pressure injected footings, cased, 30-60 ton cap., 12" diameter	50.000	V.L.F.	1,100	1,635	2,735
Pile cutoff, concrete pile with thin steel shell	1.000	Ea.		17.75	17.75
TOTAL			1,100	1,652.75	2,752.75

A1020 710	Pressure Injected Footings	COST EACH		
		MAT.	INST.	TOTAL
2200	Pressure injected footing, end bearing, 25' long, 50K load, 1 pile	500	1,150	1,650
2400	100K load, 1 pile	500	1,150	1,650
2600	2 pile cluster	1,000	2,300	3,300
2800	200K load, 2 pile cluster	1,000	2,300	3,300
3200	400K load, 4 pile cluster	2,000	4,625	6,625
3400	7 pile cluster	3,500	8,100	11,600
3800	1200K load, 6 pile cluster	4,275	8,825	13,100
4000	1600K load, 7 pile cluster	5,000	10,300	15,300
4200	50' long, 50K load, 1 pile	1,100	1,650	2,750
4400	100K load, 1 pile	1,900	1,650	3,550
4600	2 pile cluster	3,800	3,300	7,100
4800	200K load, 2 pile cluster	3,800	3,300	7,100
5000	4 pile cluster	7,600	6,625	14,225
5200	400K load, 4 pile cluster	7,600	6,625	14,225
5400	8 pile cluster	15,200	13,200	28,400
5600	800K load, 7 pile cluster	13,300	11,600	24,900
5800	1200K load, 6 pile cluster	12,200	9,900	22,100
6000	1600K load, 7 pile cluster	14,200	11,600	25,800

A10 Foundations

A1030 Slab on Grade

There are four types of Slab on Grade Systems listed: Non-industrial, Light industrial, Industrial and Heavy industrial. Each type is listed two ways: reinforced and non-reinforced. A Slab on Grade system includes three passes with a grader; 6" of compacted gravel fill; polyethylene vapor barrier; 3500 p.s.i. concrete placed by chute; bituminous fibre expansion joint; all necessary edge forms (4 uses); steel trowel finish; and sprayed-on membrane curing compound.

The Expanded System Listing shows costs on a per square foot basis. Thicknesses of the slabs range from 4" and above. Non-industrial applications are for foot traffic only with negligible abrasion. Light industrial applications are for pneumatic wheels and light abrasion. Industrial applications are for solid rubber wheels and moderate abrasion. Heavy industrial applications are for steel wheels and severe abrasion. All slabs are either shown unreinforced or reinforced with welded wire fabric.

System Components	QUANTITY	UNIT	COST PER S.F.		
			MAT.	INST.	TOTAL
SYSTEM A1030 120 2220					
SLAB ON GRADE, 4" THICK, NON INDUSTRIAL, NON REINFORCED					
Fine grade, 3 passes with grader and roller	.110	S.Y.		.65	.65
Gravel under floor slab, 4" deep, compacted	1.000	S.F.	.54	.35	.89
Polyethylene vapor barrier, standard, 6 mil	1.000	S.F.	.04	.17	.21
Concrete ready mix, regular weight, 3500 psi	.012	C.Y.	1.64		1.64
Place and vibrate concrete for slab on grade, 4" thick, direct chute	.012	C.Y.		.35	.35
Expansion joint, premolded bituminous fiber, 1/2" x 6"	.220	L.F.	.11	.37	.48
Edge forms in place for slab on grade to 6" high, 4 uses	.030	L.F.	.01	.12	.13
Cure with sprayed membrane curing compound	1.000	S.F.	.14	.11	.25
Finishing floor, monolithic steel trowel	1.000	S.F.		1.04	1.04
TOTAL			2.48	3.16	5.64

A1030 120	Plain & Reinforced	COST PER S.F.		
		MAT.	INST.	TOTAL
2220	Slab on grade, 4" thick, non industrial, non reinforced	2.48	3.16	5.64
2240	Reinforced	2.59	3.44	6.03
2260	Light industrial, non reinforced	3.07	3.85	6.92
2280	Reinforced	3.25	4.24	7.49
2300	Industrial, non reinforced	3.94	7.90	11.84
2320	Reinforced	4.12	8.30	12.42
3340	5" thick, non industrial, non reinforced	2.90	3.25	6.15
3360	Reinforced	3.08	3.64	6.72
3380	Light industrial, non reinforced	3.49	3.94	7.43
3400	Reinforced	3.67	4.33	8
3420	Heavy industrial, non reinforced	5	9.40	14.40
3440	Reinforced	5.10	9.85	14.95
4460	6" thick, non industrial, non reinforced	3.44	3.18	6.62
4480	Reinforced	3.69	3.71	7.40
4500	Light industrial, non reinforced	4.05	3.87	7.92
4520	Reinforced	4.36	4.29	8.65
4540	Heavy industrial, non reinforced	5.60	9.50	15.10
4560	Reinforced	5.85	10.05	15.90

A20 Basement Construction

A2010 Basement Excavation

1 : 1

1/2 : 1

Line of Excavation

Pricing Assumptions: Two-thirds of excavation is by 2-1/2 C.Y. wheel mounted front end loader and one-third by 1-1/2 C.Y. hydraulic excavator.

Two-mile round trip haul by 12 C.Y. tandem trucks is included for excavation wasted and storage of suitable fill from excavated soil. For excavation in clay, all is wasted and the cost of suitable backfill with two-mile haul is included.

Sand and gravel assumes 15% swell and compaction; common earth assumes 25% swell and 15% compaction; clay assumes 40% swell and 15% compaction (non-clay).

In general, the following items are accounted for in the costs in the table below.

1. Excavation for building or other structure to depth and extent indicated.
2. Backfill compacted in place.
3. Haul of excavated waste.
4. Replacement of unsuitable material with bank run gravel.

Note: Additional excavation and fill beyond this line of general excavation for the building (as required for isolated spread footings, strip footings, etc.) are included in the cost of the appropriate component systems.

System Components	QUANTITY	UNIT	COST PER S.F.		
			MAT.	INST.	TOTAL
SYSTEM A2010 110 2280					
EXCAVATE & FILL, 1000 S.F., 8' DEEP, SAND, ON SITE STORAGE					
Excavating bulk shovel, 1.5 C.Y. bucket, 150 cy/hr	.262	C.Y.		.54	.54
Excavation, front end loader, 2-1/2 C.Y.	.523	C.Y.		1.37	1.37
Haul earth, 12 C.Y. dump truck	.341	L.C.Y.		2.30	2.30
Backfill, dozer bulk push 300', including compaction	.562	C.Y.		2.53	2.53
TOTAL				6.74	6.74

A2010 110	Building Excavation & Backfill	COST PER S.F.		
		MAT.	INST.	TOTAL
2220	Excav & fill, 1000 S.F., 4' sand, gravel, or common earth, on site storage		1.15	1.15
2240	Off site storage		1.68	1.68
2260	Clay excavation, bank run gravel borrow for backfill	2.05	2.54	4.59
2280	8' deep, sand, gravel, or common earth, on site storage		6.75	6.75
2300	Off site storage		14.30	14.30
2320	Clay excavation, bank run gravel borrow for backfill	8.30	11.85	20.15
2340	16' deep, sand, gravel, or common earth, on site storage		18.40	18.40
2350	Off site storage		35.50	35.50
2360	Clay excavation, bank run gravel borrow for backfill	23	30.50	53.50
3380	4000 S.F., 4' deep, sand, gravel, or common earth, on site storage		.62	.62
3400	Off site storage		1.16	1.16
3420	Clay excavation, bank run gravel borrow for backfill	1.05	1.29	2.34
3440	8' deep, sand, gravel, or common earth, on site storage		4.66	4.66
3460	Off site storage		8	8
3480	Clay excavation, bank run gravel borrow for backfill	3.79	7.05	10.84
3500	16' deep, sand, gravel, or common earth, on site storage		11.15	11.15
3520	Off site storage		21.50	21.50
3540	Clay, excavation, bank run gravel borrow for backfill	10.15	16.85	27
4560	10,000 S.F., 4' deep, sand, gravel, or common earth, on site storage		.35	.35
4580	Off site storage		.69	.69
4600	Clay excavation, bank run gravel borrow for backfill	.64	.79	1.43
4620	8' deep, sand, gravel, or common earth, on site storage		4	4
4640	Off site storage		6	6
4660	Clay excavation, bank run gravel borrow for backfill	2.30	5.45	7.75
4680	16' deep, sand, gravel, or common earth, on site storage		9	9
4700	Off site storage		15.15	15.15
4720	Clay excavation, bank run gravel borrow for backfill	6.10	12.50	18.60

A20 Basement Construction

A2010 Basement Excavation

A2010 110	Building Excavation & Backfill	COST PER S.F.		
		MAT.	INST.	TOTAL
5740	30,000 S.F., 4' deep, sand, gravel, or common earth, on site storage		.19	.19
5760	Off site storage		.40	.40
5780	Clay excavation, bank run gravel borrow for backfill	.37	.47	.84
5800	8' deep, sand, gravel, or common earth, on site storage		3.57	3.57
5820	Off site storage		4.69	4.69
5840	Clay excavation, bank run gravel borrow for backfill	1.29	4.40	5.69
5860	16' deep, sand, gravel, or common earth, on site storage		7.65	7.65
5880	Off site storage		11	11
5900	Clay excavation, bank run gravel borrow for backfill	3.38	9.65	13.03
6910	100,000 S.F., 4' deep, sand, gravel, or common earth, on site storage		.11	.11
6920	Off site storage		.22	.22
6930	Clay excavation, bank run gravel borrow for backfill	.18	.23	.41
6940	8' deep, sand, gravel, or common earth, on site storage		3.31	3.31
6950	Off site storage		3.91	3.91
6960	Clay excavation, bank run gravel borrow for backfill	.70	3.77	4.47
6970	16' deep, sand, gravel, or common earth, on site storage		6.85	6.85
6980	Off site storage		8.65	8.65
6990	Clay excavation, bank run gravel borrow for backfill	1.82	8	9.82

A2020 Basement Walls

The Foundation Bearing Wall System includes: forms up to 16' high (four uses); 3,000 p.s.i. concrete placed and vibrated; and form removal with breaking form ties and patching walls. The wall systems list walls from 6" to 16" thick and are designed with minimum reinforcement.

Excavation and backfill are not included.

Please see the reference section for further design and cost information.

A2020 110	Walls, Cast in Place							
	WALL HEIGHT (FT.)	PLACING METHOD	CONCRETE (C.Y. per L.F.)	REINFORCING (LBS. per L.F.)	WALL THICKNESS (IN.)	COST PER L.F.		
						MAT.	INST.	TOTAL
5000	8'	direct chute	.148	6.6	6	39.50	116	155.50
5020			.199	9.6	8	48.50	122	170.50
5040			.250	12	10	57.50	122	179.50
5060			.296	14.39	12	65.50	124	189.50
5080			.347	16.19	14	68	125	193
5100			.394	19.19	16	82.50	129	211.50
5200	8'	pumped	.148	6.6	6	39.50	119	158.50
5220			.199	9.6	8	48.50	122	170.50
5240			.250	12	10	57.50	125	182.50
5260			.296	14.39	12	65.50	127	192.50
5280			.347	16.19	14	68	128	196
5300			.394	19.19	16	82.50	133	215.50
6020	10'	direct chute	.248	12	8	60.50	149	209.50
6040			.307	14.99	10	71	152	223
6060			.370	17.99	12	81.50	155	236.50
6080			.433	20.24	14	92	158	250
6100			.493	23.99	16	103	161	264
6220	10'	pumped	.248	12	8	60.50	152	212.50
6240			.307	14.99	10	71	155	226
6260			.370	17.99	12	81.50	160	241.50
6280			.433	20.24	14	92	162	254
6300			.493	23.99	16	103	166	269
7220	12'	pumped	.298	14.39	8	73	184	257
7240			.369	17.99	10	85	187	272
7260			.444	21.59	12	98	192	290
7280			.52	24.29	14	111	194	305
7300			.591	28.79	16	123	200	323
7420	12'	crane & bucket	.298	14.39	8	73	191	264
7440			.369	17.99	10	85	196	281
7460			.444	21.59	12	98	203	301
7480			.52	24.29	14	111	207	318
7500			.591	28.79	16	123	216	339
8220	14'	pumped	.347	16.79	8	85	214	299
8240			.43	20.99	10	99	218	317
8260			.519	25.19	12	115	223	338
8280			.607	28.33	14	129	227	356
8300			.69	33.59	16	144	233	377

A2020 Basement Walls

A2020 110	Walls, Cast in Place							

	WALL HEIGHT (FT.)	PLACING METHOD	CONCRETE (C.Y. per L.F.)	REINFORCING (LBS. per L.F.)	WALL THICKNESS (IN.)	COST PER L.F.		
						MAT.	INST.	TOTAL
8420	14'	crane & bucket	.347	16.79	8	85	223	308
8440			.43	20.99	10	99	229	328
8460			.519	25.19	12	115	236	351
8480			.607	28.33	14	129	242	371
8500			.69	33.59	16	144	250	394
9220	16'	pumped	.397	19.19	8	97	244	341
9240			.492	23.99	10	113	249	362
9260			.593	28.79	12	131	256	387
9280			.693	32.39	14	147	259	406
9300			.788	38.38	16	165	265	430
9420	16'	crane & bucket	.397	19.19	8	97	255	352
9440			.492	23.99	10	113	261	374
9460			.593	28.79	12	131	270	401
9480			.693	32.39	14	147	276	423
9500			.788	38.38	16	165	285	450

A2020 Basement Walls

The Wood Wall Foundations System includes: all required pressure treated wood framing members; sheathing (treated and bonded with CDX-grade glue); asphalt paper; polyethylene vapor barrier; and 3-1/2" R11 fiberglass insulation.

The Expanded System Listing shows various walls fabricated with 2" x 4", 2" x 6", 2" x 8", or 2" x 10" studs. Wall height varies from 2' to 8'. The insulation is standard, fiberglass batt type for all walls listed.

System Components	QUANTITY	UNIT	COST PER S.F.		
			MAT.	INST.	TOTAL
SYSTEM A2020 150 2000					
WOOD WALL FOUNDATION, 2" X 4" STUDS, 12" O.C., 1/2" SHEATHING, 2' HIGH					
Framing plates & studs, 2" x 4" treated	.550	B.F.	1.49	2.85	4.34
Sheathing, 1/2" thick CDX, treated	1.000	S.F.	1	1.14	2.14
Building paper, asphalt felt sheathing paper 15 lb	1.100	S.F.	.06	.19	.25
Polyethylene vapor barrier, standard, 6 mil	1.000	S.F.	.04	.17	.21
Fiberglass insulation, 3-1/2" thick, 11" wide	1.000	S.F.	.37	.56	.93
TOTAL			2.96	4.91	7.87

A2020 150	Wood Wall Foundations	COST PER S.F.		
		MAT.	INST.	TOTAL
2000	Wood wall foundation, 2" x 4" studs, 12" O.C., 1/2" sheathing, 2' high	2.96	4.91	7.87
2008	4' high	2.54	4.09	6.63
2012	6' high	2.40	3.83	6.23
2020	8' high	2.32	3.69	6.01
2500	2" x 6" studs, 12" O.C., 1/2" sheathing, 2' high	3.43	5.85	9.28
2508	4' high	2.88	4.78	7.66
2512	6' high	2.69	4.42	7.11
2520	8' high	2.60	4.25	6.85
3000	2" x 8" studs, 12" O.C., 1/2" sheathing, 2' high	4.64	5.20	9.84
3008	4' high	3.75	4.31	8.06
3016	6' high	3.44	4.01	7.45
3024	8' high	3.29	3.87	7.16
3032	16" O.C., 3/4" sheathing, 2' high	4.77	5.10	9.87
3040	4' high	3.85	4.18	8.03
3048	6' high	3.53	3.86	7.39
3056	8' high	3.37	3.71	7.08
4000	2" x 10" studs, 12" O.C., 1/2" sheathing, 2' high	5.15	5.35	10.50
4008	4' high	4.09	4.41	8.50
4012	6' high	3.75	4.10	7.85
4040	8' high	3.58	3.95	7.53
4048	16" O.C., 3/4" sheathing, 2' high	5.20	5.20	10.40
4056	4' high	4.14	4.25	8.39
4070	6' high	3.79	3.94	7.73
4080	8' high	3.60	3.77	7.37

Same Data. Simplified.

Enjoy the convenience and efficiency of accessing your costs anywhere:

- **Skip the multiplier** by setting your location
- **Quickly search,** edit, favorite and share costs
- **Stay on top of price changes** with automatic updates

Discover more at rsmeans.com/online

Geothermal System

System Components	QUANTITY	UNIT	COST EACH		
			MAT.	INST.	TOTAL
SYSTEM D3050 248 1000					
GEOTHERMAL HEAT PUMP SYSTEM 50 TON, VERTICAL LOOP, 200 LF PER TON					
Mobilization excavator	2.000	Ea.		1,706	1,706
Mobilization support crew and equipment	2.000	Ea.		680	680
Mobilization drill rig	2.000	Ea.		354	354
Drill wells 6" diameter	100.000	C.L.F.		53,800	53,800
Pipe loops 1 1/2" diameter	200.000	C.L.F.	21,200	43,200	64,400
Pipe headers 2" diameter	1600.000	L.F.	2,528	4,256	6,784
U-fittings for pipe loops	50.000	Ea.	242.50	952.50	1,195
Header tee fittings	100.000	Ea.	1,250	3,250	4,500
Header elbow fittings	10.000	Ea.	105	200	305
Excavate trench for pipe header	475.000	B.C.Y.		3,410.50	3,410.50
Backfill trench for pipe header	655.000	L.C.Y.		2,069.80	2,069.80
Compact trench for pipe header	475.000	E.C.Y.		1,282.50	1,282.50
Circulation pump 5 HP	1.000	Ea.	14,400	925	15,325
Pump control system	1.000	Ea.	1,400	1,200	2,600
Pump gauges	2.000	Ea.	40	49	89
Pump gauge fittings	2.000	Ea.	128	49	177
Pipe insulation for pump connection	12.000	L.F.	41.16	91.80	132.96
Pipe for pump connection	12.000	L.F.	219.60	422.64	642.24
Pipe fittings for pump connection	1.000	Ea.	33	214.75	247.75
Install thermostat wells	2.000	Ea.	18.20	131.20	149.40
Install gauge wells	2.000	Ea.	18.20	136.60	154.80
Thermometers, stem type, 9" case, 8" stem, 3/4" NPT	8.000	Ea.	360	858.80	1,218.80
Gauges, pressure or vacuum, 3-1/2" diameter dial	1.000	Ea.	780	310	1,090
Pipe strainer for pump	1.000	Ea.	207	315	522
Shut valve for pump	1.000	Ea.	785	335	1,120
Expansion joints for pump	2.000	Ea.	680	248	928
Heat pump 50 tons	1.000	Ea.	42,500	14,700	57,200
TOTAL			86,935.66	135,148.09	222,083.75

D3050 Terminal & Package Units

D3050 248	Geothermal Heat Pump System	COST EACH		
		MAT.	INST.	TOTAL
0990	Geothermal heat pump system, vertical loops 200' depth, 200 LF/4 gpm per ton			
1000	50 tons, vertical loops 200' depth, soft soil	87,000	135,000	222,000
1050	50 tons common soil	87,000	153,500	240,500
1075	50 tons dense soil	87,000	189,500	276,500
1100	40 tons, vertical loops 200' depth, soft soil	77,500	110,000	187,500
1150	40 tons common soil	81,500	147,000	228,500
1175	40 tons dense soil	77,500	153,000	230,500
1200	30 tons, vertical loops 200' depth, soft soil	58,000	87,000	145,000
1250	30 tons, common soil	58,000	98,000	156,000
1275	30 tons, dense soil	58,000	119,000	177,000
1300	25 tons, vertical loops 200' depth, soft soil	53,500	73,500	127,000
1350	25 tons, common soil	53,500	83,000	136,500
1375	25 tons, dense soil	53,500	100,500	154,000
1390	Geothermal heat pump system, vertical loops 250' depth, 250 LF/4 gpm per ton			
1400	50 tons, vertical loops 250' depth, soft soil	92,000	159,500	251,500
1450	50 tons common soil	92,000	182,000	274,000
1475	50 tons dense soil	92,000	227,000	319,000
1500	40 tons, vertical loops 250' depth, soft soil	81,500	129,500	211,000
1550	40 tons common soil	81,500	147,000	228,500
1575	40 tons dense soil	81,500	183,000	264,500
1600	30 tons, vertical loops 250' depth, soft soil	61,500	101,000	162,500
1650	30 tons, common soil	61,500	114,500	176,000
1675	30 tons, dense soil	61,500	141,500	203,000
1700	25 tons, vertical loops 250' depth, soft soil	56,000	86,000	142,000
1750	25 tons, common soil	56,000	97,000	153,000
1775	25 tons, dense soil	56,000	119,500	175,500
1998	Geothermal heat pump system, vertical loops 200' depth, 200 LF/4 gpm per ton			
1999	Includes pumping of wells and adding 1:1 cement grout mix with 5% waste factor			
2000	50 tons, vertical loops 200' depth, soft soil	117,500	159,000	276,500
2050	50 tons common soil	117,500	177,000	294,500
2075	50 tons dense soil	117,500	213,000	330,500
2100	40 tons, vertical loops 200' depth, soft soil	102,000	129,000	231,000
2150	40 tons common soil	106,000	166,500	272,500
2175	40 tons dense soil	102,000	172,000	274,000
2200	30 tons, vertical loops 200' depth, soft soil	76,500	101,000	177,500
2250	30 tons, common soil	76,500	112,000	188,500
2275	30 tons, dense soil	76,500	133,500	210,000
2300	25 tons, vertical loops 200' depth, soft soil	69,000	85,500	154,500
2350	25 tons, common soil	69,000	94,500	163,500
2375	25 tons, dense soil	69,000	112,500	181,500
2390	Geothermal heat pump system, vertical loops 250' depth, 250 LF/4 gpm per ton			
2391	Includes pumping of wells and adding 1:1 cement grout mix with 5% waste factor			
2400	50 tons, vertical loops 250' depth, soft soil	130,500	189,500	320,000
2450	50 tons common soil	130,500	212,000	342,500
2475	50 tons dense soil	130,500	256,500	387,000
2500	40 tons, vertical loops 250' depth, soft soil	112,000	153,500	265,500
2550	40 tons common soil	112,000	171,500	283,500
2575	40 tons dense soil	112,000	207,500	319,500
2600	30 tons, vertical loops 250' depth, soft soil	84,500	119,000	203,500
2650	30 tons, common soil	84,500	132,500	217,000
2675	30 tons, dense soil	84,500	159,500	244,000
2700	25 tons, vertical loops 250' depth, soft soil	75,500	100,500	176,000
2750	25 tons, common soil	75,500	112,000	187,500
2775	25 tons, dense soil	75,500	135,000	210,500
2998	Geothermal heat pump system, vertical loops 200' depth, 200 LF/4 gpm per ton			
2999	Includes pumping of wells and adding bentonite mix with 5% waste factor			
3000	50 tons, vertical loops 200' depth, soft soil	106,000	159,000	265,000
3050	50 tons common soil	106,000	177,000	283,000

D30 HVAC

D3050 Terminal & Package Units

D3050 248	Geothermal Heat Pump System	COST EACH		
		MAT.	INST.	TOTAL
3075	50 tons dense soil	106,000	213,000	319,000
3100	40 tons, vertical loops 200' depth, soft soil	92,500	129,000	221,500
3150	40 tons common soil	96,500	166,500	263,000
3175	40 tons dense soil	92,500	172,000	264,500
3200	30 tons, vertical loops 200' depth, soft soil	69,500	101,000	170,500
3250	30 tons, common soil	69,500	112,000	181,500
3275	30 tons, dense soil	69,500	133,500	203,000
3300	25 tons, vertical loops 200' depth, soft soil	63,000	85,500	148,500
3350	25 tons, common soil	63,000	94,500	157,500
3375	25 tons, dense soil	63,000	112,500	175,500
3390	Geothermal heat pump system, vertical loops 250' depth, 250 LF/4 gpm per ton			
3391	Includes pumping of wells and adding bentonite mix with 5% waste factor			
3400	50 tons, vertical loops 250' depth, soft soil	116,000	189,500	305,500
3450	50 tons common soil	116,000	212,000	328,000
3475	50 tons dense soil	116,000	256,500	372,500
3500	40 tons, vertical loops 250' depth, soft soil	100,500	153,500	254,000
3550	40 tons common soil	100,500	171,500	272,000
3575	40 tons dense soil	100,500	207,500	308,000
3600	30 tons, vertical loops 250' depth, soft soil	75,500	119,000	194,500
3650	30 tons, common soil	75,500	132,500	208,000
3675	30 tons, dense soil	75,500	159,500	235,000
3700	25 tons, vertical loops 250' depth, soft soil	68,000	100,500	168,500
3750	25 tons, common soil	68,000	112,000	180,000
3775	25 tons, dense soil	68,000	135,000	203,000

G10 Site Preparation

G1010 Site Clearing

System Components	QUANTITY	UNIT	COST PER ACRE		
			MAT.	INST.	TOTAL
SYSTEM G1010 120 1000					
REMOVE TREES & STUMPS UP TO 6″ DIA. BY CUT & CHIP & STUMP HAUL AWAY					
Cut & chip light trees to 6″ diameter	1.000	Acre		5,025	5,025
Grub stumps & remove, ideal conditions	1.000	Acre		1,925	1,925
TOTAL				6,950	6,950

G1010 120	Clear & Grub Site	COST PER ACRE		
		MAT.	INST.	TOTAL
0980	Clear & grub includes removal and haulaway of tree stumps			
1000	Remove trees & stumps up to 6 ″ in dia. by cut & chip & remove stumps		6,950	6,950
1100	Up to 12 ″ in diameter		11,100	11,100
1990	Removal of brush includes haulaway of cut material			
1992	Percentage of vegetation is actual plant coverage			
2000	Remove brush by saw, 4′ tall, 10 mile haul cycle		4,425	4,425

G10 Site Preparation

G1010   Site Clearing

Stripping and stock piling top soil can be done with a variety of equipment depending on the size of the job. For small jobs you can do by hand. Larger jobs can be done with a loader or skid steer. Even larger jobs can be done by a dozer.

System Components	QUANTITY	UNIT	COST PER S.Y.		
			MAT.	INST.	TOTAL
SYSTEM G1010 122 1100 **STRIP TOPSOIL AND STOCKPILE, 6 INCH DEEP BY SKID STEER** By skid steer, 6" deep, 100' haul, 101-500 S.Y.	1.000	S.Y.		3.68	3.68
TOTAL				3.68	3.68

G1010 122	Strip Topsoil	COST PER S.Y.		
		MAT.	INST.	TOTAL
1000	Strip topsoil & stockpile, 6 inch deep by hand, 50' haul		15.40	15.40
1100	100' haul, 101-500 S.Y., by skid steer		3.68	3.68
1200	100' haul, 501-900 S.Y.		2.05	2.05
1300	200' haul , 901-1100 S.Y.		2.86	2.86
1400	200' haul, 1101-4000 S.Y., by dozer		.64	.64

G10 Site Preparation

G1020 Site Demolitions and Relocations

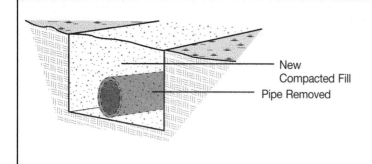

New
Compacted Fill
Pipe Removed

Demolition of utility lines includes excavation, pipe removal, backfill, compaction and one day trucking cost for loading and disposal of pipe. No waste disposal fees are included.

Demolition of utility lines @ 6' depth include hydraulic shoring.

Demolition of utility lines @ 7' to 10' depth include a trench box.

Structural fill required after removal:

14" diameter pipe:	5.15 LCY per 100 LF
16" diameter pipe:	6.72 LCY per 100 LF
18" diameter pipe:	8.53 LCY per 100 LF
24" diameter pipe:	15.1 LCY per 100 LF

System Components			COST PER L.F.		
	QUANTITY	UNIT	MAT.	INST.	TOTAL
SYSTEM G1020 205 1000					
REMOVE 4" D. I. WATER PIPE, 4' DEPTH, EXCAVATE, BACKFILL, COMPACT& HAUL					
Excavate trench 4' depth	.593	B.C.Y.		4.35	4.35
Backfill trench	.770	L.C.Y.		2.43	2.43
Compact trench	.593	E.C.Y.		3.15	3.15
Demolition of 4 inch D.I. pipe	1.000	L.F.		16.39	16.39
Truck for waste and hauling	.040	Hr.		6.10	6.10
TOTAL				32.42	32.42

G1020 205	Remove Underground Water Pipe Including Earthwork	COST PER L.F.		
		MAT.	INST.	TOTAL
1000	Remove D.I. pipe 4' depth, excavate & backfill, hauling, no dump fee, 4"		32.50	32.50
1010	6" to 12" diameter		35.50	35.50
1020	Plastic pipe, 3/4" to 4" diameter		14.40	14.40
1030	6" to 8" diameter		16.10	16.10
1040	10" to 12" diameter		20.50	20.50
1050	Concrete water pipe, 4" to 10" diameter		22.50	22.50
1060	12" diameter		27.50	27.50
1070	Welded steel water pipe, 4" to 6" diameter		29.50	29.50
1080	8" to 10" diameter		48.50	48.50
1090	Copper pipe, 3/4" to 2" diameter		15.15	15.15
1100	2-1/2" to 3" diameter		18.65	18.65
1110	4" to 6" diameter		23	23
1200	D.I. pipe 5' depth, 4" diameter		34.50	34.50
1210	6" to 12" diameter		37.50	37.50
1220	Plastic pipe, 3/4" to 4" diameter		16.25	16.25
1230	6" to 8" diameter		18.05	18.05
1240	10" to 12" diameter		22	22
1250	Concrete water pipe, 4" to 10" diameter		24.50	24.50
1260	12" diameter		30	30
1270	Welded steel water pipe, 4" to 6" diameter		31.50	31.50
1280	8" to 10" diameter		50.50	50.50
1290	Copper pipe, 3/4" to 2" diameter		17.10	17.10
1300	2-1/2" to 3" diameter		20.50	20.50
1310	4" to 6 " diameter		25	25
1400	D.I. pipe 6' depth, 4" diameter with hydraulic shoring	2.64	41.50	44.14
1410	6" to 12" diameter	2.64	44.50	47.14
1420	Plastic pipe, 3/4" to 4" diameter	2.64	23.50	26.14
1430	6" to 8" diameter	2.64	25	27.64
1440	10" to 12" diameter	2.64	29.50	32.14
1450	Concrete water pipe, 4" to 10" diameter	2.64	31.50	34.14

614

G1020 Site Demolitions and Relocations

G1020 205	Remove Underground Water Pipe Including Earthwork	COST PER L.F.		
		MAT.	INST.	TOTAL
1460	12" diameter	2.64	36.50	39.14
1470	Welded steel water pipe, 4" to 6" diameter	2.64	38.50	41.14
1480	8" to 10" diameter	2.64	57.50	60.14
1490	Copper pipe, 3/4" to 2" diameter	2.64	24	26.64
1500	2-1/2" to 3" diameter	2.64	27.50	30.14
1510	4" to 6" diameter	2.64	31.50	34.14
1600	D.I. pipe, trench 7' deep by 6' wide, with trench box, 4" diameter pipe		42.50	42.50
1610	6" to 12" diameter		46	46
1620	Plastic pipe, 3/4" to 4" diameter		26.50	26.50
1630	6" to 8" diameter		28	28
1640	10" to 12" diameter		31	31
1650	Concrete water pipe, 4" to 10" diameter		33	33
1660	12" diameter		38	38
1670	Welded steel water pipe, 4" to 6" diameter		40	40
1680	8" to 10" diameter		59.50	59.50
1690	Copper pipe, 3/4" to 2" diameter		27	27
1700	2-1/2" to 3" diameter		29	29
1710	4" to 6" diameter		33.50	33.50
1800	D.I. pipe, trench 8' deep by 6' wide, with trench box, 4" diameter pipe		46	46
1810	6" to 12" diameter		49	49
1820	Plastic pipe, 3/4" to 4" diameter		27.50	27.50
1830	6" to 8" diameter		28.50	28.50
1840	10" to 12" diameter		33.50	33.50
1850	Concrete water pipe, 4" to 10" diameter		36	36
1860	12" diameter		41	41
1870	Welded steel water pipe, 4" to 6" diameter		42.50	42.50
1880	8" to 10" diameter		62	62
1890	Copper pipe, 3/4" to 2" diameter		27.50	27.50
1900	2-1/2" to 3" diameter		32	32
1910	4" to 6" diameter		36.50	36.50
2000	D.I. pipe, trench 9' deep by 6' wide, with trench box, 4" diameter pipe		48	48
2010	6" to 12" diameter		50.50	50.50
2020	Plastic pipe, 3/4" to 4" diameter		31.50	31.50
2030	6" to 8 " diameter		32.50	32.50
2040	10" to 12" diameter		35.50	35.50
2050	Concrete water pipe, 4" to 10" diameter		37.50	37.50
2060	12" diameter		43	43
2070	Welded steel water pipe, 4" to 6" diameter		44.50	44.50
2080	8" to 10" diameter		64	64
2090	Copper pipe, 3/4" to 2" diameter		31.50	31.50
2100	2-1/2" to 3" diameter		33.50	33.50
2110	4" to 6" diameter		38.50	38.50
2200	D.I. pipe, trench 10' deep by 6' wide, with trench box, 4" diameter pipe		50	50
2210	6" to 12" diameter		53.50	53.50
2220	Plastic pipe, 3/4" to 4" diameter		34	34
2230	6" to 8" diameter		35.50	35.50
2240	10" to 12" diameter		38.50	38.50
2250	Concrete water pipe, 4" to 10" diameter		40.50	40.50
2260	12" diameter		45.50	45.50
2270	Welded steel water pipe, 4" to 6" diameter		47.50	47.50
2280	8" to 10" diameter		67	67
2290	Copper pipe, 3/4" to 2" diameter		34.50	34.50
2300	2-1/2" to 3" diameter		36.50	36.50
2310	4" to 6" diameter		41	41
2400	Concrete water pipe 5' depth, 15" to 18" diameter		39.50	39.50
2410	21" to 24" diameter		47	47
2420	D.I. water pipe, 14" to 24" diameter		46.50	46.50
2430	Plastic water pipe, 14" to 18" diameter		19.75	19.75

G10 Site Preparation

G1020 Site Demolitions and Relocations

G1020 205	Remove Underground Water Pipe Including Earthwork	COST PER L.F.		
		MAT.	INST.	TOTAL
2440	20" to 24" diameter		25	25
2500	Concrete water pipe, 6' depth, 15" to 18" diameter with hydraulic shoring	2.64	45.50	48.14
2510	21" to 24" diameter	2.64	53.50	56.14
2520	D. I. water pipe, 14" to 24" diameter	2.64	53	55.64
2530	Plastic water pipe, 14" to 18" diameter	2.64	26.50	29.14
2540	20" to 24" diameter	2.64	31	33.64
2600	Concrete water pipe, 7' deep by 6' wide, w/trench box, 15" to 18" diameter		50	50
2610	21" to 24" diameter		57.50	57.50
2620	D. I. water pipe, 14" to 24" diameter		57	57
2630	Plastic water pipe, 14" to 18" inch diameter		30	30
2640	20" to 24" diameter		35.50	35.50
2700	Concrete water pipe, 8' deep by 6' wide, w/trench box, 15" to 18" diameter		52.50	52.50
2710	21" to 24" diameter		60.50	60.50
2720	D. I. water pipe, 14" to 24" diameter		60	60
2730	Plastic water pipe, 14" to 18" diameter		32.50	32.50
2740	20" to 24" diameter		38	38
2800	Concrete water pipe, 9' deep by 6' wide, w/trench box, 15" to 18" diameter		55.50	55.50
2810	21" to 24" diameter		63.50	63.50
2820	D. I. water pipe, 14" to 24" diameter		62.50	62.50
2830	Plastic water pipe, 14" to 18" diameter		35.50	35.50
2840	20" to 24" diameter		40.50	40.50
2900	Concrete water pipe, 10' deep by 6' wide, w/trench box, 15" to 18" diameter		58.50	58.50
2910	21" to 24" diameter		65.50	65.50
2920	D. I. water pipe, 14" to 24" diameter		65.50	65.50
2930	Plastic water pipe, 14" to 18" diameter		38.50	38.50
2940	20" to 24" diameter		43.50	43.50
3000	Import stuctural fill 10 miles round trip per cubic yard	27	12.30	39.30

G10 Site Preparation

G1020 Site Demolitions and Relocations

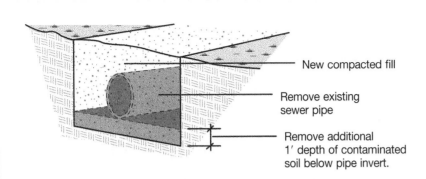

New compacted fill

Remove existing sewer pipe

Remove additional 1' depth of contaminated soil below pipe invert.

Demolition of utility lines includes excavation, pipe removal, backfill, compaction and one day trucking cost for loading and disposal of waste pipe and soil. No disposal fees are included.

Excavation is one foot deeper than pipe invert. For pipes to 12" diameter, two feet of excavated soil is waste. For pipes 15" to 18" diameter, 2.5 feet of excavated soil is waste. For pipes 21" to 24" diameter, 3 feet of excavated soil is waste.

Demolition of utility lines @ 5' to 6' depth includes hydraulic shoring.

Demolition of utility lines @ 7' to 10' depth includes trench box. Trench box requires 6' minimum trench width.

System Components	QUANTITY	UNIT	COST PER L.F. MAT.	COST PER L.F. INST.	COST PER L.F. TOTAL
SYSTEM G1020 206 1000					
REMOVE 4 INCH PLASTIC SEWER PIPE, 4' DEPTH, EXCAVATE,					
BACKFILL, COMPACT & HAUL WASTE SOIL AND PIPE					
Excavate trench 5' depth (one foot over)	.741	B.C.Y.		4.89	4.89
Backfill trench	.963	L.C.Y.		3.04	3.04
Compact trench	.741	E.C.Y.		3.93	3.93
Demolition of 4 inch plastic sewer pipe	1.000	L.F.		2.69	2.69
Truck for waste hauling	.012	Hr.		1.83	1.83
TOTAL				16.38	16.38

G1020 206	Remove Underground Sewer Pipe Including Earthwork	MAT.	INST.	TOTAL
1000	Remove plastic sewer, 4' depth, excavate, backfill, haul waste, 4" diameter		16.35	16.35
1010	6" to 8" diameter		18.05	18.05
1020	10" to 12" diameter		22	22
1030	Cast iron sewer, 4" diameter		25	25
1040	5" to 6" diameter		29	29
1050	8" to 12" diameter		32.50	32.50
1060	Concrete sewer, 4" to 10" diameter		24.50	24.50
1070	12" diameter		30	30
1100	Plastic sewer, 5' depth, with hydraulic shoring, 4" diameter	2.64	23.50	26.14
1110	6" to 8" diameter	2.64	25	27.64
1120	10" to 12" diameter	2.64	29.50	32.14
1130	Cast iron sewer, 4" diameter	2.64	31.50	34.14
1140	5" to 6" diameter	2.64	35.50	38.14
1150	8" to 12" diameter	2.64	39.50	42.14
1160	Concrete sewer, 4" to 10" inch diameter	2.64	31.50	34.14
1170	12" diameter	2.64	36.50	39.14
1200	Plastic sewer, 6' depth, with hydraulic shoring, 4" diameter	3.08	25	28.08
1210	6" to 8" diameter	3.08	26.50	29.58
1220	10" to 12" diameter	3.08	30.50	33.58
1230	Cast iron sewer, 4" diameter	3.08	33.50	36.58
1240	5" to 6" diameter	3.08	37.50	40.58
1250	8" to 12" diameter	3.08	41.50	44.58
1260	Concrete sewer, 4" to 10" diameter	3.08	33	36.08
1270	12" diameter	3.08	38.50	41.58
1300	Plastic sewer, 7' depth, with trench box, 4" diameter		28.50	28.50
1310	6" to 8" diameter		31	31

617

G10 Site Preparation

G1020 Site Demolitions and Relocations

G1020 206	Remove Underground Sewer Pipe Including Earthwork	COST PER L.F.		
		MAT.	INST.	TOTAL
1320	10" to 12" diameter		32.50	32.50
1330	Cast iron sewer, 4" diameter		35.50	35.50
1340	5" to 6" diameter		39.50	39.50
1350	8" to 12" diameter		43	43
1360	Concrete sewer, 4" to 10" diameter		35	35
1370	12" diameter		40.50	40.50
1400	Plastic sewer, 8' depth, with trench box, 4" diameter		31.50	31.50
1410	6" to 8" diameter		33.50	33.50
1420	10" to 12" diameter		37.50	37.50
1430	Cast iron sewer, 4" diameter		38.50	38.50
1440	5" to 6" diameter		42.50	42.50
1450	8" to 12" diameter		46	46
1460	Concrete sewer, 4" to 10" diameter		37.50	37.50
1470	12" diameter		43	43
1500	Plastic sewer, 9' depth, with trench box, 4" diameter		34	34
1510	6" to 8" diameter		36.50	36.50
1520	10" to 12" diameter		40.50	40.50
1530	Cast iron sewer, 4" diameter		41	41
1540	5" to 6" diameter		44.50	44.50
1550	8" to 12" diameter		48.50	48.50
1560	Concrete sewer, 4" to 10" diameter		40.50	40.50
1570	12" diameter		46	46
1600	Plastic sewer, 10' depth, with trench box, 4" diameter		37	37
1610	6" to 8" diameter		39.50	39.50
1620	10" to 12" diameter		43.50	43.50
1630	Cast iron sewer, 4" diameter		44	44
1640	5" to 6" diameter		47.50	47.50
1650	8" to 12" diameter		51.50	51.50
1660	Concrete sewer, 4" to 10" diameter		43.50	43.50
1670	12" diameter		48.50	48.50
1700	Concrete sewer, 5' depth, with hydraulic shoring, 15" to 18" diameter	2.64	49	51.64
1710	21" to 24" diameter	2.64	56.50	59.14
1720	Plastic sewer, 15" to 18" diameter	2.64	29.50	32.14
1730	21" to 24" diameter	2.64	34.50	37.14
1800	6' depth, 15" to 18" diameter	3.08	51	54.08
1810	21" to 24" diameter	3.08	60	63.08
1820	Plastic sewer, 15" to 18" diameter	3.08	32	35.08
1830	21" to 24" diameter	3.08	37.50	40.58
1900	7' depth, with trench box, 15" to 18" diameter		52.50	52.50
1910	21" to 24" diameter		60.50	60.50
1920	Plastic sewer, 15" to 18" diameter		32.50	32.50
1930	21" to 24" diameter		38	38
2000	8' depth, with trench box, 15" to 18" diameter		55.50	55.50
2010	21" to 24" diameter		63.50	63.50
2020	Plastic sewer, 15" to 18" diameter		35.50	35.50
2030	21" to 24" diameter		40.50	40.50
2100	9' depth, with trench box, 15" to 18" diameter		58.50	58.50
2110	21" to 24" diameter		65.50	65.50
2120	Plastic sewer, 15" to 18" diameter		38.50	38.50
2130	21" to 24" diameter		43.50	43.50
2200	10' depth, with trench box, 15" to 18" diameter		62	62
2210	21" to 24" diameter		69.50	69.50
2220	Plastic sewer, 15" to 18" diameter		41.50	41.50
2230	21" to 24" diameter		47	47

G10 Site Preparation

G1020 Site Demolitions and Relocations

G1020 210	Demolition Bituminous Sidewalk	COST PER S.Y.		
		MAT.	INST.	TOTAL
0990	Demolition of bituminous sidewalk & haulaway by truck			
1000	Sidewalk demolition, bituminous, 2 inch thick & trucking		8.90	8.90
1100	2-1/2 inch thick		9.50	9.50

G1020 212	Demolition Concrete Sidewalk	COST PER S.Y.		
		MAT.	INST.	TOTAL
0990	Demolition of concrete sidewalk & haulaway by truck			
1000	Sidewalk demolition, plain concrete, 4 inch thick & trucking		19.40	19.40
1100	Plain 5 inch thick		22	22
1200	Plain 6 inch thick		26	26
1300	Mesh reinforced, 4 inch thick		20.50	20.50
1400	Mesh reinforced, 5 inch thick		23.50	23.50
1500	Mesh reinforced, 6 inch thick		27.50	27.50

G10 Site Preparation

G1030 Site Earthwork

The Cut and Fill Gravel System includes: moving gravel cut from an area above the specified grade to an area below the specified grade utilizing a bulldozer and/or scraper with the addition of compaction equipment, plus a water wagon to adjust the moisture content of the soil.

The Expanded System Listing shows Cut and Fill operations with hauling distances that vary from 50' to 5000'. Lifts for compaction in the filled area vary from 6" to 12". There is no waste included in the assumptions.

System Components			COST PER C.Y.		
	QUANTITY	UNIT	EQUIP.	LABOR	TOTAL
SYSTEM G1030 105 1000					
GRAVEL CUT & FILL, 80 HP DOZER & COMPACT, 50'HAUL, 6" LIFT, 2 PASSES					
Excavating, bulk, dozer, 80 HP, 50' haul sand and gravel	1.000	C.Y.	.96	2.03	2.99
Water wagon rent per day	.003	Hr.	.19	.22	.41
Backfill, dozer, 50' haul sand and gravel	1.150	C.Y.	.46	.98	1.44
Compaction, vibrating roller, 6" lifts, 2 passes	1.000	C.Y.	.21	.31	.52
TOTAL			1.82	3.54	5.36

G1030 105	Cut & Fill Gravel	COST PER C.Y.		
		EQUIP.	LABOR	TOTAL
1000	Gravel cut & fill, 80HP dozer & roller compact, 50' haul, 6" lift, 2 passes	1.82	3.54	5.36
1050	4 passes	1.95	3.72	5.67
1100	12" lift, 2 passes	1.73	3.41	5.14
1150	4 passes	1.86	3.59	5.45
1200	150' haul, 6" lift, 2 passes	3.24	6.50	9.74
1250	4 passes	3.37	6.70	10.07
1300	12" lift, 2 passes	3.15	6.40	9.55
1350	4 passes	3.28	6.55	9.83
1400	300' haul, 6" lift, 2 passes	5.45	11.20	16.65
1450	4 passes	5.60	11.35	16.95
1500	12" lift, 2 passes	5.35	11.05	16.40
1550	4 passes	5.50	11.25	16.75
1600	105 HP dozer & roller compactor, 50' haul, 6" lift, 2 passes	2	2.65	4.65
1650	4 passes	2.13	2.83	4.96
1700	12" lift, 2 passes	1.91	2.52	4.43
1750	4 passes	2.04	2.70	4.74
1800	150' haul, 6" lift, 2 passes	3.85	5.15	9
1850	4 passes	3.98	5.30	9.28
1900	12" lift, 2 passes	3.76	5	8.76
1950	4 passes	3.89	5.20	9.09
2000	300' haul, 6" lift, 2 passes	7.10	9.50	16.60
2050	4 passes	7.25	9.65	16.90
2100	12' lift, 2 passes	7	9.35	16.35
2150	4 passes	7.15	9.55	16.70
2200	200 HP dozer & roller compactor, 150' haul, 6" lift, 2 passes	4.73	2.97	7.70
2250	4 passes	4.86	3.15	8.01
2300	12" lift, 2 passes	4.64	2.84	7.48
2350	4 passes	4.77	3.02	7.79

G10 Site Preparation

G1030 Site Earthwork

G1030 105	Cut & Fill Gravel	COST PER C.Y.		
		EQUIP.	LABOR	TOTAL
2600	300' haul, 6" lift, 2 passes	8.10	4.87	12.97
2650	4 passes	8.25	5.05	13.30
2700	12" lift, 2 passes	8.05	4.74	12.79
2750	4 passes	8.15	4.92	13.07
3000	300 HP dozer & roller compactors, 150' haul, 6" lift, 2 passes	3.52	2.02	5.54
3050	4 passes	3.65	2.20	5.85
3100	12" lift, 2 passes	3.43	1.89	5.32
3150	4 passes	3.56	2.07	5.63
3200	300' haul, 6" lift, 2 passes	6	3.22	9.22
3250	4 passes	6.15	3.40	9.55
3300	12" lift, 2 passes	5.90	3.09	8.99
3350	4 passes	6.05	3.27	9.32
4200	10 C.Y. elevating scraper & roller compacter, 1500' haul, 6" lift, 2 passes	4.55	4.41	8.96
4250	4 passes	4.96	4.90	9.86
4300	12" lift, 2 passes	4.42	4.28	8.70
4350	4 passes	4.58	4.41	8.99
4800	5000' haul, 6" lift, 2 passes	6.15	5.90	12.05
4850	4 passes	6.40	6.20	12.60
4900	12" lift, 2 passes	5.85	5.55	11.40
4950	4 passes	6.20	5.90	12.10
5000	15 C.Y. S.P. scraper & roller compactor, 1500' haul, 6" lift, 2 passes	4.91	3.25	8.16
5050	4 passes	5.15	3.55	8.70
5100	12" lift, 2 passes	4.78	3.11	7.89
5150	4 passes	4.93	3.26	8.19
5400	5000' haul, 6" lift, 2 passes	6.90	4.51	11.41
5450	4 passes	7.15	4.81	11.96
5500	12" lift, 2 passes	6.75	4.38	11.13
5550	4 passes	6.90	4.52	11.42
5600	21 C.Y. S.P. scraper & roller compactor, 1500' haul, 6" lift, 2 passes	4.09	2.60	6.69
5650	4 passes	4.36	2.90	7.26
5700	12" lift, 2 passes	3.97	2.46	6.43
5750	4 passes	4.11	2.60	6.71
6000	5000' haul, 6" lift, 2 passes	6.30	3.92	10.22
6050	4 passes	6.55	4.22	10.77
6100	12" lift, 2 passes	6.15	3.78	9.93
6150	4 passes	6.30	3.92	10.22

621

G1030 Site Earthwork

The Excavation of Sand and Gravel System balances the productivity of the Excavating equipment to the hauling equipment. It is assumed that the hauling equipment will encounter light traffic and will move up no considerable grades on the haul route. No mobilization cost is included. All costs given in these systems include a swell factor of 15%.

The Expanded System Listing shows excavation systems using backhoes ranging from 1/2 Cubic Yard capacity to 3-1/2 Cubic Yards. Power shovels indicated range from 1/2 Cubic Yard to 3 Cubic Yards. Dragline bucket rigs used range from 1/2 Cubic Yard to 3 Cubic Yards. Truck capacities indicated range from 8 Cubic Yards to 20 Cubic Yards. Each system lists the number of trucks involved and the distance (round trip) that each must travel.

System Components	QUANTITY	UNIT	COST PER C.Y.		
			EQUIP.	LABOR	TOTAL
SYSTEM G1030 110 1000					
EXCAVATE SAND & GRAVEL, 1/2 CY BACKHOE, TWO 8 CY DUMP TRUCKS 2 MRT					
Excavating, bulk, hyd. backhoe, wheel mtd., 1-1/2 CY	1.000	B.C.Y.	1.55	3.79	5.34
Haul earth, 8 CY dump truck, 2 mile round trip, 2.6 loads/hr	1.150	L.C.Y.	4.51	5.62	10.13
Spotter at earth fill dump or in cut	.020	Hr.		1.14	1.14
TOTAL			6.06	10.55	16.61

G1030 110	Excavate and Haul Sand & Gravel	COST PER C.Y.		
		EQUIP.	LABOR	TOTAL
1000	Excavate sand & gravel, 1/2 C.Y. backhoe, two 8 C.Y. dump trucks, 2 MRT	6.05	10.55	16.60
1200	Three 8 C.Y. dump trucks, 4 mile round trip	7.90	12.30	20.20
1400	Two 12 C.Y. dump trucks, 4 mile round trip	5.95	7.90	13.85
1600	3/4 C.Y. backhoe, four 8 C.Y. dump trucks, 2 mile round trip	5.40	8.15	13.55
1700	Six 8 C.Y. dump trucks, 4 mile round trip	7.65	10.95	18.60
1800	Three 12 C.Y. dump trucks, 4 mile round trip	6	6.85	12.85
1900	Two 16 C.Y. dump trailers, 3 mile round trip	4.95	5.50	10.45
2000	Two 20 C.Y. dump trailers, 4 mile round trip	4.62	5.15	9.77
2200	1-1/2 C.Y. backhoe, four 12 C.Y. dump trucks, 2 mile round trip	4.53	4.86	9.39
2300	Six 12 C.Y. dump trucks, 4 mile round trip	5.50	5.50	11
2400	Three 16 C.Y. dump trailers, 3 mile round trip	4.62	4.72	9.34
2600	Three 20 C.Y. dump trailers, 4 mile round trip	4.28	4.38	8.66
2800	2-1/2 C.Y. backhoe, six 12 C.Y. dump trucks, 2 mile round trip	4.76	5.05	9.81
2900	Eight 12 C.Y. dump trucks, 4 mile round trip	5.80	6	11.80
3000	Four 16 C.Y. dump trailers, 2 mile round trip	4.05	3.95	8
3100	Six 16 C.Y. dump trailers, 4 mile round trip	5.10	4.99	10.09
3200	Four 20 C.Y. dump trailers, 3 mile round trip	4.25	4.05	8.30
3400	3-1/2 C.Y. backhoe, eight 16 C.Y. dump trailers, 3 mile round trip	4.86	4.14	9
3500	Ten 16 C.Y. dump trailers, 4 mile round trip	5.15	4.33	9.48
3700	Six 20 C.Y. dump trailers, 2 mile round trip	3.55	2.99	6.54
3800	Nine 20 C.Y. dump trailers, 4 mile round trip	4.43	3.69	8.12
4000	1/2 C.Y. pwr. shovel, four 8 C.Y. dump trucks, 2 mile round trip	5.70	7.70	13.40
4100	Six 8 C.Y. dump trucks, 3 mile round trip	6.60	9.25	15.85
4200	Two 12 C.Y. dump trucks, 1 mile round trip	4.02	4.72	8.74
4300	Four 12 C.Y. dump trucks, 4 mile round trip	6	6.35	12.35
4400	Two 16 C.Y. dump trailers, 2 mile round trip	4.28	4.61	8.89
4500	Two 20 C.Y. dump trailers, 3 mile round trip	4.48	4.72	9.20
4600	Three 20 C.Y. dump trailers, 4 mile round trip	4.52	4.51	9.03
4800	3/4 C.Y. pwr. shovel, six 8 C.Y. dump trucks, 2 mile round trip	5.55	7.50	13.05
4900	Four 12 C.Y. dump trucks, 2 mile round trip	4.94	5.05	9.99
5000	Six 12 C.Y. dump trucks, 4 mile round trip	5.85	6.20	12.05
5100	Four 16 C.Y. dump trailers, 4 mile round trip	5.30	5	10.30
5200	Two 20 C.Y. dump trailers, 1 mile round trip	3.36	3.40	6.76
5400	1-1/2 C.Y. pwr. shovel, six 12 C.Y. dump trucks, 2 mile round trip	4.76	4.81	9.57

G1030 Site Earthwork

G1030 110	Excavate and Haul Sand & Gravel	COST PER C.Y.		
		EQUIP.	LABOR	TOTAL
5600	Four 16 C.Y. dump trailers, 1 mile round trip	3.51	3.16	6.67
5700	Six 16 C.Y. dump trailers, 3 mile round trip	4.67	4.33	9
5800	Six 20 C.Y. dump trailers, 4 mile round trip	4.34	4	8.34
6000	3 C.Y. pwr. shovel, eight 16 C.Y. dump trailers, 1 mile round trip	3.42	2.90	6.32
6200	Eight 20 C.Y. dump trailers, 2 mile round trip	3.36	2.83	6.19
6400	Twelve 20 C.Y. dump trailers, 4 mile round trip	4.26	3.55	7.81
6600	1/2 C.Y. dragline bucket, three 8 C.Y. dump trucks, 1 mile round trip	5.05	6.50	11.55
6700	Five 8 C.Y. dump trucks, 4 mile round trip	9.80	13	22.80
6800	Two 12 C.Y. dump trucks, 2 mile round trip	7.10	7.75	14.85
6900	Three 12 C.Y. dump trucks, 4 mile round trip	7.85	8.35	16.20
7000	Two 16 C.Y. dump trailers, 4 mile round trip	7.45	7.70	15.15
7200	3/4 C.Y. dragline bucket, four 8 C.Y. dump trucks, 1 mile round trip	5.40	6.90	12.30
7300	Six 8 C.Y. dump trucks, 3 mile round trip	7.80	10.30	18.10
7400	Four 12 C.Y. dump trucks, 4 mile round trip	7.15	7.90	15.05
7500	Two 16 C.Y. dump trailers, 2 mile round trip	5.55	5.75	11.30
7600	Three 16 C.Y. dump trailers, 4 mile round trip	6.45	6.35	12.80
7800	1-1/2 C.Y. dragline bucket, six 8 C.Y. dump trucks, 1 mile round trip	4.02	5	9.02
7900	Ten 8 C.Y. dump trucks, 3 mile round trip	7.25	8.95	16.20
8000	Six 12 C.Y. dump trucks, 2 mile round trip	5.30	5.30	10.60
8100	Eight 12 C.Y. dump trucks, 4 mile round trip	6.35	6.25	12.60
8200	Four 16 C.Y. dump trailers, 2 mile round trip	4.66	4.24	8.90
8300	Three 20 C.Y. dump trailers, 2 mile round trip	4.41	4.34	8.75
8400	3 C.Y. dragline bucket, eight 12 C.Y. dump trucks, 2 mile round trip	5.60	4.90	10.50
8500	Six 16 C.Y. dump trailers, 2 mile round trip	4.79	3.95	8.74
8600	Nine 16 C.Y. dump trailers, 4 mile round trip	5.75	4.75	10.50
8700	Eight 20 C.Y. dump trailers, 4 mile round trip	5.05	4.12	9.17

G10 Site Preparation

G1030 Site Earthwork

The Cut and Fill Common Earth System includes: moving common earth cut from an area above the specified grade to an area below the specified grade utilizing a bulldozer and/or scraper, with the addition of compaction equipment, plus a water wagon to adjust the moisture content of the soil.

The Expanded System Listing shows Cut and Fill operations with hauling distances that vary from 50' to 5000'. Lifts for compaction in the filled area vary from 4" to 8". There is no waste included in the assumptions.

System Components	QUANTITY	UNIT	COST PER C.Y.		
			EQUIP.	LABOR	TOTAL
SYSTEM G1030 115 1000					
EARTH CUT & FILL, 80 HP DOZER & COMPACTOR, 50' HAUL, 4" LIFT, 2 PASSES					
Excavating, bulk, dozer, 50' haul, common earth	1.000	C.Y.	1.56	3.31	4.87
Water wagon, rent per day	.008	Hr.	.50	.59	1.09
Backfill dozer, from existing stockpile, 75 H.P., 50' haul	1.310	C.Y.	.59	1.26	1.85
Compaction, roller, 4" lifts, 2 passes	1.000	C.Y.	.14	.46	.60
TOTAL			2.79	5.62	8.41

G1030 115	Cut & Fill Common Earth	COST PER C.Y.		
		EQUIP.	LABOR	TOTAL
1000	Earth cut & fill, 80 HP dozer & roller compact, 50' haul, 4" lift, 2 passes	2.79	5.60	8.40
1050	4 passes	2.47	5.10	7.55
1100	8" lift, 2 passes	2.33	4.64	6.95
1150	4 passes	2.47	5.10	7.55
1200	150' haul, 4" lift, 2 passes	4.52	9	13.50
1250	4 passes	4.76	9.80	14.55
1300	8" lift, 2 passes	4.52	9	13.50
1350	4 passes	4.76	9.80	14.55
1400	300' haul, 4" lift, 2 passes	8.10	16.15	24.50
1450	4 passes	8.55	17.55	26
1500	8" lift, 2 passes	8.10	16.15	24.50
1550	4 passes	8.55	17.55	26
1600	105 H.P. dozer and roller compactor, 50' haul, 4" lift, 2 passes	2.21	3.05	5.25
1650	4 passes	2.28	3.28	5.55
1700	8" lift, 2 passes	2.21	3.05	5.25
1750	4 passes	2.28	3.28	5.55
1800	150' haul, 4" lift, 2 passes	4.82	6.70	11.50
1850	4 passes	4.96	7.15	12.10
1900	8" lift, 2 passes	4.78	6.55	11.35
1950	4 passes	4.92	7.05	11.95
2000	300' haul, 4" lift, 2 passes	9	12.60	21.50
2050	4 passes	9.35	13.75	23
2100	8" lift, 2 passes	8.65	12.25	21
2150	4 passes	9	12.60	21.50
2200	200 H.P. dozer & roller compactor, 150' haul, 4" lift, 2 passes	5.55	3.52	9.05
2250	4 passes	5.65	3.75	9.40
2300	8" lift, 2 passes	5.55	3.41	8.95
2350	4 passes	5.55	3.52	9.05
2600	300' haul, 4" lift, 2 passes	9.85	6.35	16.20
2650	4 passes	9.95	6.80	16.75

G1030 Site Earthwork

G1030 115	Cut & Fill Common Earth	COST PER C.Y.		
		EQUIP.	LABOR	TOTAL
2700	8" lift, 2 passes	9.75	6.10	15.85
2750	4 passes	9.70	5.95	15.65
3000	300 H.P. dozer & roller compactor, 150' haul, 4" lifts, 2 passes	4.21	2.58	6.80
3050	4 passes	4.31	2.93	7.25
3100	8" lift, 2 passes	4.16	2.40	6.55
3150	4 passes	4.21	2.58	6.80
3200	300' haul, 4" lifts, 2 passes	7.20	4.11	11.30
3250	4 passes	7.30	4.46	11.75
3300	8" lifts, 2 passes	7.15	3.93	11.10
3350	4 passes	7.20	4.11	11.30
4200	10 C.Y. elevating scraper & roller compact, 1500' haul, 4" lifts, 2 passes	4.35	3.83	8.20
4250	4 passes	4.49	4.30	8.80
4300	8" lifts, 2 passes	4.28	3.60	7.90
4350	4 passes	4.35	3.83	8.20
4800	5000' haul, 4" lifts, 2 passes	5.70	4.86	10.55
4850	4 passes	5.85	5.35	11.20
4900	8" lifts, 2 passes	5.65	4.63	10.30
4950	4 passes	5.70	4.86	10.55
5000	15 C.Y. S.P. scraper & roller compact, 1500' haul, 4" lifts, 2 passes	5.25	3.17	8.40
5050	4 passes	5.35	3.46	8.80
5100	8" lifts, 2 passes	5.35	3.46	8.80
5150	4 passes	5.25	3.03	8.30
5400	5000' haul, 4" lifts, 2 passes	6.60	3.36	9.95
5450	4 passes	6.70	3.59	10.30
5500	8" lifts, 2 passes	7.20	3.97	11.15
5550	4 passes	7.20	3.91	11.10
5600	21 C.Y. S.P. scraper & roller compact, 1500' haul, 4" lifts, 2 passes	4.55	2.70	7.25
5650	4 passes	4.61	2.93	7.55
5700	8" lifts, 2 passes	4.74	2.86	7.60
5750	4 passes	4.78	2.97	7.75
6000	5000' haul, 4" lifts, 2 passes	6.50	3.42	9.90
6050	4 passes	6.60	3.77	10.35
6100	8" lifts, 2 passes	6.45	3.24	9.70
6150	4 passes	6.50	3.42	9.90

G1030 Site Earthwork

The Excavation of Common Earth System balances the productivity of the excavating equipment to the hauling equipment. It is assumed that the hauling equipment will encounter light traffic and will move up no considerable grades on the haul route. No mobilization cost is included. All costs given in these systems include a swell factor of 25% for hauling.

The Expanded System Listing shows Excavation systems using backhoes ranging from 1/2 Cubic Yard capacity to 3-1/2 Cubic Yards. Power shovels indicated range from 1/2 Cubic Yard to 3 Cubic Yards. Dragline bucket rigs range from 1/2 Cubic Yard to 3 Cubic Yards. Truck capacities range from 8 Cubic Yards to 20 Cubic Yards. Each system lists the number of trucks involved and the distance (round trip) that each must travel.

System Components	QUANTITY	UNIT	COST PER C.Y. EQUIP.	COST PER C.Y. LABOR	COST PER C.Y. TOTAL
SYSTEM G1030 120 1000					
EXCAVATE COMMON EARTH, 1/2 CY BACKHOE, TWO 8 CY DUMP TRUCKS, 1 MRT					
Excavating, bulk hyd. backhoe wheel mtd., 1/2 C.Y.	.632	B.C.Y.	.98	2.40	3.38
Hauling, 8 CY truck, cycle 0.5 mile, 20 MPH, 15 min. wait/Ld./Uld.	1.280	L.C.Y.	2.60	3.24	5.84
Spotter at earth fill dump or in cut	.016	Hr.		1.02	1.02
TOTAL			3.58	6.66	10.24

G1030 120	Excavate and Haul Common Earth	COST PER C.Y. EQUIP.	COST PER C.Y. LABOR	COST PER C.Y. TOTAL
1000	Excavate common earth, 1/2 C.Y. backhoe, two 8 C.Y. dump trucks, 1 MRT	3.58	6.65	10.23
1200	Three 8 C.Y. dump trucks, 3 mile round trip	7.30	11.40	18.70
1400	Two 12 C.Y. dump trucks, 4 mile round trip	6.60	8.70	15.30
1600	3/4 C.Y. backhoe, three 8 C.Y. dump trucks, 1 mile round trip	3.60	5.55	9.15
1700	Five 8 C.Y. dump trucks, 3 mile round trip	7.20	10.60	17.80
1800	Two 12 C.Y. dump trucks, 2 mile round trip	5.60	6.75	12.35
1900	Two 16 C.Y. dump trailers, 3 mile round trip	5.35	5.75	11.10
2000	Two 20 C.Y. dump trailers, 4 mile round trip	5.10	5.65	10.75
2200	1-1/2 C.Y. backhoe, eight 8 C.Y. dump trucks, 3 mile round trip	6.85	9.45	16.30
2300	Four 12 C.Y. dump trucks, 2 mile round trip	5	5.80	10.80
2400	Six 12 C.Y. dump trucks, 4 mile round trip	6.10	6.70	12.80
2500	Three 16 C.Y. dump trailers, 2 mile round trip	4.12	4.28	8.40
2600	Two 20 C.Y. dump trailers, 1 mile round trip	3.25	3.50	6.75
2700	Three 20 C.Y. dump trailers, 3 mile round trip	4.35	4.39	8.74
2800	2-1/2 C.Y. excavator, six 12 C.Y. dump trucks, 1 mile round trip	3.69	3.92	7.61
2900	Eight 12 C.Y. dump trucks, 3 mile round trip	5.30	5.50	10.80
3000	Four 16 C.Y. dump trailers, 1 mile round trip	3.71	3.57	7.28
3100	Six 16 C.Y. dump trailers, 3 mile round trip	5	4.85	9.85
3200	Six 20 C.Y. dump trailers, 4 mile round trip	4.65	4.47	9.12
3400	3-1/2 C.Y. backhoe, six 16 C.Y. dump trailers, 1 mile round trip	3.99	3.42	7.41
3600	Ten 16 C.Y. dump trailers, 4 mile round trip	5.70	4.85	10.55
3800	Eight 20 C.Y. dump trailers, 3 mile round trip	4.58	3.84	8.42
4000	1/2 C.Y. pwr. shovel, four 8 C.Y. dump trucks, 2 mile round trip	6.30	8.55	14.85
4100	Two 12 C.Y. dump trucks, 1 mile round trip	4.44	5.20	9.64
4200	Four 12 C.Y. dump trucks, 4 mile round trip	6.65	7.10	13.75
4300	Two 16 C.Y. dump trailers, 2 mile round trip	4.74	5.10	9.84
4400	Two 20 C.Y. dump trailers, 4 mile round trip	5.45	5.85	11.30
4800	3/4 C.Y. pwr. shovel, six 8 C.Y. dump trucks, 2 mile round trip	6.15	8.30	14.45
4900	Three 12 C.Y. dump trucks, 1 mile round trip	4.28	4.51	8.79
5000	Five 12 C.Y. dump trucks, 4 mile round trip	6.70	6.80	13.50
5100	Three 16 C.Y. dump trailers, 3 mile round trip	5.65	5.50	11.15
5200	Three 20 C.Y. dump trailers, 4 mile round trip	5.30	5.10	10.40

G1030 Site Earthwork

G1030 120	Excavate and Haul Common Earth	COST PER C.Y.		
		EQUIP.	LABOR	TOTAL
5400	1-1/2 C.Y. pwr. shovel, six 12 C.Y. dump trucks, 1 mile round trip	3.84	3.91	7.75
5500	Ten 12 C.Y. dump trucks, 4 mile round trip	6.25	6.20	12.45
5600	Six 16 C.Y. dump trailers, 3 mile round trip	5.20	4.83	10.03
5700	Four 20 C.Y. dump trailers, 2 mile round trip	3.81	3.69	7.50
5800	Six 20 C.Y. dump trailers, 4 mile round trip	4.81	4.45	9.26
6000	3 C.Y. pwr. shovel, ten 12 C.Y. dump trucks, 1 mile round trip	3.87	3.58	7.45
6200	Twelve 16 C.Y. dump trailers, 4 mile round trip	5.55	4.73	10.28
6400	Eight 20 C.Y. dump trailers, 2 mile round trip	3.71	3.12	6.83
6600	1/2 C.Y. dragline bucket, three 8 C.Y. dump trucks, 2 mile round trip	8.75	11.15	19.90
6700	Two 12 C.Y. dump trucks, 3 mile round trip	8.70	9.55	18.25
6800	Three 12 C.Y. dump trucks, 4 mile round trip	8.70	9.20	17.90
7000	Two 16 C.Y. dump trailers, 4 mile round trip	8.20	8.45	16.65
7200	3/4 C.Y. dragline bucket, four 8 C.Y. dump trucks, 1 mile round trip	4.83	6.15	10.98
7300	Six 8 C.Y. dump trucks, 3 mile round trip	8.60	11.35	19.95
7400	Two 12 C.Y. dump trucks, 1 mile round trip	5.90	6.55	12.45
7500	Four 12 C.Y. dump trucks, 4 mile round trip	7.90	8.20	16.10
7600	Two 16 C.Y. dump trailers, 2 mile round trip	6.15	6.35	12.50
7800	1-1/2 C.Y. dragline bucket, four 12 C.Y. dump trucks, 1 mile round trip	4.88	4.76	9.64
7900	Six 12 C.Y. dump trucks, 3 mile round trip	6.30	6.35	12.65
8000	Four 16 C.Y. dump trucks, 2 mile round trip	5.15	4.72	9.87
8200	Five 16 C.Y. dump trucks, 4 mile round trip	6.55	5.90	12.45
8400	3 C.Y. dragline bucket, eight 12 C.Y. dump trucks, 2 mile round trip	6.15	5.45	11.60
8500	Twelve 12 C.Y. dump trucks, 4 mile round trip	7.15	6.35	13.50
8600	Eight 16 C.Y. dump trailers, 3 mile round trip	6	4.91	10.91
8700	Six 20 C.Y. dump trailers, 2 mile round trip	4.51	3.62	8.13

G1030 Site Earthwork

The Cut and Fill Clay System includes: moving clay from an area above the specified grade to an area below the specified grade utilizing a bulldozer and compaction equipment; plus a water wagon to adjust the moisture content of the soil.

The Expanded System Listing shows Cut and Fill operations with hauling distances that vary from 50' to 5000'. Lifts for compaction in the filled areas vary from 4" to 8". There is no waste included in the assumptions.

System Components	QUANTITY	UNIT	COST PER C.Y.		
			EQUIP.	LABOR	TOTAL
SYSTEM G1030 125 1000					
CLAY CUT & FILL, 80 HP DOZER & COMPACTOR, 50' HAUL, 4" LIFTS, 2 PASSES					
Excavating , bulk, dozer, 50' haul, clay	1.000	C.Y.	1.77	3.73	5.50
Water wagon, rent per day	.004	Hr.	.25	.29	.54
Backfill dozer, from existing, stock pile, 75 H.P., 50' haul, clay	1.330	C.Y.	.69	1.46	2.15
Compaction, tamper, 4" lifts, 2 passes	245.000	S.F.	.62	.39	1.01
TOTAL			3.33	5.87	9.20

G1030 125	Cut & Fill Clay	COST PER C.Y.		
		EQUIP.	LABOR	TOTAL
1000	Clay cut & fill, 80 HP dozer & compactor, 50' haul, 4" lifts, 2 passes	3.33	5.85	9.18
1050	4 passes	4.10	6.50	10.60
1100	8" lifts, 2 passes	2.88	5.50	8.38
1150	4 passes	3.28	5.85	9.13
1200	150' haul, 4" lifts, 2 passes	5.80	11.05	16.85
1250	4 passes	6.55	11.65	18.20
1300	8" lifts, 2 passes	5.35	10.70	16.05
1350	4 passes	5.75	11	16.75
1400	300' haul, 4" lifts, 2 passes	9.70	19.30	29
1450	4 passes	10.45	19.95	30.40
1500	8" lifts, 2 passes	9.25	18.95	28.20
1550	4 passes	9.65	19.25	28.90
1600	105 HP dozer & sheepsfoot compactors, 50' haul, 4" lifts, 2 passes	3.52	4.23	7.75
1650	4 passes	4.29	4.86	9.15
1700	8" lifts, 2 passes	3.07	3.88	6.95
1750	4 passes	3.47	4.20	7.67
1800	150' haul, 4" lifts, 2 passes	6.65	8.45	15.10
1850	4 passes	7.45	9.05	16.50
1900	8" lifts, 2 passes	6.20	8.10	14.30
1950	4 passes	6.60	8.40	15
2000	300' haul, 4" lifts, 2 passes	10.30	13.35	23.65
2050	4 passes	11.10	13.95	25.05
2100	8" lifts, 2 passes	9.85	13	22.85
2150	4 passes	10.25	13.30	23.55
2200	200 HP dozer & sheepsfoot compactors, 150' haul, 4" lifts, 2 passes	8.25	4.83	13.08
2250	4 passes	9	5.45	14.45
2300	8" lifts, 2 passes	7.80	4.48	12.28
2350	4 passes	8.20	4.80	13
2600	300' haul, 4" lifts, 2 passes	13.95	8.05	22
2650	4 passes	14.75	8.70	23.45

G1030 Site Earthwork

G1030 125	Cut & Fill Clay	COST PER C.Y.		
		EQUIP.	LABOR	TOTAL
2700	8" lifts, 2 passes	13.50	7.70	21.20
2750	4 passes	13.90	8.05	21.95
3000	300 HP dozer & sheepsfoot compactors, 150' haul, 4" lifts, 2 passes	6.25	3.27	9.52
3050	4 passes	7.05	3.90	10.95
3100	8" lifts, 2 passes	5.80	2.92	8.72
3150	4 passes	6.20	3.24	9.44
3200	300' haul, 4" lifts, 2 passes	10.70	5.40	16.10
3250	4 passes	11.50	6.05	17.55
3300	8" lifts, 2 passes	10.25	5.05	15.30
3350	4 passes	10.65	5.40	16.05
4200	10 C.Y. elev. scraper & sheepsfoot rollers, 1500' haul, 4" lifts, 2 passes	10.70	6.80	17.50
4250	4 passes	10.95	7.10	18.05
4300	8" lifts, 2 passes	10.65	6.65	17.30
4350	4 passes	10.70	6.80	17.50
4800	5000' haul, 4" lifts, 2 passes	14.50	9.15	23.65
4850	4 passes	14.75	9.45	24.20
4900	8" lifts, 2 passes	14.45	9	23.45
4950	4 passes	14.50	9.15	23.65
5000	15 C.Y. SP scraper & sheepsfoot rollers, 1500' haul, 4" lifts, 2 passes	10.15	4.69	14.84
5050	4 passes	10.35	4.98	15.33
5100	8" lifts, 2 passes	10.05	4.55	14.60
5150	4 passes	10.15	4.69	14.84
5400	5000' haul, 4" lifts, 2 passes	14.45	6.65	21.10
5450	4 passes	14.65	6.95	21.60
5500	8" lifts, 2 passes	14.35	6.50	20.85
5550	4 passes	14.40	6.65	21.05
5600	21 C.Y. SP scraper & sheepsfoot rollers, 1500' haul, 4" lifts, 2 passes	8.35	3.70	12.05
5650	4 passes	8.55	4	12.55
5700	8" lifts, 2 passes	8.25	3.56	11.81
5750	4 passes	8.30	3.70	12
6000	5000' haul, 4" lifts, 2 passes	13.10	5.75	18.85
6050	4 passes	13.35	6.05	19.40
6100	8" lifts, 2 passes	13	5.60	18.60
6150	4 passes	13.10	5.75	18.85

G1030   Site Earthwork

The Excavation of Clay System balances the productivity of excavating equipment to hauling equipment. It is assumed that the hauling equipment will encounter light traffic and will move up no considerable grades on the haul route. No mobilization cost is included. All costs given in these systems include a swell factor of 40%.

The Expanded System Listing shows Excavation systems using backhoes ranging from 1/2 Cubic Yard capacity to 3-1/2 Cubic Yards. Power shovels indicated range from 1/2 Cubic Yard to 3 Cubic Yards. Dragline bucket rigs range from 1/2 Cubic Yard to 3 Cubic Yards. Truck capacities range from 8 Cubic Yards to 20 Cubic Yards. Each system lists the number of trucks involved and the distance (round trip) that each must travel.

System Components	QUANTITY	UNIT	COST PER C.Y.		
			EQUIP.	LABOR	TOTAL
SYSTEM G1030 130 1000					
EXCAVATE CLAY, 1/2 CY BACKHOE, TWO 8 CY DUMP TRUCKS, 2 MI ROUND TRIP					
Excavating bulk hyd. backhoe, wheel mtd. 1/2 C.Y.	1.000	B.C.Y.	1.55	3.79	5.34
Haul earth, 8 C.Y. dump truck, 2 mile round trip, 2.6 loads/hr	1.350	L.C.Y.	5.29	6.60	11.89
Spotter at earth fill dump or in cut	.023	Hr.		1.46	1.46
TOTAL			6.84	11.85	18.69

G1030 130	Excavate and Haul Clay	COST PER C.Y.		
		EQUIP.	LABOR	TOTAL
1000	Excavate clay, 1/2 C.Y. backhoe, two 8 C.Y. dump trucks, 2 mile round trip	6.85	11.85	18.70
1200	Three 8 C.Y. dump trucks, 4 mile round trip	9.30	14.60	23.90
1400	Two 12 C.Y. dump trucks, 4 mile round trip	7.05	9.40	16.45
1600	3/4 C.Y. backhoe, three 8 C.Y. dump trucks, 2 mile round trip	6.70	9.80	16.50
1700	Five 8 C.Y. dump trucks, 4 mile round trip	9.20	12.70	21.90
1800	Two 12 C.Y. dump trucks, 2 mile round trip	6	7.25	13.25
1900	Three 12 C.Y. dump trucks, 4 mile round trip	7.05	8.10	15.15
2000	1-1/2 C.Y. backhoe, eight 8 C.Y. dump trucks, 3 mile round trip	7.25	9.70	16.95
2100	Three 12 C.Y. dump trucks, 1 mile round trip	4.04	4.64	8.68
2200	Six 12 C.Y. dump trucks, 4 mile round trip	6.45	7.15	13.60
2300	Three 16 C.Y. dump trailers, 2 mile round trip	4.38	4.62	9
2400	Four 16 C.Y. dump trailers, 4 mile round trip	5.75	5.75	11.50
2600	Two 20 C.Y. dump trailers, 1 mile round trip	3.46	3.80	7.26
2800	2-1/2 C.Y. backhoe, eight 12 C.Y. dump trucks, 3 mile round trip	5.65	5.90	11.55
2900	Four 16 C.Y. dump trailers, 1 mile round trip	3.95	3.78	7.73
3000	Six 16 C.Y. dump trailers, 3 mile round trip	5.30	5.20	10.50
3100	Four 20 C.Y. dump trailers, 2 mile round trip	3.79	3.84	7.63
3200	Six 20 C.Y. dump trailers, 4 mile round trip	4.94	4.79	9.73
3400	3-1/2 C.Y. backhoe, ten 12 C.Y. dump trucks, 2 mile round trip	5.65	5.20	10.85
3500	Six 16 C.Y. dump trailers, 1 mile round trip	4.26	3.61	7.87
3600	Ten 16 C.Y. dump trailers, 4 mile round trip	6.05	5.15	11.20
3800	Eight 20 C.Y. dump trailers, 3 mile round trip	4.88	4.06	8.94
4000	1/2 C.Y. pwr. shovel, three 8 C.Y. dump trucks, 1 mile round trip	6.10	8.40	14.50
4100	Six 8 C.Y. dump trucks, 4 mile round trip	9.35	13.10	22.45
4200	Two 12 C.Y. dump trucks, 1 mile round trip	4.77	5.65	10.42
4400	Three 12 C.Y. dump trucks, 3 mile round trip	6.40	7.05	13.45
4600	3/4 C.Y. pwr. shovel, six 8 C.Y. dump trucks, 2 mile round trip	6.55	8.90	15.45
4700	Nine 8 C.Y. dump trucks, 4 mile round trip	9.15	12.10	21.25
4800	Three 12 C.Y. dump trucks, 1 mile round trip	4.58	4.88	9.46
4900	Four 12 C.Y. dump trucks, 3 mile round trip	6.35	6.60	12.95
5000	Two 16 C.Y. dump trailers, 1 mile round trip	4.69	4.79	9.48
5200	Three 16 C.Y. dump trailers, 3 mile round trip	6.05	5.90	11.95

G1030 Site Earthwork

G1030 130	Excavate and Haul Clay	COST PER C.Y.		
		EQUIP.	LABOR	TOTAL
5400	1-1/2 C.Y. pwr. shovel, six 12 C.Y. dump trucks, 2 mile round trip	5.60	5.70	11.30
5500	Four 16 C.Y. dump trailers, 1 mile round trip	4.14	3.77	7.91
5600	Seven 16 C.Y. dump trailers, 4 mile round trip	5.90	5.50	11.40
5800	Five 20 C.Y. dump trailers, 3 mile round trip	4.82	4.24	9.06
6000	3 C.Y. pwr. shovel, nine 16 C.Y. dump trailers, 2 mile round trip	4.55	3.86	8.41
6100	Twelve 16 C.Y. dump trailers, 4 mile round trip	5.90	5	10.90
6200	Six 20 C.Y. dump trailers, 1 mile round trip	3.63	3.09	6.72
6400	Nine 20 C.Y. dump trailers, 3 mile round trip	4.80	4.03	8.83
6600	1/2 C.Y. dragline bucket, two 8 C.Y. dump trucks, 1 mile round trip	9.65	12.30	21.95
6800	Four 8 C.Y. dump trucks, 4 mile round trip	12.50	15.85	28.35
7000	Two 12 C.Y. dump trucks, 3 mile round trip	9.40	10.30	19.70
7200	3/4 C.Y. dragline bucket, four 8 C.Y. dump trucks, 2 mile round trip	8.45	10.70	19.15
7300	Six 8 C.Y. dump trucks, 4 mile round trip	11	14.60	25.60
7400	Two 12 C.Y. dump trucks, 2 mile round trip	8.50	9.30	17.80
7500	Three 12 C.Y. dump trucks, 4 mile round trip	9.30	9.95	19.25
7600	Two 16 C.Y. dump trailers, 3 mile round trip	8.05	8.35	16.40
7800	1-1/2 C.Y. dragline bucket, five 12 C.Y. dump trucks, 3 mile round trip	7.20	6.95	14.15
7900	Three 16 C.Y. dump trailers, 1 mile round trip	5.25	4.79	10.04
8000	Five 16 C.Y. dump trailers, 4 mile round trip	7	6.30	13.30
8100	Three 20 C.Y. dump trailers, 2 mile round trip	5.25	4.80	10.05
8400	3 C.Y. dragline bucket, eight 16 C.Y. dump trailers, 3 mile round trip	6.15	5.10	11.25
8500	Six 20 C.Y. dump trailers, 2 mile round trip	4.86	3.92	8.78
8600	Eight 20 C.Y. dump trailers, 4 mile round trip	6	4.87	10.87

G10 Site Preparation

G1030 Site Earthwork

The Loading and Hauling of Rock System balances the productivity of loading equipment to hauling equipment. It is assumed that the hauling equipment will encounter light traffic and will move up no considerable grades on the haul route.

The Expanded System Listing shows Loading and Hauling systems that use either a track or wheel front-end loader. Track loaders indicated range from 1-1/2 Cubic Yards capacity to 4-1/2 Cubic Yards capacity. Wheel loaders range from 1-1/2 Cubic Yards to 5 Cubic Yards. Trucks for hauling range from 8 Cubic Yards capacity to 20 Cubic Yards capacity. Each system lists the number of trucks involved and the distance (round trip) that each must travel.

System Components	QUANTITY	UNIT	COST PER C.Y.		
			EQUIP.	LABOR	TOTAL
SYSTEM G1030 150 1000					
LOAD & HAUL ROCK, 1-1/2 C.Y. TRACK LOADER, SIX 8 C.Y. TRUCKS, 1 MRT					
Excavating bulk, F.E. loader, track mtd., 1.5 C.Y.	1.000	B.C.Y.	1.11	1.66	2.77
8 C.Y. truck, cycle 2 miles	1.650	L.C.Y.	5.71	7.11	12.82
Spotter at earth fill dump or in cut	.014	Hr.		.89	.89
TOTAL			6.82	9.66	16.48

G1030 150	Load & Haul Rock	COST PER C.Y.		
		EQUIP.	LABOR	TOTAL
1000	Load & haul rock, 1-1/2 C.Y. track loader, six 8 C.Y. trucks, 1 MRT	6.80	9.65	16.45
1200	Nine 8 C.Y. dump trucks, 3 mile round trip	9.20	12.70	21.90
1400	Six 12 C.Y. dump trucks, 4 mile round trip	8.35	9.30	17.65
1600	Three 16 C.Y. dump trucks, 2 mile round trip	5.85	6.25	12.10
2000	2-1/2 C.Y. track loader, twelve 8 C.Y. dump trucks, 3 mile round trip	9.55	12.05	21.60
2200	Five 12 C.Y. dump trucks, 1 mile round trip	5.50	5.35	10.85
2400	Eight 12 C.Y. dump trucks, 4 mile round trip	8.65	8.70	17.35
2600	Four 16 C.Y. dump trailers, 2 mile round trip	6.15	5.70	11.85
3000	3-1/2 C.Y. track loader, eight 12 C.Y. dump trucks, 2 mile round trip	7.05	6.80	13.85
3200	Five 16 C.Y. dump trucks, 1 mile round trip	5.30	4.60	9.90
3400	Seven 16 C.Y. dump trailers, 3 mile round trip	7.05	6.30	13.35
3600	Seven 20 C.Y. dump trailers, 4 mile round trip	6.60	5.85	12.45
4000	4-1/2 C.Y. track loader, nine 12 C.Y. dump trucks, 1 mile round trip	5.30	4.88	10.18
4200	Eight 16 C.Y. dump trailers, 2 mile round trip	5.80	4.98	10.78
4400	Eleven 16 C.Y. dump trailers, 4 mile round trip	7.45	6.35	13.80
4600	Seven 20 C.Y. dump trailers, 2 mile round trip	5.10	4.30	9.40
5000	1-1/2 C.Y. wheel loader, nine 8 C.Y. dump trucks, 2 mile round trip	7.15	10.20	17.35
5200	Four 12 C.Y. dump trucks, 1 mile round trip	4.87	5.70	10.57
5400	Seven 12 C.Y. dump trucks, 4 mile round trip	7.95	8.95	16.90
5600	Five 16 C.Y. dump trailers, 4 mile round trip	7	7.15	14.15
6000	3 C.Y. wheel loader, eight 12 C.Y. dump trucks, 2 mile round trip	6.30	6.55	12.85
6200	Five 16 C.Y. dump trailers, 1 mile round trip	4.61	4.34	8.95
6400	Seven 16 C.Y. dump trailers, 3 mile round trip	5.50	5.35	10.85
6600	Seven 20 C.Y. dump trailers, 4 mile round trip	5.85	5.55	11.40
7000	5 C.Y. wheel loader, twelve 12 C.Y. dump trucks, 1 mile round trip	4.69	4.59	9.28
7200	Nine 16 C.Y. dump trailers, 1 mile round trip	4.64	4.20	8.84
7400	Eight 20 C.Y. dump trailers, 1 mile round trip	3.99	3.58	7.57
7600	Twelve 20 C.Y. dump trailers, 3 mile round trip	5.40	4.75	10.15

G1030 Site Earthwork

The Excavation of Sandy Clay/Loam System balances the productivity of the excavating equipment to the hauling equipment. It is assumed that the hauling equipment will encounter light traffic and will move up no considerable grades on the haul route. No mobilization cost is included. All costs given in these systems include a swell factor of 25% for hauling.

The Expanded System Listing shows Excavation systems using backhoes ranging from 1/2 Cubic Yard capacity to 3-1/2 Cubic Yards. Truck capacities range from 8 Cubic Yards to 20 Cubic Yards. Each system lists the number of trucks involved and the distance (round trip) that each must travel.

System Components	QUANTITY	UNIT	COST PER C.Y. EQUIP.	COST PER C.Y. LABOR	COST PER C.Y. TOTAL
SYSTEM G1030 160 1000					
EXCAVATE SANDY CLAY/LOAM, 1/2 CY BACKHOE, TWO 8 CY DUMP TRUCKS, 1 MRT					
Excavating, bulk hyd. backhoe wheel mtd., 1/2 C.Y.	1.000	B.C.Y.	1.41	3.45	4.86
8 C.Y. truck, cycle 2 miles	1.300	L.C.Y.	4.50	5.60	10.10
Spotter at earth fill dump or in cut	.016	Hr.		1.02	1.02
TOTAL			5.91	10.07	15.98

G1030 160	Excavate and Haul Sandy Clay/Loam	COST PER C.Y. EQUIP.	COST PER C.Y. LABOR	COST PER C.Y. TOTAL
1000	Excavate sandy clay/loam, 1/2 C.Y. backhoe, two 8 C.Y. dump trucks, 1 MRT	5.90	10.05	15.95
1200	Three 8 C.Y. dump trucks, 3 mile round trip	7.30	11.30	18.60
1400	Two 12 C.Y. dump trucks, 4 mile round trip	6.60	8.55	15.15
1600	3/4 C.Y. backhoe, three 8 C.Y. dump trucks, 1 mile round trip	5.40	7.75	13.15
1700	Five 8 C.Y. dump trucks, 3 mile round trip	7.25	10.60	17.85
1800	Two 12 C.Y. dump trucks, 2 mile round trip	5.60	6.65	12.25
1900	Two 16 C.Y. dump trailers, 3 mile round trip	5.35	5.65	11
2000	Two 20 C.Y. dump trailers, 4 mile round trip	5.10	5.55	10.65
2200	1-1/2 C.Y. excavator, eight 8 C.Y. dump trucks, 3 mile round trip	6.85	9.45	16.30
2300	Four 12 C.Y. dump trucks, 2 mile round trip	4.99	5.70	10.70
2400	Six 12 C.Y. dump trucks, 4 mile round trip	6.10	6.65	12.75
2500	Three 16 C.Y. dump trailers, 2 mile round trip	4.08	4.19	8.25
2600	Two 20 C.Y. dump trailers, 1 mile round trip	3.19	3.39	6.60
2700	Three 20 C.Y. dump trailers, 3 mile round trip	4.31	4.30	8.60
2800	2-1/2 C.Y. excavator, six 12 C.Y. dump trucks, 1 mile round trip	3.57	3.79	7.35
2900	Eight 12 C.Y. dump trucks, 3 mile round trip	5.20	5.40	10.60
3000	Four 16 C.Y. dump trailers, 1 mile round trip	3.58	3.40	7
3100	Six 16 C.Y. dump trailers, 3 mile round trip	4.90	4.70	9.60
3200	Six 20 C.Y. dump trailers, 4 mile round trip	4.54	4.32	8.85
3400	3-1/2 C.Y. excavator, six 16 C.Y. dump trailers, 1 mile round trip	3.83	3.35	7.20
3600	Ten 16 C.Y. dump trailers, 4 mile round trip	5.55	4.82	10.35
3800	Eight 20 C.Y. dump trailers, 3 mile round trip	4.44	3.79	8.25

G1030 Site Earthwork

The Gravel Backfilling System includes: a bulldozer to place and level backfill in specified lifts; compaction equipment; and a water wagon for adjusting moisture content.

The Expanded System Listing shows Gravel Backfilling operations with bulldozers ranging from 75 H.P. to 300 H.P. The maximum hauling distance ranges from 50' to 300'. Lifts for the compaction range from 6" to 12". There is no waste included in the assumptions.

System Components	QUANTITY	UNIT	COST PER C.Y. EQUIP.	LABOR	TOTAL
SYSTEM G1030 205 1000					
GRAVEL BACKFILL, 80 HP DOZER & COMPACTORS, 50' HAUL, 6" LIFTS, 2 PASSES					
Backfilling, dozer, 80 H.P., 50' haul, sand and gravel, from stockpile	1.000	L.C.Y.	.40	.85	1.25
Water wagon rent per day	.003	Hr.	.19	.22	.41
Compaction, vibrating roller, 6" lifts, 2 passes	1.000	Hr.	.17	2.53	2.70
TOTAL			.76	3.60	4.36

G1030 205	Gravel Backfill	COST PER C.Y. EQUIP.	LABOR	TOTAL
1000	Gravel backfill, 80 HP dozer & compactors, 50' haul, 6"lifts, 2 passes	.76	3.60	4.36
1050	4 passes	.97	4.83	5.80
1100	12" lifts, 2 passes	.79	1.82	2.61
1150	4 passes	1.24	2.88	4.12
1200	150' haul, 6" lifts, 2 passes	1.16	4.44	5.60
1250	4 passes	1.37	5.65	7.02
1300	12" lifts, 2 passes	1.03	4.29	5.32
1350	4 passes	1.24	5.50	6.74
1400	300' haul, 6" lifts, 2 passes	1.55	5.25	6.80
1450	4 passes	1.76	6.50	8.26
1500	12" lifts, 2 passes	1.58	3.49	5.07
1550	4 passes	2.03	4.55	6.58
1600	105 HP dozer & vibrating compactors, 50' haul, 6" lifts, 2 passes	.88	3.44	4.32
1650	4 passes	1.09	4.67	5.76
1700	12" lifts, 2 passes	.91	1.66	2.57
1750	4 passes	1.36	2.72	4.08
1800	150' haul, 6" lifts, 2 passes	1.40	4.14	5.54
1850	4 passes	1.61	5.35	6.96
1900	12" lifts, 2 passes	1.43	2.36	3.79
1950	4 passes	1.88	3.42	5.30
2000	300' haul, 6" lifts, 2 passes	1.86	4.75	6.61
2050	4 passes	2.07	6	8.07
2100	12" lifts, 2 passes	1.89	2.97	4.86
2150	4 passes	2.34	4.03	6.37
2200	200 HP dozer & roller compactors, 150' haul, 6" lifts, 2 passes	1.75	1.29	3.04
2250	4 passes	2.01	1.62	3.63
2300	12" lifts, 2 passes	1.53	1.01	2.54
2350	4 passes	1.79	1.34	3.13
2600	300' haul, 6" lifts, 2 passes	2.46	1.69	4.15
2650	4 passes	2.72	2.02	4.74
2700	12" lifts, 2 passes	2.24	1.41	3.65
2750	4 passes	2.50	1.74	4.24

G1030 Site Earthwork

G1030 205	Gravel Backfill	COST PER C.Y.		
		EQUIP.	LABOR	TOTAL
3000	300 HP dozer & roller compactors, 150' haul, 6" lifts, 2 passes	1.28	.95	2.23
3050	4 passes	1.54	1.28	2.82
3100	12" lifts, 2 passes	1.06	.67	1.73
3150	4 passes	1.32	1	2.32
3200	300' haul, 6" lifts, 2 passes	1.69	1.15	2.84
3250	4 passes	1.95	1.48	3.43
3300	12" lifts, 2 passes	1.47	.87	2.34
3350	4 passes	1.73	1.20	2.93

G1030 Site Earthwork

The Common Earth Backfilling System includes: a bulldozer to place and level backfill in specified lifts; compaction equipment; and a water wagon for adjusting moisture content.

The Expanded System Listing shows Common Earth Backfilling with bulldozers ranging from 80 H.P. to 300 H.P. The maximum distance ranges from 50′ to 300′. Lifts for the compaction range from 4″ to 8″. There is no waste included in the assumptions.

System Components	QUANTITY	UNIT	COST PER C.Y.		
			EQUIP.	LABOR	TOTAL
SYSTEM G1030 210 1000					
EARTH BACKFILL, 80 HP DOZER & ROLLER , 50′ HAUL, 4″ LIFTS, 2 PASSES					
Backfilling, dozer 80 H.P. 50′ haul, common earth, from stockpile	1.000	L.C.Y.	.45	.96	1.41
Water wagon, rent per day	.004	Hr.	.25	.29	.54
Compaction, roller, 4″ lifts, 2 passes	.035	Hr.	1.23	4.06	5.29
TOTAL			1.93	5.31	7.24

G1030 210	Common Earth Backfill	COST PER C.Y.		
		EQUIP.	LABOR	TOTAL
1000	Earth backfill, 80 HP dozer & roller compactors, 50′ haul, 4″ lifts, 2 passes	1.93	5.30	7.23
1050	4 passes	3.15	9.35	12.50
1100	8″ lifts, 2 passes	1.33	3.34	4.67
1150	4 passes	1.93	5.30	7.23
1200	150′ haul, 4″ lifts, 2 passes	2.38	6.25	8.63
1250	4 passes	3.60	10.30	13.90
1300	8″ lifts, 2 passes	1.78	4.28	6.06
1350	4 passes	2.38	6.25	8.63
1400	300′ haul, 4″ lifts, 2 passes	2.82	7.15	9.97
1450	4 passes	4.04	11.25	15.29
1500	8″ lifts, 2 passes	2.22	5.20	7.42
1550	4 passes	2.82	7.15	9.97
1600	105 HP dozer & roller compactors, 50′ haul, 4″ lifts, 2 passes	2.05	5.10	7.15
1650	4 passes	3.27	9.15	12.42
1700	8″ lifts, 2 passes	1.45	3.14	4.59
1750	4 passes	2.05	5.10	7.15
1800	150′ haul, 4″ lifts, 2 passes	2.62	5.90	8.52
1850	4 passes	3.84	9.95	13.79
1900	8″ lifts, 2 passes	2.02	3.91	5.93
1950	4 passes	2.62	5.90	8.52
2000	300′ haul, 4″ lifts, 2 passes	3.16	6.60	9.76
2050	4 passes	4.38	10.65	15.03
2100	8″ lifts, 2 passes	2.56	4.63	7.19
2150	4 passes	3.16	6.60	9.76
2200	200 HP dozer & roller compactors, 150′ haul, 4″ lifts, 2 passes	2.40	3.46	5.86
2250	4 passes	3.05	5.80	8.85
2300	8″ lifts, 2 passes	2.08	2.30	4.38
2350	4 passes	2.40	3.46	5.86
2600	300′ haul, 4″ lifts, 2 passes	3.15	3.88	7.03
2650	4 passes	3.80	6.20	10
2700	8″ lifts, 2 passes	2.83	2.72	5.55
2750	4 passes	3.15	3.88	7.03
3000	300 HP dozer & roller compactors, 150′ haul, 4″ lift, 2 passes	1.89	3.09	4.98
3050	4 passes	2.54	5.40	7.94

G10 Site Preparation

G1030 Site Earthwork

G1030 210	Common Earth Backfill	COST PER C.Y.		
		EQUIP.	LABOR	TOTAL
3100	8" lifts, 2 passes	1.57	1.93	3.50
3150	4 passes	1.89	3.09	4.98
3200	300' haul, 4" lifts, 2 passes	2.34	3.30	5.64
3250	4 passes	2.99	5.60	8.59
3300	8" lifts, 2 passes	2.02	2.14	4.16
3350	4 passes	2.34	3.30	5.64

For customer support on your Site Work & Landscape Costs with RSMeans data, call 800.448.8182.

G10 Site Preparation

G1030 Site Earthwork

The Clay Backfilling System includes: a bulldozer to place and level backfill in specified lifts; compaction equipment; and a water wagon for adjusting moisture content.

The Expanded System Listing shows Clay Backfilling with bulldozers ranging from 75 H.P. to 300 H.P. The maximum distance ranges from 50' to 300'. Lifts for the compaction range from 4" to 8". There is no waste included in the assumptions.

System Components	QUANTITY	UNIT	COST PER C.Y.		
			EQUIP.	LABOR	TOTAL
SYSTEM G1030 215 1000					
CLAY BACKFILL, 80 HP DOZER & TAMPER, 50' HAUL, 4" LIFTS, 2 PASSES					
Backfilling, dozer 80 H.P. 50' haul, clay from stockpile	1.000	L.C.Y.	.52	1.10	1.62
Water wagon rent per day	.004	Hr.	.25	.29	.54
Compaction, tamper, 4" lifts, 2 passes	1.000	Hr.	.15	1.30	1.45
TOTAL			.92	2.69	3.61

G1030 215	Clay Backfill	COST PER C.Y.		
		EQUIP.	LABOR	TOTAL
1000	Clay backfill, 80 HP dozer & tamper compactor, 50' haul, 4" lifts, 2 passes	.92	2.69	3.61
1050	4 passes	1.07	3.99	5.06
1100	8" lifts, 2 passes	.85	2.04	2.89
1150	4 passes	.92	2.69	3.61
1200	150' haul, 4" lifts, 2 passes	1.44	3.78	5.22
1250	4 passes	1.59	5.10	6.69
1300	8" lifts, 2 passes	1.37	3.13	4.50
1350	4 passes	1.44	3.78	5.22
1400	300' haul, 4" lifts, 2 passes	1.92	4.80	6.72
1450	4 passes	2.07	6.10	8.17
1500	8" lifts, 2 passes	1.85	4.15	6
1550	4 Passes	1.92	4.80	6.72
1600	105 HP dozer & tamper compactors, 50' haul, 4" lifts, 2 passes	1.03	2.44	3.47
1650	4 passes	1.18	3.74	4.92
1700	8" lifts, 2 passes	.96	1.79	2.75
1750	4 passes	1.03	2.44	3.47
1800	150' haul, 4" lifts, 2 passes	1.67	3.28	4.95
1850	4 passes	1.82	4.58	6.40
1900	8" lifts, 2 passes	1.60	2.63	4.23
1950	4 passes	1.67	3.28	4.95
2000	300' haul, 4" lifts, 2 passes	2.28	4.11	6.39
2050	4 passes	2.43	5.40	7.83
2100	8" lifts, 2 passes	2.21	3.46	5.67
2150	4 passes	2.28	4.11	6.39
2200	200 HP dozer & sheepsfoot compactors, 150' haul, 5" lifts, 2 passes	2.57	1.64	4.21
2250	4 passes	3.09	1.97	5.06
2300	8" lifts, 2 passes	2.24	1.43	3.67
2350	4 passes	2.52	1.61	4.13
2600	300' haul, 4" lifts, 2 passes	3.38	2.09	5.47
2650	4 passes	3.90	2.42	6.32
2700	8" lifts, 2 passes	3.05	1.88	4.93
2750	4 passes	3.33	2.06	5.39

G10 Site Preparation

G1030 Site Earthwork

G1030 215	Clay Backfill	COST PER C.Y.		
		EQUIP.	LABOR	TOTAL
3000	300 HP dozer & sheepsfoot compactors, 150' haul, 4" lifts, 2 passes	2.01	1.23	3.24
3050	4 passes	2.53	1.56	4.09
3100	8" lifts, 2 passes	1.68	1.02	2.70
3150	4 passes	1.96	1.20	3.16
3200	300' haul, 4" lifts, 2 passes	2.45	1.44	3.89
3250	4 passes	2.97	1.77	4.74
3300	8" lifts, 2 passes	2.12	1.23	3.35
3350	4 passes	2.40	1.41	3.81

G10 Site Preparation

G1030 Site Earthwork

The Sandy Clay/Loam Backfilling System includes: a bulldozer to place and level backfill in specified lifts; compaction equipment; and a water wagon for adjusting moisture content.

The Expanded System Listing shows Common Earth Backfilling with bulldozers ranging from 80 H.P. to 300 H.P. The maximum distance ranges from 50' to 300'. Lifts for the compaction range from 4" to 8". There is no waste included in the assumptions.

System Components	QUANTITY	UNIT	COST PER C.Y.		
			EQUIP.	LABOR	TOTAL
SYSTEM G1030 220 1000					
SANDY CLAY BACKFILL, 80 HP DOZER & ROLLER , 50' HAUL, 4"LIFTS, 2 PASSES					
Backfilling, dozer 80 H.P. 50' haul, common earth, from stockpile	1.000	L.C.Y.	.41	.87	1.28
Water wagon, rent per day	.004	Hr.	.25	.29	.54
Compaction, roller, 4" lifts, 2 passes	.035	Hr.	1.23	4.06	5.29
TOTAL			1.89	5.22	7.11

G1030 220	Sandy Clay/Loam Backfill	COST PER C.Y.		
		EQUIP.	LABOR	TOTAL
0900	Backfill Sandy Clay/Loam using dozer and compactors			
1000	Backfill, 80 HP dozer & roller compactors, 50' haul, 4" lifts, 2 passes	1.89	5.20	7.10
1050	4 passes	3.11	9.30	12.40
1100	8" lifts, 2 passes	1.29	3.25	4.54
1150	4 passes	1.89	5.20	7.10
1200	150' haul, 4" lifts, 2 passes	2.31	6.10	8.40
1250	4 passes	3.53	10.15	13.70
1300	8" lifts, 2 passes	1.71	4.12	5.85
1350	4 passes	2.31	6.10	8.40
1400	300' haul, 4" lifts, 2 passes	2.71	6.95	9.65
1450	4 passes	3.93	11	14.95
1500	8" lifts, 2 passes	2.11	4.97	7.10
1550	4 passes	2.71	6.95	9.65
1600	105 HP dozer & roller compactors, 50' haul, 4" lifts, 2 passes	2.01	5.05	7.05
1650	4 passes	3.23	9.10	12.35
1700	8" lifts, 2 passes	1.41	3.08	4.49
1750	4 passes	2.01	5.05	7.05
1800	150' haul, 4" lifts, 2 passes	2.54	5.75	8.30
1850	4 passes	3.76	9.85	13.60
1900	8" lifts, 2 passes	1.94	3.80	5.75
1950	4 passes	2.54	5.75	8.30
2000	300' haul, 4" lifts, 2 passes	3.01	6.40	9.40
2050	4 passes	4.23	10.45	14.70
2100	8" lifts, 2 passes	2.41	4.43	6.85
2150	4 passes	3.01	6.40	9.40
2200	200 HP dozer & roller compactors, 150' haul, 4" lifts, 2 passes	2.28	3.39	5.65
2250	4 passes	2.93	5.70	8.65
2300	8" lifts, 2 passes	1.96	2.23	4.19
2350	4 passes	2.28	3.39	5.65
2600	300' haul, 4" lifts, 2 passes	2.99	3.79	6.80
2650	4 passes	3.64	6.10	9.75
2700	8" lifts, 2 passes	2.67	2.63	5.30

For customer support on your Site Work & Landscape Costs with RSMeans data, call 800.448.8182.

G1030 Site Earthwork

G1030 220	Sandy Clay/Loam Backfill	COST PER C.Y.		
		EQUIP.	LABOR	TOTAL
2750	4 passes	2.99	3.79	6.80
3000	300 HP dozer & roller compactors, 150' haul, 4" lifts, 2 passes	1.80	3.04	4.84
3050	4 passes	2.45	5.35	7.80
3100	8" lifts, 2 passes	1.48	1.88	3.36
3150	4 passes	1.80	3.04	4.84
3200	300' haul, 4" lifts, 2 passes	2.22	3.24	5.45
3250	4 passes	2.87	5.55	8.40
3300	8" lifts, 2 passes	1.90	2.08	3.98
3350	4 passes	2.22	3.24	5.45

For customer support on your Site Work & Landscape Costs with RSMeans data, call 800.448.8182.

G1030 Site Earthwork

Trenching Systems are shown on a cost per linear foot basis. The systems include: excavation; backfill and removal of spoil; and compaction for various depths and trench bottom widths. The backfill has been reduced to accommodate a pipe of suitable diameter and bedding.

The slope for trench sides varies from none to 1:1.

The Expanded System Listing shows Trenching Systems that range from 2′ to 12′ in width. Depths range from 2′ to 25′.

System Components			COST PER L.F.		
	QUANTITY	UNIT	EQUIP.	LABOR	TOTAL
SYSTEM G1030 805 1310					
TRENCHING, COMMON EARTH, NO SLOPE, 2′ WIDE, 2′ DP, 3/8 C.Y. BUCKET					
Excavation, trench, hyd. backhoe, track mtd., 3/8 C.Y. bucket	.148	B.C.Y.	.23	1.17	1.40
Backfill and load spoil, from stockpile	.153	L.C.Y.	.13	.36	.49
Compaction by vibrating plate, 6″ lifts, 4 passes	.118	E.C.Y.	.03	.43	.46
Remove excess spoil, 8 C.Y. dump truck, 2 mile roundtrip	.040	L.C.Y.	.16	.20	.36
TOTAL			.55	2.16	2.71

G1030 805	Trenching Common Earth	COST PER L.F.		
		EQUIP.	LABOR	TOTAL
1310	Trenching, common earth, no slope, 2′ wide, 2′ deep, 3/8 C.Y. bucket	.55	2.16	2.71
1320	3′ deep, 3/8 C.Y. bucket	.77	3.22	3.99
1330	4′ deep, 3/8 C.Y. bucket	.98	4.31	5.29
1340	6′ deep, 3/8 C.Y. bucket	1.28	5.55	6.83
1350	8′ deep, 1/2 C.Y. bucket	1.65	7.35	9
1360	10′ deep, 1 C.Y. bucket	3.57	8.80	12.37
1400	4′ wide, 2′ deep, 3/8 C.Y. bucket	1.29	4.28	5.57
1410	3′ deep, 3/8 C.Y. bucket	1.72	6.40	8.12
1420	4′ deep, 1/2 C.Y. bucket	1.92	7.15	9.07
1430	6′ deep, 1/2 C.Y. bucket	3.18	11.50	14.68
1440	8′ deep, 1/2 C.Y. bucket	6.65	14.95	21.60
1450	10′ deep, 1 C.Y. bucket	8.10	18.55	26.65
1460	12′ deep, 1 C.Y. bucket	10.45	23.50	33.95
1470	15′ deep, 1-1/2 C.Y. bucket	7.65	21	28.65
1480	18′ deep, 2-1/2 C.Y. bucket	12.65	29.50	42.15
1520	6′ wide, 6′ deep, 5/8 C.Y. bucket w/trench box	9	17.30	26.30
1530	8′ deep, 3/4 C.Y. bucket	11.65	22.50	34.15
1540	10′ deep, 1 C.Y. bucket	11.35	23.50	34.85
1550	12′ deep, 1-1/2 C.Y. bucket	10.30	25	35.30
1560	16′ deep, 2-1/2 C.Y. bucket	16	31.50	47.50
1570	20′ deep, 3-1/2 C.Y. bucket	20.50	37.50	58
1580	24′ deep, 3-1/2 C.Y. bucket	24	45	69
1640	8′ wide, 12′ deep, 1-1/2 C.Y. bucket w/trench box	14.35	32	46.35
1650	15′ deep, 1-1/2 C.Y. bucket	18.60	42	60.60
1660	18′ deep, 2-1/2 C.Y. bucket	23	42	65
1680	24′ deep, 3-1/2 C.Y. bucket	32.50	58.50	91
1730	10′ wide, 20′ deep, 3-1/2 C.Y. bucket w/trench box	26.50	55.50	82
1740	24′ deep, 3-1/2 C.Y. bucket	39	66.50	105.50
1780	12′ wide, 20′ deep, 3-1/2 C.Y. bucket w/trench box	41.50	70.50	112
1790	25′ deep, bucket	51.50	89.50	141
1800	1/2 to 1 slope, 2′ wide, 2′ deep, 3/8 C.Y. bucket	.77	3.22	3.99
1810	3′ deep, 3/8 C.Y. bucket	1.25	5.65	6.90
1820	4′ deep, 3/8 C.Y. bucket	1.84	8.60	10.44
1840	6′ deep, 3/8 C.Y. bucket	3.02	13.95	16.97

G1030 Site Earthwork

G1030 805	Trenching Common Earth	COST PER L.F.		
		EQUIP.	LABOR	TOTAL
1860	8' deep, 1/2 C.Y. bucket	4.75	22	26.75
1880	10' deep, 1 C.Y. bucket	12.30	31	43.30
2300	4' wide, 2' deep, 3/8 C.Y. bucket	1.50	5.35	6.85
2310	3' deep, 3/8 C.Y. bucket	2.20	8.85	11.05
2320	4' deep, 1/2 C.Y. bucket	2.70	10.90	13.60
2340	6' deep, 1/2 C.Y. bucket	5.30	20.50	25.80
2360	8' deep, 1/2 C.Y. bucket	12.95	30	42.95
2380	10' deep, 1 C.Y. bucket	18	42.50	60.50
2400	12' deep, 1 C.Y. bucket	23.50	57.50	81
2430	15' deep, 1-1/2 C.Y. bucket	21.50	61.50	83
2460	18' deep, 2-1/2 C.Y. bucket	43.50	96	139.50
2840	6' wide, 6' deep, 5/8 C.Y. bucket w/trench box	13	25.50	38.50
2860	8' deep, 3/4 C.Y. bucket	18.90	38.50	57.40
2880	10' deep, 1 C.Y. bucket	15.45	39	54.45
2900	12' deep, 1-1/2 C.Y. bucket	19.60	51	70.60
2940	16' deep, 2-1/2 C.Y. bucket	36.50	75	111.50
2980	20' deep, 3-1/2 C.Y. bucket	51	101	152
3020	24' deep, 3-1/2 C.Y. bucket	72	138	210
3100	8' wide, 12' deep, 1-1/2 C.Y. bucket w/trench box	24	58	82
3120	15' deep, 1-1/2 C.Y. bucket	35	84.50	119.50
3140	18' deep, 2-1/2 C.Y. bucket	50.50	100	150.50
3180	24' deep, 3-1/2 C.Y. bucket	80.50	151	231.50
3270	10' wide, 20' deep, 3-1/2 C.Y. bucket w/trench box	52	116	168
3280	24' deep, 3-1/2 C.Y. bucket	89	165	254
3370	12' wide, 20' deep, 3-1/2 C.Y. bucket w/trench box	75	135	210
3380	25' deep, 3-1/2 C.Y. bucket	104	193	297
3500	1 to 1 slope, 2' wide, 2' deep, 3/8 C.Y. bucket	.99	4.31	5.30
3520	3' deep, 3/8 C.Y. bucket	3.14	9.85	12.99
3540	4' deep, 3/8 C.Y. bucket	2.70	12.95	15.65
3560	6' deep, 1/2 C.Y. bucket	3.02	13.90	16.92
3580	8' deep, 1/2 C.Y. bucket	5.90	28	33.90
3600	10' deep, 1 C.Y. bucket	21	53.50	74.50
3800	4' wide, 2' deep, 3/8 C.Y. bucket	1.72	6.45	8.17
3820	3' deep, 3/8 C.Y. bucket	2.69	11.25	13.94
3840	4' deep, 1/2 C.Y. bucket	3.47	14.60	18.07
3860	6' deep, 1/2 C.Y. bucket	7.50	29.50	37
3880	8' deep, 1/2 C.Y. bucket	19.30	45.50	64.80
3900	10' deep, 1 C.Y. bucket	28	66	94
3920	12' deep, 1 C.Y. bucket	41	95.50	136.50
3940	15' deep, 1-1/2 C.Y. bucket	35.50	102	137.50
3960	18' deep, 2-1/2 C.Y. bucket	59.50	132	191.50
4030	6' wide, 6' deep, 5/8 C.Y. bucket w/trench box	17.15	34.50	51.65
4040	8' deep, 3/4 C.Y. bucket	25	47.50	72.50
4050	10' deep, 1 C.Y. bucket	22.50	57.50	80
4060	12' deep, 1-1/2 C.Y. bucket	30	77.50	107.50
4070	16' deep, 2-1/2 C.Y. bucket	57.50	118	175.50
4080	20' deep, 3-1/2 C.Y. bucket	82.50	164	246.50
4090	24' deep, 3-1/2 C.Y. bucket	120	231	351
4500	8' wide, 12' deep, 1-1/2 C.Y. bucket w/trench box	34	84.50	118.50
4550	15' deep, 1-1/2 C.Y. bucket	51	127	178
4600	18' deep, 2-1/2 C.Y. bucket	76.50	155	231.50
4650	24' deep, 3-1/2 C.Y. bucket	129	244	373
4800	10' wide, 20' deep, 3-1/2 C.Y. bucket w/trench box	77.50	177	254.50
4850	24' deep, 3-1/2 C.Y. bucket	137	258	395
4950	12' wide, 20' deep, 3-1/2 C.Y. bucket w/trench box	108	199	307
4980	25' deep, 3-1/2 C.Y. bucket	155	291	446

G10 Site Preparation

G1030 Site Earthwork

Trenching Systems are shown on a cost per linear foot basis. The systems include: excavation; backfill and removal of spoil; and compaction for various depths and trench bottom widths. The backfill has been reduced to accommodate a pipe of suitable diameter and bedding.

The slope for trench sides varies from none to 1:1.

The Expanded System Listing shows Trenching Systems that range from 2' to 12' in width. Depths range from 2' to 25'.

System Components	QUANTITY	UNIT	COST PER L.F.		
			EQUIP.	LABOR	TOTAL
SYSTEM G1030 806 1310					
TRENCHING, LOAM & SANDY CLAY, NO SLOPE, 2' WIDE, 2' DP, 3/8 C.Y. BUCKET					
Excavation, trench, hyd. backhoe, track mtd., 3/8 C.Y. bucket	.148	B.C.Y.	.21	1.08	1.29
Backfill and load spoil, from stockpile	.165	L.C.Y.	.14	.39	.53
Compaction by vibrating plate 18" wide, 6" lifts, 4 passes	.118	E.C.Y.	.03	.43	.46
Remove excess spoil, 8 C.Y. dump truck, 2 mile roundtrip	.042	L.C.Y.	.16	.21	.37
TOTAL			.54	2.11	2.65

G1030 806	Trenching Loam & Sandy Clay	COST PER L.F.		
		EQUIP.	LABOR	TOTAL
1310	Trenching, loam & sandy clay, no slope, 2' wide, 2' deep, 3/8 C.Y. bucket	.54	2.11	2.65
1320	3' deep, 3/8 C.Y. bucket	.81	3.44	4.25
1330	4' deep, 3/8 C.Y. bucket	.97	4.20	5.15
1340	6' deep, 3/8 C.Y. bucket	1.80	5	6.80
1350	8' deep, 1/2 C.Y. bucket	2.35	6.60	8.95
1360	10' deep, 1 C.Y. bucket	2.63	7.05	9.70
1400	4' wide, 2' deep, 3/8 C.Y. bucket	1.33	4.19	5.50
1410	3' deep, 3/8 C.Y. bucket	1.74	6.30	8.05
1420	4' deep, 1/2 C.Y. bucket	1.94	7.05	9
1430	6' deep, 1/2 C.Y. bucket	4.26	10.40	14.65
1440	8' deep, 1/2 C.Y. bucket	6.50	14.75	21.50
1450	10' deep, 1 C.Y. bucket	6.25	15.05	21.50
1460	12' deep, 1 C.Y. bucket	7.85	18.75	26.50
1470	15' deep, 1-1/2 C.Y. bucket	7.95	22	30
1480	18' deep, 2-1/2 C.Y. bucket	10.75	24	35
1520	6' wide, 6' deep, 5/8 C.Y. bucket w/trench box	8.75	16.95	25.50
1530	8' deep, 3/4 C.Y. bucket	11.20	22	33
1540	10' deep, 1 C.Y. bucket	10.40	22.50	33
1550	12' deep, 1-1/2 C.Y. bucket	10.15	25	35
1560	16' deep, 2-1/2 C.Y. bucket	15.65	32	47.50
1570	20' deep, 3-1/2 C.Y. bucket	18.75	37.50	56.50
1580	24' deep, 3-1/2 C.Y. bucket	23.50	46	69.50
1640	8' wide, 12' deep, 1-1/4 C.Y. bucket w/trench box	14.30	32	46.50
1650	15' deep, 1-1/2 C.Y. bucket	17.50	41	58.50
1660	18' deep, 2-1/2 C.Y. bucket	23.50	46	69.50
1680	24' deep, 3-1/2 C.Y. bucket	32	59	91
1730	10' wide, 20' deep, 3-1/2 C.Y. bucket w/trench box	32	59.50	91.50
1740	24' deep, 3-1/2 C.Y. bucket	40	73.50	114
1780	12' wide, 20' deep, 3-1/2 C.Y. bucket w/trench box	39	70.50	110
1790	25' deep, 3-1/2 C.Y. bucket	50.50	91	142
1800	1/2:1 slope, 2' wide, 2' deep, 3/8 C.Y. bucket	.75	3.15	3.90
1810	3' deep, 3/8 C.Y. bucket	1.23	5.50	6.75
1820	4' deep, 3/8 C.Y. bucket	1.81	8.40	10.20
1840	6' deep, 3/8 C.Y. bucket	4.33	12.50	16.85

644

For customer support on your Site Work & Landscape Costs with RSMeans data, call 800.448.8182.

G1030 Site Earthwork

G1030 806	Trenching Loam & Sandy Clay	COST PER L.F.		
		EQUIP.	LABOR	TOTAL
1860	8' deep, 1/2 C.Y. bucket	6.85	20	27
1880	10' deep, 1 C.Y. bucket	9	25	34
2300	4' wide, 2' deep, 3/8 C.Y. bucket	1.54	5.25	6.80
2310	3' deep, 3/8 C.Y. bucket	2.23	8.65	10.90
2320	4' deep, 1/2 C.Y. bucket	2.71	10.75	13.45
2340	6' deep, 1/2 C.Y. bucket	7.20	18.45	25.50
2360	8' deep, 1/2 C.Y. bucket	12.60	30	42.50
2380	10' deep, 1 C.Y. bucket	13.75	34.50	48.50
2400	12' deep, 1 C.Y. bucket	23	57	80
2430	15' deep, 1-1/2 C.Y. bucket	22.50	63.50	86
2460	18' deep, 2-1/2 C.Y. bucket	42.50	97.50	140
2840	6' wide, 6' deep, 5/8 C.Y. bucket w/trench box	12.65	25.50	38
2860	8' deep, 3/4 C.Y. bucket	18.20	37.50	55.50
2880	10' deep, 1 C.Y. bucket	18.70	42.50	61
2900	12' deep, 1-1/2 C.Y. bucket	19.80	51.50	71.50
2940	16' deep, 2-1/2 C.Y. bucket	36	75.50	112
2980	20' deep, 3-1/2 C.Y. bucket	49.50	102	152
3020	24' deep, 3-1/2 C.Y. bucket	70	140	210
3100	8' wide, 12' deep, 1-1/2 C.Y. bucket w/trench box	24	58.50	82.50
3120	15' deep, 1-1/2 C.Y. bucket	32.50	82	115
3140	18' deep, 2-1/2 C.Y. bucket	49	101	150
3180	24' deep, 3-1/2 C.Y. bucket	78.50	153	232
3270	10' wide, 20' deep, 3-1/2 C.Y. bucket w/trench box	63	124	187
3280	24' deep, 3-1/2 C.Y. bucket	86.50	168	255
3320	12' wide, 20' deep, 3-1/2 C.Y. bucket w/trench box	69.50	135	205
3380	25' deep, 3-1/2 C.Y. bucket w/trench box	95	181	276
3500	1:1 slope, 2' wide, 2' deep, 3/8 C.Y. bucket	.97	4.20	5.15
3520	3' deep, 3/8 C.Y. bucket	1.71	7.90	9.60
3540	4' deep, 3/8 C.Y. bucket	2.66	12.60	15.25
3560	6' deep, 1/2 C.Y. bucket	4.33	12.50	16.85
3580	8' deep, 1/2 C.Y. bucket	11.35	33.50	45
3600	10' deep, 1 C.Y. bucket	15.35	43	58.50
3800	4' wide, 2' deep, 3/8 C.Y. bucket	1.74	6.30	8.05
3820	3' deep, 1/2 C.Y. bucket	2.71	11	13.70
3840	4' deep, 1/2 C.Y. bucket	3.48	14.40	17.90
3860	6' deep, 1/2 C.Y. bucket	10.15	26.50	36.50
3880	8' deep, 1/2 C.Y. bucket	18.75	45	64
3900	10' deep, 1 C.Y. bucket	21.50	54	75.50
3920	12' deep, 1 C.Y. bucket	30.50	76.50	107
3940	15' deep, 1-1/2 C.Y. bucket	37	105	142
3960	18' deep, 2-1/2 C.Y. bucket	58.50	134	193
4030	6' wide, 6' deep, 5/8 C.Y. bucket w/trench box	16.60	34.50	51
4040	8' deep, 3/4 C.Y. bucket	25	53	78
4050	10' deep, 1 C.Y. bucket	27	62	89
4060	12' deep, 1-1/2 C.Y. bucket	29.50	78	108
4070	16' deep, 2-1/2 C.Y. bucket	56	120	176
4080	20' deep, 3-1/2 C.Y. bucket	80.50	167	248
4090	24' deep, 3-1/2 C.Y. bucket	117	234	350
4500	8' wide, 12' deep, 1-1/4 C.Y. bucket w/trench box	33.50	85	119
4550	15' deep, 1-1/2 C.Y. bucket	48	123	171
4600	18' deep, 2-1/2 C.Y. bucket	75	157	232
4650	24' deep, 3-1/2 C.Y. bucket	125	248	375
4800	10' wide, 20' deep, 3-1/2 C.Y. bucket w/trench box	93.50	189	283
4850	24' deep, 3-1/2 C.Y. bucket	133	262	395
4950	12' wide, 20' deep, 3-1/2 C.Y. bucket w/trench box	100	200	300
4980	25' deep, 3-1/2 C.Y. bucket	151	295	445

G1030 Site Earthwork

Trenching Systems are shown on a cost per linear foot basis. The systems include: excavation; backfill and removal of spoil; and compaction for various depths and trench bottom widths. The backfill has been reduced to accommodate a pipe of suitable diameter and bedding.

The slope for trench sides varies from none to 1:1.

The Expanded System Listing shows Trenching Systems that range from 2' to 12' in width. Depths range from 2' to 25'.

System Components	QUANTITY	UNIT	COST PER L.F.		
			EQUIP.	LABOR	TOTAL
SYSTEM G1030 807 1310					
TRENCHING, SAND & GRAVEL, NO SLOPE, 2' WIDE, 2' DEEP, 3/8 C.Y. BUCKET					
Excavation, trench, hyd. backhoe, track mtd., 3/8 C.Y. bucket	.148	B.C.Y.	.21	1.07	1.28
Backfill and load spoil, from stockpile	.140	L.C.Y.	.12	.33	.45
Compaction by vibrating plate 18" wide, 6" lifts, 4 passes	.118	E.C.Y.	.03	.43	.46
Remove excess spoil, 8 C.Y. dump truck, 2 mile roundtrip	.035	L.C.Y.	.14	.17	.31
TOTAL			.50	2	2.50

G1030 807	Trenching Sand & Gravel	COST PER L.F.		
		EQUIP.	LABOR	TOTAL
1310	Trenching, sand & gravel, no slope, 2' wide, 2' deep, 3/8 C.Y. bucket	.50	2	2.50
1320	3' deep, 3/8 C.Y. bucket	.75	3.32	4.07
1330	4' deep, 3/8 C.Y. bucket	.89	3.99	4.88
1340	6' deep, 3/8 C.Y. bucket	1.71	4.76	6.45
1350	8' deep, 1/2 C.Y. bucket	2.23	6.30	8.55
1360	10' deep, 1 C.Y. bucket	2.45	6.65	9.10
1400	4' wide, 2' deep, 3/8 C.Y. bucket	1.19	3.95	5.15
1410	3' deep, 3/8 C.Y. bucket	1.57	5.95	7.50
1420	4' deep, 1/2 C.Y. bucket	1.76	6.70	8.45
1430	6' deep, 1/2 C.Y. bucket	4.03	9.90	13.95
1440	8' deep, 1/2 C.Y. bucket	6.15	14	20
1450	10' deep, 1 C.Y. bucket	5.85	14.20	20
1460	12' deep, 1 C.Y. bucket	7.40	17.75	25
1470	15' deep, 1-1/2 C.Y. bucket	7.45	20.50	28
1480	18' deep, 2-1/2 C.Y. bucket	10.15	22.50	32.50
1520	6' wide, 6' deep, 5/8 C.Y. bucket w/trench box	8.25	16.05	24.50
1530	8' deep, 3/4 C.Y. bucket	10.65	21	31.50
1540	10' deep, 1 C.Y. bucket	9.80	21.50	31.50
1550	12' deep, 1-1/2 C.Y. bucket	9.50	24	33.50
1560	16' deep, 2 C.Y. bucket	14.90	30	45
1570	20' deep, 3-1/2 C.Y. bucket	17.75	35	53
1580	24' deep, 3-1/2 C.Y. bucket	22.50	43	65.50
1640	8' wide, 12' deep, 1-1/2 C.Y. bucket w/trench box	13.30	30	43.50
1650	15' deep, 1-1/2 C.Y. bucket	16.25	38.50	55
1660	18' deep, 2-1/2 C.Y. bucket	22.50	43	65.50
1680	24' deep, 3-1/2 C.Y. bucket	30	55.50	85.50
1730	10' wide, 20' deep, 3-1/2 C.Y. bucket w/trench box	30	55.50	85.50
1740	24' deep, 3-1/2 C.Y. bucket	37.50	68.50	106
1780	12' wide, 20' deep, 3-1/2 C.Y. bucket w/trench box	36.50	66	103
1790	25' deep, 3-1/2 C.Y. bucket	47	85	132
1800	1/2:1 slope, 2' wide, 2' deep, 3/8 C.Y. bucket	.70	2.99	3.69
1810	3' deep, 3/8 C.Y. bucket	1.14	5.25	6.40
1820	4' deep, 3/8 C.Y. bucket	1.68	8	9.70
1840	6' deep, 3/8 C.Y. bucket	4.12	11.95	16.05

G1030 Site Earthwork

G1030 807	Trenching Sand & Gravel	COST PER L.F.		
		EQUIP.	LABOR	TOTAL
1860	8' deep, 1/2 C.Y. bucket	6.50	19.10	25.50
1880	10' deep, 1 C.Y. bucket	8.40	23.50	32
2300	4' wide, 2' deep, 3/8 C.Y. bucket	1.38	4.96	6.35
2310	3' deep, 3/8 C.Y. bucket	2.01	8.20	10.20
2320	4' deep, 1/2 C.Y. bucket	2.46	10.15	12.60
2340	6' deep, 1/2 C.Y. bucket	6.85	17.65	24.50
2360	8' deep, 1/2 C.Y. bucket	11.95	28.50	40.50
2380	10' deep, 1 C.Y. bucket	12.95	32.50	45.50
2400	12' deep, 1 C.Y. bucket	22	54	76
2430	15' deep, 1-1/2 C.Y. bucket	21	60	81
2460	18' deep, 2-1/2 C.Y. bucket	40.50	91.50	132
2840	6' wide, 6' deep, 5/8 C.Y. bucket w/trench box	12	24.50	36.50
2860	8' deep, 3/4 C.Y. bucket	17.30	35.50	53
2880	10' deep, 1 C.Y. bucket	17.70	40	57.50
2900	12' deep, 1-1/2 C.Y. bucket	18.60	48.50	67
2940	16' deep, 2 C.Y. bucket	34	71.50	106
2980	20' deep, 3-1/2 C.Y. bucket	47	96	143
3020	24' deep, 3-1/2 C.Y. bucket	66.50	132	199
3100	8' wide, 12' deep, 1-1/4 C.Y. bucket w/trench box	22.50	55	77.50
3120	15' deep, 1-1/2 C.Y. bucket	30.50	77	108
3140	18' deep, 2-1/2 C.Y. bucket	46.50	95.50	142
3180	24' deep, 3-1/2 C.Y. bucket	74	144	218
3270	10' wide, 20' deep, 3-1/2 C.Y. bucket w/trench box	59	116	175
3280	24' deep, 3-1/2 C.Y. bucket	82	157	239
3370	12' wide, 20' deep, 3-1/2 C.Y. bucket w/trench box	66	129	195
3380	25' deep, 3-1/2 C.Y. bucket	95	181	276
3500	1:1 slope, 2' wide, 2' deep, 3/8 C.Y. bucket	1.81	5.05	6.85
3520	3' deep, 3/8 C.Y. bucket	1.57	7.50	9.05
3540	4' deep, 3/8 C.Y. bucket	2.46	12.05	14.50
3560	6' deep, 3/8 C.Y. bucket	4.13	11.95	16.10
3580	8' deep, 1/2 C.Y. bucket	10.80	32	43
3600	10' deep, 1 C.Y. bucket	14.35	40.50	55
3800	4' wide, 2' deep, 3/8 C.Y. bucket	1.58	5.95	7.55
3820	3' deep, 3/8 C.Y. bucket	2.47	10.50	12.95
3840	4' deep, 1/2 C.Y. bucket	3.17	13.65	16.80
3860	6' deep, 1/2 C.Y. bucket	9.70	25.50	35
3880	8' deep, 1/2 C.Y. bucket	17.80	42.50	60.50
3900	10' deep, 1 C.Y. bucket	20	51	71
3920	12' deep, 1 C.Y. bucket	29	72.50	102
3940	15' deep, 1-1/2 C.Y. bucket	35	99.50	135
3960	18' deep, 2-1/2 C.Y. bucket	55.50	126	182
4030	6' wide, 6' deep, 5/8 C.Y. bucket w/trench box	15.75	32.50	48.50
4040	8' deep, 3/4 C.Y. bucket	24	50.50	74.50
4050	10' deep, 1 C.Y. bucket	25.50	58.50	84
4060	12' deep, 1-1/2 C.Y. bucket	27.50	73.50	101
4070	16' deep, 2 C.Y. bucket	53.50	113	167
4080	20' deep, 3-1/2 C.Y. bucket	76	157	233
4090	24' deep, 3-1/2 C.Y. bucket	110	220	330
4500	8' wide, 12' deep, 1-1/2 C.Y. bucket w/trench box	31.50	80	112
4550	15' deep, 1-1/2 C.Y. bucket	44.50	116	161
4600	18' deep, 2-1/2 C.Y. bucket	71	148	219
4650	24' deep, 3-1/2 C.Y. bucket	118	233	350
4800	10' wide, 20' deep, 3-1/2 C.Y. bucket w/trench box	88.50	177	266
4850	24' deep, 3-1/2 C.Y. bucket	126	246	370
4950	12' wide, 20' deep, 3-1/2 C.Y. bucket w/trench box	94.50	187	282
4980	25' deep, 3-1/2 C.Y. bucket	143	277	420

G10 Site Preparation

G1030 Site Earthwork

The Pipe Bedding System is shown for various pipe diameters. Compacted bank sand is used for pipe bedding and to fill 12″ over the pipe. No backfill is included. Various side slopes are shown to accommodate different soil conditions. Pipe sizes vary from 6″ to 84″ diameter.

System Components	QUANTITY	UNIT	COST PER L.F.		
			MAT.	INST.	TOTAL
SYSTEM G1030 815 1440					
PIPE BEDDING, SIDE SLOPE 0 TO 1, 1′ WIDE, PIPE SIZE 6″ DIAMETER					
Borrow, bank sand, 2 mile haul, machine spread	.086	C.Y.	1.71	.72	2.43
Compaction, vibrating plate	.086	C.Y.		.23	.23
TOTAL			1.71	.95	2.66

G1030 815	Pipe Bedding	COST PER L.F.		
		MAT.	INST.	TOTAL
1440	Pipe bedding, side slope 0 to 1, 1′ wide, pipe size 6″ diameter	1.71	.95	2.66
1460	2′ wide, pipe size 8″ diameter	3.70	2.05	5.75
1480	Pipe size 10″ diameter	3.78	2.09	5.87
1500	Pipe size 12″ diameter	3.86	2.14	6
1520	3′ wide, pipe size 14″ diameter	6.25	3.46	9.71
1540	Pipe size 15″ diameter	6.30	3.50	9.80
1560	Pipe size 16″ diameter	6.40	3.53	9.93
1580	Pipe size 18″ diameter	6.50	3.60	10.10
1600	4′ wide, pipe size 20″ diameter	9.25	5.10	14.35
1620	Pipe size 21″ diameter	9.30	5.15	14.45
1640	Pipe size 24″ diameter	9.50	5.25	14.75
1660	Pipe size 30″ diameter	9.70	5.35	15.05
1680	6′ wide, pipe size 32″ diameter	16.60	9.15	25.75
1700	Pipe size 36″ diameter	17	9.40	26.40
1720	7′ wide, pipe size 48″ diameter	27	14.95	41.95
1740	8′ wide, pipe size 60″ diameter	33	18.20	51.20
1760	10′ wide, pipe size 72″ diameter	45.50	25.50	71
1780	12′ wide, pipe size 84″ diameter	60.50	33.50	94
2140	Side slope 1/2 to 1, 1′ wide, pipe size 6″ diameter	3.19	1.77	4.96
2160	2′ wide, pipe size 8″ diameter	5.40	3.01	8.41
2180	Pipe size 10″ diameter	5.80	3.22	9.02
2200	Pipe size 12″ diameter	6.15	3.41	9.56
2220	3′ wide, pipe size 14″ diameter	8.90	4.93	13.83
2240	Pipe size 15″ diameter	9.10	5.05	14.15
2260	Pipe size 16″ diameter	9.35	5.15	14.50
2280	Pipe size 18″ diameter	9.80	5.45	15.25
2300	4′ wide, pipe size 20″ diameter	12.95	7.15	20.10
2320	Pipe size 21″ diameter	13.25	7.35	20.60
2340	Pipe size 24″ diameter	14.10	7.80	21.90
2360	Pipe size 30″ diameter	15.60	8.65	24.25
2380	6′ wide, pipe size 32″ diameter	23	12.75	35.75
2400	Pipe size 36″ diameter	24.50	13.55	38.05
2420	7′ wide, pipe size 48″ diameter	38	21	59
2440	8′ wide, pipe size 60″ diameter	48.50	27	75.50
2460	10′ wide, pipe size 72″ diameter	66.50	37	103.50
2480	12′ wide, pipe size 84″ diameter	87	48.50	135.50
2620	Side slope 1 to 1, 1′ wide, pipe size 6″ diameter	4.67	2.58	7.25
2640	2′ wide, pipe size 8″ diameter	7.20	3.98	11.18

648

G10 Site Preparation

G1030 Site Earthwork

G1030 815	Pipe Bedding	COST PER L.F.		
		MAT.	INST.	TOTAL
2660	Pipe size 10" diameter	7.80	4.33	12.13
2680	Pipe size 12" diameter	8.50	4.72	13.22
2700	3' wide, pipe size 14" diameter	11.55	6.40	17.95
2720	Pipe size 15" diameter	11.90	6.60	18.50
2740	Pipe size 16" diameter	12.30	6.80	19.10
2760	Pipe size 18" diameter	13.15	7.25	20.40
2780	4' wide, pipe size 20" diameter	16.70	9.25	25.95
2800	Pipe size 21" diameter	17.15	9.50	26.65
2820	Pipe size 24" diameter	18.60	10.30	28.90
2840	Pipe size 30" diameter	21.50	11.90	33.40
2860	6' wide, pipe size 32" diameter	29.50	16.30	45.80
2880	Pipe size 36" diameter	32	17.70	49.70
2900	7' wide, pipe size 48" diameter	49.50	27.50	77
2920	8' wide, pipe size 60" diameter	64	35.50	99.50
2940	10' wide, pipe size 72" diameter	87.50	48.50	136
2960	12' wide, pipe size 84" diameter	114	63	177

For customer support on your Site Work & Landscape Costs with RSMeans data, call 800.448.8182.

G2010 Roadways

The Bituminous Roadway Systems are listed for pavement thicknesses between 3-1/2" and 7" and gravel bases from 3" to 22" in depth. Systems costs are expressed per linear foot for varying widths of two and multi-lane roads. Earth moving is not included. Granite curbs and line painting are added as required system components.

System Components	QUANTITY	UNIT	COST PER L.F.		
			MAT.	INST.	TOTAL
SYSTEM G2010 230 1050					
BITUM. ROADWAY, TWO LANES, 3-1/2" TH. PVMT., 3" TH. GRAVEL BASE, 24' WIDE					
Compact subgrade, 4 passes	.222	E.C.Y.		.54	.54
Bank gravel, 2 mi haul, dozer spread	.250	L.C.Y.	5.13	2.10	7.23
Compaction, granular material to 98%	.259	E.C.Y.		.18	.18
Grading, fine grade, 3 passes with grader	3.300	S.Y.		19.41	19.41
Bituminous paving, binder course, 2-1/2" thick	2.670	S.Y.	27.10	4.33	31.43
Bituminous paving, wearing course, 1" thick	2.670	S.Y.	10.76	2.54	13.30
Curbs, granite, split face, straight, 5" x 16"	2.000	L.F.	36.30	29.92	66.22
Thermoplastic, white or yellow, 4" wide, less than 6,000 L.F.	4.000	L.F.	1.84	1.20	3.04
TOTAL			81.13	60.22	141.35

G2010 230	Bituminous Roadways Gravel Base	COST PER L.F.		
		MAT.	INST.	TOTAL
1050	Bitum. roadway, two lanes, 3-1/2" th. pvmt., 3" th. gravel base, 24' wide	81	60	141
1100	28' wide	88.50	60.50	149
1150	32' wide	96	65	161
1200	4" th. gravel base, 24' wide	83	58	141
1210	28' wide	90.50	62	152.50
1220	32' wide	98	66.50	164.50
1222	36' wide	106	71	177
1224	40' wide	113	75	188
1230	6" th. gravel base, 24' wide	86.50	60	146.50
1240	28' wide	95	64	159
1250	32' wide	103	69	172
1252	36' wide	111	73.50	184.50
1254	40' wide	119	78.50	197.50
1256	8" th. gravel base, 24' wide	90.50	61	151.50
1258	28' wide	99	66.50	165.50
1260	32' wide	108	71.50	179.50
1262	36' wide	116	76	192
1264	40' wide	125	81	206
1300	9" th. gravel base, 24' wide	91.50	62	153.50
1350	28' wide	101	67.50	168.50
1400	32' wide	110	72.50	182.50
1410	10" th. gravel base, 24' wide	94	63	157
1412	28' wide	103	68	171
1414	32' wide	112	74	186
1416	36' wide	122	79	201
1418	40' wide	131	84	215
1420	12" thick gravel base, 24' wide	97.50	64.50	162
1422	28' wide	107	70.50	177.50
1424	32' wide	117	76	193
1426	36' wide	127	81.50	208.50

G2010 230	Bituminous Roadways Gravel Base	COST PER L.F.		
		MAT.	INST.	TOTAL
1428	40' wide	137	87.50	224.50
1430	15" thick gravel base, 24' wide	103	67.50	170.50
1432	28' wide	114	73.50	187.50
1434	32' wide	124	79.50	203.50
1436	36' wide	135	85.50	220.50
1438	40' wide	146	91.50	237.50
1440	18" thick gravel base, 24' wide	108	70.50	178.50
1442	28' wide	120	77	197
1444	32' wide	132	83	215
1446	36' wide	143	90	233
1448	40' wide	155	96.50	251.50
1550	4" th. pvmt., 4" th. gravel base, 24' wide	88.50	58	146.50
1600	28' wide	97	62.50	159.50
1650	32' wide	106	67	173
1652	36' wide	114	71.50	185.50
1654	40' wide	122	76.50	198.50
1700	6" th. gravel base, 24' wide	92	60	152
1710	28' wide	101	65	166
1720	32' wide	110	69.50	179.50
1722	36' wide	119	74.50	193.50
1724	40' wide	128	79	207
1726	8" th. gravel base, 24' wide	96	61.50	157.50
1728	28' wide	105	67	172
1730	32' wide	115	72	187
1732	36' wide	125	77	202
1734	40' wide	134	82	216
1800	10" th. gravel base, 24' wide	99.50	64	163.50
1850	28' wide	110	67.50	177.50
1900	32' wide	120	74.50	194.50
1902	36' wide	130	79.50	209.50
1904	40' wide	136	83.50	219.50
2050	4" th. pvmt., 5" th. gravel base, 24' wide	90.50	62	152.50
2100	28' wide	99.50	67	166.50
2150	32' wide	109	72.50	181.50
2200	8" th. gravel base, 24' wide	96.50	64	160.50
2210	28' wide	106	70	176
2220	32' wide	116	75.50	191.50
2300	12" th. gravel base, 24' wide	105	66	171
2350	28' wide	116	71.50	187.50
2400	32' wide	127	77	204
2410	36' wide	138	83	221
2412	40' wide	149	89	238
2420	15" th. gravel base, 24' wide	106	66.50	172.50
2422	28' wide	122	74.50	196.50
2424	32' wide	134	81	215
2426	36' wide	146	87	233
2428	40' wide	158	93.50	251.50
2430	18" th. gravel base, 24' wide	115	71.50	186.50
2432	28' wide	128	78.50	206.50
2434	32' wide	141	84.50	225.50
2436	36' wide	154	91.50	245.50
2438	40' wide	167	98	265
2550	4-1/2" th. pvmt., 5" th. gravel base, 24' wide	96	62.50	158.50
2600	28' wide	106	68	174
2650	32' wide	116	73	189
2700	6" th. gravel base, 24' wide	98	63.50	161.50
2710	28' wide	108	69	177
2720	32' wide	119	74	193

G2010 230	Bituminous Roadways Gravel Base	COST PER L.F.		
		MAT.	INST.	TOTAL
2800	13" th. gravel base, 24' wide	110	68.50	178.50
2850	28' wide	122	75	197
2900	32' wide	135	81.50	216.50
3050	5" th. pvmt., 6" th. gravel base, 24' wide	104	64	168
3100	28' wide	115	69.50	184.50
3150	32' wide	127	75.50	202.50
3300	14" th. gravel base, 24' wide	118	70.50	188.50
3350	28' wide	132	76.50	208.50
3400	32' wide	145	83.50	228.50
3550	5-1/2" th. pvmt., 7" th. gravel base, 24' wide	112	65.50	177.50
3600	28' wide	124	71.50	195.50
3650	32' wide	136	77.50	213.50
3800	17" th. gravel base, 24' wide	129	73.50	202.50
3850	28' wide	144	80.50	224.50
3900	32' wide	160	88	248
4050	Multi lane, 5" th. pvmt., 6" th. gravel base, 48' wide	171	98	269
4100	72' wide	239	132	371
4150	96' wide	305	165	470
4300	14" th. gravel base, 48' wide	198	111	309
4350	72' wide	278	150	428
4400	96' wide	360	191	551
4550	5-1/2" th. pvmt., 7" th. gravel base, 48' wide	186	101	287
4600	72' wide	261	136	397
4650	96' wide	335	171	506
4800	17" th. gravel base, 48' wide	221	116	337
4850	72' wide	315	159	474
4900	96' wide	405	203	608
5050	6" th. pvmt., 8" th. gravel base, 48' wide	200	103	303
5100	72' wide	282	139	421
5150	96' wide	365	175	540
5200	10" th. gravel base, 48' wide	208	106	314
5210	72' wide	293	144	437
5220	96' wide	380	182	562
5300	20" th. gravel base, 48' wide	242	121	363
5350	72' wide	345	166	511
5400	96' wide	450	212	662
5550	7" th. pvmt., 9" th. gravel base, 48' wide	226	107	333
5600	72' wide	320	145	465
5650	96' wide	415	183	598
5700	10" th. gravel base, 48' wide	231	109	340
5710	72' wide	325	148	473
5720	96' wide	425	188	613
5800	22" th. gravel base, 48' wide	272	127	399
5850	72' wide	390	175	565
5900	96' wide	505	223	728

G2010 Roadways

The Bituminous Roadway Systems are listed for pavement thicknesses between 3-1/2" and 7" and crushed stone bases from 3" to 22" in depth. Systems costs are expressed per linear foot for varying widths of two and multi-lane roads. Earth moving is not included. Granite curbs and line painting are added as required system components.

System Components	QUANTITY	UNIT	COST PER L.F. MAT.	COST PER L.F. INST.	COST PER L.F. TOTAL
SYSTEM G2010 232 1050					
BITUMINOUS ROADWAY, TWO LANES, 3-1/2" PAVEMENT					
3" THICK CRUSHED STONE BASE, 24' WIDE					
Compact subgrade, 4 passes	.222	E.C.Y.		.54	.54
3/4" crushed stonel, 2 mi haul, dozer spread	.266	C.Y.	8.11	2.22	10.33
Compaction, granular material to 98%	.259	E.C.Y.		.18	.18
Grading, fine grade, 3 passes with grader	3.300	S.Y.		19.41	19.41
Bituminous paving, binder course, 2-1/2" thick	2.670	S.Y.	27.10	4.33	31.43
Bituminous paving, wearing course, 1" thick	2.670	S.Y.	10.76	2.54	13.30
Curbs, granite, split face, straight, 5" x 16"	2.000	L.F.	36.30	29.92	66.22
Thermoplastic, white or yellow, 4" wide, less than 6,000 L.F.	4.000	L.F.	1.84	1.20	3.04
TOTAL			84.11	60.34	144.45

G2010 232	Bituminous Roadways Crushed Stone	COST PER L.F. MAT.	COST PER L.F. INST.	COST PER L.F. TOTAL
1050	Bitum. roadway, two lanes, 3-1/2" th. pvmt., 3" th. crushed stone,24' wide	84	60.50	144.50
1100	28' wide	91.50	60.50	152
1150	32' wide	99.50	70	169.50
1200	4" thick crushed stone, 24' wide	86.50	58	144.50
1210	28' wide	94.50	62	156.50
1220	32' wide	103	66.50	169.50
1222	36' wide	111	71	182
1224	40' wide	119	75	194
1230	6" thick crushed stone, 24' wide	92	59.50	151.50
1240	28' wide	101	64	165
1250	32' wide	110	69	179
1252	36' wide	119	73.50	192.50
1254	40' wide	128	78.50	206.50
1256	8" thick crushed stone, 24' wide	97.50	61	158.50
1258	28' wide	107	66.50	173.50
1260	32' wide	117	71.50	188.50
1262	36' wide	127	76	203
1264	40' wide	137	81	218
1300	9" thick crushed stone, 24' wide	100	62.50	162.50
1350	28' wide	111	67.50	178.50
1400	32' wide	121	73	194
1410	10" thick crushed stone, 24' wide	103	63	166
1412	28' wide	113	68	181
1414	32' wide	124	74	198
1416	36' wide	135	79	214
1418	40' wide	146	84	230

653

For customer support on your Site Work & Landscape Costs with RSMeans data, call 800.448.8182.

G2010 232	Bituminous Roadways Crushed Stone	COST PER L.F.		
		MAT.	INST.	TOTAL
1420	12" thick crushed stone, 24' wide	108	64.50	172.50
1422	28' wide	120	70.50	190.50
1424	32' wide	131	76.50	207.50
1426	36' wide	143	81.50	224.50
1428	40' wide	154	87.50	241.50
1430	15" thick crushed stone, 24' wide	116	67.50	183.50
1432	28' wide	129	73.50	202.50
1434	32' wide	142	79.50	221.50
1436	36' wide	155	85.50	240.50
1438	40' wide	168	91.50	259.50
1440	18" thick crushed stone, 24' wide	124	70.50	194.50
1442	28' wide	138	77	215
1444	32' wide	153	83	236
1446	36' wide	167	90	257
1448	40' wide	181	96.50	277.50
1550	4" thick pavement., 4" thick crushed stone, 24' wide	92.50	58	150.50
1600	28' wide	101	62.50	163.50
1650	32' wide	110	67	177
1652	36' wide	119	71.50	190.50
1654	40' wide	128	76.50	204.50
1700	6" thick crushed stone , 24' wide	97.50	60	157.50
1710	28' wide	107	65	172
1720	32' wide	117	69.50	186.50
1722	36' wide	127	74.50	201.50
1724	40' wide	137	79	216
1726	8" thick crushed stone , 24' wide	103	61.50	164.50
1728	28' wide	113	67	180
1730	32' wide	124	72	196
1732	36' wide	135	77	212
1734	40' wide	146	82	228
1800	10" thick crushed stone, 24' wide	109	64	173
1850	28' wide	120	68	188
1900	32' wide	132	74.50	206.50
1902	36' wide	143	79.50	222.50
1904	40' wide	148	83.50	231.50
2050	4" th. pavement, 5" thick crushed stone, 24' wide	95.50	62	157.50
2100	28' wide	105	67	172
2150	32' wide	115	72.50	187.50
2300	12" thick crushed stone, 24' wide	114	67.50	181.50
2350	28' wide	127	74	201
2400	32' wide	140	80	220
2550	4-1/2" thick pavement, 5" thick crushed stone, 24' wide	101	62.50	163.50
2600	28' wide	111	68	179
2650	32' wide	122	73.50	195.50
2800	13" thick crushed stone, 24' wide	123	69.50	192.50
2850	28' wide	137	75.50	212.50
2900	32' wide	151	82	233
3050	5" thick pavement, 6" thick crushed stone, 24' wide	109	63.50	172.50
3100	28' wide	121	69.50	190.50
3150	32' wide	133	75	208
3300	14" thick crushed stone, 24' wide	131	70.50	201.50
3350	28' wide	146	77	223
3400	32' wide	162	84	246
3550	5-1/2" thick pavement., 7" thick crushed stone, 24' wide	118	65.50	183.50
3600	28' wide	131	71.50	202.50
3650	32' wide	145	77.50	222.50
3800	17" thick crushed stone, 24' wide	145	73.50	218.50
3850	28' wide	178	84.50	262.50

G2010 232	Bituminous Roadways Crushed Stone	COST PER L.F.		
		MAT.	INST.	TOTAL
3900	32' wide	212	96.50	308.50
4050	Multi lane, 5" thick pavement, 6" thick crushed stone, 48' wide	181	97.50	278.50
4100	72' wide	253	131	384
4150	96' wide	325	164	489
4300	14" thick crushed stone, 48' wide	224	110	334
4350	72' wide	320	151	471
4400	96' wide	410	191	601
4550	5-1/2" thick pavement, 7" thick crushed stone, 48' wide	199	100	299
4600	72' wide	280	136	416
4650	96' wide	360	171	531
4800	17" thick crushed stone, 48' wide	253	117	370
4850	72' wide	360	160	520
4900	96' wide	470	203	673
5050	6" thick pavement, 8" thick crushed stone, 48' wide	215	103	318
5100	72' wide	305	140	445
5150	96' wide	395	177	572
5300	20" thick crushed stone, 48' wide	280	123	403
5350	72' wide	400	169	569
5400	96' wide	525	214	739
5550	7" thick pavement, 9" thick crushed stone, 48' wide	243	107	350
5600	72' wide	345	146	491
5650	96' wide	450	184	634
5800	22" thick crushed stone, 48' wide	315	128	443
5850	72' wide	450	177	627
5900	96' wide	590	226	816

The Parking Lot System includes: compacted bank-run gravel; fine grading with a grader and roller; and bituminous concrete wearing course. All Parking Lot systems are on a cost per car basis. There are three basic types of systems: 90° angle, 60° angle, and 45° angle. The gravel base is compacted to 98%. Final stall design and lay-out of the parking lot with precast bumpers, sealcoating and white paint is also included.

The Expanded System Listing shows the three basic parking lot types with various depths of both gravel base and wearing course. The gravel base depths range from 6″ to 10″. The bituminous paving wearing course varies from a depth of 3″ to 6″.

System Components			COST PER CAR		
	QUANTITY	UNIT	MAT.	INST.	TOTAL
SYSTEM G2020 210 1500					
PARKING LOT, 90° ANGLE PARKING, 3″ BITUMINOUS PAVING, 6″ GRAVEL BASE					
Surveying crew for layout, 4 man crew	.020	Day		54.18	54.18
Borrow, bank run gravel, haul 2 mi., spread w/dozer, no compaction	7.223	L.C.Y.	148.07	60.45	208.52
Grading, fine grade 3 passes with motor grader	43.333	S.Y.		129.36	129.36
Compact w/vibrating plate, 8″ lifts, granular mat'l. to 98%	7.223	E.C.Y.		52.08	52.08
Binder course paving 2″ thick	43.333	S.Y.	351	61.97	412.97
Wear course paving 1″ thick	43.333	S.Y.	174.63	41.16	215.79
Seal coating, petroleum resistant under 1,000 S.Y.	43.333	S.Y.	66.73	63.70	130.43
Lines on pvmt., parking stall, paint, white, 4″ wide	1.000	Ea.	5.55	4.04	9.59
Precast concrete parking bar, 6″ x 10″ x 6′-0″	1.000	Ea.	69.50	21.50	91
TOTAL			815.48	488.44	1,303.92

G2020 210	Parking Lots Gravel Base	COST PER CAR		
		MAT.	INST.	TOTAL
1500	Parking lot, 90° angle parking, 3″ bituminous paving, 6″ gravel base	815	490	1,305
1520	8″ gravel base	865	525	1,390
1540	10″ gravel base	915	565	1,480
1560	4″ bituminous paving, 6″ gravel base	1,025	515	1,540
1580	8″ gravel base	1,075	555	1,630
1600	10″ gravel base	1,125	595	1,720
1620	6″ bituminous paving, 6″ gravel base	1,400	550	1,950
1640	8″ gravel base	1,425	585	2,010
1660	10″ gravel base	1,475	625	2,100
1800	60° angle parking, 3″ bituminous paving, 6″ gravel base	815	490	1,305
1820	8″ gravel base	865	525	1,390
1840	10″ gravel base	915	565	1,480
1860	4″ bituminous paving, 6″ gravel base	1,025	515	1,540
1880	8″ gravel base	1,075	555	1,630
1900	10″ gravel base	1,125	595	1,720
1920	6″ bituminous paving, 6″ gravel base	1,400	550	1,950
1940	8″ gravel base	1,425	585	2,010
1960	10″ gravel base	1,475	625	2,100
2200	45° angle parking, 3″ bituminous paving, 6″ gravel base	835	505	1,340
2220	8″ gravel base	885	540	1,425
2240	10″ gravel base	935	580	1,515
2260	4″ bituminous paving, 6″ gravel base	1,050	530	1,580
2280	8″ gravel base	1,100	570	1,670
2300	10″ gravel base	1,150	610	1,760
2320	6″ bituminous paving, 6″ gravel base	1,425	565	1,990
2340	8″ gravel base	1,475	600	2,075
2360	10″ gravel base	1,525	640	2,165

G2020 Parking Lots

The Parking Lot System includes: compacted crushed stone; fine grading with a grader and roller; and bituminous concrete wearing course. All Parking Lot systems are on a cost per car basis. There are three basic types of systems: 90° angle, 60° angle, and 45° angle. The crushed stone is compacted to 98%. Final stall design and lay-out of the parking lot with precast bumpers, sealcoating and white paint is also included.

The Expanded System Listing shows the three basic parking lot types with various depths of both crushed stone and wearing course. The crushed stone depths range from 6″ to 10″. The bituminous paving wearing course varies from a depth of 3″ to 6″.

System Components	QUANTITY	UNIT	COST PER CAR		
			MAT.	INST.	TOTAL
SYSTEM G2020 212 1500					
PARKING LOT, 90° ANGLE PARKING, 3″ BITUMINOUS PAVING, 6″ CRUSHED STONE					
Surveying crew for layout, 4 man crew	.020	Day		54.18	54.18
Borrow, 3/4″ crushed stone, haul 2 mi., spread w/dozer, no compaction	8.645	C.Y.	263.67	72.35	336.02
Grading, fine grade 3 passes with motor grader	43.333	S.Y.		254.80	254.80
Compact w/ vibrating plate, 8″ lifts, granular mat'l. to 98%	7.222	E.C.Y.		52.07	52.07
Binder course paving, 2″ thick	43.333	S.Y.	351	61.97	412.97
Wearing course, 1″ thick	43.333	S.Y.	174.63	41.16	215.79
Seal coating, petroleum resistant under 1,000 S.Y.	43.333	S.Y.	66.73	63.70	130.43
Lines on pvmt., parking stall, paint, white, 4″ wide	1.000	Ea.	5.55	4.04	9.59
Precast concrete parking bar, 6″ x 10″ x 6′-0″	1.000	Ea.	69.50	21.50	91
TOTAL			931.08	625.77	1,556.85

G2020 212	Parking Lots Crushed Stone	COST PER CAR		
		MAT.	INST.	TOTAL
1500	Parking lot, 90° angle parking, 3″ bituminous paving, 6″ crushed stone	930	625	1,555
1520	8″ 3/4 inch crushed stone	1,025	670	1,695
1540	10″ 3/4 inch crushed stone	1,100	710	1,810
1560	4″ bituminous paving, 6″ 3/4 inch crushed stone	1,150	650	1,800
1580	8″ 3/4 inch crushed stone	1,250	695	1,945
1600	10″ 3/4 inch crushed stone	1,325	735	2,060
1620	6″ bituminous paving, 6″ 3/4 inch crushed stone	1,500	685	2,185
1640	8″ 3/4 inch crushed stone	1,600	725	2,325
1660	10″ 3/4 inch crushed stone	1,675	765	2,440
1800	60° angle parking, 3″ bituminous paving, 6″ 3/4 inch crushed stone	930	625	1,555
1820	8″ 3/4 inch crushed stone	1,025	670	1,695
1840	10″ 3/4 inch crushed stone	1,100	710	1,810
1860	4″ bituminous paving, 6″ 3/4 inch crushed stone	1,150	650	1,800
1880	8″ 3/4 inch crushed stone	1,250	695	1,945
1900	10″ 3/4 inch crushed stone	1,325	735	2,060
1920	6″ bituminous paving, 6″ 3/4 inch crushed stone	1,425	685	2,110
1940	8″ 3/4 inch crushed stone	1,600	725	2,325
1960	10″ 3/4 inch crushed stone	1,675	765	2,440
2200	45° angle parking, 3″ bituminous paving, 6″ 3/4 inch crushed stone	780	600	1,380
2220	8″ 3/4 inch crushed stone	1,050	685	1,735
2240	10″ 3/4 inch crushed stone	1,125	725	1,850
2260	4″ bituminous paving, 6″ 3/4 inch crushed stone	1,175	665	1,840
2280	8″ 3/4 inch crushed stone	1,275	710	1,985
2300	10″ 3/4 inch crushed stone	1,350	755	2,105
2320	6″ bituminous paving, 6″ 3/4 inch crushed stone	1,550	705	2,255
2340	8″ 3/4 inch crushed stone	1,625	745	2,370
2360	10″ 3/4 inch crushed stone	1,725	790	2,515

G2020 Parking Lots

System Components			COST PER EACH		
	QUANTITY	UNIT	MAT.	INST.	TOTAL
SYSTEM G2020 214 1500					
PARKING LOT, 10 CARS WITH HANDICAP AND LIGHTING					
90 ° ANGLE PARKING, 3″ BITUMINOUS PAVING, 6″ CRUSHED STONE					
Surveying crew for layout, 4 man crew	.210	Day		568.89	568.89
Excavate for poles	4.740	B.C.Y.		601.98	601.98
Concrete pole bases	.940	C.Y.	210.56	282.98	493.54
Light poles	2.000	Ea.	2,400	1,403	3,803
Backfill around poles	3.800	L.C.Y.		174.80	174.80
Compact around poles	3.800	E.C.Y.		28.96	28.96
Power feed excavation	228.000	L.F.		230.28	230.28
Undergound conduit	228.000	L.F.	323.76	925.68	1,249.44
Power feed backfill	228.000	L.F.		159.60	159.60
Borrow, 3/4″ crushed stone, haul 2 mi., spread w/dozer, no compaction	91.000	C.Y.	2,775.50	761.67	3,537.17
Grading, fine grade 3 passes with motor grader	231.000	S.Y.		1,358.28	1,358.28
Compact crushed stone	75.840	E.C.Y.		402.71	402.71
Haul asphalt to site 5 miles	101.000	L.C.Y.		787.80	787.80
Binder course paving, 2″ thick	455.000	S.Y.	3,685.50	650.65	4,336.15
Wear course paving 1″ thick	455.000	S.Y.	1,833.65	432.25	2,265.90
Seal coating pavement	455.000	S.Y.	700.70	668.85	1,369.55
Precast concrete parking bar, 6″ x 10″ x 6'-0″	10.500	Ea.	729.75	225.75	955.50
Stall parking striping	10.500	Ea.	58.28	42.43	100.71
Handicap sign	1.000	Ea.	149	150	299
400 watt HPS luminaire	2.000	Ea.	1,820	660	2,480
Safety switch for lighting feed	1.000	Ea.	65.50	228	293.50
#8 wire feed for poles	4.560	C.L.F.	136.80	417.24	554.04
TOTAL			14,889	11,161.80	26,050.80

G2020 214	Parking Lots With Handicap & Lighting	COST PER EACH		
		MAT.	INST.	TOTAL
1490	Parking Lot, 90° Angle Parking, 3″ Bituminous Paving, 6″ Crushed Stone			
1500	10 cars with one handicap & lighting	14,900	11,200	26,100
1510	20 cars with one handicap	26,600	18,900	45,500
1520	30 cars with 2 handicap	41,200	29,100	70,300
1530	40 cars with 2 handicap	55,000	38,800	93,800
1540	50 cars with 3 handicap	67,500	46,900	114,400
1550	60 cars with 3 handicap	81,500	56,500	138,000
1560	70 cars with 4 handicap	96,000	66,500	162,500
1570	80 cars with 4 handicap	107,500	74,500	182,000
1580	90 cars with 5 handicap	122,500	85,000	207,500

G2020 Parking Lots

G2020 214	Parking Lots With Handicap & Lighting	COST PER EACH		
		MAT.	INST.	TOTAL
1590	100 cars with 5 handicap	134,000	92,500	226,500
1600	110 cars with 6 handicap	148,500	103,000	251,500
1610	120 cars with 6 handicap	160,500	110,500	271,000
1690	Parking Lot, 90° Angle Parking, 3" Bituminous Paving, 8" Crushed Stone			
1700	10 cars with one handicap & lighting	15,800	11,600	27,400
1710	20 cars with one handicap	28,300	19,600	47,900
1720	30 cars with 2 handicap	43,900	30,100	74,000
1730	40 cars with 2 handicap	59,000	40,300	99,300
1740	50 cars with 3 handicap	72,000	48,800	120,800
1750	60 cars with 3 handicap	87,000	59,000	146,000
1760	70 cars with 4 handicap	102,500	69,500	172,000
1770	80 cars with 4 handicap	115,000	77,500	192,500
1780	90 cars with 5 handicap	130,500	88,000	218,500
1790	100 cars with 5 handicap	143,000	96,000	239,000
1800	110 cars with 6 handicap	158,500	107,000	265,500
1810	120 cars with 6 handicap	171,000	115,000	286,000
1890	Parking Lot, 90° Angle Parking, 3" Bituminous Paving, 10" Crushed Stone			
1900	10 cars with one handicap & lighting	16,700	12,000	28,700
1910	20 cars with one handicap	30,200	20,300	50,500
1920	30 cars with 2 handicap	46,700	31,400	78,100
1930	40 cars with 2 handicap	62,500	41,800	104,300
1940	50 cars with 3 handicap	76,500	50,500	127,000
1950	60 cars with 3 handicap	92,500	61,000	153,500
1960	70 cars with 4 handicap	109,000	72,000	181,000
1970	80 cars with 4 handicap	122,500	81,000	203,500
1980	90 cars with 5 handicap	138,500	91,500	230,000
1990	100 cars with 5 handicap	152,000	100,000	252,000
2000	110 cars with 6 handicap	168,500	111,000	279,500
2010	120 cars with 6 handicap	182,000	119,500	301,500
2190	Parking Lot, 90° Angle Parking, 4" Bituminous Paving, 6" Crushed Stone			
2200	10 cars with one handicap & lighting	17,200	11,500	28,700
2210	20 cars with one handicap	31,100	19,500	50,600
2220	30 cars with 2 handicap	48,000	29,900	77,900
2230	40 cars with 2 handicap	64,000	39,900	103,900
2240	50 cars with 3 handicap	79,000	48,300	127,300
2250	60 cars with 3 handicap	95,000	58,500	153,500
2260	70 cars with 4 handicap	112,000	69,000	181,000
2270	80 cars with 4 handicap	125,500	77,000	202,500
2280	90 cars with 5 handicap	142,500	87,000	229,500
2290	100 cars with 5 handicap	156,500	95,000	251,500
2300	110 cars with 6 handicap	173,500	106,000	279,500
2310	120 cars with 6 handicap	187,500	114,000	301,500
2390	Parking Lot, 90° Angle Parking, 4" Bituminous Paving, 8" Crushed Stone			
2400	10 cars with one handicap & lighting	18,100	11,800	29,900
2410	20 cars with one handicap	32,800	20,200	53,000
2420	30 cars with 2 handicap	50,500	31,000	81,500
2430	40 cars with 2 handicap	68,000	41,400	109,400
2440	50 cars with 3 handicap	83,500	50,000	133,500
2450	60 cars with 3 handicap	100,500	60,500	161,000
2460	70 cars with 4 handicap	118,000	71,500	189,500
2470	80 cars with 4 handicap	133,000	79,500	212,500
2480	90 cars with 5 handicap	150,500	90,500	241,000
2490	100 cars with 5 handicap	165,500	99,000	264,500
2500	110 cars with 6 handicap	183,500	110,000	293,500
2510	120 cars with 6 handicap	198,000	118,500	316,500
2590	Parking Lot, 90° Angle Parking, 4" Bituminous Paving, 10" Crushed Stone			
2600	10 cars with one handicap & lighting	19,100	12,200	31,300
2610	20 cars with one handicap	34,700	20,900	55,600

G2020 Parking Lots

G2020 214	Parking Lots With Handicap & Lighting	COST PER EACH		
		MAT.	INST.	TOTAL
2620	30 cars with 2 handicap	53,500	32,200	85,700
2630	40 cars with 2 handicap	71,500	42,900	114,400
2640	50 cars with 3 handicap	88,000	52,000	140,000
2650	60 cars with 3 handicap	106,000	63,000	169,000
2660	70 cars with 4 handicap	124,500	74,000	198,500
2670	80 cars with 4 handicap	140,500	83,000	223,500
2680	90 cars with 5 handicap	159,000	94,000	253,000
2690	100 cars with 5 handicap	174,500	103,000	277,500
2700	110 cars with 6 handicap	193,500	114,000	307,500
2710	120 cars with 6 handicap	209,000	123,000	332,000
2790	Parking Lot, 90° Angle Parking, 5" Bituminous Paving, 6" Crushed Stone			
2800	10 cars with one handicap & lighting	19,100	11,700	30,800
2810	20 cars with one handicap	34,700	19,800	54,500
2820	30 cars with 2 handicap	53,500	30,400	83,900
2830	40 cars with 2 handicap	71,500	40,600	112,100
2840	50 cars with 3 handicap	88,000	49,200	137,200
2850	60 cars with 3 handicap	106,000	59,500	165,500
2860	70 cars with 4 handicap	124,500	70,500	195,000
2870	80 cars with 4 handicap	140,500	78,500	219,000
2880	90 cars with 5 handicap	159,000	88,500	247,500
2890	100 cars with 5 handicap	175,000	97,500	272,500
2900	110 cars with 6 handicap	193,500	108,000	301,500
2910	120 cars with 6 handicap	209,000	116,500	325,500
2990	Parking Lot, 90° Angle Parking, 5" Bituminous Paving, 8" Crushed Stone			
3000	10 cars with one handicap & lighting	20,000	12,000	32,000
3010	20 cars with one handicap	36,500	20,500	57,000
3020	30 cars with 2 handicap	56,000	31,600	87,600
3030	40 cars with 2 handicap	75,000	42,200	117,200
3040	50 cars with 3 handicap	92,500	51,000	143,500
3050	60 cars with 3 handicap	111,500	61,500	173,000
3060	70 cars with 4 handicap	131,000	73,000	204,000
3070	80 cars with 4 handicap	147,500	81,000	228,500
3080	90 cars with 5 handicap	167,000	92,000	259,000
3090	100 cars with 5 handicap	183,500	101,000	284,500
3100	110 cars with 6 handicap	203,500	112,000	315,500
3110	120 cars with 6 handicap	220,000	120,500	340,500
3190	Parking Lot, 90° Angle Parking, 5" Bituminous Paving, 10" Crushed Stone			
3200	10 cars with one handicap & lighting	20,900	12,400	33,300
3210	20 cars with one handicap	38,300	21,300	59,600
3220	30 cars with 2 handicap	59,000	32,700	91,700
3230	40 cars with 2 handicap	78,500	43,700	122,200
3240	50 cars with 3 handicap	97,000	53,000	150,000
3250	60 cars with 3 handicap	117,000	64,000	181,000
3260	70 cars with 4 handicap	137,500	75,500	213,000
3270	80 cars with 4 handicap	155,000	84,500	239,500
3280	90 cars with 5 handicap	175,500	95,500	271,000
3290	100 cars with 5 handicap	193,000	104,500	297,500
3300	110 cars with 6 handicap	213,500	116,500	330,000
3310	120 cars with 6 handicap	231,000	125,000	356,000
3390	Parking Lot, 90° Angle Parking, 6" Bituminous Paving, 6" Crushed Stone			
3400	10 cars with one handicap & lighting	20,900	11,800	32,700
3410	20 cars with one handicap	38,300	20,100	58,400
3420	30 cars with 2 handicap	59,000	31,000	90,000
3430	40 cars with 2 handicap	78,500	41,300	119,800
3440	50 cars with 3 handicap	97,000	50,000	147,000
3450	60 cars with 3 handicap	116,500	60,500	177,000
3460	70 cars with 4 handicap	137,500	71,000	208,500
3470	80 cars with 4 handicap	154,500	79,500	234,000

G20 Site Improvements

G2020 Parking Lots

G2020 214	Parking Lots With Handicap & Lighting	COST PER EACH		
		MAT.	INST.	TOTAL
3480	90 cars with 5 handicap	175,500	90,000	265,500
3490	100 cars with 5 handicap	193,000	99,000	292,000
3500	110 cars with 6 handicap	213,500	109,500	323,000
3510	120 cars with 6 handicap	231,000	118,000	349,000
3590	Parking Lot, 90° Angle Parking, 6" Bituminous Paving, 8" Crushed Stone			
3600	10 cars with one handicap & lighting	21,800	12,200	34,000
3610	20 cars with one handicap	40,100	20,800	60,900
3620	30 cars with 2 handicap	61,500	32,100	93,600
3630	40 cars with 2 handicap	82,500	42,800	125,300
3640	50 cars with 3 handicap	101,500	52,000	153,500
3650	60 cars with 3 handicap	122,000	62,500	184,500
3660	70 cars with 4 handicap	143,500	74,000	217,500
3670	80 cars with 4 handicap	162,000	82,500	244,500
3680	90 cars with 5 handicap	183,500	93,500	277,000
3690	100 cars with 5 handicap	201,500	102,500	304,000
3700	110 cars with 6 handicap	223,500	114,000	337,500
3710	120 cars with 6 handicap	241,500	122,500	364,000
3790	Parking Lot, 90° Angle Parking, 6" Bituminous Paving, 10" Crushed Stone			
3800	10 cars with one handicap & lighting	22,800	12,600	35,400
3810	20 cars with one handicap	41,900	21,600	63,500
3820	30 cars with 2 handicap	64,500	33,300	97,800
3830	40 cars with 2 handicap	86,000	44,300	130,300
3840	50 cars with 3 handicap	106,000	54,000	160,000
3850	60 cars with 3 handicap	127,500	65,000	192,500
3860	70 cars with 4 handicap	150,000	76,500	226,500
3870	80 cars with 4 handicap	169,000	85,500	254,500
3880	90 cars with 5 handicap	191,500	97,000	288,500
3890	100 cars with 5 handicap	211,000	106,000	317,000
3900	110 cars with 6 handicap	233,500	118,000	351,500
3910	120 cars with 6 handicap	252,500	127,000	379,500

661

G2030 Pedestrian Paving

The Bituminous Sidewalk System includes: excavation; compacted gravel base (hand graded), bituminous surface; and hand grading along the edge of the completed walk.

The Expanded System Listing shows Bituminous Sidewalk systems with wearing course depths ranging from 1" to 2-1/2" of bituminous material. The gravel base ranges from 4" to 8". Sidewalk widths are shown ranging from 3' to 5'. Costs are on a linear foot basis.

System Components	QUANTITY	UNIT	COST PER L.F.		
			MAT.	INST.	TOTAL
SYSTEM G2030 110 1580					
BITUMINOUS SIDEWALK, 1" THICK PAVING, 4" GRAVEL BASE, 3' WIDTH					
Excavation, bulk, dozer, push 50'	.046	B.C.Y.		.11	.11
Borrow, bank run gravel, haul 2 mi., spread w/dozer, no compaction	.037	L.C.Y.	.76	.31	1.07
Compact w/vib. plate, 8" lifts	.037	E.C.Y.		.10	.10
Fine grade, area to be paved, small area	.333	S.Y.		4.05	4.05
Asphaltic concrete, 2" thick	.333	S.Y.	1.32	.81	2.13
Backfill by hand, no compaction, light soil	.006	L.C.Y.		.22	.22
TOTAL			2.08	5.60	7.68

G2030 110	Bituminous Sidewalks	COST PER L.F.		
		MAT.	INST.	TOTAL
1580	Bituminous sidewalk, 1" thick paving, 4" gravel base, 3' width	2.08	5.60	7.68
1600	4' width	2.75	6	8.75
1620	5' width	3.51	6.30	9.81
1640	6" gravel base, 3' width	2.47	5.85	8.32
1660	4' width	3.27	6.35	9.62
1680	5' width	4.11	6.90	11.01
1700	8" gravel base, 3' width	2.84	6.10	8.94
1720	4' width	3.78	6.70	10.48
1740	5' width	4.74	7.30	12.04
1800	1-1/2" thick paving, 4" gravel base, 3' width	2.81	5.85	8.66
1820	4' width	3.70	6.40	10.10
1840	5' width	4.56	7.25	11.81
1860	6" gravel base, 3' width	3.20	6.10	9.30
1880	4' width	4.15	7	11.15
1900	5' width	5.30	7.40	12.70
1920	8" gravel base, 3' width	3.50	6.60	10.10
1940	4' width	4.73	7.10	11.83
1960	5' width	5.85	8.10	13.95
2120	2" thick paving, 4" gravel base, 3' width	3.39	6.65	10.04
2140	4' width	4.51	7.35	11.86
2160	5' width	5.65	8.10	13.75
2180	6" gravel base, 3' width	3.78	6.90	10.68
2200	4' width	5.05	7.65	12.70
2240	8" gravel base, 3' width	4.15	7.10	11.25
2280	5' width	6.95	8.90	15.85
2400	2-1/2" thick paving, 4" gravel base, 3' width	4.95	7.35	12.30
2420	4' width	6.60	8.25	14.85
2460	6" gravel base, 3' width	5.35	7.60	12.95
2500	5' width	8.95	9.60	18.55
2520	8" gravel base, 3' width	5.70	7.80	13.50
2540	4' width	7.60	8.95	16.55
2560	5' width	9.60	10.05	19.65

662

For customer support on your **Site Work & Landscape Costs** with RSMeans data, call 800.448.8182.

G2030 Pedestrian Paving

The Concrete Sidewalk System includes: excavation; compacted gravel base (hand graded); forms; welded wire fabric; and 3,000 p.s.i. air-entrained concrete (broom finish).

The Expanded System Listing shows Concrete Sidewalk systems with wearing course depths ranging from 4" to 6". The gravel base ranges from 4" to 8". Sidewalk widths are shown ranging from 3' to 5'. Costs are on a linear foot basis.

System Components			COST PER L.F.		
	QUANTITY	UNIT	MAT.	INST.	TOTAL
SYSTEM G2030 120 1580					
CONCRETE SIDEWALK 4" THICK, 4" GRAVEL BASE, 3' WIDE					
Excavation, box out with dozer	.100	B.C.Y.		.23	.23
Gravel base, haul 2 miles, spread with dozer	.037	L.C.Y.	.76	.31	1.07
Compaction with vibrating plate	.037	E.C.Y.		.10	.10
Fine grade by hand	.333	S.Y.		4.11	4.11
Concrete in place including forms and reinforcing	.037	C.Y.	7.62	8.67	16.29
Backfill edges by hand	.010	L.C.Y.		.36	.36
TOTAL			8.38	13.78	22.16

G2030 120	Concrete Sidewalks	COST PER L.F.		
		MAT.	INST.	TOTAL
1580	Concrete sidewalk, 4" thick, 4" gravel base, 3' wide	8.40	13.80	22.20
1600	4' wide	11.15	16.85	28
1620	5' wide	13.95	19.95	33.90
1640	6" gravel base, 3' wide	8.75	14.05	22.80
1660	4' wide	11.70	17.20	28.90
1680	5' wide	14.60	20.50	35.10
1700	8" gravel base, 3' wide	9.15	14.30	23.45
1720	4' wide	12.20	17.50	29.70
1740	5' wide	15.20	21	36.20
1800	5" thick concrete, 4" gravel base, 3' wide	10.20	14.85	25.05
1820	4' wide	13.60	18.25	31.85
1840	5' wide	17	21.50	38.50
1860	6" gravel base, 3' wide	10.60	15.10	25.70
1880	4' wide	14.10	18.60	32.70
1900	5' wide	17.65	21	38.65
1920	8" gravel base, 3' wide	10.95	15.35	26.30
1940	4' wide	14.65	20	34.65
1960	5' wide	18.25	21.50	39.75
2120	6" thick concrete, 4" gravel base, 3' wide	11.75	15.65	27.40
2140	4' wide	15.70	19.25	34.95
2160	5' wide	19.60	23	42.60
2180	6" gravel base, 3' wide	12.15	15.90	28.05
2200	4' wide	16.20	19.60	35.80
2220	5' wide	20.50	23	43.50
2240	8" gravel base, 3' wide	12.55	16.20	28.75
2260	4' wide	16.70	19.95	36.65
2280	5' wide	21	24	45

663

G2030 Pedestrian Paving

The Plaza Systems listed include several brick and tile paving surfaces on three different bases: gravel, slab on grade and suspended slab. The system cost includes this base cost with the exception of the suspended slab. The type of bedding for the pavers depends on the base being used, and alternate bedding may be desirable. Also included in the paving costs are edging and precast grating costs and where concrete bases are involved, expansion joints.

System Components	QUANTITY	UNIT	COST PER S.F.		
			MAT.	INST.	TOTAL
SYSTEM G2030 150 2050					
PLAZA, BRICK PAVERS, 4″ X 8″ X 1-3/4″, GRAVEL BASE, STONE DUST BED					
Compact subgrade, static roller, 4 passes	.111	S.Y.		.05	.05
Bank gravel, 2 mi haul, dozer spread	.012	L.C.Y.	.25	.10	.35
Compact gravel bedding or base, vibrating plate	.012	E.C.Y.		.04	.04
Grading fine grade, 3 passes with grader	.111	S.Y.		.65	.65
Coarse washed sand bed, 1″ thick	.003	C.Y.	.06	.08	.14
Brick paver, 4″ x 8″ x 1-3/4″	4.150	Ea.	4.44	4.74	9.18
Brick edging, stood on end, 3 per L.F.	.060	L.F.	.35	.70	1.05
Concrete tree grate, 5′ square	.004	Ea.	1.89	.30	2.19
TOTAL			6.99	6.66	13.65

G2030 150	Brick & Tile Plazas	COST PER S.F.		
		MAT.	INST.	TOTAL
1050	Plaza, asphalt pavers, 6″ x 12″ x 1-1/4″, gravel base, asphalt bedding	13.70	12.25	25.95
1100	Slab on grade, asphalt bedding	16.40	13.25	29.65
1150	Suspended slab, insulated & mastic bedding	18.25	16.20	34.45
1300	6″ x 12″ x 3″, gravel base, asphalt, bedding	18.95	12.65	31.60
1350	Slab on grade, asphalt bedding	21	13.55	34.55
1400	Suspended slab, insulated & mastic bedding	23	16.50	39.50
2050	Brick pavers, 4″ x 8″ x 1-3/4″, gravel base, stone dust bedding	7	6.65	13.65
2100	Slab on grade, asphalt bedding	9.65	9.55	19.20
2150	Suspended slab, insulated & no bedding	11.50	12.50	24
2300	4″ x 8″ x 2-1/4″, gravel base, stone dust bedding	5.60	6.65	12.25
2350	Slab on grade, asphalt bedding	8.25	9.55	17.80
2400	Suspended slab, insulated & no bedding	10.10	12.50	22.60
2550	Shale pavers, 4″ x 8″ x 2-1/4″, gravel base, stone dust bedding	6.70	7.60	14.30
2600	Slab on grade, asphalt bedding	9.35	10.50	19.85
2650	Suspended slab, insulated & no bedding	6	8.80	14.80
3050	Thin set tile, 4″ x 4″ x 3/8″, slab on grade	9.45	10.60	20.05
3300	4″ x 4″ x 3/4″, slab on grade	11	7.10	18.10
3550	Concrete paving stone, 4″ x 8″ x 2-1/2″, gravel base, sand bedding	5.25	4.71	9.96
3600	Slab on grade, asphalt bedding	7.35	7	14.35
3650	Suspended slab, insulated & no bedding	4.04	5.30	9.34
3800	4″ x 8″ x 3-1/4″, gravel base, sand bedding	5.25	4.71	9.96
3850	Slab on grade, asphalt bedding	6.70	5.40	12.10
3900	Suspended slab, insulated & no bedding	4.04	5.30	9.34
4050	Concrete patio blocks, 8″ x 16″ x 2″, gravel base, sand bedding	14.50	6.10	20.60
4100	Slab on grade, asphalt bedding	16.80	8.85	25.65
4150	Suspended slab, insulated & no bedding	13.25	6.70	19.95

G2030 Pedestrian Paving

G2030 150	Brick & Tile Plazas	COST PER S.F.		
		MAT.	INST.	TOTAL
4300	16" x 16" x 2", gravel base, sand bedding	19.45	4.92	24.37
4350	Slab on grade, asphalt bedding	21	6.05	27.05
4400	Suspended slab, insulated & no bedding	18.40	6	24.40
4550	24" x 24" x 2", gravel base, sand bedding	23.50	3.75	27.25
4600	Slab on grade, asphalt bedding	26	6.50	32.50
4650	Suspended slab, insulated & no bedding	22.50	4.83	27.33
5050	Bluestone flagging, 3/4" thick irregular, gravel base, sand bedding	16.20	14.45	30.65
5100	Slab on grade, mastic bedding	22.50	19.60	42.10
5300	1" thick, irregular, gravel base, sand bedding	14.15	15.85	30
5350	Slab on grade, mastic bedding	20.50	21	41.50
5550	Flagstone, 3/4" thick irregular, gravel base, sand bedding	26.50	13.75	40.25
5600	Slab on grade, mastic bedding	33.50	18.90	52.40
5800	1-1/2" thick, random rectangular, gravel base, sand bedding	28.50	15.85	44.35
5850	Slab on grade, mastic bedding	35.50	21	56.50
6050	Granite pavers, 3-1/2" x 3-1/2" x 3-1/2", gravel base, sand bedding	26.50	14.55	41.05
6100	Slab on grade, mortar bedding	30	23	53
6300	4" x 4" x 4", gravel base, sand bedding	28	14.15	42.15
6350	Slab on grade, mortar bedding	30.50	18.90	49.40
6550	4" x 12" x 4", gravel base, sand bedding	23	13.80	36.80
6600	Slab on grade, mortar bedding	25	18.55	43.55
6800	6" x 15" x 4", gravel base, sand bedding	13.65	13.05	26.70
6850	Slab on grade, mortar bedding	15.95	17.80	33.75
7050	Limestone, 3" thick, gravel base, sand bedding	14.95	18	32.95
7100	Slab on grade, mortar bedding	17.25	22.50	39.75
7300	4" thick, gravel base, sand bedding	18.80	18.45	37.25
7350	Slab on grade, mortar bedding	21	23	44
7550	5" thick, gravel base, sand bedding	22.50	18.95	41.45
7600	Slab on grade, mortar bedding	25	23.50	48.50
8050	Slate flagging, 3/4" thick, gravel base, sand bedding	13.80	14.55	28.35
8100	Slab on grade, mastic bedding	29	21	50
8300	1" thick, gravel base, sand bedding	15.60	15.60	31.20
8350	Slab on grade, mastic bedding	22.50	21.50	44

G2030 Pedestrian Paving

The Step System has three basic types: railroad tie, cast-in-place concrete or brick with a concrete base. System elements include: gravel base compaction; and backfill.

Wood Step Systems use 6″ x 6″ railroad ties that produce 3′ to 6′ wide steps that range from 2-riser to 5-riser configurations. Cast in Place Concrete Step Systems are either monolithic or aggregate finish. They range from 3′ to 6′ in width. Concrete Steps Systems are either in a 2-riser or 5-riser configuration. Precast Concrete Step Systems are listed for 4′ to 7′ widths with both 2-riser and 5-riser models. Brick Step Systems are placed on a 12″ concrete base. The size range is the same as cost in place concrete. Costs are on a per unit basis. All systems are assumed to include a full landing at the top 4′ long.

System Components	QUANTITY	UNIT	COST PER EACH		
			MAT.	INST.	TOTAL
SYSTEM G2030 310 0960					
STAIRS, RAILROAD TIES, 6″ X 8″, 3′ WIDE, 2 RISERS					
Excavation, by hand, sandy soil	1.327	C.Y.		84.26	84.26
Borrow fill, bank run gravel	.664	C.Y.	12.35		12.35
Delivery charge	.664	C.Y.		19.22	19.22
Backfill compaction, 6″ layers, hand tamp	.664	C.Y.		16.27	16.27
Railroad ties, wood, creosoted, 6″ x 8″ x 8′-6″, C.L. lots	5.000	Ea.	307.50	190.55	498.05
Backfill by hand, no compaction, light soil	.800	C.Y.		28.80	28.80
Remove excess spoil, 12 C.Y. dump truck	1.200	L.C.Y.		8.11	8.11
TOTAL			319.85	347.21	667.06

G2030 310	Stairs	COST PER EACH		
		MAT.	INST.	TOTAL
0960	Stairs, railroad ties, 6″ x 8″, 3′ wide, 2 risers	320	345	665
0980	5 risers	755	705	1,460
1000	4′ wide, 2 risers	320	380	700
1020	5 risers	760	735	1,495
1040	5′ wide, 2 risers	385	450	835
1060	5 risers	950	900	1,850
1080	6′ wide, 2 risers	390	490	880
1100	5 risers	950	935	1,885
2520	Concrete, cast in place, 3′ wide, 2 risers	229	500	729
2540	5 risers	305	710	1,015
2560	4′ wide, 2 risers	282	655	937
2580	5 risers	365	910	1,275
2600	5′ wide, 2 risers	335	815	1,150
2620	5 risers	425	1,100	1,525
2640	6′ wide, 2 risers	380	950	1,330
2660	5 risers	490	1,275	1,765
2800	Exposed aggregate finish, 3′ wide, 2 risers	233	510	743
2820	5 risers	310	710	1,020
2840	4′ wide, 2 risers	287	655	942
2860	5 risers	375	915	1,290
2880	5′ wide, 2 risers	340	815	1,155
2900	5 risers	435	1,100	1,535
2920	6′ wide, 2 risers	390	950	1,340
2940	5 risers	505	1,275	1,780
3200	Precast, 4′ wide, 2 risers	675	365	1,040
3220	5 risers	1,075	540	1,615

G2030 Pedestrian Paving

G2030 310	Stairs	COST PER EACH		
		MAT.	INST.	TOTAL
3240	5' wide, 2 risers	760	435	1,195
3260	5 risers	1,250	580	1,830
3280	6' wide, 2 risers	840	460	1,300
3300	5 risers	1,375	625	2,000
3320	7' wide, 2 risers	1,200	490	1,690
3340	5 risers	1,975	675	2,650
4100	Brick, incl 12" conc base, 3' wide, 2 risers	380	1,275	1,655
4120	5 risers	550	1,900	2,450
4140	4' wide, 2 risers	475	1,575	2,050
4160	5 risers	680	2,350	3,030
4180	5' wide, 2 risers	570	1,875	2,445
4200	5 risers	805	2,775	3,580
4220	6' wide, 2 risers	660	2,175	2,835
4240	5 risers	940	3,175	4,115

G20 Site Improvements

G2040 Site Development

There are four basic types of Concrete Retaining Wall Systems: reinforced concrete with level backfill; reinforced concrete with sloped backfill or surcharge; unreinforced with level backfill; and unreinforced with sloped backfill or surcharge. System elements include: all necessary forms (4 uses); 3,000 p.s.i. concrete with an 8″ chute; all necessary reinforcing steel; and underdrain. Exposed concrete is patched and rubbed.

The Expanded System Listing shows walls that range in thickness from 10″ to 18″ for reinforced concrete walls with level backfill and 12″ to 24″ for reinforced walls with sloped backfill. Walls range from a height of 4′ to 20′. Unreinforced level and sloped backfill walls range from a height of 3′ to 10′.

System Components	QUANTITY	UNIT	COST PER L.F.		
			MAT.	INST.	TOTAL
SYSTEM G2040 210 1000					
CONC. RETAIN. WALL REINFORCED, LEVEL BACKFILL, 4′ HIGH					
Forms in place, cont. wall footing & keyway, 4 uses	2.000	S.F.	4.92	9.98	14.90
Forms in place, retaining wall forms, battered to 8′ high, 4 uses	8.000	SFCA	6.24	76.40	82.64
Reinforcing in place, walls, #3 to #7	.004	Ton	5	3.62	8.62
Concrete ready mix, regular weight, 3000 psi	.204	C.Y.	28.97		28.97
Placing concrete and vibrating footing con., shallow direct chute	.074	C.Y.		1.96	1.96
Placing concrete and vibrating walls, 8″ thick, direct chute	.130	C.Y.		4.64	4.64
Pipe bedding, crushed or screened bank run gravel	1.000	L.F.	2.53	1.37	3.90
Pipe, subdrainage, corrugated plastic, 4″ diameter	1.000	L.F.	.83	.84	1.67
Finish walls and break ties, patch walls	4.000	S.F.	.20	4.36	4.56
TOTAL			48.69	103.17	151.86

G2040 210	Concrete Retaining Walls	COST PER L.F.		
		MAT.	INST.	TOTAL
1000	Conc. retain. wall, reinforced, level backfill, 4′ high x 2′-2″ base, 10″ th	48.50	103	151.50
1200	6′ high x 3′-3″ base, 10″ thick	70	149	219
1400	8′ high x 4′-3″ base, 10″ thick	91	196	287
1600	10′ high x 5′-4″ base, 13″ thick	118	287	405
2200	16′ high x 8′-6″ base, 16″ thick	224	465	689
2600	20′ high x 10′-5″ base, 18″ thick	330	600	930
3000	Sloped backfill, 4′ high x 3′-2″ base, 12″ thick	59.50	107	166.50
3200	6′ high x 4′-6″ base, 12″ thick	84	153	237
3400	8′ high x 5′-11″ base, 12″ thick	112	202	314
3600	10′ high x 7′-5″ base, 16″ thick	159	300	459
3800	12′ high x 8′-10″ base, 18″ thick	208	365	573
4200	16′ high x 11′-10″ base, 21″ thick	350	515	865
4600	20′ high x 15′-0″ base, 24″ thick	545	690	1,235
5000	Unreinforced, level backfill, 3′-0″ high x 1′-6″ base	26	69.50	95.50
5200	4′-0″ high x 2′-0″ base	40	93	133
5400	6′-0″ high x 3′-0″ base	75.50	143	218.50
5600	8′-0″ high x 4′-0″ base	121	191	312
5800	10′-0″ high x 5′-0″ base	180	296	476
7000	Sloped backfill, 3′-0″ high x 2′-0″ base	31	72	103
7200	4′-0″ high x 3′-0″ base	52.50	96.50	149
7400	6′-0″ high x 5′-0″ base	110	153	263
7600	8′-0″ high x 7′-0″ base	188	207	395
7800	10′-0″ high x 9′-0″ base	290	320	610

For customer support on your Site Work & Landscape Costs with RSMeans data, call 800.448.8182.

G2040 Site Development

The Masonry Retaining Wall System includes: a reinforced concrete foundation footing for the wall with all necessary forms and 3,000 p.s.i. concrete; sand aggregate concrete blocks with steel reinforcing; solid grouting and underdrain.
The Expanded System Listing shows walls that range in thickness from 8″ to 12″ and in height from 3′-4″ to 8′.

System Components	QUANTITY	UNIT	COST PER L.F.		
			MAT.	INST.	TOTAL
SYSTEM G2040 220 1000					
REINF. CMU WALL, LEVEL FILL, 8″ THICK, 3′-4″ HIGH, 2′-4″ CONC. FTG.					
Forms in place, continuous wall footings 4 uses	1.670	S.F.	4.11	8.33	12.44
Joint reinforcing, #5 & #6 steel bars, vertical	5.590	Lb.	3.46	5.42	8.88
Concrete block, found. wall, cut joints, 8″ x 16″ x 8″ block	3.330	S.F.	11.82	22.81	34.63
Grout concrete block cores solid, 8″ thick, .258 CF/SF, pumped	3.330	S.F.	5.06	12.16	17.22
Concrete ready mix, regular wt., 3000 psi	.070	C.Y.	9.62		9.62
Pipe bedding, crushed or screened bank run gravel	1.000	L.F.	2.53	1.37	3.90
Pipe, subdrainage, corrugated plastic, 4″ diameter	1.000	L.F.	.83	.84	1.67
TOTAL			37.43	50.93	88.36

G2040 220	Masonry Retaining Walls	COST PER L.F.		
		MAT.	INST.	TOTAL
1000	Wall, reinforced CMU, level fill, 8″ thick, 3′-4″ high, 2′-4″ base	37.50	51	88.50
1200	4′-0″ high x 2′-9″ base	43	61.50	104.50
1400	4′-8″ high x 3′-3″ base	52	72.50	124.50
1600	5′-4″ high x 3′-8″ base	59.50	83	142.50
1800	6′-0″ high x 4′-2″ base	71	94.50	165.50
1840	10″ thick, 4′-0″ high x 2′-10″ base	44	62.50	106.50
1860	4′-8″ high x 3′-4″ base	53	73.50	126.50
1880	5′-4″ high x 3′-10″ base	61	83.50	144.50
1900	6′-0″ high x 4′-4″ base	72.50	95.50	168
1901	6′-0″ high x 4′-4″ base	72.50	95.50	168
1920	6′-8″ high x 4′-10″ base	80	105	185
1940	7′-4″ high x 5′-4″ base	94.50	120	214.50
2000	12″ thick, 5′-4″ high x 4′-0″ base	72.50	107	179.50
2200	6′-0″ high x 4′-6″ base	86.50	123	209.50
2400	6′-8″ high x 5′-0″ base	95.50	135	230.50
2600	7′-4″ high x 5′-6″ base	112	153	265
2800	8′-0″ high x 5′-11″ base	121	166	287

G2040 Site Development

Wood Post Retaining Wall Systems are either constructed of redwood, cedar, pressure-treated lumber or creosoted lumber. System elements include wood posts installed side-by-side to various heights; and underdrain.

The Expanded System Listing shows a wide variety of configurations with each type of wood. Costs are on a linear foot basis. Each grouping in the listing shows walls with either 4" x 4", 6" x 6", 8" x 8", or random diameter lumber.

System Components	QUANTITY	UNIT	COST PER L.F.		
			MAT.	INST.	TOTAL
SYSTEM G2040 230 1000					
WOOD POST RETAINING WALL, SIDE-BY-SIDE 4 X 4 REDWOOD, 1' HIGH					
Redwood post, columns and girts, 4" x 4"	8.000	B.F.	43.60	19.60	63.20
Pipe bedding, crushed or screened bank run gravel	1.000	L.F.	.69	.38	1.07
Pipe, subdrainage, corrugated plastic, 4" diameter	1.000	L.F.	.83	.84	1.67
TOTAL			45.12	20.82	65.94

G2040 230	Wood Post Retaining Walls	COST PER L.F.		
		MAT.	INST.	TOTAL
1000	Wood post retaining wall, side by side 4 x 4, redwood, 1' high	45	21	66
1200	2' high	89.50	40.50	130
1400	Cedar, 1' high	28.50	21	49.50
1600	2' high	56.50	40.50	97
1800	Pressure treated lumber, 1' high	10.50	18.20	28.70
2000	2' high	20	35.50	55.50
2200	Creosoted lumber, 1' high	9.40	16.10	25.50
2400	2' high	16.75	29	45.75
2600	6 x 6, redwood, 2' high	224	49	273
2800	4' high	445	97	542
3000	Cedar, 2' high	127	49	176
3200	4' high	253	97	350
3400	Pressure treated lumber, 2' high	37	40	77
3600	4' high	73	79	152
3800	Creosoted lumber, 2' high	29.50	32	61.50
4000	4' high	58.50	63	121.50
4200	8 x 8, redwood, 4' high	570	106	676
4400	6' high	1,075	198	1,273
4600	Cedar, 4' high	295	106	401
4800	6' high	555	198	753
5000	Pressure treated lumber, 4' high	120	123	243
5200	6' high	224	230	454
5400	Creosoted lumber, 4' high	97.50	100	197.50
5600	6' high	181	185	366
5800	Random diameter poles, cedar, 4' high	58	41	99
6000	6' high	86	62	148
6200	8' high	115	82	197
7000	Creosoted lumber, 4' high	22	23	45
7200	6' high	32	33.50	65.50
7400	8' high	43.50	45.50	89

G2040 Site Development

A Post and Board Retaining Wall System is constructed of one of four types of lumber: redwood, cedar, creosoted wood or pressure-treated lumber. The system includes all the elements that must go into a wall that resists lateral pressure. System elements include: posts and boards; cap; underdrain; and deadman where required.

The Expanded System Listing shows a wide variety of wall configurations with each type of lumber. The spacing of the deadman in the wall is indicated for each wall system. Wall heights vary from 3' to 10'.

System Components	QUANTITY	UNIT	COST PER L.F.		
			MAT.	INST.	TOTAL
SYSTEM G2040 240 1000					
POST & BOARD RETAINING WALL, 2″ PLANKS, 4 X 4 POSTS, 3′ HIGH					
Redwood post, 4″ x 4″	2.000	B.F.	10.90	4.90	15.80
Redwood joist and cap, 2″ x 10″	7.000	B.F.	36.60	5.84	42.44
Pipe bedding, crushed or screened bank run gravel	1.000	L.F.	1.27	.69	1.96
Pipe, subdrainage, corrugated plastic, 4″ diameter	1.000	L.F.	.83	.84	1.67
TOTAL			49.60	12.27	61.87

G2040 240	Post & Board Retaining Walls	COST PER L.F.		
		MAT.	INST.	TOTAL
1000	Post & board retaining wall, 2″ planking, redwood, 4 x 4 post, 4′ spacing, 3′ high	49.50	12.30	61.80
1200	2′ spacing, 4′ high	68.50	21	89.50
1400	6 x 6 post, 4′ spacing, 5′ high	123	26	149
1600	3′ spacing, 6′ high	174	37	211
1800	6 x 6 post with deadman, 3′ spacing, 8′ high	229	48.50	277.50
2000	8 x 8 post, 2′ spacing, 8′ high	505	95.50	600.50
2200	8 x 8 post with deadman, 2′ spacing, 9′ high	580	105	685
2400	Cedar, 4 x 4 post, 4′ spacing, 3′ high	27.50	12.85	40.35
2600	4 x 4 post with deadman, 2′ spacing, 4′ high	56.50	22.50	79
2800	6 x 6 post, 4′ spacing, 4′ high	58.50	22	80.50
3000	3′ spacing, 5′ high	85	32	117
3200	2′ spacing, 6′ high	136	51.50	187.50
3400	6 x 6 post with deadman, 3′ spacing, 6′ high	104	39.50	143.50
3600	8 x 8 post with deadman, 2′ spacing, 8′ high	290	98.50	388.50
3800	Pressure treated, 4 x 4 post, 4′ spacing, 3′ high	10.15	23.50	33.65
4000	3′ spacing, 4′ high	15.35	33.50	48.85
4200	4 x 4 post with deadman, 2′ spacing, 5′ high	34.50	46	80.50
4400	6 x 6 post, 3′ spacing, 6′ high	68	76.50	144.50
4600	2′ spacing, 7′ high	49.50	78	127.50
4800	6 x 6 post with deadman, 3′ spacing, 8′ high	54	75.50	129.50
5000	8 x 8 post with deadman, 2′ spacing, 10′ high	159	173	332
5200	Creosoted, 4 x 4 post, 4′ spacing, 3′ high	10.15	23.50	33.65
5400	3′ spacing, 4′ high	15.35	33.50	48.85
5600	4 x 4 post with deadman, 2′ spacing, 5′ high	34.50	46	80.50
5800	6 x 6 post, 3′ spacing, 6′ high	33.50	58	91.50
6000	2′ spacing, 7′ high	49.50	78	127.50
6200	6 x 6 post with deadman, 3′ spacing 8′ high	54	75.50	129.50
6400	8 x 8 post with deadman, 2′ spacing, 10′ high	159	173	332

G2040 Site Development

A Wood Tie Retaining Wall System is constructed of one of four types of lumber: redwood, cedar, creosoted wood or pressure-treated lumber. The system includes all the elements that must go into a wall that resists lateral pressure. System elements include: wood ties; threaded rod (1/2″ diameter); underdrain; and deadman where required.

The Expanded System Listing shows a wide variety of wall configurations with each type of lumber. The spacing of the deadman in the wall is indicated for each wall system. Wall heights vary from 2′ to 8′.

System Components	QUANTITY	UNIT	COST PER L.F.		
			MAT.	INST.	TOTAL
SYSTEM G2040 250 1000					
WOOD TIE WALL, ROD CONNECTOR AT 4′-0″, REDWOOD, 6″ X 6″, 2′-0″ HIGH					
Redwood post columns and girts, 6″ x 6″	12.000	B.F.	110.70	23.70	134.40
Accessories, hangers, coil threaded rods, continuous, 1/2″ diam.	.050	L.F.	.82		.82
Pipe bedding, crushed or screened bank run gravel	1.000	L.F.	1.27	.69	1.96
Pipe, subdrainage, corrugated plastic, 4″ diameter	1.000	L.F.	.83	.84	1.67
TOTAL			113.62	25.23	138.85

G2040 250	Wood Tie Retaining Walls	COST PER L.F.		
		MAT.	INST.	TOTAL
1000	Wood tie wall, rod connector at 4′-0″, redwood, 6″ x 6″, 2′-0″ high	114	25	139
1400	Deadman at 6′-0″, 4′-0″ high	263	57.50	320.50
1600	6′-0″ high	395	86	481
2000	8 x 8, deadman at 6′-0″, 4′-0″ high	390	73	463
2200	6′-0″ high	590	110	700
2400	8′-0″ high	780	146	926
3000	Cedar, 6 x 6, 2′-0″ high	65.50	25	90.50
3400	Deadman at 6′-0″, 4′-0″ high	150	57.50	207.50
3600	6′-0″ high	225	86	311
4000	8 x 8, deadman at 6′-0″, 4′-0″ high	204	73	277
4200	6′-0″ high	310	110	420
4400	8′-0″ high	410	146	556
7000	Pressure treated wood, 6 x 6, 2′-0″ high	20.50	20.50	41
7400	Deadman at 6′-0″, 4′-0″ high	45	47	92
7600	6′-0″ high	67	70.50	137.50
8000	8 x 8, deadman at 6′-0″, 4′-0″ high	84.50	85	169.50
8200	6′-0″ high	127	128	255
8400	8′-0″ high	168	169	337

G2040 Site Development

The Stone Retaining Wall System is constructed of one of four types of stone. Each of the four types is listed in terms of cost per ton. Construction is either dry set or mortar set. System elements include excavation; concrete base; crushed stone; underdrain; and backfill.

The Expanded System Listing shows five heights above grade for each type, ranging from 3' above grade to 12' above grade.

System Components	QUANTITY	UNIT	COST PER L.F.		
			MAT.	INST.	TOTAL
SYSTEM G2040 260 2400					
STONE RETAINING WALL, DRY SET, STONE AT $250/TON, 3' ABOVE GRADE					
Excavation, trench, hyd backhoe	.880	C.Y.		6.32	6.32
Strip footing, 36" x 12", reinforced	.111	C.Y.	20.09	15.70	35.79
Stone, wall material, type 1	6.550	C.F.	59.28		59.28
Setting stone wall, dry	6.550	C.F.		74.67	74.67
Stone borrow, delivered, 3/8", machine spread	.320	C.Y.	11.36	2.68	14.04
Piping, subdrainage, perforated PVC, 4" diameter	1.000	L.F.	1.84	4.53	6.37
Backfill with dozer, trench, up to 300' haul, no compaction	1.019	C.Y.		2.93	2.93
TOTAL			92.57	106.83	199.40

G2040 260	Stone Retaining Walls	COST PER L.F.		
		MAT.	INST.	TOTAL
2400	Stone retaining wall, dry set, stone at $250/ton, height above grade 3'	92.50	107	199.50
2420	Height above grade 4'	111	127	238
2440	Height above grade 6'	145	171	316
2460	Height above grade 8'	203	268	471
2480	Height above grade 10'	262	350	612
2500	Height above grade 12'	325	440	765
2600	$350/ton stone, height above grade 3'	116	107	223
2620	Height above grade 4'	142	127	269
2640	Height above grade 6'	190	171	361
2660	Height above grade 8'	271	268	539
2680	Height above grade 10'	355	350	705
2700	Height above grade 12'	440	440	880
2800	$450/ton stone, height above grade 3'	140	107	247
2820	Height above grade 4'	173	127	300
2840	Height above grade 6'	234	171	405
2860	Height above grade 8'	335	268	603
2880	Height above grade 10'	440	350	790
2900	Height above grade 12'	555	440	995
3000	$650/ton stone, height above grade 3'	187	107	294
3020	Height above grade 4'	234	127	361
3040	Height above grade 6'	320	171	491
3060	Height above grade 8'	470	268	738
3080	Height above grade 10'	620	350	970
3100	Height above grade 12'	780	440	1,220
5020	Mortar set, stone at $250/ton, height above grade 3'	92.50	94.50	187
5040	Height above grade 4'	111	111	222
5060	Height above grade 6'	145	148	293
5080	Height above grade 8'	203	230	433

G2040 Site Development

G2040 260	Stone Retaining Walls	COST PER L.F.		
		MAT.	INST.	TOTAL
5100	Height above grade 10'	262	299	561
5120	Height above grade 12'	325	375	700
5300	$350/ton stone, height above grade 3'	116	94.50	210.50
5320	Height above grade 4'	142	111	253
5340	Height above grade 6'	190	148	338
5360	Height above grade 8'	271	230	501
5380	Height above grade 10'	355	299	654
5400	Height above grade 12'	440	375	815
5500	$450/ton stone, height above grade 3'	140	94.50	234.50
5520	Height above grade 4'	173	111	284
5540	Height above grade 6'	234	148	382
5560	Height above grade 8'	335	230	565
5580	Height above grade 10'	440	299	739
5600	Height above grade 12'	555	375	930
5700	$650/ton stone, height above grade 3'	187	94.50	281.50
5720	Height above grade 4'	234	111	345
5740	Height above grade 6'	320	148	468
5760	Height above grade 8'	470	230	700
5780	Height above grade 10'	620	299	919
5800	Height above grade 12'	780	375	1,155

G2040 Site Development

The Gabion Retaining Wall Systems list three types of surcharge conditions: level, sloped and highway, for two different facing configurations: stepped and straight with batter. Costs are expressed per L.F. for heights ranging from 6' to 18' for retaining sandy or clay soil. For protection against sloughing in wet clay materials counterforts have been added for these systems. Drainage stone has been added behind the walls to avoid any additional lateral pressures.

System Components			COST PER L.F.		
	QUANTITY	UNIT	MAT.	INST.	TOTAL
SYSTEM G2040 270 1000					
GABION RET. WALL, LEVEL BACKFILL, STEPPED FACE, 4' BASE, 6' HIGH					
3' x 3' cross section gabion	2.000	Ea.	90	166.89	256.89
3' x 1' cross section gabion	1.000	Ea.	23.78	10.03	33.81
Crushed stone drainage	.220	C.Y.	7.63	2.10	9.73
TOTAL			121.41	179.02	300.43

G2040 270	Gabion Retaining Walls	COST PER L.F.		
		MAT.	INST.	TOTAL
1000	Gabion ret. wall, lvl backfill, stepped face, 4' base, 6' high, sandy soil	121	179	300
1040	Clay soil with counterforts @ 16' O.C.	181	370	551
1100	5' base, 9' high, sandy soil	217	284	501
1140	Clay soil with counterforts @ 16' O.C.	370	565	935
1200	6' base, 12' high, sandy soil	310	450	760
1240	Clay soil with counterforts @ 16' O.C.	550	1,200	1,750
1300	7'-6" base, 15' high, sandy soil	440	635	1,075
1340	Clay soil with counterforts @ 16' O.C.	1,000	1,700	2,700
1400	9' base, 18' high, sandy soil	580	890	1,470
1440	Clay soil with counterforts @ 16' O.C.	1,175	2,000	3,175
2000	Straight face w/1:6 batter, 3' base, 6' high, sandy soil	105	171	276
2040	Clay soil with counterforts @ 16' O.C.	165	365	530
2100	4'-6" base, 9' high, sandy soil	174	279	453
2140	Clay soil with counterforts @ 16' O.C.	325	560	885
2200	6' base, 12' high, sandy soil	270	445	715
2240	Clay soil with counterforts @ 16' O.C.	510	1,200	1,710
2300	7'-6" base, 15' high, sandy soil	385	640	1,025
2340	Clay soil with counterforts @ 16' O.C.	755	1,325	2,080
2400	18' high, sandy soil	495	830	1,325
2440	Clay soil with counterforts @ 16' O.C.	820	965	1,785
3000	Backfill sloped 1-1/2:1, stepped face, 4'-6" base, 6' high, sandy soil	116	192	308
3040	Clay soil with counterforts @ 16' O.C.	176	380	556
3100	6' base, 9' high, sandy soil	209	355	564
3140	Clay soil with counterforts @ 16' O.C.	360	640	1,000
3200	7'-6" base, 12' high, sandy soil	320	550	870
3240	Clay soil with counterforts @ 16' O.C.	560	1,300	1,860
3300	9' base, 15' high, sandy soil	460	800	1,260
3340	Clay soil with counterforts @ 16' O.C.	1,025	1,875	2,900
3400	10'-6" base, 18' high, sandy soil	615	1,075	1,690
3440	Clay soil with counterforts @ 16' O.C.	1,225	2,225	3,450
4000	Straight face with 1:6 batter, 4'-6" base, 6' high, sandy soil	123	194	317
4040	Clay soil with counterforts @ 16' O.C.	183	380	563
4100	6' base, 9' high, sandy soil	219	365	584
4140	Clay soil with counterforts @ 16' O.C.	370	645	1,015

675

G2040 Site Development

G2040 270	Gabion Retaining Walls	COST PER L.F.		
		MAT.	INST.	TOTAL
4200	7'-6" base, 12' high, sandy soil	335	550	885
4240	Clay soil with counterforts @ 16' O.C.	575	1,325	1,900
4300	9' base, 15' high, sandy soil	475	810	1,285
4340	Clay soil with counterforts @ 16' O.C.	1,050	1,875	2,925
4400	18' high, sandy soil	615	1,050	1,665
4440	Clay soil with counterforts @ 16' O.C.	1,225	2,200	3,425
5000	Highway surcharge, straight face, 6' base, 6' high, sandy soil	188	335	523
5040	Clay soil with counterforts @ 16' O.C.	248	525	773
5100	9' base, 9' high, sandy soil	325	590	915
5140	Clay soil with counterforts @ 16' O.C.	480	870	1,350
5200	12' high, sandy soil	605	875	1,480
5240	Clay soil with counterforts @ 16' O.C.	890	2,175	3,065
5300	12' base, 15' high, sandy soil	650	1,175	1,825
5340	Clay soil with counterforts @ 16' O.C.	1,025	1,875	2,900
5400	18' high, sandy soil	835	1,525	2,360
5440	Clay soil with counterforts @ 16' O.C.	1,450	2,625	4,075

G2040 Site Development

Wood Deck Systems are either constructed of pressure treated lumber or redwood lumber. The system includes: the deck, joists (16″ or 24″ on-center), girders, posts (8′ on-center) and railings. Decking is constructed of either 1″ x 4″ or 2″ x 6″ stock. Joists range from 2″ x 8″ to 2″ x 10″ depending on the size of the system. The size ranges hold true for both pressure treated lumber and redwood systems.

Costs are on a square foot basis.

System Components	QUANTITY	UNIT	COST PER S.F. MAT.	COST PER S.F. INST.	COST PER S.F. TOTAL
SYSTEM G2040 910 1000					
WOOD DECK, TREATED LUMBER, 2″ X 8″ JOISTS @ 16″ O.C., 2″ X 6″ DECKING					
Decking, planks, fir 2″ x 6″ treated	2.080	B.F.	1.73	5.30	7.03
Framing, joists, fir 2″ x 8″ treated	1.330	B.F.	1.33	2.83	4.16
Framing, beams, fir 2″ x 10″ treated	.133	B.F.	.12	.25	.37
Framing, post, 4″ x 4″, treated	.333	B.F.	.37	.70	1.07
Framing, railing, 2″ x 4″ lumber, treated	.667	B.F.	.63	2.38	3.01
Excavating pits by hand, heavy soil or clay	.300	C.Y.		.38	.38
Spread footing, concrete	.200	C.Y.	.45	.60	1.05
TOTAL			4.63	12.44	17.07

G2040 910	Wood Decks	COST PER S.F. MAT.	COST PER S.F. INST.	COST PER S.F. TOTAL
1000	Wood deck, treated lumber, 2″ x 8″ joists @ 16″ O.C., 2″ x 6″ decking	4.63	12.45	17.08
1004	2″ x 4″ decking	5.05	15.30	20.35
1008	1″ x 6″ decking	7.05	9.90	16.95
1012	1″ x 4″ decking	6.65	9.65	16.30
1500	2″ x 10″ joists @ 16″ O.C., 2″ x 6″ decking	4.97	13.15	18.12
1504	2″ x 4″ decking	5.40	16	21.40
1508	1″ x 6″ decking	7.40	10.60	18
1512	1″ x 4″ decking	7	10.35	17.35
1560	2″ x 10″ joists @ 24″ O.C., 2″ x 6″ decking	4.63	12.45	17.08
1564	2″ x 4″ decking	5.05	15.30	20.35
1568	1″ x 6″ decking	7.05	9.90	16.95
1572	1″ x 4″ decking	6.65	9.65	16.30
4000	Redwood lumber, 2″ x 8″ joists @ 16″ O.C., 2″ x 6″ decking	11.35	5.70	17.05
4004	2″ x 4″ decking	11.70	5.85	17.55
4008	1″ x 6″ decking	14.50	7.35	21.85
4012	1″ x 4″ decking	14.30	7.25	21.55
4500	2″ x 10″ joists @ 16″ O.C., 2″ x 6″ decking	14.75	6.25	21
4504	2″ x 4″ decking	14.90	6.60	21.50
4508	1″ x 6″ decking	15.70	7.65	23.35
4512	1″ x 4″ decking	15.50	7.55	23.05
4600	2″ x 10″ joists @ 24″ O.C., 2″ x 6″ decking	13.55	5.95	19.50
4612	2″ x 4″ decking	13.70	6.35	20.05
4616	1″ x 6″ decking	14.50	7.35	21.85
4620	1″ x 4″ decking	14.30	7.25	21.55

G20 Site Improvements

G2040 Site Development

- **Chain Link Fence**
- **Concrete Sidewalk**
- **Pool Equipment** (Building not included in price)
- **12' x 30' Swimming Pool**
- **Fence Gate**

The Swimming Pool System is a complete package. Everything from excavation to deck hardware is included in system costs. Below are three basic types of pool systems: residential, motel, and municipal. Systems elements include: excavation, pool materials, installation, deck hardware, pumps and filters, sidewalk, and fencing.

The Expanded System Listing shows three basic types of pools with a variety of finishes and basic materials. These systems are either vinyl lined with metal sides; gunite shell with a cement plaster finish or tile finish; or concrete sided with vinyl lining. Pool sizes listed here vary from 12' x 30' to 60' x 82.5'. All costs are on a per unit basis.

System Components	QUANTITY	UNIT	COST EACH MAT.	COST EACH INST.	COST EACH TOTAL
SYSTEM G2040 920 1000					
SWIMMING POOL, RESIDENTIAL, CONC. SIDES, VINYL LINED, 12' X 30'					
Swimming pool, residential, in-ground including equipment	360.000	S.F.	11,340	5,666.40	17,006.40
4" thick reinforced concrete sidewalk, broom finish, no base	400.000	S.F.	1,016	1,156	2,172
Chain link fence, residential, 4' high	124.000	L.F.	762.60	575.36	1,337.96
Fence gate, chain link, 4' high	1.000	Ea.	101	185.50	286.50
TOTAL			13,219.60	7,583.26	20,802.86

G2040 920	Swimming Pools	MAT.	INST.	TOTAL
1000	Swimming pool, residential class, concrete sides, vinyl lined, 12' x 30'	13,200	7,575	20,775
1100	16' x 32'	15,000	8,575	23,575
1200	20' x 40'	19,300	10,900	30,200
1500	Tile finish, 12' x 30'	28,800	31,900	60,700
1600	16' x 32'	29,300	33,600	62,900
1700	20' x 40'	43,800	47,300	91,100
2000	Metal sides, vinyl lined, 12' x 30'	11,400	4,925	16,325
2100	16' x 32'	12,900	5,525	18,425
2200	20' x 40'	16,700	6,950	23,650
3000	Gunite shell, cement plaster finish, 12' x 30'	22,900	13,600	36,500
3100	16' x 32'	29,000	17,900	46,900
3200	20' x 40'	40,100	17,600	57,700
4000	Motel class, concrete sides, vinyl lined, 20' x 40'	27,700	15,100	42,800
4100	28' x 60'	38,800	21,100	59,900
4500	Tile finish, 20' x 40'	51,500	54,500	106,000
4600	28' x 60'	81,500	88,500	170,000
5000	Metal sides, vinyl lined, 20' x 40'	29,000	10,800	39,800
5100	28' x 60'	38,900	14,500	53,400
6000	Gunite shell, cement plaster finish, 20' x 40'	74,500	43,700	118,200
6100	28' x 60'	111,500	65,000	176,500
7000	Municipal class, gunite shell, cement plaster finish, 42' x 75'	297,500	154,000	451,500
7100	60' x 82.5'	382,000	197,500	579,500
7500	Concrete walls, tile finish, 42' x 75'	332,000	207,000	539,000
7600	60' x 82.5'	432,000	275,000	707,000
7700	Tile finish and concrete gutter, 42' x 75'	364,000	207,000	571,000
7800	60' x 82.5'	471,000	275,000	746,000
7900	Tile finish and stainless gutter, 42' x 75'	427,000	207,000	634,000
8000	60' x 82.5'	631,500	317,500	949,000

G20 Site Improvements

G2040 Site Development

Building

Service Driveway

Landscaping

Entrance Slab

Sidewalk

Parking

Hidden Features
Utilities and Lighting
(Features do not represent actual sizes)

G2040 990	Site Development Components for Buildings	COST EACH		
		MAT.	INST.	TOTAL
0970	Assume minimal and balanced cut & fill, no rock, no demolition, no haz mat.			
0975	Lines can be adjusted linearly +/- 20% within same use & number of floors.			
0980	60,000 S.F. 2-Story Office Bldg on 3.3 Acres w/20% Green Space			
1000	Site Preparation	400	53,500	53,900
1002	Utilities	107,500	110,500	218,000
1004	Pavement	330,500	211,500	542,000
1006	Stormwater Management	332,500	66,500	399,000
1008	Sidewalks	17,800	21,000	38,800
1010	Exterior lighting	56,000	43,200	99,200
1014	Landscaping	79,500	56,500	136,000
1028	80,000 S.F. 4-Story Office Bldg on 3.5 Acres w/20% Green Space			
1030	Site preparation	410	58,000	58,410
1032	Utilities	99,500	100,500	200,000
1034	Pavement	396,000	253,500	649,500
1036	Stormwater Managmement	350,500	70,000	420,500
1038	Sidewalks	14,500	17,100	31,600
1040	Lighting	56,000	45,600	101,600
1042	Landscaping	82,000	58,500	140,500
1058	16,000 S.F. 1-Story Medical Office Bldg on 1.7 Acres w/25% Green Space			
1060	Site preparation	289	40,300	40,589
1062	Utilities	76,000	81,000	157,000
1064	Pavement	153,500	99,500	253,000
1066	Stormwater Management	162,000	32,400	194,400
1068	Sidewalks	13,000	15,400	28,400
1070	Lighting	34,300	25,100	59,400
1072	Landscaping	54,000	34,700	88,700
1098	36,000 S.F. 3-Story Apartment Bldg on 1.6 Acres w/25% Green Space			
1100	Site preparation	280	39,700	39,980

679

For customer support on your Site Work & Landscape Costs with RSMeans data, call 800.448.8182.

G2040 990	Site Development Components for Buildings	COST EACH		
		MAT.	INST.	TOTAL
1102	Utilities	70,500	73,500	144,000
1104	Pavement	155,500	100,000	255,500
1106	Stormwater Management	151,000	30,200	181,200
1108	Sidewalks	11,200	13,400	24,600
1110	Lighting	32,000	23,800	55,800
1112	Landscaping	52,000	32,900	84,900
1128	130,000 S.F. 3-Story Hospital Bldg on 5.9 Acres w/17% Green Space			
1130	Site preparation	540	76,500	77,040
1132	Utilities	144,500	145,000	289,500
1134	Pavement	673,500	430,000	1,103,500
1136	Stormwater Management	622,500	124,000	746,500
1138	Sidewalks	21,300	25,300	46,600
1140	Lighting	92,500	77,000	169,500
1142	Landscaping	114,000	89,500	203,500
1158	60,000 S.F. 1-Story Light Manufacturing Bldg on 3.9 Acres w/18% Green Space			
1160	Site preparation	440	57,000	57,440
1162	Utilities	127,500	137,500	265,000
1164	Pavement	323,000	184,000	507,000
1166	Stormwater Management	409,500	82,000	491,500
1168	Sidewalks	25,100	29,800	54,900
1170	Lighting	67,500	49,500	117,000
1172	Landscaping	86,000	63,500	149,500
1198	8,000 S.F. 1-Story Restaurant Bldg on 1.4 Acres w/26% Green Space			
1200	Site preparation	264	38,700	38,964
1202	Utilities	63,500	64,000	127,500
1204	Pavement	147,000	96,500	243,500
1206	Stormwater Management	133,000	26,600	159,600
1208	Sidewalks	9,175	10,900	20,075
1210	Lighting	28,700	21,900	50,600
1212	Landscaping	48,700	30,200	78,900
1228	20,000 S.F. 1-Story Retail Store Bldg on 2.1 Acres w/23% Green Space			
1230	Site preparation	325	47,200	47,525
1232	Utilities	84,500	89,500	174,000
1234	Pavement	199,500	129,000	328,500
1236	Stormwater Management	207,500	41,500	249,000
1238	Sidewalks	14,500	17,100	31,600
1240	Lighting	39,300	29,100	68,400
1242	Landscaping	61,500	40,800	102,300
1258	60,000 S.F. 1-Story Warehouse on 3.1 Acres w/19% Green Space			
1260	Site preparation	390	52,500	52,890
1262	Utilities	114,500	128,500	243,000
1264	Pavement	199,500	104,000	303,500
1266	Stormwater Management	319,000	64,000	383,000
1268	Sidewalks	25,100	29,800	54,900
1270	Lighting	56,500	38,400	94,900
1272	Landscaping	74,500	52,500	127,000

G2050 Landscaping

The Lawn Systems listed include different types of seeding, sodding and ground covers for flat and sloped areas. Costs are given per thousand square feet for different size jobs; residential, small commercial and large commercial. The size of the job relates to the type and productivity of the equipment being used. Components include furnishing and spreading screened loam, spreading fertilizer and limestone and mulching planted and seeded surfaces. Sloped surfaces include jute mesh or staking depending on the type of cover.

System Components	QUANTITY	UNIT	COST PER M.S.F.		
			MAT.	INST.	TOTAL
SYSTEM G2050 410 1000					
LAWN, FLAT AREA, SEEDED, TURF MIX, RESIDENTIAL					
Scarify subsoil, residential, skid steer loader w/scarifiers, 50 HP	1.000	M.S.F.		28.10	28.10
Root raking and loading, residential, no boulders	1.000	M.S.F.		35.41	35.41
Spread topsoil, skid steer loader and hand dress	18.500	C.Y.	536.50	126.17	662.67
Spread ground limestone, push spreader	110.000	S.Y.	16.50	3.30	19.80
Spread fertilizer, push spreader	110.000	S.Y.	9.90	3.30	13.20
Till topsoil, 26" rototiller	110.000	S.Y.		62.70	62.70
By hand (raking)	1.000	M.S.F.		33.50	33.50
Rolling topsoil, hand push roller	18.500	C.Y.		2.96	2.96
Seeding, turf mix, push spreader	1.000	M.S.F.	14.95	63.50	78.45
Straw, mulch	110.000	S.Y.	70.40	117.70	188.10
TOTAL			648.25	476.64	1,124.89

G2050 410	Lawns & Ground Cover	COST PER M.S.F.		
		MAT.	INST.	TOTAL
1000	Lawn, flat area, seeded, turf mix, residential	650	480	1,130
1040	Small commercial	650	520	1,170
1080	Large commercial	670	440	1,110
1200	Shade mix, residential	645	480	1,125
1240	Small commercial	645	520	1,165
1280	Large commercial	660	440	1,100
1400	Utility mix, residential	645	480	1,125
1440	Small commercial	650	520	1,170
1480	Large commercial	685	440	1,125
2000	Sod, bluegrass, residential	935	505	1,440
2040	Small commercial	880	665	1,545
2080	Large commercial	810	540	1,350
2200	Bentgrass, residential	945	500	1,445
2240	Small commercial	920	655	1,575
2280	Large commercial	900	555	1,455
2400	Ground cover, English ivy, residential	2,800	1,225	4,025
2440	Small commercial	2,800	1,475	4,275
2480	Large commercial	2,800	1,375	4,175
2600	Pachysandra, residential	4,075	6,550	10,625
2640	Small commercial	4,100	6,800	10,900
2680	Large commercial	4,100	6,675	10,775
2800	Vinca minor, residential	3,350	1,925	5,275

G20 Site Improvements

G2050 Landscaping

G2050 410	Lawns & Ground Cover	COST PER M.S.F.		
		MAT.	INST.	TOTAL
2840	Small commercial	3,375	2,175	5,550
2880	Large commercial	3,375	2,050	5,425
3000	Sloped areas, seeded, slope mix, residential	845	655	1,500
3040	Small commercial	845	650	1,495
3080	Large commercial	865	645	1,510
3200	Birdsfoot trefoil, residential	835	660	1,495
3240	Small commercial	835	650	1,485
3280	Large commercial	840	645	1,485
3400	Clover, residential	835	655	1,490
3440	Small commercial	835	650	1,485
3480	Large commercial	850	645	1,495
3600	Crown vetch, residential	985	655	1,640
3640	Small commercial	985	650	1,635
3680	Large commercial	1,025	645	1,670
3800	Wildflower, residential	835	655	1,490
3840	Small commercial	835	650	1,485
3880	Large commercial	840	645	1,485
4600	Sod, bluegrass, residential	990	1,050	2,040
4640	Small commercial	935	1,075	2,010
4680	Large commercial	870	980	1,850
4800	Bent grass, residential	1,000	570	1,570
4840	Small commercial	975	720	1,695
4880	Large commercial	955	695	1,650
5000	Ground cover, English ivy, residential	3,050	1,400	4,450
5040	Small commercial	3,050	1,600	4,650
5080	Large commercial	3,050	1,550	4,600
5200	Pachysandra, residential	4,475	7,250	11,725
5240	Small commercial	4,475	7,450	11,925
5280	Large commercial	4,475	7,400	11,875
5400	Vinca minor, residential	3,675	2,150	5,825
5440	Small commercial	2,950	2,325	5,275
5480	Large commercial	3,675	2,325	6,000

G2050 Landscaping

There are three basic types of Site Irrigation Systems: pop-up, riser mounted and quick coupling. Sprinkler heads are spray, impact or gear driven. Each system includes: the hardware for spraying the water; the pipe and fittings needed to deliver the water; and all other accessory equipment such as valves, couplings, nipples, and nozzles. Excavation heads and backfill costs are also included in the system.

The Expanded System Listing shows a wide variety of Site Irrigation Systems.

System Components	QUANTITY	UNIT	COST PER S.F. MAT.	COST PER S.F. INST.	COST PER S.F. TOTAL
SYSTEM G2050 710 1000					
SITE IRRIGATION, POP UP SPRAY, PLASTIC, 10′ RADIUS, 1000 S.F., PVC PIPE					
Excavation, chain trencher	54.000	L.F.		54.54	54.54
Pipe, fittings & nipples, PVC Schedule 40, 1″ diameter	5.000	Ea.	4.55	122.50	127.05
Fittings, bends or elbows, 4″ diameter	5.000	Ea.	62.25	277.50	339.75
Couplings PVC plastic, high pressure, 1″ diameter	5.000	Ea.	4.55	155	159.55
Valves, bronze, globe, 125 lb. rising stem, threaded, 1″ diameter	1.000	Ea.	263	40.50	303.50
Head & nozzle, pop-up spray, PVC plastic	5.000	Ea.	21.30	110	131.30
Backfill by hand with compaction	54.000	L.F.		37.80	37.80
Total cost per 1,000 S.F.			355.65	797.84	1,153.49
Total cost per S.F.			.36	.80	1.15

G2050 710	Site Irrigation	COST PER S.F. MAT.	COST PER S.F. INST.	COST PER S.F. TOTAL
1000	Site irrigation, pop up spray, 10′ radius, 1000 S.F., PVC pipe	.36	.80	1.16
1100	Polyethylene pipe	.38	.69	1.07
1200	14′ radius, 8000 S.F., PVC pipe	.13	.34	.47
1300	Polyethylene pipe	.12	.31	.43
1400	18′ square, 1000 S.F. PVC pipe	.35	.53	.88
1500	Polyethylene pipe	.35	.46	.81
1600	24′ square, 8000 S.F., PVC pipe	.11	.22	.33
1700	Polyethylene pipe	.10	.21	.31
1800	4′ x 30′ strip, 200 S.F., PVC pipe	1.69	2.25	3.94
1900	Polyethylene pipe	1.68	2.09	3.77
2000	6′ x 40′ strip, 800 S.F., PVC pipe	.45	.75	1.20
2200	Economy brass, 11′ radius, 1000 S.F., PVC pipe	.45	.76	1.21
2300	Polyethylene pipe	.45	.69	1.14
2400	14′ radius, 8000 S.F., PVC pipe	.16	.34	.50
2500	Polyethylene pipe	.15	.31	.46
2600	3′ x 28′ strip, 200 S.F., PVC pipe	1.90	2.25	4.15
2700	Polyethylene pipe	1.89	2.09	3.98
2800	7′ x 36′ strip, 1000 S.F., PVC pipe	.41	.59	1
2900	Polyethylene pipe	.41	.54	.95
3000	Hd brass, 11′ radius, 1000 S.F., PVC pipe	.49	.76	1.25
3100	Polyethylene pipe	.49	.69	1.18
3200	14′ radius, 8000 S.F., PVC pipe	.18	.34	.52
3300	Polyethylene pipe	.17	.31	.48
3400	Riser mounted spray, plastic, 10′ radius, 1000 S.F., PVC pipe	.38	.76	1.14
3500	Polyethylene pipe	.38	.69	1.07
3600	12′ radius, 5000 S.F., PVC pipe	.19	.46	.65

683

G2050 710	Site Irrigation	COST PER S.F.		
		MAT.	INST.	TOTAL
3700	Polyethylene pipe	.19	.43	.62
3800	19' square, 2000 S.F., PVC pipe	.18	.24	.42
3900	Polyethylene pipe	.17	.22	.39
4000	24' square, 8000 S.F., PVC pipe	.11	.22	.33
4100	Polyethylene pipe	.10	.21	.31
4200	5' x 32' strip, 300 S.F., PVC pipe	1.13	1.52	2.65
4300	Polyethylene pipe	1.13	1.41	2.54
4400	6' x 40' strip, 800 S.F., PVC pipe	.45	.75	1.20
4500	Polyethylene pipe	.45	.70	1.15
4600	Brass, 11' radius, 1000 S.F., PVC pipe	.43	.76	1.19
4700	Polyethylene pipe	.43	.69	1.12
4800	14' radius, 8000 S.F., PVC pipe	.15	.34	.49
4900	Polythylene pipe	.14	.31	.45
5000	Pop up gear drive stream type, plastic, 30' radius, 10,000 S.F., PVC pipe	.40	.66	1.06
5020	Polyethylene pipe	.15	.20	.35
5040	40,000 S.F., PVC pipe	.36	.66	1.02
5060	Polyethylene pipe	.13	.18	.31
5080	Riser mounted gear drive stream type, plas., 30' rad., 10,000 S.F.,PVC pipe	.39	.66	1.05
5100	Polyethylene pipe	.14	.20	.34
5120	40,000 S.F., PVC pipe	.35	.66	1.01
5140	Polyethylene pipe	.12	.18	.30
5200	Q.C. valve thread type w/impact head, brass, 75' rad, 20,000 S.F., PVC pipe	.17	.25	.42
5250	Polyethylene pipe	.07	.06	.13
5300	100,000 S.F., PVC pipe	.25	.41	.66
5350	Polyethylene pipe	.10	.09	.19
5400	Q.C. valve lug type w/impact head, brass, 75' rad, 20,000 S.F., PVC pipe	.17	.25	.42
5450	Polyethylene pipe	.07	.06	.13
5500	100,000 S.F. PVC pipe	.25	.41	.66
5550	Polyethylene pipe	.10	.09	.19
6000	Site irrigation, pop up impact type, plastic, 40' rad.,10000 S.F., PVC pipe	.25	.43	.68
6100	Polyethylene pipe	.08	.12	.20
6200	45,000 S.F., PVC pipe	.24	.47	.71
6300	Polyethylene pipe	.07	.12	.19
6400	High medium volume brass, 40' radius, 10000 S.F., PVC pipe	.26	.43	.69
6500	Polyethylene pipe	.09	.12	.21
6600	45,000 S.F. PVC pipe	.25	.47	.72
6700	Polyethylene pipe	.08	.12	.20
6800	High volume brass, 60' radius, 25,000 S.F., PVC pipe	.21	.33	.54
6900	Polyethylene pipe	.07	.08	.15
7000	100,000 S.F., PVC pipe	.20	.33	.53
7100	Polyethylene pipe	.08	.08	.16
7200	Riser mounted part/full impact type, plas., 40' rad., 10,000 S.F., PVC pipe	.25	.43	.68
7300	Polyethylene pipe	.09	.26	.35
7400	45,000 S.F., PVC pipe	.24	.47	.71
7500	Polyethylene pipe	.07	.12	.19
7600	Low medium volume brass, 40' radius, 10,000 S.F., PVC pipe	.25	.43	.68
7700	Polyethylene pipe	.08	.12	.20
7800	45,000 S.F., PVC pipe	.24	.47	.71
7900	Polyethylene pipe	.07	.12	.19
8000	Medium volume brass, 50' radius, 30,000 S.F., PVC pipe	.21	.35	.56
8100	Polyethylene pipe	.07	.09	.16
8200	70,000 S.F., PVC pipe	.31	.57	.88
8300	Polyethylene pipe	.11	.14	.25
8400	Riser mounted full only impact type, plas., 40' rad., 10,000 S.F., PVC pipe	.24	.43	.67
8500	Polyethylene pipe	.07	.12	.19
8600	45,000 S.F., PVC pipe	.23	.47	.70
8700	Polyethylene pipe	.06	.12	.18
8800	Low medium volume brass, 40' radius, 10,000 S.F., PVC pipe	.25	.43	.68

G2050 Landscaping

G2050 710	Site Irrigation	COST PER S.F.		
		MAT.	INST.	TOTAL
8900	Polyethylene pipe	.08	.12	.20
9000	45,000 S.F., PVC pipe	.24	.47	.71
9100	Polyethylene pipe	.07	.12	.19
9200	High volume brass, 80' radius, 75,000 S.F., PVC pipe	.15	.24	.39
9300	Polyethylene pipe	.01	.02	.03
9400	150,000 S.F., PVC pipe	.15	.22	.37
9500	Polyethylene pipe	.01	.02	.03
9600	Very high volume brass, 100' radius, 100,000 S.F., PVC pipe	.13	.18	.31
9700	Polyethylene pipe	.01	.02	.03
9800	200,000 S.F., PVC pipe	.13	.18	.31
9900	Polyethylene pipe	.01	.02	.03

G2050 Landscaping

Valve
Backflow Preventer
Pressure Regulator
Filter
Tubing Adapter
Drip Tubing

Drip irrigation systems can be for either lawns or individual plants. Shown below are drip irrigation systems for lawns. These drip type systems are most cost effective where there are water restrictions on use or the cost of water is very high. The supply uses PVC piping with the laterals polyethylene drip lines.

System Components	QUANTITY	UNIT	COST PER S.F.		
			MAT.	INST.	TOTAL
SYSTEM G2050 720 1000					
SITE IRRIGATION, DRIP SYSTEM, 800 S.F. (20' X 40')					
Excavate supply and header lines 4" x 18" deep	50.000	L.F.		54	54
Excavate drip lines 4" x 4" deep	560.000	L.F.		907.20	907.20
Install supply manifold (stop valve, strainer, PRV, adapters)	1.000	Ea.	26.50	166	192.50
Install 3/4" PVC supply line and risers to 4 " drip level	75.000	L.F.	24	165.75	189.75
Install 3/4" PVC fittings for supply line	33.000	Ea.	60.39	547.80	608.19
Install PVC to drip line adapters	14.000	Ea.	26.18	232.40	258.58
Install 1/4 inch drip lines with hose clamps	560.000	L.F.	61.60	308	369.60
Backfill trenches	3.180	C.Y.		146.28	146.28
Cleanup area when completed	1.000	Ea.		127	127
Total cost per 800 S.F.			198.67	2,654.43	2,853.10
Total cost per S.F.			.25	3.32	3.57

G2050 720	Site Irrigation	COST PER S.F.		
		MAT.	INST.	TOTAL
1000	Site irrigation, drip lawn watering system, 800 S.F. (20' x 40' area)	.25	3.31	3.56
1100	1000 S.F. (20' x 50' area)	.22	3	3.22
1200	1600 S.F. (40' x 40' area), 2 zone	.27	3.19	3.46
1300	2000 S.F. (40' x 50' area), 2 zone	.24	2.88	3.12
1400	2400 S.F. (60' x 40' area), 3 zone	.26	3.16	3.42
1500	3000 S.F. (60' x 50' area), 3 zone	.23	2.85	3.08
1600	3200 S.F. (2 x 40' x 40'), 4 zone	.26	3.13	3.39
1700	4000 S.F. (2 x 40' x 50'), 4 zone	.24	2.84	3.08
1800	4800 S.F. (2 x 60' x 40'), 6 zone with control	.28	3.10	3.38
1900	6000 S.F. (2 x 60' x 50'), 6 zone with control	.24	2.82	3.06
2000	6400 S.F. (4 x 40' x 40'), 8 zone with control	.27	3.16	3.43
2100	8000 S.F. (4 x 40' x 50'), 8 zone with control	.24	2.86	3.10
2200	3200 S.F. (2 x 40' x 40'), 4 zone with control	.28	3.20	3.48
2300	4000 S.F. (2 x 40' x 50'), 4 zone with control	.25	2.89	3.14
2400	2400 S.F. (60' x 40'), 3 zone with control	.28	3.26	3.54
2500	3000 S.F. (60' x 50'), 3 zone with control	.25	2.94	3.19
2600	4800 S.F. (2 x 60' x 40'), 6 zone, manual	.26	3.07	3.33
2700	6000 S.F. (2 x 60' x 50'), 6 zone, manual	.22	2.79	3.01
2800	1000 S.F. (10' x 100' area)	.13	1.87	2
2900	2000 S.F. (2 x 10' x 100' area)	.14	1.79	1.93

G2050 Landscaping

The Tree Pit Systems are listed for different heights of deciduous and evergreen trees, from 4' and above. Costs are given showing the pit size to accommodate the soil ball and the type of soil being excavated. In clay soils drainage tile has been included in addition to gravel drainage. Woven drainage fabric is applied before bark mulch is spread, assuming runoff is not harmful to roots.

System Components	QUANTITY	UNIT	COST EACH		
			MAT.	INST.	TOTAL
SYSTEM G2050 910 1220					
TREE PIT, 5' TO 6' TREE, DECIDUOUS, 2' X 1-1/4' DEEP PIT, CLAY SOIL					
Heavy soil or clay	.262	C.Y.		33.27	33.27
Fill in pit, screened gravel, 3/4" to 1/2"	.050	C.Y.	.93	2.15	3.08
Perforated PVC, schedule 40, 4" diameter	3.000	L.F.	5.52	32.70	38.22
Mix planting soil, by hand, incl. loam, peat & manure	.304	C.Y.	15.50	5.12	20.62
Prepared planting mix, by hand	.304	C.Y.		12.77	12.77
Fabric, laid in trench, polypropylene, ideal conditions	.500	S.Y.	.87	.21	1.08
Bark mulch, hand spread, 3"	.346	C.Y.	1.57	1.75	3.32
TOTAL			24.39	87.97	112.36

G2050 910	Tree Pits	COST EACH		
		MAT.	INST.	TOTAL
1000	Tree pit, 4' to 5' tree, evergreen, 2-1/2' x 1-1/2' deep pit, sandy soil	46.50	76	122.50
1020	Clay soil	43	111	154
1200	5' to 6' tree, deciduous, 2' x 1-1/4' deep pit, sandy soil	25	39	64
1220	Clay soil	24.50	88	112.50
1300	Evergreen, 3' x 1-3/4' deep pit, sandy soil	76	115	191
1320	Clay soil	67.50	166	233.50
1360	3-1/2' x 1-3/4' deep pit, sandy soil	96.50	143	239.50
1380	Clay soil	84.50	192	276.50
1400	6' to 7' tree, evergreen, 4' x 1-3/4' deep pit, sandy soil	116	167	283
1420	Clay soil	102	222	324
1600	7' to 9' tree, deciduous, 2-1/2' x 1-1/2' deep pit, sandy soil	46.50	95	141.50
1620	Clay soil	43	111	154
1800	8' to 10' tree, deciduous, 3' x 1-3/4' deep pit, sandy soil	76	115	191
1820	Clay soil	67.50	166	233.50
1900	Evergreen, 4-1/2' x 2' deep pit, sandy soil, smaller ball	172	241	413
1910	Larger ball	167	235	402
1920	Clay soil, smaller ball	146	288	434
1930	Larger ball	141	282	423
2000	10' to 12' tree, deciduous, 3-1/2' x 1-3/4' deep pit, sandy soil	96.50	141	237.50
2020	Clay soil	84.50	192	276.50

G2050 Landscaping

G2050 910	Tree Pits	COST EACH		
		MAT.	INST.	TOTAL
2060	4' x 1-3/4' deep pit, sandy soil	121	175	296
2080	Clay soil	108	238	346
2100	Evergreen, 5' x 2-1/4' deep pit, sandy soil, smaller ball	228	315	543
2110	Larger ball	221	305	526
2120	Clay soil, smaller ball	189	355	544
2130	Larger ball	181	345	526
2200	12' to 14' tree, deciduous, 4-1/2' x 2' deep pit, sandy soil, smaller ball	172	241	413
2210	Larger ball	167	235	402
2220	Clay soil, smaller ball	146	288	434
2230	Larger ball	141	282	423
2400	13' to 15' tree, evergreen, 6' x 2-1/2' deep pit, sandy soil	325	450	775
2420	Clay soil	262	470	732
2600	14' to 16' tree, deciduous, 5' x 2-1/4' deep pit, sandy soil, smaller ball	228	315	543
2610	Larger ball	221	305	526
2620	Clay soil, smaller ball	189	355	544
2630	Larger ball	181	345	526
2800	16' to 18' tree, deciduous, 6' x 2-1/2' deep pit, sandy soil	325	450	775
2820	Clay soil	262	470	732
2900	Evergreen, 7' x 3' deep pit, sandy soil	550	750	1,300
2920	Clay soil	435	730	1,165
3000	Over 18' tree, deciduous, 7' x 3' deep pit, sandy soil	550	750	1,300
3020	Clay soil	435	730	1,165
3100	Evergreen, 8' x 3-1/4' deep pit, sandy soil	765	1,050	1,815
3120	Clay soil	610	980	1,590

G30 Site Mechanical Utilities

G3010 Water Supply

The Water Service Systems are for copper service taps from 1″ to 2″ diameter into pressurized mains from 6″ to 8″ diameter. Costs are given for two offsets with depths varying from 2′ to 10′. Included in system components are excavation and backfill and required curb stops with boxes.

System Components	QUANTITY	UNIT	COST EACH MAT.	COST EACH INST.	COST EACH TOTAL
SYSTEM G3010 121 1000					
WATER SERVICE, LEAD FREE, 6″ MAIN, 1″ COPPER SERVICE, 10′ OFFSET, 2′ DEEP					
Trench excavation, 1/2 C.Y. backhoe, 1 laborer	2.220	C.Y.		15.94	15.94
Drill & tap pressurized main, 6″, 1″ to 2″ service	1.000	Ea.		465	465
Corporation stop, 1″ diameter	1.000	Ea.	108	48	156
Saddles, 3/4″ & 1″ diameter, add	1.000	Ea.	84		84
Copper tubing, type K, 1″ diameter	12.000	L.F.	127.80	52.08	179.88
Piping, curb stops, no lead, 1″ diameter	1.000	Ea.	156	48	204
Curb box, cast iron, 1/2″ to 1″ curb stops	1.000	Ea.	68.50	64.50	133
Backfill by hand, no compaction, heavy soil	2.890	C.Y.		8.33	8.33
Compaction, vibrating plate	2.220	C.Y.		21.22	21.22
TOTAL			544.30	723.07	1,267.37

G3010 121	Water Service, Lead Free	COST EACH MAT.	COST EACH INST.	COST EACH TOTAL
1000	Water service, Lead Free, 6″ main, 1″ copper service, 10′ offset, 2′ deep	545	725	1,270
1040	4′ deep	545	770	1,315
1060	6′ deep	545	985	1,530
1080	8′ deep	545	1,100	1,645
1100	10′ deep	545	1,200	1,745
1200	20′ offset, 2′ deep	545	770	1,315
1240	4′ deep	545	860	1,405
1260	6′ deep	545	1,300	1,845
1280	8′ deep	545	1,500	2,045
1300	10′ deep	545	1,725	2,270
1400	1-1/2″ copper service, 10′ offset, 2′ deep	950	790	1,740
1440	4′ deep	950	835	1,785
1460	6′ deep	950	1,050	2,000
1480	8′ deep	950	1,150	2,100
1500	10′ deep	950	1,275	2,225
1520	20′ offset, 2′ deep	950	835	1,785
1530	4′ deep	950	925	1,875
1540	6′ deep	950	1,350	2,300
1560	8′ deep	950	1,575	2,525
1580	10′ deep	950	1,800	2,750
1600	2″ copper service, 10′ offset, 2′ deep	1,250	820	2,070
1610	4′ deep	1,250	865	2,115
1620	6′ deep	1,250	1,075	2,325
1630	8′ deep	1,250	1,175	2,425
1650	10′ deep	1,250	1,300	2,550
1660	20′ offset, 2′ deep	1,250	865	2,115
1670	4′ deep	1,250	955	2,205
1680	6′ deep	1,250	1,400	2,650

G3010 Water Supply

G3010 121	Water Service, Lead Free	COST EACH		
		MAT.	INST.	TOTAL
1690	8' deep	1,250	1,600	2,850
1695	10' deep	1,250	1,825	3,075
1700	8" main, 1" copper service, 10' offset, 2' deep	545	765	1,310
1710	4' deep	545	810	1,355
1720	6' deep	545	1,025	1,570
1730	8' deep	545	1,125	1,670
1740	10' deep	545	1,250	1,795
1750	20' offset, 2' deep	545	810	1,355
1760	4' deep	545	900	1,445
1770	6' deep	545	1,325	1,870
1780	8' deep	545	1,550	2,095
1790	10' deep	545	1,775	2,320
1800	1-1/2" copper service, 10' offset, 2' deep	950	830	1,780
1810	4' deep	950	875	1,825
1820	6' deep	950	1,100	2,050
1830	8' deep	950	1,600	2,550
1840	10' deep	950	1,300	2,250
1850	20' offset, 2' deep	950	875	1,825
1860	4' deep	950	875	1,825
1870	6' deep	950	1,400	2,350
1880	8' deep	950	1,600	2,550
1890	10' deep	950	1,825	2,775
1900	2" copper service, 10' offset, 2' deep	1,250	860	2,110
1910	4' deep	1,250	905	2,155
1920	6' deep	1,250	1,125	2,375
1930	8' deep	1,250	1,225	2,475
1940	10' deep	1,250	1,325	2,575
1950	20' offset, 2' deep	1,250	905	2,155
1960	4' deep	1,250	995	2,245
1970	6' deep	1,250	1,425	2,675
1980	8' deep	1,250	1,650	2,900
1990	10' deep	1,250	1,850	3,100

G3010 Water Supply

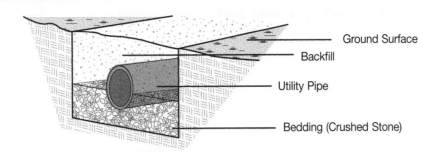

System Components	QUANTITY	UNIT	COST L.F.		
			MAT.	INST.	TOTAL
SYSTEM G3010 122 1000					
WATERLINE 4″ DUCTILE IRON INCLUDING COMMON EARTH EXCAVATION,					
BEDDING, BACKFILL AND COMPACTION					
Excavate Trench	.296	B.C.Y.		2.17	2.17
Crushed stone bedding material	.203	L.C.Y.	6.19	2.54	8.73
Compact bedding	.173	E.C.Y.		1.04	1.04
Install 4″ MJ ductile iron pipe	1.000	L.F.	46.50	18.01	64.51
Backfill trench	.160	L.C.Y.		.50	.50
Compact fill material in trench	.123	E.C.Y.		.66	.66
Dispose of excess fill material on-site	.225	L.C.Y.		2.84	2.84
TOTAL			52.69	27.76	80.45

G3010 122	Waterline (Common Earth Excavation)	COST L.F.		
		MAT.	INST.	TOTAL
0950	Waterline including common earth excavation, bedding, backfill and compaction			
1000	4″ MJ ductile iron, 4′ deep	52.50	27.50	80
1150	6 inch	62.50	32.50	95
1200	8 inch	66	37.50	103.50
1250	10 inch	85.50	42.50	128
1300	12 inch	107	45.50	152.50
1400	4″ push joint ductile iron, 4′ deep	31.50	18.80	50.30
1450	6 ″	32.50	21	53.50
1500	8 ″	43	28.50	71.50
1550	10 ″	62.50	30.50	93
1600	12 ″	66	35	101
1700	4″ PVC Class 150, 4′ deep	9.95	14.30	24.25
1750	6 ″	12.60	15.55	28.15
1800	8 ″	17	17.05	34.05
1900	4″ PVC Class 200, 4′ deep	10.45	16.20	26.65
1950	6 ″	15.05	17.80	32.85
2000	8 ″	17.20	19.85	37.05
2100	6″ PVC C900, 4′ deep	12.60	17.80	30.40
2150	8 ″	17	19.70	36.70
2160	10 ″	21	21.50	42.50
2200	12 ″	28	24.50	52.50
2300	6″ HDPE SDR 21 fusion, 4′ deep	13.30	20	33.30
2350	8 ″	18.10	22.50	40.60
2360	10 ″	19.60	23	42.60
2400	12 ″	26	26	52
2500	10″, 1/4″ wall black steel, 4′ deep	50	43.50	93.50
2550	12″	58	45	103
3000	18″ MJ ductile iron, 6′ deep with trench box	164	93.50	257.50
3050	24″	208	119	327
3100	18″ push joint ductile iron, 6′ deep with trench box	89	77.50	166.50
3200	24″	118	89	207
3300	18″ PVC C905, 6′ deep with trench box	52.50	56	108.50

691

G3010 Water Supply

G3010 122	Waterline (Common Earth Excavation)	COST L.F.		
		MAT.	INST.	TOTAL
3350	24"	72.50	65	137.50
3400	18" HDPE fusion SDR 21, 6' deep with trench box	53	73.50	126.50
3450	24"	84	87.50	171.50
3500	24" Concrete (CCP) , 6' deep with trench box	114	77.50	191.50
3600	18", 1/4" wall black steel , 6' deep with trench box	101	79	180
3650	18", 5/16" wall	135	79	214
3700	18", 3/8" wall	154	195	349
3750	24", 3/8" wall	178	227	405
3800	36" Concrete (CCP) , 8' deep with trench box	181	118	299
3900	36" PVC C905 , 8' deep with trench box	166	96.50	262.50
4000	36" 5/16" wall black steel , 8' deep with trench box	203	296	499
4050	36" 1/2" wall black steel	380	320	700
5000	4" MJ ductile iron, 5' deep	52.50	29	81.50
5005	6"	62.50	33.50	96
5010	8"	65.50	38	103.50
5015	10"	85	42.50	127.50
5020	12"	106	46	152
5100	4" push on joint ductile iron, 5' deep	31.50	19.70	51.20
5105	6"	32.50	21.50	54
5110	8"	42.50	29	71.50
5115	10"	62	31	93
5120	12"	65	34	99
5200	4" PVC Class 150, 5' deep	9.80	15.20	25
5205	6"	12.25	16.35	28.60
5210	8"	16.45	17.75	34.20
5300	4" PVC Class 200, 5' deep	10.30	17.15	27.45
5305	6"	14.70	18.60	33.30
5310	8"	16.65	20.50	37.15
5405	6" PVC C900, 5' deep	12.25	18.60	30.85
5410	8"	16.45	20.50	36.95
5415	10"	21	22.50	43.50
5420	12"	27	24.50	51.50
5505	6" HDPE SDR 21 fusion, 5' deep	12.95	21	33.95
5510	8"	17.55	23	40.55
5515	10"	19.60	24	43.60
5520	12"	25	26	51
5615	10", 1/4" wall black steel, 5' deep	49.50	43.50	93
5620	12"	57	45.50	102.50
7000	4" MJ ductile iron, 7' deep	65	56	121
7005	6"	75.50	61.50	137
7010	8"	79.50	67	146.50
7015	10"	100	72.50	172.50
7020	12"	122	76	198
7100	4" push on joint ductile iron, 7' deep	44	47.50	91.50
7105	6"	45.50	49.50	95
7110	8"	56.50	58	114.50
7115	10"	77	60.50	137.50
7120	12"	81	64	145
7200	4" PVC Class 150, 7' deep	22	42.50	64.50
7205	6"	25.50	44	69.50
7210	8"	30.50	46.50	77
7300	4" PVC Class 200, 7' deep	22.50	44.50	67
7305	6"	28	46.50	74.50
7310	8"	31	49.50	80.50
7405	6" PVC C900, 7' deep	25.50	46.50	72
7410	8"	30.50	49.50	80
7415	10"	36	52	88
7420	12"	42.50	54.50	97

G30 Site Mechanical Utilities

G3010 Water Supply

G3010 122	Waterline (Common Earth Excavation)	COST L.F.		
		MAT.	INST.	TOTAL
7505	6" HDPE SDR 21 fusion, 7' deep	26	49	75
7510	8"	31.50	51.50	83
7515	10"	34.50	53.50	88
7520	12"	41	56	97

G3010 Water Supply

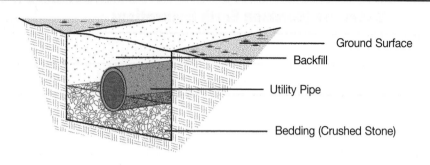

Labels on diagram: Ground Surface, Backfill, Utility Pipe, Bedding (Crushed Stone)

System Components			COST L.F.		
	QUANTITY	UNIT	MAT.	INST.	TOTAL
SYSTEM G3010 124 1000					
WATERLINE 4″ DUCTILE IRON INCLUDING LOAM & SANDY CLAY EXCAVATION,					
BEDDING, BACKFILL AND COMPACTION					
Excavate Trench	.296	B.C.Y.		2.01	2.01
Crushed stone bedding material	.203	L.C.Y.	6.19	2.54	8.73
Compact bedding	.173	E.C.Y.		1.04	1.04
Install 4″ MJ ductile iron pipe	1.000	L.F.	46.50	18.01	64.51
Backfill trench	.173	L.C.Y.		.54	.54
Compact fill material in trench	.123	E.C.Y.		.66	.66
Dispose of excess fill material on-site	.242	L.C.Y.		3.05	3.05
TOTAL			52.69	27.85	80.54

G3010 124	Waterline (Loam & Sandy Clay Excavation)	COST L.F.		
		MAT.	INST.	TOTAL
0950	Waterline including loam & sandy clay excavation, bedding, backfill & compaction			
1000	4″ MJ ductile iron, 4′ deep	52.50	28	80.50
1150	6 inch	62.50	33	95.50
1200	8 inch	66	37.50	103.50
1250	10 inch	85.50	42.50	128
1300	12 inch	107	46	153
1400	4″ push joint ductile iron, 4′ deep	31.50	18.85	50.35
1450	6 ″	32.50	21	53.50
1500	8 ″	43	28.50	71.50
1550	10 ″	62.50	31	93.50
1600	12 ″	66	34	100
1700	4″ PVC Class 150, 4′ deep	9.50	14.35	23.85
1750	6 ″	12.20	15.65	27.85
1800	8 ″	16.55	17.15	33.70
1900	4″ PVC Class 200, 4′ deep	10	16.25	26.25
1950	6 ″	14.65	17.90	32.55
2000	8 ″	16.75	19.95	36.70
2100	6″ PVC C900, 4′ deep	12.20	17.90	30.10
2150	8 ″	16.55	19.80	36.35
2200	12 ″	27.50	24.50	52
2300	6″ HDPE SDR 21 fusion, 4′ deep	12.90	20	32.90
2350	8 ″	17.65	22.50	40.15
2400	12 ″	25.50	26	51.50
2500	10″, 1/4″ wall black steel, 4′ deep	50	43.50	93.50
2550	12″	58	45.50	103.50
3000	18″ MJ ductile iron, 6′ deep with trench box	164	93.50	257.50
3050	24″	208	120	328
3100	18″ push joint ductile iron, 6′ deep with trench box	89	80.50	169.50
3200	24″	118	89.50	207.50
3300	18″ PVC C905, 6′ deep with trench box	52.50	56.50	109
3350	24″	72	65.50	137.50
3400	18″ HDPE fusion SDR 21, 6′ deep with trench box	53	73.50	126.50

G3010 Water Supply

G3010 124	Waterline (Loam & Sandy Clay Excavation)	COST L.F.		
		MAT.	INST.	TOTAL
3450	24"	83.50	88	171.50
3500	24" Concrete (CCP) , 6' deep with trench box	114	78	192
3600	18", 1/4" wall black steel , 6' deep with trench box	101	79	180
3650	18", 5/16" wall	135	79	214
3700	18", 3/8" wall	154	196	350
3750	24", 3/8" wall	178	228	406
3800	36" Concrete (CCP) , 8' deep with trench box	181	119	300
3900	36" PVC C905 , 8' deep with trench box	166	97.50	263.50
4000	36" 5/16" wall black steel , 8' deep with trench box	203	297	500
4050	36" 1/2" wall black steel	380	320	700
5000	4" MJ ductile iron, 5' deep	52.50	29	81.50
5005	6"	62.50	33.50	96
5010	8"	65.50	38	103.50
5015	10"	85	43	128
5020	12"	106	46	152
5100	4" push on joint ductile iron, 5' deep	31.50	19.70	51.20
5105	6"	32.50	22	54.50
5110	8"	42.50	29.50	72
5115	10"	62	31	93
5120	12"	65	34	99
5200	4" PVC Class 150, 5' deep	9.80	15.30	25.10
5205	6"	12.25	16.40	28.65
5210	8"	16.45	17.80	34.25
5300	4" PVC Class 200, 5' deep	10.30	17.20	27.50
5305	6"	14.70	18.65	33.35
5310	8"	16.65	20.50	37.15
5405	6" PVC C900, 5' deep	12.25	18.65	30.90
5410	8"	16.45	20.50	36.95
5415	10"	21	22.50	43.50
5420	12"	27	24.50	51.50
5505	6" HDPE SDR 21 fusion, 5' deep	12.95	21	33.95
5510	8"	17.55	23	40.55
5515	10"	19.60	24	43.60
5520	12"	25	26	51
5615	10", 1/4" wall black steel, 5' deep	49.50	44	93.50
5620	12"	57	45.50	102.50
6000	4" MJ ductile iron, 6' deep	65	52	117
6005	6"	75.50	57.50	133
6010	8"	79.50	63	142.50
6015	10"	100	68.50	168.50
6020	12"	122	72	194
6100	4" push on joint ductile iron, 6' deep	44	43	87
6105	6"	45.50	46	91.50
6110	8"	56.50	54	110.50
6115	10"	77	57	134
6120	12"	81	60.50	141.50
6200	4" PVC Class 150, 6' deep	22	39	61
6205	6"	26.50	41	67.50
6210	8"	30.50	42.50	73
6300	4" PVC Class 200, 6' deep	22.50	41	63.50
6305	6"	28	43	71
6310	8"	31	45	76
6405	6" PVC C900, 6' deep	25.50	43	68.50
6410	8"	30.50	45	75.50
6415	10"	36	47.50	83.50
6420	12"	42.50	51	93.50
6505	6" HDPE SDR 21 fusion, 6' deep	26	45.50	71.50
6510	8"	31.50	48	79.50

695

G3010 Water Supply

G3010 124	Waterline (Loam & Sandy Clay Excavation)	COST L.F.		
		MAT.	INST.	TOTAL
6515	10"	34.50	50	84.50
6520	12"	41	52.50	93.50
6615	10", 1/4" wall black steel, 6' deep	64.50	69.50	134
6620	12"	73	71.50	144.50
7000	4" MJ ductile iron, 7' deep	65	55	120
7005	6"	75.50	61	136.50
7010	8"	79.50	66	145.50
7015	10"	100	72	172
7020	12"	122	75.50	197.50
7100	4" push on joint ductile iron, 7' deep	44	46.50	90.50
7105	6"	45.50	49.50	95
7110	8"	56.50	57.50	114
7115	10"	77	60	137
7120	12"	81	63.50	144.50
7200	4" PVC Class 150, 7' deep	22	41.50	63.50
7205	6"	25.50	44	69.50
7210	8"	30.50	45.50	76
7300	4" PVC Class 200, 7' deep	22.50	43.50	66
7305	6"	28	46	74
7310	8"	31	48.50	79.50
7405	6" PVC C900, 7' deep	25.50	46	71.50
7410	8"	30.50	48.50	79
7415	10"	36	51	87
7420	12"	42.50	54	96.50
7505	6" HDPE SDR 21 fusion, 7' deep	26	48	74
7510	8"	31.50	51.50	83
7515	10"	34.50	52.50	87
7520	12"	41	56	97

G3010 Water Supply

The Fire Hydrant Systems include: four different hydrants with three different lengths of offsets, at several depths of trenching. Excavation and backfill is included with each system as well as thrust blocks as necessary and pipe bedding. Finally spreading of excess material and fine grading complete the components.

System Components	QUANTITY	UNIT	COST PER EACH		
			MAT.	INST.	TOTAL
SYSTEM G3010 410 1500					
HYDRANT, 4-1/2" VALVE SIZE, TWO WAY, 10' OFFSET, 2' DEEP					
Excavation, trench, 1 C.Y. hydraulic backhoe	3.556	C.Y.		31.25	31.25
Sleeve, 12" x 6"	1.000	Ea.	1,550	125.55	1,675.55
Ductile iron pipe, 6" diameter	8.000	L.F.	448	180.40	628.40
Gate valve, 6"	1.000	Ea.	2,650	313	2,963
Hydrant valve box, 6' long	1.000	Ea.	253	85	338
4-1/2" valve size, depth 2'-0"	1.000	Ea.	3,175	226	3,401
Thrust blocks, at valve, shoe and sleeve	1.600	C.Y.	358.40	481.66	840.06
Borrow, crushed stone, 3/8"	4.089	C.Y.	145.16	34.22	179.38
Backfill and compact, dozer, air tamped	4.089	C.Y.		80.56	80.56
Spread excess excavated material	.204	C.Y.		9.38	9.38
Fine grade, hand	400.000	S.F.		294	294
TOTAL			**8,579.56**	**1,861.02**	**10,440.58**

G3010 410	Fire Hydrants	COST PER EACH		
		MAT.	INST.	TOTAL
1000	Hydrant, 4-1/2" valve size, two way, 0' offset, 2' deep	8,125	1,275	9,400
1100	4' deep	8,150	1,300	9,450
1200	6' deep	8,700	1,350	10,050
1300	8' deep	8,075	1,425	9,500
1400	10' deep	9,450	1,500	10,950
1500	10' offset, 2' deep	8,575	1,850	10,425
1600	4' deep	8,600	2,200	10,800
1700	6' deep	9,150	2,800	11,950
1800	8' deep	8,525	3,725	12,250
1900	10' deep	9,900	5,625	15,525
2000	20' offset, 2' deep	9,125	2,475	11,600
2100	4' deep	9,150	3,000	12,150
2200	6' deep	9,700	3,925	13,625
2300	8' deep	9,100	5,225	14,325
2400	10' deep	10,500	7,025	17,525
2500	Hydrant, 4-1/2" valve size, three way, 0' offset, 2' deep	8,350	1,275	9,625
2600	4' deep	8,375	1,300	9,675
2700	6' deep	8,975	1,375	10,350
2800	8' deep	8,300	1,450	9,750
2900	10' deep	9,775	1,550	11,325
3000	10' offset, 2' deep	8,800	1,875	10,675
3100	4' deep	8,825	2,225	11,050
3200	6' deep	9,425	2,825	12,250
3300	8' deep	8,750	3,750	12,500

G3010 Water Supply

G3010 410	Fire Hydrants	COST PER EACH		
		MAT.	INST.	TOTAL
3400	10' deep	10,200	5,650	15,850
3500	20' offset, 2' deep	9,350	2,475	11,825
3600	4' deep	9,375	3,025	12,400
3700	6' deep	9,975	3,950	13,925
3800	8' deep	9,300	5,275	14,575
3900	10' deep	10,800	7,075	17,875
5000	5-1/4" valve size, two way, 0' offset, 2' deep	7,575	1,275	8,850
5100	4' deep	7,925	1,300	9,225
5200	6' deep	8,125	1,350	9,475
5300	8' deep	8,700	1,425	10,125
5400	10' deep	9,100	1,500	10,600
5500	10' offset, 2' deep	8,025	1,850	9,875
5600	4' deep	8,375	2,200	10,575
5700	6' deep	8,575	2,800	11,375
5800	8' deep	9,150	3,725	12,875
5900	10' deep	9,550	5,625	15,175
6000	20' offset, 2' deep	8,575	2,475	11,050
6100	4' deep	8,925	3,000	11,925
6200	6' deep	9,125	3,925	13,050
6300	8' deep	9,725	5,225	14,950
6400	10' deep	10,100	7,025	17,125
6500	Three way, 0' offset, 2' deep	7,750	1,275	9,025
6600	4' deep	8,125	1,300	9,425
6700	6' deep	8,350	1,375	9,725
6800	8' deep	8,975	1,450	10,425
6900	10' deep	9,400	1,550	10,950
7000	10' offset, 2' deep	8,200	1,875	10,075
7100	4' deep	8,575	2,225	10,800
7200	6' deep	8,800	2,825	11,625
7300	8' deep	9,425	3,750	13,175
7400	10' deep	9,850	5,650	15,500
7500	20' offset, 2' deep	8,775	2,475	11,250
7600	4' deep	9,150	2,725	11,875
7700	6' deep	9,350	3,950	13,300

G3020 Sanitary Sewer

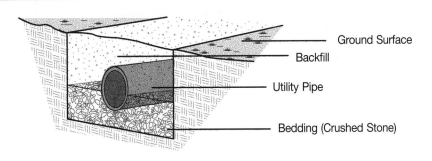

Ground Surface
Backfill
Utility Pipe
Bedding (Crushed Stone)

System Components	QUANTITY	UNIT	COST L.F.		
			MAT.	INST.	TOTAL
SYSTEM G3020 112 1000					
SEWERLINE 4″ B&S CAST IRON INCLUDING COMMON EARTH EXCAVATION,					
BEDDING, BACKFILL AND COMPACTION					
Excavate Trench	.296	B.C.Y.		2.17	2.17
Crushed stone bedding material	.203	L.C.Y.	6.19	2.54	8.73
Compact bedding	.173	E.C.Y.		1.04	1.04
Install 4″ B&S cast iron pipe	1.000	L.F.	25	21	46
Backfill trench	.160	L.C.Y.		.50	.50
Compact fill material in trench	.123	E.C.Y.		.66	.66
Dispose of excess fill material on-site	.225	L.C.Y.		2.84	2.84
TOTAL			31.19	30.75	61.94

G3020 112	Sewerline (Common Earth Excavation)	COST L.F.		
		MAT.	INST.	TOTAL
0950	Sewerline including common earth excavation, bedding, backfill and compaction			
1000	4″ B&S Cast iron, 4' deep	31	30.50	61.50
1150	6 inch	49	36	85
1200	8 inch	76	52.50	128.50
1250	10 inch	122	55.50	177.50
1300	12 inch	172	62.50	234.50
1400	4″ HDPE type S watertight, 4' deep	7.40	13.75	21.15
1450	6 ″	10.10	14.35	24.45
1500	8 ″	11.65	14.95	26.60
1550	10 ″	13.60	15.45	29.05
1600	12 ″	16.70	19.65	36.35
1700	4″ PVC, SDR 35, 20' length , 4' deep	8.05	14.25	22.30
1750	6 ″	11.15	14.95	26.10
1800	8 ″- 13' length	14.45	15.50	29.95
1900	10″	20	17.70	37.70
1950	12 ″	25.50	18.20	43.70
2000	12 ″ Concrete w/gasket class 3	27.50	24.50	52
2100	12 ″ class 4	28.50	24.50	53
2200	12 ″ class 5	27.50	24.50	52
3050	24″	47.50	52	99.50
3100	18″ PVC, SDR 35, 13' length, 6' deep with trench box	60	52	112
3150	24″	92	55	147
3200	18″ concrete w/gasket class 3, 6' deep with trench box	52	56	108
3250	24″ class 3	61.50	59	120.50
3300	18″ class 4	45.50	56	101.50
3350	24″ class 4	55.50	59	114.50
3400	18″ class 5	51.50	56	107.50
3450	24″ class 5	58	59	117
4000	30″ HDPE type S, watertight , 8' deep with trench box	69.50	72	141.50
4050	36″	85	78.50	163.50
5000	30″ concrete w/gaskets class 3, 8' deep with trench box	103	115	218
5050	36″ class 3	141	129	270

G30 Site Mechanical Utilities

G3020 Sanitary Sewer

G3020 112	Sewerline (Common Earth Excavation)	MAT.	INST.	TOTAL
5100	30" class 4	81.50	115	196.50
5150	36" class 4	102	129	231
5200	30" class 5	95.50	115	210.50
5250	36" class 5	138	129	267
5500	4" B&S Cast iron, 5' deep	31	31.50	62.50
5502	6"	49	36.50	85.50
5504	8"	75.50	53	128.50
5506	10"	121	55.50	176.50
5508	12"	171	63	234
5510	4" HDPE type S watertight, 5' deep	7.25	14.65	21.90
5512	6"	9.80	15.15	24.95
5514	8"	11.10	15.60	26.70
5516	10"	12.80	15.90	28.70
5518	12"	15.60	16.45	32.05
5520	4" PVC,SDR 35, 5' deep	7.90	15.15	23.05
5522	6"	10.80	15.75	26.55
5524	8"	13.95	16.20	30.15
5526	10"	19.20	18.15	37.35
5528	12"	24.50	18.50	43
5532	12" Concrete w/gasket class 3, 5' deep	26.50	25	51.50
5534	12" class 4	27.50	25	52.50
5536	12" class 5	26.50	25	51.50
6000	4" B&S Cast iron, 6' deep	43.50	56.50	100
6002	6"	62	63.50	125.50
6004	8"	89.50	79.50	169
6006	10"	136	83	219
6008	12"	187	90.50	277.50
6010	4" HDPE type S watertight, 6' deep	19.60	39.50	59.10
6012	6"	23	42	65
6014	8"	25	41.50	66.50
6016	10"	28	43	71
6018	12"	31.50	44	75.50
6020	4" PVC,SDR 35, 6' deep	20	40.50	60.50
6022	6"	24	42.50	66.50
6024	8"	28	42.50	70.50
6026	10"	34	45	79
6028	12"	40.50	46	86.50
6032	12" Concrete w/gasket class 3, 6' deep	42.50	53	95.50
6034	12" class 4	43.50	53	96.50
6036	12" class 5	42	53	95
6500	4" B&S Cast iron, 7' deep	43.50	59	102.50
6502	6"	62	66	128
6504	8"	89.50	82	171.50
6506	10"	136	85.50	221.50
6508	12"	187	93	280
6510	4" HDPE type S watertight, 7' deep	19.60	42	61.60
6512	6"	23	45	68
6514	8"	25	44.50	69.50
6516	10"	28	45.50	73.50
6518	12"	31.50	46.50	78
6520	4" PVC,SDR 35, 7' deep	20	42.50	62.50
6522	6"	24	45.50	69.50
6524	8"	28	45	73
6526	10"	34	47.50	81.50
6528	12"	40.50	49	89.50
6532	12" Concrete w/gasket class 3, 7' deep	42.50	55	97.50
6534	12" class 4	43.50	55	98.50
6536	12" class 5	42	55	97

G30 Site Mechanical Utilities

G3020 Sanitary Sewer

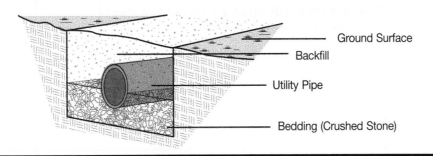

System Components	QUANTITY	UNIT	COST L.F.		
			MAT.	INST.	TOTAL
SYSTEM G3020 114 1000 **SEWERLINE 4″ B&S CAST IRON INCLUDING LOAM & SANDY CLAY EXCAVATION,** **BEDDING, BACKFILL AND COMPACTION**					
Excavate Trench	.296	B.C.Y.		2.01	2.01
Crushed stone bedding material	.203	L.C.Y.	6.19	2.54	8.73
Compact bedding	.173	E.C.Y.		1.04	1.04
Install 4″ B&S cast iron pipe	1.000	L.F.	25	21	46
Backfill trench	.173	L.C.Y.		.54	.54
Compact fill material in trench	.123	E.C.Y.		.66	.66
Dispose of excess fill material on-site	.242	L.C.Y.		3.05	3.05
TOTAL			31.19	30.84	62.03

G3020 114	Sewerline (loam & sandy clay excavation)	COST L.F.		
		MAT.	INST.	TOTAL
0950	Sewerline incl. loam & sandy clay excavation, bedding, backfill and compaction			
1000	4″ B&S Cast iron, 4′ deep	31	30.50	61.50
1150	6 inch	49	36	85
1200	8 inch	76	52.50	128.50
1250	10 inch	122	55.50	177.50
1300	12 inch	172	63	235
1400	4″ HDPE type S watertight, 4′ deep	7.40	13.80	21.20
1450	6″	10.10	14.50	24.60
1500	8″	11.65	15.05	26.70
1550	10″	13.60	15.50	29.10
1600	12″	16.70	16.25	32.95
1700	4″ PVC, SDR 35, 20′ length , 4′ deep	8.05	14.35	22.40
1750	6″	11.15	15.10	26.25
1800	8″- 13′ length	14.45	15.65	30.10
1900	10″	20	17.75	37.75
1950	12″	25.50	18.30	43.80
2000	12″ Concrete w/gasket class 3	27.50	24.50	52
2100	12″ class 4	28.50	24.50	53
2200	12″ class 5	27.50	24.50	52
3000	18″ HDPE type S watertight, 6′ deep with trench box	37.50	49.50	87
3050	24″	47.50	52	99.50
3100	18″ PVC, SDR 35, 13′ length, 6′ deep with trench box	60	52.50	112.50
3150	24″	92	55.50	147.50
3200	18″ concrete w/gasket class 3, 6′ deep with trench box	52	56	108
3250	24″ class 3	61.50	59.50	121
3300	18″ class 4	45.50	56	101.50
3350	24″ class 4	55.50	59.50	115
3400	18″ class 5	51.50	56	107.50
3450	24″ class 5	58	59.50	117.50
4000	30″ HDPE type S, watertight , 8′ deep with trench box	69.50	75	144.50
4050	36″	85	79.50	164.50
5000	30″ concrete w/gaskets class 3, 8′ deep with trench box	103	116	219
5050	36″ class 3	141	130	271
5100	30″ class 4	81.50	116	197.50

G3020 Sanitary Sewer

G3020 114	Sewerline (loam & sandy clay excavation)	COST L.F.		
		MAT.	INST.	TOTAL
5150	36" class 4	102	130	232
5200	30" class 5	95.50	116	211.50
5250	36" class 5	138	130	268
5500	4" B&S Cast iron, 5' deep	31	32	63
5502	6"	49	36.50	85.50
5504	8"	75.50	53	128.50
5506	10"	121	56	177
5508	12"	171	63	234
5510	4" HDPE type S watertight, 5' deep	7.25	14.70	21.95
5512	6"	9.80	15.25	25.05
5514	8"	11.10	15.70	26.80
5516	10"	12.80	15.95	28.75
5518	12"	15.60	16.55	32.15
5520	4" PVC, SDR 35, 5' deep	7.90	15.25	23.15
5522	6"	10.80	15.85	26.65
5524	8"	13.95	16.25	30.20
5526	10"	19.20	18.20	37.40
5528	12"	24.50	18.60	43.10
5532	12" Concrete w/gasket class 3, 5' deep	26.50	25	51.50
5534	12" class 4	27.50	25	52.50
5536	12" class 5	26.50	25	51.50
6000	4" B&S Cast iron, 6' deep	43.50	56.50	100
6002	6"	62	64	126
6004	8"	89.50	79.50	169
6006	10"	136	83	219
6008	12"	187	91	278
6010	4" HDPE type S watertight, 6' deep	19.60	39.50	59.10
6012	6"	23	42.50	65.50
6014	8"	25	42	67
6016	10"	28	43	71
6018	12"	31.50	44.50	76
6020	4" PVC,SDR 35, 6' deep	20	40	60
6022	6"	24	43	67
6024	8"	28	43	71
6026	10"	34	45.50	79.50
6028	12"	40.50	46.50	87
6032	12" Concrete w/gasket class 3, 6' deep	42.50	53	95.50
6034	12" class 4	43.50	53	96.50
6036	12" class 5	42	53	95
6500	4" B&S Cast iron, 7' deep	43.50	58	101.50
6502	6"	62	65	127
6504	8"	89.50	81	170.50
6506	10"	136	84.50	220.50
6508	12"	187	92.50	279.50
6510	4" HDPE type S watertight, 7' deep	19.60	41	60.60
6512	6"	23	44	67
6514	8"	25	43.50	68.50
6516	10"	28	44.50	72.50
6518	12"	31.50	46	77.50
6520	4" PVC,SDR 35, 7' deep	20	41.50	61.50
6522	6"	24	44.50	68.50
6524	8"	28	44	72
6526	10"	34	47	81
6528	12"	40.50	48	88.50
6532	12" Concrete w/gasket class 3, 7' deep	42.50	54	96.50
6534	12" class 4	43.50	54	97.50
6536	12" class 5	42	54	96

G30 Site Mechanical Utilities

G3020 Sanitary Sewer

The Septic System includes: a septic tank; leaching field; concrete distribution boxes; plus excavation and gravel backfill.

The Expanded System Listing shows systems with tanks ranging from 1000 gallons to 2500 gallons. Tanks are either concrete or fiberglass. Cost is on a complete unit basis.

System Components	QUANTITY	UNIT	COST EACH		
			MAT.	INST.	TOTAL
SYSTEM G3020 302 0300					
1000 GALLON TANK, LEACHING FIELD, 600 S.F., 50′ FROM BLDG.					
Mobilization	2.000	Ea.		1,706	1,706
Septic tank, precast, 1000 gallon	1.000	Ea.	1,175	283	1,458
Effluent filter	1.000	Ea.	48	83	131
Distribution box, precast, 5 outlets	1.000	Ea.	114	63.50	177.50
Flow leveler	3.000	Ea.	8.34	30.30	38.64
Sewer pipe, PVC, SDR 35, 4″ diameter	160.000	L.F.	294.40	724.80	1,019.20
Tee	1.000	Ea.	20.50	83	103.50
Elbow	2.000	Ea.	24.90	111	135.90
Viewport cap	1.000	Ea.	7.55	27.50	35.05
Filter fabric	67.000	S.Y.	115.91	28.14	144.05
Detectable marking tape	1.600	C.L.F.	15.84	5.39	21.23
Excavation	160.000	C.Y.		2,176	2,176
Backfill	133.000	L.C.Y.		420.28	420.28
Spoil	55.000	L.C.Y.		404.25	404.25
Compaction	113.000	E.C.Y.		1,020.39	1,020.39
Stone fill, 3/4″ to 1-1/2″	39.000	C.Y.	1,794	488.28	2,282.28
TOTAL			3,618.44	7,654.83	11,273.27

G3020 302	Septic Systems	COST EACH		
		MAT.	INST.	TOTAL
0300	1000 gal. concrete septic tank, 600 S.F. leaching field, 50 feet from bldg.	3,625	7,650	11,275
0301	200 feet from building	3,900	9,500	13,400
0305	750 S.F. leaching field, 50 feet from building	4,125	8,325	12,450
0306	200 feet from building	4,400	10,200	14,600
0400	1250 gal. concrete septic tank, 600 S.F. leaching field, 50 feet from bldg.	4,025	7,950	11,975
0401	200 feet from building	4,300	9,800	14,100
0405	750 S.F. leaching field, 50 feet from building	4,525	8,625	13,150
0406	200 feet from building	4,800	10,500	15,300
0410	1000 S.F. leaching field, 50 feet from building	5,425	9,900	15,325
0411	200 feet from building	5,725	11,800	17,525
0500	1500 gal. concrete septic tank, 600 S.F. leaching field, 50 feet from bldg.	4,275	8,225	12,500
0501	200 feet from building	4,525	10,100	14,625
0505	750 S.F. leaching field, 50 feet from building	4,775	8,950	13,725
0506	200 feet from building	5,050	11,800	16,850
0510	1000 S.F. leaching field, 50 feet from building	5,650	10,200	15,850
0511	200 feet from building	5,950	12,100	18,050
0515	1100 S.F. leaching field, 50 feet from building	6,000	10,700	16,700
0516	200 feet from building	6,275	12,600	18,875
0600	2000 gal concrete tank, 1200 S.F. leaching field, 50 feet from building	7,325	11,300	18,625
0601	200 feet from building	7,600	13,100	20,700

703

G30 Site Mechanical Utilities

G3020 Sanitary Sewer

G3020 302	Septic Systems	COST EACH		
		MAT.	INST.	TOTAL
0605	1500 S.F. leaching field, 50 feet from building	8,350	12,800	21,150
0606	200 feet from building	8,650	14,600	23,250
0710	2500 gal concrete tank, 1500 S.F. leaching field, 50 feet from building	8,200	13,500	21,700
0711	200 feet from building	8,500	15,400	23,900
0715	1600 S.F. leaching field, 50 feet from building	8,575	14,000	22,575
0716	200 feet from building	8,875	15,800	24,675
0720	1800 S.F. leaching field, 50 feet from building	9,275	15,000	24,275
0721	200 feet from building	9,575	16,900	26,475
1300	1000 gal. 2 compartment tank, 600 S.F. leaching field, 50 feet from bldg.	3,200	7,650	10,850
1301	200 feet from building	3,475	9,500	12,975
1305	750 S.F. leaching field, 50 feet from building	3,700	8,325	12,025
1306	200 feet from building	4,000	10,200	14,200
1400	1250 gal. 2 compartment tank, 600 S.F. leaching field, 50 feet from bldg.	3,600	7,950	11,550
1401	200 feet from building	3,875	9,800	13,675
1405	750 S.F. leaching field, 50 feet from building	4,100	8,625	12,725
1406	200 feet from building	4,375	10,500	14,875
1410	1000 S.F. leaching field, 50 feet from building	5,000	9,900	14,900
1411	200 feet from building	5,300	11,800	17,100
1500	1500 gal. 2 compartment tank, 600 S.F. leaching field, 50 feet from bldg.	4,200	8,225	12,425
1501	200 feet from building	4,450	10,100	14,550
1505	750 S.F. leaching field, 50 feet from building	4,700	8,950	13,650
1506	200 feet from building	4,975	11,800	16,775
1510	1000 S.F. leaching field, 50 feet from building	5,575	10,200	15,775
1511	200 feet from building	5,875	12,100	17,975
1515	1100 S.F. leaching field, 50 feet from building	5,925	10,700	16,625
1516	200 feet from building	6,200	12,600	18,800
1600	2000 gal 2 compartment tank, 1200 S.F. leaching field, 50 feet from bldg.	7,350	11,300	18,650
1601	200 feet from building	7,625	13,100	20,725
1605	1500 S.F. leaching field, 50 feet from building	8,375	12,800	21,175
1606	200 feet from building	8,675	14,600	23,275
1710	2500 gal 2 compartment tank, 1500 S.F. leaching field, 50 feet from bldg.	9,150	13,500	22,650
1711	200 feet from building	9,450	15,400	24,850
1715	1600 S.F. leaching field, 50 feet from building	9,525	14,000	23,525
1716	200 feet from building	9,825	15,800	25,625
1720	1800 S.F. leaching field, 50 feet from building	10,200	15,000	25,200
1721	200 feet from building	10,500	16,900	27,400

G3020 Sanitary Sewer

System Components	QUANTITY	UNIT	COST EACH		
			MAT.	INST.	TOTAL
SYSTEM G3020 730 1000					
500,000 GPD, SECONDARY DOMESTIC SEWAGE LAGOON, COMMON EARTH					
Supervision of lagoon construction	60.000	Day		39,900	39,900
Quality Control of Earthwork per day	30.000	Day		15,150	15,150
Excavate lagoon	10780.000	B.C.Y.		28,782.60	28,782.60
Finish grade slopes for seeding	40.000	M.S.F.		1,470	1,470
Spread & grade top of lagoon banks for compaction	753.000	M.S.F.		59,110.50	59,110.50
Compact lagoon banks	10725.000	E.C.Y.		8,901.75	8,901.75
Install pond liner	202.000	M.S.F.	184,830	252,500	437,330
Provide water for compaction and dust control	2156.000	E.C.Y.	2,889.04	4,700.08	7,589.12
Dispose of excess material on-site	72.000	L.C.Y.		282.96	282.96
Seed non-lagoon side of banks	40.000	M.S.F.	1,320	1,200	2,520
Place gravel on top of banks	2311.000	S.Y.	13,750.45	3,605.16	17,355.61
Provide 6 ' high fence around lagoon	1562.000	L.F.	34,364	12,105.50	46,469.50
Provide 12' wide fence gate	3.000	Opng.	1,530	1,743	3,273
Temporary Fencing	1562.000	L.F.	3,436.40	3,951.86	7,388.26
TOTAL			242,119.89	433,403.41	675,523.30

G3020 730	Secondary Sewage Lagoon	COST EACH		
		MAT.	INST.	TOTAL
0980	Lagoons, Secondary, 10 day retention, 5.1 foot depth, no mechanical aeration			
1000	500,000 GPD Secondary Domestic Sewage Lagoon, common earth,10 day retention	242,000	433,500	675,500
1100	600,000 GPD	276,500	507,500	784,000
1200	700,000 GPD	309,000	671,500	980,500
1300	800,000 GPD	342,500	614,000	956,500
1400	900,000 GPD	375,500	681,000	1,056,500
1500	1 MGD	437,000	788,000	1,225,000
1600	2 MGD	736,500	1,376,000	2,112,500
1700	3 MGD	1,183,000	2,130,500	3,313,500
2000	500,000 GPD sandy clay & loam	242,000	435,000	677,000
2100	600,000 GPD	276,500	508,500	785,000
2200	700,000 GPD	309,000	552,000	861,000
2300	800,000 GPD	342,500	616,000	958,500
2400	900,000 GPD	375,500	682,500	1,058,000
2500	1 MGD	437,000	791,000	1,228,000
2600	2 MGD	736,500	1,377,000	2,113,500
2700	3 MGD	1,183,000	2,135,000	3,318,000
3000	500,000 GPD sand & gravel	242,000	424,500	666,500
3100	600,000 GPD	276,500	495,500	772,000
3200	700,000 GPD	309,000	537,000	846,000
3300	800,000 GPD	342,500	599,500	942,000
3400	900,000 GPD	375,500	664,000	1,039,500

G30 Site Mechanical Utilities

G3020 Sanitary Sewer

G3020 730	Secondary Sewage Lagoon	COST EACH		
		MAT.	INST.	TOTAL
3500	1 MGD	437,000	770,000	1,207,000
3600	2 MGD	736,500	1,338,000	2,074,500
3700	3 MGD	1,183,000	2,074,500	3,257,500
6000	500,000 GPD, Domestic Sewage Sec. Lagoon, comm. earth, 10 day, no liner	48,300	178,500	226,800
6100	600,000 GPD	61,500	214,000	275,500
6200	700,000 GPD	65,500	224,000	289,500
6300	800,000 GPD	69,000	248,000	317,000
6400	900,000 GPD	72,500	267,500	340,000
6500	1 MGD	86,000	308,000	394,000
6600	2 MGD	114,500	526,000	640,500
6700	3 MGD	201,000	789,500	990,500
7000	500,000 GPD sandy clay & loam	57,500	182,500	240,000
7100	600,000 GPD	61,500	214,500	276,000
7200	700,000 GPD	65,500	219,500	285,000
7300	800,000 GPD	69,000	242,500	311,500
7400	900,000 GPD	72,500	268,500	341,000
7500	1 MGD	86,000	311,000	397,000
7600	2 MGD	114,500	527,000	641,500
7700	3 MGD	201,000	794,000	995,000

System Components

System Components	QUANTITY	UNIT	COST EACH		
			MAT.	INST.	TOTAL
SYSTEM G3020 740 1000					
WASTEWATER TREATMENT, AERATION, 75 CFM PREPACKAGED BLOWER SYSTEM					
Prepacked blower, 38 CFM	3.000	Ea.	10,050	2,595	12,645
Flexible coupling-2″ diameter	3.000	Ea.	108	61.50	169.50
Wafer style check valve-2″ diameter	3.000	Ea.	438	256.50	694.50
Wafer style butterfly valve-2″ diameter	3.000	Ea.	327	165	492
Wafer style butterfly valve-3″ diameter	1.000	Ea.	126	174	300
Blower Pressure Relief Valves-adjustable-3″ Diameter	1.000	Ea.	198	174	372
Silencer with polyester filter-3″ Connection	1.000	Ea.	180	174	354
Pipe-2″ diameter	30.000	L.F.	571.50	765	1,336.50
Pipe-3″ diameter	50.000	L.F.	1,775	1,400	3,175
3 x 3 x 2 Tee	3.000	Ea.	321	300	621
Pipe cap-3″ diameter	1.000	Ea.	83	46.50	129.50
Pressure gage	1.000	Ea.	20	24.50	44.50
Master control panel	1.000	Ea.	1,075	292	1,367
TOTAL			15,272.50	6,428	21,700.50

G3020 740	Blower System for Wastewater Aeration	COST EACH		
		MAT.	INST.	TOTAL
1000	75 CFM Prepackaged Blower System	15,300	6,425	21,725
1100	250 CFM	14,400	9,175	23,575
1200	450 CFM	25,500	9,700	35,200
1300	700 CFM	41,700	12,200	53,900
1400	1000 CFM	43,900	12,900	56,800

G3030 Storm Sewer

Manhole   Catch Basin

The Manhole and Catch Basin System includes: excavation with a backhoe; a formed concrete footing; frame and cover; cast iron steps and compacted backfill.

The Expanded System Listing shows manholes that have a 4', 5' and 6' inside diameter riser. Depths range from 4' to 14'. Construction material shown is either concrete, concrete block, precast concrete, or brick.

System Components	QUANTITY	UNIT	COST PER EACH		
			MAT.	INST.	TOTAL
SYSTEM G3030 210 1920					
MANHOLE/CATCH BASIN, BRICK, 4' I.D. RISER, 4' DEEP					
Excavation, hydraulic backhoe, 3/8 C.Y. bucket	14.815	B.C.Y.		130.22	130.22
Trim sides and bottom of excavation	64.000	S.F.		73.60	73.60
Forms in place, manhole base, 4 uses	20.000	SFCA	16.40	117	133.40
Reinforcing in place footings, #4 to #7	.019	Ton	23.75	24.70	48.45
Concrete, 3000 psi	.925	C.Y.	131.35		131.35
Place and vibrate concrete, footing, direct chute	.925	C.Y.		53.72	53.72
Catch basin or MH, brick, 4' ID, 4' deep	1.000	Ea.	675	1,150	1,825
Catch basin or MH steps; heavy galvanized cast iron	1.000	Ea.	19.85	15.80	35.65
Catch basin or MH frame and cover	1.000	Ea.	325	241	566
Fill, granular	12.954	L.C.Y.	349.76		349.76
Backfill, spread with wheeled front end loader	12.954	L.C.Y.		37.31	37.31
Backfill compaction, 12" lifts, air tamp	12.954	E.C.Y.		133.82	133.82
TOTAL			1,541.11	1,977.17	3,518.28

G3030 210	Manholes & Catch Basins	COST PER EACH		
		MAT.	INST.	TOTAL
1920	Manhole/catch basin, brick, 4' I.D. riser, 4' deep	1,550	1,975	3,525
1940	6' deep	2,150	2,725	4,875
1960	8' deep	2,875	3,750	6,625
1980	10' deep	3,375	4,650	8,025
3000	12' deep	4,200	5,075	9,275
3020	14' deep	5,175	7,100	12,275
3200	Block, 4' I.D. riser, 4' deep	1,325	1,575	2,900
3220	6' deep	1,825	2,250	4,075
3240	8' deep	2,425	3,100	5,525
3260	10' deep	2,875	3,850	6,725
3280	12' deep	3,650	4,900	8,550
3300	14' deep	4,550	5,975	10,525
4620	Concrete, cast-in-place, 4' I.D. riser, 4' deep	1,500	2,700	4,200
4640	6' deep	2,125	3,600	5,725
4660	8' deep	2,975	5,250	8,225
4680	10' deep	3,550	6,500	10,050
4700	12' deep	4,475	8,075	12,550
4720	14' deep	5,525	9,675	15,200
5820	Concrete, precast, 4' I.D. riser, 4' deep	1,825	1,450	3,275
5840	6' deep	2,325	1,950	4,275

G30 Site Mechanical Utilities

G3030 Storm Sewer

G3030 210	Manholes & Catch Basins	COST PER EACH		
		MAT.	INST.	TOTAL
5860	8' deep	2,975	2,775	5,750
5880	10' deep	3,550	3,400	6,950
5900	12' deep	4,475	4,100	8,575
5920	14' deep	5,525	5,350	10,875
6000	5' I.D. riser, 4' deep	2,850	1,575	4,425
6020	6' deep	3,575	2,200	5,775
6040	8' deep	4,650	2,900	7,550
6060	10' deep	5,950	3,700	9,650
6080	12' deep	7,400	4,725	12,125
6100	14' deep	8,975	5,800	14,775
6200	6' I.D. riser, 4' deep	3,975	2,050	6,025
6220	6' deep	4,825	2,700	7,525
6240	8' deep	6,150	3,800	9,950
6260	10' deep	7,700	4,825	12,525
6280	12' deep	9,400	6,100	15,500
6300	14' deep	11,300	7,425	18,725

G3030 Storm Sewer

The Headwall Systems are listed in concrete and different stone wall materials for two different backfill slope conditions. The backfill slope directly affects the length of the wing walls. Walls are listed for different culvert sizes starting at 30" diameter. Excavation and backfill are included in the system components, and are figured from an elevation 2' below the bottom of the pipe.

System Components	QUANTITY	UNIT	COST PER EACH		
			MAT.	INST.	TOTAL
SYSTEM G3030 310 2000					
HEADWALL, C.I.P. CONCRETE FOR 30" PIPE, 3' LONG WING WALLS					
Excavation, hydraulic backhoe, 3/8 C.Y. bucket	28.000	B.C.Y.		246.12	246.12
Formwork, 2 uses	157.000	SFCA	442.04	2,108.29	2,550.33
Reinforcing in place, A615 Gr 60, longer and heavier dowels, add	45.000	Lb.	30.60	84.15	114.75
Concrete, 3000 psi	2.600	C.Y.	369.20		369.20
Place concrete, spread footings, direct chute	2.600	C.Y.		150.98	150.98
Backfill, dozer	28.000	L.C.Y.		60.48	60.48
TOTAL			841.84	2,650.02	3,491.86

G3030 310	Headwalls	COST PER EACH		
		MAT.	INST.	TOTAL
2000	Headwall, 1-1/2 to 1 slope soil, C.I.P. conc, 30" pipe, 3' long wing walls	840	2,650	3,490
2020	Pipe size 36", 3'-6" long wing walls	1,050	3,175	4,225
2040	Pipe size 42", 4' long wing walls	1,275	3,725	5,000
2060	Pipe size 48", 4'-6" long wing walls	1,550	4,350	5,900
2080	Pipe size 54", 5'-0" long wing walls	1,825	5,050	6,875
2100	Pipe size 60", 5'-6" long wing walls	2,125	5,750	7,875
2120	Pipe size 72", 6'-6" long wing walls	2,850	7,375	10,225
2140	Pipe size 84", 7'-6" long wing walls	3,625	9,125	12,750
2500	$250/ton stone, pipe size 30", 3' long wing walls	595	975	1,570
2520	Pipe size 36", 3'-6" long wing walls	765	1,150	1,915
2540	Pipe size 42", 4' long wing walls	960	1,375	2,335
2560	Pipe size 48", 4'-6" long wing walls	1,175	1,625	2,800
2580	Pipe size 54", 5' long wing walls	1,425	1,900	3,325
2600	Pipe size 60", 5'-6" long wing walls	1,700	2,200	3,900
2620	Pipe size 72", 6'-6" long wing walls	2,325	2,925	5,250
2640	Pipe size 84", 7'-6" long wing walls	3,075	3,750	6,825
3000	$350/ton stone, pipe size 30", 3' long wing walls	830	975	1,805
3020	Pipe size 36", 3'-6" long wing walls	1,075	1,150	2,225
3040	Pipe size 42", 4' long wing walls	1,350	1,375	2,725
3060	Pipe size 48", 4'-6" long wing walls	1,650	1,625	3,275
3080	Pipe size 54", 5' long wing walls	2,000	1,900	3,900
3100	Pipe size 60", 5'-6" long wing walls	2,375	2,200	4,575
3120	Pipe size 72", 6'-6" long wing walls	3,275	2,925	6,200
3140	Pipe size 84", 7'-6" long wing walls	4,325	3,750	8,075
3500	$450/ton stone, pipe size 30", 3' long wing walls	1,075	975	2,050
3520	Pipe size 36", 3'-6" long wing walls	1,375	1,150	2,525
3540	Pipe size 42", 4' long wing walls	1,725	1,375	3,100
3560	Pipe size 48", 4'-6" long wing walls	2,125	1,625	3,750
3580	Pipe size 54", 5' long wing walls	2,575	1,900	4,475
3600	Pipe size 60", 5'-6" long wing walls	3,050	2,200	5,250

G3030 Storm Sewer

G3030 310	Headwalls	COST PER EACH		
		MAT.	INST.	TOTAL
3620	Pipe size 72", 6'-6" long wing walls	4,200	2,925	7,125
3640	Pipe size 84", 7'-6" long wing walls	5,525	3,750	9,275
4000	$650/ton stone, pipe size 30", 3' long wing walls	1,525	975	2,500
4020	Pipe size 36", 3'-6" long wing walls	1,975	1,150	3,125
4040	Pipe size 42", 4' long wing walls	2,475	1,375	3,850
4060	Pipe size 48", 4'-6" long wing walls	3,050	1,625	4,675
4080	Pipe size 54", 5' long wing walls	3,700	1,900	5,600
4100	Pipe size 60", 5'-6" long wing walls	4,375	2,200	6,575
4120	Pipe size 72", 6'-6" long wing walls	6,050	2,925	8,975
4140	Pipe size 84", 7'-6" long wing walls	7,950	3,750	11,700
4500	2 to 1 slope soil, C.I.P. concrete, pipe size 30", 4'-3" long wing walls	1,000	3,050	4,050
4520	Pipe size 36", 5' long wing walls	1,275	3,750	5,025
4540	Pipe size 42", 5'-9" long wing walls	1,575	4,500	6,075
4560	Pipe size 48", 6'-6" long wing walls	1,900	5,225	7,125
4580	Pipe size 54", 7'-3" long wing walls	2,250	6,100	8,350
4600	Pipe size 60", 8'-0" long wing walls	2,650	6,975	9,625
4620	Pipe size 72", 9'-6" long wing walls	3,550	9,050	12,600
4640	Pipe size 84", 11'-0" long wing walls	4,600	11,300	15,900
5000	$250/ton stone, pipe size 30", 4'-3" long wing walls	720	1,125	1,845
5020	Pipe size 36", 5' long wing walls	945	1,375	2,320
5040	Pipe size 42", 5'-9" long wing walls	1,200	1,650	2,850
5060	Pipe size 48", 6'-6" long wing walls	1,475	1,975	3,450
5080	Pipe size 54", 7'-3" long wing walls	1,800	2,325	4,125
5100	Pipe size 60", 8'-0" long wing walls	2,125	2,700	4,825
5120	Pipe size 72", 9'-6" long wing walls	2,325	3,625	5,950
5140	Pipe size 84", 11'-0" long wing walls	3,075	4,750	7,825
5500	$350/ton stone, pipe size 30", 4'-3" long wing walls	1,000	1,125	2,125
5520	Pipe size 36", 5' long wing walls	1,325	1,375	2,700
5540	Pipe size 42", 5'-9" long wing walls	1,675	1,650	3,325
5560	Pipe size 48", 6'-6" long wing walls	2,075	1,975	4,050
5580	Pipe size 54", 7'-3" long wing walls	2,525	2,325	4,850
5600	Pipe size 60", 8'-0" long wing walls	3,000	2,700	5,700
5620	Pipe size 72", 9'-6" long wing walls	4,150	3,625	7,775
5640	Pipe size 84", 11'-0" long wing walls	5,550	4,750	10,300
6000	$450/ton stone, pipe size 30", 4'-3" long wing walls	1,300	1,125	2,425
6020	Pipe size 36", 5'-0" long wing walls	1,700	1,375	3,075
6040	Pipe size 42", 5'-9" long wing walls	2,150	1,650	3,800
6060	Pipe size 48", 6'-6" long wing walls	2,675	1,975	4,650
6080	Pipe size 54", 7'-3" long wing walls	3,225	2,325	5,550
6100	Pipe size 60", 8'-0" long wing walls	3,850	2,700	6,550
6120	Pipe size 72", 9'-6" long wing walls	5,325	3,625	8,950
6140	Pipe size 84", 11'-0" long wing walls	7,100	4,750	11,850
6500	$650/ton stone, pipe size 30", 4'-3" long wing walls	1,875	1,125	3,000
6520	Pipe size 36", 5' long wing walls	2,450	1,375	3,825
6540	Pipe size 42", 5'-9" long wing walls	3,100	1,650	4,750
6560	Pipe size 48", 6'-6" long wing walls	3,825	1,975	5,800
6580	Pipe size 54", 7'-3" long wing walls	4,650	2,325	6,975
6600	Pipe size 60", 8'-0" long wing walls	5,525	2,700	8,225
6620	Pipe size 72", 9'-6" long wing walls	7,675	3,625	11,300
6640	Pipe size 84", 11'-0" long wing walls	10,200	4,750	14,950

G30 Site Mechanical Utilities

G3030 Storm Sewer

Most new project sites will require some degree of work to address the quantity and, sometimes, quality of the stormwater produced from that site. The extent of this work is determined by design professionals, based on local and federal guidelines. In general, the scope of stormwater work will be a function of the amount of impervious surface (roof and paved areas) to be constructed, the permeability of soil, and the sensitivity of the location to stormwater issues overall. Absent further information, an allowance of $3 per SF of impervious surface is advisable. Note that these costs rely on the inlet and piping of a traditional stormwater drainage system and are additive to that base work unless otherwise stated.

1050 Small Surface Retention

Small surface retention areas, or rain gardens, are typically 200 to 800 SF depressions that intercept surface storm run-off and retain and remediate (improve) water through filtration and absorption with specialized plants and fill material. These can be attractive if maintained and located in smaller areas to treat smaller flows.

1100 Large Surface Detention

Large surface detention areas collect stormwater in open basins and release runoff at a set rate. These basins make up about 10% of the impervious area of a site and must be protected by a fence. This option is the least expensive per volume of capacity, but the results are usually not attractive and it is not always allowed by local ordinance.

1150 Sub-Surface Detention, Gravel

Sub-Surface detention systems (gravel) are large areas (typically located under surface parking) in which existing earth is replaced by gravel and the voids within are utilized as a temporary storage area for storm water. Water access can be by pervious pavers or a traditional inlet system.

1200 Sub-Surface Detention, Gravel & Void Structure

Sub-Surface detention (gravel and void structure) is a hybrid solution consisting of the gravel solution above with the presence of structures that create additional volumes of storage. These typically employ inlets which can be supplemented to provide improved water quality.

1250 Sub-Surface Detention, Structural Vaults

Sub-Surface detention systems (structural vaults) are large precast enclosures providing high-volume water storage. It is the most expensive system but requires the least amount of space for a given volume of capacity.

G3030 610	Stormwater Management (costs per CF of stormwater)	MAT.	INST.	TOTAL
1050	Small Surface Retention	8.40	3.02	11.42
1100	Large Surface Detention	.27	.38	.65
1150	Sub-Surface Detention, Gravel	9.30	3.38	12.68
1200	Sub-Surface Detention, Gravel & Void Structure	8.85	1.71	10.56
1250	Sub-Surface Detention, Structural Vaults	21.50	1.99	23.49

Cost per C.F.

G3060 Fuel Distribution

System Components	QUANTITY	UNIT	COST L.F.		
			MAT.	INST.	TOTAL
SYSTEM G3060 112 1000					
GASLINE 1/2″ PE, 60 PSI COILS, COMPRESSION COUPLING, INCLUDING COMMON					
EARTH EXCAVATION, BEDDING, BACKFILL AND COMPACTION					
Excavate Trench	.170	B.C.Y.		1.25	1.25
Sand bedding material	.090	L.C.Y.	1.80	1.13	2.93
Compact bedding	.070	E.C.Y.		.42	.42
Install 1/2″ gas pipe	1.000	L.F.	.54	3.99	4.53
Backfill trench	.120	L.C.Y.		.38	.38
Compact fill material in trench	.100	E.C.Y.		.53	.53
Dispose of excess fill material on-site	.090	L.C.Y.		1.14	1.14
TOTAL			2.34	8.84	11.18

G3060 112	Gasline (Common Earth Excavation)	COST L.F.		
		MAT.	INST.	TOTAL
0950	Gasline, including common earth excavation, bedding, backfill and compaction			
1000	1/2″ diameter PE, 60 psi coils, compression coupling, SDR 11, 2′ deep	2.34	8.85	11.19
1010	4′ deep	2.34	11.35	13.69
1020	6′ deep	2.34	13.80	16.14
1050	1″ diameter PE, 60 psi coils, compression coupling, SDR 11, 2′ deep	3.05	9.30	12.35
1060	4′ deep	3.05	11.80	14.85
1070	6′ deep	3.05	14.25	17.30
1100	1 1/4″ diameter PE, 60 psi coils, compression coupling, SDR 11, 2′ deep	3.60	9.30	12.90
1110	4′ deep	3.60	11.80	15.40
1120	6′ deep	3.60	14.25	17.85
1150	2″ diameter PE, 60 psi coils, compression coupling, SDR 11, 2′ deep	4.52	9.85	14.37
1160	4′ deep	4.52	12.30	16.82
1180	6′ deep	4.52	14.80	19.32
1200	3″ diameter PE, 60 psi coils, compression coupling, SDR 11, 2′ deep	9.60	11.95	21.55
1210	4′ deep	9.60	14.45	24.05
1220	6′ deep	9.60	16.90	26.50

713

G3060 Fuel Distribution

G3060 112	Gasline (Common Earth Excavation)	COST L.F.		
		MAT.	INST.	TOTAL
1250	4" diameter PE, 60 psi 40' length, compression coupling, SDR 11, 2' deep	17.70	16.45	34.15
1260	4' deep	17.70	18.90	36.60
1270	6' deep	17.70	21.50	39.20
1300	6" diameter PE, 60 psi 40' length w/coupling, SDR 11, 2' deep	42	17.40	59.40
1310	4' deep	42	19.90	61.90
1320	6' deep	42	22.50	64.50
1350	8" diameter PE, 60 psi 40' length w/coupling, SDR 11, 4' deep	63	22.50	85.50
1360	6' deep	63	25	88
2000	1" diameter steel plain end, Schedule 40, 2' deep	7.40	15.05	22.45
2010	4' deep	7.40	17.55	24.95
2020	6' deep	7.40	20	27.40
2050	2" diameter, steel plain end, Schedule 40, 2' deep	10.55	15.75	26.30
2060	4' deep	10.55	18.30	28.85
2070	6' deep	10.55	20.50	31.05
2100	3" diameter, steel plain end, Schedule 40, 2' deep	17.35	17.75	35.10
2110	4' deep	17.35	20	37.35
2120	6' deep	17.35	22.50	39.85
2150	4" diameter, steel plain end, Schedule 40, 2' deep	22	24.50	46.50
2160	4' deep	22	27	49
2170	6' deep	22	29.50	51.50
2200	5" diameter, steel plain end, Schedule 40, 2' deep	30.50	27.50	58
2210	4' deep	30.50	30	60.50
2220	6' deep	30.50	32.50	63
2250	6" diameter, steel plain end, Schedule 40, 2' deep	36.50	32.50	69
2260	4' deep	36.50	35	71.50
2270	6' deep	36.50	37	73.50
2300	8" diameter, steel plain end, Schedule 40, 2' deep	56.50	39.50	96
2310	4' deep	56.50	42	98.50
2320	6' deep	56.50	44.50	101
2350	10" diameter, steel plain end, Schedule 40, 2' deep	129	53	182
2360	4' deep	129	56	185
2370	6' deep	129	58	187
2400	12" diameter, steel plain end, Schedule 40, 2' deep	143	65	208
2410	4' deep	143	67.50	210.50
2420	6' deep	143	70	213
2460	14" diameter, steel plain end, Schedule 40, 4' deep	153	72	225
2470	6' deep	153	75	228
2510	16" diameter, Schedule 40, 4' deep	167	77	244
2520	6' deep	167	80.50	247.50
2560	18" diameter, Schedule 40, 4' deep	215	83.50	298.50
2570	6' deep	215	86	301
2600	20" diameter, steel plain end, Schedule 40, 4' deep	330	90	420
2610	6' deep	330	93.50	423.50
2650	24" diameter, steel plain end, Schedule 40, 4' deep	380	108	488
2660	6' deep	380	111	491
3150	4" diameter, steel plain end, Schedule 80, 2' deep	47	39	86
3160	4' deep	47	41	88
3170	6' deep	47	44	91
3200	5" diameter, steel plain end, Schedule 80, 2' deep	60.50	39.50	100
3210	4' deep	60.50	42	102.50
3220	6' deep	60.50	44.50	105
3250	6" diameter, steel plain end, Schedule 80, 2' deep	97	43.50	140.50
3260	4' deep	97	46	143
3270	6' deep	97	48.50	145.50
3300	8" diameter, steel plain end, Schedule 80, 2' deep	128	49.50	177.50
3310	4' deep	128	52	180
3320	6' deep	128	54.50	182.50
3350	10" diameter, steel plain end, Schedule 80, 2' deep	191	58	249

G3060 Fuel Distribution

G3060 112	Gasline (Common Earth Excavation)	COST L.F.		
		MAT.	INST.	TOTAL
3360	4' deep	191	61	252
3370	6' deep	191	63	254
3400	12" diameter, steel plain end, Schedule 80, 2' deep	253	71	324
3410	4' deep	253	74	327
3420	6' deep	253	76	329

G4010 Electrical Distribution

GROUND SURFACE

BACKFILL

CONCRETE

CONDUIT

CONDUIT SUPPORTS

BEDDING (NATURAL)

UNDERGROUND ELECTRICAL CONDUIT

System Components	QUANTITY	UNIT	COST EACH		
			MAT.	INST.	TOTAL
SYSTEM G4010 312 1000					
2000 AMP UNDERGROUND SERVICE INCLUDING COMMON EARTH EXCAVATION,					
CONCRETE, BACKFILL AND COMPACTION					
Excavate Trench	44.440	B.C.Y.		325.75	325.75
4 inch conduit bank	108.000	L.F.	1,323	2,916	4,239
4 inch fitting	8.000	Ea.	352	488	840
4 inch bells	4.000	Ea.	20.40	244	264.40
Concrete material	16.580	C.Y.	2,354.36		2,354.36
Concrete placement	16.580	C.Y.		439.20	439.20
Backfill trench	33.710	L.C.Y.		106.52	106.52
Compact fill material in trench	25.930	E.C.Y.		137.69	137.69
Dispose of excess fill material on-site	24.070	L.C.Y.		303.76	303.76
500 kcmil power cable	18.000	C.L.F.	19,350	8,190	27,540
Wire, 600 volt stranded copper wire, type THW, 1/0	6.000	C.L.F.	1,488	1,326	2,814
TOTAL			24,887.76	14,476.92	39,364.68

G4010 312	Underground Power Feed	COST EACH		
		MAT.	INST.	TOTAL
0900	Underground electrical			
0950	Including common earth excavation, concrete, backfill and compaction			
1000	2000 Amp service, 100' length, 4' depth	24,900	14,500	39,400
1100	1600 Amp service	19,400	11,200	30,600
1200	1200 Amp service	15,200	9,275	24,475
1300	1000 Amp service	17,700	8,500	26,200
1400	800 Amp service	10,900	7,975	18,875
1500	600 Amp service	7,000	7,400	14,400

G40 Site Electrical Utilities

G4010 Electrical Distribution

System Components	QUANTITY	UNIT	COST L.F. MAT.	COST L.F. INST.	COST L.F. TOTAL
SYSTEM G4010 320 1012					
UNDERGROUND ELECTRICAL CONDUIT, 2″ DIAMETER, INCLUDING EXCAVATION,					
CONCRETE, BEDDING, BACKFILL AND COMPACTION					
Excavate Trench	.150	B.C.Y.		1.10	1.10
Base spacer	.200	Ea.	.30	2.61	2.91
Conduit	1.000	L.F.	1.42	4.06	5.48
Concrete material	.050	C.Y.	7.10		7.10
Concrete placement	.050	C.Y.		1.32	1.32
Backfill trench	.120	L.C.Y.		.38	.38
Compact fill material in trench	.100	E.C.Y.		.53	.53
Dispose of excess fill material on-site	.120	L.C.Y.		1.52	1.52
Marking tape	.010	C.L.F.	.42	.04	.46
TOTAL			9.24	11.56	20.80

G4010 320	Underground Electrical Conduit	MAT.	INST.	TOTAL
0950	Underground electrical conduit, including common earth excavation,			
0952	Concrete, bedding, backfill and compaction			
1012	2″ dia. Schedule 40 PVC, 2′ deep	9.25	11.60	20.85
1014	4′ deep	9.25	16.40	25.65
1016	6′ deep	9.25	21	30.25
1022	2 @ 2″ dia. Schedule 40 PVC, 2′ deep	10.95	18.25	29.20
1024	4′ deep	10.95	23	33.95
1026	6′ deep	10.95	27.50	38.45
1032	3 @ 2″ dia. Schedule 40 PVC, 2′ deep	12.70	25	37.70
1034	4′ deep	12.70	30	42.70
1036	6′ deep	12.70	34.50	47.20
1042	4 @ 2″ dia. Schedule 40 PVC, 2′ deep	17.20	31	48.20
1044	4′ deep	17.20	36	53.20
1046	6′ deep	17.20	40.50	57.70

717

G4010 Electrical Distribution

G4010 320	Underground Electrical Conduit	COST L.F.		
		MAT.	INST.	TOTAL
1062	6 @ 2" dia. Schedule 40 PVC, 2' deep	20.50	44.50	65
1064	4' deep	20.50	49	69.50
1066	6' deep	20.50	54	74.50
1082	8 @ 2" dia. Schedule 40 PVC, 2' deep	24	57.50	81.50
1084	4' deep	24	62	86
1086	6' deep	24	67	91
1092	9 @ 2" dia. Schedule 40 PVC, 2' deep	25.50	64	89.50
1094	4' deep	28.50	68.50	97
1096	6' deep	28.50	73.50	102
1212	3" dia. Schedule 40 PVC, 2' deep	11.95	14.25	26.20
1214	4' deep	11.95	19.15	31.10
1216	6' deep	11.95	23.50	35.45
1222	2 @ 3" dia. Schedule 40 PVC, 2' deep	13.55	23.50	37.05
1224	4' deep	13.55	28.50	42.05
1226	6' deep	13.55	32.50	46.05
1232	3 @ 3" dia. Schedule 40 PVC, 2' deep	16.55	32.50	49.05
1234	4' deep	16.55	37.50	54.05
1236	6' deep	16.55	42	58.55
1242	4 @ 3" dia. Schedule 40 PVC, 2' deep	24	42	66
1244	4' deep	24	46.50	70.50
1246	6' deep	24	51.50	75.50
1262	6 @ 3" dia. Schedule 40 PVC, 2' deep	28.50	60.50	89
1264	4' deep	28.50	65	93.50
1266	6' deep	28.50	69.50	98
1282	8 @ 3" dia. Schedule 40 PVC, 2' deep	34.50	79	113.50
1284	4' deep	34.50	83.50	118
1286	6' deep	34.50	89	123.50
1292	9 @ 3" dia. Schedule 40 PVC, 2' deep	40.50	89	129.50
1294	4' deep	41.50	93	134.50
1296	6' deep	41.50	97	138.50
1312	4" dia. Schedule 40 PVC, 2' deep	12.80	17.45	30.25
1314	4' deep	12.80	22.50	35.30
1316	6' deep	12.80	27	39.80
1322	2 @ 4" dia. Schedule 40 PVC, 2' deep	16.70	30	46.70
1324	4' deep	16.70	35.50	52.20
1326	6' deep	16.70	39.50	56.20
1332	3 @ 4" dia. Schedule 40 PVC, 2' deep	19.15	42.50	61.65
1334	4' deep	19.15	47.50	66.65
1336	6' deep	19.15	52	71.15
1342	4 @ 4" dia. Schedule 40 PVC, 2' deep	28.50	56	84.50
1344	4' deep	28.50	60	88.50
1346	6' deep	28.50	65	93.50
1362	6 @ 4" dia. Schedule 40 PVC, 2' deep	35	81	116
1364	4' deep	35	85.50	120.50
1366	6' deep	35	90.50	125.50
1382	8 @ 4" dia. Schedule 40 PVC, 2' deep	44	107	151
1384	4' deep	44	112	156
1386	6' deep	44	118	162
1392	9 @ 4" dia. Schedule 40 PVC, 2' deep	49	120	169
1394	4' deep	49	123	172
1396	6' deep	49	128	177
1412	5" dia. Schedule 40 PVC, 2' deep	14.55	21	35.55
1414	4' deep	14.55	25.50	40.05
1416	6' deep	14.55	30.50	45.05
1422	2 @ 5" dia. Schedule 40 PVC, 2' deep	20	37	57
1424	4' deep	20	41.50	61.50
1426	6' deep	20	46.50	66.50
1432	3 @ 5" dia. Schedule 40 PVC, 2' deep	25.50	53.50	79

G40 Site Electrical Utilities

G4010 Electrical Distribution

G4010 320	Underground Electrical Conduit	COST L.F.		
		MAT.	INST.	TOTAL
1434	4' deep	25.50	58.50	84
1436	6' deep	25.50	64	89.50
1442	4 @ 5" dia. Schedule 40 PVC, 2' deep	35.50	69	104.50
1444	4' deep	35.50	74	109.50
1446	6' deep	35.50	78.50	114
1462	6 @ 5" dia. Schedule 40 PVC, 2' deep	45.50	102	147.50
1464	4' deep	45.50	106	151.50
1466	6' deep	45.50	112	157.50
1482	8 @ 5" dia. Schedule 40 PVC, 2' deep	59.50	135	194.50
1484	4' deep	59.50	141	200.50
1486	6' deep	59.50	147	206.50
1492	9 @ 5" dia. Schedule 40 PVC, 2' deep	66.50	151	217.50
1494	4' deep	66.50	154	220.50
1496	6' deep	66.50	159	225.50
1512	6" dia. Schedule 40 PVC, 2' deep	17.70	25.50	43.20
1514	4' deep	17.70	30	47.70
1516	6' deep	17.70	35	52.70
1522	2 @ 6" dia. Schedule 40 PVC, 2' deep	23.50	45.50	69
1524	4' deep	23.50	50	73.50
1526	6' deep	23.50	55	78.50
1532	3 @ 6" dia. Schedule 40 PVC, 2' deep	32.50	67	99.50
1534	4' deep	32.50	72.50	105
1536	6' deep	32.50	79	111.50
1542	4 @ 6" dia. Schedule 40 PVC, 2' deep	42.50	86	128.50
1544	4' deep	42.50	91	133.50
1546	6' deep	42.50	95.50	138
1562	6 @ 6" dia. Schedule 40 PVC, 2' deep	59	128	187
1564	4' deep	59	133	192
1566	6' deep	59	140	199
1582	8 @ 6" dia. Schedule 40 PVC, 2' deep	77.50	170	247.50
1584	4' deep	77.50	178	255.50
1586	6' deep	77.50	185	262.50
1592	9 @ 6" dia. Schedule 40 PVC, 2' deep	86.50	191	277.50
1594	4' deep	86.50	195	281.50
1596	6' deep	86.50	201	287.50

G4020 Site Lighting

The Site Lighting System includes the complete unit from foundation to electrical fixtures. Each system includes: excavation; concrete base; backfill by hand; compaction with a plate compacter; pole of specified material; all fixtures; and lamps.

The Expanded System Listing shows Site Lighting Systems that use one of two types of lamps: high pressure sodium; and metal halide. Systems are listed for 400-watt and 1000-watt lamps. Pole height varies from 20′ to 40′. There are four types of poles possibly listed: aluminum, fiberglass, steel and wood.

System Components	QUANTITY	UNIT	COST EACH		
			MAT.	INST.	TOTAL
SYSTEM G4020 110 5820					
SITE LIGHTING, METAL HALIDE, 400 WATT, ALUMINUM POLE, 20′ HIGH					
Excavating, by hand, pits to 6′ deep, heavy soil	2.368	C.Y.		300.74	300.74
Concrete in place incl. forms and reinf. stl. spread footings under 1 CY	.465	C.Y.	104.16	139.98	244.14
Backfill	1.903	C.Y.		87.54	87.54
Compact, vibrating plate	1.903	C.Y.		14.50	14.50
Aluminum light pole, 20′, no concrete base	1.000	Ea.	1,200	701.50	1,901.50
Road area luminaire wire	.300	C.L.F.	9	27.45	36.45
Roadway area luminaire, metal halide 400 W	1.000	Ea.	760	330	1,090
TOTAL			2,073.16	1,601.71	3,674.87

G4020 110	Site Lighting	COST EACH		
		MAT.	INST.	TOTAL
2320	Site lighting, high pressure sodium, 400 watt, aluminum pole, 20′ high	2,225	1,600	3,825
2340	30′ high	3,325	1,950	5,275
2360	40′ high	4,100	2,525	6,625
2520	Fiberglass pole, 20′ high	1,975	1,400	3,375
2540	30′ high	2,125	1,750	3,875
2560	40′ high	3,100	2,225	5,325
2720	Steel pole, 20′ high	2,425	1,675	4,100
2740	30′ high	2,450	2,050	4,500
2760	40′ high	2,950	2,700	5,650
2920	Wood pole, 20′ high	1,300	1,550	2,850
2940	30′ high	1,400	1,950	3,350
2960	40′ high	1,775	2,400	4,175
3120	1000 watt, aluminum pole, 20′ high	2,350	1,625	3,975
3140	30′ high	3,450	1,975	5,425
3160	40′ high	4,225	2,575	6,800
3320	Fiberglass pole, 20′ high	2,100	1,425	3,525
3340	30′ high	2,250	1,775	4,025
3360	40′ high	3,225	2,250	5,475
3420	Steel pole, 20′ ttigh	2,550	1,700	4,250
3440	30′ high	2,575	2,100	4,675

G4020 Site Lighting

G4020 110	Site Lighting	COST EACH		
		MAT.	INST.	TOTAL
3460	40' high	3,075	2,725	5,800
3520	Wood pole, 20' high	1,425	1,600	3,025
3540	30' high	1,525	1,975	3,500
3560	40' high	1,875	2,425	4,300
5820	Metal halide, 400 watt, aluminum pole, 20' high	2,075	1,600	3,675
5840	30' high	3,175	1,950	5,125
5860	40' high	3,950	2,525	6,475
6120	Fiberglass pole, 20' high	1,825	1,400	3,225
6140	30' high	1,975	1,750	3,725
6160	40' high	2,950	2,225	5,175
6320	Steel pole, 20' high	2,275	1,675	3,950
6340	30' high	2,300	2,050	4,350
6360	40' high	2,800	2,700	5,500
7620	1000 watt, aluminum pole, 20' high	2,175	1,625	3,800
7640	30' high	3,275	1,975	5,250
7660	40' high	4,050	2,575	6,625
7900	Fiberglass pole, 20' high	1,925	1,425	3,350
7920	30' high	2,075	1,775	3,850
7940	40' high	3,050	2,250	5,300
8120	Steel pole, 20' high	2,375	1,700	4,075
8140	30' high	2,400	2,100	4,500
8160	40' high	2,900	2,725	5,625

For customer support on your Site Work & Landscape Costs with RSMeans data, call 800.448.8182.

Table G4020 210 Procedure for Calculating Floodlights Required for Various Footcandles
Poles should not be spaced more than 4 times the fixture mounting height for good light distribution.

Estimating Chart
Select Lamp type.

Determine total square feet.

Chart will show quantity of fixtures to provide 1 footcandle initial, at intersection of lines. Multiply fixture quantity by desired footcandle level.

Chart based on use of wide beam luminaires in an area whose dimensions are large compared to mounting height and is approximate only.

To maintain 1 footcandle over a large area use these watts per square foot:

Incandescent	0.15
Metal Halide	0.032
Mercury Vapor	0.05
High Pressure Sodium	0.024

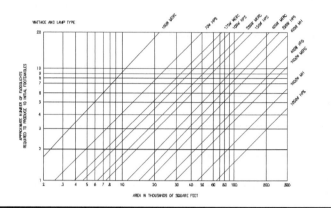

System Components	QUANTITY	UNIT	COST EACH		
			MAT.	INST.	TOTAL
SYSTEM G4020 210 0200					
LIGHT POLES, ALUMINUM, 20' HIGH, 1 ARM BRACKET					
Aluminum light pole, 20', no concrete base	1.000	Ea.	1,200	701.50	1,901.50
Bracket arm for aluminum light pole	1.000	Ea.	155	91.50	246.50
Excavation by hand, pits to 6' deep, heavy soil or clay	2.368	C.Y.		300.74	300.74
Footing, concrete incl forms, reinforcing, spread, under 1 C.Y.	.465	C.Y.	104.16	139.98	244.14
Backfill by hand	1.903	C.Y.		87.54	87.54
Compaction vibrating plate	1.903	C.Y.		14.50	14.50
TOTAL			1,459.16	1,335.76	2,794.92

G4020 210	Light Pole (Installed)	COST EACH		
		MAT.	INST.	TOTAL
0200	Light pole, aluminum, 20' high, 1 arm bracket	1,450	1,325	2,775
0240	2 arm brackets	1,625	1,325	2,950
0280	3 arm brackets	1,775	1,375	3,150
0320	4 arm brackets	1,925	1,400	3,325
0360	30' high, 1 arm bracket	2,550	1,675	4,225
0400	2 arm brackets	2,725	1,675	4,400
0440	3 arm brackets	2,875	1,725	4,600
0480	4 arm brackets	3,025	1,725	4,750
0680	40' high, 1 arm bracket	3,325	2,250	5,575
0720	2 arm brackets	3,500	2,250	5,750
0760	3 arm brackets	3,650	2,300	5,950
0800	4 arm brackets	3,800	2,300	6,100
0840	Steel, 20' high, 1 arm bracket	1,750	1,400	3,150
0880	2 arm brackets	1,850	1,400	3,250
0920	3 arm brackets	1,750	1,450	3,200
0960	4 arm brackets	1,900	1,450	3,350
1000	30' high, 1 arm bracket	1,775	1,775	3,550
1040	2 arm brackets	1,875	1,775	3,650
1080	3 arm brackets	1,775	1,825	3,600
1120	4 arm brackets	1,925	1,825	3,750
1320	40' high, 1 arm bracket	2,275	2,425	4,700
1360	2 arm brackets	2,375	2,425	4,800
1400	3 arm brackets	2,275	2,475	4,750
1440	4 arm brackets	2,425	2,475	4,900

G9020 Site Repair & Maintenance

G9020 100	Clean and Wrap Marine Piles	COST EACH		
		MAT.	INST.	TOTAL
1000	Clean & wrap 5 foot long, 8 inch diameter wood pile using nails from boat	103	109	212
1010	6 foot long	123	181	304
1020	7 foot long	144	187	331
1030	8 foot long	164	194	358
1040	9 foot long	185	201	386
1050	10 foot long	205	209	414
1060	11 foot long	226	270	496
1070	12 foot long	246	276	522
1075	13 foot long	267	283	550
1080	14 foot long	287	291	578
1090	15 foot long	310	298	608
1100	Clean & wrap 5 foot long, 10 inch diameter	128	135	263
1110	6 foot long	153	225	378
1120	7 foot long	179	234	413
1130	8 foot long	204	242	446
1140	9 foot long	230	251	481
1150	10 foot long	255	260	515
1160	11 foot long	281	345	626
1170	12 foot long	305	355	660
1175	13 foot long	330	365	695
1180	14 foot long	355	370	725
1190	15 foot long	385	380	765
1200	Clean & wrap 5 foot long, 12 inch diameter	153	161	314
1210	6 foot long	183	271	454
1220	7 foot long	214	281	495
1230	8 foot long	244	293	537
1240	9 foot long	275	305	580
1250	10 foot long	305	315	620
1260	11 foot long	335	410	745
1270	12 foot long	365	420	785
1275	13 foot long	395	430	825
1280	14 foot long	425	440	865
1290	15 foot long	460	455	915
1300	Clean & wrap 5 foot long, 13 inch diameter	165	177	342
1310	6 foot long	198	295	493
1320	7 foot long	231	305	536
1330	8 foot long	264	320	584
1340	9 foot long	297	330	627
1350	10 foot long	330	340	670
1360	11 foot long	365	440	805
1370	12 foot long	395	450	845
1375	13 foot long	430	460	890
1380	14 foot long	460	470	930
1390	15 foot long	495	485	980
1400	Clean & wrap 5 foot long, 14 inch diameter	180	190	370
1410	6 foot long	216	310	526
1420	7 foot long	252	325	577
1430	8 foot long	288	335	623
1440	9 foot long	325	350	675
1450	10 foot long	360	360	720
1460	11 foot long	395	465	860
1470	12 foot long	430	480	910
1475	13 foot long	470	490	960
1480	14 foot long	505	505	1,010
1490	15 foot long	540	520	1,060
1500	Clean & wrap 5 foot long, 8 inch diameter wood pile using straps from boat	155	105	260
1510	6 foot long	186	176	362
1520	7 foot long	217	182	399

G9020 Site Repair & Maintenance

G9020 100	Clean and Wrap Marine Piles	COST EACH		
		MAT.	INST.	TOTAL
1530	8 foot long	248	189	437
1540	9 foot long	279	194	473
1550	10 foot long	310	201	511
1560	11 foot long	340	262	602
1570	12 foot long	370	268	638
1575	13 foot long	405	274	679
1580	14 foot long	435	280	715
1590	15 foot long	465	287	752
1600	Clean & wrap 5 foot long, 10 inch diameter	193	131	324
1610	6 foot long	231	220	451
1620	7 foot long	270	228	498
1630	8 foot long	310	236	546
1640	9 foot long	345	243	588
1650	10 foot long	385	252	637
1660	11 foot long	425	335	760
1670	12 foot long	460	345	805
1675	13 foot long	500	350	850
1680	14 foot long	540	360	900
1690	15 foot long	580	370	950
1700	Clean & wrap 5 foot long, 12 inch diameter	233	156	389
1710	6 foot long	279	265	544
1720	7 foot long	325	274	599
1730	8 foot long	370	285	655
1740	9 foot long	420	294	714
1750	10 foot long	465	305	770
1760	11 foot long	510	395	905
1770	12 foot long	560	410	970
1775	13 foot long	605	420	1,025
1780	14 foot long	650	425	1,075
1790	15 foot long	700	435	1,135
1800	Clean & wrap 5 foot long, 13 inch diameter	250	171	421
1810	6 foot long	300	288	588
1820	7 foot long	350	299	649
1830	8 foot long	400	310	710
1840	9 foot long	450	320	770
1850	10 foot long	500	330	830
1860	11 foot long	550	425	975
1870	12 foot long	600	440	1,040
1875	13 foot long	650	450	1,100
1880	14 foot long	700	455	1,155
1890	15 foot long	750	465	1,215
1900	Clean & wrap 5 foot long, 14 inch diameter	270	183	453
1910	6 foot long	325	305	630
1920	7 foot long	380	315	695
1930	8 foot long	430	325	755
1940	9 foot long	485	340	825
1950	10 foot long	540	350	890
1960	11 foot long	595	455	1,050
1970	12 foot long	650	465	1,115
1975	13 foot long	700	480	1,180
1980	14 foot long	755	490	1,245
1990	15 foot long	810	500	1,310
2000	Clean & wrap 5 foot long, 12 inch diameter concrete pile from boat w/straps	233	137	370
2010	6 foot long	279	230	509
2020	7 foot long	325	239	564
2030	8 foot long	370	249	619
2040	9 foot long	420	258	678
2050	10 foot long	465	268	733

G90 Other Site Construction

G9020 Site Repair & Maintenance

G9020 100	Clean and Wrap Marine Piles	COST EACH		
		MAT.	INST.	TOTAL
2060	11 foot long	510	355	865
2070	12 foot long	560	365	925
2080	13 foot long	605	370	975
2085	14 foot long	650	380	1,030
2090	15 foot long	700	395	1,095
2100	Clean & wrap 5 foot long, 14 inch diameter	270	160	430
2110	6 foot long	325	266	591
2120	7 foot long	380	277	657
2130	8 foot long	430	288	718
2140	9 foot long	485	300	785
2150	10 foot long	540	310	850
2160	11 foot long	595	400	995
2170	12 foot long	650	410	1,060
2180	13 foot long	700	425	1,125
2185	14 foot long	755	435	1,190
2190	15 foot long	810	445	1,255

Reference Section

All the reference information is in one section, making it easy to find what you need to know . . and easy to use the data set on a daily basis. This section is visually identified by a vertical black bar on the page edges.

In this Reference Section, we've included Equipment Rental Costs, a listing of rental and operating costs; Crew Listings, a full listing of all crews and equipment, and their costs; Historical Cost Indexes for cost comparisons over time; City Cost Indexes and Location Factors for adjusting costs to the region you are in; Reference Tables, where you will find explanations, estimating information and procedures, or technical data; Change Orders, information on pricing changes to contract documents; and an explanation of all the Abbreviations in the data set.

Table of Contents

Construction Equipment Rental Costs — 729

Crew Listings — 741

Historical Cost Indexes — 778

City Cost Indexes — 779

Location Factors — 822

Reference Tables — 828
R01 General Requirements 828
R02 Existing Conditions 839
R03 Concrete.. 841
R04 Masonry 848
R05 Metals 851
R06 Wood, Plastics & Composites 858
R08 Openings.. 858
R09 Finishes.. 859
R13 Special Construction 860

Reference Tables (cont.)
R22 Plumbing 861
R23 Heating, Ventilating &
　　Air Conditioning................................... 865
R26 Electrical... 867
R31 Earthwork... 870
R32 Exterior Improvements........................ 874
R33 Utilities... 877
R34 Transportation...................................... 878

Change Orders — 879

Project Costs — 883

Abbreviations — 888

Estimating Tips

- This section contains the average costs to rent and operate hundreds of pieces of construction equipment. This is useful information when one is estimating the time and material requirements of any particular operation in order to establish a unit or total cost. Bare equipment costs shown on a unit cost line include, not only rental, but also operating costs for equipment under normal use.

Rental Costs

- Equipment rental rates are obtained from the following industry sources throughout North America: contractors, suppliers, dealers, manufacturers, and distributors.
- Rental rates vary throughout the country, with larger cities generally having lower rates. Lease plans for new equipment are available for periods in excess of six months, with a percentage of payments applying toward purchase.
- Monthly rental rates vary from 2% to 5% of the purchase price of the equipment depending on the anticipated life of the equipment and its wearing parts.
- Weekly rental rates are about 1/3 of the monthly rates, and daily rental rates are about 1/3 of the weekly rate.
- Rental rates can also be treated as reimbursement costs for contractor-owned equipment. Owned equipment costs include depreciation, loan payments, interest, taxes, insurance, storage, and major repairs.

Operating Costs

- The operating costs include parts and labor for routine servicing, such as the repair and replacement of pumps, filters, and worn lines. Normal operating expendables, such as fuel, lubricants, tires, and electricity (where applicable), are also included.
- Extraordinary operating expendables with highly variable wear patterns, such as diamond bits and blades, are excluded. These costs can be found as material costs in the Unit Price section.
- The hourly operating costs listed do not include the operator's wages.

Equipment Cost/Day

- Any power equipment required by a crew is shown in the Crew Listings with a daily cost.
- This daily cost of equipment needed by a crew includes both the rental cost and the operating cost and is based on dividing the weekly rental rate by 5 (the number of working days in the week), then adding the hourly operating cost multiplied by 8 (the number of hours in a day). This "Equipment Cost/Day" is shown in the far right column of the Equipment Rental section.
- If equipment is needed for only one or two days, it is best to develop your own cost by including components for daily rent and hourly operating costs. This is important when the listed Crew for a task does not contain the equipment needed, such as a crane for lifting mechanical heating/cooling equipment up onto a roof.

- If the quantity of work is less than the crew's Daily Output shown for a Unit Price line item that includes a bare unit equipment cost, the recommendation is to estimate one day's rental cost and operating cost for equipment shown in the Crew Listing for that line item.
- Please note, in some cases the equipment description in the crew is followed by a time period in parenthesis. For example: (daily) or (monthly). In these cases the equipment cost/day is calculated by adding the rental cost per time period to the hourly operating cost multiplied by 8.

Mobilization, Demobilization Costs

- The cost to move construction equipment from an equipment yard or rental company to the job site and back again is not included in equipment rental costs listed in the Reference Section. It is also not included in the bare equipment cost of any Unit Price line item or in any equipment costs shown in the Crew Listings.
- Mobilization (to the site) and demobilization (from the site) costs can be found in the Unit Price section.
- If a piece of equipment is already at the job site, it is not appropriate to utilize mobilization or demobilization costs again in an estimate. ■

01 54 33 | Equipment Rental

			UNIT	HOURLY OPER. COST	RENT PER DAY	RENT PER WEEK	RENT PER MONTH	EQUIPMENT COST/DAY	
10	0010	**CONCRETE EQUIPMENT RENTAL** without operators	R015433 -10						**10**
	0200	Bucket, concrete lightweight, 1/2 C.Y.	Ea.	.87	38.50	115	345	30	
	0300	1 C.Y.		.98	63.50	190	570	45.80	
	0400	1-1/2 C.Y.		1.23	60	180	540	45.85	
	0500	2 C.Y.		1.34	73.50	220	660	54.70	
	0580	8 C.Y.		6.47	93.50	280	840	107.75	
	0600	Cart, concrete, self-propelled, operator walking, 10 C.F.		2.85	175	525	1,575	127.85	
	0700	Operator riding, 18 C.F.		4.81	192	575	1,725	153.45	
	0800	Conveyer for concrete, portable, gas, 16" wide, 26' long		10.61	160	480	1,450	180.85	
	0900	46' long		10.98	175	525	1,575	192.85	
	1000	56' long		11.15	192	575	1,725	204.15	
	1100	Core drill, electric, 2-1/2 H.P., 1" to 8" bit diameter		1.56	83.50	250	750	62.50	
	1150	11 H.P., 8" to 18" cores		5.38	119	356.32	1,075	114.30	
	1200	Finisher, concrete floor, gas, riding trowel, 96" wide		9.64	153	459.60	1,375	169	
	1300	Gas, walk-behind, 3 blade, 36" trowel		2.03	96	287.50	865	73.75	
	1400	4 blade, 48" trowel		3.06	104	312.50	940	87	
	1500	Float, hand-operated (Bull float), 48" wide		.08	12.35	37	111	8.05	
	1570	Curb builder, 14 H.P., gas, single screw		14.00	253	760	2,275	263.95	
	1590	Double screw		15.00	253	760	2,275	272	
	1600	Floor grinder, concrete and terrazzo, electric, 22" path		3.03	134	401.75	1,200	104.60	
	1700	Edger, concrete, electric, 7" path		1.18	57.50	172.50	520	43.95	
	1750	Vacuum pick-up system for floor grinders, wet/dry		1.61	102	305.71	915	74.05	
	1800	Mixer, powered, mortar and concrete, gas, 6 C.F., 18 H.P.		7.39	97	291	875	117.35	
	1900	10 C.F., 25 H.P.		8.97	114	342.50	1,025	140.30	
	2000	16 C.F.		9.33	144	432.50	1,300	161.15	
	2100	Concrete, stationary, tilt drum, 2 C.Y.		7.21	80	240	720	105.70	
	2120	Pump, concrete, truck mounted, 4" line, 80' boom		29.79	287	860	2,575	410.35	
	2140	5" line, 110' boom		37.34	287	860	2,575	470.75	
	2160	Mud jack, 50 C.F. per hr.		6.43	228	685	2,050	188.45	
	2180	225 C.F. per hr.		8.52	293	880	2,650	244.15	
	2190	Shotcrete pump rig, 12 C.Y./hr.		13.93	223	670	2,000	245.40	
	2200	35 C.Y./hr.		15.75	287	860	2,575	298	
	2600	Saw, concrete, manual, gas, 18 H.P.		5.52	112	337	1,000	111.55	
	2650	Self-propelled, gas, 30 H.P.		7.87	81	242.71	730	111.50	
	2675	V-groove crack chaser, manual, gas, 6 H.P.		1.64	100	300	900	73.10	
	2700	Vibrators, concrete, electric, 60 cycle, 2 H.P.		.47	73	218.50	655	47.45	
	2800	3 H.P.		.56	73.50	221	665	48.70	
	2900	Gas engine, 5 H.P.		1.54	16.85	50.61	152	22.45	
	3000	8 H.P.		2.08	17.05	51.12	153	26.85	
	3050	Vibrating screed, gas engine, 8 H.P.		2.80	88	263.50	790	75.10	
	3120	Concrete transit mixer, 6 x 4, 250 H.P., 8 C.Y., rear discharge		50.57	70	210	630	446.55	
	3200	Front discharge		58.71	135	405	1,225	550.65	
	3300	6 x 6, 285 H.P., 12 C.Y., rear discharge		57.97	150	450	1,350	553.80	
	3400	Front discharge		60.41	170	510	1,525	585.25	
20	0010	**EARTHWORK EQUIPMENT RENTAL** without operators	R015433 -10						**20**
	0040	Aggregate spreader, push type, 8' to 12' wide	Ea.	2.59	75	225	675	65.75	
	0045	Tailgate type, 8' wide		2.54	63.50	190	570	58.30	
	0055	Earth auger, truck mounted, for fence & sign posts, utility poles		13.81	150	450	1,350	200.50	
	0060	For borings and monitoring wells		42.52	83.50	250	750	390.20	
	0070	Portable, trailer mounted		2.29	100	300	900	78.35	
	0075	Truck mounted, for caissons, water wells		85.14	150	450	1,350	771.10	
	0080	Horizontal boring machine, 12" to 36" diameter, 45 H.P.		22.70	104	312	935	244	
	0090	12" to 48" diameter, 65 H.P.		31.16	108	325	975	314.25	
	0095	Auger, for fence posts, gas engine, hand held		.45	84	251.50	755	53.85	
	0100	Excavator, diesel hydraulic, crawler mounted, 1/2 C.Y. cap.		21.66	465	1,394.28	4,175	452.10	
	0120	5/8 C.Y. capacity		28.95	610	1,833.22	5,500	598.25	
	0140	3/4 C.Y. capacity		32.56	725	2,168.88	6,500	694.25	
	0150	1 C.Y. capacity		41.09	780	2,333	7,000	795.30	

		UNIT	HOURLY OPER. COST	RENT PER DAY	RENT PER WEEK	RENT PER MONTH	EQUIPMENT COST/DAY
0200	1-1/2 C.Y. capacity	Ea.	48.44	500	1,500	4,500	687.55
0300	2 C.Y. capacity		56.41	835	2,500	7,500	951.30
0320	2-1/2 C.Y. capacity		82.39	1,350	4,027.92	12,100	1,465
0325	3-1/2 C.Y. capacity		119.76	2,000	6,000	18,000	2,158
0330	4-1/2 C.Y. capacity		151.17	3,675	11,000	33,000	3,409
0335	6 C.Y. capacity		191.81	3,225	9,650	29,000	3,464
0340	7 C.Y. capacity		174.67	3,400	10,200	30,600	3,437
0342	Excavator attachments, bucket thumbs		3.39	258	774.60	2,325	182.05
0345	Grapples		3.13	222	666.16	2,000	158.30
0346	Hydraulic hammer for boom mounting, 4000 ft lb.		13.44	890	2,670	8,000	641.50
0347	5000 ft lb.		15.90	950	2,850	8,550	697.25
0348	8000 ft lb.		23.47	1,200	3,600	10,800	907.75
0349	12,000 ft lb.		25.64	1,100	3,333	10,000	871.75
0350	Gradall type, truck mounted, 3 ton @ 15' radius, 5/8 C.Y.		43.31	835	2,500	7,500	846.50
0370	1 C.Y. capacity		59.22	835	2,500	7,500	973.80
0400	Backhoe-loader, 40 to 45 H.P., 5/8 C.Y. capacity		11.86	244	732.50	2,200	241.40
0450	45 H.P. to 60 H.P., 3/4 C.Y. capacity		17.97	117	350	1,050	213.75
0460	80 H.P., 1-1/4 C.Y. capacity		20.30	117	350	1,050	232.40
0470	112 H.P., 1-1/2 C.Y. capacity		32.89	610	1,833.22	5,500	629.75
0482	Backhoe-loader attachment, compactor, 20,000 lb.		6.42	155	464.76	1,400	144.30
0485	Hydraulic hammer, 750 ft lb.		3.67	107	320.17	960	93.40
0486	Hydraulic hammer, 1200 ft lb.		6.53	205	614.52	1,850	175.15
0500	Brush chipper, gas engine, 6" cutter head, 35 H.P.		9.14	247	740	2,225	221.10
0550	Diesel engine, 12" cutter head, 130 H.P.		23.60	340	1,020	3,050	392.80
0600	15" cutter head, 165 H.P.		26.51	415	1,239.36	3,725	459.90
0750	Bucket, clamshell, general purpose, 3/8 C.Y.		1.40	91.50	275	825	66.15
0800	1/2 C.Y.		1.51	91.50	275	825	67.10
0850	3/4 C.Y.		1.64	91.50	275	825	68.10
0900	1 C.Y.		1.70	91.50	275	825	68.55
0950	1-1/2 C.Y.		2.78	91.50	275	825	77.25
1000	2 C.Y.		2.91	91.50	275	825	78.30
1010	Bucket, dragline, medium duty, 1/2 C.Y.		.82	91.50	275	825	61.55
1020	3/4 C.Y.		.78	91.50	275	825	61.25
1030	1 C.Y.		.80	91.50	275	825	61.35
1040	1-1/2 C.Y.		1.26	91.50	275	825	65.05
1050	2 C.Y.		1.29	91.50	275	825	65.30
1070	3 C.Y.		2.07	91.50	275	825	71.60
1200	Compactor, manually guided 2-drum vibratory smooth roller, 7.5 H.P.		7.20	181	542.50	1,625	166.10
1250	Rammer/tamper, gas, 8"		2.20	48	144.59	435	46.50
1260	15"		2.62	55	165.25	495	54
1300	Vibratory plate, gas, 18" plate, 3000 lb. blow		2.12	24.50	72.81	218	31.55
1350	21" plate, 5000 lb. blow		2.61	241	722.50	2,175	165.35
1370	Curb builder/extruder, 14 H.P., gas, single screw		13.99	253	760	2,275	263.95
1390	Double screw		14.99	253	760	2,275	271.95
1500	Disc harrow attachment, for tractor		.47	82.50	246.84	740	53.15
1810	Feller buncher, shearing & accumulating trees, 100 H.P.		39.08	460	1,380	4,150	588.60
1860	Grader, self-propelled, 25,000 lb.		33.25	1,100	3,333	10,000	932.60
1910	30,000 lb.		32.76	1,325	4,000	12,000	1,062
1920	40,000 lb.		51.73	1,550	4,667	14,000	1,347
1930	55,000 lb.		66.73	1,775	5,333	16,000	1,600
1950	Hammer, pavement breaker, self-propelled, diesel, 1000 to 1250 lb.		28.31	600	1,800	5,400	586.50
2000	1300 to 1500 lb.		42.67	1,000	3,020.94	9,075	945.55
2050	Pile driving hammer, steam or air, 4150 ft lb. @ 225 bpm		12.11	500	1,500	4,500	396.90
2100	8750 ft lb. @ 145 bpm		14.30	700	2,100	6,300	534.45
2150	15,000 ft lb. @ 60 bpm		14.63	835	2,500	7,500	617.05
2200	24,450 ft lb. @ 111 bpm		15.64	965	2,900	8,700	705.15
2250	Leads, 60' high for pile driving hammers up to 20,000 ft lb.		3.66	300	900	2,700	209.25
2300	90' high for hammers over 20,000 ft lb.		5.43	540	1,620	4,850	367.45

		UNIT	HOURLY OPER. COST	RENT PER DAY	RENT PER WEEK	RENT PER MONTH	EQUIPMENT COST/DAY	
20	2350	Diesel type hammer, 22,400 ft lb.	Ea.	17.76	490	1,471.74	4,425	436.45	20
	2400	41,300 ft lb.		25.61	620	1,859.04	5,575	576.65	
	2450	141,000 ft lb.		41.20	980	2,943.48	8,825	918.35	
	2500	Vib. elec. hammer/extractor, 200 kW diesel generator, 34 H.P.		41.25	715	2,143.06	6,425	758.60	
	2550	80 H.P.		72.81	1,025	3,098.40	9,300	1,202	
	2600	150 H.P.		134.74	2,000	5,964.42	17,900	2,271	
	2700	Hydro Excavator w/EXT boom 12 C.Y., 1200 gallons		37.70	1,600	4,800	14,400	1,262	
	2800	Log chipper, up to 22" diameter, 600 H.P.		46.13	305	915	2,750	552	
	2850	Logger, for skidding & stacking logs, 150 H.P.		43.40	930	2,785	8,350	904.20	
	2860	Mulcher, diesel powered, trailer mounted		17.99	305	915	2,750	326.90	
	2900	Rake, spring tooth, with tractor		14.67	370	1,110.26	3,325	339.45	
	3000	Roller, vibratory, tandem, smooth drum, 20 H.P.		7.78	320	967	2,900	255.65	
	3050	35 H.P.		10.10	260	779.76	2,350	236.75	
	3100	Towed type vibratory compactor, smooth drum, 50 H.P.		25.20	520	1,566	4,700	514.75	
	3150	Sheepsfoot, 50 H.P.		25.56	385	1,161.90	3,475	436.90	
	3170	Landfill compactor, 220 H.P.		69.80	1,650	4,985	15,000	1,555	
	3200	Pneumatic tire roller, 80 H.P.		12.88	405	1,213.54	3,650	345.75	
	3250	120 H.P.		19.33	665	1,988.14	5,975	552.25	
	3300	Sheepsfoot vibratory roller, 240 H.P.		62.02	1,425	4,260.30	12,800	1,348	
	3320	340 H.P.		83.57	2,175	6,500	19,500	1,969	
	3350	Smooth drum vibratory roller, 75 H.P.		23.27	655	1,962.32	5,875	578.60	
	3400	125 H.P.		27.53	740	2,220.52	6,650	664.35	
	3410	Rotary mower, brush, 60", with tractor		18.73	360	1,084.44	3,250	366.75	
	3420	Rototiller, walk-behind, gas, 5 H.P.		2.13	60	180	540	53.05	
	3422	8 H.P.		2.80	132	395	1,175	101.40	
	3440	Scrapers, towed type, 7 C.Y. capacity		6.42	127	382.14	1,150	127.80	
	3450	10 C.Y. capacity		7.18	170	511.24	1,525	159.70	
	3500	15 C.Y. capacity		7.38	196	588.70	1,775	176.75	
	3525	Self-propelled, single engine, 14 C.Y. capacity		132.89	2,225	6,660	20,000	2,395	
	3550	Dual engine, 21 C.Y. capacity		140.95	2,500	7,500	22,500	2,628	
	3600	31 C.Y. capacity		187.28	3,625	10,844.40	32,500	3,667	
	3640	44 C.Y. capacity		231.98	4,650	13,942.80	41,800	4,644	
	3650	Elevating type, single engine, 11 C.Y. capacity		61.68	1,075	3,200	9,600	1,133	
	3700	22 C.Y. capacity		114.28	1,625	4,850	14,600	1,884	
	3710	Screening plant, 110 H.P. w/5' x 10' screen		21.07	645	1,933	5,800	555.15	
	3720	5' x 16' screen		26.60	1,325	4,000	12,000	1,013	
	3850	Shovel, crawler-mounted, front-loading, 7 C.Y. capacity		218.00	3,925	11,773.92	35,300	4,099	
	3855	12 C.Y. capacity		335.89	5,450	16,318.24	49,000	5,951	
	3860	Shovel/backhoe bucket, 1/2 C.Y.		2.68	73	218.95	655	65.25	
	3870	3/4 C.Y.		2.66	82	245.81	735	70.40	
	3880	1 C.Y.		2.75	91	272.66	820	76.50	
	3890	1-1/2 C.Y.		2.94	107	320.17	960	87.60	
	3910	3 C.Y.		3.43	145	433.78	1,300	114.15	
	3950	Stump chipper, 18" deep, 30 H.P.		6.88	232	697	2,100	194.45	
	4110	Dozer, crawler, torque converter, diesel 80 H.P.		25.18	335	1,000	3,000	401.40	
	4150	105 H.P.		34.23	600	1,800	5,400	633.85	
	4200	140 H.P.		41.15	720	2,166	6,500	762.40	
	4260	200 H.P.		62.97	1,675	5,000	15,000	1,504	
	4310	300 H.P.		80.49	1,875	5,600	16,800	1,764	
	4360	410 H.P.		106.42	3,200	9,630	28,900	2,777	
	4370	500 H.P.		132.98	3,900	11,670	35,000	3,398	
	4380	700 H.P.		229.47	5,475	16,421.52	49,300	5,120	
	4400	Loader, crawler, torque conv., diesel, 1-1/2 C.Y., 80 H.P.		29.45	550	1,651	4,950	565.80	
	4450	1-1/2 to 1-3/4 C.Y., 95 H.P.		30.18	695	2,091.42	6,275	659.75	
	4510	1-3/4 to 2-1/4 C.Y., 130 H.P.		47.61	965	2,900	8,700	960.90	
	4530	2-1/2 to 3-1/4 C.Y., 190 H.P.		57.61	1,175	3,540	10,600	1,169	
	4560	3-1/2 to 5 C.Y., 275 H.P.		71.19	1,525	4,595.96	13,800	1,489	
	4610	Front end loader, 4WD, articulated frame, diesel, 1 to 1-1/4 C.Y., 70 H.P.		16.58	282	846.90	2,550	302	

01 54 33 | Equipment Rental

		UNIT	HOURLY OPER. COST	RENT PER DAY	RENT PER WEEK	RENT PER MONTH	EQUIPMENT COST/DAY	
20	4620	1-1/2 to 1-3/4 C.Y., 95 H.P.	Ea.	19.94	440	1,320	3,950	423.50	**20**
	4650	1-3/4 to 2 C.Y., 130 H.P.		21.00	395	1,187.72	3,575	405.55	
	4710	2-1/2 to 3-1/2 C.Y., 145 H.P.		29.44	780	2,333	7,000	702.10	
	4730	3 to 4-1/2 C.Y., 185 H.P.		31.99	890	2,667	8,000	789.35	
	4760	5-1/4 to 5-3/4 C.Y., 270 H.P.		53.03	890	2,666	8,000	957.50	
	4810	7 to 9 C.Y., 475 H.P.		90.90	2,550	7,667	23,000	2,261	
	4870	9 to 11 C.Y., 620 H.P.		131.52	2,700	8,107.48	24,300	2,674	
	4880	Skid-steer loader, wheeled, 10 C.F., 30 H.P. gas		9.54	169	506.07	1,525	177.55	
	4890	1 C.Y., 78 H.P., diesel		18.38	420	1,265.18	3,800	400.10	
	4892	Skid-steer attachment, auger		.74	145	433.50	1,300	92.65	
	4893	Backhoe		.74	122	366.64	1,100	79.25	
	4894	Broom		.70	140	420.25	1,250	89.70	
	4895	Forks		.15	32	96	288	20.45	
	4896	Grapple		.72	88.50	265.25	795	58.80	
	4897	Concrete hammer		1.05	183	550	1,650	118.40	
	4898	Tree spade		.60	103	309.84	930	66.75	
	4899	Trencher		.65	102	305	915	66.20	
	4900	Trencher, chain, boom type, gas, operator walking, 12 H.P.		4.16	206	618.25	1,850	156.95	
	4910	Operator riding, 40 H.P.		16.64	450	1,343.75	4,025	401.85	
	5000	Wheel type, diesel, 4' deep, 12" wide		68.50	965	2,891.84	8,675	1,126	
	5100	6' deep, 20" wide		87.32	1,050	3,127.50	9,375	1,324	
	5150	Chain type, diesel, 5' deep, 8" wide		16.25	360	1,084.44	3,250	346.90	
	5200	Diesel, 8' deep, 16" wide		89.39	1,925	5,783.68	17,400	1,872	
	5202	Rock trencher, wheel type, 6" wide x 18" deep		46.98	90	270	810	429.85	
	5206	Chain type, 18" wide x 7' deep		104.38	283	850	2,550	1,005	
	5210	Tree spade, self-propelled		13.65	230	690	2,075	247.20	
	5250	Truck, dump, 2-axle, 12 ton, 8 C.Y. payload, 220 H.P.		23.88	395	1,185	3,550	428.05	
	5300	Three axle dump, 16 ton, 12 C.Y. payload, 400 H.P.		44.50	360	1,084.44	3,250	572.90	
	5310	Four axle dump, 25 ton, 18 C.Y. payload, 450 H.P.		49.85	525	1,575.02	4,725	713.80	
	5350	Dump trailer only, rear dump, 16-1/2 C.Y.		5.73	151	454.43	1,375	136.70	
	5400	20 C.Y.		6.18	170	511.24	1,525	151.70	
	5450	Flatbed, single axle, 1-1/2 ton rating		19.00	73.50	221.02	665	196.15	
	5500	3 ton rating		23.05	1,050	3,180	9,550	820.40	
	5550	Off highway rear dump, 25 ton capacity		62.67	1,475	4,389.40	13,200	1,379	
	5600	35 ton capacity		66.90	665	2,000	6,000	935.20	
	5610	50 ton capacity		83.87	1,825	5,499.66	16,500	1,771	
	5620	65 ton capacity		89.56	2,000	5,990.24	18,000	1,915	
	5630	100 ton capacity		121.24	2,950	8,830.44	26,500	2,736	
	6000	Vibratory plow, 25 H.P., walking	▼	6.77	300	900	2,700	234.20	
40	0010	**GENERAL EQUIPMENT RENTAL** without operators							**40**
	0020	Aerial lift, scissor type, to 20' high, 1200 lb. capacity, electric [R015433 -10]	Ea.	3.48	129	385.75	1,150	105	
	0030	To 30' high, 1200 lb. capacity		3.77	203	607.67	1,825	151.70	
	0040	Over 30' high, 1500 lb. capacity		5.13	243	727.50	2,175	186.60	
	0070	Articulating boom, to 45' high, 500 lb. capacity, diesel [R015433 -15]		9.92	250	750	2,250	229.35	
	0075	To 60' high, 500 lb. capacity		13.66	300	900	2,700	289.25	
	0080	To 80' high, 500 lb. capacity		16.05	900	2,702.25	8,100	668.85	
	0085	To 125' high, 500 lb. capacity		18.34	1,525	4,603.50	13,800	1,067	
	0100	Telescoping boom to 40' high, 500 lb. capacity, diesel		11.24	315	945	2,825	278.90	
	0105	To 45' high, 500 lb. capacity		12.51	320	965	2,900	293.05	
	0110	To 60' high, 500 lb. capacity		16.36	300	900	2,700	310.90	
	0115	To 80' high, 500 lb. capacity		21.27	355	1,067	3,200	383.55	
	0120	To 100' high, 500 lb. capacity		28.71	865	2,587.75	7,775	747.25	
	0125	To 120' high, 500 lb. capacity		29.16	1,450	4,348.50	13,000	1,103	
	0195	Air compressor, portable, 6.5 CFM, electric		.90	44.50	133	400	33.80	
	0196	Gasoline		.65	56	167.50	505	38.70	
	0200	Towed type, gas engine, 60 CFM		9.43	129	387.50	1,175	152.95	
	0300	160 CFM	▼	10.47	198	595	1,775	202.80	

For customer support on your Site Work & Landscape Costs with RSMeans data, call 800.448.8182.

733

		UNIT	HOURLY OPER. COST	RENT PER DAY	RENT PER WEEK	RENT PER MONTH	EQUIPMENT COST/DAY	
40	0400	Diesel engine, rotary screw, 250 CFM	Ea.	12.08	175	524	1,575	201.50	**40**
	0500	365 CFM		16.00	310	937	2,800	315.40	
	0550	450 CFM		19.95	277	832.25	2,500	326.05	
	0600	600 CFM		34.10	248	743.62	2,225	421.50	
	0700	750 CFM		34.62	435	1,306.50	3,925	538.25	
	0930	Air tools, breaker, pavement, 60 lb.		.57	81.50	245	735	53.50	
	0940	80 lb.		.56	81	242.50	730	53	
	0950	Drills, hand (jackhammer), 65 lb.		.67	68	203.50	610	46.05	
	0960	Track or wagon, swing boom, 4" drifter		54.66	1,025	3,104	9,300	1,058	
	0970	5" drifter		63.30	1,025	3,104	9,300	1,127	
	0975	Track mounted quarry drill, 6" diameter drill		101.91	1,900	5,665	17,000	1,948	
	0980	Dust control per drill		1.04	25	75.50	227	23.40	
	0990	Hammer, chipping, 12 lb.		.60	46	138	415	32.40	
	1000	Hose, air with couplings, 50' long, 3/4" diameter		.07	12	36	108	7.75	
	1100	1" diameter		.08	12.35	37	111	8	
	1200	1-1/2" diameter		.22	37.50	112.50	340	24.25	
	1300	2" diameter		.24	45	135	405	28.90	
	1400	2-1/2" diameter		.36	57.50	172.50	520	37.35	
	1410	3" diameter		.42	58.50	175	525	38.35	
	1450	Drill, steel, 7/8" x 2'		.08	12.90	38.73	116	8.40	
	1460	7/8" x 6'		.12	19.60	58.87	177	12.70	
	1520	Moil points		.03	7	21	63	4.40	
	1525	Pneumatic nailer w/accessories		.48	39.50	118	355	27.40	
	1530	Sheeting driver for 60 lb. breaker		.04	7.75	23.24	69.50	5	
	1540	For 90 lb. breaker		.13	10.50	31.50	94.50	7.35	
	1550	Spade, 25 lb.		.50	7.40	22.21	66.50	8.45	
	1560	Tamper, single, 35 lb.		.59	48.50	145.75	435	33.85	
	1570	Triple, 140 lb.		.89	61.50	184.87	555	44.05	
	1580	Wrenches, impact, air powered, up to 3/4" bolt		.43	49.50	148.25	445	33.05	
	1590	Up to 1-1/4" bolt		.58	79.50	238.50	715	52.30	
	1600	Barricades, barrels, reflectorized, 1 to 99 barrels		.03	4	12	36	2.65	
	1610	100 to 200 barrels		.02	4.41	13.22	39.50	2.85	
	1620	Barrels with flashers, 1 to 99 barrels		.03	6.40	19.16	57.50	4.10	
	1630	100 to 200 barrels		.03	5.10	15.34	46	3.30	
	1640	Barrels with steady burn type C lights		.05	8.45	25.30	76	5.45	
	1650	Illuminated board, trailer mounted, with generator		3.28	139	418.28	1,250	109.85	
	1670	Portable barricade, stock, with flashers, 1 to 6 units		.03	6.35	19.11	57.50	4.10	
	1680	25 to 50 units		.03	5.95	17.82	53.50	3.85	
	1685	Butt fusion machine, wheeled, 1.5 HP electric, 2" - 8" diameter pipe		2.63	225	675	2,025	156.05	
	1690	Tracked, 20 HP diesel, 4"-12" diameter pipe		11.23	560	1,685	5,050	426.85	
	1695	83 HP diesel, 8" - 24" diameter pipe		51.32	1,100	3,325	9,975	1,076	
	1700	Carts, brick, gas engine, 1000 lb. capacity		2.94	65	195	585	62.55	
	1800	1500 lb., 7-1/2' lift		2.92	69.50	208	625	64.95	
	1822	Dehumidifier, medium, 6 lb./hr., 150 CFM		1.19	76.50	229.28	690	55.35	
	1824	Large, 18 lb./hr., 600 CFM		2.19	585	1,750	5,250	367.55	
	1830	Distributor, asphalt, trailer mounted, 2000 gal., 38 H.P. diesel		10.99	355	1,058.62	3,175	299.65	
	1840	3000 gal., 38 H.P. diesel		12.87	380	1,136.08	3,400	330.15	
	1850	Drill, rotary hammer, electric		1.11	71.50	214	640	51.70	
	1860	Carbide bit, 1-1/2" diameter, add to electric rotary hammer		.03	41.50	125	375	25.25	
	1865	Rotary, crawler, 250 H.P.		135.77	2,300	6,868.12	20,600	2,460	
	1870	Emulsion sprayer, 65 gal., 5 H.P. gas engine		2.77	107	320.17	960	86.15	
	1880	200 gal., 5 H.P. engine		7.22	179	537.06	1,600	165.20	
	1900	Floor auto-scrubbing machine, walk-behind, 28" path		5.62	222	667	2,000	178.40	
	1930	Floodlight, mercury vapor, or quartz, on tripod, 1000 watt		.46	36.50	110	330	25.65	
	1940	2000 watt		.59	28	84.69	254	21.65	
	1950	Floodlights, trailer mounted with generator, 1 - 300 watt light		3.54	78.50	235.48	705	75.45	
	1960	2 - 1000 watt lights		4.49	87.50	262.33	785	88.35	
	2000	4 - 300 watt lights		4.24	100	299.51	900	93.85	

01 54 33 | Equipment Rental

		UNIT	HOURLY OPER. COST	RENT PER DAY	RENT PER WEEK	RENT PER MONTH	EQUIPMENT COST/DAY	
40	2005	Foam spray rig, incl. box trailer, compressor, generator, proportioner	Ea.	25.46	535	1,600.84	4,800	523.85	**40**
	2015	Forklift, pneumatic tire, rough terr, straight mast, 5000 lb, 12' lift, gas		18.59	219	655.83	1,975	279.90	
	2025	8000 lb, 12' lift		22.68	360	1,084.44	3,250	398.35	
	2030	5000 lb, 12' lift, diesel		15.41	244	733.29	2,200	269.90	
	2035	8000 lb, 12' lift, diesel		16.70	277	831.40	2,500	299.90	
	2045	All terrain, telescoping boom, diesel, 5000 lb, 10' reach, 19' lift		17.20	233	700	2,100	277.60	
	2055	6600 lb, 29' reach, 42' lift		21.04	233	700	2,100	308.30	
	2065	10,000 lb, 31' reach, 45' lift		23.03	315	950	2,850	374.20	
	2070	Cushion tire, smooth floor, gas, 5000 lb capacity		8.23	247	741.50	2,225	214.10	
	2075	8000 lb capacity		11.33	275	826.25	2,475	255.90	
	2085	Diesel, 5000 lb capacity		7.73	210	629.25	1,900	187.65	
	2090	12,000 lb capacity		12.01	400	1,194.50	3,575	335	
	2095	20,000 lb capacity		17.20	660	1,980	5,950	533.60	
	2100	Generator, electric, gas engine, 1.5 kW to 3 kW		2.57	46.50	140	420	48.55	
	2200	5 kW		3.21	91	272.50	820	80.15	
	2300	10 kW		5.91	108	322.50	970	111.80	
	2400	25 kW		7.38	405	1,210	3,625	301.10	
	2500	Diesel engine, 20 kW		9.18	229	687.50	2,075	210.95	
	2600	50 kW		15.90	370	1,110	3,325	349.20	
	2700	100 kW		28.51	445	1,340.50	4,025	496.20	
	2800	250 kW		54.19	750	2,249.33	6,750	883.40	
	2850	Hammer, hydraulic, for mounting on boom, to 500 ft lb.		2.89	93.50	279.89	840	79.10	
	2860	1000 ft lb.		4.59	139	418.28	1,250	120.40	
	2900	Heaters, space, oil or electric, 50 MBH		1.46	46.50	140	420	39.65	
	3000	100 MBH		2.71	46.50	140	420	49.70	
	3100	300 MBH		7.90	135	405	1,225	144.20	
	3150	500 MBH		13.12	200	600	1,800	224.95	
	3200	Hose, water, suction with coupling, 20' long, 2" diameter		.02	5.65	17	51	3.55	
	3210	3" diameter		.03	14.15	42.50	128	8.75	
	3220	4" diameter		.03	28	84	252	17.05	
	3230	6" diameter		.11	40.50	121.50	365	25.20	
	3240	8" diameter		.27	53.50	160	480	34.15	
	3250	Discharge hose with coupling, 50' long, 2" diameter		.01	6.50	19.50	58.50	4	
	3260	3" diameter		.01	7.35	22	66	4.50	
	3270	4" diameter		.02	21	62.50	188	12.65	
	3280	6" diameter		.06	29	87	261	17.90	
	3290	8" diameter		.24	37.50	112.50	340	24.40	
	3295	Insulation blower		.83	117	350	1,050	76.65	
	3300	Ladders, extension type, 16' to 36' long		.18	41.50	125	375	26.45	
	3400	40' to 60' long		.64	120	360.50	1,075	77.20	
	3405	Lance for cutting concrete		2.20	65	195	585	56.60	
	3407	Lawn mower, rotary, 22", 5 H.P.		1.05	38.50	115	345	31.40	
	3408	48" self-propelled		2.89	138	415	1,250	106.10	
	3410	Level, electronic, automatic, with tripod and leveling rod		1.05	37.50	112	335	30.80	
	3430	Laser type, for pipe and sewer line and grade		2.17	117	350	1,050	87.35	
	3440	Rotating beam for interior control		.90	64	192.50	580	45.70	
	3460	Builder's optical transit, with tripod and rod		.10	37.50	112	335	23.20	
	3500	Light towers, towable, with diesel generator, 2000 watt		4.25	101	303.64	910	94.75	
	3600	4000 watt		4.50	165	495	1,475	135	
	3700	Mixer, powered, plaster and mortar, 6 C.F., 7 H.P.		2.05	83.50	250	750	66.40	
	3800	10 C.F., 9 H.P.		2.24	124	372.50	1,125	92.40	
	3850	Nailer, pneumatic		.48	33.50	100.18	300	23.85	
	3900	Paint sprayers complete, 8 CFM		.85	61.50	184.87	555	43.75	
	4000	17 CFM		1.60	110	330.50	990	78.85	
	4020	Pavers, bituminous, rubber tires, 8' wide, 50 H.P., diesel		31.93	570	1,704.12	5,100	596.25	
	4030	10' wide, 150 H.P.		95.62	1,950	5,835.32	17,500	1,932	
	4050	Crawler, 8' wide, 100 H.P., diesel		87.59	2,050	6,170.98	18,500	1,935	
	4060	10' wide, 150 H.P.		103.97	2,350	7,048.86	21,100	2,241	

01 54 33 | Equipment Rental

		UNIT	HOURLY OPER. COST	RENT PER DAY	RENT PER WEEK	RENT PER MONTH	EQUIPMENT COST/DAY	
40								**40**
4070	Concrete paver, 12' to 24' wide, 250 H.P.	Ea.	87.62	1,675	5,060.72	15,200	1,713	
4080	Placer-spreader-trimmer, 24' wide, 300 H.P.		117.51	2,550	7,668.54	23,000	2,474	
4100	Pump, centrifugal gas pump, 1-1/2" diam., 65 GPM		3.92	54	162.15	485	63.80	
4200	2" diameter, 130 GPM		4.98	44	132.50	400	66.35	
4300	3" diameter, 250 GPM		5.12	56	167.50	505	74.45	
4400	6" diameter, 1500 GPM		22.24	91.50	275	825	232.90	
4500	Submersible electric pump, 1-1/4" diameter, 55 GPM		.40	36	107.50	325	24.70	
4600	1-1/2" diameter, 83 GPM		.44	43.50	130	390	29.55	
4700	2" diameter, 120 GPM		1.64	61.50	185	555	50.15	
4800	3" diameter, 300 GPM		3.03	109	327.50	985	89.75	
4900	4" diameter, 560 GPM		14.75	61.50	185	555	155	
5000	6" diameter, 1590 GPM		22.08	65	195	585	215.60	
5100	Diaphragm pump, gas, single, 1-1/2" diameter		1.13	38.50	115	345	32	
5200	2" diameter		3.98	91.50	275	825	86.80	
5300	3" diameter		4.05	95	285	855	89.40	
5400	Double, 4" diameter		6.03	95	285	855	105.25	
5450	Pressure washer 5 GPM, 3000 psi		3.87	110	330	990	96.95	
5460	7 GPM, 3000 psi		4.94	85	255	765	90.50	
5470	High pressure water jet 10 ksi		39.55	720	2,160	6,475	748.40	
5480	40 ksi		27.88	980	2,940	8,825	811.05	
5500	Trash pump, self-priming, gas, 2" diameter		3.82	108	325	975	95.55	
5600	Diesel, 4" diameter		6.68	162	485	1,450	150.40	
5650	Diesel, 6" diameter		16.85	162	485	1,450	231.80	
5655	Grout Pump		18.70	281	841.73	2,525	317.90	
5700	Salamanders, L.P. gas fired, 100,000 BTU		2.88	57.50	172.50	520	57.55	
5705	50,000 BTU		1.66	23.50	71	213	27.50	
5720	Sandblaster, portable, open top, 3 C.F. capacity		.60	132	395	1,175	83.80	
5730	6 C.F. capacity		1.00	132	395	1,175	87	
5740	Accessories for above		.14	24	71.26	214	15.35	
5750	Sander, floor		.77	73.50	220	660	50.15	
5760	Edger		.52	35	105	315	25.15	
5800	Saw, chain, gas engine, 18" long		1.75	63.50	190	570	52	
5900	Hydraulic powered, 36" long		.78	58.50	175	525	41.25	
5950	60" long		.78	65	195	585	45.25	
6000	Masonry, table mounted, 14" diameter, 5 H.P.		1.32	76.50	230	690	56.55	
6050	Portable cut-off, 8 H.P.		1.81	77.50	232.50	700	61	
6100	Circular, hand held, electric, 7-1/4" diameter		.23	13.85	41.50	125	10.10	
6200	12" diameter		.24	41	122.50	370	26.40	
6250	Wall saw, w/hydraulic power, 10 H.P.		3.29	98.50	296	890	85.50	
6275	Shot blaster, walk-behind, 20" wide		4.73	281	841.73	2,525	206.20	
6280	Sidewalk broom, walk-behind		2.24	82.50	247.87	745	67.50	
6300	Steam cleaner, 100 gallons per hour		3.34	82.50	247.87	745	76.30	
6310	200 gallons per hour		4.33	100	299.51	900	94.55	
6340	Tar Kettle/Pot, 400 gallons		16.48	127	380	1,150	207.85	
6350	Torch, cutting, acetylene-oxygen, 150' hose, excludes gases		.45	15.30	45.96	138	12.80	
6360	Hourly operating cost includes tips and gas		20.92	7	20.98	63	171.55	
6410	Toilet, portable chemical		.13	23	69.20	208	14.90	
6420	Recycle flush type		.16	28.50	85.72	257	18.45	
6430	Toilet, fresh water flush, garden hose,		.19	34	102.25	305	22	
6440	Hoisted, non-flush, for high rise		.16	28	83.66	251	18	
6465	Tractor, farm with attachment		17.37	390	1,172.50	3,525	373.40	
6480	Trailers, platform, flush deck, 2 axle, 3 ton capacity		1.69	94.50	284	850	70.30	
6500	25 ton capacity		6.23	143	428.61	1,275	135.55	
6600	40 ton capacity		8.04	203	609.35	1,825	186.20	
6700	3 axle, 50 ton capacity		8.72	225	676.48	2,025	205.05	
6800	75 ton capacity		11.08	300	898.54	2,700	268.30	
6810	Trailer mounted cable reel for high voltage line work		5.89	28.50	85	255	64.10	
6820	Trailer mounted cable tensioning rig		11.67	28.50	85	255	110.40	

For customer support on your Site Work & Landscape Costs with RSMeans data, call 800.448.8182.

01 54 33 | Equipment Rental

			UNIT	HOURLY OPER. COST	RENT PER DAY	RENT PER WEEK	RENT PER MONTH	EQUIPMENT COST/DAY	
40	6830	Cable pulling rig	Ea.	73.77	28.50	85	255	607.15	40
	6850	Portable cable/wire puller, 8000 lb max pulling capacity		3.70	120	360	1,075	101.60	
	6900	Water tank trailer, engine driven discharge, 5000 gallons		7.16	158	475.09	1,425	152.25	
	6925	10,000 gallons		9.75	215	645.50	1,925	207.10	
	6950	Water truck, off highway, 6000 gallons		71.75	835	2,504.54	7,525	1,075	
	7010	Tram car for high voltage line work, powered, 2 conductor		6.88	28.50	85	255	72.05	
	7020	Transit (builder's level) with tripod		.10	17.55	52.67	158	11.30	
	7030	Trench box, 3000 lb., 6' x 8'		.56	96.50	290.22	870	62.50	
	7040	7200 lb., 6' x 20'		.72	187	560	1,675	117.75	
	7050	8000 lb., 8' x 16'		1.08	186	557.71	1,675	120.15	
	7060	9500 lb., 8' x 20'		1.20	232	697.14	2,100	149.05	
	7065	11,000 lb., 8' x 24'		1.26	219	655.83	1,975	141.25	
	7070	12,000 lb., 10' x 20'		1.49	263	790.09	2,375	169.95	
	7100	Truck, pickup, 3/4 ton, 2 wheel drive		9.24	61.50	184.87	555	110.85	
	7200	4 wheel drive		9.48	167	500	1,500	175.85	
	7250	Crew carrier, 9 passenger		12.66	108	325	975	166.25	
	7290	Flat bed truck, 20,000 lb. GVW		15.26	133	397.63	1,200	201.60	
	7300	Tractor, 4 x 2, 220 H.P.		22.25	215	645.50	1,925	307.10	
	7410	330 H.P.		32.33	294	883.04	2,650	435.25	
	7500	6 x 4, 380 H.P.		36.09	340	1,022.47	3,075	493.25	
	7600	450 H.P.		44.23	415	1,239.36	3,725	601.75	
	7610	Tractor, with A frame, boom and winch, 225 H.P.		24.74	293	877.88	2,625	373.50	
	7620	Vacuum truck, hazardous material, 2500 gallons		12.79	310	929.52	2,800	288.25	
	7625	5,000 gallons		13.02	440	1,316.82	3,950	367.55	
	7650	Vacuum, HEPA, 16 gallon, wet/dry		.85	122	365	1,100	79.80	
	7655	55 gallon, wet/dry		.78	25.50	76.50	230	21.50	
	7660	Water tank, portable		.73	160	480.25	1,450	101.90	
	7690	Sewer/catch basin vacuum, 14 C.Y., 1500 gallons		17.31	665	1,988.14	5,975	536.15	
	7700	Welder, electric, 200 amp		3.81	33.50	100	300	50.50	
	7800	300 amp		5.55	103	310	930	106.40	
	7900	Gas engine, 200 amp		8.95	58.50	175	525	106.55	
	8000	300 amp		10.13	110	330	990	147	
	8100	Wheelbarrow, any size		.06	11.15	33.50	101	7.20	
	8200	Wrecking ball, 4000 lb.		2.50	60	180	540	56	
50	0010	HIGHWAY EQUIPMENT RENTAL without operators							50
	0050	Asphalt batch plant, portable drum mixer, 100 ton/hr.	Ea.	88.41	1,550	4,621.78	13,900	1,632	
	0060	200 ton/hr.		101.99	1,650	4,931.62	14,800	1,802	
	0070	300 ton/hr.		119.86	1,925	5,783.68	17,400	2,116	
	0100	Backhoe attachment, long stick, up to 185 H.P., 10.5' long		.37	25.50	76.43	229	18.25	
	0140	Up to 250 H.P., 12' long		.41	28.50	85.72	257	20.45	
	0180	Over 250 H.P., 15' long		.56	39	116.71	350	27.85	
	0200	Special dipper arm, up to 100 H.P., 32' long		1.16	79.50	238.58	715	56.95	
	0240	Over 100 H.P., 33' long		1.44	100	299.51	900	71.45	
	0280	Catch basin/sewer cleaning truck, 3 ton, 9 C.Y., 1000 gal.		35.39	420	1,265.18	3,800	536.15	
	0300	Concrete batch plant, portable, electric, 200 C.Y./hr.		24.18	560	1,678.30	5,025	529.15	
	0520	Grader/dozer attachment, ripper/scarifier, rear mounted, up to 135 H.P.		3.15	63.50	190.04	570	63.20	
	0540	Up to 180 H.P.		4.13	95.50	287.12	860	90.50	
	0580	Up to 250 H.P.		5.85	153	459.60	1,375	138.75	
	0700	Pvmt. removal bucket, for hyd. excavator, up to 90 H.P.		2.16	58	174.54	525	52.20	
	0740	Up to 200 H.P.		2.31	74.50	223.08	670	63.05	
	0780	Over 200 H.P.		2.52	91	273.69	820	74.90	
	0900	Aggregate spreader, self-propelled, 187 H.P.		50.60	740	2,220.52	6,650	848.90	
	1000	Chemical spreader, 3 C.Y.		3.17	96.50	290	870	83.35	
	1900	Hammermill, traveling, 250 H.P.		67.35	515	1,550	4,650	848.80	
	2000	Horizontal borer, 3" diameter, 13 H.P. gas driven		5.42	232	695	2,075	182.35	
	2150	Horizontal directional drill, 20,000 lb. thrust, 78 H.P. diesel		27.58	530	1,590	4,775	538.65	
	2160	30,000 lb. thrust, 115 H.P.		33.90	615	1,850	5,550	641.20	
	2170	50,000 lb. thrust, 170 H.P.		48.60	710	2,135	6,400	815.80	

R015433 -10

For customer support on your Site Work & Landscape Costs with RSMeans data, call 800.448.8182.

737

01 54 33 | Equipment Rental

		UNIT	HOURLY OPER. COST	RENT PER DAY	RENT PER WEEK	RENT PER MONTH	EQUIPMENT COST/DAY	
50	2190	Mud trailer for HDD, 1500 gallons, 175 H.P., gas	Ea.	25.50	175	525	1,575	309	50
	2200	Hydromulcher, diesel, 3000 gallon, for truck mounting		17.43	227	680	2,050	275.45	
	2300	Gas, 600 gallon		7.49	95	285	855	116.95	
	2400	Joint & crack cleaner, walk behind, 25 H.P.		3.16	45.50	136	410	52.45	
	2500	Filler, trailer mounted, 400 gallons, 20 H.P.		8.35	147	440	1,325	154.75	
	3000	Paint striper, self-propelled, 40 gallon, 22 H.P.		6.76	122	365	1,100	127.10	
	3100	120 gallon, 120 H.P.		19.23	380	1,140	3,425	381.80	
	3200	Post drivers, 6" I-Beam frame, for truck mounting		12.41	320	960	2,875	291.30	
	3400	Road sweeper, self-propelled, 8' wide, 90 H.P.		35.91	715	2,143.06	6,425	715.85	
	3450	Road sweeper, vacuum assisted, 4 C.Y., 220 gallons		58.28	670	2,013.96	6,050	869	
	4000	Road mixer, self-propelled, 130 H.P.		46.23	825	2,478.72	7,425	865.60	
	4100	310 H.P.		75.01	2,150	6,480.82	19,400	1,896	
	4220	Cold mix paver, incl. pug mill and bitumen tank, 165 H.P.		94.97	2,325	6,945.58	20,800	2,149	
	4240	Pavement brush, towed		3.43	100	299.51	900	87.30	
	4250	Paver, asphalt, wheel or crawler, 130 H.P., diesel		94.23	2,275	6,816.48	20,400	2,117	
	4300	Paver, road widener, gas, 1' to 6', 67 H.P.		46.66	975	2,917.66	8,750	956.80	
	4400	Diesel, 2' to 14', 88 H.P.		56.38	1,150	3,459.88	10,400	1,143	
	4600	Slipform pavers, curb and gutter, 2 track, 75 H.P.		57.83	1,250	3,769.72	11,300	1,217	
	4700	4 track, 165 H.P.		35.69	845	2,530.36	7,600	791.55	
	4800	Median barrier, 215 H.P.		58.43	1,350	4,027.92	12,100	1,273	
	4901	Trailer, low bed, 75 ton capacity		10.71	282	846.90	2,550	255.05	
	5000	Road planer, walk behind, 10" cutting width, 10 H.P.		2.45	243	730	2,200	165.65	
	5100	Self-propelled, 12" cutting width, 64 H.P.		8.26	190	570	1,700	180.05	
	5120	Traffic line remover, metal ball blaster, truck mounted, 115 H.P.		46.56	905	2,720	8,150	916.50	
	5140	Grinder, truck mounted, 115 H.P.		50.89	905	2,720	8,150	951.15	
	5160	Walk-behind, 11 H.P.		3.56	142	425	1,275	113.45	
	5200	Pavement profiler, 4' to 6' wide, 450 H.P.		216.58	1,275	3,800	11,400	2,493	
	5300	8' to 10' wide, 750 H.P.		331.58	1,325	3,975	11,900	3,448	
	5400	Roadway plate, steel, 1" x 8' x 20'		.09	61	182.50	550	37.20	
	5600	Stabilizer, self-propelled, 150 H.P.		41.14	1,025	3,100	9,300	949.10	
	5700	310 H.P.		76.18	1,300	3,900	11,700	1,389	
	5800	Striper, truck mounted, 120 gallon paint, 460 H.P.		48.74	340	1,015	3,050	592.95	
	5900	Thermal paint heating kettle, 115 gallons		7.71	61.50	185	555	98.65	
	6000	Tar kettle, 330 gallon, trailer mounted		12.27	96.50	290	870	156.20	
	7000	Tunnel locomotive, diesel, 8 to 12 ton		29.76	620	1,859.04	5,575	609.85	
	7005	Electric, 10 ton		29.25	705	2,117.24	6,350	657.40	
	7010	Muck cars, 1/2 C.Y. capacity		2.30	26.50	80.04	240	34.40	
	7020	1 C.Y. capacity		2.51	35	104.31	315	40.95	
	7030	2 C.Y. capacity		2.66	39	116.71	350	44.60	
	7040	Side dump, 2 C.Y. capacity		2.87	48	144.59	435	51.90	
	7050	3 C.Y. capacity		3.85	53	159.05	475	62.65	
	7060	5 C.Y. capacity		5.62	68.50	205.53	615	86.10	
	7100	Ventilating blower for tunnel, 7-1/2 H.P.		2.14	52.50	158.02	475	48.70	
	7110	10 H.P.		2.42	55	165.25	495	52.40	
	7120	20 H.P.		3.54	71.50	214.82	645	71.30	
	7140	40 H.P.		6.14	94.50	284.02	850	105.90	
	7160	60 H.P.		8.69	102	304.68	915	130.45	
	7175	75 H.P.		10.37	158	475.09	1,425	177.95	
	7180	200 H.P.		20.78	310	934.68	2,800	353.20	
	7800	Windrow loader, elevating		53.94	1,650	4,975	14,900	1,427	
60	0010	**LIFTING AND HOISTING EQUIPMENT RENTAL** without operators [R015433 -10]							60
	0150	Crane, flatbed mounted, 3 ton capacity	Ea.	14.41	201	604.19	1,825	236.10	
	0200	Crane, climbing, 106' jib, 6000 lb. capacity, 410 fpm [R312316 -45]		39.72	2,600	7,800	23,400	1,878	
	0300	101' jib, 10,250 lb. capacity, 270 fpm		46.43	2,275	6,800	20,400	1,731	
	0500	Tower, static, 130' high, 106' jib, 6200 lb. capacity at 400 fpm		45.16	2,250	6,715	20,100	1,704	
	0520	Mini crawler spider crane, up to 24" wide, 1990 lb. lifting capacity		12.50	550	1,652.48	4,950	430.50	
	0525	Up to 30" wide, 6450 lb. lifting capacity		14.52	655	1,962.32	5,875	508.65	
	0530	Up to 52" wide, 6680 lb. lifting capacity		23.10	800	2,401.26	7,200	665.05	

01 54 33 | Equipment Rental

		UNIT	HOURLY OPER. COST	RENT PER DAY	RENT PER WEEK	RENT PER MONTH	EQUIPMENT COST/DAY	
60								60
0535	Up to 55" wide, 8920 lb. lifting capacity	Ea.	25.79	885	2,659.46	7,975	738.25	
0540	Up to 66" wide, 13,350 lb. lifting capacity		34.92	1,375	4,131.20	12,400	1,106	
0600	Crawler mounted, lattice boom, 1/2 C.Y., 15 tons at 12' radius		36.96	830	2,483	7,450	792.30	
0700	3/4 C.Y., 20 tons at 12' radius		50.42	930	2,790	8,375	961.35	
0800	1 C.Y., 25 tons at 12' radius		67.42	985	2,950	8,850	1,129	
0900	1-1/2 C.Y., 40 tons at 12' radius		66.31	1,125	3,375	10,100	1,206	
1000	2 C.Y., 50 tons at 12' radius		88.77	1,325	4,000	12,000	1,510	
1100	3 C.Y., 75 tons at 12' radius		75.26	2,325	7,000	21,000	2,002	
1200	100 ton capacity, 60' boom		85.91	2,675	8,000	24,000	2,287	
1300	165 ton capacity, 60' boom		106.10	3,000	9,000	27,000	2,649	
1400	200 ton capacity, 70' boom		138.21	3,825	11,500	34,500	3,406	
1500	350 ton capacity, 80' boom		182.20	4,175	12,500	37,500	3,958	
1600	Truck mounted, lattice boom, 6 x 4, 20 tons at 10' radius		39.76	1,950	5,850	17,600	1,488	
1700	25 tons at 10' radius		42.73	2,325	7,000	21,000	1,742	
1800	8 x 4, 30 tons at 10' radius		45.54	2,500	7,500	22,500	1,864	
1900	40 tons at 12' radius		48.55	2,725	8,200	24,600	2,028	
2000	60 tons at 15' radius		53.69	1,650	4,950	14,900	1,419	
2050	82 tons at 15' radius		59.43	1,775	5,350	16,100	1,545	
2100	90 tons at 15' radius		66.39	1,950	5,825	17,500	1,696	
2200	115 tons at 15' radius		74.90	2,175	6,525	19,600	1,904	
2300	150 tons at 18' radius		81.09	2,700	8,100	24,300	2,269	
2350	165 tons at 18' radius		87.03	2,425	7,275	21,800	2,151	
2400	Truck mounted, hydraulic, 12 ton capacity		29.50	390	1,175	3,525	471	
2500	25 ton capacity		36.36	485	1,450	4,350	580.85	
2550	33 ton capacity		50.67	900	2,700	8,100	945.35	
2560	40 ton capacity		49.47	900	2,700	8,100	935.80	
2600	55 ton capacity		53.78	915	2,750	8,250	980.20	
2700	80 ton capacity		75.71	1,475	4,400	13,200	1,486	
2720	100 ton capacity		74.96	1,550	4,675	14,000	1,535	
2740	120 ton capacity		102.81	1,825	5,500	16,500	1,922	
2760	150 ton capacity		109.92	2,050	6,125	18,400	2,104	
2800	Self-propelled, 4 x 4, with telescoping boom, 5 ton		15.14	430	1,285	3,850	378.10	
2900	12-1/2 ton capacity		21.42	430	1,285	3,850	428.30	
3000	15 ton capacity		34.42	450	1,350	4,050	545.35	
3050	20 ton capacity		24.02	650	1,950	5,850	582.20	
3100	25 ton capacity		36.69	1,425	4,250	12,800	1,144	
3150	40 ton capacity		44.90	660	1,975	5,925	754.20	
3200	Derricks, guy, 20 ton capacity, 60' boom, 75' mast		22.74	1,425	4,250	12,800	1,032	
3300	100' boom, 115' mast		36.04	2,000	6,000	18,000	1,488	
3400	Stiffleg, 20 ton capacity, 70' boom, 37' mast		25.41	615	1,850	5,550	573.25	
3500	100' boom, 47' mast		39.32	665	2,000	6,000	714.55	
3550	Helicopter, small, lift to 1250 lb. maximum, w/pilot		99.14	2,150	6,435	19,300	2,080	
3600	Hoists, chain type, overhead, manual, 3/4 ton		.14	10.25	30.70	92	7.30	
3900	10 ton		.79	6.20	18.59	56	10	
4000	Hoist and tower, 5000 lb. cap., portable electric, 40' high		5.12	142	426	1,275	126.20	
4100	For each added 10' section, add		.12	31.50	95	285	19.95	
4200	Hoist and single tubular tower, 5000 lb. electric, 100' high		6.96	105	315	945	118.65	
4300	For each added 6'-6" section, add		.21	38.50	115	345	24.65	
4400	Hoist and double tubular tower, 5000 lb., 100' high		7.57	105	315	945	123.60	
4500	For each added 6'-6" section, add		.23	41.50	125	375	26.80	
4550	Hoist and tower, mast type, 6000 lb., 100' high		8.24	94.50	284	850	122.70	
4570	For each added 10' section, add		.13	31.50	95	285	20.05	
4600	Hoist and tower, personnel, electric, 2000 lb., 100' @ 125 fpm		17.50	25	75	225	155	
4700	3000 lb., 100' @ 200 fpm		20.02	25	75	225	175.15	
4800	3000 lb., 150' @ 300 fpm		22.22	25	75	225	192.75	
4900	4000 lb., 100' @ 300 fpm		22.98	25	75	225	198.85	
5000	6000 lb., 100' @ 275 fpm		24.70	25	75	225	212.60	
5100	For added heights up to 500', add	L.F.	.01	3.33	10	30	2.10	

For customer support on your Site Work & Landscape Costs with RSMeans data, call 800.448.8182.

739

01 54 33 | Equipment Rental

		UNIT	HOURLY OPER. COST	RENT PER DAY	RENT PER WEEK	RENT PER MONTH	EQUIPMENT COST/DAY	
60								**60**
5200	Jacks, hydraulic, 20 ton	Ea.	.05	19.65	59	177	12.20	
5500	100 ton		.40	26	78.50	236	18.90	
6100	Jacks, hydraulic, climbing w/50' jackrods, control console, 30 ton cap.		2.17	31	93	279	35.90	
6150	For each added 10' jackrod section, add		.05	5	15	45	3.40	
6300	50 ton capacity		3.48	33.50	100	300	47.85	
6350	For each added 10' jackrod section, add		.06	5	15	45	3.50	
6500	125 ton capacity		9.10	51.50	155	465	103.85	
6550	For each added 10' jackrod section, add		.61	5	15	45	7.90	
6600	Cable jack, 10 ton capacity with 200' cable		1.82	35.50	107	320	35.95	
6650	For each added 50' of cable, add		.22	15	45	135	10.75	
70	**WELLPOINT EQUIPMENT RENTAL** without operators							**70**
0010	R015433 -10							
0020	Based on 2 months rental							
0100	Combination jetting & wellpoint pump, 60 H.P. diesel	Ea.	15.67	298	895	2,675	304.35	
0200	High pressure gas jet pump, 200 H.P., 300 psi	"	33.83	275	825	2,475	435.65	
0300	Discharge pipe, 8" diameter	L.F.	.01	1.40	4.20	12.60	.90	
0350	12" diameter		.01	2.07	6.20	18.60	1.35	
0400	Header pipe, flows up to 150 GPM, 4" diameter		.01	.73	2.20	6.60	.50	
0500	400 GPM, 6" diameter		.01	1.07	3.20	9.60	.70	
0600	800 GPM, 8" diameter		.01	1.40	4.20	12.60	.95	
0700	1500 GPM, 10" diameter		.01	1.73	5.20	15.60	1.15	
0800	2500 GPM, 12" diameter		.03	2.07	6.20	18.60	1.45	
0900	4500 GPM, 16" diameter		.03	2.40	7.20	21.50	1.70	
0950	For quick coupling aluminum and plastic pipe, add		.03	9.35	28	84	5.85	
1100	Wellpoint, 25' long, with fittings & riser pipe, 1-1/2" or 2" diameter	Ea.	.07	132	395	1,175	79.55	
1200	Wellpoint pump, diesel powered, 4" suction, 20 H.P.		7.00	150	450	1,350	146	
1300	6" suction, 30 H.P.		9.39	167	500	1,500	175.15	
1400	8" suction, 40 H.P.		12.73	250	750	2,250	251.80	
1500	10" suction, 75 H.P.		18.77	265	795	2,375	309.20	
1600	12" suction, 100 H.P.		27.24	298	895	2,675	396.90	
1700	12" suction, 175 H.P.		38.98	315	950	2,850	501.80	
80	**MARINE EQUIPMENT RENTAL** without operators							**80**
0010	R015433 -10							
0200	Barge, 400 Ton, 30' wide x 90' long	Ea.	17.63	1,200	3,588.98	10,800	858.85	
0240	800 Ton, 45' wide x 90' long		22.14	1,475	4,415.22	13,200	1,060	
2000	Tugboat, diesel, 100 H.P.		29.57	238	712.63	2,150	379.10	
2040	250 H.P.		57.41	430	1,291	3,875	717.50	
2080	380 H.P.		124.99	1,300	3,873	11,600	1,774	
3000	Small work boat, gas, 16-foot, 50 H.P.		11.35	48	143.56	430	119.50	
4000	Large, diesel, 48-foot, 200 H.P.		74.68	1,375	4,105.38	12,300	1,418	

Crews - Standard

Crew No.	Bare Costs Hr.	Daily	Incl. Subs O&P Hr.	Daily	Cost Per Labor-Hour Bare Costs	Incl. O&P
Crew A-1	Hr.	Daily	Hr.	Daily	Bare Costs	Incl. O&P
1 Building Laborer	$42.10	$336.80	$63.25	$506.00	$42.10	$63.25
1 Concrete Saw, Gas Manual		111.55		122.71	13.94	15.34
8 L.H., Daily Totals		$448.35		$628.71	$56.04	$78.59
Crew A-1A	Hr.	Daily	Hr.	Daily	Bare Costs	Incl. O&P
1 Skilled Worker	$54.85	$438.80	$82.95	$663.60	$54.85	$82.95
1 Shot Blaster, 20"		206.20		226.82	25.77	28.35
8 L.H., Daily Totals		$645.00		$890.42	$80.63	$111.30
Crew A-1B	Hr.	Daily	Hr.	Daily	Bare Costs	Incl. O&P
1 Building Laborer	$42.10	$336.80	$63.25	$506.00	$42.10	$63.25
1 Concrete Saw		111.50		122.65	13.94	15.33
8 L.H., Daily Totals		$448.30		$628.65	$56.04	$78.58
Crew A-1C	Hr.	Daily	Hr.	Daily	Bare Costs	Incl. O&P
1 Building Laborer	$42.10	$336.80	$63.25	$506.00	$42.10	$63.25
1 Chain Saw, Gas, 18"		52.00		57.20	6.50	7.15
8 L.H., Daily Totals		$388.80		$563.20	$48.60	$70.40
Crew A-1D	Hr.	Daily	Hr.	Daily	Bare Costs	Incl. O&P
1 Building Laborer	$42.10	$336.80	$63.25	$506.00	$42.10	$63.25
1 Vibrating Plate, Gas, 18"		31.55		34.70	3.94	4.34
8 L.H., Daily Totals		$368.35		$540.71	$46.04	$67.59
Crew A-1E	Hr.	Daily	Hr.	Daily	Bare Costs	Incl. O&P
1 Building Laborer	$42.10	$336.80	$63.25	$506.00	$42.10	$63.25
1 Vibrating Plate, Gas, 21"		165.35		181.88	20.67	22.74
8 L.H., Daily Totals		$502.15		$687.88	$62.77	$85.99
Crew A-1F	Hr.	Daily	Hr.	Daily	Bare Costs	Incl. O&P
1 Building Laborer	$42.10	$336.80	$63.25	$506.00	$42.10	$63.25
1 Rammer/Tamper, Gas, 8"		46.50		51.15	5.81	6.39
8 L.H., Daily Totals		$383.30		$557.15	$47.91	$69.64
Crew A-1G	Hr.	Daily	Hr.	Daily	Bare Costs	Incl. O&P
1 Building Laborer	$42.10	$336.80	$63.25	$506.00	$42.10	$63.25
1 Rammer/Tamper, Gas, 15"		54.00		59.40	6.75	7.42
8 L.H., Daily Totals		$390.80		$565.40	$48.85	$70.67
Crew A-1H	Hr.	Daily	Hr.	Daily	Bare Costs	Incl. O&P
1 Building Laborer	$42.10	$336.80	$63.25	$506.00	$42.10	$63.25
1 Exterior Steam Cleaner		76.30		83.93	9.54	10.49
8 L.H., Daily Totals		$413.10		$589.93	$51.64	$73.74
Crew A-1J	Hr.	Daily	Hr.	Daily	Bare Costs	Incl. O&P
1 Building Laborer	$42.10	$336.80	$63.25	$506.00	$42.10	$63.25
1 Cultivator, Walk-Behind, 5 H.P.		53.05		58.35	6.63	7.29
8 L.H., Daily Totals		$389.85		$564.36	$48.73	$70.54
Crew A-1K	Hr.	Daily	Hr.	Daily	Bare Costs	Incl. O&P
1 Building Laborer	$42.10	$336.80	$63.25	$506.00	$42.10	$63.25
1 Cultivator, Walk-Behind, 8 H.P.		101.40		111.54	12.68	13.94
8 L.H., Daily Totals		$438.20		$617.54	$54.77	$77.19
Crew A-1M	Hr.	Daily	Hr.	Daily	Bare Costs	Incl. O&P
1 Building Laborer	$42.10	$336.80	$63.25	$506.00	$42.10	$63.25
1 Snow Blower, Walk-Behind		67.50		74.25	8.44	9.28
8 L.H., Daily Totals		$404.30		$580.25	$50.54	$72.53

Crew No.	Bare Costs Hr.	Daily	Incl. Subs O&P Hr.	Daily	Cost Per Labor-Hour Bare Costs	Incl. O&P
Crew A-2	Hr.	Daily	Hr.	Daily	Bare Costs	Incl. O&P
2 Laborers	$42.10	$673.60	$63.25	$1012.00	$43.82	$65.77
1 Truck Driver (light)	47.25	378.00	70.80	566.40		
1 Flatbed Truck, Gas, 1.5 Ton		196.15		215.76	8.17	8.99
24 L.H., Daily Totals		$1247.75		$1794.17	$51.99	$74.76
Crew A-2A	Hr.	Daily	Hr.	Daily	Bare Costs	Incl. O&P
2 Laborers	$42.10	$673.60	$63.25	$1012.00	$43.82	$65.77
1 Truck Driver (light)	47.25	378.00	70.80	566.40		
1 Flatbed Truck, Gas, 1.5 Ton		196.15		215.76		
1 Concrete Saw		111.50		122.65	12.82	14.10
24 L.H., Daily Totals		$1359.25		$1916.82	$56.64	$79.87
Crew A-2B	Hr.	Daily	Hr.	Daily	Bare Costs	Incl. O&P
1 Truck Driver (light)	$47.25	$378.00	$70.80	$566.40	$47.25	$70.80
1 Flatbed Truck, Gas, 1.5 Ton		196.15		215.76	24.52	26.97
8 L.H., Daily Totals		$574.15		$782.16	$71.77	$97.77
Crew A-3A	Hr.	Daily	Hr.	Daily	Bare Costs	Incl. O&P
1 Equip. Oper. (light)	$53.00	$424.00	$79.20	$633.60	$53.00	$79.20
1 Pickup Truck, 4x4, 3/4 Ton		175.85		193.44	21.98	24.18
8 L.H., Daily Totals		$599.85		$827.03	$74.98	$103.38
Crew A-3B	Hr.	Daily	Hr.	Daily	Bare Costs	Incl. O&P
1 Equip. Oper. (medium)	$56.75	$454.00	$84.85	$678.80	$52.85	$79.10
1 Truck Driver (heavy)	48.95	391.60	73.35	586.80		
1 Dump Truck, 12 C.Y., 400 H.P.		572.90		630.19		
1 F.E. Loader, W.M., 2.5 C.Y.		702.10		772.31	79.69	87.66
16 L.H., Daily Totals		$2120.60		$2668.10	$132.54	$166.76
Crew A-3C	Hr.	Daily	Hr.	Daily	Bare Costs	Incl. O&P
1 Equip. Oper. (light)	$53.00	$424.00	$79.20	$633.60	$53.00	$79.20
1 Loader, Skid Steer, 78 H.P.		400.10		440.11	50.01	55.01
8 L.H., Daily Totals		$824.10		$1073.71	$103.01	$134.21
Crew A-3D	Hr.	Daily	Hr.	Daily	Bare Costs	Incl. O&P
1 Truck Driver (light)	$47.25	$378.00	$70.80	$566.40	$47.25	$70.80
1 Pickup Truck, 4x4, 3/4 Ton		175.85		193.44		
1 Flatbed Trailer, 25 Ton		135.55		149.10	38.92	42.82
8 L.H., Daily Totals		$689.40		$908.94	$86.17	$113.62
Crew A-3E	Hr.	Daily	Hr.	Daily	Bare Costs	Incl. O&P
1 Equip. Oper. (crane)	$59.20	$473.60	$88.50	$708.00	$54.08	$80.92
1 Truck Driver (heavy)	48.95	391.60	73.35	586.80		
1 Pickup Truck, 4x4, 3/4 Ton		175.85		193.44	10.99	12.09
16 L.H., Daily Totals		$1041.05		$1488.23	$65.07	$93.01
Crew A-3F	Hr.	Daily	Hr.	Daily	Bare Costs	Incl. O&P
1 Equip. Oper. (crane)	$59.20	$473.60	$88.50	$708.00	$54.08	$80.92
1 Truck Driver (heavy)	48.95	391.60	73.35	586.80		
1 Pickup Truck, 4x4, 3/4 Ton		175.85		193.44		
1 Truck Tractor, 6x4, 380 H.P.		493.25		542.58		
1 Lowbed Trailer, 75 Ton		255.05		280.56	57.76	63.54
16 L.H., Daily Totals		$1789.35		$2311.36	$111.83	$144.46

For customer support on your Site Work & Landscape Costs with RSMeans data, call 800.448.8182.

741

Crews - Standard

Crew A-3G	Hr.	Daily	Hr.	Daily	Bare Costs	Incl. O&P
1 Equip. Oper. (crane)	$59.20	$473.60	$88.50	$708.00	$54.08	$80.92
1 Truck Driver (heavy)	48.95	391.60	73.35	586.80		
1 Pickup Truck, 4x4, 3/4 Ton		175.85		193.44		
1 Truck Tractor, 6x4, 450 H.P.		601.75		661.92		
1 Lowbed Trailer, 75 Ton		255.05		280.56	64.54	70.99
16 L.H., Daily Totals		$1897.85		$2430.72	$118.62	$151.92

Crew A-3H	Hr.	Daily	Hr.	Daily	Bare Costs	Incl. O&P
1 Equip. Oper. (crane)	$59.20	$473.60	$88.50	$708.00	$59.20	$88.50
1 Hyd. Crane, 12 Ton (Daily)		724.85		797.34	90.61	99.67
8 L.H., Daily Totals		$1198.45		$1505.34	$149.81	$188.17

Crew A-3I	Hr.	Daily	Hr.	Daily	Bare Costs	Incl. O&P
1 Equip. Oper. (crane)	$59.20	$473.60	$88.50	$708.00	$59.20	$88.50
1 Hyd. Crane, 25 Ton (Daily)		801.40		881.54	100.18	110.19
8 L.H., Daily Totals		$1275.00		$1589.54	$159.38	$198.69

Crew A-3J	Hr.	Daily	Hr.	Daily	Bare Costs	Incl. O&P
1 Equip. Oper. (crane)	$59.20	$473.60	$88.50	$708.00	$59.20	$88.50
1 Hyd. Crane, 40 Ton (Daily)		1272.00		1399.20	159.00	174.90
8 L.H., Daily Totals		$1745.60		$2107.20	$218.20	$263.40

Crew A-3K	Hr.	Daily	Hr.	Daily	Bare Costs	Incl. O&P
1 Equip. Oper. (crane)	$59.20	$473.60	$88.50	$708.00	$54.88	$82.03
1 Equip. Oper. (oiler)	50.55	404.40	75.55	604.40		
1 Hyd. Crane, 55 Ton (Daily)		1362.00		1498.20		
1 P/U Truck, 3/4 Ton (Daily)		142.15		156.37	94.01	103.41
16 L.H., Daily Totals		$2382.15		$2966.97	$148.88	$185.44

Crew A-3L	Hr.	Daily	Hr.	Daily	Bare Costs	Incl. O&P
1 Equip. Oper. (crane)	$59.20	$473.60	$88.50	$708.00	$54.88	$82.03
1 Equip. Oper. (oiler)	50.55	404.40	75.55	604.40		
1 Hyd. Crane, 80 Ton (Daily)		2101.00		2311.10		
1 P/U Truck, 3/4 Ton (Daily)		142.15		156.37	140.20	154.22
16 L.H., Daily Totals		$3121.15		$3779.86	$195.07	$236.24

Crew A-3M	Hr.	Daily	Hr.	Daily	Bare Costs	Incl. O&P
1 Equip. Oper. (crane)	$59.20	$473.60	$88.50	$708.00	$54.88	$82.03
1 Equip. Oper. (oiler)	50.55	404.40	75.55	604.40		
1 Hyd. Crane, 100 Ton (Daily)		2227.00		2449.70		
1 P/U Truck, 3/4 Ton (Daily)		142.15		156.37	148.07	162.88
16 L.H., Daily Totals		$3247.15		$3918.47	$202.95	$244.90

Crew A-3N	Hr.	Daily	Hr.	Daily	Bare Costs	Incl. O&P
1 Equip. Oper. (crane)	$59.20	$473.60	$88.50	$708.00	$59.20	$88.50
1 Tower Crane (monthly)		1693.00		1862.30	211.63	232.79
8 L.H., Daily Totals		$2166.60		$2570.30	$270.82	$321.29

Crew A-3P	Hr.	Daily	Hr.	Daily	Bare Costs	Incl. O&P
1 Equip. Oper. (light)	$53.00	$424.00	$79.20	$633.60	$53.00	$79.20
1 A.T. Forklift, 31' reach, 45' lift		374.20		411.62	46.77	51.45
8 L.H., Daily Totals		$798.20		$1045.22	$99.78	$130.65

Crew A-3Q	Hr.	Daily	Hr.	Daily	Bare Costs	Incl. O&P
1 Equip. Oper. (light)	$53.00	$424.00	$79.20	$633.60	$53.00	$79.20
1 Pickup Truck, 4x4, 3/4 Ton		175.85		193.44		
1 Flatbed Trailer, 3 Ton		70.30		77.33	30.77	33.85
8 L.H., Daily Totals		$670.15		$904.37	$83.77	$113.05

Crew A-3R	Hr.	Daily	Hr.	Daily	Bare Costs	Incl. O&P
1 Equip. Oper. (light)	$53.00	$424.00	$79.20	$633.60	$53.00	$79.20
1 Forklift, Smooth Floor, 8,000 Lb.		255.90		281.49	31.99	35.19
8 L.H., Daily Totals		$679.90		$915.09	$84.99	$114.39

Crew A-4	Hr.	Daily	Hr.	Daily	Bare Costs	Incl. O&P
2 Carpenters	$53.15	$850.40	$79.85	$1277.60	$50.23	$75.25
1 Painter, Ordinary	44.40	355.20	66.05	528.40		
24 L.H., Daily Totals		$1205.60		$1806.00	$50.23	$75.25

Crew A-5	Hr.	Daily	Hr.	Daily	Bare Costs	Incl. O&P
2 Laborers	$42.10	$673.60	$63.25	$1012.00	$42.67	$64.09
.25 Truck Driver (light)	47.25	94.50	70.80	141.60		
.25 Flatbed Truck, Gas, 1.5 Ton		49.04		53.94	2.72	3.00
18 L.H., Daily Totals		$817.14		$1207.54	$45.40	$67.09

Crew A-6	Hr.	Daily	Hr.	Daily	Bare Costs	Incl. O&P
1 Instrument Man	$54.85	$438.80	$82.95	$663.60	$52.92	$79.67
1 Rodman/Chainman	51.00	408.00	76.40	611.20		
1 Level, Electronic		30.80		33.88	1.93	2.12
16 L.H., Daily Totals		$877.60		$1308.68	$54.85	$81.79

Crew A-7	Hr.	Daily	Hr.	Daily	Bare Costs	Incl. O&P
1 Chief of Party	$66.05	$528.40	$99.00	$792.00	$57.30	$86.12
1 Instrument Man	54.85	438.80	82.95	663.60		
1 Rodman/Chainman	51.00	408.00	76.40	611.20		
1 Level, Electronic		30.80		33.88	1.28	1.41
24 L.H., Daily Totals		$1406.00		$2100.68	$58.58	$87.53

Crew A-8	Hr.	Daily	Hr.	Daily	Bare Costs	Incl. O&P
1 Chief of Party	$66.05	$528.40	$99.00	$792.00	$55.73	$83.69
1 Instrument Man	54.85	438.80	82.95	663.60		
2 Rodmen/Chainmen	51.00	816.00	76.40	1222.40		
1 Level, Electronic		30.80		33.88	.96	1.06
32 L.H., Daily Totals		$1814.00		$2711.88	$56.69	$84.75

Crew A-9	Hr.	Daily	Hr.	Daily	Bare Costs	Incl. O&P
1 Asbestos Foreman	$59.15	$473.20	$90.80	$726.40	$58.71	$90.14
7 Asbestos Workers	58.65	3284.40	90.05	5042.80		
64 L.H., Daily Totals		$3757.60		$5769.20	$58.71	$90.14

Crew A-10A	Hr.	Daily	Hr.	Daily	Bare Costs	Incl. O&P
1 Asbestos Foreman	$59.15	$473.20	$90.80	$726.40	$58.82	$90.30
2 Asbestos Workers	58.65	938.40	90.05	1440.80		
24 L.H., Daily Totals		$1411.60		$2167.20	$58.82	$90.30

Crew A-10B	Hr.	Daily	Hr.	Daily	Bare Costs	Incl. O&P
1 Asbestos Foreman	$59.15	$473.20	$90.80	$726.40	$58.77	$90.24
3 Asbestos Workers	58.65	1407.60	90.05	2161.20		
32 L.H., Daily Totals		$1880.80		$2887.60	$58.77	$90.24

Crew A-10C	Hr.	Daily	Hr.	Daily	Bare Costs	Incl. O&P
3 Asbestos Workers	$58.65	$1407.60	$90.05	$2161.20	$58.65	$90.05
1 Flatbed Truck, Gas, 1.5 Ton		196.15		215.76	8.17	8.99
24 L.H., Daily Totals		$1603.75		$2376.97	$66.82	$99.04

For customer support on your Site Work & Landscape Costs with RSMeans data, call 800.448.8182.

Crew No.		Bare Costs		Incl. Subs O&P		Cost Per Labor-Hour	
Crew A-10D	Hr.	Daily	Hr.	Daily	**Bare Costs**	**Incl. O&P**	
2 Asbestos Workers	$58.65	$938.40	$90.05	$1440.80	$56.76	$86.04	
1 Equip. Oper. (crane)	59.20	473.60	88.50	708.00			
1 Equip. Oper. (oiler)	50.55	404.40	75.55	604.40			
1 Hydraulic Crane, 33 Ton		945.35		1039.89	29.54	32.50	
32 L.H., Daily Totals		$2761.75		$3793.09	$86.30	$118.53	
Crew A-11	Hr.	Daily	Hr.	Daily	**Bare Costs**	**Incl. O&P**	
1 Asbestos Foreman	$59.15	$473.20	$90.80	$726.40	$58.71	$90.14	
7 Asbestos Workers	58.65	3284.40	90.05	5042.80			
2 Chip. Hammers, 12 Lb., Elec.		64.80		71.28	1.01	1.11	
64 L.H., Daily Totals		$3822.40		$5840.48	$59.73	$91.26	
Crew A-12	Hr.	Daily	Hr.	Daily	**Bare Costs**	**Incl. O&P**	
1 Asbestos Foreman	$59.15	$473.20	$90.80	$726.40	$58.71	$90.14	
7 Asbestos Workers	58.65	3284.40	90.05	5042.80			
1 Trk-Mtd Vac, 14 CY, 1500 Gal.		536.15		589.76			
1 Flatbed Truck, 20,000 GVW		201.60		221.76	11.53	12.68	
64 L.H., Daily Totals		$4495.35		$6580.73	$70.24	$102.82	
Crew A-13	Hr.	Daily	Hr.	Daily	**Bare Costs**	**Incl. O&P**	
1 Equip. Oper. (light)	$53.00	$424.00	$79.20	$633.60	$53.00	$79.20	
1 Trk-Mtd Vac, 14 CY, 1500 Gal.		536.15		589.76			
1 Flatbed Truck, 20,000 GVW		201.60		221.76	92.22	101.44	
8 L.H., Daily Totals		$1161.75		$1445.13	$145.22	$180.64	
Crew B-1	Hr.	Daily	Hr.	Daily	**Bare Costs**	**Incl. O&P**	
1 Labor Foreman (outside)	$44.10	$352.80	$66.25	$530.00	$42.77	$64.25	
2 Laborers	42.10	673.60	63.25	1012.00			
24 L.H., Daily Totals		$1026.40		$1542.00	$42.77	$64.25	
Crew B-1A	Hr.	Daily	Hr.	Daily	**Bare Costs**	**Incl. O&P**	
1 Labor Foreman (outside)	$44.10	$352.80	$66.25	$530.00	$42.77	$64.25	
2 Laborers	42.10	673.60	63.25	1012.00			
2 Cutting Torches		25.60		28.16			
2 Sets of Gases		343.10		377.41	15.36	16.90	
24 L.H., Daily Totals		$1395.10		$1947.57	$58.13	$81.15	
Crew B-1B	Hr.	Daily	Hr.	Daily	**Bare Costs**	**Incl. O&P**	
1 Labor Foreman (outside)	$44.10	$352.80	$66.25	$530.00	$46.88	$70.31	
2 Laborers	42.10	673.60	63.25	1012.00			
1 Equip. Oper. (crane)	59.20	473.60	88.50	708.00			
2 Cutting Torches		25.60		28.16			
2 Sets of Gases		343.10		377.41			
1 Hyd. Crane, 12 Ton		471.00		518.10	26.24	28.86	
32 L.H., Daily Totals		$2339.70		$3173.67	$73.12	$99.18	
Crew B-1C	Hr.	Daily	Hr.	Daily	**Bare Costs**	**Incl. O&P**	
1 Labor Foreman (outside)	$44.10	$352.80	$66.25	$530.00	$42.77	$64.25	
2 Laborers	42.10	673.60	63.25	1012.00			
1 Telescoping Boom Lift, to 60'		310.90		341.99	12.95	14.25	
24 L.H., Daily Totals		$1337.30		$1883.99	$55.72	$78.50	
Crew B-1D	Hr.	Daily	Hr.	Daily	**Bare Costs**	**Incl. O&P**	
2 Laborers	$42.10	$673.60	$63.25	$1012.00	$42.10	$63.25	
1 Small Work Boat, Gas, 50 H.P.		119.50		131.45			
1 Pressure Washer, 7 GPM		90.50		99.55	13.13	14.44	
16 L.H., Daily Totals		$883.60		$1243.00	$55.23	$77.69	

Crew No.		Bare Costs		Incl. Subs O&P		Cost Per Labor-Hour	
Crew B-1E	Hr.	Daily	Hr.	Daily	**Bare Costs**	**Incl. O&P**	
1 Labor Foreman (outside)	$44.10	$352.80	$66.25	$530.00	$42.60	$64.00	
3 Laborers	42.10	1010.40	63.25	1518.00			
1 Work Boat, Diesel, 200 H.P.		1418.00		1559.80			
2 Pressure Washers, 7 GPM		181.00		199.10	49.97	54.97	
32 L.H., Daily Totals		$2962.20		$3806.90	$92.57	$118.97	
Crew B-1F	Hr.	Daily	Hr.	Daily	**Bare Costs**	**Incl. O&P**	
2 Skilled Workers	$54.85	$877.60	$82.95	$1327.20	$50.60	$76.38	
1 Laborer	42.10	336.80	63.25	506.00			
1 Small Work Boat, Gas, 50 H.P.		119.50		131.45			
1 Pressure Washer, 7 GPM		90.50		99.55	8.75	9.63	
24 L.H., Daily Totals		$1424.40		$2064.20	$59.35	$86.01	
Crew B-1G	Hr.	Daily	Hr.	Daily	**Bare Costs**	**Incl. O&P**	
2 Laborers	$42.10	$673.60	$63.25	$1012.00	$42.10	$63.25	
1 Small Work Boat, Gas, 50 H.P.		119.50		131.45	7.47	8.22	
16 L.H., Daily Totals		$793.10		$1143.45	$49.57	$71.47	
Crew B-1H	Hr.	Daily	Hr.	Daily	**Bare Costs**	**Incl. O&P**	
2 Skilled Workers	$54.85	$877.60	$82.95	$1327.20	$50.60	$76.38	
1 Laborer	42.10	336.80	63.25	506.00			
1 Small Work Boat, Gas, 50 H.P.		119.50		131.45	4.98	5.48	
24 L.H., Daily Totals		$1333.90		$1964.65	$55.58	$81.86	
Crew B-1J	Hr.	Daily	Hr.	Daily	**Bare Costs**	**Incl. O&P**	
1 Labor Foreman (inside)	$42.60	$340.80	$64.00	$512.00	$42.35	$63.63	
1 Laborer	42.10	336.80	63.25	506.00			
16 L.H., Daily Totals		$677.60		$1018.00	$42.35	$63.63	
Crew B-1K	Hr.	Daily	Hr.	Daily	**Bare Costs**	**Incl. O&P**	
1 Carpenter Foreman (inside)	$53.65	$429.20	$80.60	$644.80	$53.40	$80.22	
1 Carpenter	53.15	425.20	79.85	638.80			
16 L.H., Daily Totals		$854.40		$1283.60	$53.40	$80.22	
Crew B-2	Hr.	Daily	Hr.	Daily	**Bare Costs**	**Incl. O&P**	
1 Labor Foreman (outside)	$44.10	$352.80	$66.25	$530.00	$42.50	$63.85	
4 Laborers	42.10	1347.20	63.25	2024.00			
40 L.H., Daily Totals		$1700.00		$2554.00	$42.50	$63.85	
Crew B-2A	Hr.	Daily	Hr.	Daily	**Bare Costs**	**Incl. O&P**	
1 Labor Foreman (outside)	$44.10	$352.80	$66.25	$530.00	$42.77	$64.25	
2 Laborers	42.10	673.60	63.25	1012.00			
1 Telescoping Boom Lift, to 60'		310.90		341.99	12.95	14.25	
24 L.H., Daily Totals		$1337.30		$1883.99	$55.72	$78.50	
Crew B-3	Hr.	Daily	Hr.	Daily	**Bare Costs**	**Incl. O&P**	
1 Labor Foreman (outside)	$44.10	$352.80	$66.25	$530.00	$47.16	$70.72	
2 Laborers	42.10	673.60	63.25	1012.00			
1 Equip. Oper. (medium)	56.75	454.00	84.85	678.80			
2 Truck Drivers (heavy)	48.95	783.20	73.35	1173.60			
1 Crawler Loader, 3 C.Y.		1169.00		1285.90			
2 Dump Trucks, 12 C.Y., 400 H.P.		1145.80		1260.38	48.23	53.05	
48 L.H., Daily Totals		$4578.40		$5940.68	$95.38	$123.76	
Crew B-3A	Hr.	Daily	Hr.	Daily	**Bare Costs**	**Incl. O&P**	
4 Laborers	$42.10	$1347.20	$63.25	$2024.00	$45.03	$67.57	
1 Equip. Oper. (medium)	56.75	454.00	84.85	678.80			
1 Hyd. Excavator, 1.5 C.Y.		687.55		756.30	17.19	18.91	
40 L.H., Daily Totals		$2488.75		$3459.11	$62.22	$86.48	

Crew No.	Bare Costs		Incl. Subs O&P		Cost Per Labor-Hour	

Crew B-3B

	Hr.	Daily	Hr.	Daily	Bare Costs	Incl. O&P
2 Laborers	$42.10	$673.60	$63.25	$1012.00	$47.48	$71.17
1 Equip. Oper. (medium)	56.75	454.00	84.85	678.80		
1 Truck Driver (heavy)	48.95	391.60	73.35	586.80		
1 Backhoe Loader, 80 H.P.		232.40		255.64		
1 Dump Truck, 12 C.Y., 400 H.P.		572.90		630.19	25.17	27.68
32 L.H., Daily Totals		$2324.50		$3163.43	$72.64	$98.86

Crew B-3C

	Hr.	Daily	Hr.	Daily	Bare Costs	Incl. O&P
3 Laborers	$42.10	$1010.40	$63.25	$1518.00	$45.76	$68.65
1 Equip. Oper. (medium)	56.75	454.00	84.85	678.80		
1 Crawler Loader, 4 C.Y.		1489.00		1637.90	46.53	51.18
32 L.H., Daily Totals		$2953.40		$3834.70	$92.29	$119.83

Crew B-4

	Hr.	Daily	Hr.	Daily	Bare Costs	Incl. O&P
1 Labor Foreman (outside)	$44.10	$352.80	$66.25	$530.00	$43.58	$65.43
4 Laborers	42.10	1347.20	63.25	2024.00		
1 Truck Driver (heavy)	48.95	391.60	73.35	586.80		
1 Truck Tractor, 220 H.P.		307.10		337.81		
1 Flatbed Trailer, 40 Ton		186.20		204.82	10.28	11.30
48 L.H., Daily Totals		$2584.90		$3683.43	$53.85	$76.74

Crew B-5

	Hr.	Daily	Hr.	Daily	Bare Costs	Incl. O&P
1 Labor Foreman (outside)	$44.10	$352.80	$66.25	$530.00	$46.57	$69.85
4 Laborers	42.10	1347.20	63.25	2024.00		
2 Equip. Oper. (medium)	56.75	908.00	84.85	1357.60		
1 Air Compressor, 250 cfm		201.50		221.65		
2 Breakers, Pavement, 60 lb.		107.00		117.70		
2 -50' Air Hoses, 1.5"		48.50		53.35		
1 Crawler Loader, 3 C.Y.		1169.00		1285.90	27.25	29.98
56 L.H., Daily Totals		$4134.00		$5590.20	$73.82	$99.83

Crew B-5A

	Hr.	Daily	Hr.	Daily	Bare Costs	Incl. O&P
1 Labor Foreman (outside)	$44.10	$352.80	$66.25	$530.00	$46.76	$70.11
6 Laborers	42.10	2020.80	63.25	3036.00		
2 Equip. Oper. (medium)	56.75	908.00	84.85	1357.60		
1 Equip. Oper. (light)	53.00	424.00	79.20	633.60		
2 Truck Drivers (heavy)	48.95	783.20	73.35	1173.60		
1 Air Compressor, 365 cfm		315.40		346.94		
2 Breakers, Pavement, 60 lb.		107.00		117.70		
8 -50' Air Hoses, 1"		64.00		70.40		
2 Dump Trucks, 8 C.Y., 220 H.P.		856.10		941.71	13.98	15.38
96 L.H., Daily Totals		$5831.30		$8207.55	$60.74	$85.50

Crew B-5B

	Hr.	Daily	Hr.	Daily	Bare Costs	Incl. O&P
1 Powderman	$54.85	$438.80	$82.95	$663.60	$52.53	$78.78
2 Equip. Oper. (medium)	56.75	908.00	84.85	1357.60		
3 Truck Drivers (heavy)	48.95	1174.80	73.35	1760.40		
1 F.E. Loader, W.M., 2.5 C.Y.		702.10		772.31		
3 Dump Trucks, 12 C.Y., 400 H.P.		1718.70		1890.57		
1 Air Compressor, 365 cfm		315.40		346.94	57.00	62.70
48 L.H., Daily Totals		$5257.80		$6791.42	$109.54	$141.49

Crew B-5C

	Hr.	Daily	Hr.	Daily	Bare Costs	Incl. O&P
3 Laborers	$42.10	$1010.40	$63.25	$1518.00	$48.84	$73.17
1 Equip. Oper. (medium)	56.75	454.00	84.85	678.80		
2 Truck Drivers (heavy)	48.95	783.20	73.35	1173.60		
1 Equip. Oper. (crane)	59.20	473.60	88.50	708.00		
1 Equip. Oper. (oiler)	50.55	404.40	75.55	604.40		
2 Dump Trucks, 12 C.Y., 400 H.P.		1145.80		1260.38		
1 Crawler Loader, 4 C.Y.		1489.00		1637.90		
1 S.P. Crane, 4x4, 25 Ton		1144.00		1258.40	59.04	64.95
64 L.H., Daily Totals		$6904.40		$8839.48	$107.88	$138.12

Crew B-5D

	Hr.	Daily	Hr.	Daily	Bare Costs	Incl. O&P
1 Labor Foreman (outside)	$44.10	$352.80	$66.25	$530.00	$46.87	$70.29
4 Laborers	42.10	1347.20	63.25	2024.00		
2 Equip. Oper. (medium)	56.75	908.00	84.85	1357.60		
1 Truck Driver (heavy)	48.95	391.60	73.35	586.80		
1 Air Compressor, 250 cfm		201.50		221.65		
2 Breakers, Pavement, 60 lb.		107.00		117.70		
2 -50' Air Hoses, 1.5"		48.50		53.35		
1 Crawler Loader, 3 C.Y.		1169.00		1285.90		
1 Dump Truck, 12 C.Y., 400 H.P.		572.90		630.19	32.80	36.07
64 L.H., Daily Totals		$5098.50		$6807.19	$79.66	$106.36

Crew B-5E

	Hr.	Daily	Hr.	Daily	Bare Costs	Incl. O&P
1 Labor Foreman (outside)	$44.10	$352.80	$66.25	$530.00	$46.87	$70.29
4 Laborers	42.10	1347.20	63.25	2024.00		
2 Equip. Oper. (medium)	56.75	908.00	84.85	1357.60		
1 Truck Driver (heavy)	48.95	391.60	73.35	586.80		
1 Water Tank Trailer, 5000 Gal.		152.25		167.47		
1 High Press. Water Jet, 40 KSI		811.05		892.16		
2 -50' Air Hoses, 1.5"		48.50		53.35		
1 Crawler Loader, 3 C.Y.		1169.00		1285.90		
1 Dump Truck, 12 C.Y., 400 H.P.		572.90		630.19	43.03	47.33
64 L.H., Daily Totals		$5753.30		$7527.47	$89.90	$117.62

Crew B-6

	Hr.	Daily	Hr.	Daily	Bare Costs	Incl. O&P
2 Laborers	$42.10	$673.60	$63.25	$1012.00	$45.73	$68.57
1 Equip. Oper. (light)	53.00	424.00	79.20	633.60		
1 Backhoe Loader, 48 H.P.		213.75		235.13	8.91	9.80
24 L.H., Daily Totals		$1311.35		$1880.72	$54.64	$78.36

Crew B-6A

	Hr.	Daily	Hr.	Daily	Bare Costs	Incl. O&P
.5 Labor Foreman (outside)	$44.10	$176.40	$66.25	$265.00	$48.36	$72.49
1 Laborer	42.10	336.80	63.25	506.00		
1 Equip. Oper. (medium)	56.75	454.00	84.85	678.80		
1 Vacuum Truck, 5000 Gal.		367.55		404.31	18.38	20.22
20 L.H., Daily Totals		$1334.75		$1854.11	$66.74	$92.71

Crew B-6B

	Hr.	Daily	Hr.	Daily	Bare Costs	Incl. O&P
2 Labor Foremen (outside)	$44.10	$705.60	$66.25	$1060.00	$42.77	$64.25
4 Laborers	42.10	1347.20	63.25	2024.00		
1 S.P. Crane, 4x4, 5 Ton		378.10		415.91		
1 Flatbed Truck, Gas, 1.5 Ton		196.15		215.76		
1 Butt Fusion Mach., 4"-12" diam.		426.85		469.54	20.86	22.94
48 L.H., Daily Totals		$3053.90		$4185.21	$63.62	$87.19

Crews - Standard

Crew B-6C

Crew B-6C	Bare Costs Hr.	Bare Costs Daily	Incl. Subs O&P Hr.	Incl. Subs O&P Daily	Cost Per Labor-Hour Bare Costs	Cost Per Labor-Hour Incl. O&P
2 Labor Foremen (outside)	$44.10	$705.60	$66.25	$1060.00	$42.77	$64.25
4 Laborers	42.10	1347.20	63.25	2024.00		
1 S.P. Crane, 4x4, 12 Ton		428.30		471.13		
1 Flatbed Truck, Gas, 3 Ton		820.40		902.44		
1 Butt Fusion Mach., 8"-24" diam.		1076.00		1183.60	48.43	53.27
48 L.H., Daily Totals		$4377.50		$5641.17	$91.20	$117.52

Crew B-6D

Crew B-6D	Bare Costs Hr.	Bare Costs Daily	Incl. Subs O&P Hr.	Incl. Subs O&P Daily	Cost Per Labor-Hour Bare Costs	Cost Per Labor-Hour Incl. O&P
0.5 Labor Foreman (outside)	$44.10	$176.40	$66.25	$265.00	$48.36	$72.49
1 Laborer	42.10	336.80	63.25	506.00		
1 Equip. Oper. (medium)	56.75	454.00	84.85	678.80		
1 Hydro Excavator, 12 C.Y.		1262.00		1388.20	63.10	69.41
20 L.H., Daily Totals		$2229.20		$2838.00	$111.46	$141.90

Crew B-7

Crew B-7	Bare Costs Hr.	Bare Costs Daily	Incl. Subs O&P Hr.	Incl. Subs O&P Daily	Cost Per Labor-Hour Bare Costs	Cost Per Labor-Hour Incl. O&P
1 Labor Foreman (outside)	$44.10	$352.80	$66.25	$530.00	$44.88	$67.35
4 Laborers	42.10	1347.20	63.25	2024.00		
1 Equip. Oper. (medium)	56.75	454.00	84.85	678.80		
1 Brush Chipper, 12", 130 H.P.		392.80		432.08		
1 Crawler Loader, 3 C.Y.		1169.00		1285.90		
2 Chain Saws, Gas, 36" Long		82.50		90.75	34.26	37.68
48 L.H., Daily Totals		$3798.30		$5041.53	$79.13	$105.03

Crew B-7A

Crew B-7A	Bare Costs Hr.	Bare Costs Daily	Incl. Subs O&P Hr.	Incl. Subs O&P Daily	Cost Per Labor-Hour Bare Costs	Cost Per Labor-Hour Incl. O&P
2 Laborers	$42.10	$673.60	$63.25	$1012.00	$45.73	$68.57
1 Equip. Oper. (light)	53.00	424.00	79.20	633.60		
1 Rake w/Tractor		339.45		373.39		
2 Chain Saws, Gas, 18"		104.00		114.40	18.48	20.32
24 L.H., Daily Totals		$1541.05		$2133.40	$64.21	$88.89

Crew B-7B

Crew B-7B	Bare Costs Hr.	Bare Costs Daily	Incl. Subs O&P Hr.	Incl. Subs O&P Daily	Cost Per Labor-Hour Bare Costs	Cost Per Labor-Hour Incl. O&P
1 Labor Foreman (outside)	$44.10	$352.80	$66.25	$530.00	$45.46	$68.21
4 Laborers	42.10	1347.20	63.25	2024.00		
1 Equip. Oper. (medium)	56.75	454.00	84.85	678.80		
1 Truck Driver (heavy)	48.95	391.60	73.35	586.80		
1 Brush Chipper, 12", 130 H.P.		392.80		432.08		
1 Crawler Loader, 3 C.Y.		1169.00		1285.90		
2 Chain Saws, Gas, 36" Long		82.50		90.75		
1 Dump Truck, 8 C.Y., 220 H.P.		428.05		470.86	37.01	40.71
56 L.H., Daily Totals		$4617.95		$6099.19	$82.46	$108.91

Crew B-7C

Crew B-7C	Bare Costs Hr.	Bare Costs Daily	Incl. Subs O&P Hr.	Incl. Subs O&P Daily	Cost Per Labor-Hour Bare Costs	Cost Per Labor-Hour Incl. O&P
1 Labor Foreman (outside)	$44.10	$352.80	$66.25	$530.00	$45.46	$68.21
4 Laborers	42.10	1347.20	63.25	2024.00		
1 Equip. Oper. (medium)	56.75	454.00	84.85	678.80		
1 Truck Driver (heavy)	48.95	391.60	73.35	586.80		
1 Brush Chipper, 12", 130 H.P.		392.80		432.08		
1 Crawler Loader, 3 C.Y.		1169.00		1285.90		
2 Chain Saws, Gas, 36" Long		82.50		90.75		
1 Dump Truck, 12 C.Y., 400 H.P.		572.90		630.19	39.59	43.55
56 L.H., Daily Totals		$4762.80		$6258.52	$85.05	$111.76

Crew B-8

Crew B-8	Bare Costs Hr.	Bare Costs Daily	Incl. Subs O&P Hr.	Incl. Subs O&P Daily	Cost Per Labor-Hour Bare Costs	Cost Per Labor-Hour Incl. O&P
1 Labor Foreman (outside)	$44.10	$352.80	$66.25	$530.00	$48.78	$73.09
2 Laborers	42.10	673.60	63.25	1012.00		
2 Equip. Oper. (medium)	56.75	908.00	84.85	1357.60		
1 Equip. Oper. (oiler)	50.55	404.40	75.55	604.40		
2 Truck Drivers (heavy)	48.95	783.20	73.35	1173.60		
1 Hyd. Crane, 25 Ton		580.85		638.93		
1 Crawler Loader, 3 C.Y.		1169.00		1285.90		
2 Dump Trucks, 12 C.Y., 400 H.P.		1145.80		1260.38	45.24	49.77
64 L.H., Daily Totals		$6017.65		$7862.81	$94.03	$122.86

Crew B-9

Crew B-9	Bare Costs Hr.	Bare Costs Daily	Incl. Subs O&P Hr.	Incl. Subs O&P Daily	Cost Per Labor-Hour Bare Costs	Cost Per Labor-Hour Incl. O&P
1 Labor Foreman (outside)	$44.10	$352.80	$66.25	$530.00	$42.50	$63.85
4 Laborers	42.10	1347.20	63.25	2024.00		
1 Air Compressor, 250 cfm		201.50		221.65		
2 Breakers, Pavement, 60 lb.		107.00		117.70		
2 -50' Air Hoses, 1.5"		48.50		53.35	8.93	9.82
40 L.H., Daily Totals		$2057.00		$2946.70	$51.42	$73.67

Crew B-9A

Crew B-9A	Bare Costs Hr.	Bare Costs Daily	Incl. Subs O&P Hr.	Incl. Subs O&P Daily	Cost Per Labor-Hour Bare Costs	Cost Per Labor-Hour Incl. O&P
2 Laborers	$42.10	$673.60	$63.25	$1012.00	$44.38	$66.62
1 Truck Driver (heavy)	48.95	391.60	73.35	586.80		
1 Water Tank Trailer, 5000 Gal.		152.25		167.47		
1 Truck Tractor, 220 H.P.		307.10		337.81		
2 -50' Discharge Hoses, 3"		9.00		9.90	19.51	21.47
24 L.H., Daily Totals		$1533.55		$2113.99	$63.90	$88.08

Crew B-9B

Crew B-9B	Bare Costs Hr.	Bare Costs Daily	Incl. Subs O&P Hr.	Incl. Subs O&P Daily	Cost Per Labor-Hour Bare Costs	Cost Per Labor-Hour Incl. O&P
2 Laborers	$42.10	$673.60	$63.25	$1012.00	$44.38	$66.62
1 Truck Driver (heavy)	48.95	391.60	73.35	586.80		
2 -50' Discharge Hoses, 3"		9.00		9.90		
1 Water Tank Trailer, 5000 Gal.		152.25		167.47		
1 Truck Tractor, 220 H.P.		307.10		337.81		
1 Pressure Washer		96.95		106.65	23.55	25.91
24 L.H., Daily Totals		$1630.50		$2220.63	$67.94	$92.53

Crew B-9D

Crew B-9D	Bare Costs Hr.	Bare Costs Daily	Incl. Subs O&P Hr.	Incl. Subs O&P Daily	Cost Per Labor-Hour Bare Costs	Cost Per Labor-Hour Incl. O&P
1 Labor Foreman (outside)	$44.10	$352.80	$66.25	$530.00	$42.50	$63.85
4 Common Laborers	42.10	1347.20	63.25	2024.00		
1 Air Compressor, 250 cfm		201.50		221.65		
2 -50' Air Hoses, 1.5"		48.50		53.35		
2 Air Powered Tampers		67.70		74.47	7.94	8.74
40 L.H., Daily Totals		$2017.70		$2903.47	$50.44	$72.59

Crew B-9E

Crew B-9E	Bare Costs Hr.	Bare Costs Daily	Incl. Subs O&P Hr.	Incl. Subs O&P Daily	Cost Per Labor-Hour Bare Costs	Cost Per Labor-Hour Incl. O&P
1 Cement Finisher	$49.95	$399.60	$73.45	$587.60	$46.02	$68.35
1 Laborer	42.10	336.80	63.25	506.00		
1 Chip. Hammers, 12 Lb., Elec.		32.40		35.64	2.02	2.23
16 L.H., Daily Totals		$768.80		$1129.24	$48.05	$70.58

Crew B-10

Crew B-10	Bare Costs Hr.	Bare Costs Daily	Incl. Subs O&P Hr.	Incl. Subs O&P Daily	Cost Per Labor-Hour Bare Costs	Cost Per Labor-Hour Incl. O&P
1 Equip. Oper. (medium)	$56.75	$454.00	$84.85	$678.80	$51.87	$77.65
.5 Laborer	42.10	168.40	63.25	253.00		
12 L.H., Daily Totals		$622.40		$931.80	$51.87	$77.65

Crew B-10A

Crew B-10A	Bare Costs Hr.	Bare Costs Daily	Incl. Subs O&P Hr.	Incl. Subs O&P Daily	Cost Per Labor-Hour Bare Costs	Cost Per Labor-Hour Incl. O&P
1 Equip. Oper. (medium)	$56.75	$454.00	$84.85	$678.80	$51.87	$77.65
.5 Laborer	42.10	168.40	63.25	253.00		
1 Roller, 2-Drum, W.B., 7.5 H.P.		166.10		182.71	13.84	15.23
12 L.H., Daily Totals		$788.50		$1114.51	$65.71	$92.88

For customer support on your Site Work & Landscape Costs with RSMeans data, call 800.448.8182.

745

Crew No.	Bare Costs		Incl. Subs O&P		Cost Per Labor-Hour	
	Hr.	Daily	Hr.	Daily	Bare Costs	Incl. O&P
Crew B-10B						
1 Equip. Oper. (medium)	$56.75	$454.00	$84.85	$678.80	$51.87	$77.65
.5 Laborer	42.10	168.40	63.25	253.00		
1 Dozer, 200 H.P.		1504.00		1654.40	125.33	137.87
12 L.H., Daily Totals		$2126.40		$2586.20	$177.20	$215.52
Crew B-10C	Hr.	Daily	Hr.	Daily	Bare Costs	Incl. O&P
1 Equip. Oper. (medium)	$56.75	$454.00	$84.85	$678.80	$51.87	$77.65
.5 Laborer	42.10	168.40	63.25	253.00		
1 Dozer, 200 H.P.		1504.00		1654.40		
1 Vibratory Roller, Towed, 23 Ton		514.75		566.23	168.23	185.05
12 L.H., Daily Totals		$2641.15		$3152.43	$220.10	$262.70
Crew B-10D	Hr.	Daily	Hr.	Daily	Bare Costs	Incl. O&P
1 Equip. Oper. (medium)	$56.75	$454.00	$84.85	$678.80	$51.87	$77.65
.5 Laborer	42.10	168.40	63.25	253.00		
1 Dozer, 200 H.P.		1504.00		1654.40		
1 Sheepsft. Roller, Towed		436.90		480.59	161.74	177.92
12 L.H., Daily Totals		$2563.30		$3066.79	$213.61	$255.57
Crew B-10E	Hr.	Daily	Hr.	Daily	Bare Costs	Incl. O&P
1 Equip. Oper. (medium)	$56.75	$454.00	$84.85	$678.80	$51.87	$77.65
.5 Laborer	42.10	168.40	63.25	253.00		
1 Tandem Roller, 5 Ton		255.65		281.21	21.30	23.43
12 L.H., Daily Totals		$878.05		$1213.02	$73.17	$101.08
Crew B-10F	Hr.	Daily	Hr.	Daily	Bare Costs	Incl. O&P
1 Equip. Oper. (medium)	$56.75	$454.00	$84.85	$678.80	$51.87	$77.65
.5 Laborer	42.10	168.40	63.25	253.00		
1 Tandem Roller, 10 Ton		236.75		260.43	19.73	21.70
12 L.H., Daily Totals		$859.15		$1192.22	$71.60	$99.35
Crew B-10G	Hr.	Daily	Hr.	Daily	Bare Costs	Incl. O&P
1 Equip. Oper. (medium)	$56.75	$454.00	$84.85	$678.80	$51.87	$77.65
.5 Laborer	42.10	168.40	63.25	253.00		
1 Sheepsfoot Roller, 240 H.P.		1348.00		1482.80	112.33	123.57
12 L.H., Daily Totals		$1970.40		$2414.60	$164.20	$201.22
Crew B-10H	Hr.	Daily	Hr.	Daily	Bare Costs	Incl. O&P
1 Equip. Oper. (medium)	$56.75	$454.00	$84.85	$678.80	$51.87	$77.65
.5 Laborer	42.10	168.40	63.25	253.00		
1 Diaphragm Water Pump, 2"		86.80		95.48		
1 -20' Suction Hose, 2"		3.55		3.90		
2 -50' Discharge Hoses, 2"		8.00		8.80	8.20	9.02
12 L.H., Daily Totals		$720.75		$1039.98	$60.06	$86.67
Crew B-10I	Hr.	Daily	Hr.	Daily	Bare Costs	Incl. O&P
1 Equip. Oper. (medium)	$56.75	$454.00	$84.85	$678.80	$51.87	$77.65
.5 Laborer	42.10	168.40	63.25	253.00		
1 Diaphragm Water Pump, 4"		105.25		115.78		
1 -20' Suction Hose, 4"		17.05		18.75		
2 -50' Discharge Hoses, 4"		25.30		27.83	12.30	13.53
12 L.H., Daily Totals		$770.00		$1094.16	$64.17	$91.18
Crew B-10J	Hr.	Daily	Hr.	Daily	Bare Costs	Incl. O&P
1 Equip. Oper. (medium)	$56.75	$454.00	$84.85	$678.80	$51.87	$77.65
.5 Laborer	42.10	168.40	63.25	253.00		
1 Centrifugal Water Pump, 3"		74.45		81.89		
1 -20' Suction Hose, 3"		8.75		9.63		
2 -50' Discharge Hoses, 3"		9.00		9.90	7.68	8.45
12 L.H., Daily Totals		$714.60		$1033.22	$59.55	$86.10

Crew No.	Bare Costs		Incl. Subs O&P		Cost Per Labor-Hour	
	Hr.	Daily	Hr.	Daily	Bare Costs	Incl. O&P
Crew B-10K						
1 Equip. Oper. (medium)	$56.75	$454.00	$84.85	$678.80	$51.87	$77.65
.5 Laborer	42.10	168.40	63.25	253.00		
1 Centr. Water Pump, 6"		232.90		256.19		
1 -20' Suction Hose, 6"		25.20		27.72		
2 -50' Discharge Hoses, 6"		35.80		39.38	24.49	26.94
12 L.H., Daily Totals		$916.30		$1255.09	$76.36	$104.59
Crew B-10L	Hr.	Daily	Hr.	Daily	Bare Costs	Incl. O&P
1 Equip. Oper. (medium)	$56.75	$454.00	$84.85	$678.80	$51.87	$77.65
.5 Laborer	42.10	168.40	63.25	253.00		
1 Dozer, 80 H.P.		401.40		441.54	33.45	36.80
12 L.H., Daily Totals		$1023.80		$1373.34	$85.32	$114.44
Crew B-10M	Hr.	Daily	Hr.	Daily	Bare Costs	Incl. O&P
1 Equip. Oper. (medium)	$56.75	$454.00	$84.85	$678.80	$51.87	$77.65
.5 Laborer	42.10	168.40	63.25	253.00		
1 Dozer, 300 H.P.		1764.00		1940.40	147.00	161.70
12 L.H., Daily Totals		$2386.40		$2872.20	$198.87	$239.35
Crew B-10N	Hr.	Daily	Hr.	Daily	Bare Costs	Incl. O&P
1 Equip. Oper. (medium)	$56.75	$454.00	$84.85	$678.80	$51.87	$77.65
.5 Laborer	42.10	168.40	63.25	253.00		
1 F.E. Loader, T.M., 1.5 C.Y.		565.80		622.38	47.15	51.87
12 L.H., Daily Totals		$1188.20		$1554.18	$99.02	$129.51
Crew B-10O	Hr.	Daily	Hr.	Daily	Bare Costs	Incl. O&P
1 Equip. Oper. (medium)	$56.75	$454.00	$84.85	$678.80	$51.87	$77.65
.5 Laborer	42.10	168.40	63.25	253.00		
1 F.E. Loader, T.M., 2.25 C.Y.		960.90		1056.99	80.08	88.08
12 L.H., Daily Totals		$1583.30		$1988.79	$131.94	$165.73
Crew B-10P	Hr.	Daily	Hr.	Daily	Bare Costs	Incl. O&P
1 Equip. Oper. (medium)	$56.75	$454.00	$84.85	$678.80	$51.87	$77.65
.5 Laborer	42.10	168.40	63.25	253.00		
1 Crawler Loader, 3 C.Y.		1169.00		1285.90	97.42	107.16
12 L.H., Daily Totals		$1791.40		$2217.70	$149.28	$184.81
Crew B-10Q	Hr.	Daily	Hr.	Daily	Bare Costs	Incl. O&P
1 Equip. Oper. (medium)	$56.75	$454.00	$84.85	$678.80	$51.87	$77.65
.5 Laborer	42.10	168.40	63.25	253.00		
1 Crawler Loader, 4 C.Y.		1489.00		1637.90	124.08	136.49
12 L.H., Daily Totals		$2111.40		$2569.70	$175.95	$214.14
Crew B-10R	Hr.	Daily	Hr.	Daily	Bare Costs	Incl. O&P
1 Equip. Oper. (medium)	$56.75	$454.00	$84.85	$678.80	$51.87	$77.65
.5 Laborer	42.10	168.40	63.25	253.00		
1 F.E. Loader, W.M., 1 C.Y.		302.00		332.20	25.17	27.68
12 L.H., Daily Totals		$924.40		$1264.00	$77.03	$105.33
Crew B-10S	Hr.	Daily	Hr.	Daily	Bare Costs	Incl. O&P
1 Equip. Oper. (medium)	$56.75	$454.00	$84.85	$678.80	$51.87	$77.65
.5 Laborer	42.10	168.40	63.25	253.00		
1 F.E. Loader, W.M., 1.5 C.Y.		423.50		465.85	35.29	38.82
12 L.H., Daily Totals		$1045.90		$1397.65	$87.16	$116.47
Crew B-10T	Hr.	Daily	Hr.	Daily	Bare Costs	Incl. O&P
1 Equip. Oper. (medium)	$56.75	$454.00	$84.85	$678.80	$51.87	$77.65
.5 Laborer	42.10	168.40	63.25	253.00		
1 F.E. Loader, W.M., 2.5 C.Y.		702.10		772.31	58.51	64.36
12 L.H., Daily Totals		$1324.50		$1704.11	$110.38	$142.01

For customer support on your Site Work & Landscape Costs with RSMeans data, call 800.448.8182.

Crew B-10U

Crew No.	Bare Costs Hr.	Daily	Incl. Subs O&P Hr.	Daily	Bare Costs	Incl. O&P
1 Equip. Oper. (medium)	$56.75	$454.00	$84.85	$678.80	$51.87	$77.65
.5 Laborer	42.10	168.40	63.25	253.00		
1 F.E. Loader, W.M., 5.5 C.Y.		957.50		1053.25	79.79	87.77
12 L.H., Daily Totals		$1579.90		$1985.05	$131.66	$165.42

Crew B-10V

Crew No.	Bare Costs Hr.	Daily	Incl. Subs O&P Hr.	Daily	Bare Costs	Incl. O&P
1 Equip. Oper. (medium)	$56.75	$454.00	$84.85	$678.80	$51.87	$77.65
.5 Laborer	42.10	168.40	63.25	253.00		
1 Dozer, 700 H.P.		5120.00		5632.00	426.67	469.33
12 L.H., Daily Totals		$5742.40		$6563.80	$478.53	$546.98

Crew B-10W

Crew No.	Bare Costs Hr.	Daily	Incl. Subs O&P Hr.	Daily	Bare Costs	Incl. O&P
1 Equip. Oper. (medium)	$56.75	$454.00	$84.85	$678.80	$51.87	$77.65
.5 Laborer	42.10	168.40	63.25	253.00		
1 Dozer, 105 H.P.		633.85		697.24	52.82	58.10
12 L.H., Daily Totals		$1256.25		$1629.04	$104.69	$135.75

Crew B-10X

Crew No.	Bare Costs Hr.	Daily	Incl. Subs O&P Hr.	Daily	Bare Costs	Incl. O&P
1 Equip. Oper. (medium)	$56.75	$454.00	$84.85	$678.80	$51.87	$77.65
.5 Laborer	42.10	168.40	63.25	253.00		
1 Dozer, 410 H.P.		2777.00		3054.70	231.42	254.56
12 L.H., Daily Totals		$3399.40		$3986.50	$283.28	$332.21

Crew B-10Y

Crew No.	Bare Costs Hr.	Daily	Incl. Subs O&P Hr.	Daily	Bare Costs	Incl. O&P
1 Equip. Oper. (medium)	$56.75	$454.00	$84.85	$678.80	$51.87	$77.65
.5 Laborer	42.10	168.40	63.25	253.00		
1 Vibr. Roller, Towed, 12 Ton		578.60		636.46	48.22	53.04
12 L.H., Daily Totals		$1201.00		$1568.26	$100.08	$130.69

Crew B-11A

Crew No.	Bare Costs Hr.	Daily	Incl. Subs O&P Hr.	Daily	Bare Costs	Incl. O&P
1 Equipment Oper. (med.)	$56.75	$454.00	$84.85	$678.80	$49.42	$74.05
1 Laborer	42.10	336.80	63.25	506.00		
1 Dozer, 200 H.P.		1504.00		1654.40	94.00	103.40
16 L.H., Daily Totals		$2294.80		$2839.20	$143.43	$177.45

Crew B-11B

Crew No.	Bare Costs Hr.	Daily	Incl. Subs O&P Hr.	Daily	Bare Costs	Incl. O&P
1 Equipment Oper. (light)	$53.00	$424.00	$79.20	$633.60	$47.55	$71.22
1 Laborer	42.10	336.80	63.25	506.00		
1 Air Powered Tamper		33.85		37.23		
1 Air Compressor, 365 cfm		315.40		346.94		
2 -50' Air Hoses, 1.5"		48.50		53.35	24.86	27.35
16 L.H., Daily Totals		$1158.55		$1577.13	$72.41	$98.57

Crew B-11C

Crew No.	Bare Costs Hr.	Daily	Incl. Subs O&P Hr.	Daily	Bare Costs	Incl. O&P
1 Equipment Oper. (med.)	$56.75	$454.00	$84.85	$678.80	$49.42	$74.05
1 Laborer	42.10	336.80	63.25	506.00		
1 Backhoe Loader, 48 H.P.		213.75		235.13	13.36	14.70
16 L.H., Daily Totals		$1004.55		$1419.93	$62.78	$88.75

Crew B-11J

Crew No.	Bare Costs Hr.	Daily	Incl. Subs O&P Hr.	Daily	Bare Costs	Incl. O&P
1 Equipment Oper. (med.)	$56.75	$454.00	$84.85	$678.80	$49.42	$74.05
1 Laborer	42.10	336.80	63.25	506.00		
1 Grader, 30,000 Lbs.		1062.00		1168.20		
1 Ripper, Beam & 1 Shank		90.50		99.55	72.03	79.23
16 L.H., Daily Totals		$1943.30		$2452.55	$121.46	$153.28

Crew B-11K

Crew No.	Bare Costs Hr.	Daily	Incl. Subs O&P Hr.	Daily	Bare Costs	Incl. O&P
1 Equipment Oper. (med.)	$56.75	$454.00	$84.85	$678.80	$49.42	$74.05
1 Laborer	42.10	336.80	63.25	506.00		
1 Trencher, Chain Type, 8' D		1872.00		2059.20	117.00	128.70
16 L.H., Daily Totals		$2662.80		$3244.00	$166.43	$202.75

Crew B-11L

Crew No.	Bare Costs Hr.	Daily	Incl. Subs O&P Hr.	Daily	Bare Costs	Incl. O&P
1 Equipment Oper. (med.)	$56.75	$454.00	$84.85	$678.80	$49.42	$74.05
1 Laborer	42.10	336.80	63.25	506.00		
1 Grader, 30,000 Lbs.		1062.00		1168.20	66.38	73.01
16 L.H., Daily Totals		$1852.80		$2353.00	$115.80	$147.06

Crew B-11M

Crew No.	Bare Costs Hr.	Daily	Incl. Subs O&P Hr.	Daily	Bare Costs	Incl. O&P
1 Equipment Oper. (med.)	$56.75	$454.00	$84.85	$678.80	$49.42	$74.05
1 Laborer	42.10	336.80	63.25	506.00		
1 Backhoe Loader, 80 H.P.		232.40		255.64	14.53	15.98
16 L.H., Daily Totals		$1023.20		$1440.44	$63.95	$90.03

Crew B-11N

Crew No.	Bare Costs Hr.	Daily	Incl. Subs O&P Hr.	Daily	Bare Costs	Incl. O&P
1 Labor Foreman (outside)	$44.10	$352.80	$66.25	$530.00	$50.14	$75.12
2 Equipment Operators (med.)	56.75	908.00	84.85	1357.60		
6 Truck Drivers (heavy)	48.95	2349.60	73.35	3520.80		
1 F.E. Loader, W.M., 5.5 C.Y.		957.50		1053.25		
1 Dozer, 410 H.P.		2777.00		3054.70		
6 Dump Trucks, Off Hwy., 50 Ton		10626.00		11688.60	199.45	219.40
72 L.H., Daily Totals		$17970.90		$21204.95	$249.60	$294.51

Crew B-11Q

Crew No.	Bare Costs Hr.	Daily	Incl. Subs O&P Hr.	Daily	Bare Costs	Incl. O&P
1 Equipment Operator (med.)	$56.75	$454.00	$84.85	$678.80	$51.87	$77.65
.5 Laborer	42.10	168.40	63.25	253.00		
1 Dozer, 140 H.P.		762.40		838.64	63.53	69.89
12 L.H., Daily Totals		$1384.80		$1770.44	$115.40	$147.54

Crew B-11R

Crew No.	Bare Costs Hr.	Daily	Incl. Subs O&P Hr.	Daily	Bare Costs	Incl. O&P
1 Equipment Operator (med.)	$56.75	$454.00	$84.85	$678.80	$51.87	$77.65
.5 Laborer	42.10	168.40	63.25	253.00		
1 Dozer, 200 H.P.		1504.00		1654.40	125.33	137.87
12 L.H., Daily Totals		$2126.40		$2586.20	$177.20	$215.52

Crew B-11S

Crew No.	Bare Costs Hr.	Daily	Incl. Subs O&P Hr.	Daily	Bare Costs	Incl. O&P
1 Equipment Operator (med.)	$56.75	$454.00	$84.85	$678.80	$51.87	$77.65
.5 Laborer	42.10	168.40	63.25	253.00		
1 Dozer, 300 H.P.		1764.00		1940.40		
1 Ripper, Beam & 1 Shank		90.50		99.55	154.54	170.00
12 L.H., Daily Totals		$2476.90		$2971.75	$206.41	$247.65

Crew B-11T

Crew No.	Bare Costs Hr.	Daily	Incl. Subs O&P Hr.	Daily	Bare Costs	Incl. O&P
1 Equipment Operator (med.)	$56.75	$454.00	$84.85	$678.80	$51.87	$77.65
.5 Laborer	42.10	168.40	63.25	253.00		
1 Dozer, 410 H.P.		2777.00		3054.70		
1 Ripper, Beam & 2 Shanks		138.75		152.63	242.98	267.28
12 L.H., Daily Totals		$3538.15		$4139.13	$294.85	$344.93

Crew B-11U

Crew No.	Bare Costs Hr.	Daily	Incl. Subs O&P Hr.	Daily	Bare Costs	Incl. O&P
1 Equipment Operator (med.)	$56.75	$454.00	$84.85	$678.80	$51.87	$77.65
.5 Laborer	42.10	168.40	63.25	253.00		
1 Dozer, 520 H.P.		3398.00		3737.80	283.17	311.48
12 L.H., Daily Totals		$4020.40		$4669.60	$335.03	$389.13

Crews - Standard

Crew No.	Bare Costs		Incl. Subs O&P		Cost Per Labor-Hour	
Crew B-11V	Hr.	Daily	Hr.	Daily	Bare Costs	Incl. O&P
3 Laborers	$42.10	$1010.40	$63.25	$1518.00	$42.10	$63.25
1 Roller, 2-Drum, W.B., 7.5 H.P.		166.10		182.71	6.92	7.61
24 L.H., Daily Totals		$1176.50		$1700.71	$49.02	$70.86
Crew B-11W	Hr.	Daily	Hr.	Daily	Bare Costs	Incl. O&P
1 Equipment Operator (med.)	$56.75	$454.00	$84.85	$678.80	$49.03	$73.47
1 Common Laborer	42.10	336.80	63.25	506.00		
10 Truck Drivers (heavy)	48.95	3916.00	73.35	5868.00		
1 Dozer, 200 H.P.		1504.00		1654.40		
1 Vibratory Roller, Towed, 23 Ton		514.75		566.23		
10 Dump Trucks, 8 C.Y., 220 H.P.		4280.50		4708.55	65.62	72.18
96 L.H., Daily Totals		$11006.05		$13981.98	$114.65	$145.65
Crew B-11Y	Hr.	Daily	Hr.	Daily	Bare Costs	Incl. O&P
1 Labor Foreman (outside)	$44.10	$352.80	$66.25	$530.00	$47.21	$70.78
5 Common Laborers	42.10	1684.00	63.25	2530.00		
3 Equipment Operators (med.)	56.75	1362.00	84.85	2036.40		
1 Dozer, 80 H.P.		401.40		441.54		
2 Rollers, 2-Drum, W.B., 7.5 H.P.		332.20		365.42		
4 Vibrating Plates, Gas, 21"		661.40		727.54	19.38	21.31
72 L.H., Daily Totals		$4793.80		$6630.90	$66.58	$92.10
Crew B-12A	Hr.	Daily	Hr.	Daily	Bare Costs	Incl. O&P
1 Equip. Oper. (crane)	$59.20	$473.60	$88.50	$708.00	$50.65	$75.88
1 Laborer	42.10	336.80	63.25	506.00		
1 Hyd. Excavator, 1 C.Y.		795.30		874.83	49.71	54.68
16 L.H., Daily Totals		$1605.70		$2088.83	$100.36	$130.55
Crew B-12B	Hr.	Daily	Hr.	Daily	Bare Costs	Incl. O&P
1 Equip. Oper. (crane)	$59.20	$473.60	$88.50	$708.00	$50.65	$75.88
1 Laborer	42.10	336.80	63.25	506.00		
1 Hyd. Excavator, 1.5 C.Y.		687.55		756.30	42.97	47.27
16 L.H., Daily Totals		$1497.95		$1970.31	$93.62	$123.14
Crew B-12C	Hr.	Daily	Hr.	Daily	Bare Costs	Incl. O&P
1 Equip. Oper. (crane)	$59.20	$473.60	$88.50	$708.00	$50.65	$75.88
1 Laborer	42.10	336.80	63.25	506.00		
1 Hyd. Excavator, 2 C.Y.		951.30		1046.43	59.46	65.40
16 L.H., Daily Totals		$1761.70		$2260.43	$110.11	$141.28
Crew B-12D	Hr.	Daily	Hr.	Daily	Bare Costs	Incl. O&P
1 Equip. Oper. (crane)	$59.20	$473.60	$88.50	$708.00	$50.65	$75.88
1 Laborer	42.10	336.80	63.25	506.00		
1 Hyd. Excavator, 3.5 C.Y.		2158.00		2373.80	134.88	148.36
16 L.H., Daily Totals		$2968.40		$3587.80	$185.53	$224.24
Crew B-12E	Hr.	Daily	Hr.	Daily	Bare Costs	Incl. O&P
1 Equip. Oper. (crane)	$59.20	$473.60	$88.50	$708.00	$50.65	$75.88
1 Laborer	42.10	336.80	63.25	506.00		
1 Hyd. Excavator, .5 C.Y.		452.10		497.31	28.26	31.08
16 L.H., Daily Totals		$1262.50		$1711.31	$78.91	$106.96
Crew B-12F	Hr.	Daily	Hr.	Daily	Bare Costs	Incl. O&P
1 Equip. Oper. (crane)	$59.20	$473.60	$88.50	$708.00	$50.65	$75.88
1 Laborer	42.10	336.80	63.25	506.00		
1 Hyd. Excavator, .75 C.Y.		694.25		763.67	43.39	47.73
16 L.H., Daily Totals		$1504.65		$1977.68	$94.04	$123.60

Crew No.	Bare Costs		Incl. Subs O&P		Cost Per Labor-Hour	
Crew B-12G	Hr.	Daily	Hr.	Daily	Bare Costs	Incl. O&P
1 Equip. Oper. (crane)	$59.20	$473.60	$88.50	$708.00	$50.65	$75.88
1 Laborer	42.10	336.80	63.25	506.00		
1 Crawler Crane, 15 Ton		792.30		871.53		
1 Clamshell Bucket, .5 C.Y.		67.10		73.81	53.71	59.08
16 L.H., Daily Totals		$1669.80		$2159.34	$104.36	$134.96
Crew B-12H	Hr.	Daily	Hr.	Daily	Bare Costs	Incl. O&P
1 Equip. Oper. (crane)	$59.20	$473.60	$88.50	$708.00	$50.65	$75.88
1 Laborer	42.10	336.80	63.25	506.00		
1 Crawler Crane, 25 Ton		1129.00		1241.90		
1 Clamshell Bucket, 1 C.Y.		68.55		75.41	74.85	82.33
16 L.H., Daily Totals		$2007.95		$2531.30	$125.50	$158.21
Crew B-12I	Hr.	Daily	Hr.	Daily	Bare Costs	Incl. O&P
1 Equip. Oper. (crane)	$59.20	$473.60	$88.50	$708.00	$50.65	$75.88
1 Laborer	42.10	336.80	63.25	506.00		
1 Crawler Crane, 20 Ton		961.35		1057.48		
1 Dragline Bucket, .75 C.Y.		61.25		67.38	63.91	70.30
16 L.H., Daily Totals		$1833.00		$2338.86	$114.56	$146.18
Crew B-12J	Hr.	Daily	Hr.	Daily	Bare Costs	Incl. O&P
1 Equip. Oper. (crane)	$59.20	$473.60	$88.50	$708.00	$50.65	$75.88
1 Laborer	42.10	336.80	63.25	506.00		
1 Gradall, 5/8 C.Y.		846.50		931.15	52.91	58.20
16 L.H., Daily Totals		$1656.90		$2145.15	$103.56	$134.07
Crew B-12K	Hr.	Daily	Hr.	Daily	Bare Costs	Incl. O&P
1 Equip. Oper. (crane)	$59.20	$473.60	$88.50	$708.00	$50.65	$75.88
1 Laborer	42.10	336.80	63.25	506.00		
1 Gradall, 3 Ton, 1 C.Y.		973.80		1071.18	60.86	66.95
16 L.H., Daily Totals		$1784.20		$2285.18	$111.51	$142.82
Crew B-12L	Hr.	Daily	Hr.	Daily	Bare Costs	Incl. O&P
1 Equip. Oper. (crane)	$59.20	$473.60	$88.50	$708.00	$50.65	$75.88
1 Laborer	42.10	336.80	63.25	506.00		
1 Crawler Crane, 15 Ton		792.30		871.53		
1 F.E. Attachment, .5 C.Y.		65.25		71.78	53.60	58.96
16 L.H., Daily Totals		$1667.95		$2157.30	$104.25	$134.83
Crew B-12M	Hr.	Daily	Hr.	Daily	Bare Costs	Incl. O&P
1 Equip. Oper. (crane)	$59.20	$473.60	$88.50	$708.00	$50.65	$75.88
1 Laborer	42.10	336.80	63.25	506.00		
1 Crawler Crane, 20 Ton		961.35		1057.48		
1 F.E. Attachment, .75 C.Y.		70.40		77.44	64.48	70.93
16 L.H., Daily Totals		$1842.15		$2348.93	$115.13	$146.81
Crew B-12N	Hr.	Daily	Hr.	Daily	Bare Costs	Incl. O&P
1 Equip. Oper. (crane)	$59.20	$473.60	$88.50	$708.00	$50.65	$75.88
1 Laborer	42.10	336.80	63.25	506.00		
1 Crawler Crane, 25 Ton		1129.00		1241.90		
1 F.E. Attachment, 1 C.Y.		76.50		84.15	75.34	82.88
16 L.H., Daily Totals		$2015.90		$2540.05	$125.99	$158.75
Crew B-12O	Hr.	Daily	Hr.	Daily	Bare Costs	Incl. O&P
1 Equip. Oper. (crane)	$59.20	$473.60	$88.50	$708.00	$50.65	$75.88
1 Laborer	42.10	336.80	63.25	506.00		
1 Crawler Crane, 40 Ton		1206.00		1326.60		
1 F.E. Attachment, 1.5 C.Y.		87.60		96.36	80.85	88.94
16 L.H., Daily Totals		$2104.00		$2636.96	$131.50	$164.81

For customer support on your Site Work & Landscape Costs with RSMeans data, call 800.448.8182.

Crew No.	Bare Costs		Incl. Subs O&P		Cost Per Labor-Hour	

Crew B-12P	Hr.	Daily	Hr.	Daily	Bare Costs	Incl. O&P
1 Equip. Oper. (crane)	$59.20	$473.60	$88.50	$708.00	$50.65	$75.88
1 Laborer	42.10	336.80	63.25	506.00		
1 Crawler Crane, 40 Ton		1206.00		1326.60		
1 Dragline Bucket, 1.5 C.Y.		65.05		71.56	79.44	87.38
16 L.H., Daily Totals		$2081.45		$2612.16	$130.09	$163.26

Crew B-12Q	Hr.	Daily	Hr.	Daily	Bare Costs	Incl. O&P
1 Equip. Oper. (crane)	$59.20	$473.60	$88.50	$708.00	$50.65	$75.88
1 Laborer	42.10	336.80	63.25	506.00		
1 Hyd. Excavator, 5/8 C.Y.		598.25		658.08	37.39	41.13
16 L.H., Daily Totals		$1408.65		$1872.08	$88.04	$117.00

Crew B-12S	Hr.	Daily	Hr.	Daily	Bare Costs	Incl. O&P
1 Equip. Oper. (crane)	$59.20	$473.60	$88.50	$708.00	$50.65	$75.88
1 Laborer	42.10	336.80	63.25	506.00		
1 Hyd. Excavator, 2.5 C.Y.		1465.00		1611.50	91.56	100.72
16 L.H., Daily Totals		$2275.40		$2825.50	$142.21	$176.59

Crew B-12T	Hr.	Daily	Hr.	Daily	Bare Costs	Incl. O&P
1 Equip. Oper. (crane)	$59.20	$473.60	$88.50	$708.00	$50.65	$75.88
1 Laborer	42.10	336.80	63.25	506.00		
1 Crawler Crane, 75 Ton		2002.00		2202.20		
1 F.E. Attachment, 3 C.Y.		114.15		125.57	132.26	145.49
16 L.H., Daily Totals		$2926.55		$3541.76	$182.91	$221.36

Crew B-12V	Hr.	Daily	Hr.	Daily	Bare Costs	Incl. O&P
1 Equip. Oper. (crane)	$59.20	$473.60	$88.50	$708.00	$50.65	$75.88
1 Laborer	42.10	336.80	63.25	506.00		
1 Crawler Crane, 75 Ton		2002.00		2202.20		
1 Dragline Bucket, 3 C.Y.		71.60		78.76	129.60	142.56
16 L.H., Daily Totals		$2884.00		$3494.96	$180.25	$218.44

Crew B-12Y	Hr.	Daily	Hr.	Daily	Bare Costs	Incl. O&P
1 Equip. Oper. (crane)	$59.20	$473.60	$88.50	$708.00	$47.80	$71.67
2 Laborers	42.10	673.60	63.25	1012.00		
1 Hyd. Excavator, 3.5 C.Y.		2158.00		2373.80	89.92	98.91
24 L.H., Daily Totals		$3305.20		$4093.80	$137.72	$170.57

Crew B-12Z	Hr.	Daily	Hr.	Daily	Bare Costs	Incl. O&P
1 Equip. Oper. (crane)	$59.20	$473.60	$88.50	$708.00	$47.80	$71.67
2 Laborers	42.10	673.60	63.25	1012.00		
1 Hyd. Excavator, 2.5 C.Y.		1465.00		1611.50	61.04	67.15
24 L.H., Daily Totals		$2612.20		$3331.50	$108.84	$138.81

Crew B-13	Hr.	Daily	Hr.	Daily	Bare Costs	Incl. O&P
1 Labor Foreman (outside)	$44.10	$352.80	$66.25	$530.00	$46.04	$69.04
4 Laborers	42.10	1347.20	63.25	2024.00		
1 Equip. Oper. (crane)	59.20	473.60	88.50	708.00		
1 Equip. Oper. (oiler)	50.55	404.40	75.55	604.40		
1 Hyd. Crane, 25 Ton		580.85		638.93	10.37	11.41
56 L.H., Daily Totals		$3158.85		$4505.34	$56.41	$80.45

Crew B-13A	Hr.	Daily	Hr.	Daily	Bare Costs	Incl. O&P
1 Labor Foreman (outside)	$44.10	$352.80	$66.25	$530.00	$48.53	$72.74
2 Laborers	42.10	673.60	63.25	1012.00		
2 Equipment Operators (med.)	56.75	908.00	84.85	1357.60		
2 Truck Drivers (heavy)	48.95	783.20	73.35	1173.60		
1 Crawler Crane, 75 Ton		2002.00		2202.20		
1 Crawler Loader, 4 C.Y.		1489.00		1637.90		
2 Dump Trucks, 8 C.Y., 220 H.P.		856.10		941.71	77.63	85.39
56 L.H., Daily Totals		$7064.70		$8855.01	$126.16	$158.13

Crew B-13B	Hr.	Daily	Hr.	Daily	Bare Costs	Incl. O&P
1 Labor Foreman (outside)	$44.10	$352.80	$66.25	$530.00	$46.04	$69.04
4 Laborers	42.10	1347.20	63.25	2024.00		
1 Equip. Oper. (crane)	59.20	473.60	88.50	708.00		
1 Equip. Oper. (oiler)	50.55	404.40	75.55	604.40		
1 Hyd. Crane, 55 Ton		980.20		1078.22	17.50	19.25
56 L.H., Daily Totals		$3558.20		$4944.62	$63.54	$88.30

Crew B-13C	Hr.	Daily	Hr.	Daily	Bare Costs	Incl. O&P
1 Labor Foreman (outside)	$44.10	$352.80	$66.25	$530.00	$46.04	$69.04
4 Laborers	42.10	1347.20	63.25	2024.00		
1 Equip. Oper. (crane)	59.20	473.60	88.50	708.00		
1 Equip. Oper. (oiler)	50.55	404.40	75.55	604.40		
1 Crawler Crane, 100 Ton		2287.00		2515.70	40.84	44.92
56 L.H., Daily Totals		$4865.00		$6382.10	$86.88	$113.97

Crew B-13D	Hr.	Daily	Hr.	Daily	Bare Costs	Incl. O&P
1 Laborer	$42.10	$336.80	$63.25	$506.00	$50.65	$75.88
1 Equip. Oper. (crane)	59.20	473.60	88.50	708.00		
1 Hyd. Excavator, 1 C.Y.		795.30		874.83		
1 Trench Box		117.75		129.53	57.07	62.77
16 L.H., Daily Totals		$1723.45		$2218.36	$107.72	$138.65

Crew B-13E	Hr.	Daily	Hr.	Daily	Bare Costs	Incl. O&P
1 Laborer	$42.10	$336.80	$63.25	$506.00	$50.65	$75.88
1 Equip. Oper. (crane)	59.20	473.60	88.50	708.00		
1 Hyd. Excavator, 1.5 C.Y.		687.55		756.30		
1 Trench Box		117.75		129.53	50.33	55.36
16 L.H., Daily Totals		$1615.70		$2099.83	$100.98	$131.24

Crew B-13F	Hr.	Daily	Hr.	Daily	Bare Costs	Incl. O&P
1 Laborer	$42.10	$336.80	$63.25	$506.00	$50.65	$75.88
1 Equip. Oper. (crane)	59.20	473.60	88.50	708.00		
1 Hyd. Excavator, 3.5 C.Y.		2158.00		2373.80		
1 Trench Box		117.75		129.53	142.23	156.46
16 L.H., Daily Totals		$3086.15		$3717.32	$192.88	$232.33

Crew B-13G	Hr.	Daily	Hr.	Daily	Bare Costs	Incl. O&P
1 Laborer	$42.10	$336.80	$63.25	$506.00	$50.65	$75.88
1 Equip. Oper. (crane)	59.20	473.60	88.50	708.00		
1 Hyd. Excavator, .75 C.Y.		694.25		763.67		
1 Trench Box		117.75		129.53	50.75	55.83
16 L.H., Daily Totals		$1622.40		$2107.20	$101.40	$131.70

Crew B-13H	Hr.	Daily	Hr.	Daily	Bare Costs	Incl. O&P
1 Laborer	$42.10	$336.80	$63.25	$506.00	$50.65	$75.88
1 Equip. Oper. (crane)	59.20	473.60	88.50	708.00		
1 Gradall, 5/8 C.Y.		846.50		931.15		
1 Trench Box		117.75		129.53	60.27	66.29
16 L.H., Daily Totals		$1774.65		$2274.68	$110.92	$142.17

Crew No.	Bare Costs		Incl. Subs O&P		Cost Per Labor-Hour	
	Hr.	Daily	Hr.	Daily	Bare Costs	Incl. O&P
Crew B-13I						
1 Laborer	$42.10	$336.80	$63.25	$506.00	$50.65	$75.88
1 Equip. Oper. (crane)	59.20	473.60	88.50	708.00		
1 Gradall, 3 Ton, 1 C.Y.		973.80		1071.18		
1 Trench Box		117.75		129.53	68.22	75.04
16 L.H., Daily Totals		$1901.95		$2414.70	$118.87	$150.92
Crew B-13J	Hr.	Daily	Hr.	Daily	Bare Costs	Incl. O&P
1 Laborer	$42.10	$336.80	$63.25	$506.00	$50.65	$75.88
1 Equip. Oper. (crane)	59.20	473.60	88.50	708.00		
1 Hyd. Excavator, 2.5 C.Y.		1465.00		1611.50		
1 Trench Box		117.75		129.53	98.92	108.81
16 L.H., Daily Totals		$2393.15		$2955.03	$149.57	$184.69
Crew B-13K	Hr.	Daily	Hr.	Daily	Bare Costs	Incl. O&P
2 Equip. Opers. (crane)	$59.20	$947.20	$88.50	$1416.00	$59.20	$88.50
1 Hyd. Excavator, .75 C.Y.		694.25		763.67		
1 Hyd. Hammer, 4000 ft-lb		641.50		705.65		
1 Hyd. Excavator, .75 C.Y.		694.25		763.67	126.88	139.56
16 L.H., Daily Totals		$2977.20		$3649.00	$186.07	$228.06
Crew B-13L	Hr.	Daily	Hr.	Daily	Bare Costs	Incl. O&P
2 Equip. Opers. (crane)	$59.20	$947.20	$88.50	$1416.00	$59.20	$88.50
1 Hyd. Excavator, 1.5 C.Y.		687.55		756.30		
1 Hyd. Hammer, 5000 ft-lb		697.25		766.98		
1 Hyd. Excavator, .75 C.Y.		694.25		763.67	129.94	142.93
16 L.H., Daily Totals		$3026.25		$3702.95	$189.14	$231.43
Crew B-13M	Hr.	Daily	Hr.	Daily	Bare Costs	Incl. O&P
2 Equip. Opers. (crane)	$59.20	$947.20	$88.50	$1416.00	$59.20	$88.50
1 Hyd. Excavator, 2.5 C.Y.		1465.00		1611.50		
1 Hyd. Hammer, 8000 ft-lb		907.75		998.52		
1 Hyd. Excavator, 1.5 C.Y.		687.55		756.30	191.27	210.40
16 L.H., Daily Totals		$4007.50		$4782.33	$250.47	$298.90
Crew B-13N	Hr.	Daily	Hr.	Daily	Bare Costs	Incl. O&P
2 Equip. Opers. (crane)	$59.20	$947.20	$88.50	$1416.00	$59.20	$88.50
1 Hyd. Excavator, 3.5 C.Y.		2158.00		2373.80		
1 Hyd. Hammer, 12,000 ft-lb		871.75		958.92		
1 Hyd. Excavator, 1.5 C.Y.		687.55		756.30	232.33	255.56
16 L.H., Daily Totals		$4664.50		$5505.03	$291.53	$344.06
Crew B-14	Hr.	Daily	Hr.	Daily	Bare Costs	Incl. O&P
1 Labor Foreman (outside)	$44.10	$352.80	$66.25	$530.00	$44.25	$66.41
4 Laborers	42.10	1347.20	63.25	2024.00		
1 Equip. Oper. (light)	53.00	424.00	79.20	633.60		
1 Backhoe Loader, 48 H.P.		213.75		235.13	4.45	4.90
48 L.H., Daily Totals		$2337.75		$3422.72	$48.70	$71.31
Crew B-14A	Hr.	Daily	Hr.	Daily	Bare Costs	Incl. O&P
1 Equip. Oper. (crane)	$59.20	$473.60	$88.50	$708.00	$53.50	$80.08
.5 Laborer	42.10	168.40	63.25	253.00		
1 Hyd. Excavator, 4.5 C.Y.		3409.00		3749.90	284.08	312.49
12 L.H., Daily Totals		$4051.00		$4710.90	$337.58	$392.57
Crew B-14B	Hr.	Daily	Hr.	Daily	Bare Costs	Incl. O&P
1 Equip. Oper. (crane)	$59.20	$473.60	$88.50	$708.00	$53.50	$80.08
.5 Laborer	42.10	168.40	63.25	253.00		
1 Hyd. Excavator, 6 C.Y.		3464.00		3810.40	288.67	317.53
12 L.H., Daily Totals		$4106.00		$4771.40	$342.17	$397.62

Crew No.	Bare Costs		Incl. Subs O&P		Cost Per Labor-M	
	Hr.	Daily	Hr.	Daily	Bare Costs	Incl. O&
Crew B-14C						
1 Equip. Oper. (crane)	$59.20	$473.60	$88.50	$708.00	$53.50	$80.
.5 Laborer	42.10	168.40	63.25	253.00		
1 Hyd. Excavator, 7 C.Y.		3437.00		3780.70	286.42	315.
12 L.H., Daily Totals		$4079.00		$4741.70	$339.92	$395.
Crew B-14F	Hr.	Daily	Hr.	Daily	Bare Costs	Incl. O&
1 Equip. Oper. (crane)	$59.20	$473.60	$88.50	$708.00	$53.50	$80.
.5 Laborer	42.10	168.40	63.25	253.00		
1 Hyd. Shovel, 7 C.Y.		4099.00		4508.90	341.58	375.
12 L.H., Daily Totals		$4741.00		$5469.90	$395.08	$455.
Crew B-14G	Hr.	Daily	Hr.	Daily	Bare Costs	Incl. O&
1 Equip. Oper. (crane)	$59.20	$473.60	$88.50	$708.00	$53.50	$80.
.5 Laborer	42.10	168.40	63.25	253.00		
1 Hyd. Shovel, 12 C.Y.		5951.00		6546.10	495.92	545.
12 L.H., Daily Totals		$6593.00		$7507.10	$549.42	$625.
Crew B-14J	Hr.	Daily	Hr.	Daily	Bare Costs	Incl. O&
1 Equip. Oper. (medium)	$56.75	$454.00	$84.85	$678.80	$51.87	$77.
.5 Laborer	42.10	168.40	63.25	253.00		
1 F.E. Loader, 8 C.Y.		2261.00		2487.10	188.42	207.
12 L.H., Daily Totals		$2883.40		$3418.90	$240.28	$284.
Crew B-14K	Hr.	Daily	Hr.	Daily	Bare Costs	Incl. O&
1 Equip. Oper. (medium)	$56.75	$454.00	$84.85	$678.80	$51.87	$77.
.5 Laborer	42.10	168.40	63.25	253.00		
1 F.E. Loader, 10 C.Y.		2674.00		2941.40	222.83	245.
12 L.H., Daily Totals		$3296.40		$3873.20	$274.70	$322.
Crew B-15	Hr.	Daily	Hr.	Daily	Bare Costs	Incl. O&
1 Equipment Oper. (med.)	$56.75	$454.00	$84.85	$678.80	$50.20	$75.
.5 Laborer	42.10	168.40	63.25	253.00		
2 Truck Drivers (heavy)	48.95	783.20	73.35	1173.60		
2 Dump Trucks, 12 C.Y., 400 H.P.		1145.80		1260.38		
1 Dozer, 200 H.P.		1504.00		1654.40	94.64	104.
28 L.H., Daily Totals		$4055.40		$5020.18	$144.84	$179.
Crew B-16	Hr.	Daily	Hr.	Daily	Bare Costs	Incl. O&
1 Labor Foreman (outside)	$44.10	$352.80	$66.25	$530.00	$44.31	$66.
2 Laborers	42.10	673.60	63.25	1012.00		
1 Truck Driver (heavy)	48.95	391.60	73.35	586.80		
1 Dump Truck, 12 C.Y., 400 H.P.		572.90		630.19	17.90	19.6
32 L.H., Daily Totals		$1990.90		$2758.99	$62.22	$86.
Crew B-17	Hr.	Daily	Hr.	Daily	Bare Costs	Incl. O&
2 Laborers	$42.10	$673.60	$63.25	$1012.00	$46.54	$69.7
1 Equip. Oper. (light)	53.00	424.00	79.20	633.60		
1 Truck Driver (heavy)	48.95	391.60	73.35	586.80		
1 Backhoe Loader, 48 H.P.		213.75		235.13		
1 Dump Truck, 8 C.Y., 220 H.P.		428.05		470.86	20.06	22.0
32 L.H., Daily Totals		$2131.00		$2938.38	$66.59	$91.8
Crew B-17A	Hr.	Daily	Hr.	Daily	Bare Costs	Incl. O&P
2 Labor Foremen (outside)	$44.10	$705.60	$66.25	$1060.00	$45.25	$68.0
6 Laborers	42.10	2020.80	63.25	3036.00		
1 Skilled Worker Foreman (out)	56.85	454.80	86.00	688.00		
1 Skilled Worker	54.85	438.80	82.95	663.60		
80 L.H., Daily Totals		$3620.00		$5447.60	$45.25	$68.0

750

For customer support on your Site Work & Landscape Costs with RSMeans data, call 800.448.8182.

Crew No.	Bare Costs		Incl. Subs O&P		Cost Per Labor-Hour	

Crew B-17B

	Hr.	Daily	Hr.	Daily	Bare Costs	Incl. O&P
2 Laborers	$42.10	$673.60	$63.25	$1012.00	$46.54	$69.76
1 Equip. Oper. (light)	53.00	424.00	79.20	633.60		
1 Truck Driver (heavy)	48.95	391.60	73.35	586.80		
1 Backhoe Loader, 48 H.P.		213.75		235.13		
1 Dump Truck, 12 C.Y., 400 H.P.		572.90		630.19	24.58	27.04
32 L.H., Daily Totals		$2275.85		$3097.72	$71.12	$96.80

Crew B-18

	Hr.	Daily	Hr.	Daily	Bare Costs	Incl. O&P
1 Labor Foreman (outside)	$44.10	$352.80	$66.25	$530.00	$42.77	$64.25
2 Laborers	42.10	673.60	63.25	1012.00		
1 Vibrating Plate, Gas, 21"		165.35		181.88	6.89	7.58
24 L.H., Daily Totals		$1191.75		$1723.89	$49.66	$71.83

Crew B-19

	Hr.	Daily	Hr.	Daily	Bare Costs	Incl. O&P
1 Pile Driver Foreman (outside)	$56.20	$449.60	$87.30	$698.40	$55.24	$84.58
4 Pile Drivers	54.20	1734.40	84.20	2694.40		
2 Equip. Oper. (crane)	59.20	947.20	88.50	1416.00		
1 Equip. Oper. (oiler)	50.55	404.40	75.55	604.40		
1 Crawler Crane, 40 Ton		1206.00		1326.60		
1 Lead, 90' High		367.45		404.19		
1 Hammer, Diesel, 22k ft-lb		436.45		480.10	31.40	34.55
64 L.H., Daily Totals		$5545.50		$7624.09	$86.65	$119.13

Crew B-19A

	Hr.	Daily	Hr.	Daily	Bare Costs	Incl. O&P
1 Pile Driver Foreman (outside)	$56.20	$449.60	$87.30	$698.40	$55.24	$84.58
4 Pile Drivers	54.20	1734.40	84.20	2694.40		
2 Equip. Oper. (crane)	59.20	947.20	88.50	1416.00		
1 Equip. Oper. (oiler)	50.55	404.40	75.55	604.40		
1 Crawler Crane, 75 Ton		2002.00		2202.20		
1 Lead, 90' High		367.45		404.19		
1 Hammer, Diesel, 41k ft-lb		576.65		634.32	46.03	50.64
64 L.H., Daily Totals		$6481.70		$8653.91	$101.28	$135.22

Crew B-19B

	Hr.	Daily	Hr.	Daily	Bare Costs	Incl. O&P
1 Pile Driver Foreman (outside)	$56.20	$449.60	$87.30	$698.40	$55.24	$84.58
4 Pile Drivers	54.20	1734.40	84.20	2694.40		
2 Equip. Oper. (crane)	59.20	947.20	88.50	1416.00		
1 Equip. Oper. (oiler)	50.55	404.40	75.55	604.40		
1 Crawler Crane, 40 Ton		1206.00		1326.60		
1 Lead, 90' High		367.45		404.19		
1 Hammer, Diesel, 22k ft-lb		436.45		480.10		
1 Barge, 400 Ton		858.85		944.74	44.82	49.31
64 L.H., Daily Totals		$6404.35		$8568.83	$100.07	$133.89

Crew B-19C

	Hr.	Daily	Hr.	Daily	Bare Costs	Incl. O&P
1 Pile Driver Foreman (outside)	$56.20	$449.60	$87.30	$698.40	$55.24	$84.58
4 Pile Drivers	54.20	1734.40	84.20	2694.40		
2 Equip. Oper. (crane)	59.20	947.20	88.50	1416.00		
1 Equip. Oper. (oiler)	50.55	404.40	75.55	604.40		
1 Crawler Crane, 75 Ton		2002.00		2202.20		
1 Lead, 90' High		367.45		404.19		
1 Hammer, Diesel, 41k ft-lb		576.65		634.32		
1 Barge, 400 Ton		858.85		944.74	59.45	65.40
64 L.H., Daily Totals		$7340.55		$9598.65	$114.70	$149.98

Crew B-20

	Hr.	Daily	Hr.	Daily	Bare Costs	Incl. O&P
1 Labor Foreman (outside)	$44.10	$352.80	$66.25	$530.00	$47.02	$70.82
1 Skilled Worker	54.85	438.80	82.95	663.60		
1 Laborer	42.10	336.80	63.25	506.00		
24 L.H., Daily Totals		$1128.40		$1699.60	$47.02	$70.82

Crew B-20A

	Hr.	Daily	Hr.	Daily	Bare Costs	Incl. O&P
1 Labor Foreman (outside)	$44.10	$352.80	$66.25	$530.00	$50.55	$75.78
1 Laborer	42.10	336.80	63.25	506.00		
1 Plumber	64.45	515.60	96.45	771.60		
1 Plumber Apprentice	51.55	412.40	77.15	617.20		
32 L.H., Daily Totals		$1617.60		$2424.80	$50.55	$75.78

Crew B-21

	Hr.	Daily	Hr.	Daily	Bare Costs	Incl. O&P
1 Labor Foreman (outside)	$44.10	$352.80	$66.25	$530.00	$48.76	$73.34
1 Skilled Worker	54.85	438.80	82.95	663.60		
1 Laborer	42.10	336.80	63.25	506.00		
.5 Equip. Oper. (crane)	59.20	236.80	88.50	354.00		
.5 S.P. Crane, 4x4, 5 Ton		189.05		207.96	6.75	7.43
28 L.H., Daily Totals		$1554.25		$2261.55	$55.51	$80.77

Crew B-21A

	Hr.	Daily	Hr.	Daily	Bare Costs	Incl. O&P
1 Labor Foreman (outside)	$44.10	$352.80	$66.25	$530.00	$52.28	$78.32
1 Laborer	42.10	336.80	63.25	506.00		
1 Plumber	64.45	515.60	96.45	771.60		
1 Plumber Apprentice	51.55	412.40	77.15	617.20		
1 Equip. Oper. (crane)	59.20	473.60	88.50	708.00		
1 S.P. Crane, 4x4, 12 Ton		428.30		471.13	10.71	11.78
40 L.H., Daily Totals		$2519.50		$3603.93	$62.99	$90.10

Crew B-21B

	Hr.	Daily	Hr.	Daily	Bare Costs	Incl. O&P
1 Labor Foreman (outside)	$44.10	$352.80	$66.25	$530.00	$45.92	$68.90
3 Laborers	42.10	1010.40	63.25	1518.00		
1 Equip. Oper. (crane)	59.20	473.60	88.50	708.00		
1 Hyd. Crane, 12 Ton		471.00		518.10	11.78	12.95
40 L.H., Daily Totals		$2307.80		$3274.10	$57.70	$81.85

Crew B-21C

	Hr.	Daily	Hr.	Daily	Bare Costs	Incl. O&P
1 Labor Foreman (outside)	$44.10	$352.80	$66.25	$530.00	$46.04	$69.04
4 Laborers	42.10	1347.20	63.25	2024.00		
1 Equip. Oper. (crane)	59.20	473.60	88.50	708.00		
1 Equip. Oper. (oiler)	50.55	404.40	75.55	604.40		
2 Cutting Torches		25.60		28.16		
2 Sets of Gases		343.10		377.41		
1 Lattice Boom Crane, 90 Ton		1696.00		1865.60	36.87	40.56
56 L.H., Daily Totals		$4642.70		$6137.57	$82.91	$109.60

Crew B-22

	Hr.	Daily	Hr.	Daily	Bare Costs	Incl. O&P
1 Labor Foreman (outside)	$44.10	$352.80	$66.25	$530.00	$49.45	$74.35
1 Skilled Worker	54.85	438.80	82.95	663.60		
1 Laborer	42.10	336.80	63.25	506.00		
.75 Equip. Oper. (crane)	59.20	355.20	88.50	531.00		
.75 S.P. Crane, 4x4, 5 Ton		283.57		311.93	9.45	10.40
30 L.H., Daily Totals		$1767.18		$2542.53	$58.91	$84.75

Crew B-22A

	Hr.	Daily	Hr.	Daily	Bare Costs	Incl. O&P
1 Labor Foreman (outside)	$44.10	$352.80	$66.25	$530.00	$48.47	$72.84
1 Skilled Worker	54.85	438.80	82.95	663.60		
2 Laborers	42.10	673.60	63.25	1012.00		
1 Equipment Operator, Crane	59.20	473.60	88.50	708.00		
1 S.P. Crane, 4x4, 5 Ton		378.10		415.91		
1 Butt Fusion Mach., 4"-12" diam.		426.85		469.54	20.12	22.14
40 L.H., Daily Totals		$2743.75		$3799.05	$68.59	$94.98

For customer support on your Site Work & Landscape Costs with RSMeans data, call 800.448.8182.

751

Crew B-22B

Crew No.	Hr.	Daily	Hr.	Daily	Bare Costs	Incl. O&P
1 Labor Foreman (outside)	$44.10	$352.80	$66.25	$530.00	$48.47	$72.84
1 Skilled Worker	54.85	438.80	82.95	663.60		
2 Laborers	42.10	673.60	63.25	1012.00		
1 Equip. Oper. (crane)	59.20	473.60	88.50	708.00		
1 S.P. Crane, 4x4, 5 Ton		378.10		415.91		
1 Butt Fusion Mach., 8"-24" diam.		1076.00		1183.60	36.35	39.99
40 L.H., Daily Totals		$3392.90		$4513.11	$84.82	$112.83

Crew B-22C

Crew No.	Hr.	Daily	Hr.	Daily	Bare Costs	Incl. O&P
1 Skilled Worker	$54.85	$438.80	$82.95	$663.60	$48.48	$73.10
1 Laborer	42.10	336.80	63.25	506.00		
1 Butt Fusion Mach., 2"-8" diam.		156.05		171.66	9.75	10.73
16 L.H., Daily Totals		$931.65		$1341.26	$58.23	$83.83

Crew B-23

Crew No.	Hr.	Daily	Hr.	Daily	Bare Costs	Incl. O&P
1 Labor Foreman (outside)	$44.10	$352.80	$66.25	$530.00	$42.50	$63.85
4 Laborers	42.10	1347.20	63.25	2024.00		
1 Drill Rig, Truck-Mounted		771.10		848.21		
1 Flatbed Truck, Gas, 3 Ton		820.40		902.44	39.79	43.77
40 L.H., Daily Totals		$3291.50		$4304.65	$82.29	$107.62

Crew B-23A

Crew No.	Hr.	Daily	Hr.	Daily	Bare Costs	Incl. O&P
1 Labor Foreman (outside)	$44.10	$352.80	$66.25	$530.00	$47.65	$71.45
1 Laborer	42.10	336.80	63.25	506.00		
1 Equip. Oper. (medium)	56.75	454.00	84.85	678.80		
1 Drill Rig, Truck-Mounted		771.10		848.21		
1 Pickup Truck, 3/4 Ton		110.85		121.94	36.75	40.42
24 L.H., Daily Totals		$2025.55		$2684.95	$84.40	$111.87

Crew B-23B

Crew No.	Hr.	Daily	Hr.	Daily	Bare Costs	Incl. O&P
1 Labor Foreman (outside)	$44.10	$352.80	$66.25	$530.00	$47.65	$71.45
1 Laborer	42.10	336.80	63.25	506.00		
1 Equip. Oper. (medium)	56.75	454.00	84.85	678.80		
1 Drill Rig, Truck-Mounted		771.10		848.21		
1 Pickup Truck, 3/4 Ton		110.85		121.94		
1 Centr. Water Pump, 6"		232.90		256.19	46.45	51.10
24 L.H., Daily Totals		$2258.45		$2941.14	$94.10	$122.55

Crew B-24

Crew No.	Hr.	Daily	Hr.	Daily	Bare Costs	Incl. O&P
1 Cement Finisher	$49.95	$399.60	$73.45	$587.60	$48.40	$72.18
1 Laborer	42.10	336.80	63.25	506.00		
1 Carpenter	53.15	425.20	79.85	638.80		
24 L.H., Daily Totals		$1161.60		$1732.40	$48.40	$72.18

Crew B-25

Crew No.	Hr.	Daily	Hr.	Daily	Bare Costs	Incl. O&P
1 Labor Foreman (outside)	$44.10	$352.80	$66.25	$530.00	$46.28	$69.41
7 Laborers	42.10	2357.60	63.25	3542.00		
3 Equip. Oper. (medium)	56.75	1362.00	84.85	2036.40		
1 Asphalt Paver, 130 H.P.		2117.00		2328.70		
1 Tandem Roller, 10 Ton		236.75		260.43		
1 Roller, Pneum. Whl., 12 Ton		345.75		380.32	30.68	33.74
88 L.H., Daily Totals		$6771.90		$9077.85	$76.95	$103.16

Crew B-25B

Crew No.	Hr.	Daily	Hr.	Daily	Bare Costs	Incl. O&P
1 Labor Foreman (outside)	$44.10	$352.80	$66.25	$530.00	$47.15	$70.70
7 Laborers	42.10	2357.60	63.25	3542.00		
4 Equip. Oper. (medium)	56.75	1816.00	84.85	2715.20		
1 Asphalt Paver, 130 H.P.		2117.00		2328.70		
2 Tandem Rollers, 10 Ton		473.50		520.85		
1 Roller, Pneum. Whl., 12 Ton		345.75		380.32	30.59	33.64
96 L.H., Daily Totals		$7462.65		$10017.08	$77.74	$104.34

Crew B-25C

Crew No.	Hr.	Daily	Hr.	Daily	Bare Costs	Incl. O&P
1 Labor Foreman (outside)	$44.10	$352.80	$66.25	$530.00	$47.32	$70.95
3 Laborers	42.10	1010.40	63.25	1518.00		
2 Equip. Oper. (medium)	56.75	908.00	84.85	1357.60		
1 Asphalt Paver, 130 H.P.		2117.00		2328.70		
1 Tandem Roller, 10 Ton		236.75		260.43	49.04	53.94
48 L.H., Daily Totals		$4624.95		$5994.73	$96.35	$124.89

Crew B-25D

Crew No.	Hr.	Daily	Hr.	Daily	Bare Costs	Incl. O&P
1 Labor Foreman (outside)	$44.10	$352.80	$66.25	$530.00	$47.54	$71.28
3 Laborers	42.10	1010.40	63.25	1518.00		
2.125 Equip. Oper. (medium)	56.75	964.75	84.85	1442.45		
.125 Truck Driver (heavy)	48.95	48.95	73.35	73.35		
.125 Truck Tractor, 6x4, 380 H.P.		61.66		67.82		
.125 Dist. Tanker, 3000 Gallon		41.27		45.40		
1 Asphalt Paver, 130 H.P.		2117.00		2328.70		
1 Tandem Roller, 10 Ton		236.75		260.43	49.13	54.05
50 L.H., Daily Totals		$4833.58		$6266.14	$96.67	$125.32

Crew B-25E

Crew No.	Hr.	Daily	Hr.	Daily	Bare Costs	Incl. O&P
1 Labor Foreman (outside)	$44.10	$352.80	$66.25	$530.00	$47.74	$71.58
3 Laborers	42.10	1010.40	63.25	1518.00		
2.250 Equip. Oper. (medium)	56.75	1021.50	84.85	1527.30		
.25 Truck Driver (heavy)	48.95	97.90	73.35	146.70		
.25 Truck Tractor, 6x4, 380 H.P.		123.31		135.64		
.25 Dist. Tanker, 3000 Gallon		82.54		90.79		
1 Asphalt Paver, 130 H.P.		2117.00		2328.70		
1 Tandem Roller, 10 Ton		236.75		260.43	49.22	54.15
52 L.H., Daily Totals		$5042.20		$6537.56	$96.97	$125.72

Crew B-26

Crew No.	Hr.	Daily	Hr.	Daily	Bare Costs	Incl. O&P
1 Labor Foreman (outside)	$44.10	$352.80	$66.25	$530.00	$46.96	$70.35
6 Laborers	42.10	2020.80	63.25	3036.00		
2 Equip. Oper. (medium)	56.75	908.00	84.85	1357.60		
1 Rodman (reinf.)	56.40	451.20	84.90	679.20		
1 Cement Finisher	49.95	399.60	73.45	587.60		
1 Grader, 30,000 Lbs.		1062.00		1168.20		
1 Paving Mach. & Equip.		2474.00		2721.40	40.18	44.20
88 L.H., Daily Totals		$7668.40		$10080.00	$87.14	$114.55

Crew B-26A

Crew No.	Hr.	Daily	Hr.	Daily	Bare Costs	Incl. O&P
1 Labor Foreman (outside)	$44.10	$352.80	$66.25	$530.00	$46.96	$70.35
6 Laborers	42.10	2020.80	63.25	3036.00		
2 Equip. Oper. (medium)	56.75	908.00	84.85	1357.60		
1 Rodman (reinf.)	56.40	451.20	84.90	679.20		
1 Cement Finisher	49.95	399.60	73.45	587.60		
1 Grader, 30,000 Lbs.		1062.00		1168.20		
1 Paving Mach. & Equip.		2474.00		2721.40		
1 Concrete Saw		111.50		122.65	41.45	45.59
88 L.H., Daily Totals		$7779.90		$10202.65	$88.41	$115.94

Crew B-26B

Crew No.	Hr.	Daily	Hr.	Daily	Bare Costs	Incl. O&P
1 Labor Foreman (outside)	$44.10	$352.80	$66.25	$530.00	$47.77	$71.55
6 Laborers	42.10	2020.80	63.25	3036.00		
3 Equip. Oper. (medium)	56.75	1362.00	84.85	2036.40		
1 Rodman (reinf.)	56.40	451.20	84.90	679.20		
1 Cement Finisher	49.95	399.60	73.45	587.60		
1 Grader, 30,000 Lbs.		1062.00		1168.20		
1 Paving Mach. & Equip.		2474.00		2721.40		
1 Concrete Pump, 110' Boom		470.75		517.83	41.74	45.91
96 L.H., Daily Totals		$8593.15		$11276.63	$89.51	$117.46

Crews - Standard

Crew No.	Bare Costs		Incl. Subs O&P		Cost Per Labor-Hour	

Crew B-26C	Hr.	Daily	Hr.	Daily	Bare Costs	Incl. O&P
1 Labor Foreman (outside)	$44.10	$352.80	$66.25	$530.00	$45.98	$68.89
6 Laborers	42.10	2020.80	63.25	3036.00		
1 Equip. Oper. (medium)	56.75	454.00	84.85	678.80		
1 Rodman (reinf.)	56.40	451.20	84.90	679.20		
1 Cement Finisher	49.95	399.60	73.45	587.60		
1 Paving Mach. & Equip.		2474.00		2721.40		
1 Concrete Saw		111.50		122.65	32.32	35.55
80 L.H., Daily Totals		$6263.90		$8355.65	$78.30	$104.45

Crew B-27	Hr.	Daily	Hr.	Daily	Bare Costs	Incl. O&P
1 Labor Foreman (outside)	$44.10	$352.80	$66.25	$530.00	$42.60	$64.00
3 Laborers	42.10	1010.40	63.25	1518.00		
1 Berm Machine		271.95		299.14	8.50	9.35
32 L.H., Daily Totals		$1635.15		$2347.15	$51.10	$73.35

Crew B-28	Hr.	Daily	Hr.	Daily	Bare Costs	Incl. O&P
2 Carpenters	$53.15	$850.40	$79.85	$1277.60	$49.47	$74.32
1 Laborer	42.10	336.80	63.25	506.00		
24 L.H., Daily Totals		$1187.20		$1783.60	$49.47	$74.32

Crew B-29	Hr.	Daily	Hr.	Daily	Bare Costs	Incl. O&P
1 Labor Foreman (outside)	$44.10	$352.80	$66.25	$530.00	$46.04	$69.04
4 Laborers	42.10	1347.20	63.25	2024.00		
1 Equip. Oper. (crane)	59.20	473.60	88.50	708.00		
1 Equip. Oper. (oiler)	50.55	404.40	75.55	604.40		
1 Gradall, 5/8 C.Y.		846.50		931.15	15.12	16.63
56 L.H., Daily Totals		$3424.50		$4797.55	$61.15	$85.67

Crew B-30	Hr.	Daily	Hr.	Daily	Bare Costs	Incl. O&P
1 Equip. Oper. (medium)	$56.75	$454.00	$84.85	$678.80	$51.55	$77.18
2 Truck Drivers (heavy)	48.95	783.20	73.35	1173.60		
1 Hyd. Excavator, 1.5 C.Y.		687.55		756.30		
2 Dump Trucks, 12 C.Y., 400 H.P.		1145.80		1260.38	76.39	84.03
24 L.H., Daily Totals		$3070.55		$3869.09	$127.94	$161.21

Crew B-31	Hr.	Daily	Hr.	Daily	Bare Costs	Incl. O&P
1 Labor Foreman (outside)	$44.10	$352.80	$66.25	$530.00	$44.71	$67.17
3 Laborers	42.10	1010.40	63.25	1518.00		
1 Carpenter	53.15	425.20	79.85	638.80		
1 Air Compressor, 250 cfm		201.50		221.65		
1 Sheeting Driver		7.35		8.09		
2 -50' Air Hoses, 1.5"		48.50		53.35	6.43	7.08
40 L.H., Daily Totals		$2045.75		$2969.89	$51.14	$74.25

Crew B-32	Hr.	Daily	Hr.	Daily	Bare Costs	Incl. O&P
1 Laborer	$42.10	$336.80	$63.25	$506.00	$53.09	$79.45
3 Equip. Oper. (medium)	56.75	1362.00	84.85	2036.40		
1 Grader, 30,000 Lbs.		1062.00		1168.20		
1 Tandem Roller, 10 Ton		236.75		260.43		
1 Dozer, 200 H.P.		1504.00		1654.40	87.59	96.34
32 L.H., Daily Totals		$4501.55		$5625.43	$140.67	$175.79

Crew B-32A	Hr.	Daily	Hr.	Daily	Bare Costs	Incl. O&P
1 Laborer	$42.10	$336.80	$63.25	$506.00	$51.87	$77.65
2 Equip. Oper. (medium)	56.75	908.00	84.85	1357.60		
1 Grader, 30,000 Lbs.		1062.00		1168.20		
1 Roller, Vibratory, 25 Ton		664.35		730.78	71.93	79.12
24 L.H., Daily Totals		$2971.15		$3762.59	$123.80	$156.77

Crew B-32B	Hr.	Daily	Hr.	Daily	Bare Costs	Incl. O&P
1 Laborer	$42.10	$336.80	$63.25	$506.00	$51.87	$77.65
2 Equip. Oper. (medium)	56.75	908.00	84.85	1357.60		
1 Dozer, 200 H.P.		1504.00		1654.40		
1 Roller, Vibratory, 25 Ton		664.35		730.78	90.35	99.38
24 L.H., Daily Totals		$3413.15		$4248.78	$142.21	$177.03

Crew B-32C	Hr.	Daily	Hr.	Daily	Bare Costs	Incl. O&P
1 Labor Foreman (outside)	$44.10	$352.80	$66.25	$530.00	$49.76	$74.55
2 Laborers	42.10	673.60	63.25	1012.00		
3 Equip. Oper. (medium)	56.75	1362.00	84.85	2036.40		
1 Grader, 30,000 Lbs.		1062.00		1168.20		
1 Tandem Roller, 10 Ton		236.75		260.43		
1 Dozer, 200 H.P.		1504.00		1654.40	58.39	64.23
48 L.H., Daily Totals		$5191.15		$6661.43	$108.15	$138.78

Crew B-33A	Hr.	Daily	Hr.	Daily	Bare Costs	Incl. O&P
1 Equip. Oper. (medium)	$56.75	$454.00	$84.85	$678.80	$52.56	$78.68
.5 Laborer	42.10	168.40	63.25	253.00		
.25 Equip. Oper. (medium)	56.75	113.50	84.85	169.70		
1 Scraper, Towed, 7 C.Y.		127.80		140.58		
1.25 Dozers, 300 H.P.		2205.00		2425.50	166.63	183.29
14 L.H., Daily Totals		$3068.70		$3667.58	$219.19	$261.97

Crew B-33B	Hr.	Daily	Hr.	Daily	Bare Costs	Incl. O&P
1 Equip. Oper. (medium)	$56.75	$454.00	$84.85	$678.80	$52.56	$78.68
.5 Laborer	42.10	168.40	63.25	253.00		
.25 Equip. Oper. (medium)	56.75	113.50	84.85	169.70		
1 Scraper, Towed, 10 C.Y.		159.70		175.67		
1.25 Dozers, 300 H.P.		2205.00		2425.50	168.91	185.80
14 L.H., Daily Totals		$3100.60		$3702.67	$221.47	$264.48

Crew B-33C	Hr.	Daily	Hr.	Daily	Bare Costs	Incl. O&P
1 Equip. Oper. (medium)	$56.75	$454.00	$84.85	$678.80	$52.56	$78.68
.5 Laborer	42.10	168.40	63.25	253.00		
.25 Equip. Oper. (medium)	56.75	113.50	84.85	169.70		
1 Scraper, Towed, 15 C.Y.		176.75		194.43		
1.25 Dozers, 300 H.P.		2205.00		2425.50	170.13	187.14
14 L.H., Daily Totals		$3117.65		$3721.43	$222.69	$265.82

Crew B-33D	Hr.	Daily	Hr.	Daily	Bare Costs	Incl. O&P
1 Equip. Oper. (medium)	$56.75	$454.00	$84.85	$678.80	$52.56	$78.68
.5 Laborer	42.10	168.40	63.25	253.00		
.25 Equip. Oper. (medium)	56.75	113.50	84.85	169.70		
1 S.P. Scraper, 14 C.Y.		2395.00		2634.50		
.25 Dozer, 300 H.P.		441.00		485.10	202.57	222.83
14 L.H., Daily Totals		$3571.90		$4221.10	$255.14	$301.51

Crew B-33E	Hr.	Daily	Hr.	Daily	Bare Costs	Incl. O&P
1 Equip. Oper. (medium)	$56.75	$454.00	$84.85	$678.80	$52.56	$78.68
.5 Laborer	42.10	168.40	63.25	253.00		
.25 Equip. Oper. (medium)	56.75	113.50	84.85	169.70		
1 S.P. Scraper, 21 C.Y.		2628.00		2890.80		
.25 Dozer, 300 H.P.		441.00		485.10	219.21	241.14
14 L.H., Daily Totals		$3804.90		$4477.40	$271.78	$319.81

For customer support on your Site Work & Landscape Costs with RSMeans data, call 800.448.8182.

753

Crew B-33F

Crew B-33F	Hr.	Daily	Hr.	Daily	Bare Costs	Incl. O&P
1 Equip. Oper. (medium)	$56.75	$454.00	$84.85	$678.80	$52.56	$78.68
.5 Laborer	42.10	168.40	63.25	253.00		
.25 Equip. Oper. (medium)	56.75	113.50	84.85	169.70		
1 Elev. Scraper, 11 C.Y.		1133.00		1246.30		
.25 Dozer, 300 H.P.		441.00		485.10	112.43	123.67
14 L.H., Daily Totals		$2309.90		$2832.90	$164.99	$202.35

Crew B-33G

Crew B-33G	Hr.	Daily	Hr.	Daily	Bare Costs	Incl. O&P
1 Equip. Oper. (medium)	$56.75	$454.00	$84.85	$678.80	$52.56	$78.68
.5 Laborer	42.10	168.40	63.25	253.00		
.25 Equip. Oper. (medium)	56.75	113.50	84.85	169.70		
1 Elev. Scraper, 22 C.Y.		1884.00		2072.40		
.25 Dozer, 300 H.P.		441.00		485.10	166.07	182.68
14 L.H., Daily Totals		$3060.90		$3659.00	$218.64	$261.36

Crew B-33H

Crew B-33H	Hr.	Daily	Hr.	Daily	Bare Costs	Incl. O&P
.5 Laborer	$42.10	$168.40	$63.25	$253.00	$52.56	$78.68
1 Equipment Operator (med.)	56.75	454.00	84.85	678.80		
.25 Equipment Operator (med.)	56.75	113.50	84.85	169.70		
1 S.P. Scraper, 44 C.Y.		4644.00		5108.40		
.25 Dozer, 410 H.P.		694.25		763.67	381.30	419.43
14 L.H., Daily Totals		$6074.15		$6973.57	$433.87	$498.11

Crew B-33J

Crew B-33J	Hr.	Daily	Hr.	Daily	Bare Costs	Incl. O&P
1 Equipment Operator (med.)	$56.75	$454.00	$84.85	$678.80	$56.75	$84.85
1 S.P. Scraper, 14 C.Y.		2395.00		2634.50	299.38	329.31
8 L.H., Daily Totals		$2849.00		$3313.30	$356.13	$414.16

Crew B-33K

Crew B-33K	Hr.	Daily	Hr.	Daily	Bare Costs	Incl. O&P
1 Equipment Operator (med.)	$56.75	$454.00	$84.85	$678.80	$52.56	$78.68
.25 Equipment Operator (med.)	56.75	113.50	84.85	169.70		
.5 Laborer	42.10	168.40	63.25	253.00		
1 S.P. Scraper, 31 C.Y.		3667.00		4033.70		
.25 Dozer, 410 H.P.		694.25		763.67	311.52	342.67
14 L.H., Daily Totals		$5097.15		$5898.88	$364.08	$421.35

Crew B-34A

Crew B-34A	Hr.	Daily	Hr.	Daily	Bare Costs	Incl. O&P
1 Truck Driver (heavy)	$48.95	$391.60	$73.35	$586.80	$48.95	$73.35
1 Dump Truck, 8 C.Y., 220 H.P.		428.05		470.86	53.51	58.86
8 L.H., Daily Totals		$819.65		$1057.66	$102.46	$132.21

Crew B-34B

Crew B-34B	Hr.	Daily	Hr.	Daily	Bare Costs	Incl. O&P
1 Truck Driver (heavy)	$48.95	$391.60	$73.35	$586.80	$48.95	$73.35
1 Dump Truck, 12 C.Y., 400 H.P.		572.90		630.19	71.61	78.77
8 L.H., Daily Totals		$964.50		$1216.99	$120.56	$152.12

Crew B-34C

Crew B-34C	Hr.	Daily	Hr.	Daily	Bare Costs	Incl. O&P
1 Truck Driver (heavy)	$48.95	$391.60	$73.35	$586.80	$48.95	$73.35
1 Truck Tractor, 6x4, 380 H.P.		493.25		542.58		
1 Dump Trailer, 16.5 C.Y.		136.70		150.37	78.74	86.62
8 L.H., Daily Totals		$1021.55		$1279.74	$127.69	$159.97

Crew B-34D

Crew B-34D	Hr.	Daily	Hr.	Daily	Bare Costs	Incl. O&P
1 Truck Driver (heavy)	$48.95	$391.60	$73.35	$586.80	$48.95	$73.35
1 Truck Tractor, 6x4, 380 H.P.		493.25		542.58		
1 Dump Trailer, 20 C.Y.		151.70		166.87	80.62	88.68
8 L.H., Daily Totals		$1036.55		$1296.24	$129.57	$162.03

Crew B-34E

Crew B-34E	Hr.	Daily	Hr.	Daily	Bare Costs	Incl. O&P
1 Truck Driver (heavy)	$48.95	$391.60	$73.35	$586.80	$48.95	$73.35
1 Dump Truck, Off Hwy., 25 Ton		1379.00		1516.90	172.38	189.61
8 L.H., Daily Totals		$1770.60		$2103.70	$221.32	$262.96

Crew B-34F

Crew B-34F	Hr.	Daily	Hr.	Daily	Bare Costs	Incl. O&P
1 Truck Driver (heavy)	$48.95	$391.60	$73.35	$586.80	$48.95	$73.35
1 Dump Truck, Off Hwy., 35 Ton		935.20		1028.72	116.90	128.59
8 L.H., Daily Totals		$1326.80		$1615.52	$165.85	$201.94

Crew B-34G

Crew B-34G	Hr.	Daily	Hr.	Daily	Bare Costs	Incl. O&P
1 Truck Driver (heavy)	$48.95	$391.60	$73.35	$586.80	$48.95	$73.35
1 Dump Truck, Off Hwy., 50 Ton		1771.00		1948.10	221.38	243.51
8 L.H., Daily Totals		$2162.60		$2534.90	$270.32	$316.86

Crew B-34H

Crew B-34H	Hr.	Daily	Hr.	Daily	Bare Costs	Incl. O&P
1 Truck Driver (heavy)	$48.95	$391.60	$73.35	$586.80	$48.95	$73.35
1 Dump Truck, Off Hwy., 65 Ton		1915.00		2106.50	239.38	263.31
8 L.H., Daily Totals		$2306.60		$2693.30	$288.32	$336.66

Crew B-34I

Crew B-34I	Hr.	Daily	Hr.	Daily	Bare Costs	Incl. O&P
1 Truck Driver (heavy)	$48.95	$391.60	$73.35	$586.80	$48.95	$73.35
1 Dump Truck, 18 C.Y., 450 H.P.		713.80		785.18	89.22	98.15
8 L.H., Daily Totals		$1105.40		$1371.98	$138.18	$171.50

Crew B-34J

Crew B-34J	Hr.	Daily	Hr.	Daily	Bare Costs	Incl. O&P
1 Truck Driver (heavy)	$48.95	$391.60	$73.35	$586.80	$48.95	$73.35
1 Dump Truck, Off Hwy., 100 Ton		2736.00		3009.60	342.00	376.20
8 L.H., Daily Totals		$3127.60		$3596.40	$390.95	$449.55

Crew B-34K

Crew B-34K	Hr.	Daily	Hr.	Daily	Bare Costs	Incl. O&P
1 Truck Driver (heavy)	$48.95	$391.60	$73.35	$586.80	$48.95	$73.35
1 Truck Tractor, 6x4, 450 H.P.		601.75		661.92		
1 Lowbed Trailer, 75 Ton		255.05		280.56	107.10	117.81
8 L.H., Daily Totals		$1248.40		$1529.28	$156.05	$191.16

Crew B-34L

Crew B-34L	Hr.	Daily	Hr.	Daily	Bare Costs	Incl. O&P
1 Equip. Oper. (light)	$53.00	$424.00	$79.20	$633.60	$53.00	$79.20
1 Flatbed Truck, Gas, 1.5 Ton		196.15		215.76	24.52	26.97
8 L.H., Daily Totals		$620.15		$849.37	$77.52	$106.17

Crew B-34M

Crew B-34M	Hr.	Daily	Hr.	Daily	Bare Costs	Incl. O&P
1 Equip. Oper. (light)	$53.00	$424.00	$79.20	$633.60	$53.00	$79.20
1 Flatbed Truck, Gas, 3 Ton		820.40		902.44	102.55	112.81
8 L.H., Daily Totals		$1244.40		$1536.04	$155.55	$192.01

Crew B-34N

Crew B-34N	Hr.	Daily	Hr.	Daily	Bare Costs	Incl. O&P
1 Truck Driver (heavy)	$48.95	$391.60	$73.35	$586.80	$52.85	$79.10
1 Equip. Oper. (medium)	56.75	454.00	84.85	678.80		
1 Truck Tractor, 6x4, 380 H.P.		493.25		542.58		
1 Flatbed Trailer, 40 Ton		186.20		204.82	42.47	46.71
16 L.H., Daily Totals		$1525.05		$2012.99	$95.32	$125.81

Crew B-34P

Crew B-34P	Hr.	Daily	Hr.	Daily	Bare Costs	Incl. O&P
1 Pipe Fitter	$65.55	$524.40	$98.10	$784.80	$56.52	$84.58
1 Truck Driver (light)	47.25	378.00	70.80	566.40		
1 Equip. Oper. (medium)	56.75	454.00	84.85	678.80		
1 Flatbed Truck, Gas, 3 Ton		820.40		902.44		
1 Backhoe Loader, 48 H.P.		213.75		235.13	43.09	47.40
24 L.H., Daily Totals		$2390.55		$3167.57	$99.61	$131.98

754

For customer support on your Site Work & Landscape Costs with RSMeans data, call 800.448.8182.

Crew No.	Bare Costs		Incl. Subs O&P		Cost Per Labor-Hour	
Crew B-34Q	Hr.	Daily	Hr.	Daily	Bare Costs	Incl. O&P
1 Pipe Fitter	$65.55	$524.40	$98.10	$784.80	$57.33	$85.80
1 Truck Driver (light)	47.25	378.00	70.80	566.40		
1 Equip. Oper. (crane)	59.20	473.60	88.50	708.00		
1 Flatbed Trailer, 25 Ton		135.55		149.10		
1 Dump Truck, 8 C.Y., 220 H.P.		428.05		470.86		
1 Hyd. Crane, 25 Ton		580.85		638.93	47.69	52.45
24 L.H., Daily Totals		$2520.45		$3318.09	$105.02	$138.25

Crew No.	Bare Costs		Incl. Subs O&P		Cost Per Labor-Hour	
Crew B-34R	Hr.	Daily	Hr.	Daily	Bare Costs	Incl. O&P
1 Pipe Fitter	$65.55	$524.40	$98.10	$784.80	$57.33	$85.80
1 Truck Driver (light)	47.25	378.00	70.80	566.40		
1 Equip. Oper. (crane)	59.20	473.60	88.50	708.00		
1 Flatbed Trailer, 25 Ton		135.55		149.10		
1 Dump Truck, 8 C.Y., 220 H.P.		428.05		470.86		
1 Hyd. Crane, 25 Ton		580.85		638.93		
1 Hyd. Excavator, 1 C.Y.		795.30		874.83	80.82	88.91
24 L.H., Daily Totals		$3315.75		$4192.93	$138.16	$174.71

Crew No.	Bare Costs		Incl. Subs O&P		Cost Per Labor-Hour	
Crew B-34S	Hr.	Daily	Hr.	Daily	Bare Costs	Incl. O&P
2 Pipe Fitters	$65.55	$1048.80	$98.10	$1569.60	$59.81	$89.51
1 Truck Driver (heavy)	48.95	391.60	73.35	586.80		
1 Equip. Oper. (crane)	59.20	473.60	88.50	708.00		
1 Flatbed Trailer, 40 Ton		186.20		204.82		
1 Truck Tractor, 6x4, 380 H.P.		493.25		542.58		
1 Hyd. Crane, 80 Ton		1486.00		1634.60		
1 Hyd. Excavator, 2 C.Y.		951.30		1046.43	97.40	107.14
32 L.H., Daily Totals		$5030.75		$6292.82	$157.21	$196.65

Crew No.	Bare Costs		Incl. Subs O&P		Cost Per Labor-Hour	
Crew B-34T	Hr.	Daily	Hr.	Daily	Bare Costs	Incl. O&P
2 Pipe Fitters	$65.55	$1048.80	$98.10	$1569.60	$59.81	$89.51
1 Truck Driver (heavy)	48.95	391.60	73.35	586.80		
1 Equip. Oper. (crane)	59.20	473.60	88.50	708.00		
1 Flatbed Trailer, 40 Ton		186.20		204.82		
1 Truck Tractor, 6x4, 380 H.P.		493.25		542.58		
1 Hyd. Crane, 80 Ton		1486.00		1634.60	67.67	74.44
32 L.H., Daily Totals		$4079.45		$5246.40	$127.48	$163.95

Crew No.	Bare Costs		Incl. Subs O&P		Cost Per Labor-Hour	
Crew B-34U	Hr.	Daily	Hr.	Daily	Bare Costs	Incl. O&P
1 Truck Driver (heavy)	$48.95	$391.60	$73.35	$586.80	$50.98	$76.28
1 Equip. Oper. (light)	53.00	424.00	79.20	633.60		
1 Truck Tractor, 220 H.P.		307.10		337.81		
1 Flatbed Trailer, 25 Ton		135.55		149.10	27.67	30.43
16 L.H., Daily Totals		$1258.25		$1707.32	$78.64	$106.71

Crew No.	Bare Costs		Incl. Subs O&P		Cost Per Labor-Hour	
Crew B-34V	Hr.	Daily	Hr.	Daily	Bare Costs	Incl. O&P
1 Truck Driver (heavy)	$48.95	$391.60	$73.35	$586.80	$53.72	$80.35
1 Equip. Oper. (crane)	59.20	473.60	88.50	708.00		
1 Equip. Oper. (light)	53.00	424.00	79.20	633.60		
1 Truck Tractor, 6x4, 450 H.P.		601.75		661.92		
1 Equipment Trailer, 50 Ton		205.05		225.56		
1 Pickup Truck, 4x4, 3/4 Ton		175.85		193.44	40.94	45.04
24 L.H., Daily Totals		$2271.85		$3009.32	$94.66	$125.39

Crew No.	Bare Costs		Incl. Subs O&P		Cost Per Labor-Hour	
Crew B-34W	Hr.	Daily	Hr.	Daily	Bare Costs	Incl. O&P
5 Truck Drivers (heavy)	$48.95	$1958.00	$73.35	$2934.00	$51.60	$77.27
2 Equip. Opers. (crane)	59.20	947.20	88.50	1416.00		
1 Equip. Oper. (mechanic)	59.15	473.20	88.40	707.20		
1 Laborer	42.10	336.80	63.25	506.00		
4 Truck Tractors, 6x4, 380 H.P.		1973.00		2170.30		
2 Equipment Trailers, 50 Ton		410.10		451.11		
2 Flatbed Trailers, 40 Ton		372.40		409.64		
1 Pickup Truck, 4x4, 3/4 Ton		175.85		193.44		
1 S.P. Crane, 4x4, 20 Ton		582.20		640.42	48.80	53.68
72 L.H., Daily Totals		$7228.75		$9428.10	$100.40	$130.95

Crew No.	Bare Costs		Incl. Subs O&P		Cost Per Labor-Hour	
Crew B-35	Hr.	Daily	Hr.	Daily	Bare Costs	Incl. O&P
1 Labor Foreman (outside)	$44.10	$352.80	$66.25	$530.00	$52.54	$78.83
1 Skilled Worker	54.85	438.80	82.95	663.60		
1 Welder (plumber)	64.45	515.60	96.45	771.60		
1 Laborer	42.10	336.80	63.25	506.00		
1 Equip. Oper. (crane)	59.20	473.60	88.50	708.00		
1 Equip. Oper. (oiler)	50.55	404.40	75.55	604.40		
1 Welder, Electric, 300 amp		106.40		117.04		
1 Hyd. Excavator, .75 C.Y.		694.25		763.67	16.68	18.35
48 L.H., Daily Totals		$3322.65		$4664.31	$69.22	$97.17

Crew No.	Bare Costs		Incl. Subs O&P		Cost Per Labor-Hour	
Crew B-35A	Hr.	Daily	Hr.	Daily	Bare Costs	Incl. O&P
1 Labor Foreman (outside)	$44.10	$352.80	$66.25	$530.00	$51.05	$76.60
2 Laborers	42.10	673.60	63.25	1012.00		
1 Skilled Worker	54.85	438.80	82.95	663.60		
1 Welder (plumber)	64.45	515.60	96.45	771.60		
1 Equip. Oper. (crane)	59.20	473.60	88.50	708.00		
1 Equip. Oper. (oiler)	50.55	404.40	75.55	604.40		
1 Welder, Gas Engine, 300 amp		147.00		161.70		
1 Crawler Crane, 75 Ton		2002.00		2202.20	38.38	42.21
56 L.H., Daily Totals		$5007.80		$6653.50	$89.42	$118.81

Crew No.	Bare Costs		Incl. Subs O&P		Cost Per Labor-Hour	
Crew B-36	Hr.	Daily	Hr.	Daily	Bare Costs	Incl. O&P
1 Labor Foreman (outside)	$44.10	$352.80	$66.25	$530.00	$48.36	$72.49
2 Laborers	42.10	673.60	63.25	1012.00		
2 Equip. Oper. (medium)	56.75	908.00	84.85	1357.60		
1 Dozer, 200 H.P.		1504.00		1654.40		
1 Aggregate Spreader		65.75		72.33		
1 Tandem Roller, 10 Ton		236.75		260.43	45.16	49.68
40 L.H., Daily Totals		$3740.90		$4886.75	$93.52	$122.17

Crew No.	Bare Costs		Incl. Subs O&P		Cost Per Labor-Hour	
Crew B-36A	Hr.	Daily	Hr.	Daily	Bare Costs	Incl. O&P
1 Labor Foreman (outside)	$44.10	$352.80	$66.25	$530.00	$50.76	$76.02
2 Laborers	42.10	673.60	63.25	1012.00		
4 Equip. Oper. (medium)	56.75	1816.00	84.85	2715.20		
1 Dozer, 200 H.P.		1504.00		1654.40		
1 Aggregate Spreader		65.75		72.33		
1 Tandem Roller, 10 Ton		236.75		260.43		
1 Roller, Pneum. Whl., 12 Ton		345.75		380.32	38.43	42.28
56 L.H., Daily Totals		$4994.65		$6624.68	$89.19	$118.30

For customer support on your Site Work & Landscape Costs with RSMeans data, call 800.448.8182.

755

Crew No.	Bare Costs		Incl. Subs O&P		Cost Per Labor-Hour	

Crew B-36B	Hr.	Daily	Hr.	Daily	Bare Costs	Incl. O&P
1 Labor Foreman (outside)	$44.10	$352.80	$66.25	$530.00	$50.53	$75.69
2 Laborers	42.10	673.60	63.25	1012.00		
4 Equip. Oper. (medium)	56.75	1816.00	84.85	2715.20		
1 Truck Driver (heavy)	48.95	391.60	73.35	586.80		
1 Grader, 30,000 Lbs.		1062.00		1168.20		
1 F.E. Loader, Crl. 1.5 C.Y.		659.75		725.73		
1 Dozer, 300 H.P.		1764.00		1940.40		
1 Roller, Vibratory, 25 Ton		664.35		730.78		
1 Truck Tractor, 6x4, 450 H.P.		601.75		661.92		
1 Water Tank Trailer, 5000 Gal.		152.25		167.47	76.63	84.29
64 L.H., Daily Totals		$8138.10		$10238.51	$127.16	$159.98

Crew B-36C	Hr.	Daily	Hr.	Daily	Bare Costs	Incl. O&P
1 Labor Foreman (outside)	$44.10	$352.80	$66.25	$530.00	$52.66	$78.83
3 Equip. Oper. (medium)	56.75	1362.00	84.85	2036.40		
1 Truck Driver (heavy)	48.95	391.60	73.35	586.80		
1 Grader, 30,000 Lbs.		1062.00		1168.20		
1 Dozer, 300 H.P.		1764.00		1940.40		
1 Roller, Vibratory, 25 Ton		664.35		730.78		
1 Truck Tractor, 6x4, 450 H.P.		601.75		661.92		
1 Water Tank Trailer, 5000 Gal.		152.25		167.47	106.11	116.72
40 L.H., Daily Totals		$6350.75		$7821.98	$158.77	$195.55

Crew B-36D	Hr.	Daily	Hr.	Daily	Bare Costs	Incl. O&P
1 Labor Foreman (outside)	$44.10	$352.80	$66.25	$530.00	$53.59	$80.20
3 Equip. Oper. (medium)	56.75	1362.00	84.85	2036.40		
1 Grader, 30,000 Lbs.		1062.00		1168.20		
1 Dozer, 300 H.P.		1764.00		1940.40		
1 Roller, Vibratory, 25 Ton		664.35		730.78	109.07	119.98
32 L.H., Daily Totals		$5205.15		$6405.78	$162.66	$200.18

Crew B-37	Hr.	Daily	Hr.	Daily	Bare Costs	Incl. O&P
1 Labor Foreman (outside)	$44.10	$352.80	$66.25	$530.00	$44.25	$66.41
4 Laborers	42.10	1347.20	63.25	2024.00		
1 Equip. Oper. (light)	53.00	424.00	79.20	633.60		
1 Tandem Roller, 5 Ton		255.65		281.21	5.33	5.86
48 L.H., Daily Totals		$2379.65		$3468.82	$49.58	$72.27

Crew B-37A	Hr.	Daily	Hr.	Daily	Bare Costs	Incl. O&P
2 Laborers	$42.10	$673.60	$63.25	$1012.00	$43.82	$65.77
1 Truck Driver (light)	47.25	378.00	70.80	566.40		
1 Flatbed Truck, Gas, 1.5 Ton		196.15		215.76		
1 Tar Kettle, T.M.		156.20		171.82	14.68	16.15
24 L.H., Daily Totals		$1403.95		$1965.98	$58.50	$81.92

Crew B-37B	Hr.	Daily	Hr.	Daily	Bare Costs	Incl. O&P
3 Laborers	$42.10	$1010.40	$63.25	$1518.00	$43.39	$65.14
1 Truck Driver (light)	47.25	378.00	70.80	566.40		
1 Flatbed Truck, Gas, 1.5 Ton		196.15		215.76		
1 Tar Kettle, T.M.		156.20		171.82	11.01	12.11
32 L.H., Daily Totals		$1740.75		$2471.99	$54.40	$77.25

Crew B-37C	Hr.	Daily	Hr.	Daily	Bare Costs	Incl. O&P
2 Laborers	$42.10	$673.60	$63.25	$1012.00	$44.67	$67.03
2 Truck Drivers (light)	47.25	756.00	70.80	1132.80		
2 Flatbed Trucks, Gas, 1.5 Ton		392.30		431.53		
1 Tar Kettle, T.M.		156.20		171.82	17.14	18.85
32 L.H., Daily Totals		$1978.10		$2748.15	$61.82	$85.88

Crew B-37D	Hr.	Daily	Hr.	Daily	Bare Costs	Incl. O&P
1 Laborer	$42.10	$336.80	$63.25	$506.00	$44.67	$67.03
1 Truck Driver (light)	47.25	378.00	70.80	566.40		
1 Pickup Truck, 3/4 Ton		110.85		121.94	6.93	7.62
16 L.H., Daily Totals		$825.65		$1194.34	$51.60	$74.65

Crew B-37E	Hr.	Daily	Hr.	Daily	Bare Costs	Incl. O&P
3 Laborers	$42.10	$1010.40	$63.25	$1518.00	$47.22	$70.77
1 Equip. Oper. (light)	53.00	424.00	79.20	633.60		
1 Equip. Oper. (medium)	56.75	454.00	84.85	678.80		
2 Truck Drivers (light)	47.25	756.00	70.80	1132.80		
4 Barrels w/ Flasher		16.40		18.04		
1 Concrete Saw		111.50		122.65		
1 Rotary Hammer Drill		51.70		56.87		
1 Hammer Drill Bit		25.25		27.77		
1 Loader, Skid Steer, 30 H.P.		177.55		195.31		
1 Conc. Hammer Attach.		118.40		130.24		
1 Vibrating Plate, Gas, 18"		31.55		34.70		
2 Flatbed Trucks, Gas, 1.5 Ton		392.30		431.53	16.51	18.16
56 L.H., Daily Totals		$3569.05		$4980.31	$63.73	$88.93

Crew B-37F	Hr.	Daily	Hr.	Daily	Bare Costs	Incl. O&P
3 Laborers	$42.10	$1010.40	$63.25	$1518.00	$43.39	$65.14
1 Truck Driver (light)	47.25	378.00	70.80	566.40		
4 Barrels w/ Flasher		16.40		18.04		
1 Concrete Mixer, 10 C.F.		140.30		154.33		
1 Air Compressor, 60 cfm		152.95		168.25		
1 -50' Air Hose, 3/4"		7.75		8.53		
1 Spade (Chipper)		8.45		9.29		
1 Flatbed Truck, Gas, 1.5 Ton		196.15		215.76	16.31	17.94
32 L.H., Daily Totals		$1910.40		$2658.60	$59.70	$83.08

Crew B-37G	Hr.	Daily	Hr.	Daily	Bare Costs	Incl. O&P
1 Labor Foreman (outside)	$44.10	$352.80	$66.25	$530.00	$44.25	$66.41
4 Laborers	42.10	1347.20	63.25	2024.00		
1 Equip. Oper. (light)	53.00	424.00	79.20	633.60		
1 Berm Machine		271.95		299.14		
1 Tandem Roller, 5 Ton		255.65		281.21	10.99	12.09
48 L.H., Daily Totals		$2651.60		$3767.96	$55.24	$78.50

Crew B-37H	Hr.	Daily	Hr.	Daily	Bare Costs	Incl. O&P
1 Labor Foreman (outside)	$44.10	$352.80	$66.25	$530.00	$44.25	$66.41
4 Laborers	42.10	1347.20	63.25	2024.00		
1 Equip. Oper. (light)	53.00	424.00	79.20	633.60		
1 Tandem Roller, 5 Ton		255.65		281.21		
1 Flatbed Truck, Gas, 1.5 Ton		196.15		215.76		
1 Tar Kettle, T.M.		156.20		171.82	12.67	13.93
48 L.H., Daily Totals		$2732.00		$3856.40	$56.92	$80.34

For customer support on your Site Work & Landscape Costs with RSMeans data, call 800.448.8182.

Crew No.	Bare Costs		Incl. Subs O&P		Cost Per Labor-Hour	
Crew B-37I	**Hr.**	**Daily**	**Hr.**	**Daily**	**Bare Costs**	**Incl. O&P**
3 Laborers	$42.10	$1010.40	$63.25	$1518.00	$47.22	$70.77
1 Equip. Oper. (light)	53.00	424.00	79.20	633.60		
1 Equip. Oper. (medium)	56.75	454.00	84.85	678.80		
2 Truck Drivers (light)	47.25	756.00	70.80	1132.80		
4 Barrels w/ Flasher		16.40		18.04		
1 Concrete Saw		111.50		122.65		
1 Rotary Hammer Drill		51.70		56.87		
1 Hammer Drill Bit		25.25		27.77		
1 Air Compressor, 60 cfm		152.95		168.25		
1 -50' Air Hose, 3/4"		7.75		8.53		
1 Spade (Chipper)		8.45		9.29		
1 Loader, Skid Steer, 30 H.P.		177.55		195.31		
1 Conc. Hammer Attach.		118.40		130.24		
1 Concrete Mixer, 10 C.F.		140.30		154.33		
1 Vibrating Plate, Gas, 18"		31.55		34.70		
2 Flatbed Trucks, Gas, 1.5 Ton		392.30		431.53	22.04	24.24
56 L.H., Daily Totals		$3878.50		$5320.71	$69.26	$95.01
Crew B-37J	**Hr.**	**Daily**	**Hr.**	**Daily**	**Bare Costs**	**Incl. O&P**
1 Labor Foreman (outside)	$44.10	$352.80	$66.25	$530.00	$44.25	$66.41
4 Laborers	42.10	1347.20	63.25	2024.00		
1 Equip. Oper. (light)	53.00	424.00	79.20	633.60		
1 Air Compressor, 60 cfm		152.95		168.25		
1 -50' Air Hose, 3/4"		7.75		8.53		
2 Concrete Mixers, 10 C.F.		280.60		308.66		
2 Flatbed Trucks, Gas, 1.5 Ton		392.30		431.53		
1 Shot Blaster, 20"		206.20		226.82	21.66	23.83
48 L.H., Daily Totals		$3163.80		$4331.38	$65.91	$90.24
Crew B-37K	**Hr.**	**Daily**	**Hr.**	**Daily**	**Bare Costs**	**Incl. O&P**
1 Labor Foreman (outside)	$44.10	$352.80	$66.25	$530.00	$44.25	$66.41
4 Laborers	42.10	1347.20	63.25	2024.00		
1 Equip. Oper. (light)	53.00	424.00	79.20	633.60		
1 Air Compressor, 60 cfm		152.95		168.25		
1 -50' Air Hose, 3/4"		7.75		8.53		
2 Flatbed Trucks, Gas, 1.5 Ton		392.30		431.53		
1 Shot Blaster, 20"		206.20		226.82	15.82	17.40
48 L.H., Daily Totals		$2883.20		$4022.72	$60.07	$83.81
Crew B-38	**Hr.**	**Daily**	**Hr.**	**Daily**	**Bare Costs**	**Incl. O&P**
1 Labor Foreman (outside)	$44.10	$352.80	$66.25	$530.00	$47.61	$71.36
2 Laborers	42.10	673.60	63.25	1012.00		
1 Equip. Oper. (light)	53.00	424.00	79.20	633.60		
1 Equip. Oper. (medium)	56.75	454.00	84.85	678.80		
1 Backhoe Loader, 48 H.P.		213.75		235.13		
1 Hyd. Hammer (1200 lb.)		175.15		192.66		
1 F.E. Loader, W.M., 4 C.Y.		789.35		868.28		
1 Pvmt. Rem. Bucket		63.05		69.36	31.03	34.14
40 L.H., Daily Totals		$3145.70		$4219.83	$78.64	$105.50
Crew B-39	**Hr.**	**Daily**	**Hr.**	**Daily**	**Bare Costs**	**Incl. O&P**
1 Labor Foreman (outside)	$44.10	$352.80	$66.25	$530.00	$44.25	$66.41
4 Laborers	42.10	1347.20	63.25	2024.00		
1 Equip. Oper. (light)	53.00	424.00	79.20	633.60		
1 Air Compressor, 250 cfm		201.50		221.65		
2 Breakers, Pavement, 60 lb.		107.00		117.70		
2 -50' Air Hoses, 1.5"		48.50		53.35	7.44	8.18
48 L.H., Daily Totals		$2481.00		$3580.30	$51.69	$74.59
Crew B-40	**Hr.**	**Daily**	**Hr.**	**Daily**	**Bare Costs**	**Incl. O&P**
1 Pile Driver Foreman (outside)	$56.20	$449.60	$87.30	$698.40	$55.24	$84.58
4 Pile Drivers	54.20	1734.40	84.20	2694.40		
2 Equip. Oper. (crane)	59.20	947.20	88.50	1416.00		
1 Equip. Oper. (oiler)	50.55	404.40	75.55	604.40		
1 Crawler Crane, 40 Ton		1206.00		1326.60		
1 Vibratory Hammer & Gen.		2271.00		2498.10	54.33	59.76
64 L.H., Daily Totals		$7012.60		$9237.90	$109.57	$144.34
Crew B-40B	**Hr.**	**Daily**	**Hr.**	**Daily**	**Bare Costs**	**Incl. O&P**
1 Labor Foreman (outside)	$44.10	$352.80	$66.25	$530.00	$46.69	$70.01
3 Laborers	42.10	1010.40	63.25	1518.00		
1 Equip. Oper. (crane)	59.20	473.60	88.50	708.00		
1 Equip. Oper. (oiler)	50.55	404.40	75.55	604.40		
1 Lattice Boom Crane, 40 Ton		2028.00		2230.80	42.25	46.48
48 L.H., Daily Totals		$4269.20		$5591.20	$88.94	$116.48
Crew B-41	**Hr.**	**Daily**	**Hr.**	**Daily**	**Bare Costs**	**Incl. O&P**
1 Labor Foreman (outside)	$44.10	$352.80	$66.25	$530.00	$43.63	$65.50
4 Laborers	42.10	1347.20	63.25	2024.00		
.25 Equip. Oper. (crane)	59.20	118.40	88.50	177.00		
.25 Equip. Oper. (oiler)	50.55	101.10	75.55	151.10		
.25 Crawler Crane, 40 Ton		301.50		331.65	6.85	7.54
44 L.H., Daily Totals		$2221.00		$3213.75	$50.48	$73.04
Crew B-42	**Hr.**	**Daily**	**Hr.**	**Daily**	**Bare Costs**	**Incl. O&P**
1 Labor Foreman (outside)	$44.10	$352.80	$66.25	$530.00	$47.46	$71.67
4 Laborers	42.10	1347.20	63.25	2024.00		
1 Equip. Oper. (crane)	59.20	473.60	88.50	708.00		
1 Equip. Oper. (oiler)	50.55	404.40	75.55	604.40		
1 Welder	57.45	459.60	90.10	720.80		
1 Hyd. Crane, 25 Ton		580.85		638.93		
1 Welder, Gas Engine, 300 amp		147.00		161.70		
1 Horz. Boring Csg. Mch.		314.25		345.68	16.28	17.91
64 L.H., Daily Totals		$4079.70		$5733.51	$63.75	$89.59
Crew B-43	**Hr.**	**Daily**	**Hr.**	**Daily**	**Bare Costs**	**Incl. O&P**
1 Labor Foreman (outside)	$44.10	$352.80	$66.25	$530.00	$46.69	$70.01
3 Laborers	42.10	1010.40	63.25	1518.00		
1 Equip. Oper. (crane)	59.20	473.60	88.50	708.00		
1 Equip. Oper. (oiler)	50.55	404.40	75.55	604.40		
1 Drill Rig, Truck-Mounted		771.10		848.21	16.06	17.67
48 L.H., Daily Totals		$3012.30		$4208.61	$62.76	$87.68
Crew B-44	**Hr.**	**Daily**	**Hr.**	**Daily**	**Bare Costs**	**Incl. O&P**
1 Pile Driver Foreman (outside)	$56.20	$449.60	$87.30	$698.40	$54.19	$83.04
4 Pile Drivers	54.20	1734.40	84.20	2694.40		
2 Equip. Oper. (crane)	59.20	947.20	88.50	1416.00		
1 Laborer	42.10	336.80	63.25	506.00		
1 Crawler Crane, 40 Ton		1206.00		1326.60		
1 Lead, 60' High		209.25		230.18		
1 Hammer, Diesel, 15K ft.-lbs.		617.05		678.76	31.75	34.93
64 L.H., Daily Totals		$5500.30		$7550.33	$85.94	$117.97
Crew B-45	**Hr.**	**Daily**	**Hr.**	**Daily**	**Bare Costs**	**Incl. O&P**
1 Equip. Oper. (medium)	$56.75	$454.00	$84.85	$678.80	$52.85	$79.10
1 Truck Driver (heavy)	48.95	391.60	73.35	586.80		
1 Dist. Tanker, 3000 Gallon		330.15		363.17		
1 Truck Tractor, 6x4, 380 H.P.		493.25		542.58	51.46	56.61
16 L.H., Daily Totals		$1669.00		$2171.34	$104.31	$135.71

For customer support on your Site Work & Landscape Costs with RSMeans data, call 800.448.8182.

757

Crews - Standard

Crew No.	Hr.	Daily	Hr.	Daily	Bare Costs	Incl. O&P
Crew B-46						
1 Pile Driver Foreman (outside)	$56.20	$449.60	$87.30	$698.40	$48.48	$74.24
2 Pile Drivers	54.20	867.20	84.20	1347.20		
3 Laborers	42.10	1010.40	63.25	1518.00		
1 Chain Saw, Gas, 36" Long		41.25		45.38	.86	.95
48 L.H., Daily Totals		$2368.45		$3608.97	$49.34	$75.19
Crew B-47	Hr.	Daily	Hr.	Daily	Bare Costs	Incl. O&P
1 Blast Foreman (outside)	$44.10	$352.80	$66.25	$530.00	$46.40	$69.57
1 Driller	42.10	336.80	63.25	506.00		
1 Equip. Oper. (light)	53.00	424.00	79.20	633.60		
1 Air Track Drill, 4"		1058.00		1163.80		
1 Air Compressor, 600 cfm		421.50		463.65		
2 -50' Air Hoses, 3"		76.70		84.37	64.84	71.33
24 L.H., Daily Totals		$2669.80		$3381.42	$111.24	$140.89
Crew B-47A	Hr.	Daily	Hr.	Daily	Bare Costs	Incl. O&P
1 Drilling Foreman (outside)	$44.10	$352.80	$66.25	$530.00	$51.28	$76.77
1 Equip. Oper. (heavy)	59.20	473.60	88.50	708.00		
1 Equip. Oper. (oiler)	50.55	404.40	75.55	604.40		
1 Air Track Drill, 5"		1127.00		1239.70	46.96	51.65
24 L.H., Daily Totals		$2357.80		$3082.10	$98.24	$128.42
Crew B-47C	Hr.	Daily	Hr.	Daily	Bare Costs	Incl. O&P
1 Laborer	$42.10	$336.80	$63.25	$506.00	$47.55	$71.22
1 Equip. Oper. (light)	53.00	424.00	79.20	633.60		
1 Air Compressor, 750 cfm		538.25		592.08		
2 -50' Air Hoses, 3"		76.70		84.37		
1 Air Track Drill, 4"		1058.00		1163.80	104.56	115.02
16 L.H., Daily Totals		$2433.75		$2979.84	$152.11	$186.24
Crew B-47E	Hr.	Daily	Hr.	Daily	Bare Costs	Incl. O&P
1 Labor Foreman (outside)	$44.10	$352.80	$66.25	$530.00	$42.60	$64.00
3 Laborers	42.10	1010.40	63.25	1518.00		
1 Flatbed Truck, Gas, 3 Ton		820.40		902.44	25.64	28.20
32 L.H., Daily Totals		$2183.60		$2950.44	$68.24	$92.20
Crew B-47G	Hr.	Daily	Hr.	Daily	Bare Costs	Incl. O&P
1 Labor Foreman (outside)	$44.10	$352.80	$66.25	$530.00	$45.33	$67.99
2 Laborers	42.10	673.60	63.25	1012.00		
1 Equip. Oper. (light)	53.00	424.00	79.20	633.60		
1 Air Track Drill, 4"		1058.00		1163.80		
1 Air Compressor, 600 cfm		421.50		463.65		
2 -50' Air Hoses, 3"		76.70		84.37		
1 Gunite Pump Rig		317.90		349.69	58.57	64.42
32 L.H., Daily Totals		$3324.50		$4237.11	$103.89	$132.41
Crew B-47H	Hr.	Daily	Hr.	Daily	Bare Costs	Incl. O&P
1 Skilled Worker Foreman (out)	$56.85	$454.80	$86.00	$688.00	$55.35	$83.71
3 Skilled Workers	54.85	1316.40	82.95	1990.80		
1 Flatbed Truck, Gas, 3 Ton		820.40		902.44	25.64	28.20
32 L.H., Daily Totals		$2591.60		$3581.24	$80.99	$111.91

Crew No.	Hr.	Daily	Hr.	Daily	Bare Costs	Incl. O&P
Crew B-48						
1 Labor Foreman (outside)	$44.10	$352.80	$66.25	$530.00	$47.59	$71.32
3 Laborers	42.10	1010.40	63.25	1518.00		
1 Equip. Oper. (crane)	59.20	473.60	88.50	708.00		
1 Equip. Oper. (oiler)	50.55	404.40	75.55	604.40		
1 Equip. Oper. (light)	53.00	424.00	79.20	633.60		
1 Centr. Water Pump, 6"		232.90		256.19		
1 -20' Suction Hose, 6"		25.20		27.72		
1 -50' Discharge Hose, 6"		17.90		19.69		
1 Drill Rig, Truck-Mounted		771.10		848.21	18.70	20.57
56 L.H., Daily Totals		$3712.30		$5145.81	$66.29	$91.89
Crew B-49	Hr.	Daily	Hr.	Daily	Bare Costs	Incl. O&P
1 Labor Foreman (outside)	$44.10	$352.80	$66.25	$530.00	$50.12	$75.61
3 Laborers	42.10	1010.40	63.25	1518.00		
2 Equip. Oper. (crane)	59.20	947.20	88.50	1416.00		
2 Equip. Oper. (oilers)	50.55	808.80	75.55	1208.80		
1 Equip. Oper. (light)	53.00	424.00	79.20	633.60		
2 Pile Drivers	54.20	867.20	84.20	1347.20		
1 Hyd. Crane, 25 Ton		580.85		638.93		
1 Centr. Water Pump, 6"		232.90		256.19		
1 -20' Suction Hose, 6"		25.20		27.72		
1 -50' Discharge Hose, 6"		17.90		19.69		
1 Drill Rig, Truck-Mounted		771.10		848.21	18.50	20.35
88 L.H., Daily Totals		$6038.35		$8444.34	$68.62	$95.96
Crew B-50	Hr.	Daily	Hr.	Daily	Bare Costs	Incl. O&P
2 Pile Driver Foremen (outside)	$56.20	$899.20	$87.30	$1396.80	$52.35	$80.15
6 Pile Drivers	54.20	2601.60	84.20	4041.60		
2 Equip. Oper. (crane)	59.20	947.20	88.50	1416.00		
1 Equip. Oper. (oiler)	50.55	404.40	75.55	604.40		
3 Laborers	42.10	1010.40	63.25	1518.00		
1 Crawler Crane, 40 Ton		1206.00		1326.60		
1 Lead, 60' High		209.25		230.18		
1 Hammer, Diesel, 15K ft.-lbs.		617.05		678.76		
1 Air Compressor, 600 cfm		421.50		463.65		
2 -50' Air Hoses, 3"		76.70		84.37		
1 Chain Saw, Gas, 36" Long		41.25		45.38	22.96	25.26
112 L.H., Daily Totals		$8434.55		$11805.73	$75.31	$105.41
Crew B-51	Hr.	Daily	Hr.	Daily	Bare Costs	Incl. O&P
1 Labor Foreman (outside)	$44.10	$352.80	$66.25	$530.00	$43.29	$65.01
4 Laborers	42.10	1347.20	63.25	2024.00		
1 Truck Driver (light)	47.25	378.00	70.80	566.40		
1 Flatbed Truck, Gas, 1.5 Ton		196.15		215.76	4.09	4.50
48 L.H., Daily Totals		$2274.15		$3336.17	$47.38	$69.50
Crew B-52	Hr.	Daily	Hr.	Daily	Bare Costs	Incl. O&P
1 Carpenter Foreman (outside)	$55.15	$441.20	$82.85	$662.80	$48.73	$72.97
1 Carpenter	53.15	425.20	79.85	638.80		
3 Laborers	42.10	1010.40	63.25	1518.00		
1 Cement Finisher	49.95	399.60	73.45	587.60		
.5 Rodman (reinf.)	56.40	225.60	84.90	339.60		
.5 Equip. Oper. (medium)	56.75	227.00	84.85	339.40		
.5 Crawler Loader, 3 C.Y.		584.50		642.95	10.44	11.48
56 L.H., Daily Totals		$3313.50		$4729.15	$59.17	$84.45
Crew B-53	Hr.	Daily	Hr.	Daily	Bare Costs	Incl. O&P
1 Equip. Oper. (light)	$53.00	$424.00	$79.20	$633.60	$53.00	$79.20
1 Trencher, Chain, 12 H.P.		156.95		172.65	19.62	21.58
8 L.H., Daily Totals		$580.95		$806.25	$72.62	$100.78

Crew No.	Bare Costs		Incl. Subs O&P		Cost Per Labor-Hour	
Crew B-54	**Hr.**	**Daily**	**Hr.**	**Daily**	**Bare Costs**	**Incl. O&P**
1 Equip. Oper. (light)	$53.00	$424.00	$79.20	$633.60	$53.00	$79.20
1 Trencher, Chain, 40 H.P.		401.85		442.04	50.23	55.25
8 L.H., Daily Totals		$825.85		$1075.64	$103.23	$134.45
Crew B-54A	**Hr.**	**Daily**	**Hr.**	**Daily**	**Bare Costs**	**Incl. O&P**
.17 Labor Foreman (outside)	$44.10	$59.98	$66.25	$90.10	$54.91	$82.15
1 Equipment Operator (med.)	56.75	454.00	84.85	678.80		
1 Wheel Trencher, 67 H.P.		1126.00		1238.60	120.30	132.33
9.36 L.H., Daily Totals		$1639.98		$2007.50	$175.21	$214.48
Crew B-54B	**Hr.**	**Daily**	**Hr.**	**Daily**	**Bare Costs**	**Incl. O&P**
.25 Labor Foreman (outside)	$44.10	$88.20	$66.25	$132.50	$54.22	$81.13
1 Equipment Operator (med.)	56.75	454.00	84.85	678.80		
1 Wheel Trencher, 150 H.P.		1324.00		1456.40	132.40	145.64
10 L.H., Daily Totals		$1866.20		$2267.70	$186.62	$226.77
Crew B-54C	**Hr.**	**Daily**	**Hr.**	**Daily**	**Bare Costs**	**Incl. O&P**
1 Laborer	$42.10	$336.80	$63.25	$506.00	$49.42	$74.05
1 Equipment Operator (med.)	56.75	454.00	84.85	678.80		
1 Wheel Trencher, 67 H.P.		1126.00		1238.60	70.38	77.41
16 L.H., Daily Totals		$1916.80		$2423.40	$119.80	$151.46
Crew B-54D	**Hr.**	**Daily**	**Hr.**	**Daily**	**Bare Costs**	**Incl. O&P**
1 Laborer	$42.10	$336.80	$63.25	$506.00	$49.42	$74.05
1 Equipment Operator (med.)	56.75	454.00	84.85	678.80		
1 Rock Trencher, 6" Width		429.85		472.83	26.87	29.55
16 L.H., Daily Totals		$1220.65		$1657.64	$76.29	$103.60
Crew B-54E	**Hr.**	**Daily**	**Hr.**	**Daily**	**Bare Costs**	**Incl. O&P**
1 Laborer	$42.10	$336.80	$63.25	$506.00	$49.42	$74.05
1 Equipment Operator (med.)	56.75	454.00	84.85	678.80		
1 Rock Trencher, 18" Width		1005.00		1105.50	62.81	69.09
16 L.H., Daily Totals		$1795.80		$2290.30	$112.24	$143.14
Crew B-55	**Hr.**	**Daily**	**Hr.**	**Daily**	**Bare Costs**	**Incl. O&P**
2 Laborers	$42.10	$673.60	$63.25	$1012.00	$43.82	$65.77
1 Truck Driver (light)	47.25	378.00	70.80	566.40		
1 Truck-Mounted Earth Auger		390.20		429.22		
1 Flatbed Truck, Gas, 3 Ton		820.40		902.44	50.44	55.49
24 L.H., Daily Totals		$2262.20		$2910.06	$94.26	$121.25
Crew B-56	**Hr.**	**Daily**	**Hr.**	**Daily**	**Bare Costs**	**Incl. O&P**
1 Laborer	$42.10	$336.80	$63.25	$506.00	$47.55	$71.22
1 Equip. Oper. (light)	53.00	424.00	79.20	633.60		
1 Air Track Drill, 4"		1058.00		1163.80		
1 Air Compressor, 600 cfm		421.50		463.65		
1 -50' Air Hose, 3"		38.35		42.19	94.87	104.35
16 L.H., Daily Totals		$2278.65		$2809.24	$142.42	$175.58

Crew No.	Bare Costs		Incl. Subs O&P		Cost Per Labor-Hour	
Crew B-57	**Hr.**	**Daily**	**Hr.**	**Daily**	**Bare Costs**	**Incl. O&P**
1 Labor Foreman (outside)	$44.10	$352.80	$66.25	$530.00	$48.51	$72.67
2 Laborers	42.10	673.60	63.25	1012.00		
1 Equip. Oper. (crane)	59.20	473.60	88.50	708.00		
1 Equip. Oper. (light)	53.00	424.00	79.20	633.60		
1 Equip. Oper. (oiler)	50.55	404.40	75.55	604.40		
1 Crawler Crane, 25 Ton		1129.00		1241.90		
1 Clamshell Bucket, 1 C.Y.		68.55		75.41		
1 Centr. Water Pump, 6"		232.90		256.19		
1 -20' Suction Hose, 6"		25.20		27.72		
20 -50' Discharge Hoses, 6"		358.00		393.80	37.78	41.56
48 L.H., Daily Totals		$4142.05		$5483.02	$86.29	$114.23
Crew B-58	**Hr.**	**Daily**	**Hr.**	**Daily**	**Bare Costs**	**Incl. O&P**
2 Laborers	$42.10	$673.60	$63.25	$1012.00	$45.73	$68.57
1 Equip. Oper. (light)	53.00	424.00	79.20	633.60		
1 Backhoe Loader, 48 H.P.		213.75		235.13		
1 Small Helicopter, w/ Pilot		2080.00		2288.00	95.57	105.13
24 L.H., Daily Totals		$3391.35		$4168.73	$141.31	$173.70
Crew B-59	**Hr.**	**Daily**	**Hr.**	**Daily**	**Bare Costs**	**Incl. O&P**
1 Truck Driver (heavy)	$48.95	$391.60	$73.35	$586.80	$48.95	$73.35
1 Truck Tractor, 220 H.P.		307.10		337.81		
1 Water Tank Trailer, 5000 Gal.		152.25		167.47	57.42	63.16
8 L.H., Daily Totals		$850.95		$1092.09	$106.37	$136.51
Crew B-59A	**Hr.**	**Daily**	**Hr.**	**Daily**	**Bare Costs**	**Incl. O&P**
2 Laborers	$42.10	$673.60	$63.25	$1012.00	$44.38	$66.62
1 Truck Driver (heavy)	48.95	391.60	73.35	586.80		
1 Water Tank Trailer, 5000 Gal.		152.25		167.47		
1 Truck Tractor, 220 H.P.		307.10		337.81	19.14	21.05
24 L.H., Daily Totals		$1524.55		$2104.09	$63.52	$87.67
Crew B-60	**Hr.**	**Daily**	**Hr.**	**Daily**	**Bare Costs**	**Incl. O&P**
1 Labor Foreman (outside)	$44.10	$352.80	$66.25	$530.00	$49.15	$73.60
2 Laborers	42.10	673.60	63.25	1012.00		
1 Equip. Oper. (crane)	59.20	473.60	88.50	708.00		
2 Equip. Oper. (light)	53.00	848.00	79.20	1267.20		
1 Equip. Oper. (oiler)	50.55	404.40	75.55	604.40		
1 Crawler Crane, 40 Ton		1206.00		1326.60		
1 Lead, 60' High		209.25		230.18		
1 Hammer, Diesel, 15K ft.-lbs.		617.05		678.76		
1 Backhoe Loader, 48 H.P.		213.75		235.13	40.11	44.12
56 L.H., Daily Totals		$4998.45		$6592.26	$89.26	$117.72
Crew B-61	**Hr.**	**Daily**	**Hr.**	**Daily**	**Bare Costs**	**Incl. O&P**
1 Labor Foreman (outside)	$44.10	$352.80	$66.25	$530.00	$44.68	$67.04
3 Laborers	42.10	1010.40	63.25	1518.00		
1 Equip. Oper. (light)	53.00	424.00	79.20	633.60		
1 Cement Mixer, 2 C.Y.		105.70		116.27		
1 Air Compressor, 160 cfm		202.80		223.08	7.71	8.48
40 L.H., Daily Totals		$2095.70		$3020.95	$52.39	$75.52
Crew B-62	**Hr.**	**Daily**	**Hr.**	**Daily**	**Bare Costs**	**Incl. O&P**
2 Laborers	$42.10	$673.60	$63.25	$1012.00	$45.73	$68.57
1 Equip. Oper. (light)	53.00	424.00	79.20	633.60		
1 Loader, Skid Steer, 30 H.P.		177.55		195.31	7.40	8.14
24 L.H., Daily Totals		$1275.15		$1840.91	$53.13	$76.70

For customer support on your Site Work & Landscape Costs with RSMeans data, call 800.448.8182.

759

Crews - Standard

Crew B-62A

Crew No.	Bare Costs Hr.	Daily	Incl. Subs O&P Hr.	Daily	Cost Per Labor-Hour Bare Costs	Incl. O&P
2 Laborers	$42.10	$673.60	$63.25	$1012.00	$45.73	$68.57
1 Equip. Oper. (light)	53.00	424.00	79.20	633.60		
1 Loader, Skid Steer, 30 H.P.		177.55		195.31		
1 Trencher Attachment		66.20		72.82	10.16	11.17
24 L.H., Daily Totals		$1341.35		$1913.72	$55.89	$79.74

Crew B-63

Crew No.	Bare Costs Hr.	Daily	Incl. Subs O&P Hr.	Daily	Cost Per Labor-Hour Bare Costs	Incl. O&P
4 Laborers	$42.10	$1347.20	$63.25	$2024.00	$44.28	$66.44
1 Equip. Oper. (light)	53.00	424.00	79.20	633.60		
1 Loader, Skid Steer, 30 H.P.		177.55		195.31	4.44	4.88
40 L.H., Daily Totals		$1948.75		$2852.91	$48.72	$71.32

Crew B-63B

Crew No.	Bare Costs Hr.	Daily	Incl. Subs O&P Hr.	Daily	Cost Per Labor-Hour Bare Costs	Incl. O&P
1 Labor Foreman (inside)	$42.60	$340.80	$64.00	$512.00	$44.95	$67.42
2 Laborers	42.10	673.60	63.25	1012.00		
1 Equip. Oper. (light)	53.00	424.00	79.20	633.60		
1 Loader, Skid Steer, 78 H.P.		400.10		440.11	12.50	13.75
32 L.H., Daily Totals		$1838.50		$2597.71	$57.45	$81.18

Crew B-64

Crew No.	Bare Costs Hr.	Daily	Incl. Subs O&P Hr.	Daily	Cost Per Labor-Hour Bare Costs	Incl. O&P
1 Laborer	$42.10	$336.80	$63.25	$506.00	$44.67	$67.03
1 Truck Driver (light)	47.25	378.00	70.80	566.40		
1 Power Mulcher (small)		221.10		243.21		
1 Flatbed Truck, Gas, 1.5 Ton		196.15		215.76	26.08	28.69
16 L.H., Daily Totals		$1132.05		$1531.38	$70.75	$95.71

Crew B-65

Crew No.	Bare Costs Hr.	Daily	Incl. Subs O&P Hr.	Daily	Cost Per Labor-Hour Bare Costs	Incl. O&P
1 Laborer	$42.10	$336.80	$63.25	$506.00	$44.67	$67.03
1 Truck Driver (light)	47.25	378.00	70.80	566.40		
1 Power Mulcher (Large)		326.90		359.59		
1 Flatbed Truck, Gas, 1.5 Ton		196.15		215.76	32.69	35.96
16 L.H., Daily Totals		$1237.85		$1647.76	$77.37	$102.98

Crew B-66

Crew No.	Bare Costs Hr.	Daily	Incl. Subs O&P Hr.	Daily	Cost Per Labor-Hour Bare Costs	Incl. O&P
1 Equip. Oper. (light)	$53.00	$424.00	$79.20	$633.60	$53.00	$79.20
1 Loader-Backhoe, 40 H.P.		241.40		265.54	30.18	33.19
8 L.H., Daily Totals		$665.40		$899.14	$83.17	$112.39

Crew B-67

Crew No.	Bare Costs Hr.	Daily	Incl. Subs O&P Hr.	Daily	Cost Per Labor-Hour Bare Costs	Incl. O&P
1 Millwright	$55.90	$447.20	$81.00	$648.00	$54.45	$80.10
1 Equip. Oper. (light)	53.00	424.00	79.20	633.60		
1 R.T. Forklift, 5,000 Lb., diesel		269.90		296.89	16.87	18.56
16 L.H., Daily Totals		$1141.10		$1578.49	$71.32	$98.66

Crew B-67B

Crew No.	Bare Costs Hr.	Daily	Incl. Subs O&P Hr.	Daily	Cost Per Labor-Hour Bare Costs	Incl. O&P
1 Millwright Foreman (inside)	$56.40	$451.20	$81.70	$653.60	$56.15	$81.35
1 Millwright	55.90	447.20	81.00	648.00		
16 L.H., Daily Totals		$898.40		$1301.60	$56.15	$81.35

Crew B-68

Crew No.	Bare Costs Hr.	Daily	Incl. Subs O&P Hr.	Daily	Cost Per Labor-Hour Bare Costs	Incl. O&P
2 Millwrights	$55.90	$894.40	$81.00	$1296.00	$54.93	$80.40
1 Equip. Oper. (light)	53.00	424.00	79.20	633.60		
1 R.T. Forklift, 5,000 Lb., diesel		269.90		296.89	11.25	12.37
24 L.H., Daily Totals		$1588.30		$2226.49	$66.18	$92.77

Crew B-68A

Crew No.	Bare Costs Hr.	Daily	Incl. Subs O&P Hr.	Daily	Cost Per Labor-Hour Bare Costs	Incl. O&P
1 Millwright Foreman (inside)	$56.40	$451.20	$81.70	$653.60	$56.07	$81.23
2 Millwrights	55.90	894.40	81.00	1296.00		
1 Forklift, Smooth Floor, 8,000 Lb.		255.90		281.49	10.66	11.73
24 L.H., Daily Totals		$1601.50		$2231.09	$66.73	$92.96

Crew B-68B

Crew No.	Bare Costs Hr.	Daily	Incl. Subs O&P Hr.	Daily	Cost Per Labor-Hour Bare Costs	Incl. O&P
1 Millwright Foreman (inside)	$56.40	$451.20	$81.70	$653.60	$59.97	$88.47
2 Millwrights	55.90	894.40	81.00	1296.00		
2 Electricians	61.35	981.60	91.35	1461.60		
2 Plumbers	64.45	1031.20	96.45	1543.20		
1 R.T. Forklift, 5,000 Lb., gas		279.90		307.89	5.00	5.50
56 L.H., Daily Totals		$3638.30		$5262.29	$64.97	$93.97

Crew B-68C

Crew No.	Bare Costs Hr.	Daily	Incl. Subs O&P Hr.	Daily	Cost Per Labor-Hour Bare Costs	Incl. O&P
1 Millwright Foreman (inside)	$56.40	$451.20	$81.70	$653.60	$59.52	$87.63
1 Millwright	55.90	447.20	81.00	648.00		
1 Electrician	61.35	490.80	91.35	730.80		
1 Plumber	64.45	515.60	96.45	771.60		
1 R.T. Forklift, 5,000 Lb., gas		279.90		307.89	8.75	9.62
32 L.H., Daily Totals		$2184.70		$3111.89	$68.27	$97.25

Crew B-68D

Crew No.	Bare Costs Hr.	Daily	Incl. Subs O&P Hr.	Daily	Cost Per Labor-Hour Bare Costs	Incl. O&P
1 Labor Foreman (inside)	$42.60	$340.80	$64.00	$512.00	$45.90	$68.82
1 Laborer	42.10	336.80	63.25	506.00		
1 Equip. Oper. (light)	53.00	424.00	79.20	633.60		
1 R.T. Forklift, 5,000 Lb., gas		279.90		307.89	11.66	12.83
24 L.H., Daily Totals		$1381.50		$1959.49	$57.56	$81.65

Crew B-68E

Crew No.	Bare Costs Hr.	Daily	Incl. Subs O&P Hr.	Daily	Cost Per Labor-Hour Bare Costs	Incl. O&P
1 Struc. Steel Foreman (inside)	$58.15	$465.20	$91.20	$729.60	$57.71	$90.53
3 Struc. Steel Workers	57.65	1383.60	90.45	2170.80		
1 Welder	57.45	459.60	90.10	720.80		
1 Forklift, Smooth Floor, 8,000 Lb.		255.90		281.49	6.40	7.04
40 L.H., Daily Totals		$2564.30		$3902.69	$64.11	$97.57

Crew B-68F

Crew No.	Bare Costs Hr.	Daily	Incl. Subs O&P Hr.	Daily	Cost Per Labor-Hour Bare Costs	Incl. O&P
1 Skilled Worker Foreman (out)	$56.85	$454.80	$86.00	$688.00	$55.52	$83.97
2 Skilled Workers	54.85	877.60	82.95	1327.20		
1 R.T. Forklift, 5,000 Lb., gas		279.90		307.89	11.66	12.83
24 L.H., Daily Totals		$1612.30		$2323.09	$67.18	$96.80

Crew B-68G

Crew No.	Bare Costs Hr.	Daily	Incl. Subs O&P Hr.	Daily	Cost Per Labor-Hour Bare Costs	Incl. O&P
2 Structural Steel Workers	$57.65	$922.40	$90.45	$1447.20	$57.65	$90.45
1 R.T. Forklift, 5,000 Lb., gas		279.90		307.89	17.49	19.24
16 L.H., Daily Totals		$1202.30		$1755.09	$75.14	$109.69

Crew B-69

Crew No.	Bare Costs Hr.	Daily	Incl. Subs O&P Hr.	Daily	Cost Per Labor-Hour Bare Costs	Incl. O&P
1 Labor Foreman (outside)	$44.10	$352.80	$66.25	$530.00	$46.69	$70.01
3 Laborers	42.10	1010.40	63.25	1518.00		
1 Equip. Oper. (crane)	59.20	473.60	88.50	708.00		
1 Equip. Oper. (oiler)	50.55	404.40	75.55	604.40		
1 Hyd. Crane, 80 Ton		1486.00		1634.60	30.96	34.05
48 L.H., Daily Totals		$3727.20		$4995.00	$77.65	$104.06

Crew B-69A

Crew No.	Bare Costs Hr.	Daily	Incl. Subs O&P Hr.	Daily	Cost Per Labor-Hour Bare Costs	Incl. O&P
1 Labor Foreman (outside)	$44.10	$352.80	$66.25	$530.00	$46.18	$69.05
3 Laborers	42.10	1010.40	63.25	1518.00		
1 Equip. Oper. (medium)	56.75	454.00	84.85	678.80		
1 Concrete Finisher	49.95	399.60	73.45	587.60		
1 Curb/Gutter Paver, 2-Track		1217.00		1338.70	25.35	27.89
48 L.H., Daily Totals		$3433.80		$4653.10	$71.54	$96.94

For customer support on your Site Work & Landscape Costs with RSMeans data, call 800.448.8182.

Crew B-69B

Crew No.	Bare Costs Hr.	Daily	Incl. Subs O&P Hr.	Daily	Cost Per Labor-Hour Bare Costs	Incl. O&P
1 Labor Foreman (outside)	$44.10	$352.80	$66.25	$530.00	$46.18	$69.05
3 Laborers	42.10	1010.40	63.25	1518.00		
1 Equip. Oper. (medium)	56.75	454.00	84.85	678.80		
1 Cement Finisher	49.95	399.60	73.45	587.60		
1 Curb/Gutter Paver, 4-Track		791.55		870.71	16.49	18.14
48 L.H., Daily Totals		$3008.35		$4185.10	$62.67	$87.19

Crew B-70

Crew No.	Bare Costs Hr.	Daily	Incl. Subs O&P Hr.	Daily	Cost Per Labor-Hour Bare Costs	Incl. O&P
1 Labor Foreman (outside)	$44.10	$352.80	$66.25	$530.00	$48.66	$72.94
3 Laborers	42.10	1010.40	63.25	1518.00		
3 Equip. Oper. (medium)	56.75	1362.00	84.85	2036.40		
1 Grader, 30,000 Lbs.		1062.00		1168.20		
1 Ripper, Beam & 1 Shank		90.50		99.55		
1 Road Sweeper, S.P., 8' wide		715.85		787.43		
1 F.E. Loader, W.M., 1.5 C.Y.		423.50		465.85	40.93	45.02
56 L.H., Daily Totals		$5017.05		$6605.44	$89.59	$117.95

Crew B-70A

Crew No.	Bare Costs Hr.	Daily	Incl. Subs O&P Hr.	Daily	Cost Per Labor-Hour Bare Costs	Incl. O&P
1 Laborer	$42.10	$336.80	$63.25	$506.00	$53.82	$80.53
4 Equip. Oper. (medium)	56.75	1816.00	84.85	2715.20		
1 Grader, 40,000 Lbs.		1347.00		1481.70		
1 F.E. Loader, W.M., 2.5 C.Y.		702.10		772.31		
1 Dozer, 80 H.P.		401.40		441.54		
1 Roller, Pneum. Whl., 12 Ton		345.75		380.32	69.91	76.90
40 L.H., Daily Totals		$4949.05		$6297.07	$123.73	$157.43

Crew B-71

Crew No.	Bare Costs Hr.	Daily	Incl. Subs O&P Hr.	Daily	Cost Per Labor-Hour Bare Costs	Incl. O&P
1 Labor Foreman (outside)	$44.10	$352.80	$66.25	$530.00	$48.66	$72.94
3 Laborers	42.10	1010.40	63.25	1518.00		
3 Equip. Oper. (medium)	56.75	1362.00	84.85	2036.40		
1 Pvmt. Profiler, 750 H.P.		3448.00		3792.80		
1 Road Sweeper, S.P., 8' wide		715.85		787.43		
1 F.E. Loader, W.M., 1.5 C.Y.		423.50		465.85	81.92	90.11
56 L.H., Daily Totals		$7312.55		$9130.49	$130.58	$163.04

Crew B-72

Crew No.	Bare Costs Hr.	Daily	Incl. Subs O&P Hr.	Daily	Cost Per Labor-Hour Bare Costs	Incl. O&P
1 Labor Foreman (outside)	$44.10	$352.80	$66.25	$530.00	$49.67	$74.42
3 Laborers	42.10	1010.40	63.25	1518.00		
4 Equip. Oper. (medium)	56.75	1816.00	84.85	2715.20		
1 Pvmt. Profiler, 750 H.P.		3448.00		3792.80		
1 Hammermill, 250 H.P.		848.80		933.68		
1 Windrow Loader		1427.00		1569.70		
1 Mix Paver, 165 H.P.		2149.00		2363.90		
1 Roller, Pneum. Whl., 12 Ton		345.75		380.32	128.41	141.26
64 L.H., Daily Totals		$11397.75		$13803.61	$178.09	$215.68

Crew B-73

Crew No.	Bare Costs Hr.	Daily	Incl. Subs O&P Hr.	Daily	Cost Per Labor-Hour Bare Costs	Incl. O&P
1 Labor Foreman (outside)	$44.10	$352.80	$66.25	$530.00	$51.51	$77.13
2 Laborers	42.10	673.60	63.25	1012.00		
5 Equip. Oper. (medium)	56.75	2270.00	84.85	3394.00		
1 Road Mixer, 310 H.P.		1896.00		2085.60		
1 Tandem Roller, 10 Ton		236.75		260.43		
1 Hammermill, 250 H.P.		848.80		933.68		
1 Grader, 30,000 Lbs.		1062.00		1168.20		
.5 F.E. Loader, W.M., 1.5 C.Y.		211.75		232.93		
.5 Truck Tractor, 220 H.P.		153.55		168.91		
.5 Water Tank Trailer, 5000 Gal.		76.13		83.74	70.08	77.09
64 L.H., Daily Totals		$7781.38		$9869.47	$121.58	$154.21

Crew B-74

Crew No.	Bare Costs Hr.	Daily	Incl. Subs O&P Hr.	Daily	Cost Per Labor-Hour Bare Costs	Incl. O&P
1 Labor Foreman (outside)	$44.10	$352.80	$66.25	$530.00	$51.39	$76.95
1 Laborer	42.10	336.80	63.25	506.00		
4 Equip. Oper. (medium)	56.75	1816.00	84.85	2715.20		
2 Truck Drivers (heavy)	48.95	783.20	73.35	1173.60		
1 Grader, 30,000 Lbs.		1062.00		1168.20		
1 Ripper, Beam & 1 Shank		90.50		99.55		
2 Stabilizers, 310 H.P.		2778.00		3055.80		
1 Flatbed Truck, Gas, 3 Ton		820.40		902.44		
1 Chem. Spreader, Towed		83.35		91.69		
1 Roller, Vibratory, 25 Ton		664.35		730.78		
1 Water Tank Trailer, 5000 Gal.		152.25		167.47		
1 Truck Tractor, 220 H.P.		307.10		337.81	93.09	102.40
64 L.H., Daily Totals		$9246.75		$11478.55	$144.48	$179.35

Crew B-75

Crew No.	Bare Costs Hr.	Daily	Incl. Subs O&P Hr.	Daily	Cost Per Labor-Hour Bare Costs	Incl. O&P
1 Labor Foreman (outside)	$44.10	$352.80	$66.25	$530.00	$51.74	$77.46
1 Laborer	42.10	336.80	63.25	506.00		
4 Equip. Oper. (medium)	56.75	1816.00	84.85	2715.20		
1 Truck Driver (heavy)	48.95	391.60	73.35	586.80		
1 Grader, 30,000 Lbs.		1062.00		1168.20		
1 Ripper, Beam & 1 Shank		90.50		99.55		
2 Stabilizers, 310 H.P.		2778.00		3055.80		
1 Dist. Tanker, 3000 Gallon		330.15		363.17		
1 Truck Tractor, 6x4, 380 H.P.		493.25		542.58		
1 Roller, Vibratory, 25 Ton		664.35		730.78	96.75	106.43
56 L.H., Daily Totals		$8315.45		$10298.08	$148.49	$183.89

Crew B-76

Crew No.	Bare Costs Hr.	Daily	Incl. Subs O&P Hr.	Daily	Cost Per Labor-Hour Bare Costs	Incl. O&P
1 Dock Builder Foreman (outside)	$56.20	$449.60	$87.30	$698.40	$55.13	$84.54
5 Dock Builders	54.20	2168.00	84.20	3368.00		
2 Equip. Oper. (crane)	59.20	947.20	88.50	1416.00		
1 Equip. Oper. (oiler)	50.55	404.40	75.55	604.40		
1 Crawler Crane, 50 Ton		1510.00		1661.00		
1 Barge, 400 Ton		858.85		944.74		
1 Hammer, Diesel, 15K ft.-lbs.		617.05		678.76		
1 Lead, 60' High		209.25		230.18		
1 Air Compressor, 600 cfm		421.50		463.65		
2 -50' Air Hoses, 3"		76.70		84.37	51.30	56.43
72 L.H., Daily Totals		$7662.55		$10149.49	$106.42	$140.97

Crew B-76A

Crew No.	Bare Costs Hr.	Daily	Incl. Subs O&P Hr.	Daily	Cost Per Labor-Hour Bare Costs	Incl. O&P
1 Labor Foreman (outside)	$44.10	$352.80	$66.25	$530.00	$45.54	$68.32
5 Laborers	42.10	1684.00	63.25	2530.00		
1 Equip. Oper. (crane)	59.20	473.60	88.50	708.00		
1 Equip. Oper. (oiler)	50.55	404.40	75.55	604.40		
1 Crawler Crane, 50 Ton		1510.00		1661.00		
1 Barge, 400 Ton		858.85		944.74	37.01	40.71
64 L.H., Daily Totals		$5283.65		$6978.14	$82.56	$109.03

Crew B-77

Crew No.	Bare Costs Hr.	Daily	Incl. Subs O&P Hr.	Daily	Cost Per Labor-Hour Bare Costs	Incl. O&P
1 Labor Foreman (outside)	$44.10	$352.80	$66.25	$530.00	$43.53	$65.36
3 Laborers	42.10	1010.40	63.25	1518.00		
1 Truck Driver (light)	47.25	378.00	70.80	566.40		
1 Crack Cleaner, 25 H.P.		52.45		57.70		
1 Crack Filler, Trailer Mtd.		154.75		170.22		
1 Flatbed Truck, Gas, 3 Ton		820.40		902.44	25.69	28.26
40 L.H., Daily Totals		$2768.80		$3744.76	$69.22	$93.62

Crew No.	Hr.	Daily	Hr.	Daily	Bare Costs	Incl. O&P
Crew B-78					Bare Costs	Incl. O&P
1 Labor Foreman (outside)	$44.10	$352.80	$66.25	$530.00	$43.29	$65.01
4 Laborers	42.10	1347.20	63.25	2024.00		
1 Truck Driver (light)	47.25	378.00	70.80	566.40		
1 Paint Striper, S.P., 40 Gallon		127.10		139.81		
1 Flatbed Truck, Gas, 3 Ton		820.40		902.44		
1 Pickup Truck, 3/4 Ton		110.85		121.94	22.05	24.25
48 L.H., Daily Totals		$3136.35		$4284.59	$65.34	$89.26
Crew B-78A	Hr.	Daily	Hr.	Daily	Bare Costs	Incl. O&P
1 Equip. Oper. (light)	$53.00	$424.00	$79.20	$633.60	$53.00	$79.20
1 Line Rem. (Metal Balls) 115 H.P.		916.50		1008.15	114.56	126.02
8 L.H., Daily Totals		$1340.50		$1641.75	$167.56	$205.22
Crew B-78B	Hr.	Daily	Hr.	Daily	Bare Costs	Incl. O&P
2 Laborers	$42.10	$673.60	$63.25	$1012.00	$43.31	$65.02
.25 Equip. Oper. (light)	53.00	106.00	79.20	158.40		
1 Pickup Truck, 3/4 Ton		110.85		121.94		
1 Line Rem.,11 H.P.,Walk Behind		113.45		124.80		
.25 Road Sweeper, S.P., 8' wide		178.96		196.86	22.40	24.64
18 L.H., Daily Totals		$1182.86		$1613.99	$65.71	$89.67
Crew B-78C	Hr.	Daily	Hr.	Daily	Bare Costs	Incl. O&P
1 Labor Foreman (outside)	$44.10	$352.80	$66.25	$530.00	$43.29	$65.01
4 Laborers	42.10	1347.20	63.25	2024.00		
1 Truck Driver (light)	47.25	378.00	70.80	566.40		
1 Paint Striper, T.M., 120 Gal.		592.95		652.25		
1 Flatbed Truck, Gas, 3 Ton		820.40		902.44		
1 Pickup Truck, 3/4 Ton		110.85		121.94	31.75	34.93
48 L.H., Daily Totals		$3602.20		$4797.02	$75.05	$99.94
Crew B-78D	Hr.	Daily	Hr.	Daily	Bare Costs	Incl. O&P
2 Labor Foremen (outside)	$44.10	$705.60	$66.25	$1060.00	$43.02	$64.61
7 Laborers	42.10	2357.60	63.25	3542.00		
1 Truck Driver (light)	47.25	378.00	70.80	566.40		
1 Paint Striper, T.M., 120 Gal.		592.95		652.25		
1 Flatbed Truck, Gas, 3 Ton		820.40		902.44		
3 Pickup Trucks, 3/4 Ton		332.55		365.81		
1 Air Compressor, 60 cfm		152.95		168.25		
1 -50' Air Hose, 3/4"		7.75		8.53		
1 Breaker, Pavement, 60 lb.		53.50		58.85	24.50	26.95
80 L.H., Daily Totals		$5401.30		$7324.51	$67.52	$91.56
Crew B-78E	Hr.	Daily	Hr.	Daily	Bare Costs	Incl. O&P
2 Labor Foremen (outside)	$44.10	$705.60	$66.25	$1060.00	$42.86	$64.38
9 Laborers	42.10	3031.20	63.25	4554.00		
1 Truck Driver (light)	47.25	378.00	70.80	566.40		
1 Paint Striper, T.M., 120 Gal.		592.95		652.25		
1 Flatbed Truck, Gas, 3 Ton		820.40		902.44		
4 Pickup Trucks, 3/4 Ton		443.40		487.74		
2 Air Compressors, 60 cfm		305.90		336.49		
2 -50' Air Hoses, 3/4"		15.50		17.05		
2 Breakers, Pavement, 60 lb.		107.00		117.70	23.80	26.18
96 L.H., Daily Totals		$6399.95		$8694.07	$66.67	$90.56

Crew No.	Hr.	Daily	Hr.	Daily	Bare Costs	Incl. O&P
Crew B-78F					Bare Costs	Incl. O&P
2 Labor Foremen (outside)	$44.10	$705.60	$66.25	$1060.00	$42.75	$64.22
11 Laborers	42.10	3704.80	63.25	5566.00		
1 Truck Driver (light)	47.25	378.00	70.80	566.40		
1 Paint Striper, T.M., 120 Gal.		592.95		652.25		
1 Flatbed Truck, Gas, 3 Ton		820.40		902.44		
7 Pickup Trucks, 3/4 Ton		775.95		853.54		
3 Air Compressors, 60 cfm		458.85		504.74		
3 -50' Air Hoses, 3/4"		23.25		25.57		
3 Breakers, Pavement, 60 lb.		160.50		176.55	25.28	27.81
112 L.H., Daily Totals		$7620.30		$10307.49	$68.04	$92.03
Crew B-79	Hr.	Daily	Hr.	Daily	Bare Costs	Incl. O&P
1 Labor Foreman (outside)	$44.10	$352.80	$66.25	$530.00	$43.53	$65.36
3 Laborers	42.10	1010.40	63.25	1518.00		
1 Truck Driver (light)	47.25	378.00	70.80	566.40		
1 Paint Striper, T.M., 120 Gal.		592.95		652.25		
1 Heating Kettle, 115 Gallon		98.65		108.52		
1 Flatbed Truck, Gas, 3 Ton		820.40		902.44		
2 Pickup Trucks, 3/4 Ton		221.70		243.87	43.34	47.68
40 L.H., Daily Totals		$3474.90		$4521.47	$86.87	$113.04
Crew B-79A	Hr.	Daily	Hr.	Daily	Bare Costs	Incl. O&P
1.5 Equip. Oper. (light)	$53.00	$636.00	$79.20	$950.40	$53.00	$79.20
.5 Line Remov. (Grinder) 115 H.P.		475.57		523.13		
1 Line Rem. (Metal Balls) 115 H.P.		916.50		1008.15	116.01	127.61
12 L.H., Daily Totals		$2028.08		$2481.68	$169.01	$206.81
Crew B-79B	Hr.	Daily	Hr.	Daily	Bare Costs	Incl. O&P
1 Laborer	$42.10	$336.80	$63.25	$506.00	$42.10	$63.25
1 Set of Gases		171.55		188.71	21.44	23.59
8 L.H., Daily Totals		$508.35		$694.71	$63.54	$86.84
Crew B-79C	Hr.	Daily	Hr.	Daily	Bare Costs	Incl. O&P
1 Labor Foreman (outside)	$44.10	$352.80	$66.25	$530.00	$43.12	$64.76
5 Laborers	42.10	1684.00	63.25	2530.00		
1 Truck Driver (light)	47.25	378.00	70.80	566.40		
1 Paint Striper, T.M., 120 Gal.		592.95		652.25		
1 Heating Kettle, 115 Gallon		98.65		108.52		
1 Flatbed Truck, Gas, 3 Ton		820.40		902.44		
3 Pickup Trucks, 3/4 Ton		332.55		365.81		
1 Air Compressor, 60 cfm		152.95		168.25		
1 -50' Air Hose, 3/4"		7.75		8.53		
1 Breaker, Pavement, 60 lb.		53.50		58.85	36.76	40.44
56 L.H., Daily Totals		$4473.55		$5891.02	$79.88	$105.20
Crew B-79D	Hr.	Daily	Hr.	Daily	Bare Costs	Incl. O&P
2 Labor Foremen (outside)	$44.10	$705.60	$66.25	$1060.00	$43.24	$64.94
5 Laborers	42.10	1684.00	63.25	2530.00		
1 Truck Driver (light)	47.25	378.00	70.80	566.40		
1 Paint Striper, T.M., 120 Gal.		592.95		652.25		
1 Heating Kettle, 115 Gallon		98.65		108.52		
1 Flatbed Truck, Gas, 3 Ton		820.40		902.44		
4 Pickup Trucks, 3/4 Ton		443.40		487.74		
1 Air Compressor, 60 cfm		152.95		168.25		
1 -50' Air Hose, 3/4"		7.75		8.53		
1 Breaker, Pavement, 60 lb.		53.50		58.85	33.90	37.29
64 L.H., Daily Totals		$4937.20		$6542.96	$77.14	$102.23

For customer support on your Site Work & Landscape Costs with RSMeans data, call 800.448.8182.

Crews - Standard

Crew No.	Bare Costs Hr.	Daily	Incl. Subs O&P Hr.	Daily	Cost Per Labor-Hour Bare Costs	Incl. O&P
Crew B-79E					Bare Costs	Incl. O&P
2 Labor Foremen (outside)	$44.10	$705.60	$66.25	$1060.00	$43.02	$64.61
7 Laborers	42.10	2357.60	63.25	3542.00		
1 Truck Driver (light)	47.25	378.00	70.80	566.40		
1 Paint Striper, T.M., 120 Gal.		592.95		652.25		
1 Heating Kettle, 115 Gallon		98.65		108.52		
1 Flatbed Truck, Gas, 3 Ton		820.40		902.44		
5 Pickup Trucks, 3/4 Ton		554.25		609.67		
2 Air Compressors, 60 cfm		305.90		336.49		
2 -50' Air Hoses, 3/4"		15.50		17.05		
2 Breakers, Pavement, 60 lb.		107.00		117.70	31.18	34.30
80 L.H., Daily Totals		$5935.85		$7912.52	$74.20	$98.91
Crew B-80					Bare Costs	Incl. O&P
1 Labor Foreman (outside)	$44.10	$352.80	$66.25	$530.00	$46.61	$69.88
1 Laborer	42.10	336.80	63.25	506.00		
1 Truck Driver (light)	47.25	378.00	70.80	566.40		
1 Equip. Oper. (light)	53.00	424.00	79.20	633.60		
1 Flatbed Truck, Gas, 3 Ton		820.40		902.44		
1 Earth Auger, Truck-Mtd.		200.50		220.55	31.90	35.09
32 L.H., Daily Totals		$2512.50		$3358.99	$78.52	$104.97
Crew B-80A					Bare Costs	Incl. O&P
3 Laborers	$42.10	$1010.40	$63.25	$1518.00	$42.10	$63.25
1 Flatbed Truck, Gas, 3 Ton		820.40		902.44	34.18	37.60
24 L.H., Daily Totals		$1830.80		$2420.44	$76.28	$100.85
Crew B-80B					Bare Costs	Incl. O&P
3 Laborers	$42.10	$1010.40	$63.25	$1518.00	$44.83	$67.24
1 Equip. Oper. (light)	53.00	424.00	79.20	633.60		
1 Crane, Flatbed Mounted, 3 Ton		236.10		259.71	7.38	8.12
32 L.H., Daily Totals		$1670.50		$2411.31	$52.20	$75.35
Crew B-80C					Bare Costs	Incl. O&P
2 Laborers	$42.10	$673.60	$63.25	$1012.00	$43.82	$65.77
1 Truck Driver (light)	47.25	378.00	70.80	566.40		
1 Flatbed Truck, Gas, 1.5 Ton		196.15		215.76		
1 Manual Fence Post Auger, Gas		53.85		59.23	10.42	11.46
24 L.H., Daily Totals		$1301.60		$1853.40	$54.23	$77.22
Crew B-81					Bare Costs	Incl. O&P
1 Laborer	$42.10	$336.80	$63.25	$506.00	$49.27	$73.82
1 Equip. Oper. (medium)	56.75	454.00	84.85	678.80		
1 Truck Driver (heavy)	48.95	391.60	73.35	586.80		
1 Hydromulcher, T.M., 3000 Gal.		275.45		303.00		
1 Truck Tractor, 220 H.P.		307.10		337.81	24.27	26.70
24 L.H., Daily Totals		$1764.95		$2412.41	$73.54	$100.52
Crew B-81A					Bare Costs	Incl. O&P
1 Laborer	$42.10	$336.80	$63.25	$506.00	$44.67	$67.03
1 Truck Driver (light)	47.25	378.00	70.80	566.40		
1 Hydromulcher, T.M., 600 Gal.		116.95		128.65		
1 Flatbed Truck, Gas, 3 Ton		820.40		902.44	58.58	64.44
16 L.H., Daily Totals		$1652.15		$2103.49	$103.26	$131.47
Crew B-82					Bare Costs	Incl. O&P
1 Laborer	$42.10	$336.80	$63.25	$506.00	$47.55	$71.22
1 Equip. Oper. (light)	53.00	424.00	79.20	633.60		
1 Horiz. Borer, 6 H.P.		182.35		200.59	11.40	12.54
16 L.H., Daily Totals		$943.15		$1340.18	$58.95	$83.76

Crew No.	Bare Costs Hr.	Daily	Incl. Subs O&P Hr.	Daily	Cost Per Labor-Hour Bare Costs	Incl. O&P
Crew B-82A					Bare Costs	Incl. O&P
2 Laborers	$42.10	$673.60	$63.25	$1012.00	$47.55	$71.22
2 Equip. Opers. (light)	53.00	848.00	79.20	1267.20		
2 Dump Truck, 8 C.Y., 220 H.P.		856.10		941.71		
1 Flatbed Trailer, 25 Ton		135.55		149.10		
1 Horiz. Dir. Drill, 20k lb. Thrust		538.65		592.51		
1 Mud Trailer for HDD, 1500 Gal.		309.00		339.90		
1 Pickup Truck, 4x4, 3/4 Ton		175.85		193.44		
1 Flatbed Trailer, 3 Ton		70.30		77.33		
1 Loader, Skid Steer, 78 H.P.		400.10		440.11	77.67	85.44
32 L.H., Daily Totals		$4007.15		$5013.31	$125.22	$156.67
Crew B-82B					Bare Costs	Incl. O&P
2 Laborers	$42.10	$673.60	$63.25	$1012.00	$47.55	$71.22
2 Equip. Opers. (light)	53.00	848.00	79.20	1267.20		
2 Dump Truck, 8 C.Y., 220 H.P.		856.10		941.71		
1 Flatbed Trailer, 25 Ton		135.55		149.10		
1 Horiz. Dir. Drill, 30k lb. Thrust		641.20		705.32		
1 Mud Trailer for HDD, 1500 Gal.		309.00		339.90		
1 Pickup Truck, 4x4, 3/4 Ton		175.85		193.44		
1 Flatbed Trailer, 3 Ton		70.30		77.33		
1 Loader, Skid Steer, 78 H.P.		400.10		440.11	80.88	88.97
32 L.H., Daily Totals		$4109.70		$5126.11	$128.43	$160.19
Crew B-82C					Bare Costs	Incl. O&P
2 Laborers	$42.10	$673.60	$63.25	$1012.00	$47.55	$71.22
2 Equip. Opers. (light)	53.00	848.00	79.20	1267.20		
2 Dump Truck, 8 C.Y., 220 H.P.		856.10		941.71		
1 Flatbed Trailer, 25 Ton		135.55		149.10		
1 Horiz. Dir. Drill, 50k lb. Thrust		815.80		897.38		
1 Mud Trailer for HDD, 1500 Gal.		309.00		339.90		
1 Pickup Truck, 4x4, 3/4 Ton		175.85		193.44		
1 Flatbed Trailer, 3 Ton		70.30		77.33		
1 Loader, Skid Steer, 78 H.P.		400.10		440.11	86.33	94.97
32 L.H., Daily Totals		$4284.30		$5318.17	$133.88	$166.19
Crew B-82D					Bare Costs	Incl. O&P
1 Equip. Oper. (light)	$53.00	$424.00	$79.20	$633.60	$53.00	$79.20
1 Mud Trailer for HDD, 1500 Gal.		309.00		339.90	38.63	42.49
8 L.H., Daily Totals		$733.00		$973.50	$91.63	$121.69
Crew B-83					Bare Costs	Incl. O&P
1 Tugboat Captain	$56.75	$454.00	$84.85	$678.80	$49.42	$74.05
1 Tugboat Hand	42.10	336.80	63.25	506.00		
1 Tugboat, 250 H.P.		717.50		789.25	44.84	49.33
16 L.H., Daily Totals		$1508.30		$1974.05	$94.27	$123.38
Crew B-84					Bare Costs	Incl. O&P
1 Equip. Oper. (medium)	$56.75	$454.00	$84.85	$678.80	$56.75	$84.85
1 Rotary Mower/Tractor		366.75		403.43	45.84	50.43
8 L.H., Daily Totals		$820.75		$1082.22	$102.59	$135.28
Crew B-85					Bare Costs	Incl. O&P
3 Laborers	$42.10	$1010.40	$63.25	$1518.00	$46.40	$69.59
1 Equip. Oper. (medium)	56.75	454.00	84.85	678.80		
1 Truck Driver (heavy)	48.95	391.60	73.35	586.80		
1 Telescoping Boom Lift, to 80'		383.55		421.90		
1 Brush Chipper, 12", 130 H.P.		392.80		432.08		
1 Pruning Saw, Rotary		26.40		29.04	20.07	22.08
40 L.H., Daily Totals		$2658.75		$3666.63	$66.47	$91.67

For customer support on your Site Work & Landscape Costs with RSMeans data, call 800.448.8182.

763

Crew No.	Bare Costs Hr.	Daily	Incl. Subs O&P Hr.	Daily	Cost Per Labor-Hour Bare Costs	Incl. O&P
Crew B-86	Hr.	Daily	Hr.	Daily	Bare Costs	Incl. O&P
1 Equip. Oper. (medium)	$56.75	$454.00	$84.85	$678.80	$56.75	$84.85
1 Stump Chipper, S.P.		194.45		213.90	24.31	26.74
8 L.H., Daily Totals		$648.45		$892.70	$81.06	$111.59
Crew B-86A	Hr.	Daily	Hr.	Daily	Bare Costs	Incl. O&P
1 Equip. Oper. (medium)	$56.75	$454.00	$84.85	$678.80	$56.75	$84.85
1 Grader, 30,000 Lbs.		1062.00		1168.20	132.75	146.03
8 L.H., Daily Totals		$1516.00		$1847.00	$189.50	$230.88
Crew B-86B	Hr.	Daily	Hr.	Daily	Bare Costs	Incl. O&P
1 Equip. Oper. (medium)	$56.75	$454.00	$84.85	$678.80	$56.75	$84.85
1 Dozer, 200 H.P.		1504.00		1654.40	188.00	206.80
8 L.H., Daily Totals		$1958.00		$2333.20	$244.75	$291.65
Crew B-87	Hr.	Daily	Hr.	Daily	Bare Costs	Incl. O&P
1 Laborer	$42.10	$336.80	$63.25	$506.00	$53.82	$80.53
4 Equip. Oper. (medium)	56.75	1816.00	84.85	2715.20		
2 Feller Bunchers, 100 H.P.		1177.20		1294.92		
1 Log Chipper, 22" Tree		552.00		607.20		
1 Dozer, 105 H.P.		633.85		697.24		
1 Chain Saw, Gas, 36" Long		41.25		45.38	60.11	66.12
40 L.H., Daily Totals		$4557.10		$5865.93	$113.93	$146.65
Crew B-88	Hr.	Daily	Hr.	Daily	Bare Costs	Incl. O&P
1 Laborer	$42.10	$336.80	$63.25	$506.00	$54.66	$81.76
6 Equip. Oper. (medium)	56.75	2724.00	84.85	4072.80		
2 Feller Bunchers, 100 H.P.		1177.20		1294.92		
1 Log Chipper, 22" Tree		552.00		607.20		
2 Log Skidders, 50 H.P.		1808.40		1989.24		
1 Dozer, 105 H.P.		633.85		697.24		
1 Chain Saw, Gas, 36" Long		41.25		45.38	75.23	82.75
56 L.H., Daily Totals		$7273.50		$9212.77	$129.88	$164.51
Crew B-89	Hr.	Daily	Hr.	Daily	Bare Costs	Incl. O&P
1 Equip. Oper. (light)	$53.00	$424.00	$79.20	$633.60	$50.13	$75.00
1 Truck Driver (light)	47.25	378.00	70.80	566.40		
1 Flatbed Truck, Gas, 3 Ton		820.40		902.44		
1 Concrete Saw		111.50		122.65		
1 Water Tank, 65 Gal.		101.90		112.09	64.61	71.07
16 L.H., Daily Totals		$1835.80		$2337.18	$114.74	$146.07
Crew B-89A	Hr.	Daily	Hr.	Daily	Bare Costs	Incl. O&P
1 Skilled Worker	$54.85	$438.80	$82.95	$663.60	$48.48	$73.10
1 Laborer	42.10	336.80	63.25	506.00		
1 Core Drill (Large)		114.30		125.73	7.14	7.86
16 L.H., Daily Totals		$889.90		$1295.33	$55.62	$80.96
Crew B-89B	Hr.	Daily	Hr.	Daily	Bare Costs	Incl. O&P
1 Equip. Oper. (light)	$53.00	$424.00	$79.20	$633.60	$50.13	$75.00
1 Truck Driver (light)	47.25	378.00	70.80	566.40		
1 Wall Saw, Hydraulic, 10 H.P.		85.50		94.05		
1 Generator, Diesel, 100 kW		496.20		545.82		
1 Water Tank, 65 Gal.		101.90		112.09		
1 Flatbed Truck, Gas, 3 Ton		820.40		902.44	94.00	103.40
16 L.H., Daily Totals		$2306.00		$2854.40	$144.13	$178.40
Crew B-89C	Hr.	Daily	Hr.	Daily	Bare Costs	Incl. O&P
1 Cement Finisher	$49.95	$399.60	$73.45	$587.60	$49.95	$73.45
1 Masonry cut-off saw, gas		61.00		67.10	7.63	8.39
8 L.H., Daily Totals		$460.60		$654.70	$57.58	$81.84

Crew No.	Bare Costs Hr.	Daily	Incl. Subs O&P Hr.	Daily	Cost Per Labor-Hour Bare Costs	Incl. O&P
Crew B-90	Hr.	Daily	Hr.	Daily	Bare Costs	Incl. O&P
1 Labor Foreman (outside)	$44.10	$352.80	$66.25	$530.00	$46.79	$70.14
3 Laborers	42.10	1010.40	63.25	1518.00		
2 Equip. Oper. (light)	53.00	848.00	79.20	1267.20		
2 Truck Drivers (heavy)	48.95	783.20	73.35	1173.60		
1 Road Mixer, 310 H.P.		1896.00		2085.60		
1 Dist. Truck, 2000 Gal.		299.65		329.62	34.31	37.74
64 L.H., Daily Totals		$5190.05		$6904.02	$81.09	$107.88
Crew B-90A	Hr.	Daily	Hr.	Daily	Bare Costs	Incl. O&P
1 Labor Foreman (outside)	$44.10	$352.80	$66.25	$530.00	$50.76	$76.02
2 Laborers	42.10	673.60	63.25	1012.00		
4 Equip. Oper. (medium)	56.75	1816.00	84.85	2715.20		
2 Graders, 30,000 Lbs.		2124.00		2336.40		
1 Tandem Roller, 10 Ton		236.75		260.43		
1 Roller, Pneum. Whl., 12 Ton		345.75		380.32	48.33	53.16
56 L.H., Daily Totals		$5548.90		$7234.35	$99.09	$129.18
Crew B-90B	Hr.	Daily	Hr.	Daily	Bare Costs	Incl. O&P
1 Labor Foreman (outside)	$44.10	$352.80	$66.25	$530.00	$49.76	$74.55
2 Laborers	42.10	673.60	63.25	1012.00		
3 Equip. Oper. (medium)	56.75	1362.00	84.85	2036.40		
1 Roller, Pneum. Whl., 12 Ton		345.75		380.32		
1 Road Mixer, 310 H.P.		1896.00		2085.60	46.70	51.37
48 L.H., Daily Totals		$4630.15		$6044.32	$96.46	$125.92
Crew B-90C	Hr.	Daily	Hr.	Daily	Bare Costs	Incl. O&P
1 Labor Foreman (outside)	$44.10	$352.80	$66.25	$530.00	$48.15	$72.17
4 Laborers	42.10	1347.20	63.25	2024.00		
3 Equip. Oper. (medium)	56.75	1362.00	84.85	2036.40		
3 Truck Drivers (heavy)	48.95	1174.80	73.35	1760.40		
3 Road Mixers, 310 H.P.		5688.00		6256.80	64.64	71.10
88 L.H., Daily Totals		$9924.80		$12607.60	$112.78	$143.27
Crew B-90D	Hr.	Daily	Hr.	Daily	Bare Costs	Incl. O&P
1 Labor Foreman (outside)	$44.10	$352.80	$66.25	$530.00	$47.22	$70.80
6 Laborers	42.10	2020.80	63.25	3036.00		
3 Equip. Oper. (medium)	56.75	1362.00	84.85	2036.40		
3 Truck Drivers (heavy)	48.95	1174.80	73.35	1760.40		
3 Road Mixers, 310 H.P.		5688.00		6256.80	54.69	60.16
104 L.H., Daily Totals		$10598.40		$13619.60	$101.91	$130.96
Crew B-90E	Hr.	Daily	Hr.	Daily	Bare Costs	Incl. O&P
1 Labor Foreman (outside)	$44.10	$352.80	$66.25	$530.00	$47.97	$71.91
4 Laborers	42.10	1347.20	63.25	2024.00		
3 Equip. Oper. (medium)	56.75	1362.00	84.85	2036.40		
1 Truck Driver (heavy)	48.95	391.60	73.35	586.80		
1 Road Mixer, 310 H.P.		1896.00		2085.60	26.33	28.97
72 L.H., Daily Totals		$5349.60		$7262.80	$74.30	$100.87
Crew B-91	Hr.	Daily	Hr.	Daily	Bare Costs	Incl. O&P
1 Labor Foreman (outside)	$44.10	$352.80	$66.25	$530.00	$50.53	$75.69
2 Laborers	42.10	673.60	63.25	1012.00		
4 Equip. Oper. (medium)	56.75	1816.00	84.85	2715.20		
1 Truck Driver (heavy)	48.95	391.60	73.35	586.80		
1 Dist. Tanker, 3000 Gallon		330.15		363.17		
1 Truck Tractor, 6x4, 380 H.P.		493.25		542.58		
1 Aggreg. Spreader, S.P.		848.90		933.79		
1 Roller, Pneum. Whl., 12 Ton		345.75		380.32		
1 Tandem Roller, 10 Ton		236.75		260.43	35.23	38.75
64 L.H., Daily Totals		$5488.80		$7324.28	$85.76	$114.44

For customer support on your Site Work & Landscape Costs with RSMeans data, call 800.448.8182.

Crew No.	Bare Costs		Incl. Subs O&P		Cost Per Labor-Hour	
	Hr.	Daily	Hr.	Daily	Bare Costs	Incl. O&P
Crew B-91B	Hr.	Daily	Hr.	Daily	Bare Costs	Incl. O&P
1 Laborer	$42.10	$336.80	$63.25	$506.00	$49.42	$74.05
1 Equipment Oper. (med.)	56.75	454.00	84.85	678.80		
1 Road Sweeper, Vac. Assist.		869.00		955.90	54.31	59.74
16 L.H., Daily Totals		$1659.80		$2140.70	$103.74	$133.79
Crew B-91C	Hr.	Daily	Hr.	Daily	Bare Costs	Incl. O&P
1 Laborer	$42.10	$336.80	$63.25	$506.00	$44.67	$67.03
1 Truck Driver (light)	47.25	378.00	70.80	566.40		
1 Catch Basin Cleaning Truck		536.15		589.76	33.51	36.86
16 L.H., Daily Totals		$1250.95		$1662.17	$78.18	$103.89
Crew B-91D	Hr.	Daily	Hr.	Daily	Bare Costs	Incl. O&P
1 Labor Foreman (outside)	$44.10	$352.80	$66.25	$530.00	$48.94	$73.34
5 Laborers	42.10	1684.00	63.25	2530.00		
5 Equip. Oper. (medium)	56.75	2270.00	84.85	3394.00		
2 Truck Drivers (heavy)	48.95	783.20	73.35	1173.60		
1 Aggreg. Spreader, S.P.		848.90		933.79		
2 Truck Tractors, 6x4, 380 H.P.		986.50		1085.15		
2 Dist. Tankers, 3000 Gallon		660.30		726.33		
2 Pavement Brushes, Towed		174.60		192.06		
2 Rollers Pneum. Whl., 12 Ton		691.50		760.65	32.33	35.56
104 L.H., Daily Totals		$8451.80		$11325.58	$81.27	$108.90
Crew B-92	Hr.	Daily	Hr.	Daily	Bare Costs	Incl. O&P
1 Labor Foreman (outside)	$44.10	$352.80	$66.25	$530.00	$42.60	$64.00
3 Laborers	42.10	1010.40	63.25	1518.00		
1 Crack Cleaner, 25 H.P.		52.45		57.70		
1 Air Compressor, 60 cfm		152.95		168.25		
1 Tar Kettle, T.M.		156.20		171.82		
1 Flatbed Truck, Gas, 3 Ton		820.40		902.44	36.94	40.63
32 L.H., Daily Totals		$2545.20		$3348.20	$79.54	$104.63
Crew B-93	Hr.	Daily	Hr.	Daily	Bare Costs	Incl. O&P
1 Equip. Oper. (medium)	$56.75	$454.00	$84.85	$678.80	$56.75	$84.85
1 Feller Buncher, 100 H.P.		588.60		647.46	73.58	80.93
8 L.H., Daily Totals		$1042.60		$1326.26	$130.32	$165.78
Crew B-94A	Hr.	Daily	Hr.	Daily	Bare Costs	Incl. O&P
1 Laborer	$42.10	$336.80	$63.25	$506.00	$42.10	$63.25
1 Diaphragm Water Pump, 2"		86.80		95.48		
1 -20' Suction Hose, 2"		3.55		3.90		
2 -50' Discharge Hoses, 2"		8.00		8.80	12.29	13.52
8 L.H., Daily Totals		$435.15		$614.18	$54.39	$76.77
Crew B-94B	Hr.	Daily	Hr.	Daily	Bare Costs	Incl. O&P
1 Laborer	$42.10	$336.80	$63.25	$506.00	$42.10	$63.25
1 Diaphragm Water Pump, 4"		105.25		115.78		
1 -20' Suction Hose, 4"		17.05		18.75		
2 -50' Discharge Hoses, 4"		25.30		27.83	18.45	20.30
8 L.H., Daily Totals		$484.40		$668.36	$60.55	$83.55
Crew B-94C	Hr.	Daily	Hr.	Daily	Bare Costs	Incl. O&P
1 Laborer	$42.10	$336.80	$63.25	$506.00	$42.10	$63.25
1 Centrifugal Water Pump, 3"		74.45		81.89		
1 -20' Suction Hose, 3"		8.75		9.63		
2 -50' Discharge Hoses, 3"		9.00		9.90	11.53	12.68
8 L.H., Daily Totals		$429.00		$607.42	$53.63	$75.93

Crew No.	Bare Costs		Incl. Subs O&P		Cost Per Labor-Hour	
Crew B-94D	Hr.	Daily	Hr.	Daily	Bare Costs	Incl. O&P
1 Laborer	$42.10	$336.80	$63.25	$506.00	$42.10	$63.25
1 Centr. Water Pump, 6"		232.90		256.19		
1 -20' Suction Hose, 6"		25.20		27.72		
2 -50' Discharge Hoses, 6"		35.80		39.38	36.74	40.41
8 L.H., Daily Totals		$630.70		$829.29	$78.84	$103.66
Crew C-1	Hr.	Daily	Hr.	Daily	Bare Costs	Incl. O&P
3 Carpenters	$53.15	$1275.60	$79.85	$1916.40	$50.39	$75.70
1 Laborer	42.10	336.80	63.25	506.00		
32 L.H., Daily Totals		$1612.40		$2422.40	$50.39	$75.70
Crew C-2	Hr.	Daily	Hr.	Daily	Bare Costs	Incl. O&P
1 Carpenter Foreman (outside)	$55.15	$441.20	$82.85	$662.80	$51.64	$77.58
4 Carpenters	53.15	1700.80	79.85	2555.20		
1 Laborer	42.10	336.80	63.25	506.00		
48 L.H., Daily Totals		$2478.80		$3724.00	$51.64	$77.58
Crew C-2A	Hr.	Daily	Hr.	Daily	Bare Costs	Incl. O&P
1 Carpenter Foreman (outside)	$55.15	$441.20	$82.85	$662.80	$51.11	$76.52
3 Carpenters	53.15	1275.60	79.85	1916.40		
1 Cement Finisher	49.95	399.60	73.45	587.60		
1 Laborer	42.10	336.80	63.25	506.00		
48 L.H., Daily Totals		$2453.20		$3672.80	$51.11	$76.52
Crew C-3	Hr.	Daily	Hr.	Daily	Bare Costs	Incl. O&P
1 Rodman Foreman (outside)	$58.40	$467.20	$87.95	$703.60	$52.65	$79.16
4 Rodmen (reinf.)	56.40	1804.80	84.90	2716.80		
1 Equip. Oper. (light)	53.00	424.00	79.20	633.60		
2 Laborers	42.10	673.60	63.25	1012.00		
3 Stressing Equipment		56.70		62.37		
.5 Grouting Equipment		122.08		134.28	2.79	3.07
64 L.H., Daily Totals		$3548.38		$5262.65	$55.44	$82.23
Crew C-4	Hr.	Daily	Hr.	Daily	Bare Costs	Incl. O&P
1 Rodman Foreman (outside)	$58.40	$467.20	$87.95	$703.60	$56.90	$85.66
3 Rodmen (reinf.)	56.40	1353.60	84.90	2037.60		
3 Stressing Equipment		56.70		62.37	1.77	1.95
32 L.H., Daily Totals		$1877.50		$2803.57	$58.67	$87.61
Crew C-4A	Hr.	Daily	Hr.	Daily	Bare Costs	Incl. O&P
2 Rodmen (reinf.)	$56.40	$902.40	$84.90	$1358.40	$56.40	$84.90
4 Stressing Equipment		75.60		83.16	4.72	5.20
16 L.H., Daily Totals		$978.00		$1441.56	$61.13	$90.10
Crew C-5	Hr.	Daily	Hr.	Daily	Bare Costs	Incl. O&P
1 Rodman Foreman (outside)	$58.40	$467.20	$87.95	$703.60	$56.25	$84.51
4 Rodmen (reinf.)	56.40	1804.80	84.90	2716.80		
1 Equip. Oper. (crane)	59.20	473.60	88.50	708.00		
1 Equip. Oper. (oiler)	50.55	404.40	75.55	604.40		
1 Hyd. Crane, 25 Ton		580.85		638.93	10.37	11.41
56 L.H., Daily Totals		$3730.85		$5371.73	$66.62	$95.92
Crew C-6	Hr.	Daily	Hr.	Daily	Bare Costs	Incl. O&P
1 Labor Foreman (outside)	$44.10	$352.80	$66.25	$530.00	$43.74	$65.45
4 Laborers	42.10	1347.20	63.25	2024.00		
1 Cement Finisher	49.95	399.60	73.45	587.60		
2 Gas Engine Vibrators		53.70		59.07	1.12	1.23
48 L.H., Daily Totals		$2153.30		$3200.67	$44.86	$66.68

For customer support on your Site Work & Landscape Costs with RSMeans data, call 800.448.8182.

765

Crew No.	Bare Costs		Incl. Subs O&P		Cost Per Labor-Hour	

Crew C-6A	**Hr.**	**Daily**	**Hr.**	**Daily**	Bare Costs	Incl. O&P
2 Cement Finishers	$49.95	$799.20	$73.45	$1175.20	$49.95	$73.45
1 Concrete Vibrator, Elec, 2 HP		47.45		52.20	2.97	3.26
16 L.H., Daily Totals		$846.65		$1227.40	$52.92	$76.71

Crew C-7	**Hr.**	**Daily**	**Hr.**	**Daily**	Bare Costs	Incl. O&P
1 Labor Foreman (outside)	$44.10	$352.80	$66.25	$530.00	$45.76	$68.48
5 Laborers	42.10	1684.00	63.25	2530.00		
1 Cement Finisher	49.95	399.60	73.45	587.60		
1 Equip. Oper. (medium)	56.75	454.00	84.85	678.80		
1 Equip. Oper. (oiler)	50.55	404.40	75.55	604.40		
2 Gas Engine Vibrators		53.70		59.07		
1 Concrete Bucket, 1 C.Y.		45.80		50.38		
1 Hyd. Crane, 55 Ton		980.20		1078.22	15.00	16.50
72 L.H., Daily Totals		$4374.50		$6118.47	$60.76	$84.98

Crew C-7A	**Hr.**	**Daily**	**Hr.**	**Daily**	Bare Costs	Incl. O&P
1 Labor Foreman (outside)	$44.10	$352.80	$66.25	$530.00	$44.06	$66.15
5 Laborers	42.10	1684.00	63.25	2530.00		
2 Truck Drivers (heavy)	48.95	783.20	73.35	1173.60		
2 Conc. Transit Mixers		1170.50		1287.55	18.29	20.12
64 L.H., Daily Totals		$3990.50		$5521.15	$62.35	$86.27

Crew C-7B	**Hr.**	**Daily**	**Hr.**	**Daily**	Bare Costs	Incl. O&P
1 Labor Foreman (outside)	$44.10	$352.80	$66.25	$530.00	$45.54	$68.32
5 Laborers	42.10	1684.00	63.25	2530.00		
1 Equipment Operator, Crane	59.20	473.60	88.50	708.00		
1 Equipment Oiler	50.55	404.40	75.55	604.40		
1 Conc. Bucket, 2 C.Y.		54.70		60.17		
1 Lattice Boom Crane, 165 Ton		2151.00		2366.10	34.46	37.91
64 L.H., Daily Totals		$5120.50		$6798.67	$80.01	$106.23

Crew C-7C	**Hr.**	**Daily**	**Hr.**	**Daily**	Bare Costs	Incl. O&P
1 Labor Foreman (outside)	$44.10	$352.80	$66.25	$530.00	$46.01	$69.03
5 Laborers	42.10	1684.00	63.25	2530.00		
2 Equipment Operators (med.)	56.75	908.00	84.85	1357.60		
2 F.E. Loaders, W.M., 4 C.Y.		1578.70		1736.57	24.67	27.13
64 L.H., Daily Totals		$4523.50		$6154.17	$70.68	$96.16

Crew C-7D	**Hr.**	**Daily**	**Hr.**	**Daily**	Bare Costs	Incl. O&P
1 Labor Foreman (outside)	$44.10	$352.80	$66.25	$530.00	$44.48	$66.76
5 Laborers	42.10	1684.00	63.25	2530.00		
1 Equip. Oper. (medium)	56.75	454.00	84.85	678.80		
1 Concrete Conveyer		204.15		224.57	3.65	4.01
56 L.H., Daily Totals		$2694.95		$3963.36	$48.12	$70.77

Crew C-8	**Hr.**	**Daily**	**Hr.**	**Daily**	Bare Costs	Incl. O&P
1 Labor Foreman (outside)	$44.10	$352.80	$66.25	$530.00	$46.72	$69.68
3 Laborers	42.10	1010.40	63.25	1518.00		
2 Cement Finishers	49.95	799.20	73.45	1175.20		
1 Equip. Oper. (medium)	56.75	454.00	84.85	678.80		
1 Concrete Pump (Small)		410.35		451.38	7.33	8.06
56 L.H., Daily Totals		$3026.75		$4353.39	$54.05	$77.74

Crew C-8A	**Hr.**	**Daily**	**Hr.**	**Daily**	Bare Costs	Incl. O&P
1 Labor Foreman (outside)	$44.10	$352.80	$66.25	$530.00	$45.05	$67.15
3 Laborers	42.10	1010.40	63.25	1518.00		
2 Cement Finishers	49.95	799.20	73.45	1175.20		
48 L.H., Daily Totals		$2162.40		$3223.20	$45.05	$67.15

Crew C-8B	**Hr.**	**Daily**	**Hr.**	**Daily**	Bare Costs	Incl. O&P
1 Labor Foreman (outside)	$44.10	$352.80	$66.25	$530.00	$45.43	$68.17
3 Laborers	42.10	1010.40	63.25	1518.00		
1 Equip. Oper. (medium)	56.75	454.00	84.85	678.80		
1 Vibrating Power Screed		75.10		82.61		
1 Roller, Vibratory, 25 Ton		664.35		730.78		
1 Dozer, 200 H.P.		1504.00		1654.40	56.09	61.69
40 L.H., Daily Totals		$4060.65		$5194.60	$101.52	$129.86

Crew C-8C	**Hr.**	**Daily**	**Hr.**	**Daily**	Bare Costs	Incl. O&P
1 Labor Foreman (outside)	$44.10	$352.80	$66.25	$530.00	$46.18	$69.05
3 Laborers	42.10	1010.40	63.25	1518.00		
1 Cement Finisher	49.95	399.60	73.45	587.60		
1 Equip. Oper. (medium)	56.75	454.00	84.85	678.80		
1 Shotcrete Rig, 12 C.Y./hr		245.40		269.94		
1 Air Compressor, 160 cfm		202.80		223.08		
4 -50' Air Hoses, 1"		32.00		35.20		
4 -50' Air Hoses, 2"		115.60		127.16	12.41	13.65
48 L.H., Daily Totals		$2812.60		$3969.78	$58.60	$82.70

Crew C-8D	**Hr.**	**Daily**	**Hr.**	**Daily**	Bare Costs	Incl. O&P
1 Labor Foreman (outside)	$44.10	$352.80	$66.25	$530.00	$47.29	$70.54
1 Laborer	42.10	336.80	63.25	506.00		
1 Cement Finisher	49.95	399.60	73.45	587.60		
1 Equipment Oper. (light)	53.00	424.00	79.20	633.60		
1 Air Compressor, 250 cfm		201.50		221.65		
2 -50' Air Hoses, 1"		16.00		17.60	6.80	7.48
32 L.H., Daily Totals		$1730.70		$2496.45	$54.08	$78.01

Crew C-8E	**Hr.**	**Daily**	**Hr.**	**Daily**	Bare Costs	Incl. O&P
1 Labor Foreman (outside)	$44.10	$352.80	$66.25	$530.00	$45.56	$68.11
3 Laborers	42.10	1010.40	63.25	1518.00		
1 Cement Finisher	49.95	399.60	73.45	587.60		
1 Equipment Oper. (light)	53.00	424.00	79.20	633.60		
1 Shotcrete Rig, 35 C.Y./hr		298.00		327.80		
1 Air Compressor, 250 cfm		201.50		221.65		
4 -50' Air Hoses, 1"		32.00		35.20		
4 -50' Air Hoses, 2"		115.60		127.16	13.48	14.83
48 L.H., Daily Totals		$2833.90		$3981.01	$59.04	$82.94

Crew C-9	**Hr.**	**Daily**	**Hr.**	**Daily**	Bare Costs	Incl. O&P
1 Cement Finisher	$49.95	$399.60	$73.45	$587.60	$46.79	$69.79
2 Laborers	42.10	673.60	63.25	1012.00		
1 Equipment Oper. (light)	53.00	424.00	79.20	633.60		
1 Grout Pump, 50 C.F./hr.		188.45		207.29		
1 Air Compressor, 160 cfm		202.80		223.08		
2 -50' Air Hoses, 1"		16.00		17.60		
2 -50' Air Hoses, 2"		57.80		63.58	14.53	15.99
32 L.H., Daily Totals		$1962.25		$2744.76	$61.32	$85.77

Crew C-10	**Hr.**	**Daily**	**Hr.**	**Daily**	Bare Costs	Incl. O&P
1 Laborer	$42.10	$336.80	$63.25	$506.00	$47.33	$70.05
2 Cement Finishers	49.95	799.20	73.45	1175.20		
24 L.H., Daily Totals		$1136.00		$1681.20	$47.33	$70.05

Crew C-10B	**Hr.**	**Daily**	**Hr.**	**Daily**	Bare Costs	Incl. O&P
3 Laborers	$42.10	$1010.40	$63.25	$1518.00	$45.24	$67.33
2 Cement Finishers	49.95	799.20	73.45	1175.20		
1 Concrete Mixer, 10 C.F.		140.30		154.33		
2 Trowels, 48" Walk-Behind		174.00		191.40	7.86	8.64
40 L.H., Daily Totals		$2123.90		$3038.93	$53.10	$75.97

Crew No.	Bare Costs		Incl. Subs O&P		Cost Per Labor-Hour	
Crew C-10C	**Hr.**	**Daily**	**Hr.**	**Daily**	**Bare Costs**	**Incl. O&P**
1 Laborer	$42.10	$336.80	$63.25	$506.00	$47.33	$70.05
2 Cement Finishers	49.95	799.20	73.45	1175.20		
1 Trowel, 48" Walk-Behind		87.00		95.70	3.63	3.99
24 L.H., Daily Totals		$1223.00		$1776.90	$50.96	$74.04
Crew C-10D	**Hr.**	**Daily**	**Hr.**	**Daily**	**Bare Costs**	**Incl. O&P**
1 Laborer	$42.10	$336.80	$63.25	$506.00	$47.33	$70.05
2 Cement Finishers	49.95	799.20	73.45	1175.20		
1 Vibrating Power Screed		75.10		82.61		
1 Trowel, 48" Walk-Behind		87.00		95.70	6.75	7.43
24 L.H., Daily Totals		$1298.10		$1859.51	$54.09	$77.48
Crew C-10E	**Hr.**	**Daily**	**Hr.**	**Daily**	**Bare Costs**	**Incl. O&P**
1 Laborer	$42.10	$336.80	$63.25	$506.00	$47.33	$70.05
2 Cement Finishers	49.95	799.20	73.45	1175.20		
1 Vibrating Power Screed		75.10		82.61		
1 Cement Trowel, 96" Ride-On		169.00		185.90	10.17	11.19
24 L.H., Daily Totals		$1380.10		$1949.71	$57.50	$81.24
Crew C-10F	**Hr.**	**Daily**	**Hr.**	**Daily**	**Bare Costs**	**Incl. O&P**
1 Laborer	$42.10	$336.80	$63.25	$506.00	$47.33	$70.05
2 Cement Finishers	49.95	799.20	73.45	1175.20		
1 Telescoping Boom Lift, to 60'		310.90		341.99	12.95	14.25
24 L.H., Daily Totals		$1446.90		$2023.19	$60.29	$84.30
Crew C-11	**Hr.**	**Daily**	**Hr.**	**Daily**	**Bare Costs**	**Incl. O&P**
1 Struc. Steel Foreman (outside)	$59.65	$477.20	$93.55	$748.40	$57.26	$88.92
6 Struc. Steel Workers	57.65	2767.20	90.45	4341.60		
1 Equip. Oper. (crane)	59.20	473.60	88.50	708.00		
1 Equip. Oper. (oiler)	50.55	404.40	75.55	604.40		
1 Lattice Boom Crane, 150 Ton		2269.00		2495.90	31.51	34.67
72 L.H., Daily Totals		$6391.40		$8898.30	$88.77	$123.59
Crew C-12	**Hr.**	**Daily**	**Hr.**	**Daily**	**Bare Costs**	**Incl. O&P**
1 Carpenter Foreman (outside)	$55.15	$441.20	$82.85	$662.80	$52.65	$79.03
3 Carpenters	53.15	1275.60	79.85	1916.40		
1 Laborer	42.10	336.80	63.25	506.00		
1 Equip. Oper. (crane)	59.20	473.60	88.50	708.00		
1 Hyd. Crane, 12 Ton		471.00		518.10	9.81	10.79
48 L.H., Daily Totals		$2998.20		$4311.30	$62.46	$89.82
Crew C-13	**Hr.**	**Daily**	**Hr.**	**Daily**	**Bare Costs**	**Incl. O&P**
1 Struc. Steel Worker	$57.65	$461.20	$90.45	$723.60	$56.08	$86.80
1 Welder	57.45	459.60	90.10	720.80		
1 Carpenter	53.15	425.20	79.85	638.80		
1 Welder, Gas Engine, 300 amp		147.00		161.70	6.13	6.74
24 L.H., Daily Totals		$1493.00		$2244.90	$62.21	$93.54
Crew C-14	**Hr.**	**Daily**	**Hr.**	**Daily**	**Bare Costs**	**Incl. O&P**
1 Carpenter Foreman (outside)	$55.15	$441.20	$82.85	$662.80	$51.36	$76.98
5 Carpenters	53.15	2126.00	79.85	3194.00		
4 Laborers	42.10	1347.20	63.25	2024.00		
4 Rodmen (reinf.)	56.40	1804.80	84.90	2716.80		
2 Cement Finishers	49.95	799.20	73.45	1175.20		
1 Equip. Oper. (crane)	59.20	473.60	88.50	708.00		
1 Equip. Oper. (oiler)	50.55	404.40	75.55	604.40		
1 Hyd. Crane, 80 Ton		1486.00		1634.60	10.32	11.35
144 L.H., Daily Totals		$8882.40		$12719.80	$61.68	$88.33

Crew No.	Bare Costs		Incl. Subs O&P		Cost Per Labor-Hour	
Crew C-14A	**Hr.**	**Daily**	**Hr.**	**Daily**	**Bare Costs**	**Incl. O&P**
1 Carpenter Foreman (outside)	$55.15	$441.20	$82.85	$662.80	$52.88	$79.39
16 Carpenters	53.15	6803.20	79.85	10220.80		
4 Rodmen (reinf.)	56.40	1804.80	84.90	2716.80		
2 Laborers	42.10	673.60	63.25	1012.00		
1 Cement Finisher	49.95	399.60	73.45	587.60		
1 Equip. Oper. (medium)	56.75	454.00	84.85	678.80		
1 Gas Engine Vibrator		26.85		29.54		
1 Concrete Pump (Small)		410.35		451.38	2.19	2.40
200 L.H., Daily Totals		$11013.60		$16359.72	$55.07	$81.80
Crew C-14B	**Hr.**	**Daily**	**Hr.**	**Daily**	**Bare Costs**	**Incl. O&P**
1 Carpenter Foreman (outside)	$55.15	$441.20	$82.85	$662.80	$52.77	$79.17
16 Carpenters	53.15	6803.20	79.85	10220.80		
4 Rodmen (reinf.)	56.40	1804.80	84.90	2716.80		
2 Laborers	42.10	673.60	63.25	1012.00		
2 Cement Finishers	49.95	799.20	73.45	1175.20		
1 Equip. Oper. (medium)	56.75	454.00	84.85	678.80		
1 Gas Engine Vibrator		26.85		29.54		
1 Concrete Pump (Small)		410.35		451.38	2.10	2.31
208 L.H., Daily Totals		$11413.20		$16947.32	$54.87	$81.48
Crew C-14C	**Hr.**	**Daily**	**Hr.**	**Daily**	**Bare Costs**	**Incl. O&P**
1 Carpenter Foreman (outside)	$55.15	$441.20	$82.85	$662.80	$50.37	$75.59
6 Carpenters	53.15	2551.20	79.85	3832.80		
2 Rodmen (reinf.)	56.40	902.40	84.90	1358.40		
4 Laborers	42.10	1347.20	63.25	2024.00		
1 Cement Finisher	49.95	399.60	73.45	587.60		
1 Gas Engine Vibrator		26.85		29.54	.24	.26
112 L.H., Daily Totals		$5668.45		$8495.14	$50.61	$75.85
Crew C-14D	**Hr.**	**Daily**	**Hr.**	**Daily**	**Bare Costs**	**Incl. O&P**
1 Carpenter Foreman (outside)	$55.15	$441.20	$82.85	$662.80	$52.62	$78.99
18 Carpenters	53.15	7653.60	79.85	11498.40		
2 Rodmen (reinf.)	56.40	902.40	84.90	1358.40		
2 Laborers	42.10	673.60	63.25	1012.00		
1 Cement Finisher	49.95	399.60	73.45	587.60		
1 Equip. Oper. (medium)	56.75	454.00	84.85	678.80		
1 Gas Engine Vibrator		26.85		29.54		
1 Concrete Pump (Small)		410.35		451.38	2.19	2.40
200 L.H., Daily Totals		$10961.60		$16278.92	$54.81	$81.39
Crew C-14E	**Hr.**	**Daily**	**Hr.**	**Daily**	**Bare Costs**	**Incl. O&P**
1 Carpenter Foreman (outside)	$55.15	$441.20	$82.85	$662.80	$51.21	$76.85
2 Carpenters	53.15	850.40	79.85	1277.60		
4 Rodmen (reinf.)	56.40	1804.80	84.90	2716.80		
3 Laborers	42.10	1010.40	63.25	1518.00		
1 Cement Finisher	49.95	399.60	73.45	587.60		
1 Gas Engine Vibrator		26.85		29.54	.31	.34
88 L.H., Daily Totals		$4533.25		$6792.34	$51.51	$77.19
Crew C-14F	**Hr.**	**Daily**	**Hr.**	**Daily**	**Bare Costs**	**Incl. O&P**
1 Labor Foreman (outside)	$44.10	$352.80	$66.25	$530.00	$47.56	$70.38
2 Laborers	42.10	673.60	63.25	1012.00		
6 Cement Finishers	49.95	2397.60	73.45	3525.60		
1 Gas Engine Vibrator		26.85		29.54	.37	.41
72 L.H., Daily Totals		$3450.85		$5097.14	$47.93	$70.79

Crew No.	Bare Costs		Incl. Subs O&P		Cost Per Labor-Hour	
Crew C-14G	Hr.	Daily	Hr.	Daily	Bare Costs	Incl. O&P
1 Labor Foreman (outside)	$44.10	$352.80	$66.25	$530.00	$46.87	$69.51
2 Laborers	42.10	673.60	63.25	1012.00		
4 Cement Finishers	49.95	1598.40	73.45	2350.40		
1 Gas Engine Vibrator		26.85		29.54	.48	.53
56 L.H., Daily Totals		$2651.65		$3921.93	$47.35	$70.03
Crew C-14H	Hr.	Daily	Hr.	Daily	Bare Costs	Incl. O&P
1 Carpenter Foreman (outside)	$55.15	$441.20	$82.85	$662.80	$51.65	$77.36
2 Carpenters	53.15	850.40	79.85	1277.60		
1 Rodman (reinf.)	56.40	451.20	84.90	679.20		
1 Laborer	42.10	336.80	63.25	506.00		
1 Cement Finisher	49.95	399.60	73.45	587.60		
1 Gas Engine Vibrator		26.85		29.54	.56	.62
48 L.H., Daily Totals		$2506.05		$3742.74	$52.21	$77.97
Crew C-14L	Hr.	Daily	Hr.	Daily	Bare Costs	Incl. O&P
1 Carpenter Foreman (outside)	$55.15	$441.20	$82.85	$662.80	$49.37	$74.03
6 Carpenters	53.15	2551.20	79.85	3832.80		
4 Laborers	42.10	1347.20	63.25	2024.00		
1 Cement Finisher	49.95	399.60	73.45	587.60		
1 Gas Engine Vibrator		26.85		29.54	.28	.31
96 L.H., Daily Totals		$4766.05		$7136.73	$49.65	$74.34
Crew C-14M	Hr.	Daily	Hr.	Daily	Bare Costs	Incl. O&P
1 Carpenter Foreman (outside)	$55.15	$441.20	$82.85	$662.80	$51.09	$76.53
2 Carpenters	53.15	850.40	79.85	1277.60		
1 Rodman (reinf.)	56.40	451.20	84.90	679.20		
2 Laborers	42.10	673.60	63.25	1012.00		
1 Cement Finisher	49.95	399.60	73.45	587.60		
1 Equip. Oper. (medium)	56.75	454.00	84.85	678.80		
1 Gas Engine Vibrator		26.85		29.54		
1 Concrete Pump (Small)		410.35		451.38	6.83	7.51
64 L.H., Daily Totals		$3707.20		$5378.92	$57.93	$84.05
Crew C-15	Hr.	Daily	Hr.	Daily	Bare Costs	Incl. O&P
1 Carpenter Foreman (outside)	$55.15	$441.20	$82.85	$662.80	$49.34	$73.79
2 Carpenters	53.15	850.40	79.85	1277.60		
3 Laborers	42.10	1010.40	63.25	1518.00		
2 Cement Finishers	49.95	799.20	73.45	1175.20		
1 Rodman (reinf.)	56.40	451.20	84.90	679.20		
72 L.H., Daily Totals		$3552.40		$5312.80	$49.34	$73.79
Crew C-16	Hr.	Daily	Hr.	Daily	Bare Costs	Incl. O&P
1 Labor Foreman (outside)	$44.10	$352.80	$66.25	$530.00	$46.72	$69.68
3 Laborers	42.10	1010.40	63.25	1518.00		
2 Cement Finishers	49.95	799.20	73.45	1175.20		
1 Equip. Oper. (medium)	56.75	454.00	84.85	678.80		
1 Gunite Pump Rig		317.90		349.69		
2 -50' Air Hoses, 3/4"		15.50		17.05		
2 -50' Air Hoses, 2"		57.80		63.58	6.99	7.68
56 L.H., Daily Totals		$3007.60		$4332.32	$53.71	$77.36
Crew C-16A	Hr.	Daily	Hr.	Daily	Bare Costs	Incl. O&P
1 Laborer	$42.10	$336.80	$63.25	$506.00	$49.69	$73.75
2 Cement Finishers	49.95	799.20	73.45	1175.20		
1 Equip. Oper. (medium)	56.75	454.00	84.85	678.80		
1 Gunite Pump Rig		317.90		349.69		
2 -50' Air Hoses, 3/4"		15.50		17.05		
2 -50' Air Hoses, 2"		57.80		63.58		
1 Telescoping Boom Lift, to 60'		310.90		341.99	21.94	24.13
32 L.H., Daily Totals		$2292.10		$3132.31	$71.63	$97.88

Crew No.	Bare Costs		Incl. Subs O&P		Cost Per Labor-Hour	
Crew C-17	Hr.	Daily	Hr.	Daily	Bare Costs	Incl. O&P
2 Skilled Worker Foremen (out)	$56.85	$909.60	$86.00	$1376.00	$55.25	$83.56
8 Skilled Workers	54.85	3510.40	82.95	5308.80		
80 L.H., Daily Totals		$4420.00		$6684.80	$55.25	$83.56
Crew C-17A	Hr.	Daily	Hr.	Daily	Bare Costs	Incl. O&P
2 Skilled Worker Foremen (out)	$56.85	$909.60	$86.00	$1376.00	$55.30	$83.62
8 Skilled Workers	54.85	3510.40	82.95	5308.80		
.125 Equip. Oper. (crane)	59.20	59.20	88.50	88.50		
.125 Hyd. Crane, 80 Ton		185.75		204.32	2.29	2.52
81 L.H., Daily Totals		$4664.95		$6977.63	$57.59	$86.14
Crew C-17B	Hr.	Daily	Hr.	Daily	Bare Costs	Incl. O&P
2 Skilled Worker Foremen (out)	$56.85	$909.60	$86.00	$1376.00	$55.35	$83.68
8 Skilled Workers	54.85	3510.40	82.95	5308.80		
.25 Equip. Oper. (crane)	59.20	118.40	88.50	177.00		
.25 Hyd. Crane, 80 Ton		371.50		408.65		
.25 Trowel, 48" Walk-Behind		21.75		23.93	4.80	5.28
82 L.H., Daily Totals		$4931.65		$7294.38	$60.14	$88.96
Crew C-17C	Hr.	Daily	Hr.	Daily	Bare Costs	Incl. O&P
2 Skilled Worker Foremen (out)	$56.85	$909.60	$86.00	$1376.00	$55.39	$83.74
8 Skilled Workers	54.85	3510.40	82.95	5308.80		
.375 Equip. Oper. (crane)	59.20	177.60	88.50	265.50		
.375 Hyd. Crane, 80 Ton		557.25		612.98	6.71	7.39
83 L.H., Daily Totals		$5154.85		$7563.27	$62.11	$91.12
Crew C-17D	Hr.	Daily	Hr.	Daily	Bare Costs	Incl. O&P
2 Skilled Worker Foremen (out)	$56.85	$909.60	$86.00	$1376.00	$55.44	$83.80
8 Skilled Workers	54.85	3510.40	82.95	5308.80		
.5 Equip. Oper. (crane)	59.20	236.80	88.50	354.00		
.5 Hyd. Crane, 80 Ton		743.00		817.30	8.85	9.73
84 L.H., Daily Totals		$5399.80		$7856.10	$64.28	$93.53
Crew C-17E	Hr.	Daily	Hr.	Daily	Bare Costs	Incl. O&P
2 Skilled Worker Foremen (out)	$56.85	$909.60	$86.00	$1376.00	$55.25	$83.56
8 Skilled Workers	54.85	3510.40	82.95	5308.80		
1 Hyd. Jack with Rods		35.90		39.49	.45	.49
80 L.H., Daily Totals		$4455.90		$6724.29	$55.70	$84.05
Crew C-18	Hr.	Daily	Hr.	Daily	Bare Costs	Incl. O&P
.125 Labor Foreman (outside)	$44.10	$44.10	$66.25	$66.25	$42.32	$63.58
1 Laborer	42.10	336.80	63.25	506.00		
1 Concrete Cart, 10 C.F.		127.85		140.63	14.21	15.63
9 L.H., Daily Totals		$508.75		$712.88	$56.53	$79.21
Crew C-19	Hr.	Daily	Hr.	Daily	Bare Costs	Incl. O&P
.125 Labor Foreman (outside)	$44.10	$44.10	$66.25	$66.25	$42.32	$63.58
1 Laborer	42.10	336.80	63.25	506.00		
1 Concrete Cart, 18 C.F.		153.45		168.79	17.05	18.75
9 L.H., Daily Totals		$534.35		$741.04	$59.37	$82.34
Crew C-20	Hr.	Daily	Hr.	Daily	Bare Costs	Incl. O&P
1 Labor Foreman (outside)	$44.10	$352.80	$66.25	$530.00	$45.16	$67.60
5 Laborers	42.10	1684.00	63.25	2530.00		
1 Cement Finisher	49.95	399.60	73.45	587.60		
1 Equip. Oper. (medium)	56.75	454.00	84.85	678.80		
2 Gas Engine Vibrators		53.70		59.07		
1 Concrete Pump (Small)		410.35		451.38	7.25	7.98
64 L.H., Daily Totals		$3354.45		$4836.85	$52.41	$75.58

For customer support on your Site Work & Landscape Costs with RSMeans data, call 800.448.8182.

Crew No.	Bare Costs		Incl. Subs O&P		Cost Per Labor-Hour	
Crew C-21	**Hr.**	**Daily**	**Hr.**	**Daily**	**Bare Costs**	**Incl. O&P**
1 Labor Foreman (outside)	$44.10	$352.80	$66.25	$530.00	$45.16	$67.60
5 Laborers	42.10	1684.00	63.25	2530.00		
1 Cement Finisher	49.95	399.60	73.45	587.60		
1 Equip. Oper. (medium)	56.75	454.00	84.85	678.80		
2 Gas Engine Vibrators		53.70		59.07		
1 Concrete Conveyer		204.15		224.57	4.03	4.43
64 L.H., Daily Totals		$3148.25		$4610.03	$49.19	$72.03
Crew C-22	**Hr.**	**Daily**	**Hr.**	**Daily**	**Bare Costs**	**Incl. O&P**
1 Rodman Foreman (outside)	$58.40	$467.20	$87.95	$703.60	$56.71	$85.34
4 Rodmen (reinf.)	56.40	1804.80	84.90	2716.80		
.125 Equip. Oper. (crane)	59.20	59.20	88.50	88.50		
.125 Equip. Oper. (oiler)	50.55	50.55	75.55	75.55		
.125 Hyd. Crane, 25 Ton		72.61		79.87	1.73	1.90
42 L.H., Daily Totals		$2454.36		$3664.32	$58.44	$87.25
Crew C-23	**Hr.**	**Daily**	**Hr.**	**Daily**	**Bare Costs**	**Incl. O&P**
2 Skilled Worker Foremen (out)	$56.85	$909.60	$86.00	$1376.00	$55.26	$83.38
6 Skilled Workers	54.85	2632.80	82.95	3981.60		
1 Equip. Oper. (crane)	59.20	473.60	88.50	708.00		
1 Equip. Oper. (oiler)	50.55	404.40	75.55	604.40		
1 Lattice Boom Crane, 90 Ton		1696.00		1865.60	21.20	23.32
80 L.H., Daily Totals		$6116.40		$8535.60	$76.45	$106.69
Crew C-23A	**Hr.**	**Daily**	**Hr.**	**Daily**	**Bare Costs**	**Incl. O&P**
1 Labor Foreman (outside)	$44.10	$352.80	$66.25	$530.00	$47.61	$71.36
2 Laborers	42.10	673.60	63.25	1012.00		
1 Equip. Oper. (crane)	59.20	473.60	88.50	708.00		
1 Equip. Oper. (oiler)	50.55	404.40	75.55	604.40		
1 Crawler Crane, 100 Ton		2287.00		2515.70		
3 Conc. Buckets, 8 C.Y.		323.25		355.57	65.26	71.78
40 L.H., Daily Totals		$4514.65		$5725.68	$112.87	$143.14
Crew C-24	**Hr.**	**Daily**	**Hr.**	**Daily**	**Bare Costs**	**Incl. O&P**
2 Skilled Worker Foremen (out)	$56.85	$909.60	$86.00	$1376.00	$55.26	$83.38
6 Skilled Workers	54.85	2632.80	82.95	3981.60		
1 Equip. Oper. (crane)	59.20	473.60	88.50	708.00		
1 Equip. Oper. (oiler)	50.55	404.40	75.55	604.40		
1 Lattice Boom Crane, 150 Ton		2269.00		2495.90	28.36	31.20
80 L.H., Daily Totals		$6689.40		$9165.90	$83.62	$114.57
Crew C-25	**Hr.**	**Daily**	**Hr.**	**Daily**	**Bare Costs**	**Incl. O&P**
2 Rodmen (reinf.)	$56.40	$902.40	$84.90	$1358.40	$45.50	$71.00
2 Rodmen Helpers	34.60	553.60	57.10	913.60		
32 L.H., Daily Totals		$1456.00		$2272.00	$45.50	$71.00
Crew C-27	**Hr.**	**Daily**	**Hr.**	**Daily**	**Bare Costs**	**Incl. O&P**
2 Cement Finishers	$49.95	$799.20	$73.45	$1175.20	$49.95	$73.45
1 Concrete Saw		111.50		122.65	6.97	7.67
16 L.H., Daily Totals		$910.70		$1297.85	$56.92	$81.12
Crew C-28	**Hr.**	**Daily**	**Hr.**	**Daily**	**Bare Costs**	**Incl. O&P**
1 Cement Finisher	$49.95	$399.60	$73.45	$587.60	$49.95	$73.45
1 Portable Air Compressor, Gas		38.70		42.57	4.84	5.32
8 L.H., Daily Totals		$438.30		$630.17	$54.79	$78.77
Crew C-29	**Hr.**	**Daily**	**Hr.**	**Daily**	**Bare Costs**	**Incl. O&P**
1 Laborer	$42.10	$336.80	$63.25	$506.00	$42.10	$63.25
1 Pressure Washer		96.95		106.65	12.12	13.33
8 L.H., Daily Totals		$433.75		$612.64	$54.22	$76.58

Crew No.	Bare Costs		Incl. Subs O&P		Cost Per Labor-Hour	
Crew C-30	**Hr.**	**Daily**	**Hr.**	**Daily**	**Bare Costs**	**Incl. O&P**
1 Laborer	$42.10	$336.80	$63.25	$506.00	$42.10	$63.25
1 Concrete Mixer, 10 C.F.		140.30		154.33	17.54	19.29
8 L.H., Daily Totals		$477.10		$660.33	$59.64	$82.54
Crew C-31	**Hr.**	**Daily**	**Hr.**	**Daily**	**Bare Costs**	**Incl. O&P**
1 Cement Finisher	$49.95	$399.60	$73.45	$587.60	$49.95	$73.45
1 Grout Pump		317.90		349.69	39.74	43.71
8 L.H., Daily Totals		$717.50		$937.29	$89.69	$117.16
Crew C-32	**Hr.**	**Daily**	**Hr.**	**Daily**	**Bare Costs**	**Incl. O&P**
1 Cement Finisher	$49.95	$399.60	$73.45	$587.60	$46.02	$68.35
1 Laborer	42.10	336.80	63.25	506.00		
1 Crack Chaser Saw, Gas, 6 H.P.		73.10		80.41		
1 Vacuum Pick-Up System		74.05		81.45	9.20	10.12
16 L.H., Daily Totals		$883.55		$1255.46	$55.22	$78.47
Crew D-1	**Hr.**	**Daily**	**Hr.**	**Daily**	**Bare Costs**	**Incl. O&P**
1 Bricklayer	$52.05	$416.40	$78.90	$631.20	$46.90	$71.10
1 Bricklayer Helper	41.75	334.00	63.30	506.40		
16 L.H., Daily Totals		$750.40		$1137.60	$46.90	$71.10
Crew D-2	**Hr.**	**Daily**	**Hr.**	**Daily**	**Bare Costs**	**Incl. O&P**
3 Bricklayers	$52.05	$1249.20	$78.90	$1893.60	$48.40	$73.31
2 Bricklayer Helpers	41.75	668.00	63.30	1012.80		
.5 Carpenter	53.15	212.60	79.85	319.40		
44 L.H., Daily Totals		$2129.80		$3225.80	$48.40	$73.31
Crew D-3	**Hr.**	**Daily**	**Hr.**	**Daily**	**Bare Costs**	**Incl. O&P**
3 Bricklayers	$52.05	$1249.20	$78.90	$1893.60	$48.18	$73.00
2 Bricklayer Helpers	41.75	668.00	63.30	1012.80		
.25 Carpenter	53.15	106.30	79.85	159.70		
42 L.H., Daily Totals		$2023.50		$3066.10	$48.18	$73.00
Crew D-4	**Hr.**	**Daily**	**Hr.**	**Daily**	**Bare Costs**	**Incl. O&P**
1 Bricklayer	$52.05	$416.40	$78.90	$631.20	$47.14	$71.17
2 Bricklayer Helpers	41.75	668.00	63.30	1012.80		
1 Equip. Oper. (light)	53.00	424.00	79.20	633.60		
1 Grout Pump, 50 C.F./hr.		188.45		207.29	5.89	6.48
32 L.H., Daily Totals		$1696.85		$2484.90	$53.03	$77.65
Crew D-5	**Hr.**	**Daily**	**Hr.**	**Daily**	**Bare Costs**	**Incl. O&P**
1 Bricklayer	52.05	416.40	78.90	631.20	52.05	78.90
8 L.H., Daily Totals		$416.40		$631.20	$52.05	$78.90
Crew D-6	**Hr.**	**Daily**	**Hr.**	**Daily**	**Bare Costs**	**Incl. O&P**
3 Bricklayers	$52.05	$1249.20	$78.90	$1893.60	$47.15	$71.45
3 Bricklayer Helpers	41.75	1002.00	63.30	1519.20		
.25 Carpenter	53.15	106.30	79.85	159.70		
50 L.H., Daily Totals		$2357.50		$3572.50	$47.15	$71.45
Crew D-7	**Hr.**	**Daily**	**Hr.**	**Daily**	**Bare Costs**	**Incl. O&P**
1 Tile Layer	$49.50	$396.00	$72.60	$580.80	$44.15	$64.75
1 Tile Layer Helper	38.80	310.40	56.90	455.20		
16 L.H., Daily Totals		$706.40		$1036.00	$44.15	$64.75
Crew D-8	**Hr.**	**Daily**	**Hr.**	**Daily**	**Bare Costs**	**Incl. O&P**
3 Bricklayers	$52.05	$1249.20	$78.90	$1893.60	$47.93	$72.66
2 Bricklayer Helpers	41.75	668.00	63.30	1012.80		
40 L.H., Daily Totals		$1917.20		$2906.40	$47.93	$72.66

For customer support on your Site Work & Landscape Costs with RSMeans data, call 800.448.8182.

769

Crew No.	Bare Costs		Incl. Subs O&P		Cost Per Labor-Hour	
Crew D-9	**Hr.**	**Daily**	**Hr.**	**Daily**	**Bare Costs**	**Incl. O&P**
3 Bricklayers	$52.05	$1249.20	$78.90	$1893.60	$46.90	$71.10
3 Bricklayer Helpers	41.75	1002.00	63.30	1519.20		
48 L.H., Daily Totals		$2251.20		$3412.80	$46.90	$71.10

Crew D-10	**Hr.**	**Daily**	**Hr.**	**Daily**	**Bare Costs**	**Incl. O&P**
1 Bricklayer Foreman (outside)	$54.05	$432.40	$81.95	$655.60	$51.76	$78.16
1 Bricklayer	52.05	416.40	78.90	631.20		
1 Bricklayer Helper	41.75	334.00	63.30	506.40		
1 Equip. Oper. (crane)	59.20	473.60	88.50	708.00		
1 S.P. Crane, 4x4, 12 Ton		428.30		471.13	13.38	14.72
32 L.H., Daily Totals		$2084.70		$2972.33	$65.15	$92.89

Crew D-11	**Hr.**	**Daily**	**Hr.**	**Daily**	**Bare Costs**	**Incl. O&P**
1 Bricklayer Foreman (outside)	$54.05	$432.40	$81.95	$655.60	$49.28	$74.72
1 Bricklayer	52.05	416.40	78.90	631.20		
1 Bricklayer Helper	41.75	334.00	63.30	506.40		
24 L.H., Daily Totals		$1182.80		$1793.20	$49.28	$74.72

Crew D-12	**Hr.**	**Daily**	**Hr.**	**Daily**	**Bare Costs**	**Incl. O&P**
1 Bricklayer Foreman (outside)	$54.05	$432.40	$81.95	$655.60	$47.40	$71.86
1 Bricklayer	52.05	416.40	78.90	631.20		
2 Bricklayer Helpers	41.75	668.00	63.30	1012.80		
32 L.H., Daily Totals		$1516.80		$2299.60	$47.40	$71.86

Crew D-13	**Hr.**	**Daily**	**Hr.**	**Daily**	**Bare Costs**	**Incl. O&P**
1 Bricklayer Foreman (outside)	$54.05	$432.40	$81.95	$655.60	$50.33	$75.97
1 Bricklayer	52.05	416.40	78.90	631.20		
2 Bricklayer Helpers	41.75	668.00	63.30	1012.80		
1 Carpenter	53.15	425.20	79.85	638.80		
1 Equip. Oper. (crane)	59.20	473.60	88.50	708.00		
1 S.P. Crane, 4x4, 12 Ton		428.30		471.13	8.92	9.82
48 L.H., Daily Totals		$2843.90		$4117.53	$59.25	$85.78

Crew D-14	**Hr.**	**Daily**	**Hr.**	**Daily**	**Bare Costs**	**Incl. O&P**
3 Bricklayers	$52.05	$1249.20	$78.90	$1893.60	$49.48	$75.00
1 Bricklayer Helper	41.75	334.00	63.30	506.40		
32 L.H., Daily Totals		$1583.20		$2400.00	$49.48	$75.00

Crew E-1	**Hr.**	**Daily**	**Hr.**	**Daily**	**Bare Costs**	**Incl. O&P**
1 Welder Foreman (outside)	$59.45	$475.60	$93.25	$746.00	$56.63	$87.52
1 Welder	57.45	459.60	90.10	720.80		
1 Equip. Oper. (light)	53.00	424.00	79.20	633.60		
1 Welder, Gas Engine, 300 amp		147.00		161.70	6.13	6.74
24 L.H., Daily Totals		$1506.20		$2262.10	$62.76	$94.25

Crew E-2	**Hr.**	**Daily**	**Hr.**	**Daily**	**Bare Costs**	**Incl. O&P**
1 Struc. Steel Foreman (outside)	$59.65	$477.20	$93.55	$748.40	$57.14	$88.49
4 Struc. Steel Workers	57.65	1844.80	90.45	2894.40		
1 Equip. Oper. (crane)	59.20	473.60	88.50	708.00		
1 Equip. Oper. (oiler)	50.55	404.40	75.55	604.40		
1 Lattice Boom Crane, 90 Ton		1696.00		1865.60	30.29	33.31
56 L.H., Daily Totals		$4896.00		$6820.80	$87.43	$121.80

Crew E-3	**Hr.**	**Daily**	**Hr.**	**Daily**	**Bare Costs**	**Incl. O&P**
1 Struc. Steel Foreman (outside)	$59.65	$477.20	$93.55	$748.40	$58.25	$91.37
1 Struc. Steel Worker	57.65	461.20	90.45	723.60		
1 Welder	57.45	459.60	90.10	720.80		
1 Welder, Gas Engine, 300 amp		147.00		161.70	6.13	6.74
24 L.H., Daily Totals		$1545.00		$2354.50	$64.38	$98.10

Crew No.	Bare Costs		Incl. Subs O&P		Cost Per Labor-Hour	
Crew E-3A	**Hr.**	**Daily**	**Hr.**	**Daily**	**Bare Costs**	**Incl. O&P**
1 Struc. Steel Foreman (outside)	$59.65	$477.20	$93.55	$748.40	$58.25	$91.37
1 Struc. Steel Worker	57.65	461.20	90.45	723.60		
1 Welder	57.45	459.60	90.10	720.80		
1 Welder, Gas Engine, 300 amp		147.00		161.70		
1 Telescoping Boom Lift, to 40'		278.90		306.79	17.75	19.52
24 L.H., Daily Totals		$1823.90		$2661.29	$76.00	$110.89

Crew E-4	**Hr.**	**Daily**	**Hr.**	**Daily**	**Bare Costs**	**Incl. O&P**
1 Struc. Steel Foreman (outside)	$59.65	$477.20	$93.55	$748.40	$58.15	$91.22
3 Struc. Steel Workers	57.65	1383.60	90.45	2170.80		
1 Welder, Gas Engine, 300 amp		147.00		161.70	4.59	5.05
32 L.H., Daily Totals		$2007.80		$3080.90	$62.74	$96.28

Crew E-5	**Hr.**	**Daily**	**Hr.**	**Daily**	**Bare Costs**	**Incl. O&P**
2 Struc. Steel Foremen (outside)	$59.65	$954.40	$93.55	$1496.80	$57.48	$89.35
5 Struc. Steel Workers	57.65	2306.00	90.45	3618.00		
1 Equip. Oper. (crane)	59.20	473.60	88.50	708.00		
1 Welder	57.45	459.60	90.10	720.80		
1 Equip. Oper. (oiler)	50.55	404.40	75.55	604.40		
1 Lattice Boom Crane, 90 Ton		1696.00		1865.60		
1 Welder, Gas Engine, 300 amp		147.00		161.70	23.04	25.34
80 L.H., Daily Totals		$6441.00		$9175.30	$80.51	$114.69

Crew E-6	**Hr.**	**Daily**	**Hr.**	**Daily**	**Bare Costs**	**Incl. O&P**
3 Struc. Steel Foremen (outside)	$59.65	$1431.60	$93.55	$2245.20	$57.38	$89.25
9 Struc. Steel Workers	57.65	4150.80	90.45	6512.40		
1 Equip. Oper. (crane)	59.20	473.60	88.50	708.00		
1 Welder	57.45	459.60	90.10	720.80		
1 Equip. Oper. (oiler)	50.55	404.40	75.55	604.40		
1 Equip. Oper. (light)	53.00	424.00	79.20	633.60		
1 Lattice Boom Crane, 90 Ton		1696.00		1865.60		
1 Welder, Gas Engine, 300 amp		147.00		161.70		
1 Air Compressor, 160 cfm		202.80		223.08		
2 Impact Wrenches		104.60		115.06	16.80	18.48
128 L.H., Daily Totals		$9494.40		$13789.84	$74.17	$107.73

Crew E-7	**Hr.**	**Daily**	**Hr.**	**Daily**	**Bare Costs**	**Incl. O&P**
1 Struc. Steel Foreman (outside)	$59.65	$477.20	$93.55	$748.40	$57.44	$89.28
4 Struc. Steel Workers	57.65	1844.80	90.45	2894.40		
1 Equip. Oper. (crane)	59.20	473.60	88.50	708.00		
1 Equip. Oper. (oiler)	50.55	404.40	75.55	604.40		
1 Welder Foreman (outside)	59.45	475.60	93.25	746.00		
2 Welders	57.45	919.20	90.10	1441.60		
1 Lattice Boom Crane, 90 Ton		1696.00		1865.60		
2 Welder, Gas Engine, 300 amp		294.00		323.40	24.88	27.36
80 L.H., Daily Totals		$6584.80		$9331.80	$82.31	$116.65

Crew E-8	**Hr.**	**Daily**	**Hr.**	**Daily**	**Bare Costs**	**Incl. O&P**
1 Struc. Steel Foreman (outside)	$59.65	$477.20	$93.55	$748.40	$57.10	$88.63
4 Struc. Steel Workers	57.65	1844.80	90.45	2894.40		
1 Welder Foreman (outside)	59.45	475.60	93.25	746.00		
4 Welders	57.45	1838.40	90.10	2883.20		
1 Equip. Oper. (crane)	59.20	473.60	88.50	708.00		
1 Equip. Oper. (oiler)	50.55	404.40	75.55	604.40		
1 Equip. Oper. (light)	53.00	424.00	79.20	633.60		
1 Lattice Boom Crane, 90 Ton		1696.00		1865.60		
4 Welder, Gas Engine, 300 amp		588.00		646.80	21.96	24.16
104 L.H., Daily Totals		$8222.00		$11730.40	$79.06	$112.79

Crew E-9

Crew No.	Bare Costs Hr.	Bare Costs Daily	Incl. Subs O&P Hr.	Incl. Subs O&P Daily	Cost Per Labor-Hour Bare Costs	Cost Per Labor-Hour Incl. O&P
2 Struc. Steel Foremen (outside)	$59.65	$954.40	$93.55	$1496.80	$57.31	$89.15
5 Struc. Steel Workers	57.65	2306.00	90.45	3618.00		
1 Welder Foreman (outside)	59.45	475.60	93.25	746.00		
5 Welders	57.45	2298.00	90.10	3604.00		
1 Equip. Oper. (crane)	59.20	473.60	88.50	708.00		
1 Equip. Oper. (oiler)	50.55	404.40	75.55	604.40		
1 Equip. Oper. (light)	53.00	424.00	79.20	633.60		
1 Lattice Boom Crane, 90 Ton		1696.00		1865.60		
5 Welder, Gas Engine, 300 amp		735.00		808.50	18.99	20.89
128 L.H., Daily Totals		$9767.00		$14084.90	$76.30	$110.04

Crew E-10

Crew No.	Bare Costs Hr.	Bare Costs Daily	Incl. Subs O&P Hr.	Incl. Subs O&P Daily	Cost Per Labor-Hour Bare Costs	Cost Per Labor-Hour Incl. O&P
1 Welder Foreman (outside)	$59.45	$475.60	$93.25	$746.00	$58.45	$91.67
1 Welder	57.45	459.60	90.10	720.80		
1 Welder, Gas Engine, 300 amp		147.00		161.70		
1 Flatbed Truck, Gas, 3 Ton		820.40		902.44	60.46	66.51
16 L.H., Daily Totals		$1902.60		$2530.94	$118.91	$158.18

Crew E-11

Crew No.	Bare Costs Hr.	Bare Costs Daily	Incl. Subs O&P Hr.	Incl. Subs O&P Daily	Cost Per Labor-Hour Bare Costs	Cost Per Labor-Hour Incl. O&P
2 Painters, Struc. Steel	$45.85	$733.60	$74.45	$1191.20	$46.70	$72.84
1 Building Laborer	42.10	336.80	63.25	506.00		
1 Equip. Oper. (light)	53.00	424.00	79.20	633.60		
1 Air Compressor, 250 cfm		201.50		221.65		
1 Sandblaster, Portable, 3 C.F.		83.80		92.18		
1 Set Sand Blasting Accessories		15.35		16.89	9.40	10.33
32 L.H., Daily Totals		$1795.05		$2661.51	$56.10	$83.17

Crew E-11A

Crew No.	Bare Costs Hr.	Bare Costs Daily	Incl. Subs O&P Hr.	Incl. Subs O&P Daily	Cost Per Labor-Hour Bare Costs	Cost Per Labor-Hour Incl. O&P
2 Painters, Struc. Steel	$45.85	$733.60	$74.45	$1191.20	$46.70	$72.84
1 Building Laborer	42.10	336.80	63.25	506.00		
1 Equip. Oper. (light)	53.00	424.00	79.20	633.60		
1 Air Compressor, 250 cfm		201.50		221.65		
1 Sandblaster, Portable, 3 C.F.		83.80		92.18		
1 Set Sand Blasting Accessories		15.35		16.89		
1 Telescoping Boom Lift, to 60'		310.90		341.99	19.11	21.02
32 L.H., Daily Totals		$2105.95		$3003.51	$65.81	$93.86

Crew E-11B

Crew No.	Bare Costs Hr.	Bare Costs Daily	Incl. Subs O&P Hr.	Incl. Subs O&P Daily	Cost Per Labor-Hour Bare Costs	Cost Per Labor-Hour Incl. O&P
2 Painters, Struc. Steel	$45.85	$733.60	$74.45	$1191.20	$44.60	$70.72
1 Building Laborer	42.10	336.80	63.25	506.00		
2 Paint Sprayer, 8 C.F.M.		87.50		96.25		
1 Telescoping Boom Lift, to 60'		310.90		341.99	16.60	18.26
24 L.H., Daily Totals		$1468.80		$2135.44	$61.20	$88.98

Crew E-12

Crew No.	Bare Costs Hr.	Bare Costs Daily	Incl. Subs O&P Hr.	Incl. Subs O&P Daily	Cost Per Labor-Hour Bare Costs	Cost Per Labor-Hour Incl. O&P
1 Welder Foreman (outside)	$59.45	$475.60	$93.25	$746.00	$56.23	$86.22
1 Equip. Oper. (light)	53.00	424.00	79.20	633.60		
1 Welder, Gas Engine, 300 amp		147.00		161.70	9.19	10.11
16 L.H., Daily Totals		$1046.60		$1541.30	$65.41	$96.33

Crew E-13

Crew No.	Bare Costs Hr.	Bare Costs Daily	Incl. Subs O&P Hr.	Incl. Subs O&P Daily	Cost Per Labor-Hour Bare Costs	Cost Per Labor-Hour Incl. O&P
1 Welder Foreman (outside)	$59.45	$475.60	$93.25	$746.00	$57.30	$88.57
.5 Equip. Oper. (light)	53.00	212.00	79.20	316.80		
1 Welder, Gas Engine, 300 amp		147.00		161.70	12.25	13.48
12 L.H., Daily Totals		$834.60		$1224.50	$69.55	$102.04

Crew E-14

Crew No.	Bare Costs Hr.	Bare Costs Daily	Incl. Subs O&P Hr.	Incl. Subs O&P Daily	Cost Per Labor-Hour Bare Costs	Cost Per Labor-Hour Incl. O&P
1 Welder Foreman (outside)	$59.45	$475.60	$93.25	$746.00	$59.45	$93.25
1 Welder, Gas Engine, 300 amp		147.00		161.70	18.38	20.21
8 L.H., Daily Totals		$622.60		$907.70	$77.83	$113.46

Crew E-16

Crew No.	Bare Costs Hr.	Bare Costs Daily	Incl. Subs O&P Hr.	Incl. Subs O&P Daily	Cost Per Labor-Hour Bare Costs	Cost Per Labor-Hour Incl. O&P
1 Welder Foreman (outside)	$59.45	$475.60	$93.25	$746.00	$58.45	$91.67
1 Welder	57.45	459.60	90.10	720.80		
1 Welder, Gas Engine, 300 amp		147.00		161.70	9.19	10.11
16 L.H., Daily Totals		$1082.20		$1628.50	$67.64	$101.78

Crew E-17

Crew No.	Bare Costs Hr.	Bare Costs Daily	Incl. Subs O&P Hr.	Incl. Subs O&P Daily	Cost Per Labor-Hour Bare Costs	Cost Per Labor-Hour Incl. O&P
1 Struc. Steel Foreman (outside)	$59.65	$477.20	$93.55	$748.40	$58.65	$92.00
1 Structural Steel Worker	57.65	461.20	90.45	723.60		
16 L.H., Daily Totals		$938.40		$1472.00	$58.65	$92.00

Crew E-18

Crew No.	Bare Costs Hr.	Bare Costs Daily	Incl. Subs O&P Hr.	Incl. Subs O&P Daily	Cost Per Labor-Hour Bare Costs	Cost Per Labor-Hour Incl. O&P
1 Struc. Steel Foreman (outside)	$59.65	$477.20	$93.55	$748.40	$57.87	$89.95
3 Structural Steel Workers	57.65	1383.60	90.45	2170.80		
1 Equipment Operator (med.)	56.75	454.00	84.85	678.80		
1 Lattice Boom Crane, 20 Ton		1488.00		1636.80	37.20	40.92
40 L.H., Daily Totals		$3802.80		$5234.80	$95.07	$130.87

Crew E-19

Crew No.	Bare Costs Hr.	Bare Costs Daily	Incl. Subs O&P Hr.	Incl. Subs O&P Daily	Cost Per Labor-Hour Bare Costs	Cost Per Labor-Hour Incl. O&P
1 Struc. Steel Foreman (outside)	$59.65	$477.20	$93.55	$748.40	$56.77	$87.73
1 Structural Steel Worker	57.65	461.20	90.45	723.60		
1 Equip. Oper. (light)	53.00	424.00	79.20	633.60		
1 Lattice Boom Crane, 20 Ton		1488.00		1636.80	62.00	68.20
24 L.H., Daily Totals		$2850.40		$3742.40	$118.77	$155.93

Crew E-20

Crew No.	Bare Costs Hr.	Bare Costs Daily	Incl. Subs O&P Hr.	Incl. Subs O&P Daily	Cost Per Labor-Hour Bare Costs	Cost Per Labor-Hour Incl. O&P
1 Struc. Steel Foreman (outside)	$59.65	$477.20	$93.55	$748.40	$57.21	$88.73
5 Structural Steel Workers	57.65	2306.00	90.45	3618.00		
1 Equip. Oper. (crane)	59.20	473.60	88.50	708.00		
1 Equip. Oper. (oiler)	50.55	404.40	75.55	604.40		
1 Lattice Boom Crane, 40 Ton		2028.00		2230.80	31.69	34.86
64 L.H., Daily Totals		$5689.20		$7909.60	$88.89	$123.59

Crew E-22

Crew No.	Bare Costs Hr.	Bare Costs Daily	Incl. Subs O&P Hr.	Incl. Subs O&P Daily	Cost Per Labor-Hour Bare Costs	Cost Per Labor-Hour Incl. O&P
1 Skilled Worker Foreman (out)	$56.85	$454.80	$86.00	$688.00	$55.52	$83.97
2 Skilled Workers	54.85	877.60	82.95	1327.20		
24 L.H., Daily Totals		$1332.40		$2015.20	$55.52	$83.97

Crew E-24

Crew No.	Bare Costs Hr.	Bare Costs Daily	Incl. Subs O&P Hr.	Incl. Subs O&P Daily	Cost Per Labor-Hour Bare Costs	Cost Per Labor-Hour Incl. O&P
3 Structural Steel Workers	$57.65	$1383.60	$90.45	$2170.80	$57.42	$89.05
1 Equipment Operator (med.)	56.75	454.00	84.85	678.80		
1 Hyd. Crane, 25 Ton		580.85		638.93	18.15	19.97
32 L.H., Daily Totals		$2418.45		$3488.53	$75.58	$109.02

Crew E-25

Crew No.	Bare Costs Hr.	Bare Costs Daily	Incl. Subs O&P Hr.	Incl. Subs O&P Daily	Cost Per Labor-Hour Bare Costs	Cost Per Labor-Hour Incl. O&P
1 Welder Foreman (outside)	$59.45	$475.60	$93.25	$746.00	$59.45	$93.25
1 Cutting Torch		12.80		14.08	1.60	1.76
8 L.H., Daily Totals		$488.40		$760.08	$61.05	$95.01

Crew E-26

Crew No.	Bare Costs Hr.	Bare Costs Daily	Incl. Subs O&P Hr.	Incl. Subs O&P Daily	Cost Per Labor-Hour Bare Costs	Cost Per Labor-Hour Incl. O&P
1 Struc. Steel Foreman (outside)	$59.65	$477.20	$93.55	$748.40	$58.91	$91.73
1 Struc. Steel Worker	57.65	461.20	90.45	723.60		
1 Welder	57.45	459.60	90.10	720.80		
.25 Electrician	61.35	122.70	91.35	182.70		
.25 Plumber	64.45	128.90	96.45	192.90		
1 Welder, Gas Engine, 300 amp		147.00		161.70	5.25	5.78
28 L.H., Daily Totals		$1796.60		$2730.10	$64.16	$97.50

For customer support on your Site Work & Landscape Costs with RSMeans data, call 800.448.8182.

771

Crew No.	Bare Costs Hr.	Daily	Incl. Subs O&P Hr.	Daily	Cost Per Labor-Hour Bare Costs	Incl. O&P
Crew E-27						
1 Struc. Steel Foreman (outside)	$59.65	$477.20	$93.55	$748.40	$57.21	$88.73
5 Struc. Steel Workers	57.65	2306.00	90.45	3618.00		
1 Equip. Oper. (crane)	59.20	473.60	88.50	708.00		
1 Equip. Oper. (oiler)	50.55	404.40	75.55	604.40		
1 Hyd. Crane, 12 Ton		471.00		518.10		
1 Hyd. Crane, 80 Ton		1486.00		1634.60	30.58	33.64
64 L.H., Daily Totals		$5618.20		$7831.50	$87.78	$122.37
Crew F-3						
4 Carpenters	$53.15	$1700.80	$79.85	$2555.20	$54.36	$81.58
1 Equip. Oper. (crane)	59.20	473.60	88.50	708.00		
1 Hyd. Crane, 12 Ton		471.00		518.10	11.78	12.95
40 L.H., Daily Totals		$2645.40		$3781.30	$66.14	$94.53
Crew F-4						
4 Carpenters	$53.15	$1700.80	$79.85	$2555.20	$53.73	$80.58
1 Equip. Oper. (crane)	59.20	473.60	88.50	708.00		
1 Equip. Oper. (oiler)	50.55	404.40	75.55	604.40		
1 Hyd. Crane, 55 Ton		980.20		1078.22	20.42	22.46
48 L.H., Daily Totals		$3559.00		$4945.82	$74.15	$103.04
Crew F-5						
1 Carpenter Foreman (outside)	$55.15	$441.20	$82.85	$662.80	$53.65	$80.60
3 Carpenters	53.15	1275.60	79.85	1916.40		
32 L.H., Daily Totals		$1716.80		$2579.20	$53.65	$80.60
Crew F-6						
2 Carpenters	$53.15	$850.40	$79.85	$1277.60	$49.94	$74.94
2 Building Laborers	42.10	673.60	63.25	1012.00		
1 Equip. Oper. (crane)	59.20	473.60	88.50	708.00		
1 Hyd. Crane, 12 Ton		471.00		518.10	11.78	12.95
40 L.H., Daily Totals		$2468.60		$3515.70	$61.72	$87.89
Crew F-7						
2 Carpenters	$53.15	$850.40	$79.85	$1277.60	$47.63	$71.55
2 Building Laborers	42.10	673.60	63.25	1012.00		
32 L.H., Daily Totals		$1524.00		$2289.60	$47.63	$71.55
Crew G-1						
1 Roofer Foreman (outside)	$48.20	$385.60	$79.55	$636.40	$43.17	$71.25
4 Roofers Composition	46.20	1478.40	76.25	2440.00		
2 Roofer Helpers	34.60	553.60	57.10	913.60		
1 Application Equipment		192.85		212.13		
1 Tar Kettle/Pot		207.85		228.63		
1 Crew Truck		166.25		182.88	10.12	11.14
56 L.H., Daily Totals		$2984.55		$4613.65	$53.30	$82.39
Crew G-2						
1 Plasterer	$48.60	$388.80	$72.55	$580.40	$44.37	$66.37
1 Plasterer Helper	42.40	339.20	63.30	506.40		
1 Building Laborer	42.10	336.80	63.25	506.00		
1 Grout Pump, 50 C.F./hr.		188.45		207.29	7.85	8.64
24 L.H., Daily Totals		$1253.25		$1800.10	$52.22	$75.00

Crew No.	Bare Costs Hr.	Daily	Incl. Subs O&P Hr.	Daily	Cost Per Labor-Hour Bare Costs	Incl. O&P
Crew G-2A						
1 Roofer Composition	$46.20	$369.60	$76.25	$610.00	$40.97	$65.53
1 Roofer Helper	34.60	276.80	57.10	456.80		
1 Building Laborer	42.10	336.80	63.25	506.00		
1 Foam Spray Rig, Trailer-Mtd.		523.85		576.24		
1 Pickup Truck, 3/4 Ton		110.85		121.94	26.45	29.09
24 L.H., Daily Totals		$1617.90		$2270.97	$67.41	$94.62
Crew G-3						
2 Sheet Metal Workers	$62.30	$996.80	$94.50	$1512.00	$52.20	$78.88
2 Building Laborers	42.10	673.60	63.25	1012.00		
32 L.H., Daily Totals		$1670.40		$2524.00	$52.20	$78.88
Crew G-4						
1 Labor Foreman (outside)	$44.10	$352.80	$66.25	$530.00	$42.77	$64.25
2 Building Laborers	42.10	673.60	63.25	1012.00		
1 Flatbed Truck, Gas, 1.5 Ton		196.15		215.76		
1 Air Compressor, 160 cfm		202.80		223.08	16.62	18.29
24 L.H., Daily Totals		$1425.35		$1980.85	$59.39	$82.54
Crew G-5						
1 Roofer Foreman (outside)	$48.20	$385.60	$79.55	$636.40	$41.96	$69.25
2 Roofers Composition	46.20	739.20	76.25	1220.00		
2 Roofer Helpers	34.60	553.60	57.10	913.60		
1 Application Equipment		192.85		212.13	4.82	5.30
40 L.H., Daily Totals		$1871.25		$2982.14	$46.78	$74.55
Crew G-6A						
2 Roofers Composition	$46.20	$739.20	$76.25	$1220.00	$46.20	$76.25
1 Small Compressor, Electric		33.80		37.18		
2 Pneumatic Nailers		54.80		60.28	5.54	6.09
16 L.H., Daily Totals		$827.80		$1317.46	$51.74	$82.34
Crew G-7						
1 Carpenter	$53.15	$425.20	$79.85	$638.80	$53.15	$79.85
1 Small Compressor, Electric		33.80		37.18		
1 Pneumatic Nailer		27.40		30.14	7.65	8.41
8 L.H., Daily Totals		$486.40		$706.12	$60.80	$88.27
Crew H-1						
2 Glaziers	$51.00	$816.00	$76.40	$1222.40	$54.33	$83.42
2 Struc. Steel Workers	57.65	922.40	90.45	1447.20		
32 L.H., Daily Totals		$1738.40		$2669.60	$54.33	$83.42
Crew H-2						
2 Glaziers	$51.00	$816.00	$76.40	$1222.40	$48.03	$72.02
1 Building Laborer	42.10	336.80	63.25	506.00		
24 L.H., Daily Totals		$1152.80		$1728.40	$48.03	$72.02
Crew H-3						
1 Glazier	$51.00	$408.00	$76.40	$611.20	$45.48	$68.67
1 Helper	39.95	319.60	60.95	487.60		
16 L.H., Daily Totals		$727.60		$1098.80	$45.48	$68.67
Crew H-4						
1 Carpenter	$53.15	$425.20	$79.85	$638.80	$49.51	$74.59
1 Carpenter Helper	39.95	319.60	60.95	487.60		
.5 Electrician	61.35	245.40	91.35	365.40		
20 L.H., Daily Totals		$990.20		$1491.80	$49.51	$74.59

Crew No.	Bare Costs		Incl. Subs O&P		Cost Per Labor-Hour	
Crew J-1	Hr.	Daily	Hr.	Daily	Bare Costs	Incl. O&P
3 Plasterers	$48.60	$1166.40	$72.55	$1741.20	$46.12	$68.85
2 Plasterer Helpers	42.40	678.40	63.30	1012.80		
1 Mixing Machine, 6 C.F.		117.35		129.09	2.93	3.23
40 L.H., Daily Totals		$1962.15		$2883.09	$49.05	$72.08
Crew J-2	Hr.	Daily	Hr.	Daily	Bare Costs	Incl. O&P
3 Plasterers	$48.60	$1166.40	$72.55	$1741.20	$47.11	$70.09
2 Plasterer Helpers	42.40	678.40	63.30	1012.80		
1 Lather	52.05	416.40	76.30	610.40		
1 Mixing Machine, 6 C.F.		117.35		129.09	2.44	2.69
48 L.H., Daily Totals		$2378.55		$3493.49	$49.55	$72.78
Crew J-3	Hr.	Daily	Hr.	Daily	Bare Costs	Incl. O&P
1 Terrazzo Worker	$49.40	$395.20	$72.45	$579.60	$44.95	$65.92
1 Terrazzo Helper	40.50	324.00	59.40	475.20		
1 Floor Grinder, 22" Path		104.60		115.06		
1 Terrazzo Mixer		161.15		177.26	16.61	18.27
16 L.H., Daily Totals		$984.95		$1347.13	$61.56	$84.20
Crew J-4	Hr.	Daily	Hr.	Daily	Bare Costs	Incl. O&P
2 Cement Finishers	$49.95	$799.20	$73.45	$1175.20	$47.33	$70.05
1 Laborer	42.10	336.80	63.25	506.00		
1 Floor Grinder, 22" Path		104.60		115.06		
1 Floor Edger, 7" Path		43.95		48.34		
1 Vacuum Pick-Up System		74.05		81.45	9.28	10.20
24 L.H., Daily Totals		$1358.60		$1926.06	$56.61	$80.25
Crew J-4A	Hr.	Daily	Hr.	Daily	Bare Costs	Incl. O&P
2 Cement Finishers	$49.95	$799.20	$73.45	$1175.20	$46.02	$68.35
2 Laborers	42.10	673.60	63.25	1012.00		
1 Floor Grinder, 22" Path		104.60		115.06		
1 Floor Edger, 7" Path		43.95		48.34		
1 Vacuum Pick-Up System		74.05		81.45		
1 Floor Auto Scrubber		178.40		196.24	12.53	13.78
32 L.H., Daily Totals		$1873.80		$2628.30	$58.56	$82.13
Crew J-4B	Hr.	Daily	Hr.	Daily	Bare Costs	Incl. O&P
1 Laborer	$42.10	$336.80	$63.25	$506.00	$42.10	$63.25
1 Floor Auto Scrubber		178.40		196.24	22.30	24.53
8 L.H., Daily Totals		$515.20		$702.24	$64.40	$87.78
Crew J-6	Hr.	Daily	Hr.	Daily	Bare Costs	Incl. O&P
2 Painters	$44.40	$710.40	$66.05	$1056.80	$45.98	$68.64
1 Building Laborer	42.10	336.80	63.25	506.00		
1 Equip. Oper. (light)	53.00	424.00	79.20	633.60		
1 Air Compressor, 250 cfm		201.50		221.65		
1 Sandblaster, Portable, 3 C.F.		83.80		92.18		
1 Set Sand Blasting Accessories		15.35		16.89	9.40	10.33
32 L.H., Daily Totals		$1771.85		$2527.11	$55.37	$78.97
Crew J-7	Hr.	Daily	Hr.	Daily	Bare Costs	Incl. O&P
2 Painters	$44.40	$710.40	$66.05	$1056.80	$44.40	$66.05
1 Floor Belt Sander		50.15		55.16		
1 Floor Sanding Edger		25.15		27.66	4.71	5.18
16 L.H., Daily Totals		$785.70		$1139.63	$49.11	$71.23

Crew No.	Bare Costs		Incl. Subs O&P		Cost Per Labor-Hour	
Crew K-1	Hr.	Daily	Hr.	Daily	Bare Costs	Incl. O&P
1 Carpenter	$53.15	$425.20	$79.85	$638.80	$50.20	$75.33
1 Truck Driver (light)	47.25	378.00	70.80	566.40		
1 Flatbed Truck, Gas, 3 Ton		820.40		902.44	51.27	56.40
16 L.H., Daily Totals		$1623.60		$2107.64	$101.47	$131.73
Crew K-2	Hr.	Daily	Hr.	Daily	Bare Costs	Incl. O&P
1 Struc. Steel Foreman (outside)	$59.65	$477.20	$93.55	$748.40	$54.85	$84.93
1 Struc. Steel Worker	57.65	461.20	90.45	723.60		
1 Truck Driver (light)	47.25	378.00	70.80	566.40		
1 Flatbed Truck, Gas, 3 Ton		820.40		902.44	34.18	37.60
24 L.H., Daily Totals		$2136.80		$2940.84	$89.03	$122.54
Crew L-1	Hr.	Daily	Hr.	Daily	Bare Costs	Incl. O&P
1 Electrician	$61.35	$490.80	$91.35	$730.80	$62.90	$93.90
1 Plumber	64.45	515.60	96.45	771.60		
16 L.H., Daily Totals		$1006.40		$1502.40	$62.90	$93.90
Crew L-2	Hr.	Daily	Hr.	Daily	Bare Costs	Incl. O&P
1 Carpenter	$53.15	$425.20	$79.85	$638.80	$46.55	$70.40
1 Carpenter Helper	39.95	319.60	60.95	487.60		
16 L.H., Daily Totals		$744.80		$1126.40	$46.55	$70.40
Crew L-3	Hr.	Daily	Hr.	Daily	Bare Costs	Incl. O&P
1 Carpenter	$53.15	$425.20	$79.85	$638.80	$57.49	$86.39
.5 Electrician	61.35	245.40	91.35	365.40		
.5 Sheet Metal Worker	62.30	249.20	94.50	378.00		
16 L.H., Daily Totals		$919.80		$1382.20	$57.49	$86.39
Crew L-3A	Hr.	Daily	Hr.	Daily	Bare Costs	Incl. O&P
1 Carpenter Foreman (outside)	$55.15	$441.20	$82.85	$662.80	$57.53	$86.73
.5 Sheet Metal Worker	62.30	249.20	94.50	378.00		
12 L.H., Daily Totals		$690.40		$1040.80	$57.53	$86.73
Crew L-4	Hr.	Daily	Hr.	Daily	Bare Costs	Incl. O&P
2 Skilled Workers	$54.85	$877.60	$82.95	$1327.20	$49.88	$75.62
1 Helper	39.95	319.60	60.95	487.60		
24 L.H., Daily Totals		$1197.20		$1814.80	$49.88	$75.62
Crew L-5	Hr.	Daily	Hr.	Daily	Bare Costs	Incl. O&P
1 Struc. Steel Foreman (outside)	$59.65	$477.20	$93.55	$748.40	$58.16	$90.61
5 Struc. Steel Workers	57.65	2306.00	90.45	3618.00		
1 Equip. Oper. (crane)	59.20	473.60	88.50	708.00		
1 Hyd. Crane, 25 Ton		580.85		638.93	10.37	11.41
56 L.H., Daily Totals		$3837.65		$5713.34	$68.53	$102.02
Crew L-5A	Hr.	Daily	Hr.	Daily	Bare Costs	Incl. O&P
1 Struc. Steel Foreman (outside)	$59.65	$477.20	$93.55	$748.40	$58.54	$90.74
2 Structural Steel Workers	57.65	922.40	90.45	1447.20		
1 Equip. Oper. (crane)	59.20	473.60	88.50	708.00		
1 S.P. Crane, 4x4, 25 Ton		1144.00		1258.40	35.75	39.33
32 L.H., Daily Totals		$3017.20		$4162.00	$94.29	$130.06

For customer support on your Site Work & Landscape Costs with RSMeans data, call 800.448.8182.

773

Crew No.	Bare Costs		Incl. Subs O&P		Cost Per Labor-Hour	
Crew L-5B	**Hr.**	**Daily**	**Hr.**	**Daily**	**Bare Costs**	**Incl. O&P**
1 Struc. Steel Foreman (outside)	$59.65	$477.20	$93.55	$748.40	$59.83	$90.82
2 Structural Steel Workers	57.65	922.40	90.45	1447.20		
2 Electricians	61.35	981.60	91.35	1461.60		
2 Steamfitters/Pipefitters	65.55	1048.80	98.10	1569.60		
1 Equip. Oper. (crane)	59.20	473.60	88.50	708.00		
1 Equip. Oper. (oiler)	50.55	404.40	75.55	604.40		
1 Hyd. Crane, 80 Ton		1486.00		1634.60	20.64	22.70
72 L.H., Daily Totals		$5794.00		$8173.80	$80.47	$113.53
Crew L-6	**Hr.**	**Daily**	**Hr.**	**Daily**	**Bare Costs**	**Incl. O&P**
1 Plumber	$64.45	$515.60	$96.45	$771.60	$63.42	$94.75
.5 Electrician	61.35	245.40	91.35	365.40		
12 L.H., Daily Totals		$761.00		$1137.00	$63.42	$94.75
Crew L-7	**Hr.**	**Daily**	**Hr.**	**Daily**	**Bare Costs**	**Incl. O&P**
2 Carpenters	$53.15	$850.40	$79.85	$1277.60	$51.16	$76.75
1 Building Laborer	42.10	336.80	63.25	506.00		
.5 Electrician	61.35	245.40	91.35	365.40		
28 L.H., Daily Totals		$1432.60		$2149.00	$51.16	$76.75
Crew L-8	**Hr.**	**Daily**	**Hr.**	**Daily**	**Bare Costs**	**Incl. O&P**
2 Carpenters	$53.15	$850.40	$79.85	$1277.60	$55.41	$83.17
.5 Plumber	64.45	257.80	96.45	385.80		
20 L.H., Daily Totals		$1108.20		$1663.40	$55.41	$83.17
Crew L-9	**Hr.**	**Daily**	**Hr.**	**Daily**	**Bare Costs**	**Incl. O&P**
1 Labor Foreman (inside)	$42.60	$340.80	$64.00	$512.00	$47.81	$72.58
2 Building Laborers	42.10	673.60	63.25	1012.00		
1 Struc. Steel Worker	57.65	461.20	90.45	723.60		
.5 Electrician	61.35	245.40	91.35	365.40		
36 L.H., Daily Totals		$1721.00		$2613.00	$47.81	$72.58
Crew L-10	**Hr.**	**Daily**	**Hr.**	**Daily**	**Bare Costs**	**Incl. O&P**
1 Struc. Steel Foreman (outside)	$59.65	$477.20	$93.55	$748.40	$58.83	$90.83
1 Structural Steel Worker	57.65	461.20	90.45	723.60		
1 Equip. Oper. (crane)	59.20	473.60	88.50	708.00		
1 Hyd. Crane, 12 Ton		471.00		518.10	19.63	21.59
24 L.H., Daily Totals		$1883.00		$2698.10	$78.46	$112.42
Crew L-11	**Hr.**	**Daily**	**Hr.**	**Daily**	**Bare Costs**	**Incl. O&P**
2 Wreckers	$42.10	$673.60	$64.80	$1036.80	$49.10	$74.33
1 Equip. Oper. (crane)	59.20	473.60	88.50	708.00		
1 Equip. Oper. (light)	53.00	424.00	79.20	633.60		
1 Hyd. Excavator, 2.5 C.Y.		1465.00		1611.50		
1 Loader, Skid Steer, 78 H.P.		400.10		440.11	58.28	64.11
32 L.H., Daily Totals		$3436.30		$4430.01	$107.38	$138.44
Crew M-1	**Hr.**	**Daily**	**Hr.**	**Daily**	**Bare Costs**	**Incl. O&P**
3 Elevator Constructors	$85.55	$2053.20	$127.00	$3048.00	$81.28	$120.65
1 Elevator Apprentice	68.45	547.60	101.60	812.80		
5 Hand Tools		50.00		55.00	1.56	1.72
32 L.H., Daily Totals		$2650.80		$3915.80	$82.84	$122.37

Crew No.	Bare Costs		Incl. Subs O&P		Cost Per Labor-Hour	
Crew M-3	**Hr.**	**Daily**	**Hr.**	**Daily**	**Bare Costs**	**Incl. O&P**
1 Electrician Foreman (outside)	$63.35	$506.80	$94.35	$754.80	$64.39	$95.86
1 Common Laborer	42.10	336.80	63.25	506.00		
.25 Equipment Operator (med.)	56.75	113.50	84.85	169.70		
1 Elevator Constructor	85.55	684.40	127.00	1016.00		
1 Elevator Apprentice	68.45	547.60	101.60	812.80		
.25 S.P. Crane, 4x4, 20 Ton		145.55		160.10	4.28	4.71
34 L.H., Daily Totals		$2334.65		$3419.41	$68.67	$100.57
Crew M-4	**Hr.**	**Daily**	**Hr.**	**Daily**	**Bare Costs**	**Incl. O&P**
1 Electrician Foreman (outside)	$63.35	$506.80	$94.35	$754.80	$63.75	$94.94
1 Common Laborer	42.10	336.80	63.25	506.00		
.25 Equipment Operator, Crane	59.20	118.40	88.50	177.00		
.25 Equip. Oper. (oiler)	50.55	101.10	75.55	151.10		
1 Elevator Constructor	85.55	684.40	127.00	1016.00		
1 Elevator Apprentice	68.45	547.60	101.60	812.80		
.25 S.P. Crane, 4x4, 40 Ton		188.55		207.41	5.24	5.76
36 L.H., Daily Totals		$2483.65		$3625.11	$68.99	$100.70
Crew Q-1	**Hr.**	**Daily**	**Hr.**	**Daily**	**Bare Costs**	**Incl. O&P**
1 Plumber	$64.45	$515.60	$96.45	$771.60	$58.00	$86.80
1 Plumber Apprentice	51.55	412.40	77.15	617.20		
16 L.H., Daily Totals		$928.00		$1388.80	$58.00	$86.80
Crew Q-1A	**Hr.**	**Daily**	**Hr.**	**Daily**	**Bare Costs**	**Incl. O&P**
.25 Plumber Foreman (outside)	$66.45	$132.90	$99.45	$198.90	$64.85	$97.05
1 Plumber	64.45	515.60	96.45	771.60		
10 L.H., Daily Totals		$648.50		$970.50	$64.85	$97.05
Crew Q-1C	**Hr.**	**Daily**	**Hr.**	**Daily**	**Bare Costs**	**Incl. O&P**
1 Plumber	$64.45	$515.60	$96.45	$771.60	$57.58	$86.15
1 Plumber Apprentice	51.55	412.40	77.15	617.20		
1 Equip. Oper. (medium)	56.75	454.00	84.85	678.80		
1 Trencher, Chain Type, 8' D		1872.00		2059.20	78.00	85.80
24 L.H., Daily Totals		$3254.00		$4126.80	$135.58	$171.95
Crew Q-2	**Hr.**	**Daily**	**Hr.**	**Daily**	**Bare Costs**	**Incl. O&P**
2 Plumbers	$64.45	$1031.20	$96.45	$1543.20	$60.15	$90.02
1 Plumber Apprentice	51.55	412.40	77.15	617.20		
24 L.H., Daily Totals		$1443.60		$2160.40	$60.15	$90.02
Crew Q-3	**Hr.**	**Daily**	**Hr.**	**Daily**	**Bare Costs**	**Incl. O&P**
1 Plumber Foreman (inside)	$64.95	$519.60	$97.20	$777.60	$61.35	$91.81
2 Plumbers	64.45	1031.20	96.45	1543.20		
1 Plumber Apprentice	51.55	412.40	77.15	617.20		
32 L.H., Daily Totals		$1963.20		$2938.00	$61.35	$91.81
Crew Q-4	**Hr.**	**Daily**	**Hr.**	**Daily**	**Bare Costs**	**Incl. O&P**
1 Plumber Foreman (inside)	$64.95	$519.60	$97.20	$777.60	$61.35	$91.81
1 Plumber	64.45	515.60	96.45	771.60		
1 Welder (plumber)	64.45	515.60	96.45	771.60		
1 Plumber Apprentice	51.55	412.40	77.15	617.20		
1 Welder, Electric, 300 amp		106.40		117.04	3.33	3.66
32 L.H., Daily Totals		$2069.60		$3055.04	$64.67	$95.47
Crew Q-5	**Hr.**	**Daily**	**Hr.**	**Daily**	**Bare Costs**	**Incl. O&P**
1 Steamfitter	$65.55	$524.40	$98.10	$784.80	$59.00	$88.30
1 Steamfitter Apprentice	52.45	419.60	78.50	628.00		
16 L.H., Daily Totals		$944.00		$1412.80	$59.00	$88.30

Crews - Standard

Crew Q-6	Hr.	Daily	Hr.	Daily	Bare Costs	Incl. O&P
2 Steamfitters	$65.55	$1048.80	$98.10	$1569.60	$61.18	$91.57
1 Steamfitter Apprentice	52.45	419.60	78.50	628.00		
24 L.H., Daily Totals		$1468.40		$2197.60	$61.18	$91.57

Crew Q-7	Hr.	Daily	Hr.	Daily	Bare Costs	Incl. O&P
1 Steamfitter Foreman (inside)	$66.05	$528.40	$98.85	$790.80	$62.40	$93.39
2 Steamfitters	65.55	1048.80	98.10	1569.60		
1 Steamfitter Apprentice	52.45	419.60	78.50	628.00		
32 L.H., Daily Totals		$1996.80		$2988.40	$62.40	$93.39

Crew Q-8	Hr.	Daily	Hr.	Daily	Bare Costs	Incl. O&P
1 Steamfitter Foreman (inside)	$66.05	$528.40	$98.85	$790.80	$62.40	$93.39
1 Steamfitter	65.55	524.40	98.10	784.80		
1 Welder (steamfitter)	65.55	524.40	98.10	784.80		
1 Steamfitter Apprentice	52.45	419.60	78.50	628.00		
1 Welder, Electric, 300 amp		106.40		117.04	3.33	3.66
32 L.H., Daily Totals		$2103.20		$3105.44	$65.72	$97.05

Crew Q-9	Hr.	Daily	Hr.	Daily	Bare Costs	Incl. O&P
1 Sheet Metal Worker	$62.30	$498.40	$94.50	$756.00	$56.08	$85.05
1 Sheet Metal Apprentice	49.85	398.80	75.60	604.80		
16 L.H., Daily Totals		$897.20		$1360.80	$56.08	$85.05

Crew Q-10	Hr.	Daily	Hr.	Daily	Bare Costs	Incl. O&P
2 Sheet Metal Workers	$62.30	$996.80	$94.50	$1512.00	$58.15	$88.20
1 Sheet Metal Apprentice	49.85	398.80	75.60	604.80		
24 L.H., Daily Totals		$1395.60		$2116.80	$58.15	$88.20

Crew Q-11	Hr.	Daily	Hr.	Daily	Bare Costs	Incl. O&P
1 Sheet Metal Foreman (inside)	$62.80	$502.40	$95.25	$762.00	$59.31	$89.96
2 Sheet Metal Workers	62.30	996.80	94.50	1512.00		
1 Sheet Metal Apprentice	49.85	398.80	75.60	604.80		
32 L.H., Daily Totals		$1898.00		$2878.80	$59.31	$89.96

Crew Q-12	Hr.	Daily	Hr.	Daily	Bare Costs	Incl. O&P
1 Sprinkler Installer	$63.25	$506.00	$94.80	$758.40	$56.92	$85.33
1 Sprinkler Apprentice	50.60	404.80	75.85	606.80		
16 L.H., Daily Totals		$910.80		$1365.20	$56.92	$85.33

Crew Q-13	Hr.	Daily	Hr.	Daily	Bare Costs	Incl. O&P
1 Sprinkler Foreman (inside)	$63.75	$510.00	$95.55	$764.40	$60.21	$90.25
2 Sprinkler Installers	63.25	1012.00	94.80	1516.80		
1 Sprinkler Apprentice	50.60	404.80	75.85	606.80		
32 L.H., Daily Totals		$1926.80		$2888.00	$60.21	$90.25

Crew Q-14	Hr.	Daily	Hr.	Daily	Bare Costs	Incl. O&P
1 Asbestos Worker	$58.65	$469.20	$90.05	$720.40	$52.77	$81.03
1 Asbestos Apprentice	46.90	375.20	72.00	576.00		
16 L.H., Daily Totals		$844.40		$1296.40	$52.77	$81.03

Crew Q-15	Hr.	Daily	Hr.	Daily	Bare Costs	Incl. O&P
1 Plumber	$64.45	$515.60	$96.45	$771.60	$58.00	$86.80
1 Plumber Apprentice	51.55	412.40	77.15	617.20		
1 Welder, Electric, 300 amp		106.40		117.04	6.65	7.32
16 L.H., Daily Totals		$1034.40		$1505.84	$64.65	$94.11

Crew Q-16	Hr.	Daily	Hr.	Daily	Bare Costs	Incl. O&P
2 Plumbers	$64.45	$1031.20	$96.45	$1543.20	$60.15	$90.02
1 Plumber Apprentice	51.55	412.40	77.15	617.20		
1 Welder, Electric, 300 amp		106.40		117.04	4.43	4.88
24 L.H., Daily Totals		$1550.00		$2277.44	$64.58	$94.89

Crew Q-17	Hr.	Daily	Hr.	Daily	Bare Costs	Incl. O&P
1 Steamfitter	$65.55	$524.40	$98.10	$784.80	$59.00	$88.30
1 Steamfitter Apprentice	52.45	419.60	78.50	628.00		
1 Welder, Electric, 300 amp		106.40		117.04	6.65	7.32
16 L.H., Daily Totals		$1050.40		$1529.84	$65.65	$95.61

Crew Q-17A	Hr.	Daily	Hr.	Daily	Bare Costs	Incl. O&P
1 Steamfitter	$65.55	$524.40	$98.10	$784.80	$59.07	$88.37
1 Steamfitter Apprentice	52.45	419.60	78.50	628.00		
1 Equip. Oper. (crane)	59.20	473.60	88.50	708.00		
1 Hyd. Crane, 12 Ton		471.00		518.10		
1 Welder, Electric, 300 amp		106.40		117.04	24.06	26.46
24 L.H., Daily Totals		$1995.00		$2755.94	$83.13	$114.83

Crew Q-18	Hr.	Daily	Hr.	Daily	Bare Costs	Incl. O&P
2 Steamfitters	$65.55	$1048.80	$98.10	$1569.60	$61.18	$91.57
1 Steamfitter Apprentice	52.45	419.60	78.50	628.00		
1 Welder, Electric, 300 amp		106.40		117.04	4.43	4.88
24 L.H., Daily Totals		$1574.80		$2314.64	$65.62	$96.44

Crew Q-19	Hr.	Daily	Hr.	Daily	Bare Costs	Incl. O&P
1 Steamfitter	$65.55	$524.40	$98.10	$784.80	$59.78	$89.32
1 Steamfitter Apprentice	52.45	419.60	78.50	628.00		
1 Electrician	61.35	490.80	91.35	730.80		
24 L.H., Daily Totals		$1434.80		$2143.60	$59.78	$89.32

Crew Q-20	Hr.	Daily	Hr.	Daily	Bare Costs	Incl. O&P
1 Sheet Metal Worker	$62.30	$498.40	$94.50	$756.00	$57.13	$86.31
1 Sheet Metal Apprentice	49.85	398.80	75.60	604.80		
.5 Electrician	61.35	245.40	91.35	365.40		
20 L.H., Daily Totals		$1142.60		$1726.20	$57.13	$86.31

Crew Q-21	Hr.	Daily	Hr.	Daily	Bare Costs	Incl. O&P
2 Steamfitters	$65.55	$1048.80	$98.10	$1569.60	$61.23	$91.51
1 Steamfitter Apprentice	52.45	419.60	78.50	628.00		
1 Electrician	61.35	490.80	91.35	730.80		
32 L.H., Daily Totals		$1959.20		$2928.40	$61.23	$91.51

Crew Q-22	Hr.	Daily	Hr.	Daily	Bare Costs	Incl. O&P
1 Plumber	$64.45	$515.60	$96.45	$771.60	$58.00	$86.80
1 Plumber Apprentice	51.55	412.40	77.15	617.20		
1 Hyd. Crane, 12 Ton		471.00		518.10	29.44	32.38
16 L.H., Daily Totals		$1399.00		$1906.90	$87.44	$119.18

Crew Q-22A	Hr.	Daily	Hr.	Daily	Bare Costs	Incl. O&P
1 Plumber	$64.45	$515.60	$96.45	$771.60	$54.33	$81.34
1 Plumber Apprentice	51.55	412.40	77.15	617.20		
1 Laborer	42.10	336.80	63.25	506.00		
1 Equip. Oper. (crane)	59.20	473.60	88.50	708.00		
1 Hyd. Crane, 12 Ton		471.00		518.10	14.72	16.19
32 L.H., Daily Totals		$2209.40		$3120.90	$69.04	$97.53

For customer support on your Site Work & Landscape Costs with RSMeans data, call 800.448.8182.

775

Crew No.	Bare Costs Hr.	Daily	Incl. Subs O&P Hr.	Daily	Cost Per Labor-Hour Bare Costs	Incl. O&P
Crew Q-23	Hr.	Daily	Hr.	Daily	Bare Costs	Incl. O&P
1 Plumber Foreman (outside)	$66.45	$531.60	$99.45	$795.60	$62.55	$93.58
1 Plumber	64.45	515.60	96.45	771.60		
1 Equip. Oper. (medium)	56.75	454.00	84.85	678.80		
1 Lattice Boom Crane, 20 Ton		1488.00		1636.80	62.00	68.20
24 L.H., Daily Totals		$2989.20		$3882.80	$124.55	$161.78
Crew R-1	Hr.	Daily	Hr.	Daily	Bare Costs	Incl. O&P
1 Electrician Foreman	$61.85	$494.80	$92.10	$736.80	$57.35	$85.39
3 Electricians	61.35	1472.40	91.35	2192.40		
2 Electrician Apprentices	49.10	785.60	73.10	1169.60		
48 L.H., Daily Totals		$2752.80		$4098.80	$57.35	$85.39
Crew R-1A	Hr.	Daily	Hr.	Daily	Bare Costs	Incl. O&P
1 Electrician	$61.35	$490.80	$91.35	$730.80	$55.23	$82.22
1 Electrician Apprentice	49.10	392.80	73.10	584.80		
16 L.H., Daily Totals		$883.60		$1315.60	$55.23	$82.22
Crew R-1B	Hr.	Daily	Hr.	Daily	Bare Costs	Incl. O&P
1 Electrician	$61.35	$490.80	$91.35	$730.80	$53.18	$79.18
2 Electrician Apprentices	49.10	785.60	73.10	1169.60		
24 L.H., Daily Totals		$1276.40		$1900.40	$53.18	$79.18
Crew R-1C	Hr.	Daily	Hr.	Daily	Bare Costs	Incl. O&P
2 Electricians	$61.35	$981.60	$91.35	$1461.60	$55.23	$82.22
2 Electrician Apprentices	49.10	785.60	73.10	1169.60		
1 Portable cable puller, 8000 lb.		101.60		111.76	3.17	3.49
32 L.H., Daily Totals		$1868.80		$2742.96	$58.40	$85.72
Crew R-1D	Hr.	Daily	Hr.	Daily	Bare Costs	Incl. O&P
1 Electrician	$61.35	$490.80	$91.35	$730.80	$55.23	$82.23
1 Electrician Apprentice	49.10	392.80	73.10	584.80		
1 Aerial lift		105.00		115.50	6.56	7.22
16 L.H., Daily Totals		$988.60		$1431.10	$61.79	$89.45
Crew R-2	Hr.	Daily	Hr.	Daily	Bare Costs	Incl. O&P
1 Electrician Foreman	$61.85	$494.80	$92.10	$736.80	$57.61	$85.84
3 Electricians	61.35	1472.40	91.35	2192.40		
2 Electrician Apprentices	49.10	785.60	73.10	1169.60		
1 Equip. Oper. (crane)	59.20	473.60	88.50	708.00		
1 S.P. Crane, 4x4, 5 Ton		378.10		415.91	6.75	7.43
56 L.H., Daily Totals		$3604.50		$5222.71	$64.37	$93.26
Crew R-3	Hr.	Daily	Hr.	Daily	Bare Costs	Incl. O&P
1 Electrician Foreman	$61.85	$494.80	$92.10	$736.80	$61.12	$91.08
1 Electrician	61.35	490.80	91.35	730.80		
.5 Equip. Oper. (crane)	59.20	236.80	88.50	354.00		
.5 S.P. Crane, 4x4, 5 Ton		189.05		207.96	9.45	10.40
20 L.H., Daily Totals		$1411.45		$2029.56	$70.57	$101.48
Crew R-4	Hr.	Daily	Hr.	Daily	Bare Costs	Incl. O&P
1 Struc. Steel Foreman (outside)	$59.65	$477.20	$93.55	$748.40	$58.79	$91.25
3 Struc. Steel Workers	57.65	1383.60	90.45	2170.80		
1 Electrician	61.35	490.80	91.35	730.80		
1 Welder, Gas Engine, 300 amp		147.00		161.70	3.67	4.04
40 L.H., Daily Totals		$2498.60		$3811.70	$62.47	$95.29

Crew No.	Bare Costs Hr.	Daily	Incl. Subs O&P Hr.	Daily	Cost Per Labor-Hour Bare Costs	Incl. O&P
Crew R-5	Hr.	Daily	Hr.	Daily	Bare Costs	Incl. O&P
1 Electrician Foreman	$61.85	$494.80	$92.10	$736.80	$53.61	$80.36
4 Electrician Linemen	61.35	1963.20	91.35	2923.20		
2 Electrician Operators	61.35	981.60	91.35	1461.60		
4 Electrician Groundmen	39.95	1278.40	60.95	1950.40		
1 Crew Truck		166.25		182.88		
1 Flatbed Truck, 20,000 GVW		201.60		221.76		
1 Pickup Truck, 3/4 Ton		110.85		121.94		
.2 Hyd. Crane, 55 Ton		196.04		215.64		
.2 Hyd. Crane, 12 Ton		94.20		103.62		
.2 Earth Auger, Truck-Mtd.		40.10		44.11		
1 Tractor w/Winch		373.50		410.85	13.44	14.78
88 L.H., Daily Totals		$5900.54		$8372.79	$67.05	$95.15
Crew R-6	Hr.	Daily	Hr.	Daily	Bare Costs	Incl. O&P
1 Electrician Foreman	$61.85	$494.80	$92.10	$736.80	$53.61	$80.36
4 Electrician Linemen	61.35	1963.20	91.35	2923.20		
2 Electrician Operators	61.35	981.60	91.35	1461.60		
4 Electrician Groundmen	39.95	1278.40	60.95	1950.40		
1 Crew Truck		166.25		182.88		
1 Flatbed Truck, 20,000 GVW		201.60		221.76		
1 Pickup Truck, 3/4 Ton		110.85		121.94		
.2 Hyd. Crane, 55 Ton		196.04		215.64		
.2 Hyd. Crane, 12 Ton		94.20		103.62		
.2 Earth Auger, Truck-Mtd.		40.10		44.11		
1 Tractor w/Winch		373.50		410.85		
3 Cable Trailers		192.30		211.53		
.5 Tensioning Rig		55.20		60.72		
.5 Cable Pulling Rig		303.57		333.93	19.70	21.67
88 L.H., Daily Totals		$6451.61		$8978.98	$73.31	$102.03
Crew R-7	Hr.	Daily	Hr.	Daily	Bare Costs	Incl. O&P
1 Electrician Foreman	$61.85	$494.80	$92.10	$736.80	$43.60	$66.14
5 Electrician Groundmen	39.95	1598.00	60.95	2438.00		
1 Crew Truck		166.25		182.88	3.46	3.81
48 L.H., Daily Totals		$2259.05		$3357.68	$47.06	$69.95
Crew R-8	Hr.	Daily	Hr.	Daily	Bare Costs	Incl. O&P
1 Electrician Foreman	$61.85	$494.80	$92.10	$736.80	$54.30	$81.34
3 Electrician Linemen	61.35	1472.40	91.35	2192.40		
2 Electrician Groundmen	39.95	639.20	60.95	975.20		
1 Pickup Truck, 3/4 Ton		110.85		121.94		
1 Crew Truck		166.25		182.88	5.77	6.35
48 L.H., Daily Totals		$2883.50		$4209.21	$60.07	$87.69
Crew R-9	Hr.	Daily	Hr.	Daily	Bare Costs	Incl. O&P
1 Electrician Foreman	$61.85	$494.80	$92.10	$736.80	$50.71	$76.24
1 Electrician Lineman	61.35	490.80	91.35	730.80		
2 Electrician Operators	61.35	981.60	91.35	1461.60		
4 Electrician Groundmen	39.95	1278.40	60.95	1950.40		
1 Pickup Truck, 3/4 Ton		110.85		121.94		
1 Crew Truck		166.25		182.88	4.33	4.76
64 L.H., Daily Totals		$3522.70		$5184.41	$55.04	$81.01
Crew R-10	Hr.	Daily	Hr.	Daily	Bare Costs	Incl. O&P
1 Electrician Foreman	$61.85	$494.80	$92.10	$736.80	$57.87	$86.41
4 Electrician Linemen	61.35	1963.20	91.35	2923.20		
1 Electrician Groundman	39.95	319.60	60.95	487.60		
1 Crew Truck		166.25		182.88		
3 Tram Cars		216.15		237.76	7.97	8.76
48 L.H., Daily Totals		$3160.00		$4568.24	$65.83	$95.17

Crew No.	Bare Costs		Incl. Subs O&P		Cost Per Labor-Hour	
Crew R-11	Hr.	Daily	Hr.	Daily	Bare Costs	Incl. O&P
1 Electrician Foreman	$61.85	$494.80	$92.10	$736.80	$58.36	$87.04
4 Electricians	61.35	1963.20	91.35	2923.20		
1 Equip. Oper. (crane)	59.20	473.60	88.50	708.00		
1 Common Laborer	42.10	336.80	63.25	506.00		
1 Crew Truck		166.25		182.88		
1 Hyd. Crane, 12 Ton		471.00		518.10	11.38	12.52
56 L.H., Daily Totals		$3905.65		$5574.98	$69.74	$99.55
Crew R-12	Hr.	Daily	Hr.	Daily	Bare Costs	Incl. O&P
1 Carpenter Foreman (inside)	$53.65	$429.20	$80.60	$644.80	$49.91	$75.30
4 Carpenters	53.15	1700.80	79.85	2555.20		
4 Common Laborers	42.10	1347.20	63.25	2024.00		
1 Equip. Oper. (medium)	56.75	454.00	84.85	678.80		
1 Steel Worker	57.65	461.20	90.45	723.60		
1 Dozer, 200 H.P.		1504.00		1654.40		
1 Pickup Truck, 3/4 Ton		110.85		121.94	18.35	20.19
88 L.H., Daily Totals		$6007.25		$8402.74	$68.26	$95.49
Crew R-13	Hr.	Daily	Hr.	Daily	Bare Costs	Incl. O&P
1 Electrician Foreman	$61.85	$494.80	$92.10	$736.80	$59.29	$88.35
3 Electricians	61.35	1472.40	91.35	2192.40		
.25 Equip. Oper. (crane)	59.20	118.40	88.50	177.00		
1 Equipment Oiler	50.55	404.40	75.55	604.40		
.25 Hydraulic Crane, 33 Ton		236.34		259.97	5.63	6.19
42 L.H., Daily Totals		$2726.34		$3970.57	$64.91	$94.54
Crew R-15	Hr.	Daily	Hr.	Daily	Bare Costs	Incl. O&P
1 Electrician Foreman	$61.85	$494.80	$92.10	$736.80	$60.04	$89.45
4 Electricians	61.35	1963.20	91.35	2923.20		
1 Equipment Oper. (light)	53.00	424.00	79.20	633.60		
1 Telescoping Boom Lift, to 40'		278.90		306.79	5.81	6.39
48 L.H., Daily Totals		$3160.90		$4600.39	$65.85	$95.84
Crew R-15A	Hr.	Daily	Hr.	Daily	Bare Costs	Incl. O&P
1 Electrician Foreman	$61.85	$494.80	$92.10	$736.80	$53.63	$80.08
2 Electricians	61.35	981.60	91.35	1461.60		
2 Common Laborers	42.10	673.60	63.25	1012.00		
1 Equip. Oper. (light)	53.00	424.00	79.20	633.60		
1 Telescoping Boom Lift, to 40'		278.90		306.79	5.81	6.39
48 L.H., Daily Totals		$2852.90		$4150.79	$59.44	$86.47
Crew R-18	Hr.	Daily	Hr.	Daily	Bare Costs	Incl. O&P
.25 Electrician Foreman	$61.85	$123.70	$92.10	$184.20	$53.85	$80.18
1 Electrician	61.35	490.80	91.35	730.80		
2 Electrician Apprentices	49.10	785.60	73.10	1169.60		
26 L.H., Daily Totals		$1400.10		$2084.60	$53.85	$80.18
Crew R-19	Hr.	Daily	Hr.	Daily	Bare Costs	Incl. O&P
.5 Electrician Foreman	$61.85	$247.40	$92.10	$368.40	$61.45	$91.50
2 Electricians	61.35	981.60	91.35	1461.60		
20 L.H., Daily Totals		$1229.00		$1830.00	$61.45	$91.50
Crew R-21	Hr.	Daily	Hr.	Daily	Bare Costs	Incl. O&P
1 Electrician Foreman	$61.85	$494.80	$92.10	$736.80	$61.36	$91.37
3 Electricians	61.35	1472.40	91.35	2192.40		
.1 Equip. Oper. (medium)	56.75	45.40	84.85	67.88		
.1 S.P. Crane, 4x4, 25 Ton		114.40		125.84	3.49	3.84
32.8 L.H., Daily Totals		$2127.00		$3122.92	$64.85	$95.21

Crew No.	Bare Costs		Incl. Subs O&P		Cost Per Labor-Hour	
Crew R-22	Hr.	Daily	Hr.	Daily	Bare Costs	Incl. O&P
.66 Electrician Foreman	$61.85	$326.57	$92.10	$486.29	$56.16	$83.62
2 Electricians	61.35	981.60	91.35	1461.60		
2 Electrician Apprentices	49.10	785.60	73.10	1169.60		
37.28 L.H., Daily Totals		$2093.77		$3117.49	$56.16	$83.62
Crew R-30	Hr.	Daily	Hr.	Daily	Bare Costs	Incl. O&P
.25 Electrician Foreman (outside)	$63.35	$126.70	$94.35	$188.70	$49.66	$74.29
1 Electrician	61.35	490.80	91.35	730.80		
2 Laborers (Semi-Skilled)	42.10	673.60	63.25	1012.00		
26 L.H., Daily Totals		$1291.10		$1931.50	$49.66	$74.29
Crew R-31	Hr.	Daily	Hr.	Daily	Bare Costs	Incl. O&P
1 Electrician	$61.35	$490.80	$91.35	$730.80	$61.35	$91.35
1 Core Drill, Electric, 2.5 H.P.		62.50		68.75	7.81	8.59
8 L.H., Daily Totals		$553.30		$799.55	$69.16	$99.94
Crew W-41E	Hr.	Daily	Hr.	Daily	Bare Costs	Incl. O&P
.5 Plumber Foreman (outside)	$66.45	$265.80	$99.45	$397.80	$55.91	$83.77
1 Plumber	64.45	515.60	96.45	771.60		
1 Laborer	42.10	336.80	63.25	506.00		
20 L.H., Daily Totals		$1118.20		$1675.40	$55.91	$83.77

Historical Cost Indexes

The table below lists both the RSMeans® historical cost index based on Jan. 1, 1993 = 100 as well as the computed value of an index based on Jan. 1, 2020 costs. Since the Jan. 1, 2020 figure is estimated, space is left to write in the actual index figures as they become available through the quarterly *RSMeans Construction Cost Indexes*.

To compute the actual index based on Jan. 1, 2020 = 100, divide the historical cost index for a particular year by the actual Jan. 1, 2020 construction cost index. Space has been left to advance the index figures as the year progresses.

Year	Historical Cost Index Jan. 1, 1993 = 100		Current Index Based on Jan. 1, 2020 = 100		Year	Historical Cost Index Jan. 1, 1993 = 100	Current Index Based on Jan. 1, 2020 = 100		Year	Historical Cost Index Jan. 1, 1993 = 100	Current Index Based on Jan. 1, 2020 = 100	
	Est.	Actual	Est.	Actual		Actual	Est.	Actual		Actual	Est.	Actual
Oct 2020*					July 2005	151.6	63.4		July 1987	87.7	36.7	
July 2020*					2004	143.7	60.1		1986	84.2	35.2	
Apr 2020*					2003	132.0	55.2		1985	82.6	34.6	
Jan 2020*	239.1		100.0	100.0	2002	128.7	53.8		1984	82.0	34.3	
July 2019		232.2	97.1		2001	125.1	52.3		1983	80.2	33.5	
2018		222.9	93.2		2000	120.9	50.6		1982	76.1	31.8	
2017		213.6	89.3		1999	117.6	49.2		1981	70.0	29.3	
2016		207.3	86.7		1998	115.1	48.1		1980	62.9	26.3	
2015		206.2	86.2		1997	112.8	47.2		1979	57.8	24.2	
2014		204.9	85.7		1996	110.2	46.1		1978	53.5	22.4	
2013		201.2	84.1		1995	107.6	45.0		1977	49.5	20.7	
2012		194.6	81.4		1994	104.4	43.7		1976	46.9	19.6	
2011		191.2	80.0		1993	101.7	42.5		1975	44.8	18.7	
2010		183.5	76.7		1992	99.4	41.6		1974	41.4	17.3	
2009		180.1	75.3		1991	96.8	40.5		1973	37.7	15.8	
2008		180.4	75.4		1990	94.3	39.4		1972	34.8	14.6	
2007		169.4	70.8		1989	92.1	38.5		1971	32.1	13.4	
2006		162.0	67.8		1988	89.9	37.6		1970	28.7	12.0	

Adjustments to Costs

The "Historical Cost Index" can be used to convert national average building costs at a particular time to the approximate building costs for some other time.

Example:

Estimate and compare construction costs for different years in the same city.

To estimate the national average construction cost of a building in 1970, knowing that it cost $900,000 in 2020:

INDEX in 1970 = 28.7

INDEX in 2020 = 239.1

Note: The city cost indexes for Canada can be used to convert U.S. national averages to local costs in Canadian dollars.

Example:

To estimate and compare the cost of a building in Toronto, ON in 2020 with the known cost of $600,000 (US$) in New York, NY in 2020:

INDEX Toronto = 115.6

INDEX New York = 137.1

$$\frac{\text{INDEX Toronto}}{\text{INDEX New York}} \times \text{Cost New York} = \text{Cost Toronto}$$

$$\frac{115.6}{137.1} \times \$600,000 = .843 \times \$600,000 = \$505,908$$

The construction cost of the building in Toronto is $505,908 (CN$).

*Historical Cost Index updates and other resources are provided on the following website:
http://info.thegordiangroup.com/RSMeans.html

Time Adjustment Using the Historical Cost Indexes:

$$\frac{\text{Index for Year A}}{\text{Index for Year B}} \times \text{Cost in Year B} = \text{Cost in Year A}$$

$$\frac{\text{INDEX 1970}}{\text{INDEX 2020}} \times \text{Cost 2020} = \text{Cost 1970}$$

$$\frac{28.7}{239.1} \times \$900,000 = .120 \times \$900,000 = \$108,000$$

The construction cost of the building in 1970 was $108,000.

For customer support on your Site Work & Landscape Costs with RSMeans data, call 800.448.8182.

How to Use the City Cost Indexes

What you should know before you begin

RSMeans City Cost Indexes (CCI) are an extremely useful tool for when you want to compare costs from city to city and region to region.

This publication contains average construction cost indexes for 731 U.S. and Canadian cities covering over 930 three-digit zip code locations, as listed directly under each city.

Keep in mind that a City Cost Index number is a percentage ratio of a specific city's cost to the national average cost of the same item at a stated time period.

In other words, these index figures represent relative construction factors (or, if you prefer, multipliers) for material and installation costs, as well as the weighted average for Total In Place costs for each CSI MasterFormat division. Installation costs include both labor and equipment rental costs. When estimating equipment rental rates only for a specific location, use 01 54 33 EQUIPMENT RENTAL COSTS in the Reference Section.

The 30 City Average Index is the average of 30 major U.S. cities and serves as a national average.

Index figures for both material and installation are based on the 30 major city average of 100 and represent the cost relationship as of July 1, 2019. The index for each division is computed from representative material and labor quantities for that division. The weighted average for each city is a weighted total of the components listed above it. It does not include relative productivity between trades or cities.

As changes occur in local material prices, labor rates, and equipment rental rates (including fuel costs), the impact of these changes should be accurately measured by the change in the City Cost Index for each particular city (as compared to the 30 city average).

Therefore, if you know (or have estimated) building costs in one city today, you can easily convert those costs to expected building costs in another city.

In addition, by using the Historical Cost Index, you can easily convert national average building costs at a particular time to the approximate building costs for some other time. The City Cost Indexes can then be applied to calculate the costs for a particular city.

Quick calculations

Location Adjustment Using the City Cost Indexes:

$$\frac{\text{Index for City A}}{\text{Index for City B}} \times \text{Cost in City B} = \text{Cost in City A}$$

Time Adjustment for the National Average Using the Historical Cost Index:

$$\frac{\text{Index for Year A}}{\text{Index for Year B}} \times \text{Cost in Year B} = \text{Cost in Year A}$$

Adjustment from the National Average:

$$\frac{\text{Index for City A}}{100} \times \text{National Average Cost} = \text{Cost in City A}$$

Since each of the other RSMeans data sets contains many different items, any *one* item multiplied by the particular city index may give incorrect results. However, the larger the number of items compiled, the closer the results should be to actual costs for that particular city.

The City Cost Indexes for Canadian cities are calculated using Canadian material and equipment prices and labor rates in Canadian dollars. Therefore, indexes for Canadian cities can be used to convert U.S. national average prices to local costs in Canadian dollars.

How to use this section

1. Compare costs from city to city.

In using the RSMeans Indexes, remember that an index number is not a fixed number but a ratio: It's a percentage ratio of a building component's cost at any stated time to the national average cost of that same component at the same time period. Put in the form of an equation:

$$\frac{\text{Specific City Cost}}{\text{National Average Cost}} \times 100 = \text{City Index Number}$$

Therefore, when making cost comparisons between cities, do not subtract one city's index number from the index number of another city and read the result as a percentage difference. Instead, divide one city's index number by that of the other city. The resulting number may then be used as a multiplier to calculate cost differences from city to city.

The formula used to find cost differences between cities for the purpose of comparison is as follows:

$$\frac{\text{City A Index}}{\text{City B Index}} \times \text{City B Cost (Known)} = \text{City A Cost (Unknown)}$$

In addition, you can use RSMeans CCI to calculate and compare costs division by division between cities using the same basic formula. (Just be sure that you're comparing similar divisions.)

2. Compare a specific city's construction costs with the national average.

When you're studying construction location feasibility, it's advisable to compare a prospective project's cost index with an index of the national average cost.

For example, divide the weighted average index of construction costs of a specific city by that of the 30 City Average, which = 100.

$$\frac{\text{City Index}}{100} = \text{\% of National Average}$$

As a result, you get a ratio that indicates the relative cost of construction in that city in comparison with the national average.

3. Convert U.S. national average to actual costs in Canadian City.

$$\frac{\text{Index for Canadian City}}{100} \times \text{National Average Cost} = \text{Cost in Canadian City in \$ CAN}$$

4. Adjust construction cost data based on a national average.
When you use a source of construction cost data which is based on a national average (such as RSMeans cost data), it is necessary to adjust those costs to a specific location.

$$\frac{\text{City Index}}{100} \times \frac{\text{Cost Based on}}{\text{National Average Costs}} = \frac{\text{City Cost}}{\text{(Unknown)}}$$

5. When applying the City Cost Indexes to demolition projects, use the appropriate division installation index. For example, for removal of existing doors and windows, use the Division 8 (Openings) index.

What you might like to know about how we developed the Indexes

The information presented in the CCI is organized according to the Construction Specifications Institute (CSI) MasterFormat 2018 classification system.

To create a reliable index, RSMeans researched the building type most often constructed in the United States and Canada. Because it was concluded that no one type of building completely represented the building construction industry, nine different types of buildings were combined to create a composite model.

The exact material, labor, and equipment quantities are based on detailed analyses of these nine building types, and then each quantity is weighted in proportion to expected usage. These various material items, labor hours, and equipment rental rates are thus combined to form a composite building representing as closely as possible the actual usage of materials, labor, and equipment in the North American building construction industry.

The following structures were chosen to make up that composite model:

1. Factory, 1 story
2. Office, 2–4 stories
3. Store, Retail
4. Town Hall, 2–3 stories
5. High School, 2–3 stories
6. Hospital, 4–8 stories
7. Garage, Parking
8. Apartment, 1–3 stories
9. Hotel/Motel, 2–3 stories

For the purposes of ensuring the timeliness of the data, the components of the index for the composite model have been streamlined. They currently consist of:

- specific quantities of 66 commonly used construction materials;
- specific labor-hours for 21 building construction trades; and
- specific days of equipment rental for 6 types of construction equipment (normally used to install the 66 material items by the 21 trades.) Fuel costs and routine maintenance costs are included in the equipment cost.

Material and equipment price quotations are gathered quarterly from cities in the United States and Canada. These prices and the latest negotiated labor wage rates for 21 different building trades are used to compile the quarterly update of the City Cost Index.

The 30 major U.S. cities used to calculate the national average are:

Atlanta, GA
Baltimore, MD
Boston, MA
Buffalo, NY
Chicago, IL
Cincinnati, OH
Cleveland, OH
Columbus, OH
Dallas, TX
Denver, CO
Detroit, MI
Houston, TX
Indianapolis, IN
Kansas City, MO
Los Angeles, CA

Memphis, TN
Milwaukee, WI
Minneapolis, MN
Nashville, TN
New Orleans, LA
New York, NY
Philadelphia, PA
Phoenix, AZ
Pittsburgh, PA
St. Louis, MO
San Antonio, TX
San Diego, CA
San Francisco, CA
Seattle, WA
Washington, DC

What the CCI does not indicate

The weighted average for each city is a total of the divisional components weighted to reflect typical usage. It does not include the productivity variations between trades or cities.

In addition, the CCI does not take into consideration factors such as the following:

- managerial efficiency
- competitive conditions
- automation
- restrictive union practices
- unique local requirements
- regional variations due to specific building codes

City Cost Indexes

DIVISION		UNITED STATES 30 CITY AVERAGE			ALABAMA ANNISTON 362			BIRMINGHAM 350 - 352			BUTLER 369			DECATUR 356			DOTHAN 363		
		MAT.	INST.	TOTAL	MAT.	INST.	TOTAL	MAT.	INST.	TOTAL	MAT.	INST.	TOTAL	MAT.	INST.	TOTAL	MAT.	INST.	TOTAL
015433	CONTRACTOR EQUIPMENT		100.0	100.0		101.9	101.9		104.7	104.7		99.5	99.5		101.9	101.9		99.5	99.5
0241, 31 - 34	SITE & INFRASTRUCTURE, DEMOLITION	100.0	100.0	100.0	90.9	89.5	89.9	91.2	94.5	93.5	105.4	84.4	90.9	84.5	88.1	87.0	102.9	84.8	90.4
0310	Concrete Forming & Accessories	100.0	100.0	100.0	85.6	68.7	71.2	90.6	69.7	72.8	82.3	68.8	70.8	90.9	62.8	66.9	90.3	69.1	72.2
0320	Concrete Reinforcing	100.0	100.0	100.0	88.8	71.2	80.3	94.2	71.3	83.1	93.9	70.5	82.6	88.1	67.7	78.2	93.9	71.3	83.0
0330	Cast-in-Place Concrete	100.0	100.0	100.0	82.3	67.7	76.9	103.8	69.3	90.9	80.3	67.8	75.6	101.9	67.2	89.0	80.3	67.7	75.6
03	CONCRETE	100.0	100.0	100.0	89.2	70.4	81.0	92.2	71.4	83.1	90.3	70.4	81.6	92.0	67.0	81.0	89.5	70.7	81.2
04	MASONRY	100.0	100.0	100.0	89.2	63.1	73.2	87.1	64.3	73.1	93.7	63.1	74.9	85.2	62.4	71.2	95.1	62.9	75.3
05	METALS	100.0	100.0	100.0	101.8	94.8	99.7	100.5	94.2	98.7	100.7	94.8	98.9	102.9	93.1	100.0	100.7	95.4	99.1
06	WOOD, PLASTICS & COMPOSITES	100.0	100.0	100.0	82.0	69.9	75.3	91.0	70.0	79.4	77.1	69.9	73.1	95.7	62.1	77.2	88.1	69.9	78.1
07	THERMAL & MOISTURE PROTECTION	100.0	100.0	100.0	96.0	63.0	81.6	94.0	67.6	82.5	96.1	65.9	82.9	93.1	64.5	80.7	96.0	65.7	82.8
08	OPENINGS	100.0	100.0	100.0	94.7	69.2	88.7	102.6	69.8	94.9	94.7	69.5	88.8	106.9	64.6	97.0	94.7	69.7	88.9
0920	Plaster & Gypsum Board	100.0	100.0	100.0	81.8	69.5	73.5	88.9	69.5	75.9	79.3	69.5	72.7	92.7	61.6	71.7	88.7	69.5	75.8
0950, 0980	Ceilings & Acoustic Treatment	100.0	100.0	100.0	77.1	69.5	72.0	84.8	69.5	74.5	77.1	69.5	72.0	83.2	61.6	68.6	77.1	69.5	72.0
0960	Flooring	100.0	100.0	100.0	82.0	69.1	78.3	99.6	69.1	91.0	86.5	69.1	81.6	89.9	67.2	84.1	91.6	69.1	85.3
0970, 0990	Wall Finishes & Painting/Coating	100.0	100.0	100.0	89.1	66.7	75.9	87.0	66.7	75.0	89.1	45.9	63.5	79.4	61.0	68.5	89.1	80.4	84.0
09	FINISHES	100.0	100.0	100.0	80.4	68.8	74.1	91.2	69.2	79.3	83.6	66.6	74.4	85.3	63.5	73.5	86.1	70.3	77.6
COVERS	DIVS. 10 - 14, 25, 28, 41, 43, 44, 46	100.0	100.0	100.0	100.0	74.0	94.2	100.0	83.6	96.3	100.0	74.0	94.2	100.0	72.9	94.0	100.0	74.0	94.2
21, 22, 23	FIRE SUPPRESSION, PLUMBING & HVAC	100.0	100.0	100.0	101.0	51.8	81.7	100.0	66.3	86.8	98.1	67.0	85.9	100.0	64.5	86.1	98.1	65.0	85.1
26, 27, 3370	ELECTRICAL, COMMUNICATIONS & UTIL.	100.0	100.0	100.0	97.7	59.9	79.1	98.4	59.7	79.3	99.8	61.7	81.0	94.5	63.8	79.4	98.6	76.1	87.5
MF2018	WEIGHTED AVERAGE	100.0	100.0	100.0	95.9	67.7	84.0	97.4	71.8	86.6	96.1	70.6	85.3	97.1	69.2	85.3	96.2	72.7	86.3

DIVISION		ALABAMA EVERGREEN 364			GADSDEN 359			HUNTSVILLE 357 - 358			JASPER 355			MOBILE 365 - 366			MONTGOMERY 360 - 361		
		MAT.	INST.	TOTAL	MAT.	INST.	TOTAL	MAT.	INST.	TOTAL	MAT.	INST.	TOTAL	MAT.	INST.	TOTAL	MAT.	INST.	TOTAL
015433	CONTRACTOR EQUIPMENT		99.5	99.5		101.9	101.9		101.9	101.9		101.9	101.9		99.5	99.5		102.0	102.0
0241, 31 - 34	SITE & INFRASTRUCTURE, DEMOLITION	105.9	84.4	91.0	89.9	89.5	89.6	84.2	89.4	87.8	89.9	89.5	89.6	97.7	85.4	89.2	95.6	89.9	91.7
0310	Concrete Forming & Accessories	79.4	68.6	70.2	83.9	69.1	71.3	90.9	65.9	69.6	88.5	64.7	68.3	89.4	68.4	71.5	92.8	69.4	72.8
0320	Concrete Reinforcing	94.0	70.4	82.6	93.5	71.3	82.8	88.1	76.3	82.4	88.1	71.3	80.0	91.5	70.5	81.3	99.7	71.4	86.0
0330	Cast-in-Place Concrete	80.3	67.6	75.6	101.9	67.8	89.2	99.2	67.8	87.5	113.0	67.8	96.1	84.2	66.9	77.7	81.5	68.7	76.7
03	CONCRETE	90.7	70.2	81.7	96.5	70.7	85.1	90.8	70.1	81.7	100.1	68.7	86.3	85.5	69.9	78.7	84.1	71.1	78.4
04	MASONRY	93.8	62.9	74.8	83.8	63.1	71.1	86.3	63.3	72.2	81.4	63.1	70.2	92.3	61.4	73.3	89.4	63.2	73.3
05	METALS	100.7	94.6	98.9	100.7	95.2	99.1	102.9	96.7	101.0	100.7	95.1	99.0	102.8	94.9	100.5	101.7	94.3	99.5
06	WOOD, PLASTICS & COMPOSITES	73.8	69.9	71.6	86.7	69.9	77.4	95.7	65.3	79.0	93.0	64.1	77.1	86.7	69.9	77.4	90.3	70.0	79.1
07	THERMAL & MOISTURE PROTECTION	96.0	65.4	82.7	93.3	66.3	81.6	93.0	65.9	81.2	93.3	64.4	80.7	95.6	65.4	82.4	94.2	66.9	82.3
08	OPENINGS	94.7	69.3	88.8	103.4	69.7	95.6	106.6	68.5	97.7	103.4	66.6	94.8	97.3	69.5	90.8	96.2	69.8	90.0
0920	Plaster & Gypsum Board	78.3	69.5	72.4	85.1	69.5	74.6	92.7	64.8	73.9	89.5	63.6	72.1	85.6	69.5	74.8	85.9	69.5	74.9
0950, 0980	Ceilings & Acoustic Treatment	77.1	69.5	72.0	79.1	69.5	72.7	85.0	64.8	71.4	79.1	63.6	68.6	83.0	69.5	73.9	84.6	69.5	74.4
0960	Flooring	84.5	69.1	80.1	86.4	69.1	81.5	89.9	69.1	84.1	88.4	69.1	82.9	91.1	69.1	84.9	91.7	69.1	85.3
0970, 0990	Wall Finishes & Painting/Coating	89.1	45.9	63.5	79.4	57.4	66.5	79.4	63.7	70.1	79.4	66.7	71.9	92.3	45.9	64.8	91.1	66.7	76.7
09	FINISHES	82.9	66.5	74.1	82.9	67.8	74.7	85.7	65.9	75.0	84.0	65.4	73.9	86.2	66.2	75.4	88.3	68.9	77.8
COVERS	DIVS. 10 - 14, 25, 28, 41, 43, 44, 46	100.0	73.9	94.2	100.0	82.9	96.2	100.0	82.5	96.1	100.0	73.4	94.1	100.0	82.5	96.1	100.0	83.3	96.3
21, 22, 23	FIRE SUPPRESSION, PLUMBING & HVAC	98.1	59.8	83.0	102.3	62.3	86.6	100.0	66.9	87.0	102.3	65.7	87.9	100.0	61.4	84.8	100.0	64.3	86.0
26, 27, 3370	ELECTRICAL, COMMUNICATIONS & UTIL.	97.1	56.8	77.2	94.5	59.7	77.4	95.4	65.9	80.8	94.2	59.9	77.3	100.4	56.8	78.9	101.1	76.1	88.8
MF2018	WEIGHTED AVERAGE	95.8	68.3	84.2	97.3	70.2	85.9	97.1	71.7	86.4	97.8	69.8	86.0	96.7	68.7	84.9	96.3	73.1	86.5

DIVISION		ALABAMA PHENIX CITY 368			SELMA 367			TUSCALOOSA 354			ALASKA ANCHORAGE 995 - 996			FAIRBANKS 997			JUNEAU 998		
		MAT.	INST.	TOTAL	MAT.	INST.	TOTAL	MAT.	INST.	TOTAL	MAT.	INST.	TOTAL	MAT.	INST.	TOTAL	MAT.	INST.	TOTAL
015433	CONTRACTOR EQUIPMENT		99.5	99.5		99.5	99.5		101.9	101.9		110.5	110.5		112.9	112.9		110.5	110.5
0241, 31 - 34	SITE & INFRASTRUCTURE, DEMOLITION	109.8	85.5	93.0	102.7	85.5	90.0	84.8	89.5	88.0	118.4	121.7	120.7	117.8	124.5	122.5	133.6	121.7	125.4
0310	Concrete Forming & Accessories	85.6	67.6	70.2	83.5	69.1	71.2	90.8	69.2	72.4	120.5	116.2	116.8	124.7	115.0	116.4	125.7	116.2	117.6
0320	Concrete Reinforcing	93.9	66.9	80.9	93.9	71.3	83.0	88.1	71.3	80.0	148.4	120.7	135.0	147.2	120.7	134.4	138.9	120.7	130.1
0330	Cast-in-Place Concrete	80.3	67.9	75.7	80.3	67.8	75.6	103.4	67.9	90.2	105.1	117.5	109.7	109.7	115.6	111.9	116.1	117.5	116.6
03	CONCRETE	93.6	69.3	82.9	89.0	70.7	81.0	92.7	70.8	83.1	107.5	116.7	111.6	101.4	115.5	107.6	114.1	116.7	115.2
04	MASONRY	93.7	63.1	74.9	97.4	63.1	76.3	85.4	63.1	71.7	171.5	119.4	139.5	181.4	117.8	142.3	162.5	119.4	136.0
05	METALS	100.6	93.7	98.6	100.6	95.2	99.0	102.1	95.3	100.1	127.3	104.4	120.5	122.6	105.4	117.5	113.1	104.4	110.6
06	WOOD, PLASTICS & COMPOSITES	81.7	67.7	74.0	78.9	69.9	73.9	95.7	69.9	81.5	112.3	113.5	112.9	126.9	112.7	119.1	121.9	113.5	117.3
07	THERMAL & MOISTURE PROTECTION	96.5	66.1	83.2	95.9	65.9	82.6	93.1	66.3	81.4	176.9	116.9	150.8	186.2	115.8	155.6	188.4	116.9	157.3
08	OPENINGS	94.7	67.4	88.3	94.7	69.7	88.8	106.6	69.7	98.0	130.8	115.8	127.3	132.8	114.7	128.6	131.0	115.8	127.5
0920	Plaster & Gypsum Board	83.0	67.3	72.4	81.1	69.5	73.3	92.7	69.5	77.1	144.1	113.8	123.7	177.5	112.9	134.0	158.2	113.8	128.3
0950, 0980	Ceilings & Acoustic Treatment	77.1	67.3	70.5	77.1	69.5	72.0	85.0	69.5	74.6	131.3	113.8	119.5	123.9	112.9	116.5	136.1	113.8	121.0
0960	Flooring	88.4	69.1	83.0	87.0	69.1	81.9	89.9	69.1	84.1	118.3	119.4	118.6	116.7	119.4	117.4	125.0	119.4	123.5
0970, 0990	Wall Finishes & Painting/Coating	89.1	79.0	83.1	89.1	66.7	75.9	79.4	66.7	71.9	113.1	119.1	116.6	111.6	116.9	114.7	108.7	119.1	114.8
09	FINISHES	85.2	68.9	76.4	83.7	68.8	75.7	85.6	68.8	76.6	127.0	117.1	121.7	128.6	116.0	121.8	128.9	117.1	122.5
COVERS	DIVS. 10 - 14, 25, 28, 41, 43, 44, 46	100.0	82.7	96.1	100.0	70.4	93.4	100.0	82.9	96.2	100.0	114.1	103.2	100.0	113.9	103.0	100.0	114.1	103.2
21, 22, 23	FIRE SUPPRESSION, PLUMBING & HVAC	98.1	64.2	84.8	98.1	65.1	85.1	100.0	67.1	87.1	100.5	107.1	103.1	100.0	107.9	103.2	100.0	107.1	103.1
26, 27, 3370	ELECTRICAL, COMMUNICATIONS & UTIL.	99.2	65.4	82.6	98.3	76.1	87.3	95.0	59.7	77.6	117.2	109.4	113.3	124.2	109.4	116.9	110.2	109.4	109.8
MF2018	WEIGHTED AVERAGE	96.8	70.8	85.8	95.9	72.5	86.0	97.1	71.4	86.3	118.5	113.1	116.2	118.8	113.0	116.3	116.8	113.1	115.2

For customer support on your Site Work & Landscape Costs with RSMeans data, call 800.448.8182.

781

ALASKA / ARIZONA

DIVISION		KETCHIKAN 999 MAT.	INST.	TOTAL	CHAMBERS 865 MAT.	INST.	TOTAL	FLAGSTAFF 860 MAT.	INST.	TOTAL	GLOBE 855 MAT.	INST.	TOTAL	KINGMAN 864 MAT.	INST.	TOTAL	MESA/TEMPE 852 MAT.	INST.	TOTAL
015433	CONTRACTOR EQUIPMENT		112.9	112.9		88.2	88.2		88.2	88.2		89.3	89.3		88.2	88.2		89.3	89.3
0241, 31 - 34	SITE & INFRASTRUCTURE, DEMOLITION	169.1	124.6	138.4	71.5	90.1	84.4	91.2	90.1	90.4	109.0	91.0	96.5	71.4	90.1	84.3	97.8	91.0	93.1
0310	Concrete Forming & Accessories	116.3	116.2	116.2	99.2	70.3	74.6	104.6	67.4	72.8	93.5	70.4	73.8	97.4	66.1	70.7	111.2	76.9	94.6
0320	Concrete Reinforcing	110.6	120.7	115.5	105.5	76.9	91.7	89.8	76.9	91.6	110.5	76.9	94.2	105.7	76.9	91.7	84.0	67.0	77.7
0330	Cast-in-Place Concrete	221.8	117.3	182.9	89.8	67.3	81.4	89.8	67.4	81.5	83.3	67.0	77.2	89.5	67.3	81.2	93.1	70.4	83.1
03	CONCRETE	170.9	116.7	147.1	93.4	70.3	83.3	114.1	69.0	94.3	101.9	70.3	88.0	93.1	68.4	82.3	93.1	70.4	83.1
04	MASONRY	188.6	119.4	146.1	96.0	59.4	73.5	96.1	59.4	73.6	101.7	59.3	75.7	96.0	59.4	73.5	101.9	59.3	75.8
05	METALS	122.8	105.4	117.7	103.0	71.6	93.8	103.5	71.7	94.2	105.8	72.5	96.0	103.7	71.6	94.3	106.2	72.6	96.3
06	WOOD, PLASTICS & COMPOSITES	117.1	113.5	115.1	101.8	73.0	85.9	108.2	68.8	86.5	92.7	73.1	81.9	96.4	67.3	80.4	96.7	73.1	83.7
07	THERMAL & MOISTURE PROTECTION	191.6	116.4	158.9	99.9	69.7	86.8	101.8	69.3	87.7	99.1	68.3	85.7	99.9	69.1	86.5	99.0	68.3	85.6
08	OPENINGS	128.6	115.8	125.6	105.8	69.3	97.3	105.9	70.1	97.6	94.9	69.4	88.9	106.0	67.9	97.1	94.9	69.4	89.0
0920	Plaster & Gypsum Board	161.0	113.8	129.2	92.8	72.5	79.1	96.5	68.2	77.5	88.2	72.5	77.6	84.2	66.6	72.4	91.7	72.5	78.7
0950, 0980	Ceilings & Acoustic Treatment	118.4	113.8	115.3	106.8	72.5	83.6	107.6	68.2	81.1	95.5	72.5	80.0	107.6	66.6	80.0	95.5	72.5	80.0
0960	Flooring	116.4	119.4	117.2	89.7	64.5	82.6	92.0	64.5	84.3	98.6	64.5	89.0	88.4	64.5	81.6	100.4	64.5	90.3
0970, 0990	Wall Finishes & Painting/Coating	111.6	119.1	116.0	87.9	58.6	70.5	87.9	58.6	70.5	90.6	58.6	71.6	87.9	58.6	70.5	90.6	58.6	71.6
09	FINISHES	129.2	117.1	122.7	91.3	68.1	78.8	94.5	65.7	78.9	96.6	68.2	81.2	89.9	64.8	76.3	96.4	68.2	81.2
COVERS	DIVS. 10 - 14, 25, 28, 41, 43, 44, 46	100.0	114.2	103.2	100.0	81.8	95.9	100.0	81.3	95.8	100.0	82.1	96.0	100.0	81.2	95.8	100.0	82.1	96.0
21, 22, 23	FIRE SUPPRESSION, PLUMBING & HVAC	98.1	107.2	101.7	97.7	76.8	89.5	100.3	77.0	91.1	96.6	76.9	88.9	97.7	76.8	89.5	100.1	77.0	91.1
26, 27, 3370	ELECTRICAL, COMMUNICATIONS & UTIL.	124.2	109.4	116.9	102.3	66.5	84.7	101.3	60.7	81.3	97.1	62.9	80.2	102.3	62.9	82.9	94.2	62.9	78.7
MF2018	WEIGHTED AVERAGE	128.4	113.4	122.1	98.3	71.8	87.1	102.4	70.5	88.9	99.5	71.4	87.6	98.3	70.4	86.5	98.7	71.5	87.2

ARIZONA / ARKANSAS

DIVISION		PHOENIX 850,853 MAT.	INST.	TOTAL	PRESCOTT 863 MAT.	INST.	TOTAL	SHOW LOW 859 MAT.	INST.	TOTAL	TUCSON 856 - 857 MAT.	INST.	TOTAL	BATESVILLE 725 MAT.	INST.	TOTAL	CAMDEN 717 MAT.	INST.	TOTAL
015433	CONTRACTOR EQUIPMENT		92.5	92.5		88.2	88.2		89.3	89.3		89.3	89.3		89.7	89.7		89.7	89.7
0241, 31 - 34	SITE & INFRASTRUCTURE, DEMOLITION	98.3	94.3	95.6	78.4	90.1	86.5	111.4	91.0	97.3	92.9	91.0	91.6	73.0	85.7	81.8	77.2	85.8	83.1
0310	Concrete Forming & Accessories	101.2	70.8	75.3	100.8	70.3	74.8	100.6	70.5	74.9	97.1	67.5	71.8	82.5	61.4	64.5	78.6	61.9	64.4
0320	Concrete Reinforcing	109.1	76.9	93.6	105.4	76.9	91.6	111.2	76.9	94.6	92.3	76.9	84.9	84.4	69.4	77.1	91.0	69.4	80.5
0330	Cast-in-Place Concrete	83.6	67.8	77.7	89.8	67.3	81.4	83.3	67.0	77.3	86.5	67.0	79.2	72.8	74.6	73.5	76.4	74.6	75.7
03	CONCRETE	95.5	70.8	84.6	99.0	70.3	86.4	104.2	70.3	89.3	91.6	69.0	81.7	73.2	68.1	70.9	75.7	68.3	72.4
04	MASONRY	94.8	62.1	74.7	96.1	59.4	73.6	101.8	59.3	75.7	88.9	59.3	70.7	94.1	60.8	73.6	104.7	60.8	77.7
05	METALS	107.8	73.7	97.8	103.5	71.6	94.2	105.6	72.5	95.9	107.0	72.6	96.9	94.7	76.0	89.2	101.1	76.1	93.8
06	WOOD, PLASTICS & COMPOSITES	102.1	72.0	85.5	103.3	73.0	86.6	101.2	73.1	85.7	96.9	69.0	81.6	91.1	63.5	75.9	89.2	64.0	75.4
07	THERMAL & MOISTURE PROTECTION	100.3	69.6	86.9	100.5	69.7	87.1	99.3	68.3	85.8	99.9	67.9	86.0	102.7	61.9	84.9	97.5	61.9	82.0
08	OPENINGS	100.7	71.4	93.9	105.9	69.3	97.4	94.2	71.1	88.8	91.4	70.2	86.5	100.0	61.0	90.9	103.1	61.3	93.4
0920	Plaster & Gypsum Board	99.4	71.2	80.4	93.3	72.5	79.3	93.5	72.5	79.4	96.3	68.2	77.4	73.7	62.9	66.5	84.7	63.5	70.4
0950, 0980	Ceilings & Acoustic Treatment	106.5	71.2	82.7	106.0	72.5	83.4	95.5	72.5	80.0	96.4	68.2	77.4	88.1	62.9	71.1	86.3	63.5	70.9
0960	Flooring	102.4	64.5	91.7	90.6	64.5	83.2	102.2	64.5	91.6	91.8	64.8	84.3	81.8	70.9	78.7	86.1	70.9	81.9
0970, 0990	Wall Finishes & Painting/Coating	95.3	57.2	72.8	87.9	58.6	70.5	90.6	58.6	71.6	91.3	58.6	71.9	92.2	54.5	69.9	88.8	54.5	68.9
09	FINISHES	101.0	68.1	83.2	92.0	68.1	79.1	98.6	68.2	82.2	94.5	65.9	79.0	76.3	62.6	68.9	79.8	62.9	70.7
COVERS	DIVS. 10 - 14, 25, 28, 41, 43, 44, 46	100.0	83.2	96.2	100.0	81.8	95.9	100.0	82.1	96.0	100.0	81.7	95.9	100.0	66.4	92.5	100.0	66.5	92.5
21, 22, 23	FIRE SUPPRESSION, PLUMBING & HVAC	100.0	78.5	91.6	100.3	76.8	91.0	96.6	76.9	88.9	100.1	74.8	90.2	96.5	51.2	78.7	96.4	57.5	81.2
26, 27, 3370	ELECTRICAL, COMMUNICATIONS & UTIL.	99.8	60.7	80.5	101.0	62.9	82.2	94.5	62.9	78.9	96.2	60.6	78.7	95.5	60.5	78.3	95.8	58.1	77.2
MF2018	WEIGHTED AVERAGE	100.6	72.3	88.6	99.9	71.3	87.8	99.7	71.5	87.8	97.7	70.2	86.1	91.6	63.6	79.8	94.0	64.7	81.6

ARKANSAS

DIVISION		FAYETTEVILLE 727 MAT.	INST.	TOTAL	FORT SMITH 729 MAT.	INST.	TOTAL	HARRISON 726 MAT.	INST.	TOTAL	HOT SPRINGS 719 MAT.	INST.	TOTAL	JONESBORO 724 MAT.	INST.	TOTAL	LITTLE ROCK 720 - 722 MAT.	INST.	TOTAL
015433	CONTRACTOR EQUIPMENT		89.7	89.7		89.7	89.7		89.7	89.7		89.7	89.7	96.7	102.4	100.7	85.3	90.3	88.8
0241, 31 - 34	SITE & INFRASTRUCTURE, DEMOLITION	72.5	85.7	81.6	77.7	85.5	83.1	77.7	85.7	83.3	80.2	85.7	84.0	85.7	61.9	65.4	93.1	62.4	66.9
0310	Concrete Forming & Accessories	78.4	61.8	64.3	95.7	61.6	66.7	86.5	61.8	65.4	76.3	61.8	64.0	81.5	65.1	73.6	90.4	69.5	80.3
0320	Concrete Reinforcing	84.4	62.1	73.6	85.4	66.3	76.2	84.0	69.4	76.9	89.3	69.4	79.7	81.5	69.4	77.0	90.4	69.4	80.3
0330	Cast-in-Place Concrete	72.9	74.6	73.5	83.3	75.5	80.4	80.8	74.5	78.5	78.1	74.6	76.8	79.4	75.8	78.0	80.7	76.9	79.3
03	CONCRETE	72.9	67.0	70.3	80.1	68.0	74.8	79.7	68.2	74.6	78.9	68.2	74.2	77.3	69.0	73.7	81.0	69.3	75.9
04	MASONRY	85.0	60.8	70.1	91.6	60.8	72.7	94.3	60.8	73.7	78.6	60.8	67.7	86.5	60.8	70.7	87.7	62.1	72.0
05	METALS	94.7	73.6	88.5	96.9	74.9	90.4	95.8	76.0	90.0	101.1	76.1	93.7	91.3	89.4	90.8	97.2	76.1	91.0
06	WOOD, PLASTICS & COMPOSITES	87.7	64.0	74.7	107.1	64.0	83.4	96.8	64.0	78.7	86.7	64.0	74.2	108.4	64.4	78.3	98.5	63.1	75.9
07	THERMAL & MOISTURE PROTECTION	103.5	61.9	85.4	103.8	61.7	85.5	103.0	61.9	85.1	97.8	61.9	82.2	105.6	60.5	95.1	95.6	60.7	87.5
08	OPENINGS	100.0	59.6	90.6	102.0	60.1	92.3	100.8	61.3	91.6	103.1	61.3	93.3	105.6	60.5	95.1	95.6	60.7	87.5
0920	Plaster & Gypsum Board	73.0	63.5	66.6	79.7	63.5	68.8	78.8	63.5	68.5	83.4	63.5	70.0	87.3	63.5	71.2	92.0	63.5	72.8
0950, 0980	Ceilings & Acoustic Treatment	88.1	63.5	71.5	89.8	63.5	72.0	89.8	63.5	72.0	86.3	63.5	70.9	92.1	63.5	72.8	87.3	63.5	71.2
0960	Flooring	78.9	70.9	76.7	88.3	70.9	83.4	84.1	70.9	80.4	85.1	70.9	81.1	59.3	70.9	62.6	89.2	81.0	86.9
0970, 0990	Wall Finishes & Painting/Coating	92.2	54.5	69.9	92.2	52.1	68.5	92.2	54.5	69.9	88.8	54.5	68.5	81.7	54.5	65.6	92.8	52.4	68.9
09	FINISHES	75.5	62.9	68.7	79.6	62.6	70.4	78.3	62.9	70.0	79.6	62.9	70.6	74.6	63.2	68.4	85.3	64.9	74.3
COVERS	DIVS. 10 - 14, 25, 28, 41, 43, 44, 46	100.0	66.1	92.4	100.0	77.7	95.0	100.0	68.5	93.0	100.0	66.2	92.5	100.0	67.0	92.6	100.0	78.2	95.1
21, 22, 23	FIRE SUPPRESSION, PLUMBING & HVAC	96.5	60.0	82.2	100.0	48.0	79.6	96.5	49.1	77.8	96.4	51.5	78.8	100.4	52.3	81.1	99.9	49.1	79.9
26, 27, 3370	ELECTRICAL, COMMUNICATIONS & UTIL.	90.0	53.9	72.2	93.1	57.0	75.3	94.2	57.5	76.1	97.7	65.2	81.7	99.0	62.3	81.0	100.1	66.7	83.7
MF2018	WEIGHTED AVERAGE	90.5	64.1	79.3	94.1	62.6	80.8	92.9	62.8	80.2	93.5	64.4	81.2	93.8	66.6	82.3	94.5	65.4	82.2

For customer support on your Site Work & Landscape Costs with RSMeans data, call 800.448.8182.

City Cost Indexes

DIVISION		ARKANSAS											CALIFORNIA					
		PINE BLUFF 716			RUSSELLVILLE 728			TEXARKANA 718			WEST MEMPHIS 723			ALHAMBRA 917 - 918			ANAHEIM 928		
		MAT.	INST.	TOTAL	MAT.	INST.	TOTAL	MAT.	INST.	TOTAL	MAT.	INST.	TOTAL	MAT.	INST.	TOTAL	MAT.	INST.	TOTAL
015433	CONTRACTOR EQUIPMENT		89.7	89.7		89.7	89.7		90.9	90.9		112.7	112.7		95.4	95.4		99.2	99.2
0241, 31 - 34	SITE & INFRASTRUCTURE, DEMOLITION	82.3	85.6	84.6	74.2	85.7	82.2	92.9	87.8	89.3	103.1	102.8	102.9	100.8	105.3	103.9	97.7	105.5	103.1
0310	Concrete Forming & Accessories	76.1	61.9	64.0	83.1	61.7	64.8	81.4	61.5	64.5	90.6	62.2	66.4	114.4	139.5	135.8	102.9	139.7	134.3
0320	Concrete Reinforcing	90.9	69.4	80.5	84.9	69.4	77.4	90.5	69.3	80.2	81.5	63.3	72.7	105.0	133.6	118.9	95.2	133.5	113.8
0330	Cast-in-Place Concrete	78.1	74.6	76.8	76.3	74.5	75.6	85.2	74.4	81.2	83.2	75.9	80.5	83.3	127.8	99.9	86.7	131.0	103.2
03	CONCRETE	79.7	68.3	74.7	76.0	68.1	72.6	77.8	68.1	73.5	83.8	68.8	77.2	94.2	133.1	111.3	96.2	134.3	112.9
04	MASONRY	110.7	60.8	80.0	90.9	60.8	72.4	92.6	60.8	73.0	75.3	60.8	66.4	106.9	139.4	126.9	78.2	138.0	115.0
05	METALS	101.9	76.1	94.3	94.7	75.9	89.1	93.8	75.8	88.5	90.4	89.1	90.0	79.9	113.2	89.7	107.7	114.2	109.6
06	WOOD, PLASTICS & COMPOSITES	86.4	64.0	74.1	92.3	64.0	76.8	93.4	64.0	77.2	100.7	64.4	80.7	99.3	138.0	120.6	97.5	138.3	119.9
07	THERMAL & MOISTURE PROTECTION	97.9	62.0	82.3	103.7	61.9	85.5	98.5	61.9	82.6	108.9	61.9	88.4	104.2	131.8	116.2	108.9	134.9	120.2
08	OPENINGS	104.3	60.8	94.1	100.0	61.3	91.0	109.2	61.3	98.0	103.2	60.1	93.1	87.8	137.4	99.4	103.9	137.6	111.8
0920	Plaster & Gypsum Board	83.1	63.5	69.9	73.7	63.5	66.8	85.7	63.5	70.7	89.2	63.5	71.9	94.5	139.3	124.6	112.7	139.3	130.6
0950, 0980	Ceilings & Acoustic Treatment	86.3	63.5	70.9	88.1	63.5	71.5	89.6	63.5	72.0	90.2	63.5	72.2	111.0	139.3	130.1	109.3	139.3	129.5
0960	Flooring	84.9	70.9	80.9	81.3	70.9	78.4	86.9	43.4	74.7	61.6	70.9	64.2	104.8	121.1	109.4	99.2	121.1	105.3
0970, 0990	Wall Finishes & Painting/Coating	88.8	52.1	67.1	92.2	54.5	69.9	88.8	54.5	68.5	81.7	54.5	65.6	103.9	120.3	113.6	92.6	120.3	109.0
09	FINISHES	79.6	62.6	70.4	76.4	62.9	69.1	81.7	57.2	68.4	75.9	63.2	69.0	101.8	134.1	119.3	99.4	134.2	118.3
COVERS	DIVS. 10 - 14, 25, 28, 41, 43, 44, 46	100.0	77.7	95.0	100.0	66.5	92.5	100.0	66.1	92.5	100.0	67.0	92.6	100.0	119.6	104.4	100.0	120.2	104.5
21, 22, 23	FIRE SUPPRESSION, PLUMBING & HVAC	100.0	51.4	80.9	96.5	50.9	78.6	100.0	54.5	82.1	96.8	64.6	84.2	96.5	130.9	110.0	99.9	130.9	112.1
26, 27, 3370	ELECTRICAL, COMMUNICATIONS & UTIL.	95.9	57.5	77.0	93.1	57.5	75.6	97.6	57.5	77.9	100.5	65.3	83.1	121.0	129.2	125.0	91.5	113.3	102.2
MF2018	WEIGHTED AVERAGE	96.0	63.6	82.3	91.7	63.1	79.6	94.8	63.3	81.5	93.3	69.8	83.3	96.7	128.6	110.2	99.3	126.6	110.9

DIVISION		CALIFORNIA																
		BAKERSFIELD 932 - 933			BERKELEY 947			EUREKA 955			FRESNO 936 - 938			INGLEWOOD 903 - 905			LONG BEACH 906 - 908		
		MAT.	INST.	TOTAL	MAT.	INST.	TOTAL	MAT.	INST.	TOTAL	MAT.	INST.	TOTAL	MAT.	INST.	TOTAL	MAT.	INST.	TOTAL
015433	CONTRACTOR EQUIPMENT		98.7	98.7		100.2	100.2		97.1	97.1		97.3	97.3		96.9	96.9		96.9	96.9
0241, 31 - 34	SITE & INFRASTRUCTURE, DEMOLITION	96.1	106.8	103.5	109.8	106.0	107.2	109.0	102.7	104.7	98.9	103.2	101.8	87.7	102.0	97.6	94.7	102.0	99.6
0310	Concrete Forming & Accessories	105.8	139.2	134.3	113.0	168.1	160.0	111.2	154.1	147.8	102.9	153.3	145.8	107.3	139.9	135.1	102.0	139.9	134.3
0320	Concrete Reinforcing	99.8	133.4	116.0	90.7	135.2	112.2	103.7	136.6	119.6	83.3	134.2	107.9	99.8	133.7	116.2	99.0	133.7	115.8
0330	Cast-in-Place Concrete	88.1	130.2	103.8	110.4	133.5	119.0	94.1	129.6	107.3	95.5	129.2	108.0	81.5	130.2	99.6	92.7	130.2	106.7
03	CONCRETE	91.7	133.7	110.2	105.1	148.2	124.0	106.8	140.8	121.7	95.5	139.9	115.0	90.9	134.2	109.9	100.5	134.2	115.3
04	MASONRY	92.1	137.5	120.0	122.7	154.9	142.5	103.3	153.4	134.1	96.9	145.0	126.5	73.1	139.5	113.9	81.7	139.5	117.2
05	METALS	102.5	112.5	105.4	109.9	116.5	111.9	107.4	116.1	109.9	102.8	115.2	106.4	88.0	114.8	95.9	87.9	114.8	95.8
06	WOOD, PLASTICS & COMPOSITES	96.9	138.3	119.7	104.3	173.4	142.3	111.2	157.3	136.6	101.2	157.3	132.1	99.0	138.4	120.7	92.5	138.4	117.8
07	THERMAL & MOISTURE PROTECTION	108.5	124.3	115.4	112.2	155.7	131.2	112.6	151.2	129.4	98.9	132.9	113.7	107.2	133.4	118.6	107.5	133.4	118.8
08	OPENINGS	92.6	134.3	102.3	91.0	161.7	107.5	103.2	143.2	112.5	94.7	145.1	106.5	87.6	137.6	99.2	87.5	137.6	99.2
0920	Plaster & Gypsum Board	94.4	139.3	124.6	113.4	175.0	154.8	117.4	158.8	145.2	90.9	158.8	136.6	100.3	139.3	126.5	96.2	139.3	125.2
0950, 0980	Ceilings & Acoustic Treatment	100.7	139.3	126.8	101.2	175.0	151.0	113.3	158.8	144.0	98.1	158.8	139.0	114.0	139.3	131.1	114.0	139.3	131.1
0960	Flooring	102.2	121.1	107.5	124.0	143.3	129.4	103.2	143.3	114.4	100.4	121.5	106.3	114.7	121.1	116.5	111.5	121.1	114.2
0970, 0990	Wall Finishes & Painting/Coating	92.5	108.6	102.0	111.6	171.3	146.9	94.1	138.6	120.4	101.3	120.7	112.8	115.1	120.3	118.2	115.1	120.3	118.2
09	FINISHES	96.4	133.0	116.2	110.1	165.4	140.0	104.6	152.0	130.3	94.6	146.3	122.6	108.2	134.3	122.4	107.2	134.3	121.9
COVERS	DIVS. 10 - 14, 25, 28, 41, 43, 44, 46	100.0	117.7	103.9	100.0	134.0	107.6	100.0	130.8	106.9	100.0	130.8	106.9	100.0	120.5	104.6	100.0	120.5	104.6
21, 22, 23	FIRE SUPPRESSION, PLUMBING & HVAC	100.1	128.8	111.4	96.7	169.6	125.3	96.4	130.6	109.9	100.1	130.1	111.9	96.1	131.0	109.8	96.1	131.0	109.8
26, 27, 3370	ELECTRICAL, COMMUNICATIONS & UTIL.	107.6	108.1	107.8	99.9	161.8	130.4	98.4	124.9	111.5	97.0	107.4	102.2	98.5	129.2	113.6	98.3	129.2	113.5
MF2018	WEIGHTED AVERAGE	98.9	124.6	109.8	103.0	151.9	123.7	102.5	134.1	115.9	98.5	129.4	111.5	93.8	128.8	108.6	95.4	128.8	109.5

DIVISION		CALIFORNIA																
		LOS ANGELES 900 - 902			MARYSVILLE 959			MODESTO 953			MOJAVE 935			OAKLAND 946			OXNARD 930		
		MAT.	INST.	TOTAL	MAT.	INST.	TOTAL	MAT.	INST.	TOTAL	MAT.	INST.	TOTAL	MAT.	INST.	TOTAL	MAT.	INST.	TOTAL
015433	CONTRACTOR EQUIPMENT		103.3	103.3		97.1	97.1		97.1	97.1		97.3	97.3		100.2	100.2		96.3	96.3
0241, 31 - 34	SITE & INFRASTRUCTURE, DEMOLITION	93.7	109.2	104.4	105.6	102.6	103.5	100.6	102.6	102.0	91.9	103.4	99.8	115.2	106.0	108.8	99.2	101.6	100.8
0310	Concrete Forming & Accessories	105.4	140.1	135.0	101.6	153.6	145.9	98.1	153.7	145.5	113.6	139.3	135.5	102.3	168.2	158.4	105.5	139.8	134.7
0320	Concrete Reinforcing	100.7	133.7	116.6	103.7	134.1	118.4	107.4	134.1	120.3	100.4	133.4	116.4	92.8	136.9	114.1	98.6	133.5	115.5
0330	Cast-in-Place Concrete	84.1	131.4	101.7	105.2	129.2	114.1	94.2	129.3	107.2	82.6	130.1	100.2	104.8	133.5	115.4	96.1	130.4	108.9
03	CONCRETE	98.3	134.7	114.3	107.1	140.0	121.6	98.2	140.1	116.6	87.4	133.8	107.8	106.3	148.5	124.8	95.5	134.1	112.5
04	MASONRY	88.6	139.5	119.9	104.2	142.9	128.0	101.5	142.9	126.9	94.9	137.4	121.0	131.5	154.9	145.9	97.9	136.8	121.8
05	METALS	94.3	115.0	100.4	106.9	114.7	109.2	103.6	114.8	106.9	100.0	113.4	104.0	105.2	117.2	108.7	98.0	113.9	102.7
06	WOOD, PLASTICS & COMPOSITES	105.6	138.6	123.7	98.3	157.3	130.8	93.9	157.3	128.8	100.8	138.4	121.5	92.8	173.4	137.2	95.6	138.4	119.1
07	THERMAL & MOISTURE PROTECTION	103.4	133.9	116.7	112.0	138.9	123.7	111.6	139.5	123.7	105.2	123.7	113.5	110.8	155.7	130.4	108.6	133.3	119.3
08	OPENINGS	98.1	136.9	107.2	102.5	145.6	112.6	101.3	147.4	112.1	89.8	134.3	100.2	91.1	162.1	107.7	92.4	137.6	102.9
0920	Plaster & Gypsum Board	95.6	139.3	125.0	110.0	158.8	142.8	112.7	158.8	143.7	99.4	139.3	126.6	107.6	175.0	152.9	94.5	139.3	124.7
0950, 0980	Ceilings & Acoustic Treatment	121.4	139.3	133.5	112.5	158.8	143.7	109.3	158.8	142.6	99.4	139.3	126.3	103.8	175.0	151.8	100.8	139.3	126.8
0960	Flooring	112.5	122.2	115.3	98.8	119.5	104.6	99.2	133.0	108.7	101.3	121.1	106.8	117.0	143.3	124.4	93.8	121.1	101.5
0970, 0990	Wall Finishes & Painting/Coating	111.6	120.3	116.7	94.1	134.1	117.8	94.1	138.6	120.4	87.2	109.6	100.5	111.6	171.3	146.9	87.2	114.8	103.6
09	FINISHES	109.4	134.7	123.1	101.7	147.4	126.4	101.0	150.2	127.7	95.0	133.2	115.6	108.1	165.4	139.1	92.5	133.7	114.8
COVERS	DIVS. 10 - 14, 25, 28, 41, 43, 44, 46	100.0	121.0	104.7	100.0	130.8	106.9	100.0	130.8	106.9	100.0	117.9	104.0	100.0	134.0	107.6	100.0	120.4	104.5
21, 22, 23	FIRE SUPPRESSION, PLUMBING & HVAC	99.9	131.0	112.1	96.4	130.0	109.6	99.9	131.7	112.4	96.5	128.8	109.2	100.2	169.6	127.5	100.0	131.0	112.2
26, 27, 3370	ELECTRICAL, COMMUNICATIONS & UTIL.	96.9	131.0	113.7	95.1	120.7	107.7	97.5	109.9	103.6	95.5	108.1	101.7	99.1	161.8	130.0	101.4	117.9	109.5
MF2018	WEIGHTED AVERAGE	98.5	129.8	111.7	101.7	131.3	114.2	100.6	130.6	113.3	95.4	124.4	107.7	103.4	152.0	123.9	98.0	126.7	110.1

For customer support on your Site Work & Landscape Costs with RSMeans data, call 800.448.8182.

783

CALIFORNIA

DIVISION		PALM SPRINGS 922 MAT.	INST.	TOTAL	PALO ALTO 943 MAT.	INST.	TOTAL	PASADENA 910-912 MAT.	INST.	TOTAL	REDDING 960 MAT.	INST.	TOTAL	RICHMOND 948 MAT.	INST.	TOTAL	RIVERSIDE 925 MAT.	INST.	TOTAL
015433	CONTRACTOR EQUIPMENT		98.1	98.1		100.2	100.2		95.4	95.4		97.1	97.1		100.2	100.2		98.1	98.1
0241, 31 - 34	SITE & INFRASTRUCTURE, DEMOLITION	89.3	103.6	99.2	106.1	106.0	106.0	97.5	105.3	102.9	122.2	102.6	108.7	114.3	106.0	108.5	96.2	103.6	101.3
0310	Concrete Forming & Accessories	99.8	136.1	130.7	100.5	168.2	158.2	103.4	139.6	134.2	105.0	153.6	146.5	115.5	167.8	160.1	103.5	139.7	134.4
0320	Concrete Reinforcing	109.0	133.5	120.9	90.7	136.7	113.0	106.0	133.6	119.3	137.8	134.1	136.0	90.7	136.7	112.9	105.9	133.5	119.3
0330	Cast-in-Place Concrete	82.7	131.0	100.6	93.4	133.5	108.4	79.1	127.8	97.2	107.2	129.3	115.4	107.4	133.3	117.1	89.9	131.0	105.2
03	CONCRETE	90.5	132.7	109.0	95.3	148.5	118.7	89.9	133.1	108.9	114.0	140.1	125.4	108.0	148.2	125.7	96.4	134.3	113.1
04	MASONRY	76.5	137.7	114.1	105.1	151.5	133.6	93.3	139.4	121.6	131.9	142.9	138.7	122.5	151.5	140.3	77.5	137.7	114.5
05	METALS	108.3	114.1	110.0	102.7	116.9	106.9	80.0	113.2	89.7	103.9	114.7	107.1	102.8	116.6	106.8	107.7	114.2	109.6
06	WOOD, PLASTICS & COMPOSITES	92.1	133.5	114.9	90.4	173.4	136.1	85.4	138.0	114.4	107.8	157.3	135.0	107.5	173.4	143.8	97.5	138.3	119.9
07	THERMAL & MOISTURE PROTECTION	108.1	134.3	119.5	110.0	155.0	129.6	103.8	131.8	116.0	128.2	138.9	132.9	110.6	154.8	129.8	109.0	134.8	120.2
08	OPENINGS	99.9	134.9	108.1	91.1	160.2	107.2	87.8	137.4	99.4	115.9	145.6	122.8	91.1	160.2	107.2	102.6	137.6	110.8
0920	Plaster & Gypsum Board	107.2	134.4	125.5	106.0	175.0	152.4	88.8	139.3	122.8	112.6	158.8	143.7	114.5	175.0	155.2	111.9	139.3	130.4
0950, 0980	Ceilings & Acoustic Treatment	106.0	134.4	125.1	102.0	175.0	151.2	111.0	139.3	130.1	148.8	158.8	155.5	102.0	175.0	151.2	114.2	139.3	131.1
0960	Flooring	101.5	121.1	107.0	115.6	143.3	123.4	98.5	121.1	104.9	87.8	128.6	99.3	126.5	143.3	131.3	102.9	121.1	108.0
0970, 0990	Wall Finishes & Painting/Coating	91.1	120.3	108.4	111.6	171.3	146.9	103.9	120.3	113.6	105.2	134.1	122.3	111.6	171.3	146.9	91.1	120.3	108.4
09	FINISHES	98.0	131.4	116.1	106.5	165.4	138.4	99.0	134.1	118.0	107.2	149.0	129.8	111.6	165.4	140.7	101.0	134.2	119.0
COVERS	DIVS. 10 - 14, 25, 28, 41, 43, 44, 46	100.0	119.6	104.4	100.0	134.0	107.6	100.0	119.6	104.4	100.0	130.8	106.9	100.0	133.8	107.5	100.0	119.3	104.3
21, 22, 23	FIRE SUPPRESSION, PLUMBING & HVAC	96.4	130.9	109.9	96.7	170.6	125.7	96.5	130.9	110.0	100.3	130.1	112.0	96.7	164.0	123.1	99.9	130.9	112.1
26, 27, 3370	ELECTRICAL, COMMUNICATIONS & UTIL.	94.6	110.8	102.6	99.0	176.6	137.2	117.5	129.2	123.2	101.3	120.8	110.9	99.6	133.2	116.2	91.3	114.4	102.7
MF2018	WEIGHTED AVERAGE	97.4	125.3	109.2	99.2	153.8	122.3	94.8	128.6	109.1	107.6	131.5	117.7	102.4	146.2	120.9	99.3	126.6	110.8

CALIFORNIA

DIVISION		SACRAMENTO 942, 956-958 MAT.	INST.	TOTAL	SALINAS 939 MAT.	INST.	TOTAL	SAN BERNARDINO 923-924 MAT.	INST.	TOTAL	SAN DIEGO 919-921 MAT.	INST.	TOTAL	SAN FRANCISCO 940-941 MAT.	INST.	TOTAL	SAN JOSE 951 MAT.	INST.	TOTAL
015433	CONTRACTOR EQUIPMENT		99.0	99.0		97.3	97.3		98.1	98.1		102.2	102.2		109.8	109.8		98.7	98.7
0241, 31 - 34	SITE & INFRASTRUCTURE, DEMOLITION	95.6	111.3	106.4	112.3	103.2	106.0	76.1	103.6	95.1	105.7	107.8	107.2	116.4	113.1	114.1	131.6	97.9	108.3
0310	Concrete Forming & Accessories	100.5	156.0	147.8	108.9	156.6	149.6	107.1	139.7	134.9	103.8	129.1	125.4	106.1	168.3	159.1	103.9	167.9	158.5
0320	Concrete Reinforcing	86.0	134.4	109.4	99.1	134.6	116.3	105.9	133.5	119.3	102.6	133.4	117.5	105.6	132.1	118.4	94.7	135.0	114.2
0330	Cast-in-Place Concrete	88.6	130.2	104.1	95.0	129.6	107.8	62.1	131.0	87.8	90.5	124.5	103.2	116.9	133.3	123.0	109.4	132.4	118.0
03	CONCRETE	97.9	141.1	116.9	105.8	141.6	121.5	71.1	134.3	98.9	100.2	127.3	112.1	117.8	148.1	131.1	131.5	151.5	143.8
04	MASONRY	106.1	145.9	130.6	95.1	148.4	127.9	83.6	137.7	116.8	86.2	133.4	115.2	140.9	154.6	149.3	101.8	121.6	107.6
05	METALS	100.8	109.8	103.4	102.7	115.8	106.5	107.7	114.2	109.6	94.8	114.1	100.5	110.7	122.4	114.2	101.8	121.6	107.6
06	WOOD, PLASTICS & COMPOSITES	87.1	160.2	127.3	100.6	160.2	133.4	101.2	138.3	121.6	95.0	126.1	112.1	95.1	173.4	138.2	106.0	173.2	143.0
07	THERMAL & MOISTURE PROTECTION	122.2	142.5	131.0	106.2	145.4	123.3	107.1	134.8	119.2	108.2	121.5	114.0	116.5	155.7	133.5	108.0	155.1	128.5
08	OPENINGS	103.9	149.0	114.4	93.5	154.4	107.7	100.0	137.6	108.7	99.8	127.8	106.3	99.5	159.0	113.3	92.6	161.5	108.7
0920	Plaster & Gypsum Board	103.4	161.5	142.5	95.4	161.8	140.1	113.3	139.3	130.8	94.4	126.7	116.1	105.8	175.0	152.4	108.3	175.0	153.2
0950, 0980	Ceilings & Acoustic Treatment	102.0	161.5	142.1	99.4	161.8	141.5	109.3	139.3	129.5	121.1	126.7	124.8	111.7	175.0	154.4	109.3	175.0	153.6
0960	Flooring	115.8	128.6	119.4	95.9	143.3	109.2	105.0	121.1	109.5	108.9	121.1	112.3	117.5	143.3	124.7	92.4	143.3	106.7
0970, 0990	Wall Finishes & Painting/Coating	107.9	134.1	123.4	88.1	171.3	137.3	99.1	120.3	108.4	101.5	118.0	111.2	109.6	174.8	148.2	94.4	171.3	139.9
09	FINISHES	105.3	150.7	129.9	94.6	157.3	128.5	99.4	134.2	118.3	106.6	126.1	117.2	109.9	165.8	140.1	99.9	165.2	135.3
COVERS	DIVS. 10 - 14, 25, 28, 41, 43, 44, 46	100.0	131.7	107.1	100.0	131.2	107.0	100.0	117.7	103.9	100.0	115.7	103.5	100.0	128.4	106.3	100.0	133.4	107.5
21, 22, 23	FIRE SUPPRESSION, PLUMBING & HVAC	100.1	131.4	112.4	96.5	137.6	112.7	96.4	130.9	110.0	100.0	129.4	111.5	100.1	182.8	132.6	99.9	170.5	127.7
26, 27, 3370	ELECTRICAL, COMMUNICATIONS & UTIL.	94.8	120.8	107.6	96.5	132.1	114.0	94.6	112.6	103.5	103.8	104.3	104.0	99.2	182.8	140.4	100.7	176.6	138.1
MF2018	WEIGHTED AVERAGE	100.8	133.0	114.4	99.1	137.2	115.2	95.0	126.3	108.2	99.7	121.7	109.0	107.4	158.4	129.0	102.6	153.5	124.2

CALIFORNIA

DIVISION		SAN LUIS OBISPO 934 MAT.	INST.	TOTAL	SAN MATEO 944 MAT.	INST.	TOTAL	SAN RAFAEL 949 MAT.	INST.	TOTAL	SANTA ANA 926-927 MAT.	INST.	TOTAL	SANTA BARBARA 931 MAT.	INST.	TOTAL	SANTA CRUZ 950 MAT.	INST.	TOTAL
015433	CONTRACTOR EQUIPMENT		97.3	97.3		100.2	100.2		99.4	99.4		98.1	98.1		97.3	97.3		98.7	98.7
0241, 31 - 34	SITE & INFRASTRUCTURE, DEMOLITION	104.4	103.4	103.7	112.1	106.0	107.9	107.6	111.2	110.1	87.8	103.6	98.7	99.1	103.4	102.1	103.9	156.9	149.0
0310	Concrete Forming & Accessories	115.3	139.7	136.1	106.3	168.2	159.0	111.5	167.8	159.5	107.4	139.7	134.9	106.1	139.7	134.8	117.1	134.6	125.6
0320	Concrete Reinforcing	100.4	133.4	116.4	90.7	136.9	113.0	91.3	135.1	112.5	109.6	133.5	121.2	98.6	133.5	115.4	117.1	134.6	125.6
0330	Cast-in-Place Concrete	102.1	130.2	112.6	104.0	133.5	115.0	121.1	132.3	125.2	79.3	131.0	98.6	95.8	130.2	108.6	108.7	131.4	117.1
03	CONCRETE	103.3	134.0	116.8	104.6	148.5	123.9	128.2	147.4	136.6	88.0	134.3	108.3	95.4	134.0	112.4	106.4	142.6	122.3
04	MASONRY	96.5	135.9	120.7	122.2	154.6	142.1	99.1	154.6	133.2	73.4	138.0	113.1	95.2	135.9	120.2	135.3	148.6	143.4
05	METALS	100.7	113.8	104.6	102.6	117.1	106.9	103.7	112.4	106.3	107.8	114.2	109.7	98.5	113.8	103.0	109.3	120.2	112.5
06	WOOD, PLASTICS & COMPOSITES	103.0	138.4	122.4	97.9	173.4	139.5	95.0	173.2	138.0	103.0	138.3	122.4	105.8	138.4	119.1	106.0	160.4	135.9
07	THERMAL & MOISTURE PROTECTION	106.4	132.6	117.8	110.4	156.4	130.4	105.0	155.4	132.5	108.4	134.9	119.9	105.8	132.6	117.5	107.6	147.9	125.1
08	OPENINGS	91.7	134.3	101.6	91.1	162.1	107.6	101.5	159.7	115.1	99.2	137.6	108.2	93.2	137.6	103.6	93.9	154.4	108.0
0920	Plaster & Gypsum Board	101.1	139.3	126.8	111.7	175.0	154.3	113.2	175.0	154.8	114.6	139.3	131.2	94.5	139.3	124.7	116.8	161.8	147.1
0950, 0980	Ceilings & Acoustic Treatment	99.4	139.3	126.3	102.0	175.0	151.2	109.3	175.0	153.6	109.3	139.3	129.5	100.8	139.3	126.8	110.9	161.8	145.2
0960	Flooring	102.0	121.1	107.4	119.5	143.3	126.2	132.6	143.3	135.6	105.6	121.1	110.0	95.0	121.1	102.4	96.6	143.3	109.7
0970, 0990	Wall Finishes & Painting/Coating	87.2	114.8	103.6	111.6	171.3	146.9	107.8	165.5	141.9	91.1	118.0	107.0	87.2	114.8	103.6	94.5	171.3	139.9
09	FINISHES	96.3	133.7	116.5	108.9	165.4	139.5	112.0	164.6	140.5	100.9	134.0	118.8	93.0	133.7	115.0	102.6	157.4	132.3
COVERS	DIVS. 10 - 14, 25, 28, 41, 43, 44, 46	100.0	126.5	105.9	100.0	134.0	107.6	100.0	133.3	107.4	100.0	120.2	104.5	100.0	119.5	104.3	100.0	131.5	107.0
21, 22, 23	FIRE SUPPRESSION, PLUMBING & HVAC	96.5	131.0	110.0	96.7	165.7	123.8	96.7	182.8	130.5	96.4	130.9	110.0	100.0	131.0	112.2	99.9	137.9	114.8
26, 27, 3370	ELECTRICAL, COMMUNICATIONS & UTIL.	95.5	110.8	103.1	99.0	168.0	133.0	96.2	124.9	110.3	94.6	111.9	103.1	94.5	113.7	103.9	99.8	132.1	115.7
MF2018	WEIGHTED AVERAGE	98.2	125.8	109.9	101.5	152.0	122.9	104.5	149.1	123.4	97.0	126.3	109.4	97.2	126.1	109.4	104.6	137.5	118.5

For customer support on your Site Work & Landscape Costs with RSMeans data, call 800.448.8182.

City Cost Indexes

		CALIFORNIA																COLORADO		
		SANTA ROSA			STOCKTON			SUSANVILLE			VALLEJO			VAN NUYS			ALAMOSA		
DIVISION		954			952			961			945			913 - 916			811		
		MAT.	INST.	TOTAL	MAT.	INST.	TOTAL	MAT.	INST.	TOTAL	MAT.	INST.	TOTAL	MAT.	INST.	TOTAL	MAT.	INST.	TOTAL
015433	CONTRACTOR EQUIPMENT		97.6	97.6		97.1	97.1		97.1	97.1		99.4	99.4		95.4	95.4		90.2	90.2
0241, 31 - 34	SITE & INFRASTRUCTURE, DEMOLITION	101.4	102.6	102.2	100.3	102.6	101.9	129.5	102.6	110.9	96.3	111.1	106.5	115.7	105.3	108.5	141.3	84.4	102.0
0310	Concrete Forming & Accessories	100.7	167.1	157.3	101.9	155.5	147.6	106.2	153.9	146.8	102.3	166.8	157.3	109.8	139.5	135.2	103.8	66.0	71.5
0320	Concrete Reinforcing	104.6	135.1	119.3	107.4	134.1	120.3	137.8	134.1	136.0	92.5	135.0	113.0	106.0	133.6	119.3	114.0	67.9	91.7
0330	Cast-in-Place Concrete	103.3	130.8	113.5	91.7	129.2	105.7	97.5	129.3	109.3	96.4	131.4	109.5	83.4	127.8	99.9	100.1	73.1	90.1
03	CONCRETE	107.1	146.9	124.6	97.3	140.9	116.5	117.0	140.2	127.2	102.4	146.6	121.8	104.4	133.1	117.0	112.5	69.4	93.6
04	MASONRY	101.8	153.5	133.6	101.4	142.9	126.9	130.4	142.9	138.1	77.4	153.5	124.1	106.9	139.4	126.9	130.1	59.5	86.7
05	METALS	108.1	117.7	110.9	103.9	114.7	107.0	102.9	114.7	106.4	103.7	111.7	106.0	79.2	113.2	89.2	105.2	78.5	97.3
06	WOOD, PLASTICS & COMPOSITES	93.6	173.0	137.3	99.5	159.9	132.8	109.6	157.6	136.0	84.8	173.2	133.4	94.1	138.0	118.3	98.5	67.4	81.4
07	THERMAL & MOISTURE PROTECTION	108.9	154.1	128.6	112.1	138.8	123.7	130.2	138.9	134.0	113.1	154.1	130.9	105.0	131.8	116.6	111.9	67.2	92.5
08	OPENINGS	100.8	161.4	114.9	101.3	147.0	112.0	116.8	145.8	123.6	103.2	159.7	116.3	87.7	137.4	99.3	96.3	68.5	89.8
0920	Plaster & Gypsum Board	109.2	175.0	153.5	112.7	161.5	145.5	113.1	159.1	144.1	107.5	175.0	152.9	92.8	139.3	124.1	80.8	66.5	71.2
0950, 0980	Ceilings & Acoustic Treatment	109.3	175.0	153.6	117.5	161.5	147.1	139.9	159.1	152.9	111.1	175.0	154.2	107.8	139.3	129.1	102.7	66.5	78.3
0960	Flooring	102.1	136.5	111.7	99.2	133.0	108.7	88.3	128.6	99.6	126.0	143.3	130.9	101.9	121.1	107.3	107.8	70.1	97.2
0970, 0990	Wall Finishes & Painting/Coating	91.1	165.5	135.2	94.1	134.1	117.8	105.2	134.1	122.3	108.7	165.5	142.3	103.9	120.3	113.6	101.7	76.9	87.0
09	FINISHES	100.0	163.0	134.1	102.6	151.3	129.0	106.7	149.2	129.7	108.4	164.4	138.7	101.2	134.1	119.0	99.8	67.5	82.3
COVERS	DIVS. 10 - 14, 25, 28, 41, 43, 44, 46	100.0	132.4	107.2	100.0	128.0	106.2	100.0	130.9	106.9	100.0	132.9	107.3	100.0	119.6	104.4	100.0	83.0	96.2
21, 22, 23	FIRE SUPPRESSION, PLUMBING & HVAC	96.4	182.2	130.1	99.9	131.7	112.4	96.8	130.1	109.9	100.2	147.2	118.7	96.5	130.9	110.0	96.5	72.4	87.0
26, 27, 3370	ELECTRICAL, COMMUNICATIONS & UTIL.	94.9	124.9	109.7	97.5	114.3	105.8	101.7	120.8	111.1	92.4	127.5	109.7	117.5	129.2	123.2	96.3	63.2	80.0
MF2018	WEIGHTED AVERAGE	101.2	148.4	121.2	100.7	131.3	113.7	107.3	131.6	117.5	100.2	141.6	117.7	97.8	128.6	110.8	103.6	70.3	89.5

		COLORADO																		
		BOULDER			COLORADO SPRINGS			DENVER			DURANGO			FORT COLLINS			FORT MORGAN		
DIVISION		803			808 - 809			800 - 802			813			805			807		
		MAT.	INST.	TOTAL	MAT.	INST.	TOTAL	MAT.	INST.	TOTAL	MAT.	INST.	TOTAL	MAT.	INST.	TOTAL	MAT.	INST.	TOTAL
015433	CONTRACTOR EQUIPMENT		92.2	92.2		90.2	90.2		99.6	99.6		90.2	90.2		92.2	92.2		92.2	92.2
0241, 31 - 34	SITE & INFRASTRUCTURE, DEMOLITION	97.1	91.4	93.2	99.2	86.7	90.6	104.1	102.5	103.0	134.3	84.4	99.9	109.6	91.2	96.9	99.3	90.9	93.5
0310	Concrete Forming & Accessories	107.0	77.4	81.8	97.3	63.5	68.5	106.1	66.4	72.2	109.9	65.7	72.2	104.5	65.8	71.5	107.6	65.7	71.9
0320	Concrete Reinforcing	108.7	68.1	89.0	107.9	67.9	88.6	107.9	70.1	89.6	114.0	67.9	91.7	108.8	68.0	89.1	108.9	68.0	89.2
0330	Cast-in-Place Concrete	113.8	73.9	98.9	116.8	73.4	100.7	121.8	73.6	103.9	115.1	73.1	99.5	128.5	72.5	107.6	111.6	72.5	97.1
03	CONCRETE	105.2	74.9	91.9	108.7	68.4	91.0	111.4	70.2	93.3	114.2	69.3	94.5	116.3	69.2	95.6	103.7	69.1	88.5
04	MASONRY	101.6	65.4	79.4	103.7	62.3	78.3	104.9	64.1	79.9	117.5	60.6	82.5	118.9	63.1	84.6	116.7	63.1	83.8
05	METALS	94.4	77.9	89.6	97.5	77.7	91.7	99.9	79.0	93.8	105.2	78.4	97.3	95.7	77.7	90.4	94.2	77.8	89.4
06	WOOD, PLASTICS & COMPOSITES	107.1	81.4	93.0	96.6	63.6	78.5	108.4	67.1	85.7	108.4	67.4	85.8	104.6	67.0	83.9	107.1	67.0	85.0
07	THERMAL & MOISTURE PROTECTION	108.7	73.2	93.2	109.4	69.7	92.1	107.1	71.7	91.7	111.9	67.5	92.6	109.0	70.6	92.3	108.6	69.6	91.6
08	OPENINGS	95.4	76.2	90.9	99.5	66.4	91.8	102.5	68.8	94.6	103.1	68.5	95.0	95.3	68.3	89.0	95.3	68.2	89.0
0920	Plaster & Gypsum Board	116.3	81.3	92.8	100.3	62.9	75.2	111.7	66.6	81.4	95.3	66.5	75.9	110.3	66.6	80.9	116.3	66.5	82.8
0950, 0980	Ceilings & Acoustic Treatment	91.9	81.3	84.8	100.0	62.9	75.0	104.4	66.6	78.9	102.7	66.5	78.3	91.9	66.6	74.8	91.9	66.5	74.8
0960	Flooring	111.2	76.7	101.5	102.2	69.0	92.9	108.4	76.7	99.5	113.1	69.0	100.7	107.3	76.7	98.7	111.7	76.7	101.8
0970, 0990	Wall Finishes & Painting/Coating	97.1	76.9	85.1	96.8	76.9	85.0	103.4	76.9	87.7	101.7	76.9	87.0	97.1	76.9	85.1	97.1	76.9	85.1
09	FINISHES	100.6	77.7	88.2	97.4	65.3	80.0	102.5	68.9	84.3	102.4	67.4	83.5	99.2	68.6	82.6	100.7	68.5	83.3
COVERS	DIVS. 10 - 14, 25, 28, 41, 43, 44, 46	100.0	84.5	96.5	100.0	82.2	96.0	100.0	86.3	96.9	100.0	82.9	96.2	100.0	82.1	96.0	100.0	82.1	96.0
21, 22, 23	FIRE SUPPRESSION, PLUMBING & HVAC	96.5	74.8	87.9	100.1	72.2	89.1	99.9	73.8	89.7	96.5	58.7	81.6	100.0	71.7	88.9	96.5	73.5	87.4
26, 27, 3370	ELECTRICAL, COMMUNICATIONS & UTIL.	99.6	80.6	90.3	103.2	73.1	88.4	105.0	80.4	92.9	95.8	50.3	73.4	99.6	80.6	90.2	100.0	78.1	89.2
MF2018	WEIGHTED AVERAGE	98.7	77.0	89.6	101.2	71.5	88.6	103.0	75.4	91.3	103.9	65.6	87.8	102.2	73.5	90.1	99.3	73.5	88.4

		COLORADO																		
		GLENWOOD SPRINGS			GOLDEN			GRAND JUNCTION			GREELEY			MONTROSE			PUEBLO		
DIVISION		816			804			815			806			814			810		
		MAT.	INST.	TOTAL	MAT.	INST.	TOTAL	MAT.	INST.	TOTAL	MAT.	INST.	TOTAL	MAT.	INST.	TOTAL	MAT.	INST.	TOTAL
015433	CONTRACTOR EQUIPMENT		93.3	93.3		92.2	92.2		93.3	93.3		92.2	92.2		91.7	91.7		90.2	90.2
0241, 31 - 34	SITE & INFRASTRUCTURE, DEMOLITION	150.9	91.9	110.2	110.4	91.4	97.3	133.7	92.1	105.0	95.9	91.2	92.7	144.0	87.9	105.3	125.2	84.5	97.1
0310	Concrete Forming & Accessories	100.6	65.5	70.7	99.7	66.7	71.5	108.7	66.6	72.8	102.4	65.8	71.2	100.1	65.5	70.6	106.1	63.7	70.0
0320	Concrete Reinforcing	112.8	68.0	91.2	108.9	68.0	89.2	113.2	67.8	91.2	108.7	68.0	89.0	112.7	67.8	91.0	109.0	67.9	89.1
0330	Cast-in-Place Concrete	100.1	72.1	89.6	111.7	73.9	97.6	110.8	73.5	96.9	107.3	72.5	94.3	100.1	72.6	89.8	99.4	73.7	89.8
03	CONCRETE	118.0	68.9	96.4	114.1	70.0	94.7	110.7	69.9	92.8	100.3	69.1	86.6	108.5	69.0	91.1	100.9	68.7	86.7
04	MASONRY	103.1	63.0	78.5	119.7	65.4	86.3	137.8	61.9	91.2	111.9	62.7	81.7	110.0	59.6	79.0	99.5	60.3	75.4
05	METALS	104.8	78.2	97.0	94.3	77.8	89.5	106.5	78.2	98.2	95.7	77.7	90.4	103.9	78.1	96.3	108.3	78.7	99.6
06	WOOD, PLASTICS & COMPOSITES	93.6	67.1	79.0	98.9	67.0	81.3	106.1	67.2	84.7	101.7	67.0	82.6	94.6	67.2	79.6	101.2	63.9	80.7
07	THERMAL & MOISTURE PROTECTION	111.8	68.2	92.8	109.8	71.6	93.2	110.9	69.3	92.8	108.2	70.4	91.8	112.0	67.2	92.5	110.4	67.8	91.8
08	OPENINGS	102.1	68.3	94.2	95.4	68.2	89.0	102.8	68.4	94.7	103.2	68.4	89.0	103.2	68.4	95.1	98.0	66.6	90.7
0920	Plaster & Gypsum Board	121.2	66.5	84.4	107.5	66.5	79.9	135.4	66.6	89.1	108.7	66.6	80.4	79.9	66.5	70.9	84.7	62.9	70.1
0950, 0980	Ceilings & Acoustic Treatment	101.9	66.5	78.0	91.9	66.5	74.8	101.9	66.6	78.1	91.9	66.6	74.8	102.7	66.5	78.3	110.0	62.9	78.3
0960	Flooring	106.9	76.7	98.4	104.9	76.7	97.0	112.4	70.1	100.5	106.2	73.1	96.9	110.4	73.1	100.0	109.1	73.1	99.0
0970, 0990	Wall Finishes & Painting/Coating	101.7	76.9	87.0	97.1	76.9	85.1	101.7	76.9	87.0	97.1	76.9	85.1	101.7	76.9	87.0	101.7	76.9	87.0
09	FINISHES	105.2	68.7	85.4	98.9	69.1	82.8	106.8	67.9	85.8	97.9	67.9	81.6	100.5	68.0	82.9	100.2	66.3	81.9
COVERS	DIVS. 10 - 14, 25, 28, 41, 43, 44, 46	100.0	82.3	96.1	100.0	82.9	96.2	100.0	83.1	96.2	100.0	82.1	96.0	100.0	82.6	96.1	100.0	82.9	96.2
21, 22, 23	FIRE SUPPRESSION, PLUMBING & HVAC	96.5	58.7	81.6	96.5	74.8	87.9	100.0	74.8	90.1	100.0	71.7	88.9	96.5	58.7	81.6	100.0	82.9	96.2
26, 27, 3370	ELECTRICAL, COMMUNICATIONS & UTIL.	93.4	52.3	73.1	100.0	80.6	90.4	95.5	54.6	75.3	99.6	80.6	90.2	95.5	50.3	73.2	96.4	63.3	80.0
MF2018	WEIGHTED AVERAGE	103.9	66.8	88.2	100.9	74.7	89.8	105.7	70.5	90.8	99.3	73.4	88.4	102.6	65.8	87.1	101.8	70.0	88.3

For customer support on your Site Work & Landscape Costs with RSMeans data, call 800.448.8182.

785

Section 1

DIVISION		COLORADO SALIDA 812			CONNECTICUT BRIDGEPORT 066			BRISTOL 060			HARTFORD 061			MERIDEN 064			NEW BRITAIN 060		
		MAT.	INST.	TOTAL	MAT.	INST.	TOTAL	MAT.	INST.	TOTAL	MAT.	INST.	TOTAL	MAT.	INST.	TOTAL	MAT.	INST.	TOTAL
015433	CONTRACTOR EQUIPMENT		91.7	91.7		95.7	95.7		95.7	95.7		99.0	99.0		96.1	96.1		95.7	95.7
0241, 31 - 34	SITE & INFRASTRUCTURE, DEMOLITION	134.0	88.2	102.4	106.1	97.8	100.3	105.1	97.7	100.0	101.6	102.4	102.2	102.8	98.5	99.8	105.3	97.7	100.1
0310	Concrete Forming & Accessories	108.6	65.9	72.2	103.4	116.7	114.7	103.4	116.6	114.6	103.5	116.8	114.8	103.1	116.6	114.6	103.9	116.6	114.7
0320	Concrete Reinforcing	112.4	67.9	90.9	116.3	144.5	130.0	116.3	144.5	130.0	111.6	144.6	127.5	116.3	144.5	130.0	116.3	144.5	130.0
0330	Cast-in-Place Concrete	114.7	72.8	99.1	108.5	127.5	115.6	101.6	127.5	111.2	106.5	128.5	114.7	97.8	127.5	108.8	103.3	127.5	112.3
03	CONCRETE	109.1	69.3	91.6	104.1	124.5	113.0	100.9	124.4	111.2	101.1	124.8	111.5	99.1	124.4	110.2	101.7	124.4	111.7
04	MASONRY	138.6	60.6	90.7	109.4	130.9	122.6	100.8	130.9	119.3	100.5	131.0	119.2	100.4	130.9	119.2	102.5	130.9	120.0
05	METALS	103.6	78.6	96.2	98.1	116.6	103.5	98.1	116.5	103.5	103.1	116.0	106.9	95.4	116.5	101.6	94.5	116.5	100.9
06	WOOD, PLASTICS & COMPOSITES	102.4	67.2	83.0	105.0	114.2	110.1	105.0	114.2	110.1	96.2	114.4	106.2	105.0	114.2	110.1	105.0	114.2	110.1
07	THERMAL & MOISTURE PROTECTION	110.8	67.5	92.0	99.7	123.5	110.0	99.8	120.6	108.9	104.4	121.1	111.7	99.8	120.6	108.9	99.8	120.7	108.9
08	OPENINGS	96.3	68.4	89.8	96.4	120.6	102.0	96.4	120.6	102.0	98.1	120.7	103.4	98.5	120.6	103.7	96.4	120.6	102.0
0920	Plaster & Gypsum Board	80.2	66.5	71.0	115.7	114.3	114.8	115.7	114.3	114.8	102.5	114.3	110.5	117.3	114.3	115.3	115.7	114.3	114.8
0950, 0980	Ceilings & Acoustic Treatment	102.7	66.5	78.3	102.0	114.3	110.3	102.0	114.3	110.3	98.8	114.3	109.3	107.5	114.3	112.1	102.0	114.3	110.3
0960	Flooring	115.9	69.9	103.0	92.6	127.0	102.3	92.6	124.5	101.6	96.9	127.0	105.4	92.6	124.5	101.6	92.6	124.5	101.6
0970, 0990	Wall Finishes & Painting/Coating	101.7	76.9	87.0	92.0	125.5	111.8	92.0	129.1	113.9	98.1	129.1	116.4	92.0	129.1	113.9	92.0	129.1	113.9
09	FINISHES	101.1	67.4	82.8	94.7	118.9	107.8	94.7	118.8	107.8	96.2	119.4	108.8	96.0	118.8	108.4	94.7	118.8	107.8
COVERS	DIVS. 10 - 14, 25, 28, 41, 43, 44, 46	100.0	82.7	96.1	100.0	113.9	103.1	100.0	113.9	103.1	100.0	114.4	103.2	100.0	114.0	103.1	100.0	113.9	103.1
21, 22, 23	FIRE SUPPRESSION, PLUMBING & HVAC	96.5	73.6	87.5	100.1	118.6	107.4	100.1	118.6	107.4	100.0	118.6	107.3	96.6	118.6	105.2	100.1	118.6	107.4
26, 27, 3370	ELECTRICAL, COMMUNICATIONS & UTIL.	95.7	63.2	79.7	93.0	105.0	98.9	93.0	103.7	98.3	92.7	109.4	100.9	92.9	105.8	99.3	93.0	103.7	98.3
MF2018	WEIGHTED AVERAGE	103.1	70.9	89.5	99.3	117.0	106.8	98.5	116.7	106.2	99.6	118.0	107.3	97.2	117.0	105.6	98.1	116.7	106.0

Section 2

DIVISION		CONNECTICUT NEW HAVEN 065			NEW LONDON 063			NORWALK 068			STAMFORD 069			WATERBURY 067			WILLIMANTIC 062		
		MAT.	INST.	TOTAL	MAT.	INST.	TOTAL	MAT.	INST.	TOTAL	MAT.	INST.	TOTAL	MAT.	INST.	TOTAL	MAT.	INST.	TOTAL
015433	CONTRACTOR EQUIPMENT		96.1	96.1		96.1	96.1		95.7	95.7		95.7	95.7		95.7	95.7		95.7	95.7
0241, 31 - 34	SITE & INFRASTRUCTURE, DEMOLITION	105.3	98.4	100.5	97.2	98.1	97.8	105.8	97.6	100.2	106.5	97.7	100.4	105.8	97.8	100.2	105.8	97.7	100.2
0310	Concrete Forming & Accessories	103.2	116.5	114.6	103.1	116.5	114.5	103.4	116.6	114.7	103.4	116.9	114.9	103.4	116.6	114.7	103.4	116.4	114.5
0320	Concrete Reinforcing	116.3	144.5	130.0	91.2	144.5	117.0	116.3	144.5	130.0	116.3	144.5	130.0	116.3	144.5	130.0	116.3	144.5	130.0
0330	Cast-in-Place Concrete	105.1	125.5	112.7	89.5	125.5	102.9	106.7	126.7	114.2	108.5	126.8	115.3	108.5	127.5	115.6	101.3	125.5	110.3
03	CONCRETE	116.3	123.7	119.6	89.2	123.7	104.4	103.3	124.1	112.4	104.1	124.3	113.2	104.1	124.4	113.0	100.6	123.5	110.8
04	MASONRY	101.1	130.9	119.4	99.3	130.9	118.7	100.6	130.1	118.7	101.4	130.1	119.0	101.4	130.9	119.5	100.6	130.9	119.3
05	METALS	94.7	116.4	101.1	94.4	116.4	100.9	98.1	116.5	103.5	98.1	116.8	103.6	98.1	116.5	103.5	97.9	116.3	103.3
06	WOOD, PLASTICS & COMPOSITES	105.0	114.2	110.1	105.0	114.2	110.1	105.0	114.2	110.1	105.0	114.2	110.1	105.0	114.2	110.1	105.0	114.2	110.1
07	THERMAL & MOISTURE PROTECTION	99.9	120.8	109.0	99.7	120.3	108.7	99.9	123.2	110.0	99.8	123.2	110.0	99.8	121.1	109.1	100.0	120.3	108.8
08	OPENINGS	96.4	120.6	102.0	98.7	120.6	103.8	96.4	120.6	102.0	96.4	120.6	102.0	96.4	120.6	102.0	98.8	120.6	103.9
0920	Plaster & Gypsum Board	115.7	114.3	114.8	115.7	114.3	114.8	115.7	114.3	114.8	115.7	114.3	114.8	115.7	114.3	114.8	115.7	114.3	114.8
0950, 0980	Ceilings & Acoustic Treatment	102.0	114.3	110.3	100.2	114.3	109.7	102.0	114.3	110.3	102.0	114.3	110.3	102.0	114.3	110.3	100.2	114.3	109.7
0960	Flooring	92.6	127.0	102.3	92.6	127.0	102.3	92.6	124.5	101.6	92.6	127.0	102.3	92.6	127.0	102.3	92.6	121.1	100.6
0970, 0990	Wall Finishes & Painting/Coating	92.0	125.5	111.8	92.0	129.1	113.9	92.0	125.5	111.8	92.0	125.5	111.8	92.0	129.1	113.9	92.0	129.1	113.9
09	FINISHES	94.7	118.9	107.8	93.9	119.3	107.6	94.7	118.5	107.6	94.8	118.9	107.8	94.6	119.3	107.9	94.5	118.3	107.3
COVERS	DIVS. 10 - 14, 25, 28, 41, 43, 44, 46	100.0	114.0	103.1	100.0	114.0	103.1	100.0	113.9	103.1	100.0	114.1	103.2	100.0	114.0	103.1	100.0	114.0	103.1
21, 22, 23	FIRE SUPPRESSION, PLUMBING & HVAC	100.1	118.6	107.4	96.6	118.6	105.2	100.1	118.6	107.4	100.1	118.6	107.4	100.1	118.6	107.4	100.1	118.4	107.3
26, 27, 3370	ELECTRICAL, COMMUNICATIONS & UTIL.	92.9	106.0	99.4	90.5	107.9	99.1	93.0	104.8	98.8	93.0	150.1	121.1	92.6	107.1	99.7	93.0	108.8	100.8
MF2018	WEIGHTED AVERAGE	99.9	117.0	107.1	95.2	117.2	104.5	98.8	116.7	106.4	99.0	123.2	109.2	98.9	117.2	106.6	98.7	117.1	106.5

Section 3

DIVISION		D.C. WASHINGTON 200 - 205			DELAWARE DOVER 199			NEWARK 197			WILMINGTON 198			FLORIDA DAYTONA BEACH 321			FORT LAUDERDALE 333		
		MAT.	INST.	TOTAL	MAT.	INST.	TOTAL	MAT.	INST.	TOTAL	MAT.	INST.	TOTAL	MAT.	INST.	TOTAL	MAT.	INST.	TOTAL
015433	CONTRACTOR EQUIPMENT		105.3	105.3		118.2	118.2		119.0	119.0		118.4	118.4		99.5	99.5		92.8	92.8
0241, 31 - 34	SITE & INFRASTRUCTURE, DEMOLITION	103.0	97.8	99.4	106.0	107.5	107.0	104.5	108.7	107.4	102.2	107.8	106.1	117.6	85.0	95.1	95.0	73.2	80.0
0310	Concrete Forming & Accessories	97.7	73.4	77.0	98.0	99.4	99.2	97.0	99.2	98.9	95.9	99.4	98.9	97.1	61.6	66.8	93.4	59.4	64.4
0320	Concrete Reinforcing	103.5	86.8	95.4	101.4	115.1	108.0	97.2	115.1	105.9	101.0	115.1	107.8	96.0	61.0	79.1	94.6	59.9	77.8
0330	Cast-in-Place Concrete	103.6	79.0	94.5	108.1	107.7	107.9	91.2	106.7	97.0	102.6	107.7	104.5	91.8	66.0	82.2	96.3	62.3	83.6
03	CONCRETE	106.0	78.6	94.0	100.4	106.0	102.9	93.4	105.7	98.8	97.7	106.0	101.4	90.5	64.8	79.2	93.4	62.3	79.7
04	MASONRY	97.6	87.8	91.6	99.0	98.5	98.7	101.0	98.5	99.5	94.4	98.5	96.9	86.4	60.9	70.7	87.6	54.4	67.2
05	METALS	103.9	95.6	101.5	102.9	121.9	108.5	104.7	123.1	110.1	103.1	121.9	108.6	101.9	88.9	98.1	96.9	88.2	94.4
06	WOOD, PLASTICS & COMPOSITES	95.4	72.2	82.6	92.3	97.6	95.2	91.0	97.4	94.5	85.7	97.6	92.2	93.2	60.1	75.0	78.3	60.6	68.6
07	THERMAL & MOISTURE PROTECTION	104.3	87.1	96.8	105.2	110.4	107.4	109.2	109.8	109.4	105.1	110.4	107.4	101.4	65.4	85.7	106.3	60.4	86.3
08	OPENINGS	100.7	74.6	94.6	91.2	97.4	95.0	91.2	107.3	94.9	88.6	107.4	93.0	92.9	59.4	85.1	94.2	59.5	86.1
0920	Plaster & Gypsum Board	104.3	71.4	82.2	96.0	97.4	96.9	96.7	97.4	97.2	97.3	97.4	97.4	92.7	59.5	70.4	107.2	60.0	75.5
0950, 0980	Ceilings & Acoustic Treatment	110.9	71.4	84.3	100.2	97.4	98.3	96.0	97.4	96.9	90.8	97.4	95.2	78.3	59.5	65.6	84.8	60.0	68.1
0960	Flooring	95.4	75.5	89.8	96.9	105.4	99.3	91.3	105.4	95.3	96.0	105.4	98.7	99.5	62.7	89.2	97.1	62.7	87.5
0970, 0990	Wall Finishes & Painting/Coating	103.7	73.1	85.6	92.6	119.1	108.3	88.2	119.1	106.4	95.3	119.1	109.4	103.8	61.4	78.7	96.5	58.3	73.9
09	FINISHES	98.5	72.8	84.6	95.9	101.6	99.0	89.7	101.5	96.1	94.2	101.6	98.2	91.5	61.1	75.1	91.3	59.4	74.1
COVERS	DIVS. 10 - 14, 25, 28, 41, 43, 44, 46	100.0	94.5	98.8	100.0	103.7	100.8	100.0	103.4	100.8	100.0	103.7	100.8	100.0	81.4	95.9	100.0	80.7	95.7
21, 22, 23	FIRE SUPPRESSION, PLUMBING & HVAC	100.0	88.4	95.4	100.0	119.4	107.6	100.3	119.3	107.7	100.1	119.4	107.7	99.9	75.6	90.4	100.0	66.7	86.3
26, 27, 3370	ELECTRICAL, COMMUNICATIONS & UTIL.	97.7	100.0	98.8	95.1	109.7	102.3	96.9	109.7	103.2	95.1	109.7	102.3	96.7	58.9	78.1	94.7	67.4	81.3
MF2018	WEIGHTED AVERAGE	101.1	87.5	95.4	99.0	109.6	103.5	98.3	109.7	103.1	97.9	109.7	102.9	97.2	69.5	85.5	96.1	66.4	83.6

786

For customer support on your Site Work & Landscape Costs with RSMeans data, call 800.448.8182.

FLORIDA

DIVISION		FORT MYERS 339, 341			GAINESVILLE 326, 344			JACKSONVILLE 320, 322			LAKELAND 338			MELBOURNE 329			MIAMI 330 - 332, 340		
		MAT.	INST.	TOTAL	MAT.	INST.	TOTAL	MAT.	INST.	TOTAL	MAT.	INST.	TOTAL	MAT.	INST.	TOTAL	MAT.	INST.	TOTAL
015433	CONTRACTOR EQUIPMENT		99.5	99.5		99.5	99.5		99.5	99.5		99.5	99.5		99.5	99.5		94.9	94.9
0241, 31 - 34	SITE & INFRASTRUCTURE, DEMOLITION	106.7	85.2	91.9	127.1	84.8	97.9	117.6	85.2	95.2	108.7	85.2	92.5	125.6	85.1	97.7	96.5	78.9	84.3
0310	Concrete Forming & Accessories	89.6	63.8	67.6	92.5	57.1	62.4	96.9	60.2	65.6	86.4	63.8	67.1	93.7	61.7	66.4	99.3	60.2	66.0
0320	Concrete Reinforcing	95.7	76.1	86.2	101.8	61.5	82.3	96.0	61.3	79.2	98.0	76.4	87.6	97.1	65.6	81.9	101.6	60.0	81.5
0330	Cast-in-Place Concrete	100.4	66.0	87.6	105.4	65.7	90.6	92.7	65.8	82.7	102.7	66.1	89.1	110.7	66.0	94.1	93.1	63.9	82.2
03	CONCRETE	94.0	68.3	82.7	101.7	62.7	84.6	90.9	64.2	79.2	95.8	68.5	83.8	102.1	65.6	86.1	91.7	63.2	79.2
04	MASONRY	82.5	58.8	67.9	99.3	60.8	75.7	86.1	58.7	69.3	96.8	61.0	74.8	84.2	60.9	69.9	88.0	55.5	68.0
05	METALS	99.0	93.6	97.4	100.7	88.4	97.1	100.4	88.6	96.9	98.9	94.7	97.7	110.6	90.6	104.7	97.1	87.6	94.3
06	WOOD, PLASTICS & COMPOSITES	75.4	63.2	68.7	87.2	54.6	69.3	93.2	58.6	74.2	71.0	63.2	66.7	88.9	60.1	73.0	92.9	60.7	75.2
07	THERMAL & MOISTURE PROTECTION	106.1	63.2	87.4	101.8	62.5	84.7	101.7	62.4	84.6	106.0	63.8	87.6	101.9	64.1	85.5	106.1	61.4	86.6
08	OPENINGS	95.5	64.6	88.3	92.5	56.5	84.1	92.9	58.8	84.9	95.5	64.7	88.3	92.2	60.5	84.8	96.6	59.5	88.0
0920	Plaster & Gypsum Board	103.3	62.7	76.0	89.6	53.8	65.6	92.7	58.0	69.3	100.2	62.7	75.0	89.6	59.5	69.4	96.2	60.0	71.9
0950, 0980	Ceilings & Acoustic Treatment	79.7	62.7	68.2	72.9	53.8	60.1	78.3	58.0	64.6	79.7	62.7	68.2	77.5	59.5	65.4	85.6	60.0	68.3
0960	Flooring	94.1	77.2	89.4	97.0	62.7	87.4	99.5	62.7	89.2	92.2	62.7	83.9	97.2	62.7	87.5	98.6	62.7	88.5
0970, 0990	Wall Finishes & Painting/Coating	101.4	63.2	78.8	103.8	63.2	79.8	103.8	63.2	79.8	101.4	63.2	78.8	103.8	60.9	90.2	100.2	58.3	75.4
09	FINISHES	91.1	66.1	77.6	90.4	58.1	72.9	91.6	60.4	74.7	90.2	63.1	75.6	91.0	63.2	76.0	90.9	59.8	74.1
COVERS	DIVS. 10 - 14, 25, 28, 41, 43, 44, 46	100.0	80.0	95.5	100.0	80.8	95.7	100.0	78.0	95.1	100.0	80.0	95.5	100.0	81.4	95.9	100.0	81.5	95.9
21, 22, 23	FIRE SUPPRESSION, PLUMBING & HVAC	98.1	59.3	82.9	98.6	63.1	84.7	99.9	63.1	85.5	98.1	59.9	83.1	99.9	74.2	89.8	100.0	63.8	85.8
26, 27, 3370	ELECTRICAL, COMMUNICATIONS & UTIL.	96.6	60.0	78.6	97.0	58.6	78.0	96.4	61.9	79.4	95.0	58.5	77.0	97.7	62.7	80.5	98.4	77.9	88.3
MF2018	WEIGHTED AVERAGE	96.5	67.7	84.3	98.8	65.8	84.8	96.9	66.6	84.1	97.1	67.6	84.6	100.1	70.3	87.5	96.7	68.0	84.6

FLORIDA

DIVISION		ORLANDO 327 - 328, 347			PANAMA CITY 324			PENSACOLA 325			SARASOTA 342			ST. PETERSBURG 337			TALLAHASSEE 323		
		MAT.	INST.	TOTAL	MAT.	INST.	TOTAL	MAT.	INST.	TOTAL	MAT.	INST.	TOTAL	MAT.	INST.	TOTAL	MAT.	INST.	TOTAL
015433	CONTRACTOR EQUIPMENT		102.0	102.0		99.5	99.5		99.5	99.5		99.5	99.5		99.5	99.5		102.0	102.0
0241, 31 - 34	SITE & INFRASTRUCTURE, DEMOLITION	115.8	89.6	97.7	131.4	85.2	99.5	131.3	85.0	99.3	119.0	85.2	95.7	110.4	85.0	92.9	110.2	89.8	96.1
0310	Concrete Forming & Accessories	101.8	61.5	67.4	96.2	66.2	70.6	94.2	61.6	66.4	93.7	63.7	68.1	92.8	60.8	65.5	99.0	62.2	67.6
0320	Concrete Reinforcing	104.9	65.6	85.9	100.3	67.4	84.4	102.8	67.4	85.7	96.6	76.2	86.7	98.0	76.0	87.4	98.4	61.5	80.5
0330	Cast-in-Place Concrete	111.6	66.7	94.9	97.4	65.9	85.7	120.4	65.3	99.9	105.1	66.1	90.6	103.8	65.8	89.7	92.3	66.7	82.8
03	CONCRETE	100.8	65.7	85.4	100.0	67.9	85.9	109.7	65.6	90.3	95.3	68.4	83.5	97.3	66.9	83.9	90.0	65.3	79.2
04	MASONRY	94.8	60.9	74.0	90.6	60.3	72.0	109.1	59.9	78.9	88.7	61.0	71.6	132.4	58.7	87.1	85.2	60.3	69.9
05	METALS	96.7	89.1	94.5	101.5	90.9	98.4	102.6	90.8	99.1	102.5	93.9	100.0	99.9	93.5	98.0	101.7	87.7	97.6
06	WOOD, PLASTICS & COMPOSITES	91.7	60.2	74.4	92.0	66.6	78.0	90.3	61.1	74.2	92.5	63.2	76.4	79.5	59.3	68.4	95.6	61.2	76.7
07	THERMAL & MOISTURE PROTECTION	109.2	66.0	90.4	102.0	64.1	85.5	101.9	63.2	85.1	100.1	63.8	84.3	106.2	62.2	87.1	96.7	64.0	82.4
08	OPENINGS	97.8	60.6	89.1	91.0	64.5	84.9	91.0	61.4	84.1	97.8	64.2	90.0	94.3	62.5	86.9	97.9	60.2	89.1
0920	Plaster & Gypsum Board	93.3	59.5	70.6	91.9	66.1	74.6	99.6	60.5	73.3	97.2	62.7	74.0	105.6	58.7	74.0	100.0	60.5	73.4
0950, 0980	Ceilings & Acoustic Treatment	89.7	59.5	69.3	77.5	66.1	69.8	77.5	60.5	66.0	83.9	62.7	69.6	81.6	58.7	66.1	86.2	60.5	68.9
0960	Flooring	95.7	62.7	86.4	99.1	72.5	91.6	95.1	62.7	86.0	103.5	54.9	89.9	96.1	60.8	86.2	97.6	61.7	87.5
0970, 0990	Wall Finishes & Painting/Coating	95.6	60.3	74.7	103.8	63.2	79.8	103.8	63.2	79.8	99.0	63.2	77.8	101.4	63.2	78.8	101.6	63.2	78.9
09	FINISHES	93.2	61.1	75.8	92.6	66.8	78.6	92.1	61.6	75.6	96.0	61.8	77.5	92.6	60.5	75.2	94.4	61.8	76.7
COVERS	DIVS. 10 - 14, 25, 28, 41, 43, 44, 46	100.0	81.7	95.9	100.0	80.6	95.7	100.0	79.7	95.5	100.0	80.0	95.5	100.0	79.6	95.4	100.0	81.5	95.9
21, 22, 23	FIRE SUPPRESSION, PLUMBING & HVAC	100.0	56.4	82.9	99.9	63.2	85.5	99.9	62.7	85.3	99.9	58.5	83.6	100.0	59.9	84.2	100.0	65.5	86.5
26, 27, 3370	ELECTRICAL, COMMUNICATIONS & UTIL.	98.4	63.1	81.0	95.5	58.6	77.3	99.0	50.7	75.2	96.9	58.5	78.0	95.0	60.6	78.0	103.7	58.6	81.4
MF2018	WEIGHTED AVERAGE	99.0	66.6	85.3	98.6	68.3	85.8	101.2	65.8	86.2	98.8	67.0	85.4	99.7	66.8	85.8	98.2	67.7	85.3

DIVISION		FLORIDA TAMPA 335 - 336, 346			WEST PALM BEACH 334, 349			GEORGIA ALBANY 317, 398			ATHENS 306			ATLANTA 300 - 303, 399			AUGUSTA 308 - 309		
		MAT.	INST.	TOTAL	MAT.	INST.	TOTAL	MAT.	INST.	TOTAL	MAT.	INST.	TOTAL	MAT.	INST.	TOTAL	MAT.	INST.	TOTAL
015433	CONTRACTOR EQUIPMENT		99.5	99.5		92.8	92.8		93.7	93.7		92.6	92.6		96.6	96.6		92.6	92.6
0241, 31 - 34	SITE & INFRASTRUCTURE, DEMOLITION	110.9	87.0	94.4	91.6	73.2	78.9	107.3	76.7	86.1	102.8	90.8	94.5	99.7	96.0	97.1	95.9	91.9	93.2
0310	Concrete Forming & Accessories	95.4	64.1	68.7	96.7	59.1	64.7	90.6	66.5	70.1	90.0	43.6	50.4	94.3	72.3	75.5	90.9	73.0	75.6
0320	Concrete Reinforcing	94.6	76.5	85.8	97.3	57.5	78.1	96.5	71.3	84.3	94.5	63.5	79.5	93.9	71.3	83.0	103.6	70.0	87.3
0330	Cast-in-Place Concrete	101.5	66.2	88.3	91.5	62.2	80.6	94.6	69.1	85.1	109.7	69.5	94.7	113.1	71.1	97.5	92.5	70.0	83.6
03	CONCRETE	95.9	68.7	83.9	90.1	61.8	77.7	86.9	70.1	79.5	99.6	57.4	81.1	101.8	72.4	88.9	92.5	72.2	83.6
04	MASONRY	88.3	61.0	71.5	87.1	52.2	65.7	92.5	68.7	77.9	75.0	77.0	76.2	87.8	68.9	76.2	88.1	68.8	76.2
05	METALS	99.0	95.0	97.8	95.9	87.0	93.3	106.2	97.4	103.6	95.3	79.1	90.6	96.2	83.5	92.5	95.0	82.6	91.4
06	WOOD, PLASTICS & COMPOSITES	83.3	63.2	72.2	83.1	60.6	70.7	80.1	66.4	72.6	92.5	35.8	61.3	97.8	74.0	84.7	93.8	75.3	83.6
07	THERMAL & MOISTURE PROTECTION	106.5	63.8	87.9	106.0	60.6	86.2	100.6	68.0	86.4	95.4	68.3	83.6	96.8	73.8	86.8	95.1	71.4	84.8
08	OPENINGS	95.5	64.7	88.3	93.8	59.0	85.7	85.7	68.5	81.7	91.3	50.2	81.7	99.4	73.5	93.3	91.4	74.0	87.3
0920	Plaster & Gypsum Board	108.2	62.7	77.6	111.9	60.0	77.0	101.1	66.0	77.5	93.0	34.4	53.6	95.0	73.6	80.6	94.0	75.0	81.2
0950, 0980	Ceilings & Acoustic Treatment	84.8	62.7	69.9	79.7	60.0	66.4	77.5	66.0	69.8	96.6	34.4	54.7	89.7	73.6	78.8	97.5	75.0	82.4
0960	Flooring	97.1	62.7	87.5	98.9	54.9	86.6	99.4	70.8	91.4	96.2	81.9	92.2	99.1	70.8	91.1	96.4	70.8	89.2
0970, 0990	Wall Finishes & Painting/Coating	101.4	63.2	78.8	96.5	58.3	73.9	88.7	94.0	91.8	92.4	94.0	93.3	96.1	97.3	96.8	92.4	90.2	91.1
09	FINISHES	93.9	63.1	77.3	91.2	57.8	73.2	91.7	69.7	79.8	95.0	53.9	72.7	95.0	74.5	83.9	94.7	74.5	83.7
COVERS	DIVS. 10 - 14, 25, 28, 41, 43, 44, 46	100.0	80.0	95.6	100.0	80.7	95.7	100.0	85.9	96.8	100.0	82.7	96.1	100.0	87.2	97.2	100.0	87.0	97.1
21, 22, 23	FIRE SUPPRESSION, PLUMBING & HVAC	100.0	59.9	84.3	98.1	58.5	82.6	100.0	68.9	87.8	96.6	66.3	84.7	100.0	69.4	88.0	100.1	69.0	87.9
26, 27, 3370	ELECTRICAL, COMMUNICATIONS & UTIL.	94.7	63.1	79.1	95.7	67.4	81.8	95.5	61.9	79.0	97.8	63.9	81.1	97.6	71.5	84.7	98.3	67.3	83.1
MF2018	WEIGHTED AVERAGE	97.6	68.4	85.3	95.1	64.1	82.0	96.5	71.9	86.1	95.6	66.9	83.5	98.2	75.0	88.4	96.0	73.9	86.6

For customer support on your Site Work & Landscape Costs with RSMeans data, call 800.448.8182.

787

GEORGIA

DIVISION		COLUMBUS 318-319			DALTON 307			GAINESVILLE 305			MACON 310-312			SAVANNAH 313-314			STATESBORO 304		
		MAT.	INST.	TOTAL	MAT.	INST.	TOTAL	MAT.	INST.	TOTAL	MAT.	INST.	TOTAL	MAT.	INST.	TOTAL	MAT.	INST.	TOTAL
015433	CONTRACTOR EQUIPMENT		93.7	93.7		107.6	107.6		92.6	92.6		103.2	103.2		96.7	96.7		95.5	95.5
0241, 31 - 34	SITE & INFRASTRUCTURE, DEMOLITION	107.2	76.7	86.2	102.6	95.8	97.9	102.5	90.7	94.3	108.6	91.0	96.4	107.0	82.0	89.7	103.7	78.0	85.9
0310	Concrete Forming & Accessories	90.5	66.6	70.1	83.2	64.5	67.3	93.3	40.7	48.5	90.3	66.6	70.1	93.4	72.0	75.2	78.5	52.2	56.1
0320	Concrete Reinforcing	96.6	71.3	84.4	94.1	60.9	78.0	94.4	63.4	79.4	97.8	71.3	85.0	104.1	70.7	88.0	93.7	71.2	82.8
0330	Cast-in-Place Concrete	94.2	68.8	84.8	106.6	68.0	92.2	115.3	68.7	98.0	93.0	68.3	83.8	98.3	69.5	87.6	109.5	68.6	94.3
03	CONCRETE	86.8	70.0	79.4	98.9	67.0	84.9	101.4	55.8	81.4	86.3	69.8	79.1	88.5	72.5	81.5	99.3	63.4	83.5
04	MASONRY	92.5	68.7	77.9	75.8	75.5	75.6	83.2	76.2	78.9	104.8	68.9	82.7	87.4	68.8	75.9	77.8	77.0	77.3
05	METALS	105.9	97.6	103.4	96.4	94.0	95.7	94.6	78.5	89.9	100.9	97.6	100.0	102.1	96.0	100.3	100.0	98.9	99.7
06	WOOD, PLASTICS & COMPOSITES	80.1	66.4	72.6	75.9	65.2	70.0	96.1	33.2	61.5	85.9	66.5	75.2	87.4	73.9	80.0	69.7	47.2	57.3
07	THERMAL & MOISTURE PROTECTION	100.5	68.9	86.8	97.3	70.1	85.5	95.4	67.7	83.3	99.0	70.6	86.6	98.5	70.4	86.3	95.9	67.1	83.4
08	OPENINGS	85.7	69.3	81.9	92.8	65.8	86.5	91.3	48.8	81.4	85.5	69.3	81.7	95.3	73.2	90.2	93.7	58.5	85.5
0920	Plaster & Gypsum Board	101.1	66.0	77.5	81.9	64.7	70.3	94.9	31.8	52.4	104.7	66.0	78.7	103.3	73.6	83.3	83.6	46.2	58.4
0950, 0980	Ceilings & Acoustic Treatment	77.5	66.0	69.8	110.2	64.7	79.5	96.6	31.8	52.9	72.8	66.0	68.3	87.3	73.6	78.0	105.8	46.2	65.6
0960	Flooring	99.4	70.8	91.4	97.0	81.9	92.8	97.8	81.9	93.3	77.9	70.8	75.9	95.6	70.8	88.6	114.6	81.9	105.4
0970, 0990	Wall Finishes & Painting/Coating	88.7	87.4	87.9	82.8	70.1	75.3	92.4	94.0	93.3	90.7	94.0	92.6	88.0	86.1	86.9	90.8	70.1	78.5
09	FINISHES	91.6	68.9	79.3	104.5	68.2	84.9	95.5	52.1	72.0	80.5	69.7	74.7	92.9	73.2	82.3	107.9	57.9	80.9
COVERS	DIVS. 10 - 14, 25, 28, 41, 43, 44, 46	100.0	85.9	96.9	100.0	85.3	96.7	100.0	35.5	85.6	100.0	86.0	96.9	100.0	87.0	97.1	100.0	83.8	96.4
21, 22, 23	FIRE SUPPRESSION, PLUMBING & HVAC	100.1	66.8	87.0	96.7	60.8	82.6	96.6	65.6	84.4	100.1	68.2	87.6	100.1	64.8	86.3	97.2	67.3	85.5
26, 27, 3370	ELECTRICAL, COMMUNICATIONS & UTIL.	95.7	67.7	81.9	106.9	62.2	84.9	97.8	70.8	84.5	94.2	62.7	78.6	99.1	63.7	81.7	98.3	63.4	81.1
MF2018	WEIGHTED AVERAGE	96.5	72.2	86.2	97.5	71.4	86.5	96.2	65.6	83.2	95.2	73.1	85.8	97.3	72.7	86.9	97.8	69.7	85.9

DIVISION		GEORGIA VALDOSTA 316			WAYCROSS 315			HAWAII HILO 967			HONOLULU 968			STATES & POSS., GUAM 969			IDAHO BOISE 836-837		
		MAT.	INST.	TOTAL	MAT.	INST.	TOTAL	MAT.	INST.	TOTAL	MAT.	INST.	TOTAL	MAT.	INST.	TOTAL	MAT.	INST.	TOTAL
015433	CONTRACTOR EQUIPMENT		93.7	93.7		93.7	93.7		97.8	97.8		99.3	99.3		160.5	160.5		94.6	94.6
0241, 31 - 34	SITE & INFRASTRUCTURE, DEMOLITION	116.9	76.7	89.1	113.7	77.0	88.3	142.5	103.1	115.3	150.0	106.6	120.0	186.3	98.5	125.7	86.8	92.0	90.4
0310	Concrete Forming & Accessories	81.7	41.6	47.5	83.5	64.0	66.8	108.2	123.8	121.5	120.2	123.7	123.2	110.5	51.5	60.2	100.2	81.1	83.9
0320	Concrete Reinforcing	98.7	59.5	79.8	98.7	59.8	79.9	143.4	127.0	135.5	166.8	127.0	147.6	253.7	27.4	144.2	107.6	79.5	94.0
0330	Cast-in-Place Concrete	92.6	68.7	83.7	104.7	68.6	91.2	180.3	123.3	159.1	145.0	123.5	137.0	156.3	95.7	133.7	90.6	96.6	92.8
03	CONCRETE	91.2	56.6	76.0	94.5	66.5	82.2	144.7	123.2	135.3	137.9	123.1	131.4	147.9	63.9	111.0	97.6	86.4	92.7
04	MASONRY	97.8	77.1	85.1	98.7	77.0	85.4	148.6	122.1	132.3	135.5	122.2	127.3	206.7	34.8	101.0	126.3	86.2	101.7
05	METALS	105.3	93.0	101.7	104.4	88.9	99.8	110.8	108.1	110.0	124.1	107.1	119.1	143.1	76.0	123.4	109.6	82.7	101.7
06	WOOD, PLASTICS & COMPOSITES	68.7	33.0	49.0	70.2	63.4	66.5	111.4	124.3	118.5	132.8	124.2	128.1	123.4	53.9	85.1	93.5	79.9	86.0
07	THERMAL & MOISTURE PROTECTION	100.8	66.6	86.0	100.6	69.1	86.3	128.7	119.3	124.6	146.3	120.0	134.9	150.7	58.2	110.4	101.4	86.6	94.9
08	OPENINGS	82.5	47.6	74.4	82.8	61.8	77.9	114.1	123.2	116.2	128.1	123.2	126.9	118.4	43.8	101.0	96.9	75.9	94.0
0920	Plaster & Gypsum Board	93.8	31.6	52.0	93.8	62.9	73.0	115.2	124.8	121.7	160.8	124.8	136.6	236.1	42.5	105.8	92.3	79.4	83.6
0950, 0980	Ceilings & Acoustic Treatment	75.9	31.6	46.0	74.0	62.9	66.5	136.6	124.8	128.6	144.8	124.8	131.3	257.2	42.5	112.4	102.8	79.4	87.0
0960	Flooring	93.0	83.5	90.3	94.3	81.9	90.8	103.8	139.3	113.8	121.3	139.3	126.4	121.2	39.5	98.2	92.9	90.7	92.3
0970, 0990	Wall Finishes & Painting/Coating	88.7	94.0	91.8	88.7	70.1	77.7	99.2	143.8	125.6	110.1	143.8	130.1	105.2	31.5	61.6	92.6	39.5	61.2
09	FINISHES	89.4	52.5	69.4	89.0	67.5	77.4	110.8	129.1	120.7	125.5	129.1	127.4	185.1	48.0	110.9	92.7	78.2	84.9
COVERS	DIVS. 10 - 14, 25, 28, 41, 43, 44, 46	100.0	82.2	96.0	100.0	85.5	96.8	100.0	113.7	103.1	100.0	113.6	103.0	100.0	66.0	92.4	100.0	88.6	97.4
21, 22, 23	FIRE SUPPRESSION, PLUMBING & HVAC	100.1	65.8	87.7	97.9	63.6	84.4	100.3	112.4	105.0	100.4	112.3	105.1	102.6	33.5	75.5	100.1	74.2	89.9
26, 27, 3370	ELECTRICAL, COMMUNICATIONS & UTIL.	94.1	57.8	76.2	97.7	63.4	80.8	108.1	124.6	116.2	109.8	124.5	117.1	158.1	35.9	97.9	97.2	68.2	82.9
MF2018	WEIGHTED AVERAGE	96.7	66.2	83.8	96.8	70.0	85.5	114.8	118.4	116.3	119.1	118.5	118.9	137.1	51.4	100.9	101.0	79.9	92.1

DIVISION		IDAHO COEUR D'ALENE 838			IDAHO FALLS 834			LEWISTON 835			POCATELLO 832			TWIN FALLS 833			ILLINOIS BLOOMINGTON 617		
		MAT.	INST.	TOTAL	MAT.	INST.	TOTAL	MAT.	INST.	TOTAL	MAT.	INST.	TOTAL	MAT.	INST.	TOTAL	MAT.	INST.	TOTAL
015433	CONTRACTOR EQUIPMENT		89.7	89.7		94.6	94.6		89.7	89.7		94.6	94.6		94.6	94.6		101.9	101.9
0241, 31 - 34	SITE & INFRASTRUCTURE, DEMOLITION	85.5	86.5	86.2	85.1	92.1	89.9	92.2	87.2	88.7	88.3	92.0	90.9	95.4	91.8	92.9	94.9	96.0	95.7
0310	Concrete Forming & Accessories	108.7	81.4	85.4	94.1	78.3	80.6	113.9	81.1	85.9	100.4	80.5	83.4	101.6	76.5	80.2	84.0	116.0	111.2
0320	Concrete Reinforcing	115.7	96.2	106.3	109.6	77.9	94.3	115.3	96.2	106.3	108.0	79.4	94.2	110.0	77.8	94.4	93.0	102.4	97.5
0330	Cast-in-Place Concrete	98.0	85.2	93.2	86.3	83.4	85.2	101.8	82.0	94.5	93.1	96.4	94.3	95.6	82.3	90.6	98.0	113.7	103.9
03	CONCRETE	103.9	85.3	95.7	89.6	80.2	85.5	107.5	84.1	97.2	96.7	86.0	92.0	104.6	79.0	93.4	91.9	113.6	101.4
04	MASONRY	128.0	87.3	103.0	121.5	87.2	100.4	128.5	84.7	101.6	124.0	86.2	100.8	126.7	82.8	99.7	110.2	121.7	117.3
05	METALS	103.0	88.1	98.6	118.1	81.6	107.4	102.4	88.2	98.3	118.2	82.0	107.6	118.2	81.4	107.4	95.2	123.1	103.4
06	WOOD, PLASTICS & COMPOSITES	96.7	79.8	87.4	87.2	76.3	81.2	102.6	79.8	90.1	93.5	79.9	86.0	94.6	76.3	84.5	81.7	113.0	99.0
07	THERMAL & MOISTURE PROTECTION	158.9	84.3	126.4	100.4	75.3	89.5	159.2	83.1	126.1	101.0	75.0	89.7	101.8	80.5	92.5	95.4	112.1	102.6
08	OPENINGS	113.8	75.9	105.0	99.8	71.3	93.2	106.7	79.6	100.4	97.6	68.8	90.9	100.6	62.4	91.7	91.3	118.7	97.7
0920	Plaster & Gypsum Board	167.0	79.4	108.1	80.3	75.7	77.2	168.9	79.4	108.7	81.9	79.4	80.2	84.0	75.7	78.4	89.1	113.5	105.5
0950, 0980	Ceilings & Acoustic Treatment	136.3	79.4	98.0	104.3	75.7	85.0	136.3	79.4	98.0	110.0	79.4	89.4	106.8	75.7	85.8	83.9	113.5	103.9
0960	Flooring	130.9	77.0	115.8	92.7	77.0	88.3	134.3	77.0	118.2	96.2	77.0	90.8	97.5	77.0	91.7	86.0	120.8	95.8
0970, 0990	Wall Finishes & Painting/Coating	111.2	73.8	89.1	92.6	39.5	61.2	111.2	71.3	87.6	92.5	39.5	61.1	92.6	39.5	61.2	87.6	134.8	115.5
09	FINISHES	158.3	79.5	115.7	90.9	74.0	81.8	159.7	78.9	116.0	93.7	75.9	84.1	94.5	72.9	82.8	86.1	119.0	103.9
COVERS	DIVS. 10 - 14, 25, 28, 41, 43, 44, 46	100.0	93.0	98.4	100.0	87.4	97.2	100.0	92.6	98.3	100.0	88.5	97.4	100.0	85.9	96.9	100.0	104.5	101.0
21, 22, 23	FIRE SUPPRESSION, PLUMBING & HVAC	99.4	85.5	94.0	101.1	81.8	93.5	100.9	85.1	94.6	100.0	73.6	89.6	100.0	72.2	89.1	96.5	105.4	100.0
26, 27, 3370	ELECTRICAL, COMMUNICATIONS & UTIL.	88.7	81.5	85.2	88.3	70.9	79.7	86.9	78.2	82.6	94.4	66.0	80.4	89.6	70.9	80.4	95.9	90.3	93.1
MF2018	WEIGHTED AVERAGE	108.1	84.4	98.1	100.5	79.9	91.8	108.2	83.5	97.8	102.0	78.4	92.0	103.2	76.8	92.1	95.0	109.5	101.1

For customer support on your Site Work & Landscape Costs with RSMeans data, call 800.448.8182.

ILLINOIS

DIVISION		CARBONDALE 629			CENTRALIA 628			CHAMPAIGN 618 - 619			CHICAGO 606 - 608			DECATUR 625			EAST ST. LOUIS 620 - 622		
		MAT.	INST.	TOTAL	MAT.	INST.	TOTAL	MAT.	INST.	TOTAL	MAT.	INST.	TOTAL	MAT.	INST.	TOTAL	MAT.	INST.	TOTAL
015433	CONTRACTOR EQUIPMENT		110.3	110.3		110.3	110.3		102.8	102.8		100.3	100.3		102.8	102.8		110.3	110.3
0241, 31 - 34	SITE & INFRASTRUCTURE, DEMOLITION	100.9	96.8	98.1	101.3	98.6	99.5	104.0	97.2	99.3	105.4	104.4	104.7	94.8	97.0	96.3	103.5	97.7	99.5
0310	Concrete Forming & Accessories	90.0	108.9	106.1	91.6	114.2	110.9	90.3	116.7	112.8	97.4	159.7	150.5	92.1	117.9	114.1	87.5	114.2	110.2
0320	Concrete Reinforcing	82.2	103.4	92.5	82.2	104.0	92.8	93.0	101.6	97.2	103.8	153.3	127.7	80.6	99.8	89.9	82.2	104.0	92.7
0330	Cast-in-Place Concrete	91.5	102.7	95.7	92.0	119.6	102.3	113.7	110.8	112.6	126.0	155.6	137.0	99.9	114.6	105.4	93.6	118.4	102.8
03	CONCRETE	79.9	107.3	91.9	80.4	115.8	96.0	104.1	112.7	107.9	108.0	156.5	129.3	90.3	114.3	100.8	81.4	115.4	96.3
04	MASONRY	75.0	111.5	97.4	75.0	122.9	104.5	133.1	125.4	128.4	105.3	163.4	141.0	70.9	122.8	102.8	75.3	123.0	104.6
05	METALS	100.9	130.4	109.6	100.9	133.5	110.5	95.2	120.3	102.6	95.9	146.8	110.8	104.9	120.7	109.5	102.1	133.5	111.3
06	WOOD, PLASTICS & COMPOSITES	87.7	105.8	97.7	90.2	111.2	101.7	88.5	114.0	102.5	100.0	158.4	132.1	88.7	116.8	104.2	84.8	111.2	99.3
07	THERMAL & MOISTURE PROTECTION	90.6	101.3	95.3	90.7	111.5	99.8	96.2	115.6	104.6	95.2	150.7	119.4	95.9	113.3	103.5	90.7	110.7	99.4
08	OPENINGS	87.3	116.3	94.1	87.3	119.2	94.8	91.8	116.9	97.7	100.9	169.3	116.9	98.2	118.2	102.9	87.4	118.4	94.6
0920	Plaster & Gypsum Board	93.7	106.2	102.1	95.0	111.6	106.2	91.0	114.5	106.8	101.3	160.0	140.8	96.5	117.4	110.6	92.2	111.6	105.3
0950, 0980	Ceilings & Acoustic Treatment	80.6	106.2	97.8	80.6	111.6	101.5	83.9	114.5	104.5	100.4	160.0	140.6	87.1	117.4	107.5	80.6	111.6	101.5
0960	Flooring	113.7	116.6	114.5	114.6	116.6	115.1	89.2	116.6	96.9	91.3	163.8	111.7	101.8	117.1	106.1	112.6	116.6	113.7
0970, 0990	Wall Finishes & Painting/Coating	103.1	105.0	104.2	103.1	111.6	108.2	87.6	114.9	103.7	96.7	170.3	140.2	95.5	117.8	108.7	103.1	111.6	108.2
09	FINISHES	92.1	110.7	102.1	92.5	114.4	104.4	88.1	116.9	103.7	95.8	162.4	131.8	92.2	118.3	106.3	91.6	114.4	103.9
COVERS	DIVS. 10 - 14, 25, 28, 41, 43, 44, 46	100.0	106.9	101.5	100.0	105.8	101.3	100.0	109.2	102.0	100.0	129.3	106.5	100.0	108.6	101.9	100.0	108.3	101.9
21, 22, 23	FIRE SUPPRESSION, PLUMBING & HVAC	96.3	106.8	100.4	96.3	97.8	96.9	96.5	107.0	100.6	99.9	137.4	114.6	99.9	98.8	99.5	99.9	99.2	99.6
26, 27, 3370	ELECTRICAL, COMMUNICATIONS & UTIL.	91.9	107.8	99.8	93.1	106.3	99.6	99.0	95.4	97.2	97.9	136.2	116.8	95.3	103.7	99.5	92.7	99.7	96.2
MF2018	WEIGHTED AVERAGE	92.5	109.6	99.7	92.8	111.2	100.5	98.5	110.5	103.6	100.1	145.6	119.3	96.6	110.1	102.3	93.8	110.4	100.9

ILLINOIS

DIVISION		EFFINGHAM 624			GALESBURG 614			JOLIET 604			KANKAKEE 609			LA SALLE 613			NORTH SUBURBAN 600 - 603		
		MAT.	INST.	TOTAL	MAT.	INST.	TOTAL	MAT.	INST.	TOTAL	MAT.	INST.	TOTAL	MAT.	INST.	TOTAL	MAT.	INST.	TOTAL
015433	CONTRACTOR EQUIPMENT		102.8	102.8		101.9	101.9		93.7	93.7		93.7	93.7		101.9	101.9		93.7	93.7
0241, 31 - 34	SITE & INFRASTRUCTURE, DEMOLITION	99.3	95.9	97.0	97.3	95.8	96.3	101.2	97.2	98.5	94.7	97.1	96.4	96.7	96.8	96.8	100.5	97.9	98.7
0310	Concrete Forming & Accessories	96.5	114.6	111.9	90.2	117.0	113.0	96.8	156.7	147.8	90.5	145.4	137.3	103.7	125.3	122.1	96.1	158.5	149.3
0320	Concrete Reinforcing	83.3	93.8	88.4	92.5	105.6	98.8	103.8	136.6	119.7	104.6	135.1	119.3	92.7	132.9	112.1	103.8	149.1	125.7
0330	Cast-in-Place Concrete	99.6	109.9	103.4	100.9	109.2	104.0	114.6	143.9	125.5	106.8	135.8	117.6	100.8	121.8	108.6	114.6	152.9	128.9
03	CONCRETE	91.0	109.9	99.3	94.8	113.0	102.8	101.0	147.9	121.6	95.0	139.7	114.7	95.6	126.1	109.0	101.0	154.1	124.3
04	MASONRY	79.1	115.3	101.3	110.4	122.4	117.8	104.5	156.7	136.6	100.6	147.0	129.1	110.4	129.7	122.3	101.2	158.8	136.6
05	METALS	101.9	114.9	105.7	95.2	123.6	103.5	94.1	134.7	106.0	94.1	133.4	105.6	95.3	142.0	109.0	95.1	141.6	108.8
06	WOOD, PLASTICS & COMPOSITES	91.0	114.0	103.6	88.3	115.4	103.2	96.6	157.7	130.2	90.0	144.7	120.1	103.8	123.2	114.5	95.4	158.2	130.0
07	THERMAL & MOISTURE PROTECTION	95.4	109.1	101.4	95.6	110.0	101.9	99.5	144.2	119.0	98.6	139.5	116.4	95.8	122.5	107.4	99.9	147.6	120.7
08	OPENINGS	93.3	114.5	98.2	91.3	115.9	97.0	99.3	161.7	113.9	92.4	156.2	107.3	91.3	133.3	101.1	99.4	167.9	115.4
0920	Plaster & Gypsum Board	96.3	114.5	108.5	91.0	115.9	107.8	93.7	159.5	138.0	92.5	146.1	128.6	98.3	124.0	115.6	97.2	160.0	139.5
0950, 0980	Ceilings & Acoustic Treatment	80.6	114.5	103.4	83.9	115.9	105.5	100.9	159.5	140.4	100.9	146.1	131.4	83.9	124.0	110.9	100.9	160.0	140.8
0960	Flooring	103.0	116.6	106.8	89.1	120.8	98.0	89.4	149.8	106.4	86.3	150.7	104.4	95.7	124.8	103.9	89.9	159.5	109.4
0970, 0990	Wall Finishes & Painting/Coating	95.5	109.3	103.7	87.6	98.5	94.0	89.0	163.0	132.8	89.0	135.7	116.6	87.6	133.0	114.5	91.0	163.9	134.1
09	FINISHES	91.3	115.1	104.2	87.4	116.3	103.1	91.6	157.4	127.2	90.0	145.8	120.2	90.4	124.7	109.0	92.2	160.1	129.0
COVERS	DIVS. 10 - 14, 25, 28, 41, 43, 44, 46	100.0	104.5	101.0	100.0	104.4	101.0	100.0	122.6	105.0	100.0	120.6	104.6	100.0	105.6	101.2	100.0	124.1	105.4
21, 22, 23	FIRE SUPPRESSION, PLUMBING & HVAC	96.4	104.5	99.6	96.5	105.7	100.1	100.0	130.0	111.8	96.5	129.6	109.5	96.5	128.4	109.0	99.9	136.2	114.1
26, 27, 3370	ELECTRICAL, COMMUNICATIONS & UTIL.	93.5	107.8	100.5	96.7	86.5	91.7	96.9	133.3	114.8	92.4	137.0	114.4	94.0	137.0	115.2	96.7	135.5	115.9
MF2018	WEIGHTED AVERAGE	95.1	109.1	101.0	95.7	108.5	101.1	98.3	138.8	115.4	95.0	134.9	111.9	95.9	127.0	109.0	98.3	142.9	117.1

ILLINOIS

DIVISION		PEORIA 615 - 616			QUINCY 623			ROCK ISLAND 612			ROCKFORD 610 - 611			SOUTH SUBURBAN 605			SPRINGFIELD 626 - 627		
		MAT.	INST.	TOTAL	MAT.	INST.	TOTAL	MAT.	INST.	TOTAL	MAT.	INST.	TOTAL	MAT.	INST.	TOTAL	MAT.	INST.	TOTAL
015433	CONTRACTOR EQUIPMENT		101.9	101.9		102.8	102.8		101.9	101.9		101.9	101.9		93.7	93.7		105.1	105.1
0241, 31 - 34	SITE & INFRASTRUCTURE, DEMOLITION	97.9	96.0	96.6	98.1	96.4	96.9	95.5	94.8	95.0	97.4	96.8	97.0	100.5	97.8	98.6	99.8	100.5	100.3
0310	Concrete Forming & Accessories	93.0	115.9	112.6	94.3	113.6	110.8	91.6	99.1	98.0	97.2	127.7	123.2	96.1	158.5	149.3	93.2	115.6	112.3
0320	Concrete Reinforcing	90.1	104.7	97.2	83.0	86.2	84.5	92.5	99.0	95.6	85.3	134.1	108.9	103.8	149.0	125.7	83.2	99.8	91.2
0330	Cast-in-Place Concrete	97.9	115.6	104.5	99.8	106.6	102.3	98.7	98.4	98.6	100.2	126.6	110.0	114.6	152.8	128.8	94.9	109.2	100.2
03	CONCRETE	92.0	114.6	102.0	90.7	107.1	97.9	92.7	100.0	95.9	92.9	129.0	108.7	101.0	154.0	124.3	88.5	111.2	98.5
04	MASONRY	109.7	121.5	116.9	98.9	113.2	107.7	110.2	100.2	104.1	85.6	142.3	120.4	101.2	158.7	136.6	80.7	122.2	106.2
05	METALS	97.9	124.2	105.6	101.9	112.6	105.1	95.2	118.5	102.1	97.9	141.5	110.7	95.1	141.4	108.7	102.5	119.1	107.3
06	WOOD, PLASTICS & COMPOSITES	96.4	113.2	105.6	88.6	114.0	102.6	90.0	97.5	94.1	96.4	123.6	111.3	95.4	158.2	130.0	90.0	114.1	103.3
07	THERMAL & MOISTURE PROTECTION	96.3	112.4	103.3	95.3	106.8	100.3	95.5	98.4	96.8	98.8	129.0	111.9	99.9	147.6	120.7	98.3	114.4	105.3
08	OPENINGS	96.6	119.5	102.0	94.0	112.1	98.2	91.3	103.3	94.1	96.6	136.5	105.9	99.4	167.9	115.4	99.2	116.4	103.2
0920	Plaster & Gypsum Board	95.5	113.7	107.7	95.0	114.5	108.1	91.0	97.6	95.5	95.5	124.4	114.9	97.2	160.0	139.5	99.9	114.5	109.7
0950, 0980	Ceilings & Acoustic Treatment	89.6	113.7	105.8	80.6	114.5	103.4	83.9	97.6	93.2	89.6	124.4	113.1	100.9	160.0	140.8	91.3	114.5	107.0
0960	Flooring	92.6	119.5	100.2	101.8	113.6	105.1	90.2	94.6	91.4	92.6	124.9	101.7	89.9	159.5	109.4	104.8	117.1	108.3
0970, 0990	Wall Finishes & Painting/Coating	87.6	134.8	115.5	95.5	111.5	105.0	87.6	95.5	92.3	87.6	142.7	120.2	91.0	163.9	134.1	98.0	111.5	106.0
09	FINISHES	90.2	118.7	105.6	90.7	114.2	103.4	87.7	97.6	93.0	90.2	128.5	110.9	92.2	160.1	129.0	98.0	115.9	107.7
COVERS	DIVS. 10 - 14, 25, 28, 41, 43, 44, 46	100.0	104.4	101.0	100.0	106.9	101.5	100.0	97.5	99.4	100.0	114.2	103.2	100.0	124.1	105.4	100.0	108.3	101.9
21, 22, 23	FIRE SUPPRESSION, PLUMBING & HVAC	100.0	100.2	100.0	96.4	102.9	98.9	96.5	97.5	96.9	100.1	116.1	106.4	99.9	136.2	114.1	99.9	103.7	101.4
26, 27, 3370	ELECTRICAL, COMMUNICATIONS & UTIL.	97.7	91.5	94.6	91.4	82.2	86.9	89.3	93.1	91.2	98.0	125.6	111.6	96.7	135.5	115.9	98.1	88.7	93.5
MF2018	WEIGHTED AVERAGE	97.6	108.8	102.3	95.7	104.2	99.3	94.6	99.5	96.7	96.6	125.5	108.8	98.3	142.8	117.1	97.5	108.3	102.1

For customer support on your Site Work & Landscape Costs with RSMeans data, call 800.448.8182.

789

City Cost Indexes

DIVISION		ANDERSON 460 MAT.	INST.	TOTAL	BLOOMINGTON 474 MAT.	INST.	TOTAL	COLUMBUS 472 MAT.	INST.	TOTAL	EVANSVILLE 476-477 MAT.	INST.	TOTAL	FORT WAYNE 467-468 MAT.	INST.	TOTAL	GARY 463-464 MAT.	INST.	TOTAL
			INDIANA																
015433	CONTRACTOR EQUIPMENT		94.5	94.5		81.5	81.5		81.5	81.5		110.3	110.3		94.5	94.5		94.5	94.5
0241, 31 - 34	SITE & INFRASTRUCTURE, DEMOLITION	98.7	89.0	92.0	86.5	87.7	87.3	83.2	87.4	86.1	91.9	115.1	107.9	99.8	88.8	92.2	99.3	92.8	94.8
0310	Concrete Forming & Accessories	94.9	77.4	80.0	100.8	82.1	84.9	95.0	80.2	82.4	94.3	78.6	80.9	92.9	73.7	76.5	95.0	111.1	108.7
0320	Concrete Reinforcing	104.9	83.0	94.3	90.4	86.9	88.7	90.8	86.9	88.9	99.0	81.0	90.3	104.9	77.9	91.9	104.9	115.8	110.2
0330	Cast-in-Place Concrete	106.3	74.9	94.6	101.3	75.7	91.7	100.8	76.5	91.8	96.8	83.2	91.7	113.0	75.2	98.9	111.1	112.2	111.5
03	CONCRETE	95.7	78.1	88.0	99.2	80.3	90.9	98.5	79.6	90.2	99.5	80.8	91.3	98.6	75.7	88.6	97.9	112.1	104.1
04	MASONRY	87.4	74.0	79.2	88.9	74.4	80.0	88.8	74.3	79.9	84.3	77.3	80.0	90.7	71.4	78.8	88.8	110.3	102.0
05	METALS	97.7	88.2	94.9	99.1	75.7	92.2	99.1	74.7	91.9	91.9	83.2	89.3	97.7	86.3	94.3	97.7	107.1	100.5
06	WOOD, PLASTICS & COMPOSITES	93.8	77.6	84.9	109.5	82.9	94.9	104.0	80.5	91.1	90.8	77.7	83.6	93.7	73.8	82.7	91.3	109.1	101.1
07	THERMAL & MOISTURE PROTECTION	109.2	75.0	94.4	95.8	77.9	88.0	95.2	78.7	88.0	100.0	81.7	92.1	108.9	76.8	94.9	107.7	106.2	107.0
08	OPENINGS	93.2	75.8	89.2	97.5	80.0	93.4	93.9	78.6	90.3	91.8	76.0	88.1	93.2	71.2	88.1	93.2	114.6	98.2
0920	Plaster & Gypsum Board	106.4	77.3	86.8	98.6	83.2	88.2	95.4	80.7	85.5	94.2	76.7	82.4	105.8	73.4	84.0	99.5	109.7	106.4
0950, 0980	Ceilings & Acoustic Treatment	89.8	77.3	81.4	77.9	83.2	81.5	77.9	80.7	79.8	81.6	76.7	78.3	89.8	73.4	78.7	89.8	109.7	103.2
0960	Flooring	93.5	75.0	88.3	99.6	83.2	95.0	94.4	83.2	91.2	94.1	71.4	87.7	93.5	71.2	87.2	93.5	110.5	98.3
0970, 0990	Wall Finishes & Painting/Coating	92.8	65.8	76.9	84.4	81.1	82.5	84.4	81.1	82.5	90.3	84.4	86.8	92.8	70.2	79.5	92.8	120.7	109.3
09	FINISHES	91.3	75.7	82.9	90.6	82.4	86.2	88.6	81.0	84.5	89.3	77.7	83.0	91.1	72.9	81.3	90.3	111.8	102.0
COVERS	DIVS. 10 - 14, 25, 28, 41, 43, 44, 46	100.0	87.4	97.2	100.0	88.3	97.4	100.0	88.0	97.3	100.0	91.2	98.0	100.0	87.2	97.1	100.0	107.9	101.8
21, 22, 23	FIRE SUPPRESSION, PLUMBING & HVAC	100.0	77.0	91.0	99.7	79.6	91.8	96.2	79.1	89.5	99.9	78.6	91.5	100.0	72.2	89.1	100.0	106.2	102.4
26, 27, 3370	ELECTRICAL, COMMUNICATIONS & UTIL.	87.7	83.2	85.5	99.9	86.6	93.3	99.1	86.7	93.0	95.7	81.4	88.7	88.4	74.9	81.7	99.5	110.1	104.7
MF2018	WEIGHTED AVERAGE	96.0	79.8	89.1	97.8	81.1	90.7	96.1	80.5	89.5	95.5	82.7	90.1	96.6	76.3	88.0	97.5	108.1	102.0

DIVISION		INDIANAPOLIS 461-462 MAT.	INST.	TOTAL	KOKOMO 469 MAT.	INST.	TOTAL	LAFAYETTE 479 MAT.	INST.	TOTAL	LAWRENCEBURG 470 MAT.	INST.	TOTAL	MUNCIE 473 MAT.	INST.	TOTAL	NEW ALBANY 471 MAT.	INST.	TOTAL
			INDIANA																
015433	CONTRACTOR EQUIPMENT		86.1	86.1		94.5	94.5		81.5	81.5		100.6	100.6		92.8	92.8		90.6	90.6
0241, 31 - 34	SITE & INFRASTRUCTURE, DEMOLITION	99.8	91.3	93.9	94.9	89.0	90.8	83.9	87.4	86.3	81.2	102.2	95.7	86.6	87.9	87.5	78.5	89.8	86.3
0310	Concrete Forming & Accessories	100.4	85.2	87.4	98.0	78.2	81.1	92.5	78.6	80.7	91.5	78.4	80.3	92.3	76.9	79.2	90.1	76.9	78.9
0320	Concrete Reinforcing	106.1	87.2	97.0	94.9	87.2	91.1	90.4	82.9	86.8	89.7	79.8	84.9	100.0	82.9	91.7	91.0	82.6	86.9
0330	Cast-in-Place Concrete	101.2	84.8	95.1	105.2	81.8	96.5	101.4	77.6	92.5	94.8	74.9	87.4	106.4	73.9	94.3	97.9	73.1	88.7
03	CONCRETE	99.0	84.8	92.8	92.5	81.6	87.7	98.7	78.7	89.9	91.8	78.0	85.7	97.4	77.6	88.7	97.3	76.8	88.3
04	MASONRY	89.8	78.7	83.0	87.1	77.2	81.0	94.0	74.0	81.7	73.8	74.2	74.1	90.5	74.1	80.4	80.2	71.1	74.6
05	METALS	94.7	76.0	89.2	94.2	89.8	92.9	97.4	73.7	90.5	93.9	86.2	91.7	100.8	88.0	97.0	96.0	82.2	91.9
06	WOOD, PLASTICS & COMPOSITES	100.1	86.0	92.4	96.8	77.4	86.1	101.2	79.2	89.1	89.1	78.3	83.2	102.6	77.1	88.6	91.2	77.6	83.7
07	THERMAL & MOISTURE PROTECTION	98.7	82.0	91.4	108.1	77.5	94.8	95.2	77.2	87.4	100.8	76.9	90.4	98.2	76.4	88.7	87.6	72.3	80.9
08	OPENINGS	103.9	81.9	98.8	88.4	76.9	85.7	92.4	76.7	88.8	93.6	75.0	89.3	91.0	75.6	87.4	91.4	76.3	87.9
0920	Plaster & Gypsum Board	96.8	85.8	89.4	111.0	77.1	88.2	93.3	79.4	83.9	71.7	78.4	76.2	94.2	77.3	82.8	92.0	77.5	82.2
0950, 0980	Ceilings & Acoustic Treatment	92.7	85.8	88.0	90.6	77.1	81.5	73.9	79.4	77.6	84.0	78.4	80.2	77.9	77.3	77.5	81.6	77.5	78.8
0960	Flooring	97.2	83.2	93.2	97.3	88.4	94.8	93.3	79.2	89.3	68.9	83.2	72.9	93.7	75.0	88.5	91.6	54.0	81.1
0970, 0990	Wall Finishes & Painting/Coating	96.5	81.1	87.4	92.8	68.8	78.6	84.4	81.4	82.6	85.2	73.0	78.0	84.4	65.8	73.4	90.3	66.3	76.1
09	FINISHES	94.9	84.6	89.4	93.1	79.1	85.5	87.2	79.2	82.8	78.8	78.9	78.8	87.9	75.4	81.1	88.5	71.4	79.2
COVERS	DIVS. 10 - 14, 25, 28, 41, 43, 44, 46	100.0	93.7	98.6	100.0	88.6	97.5	100.0	86.6	97.0	100.0	88.0	97.3	100.0	86.3	97.0	100.0	87.7	97.3
21, 22, 23	FIRE SUPPRESSION, PLUMBING & HVAC	100.1	79.9	92.2	96.5	79.3	89.7	96.2	76.2	88.3	97.0	75.9	88.7	99.7	76.9	90.7	96.4	78.6	89.4
26, 27, 3370	ELECTRICAL, COMMUNICATIONS & UTIL.	101.9	86.7	94.4	92.0	78.3	85.3	98.6	78.6	88.7	93.6	74.1	84.0	91.4	74.0	82.9	94.3	74.6	84.6
MF2018	WEIGHTED AVERAGE	98.7	83.2	92.1	94.2	81.1	88.6	95.8	78.1	88.3	92.5	79.6	87.0	96.2	78.2	88.6	93.9	77.3	86.9

DIVISION		SOUTH BEND 465-466 MAT.	INST.	TOTAL	TERRE HAUTE 478 MAT.	INST.	TOTAL	WASHINGTON 475 MAT.	INST.	TOTAL	BURLINGTON 526 MAT.	INST.	TOTAL	CARROLL 514 MAT.	INST.	TOTAL	CEDAR RAPIDS 522-524 MAT.	INST.	TOTAL
			INDIANA									IOWA							
015433	CONTRACTOR EQUIPMENT		108.2	108.2		110.3	110.3		110.3	110.3		98.8	98.8		98.8	98.8		95.7	95.7
0241, 31 - 34	SITE & INFRASTRUCTURE, DEMOLITION	97.7	94.2	95.3	93.4	75.1	80.6	93.2	115.7	108.7	98.0	91.8	93.7	84.1	83.9	83.9	102.5	83.9	86.6
0310	Concrete Forming & Accessories	95.1	76.0	78.9	95.0	77.1	79.7	96.0	80.5	82.8	96.5	94.2	94.5	95.4	85.8	90.8	95.3	80.2	88.0
0320	Concrete Reinforcing	104.1	83.6	94.2	99.0	83.0	91.3	91.6	82.6	87.2	94.7	97.8	96.2	95.4	85.8	90.8	95.3	80.2	88.0
0330	Cast-in-Place Concrete	108.2	78.4	97.1	93.7	78.3	88.0	102.0	85.0	95.6	107.3	54.8	87.8	107.3	82.2	97.9	107.6	83.4	98.6
03	CONCRETE	97.7	79.7	89.8	102.5	78.8	92.1	108.3	82.6	97.0	95.8	81.5	89.6	94.7	84.3	90.1	95.9	83.8	90.6
04	MASONRY	91.2	74.0	80.6	92.0	73.3	80.5	84.5	79.6	81.5	99.5	71.9	82.5	101.1	72.4	83.5	105.0	80.0	89.6
05	METALS	101.5	102.4	101.7	92.6	84.4	90.2	87.1	84.7	86.4	87.9	99.1	91.2	87.9	95.4	90.1	90.3	93.1	91.2
06	WOOD, PLASTICS & COMPOSITES	96.6	75.2	84.8	93.0	76.7	84.0	93.2	79.2	85.5	91.6	98.1	95.2	77.8	88.4	83.6	98.9	83.6	90.5
07	THERMAL & MOISTURE PROTECTION	103.2	78.7	92.5	100.1	79.1	91.0	100.0	83.5	92.8	104.4	77.1	92.5	104.7	78.8	93.4	105.5	80.1	94.4
08	OPENINGS	93.1	75.5	89.0	92.3	75.3	88.3	89.3	77.1	86.4	94.5	95.8	94.8	99.8	84.2	95.4	99.3	81.2	95.1
0920	Plaster & Gypsum Board	97.7	74.7	82.2	94.2	75.7	81.8	94.0	78.3	83.4	105.0	98.3	100.5	100.9	88.3	92.4	110.3	83.6	92.3
0950, 0980	Ceilings & Acoustic Treatment	94.2	74.7	81.1	81.6	75.7	77.6	76.7	78.3	77.8	99.0	98.3	98.5	99.0	88.3	91.8	101.4	83.6	89.4
0960	Flooring	90.7	86.2	89.5	94.1	75.7	88.9	95.1	83.2	91.7	93.7	69.1	86.8	88.0	79.5	85.6	107.9	84.3	101.3
0970, 0990	Wall Finishes & Painting/Coating	94.3	82.5	87.3	90.3	79.8	84.1	90.3	84.4	86.8	92.8	84.6	87.9	92.8	84.6	87.9	94.7	70.8	80.6
09	FINISHES	92.5	78.3	84.8	89.3	76.9	82.6	88.9	81.3	84.7	93.9	89.7	91.6	90.1	83.6	86.6	99.3	82.7	90.3
COVERS	DIVS. 10 - 14, 25, 28, 41, 43, 44, 46	100.0	88.7	97.5	100.0	88.7	97.5	100.0	91.8	98.2	100.0	93.4	98.5	100.0	89.1	97.8	100.0	93.4	98.5
21, 22, 23	FIRE SUPPRESSION, PLUMBING & HVAC	99.9	75.0	90.1	99.9	76.5	90.7	96.4	81.0	90.4	96.5	83.8	91.5	96.5	77.8	89.2	100.0	81.2	92.6
26, 27, 3370	ELECTRICAL, COMMUNICATIONS & UTIL.	101.3	83.9	92.7	94.1	83.5	88.8	94.5	83.6	89.2	101.1	72.2	86.9	101.8	76.9	89.5	98.8	80.1	89.5
MF2018	WEIGHTED AVERAGE	98.4	81.8	91.4	96.2	81.7	90.1	94.6	84.8	90.4	95.7	84.3	90.9	95.5	82.4	89.9	98.0	83.7	92.0

City Cost Indexes

IOWA

DIVISION		COUNCIL BLUFFS 515			CRESTON 508			DAVENPORT 527 - 528			DECORAH 521			DES MOINES 500 - 503,509			DUBUQUE 520		
		MAT.	INST.	TOTAL	MAT.	INST.	TOTAL	MAT.	INST.	TOTAL	MAT.	INST.	TOTAL	MAT.	INST.	TOTAL	MAT.	INST.	TOTAL
015433	CONTRACTOR EQUIPMENT		95.0	95.0		98.8	98.8		98.8	98.8		98.8	98.8		102.4	102.4		94.5	94.5
0241, 31 - 34	SITE & INFRASTRUCTURE, DEMOLITION	104.5	88.5	93.4	92.9	93.6	93.4	98.4	94.6	95.8	96.6	91.6	93.2	98.3	99.3	99.0	97.7	88.7	91.5
0310	Concrete Forming & Accessories	83.5	73.6	75.0	79.2	86.7	85.6	102.0	95.9	96.8	94.0	71.8	75.1	95.6	89.5	90.4	84.9	81.5	82.0
0320	Concrete Reinforcing	97.3	79.8	88.8	96.3	85.9	91.3	95.3	99.0	97.1	94.7	85.2	90.1	101.0	101.6	101.3	94.0	79.9	87.2
0330	Cast-in-Place Concrete	111.8	79.1	99.6	114.6	85.7	103.8	103.6	95.2	100.5	104.4	76.0	93.8	97.6	92.1	95.6	105.3	83.0	97.0
03	CONCRETE	98.1	77.6	89.1	97.7	86.8	92.9	94.0	96.7	95.2	93.8	76.6	86.3	91.0	92.8	91.8	92.7	82.6	88.3
04	MASONRY	106.5	76.2	87.8	99.7	79.9	87.5	102.0	92.2	96.0	120.3	69.2	88.9	87.7	89.0	88.5	105.9	68.8	83.1
05	METALS	95.2	92.8	94.5	88.4	96.0	90.6	90.3	105.5	94.8	88.0	94.2	89.9	94.1	96.8	94.9	88.9	92.4	89.9
06	WOOD, PLASTICS & COMPOSITES	76.5	72.1	74.1	69.7	88.4	80.0	98.9	95.3	96.9	88.5	70.9	78.8	87.6	88.5	88.1	78.3	82.2	80.4
07	THERMAL & MOISTURE PROTECTION	104.8	74.8	91.8	106.2	82.2	95.8	104.9	91.6	99.1	104.6	69.9	89.5	98.5	87.6	93.8	105.1	76.5	92.6
08	OPENINGS	98.4	76.8	93.3	108.6	85.9	103.3	99.3	96.7	98.7	97.4	77.9	92.9	100.8	89.6	98.2	98.4	82.4	94.6
0920	Plaster & Gypsum Board	100.9	71.8	81.3	95.8	88.3	90.8	110.3	95.4	100.2	103.7	70.3	81.2	92.8	88.3	89.8	100.9	82.1	88.3
0950, 0980	Ceilings & Acoustic Treatment	99.0	71.8	80.6	90.2	88.3	88.9	101.4	95.4	97.3	99.0	70.3	79.6	93.4	88.3	90.0	99.0	82.1	87.6
0960	Flooring	87.0	84.3	86.2	81.3	69.1	77.8	96.1	89.4	94.2	93.3	69.1	86.5	95.2	94.5	95.0	99.1	69.1	90.7
0970, 0990	Wall Finishes & Painting/Coating	89.6	59.8	72.0	83.6	84.6	84.2	92.8	90.5	91.4	92.8	84.6	87.9	91.6	84.6	87.5	93.9	78.4	84.8
09	FINISHES	91.0	73.7	81.7	85.1	83.4	84.2	95.7	94.1	94.8	93.5	72.4	82.1	93.6	90.0	91.6	94.8	78.9	86.2
COVERS	DIVS. 10 - 14, 25, 28, 41, 43, 44, 46	100.0	91.4	98.1	100.0	91.6	98.1	100.0	96.9	99.3	100.0	88.6	97.5	100.0	95.6	99.0	100.0	92.3	98.3
21, 22, 23	FIRE SUPPRESSION, PLUMBING & HVAC	100.0	72.8	89.4	96.3	79.8	89.8	100.0	93.9	97.6	96.5	74.5	87.9	99.8	86.8	94.7	100.0	75.6	90.4
26, 27, 3370	ELECTRICAL, COMMUNICATIONS & UTIL.	103.9	82.7	93.4	93.5	76.9	85.3	96.9	87.6	92.3	98.8	47.2	73.4	105.1	84.3	94.8	102.5	76.9	89.9
MF2018	WEIGHTED AVERAGE	98.9	79.2	90.6	95.7	84.2	90.8	97.1	94.5	96.0	96.4	73.7	86.8	97.2	90.2	94.3	97.2	79.9	89.9

IOWA

DIVISION		FORT DODGE 505			MASON CITY 504			OTTUMWA 525			SHENANDOAH 516			SIBLEY 512			SIOUX CITY 510 - 511		
		MAT.	INST.	TOTAL	MAT.	INST.	TOTAL	MAT.	INST.	TOTAL	MAT.	INST.	TOTAL	MAT.	INST.	TOTAL	MAT.	INST.	TOTAL
015433	CONTRACTOR EQUIPMENT		98.8	98.8		98.8	98.8		94.5	94.5		95.0	95.0		98.8	98.8		98.8	98.8
0241, 31 - 34	SITE & INFRASTRUCTURE, DEMOLITION	101.4	90.5	93.9	101.5	91.5	94.6	98.0	86.7	90.2	102.6	88.6	92.9	107.7	91.6	96.6	109.6	93.3	98.3
0310	Concrete Forming & Accessories	79.8	78.0	78.3	83.6	71.4	73.2	91.9	87.1	87.8	85.1	77.2	78.3	85.7	37.8	44.9	102.5	75.5	79.4
0320	Concrete Reinforcing	96.3	85.2	90.9	96.2	85.2	90.9	94.7	98.0	96.3	97.3	85.8	91.7	97.3	85.1	91.4	95.3	100.8	98.0
0330	Cast-in-Place Concrete	107.6	41.9	83.1	107.6	71.2	94.0	108.0	63.9	91.6	108.1	82.6	98.6	105.9	55.5	87.2	106.6	89.3	100.2
03	CONCRETE	93.1	67.6	81.9	93.3	74.8	85.2	95.4	81.5	89.3	95.6	81.5	89.4	94.7	54.1	76.8	95.3	85.6	91.0
04	MASONRY	98.6	51.9	69.9	111.6	68.0	84.8	102.3	55.2	73.3	106.0	74.5	86.6	124.6	52.2	80.1	99.2	70.0	81.2
05	METALS	88.5	94.0	90.1	88.5	94.2	90.2	87.9	99.7	91.3	94.2	95.6	94.6	88.1	93.3	89.7	90.3	101.3	93.6
06	WOOD, PLASTICS & COMPOSITES	70.1	88.4	80.2	74.0	70.9	72.3	85.7	97.9	92.4	78.3	78.5	78.4	79.1	34.2	54.4	98.9	74.4	85.4
07	THERMAL & MOISTURE PROTECTION	105.5	65.2	88.0	105.0	70.5	90.0	105.3	70.3	90.1	104.0	73.3	90.7	104.3	55.2	82.9	104.9	73.4	91.2
08	OPENINGS	102.5	75.9	96.3	94.7	77.9	90.8	98.8	92.8	97.4	90.2	76.4	87.0	95.6	46.0	84.0	99.3	81.6	95.1
0920	Plaster & Gypsum Board	95.8	88.3	90.8	95.8	70.3	78.6	101.5	98.3	99.3	100.9	78.3	85.7	100.9	32.5	54.9	110.3	73.8	85.8
0950, 0980	Ceilings & Acoustic Treatment	90.2	88.3	88.9	90.2	70.3	76.8	99.0	98.3	98.5	99.0	78.3	85.0	99.0	32.5	54.2	101.4	73.8	82.8
0960	Flooring	82.6	69.1	78.8	84.6	69.1	80.2	102.2	69.1	92.9	87.7	73.9	83.8	88.9	69.1	83.3	96.1	73.3	89.7
0970, 0990	Wall Finishes & Painting/Coating	83.6	81.3	82.2	83.6	84.6	84.2	93.9	84.6	88.4	89.6	84.6	86.6	92.8	83.6	87.3	92.8	69.1	78.8
09	FINISHES	87.0	78.0	82.1	87.6	72.1	79.2	95.9	84.8	89.9	91.1	77.1	83.5	93.3	46.4	67.9	97.2	73.8	84.5
COVERS	DIVS. 10 - 14, 25, 28, 41, 43, 44, 46	100.0	84.8	96.6	100.0	88.2	97.4	100.0	87.3	97.2	100.0	88.2	97.4	100.0	89.8	95.3	100.0	91.4	98.1
21, 22, 23	FIRE SUPPRESSION, PLUMBING & HVAC	96.3	71.4	86.6	96.3	77.7	89.0	96.5	73.1	87.3	96.5	83.9	91.6	96.5	71.5	86.7	100.0	81.2	92.6
26, 27, 3370	ELECTRICAL, COMMUNICATIONS & UTIL.	99.8	69.4	84.8	98.9	47.2	73.4	100.9	71.0	86.2	98.8	78.4	88.7	98.8	47.2	73.4	98.8	74.0	86.6
MF2018	WEIGHTED AVERAGE	95.5	73.7	86.3	95.3	74.0	86.3	96.3	78.7	88.9	96.1	81.9	90.1	96.8	62.4	82.3	97.7	81.5	90.9

IOWA / KANSAS

DIVISION		IOWA SPENCER 513			IOWA WATERLOO 506 - 507			KANSAS BELLEVILLE 669			KANSAS COLBY 677			KANSAS DODGE CITY 678			KANSAS EMPORIA 668		
		MAT.	INST.	TOTAL	MAT.	INST.	TOTAL	MAT.	INST.	TOTAL	MAT.	INST.	TOTAL	MAT.	INST.	TOTAL	MAT.	INST.	TOTAL
015433	CONTRACTOR EQUIPMENT		98.8	98.8		98.8	98.8		103.7	103.7		103.7	103.7		103.7	103.7		101.9	101.9
0241, 31 - 34	SITE & INFRASTRUCTURE, DEMOLITION	107.8	90.4	95.7	106.9	92.7	97.1	111.3	92.0	98.0	106.7	92.7	97.0	109.4	91.7	97.2	103.1	89.7	93.8
0310	Concrete Forming & Accessories	91.8	37.5	45.5	94.4	68.7	72.5	94.8	54.3	60.3	97.5	61.2	66.6	91.1	61.1	65.5	86.2	70.0	72.4
0320	Concrete Reinforcing	97.3	85.1	91.4	96.9	80.3	88.9	98.9	102.6	100.7	97.1	102.6	99.8	94.7	102.3	98.4	97.5	102.8	100.1
0330	Cast-in-Place Concrete	105.9	65.5	90.9	115.2	84.0	103.6	121.0	83.5	107.1	119.5	87.2	107.5	121.6	86.9	108.7	117.0	87.3	106.0
03	CONCRETE	95.0	57.4	78.5	99.0	77.2	89.4	110.5	74.3	94.6	108.3	78.7	95.3	109.7	78.5	96.0	103.1	82.8	94.2
04	MASONRY	124.6	52.2	80.1	99.4	75.1	84.5	86.4	58.5	69.3	100.5	64.6	78.4	110.7	59.6	79.3	91.8	66.0	76.0
05	METALS	88.1	93.0	89.5	90.7	93.0	91.4	95.6	99.2	96.7	92.8	100.0	94.9	94.2	98.6	95.5	95.3	100.2	96.8
06	WOOD, PLASTICS & COMPOSITES	85.5	34.2	57.3	87.2	64.4	74.6	93.3	51.2	70.1	96.9	57.4	75.2	88.9	57.4	71.5	84.3	68.8	75.8
07	THERMAL & MOISTURE PROTECTION	105.3	55.8	83.8	105.3	77.2	93.0	91.4	62.1	78.7	97.8	65.4	83.7	97.8	66.6	84.2	89.5	75.4	83.3
08	OPENINGS	106.7	46.0	92.5	95.1	72.9	90.0	94.1	62.5	86.7	97.9	65.9	90.4	97.8	65.9	90.4	92.1	74.4	88.0
0920	Plaster & Gypsum Board	101.5	32.5	55.1	103.8	63.6	76.7	88.7	50.0	62.6	96.8	56.3	69.6	90.5	56.3	67.5	85.5	68.1	73.8
0950, 0980	Ceilings & Acoustic Treatment	99.0	32.5	54.2	91.8	63.6	72.8	76.5	50.0	58.6	77.3	56.3	63.2	77.3	56.3	63.2	76.5	68.1	70.8
0960	Flooring	91.6	69.1	85.3	89.6	78.9	86.6	92.9	68.4	86.0	87.6	68.4	82.2	83.8	68.4	79.4	88.3	68.4	82.7
0970, 0990	Wall Finishes & Painting/Coating	92.8	57.2	71.7	83.6	81.3	82.2	89.7	57.9	70.9	96.1	57.9	73.5	96.1	57.9	73.5	89.7	57.9	70.9
09	FINISHES	94.2	42.3	66.1	90.7	70.9	80.0	86.1	55.7	69.6	84.7	60.8	71.8	82.8	60.8	70.9	83.3	67.6	74.8
COVERS	DIVS. 10 - 14, 25, 28, 41, 43, 44, 46	100.0	78.9	95.3	100.0	91.0	98.0	100.0	83.3	96.3	100.0	86.1	96.9	100.0	86.0	96.9	100.0	85.9	96.9
21, 22, 23	FIRE SUPPRESSION, PLUMBING & HVAC	96.5	71.5	86.7	99.9	80.1	92.1	96.4	70.3	86.1	96.5	71.1	86.5	100.0	71.1	88.7	96.4	73.5	87.4
26, 27, 3370	ELECTRICAL, COMMUNICATIONS & UTIL.	100.4	47.2	74.2	95.4	61.6	78.7	104.1	62.7	83.7	95.5	67.9	81.9	92.9	67.9	80.6	101.4	66.9	84.4
MF2018	WEIGHTED AVERAGE	98.2	62.3	83.0	96.8	77.4	88.6	97.9	70.7	86.4	97.3	74.0	87.4	98.6	73.3	87.9	96.1	76.5	87.8

For customer support on your Site Work & Landscape Costs with RSMeans data, call 800.448.8182.

791

KANSAS

DIVISION		FORT SCOTT 667 MAT.	INST.	TOTAL	HAYS 676 MAT.	INST.	TOTAL	HUTCHINSON 675 MAT.	INST.	TOTAL	INDEPENDENCE 673 MAT.	INST.	TOTAL	KANSAS CITY 660-662 MAT.	INST.	TOTAL	LIBERAL 679 MAT.	INST.	TOTAL
015433	CONTRACTOR EQUIPMENT		102.8	102.8		103.7	103.7		103.7	103.7		103.7	103.7		100.4	100.4		103.7	103.7
0241, 31 - 34	SITE & INFRASTRUCTURE, DEMOLITION	99.9	89.7	92.8	111.4	92.2	98.2	90.6	92.7	92.0	110.2	92.7	98.1	94.3	90.2	91.5	111.3	92.3	98.2
0310	Concrete Forming & Accessories	102.8	82.9	85.9	95.2	58.9	64.3	86.0	56.6	61.0	105.8	67.8	73.4	99.4	98.4	98.6	91.5	58.8	63.7
0320	Concrete Reinforcing	96.9	100.8	98.8	94.7	102.6	98.5	94.7	102.6	98.5	94.1	100.8	97.3	92.9	97.4	94.6	94.2	83.5	90.3
0330	Cast-in-Place Concrete	108.5	82.4	98.8	94.2	83.8	90.3	87.3	86.9	87.2	122.1	87.1	109.1	90.7	99.7	94.6	101.7	76.3	90.5
03	CONCRETE	98.4	86.6	93.2	99.8	76.5	89.6	83.0	76.5	80.2	110.8	81.3	97.8	90.7	99.7	94.6	101.7	76.3	90.5
04	MASONRY	92.7	56.1	70.2	109.5	58.6	78.2	100.2	64.5	78.3	97.8	64.5	77.3	93.5	98.1	96.3	108.2	53.6	74.6
05	METALS	95.3	99.0	96.4	92.3	99.8	94.5	92.1	99.0	94.1	92.0	98.8	94.0	103.1	106.2	104.0	92.6	98.7	94.4
06	WOOD, PLASTICS & COMPOSITES	103.8	90.5	96.5	93.9	57.4	73.8	84.0	51.4	66.0	107.3	65.9	84.5	99.6	98.5	99.0	89.6	57.4	71.9
07	THERMAL & MOISTURE PROTECTION	90.5	72.8	82.8	98.1	62.8	82.8	96.6	64.3	82.5	97.8	74.9	87.9	90.3	99.3	94.2	98.3	60.9	82.0
08	OPENINGS	92.1	85.9	90.6	97.8	65.9	90.4	97.8	62.6	89.6	95.9	70.2	89.9	93.2	97.4	94.2	97.9	65.9	90.4
0920	Plaster & Gypsum Board	91.2	90.4	90.7	94.3	56.3	68.7	89.2	50.2	63.0	105.0	65.1	78.1	84.8	98.6	94.1	91.4	56.3	67.8
0950, 0980	Ceilings & Acoustic Treatment	76.5	90.4	85.9	77.3	56.3	63.2	77.3	50.2	59.0	77.3	65.1	69.0	76.5	98.6	91.4	77.3	56.3	63.2
0960	Flooring	102.6	66.9	92.5	86.4	68.4	81.3	80.9	68.4	77.3	92.0	66.9	84.9	82.4	95.3	86.0	84.1	68.4	79.6
0970, 0990	Wall Finishes & Painting/Coating	91.3	77.5	83.1	96.1	57.9	73.5	96.1	57.9	73.5	96.1	57.9	73.5	97.1	102.7	100.4	96.1	57.9	73.5
09	FINISHES	88.5	80.7	84.3	84.5	59.3	70.9	80.3	57.3	67.9	87.3	66.1	75.8	83.6	98.3	91.5	83.7	59.3	70.5
COVERS	DIVS. 10 - 14, 25, 28, 41, 43, 44, 46	100.0	86.8	97.1	100.0	84.0	96.4	100.0	85.4	96.7	100.0	87.0	97.1	100.0	94.5	98.8	100.0	84.0	96.4
21, 22, 23	FIRE SUPPRESSION, PLUMBING & HVAC	96.4	67.4	85.0	96.5	67.4	85.1	96.5	71.1	86.5	96.5	70.4	86.3	99.9	99.1	99.6	96.5	69.3	85.8
26, 27, 3370	ELECTRICAL, COMMUNICATIONS & UTIL.	100.6	67.9	84.5	94.6	67.9	81.4	90.5	61.0	75.9	92.3	72.8	82.7	106.2	98.1	102.2	92.9	67.9	80.6
MF2018	WEIGHTED AVERAGE	96.0	77.1	88.0	96.5	72.0	86.1	92.5	71.9	83.8	97.2	76.0	88.2	97.3	93.6	95.8	96.5	71.7	86.0

KANSAS / KENTUCKY

DIVISION		SALINA 674 MAT.	INST.	TOTAL	TOPEKA 664-666 MAT.	INST.	TOTAL	WICHITA 670-672 MAT.	INST.	TOTAL	ASHLAND 411-412 MAT.	INST.	TOTAL	BOWLING GREEN 421-422 MAT.	INST.	TOTAL	CAMPTON 413-414 MAT.	INST.	TOTAL
015433	CONTRACTOR EQUIPMENT		103.7	103.7		105.0	105.0		106.7	106.7		97.4	97.4		90.6	90.6		96.8	96.8
0241, 31 - 34	SITE & INFRASTRUCTURE, DEMOLITION	99.8	92.2	94.6	98.4	94.4	95.6	96.8	98.5	97.9	114.0	78.8	89.7	78.7	89.4	86.1	87.5	90.7	89.7
0310	Concrete Forming & Accessories	87.8	61.4	65.3	96.7	66.1	70.7	92.9	57.0	62.3	87.2	89.6	89.3	86.5	78.8	79.9	89.3	80.9	82.2
0320	Concrete Reinforcing	94.1	100.5	97.2	93.5	103.8	98.5	92.4	100.4	96.3	92.5	94.2	93.3	89.7	76.8	83.5	90.6	93.0	91.7
0330	Cast-in-Place Concrete	105.6	83.8	97.5	97.6	86.4	93.4	98.7	79.5	91.5	89.1	94.6	91.1	88.4	69.7	81.4	98.6	68.1	87.2
03	CONCRETE	96.8	77.3	88.2	92.3	80.8	87.3	92.4	73.7	84.2	93.1	93.3	93.2	91.6	75.7	84.6	95.0	78.8	87.9
04	MASONRY	125.4	54.9	82.1	87.7	64.3	73.3	97.5	54.1	70.8	90.7	92.0	91.5	96.7	69.2	78.3	89.7	54.7	68.2
05	METALS	94.0	99.5	95.5	99.5	100.0	99.6	96.2	97.5	96.6	95.1	107.7	98.8	96.7	82.8	92.6	96.0	89.8	94.2
06	WOOD, PLASTICS & COMPOSITES	85.4	60.7	71.8	99.9	65.2	80.8	98.7	54.4	74.3	73.4	87.7	81.3	85.7	80.6	82.9	84.2	88.3	86.5
07	THERMAL & MOISTURE PROTECTION	97.2	62.1	82.0	94.0	74.7	85.6	96.2	60.8	80.8	91.2	88.3	90.0	87.6	77.9	83.4	100.2	67.8	86.1
08	OPENINGS	97.8	67.3	90.7	103.5	74.1	96.7	101.3	63.8	92.5	90.6	89.0	90.2	91.4	77.1	88.1	92.6	87.4	91.4
0920	Plaster & Gypsum Board	89.2	59.8	69.4	99.5	64.2	75.7	95.7	53.1	67.0	59.3	87.6	78.4	87.6	80.6	82.9	87.6	87.6	87.6
0950, 0980	Ceilings & Acoustic Treatment	77.3	59.8	65.5	87.2	64.2	71.7	88.0	53.1	64.5	77.0	87.6	84.2	81.6	80.6	80.9	81.6	87.6	85.7
0960	Flooring	82.2	68.4	78.3	98.2	68.4	89.8	94.3	68.4	87.0	73.8	81.3	75.9	89.5	61.9	81.8	91.6	64.3	83.9
0970, 0990	Wall Finishes & Painting/Coating	96.1	57.9	73.5	92.8	67.7	78.0	94.9	57.9	73.0	91.6	90.4	90.9	90.3	67.6	76.9	90.3	53.9	68.8
09	FINISHES	81.5	61.3	70.6	93.9	65.7	78.6	90.3	57.7	72.7	76.1	88.1	82.6	87.1	74.6	80.3	87.9	75.9	81.4
COVERS	DIVS. 10 - 14, 25, 28, 41, 43, 44, 46	100.0	85.3	96.7	100.0	80.3	95.6	100.0	85.1	96.7	100.0	88.2	97.4	100.0	84.4	96.5	100.0	47.8	88.4
21, 22, 23	FIRE SUPPRESSION, PLUMBING & HVAC	100.0	70.5	88.5	100.0	73.2	89.5	99.8	71.0	88.5	96.2	85.0	91.8	99.9	77.0	90.9	96.4	76.4	88.6
26, 27, 3370	ELECTRICAL, COMMUNICATIONS & UTIL.	92.7	73.8	83.4	105.1	71.9	88.7	95.9	73.8	85.0	91.9	88.3	90.1	94.6	74.7	84.8	92.1	88.2	90.2
MF2018	WEIGHTED AVERAGE	97.2	73.5	87.2	98.6	76.5	89.3	97.0	72.6	86.7	93.2	89.7	91.7	94.6	77.2	87.3	94.4	78.0	87.5

KENTUCKY

DIVISION		CORBIN 407-409 MAT.	INST.	TOTAL	COVINGTON 410 MAT.	INST.	TOTAL	ELIZABETHTOWN 427 MAT.	INST.	TOTAL	FRANKFORT 406 MAT.	INST.	TOTAL	HAZARD 417-418 MAT.	INST.	TOTAL	HENDERSON 424 MAT.	INST.	TOTAL
015433	CONTRACTOR EQUIPMENT		96.8	96.8		100.6	100.6		90.6	90.6		100.1	100.1		96.8	96.8		110.3	110.3
0241, 31 - 34	SITE & INFRASTRUCTURE, DEMOLITION	91.6	91.2	91.3	82.7	101.9	95.9	72.9	89.3	84.3	90.0	96.8	94.7	85.3	91.8	89.7	92.3	77.1	79.3
0310	Concrete Forming & Accessories	85.0	76.0	77.3	84.9	70.8	72.8	81.3	72.1	73.5	102.0	78.5	82.0	85.8	81.8	82.4	89.8	84.4	87.2
0320	Concrete Reinforcing	89.6	92.3	90.9	94.3	78.6	84.1	90.2	83.7	87.0	94.0	84.0	89.2	91.0	92.5	91.7	78.1	84.5	80.5
0330	Cast-in-Place Concrete	91.0	72.5	84.1	94.3	79.6	88.9	79.9	67.6	75.3	91.1	77.0	85.8	94.8	69.6	85.4	88.7	81.3	85.5
03	CONCRETE	84.2	78.0	81.5	93.4	76.1	85.8	83.5	73.1	79.0	87.1	79.3	83.6	91.8	79.7	86.5	88.7	81.3	85.5
04	MASONRY	81.9	60.0	68.4	104.0	72.0	84.3	77.4	61.8	67.8	78.6	74.2	75.9	88.6	56.8	69.0	96.0	78.7	85.4
05	METALS	90.9	89.3	90.5	93.9	88.1	92.2	95.9	85.3	92.7	93.1	86.4	91.1	96.0	89.8	94.2	86.8	87.0	86.9
06	WOOD, PLASTICS & COMPOSITES	72.1	77.6	75.1	81.8	68.3	74.3	80.6	73.9	76.9	101.5	77.7	88.4	80.9	88.3	85.0	88.4	75.8	81.5
07	THERMAL & MOISTURE PROTECTION	103.8	69.5	88.9	101.1	72.1	88.5	87.0	68.5	79.0	101.8	75.7	90.5	100.1	69.5	86.8	99.3	82.8	92.1
08	OPENINGS	87.3	70.5	83.4	94.5	71.8	89.2	91.4	73.0	87.1	98.9	78.4	94.1	93.0	87.2	91.6	89.6	78.5	87.0
0920	Plaster & Gypsum Board	88.4	76.6	80.5	68.2	68.0	68.1	86.3	73.7	77.8	91.6	76.6	81.5	86.3	87.6	87.2	90.9	74.8	80.0
0950, 0980	Ceilings & Acoustic Treatment	74.7	76.6	76.0	84.0	68.0	73.2	81.6	73.7	76.2	90.5	76.6	81.1	81.6	87.6	85.7	76.7	74.8	75.4
0960	Flooring	88.4	64.3	81.6	66.3	83.5	71.1	86.9	73.8	83.2	98.9	78.6	93.2	89.8	64.3	82.6	93.2	76.0	88.4
0970, 0990	Wall Finishes & Painting/Coating	87.2	62.0	72.3	85.2	71.1	76.9	90.3	68.1	77.2	94.2	91.7	92.8	90.3	53.9	68.8	90.3	86.7	88.2
09	FINISHES	83.0	72.5	77.3	77.8	72.7	75.0	85.8	71.6	78.1	93.2	80.2	86.2	87.0	76.4	81.3	87.1	77.9	82.1
COVERS	DIVS. 10 - 14, 25, 28, 41, 43, 44, 46	100.0	89.8	97.7	100.0	85.9	96.8	100.0	70.0	93.3	100.0	58.5	90.8	100.0	48.5	88.5	100.0	55.8	90.1
21, 22, 23	FIRE SUPPRESSION, PLUMBING & HVAC	96.5	74.2	87.7	97.0	76.0	88.8	96.6	77.9	89.3	99.9	81.0	92.5	96.4	77.5	89.0	96.6	79.3	89.8
26, 27, 3370	ELECTRICAL, COMMUNICATIONS & UTIL.	90.0	88.2	89.1	95.6	72.4	84.2	91.9	77.9	85.0	100.6	77.9	89.4	92.1	88.2	90.2	94.0	77.9	86.1
MF2018	WEIGHTED AVERAGE	90.8	78.0	85.4	94.3	77.8	87.3	91.3	75.6	84.7	95.4	80.3	89.0	93.8	78.7	87.4	92.2	81.9	87.8

City Cost Indexes

KENTUCKY

DIVISION		LEXINGTON 403-405 MAT.	INST.	TOTAL	LOUISVILLE 400-402 MAT.	INST.	TOTAL	OWENSBORO 423 MAT.	INST.	TOTAL	PADUCAH 420 MAT.	INST.	TOTAL	PIKEVILLE 415-416 MAT.	INST.	TOTAL	SOMERSET 425-426 MAT.	INST.	TOTAL
015433	CONTRACTOR EQUIPMENT		96.8	96.8		93.7	93.7		110.3	110.3		110.3	110.3		97.4	97.4		96.8	96.8
0241, 31 - 34	SITE & INFRASTRUCTURE, DEMOLITION	93.8	92.8	93.1	87.8	94.4	92.3	91.8	115.3	108.0	84.4	114.8	105.4	125.2	78.0	92.6	77.9	91.2	87.1
0310	Concrete Forming & Accessories	96.5	73.7	77.1	93.5	79.1	81.2	90.8	76.8	78.9	88.8	81.2	82.3	96.0	85.6	87.2	86.8	76.6	78.1
0320	Concrete Reinforcing	98.4	84.0	91.4	94.8	84.2	89.6	89.8	77.3	83.8	90.4	82.5	86.6	93.0	94.1	93.5	90.2	92.5	91.3
0330	Cast-in-Place Concrete	93.1	84.4	89.9	89.3	70.7	82.4	91.1	83.9	88.4	83.2	81.1	82.4	97.9	89.4	94.8	78.1	87.9	81.7
03	CONCRETE	86.9	79.7	83.8	85.8	77.4	82.1	100.7	79.8	91.5	93.3	81.7	88.2	106.8	89.7	99.3	78.6	83.7	80.8
04	MASONRY	80.6	71.1	74.8	78.7	72.0	74.6	88.8	78.6	82.5	91.5	79.2	83.9	88.5	83.2	85.3	83.3	63.2	71.0
05	METALS	93.3	86.7	91.3	95.5	85.5	92.5	88.3	84.7	87.3	85.3	87.5	86.0	95.0	107.5	98.7	95.9	90.0	94.2
06	WOOD, PLASTICS & COMPOSITES	87.3	71.6	78.6	88.8	80.8	84.4	86.3	75.8	80.5	84.0	81.4	82.6	83.3	87.7	85.7	81.5	77.6	79.3
07	THERMAL & MOISTURE PROTECTION	104.1	73.9	90.9	101.0	75.7	90.0	100.1	76.6	89.9	99.4	82.6	92.1	92.1	80.5	87.0	99.3	71.5	87.2
08	OPENINGS	87.5	74.6	84.5	88.8	76.8	86.0	89.6	76.7	86.6	89.0	81.1	87.1	91.2	85.7	89.9	92.1	77.1	88.6
0920	Plaster & Gypsum Board	97.7	70.4	79.4	93.7	80.6	84.9	89.3	74.8	79.5	88.6	80.6	83.2	63.1	87.6	79.6	86.3	76.6	79.8
0950, 0980	Ceilings & Acoustic Treatment	77.9	70.4	72.9	82.8	80.6	81.3	76.7	74.8	75.4	76.7	80.6	79.3	77.0	87.6	84.2	81.6	76.6	78.2
0960	Flooring	93.4	64.3	85.2	95.0	62.9	86.0	92.6	61.9	84.0	91.5	76.0	87.2	78.1	64.3	74.2	90.1	64.3	82.8
0970, 0990	Wall Finishes & Painting/Coating	87.2	79.7	82.8	93.9	68.1	78.7	90.3	86.7	88.2	90.3	73.1	80.2	91.6	91.7	91.7	90.3	68.1	77.2
09	FINISHES	86.4	72.0	78.6	90.8	74.9	82.2	87.3	74.7	80.5	86.5	79.4	82.7	78.7	82.6	80.8	86.4	73.3	79.3
COVERS	DIVS. 10 - 14, 25, 28, 41, 43, 44, 46	100.0	92.1	98.2	100.0	90.5	97.9	100.0	93.0	98.4	100.0	91.1	98.0	100.0	48.2	88.4	100.0	89.2	97.6
21, 22, 23	FIRE SUPPRESSION, PLUMBING & HVAC	100.0	76.4	90.8	99.9	80.0	92.1	99.9	77.1	91.0	96.6	78.2	89.4	96.2	82.3	90.7	96.6	74.0	87.7
26, 27, 3370	ELECTRICAL, COMMUNICATIONS & UTIL.	92.5	74.7	83.7	96.4	77.9	87.3	94.1	73.7	84.0	96.3	77.3	86.9	94.8	88.3	91.6	92.5	88.2	90.4
MF2018	WEIGHTED AVERAGE	93.1	78.1	86.7	93.9	79.6	87.8	94.7	81.0	88.9	92.4	83.2	88.6	95.8	85.4	91.4	91.6	79.5	86.5

LOUISIANA

DIVISION		ALEXANDRIA 713-714 MAT.	INST.	TOTAL	BATON ROUGE 707-708 MAT.	INST.	TOTAL	HAMMOND 704 MAT.	INST.	TOTAL	LAFAYETTE 705 MAT.	INST.	TOTAL	LAKE CHARLES 706 MAT.	INST.	TOTAL	MONROE 712 MAT.	INST.	TOTAL
015433	CONTRACTOR EQUIPMENT		90.9	90.9		91.9	91.9		89.3	89.3		89.3	89.3		88.8	88.8		90.9	90.9
0241, 31 - 34	SITE & INFRASTRUCTURE, DEMOLITION	98.7	87.7	91.1	102.1	91.2	94.6	101.2	85.7	90.5	102.5	87.9	92.4	103.2	87.1	92.0	98.7	87.6	91.1
0310	Concrete Forming & Accessories	77.8	60.0	62.6	99.0	71.8	75.8	79.0	54.9	58.5	95.6	67.8	71.9	96.4	67.8	72.0	77.4	59.4	62.1
0320	Concrete Reinforcing	92.2	54.3	73.9	93.3	54.3	74.5	92.1	54.3	73.8	93.4	54.3	74.5	93.4	54.3	74.5	91.2	54.3	73.3
0330	Cast-in-Place Concrete	89.0	66.2	80.5	88.8	73.1	82.9	87.6	62.6	78.3	87.1	66.1	79.3	91.7	66.1	82.2	89.0	64.1	79.7
03	CONCRETE	82.8	62.0	73.6	85.5	69.8	78.6	85.8	58.3	73.7	86.8	65.4	77.4	89.0	65.4	78.6	82.6	61.0	73.1
04	MASONRY	109.3	64.3	81.6	91.7	66.7	76.3	95.3	60.0	73.6	95.3	66.6	77.7	94.7	66.6	77.5	103.9	63.1	78.8
05	METALS	92.6	70.9	86.2	96.9	74.0	90.1	88.4	69.3	82.8	87.7	69.6	82.4	87.7	69.6	82.4	92.6	70.8	86.2
06	WOOD, PLASTICS & COMPOSITES	89.2	58.8	72.5	103.0	73.2	86.6	83.0	54.3	67.2	103.7	68.0	84.1	101.9	68.0	83.3	88.6	58.8	72.2
07	THERMAL & MOISTURE PROTECTION	99.0	65.6	84.5	97.7	69.2	85.3	96.7	62.9	82.0	97.3	67.4	84.3	97.1	67.2	84.1	99.0	64.8	84.1
08	OPENINGS	110.8	58.6	98.6	97.9	70.4	91.5	96.2	54.9	86.6	99.8	62.2	91.0	99.8	62.2	91.0	110.7	57.1	98.2
0920	Plaster & Gypsum Board	83.6	58.1	66.4	101.0	72.7	82.0	100.0	53.4	68.7	108.6	67.6	81.0	108.6	67.6	81.0	83.3	58.1	66.3
0950, 0980	Ceilings & Acoustic Treatment	87.1	58.1	67.6	92.9	72.7	79.3	96.9	53.4	67.6	94.4	67.6	76.3	95.3	67.6	76.6	87.1	58.1	67.6
0960	Flooring	85.1	68.2	80.4	93.0	68.2	86.0	89.9	68.2	83.8	98.2	68.2	89.8	98.2	68.2	89.8	84.7	68.2	80.1
0970, 0990	Wall Finishes & Painting/Coating	88.8	59.7	71.6	94.7	59.7	74.0	97.7	59.9	75.3	97.7	59.7	75.2	97.7	59.7	75.2	88.8	59.7	71.6
09	FINISHES	80.9	61.2	70.2	92.2	70.3	80.3	90.4	57.5	72.6	93.5	67.2	79.3	93.7	67.2	79.4	80.7	60.9	70.0
COVERS	DIVS. 10 - 14, 25, 28, 41, 43, 44, 46	100.0	79.1	95.3	100.0	84.2	96.5	100.0	80.7	95.7	100.0	83.3	96.3	100.0	83.3	96.3	100.0	78.9	95.3
21, 22, 23	FIRE SUPPRESSION, PLUMBING & HVAC	100.0	63.2	85.6	99.9	64.1	85.9	96.7	61.0	82.7	100.2	64.2	86.0	100.2	64.5	86.2	100.0	61.9	85.0
26, 27, 3370	ELECTRICAL, COMMUNICATIONS & UTIL.	94.3	61.4	78.1	101.7	58.0	80.2	97.0	69.6	83.5	98.1	63.8	81.2	97.7	65.9	82.0	95.9	57.3	76.9
MF2018	WEIGHTED AVERAGE	95.9	65.6	83.1	96.6	69.3	85.1	93.7	64.4	81.3	95.4	67.9	83.8	95.6	68.2	84.1	95.8	64.4	82.5

LOUISIANA / MAINE

DIVISION		NEW ORLEANS 700-701 MAT.	INST.	TOTAL	SHREVEPORT 710-711 MAT.	INST.	TOTAL	THIBODAUX 703 MAT.	INST.	TOTAL	AUGUSTA 043 MAT.	INST.	TOTAL	BANGOR 044 MAT.	INST.	TOTAL	BATH 045 MAT.	INST.	TOTAL
015433	CONTRACTOR EQUIPMENT		88.6	88.6		93.9	93.9		89.3	89.3		98.4	98.4		95.7	95.7		95.7	95.7
0241, 31 - 34	SITE & INFRASTRUCTURE, DEMOLITION	104.3	93.9	97.1	101.1	92.3	95.0	103.6	87.6	92.6	88.2	98.0	95.0	90.6	95.5	94.0	88.3	94.2	92.4
0310	Concrete Forming & Accessories	95.7	68.9	72.8	92.9	60.3	65.1	90.2	64.5	68.2	98.6	78.6	81.5	93.0	76.5	79.0	89.1	78.4	80.0
0320	Concrete Reinforcing	92.9	54.4	74.3	92.5	54.3	74.0	92.1	54.3	73.8	102.6	81.1	92.2	94.1	81.1	87.8	93.1	81.1	87.3
0330	Cast-in-Place Concrete	90.7	69.7	82.9	92.0	65.2	82.0	94.1	64.3	83.0	89.8	115.3	99.3	70.9	114.1	87.0	70.9	114.3	87.1
03	CONCRETE	94.4	66.6	82.2	86.2	61.7	75.4	90.6	63.2	78.5	95.5	92.3	94.1	88.5	91.0	89.6	88.6	91.0	90.1
04	MASONRY	99.8	62.6	77.0	94.5	63.4	75.4	119.9	62.8	84.8	101.0	94.1	96.8	115.8	93.7	102.2	122.5	94.1	105.0
05	METALS	99.1	62.6	88.4	96.7	70.2	88.9	88.4	69.5	82.8	105.4	92.3	101.6	93.7	93.0	93.5	92.1	92.9	92.3
06	WOOD, PLASTICS & COMPOSITES	100.3	71.1	84.2	98.0	59.7	76.9	90.5	65.5	76.8	93.5	75.6	83.6	90.7	73.0	80.9	85.1	75.4	79.8
07	THERMAL & MOISTURE PROTECTION	95.0	68.0	83.3	97.4	65.6	83.6	96.9	65.2	83.1	111.0	100.6	106.5	109.0	100.0	105.1	108.9	100.1	105.1
08	OPENINGS	98.8	64.2	90.7	104.7	56.6	93.5	100.7	56.9	90.5	102.7	78.5	97.1	96.4	77.1	91.9	96.4	78.4	92.2
0920	Plaster & Gypsum Board	96.5	70.5	79.0	91.5	58.8	69.5	101.6	65.0	77.0	109.7	74.5	86.0	116.8	71.9	86.6	112.0	74.5	86.8
0950, 0980	Ceilings & Acoustic Treatment	97.5	70.5	79.3	92.8	58.8	69.9	96.9	65.0	75.4	107.1	74.5	85.1	89.6	71.9	77.7	88.6	74.5	79.1
0960	Flooring	103.6	68.2	93.7	91.2	68.2	84.8	95.8	68.2	84.8	89.9	110.6	95.7	82.5	110.6	90.4	80.9	110.6	89.2
0970, 0990	Wall Finishes & Painting/Coating	103.6	60.6	78.2	87.7	59.7	71.1	98.9	59.9	75.8	96.0	88.9	91.8	90.6	88.9	89.6	90.6	88.9	89.6
09	FINISHES	96.6	68.2	81.2	87.7	61.5	73.5	92.6	64.8	77.6	95.8	84.7	89.8	89.7	83.0	86.1	88.3	84.6	86.3
COVERS	DIVS. 10 - 14, 25, 28, 41, 43, 44, 46	100.0	86.5	97.0	100.0	82.6	96.1	100.0	81.8	95.9	100.0	98.0	99.6	100.0	99.6	99.9	100.0	100.9	100.2
21, 22, 23	FIRE SUPPRESSION, PLUMBING & HVAC	100.1	62.5	85.4	99.9	62.8	85.3	96.7	62.5	83.2	100.0	74.3	89.9	100.3	74.1	90.0	96.7	74.3	87.9
26, 27, 3370	ELECTRICAL, COMMUNICATIONS & UTIL.	102.1	70.6	86.6	101.8	65.5	83.9	95.7	69.6	82.8	101.6	76.8	89.3	100.0	69.4	84.9	98.2	76.8	87.6
MF2018	WEIGHTED AVERAGE	99.0	68.5	86.1	97.1	66.3	84.1	96.0	67.0	83.8	100.4	85.6	94.2	97.1	83.9	91.5	95.9	85.4	91.4

For customer support on your Site Work & Landscape Costs with RSMeans data, call 800.448.8182.

793

City Cost Indexes

		MAINE																
	DIVISION	HOULTON 047			KITTERY 039			LEWISTON 042			MACHIAS 046			PORTLAND 040-041			ROCKLAND 048		
		MAT.	INST.	TOTAL	MAT.	INST.	TOTAL	MAT.	INST.	TOTAL	MAT.	INST.	TOTAL	MAT.	INST.	TOTAL	MAT.	INST.	TOTAL
015433	CONTRACTOR EQUIPMENT		95.7	95.7		95.7	95.7		95.7	95.7		95.7	95.7		99.2	99.2		95.7	95.7
0241, 31 - 34	SITE & INFRASTRUCTURE, DEMOLITION	90.2	94.2	93.0	79.5	94.3	89.7	88.2	95.5	93.3	89.5	94.2	92.8	87.4	100.6	96.5	86.0	94.2	91.7
0310	Concrete Forming & Accessories	96.7	78.4	81.1	88.8	78.7	80.2	98.2	76.6	79.8	93.8	78.4	80.7	99.8	76.8	80.2	94.9	78.4	80.9
0320	Concrete Reinforcing	94.1	81.1	87.8	89.7	81.1	85.5	115.6	81.1	98.9	94.1	81.1	87.8	106.5	81.1	94.2	94.1	81.1	87.8
0330	Cast-in-Place Concrete	70.9	113.2	86.7	70.7	114.4	87.0	72.4	114.1	87.9	70.9	114.3	87.1	85.8	115.2	96.8	72.4	114.3	88.0
03	CONCRETE	89.5	91.5	90.4	82.1	92.1	86.5	88.1	91.0	89.4	89.0	91.9	90.3	91.5	91.4	91.5	86.0	91.9	88.6
04	MASONRY	97.9	94.1	95.6	106.5	94.1	98.9	98.5	93.7	95.5	97.9	94.1	95.6	104.7	93.7	98.0	92.5	94.1	93.5
05	METALS	92.3	92.8	92.4	86.8	93.1	88.6	97.1	93.0	95.9	92.3	92.8	92.5	103.0	92.5	99.9	92.2	92.9	92.4
06	WOOD, PLASTICS & COMPOSITES	94.5	75.4	84.0	88.2	75.4	81.2	96.3	73.0	83.5	91.5	75.4	82.7	95.8	73.1	83.3	92.4	75.4	83.1
07	THERMAL & MOISTURE PROTECTION	109.1	100.1	105.1	107.8	100.1	104.4	108.7	100.0	104.9	109.0	100.1	105.1	113.6	100.6	107.9	108.6	100.1	104.9
08	OPENINGS	96.5	78.4	92.3	96.1	82.0	92.8	99.3	77.1	94.1	96.5	78.4	92.3	96.8	77.2	92.2	96.4	78.4	92.2
0920	Plaster & Gypsum Board	119.0	74.5	89.0	106.1	74.5	84.8	121.8	71.9	88.2	117.4	74.5	88.5	108.0	71.9	83.7	117.4	74.5	88.5
0950, 0980	Ceilings & Acoustic Treatment	88.6	74.5	79.1	101.8	74.5	83.4	99.3	71.9	80.8	88.6	74.5	79.1	107.4	71.9	83.5	88.6	74.5	79.1
0960	Flooring	83.7	110.6	91.2	86.6	110.6	93.4	85.2	110.6	92.3	82.9	110.6	90.7	89.6	110.6	95.5	83.3	110.6	91.0
0970, 0990	Wall Finishes & Painting/Coating	90.6	88.9	89.6	81.8	102.1	93.8	90.6	88.9	89.6	90.6	88.9	89.6	94.6	88.9	91.2	90.6	88.9	89.6
09	FINISHES	90.2	84.6	87.2	91.6	86.0	88.6	92.6	83.0	87.4	89.7	84.6	86.9	96.2	83.2	89.1	89.4	84.6	86.8
COVERS	DIVS. 10 - 14, 25, 28, 41, 43, 44, 46	100.0	97.6	99.5	100.0	100.9	100.2	100.0	99.7	99.9	100.0	97.6	99.5	100.0	100.1	100.0	100.0	100.9	100.2
21, 22, 23	FIRE SUPPRESSION, PLUMBING & HVAC	96.7	74.3	87.9	96.7	80.4	90.3	100.3	74.1	90.0	96.7	74.3	87.9	100.0	74.1	89.9	96.7	74.3	87.9
26, 27, 3370	ELECTRICAL, COMMUNICATIONS & UTIL.	101.9	76.8	89.5	90.3	76.8	83.6	101.9	76.7	89.5	101.9	76.8	89.5	105.0	76.7	91.0	101.8	76.8	89.5
MF2018	WEIGHTED AVERAGE	95.6	85.2	91.2	92.6	87.0	90.2	97.5	84.9	92.2	95.4	85.3	91.1	99.6	85.4	93.6	94.6	85.4	90.7

		MAINE			MARYLAND														
	DIVISION	WATERVILLE 049			ANNAPOLIS 214			BALTIMORE 210 - 212			COLLEGE PARK 207 - 208			CUMBERLAND 215			EASTON 216		
		MAT.	INST.	TOTAL	MAT.	INST.	TOTAL	MAT.	INST.	TOTAL	MAT.	INST.	TOTAL	MAT.	INST.	TOTAL	MAT.	INST.	TOTAL
015433	CONTRACTOR EQUIPMENT		95.7	95.7		104.5	104.5		104.1	104.1		107.8	107.8		102.4	102.4		102.4	102.4
0241, 31 - 34	SITE & INFRASTRUCTURE, DEMOLITION	90.0	94.2	92.9	100.0	93.5	95.5	100.1	97.7	98.4	100.1	93.6	95.6	92.0	90.3	90.8	98.8	87.7	91.1
0310	Concrete Forming & Accessories	88.6	78.4	79.9	101.0	78.1	81.5	98.4	76.2	79.5	85.1	76.2	77.5	92.3	81.9	83.5	90.4	73.6	76.1
0320	Concrete Reinforcing	94.1	81.1	87.8	108.6	94.9	102.0	101.2	87.3	94.5	103.5	94.0	98.9	93.7	87.4	90.6	92.9	87.1	90.1
0330	Cast-in-Place Concrete	70.9	114.3	87.1	114.1	79.4	101.2	113.8	77.1	100.1	109.4	76.4	97.2	98.0	82.9	92.4	108.8	64.8	92.4
03	CONCRETE	90.2	91.9	90.9	101.4	82.7	93.2	108.2	79.4	95.5	105.2	80.9	94.5	91.4	84.5	88.4	99.7	74.2	88.5
04	MASONRY	109.8	94.1	100.2	94.8	76.6	83.6	104.7	73.1	85.2	111.1	74.6	88.7	99.3	85.4	90.8	113.5	58.6	79.7
05	METALS	92.3	92.8	92.4	105.5	105.2	105.4	103.8	96.6	101.7	90.2	108.3	95.5	100.7	104.2	101.8	101.0	99.7	100.6
06	WOOD, PLASTICS & COMPOSITES	84.6	75.4	79.5	97.1	77.2	86.1	102.5	76.7	88.3	77.3	75.7	76.4	86.7	81.0	83.6	84.5	80.8	82.5
07	THERMAL & MOISTURE PROTECTION	109.1	100.1	105.2	102.3	83.4	94.8	102.7	80.7	93.2	105.3	81.7	95.0	100.9	81.9	92.6	101.0	73.3	89.0
08	OPENINGS	96.5	78.4	92.3	102.3	83.0	97.8	100.5	80.7	95.9	92.7	82.4	90.3	97.6	84.6	94.6	96.0	83.0	92.9
0920	Plaster & Gypsum Board	112.0	74.5	86.8	94.4	76.8	82.5	100.3	76.2	84.0	95.3	75.2	81.8	99.9	80.8	87.0	99.9	80.6	86.9
0950, 0980	Ceilings & Acoustic Treatment	88.6	74.5	79.1	89.1	76.8	80.8	100.5	76.2	84.1	115.1	75.2	88.2	99.0	80.8	86.7	99.0	80.6	86.6
0960	Flooring	80.6	110.6	89.0	90.7	79.5	87.6	92.8	75.7	88.0	87.9	78.4	85.2	85.8	93.1	87.9	85.0	75.7	82.4
0970, 0990	Wall Finishes & Painting/Coating	90.6	88.9	89.6	99.2	72.7	83.5	101.3	75.3	85.9	104.8	72.5	85.7	98.2	84.7	90.2	98.2	72.5	83.0
09	FINISHES	88.4	84.6	86.4	88.5	77.3	82.4	97.0	75.7	85.5	95.6	75.6	84.8	93.3	84.2	88.4	93.5	74.4	83.2
COVERS	DIVS. 10 - 14, 25, 28, 41, 43, 44, 46	100.0	97.5	99.4	100.0	90.3	97.8	100.0	86.8	97.1	100.0	86.7	97.0	100.0	91.3	98.1	100.0	82.9	96.2
21, 22, 23	FIRE SUPPRESSION, PLUMBING & HVAC	96.7	74.3	87.9	100.0	87.0	94.9	100.1	82.0	93.0	96.5	87.4	93.0	96.4	72.0	86.8	96.4	72.8	87.1
26, 27, 3370	ELECTRICAL, COMMUNICATIONS & UTIL.	101.9	76.8	89.5	101.6	88.9	95.3	99.6	86.9	93.3	97.3	101.9	99.5	98.2	80.8	89.6	97.7	62.3	80.3
MF2018	WEIGHTED AVERAGE	96.0	85.2	91.4	100.4	86.3	94.4	101.8	83.2	93.9	97.3	87.7	93.2	96.9	83.8	91.4	98.6	74.8	88.5

		MARYLAND														MASSACHUSETTS		
	DIVISION	ELKTON 219			HAGERSTOWN 217			SALISBURY 218			SILVER SPRING 209			WALDORF 206			BOSTON 020 - 022, 024		
		MAT.	INST.	TOTAL	MAT.	INST.	TOTAL	MAT.	INST.	TOTAL	MAT.	INST.	TOTAL	MAT.	INST.	TOTAL	MAT.	INST.	TOTAL
015433	CONTRACTOR EQUIPMENT		102.4	102.4		102.4	102.4		102.4	102.4		99.7	99.7		99.7	99.7		103.0	103.0
0241, 31 - 34	SITE & INFRASTRUCTURE, DEMOLITION	86.3	89.0	88.1	90.6	90.4	90.4	98.8	87.6	91.1	88.9	85.7	86.7	95.1	85.7	88.6	92.7	102.1	99.2
0310	Concrete Forming & Accessories	96.2	92.3	92.9	91.5	78.8	80.7	104.5	50.2	58.2	93.0	75.5	78.1	99.7	75.4	79.0	105.1	139.7	134.6
0320	Concrete Reinforcing	92.9	117.3	104.7	93.7	87.4	90.6	92.9	65.2	79.5	102.3	93.9	98.3	103.0	93.9	98.6	117.7	156.9	136.7
0330	Cast-in-Place Concrete	88.1	72.2	82.2	93.3	82.9	89.5	108.8	62.9	91.7	112.1	76.8	98.9	125.6	76.7	107.4	99.2	141.8	115.1
03	CONCRETE	84.0	90.6	86.9	87.7	83.1	85.7	100.5	59.2	82.4	102.8	80.5	93.0	113.3	80.5	98.9	105.4	142.7	121.8
04	MASONRY	98.6	67.0	79.2	105.1	85.4	93.0	112.9	55.2	77.4	110.6	75.0	88.7	94.7	75.0	82.6	112.4	146.3	133.1
05	METALS	101.0	114.0	104.8	100.9	104.3	101.9	101.0	90.9	98.1	94.8	104.3	97.6	94.8	104.3	97.5	102.8	136.1	112.6
06	WOOD, PLASTICS & COMPOSITES	91.9	101.3	97.0	85.8	76.7	80.8	102.6	51.4	74.4	84.0	75.0	79.0	90.8	75.0	82.1	103.8	140.2	123.8
07	THERMAL & MOISTURE PROTECTION	100.5	79.4	91.3	101.2	83.0	93.3	101.3	68.5	87.0	108.4	86.5	98.9	109.0	86.5	99.2	107.2	137.1	120.2
08	OPENINGS	96.0	102.1	97.4	95.9	81.2	92.5	96.2	60.8	87.9	84.9	82.1	84.2	85.5	81.5	84.5	99.9	147.3	111.0
0920	Plaster & Gypsum Board	102.8	101.7	102.0	99.9	76.4	84.1	109.4	50.4	69.7	101.9	75.2	83.9	104.8	75.2	84.8	109.0	141.0	130.6
0950, 0980	Ceilings & Acoustic Treatment	99.0	101.7	100.8	101.7	76.4	84.6	99.0	50.4	66.2	123.2	75.2	90.8	123.2	75.2	90.8	100.5	141.0	127.8
0960	Flooring	87.2	75.7	84.0	85.4	93.1	87.6	90.7	75.7	86.5	93.8	78.4	89.5	97.2	78.4	92.0	90.5	163.5	111.0
0970, 0990	Wall Finishes & Painting/Coating	98.2	72.5	83.0	98.2	72.5	83.0	98.2	72.5	83.0	112.7	72.5	88.9	112.7	72.5	88.9	94.3	160.4	133.4
09	FINISHES	93.7	88.5	90.9	93.5	80.4	86.4	96.4	56.2	74.6	95.8	75.0	84.5	97.5	75.1	85.4	98.7	147.0	124.8
COVERS	DIVS. 10 - 14, 25, 28, 41, 43, 44, 46	100.0	56.7	90.4	100.0	90.8	98.0	100.0	76.5	94.8	100.0	85.2	96.7	100.0	83.1	96.2	100.0	119.1	104.3
21, 22, 23	FIRE SUPPRESSION, PLUMBING & HVAC	96.4	78.7	89.5	99.9	83.1	93.3	96.4	70.7	86.3	96.5	87.9	93.1	96.5	87.9	93.1	100.1	127.4	110.8
26, 27, 3370	ELECTRICAL, COMMUNICATIONS & UTIL.	99.4	86.9	93.2	98.0	80.8	89.5	96.6	60.1	78.6	94.6	101.9	98.2	92.2	101.9	97.0	102.6	129.5	115.9
MF2018	WEIGHTED AVERAGE	95.9	86.2	91.8	97.4	85.3	92.3	99.0	66.9	85.4	96.5	86.7	92.4	97.3	86.6	92.8	101.9	134.0	115.5

For customer support on your Site Work & Landscape Costs with RSMeans data, call 800.448.8182.

City Cost Indexes

MASSACHUSETTS

DIVISION		BROCKTON 023 MAT.	INST.	TOTAL	BUZZARDS BAY 025 MAT.	INST.	TOTAL	FALL RIVER 027 MAT.	INST.	TOTAL	FITCHBURG 014 MAT.	INST.	TOTAL	FRAMINGHAM 017 MAT.	INST.	TOTAL	GREENFIELD 013 MAT.	INST.	TOTAL
015433	CONTRACTOR EQUIPMENT		98.1	98.1		98.1	98.1		98.9	98.9		95.7	95.7		97.4	97.4		95.7	95.7
0241, 31 - 34	SITE & INFRASTRUCTURE, DEMOLITION	91.1	98.0	95.8	81.2	98.1	92.9	90.3	98.0	95.6	82.7	97.8	93.1	79.5	97.8	92.1	86.4	97.3	93.9
0310	Concrete Forming & Accessories	100.7	121.1	118.1	98.5	120.6	117.4	100.7	120.8	117.8	94.0	118.5	114.9	101.4	121.3	118.4	92.4	117.7	114.0
0320	Concrete Reinforcing	109.7	142.7	125.7	88.0	122.3	104.6	109.7	122.3	115.8	87.9	142.5	114.3	87.9	142.8	114.4	91.5	126.5	108.4
0330	Cast-in-Place Concrete	88.1	136.6	106.2	73.2	136.4	96.7	85.2	136.9	104.4	78.4	136.5	100.0	78.4	136.7	100.1	80.5	122.2	96.0
03	CONCRETE	95.1	129.8	110.3	80.7	126.0	100.6	93.8	126.2	108.0	79.5	128.3	101.0	82.3	130.0	103.2	83.0	120.3	99.4
04	MASONRY	105.4	135.1	123.6	97.9	135.1	120.8	105.9	135.1	123.8	98.2	131.9	118.9	104.7	135.1	124.0	102.7	119.1	112.8
05	METALS	98.8	125.6	106.7	93.6	117.0	100.5	98.8	117.3	104.3	95.3	122.1	103.2	95.3	125.8	104.3	97.8	113.8	102.5
06	WOOD, PLASTICS & COMPOSITES	98.1	119.7	109.9	94.7	119.7	108.4	98.1	119.9	110.1	94.5	116.5	106.6	101.0	119.4	111.2	92.5	119.7	107.4
07	THERMAL & MOISTURE PROTECTION	103.2	128.4	114.2	102.0	125.7	112.3	103.0	124.9	112.5	103.2	122.1	111.4	103.3	128.3	114.2	103.3	109.9	106.2
08	OPENINGS	97.4	128.4	104.6	93.4	118.9	99.3	97.4	118.3	102.2	99.2	126.7	105.6	89.8	128.3	98.7	99.4	120.3	104.3
0920	Plaster & Gypsum Board	92.3	119.9	110.9	87.3	119.9	109.3	92.3	119.9	110.9	111.3	116.7	114.9	114.4	119.9	118.1	112.6	119.9	117.5
0950, 0980	Ceilings & Acoustic Treatment	104.3	119.9	114.9	86.2	119.9	108.9	104.3	119.9	114.9	93.9	116.7	109.3	93.9	119.9	111.5	103.7	119.9	114.6
0960	Flooring	85.0	162.0	106.6	82.6	162.0	104.9	83.8	162.0	105.8	86.7	162.0	107.8	88.2	162.0	108.9	85.9	135.0	99.7
0970, 0990	Wall Finishes & Painting/Coating	85.0	139.2	117.1	85.0	139.2	117.1	85.0	139.2	117.1	86.2	139.2	117.6	87.1	139.2	117.9	86.2	109.3	99.9
09	FINISHES	90.8	130.8	112.5	85.2	130.8	109.9	90.5	131.0	112.4	89.5	128.9	110.9	90.2	130.6	112.1	91.7	120.9	107.5
COVERS	DIVS. 10 - 14, 25, 28, 41, 43, 44, 46	100.0	109.7	102.2	100.0	109.7	102.2	100.0	110.2	102.3	100.0	104.2	100.9	100.0	109.3	102.1	100.0	102.9	100.7
21, 22, 23	FIRE SUPPRESSION, PLUMBING & HVAC	100.4	103.9	101.7	96.8	103.5	99.4	100.4	103.5	101.6	97.0	104.5	99.9	97.0	121.1	106.4	97.0	98.5	97.6
26, 27, 3370	ELECTRICAL, COMMUNICATIONS & UTIL.	101.3	97.7	99.5	98.5	100.1	99.3	101.2	97.7	99.4	100.6	104.1	102.3	97.4	124.4	110.7	100.5	96.1	98.4
MF2018	WEIGHTED AVERAGE	98.5	116.7	106.2	93.2	115.2	102.5	98.3	114.9	105.3	94.6	116.2	103.7	94.0	124.2	106.8	95.9	109.0	101.4

MASSACHUSETTS

DIVISION		HYANNIS 026 MAT.	INST.	TOTAL	LAWRENCE 019 MAT.	INST.	TOTAL	LOWELL 018 MAT.	INST.	TOTAL	NEW BEDFORD 027 MAT.	INST.	TOTAL	PITTSFIELD 012 MAT.	INST.	TOTAL	SPRINGFIELD 010 - 011 MAT.	INST.	TOTAL
015433	CONTRACTOR EQUIPMENT		98.1	98.1		98.1	98.1		95.7	95.7		98.9	98.9		95.7	95.7		95.7	95.7
0241, 31 - 34	SITE & INFRASTRUCTURE, DEMOLITION	87.4	98.1	94.8	91.6	98.0	96.0	90.6	98.1	95.8	88.7	98.0	95.1	91.6	96.7	95.2	91.1	97.1	95.2
0310	Concrete Forming & Accessories	92.7	120.6	116.5	102.5	121.6	118.8	99.0	122.6	119.1	100.7	120.8	117.9	99.0	102.9	102.3	99.3	117.7	115.0
0320	Concrete Reinforcing	88.0	122.3	104.6	108.8	148.3	127.9	109.7	148.1	128.2	109.7	122.4	115.8	90.8	114.1	102.1	109.7	126.5	117.8
0330	Cast-in-Place Concrete	80.4	136.4	101.3	90.7	136.8	107.9	82.4	138.5	103.3	75.1	136.9	98.1	90.0	113.4	98.7	85.8	122.2	99.4
03	CONCRETE	86.7	126.0	104.0	96.2	131.1	111.5	88.0	131.9	107.3	89.1	126.3	105.4	89.0	108.3	97.5	89.5	120.2	103.0
04	MASONRY	104.3	135.1	123.2	110.1	135.9	125.9	97.3	134.9	120.4	104.2	135.7	123.6	98.0	111.5	106.3	97.6	119.1	110.8
05	METALS	95.1	117.0	101.5	98.2	128.4	107.1	98.2	125.0	106.1	98.8	117.4	104.3	98.0	108.2	101.0	100.9	113.7	104.6
06	WOOD, PLASTICS & COMPOSITES	88.0	119.7	105.4	101.5	119.7	111.5	100.8	119.7	111.2	98.1	119.9	110.1	100.8	102.5	101.8	100.8	119.7	111.2
07	THERMAL & MOISTURE PROTECTION	102.5	125.7	112.6	104.0	128.6	114.7	103.7	128.0	114.3	102.9	125.1	112.6	103.8	104.3	104.0	103.7	109.9	106.4
08	OPENINGS	93.9	118.9	99.8	93.4	129.9	101.9	100.4	129.9	107.3	97.4	123.0	103.3	100.4	107.6	102.1	100.4	120.3	105.1
0920	Plaster & Gypsum Board	83.2	119.9	107.9	117.3	119.9	119.1	117.3	119.9	119.1	92.3	119.9	110.9	117.3	102.3	107.2	117.3	119.9	119.1
0950, 0980	Ceilings & Acoustic Treatment	96.2	119.9	112.2	105.5	119.9	115.2	105.5	119.9	115.2	104.3	119.9	114.9	105.5	102.3	103.4	105.5	119.9	115.2
0960	Flooring	80.0	162.0	103.0	88.8	162.0	109.4	88.8	162.0	109.4	83.8	162.0	105.8	89.1	129.8	100.6	88.2	135.0	101.3
0970, 0990	Wall Finishes & Painting/Coating	85.0	139.2	117.1	86.3	139.2	117.6	86.2	139.2	117.6	85.0	139.2	117.1	86.2	109.3	99.9	87.0	109.3	100.2
09	FINISHES	86.3	130.8	110.4	93.8	130.8	113.8	93.7	131.6	114.2	90.4	131.0	112.3	93.8	108.6	101.8	93.6	120.9	108.4
COVERS	DIVS. 10 - 14, 25, 28, 41, 43, 44, 46	100.0	109.7	102.2	100.0	109.8	102.2	100.0	110.8	102.4	100.0	110.2	102.3	100.0	99.5	99.9	100.0	102.9	100.7
21, 22, 23	FIRE SUPPRESSION, PLUMBING & HVAC	100.4	104.2	101.9	100.1	119.7	107.8	100.1	121.2	108.4	100.4	103.4	101.5	100.1	94.2	97.8	100.0	98.5	99.5
26, 27, 3370	ELECTRICAL, COMMUNICATIONS & UTIL.	98.9	97.7	98.3	99.7	122.9	111.1	100.1	119.3	109.6	102.1	97.7	99.9	100.1	96.1	98.2	100.1	93.0	96.6
MF2018	WEIGHTED AVERAGE	95.7	115.0	103.8	98.4	124.2	109.3	97.5	123.8	108.6	97.7	115.2	105.1	97.7	102.6	99.7	98.2	108.5	102.5

MASSACHUSETTS / MICHIGAN

DIVISION		WORCESTER 015 - 016 MAT.	INST.	TOTAL	ANN ARBOR 481 MAT.	INST.	TOTAL	BATTLE CREEK 490 MAT.	INST.	TOTAL	BAY CITY 487 MAT.	INST.	TOTAL	DEARBORN 481 MAT.	INST.	TOTAL	DETROIT 482 MAT.	INST.	TOTAL
015433	CONTRACTOR EQUIPMENT		95.7	95.7		109.1	109.1		96.1	96.1		109.1	109.1		109.1	109.1		97.8	97.8
0241, 31 - 34	SITE & INFRASTRUCTURE, DEMOLITION	91.0	97.9	95.7	81.0	92.2	88.7	93.6	81.7	85.3	72.5	91.7	85.8	80.8	92.3	88.7	98.2	101.2	100.2
0310	Concrete Forming & Accessories	99.6	118.5	115.7	96.3	104.1	103.0	96.0	77.3	80.0	96.4	81.5	83.7	96.2	104.7	103.5	100.4	105.3	104.6
0320	Concrete Reinforcing	109.7	155.5	131.9	101.2	103.5	102.3	98.5	80.9	90.0	101.2	102.7	101.9	101.2	103.6	102.3	101.4	104.8	103.0
0330	Cast-in-Place Concrete	85.3	136.5	104.4	88.3	96.4	91.3	86.1	91.0	87.9	84.5	83.5	84.1	86.4	97.2	90.4	104.5	101.3	103.3
03	CONCRETE	89.3	130.6	107.5	89.8	102.2	95.2	85.7	82.6	84.3	88.0	87.3	87.7	88.9	102.8	95.0	103.1	103.2	103.2
04	MASONRY	97.2	131.9	118.5	99.6	96.4	97.6	99.1	78.4	86.4	99.2	77.7	86.0	99.5	98.2	98.7	102.6	99.8	100.9
05	METALS	100.9	127.4	108.7	102.7	115.7	106.5	103.7	83.7	97.9	103.3	114.0	106.4	102.8	115.8	106.6	104.3	94.3	101.3
06	WOOD, PLASTICS & COMPOSITES	101.2	116.5	109.6	89.2	106.6	98.8	88.4	75.6	81.3	89.2	81.9	85.2	89.2	106.6	98.8	97.1	106.7	102.4
07	THERMAL & MOISTURE PROTECTION	103.8	122.1	111.8	105.3	98.2	102.2	96.1	79.0	88.7	102.7	81.0	93.2	103.6	100.7	102.3	102.6	102.9	102.7
08	OPENINGS	100.4	130.2	107.4	94.2	100.8	95.7	86.5	74.1	83.6	94.2	83.5	91.7	94.2	101.1	95.8	97.2	101.9	98.3
0920	Plaster & Gypsum Board	117.3	116.7	116.9	107.4	106.5	106.8	91.7	71.7	78.2	107.4	81.1	89.7	107.4	106.5	106.8	104.1	106.5	105.7
0950, 0980	Ceilings & Acoustic Treatment	105.5	116.7	113.0	84.8	106.5	99.4	78.7	71.7	74.0	85.7	81.1	82.6	84.8	106.5	99.4	96.9	106.5	103.4
0960	Flooring	88.8	154.9	107.4	91.2	107.2	95.7	91.5	70.4	85.6	91.2	77.0	87.2	90.6	101.6	93.6	95.7	106.7	98.8
0970, 0990	Wall Finishes & Painting/Coating	86.2	139.2	117.6	82.7	95.5	90.3	82.3	76.3	78.7	82.7	79.1	80.6	82.7	93.8	89.3	93.1	100.1	97.3
09	FINISHES	93.7	127.5	112.0	90.8	104.3	98.1	84.7	75.2	79.6	90.6	79.9	84.8	90.6	103.5	97.6	97.7	105.4	101.9
COVERS	DIVS. 10 - 14, 25, 28, 41, 43, 44, 46	100.0	104.2	100.9	100.0	93.7	98.6	100.0	93.7	98.6	100.0	88.0	97.3	100.0	94.1	98.7	100.0	102.4	100.5
21, 22, 23	FIRE SUPPRESSION, PLUMBING & HVAC	100.1	104.7	101.9	100.0	91.3	96.6	100.1	82.6	93.2	100.0	78.6	91.6	100.0	100.0	100.0	99.9	102.7	101.0
26, 27, 3370	ELECTRICAL, COMMUNICATIONS & UTIL.	100.1	104.1	102.1	97.3	103.0	100.1	96.6	77.7	87.3	96.4	82.5	89.5	97.3	96.1	96.7	99.9	103.5	101.6
MF2018	WEIGHTED AVERAGE	98.1	117.0	106.1	97.1	99.8	98.2	95.5	80.4	89.1	96.6	85.3	91.8	96.9	100.9	98.6	100.7	102.0	101.3

For customer support on your Site Work & Landscape Costs with RSMeans data, call 800.448.8182.

795

MICHIGAN

DIVISION		FLINT 484 - 485			GAYLORD 497			GRAND RAPIDS 493, 495			IRON MOUNTAIN 498 - 499			JACKSON 492			KALAMAZOO 491		
		MAT.	INST.	TOTAL	MAT.	INST.	TOTAL	MAT.	INST.	TOTAL	MAT.	INST.	TOTAL	MAT.	INST.	TOTAL	MAT.	INST.	TOTAL
015433	CONTRACTOR EQUIPMENT		109.1	109.1		103.6	103.6		98.3	98.3		91.1	91.1		103.6	103.6		96.1	96.1
0241, 31 - 34	SITE & INFRASTRUCTURE, DEMOLITION	70.2	91.8	85.1	87.4	79.6	82.0	92.9	85.9	88.1	96.4	88.5	90.9	111.0	81.2	90.4	93.9	81.6	85.4
0310	Concrete Forming & Accessories	99.1	81.8	84.4	95.0	73.9	77.0	95.6	75.7	78.6	86.6	78.6	79.8	91.6	82.3	83.7	96.0	77.1	79.9
0320	Concrete Reinforcing	101.2	103.1	102.1	91.9	93.0	92.4	103.4	80.7	92.4	91.6	85.0	88.4	89.3	103.1	96.0	98.5	76.2	87.8
0330	Cast-in-Place Concrete	88.9	85.8	87.8	85.8	78.4	83.1	90.4	89.0	89.9	101.7	67.6	89.0	77.8	89.8	83.1	88.9	81.7	85.7
03	CONCRETE	90.2	88.4	89.4	82.6	80.5	81.6	89.7	81.1	85.9	90.9	76.6	84.6	82.8	88.3	85.1	89.3	79.5	85.1
04	MASONRY	99.7	85.9	91.2	109.7	72.7	87.0	92.0	74.1	81.0	95.4	79.7	85.7	89.2	86.9	87.8	97.6	78.4	85.8
05	METALS	102.8	114.6	106.3	105.5	110.8	107.0	100.8	83.2	95.6	104.8	91.1	100.8	105.7	112.7	107.8	103.7	81.9	97.3
06	WOOD, PLASTICS & COMPOSITES	92.6	80.7	86.1	81.9	73.4	77.2	92.2	73.9	82.2	78.2	78.9	78.6	80.8	79.8	80.2	88.4	75.6	81.3
07	THERMAL & MOISTURE PROTECTION	103.0	84.5	95.0	94.7	74.7	86.0	98.0	71.8	86.6	98.4	74.8	88.1	94.0	88.3	91.5	96.1	79.0	88.7
08	OPENINGS	94.2	82.6	91.5	86.0	78.7	84.3	101.8	73.8	95.2	92.5	69.8	87.2	85.2	85.5	85.3	86.5	73.5	83.4
0920	Plaster & Gypsum Board	109.0	79.8	89.4	91.3	71.8	78.2	99.8	70.2	79.8	51.3	79.0	69.9	90.1	78.3	82.1	91.7	71.7	78.2
0950, 0980	Ceilings & Acoustic Treatment	84.8	79.8	81.5	77.1	71.8	73.5	91.1	70.2	77.0	76.0	79.0	78.0	77.1	78.3	77.9	78.7	71.7	74.0
0960	Flooring	91.2	86.0	89.7	84.9	85.6	85.1	95.7	76.1	90.2	102.0	91.0	98.9	83.4	78.7	82.1	91.5	70.4	85.6
0970, 0990	Wall Finishes & Painting/Coating	82.7	78.3	80.1	78.7	81.6	80.4	89.3	76.9	82.0	94.8	69.6	79.9	78.7	93.8	87.7	82.3	76.3	78.7
09	FINISHES	90.1	81.6	85.5	84.1	76.2	79.8	91.4	75.2	82.6	84.6	80.0	82.1	85.2	81.9	83.4	84.7	75.2	79.6
COVERS	DIVS. 10 - 14, 25, 28, 41, 43, 44, 46	100.0	88.8	97.5	100.0	77.9	95.1	100.0	93.0	98.4	100.0	85.9	96.8	100.0	91.6	98.1	100.0	93.7	98.6
21, 22, 23	FIRE SUPPRESSION, PLUMBING & HVAC	100.0	83.3	93.4	96.8	79.0	89.8	100.0	79.6	92.0	96.7	83.9	91.7	96.8	85.7	92.4	100.1	78.2	91.5
26, 27, 3370	ELECTRICAL, COMMUNICATIONS & UTIL.	97.3	89.1	93.2	94.5	76.3	85.5	102.2	82.9	92.7	100.8	80.7	90.9	98.3	103.0	100.6	96.4	74.1	85.4
MF2018	WEIGHTED AVERAGE	96.8	88.6	93.3	94.5	80.6	88.6	97.9	79.9	90.3	96.5	81.7	90.2	94.0	90.7	92.6	95.8	78.6	88.6

MICHIGAN / MINNESOTA

DIVISION		LANSING 488 - 489			MUSKEGON 494			ROYAL OAK 480, 483			SAGINAW 486			TRAVERSE CITY 496			BEMIDJI 566		
		MAT.	INST.	TOTAL	MAT.	INST.	TOTAL	MAT.	INST.	TOTAL	MAT.	INST.	TOTAL	MAT.	INST.	TOTAL	MAT.	INST.	TOTAL
015433	CONTRACTOR EQUIPMENT		111.0	111.0		96.1	96.1		89.8	89.8		109.1	109.1		91.1	91.1		97.7	97.7
0241, 31 - 34	SITE & INFRASTRUCTURE, DEMOLITION	90.7	96.5	94.7	91.5	81.7	84.8	84.8	92.6	90.2	73.5	91.7	86.1	81.8	87.5	85.7	95.1	95.1	95.1
0310	Concrete Forming & Accessories	93.8	74.9	77.7	96.3	79.4	81.9	92.3	105.2	103.2	96.3	79.7	82.1	86.6	72.4	74.5	88.0	82.8	83.6
0320	Concrete Reinforcing	104.8	102.9	103.8	99.3	80.7	90.3	91.7	104.8	98.0	101.2	102.7	101.9	93.1	80.2	86.9	97.4	101.8	99.5
0330	Cast-in-Place Concrete	97.9	85.1	93.1	85.8	90.5	87.5	77.3	97.5	84.8	87.3	83.4	85.8	79.3	76.7	78.3	101.8	104.5	102.8
03	CONCRETE	91.4	84.9	88.5	84.2	83.3	83.8	77.1	101.6	87.9	89.3	86.5	88.1	74.8	76.1	75.4	91.0	95.0	92.8
04	MASONRY	94.5	84.6	88.4	96.3	76.2	83.9	93.2	99.5	97.1	101.2	77.7	86.7	93.5	74.8	82.0	98.9	102.4	101.0
05	METALS	101.6	112.7	104.9	101.5	83.1	96.1	106.3	92.2	102.2	102.8	113.8	106.0	104.8	88.9	100.1	91.2	117.9	99.0
06	WOOD, PLASTICS & COMPOSITES	87.0	71.2	78.3	85.4	78.0	81.3	85.0	106.6	96.9	85.9	79.7	82.5	78.2	71.9	74.8	68.8	77.0	73.3
07	THERMAL & MOISTURE PROTECTION	102.0	83.8	94.1	95.0	73.5	85.7	101.2	100.9	101.1	103.9	80.9	93.9	97.3	74.2	87.3	106.3	90.7	99.5
08	OPENINGS	102.7	77.2	96.8	85.8	76.5	83.6	94.0	101.3	95.7	92.4	82.2	90.0	92.5	65.4	86.2	99.7	101.3	100.1
0920	Plaster & Gypsum Board	100.1	70.2	79.9	71.8	74.2	73.4	104.8	106.5	106.0	107.4	78.8	88.2	51.3	71.8	65.1	104.3	76.8	85.8
0950, 0980	Ceilings & Acoustic Treatment	83.9	70.2	74.6	78.7	74.2	75.6	84.1	106.5	99.2	84.8	78.8	80.8	76.0	71.8	73.2	128.8	76.8	93.7
0960	Flooring	97.2	78.7	92.0	90.2	72.3	85.1	88.2	106.7	93.4	91.2	77.0	87.2	102.0	85.6	97.4	90.7	91.4	90.9
0970, 0990	Wall Finishes & Painting/Coating	91.1	76.9	82.7	80.5	79.7	80.0	84.2	93.8	89.9	82.7	79.1	80.6	94.8	39.4	62.1	89.6	105.8	99.2
09	FINISHES	90.2	74.6	81.7	81.1	77.1	78.9	89.7	104.3	97.6	90.5	78.6	84.0	83.5	71.0	76.7	98.2	86.0	91.6
COVERS	DIVS. 10 - 14, 25, 28, 41, 43, 44, 46	100.0	88.8	97.5	100.0	94.5	98.8	100.0	100.0	100.0	100.0	87.7	97.3	100.0	84.4	96.5	100.0	95.6	99.0
21, 22, 23	FIRE SUPPRESSION, PLUMBING & HVAC	99.9	83.7	93.5	99.9	83.4	93.4	96.7	101.4	98.5	100.0	78.3	91.5	96.7	78.9	89.7	96.7	84.8	92.0
26, 27, 3370	ELECTRICAL, COMMUNICATIONS & UTIL.	99.1	87.6	93.4	96.8	73.0	85.1	99.4	103.1	101.3	95.4	84.5	90.0	96.2	76.2	86.4	105.0	100.7	102.8
MF2018	WEIGHTED AVERAGE	98.0	86.8	93.3	94.3	80.0	88.3	95.1	100.3	97.3	96.5	85.1	91.7	93.4	77.7	86.8	96.7	95.4	96.1

MINNESOTA

DIVISION		BRAINERD 564			DETROIT LAKES 565			DULUTH 556 - 558			MANKATO 560			MINNEAPOLIS 553 - 555			ROCHESTER 559		
		MAT.	INST.	TOTAL	MAT.	INST.	TOTAL	MAT.	INST.	TOTAL	MAT.	INST.	TOTAL	MAT.	INST.	TOTAL	MAT.	INST.	TOTAL
015433	CONTRACTOR EQUIPMENT		100.2	100.2		97.7	97.7		102.9	102.9		100.2	100.2		108.4	108.4		100.7	100.7
0241, 31 - 34	SITE & INFRASTRUCTURE, DEMOLITION	96.0	99.9	98.7	93.3	95.4	94.7	98.2	101.7	100.6	92.6	99.4	97.3	99.3	114.0	111.8	99.2	94.8	95.4
0310	Concrete Forming & Accessories	88.7	83.6	84.4	84.9	82.8	83.1	98.4	94.6	95.2	96.9	96.6	96.6	96.1	109.4	102.5	98.9	108.8	103.7
0320	Concrete Reinforcing	96.2	102.2	99.1	97.4	101.8	99.6	102.6	102.0	102.3	96.0	109.2	102.4	96.1	109.4	102.5	94.8	95.3	95.0
0330	Cast-in-Place Concrete	110.7	109.1	110.1	98.9	107.5	102.1	94.3	97.7	95.6	101.9	99.5	101.0	95.6	115.3	102.9	94.8	95.3	95.0
03	CONCRETE	94.8	97.1	95.8	88.6	96.1	91.9	95.3	98.1	96.5	90.5	100.8	95.0	97.6	114.3	104.9	91.4	98.6	94.6
04	MASONRY	122.5	116.3	118.7	122.6	109.8	114.8	95.3	105.7	101.7	111.3	110.8	111.0	114.5	119.5	117.6	102.9	103.5	103.3
05	METALS	92.2	118.3	99.9	91.1	117.5	98.9	100.2	119.1	105.8	92.1	121.6	100.7	98.7	124.5	106.3	99.3	123.1	106.3
06	WOOD, PLASTICS & COMPOSITES	84.9	74.2	79.1	65.8	74.4	70.6	93.9	91.9	92.8	94.5	95.4	95.0	96.1	110.8	104.2	95.2	92.9	93.9
07	THERMAL & MOISTURE PROTECTION	104.7	102.4	103.7	106.1	99.4	103.2	97.9	102.2	99.8	105.1	93.4	100.1	101.2	115.9	107.6	105.3	90.0	98.6
08	OPENINGS	86.4	99.8	89.5	99.9	99.9	99.9	103.3	104.6	103.6	90.9	112.2	95.9	99.3	121.5	104.5	98.6	112.5	101.8
0920	Plaster & Gypsum Board	91.5	74.1	79.8	103.7	74.1	83.8	94.5	92.1	92.9	95.6	95.9	95.8	101.1	111.2	107.9	103.2	93.2	96.5
0950, 0980	Ceilings & Acoustic Treatment	58.8	74.1	69.1	128.8	74.1	91.9	89.2	92.1	91.1	58.8	95.9	83.8	103.1	111.2	108.6	93.9	93.2	93.4
0960	Flooring	89.6	85.6	88.4	89.5	85.6	88.4	93.2	123.9	101.8	91.4	81.8	88.7	100.6	114.6	104.5	92.3	81.8	89.3
0970, 0990	Wall Finishes & Painting/Coating	84.0	105.8	96.9	89.6	83.4	85.9	86.0	104.4	96.9	95.0	103.7	100.2	101.3	123.8	114.6	85.3	98.2	92.9
09	FINISHES	82.1	85.2	83.8	97.6	82.5	89.4	89.4	100.2	95.3	83.7	94.9	89.7	98.9	115.0	107.6	90.0	93.1	91.7
COVERS	DIVS. 10 - 14, 25, 28, 41, 43, 44, 46	100.0	97.3	99.4	100.0	97.1	99.4	100.0	94.6	98.8	100.0	97.5	99.4	100.0	105.8	101.3	100.0	101.8	100.4
21, 22, 23	FIRE SUPPRESSION, PLUMBING & HVAC	96.0	88.6	93.1	96.7	87.5	93.1	99.8	94.5	97.7	96.0	84.6	91.5	100.0	110.6	104.2	100.0	91.6	96.7
26, 27, 3370	ELECTRICAL, COMMUNICATIONS & UTIL.	102.7	101.7	102.2	104.8	71.1	88.2	100.3	101.7	101.0	109.1	93.1	101.2	107.2	110.3	108.7	103.0	93.1	98.1
MF2018	WEIGHTED AVERAGE	95.6	98.6	96.8	97.4	92.4	95.3	98.6	101.3	99.7	95.7	98.5	96.9	100.7	114.1	106.3	98.3	98.7	98.5

City Cost Indexes

MINNESOTA / MISSISSIPPI

DIVISION		SAINT PAUL 550 - 551 MAT.	INST.	TOTAL	ST. CLOUD 563 MAT.	INST.	TOTAL	THIEF RIVER FALLS 567 MAT.	INST.	TOTAL	WILLMAR 562 MAT.	INST.	TOTAL	WINDOM 561 MAT.	INST.	TOTAL	BILOXI 395 MAT.	INST.	TOTAL
015433	CONTRACTOR EQUIPMENT		102.9	102.9		100.2	100.2		97.7	97.7		100.2	100.2		100.2	100.2		100.1	100.1
0241, 31 - 34	SITE & INFRASTRUCTURE, DEMOLITION	96.5	102.5	100.7	91.4	100.8	97.9	94.0	94.9	94.7	90.6	99.7	96.9	84.5	98.9	94.4	105.2	85.9	91.9
0310	Concrete Forming & Accessories	98.6	117.0	114.3	86.0	113.5	109.5	88.8	81.9	82.9	85.8	86.8	86.7	90.1	82.2	83.4	92.8	64.1	68.4
0320	Concrete Reinforcing	105.4	109.9	107.6	96.2	109.4	102.6	97.8	101.6	99.6	95.8	109.2	102.3	95.8	108.1	101.8	93.4	48.8	71.8
0330	Cast-in-Place Concrete	99.9	116.3	106.0	97.7	115.0	104.1	100.9	81.3	93.6	99.2	79.5	91.9	85.8	83.3	84.8	114.2	66.3	96.4
03	CONCRETE	95.3	116.1	104.4	86.5	114.0	98.5	89.8	86.5	88.3	86.3	89.5	87.7	77.6	88.5	82.4	94.4	64.0	81.0
04	MASONRY	105.3	127.4	118.9	107.3	121.8	116.2	98.9	102.3	101.0	112.1	114.5	113.6	122.0	87.9	101.0	90.4	62.8	73.4
05	METALS	100.1	125.0	107.4	92.9	123.1	101.8	91.3	116.7	98.7	92.0	121.5	100.7	91.9	119.5	100.0	90.3	84.5	88.6
06	WOOD, PLASTICS & COMPOSITES	95.0	114.2	105.6	82.1	110.3	97.6	70.0	77.0	73.8	81.8	79.9	80.8	86.3	79.9	82.8	93.1	64.5	77.3
07	THERMAL & MOISTURE PROTECTION	102.0	120.5	110.1	104.9	111.7	107.9	107.1	88.1	98.8	104.7	101.0	103.1	104.6	83.0	95.2	100.0	63.1	83.9
08	OPENINGS	98.2	124.3	104.3	91.4	122.1	98.5	99.7	101.3	100.1	88.4	102.8	91.7	91.9	102.8	94.4	96.3	55.0	86.7
0920	Plaster & Gypsum Board	94.5	115.0	108.3	91.5	111.2	104.7	104.3	76.8	85.8	91.5	79.9	83.7	91.5	79.9	83.7	103.8	64.0	77.0
0950, 0980	Ceilings & Acoustic Treatment	91.6	115.0	107.4	58.8	111.2	94.1	128.8	76.8	93.7	58.8	79.9	73.0	58.8	79.9	73.0	86.4	64.0	71.3
0960	Flooring	92.4	122.8	101.0	86.4	120.9	96.1	90.4	85.6	89.0	87.8	85.6	87.2	90.3	85.6	89.0	94.0	65.0	85.8
0970, 0990	Wall Finishes & Painting/Coating	92.1	129.1	114.0	95.0	123.8	112.0	89.6	83.4	85.9	89.6	83.4	85.9	89.6	103.7	98.0	86.0	46.9	62.8
09	FINISHES	91.1	119.1	106.3	81.5	115.9	100.1	98.1	82.6	89.7	81.6	85.3	83.6	81.8	84.8	83.4	89.4	62.8	75.0
COVERS	DIVS. 10 - 14, 25, 28, 41, 43, 44, 46	100.0	108.7	101.9	100.0	103.2	100.7	100.0	95.5	99.0	100.0	97.3	99.4	100.0	93.7	98.6	100.0	71.9	93.7
21, 22, 23	FIRE SUPPRESSION, PLUMBING & HVAC	99.9	116.7	106.5	99.5	110.5	103.8	96.7	84.4	91.9	96.0	101.6	98.2	96.0	81.0	90.1	100.0	55.3	82.4
26, 27, 3370	ELECTRICAL, COMMUNICATIONS & UTIL.	102.9	115.1	108.9	102.7	115.1	108.8	102.1	71.0	86.8	102.7	83.7	93.4	109.1	93.1	101.2	101.0	53.7	77.7
MF2018	WEIGHTED AVERAGE	99.0	117.6	106.9	95.1	114.2	103.2	96.3	89.3	93.3	94.0	98.0	95.7	94.3	91.2	93.0	96.2	63.9	82.6

MISSISSIPPI

DIVISION		CLARKSDALE 386 MAT.	INST.	TOTAL	COLUMBUS 397 MAT.	INST.	TOTAL	GREENVILLE 387 MAT.	INST.	TOTAL	GREENWOOD 389 MAT.	INST.	TOTAL	JACKSON 390 - 392 MAT.	INST.	TOTAL	LAUREL 394 MAT.	INST.	TOTAL
015433	CONTRACTOR EQUIPMENT		100.1	100.1		100.1	100.1		100.1	100.1		100.1	100.1		102.6	102.6		100.1	100.1
0241, 31 - 34	SITE & INFRASTRUCTURE, DEMOLITION	101.4	84.4	89.7	103.9	85.2	91.0	107.5	86.1	92.7	104.5	84.1	90.4	101.1	90.6	93.9	109.6	84.5	92.2
0310	Concrete Forming & Accessories	83.8	43.9	49.8	81.9	46.3	51.6	80.7	62.7	65.4	92.5	44.3	51.4	92.0	64.6	68.6	82.0	59.5	62.8
0320	Concrete Reinforcing	104.2	66.5	85.9	99.9	66.6	83.8	104.7	66.7	86.3	104.2	66.5	86.0	104.0	51.7	78.7	100.6	32.4	67.6
0330	Cast-in-Place Concrete	104.1	58.8	87.2	116.5	61.1	95.9	107.2	65.7	91.8	112.0	58.9	92.2	99.6	66.5	87.3	113.9	60.1	93.9
03	CONCRETE	93.8	55.1	76.8	95.8	57.1	78.8	99.3	66.2	84.8	100.1	55.4	80.4	89.2	64.7	78.4	97.9	56.9	79.9
04	MASONRY	89.1	50.8	65.6	115.4	53.6	77.4	132.2	61.4	88.7	89.8	50.7	65.8	95.7	61.5	74.7	111.5	49.4	73.4
05	METALS	92.5	87.9	91.1	87.4	90.4	88.3	93.6	91.2	92.8	92.5	87.8	91.1	96.9	84.6	93.3	87.5	76.0	84.1
06	WOOD, PLASTICS & COMPOSITES	79.4	44.4	60.1	78.8	45.2	60.3	76.2	62.8	68.8	91.8	44.4	65.7	96.3	65.5	79.4	79.9	64.5	71.4
07	THERMAL & MOISTURE PROTECTION	97.7	52.0	77.8	100.0	55.4	80.6	98.2	62.1	82.5	98.1	54.2	79.0	98.3	62.9	82.9	100.2	56.5	81.2
08	OPENINGS	95.1	48.2	84.1	95.9	48.6	84.9	94.8	58.1	86.2	95.1	48.2	84.1	99.5	56.0	89.4	93.2	51.8	83.5
0920	Plaster & Gypsum Board	89.9	43.3	58.6	94.3	44.2	60.6	89.6	62.3	71.2	100.6	43.3	62.1	90.7	64.9	73.3	94.3	64.0	73.9
0950, 0980	Ceilings & Acoustic Treatment	82.3	43.3	56.0	81.5	44.2	56.3	85.0	62.3	69.7	82.3	43.3	56.0	89.7	64.9	73.0	81.5	64.0	69.7
0960	Flooring	97.8	45.0	83.0	87.9	65.0	81.4	96.2	65.0	87.4	103.5	65.0	92.7	92.1	65.0	84.5	86.6	65.0	80.5
0970, 0990	Wall Finishes & Painting/Coating	94.6	46.9	66.4	86.0	46.9	62.8	94.6	56.7	72.1	94.6	46.9	66.4	89.9	56.7	70.3	86.0	46.9	62.8
09	FINISHES	90.2	43.9	65.1	85.4	49.1	65.8	90.8	62.5	75.5	93.7	48.0	68.9	89.9	64.1	75.9	85.6	60.5	72.0
COVERS	DIVS. 10 - 14, 25, 28, 41, 43, 44, 46	100.0	48.0	88.4	100.0	49.0	88.6	100.0	71.3	93.6	100.0	48.0	88.4	100.0	71.9	93.7	100.0	34.2	85.3
21, 22, 23	FIRE SUPPRESSION, PLUMBING & HVAC	98.4	51.3	79.9	98.0	53.1	80.4	100.0	57.4	83.3	98.4	51.7	80.1	100.0	59.1	83.9	98.1	47.0	78.0
26, 27, 3370	ELECTRICAL, COMMUNICATIONS & UTIL.	96.8	41.6	69.6	98.6	55.1	77.2	96.8	55.9	76.6	96.8	39.0	68.3	102.7	55.9	79.6	100.1	57.9	79.3
MF2018	WEIGHTED AVERAGE	95.3	55.2	78.4	95.9	59.1	80.4	98.7	65.5	84.7	96.6	55.5	79.2	97.3	65.6	83.9	96.1	57.8	79.9

MISSISSIPPI / MISSOURI

DIVISION		MCCOMB 396 MAT.	INST.	TOTAL	MERIDIAN 393 MAT.	INST.	TOTAL	TUPELO 388 MAT.	INST.	TOTAL	BOWLING GREEN 633 MAT.	INST.	TOTAL	CAPE GIRARDEAU 637 MAT.	INST.	TOTAL	CHILLICOTHE 646 MAT.	INST.	TOTAL
015433	CONTRACTOR EQUIPMENT		100.1	100.1		100.1	100.1		100.1	100.1		107.1	107.1		107.1	107.1		101.5	101.5
0241, 31 - 34	SITE & INFRASTRUCTURE, DEMOLITION	96.3	84.3	88.0	100.7	86.1	90.7	98.9	84.3	88.8	87.6	90.0	89.3	89.3	89.9	89.7	101.2	88.4	92.3
0310	Concrete Forming & Accessories	81.9	45.5	50.9	79.4	62.8	65.2	81.2	45.9	51.1	96.0	94.0	94.3	88.9	82.0	83.0	86.2	95.7	94.3
0320	Concrete Reinforcing	101.2	34.3	68.8	99.9	51.6	76.6	101.9	66.5	84.8	93.4	96.5	94.9	94.6	79.6	87.3	100.4	102.6	101.5
0330	Cast-in-Place Concrete	101.1	58.4	85.2	108.2	66.0	92.5	104.1	67.9	90.6	90.7	95.5	92.5	89.8	86.6	88.6	91.7	86.0	89.6
03	CONCRETE	85.9	50.2	70.2	90.1	63.8	78.6	93.4	59.2	78.4	91.2	96.4	93.5	90.4	85.0	88.0	95.0	94.3	94.7
04	MASONRY	116.7	50.3	75.9	90.0	62.1	72.9	121.7	53.3	79.6	109.6	97.9	102.4	105.9	80.5	90.3	99.2	92.2	94.9
05	METALS	87.6	75.6	84.1	88.5	85.6	87.7	92.4	87.9	91.1	91.8	117.7	99.4	92.9	109.4	97.8	85.1	110.6	92.5
06	WOOD, PLASTICS & COMPOSITES	78.8	46.8	61.2	76.2	62.8	68.8	76.7	45.2	59.4	96.7	94.4	95.4	89.2	80.0	84.1	94.0	97.0	95.7
07	THERMAL & MOISTURE PROTECTION	99.4	54.2	79.8	99.6	62.4	83.4	97.6	53.9	78.6	96.4	99.2	97.6	95.9	85.3	91.3	92.7	92.7	92.7
08	OPENINGS	96.0	41.3	83.2	95.6	54.8	86.1	95.0	51.4	84.9	98.7	98.2	98.5	98.6	76.1	93.4	86.0	96.2	88.3
0920	Plaster & Gypsum Board	94.3	45.8	61.7	94.3	62.3	72.8	89.6	44.2	59.0	101.7	94.6	96.9	101.1	79.8	86.8	98.2	96.9	97.3
0950, 0980	Ceilings & Acoustic Treatment	81.5	45.8	57.4	82.4	62.3	68.8	82.3	44.2	56.6	88.9	94.6	92.8	88.9	79.8	82.8	87.4	96.9	93.8
0960	Flooring	87.9	65.0	81.4	86.5	65.0	80.5	96.5	65.0	87.6	95.7	96.9	96.0	92.4	84.5	90.2	94.9	99.7	96.2
0970, 0990	Wall Finishes & Painting/Coating	86.0	46.9	62.8	86.0	46.9	62.8	94.6	45.2	65.4	95.4	106.3	101.8	95.4	67.1	78.6	94.3	103.2	99.6
09	FINISHES	84.8	49.2	65.5	84.8	61.6	72.3	89.7	48.9	67.6	98.2	95.7	96.8	97.0	79.8	87.7	97.1	97.6	97.3
COVERS	DIVS. 10 - 14, 25, 28, 41, 43, 44, 46	100.0	50.9	89.1	100.0	71.5	93.6	100.0	49.0	88.6	100.0	81.4	95.8	100.0	94.3	98.7	100.0	81.3	95.8
21, 22, 23	FIRE SUPPRESSION, PLUMBING & HVAC	98.0	50.7	79.5	100.0	57.6	83.3	98.5	52.7	80.5	96.5	99.5	97.7	100.0	98.7	99.5	96.6	99.9	97.9
26, 27, 3370	ELECTRICAL, COMMUNICATIONS & UTIL.	97.3	56.0	76.9	100.1	55.6	78.2	96.5	55.0	76.1	99.2	76.8	88.2	99.2	98.5	98.9	94.9	75.8	85.5
MF2018	WEIGHTED AVERAGE	94.4	55.8	78.0	94.6	64.5	81.9	96.6	59.0	80.7	96.4	95.5	96.0	97.0	91.2	94.5	93.7	93.8	93.7

For customer support on your Site Work & Landscape Costs with RSMeans data, call 800.448.8182.

797

MISSOURI

DIVISION		COLUMBIA 652 MAT.	INST.	TOTAL	FLAT RIVER 636 MAT.	INST.	TOTAL	HANNIBAL 634 MAT.	INST.	TOTAL	HARRISONVILLE 647 MAT.	INST.	TOTAL	JEFFERSON CITY 650 - 651 MAT.	INST.	TOTAL	JOPLIN 648 MAT.	INST.	TOTAL
015433	CONTRACTOR EQUIPMENT		110.3	110.3		107.1	107.1		107.1	107.1		101.5	101.5		113.1	113.1		104.8	104.8
0241, 31 - 34	SITE & INFRASTRUCTURE, DEMOLITION	94.2	93.8	93.9	90.2	89.7	89.8	85.4	89.8	88.4	92.9	89.5	90.6	94.4	98.7	97.4	102.0	92.6	95.5
0310	Concrete Forming & Accessories	84.4	81.0	81.5	102.4	88.3	90.4	94.3	82.5	84.3	83.4	99.0	96.7	96.3	78.6	81.2	97.4	74.2	77.7
0320	Concrete Reinforcing	85.6	93.3	89.3	94.6	102.9	98.6	92.9	96.5	94.6	100.0	110.9	105.3	92.4	93.3	92.9	103.7	84.3	94.3
0330	Cast-in-Place Concrete	88.6	84.8	87.2	93.7	90.6	92.5	85.9	93.5	88.7	93.9	99.7	96.0	93.9	84.1	90.2	99.3	75.5	90.4
03	CONCRETE	80.8	86.3	83.2	94.1	93.3	93.8	87.6	90.5	88.9	91.1	102.0	95.9	87.5	84.8	86.3	93.9	77.7	86.7
04	MASONRY	136.9	86.4	105.9	106.5	76.4	88.0	101.6	94.6	97.3	94.0	99.4	97.3	96.0	86.5	90.1	92.4	80.2	84.9
05	METALS	96.3	115.4	101.9	91.7	119.2	99.8	91.8	117.2	99.3	85.4	115.1	94.1	95.8	114.0	101.1	87.9	98.8	91.1
06	WOOD, PLASTICS & COMPOSITES	80.3	78.6	79.3	105.2	89.6	96.6	94.9	81.1	87.3	90.4	98.7	95.0	97.1	75.6	85.2	106.1	73.5	88.2
07	THERMAL & MOISTURE PROTECTION	90.6	86.1	88.7	96.6	91.0	94.2	96.2	93.4	95.0	91.8	100.8	95.7	97.5	86.1	92.5	91.9	81.2	87.3
08	OPENINGS	94.4	81.7	91.4	98.6	97.4	98.4	98.7	83.7	95.2	85.7	103.2	89.8	94.5	80.0	91.1	86.9	76.3	84.5
0920	Plaster & Gypsum Board	84.7	78.2	80.3	107.7	89.7	95.6	101.4	81.0	87.7	94.1	98.6	97.1	96.2	74.9	81.9	105.4	72.7	83.4
0950, 0980	Ceilings & Acoustic Treatment	88.5	78.2	81.5	88.9	89.7	89.4	88.9	81.0	83.6	87.4	98.6	95.0	93.3	74.9	80.9	88.3	72.7	77.8
0960	Flooring	90.2	98.6	92.5	98.9	84.5	94.9	95.0	96.9	95.5	90.3	100.8	93.2	97.6	71.5	90.3	120.1	72.2	106.6
0970, 0990	Wall Finishes & Painting/Coating	93.1	81.9	86.4	95.4	71.7	81.4	95.4	94.1	94.6	98.6	107.8	104.0	91.2	81.9	85.7	93.9	75.9	83.2
09	FINISHES	84.0	83.5	83.7	100.0	85.1	92.0	97.7	85.3	91.0	94.8	100.2	97.7	91.5	77.1	83.7	103.8	74.4	87.9
COVERS	DIVS. 10 - 14, 25, 28, 41, 43, 44, 46	100.0	93.6	98.6	100.0	94.0	98.7	100.0	78.8	95.3	100.0	82.7	96.2	100.0	93.6	98.6	100.0	79.5	95.4
21, 22, 23	FIRE SUPPRESSION, PLUMBING & HVAC	100.0	96.9	98.8	96.5	96.8	96.6	96.5	97.7	97.0	96.6	100.8	98.2	100.0	96.9	98.8	100.1	71.1	88.5
26, 27, 3370	ELECTRICAL, COMMUNICATIONS & UTIL.	97.8	81.4	89.7	103.8	98.5	101.2	98.0	76.8	87.5	101.5	101.2	101.4	103.2	81.4	92.4	92.9	66.1	79.7
MF2018	WEIGHTED AVERAGE	96.1	90.7	93.8	97.3	94.2	96.0	95.3	91.6	93.7	93.2	100.8	96.4	96.4	89.9	93.7	95.1	77.7	87.7

MISSOURI

DIVISION		KANSAS CITY 640 - 641 MAT.	INST.	TOTAL	KIRKSVILLE 635 MAT.	INST.	TOTAL	POPLAR BLUFF 639 MAT.	INST.	TOTAL	ROLLA 654 - 655 MAT.	INST.	TOTAL	SEDALIA 653 MAT.	INST.	TOTAL	SIKESTON 638 MAT.	INST.	TOTAL
015433	CONTRACTOR EQUIPMENT		104.4	104.4		97.4	97.4		99.6	99.6		110.3	110.3		100.4	100.4		99.6	99.6
0241, 31 - 34	SITE & INFRASTRUCTURE, DEMOLITION	94.6	98.6	97.3	88.8	84.9	86.1	76.1	88.5	84.7	92.9	94.1	93.8	92.1	89.3	90.2	79.6	89.1	86.2
0310	Concrete Forming & Accessories	97.2	103.5	102.6	86.9	78.4	79.6	87.4	79.3	80.5	91.6	94.0	93.6	89.6	78.2	79.9	88.2	79.4	80.7
0320	Concrete Reinforcing	98.5	111.1	104.6	93.8	83.3	88.7	96.7	76.0	86.7	86.0	93.4	89.6	84.8	110.3	97.1	96.0	76.0	86.4
0330	Cast-in-Place Concrete	97.3	102.6	99.3	93.6	82.2	89.4	72.0	83.5	76.3	90.6	94.7	92.1	94.7	81.3	89.7	77.0	83.5	79.4
03	CONCRETE	93.8	104.9	98.7	106.9	81.8	95.9	81.5	81.4	81.5	82.5	95.6	88.2	95.7	86.1	91.5	85.2	81.5	83.6
04	MASONRY	99.6	103.3	101.9	113.2	84.8	95.8	104.5	74.1	85.8	110.8	85.3	95.2	117.1	81.6	95.3	104.2	74.1	85.7
05	METALS	94.9	112.8	100.2	91.5	101.0	94.3	92.1	97.8	93.7	95.7	115.9	101.6	94.5	112.8	99.9	92.4	97.9	94.0
06	WOOD, PLASTICS & COMPOSITES	100.1	103.8	102.2	82.2	76.5	79.1	81.3	80.0	80.6	88.0	95.6	92.2	81.6	76.7	78.9	82.9	80.0	81.3
07	THERMAL & MOISTURE PROTECTION	91.8	105.2	97.7	102.2	90.8	97.3	100.5	83.0	92.9	90.9	92.9	91.8	96.1	87.8	92.5	100.7	82.1	92.6
08	OPENINGS	96.5	106.0	98.7	104.1	78.8	98.2	105.0	75.2	98.1	94.4	91.1	93.6	99.4	87.0	96.5	105.0	75.2	98.1
0920	Plaster & Gypsum Board	100.9	104.0	103.0	96.3	76.2	82.8	96.6	79.8	85.3	86.9	95.6	92.8	80.5	76.2	77.6	98.5	79.8	85.9
0950, 0980	Ceilings & Acoustic Treatment	90.5	104.0	99.6	87.3	76.2	79.8	88.9	79.8	82.8	88.5	95.6	93.3	88.5	76.2	80.2	88.9	79.8	82.8
0960	Flooring	95.9	101.8	97.6	73.8	96.4	80.2	87.3	84.5	86.5	93.8	96.4	94.5	72.5	72.9	72.6	87.8	84.5	86.9
0970, 0990	Wall Finishes & Painting/Coating	97.2	107.8	103.5	91.3	78.3	83.6	90.6	67.1	76.7	93.1	90.2	91.4	93.1	103.2	99.1	90.6	67.1	76.7
09	FINISHES	98.1	103.6	101.1	97.4	81.1	88.6	96.9	78.4	86.9	85.4	94.0	90.1	82.9	79.1	80.8	97.5	78.7	87.3
COVERS	DIVS. 10 - 14, 25, 28, 41, 43, 44, 46	100.0	98.4	99.7	100.0	78.3	95.2	100.0	92.2	98.3	100.0	96.7	99.3	100.0	87.9	97.3	100.0	92.2	98.3
21, 22, 23	FIRE SUPPRESSION, PLUMBING & HVAC	100.0	103.4	101.4	96.6	97.1	96.8	96.6	94.9	95.9	96.5	98.6	97.3	96.4	94.9	95.8	96.6	95.0	95.9
26, 27, 3370	ELECTRICAL, COMMUNICATIONS & UTIL.	103.3	101.2	102.2	98.2	76.8	87.6	98.4	98.4	98.4	96.4	79.4	88.0	97.2	101.2	99.2	97.5	98.4	97.9
MF2018	WEIGHTED AVERAGE	97.9	103.8	100.4	98.9	86.5	93.7	95.1	87.7	92.0	94.1	94.3	94.2	96.4	91.5	94.3	95.7	87.8	92.3

MISSOURI / MONTANA

DIVISION		SPRINGFIELD 656 - 658 MAT.	INST.	TOTAL	ST. JOSEPH 644 - 645 MAT.	INST.	TOTAL	ST. LOUIS 630 - 631 MAT.	INST.	TOTAL	BILLINGS 590 - 591 MAT.	INST.	TOTAL	BUTTE 597 MAT.	INST.	TOTAL	GREAT FALLS 594 MAT.	INST.	TOTAL
015433	CONTRACTOR EQUIPMENT		102.8	102.8		101.5	101.5		108.6	108.6		97.9	97.9		97.7	97.7		97.7	97.7
0241, 31 - 34	SITE & INFRASTRUCTURE, DEMOLITION	94.4	91.4	92.4	96.7	87.4	90.3	94.4	97.8	96.7	91.8	93.5	93.0	97.3	93.3	94.5	101.0	93.3	95.7
0310	Concrete Forming & Accessories	97.7	75.4	78.7	96.3	89.1	90.2	99.5	102.5	102.1	99.3	68.3	72.9	86.5	68.2	70.9	99.2	68.2	72.8
0320	Concrete Reinforcing	82.5	92.9	87.5	97.3	110.4	103.6	86.5	105.2	95.5	98.0	81.8	90.2	106.3	81.7	94.4	98.0	81.8	90.2
0330	Cast-in-Place Concrete	96.3	73.7	87.9	92.2	95.5	93.4	101.3	102.5	101.7	118.7	72.7	101.6	130.6	72.5	109.0	138.1	72.5	113.7
03	CONCRETE	92.1	79.0	86.3	90.3	96.0	92.8	97.1	104.0	100.1	98.3	73.1	87.2	101.9	73.0	89.2	106.9	73.0	92.0
04	MASONRY	89.1	79.9	83.5	95.3	88.7	91.2	91.1	107.1	100.9	125.8	83.8	100.0	120.9	82.7	97.4	125.4	83.8	99.8
05	METALS	100.7	101.7	101.0	91.5	113.8	98.0	97.4	119.0	103.8	107.0	89.6	101.9	101.2	89.4	97.7	104.4	89.5	100.0
06	WOOD, PLASTICS & COMPOSITES	89.4	75.0	81.5	106.1	88.7	96.5	97.6	101.0	99.5	90.1	64.6	76.1	76.8	64.6	70.1	91.4	64.6	76.6
07	THERMAL & MOISTURE PROTECTION	94.6	75.5	86.3	92.2	88.7	90.7	94.3	105.5	99.2	108.3	73.7	93.2	107.8	73.3	92.8	108.5	73.7	93.4
08	OPENINGS	101.7	84.3	97.6	89.9	96.7	91.5	100.3	105.8	101.6	98.6	65.9	91.0	96.9	65.9	89.7	99.8	65.9	91.9
0920	Plaster & Gypsum Board	87.8	74.4	78.8	106.9	88.3	94.4	106.5	101.5	103.1	118.3	64.0	81.8	118.8	64.0	81.9	128.7	64.0	85.2
0950, 0980	Ceilings & Acoustic Treatment	88.5	74.4	79.0	94.8	88.3	90.4	90.5	101.5	97.9	96.6	64.0	74.6	104.1	64.0	77.0	105.7	64.0	77.6
0960	Flooring	92.8	71.5	86.8	99.4	99.0	99.3	98.0	99.1	98.3	88.1	80.8	86.0	85.4	90.9	87.0	91.9	80.8	88.8
0970, 0990	Wall Finishes & Painting/Coating	87.6	77.6	81.6	94.3	107.8	102.3	96.4	106.3	102.2	93.8	91.0	92.2	92.3	67.8	77.8	92.3	91.0	91.5
09	FINISHES	87.6	77.6	82.2	100.3	92.8	96.3	100.5	101.9	101.3	92.5	72.3	81.6	93.0	71.8	81.5	96.8	72.3	83.5
COVERS	DIVS. 10 - 14, 25, 28, 41, 43, 44, 46	100.0	90.5	97.9	100.0	94.0	98.7	100.0	102.8	100.6	100.0	91.6	98.1	100.0	91.6	98.1	100.0	91.6	98.1
21, 22, 23	FIRE SUPPRESSION, PLUMBING & HVAC	100.0	69.4	88.0	100.1	87.5	95.2	100.0	105.5	102.2	100.0	76.4	90.7	100.0	71.2	88.7	100.0	71.2	88.7
26, 27, 3370	ELECTRICAL, COMMUNICATIONS & UTIL.	101.3	68.8	85.3	101.5	75.8	88.9	101.8	98.5	100.2	102.3	75.2	88.9	109.2	71.3	90.5	101.7	70.8	86.5
MF2018	WEIGHTED AVERAGE	97.6	79.0	89.7	96.1	90.9	93.9	98.8	104.5	101.2	101.6	78.4	91.8	101.5	76.5	90.9	102.9	76.6	91.8

City Cost Indexes

MONTANA

DIVISION		HAVRE 595 MAT.	INST.	TOTAL	HELENA 596 MAT.	INST.	TOTAL	KALISPELL 599 MAT.	INST.	TOTAL	MILES CITY 593 MAT.	INST.	TOTAL	MISSOULA 598 MAT.	INST.	TOTAL	WOLF POINT 592 MAT.	INST.	TOTAL
015433	CONTRACTOR EQUIPMENT		97.7	97.7		99.9	99.9		97.7	97.7		97.7	97.7		97.7	97.7		97.7	97.7
0241, 31 - 34	SITE & INFRASTRUCTURE, DEMOLITION	104.2	93.1	96.6	91.1	96.8	95.0	87.5	93.3	91.5	93.5	93.2	93.3	80.5	93.3	89.3	110.2	93.2	98.4
0310	Concrete Forming & Accessories	79.6	67.2	69.0	100.2	68.3	73.0	89.7	68.3	71.4	97.7	67.3	71.8	89.7	68.4	71.5	90.4	67.3	70.7
0320	Concrete Reinforcing	107.1	78.9	93.5	113.0	81.8	97.9	109.1	85.8	97.8	106.7	78.9	93.3	108.1	85.8	97.3	108.2	78.2	93.7
0330	Cast-in-Place Concrete	140.7	71.2	114.9	102.1	73.3	91.4	113.4	72.5	98.2	124.2	71.3	104.5	96.2	72.6	87.4	139.1	70.2	113.5
03	CONCRETE	109.8	71.6	93.0	94.7	73.3	85.3	91.7	73.7	83.8	98.8	71.7	86.9	80.5	73.8	77.6	113.2	71.2	94.8
04	MASONRY	121.9	80.6	96.5	114.2	82.7	94.8	119.6	82.7	96.9	127.6	80.6	98.7	144.5	82.7	106.5	128.9	80.6	99.2
05	METALS	97.3	88.2	94.6	103.2	88.5	98.9	97.2	90.8	95.3	96.5	88.4	94.1	97.7	91.0	95.7	96.6	88.1	94.1
06	WOOD, PLASTICS & COMPOSITES	68.1	64.6	66.2	93.8	64.7	77.8	80.2	64.6	71.6	88.4	64.6	75.3	80.2	64.6	71.6	79.5	64.6	71.3
07	THERMAL & MOISTURE PROTECTION	108.3	66.2	90.0	103.1	73.9	90.4	107.4	73.6	92.7	107.8	67.7	90.3	106.9	75.6	93.3	108.9	67.5	90.9
08	OPENINGS	97.4	65.2	89.9	96.3	65.9	89.2	97.4	66.8	90.3	96.9	65.2	89.5	96.9	66.8	89.9	96.9	65.1	89.5
0920	Plaster & Gypsum Board	114.1	64.0	80.4	113.5	64.0	80.2	118.8	64.0	81.9	127.8	64.0	84.9	118.8	64.0	81.9	121.8	64.0	82.9
0950, 0980	Ceilings & Acoustic Treatment	104.1	64.0	77.0	107.9	64.0	78.3	104.1	64.0	77.0	101.6	64.0	76.3	104.1	64.0	77.0	101.6	64.0	76.3
0960	Flooring	83.0	90.9	85.3	95.9	90.9	94.5	87.1	90.9	88.2	91.8	90.9	91.6	87.1	90.9	88.2	88.4	90.9	89.1
0970, 0990	Wall Finishes & Painting/Coating	92.3	67.8	77.8	97.6	67.8	79.9	92.3	91.0	91.5	92.3	67.8	77.8	92.3	91.0	91.5	92.3	67.8	77.8
09	FINISHES	92.3	71.3	80.9	100.1	71.8	84.8	92.9	74.3	82.8	95.4	71.3	82.3	92.3	74.3	82.6	95.0	71.3	82.1
COVERS	DIVS. 10 - 14, 25, 28, 41, 43, 44, 46	100.0	88.8	97.5	100.0	91.9	98.2	100.0	89.5	97.7	100.0	88.9	97.5	100.0	92.4	98.3	100.0	88.9	97.5
21, 22, 23	FIRE SUPPRESSION, PLUMBING & HVAC	96.5	68.9	85.6	100.0	71.2	88.7	96.5	69.4	85.8	96.5	75.2	88.1	100.0	71.2	88.7	96.5	75.2	88.1
26, 27, 3370	ELECTRICAL, COMMUNICATIONS & UTIL.	101.7	69.2	85.7	109.0	70.8	90.2	106.0	68.1	87.3	101.7	74.4	88.3	107.0	70.2	88.9	101.7	74.4	88.3
MF2018	WEIGHTED AVERAGE	100.4	74.8	89.6	100.9	76.7	90.7	98.1	76.2	88.9	99.2	77.0	89.8	98.6	77.1	89.5	101.4	76.9	91.1

NEBRASKA

DIVISION		ALLIANCE 693 MAT.	INST.	TOTAL	COLUMBUS 686 MAT.	INST.	TOTAL	GRAND ISLAND 688 MAT.	INST.	TOTAL	HASTINGS 689 MAT.	INST.	TOTAL	LINCOLN 683 - 685 MAT.	INST.	TOTAL	MCCOOK 690 MAT.	INST.	TOTAL
015433	CONTRACTOR EQUIPMENT		95.3	95.3		101.9	101.9		101.9	101.9		101.9	101.9		105.0	105.0		101.9	101.9
0241, 31 - 34	SITE & INFRASTRUCTURE, DEMOLITION	98.2	96.2	96.8	101.8	90.6	94.1	106.8	90.7	95.7	105.4	90.6	95.2	94.9	95.7	95.4	101.5	90.6	94.0
0310	Concrete Forming & Accessories	87.0	55.3	60.0	96.5	75.0	78.2	96.1	68.8	72.9	99.3	72.5	76.5	94.4	75.5	78.3	92.3	55.9	61.2
0320	Concrete Reinforcing	107.4	87.4	97.7	100.4	85.9	93.4	99.8	76.1	88.4	99.8	76.1	88.4	97.6	76.4	87.3	100.1	76.3	88.6
0330	Cast-in-Place Concrete	108.1	81.8	98.3	107.3	81.8	97.8	113.6	77.7	100.2	113.6	74.0	98.9	87.2	83.1	85.6	116.9	73.9	100.9
03	CONCRETE	114.3	70.9	95.2	100.2	80.4	91.5	104.9	74.5	91.5	105.1	74.9	91.8	87.5	79.4	83.9	103.7	67.4	87.7
04	MASONRY	106.0	77.4	88.4	111.1	77.4	90.4	104.3	73.8	85.6	112.8	74.0	88.9	93.6	78.2	84.1	101.9	77.4	86.8
05	METALS	99.1	85.2	95.0	91.4	98.0	93.3	93.2	93.7	93.3	94.0	93.7	93.9	95.0	93.3	94.5	93.9	93.8	93.9
06	WOOD, PLASTICS & COMPOSITES	82.6	48.7	64.0	94.3	74.9	83.6	93.4	66.3	78.5	97.2	72.1	83.4	99.0	75.0	85.8	90.7	49.7	68.1
07	THERMAL & MOISTURE PROTECTION	100.7	75.8	89.9	101.6	80.4	92.4	101.8	78.3	91.5	101.8	77.6	91.3	100.1	81.1	91.8	96.0	76.1	87.3
08	OPENINGS	90.8	58.1	83.2	91.3	73.7	87.2	91.3	66.8	85.6	91.3	69.8	86.3	105.4	68.8	96.9	91.2	55.8	83.0
0920	Plaster & Gypsum Board	79.8	47.5	58.1	93.4	74.3	80.6	92.5	65.5	74.3	94.4	71.4	78.9	102.7	74.3	83.6	91.0	48.4	62.4
0950, 0980	Ceilings & Acoustic Treatment	88.6	47.5	60.9	83.4	74.3	77.3	83.4	65.5	71.3	83.4	71.4	75.3	98.3	74.3	82.1	83.9	48.4	60.0
0960	Flooring	91.4	87.3	90.2	84.0	87.3	84.9	83.7	78.9	82.4	85.0	81.3	84.0	96.3	86.6	93.6	90.0	87.3	89.2
0970, 0990	Wall Finishes & Painting/Coating	155.8	51.7	94.2	74.9	60.1	66.1	74.9	62.5	67.6	74.9	60.1	66.1	94.4	79.3	85.5	87.6	45.0	62.4
09	FINISHES	90.8	59.2	73.7	84.9	75.5	79.8	84.9	69.3	76.5	85.6	72.7	78.6	95.6	77.9	86.0	88.4	59.1	72.5
COVERS	DIVS. 10 - 14, 25, 28, 41, 43, 44, 46	100.0	83.9	96.4	100.0	83.5	96.3	100.0	89.7	97.7	100.0	83.1	96.2	100.0	91.0	98.0	100.0	84.1	96.5
21, 22, 23	FIRE SUPPRESSION, PLUMBING & HVAC	96.7	75.0	88.2	96.6	75.3	88.2	100.1	79.5	92.0	96.6	74.6	87.9	100.0	79.6	92.0	96.5	75.2	88.1
26, 27, 3370	ELECTRICAL, COMMUNICATIONS & UTIL.	93.6	65.6	79.8	97.3	81.4	89.5	95.9	65.7	81.0	95.3	79.4	87.5	109.2	65.7	87.8	95.6	65.7	80.9
MF2018	WEIGHTED AVERAGE	98.7	73.3	87.9	96.0	80.7	89.6	97.4	76.8	88.7	97.1	78.2	89.1	98.3	79.7	90.5	96.3	73.1	86.5

NEBRASKA / NEVADA

DIVISION		NORFOLK 687 MAT.	INST.	TOTAL	NORTH PLATTE 691 MAT.	INST.	TOTAL	OMAHA 680 - 681 MAT.	INST.	TOTAL	VALENTINE 692 MAT.	INST.	TOTAL	CARSON CITY 897 MAT.	INST.	TOTAL	ELKO 898 MAT.	INST.	TOTAL
015433	CONTRACTOR EQUIPMENT		91.7	91.7		101.9	101.9		94.2	94.2		95.0	95.0		97.9	97.9		94.6	94.6
0241, 31 - 34	SITE & INFRASTRUCTURE, DEMOLITION	84.1	89.7	88.0	102.9	90.4	94.3	89.1	93.9	92.4	86.3	95.2	92.4	83.3	97.4	93.0	67.8	91.3	84.0
0310	Concrete Forming & Accessories	83.2	74.1	75.5	94.8	75.2	78.1	93.1	76.7	79.1	83.5	53.1	57.6	106.1	76.3	80.7	111.4	96.3	98.6
0320	Concrete Reinforcing	100.5	66.5	84.1	99.5	76.2	88.3	100.8	76.5	89.0	100.1	66.3	83.7	115.8	116.1	116.0	124.5	114.5	119.7
0330	Cast-in-Place Concrete	107.9	71.3	94.3	116.9	60.6	96.0	91.4	78.6	86.7	103.1	54.0	84.8	98.0	82.2	92.1	95.0	72.4	86.6
03	CONCRETE	98.8	72.4	87.2	103.8	71.4	89.6	89.8	77.8	84.5	101.2	56.9	81.7	100.1	85.3	94.1	98.5	90.8	95.1
04	MASONRY	117.1	77.3	92.7	90.4	76.9	82.1	93.0	80.0	85.0	101.4	76.8	86.3	118.2	68.8	87.8	125.0	67.7	89.8
05	METALS	94.8	80.4	90.6	93.1	93.2	93.2	95.0	83.9	91.7	105.1	79.4	97.5	111.4	94.1	106.3	115.0	93.6	108.7
06	WOOD, PLASTICS & COMPOSITES	78.4	74.4	76.2	92.7	76.9	84.0	93.4	76.6	84.2	77.5	47.4	61.0	93.6	74.4	83.0	104.7	102.3	103.4
07	THERMAL & MOISTURE PROTECTION	101.2	78.3	91.2	95.9	78.4	88.3	96.7	80.8	89.8	96.5	73.9	86.7	118.1	79.1	101.2	114.2	73.6	96.5
08	OPENINGS	92.7	68.7	87.1	90.6	72.1	86.3	99.8	76.8	94.4	92.7	53.4	83.6	101.1	77.6	95.6	102.5	92.6	100.2
0920	Plaster & Gypsum Board	93.2	74.3	80.5	91.0	76.4	81.2	102.9	76.4	85.1	92.5	46.6	61.6	100.8	73.7	82.6	106.9	102.5	103.9
0950, 0980	Ceilings & Acoustic Treatment	97.6	74.3	81.9	83.9	76.4	78.8	101.0	76.4	84.4	100.4	46.6	64.1	96.0	73.7	80.9	94.5	102.5	99.9
0960	Flooring	104.1	87.3	99.4	91.1	78.9	87.7	100.6	87.3	96.9	116.8	87.3	108.5	99.5	67.1	90.4	103.0	67.1	92.9
0970, 0990	Wall Finishes & Painting/Coating	132.5	60.1	89.7	87.6	58.9	70.6	104.2	61.0	78.6	155.3	61.0	99.5	95.8	79.3	86.1	94.2	79.3	85.4
09	FINISHES	101.8	75.2	87.4	88.7	74.8	81.2	100.4	76.9	87.7	109.1	59.3	82.1	96.5	74.0	84.3	94.0	90.4	92.0
COVERS	DIVS. 10 - 14, 25, 28, 41, 43, 44, 46	100.0	85.1	96.7	100.0	61.1	91.3	100.0	90.2	97.8	100.0	56.1	90.2	100.0	100.2	100.0	100.0	91.0	98.0
21, 22, 23	FIRE SUPPRESSION, PLUMBING & HVAC	96.3	74.3	87.7	100.0	73.7	89.6	100.0	80.3	92.3	96.1	73.5	87.2	100.0	77.0	91.0	98.3	76.9	89.9
26, 27, 3370	ELECTRICAL, COMMUNICATIONS & UTIL.	96.2	81.4	88.9	93.9	65.7	80.0	103.9	83.7	94.0	91.4	65.7	78.7	102.2	88.3	95.3	98.8	88.3	93.6
MF2018	WEIGHTED AVERAGE	97.4	77.5	89.0	96.3	75.6	87.6	97.6	81.5	90.8	98.6	69.2	86.2	102.9	82.5	94.3	102.3	85.2	95.0

For customer support on your Site Work & Landscape Costs with RSMeans data, call 800.448.8182.

799

NEVADA / NEW HAMPSHIRE

	DIVISION	ELY 893			LAS VEGAS 889 - 891			RENO 894 - 895			CHARLESTON 036			CLAREMONT 037			CONCORD 032 - 033		
		MAT.	INST.	TOTAL	MAT.	INST.	TOTAL	MAT.	INST.	TOTAL	MAT.	INST.	TOTAL	MAT.	INST.	TOTAL	MAT.	INST.	TOTAL
015433	CONTRACTOR EQUIPMENT		94.6	94.6		94.6	94.6		94.6	94.6		95.7	95.7		95.7	95.7		98.4	98.4
0241, 31 - 34	SITE & INFRASTRUCTURE, DEMOLITION	73.4	92.5	86.6	76.6	95.5	89.7	73.4	92.6	86.7	80.6	96.1	91.3	74.6	96.1	89.5	88.7	101.2	97.3
0310	Concrete Forming & Accessories	104.3	101.6	102.0	105.2	105.6	105.5	100.6	76.3	79.9	86.6	83.1	83.6	92.1	83.2	84.5	97.2	94.8	95.1
0320	Concrete Reinforcing	123.1	114.7	119.0	113.5	125.9	119.5	116.4	124.2	120.1	89.7	88.6	89.1	89.7	88.6	89.1	100.9	88.7	95.0
0330	Cast-in-Place Concrete	102.0	95.2	99.4	98.7	108.2	102.2	107.9	81.1	97.9	84.7	116.1	96.4	77.8	116.1	92.1	100.9	117.8	107.2
03	CONCRETE	107.5	101.2	104.7	102.4	109.5	105.5	106.7	86.3	97.8	90.4	95.8	92.7	82.8	95.8	88.5	96.8	101.6	98.9
04	MASONRY	130.5	74.8	96.3	117.4	106.2	110.5	124.2	68.7	90.1	89.9	99.5	95.8	89.9	99.5	95.8	97.8	101.4	100.0
05	METALS	115.0	96.5	109.5	124.1	103.9	118.2	116.7	97.4	111.1	94.2	92.1	93.5	94.2	92.1	93.6	101.3	91.8	98.5
06	WOOD, PLASTICS & COMPOSITES	94.8	104.4	100.1	92.4	102.3	97.8	88.4	74.3	80.6	86.6	79.4	82.6	92.4	79.4	85.3	93.3	93.6	93.5
07	THERMAL & MOISTURE PROTECTION	114.7	92.2	104.9	129.4	101.9	117.4	114.2	78.3	98.6	107.3	106.5	107.0	107.1	106.5	106.8	111.3	109.5	110.5
08	OPENINGS	102.4	93.8	100.4	101.4	109.6	103.3	100.2	79.2	95.3	96.1	80.6	92.5	97.2	80.6	93.3	97.0	88.5	95.1
0920	Plaster & Gypsum Board	101.2	104.6	103.5	93.9	102.5	99.7	89.0	73.7	78.7	106.1	78.5	87.5	107.0	78.5	87.9	108.7	93.1	98.2
0950, 0980	Ceilings & Acoustic Treatment	94.5	104.6	101.3	101.0	102.5	102.0	97.8	73.7	81.5	101.8	78.5	86.1	101.8	78.5	86.1	106.1	93.1	97.4
0960	Flooring	100.5	67.1	91.1	92.0	104.1	95.4	97.3	67.1	88.8	85.1	113.1	92.9	87.6	113.1	94.7	96.4	113.1	101.1
0970, 0990	Wall Finishes & Painting/Coating	94.2	112.0	104.7	96.8	119.4	110.2	94.2	79.3	85.4	81.8	90.0	86.6	81.8	90.0	86.6	95.9	90.0	92.4
09	FINISHES	92.9	97.0	95.1	90.8	106.4	99.3	90.8	73.9	81.7	90.2	88.5	89.3	90.6	88.5	89.5	95.9	97.4	96.7
COVERS	DIVS. 10 - 14, 25, 28, 41, 43, 44, 46	100.0	70.7	93.5	100.0	105.1	101.1	100.0	99.4	100.0	100.0	88.5	97.4	100.0	88.5	97.4	100.0	109.5	102.1
21, 22, 23	FIRE SUPPRESSION, PLUMBING & HVAC	98.3	96.2	97.4	100.1	103.0	101.3	100.0	77.1	91.0	96.7	82.2	91.0	96.7	82.2	91.0	100.0	87.2	95.0
26, 27, 3370	ELECTRICAL, COMMUNICATIONS & UTIL.	99.0	93.0	96.0	103.5	107.8	105.6	99.4	88.3	93.9	91.4	75.2	83.4	91.4	75.2	83.4	90.2	75.2	82.8
MF2018	WEIGHTED AVERAGE	103.7	93.2	99.2	105.0	105.1	105.0	103.6	82.6	94.7	94.1	88.4	91.7	93.1	88.4	91.1	98.0	93.2	96.0

NEW HAMPSHIRE / NEW JERSEY

	DIVISION	KEENE 034			LITTLETON 035			MANCHESTER 031			NASHUA 030			PORTSMOUTH 038			ATLANTIC CITY 082, 084		
		MAT.	INST.	TOTAL	MAT.	INST.	TOTAL	MAT.	INST.	TOTAL	MAT.	INST.	TOTAL	MAT.	INST.	TOTAL	MAT.	INST.	TOTAL
015433	CONTRACTOR EQUIPMENT		95.7	95.7		95.7	95.7		99.0	99.0		95.7	95.7		95.7	95.7		93.3	93.3
0241, 31 - 34	SITE & INFRASTRUCTURE, DEMOLITION	88.0	96.2	93.6	74.7	95.2	88.8	87.8	101.3	97.1	89.5	96.3	94.2	83.3	97.0	92.7	88.2	99.7	96.1
0310	Concrete Forming & Accessories	90.7	83.4	84.5	102.4	77.3	81.0	97.5	95.1	95.5	98.5	94.9	95.4	87.9	94.1	93.2	114.8	142.0	138.0
0320	Concrete Reinforcing	89.7	88.6	89.2	90.4	88.6	89.0	108.5	88.7	98.9	112.0	88.7	100.7	89.7	88.7	89.2	81.1	138.0	108.6
0330	Cast-in-Place Concrete	85.1	116.2	96.7	76.3	107.5	87.9	99.2	118.3	106.3	80.5	117.4	94.3	76.3	117.2	91.5	78.5	137.1	100.3
03	CONCRETE	89.9	95.9	92.6	82.2	90.1	85.7	97.0	101.9	99.2	90.0	101.6	95.1	82.4	101.2	90.7	87.7	138.1	109.8
04	MASONRY	92.5	99.5	96.8	100.9	84.1	90.6	94.4	101.4	98.7	94.4	101.4	98.7	90.2	99.8	96.1	104.1	140.0	126.1
05	METALS	94.8	92.5	94.2	94.9	92.2	94.1	102.0	92.2	99.1	100.5	92.7	98.2	96.4	94.3	95.8	98.8	115.0	103.6
06	WOOD, PLASTICS & COMPOSITES	90.9	79.4	84.6	101.9	79.4	89.5	93.6	93.7	93.7	100.2	93.6	96.6	87.8	93.6	91.0	122.8	143.7	134.3
07	THERMAL & MOISTURE PROTECTION	107.9	106.4	107.2	107.2	99.5	103.8	112.5	109.5	111.2	108.3	108.9	108.6	107.8	108.1	107.9	102.1	134.0	116.0
08	OPENINGS	94.8	84.1	92.3	98.1	80.6	94.0	95.7	93.3	95.2	99.0	91.3	97.2	99.6	81.9	95.5	96.4	139.2	106.4
0920	Plaster & Gypsum Board	106.4	78.5	87.6	120.6	78.5	92.3	99.7	93.1	95.3	115.6	93.1	100.5	106.1	93.1	97.4	114.9	144.6	134.9
0950, 0980	Ceilings & Acoustic Treatment	101.8	78.5	86.1	101.8	78.5	86.1	101.2	93.1	95.8	114.1	93.1	96.3	102.7	93.1	96.3	96.4	144.6	128.9
0960	Flooring	87.2	113.1	94.5	97.9	113.1	102.2	93.8	113.1	99.2	91.0	113.1	97.2	85.2	113.1	93.0	97.5	162.5	115.8
0970, 0990	Wall Finishes & Painting/Coating	81.8	103.4	94.6	81.8	90.0	86.6	95.3	119.0	109.3	81.8	102.2	93.9	81.8	102.2	93.9	81.9	141.8	117.4
09	FINISHES	92.1	89.9	90.9	95.5	84.6	89.6	94.8	100.6	97.9	96.8	98.7	97.8	91.1	98.2	95.2	93.5	147.2	122.6
COVERS	DIVS. 10 - 14, 25, 28, 41, 43, 44, 46	100.0	94.7	98.8	100.0	95.6	99.0	100.0	109.7	102.2	100.0	109.4	102.1	100.0	108.7	101.9	100.0	114.9	103.3
21, 22, 23	FIRE SUPPRESSION, PLUMBING & HVAC	96.7	82.3	91.0	96.7	74.1	87.8	100.1	87.3	95.1	100.2	87.3	95.1	100.2	86.3	94.7	99.7	133.1	112.8
26, 27, 3370	ELECTRICAL, COMMUNICATIONS & UTIL.	91.4	75.2	83.4	92.2	49.1	71.0	93.9	79.3	86.7	93.2	79.3	86.4	91.7	75.3	83.6	94.2	143.4	118.5
MF2018	WEIGHTED AVERAGE	94.5	89.0	92.2	94.4	80.2	88.4	98.2	94.5	96.6	97.5	93.7	95.9	94.9	92.4	93.8	96.8	133.2	112.2

NEW JERSEY

	DIVISION	CAMDEN 081			DOVER 078			ELIZABETH 072			HACKENSACK 076			JERSEY CITY 073			LONG BRANCH 077		
		MAT.	INST.	TOTAL	MAT.	INST.	TOTAL	MAT.	INST.	TOTAL	MAT.	INST.	TOTAL	MAT.	INST.	TOTAL	MAT.	INST.	TOTAL
015433	CONTRACTOR EQUIPMENT		93.3	93.3		95.7	95.7		95.7	95.7		95.7	95.7		93.3	93.3		92.9	92.9
0241, 31 - 34	SITE & INFRASTRUCTURE, DEMOLITION	89.3	99.3	96.2	103.8	101.4	102.2	108.9	101.4	103.7	104.8	101.4	102.5	95.2	101.4	99.5	100.0	100.5	100.4
0310	Concrete Forming & Accessories	105.0	137.3	132.5	96.4	148.3	140.6	108.0	148.5	142.5	96.4	148.3	140.7	100.6	148.4	141.3	101.1	139.0	133.4
0320	Concrete Reinforcing	106.5	128.5	117.1	75.3	156.5	114.6	75.3	156.5	114.6	75.3	156.5	114.6	97.7	156.5	126.2	75.3	156.1	114.4
0330	Cast-in-Place Concrete	76.1	133.8	97.6	81.3	131.5	100.0	69.8	140.3	96.1	79.4	140.3	102.1	63.4	131.6	88.8	70.5	133.8	94.0
03	CONCRETE	88.2	133.2	108.0	86.1	142.3	110.8	83.3	145.5	110.6	84.5	145.4	111.2	81.0	142.2	107.9	85.1	138.6	108.6
04	MASONRY	94.2	135.0	119.3	91.2	145.6	124.7	107.2	145.6	130.8	95.3	145.6	122.4	85.3	145.6	122.4	99.6	135.5	121.6
05	METALS	104.7	111.0	106.5	96.5	125.1	104.9	98.0	125.3	106.0	96.5	125.1	104.9	102.4	122.2	108.2	96.6	121.0	103.7
06	WOOD, PLASTICS & COMPOSITES	110.1	137.9	125.4	94.9	148.3	124.3	109.7	148.3	131.0	94.9	148.3	124.3	95.8	148.3	124.7	96.9	139.3	120.2
07	THERMAL & MOISTURE PROTECTION	102.0	133.0	115.5	103.7	142.8	120.7	104.0	144.1	121.4	103.5	136.7	117.9	103.2	142.8	120.4	103.4	129.0	114.5
08	OPENINGS	98.4	132.0	106.3	102.0	145.1	112.1	100.4	145.1	110.8	99.8	145.1	110.4	98.5	145.1	109.4	94.6	138.5	104.8
0920	Plaster & Gypsum Board	110.9	138.6	129.6	111.5	149.4	137.0	119.0	149.4	139.5	111.5	149.4	137.0	114.6	149.4	138.0	113.7	140.2	131.5
0950, 0980	Ceilings & Acoustic Treatment	106.3	138.6	128.1	92.3	149.4	130.8	94.1	149.4	131.4	92.3	149.4	130.8	102.2	149.4	134.0	92.3	140.2	124.6
0960	Flooring	93.7	162.5	113.0	83.7	188.8	113.2	88.7	188.8	116.8	83.7	188.8	113.2	84.7	188.8	114.0	84.9	176.9	110.8
0970, 0990	Wall Finishes & Painting/Coating	81.9	141.8	117.4	83.8	148.6	122.1	83.8	148.6	122.1	83.8	148.6	122.1	83.9	148.6	122.2	83.9	141.8	118.2
09	FINISHES	93.7	143.7	120.8	90.0	156.2	125.8	93.4	156.2	127.4	89.9	156.2	125.8	92.4	156.6	127.1	91.0	147.2	121.4
COVERS	DIVS. 10 - 14, 25, 28, 41, 43, 44, 46	100.0	114.3	103.2	100.0	131.7	107.1	100.0	131.7	107.1	100.0	131.7	107.1	100.0	131.7	107.1	100.0	114.5	103.2
21, 22, 23	FIRE SUPPRESSION, PLUMBING & HVAC	100.0	126.3	110.3	99.7	135.7	113.8	100.0	137.1	114.6	99.7	135.7	113.8	100.0	135.7	114.0	99.7	131.1	112.1
26, 27, 3370	ELECTRICAL, COMMUNICATIONS & UTIL.	98.4	132.1	115.0	96.0	138.5	116.9	96.6	138.5	117.2	96.0	142.3	118.8	100.5	142.3	121.1	95.7	129.9	112.6
MF2018	WEIGHTED AVERAGE	98.0	127.8	110.6	96.4	137.5	113.7	97.5	138.3	114.7	96.1	138.3	114.0	96.5	137.8	114.0	95.8	130.9	110.7

For customer support on your Site Work & Landscape Costs with RSMeans data, call 800.448.8182.

NEW JERSEY

DIVISION		NEW BRUNSWICK 088 - 089			NEWARK 070 - 071			PATERSON 074 - 075			POINT PLEASANT 087			SUMMIT 079			TRENTON 085 - 086		
		MAT.	INST.	TOTAL	MAT.	INST.	TOTAL	MAT.	INST.	TOTAL	MAT.	INST.	TOTAL	MAT.	INST.	TOTAL	MAT.	INST.	TOTAL
015433	CONTRACTOR EQUIPMENT		92.9	92.9		98.3	98.3		95.7	95.7		92.9	92.9		95.7	95.7		97.0	97.0
0241, 31 - 34	SITE & INFRASTRUCTURE, DEMOLITION	101.1	100.9	100.9	110.0	106.2	107.4	106.6	101.4	103.0	102.6	100.5	101.2	106.3	101.4	102.9	88.6	105.4	100.2
0310	Concrete Forming & Accessories	108.5	148.0	142.2	100.7	148.7	141.6	98.3	148.3	140.9	102.4	139.1	133.6	99.0	148.5	141.2	103.5	138.8	133.6
0320	Concrete Reinforcing	82.1	156.1	117.9	96.2	156.6	125.4	97.7	156.5	126.2	82.1	156.1	117.9	75.3	156.5	114.6	107.7	116.2	111.8
0330	Cast-in-Place Concrete	97.0	136.9	111.8	89.8	140.9	108.8	80.8	140.2	102.9	97.0	135.8	111.5	67.5	140.3	94.6	95.6	134.4	110.0
03	CONCRETE	104.2	143.8	121.6	93.1	145.7	116.2	88.8	145.3	113.7	103.9	139.4	119.5	80.6	145.5	109.1	97.4	132.1	112.6
04	MASONRY	102.5	140.6	125.9	94.9	145.7	126.1	91.7	145.6	124.9	91.4	135.5	118.5	93.8	145.6	125.6	95.4	135.5	120.0
05	METALS	98.9	121.4	105.5	104.2	124.7	110.3	97.5	125.1	105.6	98.9	121.0	105.4	96.5	125.3	104.9	104.3	108.8	105.6
06	WOOD, PLASTICS & COMPOSITES	115.7	148.3	133.6	97.8	148.4	125.6	97.5	148.3	125.5	107.3	139.3	124.9	98.7	148.3	126.0	101.6	139.4	122.4
07	THERMAL & MOISTURE PROTECTION	102.5	138.2	118.0	105.2	144.7	122.4	103.8	136.7	118.1	102.5	131.3	115.0	104.1	144.1	121.5	102.9	135.2	117.0
08	OPENINGS	91.6	145.1	104.1	100.5	145.1	110.9	105.0	145.1	114.3	93.4	141.0	104.5	106.1	145.1	115.2	96.9	130.0	104.6
0920	Plaster & Gypsum Board	113.0	149.4	137.5	106.2	149.4	135.3	114.6	149.4	138.0	107.9	140.2	129.6	113.7	149.4	137.7	101.6	140.2	127.5
0950, 0980	Ceilings & Acoustic Treatment	96.4	149.4	132.2	103.0	149.4	134.3	102.2	149.4	134.0	96.4	140.2	125.9	92.3	149.4	130.8	101.2	140.2	127.5
0960	Flooring	95.1	188.8	121.4	96.7	188.8	122.6	84.7	188.8	114.0	92.5	162.5	112.2	84.9	188.8	114.1	100.1	176.9	121.7
0970, 0990	Wall Finishes & Painting/Coating	81.9	148.6	121.4	94.1	148.6	126.3	83.8	148.6	122.1	81.9	141.8	117.4	83.8	148.6	122.1	94.9	141.8	122.7
09	FINISHES	93.8	156.2	127.6	97.2	156.7	129.4	92.5	156.2	127.0	92.3	144.7	120.7	91.0	156.2	126.3	97.0	147.3	124.2
COVERS	DIVS. 10 - 14, 25, 28, 41, 43, 44, 46	100.0	131.5	107.0	100.0	131.8	107.1	100.0	131.7	107.1	100.0	111.6	102.6	100.0	131.7	107.1	100.0	114.7	103.3
21, 22, 23	FIRE SUPPRESSION, PLUMBING & HVAC	99.7	133.8	113.1	100.0	139.2	115.4	100.0	135.5	114.0	99.7	131.1	112.0	99.8	137.1	114.4	100.1	130.8	112.1
26, 27, 3370	ELECTRICAL, COMMUNICATIONS & UTIL.	94.8	133.7	114.0	105.0	142.3	123.4	100.5	138.5	119.2	94.2	129.9	111.8	96.6	138.5	117.2	102.6	128.7	115.4
MF2018	WEIGHTED AVERAGE	98.7	135.6	114.3	100.3	139.7	117.0	98.0	137.7	114.8	98.1	130.8	111.9	96.4	138.3	114.1	99.7	128.9	112.0

DIVISION		NEW JERSEY VINELAND 080, 083			NEW MEXICO ALBUQUERQUE 870 - 872			CARRIZOZO 883			CLOVIS 881			FARMINGTON 874			GALLUP 873		
		MAT.	INST.	TOTAL	MAT.	INST.	TOTAL	MAT.	INST.	TOTAL	MAT.	INST.	TOTAL	MAT.	INST.	TOTAL	MAT.	INST.	TOTAL
015433	CONTRACTOR EQUIPMENT		93.3	93.3		107.5	107.5		107.5	107.5		107.5	107.5		107.5	107.5		107.5	107.5
0241, 31 - 34	SITE & INFRASTRUCTURE, DEMOLITION	92.5	99.4	97.3	91.1	97.5	95.5	109.7	97.5	101.3	96.6	97.5	97.2	97.3	97.5	97.5	107.2	97.5	100.5
0310	Concrete Forming & Accessories	99.2	137.6	131.9	101.0	66.2	71.3	98.6	66.2	71.0	98.6	66.1	70.9	101.1	66.2	71.4	101.1	66.2	71.4
0320	Concrete Reinforcing	81.1	131.2	105.3	99.9	71.5	86.1	116.0	71.5	94.5	117.3	71.4	95.1	109.2	71.5	90.9	104.5	71.5	88.5
0330	Cast-in-Place Concrete	84.6	133.9	103.0	90.5	70.1	82.9	94.8	70.1	85.6	94.7	70.1	85.5	91.3	70.1	83.4	86.0	70.1	80.1
03	CONCRETE	92.2	133.8	110.5	92.9	69.8	82.7	116.7	69.8	96.1	104.3	69.7	89.1	96.4	69.8	84.7	103.3	69.8	88.6
04	MASONRY	92.6	135.5	118.9	106.0	60.5	78.1	108.5	60.5	79.0	108.5	60.5	79.0	113.8	60.5	81.1	101.6	60.5	76.4
05	METALS	98.8	112.6	102.8	108.1	90.6	102.9	106.6	90.6	101.9	106.3	90.5	101.6	105.7	90.6	101.3	104.8	90.6	100.7
06	WOOD, PLASTICS & COMPOSITES	103.7	137.9	122.5	102.1	67.2	82.9	93.5	67.2	79.0	93.5	67.2	79.0	102.2	67.2	82.9	102.2	67.2	82.9
07	THERMAL & MOISTURE PROTECTION	101.9	132.2	115.1	101.0	72.8	88.7	106.3	72.8	91.7	105.0	72.8	91.0	101.3	72.8	88.9	102.5	72.8	89.6
08	OPENINGS	92.9	134.9	102.7	98.5	67.0	91.1	96.5	67.0	89.6	96.6	67.0	89.7	100.8	67.0	92.9	100.8	67.0	93.0
0920	Plaster & Gypsum Board	106.4	138.6	128.1	113.0	66.1	81.5	80.2	66.1	70.7	80.2	66.1	70.7	99.4	66.1	77.0	99.4	66.1	77.0
0950, 0980	Ceilings & Acoustic Treatment	96.4	138.6	124.9	99.8	66.1	77.1	102.7	66.1	78.1	102.7	66.1	78.1	98.6	66.1	76.7	98.6	66.1	76.7
0960	Flooring	91.7	162.5	111.6	88.9	66.8	82.7	97.2	66.8	88.7	97.2	66.8	88.7	90.4	66.8	83.8	90.4	66.8	83.8
0970, 0990	Wall Finishes & Painting/Coating	81.9	141.8	117.4	97.8	52.6	71.1	92.6	52.6	68.9	92.6	52.6	68.9	92.1	52.6	68.7	92.1	52.6	68.7
09	FINISHES	91.1	143.9	119.6	92.3	64.7	77.4	94.2	64.7	78.3	92.9	64.7	77.7	90.7	64.7	76.6	92.1	64.7	77.3
COVERS	DIVS. 10 - 14, 25, 28, 41, 43, 44, 46	100.0	114.5	103.2	100.0	85.1	96.7	100.0	85.1	96.7	100.0	85.1	96.7	100.0	85.1	96.7	100.0	85.1	96.7
21, 22, 23	FIRE SUPPRESSION, PLUMBING & HVAC	99.7	126.6	110.3	100.3	69.0	88.0	97.9	69.0	86.5	97.9	68.6	86.4	100.2	69.0	88.0	98.1	69.0	86.6
26, 27, 3370	ELECTRICAL, COMMUNICATIONS & UTIL.	94.2	143.3	118.5	87.9	69.5	78.8	90.1	69.5	80.0	87.9	69.5	78.8	85.9	69.5	77.8	85.3	69.5	77.5
MF2018	WEIGHTED AVERAGE	96.3	129.8	110.5	98.6	72.5	87.6	101.6	72.5	89.3	99.3	72.4	87.9	99.0	72.5	87.8	99.0	72.5	87.8

NEW MEXICO

DIVISION		LAS CRUCES 880			LAS VEGAS 877			ROSWELL 882			SANTA FE 875			SOCORRO 878			TRUTH/CONSEQUENCES 879		
		MAT.	INST.	TOTAL	MAT.	INST.	TOTAL	MAT.	INST.	TOTAL	MAT.	INST.	TOTAL	MAT.	INST.	TOTAL	MAT.	INST.	TOTAL
015433	CONTRACTOR EQUIPMENT		83.5	83.5		107.5	107.5		107.5	107.5		110.4	110.4		107.5	107.5		83.5	83.5
0241, 31 - 34	SITE & INFRASTRUCTURE, DEMOLITION	97.8	77.5	83.8	96.2	97.5	97.1	99.0	97.5	98.0	99.1	102.8	101.6	92.8	97.5	96.1	112.5	77.5	88.3
0310	Concrete Forming & Accessories	95.4	65.2	69.7	101.1	66.2	71.4	98.6	66.2	71.0	100.0	66.3	71.2	101.1	66.2	71.4	98.8	65.1	70.1
0320	Concrete Reinforcing	113.1	71.2	92.9	106.3	71.5	89.5	117.3	71.5	95.1	99.3	71.5	85.9	108.4	71.5	90.5	101.7	71.3	87.0
0330	Cast-in-Place Concrete	89.5	63.1	79.7	88.8	70.1	81.9	94.7	70.1	85.6	94.3	70.9	85.6	87.0	70.1	80.7	95.4	63.1	83.4
03	CONCRETE	82.1	66.5	75.3	94.0	69.8	83.4	105.3	69.8	89.7	92.9	70.0	82.9	92.9	69.8	82.8	86.1	66.4	77.4
04	MASONRY	104.0	60.2	77.1	101.9	60.5	76.5	119.5	60.5	83.3	95.0	60.6	73.9	101.8	60.5	76.4	98.7	60.2	75.0
05	METALS	105.0	83.0	98.5	104.5	90.6	100.4	107.5	90.6	102.6	101.6	89.6	98.1	104.8	90.6	100.7	104.4	83.0	98.1
06	WOOD, PLASTICS & COMPOSITES	82.9	66.2	73.7	102.2	67.2	82.9	93.5	67.2	79.0	99.0	67.2	81.5	102.2	67.2	82.9	93.5	66.2	78.5
07	THERMAL & MOISTURE PROTECTION	92.2	68.2	81.7	100.8	72.8	88.6	105.2	72.8	91.1	103.5	73.7	90.5	100.8	72.8	88.6	90.1	68.2	80.5
08	OPENINGS	92.0	66.4	86.1	97.3	67.0	90.3	96.4	67.0	89.6	98.8	67.0	91.4	97.2	67.0	90.1	90.6	66.4	85.0
0920	Plaster & Gypsum Board	78.5	66.1	70.2	99.4	66.1	77.0	80.2	66.1	70.7	114.1	66.1	81.8	99.4	66.1	77.0	100.6	66.1	77.4
0950, 0980	Ceilings & Acoustic Treatment	87.3	66.1	73.0	98.6	66.1	76.7	102.7	66.1	78.1	95.2	66.1	75.6	98.6	66.1	76.7	86.2	66.1	72.7
0960	Flooring	128.1	66.8	110.9	90.4	66.8	83.8	97.2	66.8	88.7	100.9	66.8	91.3	90.4	66.8	83.8	119.2	66.8	104.5
0970, 0990	Wall Finishes & Painting/Coating	81.8	52.6	64.5	92.1	52.6	68.7	92.6	52.6	68.9	99.8	52.6	71.9	92.1	52.6	68.7	84.5	52.6	65.6
09	FINISHES	103.0	64.0	81.9	90.5	64.7	76.6	93.0	64.7	77.7	98.2	64.8	80.1	90.4	64.7	76.5	102.4	64.0	81.6
COVERS	DIVS. 10 - 14, 25, 28, 41, 43, 44, 46	100.0	82.7	96.1	100.0	85.1	96.7	100.0	85.1	96.7	100.0	85.2	96.7	100.0	85.1	96.7	100.0	82.6	96.1
21, 22, 23	FIRE SUPPRESSION, PLUMBING & HVAC	100.4	68.7	88.0	98.1	69.0	86.6	100.0	69.0	87.8	100.3	69.0	88.0	98.1	69.0	86.6	98.0	68.7	86.5
26, 27, 3370	ELECTRICAL, COMMUNICATIONS & UTIL.	90.1	83.8	87.0	87.4	69.5	78.6	89.2	69.5	79.5	100.3	69.5	85.2	85.7	69.5	77.7	89.5	69.5	79.6
MF2018	WEIGHTED AVERAGE	96.9	71.4	86.1	97.2	72.5	86.8	100.8	72.5	88.9	99.0	72.9	88.0	96.8	72.5	86.6	96.6	69.4	85.1

For customer support on your Site Work & Landscape Costs with RSMeans data, call 800.448.8182.

801

		NEW MEXICO			NEW YORK													
		TUCUMCARI			ALBANY			BINGHAMTON			BRONX			BROOKLYN			BUFFALO	
	DIVISION	884			120 - 122			137 - 139			104			112			140 - 142	
		MAT.	INST.	TOTAL	MAT.	INST.	TOTAL	MAT.	INST.	TOTAL	MAT.	INST.	TOTAL	MAT.	INST.	TOTAL	MAT.	INST.	TOTAL
015433	CONTRACTOR EQUIPMENT		107.5	107.5		115.6	115.6		117.6	117.6		104.4	104.4		109.4	109.4		100.5	100.5
0241, 31 - 34	SITE & INFRASTRUCTURE, DEMOLITION	96.2	97.5	97.1	80.0	105.1	97.4	94.3	89.3	90.8	97.3	109.7	105.9	118.1	120.6	119.8	97.3	102.3	100.7
0310	Concrete Forming & Accessories	98.6	66.1	70.9	97.3	107.3	105.8	99.4	93.4	94.3	95.7	189.1	175.3	105.1	189.1	176.7	101.3	117.1	114.7
0320	Concrete Reinforcing	115.0	71.4	93.9	101.3	115.2	108.0	95.8	106.3	100.9	94.8	182.0	136.9	97.2	242.5	167.5	109.1	123.5	114.4
0330	Cast-in-Place Concrete	94.7	70.1	85.5	81.0	116.4	94.2	106.3	106.9	106.5	85.6	172.0	117.8	108.3	170.5	131.5	104.9	118.7	111.0
03	CONCRETE	103.6	69.7	88.7	87.7	112.8	98.7	94.3	102.6	97.9	88.0	181.1	128.9	106.4	189.9	143.1	104.9	118.7	111.0
04	MASONRY	119.9	60.5	83.4	88.0	117.9	106.4	104.3	104.9	104.7	88.8	188.9	150.3	114.8	188.8	160.3	112.6	122.2	118.5
05	METALS	106.3	90.5	101.6	102.2	125.0	108.9	95.0	133.6	106.3	86.4	174.6	112.3	103.1	175.0	124.2	97.0	107.5	100.1
06	WOOD, PLASTICS & COMPOSITES	93.5	67.2	79.0	94.6	104.0	99.8	101.7	90.2	95.4	97.2	188.6	147.5	103.7	188.3	150.3	98.6	115.8	108.1
07	THERMAL & MOISTURE PROTECTION	105.0	72.8	91.0	105.4	110.4	107.6	109.2	94.8	102.9	102.5	168.8	131.3	109.3	168.2	134.9	102.5	111.7	106.5
08	OPENINGS	96.4	67.0	89.5	96.4	103.8	98.1	90.4	93.7	91.2	92.3	197.7	116.9	87.8	197.5	113.4	99.8	111.4	102.5
0920	Plaster & Gypsum Board	80.2	66.1	70.7	97.8	103.9	101.9	109.2	89.7	96.1	102.8	191.0	162.1	103.3	191.0	162.3	105.8	116.1	112.7
0950, 0980	Ceilings & Acoustic Treatment	102.7	66.1	78.1	97.1	103.9	101.7	96.6	89.7	92.0	89.1	191.0	157.8	91.8	191.0	158.7	104.2	116.1	112.2
0960	Flooring	97.2	66.8	88.7	90.9	110.3	96.4	103.1	102.4	102.9	99.9	182.5	123.1	110.2	182.5	130.5	96.3	116.6	102.0
0970, 0990	Wall Finishes & Painting/Coating	92.6	52.6	68.9	95.9	104.4	100.9	89.3	106.3	99.4	106.2	167.4	142.4	116.0	167.4	146.4	96.9	117.8	109.3
09	FINISHES	92.8	64.7	77.6	90.6	107.1	99.5	93.6	95.9	94.8	98.0	186.2	145.7	107.1	186.0	149.8	102.0	117.6	110.5
COVERS	DIVS. 10 - 14, 25, 28, 41, 43, 44, 46	100.0	85.1	96.7	100.0	103.8	100.9	100.0	96.6	99.2	100.0	141.5	109.3	100.0	140.8	109.1	100.0	106.5	101.4
21, 22, 23	FIRE SUPPRESSION, PLUMBING & HVAC	97.9	68.6	86.4	100.1	110.3	104.1	100.6	97.0	99.2	100.3	177.9	130.7	99.8	177.9	130.5	100.0	102.4	100.9
26, 27, 3370	ELECTRICAL, COMMUNICATIONS & UTIL.	90.1	69.5	80.0	100.0	108.4	104.1	98.7	101.0	99.8	90.6	184.5	136.9	98.5	184.5	140.9	99.5	105.3	102.4
MF2018	WEIGHTED AVERAGE	99.9	72.4	88.3	96.8	111.2	102.8	97.3	101.5	99.1	93.8	175.3	128.3	101.9	177.3	133.8	100.8	110.3	104.8

		NEW YORK																
		ELMIRA			FAR ROCKAWAY			FLUSHING			GLENS FALLS			HICKSVILLE			JAMAICA	
	DIVISION	148 - 149			116			113			128			115, 117, 118			114	
		MAT.	INST.	TOTAL	MAT.	INST.	TOTAL	MAT.	INST.	TOTAL	MAT.	INST.	TOTAL	MAT.	INST.	TOTAL	MAT.	INST.	TOTAL
015433	CONTRACTOR EQUIPMENT		120.0	120.0		109.4	109.4		109.4	109.4		112.6	112.6		109.4	109.4		109.4	109.4
0241, 31 - 34	SITE & INFRASTRUCTURE, DEMOLITION	97.4	89.6	92.0	121.2	120.6	120.8	121.3	120.6	120.8	71.4	99.6	90.9	111.1	119.1	116.6	115.5	120.6	119.0
0310	Concrete Forming & Accessories	85.0	96.3	94.6	92.2	189.1	174.8	95.9	189.1	175.3	81.5	99.9	97.2	88.8	157.2	147.1	95.9	189.1	175.3
0320	Concrete Reinforcing	98.6	105.3	101.8	97.2	242.5	167.5	98.9	242.5	168.4	96.9	112.7	104.5	97.2	174.4	134.6	97.2	242.5	167.5
0330	Cast-in-Place Concrete	98.8	106.4	101.6	117.2	170.5	137.1	117.2	170.5	137.1	78.1	111.6	90.6	99.5	161.5	122.6	108.3	170.5	131.5
03	CONCRETE	90.9	103.7	96.5	112.8	189.9	146.7	113.2	189.9	146.9	81.2	107.4	92.8	98.2	161.8	126.2	105.8	189.9	142.8
04	MASONRY	101.4	106.7	104.6	118.6	188.8	161.8	112.8	188.8	159.5	93.4	112.6	105.2	109.0	174.4	149.2	117.2	188.8	161.2
05	METALS	95.5	135.8	107.3	103.2	175.0	124.3	103.2	175.0	124.3	95.5	124.5	104.0	104.7	172.4	124.6	103.2	175.0	124.3
06	WOOD, PLASTICS & COMPOSITES	84.1	94.4	89.8	87.1	188.3	142.8	91.7	188.3	144.9	82.7	96.1	90.1	83.7	154.0	122.4	91.7	188.3	144.9
07	THERMAL & MOISTURE PROTECTION	108.9	95.2	102.9	109.2	168.3	134.9	109.3	168.3	134.9	98.6	107.1	102.3	108.9	157.8	130.2	109.1	168.3	134.8
08	OPENINGS	96.7	95.7	96.4	86.6	197.5	112.5	86.6	197.5	112.5	90.0	99.1	92.1	87.0	178.6	108.3	86.6	197.5	112.5
0920	Plaster & Gypsum Board	99.5	94.2	96.0	91.7	191.0	158.5	94.2	191.0	159.3	91.0	96.0	94.4	91.4	155.6	134.6	94.2	191.0	159.3
0950, 0980	Ceilings & Acoustic Treatment	106.0	94.2	98.1	81.2	191.0	155.2	81.2	191.0	155.2	86.4	96.0	92.9	80.3	155.6	131.1	81.2	191.0	155.2
0960	Flooring	87.6	102.4	91.8	105.0	182.5	126.8	106.5	182.5	127.9	82.3	107.7	89.5	104.0	179.7	125.2	106.5	182.5	127.9
0970, 0990	Wall Finishes & Painting/Coating	97.2	93.3	94.9	116.0	167.4	146.4	116.0	167.4	146.4	89.3	104.4	98.2	116.0	167.4	146.4	116.0	167.4	146.4
09	FINISHES	94.2	96.9	95.7	102.4	186.0	147.6	103.2	186.0	148.0	84.0	101.1	93.2	101.0	161.6	133.8	102.7	186.0	147.8
COVERS	DIVS. 10 - 14, 25, 28, 41, 43, 44, 46	100.0	97.6	99.5	100.0	140.8	109.1	100.0	140.8	109.1	100.0	96.9	99.3	100.0	132.1	107.1	100.0	140.8	109.1
21, 22, 23	FIRE SUPPRESSION, PLUMBING & HVAC	96.6	95.5	96.2	96.2	177.9	128.3	96.2	177.9	128.3	96.7	109.3	101.6	99.8	161.0	123.8	96.2	177.9	128.3
26, 27, 3370	ELECTRICAL, COMMUNICATIONS & UTIL.	96.4	104.1	100.2	105.3	184.5	144.4	105.3	184.5	144.4	94.3	105.0	99.6	98.0	141.0	119.2	97.0	184.5	140.1
MF2018	WEIGHTED AVERAGE	96.3	102.5	98.9	102.2	177.3	133.9	102.1	177.3	133.9	92.0	107.3	98.5	99.9	157.1	124.1	100.3	177.3	132.8

		NEW YORK																
		JAMESTOWN			KINGSTON			LONG ISLAND CITY			MONTICELLO			MOUNT VERNON			NEW ROCHELLE	
	DIVISION	147			124			111			127			105			108	
		MAT.	INST.	TOTAL	MAT.	INST.	TOTAL	MAT.	INST.	TOTAL	MAT.	INST.	TOTAL	MAT.	INST.	TOTAL	MAT.	INST.	TOTAL
015433	CONTRACTOR EQUIPMENT		90.6	90.6		109.4	109.4		109.4	109.4		109.4	109.4		104.4	104.4		104.4	104.4
0241, 31 - 34	SITE & INFRASTRUCTURE, DEMOLITION	98.8	89.9	92.6	139.6	115.9	123.2	119.2	120.6	120.2	134.9	115.8	121.7	102.7	105.4	104.5	102.2	105.3	104.3
0310	Concrete Forming & Accessories	85.1	90.1	89.3	82.7	130.7	123.7	100.0	189.1	176.0	89.3	130.8	124.7	86.7	138.8	131.1	100.5	135.3	130.1
0320	Concrete Reinforcing	98.9	111.2	104.8	97.3	160.4	127.8	97.2	242.5	167.5	96.5	160.4	127.4	93.5	180.8	135.7	93.6	180.6	135.7
0330	Cast-in-Place Concrete	102.4	105.2	103.4	105.5	146.3	120.7	111.8	170.5	133.7	98.7	146.3	116.5	95.6	150.0	115.8	95.6	149.7	115.7
03	CONCRETE	93.9	99.0	96.2	102.4	140.8	119.3	109.0	189.9	144.5	97.3	140.8	116.4	97.2	150.4	120.6	96.4	148.7	119.4
04	MASONRY	110.2	104.6	106.8	108.1	155.4	137.2	111.7	188.8	159.1	100.8	155.4	134.3	94.6	156.4	132.6	94.6	156.4	132.6
05	METALS	93.0	101.9	95.6	104.0	135.9	113.4	103.1	175.0	124.2	104.0	136.0	113.4	86.2	169.8	110.7	86.4	169.0	110.7
06	WOOD, PLASTICS & COMPOSITES	82.9	86.2	84.7	83.9	124.0	106.0	97.8	188.3	147.6	90.4	124.0	108.9	87.6	132.4	112.3	104.5	128.7	117.8
07	THERMAL & MOISTURE PROTECTION	108.5	95.4	102.8	121.5	145.0	131.7	109.2	168.3	134.9	121.1	145.0	131.5	103.4	147.3	122.5	103.5	144.7	121.4
08	OPENINGS	96.5	93.0	95.7	92.1	142.5	103.9	86.6	197.5	112.5	88.0	142.4	100.7	92.3	168.4	110.1	92.4	166.3	109.6
0920	Plaster & Gypsum Board	89.2	85.8	86.9	91.0	124.9	113.8	99.3	191.0	161.0	91.7	124.8	114.0	98.3	133.2	121.8	111.0	129.3	123.3
0950, 0980	Ceilings & Acoustic Treatment	102.7	85.8	91.3	77.5	124.9	109.4	81.2	191.0	155.2	77.5	124.8	109.4	87.5	133.2	118.3	87.5	129.3	115.7
0960	Flooring	90.3	102.4	93.7	98.6	160.5	116.0	108.2	182.5	129.1	101.0	160.5	117.8	91.2	179.7	116.1	99.2	164.0	117.4
0970, 0990	Wall Finishes & Painting/Coating	98.8	100.6	99.9	116.6	130.9	125.0	116.0	167.4	146.4	116.6	123.4	120.6	104.6	167.4	141.8	104.6	167.4	141.8
09	FINISHES	93.2	92.5	92.9	97.9	135.4	118.2	104.1	186.0	148.4	98.3	134.6	117.9	95.0	148.0	123.7	98.9	142.6	122.5
COVERS	DIVS. 10 - 14, 25, 28, 41, 43, 44, 46	100.0	96.8	99.3	100.0	121.2	104.7	100.0	140.8	109.1	100.0	121.3	104.7	100.0	129.2	106.5	100.0	113.0	102.9
21, 22, 23	FIRE SUPPRESSION, PLUMBING & HVAC	96.5	93.6	95.4	96.6	134.6	111.5	99.8	177.9	130.5	96.6	138.9	113.2	96.8	145.9	116.1	96.8	145.6	116.0
26, 27, 3370	ELECTRICAL, COMMUNICATIONS & UTIL.	95.4	94.9	95.1	95.7	116.3	105.9	97.5	184.5	140.4	95.7	116.3	105.9	88.8	170.6	129.1	88.8	141.7	114.9
MF2018	WEIGHTED AVERAGE	96.4	96.0	96.3	100.6	133.8	114.6	101.6	177.3	133.6	99.1	134.6	114.1	94.1	150.6	118.0	94.4	144.8	115.7

NEW YORK

DIVISION		NEW YORK 100-102 MAT.	INST.	TOTAL	NIAGARA FALLS 143 MAT.	INST.	TOTAL	PLATTSBURGH 129 MAT.	INST.	TOTAL	POUGHKEEPSIE 125-126 MAT.	INST.	TOTAL	QUEENS 110 MAT.	INST.	TOTAL	RIVERHEAD 119 MAT.	INST.	TOTAL
015433	CONTRACTOR EQUIPMENT		104.6	104.6		90.6	90.6		94.5	94.5		109.4	109.4		109.4	109.4		109.4	109.4
0241, 31-34	SITE & INFRASTRUCTURE, DEMOLITION	106.6	111.9	110.3	100.9	90.8	94.0	109.3	96.7	100.6	135.9	114.8	121.3	114.3	120.6	118.7	112.1	118.6	116.6
0310	Concrete Forming & Accessories	104.3	188.9	176.4	85.0	113.1	108.9	86.8	91.4	90.7	82.7	167.6	155.1	89.0	189.1	174.3	93.2	156.2	146.9
0320	Concrete Reinforcing	100.3	178.3	138.0	97.5	112.0	104.5	101.4	112.0	106.5	97.3	160.0	127.6	98.9	242.5	168.4	99.1	215.6	155.5
0330	Cast-in-Place Concrete	98.9	172.2	126.2	106.0	124.1	112.7	95.5	100.6	97.4	102.2	135.4	114.5	102.9	170.5	128.1	101.2	159.7	123.0
03	CONCRETE	101.6	180.3	136.2	96.2	116.1	105.0	94.3	98.0	95.9	99.7	153.6	123.4	101.5	189.9	140.3	98.7	166.5	128.5
04	MASONRY	102.1	188.8	155.4	117.5	125.1	122.2	88.3	97.4	93.9	100.6	139.5	124.5	106.0	188.8	156.9	114.3	173.3	150.6
05	METALS	97.5	171.0	119.1	95.6	102.4	97.6	99.8	99.5	99.7	104.0	135.6	113.3	103.1	175.0	124.2	105.1	160.6	121.4
06	WOOD, PLASTICS & COMPOSITES	102.0	188.6	149.7	82.8	108.2	96.8	88.8	88.2	88.5	83.9	179.3	136.4	83.8	188.3	141.3	88.7	154.0	124.6
07	THERMAL & MOISTURE PROTECTION	105.2	170.0	133.4	108.6	110.0	109.2	116.0	97.2	107.8	121.4	144.1	131.3	108.8	168.3	134.7	110.0	154.8	129.5
08	OPENINGS	96.0	197.5	119.7	96.5	104.7	98.4	97.4	94.7	96.8	92.1	170.5	110.4	86.6	197.5	112.5	87.0	172.1	106.8
0920	Plaster & Gypsum Board	108.6	191.0	164.0	89.2	108.4	102.1	108.2	87.4	94.2	91.0	181.7	152.0	91.4	191.0	158.4	92.5	155.6	135.0
0950, 0980	Ceilings & Acoustic Treatment	104.5	191.0	162.8	102.7	108.4	106.6	107.3	87.4	93.9	77.5	181.7	147.7	81.2	191.0	155.2	81.1	155.6	131.4
0960	Flooring	101.3	182.5	124.1	90.3	112.1	96.4	104.2	105.1	104.4	98.6	158.6	115.5	104.0	182.5	126.0	105.0	166.9	122.4
0970, 0990	Wall Finishes & Painting/Coating	102.7	167.4	141.0	98.8	111.5	106.3	111.2	95.1	101.7	116.6	123.4	120.6	116.0	167.4	146.4	116.0	167.4	146.4
09	FINISHES	102.9	186.2	148.0	93.3	112.9	103.9	95.0	93.5	94.2	97.7	164.2	133.7	101.3	186.0	147.1	101.5	158.6	132.4
COVERS	DIVS. 10-14, 25, 28, 41, 43, 44, 46	100.0	141.5	109.3	100.0	102.7	100.6	100.0	92.7	98.4	100.0	116.5	103.7	100.0	140.8	109.1	100.0	120.1	104.5
21, 22, 23	FIRE SUPPRESSION, PLUMBING & HVAC	100.2	178.3	130.9	96.5	105.1	99.9	96.6	100.3	98.1	96.6	120.3	105.9	99.8	177.9	130.5	100.0	155.9	121.9
26, 27, 3370	ELECTRICAL, COMMUNICATIONS & UTIL.	96.8	184.5	140.0	94.1	98.9	96.4	92.1	91.7	91.9	95.7	121.2	108.3	98.0	184.5	140.6	99.5	134.9	116.9
MF2018	WEIGHTED AVERAGE	99.7	175.2	131.6	97.4	107.5	101.6	97.0	96.6	96.8	99.8	136.8	115.4	100.0	177.3	132.6	100.6	153.5	122.9

NEW YORK

DIVISION		ROCHESTER 144-146 MAT.	INST.	TOTAL	SCHENECTADY 123 MAT.	INST.	TOTAL	STATEN ISLAND 103 MAT.	INST.	TOTAL	SUFFERN 109 MAT.	INST.	TOTAL	SYRACUSE 130-132 MAT.	INST.	TOTAL	UTICA 133-135 MAT.	INST.	TOTAL
015433	CONTRACTOR EQUIPMENT		117.6	117.6		112.6	112.6		104.4	104.4		104.4	104.4		112.6	112.6		112.6	112.6
0241, 31-34	SITE & INFRASTRUCTURE, DEMOLITION	88.8	105.1	100.0	81.4	100.1	94.3	107.0	109.7	108.9	99.4	103.7	102.4	93.0	98.5	96.8	71.9	98.1	90.0
0310	Concrete Forming & Accessories	105.8	100.3	101.2	96.4	107.1	105.5	86.1	189.3	174.1	94.4	140.8	133.9	98.5	89.1	90.5	99.5	86.3	88.2
0320	Concrete Reinforcing	101.4	105.0	103.1	95.9	115.1	105.2	94.8	216.0	153.4	93.6	152.0	121.9	96.8	106.1	101.3	96.8	100.3	98.5
0330	Cast-in-Place Concrete	97.8	103.3	99.9	94.4	115.4	102.2	95.6	172.1	124.1	92.7	138.9	109.9	98.7	102.3	100.0	90.3	101.0	94.3
03	CONCRETE	99.5	103.3	101.2	94.9	112.4	102.6	99.1	186.4	137.5	93.8	141.3	114.7	97.6	98.1	97.8	95.8	95.4	95.6
04	MASONRY	93.6	103.4	99.7	90.7	117.9	107.4	100.9	188.9	154.9	94.5	142.2	123.8	97.0	101.0	99.5	88.4	99.4	95.2
05	METALS	104.6	119.3	108.9	99.7	126.1	107.4	84.6	174.8	111.1	84.6	131.9	98.5	98.6	119.8	104.9	96.7	117.5	102.9
06	WOOD, PLASTICS & COMPOSITES	108.4	100.2	103.9	100.1	103.8	102.1	86.2	188.6	142.6	96.9	142.5	122.0	98.2	86.0	91.5	98.2	82.7	89.7
07	THERMAL & MOISTURE PROTECTION	116.4	99.8	109.2	100.1	109.8	104.3	102.9	168.8	131.5	103.4	140.9	119.7	104.0	94.1	99.7	92.7	94.0	93.3
08	OPENINGS	100.7	99.3	100.3	95.3	103.7	97.3	92.3	197.7	116.9	92.4	150.7	106.0	92.4	89.5	91.8	95.2	86.3	93.1
0920	Plaster & Gypsum Board	103.8	100.1	101.3	100.3	103.9	102.7	98.4	191.0	160.7	102.1	143.5	130.0	96.8	85.6	89.3	96.8	82.2	87.0
0950, 0980	Ceilings & Acoustic Treatment	100.4	100.1	100.2	92.8	103.9	100.3	89.1	191.0	157.8	87.5	143.5	125.2	96.6	85.6	89.2	96.6	82.2	86.9
0960	Flooring	90.4	105.0	94.5	89.6	110.3	95.4	95.1	182.5	119.6	95.0	171.6	116.5	92.0	92.3	92.1	90.0	92.4	90.7
0970, 0990	Wall Finishes & Painting/Coating	97.7	98.4	98.1	89.3	104.4	98.2	106.2	167.4	142.4	104.6	126.9	117.8	91.9	98.7	95.9	85.6	98.7	93.4
09	FINISHES	95.7	101.4	98.8	89.2	107.0	98.9	96.7	186.2	145.2	96.3	145.8	123.1	92.4	90.0	91.1	90.8	87.9	89.2
COVERS	DIVS. 10-14, 25, 28, 41, 43, 44, 46	100.0	98.9	99.7	100.0	103.5	100.8	100.0	141.5	109.3	100.0	114.7	103.3	100.0	95.1	98.9	100.0	89.8	97.7
21, 22, 23	FIRE SUPPRESSION, PLUMBING & HVAC	99.9	88.1	95.3	100.2	106.2	102.6	100.3	177.9	130.8	96.8	123.7	107.4	100.3	94.4	98.0	100.3	92.5	97.2
26, 27, 3370	ELECTRICAL, COMMUNICATIONS & UTIL.	99.5	89.9	94.8	98.7	108.4	103.5	90.6	184.5	136.9	94.8	114.6	104.6	98.6	101.0	99.8	96.6	101.0	98.8
MF2018	WEIGHTED AVERAGE	100.3	99.2	99.8	97.0	109.9	102.4	95.6	176.1	129.6	94.1	130.2	109.4	97.8	98.3	98.0	96.0	96.6	96.2

NEW YORK / NORTH CAROLINA

DIVISION		WATERTOWN 136 MAT.	INST.	TOTAL	WHITE PLAINS 106 MAT.	INST.	TOTAL	YONKERS 107 MAT.	INST.	TOTAL	ASHEVILLE 287-288 MAT.	INST.	TOTAL	CHARLOTTE 281-282 MAT.	INST.	TOTAL	DURHAM 277 MAT.	INST.	TOTAL
015433	CONTRACTOR EQUIPMENT		112.6	112.6		104.4	104.4		104.4	104.4		99.0	99.0		100.4	100.4		104.1	104.1
0241, 31-34	SITE & INFRASTRUCTURE, DEMOLITION	79.5	98.6	92.7	97.1	105.4	102.8	104.5	105.4	105.1	96.4	77.8	83.5	98.6	81.9	87.1	98.5	86.3	90.1
0310	Concrete Forming & Accessories	85.3	93.0	91.9	99.2	149.3	141.9	99.4	149.1	141.8	90.5	61.4	65.7	96.6	61.5	66.6	96.2	61.4	66.5
0320	Concrete Reinforcing	97.5	106.1	101.6	93.6	180.8	135.8	97.3	180.8	137.7	95.4	66.0	81.2	99.6	66.8	83.7	104.4	66.8	86.2
0330	Cast-in-Place Concrete	105.1	104.7	105.0	84.9	150.0	109.1	95.0	150.0	115.4	105.9	71.5	93.1	108.0	71.6	94.4	106.3	71.2	93.2
03	CONCRETE	108.8	100.7	105.3	87.6	155.2	117.3	96.6	155.1	122.3	91.1	67.4	80.7	91.2	67.5	80.8	95.2	67.4	83.0
04	MASONRY	89.7	104.8	99.0	93.9	156.4	132.3	97.1	156.4	133.6	84.9	64.4	72.3	88.6	64.4	73.8	84.3	64.4	72.0
05	METALS	96.8	119.8	103.6	86.1	169.8	110.7	94.1	169.9	116.4	101.2	89.9	97.9	102.1	88.9	98.2	117.9	90.2	109.7
06	WOOD, PLASTICS & COMPOSITES	80.3	89.9	85.6	102.4	146.7	126.8	102.3	146.3	126.5	87.6	59.0	71.9	89.9	59.0	72.9	90.6	59.0	73.2
07	THERMAL & MOISTURE PROTECTION	93.0	97.2	94.8	103.2	148.7	123.0	103.5	149.3	123.4	99.1	63.8	83.8	93.7	64.5	81.0	102.8	63.8	85.8
08	OPENINGS	95.2	93.8	94.8	92.4	176.2	112.0	96.0	176.5	114.8	89.0	59.8	82.2	99.0	60.0	89.9	95.6	60.0	87.3
0920	Plaster & Gypsum Board	87.6	89.6	89.0	105.3	147.8	133.9	108.7	147.4	134.7	102.2	57.8	72.3	98.1	57.8	71.0	89.9	57.8	68.3
0950, 0980	Ceilings & Acoustic Treatment	96.6	89.6	91.9	87.5	147.8	128.2	103.7	147.4	133.2	79.4	57.8	64.9	83.5	57.8	66.2	82.8	57.8	66.0
0960	Flooring	83.1	92.4	85.7	97.5	179.7	120.6	96.9	182.5	121.0	94.9	66.8	87.0	95.3	66.8	87.3	102.0	66.8	92.1
0970, 0990	Wall Finishes & Painting/Coating	85.6	93.0	90.0	104.6	167.4	141.8	104.6	167.4	141.8	102.5	57.1	75.7	95.9	57.1	72.9	104.4	57.1	76.4
09	FINISHES	88.2	92.4	90.5	97.1	156.4	129.2	101.0	156.7	131.1	88.8	61.6	74.1	88.3	61.6	73.9	90.4	61.6	74.8
COVERS	DIVS. 10-14, 25, 28, 41, 43, 44, 46	100.0	96.4	99.2	100.0	127.5	106.1	100.0	130.7	106.8	100.0	82.5	96.1	100.0	82.5	96.1	100.0	82.5	96.1
21, 22, 23	FIRE SUPPRESSION, PLUMBING & HVAC	100.3	87.8	95.4	100.4	145.9	118.3	100.4	146.0	118.3	100.4	61.3	85.0	99.9	63.1	85.5	100.5	61.3	85.1
26, 27, 3370	ELECTRICAL, COMMUNICATIONS & UTIL.	98.6	90.2	94.4	88.8	170.6	129.1	94.9	170.6	132.2	100.8	57.8	79.6	100.0	60.3	80.4	96.5	57.4	77.2
MF2018	WEIGHTED AVERAGE	97.7	96.8	97.4	93.8	152.8	118.8	97.9	153.0	121.2	96.4	66.6	83.8	97.4	67.6	84.8	100.1	67.2	86.2

NORTH CAROLINA

DIVISION		ELIZABETH CITY 279			FAYETTEVILLE 283			GASTONIA 280			GREENSBORO 270, 272 - 274			HICKORY 286			KINSTON 285		
		MAT.	INST.	TOTAL	MAT.	INST.	TOTAL	MAT.	INST.	TOTAL	MAT.	INST.	TOTAL	MAT.	INST.	TOTAL	MAT.	INST.	TOTAL
015433	CONTRACTOR EQUIPMENT		108.0	108.0		104.1	104.1		99.0	99.0		104.1	104.1		104.1	104.1		104.1	104.1
0241, 31 - 34	SITE & INFRASTRUCTURE, DEMOLITION	102.7	87.9	92.4	95.6	86.1	89.0	96.1	78.1	83.7	98.3	86.4	90.1	95.1	85.1	88.2	93.9	84.9	87.7
0310	Concrete Forming & Accessories	82.6	63.6	66.4	90.1	60.1	64.5	96.4	61.4	66.6	96.0	61.3	66.5	87.1	61.2	65.0	83.6	59.9	63.4
0320	Concrete Reinforcing	102.3	69.8	86.6	99.1	66.8	83.4	95.8	66.8	81.7	103.2	66.8	85.6	95.4	66.7	81.5	94.9	66.7	81.3
0330	Cast-in-Place Concrete	106.5	71.7	93.5	111.1	69.3	95.5	103.6	71.2	91.5	105.5	71.2	92.7	105.9	71.2	93.0	102.2	69.2	90.0
03	CONCRETE	95.4	69.1	83.8	93.1	66.2	81.3	89.8	67.5	80.0	94.7	67.4	82.7	90.9	67.3	80.5	88.1	66.1	78.4
04	MASONRY	95.4	61.0	74.2	88.3	61.0	71.5	89.4	64.4	74.0	81.8	64.4	71.1	74.6	64.4	68.3	81.2	61.0	68.8
05	METALS	103.5	92.3	100.2	122.2	90.2	112.8	101.8	90.2	98.4	110.0	90.2	104.1	101.3	90.0	98.0	100.0	90.0	97.1
06	WOOD, PLASTICS & COMPOSITES	75.1	63.4	68.7	86.5	59.0	71.4	95.4	59.0	75.4	90.2	59.0	73.1	82.4	59.0	69.6	79.3	59.0	68.2
07	THERMAL & MOISTURE PROTECTION	102.0	62.5	84.8	98.6	62.3	82.8	99.3	63.8	83.9	102.6	63.8	85.7	99.5	63.8	84.0	99.3	62.3	83.2
08	OPENINGS	92.9	63.1	86.0	89.1	60.0	82.3	92.3	60.0	84.8	95.6	60.0	87.3	89.0	60.0	82.3	89.1	60.0	82.3
0920	Plaster & Gypsum Board	84.4	61.7	69.1	105.8	57.8	73.5	107.9	57.8	74.2	91.1	57.8	68.7	102.2	57.8	72.3	102.2	57.8	72.3
0950, 0980	Ceilings & Acoustic Treatment	82.8	61.7	68.5	81.9	57.8	65.7	83.5	57.8	66.2	82.8	57.8	66.0	79.4	57.8	64.9	83.5	57.8	66.2
0960	Flooring	94.1	66.8	86.4	95.1	66.8	87.1	98.0	66.8	89.2	102.0	66.8	92.1	94.8	66.8	86.9	92.2	66.8	85.0
0970, 0990	Wall Finishes & Painting/Coating	104.4	57.1	76.4	102.5	57.1	75.7	102.5	57.1	75.7	104.4	57.1	76.4	102.5	57.1	75.7	102.5	57.1	75.7
09	FINISHES	87.7	63.4	74.6	89.7	60.8	74.0	91.1	61.6	75.2	90.6	61.6	74.9	88.9	61.6	74.1	88.7	60.8	73.6
COVERS	DIVS. 10 - 14, 25, 28, 41, 43, 44, 46	100.0	45.6	96.8	100.0	81.4	95.8	100.0	82.5	96.1	100.0	79.8	95.5	100.0	82.5	96.1	100.0	81.3	95.8
21, 22, 23	FIRE SUPPRESSION, PLUMBING & HVAC	96.9	58.6	81.9	100.2	59.5	84.2	100.4	60.1	84.6	100.4	61.3	85.0	96.9	60.2	82.5	96.9	58.3	81.7
26, 27, 3370	ELECTRICAL, COMMUNICATIONS & UTIL.	96.2	64.6	80.6	100.6	57.4	79.3	100.3	60.4	80.6	95.7	57.8	77.0	98.6	60.4	79.8	98.4	55.8	77.4
MF2018	WEIGHTED AVERAGE	96.9	68.3	84.8	100.2	66.1	85.8	97.0	66.8	84.2	98.5	67.2	85.3	94.7	67.3	83.1	94.4	65.5	82.2

NORTH CAROLINA / NORTH DAKOTA

DIVISION		MURPHY 289			RALEIGH 275 - 276			ROCKY MOUNT 278			WILMINGTON 284			WINSTON-SALEM 271			BISMARCK 585		
		MAT.	INST.	TOTAL	MAT.	INST.	TOTAL	MAT.	INST.	TOTAL	MAT.	INST.	TOTAL	MAT.	INST.	TOTAL	MAT.	INST.	TOTAL
015433	CONTRACTOR EQUIPMENT		99.0	99.0		106.5	106.5		104.1	104.1		99.0	99.0		104.1	104.1		99.9	99.9
0241, 31 - 34	SITE & INFRASTRUCTURE, DEMOLITION	97.4	76.3	82.8	98.8	90.8	93.3	100.7	86.3	90.8	97.4	77.6	83.7	98.7	86.4	90.2	99.8	97.8	98.4
0310	Concrete Forming & Accessories	97.0	59.8	65.3	96.9	61.0	66.3	88.5	60.9	64.9	91.8	60.1	64.8	97.9	61.4	66.8	110.6	76.6	81.6
0320	Concrete Reinforcing	94.9	64.3	80.1	105.4	66.8	86.7	102.3	66.8	85.1	96.1	66.8	81.9	103.2	66.8	85.6	92.6	98.3	95.3
0330	Cast-in-Place Concrete	109.6	69.2	94.6	109.3	71.3	95.1	104.2	70.5	91.6	105.5	69.3	92.0	108.1	71.2	94.4	106.3	86.6	98.9
03	CONCRETE	93.9	65.6	81.4	93.9	67.2	82.2	95.9	66.9	83.2	91.1	66.2	80.1	96.0	67.4	83.5	93.3	84.4	89.4
04	MASONRY	77.3	61.0	67.3	80.5	63.1	69.8	74.4	63.1	67.4	75.1	61.0	66.4	82.0	64.4	71.2	105.3	83.4	91.8
05	METALS	99.0	89.1	96.1	102.2	89.1	98.3	102.6	90.2	99.0	100.7	90.2	97.6	107.0	90.2	102.1	95.6	94.2	95.2
06	WOOD, PLASTICS & COMPOSITES	96.1	58.9	75.6	90.9	59.2	73.4	82.0	59.0	69.4	89.2	59.0	72.6	90.2	59.0	73.1	106.6	72.7	88.0
07	THERMAL & MOISTURE PROTECTION	99.3	62.3	83.2	97.2	63.8	82.7	102.5	63.2	85.4	99.1	62.3	83.1	102.6	63.8	85.7	109.3	86.0	99.2
08	OPENINGS	89.0	59.4	82.1	97.6	60.1	88.8	92.2	60.0	84.7	89.1	60.0	82.3	95.6	60.0	87.3	103.7	81.9	98.6
0920	Plaster & Gypsum Board	106.9	57.7	73.8	86.5	57.8	67.2	85.8	57.8	67.0	103.7	57.8	72.8	91.1	57.8	68.7	102.4	72.2	82.1
0950, 0980	Ceilings & Acoustic Treatment	79.4	57.7	64.8	83.5	57.8	66.2	80.4	57.8	65.2	81.9	57.8	65.7	82.8	57.8	66.0	108.5	72.2	84.1
0960	Flooring	98.3	66.8	89.4	96.8	66.8	88.4	97.6	66.8	88.9	95.5	66.8	87.5	102.0	66.8	92.1	86.9	53.7	77.6
0970, 0990	Wall Finishes & Painting/Coating	102.5	57.1	75.7	97.8	57.1	73.7	104.4	57.1	76.4	102.5	57.1	75.7	104.4	57.1	76.4	90.9	59.8	72.5
09	FINISHES	90.6	60.7	74.4	90.0	61.4	74.5	88.6	61.3	73.8	89.6	60.8	74.0	90.6	61.6	74.9	96.1	71.1	82.5
COVERS	DIVS. 10 - 14, 25, 28, 41, 43, 44, 46	100.0	81.3	95.8	100.0	82.3	96.1	100.0	82.1	96.0	100.0	81.4	95.8	100.0	82.5	96.1	100.0	94.8	98.8
21, 22, 23	FIRE SUPPRESSION, PLUMBING & HVAC	96.9	58.3	81.7	100.0	60.6	84.5	96.9	59.5	82.2	100.4	59.5	84.3	100.4	61.3	85.0	99.9	77.0	90.9
26, 27, 3370	ELECTRICAL, COMMUNICATIONS & UTIL.	101.7	55.8	79.1	99.1	56.3	78.0	98.0	57.4	78.0	101.1	55.8	78.8	95.7	57.8	77.0	98.7	73.4	86.3
MF2018	WEIGHTED AVERAGE	95.5	64.7	82.5	97.4	67.0	84.5	96.0	66.6	83.6	96.0	65.2	83.0	98.2	67.3	85.2	98.9	81.6	91.6

NORTH DAKOTA

DIVISION		DEVILS LAKE 583			DICKINSON 586			FARGO 580 - 581			GRAND FORKS 582			JAMESTOWN 584			MINOT 587		
		MAT.	INST.	TOTAL	MAT.	INST.	TOTAL	MAT.	INST.	TOTAL	MAT.	INST.	TOTAL	MAT.	INST.	TOTAL	MAT.	INST.	TOTAL
015433	CONTRACTOR EQUIPMENT		97.7	97.7		97.7	97.7		99.9	99.9		97.7	97.7		97.7	97.7		97.7	97.7
0241, 31 - 34	SITE & INFRASTRUCTURE, DEMOLITION	107.5	93.7	98.0	115.7	93.7	100.5	101.3	97.6	98.8	111.5	93.7	99.2	106.5	93.7	97.7	108.9	94.1	98.7
0310	Concrete Forming & Accessories	107.0	71.6	76.8	95.5	71.5	75.1	99.4	72.1	76.1	99.7	71.6	75.7	97.1	71.6	75.4	95.1	71.8	75.2
0320	Concrete Reinforcing	94.8	98.0	96.4	95.7	98.2	96.9	95.7	98.7	97.2	93.4	98.0	95.6	95.4	98.7	97.0	96.6	98.1	97.4
0330	Cast-in-Place Concrete	126.2	81.6	109.6	114.1	81.5	102.0	102.5	85.6	96.2	114.1	81.5	102.0	124.6	81.6	108.6	114.1	81.8	102.1
03	CONCRETE	104.4	80.4	93.8	103.5	80.4	93.3	98.3	82.1	91.2	100.9	80.3	91.8	102.9	80.5	93.1	99.4	80.6	91.1
04	MASONRY	118.0	79.4	94.2	120.1	81.4	96.4	102.7	90.0	94.9	112.1	79.2	91.9	130.9	89.9	105.7	110.1	80.7	92.1
05	METALS	94.7	94.1	94.5	94.6	94.0	94.4	100.2	94.8	98.6	94.6	93.8	94.4	94.6	94.9	94.7	94.9	94.8	94.9
06	WOOD, PLASTICS & COMPOSITES	99.0	67.5	81.7	85.6	67.5	75.7	95.9	67.6	80.4	90.4	67.5	77.8	87.7	67.5	76.6	85.3	67.5	75.5
07	THERMAL & MOISTURE PROTECTION	107.1	82.3	96.3	107.7	83.7	97.2	104.1	87.4	96.8	107.3	83.0	96.7	106.9	85.7	97.7	107.0	83.9	97.0
08	OPENINGS	100.9	79.0	95.8	100.9	79.0	95.8	101.0	79.1	95.9	99.5	79.0	94.7	100.9	79.0	95.8	99.7	79.0	94.8
0920	Plaster & Gypsum Board	118.2	67.0	83.8	109.1	67.0	80.8	101.1	67.0	78.2	110.3	67.0	81.2	110.0	67.0	81.1	109.1	67.0	80.8
0950, 0980	Ceilings & Acoustic Treatment	105.6	67.0	79.6	105.6	67.0	79.6	95.4	67.0	76.3	105.6	67.0	79.6	105.6	67.0	79.6	105.6	67.0	79.6
0960	Flooring	93.8	53.7	82.5	87.4	53.7	77.9	101.2	53.7	87.9	89.2	53.7	79.2	88.0	53.7	78.4	87.1	53.7	77.7
0970, 0990	Wall Finishes & Painting/Coating	86.6	56.4	68.7	86.6	56.4	68.7	93.1	67.4	77.9	86.6	66.0	74.4	86.6	56.4	68.7	86.6	57.5	69.4
09	FINISHES	95.2	66.8	79.8	94.0	66.8	79.3	96.2	68.1	81.0	93.2	67.8	79.5	92.3	66.8	78.5	92.1	66.9	78.5
COVERS	DIVS. 10 - 14, 25, 28, 41, 43, 44, 46	100.0	87.2	97.2	100.0	87.2	97.2	100.0	93.6	98.6	100.0	87.2	97.2	100.0	87.2	97.2	100.0	93.3	98.5
21, 22, 23	FIRE SUPPRESSION, PLUMBING & HVAC	96.6	78.3	89.4	96.6	73.3	87.4	100.0	74.1	89.8	100.1	72.3	89.2	96.6	72.3	87.0	100.1	72.1	89.1
26, 27, 3370	ELECTRICAL, COMMUNICATIONS & UTIL.	94.3	68.3	81.5	101.9	67.9	85.1	98.5	69.4	84.2	97.3	68.3	83.0	94.3	68.3	81.5	100.1	72.6	86.6
MF2018	WEIGHTED AVERAGE	99.2	78.8	90.6	99.9	77.9	90.6	99.7	80.3	91.5	99.4	77.6	90.2	99.2	78.8	90.6	99.3	78.6	90.5

For customer support on your Site Work & Landscape Costs with RSMeans data, call 800.448.8182.

		NORTH DAKOTA			OHIO													
		WILLISTON			AKRON			ATHENS			CANTON			CHILLICOTHE			CINCINNATI		
DIVISION		588			442 - 443			457			446 - 447			456			451 - 452		
		MAT.	INST.	TOTAL	MAT.	INST.	TOTAL	MAT.	INST.	TOTAL	MAT.	INST.	TOTAL	MAT.	INST.	TOTAL	MAT.	INST.	TOTAL
015433	CONTRACTOR EQUIPMENT		97.7	97.7		88.7	88.7		84.9	84.9		88.7	88.7		95.5	95.5		95.9	95.9
0241, 31 - 34	SITE & INFRASTRUCTURE, DEMOLITION	109.2	91.5	97.0	96.2	94.1	94.8	107.3	85.3	92.1	96.3	93.9	94.7	93.8	94.9	94.6	90.0	98.2	95.7
0310	Concrete Forming & Accessories	101.7	71.3	75.8	102.8	82.8	85.8	95.3	78.9	81.4	102.8	74.3	78.5	98.0	81.5	84.0	101.5	80.2	83.3
0320	Concrete Reinforcing	97.5	98.1	97.8	94.7	91.0	92.9	86.9	89.2	88.0	94.7	75.0	85.2	84.0	88.8	86.3	89.1	77.8	83.6
0330	Cast-in-Place Concrete	114.1	81.4	101.9	102.6	89.0	97.6	111.2	95.8	105.5	103.6	87.2	97.5	100.9	92.4	97.8	96.4	77.5	89.3
03	CONCRETE	100.8	80.2	91.7	99.8	85.9	93.7	101.0	86.3	94.6	100.3	78.7	90.8	95.6	86.9	91.8	94.7	79.0	87.8
04	MASONRY	105.1	80.7	90.1	90.9	89.4	90.0	73.6	97.2	88.1	91.5	80.7	84.9	80.2	90.5	86.5	83.0	79.8	81.0
05	METALS	94.8	93.6	94.4	97.5	80.2	92.4	97.1	80.5	92.3	97.5	73.7	90.5	89.4	89.8	89.5	91.5	81.7	88.6
06	WOOD, PLASTICS & COMPOSITES	92.1	67.5	78.6	105.9	81.1	92.2	87.4	74.0	80.0	106.2	71.8	87.3	100.3	78.0	88.0	102.3	80.1	90.1
07	THERMAL & MOISTURE PROTECTION	107.3	83.5	96.9	104.9	91.1	98.9	102.9	91.5	98.0	106.0	87.2	97.8	104.7	88.4	97.6	102.7	80.9	93.2
08	OPENINGS	101.0	79.0	95.8	107.5	82.7	101.7	96.2	75.8	91.5	101.3	69.8	93.9	88.1	78.0	85.7	98.3	76.4	93.2
0920	Plaster & Gypsum Board	110.3	67.0	81.2	98.3	80.5	86.3	93.1	73.2	79.7	99.2	71.0	80.2	96.6	77.8	83.9	96.1	79.9	85.2
0950, 0980	Ceilings & Acoustic Treatment	105.6	67.0	79.6	90.6	80.5	83.8	105.9	73.2	83.8	90.6	71.0	77.4	98.7	77.8	84.6	93.6	79.9	84.3
0960	Flooring	90.1	53.7	79.9	91.7	83.3	89.3	120.7	75.3	108.0	91.8	74.1	86.8	97.5	75.3	91.3	98.7	78.5	93.0
0970, 0990	Wall Finishes & Painting/Coating	86.6	56.4	68.7	98.3	90.9	93.9	100.1	89.1	93.6	98.3	73.9	83.9	97.4	89.1	92.5	97.0	71.7	82.0
09	FINISHES	93.4	66.8	79.0	95.0	83.6	88.8	99.4	78.7	88.2	95.2	73.4	83.4	96.5	80.4	87.8	95.9	79.1	86.8
COVERS	DIVS. 10 - 14, 25, 28, 41, 43, 44, 46	100.0	87.2	97.1	100.0	90.3	97.8	100.0	87.9	97.3	100.0	88.4	97.4	100.0	86.8	97.1	100.0	88.4	97.4
21, 22, 23	FIRE SUPPRESSION, PLUMBING & HVAC	96.6	73.2	87.4	100.0	87.8	95.2	96.5	81.2	90.5	100.0	78.7	91.6	97.0	92.4	95.2	99.9	76.9	90.9
26, 27, 3370	ELECTRICAL, COMMUNICATIONS & UTIL.	97.7	68.7	83.4	98.6	81.8	90.4	97.9	89.7	93.9	97.9	85.0	91.5	97.2	89.7	93.5	96.3	74.2	85.4
MF2018	WEIGHTED AVERAGE	98.4	77.7	89.6	99.4	86.0	93.8	97.1	84.8	91.9	98.9	79.9	90.9	94.3	88.5	91.8	96.1	80.0	89.3

		OHIO																
		CLEVELAND			COLUMBUS			DAYTON			HAMILTON			LIMA			LORAIN		
DIVISION		441			430 - 432			453 - 454			450			458			440		
		MAT.	INST.	TOTAL	MAT.	INST.	TOTAL	MAT.	INST.	TOTAL	MAT.	INST.	TOTAL	MAT.	INST.	TOTAL	MAT.	INST.	TOTAL
015433	CONTRACTOR EQUIPMENT		92.0	92.0		92.3	92.3		88.9	88.9		95.5	95.5		88.3	88.3		88.7	88.7
0241, 31 - 34	SITE & INFRASTRUCTURE, DEMOLITION	94.9	96.2	95.8	101.8	92.3	95.3	90.0	94.2	92.9	90.0	94.5	93.1	101.1	84.9	89.9	95.5	94.3	94.7
0310	Concrete Forming & Accessories	102.3	90.2	92.0	98.7	78.1	81.1	99.9	77.2	80.6	99.9	77.4	80.7	95.3	76.3	79.1	102.8	74.6	78.7
0320	Concrete Reinforcing	95.2	91.4	93.4	99.9	78.7	89.7	89.1	77.4	83.3	89.1	77.2	83.3	86.9	79.3	83.2	94.7	91.3	93.1
0330	Cast-in-Place Concrete	99.5	97.2	98.6	102.1	80.9	94.2	86.5	80.6	84.3	92.6	80.9	88.3	102.1	89.5	97.4	97.7	91.4	95.4
03	CONCRETE	100.1	92.4	96.7	98.6	79.2	90.0	87.0	78.6	83.3	89.9	79.1	85.1	93.9	81.5	88.5	97.5	83.1	91.2
04	MASONRY	96.9	97.9	97.5	91.4	85.2	87.6	79.1	76.4	77.4	79.5	81.2	80.5	98.8	78.1	86.1	87.8	93.1	91.0
05	METALS	99.1	83.3	94.5	98.1	79.3	92.5	90.9	76.8	86.7	90.9	85.6	89.4	97.2	80.1	92.2	98.1	81.6	93.3
06	WOOD, PLASTICS & COMPOSITES	99.8	88.5	93.6	98.5	77.7	87.0	104.2	77.1	89.3	103.0	77.1	88.7	87.3	75.1	80.6	105.9	70.1	86.2
07	THERMAL & MOISTURE PROTECTION	102.0	96.9	99.8	93.6	83.8	89.3	108.8	79.5	96.1	104.8	79.9	94.0	102.4	83.7	94.3	105.9	91.0	99.4
08	OPENINGS	100.3	86.4	97.0	98.3	74.0	92.7	95.0	74.6	90.3	92.8	74.8	88.6	96.3	73.7	91.0	101.3	76.6	95.5
0920	Plaster & Gypsum Board	97.9	88.1	91.3	94.1	77.0	82.6	98.1	76.9	83.9	98.1	76.9	83.9	93.1	74.3	80.5	98.3	69.2	78.7
0950, 0980	Ceilings & Acoustic Treatment	86.3	88.1	87.5	93.5	77.0	82.4	99.6	76.9	84.3	98.7	76.9	84.0	105.9	74.3	84.6	90.6	69.2	76.2
0960	Flooring	93.1	89.8	92.2	95.1	75.3	89.5	101.2	72.4	93.1	98.5	78.5	92.9	119.8	77.2	107.8	91.8	89.8	91.2
0970, 0990	Wall Finishes & Painting/Coating	101.2	90.9	95.1	99.5	79.6	87.7	97.4	70.7	81.6	97.4	71.3	82.0	100.2	76.5	86.2	98.3	90.9	93.9
09	FINISHES	95.3	90.1	92.5	95.0	77.5	85.5	97.7	75.4	85.6	96.7	76.9	86.0	98.7	76.0	86.4	95.0	78.0	85.8
COVERS	DIVS. 10 - 14, 25, 28, 41, 43, 44, 46	100.0	96.4	99.2	100.0	88.2	97.4	100.0	85.1	96.7	100.0	85.3	96.7	100.0	84.4	96.5	100.0	90.9	98.0
21, 22, 23	FIRE SUPPRESSION, PLUMBING & HVAC	100.0	89.9	96.0	100.0	84.1	93.7	100.7	81.2	93.0	100.5	74.9	90.5	96.5	91.7	94.6	100.0	88.4	95.4
26, 27, 3370	ELECTRICAL, COMMUNICATIONS & UTIL.	98.3	92.7	95.5	99.9	80.2	90.2	94.9	75.7	85.4	95.2	75.5	85.5	98.2	75.7	87.1	98.0	78.2	88.2
MF2018	WEIGHTED AVERAGE	99.1	91.6	95.9	98.4	82.0	91.4	94.8	79.2	88.2	94.8	79.5	88.3	97.2	81.7	90.7	98.4	84.7	92.6

		OHIO																
		MANSFIELD			MARION			SPRINGFIELD			STEUBENVILLE			TOLEDO			YOUNGSTOWN		
DIVISION		448 - 449			433			455			439			434 - 436			444 - 445		
		MAT.	INST.	TOTAL	MAT.	INST.	TOTAL	MAT.	INST.	TOTAL	MAT.	INST.	TOTAL	MAT.	INST.	TOTAL	MAT.	INST.	TOTAL
015433	CONTRACTOR EQUIPMENT		88.7	88.7		88.7	88.7		88.9	88.9		92.4	92.4		92.4	92.4		88.7	88.7
0241, 31 - 34	SITE & INFRASTRUCTURE, DEMOLITION	91.7	94.0	93.3	95.7	90.6	92.2	90.3	94.2	93.0	141.5	98.2	111.6	99.8	90.9	93.6	96.1	93.9	94.6
0310	Concrete Forming & Accessories	92.1	73.5	76.3	96.6	79.4	81.9	99.9	77.1	80.5	97.6	79.0	81.7	100.3	84.9	87.1	102.8	76.6	80.5
0320	Concrete Reinforcing	86.2	79.0	82.7	92.1	79.0	85.8	89.1	79.1	84.2	89.7	94.9	92.2	99.9	82.9	91.7	94.7	85.2	90.1
0330	Cast-in-Place Concrete	95.1	87.8	92.3	88.5	88.1	88.3	88.8	80.3	85.6	96.1	89.3	93.6	96.9	90.2	94.4	101.7	87.2	96.3
03	CONCRETE	92.1	79.3	86.4	85.4	82.2	84.0	88.1	78.5	83.9	89.4	84.9	87.4	93.2	86.4	90.2	99.4	81.5	91.5
04	MASONRY	89.9	88.9	89.3	92.7	89.9	90.9	79.3	75.9	77.2	80.7	91.1	87.1	98.9	91.8	94.5	91.1	86.6	88.4
05	METALS	98.4	76.0	91.8	97.1	78.9	91.7	90.9	76.8	86.7	93.6	81.8	90.1	97.8	85.3	94.2	97.5	77.9	91.8
06	WOOD, PLASTICS & COMPOSITES	92.8	70.1	80.3	92.8	77.5	84.4	105.6	77.1	89.9	88.1	76.4	81.6	96.9	83.9	89.8	105.9	74.6	88.7
07	THERMAL & MOISTURE PROTECTION	104.1	88.4	97.3	90.5	88.8	89.8	108.7	79.2	95.9	102.0	87.4	95.7	91.0	91.2	91.1	106.1	88.2	98.3
08	OPENINGS	102.3	71.0	95.0	91.2	75.1	87.5	93.2	74.6	88.9	91.3	78.4	88.3	93.8	81.3	90.9	101.3	75.2	95.2
0920	Plaster & Gypsum Board	90.9	69.2	76.3	95.3	77.0	83.0	98.1	76.9	83.9	93.5	75.4	81.3	97.2	83.6	88.0	98.3	73.9	81.9
0950, 0980	Ceilings & Acoustic Treatment	91.4	69.2	76.4	99.5	77.0	84.3	99.6	76.9	84.3	96.7	75.4	82.3	99.5	83.6	88.8	90.6	73.9	79.3
0960	Flooring	86.6	92.7	88.3	94.4	92.7	93.9	101.2	72.4	93.1	122.5	92.7	114.1	94.8	94.4	94.7	91.8	90.0	91.3
0970, 0990	Wall Finishes & Painting/Coating	98.3	77.8	86.2	104.0	77.8	88.5	97.4	70.7	81.6	117.8	87.1	99.6	104.0	87.0	94.0	98.3	80.0	87.5
09	FINISHES	92.4	76.6	83.8	96.3	81.2	88.1	97.7	75.3	85.6	114.5	81.6	96.7	97.0	86.6	91.4	95.1	78.8	86.3
COVERS	DIVS. 10 - 14, 25, 28, 41, 43, 44, 46	100.0	88.3	97.4	100.0	85.9	96.8	100.0	84.9	96.6	100.0	86.2	96.9	100.0	91.3	98.1	100.0	88.5	97.4
21, 22, 23	FIRE SUPPRESSION, PLUMBING & HVAC	96.5	87.5	92.9	96.5	93.1	95.1	100.7	80.9	92.9	96.9	91.3	94.7	100.0	93.6	97.5	100.0	83.9	93.7
26, 27, 3370	ELECTRICAL, COMMUNICATIONS & UTIL.	95.7	90.8	93.3	94.0	90.8	92.4	94.9	80.2	87.7	88.7	106.4	97.4	100.0	103.4	101.7	98.0	74.3	86.3
MF2018	WEIGHTED AVERAGE	96.4	84.2	91.3	94.3	86.7	91.1	94.8	79.7	88.4	96.1	90.0	93.5	97.6	91.2	94.9	98.7	81.9	91.6

City Cost Indexes

		OHIO			OKLAHOMA															
		ZANESVILLE			ARDMORE			CLINTON			DURANT			ENID			GUYMON		
	DIVISION	437 - 438			734			736			747			737			739		
		MAT.	INST.	TOTAL	MAT.	INST.	TOTAL	MAT.	INST.	TOTAL	MAT.	INST.	TOTAL	MAT.	INST.	TOTAL	MAT.	INST.	TOTAL
015433	CONTRACTOR EQUIPMENT		88.7	88.7		80.7	80.7		79.9	79.9		79.9	79.9		79.9	79.9		79.9	79.9
0241, 31 - 34	SITE & INFRASTRUCTURE, DEMOLITION	98.9	90.5	93.1	96.9	92.6	94.0	98.1	91.2	93.3	93.4	88.6	90.1	100.0	91.2	93.9	102.4	90.5	94.2
0310	Concrete Forming & Accessories	93.7	78.3	80.6	88.0	55.4	60.2	86.8	55.4	60.1	82.9	55.0	59.1	89.9	55.6	60.7	92.6	55.1	60.7
0320	Concrete Reinforcing	91.6	92.6	92.1	79.5	67.7	73.8	80.0	67.7	74.0	88.6	63.6	76.5	79.5	67.7	73.8	80.0	63.3	71.9
0330	Cast-in-Place Concrete	93.2	86.5	90.7	96.6	70.5	86.9	93.4	70.5	84.9	88.6	70.3	81.8	93.4	70.6	84.9	93.4	70.0	84.7
03	CONCRETE	89.2	83.5	86.7	86.7	62.8	76.2	86.2	62.8	75.9	84.1	61.8	74.3	86.6	62.9	76.2	89.4	61.7	77.2
04	MASONRY	90.7	86.6	88.2	95.2	57.0	71.8	119.3	57.0	81.0	88.6	62.1	72.3	101.3	57.0	74.1	97.5	55.4	71.7
05	METALS	98.5	84.2	94.3	97.3	62.2	87.0	97.4	62.2	87.0	92.7	60.6	83.2	98.8	62.4	88.1	97.8	59.3	86.5
06	WOOD, PLASTICS & COMPOSITES	88.2	77.5	82.3	97.4	54.4	73.7	96.5	54.4	73.4	87.8	54.4	69.5	99.7	54.4	74.8	102.9	54.4	76.2
07	THERMAL & MOISTURE PROTECTION	90.6	84.5	88.0	100.7	64.8	85.1	100.8	64.8	85.1	96.4	64.6	82.6	100.9	64.8	85.2	101.3	62.2	84.3
08	OPENINGS	91.2	79.0	88.4	103.7	55.0	92.3	103.7	55.0	92.3	96.0	54.0	86.2	104.9	55.9	93.5	103.8	54.0	92.2
0920	Plaster & Gypsum Board	91.2	77.0	81.7	90.5	53.7	65.8	90.2	53.7	65.7	78.5	53.7	61.8	91.1	53.7	66.0	91.1	53.7	66.0
0950, 0980	Ceilings & Acoustic Treatment	99.5	77.0	84.3	89.1	53.7	65.3	89.1	53.7	65.3	83.0	53.7	63.3	89.1	53.7	65.3	89.1	53.7	65.3
0960	Flooring	92.6	75.3	87.7	82.7	57.3	75.6	81.7	57.3	74.9	89.7	51.7	79.0	83.3	72.9	80.4	84.8	57.3	77.1
0970, 0990	Wall Finishes & Painting/Coating	104.0	89.1	95.2	86.4	44.8	61.8	86.4	44.8	61.8	92.8	44.8	64.4	86.4	44.8	61.8	86.4	42.2	60.3
09	FINISHES	95.4	78.2	86.1	82.7	53.5	66.9	82.5	53.5	66.8	83.2	52.4	66.5	83.2	56.7	68.8	84.0	54.5	68.0
COVERS	DIVS. 10 - 14, 25, 28, 41, 43, 44, 46	100.0	83.3	96.3	100.0	79.4	95.4	100.0	79.4	95.4	100.0	79.3	95.4	100.0	79.4	95.4	100.0	79.4	95.4
21, 22, 23	FIRE SUPPRESSION, PLUMBING & HVAC	96.5	89.8	93.9	96.5	67.5	85.1	96.5	67.5	85.1	96.6	67.4	85.2	100.0	67.5	87.2	96.5	63.6	83.6
26, 27, 3370	ELECTRICAL, COMMUNICATIONS & UTIL.	94.1	89.7	92.0	95.0	71.0	83.2	95.9	71.0	83.7	97.3	69.2	83.4	95.9	71.0	83.7	97.6	65.6	81.8
MF2018	WEIGHTED AVERAGE	94.9	85.7	91.0	95.2	65.6	82.7	96.4	65.5	83.3	93.1	65.1	81.3	97.0	66.0	83.9	96.3	63.3	82.4

		OKLAHOMA																	
		LAWTON			MCALESTER			MIAMI			MUSKOGEE			OKLAHOMA CITY			PONCA CITY		
	DIVISION	735			745			743			744			730 - 731			746		
		MAT.	INST.	TOTAL	MAT.	INST.	TOTAL	MAT.	INST.	TOTAL	MAT.	INST.	TOTAL	MAT.	INST.	TOTAL	MAT.	INST.	TOTAL
015433	CONTRACTOR EQUIPMENT		80.7	80.7		79.9	79.9		90.9	90.9		90.9	90.9		85.6	85.6		79.9	79.9
0241, 31 - 34	SITE & INFRASTRUCTURE, DEMOLITION	96.2	92.7	93.7	86.8	90.5	89.4	88.1	88.1	88.1	88.7	87.9	88.1	94.2	98.0	96.8	93.9	90.9	91.8
0310	Concrete Forming & Accessories	92.5	55.6	61.1	81.1	41.4	47.3	93.2	55.2	60.8	97.3	54.8	61.1	92.8	63.8	68.0	89.0	55.4	60.3
0320	Concrete Reinforcing	79.7	67.7	73.9	88.3	63.3	76.2	86.9	67.7	77.6	87.7	62.8	75.7	87.9	67.8	78.2	87.7	67.7	78.0
0330	Cast-in-Place Concrete	90.3	70.6	83.0	77.7	70.0	74.8	81.4	71.4	77.7	82.4	71.1	78.2	90.3	73.4	84.0	91.0	70.5	83.4
03	CONCRETE	83.0	62.9	74.2	75.3	55.4	66.6	79.7	64.0	72.8	81.2	62.8	73.1	86.1	67.6	78.0	86.1	62.7	75.8
04	MASONRY	97.4	57.0	72.6	106.1	56.9	75.9	91.3	57.1	70.3	108.1	49.2	71.9	99.8	57.4	73.8	84.1	57.0	67.5
05	METALS	102.5	62.3	90.7	92.6	59.4	82.9	92.6	77.0	88.0	94.1	73.9	88.2	95.7	63.7	86.3	92.6	62.1	83.6
06	WOOD, PLASTICS & COMPOSITES	102.0	54.4	75.8	85.5	36.2	58.4	100.0	54.6	75.0	104.3	54.6	76.9	92.5	65.1	77.4	95.6	54.4	72.9
07	THERMAL & MOISTURE PROTECTION	100.7	64.8	85.1	96.0	61.2	80.9	96.5	64.1	82.4	96.8	61.1	81.2	92.1	66.9	81.1	96.6	65.0	82.9
08	OPENINGS	106.6	55.9	94.8	96.0	43.9	83.9	96.0	55.0	86.5	97.2	54.0	87.1	100.0	61.8	91.1	96.0	55.0	86.4
0920	Plaster & Gypsum Board	92.8	53.7	66.5	77.6	35.0	48.9	84.2	53.7	63.7	86.4	53.7	64.4	94.7	64.6	74.4	82.9	53.7	63.3
0950, 0980	Ceilings & Acoustic Treatment	96.4	53.7	67.6	83.0	35.0	50.6	83.0	53.7	63.3	92.7	53.7	66.4	87.4	64.6	72.0	83.0	53.7	63.3
0960	Flooring	85.0	72.9	81.6	88.6	57.3	79.8	95.7	51.7	83.3	98.2	33.2	79.9	86.4	72.9	82.6	92.8	57.3	82.8
0970, 0990	Wall Finishes & Painting/Coating	86.4	44.8	61.8	92.8	42.2	62.9	92.8	42.2	62.9	92.8	42.2	62.9	89.5	44.8	63.1	92.8	44.8	64.4
09	FINISHES	84.9	56.7	69.6	82.2	42.4	60.7	85.1	52.2	67.3	88.1	48.8	66.9	86.3	63.1	73.7	84.9	54.4	68.4
COVERS	DIVS. 10 - 14, 25, 28, 41, 43, 44, 46	100.0	79.4	95.4	100.0	77.4	95.0	100.0	76.3	94.7	100.0	76.3	94.7	100.0	80.9	95.8	100.0	79.4	95.4
21, 22, 23	FIRE SUPPRESSION, PLUMBING & HVAC	100.0	67.5	87.2	96.6	63.5	83.6	96.6	63.6	83.7	100.1	61.3	84.9	99.9	67.8	87.3	96.6	63.6	83.7
26, 27, 3370	ELECTRICAL, COMMUNICATIONS & UTIL.	97.6	69.2	83.6	95.8	67.0	81.6	97.1	67.0	82.3	95.4	67.2	81.5	103.4	71.1	87.5	95.4	65.6	80.7
MF2018	WEIGHTED AVERAGE	97.3	65.8	84.0	92.4	60.5	78.9	92.7	65.1	81.1	95.0	62.9	81.4	96.4	68.7	84.7	93.1	64.0	80.8

		OKLAHOMA												OREGON					
		POTEAU			SHAWNEE			TULSA			WOODWARD			BEND			EUGENE		
	DIVISION	749			748			740 - 741			738			977			974		
		MAT.	INST.	TOTAL	MAT.	INST.	TOTAL	MAT.	INST.	TOTAL	MAT.	INST.	TOTAL	MAT.	INST.	TOTAL	MAT.	INST.	TOTAL
015433	CONTRACTOR EQUIPMENT		89.7	89.7		79.9	79.9		90.9	90.9		79.9	79.9		97.3	97.3		97.3	97.3
0241, 31 - 34	SITE & INFRASTRUCTURE, DEMOLITION	74.8	83.8	81.0	96.9	90.9	92.7	94.8	87.2	89.5	98.5	91.2	93.5	105.9	98.5	100.8	96.6	98.5	97.9
0310	Concrete Forming & Accessories	87.0	54.9	59.7	82.8	55.3	59.4	97.4	56.0	62.1	86.9	43.7	50.1	107.5	96.6	98.2	104.2	96.6	97.7
0320	Concrete Reinforcing	88.7	67.7	78.5	87.7	65.4	76.9	88.0	67.7	78.2	79.5	67.7	73.8	91.0	112.8	101.5	94.9	112.8	103.5
0330	Cast-in-Place Concrete	81.4	71.3	77.7	93.9	70.5	85.2	89.7	73.2	83.6	93.4	70.5	84.9	117.6	100.6	111.3	113.9	100.6	108.9
03	CONCRETE	81.9	63.8	73.9	87.7	62.3	76.6	86.2	65.0	76.9	86.5	57.5	73.7	110.5	100.3	106.0	101.3	100.3	100.9
04	MASONRY	91.6	57.1	70.4	107.5	57.0	76.5	92.2	57.1	70.6	90.8	57.0	70.0	105.1	100.6	102.3	102.1	100.6	101.1
05	METALS	92.6	76.7	87.9	92.5	61.3	83.3	97.0	77.1	91.2	97.4	62.1	87.0	108.0	96.0	104.5	108.7	95.9	105.0
06	WOOD, PLASTICS & COMPOSITES	92.4	54.6	71.6	87.7	54.4	69.4	103.5	55.6	77.2	96.6	38.7	64.8	99.6	96.1	97.7	95.4	96.1	95.8
07	THERMAL & MOISTURE PROTECTION	96.6	64.1	82.5	96.6	63.9	82.4	96.8	64.3	82.7	100.9	63.3	84.5	118.3	100.2	110.4	117.4	102.7	111.0
08	OPENINGS	96.0	55.0	86.5	96.0	54.4	86.3	98.9	56.2	88.9	103.7	46.3	90.3	96.9	100.0	97.6	97.2	100.0	97.8
0920	Plaster & Gypsum Board	81.3	53.7	62.8	78.5	53.7	61.8	86.4	54.8	65.1	90.2	37.6	54.8	122.7	95.9	104.6	120.8	95.9	104.0
0950, 0980	Ceilings & Acoustic Treatment	83.0	53.7	63.3	83.0	53.7	63.3	92.7	54.8	67.1	89.1	37.6	54.4	85.7	95.9	92.6	86.6	95.9	92.9
0960	Flooring	92.2	54.6	81.6	89.7	57.3	80.6	97.0	61.6	87.1	81.7	54.6	74.1	105.8	105.7	105.7	104.2	105.7	104.6
0970, 0990	Wall Finishes & Painting/Coating	92.8	44.8	64.4	92.8	42.2	62.9	92.8	51.3	68.2	86.4	44.8	61.8	98.6	77.3	86.0	98.6	69.6	81.4
09	FINISHES	82.9	52.5	66.4	83.5	53.2	67.1	88.0	55.8	70.5	82.6	43.7	61.5	100.1	95.9	97.8	98.6	95.0	96.7
COVERS	DIVS. 10 - 14, 25, 28, 41, 43, 44, 46	100.0	79.6	95.5	100.0	79.4	95.4	100.0	76.4	94.7	100.0	77.7	95.0	100.0	102.0	100.4	100.0	102.0	100.4
21, 22, 23	FIRE SUPPRESSION, PLUMBING & HVAC	96.6	63.6	83.7	96.6	67.5	85.2	100.1	63.7	85.8	96.5	67.5	85.1	96.6	105.2	100.0	100.1	98.6	99.5
26, 27, 3370	ELECTRICAL, COMMUNICATIONS & UTIL.	95.5	67.0	81.5	97.4	71.0	84.4	97.3	67.0	82.4	97.4	71.0	84.4	102.0	99.8	100.9	100.7	99.8	100.2
MF2018	WEIGHTED AVERAGE	92.3	64.9	80.7	94.5	65.2	82.2	95.9	65.8	83.2	95.3	62.9	81.6	102.6	100.2	101.6	101.7	98.7	100.5

City Cost Indexes

OREGON

DIVISION		KLAMATH FALLS 976 MAT.	INST.	TOTAL	MEDFORD 975 MAT.	INST.	TOTAL	PENDLETON 978 MAT.	INST.	TOTAL	PORTLAND 970-972 MAT.	INST.	TOTAL	SALEM 973 MAT.	INST.	TOTAL	VALE 979 MAT.	INST.	TOTAL
015433	CONTRACTOR EQUIPMENT		97.3	97.3		97.3	97.3		94.9	94.9		97.3	97.3		99.9	99.9		94.9	94.9
0241, 31 - 34	SITE & INFRASTRUCTURE, DEMOLITION	109.8	98.4	102.0	104.0	98.4	100.2	103.0	91.9	95.4	99.1	98.5	98.7	92.6	102.1	99.2	90.6	91.6	91.5
0310	Concrete Forming & Accessories	100.7	96.3	97.0	99.7	96.3	96.8	101.2	96.6	97.3	105.3	96.8	98.1	105.8	96.8	98.1	107.5	95.3	97.1
0320	Concrete Reinforcing	91.0	112.7	101.5	92.6	112.7	102.3	90.3	112.8	101.2	95.6	112.8	103.9	102.7	112.8	107.6	88.1	112.5	99.9
0330	Cast-in-Place Concrete	117.6	97.0	110.0	117.6	100.4	111.2	118.4	97.5	110.7	117.0	100.6	110.9	107.6	101.7	105.4	93.2	97.8	94.9
03	CONCRETE	113.5	98.9	107.1	107.8	100.1	104.4	93.7	99.3	96.2	103.0	100.4	101.8	98.7	100.7	99.6	77.8	98.8	87.0
04	MASONRY	119.1	100.6	107.7	99.0	100.6	100.0	109.9	100.6	104.2	104.0	100.6	101.9	108.2	100.6	103.5	107.7	100.6	103.4
05	METALS	108.0	95.6	104.4	108.3	95.6	104.6	116.3	96.4	110.4	110.1	96.1	106.0	117.1	95.4	110.8	116.2	95.1	110.0
06	WOOD, PLASTICS & COMPOSITES	90.4	96.1	93.5	89.3	96.1	93.0	92.3	96.2	94.5	96.4	96.1	96.3	90.3	96.3	93.6	100.9	96.2	98.3
07	THERMAL & MOISTURE PROTECTION	118.5	96.8	109.1	118.1	96.1	108.5	110.7	95.4	104.1	117.3	100.2	109.9	114.1	100.9	108.3	110.1	91.4	102.0
08	OPENINGS	96.9	100.0	97.6	99.6	100.0	99.7	93.2	100.0	94.8	95.1	100.0	96.3	102.6	100.0	102.0	93.2	89.0	92.2
0920	Plaster & Gypsum Board	117.1	95.9	102.8	116.5	95.9	102.6	102.4	95.9	98.0	120.3	95.9	103.9	115.4	95.9	102.3	109.0	95.9	100.2
0950, 0980	Ceilings & Acoustic Treatment	93.0	95.9	95.0	98.5	95.9	96.7	61.9	95.9	84.8	88.5	95.9	93.5	96.0	95.9	95.9	61.9	95.9	84.8
0960	Flooring	102.8	105.7	103.6	102.2	105.7	103.2	70.8	105.7	80.6	101.9	105.7	102.9	107.4	105.7	106.9	72.9	105.7	82.1
0970, 0990	Wall Finishes & Painting/Coating	98.6	65.2	78.8	98.6	65.2	78.8	88.3	77.3	81.8	98.4	77.3	85.9	97.3	75.2	84.2	88.3	77.3	81.8
09	FINISHES	100.3	94.6	97.2	100.4	94.6	97.3	70.4	96.0	84.2	98.2	95.9	97.0	99.7	95.8	97.6	71.0	96.0	84.5
COVERS	DIVS. 10 - 14, 25, 28, 41, 43, 44, 46	100.0	101.9	100.4	100.0	101.9	100.4	100.0	96.5	99.2	100.0	102.1	100.5	100.0	102.4	100.5	100.0	102.2	100.5
21, 22, 23	FIRE SUPPRESSION, PLUMBING & HVAC	96.6	105.2	100.0	100.1	105.2	102.1	98.7	112.0	104.0	100.1	111.5	104.6	100.1	106.9	102.8	98.7	71.1	87.9
26, 27, 3370	ELECTRICAL, COMMUNICATIONS & UTIL.	100.7	82.6	91.8	104.1	82.6	93.5	92.8	96.8	94.8	100.9	108.7	104.7	109.3	99.8	104.6	92.8	68.1	80.6
MF2018	WEIGHTED AVERAGE	103.5	97.3	100.9	103.3	97.4	100.8	98.6	100.3	99.3	102.1	102.8	102.4	104.4	100.8	102.9	96.2	87.0	92.3

PENNSYLVANIA

DIVISION		ALLENTOWN 181 MAT.	INST.	TOTAL	ALTOONA 166 MAT.	INST.	TOTAL	BEDFORD 155 MAT.	INST.	TOTAL	BRADFORD 167 MAT.	INST.	TOTAL	BUTLER 160 MAT.	INST.	TOTAL	CHAMBERSBURG 172 MAT.	INST.	TOTAL
015433	CONTRACTOR EQUIPMENT		112.6	112.6		112.6	112.6		110.7	110.7		112.6	112.6		112.6	112.6		111.8	111.8
0241, 31 - 34	SITE & INFRASTRUCTURE, DEMOLITION	91.7	97.2	95.5	95.1	97.1	96.5	103.6	94.8	97.6	90.5	96.0	94.3	86.3	97.8	94.2	86.3	95.3	92.5
0310	Concrete Forming & Accessories	97.9	108.2	106.7	83.7	82.6	82.8	82.0	81.2	81.3	85.8	96.5	94.9	85.1	94.1	92.8	88.0	76.8	78.5
0320	Concrete Reinforcing	96.8	113.3	104.8	93.8	105.7	99.6	93.0	105.7	99.1	95.8	105.8	100.6	94.4	119.3	106.5	94.5	112.2	103.1
0330	Cast-in-Place Concrete	89.4	99.9	93.3	99.7	86.9	94.9	109.1	85.9	100.5	95.2	90.6	93.5	88.0	96.2	91.0	93.0	93.8	93.3
03	CONCRETE	92.3	107.0	98.8	87.5	89.5	88.4	102.2	88.5	96.2	94.2	97.0	95.5	79.5	100.3	88.7	97.1	90.5	94.2
04	MASONRY	92.1	93.1	92.7	95.7	83.4	88.1	107.1	80.2	90.6	93.1	82.0	86.3	97.6	93.0	94.8	93.8	79.1	84.8
05	METALS	98.9	120.5	105.3	92.9	114.8	99.3	102.2	113.4	105.5	96.7	113.2	101.6	92.6	121.7	101.2	98.3	118.8	104.3
06	WOOD, PLASTICS & COMPOSITES	97.6	111.5	105.3	76.7	81.8	79.5	79.4	81.7	80.7	82.7	101.2	92.9	78.1	94.1	86.9	84.3	75.6	79.5
07	THERMAL & MOISTURE PROTECTION	104.0	107.8	105.7	102.7	90.0	97.2	98.3	87.4	93.6	103.8	88.5	97.2	102.2	94.4	98.8	96.1	83.1	90.4
08	OPENINGS	92.4	107.3	95.9	86.3	85.6	86.1	93.4	85.6	91.6	92.4	95.3	93.1	86.2	99.3	89.3	89.2	81.4	87.4
0920	Plaster & Gypsum Board	94.9	111.8	106.3	86.3	81.2	82.9	97.5	81.2	86.6	87.1	101.2	96.6	86.3	93.9	91.4	108.9	74.9	86.0
0950, 0980	Ceilings & Acoustic Treatment	88.5	111.8	104.2	90.9	81.2	84.4	101.1	81.2	87.7	91.0	101.2	97.9	91.8	93.9	93.2	96.0	74.9	81.8
0960	Flooring	92.0	95.1	92.8	85.3	98.9	89.2	95.0	103.7	97.5	86.2	103.7	91.1	86.3	105.3	91.6	90.5	78.0	87.0
0970, 0990	Wall Finishes & Painting/Coating	91.9	101.8	97.8	87.2	106.2	98.4	97.6	105.7	102.4	91.9	105.7	100.0	87.2	106.2	98.4	92.0	100.7	97.1
09	FINISHES	90.6	105.8	98.8	88.2	87.5	87.8	100.6	87.4	93.5	88.4	99.4	94.3	88.1	97.2	93.0	90.8	78.8	84.3
COVERS	DIVS. 10 - 14, 25, 28, 41, 43, 44, 46	100.0	99.6	99.9	100.0	94.5	98.8	100.0	93.2	98.5	100.0	96.1	99.1	100.0	97.3	99.4	100.0	92.7	98.4
21, 22, 23	FIRE SUPPRESSION, PLUMBING & HVAC	100.3	114.5	105.9	99.8	83.6	93.4	96.6	84.4	91.8	96.8	90.6	94.4	96.3	95.4	96.0	96.7	89.7	94.0
26, 27, 3370	ELECTRICAL, COMMUNICATIONS & UTIL.	98.0	95.8	96.9	88.7	107.7	98.0	94.2	107.7	100.9	92.1	107.7	99.8	89.2	107.7	98.3	92.4	85.8	89.2
MF2018	WEIGHTED AVERAGE	96.7	105.8	100.6	93.3	92.8	93.1	98.8	92.1	96.0	94.9	97.1	95.8	91.3	100.6	95.2	95.1	89.4	92.7

PENNSYLVANIA

DIVISION		DOYLESTOWN 189 MAT.	INST.	TOTAL	DUBOIS 158 MAT.	INST.	TOTAL	ERIE 164 - 165 MAT.	INST.	TOTAL	GREENSBURG 156 MAT.	INST.	TOTAL	HARRISBURG 170 - 171 MAT.	INST.	TOTAL	HAZLETON 182 MAT.	INST.	TOTAL
015433	CONTRACTOR EQUIPMENT		91.5	91.5		110.7	110.7		112.6	112.6		110.7	110.7		114.7	114.7		112.6	112.6
0241, 31 - 34	SITE & INFRASTRUCTURE, DEMOLITION	104.8	85.9	91.7	108.4	95.2	99.3	92.1	97.4	95.7	99.6	96.7	97.6	86.6	100.3	96.1	85.0	97.1	93.3
0310	Concrete Forming & Accessories	82.9	126.8	120.3	81.6	83.7	83.4	97.2	85.8	87.5	87.8	88.6	88.5	100.5	85.7	87.9	80.6	88.8	87.6
0320	Concrete Reinforcing	93.6	150.9	121.3	92.3	119.4	105.4	95.8	107.3	101.4	92.3	119.0	105.2	103.4	111.0	107.1	94.0	113.1	103.2
0330	Cast-in-Place Concrete	84.5	129.7	101.3	105.2	94.1	101.1	98.0	88.7	94.5	101.3	95.7	99.2	91.7	96.9	93.6	84.5	95.6	88.6
03	CONCRETE	88.0	131.2	107.0	104.4	94.7	100.1	86.4	91.9	88.8	97.2	97.5	97.4	91.9	95.3	93.4	85.0	96.7	90.2
04	MASONRY	95.4	134.9	119.7	107.8	93.4	98.9	85.0	86.8	86.1	117.5	89.1	100.0	88.8	83.9	85.8	104.7	89.5	95.4
05	METALS	96.5	123.9	104.5	102.2	118.8	107.1	93.1	115.6	99.7	102.1	119.7	107.3	105.1	117.4	108.7	98.7	119.7	104.8
06	WOOD, PLASTICS & COMPOSITES	78.5	125.8	104.5	78.3	81.7	80.2	94.0	84.4	88.7	85.3	87.1	86.3	101.8	85.3	92.7	77.1	87.0	82.6
07	THERMAL & MOISTURE PROTECTION	101.1	131.7	114.4	98.7	92.9	96.1	103.2	88.6	96.8	98.2	92.2	95.6	98.9	100.4	99.6	103.3	100.7	102.2
08	OPENINGS	94.8	133.2	103.7	93.4	88.7	92.3	86.4	88.0	86.8	93.4	95.5	93.9	100.9	86.5	97.5	93.0	93.3	93.0
0920	Plaster & Gypsum Board	85.4	126.5	113.0	96.3	81.2	86.2	94.9	83.9	87.5	98.3	86.8	90.5	113.8	84.7	94.2	85.8	86.6	86.4
0950, 0980	Ceilings & Acoustic Treatment	87.7	126.5	113.8	101.1	81.2	87.7	88.5	83.9	85.4	100.3	86.8	91.2	104.0	84.7	91.0	89.4	86.6	87.5
0960	Flooring	76.2	139.3	93.9	94.8	103.7	97.3	92.2	98.9	94.1	98.4	78.0	92.7	94.7	90.4	93.5	83.5	89.3	85.1
0970, 0990	Wall Finishes & Painting/Coating	91.5	143.7	122.4	97.6	106.2	102.7	98.2	92.3	94.7	97.6	106.2	102.7	95.8	84.8	89.3	91.9	104.1	99.1
09	FINISHES	81.9	130.5	108.2	100.9	88.8	94.4	91.6	88.4	89.9	101.3	88.1	94.1	96.0	85.8	90.5	86.6	90.2	88.6
COVERS	DIVS. 10 - 14, 25, 28, 41, 43, 44, 46	100.0	113.0	102.9	100.0	95.2	98.9	100.0	96.0	99.1	100.0	96.4	99.2	100.0	95.4	99.0	100.0	95.5	99.0
21, 22, 23	FIRE SUPPRESSION, PLUMBING & HVAC	96.3	132.5	110.6	96.6	87.5	93.0	99.8	93.6	97.4	96.6	87.3	92.9	100.1	91.7	96.8	96.8	95.5	96.3
26, 27, 3370	ELECTRICAL, COMMUNICATIONS & UTIL.	91.6	132.4	111.7	94.8	107.7	101.1	90.3	93.1	91.7	94.8	107.7	101.2	99.7	85.8	92.9	92.9	90.0	91.5
MF2018	WEIGHTED AVERAGE	94.0	127.1	108.0	99.3	96.0	97.9	93.2	93.9	93.5	98.7	96.3	97.7	98.7	93.0	96.3	94.4	95.9	95.0

For customer support on your Site Work & Landscape Costs with RSMeans data, call 800.448.8182.

807

City Cost Indexes

PENNSYLVANIA

DIVISION		INDIANA 157 MAT.	INST.	TOTAL	JOHNSTOWN 159 MAT.	INST.	TOTAL	KITTANNING 162 MAT.	INST.	TOTAL	LANCASTER 175-176 MAT.	INST.	TOTAL	LEHIGH VALLEY 180 MAT.	INST.	TOTAL	MONTROSE 188 MAT.	INST.	TOTAL
015433	CONTRACTOR EQUIPMENT		110.7	110.7		110.7	110.7		112.6	112.6		111.8	111.8		112.6	112.6		112.6	112.6
0241, 31 - 34	SITE & INFRASTRUCTURE, DEMOLITION	97.8	95.8	96.4	104.1	96.0	98.5	88.8	97.4	94.7	78.4	95.6	90.3	88.9	97.5	94.8	87.5	97.2	94.2
0310	Concrete Forming & Accessories	82.6	92.9	91.4	81.6	82.5	82.4	85.1	89.7	89.0	90.0	86.4	86.9	91.7	110.3	107.6	81.5	89.3	88.2
0320	Concrete Reinforcing	91.7	119.4	105.1	93.0	118.9	105.5	94.4	119.2	106.4	94.1	111.0	102.3	94.0	108.0	100.8	98.4	116.2	107.0
0330	Cast-in-Place Concrete	99.3	94.4	97.5	110.1	86.7	101.4	91.5	94.7	92.7	79.1	96.7	85.7	91.5	99.2	94.3	89.7	93.0	90.9
03	CONCRETE	94.7	99.1	96.6	103.2	91.6	98.1	82.1	97.7	88.9	85.2	95.6	89.8	91.4	106.9	98.2	90.1	96.6	93.0
04	MASONRY	103.8	93.5	97.5	104.7	83.7	91.8	100.1	88.8	93.1	99.6	86.7	91.7	92.1	96.5	94.8	92.0	91.5	91.7
05	METALS	102.3	119.9	107.4	102.2	117.9	106.8	92.7	120.6	100.9	98.3	118.7	104.3	98.6	119.0	104.6	96.8	120.4	103.7
06	WOOD, PLASTICS & COMPOSITES	80.1	94.0	87.7	78.3	81.7	80.2	78.1	89.3	84.3	87.4	85.1	86.2	89.1	112.6	102.0	77.8	87.0	82.9
07	THERMAL & MOISTURE PROTECTION	98.1	94.2	96.4	98.4	88.5	94.1	102.3	92.4	98.0	95.4	101.0	97.8	103.7	111.9	107.3	103.3	91.0	97.9
08	OPENINGS	93.4	95.4	93.9	93.4	88.7	92.3	86.3	96.7	88.7	89.2	86.4	88.5	92.9	106.6	96.1	89.8	91.9	90.3
0920	Plaster & Gypsum Board	97.8	93.9	95.2	96.1	81.2	86.1	86.3	89.0	88.1	111.8	84.7	93.6	88.0	112.9	104.7	86.2	86.6	86.5
0950, 0980	Ceilings & Acoustic Treatment	101.1	93.9	96.2	100.3	81.2	87.5	91.8	89.0	89.9	96.0	84.7	88.4	89.4	112.9	105.2	91.0	86.6	88.1
0960	Flooring	95.6	103.7	97.8	94.8	78.0	90.1	86.3	103.7	91.2	91.5	95.7	92.7	88.9	94.1	90.4	84.1	103.7	89.6
0970, 0990	Wall Finishes & Painting/Coating	97.6	106.2	102.7	97.6	106.2	102.7	87.2	106.2	98.4	92.0	86.6	88.8	91.9	101.0	97.3	91.9	104.1	99.1
09	FINISHES	100.4	96.1	98.1	100.3	83.9	91.4	88.3	93.4	91.0	90.4	87.3	88.9	88.8	106.5	98.4	87.4	92.7	90.3
COVERS	DIVS. 10 - 14, 25, 28, 41, 43, 44, 46	100.0	96.5	99.2	100.0	94.4	98.8	100.0	96.2	99.1	100.0	95.7	99.0	100.0	102.5	100.6	100.0	96.2	99.2
21, 22, 23	FIRE SUPPRESSION, PLUMBING & HVAC	96.6	85.4	92.2	96.6	78.9	89.6	96.3	90.7	94.1	96.7	92.7	95.1	96.8	117.6	104.9	96.8	94.8	96.0
26, 27, 3370	ELECTRICAL, COMMUNICATIONS & UTIL.	94.8	107.7	101.2	94.8	107.7	101.1	88.7	107.7	98.1	93.7	93.3	93.5	92.9	129.4	110.9	92.1	95.2	93.6
MF2018	WEIGHTED AVERAGE	97.6	97.7	97.7	98.9	91.9	95.9	91.8	97.9	94.4	93.8	94.5	94.1	94.9	111.6	102.0	93.8	96.8	95.1

PENNSYLVANIA

DIVISION		NEW CASTLE 161 MAT.	INST.	TOTAL	NORRISTOWN 194 MAT.	INST.	TOTAL	OIL CITY 163 MAT.	INST.	TOTAL	PHILADELPHIA 190-191 MAT.	INST.	TOTAL	PITTSBURGH 150-152 MAT.	INST.	TOTAL	POTTSVILLE 179 MAT.	INST.	TOTAL
015433	CONTRACTOR EQUIPMENT		112.6	112.6		97.5	97.5		112.6	112.6		99.8	99.8		99.7	99.7		111.8	111.8
0241, 31 - 34	SITE & INFRASTRUCTURE, DEMOLITION	86.7	97.7	94.3	97.3	95.1	95.8	85.2	96.3	92.9	99.8	101.1	100.7	103.7	96.1	98.5	81.3	95.8	91.3
0310	Concrete Forming & Accessories	85.1	93.6	92.3	83.7	125.3	119.1	85.1	92.7	91.6	99.8	142.4	136.2	97.5	97.5	97.5	82.0	87.7	86.8
0320	Concrete Reinforcing	93.3	100.2	96.7	95.9	150.8	122.5	94.4	94.9	94.7	100.4	143.1	121.0	93.5	122.9	107.7	93.4	114.1	103.4
0330	Cast-in-Place Concrete	88.8	94.3	90.8	86.1	127.4	101.5	86.3	94.4	89.3	89.8	135.6	106.8	108.6	101.2	105.8	84.3	96.4	88.8
03	CONCRETE	79.9	96.2	87.0	88.0	129.7	106.3	78.3	95.0	85.6	99.4	139.1	116.8	103.8	103.1	103.5	88.7	96.7	92.2
04	MASONRY	97.3	92.6	94.4	108.3	131.2	122.4	96.5	88.7	91.7	100.2	139.7	124.5	100.4	101.4	101.0	93.3	87.6	89.8
05	METALS	92.7	114.3	99.0	100.3	124.0	107.3	92.7	113.2	98.7	103.8	125.3	110.1	103.7	107.4	104.8	98.6	120.4	105.0
06	WOOD, PLASTICS & COMPOSITES	78.1	94.1	86.9	76.2	125.7	103.4	78.1	94.1	86.9	99.1	143.5	123.6	99.6	96.7	98.0	77.2	85.1	81.6
07	THERMAL & MOISTURE PROTECTION	102.2	92.0	97.8	108.7	130.1	118.0	102.1	90.7	97.1	103.4	139.7	119.2	98.6	99.3	98.9	95.6	99.6	97.3
08	OPENINGS	86.3	91.4	87.5	86.5	133.1	97.4	86.3	94.9	88.3	97.6	144.9	108.6	96.8	102.4	98.1	89.2	93.0	90.1
0920	Plaster & Gypsum Board	86.3	93.9	91.4	84.5	126.5	112.7	86.3	93.9	91.4	103.0	144.6	131.0	97.5	96.4	96.7	106.4	84.7	91.8
0950, 0980	Ceilings & Acoustic Treatment	91.8	93.9	93.2	91.1	126.5	114.9	91.8	93.9	93.2	104.2	144.6	131.5	95.2	96.4	96.0	96.0	84.7	88.4
0960	Flooring	86.3	105.3	91.6	86.7	139.3	101.5	86.3	103.7	91.2	97.8	156.2	114.2	103.1	107.3	104.3	87.5	103.7	92.1
0970, 0990	Wall Finishes & Painting/Coating	87.2	106.2	98.4	89.4	143.7	121.5	87.2	106.2	98.4	97.2	152.6	130.0	100.7	111.3	107.0	92.0	104.1	99.1
09	FINISHES	88.2	96.9	92.9	85.3	129.5	109.2	88.0	96.1	92.4	98.3	146.5	124.4	102.4	100.3	101.2	89.2	91.3	90.3
COVERS	DIVS. 10 - 14, 25, 28, 41, 43, 44, 46	100.0	97.3	99.4	100.0	109.1	102.0	100.0	96.7	99.3	100.0	119.2	104.3	100.0	103.0	100.7	100.0	97.5	99.4
21, 22, 23	FIRE SUPPRESSION, PLUMBING & HVAC	96.3	96.8	96.5	96.6	130.6	110.0	96.3	94.2	95.5	100.1	141.2	116.3	100.0	99.8	99.9	96.7	95.8	96.3
26, 27, 3370	ELECTRICAL, COMMUNICATIONS & UTIL.	89.2	98.2	93.7	91.9	142.0	116.6	90.9	107.7	99.2	98.8	159.0	128.4	97.1	111.3	104.1	92.0	91.3	91.6
MF2018	WEIGHTED AVERAGE	91.4	97.8	94.1	94.7	128.0	108.8	91.3	97.8	94.0	100.2	138.7	116.5	100.7	102.7	101.5	93.7	96.1	94.7

PENNSYLVANIA

DIVISION		READING 195-196 MAT.	INST.	TOTAL	SCRANTON 184-185 MAT.	INST.	TOTAL	STATE COLLEGE 168 MAT.	INST.	TOTAL	STROUDSBURG 183 MAT.	INST.	TOTAL	SUNBURY 178 MAT.	INST.	TOTAL	UNIONTOWN 154 MAT.	INST.	TOTAL
015433	CONTRACTOR EQUIPMENT		119.0	119.0		112.6	112.6		111.8	111.8		112.6	112.6		112.6	112.6		110.7	110.7
0241, 31 - 34	SITE & INFRASTRUCTURE, DEMOLITION	101.8	107.7	105.9	92.2	97.2	95.7	82.7	95.7	91.7	86.6	97.3	94.0	92.6	96.7	95.4	98.3	96.5	97.1
0310	Concrete Forming & Accessories	98.8	86.2	88.0	98.0	86.6	88.3	84.3	82.8	83.0	86.5	90.5	89.9	93.1	84.8	86.0	76.5	93.9	91.4
0320	Concrete Reinforcing	97.2	148.6	122.1	96.8	116.3	106.2	95.1	105.8	100.3	97.1	116.4	106.4	96.0	112.3	103.9	92.3	119.5	105.5
0330	Cast-in-Place Concrete	77.2	97.0	84.6	93.4	93.2	93.3	90.0	87.0	88.9	88.0	94.5	90.4	92.2	94.9	93.2	99.3	95.6	97.9
03	CONCRETE	87.1	102.1	93.7	94.1	95.4	94.7	94.6	89.7	92.5	88.8	97.7	92.7	92.0	94.5	93.1	94.4	100.0	96.9
04	MASONRY	97.5	91.5	93.8	92.5	93.8	93.3	97.8	83.9	89.3	90.3	96.3	94.0	93.6	81.0	85.8	119.3	95.4	104.6
05	METALS	100.6	133.0	111.3	100.9	120.5	106.6	96.6	115.1	102.0	98.7	120.9	105.2	98.3	119.4	104.5	102.0	120.2	107.3
06	WOOD, PLASTICS & COMPOSITES	93.7	82.4	87.5	97.6	83.3	89.8	84.5	81.8	83.0	83.5	87.0	85.5	85.4	85.1	85.2	72.8	94.0	84.4
07	THERMAL & MOISTURE PROTECTION	109.0	103.3	106.6	103.9	91.3	98.4	102.9	97.6	100.6	103.5	91.3	98.2	96.8	92.7	95.0	98.0	95.1	96.7
08	OPENINGS	90.9	99.4	92.9	92.4	89.8	91.8	89.6	85.6	88.7	93.0	94.5	93.3	89.3	89.9	89.4	93.3	99.3	94.7
0920	Plaster & Gypsum Board	94.1	81.9	85.9	96.8	82.8	87.4	87.8	81.2	83.4	86.7	86.6	86.6	105.9	84.7	91.6	94.2	93.9	94.0
0950, 0980	Ceilings & Acoustic Treatment	83.7	81.9	82.5	96.6	82.8	87.3	88.5	81.2	83.6	87.7	86.6	87.0	93.6	84.7	87.6	100.3	93.9	96.0
0960	Flooring	91.3	97.9	93.2	92.0	92.9	92.2	89.4	97.9	91.8	86.6	94.1	88.8	88.3	103.7	92.6	92.0	103.7	95.3
0970, 0990	Wall Finishes & Painting/Coating	88.2	101.8	96.3	91.9	108.3	101.6	91.9	106.2	100.4	91.9	104.1	99.1	92.0	104.1	99.1	97.6	110.0	104.9
09	FINISHES	86.8	88.4	87.7	92.4	89.3	90.7	86.8	87.3	87.6	87.5	91.8	89.8	90.2	89.6	89.9	98.7	97.0	97.8
COVERS	DIVS. 10 - 14, 25, 28, 41, 43, 44, 46	100.0	95.8	99.1	100.0	94.6	98.8	100.0	94.6	98.8	100.0	97.2	99.4	100.0	93.1	98.5	100.0	97.2	99.4
21, 22, 23	FIRE SUPPRESSION, PLUMBING & HVAC	100.3	106.8	102.8	100.3	96.0	98.6	96.8	91.4	94.7	96.8	97.9	97.2	96.7	87.4	93.0	96.6	91.7	94.6
26, 27, 3370	ELECTRICAL, COMMUNICATIONS & UTIL.	97.8	91.3	94.6	98.0	95.2	96.6	91.3	107.7	99.4	92.9	141.8	117.0	92.4	91.9	92.1	91.8	107.7	99.7
MF2018	WEIGHTED AVERAGE	96.5	101.5	98.6	97.4	96.5	97.1	94.5	94.6	94.5	94.3	104.7	98.7	94.6	92.7	93.8	97.8	99.8	98.6

For customer support on your Site Work & Landscape Costs with RSMeans data, call 800.448.8182.

PENNSYLVANIA

DIVISION		WASHINGTON 153 MAT.	INST.	TOTAL	WELLSBORO 169 MAT.	INST.	TOTAL	WESTCHESTER 193 MAT.	INST.	TOTAL	WILKES-BARRE 186 - 187 MAT.	INST.	TOTAL	WILLIAMSPORT 177 MAT.	INST.	TOTAL	YORK 173 - 174 MAT.	INST.	TOTAL
015433	CONTRACTOR EQUIPMENT		110.7	110.7		112.6	112.6		97.5	97.5		112.6	112.6		112.6	112.6		111.8	111.8
0241, 31 - 34	SITE & INFRASTRUCTURE, DEMOLITION	98.3	96.8	97.3	94.0	96.6	95.8	103.4	96.1	98.4	84.5	97.2	93.3	83.7	96.8	92.8	82.4	95.6	91.5
0310	Concrete Forming & Accessories	82.8	94.0	92.4	85.2	84.2	84.3	90.0	126.6	121.2	89.1	85.3	85.8	90.1	85.6	86.3	85.1	86.3	86.1
0320	Concrete Reinforcing	92.3	119.5	105.5	95.1	116.1	105.3	95.0	150.8	122.0	95.8	116.3	105.7	95.3	110.9	102.8	96.0	111.0	103.3
0330	Cast-in-Place Concrete	99.3	95.9	98.0	94.4	88.2	92.1	95.4	129.5	108.1	84.5	93.0	87.6	78.0	90.0	82.5	84.7	96.7	89.2
03	CONCRETE	94.8	100.1	97.2	96.9	92.5	95.0	95.8	131.0	111.3	85.8	94.7	89.7	80.2	92.9	85.8	89.8	95.6	92.4
04	MASONRY	103.6	95.5	98.6	98.7	81.4	88.1	102.8	134.9	122.5	105.0	93.5	97.9	85.9	86.1	86.0	94.5	86.7	89.7
05	METALS	101.9	120.6	107.4	96.6	119.7	103.4	100.3	124.0	107.3	96.8	120.4	103.7	98.3	118.1	104.1	100.0	118.6	105.5
06	WOOD, PLASTICS & COMPOSITES	80.2	94.0	87.8	82.1	84.4	83.4	82.8	125.7	106.4	85.9	81.7	83.6	81.9	85.1	83.7	80.7	85.1	83.1
07	THERMAL & MOISTURE PROTECTION	98.1	95.1	96.8	104.1	86.0	96.2	109.1	127.0	116.9	103.3	90.9	97.9	96.1	88.3	92.7	95.6	101.0	98.0
08	OPENINGS	93.3	99.3	94.7	92.4	90.0	91.8	86.5	133.1	97.4	89.8	88.9	89.6	89.3	89.6	89.4	89.2	86.4	88.5
0920	Plaster & Gypsum Board	97.7	93.9	95.1	86.5	83.9	84.8	85.2	126.5	112.9	87.4	81.2	83.2	106.4	84.7	91.8	106.7	84.7	91.9
0950, 0980	Ceilings & Acoustic Treatment	100.3	93.9	96.0	88.5	83.9	85.4	91.1	126.5	114.9	91.0	81.2	84.4	96.0	84.7	88.4	95.1	84.7	88.1
0960	Flooring	95.7	103.7	97.9	85.8	103.7	90.9	89.6	139.3	103.6	87.7	92.9	89.1	87.2	86.8	87.1	88.9	95.7	90.8
0970, 0990	Wall Finishes & Painting/Coating	97.6	110.0	104.9	91.9	104.1	99.1	89.4	143.7	121.5	91.9	101.8	97.8	92.0	101.8	97.8	92.0	84.8	87.7
09	FINISHES	100.3	97.0	98.5	88.1	89.6	88.9	86.7	130.5	110.4	88.4	87.6	88.0	89.7	86.7	88.0	89.5	87.1	88.2
COVERS	DIVS. 10 - 14, 25, 28, 41, 43, 44, 46	100.0	97.2	99.4	100.0	96.2	99.2	100.0	115.3	103.4	100.0	94.3	98.7	100.0	95.5	99.0	100.0	95.6	99.0
21, 22, 23	FIRE SUPPRESSION, PLUMBING & HVAC	96.6	93.2	95.2	96.8	87.4	93.1	96.6	132.6	110.7	96.8	96.6	96.7	96.7	92.2	94.9	100.2	92.7	97.3
26, 27, 3370	ELECTRICAL, COMMUNICATIONS & UTIL.	94.2	107.7	100.9	92.1	95.2	93.6	91.7	132.4	111.8	92.9	90.0	91.5	92.8	83.4	88.2	93.7	82.9	88.4
MF2018	WEIGHTED AVERAGE	97.5	100.2	98.6	95.5	92.9	94.4	95.8	127.9	109.3	94.0	95.5	94.7	92.5	92.3	92.4	95.2	93.0	94.3

DIVISION		PUERTO RICO SAN JUAN 009 MAT.	INST.	TOTAL	RHODE ISLAND NEWPORT 028 MAT.	INST.	TOTAL	PROVIDENCE 029 MAT.	INST.	TOTAL	SOUTH CAROLINA AIKEN 298 MAT.	INST.	TOTAL	BEAUFORT 299 MAT.	INST.	TOTAL	CHARLESTON 294 MAT.	INST.	TOTAL
015433	CONTRACTOR EQUIPMENT		87.3	87.3		97.9	97.9		100.1	100.1		103.7	103.7		103.7	103.7		103.7	103.7
0241, 31 - 34	SITE & INFRASTRUCTURE, DEMOLITION	81.4	87.9	85.9	86.3	98.0	94.4	89.4	102.2	98.3	128.3	85.2	98.5	123.4	84.7	96.7	108.2	85.2	92.3
0310	Concrete Forming & Accessories	105.9	21.6	34.0	100.7	124.7	121.1	99.9	124.7	121.0	94.0	64.6	68.9	93.1	38.9	46.9	92.3	64.9	69.0
0320	Concrete Reinforcing	114.9	18.7	68.4	109.7	127.1	118.1	103.7	127.1	115.0	94.3	60.6	78.0	93.4	66.7	80.5	93.2	67.3	80.7
0330	Cast-in-Place Concrete	70.9	32.7	56.6	71.7	118.5	89.1	87.5	119.0	99.2	92.9	66.9	83.2	92.9	66.8	83.2	109.1	67.2	93.5
03	CONCRETE	89.8	26.1	61.8	87.5	122.4	102.9	91.8	122.5	105.3	101.0	66.4	85.8	98.4	55.7	79.7	94.9	67.8	83.0
04	MASONRY	110.0	24.2	57.3	99.2	123.1	113.8	103.4	123.1	115.5	79.9	64.8	70.6	94.9	64.8	76.4	96.4	67.5	78.6
05	METALS	116.8	40.6	94.4	98.8	116.7	104.1	103.2	115.9	106.9	102.6	89.1	98.7	102.7	90.7	99.2	104.6	92.1	100.9
06	WOOD, PLASTICS & COMPOSITES	145.4	20.5	76.6	98.0	124.0	112.3	98.4	124.0	112.5	89.7	66.8	77.1	88.4	32.5	57.6	87.2	66.8	76.0
07	THERMAL & MOISTURE PROTECTION	134.1	27.7	87.8	102.6	123.0	111.5	108.3	123.5	114.9	98.1	64.2	83.4	97.8	56.2	79.7	96.8	66.1	83.5
08	OPENINGS	96.8	19.1	78.7	97.4	124.4	103.7	99.0	124.4	104.9	94.0	62.5	86.7	94.1	44.9	82.6	97.6	65.1	90.0
0920	Plaster & Gypsum Board	76.4	25.2	48.0	91.4	124.3	113.6	104.9	124.3	118.0	83.4	65.8	71.6	87.0	30.5	49.0	88.1	65.8	73.1
0950, 0980	Ceilings & Acoustic Treatment	158.8	19.3	64.7	90.3	124.3	113.2	104.1	124.1	117.7	78.8	65.8	70.1	82.1	30.5	47.3	82.1	65.8	71.1
0960	Flooring	117.8	29.7	93.1	83.8	127.3	96.0	86.4	127.3	97.9	97.5	89.8	95.3	99.0	75.8	92.5	98.6	81.9	93.9
0970, 0990	Wall Finishes & Painting/Coating	90.2	24.6	51.4	85.0	118.9	105.0	90.8	118.9	107.4	95.2	66.0	77.9	95.2	56.9	72.5	95.2	70.0	80.3
09	FINISHES	106.0	23.2	61.2	87.7	125.1	107.9	93.7	125.1	110.7	89.0	69.7	78.5	90.1	46.0	66.2	88.2	68.9	77.8
COVERS	DIVS. 10 - 14, 25, 28, 41, 43, 44, 46	100.0	26.6	83.6	100.0	110.2	102.3	100.0	110.3	102.3	100.0	70.3	93.4	100.0	80.1	95.6	100.0	70.3	93.4
21, 22, 23	FIRE SUPPRESSION, PLUMBING & HVAC	88.5	18.8	61.1	100.4	114.8	106.0	100.2	114.8	105.9	96.9	55.2	80.5	96.9	55.6	80.7	100.4	58.4	83.9
26, 27, 3370	ELECTRICAL, COMMUNICATIONS & UTIL.	84.4	25.0	55.2	102.1	95.5	98.9	102.3	95.5	99.0	96.2	61.9	79.3	99.7	30.9	65.8	98.1	58.9	78.8
MF2018	WEIGHTED AVERAGE	98.5	29.8	69.4	97.0	114.7	104.5	99.3	115.0	105.9	97.6	67.1	84.7	98.3	57.5	81.1	98.7	68.2	85.8

DIVISION		SOUTH CAROLINA COLUMBIA 290 - 292 MAT.	INST.	TOTAL	FLORENCE 295 MAT.	INST.	TOTAL	GREENVILLE 296 MAT.	INST.	TOTAL	ROCK HILL 297 MAT.	INST.	TOTAL	SPARTANBURG 293 MAT.	INST.	TOTAL	SOUTH DAKOTA ABERDEEN 574 MAT.	INST.	TOTAL
015433	CONTRACTOR EQUIPMENT		105.6	105.6		103.7	103.7		103.7	103.7		103.7	103.7		103.7	103.7		97.7	97.7
0241, 31 - 34	SITE & INFRASTRUCTURE, DEMOLITION	105.9	89.5	94.6	117.6	84.9	95.1	112.6	85.4	93.8	110.0	84.3	92.2	112.4	85.4	93.8	99.2	93.4	95.2
0310	Concrete Forming & Accessories	93.2	64.9	69.1	81.1	64.8	67.2	91.8	64.9	68.8	90.1	64.0	67.9	94.4	64.9	69.2	98.9	75.7	79.1
0320	Concrete Reinforcing	98.3	67.3	83.3	92.8	67.2	80.4	92.8	64.4	79.1	93.5	65.8	80.1	92.8	67.3	80.4	101.0	71.4	86.7
0330	Cast-in-Place Concrete	110.3	67.5	94.4	92.9	67.0	83.2	92.9	67.2	83.3	92.8	66.4	83.0	92.9	67.2	83.3	109.5	78.8	98.1
03	CONCRETE	93.2	67.8	82.1	93.3	67.6	82.0	92.0	67.3	81.2	90.1	66.8	79.9	92.2	67.8	81.5	98.7	76.7	89.1
04	MASONRY	89.9	67.5	76.1	79.9	67.4	72.2	77.9	67.5	71.5	101.6	64.0	78.5	79.9	67.5	72.3	109.4	73.3	87.2
05	METALS	101.6	91.0	98.5	103.5	91.4	99.9	103.5	91.7	100.0	102.7	90.5	99.1	103.5	92.1	100.1	93.8	82.1	90.4
06	WOOD, PLASTICS & COMPOSITES	88.6	66.9	76.7	74.8	66.8	70.4	86.8	66.8	75.8	85.4	66.8	75.2	90.6	66.8	77.5	97.6	74.2	84.7
07	THERMAL & MOISTURE PROTECTION	93.3	66.3	81.5	97.2	66.1	83.7	97.1	66.1	83.6	96.9	59.2	80.5	97.1	66.1	83.6	103.1	77.2	91.8
08	OPENINGS	99.7	65.1	91.7	94.1	65.1	87.3	94.0	64.8	87.2	94.1	63.6	87.0	94.1	65.1	87.3	97.2	62.2	89.0
0920	Plaster & Gypsum Board	88.7	65.8	73.3	78.2	65.8	69.9	82.1	65.8	71.2	81.8	65.8	71.1	84.7	65.8	72.0	110.1	73.9	85.7
0950, 0980	Ceilings & Acoustic Treatment	86.2	65.8	72.5	79.6	65.8	70.3	78.8	65.8	70.1	78.8	65.8	70.1	78.8	65.8	70.1	98.8	73.9	82.0
0960	Flooring	90.7	80.4	87.8	89.9	80.4	87.2	96.3	80.4	91.8	95.2	75.8	89.7	97.6	80.4	92.8	91.1	43.8	77.8
0970, 0990	Wall Finishes & Painting/Coating	93.9	70.0	79.7	95.2	70.0	80.3	95.2	70.0	80.3	95.2	66.0	77.9	95.2	70.0	80.3	89.8	33.4	56.4
09	FINISHES	86.5	68.7	76.8	85.0	68.7	76.2	86.9	68.7	77.0	86.3	67.1	75.9	87.6	68.7	77.4	92.4	65.9	78.1
COVERS	DIVS. 10 - 14, 25, 28, 41, 43, 44, 46	100.0	70.5	93.4	100.0	70.3	93.4	100.0	70.3	93.4	100.0	70.0	93.3	100.0	70.3	93.4	100.0	84.8	96.6
21, 22, 23	FIRE SUPPRESSION, PLUMBING & HVAC	100.0	58.3	83.6	100.4	58.2	83.9	100.4	58.2	83.9	96.9	54.8	80.4	100.4	58.2	83.9	100.0	54.2	82.0
26, 27, 3370	ELECTRICAL, COMMUNICATIONS & UTIL.	98.5	62.8	80.9	96.1	62.8	79.7	98.2	59.8	79.3	98.2	59.9	79.3	98.2	59.8	79.3	101.1	65.3	83.4
MF2018	WEIGHTED AVERAGE	97.6	68.9	85.5	96.9	68.6	84.9	97.0	68.2	84.8	96.7	66.4	83.9	97.2	68.3	85.0	98.6	70.1	86.5

For customer support on your Site Work & Landscape Costs with RSMeans data, call 800.448.8182.

809

City Cost Indexes

SOUTH DAKOTA

DIVISION		MITCHELL 573 MAT.	INST.	TOTAL	MOBRIDGE 576 MAT.	INST.	TOTAL	PIERRE 575 MAT.	INST.	TOTAL	RAPID CITY 577 MAT.	INST.	TOTAL	SIOUX FALLS 570-571 MAT.	INST.	TOTAL	WATERTOWN 572 MAT.	INST.	TOTAL
015433	CONTRACTOR EQUIPMENT		97.7	97.7		97.7	97.7		99.9	99.9		97.7	97.7		100.9	100.9		97.7	97.7
0241, 31 - 34	SITE & INFRASTRUCTURE, DEMOLITION	95.8	93.2	94.0	95.7	93.2	94.0	97.2	96.3	96.6	97.5	92.8	94.3	92.5	98.7	96.8	95.6	93.2	94.0
0310	Concrete Forming & Accessories	97.9	44.2	52.1	88.2	44.6	51.0	99.7	45.6	53.6	106.9	56.6	64.0	103.3	78.1	81.8	84.6	74.5	76.0
0320	Concrete Reinforcing	100.4	70.5	86.0	103.0	70.5	87.3	97.8	100.2	99.0	94.4	100.4	97.3	106.6	100.6	103.7	97.6	70.5	84.5
0330	Cast-in-Place Concrete	106.4	52.3	86.3	106.4	77.2	95.6	103.1	77.8	93.7	105.6	77.2	95.0	88.3	78.8	84.7	106.4	77.2	95.5
03	CONCRETE	96.5	53.1	77.4	96.2	62.0	81.2	91.8	67.4	81.1	95.9	72.3	85.5	88.6	82.7	86.0	95.3	75.5	86.6
04	MASONRY	97.8	75.5	84.1	106.5	72.8	85.8	106.5	72.6	85.7	106.0	74.0	86.3	91.1	75.6	81.5	132.0	73.7	96.2
05	METALS	92.9	81.6	89.6	92.9	82.0	89.7	96.3	89.5	94.3	95.5	91.2	94.2	95.3	92.6	94.5	92.9	81.9	89.6
06	WOOD, PLASTICS & COMPOSITES	96.6	33.7	62.0	84.4	33.4	56.4	104.2	34.0	65.6	102.2	48.0	72.3	101.6	76.5	87.8	80.5	74.2	77.0
07	THERMAL & MOISTURE PROTECTION	102.8	72.9	89.8	102.9	75.1	90.8	104.1	72.7	90.4	103.6	78.5	92.7	103.2	83.3	94.6	102.6	76.7	91.3
08	OPENINGS	96.3	39.6	83.1	98.9	39.0	84.9	99.8	58.4	90.2	101.0	66.1	92.9	104.2	81.8	99.0	96.3	61.5	88.2
0920	Plaster & Gypsum Board	109.0	32.2	57.3	102.2	32.0	54.9	108.1	32.5	57.2	109.8	46.9	67.5	103.6	76.1	85.1	100.5	73.9	82.6
0950, 0980	Ceilings & Acoustic Treatment	95.5	32.2	52.8	98.8	32.0	53.7	98.6	32.5	54.0	100.4	46.9	64.3	97.7	76.1	83.1	95.5	73.9	80.9
0960	Flooring	90.7	43.8	77.6	86.5	46.4	75.3	98.2	31.2	79.4	90.5	72.7	85.5	93.9	75.0	88.6	85.2	43.8	73.6
0970, 0990	Wall Finishes & Painting/Coating	89.8	36.6	58.3	89.8	37.5	58.8	97.1	95.6	96.2	89.8	95.6	93.2	100.3	95.6	97.5	89.8	33.4	56.4
09	FINISHES	91.4	41.1	64.2	89.9	41.6	63.7	98.4	45.2	69.6	92.5	61.8	75.9	96.4	79.1	87.1	88.6	64.7	75.7
COVERS	DIVS. 10 - 14, 25, 28, 41, 43, 44, 46	100.0	82.0	96.0	100.0	81.9	96.0	100.0	82.3	96.1	100.0	83.5	96.3	100.0	87.0	97.1	100.0	85.8	96.8
21, 22, 23	FIRE SUPPRESSION, PLUMBING & HVAC	96.5	49.0	77.8	96.5	69.0	85.7	100.0	76.8	90.9	100.0	76.8	90.9	99.9	70.2	88.2	96.5	52.1	79.0
26, 27, 3370	ELECTRICAL, COMMUNICATIONS & UTIL.	99.4	63.4	81.7	101.1	39.8	70.9	103.9	47.7	76.2	97.6	47.7	73.0	102.6	63.4	83.3	98.6	63.4	81.2
MF2018	WEIGHTED AVERAGE	96.3	60.9	81.3	96.9	62.9	82.5	99.0	68.5	86.1	98.3	72.0	87.2	97.7	78.4	89.5	97.3	69.0	85.4

TENNESSEE

DIVISION		CHATTANOOGA 373-374 MAT.	INST.	TOTAL	COLUMBIA 384 MAT.	INST.	TOTAL	COOKEVILLE 385 MAT.	INST.	TOTAL	JACKSON 383 MAT.	INST.	TOTAL	JOHNSON CITY 376 MAT.	INST.	TOTAL	KNOXVILLE 377-379 MAT.	INST.	TOTAL
015433	CONTRACTOR EQUIPMENT		104.7	104.7		99.5	99.5		99.5	99.5		105.5	105.5		98.8	98.8		98.8	98.8
0241, 31 - 34	SITE & INFRASTRUCTURE, DEMOLITION	106.6	93.8	97.8	91.5	84.6	86.7	97.3	81.6	86.5	100.9	94.0	96.2	113.6	81.1	91.2	92.6	83.8	86.5
0310	Concrete Forming & Accessories	95.8	57.3	63.0	80.9	62.4	65.1	81.1	31.9	39.2	87.5	41.0	47.8	82.7	57.5	61.2	94.3	61.7	66.5
0320	Concrete Reinforcing	101.6	65.1	83.9	91.4	65.0	78.6	91.4	64.6	78.4	91.3	67.2	79.7	102.2	61.9	82.7	101.6	61.8	82.3
0330	Cast-in-Place Concrete	97.5	64.1	85.1	92.8	63.8	82.0	102.4	56.7	87.2	102.7	66.0	89.0	78.4	58.0	70.8	91.6	64.0	81.3
03	CONCRETE	93.1	60.3	79.8	92.4	65.1	80.4	102.4	48.7	78.8	93.2	56.5	77.0	102.7	60.2	84.0	90.7	64.2	79.0
04	MASONRY	98.6	56.3	72.6	111.4	55.5	77.0	106.7	39.4	65.3	111.7	44.4	70.4	111.0	43.2	69.3	76.3	50.7	60.6
05	METALS	93.0	88.1	91.6	93.5	88.9	92.1	93.5	87.6	91.8	96.0	89.5	94.1	90.3	86.4	89.2	93.4	86.5	91.4
06	WOOD, PLASTICS & COMPOSITES	106.5	56.3	78.9	67.8	63.4	65.4	68.0	29.5	46.8	82.9	39.7	59.1	78.7	62.6	69.8	93.1	62.6	76.4
07	THERMAL & MOISTURE PROTECTION	99.2	61.7	82.9	95.6	62.9	81.4	96.1	52.6	77.2	98.0	55.5	79.5	95.3	55.6	78.0	92.8	60.8	78.9
08	OPENINGS	102.0	57.5	91.6	91.0	53.3	82.2	91.1	35.5	78.1	97.7	44.3	85.2	98.3	59.5	89.3	95.7	55.2	86.2
0920	Plaster & Gypsum Board	80.0	55.6	63.6	87.4	62.9	70.9	87.4	28.0	47.4	90.0	38.5	55.3	101.0	62.1	74.8	109.1	62.1	77.5
0950, 0980	Ceilings & Acoustic Treatment	97.9	55.6	69.4	75.8	62.9	67.1	75.8	28.0	43.6	83.9	38.5	53.3	93.5	62.1	72.3	94.3	62.1	72.6
0960	Flooring	98.3	55.2	86.2	81.7	53.9	73.9	81.7	50.0	72.8	81.0	54.2	73.5	93.3	43.9	79.4	98.4	49.7	84.7
0970, 0990	Wall Finishes & Painting/Coating	98.0	59.5	75.2	85.0	56.5	68.1	85.0	56.5	68.1	87.0	56.5	68.9	94.9	58.7	73.5	94.9	57.6	72.8
09	FINISHES	94.4	56.6	73.9	86.2	60.1	72.1	86.8	36.2	59.4	86.1	43.4	63.0	98.5	55.3	75.1	91.4	58.9	73.8
COVERS	DIVS. 10 - 14, 25, 28, 41, 43, 44, 46	100.0	67.8	92.8	100.0	68.0	92.9	100.0	35.9	85.7	100.0	62.0	91.5	100.0	75.8	94.6	100.0	79.3	95.4
21, 22, 23	FIRE SUPPRESSION, PLUMBING & HVAC	100.1	59.3	84.1	97.8	73.6	88.3	97.8	67.1	85.8	100.1	61.4	84.9	100.0	55.5	82.5	100.0	60.5	84.5
26, 27, 3370	ELECTRICAL, COMMUNICATIONS & UTIL.	100.8	82.9	92.0	94.0	50.6	72.6	95.5	61.3	78.6	100.5	65.7	83.3	91.5	40.8	66.5	97.0	54.9	76.2
MF2018	WEIGHTED AVERAGE	98.0	68.1	85.3	94.8	66.5	82.8	96.2	56.9	79.6	97.6	61.4	82.3	98.1	58.6	81.4	94.8	63.7	81.6

TENNESSEE / TEXAS

DIVISION		MCKENZIE 382 MAT.	INST.	TOTAL	MEMPHIS 375, 380-381 MAT.	INST.	TOTAL	NASHVILLE 370-372 MAT.	INST.	TOTAL	ABILENE 795-796 MAT.	INST.	TOTAL	AMARILLO 790-791 MAT.	INST.	TOTAL	AUSTIN 786-787 MAT.	INST.	TOTAL
015433	CONTRACTOR EQUIPMENT		99.5	99.5		102.0	102.0		105.5	105.5		90.9	90.9		93.9	93.9		92.5	92.5
0241, 31 - 34	SITE & INFRASTRUCTURE, DEMOLITION	96.9	81.8	86.5	91.1	95.4	94.0	102.8	98.1	99.6	91.9	87.4	88.8	91.3	91.3	91.3	95.5	90.1	91.7
0310	Concrete Forming & Accessories	88.4	34.1	42.1	94.6	64.7	69.1	95.9	68.1	72.2	94.0	59.2	64.3	95.6	51.7	58.2	95.3	53.0	59.3
0320	Concrete Reinforcing	91.5	65.5	78.9	110.9	66.0	89.2	107.1	66.8	87.6	88.5	51.7	70.7	95.3	49.6	73.2	86.8	47.2	67.6
0330	Cast-in-Place Concrete	102.9	57.9	86.2	96.5	76.7	89.1	88.2	69.0	81.1	84.6	64.9	77.3	83.2	65.7	76.7	94.0	65.9	83.5
03	CONCRETE	101.0	50.2	78.7	96.5	70.2	84.9	92.9	69.4	82.6	84.5	60.7	74.1	87.1	57.2	74.0	89.4	57.4	75.4
04	MASONRY	110.5	45.1	70.3	92.6	56.3	70.3	87.7	55.9	68.2	99.9	60.6	75.8	97.6	60.2	74.6	91.7	60.7	72.7
05	METALS	93.5	86.2	91.9	87.2	82.5	85.8	101.6	85.1	96.7	107.1	70.4	96.3	101.9	68.7	92.1	100.7	67.4	90.9
06	WOOD, PLASTICS & COMPOSITES	76.6	31.7	51.9	98.9	65.4	80.4	100.1	70.1	83.6	99.2	60.8	78.1	102.6	51.0	74.2	92.9	52.6	70.7
07	THERMAL & MOISTURE PROTECTION	96.1	53.6	77.6	93.0	66.4	81.4	96.9	65.0	83.0	98.3	63.4	83.1	100.1	62.3	83.6	95.1	63.8	81.5
08	OPENINGS	91.1	37.2	78.5	99.5	61.3	90.6	101.0	67.5	93.2	101.8	56.7	91.2	103.2	50.8	91.0	101.1	49.6	89.1
0920	Plaster & Gypsum Board	90.9	30.3	50.1	90.9	64.5	73.1	95.9	69.3	78.0	88.2	60.1	69.3	100.7	50.0	66.6	88.9	51.6	63.8
0950, 0980	Ceilings & Acoustic Treatment	75.8	30.3	45.1	90.8	64.5	73.0	94.1	69.3	77.4	99.2	60.1	72.9	100.7	50.0	66.5	93.3	51.6	65.2
0960	Flooring	84.5	53.9	75.9	97.8	55.2	85.9	97.1	57.5	86.0	93.2	67.6	86.0	93.7	63.8	85.3	94.1	63.8	85.6
0970, 0990	Wall Finishes & Painting/Coating	85.0	44.2	60.8	93.2	59.3	73.1	100.7	68.6	81.7	92.8	51.7	68.5	91.0	51.7	67.7	95.7	44.2	65.2
09	FINISHES	88.0	37.5	60.7	95.9	61.9	77.5	97.8	66.1	80.6	86.1	60.1	72.0	92.6	53.5	71.4	91.1	53.6	70.8
COVERS	DIVS. 10 - 14, 25, 28, 41, 43, 44, 46	100.0	24.0	83.1	100.0	80.5	95.7	100.0	81.1	95.8	100.0	78.9	95.3	100.0	72.7	93.9	100.0	76.5	94.8
21, 22, 23	FIRE SUPPRESSION, PLUMBING & HVAC	97.8	58.6	82.4	100.1	70.3	88.4	99.9	75.1	90.2	100.1	52.1	81.3	99.9	51.1	80.7	99.9	58.7	83.8
26, 27, 3370	ELECTRICAL, COMMUNICATIONS & UTIL.	95.3	55.4	75.6	102.9	63.6	83.5	93.5	63.2	78.6	97.9	54.0	76.3	100.9	58.9	80.2	99.5	57.2	77.6
MF2018	WEIGHTED AVERAGE	96.3	55.1	78.9	96.6	69.8	85.3	98.0	71.8	86.9	97.8	61.4	82.4	98.3	60.0	82.1	97.2	61.4	82.1

For customer support on your Site Work & Landscape Costs with RSMeans data, call 800.448.8182.

TEXAS

DIVISION		BEAUMONT 776-777			BROWNWOOD 768			BRYAN 778			CHILDRESS 792			CORPUS CHRISTI 783-784			DALLAS 752-753		
		MAT.	INST.	TOTAL	MAT.	INST.	TOTAL	MAT.	INST.	TOTAL	MAT.	INST.	TOTAL	MAT.	INST.	TOTAL	MAT.	INST.	TOTAL
015433	CONTRACTOR EQUIPMENT		95.3	95.3		90.9	90.9		95.3	95.3		90.9	90.9		100.6	100.6		106.9	106.9
0241, 31 - 34	SITE & INFRASTRUCTURE, DEMOLITION	90.6	91.8	91.4	101.3	87.3	91.7	81.5	91.9	88.7	101.8	85.8	90.8	144.2	82.8	101.8	106.0	97.6	100.2
0310	Concrete Forming & Accessories	102.3	54.6	61.7	97.6	58.7	64.4	82.1	58.2	61.7	92.6	58.6	63.6	98.4	52.2	59.0	95.0	62.3	67.1
0320	Concrete Reinforcing	85.7	62.0	74.2	81.9	51.1	67.0	87.9	51.2	70.2	88.7	51.0	70.5	79.8	50.9	65.8	91.4	52.3	72.5
0330	Cast-in-Place Concrete	90.7	65.2	81.2	96.8	58.6	82.6	72.6	59.6	67.8	86.7	58.6	76.1	114.1	67.1	96.6	92.6	71.4	84.7
03	CONCRETE	91.0	60.7	77.7	90.3	58.2	76.2	75.8	58.5	68.2	93.4	58.2	77.9	95.2	59.1	79.4	93.3	65.3	81.0
04	MASONRY	102.9	61.7	77.6	122.8	60.6	84.6	143.3	60.6	92.5	103.5	60.2	76.9	81.4	60.7	68.7	99.5	60.2	75.3
05	METALS	99.4	76.2	92.6	102.6	69.8	92.9	99.5	72.1	91.4	104.4	69.6	94.2	97.3	84.4	93.5	101.7	82.6	96.1
06	WOOD, PLASTICS & COMPOSITES	113.6	54.1	80.9	105.8	60.8	81.0	78.6	59.3	68.0	98.6	60.8	77.8	115.6	51.5	80.3	100.8	63.6	80.3
07	THERMAL & MOISTURE PROTECTION	92.8	64.3	80.4	92.0	61.9	78.9	85.2	63.8	75.9	99.0	60.4	82.2	97.7	62.2	82.2	87.3	67.2	78.6
08	OPENINGS	95.0	55.0	85.6	100.4	56.6	90.2	95.7	56.0	86.4	96.1	56.6	86.9	101.3	50.0	89.4	96.5	59.5	87.9
0920	Plaster & Gypsum Board	101.0	53.2	68.9	82.6	60.1	67.5	88.2	58.7	68.3	87.6	60.1	69.1	96.5	50.5	65.5	92.7	62.5	72.4
0950, 0980	Ceilings & Acoustic Treatment	105.3	53.2	70.2	86.6	60.1	68.7	96.7	58.7	71.1	96.8	60.1	72.1	98.8	50.5	66.2	102.4	62.5	75.5
0960	Flooring	116.9	74.6	105.0	75.7	54.6	69.8	86.7	67.0	81.1	91.6	54.6	81.2	107.4	67.7	96.3	94.2	69.4	87.2
0970, 0990	Wall Finishes & Painting/Coating	94.9	53.9	70.6	92.8	51.7	68.4	92.2	57.4	71.6	92.8	51.7	68.5	110.1	44.2	71.1	96.1	54.8	71.6
09	FINISHES	98.0	57.8	76.2	77.6	57.4	66.7	85.4	59.4	71.3	86.2	57.4	70.6	101.9	53.8	75.9	97.9	62.7	78.8
COVERS	DIVS. 10 - 14, 25, 28, 41, 43, 44, 46	100.0	77.8	95.0	100.0	73.4	94.1	100.0	77.0	94.9	100.0	73.4	94.1	100.0	77.8	95.0	100.0	82.5	96.1
21, 22, 23	FIRE SUPPRESSION, PLUMBING & HVAC	100.0	63.0	85.5	96.5	49.5	78.0	96.5	62.4	83.1	96.6	51.7	79.0	100.0	48.7	79.9	99.9	62.3	85.1
26, 27, 3370	ELECTRICAL, COMMUNICATIONS & UTIL.	100.1	63.4	82.0	98.6	46.1	72.7	97.8	63.4	80.8	97.9	55.8	77.1	94.0	63.4	78.9	95.2	63.2	79.4
MF2018	WEIGHTED AVERAGE	97.9	65.5	84.2	97.4	58.7	81.1	95.2	64.9	82.4	97.6	60.3	81.8	98.9	61.4	83.1	98.2	68.0	85.4

TEXAS

DIVISION		DEL RIO 788			DENTON 762			EASTLAND 764			EL PASO 798 - 799, 885			FORT WORTH 760 - 761			GALVESTON 775		
		MAT.	INST.	TOTAL	MAT.	INST.	TOTAL	MAT.	INST.	TOTAL	MAT.	INST.	TOTAL	MAT.	INST.	TOTAL	MAT.	INST.	TOTAL
015433	CONTRACTOR EQUIPMENT		89.9	89.9		100.2	100.2		90.9	90.9		93.9	93.9		93.9	93.9		107.9	107.9
0241, 31 - 34	SITE & INFRASTRUCTURE, DEMOLITION	122.2	85.5	96.9	101.4	79.7	86.4	104.0	85.5	91.2	89.7	92.7	91.8	97.4	92.7	94.2	108.4	90.6	96.1
0310	Concrete Forming & Accessories	95.1	51.0	57.5	105.0	58.8	65.7	98.4	58.6	64.4	95.6	56.7	62.5	98.2	59.5	65.2	90.9	55.8	61.0
0320	Concrete Reinforcing	80.4	47.6	64.5	83.2	51.1	67.7	82.1	51.1	67.1	94.2	51.1	73.4	88.1	50.9	70.1	87.5	59.7	74.0
0330	Cast-in-Place Concrete	122.9	58.4	98.9	74.9	59.7	69.2	102.3	58.5	86.0	73.9	68.0	71.7	89.3	65.9	80.6	96.4	60.6	83.1
03	CONCRETE	115.8	54.0	88.6	70.6	59.7	65.8	94.8	58.1	78.7	82.7	60.5	72.9	86.1	61.0	75.1	92.1	60.3	78.1
04	MASONRY	95.4	60.5	74.0	131.0	59.3	86.9	92.7	60.6	73.0	91.6	62.7	73.8	92.5	59.3	72.1	99.0	60.7	75.5
05	METALS	97.2	67.2	88.4	102.1	84.9	97.1	102.4	69.6	92.7	102.1	69.4	92.5	104.3	69.5	94.1	101.2	91.4	98.3
06	WOOD, PLASTICS & COMPOSITES	96.1	50.4	71.0	117.8	60.9	86.5	113.0	60.8	84.3	90.2	57.1	72.0	99.6	60.9	78.3	95.3	55.9	73.6
07	THERMAL & MOISTURE PROTECTION	94.7	62.0	80.5	90.0	62.4	78.0	92.4	61.9	79.1	91.9	64.4	80.0	89.7	64.4	78.7	84.6	64.4	75.8
08	OPENINGS	96.6	47.8	85.2	118.7	56.2	104.2	70.6	56.6	67.4	96.3	52.8	86.2	101.2	56.6	90.8	99.7	56.0	89.5
0920	Plaster & Gypsum Board	93.3	49.5	63.8	87.3	60.1	69.0	82.6	60.1	67.5	97.3	56.2	69.6	89.8	60.1	69.8	94.8	55.0	68.0
0950, 0980	Ceilings & Acoustic Treatment	95.3	49.5	64.4	90.7	60.1	70.1	86.6	60.1	68.7	94.2	56.2	68.5	92.7	60.1	70.7	100.8	55.0	69.9
0960	Flooring	91.9	54.6	81.4	71.9	54.6	67.1	95.9	54.6	84.3	97.3	69.7	89.5	94.5	64.5	86.0	102.4	67.0	92.5
0970, 0990	Wall Finishes & Painting/Coating	97.8	42.6	65.1	103.3	48.9	71.1	94.2	51.7	69.0	98.6	49.6	69.6	101.2	51.8	72.0	104.6	53.7	74.5
09	FINISHES	94.5	50.3	70.6	76.3	57.2	66.0	84.1	57.4	69.6	92.3	58.1	73.8	89.6	59.5	73.3	95.4	56.9	74.6
COVERS	DIVS. 10 - 14, 25, 28, 41, 43, 44, 46	100.0	73.6	94.1	100.0	79.2	95.4	100.0	71.7	93.7	100.0	76.6	94.8	100.0	79.2	95.4	100.0	78.3	95.2
21, 22, 23	FIRE SUPPRESSION, PLUMBING & HVAC	96.5	59.4	81.9	96.5	55.8	80.5	96.5	49.5	78.0	99.8	65.3	86.3	99.9	55.4	82.4	96.5	62.5	83.2
26, 27, 3370	ELECTRICAL, COMMUNICATIONS & UTIL.	95.9	59.8	78.1	101.0	58.7	80.1	98.5	58.6	78.8	97.5	50.5	74.3	100.2	58.6	79.7	99.6	63.5	81.8
MF2018	WEIGHTED AVERAGE	99.7	60.4	83.1	97.3	62.9	82.7	94.2	60.3	79.8	95.9	63.6	82.3	97.6	62.9	82.9	97.7	66.5	84.5

TEXAS

DIVISION		GIDDINGS 789			GREENVILLE 754			HOUSTON 770 - 772			HUNTSVILLE 773			LAREDO 780			LONGVIEW 756		
		MAT.	INST.	TOTAL	MAT.	INST.	TOTAL	MAT.	INST.	TOTAL	MAT.	INST.	TOTAL	MAT.	INST.	TOTAL	MAT.	INST.	TOTAL
015433	CONTRACTOR EQUIPMENT		89.9	89.9		100.4	100.4		103.3	103.3		95.3	95.3		89.9	89.9		91.8	91.8
0241, 31 - 34	SITE & INFRASTRUCTURE, DEMOLITION	106.8	85.8	92.3	96.4	82.4	86.7	107.7	95.5	99.3	96.8	91.6	93.2	100.2	86.2	90.5	93.7	88.6	90.2
0310	Concrete Forming & Accessories	92.6	51.1	57.3	86.9	57.1	61.5	96.0	58.9	64.4	89.2	54.1	59.2	95.1	52.1	58.5	82.9	58.8	62.3
0320	Concrete Reinforcing	80.9	48.4	65.2	91.9	51.1	72.2	87.3	52.2	70.3	88.2	51.1	70.3	80.4	50.9	66.1	90.9	50.7	71.5
0330	Cast-in-Place Concrete	104.1	58.7	87.2	92.1	59.6	80.0	89.6	68.9	81.9	99.9	59.5	84.9	87.8	65.0	79.3	107.1	58.5	89.0
03	CONCRETE	92.6	54.3	75.8	86.7	58.7	74.4	91.4	62.5	78.7	98.7	56.6	80.2	87.3	57.4	74.2	102.7	58.1	83.1
04	MASONRY	102.2	60.6	76.7	158.6	59.3	97.5	98.0	63.1	76.6	141.3	60.6	91.7	88.9	60.6	71.5	154.2	59.2	95.8
05	METALS	96.7	68.2	88.4	99.2	83.2	94.5	103.9	77.3	96.1	99.4	71.8	91.3	99.6	69.4	90.7	91.8	68.4	84.9
06	WOOD, PLASTICS & COMPOSITES	95.3	50.4	70.6	87.2	58.4	71.4	100.6	59.2	77.8	88.1	54.1	69.4	96.1	51.4	71.5	81.2	60.8	70.0
07	THERMAL & MOISTURE PROTECTION	95.0	62.7	81.0	87.3	61.5	76.1	86.3	68.5	78.5	86.5	63.2	76.4	93.8	63.0	80.4	89.1	60.9	76.8
08	OPENINGS	95.8	48.7	84.8	92.2	55.3	83.6	106.6	56.4	94.9	95.7	52.5	85.6	96.7	49.8	85.8	83.4	56.2	77.0
0920	Plaster & Gypsum Board	92.4	49.5	63.5	82.8	57.5	65.8	94.0	58.0	69.8	92.3	53.2	66.0	94.1	50.5	64.7	81.7	60.1	67.2
0950, 0980	Ceilings & Acoustic Treatment	95.3	49.5	64.4	98.1	57.5	70.7	97.9	58.0	71.0	96.7	53.2	67.4	98.6	50.5	66.1	93.2	60.1	70.9
0960	Flooring	91.7	54.6	81.2	87.7	54.6	78.3	103.5	71.2	94.4	91.3	54.6	81.0	91.8	63.8	83.9	93.7	54.6	82.7
0970, 0990	Wall Finishes & Painting/Coating	97.8	44.2	66.1	90.4	51.7	67.5	102.1	58.6	76.3	92.2	57.4	71.6	97.8	44.2	66.1	82.1	48.9	62.4
09	FINISHES	93.1	50.5	70.0	92.3	56.1	72.7	102.7	60.8	80.1	88.3	54.1	69.8	93.4	52.9	71.5	98.1	57.2	76.0
COVERS	DIVS. 10 - 14, 25, 28, 41, 43, 44, 46	100.0	70.3	93.4	100.0	76.5	94.8	100.0	83.1	96.2	100.0	69.8	93.3	100.0	75.7	94.6	100.0	79.0	95.3
21, 22, 23	FIRE SUPPRESSION, PLUMBING & HVAC	96.5	61.4	82.7	96.5	58.0	81.3	100.0	65.3	86.3	96.5	62.4	83.1	100.0	58.9	83.8	96.4	57.5	81.1
26, 27, 3370	ELECTRICAL, COMMUNICATIONS & UTIL.	93.1	55.6	74.6	92.1	58.7	75.6	101.8	67.8	85.1	97.8	59.1	78.7	96.0	57.1	76.8	92.6	50.9	72.0
MF2018	WEIGHTED AVERAGE	96.1	60.4	81.0	97.4	62.9	82.8	100.4	68.2	86.8	98.7	62.8	83.6	96.4	61.1	81.5	97.5	61.0	82.1

For customer support on your Site Work & Landscape Costs with RSMeans data, call 800.448.8182.

811

City Cost Indexes

TEXAS

DIVISION		LUBBOCK 793-794 MAT.	INST.	TOTAL	LUFKIN 759 MAT.	INST.	TOTAL	MCALLEN 785 MAT.	INST.	TOTAL	MCKINNEY 750 MAT.	INST.	TOTAL	MIDLAND 797 MAT.	INST.	TOTAL	ODESSA 797 MAT.	INST.	TOTAL
015433	CONTRACTOR EQUIPMENT		102.9	102.9		91.8	91.8		100.7	100.7		100.4	100.4		102.9	102.9		90.9	90.9
0241, 31 - 34	SITE & INFRASTRUCTURE, DEMOLITION	115.0	86.7	95.5	88.6	90.0	89.5	149.0	82.9	103.3	92.8	82.4	85.6	117.7	85.8	95.6	92.1	88.1	89.3
0310	Concrete Forming & Accessories	92.9	53.0	58.9	85.9	54.1	58.8	99.1	51.2	58.3	86.0	57.1	61.4	96.4	59.1	64.6	93.9	59.2	64.3
0320	Concrete Reinforcing	89.7	51.8	71.4	92.6	66.7	80.1	80.0	50.9	65.9	91.9	51.1	72.2	90.7	51.7	71.8	88.5	51.6	70.7
0330	Cast-in-Place Concrete	84.7	68.2	78.6	95.8	58.3	81.8	124.0	59.7	100.1	86.3	59.6	76.4	90.0	65.9	81.1	84.6	64.9	77.3
03	CONCRETE	83.1	60.1	73.0	95.1	58.6	79.0	102.8	56.2	82.3	82.1	58.8	71.9	87.1	62.1	76.1	84.5	60.7	74.1
04	MASONRY	99.2	61.0	75.7	117.3	60.5	82.4	95.1	60.7	74.0	170.8	59.3	102.3	116.2	60.2	81.8	99.9	60.2	75.5
05	METALS	110.9	85.8	103.5	99.4	73.3	91.7	97.1	84.2	93.3	99.2	83.3	94.5	109.2	85.4	102.2	106.4	70.4	95.8
06	WOOD, PLASTICS & COMPOSITES	98.6	52.1	73.0	88.6	54.2	69.6	114.3	50.5	79.2	86.1	58.4	70.9	103.0	60.9	79.8	99.2	60.8	78.1
07	THERMAL & MOISTURE PROTECTION	88.2	63.5	77.5	88.8	60.0	76.3	97.6	62.9	82.5	87.1	61.5	76.0	88.5	63.3	77.5	98.3	62.7	82.8
08	OPENINGS	108.6	52.4	95.5	63.3	56.6	61.7	100.4	48.6	88.3	92.2	55.3	83.6	110.0	56.7	97.6	101.8	56.7	91.2
0920	Plaster & Gypsum Board	88.4	51.1	63.3	80.6	53.2	62.2	97.5	49.5	65.2	82.8	57.5	65.8	90.0	60.1	69.9	88.2	60.1	69.3
0950, 0980	Ceilings & Acoustic Treatment	100.0	51.1	67.0	88.3	53.2	64.7	98.6	49.5	65.5	98.1	57.5	70.7	97.6	60.1	72.3	99.2	60.1	72.9
0960	Flooring	87.5	69.0	82.3	126.2	54.6	106.1	106.9	72.8	97.3	87.2	54.6	78.1	88.6	63.8	81.7	93.2	63.8	84.9
0970, 0990	Wall Finishes & Painting/Coating	103.4	53.7	74.0	82.1	51.7	64.1	110.1	42.6	70.2	90.4	51.7	67.5	103.4	51.7	72.8	92.8	51.7	68.5
09	FINISHES	88.5	55.5	70.6	106.5	53.9	78.0	102.4	54.1	76.3	91.9	56.1	72.5	89.0	59.4	72.9	86.1	59.3	71.6
COVERS	DIVS. 10 - 14, 25, 28, 41, 43, 44, 46	100.0	78.3	95.2	100.0	70.1	93.3	100.0	75.8	94.6	100.0	79.1	95.3	100.0	73.8	94.1	100.0	73.5	94.1
21, 22, 23	FIRE SUPPRESSION, PLUMBING & HVAC	99.7	53.1	81.4	96.4	58.1	81.4	96.5	48.7	77.7	96.5	60.2	82.2	96.1	47.4	77.0	100.1	52.5	81.4
26, 27, 3370	ELECTRICAL, COMMUNICATIONS & UTIL.	96.7	60.1	78.6	93.9	61.4	77.8	93.8	31.6	63.1	92.2	58.7	75.6	96.7	58.9	78.1	98.0	58.9	78.7
MF2018	WEIGHTED AVERAGE	99.2	62.9	83.9	94.7	62.6	81.1	99.7	56.5	81.4	97.2	63.5	82.9	99.7	62.2	83.8	97.7	61.8	82.6

TEXAS

DIVISION		PALESTINE 758 MAT.	INST.	TOTAL	SAN ANGELO 769 MAT.	INST.	TOTAL	SAN ANTONIO 781-782 MAT.	INST.	TOTAL	TEMPLE 765 MAT.	INST.	TOTAL	TEXARKANA 755 MAT.	INST.	TOTAL	TYLER 757 MAT.	INST.	TOTAL
015433	CONTRACTOR EQUIPMENT		91.8	91.8		90.9	90.9		94.1	94.1		90.9	90.9		91.8	91.8		91.8	91.8
0241, 31 - 34	SITE & INFRASTRUCTURE, DEMOLITION	94.2	88.8	90.5	97.8	87.7	90.8	101.5	93.7	96.1	86.7	87.6	87.3	83.4	91.1	88.7	92.9	88.7	90.0
0310	Concrete Forming & Accessories	77.5	58.8	61.5	97.9	51.4	58.2	94.7	52.6	58.8	101.2	51.2	58.5	93.0	59.4	64.3	87.6	58.9	63.1
0320	Concrete Reinforcing	90.2	51.0	71.3	81.8	51.6	67.2	89.6	47.7	69.3	82.0	50.8	66.9	90.1	50.8	71.1	90.9	50.8	71.5
0330	Cast-in-Place Concrete	87.5	58.5	76.7	91.4	64.9	81.5	85.7	68.3	79.2	74.9	58.6	68.8	88.2	64.9	79.5	105.2	58.6	87.8
03	CONCRETE	97.1	58.1	80.0	86.1	57.2	73.4	88.0	58.0	74.8	73.4	54.7	65.2	88.0	60.6	75.9	102.3	58.2	82.9
04	MASONRY	111.9	59.2	79.5	119.4	60.6	83.3	89.7	60.8	71.9	130.1	60.6	87.4	175.3	59.2	103.9	164.4	59.2	99.7
05	METALS	99.1	68.5	90.1	102.7	70.3	93.2	102.0	65.4	91.3	102.4	69.2	92.7	91.7	69.0	85.0	99.0	68.5	90.1
06	WOOD, PLASTICS & COMPOSITES	79.3	60.8	69.1	106.1	50.4	75.5	98.1	51.7	72.5	115.7	50.4	79.8	93.7	60.8	75.6	90.3	60.8	74.1
07	THERMAL & MOISTURE PROTECTION	89.4	61.8	77.4	91.8	62.6	79.1	86.8	65.6	77.5	91.3	61.7	78.4	88.6	63.0	77.5	89.2	61.8	77.3
08	OPENINGS	63.3	56.6	61.7	100.5	51.0	88.9	102.1	49.5	89.8	67.3	50.8	63.4	83.3	56.5	77.1	63.2	56.5	61.7
0920	Plaster & Gypsum Board	79.0	60.1	66.3	82.6	49.5	60.3	93.3	50.5	64.5	82.6	49.5	60.3	86.8	60.1	68.8	80.6	60.1	66.8
0950, 0980	Ceilings & Acoustic Treatment	88.3	60.1	69.3	86.6	49.5	61.6	95.2	50.5	65.1	86.6	49.5	61.6	93.2	60.1	70.9	88.3	60.1	69.3
0960	Flooring	116.4	54.6	99.0	75.8	54.6	69.9	101.2	67.7	91.8	97.5	54.6	85.4	102.7	63.8	91.7	128.7	54.6	107.8
0970, 0990	Wall Finishes & Painting/Coating	82.1	51.7	64.1	92.8	51.7	68.4	99.4	44.2	66.7	94.2	44.2	64.6	82.1	51.7	64.1	82.1	51.7	64.1
09	FINISHES	104.0	57.5	78.8	77.4	51.3	63.3	104.2	54.0	77.0	83.3	50.5	65.5	100.9	59.3	78.4	107.5	57.5	80.4
COVERS	DIVS. 10 - 14, 25, 28, 41, 43, 44, 46	100.0	73.5	94.1	100.0	76.0	94.7	100.0	78.0	95.1	100.0	72.3	93.8	100.0	79.1	95.3	100.0	79.0	95.3
21, 22, 23	FIRE SUPPRESSION, PLUMBING & HVAC	96.4	56.7	80.8	96.5	52.2	79.1	100.0	59.8	84.2	96.5	53.2	79.5	96.4	55.5	80.3	96.4	58.2	81.4
26, 27, 3370	ELECTRICAL, COMMUNICATIONS & UTIL.	90.1	48.4	69.5	102.8	54.6	79.1	97.5	59.8	78.9	99.8	52.0	76.2	93.7	54.5	74.4	92.6	53.7	73.4
MF2018	WEIGHTED AVERAGE	94.2	60.4	79.9	97.1	59.4	81.2	98.3	62.3	83.1	92.5	58.5	78.2	96.7	62.0	82.1	97.8	61.6	82.5

TEXAS / UTAH

DIVISION		VICTORIA 779 MAT.	INST.	TOTAL	WACO 766-767 MAT.	INST.	TOTAL	WAXAHACHIE 751 MAT.	INST.	TOTAL	WHARTON 774 MAT.	INST.	TOTAL	WICHITA FALLS 763 MAT.	INST.	TOTAL	LOGAN 843 MAT.	INST.	TOTAL
015433	CONTRACTOR EQUIPMENT		106.6	106.6		90.9	90.9		100.4	100.4		107.9	107.9		90.9	90.9		93.9	93.9
0241, 31 - 34	SITE & INFRASTRUCTURE, DEMOLITION	113.4	88.0	95.8	95.1	88.0	90.2	94.7	82.5	86.3	118.7	90.0	98.9	95.8	88.1	90.5	99.8	89.6	92.7
0310	Concrete Forming & Accessories	90.6	51.3	57.1	99.8	59.2	65.2	86.0	59.0	63.0	85.7	53.1	57.9	99.8	59.2	65.2	103.5	67.2	72.6
0320	Concrete Reinforcing	84.0	50.9	68.0	81.6	47.2	65.0	91.9	51.1	72.2	87.4	50.9	69.7	81.6	51.7	67.1	108.0	84.4	96.6
0330	Cast-in-Place Concrete	108.3	60.5	90.5	81.0	64.9	75.0	91.1	59.8	79.5	111.3	59.5	92.0	86.5	65.0	78.5	86.4	74.3	81.9
03	CONCRETE	99.7	56.7	80.8	79.6	60.0	71.0	85.8	59.7	74.3	103.8	57.1	83.3	82.2	62.0	72.8	105.4	73.2	91.2
04	MASONRY	116.9	60.7	82.4	90.9	60.6	72.3	159.2	59.3	97.8	100.3	60.6	75.9	91.4	60.1	72.2	109.5	60.5	79.4
05	METALS	99.4	87.8	96.0	104.8	68.4	94.1	99.2	83.6	94.6	101.2	87.4	97.1	104.7	70.4	94.6	109.5	82.5	101.5
06	WOOD, PLASTICS & COMPOSITES	98.4	50.5	72.1	113.9	60.8	84.7	86.1	61.0	72.3	88.4	52.7	68.8	113.9	60.8	84.7	85.9	67.1	75.5
07	THERMAL & MOISTURE PROTECTION	88.0	61.1	76.3	92.2	63.7	79.8	87.2	61.6	76.1	85.0	63.7	75.7	92.2	62.9	79.5	103.0	68.8	88.1
08	OPENINGS	99.5	51.1	88.2	78.4	55.6	73.1	92.2	56.7	83.9	99.7	52.3	88.6	78.4	56.7	73.4	92.2	66.1	86.1
0920	Plaster & Gypsum Board	90.9	49.5	63.0	82.7	60.1	67.5	83.2	60.1	67.7	90.4	51.7	64.4	82.7	60.1	67.5	80.6	66.2	70.9
0950, 0980	Ceilings & Acoustic Treatment	101.6	49.5	66.5	87.4	60.1	69.0	99.7	60.1	73.0	100.8	51.7	67.7	87.4	60.1	69.0	104.3	66.2	78.7
0960	Flooring	100.7	54.6	87.7	96.8	63.8	87.5	87.2	54.6	78.1	99.5	66.6	90.2	97.4	71.9	90.2	97.2	60.5	86.9
0970, 0990	Wall Finishes & Painting/Coating	104.8	57.4	76.8	94.2	51.7	69.0	90.4	51.7	67.5	104.6	55.7	75.7	96.1	51.7	69.8	92.6	60.7	73.7
09	FINISHES	92.4	52.0	70.6	83.7	59.3	70.5	92.5	57.6	73.6	94.7	55.2	73.3	84.1	60.9	71.5	93.9	64.9	78.2
COVERS	DIVS. 10 - 14, 25, 28, 41, 43, 44, 46	100.0	69.7	93.2	100.0	78.9	95.3	100.0	79.4	95.4	100.0	70.0	93.3	100.0	73.5	94.1	100.0	84.4	96.5
21, 22, 23	FIRE SUPPRESSION, PLUMBING & HVAC	96.5	62.3	83.1	100.0	58.7	83.8	96.5	58.0	81.4	96.5	61.9	82.9	100.0	52.1	81.2	100.0	69.2	87.9
26, 27, 3370	ELECTRICAL, COMMUNICATIONS & UTIL.	104.9	54.0	79.8	103.0	54.1	78.9	92.1	58.7	75.6	103.9	59.2	81.9	105.0	56.0	80.8	95.2	71.3	83.4
MF2018	WEIGHTED AVERAGE	99.7	62.9	84.2	94.4	62.4	80.9	97.2	63.5	83.0	99.8	64.3	84.8	95.0	61.6	80.9	100.9	71.7	88.5

For customer support on your Site Work & Landscape Costs with RSMeans data, call 800.448.8182.

UTAH / VERMONT

DIVISION		OGDEN 842, 844 MAT.	INST.	TOTAL	PRICE 845 MAT.	INST.	TOTAL	PROVO 846 - 847 MAT.	INST.	TOTAL	SALT LAKE CITY 840 - 841 MAT.	INST.	TOTAL	BELLOWS FALLS 051 MAT.	INST.	TOTAL	BENNINGTON 052 MAT.	INST.	TOTAL
015433	CONTRACTOR EQUIPMENT		93.9	93.9		92.9	92.9		92.9	92.9		93.8	93.8		95.7	95.7		95.7	95.7
0241, 31 - 34	SITE & INFRASTRUCTURE, DEMOLITION	87.3	89.6	88.9	97.1	87.4	90.4	95.9	88.0	90.5	87.0	89.5	88.7	86.8	96.1	93.3	86.2	96.1	93.1
0310	Concrete Forming & Accessories	103.5	67.2	72.6	106.0	64.2	70.4	105.3	67.2	72.9	105.8	67.2	72.9	96.5	101.7	100.9	94.0	101.6	100.4
0320	Concrete Reinforcing	107.6	84.4	96.4	115.8	84.0	100.4	116.8	84.4	101.1	110.0	84.4	97.6	85.1	86.3	85.7	85.1	86.3	85.6
0330	Cast-in-Place Concrete	87.7	74.3	82.7	86.5	71.6	80.9	86.5	74.3	81.9	95.8	74.3	87.8	89.6	115.9	99.4	89.6	115.8	99.4
03	CONCRETE	94.4	73.2	85.1	106.6	70.8	90.8	104.9	73.2	91.0	114.2	73.2	96.2	88.7	103.7	95.3	88.5	103.6	95.2
04	MASONRY	104.0	60.5	77.3	115.1	65.1	84.4	115.3	60.5	81.6	118.1	60.5	82.7	96.3	86.2	90.1	105.1	86.2	93.5
05	METALS	110.0	82.5	101.9	106.5	82.5	99.5	107.5	82.5	100.1	113.7	82.5	104.5	95.2	91.1	94.0	95.1	90.9	93.9
06	WOOD, PLASTICS & COMPOSITES	85.9	67.1	75.5	89.2	64.0	75.3	87.5	67.1	76.3	87.8	67.1	76.4	102.8	106.8	105.0	100.1	106.8	103.8
07	THERMAL & MOISTURE PROTECTION	101.8	68.8	87.4	104.8	68.3	88.9	104.9	68.8	89.2	110.0	68.8	92.1	100.9	89.1	95.8	100.9	89.1	95.8
08	OPENINGS	92.2	66.1	86.1	96.1	76.3	91.5	96.1	66.1	89.1	94.1	66.1	87.5	100.1	97.5	99.5	100.1	97.5	99.5
0920	Plaster & Gypsum Board	80.6	66.2	70.9	83.4	63.1	69.8	81.2	66.2	71.1	90.9	66.2	74.3	108.9	106.7	107.4	107.9	106.7	107.1
0950, 0980	Ceilings & Acoustic Treatment	104.3	66.2	78.7	104.3	63.1	76.5	104.3	66.2	78.7	97.1	66.2	76.3	97.2	106.7	103.6	97.2	106.7	103.6
0960	Flooring	95.2	60.5	85.4	98.4	56.1	86.5	98.2	60.5	87.6	99.0	60.5	88.1	92.0	98.9	94.0	91.2	98.9	93.4
0970, 0990	Wall Finishes & Painting/Coating	92.6	60.7	73.7	92.6	60.7	73.7	92.6	60.7	73.7	95.7	60.7	75.0	86.4	101.5	95.3	86.4	101.5	95.3
09	FINISHES	92.0	64.9	77.3	95.0	61.7	77.0	94.5	64.9	78.5	93.7	64.9	78.1	90.6	101.8	96.7	90.3	101.8	96.5
COVERS	DIVS. 10 - 14, 25, 28, 41, 43, 44, 46	100.0	84.4	96.5	100.0	81.6	95.9	100.0	84.4	96.5	100.0	84.4	96.5	100.0	91.8	98.2	100.0	91.7	98.2
21, 22, 23	FIRE SUPPRESSION, PLUMBING & HVAC	100.0	69.2	87.9	98.1	63.2	84.4	100.0	69.2	87.9	100.1	69.2	88.0	96.7	87.6	93.1	96.7	87.6	93.1
26, 27, 3370	ELECTRICAL, COMMUNICATIONS & UTIL.	95.5	71.3	83.6	100.2	68.8	84.7	95.7	68.6	82.3	98.1	71.3	84.9	108.3	79.5	94.1	108.3	54.5	81.8
MF2018	WEIGHTED AVERAGE	98.8	71.7	87.4	101.4	70.0	88.1	101.2	71.2	88.5	103.5	71.7	90.1	96.7	92.1	94.8	97.0	88.6	93.4

VERMONT

DIVISION		BRATTLEBORO 053 MAT.	INST.	TOTAL	BURLINGTON 054 MAT.	INST.	TOTAL	GUILDHALL 059 MAT.	INST.	TOTAL	MONTPELIER 056 MAT.	INST.	TOTAL	RUTLAND 057 MAT.	INST.	TOTAL	ST. JOHNSBURY 058 MAT.	INST.	TOTAL
015433	CONTRACTOR EQUIPMENT		95.7	95.7		99.0	99.0		95.7	95.7		99.0	99.0		95.7	95.7		95.7	95.7
0241, 31 - 34	SITE & INFRASTRUCTURE, DEMOLITION	87.8	96.1	93.5	92.6	101.0	98.4	85.8	91.9	90.0	89.7	100.8	97.3	90.4	96.1	94.3	85.9	95.2	92.3
0310	Concrete Forming & Accessories	96.7	101.7	101.0	99.4	80.2	83.0	94.6	95.1	95.1	97.3	101.3	100.7	97.0	80.0	82.5	93.5	95.6	95.2
0320	Concrete Reinforcing	84.1	86.3	85.2	107.3	86.2	97.1	85.8	86.1	86.0	101.6	86.2	94.2	106.4	86.1	96.6	84.1	86.2	85.1
0330	Cast-in-Place Concrete	92.5	115.9	101.2	109.6	116.7	112.2	86.9	107.2	94.4	104.5	116.7	109.0	87.8	115.6	98.2	86.9	107.3	94.5
03	CONCRETE	90.7	103.7	96.4	101.7	94.2	98.4	86.2	97.7	91.3	98.5	103.7	100.8	91.4	93.8	92.5	85.9	97.9	91.2
04	MASONRY	104.6	86.2	93.3	103.9	86.4	93.1	104.8	71.5	84.4	101.4	86.4	92.2	87.2	86.3	86.7	131.1	71.5	94.5
05	METALS	95.1	91.1	93.9	102.8	89.8	99.0	95.2	90.0	93.7	101.2	89.9	97.9	101.1	90.3	97.9	95.2	90.4	93.8
06	WOOD, PLASTICS & COMPOSITES	103.1	106.8	105.1	99.2	78.5	87.8	99.0	106.8	103.3	96.0	106.8	102.0	103.2	78.3	89.5	94.1	106.8	101.0
07	THERMAL & MOISTURE PROTECTION	101.0	89.1	95.8	107.4	86.6	98.3	100.7	82.4	92.8	109.4	89.7	100.8	101.1	86.1	94.6	100.6	82.4	92.7
08	OPENINGS	100.1	97.8	99.6	102.6	78.4	96.9	100.1	94.0	98.7	103.1	94.1	101.0	103.2	78.3	97.4	100.1	94.0	98.7
0920	Plaster & Gypsum Board	108.9	106.7	107.4	109.9	77.4	88.1	114.5	106.7	109.2	110.4	106.7	107.9	109.4	77.4	87.9	115.8	106.7	109.6
0950, 0980	Ceilings & Acoustic Treatment	97.2	106.7	103.6	103.9	77.4	86.0	97.2	106.7	103.6	104.9	106.7	106.1	103.2	77.4	85.8	97.2	106.7	103.6
0960	Flooring	92.2	98.9	94.1	101.8	98.9	101.0	95.0	98.9	96.1	98.5	98.9	98.6	92.0	98.9	94.0	98.3	98.9	98.5
0970, 0990	Wall Finishes & Painting/Coating	86.4	101.5	95.3	98.0	88.3	92.3	86.4	88.3	87.5	97.8	88.3	92.2	86.4	88.3	87.5	86.4	88.3	87.5
09	FINISHES	90.8	101.8	96.8	99.1	83.8	90.8	92.2	96.7	94.6	97.9	100.6	99.4	92.0	83.7	87.5	93.3	96.7	95.2
COVERS	DIVS. 10 - 14, 25, 28, 41, 43, 44, 46	100.0	91.8	98.2	100.0	89.1	97.6	100.0	86.7	97.0	100.0	92.2	98.3	100.0	88.6	97.5	100.0	86.8	97.0
21, 22, 23	FIRE SUPPRESSION, PLUMBING & HVAC	96.7	87.6	93.1	100.0	68.3	87.6	96.7	60.5	82.5	96.5	68.3	85.4	100.2	68.2	87.7	96.7	60.5	82.5
26, 27, 3370	ELECTRICAL, COMMUNICATIONS & UTIL.	108.3	79.5	94.1	107.6	53.6	81.0	108.3	54.5	81.8	108.1	53.6	81.3	108.4	53.6	81.4	108.3	54.5	81.8
MF2018	WEIGHTED AVERAGE	97.4	92.1	95.2	101.9	79.8	92.5	96.8	79.0	89.3	100.2	84.4	93.5	98.9	79.4	90.7	98.1	79.3	90.2

VERMONT / VIRGINIA

DIVISION		WHITE RIVER JCT. 050 MAT.	INST.	TOTAL	ALEXANDRIA 223 MAT.	INST.	TOTAL	ARLINGTON 222 MAT.	INST.	TOTAL	BRISTOL 242 MAT.	INST.	TOTAL	CHARLOTTESVILLE 229 MAT.	INST.	TOTAL	CULPEPER 227 MAT.	INST.	TOTAL
015433	CONTRACTOR EQUIPMENT		95.7	95.7		105.3	105.3		104.1	104.1		104.1	104.1		108.1	108.1		104.1	104.1
0241, 31 - 34	SITE & INFRASTRUCTURE, DEMOLITION	90.3	95.2	93.7	113.7	89.0	96.7	123.9	87.0	98.5	108.2	86.0	92.8	112.7	88.3	95.9	111.4	86.5	94.2
0310	Concrete Forming & Accessories	91.4	96.1	95.4	91.5	71.9	74.8	90.5	71.6	74.4	86.6	64.1	67.5	85.1	61.7	65.2	82.4	71.0	72.7
0320	Concrete Reinforcing	85.1	86.2	85.6	86.6	85.0	85.9	97.8	85.1	91.6	97.7	83.7	91.0	97.1	73.0	85.4	97.8	84.9	91.6
0330	Cast-in-Place Concrete	92.5	108.1	98.3	108.1	77.7	96.8	105.2	77.2	94.8	104.8	44.5	82.3	109.0	75.5	96.5	107.7	74.4	95.3
03	CONCRETE	92.6	98.4	95.2	99.9	77.7	90.1	104.5	77.4	92.6	100.8	62.4	83.9	101.3	70.2	87.7	98.4	76.0	88.6
04	MASONRY	117.2	72.9	90.0	89.8	73.7	79.9	102.3	72.6	84.0	92.3	63.5	74.6	116.9	55.3	79.0	105.1	70.8	84.1
05	METALS	95.2	90.4	93.8	106.0	100.8	104.5	104.6	101.8	103.8	103.4	97.8	101.8	103.7	95.8	101.4	103.7	99.6	102.5
06	WOOD, PLASTICS & COMPOSITES	97.1	106.8	102.4	91.0	70.1	79.5	87.5	70.1	78.0	80.2	69.8	74.5	78.7	59.8	68.3	77.4	70.1	73.4
07	THERMAL & MOISTURE PROTECTION	101.1	83.0	93.2	102.5	81.1	93.2	104.6	80.5	94.1	103.9	58.8	84.3	103.5	69.7	88.8	103.7	78.9	92.9
08	OPENINGS	100.1	94.0	98.7	95.4	74.3	90.5	93.7	74.3	89.2	96.4	67.9	89.7	94.8	65.7	88.0	95.1	73.8	90.1
0920	Plaster & Gypsum Board	106.6	106.7	106.7	98.0	69.2	78.6	94.8	69.2	77.6	91.1	68.9	76.2	91.1	57.9	68.8	91.3	69.2	76.4
0950, 0980	Ceilings & Acoustic Treatment	97.2	106.7	103.6	92.4	69.2	76.8	90.8	69.2	76.2	90.0	68.9	75.8	90.0	57.9	68.4	90.8	69.2	76.2
0960	Flooring	90.2	98.9	92.6	97.1	77.4	91.5	95.6	75.7	90.0	92.3	55.7	82.0	90.8	55.7	80.9	90.8	74.4	86.2
0970, 0990	Wall Finishes & Painting/Coating	86.4	88.3	87.5	115.8	76.0	92.3	115.8	73.1	90.6	102.1	55.3	74.4	102.1	58.6	76.3	115.8	73.1	90.6
09	FINISHES	90.1	97.1	93.9	93.8	72.6	82.3	93.8	71.6	81.8	90.6	62.3	75.3	90.1	59.4	73.5	90.8	71.2	80.2
COVERS	DIVS. 10 - 14, 25, 28, 41, 43, 44, 46	100.0	87.2	97.1	100.0	86.6	97.0	100.0	84.4	96.5	100.0	75.1	94.5	100.0	83.2	96.2	100.0	83.6	96.3
21, 22, 23	FIRE SUPPRESSION, PLUMBING & HVAC	96.7	61.2	82.8	100.4	86.2	94.8	100.4	85.6	94.6	96.8	46.0	76.9	96.8	69.1	86.0	96.8	85.3	92.3
26, 27, 3370	ELECTRICAL, COMMUNICATIONS & UTIL.	108.3	54.3	81.7	96.6	100.2	98.4	94.4	100.2	97.2	96.2	36.3	66.7	96.2	70.4	83.5	98.5	96.9	97.7
MF2018	WEIGHTED AVERAGE	98.2	79.8	90.4	99.6	84.8	93.3	100.4	84.2	93.6	98.2	59.3	81.8	99.4	71.0	87.4	98.7	83.0	92.1

For customer support on your Site Work & Landscape Costs with RSMeans data, call 800.448.8182.

813

City Cost Indexes

VIRGINIA

DIVISION		FAIRFAX 220-221 MAT.	INST.	TOTAL	FARMVILLE 239 MAT.	INST.	TOTAL	FREDERICKSBURG 224-225 MAT.	INST.	TOTAL	GRUNDY 246 MAT.	INST.	TOTAL	HARRISONBURG 228 MAT.	INST.	TOTAL	LYNCHBURG 245 MAT.	INST.	TOTAL
015433	CONTRACTOR EQUIPMENT		104.1	104.1		108.1	108.1		104.1	104.1		104.1	104.1		104.1	104.1		104.1	104.1
0241, 31 - 34	SITE & INFRASTRUCTURE, DEMOLITION	122.4	86.7	97.8	108.5	87.3	93.9	110.9	86.4	93.9	106.0	85.3	91.7	119.3	85.0	95.6	106.8	86.7	92.9
0310	Concrete Forming & Accessories	85.1	63.8	67.0	98.9	60.3	66.0	85.1	65.4	68.3	89.6	38.2	45.8	81.4	56.6	60.2	86.6	70.5	72.9
0320	Concrete Reinforcing	97.8	90.3	94.2	92.9	67.9	80.8	98.5	80.1	89.6	96.4	44.1	71.1	97.8	57.3	78.2	97.1	84.5	91.0
0330	Cast-in-Place Concrete	105.2	78.5	95.2	106.8	83.9	98.3	106.8	74.2	94.7	104.8	50.9	84.7	102.0	57.1	82.3	99.2	75.8	88.9
03	CONCRETE	104.2	75.2	91.5	101.8	71.5	88.5	98.1	72.7	86.9	94.9	45.8	75.8	101.4	57.9	74.6	109.3	63.9	81.4
04	MASONRY	102.2	73.6	84.6	101.8	50.0	69.9	104.0	66.7	81.1	94.9	55.4	70.6	101.4	57.9	74.6	109.3	63.9	81.4
05	METALS	103.8	102.5	103.4	101.3	90.8	98.2	103.8	98.0	102.1	103.4	75.2	95.1	103.7	87.2	98.9	103.6	99.8	102.5
06	WOOD, PLASTICS & COMPOSITES	80.2	59.2	68.6	93.0	59.8	74.7	80.2	64.6	71.6	83.1	30.6	54.2	76.3	59.2	66.9	80.2	71.5	75.4
07	THERMAL & MOISTURE PROTECTION	104.4	79.9	93.7	105.2	69.4	89.6	103.7	76.3	91.8	103.9	57.3	83.6	104.1	69.6	89.1	103.7	70.4	89.2
08	OPENINGS	93.7	69.6	88.0	94.3	54.9	85.1	94.8	68.5	88.7	96.4	30.1	80.9	95.0	57.0	86.2	95.0	68.8	88.9
0920	Plaster & Gypsum Board	91.3	57.9	68.9	103.1	57.9	72.7	91.3	63.6	72.7	91.1	28.5	49.0	91.1	57.9	68.8	91.1	70.7	77.4
0950, 0980	Ceilings & Acoustic Treatment	90.8	57.9	68.6	86.5	57.9	67.2	90.8	63.6	72.4	90.0	28.5	48.6	90.0	57.9	68.4	90.0	70.7	77.0
0960	Flooring	92.5	76.7	88.1	95.6	50.5	82.9	92.5	72.4	86.9	93.5	28.2	75.2	90.6	74.4	86.1	92.3	65.4	84.7
0970, 0990	Wall Finishes & Painting/Coating	115.8	73.6	90.8	99.6	57.2	74.5	115.8	57.0	81.0	102.1	31.2	60.2	115.8	48.3	75.9	102.1	55.3	74.4
09	FINISHES	92.4	65.7	77.9	90.6	57.7	72.8	91.3	65.0	77.1	90.7	33.9	60.0	91.3	58.9	73.7	90.4	67.8	78.2
COVERS	DIVS. 10 - 14, 25, 28, 41, 43, 44, 46	100.0	84.2	96.5	100.0	77.1	94.9	100.0	78.3	95.2	100.0	73.6	94.1	100.0	62.8	91.7	100.0	78.8	95.3
21, 22, 23	FIRE SUPPRESSION, PLUMBING & HVAC	96.8	88.1	93.4	96.9	52.0	79.3	96.8	79.9	90.2	96.8	65.7	84.6	96.8	65.4	84.5	96.8	69.5	86.1
26, 27, 3370	ELECTRICAL, COMMUNICATIONS & UTIL.	97.3	101.1	99.1	91.2	68.3	79.9	94.5	91.8	93.2	96.2	68.3	82.5	96.4	86.4	91.5	97.2	67.8	82.7
MF2018	WEIGHTED AVERAGE	99.5	83.6	92.8	97.9	65.3	84.1	98.3	78.7	90.0	98.1	58.8	81.5	99.0	68.8	86.3	98.7	73.8	88.2

VIRGINIA

DIVISION		NEWPORT NEWS 236 MAT.	INST.	TOTAL	NORFOLK 233-235 MAT.	INST.	TOTAL	PETERSBURG 238 MAT.	INST.	TOTAL	PORTSMOUTH 237 MAT.	INST.	TOTAL	PULASKI 243 MAT.	INST.	TOTAL	RICHMOND 230-232 MAT.	INST.	TOTAL
015433	CONTRACTOR EQUIPMENT		108.2	108.2		110.2	110.2		108.1	108.1		108.0	108.0		104.1	104.1		109.6	109.6
0241, 31 - 34	SITE & INFRASTRUCTURE, DEMOLITION	107.6	88.4	94.3	106.9	93.8	97.9	111.1	88.4	95.4	106.2	88.2	93.8	105.4	85.7	91.8	103.4	92.8	96.1
0310	Concrete Forming & Accessories	98.1	61.1	66.6	103.6	61.2	67.5	91.8	61.8	66.2	87.8	61.3	65.2	89.6	40.4	47.7	94.2	61.9	66.7
0320	Concrete Reinforcing	92.8	66.2	79.9	101.2	66.2	84.3	92.4	73.6	83.3	92.4	66.2	79.7	96.4	83.8	90.3	102.1	73.6	88.3
0330	Cast-in-Place Concrete	103.8	64.5	89.2	111.0	76.1	98.0	110.2	76.6	97.7	102.8	65.6	89.0	104.8	82.4	96.5	95.3	76.6	88.4
03	CONCRETE	98.9	65.0	84.0	102.1	69.0	87.6	104.5	70.8	89.7	97.7	65.5	83.6	99.3	64.7	84.1	94.5	70.7	84.0
04	MASONRY	97.3	55.0	71.3	97.1	55.0	71.3	110.2	64.6	82.2	103.0	55.6	73.9	87.9	54.3	67.3	93.3	64.7	75.7
05	METALS	103.8	93.1	100.6	105.3	91.8	101.3	101.4	96.7	100.0	102.7	93.1	99.9	103.5	96.0	101.3	105.9	95.3	102.8
06	WOOD, PLASTICS & COMPOSITES	91.8	59.8	74.2	99.3	59.7	77.5	83.2	59.5	70.2	79.6	59.8	68.7	83.1	32.0	55.0	93.7	59.7	75.0
07	THERMAL & MOISTURE PROTECTION	105.1	67.7	88.9	104.4	70.0	89.5	105.1	72.3	90.9	105.1	68.0	89.0	103.9	66.6	87.6	101.0	72.9	88.8
08	OPENINGS	94.6	60.1	86.6	95.2	60.1	87.0	94.0	65.6	87.4	94.7	56.0	85.7	96.4	42.9	83.9	101.7	65.7	93.3
0920	Plaster & Gypsum Board	104.3	57.9	73.1	98.5	57.9	71.2	97.8	57.7	70.8	98.6	57.9	71.2	91.1	30.0	50.0	99.5	57.9	71.5
0950, 0980	Ceilings & Acoustic Treatment	91.4	57.9	68.8	88.8	57.9	68.0	88.2	57.7	67.6	91.4	57.9	68.8	90.0	30.0	49.6	89.5	57.9	68.2
0960	Flooring	95.6	51.1	83.1	96.2	51.1	83.5	91.7	71.7	86.1	88.9	51.7	78.4	93.5	55.7	82.9	94.0	71.7	87.8
0970, 0990	Wall Finishes & Painting/Coating	99.6	59.7	76.0	101.2	59.7	76.6	99.6	58.6	75.3	99.6	59.7	76.0	102.1	48.3	70.3	91.9	62.5	76.0
09	FINISHES	91.5	58.4	73.6	90.5	58.3	73.1	89.3	62.5	74.8	88.8	58.6	72.5	90.7	40.9	63.8	100.0	62.5	76.0
COVERS	DIVS. 10 - 14, 25, 28, 41, 43, 44, 46	100.0	83.4	96.3	100.0	83.2	96.3	100.0	83.1	96.2	100.0	83.6	96.3	100.0	73.2	94.0	100.0	83.0	96.2
21, 22, 23	FIRE SUPPRESSION, PLUMBING & HVAC	100.4	66.8	87.2	100.1	66.8	87.0	96.9	69.1	86.0	100.4	67.1	87.4	96.8	66.0	84.7	100.0	69.1	87.9
26, 27, 3370	ELECTRICAL, COMMUNICATIONS & UTIL.	93.6	64.2	79.1	98.0	60.7	79.6	93.8	70.5	82.3	92.1	60.7	76.7	96.2	83.4	89.9	99.2	72.8	88.1
MF2018	WEIGHTED AVERAGE	98.8	68.3	85.9	99.9	68.7	86.7	98.7	72.6	87.7	98.3	67.8	85.4	97.8	67.1	84.8	99.2	72.8	88.1

VIRGINIA / WASHINGTON

DIVISION		ROANOKE 240-241 MAT.	INST.	TOTAL	STAUNTON 244 MAT.	INST.	TOTAL	WINCHESTER 226 MAT.	INST.	TOTAL	CLARKSTON 994 MAT.	INST.	TOTAL	EVERETT 982 MAT.	INST.	TOTAL	OLYMPIA 985 MAT.	INST.	TOTAL
015433	CONTRACTOR EQUIPMENT		104.1	104.1		108.1	108.1		104.1	104.1		90.2	90.2		99.3	99.3		101.6	101.6
0241, 31 - 34	SITE & INFRASTRUCTURE, DEMOLITION	106.0	86.8	92.7	109.3	87.3	94.1	118.1	86.4	96.2	98.1	86.1	89.8	111.6	104.6	105.7	101.6	104.3	103.9
0310	Concrete Forming & Accessories	95.7	71.2	74.8	89.3	61.4	65.5	83.6	66.4	68.9	107.0	64.2	70.5	109.2	114.5	111.8	119.5	114.4	117.0
0320	Concrete Reinforcing	97.5	84.6	91.2	97.1	83.8	90.7	97.1	80.2	88.9	108.4	95.9	102.3	102.6	109.4	105.1	95.3	109.8	100.7
0330	Cast-in-Place Concrete	119.0	86.2	106.8	109.0	85.5	100.2	105.2	63.9	89.8	82.3	82.9	82.5	95.1	107.4	100.5	93.7	107.2	99.7
03	CONCRETE	103.5	80.2	93.3	100.7	75.2	89.5	101.5	69.5	87.4	87.6	76.6	82.8	95.1	107.4	100.5	93.7	107.2	99.7
04	MASONRY	96.3	65.9	77.6	105.0	55.3	74.5	98.7	69.8	80.9	97.8	91.7	94.0	121.0	101.8	109.2	115.3	101.7	107.0
05	METALS	105.8	100.2	104.2	103.6	96.0	101.4	103.8	97.9	102.1	92.8	87.3	91.2	116.8	98.4	111.4	117.8	96.1	111.4
06	WOOD, PLASTICS & COMPOSITES	91.9	71.5	80.7	83.1	59.8	70.3	78.7	64.8	71.1	104.4	58.5	79.2	107.5	104.7	105.9	97.5	104.8	101.5
07	THERMAL & MOISTURE PROTECTION	103.6	76.0	91.6	103.4	67.5	87.8	104.2	75.9	91.9	159.7	82.9	126.3	112.4	105.9	109.6	108.3	103.0	106.0
08	OPENINGS	95.4	68.8	89.2	95.0	58.5	86.5	96.5	68.3	89.9	116.4	65.0	104.4	107.0	107.3	106.3	109.6	104.8	108.5
0920	Plaster & Gypsum Board	98.0	70.7	79.6	91.1	57.9	68.8	91.3	63.8	72.8	154.4	57.3	89.1	112.7	104.9	107.5	106.5	104.9	105.4
0950, 0980	Ceilings & Acoustic Treatment	92.4	70.7	77.8	90.0	57.9	68.4	90.8	63.8	72.6	103.8	57.3	72.4	101.2	104.9	103.7	98.7	104.9	102.9
0960	Flooring	97.1	67.3	88.7	93.1	33.1	76.2	92.0	74.4	87.1	85.4	77.5	83.2	109.9	96.0	106.0	97.6	96.0	97.1
0970, 0990	Wall Finishes & Painting/Coating	102.1	55.3	74.4	102.1	30.7	59.8	115.8	77.6	93.2	84.4	72.1	77.2	97.6	95.2	96.2	99.6	93.1	95.8
09	FINISHES	92.7	68.5	79.6	90.6	52.6	70.0	91.8	68.3	79.1	108.2	65.8	85.2	105.4	101.6	103.4	94.9	101.5	98.5
COVERS	DIVS. 10 - 14, 25, 28, 41, 43, 44, 46	100.0	79.2	95.4	100.0	78.2	95.1	100.0	82.7	96.1	100.0	89.3	97.6	100.0	103.7	100.8	100.0	99.9	100.0
21, 22, 23	FIRE SUPPRESSION, PLUMBING & HVAC	100.4	67.9	87.6	96.8	59.3	82.1	96.8	81.4	90.8	96.7	80.6	90.4	100.0	105.2	102.0	99.9	100.6	100.2
26, 27, 3370	ELECTRICAL, COMMUNICATIONS & UTIL.	96.2	60.8	78.8	95.2	68.4	82.0	94.9	91.8	93.4	89.4	95.6	92.5	107.0	104.1	105.6	105.4	97.7	101.6
MF2018	WEIGHTED AVERAGE	100.1	73.6	88.9	98.6	67.9	85.6	98.9	79.5	90.7	99.1	81.9	91.8	105.1	104.0	104.6	104.0	101.8	103.1

City Cost Indexes

WASHINGTON

DIVISION		RICHLAND 993 MAT.	INST.	TOTAL	SEATTLE 980 - 981, 987 MAT.	INST.	TOTAL	SPOKANE 990 - 992 MAT.	INST.	TOTAL	TACOMA 983 - 984 MAT.	INST.	TOTAL	VANCOUVER 986 MAT.	INST.	TOTAL	WENATCHEE 988 MAT.	INST.	TOTAL
015433	CONTRACTOR EQUIPMENT		90.2	90.2		102.4	102.4		90.2	90.2		99.3	99.3		95.8	95.8		99.3	99.3
0241, 31 - 34	SITE & INFRASTRUCTURE, DEMOLITION	100.3	86.3	90.6	97.8	108.9	105.5	99.7	86.3	90.4	95.2	105.1	102.1	105.2	93.1	96.8	104.0	102.8	103.2
0310	Concrete Forming & Accessories	107.1	80.0	84.0	107.4	109.5	109.2	112.0	79.5	84.3	102.3	104.1	103.9	102.7	95.7	96.7	104.0	76.5	80.5
0320	Concrete Reinforcing	103.9	95.4	99.8	109.2	117.3	113.1	104.6	95.3	100.1	107.8	114.3	110.9	108.8	114.8	111.7	108.8	95.9	102.5
0330	Cast-in-Place Concrete	82.5	84.4	83.2	105.9	113.9	108.9	85.7	84.2	85.2	105.5	108.8	106.7	117.8	99.9	111.2	107.7	90.8	101.4
03	CONCRETE	87.2	84.2	85.9	102.8	111.8	106.7	89.2	83.9	86.8	97.0	106.9	101.4	107.2	100.2	104.1	105.3	84.9	96.3
04	MASONRY	98.8	83.7	89.5	123.5	107.2	113.5	99.4	83.7	89.7	117.6	100.2	106.9	118.8	105.5	110.6	120.9	94.1	104.4
05	METALS	93.2	87.1	91.4	116.1	101.9	111.9	95.4	86.6	92.8	118.7	96.5	112.2	116.0	99.3	111.1	116.0	87.3	107.6
06	WOOD, PLASTICS & COMPOSITES	104.6	77.9	89.9	103.4	108.7	106.3	113.5	77.9	93.9	97.1	104.7	101.2	89.5	94.9	92.5	98.7	73.3	84.8
07	THERMAL & MOISTURE PROTECTION	161.1	84.4	127.7	111.6	110.8	111.3	157.4	84.7	125.8	112.1	102.1	107.8	111.8	101.6	107.4	111.6	88.3	101.5
08	OPENINGS	114.4	74.7	105.2	106.1	110.7	107.2	115.0	74.7	105.6	106.7	104.7	106.2	102.8	101.5	102.5	106.2	73.0	98.5
0920	Plaster & Gypsum Board	154.4	77.2	102.5	111.2	109.1	109.8	144.2	77.2	99.1	110.6	104.9	106.8	109.0	95.2	99.7	113.9	72.7	86.2
0950, 0980	Ceilings & Acoustic Treatment	109.2	77.2	87.6	106.6	109.1	108.3	105.6	77.2	86.4	104.5	104.9	104.8	102.4	95.2	97.5	97.8	72.7	80.9
0960	Flooring	85.7	77.5	83.4	109.5	103.4	107.8	84.8	77.5	82.7	102.2	96.0	100.5	107.4	107.1	107.3	105.3	77.5	97.5
0970, 0990	Wall Finishes & Painting/Coating	84.4	75.1	78.9	107.1	95.8	100.4	84.6	76.4	79.8	97.6	93.1	95.0	100.3	79.9	88.2	97.6	70.9	81.8
09	FINISHES	109.5	78.1	92.5	108.9	106.4	107.6	107.3	78.2	91.6	103.6	101.4	102.4	101.3	95.6	98.3	104.7	74.9	88.5
COVERS	DIVS. 10 - 14, 25, 28, 41, 43, 44, 46	100.0	92.2	98.3	100.0	105.2	101.2	100.0	92.2	98.3	100.0	99.6	99.9	100.0	101.5	100.3	100.0	91.2	98.0
21, 22, 23	FIRE SUPPRESSION, PLUMBING & HVAC	100.3	108.6	103.6	99.9	123.0	109.0	100.3	83.7	93.8	100.0	100.6	100.2	100.0	110.7	104.2	96.5	92.0	94.7
26, 27, 3370	ELECTRICAL, COMMUNICATIONS & UTIL.	87.2	92.8	90.0	106.2	118.9	112.5	85.4	77.1	81.3	106.8	97.7	102.3	113.1	107.3	110.2	107.7	94.5	101.2
MF2018	WEIGHTED AVERAGE	99.8	90.0	95.6	106.3	112.5	108.9	100.1	82.4	92.6	105.4	101.3	103.7	106.6	102.8	105.0	105.6	88.7	98.5

DIVISION		WASHINGTON YAKIMA 989 MAT.	INST.	TOTAL	WEST VIRGINIA BECKLEY 258 - 259 MAT.	INST.	TOTAL	BLUEFIELD 247 - 248 MAT.	INST.	TOTAL	BUCKHANNON 262 MAT.	INST.	TOTAL	CHARLESTON 250 - 253 MAT.	INST.	TOTAL	CLARKSBURG 263 - 264 MAT.	INST.	TOTAL
015433	CONTRACTOR EQUIPMENT		99.3	99.3		104.1	104.1		104.1	104.1		104.1	104.1		106.8	106.8		104.1	104.1
0241, 31 - 34	SITE & INFRASTRUCTURE, DEMOLITION	97.6	103.8	101.9	100.8	87.4	91.6	99.9	87.4	91.3	106.0	87.4	93.1	100.4	93.0	95.3	106.8	87.4	93.4
0310	Concrete Forming & Accessories	102.9	99.2	99.8	85.0	87.1	86.8	86.6	87.1	87.0	86.1	83.7	84.0	96.0	88.6	89.7	83.6	83.8	83.7
0320	Concrete Reinforcing	108.3	95.4	102.1	98.0	87.7	93.1	96.0	82.5	89.5	96.7	82.4	89.8	106.1	87.0	96.9	96.7	95.3	96.0
0330	Cast-in-Place Concrete	112.7	83.0	101.7	99.2	92.5	96.7	102.5	92.5	98.8	102.1	92.0	98.4	95.9	96.6	96.2	111.9	87.6	102.9
03	CONCRETE	101.8	92.4	97.7	91.8	90.2	91.1	95.1	89.3	92.6	98.2	87.5	93.5	90.7	92.2	91.4	102.2	88.2	96.1
04	MASONRY	110.3	81.0	92.3	91.0	89.0	89.8	89.5	89.0	89.2	100.6	85.8	91.5	88.4	90.7	89.8	104.1	85.8	92.9
05	METALS	116.7	87.3	108.0	98.2	103.3	99.7	103.6	101.5	103.0	103.8	101.6	103.2	96.8	102.5	98.5	103.8	106.0	104.5
06	WOOD, PLASTICS & COMPOSITES	97.4	104.7	101.4	80.2	87.4	84.2	82.1	87.4	85.0	81.4	83.0	82.3	89.3	88.2	88.7	78.0	83.0	80.8
07	THERMAL & MOISTURE PROTECTION	112.3	86.1	100.9	105.0	86.5	96.9	103.6	86.5	96.1	104.0	86.6	96.4	101.2	88.1	95.5	103.8	86.3	96.2
08	OPENINGS	106.2	100.5	104.9	94.5	83.0	91.8	96.9	81.8	93.4	96.9	79.4	92.8	94.9	83.4	92.2	96.9	82.1	93.4
0920	Plaster & Gypsum Board	110.4	104.9	106.7	93.5	87.0	89.1	90.2	87.0	88.0	90.8	82.5	85.2	98.7	87.6	91.2	88.6	82.5	84.5
0950, 0980	Ceilings & Acoustic Treatment	99.6	104.9	103.2	78.1	87.0	84.1	87.5	87.0	87.2	90.0	82.5	84.9	89.6	87.6	88.2	90.0	82.5	84.9
0960	Flooring	103.2	77.5	96.0	91.0	98.9	93.2	90.1	98.9	92.6	89.8	95.2	91.3	97.1	98.9	97.6	88.8	95.2	90.6
0970, 0990	Wall Finishes & Painting/Coating	97.6	76.4	85.1	89.3	92.1	91.0	102.1	89.8	94.8	102.1	88.0	93.7	88.6	91.9	90.5	102.1	88.0	93.7
09	FINISHES	103.1	93.3	97.8	86.3	90.2	88.4	88.7	89.9	89.4	89.7	86.1	87.8	92.8	91.0	91.9	89.0	86.1	87.4
COVERS	DIVS. 10 - 14, 25, 28, 41, 43, 44, 46	100.0	96.1	99.1	100.0	90.3	97.8	100.0	90.3	97.8	100.0	90.6	97.9	100.0	92.5	98.3	100.0	90.6	97.9
21, 22, 23	FIRE SUPPRESSION, PLUMBING & HVAC	100.0	107.4	102.9	97.1	80.3	90.5	96.8	84.0	91.8	96.8	89.7	94.0	100.1	81.4	92.8	96.8	89.8	94.1
26, 27, 3370	ELECTRICAL, COMMUNICATIONS & UTIL.	109.9	92.9	101.5	92.9	84.1	88.6	95.2	84.1	89.8	96.6	89.3	93.0	98.0	87.1	92.6	96.6	89.3	93.0
MF2018	WEIGHTED AVERAGE	105.6	95.4	101.3	95.2	87.7	92.0	97.0	88.1	93.2	98.3	88.8	94.3	96.5	89.4	93.5	98.9	89.5	94.9

WEST VIRGINIA

DIVISION		GASSAWAY 266 MAT.	INST.	TOTAL	HUNTINGTON 255 - 257 MAT.	INST.	TOTAL	LEWISBURG 249 MAT.	INST.	TOTAL	MARTINSBURG 254 MAT.	INST.	TOTAL	MORGANTOWN 265 MAT.	INST.	TOTAL	PARKERSBURG 261 MAT.	INST.	TOTAL
015433	CONTRACTOR EQUIPMENT		104.1	104.1		104.1	104.1		104.1	104.1		104.1	104.1		104.1	104.1		104.1	104.1
0241, 31 - 34	SITE & INFRASTRUCTURE, DEMOLITION	103.5	87.4	92.4	105.4	88.5	93.7	115.5	87.4	96.1	104.7	88.0	93.2	100.9	88.2	92.1	109.3	88.4	94.9
0310	Concrete Forming & Accessories	85.5	86.9	86.7	96.0	89.3	90.3	83.9	86.6	86.2	85.0	79.5	80.3	83.9	83.8	83.9	88.0	86.0	86.3
0320	Concrete Reinforcing	96.7	91.6	94.2	99.5	93.1	96.4	96.7	82.4	89.8	98.0	95.1	96.6	96.7	95.3	96.0	96.0	82.3	89.4
0330	Cast-in-Place Concrete	106.9	94.0	102.1	108.2	94.0	102.9	102.5	92.3	98.7	104.0	88.3	98.1	102.1	92.0	98.4	104.4	89.0	98.7
03	CONCRETE	98.6	91.3	95.4	96.8	92.6	95.0	105.1	89.0	98.0	95.3	86.5	91.4	94.9	89.8	92.7	100.3	87.5	94.7
04	MASONRY	105.0	87.9	94.5	91.7	93.4	92.7	94.3	89.0	91.1	93.1	83.9	87.5	122.6	85.8	100.0	80.1	85.2	83.2
05	METALS	103.8	104.8	104.1	100.6	105.7	102.1	103.7	101.0	102.9	98.6	104.4	100.3	103.9	106.0	104.5	104.5	101.4	103.6
06	WOOD, PLASTICS & COMPOSITES	80.4	87.4	84.3	90.8	88.7	89.6	78.3	87.4	83.3	80.2	78.6	79.3	78.3	83.0	80.9	81.5	85.6	83.8
07	THERMAL & MOISTURE PROTECTION	103.6	87.9	96.8	105.2	88.4	97.9	104.6	86.4	96.7	105.3	80.8	94.6	103.7	86.6	96.2	103.9	86.7	96.4
08	OPENINGS	95.2	83.9	92.6	93.7	84.8	91.7	96.9	81.8	93.4	96.2	73.8	91.0	98.0	82.1	94.3	95.8	80.4	92.2
0920	Plaster & Gypsum Board	89.9	87.0	87.9	99.5	88.3	91.9	88.6	87.0	87.5	93.5	77.9	83.0	88.6	82.5	84.5	91.1	85.1	87.1
0950, 0980	Ceilings & Acoustic Treatment	90.0	87.0	88.0	78.1	88.3	85.0	90.0	87.0	88.0	78.1	77.9	78.0	90.0	82.5	84.9	90.0	85.1	86.7
0960	Flooring	89.6	98.9	92.2	98.4	103.3	99.7	88.9	98.9	91.7	91.0	98.9	93.2	88.9	95.2	90.7	92.6	91.9	92.4
0970, 0990	Wall Finishes & Painting/Coating	102.1	91.9	96.0	89.3	89.8	89.6	102.1	64.5	79.8	89.3	83.7	86.0	102.1	88.0	93.7	102.1	83.7	91.2
09	FINISHES	89.2	89.9	89.6	89.3	92.0	90.8	90.1	87.2	88.5	86.6	83.0	84.6	88.6	86.1	87.3	90.6	86.8	88.6
COVERS	DIVS. 10 - 14, 25, 28, 41, 43, 44, 46	100.0	91.1	98.0	100.0	92.3	98.3	100.0	65.4	92.3	100.0	81.4	95.8	100.0	83.5	96.3	100.0	91.3	98.1
21, 22, 23	FIRE SUPPRESSION, PLUMBING & HVAC	96.8	83.4	91.6	100.6	90.6	96.7	96.8	80.2	90.3	97.1	80.9	90.8	96.8	89.8	94.1	100.3	88.3	95.6
26, 27, 3370	ELECTRICAL, COMMUNICATIONS & UTIL.	96.6	87.1	91.9	96.6	89.6	93.1	92.8	84.1	88.5	98.6	76.9	87.9	96.7	89.3	93.1	96.7	88.3	92.5
MF2018	WEIGHTED AVERAGE	98.3	89.0	94.3	97.8	92.2	95.4	98.7	86.1	93.4	96.7	84.1	91.4	98.8	89.6	94.9	98.6	88.6	94.4

For customer support on your Site Work & Landscape Costs with RSMeans data, call 800.448.8182.

815

West Virginia / Wisconsin

		PETERSBURG 268			ROMNEY 267			WHEELING 260			BELOIT 535			EAU CLAIRE 547			GREEN BAY 541 - 543		
DIVISION		MAT.	INST.	TOTAL	MAT.	INST.	TOTAL	MAT.	INST.	TOTAL	MAT.	INST.	TOTAL	MAT.	INST.	TOTAL	MAT.	INST.	TOTAL
015433	CONTRACTOR EQUIPMENT		104.1	104.1		104.1	104.1		104.1	104.1		99.0	99.0		99.9	99.9		97.7	97.7
0241, 31 - 34	SITE & INFRASTRUCTURE, DEMOLITION	100.2	88.3	92.0	103.0	88.3	92.9	109.9	88.2	94.9	97.1	101.7	100.3	96.0	99.0	98.1	99.5	95.3	96.6
0310	Concrete Forming & Accessories	87.0	86.4	86.5	83.3	86.8	86.3	89.7	83.8	84.7	99.8	96.2	96.7	98.5	94.7	95.3	108.2	97.5	99.1
0320	Concrete Reinforcing	96.0	91.4	93.8	96.7	95.4	96.1	95.5	95.3	95.4	95.0	137.1	115.4	94.2	114.2	103.9	92.3	109.4	100.5
0330	Cast-in-Place Concrete	102.1	91.8	98.3	106.9	83.5	98.2	104.4	92.0	99.8	105.8	100.2	103.7	99.9	94.8	98.0	103.3	99.9	102.1
03	CONCRETE	95.0	90.2	92.9	98.4	88.2	93.9	100.3	89.8	95.7	96.7	104.7	100.2	93.1	98.4	95.4	96.2	100.6	98.2
04	MASONRY	96.1	85.8	89.8	94.7	87.9	90.5	103.4	85.7	92.5	99.3	100.3	99.9	91.4	96.1	94.3	122.7	97.8	107.4
05	METALS	103.9	104.1	104.0	104.0	105.9	104.5	104.7	106.1	105.1	98.6	112.0	102.5	93.8	105.8	97.3	96.4	104.2	98.7
06	WOOD, PLASTICS & COMPOSITES	82.3	87.4	85.1	77.5	87.4	82.9	83.1	83.0	83.1	96.9	94.1	95.4	100.6	94.2	97.1	107.8	97.2	101.9
07	THERMAL & MOISTURE PROTECTION	103.7	82.5	94.5	103.8	83.8	95.1	104.1	86.7	96.6	105.4	93.4	100.2	104.0	94.6	99.9	106.1	97.4	102.3
08	OPENINGS	98.1	83.6	94.7	98.0	84.5	94.8	96.6	82.1	93.2	98.1	108.3	100.5	103.1	99.3	102.2	99.1	101.2	99.6
0920	Plaster & Gypsum Board	90.8	87.0	88.2	88.3	87.0	87.4	91.1	82.5	85.3	91.6	94.5	93.5	105.5	94.5	98.1	103.3	97.5	99.4
0950, 0980	Ceilings & Acoustic Treatment	90.0	87.0	88.0	90.0	87.0	88.0	90.0	82.5	84.9	84.4	94.5	91.2	90.9	94.5	93.3	83.6	97.5	93.0
0960	Flooring	90.6	95.2	91.9	88.7	98.9	91.6	93.5	95.2	94.0	94.8	118.5	101.4	80.5	106.5	87.8	96.0	114.8	101.3
0970, 0990	Wall Finishes & Painting/Coating	102.1	88.0	93.7	102.1	88.0	93.7	102.1	88.0	93.7	95.0	103.6	100.1	83.3	76.4	79.2	92.0	80.0	84.9
09	FINISHES	89.4	88.7	89.0	88.7	89.5	89.1	90.9	86.1	88.3	92.3	100.7	96.9	87.3	94.9	91.4	91.4	99.1	95.6
COVERS	DIVS. 10 - 14, 25, 28, 41, 43, 44, 46	100.0	55.0	90.0	100.0	91.0	98.0	100.0	84.4	96.5	100.0	98.6	99.7	100.0	94.1	98.7	100.0	95.7	99.0
21, 22, 23	FIRE SUPPRESSION, PLUMBING & HVAC	96.8	83.7	91.7	96.8	83.6	91.6	100.4	89.8	96.2	99.8	95.5	98.1	100.0	88.2	95.4	100.2	83.4	93.6
26, 27, 3370	ELECTRICAL, COMMUNICATIONS & UTIL.	99.7	76.9	88.4	99.0	76.9	88.1	94.1	89.3	91.7	100.6	84.3	92.5	104.3	84.4	94.5	99.0	81.2	90.2
MF2018	WEIGHTED AVERAGE	98.0	85.9	92.9	98.3	87.2	93.6	99.6	89.6	95.4	98.6	98.9	98.7	97.5	94.1	96.1	99.4	93.4	96.9

Wisconsin

		KENOSHA 531			LA CROSSE 546			LANCASTER 538			MADISON 537			MILWAUKEE 530, 532			NEW RICHMOND 540		
DIVISION		MAT.	INST.	TOTAL	MAT.	INST.	TOTAL	MAT.	INST.	TOTAL	MAT.	INST.	TOTAL	MAT.	INST.	TOTAL	MAT.	INST.	TOTAL
015433	CONTRACTOR EQUIPMENT		97.1	97.1		99.9	99.9		99.0	99.0		101.4	101.4		90.2	90.2		100.2	100.2
0241, 31 - 34	SITE & INFRASTRUCTURE, DEMOLITION	103.0	98.4	99.9	89.8	99.0	96.1	96.0	101.1	99.5	94.7	105.3	102.0	93.6	96.0	95.2	93.3	99.6	97.7
0310	Concrete Forming & Accessories	108.7	106.1	106.5	85.2	94.5	93.2	99.1	95.2	95.7	103.8	95.8	97.0	102.9	113.9	112.3	93.6	92.5	92.6
0320	Concrete Reinforcing	94.8	110.5	102.4	93.9	106.2	99.8	96.3	106.2	101.1	97.8	106.4	101.9	99.9	115.1	107.2	91.3	114.1	102.3
0330	Cast-in-Place Concrete	115.2	102.7	110.5	89.8	96.5	92.3	105.2	99.7	103.2	101.4	100.3	100.7	93.9	112.1	100.7	104.0	102.7	103.5
03	CONCRETE	101.5	105.5	103.3	84.6	97.5	90.3	96.4	98.8	97.4	96.3	99.2	97.6	94.5	112.5	102.4	90.8	100.1	94.9
04	MASONRY	97.1	108.0	103.8	90.5	96.1	93.9	99.4	100.1	99.9	97.4	99.1	98.5	102.3	117.2	111.4	117.9	98.4	105.9
05	METALS	99.6	103.3	100.7	93.7	102.8	96.4	96.0	100.3	97.3	102.8	100.4	102.1	97.5	97.5	97.5	94.1	105.4	97.4
06	WOOD, PLASTICS & COMPOSITES	103.9	105.7	104.9	85.2	94.2	90.2	96.1	94.1	95.0	99.4	94.3	96.6	101.2	112.3	107.3	89.5	90.2	89.9
07	THERMAL & MOISTURE PROTECTION	105.5	104.0	104.9	103.4	93.5	99.1	105.2	93.7	100.2	106.3	100.5	103.7	104.8	112.7	108.2	104.9	100.2	102.9
08	OPENINGS	92.5	105.4	95.5	103.1	92.0	100.5	94.0	91.9	93.5	101.6	98.9	100.9	103.0	111.5	105.0	88.6	90.3	89.0
0920	Plaster & Gypsum Board	80.2	106.4	97.8	100.9	94.5	96.6	90.9	94.5	93.3	101.7	94.5	96.8	96.4	112.6	107.3	91.7	90.5	90.9
0950, 0980	Ceilings & Acoustic Treatment	84.4	106.4	99.2	90.0	94.5	93.0	81.1	94.5	90.1	86.5	94.5	91.9	93.1	112.6	106.3	57.1	90.5	79.6
0960	Flooring	112.8	114.8	113.4	74.7	111.6	85.1	94.4	111.0	99.0	94.3	111.0	99.0	101.7	118.6	106.4	89.7	114.5	96.7
0970, 0990	Wall Finishes & Painting/Coating	106.4	118.7	113.7	83.3	81.7	82.4	95.0	97.7	96.6	94.9	103.6	100.0	104.5	124.3	116.3	95.0	81.7	87.1
09	FINISHES	97.1	109.4	103.7	84.4	96.5	90.9	91.4	98.4	95.2	93.7	99.3	96.7	99.5	116.1	108.5	82.5	95.2	89.4
COVERS	DIVS. 10 - 14, 25, 28, 41, 43, 44, 46	100.0	97.3	99.4	100.0	92.8	98.4	100.0	84.2	96.5	100.0	96.8	99.3	100.0	105.2	101.2	100.0	95.1	98.9
21, 22, 23	FIRE SUPPRESSION, PLUMBING & HVAC	100.0	95.9	98.4	100.0	88.1	95.3	96.3	89.2	93.5	99.8	95.3	98.0	99.8	107.0	102.6	96.0	89.0	93.2
26, 27, 3370	ELECTRICAL, COMMUNICATIONS & UTIL.	100.9	97.1	99.0	104.6	84.4	94.7	100.3	84.9	92.7	101.5	94.0	97.8	100.2	102.4	101.3	102.4	84.9	93.7
MF2018	WEIGHTED AVERAGE	99.4	102.0	100.5	95.9	93.5	94.9	96.7	94.3	95.7	99.7	98.1	99.0	99.2	107.9	102.9	95.4	94.6	95.0

Wisconsin

		OSHKOSH 549			PORTAGE 539			RACINE 534			RHINELANDER 545			SUPERIOR 548			WAUSAU 544		
DIVISION		MAT.	INST.	TOTAL	MAT.	INST.	TOTAL	MAT.	INST.	TOTAL	MAT.	INST.	TOTAL	MAT.	INST.	TOTAL	MAT.	INST.	TOTAL
015433	CONTRACTOR EQUIPMENT		97.7	97.7		99.0	99.0		99.0	99.0		97.7	97.7		100.2	100.2		97.7	97.7
0241, 31 - 34	SITE & INFRASTRUCTURE, DEMOLITION	91.0	95.3	94.0	86.9	100.9	96.6	96.9	102.1	100.5	103.2	95.3	97.7	90.1	99.3	96.5	86.9	95.8	93.0
0310	Concrete Forming & Accessories	90.7	94.7	94.1	91.3	95.7	95.1	100.1	106.1	105.2	88.3	94.6	93.7	91.6	88.6	89.0	90.1	97.4	96.3
0320	Concrete Reinforcing	92.4	109.3	100.6	96.4	106.4	101.2	95.0	110.5	102.5	92.6	106.4	99.3	91.3	110.2	100.4	92.6	106.2	99.2
0330	Cast-in-Place Concrete	95.8	98.0	96.6	90.3	98.3	93.3	103.8	102.5	103.3	108.6	97.9	104.6	98.0	98.3	98.1	89.3	92.4	90.5
03	CONCRETE	86.2	98.6	91.7	84.2	98.7	90.6	95.8	105.4	100.0	98.0	98.1	98.1	85.5	96.7	90.2	81.4	97.4	88.4
04	MASONRY	104.7	97.8	100.4	98.2	100.2	99.4	99.3	108.0	104.6	121.3	97.6	106.8	117.1	100.4	106.8	104.2	97.6	100.1
05	METALS	94.3	103.5	97.0	96.7	104.3	98.9	92.0	103.3	101.1	94.2	102.5	96.6	95.1	104.6	97.9	94.0	102.9	96.6
06	WOOD, PLASTICS & COMPOSITES	86.7	94.2	90.9	85.8	94.1	90.4	97.2	105.7	101.9	84.2	94.2	89.7	87.9	85.8	86.7	86.2	97.2	92.2
07	THERMAL & MOISTURE PROTECTION	105.0	83.3	95.6	104.5	100.1	102.6	105.5	102.7	104.3	106.0	83.1	96.0	104.5	95.9	100.8	104.8	95.1	100.6
08	OPENINGS	95.6	96.2	95.7	94.1	99.7	95.4	98.1	105.4	99.8	95.6	91.9	94.7	88.0	91.9	88.9	95.8	97.1	96.1
0920	Plaster & Gypsum Board	90.7	94.5	93.2	83.8	94.5	91.0	91.6	106.4	101.5	90.7	94.5	93.2	92.1	86.0	88.0	90.7	97.5	95.3
0950, 0980	Ceilings & Acoustic Treatment	83.6	94.5	90.9	83.6	94.5	90.9	84.4	106.4	99.2	83.6	94.5	90.9	58.8	86.0	77.1	83.6	97.5	93.0
0960	Flooring	87.8	114.8	95.4	90.5	114.5	97.3	94.8	114.8	100.4	87.2	114.5	94.9	90.8	123.1	99.9	87.7	114.8	95.2
0970, 0990	Wall Finishes & Painting/Coating	89.6	104.3	98.3	95.0	97.7	96.6	95.0	115.9	107.4	89.6	80.0	83.9	84.0	105.8	96.9	89.6	83.2	85.8
09	FINISHES	86.5	100.0	93.8	89.2	99.4	94.7	92.3	109.1	101.4	87.4	97.3	92.7	82.1	96.4	89.8	86.2	99.4	93.3
COVERS	DIVS. 10 - 14, 25, 28, 41, 43, 44, 46	100.0	85.6	96.8	100.0	85.6	96.8	100.0	97.3	99.4	100.0	85.9	96.8	100.0	93.9	98.6	100.0	95.7	99.0
21, 22, 23	FIRE SUPPRESSION, PLUMBING & HVAC	96.7	83.1	91.3	96.3	95.8	96.1	99.8	96.0	98.3	96.7	88.6	93.5	96.0	91.1	94.1	96.7	88.8	93.6
26, 27, 3370	ELECTRICAL, COMMUNICATIONS & UTIL.	103.0	79.1	91.2	104.0	94.0	99.1	100.4	98.0	99.2	102.3	78.1	90.4	107.2	98.5	102.9	104.1	78.1	91.3
MF2018	WEIGHTED AVERAGE	95.4	91.9	93.9	95.2	98.0	96.4	98.7	102.3	100.2	97.9	92.2	95.5	95.1	96.5	95.7	94.7	93.4	94.2

City Cost Indexes

WYOMING

DIVISION		CASPER 826 MAT.	INST.	TOTAL	CHEYENNE 820 MAT.	INST.	TOTAL	NEWCASTLE 827 MAT.	INST.	TOTAL	RAWLINS 823 MAT.	INST.	TOTAL	RIVERTON 825 MAT.	INST.	TOTAL	ROCK SPRINGS 829-831 MAT.	INST.	TOTAL
015433	CONTRACTOR EQUIPMENT		97.3	97.3		94.6	94.6		94.6	94.6		94.6	94.6		94.6	94.6		94.6	94.6
0241, 31-34	SITE & INFRASTRUCTURE, DEMOLITION	97.8	94.7	95.7	91.6	90.1	90.6	83.6	89.3	87.6	97.5	89.3	91.9	91.1	89.3	89.8	87.4	89.5	88.9
0310	Concrete Forming & Accessories	101.3	53.6	60.6	102.5	64.6	70.2	93.7	70.5	74.0	97.6	70.8	74.7	92.7	60.3	65.0	99.7	64.7	69.9
0320	Concrete Reinforcing	111.1	83.4	97.7	106.4	83.4	95.3	114.5	83.2	99.3	114.2	83.2	99.2	115.2	83.2	99.7	115.2	81.6	99.0
0330	Cast-in-Place Concrete	103.8	77.8	94.1	97.9	77.4	90.3	99.0	72.8	89.2	99.0	72.9	89.3	99.0	72.8	89.2	99.0	75.0	90.0
03	CONCRETE	103.3	67.9	87.7	100.9	72.8	88.6	101.1	73.8	89.1	116.0	74.0	97.5	110.1	69.2	92.1	99.0	75.0	90.0
04	MASONRY	102.0	64.6	79.0	105.3	65.2	80.7	102.2	59.3	75.8	102.2	59.3	75.8	102.2	59.3	75.8	101.6	71.7	88.4
05	METALS	102.8	80.1	96.1	105.3	80.8	98.1	101.4	79.9	95.1	101.5	80.1	95.2	101.6	79.9	95.2	164.5	61.8	101.4
06	WOOD, PLASTICS & COMPOSITES	95.2	47.8	69.1	91.8	62.6	75.7	82.4	72.7	77.1	86.2	72.7	78.8	81.4	58.7	68.9	91.3	62.6	75.5
07	THERMAL & MOISTURE PROTECTION	112.7	65.9	92.3	107.2	68.0	90.2	108.8	63.9	89.3	110.3	74.9	94.9	109.7	67.7	91.4	108.9	69.5	91.8
08	OPENINGS	109.3	59.4	97.7	107.2	67.7	98.0	111.4	72.3	102.3	111.1	72.3	102.0	111.3	64.6	100.4	111.8	66.5	101.2
0920	Plaster & Gypsum Board	101.7	46.2	64.4	90.6	61.7	71.2	87.6	72.0	77.1	87.9	72.0	77.2	87.6	57.7	67.5	100.2	61.7	74.3
0950, 0980	Ceilings & Acoustic Treatment	105.3	46.2	65.5	98.5	61.7	73.7	100.3	72.0	81.2	100.3	72.0	81.2	100.3	57.7	71.6	100.3	61.7	74.3
0960	Flooring	102.2	73.8	94.2	100.3	73.8	92.8	94.2	44.5	80.2	97.0	44.5	82.3	93.6	60.3	84.2	100.3	61.7	74.3
0970, 0990	Wall Finishes & Painting/Coating	95.7	58.7	73.8	97.9	58.7	74.7	94.6	60.5	74.4	94.6	60.5	74.4	94.6	60.5	74.4	99.3	59.9	88.3
09	FINISHES	99.6	56.3	76.2	96.3	65.1	79.4	91.4	64.6	76.9	93.6	64.6	77.9	92.0	59.5	74.4	94.6	60.5	74.4
COVERS	DIVS. 10-14, 25, 28, 41, 43, 44, 46	100.0	92.5	98.3	100.0	87.4	97.2	100.0	91.8	98.2	100.0	91.8	98.2	100.0	77.3	94.9	100.0	86.9	97.1
21, 22, 23	FIRE SUPPRESSION, PLUMBING & HVAC	100.0	70.5	88.4	100.1	73.3	89.6	98.2	71.8	87.8	98.2	71.9	87.8	98.2	71.8	87.8	100.0	73.9	89.8
26, 27, 3370	ELECTRICAL, COMMUNICATIONS & UTIL.	94.1	60.3	77.5	95.2	69.0	82.3	94.0	62.3	78.4	94.0	62.3	78.4	94.0	59.1	76.8	92.5	66.0	79.4
MF2018	WEIGHTED AVERAGE	101.5	68.9	87.7	101.1	72.7	89.1	99.6	71.1	87.5	102.0	71.5	89.1	101.0	68.6	87.3	103.4	71.4	89.9

WYOMING / CANADA

DIVISION		SHERIDAN 828 MAT.	INST.	TOTAL	WHEATLAND 822 MAT.	INST.	TOTAL	WORLAND 824 MAT.	INST.	TOTAL	YELLOWSTONE NAT'L PA 821 MAT.	INST.	TOTAL	BARRIE, ONTARIO MAT.	INST.	TOTAL	BATHURST, NEW BRUNSWICK MAT.	INST.	TOTAL
015433	CONTRACTOR EQUIPMENT		94.6	94.6		94.6	94.6		94.6	94.6		94.6	94.6		100.1	100.1		99.7	99.7
0241, 31-34	SITE & INFRASTRUCTURE, DEMOLITION	91.5	90.1	90.5	88.1	89.3	88.9	85.5	89.3	88.1	85.6	89.3	88.1	117.0	94.7	101.6	105.1	90.6	95.1
0310	Concrete Forming & Accessories	100.5	61.6	67.4	95.4	48.4	55.3	95.5	61.1	66.1	95.5	61.1	66.2	125.6	82.3	88.7	106.1	57.8	64.9
0320	Concrete Reinforcing	115.2	83.2	99.7	114.5	82.6	99.1	115.2	83.2	99.7	117.2	82.3	100.3	175.2	86.9	132.5	140.5	57.5	100.4
0330	Cast-in-Place Concrete	102.3	77.4	93.1	103.3	72.7	91.9	99.0	72.7	89.2	99.0	72.8	89.2	157.7	82.1	129.6	114.0	56.2	92.5
03	CONCRETE	109.9	71.4	93.0	106.3	63.7	87.6	101.3	69.5	87.3	101.6	69.4	87.4	138.9	83.5	114.5	110.3	58.3	87.4
04	MASONRY	102.5	62.0	77.6	102.6	51.2	71.0	102.2	59.3	75.8	102.2	59.3	75.8	165.6	89.3	118.7	159.8	57.4	96.9
05	METALS	105.1	80.5	97.8	101.4	79.1	94.8	101.6	79.8	95.2	102.2	79.1	95.4	111.1	91.5	105.3	114.1	73.6	102.2
06	WOOD, PLASTICS & COMPOSITES	93.3	58.7	74.3	84.2	43.0	61.5	84.3	60.0	70.9	84.3	60.0	70.9	116.5	80.8	96.9	97.5	57.9	75.7
07	THERMAL & MOISTURE PROTECTION	110.1	66.7	91.2	109.1	58.1	86.9	108.9	64.8	89.7	108.3	64.8	89.4	115.3	86.1	102.6	111.8	58.4	88.6
08	OPENINGS	112.0	64.6	101.0	110.0	55.9	97.4	111.6	65.3	100.8	104.5	64.8	95.2	90.5	80.9	88.3	84.0	51.2	76.3
0920	Plaster & Gypsum Board	111.2	57.7	75.2	87.6	41.5	56.6	87.6	58.9	68.3	87.8	58.9	68.4	153.5	80.3	104.3	124.2	56.7	78.8
0950, 0980	Ceilings & Acoustic Treatment	103.9	57.7	72.7	100.3	41.5	60.0	101.1	58.9	72.7	101.1	58.9	72.7	94.3	80.3	84.9	116.8	56.7	76.3
0960	Flooring	97.9	60.3	87.4	95.6	43.8	81.0	95.6	44.5	81.3	95.6	44.5	81.3	119.3	87.3	110.3	99.6	41.5	83.3
0970, 0990	Wall Finishes & Painting/Coating	96.8	58.7	74.2	94.6	58.4	73.2	94.6	60.5	74.4	94.6	60.5	74.4	105.6	83.8	92.7	111.3	47.9	73.8
09	FINISHES	99.2	60.4	78.2	92.1	46.7	67.5	91.8	57.1	73.0	92.0	57.2	73.2	111.3	83.5	96.2	107.7	53.8	78.5
COVERS	DIVS. 10-14, 25, 28, 41, 43, 44, 46	100.0	93.4	98.5	100.0	81.4	95.9	100.0	76.1	94.7	100.0	76.2	94.7	139.2	65.0	122.7	131.1	57.9	114.8
21, 22, 23	FIRE SUPPRESSION, PLUMBING & HVAC	98.2	73.3	88.4	98.2	71.8	87.8	98.2	71.8	87.8	98.2	71.8	87.8	103.7	93.5	99.7	103.8	64.8	88.5
26, 27, 3370	ELECTRICAL, COMMUNICATIONS & UTIL.	96.1	59.1	77.9	94.0	76.6	85.4	94.0	76.6	85.4	93.1	86.1	89.7	117.0	82.9	100.2	113.5	55.9	85.1
MF2018	WEIGHTED AVERAGE	102.5	70.2	88.8	100.3	67.1	86.2	99.7	70.7	87.5	99.0	71.9	87.6	116.5	87.1	104.1	110.7	62.3	90.2

CANADA

DIVISION		BRANDON, MANITOBA MAT.	INST.	TOTAL	BRANTFORD, ONTARIO MAT.	INST.	TOTAL	BRIDGEWATER, NOVA SCOTIA MAT.	INST.	TOTAL	CALGARY, ALBERTA MAT.	INST.	TOTAL	CAP-DE-LA-MADELEINE, QUEBEC MAT.	INST.	TOTAL	CHARLESBOURG, QUEBEC MAT.	INST.	TOTAL
015433	CONTRACTOR EQUIPMENT		101.5	101.5		99.7	99.7		99.4	99.4		127.1	127.1		100.2	100.2		100.2	100.2
0241, 31-34	SITE & INFRASTRUCTURE, DEMOLITION	126.4	92.9	103.3	116.9	94.8	101.6	101.1	92.3	95.0	125.5	118.4	120.6	96.9	93.7	94.7	96.9	93.7	94.7
0310	Concrete Forming & Accessories	145.4	66.1	77.8	127.2	89.1	94.7	99.7	67.9	72.6	125.0	94.9	99.3	132.9	79.2	87.1	132.9	79.2	87.1
0320	Concrete Reinforcing	170.2	54.5	114.3	163.7	85.6	125.9	139.3	47.6	95.0	136.4	82.0	110.1	139.3	72.4	107.0	139.3	72.4	107.0
0330	Cast-in-Place Concrete	109.4	70.2	94.8	130.7	101.8	119.9	134.3	54.6	109.2	136.4	104.3	124.4	105.5	87.5	98.8	105.5	87.5	98.8
03	CONCRETE	118.1	66.6	95.5	121.8	93.2	109.3	120.1	65.1	95.9	126.3	96.8	113.3	107.5	81.5	96.1	107.5	81.5	96.1
04	MASONRY	212.8	60.7	119.3	166.1	93.4	121.4	161.8	65.6	102.6	212.4	89.8	137.1	162.4	76.3	109.5	162.4	76.3	109.5
05	METALS	127.2	78.2	112.8	112.3	92.0	106.3	111.6	76.0	101.1	131.0	102.4	122.6	110.7	85.0	103.1	110.7	85.0	103.1
06	WOOD, PLASTICS & COMPOSITES	150.5	66.8	104.4	120.3	87.7	102.4	89.6	67.4	77.4	102.6	94.4	98.1	131.6	79.0	102.7	131.6	79.0	102.7
07	THERMAL & MOISTURE PROTECTION	128.4	68.7	102.4	121.6	91.4	108.5	116.1	67.6	95.0	132.0	98.0	117.2	114.8	84.1	101.5	114.8	84.1	101.5
08	OPENINGS	100.3	59.9	90.8	87.7	86.6	87.5	82.4	61.3	77.5	82.3	83.7	82.6	89.1	72.3	85.2	89.1	72.3	85.2
0920	Plaster & Gypsum Board	115.4	65.6	81.9	115.8	87.4	96.7	124.3	66.5	85.4	126.4	93.3	104.1	147.6	78.4	101.0	147.6	78.4	101.0
0950, 0980	Ceilings & Acoustic Treatment	123.6	65.6	84.6	103.5	87.4	92.7	103.5	66.5	78.6	149.2	93.3	111.5	103.5	78.4	86.6	103.5	78.4	86.6
0960	Flooring	131.2	61.9	111.7	113.6	87.3	106.2	95.7	58.9	85.4	118.1	84.8	108.7	113.6	86.0	105.8	113.6	86.0	105.8
0970, 0990	Wall Finishes & Painting/Coating	117.9	54.1	80.1	110.3	92.0	99.5	110.3	59.3	80.1	113.7	106.4	109.4	110.3	83.6	94.5	110.3	83.6	94.5
09	FINISHES	122.9	64.3	91.2	108.7	89.3	98.2	103.8	65.5	83.1	123.6	94.8	108.0	112.2	81.2	95.4	112.2	81.2	95.4
COVERS	DIVS. 10-14, 25, 28, 41, 43, 44, 46	131.1	60.1	115.3	131.1	66.9	116.8	131.1	60.8	115.5	131.1	92.7	122.6	131.1	75.6	118.8	131.1	75.6	118.8
21, 22, 23	FIRE SUPPRESSION, PLUMBING & HVAC	104.0	78.9	94.1	103.8	96.3	100.9	103.8	79.5	94.3	105.1	90.2	99.2	104.2	84.2	96.4	104.2	84.2	96.4
26, 27, 3370	ELECTRICAL, COMMUNICATIONS & UTIL.	113.9	63.1	88.8	111.5	82.5	97.2	115.8	58.7	87.6	107.1	93.9	100.6	110.2	65.2	88.0	110.2	65.2	88.0
MF2018	WEIGHTED AVERAGE	120.6	70.5	99.4	113.1	90.7	103.6	111.3	70.3	94.0	119.7	95.6	109.5	110.6	80.0	97.7	110.6	80.0	97.7

For customer support on your Site Work & Landscape Costs with RSMeans data, call 800.448.8182.

817

CANADA

DIVISION		CHARLOTTETOWN, PRINCE EDWARD ISLAND			CHICOUTIMI, QUEBEC			CORNER BROOK, NEWFOUNDLAND			CORNWALL, ONTARIO			DALHOUSIE, NEW BRUNSWICK			DARTMOUTH, NOVA SCOTIA		
		MAT.	INST.	TOTAL	MAT.	INST.	TOTAL	MAT.	INST.	TOTAL	MAT.	INST.	TOTAL	MAT.	INST.	TOTAL	MAT.	INST.	TOTAL
015433	CONTRACTOR EQUIPMENT		116.8	116.8		100.3	100.3		100.0	100.0		99.7	99.7		99.7	99.7		98.9	98.9
0241, 31 - 34	SITE & INFRASTRUCTURE, DEMOLITION	134.6	101.5	111.7	102.0	93.6	96.2	130.4	90.5	102.8	115.1	94.3	100.7	101.0	90.6	93.8	117.9	92.0	100.0
0310	Concrete Forming & Accessories	123.7	53.1	63.5	134.9	89.8	96.4	121.9	75.6	82.4	125.0	82.4	88.7	105.9	58.0	65.1	111.8	67.8	74.3
0320	Concrete Reinforcing	156.7	46.6	103.5	106.0	94.5	100.4	156.0	48.8	104.2	163.7	85.3	125.8	142.5	57.6	101.4	163.2	47.5	107.3
0330	Cast-in-Place Concrete	152.1	57.6	116.9	107.4	94.7	102.7	127.3	63.6	103.6	117.5	92.8	108.3	110.5	56.3	90.3	122.9	66.8	102.0
03	CONCRETE	136.8	55.4	101.0	101.6	92.4	97.6	149.1	67.5	113.3	115.7	87.0	103.1	113.1	58.4	89.0	131.9	65.0	102.5
04	MASONRY	187.9	54.9	106.2	161.9	89.8	117.6	208.6	74.3	126.1	165.0	85.6	116.2	163.3	57.4	98.3	222.6	65.6	126.1
05	METALS	136.4	79.6	119.8	113.9	91.9	107.4	127.7	75.2	112.2	112.1	90.7	105.8	105.6	73.7	96.2	128.1	75.7	112.7
06	WOOD, PLASTICS & COMPOSITES	104.3	52.5	75.8	131.8	90.3	108.9	128.1	81.4	102.4	118.5	81.6	98.2	95.8	57.9	74.9	115.9	67.4	89.2
07	THERMAL & MOISTURE PROTECTION	137.8	57.4	102.8	112.3	96.4	105.4	133.0	65.9	103.8	121.4	86.2	106.1	121.0	58.4	93.8	130.1	67.6	102.9
08	OPENINGS	85.7	45.8	76.4	87.8	76.3	85.2	106.3	66.6	97.0	89.1	80.6	87.1	85.5	51.2	77.5	91.3	61.3	84.3
0920	Plaster & Gypsum Board	126.3	50.5	75.3	143.9	89.9	107.6	151.2	80.9	103.9	174.9	81.1	111.8	129.5	56.7	80.5	146.6	66.5	92.7
0950, 0980	Ceilings & Acoustic Treatment	128.9	50.5	76.1	116.0	89.9	98.4	124.6	80.9	95.1	107.6	81.1	89.8	106.6	56.7	72.9	131.1	66.5	87.5
0960	Flooring	115.1	55.3	98.3	115.7	86.0	107.3	112.8	49.5	95.0	113.6	86.0	105.8	101.9	63.4	91.1	107.6	58.9	93.9
0970, 0990	Wall Finishes & Painting/Coating	117.3	39.9	71.5	111.3	104.1	107.1	117.8	56.5	81.5	110.3	85.7	95.7	113.8	47.9	74.8	117.8	59.3	83.2
09	FINISHES	119.8	52.3	83.3	114.6	91.6	102.1	122.6	70.2	94.2	117.3	83.6	99.0	108.4	58.2	81.2	120.3	65.5	90.7
COVERS	DIVS. 10 - 14, 25, 28, 41, 43, 44, 46	131.1	58.4	114.9	131.1	81.5	120.1	131.1	60.6	115.4	131.1	64.6	116.3	131.1	57.8	114.8	131.1	60.8	115.5
21, 22, 23	FIRE SUPPRESSION, PLUMBING & HVAC	104.1	59.0	86.4	103.8	81.0	94.9	104.0	65.9	89.0	104.2	94.3	100.3	103.9	64.8	88.5	104.0	79.5	94.4
26, 27, 3370	ELECTRICAL, COMMUNICATIONS & UTIL.	110.0	47.5	79.2	109.9	84.1	97.2	110.9	51.5	81.6	111.6	83.4	97.7	113.9	52.8	83.8	115.0	58.7	87.2
MF2018	WEIGHTED AVERAGE	121.3	60.2	95.5	110.4	87.6	100.7	124.7	68.4	100.9	113.1	87.3	102.2	110.2	62.5	90.0	121.5	70.2	99.8

CANADA

DIVISION		EDMONTON, ALBERTA			FORT MCMURRAY, ALBERTA			FREDERICTON, NEW BRUNSWICK			GATINEAU, QUEBEC			GRANBY, QUEBEC			HALIFAX, NOVA SCOTIA		
		MAT.	INST.	TOTAL	MAT.	INST.	TOTAL	MAT.	INST.	TOTAL	MAT.	INST.	TOTAL	MAT.	INST.	TOTAL	MAT.	INST.	TOTAL
015433	CONTRACTOR EQUIPMENT		129.1	129.1		101.9	101.9		113.9	113.9		100.2	100.2		100.2	100.2		115.6	115.6
0241, 31 - 34	SITE & INFRASTRUCTURE, DEMOLITION	125.1	121.8	122.8	121.7	95.4	103.5	116.0	100.8	105.5	96.7	93.7	94.6	97.2	93.7	94.8	102.8	103.6	103.4
0310	Concrete Forming & Accessories	128.3	94.9	99.8	125.0	87.5	93.1	126.9	58.6	68.7	132.9	79.1	87.0	132.9	79.0	87.0	124.1	87.7	93.0
0320	Concrete Reinforcing	136.0	82.0	109.9	151.5	81.9	117.8	141.0	57.8	100.8	147.4	72.4	111.1	147.4	72.4	111.1	156.3	77.9	118.3
0330	Cast-in-Place Concrete	145.8	104.3	130.4	173.5	98.2	145.5	112.7	57.5	92.1	104.0	87.5	97.9	107.5	87.4	100.0	99.4	85.1	94.0
03	CONCRETE	130.8	96.8	115.9	139.8	90.6	118.2	116.9	59.7	91.8	107.9	81.4	96.2	109.5	81.4	97.1	111.9	86.1	100.6
04	MASONRY	212.7	89.8	137.2	206.1	86.3	132.5	189.1	58.8	109.0	162.3	76.3	109.5	162.6	76.3	109.6	185.4	86.4	124.6
05	METALS	134.6	102.4	125.1	139.8	90.9	125.5	136.6	83.3	121.0	110.7	84.8	103.1	110.9	84.8	103.2	136.3	99.2	125.4
06	WOOD, PLASTICS & COMPOSITES	101.6	94.4	97.6	114.5	86.7	99.2	106.5	58.2	79.9	131.6	79.0	102.7	131.6	79.0	102.7	103.7	87.8	95.0
07	THERMAL & MOISTURE PROTECTION	143.5	98.0	123.7	129.8	92.6	113.6	137.4	59.4	103.5	114.8	84.1	101.5	114.8	82.6	100.8	137.4	88.7	116.2
08	OPENINGS	80.9	83.7	81.5	89.1	79.5	86.9	88.0	50.3	79.2	89.1	67.9	84.2	89.1	67.9	84.2	88.9	79.2	86.7
0920	Plaster & Gypsum Board	129.8	93.3	105.2	116.6	86.1	96.1	130.3	56.7	80.8	114.4	78.4	90.2	114.4	78.4	90.2	117.8	87.1	97.1
0950, 0980	Ceilings & Acoustic Treatment	144.5	93.3	110.0	112.5	86.1	94.7	137.2	56.7	82.9	103.5	78.4	86.6	103.5	78.4	86.6	139.8	87.1	104.2
0960	Flooring	121.8	84.8	111.4	113.6	84.8	105.5	115.6	66.2	101.7	113.6	86.0	105.8	113.6	86.0	105.8	109.5	81.5	101.6
0970, 0990	Wall Finishes & Painting/Coating	115.6	107.4	110.8	110.4	88.3	97.3	118.8	61.1	84.6	110.3	83.6	94.5	110.3	83.6	94.5	118.4	90.0	102.1
09	FINISHES	123.6	94.9	108.0	111.8	87.4	98.6	125.1	60.4	90.1	107.8	81.2	93.4	107.8	81.2	93.4	117.7	87.5	101.3
COVERS	DIVS. 10 - 14, 25, 28, 41, 43, 44, 46	131.1	93.6	122.8	131.1	90.1	122.0	131.1	58.5	115.0	131.1	75.6	118.8	131.1	75.6	118.8	131.1	68.9	117.3
21, 22, 23	FIRE SUPPRESSION, PLUMBING & HVAC	104.9	90.4	99.2	104.2	93.2	99.9	104.2	73.7	92.2	104.2	84.2	96.3	103.8	84.2	96.1	104.9	82.3	96.0
26, 27, 3370	ELECTRICAL, COMMUNICATIONS & UTIL.	115.0	93.9	104.6	105.4	77.5	91.7	112.2	69.7	91.2	110.2	65.2	88.0	110.9	65.2	88.4	113.3	90.6	102.1
MF2018	WEIGHTED AVERAGE	121.8	95.9	110.9	121.9	88.4	107.7	119.3	69.0	98.0	110.3	79.8	97.4	110.5	79.7	97.5	117.9	88.0	105.2

CANADA

DIVISION		HAMILTON, ONTARIO			HULL, QUEBEC			JOLIETTE, QUEBEC			KAMLOOPS, BRITISH COLUMBIA			KINGSTON, ONTARIO			KITCHENER, ONTARIO		
		MAT.	INST.	TOTAL	MAT.	INST.	TOTAL	MAT.	INST.	TOTAL	MAT.	INST.	TOTAL	MAT.	INST.	TOTAL	MAT.	INST.	TOTAL
015433	CONTRACTOR EQUIPMENT		115.5	115.5		100.2	100.2		100.2	100.2		103.3	103.3		101.9	101.9		101.6	101.6
0241, 31 - 34	SITE & INFRASTRUCTURE, DEMOLITION	105.8	107.0	106.6	96.7	93.7	94.6	97.3	93.7	94.8	120.1	97.1	104.3	115.1	98.0	103.3	94.2	99.8	98.0
0310	Concrete Forming & Accessories	130.0	94.7	99.9	132.9	79.1	87.0	132.9	79.2	87.1	124.3	82.3	88.5	125.1	82.5	88.7	117.3	88.0	92.3
0320	Concrete Reinforcing	136.9	103.3	120.7	147.4	72.4	111.1	139.3	72.4	107.0	109.3	75.8	93.1	163.7	85.3	125.8	96.4	103.1	99.6
0330	Cast-in-Place Concrete	107.6	102.1	105.6	104.0	87.5	97.9	108.4	87.5	100.6	93.8	91.8	93.1	117.5	92.8	108.3	111.7	97.2	106.3
03	CONCRETE	110.2	99.3	105.4	107.9	81.4	96.2	108.8	81.5	96.8	117.2	85.0	103.0	117.5	87.0	104.1	97.1	94.1	95.8
04	MASONRY	173.2	100.5	128.5	162.3	76.3	109.5	162.7	76.3	109.6	169.4	84.0	116.9	171.9	85.7	118.9	142.2	98.6	115.4
05	METALS	134.7	106.4	126.4	110.9	84.8	103.2	110.9	85.0	103.3	112.8	87.1	105.3	113.5	90.7	106.8	124.1	98.0	116.5
06	WOOD, PLASTICS & COMPOSITES	111.4	94.0	101.8	131.6	79.0	102.7	131.6	79.0	102.7	102.2	80.7	90.3	118.5	81.7	98.3	110.3	85.9	96.9
07	THERMAL & MOISTURE PROTECTION	134.0	101.5	119.8	114.8	84.1	101.5	114.8	84.1	101.5	131.2	81.7	109.7	121.4	87.3	106.6	116.4	98.4	108.5
08	OPENINGS	85.3	93.2	87.1	89.1	67.9	84.2	89.1	72.3	85.2	86.1	78.5	84.3	89.1	80.3	87.1	80.0	87.1	81.7
0920	Plaster & Gypsum Board	131.0	93.1	105.5	114.4	78.4	90.2	147.6	78.4	101.0	101.1	79.8	86.8	177.6	81.2	112.8	107.2	85.6	92.6
0950, 0980	Ceilings & Acoustic Treatment	139.9	93.1	108.4	103.5	78.4	86.6	103.5	78.4	86.6	103.5	79.8	87.5	119.0	81.2	93.5	107.9	85.6	92.9
0960	Flooring	115.3	95.0	109.6	113.6	86.0	105.8	113.6	86.0	105.8	112.5	49.2	94.7	113.6	86.0	105.8	101.3	95.1	99.6
0970, 0990	Wall Finishes & Painting/Coating	113.0	104.1	107.8	110.3	83.6	94.5	110.3	83.6	94.5	110.3	76.4	90.3	110.3	79.2	91.9	104.4	93.9	98.2
09	FINISHES	121.0	95.6	107.3	107.8	81.2	93.4	112.2	81.2	95.4	108.3	75.8	90.7	119.9	82.9	99.9	102.1	89.4	95.2
COVERS	DIVS. 10 - 14, 25, 28, 41, 43, 44, 46	131.1	92.5	122.5	131.1	75.6	118.8	131.1	75.6	118.8	131.1	84.4	120.7	131.1	64.6	116.3	131.1	89.6	121.9
21, 22, 23	FIRE SUPPRESSION, PLUMBING & HVAC	105.2	92.1	100.0	103.8	84.2	96.1	103.8	84.2	96.1	103.8	87.5	97.4	104.2	94.4	100.4	104.2	91.3	99.1
26, 27, 3370	ELECTRICAL, COMMUNICATIONS & UTIL.	108.5	103.6	106.1	112.3	65.2	89.1	110.9	65.2	88.4	114.3	74.2	94.5	111.6	82.1	97.1	109.6	99.8	104.8
MF2018	WEIGHTED AVERAGE	116.3	98.8	108.9	110.5	79.8	97.5	110.8	80.0	97.8	113.0	83.5	100.5	114.1	87.4	102.8	108.6	94.6	102.7

For customer support on your Site Work & Landscape Costs with RSMeans data, call 800.448.8182.

City Cost Indexes

CANADA

DIVISION		LAVAL, QUEBEC			LETHBRIDGE, ALBERTA			LLOYDMINSTER, ALBERTA			LONDON, ONTARIO			MEDICINE HAT, ALBERTA			MONCTON, NEW BRUNSWICK		
		MAT.	INST.	TOTAL	MAT.	INST.	TOTAL	MAT.	INST.	TOTAL	MAT.	INST.	TOTAL	MAT.	INST.	TOTAL	MAT.	INST.	TOTAL
015433	CONTRACTOR EQUIPMENT		100.2	100.2		101.9	101.9		101.9	101.9		116.7	116.7		101.9	101.9		99.7	99.7
0241, 31 - 34	SITE & INFRASTRUCTURE, DEMOLITION	97.2	93.9	94.9	114.5	96.0	101.7	114.5	95.4	101.3	103.5	106.8	105.8	113.2	95.5	101.0	104.5	92.5	96.2
0310	Concrete Forming & Accessories	133.2	81.1	88.8	126.4	87.6	93.3	124.6	78.3	85.1	129.1	89.1	95.0	126.4	78.2	85.3	106.1	66.1	72.0
0320	Concrete Reinforcing	147.4	74.4	112.1	151.5	81.9	117.8	151.5	81.8	117.8	126.7	102.1	114.8	151.5	81.8	117.8	140.5	69.8	106.3
0330	Cast-in-Place Concrete	107.5	89.4	100.7	130.1	98.2	118.2	120.7	94.6	111.0	120.2	100.8	113.0	120.7	94.6	111.0	109.8	68.0	94.3
03	CONCRETE	109.5	83.4	98.0	120.0	90.6	107.1	115.5	85.2	102.2	114.6	96.2	106.5	115.7	85.2	102.3	108.4	68.5	90.9
04	MASONRY	162.6	78.4	110.8	180.1	86.3	122.4	161.6	80.1	111.6	183.8	99.5	132.0	161.6	80.1	111.6	159.4	74.6	107.3
05	METALS	110.7	85.9	103.5	134.1	91.0	121.5	112.8	90.8	106.4	134.7	106.8	126.5	113.0	90.8	106.5	114.1	86.8	106.1
06	WOOD, PLASTICS & COMPOSITES	131.7	81.1	103.9	118.0	86.7	100.8	114.5	77.4	94.1	114.8	86.8	99.4	118.0	77.4	95.7	97.5	65.0	79.6
07	THERMAL & MOISTURE PROTECTION	115.4	86.0	102.6	127.1	92.6	112.1	123.5	88.2	108.1	129.9	99.0	116.5	130.5	88.2	112.1	116.6	72.4	97.4
08	OPENINGS	89.1	69.8	84.6	89.1	79.5	86.9	89.1	74.4	85.7	81.6	87.9	83.1	89.1	74.4	85.7	84.0	61.9	78.8
0920	Plaster & Gypsum Board	114.7	80.5	91.7	106.4	86.1	92.7	102.1	76.6	84.9	134.4	85.8	101.7	104.3	76.6	85.6	124.2	64.1	83.7
0950, 0980	Ceilings & Acoustic Treatment	103.5	80.5	88.0	112.5	86.1	94.7	103.5	76.6	85.3	138.5	85.8	102.9	103.5	76.6	85.3	116.8	64.1	81.2
0960	Flooring	113.6	88.4	106.5	113.6	84.8	105.5	113.6	84.8	105.5	109.9	95.1	105.7	113.6	84.8	105.5	99.6	65.6	90.1
0970, 0990	Wall Finishes & Painting/Coating	110.3	85.9	95.8	110.2	96.4	102.1	110.4	75.1	89.5	113.5	100.5	105.8	110.2	75.1	89.4	111.3	83.6	94.9
09	FINISHES	107.8	83.3	94.5	109.6	88.3	98.0	107.4	78.9	92.0	121.5	91.0	105.0	107.5	78.9	92.0	107.7	67.5	86.0
COVERS	DIVS. 10 - 14, 25, 28, 41, 43, 44, 46	131.1	77.5	119.2	131.1	89.2	121.8	131.1	87.1	121.3	131.1	91.6	122.3	131.1	86.1	121.1	131.1	61.0	115.5
21, 22, 23	FIRE SUPPRESSION, PLUMBING & HVAC	104.3	86.5	97.3	104.1	90.1	98.6	104.2	90.0	98.6	105.1	89.2	98.8	103.8	86.8	97.2	103.8	73.7	92.0
26, 27, 3370	ELECTRICAL, COMMUNICATIONS & UTIL.	110.9	67.0	89.3	107.1	77.5	92.5	104.6	77.5	91.2	103.5	102.0	102.8	104.6	77.5	91.2	117.3	88.5	103.1
MF2018	WEIGHTED AVERAGE	110.7	81.6	98.4	117.0	87.9	104.7	111.5	84.7	100.2	116.3	96.5	108.0	111.7	84.0	100.0	110.9	76.0	96.2

CANADA

DIVISION		MONTREAL, QUEBEC			MOOSE JAW, SASKATCHEWAN			NEW GLASGOW, NOVA SCOTIA			NEWCASTLE, NEW BRUNSWICK			NORTH BAY, ONTARIO			OSHAWA, ONTARIO		
		MAT.	INST.	TOTAL	MAT.	INST.	TOTAL	MAT.	INST.	TOTAL	MAT.	INST.	TOTAL	MAT.	INST.	TOTAL	MAT.	INST.	TOTAL
015433	CONTRACTOR EQUIPMENT		117.7	117.7		98.3	98.3		98.9	98.9		99.7	99.7		99.3	99.3		101.6	101.6
0241, 31 - 34	SITE & INFRASTRUCTURE, DEMOLITION	113.0	105.5	107.8	114.4	90.0	97.5	111.9	92.0	98.1	105.1	90.6	95.1	126.8	93.6	103.9	104.6	99.6	101.2
0310	Concrete Forming & Accessories	130.9	90.4	96.4	109.6	54.6	62.7	111.8	67.8	74.3	106.1	58.0	65.1	146.2	80.0	89.8	122.8	91.3	96.0
0320	Concrete Reinforcing	126.8	94.6	111.3	106.9	61.4	84.9	156.0	47.6	103.6	140.5	57.6	100.4	184.7	84.8	136.4	152.7	103.7	129.0
0330	Cast-in-Place Concrete	124.4	97.4	114.3	117.1	63.7	97.2	122.9	66.8	102.0	114.0	56.3	92.6	118.4	80.1	104.1	129.2	106.4	120.7
03	CONCRETE	116.7	94.3	106.8	101.4	60.0	83.2	131.0	65.0	102.0	110.3	58.4	87.5	132.7	81.4	110.2	118.3	98.8	109.8
04	MASONRY	178.8	89.8	124.1	160.3	56.4	96.4	208.2	65.6	120.5	159.8	57.4	96.9	214.4	81.9	133.0	145.1	101.9	118.6
05	METALS	144.3	102.2	131.9	109.6	75.0	99.4	125.6	75.7	111.0	114.1	73.8	102.3	126.5	90.2	115.8	114.2	98.6	109.6
06	WOOD, PLASTICS & COMPOSITES	111.3	90.8	100.0	99.4	53.2	74.0	115.9	67.4	89.2	97.5	57.9	75.7	153.5	80.2	113.2	117.6	89.4	102.1
07	THERMAL & MOISTURE PROTECTION	125.9	97.2	113.4	114.1	60.1	90.6	130.1	67.6	102.9	116.6	58.4	91.3	136.4	82.7	113.0	117.3	104.1	111.6
08	OPENINGS	87.4	78.7	85.4	85.3	51.4	77.4	91.3	61.3	84.3	84.0	51.2	76.3	98.6	78.2	93.8	85.0	90.2	86.2
0920	Plaster & Gypsum Board	126.2	89.9	101.8	98.2	51.9	67.1	145.0	66.5	92.2	124.2	56.7	78.8	140.6	79.8	99.7	110.7	89.2	96.2
0950, 0980	Ceilings & Acoustic Treatment	146.1	89.9	108.2	103.5	51.9	68.7	123.8	66.5	85.2	116.8	56.7	76.3	123.8	79.8	94.1	103.9	89.2	94.0
0960	Flooring	112.4	90.5	106.2	103.6	53.7	89.6	107.6	58.9	93.9	99.6	63.4	89.5	131.2	86.0	118.5	104.2	97.6	102.4
0970, 0990	Wall Finishes & Painting/Coating	115.0	104.1	108.5	110.3	61.0	81.1	117.8	59.3	83.2	111.3	47.9	73.8	117.8	85.0	98.4	104.4	108.1	106.6
09	FINISHES	121.6	92.8	106.0	103.7	54.8	77.2	118.7	65.5	89.9	107.7	58.2	80.9	125.9	81.9	102.1	103.2	93.7	98.1
COVERS	DIVS. 10 - 14, 25, 28, 41, 43, 44, 46	131.1	82.8	120.4	131.1	57.5	114.7	131.1	60.8	115.5	131.1	57.9	114.8	131.1	63.4	116.0	131.1	90.4	122.0
21, 22, 23	FIRE SUPPRESSION, PLUMBING & HVAC	105.1	81.2	95.7	104.2	70.3	90.9	104.0	79.5	94.4	103.8	64.8	88.5	104.0	92.4	99.4	104.2	91.9	99.3
26, 27, 3370	ELECTRICAL, COMMUNICATIONS & UTIL.	107.9	84.1	96.2	113.4	56.3	85.3	111.3	58.7	85.4	112.9	55.9	84.8	112.6	83.3	98.1	110.6	103.8	107.3
MF2018	WEIGHTED AVERAGE	119.1	90.1	106.9	109.0	63.9	89.9	119.6	70.2	98.7	110.7	62.9	90.5	122.5	85.2	106.8	110.8	97.2	105.0

CANADA

DIVISION		OTTAWA, ONTARIO			OWEN SOUND, ONTARIO			PETERBOROUGH, ONTARIO			PORTAGE LA PRAIRIE, MANITOBA			PRINCE ALBERT, SASKATCHEWAN			PRINCE GEORGE, BRITISH COLUMBIA		
		MAT.	INST.	TOTAL	MAT.	INST.	TOTAL	MAT.	INST.	TOTAL	MAT.	INST.	TOTAL	MAT.	INST.	TOTAL	MAT.	INST.	TOTAL
015433	CONTRACTOR EQUIPMENT		117.8	117.8		100.1	100.1		99.7	99.7		101.9	101.9		98.3	98.3		103.3	103.3
0241, 31 - 34	SITE & INFRASTRUCTURE, DEMOLITION	107.0	107.2	107.1	117.0	94.6	101.5	116.9	94.2	101.3	115.4	93.2	100.1	110.2	90.1	96.3	123.4	97.1	105.3
0310	Concrete Forming & Accessories	124.8	88.4	93.7	125.6	78.7	85.6	127.2	81.0	87.8	126.6	65.7	74.7	109.6	54.4	62.5	115.1	77.5	83.0
0320	Concrete Reinforcing	137.0	102.0	120.1	175.2	86.9	132.5	163.7	85.3	125.8	151.5	54.5	104.6	111.7	61.4	87.3	109.3	75.8	93.1
0330	Cast-in-Place Concrete	122.3	103.0	115.1	157.7	76.6	127.5	130.7	81.8	112.5	120.7	69.8	101.8	106.2	63.6	90.3	117.6	91.8	108.0
03	CONCRETE	116.6	96.7	107.9	138.9	79.9	113.0	121.8	82.5	104.6	109.5	66.2	90.5	96.9	59.9	80.7	127.5	82.8	107.9
04	MASONRY	173.2	99.4	127.8	165.6	87.1	117.3	166.1	88.0	118.1	164.8	59.7	100.2	159.5	56.4	96.1	171.4	84.0	117.7
05	METALS	135.5	108.9	127.7	111.1	91.3	105.3	112.3	90.8	106.0	113.0	78.4	102.8	109.7	74.8	99.4	112.8	87.2	105.3
06	WOOD, PLASTICS & COMPOSITES	108.4	85.7	95.9	116.5	77.3	94.9	120.3	79.0	97.6	118.0	66.8	89.8	99.4	53.2	74.0	102.2	74.1	86.7
07	THERMAL & MOISTURE PROTECTION	137.3	99.7	121.0	115.3	83.5	101.5	121.6	88.0	107.0	114.6	68.2	94.4	114.0	59.0	90.1	124.5	81.0	105.6
08	OPENINGS	90.3	87.5	89.7	90.5	77.6	87.5	87.7	79.9	85.9	89.1	59.9	82.3	84.3	51.4	76.6	86.1	74.9	83.5
0920	Plaster & Gypsum Board	129.1	84.6	99.1	153.5	76.7	101.8	115.8	78.5	90.7	104.0	65.6	78.2	98.2	51.9	67.1	101.1	73.0	82.5
0950, 0980	Ceilings & Acoustic Treatment	142.3	84.6	103.4	94.3	76.7	82.4	103.5	78.5	86.7	103.5	65.6	78.0	103.5	51.9	68.7	103.5	73.0	83.0
0960	Flooring	108.3	90.6	103.4	119.3	87.3	110.3	113.6	86.0	105.8	113.6	61.9	99.1	103.6	53.7	89.6	109.0	67.4	97.3
0970, 0990	Wall Finishes & Painting/Coating	113.5	95.9	103.1	105.6	83.8	92.7	110.3	87.1	96.6	110.4	54.1	77.1	110.3	52.0	75.8	110.3	76.4	90.3
09	FINISHES	122.3	89.0	104.3	111.3	80.8	94.8	108.7	82.6	94.5	107.4	64.1	84.0	103.7	53.8	76.7	107.2	75.0	89.8
COVERS	DIVS. 10 - 14, 25, 28, 41, 43, 44, 46	131.1	89.5	121.9	139.2	63.9	122.4	131.1	64.8	116.3	131.1	59.8	115.2	131.1	57.5	114.7	131.1	83.7	120.6
21, 22, 23	FIRE SUPPRESSION, PLUMBING & HVAC	105.1	90.7	99.4	103.7	92.4	99.3	103.8	95.8	100.7	103.8	78.4	93.8	104.2	63.4	88.2	103.8	87.5	97.4
26, 27, 3370	ELECTRICAL, COMMUNICATIONS & UTIL.	104.4	101.9	103.2	118.5	82.1	100.6	111.5	82.9	97.4	113.0	54.4	84.2	113.4	56.3	85.3	111.5	74.2	93.1
MF2018	WEIGHTED AVERAGE	117.5	96.8	108.7	116.7	85.4	103.4	113.1	87.1	102.1	111.6	69.1	93.6	108.2	62.2	88.8	114.0	82.8	100.8

For customer support on your Site Work & Landscape Costs with RSMeans data, call 800.448.8182.

819

City Cost Indexes

CANADA

	DIVISION	QUEBEC CITY, QUEBEC MAT.	INST.	TOTAL	RED DEER, ALBERTA MAT.	INST.	TOTAL	REGINA, SASKATCHEWAN MAT.	INST.	TOTAL	RIMOUSKI, QUEBEC MAT.	INST.	TOTAL	ROUYN-NORANDA, QUEBEC MAT.	INST.	TOTAL	SAINT HYACINTHE, QUEBEC MAT.	INST.	TOTAL
015433	CONTRACTOR EQUIPMENT		117.4	117.4		101.9	101.9		128.3	128.3		100.2	100.2		100.2	100.2		100.2	100.2
0241, 31 - 34	SITE & INFRASTRUCTURE, DEMOLITION	114.1	104.7	107.6	113.2	95.5	101.0	126.4	119.6	121.7	97.0	93.6	94.6	96.7	93.7	94.6	97.2	93.7	94.8
0310	Concrete Forming & Accessories	127.4	90.5	95.9	139.7	78.2	87.3	131.0	91.5	97.4	132.9	89.8	96.2	132.9	79.1	87.0	132.9	79.1	87.0
0320	Concrete Reinforcing	124.2	94.6	109.9	151.5	81.8	117.8	123.8	88.8	106.9	105.1	94.5	100.0	147.4	72.4	111.1	147.4	72.4	111.1
0330	Cast-in-Place Concrete	131.6	97.0	118.7	120.7	94.6	111.0	153.8	97.7	132.9	109.4	94.7	103.9	104.0	87.5	97.9	107.5	87.5	100.0
03	CONCRETE	119.4	94.3	108.4	116.5	85.2	102.7	131.3	94.0	114.9	105.0	92.4	99.5	107.9	81.4	96.2	109.5	81.4	97.1
04	MASONRY	173.6	89.8	122.1	161.6	80.1	111.6	205.2	86.7	132.4	162.1	89.8	117.6	162.3	76.3	109.5	162.6	76.3	109.6
05	METALS	138.0	103.3	127.8	113.0	90.8	106.5	139.1	101.4	128.0	110.4	91.8	104.9	110.9	84.8	103.2	110.9	84.8	103.2
06	WOOD, PLASTICS & COMPOSITES	114.2	90.7	101.3	118.0	77.4	95.7	108.0	92.2	99.3	131.6	90.3	108.9	131.6	79.0	102.7	131.6	79.0	102.7
07	THERMAL & MOISTURE PROTECTION	125.1	97.2	112.9	141.4	88.2	118.3	143.8	86.6	118.9	114.8	96.4	106.8	114.8	84.1	101.5	115.2	84.1	101.7
08	OPENINGS	88.4	86.0	87.8	89.1	74.4	85.7	87.9	79.8	86.1	88.7	76.3	85.8	89.1	67.9	84.2	89.1	67.9	84.2
0920	Plaster & Gypsum Board	128.5	89.9	102.6	104.3	76.6	85.6	140.8	91.2	107.4	147.4	89.9	108.7	114.2	78.4	90.1	114.2	78.4	90.1
0950, 0980	Ceilings & Acoustic Treatment	131.7	89.9	103.5	103.5	76.6	85.3	154.3	91.2	111.7	102.7	89.9	94.1	102.7	78.4	86.3	102.7	78.4	86.3
0960	Flooring	108.9	90.5	103.8	116.0	84.8	107.2	123.6	95.1	115.6	114.8	86.0	106.7	113.6	86.0	105.8	113.6	86.0	105.8
0970, 0990	Wall Finishes & Painting/Coating	120.8	104.1	111.0	110.2	75.1	89.4	118.0	89.0	100.9	113.4	104.1	107.9	110.3	83.6	94.5	110.3	83.6	94.5
09	FINISHES	117.4	92.7	104.0	108.2	78.9	92.4	130.9	92.9	110.3	112.6	91.6	101.2	107.6	81.2	93.3	107.6	81.2	93.3
COVERS	DIVS. 10 - 14, 25, 28, 41, 43, 44, 46	131.1	82.5	120.3	131.1	86.1	121.1	131.1	70.9	117.7	131.1	81.5	120.1	131.1	75.6	118.8	131.1	75.6	118.8
21, 22, 23	FIRE SUPPRESSION, PLUMBING & HVAC	105.1	81.2	95.7	103.8	86.8	97.2	104.8	86.9	97.8	103.8	81.0	94.9	103.8	84.2	96.1	100.5	84.2	94.1
26, 27, 3370	ELECTRICAL, COMMUNICATIONS & UTIL.	114.7	84.1	99.6	104.6	77.5	91.2	110.8	91.5	101.3	110.9	84.1	97.7	110.9	65.2	88.4	111.5	65.2	88.7
MF2018	WEIGHTED AVERAGE	118.7	90.4	106.7	112.1	84.0	100.2	123.1	92.4	110.2	110.2	87.6	100.6	110.3	79.8	97.4	109.8	79.8	97.1

CANADA

	DIVISION	SAINT JOHN, NEW BRUNSWICK MAT.	INST.	TOTAL	SARNIA, ONTARIO MAT.	INST.	TOTAL	SASKATOON, SASKATCHEWAN MAT.	INST.	TOTAL	SAULT STE MARIE, ONTARIO MAT.	INST.	TOTAL	SHERBROOKE, QUEBEC MAT.	INST.	TOTAL	SOREL, QUEBEC MAT.	INST.	TOTAL
015433	CONTRACTOR EQUIPMENT		99.7	99.7		99.7	99.7		98.4	98.4		99.7	99.7		100.2	100.2		100.2	100.2
0241, 31 - 34	SITE & INFRASTRUCTURE, DEMOLITION	105.2	92.5	96.4	115.3	94.3	100.9	112.7	92.8	98.5	115.1	86.2	90.5	97.2	93.7	94.8	97.3	93.7	94.8
0310	Concrete Forming & Accessories	126.2	63.4	72.7	125.9	87.6	93.3	109.7	90.7	96.5	105.1	85.4	95.6	132.9	79.1	87.0	132.9	79.2	87.1
0320	Concrete Reinforcing	140.5	69.8	106.3	116.5	86.7	102.1	114.2	88.7	101.9	108.3	80.5	98.0	147.4	72.4	111.1	139.3	72.4	107.0
0330	Cast-in-Place Concrete	112.4	68.0	95.9	120.6	94.2	110.8	113.5	93.1	105.9	97.4	84.7	91.8	107.5	87.5	100.0	108.4	87.5	100.6
03	CONCRETE	110.9	67.3	91.7	111.1	90.0	101.8	101.3	91.2	96.9	97.4	84.7	91.8	109.5	81.4	97.1	108.8	81.5	96.8
04	MASONRY	180.8	74.4	115.4	177.6	90.5	124.1	172.1	86.6	119.5	162.9	90.1	118.2	162.6	76.3	109.6	162.7	76.3	109.6
05	METALS	114.0	86.8	106.0	112.3	91.2	106.1	106.5	88.6	101.2	111.5	95.2	106.7	110.7	84.8	103.1	110.9	85.0	103.3
06	WOOD, PLASTICS & COMPOSITES	120.7	61.4	88.0	119.4	86.7	101.4	96.0	91.3	93.4	106.5	87.9	96.3	131.6	79.0	102.7	131.6	79.0	102.7
07	THERMAL & MOISTURE PROTECTION	116.9	72.2	97.5	121.7	91.5	108.6	117.0	84.7	103.0	120.4	87.3	106.0	114.8	84.1	101.5	114.8	84.1	101.5
08	OPENINGS	83.9	58.9	78.0	90.5	83.5	88.8	85.3	79.4	83.9	82.4	86.1	83.2	89.1	67.9	84.2	89.1	72.3	85.2
0920	Plaster & Gypsum Board	137.9	60.3	85.7	143.2	86.4	105.0	115.6	91.2	99.1	107.1	87.6	94.0	114.2	78.4	90.1	147.4	78.4	101.0
0950, 0980	Ceilings & Acoustic Treatment	121.7	60.3	80.3	109.2	86.4	93.9	124.5	91.2	102.0	103.5	87.6	92.8	102.7	78.4	86.3	102.7	78.4	86.3
0960	Flooring	111.4	65.6	98.5	113.6	94.1	108.1	106.5	95.1	103.3	106.5	92.1	102.4	113.6	86.0	105.8	113.6	86.0	105.8
0970, 0990	Wall Finishes & Painting/Coating	111.3	83.6	94.9	110.3	98.5	103.3	113.8	89.0	99.2	110.3	91.3	99.1	110.3	83.6	94.5	110.3	83.6	94.5
09	FINISHES	114.0	65.4	87.7	113.5	90.2	100.9	112.1	92.3	101.4	104.7	88.2	95.8	107.6	81.2	93.3	112.0	81.2	95.3
COVERS	DIVS. 10 - 14, 25, 28, 41, 43, 44, 46	131.1	61.0	115.5	131.1	66.1	116.6	131.1	69.3	117.4	131.1	87.9	121.5	131.1	75.6	118.8	131.1	75.6	118.8
21, 22, 23	FIRE SUPPRESSION, PLUMBING & HVAC	103.8	75.1	92.6	103.8	101.5	102.9	104.4	86.7	97.5	103.8	90.5	98.6	104.2	84.2	96.3	103.8	84.2	96.1
26, 27, 3370	ELECTRICAL, COMMUNICATIONS & UTIL.	119.9	88.5	104.4	114.7	85.0	100.0	114.6	91.4	103.2	113.2	83.3	98.5	110.9	65.2	88.4	110.9	65.2	88.4
MF2018	WEIGHTED AVERAGE	113.2	75.7	97.3	113.2	91.3	103.9	109.9	88.5	100.9	108.6	88.7	100.2	110.6	79.8	97.6	110.8	80.0	97.8

CANADA

	DIVISION	ST. CATHARINES, ONTARIO MAT.	INST.	TOTAL	ST JEROME, QUEBEC MAT.	INST.	TOTAL	ST. JOHN'S, NEWFOUNDLAND MAT.	INST.	TOTAL	SUDBURY, ONTARIO MAT.	INST.	TOTAL	SUMMERSIDE, PRINCE EDWARD ISLAND MAT.	INST.	TOTAL	SYDNEY, NOVA SCOTIA MAT.	INST.	TOTAL
015433	CONTRACTOR EQUIPMENT		99.5	99.5		100.2	100.2		123.8	123.4		99.5	99.5		98.8	98.8		98.9	98.9
0241, 31 - 34	SITE & INFRASTRUCTURE, DEMOLITION	94.2	96.1	95.6	96.7	93.7	94.6	115.4	111.8	112.9	94.4	95.7	95.3	121.8	89.4	99.5	108.0	92.0	96.9
0310	Concrete Forming & Accessories	115.3	93.8	97.0	132.9	79.1	87.0	125.3	85.2	91.1	111.0	88.9	92.1	112.0	52.5	61.3	111.8	67.8	74.3
0320	Concrete Reinforcing	97.2	103.2	100.1	147.4	72.4	111.1	172.1	82.4	128.7	98.0	101.4	99.7	154.0	46.5	102.0	156.0	47.6	103.6
0330	Cast-in-Place Concrete	106.7	98.7	103.7	104.0	87.5	97.9	137.1	98.5	122.8	107.6	95.3	103.0	116.7	54.7	93.6	94.8	66.8	84.4
03	CONCRETE	94.8	97.2	95.8	107.9	81.4	96.2	132.7	90.3	114.1	95.0	93.5	94.3	139.1	53.4	101.5	118.1	65.0	94.8
04	MASONRY	141.8	100.8	116.6	162.3	76.3	109.5	200.1	89.9	132.4	141.9	94.9	113.0	207.4	54.8	113.7	205.8	65.9	119.6
05	METALS	114.2	97.9	109.4	110.9	84.8	103.2	137.9	99.1	126.5	113.6	97.1	108.8	125.6	69.2	109.1	125.6	75.7	111.0
06	WOOD, PLASTICS & COMPOSITES	107.8	93.2	99.8	131.6	79.0	102.7	106.9	83.4	94.0	103.5	87.9	94.9	116.4	51.9	80.9	115.9	67.4	89.2
07	THERMAL & MOISTURE PROTECTION	116.4	102.0	110.1	114.8	84.1	101.5	140.2	94.7	120.4	115.7	96.3	107.3	129.4	57.3	98.0	130.1	67.6	102.9
08	OPENINGS	79.5	91.4	82.3	89.1	67.9	84.2	85.4	74.3	82.8	80.2	86.7	81.7	102.5	45.4	89.2	91.3	61.3	84.3
0920	Plaster & Gypsum Board	100.8	93.1	95.6	114.2	78.4	90.1	150.0	82.1	104.3	100.0	87.6	91.7	145.9	50.5	81.8	145.0	66.5	92.2
0950, 0980	Ceilings & Acoustic Treatment	103.9	93.1	96.6	102.7	78.4	86.3	143.7	82.1	102.2	98.2	87.6	91.0	123.8	50.5	74.4	123.8	66.5	85.2
0960	Flooring	100.2	91.9	97.8	113.6	86.0	105.8	116.8	52.2	98.7	98.3	92.1	96.6	107.6	55.3	92.9	107.6	58.9	93.9
0970, 0990	Wall Finishes & Painting/Coating	104.4	104.1	104.2	110.3	83.6	94.5	120.8	98.6	107.7	104.4	95.3	99.0	117.8	39.9	71.7	117.8	59.3	83.2
09	FINISHES	100.1	94.6	97.1	107.6	81.2	93.3	129.2	80.4	102.8	98.3	89.9	93.7	119.8	51.9	83.1	118.7	65.5	89.9
COVERS	DIVS. 10 - 14, 25, 28, 41, 43, 44, 46	131.1	69.7	117.4	131.1	75.6	118.8	131.1	68.6	117.2	131.1	89.6	121.9	131.1	57.0	114.6	131.1	60.8	115.5
21, 22, 23	FIRE SUPPRESSION, PLUMBING & HVAC	104.2	91.1	99.0	103.8	84.2	96.1	104.8	83.0	96.2	103.6	89.4	98.1	104.0	58.9	86.2	111.3	58.7	85.4
26, 27, 3370	ELECTRICAL, COMMUNICATIONS & UTIL.	111.3	101.8	106.6	111.6	65.2	88.7	115.6	81.4	98.7	109.5	102.1	105.9	110.2	47.4	79.2	117.8	70.2	97.7
MF2018	WEIGHTED AVERAGE	106.6	95.6	101.9	110.4	79.8	97.4	122.6	87.4	107.7	106.1	93.6	100.8	121.9	57.9	94.9	117.8	70.2	97.7

For customer support on your Site Work & Landscape Costs with RSMeans data, call 800.448.8182.

		CANADA																		
	DIVISION	THUNDER BAY, ONTARIO			TIMMINS, ONTARIO			TORONTO, ONTARIO			TROIS RIVIERES, QUEBEC			TRURO, NOVA SCOTIA			VANCOUVER, BRITISH COLUMBIA		
		MAT.	INST.	TOTAL	MAT.	INST.	TOTAL	MAT.	INST.	TOTAL	MAT.	INST.	TOTAL	MAT.	INST.	TOTAL	MAT.	INST.	TOTAL
015433	CONTRACTOR EQUIPMENT		99.5	99.5		99.7	99.7		117.5	117.5		99.7	99.7		99.4	99.4		135.3	135.3
0241, 31 - 34	SITE & INFRASTRUCTURE, DEMOLITION	98.8	96.0	96.9	116.9	93.9	101.0	108.4	107.7	107.9	108.2	93.4	97.9	101.3	92.3	95.1	117.0	122.8	121.0
0310	Concrete Forming & Accessories	122.8	92.5	97.0	127.2	80.0	87.0	128.7	102.5	106.3	153.7	79.1	90.1	99.7	67.9	72.6	131.0	91.6	97.4
0320	Concrete Reinforcing	86.9	102.6	94.5	163.7	84.8	125.6	136.9	103.9	120.9	156.0	72.4	115.6	139.3	47.6	95.0	132.5	95.2	114.5
0330	Cast-in-Place Concrete	117.5	98.0	110.2	130.7	80.2	111.9	108.4	113.7	110.4	98.2	87.4	94.2	135.7	66.9	110.1	123.5	90.6	111.3
03	CONCRETE	101.9	96.2	99.4	121.8	81.5	104.1	110.5	107.1	109.0	120.3	81.4	103.2	120.8	65.1	96.3	119.7	93.0	108.0
04	MASONRY	142.4	100.5	116.7	166.1	81.9	114.4	173.3	108.7	133.6	209.8	76.3	127.8	161.9	65.6	102.7	165.3	85.6	116.3
05	METALS	114.2	97.0	109.1	112.3	90.4	105.9	135.3	109.4	127.7	124.9	84.7	113.1	111.6	76.0	101.1	133.5	109.8	126.6
06	WOOD, PLASTICS & COMPOSITES	117.6	91.3	103.1	120.3	80.3	98.3	114.8	100.5	107.0	168.2	79.0	119.1	89.6	67.4	77.4	110.3	91.0	99.7
07	THERMAL & MOISTURE PROTECTION	116.6	99.3	109.1	121.6	82.7	104.7	135.6	109.4	124.2	128.6	84.1	109.2	116.1	67.6	95.0	141.7	88.6	118.6
08	OPENINGS	78.9	89.8	81.4	87.7	78.2	85.5	83.9	98.9	87.4	100.3	72.3	93.7	82.4	61.3	77.5	84.5	88.7	85.5
0920	Plaster & Gypsum Board	128.0	91.2	103.2	115.8	79.8	91.6	129.1	99.9	109.5	174.3	78.4	109.8	124.3	66.5	85.4	127.2	89.7	102.0
0950, 0980	Ceilings & Acoustic Treatment	98.2	91.2	93.4	103.5	79.8	87.5	139.9	99.9	112.9	122.2	78.4	92.6	103.5	66.5	78.6	151.0	89.7	109.7
0960	Flooring	104.2	98.5	102.6	113.6	86.0	105.8	109.1	100.8	106.8	131.2	86.0	118.5	95.7	58.9	85.4	124.7	88.5	114.6
0970, 0990	Wall Finishes & Painting/Coating	104.4	96.2	99.5	110.3	85.0	95.3	109.8	108.1	108.8	117.8	83.6	97.6	110.3	59.3	80.1	113.7	100.3	105.8
09	FINISHES	104.1	93.9	98.6	108.7	81.9	94.2	117.8	102.8	109.7	129.3	81.1	103.2	103.8	65.5	83.1	128.0	92.1	108.6
COVERS	DIVS. 10 - 14, 25, 28, 41, 43, 44, 46	131.1	69.7	117.4	131.1	63.4	116.0	131.1	95.4	123.2	131.1	75.6	118.8	131.1	60.8	115.5	131.1	90.8	122.1
21, 22, 23	FIRE SUPPRESSION, PLUMBING & HVAC	104.2	91.2	99.1	103.8	92.4	99.3	105.1	99.6	103.0	104.0	84.2	96.2	103.8	79.5	94.3	105.2	87.4	98.2
26, 27, 3370	ELECTRICAL, COMMUNICATIONS & UTIL.	109.6	101.2	105.4	113.2	83.3	98.5	106.8	103.8	105.3	111.6	65.2	88.8	110.5	58.7	85.0	108.7	78.6	93.9
MF2018	WEIGHTED AVERAGE	107.8	95.0	102.4	113.3	85.3	101.4	116.0	104.2	111.0	120.2	79.9	103.2	110.8	70.3	93.7	117.9	92.4	107.2

		CANADA																		
	DIVISION	VICTORIA, BRITISH COLUMBIA			WHITEHORSE, YUKON			WINDSOR, ONTARIO			WINNIPEG, MANITOBA			YARMOUTH, NOVA SCOTIA			YELLOWKNIFE, NWT		
		MAT.	INST.	TOTAL	MAT.	INST.	TOTAL	MAT.	INST.	TOTAL	MAT.	INST.	TOTAL	MAT.	INST.	TOTAL	MAT.	INST.	TOTAL
015433	CONTRACTOR EQUIPMENT		106.3	106.3		132.8	132.8		99.5	99.5		125.5	125.5		98.9	98.9		130.7	130.7
0241, 31 - 34	SITE & INFRASTRUCTURE, DEMOLITION	124.3	101.3	108.4	135.7	119.6	124.6	90.8	96.0	94.4	113.7	115.3	114.8	111.7	92.0	98.1	146.3	123.6	130.6
0310	Concrete Forming & Accessories	114.9	88.9	92.7	135.0	56.7	68.2	122.8	90.6	95.3	130.4	65.4	75.0	111.8	67.8	74.3	137.3	75.8	84.8
0320	Concrete Reinforcing	111.7	95.0	103.6	167.7	63.3	117.2	95.2	101.9	98.5	127.3	61.0	95.3	156.0	47.6	103.6	138.6	65.8	103.4
0330	Cast-in-Place Concrete	117.6	87.0	106.2	152.4	71.2	122.2	109.3	99.7	105.7	143.6	72.0	116.9	121.6	66.8	101.2	174.9	87.6	142.4
03	CONCRETE	129.4	89.5	111.9	144.0	64.8	109.1	96.2	95.9	96.0	127.0	68.4	101.3	130.4	65.0	101.7	150.7	79.5	119.4
04	MASONRY	175.0	85.5	120.0	246.5	58.1	130.7	141.9	100.1	116.3	194.6	65.3	115.2	208.1	65.6	120.5	234.9	68.9	132.9
05	METALS	107.7	92.2	103.2	145.1	90.5	129.1	114.2	97.3	109.2	141.6	88.0	125.8	125.6	75.7	111.0	140.5	92.7	126.4
06	WOOD, PLASTICS & COMPOSITES	100.0	88.3	93.6	121.3	55.3	85.0	117.6	89.0	101.9	107.7	66.0	84.8	115.9	67.4	89.2	130.1	77.1	100.9
07	THERMAL & MOISTURE PROTECTION	126.9	86.3	109.2	145.8	63.3	109.9	116.4	99.1	108.9	136.1	70.3	107.5	130.1	67.6	102.9	143.7	78.8	115.5
08	OPENINGS	86.8	82.9	85.9	99.5	53.0	88.6	78.7	88.9	81.1	87.5	59.7	81.0	91.3	61.3	84.3	93.9	65.5	87.3
0920	Plaster & Gypsum Board	109.1	87.7	94.7	178.3	53.0	94.0	111.4	88.8	96.2	133.0	64.2	86.7	145.0	66.5	92.2	192.2	75.5	113.7
0950, 0980	Ceilings & Acoustic Treatment	105.8	87.7	93.6	169.3	53.0	90.9	98.2	88.8	91.8	145.3	64.2	90.6	123.8	66.5	85.2	169.0	75.5	105.9
0960	Flooring	112.0	67.4	99.5	122.4	55.2	103.5	104.2	95.9	101.9	115.9	68.5	102.6	107.6	58.9	93.9	121.9	82.9	111.0
0970, 0990	Wall Finishes & Painting/Coating	113.8	100.3	105.8	120.7	53.1	80.7	104.4	97.1	100.1	113.9	53.5	78.1	117.8	59.3	83.2	123.3	75.9	95.2
09	FINISHES	110.7	86.8	97.8	143.7	55.8	96.1	101.6	92.1	96.5	126.9	65.0	93.4	118.7	65.5	89.9	147.5	76.6	109.1
COVERS	DIVS. 10 - 14, 25, 28, 41, 43, 44, 46	131.1	65.8	116.6	131.1	60.6	115.4	131.1	69.2	117.3	131.1	63.8	116.1	131.1	60.8	115.5	131.1	63.9	116.1
21, 22, 23	FIRE SUPPRESSION, PLUMBING & HVAC	103.9	88.7	97.9	104.3	71.6	91.5	104.2	90.7	98.9	105.2	64.6	89.2	104.0	79.5	94.4	104.8	88.5	98.4
26, 27, 3370	ELECTRICAL, COMMUNICATIONS & UTIL.	112.8	82.1	97.7	130.5	57.1	94.3	114.5	100.4	107.6	113.1	63.4	88.6	111.3	58.7	85.4	125.8	77.4	102.0
MF2018	WEIGHTED AVERAGE	114.1	87.6	102.9	132.2	69.2	105.6	107.1	94.4	101.8	122.0	71.2	100.5	119.5	70.2	98.7	131.4	83.2	111.0

For customer support on your Site Work & Landscape Costs with RSMeans data, call 800.448.8182.

821

Costs shown in RSMeans cost data publications are based on national averages for materials and installation. To adjust these costs to a specific location, simply multiply the base cost by the factor and divide by 100 for that city. The data is arranged alphabetically by state and postal zip code numbers. For a city not listed, use the factor for a nearby city with similar economic characteristics.

STATE/ZIP	CITY	MAT.	INST.	TOTAL
ALABAMA				
350-352	Birmingham	97.4	71.8	86.6
354	Tuscaloosa	97.1	71.4	86.3
355	Jasper	97.8	69.8	86.0
356	Decatur	97.1	69.2	85.3
357-358	Huntsville	97.1	71.7	86.4
359	Gadsden	97.3	70.2	85.9
360-361	Montgomery	96.3	73.1	86.5
362	Anniston	95.9	67.7	84.0
363	Dothan	96.2	72.7	86.3
364	Evergreen	95.8	68.3	84.2
365-366	Mobile	96.7	68.7	84.9
367	Selma	95.9	72.5	86.0
368	Phenix City	96.8	70.8	85.8
369	Butler	96.1	70.6	85.3
ALASKA				
995-996	Anchorage	118.5	113.1	116.2
997	Fairbanks	118.8	113.0	116.3
998	Juneau	116.8	113.1	115.2
999	Ketchikan	128.4	113.4	122.1
ARIZONA				
850,853	Phoenix	100.6	72.3	88.6
851,852	Mesa/Tempe	98.7	71.5	87.2
855	Globe	99.5	71.4	87.6
856-857	Tucson	97.7	70.2	86.1
859	Show Low	99.7	71.5	87.8
860	Flagstaff	102.4	70.5	88.9
863	Prescott	99.9	71.3	87.8
864	Kingman	98.3	70.4	86.5
865	Chambers	98.3	71.8	87.1
ARKANSAS				
716	Pine Bluff	96.0	63.6	82.3
717	Camden	94.0	64.7	81.6
718	Texarkana	94.8	63.3	81.5
719	Hot Springs	93.5	64.4	81.2
720-722	Little Rock	94.5	65.4	82.2
723	West Memphis	93.3	69.8	83.3
724	Jonesboro	93.8	66.6	82.3
725	Batesville	91.6	63.6	79.8
726	Harrison	92.9	62.8	80.2
727	Fayetteville	90.5	64.1	79.3
728	Russellville	91.7	63.1	79.6
729	Fort Smith	94.1	62.6	80.8
CALIFORNIA				
900-902	Los Angeles	98.5	129.8	111.7
903-905	Inglewood	93.8	128.8	108.6
906-908	Long Beach	95.4	128.8	109.5
910-912	Pasadena	94.8	128.6	109.1
913-916	Van Nuys	97.8	128.6	110.8
917-918	Alhambra	96.7	128.6	110.2
919-921	San Diego	99.7	121.7	109.0
922	Palm Springs	97.4	125.3	109.2
923-924	San Bernardino	95.0	126.3	108.2
925	Riverside	99.3	126.6	110.8
926-927	Santa Ana	97.0	126.3	109.4
928	Anaheim	99.3	126.6	110.9
930	Oxnard	98.0	126.7	110.1
931	Santa Barbara	97.2	126.1	109.4
932-933	Bakersfield	98.9	124.6	109.8
934	San Luis Obispo	98.2	125.8	109.9
935	Mojave	95.4	124.4	107.7
936-938	Fresno	98.5	129.4	111.5
939	Salinas	99.1	137.2	115.2
940-941	San Francisco	107.4	158.4	129.0
942,956-958	Sacramento	100.8	133.0	114.4
943	Palo Alto	99.2	153.8	122.3
944	San Mateo	101.5	152.0	122.9
945	Vallejo	100.2	141.6	117.7
946	Oakland	103.4	152.0	123.9
947	Berkeley	103.0	151.9	123.7
948	Richmond	102.4	146.2	120.9
949	San Rafael	104.5	149.1	123.4
950	Santa Cruz	104.6	137.5	118.5
CALIFORNIA (CONT'D)				
951	San Jose	102.6	153.5	124.2
952	Stockton	100.7	131.3	113.7
953	Modesto	100.6	130.6	113.3
954	Santa Rosa	101.2	148.4	121.2
955	Eureka	102.5	134.1	115.9
959	Marysville	101.7	131.3	114.2
960	Redding	107.6	131.5	117.7
961	Susanville	107.3	131.6	117.5
COLORADO				
800-802	Denver	103.0	75.4	91.3
803	Boulder	98.7	77.0	89.6
804	Golden	100.9	74.7	89.8
805	Fort Collins	102.2	73.5	90.1
806	Greeley	99.3	73.4	88.4
807	Fort Morgan	99.3	73.5	88.4
808-809	Colorado Springs	101.2	71.5	88.6
810	Pueblo	101.8	70.0	88.3
811	Alamosa	103.6	70.3	89.5
812	Salida	103.1	70.9	89.5
813	Durango	103.9	65.6	87.8
814	Montrose	102.6	65.8	87.1
815	Grand Junction	105.7	70.5	90.8
816	Glenwood Springs	103.9	66.8	88.2
CONNECTICUT				
060	New Britain	98.1	116.7	106.0
061	Hartford	99.6	118.0	107.3
062	Willimantic	98.7	117.1	106.5
063	New London	95.2	117.2	104.5
064	Meriden	97.2	117.0	105.6
065	New Haven	99.9	117.0	107.1
066	Bridgeport	99.3	117.0	106.8
067	Waterbury	98.9	117.2	106.6
068	Norwalk	98.8	116.7	106.4
069	Stamford	99.0	123.2	109.2
D.C.				
200-205	Washington	101.1	87.5	95.4
DELAWARE				
197	Newark	98.3	109.7	103.1
198	Wilmington	97.9	109.7	102.9
199	Dover	99.0	109.6	103.5
FLORIDA				
320,322	Jacksonville	96.9	66.6	84.1
321	Daytona Beach	97.2	69.5	85.5
323	Tallahassee	98.2	67.7	85.3
324	Panama City	98.6	68.3	85.8
325	Pensacola	101.2	65.8	86.2
326,344	Gainesville	98.8	65.8	84.8
327-328,347	Orlando	99.0	66.6	85.3
329	Melbourne	100.1	70.3	87.5
330-332,340	Miami	96.7	68.0	84.6
333	Fort Lauderdale	96.1	66.4	83.6
334,349	West Palm Beach	95.1	64.1	82.0
335-336,346	Tampa	97.6	68.4	85.3
337	St. Petersburg	99.7	66.8	85.8
338	Lakeland	97.1	67.6	84.6
339,341	Fort Myers	96.5	67.7	84.3
342	Sarasota	98.8	67.0	85.4
GEORGIA				
300-303,399	Atlanta	98.2	75.0	88.4
304	Statesboro	97.8	69.7	85.9
305	Gainesville	96.2	65.6	83.2
306	Athens	95.6	66.9	83.5
307	Dalton	97.5	71.4	86.5
308-309	Augusta	96.0	73.9	86.6
310-312	Macon	95.2	73.1	85.8
313-314	Savannah	97.3	72.7	86.9
315	Waycross	96.8	70.0	85.5
316	Valdosta	96.7	66.2	83.8
317,398	Albany	96.5	71.9	86.1
318-319	Columbus	96.5	72.2	86.2

Location Factors - Commercial

STATE/ZIP	CITY	MAT.	INST.	TOTAL
HAWAII				
967	Hilo	114.8	118.4	116.3
968	Honolulu	119.1	118.5	118.9
STATES & POSS.				
969	Guam	137.1	51.4	100.9
IDAHO				
832	Pocatello	102.0	78.4	92.0
833	Twin Falls	103.2	76.8	92.1
834	Idaho Falls	100.5	79.9	91.8
835	Lewiston	108.2	83.5	97.8
836-837	Boise	101.0	79.9	92.1
838	Coeur d'Alene	108.1	84.4	98.1
ILLINOIS				
600-603	North Suburban	98.3	142.9	117.1
604	Joliet	98.3	138.8	115.4
605	South Suburban	98.3	142.8	117.1
606-608	Chicago	100.1	145.6	119.3
609	Kankakee	95.0	134.9	111.9
610-611	Rockford	96.6	125.5	108.8
612	Rock Island	94.6	99.5	96.7
613	La Salle	95.9	127.0	109.0
614	Galesburg	95.7	108.5	101.1
615-616	Peoria	97.6	108.8	102.3
617	Bloomington	95.0	109.5	101.1
618-619	Champaign	98.5	110.5	103.6
620-622	East St. Louis	93.8	110.4	100.9
623	Quincy	95.7	104.2	99.3
624	Effingham	95.1	109.1	101.0
625	Decatur	96.6	110.1	102.3
626-627	Springfield	97.5	108.3	102.1
628	Centralia	92.8	111.2	100.5
629	Carbondale	92.5	109.6	99.7
INDIANA				
460	Anderson	96.0	79.8	89.1
461-462	Indianapolis	98.7	83.2	92.1
463-464	Gary	97.5	108.1	102.0
465-466	South Bend	98.4	81.8	91.4
467-468	Fort Wayne	96.6	76.3	88.0
469	Kokomo	94.2	81.1	88.6
470	Lawrenceburg	92.5	79.6	87.0
471	New Albany	93.9	77.3	86.9
472	Columbus	96.1	80.5	89.5
473	Muncie	96.2	78.2	88.6
474	Bloomington	97.8	81.1	90.7
475	Washington	94.6	84.8	90.4
476-477	Evansville	95.5	82.7	90.1
478	Terre Haute	96.2	81.7	90.1
479	Lafayette	95.8	78.1	88.3
IOWA				
500-503,509	Des Moines	97.2	90.2	94.3
504	Mason City	95.3	74.0	86.3
505	Fort Dodge	95.5	73.7	86.3
506-507	Waterloo	96.8	77.4	88.6
508	Creston	95.7	84.2	90.8
510-511	Sioux City	97.7	81.5	90.9
512	Sibley	96.8	62.4	82.3
513	Spencer	98.2	62.3	83.0
514	Carroll	95.5	82.4	89.9
515	Council Bluffs	98.9	79.2	90.6
516	Shenandoah	96.1	81.9	90.1
520	Dubuque	97.2	79.9	89.9
521	Decorah	96.4	73.7	86.8
522-524	Cedar Rapids	98.0	83.7	92.0
525	Ottumwa	96.3	78.7	88.9
526	Burlington	95.7	84.3	90.9
527-528	Davenport	97.1	94.5	96.0
KANSAS				
660-662	Kansas City	97.3	98.6	97.8
664-666	Topeka	98.6	76.5	89.3
667	Fort Scott	96.0	77.1	88.0
668	Emporia	96.1	76.5	87.8
669	Belleville	97.9	70.7	86.4
670-672	Wichita	97.0	72.6	86.7
673	Independence	97.2	76.0	88.2
674	Salina	97.2	73.5	87.2
675	Hutchinson	92.5	71.9	83.8
676	Hays	96.5	72.0	86.1
677	Colby	97.3	74.0	87.4

STATE/ZIP	CITY	MAT.	INST.	TOTAL
KANSAS (CONT'D)				
678	Dodge City	98.6	73.3	87.9
679	Liberal	96.5	71.7	86.0
KENTUCKY				
400-402	Louisville	93.9	79.6	87.8
403-405	Lexington	93.1	78.1	86.7
406	Frankfort	95.4	80.3	89.0
407-409	Corbin	90.8	78.0	85.4
410	Covington	94.3	77.8	87.3
411-412	Ashland	93.2	89.7	91.7
413-414	Campton	94.4	78.0	87.5
415-416	Pikeville	95.8	85.4	91.4
417-418	Hazard	93.8	78.7	87.4
420	Paducah	92.4	83.2	88.6
421-422	Bowling Green	94.6	77.2	87.3
423	Owensboro	94.7	81.0	88.9
424	Henderson	92.2	81.9	87.8
425-426	Somerset	91.6	79.5	86.5
427	Elizabethtown	91.3	75.6	84.7
LOUISIANA				
700-701	New Orleans	99.0	68.5	86.1
703	Thibodaux	96.0	67.0	83.8
704	Hammond	93.7	64.4	81.3
705	Lafayette	95.4	67.9	83.8
706	Lake Charles	95.6	68.2	84.1
707-708	Baton Rouge	96.6	69.3	85.1
710-711	Shreveport	97.1	66.3	84.1
712	Monroe	95.8	64.4	82.5
713-714	Alexandria	95.9	65.6	83.1
MAINE				
039	Kittery	92.6	87.0	90.2
040-041	Portland	99.6	85.4	93.6
042	Lewiston	97.5	84.9	92.2
043	Augusta	100.4	85.6	94.2
044	Bangor	97.1	83.9	91.5
045	Bath	95.9	85.4	91.4
046	Machias	95.4	85.3	91.1
047	Houlton	95.6	85.2	91.2
048	Rockland	94.6	85.4	90.7
049	Waterville	96.0	85.2	91.4
MARYLAND				
206	Waldorf	97.3	86.6	92.8
207-208	College Park	97.3	87.7	93.2
209	Silver Spring	96.5	86.7	92.4
210-212	Baltimore	101.8	83.2	93.9
214	Annapolis	100.4	86.3	94.4
215	Cumberland	96.9	83.8	91.4
216	Easton	98.6	74.8	88.5
217	Hagerstown	97.4	85.3	92.3
218	Salisbury	99.0	66.9	85.4
219	Elkton	95.9	86.2	91.8
MASSACHUSETTS				
010-011	Springfield	98.2	108.5	102.5
012	Pittsfield	97.7	102.6	99.7
013	Greenfield	95.9	109.0	101.4
014	Fitchburg	94.6	116.2	103.7
015-016	Worcester	98.1	117.0	106.1
017	Framingham	94.0	124.2	106.8
018	Lowell	97.5	123.8	108.6
019	Lawrence	98.4	124.2	109.3
020-022, 024	Boston	101.9	134.0	115.5
023	Brockton	98.5	116.7	106.2
025	Buzzards Bay	93.2	115.2	102.5
026	Hyannis	95.7	115.0	103.8
027	New Bedford	97.7	115.2	105.1
MICHIGAN				
480,483	Royal Oak	95.1	100.3	97.3
481	Ann Arbor	97.1	99.8	98.2
482	Detroit	100.7	102.0	101.3
484-485	Flint	96.8	88.6	93.3
486	Saginaw	96.5	85.1	91.7
487	Bay City	96.6	85.3	91.8
488-489	Lansing	98.0	86.8	93.3
490	Battle Creek	95.5	80.4	89.1
491	Kalamazoo	95.8	78.6	88.6
492	Jackson	94.0	90.7	92.6
493,495	Grand Rapids	97.9	79.9	90.3
494	Muskegon	94.3	80.0	88.3

For customer support on your Site Work & Landscape Costs with RSMeans data, call 800.448.8182.

823

Location Factors - Commercial

STATE/ZIP	CITY	MAT.	INST.	TOTAL
MICHIGAN (CONT'D)				
496	Traverse City	93.4	77.7	86.8
497	Gaylord	94.5	80.6	88.6
498-499	Iron Mountain	96.5	81.7	90.2
MINNESOTA				
550-551	Saint Paul	99.0	117.6	106.9
553-555	Minneapolis	100.7	114.1	106.3
556-558	Duluth	98.6	101.3	99.7
559	Rochester	98.3	98.7	98.5
560	Mankato	95.7	98.5	96.9
561	Windom	94.3	91.2	93.0
562	Willmar	94.0	98.0	95.7
563	St. Cloud	95.1	114.2	103.2
564	Brainerd	95.6	98.6	96.8
565	Detroit Lakes	97.4	92.4	95.3
566	Bemidji	96.7	95.4	96.1
567	Thief River Falls	96.3	89.3	93.3
MISSISSIPPI				
386	Clarksdale	95.3	55.2	78.4
387	Greenville	98.7	65.5	84.7
388	Tupelo	96.6	59.0	80.7
389	Greenwood	96.6	55.5	79.2
390-392	Jackson	97.3	65.6	83.9
393	Meridian	94.6	64.5	81.9
394	Laurel	96.1	57.8	79.9
395	Biloxi	96.2	63.9	82.6
396	McComb	94.4	55.8	78.0
397	Columbus	95.9	59.1	80.4
MISSOURI				
630-631	St. Louis	98.8	104.5	101.2
633	Bowling Green	96.4	95.5	96.0
634	Hannibal	95.3	91.6	93.7
635	Kirksville	98.9	86.5	93.7
636	Flat River	97.3	94.2	96.0
637	Cape Girardeau	97.0	91.2	94.5
638	Sikeston	95.7	87.8	92.3
639	Poplar Bluff	95.1	87.7	92.0
640-641	Kansas City	97.9	103.8	100.4
644-645	St. Joseph	96.1	90.9	93.9
646	Chillicothe	93.7	93.8	93.7
647	Harrisonville	93.2	100.8	96.4
648	Joplin	95.1	77.7	87.7
650-651	Jefferson City	96.4	89.9	93.7
652	Columbia	96.1	90.7	93.8
653	Sedalia	96.4	91.5	94.3
654-655	Rolla	94.1	94.3	94.2
656-658	Springfield	97.6	79.0	89.7
MONTANA				
590-591	Billings	101.6	78.4	91.8
592	Wolf Point	101.4	76.9	91.1
593	Miles City	99.2	77.0	89.8
594	Great Falls	102.9	76.6	91.8
595	Havre	100.4	74.8	89.6
596	Helena	100.9	76.7	90.7
597	Butte	101.5	76.5	90.9
598	Missoula	98.6	77.1	89.5
599	Kalispell	98.1	76.2	88.9
NEBRASKA				
680-681	Omaha	97.6	81.5	90.8
683-685	Lincoln	98.3	79.7	90.5
686	Columbus	96.0	80.7	89.6
687	Norfolk	97.4	77.5	89.0
688	Grand Island	97.4	76.8	88.7
689	Hastings	97.1	78.2	89.1
690	McCook	96.3	73.1	86.5
691	North Platte	96.3	75.6	87.6
692	Valentine	98.6	69.2	86.2
693	Alliance	98.7	73.3	87.9
NEVADA				
889-891	Las Vegas	105.0	105.1	105.0
893	Ely	103.7	93.2	99.2
894-895	Reno	103.6	82.6	94.7
897	Carson City	102.9	82.5	94.3
898	Elko	102.3	85.2	95.0
NEW HAMPSHIRE				
030	Nashua	97.5	93.7	95.9
031	Manchester	98.2	94.5	96.6

STATE/ZIP	CITY	MAT.	INST.	TOTAL
NEW HAMPSHIRE (CONT'D)				
032-033	Concord	98.0	93.2	96.0
034	Keene	94.5	89.0	92.2
035	Littleton	94.4	80.2	88.4
036	Charleston	94.1	88.4	91.7
037	Claremont	93.1	88.4	91.1
038	Portsmouth	94.9	92.4	93.8
NEW JERSEY				
070-071	Newark	100.3	139.7	117.0
072	Elizabeth	97.5	138.3	114.7
073	Jersey City	96.5	137.8	114.0
074-075	Paterson	98.0	137.7	114.8
076	Hackensack	96.1	138.3	114.0
077	Long Branch	95.8	130.9	110.7
078	Dover	96.4	137.5	113.7
079	Summit	96.4	138.3	114.1
080,083	Vineland	96.3	129.8	110.5
081	Camden	98.0	127.8	110.6
082,084	Atlantic City	96.8	133.2	112.2
085-086	Trenton	99.7	128.9	112.0
087	Point Pleasant	98.1	130.8	111.9
088-089	New Brunswick	98.7	135.6	114.3
NEW MEXICO				
870-872	Albuquerque	98.6	72.5	87.6
873	Gallup	99.0	72.5	87.8
874	Farmington	99.0	72.5	87.8
875	Santa Fe	99.0	72.9	88.0
877	Las Vegas	97.2	72.5	86.8
878	Socorro	96.8	72.5	86.6
879	Truth/Consequences	96.6	69.4	85.1
880	Las Cruces	96.9	71.4	86.1
881	Clovis	99.3	72.4	87.9
882	Roswell	100.8	72.5	88.9
883	Carrizozo	101.6	72.5	89.3
884	Tucumcari	99.9	72.4	88.3
NEW YORK				
100-102	New York	99.7	175.2	131.6
103	Staten Island	95.6	176.1	129.6
104	Bronx	93.8	175.3	128.3
105	Mount Vernon	94.1	150.6	118.0
106	White Plains	93.8	152.8	118.8
107	Yonkers	97.9	153.0	121.2
108	New Rochelle	94.4	144.8	115.7
109	Suffern	94.1	130.2	109.4
110	Queens	100.0	177.3	132.6
111	Long Island City	101.6	177.3	133.6
112	Brooklyn	101.9	177.3	133.8
113	Flushing	102.1	177.3	133.9
114	Jamaica	100.3	177.3	132.8
115,117,118	Hicksville	99.9	157.1	124.1
116	Far Rockaway	102.2	177.3	133.9
119	Riverhead	100.6	153.5	122.9
120-122	Albany	96.8	111.2	102.8
123	Schenectady	97.0	109.9	102.4
124	Kingston	100.6	133.8	114.6
125-126	Poughkeepsie	99.8	136.8	115.4
127	Monticello	99.1	134.6	114.1
128	Glens Falls	92.0	107.3	98.5
129	Plattsburgh	97.0	96.6	96.8
130-132	Syracuse	97.8	98.3	98.0
133-135	Utica	96.0	96.6	96.2
136	Watertown	97.7	96.8	97.4
137-139	Binghamton	97.3	101.5	99.1
140-142	Buffalo	100.8	110.3	104.8
143	Niagara Falls	97.4	107.5	101.6
144-146	Rochester	100.3	99.2	99.8
147	Jamestown	96.4	96.0	96.3
148-149	Elmira	96.3	102.5	98.9
NORTH CAROLINA				
270,272-274	Greensboro	98.5	67.2	85.3
271	Winston-Salem	98.2	67.3	85.2
275-276	Raleigh	97.4	67.0	84.5
277	Durham	100.1	67.2	86.2
278	Rocky Mount	96.0	66.6	83.6
279	Elizabeth City	96.9	68.3	84.8
280	Gastonia	97.0	66.8	84.2
281-282	Charlotte	97.4	67.6	84.8
283	Fayetteville	100.2	66.1	85.8
284	Wilmington	96.0	65.2	83.0
285	Kinston	94.4	65.5	82.2

For customer support on your Site Work & Landscape Costs with RSMeans data, call 800.448.8182.

STATE/ZIP	CITY	MAT.	INST.	TOTAL		STATE/ZIP	CITY	MAT.	INST.	TOTAL
NORTH CAROLINA (CONT'D)						**PENNSYLVANIA (CONT'D)**				
286	Hickory	94.7	67.3	83.1		177	Williamsport	92.5	92.3	92.4
287-288	Asheville	96.4	66.6	83.8		178	Sunbury	94.6	92.7	93.8
289	Murphy	95.5	64.7	82.5		179	Pottsville	93.7	96.1	94.7
						180	Lehigh Valley	94.9	111.6	102.0
NORTH DAKOTA						181	Allentown	96.7	105.8	100.6
580-581	Fargo	99.7	80.3	91.5		182	Hazleton	94.4	95.9	95.0
582	Grand Forks	99.4	77.6	90.2		183	Stroudsburg	94.3	104.7	98.7
583	Devils Lake	99.2	78.8	90.6		184-185	Scranton	97.4	96.5	97.1
584	Jamestown	99.2	78.8	90.6		186-187	Wilkes-Barre	94.0	95.5	94.7
585	Bismarck	98.9	81.6	91.6		188	Montrose	93.8	96.8	95.1
586	Dickinson	99.9	77.9	90.6		189	Doylestown	94.0	127.1	108.0
587	Minot	99.3	78.6	90.5		190-191	Philadelphia	100.2	138.7	116.5
588	Williston	98.4	77.7	89.6		193	Westchester	95.8	127.9	109.3
						194	Norristown	94.7	128.0	108.8
OHIO						195-196	Reading	96.5	101.5	98.6
430-432	Columbus	98.4	82.0	91.4						
433	Marion	94.3	86.7	91.1		**PUERTO RICO**				
434-436	Toledo	97.6	91.2	94.9		009	San Juan	98.5	29.8	69.4
437-438	Zanesville	94.9	85.7	91.0						
439	Steubenville	96.1	90.0	93.5		**RHODE ISLAND**				
440	Lorain	98.4	84.7	92.6		028	Newport	97.0	114.7	104.5
441	Cleveland	99.1	91.6	95.9		029	Providence	99.3	115.0	105.9
442-443	Akron	99.4	86.0	93.8						
444-445	Youngstown	98.7	81.9	91.6		**SOUTH CAROLINA**				
446-447	Canton	98.9	79.9	90.9		290-292	Columbia	97.6	68.9	85.5
448-449	Mansfield	96.4	84.2	91.3		293	Spartanburg	97.2	68.3	85.0
450	Hamilton	94.8	79.5	88.3		294	Charleston	98.7	68.2	85.8
451-452	Cincinnati	96.1	80.0	89.3		295	Florence	96.9	68.6	84.9
453-454	Dayton	94.8	79.2	88.2		296	Greenville	97.0	68.2	84.8
455	Springfield	94.8	79.7	88.4		297	Rock Hill	96.7	66.4	83.9
456	Chillicothe	94.3	88.5	91.8		298	Aiken	97.6	67.1	84.7
457	Athens	97.1	84.8	91.9		299	Beaufort	98.3	57.5	81.1
458	Lima	97.2	81.7	90.7						
						SOUTH DAKOTA				
OKLAHOMA						570-571	Sioux Falls	97.7	78.4	89.5
730-731	Oklahoma City	96.4	68.7	84.7		572	Watertown	97.3	69.0	85.4
734	Ardmore	95.2	65.6	82.7		573	Mitchell	96.3	60.9	81.3
735	Lawton	97.3	65.8	84.0		574	Aberdeen	98.6	70.1	86.5
736	Clinton	96.4	65.5	83.3		575	Pierre	99.0	68.5	86.1
737	Enid	97.0	66.0	83.9		576	Mobridge	96.9	62.9	82.5
738	Woodward	95.3	62.9	81.6		577	Rapid City	98.3	72.0	87.2
739	Guymon	96.3	63.3	82.4						
740-741	Tulsa	95.9	65.8	83.2		**TENNESSEE**				
743	Miami	92.7	65.1	81.1		370-372	Nashville	98.0	71.8	86.9
744	Muskogee	95.0	62.9	81.4		373-374	Chattanooga	98.0	68.1	85.3
745	McAlester	92.4	60.5	78.9		375,380-381	Memphis	96.6	69.8	85.3
746	Ponca City	93.1	64.0	80.8		376	Johnson City	98.1	58.6	81.4
747	Durant	93.1	65.1	81.3		377-379	Knoxville	94.8	63.7	81.6
748	Shawnee	94.5	65.2	82.2		382	McKenzie	96.3	55.1	78.9
749	Poteau	92.3	64.9	80.7		383	Jackson	97.6	61.4	82.3
						384	Columbia	94.8	66.5	82.8
OREGON						385	Cookeville	96.2	56.9	79.6
970-972	Portland	102.1	102.8	102.4						
973	Salem	104.4	100.8	102.9		**TEXAS**				
974	Eugene	101.7	98.7	100.5		750	McKinney	97.2	63.5	82.9
975	Medford	103.3	97.4	100.8		751	Waxahachie	97.2	63.5	83.0
976	Klamath Falls	103.5	97.3	100.9		752-753	Dallas	98.2	68.0	85.4
977	Bend	102.6	100.2	101.6		754	Greenville	97.4	62.9	82.8
978	Pendleton	98.6	100.3	99.3		755	Texarkana	96.7	62.0	82.1
979	Vale	96.2	87.0	92.3		756	Longview	97.5	61.0	82.1
						757	Tyler	97.8	61.6	82.5
PENNSYLVANIA						758	Palestine	94.2	60.4	79.9
150-152	Pittsburgh	100.7	102.7	101.5		759	Lufkin	94.7	62.6	81.1
153	Washington	97.5	100.2	98.6		760-761	Fort Worth	97.6	62.9	82.9
154	Uniontown	97.8	99.8	98.6		762	Denton	97.3	62.9	82.7
155	Bedford	98.8	92.1	96.0		763	Wichita Falls	95.0	61.6	80.9
156	Greensburg	98.7	96.3	97.7		764	Eastland	94.2	60.3	79.8
157	Indiana	97.6	97.7	97.7		765	Temple	92.5	58.5	78.2
158	Dubois	99.3	96.0	97.9		766-767	Waco	94.4	62.4	80.9
159	Johnstown	98.9	91.9	95.9		768	Brownwood	97.4	58.7	81.1
160	Butler	91.3	100.6	95.2		769	San Angelo	97.1	59.4	81.2
161	New Castle	91.4	97.8	94.1		770-772	Houston	100.4	68.2	86.8
162	Kittanning	91.8	97.9	94.4		773	Huntsville	98.7	62.8	83.6
163	Oil City	91.3	97.8	94.0		774	Wharton	99.8	64.3	84.8
164-165	Erie	93.2	93.9	93.5		775	Galveston	97.7	66.5	84.5
166	Altoona	93.3	92.8	93.1		776-777	Beaumont	97.9	65.5	84.2
167	Bradford	94.9	97.1	95.8		778	Bryan	95.2	64.9	82.4
168	State College	94.5	94.6	94.5		779	Victoria	99.7	62.9	84.2
169	Wellsboro	95.5	92.9	94.4		780	Laredo	96.4	61.1	81.5
170-171	Harrisburg	98.7	93.0	96.3		781-782	San Antonio	98.3	62.3	83.1
172	Chambersburg	95.1	89.4	92.7		783-784	Corpus Christi	98.9	61.4	83.1
173-174	York	95.2	93.0	94.3		785	McAllen	99.7	56.5	81.4
175-176	Lancaster	93.8	94.5	94.1		786-787	Austin	97.2	61.4	82.1

For customer support on your Site Work & Landscape Costs with RSMeans data, call 800.448.8182.

825

Location Factors - Commercial

STATE/ZIP	CITY	MAT.	INST.	TOTAL
TEXAS (CONT'D)				
788	Del Rio	99.7	60.4	83.1
789	Giddings	96.1	60.4	81.0
790-791	Amarillo	98.2	60.0	82.1
792	Childress	97.6	60.3	81.8
793-794	Lubbock	99.2	62.9	83.9
795-796	Abilene	97.8	61.4	82.4
797	Midland	99.7	62.2	83.8
798-799,885	El Paso	95.9	63.6	82.3
UTAH				
840-841	Salt Lake City	103.5	71.7	90.1
842,844	Ogden	98.8	71.7	87.4
843	Logan	100.9	71.7	88.5
845	Price	101.4	70.0	88.1
846-847	Provo	101.2	71.2	88.5
VERMONT				
050	White River Jct.	98.2	79.8	90.4
051	Bellows Falls	96.7	92.1	94.8
052	Bennington	97.0	88.6	93.4
053	Brattleboro	97.4	92.1	95.2
054	Burlington	101.9	79.8	92.5
056	Montpelier	100.2	84.4	93.5
057	Rutland	98.9	79.4	90.7
058	St. Johnsbury	98.1	79.3	90.2
059	Guildhall	96.8	79.0	89.3
VIRGINIA				
220-221	Fairfax	99.5	83.6	92.8
222	Arlington	100.4	84.2	93.6
223	Alexandria	99.6	84.8	93.3
224-225	Fredericksburg	98.3	78.7	90.0
226	Winchester	98.9	79.5	90.7
227	Culpeper	98.7	83.0	92.1
228	Harrisonburg	99.0	68.8	86.3
229	Charlottesville	99.4	71.0	87.4
230-232	Richmond	99.2	72.8	88.1
233-235	Norfolk	99.9	68.7	86.7
236	Newport News	98.8	68.3	85.9
237	Portsmouth	98.3	67.8	85.4
238	Petersburg	98.7	72.6	87.7
239	Farmville	97.9	65.3	84.1
240-241	Roanoke	100.1	73.6	88.9
242	Bristol	98.2	59.3	81.8
243	Pulaski	97.8	67.1	84.8
244	Staunton	98.6	67.9	85.6
245	Lynchburg	98.7	73.8	88.2
246	Grundy	98.1	58.8	81.5
WASHINGTON				
980-981,987	Seattle	106.3	112.5	108.9
982	Everett	105.1	104.0	104.6
983-984	Tacoma	105.4	101.3	103.7
985	Olympia	104.0	101.8	103.1
986	Vancouver	106.6	102.8	105.0
988	Wenatchee	105.6	88.7	98.5
989	Yakima	105.6	95.4	101.3
990-992	Spokane	100.1	82.4	92.6
993	Richland	99.8	90.0	95.6
994	Clarkston	99.1	81.9	91.8
WEST VIRGINIA				
247-248	Bluefield	97.0	88.1	93.2
249	Lewisburg	98.7	86.1	93.4
250-253	Charleston	96.5	89.4	93.5
254	Martinsburg	96.7	84.1	91.4
255-257	Huntington	97.8	92.2	95.4
258-259	Beckley	95.2	87.7	92.0
260	Wheeling	99.6	89.6	95.4
261	Parkersburg	98.6	88.6	94.4
262	Buckhannon	98.3	88.8	94.3
263-264	Clarksburg	98.9	89.5	94.9
265	Morgantown	98.8	89.6	94.9
266	Gassaway	98.3	89.0	94.3
267	Romney	98.3	87.2	93.6
268	Petersburg	98.0	85.9	92.9
WISCONSIN				
530,532	Milwaukee	99.2	107.9	102.9
531	Kenosha	99.4	102.0	100.5
534	Racine	98.7	102.3	100.2
535	Beloit	98.6	98.9	98.7
537	Madison	99.7	98.1	99.0

STATE/ZIP	CITY	MAT.	INST.	TOTAL
WISCONSIN (CONT'D)				
538	Lancaster	96.7	94.3	95.7
539	Portage	95.2	98.0	96.4
540	New Richmond	95.4	94.6	95.0
541-543	Green Bay	99.4	93.4	96.9
544	Wausau	94.7	93.4	94.2
545	Rhinelander	97.9	92.2	95.5
546	La Crosse	95.9	93.5	94.9
547	Eau Claire	97.5	94.1	96.1
548	Superior	95.1	96.5	95.7
549	Oshkosh	95.4	91.9	93.9
WYOMING				
820	Cheyenne	101.1	72.7	89.1
821	Yellowstone Nat'l Park	99.0	71.9	87.6
822	Wheatland	100.3	67.1	86.2
823	Rawlins	102.0	71.5	89.1
824	Worland	99.7	70.7	87.5
825	Riverton	101.0	68.6	87.3
826	Casper	101.5	68.9	87.7
827	Newcastle	99.6	71.1	87.5
828	Sheridan	102.5	70.2	88.8
829-831	Rock Springs	103.4	71.4	89.9
CANADIAN FACTORS (reflect Canadian currency)				
ALBERTA				
	Calgary	119.7	95.6	109.5
	Edmonton	121.8	95.9	110.9
	Fort McMurray	121.9	88.4	107.7
	Lethbridge	117.0	87.9	104.7
	Lloydminster	111.5	84.7	100.2
	Medicine Hat	111.7	84.0	100.0
	Red Deer	112.1	84.0	100.2
BRITISH COLUMBIA				
	Kamloops	113.0	83.5	100.5
	Prince George	114.0	82.8	100.8
	Vancouver	117.9	92.4	107.2
	Victoria	114.1	87.6	102.9
MANITOBA				
	Brandon	120.6	70.5	99.4
	Portage la Prairie	111.6	69.1	93.6
	Winnipeg	122.0	71.2	100.5
NEW BRUNSWICK				
	Bathurst	110.7	62.3	90.2
	Dalhousie	110.2	62.5	90.0
	Fredericton	119.3	69.0	98.0
	Moncton	110.9	76.0	96.2
	Newcastle	110.7	62.9	90.5
	Saint John	113.2	75.7	97.3
NEWFOUNDLAND				
	Corner Brook	124.7	68.4	100.9
	St. John's	122.6	87.4	107.7
NORTHWEST TERRITORIES				
	Yellowknife	131.4	83.2	111.0
NOVA SCOTIA				
	Bridgewater	111.3	70.3	94.0
	Dartmouth	121.5	70.2	99.8
	Halifax	117.9	88.0	105.2
	New Glasgow	119.6	70.2	98.7
	Sydney	117.8	70.2	97.7
	Truro	110.8	70.3	93.7
	Yarmouth	119.5	70.2	98.7
ONTARIO				
	Barrie	116.5	87.1	104.1
	Brantford	113.1	90.7	103.6
	Cornwall	113.1	87.3	102.2
	Hamilton	116.3	98.8	108.9
	Kingston	114.1	87.4	102.8
	Kitchener	108.6	94.6	102.7
	London	116.3	96.5	108.0
	North Bay	122.5	85.2	106.8
	Oshawa	110.8	97.2	105.0
	Ottawa	117.5	96.8	108.7
	Owen Sound	116.7	85.4	103.4
	Peterborough	113.1	87.1	102.1
	Sarnia	113.2	91.3	103.9

For customer support on your Site Work & Landscape Costs with RSMeans data, call 800.448.8182.

Location Factors - Commercial

STATE/ZIP	CITY	MAT.	INST.	TOTAL
ONTARIO (CONT'D)				
	Sault Ste. Marie	108.6	88.7	100.2
	St. Catharines	106.6	95.6	101.9
	Sudbury	106.1	93.6	100.8
	Thunder Bay	107.8	95.0	102.4
	Timmins	113.3	85.3	101.4
	Toronto	116.0	104.2	111.0
	Windsor	107.1	94.4	101.8
PRINCE EDWARD ISLAND				
	Charlottetown	121.3	60.2	95.5
	Summerside	121.9	57.9	94.9
QUEBEC				
	Cap-de-la-Madeleine	110.6	80.0	97.7
	Charlesbourg	110.6	80.0	97.7
	Chicoutimi	110.4	87.6	100.7
	Gatineau	110.3	79.8	97.4
	Granby	110.5	79.7	97.5
	Hull	110.5	79.8	97.5
	Joliette	110.8	80.0	97.8
	Laval	110.7	81.6	98.4
	Montreal	119.1	90.1	106.9
	Quebec City	118.7	90.4	106.7
	Rimouski	110.2	87.6	100.6
	Rouyn-Noranda	110.3	79.8	97.4
	Saint-Hyacinthe	109.8	79.8	97.1
	Sherbrooke	110.6	79.8	97.6
	Sorel	110.8	80.0	97.8
	Saint-Jerome	110.4	79.8	97.4
	Trois-Rivieres	120.2	79.9	103.2
SASKATCHEWAN				
	Moose Jaw	109.0	63.9	89.9
	Prince Albert	108.2	62.2	88.8
	Regina	123.1	92.4	110.2
	Saskatoon	109.9	88.5	100.9
YUKON				
	Whitehorse	132.2	69.2	105.6

For customer support on your Site Work & Landscape Costs with RSMeans data, call 800.448.8182.

827

R011105-05 Tips for Accurate Estimating

1. Use pre-printed or columnar forms for orderly sequence of dimensions and locations and for recording telephone quotations.

2. Use only the front side of each paper or form except for certain pre-printed summary forms.

3. Be consistent in listing dimensions: For example, length x width x height. This helps in rechecking to ensure that, the total length of partitions is appropriate for the building area.

4. Use printed (rather than measured) dimensions where given.

5. Add up multiple printed dimensions for a single entry where possible.

6. Measure all other dimensions carefully.

7. Use each set of dimensions to calculate multiple related quantities.

8. Convert foot and inch measurements to decimal feet when listing. Memorize decimal equivalents to .01 parts of a foot (1/8″ equals approximately .01′).

9. Do not "round off" quantities until the final summary.

10. Mark drawings with different colors as items are taken off.

11. Keep similar items together, different items separate.

12. Identify location and drawing numbers to aid in future checking for completeness.

13. Measure or list everything on the drawings or mentioned in the specifications.

14. It may be necessary to list items not called for to make the job complete.

15. Be alert for: Notes on plans such as N.T.S. (not to scale); changes in scale throughout the drawings; reduced size drawings; discrepancies between the specifications and the drawings.

16. Develop a consistent pattern of performing an estimate. For example:
    a. Start the quantity takeoff at the lower floor and move to the next higher floor.
    b. Proceed from the main section of the building to the wings.
    c. Proceed from south to north or vice versa, clockwise or counterclockwise.
    d. Take off floor plan quantities first, elevations next, then detail drawings.

17. List all gross dimensions that can be either used again for different quantities, or used as a rough check of other quantities for verification (exterior perimeter, gross floor area, individual floor areas, etc.).

18. Utilize design symmetry or repetition (repetitive floors, repetitive wings, symmetrical design around a center line, similar room layouts, etc.). Note: Extreme caution is needed here so as not to omit or duplicate an area.

19. Do not convert units until the final total is obtained. For instance, when estimating concrete work, keep all units to the nearest cubic foot, then summarize and convert to cubic yards.

20. When figuring alternatives, it is best to total all items involved in the basic system, then total all items involved in the alternates. Therefore you work with positive numbers in all cases. When adds and deducts are used, it is often confusing whether to add or subtract a portion of an item; especially on a complicated or involved alternate.

For customer support on your Site Work & Landscape Costs with RSMeans data, call 800.448.8182.

R011105-50   Metric Conversion Factors

Description: This table is primarily for converting customary U.S. units in the left hand column to SI metric units in the right hand column. In addition, conversion factors for some commonly encountered Canadian and non-SI metric units are included.

	If You Know		Multiply By		To Find
Length	Inches	x	25.4[a]	=	Millimeters
	Feet	x	0.3048[a]	=	Meters
	Yards	x	0.9144[a]	=	Meters
	Miles (statute)	x	1.609	=	Kilometers
Area	Square inches	x	645.2	=	Square millimeters
	Square feet	x	0.0929	=	Square meters
	Square yards	x	0.8361	=	Square meters
Volume	Cubic inches	x	16,387	=	Cubic millimeters
(Capacity)	Cubic feet	x	0.02832	=	Cubic meters
	Cubic yards	x	0.7646	=	Cubic meters
	Gallons (U.S. liquids)[b]	x	0.003785	=	Cubic meters[c]
	Gallons (Canadian liquid)[b]	x	0.004546	=	Cubic meters[c]
	Ounces (U.S. liquid)[b]	x	29.57	=	Milliliters[c, d]
	Quarts (U.S. liquid)[b]	x	0.9464	=	Liters[c, d]
	Gallons (U.S. liquid)[b]	x	3.785	=	Liters[c, d]
Force	Kilograms force[d]	x	9.807	=	Newtons
	Pounds force	x	4.448	=	Newtons
	Pounds force	x	0.4536	=	Kilograms force[d]
	Kips	x	4448	=	Newtons
	Kips	x	453.6	=	Kilograms force[d]
Pressure,	Kilograms force per square centimeter[d]	x	0.09807	=	Megapascals
Stress,	Pounds force per square inch (psi)	x	0.006895	=	Megapascals
Strength	Kips per square inch	x	6.895	=	Megapascals
(Force per unit area)	Pounds force per square inch (psi)	x	0.07031	=	Kilograms force per square centimeter[d]
	Pounds force per square foot	x	47.88	=	Pascals
	Pounds force per square foot	x	4.882	=	Kilograms force per square meter[d]
Flow	Cubic feet per minute	x	0.4719	=	Liters per second
	Gallons per minute	x	0.0631	=	Liters per second
	Gallons per hour	x	1.05	=	Milliliters per second
Bending	Inch-pounds force	x	0.01152	=	Meter-kilograms force[d]
Moment	Inch-pounds force	x	0.1130	=	Newton-meters
Or Torque	Foot-pounds force	x	0.1383	=	Meter-kilograms force[d]
	Foot-pounds force	x	1.356	=	Newton-meters
	Meter-kilograms force[d]	x	9.807	=	Newton-meters
Mass	Ounces (avoirdupois)	x	28.35	=	Grams
	Pounds (avoirdupois)	x	0.4536	=	Kilograms
	Tons (metric)	x	1000	=	Kilograms
	Tons, short (2000 pounds)	x	907.2	=	Kilograms
	Tons, short (2000 pounds)	x	0.9072	=	Megagrams[e]
Mass per	Pounds mass per cubic foot	x	16.02	=	Kilograms per cubic meter
Unit	Pounds mass per cubic yard	x	0.5933	=	Kilograms per cubic meter
Volume	Pounds mass per gallon (U.S. liquid)[b]	x	119.8	=	Kilograms per cubic meter
	Pounds mass per gallon (Canadian liquid)[b]	x	99.78	=	Kilograms per cubic meter
Temperature	Degrees Fahrenheit	(F-32)/1.8		=	Degrees Celsius
	Degrees Fahrenheit	(F+459.67)/1.8		=	Degrees Kelvin
	Degrees Celsius	C+273.15		=	Degrees Kelvin

[a] The factor given is exact
[b] One U.S. gallon = 0.8327 Canadian gallon
[c] 1 liter = 1000 milliliters = 1000 cubic centimeters
  1 cubic decimeter = 0.001 cubic meter

[d] Metric but not SI unit
[e] Called "tonne" in England and
  "metric ton" in other metric countries

R011105-60   Weights and Measures

Measures of Length
1 Mile = 1760 Yards = 5280 Feet
1 Yard = 3 Feet = 36 inches
1 Foot = 12 Inches
1 Mil = 0.001 Inch
1 Fathom = 2 Yards = 6 Feet
1 Rod = 5.5 Yards = 16.5 Feet
1 Hand = 4 Inches
1 Span = 9 Inches
1 Micro-inch = One Millionth Inch or 0.000001 Inch
1 Micron = One Millionth Meter + 0.00003937 Inch

Surveyor's Measure
1 Mile = 8 Furlongs = 80 Chains
1 Furlong = 10 Chains = 220 Yards
1 Chain = 4 Rods = 22 Yards = 66 Feet = 100 Links
1 Link = 7.92 Inches

Square Measure
1 Square Mile = 640 Acres = 6400 Square Chains
1 Acre = 10 Square Chains = 4840 Square Yards =
   43,560 Sq. Ft.
1 Square Chain = 16 Square Rods = 484 Square Yards =
   4356 Sq. Ft.
1 Square Rod = 30.25 Square Yards = 272.25 Square Feet = 625 Square
   Lines
1 Square Yard = 9 Square Feet
1 Square Foot = 144 Square Inches
An Acre equals a Square 208.7 Feet per Side

Cubic Measure
1 Cubic Yard = 27 Cubic Feet
1 Cubic Foot = 1728 Cubic Inches
1 Cord of Wood = 4 x 4 x 8 Feet = 128 Cubic Feet
1 Perch of Masonry = 16½ x 1½ x 1 Foot = 24.75 Cubic Feet

Avoirdupois or Commercial Weight
1 Gross or Long Ton = 2240 Pounds
1 Net or Short Ton = 2000 Pounds
1 Pound = 16 Ounces = 7000 Grains
1 Ounce = 16 Drachms = 437.5 Grains
1 Stone = 14 Pounds

Power
1 British Thermal Unit per Hour = 0.2931 Watts
1 Ton (Refrigeration) = 3.517 Kilowatts
1 Horsepower (Boiler) = 9.81 Kilowatts
1 Horsepower (550 ft-lb/s) = 0.746 Kilowatts

Shipping Measure
For Measuring Internal Capacity of a Vessel:
   1 Register Ton = 100 Cubic Feet

For Measurement of Cargo:
   Approximately 40 Cubic Feet of Merchandise is considered a Shipping
   Ton, unless that bulk would weigh more than 2000 Pounds, in which case
   Freight Charge may be based upon weight.

40 Cubic Feet = 32.143 U.S. Bushels = 31.16 Imp. Bushels

Liquid Measure
1 Imperial Gallon = 1.2009 U.S. Gallon = 277.42 Cu. In.
1 Cubic Foot = 7.48 U.S. Gallons

R011110-10   Architectural Fees

Tabulated below are typical percentage fees by project size, for good professional architectural service. Fees may vary from those listed depending upon degree of design difficulty and economic conditions in any particular area.

Rates can be interpolated horizontally and vertically. Various portions of the same project requiring different rates should be adjusted proportionately. For alterations, add 50% to the fee for the first $500,000 of project cost and add 25% to the fee for project cost over $500,000.

Architectural fees tabulated below include Structural, Mechanical and Electrical Engineering Fees. They do not include the fees for special consultants such as kitchen planning, security, acoustical, interior design, etc

Civil Engineering fees are included in the Architectural fee for project sites requiring minimal design such as city sites. However, separate Civil Engineering fees must be added when utility connections require design, drainage calculations are needed, stepped foundations are required, or provisions are required to protect adjacent wetlands.

Building Types	Total Project Size in Thousands of Dollars						
	100	250	500	1,000	5,000	10,000	50,000
Factories, garages, warehouses, repetitive housing	9.0%	8.0%	7.0%	6.2%	5.3%	4.9%	4.5%
Apartments, banks, schools, libraries, offices, municipal buildings	12.2	12.3	9.2	8.0	7.0	6.6	6.2
Churches, hospitals, homes, laboratories, museums, research	15.0	13.6	12.7	11.9	9.5	8.8	8.0
Memorials, monumental work, decorative furnishings	—	16.0	14.5	13.1	10.0	9.0	8.3

R011110-30   Engineering Fees

Typical **Structural Engineering Fees** based on type of construction and total project size. These fees are included in Architectural Fees.

Type of Construction	Total Project Size (in thousands of dollars)			
	$500	$500-$1,000	$1,000-$5,000	Over $5000
Industrial buildings, factories & warehouses	Technical payroll times 2.0 to 2.5	1.60%	1.25%	1.00%
Hotels, apartments, offices, dormitories, hospitals, public buildings, food stores		2.00%	1.70%	1.20%
Museums, banks, churches and cathedrals		2.00%	1.75%	1.25%
Thin shells, prestressed concrete, earthquake resistive		2.00%	1.75%	1.50%
Parking ramps, auditoriums, stadiums, convention halls, hangars & boiler houses		2.50%	2.00%	1.75%
Special buildings, major alterations, underpinning & future expansion	↓	Add to above 0.5%	Add to above 0.5%	Add to above 0.5%

For complex reinforced concrete or unusually complicated structures, add 20% to 50%.

Typical **Mechanical and Electrical Engineering Fees** are based on the size of the subcontract. The fee structure for both is shown below. These fees are included in Architectural Fees.

Type of Construction	Subcontract Size							
	$25,000	$50,000	$100,000	$225,000	$350,000	$500,000	$750,000	$1,000,000
Simple structures	6.4%	5.7%	4.8%	4.5%	4.4%	4.3%	4.2%	4.1%
Intermediate structures	8.0	7.3	6.5	5.6	5.1	5.0	4.9	4.8
Complex structures	10.1	9.0	9.0	8.0	7.5	7.5	7.0	7.0

For renovations, add 15% to 25% to applicable fee.

R012153-60   Security Factors

Contractors entering, working in, and exiting secure facilities often lose productive time during a normal workday. The recommended allowances in this section are intended to provide for the loss of productivity by increasing labor costs. Note that different costs are associated with searches upon entry only and searches upon entry and exit. Time spent in a queue is unpredictable and not part of these allowances. Contractors should plan ahead for this situation.

Security checkpoints are designed to reflect the level of security required to gain access or egress. An extreme example is when contractors, along with any materials, tools, equipment, and vehicles, must be physically searched and have all materials, tools, equipment, and vehicles inventoried and documented prior to both entry and exit.

Physical searches without going through the documentation process represent the next level and take up less time.

Electronic searches—passing through a detector or x-ray machine with no documentation of materials, tools, equipment, and vehicles—take less time than physical searches.

Visual searches of materials, tools, equipment, and vehicles represent the next level of security.

Finally, access by means of an ID card or displayed sticker takes the least amount of time.

Another consideration is if the searches described above are performed each and every day, or if they are performed only on the first day with access granted by ID card or displayed sticker for the remainder of the project. The figures for this situation have been calculated to represent the initial check-in as described and subsequent entry by ID card or displayed sticker for up to 20 days on site. For the situation described above, where the time period is beyond 20 days, the impact on labor cost is negligible.

There are situations where tradespeople must be accompanied by an escort and observed during the work day. The loss of freedom of movement will slow down productivity for the tradesperson. Costs for the observer have not been included. Those costs are normally born by the owner.

R012909-80   Sales Tax by State

State sales tax on materials is tabulated below (5 states have no sales tax). Many states allow local jurisdictions, such as a county or city, to levy additional sales tax.

Some projects may be sales tax exempt, particularly those constructed with public funds.

State	Tax (%)	State	Tax (%)	State	Tax (%)	State	Tax (%)
Alabama	4	Illinois	6.25	Montana	0	Rhode Island	7
Alaska	0	Indiana	7	Nebraska	5.5	South Carolina	6
Arizona	5.6	Iowa	6	Nevada	6.85	South Dakota	4.5
Arkansas	6.5	Kansas	6.5	New Hampshire	0	Tennessee	7
California	7.25	Kentucky	6	New Jersey	6.625	Texas	6.25
Colorado	2.9	Louisiana	4.45	New Mexico	5.125	Utah	4.85
Connecticut	6.35	Maine	5.5	New York	4	Vermont	6
Delaware	0	Maryland	6	North Carolina	4.75	Virginia	4.3
District of Columbia	6	Massachusetts	6.25	North Dakota	5	Washington	6.5
Florida	6	Michigan	6	Ohio	5.75	West Virginia	6
Georgia	4	Minnesota	6.875	Oklahoma	4.5	Wisconsin	5
Hawaii	4	Mississippi	7	Oregon	0	Wyoming	4
Idaho	6	Missouri	4.225	Pennsylvania	6	Average	5.06 %

Sales Tax by Province (Canada)

GST - a value-added tax, which the government imposes on most goods and services provided in or imported into Canada. PST - a retail sales tax, which five of the provinces impose on the prices of most goods and some services. QST - a value-added tax, similar to the federal GST, which Quebec imposes. HST - Three provinces have combined their retail sales taxes with the federal GST into one harmonized tax.

Province	PST (%)	QST (%)	GST (%)	HST (%)
Alberta	0	0	5	0
British Columbia	7	0	5	0
Manitoba	7	0	5	0
New Brunswick	0	0	0	15
Newfoundland	0	0	0	15
Northwest Territories	0	0	5	0
Nova Scotia	0	0	0	15
Ontario	0	0	0	13
Prince Edward Island	0	0	0	15
Quebec	9.975	0	5	14.975
Saskatchewan	6	0	5	0
Yukon	0	0	5	0

R012909-80   Unemployment Taxes and Social Security Taxes

State unemployment tax rates vary not only from state to state, but also with the experience rating of the contractor. The federal unemployment tax rate is 6.0% of the first $7,000 of wages. This is reduced by a credit of up to 5.4% for timely payment to the state. The minimum federal unemployment tax is 0.6% after all credits.

Social security (FICA) for 2020 is estimated at time of publication to be 7.65% of wages up to $137,100.

R012909-86   Unemployment Tax by State

Information is from the U.S. Department of Labor, state unemployment tax rates.

State	Tax (%)	State	Tax (%)	State	Tax (%)	State	Tax (%)
Alabama	6.80	Illinois	6.93	Montana	6.30	Rhode Island	9.49
Alaska	5.4	Indiana	7.4	Nebraska	5.4	South Carolina	5.46
Arizona	12.76	Iowa	7.5	Nevada	5.4	South Dakota	9.35
Arkansas	14.3	Kansas	7.6	New Hampshire	7.5	Tennessee	10.0
California	6.2	Kentucky	9.3	New Jersey	5.8	Texas	6.5
Colorado	8.15	Louisiana	6.2	New Mexico	5.4	Utah	7.1
Connecticut	6.8	Maine	5.46	New York	9.1	Vermont	7.7
Delaware	8.20	Maryland	7.50	North Carolina	5.76	Virginia	6.21
District of Columbia	7	Massachusetts	12.65	North Dakota	10.74	Washington	7.73
Florida	5.4	Michigan	10.3	Ohio	9	West Virginia	8.5
Georgia	5.4	Minnesota	9.1	Oklahoma	5.5	Wisconsin	12.0
Hawaii	5.6	Mississippi	5.6	Oregon	5.4	Wyoming	8.8
Idaho	5.4	Missouri	8.37	Pennsylvania	11.03	Median	7.40 %

R012909-90   Overtime

One way to improve the completion date of a project or eliminate negative float from a schedule is to compress activity duration times. This can be achieved by increasing the crew size or working overtime with the proposed crew.

To determine the costs of working overtime to compress activity duration times, consider the following examples. Below is an overtime efficiency and cost chart based on a five, six, or seven day week with an eight through twelve hour day. Payroll percentage increases for time and one half and double times are shown for the various working days.

Days per Week	Hours per Day	Production Efficiency					Payroll Cost Factors	
		1st Week	2nd Week	3rd Week	4th Week	Average 4 Weeks	@ 1-1/2 Times	@ 2 Times
5	8	100%	100%	100%	100%	100%	1.000	1.000
	9	100	100	95	90	96	1.056	1.111
	10	100	95	90	85	93	1.100	1.200
	11	95	90	75	65	81	1.136	1.273
	12	90	85	70	60	76	1.167	1.333
6	8	100	100	95	90	96	1.083	1.167
	9	100	95	90	85	93	1.130	1.259
	10	95	90	85	80	88	1.167	1.333
	11	95	85	70	65	79	1.197	1.394
	12	90	80	65	60	74	1.222	1.444
7	8	100	95	85	75	89	1.143	1.286
	9	95	90	80	70	84	1.183	1.365
	10	90	85	75	65	79	1.214	1.429
	11	85	80	65	60	73	1.240	1.481
	12	85	75	60	55	69	1.262	1.524

R013113-40   Builder's Risk Insurance

Builder's risk insurance is insurance on a building during construction. Premiums are paid by the owner or the contractor. Blasting, collapse and underground insurance would raise total insurance costs.

For customer support on your Site Work & Landscape Costs with RSMeans data, call 800.448.8182.

833

R013113-50   General Contractor's Overhead

There are two distinct types of overhead on a construction project: Project overhead and main office overhead. Project overhead includes those costs at a construction site not directly associated with the installation of construction materials. Examples of project overhead costs include the following:

1. Superintendent
2. Construction office and storage trailers
3. Temporary sanitary facilities
4. Temporary utilities
5. Security fencing
6. Photographs
7. Cleanup
8. Performance and payment bonds

The above project overhead items are also referred to as general requirements and therefore are estimated in Division 1. Division 1 is the first division listed in the CSI MasterFormat but it is usually the last division estimated. The sum of the costs in Divisions 1 through 49 is referred to as the sum of the direct costs.

All construction projects also include indirect costs. The primary components of indirect costs are the contractor's main office overhead and profit. The amount of the main office overhead expense varies depending on the following:

1. Owner's compensation
2. Project managers' and estimators' wages
3. Clerical support wages
4. Office rent and utilities
5. Corporate legal and accounting costs
6. Advertising
7. Automobile expenses
8. Association dues
9. Travel and entertainment expenses

These costs are usually calculated as a percentage of annual sales volume. This percentage can range from 35% for a small contractor doing less than $500,000 to 5% for a large contractor with sales in excess of $100 million.

R013113-55   Installing Contractor's Overhead

Installing contractors (subcontractors) also incur costs for general requirements and main office overhead.

Included within the total incl. overhead and profit costs is a percent mark-up for overhead that includes:

1. Compensation and benefits for office staff and project managers
2. Office rent, utilities, business equipment, and maintenance
3. Corporate legal and accounting costs

4. Advertising
5. Vehicle expenses (for office staff and project managers)
6. Association dues
7. Travel, entertainment
8. Insurance
9. Small tools and equipment

R013113-60   Workers' Compensation Insurance Rates by Trade

The table below tabulates the national averages for workers' compensation insurance rates by trade and type of building. The average "Insurance Rate" is multiplied by the "% of Building Cost" for each trade. This produces the "Workers' Compensation" cost by % of total labor cost, to be added for each trade by building type to determine the weighted average workers' compensation rate for the building types analyzed.

Trade	Insurance Rate (% Labor Cost) Range	Average	% of Building Cost Office Bldgs.	Schools & Apts.	Mfg.	Workers' Compensation Office Bldgs.	Schools & Apts.	Mfg.
Excavation, Grading, etc.	2.3 % to 16.9%	9.6%	4.8%	4.9%	4.5%	0.46%	0.47%	0.43%
Piles & Foundations	3.3 to 28.0	15.7	7.1	5.2	8.7	1.11	0.82	1.37
Concrete	3.1 to 25.8	14.4	5.0	14.8	3.7	0.72	2.13	0.53
Masonry	3.3 to 52.1	27.7	6.9	7.5	1.9	1.91	2.08	0.53
Structural Steel	4.4 to 31.8	18.1	10.7	3.9	17.6	1.94	0.71	3.19
Miscellaneous & Ornamental Metals	2.9 to 21.4	12.2	2.8	4.0	3.6	0.34	0.49	0.44
Carpentry & Millwork	3.4 to 29.1	16.2	3.7	4.0	0.5	0.60	0.65	0.08
Metal or Composition Siding	4.8 to 124.6	64.7	2.3	0.3	4.3	1.49	0.19	2.78
Roofing	4.8 to 110.4	57.6	2.3	2.6	3.1	1.32	1.50	1.79
Doors & Hardware	3.1 to 29.1	16.1	0.9	1.4	0.4	0.14	0.23	0.06
Sash & Glazing	3.9 to 21.6	12.8	3.5	4.0	1.0	0.45	0.51	0.13
Lath & Plaster	2.7 to 30.1	16.4	3.3	6.9	0.8	0.54	1.13	0.13
Tile, Marble & Floors	2.0 to 19.0	10.5	2.6	3.0	0.5	0.27	0.32	0.05
Acoustical Ceilings	1.7 to 29.7	15.7	2.4	0.2	0.3	0.38	0.03	0.05
Painting	3.3 to 37.4	20.3	1.5	1.6	1.6	0.30	0.32	0.32
Interior Partitions	3.4 to 29.1	16.2	3.9	4.3	4.4	0.63	0.70	0.71
Miscellaneous Items	1.9 to 97.7	10.3	5.2	3.7	9.7	0.54	0.38	1.00
Elevators	1.3 to 9.0	5.1	2.1	1.1	2.2	0.11	0.06	0.11
Sprinklers	1.8 to 14.6	8.2	0.5	—	2.0	0.04	—	0.16
Plumbing	1.4 to 13.3	7.4	4.9	7.2	5.2	0.36	0.53	0.38
Heat., Vent., Air Conditioning	2.8 to 15.8	9.3	13.5	11.0	12.9	1.26	1.02	1.20
Electrical	1.7 to 10.7	6.2	10.1	8.4	11.1	0.63	0.52	0.69
Total	1.3 % to 124.6%	—	100.0%	100.0%	100.0%	15.54%	14.79%	16.13%

Overall Weighted Average   15.49%

Workers' Compensation Insurance Rates by States

The table below lists the weighted average Workers' Compensation base rate for each state with a factor comparing this with the national average of 9.9%.

State	Weighted Average	Factor	State	Weighted Average	Factor	State	Weighted Average	Factor
Alabama	12.4%	126	Kentucky	10.5%	108	North Dakota	6.4%	65
Alaska	8.6	88	Louisiana	17.5	178	Ohio	5.5	56
Arizona	7.7	79	Maine	8.0	81	Oklahoma	7.4	76
Arkansas	5.1	52	Maryland	9.7	99	Oregon	7.5	76
California	19.6	200	Massachusetts	8.8	90	Pennsylvania	17.2	175
Colorado	6.2	63	Michigan	6.3	64	Rhode Island	10.0	102
Connecticut	12.4	126	Minnesota	13.4	137	South Carolina	16.2	165
Delaware	11.8	120	Mississippi	9.2	94	South Dakota	7.6	77
District of Columbia	7.7	78	Missouri	12.0	122	Tennessee	6.0	61
Florida	8.4	86	Montana	7.1	73	Texas	4.9	50
Georgia	30.0	307	Nebraska	11.7	120	Utah	5.7	58
Hawaii	8.4	86	Nevada	7.6	77	Vermont	8.9	91
Idaho	7.8	79	New Hampshire	8.8	90	Virginia	6.7	69
Illinois	17.5	179	New Jersey	14.2	145	Washington	7.3	75
Indiana	3.1	32	New Mexico	13.0	132	West Virginia	4.2	42
Iowa	9.9	101	New York	15.6	159	Wisconsin	11.2	115
Kansas	5.6	57	North Carolina	11.6	119	Wyoming	4.8	49

Weighted Average for U.S. is   9.9% of payroll = 100%

The weighted average skilled worker rate for 35 trades is 9.8%. For bidding purposes, apply the full value of Workers' Compensation directly to total labor costs, or if labor is 38%, materials 42% and overhead and profit 20% of total cost, carry 38/80 x 9.8% = 4.66% of cost (before overhead and profit)

into overhead. Rates vary not only from state to state but also with the experience rating of the contractor.

Rates are the most current available at the time of publication.

R015423-10   Steel Tubular Scaffolding

On new construction, tubular scaffolding is efficient up to 60' high or five stories. Above this it is usually better to use a hung scaffolding if construction permits. Swing scaffolding operations may interfere with tenants. In this case, the tubular is more practical at all heights.

In repairing or cleaning the front of an existing building the cost of tubular scaffolding per S.F. of building front increases as the height increases above the first tier. The first tier cost is relatively high due to leveling and alignment.

The minimum efficient crew for erecting and dismantling is three workers. They can set up and remove 18 frame sections per day up to 5 stories high. For 6 to 12 stories high, a crew of four is most efficient. Use two or more on top and two on the bottom for handing up or hoisting. They can also set up and remove 18 frame sections per day. At 7' horizontal spacing, this will run about 800 S.F. per day of erecting and dismantling. Time for placing and removing planks must be added to the above. A crew of three can place and remove 72 planks per day up to 5 stories. For over 5 stories, a crew of four can place and remove 80 planks per day.

The table below shows the number of pieces required to erect tubular steel scaffolding for 1000 S.F. of building frontage. This area is made up of a scaffolding system that is 12 frames (11 bays) long by 2 frames high.

For jobs under twenty-five frames, add 50% to rental cost. Rental rates will be lower for jobs over three months duration. Large quantities for long periods can reduce rental rates by 20%.

Description of Component	Number of Pieces for 1000 S.F. of Building Front	Unit
5' Wide Standard Frame, 6'-4" High	24	Ea.
Leveling Jack & Plate	24	
Cross Brace	44	
Side Arm Bracket, 21"	12	
Guardrail Post	12	
Guardrail, 7' section	22	
Stairway Section	2	
Stairway Starter Bar	1	
Stairway Inside Handrail	2	
Stairway Outside Handrail	2	
Walk-Thru Frame Guardrail	2	

Scaffolding is often used as falsework over 15' high during construction of cast-in-place concrete beams and slabs. Two foot wide scaffolding is generally used for heavy beam construction. The span between frames depends upon the load to be carried with a maximum span of 5'.

Heavy duty shoring frames with a capacity of 10,000#/leg can be spaced up to 10' O.C. depending upon form support design and loading.

Scaffolding used as horizontal shoring requires less than half the material required with conventional shoring.

On new construction, erection is done by carpenters.

Rolling towers supporting horizontal shores can reduce labor and speed the job. For maintenance work, catwalks with spans up to 70' can be supported by the rolling towers.

R015433-10   Contractor Equipment

Rental Rates shown elsewhere in the data set pertain to late model high quality machines in excellent working condition, rented from equipment dealers. Rental rates from contractors may be substantially lower than the rental rates from equipment dealers depending upon economic conditions; for older, less productive machines, reduce rates by a maximum of 15%. Any overtime must be added to the base rates. For shift work, rates are lower. Usual rule of thumb is 150% of one shift rate for two shifts; 200% for three shifts.

For periods of less than one week, operated equipment is usually more economical to rent than renting bare equipment and hiring an operator.

Costs to move equipment to a job site (mobilization) or from a job site (demobilization) are not included in rental rates, nor in any Equipment costs on any Unit Price line items or crew listings. These costs can be found elsewhere. If a piece of equipment is already at a job site, it is not appropriate to utilize mob/demob costs in an estimate again.

Rental rates vary throughout the country with larger cities generally having lower rates. Lease plans for new equipment are available for periods in excess of six months with a percentage of payments applying toward purchase.

Rental rates can also be treated as reimbursement costs for contractor-owned equipment. Owned equipment costs include depreciation, loan payments, interest, taxes, insurance, storage, and major repairs.

Monthly rental rates vary from 2% to 5% of the cost of the equipment depending on the anticipated life of the equipment and its wearing parts. Weekly rates are about 1/3 the monthly rates and daily rental rates are about 1/3 the weekly rates.

The hourly operating costs for each piece of equipment include costs to the user such as fuel, oil, lubrication, normal expendables for the equipment, and a percentage of the mechanic's wages chargeable to maintenance. The hourly operating costs listed do not include the operator's wages.

The daily cost for equipment used in the standard crews is figured by dividing the weekly rate by five, then adding eight times the hourly operating cost to give the total daily equipment cost, not including the operator. This figure is in the right hand column of the Equipment listings under Equipment Cost/Day.

Pile Driving rates shown for pile the hammer and extractor do not include leads, cranes, boilers or compressors. Vibratory pile driving requires an added field specialist during set-up and pile driving operation for the electric model. The hydraulic model requires a field specialist for set-up only. Up to 125 reuses of sheet piling are possible using vibratory drivers. For normal conditions, crane capacity for hammer type and size is as follows.

Crane Capacity	Hammer Type and Size		
	Air or Steam	Diesel	Vibratory
25 ton	to 8,750 ft.-lb.		70 H.P.
40 ton	15,000 ft.-lb.	to 32,000 ft.-lb.	170 H.P.
60 ton	25,000 ft.-lb.		300 H.P.
100 ton		112,000 ft.-lb.	

Cranes should be specified for the job by size, building and site characteristics, availability, performance characteristics, and duration of time required.

Backhoes & Shovels rent for about the same as equivalent size cranes but maintenance and operating expenses are higher. The crane operator's rate must be adjusted for high boom heights. Average adjustments: for 150' boom add 2% per hour; over 185', add 4% per hour; over 210', add 6% per hour; over 250', add 8% per hour and over 295', add 12% per hour.

Tower Cranes of the climbing or static type have jibs from 50' to 200' and capacities at maximum reach range from 4,000 to 14,000 pounds. Lifting capacities increase up to maximum load as the hook radius decreases.

Typical rental rates, based on purchase price, are about 2% to 3% per month.

Erection and dismantling run between 500 and 2000 labor hours. Climbing operation takes 10 labor hours per 20' climb. Crane dead time is about 5 hours per 40' climb. If crane is bolted to side of the building add cost of ties and extra mast sections. Climbing cranes have from 80' to 180' of mast while static cranes have 80' to 800' of mast.

Truck Cranes can be converted to tower cranes by using tower attachments. Mast heights over 400' have been used.

A single 100' high material **Hoist and Tower** can be erected and dismantled in about 400 labor hours; a double 100' high hoist and tower in about 600 labor hours. Erection times for additional heights are 3 and 4 labor hours

per vertical foot respectively up to 150', and 4 to 5 labor hours per vertical foot over 150' high. A 40' high portable Buck hoist takes about 160 labor hours to erect and dismantle. Additional heights take 2 labor hours per vertical foot to 80' and 3 labor hours per vertical foot for the next 100'. Most material hoists do not meet local code requirements for carrying personnel.

A 150' high **Personnel Hoist** requires about 500 to 800 labor hours to erect and dismantle. Budget erection time at 5 labor hours per vertical foot for all trades. Local code requirements or labor scarcity requiring overtime can add up to 50% to any of the above erection costs.

Earthmoving Equipment: The selection of earthmoving equipment depends upon the type and quantity of material, moisture content, haul distance, haul road, time available, and equipment available. Short haul cut and fill operations may require dozers only, while another operation may require excavators, a fleet of trucks, and spreading and compaction equipment. Stockpiled material and granular material are easily excavated with front end loaders. Scrapers are most economically used with hauls between 300' and 1-1/2 miles if adequate haul roads can be maintained. Shovels are often used for blasted rock and any material where a vertical face of 8' or more can be excavated. Special conditions may dictate the use of draglines, clamshells, or backhoes. Spreading and compaction equipment must be matched to the soil characteristics, the compaction required and the rate the fill is being supplied.

R015433-15   Heavy Lifting

Hydraulic Climbing Jacks

The use of hydraulic heavy lift systems is an alternative to conventional type crane equipment. The lifting, lowering, pushing, or pulling mechanism is a hydraulic climbing jack moving on a square steel jackrod from 1-5/8" to 4" square, or a steel cable. The jackrod or cable can be vertical or horizontal, stationary or movable, depending on the individual application. When the jackrod is stationary, the climbing jack will climb the rod and push or pull the load along with itself. When the climbing jack is stationary, the jackrod is movable with the load attached to the end and the climbing jack will lift or lower the jackrod with the attached load. The heavy lift system is normally operated by a single control lever located at the hydraulic pump.

The system is flexible in that one or more climbing jacks can be applied wherever a load support point is required, and the rate of lift synchronized.

Economic benefits have been demonstrated on projects such as: erection of ground assembled roofs and floors, complete bridge spans, girders and trusses, towers, chimney liners and steel vessels, storage tanks, and heavy machinery. Other uses are raising and lowering offshore work platforms, caissons, tunnel sections and pipelines.

R015436-50   Mobilization

Costs to move rented construction equipment to a job site from an equipment dealer's or contractor's yard (mobilization) or off the job site (demobilization) are not included in the rental or operating rates, nor in the equipment cost on a unit price line or in a crew listing. These costs can be found consolidated in the Mobilization section of the data and elsewhere in particular site work sections. If a piece of equipment is already on the job site, it is not appropriate to include mob/demob costs in a new estimate that requires use of that equipment. The following table identifies approximate sizes of rented construction equipment that would be hauled on a towed trailer. Because this listing is not all-encompassing, the user can infer as to what size trailer might be required for a piece of equipment not listed.

3-ton Trailer	20-ton Trailer	40-ton Trailer	50-ton Trailer
20 H.P. Excavator	110 H.P. Excavator	200 H.P. Excavator	270 H.P. Excavator
50 H.P. Skid Steer	165 H.P. Dozer	300 H.P. Dozer	Small Crawler Crane
35 H.P. Roller	150 H.P. Roller	400 H.P. Scraper	500 H.P. Scraper
40 H.P. Trencher	Backhoe	450 H.P. Art. Dump Truck	500 H.P. Art. Dump Truck

R024119-10   Demolition Defined

Whole Building Demolition - Demolition of the whole building with no concern for any particular building element, component, or material type being demolished. This type of demolition is accomplished with large pieces of construction equipment that break up the structure, load it into trucks and haul it to a disposal site, but disposal or dump fees are not included. Demolition of below-grade foundation elements, such as footings, foundation walls, grade beams, slabs on grade, etc., is not included. Certain mechanical equipment containing flammable liquids or ozone-depleting refrigerants, electric lighting elements, communication equipment components, and other building elements may contain hazardous waste, and must be removed, either selectively or carefully, as hazardous waste before the building can be demolished.

Foundation Demolition - Demolition of below-grade foundation footings, foundation walls, grade beams, and slabs on grade. This type of demolition is accomplished by hand or pneumatic hand tools, and does not include saw cutting, or handling, loading, hauling, or disposal of the debris.

Gutting - Removal of building interior finishes and electrical/mechanical systems down to the load-bearing and sub-floor elements of the rough building frame, with no concern for any particular building element, component, or material type being demolished. This type of demolition is accomplished by hand or pneumatic hand tools, and includes loading into trucks, but not hauling, disposal or dump fees, scaffolding, or shoring. Certain mechanical equipment containing flammable liquids or ozone-depleting refrigerants, electric lighting elements, communication equipment components, and other building elements may contain hazardous waste, and must be removed, either selectively or carefully, as hazardous waste, before the building is gutted.

Selective Demolition - Demolition of a selected building element, component, or finish, with some concern for surrounding or adjacent elements, components, or finishes (see the first Subdivision (s) at the beginning of appropriate Divisions). This type of demolition is accomplished by hand or pneumatic hand tools, and does not include handling, loading, storing, hauling, or disposal of the debris, scaffolding, or shoring. "Gutting" methods may be used in order to save time, but damage that is caused to surrounding or adjacent elements, components, or finishes may have to be repaired at a later time.

Careful Removal - Removal of a piece of service equipment, building element or component, or material type, with great concern for both the removed item and surrounding or adjacent elements, components or finishes. The purpose of careful removal may be to protect the removed item for later re-use, preserve a higher salvage value of the removed item, or replace an item while taking care to protect surrounding or adjacent elements, components, connections, or finishes from cosmetic and/or structural damage. An approximation of the time required to perform this type of removal is 1/3 to 1/2 the time it would take to install a new item of like kind. This type of removal is accomplished by hand or pneumatic hand tools, and does not include loading, hauling, or storing the removed item, scaffolding, shoring, or lifting equipment.

Cutout Demolition - Demolition of a small quantity of floor, wall, roof, or other assembly, with concern for the appearance and structural integrity of the surrounding materials. This type of demolition is accomplished by hand or pneumatic hand tools, and does not include saw cutting, handling, loading, hauling, or disposal of debris, scaffolding, or shoring.

Rubbish Handling - Work activities that involve handling, loading or hauling of debris. Generally, the cost of rubbish handling must be added to the cost of all types of demolition, with the exception of whole building demolition.

Minor Site Demolition - Demolition of site elements outside the footprint of a building. This type of demolition is accomplished by hand or pneumatic hand tools, or with larger pieces of construction equipment, and may include loading a removed item onto a truck (check the Crew for equipment used). It does not include saw cutting, hauling or disposal of debris, and, sometimes, handling or loading.

R024119-20   Dumpsters

Dumpster rental costs on construction sites are presented in two ways.

The cost per week rental includes the delivery of the dumpster; its pulling or emptying once per week, and its final removal. The assumption is made that the dumpster contractor could choose to empty a dumpster by simply bringing in an empty unit and removing the full one. These costs also include the disposal of the materials in the dumpster.

The Alternate Pricing can be used when actual planned conditions are not approximated by the weekly numbers. For example, these lines can be used when a dumpster is needed for 4 weeks and will need to be emptied 2 or 3 times per week. Conversely the Alternate Pricing lines can be used when a dumpster will be rented for several weeks or months but needs to be emptied only a few times over this period.

R024119-30   Rubbish Handling Chutes

To correctly estimate the cost of rubbish handling chute systems, the individual components must be priced separately. First choose the size of the system; a 30-inch diameter chute is quite common, but the sizes range from 18 to 36 inches in diameter. The 30-inch chute comes in a standard weight and two thinner weights. The thinner weight chutes are sometimes chosen for cost savings, but they are more easily damaged.

There are several types of major chute pieces that make up the chute system. The first component to consider is the top chute section (top intake hopper) where the material is dropped into the chute at the highest point. After determining the top chute, the intermediate chute pieces called the regular chute sections are priced. Next, the number of chute control door sections (intermediate intake hoppers) must be determined. In the more complex systems, a chute control door section is provided at each floor level. The last major component to consider is bolt down frames; these are usually provided at every other floor level.

There are a number of accessories to consider for safe operation and control. There are covers for the top chute and the chute door sections. The top chute can have a trough that allows for better loading of the chute. For the safest operation, a chute warning light system can be added that will warn the other chute intake locations not to load while another is being used. There are dust control devices that spray a water mist to keep down the dust as the debris is loaded into a Dumpster. There are special breakaway cords that are used to prevent damage to the chute if the Dumpster is removed without disconnecting from the chute. There are chute liners that can be installed to protect the chute structure from physical damage from rough abrasive materials. Warning signs can be posted at each floor level that is provided with a chute control door section.

In summary, a complete rubbish handling chute system will include one top section, several intermediate regular sections, several intermediate control door (intake hopper) sections and bolt down frames at every other floor level starting with the top floor. If so desired, the system can also include covers and a light warning system for a safer operation. The bottom of the chute should always be above the Dumpster and should be tied off with a breakaway cord to the Dumpster.

R026510-20 Underground Storage Tank Removal

Underground Storage Tank Removal can be divided into two categories: Non-Leaking and Leaking. Prior to removing an underground storage tank, tests should be made, with the proper authorities present, to determine whether a tank has been leaking or the surrounding soil has been contaminated.

To safely remove Liquid Underground Storage Tanks:
1. Excavate to the top of the tank.
2. Disconnect all piping.
3. Open all tank vents and access ports.
4. Remove all liquids and/or sludge.
5. Purge the tank with an inert gas.
6. Provide access to the inside of the tank and clean out the interior using proper personal protective equipment (PPE).
7. Excavate soil surrounding the tank using proper PPE for on-site personnel.
8. Pull and properly dispose of the tank.
9. Clean up the site of all contaminated material.
10. Install new tanks or close the excavation.

R031113-10   Wall Form Materials

Aluminum Forms

Approximate weight is 3 lbs. per S.F.C.A. Standard widths are available from 4" to 36" with 36" most common. Standard lengths of 2', 4', 6' to 8' are available. Forms are lightweight and fewer ties are needed with the wider widths. The form face is either smooth or textured.

Metal Framed Plywood Forms

Manufacturers claim over 75 reuses of plywood and over 300 reuses of steel frames. Many specials such as corners, fillers, pilasters, etc. are available. Monthly rental is generally about 15% of purchase price for first month and 9% per month thereafter with 90% of rental applied to purchase for the first month and decreasing percentages thereafter. Aluminum framed forms cost 25% to 30% more than steel framed.

After the first month, extra days may be prorated from the monthly charge. Rental rates do not include ties, accessories, cleaning, loss of hardware or freight in and out. Approximate weight is 5 lbs. per S.F. for steel; 3 lbs. per S.F. for aluminum.

Forms can be rented with option to buy.

Plywood Forms, Job Fabricated

There are two types of plywood used for concrete forms.

1. Exterior plyform is completely waterproof. This is face oiled to facilitate stripping. Ten reuses can be expected with this type with 25 reuses possible.
2. An overlaid type consists of a resin fiber fused to exterior plyform. No oiling is required except to facilitate cleaning. This is available in both high density (HDO) and medium density overlaid (MDO). Using HDO, 50 reuses can be expected with 200 possible.

Plyform is available in 5/8" and 3/4" thickness. High density overlaid is available in 3/8", 1/2", 5/8" and 3/4" thickness.

5/8" thick is sufficient for most building forms, while 3/4" is best on heavy construction.

Plywood Forms, Modular, Prefabricated

There are many plywood forming systems without frames. Most of these are manufactured from 1-1/8" (HDO) plywood and have some hardware attached. These are used principally for foundation walls 8' or less high. With care and maintenance, 100 reuses can be attained with decreasing quality of surface finish.

Steel Forms

Approximate weight is 6-1/2 lbs. per S.F.C.A. including accessories. Standard widths are available from 2" to 24", with 24" most common. Standard lengths are from 2' to 8', with 4' the most common. Forms are easily ganged into modular units.

Forms are usually leased for 15% of the purchase price per month prorated daily over 30 days.

Rental may be applied to sale price, and usually rental forms are bought. With careful handling and cleaning 200 to 400 reuses are possible.

Straight wall gang forms up to 12' x 20' or 8' x 30' can be fabricated. These crane handled forms usually lease for approx. 9% per month.

Individual job analysis is available from the manufacturer at no charge.

R031113-40   Forms for Reinforced Concrete

Design Economy

Avoid many sizes in proportioning beams and columns.

From story to story avoid changing column dimensions. Gain strength by adding steel or using a richer mix. If a change in size of column is necessary, vary one dimension only to minimize form alterations. Keep beams and columns the same width.

From floor to floor in a multi-story building vary beam depth, not width, as that will leave the slab panel form unchanged. It is cheaper to vary the strength of a beam from floor to floor by means of a steel area than by 2" changes in either width or depth.

Cost Factors

Material includes the cost of lumber, cost of rent for metal pans or forms if used, nails, form ties, form oil, bolts and accessories.

Labor includes the cost of carpenters to make up, erect, remove and repair, plus common labor to clean and move. Having carpenters remove forms minimizes repairs.

Improper alignment and condition of forms will increase finishing cost. When forms are heavily oiled, concrete surfaces must be neutralized before finishing. Special curing compounds will cause spillages to spall off in first frost. Gang forming methods will reduce costs on large projects.

Materials Used

Boards are seldom used unless their architectural finish is required. Generally, steel, fiberglass and plywood are used for contact surfaces. Labor on plywood is 10% less than with boards. The plywood is backed up with

2 x 4's at 12" to 32" O.C. Walers are generally 2 - 2 x 4's. Column forms are held together with steel yokes or bands. Shoring is with adjustable shoring or scaffolding for high ceilings.

Reuse

Floor and column forms can be reused four or possibly five times without excessive repair. Remember to allow for 10% waste on each reuse.

When modular sized wall forms are made, up to twenty uses can be expected with exterior plyform.

When forms are reused, the cost to erect, strip, clean and move will not be affected. 10% replacement of lumber should be included and about one hour of carpenter time for repairs on each reuse per 100 S.F.

The reuse cost for certain accessory items normally rented on a monthly basis will be lower than the cost for the first use.

After the fifth use, new material required plus time needed for repair prevent the form cost from dropping further; it may go up. Much depends on care in stripping, the number of special bays, changes in beam or column sizes and other factors.

Costs for multiple use of formwork may be developed as follows:

2 Uses	3 Uses	4 Uses
$\dfrac{(\text{1st Use + Reuse})}{2} = \text{avg. cost/2 uses}$	$\dfrac{(\text{1st Use + 2 Reuses})}{3} = \text{avg. cost/3 uses}$	$\dfrac{(\text{1st use + 3 Reuses})}{4} = \text{avg. cost/4 uses}$

R031113-60   Formwork Labor-Hours

Item	Unit	Hours Required Fabricate	Hours Required Erect & Strip	Hours Required Clean & Move	Total Hours 1 Use	Multiple Use 2 Use	Multiple Use 3 Use	Multiple Use 4 Use
Beam and Girder, interior beams, 12" wide	100 S.F.	6.4	8.3	1.3	16.0	13.3	12.4	12.0
Hung from steel beams		5.8	7.7	1.3	14.8	12.4	11.6	11.2
Beam sides only, 36" high		5.8	7.2	1.3	14.3	11.9	11.1	10.7
Beam bottoms only, 24" wide		6.6	13.0	1.3	20.9	18.1	17.2	16.7
Box out for openings		9.9	10.0	1.1	21.0	16.6	15.1	14.3
Buttress forms, to 8' high		6.0	6.5	1.2	13.7	11.2	10.4	10.0
Centering, steel, 3/4" rib lath			1.0		1.0			
3/8" rib lath or slab form			0.9		0.9			
Chamfer strip or keyway	100 L.F.		1.5		1.5	1.5	1.5	1.5
Columns, fiber tube 8" diameter			20.6		20.6			
12"			21.3		21.3			
16"			22.9		22.9			
20"			23.7		23.7			
24"			24.6		24.6			
30"			25.6		25.6			
Columns, round steel, 12" diameter			22.0		22.0	22.0	22.0	22.0
16"			25.6		25.6	25.6	25.6	25.6
20"			30.5		30.5	30.5	30.5	30.5
24"			37.7		37.7	37.7	37.7	37.7
Columns, plywood 8" x 8"	100 S.F.	7.0	11.0	1.2	19.2	16.2	15.2	14.7
12" x 12"		6.0	10.5	1.2	17.7	15.2	14.4	14.0
16" x 16"		5.9	10.0	1.2	17.1	14.7	13.8	13.4
24" x 24"		5.8	9.8	1.2	16.8	14.4	13.6	13.2
Columns, steel framed plywood 8" x 8"			10.0	1.0	11.0	11.0	11.0	11.0
12" x 12"			9.3	1.0	10.3	10.3	10.3	10.3
16" x 16"			8.5	1.0	9.5	9.5	9.5	9.5
24" x 24"			7.8	1.0	8.8	8.8	8.8	8.8
Drop head forms, plywood		9.0	12.5	1.5	23.0	19.0	17.7	17.0
Coping forms		8.5	15.0	1.5	25.0	21.3	20.0	19.4
Culvert, box			14.5	4.3	18.8	18.8	18.8	18.8
Curb forms, 6" to 12" high, on grade		5.0	8.5	1.2	14.7	12.7	12.1	11.7
On elevated slabs		6.0	10.8	1.2	18.0	15.5	14.7	14.3
Edge forms to 6" high, on grade	100 L.F.	2.0	3.5	0.6	6.1	5.6	5.4	5.3
7" to 12" high	100 S.F.	2.5	5.0	1.0	8.5	7.8	7.5	7.4
Equipment foundations		10.0	18.0	2.0	30.0	25.5	24.0	23.3
Flat slabs, including drops		3.5	6.0	1.2	10.7	9.5	9.0	8.8
Hung from steel		3.0	5.5	1.2	9.7	8.7	8.4	8.2
Closed deck for domes		3.0	5.8	1.2	10.0	9.0	8.7	8.5
Open deck for pans		2.2	5.3	1.0	8.5	7.9	7.7	7.6
Footings, continuous, 12" high		3.5	3.5	1.5	8.5	7.3	6.8	6.6
Spread, 12" high		4.7	4.2	1.6	10.5	8.7	8.0	7.7
Pile caps, square or rectangular		4.5	5.0	1.5	11.0	9.3	8.7	8.4
Grade beams, 24" deep		2.5	5.3	1.2	9.0	8.3	8.0	7.9
Lintel or Sill forms		8.0	17.0	2.0	27.0	23.5	22.3	21.8
Spandrel beams, 12" wide		9.0	11.2	1.3	21.5	17.5	16.2	15.5
Stairs			25.0	4.0	29.0	29.0	29.0	29.0
Trench forms in floor		4.5	14.0	1.5	20.0	18.3	17.7	17.4
Walls, Plywood, at grade, to 8' high		5.0	6.5	1.5	13.0	11.0	9.7	9.5
8' to 16'		7.5	8.0	1.5	17.0	13.8	12.7	12.1
16' to 20'		9.0	10.0	1.5	20.5	16.5	15.2	14.5
Foundation walls, to 8' high		4.5	6.5	1.0	12.0	10.3	9.7	9.4
8' to 16' high		5.5	7.5	1.0	14.0	11.8	11.0	10.6
Retaining wall to 12' high, battered		6.0	8.5	1.5	16.0	13.5	12.7	12.3
Radial walls to 12' high, smooth		8.0	9.5	2.0	19.5	16.0	14.8	14.3
2' chords		7.0	8.0	1.5	16.5	13.5	12.5	12.0
Prefabricated modular, to 8' high		—	4.3	1.0	5.3	5.3	5.3	5.3
Steel, to 8' high		—	6.8	1.2	8.0	8.0	8.0	8.0
8' to 16' high		—	9.1	1.5	10.6	10.3	10.2	10.2
Steel framed plywood to 8' high		—	6.8	1.2	8.0	7.5	7.3	7.2
8' to 16' high		—	9.3	1.2	10.5	9.5	9.2	9.0

R032110-10   Reinforcing Steel Weights and Measures

Bar Designation No.**	Nominal Weight Lb./Ft.	U.S. Customary Units Nominal Dimensions*			SI Units Nominal Dimensions*			
		Diameter in.	Cross Sectional Area, in.²	Perimeter in.	Nominal Weight kg/m	Diameter mm	Cross Sectional Area, cm²	Perimeter mm
3	.376	.375	.11	1.178	.560	9.52	.71	29.9
4	.668	.500	.20	1.571	.994	12.70	1.29	39.9
5	1.043	.625	.31	1.963	1.552	15.88	2.00	49.9
6	1.502	.750	.44	2.356	2.235	19.05	2.84	59.8
7	2.044	.875	.60	2.749	3.042	22.22	3.87	69.8
8	2.670	1.000	.79	3.142	3.973	25.40	5.10	79.8
9	3.400	1.128	1.00	3.544	5.059	28.65	6.45	90.0
10	4.303	1.270	1.27	3.990	6.403	32.26	8.19	101.4
11	5.313	1.410	1.56	4.430	7.906	35.81	10.06	112.5
14	7.650	1.693	2.25	5.320	11.384	43.00	14.52	135.1
18	13.600	2.257	4.00	7.090	20.238	57.33	25.81	180.1

* The nominal dimensions of a deformed bar are equivalent to those of a plain round bar having the same weight per foot as the deformed bar.
** Bar numbers are based on the number of eighths of an inch included in the nominal diameter of the bars.

R032110-70   Bend, Place and Tie Reinforcing

Placing and tying by rodmen for footings and slabs run from nine hrs. per ton for heavy bars to fifteen hrs. per ton for light bars. For beams, columns, and walls, production runs from eight hrs. per ton for heavy bars to twenty hrs. per ton for light bars. The overall average for typical reinforced concrete buildings is about fourteen hrs. per ton. These production figures include the time for placing accessories and usual inserts, but not their material cost (allow 15% of the cost of delivered bent rods). Equipment

handling is necessary for the larger-sized bars so that installation costs for the very heavy bars will not decrease proportionately.

Installation costs for splicing reinforcing bars include allowance for equipment to hold the bars in place while splicing as well as necessary scaffolding for iron workers.

R032110-80   Shop-Fabricated Reinforcing Steel

The material prices for reinforcing, shown in the unit cost sections of the data set, are for 50 tons or more of shop-fabricated reinforcing steel and include:
1. Mill base price of reinforcing steel
2. Mill grade/size/length extras
3. Mill delivery to the fabrication shop
4. Shop storage and handling
5. Shop drafting/detailing
6. Shop shearing and bending
7. Shop listing
8. Shop delivery to the job site

Both material and installation costs can be considerably higher for small jobs consisting primarily of smaller bars, while material costs may be slightly lower for larger jobs.

R032205-30   Common Stock Styles of Welded Wire Fabric

This table provides some of the basic specifications, sizes, and weights of welded wire fabric used for reinforcing concrete.

	New Designation	Old Designation		Steel Area per Foot				Approximate Weight	
	Spacing — Cross Sectional Area (in.) — (Sq. in. 100)	Spacing — Wire Gauge (in.) — (AS & W)		Longitudinal		Transverse		per 100 S.F.	
				in.	cm	in.	cm	lbs	kg
Rolls	6 x 6 — W1.4 x W1.4	6 x 6 — 10 x 10		.028	.071	.028	.071	21	9.53
	6 x 6 — W2.0 x W2.0	6 x 6 — 8 x 8	1	.040	.102	.040	.102	29	13.15
	6 x 6 — W2.9 x W2.9	6 x 6 — 6 x 6		.058	.147	.058	.147	42	19.05
	6 x 6 — W4.0 x W4.0	6 x 6 — 4 x 4		.080	.203	.080	.203	58	26.91
	4 x 4 — W1.4 x W1.4	4 x 4 — 10 x 10		.042	.107	.042	.107	31	14.06
	4 x 4 — W2.0 x W2.0	4 x 4 — 8 x 8	1	.060	.152	.060	.152	43	19.50
	4 x 4 — W2.9 x W2.9	4 x 4 — 6 x 6		.087	.227	.087	.227	62	28.12
	4 x 4 — W4.0 x W4.0	4 x 4 — 4 x 4		.120	.305	.120	.305	85	38.56
Sheets	6 x 6 — W2.9 x W2.9	6 x 6 — 6 x 6		.058	.147	.058	.147	42	19.05
	6 x 6 — W4.0 x W4.0	6 x 6 — 4 x 4		.080	.203	.080	.203	58	26.31
	6 x 6 — W5.5 x W5.5	6 x 6 — 2 x 2	2	.110	.279	.110	.279	80	36.29
	4 x 4 — W1.4 x W1.4	4 x 4 — 4 x 4		.120	.305	.120	.305	85	38.56

NOTES: 1. Exact W—number size for 8 gauge is W2.1
        2. Exact W—number size for 2 gauge is W5.4

The above table was compiled with the following excerpts from the WRI Manual of Standard Practices, 7th Edition, Copyright 2006. Reproduced with permission of the Wire Reinforcement Institute, Inc.:

1. Chapter 3, page 7, Table 1 Common Styles of Metric Wire Reinforcement (WWR) With Equivalent US Customary Units
2. Chapter 6, page 19, Table 5 Customary Units
3. Chapter 6, Page 23, Table 7 Customary Units (in.) Welded Plain Wire Reinforcement
4. Chapter 6, Page 25 Table 8 Wire Size Comparison
5. Chapter 9, Page 30, Table 9 Weight of Longitudinal Wires Weight (Mass) Estimating Tables
6. Chapter 9, Page 31, Table 9M Weight of Longitudinal Wires Weight (Mass) Estimating Tables
7. Chapter 9, Page 32, Table 10 Weight of Transverse Wires Based on 62″ lengths of transverse wire (60″ width plus 1″ overhand each side)
8. Chapter 9, Page 33, Table 10M Weight of Transverse Wires

R033053-60   Maximum Depth of Frost Penetration in Inches

THIS MAP IS REASONABLY ACCURATE FOR MOST PARTS
OF THE UNITED STATES BUT IS NECESSARILY HIGHLY
GENERALIZED, AND CONSEQUENTLY NOT TOO ACCURATE IN
MOUNTAINOUS REGIONS, PARTICULARLY IN THE ROCKIES.

R033105-20 Materials for One C.Y. of Concrete

This is an approximate method of figuring quantities of cement, sand and coarse aggregate for a field mix with waste allowance included.

With crushed gravel as coarse aggregate, to determine barrels of cement required, divide 10 by total mix; that is, for 1:2:4 mix, 10 divided by 7 = 1-3/7 barrels.

If the coarse aggregate is crushed stone, use 10-1/2 instead of 10 as given for gravel.

To determine tons of sand required, multiply barrels of cement by parts of sand and then by 0.2; that is, for the 1:2:4 mix, as above, 1-3/7 x 2 x .2 = .57 tons.

Tons of crushed gravel are in the same ratio to tons of sand as parts in the mix, or 4/2 x .57 = 1.14 tons.

1 bag cement = 94#	1 C.Y. sand or crushed gravel = 2700#	1 C.Y. crushed stone = 2575#
4 bags = 1 barrel	1 ton sand or crushed gravel = 20 C.F.	1 ton crushed stone = 21 C.F.

Average carload of cement is 692 bags; of sand or gravel is 56 tons.

Do not stack stored cement over 10 bags high.

R033105-65 Field-Mix Concrete

Presently most building jobs are built with ready-mixed concrete except at isolated locations and some larger jobs requiring over 10,000 C.Y. where land is readily available for setting up a temporary batch plant.

The most economical mix is a controlled mix using local aggregate proportioned by trial to give the required strength with the least cost of material.

R033105-70 Placing Ready-Mixed Concrete

For ground pours allow for 5% waste when figuring quantities.

Prices in the front of the data set assume normal deliveries. If deliveries are made before 8 A.M. or after 5 P.M. or on Saturday afternoons add 30%. Negotiated discounts for large volumes are not included in prices in front of the data set.

For the lower floors without truck access, concrete may be wheeled in rubber-tired buggies, conveyer handled, crane handled or pumped. Pumping is economical if there is top steel. Conveyers are more efficient for thick slabs.

At higher floors the rubber-tired buggies may be hoisted by a hoisting tower and wheeled to the location. Placement by a conveyer is limited to three floors and is best for high-volume pours. Pumped concrete is best when the building has no crane access. Concrete may be pumped directly as high as thirty-six stories using special pumping techniques. Normal maximum height is about fifteen stories.

The best pumping aggregate is screened and graded bank gravel rather than crushed stone.

Pumping downward is more difficult than pumping upward. The horizontal distance from pump to pour may increase preparation time prior to pour. Placing by cranes, either mobile, climbing or tower types, continues as the most efficient method for high-rise concrete buildings.

R040513-10 Cement Mortar (material only)

Type N - 1:1:6 mix by volume. Use everywhere above grade except as noted below. - 1:3 mix using conventional masonry cement which saves handling two separate bagged materials.

Type M - 1:1/4:3 mix by volume, or 1 part cement, 1/4 (10% by wt.) lime, 3 parts sand. Use for heavy loads and where earthquakes or hurricanes may occur. Also for reinforced brick, sewers, manholes and everywhere below grade.

Mix Proportions by Volume and Compressive Strength of Mortar

Where Used	Mortar Type	Portland Cement	Masonry Cement	Hydrated Lime	Masonry Sand	Compressive Strength @ 28 days
Plain Masonry	M	1	1	—	6	
		1	—	1/4	3	2500 psi
	S	1/2	1	—	4	
		1	—	1/4 to 1/2	4	1800 psi
	N	—	1	—	3	
		1	—	1/2 to 1-1/4	6	750 psi
	O	—	1	—	3	
		1	—	1-1/4 to 2-1/2	9	350 psi
	K	1	—	2-1/2 to 4	12	75 psi
Reinforced Masonry	PM	1	1	—	6	2500 psi
	PL	1	—	1/4 to 1/2	4	2500 psi

Note: The total aggregate should be between 2.25 to 3 times the sum of the cement and lime used.

The labor cost to mix the mortar is included in the productivity and labor cost of unit price lines in unit cost sections for brickwork, blockwork and stonework.

The material cost of mixed mortar is included in the material cost of those same unit price lines and includes the cost of renting and operating a 10 C.F. mixer at the rate of 200 C.F. per day.

There are two types of mortar color used. One type is the inert additive type with about 100 lbs. per M brick as the typical quantity required. These colors are also available in smaller-batch-sized bags (1 lb. to 15 lb.) which can be placed directly into the mixer without measuring. The other type is premixed and replaces the masonry cement. Dark green color has the highest cost.

R040519-50 Masonry Reinforcing

Horizontal joint reinforcing helps prevent wall cracks where wall movement may occur and in many locations is required by code. Horizontal joint reinforcing is generally not considered to be structural reinforcing and an unreinforced wall may still contain joint reinforcing.

Reinforcing strips come in 10' and 12' lengths and in truss and ladder shapes, with and without drips. Field labor runs between 2.7 to 5.3 hours per 1000 L.F. for wall thicknesses up to 12".

The wire meets ASTM A82 for cold drawn steel wire and the typical size is 9 ga. sides and ties with 3/16" diameter also available. Typical finish is mill galvanized with zinc coating at .10 oz. per S.F. Class I (.40 oz. per S.F.) and Class III (.80. oz. per S.F.) are also available, as is hot dipped galvanizing at 1.50 oz. per S.F.

R042110-10   Economy in Bricklaying

Have adequate supervision. Be sure bricklayers are always supplied with materials so there is no waiting. Place experienced bricklayers at corners and openings.

Use only screened sand for mortar. Otherwise, labor time will be wasted picking out pebbles. Use seamless metal tubs for mortar as they do not leak or catch the trowel. Locate stack and mortar for easy wheeling.

Have brick delivered for stacking. This makes for faster handling, reduces chipping and breakage, and requires less storage space. Many dealers will deliver select common in 2' x 3' x 4' pallets or face brick packaged. This affords quick handling with a crane or forklift and easy tonging in units of ten, which reduces waste.

Use wider bricks for one wythe wall construction. Keep scaffolding away from the wall to allow mortar to fall clear and not stain the wall.

On large jobs develop specialized crews for each type of masonry unit.

Consider designing for prefabricated panel construction on high rise projects.

Avoid excessive corners or openings. Each opening adds about 50% to the labor cost for area of opening.

Bolting stone panels and using window frames as stops reduce labor costs and speed up erection.

R042110-20   Common and Face Brick

Common building brick manufactured according to ASTM C62 and facing brick manufactured according to ASTM C216 are the two standard bricks available for general building use.

Building brick is made in three grades: SW, where high resistance to damage caused by cyclic freezing is required; MW, where moderate resistance to cyclic freezing is needed; and NW, where little resistance to cyclic freezing is needed. Facing brick is made in only the two grades SW and MW. Additionally, facing brick is available in three types: FBS, for general use; FBX, for general use where a higher degree of precision and lower permissible variation in size than FBS are needed; and FBA, for general use to produce characteristic architectural effects resulting from non-uniformity in size and texture of the units.

In figuring the material cost of brickwork, an allowance of 25% mortar waste and 3% brick breakage was included. If bricks are delivered palletized

with 280 to 300 per pallet, or packaged, allow only 1-1/2% for breakage. Packaged or palletized delivery is practical when a job is big enough to have a crane or other equipment available to handle a package of brick. This is so on all industrial work but not always true on small commercial buildings.

The use of buff and gray face is increasing, and there is a continuing trend to the Norman, Roman, Jumbo and SCR brick.

Common red clay brick for backup is not used that often. Concrete block is the most usual backup material with occasional use of sand lime or cement brick. Building brick is commonly used in solid walls for strength and as a fire stop.

Brick panels built on the ground and then crane erected to the upper floors have proven to be economical. This allows the work to be done under cover and without scaffolding.

R042110-50   Brick, Block & Mortar Quantities

Running Bond							For Other Bonds Standard Size Add to S.F. Quantities in Table to Left		
Number of Brick per S.F. of Wall - Single Wythe with 3/8" Joints					C.F. of Mortar per M Bricks, Waste Included				
Type Brick	Nominal Size (incl. mortar) L H W		Modular Coursing	Number of Brick per S.F.	3/8" Joint	1/2" Joint	Bond Type	Description	Factor
Standard	8 x 2-2/3 x 4		3C=8"	6.75	8.1	10.3	Common	full header every fifth course	+20%
Economy	8 x 4 x 4		1C=4"	4.50	9.1	11.6		full header every sixth course	+16.7%
Engineer	8 x 3-1/5 x 4		5C=16"	5.63	8.5	10.8	English	full header every second course	+50%
Fire	9 x 2-1/2 x 4-1/2		2C=5"	6.40	550 # Fireclay	—	Flemish	alternate headers every course	+33.3%
Jumbo	12 x 4 x 6 or 8		1C=4"	3.00	22.5	29.2		every sixth course	+5.6%
Norman	12 x 2-2/3 x 4		3C=8"	4.50	11.2	14.3	Header = W x H exposed		+100%
Norwegian	12 x 3-1/5 x 4		5C=16"	3.75	11.7	14.9	Rowlock = H x W exposed		+100%
Roman	12 x 2 x 4		2C=4"	6.00	10.7	13.7	Rowlock stretcher = L x W exposed		+33.3%
SCR	12 x 2-2/3 x 6		3C=8"	4.50	21.8	28.0	Soldier = H x L exposed		—
Utility	12 x 4 x 4		1C=4"	3.00	12.3	15.7	Sailor = W x L exposed		-33.3%

Concrete Blocks Nominal Size		Approximate Weight per S.F.		Blocks per 100 S.F.	Mortar per M block, waste included	
		Standard	Lightweight		Partitions	Back up
2"	x 8" x 16"	20 PSF	15 PSF	113	27 C.F.	36 C.F.
4"		30	20		41	51
6"		42	30		56	66
8"		55	38		72	82
10"		70	47		87	97
12"		85	55		102	112

Brick & Mortar Quantities
©Brick Industry Association. 2009 Feb. Technical Notes on
Brick Construction 10:
     Dimensioning and Estimating Brick Masonry. Reston (VA): BIA. Table 1
     Modular Brick Sizes and Table 4 Quantity Estimates for Brick Masonry.

R042210-20   Concrete Block

The material cost of special block such as corner, jamb and head block can be figured at the same price as ordinary block of equal size. Labor on specials is about the same as equal-sized regular block.

Bond beams and 16″ high lintel blocks are more expensive than regular units of equal size. Lintel blocks are 8″ long and either 8″ or 16″ high.

Use of a motorized mortar spreader box will speed construction of continuous walls.

Hollow non-load-bearing units are made according to ASTM C129 and hollow load-bearing units according to ASTM C90.

R050516-30 Coating Structural Steel

On field-welded jobs, the shop-applied primer coat is necessarily omitted. All painting must be done in the field and usually consists of red oxide rust inhibitive paint or an aluminum paint. The table below shows paint coverage and daily production for field painting.

See Division 09 97 13.23 for hot-dipped galvanizing and for field-applied cold galvanizing and other paints and protective coatings.

See Division 05 01 10.51 for steel surface preparation treatments such as wire brushing, pressure washing and sand blasting.

Type Construction	Surface Area per Ton	Coat	One Gallon Covers		In 8 Hrs. Person Covers		Average per Ton Spray	
			Brush	Spray	Brush	Spray	Gallons	Labor-hours
Light Structural	300 S.F. to 500 S.F.	1st	500 S.F.	455 S.F.	640 S.F.	2000 S.F.	0.9 gals.	1.6 L.H.
		2nd	450	410	800	2400	1.0	1.3
		3rd	450	410	960	3200	1.0	1.0
Medium	150 S.F. to 300 S.F.	All	400	365	1600	3200	0.6	0.6
Heavy Structural	50 S.F. to 150 S.F.	1st	400	365	1920	4000	0.2	0.2
		2nd	400	365	2000	4000	0.2	0.2
		3rd	400	365	2000	4000	0.2	0.2
Weighted Average	225 S.F.	All	400	365	1350	3000	0.6	0.6

R050521-20 Welded Structural Steel

Usual weight reductions with welded design run 10% to 20% compared with bolted or riveted connections. This amounts to about the same total cost compared with bolted structures since field welding is more expensive than bolts. For normal spans of 18′ to 24′ figure 6 to 7 connections per ton.

Trusses — For welded trusses add 4% to weight of main members for connections. Up to 15% less steel can be expected in a welded truss compared to one that is shop bolted. Cost of erection is the same whether shop bolted or welded.

General — Typical electrodes for structural steel welding are E6010, E6011, E60T and E70T. Typical buildings vary between 2# to 8# of weld rod per ton of steel. Buildings utilizing continuous design require about three times as much welding as conventional welded structures. In estimating field erection by welding, it is best to use the average linear feet of weld per ton to arrive at the welding cost per ton. The type, size and position of the weld will have a direct bearing on the cost per linear foot. A typical field welder will deposit 1.8# to 2# of weld rod per hour manually. Using semiautomatic methods can increase production by as much as 50% to 75%.

R050523-10 High Strength Bolts

Common bolts (A307) are usually used in secondary connections (see Division 05 05 23.10).

High strength bolts (A325 and A490) are usually specified for primary connections such as column splices, beam and girder connections to columns, column bracing, connections for supports of operating equipment or of other live loads which produce impact or reversal of stress, and in structures carrying cranes of over 5-ton capacity.

Allow 20 field bolts per ton of steel for a 6 story office building, apartment house or light industrial building. For 6 to 12 stories allow 25 bolts per ton, and above 12 stories, 30 bolts per ton. On power stations, 20 to 25 bolts per ton are needed.

R051223-10 Structural Steel

The bare material prices for structural steel, shown in the unit cost sections of the data set, are for 100 tons of shop-fabricated structural steel and include:

1. Mill base price of structural steel
2. Mill scrap/grade/size/length extras
3. Mill delivery to a metals service center (warehouse)
4. Service center storage and handling
5. Service center delivery to a fabrication shop
6. Shop storage and handling
7. Shop drafting/detailing
8. Shop fabrication
9. Shop coat of primer paint
10. Shop listing
11. Shop delivery to the job site

In unit cost sections of the data set that contain items for field fabrication of steel components, the bare material cost of steel includes:

1. Mill base price of structural steel
2. Mill scrap/grade/size/length extras
3. Mill delivery to a metals service center (warehouse)
4. Service center storage and handling
5. Service center delivery to the job site

For customer support on your Site Work & Landscape Costs with RSMeans data, call 800.448.8182.

851

R051223-20   Steel Estimating Quantities

One estimate on erection is that a crane can handle 35 to 60 pieces per day. Say the average is 45. With usual sizes of beams, girders, and columns, this would amount to about 20 tons per day. The type of connection greatly affects the speed of erection. Moment connections for continuous design slow down production and increase erection costs.

Short open web bar joists can be set at the rate of 75 to 80 per day, with 50 per day being the average for setting long span joists.

After main members are calculated, add the following for usual allowances: base plates 2% to 3%; column splices 4% to 5%; and miscellaneous details 4% to 5%, for a total of 10% to 13% in addition to main members.

The ratio of column to beam tonnage varies depending on type of steels used, typical spans, story heights and live loads.

It is more economical to keep the column size constant and to vary the strength of the column by using high strength steels. This also saves floor space. Buildings have recently gone as high as ten stories with 8″ high strength columns. For light columns under W8X31 lb. sections, concrete filled steel columns are economical.

High strength steels may be used in columns and beams to save floor space and to meet head room requirements. High strength steels in some sizes sometimes require long lead times.

Round, square and rectangular columns, both plain and concrete filled, are readily available and save floor area, but are higher in cost per pound than rolled columns. For high unbraced columns, tube columns may be less expensive.

Below are average minimum figures for the weights of the structural steel frame for different types of buildings using A36 steel, rolled shapes and simple joints. For economy in domes, rise to span ratio = .13. Open web joist framing systems will reduce weights by 10% to 40%. Composite design can reduce steel weight by up to 25% but additional concrete floor slab thickness may be required. Continuous design can reduce the weights up to 20%. There are many building codes with different live load requirements and different structural requirements, such as hurricane and earthquake loadings, which can alter the figures.

Structural Steel Weights per S.F. of Floor Area									
Type of Building	No. of Stories	Avg. Spans	L.L. #/S.F.	Lbs. Per S.F.	Type of Building	No. of Stories	Avg. Spans	L.L. #/S.F.	Lbs. Per S.F.
Steel Frame Mfg.	1	20′x20′ 30′x30′ 40′x40′	40	8 13 18	Apartments	2-8 9-25	20′x20′	40	8 14
Parking garage	4	Various	80	8.5	Office	to 10 20 30 over 50	Various	80	10 18 26 35
Domes (Schwedler)*	1	200′ 300′	30	10 15					

R051223-25   Common Structural Steel Specifications

ASTM A992 (formerly A36, then A572 Grade 50) is the all-purpose carbon grade steel widely used in building and bridge construction.

The other high-strength steels listed below may each have certain advantages over ASTM A992 structural carbon steel, depending on the application. They have proven to be economical choices where, due to lighter members, the reduction of dead load and the associated savings in shipping cost can be significant.

ASTM A588 atmospheric weathering, high-strength, low-alloy steels can be used in the bare (uncoated) condition, where exposure to normal atmosphere causes a tightly adherant oxide to form on the surface, protecting the steel from further oxidation. ASTM A242 corrosion-resistant, high-strength, low-alloy steels have enhanced atmospheric corrosion resistance of at least two times that of carbon structural steels with copper, or four times that of carbon structural steels without copper. The reduction or elimination of maintenance resulting from the use of these steels often offsets their higher initial cost.

Steel Type	ASTM Designation	Minimum Yield Stress in KSI	Shapes Available
Carbon	A36	36	All structural shape groups, and plates & bars up through 8" thick
	A529	50	Structural shape group 1, and plates & bars up through 2" thick
High-Strength Low-Alloy Quenched & Self-Tempered	A913	50	All structural shape groups
		60	
		65	
		70	
High-Strength Low-Alloy Columbium-Vanadium	A572	42	All structural shape groups, and plates & bars up through 6" thick
		50	All structural shape groups, and plates & bars up through 4" thick
		55	Structural shape groups 1 & 2, and plates & bars up through 2" thick
		60	Structural shape groups 1 & 2, and plates & bars up through 1-1/4" thick
		65	Structural shape group 1, and plates & bars up through 1-1/4" thick
High-Strength Low-Alloy Columbium-Vanadium	A992	50	All structural shape groups
Weathering High-Strength Low-Alloy	A242	42	Structural shape groups 4 & 5, and plates & bars over 1-1/2" up through 4" thick
		46	Structural shape group 3, and plates & bars over 3/4" up through 1-1/2" thick
		50	Structural shape groups 1 & 2, and plates & bars up through 3/4" thick
Weathering High-Strength Low-Alloy	A588	42	Plates & bars over 5" up through 8" thick
		46	Plates & bars over 4" up through 5" thick
		50	All structural shape groups, and plates & bars up through 4" thick
Quenched and Tempered Low-Alloy	A852	70	Plates & bars up through 4" thick
Quenched and Tempered Alloy	A514	90	Plates & bars over 2-1/2" up through 6" thick
		100	Plates & bars up through 2-1/2" thick

R051223-30   High Strength Steels

The mill price of high strength steels may be higher than A992 carbon steel, but their proper use can achieve overall savings through total reduced weights. For columns with L/r over 100, A992 steel is best; under 100, high strength steels are economical. For heavy columns, high strength steels are economical when cover plates are eliminated. There is no economy using high strength steels for clip angles or supports or for beams where deflection governs. Thinner members are more economical than thick.

The per ton erection and fabricating costs of the high strength steels will be higher than for A992 since the same number of pieces, but less weight, will be installed.

R051223-35 Common Steel Sections

The upper portion of this table shows the name, shape, common designation and basic characteristics of commonly used steel sections. The lower portion explains how to read the designations used for the above illustrated common sections.

Shape & Designation	Name & Characteristics	Shape & Designation	Name & Characteristics
W	W Shape — Parallel flange surfaces	MC	Miscellaneous Channel — Infrequently rolled by some producers
S	American Standard Beam (I Beam) — Sloped inner flange	L	Angle — Equal or unequal legs, constant thickness
M	Miscellaneous Beams — Cannot be classified as W, HP or S; infrequently rolled by some producers	T	Structural Tee — Cut from W, M or S on center of web
C	American Standard Channel — Sloped inner flange	HP	Bearing Pile — Parallel flanges and equal flange and web thickness

Common drawing designations follow:

W Shape
- W 18 x 35
  - Weight in Pounds per Foot
  - Nominal Depth in Inches (Actual 17-3/4")

American Standard Beam
- S 12 x 31.8
  - Weight in Pounds per Foot
  - Depth in Inches

Miscellaneous Beam
- M 8 x 6.5
  - Weight in Pounds per Foot
  - Depth in Inches

American Standard Channel
- C 8 x 11.5
  - Weight in Pounds per Foot
  - Depth in Inches

Miscellaneous Channel
- MC 8 x 22.8
  - Weight in Pounds per Foot
  - Depth in Inches

Angle
- L 6 x 3-1/2 x 3/8
  - Length of Long Leg in Inches
  - Thickness of Each Leg in Inches
  - Length of Other Leg in Inches

Tee Cut from W16 x 100
- WT 8 x 50
  - Weight in Pounds per Foot
  - Nominal Depth in Inches (Actual 8-1/2")

Tee Cut from S12 x 35
- ST 6 x 17.5
  - Weight in Pounds per Foot
  - Depth in Inches (Actual 6-1/4")

Tee Cut from M10 x 9
- MT 5 x 4.5
  - Weight in Pounds per Foot
  - Depth in Inches

Bearing Pile
- HP 12 x 84
  - Weight in Pounds per Foot
  - Nominal Depth in Inches (Actual 12-1/4")

R051223-45   Installation Time for Structural Steel Building Components

The following tables show the expected average installation times for various structural steel shapes. Table A presents installation times for columns, Table B for beams, Table C for light framing and bolts, and Table D for structural steel for various project types.

Table A

Description	Labor-Hours	Unit
Columns		
Steel, Concrete Filled		
3-1/2" Diameter	.933	Ea.
6-5/8" Diameter	1.120	Ea.
Steel Pipe		
3" Diameter	.933	Ea.
8" Diameter	1.120	Ea.
12" Diameter	1.244	Ea.
Structural Tubing		
4" x 4"	.966	Ea.
8" x 8"	1.120	Ea.
12" x 8"	1.167	Ea.
W Shape 2 Tier		
W8 x 31	.052	L.F.
W8 x 67	.057	L.F.
W10 x 45	.054	L.F.
W10 x 112	.058	L.F.
W12 x 50	.054	L.F.
W12 x 190	.061	L.F.
W14 x 74	.057	L.F.
W14 x 176	.061	L.F.

Table B

Description	Labor-Hours	Unit	Labor-Hours	Unit
Beams, W Shape				
W6 x 9	.949	Ea.	.093	L.F.
W10 x 22	1.037	Ea.	.085	L.F.
W12 x 26	1.037	Ea.	.064	L.F.
W14 x 34	1.333	Ea.	.069	L.F.
W16 x 31	1.333	Ea.	.062	L.F.
W18 x 50	2.162	Ea.	.088	L.F.
W21 x 62	2.222	Ea.	.077	L.F.
W24 x 76	2.353	Ea.	.072	L.F.
W27 x 94	2.581	Ea.	.067	L.F.
W30 x 108	2.857	Ea.	.067	L.F.
W33 x 130	3.200	Ea.	.071	L.F.
W36 x 300	3.810	Ea.	.077	L.F.

Table C

Description	Labor-Hours	Unit
Light Framing		
Angles 4" and Larger	.055	lbs.
Less than 4"	.091	lbs.
Channels 8" and Larger	.048	lbs.
Less than 8"	.072	lbs.
Cross Bracing Angles	.055	lbs.
Rods	.034	lbs.
Hanging Lintels	.069	lbs.
High Strength Bolts in Place		
3/4" Bolts	.070	Ea.
7/8" Bolts	.076	Ea.

Table D

Description	Labor-Hours	Unit	Labor-Hours	Unit
Apartments, Nursing Homes, etc.				
1-2 Stories	4.211	Piece	7.767	Ton
3-6 Stories	4.444	Piece	7.921	Ton
7-15 Stories	4.923	Piece	9.014	Ton
Over 15 Stories	5.333	Piece	9.209	Ton
Offices, Hospitals, etc.				
1-2 Stories	4.211	Piece	7.767	Ton
3-6 Stories	4.741	Piece	8.889	Ton
7-15 Stories	4.923	Piece	9.014	Ton
Over 15 Stories	5.120	Piece	9.209	Ton
Industrial Buildings				
1 Story	3.478	Piece	6.202	Ton

R051223-80   Dimensions and Weights of Sheet Steel

Gauge No.	Approximate Thickness				Weight		
	Inches (in fractions)	Inches (in decimal parts)		Millimeters	per S.F. in Ounces	per S.F. in Lbs.	per Square Meter in Kg.
	Wrought Iron	Wrought Iron	Steel	Steel			
0000000	1/2"	.5	.4782	12.146	320	20.000	97.650
000000	15/32"	.46875	.4484	11.389	300	18.750	91.550
00000	7/16"	.4375	.4185	10.630	280	17.500	85.440
0000	13/32"	.40625	.3886	9.870	260	16.250	79.330
000	3/8"	.375	.3587	9.111	240	15.000	73.240
00	11/32"	.34375	.3288	8.352	220	13.750	67.130
0	5/16"	.3125	.2989	7.592	200	12.500	61.030
1	9/32"	.28125	.2690	6.833	180	11.250	54.930
2	17/64"	.265625	.2541	6.454	170	10.625	51.880
3	1/4"	.25	.2391	6.073	160	10.000	48.820
4	15/64"	.234375	.2242	5.695	150	9.375	45.770
5	7/32"	.21875	.2092	5.314	140	8.750	42.720
6	13/64"	.203125	.1943	4.935	130	8.125	39.670
7	3/16"	.1875	.1793	4.554	120	7.500	36.320
8	11/64"	.171875	.1644	4.176	110	6.875	33.570
9	5/32"	.15625	.1495	3.797	100	6.250	30.520
10	9/64"	.140625	.1345	3.416	90	5.625	27.460
11	1/8"	.125	.1196	3.038	80	5.000	24.410
12	7/64"	.109375	.1046	2.657	70	4.375	21.360
13	3/32"	.09375	.0897	2.278	60	3.750	18.310
14	5/64"	.078125	.0747	1.897	50	3.125	15.260
15	9/128"	.0713125	.0673	1.709	45	2.813	13.730
16	1/16"	.0625	.0598	1.519	40	2.500	12.210
17	9/160"	.05625	.0538	1.367	36	2.250	10.990
18	1/20"	.05	.0478	1.214	32	2.000	9.765
19	7/160"	.04375	.0418	1.062	28	1.750	8.544
20	3/80"	.0375	.0359	.912	24	1.500	7.324
21	11/320"	.034375	.0329	.836	22	1.375	6.713
22	1/32"	.03125	.0299	.759	20	1.250	6.103
23	9/320"	.028125	.0269	.683	18	1.125	5.490
24	1/40"	.025	.0239	.607	16	1.000	4.882
25	7/320"	.021875	.0209	.531	14	.875	4.272
26	3/160"	.01875	.0179	.455	12	.750	3.662
27	11/640"	.0171875	.0164	.417	11	.688	3.357
28	1/64"	.015625	.0149	.378	10	.625	3.052

R053100-10   Decking Descriptions

General - All Deck Products

A steel deck is made by cold forming structural grade sheet steel into a repeating pattern of parallel ribs. The strength and stiffness of the panels are the result of the ribs and the material properties of the steel. Deck lengths can be varied to suit job conditions, but because of shipping considerations, are usually less than 40 feet. Standard deck width varies with the product used but full sheets are usually 12", 18", 24", 30", or 36". The deck is typically furnished in a standard width with the ends cut square. Any cutting for width, such as at openings or for angular fit, is done at the job site.

The deck is typically attached to the building frame with arc puddle welds, self-drilling screws, or powder or pneumatically driven pins. Sheet to sheet fastening is done with screws, button punching (crimping), or welds.

Composite Floor Deck

After installation and adequate fastening, a floor deck serves several purposes. It (a) acts as a working platform, (b) stabilizes the frame, (c) serves as a concrete form for the slab, and (d) reinforces the slab to carry the design loads applied during the life of the building. Composite decks are distinguished by the presence of shear connector devices as part of the deck. These devices are designed to mechanically lock the concrete and deck together so that the concrete and the deck work together to carry subsequent floor loads. These shear connector devices can be rolled-in embossments, lugs, holes, or wires welded to the panels. The deck profile can also be used to interlock concrete and steel.

Composite deck finishes are either galvanized (zinc coated) or phosphatized/painted. Galvanized deck has a zinc coating on both the top and bottom surfaces. The phosphatized/painted deck has a bare (phosphatized) top surface that will come into contact with the concrete. This bare top surface can be expected to develop rust before the concrete is placed. The bottom side of the deck has a primer coat of paint.

A composite floor deck is normally installed so the panel ends do not overlap on the supporting beams. Shear lugs or panel profile shapes often prevent a tight metal to metal fit if the panel ends overlap; the air gap caused by overlapping will prevent proper fusion with the structural steel supports when the panel end laps are shear stud welded.

Adequate end bearing of the deck must be obtained as shown on the drawings. If bearing is actually less in the field than shown on the drawings, further investigation is required.

Roof Deck

A roof deck is not designed to act compositely with other materials. A roof deck acts alone in transferring horizontal and vertical loads into the building frame. Roof deck rib openings are usually narrower than floor deck rib openings. This provides adequate support of the rigid thermal insulation board.

A roof deck is typically installed to endlap approximately 2" over supports. However, it can be butted (or lapped more than 2") to solve field fit problems. Since designers frequently use the installed deck system as part of the horizontal bracing system (the deck as a diaphragm), any fastening substitution or change should be approved by the designer. Continuous perimeter support of the deck is necessary to limit edge deflection in the finished roof and may be required for diaphragm shear transfer.

Standard roof deck finishes are galvanized or primer painted. The standard factory applied paint for roof decks is a primer paint and is not intended to weather for extended periods of time. Field painting or touching up of abrasions and deterioration of the primer coat or other protective finishes is the responsibility of the contractor.

Cellular Deck

A cellular deck is made by attaching a bottom steel sheet to a roof deck or composite floor deck panel. A cellular deck can be used in the same manner as a floor deck. Electrical, telephone, and data wires are easily run through the chase created between the deck panel and the bottom sheet.

When used as part of the electrical distribution system, the cellular deck must be installed so that the ribs line up and create a smooth cell transition at abutting ends. The joint that occurs at butting cell ends must be taped or otherwise sealed to prevent wet concrete from seeping into the cell. Cell interiors must be free of welding burrs, or other sharp intrusions, to prevent damage to wires.

When used as a roof deck, the bottom flat plate is usually left exposed to view. Care must be maintained during erection to keep good alignment and prevent damage.

A cellular deck is sometimes used with the flat plate on the top side to provide a flat working surface. Installation of the deck for this purpose requires special methods for attachment to the frame because the flat plate, now on the top, can prevent direct access to the deck material that is bearing on the structural steel. It may be advisable to treat the flat top surface to prevent slipping.

A cellular deck is always furnished galvanized or painted over galvanized.

Form Deck

A form deck can be any floor or roof deck product used as a concrete form. Connections to the frame are by the same methods used to anchor floor and roof decks. Welding washers are recommended when welding a deck that is less than 20 gauge thickness.

A form deck is furnished galvanized, prime painted, or uncoated. A galvanized deck must be used for those roof deck systems where a form deck is used to carry a lightweight insulating concrete fill.

Wood, Plastics & Comp. | R0611 Wood Framing

R061110-30 Lumber Product Material Prices

The price of forest products fluctuates widely from location to location and from season to season depending upon economic conditions. The bare material prices in the unit cost sections of the data set show the National Average material prices in effect Jan. 1 of this data year. It must be noted that lumber prices in general may change significantly during the year.

Availability of certain items depends upon geographic location and must be checked prior to firm-price bidding.

Wood, Plastics & Comp. | R0616 Sheathing

R061636-20 Plywood

There are two types of plywood used in construction: interior, which is moisture-resistant but not waterproofed, and exterior, which is waterproofed.

The grade of the exterior surface of the plywood sheets is designated by the first letter: A, for smooth surface with patches allowed; B, for solid surface with patches and plugs allowed; C, which may be surface plugged or may have knot holes up to 1″ wide; and D, which is used only for interior type plywood and may have knot holes up to 2-1/2″ wide. "Structural Grade" is specifically designed for engineered applications such as box beams. All CC & DD grades have roof and floor spans marked on them.

Underlayment-grade plywood runs from 1/4″ to 1-1/4″ thick. Thicknesses 5/8″ and over have optional tongue and groove joints which eliminate the need for blocking the edges. Underlayment 19/32″ and over may be referred to as Sturd-i-Floor.

The price of plywood can fluctuate widely due to geographic and economic conditions.

Typical uses for various plywood grades are as follows:

AA-AD Interior — cupboards, shelving, paneling, furniture

BB Plyform — concrete form plywood

CDX — wall and roof sheathing

Structural — box beams, girders, stressed skin panels

AA-AC Exterior — fences, signs, siding, soffits, etc.

Underlayment — base for resilient floor coverings

Overlaid HDO — high density for concrete forms & highway signs

Overlaid MDO — medium density for painting, siding, soffits & signs

303 Siding — exterior siding, textured, striated, embossed, etc.

Openings | R0813 Metal Doors

R081313-20 Steel Door Selection Guide

Standard steel doors are classified into four levels, as recommended by the Steel Door Institute in the chart below. Each of the four levels offers a range of construction models and designs to meet architectural requirements for preference and appearance, including full flush, seamless, and stile & rail. Recommended minimum gauge requirements are also included.

For complete standard steel door construction specifications and available sizes, refer to the Steel Door Institute Technical Data Series, ANSI A250.8-98 (SDI-100), and ANSI A250.4-94 Test Procedure and Acceptance Criteria for Physical Endurance of Steel Door and Hardware Reinforcements.

Level		Model	Construction	For Full Flush or Seamless		
				Min. Gauge	Thickness (in)	Thickness (mm)
I	Standard Duty	1	Full Flush			
		2	Seamless	20	0.032	0.8
II	Heavy Duty	1	Full Flush			
		2	Seamless	18	0.042	1.0
III	Extra Heavy Duty	1	Full Flush			
		2	Seamless			
		3	*Stile & Rail	16	0.053	1.3
IV	Maximum Duty	1	Full Flush			
		2	Seamless	14	0.067	1.6

*Stiles & rails are 16 gauge; flush panels, when specified, are 18 gauge

For customer support on your Site Work & Landscape Costs with RSMeans data, call 800.448.8182.

R092910-10   Levels of Gypsum Board Finish

In the past, contract documents often used phrases such as "industry standard" and "workmanlike finish" to specify the expected quality of gypsum board wall and ceiling installations. The vagueness of these descriptions led to unacceptable work and disputes.

In order to resolve this problem, four major trade associations concerned with the manufacture, erection, finish, and decoration of gypsum board wall and ceiling systems developed an industry-wide *Recommended Levels of Gypsum Board Finish.*

The finish of gypsum board walls and ceilings for specific final decoration is dependent on a number of factors. A primary consideration is the location of the surface and the degree of decorative treatment desired. Painted and unpainted surfaces in warehouses and other areas where appearance is normally not critical may simply require the taping of wallboard joints and 'spotting' of fastener heads. Blemish-free, smooth, monolithic surfaces often intended for painted and decorated walls and ceilings in habitated structures, ranging from single-family dwellings through monumental buildings, require additional finishing prior to the application of the final decoration.

Other factors to be considered in determining the level of finish of the gypsum board surface are (1) the type of angle of surface illumination (both natural and artificial lighting), and (2) the paint and method of application or the type and finish of wallcovering specified as the final decoration. Critical lighting conditions, gloss paints, and thin wall coverings require a higher level of gypsum board finish than heavily textured surfaces which are subsequently painted or surfaces which are to be decorated with heavy grade wall coverings.

The following descriptions were developed by the Association of the Wall and Ceiling Industries-International (AWCI), Ceiling & Interior Systems Construction Association (CISCA), Gypsum Association (GA), and Painting and Decorating Contractors of America (PDCA) as a guide.

Level 0: Used in temporary construction or wherever the final decoration has not been determined. Unfinished. No taping, finishing or corner beads are required. Also could be used where non-predecorated panels will be used in demountable-type partitions that are to be painted as a final finish.

Level 1: Frequently used in plenum areas above ceilings, in attics, in areas where the assembly would generally be concealed, or in building service corridors and other areas not normally open to public view. Some degree of sound and smoke control is provided; in some geographic areas, this level is referred to as "fire-taping," although this level of finish does not typically meet fire-resistant assembly requirements. Where a fire resistance rating is required for the gypsum board assembly, details of construction should be in accordance with reports of fire tests of assemblies that have met the requirements of the fire rating acceptable.

All joints and interior angles shall have tape embedded in joint compound. Accessories are optional at specifier discretion in corridors and other areas with pedestrian traffic. Tape and fastener heads need not be covered with joint compound. Surface shall be free of excess joint compound. Tool marks and ridges are acceptable.

Level 2: It may be specified for standard gypsum board surfaces in garages, warehouse storage, or other similar areas where surface appearance is not of primary importance.

All joints and interior angles shall have tape embedded in joint compound and shall be immediately wiped with a joint knife or trowel, leaving a thin coating of joint compound over all joints and interior angles. Fastener heads and accessories shall be covered with a coat of joint compound. Surface shall be free of excess joint compound. Tool marks and ridges are acceptable.

Level 3: Typically used in areas receiving heavy texture (spray or hand applied) finishes before final painting, or where commercial-grade (heavy duty) wall coverings are to be applied as the final decoration. This level of finish should not be used where smooth painted surfaces or where lighter weight wall coverings are specified. The prepared surface shall be coated with a drywall primer prior to the application of final finishes.

All joints and interior angles shall have tape embedded in joint compound and shall be immediately wiped with a joint knife or trowel, leaving a thin coating of joint compound over all joints and interior angles. One additional coat of joint compound shall be applied over all joints and interior angles. Fastener heads and accessories shall be covered with two separate coats of joint compound. All joint compounds shall be smooth and free of tool marks and ridges. The prepared surface shall be covered with a drywall primer prior to the application of the final decoration.

Level 4: This level should be used where residential grade (light duty) wall coverings, flat paints, or light textures are to be applied. The prepared surface shall be coated with a drywall primer prior to the application of final finishes. Release agents for wall coverings are specifically formulated to minimize damage if coverings are subsequently removed.

The weight, texture, and sheen level of the wall covering material selected should be taken into consideration when specifying wall coverings over this level of drywall treatment. Joints and fasteners must be sufficiently concealed if the wall covering material is lightweight, contains limited pattern, has a glossy finish, or has any combination of these features. In critical lighting areas, flat paints applied over light textures tend to reduce joint photographing. Gloss, semi-gloss, and enamel paints are not recommended over this level of finish.

All joints and interior angles shall have tape embedded in joint compound and shall be immediately wiped with a joint knife or trowel, leaving a thin coating of joint compound over all joints and interior angles. In addition, two separate coats of joint compound shall be applied over all flat joints and one separate coat of joint compound applied over interior angles. Fastener heads and accessories shall be covered with three separate coats of joint compound. All joint compounds shall be smooth and free of tool marks and ridges. The prepared surface shall be covered with a drywall primer like Sheetrock® First Coat prior to the application of the final decoration.

Level 5: The highest quality finish is the most effective method to provide a uniform surface and minimize the possibility of joint photographing and of fasteners showing through the final decoration. This level of finish is required where gloss, semi-gloss, or enamel is specified; when flat joints are specified over an untextured surface; or where critical lighting conditions occur. The prepared surface shall be coated with a drywall primer prior to the application of the final decoration.

All joints and interior angles shall have tape embedded in joint compound and be immediately wiped with a joint knife or trowel, leaving a thin coating of joint compound over all joints and interior angles. Two separate coats of joint compound shall be applied over all flat joints and one separate coat of joint compound applied over interior angles. Fastener heads and accessories shall be covered with three separate coats of joint compound.

A thin skim coat of joint compound shall be trowel applied to the entire surface. Excess compound is immediately troweled off, leaving a film or skim coating of compound completely covering the paper. As an alternative to a skim coat, a material manufactured especially for this purpose may be applied such as Sheetrock® Tuff-Hide primer surfacer. The surface must be smooth and free of tool marks and ridges. The prepared surface shall be covered with a drywall primer prior to the application of the final decoration.

R131113-20   Swimming Pools

Pool prices given per square foot of surface area include pool structure, filter and chlorination equipment, pumps, related piping, ladders/steps, maintenance kit, skimmer and vacuum system. Decks and electrical service to equipment are not included.

Residential in-ground pool construction can be divided into two categories: vinyl lined and gunite. Vinyl lined pool walls are constructed of different materials including wood, concrete, plastic or metal. The bottom is often graded with sand over which the vinyl liner is installed. Vermiculite or soil cement bottoms may be substituted for an added cost.

Gunite pool construction is used both in residential and municipal installations. These structures are steel reinforced for strength and finished with a white cement limestone plaster.

Municipal pools will have a higher cost because plumbing codes require more expensive materials, chlorination equipment and higher filtration rates.

Municipal pools greater than 1,800 S.F. require gutter systems to control waves. This gutter may be formed into the concrete wall. Often a vinyl/stainless steel gutter or gutter/wall system is specified, which will raise the pool cost.

Competition pools usually require tile bottoms and sides with contrasting lane striping, which will also raise the pool cost.

R133113-10   Air Supported Structures

Air supported structures are made from fabrics that can be classified into two groups: temporary and permanent. Temporary fabrics include nylon, woven polyethylene, vinyl film, and vinyl coated dacron. These have lifespans that range from five to fifteen plus years. The cost per square foot includes a fabric shell, tension cables, primary and back-up inflation systems and doors. The lower cost structures are used for construction shelters, bulk storage and pond covers. The more expensive are used for recreational structures and warehouses.

Permanent fabrics are teflon coated fiberglass. The life of this structure is twenty plus years. The high cost limits its application to architectural designed structures which call for a clear span covered area, such as stadiums and convention centers. Both temporary and permanent structures are available in translucent fabrics which eliminate the need for daytime lighting.

Areas to be covered vary from 10,000 S.F. to any area up to 1000 feet wide by any length. Height restrictions range from a maximum of 1/2 of the

width to a minimum of 1/6 of the width. Erection of even the largest of the temporary structures requires no more than a week.

Centrifugal fans provide the inflation necessary to support the structure during application of live loads. Airlocks are usually used at large entrances to prevent loss of static pressure. Some manufacturers employ propeller fans which generate sufficient airflow (30,000 CFM) to eliminate the need for airlocks. These fans may also be automatically controlled to resist high wind conditions, regulate humidity (air changes), and provide cooling and heat.

Insulation can be provided with the addition of a second or even third interior liner, creating a dead air space with an "R" value of four to nine. Some structures allow for the liner to be collapsed into the outer shell to enable the internal heat to melt accumulated snow. For cooling or air conditioning, the exterior face of the liner can be aluminized to reflect the sun's heat.

R133419-10   Pre-Engineered Steel Buildings

These buildings are manufactured by many companies and normally erected by franchised dealers throughout the U.S. The four basic types are: Rigid Frames, Truss type, Post and Beam and the Sloped Beam type. The most popular roof slope is a low pitch of 1" in 12". The minimum economical area of these buildings is about 3000 S.F. of floor area. Bay sizes are usually 20' to 24' but can go as high as 30' with heavier girts and purlins. Eave heights are usually 12' to 24' with 18' to 20' most typical.

Material prices shown in the Unit Price section are bare costs for the building shell only and do not include floors, foundations, anchor bolts, interior finishes or utilities. Costs assume at least three bays of 24' each, a 1" in 12" roof slope, and they are based on a 30 psf roof load and a 20 psf wind

load and no unusual requirements. Wind load is a function of wind speed, building height, and terrain characteristics; this should be determined by a registered structural engineer. Costs include the structural frame, 26 ga. non-insulated colored corrugated or ribbed roofing and siding panels, fasteners, closures, trim and flashing but no allowance for insulation, doors, windows, skylights, gutters or downspouts. Very large projects would generally cost less for materials than the prices shown. For roof panel substitutions and wall panel substitutions, see appropriate Unit Price sections.

Conditions at the site, weather, shape and size of the building, and labor availability will affect the erection cost of the building.

R133423-30   Dome Structures

Steel — The four types are Lamella, Schwedler, Arch and Geodesic. For maximum economy, the rise should be about 15 to 20% of the diameter. Most common diameters are in the 200' to 300' range. Lamella domes weigh about 5 P.S.F. of floor area less than Schwedler domes. The Schwedler dome weight in lbs. per S.F. approaches .046 times the diameter. Domes below 125' in diameter weigh .07 times the diameter and the cost per ton of steel is higher. See R051223-20 for estimating weight.

Wood — Small domes are of sawn lumber. Larger ones are laminated. In larger sizes, triaxial and triangular cost about the same; radial domes cost more. Radial domes are economical in the 60' to 70' diameter range. The most economical range of all types is 80' to 200' diameters. Diameters can

run over 400'. All costs are quoted above the foundation. Prices include 2" decking and a tension tie ring in place.

Plywood — Stock prefab geodesic domes are available with diameters from 24' to 60'.

Fiberglass — Aluminum framed translucent sandwich panels with spans from 5' to 45' are commercially available.

Aluminum — Stressed skin aluminum panels form geodesic domes with spans ranging from 82' to 232'. An aluminum space truss, triangulated or nontriangulated, with aluminum or clear acrylic closure panels can be used for clear spans of 40' to 415'.

R220102-20   Labor Adjustment Factors

Labor Adjustment Factors are provided for Divisions 21, 22, and 23 to assist the mechanical estimator account for the various complexities and special conditions of any particular project. While a single percentage has been entered on each line of Division 22 01 02.20, it should be understood that these are just suggested midpoints of ranges of values commonly used by mechanical estimators. They may be increased or decreased depending on the severity of the special conditions.

The group for "existing occupied buildings" has been the subject of requests for explanation. Actually there are two stages to this group: buildings that are existing and "finished" but unoccupied, and those that also are occupied. Buildings that are "finished" may result in higher labor costs due to the

workers having to be more careful not to damage finished walls, ceilings, floors, etc. and may necessitate special protective coverings and barriers. Also corridor bends and doorways may not accommodate long pieces of pipe or larger pieces of equipment. Work above an already hung ceiling can be very time consuming. The addition of occupants may force the work to be done on premium time (nights and/or weekends), eliminate the possible use of some preferred tools such as pneumatic drivers, powder charged drivers, etc. The estimator should evaluate the access to the work area and just how the work is going to be accomplished to arrive at an increase in labor costs over "normal" new construction productivity.

R220523-80   Valve Materials

VALVE MATERIALS

Bronze:
Bronze is one of the oldest materials used to make valves. It is most commonly used in hot and cold water systems and other non-corrosive services. It is often used as a seating surface in larger iron body valves to ensure tight closure.

Carbon Steel:
Carbon steel is a high strength material. Therefore, valves made from this metal are used in higher pressure services, such as steam lines up to 600 psi at 850° F. Many steel valves are available with butt-weld ends for economy and are generally used in high pressure steam service as well as other higher pressure non-corrosive services.

Forged Steel:
Valves from tough carbon steel are used in service up to 2000 psi and temperatures up to 1000° F in gate, globe and check valves.

Iron:
Valves are normally used in medium to large pipe lines to control non-corrosive fluid and gases, where pressures do not exceed 250 psi at 450° F or 500 psi cold water, oil or gas.

Stainless Steel:
Developed steel alloys can be used in over 90% corrosive services.

Plastic PVC:
This is used in a great variety of valves generally in high corrosive services with lower temperatures and pressures.

VALVE SERVICE PRESSURES

Pressure ratings on valves provide an indication of the safe operating pressure for a valve at some elevated temperature. This temperature is dependent upon the materials used and the fabrication of the valve. When specific data is not available, a good "rule-of-thumb" to follow is the temperature of saturated steam on the primary rating indicated on the valve body. Example: The valve has the number 150S printed on the side indicating 150 psi and hence, a maximum operating temperature of 367° F (temperature of saturated steam and 150 psi).

DEFINITIONS

1. "WOG" – Water, oil, gas (cold working pressures).
2. "SWP" – Steam working pressure.
3. 100% area (full port) – means the area through the valve is equal to or greater than the area of standard pipe.
4. "Standard Opening" – means that the area through the valve is less than the area of standard pipe and therefore these valves should be used only where restriction of flow is unimportant.
5. "Round Port" – means the valve has a full round opening through the plug and body, of the same size and area as standard pipe.
6. "Rectangular Port" – valves have rectangular shaped ports through the plug body. The area of the port is either equal to 100% of the area of standard pipe, or restricted (standard opening). In either case it is clearly marked.
7. "ANSI" – American National Standards Institute.

R220523-90   Valve Selection Considerations

INTRODUCTION: In any piping application, valve performance is critical. Valves should be selected to give the best performance at the lowest cost.

The following is a list of performance characteristics generally expected of valves.

1. Stopping flow or starting it.
2. Throttling flow (Modulation).
3. Flow direction changing.
4. Checking backflow (Permitting flow in only one direction).
5. Relieving or regulating pressure.

In order to properly select the right valve, some facts must be determined.

A. What liquid or gas will flow through the valve?

B. Does the fluid contain suspended particles?

C. Does the fluid remain in liquid form at all times?

D. Which metals does fluid corrode?

E. What are the pressure and temperature limits? (As temperature and pressure rise, so will the price of the valve.)

F. Is there constant line pressure?

G. Is the valve merely an on-off valve?

H. Will checking of backflow be required?

I. Will the valve operate frequently or infrequently?

Valves are classified by design type into such classifications as gate, globe, angle, check, ball, butterfly and plug. They are also classified by end connection, stem, pressure restrictions and material such as bronze, cast iron, etc. Each valve has a specific use. A quality valve used correctly will provide a lifetime of trouble-free service, but a high quality valve installed in the wrong service may require frequent attention.

STEM TYPES
(O.S. & Y)—Rising Stem-Outside Screw and Yoke

Offers a visual indication of whether the valve is open or closed. Recommended where high temperatures, corrosives, and solids in the line might cause damage to inside-valve stem threads. The stem threads are engaged by the yoke bushing so the stem rises through the hand wheel as it is turned.

(R.S.)—Rising Stem-Inside Screw

Adequate clearance for operation must be provided because both the hand wheel and the stem rise.
The valve wedge position is indicated by the position of the stem and hand wheel.

(N.R.S.)—Non-Rising Stem-Inside Screw

A minimum clearance is required for operating this type of valve. Excessive wear or damage to stem threads inside the valve may be caused by heat, corrosion, and solids. Because the hand wheel and stem do not rise, wedge position cannot be visually determined.

VALVE TYPES
Gate Valves

Provide full flow, minute pressure drop, minimum turbulence and minimum fluid trapped in the line.
They are normally used where operation is infrequent.

Globe Valves

Globe valves are designed for throttling and/or frequent operation with positive shut-off. Particular attention must be paid to the several types of seating materials available to avoid unnecessary wear. The seats must be compatible with the fluid in service and may be composition or metal. The configuration of the globe valve opening causes turbulence which results in increased resistance. Most bronze globe valves are rising stem-inside screw, but they are also available on O.S. & Y.

Angle Valves

The fundamental difference between the angle valve and the globe valve is the fluid flow through the angle valve. It makes a 90° turn and offers less resistance to flow than the globe valve while replacing an elbow. An angle valve thus reduces the number of joints and installation time.

Check Valves

Check valves are designed to prevent backflow by automatically seating when the direction of fluid is reversed.
Swing check valves are generally installed with gate valves, as they provide comparable full flow. Usually recommended for lines where flow velocities are low and should not be used on lines with pulsating flow. Recommended for horizontal installation or in vertical lines only where flow is upward.

Lift Check Valves

These are commonly used with globe and angle valves since they have similar diaphragm seating arrangements and are recommended for preventing backflow of steam, air, gas and water, and on vapor lines with high flow velocities. For horizontal lines, horizontal lift checks should be used and vertical lift checks for vertical lines.

Ball Valves

Ball valves are light and easily installed, yet because of modern elastomeric seats, provide tight closure. Flow is controlled by rotating up to 90° a drilled ball which fits tightly against resilient seals. This ball seats with flow in either direction, and valve handle indicates the degree of opening. Recommended for frequent operation readily adaptable to automation, ideal for installation where space is limited.

Butterfly Valves

Butterfly valves provide bubble-tight closure with excellent throttling characteristics. They can be used for full-open, closed and for throttling applications.

The butterfly valve consists of a disc within the valve body which is controlled by a shaft. In its closed position, the valve disc seals against a resilient seat. The disc position throughout the full 90° rotation is visually indicated by the position of the operator.

A butterfly valve is only a fraction of the weight of a gate valve and requires no gaskets between flanges in most cases. Recommended for frequent operation and adaptable to automation where space is limited.

Wafer and lug type bodies, when installed between two pipe flanges, can be easily removed from the line. The pressure of the bolted flanges holds the valve in place.
Locating lugs makes installation easier.

Plug Valves

Lubricated plug valves, because of the wide range of services to which they are adapted, may be classified as all purpose valves. They can be safely used at all pressure and vacuums, and at all temperatures up to the limits of available lubricants. They are the most satisfactory valves for the handling of gritty suspensions and many other destructive, erosive, corrosive and chemical solutions.

For customer support on your Site Work & Landscape Costs with RSMeans data, call 800.448.8182.

863

R221113-50   Pipe Material Considerations

1. Malleable fittings should be used for gas service.
2. Malleable fittings are used where there are stresses/strains due to expansion and vibration.
3. Cast fittings may be broken as an aid to disassembling heating lines frozen by long use, temperature and minerals.
4. A cast iron pipe is extensively used for underground and submerged service.
5. Type M (light wall) copper tubing is available in hard temper only and is used for nonpressure and less severe applications than K and L.

6. Type L (medium wall) copper tubing, available hard or soft for interior service.
7. Type K (heavy wall) copper tubing, available in hard or soft temper for use where conditions are severe. For underground and interior service.
8. Hard drawn tubing requires fewer hangers or supports but should not be bent. Silver brazed fittings are recommended, but soft solder is normally used.
9. Type DMV (very light wall) copper tubing designed for drainage, waste and vent plus other non-critical pressure services.

Domestic/Imported Pipe and Fittings Costs

The prices shown in this publication for steel/cast iron pipe and steel, cast iron, and malleable iron fittings are based on domestic production sold at the normal trade discounts. The above listed items of foreign manufacture may be available at prices 1/3 to 1/2 of those shown. Some imported items after minor machining or finishing operations are being sold as domestic to further complicate the system.

Caution: Most pipe prices in this data set also include a coupling and pipe hangers which for the larger sizes can add significantly to the per foot cost and should be taken into account when comparing "book cost" with the quoted supplier's cost.

Fig. R238313-11

R238313-10   Heat Trace Systems

Before you can determine the cost of a HEAT TRACE installation the method of attachment must be established. There are four (4) common methods:

1. Cable is simply attached to the pipe with polyester tape every 12'.
2. Cable is attached with a continuous cover of 2" wide aluminum tape.
3. Cable is attached with factory extruded heat transfer cement and covered with metallic raceway with clips every 10'.
4. Cable is attached between layers of pipe insulation using either clips or polyester tape.

Example: Components for method 3 must include:
   A. Heat trace cable by voltage and watts per linear foot.
   B. Heat transfer cement, 1 gallon per 60 linear feet of cover.
   C. Metallic raceway by size and type.
   D. Raceway clips by size of pipe.

When taking off linear foot lengths of cable add the following for each valve in the system. (E)

In all of the above methods each component of the system must be priced individually.

SCREWED OR WELDED VALVE:			FLANGED VALVE:			BUTTERFLY VALVES:		
1/2"	=	6"	1/2"	=	1' -0"	1/2"	=	0'
3/4"	=	9"	3/4"	=	1' -6"	3/4"	=	0'
1"	=	1' -0"	1"	=	2' -0"	1"	=	1' -0"
1-1/2"	=	1' -6"	1-1/2"	=	2' -6"	1-1/2"	=	1' -6"
2"	=	2'	2"	=	2' -6"	2"	=	2' -0"
2-1/2"	=	2' -6"	2-1/2"	=	3' -0"	2-1/2"	=	2' -6"
3"	=	2' -6"	3"	=	3' -6"	3"	=	2' -6"
4"	=	4' -0"	4"	=	4' -0"	4"	=	3' -0"
6"	=	7' -0"	6"	=	8' -0"	6"	=	3' -6"
8"	=	9' -6"	8"	=	11' -0"	8"	=	4' -0"
10"	=	12' -6"	10"	=	14' -0"	10"	=	4' -0"
12"	=	15' -0"	12"	=	16' -6"	12"	=	5' -0"
14"	=	18' -0"	14"	=	19' -6"	14"	=	5' -6"
16"	=	21' -6"	16"	=	23' -0"	16"	=	6' -0"
18"	=	25' -6"	18"	=	27' -0"	18"	=	6' -6"
20"	=	28' -6"	20"	=	30' -0"	20"	=	7' -0"
24"	=	34' -0"	24"	=	36' -0"	24"	=	8' -0"
30"	=	40' -0"	30"	=	42' -0"	30"	=	10' -0"

R238313-10 Heat Trace Systems (cont.)

Add the following quantities of heat transfer cement to linear foot totals for each valve:

Nominal Valve Size	Gallons of Cement per Valve
1/2"	0.14
3/4"	0.21
1"	0.29
1-1/2"	0.36
2"	0.43
2-1/2"	0.70
3"	0.71
4"	1.00
6"	1.43
8"	1.48
10"	1.50
12"	1.60
14"	1.75
16"	2.00
18"	2.25
20"	2.50
24"	3.00
30"	3.75

The following must be added to the list of components to accurately price HEAT TRACE systems:

1. Expediter fitting and clamp fasteners (F)
2. Junction box and nipple connected to expediter fitting (G)
3. Field installed terminal blocks within junction box
4. Ground lugs
5. Piping from power source to expediter fitting
6. Controls
7. Thermostats
8. Branch wiring
9. Cable splices
10. End of cable terminations
11. Branch piping fittings and boxes

Deduct the following percentages from labor if cable lengths in the same area exceed:

150' to 250'	10%	351' to 500'	20%
251' to 350'	15%	Over 500'	25%

Add the following percentages to labor for elevated installations:

15' to 20' high	10%	31' to 35' high	40%
21' to 25' high	20%	36' to 40' high	50%
26' to 30' high	30%	Over 40' high	60%

R238313-20 Spiral-Wrapped Heat Trace Cable (Pitch Table)

In order to increase the amount of heat, occasionally a heat trace cable is wrapped in a spiral fashion around a pipe, increasing the number of feet of heater cable per linear foot of pipe.

Engineers first determine the heat loss per foot of pipe (based on the insulating material, its thickness, and the temperature differential across it). A ratio is then calculated by the formula:

$$\text{Feet of Heat Trace per Foot of Pipe} = \frac{\text{Watts/Foot of Heat Loss}}{\text{Watts/Foot of the Cable}}$$

The linear distance between wraps (pitch) is then taken from a chart or table. Generally, the pitch is listed on a drawing leaving the estimator to calculate the total length of heat tape required. An approximation may be taken from this table.

Pitch In Inches	1	1¼	1½	2	2½	3	4	6	8	10	12	14	16	18	20	24
3.5	1.80															
4	1.65															
5	1.46	1.60	1.80													
6	1.34	1.45	1.55	1.75												
7	1.25	1.35	1.43	1.57	1.75											
8	1.20	1.28	1.34	1.45	1.60	1.80										
9	1.16	1.23	1.28	1.37	1.51	1.68										
10	1.13	1.19	1.24	1.32	1.44	1.57	1.82									
15	1.06	1.08	1.10	1.15	1.21	1.29	1.42	1.78								
20	1.04	1.05	1.06	1.08	1.13	1.17	1.25	1.49	1.73							
25		1.04	1.04	1.06	1.08	1.11	1.17	1.33	1.51	1.72						
30				1.04	1.05	1.07	1.12	1.24	1.37	1.54	1.70					
35						1.06	1.09	1.17	1.28	1.42	1.54	1.80				
40						1.05	1.07	1.14	1.22	1.33	1.44	1.64	1.78			
50							1.05	1.09	1.15	1.28	1.33	1.52	1.64	1.75	1.64	1.83
60								1.06	1.11	1.16	1.21	1.35	1.44	1.53	1.46	1.62
70								1.05	1.08	1.12	1.17	1.25	1.31	1.39	1.35	1.47
80									1.06	1.09	1.13	1.19	1.24	1.30	1.28	1.38
90									1.04	1.06	1.10	1.15	1.19	1.24	1.23	1.32
100										1.05	1.08	1.13	1.16	1.19	1.19	1.23

Feet of Heat Trace Per Foot of Pipe — Nominal Pipe Size in Inches

Note: Common practice would normally limit the lower end of the table to 5% of additional heat. Above 80% an engineer would likely opt for two (2) parallel cables.

R260519-93 Metric Equivalent, Wire

United States — Size AWG or kcmil	United States — Area Cir. Mils. (cmil) mm²	European — Size mm²	European — Area Cir. Mils.
18	1620/.82	.75	1480
16	2580/1.30	1.0	1974
14	4110/2.08	1.5	2961
12	6530/3.30	2.5	4935
10	10,380/5.25	4	7896
8	16,510/8.36	6	11,844
6	26,240/13.29	10	19,740
4	41,740/21.14	16	31,584
3	52,620/26.65	25	49,350
2	66,360/33.61	–	–
1	83,690/42.39	35	69,090
1/0	105,600/53.49	50	98,700
2/0	133,100/67.42	–	–
3/0	167,800/85.00	70	138,180
4/0	211,600/107.19	95	187,530
250	250,000/126.64	120	236,880
300	300,000/151.97	150	296,100
350	350,000/177.30	–	–
400	400,000/202.63	185	365,190
500	500,000/253.29	240	473,760
600	600,000/303.95	300	592,200
700	700,000/354.60	–	–
750	750,000/379.93	–	–

R260533-22 Conductors in Conduit

The table below lists the maximum number of conductors for various sized conduit using THW, TW or THWN insulations.

Copper Wire Size	1/2" TW	1/2" THW	1/2" THWN	3/4" TW	3/4" THW	3/4" THWN	1" TW	1" THW	1" THWN	1-1/4" TW	1-1/4" THW	1-1/4" THWN	1-1/2" TW	1-1/2" THW	1-1/2" THWN	2" TW	2" THW	2" THWN	2-1/2" TW	2-1/2" THW	2-1/2" THWN	3" THW	3" THWN	3-1/2" THW	3-1/2" THWN	4" THW	4" THWN
#14	9	6	13	15	10	24	25	16	39	44	29	69	60	40	94	99	65	154	142	93		143		192			
#12	7	4	10	12	8	18	19	13	29	35	24	51	47	32	70	78	53	114	111	76	164	117		157			
#10	5	4	6	9	6	11	15	11	18	26	19	32	36	26	44	60	43	73	85	61	104	95	160	127		163	
#8	2	1	3	4	3	5	7	5	9	12	10	16	17	13	22	28	22	36	40	32	51	49	79	66	106	85	136
#6		1	1		2	4		4	6		7	11		10	15		16	26		23	37	36	57	48	76	62	98
#4		1	1		1	2		3	4		5	7		7	9		12	16		17	22	27	35	36	47	47	60
#3		1	1		1	1		2	3		4	6		6	8		10	13		15	19	23	29	31	39	40	51
#2		1	1		1	1		2	3		4	5		5	7		9	11		13	16	20	25	27	33	34	43
#1					1	1		1	1		3	3		4	5		6	8		9	12	14	18	19	25	25	32
1/0					1	1		1	1		2	3		3	4		5	7		8	10	12	15	16	21	21	27
2/0					1	1		1	1		1	2		3	3		5	6		7	8	10	13	14	17	18	22
3/0					1	1		1	1		1	1		2	3		4	5		6	7	9	11	12	14	15	18
4/0						1		1	1		1	1		1	2		3	4		5	6	7	9	10	12	13	15
250 kcmil								1	1		1	1		1	1		2	3		4	4	6	7	8	10	10	12
300								1	1		1	1		1	1		2	3		3	4	5	6	7	8	9	11
350									1		1	1		1	1		1	2		3	3	4	5	6	7	8	9
400											1	1		1	1		1	1		2	3	4	5	5	6	7	8
500											1	1		1	1		1	1		1	2	3	4	4	5	6	7
600														1	1		1	1		1	1	3	3	4	4	5	5
700														1	1		1	1		1	1	2	3	3	4	4	5
750														1	1		1	1		1	1	2	2	3	3	4	4

For customer support on your Site Work & Landscape Costs with RSMeans data, call 800.448.8182.

867

R260590-05 Typical Overhead Service Entrance

R262716-40   Standard Electrical Enclosure Types

NEMA Enclosures

Electrical enclosures serve two basic purposes: they protect people from accidental contact with enclosed electrical devices and connections, and they protect the enclosed devices and connections from specified external conditions. The National Electrical Manufacturers Association (NEMA) has established the following standards. Because these descriptions are not intended to be complete representations of NEMA listings, consultation of NEMA literature is advised for detailed information.

The following definitions and descriptions pertain to NONHAZARDOUS locations.

NEMA   Type 1: General purpose enclosures intended for use indoors, primarily to prevent accidental contact of personnel with the enclosed equipment in areas that do not involve unusual conditions.

NEMA   Type 2: Dripproof indoor enclosures intended to protect the enclosed equipment against dripping noncorrosive liquids and falling dirt.

NEMA   Type 3: Dustproof, raintight and sleet-resistant (ice-resistant) enclosures intended for use outdoors to protect the enclosed equipment against wind-blown dust, rain, sleet, and external ice formation.

NEMA   Type 3R: Rainproof and sleet-resistant (ice-resistant) enclosures which are intended for use outdoors to protect the enclosed equipment against rain. These enclosures are constructed so that the accumulation and melting of sleet (ice) will not damage the enclosure and its internal mechanisms.

NEMA   Type 3S: Enclosures intended for outdoor use to provide limited protection against wind-blown dust, rain, and sleet (ice) and to allow operation of external mechanisms when ice-laden.

NEMA   Type 4: Watertight and dust-tight enclosures intended for use indoors and out – to protect the enclosed equipment against splashing water, seepage of water, falling or hose-directed water, and severe external condensation.

NEMA   Type 4X: Watertight, dust-tight, and corrosion-resistant indoor and outdoor enclosures featuring the same provisions as Type 4 enclosures, plus corrosion resistance.

NEMA   Type 5: Indoor enclosures intended primarily to provide limited protection against dust and falling dirt.

NEMA   Type 6: Enclosures intended for indoor and outdoor use – primarily to provide limited protection against the entry of water during occasional temporary submersion at a limited depth.

NEMA   Type 6R: Enclosures intended for indoor and outdoor use – primarily to provide limited protection against the entry of water during prolonged submersion at a limited depth.

NEMA   Type 11: Enclosures intended for indoor use – primarily to provide, by means of oil immersion, limited protection to enclosed equipment against the corrosive effects of liquids and gases.

NEMA   Type 12: Dust-tight and driptight indoor enclosures intended for use indoors in industrial locations to protect the enclosed equipment against fibers, flyings, lint, dust, and dirt, as well as light splashing, seepage, dripping, and external condensation of noncorrosive liquids.

NEMA   Type 13: Oil-tight and dust-tight indoor enclosures intended primarily to house pilot devices, such as limit switches, foot switches, push buttons, selector switches, and pilot lights, and to protect these devices against lint and dust, seepage, external condensation, and sprayed water, oil, and noncorrosive coolant.

The following definitions and descriptions pertain to HAZARDOUS, or CLASSIFIED, locations:

NEMA   Type 7: Enclosures intended for use in indoor locations classified as Class 1, Groups A, B, C, or D, as defined in the National Electrical Code.

NEMA   Type 9: Enclosures intended for use in indoor locations classified as Class 2, Groups E, F, or G, as defined in the National Electrical Code.

R312316-40   Excavating

The selection of equipment used for structural excavation and bulk excavation or for grading is determined by the following factors.

1. Quantity of material
2. Type of material
3. Depth or height of cut
4. Length of haul
5. Condition of haul road
6. Accessibility of site
7. Moisture content and dewatering requirements
8. Availability of excavating and hauling equipment

Some additional costs must be allowed for hand trimming the sides and bottom of concrete pours and other excavation below the general excavation.

Number of B.C.Y. per truck = 1.5 C.Y. bucket x 8 passes = 12 loose C.Y.

$$= 12 \times \frac{100}{118} = 10.2 \text{ B.C.Y. per truck}$$

Truck Haul Cycle:

Load truck, 8 passes	=	4 minutes
Haul distance, 1 mile	=	9 minutes
Dump time	=	2 minutes
Return, 1 mile	=	7 minutes
Spot under machine	=	1 minute
		23 minute cycle

Add the mobilization and demobilization costs to the total excavation costs. When equipment is rented for more than three days, there is often no mobilization charge by the equipment dealer. On larger jobs outside of urban areas, scrapers can move earth economically provided a dump site or fill area and adequate haul roads are available. Excavation within sheeting bracing or cofferdam bracing is usually done with a clamshell and production

When planning excavation and fill, the following should also be considered.

1. Swell factor
2. Compaction factor
3. Moisture content
4. Density requirements

A typical example for scheduling and estimating the cost of excavation of a 15′ deep basement on a dry site when the material must be hauled off the site is outlined below.

Assumptions:

1. Swell factor, 18%
2. No mobilization or demobilization
3. Allowance included for idle time and moving on job
4. No dewatering, sheeting, or bracing
5. No truck spotter or hand trimming

Fleet Haul Production per day in B.C.Y.

$$4 \text{ trucks} \times \frac{50 \text{ min. hour}}{23 \text{ min. haul cycle}} \times 8 \text{ hrs.} \times 10.2 \text{ B.C.Y.}$$

$$= 4 \times 2.2 \times 8 \times 10.2 = 718 \text{ B.C.Y./day}$$

is low, since the clamshell may have to be guided by hand between the bracing. When excavating or filling an area enclosed with a wellpoint system, add 10% to 15% to the cost to allow for restricted access. When estimating earth excavation quantities for structures, allow work space outside the building footprint for construction of the foundation and a slope of 1:1 unless sheeting is used.

R312316-45 Excavating Equipment

The table below lists theoretical hourly production in C.Y./hr. bank measure for some typical excavation equipment. Figures assume 50 minute hours, 83% job efficiency, 100% operator efficiency, 90° swing and properly sized hauling units, which must be modified for adverse digging and loading conditions. Actual production costs in the front of the data set average about 50% of the theoretical values listed here.

Equipment	Soil Type	B.C.Y. Weight	% Swell	1 C.Y.	1-1/2 C.Y.	2 C.Y.	2-1/2 C.Y.	3 C.Y.	3-1/2 C.Y.	4 C.Y.
Hydraulic Excavator "Backhoe" 15' Deep Cut	Moist loam, sandy clay	3400 lb.	40%	85	125	175	220	275	330	380
	Sand and gravel	3100	18	80	120	160	205	260	310	365
	Common earth	2800	30	70	105	150	190	240	280	330
	Clay, hard, dense	3000	33	65	100	130	170	210	255	300
Power Shovel Optimum Cut (Ft.)	Moist loam, sandy clay	3400	40	170 (6.0)	245 (7.0)	295 (7.8)	335 (8.4)	385 (8.8)	435 (9.1)	475 (9.4)
	Sand and gravel	3100	18	165 (6.0)	225 (7.0)	275 (7.8)	325 (8.4)	375 (8.8)	420 (9.1)	460 (9.4)
	Common earth	2800	30	145 (7.8)	200 (9.2)	250 (10.2)	295 (11.2)	335 (12.1)	375 (13.0)	425 (13.8)
	Clay, hard, dense	3000	33	120 (9.0)	175 (10.7)	220 (12.2)	255 (13.3)	300 (14.2)	335 (15.1)	375 (16.0)
Drag Line Optimum Cut (Ft.)	Moist loam, sandy clay	3400	40	130 (6.6)	180 (7.4)	220 (8.0)	250 (8.5)	290 (9.0)	325 (9.5)	385 (10.0)
	Sand and gravel	3100	18	130 (6.6)	175 (7.4)	210 (8.0)	245 (8.5)	280 (9.0)	315 (9.5)	375 (10.0)
	Common earth	2800	30	110 (8.0)	160 (9.0)	190 (9.9)	220 (10.5)	250 (11.0)	280 (11.5)	310 (12.0)
	Clay, hard, dense	3000	33	90 (9.3)	130 (10.7)	160 (11.8)	190 (12.3)	225 (12.8)	250 (13.3)	280 (12.0)

Equipment	Soil Type	B.C.Y. Weight	% Swell	Wheel Loaders				Track Loaders		
				3 C.Y.	4 C.Y.	6 C.Y.	8 C.Y.	2-1/4 C.Y.	3 C.Y.	4 C.Y.
Loading Tractors	Moist loam, sandy clay	3400	40	260	340	510	690	135	180	250
	Sand and gravel	3100	18	245	320	480	650	130	170	235
	Common earth	2800	30	230	300	460	620	120	155	220
	Clay, hard, dense	3000	33	200	270	415	560	110	145	200
	Rock, well-blasted	4000	50	180	245	380	520	100	130	180

For customer support on your Site Work & Landscape Costs with RSMeans data, call 800.448.8182.

871

R312319-90 Wellpoints

A single stage wellpoint system is usually limited to dewatering an average 15' depth below normal ground water level. Multi-stage systems are employed for greater depth with the pumping equipment installed only at the lowest header level. Ejectors with unlimited lift capacity can be economical when two or more stages of wellpoints can be replaced or when horizontal clearance is restricted, such as in deep trenches or tunneling projects, and where low water flows are expected. Wellpoints are usually spaced on 2-1/2' to 10' centers along a header pipe. Wellpoint spacing, header size, and pump size are all determined by the expected flow as dictated by soil conditions.

In almost all soils encountered in wellpoint dewatering, the wellpoints may be jetted into place. Cemented soils and stiff clays may require sand wicks about 12" in diameter around each wellpoint to increase efficiency and eliminate weeping into the excavation. These sand wicks require 1/2 to 3 C.Y. of washed filter sand and are installed by using a 12" diameter steel casing and hole puncher jetted into the ground 2' deeper than the wellpoint. Rock may require predrilled holes.

Labor required for the complete installation and removal of a single stage wellpoint system is in the range of 3/4 to 2 labor-hours per linear foot of header, depending upon jetting conditions, wellpoint spacing, etc.

Continuous pumping is necessary except in some free draining soil where temporary flooding is permissible (as in trenches which are backfilled after each day's work). Good practice requires provision of a stand-by pump during the continuous pumping operation.

Systems for continuous trenching below the water table should be installed three to four times the length of expected daily progress to ensure uninterrupted digging, and header pipe size should not be changed during the job.

For pervious free draining soils, deep wells in place of wellpoints may be economical because of lower installation and maintenance costs. Daily production ranges between two to three wells per day, for 25' to 40' depths, to one well per day for depths over 50'.

Detailed analysis and estimating for any dewatering problem is available at no cost from wellpoint manufacturers. Major firms will quote "sufficient equipment" quotes or their affiliates will offer lump sum proposals to cover complete dewatering responsibility.

Description for 200' System with 8" Header		Quantities
Equipment & Material	Wellpoints 25' long, 2" diameter @ 5' O.C.	40 Each
	Header pipe, 8" diameter	200 L.F.
	Discharge pipe, 8" diameter	100 L.F.
	8" valves	3 Each
	Combination jetting & wellpoint pump (standby)	1 Each
	Wellpoint pump, 8" diameter	1 Each
	Transportation to and from site	1 Day
	Fuel for 30 days x 60 gal./day	1800 Gallons
	Lubricants for 30 days x 16 lbs./day	480 Lbs.
	Sand for points	40 C.Y.
Labor	Technician to supervise installation	1 Week
	Labor for installation and removal of system	300 Labor-hours
	4 Operators straight time 40 hrs./wk. for 4.33 wks.	693 Hrs.
	4 Operators overtime 2 hrs./wk. for 4.33 wks.	35 Hrs.

R312323-30 Compacting Backfill

Compaction of fill in embankments, around structures, in trenches, and under slabs is important to control settlement. Factors affecting compaction are:

1. Soil gradation
2. Moisture content
3. Equipment used
4. Depth of fill per lift
5. Density required

Production Rate:

$$\frac{1.75' \text{ plate width x 50 F.P.M. x 50 min./hr. x .67' lift}}{27 \text{ C.F. per C.Y.}} = 108.5 \text{ C.Y./hr.}$$

Production Rate for 4 Passes:

$$\frac{108.5 \text{ C.Y.}}{4 \text{ passes}} = 27.125 \text{ C.Y./hr. x 8 hrs.} = 217 \text{ C.Y./day}$$

Example:

Compact granular fill around a building foundation using a 21" wide x 24" vibratory plate in 8" lifts. Operator moves at 50 F.P.M. working a 50 minute hour to develop 95% Modified Proctor Density with 4 passes.

R314116-40 Wood Sheet Piling

Wood sheet piling may be used for depths to 20′ where there is no ground water. If moderate ground water is encountered Tongue & Groove sheeting will help to keep it out. When considerable ground water is present, steel sheeting must be used.

For estimating purposes on trench excavation, sizes are as follows:

Depth	Sheeting	Wales	Braces	B.F. per S.F.
To 8′	3 x 12's	6 x 8's, 2 line	6 x 8's, @ 10′	4.0 @ 8′
8′ x 12′	3 x 12's	10 x 10's, 2 line	10 x 10's, @ 9′	5.0 average
12′ to 20′	3 x 12's	12 x 12's, 3 line	12 x 12's, @ 8′	7.0 average

Sheeting to be toed in at least 2′ depending upon soil conditions. A five person crew with an air compressor and sheeting driver can drive and brace 440 SF/day at 8′ deep, 360 SF/day at 12′ deep, and 320 SF/day at 16′ deep.

For normal soils, piling can be pulled in 1/3 the time to install. Pulling difficulty increases with the time in the ground. Production can be increased by high pressure jetting.

R314116-45 Steel Sheet Piling

Limiting weights are 22 to 38#/S.F. of wall surface with 27#/S.F. average for usual types and sizes. (Weights of piles themselves are from 30.7#/L.F. to 57#/L.F. but they are 15″ to 21″ wide.) Lightweight sections 12″ to 28″ wide from 3 ga. to 12 ga. thick are also available for shallow excavations. Piles may be driven two at a time with an impact or vibratory hammer (use vibratory to pull) hung from a crane without leads. A reasonable estimate of the life of steel sheet piling is 10 uses with up to 125 uses possible if a vibratory hammer is used. Used piling costs from 50% to 80% of new piling depending on location and market conditions. Sheet piling and H piles can be rented for about 30% of the delivered mill price for the first month and 5% per month thereafter. Allow 1 labor-hour per pile for cleaning and trimming after driving. These costs increase with depth and hydrostatic head. Vibratory drivers are faster in wet granular soils and are excellent for pile extraction. Pulling difficulty increases with the time in the ground and may cost more than driving. It is often economical to abandon the sheet piling, especially if it can be used as the outer wall form. Allow about 1/3 additional length or more for toeing into ground. Add bracing, waler and strut costs. Waler costs can equal the cost per ton of sheeting.

R314513-90 Vibroflotation and Vibro Replacement Soil Compaction

Vibroflotation is a proprietary system of compacting sandy soils in place to increase relative density to about 70%. Typical bearing capacities attained will be 6000 psf for saturated sand and 12,000 psf for dry sand. Usual range is 4000 to 8000 psf capacity. Costs in the front of the data set are for a vertical foot of compacted cylinder 6′ to 10′ in diameter.

Vibro replacement is a proprietary system of improving cohesive soils in place to increase bearing capacity. Most silts and clays above or below the water table can be strengthened by installation of stone columns.

The process consists of radial displacement of the soil by vibration. The created hole is then backfilled in stages with coarse granular fill which is thoroughly compacted and displaced into the surrounding soil in the form of a column.

The total project cost would depend on the number and depth of the compacted cylinders. The installing company guarantees relative soil density of the sand cylinders after compaction and the bearing capacity of the soil after the replacement process. Detailed estimating information is available from the installer at no cost.

R316326-60   Caissons

The three principal types of caissons are:

(1) **Belled Caissons,** which except for shallow depths and poor soil conditions, are generally recommended. They provide more bearing than shaft area. Because of its conical shape, no horizontal reinforcement of the bell is required.

(2) **Straight Shaft Caissons** are used where relatively light loads are to be supported by caissons that rest on high value bearing strata. While the shaft is larger in diameter than for belled types this is more than offset by the savings in time and labor.

(3) **Keyed Caissons** are used when extremely heavy loads are to be carried. A keyed or socketed caisson transfers its load into rock by a combination of end-bearing and shear reinforcing of the shaft. The most economical shaft often consists of a steel casing, a steel wide flange core and concrete. Allowable compressive stresses of .225 f'c for concrete, 16,000 psi for the wide flange core, and 9,000 psi for the steel casing are commonly used. The usual range of shaft diameter is 18″ to 84″. The number of sizes specified for any one project should be limited due to the problems of casing and auger storage. When hand work is to be performed, shaft diameters should not be less than 32″. When inspection of borings is required a minimum shaft diameter of 30″ is recommended. Concrete caissons are intended to be poured against earth excavation so permanent forms, which add to cost, should not be used if the excavation is clean and the earth is sufficiently impervious to prevent excessive loss of concrete.

Soil Conditions for Belling		
Good	Requires Handwork	Not Recommended
Clay	Hard Shale	Silt
Sandy Clay	Limestone	Sand
Silty Clay	Sandstone	Gravel
Clayey Silt	Weathered Mica	Igneous Rock
Hard-pan		
Soft Shale		
Decomposed Rock		

Exterior Improvements   R3292 Turf & Grasses

R329219-50   Seeding

The type of grass is determined by light, shade and moisture content of soil plus intended use. Fertilizer should be disked 4″ before seeding. For steep slopes disk five tons of mulch and lay two tons of hay or straw on surface per acre after seeding. Surface mulch can be staked, lightly disked or tar emulsion sprayed. Material for mulch can be wood chips, peat moss, partially rotted hay or straw, wood fibers and sprayed emulsions. Hemp seed blankets with fertilizer are also available. For spring seeding, watering is necessary. Late fall seeding may have to be reseeded in the spring. Hydraulic seeding, power mulching, and aerial seeding can be used on large areas.

R329343-10   Plant Spacing Chart

This chart may be used when plants are to be placed equidistant from each other, staggering their position in each row.

Plant Spacing (Inches)	Row Spacing (Inches)	Plants Per (CSF)	Plant Spacing (Feet)	Row Spacing (Feet)	Plants Per (MSF)
			4	3.46	72
6	5.20	462	5	4.33	46
8	6.93	260	6	5.20	32
10	8.66	166	8	6.93	18
12	10.39	115	10	8.66	12
15	12.99	74	12	10.39	8
18	15.59	51.32	15	12.99	5.13
21	18.19	37.70	20	17.32	2.89
24	20.78	28.87	25	21.65	1.85
30	25.98	18.48	30	25.98	1.28
36	31.18	12.83	40	34.64	0.72

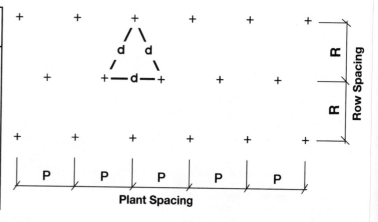

R329343-20   Trees and Plants by Environment and Purposes

Dry, Windy, Exposed Areas
Barberry
Junipers, all varieties
Locust
Maple
Oak
Pines, all varieties
Poplar, Hybrid
Privet
Spruce, all varieties
Sumac, Staghorn

Lightly Wooded Areas
Dogwood
Hemlock
Larch
Pine, White
Rhododendron
Spruce, Norway
Redbud

Total Shade Areas
Hemlock
Ivy, English
Myrtle
Pachysandra
Privet
Spice Bush
Yews, Japanese

Cold Temperatures of Northern U.S. and Canada
Arborvitae, American
Birch, White
Dogwood, Silky
Fir, Balsam
Fir, Douglas
Hemlock
Juniper, Andorra
Juniper, Blue Rug
Linden, Little Leaf
Maple, Sugar
Mountain Ash
Myrtle
Olive, Russian

Pine, Mugho
Pine, Ponderosa
Pine, Red
Pine, Scotch
Poplar, Hybrid
Privet
Rosa Rugosa
Spruce, Dwarf Alberta
Spruce, Black Hills
Spruce, Blue
Spruce, Norway
Spruce, White, Engelman
Yellow Wood

Wet, Swampy Areas
American Arborvitae
Birch, White
Black Gum
Hemlock
Maple, Red
Pine, White
Willow

Poor, Dry, Rocky Soil
Barberry
Crownvetch
Eastern Red Cedar
Juniper, Virginiana
Locust, Black
Locust, Bristly
Locust, Honey
Olive, Russian
Pines, all varieties
Privet
Rosa Rugosa
Sumac, Staghorn

Seashore Planting
Arborvitae, American
Juniper, Tamarix
Locust, Black
Oak, White
Olive, Russian
Pine, Austrian
Pine, Japanese Black

Pine, Mugho
Pine, Scotch
Privet, Amur River
Rosa Rugosa
Yew, Japanese

City Planting
Barberry
Fir, Concolor
Forsythia
Hemlock
Holly, Japanese
Ivy, English
Juniper, Andorra
Linden, Little Leaf
Locust, Honey
Maple, Norway, Silver
Oak, Pin, Red
Olive, Russian
Pachysandra
Pine, Austrian
Pine, White
Privet
Rosa Rugosa
Sumac, Staghorn
Yew, Japanese

Bonsai Planting
Azaleas
Birch, White
Ginkgo
Junipers
Pine, Bristlecone
Pine, Mugho
Spruce, Engelmann
Spruce, Dwarf Alberta

Street Planting
Linden, Little Leaf
Oak, Pin
Ginkgo

Fast Growth
Birch, White
Crownvetch
Dogwood, Silky

Fir, Douglas
Juniper, Blue Pfitzer
Juniper, Blue Rug
Maple, Silver
Olive, Autumn
Pines, Austrian, Ponderosa, Red
   Scotch and White
Poplar, Hybrid
Privet
Spruce, Norway
Spruce, Serbian
Texus, Cuspidata, Hicksi
Willow

Dense, Impenetrable Hedges
Field Plantings:
   Locust, Bristly,
   Olive, Autumn
   Sumac
Residential Area:
   Barberry, Red or Green
   Juniper, Blue Pfitzer
   Rosa Rugosa

Food for Birds
Ash, Mountain
Barberry
Bittersweet
Cherry, Manchu
Dogwood, Silky
Honeysuckle, Rem Red
Hawthorn
Oaks
Olive, Autumn, Russian
Privet
Rosa Rugosa
Sumac

Erosion Control
Crownvetch
Locust, Bristly
Willow

R329343-30 Zones of Plant Hardiness

R331113-80   Piping Designations

There are several systems currently in use to describe pipe and fittings. The following paragraphs will help to identify and clarify classifications of piping systems used for water distribution.

Piping may be classified by schedule. Piping schedules include 5S, 10S, 10, 20, 30, Standard, 40, 60, Extra Strong, 80, 100, 120, 140, 160 and Double Extra Strong. These schedules are dependent upon the pipe wall thickness. The wall thickness of a particular schedule may vary with pipe size.

Ductile iron pipe for water distribution is classified by Pressure Classes such as Class 150, 200, 250, 300 and 350. These classes are actually the rated water working pressure of the pipe in pounds per square inch (psi). The pipe in these pressure classes is designed to withstand the rated water working pressure plus a surge allowance of 100 psi.

The American Water Works Association (AWWA) provides standards for various types of **plastic pipe.** C-900 is the specification for polyvinyl chloride (PVC) piping used for water distribution in sizes ranging from 4″ through 12″. C-901 is the specification for polyethylene (PE) pressure pipe, tubing and fittings used for water distribution in sizes ranging from 1/2″ through 3″. C-905 is the specification for PVC piping sizes 14″ and greater.

PVC pressure-rated pipe is identified using the standard dimensional ratio (SDR) method. This method is defined by the American Society for Testing and Materials (ASTM) Standard D 2241. This pipe is available in SDR numbers 64, 41, 32.5, 26, 21, 17, and 13.5. A pipe with an SDR of 64 will have the thinnest wall while a pipe with an SDR of 13.5 will have the thickest wall. When the pressure rating (PR) of a pipe is given in psi, it is based on a line supplying water at 73 degrees F.

The National Sanitation Foundation (NSF) seal of approval is applied to products that can be used with potable water. These products have been tested to ANSI/NSF Standard 14.

Valves and strainers are classified by American National Standards Institute (ANSI) Classes. These Classes are 125, 150, 200, 250, 300, 400, 600, 900, 1500 and 2500. Within each class there is an operating pressure range dependent upon temperature. Design parameters should be compared to the appropriate material dependent, pressure-temperature rating chart for accurate valve selection.

R337119-30   Concrete for Conduit Encasement

The table below lists C.Y. of concrete for 100 L.F. of trench. Conduits separation center to center should meet 7.5″ (N.E.C.).

Number of Conduits	1	2	3	4	6	8	9	Number of Conduits
Trench Dimension	11.5″ x 11.5″	11.5″ x 19″	11.5″ x 27″	19″ x 19″	19″ x 27″	19″ x 38″	27″ x 27″	Trench Dimension
Conduit Diameter 2.0″	3.29	5.39	7.64	8.83	12.51	17.66	17.72	Conduit Diameter 2.0″
2.5″	3.23	5.29	7.49	8.62	12.19	17.23	17.25	2.5″
3.0″	3.15	5.13	7.24	8.29	11.71	16.59	16.52	3.0″
3.5″	3.08	4.97	7.02	7.99	11.26	15.98	15.84	3.5″
4.0″	2.99	4.80	6.76	7.65	10.74	15.30	15.07	4.0″
5.0″	2.78	4.37	6.11	6.78	9.44	13.57	13.12	5.0″
6.0″	2.52	3.84	5.33	5.74	7.87	11.48	10.77	6.0″

R347216-10 Single Track R.R. Siding

The costs for a single track RR siding in the Unit Price section include the components shown in the table below.

Description of Component	Qty. per L.F. of Track	Unit
Ballast, 1-1/2" crushed stone	.667	C.Y.
6" x 8" x 8'-6" Treated timber ties, 22" O.C.	.545	Ea.
Tie plates, 2 per tie	1.091	Ea.
Track rail	2.000	L.F.
Spikes, 6", 4 per tie	2.182	Ea.
Splice bars w/ bolts, lock washers & nuts, @ 33' O.C.	.061	Pair
Crew B-14 @ 57 L.F./Day	.018	Day

R347216-20 Single Track, Steel Ties, Concrete Bed

The costs for a R.R. siding with steel ties and a concrete bed in the Unit Price section include the components shown in the table below.

Description of Component	Qty. per L.F. of Track	Unit
Concrete bed, 9' wide, 10" thick	.278	C.Y.
Ties, W6x16 x 6'-6" long, @ 30" O.C.	.400	Ea.
Tie plates, 4 per tie	1.600	Ea.
Track rail	2.000	L.F.
Tie plate bolts, 1", 8 per tie	3.200	Ea.
Splice bars w/bolts, lock washers & nuts, @ 33' O.C.	.061	Pair
Crew B-14 @ 22 L.F./Day	.045	Day

Change Orders

Change Order Considerations

A change order is a written document usually prepared by the design professional and signed by the owner, the architect/engineer, and the contractor. A change order states the agreement of the parties to: an addition, deletion, or revision in the work; an adjustment in the contract sum, if any; or an adjustment in the contract time, if any. Change orders, or "extras", in the construction process occur after execution of the construction contract and impact architects/ engineers, contractors, and owners.

Change orders that are properly recognized and managed can ensure orderly, professional, and profitable progress for everyone involved in the project. There are many causes for change orders and change order requests. In all cases, change orders or change order requests should be addressed promptly and in a precise and prescribed manner. The following paragraphs include information regarding change order pricing and procedures.

The Causes of Change Orders

Reasons for issuing change orders include:

- Unforeseen field conditions that require a change in the work
- Correction of design discrepancies, errors, or omissions in the contract documents
- Owner-requested changes, either by design criteria, scope of work, or project objectives
- Completion date changes for reasons unrelated to the construction process
- Changes in building code interpretations, or other public authority requirements that require a change in the work
- Changes in availability of existing or new materials and products

Procedures

Properly written contract documents must include the correct change order procedures for all parties—owners, design professionals, and contractors—to follow in order to avoid costly delays and litigation.

Being "in the right" is not always a sufficient or acceptable defense. The contract provisions requiring notification and documentation must be adhered to within a defined or reasonable time frame.

The appropriate method of handling change orders is by a written proposal and acceptance by all parties involved. Prior to starting work on a project, all parties should identify their

authorized agents who may sign and accept change orders, as well as any limits placed on their authority.

Time may be a critical factor when the need for a change arises. For such cases, the contractor might be directed to proceed on a "time and materials" basis, rather than wait for all paperwork to be processed—a delay that could impede progress. In this situation, the contractor must still follow the prescribed change order procedures including, but not limited to, notification and documentation.

Lack of documentation can be very costly, especially if legal judgments are to be made, and if certain field personnel are no longer available. For time and material change orders, the contractor should keep accurate daily records of all labor and material allocated to the change.

Owners or awarding authorities who do considerable and continual building construction (such as the federal government) realize the inevitability of change orders for numerous reasons, both predictable and unpredictable. As a result, the federal government, the American Institute of Architects (AIA), the Engineers Joint Contract Documents Committee (EJCDC), and other contractor, legal, and technical organizations have developed standards and procedures to be followed by all parties to achieve contract continuance and timely completion, while being financially fair to all concerned.

Pricing Change Orders

When pricing change orders, regardless of their cause, the most significant factor is when the change occurs. The need for a change may be perceived in the field or requested by the architect/engineer *before* any of the actual installation has begun, or may evolve or appear *during* construction when the item of work in question is partially installed. In the latter cases, the original sequence of construction is disrupted, along with all contiguous and supporting systems. Change orders cause the greatest impact when they occur *after* the installation has been completed and must be uncovered, or even replaced. Post-completion changes may be caused by necessary design changes, product failure, or changes in the owner's requirements that are not discovered until the building or the systems begin to function.

Specified procedures of notification and record keeping must be adhered to and enforced regardless of the stage of construction: *before, during,* or *after* installation. Some bidding documents anticipate change orders by requiring that unit prices including overhead and profit percentages—for additional as well as deductible changes—be listed. Generally these unit prices do not fully take into account the ripple effect, or impact on other trades, and should be used for general guidance only.

When pricing change orders, it is important to classify the time frame in which the change occurs. There are two basic time frames for change orders: *pre-installation change orders,* which occur before the start of construction, and *post-installation change orders,* which involve reworking after the original installation. Change orders that occur between these stages may be priced according to the extent of work completed using a combination of techniques developed for pricing *pre-* and *post-installation* changes.

Factors To Consider When Pricing Change Orders

As an estimator begins to prepare a change order, the following questions should be reviewed to determine their impact on the final price.

General

- *Is the change order work* pre-installation *or* post-installation?

  Change order work costs vary according to how much of the installation has been completed. Once workers have the project scoped in their minds, even though they have not started, it can be difficult to refocus. Consequently they may spend more than the normal amount of time understanding the change. Also, modifications to work in place, such as trimming or refitting, usually take more time than was initially estimated. The greater the amount of work in place, the more reluctant workers are to change it. Psychologically they may resent the change and as a result the rework takes longer than normal. Post-installation change order estimates must include demolition of existing work as required to accomplish the change. If the work is performed at a later time, additional obstacles, such as building finishes, may be present which must be protected. Regardless of whether the change occurs pre-installation or post-installation, attempt to isolate the identifiable factors and price them separately. For example, add shipping costs that may be required pre-installation or any demolition required post-installation. Then analyze the potential impact on productivity of psychological and/or learning curve factors and adjust the output rates accordingly. One approach is to break down the typical workday into segments and quantify the impact on each segment.

Change Order Installation Efficiency

The labor-hours expressed (for new construction) are based on average installation time, using an efficiency level. For change order situations, adjustments to this efficiency level should reflect the daily labor-hour allocation for that particular occurrence.

- *Will the change substantially delay the original completion date?*

  A significant change in the project may cause the original completion date to be extended. The extended schedule may subject the contractor to new wage rates dictated by relevant labor contracts. Project supervision and other project overhead must also be extended beyond the original completion date. The schedule extension may also put installation into a new weather season. For example, underground piping scheduled for October installation was delayed until January. As a result, frost penetrated the trench area, thereby changing the degree of difficulty of the task. Changes and delays may have a ripple effect throughout the project. This effect must be analyzed and negotiated with the owner.

- *What is the net effect of a deduct change order?*

  In most cases, change orders resulting in a deduction or credit reflect only bare costs. The contractor may retain the overhead and profit based on the original bid.

Materials

- *Will you have to pay more or less for the new material, required by the change order, than you paid for the original purchase?*

  The same material prices or discounts will usually apply to materials purchased for change orders as new construction. In some

instances, however, the contractor may forfeit the advantages of competitive pricing for change orders. Consider the following example:

A contractor purchased over $20,000 worth of fan coil units for an installation and obtained the maximum discount. Some time later it was determined the project required an additional matching unit. The contractor has to purchase this unit from the original supplier to ensure a match. The supplier at this time may not discount the unit because of the small quantity, and he is no longer in a competitive situation. The impact of quantity on purchase can add between 0% and 25% to material prices and/or subcontractor quotes.

- *If materials have been ordered or delivered to the job site, will they be subject to a cancellation charge or restocking fee?*

  Check with the supplier to determine if ordered materials are subject to a cancellation charge. Delivered materials not used as a result of a change order may be subject to a restocking fee if returned to the supplier. Common restocking charges run between 20% and 40%. Also, delivery charges to return the goods to the supplier must be added.

Labor

- *How efficient is the existing crew at the actual installation?*

  Is the same crew that performed the initial work going to do the change order? Possibly the change consists of the installation of a unit identical to one already installed; therefore, the change should take less time. Be sure to consider this potential productivity increase and modify the productivity rates accordingly.

- *If the crew size is increased, what impact will that have on supervision requirements?*

  Under most bargaining agreements or management practices, there is a point at which a working foreman is replaced by a nonworking foreman. This replacement increases project overhead by adding a nonproductive worker. If additional workers are added to accelerate the project or to perform changes while maintaining the schedule, be sure to add additional supervision time if warranted. Calculate the

hours involved and the additional cost directly if possible.

- *What are the other impacts of increased crew size?*

  The larger the crew, the greater the potential for productivity to decrease. Some of the factors that cause this productivity loss are: overcrowding (producing restrictive conditions in the working space) and possibly a shortage of any special tools and equipment required. Such factors affect not only the crew working on the elements directly involved in the change order, but other crews whose movements may also be hampered. As the crew increases, check its basic composition for changes by the addition or deletion of apprentices or nonworking foreman, and quantify the potential effects of equipment shortages or other logistical factors.

- *As new crews, unfamiliar with the project, are brought onto the site, how long will it take them to become oriented to the project requirements?*

  The orientation time for a new crew to become 100% effective varies with the site and type of project. Orientation is easiest at a new construction site and most difficult at existing, very restrictive renovation sites. The type of work also affects orientation time. When all elements of the work are exposed, such as concrete or masonry work, orientation is decreased. When the work is concealed or less visible, such as existing electrical systems, orientation takes longer. Usually orientation can be accomplished in one day or less. Costs for added orientation should be itemized and added to the total estimated cost.

- *How much actual production can be gained by working overtime?*

  Short term overtime can be used effectively to accomplish more work in a day. However, as overtime is scheduled to run beyond several weeks, studies have shown marked decreases in output. The following chart shows the effect of long term overtime on worker efficiency. If the anticipated change requires extended overtime to keep the job on schedule, these factors can be used as a guide to predict the impact on time and cost. Add project overhead, particularly supervision, that may also be incurred.

For customer support on your Site Work & Landscape Costs with RSMeans data, call 800.448.8182.

Days per Week	Hours per Day	Production Efficiency					Payroll Cost Factors	
		1st Week	2nd Week	3rd Week	4th Week	Average 4 Weeks	@ 1-1/2 Times	@ 2 Times
5	8	100%	100%	100%	100%	100%	100%	100%
	9	100	100	95	90	96.25	105.6	111.1
	10	100	95	90	85	92.50	110.0	120.0
	11	95	90	75	65	81.25	113.6	127.3
	12	90	85	70	60	76.25	116.7	133.3
6	8	100	100	95	90	96.25	108.3	116.7
	9	100	95	90	85	92.50	113.0	125.9
	10	95	90	85	80	87.50	116.7	133.3
	11	95	85	70	65	78.75	119.7	139.4
	12	90	80	65	60	73.75	122.2	144.4
7	8	100	95	85	75	88.75	114.3	128.6
	9	95	90	80	70	83.75	118.3	136.5
	10	90	85	75	65	78.75	121.4	142.9
	11	85	80	65	60	72.50	124.0	148.1
	12	85	75	60	55	68.75	126.2	152.4

Effects of Overtime

Caution: Under many labor agreements, Sundays and holidays are paid at a higher premium than the normal overtime rate.

The use of long-term overtime is counterproductive on almost any construction job; that is, the longer the period of overtime, the lower the actual production rate. Numerous studies have been conducted, and while they have resulted in slightly different numbers, all reach the same conclusion. The figure above tabulates the effects of overtime work on efficiency.

As illustrated, there can be a difference between the *actual* payroll cost per hour and the *effective* cost per hour for overtime work. This is due to the reduced production efficiency with the increase in weekly hours beyond 40. This difference between actual and effective cost results from overtime work over a prolonged period. Short-term overtime work does not result in as great a reduction in efficiency and, in such cases, effective cost may not vary significantly from the actual payroll cost. As the total hours per week are increased on a regular basis, more time is lost due to fatigue, lowered morale, and an increased accident rate.

As an example, assume a project where workers are working 6 days a week, 10 hours per day. From the figure above (based on productivity studies), the average effective productive hours over a 4-week period are:

$$0.875 \times 60 = 52.5$$

Depending upon the locale and day of week, overtime hours may be paid at time and a half or double time. For time and a half, the overall (average) *actual* payroll cost (including regular and overtime hours) is determined as follows:

$$\frac{40 \text{ reg. hrs.} + (20 \text{ overtime hrs.} \times 1.5)}{60 \text{ hrs.}} = 1.167$$

Based on 60 hours, the payroll cost per hour will be 116.7% of the normal rate at 40 hours per week. However, because the effective production (efficiency) for 60 hours is reduced to the equivalent of 52.5 hours, the effective cost of overtime is calculated as follows:

For time and a half:

$$\frac{40 \text{ reg. hrs.} + (20 \text{ overtime hrs.} \times 1.5)}{52.5 \text{ hrs.}} = 1.33$$

The installed cost will be 133% of the normal rate (for labor).

Thus, when figuring overtime, the actual cost per unit of work will be higher than the apparent overtime payroll dollar increase, due to the reduced productivity of the longer work week. These efficiency calculations are true only for those cost factors determined by hours worked. Costs that are applied weekly or monthly, such as equipment rentals, will not be similarly affected.

Equipment

- *What equipment is required to complete the change order?*

Change orders may require extending the rental period of equipment already on the job site, or the addition of special equipment brought in to accomplish the change work. In either case, the additional rental charges and operator labor charges must be added.

Summary

The preceding considerations and others you deem appropriate should be analyzed and applied to a change order estimate. The impact of each should be quantified and listed on the estimate to form an audit trail.

Change orders that are properly identified, documented, and managed help to ensure the orderly, professional, and profitable progress of the work. They also minimize potential claims or disputes at the end of the project.

Back by customer demand!

You asked and we listened. For customer convenience and estimating ease, we have made the 2020 Project Costs available for download at **RSMeans.com/2020books**. You will also find sample estimates, an RSMeans data overview video, and a book registration form to receive quarterly data updates throughout 2020.

Estimating Tips

- The cost figures available in the download were derived from hundreds of projects contained in the RSMeans database of completed construction projects. They include the contractor's overhead and profit. The figures have been adjusted to January of the current year.

- These projects were located throughout the U.S. and reflect a tremendous variation in square foot (S.F.) costs. This is due to differences, not only in labor and material costs, but also in individual owners' requirements. For instance, a bank in a large city would have different features than one in a rural area. This is true of all the different types of buildings analyzed. Therefore, caution should be exercised when using these Project Costs. For example, for courthouses, costs in the database are local courthouse costs and will not apply to the larger, more elaborate federal courthouses.

- None of the figures "go with" any others. All individual cost items were computed and tabulated separately. Thus, the sum of the median figures for plumbing, HVAC, and electrical will not normally total up to the total mechanical and electrical costs arrived at by separate analysis and tabulation of the projects.

- Each building was analyzed as to total and component costs and percentages. The figures were arranged in ascending order with the results tabulated as shown. The 1/4 column shows that 25% of the projects had lower costs and 75% had higher. The 3/4 column shows that 75% of the projects had lower costs and 25% had higher. The median column shows that 50% of the projects had lower costs and 50% had higher.

- Project Costs are useful in the conceptual stage when no details are available. As soon as details become available in the project design, the square foot approach should be discontinued and the project should be priced as to its particular components. When more precision is required, or for estimating the replacement cost of specific buildings, the current edition of *Square Foot Costs with RSMeans data* should be used.

- In using the figures in this section, it is recommended that the median column be used for preliminary figures if no additional information is available. The median figures, when multiplied by the total city construction cost index figures (see City Cost Indexes) and then multiplied by the project size modifier at the end of this section, should present a fairly accurate base figure, which would then have to be adjusted in view of the estimator's experience, local economic conditions, code requirements, and the owner's particular requirements. There is no need to factor in the percentage figures, as these should remain constant from city to city.

- The editors of this data would greatly appreciate receiving cost figures on one or more of your recent projects, which would then be included in the averages for next year. All cost figures received will be kept confidential, except that they will be averaged with other similar projects to arrive at square foot cost figures for next year.

See the website above for details and the discount available for submitting one or more of your projects.

Same Data. Simplified.

Enjoy the convenience and efficiency of accessing your costs anywhere:

- **Skip the multiplier** by setting your location
- **Quickly search,** edit, favorite and share costs
- **Stay on top of price changes** with automatic updates

Discover more at rsmeans.com/online

50 17 00 | Project Costs	UNIT	UNIT COSTS			% OF TOTAL			
		1/4	MEDIAN	3/4	1/4	MEDIAN	3/4	
0000 **Auto Sales with Repair**	S.F.							01
0100 Architectural		106	119	128	58%	64%	67%	
0200 Plumbing		8.90	9.30	12.40	4.84%	5.20%	6.80%	
0300 Mechanical		11.90	15.95	17.60	6.40%	8.70%	10.15%	
0400 Electrical		18.30	22.50	28.50	9.05%	11.70%	15.90%	
0500 Total Project Costs		178	186	191				
0000 **Banking Institutions**	S.F.							02
0100 Architectural		160	197	239	59%	65%	69%	
0200 Plumbing		6.45	9	12.50	2.12%	3.39%	4.19%	
0300 Mechanical		12.80	17.70	21	4.41%	5.10%	10.75%	
0400 Electrical		31.50	38	58	10.45%	13.05%	15.90%	
0500 Total Project Costs		266	299	370				
0000 **Court House**	S.F.							03
0100 Architectural		84.50	166	166	54.50%	58.50%	58.50%	
0200 Plumbing		3.19	3.19	3.19	2.07%	2.07%	2.07%	
0300 Mechanical		19.95	19.95	19.95	12.95%	12.95%	12.95%	
0400 Electrical		25.50	25.50	25.50	16.60%	16.60%	16.60%	
0500 Total Project Costs		154	284	284				
0000 **Data Centers**	S.F.							04
0100 Architectural		191	191	191	68%	68%	68%	
0200 Plumbing		10.45	10.45	10.45	3.71%	3.71%	3.71%	
0300 Mechanical		26.50	26.50	26.50	9.45%	9.45%	9.45%	
0400 Electrical		25.50	25.50	25.50	9%	9%	9%	
0500 Total Project Costs		281	281	281				
0000 **Detention Centers**	S.F.							05
0100 Architectural		177	187	198	52%	53%	60.50%	
0200 Plumbing		18.65	22.50	27.50	5.15%	7.10%	7.25%	
0300 Mechanical		23.50	34	40.50	7.55%	9.50%	13.80%	
0400 Electrical		39	46	60	10.90%	14.85%	17.95%	
0500 Total Project Costs		299	315	370				
0000 **Fire Stations**	S.F.							06
0100 Architectural		100	129	187	49%	54.50%	61.50%	
0200 Plumbing		10.75	13.95	16.95	4.67%	5.60%	6.30%	
0300 Mechanical		14.95	21	29.50	6.10%	8.45%	10.20%	
0400 Electrical		23	29.50	35.50	10.75%	12.75%	14.95%	
0500 Total Project Costs		210	238	320				
0000 **Gymnasium**	S.F.							07
0100 Architectural		88.50	116	116	57%	64.50%	64.50%	
0200 Plumbing		2.17	7.10	7.10	1.58%	3.48%	3.48%	
0300 Mechanical		3.32	30	30	2.42%	14.65%	14.65%	
0400 Electrical		10.90	21	21	7.95%	10.35%	10.35%	
0500 Total Project Costs		137	204	204				
0000 **Hospitals**	S.F.							08
0100 Architectural		107	176	191	43%	47.50%	48%	
0200 Plumbing		7.85	15.05	32.50	6%	7.45%	7.65%	
0300 Mechanical		52	58.50	76.50	14.20%	17.95%	23.50%	
0400 Electrical		23.50	47.50	61.50	10.95%	13.75%	16.85%	
0500 Total Project Costs		250	375	405				
0000 **Industrial Buildings**	S.F.							09
0100 Architectural		45	71.50	232	46%	54%	56.50%	
0200 Plumbing		1.74	6.55	13.25	2%	3.06%	6.30%	
0300 Mechanical		4.81	9.15	43.50	4.77%	5.55%	14.80%	
0400 Electrical		7.35	8.40	70	7.85%	13.55%	16.20%	
0500 Total Project Costs		80	104	430				
0000 **Medical Clinics & Offices**	S.F.							10
0100 Architectural		89.50	121	161	48.50%	55.50%	62.50%	
0200 Plumbing		8.90	13.15	21	4.47%	6.60%	8.65%	
0300 Mechanical		13.95	22.50	44	7.80%	10.95%	16.10%	
0400 Electrical		20	27	37.50	9.70%	11.65%	14.05%	
0500 Total Project Costs		166	217	292				

50 17 00 | Project Costs

			UNIT	UNIT COSTS			% OF TOTAL		
				1/4	MEDIAN	3/4	1/4	MEDIAN	3/4
11	0000	**Mixed Use**	S.F.						
	0100	Architectural		91.50	129	211	48.50%	57%	62.50%
	0200	Plumbing		6.15	9.35	11.80	3.23%	3.44%	4.18%
	0300	Mechanical		16.40	24.50	47.50	6.10%	13.75%	18.85%
	0400	Electrical		16	26.50	41.50	8.30%	11.40%	14%
	0500	Total Project Costs		189	216	345			
12	0000	**Multi-Family Housing**	S.F.						
	0100	Architectural		76	114	170	56.50%	62%	66.50%
	0200	Plumbing		6.55	11.55	15.05	5.30%	6.85%	8%
	0300	Mechanical		7	9.45	27	4.92%	6.90%	10.40%
	0400	Electrical		9.90	15.45	19.85	6.20%	7.90%	10.25%
	0500	Total Project Costs		111	208	271			
13	0000	**Nursing Home & Assisted Living**	S.F.						
	0100	Architectural		71.50	94	118	51.50%	55.50%	63.50%
	0200	Plumbing		7.75	11.60	12.80	6.25%	7.40%	8.80%
	0300	Mechanical		6.35	9.35	18.30	4.04%	6.70%	9.55%
	0400	Electrical		10.50	16.55	23	7%	10.75%	13.10%
	0500	Total Project Costs		122	160	192			
14	0000	**Office Buildings**	S.F.						
	0100	Architectural		94.50	129	177	56%	61%	69%
	0200	Plumbing		5.10	8	15.05	2.65%	3.56%	5.85%
	0300	Mechanical		11	17	26	5.60%	8.20%	11.10%
	0400	Electrical		12.75	21.50	34	7.75%	10%	12.70%
	0500	Total Project Costs		161	200	282			
15	0000	**Parking Garage**	S.F.						
	0100	Architectural		31.50	38.50	40.50	70%	79%	88%
	0200	Plumbing		1.04	1.09	2.04	2.05%	2.70%	2.83%
	0300	Mechanical		.81	1.24	4.72	2.11%	3.62%	3.81%
	0400	Electrical		2.78	3.05	6.35	5.30%	6.35%	7.95%
	0500	Total Project Costs		38.50	47	51			
16	0000	**Parking Garage/Mixed Use**	S.F.						
	0100	Architectural		103	112	114	61%	62%	65.50%
	0200	Plumbing		3.30	4.32	6.60	2.47%	2.72%	3.66%
	0300	Mechanical		14.10	15.85	23	7.80%	13.10%	13.60%
	0400	Electrical		14.75	21	22	8.20%	12.65%	18.15%
	0500	Total Project Costs		168	175	181			
17	0000	**Police Stations**	S.F.						
	0100	Architectural		116	130	164	49%	56.50%	61%
	0200	Plumbing		15.30	18.40	18.50	5.05%	5.55%	9.05%
	0300	Mechanical		34.50	48.50	50.50	13%	14.55%	16.55%
	0400	Electrical		26	28.50	30.50	9.15%	12.10%	14%
	0500	Total Project Costs		217	267	305			
18	0000	**Police/Fire**	S.F.						
	0100	Architectural		113	113	345	55.50%	66%	68%
	0200	Plumbing		9.05	9.35	34.50	5.45%	5.50%	5.55%
	0300	Mechanical		13.85	22	79	8.35%	12.70%	12.80%
	0400	Electrical		15.75	20	90.50	9.50%	11.75%	14.55%
	0500	Total Project Costs		166	171	625			
19	0000	**Public Assembly Buildings**	S.F.						
	0100	Architectural		115	159	238	57.50%	61.50%	66%
	0200	Plumbing		6.10	8.90	13.40	2.60%	3.36%	4.79%
	0300	Mechanical		12.85	23	35.50	6.55%	8.95%	12.45%
	0400	Electrical		19	26	41	8.60%	10.75%	13%
	0500	Total Project Costs		185	253	375			
20	0000	**Recreational**	S.F.						
	0100	Architectural		106	168	236	53.50%	60%	66%
	0200	Plumbing		7.80	14.75	21.50	3.08%	5%	6.85%
	0300	Mechanical		13.15	20	31.50	5.15%	6.95%	11.70%
	0400	Electrical		15.20	26.50	39.50	7.35%	8.95%	10.75%
	0500	Total Project Costs		168	281	445			

For customer support on your Site Work & Landscape Costs with RSMeans data, call 800.448.8182.

885

50 17 | Project Costs

		50 17 00 | Project Costs	UNIT	UNIT COSTS			% OF TOTAL		
				1/4	MEDIAN	3/4	1/4	MEDIAN	3/4
21	0000	**Restaurants**	S.F.						
	0100	Architectural		126	193	248	59%	60%	63.50%
	0200	Plumbing		13.80	31.50	40	7.35%	7.75%	8.95%
	0300	Mechanical		14.85	17.55	37	6.50%	8.15%	11.15%
	0400	Electrical		14.80	24	48.50	7.10%	10.30%	11.60%
	0500	Total Project Costs		208	305	415			
22	0000	**Retail**	S.F.						
	0100	Architectural		56	86	127	60%	62%	64.50%
	0200	Plumbing		7	9.25	12.10	5.05%	6.70%	9%
	0300	Mechanical		6.55	9.25	17.05	5.70%	6.20%	10.20%
	0400	Electrical		10.45	18.95	31.50	8.05%	11.25%	12.45%
	0500	Total Project Costs		85.50	114	186			
23	0000	**Schools**	S.F.						
	0100	Architectural		97	125	164	52.50%	56%	61%
	0200	Plumbing		7.75	10.60	16.05	3.85%	4.82%	7.25%
	0300	Mechanical		18.90	26.50	38.50	9.50%	12.35%	15.10%
	0400	Electrical		17.85	25	32.50	9.45%	11.45%	13.30%
	0500	Total Project Costs		169	227	300			
24	0000	**University, College & Private School Classroom & Admin Buildings**	S.F.						
	0100	Architectural		124	153	192	50.50%	55%	59.50%
	0200	Plumbing		7.05	10.95	15.45	2.74%	4.30%	6.35%
	0300	Mechanical		26.50	38.50	46	10.10%	12.15%	14.70%
	0400	Electrical		19.95	28	34	7.65%	9.50%	11.55%
	0500	Total Project Costs		205	284	375			
25	0000	**University, College & Private School Dormitories**	S.F.						
	0100	Architectural		81	142	151	54.50%	65%	68.50%
	0200	Plumbing		10.70	15.10	22.50	6.45%	7.30%	9.15%
	0300	Mechanical		4.79	20.50	32.50	4.13%	9%	12.05%
	0400	Electrical		5.70	19.75	30	4.75%	7.35%	12.30%
	0500	Total Project Costs		119	227	268			
26	0000	**University, College & Private School Science, Eng. & Lab Buildings**	S.F.						
	0100	Architectural		143	164	193	50.50%	56.50%	58%
	0200	Plumbing		9.60	15.15	26.50	3.29%	3.95%	8.40%
	0300	Mechanical		43.50	68.50	70	12.75%	19.40%	23.50%
	0400	Electrical		29	33.50	38.50	9%	11.55%	13.15%
	0500	Total Project Costs		290	315	365			
27	0000	**University, College & Private School Student Union Buildings**	S.F.						
	0100	Architectural		110	289	289	54.50%	54.50%	59.50%
	0200	Plumbing		16.65	16.65	25	3.13%	4.27%	11.45%
	0300	Mechanical		31.50	51	51	9.60%	9.60%	14.55%
	0400	Electrical		27.50	48	48	9.05%	12.80%	13.15%
	0500	Total Project Costs		217	530	530			
28	0000	**Warehouses**	S.F.						
	0100	Architectural		47	73.50	174	61.50%	67.50%	72%
	0200	Plumbing		2.45	5.25	10.10	2.82%	3.72%	5%
	0300	Mechanical		2.91	16.55	26	4.56%	8.20%	10.70%
	0400	Electrical		5.30	19.85	33	7.50%	10.10%	18.30%
	0500	Total Project Costs		70.50	125	243			

Square Foot Project Size Modifier

One factor that affects the S.F. cost of a particular building is the size. In general, for buildings built to the same specifications in the same locality, the larger building will have the lower S.F. cost. This is due mainly to the decreasing contribution of the exterior walls plus the economy of scale usually achievable in larger buildings. The Area Conversion Scale shown below will give a factor to convert costs for the typical size building to an adjusted cost for the particular project.

Example: Determine the cost per S.F. for a 152,600 S.F. Multi-family housing.

$$\frac{\text{Proposed building area} = 152,600 \text{ S.F.}}{\text{Typical size from below} = 49,900 \text{ S.F.}} = 2.00$$

Enter Area Conversion Scale at 2.0, intersect curve, read horizontally the appropriate cost multiplier of .94. Size adjusted cost becomes .94 x $208.00 = $195.52 based on national average costs.

Note: For Size Factors less than .50, the Cost Multiplier is 1.1
 For Size Factors greater than 3.5, the Cost Multiplier is .90

The Square Foot Base Size lists the median costs, most typical project size in our accumulated data, and the range in size of the projects.

The Size Factor for your project is determined by dividing your project area in S.F. by the typical project size for the particular Building Type. With this factor, enter the Area Conversion Scale at the appropriate Size Factor and determine the appropriate Cost Multiplier for your building size.

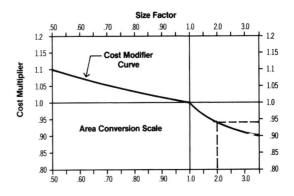

System	Median Cost (Total Project Costs)	Typical Size Gross S.F. (Median of Projects)	Typical Range (Low – High) (Projects)
Auto Sales with Repair	$186.00	24,900	4,700 – 29,300
Banking Institutions	299.00	9,300	3,300 – 38,100
Court House	284.00	47,600	24,700 – 70,500
Data Centers	281.00	14,400	14,400 – 14,400
Detention Centers	315.00	37,800	12,300 – 183,300
Fire Stations	238.00	13,100	6,300 – 49,600
Gymnasium	204.00	52,400	22,800 – 82,000
Hospitals	375.00	87,100	22,400 – 410,300
Industrial Buildings	104.00	22,100	5,100 – 200,600
Medical Clinics & Offices	217.00	20,900	2,300 – 327,000
Mixed Use	216.00	28,500	7,200 – 188,900
Multi-Family Housing	208.00	49,900	2,500 – 1,161,500
Nursing Home & Assisted Living	160.00	38,200	1,500 – 242,600
Office Buildings	200.00	20,500	1,100 – 930,000
Parking Garage	47.00	151,800	99,900 – 287,000
Parking Garage/Mixed Use	175.00	254,200	5,300 – 318,000
Police Stations	267.00	28,500	15,400 – 88,600
Police/Fire	171.00	44,300	8,600 – 50,300
Public Assembly Buildings	253.00	21,000	2,200 – 235,300
Recreational	281.00	28,800	1,000 – 223,800
Restaurants	305.00	6,000	4,000 – 42,000
Retail	114.00	28,700	5,200 – 84,300
Schools	227.00	70,600	1,300 – 410,800
University, College & Private School Classroom & Admin Buildings	284.00	48,300	9,400 – 196,200
University, College & Private School Dormitories	227.00	28,900	1,500 – 126,900
University, College & Private School Science, Eng. & Lab Buildings	315.00	39,800	5,300 – 117,600
University, College & Private School Student Union Buildings	530.00	48,700	42,100 – 50,000
Warehouses	125.00	10,400	600 – 303,800

A	Area Square Feet; Ampere	Brk., brk	Brick	Csc	Cosecant
AAFES	Army and Air Force Exchange Service	brkt	Bracket	C.S.F.	Hundred Square Feet
		Brs.	Brass	CSI	Construction Specifications Institute
ABS	Acrylonitrile Butadiene Stryrene; Asbestos Bonded Steel	Brz.	Bronze		
		Bsn.	Basin	CT	Current Transformer
A.C., AC	Alternating Current; Air-Conditioning; Asbestos Cement; Plywood Grade A & C	Btr.	Better	CTS	Copper Tube Size
		BTU	British Thermal Unit	Cu	Copper, Cubic
		BTUH	BTU per Hour	Cu. Ft.	Cubic Foot
		Bu.	Bushels	cw	Continuous Wave
ACI	American Concrete Institute	BUR	Built-up Roofing	C.W.	Cool White; Cold Water
ACR	Air Conditioning Refrigeration	BX	Interlocked Armored Cable	Cwt.	100 Pounds
ADA	Americans with Disabilities Act	°C	Degree Centigrade	C.W.X.	Cool White Deluxe
AD	Plywood, Grade A & D	c	Conductivity, Copper Sweat	C.Y.	Cubic Yard (27 cubic feet)
Addit.	Additional	C	Hundred; Centigrade	C.Y./Hr.	Cubic Yard per Hour
Adh.	Adhesive	C/C	Center to Center, Cedar on Cedar	Cyl.	Cylinder
Adj.	Adjustable	C-C	Center to Center	d	Penny (nail size)
af	Audio-frequency	Cab	Cabinet	D	Deep; Depth; Discharge
AFFF	Aqueous Film Forming Foam	Cair.	Air Tool Laborer	Dis., Disch.	Discharge
AFUE	Annual Fuel Utilization Efficiency	Cal.	Caliper	Db	Decibel
AGA	American Gas Association	Calc	Calculated	Dbl.	Double
Agg.	Aggregate	Cap.	Capacity	DC	Direct Current
A.H., Ah	Ampere Hours	Carp.	Carpenter	DDC	Direct Digital Control
A hr.	Ampere-hour	C.B.	Circuit Breaker	Demob.	Demobilization
A.H.U., AHU	Air Handling Unit	C.C.A.	Chromate Copper Arsenate	d.f.t.	Dry Film Thickness
A.I.A.	American Institute of Architects	C.C.F.	Hundred Cubic Feet	d.f.u.	Drainage Fixture Units
AIC	Ampere Interrupting Capacity	cd	Candela	D.H.	Double Hung
Allow.	Allowance	cd/sf	Candela per Square Foot	DHW	Domestic Hot Water
alt., alt	Alternate	CD	Grade of Plywood Face & Back	DI	Ductile Iron
Alum.	Aluminum	CDX	Plywood, Grade C & D, exterior glue	Diag.	Diagonal
a.m.	Ante Meridiem			Diam., Dia	Diameter
Amp.	Ampere	Cefi.	Cement Finisher	Distrib.	Distribution
Anod.	Anodized	Cem.	Cement	Div.	Division
ANSI	American National Standards Institute	CF	Hundred Feet	Dk.	Deck
		C.F.	Cubic Feet	D.L.	Dead Load; Diesel
APA	American Plywood Association	CFM	Cubic Feet per Minute	DLH	Deep Long Span Bar Joist
Approx.	Approximate	CFRP	Carbon Fiber Reinforced Plastic	dlx	Deluxe
Apt.	Apartment	c.g.	Center of Gravity	Do.	Ditto
Asb.	Asbestos	CHW	Chilled Water; Commercial Hot Water	DOP	Dioctyl Phthalate Penetration Test (Air Filters)
A.S.B.C.	American Standard Building Code				
Asbe.	Asbestos Worker	C.I., CI	Cast Iron	Dp., dp	Depth
ASCE	American Society of Civil Engineers	C.I.P., CIP	Cast in Place	D.P.S.T.	Double Pole, Single Throw
A.S.H.R.A.E.	American Society of Heating, Refrig. & AC Engineers	Circ.	Circuit	Dr.	Drive
		C.L.	Carload Lot	DR	Dimension Ratio
ASME	American Society of Mechanical Engineers	CL	Chain Link	Drink.	Drinking
		Clab.	Common Laborer	D.S.	Double Strength
ASTM	American Society for Testing and Materials	Clam	Common Maintenance Laborer	D.S.A.	Double Strength A Grade
		C.L.F.	Hundred Linear Feet	D.S.B.	Double Strength B Grade
Attchmt.	Attachment	CLF	Current Limiting Fuse	Dty.	Duty
Avg., Ave.	Average	CLP	Cross Linked Polyethylene	DWV	Drain Waste Vent
AWG	American Wire Gauge	cm	Centimeter	DX	Deluxe White, Direct Expansion
AWWA	American Water Works Assoc.	CMP	Corr. Metal Pipe	dyn	Dyne
Bbl.	Barrel	CMU	Concrete Masonry Unit	e	Eccentricity
B&B, BB	Grade B and Better; Balled & Burlapped	CN	Change Notice	E	Equipment Only; East; Emissivity
		Col.	Column	Ea.	Each
B&S	Bell and Spigot	CO₂	Carbon Dioxide	EB	Encased Burial
B.&W.	Black and White	Comb.	Combination	Econ.	Economy
b.c.c.	Body-centered Cubic	comm.	Commercial, Communication	E.C.Y	Embankment Cubic Yards
B.C.Y.	Bank Cubic Yards	Compr.	Compressor	EDP	Electronic Data Processing
BE	Bevel End	Conc.	Concrete	EIFS	Exterior Insulation Finish System
B.F.	Board Feet	Cont., cont	Continuous; Continued, Container	E.D.R.	Equiv. Direct Radiation
Bg. cem.	Bag of Cement	Corkbd.	Cork Board	Eq.	Equation
BHP	Boiler Horsepower; Brake Horsepower	Corr.	Corrugated	EL	Elevation
		Cos	Cosine	Elec.	Electrician; Electrical
B.I.	Black Iron	Cot	Cotangent	Elev.	Elevator; Elevating
bidir.	bidirectional	Cov.	Cover	EMT	Electrical Metallic Conduit; Thin Wall Conduit
Bit., Bitum.	Bituminous	C/P	Cedar on Paneling		
Bit., Conc.	Bituminous Concrete	CPA	Control Point Adjustment	Eng.	Engine, Engineered
Bk.	Backed	Cplg.	Coupling	EPDM	Ethylene Propylene Diene Monomer
Bkrs.	Breakers	CPM	Critical Path Method		
Bldg., bldg	Building	CPVC	Chlorinated Polyvinyl Chloride	EPS	Expanded Polystyrene
Blk.	Block	C.Pr.	Hundred Pair	Eqhv.	Equip. Oper., Heavy
Bm.	Beam	CRC	Cold Rolled Channel	Eqlt.	Equip. Oper., Light
Boil.	Boilermaker	Creos.	Creosote	Eqmd.	Equip. Oper., Medium
bpm	Blows per Minute	Crpt.	Carpet & Linoleum Layer	Eqmm.	Equip. Oper., Master Mechanic
BR	Bedroom	CRT	Cathode-ray Tube	Eqol.	Equip. Oper., Oilers
Brg., brng.	Bearing	CS	Carbon Steel, Constant Shear Bar Joist	Equip.	Equipment
Brhe.	Bricklayer Helper			ERW	Electric Resistance Welded
Bric.	Bricklayer				

.S.	Energy Saver
Est.	Estimated
su	Electrostatic Units
.W.	Each Way
WT	Entering Water Temperature
xcav.	Excavation
xcl	Excluding
xp., exp	Expansion, Exposure
xt., ext	Exterior; Extension
xtru.	Extrusion
	Fiber Stress
F	Fahrenheit; Female; Fill
ab., fab	Fabricated; Fabric
BGS	Fiberglass
C.	Footcandles
c.c.	Face-centered Cubic
c.	Compressive Stress in Concrete; Extreme Compressive Stress
E.	Front End
FEP	Fluorinated Ethylene Propylene (Teflon)
G.	Flat Grain
HA.	Federal Housing Administration
Fig.	Figure
Fin.	Finished
FIPS	Female Iron Pipe Size
Fixt.	Fixture
FJP	Finger jointed and primed
Fl. Oz.	Fluid Ounces
Flr.	Floor
Flrs.	Floors
FM	Frequency Modulation; Factory Mutual
Fmg.	Framing
FM/UL	Factory Mutual/Underwriters Labs
Fdn.	Foundation
FNPT	Female National Pipe Thread
Fori.	Foreman, Inside
Foro.	Foreman, Outside
Fount.	Fountain
fpm	Feet per Minute
FPT	Female Pipe Thread
Fr	Frame
FR.	Fire Rating
FRK	Foil Reinforced Kraft
FSK	Foil/Scrim/Kraft
FRP	Fiberglass Reinforced Plastic
FS	Forged Steel
FSC	Cast Body; Cast Switch Box
Ft., ft	Foot; Feet
Ftng.	Fitting
Ftg.	Footing
Ft lb.	Foot Pound
Furn.	Furniture
FVNR	Full Voltage Non-Reversing
FVR	Full Voltage Reversing
FXM	Female by Male
Fy.	Minimum Yield Stress of Steel
g	Gram
G	Gauss
Ga.	Gauge
Gal., gal.	Gallon
Galv., galv	Galvanized
GC/MS	Gas Chromatograph/Mass Spectrometer
Gen.	General
GFI	Ground Fault Interrupter
GFRC	Glass Fiber Reinforced Concrete
Glaz.	Glazier
GPD	Gallons per Day
gpf	Gallon per Flush
GPH	Gallons per Hour
gpm, GPM	Gallons per Minute
GR	Grade
Gran.	Granular
Grnd.	Ground
GVW	Gross Vehicle Weight
GWB	Gypsum Wall Board

H	High Henry
HC	High Capacity
H.D., HD	Heavy Duty; High Density
H.D.O.	High Density Overlaid
HDPE	High Density Polyethylene Plastic
Hdr.	Header
Hdwe.	Hardware
H.I.D., HID	High Intensity Discharge
Help.	Helper Average
HEPA	High Efficiency Particulate Air Filter
Hg	Mercury
HIC	High Interrupting Capacity
HM	Hollow Metal
HMWPE	High Molecular Weight Polyethylene
HO	High Output
Horiz.	Horizontal
H.P., HP	Horsepower; High Pressure
H.P.F.	High Power Factor
Hr.	Hour
Hrs./Day	Hours per Day
HSC	High Short Circuit
Ht.	Height
Htg.	Heating
Htrs.	Heaters
HVAC	Heating, Ventilation & Air-Conditioning
Hvy.	Heavy
HW	Hot Water
Hyd.; Hydr.	Hydraulic
Hz	Hertz (cycles)
I.	Moment of Inertia
IBC	International Building Code
I.C.	Interrupting Capacity
ID	Inside Diameter
I.D.	Inside Dimension; Identification
I.F.	Inside Frosted
I.M.C.	Intermediate Metal Conduit
In.	Inch
Incan.	Incandescent
Incl.	Included; Including
Int.	Interior
Inst.	Installation
Insul., insul	Insulation/Insulated
I.P.	Iron Pipe
I.P.S., IPS	Iron Pipe Size
IPT	Iron Pipe Threaded
I.W.	Indirect Waste
J	Joule
J.I.C.	Joint Industrial Council
K	Thousand; Thousand Pounds; Heavy Wall Copper Tubing, Kelvin
K.A.H.	Thousand Amp. Hours
kcmil	Thousand Circular Mils
KD	Knock Down
K.D.A.T.	Kiln Dried After Treatment
kg	Kilogram
kG	Kilogauss
kgf	Kilogram Force
kHz	Kilohertz
Kip	1000 Pounds
KJ	Kilojoule
K.L.	Effective Length Factor
K.L.F.	Kips per Linear Foot
Km	Kilometer
KO	Knock Out
K.S.F.	Kips per Square Foot
K.S.I.	Kips per Square Inch
kV	Kilovolt
kVA	Kilovolt Ampere
kVAR	Kilovar (Reactance)
KW	Kilowatt
KWh	Kilowatt-hour
L	Labor Only; Length; Long; Medium Wall Copper Tubing
Lab.	Labor
lat	Latitude

Lath.	Lather
Lav.	Lavatory
lb.; #	Pound
L.B., LB	Load Bearing; L Conduit Body
L. & E.	Labor & Equipment
lb./hr.	Pounds per Hour
lb./L.F.	Pounds per Linear Foot
lbf/sq.in.	Pound-force per Square Inch
L.C.L.	Less than Carload Lot
L.C.Y.	Loose Cubic Yard
Ld.	Load
LE	Lead Equivalent
LED	Light Emitting Diode
L.F.	Linear Foot
L.F. Hdr	Linear Feet of Header
L.F. Nose	Linear Foot of Stair Nosing
L.F. Rsr	Linear Foot of Stair Riser
Lg.	Long; Length; Large
L & H	Light and Heat
LH	Long Span Bar Joist
L.H.	Labor Hours
L.L., LL	Live Load
L.L.D.	Lamp Lumen Depreciation
lm	Lumen
lm/sf	Lumen per Square Foot
lm/W	Lumen per Watt
LOA	Length Over All
log	Logarithm
L-O-L	Lateralolet
long.	Longitude
L.P., LP	Liquefied Petroleum; Low Pressure
L.P.F.	Low Power Factor
LR	Long Radius
L.S.	Lump Sum
Lt.	Light
Lt. Ga.	Light Gauge
L.T.L.	Less than Truckload Lot
Lt. Wt.	Lightweight
L.V.	Low Voltage
M	Thousand; Material; Male; Light Wall Copper Tubing
M^2CA	Meters Squared Contact Area
m/hr.; M.H.	Man-hour
mA	Milliampere
Mach.	Machine
Mag. Str.	Magnetic Starter
Maint.	Maintenance
Marb.	Marble Setter
Mat; Mat'l.	Material
Max.	Maximum
MBF	Thousand Board Feet
MBH	Thousand BTU's per hr.
MC	Metal Clad Cable
MCC	Motor Control Center
M.C.F.	Thousand Cubic Feet
MCFM	Thousand Cubic Feet per Minute
M.C.M.	Thousand Circular Mils
MCP	Motor Circuit Protector
MD	Medium Duty
MDF	Medium-density fibreboard
M.D.O.	Medium Density Overlaid
Med.	Medium
MF	Thousand Feet
M.F.B.M.	Thousand Feet Board Measure
Mfg.	Manufacturing
Mfrs.	Manufacturers
mg	Milligram
MGD	Million Gallons per Day
MGPH	Million Gallons per Hour
MH, M.H.	Manhole; Metal Halide; Man-Hour
MHz	Megahertz
Mi.	Mile
MI	Malleable Iron; Mineral Insulated
MIPS	Male Iron Pipe Size
mj	Mechanical Joint
m	Meter
mm	Millimeter
Mill.	Millwright
Min., min.	Minimum, Minute

Misc.	Miscellaneous	PCM	Phase Contrast Microscopy	SBS	Styrene Butadiere Styrene
ml	Milliliter, Mainline	PDCA	Painting and Decorating	SC	Screw Cover
M.L.F.	Thousand Linear Feet		Contractors of America	SCFM	Standard Cubic Feet per Minute
Mo.	Month	P.E., PE	Professional Engineer;	Scaf.	Scaffold
Mobil.	Mobilization		Porcelain Enamel;	Sch., Sched.	Schedule
Mog.	Mogul Base		Polyethylene; Plain End	S.C.R.	Modular Brick
MPH	Miles per Hour	P.E.C.I.	Porcelain Enamel on Cast Iron	S.D.	Sound Deadening
MPT	Male Pipe Thread	Perf.	Perforated	SDR	Standard Dimension Ratio
MRGWB	Moisture Resistant Gypsum	PEX	Cross Linked Polyethylene	S.E.	Surfaced Edge
	Wallboard	Ph.	Phase	Sel.	Select
MRT	Mile Round Trip	P.I.	Pressure Injected	SER, SEU	Service Entrance Cable
ms	Millisecond	Pile.	Pile Driver	S.F.	Square Foot
M.S.F.	Thousand Square Feet	Pkg.	Package	S.F.C.A.	Square Foot Contact Area
Mstz.	Mosaic & Terrazzo Worker	Pl.	Plate	S.F. Flr.	Square Foot of Floor
M.S.Y.	Thousand Square Yards	Plah.	Plasterer Helper	S.F.G.	Square Foot of Ground
Mtd., mtd., mtd	Mounted	Plas.	Plasterer	S.F. Hor.	Square Foot Horizontal
Mthe.	Mosaic & Terrazzo Helper	plf	Pounds Per Linear Foot	SFR	Square Feet of Radiation
Mtng.	Mounting	Pluh.	Plumber Helper	S.F. Shlf.	Square Foot of Shelf
Mult.	Multi; Multiply	Plum.	Plumber	S4S	Surface 4 Sides
MUTCD	Manual on Uniform Traffic Control	Ply.	Plywood	Shee.	Sheet Metal Worker
	Devices	p.m.	Post Meridiem	Sin.	Sine
M.V.A.	Million Volt Amperes	Pntd.	Painted	Skwk.	Skilled Worker
M.V.A.R.	Million Volt Amperes Reactance	Pord.	Painter, Ordinary	SL	Saran Lined
MV	Megavolt	pp	Pages	S.L.	Slimline
MW	Megawatt	PP, PPL	Polypropylene	Sldr.	Solder
MXM	Male by Male	P.P.M.	Parts per Million	SLH	Super Long Span Bar Joist
MYD	Thousand Yards	Pr.	Pair	S.N.	Solid Neutral
N	Natural; North	P.E.S.B.	Pre-engineered Steel Building	SO	Stranded with oil resistant inside
nA	Nanoampere	Prefab.	Prefabricated		insulation
NA	Not Available; Not Applicable	Prefin.	Prefinished	S-O-L	Socketolet
N.B.C.	National Building Code	Prop.	Propelled	sp	Standpipe
NC	Normally Closed	PSF, psf	Pounds per Square Foot	S.P.	Static Pressure; Single Pole; Self-
NEMA	National Electrical Manufacturers	PSI, psi	Pounds per Square Inch		Propelled
	Assoc.	PSIG	Pounds per Square Inch Gauge	Spri.	Sprinkler Installer
NEHB	Bolted Circuit Breaker to 600V.	PSP	Plastic Sewer Pipe	spwg	Static Pressure Water Gauge
NFPA	National Fire Protection Association	Pspr.	Painter, Spray	S.P.D.T.	Single Pole, Double Throw
NLB	Non-Load-Bearing	Psst.	Painter, Structural Steel	SPF	Spruce Pine Fir; Sprayed
NM	Non-Metallic Cable	P.T.	Potential Transformer		Polyurethane Foam
nm	Nanometer	P. & T.	Pressure & Temperature	S.P.S.T.	Single Pole, Single Throw
No.	Number	Ptd.	Painted	SPT	Standard Pipe Thread
NO	Normally Open	Ptns.	Partitions	Sq.	Square; 100 Square Feet
N.O.C.	Not Otherwise Classified	Pu	Ultimate Load	Sq. Hd.	Square Head
Nose.	Nosing	PVC	Polyvinyl Chloride	Sq. In.	Square Inch
NPT	National Pipe Thread	Pvmt.	Pavement	S.S.	Single Strength; Stainless Steel
NQOD	Combination Plug-on/Bolt on	PRV	Pressure Relief Valve	S.S.B.	Single Strength B Grade
	Circuit Breaker to 240V.	Pwr.	Power	sst, ss	Stainless Steel
N.R.C., NRC	Noise Reduction Coefficient/	Q	Quantity Heat Flow	Sswk.	Structural Steel Worker
	Nuclear Regulator Commission	Qt.	Quart	Sswl.	Structural Steel Welder
N.R.S.	Non Rising Stem	Quan., Qty.	Quantity	St.; Stl.	Steel
ns	Nanosecond	Q.C.	Quick Coupling	STC	Sound Transmission Coefficient
NTP	Notice to Proceed	r	Radius of Gyration	Std.	Standard
nW	Nanowatt	R	Resistance	Stg.	Staging
OB	Opposing Blade	R.C.P.	Reinforced Concrete Pipe	STK	Select Tight Knot
OC	On Center	Rect.	Rectangle	STP	Standard Temperature & Pressure
OD	Outside Diameter	recpt.	Receptacle	Stpi.	Steamfitter, Pipefitter
O.D.	Outside Dimension	Reg.	Regular	Str.	Strength; Starter; Straight
ODS	Overhead Distribution System	Reinf.	Reinforced	Strd.	Stranded
O.G.	Ogee	Req'd.	Required	Struct.	Structural
O.H.	Overhead	Res.	Resistant	Sty.	Story
O&P	Overhead and Profit	Resi.	Residential	Subj.	Subject
Oper.	Operator	RF	Radio Frequency	Subs.	Subcontractors
Opng.	Opening	RFID	Radio-frequency Identification	Surf.	Surface
Orna.	Ornamental	Rgh.	Rough	Sw.	Switch
OSB	Oriented Strand Board	RGS	Rigid Galvanized Steel	Swbd.	Switchboard
OS&Y	Outside Screw and Yoke	RHW	Rubber, Heat & Water Resistant;	S.Y.	Square Yard
OSHA	Occupational Safety and Health		Residential Hot Water	Syn.	Synthetic
	Act	rms	Root Mean Square	S.Y.P.	Southern Yellow Pine
Ovhd.	Overhead	Rnd.	Round	Sys.	System
OWG	Oil, Water or Gas	Rodm.	Rodman	t.	Thickness
Oz.	Ounce	Rofc.	Roofer, Composition	T	Temperature; Ton
P.	Pole; Applied Load; Projection	Rofp.	Roofer, Precast	Tan	Tangent
p.	Page	Rohe.	Roofer Helpers (Composition)	T.C.	Terra Cotta
Pape.	Paperhanger	Rots.	Roofer, Tile & Slate	T & C	Threaded and Coupled
P.A.P.R.	Powered Air Purifying Respirator	R.O.W.	Right of Way	T.D.	Temperature Difference
PAR	Parabolic Reflector	RPM	Revolutions per Minute	TDD	Telecommunications Device for
P.B., PB	Push Button	R.S.	Rapid Start		the Deaf
Pc., Pcs.	Piece, Pieces	Rsr	Riser	T.E.M.	Transmission Electron Microscopy
P.C.	Portland Cement; Power Connector	RT	Round Trip	temp	Temperature, Tempered, Temporary
P.C.F.	Pounds per Cubic Foot	S.	Suction; Single Entrance; South	TFFN	Nylon Jacketed Wire

Abbreviations

TFE	Tetrafluoroethylene (Teflon)	U.L., UL	Underwriters Laboratory	w/	With
T. & G.	Tongue & Groove;	Uld.	Unloading	W.C., WC	Water Column; Water Closet
	Tar & Gravel	Unfin.	Unfinished	W.F.	Wide Flange
Th., Thk.	Thick	UPS	Uninterruptible Power Supply	W.G.	Water Gauge
Thn.	Thin	URD	Underground Residential	Wldg.	Welding
Thrded	Threaded		Distribution	W. Mile	Wire Mile
Tilf.	Tile Layer, Floor	US	United States	W-O-L	Weldolet
Tilh.	Tile Layer, Helper	USGBC	U.S. Green Building Council	W.R.	Water Resistant
THHN	Nylon Jacketed Wire	USP	United States Primed	Wrck.	Wrecker
THW.	Insulated Strand Wire	UTMCD	Uniform Traffic Manual For Control	WSFU	Water Supply Fixture Unit
THWN	Nylon Jacketed Wire		Devices	W.S.P.	Water, Steam, Petroleum
T.L., TL	Truckload	UTP	Unshielded Twisted Pair	WT., Wt.	Weight
T.M.	Track Mounted	V	Volt	WWF	Welded Wire Fabric
Tot.	Total	VA	Volt Amperes	XFER	Transfer
T-O-L	Threadolet	VAT	Vinyl Asbestos Tile	XFMR	Transformer
tmpd	Tempered	V.C.T.	Vinyl Composition Tile	XHD	Extra Heavy Duty
TPO	Thermoplastic Polyolefin	VAV	Variable Air Volume	XHHW	Cross-Linked Polyethylene Wire
T.S.	Trigger Start	VC	Veneer Core	XLPE	Insulation
Tr.	Trade	VDC	Volts Direct Current	XLP	Cross-linked Polyethylene
Transf.	Transformer	Vent.	Ventilation	Xport	Transport
Trhv.	Truck Driver, Heavy	Vert.	Vertical	Y	Wye
Trlr	Trailer	V.F.	Vinyl Faced	yd	Yard
Trlt.	Truck Driver, Light	V.G.	Vertical Grain	yr	Year
TTY	Teletypewriter	VHF	Very High Frequency	Δ	Delta
TV	Television	VHO	Very High Output	%	Percent
T.W.	Thermoplastic Water Resistant	Vib.	Vibrating	~	Approximately
	Wire	VLF	Vertical Linear Foot	Ø	Phase; diameter
UCI	Uniform Construction Index	VOC	Volatile Organic Compound	@	At
UF	Underground Feeder	Vol.	Volume	#	Pound; Number
UGND	Underground Feeder	VRP	Vinyl Reinforced Polyester	<	Less Than
UHF	Ultra High Frequency	W	Wire; Watt; Wide; West	>	Greater Than
U.I.	United Inch			Z	Zone

A

Abandon catch basin 28
ABC extinguisher portable 182
   extinguisher wheeled 182
Abrasive aluminum oxide 52
   floor 82
   silicon carbide 52, 82
ABS DWV pipe 225
Absorber shock 241
Absorption testing 14
Access door and panel 163
   door basement 85
   door floor 163
   road and parking area 21
Accessory anchor bolt 67
   dock 558
   masonry 96
   rebar 71
   reinforcing 71
   steel wire rope 120
Acid etch finish concrete 82
   proof floor 174
   resistant pipe 257
Acoustical booth 208
   ceiling 174
   enclosure 208
Acrylic coatings epoxy 384
Acrylic emulsion 384
   sign 180
Activated sludge treatment cell . . 567
ADA compliant drinking fount. . . 255
Adhesive EPDM 143
   neoprene 143
   PVC 143
   roof 151
Adjustable jack post 116
Adjustment factor 10
   factor labor 212
Admixture cement 82
   concrete 52
Aerate lawn 366
Aeration blower 707
   equipment 563
   sewage 562
Aerator 256
Aerial photography 14
   seeding 405
   survey 26
Agent form release 53
Aggregate exposed 82, 369
   lightweight 53
   masonry 93
   spreader 730, 737
   testing 14
Air compressor 733
   compressor portable 733
   conditioner removal 260
   diffuser system 563, 566
   entraining agent 52
   hose 734
   lock 204
   process piping 564
   release & vacuum valve 493
   release valve 493
   release valve sewer 495
   spade 734
   supported building 203
   supported storage tank cover . 203
   supported structures 860
Air-compressor mobilization 289
Aircraft cable 123
Airplane hangar 209
Airport compaction 333
   painted marking 382, 383
   thermoplastic marking 383

Alloy steel chain 132
Altitude valve 492
Aluminum bench 398
   bulkhead 555
   coping 109
   curtain wall glazed 165
   dome 205
   door 164
   downspout 152
   drip edge 154
   edging 468
   entrance 165
   expansion joint 154
   fence 386, 387
   flagpole 184
   flashing 148, 152
   foil 146
   gangway 557
   grating 127
   gutter 152
   joist shoring 63
   ladder 557
   light pole 722
   mesh grating 129
   oxide abrasive 52
   pipe 501, 504
   plank grating 128
   roof 147
   roof panel 147
   salvage 213
   sash 165
   screw 148
   service entrance cable 268
   sheet metal 152
   sheeting 555
   shelter 209
   shielded cable 267
   siding 148
   siding panel 148
   sign 180
   storefront 165
   tile 147
   trench cover 130
   weld rod 116
   window 165
   window demolition 160
American elm 447
Analysis petrographic 14
   sieve 14
Anchor bolt 67, 68, 94
   bolt accessory 67
   bolt screw 72
   bolt sleeve 67
   bolt template 67
   brick 94
   buck 94
   channel slot 95
   chemical 87, 113
   dovetail 70
   epoxy 87, 113
   expansion 113
   lead screw 113
   machinery 71
   masonry 94
   partition 95
   pipe conduit casing 533
   plastic screw 113
   pole 535
   rigid 95
   rock 65
   screw 72, 113
   steel 95
   stone 95
   toggle bolt 113
   wedge 113
Angle corner 117

curb edging 117
   valve bronze 213
Arch climber 190
   culvert oval 508
   radial 138
Architectural fee 10, 830
Area window well 184
Arrow 381
Articulating boom lift 733
Artificial grass surfacing 384
   plant 194
   turf 384
Ashlar stone 106
   veneer 104
Asphalt base sheet 151
   block 378
   block floor 378
   coating 142
   concrete 288
   curb 380
   cutting 44
   distributor 734
   felt 151
   flashing 152
   flood coat 151
   grinding 361
   mix design 14
   pavement 384
   paver 735
   paving 372
   paving cold reused 361
   paving cold-mix 375
   paving hot reused 361
   plant portable 737
   roof shingle 146
   rubberized 364
   sealcoat rubberized 360
   shingle 146
   sidewalk 369
   soil stabilization 335
   stabilization 335
   surface treatment 360
   tile paving 664
Asphaltic binder 372
   concrete 372
   concrete paving 375
   emulsion 360, 402
   pavement 372
   wearing course 372
Athletic backstop 388
   field seeding 405
   paving 384
   pole 190
   post 191
   screening demo 38
   surface demo 36
Athletic/playground
equipment demo 38
Attachment ripper 737
Attachments skid-steer 733
Auger boring 27
   earth 730
   hole 26
Augering 289
Auto park drain 250
Automotive equipment 188
   spray painting booth 188
Autoscrub machine 734
Awning canvas 183
   fabric 183
   window 183
   window aluminum 165
   window metal-clad 166

B

B and S pipe gasket 248
Backer rod 155
   rod polyethylene 67
Backerboard 143
Backfill 308, 310
   clay 638
   compaction 872
   dozer 308
   earth 636
   gravel 634
   planting pit 404
   sandy clay/loam 640
   structural 308
   trench 294, 297
Backflow preventer 240
Backhoe 301, 731
   bucket 732
   excavation 622, 626, 630, 633
   extension 737
   trenching 642, 644, 646
Backstop athletic 388
   baseball 388
   basketball 388
   chain link 388
   handball court 191
   squash court 191
   tennis court 388
Backwater valve 253
Balance excavation 624, 625
Balanced excavation . . 620, 621, 628,
                       629
Ball check valve 215
   check valve plastic 215
   valve plastic 215
   wrecking 737
Ballast high intensity discharge . . 279
   lighting 279
   railroad 543
   replacement 279
Bank run gravel 288
Bar chair reinforcing 71
   grating tread 131
   masonry reinforcing 96
   tie reinforcing 71
Barbed wire fence 386
Barge construction 740
   driven pile 348
   mobilization 138
Bark mulch 402
   mulch redwood 402
Barrels flasher 734
   reflectorized 734
Barricade 22, 545
   flasher 734
   tape 22
Barrier and enclosure 21
   crash 546
   delineator 546
   dust 21
   highway sound traffic 399
   impact 546
   median 545
   moisture 144
   parking 381
   security 189
   slipform paver 738
   weather 145
   weed 402
Base course 370, 371
   course drainage layer 370
   gravel 371
   road 371
   screed 72
   sheet 151

sheet asphalt 151
stabilization 335
stabilizer 738
stone 370
transformer 280
wall 175
Baseball backstop 388
scoreboard 190
Basement bulkhead stair 85
Base plate scaffold 19
shoring 19
Basket street 198
weave fence vinyl 392
Basketball backstop 388
Basket weave fence 393
Batch trial 15
Bathtub removal 213
Beach stone 370
Bead blast demo 170
Beam and girder formwork 54
bond 94, 100
bottom formwork 54
concrete 76
concrete placement 79
concrete placement grade 80
fireproofing 155
formwork side 54
precast 85
precast concrete 85
precast concrete tee 85
reinforcing 73
side formwork 54
soldier 343
test 15
Bearing pad 115
Bed fertilizer flower 366
maintenance flower 366
Bedding brick 378
pipe 310
placement joint 83
Belgian block 380
Bell & spigot pipe 243
caisson 356
Belt material handling 560
Bench 397
aluminum 398
fiberglass 397
greenhouse 206
park 397
planter 397
player 398
wood 398
Bender duct 539
Bentonite 144, 344
grout mix 342
panel waterproofing 144
Berm pavement 380
road 380
Bevel siding 149
Bicycle rack 190, 197
Bimodal MDPE, ASTM D2513 ... 513
Bin retaining wall metal 395
Binder asphaltic 372
Birch hollow core door 163
Bit core drill 89
Bitum concrete plant
mixed recycled 288
Bituminous block 378
coating 142, 472
cold patch 367
concrete curb 380
parking lot 656
paver 735
paving 650, 653, 656
roadway 650
waterproofing 585

Bituminous-stabilized
base course 372
Black steel pipe 488
Blanket curing 84
Blast demo bead 170
floor shot 170
Blaster shot 736
Blasting 299
cap 299
mat 299
Bleacher 206
outdoor 206
stadium 206
Block asphalt 378
Belgian 380
bituminous 378
Block, brick and mortar 849
Block concrete 99-103, 850
concrete bond beam 100
concrete exterior 102
decorative concrete 101
granite 379
grooved 101
ground face 101
hexagonal face 102
insulation 100
insulation insert concrete 100
lightweight 103
lintel 103
manhole 478
manhole/catch basin 708
partition 103
profile 101
scored 102
scored split face 101
slump 101
split rib 101
thrust 489
wall removal 40
Blower 707
control panel 565
control single phase duplex .. 565
control single phase simplex .. 565
control three phase duplex ... 566
control three phase simplex .. 565
coupling 564
insulation 735
prepackaged 563
pressure relief valve 564
rotary lobe 563
Bluegrass sod 406
Bluestone 105
sidewalk 370
step 370
Board & batten siding 149
directory 180
drain 144
fence 393
insulation 144
insulation foam 145
siding wood 149
Boat slip 556
work 740
Boiler demolition 260
mobilization 289
removal 260
Bollard 116
light 283
pipe 381
security 544
Bolt & gasket set 221
& hex nut steel 114
anchor 67, 68, 94
remove 112
steel 136
Bond 13

beam 94, 100
performance 13
Bonding agent 52
surface 93
Boom lift articulating 733
lift telescoping 733
Boot pile 346
rubber 479
Booth acoustical 208
painting 188
portable 208
ticket 208
Bored pile 355
Borer horizontal 737
Boring and exploratory drilling .. 26
auger 27
cased 26
horizontal 477
machine horizontal 730
service 478
Borosilicate pipe 256
Borrow 288
clay 310
loading and/or spreading 309
rock 310
select 310
Bottle vial 49
Boulder excavation 299
Boundary and survey marker ... 26
Bowstring truss 138
Box Buffalo 480
culvert 505
culvert demo 31
culvert formwork 55
distribution 500
electrical 274
electrical pull 274
fence shadow 394
flow leveler distribution 500
steel mesh 341
storage 18
trench 344, 737
utility 479
Brace cross 118
Bracing shoring 19
Bracket scaffold 19
Brass hinge 167
salvage 213
Brazed connection 271
Break tie concrete 82
Breaker pavement 734
Breakwater bulkhead residential . 555
Brick 99
anchor 94
bedding 378
bedding mortar 369
Brick, block and mortar 849
Brick cart 734
catch basin 478
common 98
common building 98
demolition 92, 170
economy 98
edging 468
engineer 99
face 98, 849
floor 174, 378
flooring miscellaneous 174
manhole 478
paving 378, 664
removal 28
sand 53
stair 104
step 370
testing 15
veneer 97

veneer thin 98
wall 104
Bricklaying 849
Bridge 399
concrete 399
deck 398
foundation 398
girder 398
highway 398
pedestrian 399
pedestrian demo 41
railing 398
sidewalk 20
Broadcast stolen 406
Bronze angle valve 213
globe valve 214
letter 180
plaque 180
swing check valve 213
valve 213
Broom finish concrete 81
sidewalk 736
Brown coat 175
Brownstone 107
Brush chipper 731
clearing 290
cutter 731, 732
mowing 290
Bubbler 255
drinking 255
Buck anchor 94
Bucket backhoe 732
clamshell 731
concrete 730
dragline 731
excavator 737
pavement 737
Buffalo box 480
Buggy concrete 81, 730
Builder's risk insurance 833
Building air-supported 203
component deconstruction 45
deconstruction 44
demo footing & foundation ... 40
demolition 40
demolition selective 41
excavation and backfill 602
greenhouse 205
hangar 209
model 10
moving 47
paper 146
permit 14
portable 18
prefabricated 205
relocation 47
shoring 341
site development 679, 680
slab 600
subdrain 584
temporary 18
tension 204
Built-up roof 151
roofing component 151
roofing system 151
Bulb end pile 599
Bulk bank measure excavating .. 301
dozer excavating 304
dragline excavation 305
drilling and blasting 299
scraper excavation 306
storage dome 208
Bulkhead aluminum 555
canal 555
door 163
marine 555

moveable 202
   residential 555
Bulkhead/cellar door 163
Bull run valve 500
Bulldozer 732
Bull float concrete 730
Bullnose block 102
Bumper car 381
   dock 189, 558
   door 167
   parking 381
   railroad 543
   wall 167
Buncher feller 731
Bunk house trailer 18
Burial cable direct 269
   cell 47
Buried conduit PVC 536
Burlap curing 84
   rubbing concrete 82
Bush hammer 82
   hammer finish concrete 82
Butt fusion machine 734
Butterfly valve 260, 491
   valve gear operated 565
   valve grooved joint 223
   valve lever operated 565
Butyl expansion joint 154
   rubber 155
   waterproofing 143

C

Cable aircraft 123
   aluminum shielded 267
   copper shielded 266
   guide rail 545
   jack 740
   medium-voltage 266
   pulling 737
   railing 134
   safety railing 123
   sheathed nonmetallic 268
   sheathed Romex 268
   tensioning 736
   trailer 736
   underground feeder 268
   underground URD 268
Cable/wire puller 737
Cactus 435
Caisson 598
   bell 356
   concrete 355
   displacement 357
   foundation 355
Caissons 874
Calcium chloride 52, 335
   silicate insulated system . . . 533
   silicate insulation 533
Canal bulkhead 555
   gate 553
Canopy 183
   entry 183
   metal 183
   patio/deck 183
Cant roof 151
Cantilever retaining wall 395
Canvas awning 183
Cap blasting 299
   concrete placement pile 80
   dowel 74
   end 502
   HDPE piping 488
   pile 78
   pipe 483, 497, 502

service entrance 269
Car bumper 381
   tram 737
Carborundum rub finish 82
Carpet removal 170
   tile removal 170
Carport 184
Cart brick 734
   concrete 81, 730
   mounted extinguisher 182
Cased boring 26
Casement window 165, 166
   window metal-clad 166
Casing anchor pipe conduit . . . 533
Cast-in-place concrete 76, 79
   in place concrete curb 379
   in place pile 357
   in place pile add 357
   iron bench 398
   iron fitting 244, 496
   iron grate cover 130
   iron manhole cover 509
   iron pipe 243, 496
   iron pipe fitting 244
   iron pipe gasket 247
   iron pull box 274
Caster scaffold 19
Casting construction 132
Cast iron column base 132
   weld rod 116
Catch basin 508
   basin and manhole 709
   basin brick 478
   basin cleaner 737
   basin cover remove/replace . . 509
   basin precast 478
   basin removal 28
   basin vacuum 737
   basin/manhole 708
Cathodic protection 209
Catwalk scaffold 20
Caulking 157
   lead 243
   oakum 243
   polyurethane 157
   sealant 156
Cavity truss reinforcing 96
   wall grout 94
   wall insulation 145
   wall reinforcing 96
Cedar fence 394
   pole wall 670
   post 136, 139
   post and board wall 671
   siding 149
Ceiling 174
   acoustical 174
   demolition 170
   painting 176
   suspended 174
   tile 174
Cell activated sludge treatment . . 567
Cellar door 163
Cellular concrete 77
   decking 125
Cement admixture 82
   color 93
   duct 539
   grout 94, 342
   liner 476
   masonry 93
   masonry unit 101-103
   mortar 848
   parging 142
   Portland 53
   PVC conduit 274

soil 335
   soil stabilization 335
   stabilization 335
   testing 15
   underlayment self-leveling . . 87
Cementitious coating 585
   waterproofing 144
Cementitious/metallic slurry . . 585
Center bulb waterstop 66
   meter 276
Centrifugal pump 736
Ceramic disc air diffuser system . 566
   mulch 402
   tile 173
   tile demolition 170
   tile waterproofing membrane . . 173
   tiling 173
Certification welding 16
Chain core rope 189
   fall hoist 739
   hoist 560
   hoist door 164
   link backstop 388
   link barbed wire 391
   link fabric 391
   link fence 23, 387, 393
   link fence & gate demo 34
   link fence high-security 390
   link gate 387
   link gate and fence 385
   link gate industrial 391
   link gate internal 391
   link gate residential 391
   link industrial fence 386
   link line post 390
   link terminal post 390
   link top rail 391
   link transom 392
   saw 736
   steel 132
   trencher 297, 733
Chair lifeguard 202
Chamber infiltration 501
   silencer 564
Chamfer strip 63
Channel curb edging 117
   grating tread 131
   siding 149
   slot 95
   slot anchor 95
Channelizing traffic 546
Charge powder 115
Check swing valve 213
   valve 493
   valve ball 215
   valve flanged cast iron 492
   valve flanged steel 565
   valve steel 261
   valve wafer style 564
Checkered plate 130
   plate cover 130
   plate platform 130
Chemical anchor 87, 113
   dry extinguisher 182
   spreader 737
   termite control 334
   toilet 736
   water closet 562
Chilled water distribution 530
   water pipe underground 531
Chilled/HVAC
hot water distribution 530
Chimney demolition 92
   foundation 76
Chip quartz 53
   silica 53

white marble 53
   wood 402
Chipper brush 731
   log 732
   stump 732
Chipping hammer 734
   stump 290
Chloride accelerator concrete . . . 79
Chlorination system 256
Chlorosulfunated polyethylene . . 47
Chute system 42
C.I.P. pile 586, 587
   wall 603
Circuit-breaker 277
   explosion proof 277
Circular saw 736
Circulating pump 263
Clamp ground rod 270
   pipe 472
   pipe repair 473
   ring 484
   water pipe ground 270
Clamshell 301, 555
   bucket 731
Clapboard painting 175
Clay 630, 638
   borrow 310
   cut & fill 628
   excavation 630
   fire 108
   pipe 499, 501
   tennis court 384
   tile 147
   tile coping 109
Clean concrete pile 550
   control joint 66
   ditch 307
   steel pile 551
   tank 48
   wood pile 550
Cleaner catch basin 737
   crack 738
   steam 736
Cleaning internal pipe 475
   marine pier pile 550
   metal 112
   up 24
Cleanout floor 216
   pipe 216
   PVC 216
   tee 216
Clear and grub 612
   and grub site 289
   grub 289
Clearing brush 290
   selective 290
   site 289
Cleat dock 557
Clevis hook 132
Climber arch 190
Climbing crane 738
   hydraulic jack 740
Clip wire rope 120
Closed deck grandstand 206
Closure trash 198
Cloth hardware 393
CMU 99
Coarse washed sand bed 370
Coat brown 175
   scratch 175
   tack 360
Coating 178
   & paint 177
   bituminous 142, 472
   epoxy 82, 472
   exterior steel 178

Index

intumescent 177
polyethylene 472
reinforcing 74
rubber 144
silicone 144
spray 142
structural steel 851
trowel 142
water repellent 144
waterproofing 142
Coats seal 375
Cofferdam 343
excavation 302
Cohesion testing 14
Coil bolt formwork 64
tie formwork 64
CO₂ extinguisher 182
Cold laid asphalt 375
milling asphalt paving 361
mix paver 738
patch 361
planing 361
recycling 361
Cold-mix asphalt paving 375
Collection concrete pipe sewage . 496
plastic pipe sewage 496
sewage/drainage 505
Color floor 82
Coloring concrete 53
Column 139
base cast-iron 132
brick 98
concrete 76
concrete placement 80
fireproof 155
footing 581
formwork 54
framing 136
lally 116
lightweight 116
plywood formwork 55
reinforcing 73
round fiber tube formwork 55
round fiberglass formwork 54
steel framed formwork 55
structural 116
structural shape 117
tie 95
wood 136
Command dog 23
Commercial 4′ H fence 386
greenhouse 205
graywater 253
half glass door 161
Commissioning 24
Common brick 98
building brick 98
earth 636
Compact fill 311
Compacted concrete roller 84
Compaction 332, 333
airport 333
backfill 872
earth 332
soil 308
structural 333
test Proctor 15
vibroflotation 873
Compactor earth 731
landfill 732
plate 332
sheepsfoot 332
tamper 333
Compensation workers 12
Compliant drinking fount. ADA . 255
Composite door 162

metal deck 124
Composition flooring removal . . . 170
Compound dustproofing 52
Compressed air equipment
automotive 188
Compressive strength 15
Compressor air 733
Concrete 847
acid etch finish 82
admixture 52
asphalt 288
asphaltic 372
beam 76
block 99-103, 850
block back-up 100
block bond beam 100
block decorative 101
block demolition 41
block exterior 102
block foundation 102
block grout 94
block insulation 145
block insulation insert 100
block lintel 103
block partition 103
block planter 378
break tie 82
bridge 399
broom finish 81
bucket 730
buggy 81, 730
bull float 730
burlap rubbing 82
bush hammer finish 82
caisson 355, 598
cart 81, 730
cast-in-place 76, 79
catch basin 478
cellular 77
chloride accelerator 79
coloring 53
column 76
conveyer 730
coping 109
core drilling 88
crane and bucket 80
cribbing 395
culvert 506
curb 379
curing 84, 377
curing compound 84
cutout 41
cylinder 15
cylinder pipe 482
dampproofing 54
demo 27
demolition 41, 52
direct chute 80
distribution box 500
drilling 89
encapsulated billet 556
equipment rental 730
fiber reinforcing 79
field mix 847
finish 377
flexural 77
float finish 81, 82
floor edger 730
floor grinder 730
floor patching 52
footing 77, 580, 581
form liner 60
forms 841, 842
foundation 603, 604
furring 137
granolithic finish 86

grout 94
hand hole 538
hand mix 78
hand trowel finish 81
hardener 82
headwall 507
hydrodemolition 27
impact drilling 89
inspection 15
integral color 52
integral topping 86
integral waterproofing 53
lance 735
lightweight 77, 79
machine trowel finish 81
manhole 478, 538
materials 847
median 380
membrane curing 84
mix design 15
mixer 730
non-chloride accelerator 79
pan tread 81
paper curing 84
patio block 378
pavement highway 377
paver 736
pier and shaft drilled 357
pile 586-588
pile cap 582
pile clean 550
pile prestressed 345
pipe 506
pipe gasket 507
pipe removal 28
placement beam 79
placement column 80
placement footing 80
placement slab 80
placement wall 80
placing 79, 847
plant mixed recycled bitum . . 288
plantable paver precast 378
planter 397
pontoon 556
post 545
pressure grouting 342
pump 730
pumped 80
ready mix 79
receptacle 198
reinforcing 73
removal 27
retaining wall 395, 668
retaining wall segmental 395
retarder 79
revetment 555
roller compacted 83
sand 53
sandblast finish 82
saw 730
saw blade 88
scarify 170
seawall 555
septic tank 499
shingle 147
short load 79
sidewalk 369, 663
slab 600
slab conduit 274
slab cutting 88
slab raceway 275
slab sidewalk 663
slab X-ray 16
small quantity 78
splash block 369

spreader 736
stair 78
stair finish 82
step 666
tank 495
tie railroad 543
tile 147
topping 86
track cross tie 543
trowel 730
truck 730
truck holding 79
unit paving slab precast 378
utility vault 479
vertical shoring 63
vibrator 730
volumetric mixed 79
wall 77, 578, 603
wall demolition 41
wall panel tilt-up 86
wall patching 52
water reducer 79
water storage tank 495
wheeling 81
winter 79
Concrete-filled steel pile 354
Conductor 268
& grounding 266, 268, 270
Conductors 867
Conduit concrete slab 274
electrical 272
encasement 877
fitting pipe 532
high installation 274
in slab PVC 274
in trench electrical 275
in trench steel 275
pipe prefabricated 534
prefabricated pipe 531, 533
preinsulated pipe 531
PVC 272
PVC buried 536
PVC fitting 272
rigid in slab 275
Cone traffic 22
Connection brazed 271
temporary 17
utility 479
wire weld 270
Connector dock 557
timber 136
Consolidation test 15
Construction aid 18
barge 740
casting 132
management fee 10
photography 14
sump hole 307
temporary 21
time 831
Contained pipe 235
Containment hazardous waste . . . 47
unit dust 22
Contaminated soil 48
Contingency 10
Continuous footing 580
Contractor scale 186
Control erosion 334
joint 66, 96
joint clean 66
joint PVC 96
joint rubber 96
joint sawn 66
panel blower 565
Conveyor 560
material handling 560

Coping	106, 109
aluminum	109
clay tile	109
concrete	109
terra cotta	109
Copper fitting	217
ground wire	270
gutter	152
pipe	217, 488
salvage	213
shielded cable	266
tube	217, 488
tube fitting	217
Core drill	15, 730
drill bit	89
drilling concrete	88
testing	15
Corner guard	117
Corrosion resistance	472
resistant fitting	257
resistant pipe	256, 257
resistant pipe fitting	257
Corrugated metal pipe	501, 503
pipe	501
roof tile	147
siding	147-149
steel pipe	501
tubing	502
Cost mark-up	13
Coupling blower	564
plastic	229
Course drainage layer base	370
obstacle	190
wearing	372
Court air supported	204
paddle tennis	191
surfacing	384
Cover aluminum trench	130
cast iron grate	130
checkered plate	130
ground	407, 436
manhole	509
pool	202
stadium	204
stair tread	23
tank	203
trench	130
walkway	184
CPVC pipe	225
Crack cleaner	738
filler	738
filling	360
repair asphalt paving	363
Crane	560
and bucket concrete	80
climbing	738
crawler	739
crew daily	18
hydraulic	739
mobilization	289
rail	560
scale	185
tower	19, 738
truck mounted	739
Crash barrier	546
Crawler crane	739
drill rotary	734
shovel	732
Creosoted wood post wall	670
Crew daily crane	18
forklift	18
survey	24
Cribbing concrete	395
Cross	490
brace	118
Crushed stone	288

stone sidewalk	370
Cultured stone	109
Culvert box	505
concrete	506
end	506
headwall	710
pipe	507
reinforced	506
Curb	379
and gutter	379
asphalt	380
box	489
builder	730
concrete	379
edging	117, 380
edging angle	117
edging channel	117
extruder	731
granite	106, 380
inlet	380, 509
precast	380
removal	27
seal	361
slipform paver	738
Cured-in-place pipe	476
Curing blanket	84
compound concrete	84
concrete	84, 377
concrete membrane	84
concrete paper	84
concrete water	84
paper	145
Curtain wall	165
wall glazed aluminum	165
Custom architectural door	162
Cut & fill clay	628
& fill earth	624
& fill gravel	620
drainage ditch	307
in sleeve	490
in valve	490
pipe groove	224
stone curb	380
Cutoff pile	289
Cutout demolition	41
slab	41
Cutter brush	731, 732
Cutting asphalt	44
concrete floor saw	88
concrete slab	88
concrete wall saw	88
masonry	44
steel	114
torch	114, 736
wood	44
Cylinder concrete	15
lockset	166

D

Daily crane crew	18
Dam expansion	398
Dampproofing	144
Dasher hockey	203
ice rink	203
Dead lock	166
Deadbolt	166
Deciduous tree	437
Deck drain	250
metal floor	125
slab form	125
Decking	857
cellular	125
floor	124, 125
form	125

roof	124
Deconstruction building	44
building component	45
material handling	46
Decorative block	101
fence	392
Dehumidifier	734
Delineator	181
barrier	546
Delivery charge	20
Demo athletic screening	38
athletic surface	36
athletic/playground equipment	38
box culvert	31
bridge pedestrian	41
chain link fence & gate	34
concrete	27
electric duct & fitting	32
electric duct bank	33
footing & foundation building	40
gasoline containment piping	32
geodesic dome	200
lawn sprinkler system	37
lightning protection	200
manhole & catch basin	31
metal drainage piping	30
misc metal fence & gate	34
modular playground equip.	38
natural gas PE pipe	32
natural gas steel pipe	32
natural gas valve & fitting	32
parking appurtenance	37
pile	35
pre-engineered steel building	200
radio tower	36
retaining wall	37
rip-rap & rock lining	34
septic tank & component	31
silo	200
site furnishing	38
sod edging planter & guy	39
specialty	180
steel pipe with insulation	31
storage tank	200
tension structure	201
traffic light system	39
utility pole & cross arm	36
utility valve	29
vinyl fence & gate	34
walk or step or paver	36
water & sewer piping & fitting	29
water well	35
wood fence & gate	34
Demolish remove pavement and curb	27
Demolition	839
bituminous sidewalk	619
boiler	260
brick	92, 170
ceiling	170
ceramic tile	170
chimney	92
concrete	41, 52
concrete block	41
concrete sidewalk	619
concrete wall	41
disposal	42
door	160
drywall	170
ductwork	260
electrical	266
explosive/implosive	40
fencing	33
fireplace	92
flooring	170
glass	160

gutting	44
hammer	731
highway guard	27
HVAC	260
implosive	40
lath	170
masonry	41, 92, 170
metal	112
metal stud	171
pavement	27
pipe	614, 617
plaster	170, 171
plumbing	213
plywood	170, 171
saw cutting	44
selective building	41
shore protection	35
site	27
steel	112
steel window	160
terrazzo	170
tile	170
torch cutting	44
utility material	29
wall	170
wall and partition	170
window	160
window & door	160
wood	170
wood framing	136
Dental metal interceptor	251
Derail railroad	543
Derrick crane guyed	739
crane stiffleg	739
Detection tape	479
Detector check valve	493
meter water supply	239
Detour sign	23
Device & box wiring	274
Dewater	307, 308
pumping	308
Dewatering	307, 872
equipment	740
Diamond plate tread	131
Diaphragm pump	736
Diesel generator	278
generator set	278
hammer	732
tugboat	740
Diesel-engine-driven generator	278
Diffuser system air	563, 566
Dimensions and weights of sheet steel	856
Direct burial cable	269
burial PVC	536
chute concrete	80
Directional drill horizontal	737
drilling	477
sign	181
Directory	180
board	180
Disc harrow	731
Discharge hose	735
Dispenser ticket	188
Displacement caisson	357
Disposal	41
field	500
or salvage value	44
Distribution box	500
box concrete	500
box flow leveler	500
box HDPE	500
pipe water	483, 485
valve & meter gas	529
Distributor asphalt	734
Ditch clean	307

Ditching 307, 344
Diving board 202
  stand 202
Dock accessory 558
  accessory floating 557
  accessory jetties 557
  bumper 189, 558
  cleat 557
  connector 557
  floating 556
  galvanized 556
  ladder 557
  loading 189
  pile guide 557
  truck 189
  utility 557
  wood 137, 556
Dog command 23
Dogwood 439
Dome 208
  geodesic 205
  structures 860
Door air lock 204
  aluminum 164
  and panel access 163
  and window material 160
  birch hollow core 163
  bulkhead 163
  bulkhead/cellar 163
  bumper 167
  chain hoist 164
  commercial floor 163
  commercial half glass 161
  composite 162
  custom architectural 162
  demolition 160
  fire 162
  frame grout 94
  glass 164
  hollow metal 161
  labeled 162
  metal 164
  metal fire 162
  overhead 164
  overhead commercial 164
  patio 163
  plastic laminate 162
  prefinished 162
  removal 160
  revolving 204
  sectional 164
  sliding glass 163
  sliding glass vinyl-clad 163
  sliding vinyl-clad 163
  sliding wood 163
  special 163
  steel 161, 858
  stop 167
  swinging glass 164
  wood 162
  wood fire 162
Double-hung wood window 166
  wall pipe 231, 510
Dovetail anchor 70
Dowel 95
  cap 74
  reinforcing 74
  sleeve 74
Downspout 152
  aluminum 152
  steel 152
Dozer 732
  backfill 308, 634-636, 638-641
  cut & fill . . 620, 621, 624, 625, 628,
                          629
  excavation 304

ripping 300
Dozing ripped material 300
Dragline 301, 554
  bucket 731
  excavation 623, 627, 631
  excavation bulk 305
Drain board 144
  deck 250
  facility trench 253
  floor 251
  geotextile 502
  landscape 508
  pipe 307
  sanitary 250
  sediment bucket 251
  site 508
  trench 253
Drainage & sewage piping 505
  ditch cut 307
  field 500
  flexible 502
  grate 508
  mat 502
  pipe . . 256, 257, 497, 498, 503, 505
  site 563
  storm 510
Dredge 554
  hydraulic 554
Drill auger 27
  core 15, 730
  earth 26, 355
  main 489
  quarry 734
  rig 26
  rig mobilization 289
  rock 26, 299
  steel 734
  tap main 489
  track 734
  wood 136, 396
Drilled concrete pier and shaft . . 357
  concrete pier uncased 357
  micropile metal pipe 358
  pier 356
Drilling and blasting bulk 299
  and blasting rock 299
  concrete 89
  concrete impact 89
  directional 477
  horizontal 477
  quarry 299
  rig 730
  rock bolt 299
  steel 114
Drinking bubbler 255
  fountain 255
  fountain deck rough-in 255
  fountain floor rough-in 255
  fountain wall rough-in 255
Drip edge 154
  edge aluminum 154
  irrigation 399
  irrigation subsurface 399
Driven pile 345
Driver post 738
  sheeting 734
Driveway 369
  removal 27
Drum polyethylene 556
Dry fall painting 176
  stone wall 673
Drywall 172
  cutout 42
  demolition 170
  finish 859
  gypsum 172

partition 171
  removal 170
Duck tarpaulin 21
Duct and manhole undrgnd. 536
  bender 539
  cement 539
  furnace 264
  underground 538
  utility 538
Ductile iron fitting 483
  iron fitting mech joint 484
  iron pipe 483
Ductwork demolition 260
Dumbbell waterstop 66
Dump charge 44
  truck 733
Dumpster 43
Dumpsters 839
Dump truck off highway 733
Dust barrier 21
  containment unit 22
Dustproofing 82
  compound 52
DWV pipe ABS 225
  PVC pipe 225, 498

E

Earth auger 730
  compaction 332
  compactor 731
  drill 26, 355
  excavation 626, 633
  moving . . . 620, 622, 624, 628, 632,
            634, 636, 638, 640
  retaining 668
  rolling 333
  scraper 732
  vibrator 294, 308, 333
Earthwork 292
  equipment 308
  equipment rental 730
Economy brick 98
Edge drip 154
Edger concrete floor 730
Edging 468
  aluminum 468
  curb 117, 380
  precast 468
  wood 469
Efflorescence testing 15
Effluent-filter septic system 500
Ejector pump 252
Elastomeric sheet waterproofing . 143
  waterproofing 143
Elbow HDPE piping 487
Electric duct & fitting demo 32
  duct bank demo 33
  fixture 283, 285
  fixture exterior 281
  generator 735
  hoist 560
  lamp 285
  panelboard 276
  switch 277
  utility 538
Electrical & telephone
underground 538
  box 274
  conduit 272, 717
  demolition 266
  pole 536
Electricity temporary 16
Electrofusion welded PP pipe . . . 258
Electrolytic grounding 271

Elevated conduit 274
  installation add 212
  pipe add 212
  slab formwork 56
  slab reinforcing 73
  water storage tank 494
  water tank 494
Elevating scraper 306
Elm American 447
Employer liability 12
Emulsion acrylic 384
  asphaltic 360, 402
  pavement 360
  penetration 372
  sprayer 734
Encasement pile 346
Enclosure acoustical 208
  swimming pool 206
End cap 502
  culvert 506
Energy distribution pipe steam . . 533
Engineer brick 99
Engineering fee 831
Entrance aluminum 165
  and storefront 164
  cable service 269
  floor mat 194
Entry canopy 183
EPDM adhesive 143
Epoxy acrylic coatings 384
Epoxy anchor 87, 113
  coated reinforcing 74
  coating 82, 472
  fiberglass wound pipe 257
  grout 342
  resin fitting 226
  welded wire 75
Equipment 554, 730
  aeration 563
  automotive 188
  earthwork 308
  foundation formwork 56
  ice skating 203
  insurance 12
  loading dock 189
  lube 188
  lubrication 188
  mobilization 357
  pad 77
  parking 188
  parking collection 188
  parking control 188
  parking gate 188
  parking ticket 188
  playfield 190
  playground 190, 388
  refrigeration 203
  rental 730, 837, 838
  rental concrete 730
  rental earthwork 730
  rental general 733
  rental highway 737
  rental lifting 738
  rental marine 740
  rental wellpoint 740
  swimming pool 202
  waste handling 192
Erosion control 334
  control synthetic 334
Estimating 828
Excavate & haul . . 622, 623, 626, 627,
               630, 631, 633
Excavating bulk bank measure . . 301
  bulk dozer 304
  equipment 871
  large volume project 302

Index

trench 292
utility trench 297
utility trench plow 298
Excavation . . 292, 301, 620, 622, 626,
628, 630, 633, 870
& backfill building 601, 602
and backfill 601, 602
boulder 299
bulk dragline 305
bulk scraper 306
cofferdam 302
dozer 304
hand 294, 298, 308
machine 308
planting pit 404
scraper 306
septic tank 500
structural 298
tractor 294
trench 293, 297, 642-644, 646
Excavator bucket 737
hydraulic 730
Excavator hydro 732
Expansion anchor 113
dam 398
joint 67, 154
joint assembly 157
joint butyl 154
joint neoprene 154
joint roof 154
shield 113
Expense office 12
Explosion proof circuit breaker . . 277
Explosive 299
Explosive/implosive demolition . . 40
Exposed aggregate 82, 369
Extension backhoe 737
ladder 735
Exterior concrete block 102
fixture LED 281
fixture lighting 281
floodlamp 284
LED fixture 282
lighting 720
lighting fixture 281
plaster 172
siding painting 175
signage 180
steel coating 178
tile 173
Extinguisher cart mounted 182
chemical dry 182
CO$_2$ 182
fire 182
installation 182
portable ABC 182
pressurized fire 182
standard 182
wheeled ABC 182
Extra work 12
Extruder curb 731
Eye bolt screw 72

F

Fabric awning 183
flashing 152
pillow water tank 495
polypropylene 502
stabilization 371
stile 167
structure 203
waterproofing 143
welded wire 75
wire 393

Face brick 98, 849
Facility trench drain 253
Facing panel 107
stone 106
Factor 10
security 11
Fall painting dry 176
Farm type siding 148
Faucet & fitting 255
Fee architectural 10
Feeder cable underground 268
Feller buncher 731
Felt asphalt 151
waterproofing 143
Fence aluminum 386, 387
and gate 385
board 393
cedar 394
chain link 23, 393
chain link gate & post 387
chain link industrial 386
chain link residential 386
commercial 4' H 386
decorative 392
fabric & accessory 388
helical topping 393
mesh 386
metal 386
misc metal 393
open rail 394
plywood 23
recycled plastic 392
safety 23
snow 48
steel 386, 392
temporary 23
tennis court 387
tubular 392
vinyl 392
wire 23, 393
wood 393
wood rail 394
wrought iron 386
Fencing demolition 33
wire 393
Fender 557
Fertilize lawn 368
Fertilizer 404, 406
shrub bed 367
tree 367
Fertilizing 368
Fiber reinforcing concrete 79
steel 76
synthetic 76
Fiberglass bench 397
flagpole 184
insulation 144
light pole 280
panel 21, 148
planter 397
rainwater storage tank 254
tank 262
trash receptacle 198
trench drain 253
underground storage tank 243
waterproofing 143
Field disposal 500
drainage 500
mix concrete 847
office 18
office expense 18
personnel 12, 13
sample 49
seeding 404
Fieldstone 104
Fill 308, 310

by borrow & utility bedding . . 310
floor 77
flowable 79
gravel 288, 310
Filler crack 738
joint 155, 360
Fillet welding 114
Filling crack 360
Film polyethylene 402
Filter rainwater 562
stone 341
swimming pool 256
Filtration equipment 201
Fine grade 292, 404
Finish concrete 377
grading 292
lime 93
Finishing floor concrete 81
wall concrete 82
Fire brick 108
clay 108
door 162
door metal 162
door wood 162
extinguisher 182
extinguisher portable 182
hydrant 493, 697
hydrant remove 28
retardant pipe 257
Firebrick 108
Fireplace box 108
demolition 92
masonry 108
Fireproofing 155
plaster 155
sprayed cementitious 155
Fitness trail 190
Fitting cast iron 244
cast iron pipe 219, 244
conduit PVC 272
copper 217
copper pipe 217
corrosion resistant 257
ductile iron 483
DWV pipe 228
epoxy resin 226
glass 256
grooved joint 222
HDPE 231
malleable iron 220
malleable iron pipe 221
pipe 485, 532, 533
pipe conduit 532
plastic 228
polypropylene 258
PVC 227, 486
PVC duct 539
steel 219, 221
subdrainage 502
underground duct 539
waterstop 66
Fixed dock jetties 137
end caisson pile 355
vehicle delineator 546
Fixture landscape 282
lantern 279
LED exterior 281
lighting 281
plumbing 251, 255, 263, 482
removal 213
residential 279
sodium high pressure 284
Flagging 174, 379
slate 379
Flagpole 184
aluminum 184

fiberglass 184
foundation 185
steel internal halyard 184
structure mounted 185
wall 185
wood 185
Flagstone paving 665
Flange gasket 221
grooved joint 223
PVC 230
steel 223
tie 95
Flap gate 553
Flasher barrels 734
barricade 734
Flashing 152
aluminum 148, 152
asphalt 152
fabric 152
masonry 152
sheet metal 152
Flatbed truck 733, 737
truck crane 738
Flexible drainage 502
pavement repair 361
Flexible paving surface
treatment 360
Flexural concrete 77
Float finish concrete 81, 82
glass 168
Floater equipment 12
Floating dock 556
dock accessory 557
dock jetties 556
wood pier 556
Floodlamp exterior 284
Floodlight LED 285
pole mounted 284
trailer 734
tripod 734
Floor 174
abrasive 82
acid proof 174
asphalt block 378
brick 174, 378
cleanout 216
color 82
concrete finishing 81
decking 124, 125
door commercial 163
drain 251
fill 77
fill lightweight 77
flagging 379
framing removal 42
grating aluminum 128
grating steel 128
hardener 53
marble 107
mat 194
mat entrance 194
plate stair 125
quarry tile 173
removal 40
rubber 384
sander 736
sealer 53
slate 107
tile or terrazzo base 175
topping 82, 86
vinyl 175
Flooring demolition 170
masonry 174
miscellaneous brick 174
Flotation polystyrene 556
Flowable fill 79

For customer support on your Site Work & Landscape Costs with RSMeans data, call 800.448.8182.

Flower bed clean-up 366
  bed cultivate 366
  bed fertilizer 366
  bed maintenance 366
  bed spring prepare 366
  bed weed 366
Flush tube framing 164
Flying truss shoring 63
Foam board insulation 145
  core DWV ABS pipe 225
  insulation 145
  spray rig 735
Fog seal 360
Foil aluminum 146
Folding gate 182
Football goal post 191
  scoreboard 190
Footing concrete 77, 580
  concrete placement 80
  formwork continuous 57
  formwork spread 57
  keyway 57
  keyway formwork 57
  pressure injected 599
  reinforcing 73
  removal 40
  spread 77
Forged valve 261
Forklift 735
  crew 18
Form decking 125
  liner concrete 60
  material 841
  release agent 53
Formblock 101
Forms 842
  concrete 842
Formwork beam and girder 54
  beam bottom 54
  beam side 54
  box culvert 55
  coil bolt 64
  coil tie 64
  column 54
  column plywood 55
  column round fiber tube 55
  column round fiberglass 54
  column steel framed 55
  continuous footing 57
  elevated slab 56
  equipment foundation 56
  footing keyway 57
  gang wall 59
  gas station 57
  gas station island 57
  grade beam 57
  hanger accessory 63
  insert concrete 70
  insulating concrete 60
  labor hours 843
  light base 57
  mat foundation 57
  oil 65
  patch 65
  pencil rod 65
  pile cap 57
  plywood 55
  radial wall 59
  retaining wall 59
  she-bolt 65
  shoring concrete 63
  side beam 54
  sign base 57
  slab blockout 58
  slab box out opening 56
  slab bulkhead 56, 58

slab curb 56, 58
slab edge 56, 58
slab flat plate 56
slab haunch 78
slab on grade 58
slab thickened edge 78
slab trench 58
slab turndown 78
slab void 56, 58
slab with drop panel 56
sleeve 63
snap tie 64
spread footing 57
stair 62
stake 65
steel framed plywood 59
taper tie 65
tie cone 65
upstanding beam 54
waler holder 65
wall 58
wall accessory 64
wall box out 58
wall brick shelf 58
wall bulkhead 58
wall buttress 58
wall corbel 58
wall lintel 59
wall pilaster 59
wall plywood 58
wall prefab plywood 59
wall sill 59
wall steel framed 59
Foundation . 578, 580, 590, 592, 594,
         595, 597, 598, 605
  backfill 602
  beam 597
  caisson 355
  chimney 76
  concrete block 102
  dampproofing 585
  excavation 601, 602
  flagpole 185
  mat 77
  mat concrete placement 80
  pile 357
  pole 535
  scale pit 185
  underdrain 584
  underpin 343
  wall 102, 603, 604
  waterproofing 585
Fountain 202
  drinking 255
  lighting 256
  outdoor 202
  pump 256
  water pump 256
  yard 202
Frame grating 130
  metal 161
  scaffold 19
  shoring 63
  steel 161
  trench grating 130
  welded 161
Framing column 136
  laminated 138
  light 136
  lightweight 117
  lightweight angle 117
  lightweight channel 117
  pipe support 118
  sleeper 137
  slotted channel 118
  steel 119

treated lumber 137
  tube 164
  wood 136
Friction pile 346, 357
Front end loader 301
Frost penetration 846
Fruit and nut tree 467
Fuel tank 263
Furnace duct 264
Furnishing site 197
Furring concrete 137
  masonry 137
  wall 137
  wood 137
Fused pipe 487
Fusion machine butt 734

G

Gabion box 341
  retaining wall stone 396
  wall 675
Galley septic 500
Galvanized dock 556
  plank grating 129
  receptacle 198
  reinforcing 74
  roof 148
  welded wire 75
Galvanizing 178
  lintel 118
  metal in field 178
  metal in shop 112
Gang wall formwork 59
Garage 208
  residential 208
Garden house 208
Gas distribution valve & meter .. 529
  generator set 278
  line 713
  pipe 511, 513
  service steel piping 511
  station formwork 57
  station island formwork ... 57
  station tank 263
  stop 529
Gas-fired duct heater 264
  heating 264
Gasket & bolt set 222
  b and S pipe 248
  cast iron pipe 247
  concrete pipe 507
  flange 222
  joint pipe 243
  joint pipe fitting 246
Gasoline containment
piping demo 32
  generator 278
  piping 510
Gate box 491
  canal 553
  chain link 387
  fence 386
  flap 553
  folding 182
  knife 554
  parking 188
  security 182
  slide 386, 554
  sluice 553
  swing 386
  valve 260, 261, 492, 493
  valve grooved-joint 224
  valve soldered 214
General contractor's overhead .. 834

equipment rental 733
  fill 310
Generator diesel 278
  diesel-engine-driven 278
  electric 735
  emergency 278
  gasoline 278
  natural gas 278
  set 278
Geodesic dome 205
  dome demo 200
  hemisphere greenhouse ... 206
Geo-grid 395
Geotextile drain 502
  soil stabilization 335
  stabilization 335
  subsurface drainage 502
Geothermal heat pump system .. 609,
                    610
Girder bridge 398
  reinforcing 73
Girt steel 119
Gland seal pipe conduit 532
Glass curtain wall 165
  demolition 160
  door 164
  door sliding 163
  door swinging 164
  fiber rod reinforcing 74
  fitting 256
  float 168
  insulating 168
  laminated 168
  mirror 168
  pipe 256
  pipe fitting 256
  reduce heat transfer ... 168
  window wall 164
Glazed aluminum curtain wall ... 165
  brick 99
Glazing 168
Globe valve 261
  valve bronze 214
Glued laminated 138
Goal post 191
  football 191
  soccer 191
Golf shelter 190
Gore line 382, 383
Gradall 301, 731
Grade beam 597
  beam concrete placement 80
  beam formwork 57
  fine 292, 404
Grader motorized 731
  ripping 300
Grading 291, 316
  and seeding 404
  maintenance 542
  rough 292
  site 292
  slope 292
Graffiti-resistant treatment ... 176
Grandstand 206
Granite 105
  block 379
  building 105
  chip 402
  curb 106, 380
  edging 468
  Indian 380
  paver 106, 379
  paving 665
  paving block 379
  reclaimed or antique 106
  sidewalk 379

Index

Granolithic finish concrete 86
Grass lawn 404
  ornamental 412
  seed 404
  surfacing artificial 384
Grate drainage 508
  pipe 508
  precast 369
  tree 369
Grating aluminum 127
  aluminum floor 128
  aluminum mesh 129
  aluminum plank 128
  area wall 184
  area way 184
  frame 130
  galvanized plank 129
  plank 129
  stainless bar 129
  stainless plank 129
  stair 126
  steel 128
  steel floor 128
  steel mesh 129
Gravel 634
  base 371
  excavation 622
  fill 288, 310
  moving 620
  pack well 480
  pea 402
  roof 53
  stop 152
Gravity retaining wall 395
  wall 668
Grease interceptor 251
  trap 251
Greenhouse 205
  air-supported 204
  commercial 205
  cooling 205
  geodesic hemisphere 206
  residential 205
Graywater commercial 253
  lift station 253
  recovery system resi 252
  residential 252
  system component 253
Grille painting 175
Grinder concrete floor 730
  system pump 251
Grinding asphalt 361
Grit removal and
handling equipment 562
Groove cut labor steel 224
  roll labor steel 224
Grooved block 101
  joint fitting 222
  joint flange 223
  joint pipe 222
Grooving pavement 361
Ground box hydrant 241
  clamp water pipe 270
  cover 407, 436, 681, 682
  face block 101
  insulated wire 271
  post hydrant 242
  rod 269
  rod clamp 270
  rod wire 271
  socket 190
  water monitoring 16
  wire 270
Grounding 269
  & conductor 266, 268, 270
  electrolytic 271

wire brazed 271
Grout 94
  cavity wall 94
  cement 94, 342
  concrete 94
  concrete block 94
  door frame 94
  epoxy 342
  metallic non-shrink 87
  non-metallic non-shrink 87
  pump 736
  topping 87
  wall 94
Grouted pile 358
  soil nailing 336
Guard corner 117
  house 208
  service 23
  snow 155
  tree 469
  wall 117
Guardrail scaffold 19
  temporary 22
Guide rail 545
  rail cable 545
  rail removal 28
  rail timber 545
  rail vehicle 545
Guide/guard rail 545
Gunite 83
  dry-mix 83
  mesh 83
Gutter 153
  aluminum 152
  copper 152
  lead coated copper 153
  monolithic 380
  stainless 153
  steel 153
  vinyl 153
  wood 153
Gutting 44
  demolition 44
Guyed derrick crane 739
  tower 539
Guying power pole 535
Gypsum block demolition 41
  board system 171
  drywall 172
  partition NLB 171
  plaster 171

H

H pile 345, 592
  pile steel 592
Hair interceptor 251
Hammer bush 82
  chipping 734
  demolition 731
  diesel 732
  drill rotary 734
  hydraulic 731, 735
  pile 731
  pile mobilization 289
  vibratory 732
Hammermill 737
Hand clearing 290
  excavation 294, 298, 308
  hole 479
  hole concrete 538
  trowel finish concrete 81
Handball court backstop 191
Handball/squash court 191
Handicap ramp 77

Handling material 24, 560
  waste 192
Hangar 209
Hanger accessory formwork 63
Hanging lintel 118
Hardboard receptacle 198
Hardener concrete 82
  floor 53
Hardware 166
  cloth 393
Harrow disc 731
Haul earth 602
Hauling 311
  cycle 311
Haunch slab 78
Hay 402
Hazardous waste cleanup 49
  waste containment 47
  waste disposal 49
  waste handling 192
HDPE 47
  distribution box 500
  drainage and sewage pipe 505
  fitting 231
  infiltration chamber 500, 510
  pipe 231
  pipe dual wall 235
  pipe liner 477
  piping 487
  piping cap 488
  piping elbow 487
  piping tee 488
Header pipe wellpoint 740
Headwall 507
  concrete 507
  precast 507
  sitework 710
Hearth 108
Heat greenhouse 206
  temporary 54
Heated pad 54
Heater gas-fired duct 264
  space 735
Heating gas-fired 264
  kettle 738
  subsoil 203
Heavy construction 399
  duty shoring 19
  lifting 837
  rail track 543
Helicopter 739
Hex bolt steel 114
Hexagonal face block 102
High chair reinforcing 71
  installation conduit 274
  intensity discharge ballast 279
  intensity discharge lamp ... 285
  pressure HDPE, ASTM D2513 . 520
  strength bolt 851
  strength decorative block 101
  strength steel 853
High-security chain link fence ... 390
Highway bridge 398
  equipment rental 737
  guard demolition 27
  pavement 650-655
  paver 379
  paving 398
  sign 181
Hinge 167
  brass 167
  paumelle 167
  steel 167
  surface 167
Hockey dasher 203
Hoist 560

chain 560
  chain fall 739
  electric 560
  overhead 560
  personnel 739
  tower 739
Holder screed 72
Holding tank 562
Hollow metal door 161
  metal frame 161
  precast concrete plank 84
Honed block 102
Honeylocust 441
Hook clevis 132
Horizontal borer 737
  boring 477
  boring machine 730
  directional drill 737
  drilling 477
  timber wall 672
Hose air 734
  bibb sillcock 255
  discharge 735
  suction 735
  water 735
House garden 208
  guard 208
Housewrap 145
Humidification equipment 206
Humus peat 402
HVAC demolition 260
Hydrant 494
  fire 493
  ground box 241
  ground post 242
  removal 28
  remove fire 28
  wall 241
  water 241
Hydrated lime 93
Hydraulic crane 739
  dredge 554
  excavator 730
  hammer 731, 735
  jack 740
  jack climbing 740
  jacking 837
  pressure test 261
  rock breaking 300
  seeding 404
  sluice gate 553
  structure 553, 554
Hydro excavator 732
Hydro seeding 406
Hydrodemolition 27
  concrete 27
Hydroexcavation 27
Hydromulcher 738
Hydronic energy underground ... 533

I

Ice melting compound 367
  rink dasher 203
  skating equipment 203
Impact barrier 546
  wrench 734
Implosive demolition 40
Incandescent exterior lamp 285
  fixture 279
  interior lamp 285
Incinerator municipal 192
Indian granite 380
Indicator valve post 494
Induction lighting 281

For customer support on your Site Work & Landscape Costs with RSMeans data, call 800.448.8182.

Industrial chain link gate 391
railing 127
Inert gas 48
Infiltration chamber 501
chamber HDPE 500, 510
Inlet curb 380, 509
Insecticide 334
Insert concrete formwork 70
Inspection internal pipe 475
technician 15
Installation add elevated 212
extinguisher 182
Insulated pipe 530
precast wall panel 85
steel pipe 530
Insulating concrete formwork . . . 60
glass 168
Insulation 144
blower 735
board 144
fiberglass 144
foam 145
insert concrete block 100
isocyanurate 145
masonry 145
polystyrene 145
rigid 144, 145
subsoil 203
vapor barrier 146
wall 144
Insurance 12, 833
builder risk 12
equipment 12
public liability 12
Integral color concrete 52
topping concrete 86
waterproofing 54
waterproofing concrete 53
Interceptor 251
grease 251
hair 251
metal recovery 251
oil 251
Interior paint wall & ceiling 176
plant 194
planter 195
Interlocking paving 664
precast concrete unit 378
Internal chain link gate 391
cleaning pipe 475
Intumescent coating 177
Invert manhole 509
Iron alloy mechanical joint pipe . 257
body valve 260
grate 369
Ironspot brick 174
Irrigation drip 399
system sprinkler 400
Isocyanurate insulation 145

J

Jack cable 740
hydraulic 740
post adjustable 116
screw 341
Jackhammer 299, 734
Jacking 477
Jet water system 482
Jetty dock accessory 557
fixed dock 137
floating dock 556
pier 137
Joint assembly expansion 157
bedding placement 83

control 66, 96
expansion 67, 154
filler 155, 360
push-on 483
reinforcing 96
restraint 484, 487
roof 154
sealant replacement 142
sealer 155, 156
Joist longspan 123
open web bar 124
Jumbo brick 99
Jute mesh 334

K

Kennel fence 393
Kettle heating 738
tar 736, 738
Keyway footing 57
King brick 99
Kiosk 208
Knife gate 554
Kraft paper 146

L

Labeled door 162
Labor adjustment factor 212
formwork 843
modifier 212
pipe groove roll 224
Laboratory analytical service 49
Ladder aluminum 557
dock 557
extension 735
inclined metal 126
monkey 190
reinforcing 96
rolling 20
ship 126
swimming pool 202
vertical metal 126
Lag screw 115
Lagging 343
Lally column 116
Laminated framing 138
glass 168
glued 138
veneer member 138
Lamp electric 285
high intensity discharge 285
incandescent exterior 285
incandescent interior 285
mercury vapor 285
metal halide 285
post 133, 280
quartz line 285
sodium high pressure 285
sodium low pressure 285
Lance concrete 735
Landfill compactor 732
Landing metal pan 125
Landscape drain 508
fee 10
fixture 282
layout 437
LED fixture 283
light 283
lighting 282
Landscaping 875
timber step 666
timber wall 672
Lantern fixture 279

Large volume earth project 302
Laser level 735
Latch set 166
Lateral arm awning retractable . . . 183
Latex caulking 156
Lath demolition 170
metal 171
Lava stone 106
Lavatory removal 213
Lawn aerate 366
bed preparation 402
edging 366
fertilize 368
grass 404
maintenance 366
mower 735
seed 367, 404
seeding 681
sodding 681, 682
sprinkler 683
sprinkler system demo 37
trim 368
water 369
Layout landscape 437
Leaching field 703
field chamber 500
pit 500
Lead caulking 243
coated copper gutter 153
salvage 213
screw anchor 113
Leads pile 731
Lean-to type greenhouse 205
LED fixture exterior 282
floodlight 285
lighting parking 281
luminaire roadway 281
luminaire walkway 284
Letter sign 180
Level laser 735
Leveling jack shoring 19
Liability employer 12
insurance 834
Lifeguard chair 202
Lift scissor 733
slab 78
station greywater 253
station packaged 499
Lifting equipment rental 738
Light base formwork 57
bollard 283
framing 136
landscape 283
pole 722
pole aluminum 280
pole fiberglass 280
pole steel 280
pole wood 280
post 279
shield 545
temporary 17
tower 735
underwater 202
Lighting 285
ballast 279
exterior fixture 281
fixture 281
fixture exterior 281
fountain 256
induction 281
landscape 282
pole 280
residential 279
temporary 17
underwater 283
Lightning protection demo 200

Lightweight aggregate 53
angle framing 117
block 103
channel framing 117
column 116
concrete 77, 79
floor fill 77
framing 117
natural stone 106
Lime 93
finish 93
hydrated 93
soil stabilization 335
Limestone 106, 404
coping 109
paver 370
Line cover protective device 280
gore 382, 383
pole 535
pole & fixture 535
pole transmission 535
remover traffic 738
Liner cement 476
pipe 476
pond 288
Link barbed wire chain 391
fabric chain 391
gate and fence chain 385
line post chain 390
terminal post chain 390
top rail chain 391
transom chain 392
Lintel 106
block 103
concrete block 103
galvanizing 118
hanging 118
precast concrete 86
steel 118
Load & haul rock 632
test pile 289
Loader front end 301
skid-steer 733
tractor 732
wheeled 732
windrow 738
Loading dock 189
dock equipment 189
rock 632
Loam 288, 291, 310, 403
spread 403
Lock tubular 166
Lockset cylinder 166
Locomotive tunnel 738
Log chipper 732
skidder 732
Longspan joist 123
Low bed trailer 738
Lube equipment 188
Lubricated valve 530
Lubrication equipment 188
Lumber foundation 605
product prices 858
Luminaire roadway 281
walkway 283, 284

M

Macadam 372
penetration 372
Machine auto-scrub 734
excavation 308
screw 115
trowel finish concrete 81
welding 737

Index

Machinery anchor 71
Magnetic particle test 15
Magnolia 443
Main office expense 12
Maintenance grading 542
   grading roadway 542
   lawn 366
   railroad 543
   railroad track 543
   shrub 367
   site 366
   walk 367
Mall front 165
Malleable iron fitting 220
   iron pipe fitting 220
Management fee construction 10
   stormwater 509
Manhole 478
   & catch basin demo 31
   brick 478
   concrete 478, 538
   cover 509
   electric service 538
   frame and cover 509
   invert 509
   raise 509
   removal 28
   step 478
Manual transfer switch 279
Maple tree 438
Marble 106
   chip 402
   chip white 53
   coping 109
   floor 107
   stair 107
Marina small boat 556
Marine bulkhead 555
   equipment rental 740
   pier pile cleaning 550
   pier pile wrapping 552
   piling 137
   recycled plastic/steel pile 349
   wood pile 346
Marker boundary and survey 26
Mark-up cost 12
Mason scaffold 19
Masonry accessory 96
   aggregate 93
   anchor 94
   cement 93
   color 93
   cutting 44
   demolition 41, 92, 170
   fireplace 108
   flashing 152
   flooring 174
   furring 137
   insulation 145
   manhole 478
   patching 92
   pointing 92
   reinforcing 96, 848
   removal 28, 92
   restoration 93
   saw 736
   selective demolition 92
   step 370
   testing 15
   toothing 42
   wall 397
   wall tie 94
   waterproofing 94
Mat blasting 299
   concrete placement foundation . 80
   drainage 502

floor 194
foundation 77
   foundation formwork 57
Material door and window 160
   handling 24, 560
   handling belt 560
   handling conveyor 560
   handling deconstruction 46
   handling system 560
   removal/salvage 44
Materials concrete 847
Mechanical 861
   dredging 554
   equipment demolition 260
   seeding 404
Median barrier 545
   concrete 380
   precast 545
Medium-voltage cable 266
Melting compound ice 367
Membrane roofing 151
   waterproofing 143, 363
Mercury vapor lamp 285
Mesh fence 386
   gunite 83
Metal bin retaining wall 395
   canopy 183
   cleaning 112
   deck composite 124
   demolition 112
   door 164
   drainage piping demo 30
   fence 386
   floor deck 125
   frame 161
   halide lamp 285
   in field galvanizing 178
   in field painting 178
   in shop galvanizing 112
   interceptor dental 251
   ladder inclined 126
   lath 171
   paint 178
   paint & protective coating . . . 112
   pan landing 125
   pan stair 125
   parking bumper 381
   pipe 503
   pipe removal 213
   plate stair 125
   pressure washing 112
   recovery interceptor 251
   roof 147, 148
   sandblasting 112
   sash 165
   screen 165
   siding 147, 148
   sign 180
   steam cleaning 112
   stud 171
   stud demolition 171
   water blasting 112
   wire brushing 112
Metal-clad dbl.-hung window . . . 166
Metallic hardener 82
   non-shrink grout 87
Meter center 276
   pit 479
   socket 276
   water supply 239
   water supply detector 239
   water supply domestic 239
Metric conversion factors 829
Micropile metal pipe drilled 358
Microtunneling 477, 478
Mill base price reinforcing 72

Milling pavement 361
Minor site demolition 28
Mirror 168
   glass 168
   wall 168
Misc metal fence & gate demo . . 34
Miscellaneous painting 175
Mix asphaltic base course plant . 371
   design asphalt 14
   planting pit 404
Mixed bituminous conc. plant . . . 288
Mixer concrete 730
   mortar 730, 735
   plaster 735
   road 738
Mixing windrow 361
Mobilization 289, 356
   air-compressor 289
   barge 138
   equipment 343, 357
   or demobilization 20
Model building 10
Modifier labor 212
Modular playground 190
   playground equipment demo . 38
Modulus of elasticity 15
Moil point 734
Moisture barrier 144
   content test 15
Monitoring sampling 49
Monkey ladder 190
Monolithic gutter 380
Monument survey 26
Mortar 93
   admixture 94
Mortar, brick and block 849
Mortar cement 848
   masonry cement 93
   mixer 730, 735
   pigment 93
   Portland cement 93
   restoration 93
   sand 93
   set stone wall 673
   testing 15
Moss peat 402
Motorized grader 731
   roof 206
Moveable bulkhead 202
Moving building 47
   shrub 469
   structure 47
   tree 469
Mower lawn 735
Mowing 368
   brush 290
Muck car tunnel 738
Mud pump 730
   trailer 738
Mulch bark 402
   ceramic 402
   stone 402
Mulcher power 732
Mulching 402
Municipal incinerator 192
Mylar tarpaulin 21

N

Nail stake 65
Nailer pneumatic 734, 735
Natural gas generator 278
   gas PE pipe demo 32
   gas steel pipe demo 32
   gas valve & fitting demo 32

NEMA enclosures 869
Neoprene adhesive 143
   expansion joint 154
   waterproofing 143
Net safety 18
   tennis court 385
No hub pipe 244
   hub pipe fitting 248
Non-chloride accelerator conc. . . 79
Nondestructive pipe testing 261
Non destructive testing 15
Non-metallic non-shrink grout . . 87
Non-template hinge 167
Norwegian brick 99
Nut remove 112

O

Oakum caulking 243
Observation well 480
Obstacle course 190
Off highway dump truck 733
Office & storage space 18
   expense 12
   field 18
   trailer 18
Oil formwork 65
   interceptor 251
Oil-filled transformer 276
Oil/water separator 563
Omitted work 13
One piece astragal 167
Open rail fence 394
   web bar joist 124
Opening roof frame 118
Ornamental aluminum rail 134
   glass rail 134
   grass 412
   railing 134
   steel rail 134
   wrought iron rail 134
Outdoor bleacher 206
   fountain 202
Outrigger wall pole 185
Oval arch culvert 508
Overhaul 44
Overhead & profit 13
   commercial door 164
   contractor 834
   door 164
   hoist 560
Overlapping astragal 167
Overlay face door 162
   pavement 360
Overpass 399
Overtime 11, 13
Oxygen lance cutting 44

P

Packaged lift station 499
   pumping station 499
Pad bearing 115
   equipment 77
   heated 54
   vibration 115
Paddle tennis court 191
Paint 177
   & coating 175
   & protective coating metal . . . 112
   and protective coating 112
   exterior 177
   floor concrete 175
   floor wood 175

interior miscellaneous 175
metal 178
miscellaneous 178
sprayer 735
striper 738
wall & ceiling interior 176
Painted marking airport 382, 383
pavement marking 381
Painting booth 188
ceiling 176
clapboard 175
exterior siding 175
grille 175
metal in field 178
miscellaneous 175
parking stall 382
pavement 381
pavement letter 382
pipe 175
reflective 381
siding 175
steel 178
stucco 175
swimming pool 201
temporary road 381
tennis court 385
thermoplastic 381
truss 176
Pan stair metal 125
tread concrete 81
Panel facing 107
fiberglass 21, 148
sandwich 148
sound dampening gypsum . . . 172
steel roofing 148
Panelboard 276
electric 276
w/circuit-breaker 276
Paneling cutout 42
Paper building 146
sheathing 146
Parging cement 142
Park bench 397
Parking appurtenance demo 37
barrier 381
barrier precast 381
bumper 381
bumper metal 381
bumper plastic 381
bumper wood 381
collection equipment 188
control equipment 188
equipment 188
gate 188
gate equipment 188
LED lighting 281
lot 656, 657
lot crushed stone 657
lot gravel base 656
lot paving 375
lot with lighting 658-661
marking pavement 382
stall painting 382
ticket equipment 188
Partition 182
anchor 95
block 103
concrete block 103
drywall 171
NLB gypsum 171
plaster 171
wall 171
Patch core hole 15
formwork 65
repair asphalt pavement 362
Patching asphalt 288

concrete floor 52
concrete wall 52
masonry 92
rigid pavement 364
Patio 369, 370
block concrete 378
door 163
Patio/deck canopy 183
Paumelle hinge 167
Pavement 378
asphalt 384
asphaltic 372
berm 380
breaker 734
bucket 737
demolition 27
emulsion 360
grooving 361
highway concrete 377
letter painting 382
line & marking removal 36
marking 381
marking painted 381
marking plastic 382
milling 361
overlay 360
painting 381
parking marking 382
planer 738
profiler 738
profiling 361
pulverization 361
reclamation 361
recycled 288
recycling 361
removal 28
repair flexible 361
repair rigid 364
replacement 372
replacement rigid 364
sealer 385
slate 107
widener 738
Paver asphalt 735
bituminous 735
cold mix 738
concrete 736
floor 174
granite 379
highway 379
roof 155
shoulder 738
thinset 370
tile exposed aggregate 378
Paving asphalt 372
asphaltic concrete 375
athletic 384
block granite 379
brick 378
cleaning 360
cold reused asphalt 361
highway 398
hot reused asphalt 361
parking lot 375
Paving surface trtmnt. flexible . . . 360
Pea gravel 402
stone 53
Peat humus 402
moss 402
Pedestal pile 599
Pedestrian bridge 399
paving 666
Pencil rod formwork 65
Penetration macadam 372
test 14
Perennial 414

Perforated aluminum pipe 501
pipe 501
PVC pipe 501
Performance bond 13
Perimeter drain 584
Perlite insulation 145
Permit building 14
Personnel field 12, 13
hoist 739
Petrographic analysis 14
PEX pipe 226
tubing 226
Photography 14
aerial 14
construction 14
time lapse 14
Photoluminescent signage 181
Picket fence 394
fence vinyl 392
railing 127
Pickup truck 737
Pick-up vacuum 730
Pier brick 98
Pigment mortar 93
Pile 357, 590, 594, 595, 599
barge driven 348
boot 346
bored 355
caisson 598
cap 78, 582, 583
cap concrete placement 80
cap formwork 57
cast-in-place 586, 587
C.I.P. concrete 586, 587
concrete 586, 587
concrete filled pipe 590, 591
cutoff 289
demo 35
driven 345
driving 837, 838
driving mobilization 289
encasement 346
foundation 357
friction 346, 357
grouted 358
H 343, 345
hammer 731
high strength 342
leads 731
lightweight 342
load test 289
mobilization hammer 289
pin 358
pipe 357
point 345, 346, 355
point heavy duty 355
precast 345
precast concrete 588
prestressed 345, 357
prestressed concrete 588
recycled plastic and steel 349
shore driven 346
sod 404
splice 345, 346, 354
steel 345
steel H 592
steel pipe 590, 591
steel sheet 341
steel shell 587
step tapered 345
step-tapered steel 594
supported dock 137
testing 289
timber 346
treated 346
treated wood 595

wood 346
wood sheet 342
wrap concrete 552
wrap steel 552
wrap wood 552
wrapping 723
Piling marine 137
sheet 341, 873
special cost 289
Pillow tank 495
Pin pile 358
powder 115
Pine shelving 183
siding 149
straw 402
Pipe & fitting backflow
preventer 240
& fitting backwater valve 253
& fitting bronze 215
& fitting cast iron 216, 244
& fitting copper 217
& fitting DWV 228
& fitting grooved joint 222
& fitting hydrant 241
& fitting iron body 261
& fitting malleable iron 221
& fitting polypropylene 257
& fitting PVC 225
& fitting sanitary drain 250
& fitting steel 219
& fitting steel cast iron 220
& fitting turbine water meter . . 240
acid resistant 257
add elevated 212
aluminum 501, 504
and fittings 864
bedding 310, 648, 649
bedding trench 298
black steel 511
bollard 381
bursting 476
cap 483, 497, 502
cast iron 243
clamp 472
clay 499, 501
cleaning internal 475
cleanout 216
concrete 506
concrete cylinder 482
conduit gland seal 532
conduit prefabricated 531, 533
conduit preinsulated 531
conduit system 531, 534
contained 235
copper 217, 488
corrosion resistant 257
corrugated 501
corrugated metal 501, 503
CPVC 225
culvert 507
culvert system 507
cured-in-place 476
double wall 231, 510
drain 307
drainage . . 256, 257, 497, 498, 503,
505
dual wall HDPE 235
ductile iron 483
DWV PVC 225, 498
epoxy fiberglass wound 257
fire retardant 257
fitting 485, 532, 533
fitting cast iron 219, 244
fitting copper 217
fitting corrosion resistant 257
fitting DWV 228

fitting gasket joint	246
fitting glass	256
fitting malleable iron	221
fitting no hub	248
fitting plastic	226, 229, 230, 485
fitting soil	244
foam core DWV ABS	225
freezing	473
fused	487
gas	511, 513
glass	256
grate	508
groove cut	224
groove roll labor	224
grooved joint	222
HDPE	231
inspection internal	475
insulated	530
insulated steel	530
internal cleaning	475
iron alloy mechanical joint	257
liner	476
liner HDPE	477
lining	477
metal	503
no hub	244
painting	175
perforated	501
perforated aluminum	501
PEX	226
pile	357
plastic	224, 225
plastic FRP	224
polyethylene	484, 513
polypropylene	257
prefabricated	531, 533
preinsulated	531, 533
proxylene	257
PVC	225, 485, 497
rail aluminum	126
rail galvanized	126
rail stainless	126
rail steel	126
rail wall	127
railing	126
reinforced concrete	506
relay	307
removal	28
removal metal	213
removal plastic	213
repair	472
repair clamp	473
rodding	475
sewage	497, 503, 506
sewage collection PVC	497
shock absorber	241
single hub	243
sleeve plastic	63
sleeve sheet metal	63
sleeve steel	64
soil	243
steel	218, 488, 504, 511
subdrainage	501
subdrainage plastic	501
support framing	118
tee	223
testing	261
testing nondestructive	261
vitrified clay	498, 501
water	483
water distribution	488
wrapping	472
X-ray	262
Pipe bursting	476
Piping air process	564
Piping designations	877

Piping drainage & sewage	505
excavation	642, 644, 646
gas service polyethylene	513
gas service steel	511
gasoline	510
HDPE	487
sewage pipe vitrified clay	498
storm drainage	503
subdrainage vitrified clay	501
Pit excavation	298
leaching	500
meter	479
scale	185
sump	307
test	26
Pitch emulsion tar	360
Placing concrete	79, 847
reinforcment	844
Plain & reinf. slab on grade	600
Planer pavement	738
Planing cold	361
Plank grating	129
hollow precast concrete	84
precast concrete roof	84
precast concrete slab	84
scaffolding	20
Plant and bulb transplanting	469
and planter	194
artificial	194
bed preparation	404
hardiness	876
interior	194
mix asphaltic base course	371
mixed bitum conc. recycled	288
mixed bituminous concrete	288
screening	732
spacing chart	875
stolen	406
Planter	195, 397
bench	397
concrete	397
fiberglass	397
interior	195
Planting	469
pit	687, 688
shrub	436
tree	436
Plant-mix asphalt paving	372
Plaque bronze	180
Plaster	171
cutout	42
demolition	170, 171
gypsum	171
mixer	735
partition	171
thin coat	172
wall	171
Plasterboard	172
Plastic and steel pile recycled	349
ball check valve	215
ball valve	215
coupling	229
fence recycled	392
fitting	228
FRP pipe	224
laminate door	162
parking bumper	381
pavement marking	382
pipe	224, 225
pipe fitting	226, 229, 230, 485
pipe polyethylene	231
pipe removal	213
pipe weld	234
screw anchor	113
sign	180
trench drain	253

valve	215
Plate checkered	130
compactor	332
roadway	738
steel	118
stiffener	118
vibrating	332
Platform checkered plate	130
framing	136
tennis	191
trailer	736
Player bench	398
Playfield equipment	190
Playground equipment	190, 388
modular	190
protective surfacing	384
slide	190
whirler	190
Plaza brick & tile	664, 665
paving	664
Plow vibrator	733
Plug valve	530
wall	97
Plumbing	240
demolition	213
fixture	251, 255, 263, 482
fixture removal	213
Plywood	858
demolition	170, 171
fence	23
formwork	55
formwork steel framed	60
sidewalk	23
siding	149
sign	181
Pneumatic nailer	734, 735
Point heavy duty pile	355
moil	734
pile	345, 346, 355
Pointing masonry	92
Poisoning soil	334
Pole & fixture line	535
aluminum light	280
anchor	535
athletic	190
crossarm	535, 536
electrical	536
fiberglass light	280
foundation	535
lighting	280
portable decorative	189
steel electrical utility	535
steel light	280
telephone	536
utility	280
wood	536
wood electrical utility	536
wood light	280
Polyethylene backer rod	67
coating	472
drum	556
film	402
pipe	484, 513, 683, 686
plastic pipe	231
pool cover	202
septic tank	500
tarpaulin	21
waterproofing	143
Polymer trench drain	253
Polypropylene fabric	502
fitting	258
pipe	257
siding	150
Polystyrene insulation	145
Polyurethane caulking	157
Polyvinyl chloride (PVC)	47

tarpaulin	21
Pond liner	288
Pontoon concrete	556
Pool accessory	202
concrete side/vinyl lined	678
cover	202
cover polyethylene	202
filtration swimming	256
gunite shell	678
swimming	201
Poplar	444
Porcelain tile	173
Portable air compressor	733
asphalt plant	737
booth	208
building	18
fire extinguisher	182
scale	185
Portland cement	53
Post and board retaining wall	671
and panel signage	181
athletic	191
cedar	136, 139
concrete	545
driver	738
indicator valve	494
lamp	133, 280
light	279
pedestrian traffic control	189
recreational	190
redwood	136
shore	63
sign	181
tennis court	385
wood	136
Post-tensioned slab on grade	75
Potable-water storage tank	243
Powder actuated tool	115
charge	115
pin	115
Power mulcher	732
pole guying	535
temporary	16, 17
trowel	730
Preblast survey	299
Precast beam	85
bridge	399
catch basin	478
concrete beam	85
concrete lintel	86
concrete pile	588
concrete plantable paver	378
concrete roof plank	84
concrete slab plank	84
concrete stair	85
concrete tee beam	85
concrete unit paving slab	378
concrete wall	85
concrete wall panel tilt-up	86
coping	109
curb	380
edging	468
grate	369
headwall	507
median	545
parking barrier	381
paving	664
pile	345
septic tank	499
wall panel	85
wall panel insulated	85
Pre-engineered steel building	206
steel building demo	200
steel buildings	860
Prefabricated building	205
pipe	531, 533

Prefinished door 162
Preformed roof panel 147
   roofing & siding 147
Preinsulated pipe 531, 533
Prepackaged blower 563
Preparation lawn bed 402
   plant bed 404
Preservation topsoil 369
Pressure grouting cement 342
   injected footing 599
   reducing valve water 215
   regulator valve 529
   relief valve 214
   relief valve adjustable 564
   relief valve preset 564
   relief valve weight loaded 564
   switch well water 482
   test hydraulic 261
   treated lumber 136
   valve relief 214
   washer 736
   washing metal 112
Pressurized fire extinguisher 182
Prestressed concrete pile . . 345, 588
   pile 345, 357
Prestressing steel 75
Preventer backflow 240
Prices lumber products 858
Prime coat 372
Primer steel 178
Privacy fence vinyl 392
   slat 386
Proctor compaction test 15
Product piping 510
Profile block 101
Profiler pavement 738
Profiling pavement 361
Progress schedule 14
Project overhead 13
   sign 23
Promenade drain 250
Property line survey 26
Protection slope 334
   stile 167
   temporary 24
   termite 334
   winter 21, 54
Proxylene pipe 257
Prune shrub 369
   tree 369
Pruning 369
   tree 368
P&T relief valve 214
Pull box electrical 274
Pulverization pavement 361
Pump 263, 480
   centrifugal 736
   circulating 263
   concrete 730
   contractor 307
   diaphragm 736
   fountain 256
   general utility 242
   grinder system 251
   grout 736
   in-line centrifugal 263
   mud 730
   operator 308
   sewage ejector 252
   shallow well 482
   shotcrete 730
   submersible 254, 480, 736
   sump 254
   trash 736
   water 263, 480, 736
   water supply well 482

wellpoint 740
Pumped concrete 80
Pumping 307
   dewater 308
   station 499
Purlin steel 119
Push-on joint 483
PVC adhesive 143
   cleanout 216
   conduit 272
   conduit cement 274
   conduit fitting 272
   conduit in slab 274
   control joint 96
   direct burial 536
   duct fitting 539
   fitting 227, 486
   flange 230
   pipe 225, 485, 497, 683, 686
   pipe perforated 501
   sheet 143
   swing check valve 495
   underground duct 538
   union 230
   waterstop 66
   well casing 481

Q

Quarry drill 734
   drilling 299
   tile 173
Quartz 402
   chip 53
   line lamp 285
Quoin 106

R

Raceway 272, 273
   concrete slab 275
Rack bicycle 190, 197
Radial arch 138
   wall formwork 59
Radio tower 539
   tower demo 36
Radiography test 16
Rail aluminum pipe 126
   crane 560
   galvanized pipe 126
   guide 545
   guide/guard 545
   ornamental aluminum 134
   ornamental glass 134
   ornamental steel 134
   ornamental wrought iron 134
   stainless pipe 126
   steel pipe 126
   wall pipe 127
Railing bridge 398
   cable 134
   industrial 127
   ornamental 134
   picket 127
   pipe 126
Railroad ballast 543
   bumper 543
   concrete tie 543
   derail 543
   maintenance 543
   relay rail 546
   siding 546, 878
   tie 370, 468
   tie retaining wall 672

tie step 370
timber switch tie 543
timber tie 543
track accessory 543
track heavy rail 543
track maintenance 543
track material 543
track removal 29
turnout 547
wheel stop 543
wood tie 546
Rainwater filter 562
   storage tank fiberglass 254
Raise manhole 509
   manhole frame 509
Raising concrete 343
Rake topsoil 402
   tractor 732
Rammer tamper 333
Rammer/tamper 731
Ramp handicap 77
   temporary 21
Ratio water cement 15
Razor wire 393
Ready mix concrete 79, 847
Rebar accessory 71
Receptacle concrete 198
   galvanized 198
   trash 198
Recirculating chemical toilet . . . 562
Reclaimed or antique granite . . . 106
Reclamation pavement 361
Recovery system resi graywater . . 252
Recreational post 190
Recycled pavement 288
   plant mixed bitum concrete . . 288
   plastic and steel pile 349
   plastic/steel pile marine 349
Recycling cold 361
   pavement 361
Reduce heat transfer glass 168
Redwood bark mulch 402
   deck 677
   post 136
   post and board wall 671
   post wall 670
   siding 149
Refinish textured ceiling 170
Reflective painting 381
   sign 181
Reflectorized barrels 734
Refrigerant removal 260
Refrigeration equipment 203
Reinforced beam 597
   concrete pipe 506
   culvert 506
Reinforcement welded wire 845
Reinforcing 844
   accessory 71
   bar chair 71
   bar tie 71
   beam 73
   chair subgrade 72
   coating 74
   column 73
   concrete 73
   dowel 74
   elevated slab 73
   epoxy- coated 74
   footing 73
   galvanized 74
   girder 73
   glass fiber rod 74
   high chair 71
   joint 96
   ladder 96

masonry 96
mill base price 72
shop extra 72
slab 73
sorting 73
spiral 73
steel 844
steel fiber 76
synthetic fiber 76
testing 15
tie wire 72
truss 96
wall 73
Relay pipe 307
   rail railroad 546
Release valve sewer air 495
Relief valve P&T 214
   valve self-closing 214
   valve sewer 495
   valve temperature 214
Relining sewer 476
Removal air conditioner 260
   bathtub 213
   block wall 40
   boiler 260
   catch basin 28
   concrete 27
   concrete pipe 28
   curb 27
   driveway 27
   fixture 213
   floor 40
   guide rail 28
   hydrant 28
   lavatory 213
   masonry 28, 92
   pavement 28
   pavement line & marking 36
   pipe 28
   plumbing fixture 213
   railroad track 29
   refrigerant 260
   sidewalk 28
   sign 180
   sink 213
   sod 404
   steel pipe 28
   stone 28
   stump 291
   tank 48
   tree 289, 291, 469
   urinal 213
   utility line 28
   water closet 213
   water fountain 213
   water heater 213
   window 160
Removal/salvage material 44
Remove bolt 112
   nut 112
   rock 403
   topsoil 402
Remove/replace catch
basin cover 509
Rendering 10
Renovation tread cover 132
Rental equipment 730
Repair asphalt pavement patch . . 362
   asphalt paving crack 363
   pipe 472
Replacement joint sealant 142
   pavement 372
Reservoir liner 288
Residential bulkhead 555
   chain link gate 391
   fixture 279

garage 208
greenhouse 205
graywater 252
gutting 44
lighting 279
pool 678
Resilient base 175
pavement 384
Resistance corrosion 472
Restoration masonry 93
mortar 93
Retaining wall 78, 397
wall & footing masonry 669
wall cast concrete 395
wall concrete segmental 395
wall demo 37
wall formwork 59
wall segmental 395
wall stone 397
wall stone gabion 396
wall timber 396
wall wood 670
Retarder concrete 79
vapor 145, 146
Retractable lateral arm awning . . . 183
Revetment concrete 555
Revolving door 204
Ribbed waterstop 66
Ridge cap 148
Rig drill 26
Rigid anchor 95
conduit in trench 275
in slab conduit 275
insulation 144, 145
pavement patching 364
pavement repair 364
pavement replacement 364
paving surface treatment 363
Ring clamp 484
Ripper attachment 737
Ripping 300
dozer 300
grader 300
Rip-rap & rock lining demo 34
Riprap and rock lining 341
Riser pipe wellpoint 740
rubber 175
River stone 53, 370
Road base 371
berm 380
mixer 738
pavement 650, 653
sign 181
sweeper 738
temporary 21
Roadway LED luminaire 281
luminaire 281
maintenance grading 542
plate 738
Rock anchor 65
bolt drilling 299
borrow 310
breaking hydraulic 300
drill 26, 299
removal 299
remove 403
salt 367
trencher 733
Rod backer 155
ground 269
tie 118, 342
weld 116
Roll roof 151
Roller compacted concrete 83
compaction 308
sheepsfoot 333, 732

static 333
tandem 732
vibrating 332
vibratory 732
Rolling earth 333
ladder 20
topsoil 403
tower scaffold 20
Romex copper 268
Roof adhesive 151
aluminum 147
beam 138
built-up 151
cant 151
clay tile 147
decking 124
expansion joint 154
fiberglass 148
frame opening 118
framing removal 42
gravel 53
joint 154
metal 147, 148
metal tile 147
panel aluminum 147
panel preformed 147
paver 155
paver and support 155
roll 151
shingle asphalt 146
specialty prefab 152
steel 148
tile 147
truss 138
Roofing & siding preformed 147
membrane 151
system built-up 151
Root raking loading 403
Rope decorative 189
steel wire 122
Rotary crawler drill 734
hammer drill 734
lobe blower 563
Rototiller 732
Rough grading 292
grading site 292
stone wall 104
Rough-in drinking fount. deck . . 255
drinking fountain floor 255
drinking fountain wall 255
Round wood 370
Rub finish carborundum 82
Rubber base 175
boot 479
coating 144
control joint 96
floor 384
pavement 384
riser 175
threshold 167
waterproofing 143
waterstop 66
Rubberized asphalt 364
asphalt sealcoat 360
Rubbing wall 82
Rubbish handling 42
handling chutes 839
Rumble strip 546
Run gravel bank 288
valve bull 500
Running track surfacing 384
Rupture testing 15
Russian Olive 440

S

Safety fence 23
net 18
railing cable 123
switch 277
Salamander 736
Sales tax 11, 832, 833
Salt rock 367
Salvage or disposal value 44
Sample field 49
Sand 93
backfill 648
brick 53
concrete 53
excavation 622
fill 288
screened 93
Sandblast finish concrete 82
Sandblasting equipment 736
metal 112
Sander floor 736
Sandstone 107
flagging 379
Sandwich panel 148
Sandy clay/loam 640
Sanitary drain 250
tee 248
Sash aluminum 165
metal 165
security 165
steel 165
Saw blade concrete 88
chain 736
circular 736
concrete 730
cutting concrete floor 88
cutting concrete wall 88
cutting demolition 44
cutting slab 88
masonry 736
Sawn control joint 66
Scaffold baseplate 19
bracket 19
caster 19
catwalk 20
frame 19
guardrail 19
mason 19
rolling tower 20
specialty 20
stairway 19
wood plank 19
Scaffolding 836
plank 20
tubular 19
Scale 185
commercial 185
contractor 186
crane 185
pit 185
portable 185
truck 185, 186
Scarify 403
concrete 170
subsoil 403
Schedule 14
progress 14
Scissor gate 182
lift 733
Scoreboard baseball 190
football 190
Scored block 102
split face block 101
Scraper cut & fill 621, 625, 629
earth 732

elevating 306
excavation 306
self propelled 306
Scratch coat 175
Screed base 72
Screed, gas engine,
8HP vibrating 730
Screed holder 72
Screen metal 165
security 165
Screened loam 402
sand 93
topsoil 403
Screening plant 732
Screw aluminum 148
anchor 72, 113
anchor bolt 72
eye bolt 72
jack 341
lag 115
machine 115
Seal coats 375, 376
curb 361
fog 360
pavement 360
security 167
Sealant 142
caulking 156
replacement joint 142
Sealcoat 360
Sealer floor 53
joint 155, 156
pavement 385
Sealing crack asphalt paving 363
foundation 585
Seawall concrete 555
Secondary sewage lagoon 705
treatment plant 562
Sectional door 164
Security barrier 189
bollard 544
factor 11
gate 182
planter 544
sash 165
screen 165
seal 167
vehicle barrier 544
Sediment bucket drain 251
Seeding 404, 874
athletic field 405
birdsfoot trefoil 405
bluegrass 405
clover 405
crown vetch 405
fescue 405
hydro 405
rye 405
shade mix 405
slope mix 405
turf mix 405
utility mix 405
wildflower 406
See-saw 190
Segmental conc. retaining wall . . 395
retaining wall 395
Select borrow 310
Selective clearing 290
demolition fencing 33
demolition masonry 92
tree removal 290
Self-closing relief valve 214
Self-propelled scraper 306
Self-supporting tower 539
Separator oil/water 563
soil 468

Septic galley	500
system	499, 703
system chamber	501
system effluent-filter	500
tank	499, 703
tank & component demo	31
tank concrete	499
tank polyethylene	500
tank precast	499
trench	703
Service boring	478
entrance cable aluminum	268
entrance cap	269
laboratory analytical	49
station equipment	188
tap	689
Setting stone	397
Sewage aeration	562
collection concrete pipe	496
collection plastic pipe	496
collection PVC pipe	497
collection valve	495
collection vent	496
ejector pump	252
municipal waste water	562
pipe	497, 503, 506
pipe vitrified clay piping	498
treatment plant	562
vent	496
Sewage/drainage collection	505
Sewer relief valve	495
relining	476
Sewerline	699, 701
common earth excavate	699, 700
loam/sandy clay excavate	701, 702
Shadow box fence	394
Shale paver	370
Shear test	15
Sheathed nonmetallic cable	268
Romex cable	268
Sheathing paper	146
She-bolt formwork	65
Sheepsfoot compactor	332
roller	333, 732
Sheet metal aluminum	152
metal flashing	152
metal screw	148
piling	341, 873
piling seawall	555
steel pile	345
Sheeting	341
aluminum	555
driver	734
open	344
tie-back	344
wale	342
wood	307, 342, 873
Shelter aluminum	209
golf	190
temporary	54
Shelving	183
pine	183
wood	183
Shield expansion	113
light	545
Shielded aluminum cable	267
copper cable	266
Shift work	11
Shingle	146
asphalt	146
concrete	147
strip	146
Ship ladder	126
Shock absorber	241
absorber pipe	241
Shop extra reinforcing	72

Shore driven pile	346
post	63
protection demolition	35
service	557
Shoring	341, 342
aluminum joist	63
baseplate	19
bracing	19
concrete formwork	63
concrete vertical	63
flying truss	63
frame	63
heavy duty	19
leveling jack	19
slab	20
steel beam	63
temporary	342
wood	344
Short load concrete	79
Shot blast floor	170
blaster	736
Shotcrete	83
pump	730
Shoulder paver	738
Shovel	301
crawler	732
excavation	622, 623, 626, 627, 630
Shrinkage test	15
Shrub bed fertilizer	367
maintenance	367
moving	469
planting	436
prune	369
temperate zones 2 – 6	419
warm temperate	431
weed	367
Sidewalk	174, 378, 663
asphalt	369
bituminous	662
bridge	20
broom	736
concrete	369
driveway and patio	369
removal	28
steam clean	367
temporary	23
Siding aluminum	148
bevel	149
cedar	149
fiberglass	148
metal	147, 148
painting	175
panel aluminum	148
plywood	149
polypropylene	150
railroad	546, 878
redwood	149
steel	149
vinyl	149
wood	149
wood board	149
Sieve analysis	14
Sign	23, 180, 181
acrylic	180
aluminum	180
base formwork	57
detour	23
directional	181
highway	181
letter	180
metal	180
plastic	180
plywood	181
post	181
project	23
reflective	181

removal	180
road	181
stainless	180
street	180
traffic	181
Signage exterior	180
photoluminescent	181
post and panel	181
Signal system traffic	544
Silencer	563
chamber	564
with paper filter	563
with polyester filter	564
Silica chip	53
Silicon carbide abrasive	52, 82
Silicone	156
coating	144
water repellent	144
Sill	106
stone	105, 108
Sillcock hose bibb	255
Silo	209
demo	200
Simulated stone	109
Single hub pipe	243
hung aluminum window	165
Sink removal	213
Site clear and grub	289
clearing	289
demolition	27, 614, 617, 619
drain	508
drainage	563
furnishing	197
furnishing demo	38
grading	292
improvement	397, 399
irrigation	683, 686
lighting	720, 721
maintenance	366
preparation	26
rough grading	292
utility	648
Sitework	601, 602
manhole	708
Skidder log	732
Skid-steer attachments	733
loader	733
Slab blockout formwork	58
box out opening formwork	56
bulkhead formwork	56, 58
concrete	600
concrete placement	80
curb formwork	56, 58
cutout	41
edge formwork	56, 58
flat plate formwork	56
haunch	78
haunch formwork	78
lift	78
on grade	77, 600
on grade formwork	58
on grade post-tensioned	75
on grade removal	28
reinforcing	73
saw cutting	88
shoring	20
stamping	83
textured	78
thickened edge	78
thickened edge formwork	78
trench formwork	58
turndown	78
turndown formwork	78
void formwork	56, 58
with drop panel formwork	56
Slabjacking	343

Slab-on-grade sidewalk	663
Slat privacy	386
Slate	107
flagging	379
pavement	107
paving	665
sidewalk	379
stair	108
Sleeper	137
framing	137
Sleeve	490
anchor bolt	67
and tap	484
cut-in	490
dowel	74
formwork	63
plastic pipe	63
sheet metal pipe	63
steel pipe	64
tapping	490
Slide gate	386, 554
playground	190
swimming pool	202
Sliding glass door	163
glass vinyl-clad door	163
vinyl-clad door	163
window	166
window aluminum	165
wood door	163
Slipform paver barrier	738
paver curb	738
Slope grading	292
protection	334
protection rip-rap	341
Slot channel	95
Slotted channel framing	118
pipe	507
Sludge treatment cell	567
treatment cell activated	567
Sluice gate	553
gate hydraulic	553
Slump block	101
Slurry trench	344
Snap-tie formwork	64
Snow fence	48
guard	155
removal	367
Snowplowing	367
Soccer goal post	191
Socket ground	190
meter	276
wire rope	121
Sod	406
edging planter & guy demo	39
tennis court	384
Sodding	406
Sodium high pressure fixture	284
high pressure lamp	285
low pressure fixture	281
low pressure lamp	285
Soil cement	335
compaction	308
decontamination	48
pipe	243
poisoning	334
sample	27
separator	468
stabilization	335
stabilization asphalt	335
stabilization cement	335
stabilization geotextile	335
stabilization lime	335
tamping	308
treatment	334
Soldier beam	343
Sorting reinforcing	73

Index

Space heater 735
  office & storage 18
Spade air 734
  tree 733
Spanish roof tile 147
Special door 163
Specialty demo 180
  scaffold 20
Specific gravity 15
  gravity testing 14
Speed bump 545
Spiral reinforcing 73
  stair 133
Splashblock concrete 369
Splice pile 345, 346, 354
Split rib block 101
Spotter 316
Spray coating 142
  painting booth automotive 188
  rig foam 735
Sprayed cementitious
fireproofing 155
Sprayer emulsion 734
  paint 735
Spread footing 77, 581
  footing formwork 57
  loam 403
  park soil 403
  soil conditioner 403
  stone 370
  topsoil 403
Spreader aggregate 730, 737
  chemical 737
  concrete 736
Spring bronze weatherstrip 167
  hinge 167
Sprinkler irrigation system 400
  underground 400
Squash court backstop 191
Stabilization asphalt 335
  base 335
  cement 335
  fabric 371
  geotextile 335
  soil 335
Stabilizer base 738
Stacked bond block 102
Stadium 206
  bleacher 206
  cover 204
Stainless bar grating 129
  gutter 153
  plank grating 129
  sign 180
  weld rod 116
Stair basement bulkhead 85
  brick 104
  concrete 78
  finish concrete 82
  floor plate 125
  formwork 62
  grating 126
  marble 107
  metal pan 125
  metal plate 125
  pan tread 81
  precast concrete 85
  railroad tie 370, 666, 667
  riser vinyl 175
  slate 108
  spiral 133
  temporary protection 23
  tread 105, 131
  tread and riser 175
  tread insert 64
Stairway scaffold 19

Stake formwork 65
  nail 65
  subgrade 72
Stamping slab 83
  texture 83
Standard extinguisher 182
Standpipe steel 495
Static roller 333
Station transfer 192
  weather 210
Steam cleaner 736
  cleaning metal 112
  energy distribution pipe 533
  pipe underground 531
Steel anchor 95
  beam shoring 63
  bin wall 395
  bolt 136
  bolt & hex nut 114
  bridge 398, 399
  building components 855
  building pre-engineered 206
  chain 132
  conduit in slab 275
  conduit in trench 275
  corner guard 117
  cutting 114
  demolition 112
  dome 208
  door 161, 858
  downspout 152
  drill 734
  drilling 114
  drum receptacle 198
  edging 370, 469
  estimating 852
  fence 386, 392
  fiber 76
  fiber reinforcing 76
  fitting 219, 221
  flange 223
  frame 161
  framing 119
  girt 119
  grating 128
  gutter 153
  H pile 592
  hex bolt 114
  hinge 167
  light pole 722
  lintel 118
  member structural 119
  mesh box 341
  mesh grating 129
  painting 178
  pile 345
  pile clean 551
  pile concrete-filled 354
  pile sheet 345
  pile step-tapered 594
  pipe 218, 488, 504, 511
  pipe corrugated 501
  pipe pile 590, 591
  pipe removal 28
  pipe with insulation demo 31
  plate 118
  prestressing 75
  primer 178
  project structural 119
  purlin 119
  reinforcing 844
  roof 148
  roofing panel 148
  salvage 213
  sash 165
  sections 854

sheet pile 341
sheet piling seawall 555
shell pile 586, 587
siding 149
silo 209
standpipe 495
step-tapered pile 594
structural 851
tower 539
underground duct 538
underground storage tank 262
valve 261
water storage tank 494
water tank ground level 494
weld rod 116
well casing 481
well screen 481
window 165
window demolition 160
wire rope 122
wire rope accessory 120
Step 181
  bluestone 370
  brick 370
  manhole 478
  masonry 370
  railroad tie 370
  stone 106
  tapered pile 345
  tapered steel pile 594
Stiffener plate 118
Stiffleg derrick crane 739
Stile fabric 167
  protection 167
Stockade fence 394
Stockpile borrow 632
Stockpiling of soil 291
Stolen broadcast 406
  plant 406
Stolonizing 406
Stone anchor 95
  ashlar 106
  base 370
  crushed 288
  cultured 109
  curb cut 380
  fill 288
  filter 341
  floor 107
  gabion retaining wall 396
  mulch 402
  paver 106, 379
  pea 53
  removal 28
  retaining wall 397, 673
  river 53, 370
  setting 397
  sill 105, 108
  simulated 109
  step 106
  stool 108
  tread 105
  wall 104, 397
Stool stone 108
Stop gas 529
  gravel 152
Storage box 18
  dome 208
  dome bulk 208
  tank 262, 495
  tank cover 203
  tank demo 200
  tank fiberglass rainwater 254
  tank fiberglass underground . . 243
  tank potable-water 243
  tank steel underground 262

tank underground 262
Storefront aluminum 165
Storm drainage 510
  drainage manhole frame 478
  drainage piping 503
Stormwater management . . . 509, 712
Straw 402
  pine 402
Street basket 198
  sign 180
Streetlight 281
Strength compressive 15
Strip chamfer 63
  footing 77, 580
  rumble 546
  shingle 146
  soil 291
  topsoil 613
Striper paint 738
Striping thermoplastic 382
Stripping topsoil 291
Structural backfill 308
  column 116
  compaction 333
  excavation 298
  framing 853
  shape column 117
  soil mixing 402
  steel 851
  steel member 119
  steel project 119
  welding 114
Structure fabric 203
  hydraulic 553, 554
  moving 47
  tension 204
Stucco 172
  painting 175
Stud metal 171
  wood 171
Stump chipper 732
  chipping 290
  removal 291
Subcontractor O&P 13
Subdrainage fitting 502
  pipe 501
  plastic pipe 501
  system 501
  vitrified clay piping 501
Subgrade reinforcing chair 72
  stake 72
Submersible pump 254, 480, 736
  sump pump 254
Subsoil heating 203
Substructure 600
Subsurface drip irrigation 399
  exploration 26
Suction hose 735
Sump hole construction 307
  pit 307
  pump 254
  pump submersible 254
Supply black steel pipe water . . . 488
  concrete pipe water 482
  copper pipe water 488
  ductile iron pipe water 483
  HDPE water 487
  polyethylene pipe water 484
Support framing pipe 118
Surface bonding 93
  hinge 167
  treatment 361
Surfacing 360, 363
  court 384
  synthetic 384
Survey aerial 26

For customer support on your Site Work & Landscape Costs with RSMeans data, call 800.448.8182.

Index

crew	24	Taper-tie formwork	65	well	480
monument	26	Tapping cross and sleeve	489	Testing	14
preblast	299	main	489	pile	289
property line	26	sleeve	490	pipe	261
stake	24	valve	490	sulfate soundness	15
topographic	26	Tar kettle	736, 738	tank	16
Suspended ceiling	174	paper	402	Texture stamping	83
Sweep drive	367	pitch emulsion	360	Textured ceiling refinish	170
walk	367	Tarpaulin	21	slab	78
Sweeper road	738	duck	21	Thermoplastic marking airport	383
Swell testing	14	Mylar	21	painting	381
Swimming pool	201	polyethylene	21	striping	382
pool blanket	202	polyvinyl	21	Thickened edge slab	78
pool enclosure	206	Tax	11	Thimble wire rope	120
pool equipment	202	sales	11	Thin brick veneer	98
pool filter	256	Social Security	11	Thin coat plaster	172
pool filtration	256	unemployment	11	Thinning tree	290
pool ladder	202	Technician inspection	15	Thinset paver	370
pool painting	201	Tee cleanout	216	Thin-set tile	173
pool residential	678	HDPE piping	488	Threshold	167
pools	860	pipe	223	stone	107
Swing	190	precast concrete beam	85	Thrust block	489
check steel valve	261	Telephone manhole	538	Ticket booth	208
check valve	213, 260	pole	536	dispenser	188
check valve bronze	213	Telescoping boom lift	733	Tie adjustable wall	94
gate	386	Temperature relief valve	214	back sheeting	344
Switch electric	277	Tempered glass greenhouse	205	column	95
general duty	277	Tempering valve	215	cone formwork	65
safety	277	valve water	215	flange	95
well water pressure	482	Template anchor bolt	67	railroad	370, 468
Synthetic erosion control	334	Temporary barricade	22	rod	118, 342
fiber	76	building	18	wall	94
fiber reinforcing	76	connection	17	wire	95
surfacing	384	construction	17, 21	wire reinforcing	72
System chamber septic	501	electricity	16	Tile	173
component graywater	253	facility	22	aluminum	147
grinder pump	251	fence	23	ceiling	174
light pole	722	guardrail	22	ceramic	173
pipe conduit	531, 534	heat	54	clay	147
septic	499	light	17	concrete	147
sitework catch basin	708	lighting	17	demolition	170
sitework swimming pool	678	power	16, 17	exposed aggregate paver	378
sitework trenching	642, 644, 646	protection	24	exterior	173
subdrainage	501	ramp	21	or terrazzo base floor	175
		road	21	porcelain	173
		road painting	381	quarry	173
T		shelter	54	roof	147
		shoring	342	roof metal	147
Tack coat	360	tape	381	vinyl composition	175
Tactile warning surfacing	383	toilet	736	waterproofing membrane	
Tamper	294, 734	utility	17	ceramic	173
compactor	333	watchman	23	Tiling ceramic	173
rammer	333	Tennis court air-supported	204	Tilling topsoil	403
Tamping soil	308	court backstop	388	Tilt-up concrete wall panel	86
Tandem roller	732	court clay	384	precast concrete wall panel	86
Tank clean	48	court fence	387, 393	Timber connector	136
cover	203	court fence and gate	387	guide rail	545
disposal	48	court net	385	pile	346
fiberglass	262	court painting	385	retaining wall	396
fiberglass rainwater storage	254	court post	385	switch tie	543
fiberglass underground		court sod	384	switch tie railroad	543
storage	243	court surfacing	384	tie railroad	543
holding	562	Tensile test	15	Time lapse photography	14
pillow	495	Tension structure	204	Toggle-bolt anchor	113
potable-water storage	243	structure demo	201	Toilet	562
removal	48	Termite control chemical	334	chemical	736
septic	499	protection	334	partition removal	171
steel underground storage	262	Terra cotta coping	109	temporary	736
storage	262, 495	cotta demolition	41	Tool powder actuated	115
testing	16	Terrazzo demolition	170	Toothing masonry	42
water	262, 495, 737	Test beam	15	Top dressing	404
Tap and sleeve	484	load pile	289	Topographic survey	26
Tape barricade	22	moisture content	15	Topographical survey	26
detection	479	pile load	289	Topping concrete	86
temporary	381	pit	26	floor	82, 86
underground	479	ultrasonic	16	grout	87

Topsoil	288, 291, 310
placement and grading	404
preservation	369
remove	402
screened	403
stripping	291
Torch cutting	114, 736
cutting demolition	44
Tower crane	19, 738, 837, 838
hoist	739
light	735
radio	539
Track accessory	543
accessory railroad	543
drill	734
heavy rail railroad	543
material railroad	543
Tractor loader	732
rake	732
truck	737
Traffic barrier highway sound	399
channelizing	546
cone	22
light system demo	39
line	381
line remover	738
sign	181
signal system	544
Trail fitness	190
Trailer bunk house	18
floodlight	734
low bed	738
mud	738
office	18
platform	736
truck	736
water	737
Tram car	737
Transfer station	192
switch manual	279
Transformer base	280
oil-filled	276
weight	868
Transmission line pole	535
Transplanting plant and bulb	469
Trap grease	251
rock surface	82
Trash closure	198
pump	736
receptacle	198
receptacle fiberglass	198
Tread and riser stair	175
bar grating	131
channel grating	131
cover renovation	132
diamond plate	131
insert stair	64
stair	105, 131
stair pan	81
stone	107, 108
vinyl	175
Treated lumber framing	137
pile	137, 346
wood pile	595
wood post and board wall	671
Treatment cell activated sludge	567
cell sludge	567
plant secondary	562
plant sewage	562
Tree & palm	463
conifer	448
deciduous	437
fertilizer	367
fruit and nut	467
grate	369
guard	469

guying system 469
maintenance 367
maple 438
moving 469
pit 687
planting 436
prune 369
pruning 368
removal 289, 291, 469
removal selective 290
spade 733
thinning 290
water 369
Trench backfill ... 294, 297, 648, 649
box 344, 737
cover 130
disposal field 500
drain 253
drain fiberglass 253
drain plastic 253
drain polymer 253
excavating 292
excavation 293, 297
grating frame 130
slurry 344
utility 297
Trencher chain 297, 733
rock 733
wheel 733
Trenching 299, 307, 643
common earth 642, 643
loam & sandy clay ... 644, 645
sand & gravel 646, 647
Trial batch 15
Trim lawn 368
Tripod floodlight 734
Trowel coating 142
concrete 730
power 730
Truck concrete 730
crane flatbed 738
dock 189
dump 733
flatbed 733, 737
holding concrete 79
loading 298
mounted crane 739
pickup 737
scale 185, 186
tractor 737
trailer 736
vacuum 737
winch 737
Truss bowstring 138
painting 176
reinforcing 96
roof 138
varnish 176
Tube copper 217, 488
fitting copper 217
framing 164
Tubing copper 217
corrugated 502
PEX 226
Tubular fence 392
lock 166
scaffolding 19
Tugboat diesel 740
Tunnel locomotive 738
muck car 738
ventilator 738
Turbine water meter 240
Turf artificial 384
Turnbuckle wire rope 121
Turndown slab 78
Turnout railroad 547

Turnstile 189
TV inspection sewer pipeline ... 475
Two piece astragal 167
Tying wire 63

U

UF feeder cable underground ... 268
underground cable 269
Ultrasonic test 16
Uncased drilled concrete pier ... 357
Underdrain 478
Underground cable UF 269
chilled water pipe 531
duct 538
duct and manhole 536
duct fitting 539
electrical & telephone 538
electrical conduit 717
hydronic energy 533
piping 648
power feed 716
sewer pipe removal ... 617, 618
sprinkler 400
steam pipe 531
storage tank 262
storage tank fiberglass 243
storage tank removal 840
storage tank steel 262
tape 479
UF feeder cable 269
water pipe removal ... 614-616
Underlayment self-leveling
cement 87
Underpin foundation 343
Underwater light 202
lighting 283
Unemployment tax 11
Union PVC 230
Upstanding beam formwork 54
URD cable underground 268
Urinal removal 213
Utility accessory 479
box 479
connection 479
duct 538
electric 538
excavation 642, 644, 646
line removal 28
material demolition 29
pole 280
pole & cross arm demo 36
pole steel electrical 535
pole wood electrical 536
sitework 538
structure 708
temporary 17
trench 297
trench excavating 297
trench plow excavating 298
valve demo 29

V

Vacuum catch basin 737
pick-up 730
truck 737
wet/dry 737
Valve 260, 492
air release 493
air release & vacuum 493
altitude 492
backwater 253
ball check 215

blower pressure relief 564
box 491
bronze 213
bronze angle 213
butterfly 260, 491
check 493
cut-in 490
forged 261
gate 260, 261, 492, 493
globe 261
grooved-joint butterfly 223
grooved-joint gate 224
iron body 260
lubricated 530
materials 861-863
plastic 215
plug 530
pressure regulator 529
PVC swing check 495
relief pressure 214
selection considerations 861
sewage collection 495
soldered gate 214
steel 261
swing check 213, 260
swing check steel 261
tapping 490
tempering 215
water pressure 215
water utility distribution ... 491
Vapor barrier 146
retarder 145, 146
Varnish truss 176
VCT removal 170
Vehicle barrier security 544
guide rail 545
Veneer ashlar 104
brick 97
granite 105
member laminated 138
Vent sewage 496
sewage collection 496
Ventilator tunnel 738
Vertical metal ladder 126
pole wall 670
post wall 670
Very low density polyethylene ... 47
Vial bottle 49
Vibrating plate 332
roller 332
screed, gas engine, 8HP 730
Vibration pad 115
Vibrator concrete 730
earth 294, 308, 333
plate 308
plow 733
Vibratory hammer 732
roller 732
Vibroflotation 343
compaction 873
Vinyl-clad premium window 166
coated fence 387
composition tile 175
fence 392
fence & gate demo 34
floor 175
gutter 153
siding 149
stair riser 175
stair tread 175
tread 175
Vitrified clay pipe 498, 501
Volume earth project large ... 302

W

Wale 342
sheeting 342
Waler holder formwork 65
Walk 369
maintenance 367
or step or paver demo 36
Walkway cover 184
LED luminaire 284
luminaire 283, 284
Wall accessory formwork 64
aluminum 148
and partition demolition ... 170
base 175
box out formwork 58
brick 104
brick shelf formwork 58
bulkhead formwork 58
bumper 167
buttress formwork 58
canopy 183
cast concrete retaining 395
cast in place 603
ceramic tile 173
concrete 77, 578
concrete finishing 82
concrete placement 80
concrete segmental retaining .. 395
corbel formwork 58
curtain 165
cutout 41
demolition 170
flagpole 185
forms 841
formwork 58
foundation 102, 578, 579, 603
framing removal 42
furring 137
grout 94
guard 117
hydrant 241
insulation 144
lintel formwork 59
masonry 397
masonry retaining 669
mirror 168
panel precast 85
partition 171
pilaster formwork 59
plaster 171
plug 97
plywood formwork 58
precast concrete 85
prefab plywood formwork 59
reinforcing 73
retaining 78, 397
rubbing 82
sill formwork 59
steel bin 395
steel framed formwork 59
stone 104, 397
stone gabion retaining 396
stucco 172
stud 171
tie 94
tie masonry 94
window 164
Wardrobe 183
Warehouse 203
Warning surfacing tactile ... 383
Washer pressure 736
Waste cleanup hazardous ... 49
disposal hazardous 49
handling 192
handling equipment 192

Index

Wastewater treatment system 562
Watchman service 23
  temporary 23
Water & sewer piping
 & fitting demo 29
  blasting metal 112
  cement ratio 15
  closet chemical 562
  closet removal 213
  curing concrete 84
  distribution chilled 530
  distribution pipe 483, 485, 488
  effect testing 14
  fountain removal 213
  hammer arrester 241
  heater removal 213
  hose 735
  hydrant 241
  meter turbine 240
  pipe 483
  pipe ground clamp 270
  pressure reducing valve 215
  pressure relief valve 215
  pressure switch well 482
  pressure valve 215
  pump 263, 480, 736
  pump fountain 256
  pumping 307
  reducer concrete 79
  repellent coating 144
  repellent silicone 144
  service lead free 689
  supply black steel pipe 488
  supply concrete pipe 482
  supply copper pipe 488
  supply detector meter 239
  supply domestic meter 239
  supply ductile iron pipe 483
  supply HDPE 487
  supply meter 239
  supply polyethylene pipe 484
  supply PVC pipe 485
  supply well pump 482
  tank 262, 495, 737
  tank elevated 494
  tank ground-level steel 494
  tempering valve 215
  trailer 737
  treatment 562
  tree 369
  utility distribution valve 491
  well 480
  well demo 35
Watering 369
Waterline 691, 694
  common earth excavate ... 691-693
  loam/sandy clay excavate .. 694-696
Waterproofing 144
  butyl 143
  cementitious 144
  coating 142
  elastomeric 143
  elastomeric sheet 143
  foundation 585

integral 54
masonry 94
membrane 143, 363
neoprene 143
rubber 143
Waterstop center bulb 66
  dumbbell 66
  fitting 66
  PVC 66
  ribbed 66
  rubber 66
Wearing course 372
Weather barrier 145
  barrier or wrap 145
  station 210
Weatherstrip spring bronze 167
  window 168
  zinc 167
Weatherstripping 167
Wedge anchor 113
Weed barrier 402
  shrub 367
Weeding 369
Weights and measures 830
Weld plastic pipe 234
  rod 116
  rod aluminum 116
  rod cast iron 116
  rod stainless 116
  rod steel 116
  X-ray 262
Welded frame 161
  structural steel 851
  wire epoxy 75
  wire fabric 75, 845
  wire galvanized 75
Welding certification 16
  fillet 114
  machine 737
  structural 114
Well 308
  & accessory 480
  area window 184
  casing PVC 481
  casing steel 481
  gravel pack 480
  pump 480
  pump shallow 482
  screen steel 481
  water 480
  water pressure switch 482
Wellpoint 308
  discharge pipe 740
  equipment rental 740
  header pipe 740
  pump 740
  riser pipe 740
Wellpoints 872
Wet/dry vacuum 737
Wheel stop railroad 543
  trencher 733
Wheelbarrow 737
Wheeled loader 732
Whirler playground 190

Widener pavement 738
Winch truck 737
Window 165
  & door demolition 160
  aluminum 165
  aluminum awning 165
  aluminum sliding 165
  awning 183
  casement 165, 166
  demolition 160
  double-hung wood 166
  metal-clad awning 166
  metal-clad casement 166
  metal-clad double-hung 166
  removal 160
  single-hung aluminum 165
  sliding 166
  steel 165
  vinyl-clad premium 166
  wall 164
  wall aluminum 164
  weatherstrip 168
  well area 184
  wood 166
  wood double-hung 166
Windrow loader 738
  mixing 361
Winter concrete 79
  protection 21, 54
Wire 867
  brushing metal 112
  copper ground 270
  fabric welded 75
  fence 23, 393
  fencing 393
  ground 270
  ground insulated 271
  ground rod 271
  mesh 393
  razor 393
  reinforcement 845
  rope clip 120
  rope socket 121
  rope thimble 120
  rope turnbuckle 121
  tie 95
  tying 63
Wiring device & box 274
Wood bench 398
  bridge 399
  chip 402
  column 136
  cutting 44
  deck 677
  demolition 170
  dock 137, 556
  dome 208
  door 162
  double-hung window 166
  edging 469
  electrical utility pole 536
  fence 393
  fence & gate demo 34
  flagpole 185

foundation 605
framing 136
framing demolition 136
furring 137
gutter 153
parking bumper 381
pier floating 556
pile 346
pile clean 550
pile marine 346
pile treated 595
plank scaffold 19
planter 397
pole 536
post 136
post retaining wall 670
rail fence 394
round 370
sheet piling 873
sheeting 307, 342, 873
shelving 183
shoring 344
sidewalk 370
siding 149
stud 171
tie railroad 546
tie retaining wall 672
wall 605
wall foundation 605
window 166
window demolition 160
Work boat 740
  extra 12
Workers' compensation 12, 835
Wrap concrete pile 552
  steel pile 552
  wood pile 552
Wrapping marine pier pile 552
  pipe 472
Wrecking ball 737
Wrench impact 734
Wrought iron fence 386

X

X-ray concrete slab 16
  pipe 262
  weld 262

Y

Yard fountain 202

Z

Zinc weatherstrip 167

Division Notes

	CREW	DAILY OUTPUT	LABOR-HOURS	UNIT	BARE COSTS				TOTAL INCL O&P
					MAT.	LABOR	EQUIP.	TOTAL	

Division Notes

	CREW	DAILY OUTPUT	LABOR-HOURS	UNIT	BARE COSTS				TOTAL INCL O&P
					MAT.	LABOR	EQUIP.	TOTAL	

Division Notes

		CREW	DAILY OUTPUT	LABOR-HOURS	UNIT	BARE COSTS				TOTAL INCL O&P
						MAT.	LABOR	EQUIP.	TOTAL	

Division Notes

	CREW	DAILY OUTPUT	LABOR-HOURS	UNIT	BARE COSTS				TOTAL INCL O&P
					MAT.	LABOR	EQUIP.	TOTAL	

Division Notes

	CREW	DAILY OUTPUT	LABOR-HOURS	UNIT	BARE COSTS				TOTAL INCL O&P
					MAT.	LABOR	EQUIP.	TOTAL	
	CREW	DAILY OUTPUT	LABOR-HOURS	UNIT	BARE COSTS				TOTAL INCL O&P
					MAT.	LABOR	EQUIP.	TOTAL	

Division Notes

	CREW	DAILY OUTPUT	LABOR-HOURS	UNIT	BARE COSTS				TOTAL INCL O&P
					MAT.	LABOR	EQUIP.	TOTAL	

Division Notes

	CREW	DAILY OUTPUT	LABOR-HOURS	UNIT	BARE COSTS				TOTAL INCL O&P
					MAT.	LABOR	EQUIP.	TOTAL	

Division Notes

			CREW	DAILY OUTPUT	LABOR-HOURS	UNIT	BARE COSTS				TOTAL INCL O&P
							MAT.	LABOR	EQUIP.	TOTAL	

Cost Data Selection Guide

The following table provides definitive information on the content of each cost data publication. The number of lines of data provided in each unit price or assemblies division, as well as the number of crews, is listed for each data set. The presence of other elements such as reference tables, square foot models, equipment rental costs, historical cost indexes, and city cost indexes, is also indicated. You can use the table to help select the RSMeans data set that has the quantity and type of information you most need in your work.

Unit Cost Divisions	Building Construction	Mechanical	Electrical	Commercial Renovation	Square Foot	Site Work Landsc.	Green Building	Interior	Concrete Masonry	Open Shop	Heavy Construction	Light Commercial	Facilities Construction	Plumbing	Residential
1	609	444	465	564	0	533	198	365	495	608	550	310	1078	450	217
2	754	278	87	710	0	970	181	397	219	753	737	479	1197	285	274
3	1745	341	232	1265	0	1537	1043	355	2274	1745	1930	538	2028	317	445
4	960	22	0	920	0	724	180	613	1158	928	614	532	1175	0	446
5	1890	158	155	1094	0	853	1788	1107	729	1890	1026	980	1907	204	746
6	2462	18	18	2121	0	110	589	1544	281	2458	123	2151	2135	22	2671
7	1593	215	128	1633	0	580	761	532	523	1590	26	1326	1693	227	1046
8	2140	80	3	2733	0	255	1138	1813	105	2142	0	2328	2966	0	1552
9	2125	86	45	1943	0	313	464	2216	424	2062	15	1779	2379	54	1544
10	1088	17	10	684	0	232	32	898	136	1088	34	588	1179	237	224
11	1096	199	166	540	0	135	56	924	29	1063	0	230	1116	162	108
12	539	0	2	297	0	219	147	1546	14	506	0	272	1565	23	216
13	740	149	157	252	0	365	124	250	77	716	266	109	756	115	103
14	273	36	0	223	0	0	0	257	0	273	0	12	293	16	6
21	127	0	41	37	0	0	0	293	0	127	0	121	665	685	259
22	1165	7543	160	1226	0	2010	1061	849	20	1154	2109	875	7505	9400	719
23	1170	6906	546	940	0	157	865	775	38	1153	98	887	5143	1919	486
25	0	0	0	14	0	0	0	0	0	0	0	0	0	0	0
26	1513	491	10465	1293	0	860	646	1159	55	1439	649	1361	10246	399	636
27	95	0	448	102	0	0	0	71	0	95	39	67	389	0	56
28	143	79	223	124	0	0	28	97	0	127	0	70	209	57	41
31	1511	733	610	807	0	3263	286	7	1216	1456	3280	607	1568	660	616
32	896	49	8	937	0	4523	408	417	361	867	1941	486	1800	140	533
33	1255	1088	565	260	0	3078	33	0	241	532	3213	135	1726	2101	161
34	107	0	47	4	0	190	0	0	31	62	221	0	136	0	0
35	18	0	0	0	0	327	0	0	0	18	442	0	84	0	0
41	63	0	0	34	0	8	0	22	0	62	31	0	69	14	0
44	75	79	0	0	0	0	0	0	0	0	0	0	75	75	0
46	23	16	0	0	0	274	261	0	0	23	264	0	33	33	0
48	8	0	36	2	0	0	21	0	0	8	15	8	21	0	8
Totals	26183	19027	14631	20759	0	21516	10310	16507	8426	24945	17623	16251	51136	17595	13113

Assem Div	Building Construction	Mechanical	Electrical	Commercial Renovation	Square Foot	Site Work Landscape	Assemblies	Green Building	Interior	Concrete Masonry	Heavy Construction	Light Commercial	Facilities Construction	Plumbing	Asm Div	Residential
A		15	0	188	164	577	598	0	0	536	571	154	24	0	1	378
B		0	0	848	2554	0	5661	56	329	1976	368	2094	174	0	2	211
C		0	0	647	954	0	1334	0	1641	146	0	844	251	0	3	591
D		1057	941	712	1858	72	2538	330	824	0	0	1345	1104	1088	4	851
E		0	0	85	261	0	301	0	5	0	0	258	5	0	5	391
F		0	0	0	114	0	143	0	0	0	0	114	0	0	6	357
G		527	447	318	312	3378	792	0	0	535	1349	205	293	677	7	307
															8	760
															9	80
															10	0
															11	0
															12	0
Totals		1599	1388	2798	6217	4027	11367	386	2799	3193	2288	5014	1851	1765		3926

Reference Section	Building Construction Costs	Mechanical	Electrical	Commercial Renovation	Square Foot	Site Work Landscape	Assem.	Green Building	Interior	Concrete Masonry	Open Shop	Heavy Construction	Light Commercial	Facilities Construction	Plumbing	Resi.
Reference Tables	yes	yes	yes	yes	no	yes	yes	yes	yes	yes	yes	yes	yes	yes	yes	yes
Models					111			25					50			28
Crews	582	582	582	561		582		582	582	582	560	582	560	561	582	560
Equipment Rental Costs	yes	yes	yes	yes		yes		yes	yes	yes	yes	yes	yes	yes	yes	yes
Historical Cost Indexes	yes	yes	yes	yes	yes	yes	yes	yes	yes	yes	yes	yes	yes	yes	yes	no
City Cost Indexes	yes	yes	yes	yes	yes	yes	yes	yes	yes	yes	yes	yes	yes	yes	yes	yes

2020 Seminar Schedule 📞 877.620.6245

Note: call for exact dates, locations, and details as some cities are subject to change.

Location	Dates	Location	Dates
Seattle, WA	January and August	San Francisco, CA	June
Dallas/Ft. Worth, TX	January	Bethesda, MD	June
Austin, TX	February	Dallas, TX	September
Jacksonville, FL	February	Raleigh, NC	October
Anchorage, AK	March and September	Baltimore, MD	November
Las Vegas, NV	March	Orlando, FL	November
Washington, D.C.	April and September	San Diego, CA	December
Charleston, SC	April	San Antonio, TX	December
Toronto	May		
Denver, CO	May		

Gordian also offers a suite of online RSMeans data self-paced offerings.
Check our website at RSMeans.com/products/training.aspx for more information.

Facilities Construction Estimating

In this two-day course, professionals working in facilities management can get help with their daily challenges to establish budgets for all phases of a project.

Some of what you'll learn:
- Determining the full scope of a project
- Understanding of Means data and what is included in prices
- Identifying appropriate factors to be included in your estimate
- Creative solutions to estimating issues
- Organizing estimates for presentation and discussion
- Special estimating techniques for repair/remodel and maintenance projects
- Appropriate use of contingency, city cost indexes, and reference notes
- Techniques to get to the correct estimate quickly

Who should attend: facility managers, engineers, contractors, facility tradespeople, planners, and project managers.

Mechanical & Electrical Estimating

This two-day course teaches attendees how to prepare more accurate and complete mechanical/electrical estimates, avoid the pitfalls of omission and double-counting, and understand the composition and rationale within the RSMeans mechanical/electrical database.

Some of what you'll learn:
- The unique way mechanical and electrical systems are interrelated
- M&E estimates—conceptual, planning, budgeting, and bidding stages
- Order of magnitude, square foot, assemblies, and unit price estimating
- Comparative cost analysis of equipment and design alternatives

Who should attend: architects, engineers, facilities managers, mechanical and electrical contractors, and others who need a highly reliable method for developing, understanding, and evaluating mechanical and electrical contracts.

Construction Cost Estimating: Concepts and Practice

This one or two day introductory course to improve estimating skills and effectiveness starts with the details of interpreting bid documents and ends with the summary of the estimate and bid submission.

Some of what you'll learn:
- Using the plans and specifications to create estimates
- The takeoff process—deriving all tasks with correct quantities
- Developing pricing using various sources; how subcontractor pricing fits in
- Summarizing the estimate to arrive at the final number
- Formulas for area and cubic measure, adding waste and adjusting productivity to specific projects
- Evaluating subcontractors' proposals and prices
- Adding insurance and bonds
- Understanding how labor costs are calculated
- Submitting bids and proposals

Who should attend: project managers, architects, engineers, owners' representatives, contractors, and anyone who's responsible for budgeting or estimating construction projects.

Assessing Scope of Work for Facilities Construction Estimating

This two-day practical training program addresses the vital importance of understanding the scope of projects in order to produce accurate cost estimates for facilities repair and remodeling.

Some of what you'll learn:
- Discussions of site visits, plans/specs, record drawings of facilities, and site-specific lists
- Review of CSI divisions, including means, methods, materials, and the challenges of scoping each topic
- Exercises in scope identification and scope writing for accurate estimating of projects
- Hands-on exercises that require scope, take-off, and pricing

Who should attend: corporate and government estimators, planners, facility managers, and others who need to produce accurate project estimates.

Practical Project Management for Construction Professionals

In this two-day course, acquire the essential knowledge and develop the skills to effectively and efficiently execute the day-to-day responsibilities of the construction project manager.

Some of what you'll learn:
- General conditions of the construction contract
- Contract modifications: change orders and construction change directives
- Negotiations with subcontractors and vendors
- Effective writing: notification and communications
- Dispute resolution: claims and liens

Who should attend: architects, engineers, owners' representatives, and project managers.

Maintenance & Repair Estimating for Facilities

This two-day course teaches attendees how to plan, budget, and estimate the cost of ongoing and preventive maintenance and repair for existing buildings and grounds.

Some of what you'll learn:
- The most financially favorable maintenance, repair, and replacement scheduling and estimating
- Auditing and value engineering facilities
- Preventive planning and facilities upgrading
- Determining both in-house and contract-out service costs
- Annual, asset-protecting M&R plan

Who should attend: facility managers, maintenance supervisors, buildings and grounds superintendents, plant managers, planners, estimators, and others involved in facilities planning and budgeting.

Life Cycle Cost Estimating for Facilities Asset Managers

Life Cycle Cost Estimating will take the attendee through choosing the correct RSMeans database to use and then correctly applying RSMeans data to their specific life cycle application. Conceptual estimating through RSMeans' new building models, conceptual estimating of major existing building projects through RSMeans' renovation models, pricing specific renovation elements, estimating repair, replacement and preventive maintenance costs today and forward up to 30 years will be covered.

Some of what you'll learn:
- Cost implications of managing assets
- Planning projects and initial & life cycle costs
- How to use RSMeans data online

Who should attend: facilities owners and managers and anyone involved in the financial side of the decision making process in the planning, design, procurement, and operation of facilities real assets.

Please bring a laptop with ability to access the internet.

Building Systems and the Construction Process

This one-day course was written to assist novices and those outside the industry in obtaining a solid understanding of the construction process - from both a building systems and construction administration approach.

Some of what you'll learn:
- Various systems used and how components come together to create a building
- Start with foundation and end with the physical systems of the structure such as HVAC and Electrical
- Focus on the process from start of design through project closeout

This training session requires you to bring a laptop computer to class.

Who should attend: building professionals or novices to help make the crossover to the construction industry; suited for anyone responsible for providing high level oversight on construction projects.

Training for our Online Estimating Solution

Construction estimating is vital to the decision-making process at each state of every project. Our online solution works the way you do. It's systematic, flexible and intuitive. In this one-day class you will see how you can estimate any phase of any project faster and better.

Some of what you'll learn:
- Customizing our online estimating solution
- Making the most of RSMeans "Circle Reference" numbers
- How to integrate your cost data
- Generating reports, exporting estimates to MS Excel, sharing, collaborating and more

Also offered as a self-paced or on-site training program!

Training for our CD Estimating Solution

This one-day course helps users become more familiar with the functionality of the CD. Each menu, icon, screen, and function found in the program is explained in depth. Time is devoted to hands-on estimating exercises.

Some of what you'll learn:
- Searching the database using all navigation methods
- Exporting RSMeans data to your preferred spreadsheet format
- Viewing crews, assembly components, and much more
- Automatically regionalizing the database

This training session requires you to bring a laptop computer to class.

When you register for this course you will receive an outline for your laptop requirements.

Also offered as a self-paced or on-site training program!

Site Work Estimating with RSMeans data

This one-day program focuses directly on site work costs. Accurately scoping, quantifying, and pricing site preparation, underground utility work, and improvements to exterior site elements are often the most difficult estimating tasks on any project. Some of what you'll learn:
- Evaluation of site work and understanding site scope including: site clearing, grading, excavation, disposal and trucking of materials, backfill and compaction, underground utilities, paving, sidewalks, and seeding & planting.
- Unit price site work estimates—Correct use of RSMeans site work cost data to develop a cost estimate.
- Using and modifying assemblies—Save valuable time when estimating site work activities using custom assemblies.

Who should attend: Engineers, contractors, estimators, project managers, owner's representatives, and others who are concerned with the proper preparation and/or evaluation of site work estimates.

Please bring a laptop with ability to access the internet.

Facilities Estimating Using the CD

This two-day class combines hands-on skill-building with best estimating practices and real-life problems. You will learn key concepts, tips, pointers, and guidelines to save time and avoid cost oversights and errors.

Some of what you'll learn:
- Estimating process concepts
- Customizing and adapting RSMeans cost data
- Establishing scope of work to account for all known variables
- Budget estimating: when, why, and how
- Site visits: what to look for and what you can't afford to overlook
- How to estimate repair and remodeling variables

This training session requires you to bring a laptop computer to class.

Who should attend: facility managers, architects, engineers, contractors, facility tradespeople, planners, project managers, and anyone involved with JOC, SABRE or IDIQ.

Registration Information

Register early to save up to $100!!!
Register 45+ days before the date of a class and save $50 off each class. This savings cannot be combined with any other promotion or discounting of the regular price of classes!

How to register
By Phone
Register by phone at 877.620.6245

Online
Register online at
RSMeans.com/products/seminars.aspx

Note: Purchase Orders or Credits Cards are required to register.

Two-day seminar registration fee - $1,200*.

One-Day Construction Cost Estimating or Building Systems and the Construction Process - $765*.

Government pricing
All federal government employees save off the regular seminar price. Other promotional discounts cannot be combined with the government discount. Call 781.422.5115 for government pricing.

CANCELLATION POLICY:
If you are unable to attend a seminar, substitutions may be made at any time before the session starts by notifying the seminar registrar at 781.422.5115 or your sales representative.
If you cancel twenty-one (21) days or more prior to the seminar, there will be no penalty and your registration fees will be refunded. These cancellations must be received by the seminar registrar or your sales representative and will be confirmed to be eligible for cancellation.
If you cancel fewer than twenty-one (21) days prior to the seminar, you will forfeit the registration fee.
In the unfortunate event of an RSMeans cancellation, RSMeans will work with you to reschedule your attendance in the same seminar at a later date or will fully refund your registration fee. RSMeans cannot be responsible for any non-refundable travel expenses incurred by you or another as a result of your registration, attendance at, or cancellation of an RSMeans seminar.
Any on-demand training modules are not eligible for cancellation, substitution, transfer, return or refund.

AACE approved courses
Many seminars described and offered here have been approved for 14 hours (1.4 recertification credits) of credit by the AACE International Certification Board toward meeting the continuing education requirements for recertification as a Certified Cost Engineer/Certified Cost Consultant.

AIA Continuing Education
We are registered with the AIA Continuing Education System (AIA/CES) and are committed to developing quality learning activities in accordance with the CES criteria. Many seminars meet the AIA/CES criteria for Quality Level 2. AIA members may receive 14 learning units (LUs) for each two-day RSMeans course

Daily course schedule
The first day of each seminar session begins at 8:30 a.m. and ends at 4:30 p.m. The second day begins at 8:00 a.m. and ends at 4:00 p.m. Participants are urged to bring a hand-held calculator since many actual problems will be work out in each session.

Continental breakfast
Your registration includes the cost of a continental breakfast and a morning an afternoon refreshment break. These informal segments allow you to discuss topics of mutual interest with other seminar attendees. (You are free to make your own lunch and dinner arrangements.)

Hotel/transportation arrangements
We arrange to hold a block of rooms at most host hotels. To take advantage of special group rates when making your reservation, be sure to mention that you are attending the RSMeans Institute data seminar. You are, of course, free to stay at the lodging place of your choice. (Hotel reservations and transportation arrangements should be made directly by seminar attendees.)

Important
Class sizes are limited, so please register as soon as possible.

*Note: Pricing subject to change.